Encyclopedia of Fluid Mechanics

VOLUME 1

Flow Phenomena and Measurement

Gulf Publishing Company
Book Division
Houston, London, Paris, Tokyo

Encyclopedia of Fluid Mechanics

VOLUME 1

Flow Phenomena and Measurement

N. P. Cheremisinoff, Editor

in collaboration with—

G. Akay
P. H. Alfredsson
L. Baker
J. Baldyga
W. H. Bell
N. S. Berman
D. L. Book
J. R. Bourne
I. P. Castro
R. P. Chhabra
W. L. Chow
W. T. Clark
E. L. Cussler
E. Dick
S. K.-Djurdjevic
A. Dudukovic
T. Z. Fahidy
F. D. Fan
K. Gotoh
W. Hayduk
I. Inoue
A. V. Johansson
K. Kataoka
H. Knapp
J. Koplik
D. Kumar
C. Kuroda
J. Lee
Y. S. Lee
M. F. Letelier S.
S. P. Lin
S. L. Liu
J. F. Lyness
Y. Matsuzaki
D. L. McCullum
W. R. C. Myers
A. Nakayama
K. D. P. Nigam
T. Nishimura
K. Ogawa
M. S. Quraishi
A. K. Saxena
A. K. Sen
A. V. Shenoy
N. Shiragami
L. C. Smith
C. C. S. Song
S. Tavoularis
C. H. Thomson
C. Y. Wang
M. Yamada
C. T. Yang
F. Zdanski
X. H. Zhou

Encyclopedia of Fluid Mechanics

VOLUME 1

Flow Phenomena and Measurement

Library of Congress Cataloging in Publication Data

(Revised for volume 2)
Main entry under title;

Encyclopedia of fluid mechanics.

Includes indexes.
Contents: v. 1. Flow phenomena and measurement—
v. 2. Dynamics of single-fluid flows and mixing.
1. Fluid mechanics—Dictionaries. I. Cheremisinoff, Nicholas P.
TA357.E53 1986 620.1′06 85-9742

ISBN 0-87201-513-0 (v. 1)
ISBN 0-87201-514-9 (v. 2)

ISBN 0-87201-513-0

CONTENTS

SECTION II: FLOW DYNAMICS AND FRICTIONAL BEHAVIOR

SECTION III: FLOW AND TURBULENCE MEASUREMENT

CONTRIBUTORS TO THIS VOLUME

G. Akay, School of Industrial Science, Cranfield Institute of Technology, Cranfield, Bedford, United Kingdom.

P. H. Alfredsson, Department of Mechanics, The Royal Institute of Technology, Stockholm, Sweden.

L. Baker, Mission Research Corp., Albuquerque, New Mexico, USA.

J. Baldyga, Instytut Inzynierii Chemicznej Politechniki Warszawskiej, Warsaw, Poland.

W. H. Bell, Institute of Ocean Sciences, Sidney, British Columbia, Canada.

N. S. Berman, Department of Chemical and Bio-Engineering, Arizona State University, Tempe, Arizona, USA.

D. L. Book, Laboratory for Computational Physics, Naval Research Laboratory, Washington, DC, USA.

J. R. Bourne, Technisch-chemisches Laboratorium ETH, Zuerich, Switzerland.

I. P. Castro, Department of Mechanical Engineering, University of Surrey, Guildford Surrey, United Kingdom.

N. P. Cheremisinoff, Exxon Chemical Co., Linden, New Jersey, USA.

R. P. Chhabra, Department of Chemical Engineering, University College of Swansea, Swansea, United Kingdom.

W. L. Chow, Department of Mechanical and Industrial Engineering, University of Illinois at Urbana-Champaign, Urbana, Illinois, USA.

W. T. Clark, Ramapo Instrument Co., Inc., Montville, New Jersey, USA.

E. L. Cussler, Department of Chemical Engineering, University of Minnesota, Minneapolis, Minnesota, USA.

E. Dick, Department of Machinery, State University of Ghent, Ghent, Belgium.

S. K.-Djurdjevic, Faculty of Technology and Metallurgy, University of Beograd, Beograd, Yugoslavia.

A. Dudukovic, Faculty of Technology, University of Novi Sad, Novi Sad, Yugoslavia.

T. Z. Fahidy, Department of Chemical Engineering, University of Waterloo, Waterloo, Ontario, Canada.

F. D. Fan, Northwestern Polytechnical University, Xi'an, China.

K. Gotoh, Research Institute for Mathematical Sciences, Kyoto University, Kyoto, Japan.

W. Hayduk, Department of Chemical Engineering, University of Ottawa, Ottawa, Canada.

I. Inoue, Department of Chemical Engineering, Tokyo Institute of Technology, Wako, Saitama, Japan.

A. V. Johansson, Department of Mechanics, The Royal Institute of Technology, Stockholm, Sweden.

K. Kataoka, Department of Chemical Engineering, Kobe University, Kobe, Japan.

H. Knapp, Technical University of Berlin, Berlin, Federal Republic of W. Germany.

J. Koplik, Schlumberg-Doll Research, Ridgefield, Connecticut, USA.

D. Kumar, Tehri Dam Circle IV, Pragatipuram, Near-Rishikesh, India.

C. Kuroda, Department of Chemical Engineering, Tokyo Institute of Technology, Tokyo, Japan.

J. Lee, Air Force Wright Aeronautical Laboratories, Wright-Patterson Air Force Base, Ohio, USA.

Y. S. Lee, Westinghouse Electric Corp., Pittsburgh, Pennsylvania, USA.

M. F. Letelier S., Department of Mechanical Engineering, University of Santiago, Chile.

S. P. Lin, Department of Mechanical and Industrial Engineering, Clarkson University, Potsdam, New York, USA.

S. L. Liu, Northwestern Polytechnical University, Xi'an, China.

J. F. Lyness, Department of Civil Engineering, University of Ulster, Newtownabbey, County Antrim, Northern Ireland.

Y. Matsuzaki, National Aerospace Laboratory, Jindaiji, Chofu, Tokyo, Japan.

D. L. McCullum, Department of Chemical Engineering, University of Minnesota, Minneapolis, Minnesota, USA.

W. R. C. Myers, Department of Civil Engineering, University of Ulster, Newtownabbey, County Antrim, Northern Ireland.

A. Nakayama, Department of Mechanical Engineering, Shizuoka University, Tohoku, Japan.

K. D. P. Nigam, Department of Chemical Engineering, Indian Institute of Technology, New Delhi, India.

T. Nishimura, Hiroshima University, Hiroshima, Japan.

K. Ogawa, Department of Chemical Engineering, Tokyo Institute of Technology, Tokyo, Japan.

M. S. Quraishi, Department of Chemical Engineering, University of Waterloo, Waterloo, Ontario, Canada.

A. K. Saxena, Indian Institute of Petroleum, C.S.I.R. Complex, New Delhi, India.

A. K. Sen, Department of Mathematical Sciences, Purdue School of Science at Indianapolis, Indiana, USA.

A. V. Shenoy, Chemical Engineering Division, National Chemical Laboratory, Pune, India.

N. Shiragami, Chemical Engineering Laboratory, Institute of Physical and Chemical Research, Wako, Saitama, Japan.

L. C. Smith, Westinghouse Electric Corp., Pittsburgh, Pennsylvania, USA.

C. C. S. Song, St. Anthony Falls Hydraulic Laboratory, University of Minnesota, Minneapolis, Minnesota, USA.

S. Tavoularis, Department of Chemical Engineering, University of Ottawa, Ottawa, Canada.

G. H. Thomson, Phillips Petroleum Company, Bartlesville, Oklahoma, USA.

C. Y. Wang, Department of Mathematics, Michigan State University, East Lansing, Michigan, USA.

M. Yamada, Department of Physics, Kyoto University, Kyoto, Japan.

C. T. Yang, U. S. Department of the Interior, Bureau of Reclamation, Engineering and Research Center, Denver, Colorado, USA.

F. Zdanski, Faculty of Technology and Metallurgy, University of Beograd, Beograd, Yugoslavia.

X. H. Zhou, Northwestern Polytechnical University, Xi'an, China.

ABOUT THE EDITOR

Nicholas P. Cheremisinoff heads the product development group in the Elastomers Technology Division of Exxon Chemical Company. Previously, he led the Reactor and Fluid Dynamics Modeling Group at Exxon Research and Engineering Company. He received his B.S., M.S., and Ph.D. degrees in chemical engineering from Clarkson College of Technology, and he is also a member of a number of professional societies including AIChE, Tau Beta Pi, and Sigma Xi.

PREFACE

Fluid mechanics plays a dominant role in transport processes that form the foundation of engineering designs aimed at producing countless products to improve the quality of life and the environment. The extent and diversity of the subject is enormous. The number of technical articles and review papers written in the last two decades is all but staggering and well beyond what the average practicing engineer or researcher may hope to consult.

Although there are many excellent books that deal with specialized topics of the overall subject, they do not, even collectively, fulfill the technical community's needs for a theoretical background and state-of-the-art understanding and working knowledge. The intent of the *Encyclopedia of Fluid Mechanics* is to provide an integrated presentation of the extensive and widely scattered information relating to all types of flow phenomena of practical interest. The work represents the efforts of hundreds of researchers and engineers from around the world, who have collectively combined their expertise in this multivolume series.

Volume I, *Flow Phenomena and Measurement*, provides a detailed treatment of single fluid flow behavior and measurement principles and techniques. A rigorous presentation of the principles of momentum and energy transport is given to provide background for later volumes dealing with the flow of complex mixtures.

The volume is divided into three sections. Section I, "Transport Properties and Flow Instability," contains nine chapters that address physical and transport properties, diffusion, micro- and macro-mixing phenomena, and turbulence structure and initiation. Detailed analytical treatment of special flow instabilities such as Rayleigh-Taylor instabilities are included.

Section II, "Flow Dynamics and Frictional Behavior," contains 26 chapters that address the theory and practical concepts of flow and turbulence, flow through irregular geometries, and non-Newtonian behavior. Transonic flows, stability of immersed objects, and principles of aerodynamics are covered, and a state-of-the-art review of the theory of minimum energy and energy dissipation is presented. The section provides a well-balanced presentation of phenomenological and theoretical aspects of flow dynamics for both the researcher and practicing engineer requiring descriptive knowledge.

Section III, "Flow and Turbulence Measurement," contains six chapters covering experimental techniques and industrial methods/instrumentation for studying flow behavior on micro- and macro-scales. Many of the techniques discussed are applicable to multiphase flows treated in later volumes.

This volume is structured to provide a balanced presentation of both theoretical and engineering-oriented information. To ensure the highest degree of reliability, the services of many specialists were enlisted. This first volume represents the efforts of 55 experts. Additionally, it brings experience and opinions of scores of engineers and researchers who aided with advice and suggestions in reviewing and refereeing the material presented. Each contributor is to be regarded as responsible for the statements and recommendations in his chapter. These individuals are to be congratulated for devoting their time and efforts to producing this volume. Without their efforts this work could not have become a reality. Special thanks go to Gulf Publishing Company for producing this series.

Nicholas P. Cheremisinoff

ENCYCLOPEDIA OF FLUID MECHANICS

VOLUME 1: FLOW PHENOMENA AND MEASUREMENT

Transport Properties and Flow Instability
Flow Dynamics and Frictional Behavior
Flow and Turbulence Measurement

VOLUME 2: DYNAMICS OF SINGLE-FLUID FLOWS AND MIXING

Channel and Free Surface Flows
Mixing Phenomena and Practices
Fluid Transport Equipment

VOLUME 3: GAS-LIQUID FLOWS

Properties of Dispersed and Atomized Flows
Flow Regimes, Hold-Up, and Pressure Drop
Reactors and Industrial Applications

VOLUME 4: SOLIDS AND GAS-SOLIDS FLOWS

Properties of Particulates and Powders
Particle-Gas Flows
Fluidization and Industrial Applications
Particulate Capture and Classification

VOLUME 5: SLURRY FLOW TECHNOLOGY

Slurry and Suspension Flow Properties
Unit Operations of Slurry Flows

VOLUME 6: COMPLEX FLOW PHENOMENA AND MODELING

Special Topics in Complex and Multiphase Flows
Transport Phenomena in the Environment
Flow Simulation and Modeling

SECTION I

TRANSPORT PROPERTIES AND FLOW INSTABILITY

CONTENTS

CHAPTER 1

GENERALIZED EQUATIONS OF STATE FOR PROCESS DESIGN

Helmut Knapp

Institute of Thermodynamics and Plant Design
Technical University of Berlin
Federal Republic of West Germany

CONTENTS

INTRODUCTION

When the process of a production plant is conceived, designed, revised, analyzed, or optimized, accurate and reliable information is required about the properties of all materials handled in various operations. Beginning with the specifications for feed and products, the process engineer must calculate mass and energy balances, changes of state, phase equilibria, etc.

The basis of equipment design is the knowledge of the operating conditions such as temperature, pressure, composition, transfer of mass, heat, and work at all important points in the process. The transfer surfaces required for heat and mass transfer can then be determined.

As an example, consider a heat exchanger where one stream is cooled and partially condensed while a countercurrent stream is heated and partially evaporated. The required surface can be determined if the cooling and heating curves (i.e., the Q–T or H–T diagrams) and the heat transfer coefficients are known.

As another example, consider a distillation column where heat and mass is transferred between the vapor and the liquid phase. The number of theoretical stages can be determined if the condition of the final thermodynamic equilibrium is known.

All these calculations require information about the thermodynamic properties such as density, heat capacity, enthalpy, heat of vaporization, distribution coefficients in phase equilibrium and heats of reaction.

The engineer engaged in process design and in need of the information should consider all possible methods and all available data:

- Experimental data, often correlated by empirical and graphical methods.
- Reliable and proven equations describing the thermodynamic properties.
- Plausible, mostly simplified, physical, or chemical models as a basis for proposing adequate correlations.
- General thermodynamic laws.

In this chapter a review is given of generalized methods that can be used to predict the required information on volumetric and caloric properties of a great number of technically important substances and mixtures thereof.

- The effect of pressure at any temperature on all thermodynamic properties can be calculated with an equation of state.
- The principle of corresponding states allows the use of generalized equations of state. Only a few parameters are needed to characterize a pure substance or a mixture.
- Electronic computers can be easily used as the information is available in the form of analytical equations.

The application of the method and its advantages are limited to fluid materials consisting of so-called normal fluids such as noble gases, nitrogen, oxygen, carbon monoxide, hydrocarbons, and certain hydrocarbon derivatives. Carbon dioxide, hydrogen sulfide, hydrogen and—with reservations—some polar substances can also be included.

At the end of the chapter references are listed offering more detailed information for each paragraph.

EQUATIONS OF STATE

Equations of states (EQS) are material equations that describe the PVT behavior of pure substances or the PVTX behavior of mixtures in the fluid state. Background reading can be found in References 1 through 3.

There are various analytical forms of EQS for pure substances

Example for a pure substance:

$$p = p(T, v, A_{ik})$$

$$\text{or} \quad v = v(T, p, A_{ik})$$

$$\text{or} \quad z = z(T, p, A_{ik}) \qquad k = 1, 2, \ldots$$

$$\text{or} \quad z = z(T, v, A_{ik})$$

$$\text{or} \quad z = z(T_r, v_r, A_k)$$

where A_{ik} = specific coefficient for substance i, k = 1, 2, . . .
A_k = general coefficient, k = 1, 2, . . .
n = number of moles in a system
p = pressure
R = universal gas constant, 8.314 J/mol K
T = temperature
T_r = reduced temperature, T/T_c
T_c, v_c = critical properties

V = total volume of a system
v = molar volume, V/n
v_r = reduced volume, v/v_c
x = concentration of a component in a mixture, e.g., mole fraction

PRESENTATION OF PVT PROPERTIES

The information on the PVT properties of a pure substance can be presented in various formats

- In tables: for example, Table 1, a steam table of water,
- In diagrams: for example, Figure 1, a schematic p-v diagram with isotherms; Figure 2, a T-v diagram of H_2O with isobars; Figure 3, a schematic p-T diagram with isochores

Table 1
Properties of Water and Steam

T °C	p bar	v^L m^3/kg	v^V m^3/kg
0	0.0061	0.00100	206.29
50	0.1233	0.00101	12.045
100	1.0132	0.00104	1.673
200	15.550	0.00115	0.127
300	85.92	0.00140	0.021
374.15	221.20	0.0031	0.0031

T = temperature; p = pressure; v^L = specific volume of saturated liquid; v^V = specific volume of saturated vapor.

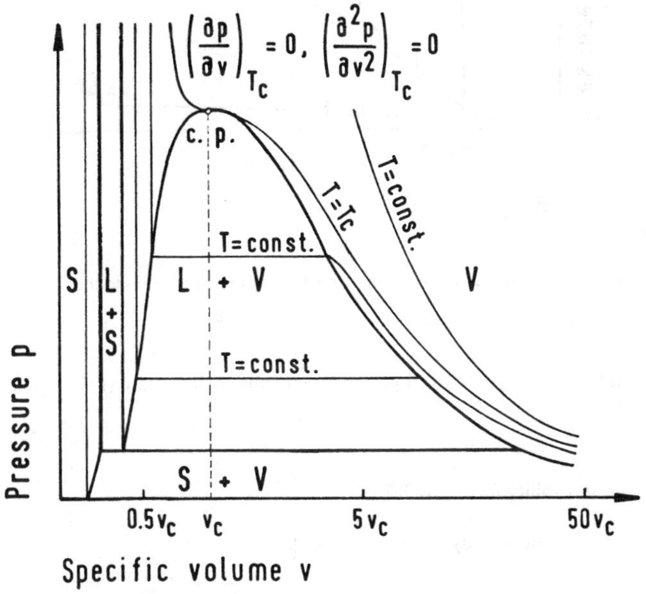

Figure 1. Typical, schematic diagram showing pressure vs. specific volume with isotherms and boundaries of one- and two-phase areas.

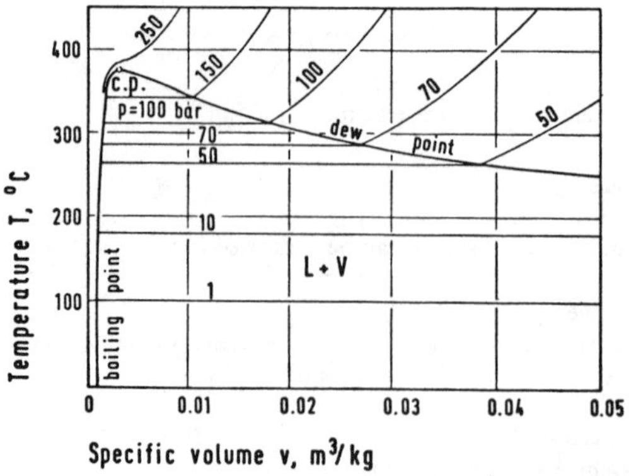

Figure 2. Diagram showing temperature vs. specific volume of water with isobars and saturation curves for vapor and liquid.

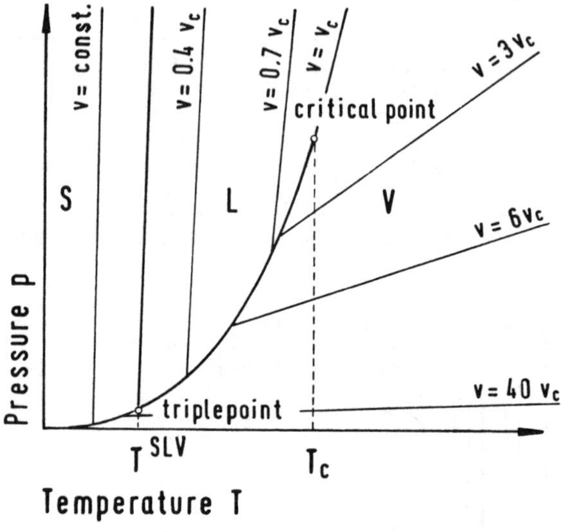

Figure 3. Schematic diagram showing pressure vs. temperature with isochores and coexistence curves.

- By multivariable mathematical functions, so-called equations of state (EQS); for example, EQS for the ideal gas state:

$$pv = TR \qquad \text{or} \qquad z = 1 \tag{1}$$

where z = compressibility factor, pv/RT

or EQS for a real gas proposed by van der Waals [1]:

$$\left(p + \frac{a}{v^2}\right)(v - b) = RT \qquad \text{or} \qquad z = \frac{v}{v - b} - \frac{a}{RTv} \tag{2}$$

where a, b = specific constants of a substance

The corrective term a/v^2 accounts for the effect of the forces acting between the molecules, and b accounts for the "hard" volume occupied by the molecules. In the equation for compressibility, the first term is often called the "repulsive" and the second the "attractive" term. These meaningful corrections give a simple, qualitively correct, but quantitatively less accurate analytical description of the behavior of a real gas.

Example: Virial EQS

$$z = 1 + B(T)\rho + C(T)\rho^2 + \cdots \tag{3}$$

where ρ = molar density, $1/v$
$B(T)$ = second virial coefficient

In the virial EQS proposed by Kamerlingh Onnes [5] corrections are made for the effect of interaction between two, three, and more molecules. The coefficients for pure substances depend only on temperature and can be calculated by the methods of statistical mechanics based on knowledge about the interaction energy of the molecules. The virial EQS is theoretically important but fails at higher densities.

Example: EQS with "Simple" and "Reference" Fluids

$$z = z^{(0)} + \omega z^{(1)} \tag{4}$$

where $z^{(0)}$ = compressibility of simple fluid
$z^{(r)}$ = compressibility of reference fluid
$z^{(1)}$ = corrective term, $(z^{(r)} - z^{(0)})/\omega^{(r)}$
ω = accentric factor

According to Pitzer's [6] suggestion, the compressibility consists of a term for simple spherical molecules and a corrective term that accounts for the effect of the nonspherical shape of the molecules. $z^{(1)}$ can be calculated as the difference between the compressibility of a reference and of a simple fluid.

Example: Augmented Rigid Body EQS

$$z = z^{\text{repulsive}} - z^{\text{attractive}} \tag{5}$$

where $$z^{\text{rep}} = \frac{1 + \rho^* + \rho^{*2} - \rho^{*3}}{(1 - \rho^*)^3} \tag{5a}$$

$$z^{\text{attr}} = \frac{f_1(\rho^*)}{T} + \frac{f_2(\rho^*)}{T^2} \tag{5b}$$

with ρ^* = reduced density, ρv^*
v^* = volume of hard sphere molecules
f_1, f_2 = functions of reduced density

As a result of computer calculations for the behavior of hard bodies, simple analytical equations were developed (e.g., Equation 5a by Carnahan and Starling [17]). The attractive term is developed in powers of $(1/T)$, modeling the results of perturbation theory.

Example: Theoretical Nonanalytical EQS

$$z = 1 + \rho\left(\frac{\partial A^+}{\partial \rho}\right)_T \tag{6}$$

where A^+ = molar residual Helmholtz free energy

Starting with our knowledge of the properties of molecules and the laws governing the motion and interaction of molecules, it is possible with the help of statistical mechanics to calculate macroscopic properties such as the free energy. For process calculations, however, this theoretical approach is too complicated, too time consuming, and too inaccurate.

EQUATION OF STATE FOR MIXTURES

The materials treated in chemical production plants are sometimes pure substances but are usually mixtures of many components. Thermodynamic properties of multicomponent systems depend not only on temperature and pressure but also on composition. The PVTX behavior of the mixture can be described by an equation of state for a fluid whose coefficients depend on the composition.

Example: Structure of EQS for Mixtures

$$z = z(T, v, A_{mk}) \tag{7}$$

where A_{mk} = coefficients for a specific mixture of constant composition, k = 1, 2, . . .

The coefficients could be found by fitting to experimental PVTX data or—in lack of sufficient experimental dáta—more conveniently by combining the coefficients of the pure components in the appropriate manner.

Example: Structure of Mixing Rule

$$A_{mk} = A_{mk}(A_{ik}, x_i, k_{ij}) \tag{8}$$

The mixing and combination rules are empirical, but do have a theoretical basis.

Example: Mixing Rules and Combination Rule for the Constants a and b in Equation 9

$$a_m = \sum \sum x_i x_j a_{ij} \qquad a_{ij} = \sqrt{a_i a_j} \tag{9}$$

$$b_m = \sum x_i b_i \tag{10}$$

Liquid mixtures of strongly interacting substances (e.g. by chemical association or solvation) can be better described with models for excess functions. The equation of state fails mainly because simple mixing rules are inadequate. More sophisticated mixing rules, however, often improve the accuracy of the calculations.

MODIFICATIONS OF EQUATIONS OF STATE

During the past century hundreds of modifications to the basic equations and to the mixing rules have been proposed in order to improve the accuracy and to extend the range of validity. The development is characterized by increasingly complicated structures—especially after the advent of fast electronic computers. Most of the equations have been arranged empirically and the numerable coefficients (sometimes more than fifty) are determined by fitting to all available experimental data using the criterion of minimum deviation between experimental and calculated data. The sets of coefficients are not unique. For the description of mixtures, mixing rules must be given for each coefficient.

GENERALIZED EQUATIONS OF STATE (GEQS)

Experience indicates and molecular theory explains that many fluids, pure substances or mixtures, have a similar PVT behavior, i.e., isotherms in p–v diagrams or vapor pressure curves look alike. According to the principle of corresponding states, proposed originally by van der Waals [8], a fluid can be characterized by the coordinates of one outstanding point on the PVT surface, such as the critical point or the triple point. The critical point is preferable as it contains more "information" about the fluid state than the triple point.

In accordance with a few "general" rules the coefficients of an equation of state can be calculated for a specific substance from the coordinates of the "correspondence point." It is also possible to introduce reduced variables, e.g., $T_r = T/T_c$, $p_r = p/p_c$ and $v = v/_c$ into an equation of state which then has a "universal" set of coefficients, once and for all fitted to a "reference" fluid. A mixture can be represented by "pseudo critical" coordinates calculated with a few mixing rules.

Example: GEQS with Specific Coefficients for Pure Components or Mixtures

$$z = z(T, v, A_{ik}) \tag{11}$$

where $A_{ik} = A_{ik}(T_{ci}, p_{ci}, v_{ci} \text{ or } \omega_i)$ (12)

If the GEQS describes the PVTX behavior of a multicomponent fluid, mixing rules are required for all k coefficients, A_{ik}.

Example: GEQS with Reduced Variables and "Universal" Coefficients

$$z = z(T_r, v_r, \omega, A_k) \tag{13}$$

where A_k = "universal" constants

For multicomponent fluids pseudocritical values are calculated for all correspondence parameters with empirical mixing and combination rules, e.g.,

$$T_{cM} = \sum\sum x_i x_j T_{cij} \quad \text{with} \quad T_{cij} = \sqrt{T_{ci}T_{cj}}\,(1 - k_{ij}) \tag{14}$$

$$v_{cM} = \sum\sum x_i x_j v_{cij} \quad \text{with} \quad v_{cij} = (1/8)(v_{ci}^{1/3} + v_{cj}^{1/3})^3 \tag{15}$$

$$\omega_{cM} = \sum x_i \omega_i \tag{16}$$

The reader may refer to References 9 and 10 for additional discussions.

CHOICE OF EQUATION OF STATE

It is important to consider the purpose for which an equation of state is to be used (see Table 2).

Table 2
Input and Output of Various Types of Equations of State

Information	Comp.	Result	Type
$\begin{array}{ccc} H & & H \\ | & & | \\ H-C & - & C-H \\ | & & | \\ H & & H \end{array}$	1	Estimate PVT	EQS: VdW
Few $\rho(T, p)$ Few $p^{LV}(T)$	1	Calculate PVT	EQS: VdW
$>100 \begin{cases} PVT \\ caloric \end{cases}$ data	1	Calculate PVT, $h - h^0$ Construct table or diagram	Empirical virial EQS with fitted coeff. A_{ik}, $k = 20 - 60$
T_c, p_c, v_c or ω	1	Calculate PVT, $h - h^0$	GEQS with calculated coeff. $A_{ik}(T_{ci}, p_{ci}, v_{ci},$ or $\omega_i)$
T_c, p_c, v_c or ω VLE $\rightarrow k_{ij}$	2, 3, ...	Calculate PVT, $h - h^0$ VLE, LLE	GEQS: VdW or virial + mix. rules $A_{mk}(A_{ik}, k_{ij}, x)$

Info—Input information that is or should be available. Comp.—Number of components in the fluid. Result—Information that is requested and can be calculated with the equation. Type—Equation that is recommended for the specified in- and output.

Case 1

Sometimes a pure substance such as water, nitrogen, methane, or ammonia has been thoroughly investigated, i.e., many pvT data, vapor pressures, heat capacities, velocities of sound have been measured. If the properties of the substance should be represented in tables or diagrams within the experimental accuracy, it is practical to interpolate the data with an equation of state containing many adjustable coefficients (often more than 40). A smaller set of coefficients might be sufficient if different sets are used for interpolations in limited areas of the p–T field.

Case 2

Only few experimental points (e.g., a few vapor pressures or densities) are known. Sometimes only the chemical structural formula is known. A simple but "safe" equation of state, mostly of the van der Waals type, allows the estimation of PVT properties.

Case 3

If process calculations for a great variety of multicomponent mixtures must be done, it is most practical to use a generalized equation of state. Only a few specific parameters for each substance and only one or two parameters for each binary combination is required.

Further discussions will concern Case 3.

THERMODYNAMIC FUNCTIONS

EQS give information on how the specific volume or the density depend on temperature, pressure, and composition. In addition, however, many more thermodynamic properties can be calculated from EQS by referring to general thermodynamic relations.

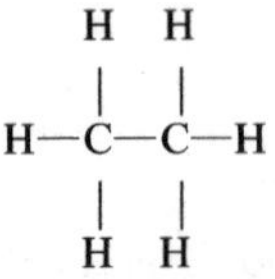

Example: Enthalpy Departure $h - h^0$

The enthalpy of a substance in the state of an ideal gas depends only on temperature:

$$h^0(T, p^0) = h^0(T^+, p^0) + \int_{T^+}^{T} c_p^0(T)\, dT \tag{17}$$

where h^0 = standard enthalpy ideal gas state
T^+ = arbitrary reference temperature
p^0 = pressure in standard state, e.g., $p^0 = 0$
$c_p^0(T)$ = molar heat capacity in ideal gas state as function of temperature, listed in standard thermophysical table

$$c_p^0(T) = \sum C_m T^m \tag{18}$$

The effect of pressure on the enthalpy, the so-called enthalpy departure $h - h^0$, can be calculated if the PVTX behavior of the fluid is known:

$$h - h^0 = \int_{p^0}^{p} \left(\frac{\partial h}{\partial p}\right)_{T,x} dp \tag{19}$$

$$= \int_{p^0}^{p} \left[v - T\left(\frac{\partial v}{\partial T}\right)_{p,x} \right] dp \tag{19a}$$

$$= \int_{\infty}^{v} \left[T\left(\frac{\partial p}{\partial T}\right)_{v,x} - p \right] dv + pv - RT \tag{19b}$$

Useful equations of state are structured such that the differentiation and integration can be carried out analytically.

The knowledge required to calculate HTPX is therefore (1) $c_p^0(T)$ and (2) $z(T, v, x)$–see Figure 4.

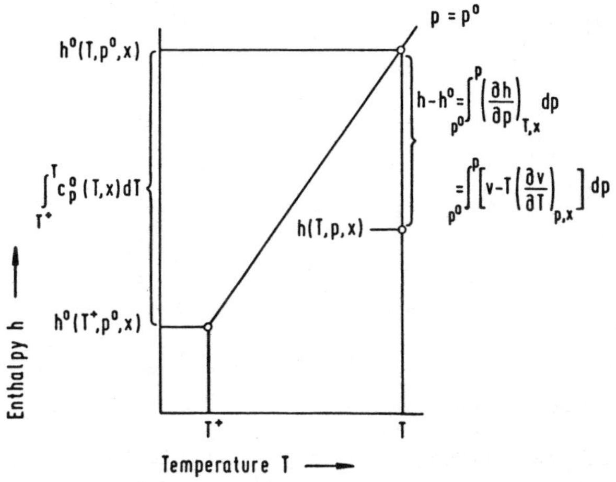

Figure 4. Enthalpy-temperature diagram of fluid with constant composition showing the isobar $p = p^0$ for the standard state (ideal gas) and the isothermal pressure effect at one temperature.

The enthalpy or the molar heat capacity of an ideal gas mixture can be calculated with ideal mixing rules:

$$h^o(T, p^o, x) = \sum x_i h_i^o(T) \tag{20}$$

$$c_p^o(T, x) = \sum x_i c_{p,i}^o(T) \tag{21}$$

Example: Entropy Departure $s - s^0$

The effect of pressure on the entropy can be calculated in accordance with a so-called Maxwell relation by referring to an EQS:

$$\left(\frac{\partial s}{\partial p}\right)_{T,x} = -\left(\frac{\partial v}{\partial T}\right)_{p,x} \tag{22}$$

The entropy in the standard state (ideal gas at $p^0 = 0.1013$ MPa) at the reference temperature $T^+ = 25°C = 298.15$ K is usually listed. The entropy of a mixture can then be calculated:

$$s(T, p, x) = \sum x_i s_i^0(T^+, p^0) + R \sum x_i \ln x_i + \int_{T^+}^{T} \sum x_i c_{p,i}^0(T) d \ln T - \int_{p^0}^{p} \left(\frac{\partial v}{\partial T}\right)_{p,x} dp \tag{23}$$

Example: Fugacity of a Pure Substance or of a Component in a Mixture

Knowledge of the fugacity helps to find important information, e.g., vapor pressures, distribution coefficients, or K-values for components in coexisting phases. Thermodynamic equilibrium in coexisting phases is characterized by equal temperature, pressure and fugacity of each component in all phases.

The fugacity of a single- or multi-component fluid, $f = \varphi p$, and of a component i for a mixture, $f_i = \varphi_i p_i$, can be calculated provided an EQS is known:

$$RT \ln \varphi = \frac{p}{p_0}\left(v - \frac{RT}{p}\right) dp \tag{24a}$$

$$= \int_{\infty}^{v} \left(\frac{RT}{v} - p\right) dv + pv - RT - RT \ln z \tag{24b}$$

$$RT \ln \varphi_i = \int_{p^0}^{p} \left[\left(\frac{\partial V}{\partial n_i}\right)_{T,p,nj \neq i} - \frac{RT}{p}\right] dp \tag{25a}$$

$$= \int_{\infty}^{v} \left[\frac{RT}{v} - n\left(\frac{\partial p}{\partial n_i}\right)_{T,p,nj \neq i}\right] dv - RT \ln z \tag{25b}$$

where φ = fugacity coefficient, $\varphi = 1$ for $p = 0$
φ_i = fugacity coefficient of component i, $\varphi_i = 1$ for $p = 0$
p_i = partial pressure, $x_i p$

Example: Vapor Pressure of a Pure Substance

If vapor and liquid coexist, the temperature, pressure and the chemical potential μ or the fugacity ($f = \varphi p$) and hence the fugacity coefficient φ in both phases are equal

$$f^L(T, p) = f^V(T, p) \tag{26a}$$

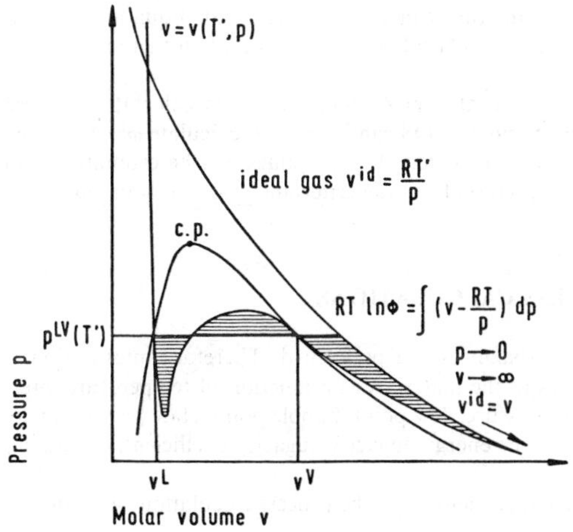

Figure 5. Schematic pressure-volume diagram with subcritical isotherm $T = T' < T_c$ for the ideal gas and for a real fluid. The shaded area represents $\int(v^{real} - v^{ideal})\, dp$ and is an indication for the "non-ideality" of the fluid.

or

$$\varphi^L(T, p) = \varphi^V(T, p) \tag{26b}$$

For a given temperature $T = T'$ a pressure must be found where the EQS has two solutions v^L and v^V for which the fugacities φ^L and φ^V calculated with Equation 24 are equal. As shown in Figure 5 the equilibrium condition $\varphi_L = \varphi^V$ means that the shaded areas between $v^L < v < v^V$ are equal (Maxwell criterion).

Example: Phase Equilibrium in Multicomponent Systems

When mixtures are cooled or heated or separated it is important to know dew or bubble points or compositions and fractions of all coexisting phases. The problems can be solved if the distribution coefficients or the K-values of all components in a mixture are known.

In the thermodynamic equilibrium the fugacities of each component in all coexisting phases are equal.

Example: Conditions for Vapor-Liquid Equilibrium

$$f_i^L = f_i^V \tag{27a}$$

$$\varphi_i^L x_i p = \varphi_i^V y_i p \tag{27b}$$

and

$$K_i = y_i/x_i = \varphi_i^L/\varphi_i^V \tag{28}$$

where x_i = mole fraction, φ_i^L = fugacity coefficient } of component i in liquid

y_i = mole fraction, φ_i^V = fugacity coefficient } of component i in vapor

As fugacity coefficients of components in a mixture can be calculated with Equation 25, it is possible to calculate K-values. The procedures are all "trial and error," and often there are several stacked iterative loops.

The bubble point pressure of a liquid mixture at a given temperature is calculated by first assuming a pressure. With T, p, x, Equation 25 and an EQS can be used to calculate φ_i^L. Assuming $\varphi_i^V = 1$, the vapor mole fractions $y_i = x_i/\varphi_i^L$ are calculated. T, p, y_i values can be estimated from φ_i^V with φ_i^L obtained from $y_i = x_i\varphi_i^V/\varphi_i^L$. The deviation from the criterium $\sum y_i = 1$ is an indication for the next iteration of the pressure.

GENERAL PROCEDURES FOR PROCESS CALCULATIONS

In basic process design, energy- and mass-balances are performed. Therefore information is required about the thermodynamic properties of the materials as a function of temperature, pressure, and composition such as density, vapor pressure, dew point, bubble point, state (vapor, liquid, multiphase), heat capacity, enthalpy, entropy, free energy, fugacity, fugacity coefficient, or equilibrium constant for phase or chemical equilibria.

Nowadays it is popular, most convenient, and efficient to do the process calculations with the aid of electronic computers.

It is possible and opportune to calculate volumetric and caloric properties of a fluid material for the state of an ideal gas first and then correct for the effect of pressure.

The properties in the state of an ideal gas can be calculated with two material equations:

$$pV = nRT \tag{29}$$

where $n = m/M$

$$c_p^0(T) = C_0 + C_1T + C_2T^2 + \cdots \tag{30}$$

For each specific substance the values of M, C_0, C_1, C_2 and the values of h_{293}^0 and s_{298}^0 must be known and stored. For mixtures, ideal mixing rules are valid.

The effect of pressure can be calculated in accordance with Equations 19, 22, 24, and 25. The only material equation required for specific substances or mixtures is an equation of state including mixing rules, viz.

$z = z(T, p, A_{ik})$ for pure substances

or

$z = z(T, p, A_{mk})$ for mixtures

For each specific substance or mixture the coefficients in the equation of state and in the mixing and combination rules must be known and stored. Generalized equations of state GEQS are very practical because only a few specific parameters (e.g., T_c, p_c, v_c, or ω) for each substance and often only one specific parameter (k_{ij}) for each binary combination need be known and stored.

EXAMPLES OF GEQS

Four frequently used and "popular" GEQS were selected. These equations all require 3 parameters to characterize a pure substance. One binary parameter is required to achieve better accuracy in the description of binary or multicomponent mixtures. A detailed review is given by Oellrich et al. [9] and Knapp et al. [10].

Example: RKS Redlich, Kwong, and Soave

A considerable improvement of the accuracy of the van der Waals EQS could be attained by a modification of Redlich and Kwong [11].

Soave [12] has generalized the RK-EQS by suggesting rules for determining a and b from T_c, p_c, and ω.

$$z = \frac{v}{v-b} - \frac{a}{RT(v+b)} \tag{31}$$

where $a = 0.42747\alpha R^2T_c^2/p_c$ (31a)

$$\alpha = [1 + m(1 - \sqrt{T/T_c})]^2 \tag{31b}$$

$$m = 0.48 + 1.574\omega - 0.176\omega^2 \tag{31c}$$

$$b = 0.0866\ RT_c/p_c \tag{31d}$$

For mixtures, a and b are calculated with mixing rules

$$b = \sum x_i b_i \tag{32a}$$

$$a = \sum x_i x_j a_{ij} \tag{32b}$$

and the combination rule

$$a_{ij} = \sqrt{a_i a_j}\,(1 - k_{ij}) \tag{32c}$$

where $k_{ij} = 0$ for $i = j$
k_{ij} for $i \neq j$ can be found by fitting to experimental VLE data

The structure of useful GEQS permits the closed differentiation and integration when calculating departure functions. Real gas properties can be calculated with RKS as follows:

$$h - h_0 = \frac{1}{b}\left(T\frac{da}{dT} - a\right)\ln\frac{v+b}{v} + pv - RT \tag{33}$$

$$\ln\varphi = z - 1 - \ln\frac{p(v-b)}{RT} - \frac{p(v-b)}{b\,RT} - \frac{a}{b\,RT}\left(2\sqrt{\frac{a_i}{a}} - \frac{b_i}{b}\right)\ln\frac{v+b}{v} \tag{34}$$

Example: PR (Peng and Robinson)

The equation was proposed by Peng and Robinson [13] as a modification of the attractive term of the van der Waals EQS:

$$z = \frac{v}{v-b} - \frac{av}{RT(v(v+b) + b(v-b))} \tag{35}$$

Parameters a and b for pure substances can be determined according to general rules

$$a = 0.45724R^2T_c^2\alpha/p_c \tag{35a}$$

$$\alpha = [1 + \kappa(1 - \sqrt{T/T_c})]^2 \tag{35b}$$

$$\kappa = 0.37464 + 1.54226\omega - 0.26992\omega^2 \tag{35c}$$

$$b = 0.07780RT_c/p_c \tag{35d}$$

The mixing and combination rules are the same as for RKS. Departure functions are also closed expressions and can be found in literature.

Example: BWRS (Benedict, Webb, Rubin, and Starling)

Benedict, Webb and Rubin [14] have developed an empirical virial equation with eight coefficients. Sets of coefficients were determined by fitting to volumetric and caloric data of individual substances, mostly hydrocarbons. Starling [15] extended the EQS to 11 coefficients and proposed general rules for determining these coefficients from T_c, ρ_c, and ω.

The EQS gives the pressure as a function of temperature and density:

$$p = RT\rho + \left(B_0RT - A_0 - \frac{C_0}{T^2} + \frac{D_0}{T^3} - \frac{E_0}{T^4}\right)\rho^2 + \left(bRT - a - \frac{d}{T}\right)\rho^3$$
$$+\alpha\left(a + \frac{d}{T}\right)\rho^6 + \frac{c}{T^2}\rho^3(1 + \gamma\rho^2)e^{-\gamma\rho^2} \tag{36}$$

It can be rewritten for the compressibility factor $z = pv/RT$ as a function of temperature and molar volume.

$$z = 1 + \frac{B}{v} + \frac{C}{v^2} + \frac{D}{v^5} + \frac{C'}{v^2}\left(1 + \frac{\gamma}{v^2}\right)e^{-\gamma/v^2} \tag{37}$$

where B, C, D, C′ are abbreviations for the expressions in Equation 36.

The 11 coefficients $B_0 \cdots \gamma$ can be determined by using the following 11 rules from T_c, ρ_c, and ω for individual substances.

$$\rho_c B_0 = A_1 + B_1\omega$$

$$\rho_c A_0/(RT_c) = A_2 + B_2\omega$$

$$\rho_c C_0/(RT_c^3) = A_3 + B_3\omega$$

$$\rho_c^2 = A_4 + B_4\omega$$

$$\rho_c^2 b = A_5 + B_5\omega$$

$$\rho_c^2 a/(RT_c) = A_6 + B_6\omega$$

$$\rho_c^3 \alpha = A_7 + B_7\omega$$

$$\rho_c^2 c/(RT_c^3) = A_8 + B_8\omega$$

$$\rho_c D_0/(RT_c^4) = A_9 + B_9\omega$$

$$\rho_c^2 d/(RT_c^2) = A_{10} + B_{10}\omega$$

$$\rho_c E_0/(RT_c^5) = A_{11} + B_{11}\omega \exp(-3.8\omega)$$

The 2 sets of 11 general constants A_j and B_j were found by simultaneous fitting to volumetric and caloric data of alkanes.

Eleven different mixing rules are given to calculate the eleven coefficients of a mixture.

Example: LKP (Lee, Kesler, and Plöcker)

The volumetric and thermodynamic functions correlated by Pitzer and co-workers [6] in accordance with the 3-parameter-corresponding-states principle originally presented in tables were presented in an analytical form by Lee and Kesler [16]:

$$z = z^{(0)} + \frac{\omega}{\omega^{(r)}}(z^{(r)} - z^{(0)}) \tag{38}$$

The compressibility factors of both a simple and a reference fluid are given as functions of reduced variables $T_r = T/T_c$, $p_r = p/p_c$, and $v_r = p_c v/(RT_c)$.

$$z = pv/RT = p_r v_r/T_r = 1 + \frac{B}{v_r} + \frac{C}{v_r^2} + \frac{D}{v_r^5} + \frac{C_4}{T_r^3 v_r^2}\left(\beta + \frac{\gamma}{v_r^2}\right)e^{-\gamma/v_r^2} \tag{39}$$

where $B = b_1 - \frac{b_2}{T_r} - \frac{b_3}{T_r^2} - \frac{b_4}{T_r^3}$

$$C = c_1 - \frac{c_2}{T_r} + \frac{c_3}{T_r^3}$$

$$D = d_1 + \frac{d_2}{T_r}$$

Experimental data for argon, krypton, and methane were used for the optimization of the 12 "general" constants $b_1, b_2, \ldots, \gamma$ for the simple fluid, and experimental data for octane were used as the reference fluid.

Mixtures are characterized by pseudocritical properties in accordance with three mixing rules suggested by Plöcker, Knapp, and Prausnitz [17]:

$$T_{cm} = \left(\frac{1}{v_{cm}}\right)^{\eta} \sum\sum x_i x_j (v_{cij})^{\eta} T_{cij} \tag{40}$$

where $T_{cij} = \sqrt{T_{ci}T_{cj}}\, k_{ij}$ (40a)
$k_{ii} = k_{ij} = 1$

k_{ij} for $i \neq j$ can be found by fitting to binary VLE data. It can also be correlated for certain groups of substances.

$$v_{cm} = \sum\sum x_i x_j v_{cij} \tag{40b}$$

where $v_{cij} = \frac{1}{8}(v_{ci}^{1/3} + v_{cj}^{1/3})^3$ (40c)

$$v_{ci} = z_{ci}RT_{ci}/p_{ci} \tag{40d}$$

$$z_{ci} = 0.2905 - 0.085\omega_i \tag{40e}$$

$$p_{cm} = RT_{cm}z_{cm}/v_{cm} \tag{40f}$$

$$z_{cm} = 0.2905 - 0.085\omega_m \tag{40g}$$

$$\omega_m = \sum x_i\omega_i \tag{40h}$$

The optimal value of η was empirically found to be 0.25.

The isothermal departure functions showing the effect of pressure on the thermodynamic properties of a fluid can also be calculated in accordance with Equations 19 through 25.

RESULTS OF PROCESS CALCULATIONS

It might be interesting to give a few examples illustrating the application of GEQS for the calculation of thermodynamic properties. The accuracy of results can be presented by showing experimental and calculated data on diagrams.

Example: PVT Properties of Pure Methane

In Figure 6 three different generalized equations of state show good accuracy at low densities. At higher densities only the LK-GEQS represents experimental data well. It is not surprising that the 11 constants for the simple fluid were fitted to methane data.

Example: Vapor Pressure of Pure Ethane

In Figure 7 the original experimental work was done by Loomis and Walters [18]. The results can be represented by the Antoine equation with a standard deviation of 1.5 Torr. The coefficients of the equation are listed in a standard handbook [19]. It is obvious that the RKS–GEQS can

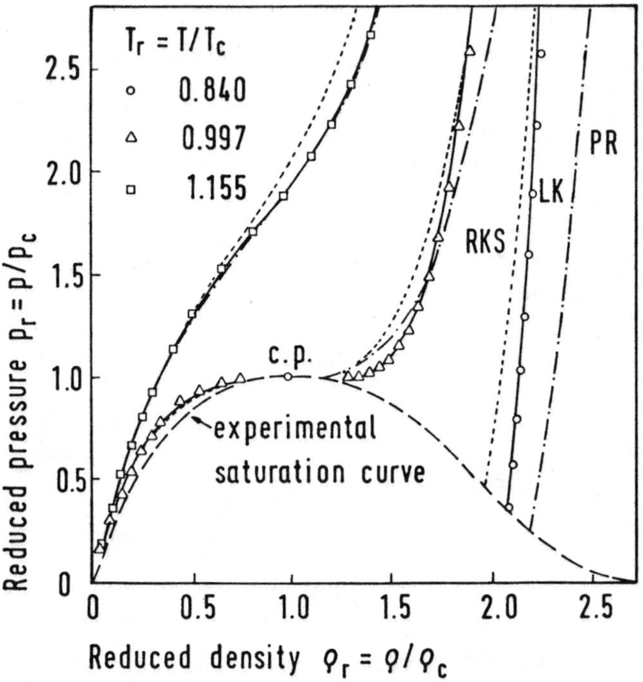

Figure 6. Diagram showing reduced pressure $p_r = p/p_c$ vs. reduced density $\rho_r = \rho/\rho_c$ for pure methane at three reduced temperatures $T_r = T/T_c$. The points are experimental, isotherms were calculated with three generalized equations of state: RKS equation (31), PR equation (35), and LK equation (38).

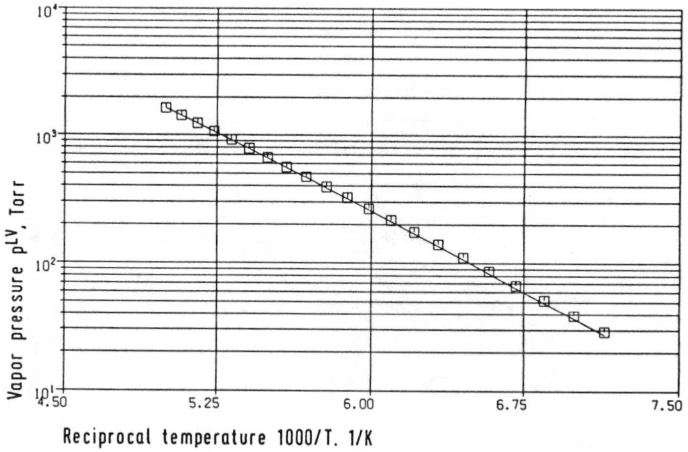

Figure 7. Vapor pressure vs. reciprocal temperature for pure ethane. The curve is based on experimental data interpolated with the Antoine equation. The points are calculated with a bubble point procedure on the basis of RKS-GEQS.

reproduce the vapor pressure satisfactorily although only 2 critical data and the acentric factor must be known to characterize the properties of the pure fluid.

Example: Vapor-Liquid-Phase Equilibrium

The design of separation processes is based on our knowledge about phase-equilibria, i.e., about the K-values. It is important to judge the accuracy of a correlation. In Figure 8 experimental data and calculated isotherms are compared.

Example: Enthalpy-Temperature Diagram

For the design of heat exchangers, Q–T or H–T diagrams contain the information required by the process engineer (see Figure 9). At temperatures below the dew point it is first necessary to determine the amount and the composition of the vapor and liquid fraction. The enthalpy of the two-phase mixture depends mainly on the liquid/vapor ratio i.e., on the result of the flash calculation preceding the enthalpy calculation.

Example: Coefficient of Performance of Heat-Pump Cycles

For process studies for heat-pump operation the coefficient of performance should be known, i.e., the energy offered in the condenser as heat to the energy required in the compressor. By combining all unit-operations of the cycle in a computer program, characteristic curves can be produced provided the caloric properties of the working fluid are known. When a GEQS is used for the PVT properties of the fluid, the results of the calculations can be compared with results obtained with information on h–p diagrams. The deviations in the example shown in Figure 10 were 3%–5%; however, they can also be due to inaccuracies of the diagram.

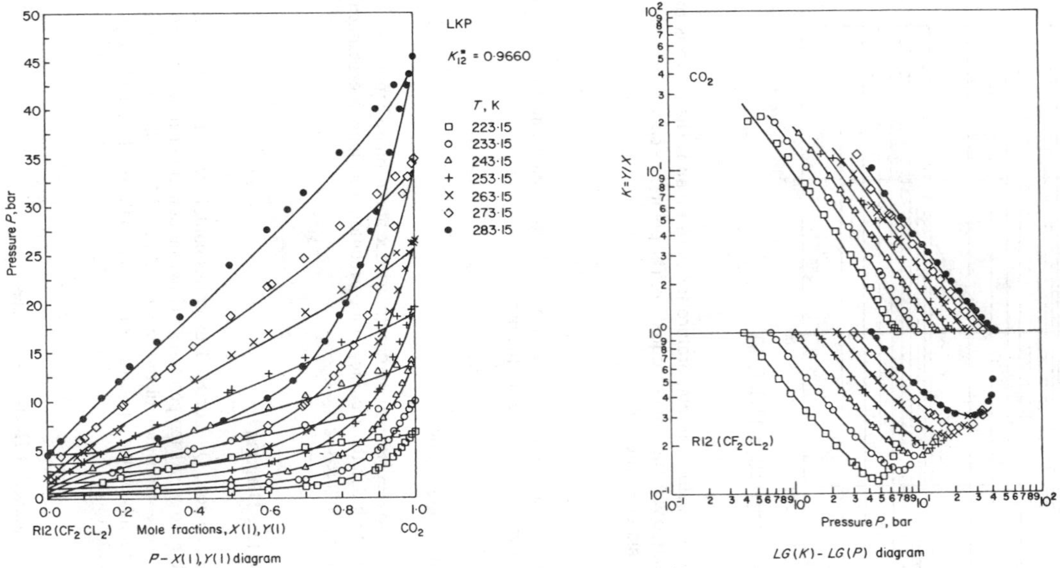

Figure 8. Diagram showing dew- and bubble-point pressure vs. vapor and liquid composition at various temperatures and diagram showing K-values vs. pressure at various temperatures for the binary mixture CO_2-CF_2Cl_2. The points are experimental data taken by Dorau et al [12]; the isotherms were calculated with a flash procedure. The PVTX properties were represented with LKP-EQS.

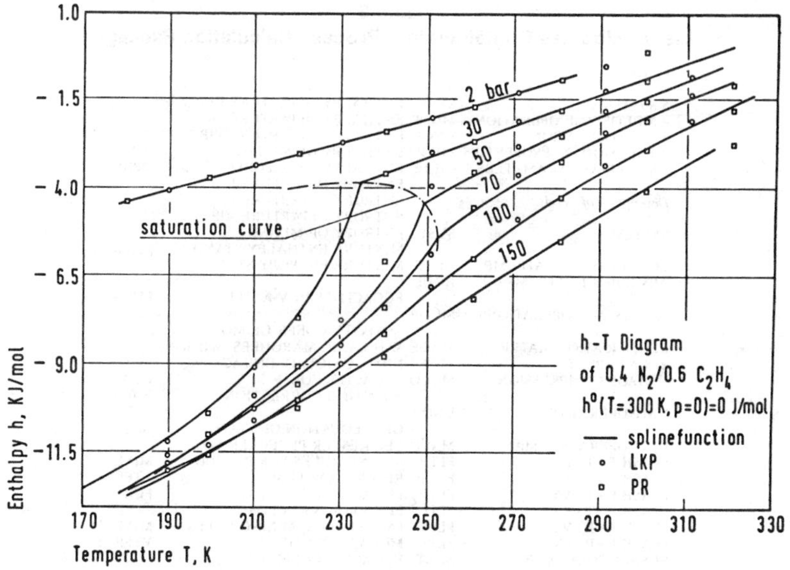

Figure 9. Enthalpy-temperature diagram for a nitrogen-ethylene mixture. The isobars are experimental data taken by Schmid [21] and interpolated by a spline function. The points were calculated with Equation 19 for enthalpy departures on the basis of two GEQS: PR equation (35) and LKP equation (38-40).

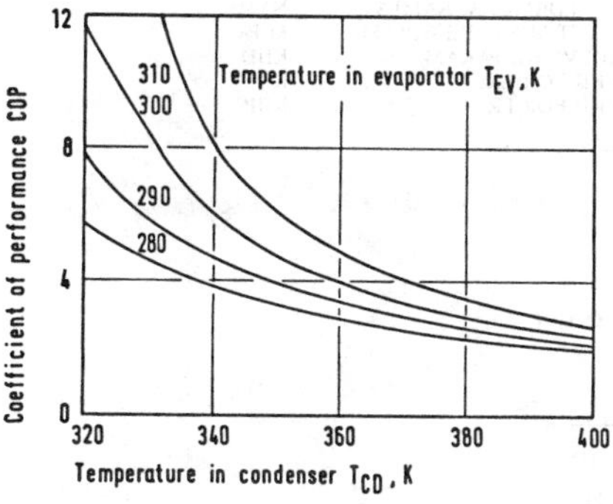

Figure 10. Diagram showing the coefficient of performance of a compression heat pump cycle as a function of condenser and evaporator temperatures. The working fluid is R11, the isentropic efficiency of the compressor 0.75, and the PVT properties of R11 were represented with LKP-EQS.

ADDITIONAL COMMENTS

All the required procedures, most of the thermodynamic relations, many useful correlations for material properties, and many characteristic coefficients are included in process compilers or process calculation packages. Process calculation software is available for small and large electronic computers [22, 23].

The structure of a typical process calculation package can be demonstrated by explaining the contents of a package developed at the Institute of Thermodynamics and Plant Design, TU Berlin/West.

Table 3
List of Modules Contained in a Process Calculation Package

Levels			
5 SYSTEMS	SYST	ENTHALPY DEPARTURE LKP	
4 MULTISTAGE OPERATION	MUOP	ENTHALPY DEPARTURE PR	
3 UNIT OPERATION	UNOP	ENTHALPY DEPARTURE VIR	
2 THERMODYN. PROPERTIES	THEP	ENTHALPY OF MIXING	
1 MATERIAL PARAMETERS	CHAP	ENTHALPY OF VAPORIZ.	DPHS
		ENTROPY DEPARTURE LKP	
Description of Levels.		ENTROPY DEPARTURE PR	
		ENTROPY DEPARTURE VIR	
SYSTEMS	SYST	ENTROPY OF MIXING	
		MIXTURE ENTHALPY STAND.	COTX
COMPRESSOR HEATPUMP	HPCC	MIXTURE ENTROPY STAND.	
ABSORBER HEATPUMP	HPAB		
		FUGACITIES IN V AND L	FUGA
MULTISTAGE OPERATIONS	MUOP		
		ACTIVITY COEFF. GE MODELS	
RECTIFICATION MATRIC	MADE	VAN LAAR, MARGULES, WILSON,	
Y/X-NTS PLOT	STYS	NRTL, UNIQUAC, UNIFAC	ACTI
N-STAGE COMPRESSION	MUCO	FUGACITY COEFF. IN V	FCIV
		POYNTING CORRECTION	POYN
UNIT OPERATION	UNOP		
		GEN. EQUATION OF STATE	GEQS
UNIT OPERATION MAN.	MAUO	LEE KESLER PLOECKER EQS	
FLASH F+T+P	FLTP	MIXING RULES FOR PSEUDO	MIXL
FLASH F+Q+P	FLQP	REDUCED VOLUME	VTPL
FLASH F+T+VY	FLTV	FUGACITY COEFF. LKP	FCIL
FLASH F+T+LX	FLTL	PENG ROBINSON EQS	
FLASH F+P+VY	FLPV	PARAM. A+B, MIXING RULES	MIXP
FLASH F+P+LX	FLPL	MOLAR VOLUME	VTPP
BUBBLE POINT X+T	BUBT	FUGACITY COEFF. PR	FCIP
BUBBLE POINT X+P	BUBP		
DEW POINT T+X	DEWT	COMPONENT PROPERTIES	COPA
DEW POINT P+Y	DEWP	VAPOR PRESSURE	VAPP
ISOBARIC HEAT OR COOL	QTPX	LIQUID VOLUME	VOLT
ADIABATIC COMPRESSION	ADCO	SECOND VIRIAL COEFF.	BMTX
ADIABATIC EXPANSION	ADEX		
ISENTHALPIC THROTTLING	IREX	MATERIAL PARAMETERS	CHAP
		MANAGEMENT FOR PARAM:	MAPA
THERMODYN. PROPERTIES	THEP	REQUEST PARAMETERS	PUDA
		RETRIEVE NAMES	NAME
THERMODYN. EQUILIBRIA	THEQ	ARRANGE SUBSTANCES	NORG
VLE CONSTANT K=Y/X	KVLE	M, TLV, TC, PC, VC, OMEGA	PURE
CHEM. REACT. CONSTANT	KCHE	CHEM. FORMULA, NAMES	NAMS
		REQUEST MIXTURE PARAM.	MIPA
CALORIC PROPERTIES	CALO	RETRIEVE KIJ PARAM.	KIJD
		4000 KIJ FOR LKP	KIJL
MIXTURE ENTHALPY MIXTURE ENTROPY	MACA	220 KIJ FOR PR	KIJP

BPBP = Berliner Prozeß Berechnungspaket

Program modules are arranged in a hierarchical order

Level	Description	Code
4	Multistage Operations	MUOP
3	Single Stage Unit Operations	UNOP
2	Thermodynamic Properties	THEP
1	Material Parameters	CHAP

In Table 3 all modules of the four levels are listed. In this table the significance of GEQS can be easily recognized.

Example: Q–T Diagram of a Process Stream

In this example refer to Figures 11 and 12. Module QTPX in level 3 will provide a management program for the calculation of the enthalpy of a mixture of specified composition as a function

```
Example for the calculation of a cooling curve:   QTPX
===============================================

The calculation can be done for any specified temperature-intervals.

         (Heat-exchanger)                                  (Phase-seperator)


INPUT: T           T-intervals          T                ********
       P                                                 *      *
       F                                                 *      *
       z  ***************************                    *      *
          *                         *                    *      *
  =======>*=========================*===============>*      *=====>
          *                         *                    *      *
          ***************************                    *______*
                                                         *      *
                                                         *      *
                                                         ********
OUTPUT:                 +/- Q                                L       V
                                                             x       y

                        UNITOPERATION: QTPX
 ============================================================

         SUBSTANCES OF THE 5-COMPONENT-MIXTURE

          NUMBER     COMP.NR.  NAME                           FORMULA

            1          116     NITROGEN                       N2
            2           22     METHANE                        CH4
            3           23     ETHANE                         C2H6
            4           24     PROPANE                        C3H8
            5           25     N-BUTANE                       C4H10


      THE CALCULATION WAS DONE WITH PR-EQS


   T (K)      P (BAR)     V/F (-)

  300.00       40.000      1.0000


      Z(I)       F*Z       X(I)       L*X       Y(I)       V*Y       K(I)

 1 .500E-01 .500E-010.        0.         .500E-01 .500E-010.
 2 .860E+00 .860E+000.        0.         .860E+00 .860E+000.
 3 .600E-01 .600E-010.        0.         .600E-01 .600E-010.
 4 .200E-01 .200E-010.        0.         .200E-01 .200E-010.
 5 .100E-01 .100E-010.        0.         .100E-01 .100E-010.
           ---------        ---------             ---------
           .100E+01         .100E-05              .100E+01

          M (G/MOL)    V (CCM/MOL)    H (J/MOL)   Q (J)

F         .1847E+02     .5627E+03     .9452E+04    0.
L        0.            0.            -.1370E+05
V         .1847E+02     .5627E+03     .9452E+04
```

Figure 11. Printout of results of module QTPX. The calculations can be done at specified temperature intervals. However, only three points at T = 300,200, and 130 K are printed as examples.

```
      THE CALCULATION WAS DONE WITH PR-EQS

   T (K)      P (BAR)      V/F (-)

  200.00       40.000        .8129

     Z(I)      F*Z        X(I)       L*X        Y(I)       V*Y        K(I)

 1 .500E-01 .500E-01 .124E-01 .232E-02 .587E-01 .477E-01 .472E+01
 2 .860E+00 .860E+00 .662E+00 .124E+00 .906E+00 .736E+00 .137E+01
 3 .600E-01 .600E-01 .183E+00 .343E-01 .317E-01 .257E-01 .173E+00
 4 .200E-01 .200E-01 .911E-01 .170E-01 .363E-02 .295E-02 .399E-01
 5 .100E-01 .100E-01 .514E-01 .962E-02 .469E-03 .381E-03 .912E-02
           ---------           --------           ---------
            .100E+01            .187E-00           .813E+00

          M (G/MOL)    V (CCM/MOL)      H (J/MOL)     Q (J)

F          .1847E+02     .2270E+03      .3424E+04     .6028E+04
L          .2348E+02     .5287E+02     -.2561E+04
V          .1731E+02     .2671E+03      .4802E+04

      THE CALCULATION WAS DONE WITH PR-EQS

   T (K)      P (BAR)      V/F (-)

  130.00       40.000        .0000

     Z(I)      F*Z        X(I)       L*X        Y(I)       V*Y        K(I)

 1 .500E-01 .500E-01 .500E-01 .500E-010.             0.         0.
 2 .860E+00 .860E+00 .860E+00 .860E+000.             0.         0.
 3 .600E-01 .600E-01 .600E-01 .600E-010.             0.         0.
 4 .200E-01 .200E-01 .200E-01 .200E-010.             0.         0.
 5 .100E-01 .100E-01 .100E-01 .200E-010.             0.         0.
           ---------           --------           ---------
            .100E+01            .100E+01           .100E-05

          M (G/MOL)    V (CCM/MOL)      H (J/MOL)     Q (J)

F          .1847E+02     .3574E+02     -.4918E+04     .1437E+05
L          .1847E+02     .3574E+02     -.4918E+04
V        0.            0.              -.6388E+04
```

Figure 12. (Continued)

of temperature, in specified temperature intervals at a specified pressure. If the option PR–EQS is selected, the enthalpy departure $h-h^0$, i.e., module HDEP in level 2, will be calculated with the chosen GEQS. Modules EQSL, MIXL, VTPL, and FCIL will be called in combination with FLTP to find the state of the mixture at each process point. MACA will call the modules for the standard enthalpy of the mixture and add the enthalpy departure. From level 1 the modules MAPA, PURE, MIPA, and KIJL will supply all parameters that are needed in level 2 as coefficients of the specified component of the mixture.

Figures 11 and 12 show printouts of QTPX for a 5-component mixture being cooled from 300° to 130°K.

References 19, 22, and 23 provide further headings.

Generalized equations of state can be very useful in describing analytically the PVTX behavior of pure substances and of mixtures. GEQS not only give information about the specific or molar

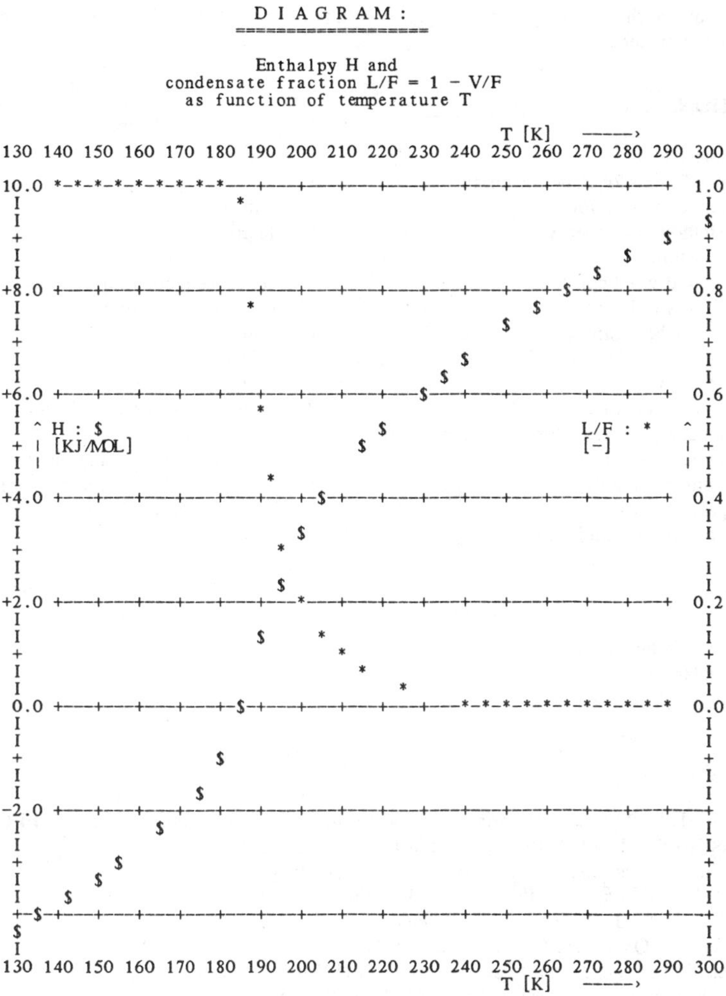

Figure 12. Print-plot showing the enthalpy H and the condensate fraction L/F (liquid/feed) as a function of temperature for natural gas with a composition specified in Figure 11. The print plot illustrates the results of a QTPX calculation.

volume as a function of temperature, pressure, and composition but also help to calculate departure functions, i.e., the effect of pressure on thermodynamic properties such as enthalpy, entropy, and fugacity.

Beginning with our knowledge about the property in the state of the ideal gas, all important thermodynamic properties at higher pressures (i.e., higher densities) can be calculated as long as the validity of the GEQS can be accepted. In many cases the PVTX information of the GEQS might not be very exact, but it is the only available information. The responsible process engineer has to be aware that at critical process points results of predictive methods must be checked by experimental evidence.

NOTATION

A_k	coefficient
A_{mk}	coefficients for a specific mixture of constant composition
A^+	Helholtz free energy
a, b	constants
B(T)	second virial constant
C(T)	third virial constant
C_P^0	molar heat capacity in ideal gas state
f_1, f_2	functions of reduced densities
h	enthalpy
h^0	standard enthalpy for ideal gas
k_{ij}	binary parameter for the combination of component i and j
n	number of moles
p_i	partial pressure
p^0	pressure at standard state
p	pressure
R	universal gas constant
s	entropy
T	temperature
T_r	reduced temperature
T^+	reference temperature
T_c	critical temperature
V	volume
v	molar volume
v_r	reduced volume
v^*	volume of spherical molecules
x	concentration
x_i	mole fraction of component i in liquid
y_i	mole fraction of component i in vapor
z	compressibility factor

Greek Symbols

γ	specific heat ratio
ρ	molar density
ρ^*	reduced density
φ	fugacity coefficient
ω	acentric factor

REFERENCES

1. Leland, T. W., "Phase Equilibria and Fluid Properties in the Chemical Industry," EFCE Publ. Series No. 11. (1980) Dechema, Frankfurt-Main, FRG, 281–332.
2. Martin, J. J., *Ind. Eng. Chem. Fundam.* Vol. 18, (1979), p. 81.
3. Vidal, J., *Fluid Phase Equilibria*, Vol. 13 (1983), pp. 15–34.
4. van der Waals, J. D., Dissertation, Leiden, (1823).
5. Kamerlingh Onnes, H., *Comm. Phys. Lab. Leiden*, Vol. 71 and 74 (1901).
6. Pitzer, K. S., et al., *J. Am. Chem. Soc.*, Vol. 77 (1955), p. 3433.
7. Carnahan, N. F., and Starling, K. F., *AIChE j.*, Vol. 18 (1972), p. 184.
8. van der Waals, J. D., Verband. Kon. Akad. Wetensch., Amsterdam 20 (1880).
9. Oellrich, L., et al., *Intern. Chem. Eng.*, Vol. 21, No. 1 (1981), pp. 1–17.
10. Knapp, H., et al., "Vapor-Liquid Equilibria for Mixtures of Low Boiling Substances", Chemistry Data Series VI, Dechema, Frankfurt-Main, FRG., (1982).
11. Redlich, O., and Kwong, J. N. S., *Chem. Reviews*, Vol. 44 (1949), p. 233.
12. Soave, G., *Chem. Eng. Sci.*, Vol. 27 (1972), p. 1197.
13. Peng, D. Y., and Robinson, D. B., *Ind. Eng. Chem. Fundam.*, Vol. 15 (1976), p. 59.
14. Benedict, M., Webb, R. B., Rubin, L. L., *J. Chem. Phys.*, Vol. 8 (1940), p. 314.
15. Starling, K. F., *Hydroc. Proc.*, Vol. 50 (1971), p. 101; and subsequent articles in *Hydroc. Proc.*
16. Lee, B. L., and Kesler, M. G., *AIChE J.*, Vol. 21 (1975), pp. 510, 1040.
17. Plöcker, U., Knapp, H., and Prausnitz, J. M., *Ind. Eng. Chem. Proc Des. Dev.*, Vol. 17 (1978), p. 324.
18. Loomis, A. G., and Walter, J. E., *J. Am. Chem. Soc.*, Vol. 48 (1926), p. 2051.
19. Reid, J. C., Prausnitz, J. M., Sherwood, T. K., *The Properties of Gases and Liquids*, 3rd ed., McGraw Hill Inc., New York (1977).

20. Dorau, W., Al-Wakeel, I. M., and Knapp, H., *Cryogenics*, Vol. 1 (1983), p. 29–36.
21. Schmid, O., Disseration, TU-Berlin, (1980).
22. Eckermann, R., "Phase Equilibria and Fluid Properties in the Chemical Industry," EFCE Publ. Series No. 11. (1980), Dechema, Frankfurt-Main, FRG. 739–758.
23. Neumann, K. K., and Futterer, E., "Phase Equilibria and Fluid Properties in the Chemical Industry," EFCE Publ. Series No. 11, (1980), Dechema, Frankfurt-Main, FRG. 807–820.
24. Curl, R. F., and Pitzer, K. S., *Ind. Eng. Chem.*, Vol. 50 (1958), p. 265.
25. Starling, K. F., *Fluid Thermodynamic Properties for Light Petroleum Systems*, Gulf Publ. Comp., Houston (1973).

CHAPTER 2

DENSITY CORRELATIONS FOR SATURATED AND COMPRESSED LIQUIDS AND LIQUID MIXTURES

George H. Thomson

Technical Systems Development
Phillips Petroleum Co.
Bartlesville, Oklahoma, USA

CONTENTS

INTRODUCTION

Uses of Liquid Density

Density is the basic relationship between the mass of a substance and its volume. It is widely used in science and engineering to convert mass into volume and vice versa. Density is also used in the calculation of values of many physical and thermodynamic properties because it reflects the intermolecular forces in matter. Engineers use densities to size pumps, tanks, and pipes, and to calculate the mass of liquid flowing through a pipe from its volume. The use of this sort of calculation to determine the mass of liquid that has passed from the seller to the buyer of a product is called custody transfer, and is often the basis for the exchange of considerable sums of money.

Definition

Density, ρ, is defined as the mass (or, loosely, the weight) of a given quantity of a substance divided by the volume of that quantity. The traditional scientific unit is grams per cubic centimeter. Engineers often use pounds per cubic foot, and the SI unit is kilograms per cubic meter. Specific volume is often used instead of density. It is the reciprocal of density and has such units as cubic feet per pound. Molar volumes are widely used in chemistry and chemical engineering. Typical dimensions are cubic feet per pound mole and liters per gram mole. A number of other units of density and specific volume are used in various branches of science and engineering. Only two conversion ratios will be given here:

$$1\ \text{g/cm}^3 = 62.42797\ \text{lb/ft}^3$$
$$= 1{,}000\ \text{kg/m}^3$$

Specific gravity is defined as the density of a liquid at some temperature t divided by the density of water at temperature t′. The two temperatures may be the same. Thus, specific gravity 60°F/60°F means that both densities are at 60°F (15°C, 20°C, and 25°C are also common reference temperatures). The symbol d_4^{20} indicating the density of a liquid at 20°C compared to the density of water at 4°C, really describes a specific gravity. The temperature 4°C is used because the density of water at that temperature is very close to one gram per cubic centimeter. API gravity, commonly used in the petroleum industry, has units of degrees API, and is defined as (141.5/SG)—131.5, where SG is the 60°F specific gravity of the liquid (60°F/60°F). A gravity of 10° API corresponds to a specific gravity of one, but the scales increase in opposite directions, i.e., a denser liquid has a higher specific gravity but a lower API gravity.

A discussion of liquid densities may be divided into two parts, one dealing with the densities of saturated liquids and the other dealing with compressed liquids. A saturated liquid is a liquid in equilibrium with its vapor at a given temperature. For a pure liquid this means that the vapor over the liquid has the same composition as the liquid (no air is present) and a pressure equal to the vapor pressure of the liquid at that temperature. For a mixture this means that the liquid is in equilibrium with vapor of equilibrium composition (usually not the same composition as that of the liquid) at a pressure equal to the bubble-point pressure of the liquid at the given temperature. The bubble-point pressure is the pressure at which the first bubble of vapor appears at a given temperature. Densities of liquids at these conditions are sometimes called orthobaric or bubble-point densities.

A compressed liquid is contained by a pressure greater than the vapor pressure or the bubble-point pressure of the liquid at the prevailing temperature. These liquids are sometimes called subcooled liquids and have subcooled densities. The effect of the composition and the pressure of the vapor on the liquid density is small when the temperature is well below the normal boiling point of the liquid.

Another complication is that some liquids are polar. Densities of polar liquids, like many of their properties, do not exhibit the same temperature and pressure behavior that densities of nonpolar liquids do. For this reason density correlations have been devised specifically for polar liquids. These correlations may or may not describe densities of nonpolar liquids well, but they are always more complex, usually including one or more additional parameters derived specifically for polar substances. Polar liquids are described in more detail under "Polar Liquids" later in this chapter.

Expected Accuracy of Correlations

The apparent accuracy of any correlation depends on the set of experimental data against which it is tested. A correlation for nonpolar liquids will predict densities of alcohols poorly; a correlation for polar liquids may not predict densities of hydrocarbons well. Although references to, and data bases containing, experimental data are sometimes shared, to a considerable extent each evaluator tests against his own data base. For this reason magnitudes of errors and sometimes even the ranking of correlations will vary from one comparison to another. For example, one correlation gave 0.58% average absolute error when tested against hydrocarbons and 2.43% when tested against alcohols. This is partly because far more accurate values of required parameters, such as critical constants and acentric factors, have been reported for hydrocarbons than for other classes of organic compounds. Another reason is that correlations for nonpolar substances, such as hydrocarbons, are simpler and more accurate than correlations for polar substances.

Tables 1 and 2 give approximate magnitudes of the errors to be expected when the correlations discussed in this chapter are applied to the specified type of liquid. These figures are somewhat misleading, however, in that they imply that compressed liquid density correlations for nonpolar mixtures are more accurate than saturated liquid density correlations for the same mixtures. This is probably because the compressed density correlations were tested against smaller, less varied data bases.

Table 1
Expected Percentage Errors for Saturated Liquid Density Correlations

Type of liquid	Pure, nonpolar	Pure, polar	Nonpolar mixtures	Polar mixtures
Expected Accuracy	0.4–0.8	0.4–1.2	1.4–2.5	1.5–4.0

Table 2
Expected Percentage Errors for Compressed Liquid Density Correlations

Type of liquid	Pure, nonpolar	Pure, polar	Nonpolar mixtures	Polar mixtures
Expected Accuracy	0.5–1.3	2.0–3.0	1.0–2.0	7.0 up

General References

Liquid density correlations have been reviewed by Reid and Sherwood [1], Reid, et al. [2], Spencer and Danner [3], Spencer and Adler [4], and Rea, et al. [5]. T. E. Daubert and R. P. Danner of the Chemical Engineering Department at Pennsylvania State University have been evaluating liquid density correlations for the API Technical Data Book-Petroleum Refining and the American Institute of Chemical Engineers' Design Institute for Physical Properties for some time, but their results are confidential to members of the API and supporters of the DIPPR project.

GENERAL PRINCIPLES

Basic Variables

The object of any liquid density correlation is to describe how the density of a liquid changes as the temperature, pressure, composition, or any combination of these variables changes. The change in density with temperature is determined by the coefficient of thermal expansion of the liquid:

$$\alpha = \frac{1}{V}\left(\frac{\partial V}{\partial T}\right)_{P,x} \tag{1}$$

Similarly, the change in density with pressure depends on the compressibility of the liquid:

$$\beta = -\frac{1}{V}\left(\frac{\partial V}{\partial P}\right)_{T,x} \tag{2}$$

The change in density with composition is related to the partial molar volumes of the components of the mixture, $\partial V/\partial x_i$. For one mole of a binary mixture,

$$V = x_1\left(\frac{\partial V}{\partial x_1}\right)_{T,P,x_2} + x_2\left(\frac{\partial V}{\partial x_2}\right)_{T,P,x_1} \tag{3}$$

For the simplest case, the ideal mixture,

$$V = \sum_i x_i v_i \tag{4}$$

and

$$\left(\frac{\partial v}{\partial x_i}\right)_{T,P,x_j} = \bar{V}_i \tag{5}$$

where x_i are the mole fractions of the components, v_i are the molar volumes of the pure substances, and $\bar{V}_i$ are their partial molar volumes in the mixture.

These three quantities, α, β, and the partial molar volumes change in different ways as temperature, pressure, and composition change, so the final form of a liquid density correlation depends strongly on the functional forms chosen to represent these changes, the assumptions that are made concerning them, and the details of the physical models used. Traditionally, however, correlations for liquid densities have been empirical rather than theoretical. Because the liquid state is more difficult to describe than the gaseous state or the solid state, theoretical liquid density correlations have been too inaccurate, too complicated, too difficult to use, too limited in temperature, pressure, or composition range, or too limited in the number or type of components allowed to be useful. This situation seems to be changing fairly rapidly, however, since there have been significant recent advances in the application of the theory of corresponding states and renormalization group theory, as well as statistical mechanics in general, to the liquid state.

The effects of changes in temperature, pressure, and composition are often treated separately, so that we have saturated liquid density equations that attempt to describe the behavior of pure liquids or mixtures along the saturation (liquid-vapor equilibrium) line as the temperature changes. The effect of changes in composition is often calculated at this point. Many compressed liquid density correlations then start with the saturated liquid at the temperature of interest and attempt to describe the effect of pressure on the density of that liquid while the temperature remains constant.

Factor Models

Liquid density correlations may be classified as compositional or non-compositional, depending on whether composition enters into the correlation explicitly. Non-compositional correlations are often used for complex mixtures, such as crude oils and petroleum products (gasolines, diesel fuels, etc.). Hankinson [6] has called these correlations "factor models." Factor models give the density of the liquid at the required temperature or pressure as a function of a reference density for that liquid. The reference density is usually a measured value and is often corrected to some reference temperature such as 60°F. The use of a reference density automatically accounts for the composition of the liquid so that the factor model has to account only for changes in temperature or pressure. Volume correction factors and compressibility factors are used for these purposes. Two specific factor models will be discussed under "Crude Oils and Petroleum Products" later in this chapter.

Compositional Models

Saturated Liquid Density Correlations

Simple functions. The earliest liquid density correlations were probably fairly simple equations representing the change in density or specific volume with temperature. Liquid densities are often almost linear functions of temperature over small ranges of temperature far from the critical point of the liquid. As the temperature approaches the critical temperature, however, the density decreases sharply. This behavior, almost linear at low temperatures with a sharp decrease at high temperatures, is difficult to fit with a simple mathematical function, so correlations of this kind were not very successful.

Corresponding states correlations. Many saturated density correlations are based on the law of corresponding states. In its original form this law stated that substances that were in "corresponding states" had the same physical and thermodynamic properties. Pure substances are in corresponding states when they have the same values of any two of the three variables—reduced temperature, reduced pressure, and reduced volume. Reduced temperature, T_r, is the ratio of the absolute temperature of a substance to its absolute critical temperature. Similarly, $P_r = P/P_c$, $V_r = V/V_c$, and $\rho_r = \rho/\rho_c$. Any two of these quantities are sufficient to specify the state of a pure substance, because its density and specific volume depend only on the temperature and the pressure.

Pitzer et al. [7] extended the law of corresponding states with the addition of a third parameter, the acentric factor, defined by the equation

$$\omega = -\log P_r(\text{at } T_r = 0.7) - 1.000 \tag{6}$$

Values of ω range from roughly zero to perhaps 1.5 or 2 for different liquids. The acentric factor was to account for the difference in behavior of spherical and non-spherical molecules. This model is often called the extended law of corresponding states or the three-parameter corresponding states model. Saturated liquid density correlations of this type are often cast in the form

$$V_r = V_r^0 + \omega V_r^1 \tag{7}$$

where both V_r^0 and V_r^1 are functions of T_r, and are independent of the nature of the liquid. V_r^0 is supposed to describe the temperature-volume behavior of spherical molecules, while ωV_r^1 describes the difference in behavior of spherical and non-spherical molecules. Although there is a theoretical base for the law of corresponding states, the forms of the temperature functions used for V_r^0 and V_r^1 are usually empirical.

A simple illustration of the hierarchy described above is given by the following equations. Here, f(x) and g(x) are unspecified functions of x, V can be replaced with $1/\rho$, and V_r with $1/\rho_r$ or its equivalent, ρ_c/ρ.

$V = f(T)$, volume or density expressed as a function of temperature
$V_r = f(T_r)$, simple or two-parameter corresponding states
$V_r = f(T_r) + \omega g(T_r)$, extended or three-parameter corresponding states

Two other extended corresponding states methods have been used for correlation of densities as well as other properties of liquids. The first has been called an extended corresponding states conformal liquid model. The second is the two-fluid corresponding states procedure. These are discussed later in this chapter under "Extended Corresponding States Conformal Liquid Method" and "Two-Fluid Corresponding States Method."

A reduced volume or reduced density calculated using any correlation must be multiplied by the critical volume or the critical density in order to obtain a saturated density. Critical volume, however, is usually known less accurately than critical temperature or critical pressure, hence a correlation for reduced density will probably be less accurate and less widely applicable than one for reduced temperature or reduced pressure. To avoid this problem the critical volume, for example, is often replaced with some other parameter whose value is close to that of V_c, but not equal to it. The value of this new parameter is determined by fitting the correlation to experimental liquid densities. This procedure replaces a relatively inaccurate critical parameter with a parameter regressed from more accurate experimental data. This approach is not entirely adequate, however, because substances behave differently very close to their critical points. The theory of the critical state has been extended significantly with the development of renormalization group theory.

Compressed Liquid Density Correlations

Many compressed liquid density correlations express the compressed density as a deviation from a saturated density:

$$V = V_s[1 - f(P)] \tag{8}$$

Here, f(P) is some function of pressure which may be based on the compressibility of the liquid or may be an empirical function. The accuracy of the compressed densities obtained using a correlation of this type depends not only on the accuracy of the function used for f(P), but also on the accuracy, or at least the suitability, of the correlation used for V_s. It is possible that a particular V_s equation and a particular f(P) could give fairly large errors if used with other V_s and f(P) equations, but smaller errors if used together. The importance of the saturated liquid density correlation is shown by one author's finding that the errors obtained using a particular compressed density equation dropped from 2.98% to 1.39% when one saturated density correlation was substituted for another. Regardless of all these factors, however, compressed liquid density correlations are less accurate than saturated density correlations.

Properties of Correlating Equations and Parameters

A correlation that contains only one or two specific constants for each substance, such as the Rackett equation with its Z_{RA} (see "Saturated Liquid Density Correlations" later in this chapter) has an advantage over correlations that require several constants for each substance in that, obviously, there are fewer constants whose values must be determined. Correlations that require several constants for each substance, such as the Hankinson-Thomson correlation (discussed later in this chapter) have the advantage, however, that they are more flexible and can be "adjusted" to fit density-temperature curves of different shapes. At the same time, it is desirable that the correlating parameters have some physical significance, and, generally, the more parameters used the less their physical meaning. It is often desirable, however, to model the entire P-V-T surface of one substance quite accurately so that it may be used as the reference fluid in the extended corresponding states conformal liquid method, which is discussed later in this chapter, for example. In this case thirty to forty constants are often used and there is no pretense that they have any physical significance. They are sometimes designated as "curve-fit constants."

There is another sort of trade-off in the matter of physical significance of parameters. The critical constants of a pure substance define its properties at the point where the properties of the liquid are identical with the properties of the vapor. A critical compressibility factor or a critical volume that is optimized to reproduce liquid densities is no longer a true critical constant and should be given a different name and symbol. Values of acentric factors, however, are often determined from values of physical properties, such as vapor pressures or vapor-liquid equilibria. Now it becomes necessary to decide whether one prefers one set of acentric factors that reproduces several types of physical property data fairly well or a separate set of acentric factors for each property. Examples of these two alternatives are the Hankinson-Thomson correlation (discussed later in this chapter) in which acentric factors determined from vapor-liquid equilibrium calculations are used to predict densities, and Teja's method (also discussed later) in which the usual acentric factors calculated from vapor pressure measurements are replaced with acentric factors optimized to reproduce densities of liquified natural gas (LNG).

Mixtures

In principle one could calculate the density of a mixture by combining the densities of the pure components in some way. A simple way to do this is to assume that the mixture is ideal, that is, that the volume of the mixture is the mole fraction average of the volumes of the components, so that $V_m = \sum_i x_i V_i$. In fact real liquids usually behave differently. Small molecules squeeze in between larger molecules, for example, or polar molecules arrange themselves in a rather loosely-packed structure. A more useful approach is to assume that there is a hypothetical pure liquid whose properties are the same as those of the mixture, and whose characterizing parameters, critical properties, and acentric factor, for example, can be calculated from the corresponding parameters of the pure components using some sort of mixing rules. The pure-component characterizing parameters are almost always combined in pairs. An example is

$$T_{cij} = (T_{ci} T_{cj})^{1/2} \tag{9}$$

where i refers to one component and j to the other. These pair parameters are then combined to give the mixture characterizing parameters using some mixing rule, such as

$$T_{cm} = \sum_i \sum_j x_i x_j T_{cij} \tag{10}$$

The simplest mixing rules, sometimes called Kay's rules, are just mole fraction averages of the pure-component parameters, for example, $\omega_m = \sum_i x_i \omega_i$. Although there is some theoretical basis for selecting mixing rules, a common approach seems to be to guess and try.

Interaction Parameters

It is often desirable to fit experimental liquid mixture densities more accurately than is possible with most liquid density correlations in their pristine form. This is usually done by introducing interaction parameters that account for specific intermolecular interactions between the components of a mixture. Interaction parameters are obtained by fitting the equations for the model under consideration to one or more experimental densities and calculating a value for the interaction parameter that makes the model fit more accurately. These interaction parameters are almost always binary parameters. This means that they are calculated from densities of binary mixtures and compensate for the interaction between two different kinds of molecules.

One common way of using interaction parameters is as multipliers as in the extended corresponding states conformal liquid correlation:

$$V_{cij}T_{cij} = \xi_{ij}(V_{ci}T_{ci}V_{cj}T_{cj}) \tag{11}$$

where ξ_{ij} is the interaction parameter. Another common formulation is multiplication by the term $(1 - k_{ij})$ as in

$$T_{cm} = \sum_i \sum_j x_i x_j (V_{ci}T_{ci}V_{cj}T_{cj})^{1/2}(1 - k_{ij}) \tag{12}$$

where k_{ij} is the interaction parameter. In the first formulation the values of the interaction parameter are usually close to one; in the second they are usually quite small so that the value of the term $1 - k_{ij}$ is close to one.

Polar Liquids

It is often convenient to classify liquids as polar or nonpolar. A polar liquid is one whose molecules have a more positive part and a more negative part. A more formal definition is that the center of positive charge of the molecule does not coincide with the center of negative charge. Organic nitriles like acetonitrile, CH_3CN, are polar because the CN group is quite negative. Polar molecules tend to align themselves positive end to negative end. This alignment makes their densities different from those of nonpolar molecules. An additional complication is that some polar molecules form hydrogen bonds. In a hydrogen bond a hydrogen atom chemically bonded to an electron-attracting atom, such as an oxygen, nitrogen, fluorine, or chlorine atom, in one molecule is weakly attracted to a similar atom in a different molecule. Hydrogen bonds are intermolecular bonds, which are relatively weak compared to normal chemical bonds, but are strong enough to influence the physical properties of the substance in which they occur, particularly at lower temperatures. Pure liquids containing hydrogen bonds, such as water and the alcohols, are often called self-associating liquids.

Equations of State

Equations of state are representations of the pressure-volume-temperature properties of a substance. The simplest is the ideal gas equation of state, $PV = nRT$. Some other common equations of state are the Benedict-Webb-Rubin, the Soave-Redlich-Kwong, and the Peng-Robinson equations. Equations of state are widely used to calculate physical and thermodynamic properties of gases and liquids. Although these equations are often used to calculate densities of gases and sometimes densities of liquids, they generally give considerably less accurate liquid densities than the correlations discussed here. They seldom give errors smaller than three to five percent, while most good liquid density correlations give errors considerably smaller than three percent.

Trends

As industry becomes more concerned with heavier (higher boiling) materials, such as heavy crude oils, shale oils, and tar sand oils, engineers have attempted to extend old correlations to these new materials, One result has been the recognition that the critical state is not a possible real condition for these materials because they decompose ("crack") before they reach their critical temperatures. This has lead to the treatment of critical constants as adjustable parameters optimized to reproduce one or more kinds of physical property data, such as densities and liquid-vapor equilibrium data, to the development of new correlations for these adjustable critical constants, which give better predictions for the new materials, and to the development of physical property correlations based on other, more realistic parameters.

CRUDE OILS AND PETROLEUM PRODUCTS

Effect of Temperature

The effect of temperature on the densities of crude oils and petroleum products is described by the factor model on which API Standard 2540, IP Standard 200, ASTM Standard D1250, and ISO/R91 are based [8]. This model was developed using the results of extensive, precise measurements of densities of domestic crude oils, foreign crude oils, and petroleum products made at the National Bureau of Standards. The model is partly compositional in that the materials to which it applies are divided into five classes: crude oils, gasolines, jet fuels, fuel oils, and lubricating oils, each of which has a different coefficient of thermal expansion. The equations for the model are

$$\frac{\rho}{\rho_T} = \exp[-\alpha_T \Delta t(1.0 + 0.8\alpha_T \Delta t)] \quad (13)$$

$$\alpha_T = (K_0 + K_1\rho_T)/\rho_T^2 \quad (14)$$

In these equations ρ is the desired density at some temperature, t, ρ_t is the reference density at temperature T, Δt is t − T, and K_0 and K_1 are constants. All temperatures are in degrees Fahrenheit. Specific extension of the model has been made for lubricating oils and gasoline-alcohol mixtures, and provision has been made for application of the model to other materials, such as coal liquids and shale oils, which have different coefficients of thermal expansion. The model and its development have been reviewed by Hankinson, et al. [9].

Effect of Pressure

The factor model that describes the effect of pressure on crude oils and petroleum products is given in API Standard 1101, "Measurement of Petroleum Liquid Hydrocarbons by Positive Displacement Meter" [10]. This equation gives volume under pressure, V_h, as a function of saturated or equilibrium volume, V_e, the higher pressure P_h, and the saturation or equilibrium pressure, P_e. The pressures are in pounds per square inch gauge. The equation is

$$V_h = V_e[1 - (P_h - P_e)F] \quad (15)$$

Using terms defined in Equation 8, the equation may be written

$$V = V_s[1 - f(P)] \quad (16)$$

which shows more clearly that it is one of the correlations that give compressed volumes as a deviation from saturated volume. Values of the function F are obtained either from graphs giving them as a function of API gravity and Fahrenheit temperature, or from a table corresponding to the graphs. F is called a compressibility factor and has the dimensions percent per 1,000 psi pressure difference. New values of F based on new density data were approved in 1983.

Effect of Composition

The reference density used in factor models accounts for all composition and mixing effects.

SATURATED LIQUID DENSITY CORRELATIONS

Rackett Equation

The Rackett equation is an empirical correlation relating the reduced volume, reduced temperature, and critical compressibility factor of a liquid. Rackett [11] wrote the equation

$$\log \frac{V_f}{V_c} = (1 - T_r)^{2/7} \log Z_c \tag{17}$$

v_f is the saturated specific volume of the liquid. It has been claimed that this relatively simple equation can reproduce the densities of pure, saturated, nonpolar liquids accurately from their triple points to their critical points. Spencer and Danner [3, 12] improved the Rackett equation by making Z_c an adjustable parameter which they called Z_{RA} and by simplifying Rackett's mixing rules. Using a Z_{RA} regressed from measured densities or even calculated from one density point increases the accuracy of the correlation because the critical constant, Z_c, is replaced with a constant optimized to reproduce the liquid density, and because accurate liquid densities are much easier to find than accurate values of Z_c. Values of Z_{RA} for many liquids have been published. Spencer and Danner [3] wrote their modification of the Rackett equation:

$$\frac{1}{\rho_s} = \left(\frac{RT_c}{P_c}\right) Z_{RA}^{[1+(1-T_r)^{2/7}]} \tag{18}$$

and suggested these mixing rules:

$$V_s = R \sum_i \left(\frac{x_i T_{ci}}{P_{ci}}\right) Z_{RAm}^{[1+(1-T_r)^{2/7}]} \tag{19}$$

$$Z_{RAm} = \sum_i x_i Z_{RAi} \tag{20}$$

$$T_{cm} = \sum_i \sum_j \phi_i \phi_j T_{cij} \tag{21}$$

$$\phi_i = x_i V_{ci} \Big/ \sum_i x_i V_{ci} \tag{22}$$

$$T_{cij} = (T_{ci} T_{cj})^{1/2}(1 - k_{ij}) \tag{23}$$

$$k_{ij} = 1 - \left[\frac{2(V_{ci}^{1/3} V_{cj}^{1/3})^{1/2}}{V_{ci}^{1/3} + V_{cj}^{1/3}}\right]^3 \tag{24}$$

Spencer and Danner [3] found that the modified Rackett equation gave 0.50 average absolute percent error for 2,795 density points for hydrocarbons, 0.60% for 652 density points for other organic compounds, and 0.74% for 148 density points for inorganic compounds. Using a data base consisting of 4,720 points, roughly half for hydrocarbons and half for "organics," alcohols, acids, and inorganics, Spencer and Adler [4] determined that the modified Rackett equation gave an average absolute error of about 0.58%. Spencer and Danner [12] found that the modified Rackett equation gave an average absolute error of 1.76% when tested against 692 density points for hydrocarbon mixtures and 2.91% when tested against 273 density points for mixtures containing hydrogen, hydrogen sulfide, or carbon dioxide with a hydrocarbon. The modified Rackett equation is recommended in the API Technical Data Book-Petroleum Refining as one of two alternative methods for calculating densities of pure, nonpolar, saturated liquids and nonpolar, saturated mixtures.

Hankinson-Thomson Correlation

Hankinson and Thomson [13] modified an extended corresponding states correlation developed by Gunn and Yamada (14) to give an accurate method for estimating saturated liquid densities. They called their method COSTALD for Corresponding States Liquid Density. The basic equation is

$$V_s/V^* = V_r^0(1 - \omega_{SRK} V_r^{(\delta)}) \tag{25}$$

where V^* is a characteristic volume analogous to the critical volume, $V_r^{(0)}$ is the spherical molecule function, and $V_r^{(\delta)}$ is a function which, when multiplied by V^* and by ω_{SRK}, gives the deviation function for non-spherical molecules. ω_{SRK} is a modified acentric factor based on the Soave-Redlich-Kwong equation of state. $V_r^{(0)}$ and $V_r^{(\delta)}$ are both functions of reduced temperature only:

$$V_r^0 = 1 + a(1 - T_r)^{1/3} + b(1 - T_r)^{2/3} + c(1 - T_r) + d(1 - T_r)^{4/3} \tag{26}$$

$$V_r^{(\delta)} = \frac{e + fT_r + gT_r^2 + hT_r^3}{T_r - 1.00001} \tag{27}$$

V^* is usually regressed by fitting these three equations to experimental densities using known values of T_r and ω_{SRK}, but it can be calculated from a single density point, or estimated from a generalized correlation of the form

$$V^* = \frac{RT_c}{P_c}(k_1 + k_2\omega_{SRK} + k_3\omega_{SRK}^2) \tag{28}$$

Values of V^* and ω_{SRK} for 200 compounds were given in the original paper, and values for almost 200 additional compounds have been obtained more recently.

Hankinson and Thomson recommended these mixing rules:

$$T_{cm} = \frac{\sum_i \sum_j x_i x_j V_{ij}^* T_{cij}}{V_m^*} \tag{29}$$

$$V_m^* = \frac{1}{4}\left[\sum_i x_i V_i^* + 3\left(\sum_i x_i V_i^{*2/3}\right)\left(\sum_i x_i V_i^{*1/3}\right)\right] \tag{30}$$

$$V_{ij}^* T_{cij} = (V_i^* T_{ci} V_j^* T_{cj})^{1/2} \tag{31}$$

$$\omega_m = \sum_i x_i \omega_{SRKi} \tag{32}$$

Hankinson and Thomson found that their correlation gave an average absolute error of 0.37% when tested against a data base of 2,657 data points for 97 pure hydrocarbons and 1,851 data points for 103 other compounds (mostly nonpolar or slightly polar), and 1.40% for 2,994 mixture data points. The Hankinson-Thomson correlation is one of two alternative methods recommended in the API Technical Data Book-Petroleum Refining for calculation of densities of pure, nonpolar, saturated liquids and nonpolar, saturated mixtures.

Extended Corresponding States Conformal Liquid Method

Mollerup [15] and Ely and Hanley [16] used a generalization of Pitzer's extended corresponding states model to calculate densities and transport properties of nonpolar liquids and mixtures. Unfortunately, both the three-parameter corresponding states model and this generalization are called extended corresponding states models. In using this model it is assumed that there is a hypothetical pure fluid whose properties are the same as the properties of the pure liquid or mixture under consideration, that is, that the liquids are conformal or that their intermolecular potential functions have the same mathematical form. The properties of this hypothetical fluid are then calculated from the properties of a reference fluid using the extended corresponding states approach.

In the classical corresponding states theory two substances i and j at temperatures T_i and T_j with volumes V_i and V_j have the same properties when they have the same reduced temperature and reduced volume:

$$T_i/T_{ci} = T_j/T_{cj} \qquad \text{and} \qquad V_i/V_{ci} = V_j/V_{cj} \tag{33}$$

In the generalized approach two scale factors are defined:

$$f_{i,j} = T_{ci}/T_{cj} \qquad \text{and} \qquad h_{i,j} = V_{ci}/V_{cj} \tag{34}$$

so that the temperature of substance j which corresponds with that of substance i is simply $T_j = T_i/f_{ij}$.

The third parameter in the generalized extended corresponding states method is still the acentric factor, which occurs in the shape factors θ_{ij} and ϕ_{ij}. The shape factors are functions of T_c, V_c, and ω, and multiply the scale factors:

$$f_{ij} = (T_{ci}/T_{cj})\theta_{ij} \qquad \text{and} \qquad h_{ij} = (V_{ci}/V_{cj})\phi_{ij} \tag{35}$$

A somewhat simplified outline of the procedure used to calculate densities using this method is the following: it is desired to calculate the density of a pure liquid or mixture x at temperature T and pressure P. First, the corresponding temperature T_o and pressure P_o of the hypothetical fluid are calculated from

$$T_o = \frac{T}{f_{xo}} \qquad \text{and} \qquad P_o = P\left(\frac{h_{xo}}{f_{xo}}\right) \tag{36}$$

where f_{xo} and h_{xo} are the complete scale factors given in Equations (35). Next, the density of the reference fluid, ρ_o, is found at T_o and P_o. In practice one would have an equation giving the density of the reference fluid as a function of temperature and pressure, or, better, an equation for the entire P-V-T surface of the reference fluid, such as those generated at the National Bureau of Standards. In principle, however, one could look up ρ_o in a table at the appropriate values of T_o and P_o. Finally, the required density is obtained from

$$\rho = \rho_o/h_{xo} \tag{37}$$

Ely and Hanley's mixing rules are

$$f_{ijo} = (f_{io}f_{jo})^{1/2}(1 - k_{ij}) \tag{38}$$

$$h_{ijo} = 1/8(h_{io}^{1/3} + h_{jo}^{1/3})^3(1 - \ell_{ij}) \tag{39}$$

where k_{ij} and ℓ_{ij} are interaction parameters, and

$$f_{m,o} = (1/h_{m,o}) \sum_i \sum_j x_i x_j f_{ijo} h_{ijo} \tag{40}$$

$$h_{m,o} = \sum_i \sum_j x_i x_j h_{ijo} \tag{41}$$

for the mixture. Mollerup used interaction parameters, but Ely and Hanley set them to zero. Without going into detail, Ely and Hanley say that the errors in their predicted densities for a data set consisting almost entirely of hydrocarbons and their mixtures are generally less than 1%. This figure apparently applies to compressed as well as saturated densities. Mollerup reported errors in volume as low as 0.01% with many errors less than 0.1% for liquified natural gas. However, he used interaction parameters. If his equation was tested on the same basis as the others reported here, without interaction parameters, the errors would probably be larger than 1% in many cases.

In principle, the generalized extended corresponding states method applies to nonpolar liquids and mixtures over the whole temperature and pressure range from dilute gas to compressed liquid. The need for a reference fluid whose properties are known accurately over the temperature-pressure range of interest is a distinct disadvantage of this method. There are only a few liquids whose P-V-T surfaces have been modeled accurately enough for them to serve as reference fluids in either the extended corresponding states conformal liquid method or the two-fluid corresponding states method (see "Two-Fluid Corresponding States Method" later in this chapter). Almost all of these liquids are low-molecular-weight hydrocarbons, namely methane, ethane, and propane. While these compounds would make good reference fluids for liquids, such as liquified natural gas, they would be less suitable as reference fluids for the higher-molecular-weight hydrocarbons and for other kinds of organic compounds. The physical properties of the members of a homologous series of organic compounds change quite rapidly over the first four to six members of the series. For this reason propane would probably not be a very good reference fluid for hexane, for example, and vice versa. Also, since the physical properties of some kinds of organic compounds differ considerably from the properties of other kinds, the linear alkanes (straight-chain hydrocarbons with no double or triple bonds) such as methane, ethane, and propane, might serve as reasonably good reference fluids for ethers, but probably not for alcohols. Of course, selection of a reference fluid quite similar to the liquid under consideration, if one were available, would help to ensure accurate predictions. Ely and Hanley say it is easy to change reference fluids, and are themselves changing their reference fluid from methane to propane. The results of this change have not been published yet. There seem to be some convergence problems with Ely and Hanley's computer program. This is presumably due to the programming and not to the nature of the correlation.

Hall's Method

Hall [17] has derived a saturated liquid density correlation from renormalization group theory which, he claims, will reproduce the densities of pure, saturated liquids to within experimental accuracy from their triple points to their critical points.

Joffe-Zudkevitch Correlation

Joffe and Zudkevitch [18] reported an equation for reduced densities of saturated polar liquids:

$$\rho_r = 1 + 0.85(1 - T_r) + (1.6916 + 0.9846\psi)(1 - T_r)^{1/3} \tag{42}$$

where ψ is a "correlation parameter." They found that for nonpolar liquids ψ is a constant, different for each liquid, but independent of temperature. ψ is a function of temperature for polar liquids, however. The ψ–T curve is often close to a straight line, which may increase or decrease with temperature, but it may also be a curve. Joffe and Zudkevitch found that, for most substances they tested, ψ was roughly a linear function of reduced temperature for values of T_r between 0.4 and 0.85. If neither values for ψ at various temperatures, nor for S, the slope of the ψ–T line are given, they may be calculated using two reference densities. Equation (42) is used to calculate two values of ψ using the two reference densities. S is then calculated using the equation for a straight line.

$$S = (\psi_2 - \psi_1)/(T_{r2} - T_{r1}) \tag{43}$$

Given S and a value of ψ at any reduced temperature, a value for ψ at any other reduced temperature is easily calculated. Joffe and Zudkevitch recommend that, if T_r is less than 0.4 it be set equal to 0.4, and if it is larger than 0.85 it be set equal to 0.85, and that the values of ψ calculated at $T_r = 0.4$ and $T_r = 0.85$ be used below and above those reduced temperatures.

As Joffe and Zudkevitch use their correlation for polar compounds, it is a four-parameter corresponding states correlation which requires the parameter S. If ψ is independent of temperature, however, it becomes a three parameter (T_c, ρ_c, and ψ) correlation.

Joffe and Zudkevitch tested their correlation against 274 data points for 24 polar compounds. The overall average error for their comparisons is 0.58%. Spencer and Adler [4] obtained 0.74% error in testing this correlation against 2,455 density data points for 75 hydrocarbons and 0.76% error for 407 data points for 14 "organics." It is interesting that three of the four compounds for which Joffe and Zudkevitch obtained the largest errors are first members of a series—dimethyl ether, methanol, and acetonitrile. The fourth was water. Joffe and Zudkevitch comment that they made an extensive comparison to show that ψ is independent of temperature for nonpolar substances and concluded that their correlation did extremely well for these substances. They did not, however give details of these comparisons. Joffe and Zudkevitch also stated that comparisons for mixtures would be published, but that work does not seem to have appeared.

Conclusions

Spencer and Danner's modification of the Rackett equation is simple and quite accurate over a wide temperature range. Parameters have been published for many compounds. It is less accurate near the critical point and for polar compounds.

The Hankinson-Thomson correlation is slightly more complicated than the Rackett equation. It is also quite accurate over wide temperature ranges and the necessary parameters have been published for many substances. It is also less accurate close to the critical point and for polar compounds. The Hankinson-Thomson method and Spencer and Danner's modification of the Rackett equation are about equally accurate for pure, saturated liquids, but the former is more accurate for saturated mixtures. This correlation also has several other useful features. It predicts the densities of liquified natural gas well, it fits the densities of crude oils and petroleum fractions quite well [19], and it has been suggested that it can fit the densities of aqueous solutions well. In addition, Hankinson and Thomson showed that, given good values of T_c and ω_{SRK}, their correlation can be used to give good estimates of the critical volumes of nonpolar liquids.

The generalized extended corresponding states correlation is more complicated than the Rackett equation or the Hankinson-Thomson correlation and requires accurate densities for the reference fluid. In principle, it is applicable over a very wide range of temperatures and pressures, and parameters are available for many compounds. It does not give accurate predictions for polar liquids. It is less accurate than the Rackett equation or the Hankinson-Thomson correlation now, but use of a different reference liquid may change that situation.

Hall's method seems promising, but no evaluations have been published.

The Joffe-Zudkevitch method is the best method for polar liquids, and is fairly accurate for nonpolar liquids, but it requires an additional parameter that has been calculated for only a few liquids. It is limited to a slightly smaller temperature range than the other correlations. It is claimed to give accurate densities for mixtures, but the evidence does not seem to have been published.

COMPRESSED LIQUID DENSITY CORRELATIONS

Lu Chart

Lu [20] developed a graph relating a combination of properties including the molar volume and critical constants of a liquid to its reduced temperature and pressure:

$$\frac{V_c^{1.77}}{V}\left(\frac{P_c}{RT_c}\right)^{0.77} = K = f(T_r, P_r) \tag{44}$$

Multiplying both sides of the equation by V gives

$$VK = V_c^{1.77}\left(\frac{P_c}{RT_c}\right)^{0.77} \tag{45}$$

For a pure liquid the right hand side of this equation is constant, so that the equation can be simplified to

$$VK = \text{constant} \tag{46}$$

Lu then evaluated the generalized K functions from his graph at T_1 and P_1 corresponding to a known volume V_1 and at T_2 and P_2 corresponding to the desired volume V_2. Then $V_1K_1 = V_2K_2$, or

$$V_2 = V_1(K_1/K_2) \tag{47}$$

Since Lu gave a K line on his graph for the saturated liquid as well as for compressed liquids, his correlation can be used to estimate one saturated density from another, a compressed density from a saturated density, a saturated density from a compressed density, or one compressed density from another. Rea et al. [5] converted Lu's graph into a series of equations giving K as a cubic equation in T_r. The coefficient of each term of this cubic equation is a fourth degree polynomial in P_r.

Lu's method is not as accurate for saturated liquid densities as are other correlations previously described, but it is fairly accurate for a compressed density correlation. Using their equations to represent Lu's chart, Rea et al. [5] found 1.09% average absolute error for 2,590 compressed densities of 32 compounds of which 30 were hydrocarbons. Errors were somewhat larger for hydrogen sulfide and carbon dioxide.

Lu's method requires one reference density for the compound under consideration, but this is usually not a problem. Although Spencer and Danner [12] tested the Lu equation with several different sets of mixing rules for saturated mixtures, the method does not seem to have been applied to compressed mixtures.

Thomson-Brobst-Hankinson Correlation

Thomson, Brobst, and Hankinson [21] extended the Tait [22] equation

$$V = V_s\left(1 - C\ln\frac{B + P}{B + P_s}\right) \tag{48}$$

by making it temperature-dependent and compositional. The temperature dependence was introduced by expressing B as a polynomial in $1 - T_r$. The composition dependence was added by making two constants functions of a modified acentric factor, ω_{SRK}. Their equation for B is

$$B/P_c = -1 + a(1 - T_r)^{1/3} + b(1 - T_r)^{2/3} + d(1 - T_r) + e(1 - T_r)^{4/3} \tag{49}$$

The constant C in Equation 48 and e in Equation 49 were defined as

$$C = j + k\omega_{SRK} \tag{50}$$

$$e = \exp(f + g\omega_{SRK} + h\omega_{SRK}^2) \tag{51}$$

The constants a, b, d, e, f, g, h, j, and k are the same for all liquids. In applying their correlation to pure liquids, Thomson, et al. calculated the liquid saturation pressures from a vapor pressure equation using specific constants for each liquid. For mixture saturation pressures they used a modification of Riedel's generalized vapor pressure equation. The other mixture parameters were calculated using the mixing rules given for their saturated density model (see "Hankinson-Thomson Correlation"). Although any one of a number of saturated liquid density correlations may be used to calculate V_s, the Hankinson-Thomson correlation seems to give the most accurate results. Thomson et al. claim an average absolute error of 0.446% for 6,338 data points for pure, nonpolar liquids; 2.57% for 1,352 data points for pure, polar liquids; and 1.61% for all mixtures. The correlation does not predict the densities of compressed, polar mixtures well, however. It has been suggested that saturated densities may be more accurate than compressed densities for these mixtures.

Perhaps the major limitation of the Thomson-Brobst-Hankinson correlation is the use of Riedel's vapor pressure equation for mixtures. To date, however, this correlation gives better predictions for a wide variety of pure liquids and mixtures than any other correlation available. The Thomson-Brobst-Hankinson method is recommended in the API Technical Data Book-Petroleum Refining for calculation of densities of pure, compressed, nonpolar liquids and nonpolar, compressed mixtures.

Extended Corresponding States Conformal Liquid Method

As stated earlier this correlation applies to compressed as well as to saturated liquids. Its accuracy seems to be comparable with that of Lu's method for pure, nonpolar liquids and their mixtures. It is less accurate when applied to polar liquids.

Conclusions

Lu's method is simple and fairly accurate when applied to pure, nonpolar liquids. It requires a reference density and is less accurate for polar liquids. It has not been applied to mixtures.

The Thomson-Brobst-Hankinson method is the most accurate method available for densities of compressed, nonpolar liquids and mixtures, but is less accurate when applied to polar liquids and mixtures. When possible, a bubble-point pressure from a flash or similar calculation should be used to calculate P_s for mixtures.

The extended corresponding states conformal liquid approach shows promise, but is less accurate than the Thomson-Brobst-Hankinson correlation at present.

CORRELATIONS FOR LIQUIFIED NATURAL GAS AND LIQUIFIED PETROLEUM GAS

Introduction

Because of their commercial importance and their transport and sale in large quantities, special emphasis has been placed on the development of accurate liquid density correlations for liquified natural gas (LNG) and liquified petroleum gas (LPG). Development of correlations for LNG is easier than it is for other mixtures for four reasons. First, accurate results are required over a fairly small temperature range about 90 to 150 K (−298 to −190°F). Second, LNG contains only a small number of relatively simple components, the paraffins methane through normal and isopentane, nitrogen, and sometimes carbon dioxide. Third, the physical constants of all of these substances are known accurately. Fourth, none of these LNG components is polar. The difficulty is the accuracy required. The densities of LNG mixtures can be measured to about 0.1%, so density correlations should have the same accuracy. It is meaningless, of course, to assign to a correlation an accuracy greater than that of the data against which it is tested.

Liquid density models used for custody transfer of LNG's usually contain interaction parameters derived from measured densities of binary mixtures (see "Interaction Parameters" earlier in this chapter).

McCarty's Comparisons

McCarty [23] tested four LNG density correlations against accurate LNG densities measured at the National Bureau of Standards in Boulder, Colorado. The four correlations were the extended corresponding states conformal liquid model previously discussed, a hard sphere model, his extension of Klosek and McKinley's empirical correlation, and a cell model. McCarty found that the extended corresponding states conformal liquid model gave 0.04% average absolute error. The cell model gave 0.07%, the hard sphere model was third with 0.18%, and McCarty's revision of the Klosek-McKinley recipe last with 0.23% average absolute error. McCarty found that the extended corresponding states correlation gave good results throughout the range of temperatures, pressures, and compositions covered by his comparisons, and apparently did not have the convergence problems that others have encountered using this method. He suggested that neither the cell model nor the hard sphere model be used for nitrogen-containing mixtures at or above 120 K (about −244°F.), and suggested even more restricted temperature and composition ranges for his revision of the Klosek-McKinley method.

Mollerup's Results

Mollerup [15] also tested the accuracy of the extended corresponding states model for LNG components and mixtures. He obtained an average absolute error of 0.10% for 465 points. One hundred twenty-two of these points, however, were for pure, saturated methane for which he obtained 0.03% error.

Hankinson-Thomson Correlation

In testing the Hankinson-Thomson correlation with interaction parameters against the National Bureau of Standards data, Hankinson, Coker, and Thomson [24] found an average absolute error of 0.08% for the forty density points for mixtures McCarty designated as "LNG-like," and 0.20% for the whole set of data. Better mixing rules or better interaction parameters might improve their results.

Two-Fluid Corresponding States Method

Teja [25] used a two-fluid corresponding states approach to calculate LNG densities. His model is similar to the extended corresponding states method except that the properties of the liquid of interest are obtained by interpolation between or extrapolation from the properties of two reference liquids, and that Teja did not use shape factors. The interpolation or extrapolation is made using acentric factors. Teja's equation is

$$V_s\left(\frac{P_c}{RT_c}\right) = \left(\frac{V_sP_c}{RT_c}\right)^{r_1} + \frac{\omega - \omega^{r_1}}{(\omega^{r_2} - \omega^{r_1})}\left[\left(\frac{V_sP_c}{RT_c}\right)^{r_2} - \left(\frac{V_sP_c}{RT_c}\right)^{r_1}\right] \tag{52}$$

where the superscripts r_1 and r_2 indicate the two reference liquids. As Teja points out, this is a convenient approach when the critical properties of the liquids are known accurately as are those of LNG components. It is necessary to know the densities of the reference liquids accurately, however. Teja used methane and n-butane as reference liquids. He also used modified acentric factors for the pure components. Each was found by using equation [52] to calculate an acentric factor from a known liquid density. Teja states that the differences between his acentric factors and those recommended by Passut and Danner [26] are not large. While the absolute differences are not, the relative differences are. His acentric factor for methane differs from Passut and Danner's value by 40% (0.012 vs. 0.0072), for ethane by 20% (0.114 vs. 0.0908), and for propane by 10% (0.162 vs. 0.1454).

For mixtures Teja used the following mixing rules:

$$V_{cm}T_{cm} = \sum_i \sum_j x_i x_j V_{cij} T_{cij} \tag{53}$$

$$V_{cm} = \sum_i \sum_j x_i x_j V_{cij} \tag{54}$$

$$Z_{cm} = \sum_i x_i Z_{ci} \tag{55}$$

$$\omega_m = \sum_i x_i \omega_i \tag{56}$$

with

$$V_{cij}T_{cij} = \xi_{ij}(V_{ci}T_{ci}V_{cj}T_{cj})^{1/2} \tag{57}$$

and

$$V_{cij} = (\eta_{ij}/8)(V_{ci}^{1/3} + V_{cj}^{1/3})^3 \tag{58}$$

where ξ_{ij} and η_{ij} are interaction parameters.

Teja tested his equation against densities for pure liquids; for binary mixtures using one interaction parameter for some of them; for three ternary mixtures, using interaction parameters for the methane-ethane and methane-nitrogen pairs; and against the density calculated for a six-component mixture using the extended corresponding states method. The average absolute errors were 0.15%, 0.046%, 0.05%, and 0.13%, respectively.

Conclusions

The extended corresponding states method and the cell model are the most accurate methods for calculating LNG densities. Neither is really easy to use and the cell model is the more complicated of the two. The extended corresponding states model requires accurate densities for the reference fluid. The Hankinson-Thomson method is easy to use and predicts densities for many LNG mixtures well. Teja's method is easy to use and gives accurate densities if the critical densities, critical compressibility factors, and saturated liquid densities of the two reference liquids are known accurately and interaction parameters and special acentric factors are available.

SUMMARY

In this chapter some of the basic variables required in liquid density correlations were outlined, the term "factor models" was explained, general aspects of compositional models were discussed, and comments were made concerning the properties of correlating equations, mixtures, polar liquids, equations of state, and present trends in the development of liquid density correlations. Two factor models for crude oils and petroleum products were discussed and saturated liquid density correlations were reviewed. The conclusions were that the modified Rackett equation and the Hankinson-Thomson correlation are the most accurate methods for predicting densities of pure, nonpolar liquids and mixtures, while the Joffe-Zudkevitch method is the most accurate for pure, polar liquids. The conclusions concerning the compressed liquid density correlations reviewed were that the Thomson-Brobst-Hankinson correlation is the most accurate. For the density correlations for liquified natural gas, it was concluded that the cell model is quite accurate, but quite complicated; the extended corresponding states model is also quite accurate, but requires accurate densities for the reference fluid; the Hankinson-Thomson correlation is quite accurate for many LNG mixtures; and Teja's method is accurate if special interaction parameters and acentric factors are used.

NOTATION

C	constant, see Equation 50	S	slope of ψ–T line
F	compressibility factor	T_r	reduced temperature
f_{ij}	scale factor	T	temperature
h_{ij}	scale factor	t	temperature
k_{ij}	interaction parameter	V*	characteristic volume
k_0, k_1	constants in Equation 14	V_c	critical volume
l_{ij}	interaction parameter	V_e	equilibrium volume
P_c	critical pressure	V	volume
P_e	equilibrium pressure	V_s	saturation volume
P_r	reduced pressure	x_i	mole fraction
P	pressure	Z	Rackett parameter, see Equation 17
R	universal gas law constant		

Greek Symbols

α	coefficient of thermal expansion	ρ	density
β	compressibility coefficient	ϕ_i	mole ratio
η_{ij}	interaction parameter in Equation 58	ψ	correlation parameter
θ_{ij}, ϕ_{ij}	shape factors	ω	acentric factor
ξ_{ij}	interaction parameter, Equation 57		

ACKNOWLEDGMENTS

The author is indebted to R. W. Hankinson, Process Division, Corporate Engineering, Phillips Petroleum Co., for ideas concerning liquid densities and their correlations. As far as I know, he is responsible for the concept of factor models. I have also used several ideas from the paper "Calculation of Liquid Densities and Their Mixtures" by D. K. Petree, G. H. Thomson, and R. W. Hankinson, which was presented at the North Sea Flow Metering Workshop in Stavanger, Norway, September, 1983.

REFERENCES

1. Reid, R. C. and Sherwood, T. K. *The Properties of Gases and Liquids*, Second Edition, McGraw-Hill, New York, 1966.
2. Reid, R. C., Prausnitz, J. M., and Sherwood, T. K. *The Properties of Gases and Liquids*, Third Edition, McGraw-Hill, New York, 1977.
3. Spencer, C. F., and Danner, R. P. *J. Chem. Eng. Data*, 17(2), 236 (1972).
4. Spencer, C. F., and Adler, S. B. *J. Chem. Eng. Data*, 23(1), 82 (1978).
5. Rea, H. E., Spencer, C. F., and Danner, R. P. *J. Chem. Eng. Data*, 18(2), 227 (1973).
6. Hankinson, R. W., Phillips Petroleum Co., Bartlesville, OK., Personal Communication–"Factor Models" (1983).
7. Pitzer, K. S., Lippmann, D. Z., Curl, R. F., Jr., Huggins, C. M., and Petersen, D. E. *J. Am. Chem. Soc.* 77, 3433 (1955).
8. "Manual of Petroleum Measurement Standards", Vol. X, Chapter 11.1, American Petroleum Institute, Washington, D. C., 1980.
9. Hankinson, R. W., Segers, R. G., Buck, T. K., and Gielzecki, F. P. *Oil Gas J.* 77(52), 66, Dec. 24, 1979.
10. "Measurement of Petroleum Liquid Hydrocarbons by Positive Displacement Meter", API Standard 1101, American Petroleum Institute, Washington, D. C., 1960.
11. Rackett, H. G. *J. Chem. Eng.* Data, 15(4), 514 (1970).
12. Spencer, C. F., and Danner, R. P. *J. Chem. Eng. Data*, 18(2), 230 (1973).

13. Hankinson, R. W. and Thomson, G. H. *AIChE J.* 25(4), 653 (1979).
14. Gunn, R. D. and Yamada, T. *AIChE J.* 17. 1341 (1971).
15. Mollerup, J. *Ber. Bunsenges. Phys. Chem.* 81(10), 1015 (1977).
16. Ely, J. F. and Hanley, H. M. NBS Technical Note 1039, National Bureau of Standards, Boulder, CO, 1981.
17. Hall, K. R. Texas A. & M. University, College Station, TX, Personal Communication -"Saturated Liquid Density Model" (1983).
18. Joffe, J. and Zudkevitch, D. AIChE Symp. Ser., 70(140), 22 (1974).
19. Robinson, E. R. *Hydrocarbon Processing,* May, 1983, p. 115.
20. Lu, B. C. -Y. *Chemical Engineering,* 66(9), 137 (1959).
21. Thomson, G. H., Brobst, K. R., and Hankinson, R. W. *AIChE J.* 28(4), 671 (1982).
22. Tait, P. G. "Physics and Chemistry of the Voyage of *H. M. S. Challenger*" II, Part IV, S. P. LXI, H. M. S. O., London, 1888.
23. McCarty, R. D. NBS Technical Note 1030, National Bureau of Standards, Boulder, CO, 1980.
24. Hankinson, R. W., Coker, T. A., and Thomson, G. H. *Hydrocarbon Processing,* April, 1982, p. 207.
25. Teja, A. S. *AIChE J.* 26(3), 337 (1980).
26. Passut, C. A. and Danner, R. P. *Ind. Eng. Chem. Process Des. Develop.* 12(3), 365 (1973).

CHAPTER 3

CORRELATIONS FOR MOLECULAR DIFFUSIVITIES IN LIQUIDS

W. Hayduk

Department of Chemical Engineering
University of Ottawa
Ottawa, Ontario, Canada

CONTENTS

INTRODUCTION

The separation processes most frequently used for liquid solutions are known as mass transfer operations. They all share common elements; in all of them mass transfer occurs between two phases separated by an interface, through which material passes by the mechanism of molecular diffusion. Whereas distillation involves the vapor and liquid phases, absorption the gas and liquid phases, and liquid extraction two immiscible liquid phases, leaching involves the solid and liquid phases. In every case, however, a concentration difference is the cause or driving force for the molecular diffusion of the one or more components being removed or concentrated in the process.

The equilibrium concentration that would be reached if a sufficient period of time were allowed to elapse is also of primary significance. The actual deviation from the equilibrium concentration determines, to a large extent, the rate of mass transfer between the phases. The equipment utilized for mass transfer frequently entails the provision for mixing to enhance the rate of transfer. In some cases, the thickness of the more quiescent liquid, especially near a solid interface, is therefore reduced or, in other cases, the exposed surface is renewed more rapidly or the interfacial area is increased by creating a greater number of bubbles or droplets for those phases separated by mobile interfaces.

It appears axiomatic that it is impossible to produce a degree of turbulence so intense that the effect of molecular diffusion is no longer significant. Even at highly turbulent interfaces, where the interface is renewed rapidly and continuously so that transient diffusion may be considered to occur only for a fraction of a second before the surface is again renewed, the transient rate of molecular diffusion, or the molecular diffusion coefficient; determines the rate of mass transfer.

The separation process, adsorption, is being used to an ever increasing degree especially for water treatment or purification. Molecular diffusion is the mechanism of transfer of contaminant through the intricacies of the minute pore structure to the internal surface of the sorbent. Regardless of the intensity of the flow surrounding the granules or particles of the sorbent, the removal rate of the undesired species is determined by the relatively slow process of molecular diffusion through the pores. Molecular diffusion is the determining factor in the rate of mass transfer for all common mass transfer operations. Thus, a knowledge of the rates of molecular diffusion is invariably required in the design or evaluation of the performance of mass transfer equipment.

METHODS FOR MEASURING DIFFUSIVITIES IN LIQUIDS

The following discussion is meant to be a brief overview of the experimental methods for the determination of molecular diffusivities of dissolved substances in liquids. A survey of this subject may be found in Ertl et al. [1].

Diaphragm Cell

Molecular diffusivities have been experimentally determined using a variety of methods, ingenious and often complex, reflecting the difficulties involved. The most common method for the determination of molecular diffusivities of dissolved liquid or solid substances in liquids has been one utilizing the diaphragm cell. In this method a diaphragm, consisting of a porous material having capillary-size passages through it, is positioned in a cell separating different concentrations of solute, which occupy chambers above and below it. The passages through the diaphragm, although circuitous and irregular in cross-sectional area, connect the two chambers. The passages through the diaphragm are sufficiently small so that the liquid confined therein experiences no convection, only molecular diffusion of the dissolved species. If the concentrations in both cells can be accurately determined before and after a period of essentially steady-state diffusion, the molecular diffusivity of the solute can be calculated.

The accurate determination of the diffusivity, however, is contingent on at least three additional underlying conditions. These include a knowledge of the true concentrations of solute at either edge of the diaphragm; this usually requires some degree of stirring in each chamber to ensure that each has essentially a uniform concentration. Because the true length of the circuitous pores is unknown, it is necessary to calibrate the diaphragm cell by making measurements for a solute whose diffusivity is accurately known. A sufficiently long period of time must be initially provided so that a steady-state concentration profile may be established within the diaphragm passages. The necessary material balance does not allow for changes of concentration within the diaphragm.

In calculating diffusivity it may be necessary to consider the volume occupied by the diffusing species as it enters the opposite cell. The minute displacement of the cell contents backward through the diaphragm as a result of the volumetric displacement of entering solute may not be completely negligible. Many variations in design of the diaphragm cell have been used for many different types of diffusivity measurements. A recent innovation is one by Asfour and Dullien [2]. In at least two instances a diaphragm cell has been constructed with capillaries of known dimensions by Ross and Hildebrand [3] and Hayduk and Ioakimidis [4]; capillary tubes were imbedded in a supporting matrix or drilled in a plate. Such cells require no prior calibration since the diffusion path length is known.

Transient Effusion from an Open-Ended Capillary

A method utilizing individual capillary-sized tubes for eliminating convection within them has also proved successful. The uniform barrel of a 1-mm diameter micro-syringe was employed as an open-ended capillary, for example, by Witherspoon and Saraf [5]. When filled with a solution of known concentration and immersed in the corresponding solvent, transient effusion of solute from the capillary occurs. Using the mathematical expression for the average concentration of solute remaining in the capillary after some elapsed time, the diffusivity may be calculated. When compared with the diaphragm cell method, the open-ended capillary cell method requires much more time, taking up to five days for a single measurement. It also requires accurate temperature control to prevent volume changes in the capillary, as well as an accurate micro-analytical technique for measuring the solute concentration in the small volume of solution; it is an absolute method, however, in that it requires no prior calibration.

Interferometric, Optical or Laser Methods

The objective in the interferometric method is to provide an instantaneous optical anlaysis of the concentration gradient of solution in a diffusion cell during transient diffusion, making use of the changes in optical properties of the solution with concentration. Thus, the cell must be constructed of special glass or have optical windows or slits for the transmission of light or laser beams in the region of the cell where the concentration is changing.

The initial condition for an experiment requires a sharp, step change in concentration from pure solvent to a constant solute concentration within the optical path. Various devices are used for this purpose including withdrawing liquid through a needle until the desired initial concentration step is attained, or by using a removable separating membrane or sliding valve separating solvent and solution. On passing through the changing concentration, the light produces interference fringes or lines when photographed, the spacing of which is related to the wavelength of the light and concentration gradient in the solution. These interference lines are generally very close together so that they require magnification for interpretation. Considerable exactitude is required in using these methods for measurement of diffusivities, but once a suitable apparatus and system of operation have been developed, they are highly accurate.

Taylor Axial Dispersion

The axial dispersion of solute within a fluid in laminar flow when a momentary increase in concentration is made, has been related to the molecular diffusion coefficient of the solute. This phenomenon was originally described by Taylor [6] and subsequently termed "Taylor axial dispersion." When a pulse of solution of higher concentration is added within a short time interval to a fluid in laminar flow, an approximately parabolic band of solute develops within the fluid. The concentration band spreads at a rate related to the solute diffusivity, average fluid velocity, and length of tube. The resulting axial distribution of the average concentration is found to be gaussian about a plane moving at the average velocity. Thus, the diffusivity is tractable from a continuous or frequently repeated analysis of the tube effluent. For accurate diffusivities, the tube size and variance of the concentration-time curve must be accurately known.

A thorough discussion of Taylor dispersion may be found in Sherwood, Pigford, and Wilke [7].

Dissolved Gases

Some twenty years ago Himmelblau [8] reviewed methods of diffusion of dissolved gases in liquids, including those utilizing the diaphragm cell and interferometric techniques suitable with

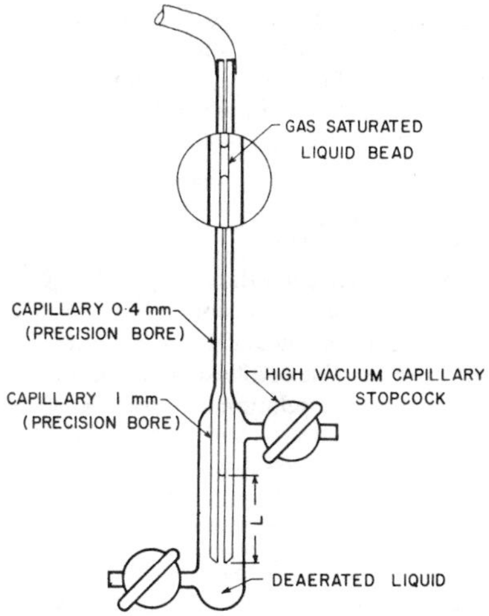

Figure 1. Details of diffusion cell.

some adaptation for both gas and liquid solutes. In addition, experiments involving the laminar flow of solvent in different devices, in each case exposing an interface for gas absorption, and a device using a capillary tube were also discussed. Since that time, the application of these methods has been expanded and at least one new method has been introduced. That method involves measuring the dissolution rate of a miniature gas bubble suspended in a solvent.

A volumetric method using the simultaneous absorption of gas at a liquid interface and diffusion through a column of liquid confined at the end of a capillary has been used successfully for a number of gases [9, 10]. A diagram of the diffusion cell is shown in Figure 1. In one sequence of experiments it was shown that a capillary bore of no more than 1 mm in diameter and a temperature control of $\pm 0.02°C$ was required to reduce the effect of convection in the liquid to a degree that the major mechanism of transfer through the confined liquid was by molecular diffusion; otherwise, the diffusion rate was enhanced by convection. Several different methods have been devised for determining the diffusivities of gases entailing the laminar flow of liquid solvents for the absorption of gases. In these methods, if the rate of absorption can be independently measured and the velocity or velocity profile in the liquid accurately described, the rate of molecular diffusion into the exposed liquid surface can be related to the parameters involved. These methods include laminar flow of solvent in the form of a jet of circular cross-section [11, 12], a wetted-wall cylinder [13, 14], a thin film over a sphere [15], and a thin film flowing radially over a circular flat plate [16]. The appropriate mathematical development has posed a challenging problem in using laminar flow methods for measuring diffusivities of dissolved gases [17, 18]. Finally, a novel method for determining diffusivities of dissolved gases entails measuring the shrinkage rate of a miniature bubble of gas held by a thread in a quiescent solvent [19]. The change in size of the bubble can be measured accurately enough to permit determination of the diffusivity from a known gas solubility and allowing for the effects of surface tension. The nature of the interface and the correct mathematical solution to the problem are the subjects of discussion, however [20].

DEFINITION OF BASIC PARAMETERS DESCRIBING MOLECULAR DIFFUSION

Fick's first law is customarily used to describe the rate of diffusion of a dilute component in a two-component solution. The term mutual diffusivity is usually used in describing the rate of counterdiffusion of two components in solutions in which neither component is dilute. A single mutual diffusivity describes the diffusion rates of both components at the particular concentration. Such diffusivities are usually highly concentration dependent so that a detailed knowledge of the concentration dependence of the diffusivity is required for the determination of diffusion rates. The discussion here is limited to diffusivities of solutes at low concentrations only, or as they are frequently called, diffusivities at infinite dilution. Fick's first law for unidirectional flow is most frequently written in terms of the molar flux and concentration of the diffusing species:

$$J_A = -D_{AB}\frac{\partial C_A}{\partial x} \quad (1)$$

In the application of Fick's first law it may be considered that the molecular diffusivity is the factor that relates the flux to the concentration gradient. It is somewhat of a misconception to speak of a diffusivity that refers to a single solute concentration, because diffusion always entails a concentration difference along the diffusion path. Usually, a constant average value of diffusivity is considered to apply for concentrations within some narrow range, however. The relationship describing the diffusion rate of the *major* component through the solution is:

$$J_B = -D_{BA}\frac{\partial C_B}{\partial x} \quad (2)$$

For a solution of constant total molar concentration it can be shown that the molecular diffusivity of both components at the particular concentration is one and the same. If,

$$C_A + C_B = C \quad (3)$$

then,

$$D_{AB} = D_{BA} = D \quad (4)$$

It can be easily shown that for only a fraction of two component liquid solutions is the molar concentration constant. Hence, an assumption that Fick's law with the flux and concentration expressed in *molar* units can accurately describe the rate of diffusion, may be erroneous. Fick's law in one of the alternate forms is often more accurate for application to liquids; it may be expressed in terms of *mass* fluxes and concentrations.

$$j_A = -D_{AB}\frac{\partial c_A}{\partial x} \quad (5)$$

In this case, a necessary condition for the accurate application of Equation 5 is that the total mass concentration of the solution be constant. If,

$$c_A + c_B = \rho \quad (6)$$

then,

$$D_{AB} = D_{BA} = D \quad (7)$$

The total mass concentration, representing a solution density, is usually more likely to be nearly constant than the total molar concentration; hence, Equation 5 is more often a useful one for analysis of rates of molecular diffusion in liquids than is Equation 2.

Although Fick's law can be used to describe the molecular diffusion of a component in a diffusion process, there often is some motion of the bulk of the liquid during diffusion, even for an apparently quiescent liquid. The equation expressing the combination of bulk flow and molecular diffusion is sometimes termed the "general diffusion equation." It is written in terms of n, the total or net mass flux, representing mass or bulk flow of the solution:

$$n = n_A + n_B \quad (8)$$

$$n_A = (nc_A) + (j_A) \tag{9}$$

In Equation 9, (nc_A) represents the flux of the dilute component as a result of motion or bulk flow of solution, whereas j_A represents the flux resulting from the molecular diffusion process as expressed by Fick's law, vectorially added. The bulk flow term is concerned with the small displacement of the solution as a whole, which arises as a result of the diffusion process. An example is offered in further explanation. The rate of diffusion of a slightly soluble solid dissolving in a liquid can be mathematically described using the general diffusion equation. The rate of solution depends on the rate of diffusion of the dissolved solid through a film of liquid adjacent to the solid-liquid interface. As the solid dissolves, the interface representing the boundary of the solid, recedes. Because the liquid moves to take the place of the space initially occupied by the solid, bulk flow must occur. It is apparent that the volumetric rate of solution of solid is equal in magnitude but opposite in direction to the volumetric rate of bulk flow. An expression for the related volumetric displacements can be written in terms of the (constant) solution density and the density of the solid:

$$\frac{n_A}{\rho_{A,s}} = -\frac{n}{\rho} \tag{10}$$

The general diffusion equation can then be readily solved because both the bulk flow and Fick's law terms are expressed as a function of the rate of molecular diffusion.

An analysis of the type described is usually required for an accurate mathematical formulation for processes involving molecular diffusion. A detailed discussion of diffusion fluxes has been given by Bird et al. [21]. Various summaries and interpretations of this subject may be found in the books on "Mass Transfer."

CORRELATIONS FOR PREDICTION OF MOLECULAR DIFFUSIVITIES IN DILUTE SOLUTIONS

Because a theory of the liquid state suitable for predicting properties such as diffusivity from fundamental principles is not yet available, correlations for diffusivity are generally empirical and are based on properties that can be readily measured or calculated. Particularly for the solvent, the properties considered to be relevant are the viscosity, molar volume, molecular weight, latent heat of vaporization, and some measure of the degree of association of these polar solvents with molecular associating tendencies. In addition to these, the surface tension, or parachor, and the radius of gyration have been considered to affect the diffusion of solute. Except for viscosity and association factor, the same properties have also been considered to influence the rate of diffusion when applied to the solute. Because of their fundamental nature, the solvent viscosity and molecular size may be expected to influence solute diffusivity; but the possible effects of solvent properties such as latent heat, radius of gyration, and the parachor, is not as apparent.

One may visualize that in liquids of high viscosity or surface tension (parachor), the molecular forces between molecules may likewise be high, thus inhibiting the motion of solute molecules through them. Similarly, it is speculated that the forces between molecules must be overcome when vaporizing a liquid; that is, the energy or latent heat of vaporization is a measure of the intermolecular forces. Related properties for liquids are the various measures of molecular size, such as molecular (or molar) volume. Molecular size has also been described by the radius of gyration, by the collision diameter, particularly for gases, or simply by the molecular weight. Any property that is in some way related to the intermolecular forces between liquid molecules has also been considered to affect the resistance to motion of solute molecules through them and hence the diffusion rate. The extent of solvent association which, in effect, changes the molecular size undoubtedly has a strong influence on molecular diffusivity; however, the effect of solvent association on solute diffusivity has generally been only superficially dealt with if at all. Certain properties of solute molecules, especially those associated with molecular *size*, such as the collision diameter for gases, or simply the molar volume, would be expected to influence the diffusion rate. The solute molecular *shape* has also been considered in determining its diffusion rate; a spherical molecule, exposing a larger cross-sectional area, might be expected to diffuse at a lower rate than one of the same mass but elongated in shape, as observed for the straight-chained, alkane hydrocarbons. Various combinations of the previously mentioned properties for both solute and solvent have been utilized by researchers in providing diffusivities to an accuracy approaching the accuracy of the diffusivity measurements themselves for a certain limited solute-solvent system; this may be contrasted with correlations which yield predicted diffusivities with a probable error of about 15% and a maximum error of up to 50% for other solute-solvent systems. A review of theories for diffusion in liquids may be found in Ghai et al. [22] and Ertl et al. [1].

Available empirical correlations for prediction of binary diffusion coefficients of non-electrolytes in liquids at low concentrations will be discussed under four headings:

- General correlations for all types of solutes, gas, liquid or solid, in all types of liquids, polar and non-polar.
- Correlations restricted to non-polar solvents.
- Correlations for water as solvent.
- Correlations for which both solute and solvents are composed of normal alkane hydrocarbons.

Based on a recent examination of the available correlations and extensive compilations of diffusivity data [23], the correlations yielding the lowest average error will be recommended for each type of solvent or solute-solvent classification. The purpose for making the different classifications is that correlations devised for a limited amount of data for chemically similar solvents usually yield a *lower* error of prediction, or are simpler in form, than those for a general correlation that encompasses a host of different solvent and solute properties.

General Correlations for Polar and Non-Polar Systems

The most general and best-known equation is that of Wilke and Chang [24], which was intended for all solutes but contains a parameter to account for molecular association in water and other highly associated solvents:

$$D^{\circ}_{AB} = 7.4(10^{-8})(\phi M)^{0.5} T \mu_B^{-1} V_A^{-0.6} \tag{11}$$

Because of its generality and relative simplicity, it cannot be used to predict diffusivities of high accuracy for all different types of molecular interactions, although the degree of success, even with associating solvents, is notable.

Another general correlation was developed by King et al. [25] utilizing latent heats of vaporization as a measure of the intermolecular forces involved:

$$D^{\circ}_{AB} \mu_B T^{-1} = 4.4(10^{-8})(V_B/V_A)^{1/6}(\Delta H_B/\Delta H_A)^{0.5} \tag{12}$$

Of those tested, Equation 12 was shown by Reid et al. [26] to give the smallest deviation from experimental results for a number of solutes in several selected organic solvents. It was found to give somewhat smaller deviations than the Wilke-Chang correlation.

A recent general correlation developed by Tyn and Calus [27] utilized the parachor as a measure of molecular interaction. This correlation, which was published recently enough to prevent its being discussed in the previously mentioned review, is as follows:

$$D^{\circ}_{AB} = 8.93(10^{-8}) T \mu_B^{-1} V_A^{1/6} V_B^{-1/3} (P_B/P_A)^{0.6} \tag{13}$$

Equation 13 was considered by the authors to be the best available general correlation for all solute-solvent systems and a significant improvement over both the Wilke-Chang [24] and King et al. [25] equations. To be used in conjunction with Equation 13 were a number of specific rules. The first was that water molecules should be treated as dimers in that the molar volume and parachor values should be doubled. Organic acid solutes should likewise be treated as dimers except when diffusing in water, methanol, and butanol; then they should be treated as monomers. The final rule was that for diffusion of non-associating solutes in monohydroxy alcohols, a variable multiplying factor was to be used for the solvent molar volume and parachor that was related to the solvent viscosity:

$$f = 8\mu_B \tag{14}$$

For use in Equation 13, parachor values were obtained from a compilation by Quayle [28] or calculated by the addition method of Sugden [29] also discussed by Reid et al. [26]. Since the parachor is related to the surface tension of liquids, it was considered to be a more sensitive indicator of intermolecular forces than the latent heat of vaporization, particularly for polar compounds.

A more recent correlation using the same parameters and rules of application as those of Tyn and Calus [27] has been developed by Hayduk and Minhas [23]:

$$D^{\circ}_{AB} = 1.55(10^{-8})T^{1.29}\mu_B^{-0.92}V_B^{-0.23}P_B^{0.5}P_A^{-0.42} \tag{15}$$

A comparison of the four general equations for the prediction of diffusivity in non-electrolyte solutions is shown in Table 1. In the original comparison some 756 data points were selected as being the most recent or reliable (when several were available for one system) and to cover all types of solute-solvent interactions. The same comparison was not extended to the King et al. correlation [25] because the data for the latent heat of vaporization were not available for many of the solutes and solvents, especially for gaseous and solid solutes.

Both correlations utilizing parachor values yield a significantly lower average error than that obtained for the Wilke-Chang equation [24]. Equation 15 is only a slight improvement over Equation 13 so that either one may be used with essentially equal effectiveness. Since parachor values are readily available for many substances or can be estimated if required, the correlations involving the parachor Equations 13 and 15 are considered to be the best currently available as general

Table 1
Comparison of Correlations for Liquid Diffusivities at Infinite Dilution [23]

	% Average Error	Number of Data
General correlation for polar and non-polar substances:		
Wilke-Chang, Equation 1, for $\phi = 2.26$ for water solvent	16.1	756
Tyn-Calus, Equation 13, using parachor	14.2	756
Hayduk-Minhas, Equation 15, using parachor	13.9	756
Correlations tested with data for non-polar solvents only:		
Tyn-Calus	14.1	254
New, Equation 15 using parachor	13.4	254
Correlations tested for water as solvent:		
Tyn-Calus, Equation 13	13.3	237
New, Equation 15	12.8	237
Wilke-Chang with $\phi = 2.26$ Equation 1	10.4	237
Hayduk-Laudie, Equation 17	10.3	237
New, Equation 18	9.4	237
Correlations tested for normal paraffin solvents and solutes:		
Wilke-Chang, Equation 1	13.3	58
New, Equation 15	12.7	58
Tyn-Calus	12.5	58
New, Equation 19	3.4	58

correlations. When a diffusion coefficient is to be predicted for a specific solvent or solute-solvent system as detailed in section B, C, and D, more accurate correlations may be available.

Sample Prediction for Estimating the Diffusivity of Carbon Dioxide in Ethanol at 17°C

Using the recommended equation:

$$D^{\circ}_{AB} = 1.55(10^{-8})T^{1.29}\mu_B^{-0.92}V_B^{-0.23}P_B^{0.5}P_A^{-0.42} \quad (15)$$

and the correction factor for solvent molar volume and solvent parachor:

$$f = 8\mu_B \quad (14)$$

At 17°C, viscosity of ethanol is 1.26 cp:

$$f = 8(1.26)$$

The parachor value for ethanol may be obtained from Table 2:

$$P_B = 126.8$$

$$P'_B = 8(1.26)126.8 = 1278.1$$

$$V_B = 62.5 \text{ (Table 6)}$$

$$V'_B = 8(1.26)60.8 = 630.0$$

For carbon dioxide, the parachor may also be obtained from Table 2:

$$P_A = 77.5$$

Hence,

$$D^{\circ}_{AB}(17°C) = 1.55(10^{-8})(290.15)^{1.29}(1.26)^{-0.92}(612.9)^{-0.23}(1278.1)^{0.5}(77.5)^{-0.42}$$
$$= 2.46(10^{-5}) \text{ cm}^2/\text{s}$$

Reported diffusivity for carbon dioxide in ethanol at 17°C is $3.20(10^{-5})$ cm²/s (Reid et al. [26]).

Deviation: $(D_{calc} - D_{exp})100/D_{exp} = -23.1\%$

Also using the Tyn and Calus equation:

$$D^{\circ}_{AB} = 8.93(10^{-8})T\mu_B^{-1}V_A^{1/6}V_B^{-1/3}(P_B/P_A)^{0.6} \quad (13)$$

Carbon dioxide molar volume estimated at (see Table 6):

$$V_A = 34.0$$

Hence,

$$D^{\circ}_{AB} = 8.93(10^{-8})(290.15)(1.26)^{-1}(34.0)^{1/6}(630)^{-1/3}(1278.1/77.5)^{0.6}$$
$$= 2.32(10^{-5}) \text{ cm}^2/\text{s}$$

Deviation: $(D_{calc} - D_{exp})100/D_{exp} = -27.5$

Table 2
Parachors for Organic Compounds

Compounds Containing Hydrogen and Carbon Only:

Formula	Name	Parachor	Formula	Name	Parachor
CH_4	methane	72.6	$C_{12}H_{26}$	dodecane	510.1
C_2H_6	ethane	110.5	$C_{16}H_{34}$	hexadecane	670.6
C_3H_8	propane	150.8	$C_{32}H_{66}$	dotriacontane	1322
C_4H_{10}	butane	190.3			
C_5H_{12}	pentane	231.0	C_2H_4	ethylene	99.5
C_6H_{14}	hexane	270.7	C_2H_2	acetylene	88.6
C_7H_{16}	heptane	311.3	C_6H_6	benzene	205.7
C_8H_{18}	octane	351.1	C_7H_8	toluene	245.7
			C_6H_{12}	cyclohexene	242.1
			C_3H_8	propylene	139.9
			C_8H_{10}	ethylbenzene	284.3
			C_8H_{10}	o-xylene	285.4

Compounds Containing Oxygen:

Formula	Name	Parachor	Formula	Name	Parachor
CO	carbon monoxide	61.6	C_2H_6O	methyl ether	136.0
CO_2	carbon dioxide	77.5	$C_4H_8O_2$	1,4-dioxane	202.0
CH_4O	methanol	88.8	$C_4H_{10}O$	ethyl ether	211.9
C_2H_6O	ethanol	126.8	$C_6H_{14}O$	n-proply ether	290.9
C_3H_8O	propanol	165.0			
C_3H_8O	2-propanol	164.4	C_3H_6O	methyl acetate	177.1
$C_6H_{12}O$	cyclohexanol	255.5	$C_4H_8O_2$	ethyl acetate	216.0
$C_2H_6O_2$	ethylene glycol	148.9	$C_5H_{10}O_2$	propyl acetate	255.3
C_6H_6O	phenol	222.0	$C_8H_8O_2$	methyl benzoate	310.4
C_7H_8O	o-cresol	275.5	$C_9H_{10}O_2$	ethyl benzoate	347.6
$C_4H_{10}O$	butanol	186.1	$C_5H_{10}O_2$	isopropyl acetate	254.4
C_2H_4O	ethylene oxide	112.6	C_4H_4O	furan	160.4
$C_2H_4O_2$	acetic acid	131.2	C_3H_60	acetone	161.7
$C_3H_6O_2$	propionic acid	169.0	C_4H_8O	methyl ethyl ketone	198.8
CH_2O_2	formic acid	94.0	$C_6H_{10}O$	cyclohexanone	253.0
			C_8H_8O	acetophenone	293.8
C_7H_6O	benzaldehyge	256.2			

Compounds Containing Oxygen:

Formula	Name	Parachor	Formula	Name	Parachor
CH_3O_2N	nitromethane	132.2	C_8H_9NO	acetanilide	321.8
$C_2H_5O_2N$	nitroethane	170.7	CH_3ON	formamide	107.2
$C_3H_7O_2N$	1-nitroethane	208.1	C_2H_5ON	acetamide	148.0
$C_6H_5O_2N$	nitrobenzene	262.6	$C_4H_{11}N$	diethylamine	217.5
$C_7H_7O_2N$	o-nitrotoluene	297.7	C_2H_7N	dimethylamine	136.6
$C_6H_5O_3N$	o-nitrophenol	273.5	C_7H_9N	benzylamine	273.7
			$C_6H_{13}N$	cyclohexylamine	298.6
C_7H_5N	benzonitrile	256.8	C_6H_7N	aniline	235.2
C_2H_3N	acetonitrile	122.2	C_5H_5N	pyridine	197.4
C_3H_5N	propionitrile	160.5	CH_5N	methylamine	95.9
			C_2H_7N	ethylamine	139.9
			C_7H_9N	o-toluidine	272.0

Compounds Containing Sulfur or Halogens:

Formula	Name	Parachor	Formula	Name	Parachor
CS_2	Carbon disulfide	143.6	CCl_4	carbon tetrachloride	219.9
C_2H_6S	dimethyl sulfide	163.0	$CHCl_3$	chloroform	183.4
$C_4H_{10}S_2$	dimethyl disulfide	290.5	C_2H_5Cl	chloroethane	151.6
$C_2H_6S_2$	dimethyl disulfide	213.6	C_6H_5Cl	chlorobenzene	244.5
C_2H_3NS	methyl thiocyanate	168.5	C_6H_5Br	bromobenzene	257.8
C_3H_5NS	ethyl thiocyanate	206.8			

Table 3
Structural Contributions for Calculation of the Parachor

Carbon-Hydrogen:	
C	9.0
H	15.5
CH_3—	55.5
—CH_2—	40.0
CH_3—$CH(CH_3)$—	133.3
CH_3—CH_2—$CH(CH_3)$—	171.9
CH_3—CH_2—CH_2—$CH(CH_3)$—	211.7
CH_3—$CH(CH_3)$—CH_2—	173.3
CH_3CH_2—$CH(C_2H_5)$—	209.5
CH_3—$C(CH_3)_2$—	170.4
CH_3—CH_2—$C(CH_3)_2$—	207.5
CH_3—$CH(CH_3)$—$CH(CH_3)$—	207.9
CH_3—$CH(CH_3)$—$C(CH_3)_2$—	243.5
C_6H_5—	189.6
Special Groups:	
—COO—	63.8
—COOH	73.8
—OH	29.8
—NH_2	42.5
—O—	20.0
—NO_2 (nitrite)	74
—NO_3 (nitrate)	93
—CO (NH_2)	91.7
=O (ketone)	
Three carbon atoms	22.3
Four carbon atoms	20.0
Five carbon atoms	18.5
Six carbon atoms	17.3
—CHO	66
O (not noted above)	20
N (not noted above)	17.5
S	49.1
P	40.5
F	26.1
Cl	55.2
Br	68.0
I	90.3
Ethylenic Bond:	
Terminal	19.1
2,2 position	17.7
3,4 position	16.3
Triple bond	40.6
Ring Closure:	
Three-membered	12.5
Four-membered	6.0
Five-membered	3.0
Six-membered	0.8

If $n \geq 12$ in $(—CH_2—)_n$, increase increment to 40.3.

Correlations for Non-Polar Solvents

Equations for predicting molecular diffusivity should be more accurate when based only on data for non-polar solvents. When the average errors obtained by Equations 13 and 15 and using data for non-polar solvents only were compared with those obtained when data for polar solvents were included, a reduction in the average error was indeed reported (Table 1). A comparison for some 254 non-polar solvents yielded average errors of 14.1% and 13.4% for Equations 13 and 15, respectively. Equation 15 is recommended not only because it yields the lowest average error when tested with diffusivities for non-polar solvents, but also because it is nearly equally successful for predicting diffusivities in polar solvents as well. Hence the same equation can be used as a general correlation and for non-polar solvents with some confidence.

The empirical nature of Equation 15 is indicated by the fractional indices for all the variables. In the presentation of several empirical equations [23], no attempt was made to attach any theoretical significance to the dependence of the molecular diffusivity on the various parameters. The method for evaluating the indices was indicated to involve a trial-and-error procedure in which the form of the equation was specified and the constants were evaluated using essentially a non-linear regression analysis of the data involved. Each of the resulting equations is the product of a number of trials involving many forms of tentative correlating equations of varying complexity. Hence the equations can be considered to be a statistical representation of the available data with the high probability that it can be used to predict diffusivities with equivalent accuracy.

Sample Prediction for Non-Polar Solvent, Hexane in Carbon Tetrachloride at 25°C

Using the recommended equation:

$$D^{\circ}_{AB} = 1.55(10^{-8})T^{1.29}\mu_B^{-0.92}V_B^{-0.23}P_B^{0.5}P_A^{-0.42} \tag{15}$$

Parachor values for hexane and carbon tetrachloride from Table 2:

$P_A = 270.7$

$P_B = 219.9$

Molar volumes at the normal boiling point:

$V_A = 141.0$ (hexane; not required)

$V_B = 102.0$ (carbon tetrachloride)

$\mu_B = 0.906$ cP.

Hence,

$$D^{\circ}_{AB} = 1.55(10^{-8})(298.15)^{1.29}(0.906)^{-0.92}(102)^{-0.23}(219.9)^{0.5}(270.7)^{-0.42}$$
$$= 1.29(10^{-5})\ \text{cm}^2/\text{s}$$

Experimental value as listed in Hayduk and Buckley [30]:

$D_{exp} = 1.50(10^{-5})$ cm²/s

Deviation: $(D_{calc} - D_{exp})100/D_{exp} = -14.0\%$

Correlations for Non-Electrolytes in Aqueous Solution

A number of correlations are available for the diffusivities of non-electrolytes in water; of these seven will be mentioned and compared. The Wilke-Chang and Tyn and Calus [27] equations were formulated as general equations for predicting diffusivities in a range of solvents including water. In addition to those is the Othmer-Thakar [31] equation and the slight variation of it proposed by Hayduk and Laudie [32] specifically for water as solvent. The last two are shown in the order mentioned:

$$D^{\circ}_{AB} = 14.0(10^{-5})\mu_B^{-1.1}V_A^{-0.6} \tag{16}$$

$$D^{\circ}_{AB} = 13.26(10^{-5})\mu_B^{-1.14}V_A^{-0.589} \tag{17}$$

More than ten years ago a compilation of some 89 solutes in water, reported in the literature between 1950 and 1973, were used to test the relative effectiveness of the four correlations mentioned above [32]. For those data, the Wilke-Chang equation predicted diffusivities that were an average of 6.9% too high when compared with the actual data. The correlation was not tested for other solvents at that time so that there appeared to be no justification for recommending a change in the constant in the equation. The *association parameter* was specific to water however, so that predicted diffusivities in water could be suitably adjusted by reducing the value of the parameter from 2.6 to 2.26. When again tested with the data, that adjustment to the association parameter of the Wilke-Chang equation yielded a low average error of 5.8%, nearly identical to those obtained for the Othmer-Thakar and Hayduk-Laudie equations. The result of that comparison for diffusivity at 25°C is

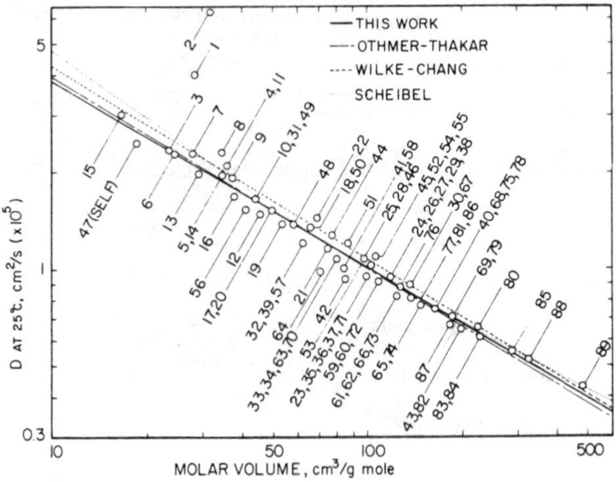

Figure 2. Correlation of diffusivity in water as a function of solute molar volume at the normal boiling point.

graphically reproduced in Figure 2. The tabulated values and sources are shown in Table 4. That the Hayduk-Laudie equation, based on more recent data, yielded nearly the same constants as the Othmer-Thakar equation, developed on the basis of data reported prior to 1950, was considered significant. As a consequence of the comparison of the two correlations, it was considered that data published prior to 1950 were as reliable as those published between 1950 and 1973, and that the Othmer-Thakar equation was a highly successful correlation in its original form. In that same survey [32] a comparison of whether the true solute molar volumes yield more accurate predicted diffusivities than those based on LeBas molar volume was also made. To determine which method of expressing molar volumes was more effective, diffusivities were estimated using molar volumes determined by *both* methods when this was possible, and average errors were calculated in each case. As anomalous as it may seem, it was reported that when molar volumes estimated by the LeBas method were used in three of the equations, the average errors were 0.1% lower than they were for the same equations using the true solute molar volumes, although the difference was not considered significant. It was concluded that when using equations involving solute molar volumes, the estimated LeBas molar volumes and actual ones based on solute densities may be expected to give essentially equivalent predicted diffusivities.

Recently, all the available data for diffusivities of dilute solutes in water were again analyzed and correlated, this time by Hayduk and Minhas [23]. All the available data were used, including that reported before 1950 as well as that reported since 1973 and which comprised some 237 different values. In addition to the four correlations mentioned earlier, three more were examined. These were the general correlation for all solvents of Tyn and Calus [27] (Equation 13), the general equation developed for all solvents utilizing the parachor (Equation 15), and finally a new correlation for diffusivities in water only. The last-mentioned equation follows:

$$D^{\circ}_{AB} = (1.25V_A^{-0.19} - 0.365)10^{-8}\mu_B^{(9.58/V_A - 1.12)}T^{1.52} \tag{18}$$

Again diffusivities were predicted by means of the correlations and compared with the reported diffusivities, the comparison made on the basis of an average error of estimation. Probably because of the much large number of data examined, the average errors reported for all correlations were considerably larger than reported previously [32]. It is considered that the relative success of the correlations is indicated by the relative reduction in average error. The new correlation was the result

Table 4
Diffusivities in Water at 25°C and Sources

No.	Substance	V_2 LeBas	V_2 Actual	$D \times 10^5$	Sources for diffusivities
1.	H_2	14.3	28.5[a]	4.40	Ferrell and Himmelblau, [37]
2.	He	—	31.9[a]	6.28	Ferrell and Himmelblau, [37]
3.	NO	23.6	23.6[b]	2.34	Wise and Houghton, [38]
4.	N_2O	36.4	36.0[a]	2.10	Thomas and Adams, [39]
5.	N_2	31.2	34.7[a]	1.99	Ferrell and Himmelblau, [37]
6.	NH_3	25.8	24.5[a]	2.28	Himmelblau, [8]
7.	O_2	25.6	27.9[a]	2.29	Ferrell and Himmelblau, [37]
8.	CO	30.7	34.5[a]	2.30	Wise and Houghton, [38]
9.	CO_2	34.0	37.3[a]	1.92	Ferrell and Himmelblau, [37]
10.	SO_2	44.8	43.8[a]	1.59	Groothuis and Kramers, [40]
11.	H_2S	32.9	35.2[a]	2.10	Himmelblau, [8]
12.	Cl_2	48.4	45.5[a]	1.48	Vivian and Peaceman, [41]
13.	Ar	—	29.2[a]	1.98	Baird and Davidson, [42]
14.	Kr	—	34.9[c]	1.92	Wise and Houghton, [38]
15.	Ne	—	16.8[c]	3.01	Houghton et al, [43]
16.	CH_4	29.6	37.7[b]	1.67	Witherspoon and Bonoli, [44]
17.	CH_3Cl	47.5	50.6[a]	1.49	Chiang and Toor, [45]
18.	C_2H_3Cl	62.3	64.5	1.34	Hayduk and Laudie, [32]
19.	C_2H_6	51.8	53.5[c]	1.38	Witherspoon and Bonoli, [44]
20.	C_2H_4	44.4	49.4[b]	1.55	Huq and Wood, [46]
21.	C_3H_8	74.0	74.5[b]	1.16	Witherspoon and Bonoli, [44]
22.	C_3H_6	66.6	69.0[a]	1.44	Vivian and King, [47]
23.	C_4H_{10}	96.2	96.6[d]	0.97	Witherspoon and Saraf, [5]
24.	Pentane	118	118[d]	0.97	Witherspoon and Bonoli, [44]
25.	C-pentane	99.5	97.9[e]	1.04	Bonoli and Witherspoon, [48]
26.	C-hexane	118	117[b]	0.90	Bonoli and Witherspoon, [48]
27.	M-c-pentane	112	120[e]	0.93	Bonoli and Witherspoon, [48]
28.	Benzene	96.0	96.5[b]	1.09	Ratcliff and Reid, [49]
29.	Toluene	118	118[d]	0.95	Bonoli and Witherspoon, [48]
30.	E-benzene	140	140[d]	0.90	Bonoli and Witherspoon, [48]
31.	Methanol	37.0	42.5[b]	1.66	Gary-Bobo and Weber, [50]
32.	Ethanol	59.2	62.6[e]	1.24	Hammond and Stokes, [51]
33.	Propanol	81.4	81.8[b]	1.12	Gary-Bobo and Weber, [50]
34.	I-propanol	81.4	81.8[e]	1.08	Gary-Bobo and Weber, [50]
35.	Butanol	104	101[e]	0.98	Lyons and Sandquist, [52]
36.	I-butanol	104	101[e]	0.93	Garner and Marchant, [53]
37.	N-butanol	104	102[e]	0.97	Gary-Bobo and Weber, [50]
38.	Benzyl alcohol	126	121[e]	0.93	Garner and Marchant, [53]
39.	E-glycol	66.6	63.5[f]	1.16	Byers and King, [54]
40.	Tri-e-glycol	170	163[f]	0.76	Merliss and Colver, [55]
41.	Pro-glycol	88.8	84.5[f]	1.00	Garner and Marchant, [53]
42.	Glycerol	96.2	85.1[e]	0.93	Garner and Marchant [53]
43.	Ethane-hex-diol	200	197[g]	0.64	Garner and Marchant, [53]
44.	Acetone	74.0	77.5[b]	1.28	Anderson et al, [56]
45.	E-acetate	111	106[b]	1.12	Chandrasekaran and King, [57]
46.	Furfural	95.7	97.5[e]	1.12	Lewis, [58]
47.	Water (self)	18.9	18.8	2.45	Robinson and Stokes, [59]

(Continued)

Table 4 (Continued)

No.	Substance	V_2 LeBas	V_2 Actual	D × 10^5	Sources for diffusivities
48.	Urea	58.0	^{h}	1.38	Gosting and Akeley, [60]
49.	Formamide	43.8	46.6^e	1.67	Gary-Bobo and Weber, [50]
50.	Acetamide	66.0	67.1^g	1.32	Gary-Bobo and Weber, [50]
51.	Propionamide	88.2	87.4^e	1.20	Gary-Bobo and Weber, [50]
52.	Butyramide	110	109^g	1.07	Gary-Bobo and Weber, [50]
53.	I-butyramide	110	103^g	1.02	Gary-Bobo and Weber, [50]
54.	Diethylamine	112	109^b	1.11	Lewis, [58]
55.	Aniline	110	106^e	1.05	Lewis, [58]
56.	Formic acid	41.6	41.1^e	1.52	Bidstrup and Geankoplis, [60]
57.	Acetic acid	63.8	64.1^b	1.19	Lewis, [58]
58.	Propionic acid	86.0	85.3^e	1.01	Albery et al, [61]
59.	Butyric acid	108	108^e	0.92	Bidstrup and Geankoplis, [60]
60.	I-butyric acid	108	109^b	0.95	Albery et al., [61]
61.	Valeric acid	130	129^e	0.82	Bidstrup and Geankoplis, [60]
62.	Tri-m-acetic acid	130	127^g	0.82	Albery et al., [61]
63.	Chloracetic acid	81.7	79.6^e	1.04	Garland and Stockmayer, [62]
64.	Glycolic acid	71.2	^{h}	0.98	Albery et al., [61]
65.	Caproic acid	153	^{h}	0.78	Bidstrup and Geankoplis, [60]
66.	Succinic acid	120	^{h}	0.86	Albery et al., [61]
67.	Glutaric acid	142	^{h}	0.79	Albery et al., [61]
68.	Adipic acid	165	^{h}	0.74	Albery et al., [61]
69.	Pimelic acid	187	^{h}	0.71	Albery et al., [61]
70.	Glycine	78.0	^{h}	1.06	Longsworth, [63, 64]
71.	Alanine	100	^{h}	0.91	Longsworth, [64]
72.	Serine	108	^{h}	0.88	Longsworth, [64]
73.	Aminobutyric acid	122	^{h}	0.83	Longsworth, [64]
74.	Valine	145	^{h}	0.77	Longsworth, [64]
75.	Leucine	167	^{h}	0.73	Longsworth, [64]
76.	Proline	127	^{h}	0.88	Longsworth, [64]
77.	Hydroxy-proline	135	^{h}	0.83	Longsworth, [64]
78.	Histidine	168	^{h}	0.73	Longsworth, [64]
79.	Phenylalanine	189	^{h}	0.705	Longsworth, [64]
80.	Tryptophan	223	^{h}	0.660	Longsworth, [64]
81.	Diglycine	138	^{h}	0.791	Longsworth, [64]
82.	Triglycine	198	^{h}	0.665	Longsworth, [64]
83.	Glycyl-leucine	227	^{h}	0.623	Longsworth, [64]
84.	Leucyl-glycine	227	^{h}	0.613	Longsworth, [64]
85.	Leucyl-glycyl-glycine	287	^{h}	0.551	Longsworth, [64]
86.	Amino benzoic acid	144	^{h}	0.840	Longsworth, [64, 65]
87.	Dextrose	166	^{h}	0.675	Gladden and Dole, [66]
88.	Sucrose	325	^{h}	0.524	Henrion, [67]
89.	Raffinose	480	^{h}	0.434	Longsworth, [64]

Sources for Molar volume or density data: a. Himmelblau, [8]; b. Reid and Sherwood, [34]; c. Bell, [68]; d. Rossini, [69]; e. Timmermans, [70]; f. Gallant, [71]; g. Weast, [72]; h. substance sublimes or decomposes.

of testing some six variations of Equation 17 utilizing the non-linear regression technique. The slightly more complex form of the equation, Equation 18, gave an average deviation of 9.4% as compared with 10.3% for Equation 17 and 10.4% for the Wilke-Chang equation using an association factor of 2.26. The Othmer-Thakar equation was considered equivalent to Equation 17. The other equations yielded somewhat higher average errors as shown in Table 1.

Equation 18 is considered a significant improvement for predicting diffusivities in water and is recommended. Only slightly higher average errors would result if Equations 16 or 17 were used. It is considered of interest that the recent equations developed for water yield average errors smaller than the average errors for the comparable general correlations for all solvents combined, and even those for the non-polar solvents.

Sample Prediction in Aqueous Solution, Vinyl Chloride in Water at 50°C

Using the recommended equation:

$$D^{\circ}_{AB} = (1.25V_A^{-0.19} - 0.365)10^{-8}\mu_B^{(9.58/V_A - 1.12)}T^{1.52} \quad (18)$$

The molar volume at the normal boiling point for vinyl chloride is estimated using the LeBas method [26]:

$$V_A = 62.3$$

$$\mu_B(50°C) = 0.5468cP$$

Then:

$$D^{\circ}_{AB} = [1.25(62.3^{-0.19}) - 0.365]10^{-8}0.5468^{(9.58/62.3 - 1.12)}2.98^{1.52}$$
$$= 2.12(10^{-5})\ cm^2/s$$

Experimentally determined diffusivity at 50°C [32] is $2.42(10^{-5})\ cm^2/s$.

Deviation: $(D_{calc} - D_{exp})100/D_{exp} = -12.4\%$

Using another equation:

$$D^{\circ}_{AB} = 13.26(10^{-5})\mu_B^{-1.14}V_A^{-0.589} \quad (17)$$

$$D_{calc} = 13.26(10^{-5})/[0.5468^{1.14}(62.3)^{0.589}]$$
$$= 2.31(10^{-5})\ cm^2/s$$

Deviation $= -4.5\%$

In this case the older equation gives a better estimate of the diffusivity.

Correlations for Alkane Solutes in Alkane Solvents

A viscosity-diffusivity "map" for normal alkanes [4], which graphically indicates the relationship between diffusivity versus solvent viscosity at 25°C for solutes of different carbon content, was used as a basis for correlating diffusivities in normal alkane systems. The "map" is reproduced in Figure 3. Diffusivities were obtained from the graph corresponding to a particular viscosity, 1 mPa·s, (cP). It was found that diffusivities at one solvent viscosity could be correlated as a function of solute

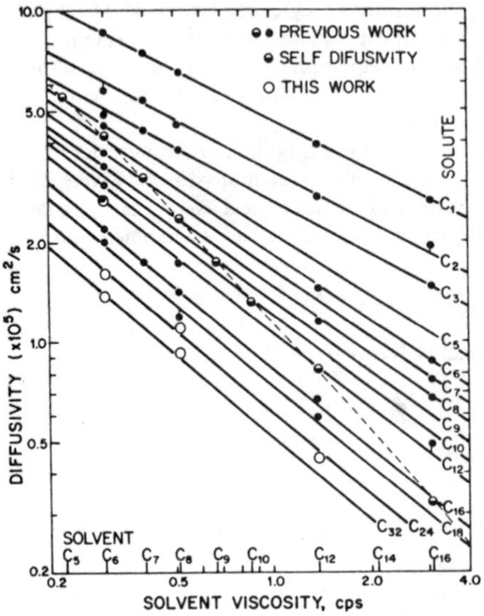

Figure 3. Diffusivity-viscosity "map" for n-paraffin solutions at 25°C.

Table 5
Diffusivity Data at 25°C in N-Alkane Solutions

	10^5D^0, cm^2/s in Solvents Indicated						
Solute	**C_6**	**C_7**	**C_8**	**C_9, C_{10}**	**C_{12}**	**C_{16}**	**Ref.**
C_1	8.64	7.52	6.49		3.94	2.66	h
C_2	5.79	5.44	4.57		2.73	1.95	i
C_3	4.87	4.40	3.83			1.48	j
C_5	4.59						b
C_6	4.13				1.39		f
	4.21[a]				1.42[d]	0.85[c]	
C_7	2.78[b]	3.12[a]				0.74[d]	
C_8	3.47[b]		2.36[e]		1.14[e]	0.67[f]	
C_9				1.70			a
C_{10}	3.02[b]			1.31[a]			
C_{12}	2.74[b]					0.49[g]	
	2.71				0.84	0.57	f
			1.72		0.81		e
C_{16}	2.21[c]	1.78[d]	1.43[f]		0.67[g]		
C_{18}	2.01[b]		1.20[e]		0.59[e]		
C_{24}	1.59[k]		1.09[k]		0.445[k]		
C_{32}	1.36[k]		0.916[k]				

[a] *Douglas and McCall [73]*
[b] *Bidlack et al. [74]*
[c] *Bidlack and Anderson [74]*
[d] *Bidlack and Anderson [74]*
[e] *Van Geet and Adamson [75]*
[f] *Shieh and Lyons [76]*
[g] *Kett [76]*
[h] *Hayduk and Buckley [30]*
[i] *Hayduk and Cheng [77]*
[j] *Hayduk et al. [10]*
[k] *Hayduk and Ioakimidis [4]*

molar volume. The slopes of the lines were read and related again to the solute molar volume. Thus, a basic form of the correlation was obtained. The effect of temperature was determined by considering the index of temperature to be variable. The final form of the equation for normal alkane solutions is as follows:

$$D^{\circ}_{AB} = 13.3(10^{-8})T^{1.47}\mu_B^{(10.2/V_A - 0.791)}V_A^{-0.71} \tag{19}$$

Equation 19 was developed utilizing some 58 data points for normal alkane solutes from C_5 to C_{32} and for normal alkane solvents from C_5 to C_{16}. Data utilized included those for self-diffusivities as well as those listed for different temperatures by Tyn and Calus [27]. These data are listed in Table 5. After several trials using equations of different form, it was found that Equation 19 yielded the lowest average deviation from the experimental results of 3.4%. The low average error is considered exceptional since the diffusivity data were obtained by a number of workers using several different methods for which the inherent experimental error could easily have approached the average deviation in correlation. This good correlation would appear to confirm that the mechanism for diffusion is similar for all long-chained alkane solutes in normal alkane solutions. Furthermore, there appears to be an interaction between the solute molar volume and the solvent viscosity since the effect of solvent viscosity on the diffusion rate is different depending on how large the diffusing molecules are. This latter observation may be made from an inspection of the complex index of the viscosity in Equation 19 or by inspecting the changing slope of the diffusivity-viscosity relationship reproduced in Figure 4, for solutes of various carbon numbers.

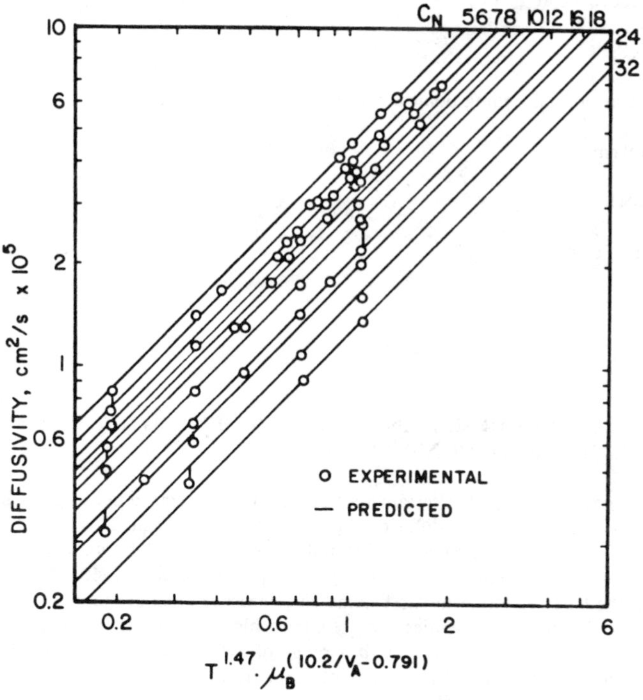

Figure 4. Diffusivities at infinite dilution in n-paraffin solutions versus the factor $T^{1.47}\mu_B^{(10.2/V_A - 0.791)}$ at temperatures ranging from 0 to 100°C.

No correlations utilizing parachor values have been reported specifically for normal alkane systems; hence it is not possible to indicate conclusively whether or not such correlations would improve accuracy. Some improvement appears possible if such a correlation were developed, simply because using the parachor improved the accuracy of several other diffusivity correlations. In that case, however, such improvements, if any, would probably be small because the recommended correlation already yields probable errors of the order of the accuracy of the measurements themselves.

The correlation for dilute solutions of normal alkane solutes in normal alkane solvents was considered successful not only because it accurately represented the consistent data available for these systems, but also because self-diffusivities of normal alkane solvents were correlated equally well when considered as solutes diffusing through solvents of the same molecular weight. The consistency of the correlation in expressing self-diffusivities is considered significant by comparison with some other correlations in that self-diffusivities are not always correlated in this way. Hence, this correlation appears highly self-consistent. Finally, it is observed that there is probably no other series of chemicals exhibiting such a wide range in molecular size as the normal alkanes, but which still belong to the same homologous series.

Sample Prediction in Alkane System, Hexane in Dodecane at 25°C

Using the recommended equation:

$$D^{\circ}_{AB} = 13.3(10^{-8})T^{1.47}\mu_B^{(10.2/V_A - 0.791)}V_A^{-0.71} \tag{19}$$

Viscosity of dodecane at 25°C, μ_B:

$$\mu_B = 1.378\text{cP}$$

Molar volume at normal boiling point for hexane, V_A:

$$V_A = 141.0\ \text{cm}^3/\text{mole}$$

$$D^{\circ}_{AB} = 1.365(10^{-5}\ \text{cm}^2/\text{s})$$

Experimentally determined [76],

$$D_{exp} = 1.39(10^{-5})\ \text{cm}^2/\text{s}$$

Deviation: $(D_{calc} - D_{exp})100/D_{exp} = -1.7\%$

EFFECT OF MOLECULAR SIZE AND SHAPE ON DIFFUSIVITY IN DILUTE LIQUID SOLUTIONS

It is of theoretical interest and in certain applications it is of practical use to know how the solute molecular shape and size affect the diffusion rate in a particular solvent. This subject was addressed some time ago by Hayduk and Buckley [30] and some useful results obtained. A comparison of the various diffusion coefficients in two non-associating solvents, carbon tetrachloride and hexane, was made on the basis of solute molar volume as a measure of solute molecular size. Over fifty data points were chosen for diffusivities in carbon tetrachloride of solutes ranging in molecular size from hydrogen to dotriacontane (C_{32}) with a range of diffusivity values of over twenty-fold. In addition, as many reliable data for hexane solvent were also compared. Figure 5 shows these comparisons.

For comparison, molar volumes were experimentally measured or estimated by either the LeBas incremental method, or, for a number of the lower molecular weight substances for which critical volumes and critical pressures were available, by means of the following Benson equation [33]

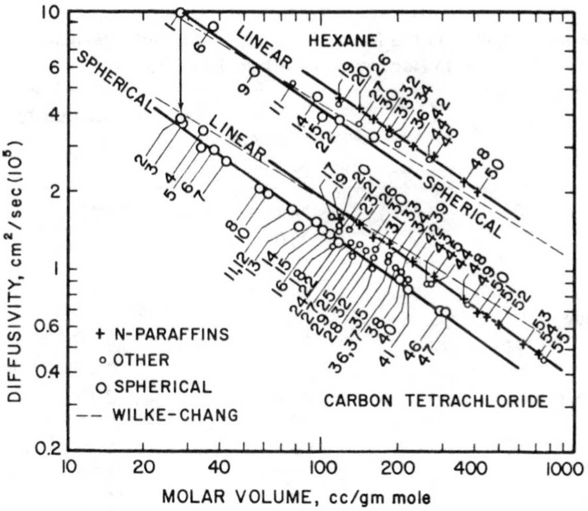

Figure 5. Diffusivities at 25°C in carbon tetrachloride and hexane as a function of molar volume at the normal boiling point and shape of solute molecules compared with diffusivities estimated using the Wilke-Chang equation.

recommended by Reid and Sherwood [34]:

$$V_c/V = 0.422 \log P_c + 1.981 \tag{20}$$

A comparison of molar volumes determined by the LeBas method and Benson equation with those experimentally measured indicated that the Benson equation gives a better estimate than the LeBas method, especially for small molecules. While the Wilke-Chang line of slope 0.6 in Figure 5 fairly represents the diffusivities in hexane, it poorly represents carbon tetrachloride. Parallel lines of slope 0.70 much better represent the data for diffusivities in carbon tetrachloride, where one of the lines represents diffusivities of essentially spherical molecules and the other for diffusivities of essentially linear molecules. Molecules such as O_2, Ar, CH_4, CF_4, SF_6 were considered spherical while others such as ethane, hexachloro-ethane, benzene, phenol, tetrahydrofuran, methanol, cyclohexane or triphenyl methane were only considered to approach sphericity.

The diffusivities for linear paraffins from pentane to dotricontane showed distinctly separate lines in both hexane and carbon tetrachloride solvents, which were approximately 30% higher than for substances essentially spherical in molecular structure. Furthermore, hexadecanol and palmitic acid dimer, with diffusivities that correspond closely with those for hexadecane, and dotricontane, respectively, indicate that the particular diffusivity behavior is more a function of molecular shape than chemical nature. Substances such as isooctane, biphenyl, benzoic acid dimer, anthracene, phenanthrene, and branched paraffins appear to have diffusivities approximately 15% greater than for essentially spherical molecules of the same molar volume. Similar trends for diffusivities in hexane are observed. The data sources for Figure 5 are given in the original reference and in Table 6.

No recent work has been found that further elucidates the effect of molecular shape on molecular diffusivity. It may be concluded, however, that the solute molecular shape cannot be ignored in correlations for diffusivity without loss in accuracy. One may speculate that while elongated molecules appear generally to diffuse at a higher rate than spherical molecules of the same molar volume, this may not be true in solvents of highly differing chemical nature. It has been frequently postulated that dissolved long-chained *polymer* molecules uncoil or extend most completely in "good" solvents, whereas they tend to assume a more compact, coiled-up or spherical shape in a poor (or theta)

Table 6
Molar Volumes at the Normal Boiling Point, and Diffusivities at 25° in Hexane and Carbon Tetrachloride

Solute	Molar volume (cm^3/g mole)	Diffusivity × (10^5) cm^2/sec	
		in CCl_4	in Hexane
1. Hydrogen	28.4[c]	9.75[d]	
2. Oxygen	28.0[c]	3.82[e]	
3. Argon	28.5[c]	3.63[d]	
4. Nitrogen	34.6[c]	3.42[d]	
5. Carbon dioxide	34.0[b]	2.95[e]	
6. Methane	37.7[c]	2.89[d]	8.64[p]
7. Methanol	42.5[c]	2.61[f]	
8. Carbon tetrafluoride	57.4[b]	2.04[d]	
9. Ethane	55.1[c]		5.79[o]
10. Ethanol	62.5[c]	1.95[f]	
11. Acetone	77.5[c]	1.70[g]	5.26[l]
12. Sulfur hexafluoride	77.8[a]	1.71[e]	
13. Tetrahydrofuran	82.9[b]	1.47[f]	
14. Benzene	96.5[c]	1.54[h]	4.64[l]
15. Carbon tetrachloride	102.0[c]	1.41[i]	3.86[l]
16. Phenol	109.0[a]	1.37[f]	
17. Aniline	111.0[a]	1.58[j]	
18. Dimethylacetamide	112.0[a]	1.23[f]	
19. Pentane	118.0[a]	1.57[k]	4.59[l]
20. 2-Methylbutane	118.0[a]	1.49[l]	4.40[l]
21. Toluene	117.0[a]	1.40[f]	
22. Cyclohexane	118.0[a]	1.27[m]	3.77[l]
23. Acetic acid (dimer)	128.0[a]	1.42[f]	
24. Benzyl alcohol	131.0[a]	1.28[f]	
25. Dimethyl propionamide	134.0[a]	1.14[f]	
26. Hexane	141.0[a]	1.50[k]	4.21[n]
27. 2-2 dimethyl butane	141.0[a]	1.25[l]	3.63[l]
28. Hexachloro ethane	159.0[a]	1.01[f]	
29. Napthalene	148.0[a]	1.20[f]	
30. Heptane	162.0[a]	1.34[k]	3.78[l]
31. Mesitylene	163.0[a]	1.19[f]	
32. Tetralin	162.0[a]	1.10[l]	3.27[l]
33. Octane	185.0[a]	1.26[k]	3.47[l]
34. Iso octane	185.0[a]	1.13[l]	3.38[l]
35. Biphenyl	185.0[a]	1.07[f]	
36. Phenanthrene	197.0[a]	1.03[l]	3.08[l]
37. Anthracene	197.0[a]	1.03[f]	
38. Hexachlorobenzene	203.0[a]	0.922[f]	
39. Biphenyl methane	207.0[a]	0.985[f]	
40. Benzohydrol	219.0[a]	0.918[f]	
41. Hexachlorocyclohane	220.0[a]	0.843[f]	
42. Decane	229.0[a]	1.09[kl]	3.02[l]
43. Benzoic acid (dimer)	261.0[a]	0.882[f]	
44. 2-Methyl propene (trimer)	266.0[a]	0.884[l]	2.68[l]
45. Dodecane	274.0[a]	0.954[f]	2.74[l]

(Continued)

Table 6 (Continued)

Solute	Molar volume (cm^3/g mole)	Diffusivity × (10^5) cm^2/sec in CCl_4	in Hexane
46. Triphenyl methane	295.0[a]	0.694[f]	
47. Triphenylmethanol	307.0[a]	0.687[f]	
48. Hexadecane	363.0[a]	0.780[f]	2.21[l]
49. Hexadecanol	375.0[a]	0.741[f]	
50. Octadecane	407.0[a]	0.690[k]	2.01[l]
51. Eicosane	451.0[a]	0.664[f]	
52. Docosane	496.0[a]	0.620[f]	
53. Octacosane	629.0[a]	0.528[f]	
54. Dotricontane	718.0[a]	0.479[f]	
55. Palmitic acid (dimer)	749.0[a]	0.448[f]	

[a] *LeBas incremental method*
[b] *Benson equation*
[c] *Experimental*
[d] *Ross and Hildebrand [3]*
[e] *Nakanishi et al. [78]*
[f] *Longsworth [79]*
[g] *Anderson and Babb [80]*
[h] *Cadwell and Babb [81]*
[i] *Watts et al. [82]*
[j] *Rao and Bennett [83]*
[k] *Dewan and Van Holde [84]*
[l] *Bidlack et al. [74]*
[m] *Hammond and Stokes [51]*
[n] *Douglas and McCall [73]*
[o] *Hayduk and Cheng [77]*
[p] *Hayduk and Buckley [30]*

solvent. If it is postulated that this is also true for much smaller molecules, then the diffusivity might be expected to vary by some 30% depending on whether the solvent was a good, or a poor solvent for the particular solute. These aspects may, in part, explain the difficulty in achieving accurate diffusivity correlations involving solute molecules of various shapes.

THE PARACHOR

The parachor is a property related to surface tension that has only recently been used to correlate molecular diffusivities, even though it was originally considered to be a "true measure of molecular volume" [29]. Its successful use for such correlations suggests that it is a better measure of molecular size for describing molecular motion than the molar volume at the normal boiling point, or simply the molar volume. The parachor is related to the liquid surface tension, molecular weight, and densities of the liquid and vapor phases [29]:

$$P = \sigma^{0.25} M(\rho_L - \rho_V)^{-1} \tag{21}$$

In Equation 21 the units customarily used are dynes/cm for surface tension, and g/cm^3 for the densities; hence it is a dimensional parameter. The density divided by the molecular weight (ρ/M) may be considered a molar concentration and is sometimes indicated as such [26]. The vapor density is frequently considered negligible compared with the liquid density and is omitted. The resulting expression is:

$$\rho = \sigma^{0.25} M(\rho_L)^{-1} \tag{22}$$

An extensive listing of parachor values is available for some 1,600 organic compounds as compiled by Quayle [28]. A much abreviated list is included in Table 2. If a required parachor is not otherwise available, a simple procedure may be used for its estimation as originally suggested by Sugden [35] and modified by Quayle [28]; it is also presented in this section. It is useful to know that for most unassociated liquids the parachor is subject to only minor variations with temperature.

Estimation of Parachor by Sugden-Quayle Method

Based essentially on the contribution of atoms, or recurring structures in organic molecules, an additive method was devised by Sugden [29] and improved by Quayle [28] for estimating parachor. An important component of this development was determining the contribution of the methylene ($—CH_2—$) unit to the parachor of many organic compounds. By comparing the parachor of many substances, the contribution of the methylene group and many other different components of molecular structure was established. Hence a list of the different structural contributions for calculating the parachor was evolved. Utilizing this list, once the structure of any compound is known, the parachor can be estimated. The structural contributions for the calculation of parachor are listed in Table 3.

It is of interest that the parachor for water, based on its surface tension of 72.8 dynes/cm at 20°C is 52.7. The value obtained by using the structural increments for 0 and H is 51.0, only 3% in error. Examples showing the procedure for parachor estimation (and surface tension) are given by Lyman et al. [36] and Reid et al. [26].

The abbreviated list of parachors for organic compounds is given in Table 2, abstracted from the list of Quayle [28]. There are many isomers among the organic compounds. While there is some variation with the number and position of the side groups, the parachor seems to be more dependent on the total number of atoms, such as carbon and hydrogen, present. Hence, the many isomers are not shown in this list. Nor are many special compounds such as the siloxanes, for example. For the much more detailed list, the original reference by Quayle [28] should be consulted.

CONCLUSION

Diffusivity data are often inaccurate. If one wishes to find the diffusivity of hydrogen in water; for example, he will discover a wide range of values differing by more than a *factor* of two. The difficulties associated with the measurement of diffusivity are generally reflected in inaccuracies of the results. Consequently, correlations for diffusivity, which can only be as reliable as the data on which they are based, are also inaccurate. Hence, it would appear that a significant improvement in accuracy of data is required before significantly better correlations will be forthcoming.

Diffusivity data for dilute liquid solutions may be found in many sources. The major listings of data are those of Ertl et al. [1] and Tyn and Calus [27]. In both cases the detailed lists and references are available only as supplementary material, which may be ordered and purchased from the appropriate sources; the data are too copious for reproduction here, however. Some diffusivity data for aqueous solutions (Table 4) solutions of alkanes (Table 5) and those in carbon tetrachloride and hexane for a temperature of 25°C (Table 6) are included for reference. In no case are the compilations exhaustive.

NOTATION

- c_i mass concentration of component i, g/cm^3
- c total mass concentration of solution, (ρ), g/cm^3
- C_i molar concentration of component i, $moles/cm^3$
- C total molar concentration, $moles/cm^3$
- D°_{AB} diffusivity of A in B at infinite dilution, cm^2/s
- f self-associating factor defined by Equation 14.
- j_i mass flux of component i resulting from molecular diffusion, $g/(cm^2s)$
- J_i molar flux of component i resulting from molecular diffusion moles/(cm^2s)
- M molar mass, g/mole
- n_i total mass flux of component i resulting from molecular diffusion and bulk (flow, $g/(cm^2s)$
- n total mass flux of solution, $g/(cm^2s)$
- P, P_i parachor, defined by Equation 21
- T absolute temperature, K
- V, V_i liquid molar volume at the normal boiling point, cm^3/mole
- V_c molar volume at the critical temperature, cm^3/mole
- x position variable along the diffusion path, cm

Greek Symbols

ΔH_i latent heat of vaporization, cal/mol
μ_B solvent viscosity, cP
ρ liquid or solution density, g/cm^3
$\rho_{A,s}$ Density of solid, g/cm^3
σ Surface tension, g/cm^3
ϕ association parameter in Wilke-Chang equation

Subscript

i component A or B

REFERENCES

1. Ertl, H., Ghai, R. K., and Dullien, F. A. L. *AIChE J.*, 20, 1 (1974).
2. Asfour, A. A. and Dullien, F. A. L. *AIChE J.*, 29, 347 (1983).
3. Ross, M. and Hildebrand, J. H. *J. Chm. Phy.*, 40, 2397 (1964).
4. Hayduk, W. and Ioakimidis, S. *J. Chem. Eng. Data*, 21, 255 (1976).
5. Witherspoon, P. A. and Bonoli, L. *Ind. Eng. Chem. Fund.*, 8, 589 (1969).
6 Taylor, G. J., *Proc. Roy. Soc.* (London), A219, 186 (1953).
7. Sherwood, T. K., Pigford, R. L. and Wilke, C. R. *Mass Transfer*, McGraw-Hill, New York (1975).
8. Himmelblau, P. M., *Chem. Rev.*, 64, 527 (1964).
9. Malik, V. K. and Hayduk, W. *Can. J. Chem. Eng.*, 46, 462 (1968).
10. Hayduk, W., Castaneda, R., Bromfield H. and Perras, R. R. *AIChE J.*, 19, 859 (1973).
11. Unver, A. A. and Himmelblau, D. M. *J. Chem. Eng. Data*, 9, 428 (1964).
12. Zhadi, I. and Turner, C. D. *Chem. Eng. Sci.*, 25, 517 (1970).
13. Emmert, R. E. and Pigford, R. L. *Chem. Eng. Prog.*, 50, 87 (1954).
14. Mazarie, A. F. and Sandall, O. C. *AIChE J.*, 26, 154 (1980).
15. Davidson, J. F. and Cullen, E. J. *Trans. Inst. Chem. Engrs.*, 35, 51 (1957).
16. Sovova, H. and Prochazka, J. *Chem. Eng. Sci.*, 31, 1091 (1976).
17. Duda, J. L. and Vrentas, J. S. *AIChE J.*, 14, 286 (1968).
18. Olbrich, W. E. and Wild, J. D. *Chem. Eng. Sci.*, 24, 25 (1969).
19. Krieger, I. M., Mulholland, G. W., and Dickey, C. S. *J. Phys. Chem.*, 71, 1123, (1967).
20. Hanika, J., Sporka, K. and Ruzicka, V. *Coll. Czech. Chem. Comm.*, 36, 1338 (1971).
21. Bird, R. B., Stewart, W. E., and Lightfoot, E. N. *Transport Phenomena*, John Wiley, NY, Chap. 16, (1960).
22. Ghai, R. K., Ertl, H., and Dullien, F. A. L. *AIChE J.*, 19, 881 (1973).
23. Hayduk, W. and Minhas B. S. *Can. J. Chem. Eng.*, 60, 295 (1982).
24. Wilke, C. R. and Chang, P. *AIChE J.*, 1, 264 (1955).
25. King, C. J., Hsueh, L., and Mao, K. W. *J. Chem. Eng. Data*, 10, 348 (1965).
26. Reid, R., Prausnitz, J. M. and Sherwood, T. K. *The Properties of Gases and Liquids*, Third Edition, McGraw-Hill, New York, NY, Chap. 11, (1977).
27. Tyn, M. T. and Calus, W. F. *J. Chem. Eng. Data*, 20, 106 (1975).
28. Quayle, O. R., *Chem. Rev.*, 53, 439 (1953).
29. Sugden, S., *The Parachor and Valency*, Routledge, London (1930).
30. Hayduk, W. and Buckley, W. K. *Chem. Eng. Sci.*, 27, 1997 (1972).
31. Othmer, D. F. and Thakar, M. S. *Ind. Eng. Chem.*, 45, 589 (1953).
32. Hayduk, W. and Laudie, H. *AIChE J.*, 20, 611 (1974).
33. Benson, S. W., *J. Phys. Sci. Colloid. Chem.*, 52, 1060 (1948).
34. Reid, R. C. and Sherwood, T. K. *The Properties of Gases and Liquids*, Second Edition, p. 88, Chap. 11, McGraw-Hill, NY (1966).
35. Sugden, S., *J. Chem. Sci.*, 1550 (1935).
36. Lyman, W. J., Reehl, W. F., and Rosenblatt, D. H. *Chemical Property Estimation Methods*, McGraw-Hill, Chap. 20 (1982).

37. Ferrel, R. T. and Himmelblau, D. M. *J. Chem. Eng. Data*, 12, 111 (1967a).
38. Wise, D. L. and Houghton, G. *Chem. Eng. Sci.*, 23, 1211 (1968).
39. Thomas, W. J. and Adams, M. J. *Trans. Far. Soc.*, 61, 668 (1965).
40. Groothuis, H. and Kramers, H. *Chem. Eng. Sci.*, 55, 17 (1955).
41. Vivian, J. E. and Peaceman, D. W. *Chem. Eng. Sci.*, 2, 437 (1956).
42. Baird, M. H. I. and Davidson, J. F. *Chem. Eng. Sci.*, 17, 473 (1962).
43. Houghton, G., Ritchie, P. D. and Thomson, J. A. *Chem. Eng. Sci.*, 17, 221 (1962).
44. Witherspoon, P. A. and Bonoli, L. *Ind. Eng. Chem. Fund.*, 8, 589 (1969).
45. Chiang, S. H. and Toor, H. L. *Ind. Eng. Chem. Fund.*, 6, 539 (1960).
46. Huq, A. and Wood, T. *J. Chem. Eng. Data*, 13, 256 (1968).
47. Vivian, J. E. and King, C. J. *AIChE J.*, 10, 220 (1964).
48. Bonoli, L. and Witherspoon, P. A. *J. Phys. Chem.*, 72, 2532 (1968).
49. Ratcliff, G. A. and Reid, K. J. *Trans. Inst. Chem. Engrs.*, 39, 423 (1961).
50. Gary-Bobo, C. M. and Weber, H. W. *J. Phys. Chem.*, 73, 1155 (1969).
51. Hammond, B. R. and Stokes, R. H. *Trans. Far. Soc.*, 49, 890 (1953).
52. Lyons, P. A. and Sandquist, C. L. *J. Am. Chem. Soc.*, 75, 3890 (1953).
53. Garner, F. H. and Marchant, P. J. M. *Trans. Inst. Chem. Engrs.*, 39, 397 (1961).
54. Byers, C. H. and King C. J. *Trans. Inst. Chem. Engrs.* 70, 2499 (1966).
55. Merliss, F. E. and Colver, C. P. *J. Chem. Eng. Data*, 14, 149 (1969).
56. Anderson, D. K., Hall, J. R., and Babb, A. L. *J. Phys. Chem.*, 62, 404 (1958).
57. Chandrasekaran, S. K. and King, C. J. *AIChE J.*, 18, 513 (1972).
58. Lewis, J. B., *J. Appl. Chem.*, 5, 228 (1955).
59. Robinson, R. A. and Stokes, R. H. *Electrolyte Solutions*, p. 316, Academic Press, NY (1955).
60. Bidstrup, D. E. and Geankoplis, C. J. *J. Chem. Eng. Data*, 8, 170 (1963).
61. Albery, W. J., Greenwood, A. R., and Kibble, R. K. *Trans. Far. Soc.*, 63, 360 (1967).
62. Garland, C. W. and Stockmayer, W. H. *J. Phys. Chem.*, 69, 2469 (1965).
63. Longsworth, L. G., *J. Am. Chem. Soc.*, 74, 4155 (1952).
64. Longsworth, L. G., *J. Am. Chem. Soc.*, 75, 5705 (1953).
65. Longsworth, L. G., *J. Phys. Chem.*, 58, 770 (1954).
66. Gladden, J. K. and Dole, M. *J. Am. Chem. Soc.*, 75, 3900 (1953).
67. Henrion, P. N., *Trans. Far. Soc.*, 60, 72 (1964).
68. Bell, J. H., *Cryogenic Engineering*, p. 384, Prentice-Hall, Englewood Cliffs, NJ (1963).
69. Rossini, F. D., (ed), *Selected Values of Physical and Thermodynamic Properties of Hydrocarbons and Related Compounds*, Carnegie Press, Pittsburgh, PA (1953).
70. Timmermans, J., *Physico-Chemical Constants of Pure Organic Compounds*, Elsevier, NY (1950).
71. Gallant, R. W., Hydrocarbon Process, 46, 183, 201 (1967).
72. Weast, R. C., (ed), *Handbook of Chemistry and Physics*, 48, F-36, Chemical Rubber, Cleveland, OH (1967–68).
73. Douglass, D. C. and McCall, D. W. *J. Chem. Phys.*, 62, 1102 (1958).
74. Bidlack, D. L., Kett, T. K., Kelly, C. M., and Anderson, D. K. *J. Chem. Data*, 14, 342 (1969).
75. VanGeet, A. L. and Adamson, A. W. *J. Phys. Chem.*, 68, 238 (1964).
76. *International Critical Tables, Coefficients of Diffusion in Liquids*, Vol. V., p. 63, McGraw-Hill, NY (1929).
77. Hayduk, W. and Cheng, S. C. *Chem. Eng. Sci.*, 26, 635 (1971).
78. Nakanishi, K., Voigt, E. M., and Hildebrand, J. H. *J. Chem. Phys.*, 42, 1860 (1965).
79. Longsworth, L. G., *J. Coll. Interf. Sci.*, 22, 3 (1966).
80. Anderson, D. K. and Babb, A. L. *J. Phys. Chem.*, 66, 899 (1962).
81. Caldwell, C. S. and Babb, A. L. *J. Phys. Chem.*, 60, 51 (1956).
82. Watts, H., Alder, B. J., and Hildebrand, J. H. *J. Chem. Phys.*, 23, 659 (1955).
83. Rao, S. S. and Bennett, C. D. *AIChE. J.*, 17, 75 (1971).
84. Dewan, R. K. and VanHolde, K. E. *J. Chem. Phys.*, 39, 1820 (1963).
85. Ferrel, R. T. and Himmelblau, D. M. *AIChE J.*, 13, 702 (1967b).
86. Komiyama, H. and Smith, J. M. *J. Chem. Eng. Data*, 19, 384 (1974).
87. Tang, Y. P. and Himmelblau, D. M. *AIChE J.*, 11, 54 (1965).

CHAPTER 4

DIFFUSION IN LIQUIDS

D. L. Foster and E. L. Cussler

Department of Chemical Engineering
University of Minnesota
Minneapolis, Minnesota, USA

CONTENTS

INTRODUCTION

Diffusion is caused by random molecular motion that leads to complete mixing. In liquids, diffusion is slow, about 0.4 cm in an hour. As a result, it is frequently rate limiting and hence important. Diffusion processes in liquids may often be accelerated by agitation. While agitation is a macroscopic process, moving portions of fluid over large distances, diffusion mixes adjacent fluid layers on a microscopic scale.

Diffusion processes are analyzed by one of two different mathematical models. The more fundamental model, Fick's law, expresses the linear mass flux per unit cross sectional area j_1 as proportional to the concentration gradient,

$$j_1 = -D\frac{dc_1}{dz} \qquad (1)$$

where the proportionality constant D is called the diffusion coefficient. In liquids D is on the order of 10^{-5} cm^2/sec. The second model, an engineering idea, linearizes the concentration gradient in a small part of the system's volume near its boundaries. As a result, the mass flux at this boundary, N_1, is proportional to a concentration difference,

$$N_1 = k(c_{1i} - c_1) \tag{2}$$

where c_{1i} and c_1 are the concentrations at the interface and in the bulk solution respectively; and k, the mass transfer coefficient, is about 10^{-3} cm/sec. The mass transfer coefficient is often useful for practical situations. The diffusion coefficient is more general and fundamental.

Although both of these models are important, this section emphasizes the more basic and scientific diffusion model. In the first subsection, we present the simplest and most frequently encountered cases of a dilute solute diffusing across a thin film and into a thick slab. These cases are then extended to concentrated solutions. In the second subsection, we review the values of diffusion coefficients and the means of obtaining these. Interactions of solutes with other solutes, solvents, and boundaries are discussed in the third subsection. The fourth subsection contains comments on diffusion in multicomponent mixtures. Finally, we briefly describe how the mass transfer model is applied.

FUNDAMENTALS OF DIFFUSION

Most diffusion problems occur in dilute solution. Accordingly, we consider first diffusion in dilute solution because it is common and easier to understand. Then we consider concentrated solutions.

Dilute Binary Solutions

Our present picture of diffusion comes from the work of Adolf Fick, who hypothesized that diffusion can be described with the same mathematical equations as Fourier's law for heat conduction. Using this hypothesis, Fick developed the laws of diffusion by means of analogies with Fourier's work [1]. He defined a total one-dimensional flux J_1 as

$$J_1 = Aj_1 = -AD\frac{\partial c_1}{\partial Z} \tag{3}$$

where A is the area across which diffusion occurs, j_1 is the flux per unit area, c_1 is concentration, z is distance and D is the diffusion coefficient. This equation, now known as Fick's law, can be restated in other useful forms.

For one-dimensional diffusion in Cartesian coordinates:

$$-j_1 = D\frac{dc_1}{dz} \tag{3a}$$

For radial diffusion in cylindrical and spherical coordinates:

$$-j_1 = D\frac{dc_1}{dr} \tag{3b}$$

Fick's law can be combined with a mass balance and solved to find the concentration and the flux as a function of position and time. Two of these solutions have particular value, for they bracket almost all observed behavior for diffusion in liquids. These two cases are diffusion across a thin film and diffusion into a thick slab. Because they are so important, they are discussed in detail in the following sections.

Steady Diffusion Across a Thin Film

Consider steady diffusion across the thin film illustrated schematically in Figure 1. On each side of the film, there is a well mixed solution of one solute, species "1." Both these solutions are dilute.

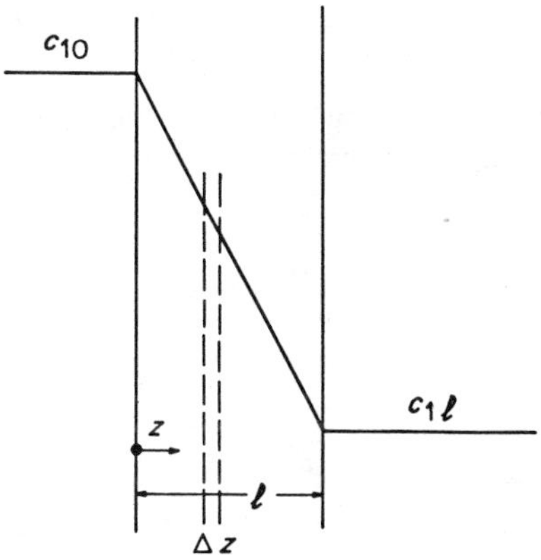

Figure 1. Diffusion across a thin film. This is the simplest and most important diffusion problem. Note that the concentration profile is independent of the diffusion coefficient.

The solute diffuses from the fixed higher concentration, located at $z \leq 0$ on the left hand side of the film, into the fixed, less concentrated solution located at $z \geq \ell$ on the right hand side.

We want to find the solute concentration profile and the flux across this film. To do this, we first write a mass balance on a thin layer Δz, located at some arbitrary position z within the thin film. The mass balance in this layer is

$$\begin{bmatrix}\text{solute}\\ \text{accumulation}\end{bmatrix} = \begin{bmatrix}\text{rate of diffusion}\\ \text{into the layer at z}\end{bmatrix} - \begin{bmatrix}\text{rate of diffusion out}\\ \text{of the layer at z} + \Delta\text{z}\end{bmatrix}$$

Because the process is in steady state, the accumulation is zero. The diffusion rate is the diffusion flux times the film's area A:

$$0 = A\left(j_1\big|_z - j_1\big|_{z+\Delta z}\right) \tag{4}$$

Dividing this equation by the layer's volume (A Δz) and letting Δz become small, we find

$$0 = \frac{d}{dz} j_1 \tag{5}$$

Combining this equation with Fick's law,

$$-j_1 = D\frac{dc_1}{dz} \tag{6}$$

we find for a constant diffusion coefficient D

$$0 = D\frac{d^2}{dz^2} c_1 \tag{7}$$

This differential equation is subject to two boundary conditions

$$\begin{aligned} z = 0 \qquad & c_1 = c_{10} \\ z = \ell \qquad & c_1 = c_{1\ell} \end{aligned} \tag{8}$$

Again, because this system is in steady state, the concentrations c_{10} and $c_{1\ell}$ are independent of time. Physically, this means that the volumes of the adjacent solutions must be much greater than the volume of the film.

The desired concentration profile and flux are now easily found. We integrate the differential equation and use the boundary conditions to find the concentration profile

$$c_1 = c_{10} + (c_{1\ell} - c_{10})\frac{z}{\ell} \tag{9}$$

The flux is found by differentiating this profile:

$$j_1 = -D\frac{dc_1}{dz} = \frac{D}{\ell}(c_{10} - c_{1\ell}) \tag{10}$$

Because the system is in steady state, the flux is a constant.

The difficulty with this result is that it is mathematically simple, so that many investigators decide that it has little content. This is not true. While simple, it is the single most important result for diffusion, basic to much of what follows.

As an example of the subtle complexity of this equation, consider the concentration profile and the flux for a single solute diffusing across a thin membrane. As in the preceding case of a film, the membrane separates two well-stirred solutions. Unlike the film, the membrane is chemically different from these solutions. Thus, the only change is that the mass balance is subject to the boundary conditions:

$$\begin{aligned} z &= 0 \qquad c_1 = HC_{10} \\ z &= \ell \qquad c_1 = HC_{1\ell} \end{aligned} \tag{11}$$

where H is a partition coefficient, the concentration in the membrane divided by that in the adjacent solution. The partition coefficient is an equilibrium property, so its use implies that equilibrium exists across the membrane interfaces. The concentration profile that results from these relations is

$$c_1 = HC_{10} + H(C_{1\ell} - C_{10})\frac{z}{\ell} \tag{12}$$

This result suggests concentration profiles like those in Figure 2, which contain sudden discontinuities at the interface. If the solute is more soluble in the membrane than in the surrounding solutions, then the concentration increases. If the solute is less soluble in the membrane, then its concentration drops. Either case produces enigmas. For example, at the left hand side of the membrane in Figure 2a, solute diffuses from the solution at C_{10} into the higher concentration within the membrane.

This apparent quandary is resolved when one thinks carefully about the solute's diffusion. Diffusion often can occur from a region of low concentration into a region of high concentration; indeed, this is the basis of many liquid-liquid extractions. Thus, the jumps in concentration in Figure 2 are graphical accidents that result from using the same scale to represent concentrations inside and outside the membrane.

This type of diffusion can also be described in terms of the solute's energy, or more exactly, in terms of its chemical potential. The solute's chemical potential does not change across the membrane's interface because equilibrium exists there. Moreover, this potential, which drops smoothly with concentration as shown in Figure 2c, is the driving force responsible for the diffusion. However, flux equations based on chemical potential turn out to be inconvenient to use.

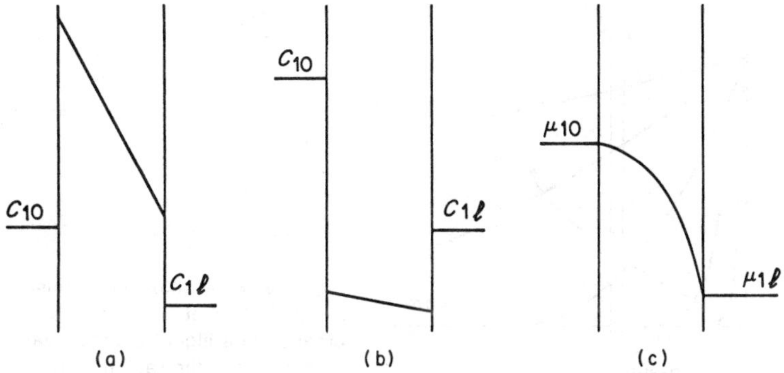

Figure 2. Concentration profiles across thin membrances. In (a), the solute is more soluble in the membrane than in the adjacent solutions; in (b) it is less so. Both cases correspond to a chemical potential gradient like that in (c).

The flux across the thin membrane can be found by combining the foregoing concentration profile with Fick's law to obtain,

$$j_1 = \left[\frac{DH}{\ell}\right](C_{10} - C_{1\ell}) \tag{13}$$

The quantity in square brackets in this equation, called the permeability, is often reported experimentally. Sometimes, this same term is called the permeability per unit length. Because the partition coefficient, H, is found to vary more widely than the diffusion coefficient, D, differences in diffusion coefficient tend to be much less important than the differences in solubility.

Unsteady Diffusion in a Semi-Infinite Slab

As previously mentioned above, the two most important cases of calculations made with Fick's law are steady diffusion across a film and unsteady diffusion into a thick slab. The first important case, the thin film, is discussed in the preceding paragraphs. The second important case, the thick slab, is discussed next.

We first consider a semi-infinite slab like that in Figure 3. This slab starts at an interface and extends a long way. We want to find how the concentration and flux within the semi-infinite slab vary as a result of an abrupt concentration change at its interface. In other words, we seek the concentration and flux as functions of position and time.

To make this calculation, we first write a mass balance on a thin volume of A Δz within the slab.

$$\frac{\partial}{\partial t}[A\,\Delta z\,c_1] = A\left(j_1\Big|_z - j_1\Big|_{z+\Delta z}\right) \tag{14}$$

Dividing by A Δz, we find

$$\frac{\partial c_1}{\partial t} = -\frac{\partial j_1}{\partial z} \tag{15}$$

Combining with Fick's law,

$$\frac{\partial c_1}{\partial t} = D\frac{\partial^2 c_1}{\partial z^2} \tag{16}$$

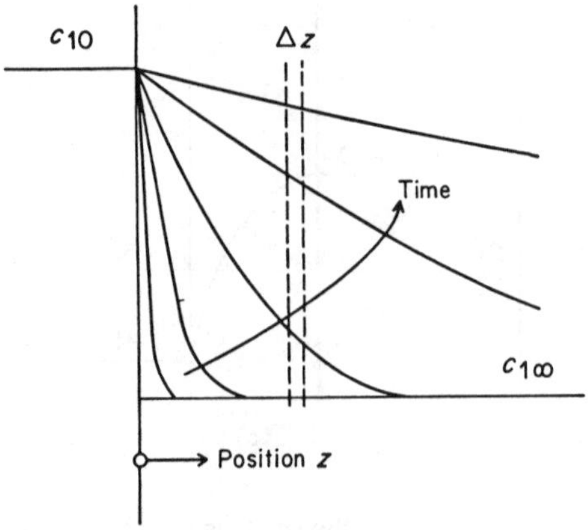

Figure 3. Free diffusion. In this case, the concentration at the left is suddenly increased to a higher constant value. Diffusion occurs in the region to the right. This case and that in Figure 1 are basic to most diffusion problems.

The boundary conditions here are

$$\begin{aligned} t < 0 \quad & \text{all } z \quad c_1 = c_{1\infty} \\ t \geq 0 \quad & z = 0 \quad c_1 = c_{10} \\ & z = \infty \quad c_1 = c_{1\infty} \end{aligned} \tag{17}$$

Solving yields the concentration profile

$$\frac{c_1 - c_{10}}{c_{1\infty} - c_{10}} = \operatorname{erf} \zeta \tag{18}$$

where ζ equals $(z/\sqrt{4Dt})$ and $\operatorname{erf} \zeta \; (= (2/\sqrt{\pi}) \int_0^\zeta e^{-\eta^2}\, d\eta)$ is the error function. Combining this solution with Fick's law gives the flux:

$$j_1 = -D\frac{\partial c}{\partial z} = \sqrt{D/\pi t}\, e^{-z^2/4Dt}(c_{10} - c_{1\infty}) \tag{19}$$

The most important limit of this result is the interfacial flux where $z = 0$

$$j_1\Big|_{z=0} = \sqrt{D/\pi t}\,(c_{10} - c_{1\infty}) \tag{20}$$

We should stress that this is the flux at a particular time t. The average flux from zero time to some particular time is easily shown to be twice this result.

Because this result is mathematically more sophisticated than that for the thin film, fewer need to be convinced of its importance. Still, we can illustrate its utility by a simple example.

As this example, we consider diffusion into a slab containing immobile reactive sites. We imagine that the sites react rapidly and reversibly with any diffusing solute; this rapid reversible reaction is characterized by an equilibrium constant K. In this case, the interfacial flux is

$$j_1\Big|_{z=0} = \sqrt{\frac{D(1 + K)}{\pi t}}\,(c_{10} - c_{1\infty}) \tag{21}$$

When K is zero, there is no reaction and the flux is the same as before. When K is very large the flux is dramatically increased. This result is important in dyeing. Interestingly, the existence of a similar rapid chemical reaction in a thin film has no effect on the flux.

Table 1
Comparison of Notation in References

Variable	Notation Here	Crank's Notation	Carslaw and Jaeger Analogue
Time	t	t	t
Position	x, y, z, r	x, y, z, r	x, y, z, r
Concentration	c_1	C	Temperature v
Concentration at boundary	$c_{10}, c_{1\ell}, \ldots$	C_1, C_0	Temperature at boundary ϕ
Binary diffusion coefficient	D	D	"Thermometric conductivity" κ
Flux relative to reference velocity	$\underset{\sim}{j}_1$	F	Heat flux f
Flux relative to fixed coordinates	$\underset{\sim}{n}_1$	F	Heat flux f
Flux at boundary	N_1 or $n_1\|_{z=\ell}$	—	Heat flux at boundary F_0
Total amount diffusing from time 0 to t	M_t	M_t	—

The example of a thin film and a semi-infinite slab are the two most important cases of diffusion in dilute solutions. Frequently, other diffusion problems are described by the same differential equations, but have different geometries or different boundary conditions. Solutions to these problems are tabulated in Crank's *Mathematics of Diffusion* [2] and Carslaw and Jaeger's *Conduction of Heat in Solids* [3]. The notation used here is compared with these references in Table 1.

As previously mentioned, the cases of a "thin film" and a "semi-infinite slab" bracket observed behavior, providing bounds for all other cases. A criterion to determine which of the two cases is more closely approached is the magnitude of the dimensionless variable ζ, defined by

$$\zeta = \frac{(\text{characteristic length})^2}{\sqrt{4Dt}} \tag{22}$$

This variable appears in the concentration profile for the semi-infinite slab above. If in a new problem $\zeta \gg 1$, then the physical situation is similar to that of the semi-infinite slab. If in a new problem $\zeta \ll 1$, then the situation will be closer to a thin film. If ζ is approximately one, we must use a more complex analysis like that given in the literature.

Concentrated Binary Solutions

Diffusion in concentrated solutions can require a more complicated analysis than that needed in dilute solution. This is because diffusion will always produce convection. In dilute solutions, the diffusion-engendered convection is vanishingly small. However, in concentrated solutions, it can be large. In this section, we first explore why this convection occurs, and then present its mathematical description.

How Diffusion Causes Convection

To see how diffusion and convection are coupled, consider a volume of liquid benzene connected by a vapor-filled capillary tube to a large volume of air, as shown in Figure 4. At room temperature benzene evaporates as a dilute vapor, which slowly diffuses into the air. This slow diffusion can be described with the form of Fick's law given above. Such a description leads to the concentration

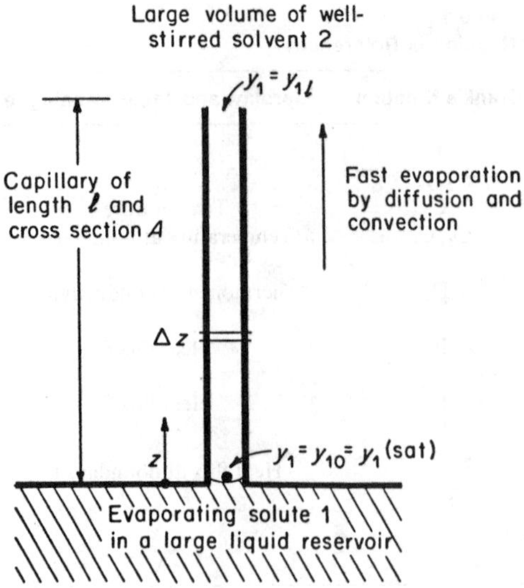

Figure 4. Fast evaporation in a thin capillary. This problem is analogous to that shown in Figure 1 but for a concentrated solution.

profile

$$c_1 = c_{10} + (c_{1\ell} - c_{10})\frac{z}{\ell} \tag{23}$$

This implies that doubling the concentration difference will double the flux.

However, when the benzene is at its boiling point, its flux up the tube is not due to diffusion at all. Instead, its flux is due to the evaporation rate, which in turn reflects the heat added to vaporize the benzene. The concentration now is not described by the linear profile just given. This flux is now due to convection.

At intermediate temperatures, we must describe the benzene flux in terms of both diffusion and convection. The total benzene flux n_1 is thus taken to be

$$\underset{\sim}{n}_1 = \underset{\sim}{j}_1^a + c_1\underset{\sim}{v}^a \tag{24}$$

where $\underset{\sim}{j}_1^a$ is the diffusion flux, c_1 is the local concentration, and $\underset{\sim}{v}^a$ is a convective "reference" velocity. A variety of reference velocities have been suggested in the literature, leading to the different equations shown in Table 2. We must choose between these equations. In most cases, our best choice is to choose $\underset{\sim}{v}^a$ so that it is zero as frequently as possible. For ideal gases, the mole average velocity and the volume average velocity are zero. For liquids, the mass average velocity and the volume average velocity are close to zero. Thus, the volume average velocity is the single most useful choice.

The Film and the Slab Revisited

In the earlier sections of this article, we reviewed in detail the two cases of thin film and a semi-infinite slab, for these tend to bracket observed behavior. However, our derivation was for dilute systems without convection. We now must estimate the changes in these results for concentrated solution.

The changes in these results depend on the volume average velocity of the system. In many cases, including almost all experimental methods of measuring diffusion coefficients, this velocity is zero. The results in the dilute solution then apply to concentrated solutions as well.

Table 2
Different Forms of the Diffusion Equation

Choice	Total Flux (diffusion + convection)	Diffusion Equation	Reference Velocity	Where Best Used
Mass	$\mathbf{n}_1 = \mathbf{j}_1^m + c_1\mathbf{v}$	$\mathbf{j}_1^m = \rho_1(\mathbf{v}_1 - \mathbf{v}_1) = -D\rho\,\nabla\omega_1$	$\mathbf{v} = \omega_1\mathbf{v}_1 + \omega_2\mathbf{v}_2$; $\rho\mathbf{v} = \mathbf{n}_1 + \mathbf{n}_2$	Constant density liquids; coupled mass and momentum transport
Molar	$\mathbf{n}_1 = \mathbf{j}_1^* + c_1\mathbf{v}^*$	$\mathbf{j}_1^* = c_1(\mathbf{v}_1 - \mathbf{v}^*) = -Dc\,\nabla y_1$	$\mathbf{v}^* = y_1\mathbf{v}_1 + y_2\mathbf{v}_2$; $c\mathbf{v}^* = n_1 + n_2$	Ideal gases where the molar concentration c is constant
Volume	$\mathbf{n}_1 = \mathbf{j}_1 + c_1\mathbf{v}^0$	$\mathbf{j}_1 = c_1(\mathbf{v}_1 - \mathbf{v}^0) = -D\,\nabla c_1$	$\mathbf{v}^0 = c_1\bar{V}_1\mathbf{v}_1 + c_2\bar{V}_2\mathbf{v}_2 = \bar{V}_1 n_1 + \bar{V}_2 n_2$	Best overall; good for constant density liquids and for ideal gases; may use either mass or mole concentration
Solvent	$\mathbf{n}_1 = \mathbf{j}_1^{(2)} + c_1\mathbf{v}_2$	$\mathbf{j}_1^{(2)} = c_1(\mathbf{v}_1 - \mathbf{v}_2) = -D_1\,\nabla c_1$	$\mathbf{v}_2$	Rare except for some membranes; note that $D_1 \neq D_2 \neq D$
Stefan-Maxwell		$\nabla y_1 = \frac{y_1 y_2}{D}(\mathbf{v}_2 - \mathbf{v}_1)$	None	Frequent theoretical result; difficult to use in practice

In some cases, however, the volume average velocity is *not* zero. The most common of these cases occurs when a solute is transferred through a stagnant solvent. This is the case for the example given in Figure 4, where benzene moves through stagnant air. The results are especially simple for ideal gases. If the solute moves through a thin film, the concentration profile is

$$\frac{c - c_1}{c - c_{10}} = \left(\frac{c - c_{1\ell}}{c - c_{10}}\right)^{z/\ell} \tag{25}$$

where c is the total molar concentration. The total flux in this case is

$$n_1 = \frac{Dc}{\ell} \ln\left(\frac{c - c_{1\ell}}{c - c_{10}}\right) \tag{26}$$

Note that doubling the concentration difference does not double n_1. The diffusion flux is found to vary with z:

$$j_1 = -D\frac{dc_1}{dz} = D\left(\frac{c - c_{10}}{\ell}\right)\left(\frac{c - c_{1\ell}}{c - c_{10}}\right)^{z/\ell} \ln\left(\frac{c - c_{1\ell}}{c - c_{10}}\right) \tag{27}$$

These three equations reduce to those given earlier for dilute diffusion.

Similarly, if the solute moves into a semi-infinite slab of stagnant solution, the concentration profile is given by

$$\left(\frac{c_1 - c_{10}}{c_{1\infty} - c_{10}}\right) = \frac{1 - \text{erf}(\zeta - \Phi)}{1 + \text{erf}\,\Phi} \tag{28}$$

where ζ again equals $z/\sqrt{4Dt}$, erf is the error function, and

$$\Phi = -\frac{1}{2}\left[\frac{\bar{V}_1(dc_1/d\zeta)}{1 - c_1\bar{V}_1}\right]_{\zeta=0} \tag{29}$$

Physically, Φ is an interfacial velocity for which tabulated values are available. The interfacial flux is

$$n_1\Big|_{z=0} = \sqrt{D/\pi t}\left[\frac{1}{1 - \bar{V}_1 c_1(\text{sat})}\right]\left[\frac{e^{-\phi^2}}{1 + \text{erf}\,\Phi}\right](c_{10} - c_{1\infty}) \tag{30}$$

For a dilute solution, Φ equals zero and the concentration profile and the flux again reduce to the dilute limits.

Thus, the description of diffusion in concentrated solutions may be more complicated than the same description in dilute solutions. The complexities occur when the diffusion engenders a bulk flow; this most frequently is significant when a concentrated solute diffuses through a stagnant solvent. Under these conditions, it is usually most convenient to describe diffusion relative to the volume average velocity. The concentration profiles and the fluxes can then be calculated for cases like the thin film and the semi-infinite slab.

Multicomponent Diffusion

Whether the solution is concentrated or dilute diffusion processes may include the transport of many solutes. Multicomponent diffusion effects are minor in dilute solution, but can be significant in non-ideal, concentrated solutions. Table 3 gives examples of systems showing multicomponent effects.

Multicomponent diffusion is most simply described by generalizing Fick's law for binary diffusion to an n-component system [4]:

$$-\underset{\sim}{j}_i = c_i(\underset{\sim}{v}^0 - \underset{\sim}{v}^i) = \sum_{j=1}^{n-1} D_{ij}\,\nabla c_j \tag{31}$$

where D_{ij} are multicomponent diffusion coefficients. Thus, for an n-component system there are $(n-1)^2$ diffusion coefficients. The diagonal terms, i.e., the D_{ii}, are called "main-term" diffusion coefficients and often are approximately equal to the binary diffusion coefficients. The off-diagonal terms, i.e. the $D_{ij,i\neq j}$, are called "cross-term" diffusion coefficients. They are usually smaller in magnitude than the "main-term" coefficients, and are not symmetric ($D_{ij,i\neq j} \neq D_{ji,i\neq j}$).

The multicomponent flux equation can be derived from irreversible thermodynamics. The three primary assumptions in this derivation are: (1) thermodynamic variables like chemical potential can be defined for a differential volume that is not in equilibrium; (2) there is a linear relationship between forces and fluxes; (3) the coefficients of the linear relationship between forces and fluxes are related via chemical potential derivatives.

Table 3
Systems with Large Multicomponent Effects

Type of System	Examples
Solutes of very different sizes	hydrogen–methane–argon [5] polystyrene–cyclohexane–toluene [6]
Solutes in highly nonideal solutions	mannitol–sucrose–water [7] acetic acid–chloroform–water [8]
Concentrated electrolytes	sodium sulfate–sulfuric acid–water [9] sodium chloride–polyacrylic acid–water [10]
Concentrated alloys	zinc–cadmium–silver [11]
Membranes with mobile carriers	sodium chloride–hydrochloric acid–monensin as a mobile carrier [12] oxygen–carbon dioxide–hemoglobin as a mobile carrier [13]

The concentration profiles for the multicomponent flux equation can usually be calculated from the binary solution profiles. The ternary case that follows is mathematically the simplest and easiest to understand. In general, the solution to a binary diffusion problem that has a concentration difference varying with position and time may be written as

$$\Delta c_1 = \Delta c_{10} F(D)$$

where Δc_{10} is a concentration difference containing initial and boundary conditions and F(D) is the explicit function of position and time. The analogous ternary diffusion problem has the solution

$$\begin{aligned} \Delta c_1 &= P_{11} F(\sigma_1) + P_{12} F(\sigma_2) \\ \Delta c_2 &= P_{21} F(\sigma_1) + P_{22} F(\sigma_2) \end{aligned} \tag{32}$$

where F(D) is the solution to the binary problem, and the values of σ_i and P_{ij} are:

Eigenvalues

$$\sigma_1 = \tfrac{1}{2}\left(D_{11} + D_{22} + \sqrt{(D_{11} - D_{22})^2 + 4D_{12}D_{21}}\right)$$

$$\sigma_2 = \tfrac{1}{2}\left(D_{11} + D_{22} + \sqrt{(D_{11} - D_{22})^2 + 4D_{12}D_{21}}\right)$$

Weighting Factors

$$P_{11} = \left(\frac{D_{11} - \sigma_2}{\sigma_1 - \sigma_2}\right) \Delta c_{10} + \left(\frac{D_{12}}{\sigma_1 - \sigma_2}\right) \Delta c_{20}$$

$$P_{12} = \left(\frac{D_{11} - \sigma_1}{\sigma_2 - \sigma_1}\right) \Delta c_{10} + \left(\frac{D_{12}}{\sigma_2 - \sigma_1}\right) \Delta c_{20}$$

$$P_{21} = \left(\frac{D_{21}}{\sigma_1 - \sigma_2}\right) \Delta c_{20} + \left(\frac{D_{22} - \sigma_2}{\sigma_1 - \sigma_2}\right) \Delta c_{10}$$

$$P_{22} = \left(\frac{D_{21}}{\sigma_2 - \sigma_1}\right) \Delta c_{20} + \left(\frac{D_{22} - \sigma_1}{\sigma_2 - \sigma_1}\right) \Delta c_{10}$$

As an example, we consider the semi-infinite slab. The binary solution may be written as

$$(c_1 - c_{10}) = (c_{1\infty} - c_{10})\,\mathrm{erf}\left(\frac{z}{\sqrt{4Dt}}\right) \tag{33}$$

The analogous ternary solution is, by comparison,

$$c_1 - c_{10} = \left[\left(\frac{D_{11} - \sigma_2}{\sigma_1 - \sigma_2}\right)(c_{1\infty} - c_{10}) + \left(\frac{D_{12}}{\sigma_1 - \sigma_2}\right)(c_{2\infty} - c_{20})\right]\mathrm{erf}\frac{z}{\sqrt{4\sigma_1 t}}$$
$$+\left[\left(\frac{D_{11} - \sigma_1}{\sigma_2 - \sigma_1}\right)(c_{1\infty} - c_{10}) + \left(\frac{D_{12}}{\sigma_2 - \sigma_1}\right)(c_{2\infty} - c_{20})\right]\mathrm{erf}\frac{z}{\sqrt{4\sigma_2 t}} \tag{34}$$

The fluxes may be found in the same manner. Thus, ternary diffusion is almost always a close mathematical parallel to the binary case.

VALUES OF LIQUID DIFFUSION COEFFICIENTS

Liquid diffusion coefficients are best obtained by experimental measurement because no universal theory permits their accurate calculation. Experimentally obtained values of diffusion coefficients are given in Tables 4 and 5 [14, 16]. The tables show that diffusion coefficient in liquids cluster around 10^{-5} cm^2/sec. Because these coefficients are smaller than in gases and frequently slower than chemical rate constants, diffusion in liquids frequently limits the rates of processes like gas absorption, acid-base reaction, and corrosion.

Table 4
Diffusion Coefficients at Infinite Dilution in Water at 25°C [14, 15]

Solute	D($\cdot 10^{-5}$ cm^2/sec)	Solute	D($\cdot 10^{-5}$ cm^2/sec)
Argon	2.00	Acetylene	0.88
Air	2.00	Methanol	0.85
Bromine	1.18	Ethanol	0.84
Carbon dioxide	1.92	1-propanol	0.87
Carbon monoxide	2.03	2-propanol	0.87
Chlorine	1.25	n-Butanol	0.77
Ethane	1.20	benzyl alcohol	0.821
Ethylene	1.87	Formic acid	1.50
Helium	6.28	Acetic acid	1.21
Hydrogen	4.50	Propionic acid	1.06
Methane	1.49	Benzoic acid	1.00
Nitric oxide	2.07	Glycine	1.06
Nitrogen	1.88	Valine	0.83
Oxygen	2.10	Acetone	1.16
Propane	0.97	Urea	$(1.380 - 0.0782c_1 + 0.00464c_1)$
Ammonia	1.64	Sucrose	$(0.5228 - 0.265c_1)$
Benzene	1.02	Qualbumin	0.078
Hydrogen sulfide	1.41	Hemoglobin	0.069
Sulfuric acid	1.73	Urease	0.035
Nitric acid	2.6	Fibrinogen	0.020

Table 5
Diffusion Coefficients at Infinite Dilution in Nonaqueous Liquids [16]

Solute	Solvent	D($\cdot 10^{-5}$ cm^2/sec)
Acetone	Chloroform	2.35
Benzene		2.89
n-Butyl acetate		1.71
Ethyl alcohol (15°)		2.20
Ethyl ether		2.14
Ethyl acetate		2.02
Methyl ethyl ketone		2.13
Acetic acid	Benzene	2.09
Aniline		1.96
Benzoic acid		1.38
Cyclohexane		2.09
Ethyl alcohol (15°)		2.25
n-Heptane		2.10
Methyl ethyl ketone (30°)		2.09
Oxygen (29.6°)		2.89
Toluene		1.85
Acetic acid	Acetone	3.31
Benzoic acid		2.62
Nitrobenzene (20°)		2.94
Water		4.56
Carbon tetrachloride	n-Hexane	3.70
Dodecane		2.73
n-Hexane		4.21
Methyl ethyl ketone (30°)		3.74
Propane		4.87
Toluene		4.21
Benzene	Ethyl alcohol	1.81
Camphor (20°)		0.70
Iodine		1.32
Iodobenzene (20°)		1.00
Oxygen (29.6°)		2.64
Water		1.24
Carbon tetrachloride		1.50
Benzene	n-Butyl alcohol	0.988
Biphenyl		0.627
p-Dichlorobenzene		0.817
Propane		1.57
Water		0.56
Acetone (20°)	Ethyl acetate	3.18
Methyl ethyl ketone (30°)		2.93
Nitrobenzene (20°)		2.25
Water		3.20
Benzene	n-Heptane	3.40

The Stokes-Einstein Equation

Since experimental measurements of diffusion coefficients are difficult, estimates are often useful. The most common basis for these estimates is the Stokes-Einstein equation. This gives estimates to within ±20% and is a standard by which other correlations are judged [16]. The Stokes-Einstein equation, derived by assuming that a rigid solute sphere diffuses in a continuum of solvent, is

$$D = \frac{kT}{f} = \frac{kT}{6\pi\mu R_0} \tag{35}$$

where f is the frictional coefficient of the solute, k is Boltzmann's constant, T is the temperature, μ is the solvent viscosity, and R_0 is the solute radius. The inverse dependence on viscosity is accurate if the solute size is at least five times the solvent size and if the viscosity is less than about 3 centipoise. In viscous solvents, diffusion varies with the −2/3 power of viscosity. For nonspherical solutes, R_0 is an average over the shape. These averages are known for ellipsoids.

Alternative Correlations

Many correlations have been developed for cases in which solute and solvent are similar in size. Five of these are shown in Table 6; others are given by Reid [16]. All of these are similar to the Stokes-Einstein equation.

However, all these correlations including the Stokes-Einstein equation are limited to dilute solutions. In concentrated solutions the diffusion coefficient can vary with concentration. This concentration dependence may be estimated in two different ways. First, the Stokes-Einstein equation may be extended to give [18, 19]

$$D = \frac{kT}{6\pi\mu R_0}(1 + 1.5\phi_1 + \cdots) \tag{36}$$

where ϕ_1 is the volume fraction of solute. The variation predicted by this relation is often smaller than those observed experimentally.

The second, more successful, method for estimating the concentration dependent diffusion is empirical:

$$D = D_0\left(1 + \frac{\partial \ln \gamma_1}{\partial \ln c_1}\right)^{\eta} \tag{37}$$

where D_0 is a new transport coefficient, estimated from dilute solution, γ_1 is the activity coefficient, and η is a parameter equal for all solutes. Values of D_0 were originally based on an arithmetic average

$$D_0 = x_1 D_0(x_1 = 1) + x_2 D_0(x_2 = 1) \tag{38}$$

More recent theories show that a geometric average is more accurate:

$$D_0 = [D_0(x_1 = 1)]^{x_1}[D_0(x_2 = 1)]^{x_2} \tag{39}$$

The quantity η was originally taken as equal to one, based on extensions of Einstein's original derivation. More recent estimates, based on scaling theories near the critical point, suggest an exponent of 0.62. Which exponent is better is currently unclear.

Table 6
Alternatives to Stokes-Einstein Equation for Diffusion in Liquids[a,b] [17]

Authors	Origin	Basic equation	Viscosity variation	Solute size variation	Remarks
Sutherland (1905)	Parallel to Stokes-Einstein, but "no stick" at sphere	$D = \frac{k_B T}{4\pi\mu R_0}$	μ^{-1}	R_0^{-1}	Always mentioned but rarely used
Glasstone et al. (1941)	Diffusion as a rate process	$D = \frac{k_B T}{2\mu R_0}$	μ^{-1}	R_0^{-1}	Smaller coefficient; closer to some experimental results
Scheibel (1954)[c,d]	Empirical	$D = \frac{AT}{\mu(\bar{V}_1)^{1/3}}\left[1 + \left(\frac{3\bar{V}_2}{\bar{V}_1}\right)^{2/3}\right]$	μ^{-1}	Equivalent to R_0^{-1} for large solutes and R_0^{-3} for small ones	Variation with solute size is the interesting feature of this equation
Wilke & Chang (1955)[c,e]	Empirical	$D = \frac{7.4 \cdot 10^{-8}(\phi\bar{M}_2)^{1/2} T}{\mu_2 \bar{V}_1^{0.6}}$	μ^{-1}	Equivalent to $R_0^{-1.8}$	Factor ϕ for solute-solvent interaction
King et al. (1965)[c]	Empirical	$D = 4.4 \cdot 10^{-8} \frac{T}{\mu_2} \cdot \left(\frac{\bar{V}_2}{\bar{V}_1}\right)^{1/6}\left(\frac{\Delta H_{vap \cdot 2}}{\Delta H_{vap \cdot 1}}\right)$		$R_0^{-0.5}$	Not suitable for viscous solvent or aqueous systems

[a] *The subscripts 1 and 2 indicate the solute and solvent, respectively.*
[b] *These relations are accurate within about 10% for water and 20% for most organics, but they are often inaccurate for alcohols and other hydrogen-bounded solvents.*
[c] *Specific units implied are* $D[=]\ cm^2/sec$; $T[=]\ ^\circ K$; $\mu[=]\ 10^{-2}\ g/cm\text{-}sec$; $V_i[=]\ cm^3/g\text{-}mol$.
[d] *The* $\bar{V}$ *are the molar volumes at the boiling points. The constant A equals* $8.2 \cdot 10^{-8}$, *except as follows:* $25.2 \cdot 10^{8}$ *for water if* $\bar{V}_1 < \bar{V}_2$; $18.9 \cdot 10^{-8}$ *for benzene when* $\bar{V}_1 < 2V_2$; $17.5 \cdot 10^{-8}$ *for others if* $\bar{V}_1 < 2.5\bar{V}_2$.
[e] *The factor* ϕ *has the following values: 2.26 for water, 1.9 for methanol, 1.5 for ethanol, and 1.0 for non-hydrogen-bonded solvents.*

Table 7
Comparison of Methods for Measuring Diffusion Coefficients

Methods	Nature of Diffusion	Apparatus Expense	Apparatus Construction	Concentration Difference Required	Method of Obtaining Data	Overall Value	Accuracy	Reference
Diaphragm cell	Pseudy-steady state	Small	Easy	Large	Concentration at known time; requires chemical analysis	Simple equipment; occasional erratic results	±10–0.2%	[20, 21]
Taylor dispersion	Decay of a pulse	Moderate	Easy	Average	Refractive index vs. time at known position	Excellent for dilute solutions	±1%	[22, 23]
Accurate Methods:								
Guoy interferometer	Unsteady in infinite cell	Large	Moderate	Small	Refractive index gradient vs. position and time	Excellent data at great effort	±0.1%	[24]
Rayleigh interferometer	Unsteady in infinite cell	Large	Difficult	Small	Refractive index vs. position and time	Best for concentration dependent diffusion	±0.1%	[25, 26]
Bryngdahl interferometer	Unsteady in infinite cell	Large	Difficult	Very Small	Refractive index difference vs. time	Suitable for dilute solutions	±0.1%	[27]
Other Methods:								
Laser Doppler light scattering	Homogeneous solution	Large	Difficult	None	Scattered liquid measurements with photomultiplier tube	Suitable for large solutes	±2%	[28, 29]
Spinning disc	Dissolution of solid or liquid	Small	Easy	Large	Concentration vs. time; requires chemical analysis	Requires diffusion controlled dissolution	±5%	[30, 31]
Steady State Methods	Steady diffusion across known length	Moderate	Moderate	Large	Small concentration changes; requires analysis	Difficult experiments, easy analysis	±5%	[32]
Spin-Echo NMR	Unsteady in infinite cell	Large	Moderate	Moderate	Amplitude of signal from NMR pulses	Used for self-diffusion coefficients, emulsions	±2–5%	[33]

Measurement of Diffusion Coefficients

Experimental measurements are required for accurate values of diffusion coefficients in liquids. The most useful methods for these measurements are shown in Table 7. Some of these methods are discussed in more detail below, roughly in order of overall value.

Diaphragm Cell

The most useful method for measuring diffusion in liquids is the diaphragm cell. It is cheap and potentially accurate, but can be hard to use. This cell consists of two compartments separated by a membrane or by a glass frit (Figure 5). Each compartment is stirred with a rotating magnet. Initially each compartment is filled with a solution of different concentration. When the experiment is complete, the concentrations of the two solutions are measured. The diffusion coefficient D is then calculated from the equation

$$D = \frac{1}{\beta t} \ln \left\{ \frac{(c_{1,\text{lower}} - c_{1,\text{upper}})_{\text{initial}}}{(c_{1,\text{lower}} - c_{1,\text{upper}})_{\text{finial}}} \right\} \quad (40)$$

where t is the final time when the experiment was stopped, c_1 is the solute concentration and β (in cm^{-2}) is a calibration constant that must be found experimentally, usually with KCl-water or urea-water systems [34]. Diffusion in the diaphragm cell should always take place vertically to avoid free convection, i.e. the diaphragm should be in the horizontal plane.

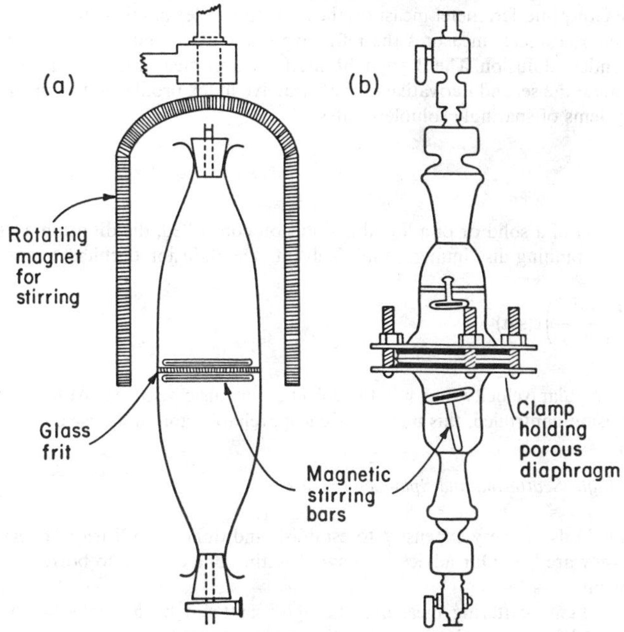

Figure 5. Diaphragm cell. The cell on the left, which uses a porous glass frit is more accurate than that on the right, which uses filter paper as a diaphragm. However, the cell with the glass frit requires much longer experiments.

Taylor Dispersion

A second method, useful especially for dilute solution, is Taylor dispersion. The equipment for Taylor dispersion takes more time and money to build than diaphragm cells, but is much easier to use. The basic apparatus is a long tube through which solvent is slowly pumped in laminar flow. A sharp pulse of solute is injected into the tube. The concentration profile caused by this pulse is measured with a refractometer as it comes out the other end of the tube. The concentration profile found in this apparatus is analyzed with the equation

$$c_1 = \frac{M}{\pi R_0^2} \frac{e^{-z^2/4Et}}{\sqrt{4\pi Et}} \tag{41}$$

where M is the total solute injected, R_0 is the tube radius, and E is a dispersion coefficient. This dispersion coefficient is given by

$$E = \frac{(vR_0)^2}{48D} \tag{42}$$

where v is the solvent velocity.

Interferometers

Though expensive to build and tedious to operate, interferometers provide values of diffusion coefficients with very high accuracy by measuring an unsteady-state refractive index profile. More specifically, the Gouy interferometer measures the refractive-index gradient and is the most accurate. The Rayleigh interferometer measures the refractive index itself, and is especially useful for concentration dependent diffusion. The Bryngdahl interferometer measures refractive index differences which approximate the second derivative of the refractive index profile, and is much harder to use. It is best for systems of sparingly soluble solutes.

Spinning Disc

If the dissolution of a solid or of a liquid is diffusion controlled, the diffusion coefficient can be found by using a spinning disc immersed in a solvent. The diffusion coefficient can be found from

$$j_1 = 0.62\left(\frac{D^{2/3}\omega^{1/2}}{\nu^{1/6}}\right) c(\text{sat}) \tag{43}$$

where ω is the angular velocity and ν is the solvent's intrinsic viscosity. Although dissolution is not always diffusion controlled, this method is cheap, reliable, and underused.

Laser-Doppler Light Scattering and Spin-Echo NMR

Both these methods are very expensive to establish and difficult to learn to operate. Once they are operating, they are fast. Our advice is to avoid setting one up, and to borrow or rent time on existing equipment.

Laser-Doppler light scattering measures the light scattered by a single solute particle. This method is best used for large solutes because they scatter more light than do small solutes. Since the light scattered by these particles can be sensitively analyzed, both translational and rotational diffusion can be calculated.

Spin-echo NMR measures the diffusion coefficients of solutes whose nuclei have magnetic moments. These coefficients are calculated from the altered magnetic moment after a series of pulses in magnetic field. The technique has been applied to the study of self-diffusion and of restricted diffusion in systems such as emulsions and foams.

This concludes our discussion of diffusion coefficients. We now turn to molecular information, which can be inferred from these coefficients.

INFORMATION IMPLIED BY DIFFUSION COEFFICIENTS

In the previous sections, we have discussed the mathematical description of diffusion, both for binary and multicomponent solutions. We have presented values of diffusion coefficients of a wide variety of solutes in both water and organic liquids, and have described how diffusion coefficients can be approximately estimated or accurately measured.

In these previous sections, we have implied that the diffusion coefficients are a means to an end. This end can be a concentration profile or an interfacial flux. However, the diffusion coefficient itself has not been of interest.

In this section, we want to see what the diffusion coefficient implies about molecular interactions. This means that our focus is the diffusion coefficients themselves, and hence is the antithesis of our earlier attitude. We begin with strong electrolytes, and then consider associating solutes, including weak electrolytes. We conclude with hydration and with solute-boundary interactions.

Strong Electrolytes

When a strong electrolyte like sodium chloride diffuses in water, it diffuses as free ions. This diffusion may be described by a single diffusion coefficient because the separate ions are electrostatically coupled, diffusing at the same rate.

The flux of ion "1" of a 1–1 electrolyte in dilute solution is given by [17]

$$-\underset{\sim}{j}_1 = \frac{2D_1D_2}{D_1 + D_2}\nabla c_1 - \frac{D_1}{D_1 + D_2}(|z|\underset{\sim}{i}) \tag{44}$$

where $|z|$ is the magnitude of the ionic charge, $\underset{\sim}{i}$ is the current density, and D_1 and D_2 are the ionic diffusion coefficients. Values of these ionic diffusion coefficients are given in Table 8 [35]. There

Table 8
Ionic Diffusion Coefficients in Water at 25°C

Cation$_i$	D_i[a]	Anion$_i$	D_i[a]
H^+	9.31	OH^-	5.28
Li^+	1.03	F^-	1.47
Na^+	1.33	Cl^-	2.03
K^+	1.96	Br^-	2.08
Rb^+	2.07	I^-	2.05
Cs^+	2.06	NO_3^-	1.90
Ag^+	1.65	CH_3COO^-	1.09
NH_4^+	1.96	$CH_3CH_2COO^-$	0.95
$N(C_4H_9)_4^+$	0.52	$B(C_6H_5)_4^-$	0.53
Ca^{2+}	0.79	SO_4^{2-}	1.06
Mg^{2+}	0.71	CO_3^{2-}	0.92
La^{3+}	0.62	$Fe(CN)_6^{3-}$	0.98

[a] *Values at infinite dilution in 10^{-5} cm^2/sec. Calculated from data in Robinson and Stokes [38].*

are two useful limits of this relation. First, when there is no current,

$$\underset{\sim}{j_1} = \underset{\sim}{j_2} = -\left[\frac{2}{1/D_1 + 1/D_2}\right]\nabla c_1 \tag{45}$$

where the quantity in square brackets is the diffusion coefficient of the electrolyte. This coefficient is a harmonic average of the ionic values, and hence is dominated by the slower ion.

The second useful limit occurs when the solution is well mixed so there is no concentration gradient

$$\underset{\sim}{j_1} = \left[\frac{D_1}{D_1 + D_2}\right]|z|\underset{\sim}{i} \tag{46}$$

where the quantity in square brackets is called the transference number or the transport number. In physical terms, the transference number of ion "1" is the fraction of the current carried by ion "1." This number implies an arithmetic average, which is the same average that occurs in conductance experiments.

The results for a non 1–1 electrolyte are more complicated than those for a 1–1 electrolyte. In particular, for diffusion in dilute solution with no current,

$$-j_T = D\,\nabla c_T = \left[\frac{|z_1| + |z_2|}{|z_2|/D_1 + |z_1|/D_2}\right]\nabla c_T \tag{47}$$

where:

$$\begin{aligned} \underset{\sim}{j_T} &= \underset{\sim}{j_1}/|z_2| = \underset{\sim}{j_2}/|z_1| \\ c_T &= c_1/|z_2| = c_2/|z_1| \end{aligned} \tag{48}$$

Like the 1–1 electrolyte, the final value is a harmonic average, although now it is weighted by the change.

Associating Solutes

Weak electrolytes only partially dissociate; detergents can associate to form micelles and dyes may aggregate. These associations often dramatically affect diffusion coefficients.

We first consider a weak 1–1 electrolyte that diffuses steadily both as ions and as ion pairs. The cations and anions diffuse at the same rate, but this rate is different from that at which the ion pairs diffuse. Moreover, as diffusion proceeds, the fraction of solute present as ions or as ion pairs changes. Under these conditions, the total solute flux J_T may be shown to be

$$-j_T = \left[\frac{D_1 + D_2(\sqrt{1 + 8Kc_T} - 1)}{\sqrt{1 + 8Kc_T}}\right]\frac{dc_T}{dz} \tag{49}$$

where K is the association constant of the weak electrolyte, c_T is the total solute concentration and D_1 and D_2 are the diffusion coefficients of the ions and ion pairs, respectively. The quantity in brackets is the apparent diffusion coefficient D of the weak electrolyte, which varies with concentration. At low c_T, this coefficient equals D_1; at high c_T, it equals D_2.

Solutes other than weak electrolytes also associate. This association is abrupt for many detergents, but can be gradual for other solutes like organic dyes. We next discuss these two cases.

Detergent molecules have a polar end and a nonpolar end. In water, the molecules may aggregate so that the nonpolar parts of many molecules are together. Such aggregates are called "micelles." This aggregation often occurs abruptly, at a particular concentration called the "critical micelle concentration" or CMC. Below this concentration, the detergent diffuses with a diffusion coefficient D_1, characteristic of the monomer. Above this concentration the diffusion involves parallel transport of monomer and micelles. The total detergent flux J_T at steady state is

$$-j_T = \left[D_m + \frac{D_1(nK)^{-1/n}}{n(c_T - c_{CMC})}\right]\frac{dc_T}{dz} \tag{50}$$

where D_m is the micelle diffusion coefficient, c_T is the total solute concentration and K is the equilibrium constant for the micelle forming reaction. Again, the quantity in square brackets is the detergent's diffusion coefficient. This equation is accurate for nonionic detergents and ionic detergents at high ionic strength where electrostatic effects are not major. Under other conditions, the results are more complicated.

In contrast to detergents, solutes like dyes associate slowly, one molecule at a time. Thus a solution of dyes contains monomers, dimers, trimers, etc. in ever decreasing proportion. In contrast, a detergent solution contains mostly monomers and very large aggregates.

In many cases, each step of the dye aggregation is described by the same association constant:

$$c_i = Kc_{i-1}c_1 \tag{51}$$

where K is an equilibrium constant independent of the size of the aggregate. Such aggregation is called "isodesmmic." The total solute flux is now given by

$$\begin{aligned} -j_T = [D_1 - Kc_T(4D_1 - 4D_2) + K^2c_T^2(15D_1 - 24D_2 + 9D_3) \\ - K^3c_T^3(56D_1 - 112D_2 - 72D_3 + 16D_4) + \cdots] \frac{dc_T}{dz} \end{aligned} \tag{52}$$

where the D_i are the diffusion coefficients of i-mers. Again, the quantity in square brackets is the apparent diffusion coefficient; note that it is independent of the solute concentration only if the D_i are equal.

Solvation

Often, solvent and solute react to form new species in solution. This is especially true for water, where ions and other solutes are routinely hydrated. When we study diffusion in such a system, we implicitly study the transport of these solvated species, and not of the solutes themselves.

The diffusion coefficient of such a solute can be either increased or decreased by this solvation. To see how these effects can occur, we return to the Stokes-Einstein equation written for a non-ideal solution:

$$-\underset{\sim}{j_1} = \left[\frac{kT}{6\pi\mu R_0}\right]\left(1 + \frac{\partial \ln \gamma_1}{\partial \ln c_1}\right)\nabla c_1 \tag{53}$$

where γ_1 is an activity coefficient. As before, quantity in square brackets is the apparent diffusion coefficient. Solvation affects both R_0 and γ_1. It will make R_0 larger, thus reducing the diffusion coefficient. Its affect on γ_1 is harder to estimate. To do so, we assume that the solute's activity $C_1\gamma_1$ equals the solute's true mole fraction after solvation:

$$\begin{aligned} c_1\gamma_1 &= \frac{\text{number of solute molecules}}{(1-n)(\text{number of solute molecules}) + (\text{total number of water molecules})} \\ &= \frac{c_1}{(1-n)c_1 + c_2} \end{aligned} \tag{54}$$

where n is the "hydration number" (the number of solvent molecules bound to a solute). Values for several solutes in water are given in Table 9. The concentration of solute molecules, c_1 and of water molecules, c_2 are related by their partial volumes:

$$c_1\bar{V}_1 + c_2\bar{V}_2 = 1 \tag{55}$$

Table 9
Hydration Numbers Found by Various Methods

Ion	Observed Diffusion Coefficient at Infinite Dilution[a]	Hydration Numbers from diffusion's Concentration Dependence[b]	Hydration Numbers from Activity Coefficients[c]	Hydration Numbers from Transference Methods[c]
H^+	9.33	—	4	1
Li^+	1.03	2.8	4	14
Na^+	1.34	1.2	3	8
K^+	1.96	0.9	1	5
Cl^-	2.03	0	1	4
Br^-	2.08	0.2	1	5
I^-	2.04	0.7	2	2

[a] *10^{-5} cm²/sec*
[b] *Data of Robinson and Stokes [35]*
[c] *Data of Hinton and Amis [36]*

so that the diffusion coefficient from the flux equation becomes for dilute solutions

$$D = \frac{kT}{6\pi\mu R_0}\left[1 - \frac{(1 - n - \bar{V}_1/\bar{V}_2)c_1}{1/\bar{V}_2 + (1 - n - \bar{V}_1/\bar{V}_2)c_2}\right] \doteq \frac{kT}{6\pi\mu R_0}[1 + n\bar{V}_2 c_1 + \cdots] \tag{56}$$

Thus, solvation decreases the diffusion coefficient at infinite dilution but increases it as the solute's concentration increases.

Solute-Boundary Interactions

In many cases, solutes diffuse not through homogeneous material, but through pores in a less permeable matrix. This diffusion represents an average over the pores and the matrix, an average that must include both the chemical properties and the particular geometry. Because this case is common, it has been studied extensively.

Three special cases of diffusion in porous media merit discussion here. The first is *tortuosity*, an empirical concept used to describe diffusion through the fluid-filled pores of an impermeable solid matrix. This phenomena is frequently described with an effective diffusion coefficient D_{eff} [37]:

$$D_{eff} = D/\tau \tag{57}$$

where D is the diffusion coefficient within the pores and τ is a tortuosity which accounts for the pores. Tortuosities may be defined as

$$\tau = a^2/\epsilon \tag{58}$$

where a is the actual pore length per solid length and ϵ is the void fraction. These tortuosities usually range in value between 2 and 6.

The second and third special cases are not empirical, but have a definite physical basis. In the second case, the porous composite consists of periodically spaced spheres. Diffusion occurs in both the interstitial region between the spheres and through the spheres. The effective diffusion coefficient in this medium can be calculated exactly [38]:

$$\frac{D_{eff}}{D} = \frac{2/D_s + 1/D - 2\phi_s(1/D_s - 1/D)}{2/D_s + 1/D + \phi_s(1/D_s - 1/D)} \tag{59}$$

where D is the diffusion coefficient in the interstitial space, D_s is the diffusion coefficient through the spheres and ϕ_s is the volume fraction of the spheres in the composite material. The diffusion does not depend on the size of the spheres but only on their volume fraction.

In the third case, the porous composite contains liquid-filled cylindrical pores in an impermeable solid. Diffusion of a solute in these pores is affected by the viscous drag caused by the adjacent walls. The diffusion coefficient is now [39]

$$\frac{D}{D_0} = 1 + \frac{9}{8}\lambda \ln \lambda - 1.54\lambda + 0(\lambda^2) \tag{60}$$

where $\lambda(= 2R_0/d)$ is the solute diameter divided by the pore diameter. This result is accurate to around 2% for λ smaller than 0.2 [40].

This concludes the description of solute-boundary interactions. In this and the earlier parts of this section, we have tried to show how diffusion coefficients reflect information about the molecular process. In the cases considered, this reflection originates from electrostatics, from solute association or solvation, and from a porous media. These different situations alter the diffusion coefficient in ways that suggest which molecular processes are important.

INTERFACIAL MASS TRANSFER

In the earlier parts of this chapter, we treated diffusion as described by Fick's law. This law has a strong scientific basis, and can in principle give a good description of diffusion in any situation. However, in many cases, this law can be very difficult to apply. This is especially true for diffusion across interfaces, from one phase into another. Examples include gas absorption, liquid–liquid extracting, and leaching.

As a result, a second unnamed model often provides a more convenient description of diffusion than does Fick's law. This model is described in this section. While it was developed in engineering, it has major potential value in other fields, even though it is rarely used there.

In this alternative model, changes in concentration are limited to the small part of the system's volume near its boundaries. In other words, all phases are well mixed except for a thin interfacial region. Diffusion across this interfacial region is desribed with a "mass transfer coefficient" rather than with a diffusion coefficient. Such mass transfer coefficients are basic to the design of industrial processes.

Mass Transfer Coefficients

The amount of mass transferred across some interface into a well-mixed solution is assumed proportional to the product of the concentration difference and an interfacial area:

(amount of mass transferred) = k(interfacial area)(concentration difference)

If we divide this equation by the interfacial area

$$N_1 = k(c_{1i} - c_1) \tag{61}$$

where N_1 is the flux at the interface, k is the mass transfer coefficient, and c_{1i} and c_1 are the concentrations at the interface and in the well mixed solution, respectively. Most commonly the concentration difference is defined locally, at a single position and time in the system. In some industrial situations, the difference is defined in other ways. We will discuss only the local definition here.

Mass transfer coefficients can be calculated from empirical correlations reported in the literature. These correlations are usually reported in terms of dimensionless groups. Common dimensionless groups are given in Table 10; their definition implies a specific physical system. Correlations of mass transfer coefficients are shown in Table 11. These correlations are typically accurate to within 20%.

Table 10
Significance of Common Dimensionless Groups

Group[a]	Physical meaning	Used in
Sherwood number $\frac{k\ell}{D}$	$\frac{\text{mass transfer velocity}}{\text{diffusion velocity}}$	Usual dependent variable
Stanton number $\frac{k}{v^0}$	$\frac{\text{mass transfer velocity}}{\text{flow velocity}}$	Occasional dependent variable
Schmidt number $\frac{\nu}{D}$	$\frac{\text{diffusivity of momentum}}{\text{diffusivity of mass}}$	Correlation of gas and liquid data
Lewis number $\frac{\alpha}{D}$	$\frac{\text{diffusivity of energy}}{\text{diffusivity of mass}}$	Simultaneous heat and mass transfer
Reynolds number $\frac{\ell v^0}{\nu}$	$\frac{\text{inertial forces}}{\text{viscous forces}}$ or	Forced convection
Peclet number $\frac{v^0\ell}{D}$	$\frac{\text{flow velocity}}{\text{diffusion velocity}}$	Correlations of gas or liquid data
Second Damköhler number $\frac{\kappa\ell^2}{D}$	$\frac{\text{reaction velocity}}{\text{diffusion velocity}}$	Correlations involving reactions

[a] *The symbols and their dimensions are as follows:*

D	*diffusion coefficient* (L^2/t)
g	*acceleration due to gravity* (L/t^2)
k	*mass transfer coefficient* (L/t)
ℓ	*characteristic length* (L)
v^0	*fluid velocity* (L/t)
α	*thermal diffusivity* (L^2/t)
κ	*first-order reaction rate constant* (t^{-1})
ν	*kinematic viscosity* (L^2/t)
$\Delta\rho/\rho$	*fractional density change*

As an example of the use of mass transfer coefficients, we consider a solid disk of benzoic acid 2.5 cm in diameter spinning at 20 rpm and 25°C. We want to know how fast will it dissolve in a large volume of water. We do know that the diffusion coefficient is 1.00×10^{-5} cm²/sec and the solubility of benzoic acid in water is 0.003 g/cm³.

The dissolution rate of benzoic acid is

$$N_1 = kc_1(\text{sat}) \tag{62}$$

where $c_1(\text{sat})$ is the concentration of benzoic acid in water at saturation. From Table 11, the mass transfer coefficient is

$$k = 0.62D\left(\frac{\omega}{\nu}\right)^{1/2}\left(\frac{\nu}{D}\right)^{1/3}$$

$$= 0.62(1.00 \times 10^{-5}\ \text{cm}^2/\text{sec})\left(\frac{(20/60)(2\pi/\text{sec})}{0.01\ \text{cm}^2/\text{sec}}\right)^{1/2}\left(\frac{0.01\ \text{cm}^2/\text{sec}}{1.00 \times 10^{-5}\ \text{cm}^2/\text{sec}}\right)^{1/3}$$

$$= 0.90 \times 10^{-3}\ \text{cm/sec}.$$

Table 11
A Selection of Mass Transfer Correlation[a,b]

Physical Situation	Basic Equation[c]	Key Variables	Remarks
Solid Interfaces			
Membrane	$\frac{k\ell}{D} = 1$	ℓ = membrane thickness	Often applied even where membrane is hypothetical
Laminar flow along flat plate	$\frac{kz}{D} = 0.323\left(\frac{zv^0}{\nu}\right)^{1/2}\left(\frac{\nu}{D}\right)^{1/3}$	z = distance from start of plate v^0 = bulk velocity	Solid theoretical foundation, which is unusual
Turbulent flow through horizontal slit	$\frac{kd}{D} = 0.026\left(\frac{dv^0}{\nu}\right)^{0.8}\left(\frac{\nu}{D}\right)^{1/3}$	v^0 = average velocity in slit d = $(2/\pi)$(slit width)	Mass transfer here is identical with a pipe of equal wetted perimeter
Turbulent flow through circular pipe	$\frac{kd}{D} = 0.026\left(\frac{dv^0}{\nu}\right)^{0.8}\left(\frac{\nu}{D}\right)^{1/3}$	v^0 = average velocity in pipe d = pipe diameter	Same as slit, because only wall region is involved
Laminar flow through circular pipe(d)	$\frac{kd}{D} = 1.86\left(\frac{dv^0}{D}\right)^{1/3}$	d = pipe diameter L = pipe length v^0 = average velocity in pipe	Not reliable when $(dv/D) < 10$ because of free convection
Forced convection around a solid sphere	$\frac{kd}{D} = 2.0 + 0.6\left(\frac{dv^0}{\nu}\right)^{1/2}\left(\frac{\nu}{D}\right)^{1/3}$	d = sphere diameter v^0 = velocity of sphere	Very difficult to reach $(kd/D) = 2$ experimentally. No sudden laminar-turbulent transition
Free convection around a solid sphere	$\frac{kd}{D} = 2.0 + 0.6\left(\frac{d^3[\Delta\rho]g}{\rho\nu^2}\right)^{1/4}\left(\frac{\nu}{D}\right)^{1/3}$	d = sphere diameter g = gravitational acceleration	For a 1 cm sphere in water, free convection is important when $\Delta\rho = 10^{-9}$ g/cm^3
Spinning disc	$\frac{kd}{D} = 0.62\left(\frac{d^2\omega}{\nu}\right)^{1/2}\left(\frac{\nu}{D}\right)^{1/3}$	d = disc diameter ω = disc rotation (radians/time)	Valid for Reynolds numbers between 100 and 20,000

(Continued)

Table 11 (Continued)

Physical Situation	Basic Equation[c]	Key Variables	Remarks
Flow normal to capillary	$\frac{kd}{D} = f\left(\frac{dv^0}{\nu}, \frac{\nu}{D}\right)$	d = tube diameter v^0 = average velocity	Large number of correlations with different exponents found by analogy with heat transfer
Packed beds	$\frac{k}{v^0} = 1.17\left(\frac{dv^0}{\nu}\right)^{-0.42}\left(\frac{\nu}{D}\right)^{-0.67}$	d = particle diameter v^0 = superficial velocity	The superficial velocity is that which would exist without packing
Fluid-Fluid Interfaces			
Drops or bubbles in stirred solutions	$\frac{kL}{D} = 0.13\left[\frac{L^4(P/V)}{\rho\nu^3}\right]^{1/4}\left(\frac{\nu}{D}\right)^{1/3}$	L = stirrer length P/V = power per volume	Correlation vs. power per volume are common for dispersions
Large drops in unstirred solution	$\frac{kd}{D} = 0.42\left(\frac{d^3\,\Delta\rho g}{\rho\nu^2}\right)^{1/3}\left(\frac{\nu}{D}\right)^{1/2}$	d = bubble diameter $\Delta\rho$ = density difference between bubble and surrounding fluid	Large is defined as ~0.3 cm diameter
Small drops of pure solute in unstirred solution	$\frac{kd}{D} = 1.13\left(\frac{dv^0}{D}\right)^{1/2}$	d = bubble diameter v^0 = bubble velocity	These behave like rigid spheres
Falling films	$\frac{kz}{D} = 0.69\left(\frac{zv^0}{D}\right)^{1/2}$	z = position along film v^0 = average film velocity	Frequently embroidered and embellished

[a] *The symbols used include: ρ is the fluid density; ν is the kinematic viscosity; D is the diffusion coefficient of the material being transferred; and k is the local mass transfer coefficient. Other symbols are defined for the specific situation.*

[b] *This table is abstracted from Calderbank [11], McCabe and Smith [42], H. Schlichting [43], Sherwood, et al., [15], and Treybal [44].*

[c] *The dimensionless groups are defined as follows: (dv/ν) and $(d^2\omega/\nu)$ are the Reynolds number; ν/D is the Schmidt number; $(d^3\,\Delta\rho g/\rho\nu^2)$ is the Grashöf number; kd/D is the Sherwood number; and k/v is the Stanton number.*

[d] *The mass transfer coefficient given here is the value averaged over the length.*

Thus, the flux is

$$N_1 = (0.90 \times 10^{-3}\ \text{cm/sec})(0.003\ \text{g/cm}^3)$$
$$= 2.7 \times 10^{-6}\ \text{g/cm}^2\cdot\text{sec}$$

Note that doubling the diffusion coefficient increases the mass transfer about 60%.

Mass Transfer Across Interfaces

The correlations of mass transfer coefficients in Table 11 provide a quick, accurate method for estimating mass transfer coefficients. However, the coefficients calculated are for one phase, i.e., for transfer from the interface into the bulk in a single phase. A good example is that given directly above, where the mass transfer occurs from a solid disc into the bulk solution.

In many cases, mass transfer occurs *across* an interface, from one phase into another. Here, there are two mass transfer coefficients, one from one bulk solution to the interface, and the second from the interface into the other bulk solution. Moreover, equal concentrations in the two bulk phases does often not imply equilibrium: a partition coefficient must be involved.

To describe mass transfer across an interface, we must combine two mass transfer coefficients and a partition coefficient. This combination is described in terms of a flux equation like that used before:

$$N_1 = K\ \Delta c_1 \tag{63}$$

where N_1 is the solute flux relative to the interface, K is called an "overall mass transfer coefficient" and Δc_1 is an appropriate concentration difference. This concentration difference can be subtle. For example, if a benzene solution of bromine is brought into contact with an aqueous solution of bromine, one appropriate concentration difference is the actual concentration of bromine in benzene minus that concentration of bromine in benzene, which would exist if the benzene solution were in equilibrium with the actual concentration in water:

$$N_1 = K[c_1(\text{in benzene}) - Hc_1(\text{in water})] \tag{64}$$

Here H is the partition coefficient, equal to the equilibrium concentration in benzene divided by the equilibrium concentration in water. Obviously, the overall mass transfer coefficient K is an average of the local mass transfer coefficients k. As an example, we consider the same benzene-water interface. The flux in the benzene phase is

$$N_1 = k(c_{10} - c_{1i}) \tag{65}$$

where k is the mass transfer coefficient from the bulk benzene to the benzene-water interface. The flux in the aqueous phase is

$$N_1 = k'(c'_{1i} - c'_{10}) \tag{66}$$

where k' is the mass transfer coefficient from the interface into the bulk water. The interfacial concentrations are in equilibrium so that

$$c_{1i} = Hc'_{1i} \tag{67}$$

Removing for the interfacial concentration, we find the flux

$$N_1 = K[c_1 - Hc'_1] \tag{68}$$

we can find the overall mass transfer coefficient K′ based on a driving force in the benzene:

$$K = \frac{1}{1/k + H/k'} \tag{69}$$

Thus, the overall mass transfer coefficient is a harmonic average of the individual coefficients, but weighted with a partition coefficient.

NOTATION

A	area
a	pore length per solid length
c_i	concentration
D	diffusion coefficient
D_{eff}	effective diffusion coefficient
D_0	transport coefficient
E	dispersion coefficient
f	functional coefficient
H	partition coefficient
j_i	linear mass flux per unit area
$\underset{\sim}{j}_i^a$	diffusion flux
K	association constant; equilibrium constant
k	Boltzmann's constant
k	mass transfer coefficient
l	length
M	solute mass
N_1	flux at interface
N_1	mass flux
n_1	total flux
P	pressure
R_0	solute radius
T	temperature
t	time
$\underset{\sim}{v}^a$	convective reference velocity
z	coordinate

Greek Symbols

β	calibration constant
γ_1	activity coefficient
ϵ	void fraction
ξ	function
η	parameter
λ	solute diameter divided by pore diameter
μ_0	solvent viscosity
v	intrinsic velocity
τ	tortuosity
ϕ	interfacial velocity
ϕ_1	volume fraction of solute
ϕ_s	volume fraction
ω	angular velocity

REFERENCES

1. Fourier, J. B. J., *Théorie analytique de la chaleur*. Paris, 1822.
2. Crank, J., *The Mathematics of Diffusion*, second ed., Oxford: Clarendon Press, 1975.
3. Carslaw, H. S., and Jaeger, J. C., *The Conduction of Heat in Solids*, second ed., Oxford: Clarendon Press, 1959.
4. Onsager, L., "Theories and Problems of Liquid Diffusion." *New York Academy of Sciences Annals*, Vol. 46, 1945, pp. 241–286.
5. Arnold, K. R., and Toor, H. L., "Unsteady Diffusion in Ternary Gas Mixtures." *American Institute of Chemical Engineers Journal*, Vol. 13, 1967, pp. 909–914.
6. Cussler, E. L., and Lightfoot, E. N., "Multicomponent Diffusion Involving High Polymers. I. Diffusion of Monodisperse Polystyrene in Mixed Solvents." *Journal of Physical Chemistry*, Vol. 69, 1965, pp. 1135–1144.
7. Ellerton, H. D., and Dunlop, P. J., "Diffusion and Frictional Coefficients for Four Compositions of the System Water-Sucrose-Mannitol at 25°. Test of the Onsager Reciprocal Relation." *Journal of Physical Chemistry*, Vol. 71, 1967, pp. 1291–1297.
8. Vitagliano, V., Laurentino, R., and Constantino, L., "Diffusion in a Ternary System with Strong Interacting Flows." *Journal of Physical Chemistry*, Vol. 73, 1969, pp. 2456–2457.

9. Wendt, R. P., "Studies of Isothermal Diffusion at 25° in the System Water-Sodium Sulfate-Sulfuric Acid and Tests of the Onsager Relation." *Journal of Physical Chemistry*, Vol. 66, 1962, pp. 1279–1288.
10. Vermeulen, T., Clazie, R. N., and Klein, G., OSW RD Progress Report No. 326, 1967.
11. Carlson, P. T., Dayananda, M. A., and Grace, R. E., "Diffusion in Ternary Ag-Zn-Cd Solid Solutions." *Metallurgical Transactions*, Vol. 3, 1972, pp. 819–828.
12. Cussler, E. L., Evans, D. F., and Matesich, M. A., "Theoretical and Experimental Basis for a Specific Countertransport System in Membranes." *Science*, Vol. 172, 1971. pp. 377–379.
13. Keller, K. H., and Friedlander, S. K., "The Steady-State Transport of Oxygen through Hemoglobin Solutions." *Journal of General Physiology*, Vol. 49, March 1966, pp. 663–679.
14. Cussler, E. L., *Multicomponent Diffusion*. Amsterdam: Elsevier, 1976.
15. Sherwood, T. K., Pigford, R. L., and Wilke, C. R., *Mass Transfer*. New York: McGraw-Hill, 1975.
16. Reid, R. C., Sherwood, T. K., and Prausnitz, J. M., *Properties of Gases and Liquids*, third ed., New York: McGraw Hill, 1977.
17. Cussler, E. L., *Diffusion, Mass Transfer in Fluid Systems*. New York: Cambridge University Press, 1984.
18. Batchelor, G. K., "Brownian Diffusion of Particles with Hydrodynamic Interaction." *Journal of Fluid Mechanics*, Vol. 71, 1976, pp. 1–30.
19. Reed, C. C., and Anderson, J. L., "Hindered Settling of a Suspension at Low Reynolds Number." *American Institute of Chemical Engineers Journal*, Vol. 26, 1980. pp. 816–824.
20. Stokes, R. H., "The Diffusion Coefficient of Eight Uni-univalent Electrolytes in Aqueous Solution at 25°." *Journal of the American Chemical Society*, Vol. 72, 1950, pp. 2243–2247.
21. Mills, R., Woolf, L. A., and Watts, R. O., "Simple Procedures for Diaphragm-Cell Diffusion Studies." *American Institute of Chemical Engineers Journal*, Vol. 14, 1968, pp. 671–681.
22. Quano, A. C., "Diffusion in Liquid Systems. I. A Simple and Fast Method of Measuring Diffusion Constants." *Industrial and Engineering Chemistry Fundamentals*, Vol. 11, 1972, pp. 268–271.
23. Maynard, V. R., and Grushka, E., "Measurement of Diffusion Coefficients by Gas-Chromatography Broadening Techniques: A Review." In *Advances in Chromatography*, Vol. 12, J. C. Giddings *et al.*, Eds., New York: Dekker, 1975, pp. 99–139.
24. Gosting, L. J., "Measurement and Interpretation of Diffusion Coefficients of Proteins." *Advances in Protein Chemistry*, Vol. 11, 1956, pp. 429–548.
25. Longsworth, L. G., "Tests of Flowing Junction Diffusion Cells with Interference Methods." *Review of Scientific Instruments*, Vol. 21, 1950, pp. 524–528.
26. Creeth, J. M., and Gosting, L. J., "Studies of Free Diffusion in Liquids with the Rayleigh Method. II. An Analysis for Systems Containing Two Solutes." *Journal of Physical Chemistry*, Vol. 62, January 1958, pp. 58–65.
27. Bryngdahl, O., "The Interferometric Method for the Determination of Diffusion Coefficients of Very Dilute Solutions." *Acta Chemica Scandinavica*. Vol. 11, 1957, pp. 1017–1033.
28. Bloomfield, V. A., "Quasi-Elastic Light Scattering Applications in Biochemistry and Biology." *Annual Review of Biophysics and Bioengineering*, Vol. 10, 1981, pp. 421–450.
29. Berne, B. J., and Pecora, R., *Dynamic Light Scattering*. New York: Wiley, 1976.
30. Levich, V., *Physicochemical Hydrodynamics*. New Jersey: Prentice-Hall, 1962.
31. Chan, A. F., Evans, D. F., and Cussler, E. L., "Explaining Solubilization Kinetics." *American Institute of Chemical Engineers Journal*, Vol. 22, November, 1976, pp. 1006–1012.
32. Rao, S. S., and Bennett, C. O., "Steady State Technique for Measuring Fluxes and Diffusivities in Binary Liquid Systems." *American Institute of Chemical Engineers Journal*, Vol. 17, 1971, pp. 75–81.
33. Dunlop, P. J., Steele, B. J., and Lane, J. E., "Experimental Methods for Studying Diffusion in Liquids, Gases and Solids." In *Physical Methods of Chemistry*, Part IV, A. Weissberger and B. W. Tossiter, Eds., New York: Wiley, 1972, pp. 207–337.
34. Gostings, L. J., and Akeley, D. F., "A Study of the Diffusion of Urea in Water at 25° with the Gouy Interference Method." *Journal of the American Chemical Society*, Vol. 74, April 20, 1952, pp. 2058–2060.

35. Robinson, R. A., and Stokes, R. H., *Electrolyte Solutions.* London: Butterworth, 1960.
36. Hinton, J. F., and Amis, E. S., "Solvation Number of Ions." *Chemical Reviews*, Vol. 71, 1971, pp. 627–674.
37. Geankopolis, C., *Mass Transfer Phenomena*, New York: Holt, Rinehart and Winston, 1972, pp. 149ff.
38. Maxwell, J. C., *A Treatise on Electricity and Magnetism*, Vol. 1: Oxford: Clarendon Press, 1873. p. 365.
39. Gajdos, L. J., and Brenner, H., "Field-Flow Fractionation: Extensions to Nonspherical Particles and Wall Effects." *Separation Sciences and Technology*, Vol. 13, 1978, pp. 215–240.
40. Quinn, J. A., Anderson, J. L., Ho, W. S., and Petzny, W. J., "Model Pores of Molecular Dimension." *Biophysical Journal*, Vol. 12, 1972, pp. 990–1007.
41. Calderbank, P. H., "Mass Transfer." In *Mixing*, V. Vhl and J. Gray, Eds., New York: Academic Press, 1967, pp. 2–111.
42. McCabe, W. L., and Smith, J. C., *Unit Operations of Chemical Engineering*, third ed., New York: McGraw-Hill, 1975.
43. Schlichting, H., *Boundary Layer Theory*, seventh ed., New York: McGraw-Hill, 1979.
44. Treybal, R. E., *Mass Transfer Operations*, third ed., New York: McGraw-Hill 1980.

Bibliography

Cussler, E. L., *Diffusion: Mass Transfer in Fluid Systems.* New York: Cambridge University Press 1984.

Reid, R. C., Sherwood, T. K., and Prausnitz, J. M., *Properties of Gases and Liquids*, third ed., New York: McGraw Hill, 1977.

Jost, W., *Diffusion in Solids, Liquids and Gases*, New York: Academic Press, 1952.

CHAPTER 5

MIXING AND ISOTROPY IN HOMOGENEOUS TURBULENCE

Jon Lee

Flight Dynamic Laboratory
Wright-Patterson AFB, OH 45433

CONTENTS

INTRODUCTION

In the statistical theory of turbulence, ergodicity is assumed to be exhibited by the Navier-Stokes equations as required by classical statistical mechanics. For homogeneous turbulence, one further assumes mixing in phase space, which implies ergodicity, and isotropy, which imposes symmetries on the statistical functions. In the inviscid case to be investigated here, mixing gives rise to energy equipartition, which is a prerequisite for isotropy. In fact, ergodicity is a plausible assumption in that total energy and helicity are the only isolating constants of motion, when sufficiently many triad (three-Fourier-mode) interactions are included in a truncated Fourier representation of the three-dimensional Navier-Stokes equations [1]. As an example, Orszag's [2] five-mode turbulence model develops random trajectories and exhibits mixing (and *a fortiori* ergodicity), when there is no isolating constant of motion other than energy.

We shall attempt to show that trajectory of the Navier-Stokes equations can develop mixing in phase space and exhibit isotropy of the covariance spectral tensor. In a homogeneous field with no mean flow, the nonlinear term of the incompressible Navier-Stokes equations can be decomposed into enumerable triad-interactions over infinitely coupled wavevectors. By singling out a typical triad-interaction, it is found that its dynamical system, called the fundamental triad-interaction, is not ergodic, let alone mixing, on the constant energy-helicity surface, nor will it develop isotropy of the covariance spectral tensor after a long evolution time. This is due to the presence of extraneous constants of motion besides energy and helicity. Surprisingly, some of them are not strictly invariant, yet they can restrict the trajectory flow just as the true constants of motion. Based on an earlier result [1], one may conjecture that more and more triad-interactions added to the dynamical system would annihilate the extraneous constants of motion, thereby engendering mixing in phase space. We shall therefore follow through this conjecture by examining a sequence of dynamical systems with increasingly many triad-interactions. To obtain such a sequence, the wavevector space will be truncated to obtain only those wavevectors lying in a sphere of specified radius—the spherical truncation. The lowest-order truncation involves 22 triad-interactions over 13 wavevectors lying in a sphere of wavenumber radius $\sqrt{3}$. We shall examine here several of the higher-order truncated systems extending up to the spherical truncation of wavenumber radius 4.

In the beginning, we had intended to provide a direct numerical test for ergodicity and mixing. However, this has proven untenable because the true trajectory cannot be computed for a sufficiently long time for such tests. Evidently, the culprit here is trajectory instability, whereby the two initially nearby trajectories break away from each other exponentially after a short threshold time. Unfortunately, a numerical consequence is the accumulation of round-off errors, which then impedes trajectory computation over a time period longer than several characteristic correlation times of the smallest eddy. The trajectory instability can best be quantified by an effective evolution time, beyond which one can no longer guarantee the accuracy of trajectory computation. This has been suggested by the two extremes. The effective evolution time is practically infinite in the absence of trajectory instability, whereas it is in the order of a typical characteristic correlation time when the system is mixing. Theoretically speaking, trajectory instability is exhibited by K-systems; hence it is a stronger dynamical property than mixing. In the hierarchy of dynamical properties, mixing implies ergodicity, and isotropy of the covariance spectral tensor follows from the mixing property.

The computability problem as manifested by trajectory instability was first reported by Birkhoff and Fisher [3] who observed irregular evolution of the initially smooth vortex sheets modelled by discrete point vortices. Later, it reappears as the predictability or internal-error-growth problem in numerical weather forecasting [4, 5]. Relevant is the more recent evidence of trajectory instability in a variety of model flow problems [6, 7, 8]. Although trajectory instability is detrimental to the numerical trajectory evolution, it is just the ingradient that we need for the statistical turbulence formulation. This is because trajectory instability engenders the loss of initial information, thereby providing a logical link between the macroscopic irreversibility and dynamic reversibility.

Numerical details of the lowest-order truncation are presented later to demonstrate the sensitive dependence upon initial conditions, random trajectory, ergodic behavior of modal energies, isotropy of the spectral tensor, decay of autocorrelation, and positive Kolmogorov entropy. The numerical result of the present investigation is this. For the lowest-order truncation the effective evolution time is only 110. The effective evolution time decreases monotonically with the increasing order of spherical truncation. It is therefore concluded that the three-dimensional Navier-Stokes system develops mixing as more and more triad-interaction are included.

In the absence of mean flow motion and boundary effects, the trajectory instability that we have been speaking of is solely attributed to nonlinearity of the Navier-Stokes equations. It is what Liepmann [9] called a dynamical instability of rapidly increasing three-dimensional perturbations, in contrast to laminar or viscous instability of the onset of turbulence.

DYNAMIC EQUATIONS

For the homogeneous field with no mean flow, it is most convenient to Fourier analyze the three-dimensional velocity field $\underset{\sim}{U}(\underset{\sim}{x}, t)$ in a cyclic box of side L

$$U_i(\underset{\sim}{x}, t) = (2\pi/L)^{3/2} \sum_{k \neq 0} U_i(\underset{\sim}{k}, t) \exp(i\underset{\sim}{k} \cdot \underset{\sim}{x}), \qquad (i = 1, 2, 3) \tag{1}$$

where the wave vector $\underset{\sim}{k} = (2\pi/L)\underset{\sim}{n}$ ($n_i = 0, \pm 1, \pm 2, \ldots$). Rather than dealing directly with the Fourier amplitudes $U_i(\underset{\sim}{k}, t)$, the triad-interaction representation involves decomposing $U_i(\underset{\sim}{k}, t)$ into a sum of the divergenceless and curlless parts (the Helmholtz theorem). Only the divergenceless part is present in the incompressible flow field, and that part will be spanned by the polarization vectors to be described presently.

Polarization-Vector Expansion

The incompressible fluid motion may be represented by

$$U_i(\underset{\sim}{k}, t) = \sum_{\mu=1,2} \epsilon_i^\mu(\underset{\sim}{k}, \xi) u^\mu(\underset{\sim}{k}, t) \tag{2}$$

Here the polarization vectors $\underset{\sim}{\epsilon}^\mu(\underset{\sim}{k}, \xi)$ obey [10]

$$\underset{\sim}{k} \cdot \underset{\sim}{\epsilon}^\mu(\underset{\sim}{k}, \xi) = 0 \tag{3a}$$

$$\underset{\sim}{\epsilon}^\mu(k, \xi) \cdot \underset{\sim}{\epsilon}^\nu(\underset{\sim}{k}, \xi) = \delta_{\mu\nu} \tag{3b}$$

$$\sum_{\mu=1,2} \epsilon_i^\mu(\underset{\sim}{k}, \xi)\epsilon_j^\mu(\underset{\sim}{k}, \xi) = P_{ij}(\underset{\sim}{k}) \, (= \delta_{ij} - k_i k_j/k^2) \tag{3c}$$

for any value of the parameter ξ in $[0, 2\pi]$. The presence of ξ will be explained shortly. In words, Equation 3a states that the polarization vectors lie in a plane normal to the wave vector $\underset{\sim}{k}$, Equation 3b is the orthonormality of polarization vectors, and Equation 3c is an identity of the orthonormal triple vectors $[\underset{\sim}{\epsilon}^1(\underset{\sim}{k}, \xi), \underset{\sim}{\epsilon}^2(\underset{\sim}{k}, \xi), \underset{\sim}{k}/k]$ in Euclidean 3-space (E_3, for short).

For a given wave vector $\underset{\sim}{K}$, the general expression for $\underset{\sim}{\epsilon}^\mu(\underset{\sim}{K}, \xi)$ can be obtained as follows: First, align $\underset{\sim}{K}$ along with one of the fixed coordinate axes, say the 1 axis, and hence $\underset{\sim}{K} = (K, 0, 0)$. Then, by virtue of Equation 3a any orthonormal vectors in the 2–3 plane are candidates for polarization vectors, i.e.,

$$\epsilon^1(\underset{\sim}{K}, \xi) = \begin{bmatrix} 0 \\ \cos\xi \\ \sin\xi \end{bmatrix}, \qquad \underset{\sim}{\epsilon}^2(\underset{\sim}{K}, \xi) = \begin{bmatrix} 0 \\ -\sin\xi \\ \cos\xi \end{bmatrix} \tag{4}$$

Note that ξ denotes the arbitrary rotation of $\underset{\sim}{\epsilon}^\mu(\underset{\sim}{K}, \xi)$ about the 1 axis. Let us now introduce a matrix

$$R = \begin{bmatrix} \cos\theta_k \cos\eta_k & -\sin\theta_k & -\cos\theta_k \sin\eta_k \\ \sin\theta_k \cos\eta_k & \cos\theta_k & -\sin\theta_k \sin\eta_k \\ \sin\eta_k & 0 & \cos\eta_k \end{bmatrix}$$

where $\cos\theta_k = K_1/K'$, $\sin\theta_k = K_2/K'$, $\cos\eta_k = K'/K$, $\sin\eta_k = K_3/K$. Here, we shall assume $K' = (K_1 + K_2)^{1/2}$ to be nonzero, for the zero wave vector is excluded from the expansion (Equation 1).

It is simple to check that the matrix R rotates the vector (K, 0, 0) into the actual configuration; i.e.,

$$R\begin{bmatrix} K \\ 0 \\ 0 \end{bmatrix} = \begin{bmatrix} K_1 \\ K_2 \\ K_3 \end{bmatrix}$$

Therefore, the premultiplication of Equation 4 by R would yield the polarization vectors associated with (K_1, K_2, K_3) (see Figure 1).

$$\underset{\sim}{\epsilon}^1(\underset{\sim}{K}, \xi) = \begin{bmatrix} -\sin\theta_k \cos\xi - \cos\theta_k \sin\eta_k \sin\xi \\ \cos\theta_k \cos\xi - \sin\theta_k \sin\eta_k \sin\xi \\ \cos\eta_k \sin\xi \end{bmatrix}$$

$$\underset{\sim}{\epsilon}^2(\underset{\sim}{K}, \xi) = \begin{bmatrix} \sin\theta_k \sin\xi - \cos\theta_k \sin\eta_k \cos\xi \\ -\cos\theta_k \sin\xi - \sin\theta_k \sin\eta_k \cos\xi \\ \cos\eta_k \cos\xi \end{bmatrix} \tag{5}$$

Note that for $\xi = 0$, Equation 5 becomes

$$\underset{\sim}{\epsilon}^1(\underset{\sim}{K}, 0) = \begin{bmatrix} -\sin\theta_k \\ \cos\theta_k \\ 0 \end{bmatrix}, \qquad \underset{\sim}{\epsilon}^2(\underset{\sim}{K}, 0) = \begin{bmatrix} -\cos\theta_k \sin\eta_k \\ -\sin\theta_k \sin\eta_k \\ \cos\eta_k \end{bmatrix} \tag{6}$$

which are the particular choice used by Herring [11] and Lee [12].

As may be verified directly, the triple vector $[\underset{\sim}{\epsilon}^1(\underset{\sim}{k}, \xi), \underset{\sim}{\epsilon}^2(\underset{\sim}{k}, \xi), \underset{\sim}{k}/k]$ forms the right-handed coordinate system, whence

$$\underset{\sim}{\epsilon}^1(\underset{\sim}{k}, \xi) = -(\underset{\sim}{k}/k) \times \underset{\sim}{\epsilon}^2(\underset{\sim}{k}, \xi)$$

$$\underset{\sim}{\epsilon}^2(\underset{\sim}{k}, \xi) = (\underset{\sim}{k}/k) \times \underset{\sim}{\epsilon}^1(\underset{\sim}{k}, \xi) \tag{7}$$

Furthermore, the triple vector obeys Equation 3 for any ξ; that is, the polarization vectors may be rotated arbitrarily in a plane normal to the wave vector $\underset{\sim}{k}$. This is because the zero divergence, which is orthogonality with the wave vector $\underset{\sim}{k}$, is satisfied by any vector in a plane normal to $\underset{\sim}{k}$.

One can formally show that $\underset{\sim}{\epsilon}^{\mu}(\underset{\sim}{k}, \xi)$ are the eigenvectors of $P_{ij}(\underset{\sim}{k})$ for the nonzero eigenvalues. Consider the eigenvalue problem

$$P_{ij}(\underset{\sim}{k})e_j(\underset{\sim}{k}) = e_i(\underset{\sim}{k}) \tag{8}$$

The eigenvalues of $P_{ij}(\underset{\sim}{k})$ are $\lambda_1 = \lambda_2 = 1$ and $\lambda_3 = 0$. The eigenvector for $\lambda_3 = 0$ is $\underset{\sim}{e}^3 = \underset{\sim}{k}$, but for

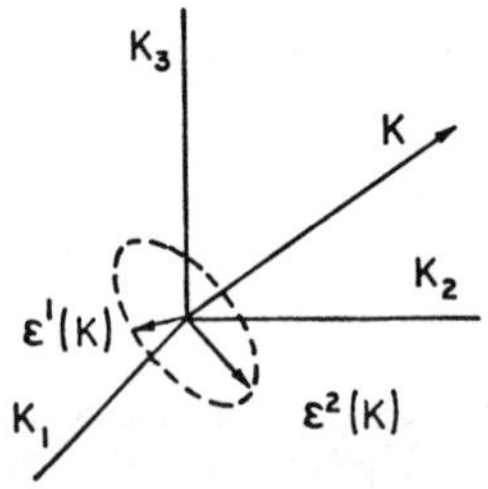

Figure 1. Polarization vectors.

the other eigenvalues, Equation 8 reduces to a single equation

$$\underset{\sim}{k} \cdot \underset{\sim}{\epsilon}^{\mu}(\underset{\sim}{k}) = 0 \qquad (\mu = 1, 2)$$

which is nothing but Equation 3a. Hence, the subspace spanned by the eigenvectors is normal to $\underset{\sim}{k}$, yet the choice of eigenvectors is left unspecified. We may also consider linear combinations of $\underset{\sim}{\epsilon}^{\mu}(k, \xi)$ as follows;

$$\underset{\sim}{Q}_1(\underset{\sim}{k}, \xi) = 2^{-1/2}[-\underset{\sim}{\epsilon}^1(\underset{\sim}{k}, \xi) - i\underset{\sim}{\epsilon}^2(\underset{\sim}{k}, \xi)]$$

$$\underset{\sim}{Q}_{-1}(\underset{\sim}{k}, \xi) = 2^{-1/2}[\underset{\sim}{\epsilon}^1(\underset{\sim}{k}, \xi) - i\underset{\sim}{\epsilon}^2(\underset{\sim}{k}, \xi)]$$

With the choice of $\xi = \sin^{-1}(k_1/(k_1 + k_2)^{1/2})$, they reduce to the $\underset{\sim}{Q}_\lambda(\underset{\sim}{k})$ vectors of Moses [13]. It is simple to check using Equation 7 that

$$\underset{\sim}{k} \times \underset{\sim}{Q}_\lambda(\underset{\sim}{k}, \xi) = -i\lambda k \underset{\sim}{Q}_\lambda(\underset{\sim}{k}, \xi), \qquad (\lambda = \pm 1)$$

Since $\underset{\sim}{\nabla}$x goes to $\underset{\sim}{k}$x in the Fourier space, Moses calls $\underset{\sim}{Q}_\lambda(\underset{\sim}{k})$ the eigenvectors of the curl operator with the eigenvalues $\lambda = \pm 1$.

Triad-Interaction Representation

Let us introduce Equations 1 and 2 into the incompressible Navier-Stokes equations. The ensuing equations for $u^{\mu}(\underset{\sim}{k}, t)$ are the triad-interaction representation of three-dimensional flow $(\mu, \lambda, \rho = 1, 2)$

$$(\partial/\partial t + \nu k^2)u^{\mu}(\underset{\sim}{k}, t) = -i(2\pi/L)^{3/2} \sum_{\lambda,\rho} \sum_{\underset{\sim}{k}=\underset{\sim}{p}+\underset{\sim}{q}} \phi^{\mu|\lambda,\rho}_{\underset{\sim}{k}|\underset{\sim}{p},\underset{\sim}{q}}(\xi) u^{\lambda}(\underset{\sim}{p}, t) u^{\rho}(\underset{\sim}{q}, t) \tag{9}$$

where $\phi^{\mu|\lambda,\rho}_{\underset{\sim}{k}|\underset{\sim}{p},\underset{\sim}{q}} = [\underset{\sim}{k} \cdot \underset{\sim}{\epsilon}^{\rho}(\underset{\sim}{q}, \xi)][\underset{\sim}{\epsilon}^{\mu}(\underset{\sim}{k}, \xi) \cdot \underset{\sim}{\epsilon}^{\lambda}(\underset{\sim}{p}, \xi)]$ and ν is the kinematic viscosity. It must be pointed out that the quadratic term in Equation 9 reflects only the inertia term $[(\underset{\sim}{U} \cdot \underset{\sim}{\nabla})\underset{\sim}{U}]$; the pressure term does not appear explicitly in the triad-interaction representation just as in the vorticity equation. In fact, the triad-interaction representation is equivalent to the vorticity form of homogeneous flow [14]. Unlike the Fourier amplitude $U_i(\underset{\sim}{k}, t)$ with three components $(i = 1, 2, 3)$, $u^{\mu}(\underset{\sim}{k}, t)$ has only two components $(\mu = 1, 2)$; a not insignificant reduction in the dynamic variables. To preserve the same form of the reality requirement $u^{\mu *}(\underset{\sim}{k}) = u^{\mu}(-\underset{\sim}{k})$ as the original $\underset{\sim}{U}^*(\underset{\sim}{k}) = \underset{\sim}{U}(-\underset{\sim}{k})$, we shall demand that the polarization vectors are unaffected by wave vector reversal $\underset{\sim}{\epsilon}^{\mu}(\underset{\sim}{k}, \xi) = \epsilon^{\mu}(-\underset{\sim}{k}, \xi)$. Consequently, we find that $\bar{\phi}^{\mu|\lambda,\rho}_{\underset{\sim}{k}|\underset{\sim}{p},\underset{\sim}{q}}(\xi) = -\bar{\phi}^{\mu|\lambda,\rho}_{-\underset{\sim}{k}|-\underset{\sim}{p},-\underset{\sim}{q}}(\xi)$, which assures equality of the evolution of $u^{\mu *}(\underset{\sim}{k}, t)$and $u^{\mu}(-\underset{\sim}{k}, t)$ for all t, hence is the dynamic reality requirement. Upon reversing the directions of p and q, we can put Equation 9 in a symmetric form

$$(\partial/\partial t + \nu k^2)u^{\mu}(\underset{\sim}{k}, t) = -i(2\pi/L)^{3/2} \sum_{\lambda,\rho} \sum_{\underset{\sim}{k}+\underset{\sim}{p}+\underset{\sim}{q}=0} \bar{\phi}^{\mu|\lambda,\rho}_{\underset{\sim}{k}|\underset{\sim}{p},\underset{\sim}{q}}(\xi) u^{\lambda *}(\underset{\sim}{p}, t) u^{\rho *}(\underset{\sim}{q}, t) \tag{10}$$

where the symmetrized coupling coefficient is given by

$$\bar{\phi}^{\mu|\lambda,\rho}_{\underset{\sim}{k}|\underset{\sim}{p},\underset{\sim}{q}}(\xi) = \tfrac{1}{2}[\phi^{\mu|\lambda,\rho}_{\underset{\sim}{k}|\underset{\sim}{p},\underset{\sim}{q}}(\xi) + \phi^{\mu|\rho,\lambda}_{\underset{\sim}{k}|\underset{\sim}{q},\underset{\sim}{p}}(\xi)]$$

The coupling coefficients satisfy the two conservation constraints over the triad wave vector $\underset{\sim}{K} + \underset{\sim}{P} + \underset{\sim}{Q} = 0$. The first is

$$\bar{\phi}^{\mu|\lambda,\rho}_{\underset{\sim}{K}|\underset{\sim}{P},\underset{\sim}{Q}}(\xi) + \bar{\phi}^{\lambda|\rho,\mu}_{\underset{\sim}{P}|\underset{\sim}{Q},\underset{\sim}{K}}(\xi) + \bar{\phi}^{\rho|\mu,\lambda}_{\underset{\sim}{Q}|\underset{\sim}{K},\underset{\sim}{P}}(\xi) = 0 \tag{11}$$

for the energy conservation, and the second is

$$(-1)^{\mu}K\,\bar{\phi}^{\mu|\lambda,\rho}_{\mathbf{K}|\mathbf{P},\mathbf{Q}}(\xi) + (-1)^{3-\lambda}P\,\bar{\phi}^{3-\lambda|\rho,3-\mu}_{\mathbf{P}|\mathbf{Q},\mathbf{K}}(\xi) + (-1)^{3-\rho}Q\,\bar{\phi}^{3-\rho|3-\mu,}_{\mathbf{Q}|\mathbf{K},\mathbf{P}}(\xi) = 0 \qquad (12)$$

from which follows the helicity conservation. Total energy and helicity are conserved not only for any ξ, but for arbitrarily many triad wave vectors coupled according to the Navier-Stokes equations.

TRUNCATION

From the outset we shall restrict ourselves to the inviscid case ($\nu = 0$). For definiteness, the infinite set Equation 10 of equations must be truncated at a level appropriate for computation. To this end, let us set $L = 2\pi$, so that $\mathbf{k} = \mathbf{n}$. In other words, the wave vectors have only the integer components and the wave vector space is now a three-dimensional lattice with integer coordinates.

Fundamental Triad-Interaction System

To illustrate the basic building block of triad interaction, we shall single out a typical triad wave vector $\mathbf{k}_1 + \mathbf{k}_2 + \mathbf{k}_3 = 0$ and then suppress all but the interaction terms involving this triad

$$\begin{aligned}
\dot{u}^{\mu}(\mathbf{k}_1) &= -i\sum_{\lambda,\rho}\bar{\phi}^{\mu|\lambda,\rho}_{\mathbf{k}_1|\mathbf{k}_2,\mathbf{k}_3}(\xi)u^{\lambda *}(\mathbf{k}_2)u^{\rho *}(\mathbf{k}_3)\\
\dot{u}^{\mu}(\mathbf{k}_2) &= -i\sum_{\lambda,\rho}\bar{\phi}^{\mu|\lambda,\rho}_{\mathbf{k}_2|\mathbf{k}_3,\mathbf{k}_1}(\xi)u^{\lambda *}(\mathbf{k}_3)u^{\rho *}(\mathbf{k}_1)\\
\dot{u}^{\mu}(\mathbf{k}_3) &= -i\sum_{\lambda,\rho}\bar{\phi}^{\mu|\lambda,\rho}_{\mathbf{k}_3|\mathbf{k}_1,\mathbf{k}_2}(\xi)u^{\lambda *}(\mathbf{k}_1)u^{\rho *}(\mathbf{k}_2)
\end{aligned} \qquad (13)$$

where the overhead dot denotes $\partial/\partial t$. This is the fundamental triad-interaction system, which is fully endowed with the essential properties of inviscid flow, such as incompressibility, energy and helicity conservations, and measure-invariance under time evolution.

Let us simplify the notation by writing $u^{\mu}_i(t) = u^{\mu}(\mathbf{k}_i, t)$. It is often more advantageous to resolve the motion of $u^{\mu}_i(t)$ by the polar representation

$$u^{\mu}_i(t) = R^{\mu}_i(t)\exp(i2\pi\omega^{\mu}_i(t)) \qquad (14)$$

into the action $J^{\mu}_i = (R^{\mu}_i)^2$ and angle ω^{μ}_i. Because of the periodicity $\omega^{\mu}_i = [\omega^{\mu}_i, \text{mod}(1)]$, points that are identical in rectangular representation may have an angle difference of integer multiples. Much insight may be gained by recasting Equation 13 into polar form

$$\begin{aligned}
\dot{R}^{\mu}_1 &= -\sum_{\lambda,\rho}\bar{\phi}^{\mu|\lambda,\rho}_{\mathbf{k}_1|\mathbf{k}_2,\mathbf{k}_3}(\xi)R^{\lambda}_2R^{\rho}_3\sin(2\pi\Omega^{\mu\lambda\rho}_{123})\\
\dot{R}^{\mu}_2 &= -\sum_{\lambda,\rho}\bar{\phi}^{\mu|\lambda,\rho}_{\mathbf{k}_2|\mathbf{k}_3,\mathbf{k}_1}(\xi)R^{\lambda}_3R^{\rho}_1\sin(2\pi\Omega^{\mu\lambda\rho}_{231})\\
\dot{R}^{\mu}_3 &= -\sum_{\lambda,\rho}\bar{\phi}^{\mu|\lambda,\rho}_{\mathbf{k}_3|\mathbf{k}_1,\mathbf{k}_2}(\xi)R^{\lambda}_1R^{\rho}_2\sin(2\pi\Omega^{\mu\lambda\rho}_{312})\\
\dot{\omega}^{\mu}_1 &= -(2\pi R^{\mu}_1)^{-1}\sum_{\lambda,\rho}\bar{\phi}^{\mu|\lambda,\rho}_{\mathbf{k}_1|\mathbf{k}_2,\mathbf{k}_3}(\xi)R^{\lambda}_2R^{\rho}_3\cos(2\pi\Omega^{\mu\lambda\rho}_{123})\\
\dot{\omega}^{\mu}_2 &= -(2\pi R^{\mu}_2)^{-1}\sum_{\lambda,\rho}\bar{\phi}^{\mu|\lambda,\rho}_{\mathbf{k}_2|\mathbf{k}_3,\mathbf{k}_1}(\xi)R^{\lambda}_3R^{\rho}_1\cos(2\pi\Omega^{\mu\lambda\rho}_{231})\\
\dot{\omega}^{\mu}_3 &= -(2\pi R^{\mu}_3)^{-1}\sum_{\lambda,\rho}\bar{\phi}^{\mu|\lambda,\rho}_{\mathbf{k}_3|\mathbf{k}_1,\mathbf{k}_2}(\xi)R^{\lambda}_1R^{\rho}_2\cos(2\pi\Omega^{\mu\lambda\rho}_{312})
\end{aligned} \qquad (15)$$

where $\Omega^{\mu\lambda\rho}_{123} = \omega^{\mu}_{1} + \omega^{\lambda}_{2} + \omega^{\rho}_{3}$, and $\Omega^{\mu\lambda\rho}_{231}$ and $\Omega^{\mu\lambda\rho}_{312}$ are similarly defined. In this representation, the motion of each mode is resolved into the amplitude and angle. Hence, Equation 15 describes the nonlinear dynamics of six coupled harmonic oscillators. Let us now write out Equation 15 in detail

$$
\begin{aligned}
\begin{bmatrix} \dot{R}^1_1 \\ \dot{R}^1_2 \\ \dot{R}^1_3 \\ \dot{R}^2_1 \\ \dot{R}^2_2 \\ \dot{R}^2_3 \end{bmatrix}
= &- \begin{bmatrix} \phi_{11}R^1_2R^1_3 \\ \phi_{12}R^1_3R^1_1 \\ \phi_{13}R^1_1R^1_2 \\ 0 \\ 0 \\ 0 \end{bmatrix} \sin(2\pi\Omega_5)
- \begin{bmatrix} \phi_{21}R^1_2R^2_3 \\ \phi_{22}R^2_3R^1_1 \\ 0 \\ 0 \\ 0 \\ \phi_{23}R^1_1R^1_2 \end{bmatrix} \sin(2\pi\Omega_3)
- \begin{bmatrix} \phi_{31}R^2_2R^1_3 \\ 0 \\ \phi_{33}R^1_1R^2_2 \\ 0 \\ \phi_{32}R^1_3R^1_1 \\ 0 \end{bmatrix} \sin(2\pi\Omega_2) \\
&- \begin{bmatrix} \phi_{41}R^2_2R^2_3 \\ 0 \\ 0 \\ 0 \\ \phi_{42}R^2_3R^1_1 \\ \phi_{43}R^1_1R^2_2 \end{bmatrix} \sin(2\pi\Omega_8)
- \begin{bmatrix} 0 \\ \phi_{52}R^1_3R^2_1 \\ \phi_{53}R^2_1R^1_2 \\ \phi_{51}R^1_2R^1_3 \\ 0 \\ 0 \end{bmatrix} \sin(2\pi\Omega_1)
- \begin{bmatrix} 0 \\ \phi_{62}R^2_3R^2_1 \\ 0 \\ \phi_{61}R^1_2R^2_3 \\ 0 \\ \phi_{63}R^2_1R^1_2 \end{bmatrix} \sin(2\pi\Omega_7) \\
&- \begin{bmatrix} 0 \\ 0 \\ \phi_{73}R^2_1R^2_2 \\ \phi_{71}R^2_2R^1_3 \\ \phi_{72}R^1_3R^2_1 \\ 0 \end{bmatrix} \sin(2\pi\Omega_6)
- \begin{bmatrix} 0 \\ 0 \\ 0 \\ \phi_{81}R^2_2R^2_3 \\ \phi_{82}R^2_3R^2_1 \\ \phi_{83}R^2_1R^2_2 \end{bmatrix} \sin(2\pi\Omega_4)
\end{aligned}
\tag{16a}
$$

$$
2\pi \begin{bmatrix} \dot{\omega}^1_1R^1_1 \\ \dot{\omega}^1_2R^1_2 \\ \cdots \\ \cdots \\ \dot{\omega}^2_3R^2_3 \end{bmatrix} = \text{The same 8 columns of Equation 16a but with } \sin(2\pi\Omega_n) \text{ being replaced by } \cos(2\pi\Omega_n). \tag{16b}
$$

Here,

$$
\begin{aligned}
&\Omega_1 = \omega^2_1 + \omega^1_2 + \omega^1_3, \quad &&\Omega_2 = \omega^1_1 + \omega^2_2 + \omega^1_3, \quad &&\Omega_3 = \omega^1_1 + \omega^1_2 + \omega^2_3, \\
&\Omega_4 = \omega^2_1 + \omega^2_2 + \omega^2_3, \quad &&\Omega_5 = \omega^1_1 + \omega^1_2 + \omega^1_3, \quad &&\Omega_6 = \omega^2_1 + \omega^2_2 + \omega^1_3, \\
&\Omega_7 = \omega^2_1 + \omega^1_2 + \omega^2_3, \quad &&\Omega_8 = \omega^1_1 + \omega^2_2 + \omega^2_3
\end{aligned}
$$

Also, the coupling coefficients have been abbreviated by ϕ_{ij}

$$(\bar{\phi}^{1|1,1}_{\underline{k}_1|\underline{k}_2,\underline{k}_3}, \bar{\phi}^{1|1,1}_{\underline{k}_2|\underline{k}_3,\underline{k}_1}, \bar{\phi}^{1|1,1}_{\underline{k}_3|\underline{k}_1,\underline{k}_2}) = (\phi_{11}, \phi_{12}, \phi_{13})$$

$$(\bar{\phi}^{1|1,2}_{\underline{k}_1|\underline{k}_2,\underline{k}_3}, \bar{\phi}^{1|2,1}_{\underline{k}_2|\underline{k}_3,\underline{k}_1}, \bar{\phi}^{2|1,1}_{\underline{k}_3|\underline{k}_1,\underline{k}_2}) = (\phi_{21}, \phi_{22}, \phi_{23})$$

$$(\bar{\phi}^{1|2,1}_{\underline{k}_1|\underline{k}_2,\underline{k}_3}, \bar{\phi}^{2|1,1}_{\underline{k}_2|\underline{k}_3,\underline{k}_1}, \bar{\phi}^{1|1,2}_{\underline{k}_3|\underline{k}_1,\underline{k}_2}) = (\phi_{31}, \phi_{32}, \phi_{33})$$

$$(\bar{\phi}^{1|2,2}_{\underline{k}_1|\underline{k}_2,\underline{k}_3}, \bar{\phi}^{2|2,1}_{\underline{k}_2|\underline{k}_3,\underline{k}_1}, \bar{\phi}^{2|1,2}_{\underline{k}_3|\underline{k}_1,\underline{k}_2}) = (\phi_{41}, \phi_{42}, \phi_{43})$$

$$(\bar{\phi}^{2|1,1}_{\underline{k}_1|\underline{k}_2,\underline{k}_3}, \bar{\phi}^{1|1,2}_{\underline{k}_2|\underline{k}_3,\underline{k}_1}, \bar{\phi}^{1|2,1}_{\underline{k}_3|\underline{k}_1,\underline{k}_2}) = (\phi_{51}, \phi_{52}, \phi_{53})$$

$$(\bar{\phi}^{2|1,2}_{\underline{k}_1|\underline{k}_2,\underline{k}_3}, \bar{\phi}^{1|2,2}_{\underline{k}_2|\underline{k}_3,\underline{k}_1}, \bar{\phi}^{2|2,1}_{\underline{k}_3|\underline{k}_1,\underline{k}_2}) = (\phi_{61}, \phi_{62}, \phi_{63})$$

$$(\bar{\phi}^{2|2,1}_{\underset{\sim}{k}_1|\underset{\sim}{k}_2,\underset{\sim}{k}_3}, \bar{\phi}^{2|1,2}_{\underset{\sim}{k}_2|\underset{\sim}{k}_3,\underset{\sim}{k}_1}, \bar{\phi}^{1|2,2}_{\underset{\sim}{k}_3|\underset{\sim}{k}_1,\underset{\sim}{k}_2}) = (\phi_{71}, \phi_{72}, \phi_{73})$$

$$(\bar{\phi}^{2|2,2}_{\underset{\sim}{k}_1|\underset{\sim}{k}_2,\underset{\sim}{k}_3}, \bar{\phi}^{2|2,2}_{\underset{\sim}{k}_2|\underset{\sim}{k}_3,\underset{\sim}{k}_1}, \bar{\phi}^{2|2,2}_{\underset{\sim}{k}_3|\underset{\sim}{k}_1,\underset{\sim}{k}_2}) = (\phi_{81}, \phi_{82}, \phi_{83})$$

In terms of ϕ_{ij}, the constraints (Equations 11 and 12) take the respective forms

$$\phi_{i1} + \phi_{i2} + \phi_{i3} = 0, \qquad (i = 1 - 8)$$

and

$$-k_1\phi_{11}+k_2\phi_{72}+k_3\phi_{63}=0, \qquad -k_1\phi_{21}+k_2\phi_{82}-k_3\phi_{53}=0, \qquad -k_1\phi_{31}-k_2\phi_{52}+k_3\phi_{83}=0,$$

$$-k_1\phi_{41}-k_2\phi_{62}-k_3\phi_{73}=0, \qquad k_1\phi_{51}+k_2\phi_{32}+k_3\phi_{23}=0, \qquad k_1\phi_{61}+k_2\phi_{42}-k_3\phi_{13}=0,$$

$$k_1\phi_{71}-k_2\phi_{12}+k_3\phi_{43}=0, \qquad k_1\phi_{81}-k_2\phi_{22}-k_3\phi_{33}=0$$

The explicit representation (Equation 16) suggests an important invariant set of the fundamental triad-interaction system. Suppose that we initially choose the ω_i^μ as follows:

$$S = \{\text{all } \omega_i^\mu \text{ such that } \Omega_n = \pm\tfrac{1}{4}, \pm\tfrac{3}{4}, \pm\tfrac{5}{4}, \ldots (n = 1 - 8)\} \tag{17}$$

It is readily seen that for such Ω_n the right-hand side of Equation 16b is identically zero and will remain so for all times. Hence Equation 17 is an invariant set. The corresponding motion is then governed solely by Equation 16a with $\sin(2\pi\Omega_n) = \pm 1$. An important subset of Equation 17 to be discussed later is

$$S' = \{\text{all } \omega_1^\mu = \omega_2^\mu \text{ such that } \Omega_n = \pm\tfrac{1}{4}, \pm\tfrac{3}{4}, \pm\tfrac{5}{4}, \ldots (n = 1 - 8)\} \tag{18}$$

This subset represents part of R-space that is independent of helicity.

Low-Order Systems of the Spherical Truncation

For a systematic truncation we must include all the triad wave vectors contained in a prescribed wave vector domain. The lowest-order truncation involves the wave vector lattice of $(\pm 1) \times (\pm 1) \times (\pm 1)$ as shown in Figure 2. Of the 26 wave vectors (excluding the zero vector), only a half of them need to be considered, for the remaining half are redundant due to the reality requirement. Without loss of generality, we shall therefore choose the 13 wave vectors lying on or above the 2–3 plane as indicated by the solid dots in Figure 2. After reordering the 13 wave vectors

$$\underset{\sim}{k}_1 = \begin{bmatrix}0\\0\\1\end{bmatrix}, \quad \underset{\sim}{k}_2 = \begin{bmatrix}0\\1\\-1\end{bmatrix}, \quad \underset{\sim}{k}_3 = \begin{bmatrix}0\\1\\0\end{bmatrix}, \quad \underset{\sim}{k}_4 = \begin{bmatrix}0\\1\\1\end{bmatrix}, \quad \underset{\sim}{k}_5 = \begin{bmatrix}1\\-1\\0\end{bmatrix},$$

$$\underset{\sim}{k}_6 = \begin{bmatrix}1\\0\\-1\end{bmatrix}, \quad \underset{\sim}{k}_7 = \begin{bmatrix}1\\0\\0\end{bmatrix}, \quad \underset{\sim}{k}_8 = \begin{bmatrix}1\\0\\1\end{bmatrix}, \quad \underset{\sim}{k}_9 = \begin{bmatrix}1\\1\\0\end{bmatrix}, \quad \underset{\sim}{k}_{10} = \begin{bmatrix}1\\-1\\-1\end{bmatrix},$$

$$\underset{\sim}{k}_{11} = \begin{bmatrix}1\\-1\\1\end{bmatrix}, \quad \underset{\sim}{k}_{12} = \begin{bmatrix}1\\1\\-1\end{bmatrix}, \quad \underset{\sim}{k}_{13} = \begin{bmatrix}1\\1\\1\end{bmatrix}$$

enumeration of Equation 10 over these wave vectors give a closed set of equations summarized in the following skeleton form:

$$
\begin{aligned}
\dot{\mathbf{u}}^{\mu}(\mathbf{k}, t) = -i \Bigg\{ &\begin{bmatrix} \mathbf{k}_1 | \mathbf{k}_2, -\mathbf{k}_3 \\ \mathbf{k}_2 | -\mathbf{k}_3, \mathbf{k}_1 \\ \mathbf{k}_3 | -\mathbf{k}_1, -\mathbf{k}_2 \end{bmatrix} + \begin{bmatrix} \mathbf{k}_1 | \mathbf{k}_3, -\mathbf{k}_4 \\ \mathbf{k}_3 | -\mathbf{k}_4, \mathbf{k}_1 \\ \mathbf{k}_4 | -\mathbf{k}_1, -\mathbf{k}_3 \end{bmatrix} + \begin{bmatrix} \mathbf{k}_1 | \mathbf{k}_6, -\mathbf{k}_7 \\ \mathbf{k}_6 | -\mathbf{k}_7, \mathbf{k}_1 \\ \mathbf{k}_7 | -\mathbf{k}_1, -\mathbf{k}_6 \end{bmatrix} + \begin{bmatrix} \mathbf{k}_1 | \mathbf{k}_7, -\mathbf{k}_8 \\ \mathbf{k}_7 | -\mathbf{k}_8, \mathbf{k}_1 \\ \mathbf{k}_8 | -\mathbf{k}_1, -\mathbf{k}_7 \end{bmatrix} \\
&+ \begin{bmatrix} \mathbf{k}_1 | -\mathbf{k}_5, \mathbf{k}_{10} \\ \mathbf{k}_5 | -\mathbf{k}_{10}, -\mathbf{k}_1 \\ \mathbf{k}_{10} | \mathbf{k}_1, -\mathbf{k}_5 \end{bmatrix} + \begin{bmatrix} \mathbf{k}_1 | \mathbf{k}_5, -\mathbf{k}_{11} \\ \mathbf{k}_5 | -\mathbf{k}_{11}, \mathbf{k}_1 \\ \mathbf{k}_{11} | -\mathbf{k}_1, -\mathbf{k}_5 \end{bmatrix} + \begin{bmatrix} \mathbf{k}_1 | \mathbf{k}_9, -\mathbf{k}_{12} \\ \mathbf{k}_9 | -\mathbf{k}_{12}, -\mathbf{k}_1 \\ \mathbf{k}_{12} | \mathbf{k}_1, -\mathbf{k}_9 \end{bmatrix} + \begin{bmatrix} \mathbf{k}_1 | \mathbf{k}_9, -\mathbf{k}_{13} \\ \mathbf{k}_9 | -\mathbf{k}_{13}, \mathbf{k}_1 \\ \mathbf{k}_{13} | -\mathbf{k}_1, -\mathbf{k}_9 \end{bmatrix} \\
&+ \begin{bmatrix} \mathbf{k}_2 | \mathbf{k}_5, -\mathbf{k}_6 \\ \mathbf{k}_5 | -\mathbf{k}_6, \mathbf{k}_2 \\ \mathbf{k}_6 | -\mathbf{k}_2, -\mathbf{k}_5 \end{bmatrix} + \begin{bmatrix} \mathbf{k}_2 | \mathbf{k}_8, -\mathbf{k}_9 \\ \mathbf{k}_8 | -\mathbf{k}_9, \mathbf{k}_2 \\ \mathbf{k}_9 | -\mathbf{k}_2, -\mathbf{k}_8 \end{bmatrix} + \begin{bmatrix} \mathbf{k}_2 | -\mathbf{k}_7, \mathbf{k}_{11} \\ \mathbf{k}_7 | -\mathbf{k}_{11}, -\mathbf{k}_2 \\ \mathbf{k}_{11} | \mathbf{k}_2, -\mathbf{k}_7 \end{bmatrix} + \begin{bmatrix} \mathbf{k}_2 | \mathbf{k}_7, -\mathbf{k}_{12} \\ \mathbf{k}_7 | -\mathbf{k}_{12}, \mathbf{k}_2 \\ \mathbf{k}_{12} | -\mathbf{k}_2, -\mathbf{k}_7 \end{bmatrix} \\
&+ \begin{bmatrix} \mathbf{k}_3 | \mathbf{k}_5, -\mathbf{k}_7 \\ \mathbf{k}_5 | -\mathbf{k}_7, \mathbf{k}_3 \\ \mathbf{k}_7 | -\mathbf{k}_3, -\mathbf{k}_5 \end{bmatrix} + \begin{bmatrix} \mathbf{k}_3 | \mathbf{k}_7, -\mathbf{k}_9 \\ \mathbf{k}_7 | -\mathbf{k}_9, \mathbf{k}_3 \\ \mathbf{k}_9 | -\mathbf{k}_3, -\mathbf{k}_7 \end{bmatrix} + \begin{bmatrix} \mathbf{k}_3 | -\mathbf{k}_6, \mathbf{k}_{10} \\ \mathbf{k}_6 | -\mathbf{k}_{10}, -\mathbf{k}_3 \\ \mathbf{k}_{10} | \mathbf{k}_3, -\mathbf{k}_6 \end{bmatrix} + \begin{bmatrix} \mathbf{k}_3 | -\mathbf{k}_8, \mathbf{k}_{11} \\ \mathbf{k}_8 | -\mathbf{k}_{11}, -\mathbf{k}_3 \\ \mathbf{k}_{11} | \mathbf{k}_3, -\mathbf{k}_8 \end{bmatrix} \\
&+ \begin{bmatrix} \mathbf{k}_3 | \mathbf{k}_6, -\mathbf{k}_{12} \\ \mathbf{k}_6 | -\mathbf{k}_{12}, \mathbf{k}_3 \\ \mathbf{k}_{12} | -\mathbf{k}_3, -\mathbf{k}_6 \end{bmatrix} + \begin{bmatrix} \mathbf{k}_3 | \mathbf{k}_8, -\mathbf{k}_{13} \\ \mathbf{k}_8 | -\mathbf{k}_{13}, \mathbf{k}_3 \\ \mathbf{k}_{13} | -\mathbf{k}_3, -\mathbf{k}_8 \end{bmatrix} + \begin{bmatrix} \mathbf{k}_4 | \mathbf{k}_5, -\mathbf{k}_8 \\ \mathbf{k}_5 | -\mathbf{k}_8, \mathbf{k}_4 \\ \mathbf{k}_8 | -\mathbf{k}_4, -\mathbf{k}_5 \end{bmatrix} + \begin{bmatrix} \mathbf{k}_4 | \mathbf{k}_6, -\mathbf{k}_9 \\ \mathbf{k}_6 | -\mathbf{k}_9, \mathbf{k}_4 \\ \mathbf{k}_9 | -\mathbf{k}_4, -\mathbf{k}_6 \end{bmatrix} \\
&+ \begin{bmatrix} \mathbf{k}_4 | -\mathbf{k}_7, \mathbf{k}_{10} \\ \mathbf{k}_7 | -\mathbf{k}_{10}, -\mathbf{k}_4 \\ \mathbf{k}_{10} | \mathbf{k}_4, -\mathbf{k}_7 \end{bmatrix} + \begin{bmatrix} \mathbf{k}_4 | \mathbf{k}_7, -\mathbf{k}_{13} \\ \mathbf{k}_7 | -\mathbf{k}_{13}, \mathbf{k}_4 \\ \mathbf{k}_{13} | -\mathbf{k}_4, -\mathbf{k}_7 \end{bmatrix} \Bigg\}
\end{aligned} \tag{19}
$$

where the column vector $\mathbf{u}^{\mu}(\mathbf{k}) = [u^{\mu}(\mathbf{k}_1), u^{\mu}(\mathbf{k}_2), \ldots, u^{\mu}(\mathbf{k}_{13})]$ subsumes all flow variables of 13 wave vectors.

Each column vector of the right-hand side has 13 entries; shown explicitly in Equation 19 are the nonzero entries, compressing out all zeros. Since the fundamental triad-interaction Equation 13 is uniquely characterized by the three sets of triad wavevectors which appear as a subscript of $\bar{\phi}$, each column vector of Equation 19 does indeed represent a fundamental triad-interaction. For instance, the first column vector signifies in full

$$
\begin{bmatrix}
\sum_{\lambda,\rho} \bar{\phi}^{\mu|\lambda,\rho}_{\mathbf{k}_1|\mathbf{k}_2,-\mathbf{k}_3}(\xi) u^{\lambda *}(\mathbf{k}_2) u^{\rho *}(-\mathbf{k}_3) \\
\sum_{\lambda,\rho} \bar{\phi}^{\mu|\lambda,\rho}_{\mathbf{k}_2|-\mathbf{k}_3,\mathbf{k}_1}(\xi) u^{\lambda *}(-\mathbf{k}_3) u^{\rho *}(\mathbf{k}_1) \\
\sum_{\lambda,\rho} \bar{\phi}^{\mu|\lambda,\rho}_{\mathbf{k}_3|-\mathbf{k}_1,-\mathbf{k}_2}(\xi) u^{\lambda *}(-\mathbf{k}_1) u^{\rho *}(-\mathbf{k}_2) \\
0 \\
\vdots \\
0
\end{bmatrix}
$$

and the expression for the remaining column vectors can be written down similarly.

By extending the polar transformation (Equation 14) to the present 13 wave vectors, the dynamic equation (Equation 19) can be expressed in terms of R_i^μ and ω_i^μ:

$$\begin{bmatrix} \dot{R}_1^\mu \\ \dot{R}_2^\mu \\ \dot{R}_3^\mu \\ \vdots \end{bmatrix} = \sum_{\lambda,\rho} \left\{ \begin{bmatrix} -\bar{\phi}^{\mu|\lambda,\rho}_{\underset{\sim}{k}_1|\underset{\sim}{k}_2,-\underset{\sim}{k}_3}(\xi) R_2^\lambda R_3^\rho \sin 2\pi(\omega_1^\mu + \omega_2^\lambda - \omega_3^\rho) \\ -\bar{\phi}^{\mu|\lambda,\rho}_{\underset{\sim}{k}_2|-\underset{\sim}{k}_3,\underset{\sim}{k}_1}(\xi) R_3^\lambda R_1^\rho \sin 2\pi(\omega_1^\rho + \omega_2^\mu - \omega_3^\lambda) \\ -\bar{\phi}^{\mu|\lambda,\rho}_{\underset{\sim}{k}_3|-\underset{\sim}{k}_1,-\underset{\sim}{k}_2}(\xi) R_1^\lambda R_2^\rho \sin 2\pi(\omega_1^\lambda + \omega_2^\rho - \omega_3^\mu) \\ 0 \\ \vdots \\ 0 \end{bmatrix} + \cdots \right\}$$

$$\begin{bmatrix} \dot{\omega}_1^\mu \\ \dot{\omega}_1^\mu \\ \dot{\omega}_3^\mu \\ \vdots \end{bmatrix} = \sum_{\lambda,\rho} \left\{ \begin{bmatrix} -(2\pi R_1^\mu)^{-1} \bar{\phi}^{\mu|\lambda,\rho}_{\underset{\sim}{k}_1|\underset{\sim}{k}_2,-\underset{\sim}{k}_3}(\xi) R_2^\lambda R_3^\rho \cos 2\pi(\omega_1^\mu + \omega_2^\lambda - \omega_3^\rho) \\ -(2\pi R_2^\mu)^{-1} \bar{\phi}^{\mu|\lambda,\rho}_{\underset{\sim}{k}_2|-\underset{\sim}{k}_3,\underset{\sim}{k}_1}(\xi) R_3^\lambda R_1^\rho \cos 2\pi(\omega_1^\rho + \omega_2^\mu - \omega_3^\lambda) \\ -(2\pi R_3^\mu)^{-1} \bar{\phi}^{\mu|\lambda,\rho}_{\underset{\sim}{k}_3|-\underset{\sim}{k}_1,-\underset{\sim}{k}_2}(\xi) R_1^\lambda R_2^\rho \cos 2\pi(\omega_1^\lambda + \omega_2^\rho - \omega_3^\mu) \\ 0 \\ \vdots \\ 0 \end{bmatrix} + \cdots \right\} \tag{20}$$

Only the terms corresponding to the first right-hand side column vector of Equation 19 are shown here explicitly; the three dots represent the remaining 21 terms.

The invariant sets S and S′ of a single triad-interaction can be extended to incorporate all 22 triad-interactions. This entails replacing Ω_n in Equation 17 by $\omega_i^\mu \pm \omega_j^\lambda \pm \omega_m^\rho$, which are the arguments of sines and cosines. That is,

$$S = \{\text{all } \omega_i^\mu \text{ such that } \omega_1^\mu + \omega_2^\lambda - \omega_3^\rho, \omega_1^\rho + \omega_2^\mu - \omega_3^\lambda, \omega_1^\lambda + \omega_2^\rho - \omega_3^\mu, \ldots = \pm\tfrac{1}{4}, \pm\tfrac{3}{4}, \pm\tfrac{5}{4}, \ldots\}$$

And, the subset (Equation 18) becomes

$$S' = \{\text{all } \omega_i^1 = \omega_i^2 \text{ such that } \omega_1^1 + \omega_2^1 - \omega_3^1, \omega_1^1 + \omega_3^1 - \omega_4^1, \ldots = \pm\tfrac{1}{4}, \pm\tfrac{3}{4}, \pm\tfrac{5}{4}, \ldots\}$$

When the wave vector lattice is larger than that of Figure 2, it becomes all but impractical to enumerate Equation 10 by hand. However, there is no need to despair, because enumeration Equation 10 by hand. However, there is no need to despair, because enumeration can readily be carried out with the aid of a computer, thereby systematically sorting out the triad wave vectors allowed

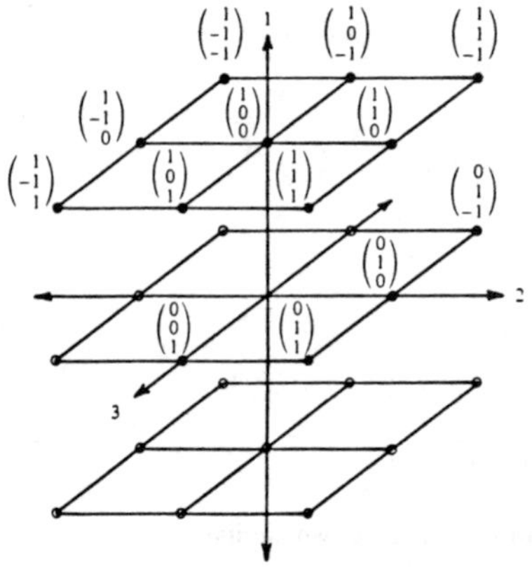

Figure 2. Three-dimensional wavevector lattice of $(\pm 1) \times (\pm 1) \times (\pm 1)$.

Table 1
Some Lower-Order Spherical Truncations

Upper k^2	Wave number radius	Wave vectors	Triad interactions or column vectors	Notation
3	1.73	13	22	D(13)
5	2.24	28	106	D(28)
7	2.56	40	242	D(40)
9	3	61	549	D(61)
11	3.32	85	1059	D(85)
16	4	128	2522	D(128)
25	5	257	10186	—
36	6	462	33152	—

in a wave vector lattice of given truncation order. It must be noticed that the 13 wave vectors of Figure 2 all lie in a sphere of wave number radius $k = \sqrt{3}$, and no other wave vectors exist in that sphere. Hence, Equation 19 represents the spherical truncation of wave number radius $k = \sqrt{3}$. Summarized in Table 1 are the pertinent results of spherical truncation up to the wave number radius $k = 6$.

As seen from Table 1, the numbers of wave vectors and, particularly, of fundamental triad-interactions increase very rapidly with the truncation order. Even for a relatively low order truncation of wave number radius $k = 6$, compared with Orszag and Patterson's [15] simulation code of $k = 16$, the 462 wave vectors give rise to 924 (462×2) dynamic variables. Hence, the corresponding system would have 924 complex differential equations with 33,152 right-hand-side column vectors; a non-trivial numerical task. For future reference, the first six truncated systems of Table 1 will be denoted by D(N), where N is the number of wave vectors retained. In this notation, the fundamental triad-interaction system is D(3).

THEORETICAL BACKGROUND

Because of many degrees of freedom, the investigation of D(N) must rely heavily on numerical analyses. It is therefore essential that a theoretical framework exists to guide us in what to compute and how to interpret it in support (or rejection) of certain dynamical propositions. To provide such theoretical guidance, we shall in this chapter review some basic results of classical statistical mechanics and modern dynamical systems, which are relevant to the present conservative (inviscid) system. (see Landford's [16] review for a conceptual discussion on conservative systems, as well as the strange attractor of fluid dynamics.) To begin with, let us split u_i^μ into the real and imaginary parts, $u_i^\mu = v_i^\mu + iw_i^\mu$. Since the representation of v_i^μ and w_i^μ is equivalent to the polar representation of J_i^μ and ω_i^μ, we shall use at will either one of the representations, as dictated by expediency.

Energy and Helicity Conservations

For a given $\underset{\sim}{k}_i$, consider the Hermitian matrix

$$\begin{bmatrix} u_i^1 u_i^{1*} & i u_i^1 u_i^{2*} \\ -i u_i^2 u_i^{1*} & u_i^2 u_i^{2*} \end{bmatrix} \tag{21}$$

It has been shown [1] that the trace is modal energy $E_i = |u_i^1|^2 + |u_i^2|^2$ and the sum of the off-diagonals times k_i is modal helicity $H_i = ik_i(u_i^1 u_i^{2*} - u_i^2 u_i^{1*})$, ignoring any constant factor. The total energy and helicity are then

$$E = \sum_{i=1}^{N} \sum_{\mu=1,2} |u_i^\mu|^2 = \sum_{i=1}^{N} \sum_{\mu=1,2} J_i^\mu$$

$$H = i \sum_{i=1}^{N} k_i(u_i^1 u_i^{2*} - u_i^2 u_i^{1*}) = 2 \sum_{i=1}^{N} k_i R_i^1 R_i^2 \sin 2\pi(\omega_i^1 - \omega_i^2)$$

in both representations.

Now, write the sum of all four elements of Equation 21 in a Hermitian form $U_i^+ A U_i$, where U_i is the column vector (u_i^1, u_i^2), $A = \left[\begin{smallmatrix} 1 & i \\ -i & 1 \end{smallmatrix}\right]$, and + denotes the transjugate. Since the eigenvalues of A are 0 and 2, $U_i^+ A U_i$ is positive semi-definite. Note that $U_i^+ A^+ U_i$ is also positive semi-definite, for A and A^+ have the same eigenvalues. Yet, $U_i^+ A U_i$ and $U_i^+ A^+ U_i$ have different signs for their imaginary part; hence, we infer from these quadratics that

$$|u_i^1|^2 + |u_i^2|^2 \gg |i(u_i^1 u_i^{2*} - u_i^2 u_i^{1*})|$$

Or, in polar representation

$$J_i^1 + J_i^2 \gg 2R_i^1 R_i^2 |\sin 2\pi(\omega_i^1 - \omega_i^2)| \tag{22}$$

Multiplying both sides by k_i and summing over i, we have

$$\sum_{i=1}^{N} k_i E_i \gg |H|$$

Its statistical form in isotropic field is $kE(k) > |H(k)|$, where E(k) and H(k) are respectively the isotropic energy and helicity spectra [17, 18].

Classical Liouville Theorem

In terms of the real and imaginary parts v_i^μ and w_i^μ, we find from Equation 10 the incompressibility of phase space

$$\sum_{i=1}^{N} \sum_{\mu=1,2} (\partial \dot{v}_i^\mu / \partial v_i^\mu + \partial \dot{w}_i^\mu / \partial w_i^\mu) = 0$$

hence the volume element in phase space is an integral invariant. Or

$$\sum_{i=1}^{N} \sum_{\mu=1,2} (\partial \dot{J}_i^\mu / \partial J_i^\mu + \partial \dot{\omega}_i^\mu / \partial \omega_i^\mu) = 0$$

in action-angle variables (but not in R_i^μ and ω_i^μ, for they are not canonical variables). This is the content of the classical Liouville theorem, stating the preservation of measure under the time evolution. Clearly, the Liouville theorem will not hold in the presence of viscous damping; hence, it is a property of inviscid flow (conservative) systems. It was more than 30 years ago that Lee [19] showed the Navier-Stokes equations in Fourier space obey the classical Liouville theorem. Although the measure-invariance is necessary for ergodicity and mixing, it is too presumptuous to infer ergodicity and, particularly, energy equipartition from the Liouville theorem alone (as is often implied in the literature).

There are several technical comments. First, dynamical systems with an invariant measure return arbitrarily close to the initial points (recurrence theorem [20]). However, the recurrence time (frequency of return), which is not given by the theorem, may be extremely long for a large degree-of-freedom system [21]. Hence, the return to initial state is practically inconceivable for D(N) even under the assumption of no chaos. Secondly, all recurrence motion are central motions, but the converse is not true [22]. Lastly, because of the measure invariance, the trajectory of an inviscid flow will not wind down to a submanifold of phase space. Hence, one does not usually speak of an attractor of conservative systems, although an attractor may be defined to encompass the whole state space [23].

Ergodicity and Mixing

For each $\underset{\sim}{k}_i$ there are two components $u_i^\mu(t)$ for $\mu = 1$ and 2, each of which in turn splits to either v_i^μ and w_i^μ or R_i^μ and ω_i^μ. We can therefore span the phase space of D(N) by the coordinates $\underset{\sim}{x} = \{x_1, x_2, \ldots, x_{4N}\}$ in E_{4N}. First of all, the trajectory of D(N) must lie on the surface of constant energy and helicity. Because of the measure invariance, the initial volume element of, say, a unit (4N − 2)-dimensional sphere will spread over the constant energy-helicity surface, but without suffering any volume change. Ergodicity and mixing are the criteria on the shape of spreading volume element and the manner in which it proceeds. For phase function $f(\underset{\sim}{x})$ along a trajectory, one can compute the time average over a period t by $\bar{f}(x, t) = t^{-1} \int_0^t \underset{\sim}{f}(x, s)\,ds$. According to Birkhoff's theorem [24] this average over a sufficiently long time, i.e., $\lim_{t\to\infty} \bar{f}(\underset{\sim}{x}, t)$, approaches a constant value $\bar{f}(\underset{\sim}{x})$ for almost all initial conditions, when the trajectory is restricted to the invariant part of phase space. After the trajectory has traversed every extension of the phase space with equal frequencies, it is reasonable to expect that $\bar{f}(\underset{\sim}{x})$ should be close to that obtained by averaging $f(\underset{\sim}{x})$ over all possible extensions of the phase space. We shall denote the phase average symbolically by $\langle f(\underset{\sim}{x})\rangle = \int_s f(\underset{\sim}{x})\,d\mu(S)$, where $\mu(S)$ is a normalized measure of the constant energy-helicity surface S. The system is said to be ergodic if and only if

$$\lim_{t\to\infty} \bar{f}(\underset{\sim}{x}, t) = \langle f(\underset{\sim}{x})\rangle \tag{23}$$

Clearly, if the trajectory is restricted to part of the phase space or if the phase space is split into two invariant parts of nonzero measure (metric decomposability), then it is not possible to guarantee ergodicity. Hence, the condition of metric indecomposability has long been used synonymously with ergodicity.

The numerical testing of ergodicity, however, runs into two difficulties. First, the equality (Equation 23) should be checked out for all phase functions. Should there be found a phase function violating Equation 23, the ergodic claim is nullified. In practice, ergodicity is understood to have been claimed for a certain class of measureable phase functions of physical importance (see [25] and [7] for the choice of phase functions for two-dimensional turbulence). The second difficulty is that the ergodic theorem is a statement about the equilibrium trajectory in the limit as $t \to \infty$. One may then ask why can't we evolve D(N) over a sufficiently long time to test Equation 23 in a sort of numerical quasi-equilibrium sense. Unfortunately, the answer is negative because of the emergence of a random or chaotic trajectory. Consequently, numerical trajectory evolution is restricted to several characteristic correlation times of a typical dynamic variable. Complete chaos is referred to as mixing in the phase space, which has the everyday analogy of mixing cream into a cup of coffee. Quantitatively, for any measureable functions $f(\underset{\sim}{x})$ and $g(\underset{\sim}{x})$ the system is said to be (strong) mixing if [26]

$$\lim_{t\to\infty} \langle f(\underset{\sim}{x}, t)g(\underset{\sim}{x})\rangle = \langle f(\underset{\sim}{x})\rangle\langle g(\underset{\sim}{x})\rangle \tag{24}$$

Intuitively, in an ergodic system the trajectory covers every extension of the phase space with equal frequencies. However, when it does so in a random fashion the system is mixing. Mixing, therefore, implies ergodicity, but not *vise versa*.

Ergodic Definition of Khinchin

As pointed out earlier, ergodicity and metric indecomposability have long been defined in terms of each other. Khinchin has therefore attempted to define ergodicity without resorting to the metric indecomposability. Let us begin by assuming that

$$\langle f(\underset{\sim}{x})\rangle = 0 \tag{25}$$

This zero-mean condition is subject to verification, although it is plausible for $f(\underset{\sim}{x})$, which is a function of a single Fourier mode. Define a correlation function with respect to the zero initial time:

$$\rho(t) = \langle f(\underset{\sim}{x}, t) f(\underset{\sim}{x}, 0)\rangle / \langle f^2(\underset{\sim}{x})\rangle$$

Then Khinchin's theorem states that if $\rho(t) \to 0$ for $t \to \infty$, the function $f(\underset{\sim}{x})$ is ergodic [24]. In fact, the claim of this theorem is too modest; it further implies mixing. To see this, we write Equation 24 for f = g:

$$\lim_{t \to \infty} \langle f(\underset{\sim}{x}, t) f(\underset{\sim}{x}, 0)\rangle = \langle f(\underset{\sim}{x})\rangle \langle f(\underset{\sim}{x})\rangle = 0 \tag{26}$$

The second equality follows from Equation 25. (The derivation of Equation 26 is also found in Lebowitz [27].) Since Equation 26 is precisely the ergodic condition of Khinchin, his theorem may be considered more correctly as a mixing than an ergodic theorem.

Since ergodicity does not imply mixing, one would never test for mixing unless $f(\underset{\sim}{x})$ is known or suspected to be ergodic. In other words, the establishment of ergodicity precedes the test for mixing. Since ergodicity means that phase and time averages are the same, we may recast $\rho(t)$ into the auto-correlation of the phase function evolved over T:

$$\rho(T, \tau) = \int_0^{T-\tau} f(\underset{\sim}{x}, s+\tau) f(\underset{\sim}{x}, s)\, ds \Big/ \int_0^{T-\tau} f^2(\underset{\sim}{x}, s)\, ds \tag{27}$$

defined similarly to that of Kells and Orszag [25] for two-dimensional turbulence. Whence the content of Khinchin's theorem is

$$\lim_{\tau \to \infty} \rho(T, \tau) = 0$$

for $T > \tau$ sufficiently large for good data sampling.

Microcanonical Distribution

We shall now express the normalized measure $\mu(S)$ expressed earlier in terms of the phase-space coordinates. To this end, we write the modal energy and helicity in the Hermitian quadratic form:

$$E_i = U_i^+ I U_i, \qquad H_i = U_i^+ B_i U_i$$

where I is the unit matrix and

$$B_i = \begin{bmatrix} 0 & -ik_i \\ ik_i & 0 \end{bmatrix}$$

By the transformation

$$U_i = 2^{-1/2} \begin{bmatrix} 1 & i \\ i & 1 \end{bmatrix} Y_i$$

the E_i and H_i become diagonal (though E_i already is) in $Y_i = (y_i^1, y_i^2)$. Extending the similarity transformation to the sums of E_i and H_i, we find that

$$E = \sum_{i=1}^{N} |y_i^1|^2 + |y_i^2|^2, \qquad H = \sum_{i=1}^{N} k_i(|y_i^1|^2 - |y_i^2|^2)$$

Now, rearranging the real and imaginary parts of y_i in a sequence of 4N variables, and identifying this sequence with the $\underset{\sim}{x}$-coordinates already introduced, the total energy and helicity may be put in the real quadratic form:

$$E(\underset{\sim}{x}) = \sum_{i=1}^{4N} x_i^2, \qquad H(\underset{\sim}{x}) = \sum_{i=1}^{4N} c_i x_i^2.$$

Here c_i are not all distinct, and exactly half of them are negative.

Since the energy-helicity surface is of the same type of quadratic constraint as the energy-entrophy surface, by parroting the result of two-dimensional turbulence [28, 29] we can at once extend the classical Khinchin theorem to three-dimensional turbulence:

$$\frac{\partial^2}{\partial E\,\partial H}\int_V f(\underset{\sim}{x})\,d\underset{\sim}{x} = \int_S f(\underset{\sim}{x})\frac{d\Sigma}{|\text{grad } E|\,|\text{grad } H|\sin\theta} \tag{28}$$

Here θ is the angle between grad E and grad H, and $d\Sigma$ is a differential element of S. The parametrization of $d\Sigma$ in terms of $4N - 2$ coordinates may follow the procedure of two-dimensional turbulence. To complete the discussion of the right-hand side, however, it is necessary to show that S is smooth and connected (as was done by Glaz [28]) for the energy-enstropy surface), which will not be pursued here. For the left-hand side V is the intersection of two phase spaces enclosed respectively by the constant energy and helicity surfaces. By extending the $\underset{\sim}{x}$-integration over all phase space with the use of unit step function, and alternative form of Equation 28 is obtained

$$\int f(\underset{\sim}{x})\,\delta(E - E(\underset{\sim}{x}))\,\delta(H - H(\underset{\sim}{x}))\,d\underset{\sim}{x} = \int_S f(\underset{\sim}{x})\frac{d\Sigma}{|\text{grad } E|\,|\text{grad } H|\sin\theta}$$

We have thus established equivalence of the microcanonical distributions $\delta(E - E(\underset{\sim}{x}))\,\delta(H - H(\underset{\sim}{x}))$ over all the $\underset{\sim}{x}$-space and $1/|\text{grad } E|\,|\text{grad } H|\sin\theta$ on the energy-helicity surface S. For instance, in terms of the latter distribution, the normalized measure $\mu(S)$ gives rise to

$$d\mu(S) = \frac{d\Sigma}{\Omega(S)|\text{grad } E|\,|\text{grad } H|\sin\theta}$$

where $\Omega(S) = \int_S d\Sigma/|\text{grad } E|\,|\text{grad } H|\sin\theta$ is the so-called structural function [24].

Canonical Distribution

Logically speaking, the canonical distribution is the asymptotic form of the *a priori* probability density $d\Sigma/|\text{grad } E|\,|\text{grad } H|\sin\theta$ on the energy-helicity surface in the limit as $N \to \infty$. The Gaussian distribution then emerges as a consequence of the central-limit theorem [24]. More provincially, however, we can write down the Gaussian canonical distribution as a function of the sum of constants of motion $C_1E_i + C_2H_i$, where C_1 and C_2 are constants [12]

$$\prod_{i=1}^{N}\frac{(C_1^2 - C_2^2k_i^2)^{1/2}}{\pi^2}\exp\{-C_1(|u_i^1|^2 + |u_i^2|^2) - iC_2k_i(u_i^1u_i^{2*} - u_i^2u_i^{1*})\}.$$

In fact, this is an equilibrium solution of the Liouville equation and, moreover, is a stable equilibrium distribution [18]. We first note the difference between the microcanonical and canonical distributions: the former is an *a priori* distribution on the constant energy-helicity surface S, whereas the latter is a Gaussian distribution extending over all phase space. How can they then be related? To see this, recall the alternative expression of the microcanonical distribution $\delta(E - E(\underset{\sim}{x}))\,\delta(H - H(\underset{\sim}{x}))$, which is peaked infinitely sharply on the submanifold S. Although a Gaussian distribution cannot peak out as sharply as the Dirac delta function, when we consider only a single component of u_i^μ, the canonical distribution becomes more pronouncedly peaked about S as there are more and more components of u_i^μ included in the exponent. This is how the canonical (Gaussian) distribution can approximate the microcanonical (delta-function) distribution closely as N becomes large.

Averaging over the canonical distribution, we find that [1]:

$$\langle |u_i^1|^2 \rangle = \langle |u_i^2|^2 \rangle = \frac{C_1}{C_1^2 - C_2^2 k_i^2} \tag{29a}$$

$$\langle u_i^2 u_i^{1*} \rangle = -\langle u_i^1 u_i^{2*} \rangle = \frac{iC_2 k_i}{C_1^2 - C_2^2 k_i^2} \tag{29b}$$

Let us examine the consequence of Equation 29. First, when helicity is zero $C_2 = 0$, hence energy equipartition follows

$$\langle |u_i^1|^2 \rangle = \langle u_i^2|^2 \rangle = 1/C_1 \qquad \text{for } i = 1 \text{ and } 2 \tag{30}$$

Secondly, in general $C_2 \neq 0$ for nonzero helicity. Then Equation 29a implies the equality of modal energy distribution for u_i^1 and u_i^2, although there is no equipartition among i. Thirdly, Equation 29b gives the equilibrium helicity distribution $2C_2k_i^2/(C_1^2 - C_2^2k_i^2)$. Lastly, since $\langle u_i^1 u_i^{2*} \rangle$ and $\langle u_i^2 u_i^{1*} \rangle$ are purely imaginary, Equation 29b implies the vanishing of the real part of the reflexional asymmetry, which is a condition for isotropy to be discussed presently.

Isotropy of the Covariance Spectral Tensor

Isotropy imposes certain symmetries on the spectral tensor $\Phi_{ij}(\underset{\sim}{k}) = \langle U_i^*(\underset{\sim}{k})U_j(\underset{\sim}{k}) \rangle$ (suppressing the time arguments), where the angle brackets now refer to ensemble average. Expressing the spectral tensor in terms of Equation 2 yields

$$\Phi_{ij}(\underset{\sim}{k}) = \sum_{\mu} \epsilon_i^\mu(\underset{\sim}{k}, \xi)\epsilon_k^\mu(\underset{\sim}{k}, \xi)U^{\mu\mu}(\underset{\sim}{k}) + \sum_{\mu \neq v} \epsilon_i^\mu = (\underset{\sim}{k}, \xi)\epsilon_j^v(\underset{\sim}{k}, \xi)U^{\mu v}(\underset{\sim}{k})$$

The covariance matrix $U^{\mu v}(\underset{\sim}{k}) = \langle u^{\mu *}(\underset{\sim}{k})u^v(\underset{\sim}{k}) \rangle$ which is by definition Hermitian may be written out in detail

$$U^{\mu v}(\underset{\sim}{k}) = \begin{bmatrix} U_1(\underset{\sim}{k}, \xi) + U_2(\underset{\sim}{k}; \xi) & ikU_3(\underset{\sim}{k}, \xi) + U_4(\underset{\sim}{k}, \xi) \\ -ikU_3(\underset{\sim}{k}, \xi) + U_4(\underset{\sim}{k}, \xi) & U_1(\underset{\sim}{k}, \xi) - U_2(\underset{\sim}{k}, \xi) \end{bmatrix} \tag{31}$$

Here $U_i(\underset{\sim}{k}, \xi)$ are all real, and we are emphasizing the explicit dependence upon ξ. With the use of Equation 31, the spectral tensor becomes in detail

$$\begin{aligned}\Phi_{ij}(\underset{\sim}{k}) = {} & \sum_{\mu} \epsilon_i^\mu(\underset{\sim}{k}, \xi)\epsilon_j^\mu(\underset{\sim}{k}, \xi)U_1(\underset{\sim}{k}, \xi) + [\epsilon_i^1(\underset{\sim}{k}, \xi)\epsilon_j^1(\underset{\sim}{k}, \xi) - \epsilon_i^2(\underset{\sim}{k}, \xi)\epsilon_j^2(\underset{\sim}{k}, \xi)]U_2(\underset{\sim}{k}, \xi) \\ & + i[\epsilon_i^1(\underset{\sim}{k}, \xi)\epsilon_j^2(\underset{\sim}{k}, \xi) - \epsilon_i^2(\underset{\sim}{k}, \xi)\epsilon_j^1(\underset{\sim}{k}, \xi)]kU_3(\underset{\sim}{k}, \xi) + \sum_{\mu \neq v} \epsilon_i^\mu(\underset{\sim}{k}, \xi)\epsilon_j^v(\underset{\sim}{k}, \xi)U_4(\underset{\sim}{k}, \xi)\end{aligned}$$

The tensor coefficients may be computed explicitly using the polarization vectors (Equation 5) to yield

$$\begin{aligned}\Phi_{ij}(\underset{\sim}{k}) = {} & P_{ij}(\underset{\sim}{k})U_1(\underset{\sim}{k}, \xi) + ie_{ijm}k_m U_3(\underset{\sim}{k}, \xi) \\ & + \mathrm{Re}[\Pi_{ij} \exp(i2\xi)]U_2(\underset{\sim}{k}, \xi) - \mathrm{Im}[\Pi_{ij} \exp(i2\xi)]U_4(\underset{\sim}{k}, \xi)\end{aligned} \tag{32}$$

where e_{ijm} is the alternating tensor. The components of Π_{ij} are given by

$$\Pi_{ij} = \begin{bmatrix} C^2 & -CD & iC \cos \eta_k \\ & D^2 & -iD \cos \eta_k \\ \text{(symmetric)} & & -\cos^2 \eta_k \end{bmatrix}$$

where $C = \sin\theta_k - i\cos\theta_k \sin\eta_k$ and $D = \cos\theta_k + i\sin\theta_k \sin\eta_k$. The isotropy requirements on covariance tensor are then

$$U_i(\underset{\sim}{k}, \xi) = U_i(\underset{\sim}{k}), \qquad (i = 1 \text{ and } 3) \tag{33a}$$

$$U_i(\underset{\sim}{k}) = U_i(k), \qquad (i = 1 \text{ and } 3) \tag{33b}$$

$$U_2(\underset{\sim}{k}, \xi) = 0 \tag{33c}$$

$$U_4(\underset{\sim}{k}, \xi) = 0 \tag{33d}$$

In words, Equation 33a states that U_1 and U_3 are independent of ξ; Equation 33b is the spherical symmetry; Equation 33c implies rotational symmetry (no rotational asymmetry); and Equation 33d is the disappearance of the real part of the reflexional asymmetry. The implication of Equation 33 can best be brought out by imposing it on Equation 32. Identifying $U_1(k)$ and $U_3(k)$ with the energy and helicity spectra, $E(k) = 4\pi k^2 U_1(k)$ and $H(k) = 8\pi k^4 U_3(k)$, we recover the spectral tensor in a familiar form

$$\Phi_{ij}(\underset{\sim}{k}) = P_{ij}(\underset{\sim}{k})E(k)/4\pi k^2 + ie_{ijm}k_m H(k)/8\pi k^4$$

which is Moffatt's [30] spectral tensor of pseudo-isotropic turbulence or, more precisely, the hemitropic turbulence. Since helicity is a constant of motion, it can be chosen at will. For instance, the usual isotropic form $\Phi_{ij}(\underset{\sim}{k}) = P_{ij}(\underset{\sim}{k})E(k)/4\pi k^2$ is obtained under zero helicity. We have thus shown that conditions (Equation 33) are necessary for the isotropic $\Phi_{ij}(\underset{\sim}{k})$.

In inviscid flow, both the spherical and rotational symmetries are realized by the energy equipartition. The reflexional asymmetry given by

$$U_4 = \sum_{i=1}^{N} (u_i^1 u_i^{2*} + u_i^2 u_i^{1*}) = 2 \sum_{i=1}^{N} R_i^1 R_i^2 \cos 2\pi(\omega_i^1 - \omega_i^2) \tag{34}$$

must be tested for its persistence/disappearance under the time evolution. According to Equation 29b, the phase-average $\langle U_4 \rangle$ vanishes identically. Hence, isotropy of the covariance tensor is indeed consistent with the canonical distribution.

Kolmogorov Entropy

Mixing is engendered by the trajectory instability, whereby the two nearby trajectories become separated exponentially in phase space with the evolution time; hence, it is also called the exponential orbit separation [31]. Consider an open set of initial states. If the system is mixing, the initial open set will spread out randomly and eventually cover the entire energy-helicity surface in the limit as $t \to \infty$. In fact, the open set of trajectories does not grow in actual size owing to measure invariance. It is the mutual distance of trajectories that becomes large as evolution proceeds. Trajectory instability can be quantified by the Kolmogorov entropy adapted from the one proposed by Benettin, Galgani and Strelcyn [32]. To define such a metric entropy, however, calls for explicit computation of trajectories of initial proximity. Since the trajectory is dependent on the initial condition I, the parameter ξ, and the initial T_0 and final evolution time T, we shall adopt the notation $D(N, I, \xi, T_0, T)$ to include all the parameters. (Note that D(N) is still reserved for a generic notation when the parameter values are of no direct concern.)

Let us denote by U the trajectory of $D(N, I, \xi, T_0, T)$ on the constant energy-helicity surface. As mentioned earlier, the representation of U may consist of the real and imaginary components v_i^μ and w_i^μ, respectively. To speak precisely of the distance of another trajectory U′ with the corresponding components $v_i^{\mu\prime}$ and $w_i^{\mu\prime}$, from U we introduce a Euclidean norm

$$\|U - U'\| = \left[\sum_{i=1}^{N} \sum_{\mu=1,2} (v_i^\mu - v_i^{\mu\prime})^2 + (w_i^\mu - w_i^{\mu\prime})^2 \right]^{1/2} \tag{35}$$

At T_0 we consider an initial condition i_0 that is removed from I by a small distance d, i.e., $\|I - i_0\| = d$, and evolve the trajectory from it over a pre-assigned time interval $\Delta\tau$. At the end of the first time interval $(0, \Delta\tau)$, we then compute the trajectory distance by

$$d_1 = \|D(N, I, \xi, 0, \Delta\tau) - D(N, i_0, \xi, 0, \Delta\tau)\|$$

Note that the trajectories of $D(N, I, \xi, 0, \Delta\tau)$ and $D(N, i_0, \xi, 0, \Delta\tau)$ do not necessarily lie on the same energy-helicity surface. Now, for the second time interval $(\Delta\tau, 2\,\Delta\tau)$, we specify a new initial point i_1 that is again a distance d away from the U at $T = \Delta\tau$, i.e., $D(N, I, \xi, 0, \Delta\tau)$. After evolving the trajectory from i_1 over $\Delta\tau$, we compute the trajectory separation by

$$d_2 = \|D(N, I, \xi, 0, 2\,\Delta\tau) - D(N, i_1, \xi, 0, \Delta\tau)\|$$

for the second time interval $(\Delta\tau, 2\,\Delta\tau)$. In a similar fashion, the trajectory separation of successive time intervals can be computed to yield a sequence of positive numbers $\{d_j\}$ for $j = 1, 2, 3, \ldots$. Figure 3 schematizes the computation of $\{d_j\}$, which differs from Figure 1 of Benettin et al. [32] in that, for instance, i_1 is not in general required to lie on the dotted-line segment connecting $D(N, I, \xi, 0, \Delta\tau)$ and $D(N, i_0, \xi, 0, \Delta\tau)$. As will be discussed later, however, this difference is irrelevant. Benettin et al. [32] defined an entropy-like quantity by

$$k_n(\Delta\tau, U, d) = \frac{1}{n\,\Delta\tau} \sum_{j=1}^{n} \ln(d_j/d)$$

They have observed from the numerical study of a Hamiltonian system (e.g., the Hénon-Heiles [33] model) that when U is in region of the phase space exhibiting chaos (i) the long-time limit of summation settles down to a constant value, i.e., $\lim_{n\to\infty} k_n(\Delta\tau, U, d) = k(\Delta\tau, U, d)$ for suitably chosen $\Delta\tau$ and d, (ii) $k(\Delta\tau, U, d)$ is independent of $\Delta\tau$, and (iii) $k(\Delta\tau, U, d)$ is independent of d also. Under these conditions we may identify $k(\Delta\tau, U, d)$ with the Kolmogorov entropy (Lyapunov characteristic numbers). A dynamical system with positive Kolmogorov entropy is called a "K-system," which is a stronger characterization than mixing [31].

For non-chaotic trajectories (including a quasi-periodic motion), on the other hand, $\lim_{n\to\infty} k_n(\Delta\tau, U, d)$ is non-positive. Hence, the Kolmogorov entropy can provide a means for discerning stochastic islands of the phase space from the ordered (non-random) region.

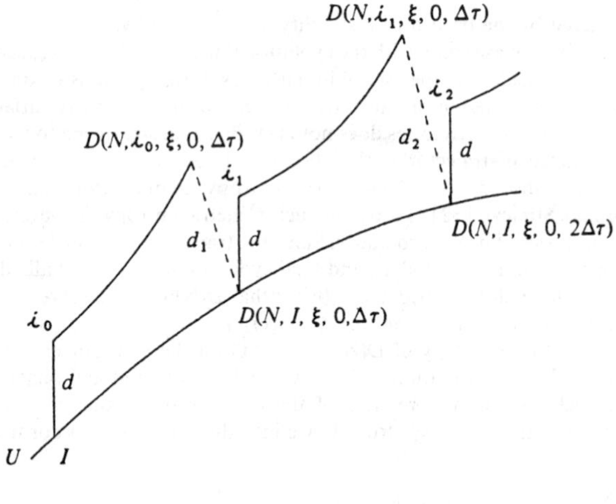

Figure 3. Computational schematics for Kolmogorov entropy.

Trajectory Instability

Trajectory instability, to be distinguished from numerical instability, has been encountered for a long time in fluid dynamics, although it has only recently been identified as such. The first evidence was reported more than 20 years ago by Bilkhoff and Fisher [3], who observed irregular evolution of the initially smooth vortex sheets modeled by discrete point vortices. Their work, however, did not receive due attention because the idea of an irregular fluid trajectory was incompatible with computational fluid dynamics. Nevertheless, trajectory instability has resurfaced in the late 1960s, but now in the numerical prediction of meteorological flows [4, 5]. It is the so-called predictability or internal-error-growth problem in numerical weather forecasting, the viability of which is known to be limited to only a few days [34], However, considerable evidence accumulated in recent years clearly indicates that trajectory instability manifested by a limited evolution time is not an isolated incident. In fact, it occurs in a variety of model flow problems, such as the Bénard convection problem [6], the inviscid Burger's model [8] and two-dimensional inviscid turbulence [7].

As a matter of fact, trajectory instability manifests itself in a dichotomic manner. On the one hand, it gives rise to the sensitive dependence of D(N) on the initial condition. Since this implies the loss of initial information, D(N) can exhibit macroscopic irreversibility of a trajectory motion that is dynamically reversible; an ingradient necessary for statistical mechanics. It must be pointed out Ruelle [35] has proposed the asymptotic decay of the time-correlation function defined exactly as in Equation 26 as a practical test for the sensitive dependence of dynamical systems on the initial condition.

On the other hand, one cannot compute the trajectory over a long time period because of the accumulation of round-off errors. Note that digital computers have a finite accuracy, and numerical trajectory computation is, at best, an attempt to approximate the true orbit (integral curve) with a finite number of digits. At any time step of integration, therefore, the difference in numerical and true orbits, however small it may be, will grow exponentially by the very definition of trajectory instability. The computed orbit will eventually depart from the true trajectory drastically; hence, it is called a pseudo-orbit.

For the Anosov system (Y-system), however, the dichotomy of trajectory instability can be conciliated, and can thereby provide a logical link between dynamics and statistics. Benettin, Casartelli, Galgani, Giorgilli and Strelcyn [36] have shown that for Anosov systems time averages computed from a psuedo-orbit are identical to those computed from the true orbit, although the pseudo-orbit may be completely different from the true orbit. This is based on a theorem of Anosov and Bowen which states roughly that near a pseudo-orbit of an Anosov system there are (infinitely) many true trajectories (originating, of course, from some other initial states), all of which would give rise to the same time average over a long time. Intuitively speaking, since the Anosov system is a model of a completely unstable system, one may expect that the deviation of the pseudo-orbit from the true orbit is so random that its time-average effect is zero due to cancellations.

Since the claim of Benettin et al. [36] has also be verified on stochastic systems which are not strictly of Anosov's type, we shall assume that their work is applicable to our system D(N). This then enables us to compute time averages much beyond the evolution time for the true trajectory.

Summing up, trajectory instability presents us with a choice: either to formulate the turbulence problem for long-time statistics, or to simulate it numerically but only for a short-time dynamics. We therefore come to echo Prigogine's [37] view that statistics begins where dynamics stops.

FUNDAMENTAL TRIAD INTERACTION

Preliminary to the investigation of D(N), we shall present here some pertinent dynamical results of the fundamental triad-interaction system formulated earlier. To be specific, the triad wave vector must be specified. We shall consider the following two of $(\pm 3) \times (\pm 3) \times (\pm 3)$ wave vector lattice:

$$W_1 = \left[\begin{pmatrix}1\\1\\1\end{pmatrix}, \begin{pmatrix}1\\1\\2\end{pmatrix}, \begin{pmatrix}-2\\-2\\-2\end{pmatrix}\right] \quad \text{and} \quad W_2 = \left[\begin{pmatrix}1\\1\\1\end{pmatrix}, \begin{pmatrix}1\\2\\1\end{pmatrix}, \begin{pmatrix}-2\\-3\\-2\end{pmatrix}\right]$$

Though the choice is arbitrary, they are basically of the different types of triad wave vector. Note that all three vectors of W_1 are in a plane that contains the 3 axis and intersects the 1–2 plane diagonally. Hence, this special geometry can give rise to certain additional symmetries in coupling coefficients. In particular, $\phi_{ij}(\xi)$ exhibit a peculiar pattern for $\xi = 0$ and $\pi/4$. Under W_1 all but the following coupling coefficients are identically zero for $\xi = 0$

$$[\phi_{21}(0), \phi_{22}(0), \phi_{23}(0)] = (-\sqrt{2}, \sqrt{2}, 0)/2\sqrt{17}$$

$$[\phi_{31}(0), \phi_{32}(0), \phi_{33}(0)] = (\sqrt{2}, 0, -\sqrt{2})/2\sqrt{6}$$

$$[\phi_{51}(0), \phi_{52}(0), \phi_{53}(0)] = (0, -\sqrt{2}, \sqrt{2})/2\sqrt{3}$$

$$[\phi_{81}(0), \phi_{82}(0), \phi_{83}(0)] = (-11, 14, -3)/2\sqrt{17 \times 9} \tag{36}$$

And, the coupling coefficients for $\xi = \pi/4$ obey

$$\phi_{ij}(\pi/4) = \phi_{(9-i),j}(\pi/4), \qquad (i = 1-4) \tag{37}$$

in addition to $\phi_{ij}(\xi) = \phi_{ij}(\xi + 2\pi)$, $\phi_{ij}(\xi) = -\phi_{ij}(\xi + \pi)$, $\phi_{ij}(\xi) = -\phi_{(9-i),j}(\xi + \pi/2)$, $(i = 1, 4, 6, 7)$ and $\phi_{ij}(\xi) = \phi_{(9-i),j}(\xi + \pi/2)$, $(i = 2, 3, 5, 8)$ which follow from the periodicity of polarization vectors. For future reference, we shall make note of the fact that in magnitude-wise [38]:

$$\phi_{11}(\pi/4) \simeq \phi_{12}(\pi/4) \gg \phi_{13}(\pi/4)$$

$$\phi_{32}(\pi/4) \simeq \phi_{33}(\pi/4) \gg \phi_{31}(\pi/4)$$

$$\phi_{41}(\pi/4) \simeq \phi_{43}(\pi/4) \gg \phi_{42}(\pi/4) \tag{38}$$

Under arbitrary triad wave vector, $\phi_{ij}(\xi)$ are typically nonzero, nor do they exhibit special symmetry like Equation 37. For instance, none of $\phi_{ij}(0)$ of W_2 are zero in contrast to Equation 36.

We shall defer the full discussion on numerical trajectory computation until later. However, it suffices to point out that, although Equations 13 and 15 are theoretically equivalent, the polar representation is not amenable to numerical integration because of the factor $(R_i^\mu)^{-1}$ in the equation for ω_i^μ. When R_i^μ becomes small at some time, the ω_i^μ-equation will violate a Lipschitz condition there. (This can also happen in a Hamiltonian system, as observed by Jackson [39].) Hence, the consequence is the loss of uniqueness [40], which is to be expected from the periodicity of polar representation. The results of rectangular representation can then be expressed in action-angle variables in a straightforward manner. First, J_i^μ is the squared magnitude $|u_i^\mu|^2$. Secondly, ω_i^μ is computed by the convention that the positive (negative) angle is referred to the clockwise (counter-clockwise) rotation.

Reduced System

As may be seen from Equation 6, the choice of $\xi = 0$ corresponds to a particular rotation in which $\underset{\sim}{\epsilon}^1(\underset{\sim}{k})$ is normal to the plane of $\underset{\sim}{k}$ and k_3 axis, and $\epsilon^1(\underset{\sim}{k})$ is on that plane. Then in view of Equation 36, the dynamic equation (Equation 16) simplifies greatly under W_1.

$$\begin{bmatrix} \dot{R}_1^1 \\ \dot{R}_2^1 \\ \dot{R}_3^1 \end{bmatrix} = -\Psi_s \begin{bmatrix} R_1^1 \\ R_2^1 \\ R_3^1 \end{bmatrix}, \qquad \begin{bmatrix} \dot{\omega}_1^1 R_1^1 \\ \dot{\omega}_2^1 R_2^1 \\ \dot{\omega}_3^1 R_3^1 \end{bmatrix} = -(\Psi_c/2\pi) \begin{bmatrix} R_1^1 \\ R_2^1 \\ R_3^1 \end{bmatrix} \tag{39a}$$

$$\begin{bmatrix} \dot{R}_1^2 \\ \dot{R}_2^2 \\ \dot{R}_3^2 \end{bmatrix} = -\begin{bmatrix} \phi_{81} R_2^2 R_3^2 \\ \phi_{82} R_3^2 R_1^2 \\ \phi_{83} R_1^2 R_2^2 \end{bmatrix} \sin(2\pi\Omega_4), \qquad 2\pi \begin{bmatrix} \dot{\omega}_1^2 \\ \dot{\omega}_2^2 \\ \dot{\omega}_3^2 \end{bmatrix} = -\begin{bmatrix} \phi_{81} R_2^2 R_3^2/R_1^2 \\ \phi_{82} R_3^2 R_1^2/R_2^2 \\ \phi_{83} R_1^2 R_2^2/R_3^2 \end{bmatrix} \cos(2\pi\Omega_4) \tag{39b}$$

where $\Psi_s = \begin{bmatrix} 0 & \phi_{21}R_3^2\sin(2\pi\Omega_3) & \phi_{31}R_2^2\sin(2\pi\Omega_2) \\ & 0 & \phi_{42}R_1^2\sin(2\pi\Omega_1) \\ \text{(antisymmetric)} & & 0 \end{bmatrix}$

and the matrix Ψ_c is the same as Ψ_s if $\sin(2\pi\Omega_n)$ are replaced by $\cos(2\pi\Omega_n)$. Note that the systems (Equations 39a and b) are uncoupled; the former is merely driven by the latter through Ψ.

The Uncoupled System (Equation 39b)

This system turns out to be a fundamental triad-interaction of two-dimensional flow. There are three known isolating constants of motion:

$$\tfrac{1}{2}(J_1^2 + J_2^2 + J_3^2) = E \tag{40a}$$

$$(\phi_{83} - \phi_{82})J_1^2 + (\phi_{81} - \phi_{83})J_2^2 + (\phi_{82} - \phi_{81})J_3^2 = C_1 \tag{40b}$$

$$R_1^2R_2^2R_3^2\cos(2\pi\Omega_4) = C_2 \tag{40c}$$

First, Equation 40a is the energy conservation. Secondly, the mutual orthogonality of three vectors (1, 1, 1), $(\phi_{81}, \phi_{82}, \phi_{83})$, and $[(\phi_{83} - \phi_{82}), (\phi_{81} - \phi_{83}), (\phi_{82} - \phi_{81})]$ yields Equation 40b, which is the enstrophy conservation in two-dimensional flow. Finally, Equation 40c is obtained by combining Equation 39b, $2\pi\dot{\Omega}_4\tan(2\pi\Omega_4) = \sum_{i=1}^{3}\dot{R}_i^2/R_i^2$, and then integrating in phase space [38, 41].

For the invariant set (Equation 17), which presently reduces to $S = \{\text{all } \omega_i^2 \text{ such that } \Omega_4 = \pm\tfrac{1}{4}, \pm\tfrac{3}{4}, \ldots\}$, Equation 39b degenerates to

$$\dot{R}_1^2 = -\phi_{81}R_2^2R_3^2, \qquad \dot{R}_2^2 = -\phi_{82}R_3^2R_1^2, \qquad \dot{R}_3^2 = -\phi_{83}R_1^2R_2^2 \tag{41}$$

This is the classical Euler equation for a rigid body moving with one point fixed under no external forces [42]. In parallel to the analytical solution of Equation 41, one may also express the solution of Equation 39b in terms of Jacobian elliptic functions: $R_1^2 = b_1\mathrm{cn}[\lambda(t)]$, $R_2^2 = b_2\,\mathrm{dn}[\lambda(t)]$, and $R_3^2 = b_3\mathrm{sn}[\lambda(t)]$. However, the determination of b_i and $\lambda(t)$ will be much more complicated than the classical case. Instead, it is more illuminating to discuss the trajectory behavior in qualitative terms. From the initial condition $I_1 = \{R_i^2 = 1 \text{ and } \omega_i^2 = \tfrac{1}{8}(i = 1 - 3)\}$, we have evolved Equation 39b over the time interval $T = 300$, with the results presented in Figure 4. On the energy sphere (Figure 4a), the periodic R_i lies in the intersection with the paraboloid C_1. Since $R_i \geq 0$ by definition, the trajectory of Figure 4a is restricted to the octant with the positive coordinates. Indeed, the role of C_2 is to delimit the trajectory to the positive octant. The phase plots of $\mathrm{Re}(u_i^2)$ vs $\mathrm{Im}(u_i^2)$ are apparently not closed in Figures 4(b–d), thereby indicating aperiodic motion.

To characterize the trajectory more qualitatively, let us consider a special invariant set of Equation 39b.

$$T = \{R_i^2 = \text{a constant and all } \omega_i^2 \text{ such that } \Omega_4 = 0\}$$

Note that Ω_4 then remains zero for all times. With the use of Equation 40c, the ω_i^2-equation may be put in the form

$$\begin{bmatrix} \dot{\omega}_1^2 \\ \dot{\omega}_2^2 \\ \dot{\omega}_3^2 \end{bmatrix} = -(C_2/2\pi)\begin{bmatrix} \phi_{81}/J_1^2 \\ \phi_{82}/J_2^2 \\ \phi_{83}/J_3^2 \end{bmatrix}$$

In terms of the frequency $f_i = \omega_i^2/t$, we find that $f_1{:}f_2{:}f_3 = \phi_{81}{:}\phi_{82}{:}\phi_{83}$ for this invariant set. But, because of $\phi_{81} + \phi_{82} + \phi_{83} = 0$, the frequencies are rationally dependent. Hence, the set T represents a periodic motion on a three-dimensional torus. Since the motion of R_i^2 is in general periodic, one

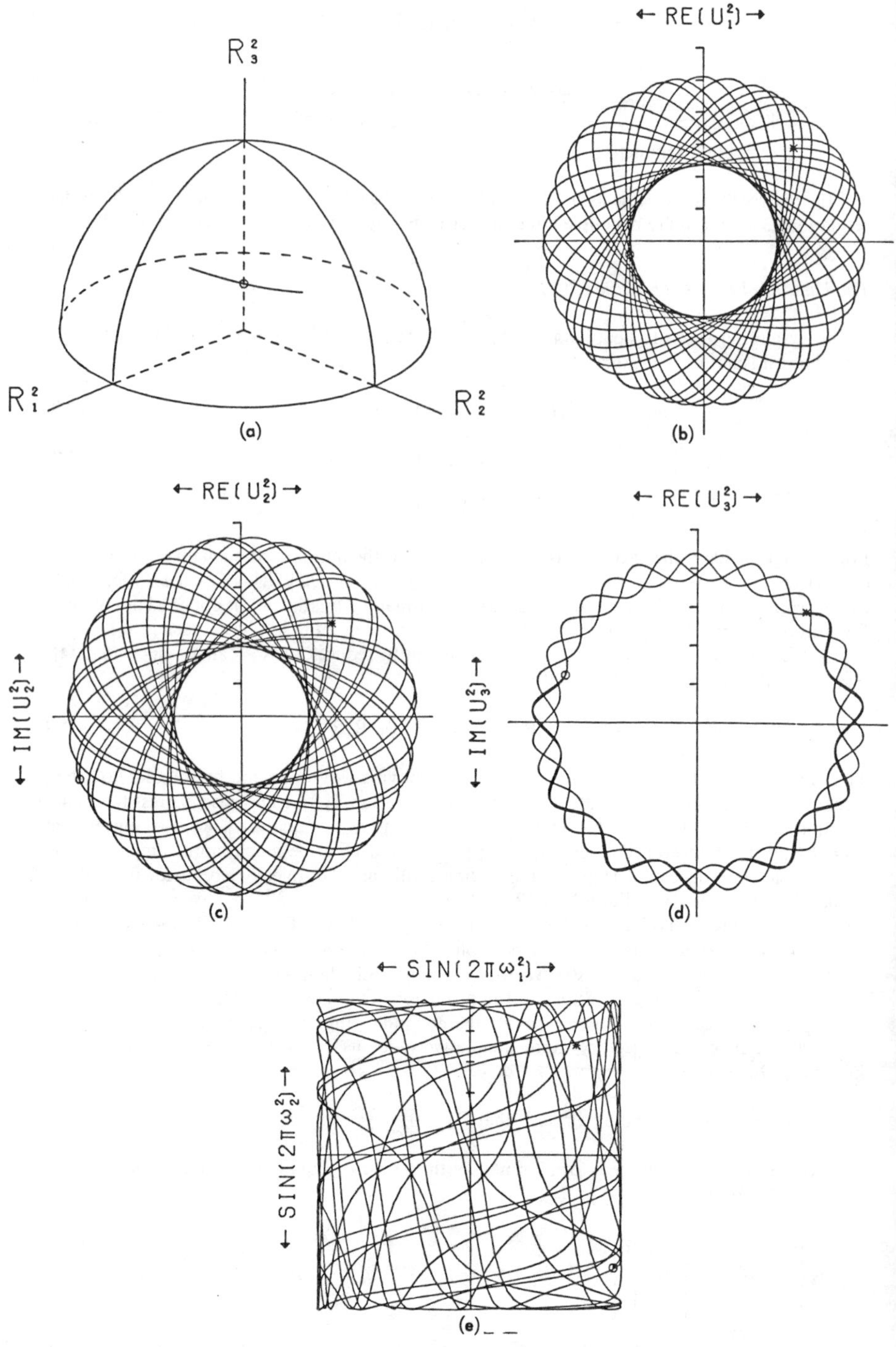

Figure 4. Time evolution of Equation 39b from the initial condition I_1 over time T = 300. o, initial point; *, final point. Energy sphere (a). Phase plots of u_1^2 (b); u_2^2 (c); u_3^2 (d). Lissajous figure (e).

may visualize the typical manifold of Equation 39b as a three-dimensional torus but with the periodically varying configuration. (For instance, a wavy doughnut in the case of two-dimensions with periodically varying meridian and parallel.) The motion on such a torus is periodic (quasi-periodic) if the frequencies are commensurable (incommensurable). A typical Lissajous figure of $\sin(2\pi\omega_1^2)$ vs $\sin(2\pi\omega_2^2)$ is not closed in Figure 4(e), thereby implying incommensurable frequencies.

To sum up, the motion of Equation 39b is ergodic on a three-dimensional torus with periodically varying configuration. The invariant sets S and T are, however, exceptional.

The Complete System (Equation 39)

As shown previously [1], the system (Equation 39) has constants of motion of the helicity form

$$\sum_{i=1}^{3} iR_i^1R_i^2\{\alpha_i^1 \exp[i2\pi(\omega_i^1 - \omega_i^2)] - \alpha_i^2 \exp[i2\pi(\omega_i^1 - \omega_i^2)]\} = \text{constant} \tag{42}$$

where the vector $\underset{\sim}{\alpha} = (\alpha_1^1, \alpha_2^1, \alpha_3^1, \alpha_1^2, \alpha_2^2, \alpha_3^2)$ satisfies

$$\begin{bmatrix} \phi_{21} & 0 & \phi_{53} & 0 & -\phi_{82} & 0 \\ \phi_{31} & \phi_{52} & 0 & 0 & 0 & -\phi_{83} \\ 0 & \phi_{22} & \phi_{33} & -\phi_{81} & 0 & 0 \end{bmatrix} \underset{\sim}{\alpha} = 0$$

Since the coefficient matrix has the rank 3, there are two constants of the form (Equation 42) besides helicity.

We have evolved Equation 39 from the initial condition $I_2 = \{R_i^\mu = 1$ and $\omega_i^\mu = \frac{1}{8}(\mu = 1, 2$ and $i = 1 - 3)\}$, which is a simple extension of I_1. Of course, the trajectory of $\mu = 2$ is the same as that of Figure 4. In contrast to the periodic Ω_4, the time histories of $\sin(2\pi\Omega_n)$ shown in Figures 5A–C

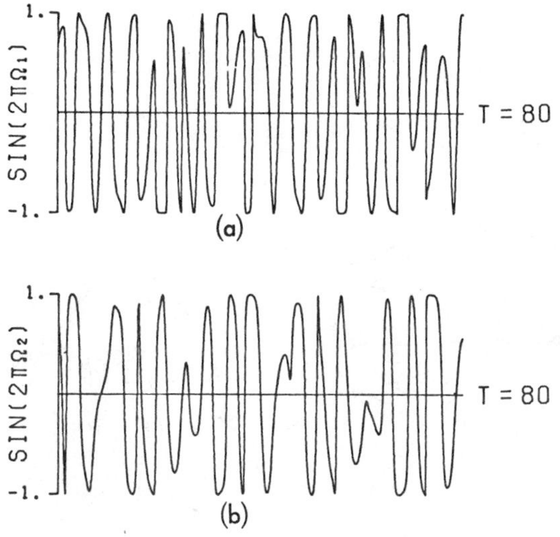

Figure 5. Time evolution of Equation 39 from the initial condition I_2 over time T = 300. o, initial point; $*$, final point. Time histories of Ω_1 (a); Ω_2 (b); Ω_3 (c). Phase plots of u_1^1 (d); u_2^1 (e); u_3^1 (f). Energy sphere (g).

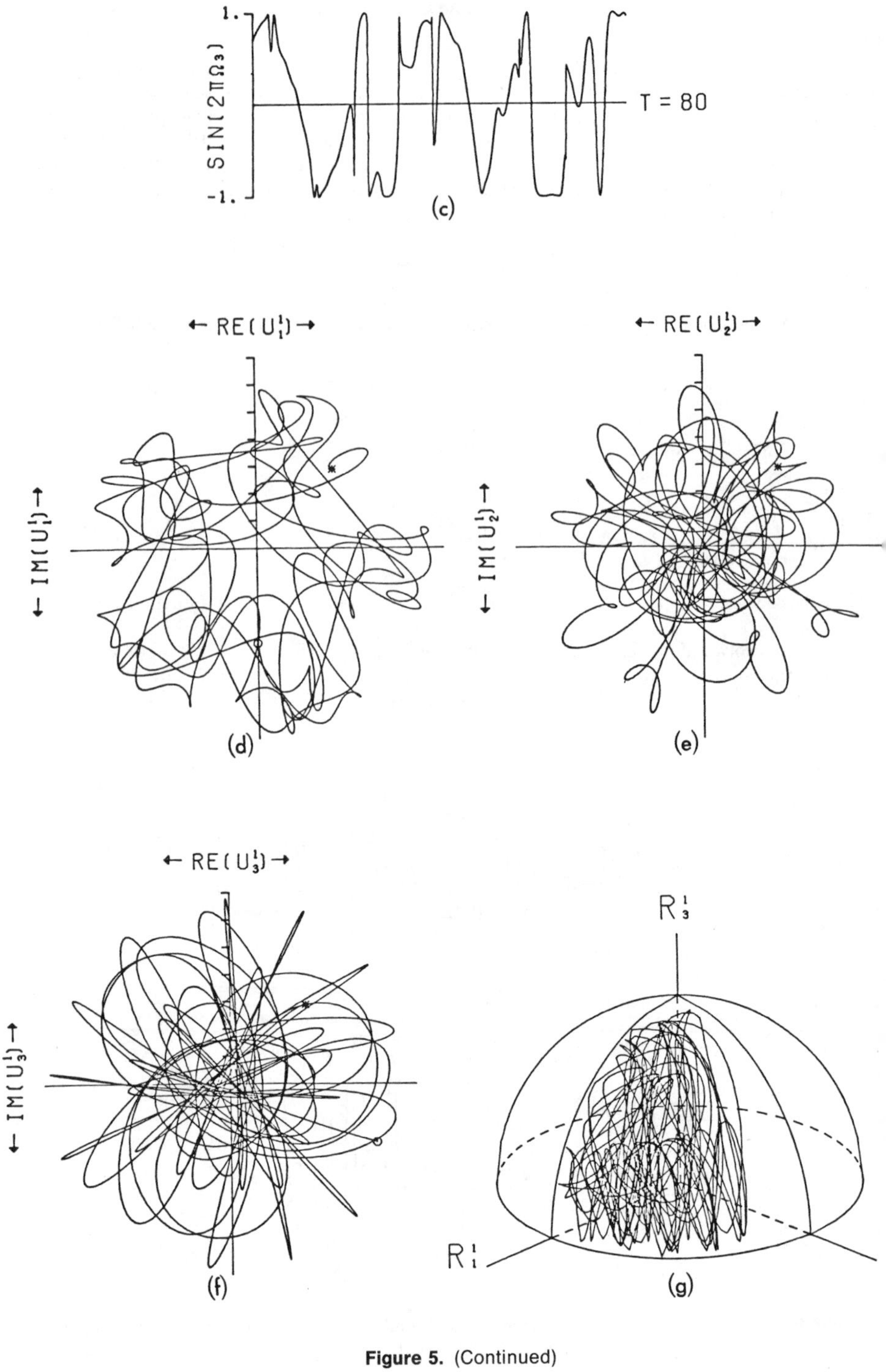

Figure 5. (Continued)

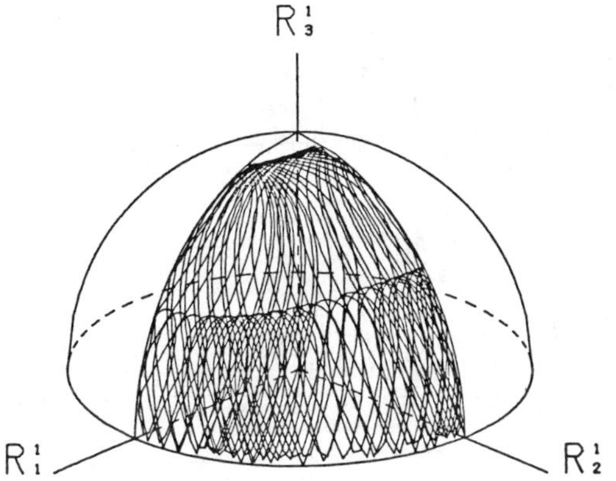

Figure 6. Energy sphere. Time evolution of Equation 39 from the initial condition I_3 over time $T = 600$.

develop irregular pattern. This is indeed a surprise. Since $\sin(2\pi\Omega_n)$ and $\cos(2\pi\Omega_n)$ enter into Ψ_s and Ψ_c, respectively, the phase plots in Figures 5(d–f) also become chaotic. Because of the helicity, one cannot expect the trajectory flow to cover the entire energy sphere [Figure 5(g)]. To suppress the helicity effect, we shall however consider the initial condition $I_3 = \{R_i^\mu = 1$ and $\omega_i^\mu = \frac{1}{12}(\mu = 1, 2$ and $i = 1 - 3)\}$ which is of the invariant set (Equation 18). As shown in Figure 6, the trajectory evolved from I_3 seems to wander all over the energy sphere with $R_i \geq 0$. This is an indication of the absence of constants of motion besides energy in this reduced manifold, though a small region near the North Pole is still inaccessible.

Summing up, the system (Equation 39b) is ergodic on a three-dimensional torus with periodically varying configuration. On the other hand, the trajectory of Equation 39a driven by Equation 39b covers almost all the energy sphere of R_i^1 when helicity-like constants (Equation 42) are suppressed. As a whole, due to the existence of extraneous constants of motion, the system (Equation 39) cannot be ergodic on the constant energy-helicity surface. Nevertheless, it is the first and simplest model to exhibit chaotic motion by retaining only four column vectors in Equation 16.

Non-ergodicity of the Fundamental Triad-Interaction System

The existence of extraneous constants of motion has precluded once and for all the possibility of Equation 39 being ergodic on the energy-helicity surface. Although a formal algebraic procedure exists for searching out quadratic constants of motion [1], it becomes inoperative when a constant of motion in hand is not strictly constant, but behaves like an isolating constant. We shall therefore attempt to infer ergodic/nonergodic behavior of fundamental triad-interaction system directly from the trajectory covering of constant energy-helicity surface.

As for a more efficient notation, let us denote the fundamental triad-interaction by TI(W, ξ, I, T), where the wave vector W is W_1 or W_2, and the remaining parameters are the same as before. Note that the initial time T_0 has been left out, for it is always zero in the present discussion. Typical results are summarized here of evolving TI(W_i, ξ, I, T) for several of selected and random values of ξ.

The TI(W_1, $\pi/4$, I, T) System

Due to the symmetry (Equation 37), the equations for u_i^1 and u_i^2 have the same ϕ_{ij}. Hence, the trajectories for $\mu = 1$ and 2 are the same if they are so initially. Except for this degenerate case,

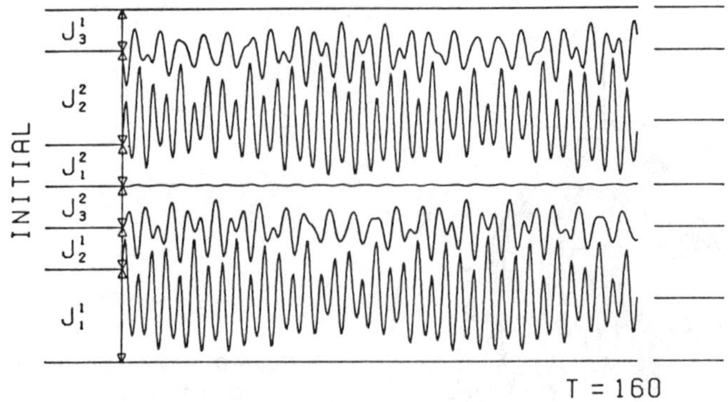

Figure 7. Action. Time evolution of TI(W_1, $\pi/4$, I_4, 160).

this system has the following constants of motion besides energy and helicity:

$$\sum_{i=1}^{3} R_i^1 R_i^2 \cos[2\pi(\omega_i^1 - \omega_i^2)] = U_4 \tag{43a}$$

$$J_1^1 + J_2^1 + J_3^2 \simeq C_3 \qquad \text{or} \qquad J_1^2 + J_2^2 + J_3^1 \simeq C_4 \tag{43b}$$

Note that Equation 43a follows from the special property of $\phi_{ij}(\pi/4)$; $\phi_{11} + \phi_{22} + \phi_{33} = 0$, $\phi_{21} + \phi_{12} + \phi_{43} = 0$, $\phi_{31} + \phi_{42} + \phi_{13} = 0$, and $\phi_{41} + \phi_{32} + \phi_{23} = 0$. We then indicate by $\simeq$ that Equation 43b is approximately constant, as may be verified by using Equation 38. Hence, C_3 and C_4 are called the quasi-constants of motion. Figure 7 depicts the time histories of J_i^μ generated by TI(W_1, $\pi/4$, I_4, 160), where $I_4 = \{R_1^1 = R_2^2 = 1.5$, all other $R_i^\mu = 1$, and $\omega_i^\mu = 1/8(\mu = 1, 2$ and $i = 1 - 3)\}$. The quasi-constancy of C_3 and C_4 is quite evident. Also indicated in the figure are the time-averaged actions tending toward $\bar{J}_i^1 \simeq \bar{J}_i^2$.

The TI($W_1\pi/8$, I, T) System

Unlike the previous systems under $\xi = 0$ and $\pi/4$, we have no *a priori* knowledge of the extraneous constants of motion. First, the phase plots of TI(W_1, $\pi/8$, I_2, T) in Figures 8A–F should be compared with the counterparts of TI(W_1, 0, I_2, T) [Figures 4(B–D) and 5(D–F)]. Since the phase plots for $\mu = 2$ are confined to an annular region, one suspects the presence of extraneous constants of motion. To suppress the helicity effect, we have evolved this system from the initial condition I_3. Now, the energy surface is E_6. To sample certain E_3 projections of it, we first relabel $R_1^1, R_2^1, \ldots, R_3^2$ as R_i $(i = 1 - 6)$, and then define the polar and azimuthal angles, $\theta_{ijk} = \tan^{-1}(R_i/R_j)$ and $\xi_{ijk} = \tan^{-1}(R_j/R_k \cos\theta_{ijk})$, for a distinct triplet $(i \neq j \neq k)$. For a trajectory wandering everywhere on the energy surface, one may expect the graph of θ_{ijk} vs ξ_{ijk} to fill up almost every point of the square of side $\pi/2$. Since the squares of Figure 9 are only partially traversed, it is concluded that there are extraneous (quasi- or strict) constants which restrict the trajectory flow to part of the energy surface that is unaffected by helicity.

We have further investigated TI(W_1, ξ, I, T) for randomly chosen parameter values; $\xi_1 = 0.35422\pi$, $\xi_2 = 0.69528\pi$, $\xi_3 = 0.94236\pi$, $\xi_4 = 1.1793\pi$, and $\xi_5 = 1.7934\pi$. For ξ_2 and ξ_5 the system has U_4 as a quasi-constant of motion; whereas, for the remaining ξ_i (i = 1, 3, and 4) the trajectory flow is restricted to a part of the constant energy-helicity surface (as evidenced by a partial covering of the graphs like Figure 9).

We shall continue to examine the fundamental triad-interaction under another triad wave vector W_2.

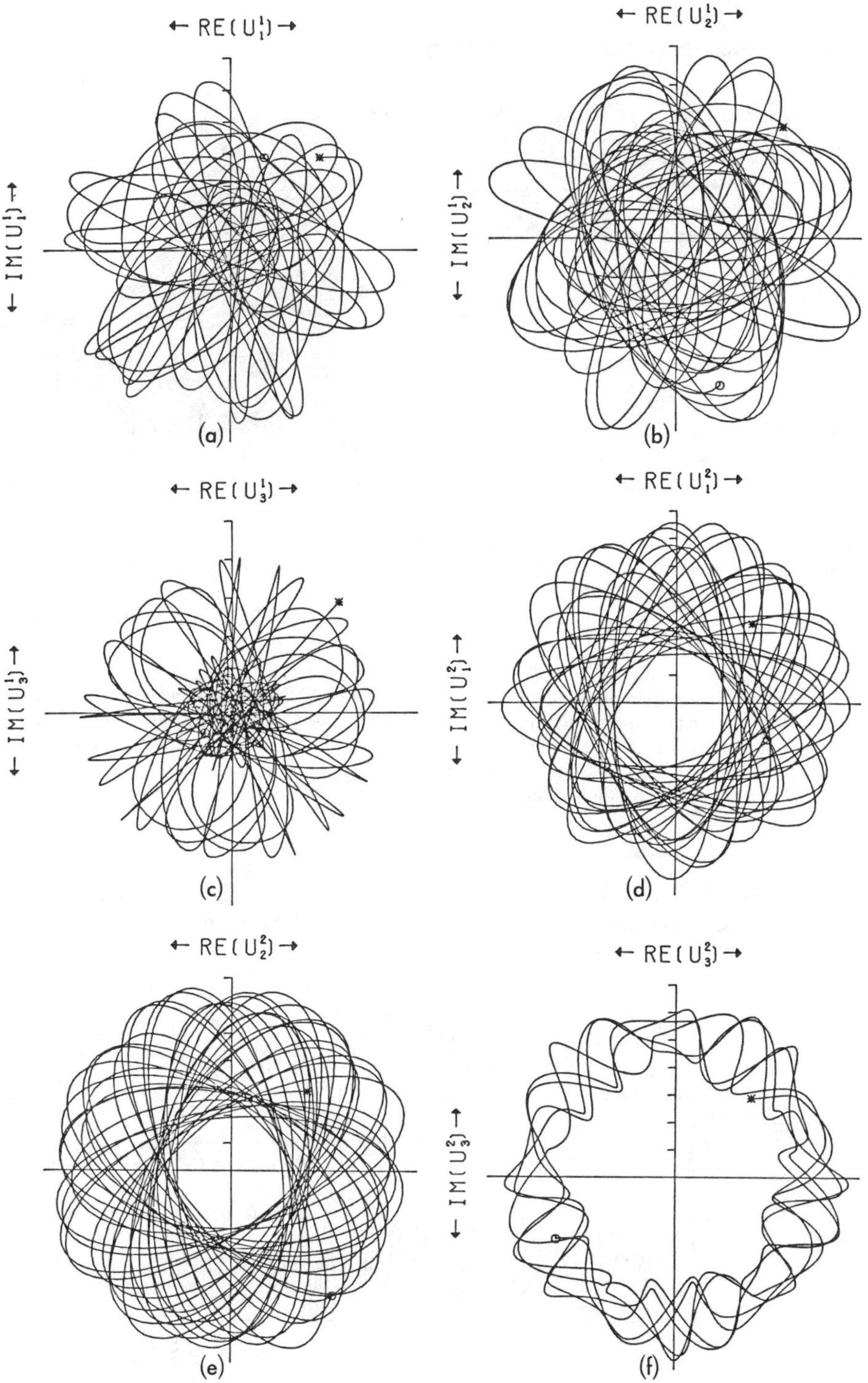

Figure 8. Time evolution of TI(W_1, $\pi/8$, I_2, 300). o, initial point; $*$, final point. Phase plots of u_1^1 (a); u_2^1 (b); u_3^1 (c); u_1^2 (d); u_2^2 (e); u_3^2 (f).

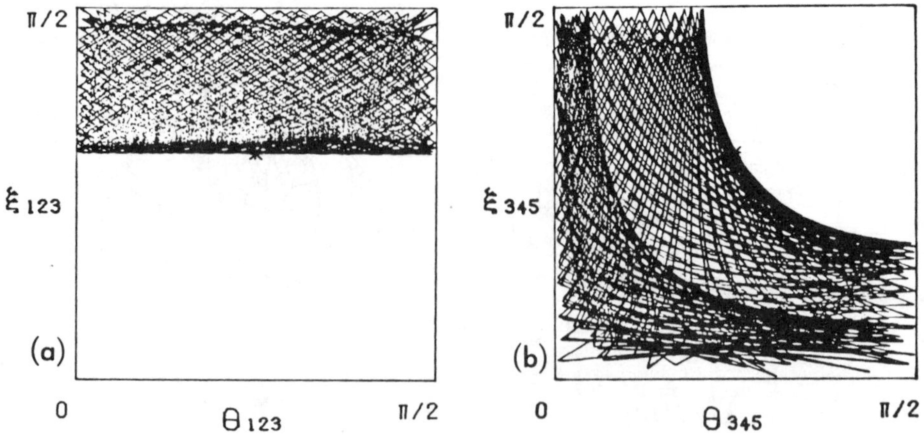

Figure 9. Three-dimensional projection of the energy surface of TI(W_1, $\pi/8$, I_3, 600). Graphs of θ_{123} vs ξ_{123} (a); θ_{345} vs ξ_{345} (b).

The TI(W_2, 0, I, T) System

In contrast to Equation 36, the coupling coefficients are typically nonzero under W_2. Time histories of action are presented in Figure 10 for TI(W_2, 0, I_2, 280). Overtly stretching the meaning of quasi-constancy, it is suggested that $J_1^1 + J_2^1 + J_3^1 \simeq C_5$ and $J_1^2 + J_2^2 + J_3^2 \simeq C_6$ are quasi-constants of motion. This is because the time averages of C_5 and C_6 are nearly the same. Recall that C_5 and C_6 were the strict constant under W_1.

The TI(W_2, I, T) System

Unlike TI(W_1, $\pi/4$, I, T), this system can evolve from the identical initial conditions for u_i^1 and u_i^2 without resulting in redundant trajectories for $\mu = 1$ and 2. Here, U_4 plays the role of quasi-constant, whereas it was the strict constant of motion under W_1.

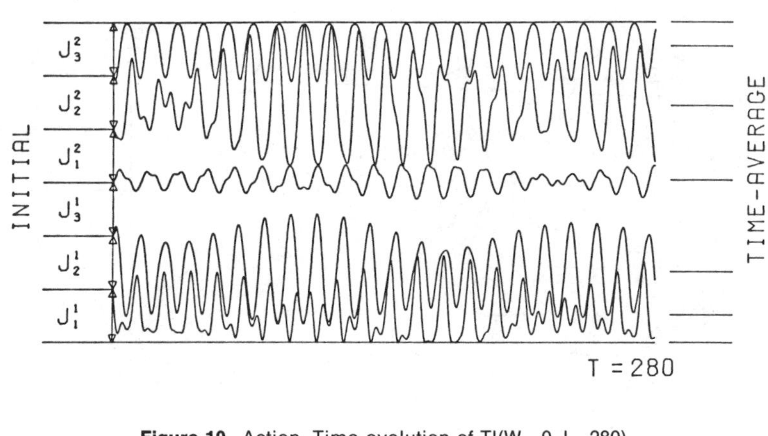

Figure 10. Action. Time evolution of TI(W_1, 0, I_2, 280).

To sum up, under W_1 and W_2 the fundamental triad-interaction system is not ergodic, for the trajectory flow is restricted to part of the constant energy-helicity surface. A surprise, however, here is the occurrence of quasi-constants of motion that can constrain the trajectory flow in phase space just as the strict isolating constant.

Lyapunov Instability

As evidenced by Figures 5 and 8, TI(W, ξ, I, T) develops a chaotic motion for most of ξ values that we have considered here. As already discussed, a chaotic orbit exhibits trajectory instability, which manifests itself by the exponential separation of initially nearby trajectories. For a graphical demonstration of this, let us construct a set of initial states clustered around I_2 as follows: First, consider a small circle centered around $R_i^\mu = 1$ on the energy surface; the solid angle subtended by the circle is $\pi/40$ in E_6. We then choose the amplitudes of each initial state so that the vector R_i^μ lies on that circle. Second, randomly choose ω_i^1 of each initial state in the range $[\frac{1}{8} - 0.01, \frac{1}{8} + 0.01]$ and set $\omega_i^2 = \omega_i^1$. A set of 40 such initial states including I_2 will be denoted by I_2. After evolving I_2 for a short time, we have compared in Figure 11 the phase plots of Re(u_1^1) vs Im(u_1^1) at the initial and final times. Also shown in the figure is TI(W_1, $\pi/8$, I_2, 80) which serves as the center of Lyapunov tube. Scattered transversal points in the final phase plot signify tendency for the trajectories to

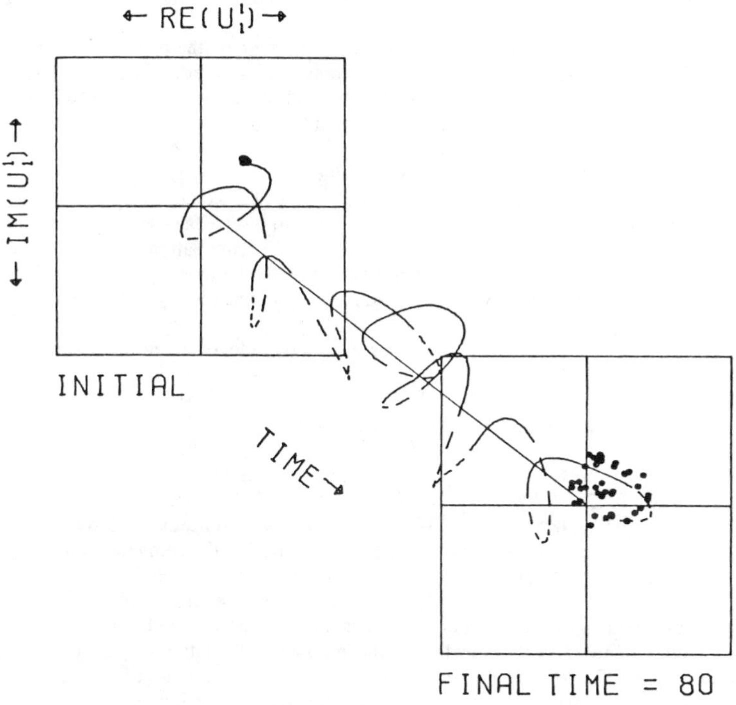

Figure 11. Lyapunov instability exhibited by the set I_2 of 40 initial data points about I_2. The center of Lyapunov tube is traced by TI(W_1, $\pi/8$, I_2, 80), which is denoted by solid-line when Im(u_1^1) $\gg 0$ and dotted-line Im(u_1^1) < 0.

break away from each other after a very short evolution. As will be seen later, trajectory instability is one of most important features of D(N).

EFFECTIVE EVOLUTION TIME

Earlier the numerical integration was performed by the differential-equation solver ODE developed and documented by Shampine and Gordon [44]. At the risk of misquoting, let us venture to say that the ODE is a variable-step, variable-order Adams-Bashford predictor and Adams-Moulton corrector code. Since the internal working of ODE is of no direct concern, we shall restrict ourselves here to the use of it as a black box, but communicating only through the input/return flag (IFLAG). Both the relative and absolute error tolerances, RELERR and ABSERR, are used to control the local error by

$$\text{Local error} = \text{RELERR} \times |\text{Approximated } u_i^\mu| + \text{ABSERR}$$

Since $|u_i^\mu|$ is usually <1, we shall always set RELERR $= 0$ and specify only the absolute error tolerance. By successively reducing ABSERR by the factor of 0.1, i.e., ABSERR $= 10^{-7}$, 10^{-8}, etc, it was found that on CDC CYBER 74/175 the smallest ABSERR that ODE can cope with is 10^{-13}. (In other words, ODE will automatically boost the error tolerance when ABSERR $= 10^{-14}$ is imposed.) That this is perhaps the smallest ABSERR may be seen from the unit round-off error of CDC single precision, which Shampine and Gordon have estimated to be 7.1×10^{-15}. Hence, all numerical experiments were carried out with RELERR $= 0$ and ABSERR $= 10^{-13}$, unless otherwise stated.

Shampine and Gordon [44] proposed a global error estimate by the reintegration, whereby one reintegrates the problem with the reduced error tolerance by an order of magnitude and compares the results (also adopted by Glaz [7] for two-dimensional turbulence). However, this is not really relevant to our application of the open-ended evolution time. Rather, the pertinent question is how long can we evolve D(N) before the accuracy of trajectory computation is in doubt. This question can best be dealt with by the forward-backward time integration, whereby one integrates the problem in forward time up to a predetermined value and then in backward time to recover the initial condition. (Note that ODE is well suited for this; one simply has to reverse the sign of the time step, DT $=$ $-$DT, for the backward time integration.) When there is no accumulation of round-off errors, one should recover the initial data, of course, within the order of ABSERR, independent of the integration time period. Therefore, the recovery of initial data after a forward-backward time integration can provide an overall measure of computational accuracy.

For the test of initial data recovery, D(13) was integrated in forward and backward times under the initial condition I(13) defined by

$$I(n) = \{R_i^\mu = (2n)^{-1/2}, \omega_i^\mu = \tfrac{1}{8}(\mu = 1, 2; i = 1, 2, \ldots, n)\}$$

The interpretation of I(n) will be given later. The forward-backward time integration was carried out for three T $= 110$, 120, and 130, but all under $\xi = 3\pi/8$ and ABSERR $= 10^{-13}$. The result of forward time evolution D(13, I(13), $3\pi/8$, 0, T) followed by backward time integration D(13, D(13, I(13), $3\pi/8$, 0, T), $3\pi/8$, T, 0) can best be summarized by superimposing the recovered initial data on the imposed initial data. Since I(13) has equal action and angle for all modes, it is simply represented by a single point, denoted by * in Figure 12, in the phase plot of Re(u_k^μ) vs Im(u_k^μ). For T $= 110$ the recovery of initial data appears quite good; at least as shown graphically in Figure 12A; the recovered initial data points, denoted by o, coincide with the imposed initial data point. Quantitatively, however, the deviations of recovered R_i^μ and ω_i^μ from the respective initial values $\sqrt{\frac{1}{26}}$ and $\frac{1}{8}$ do not exceed $\pm 0.75\%$; hence, this maximum deviation will hereinafter be adopted as a criterion of good recovery. With a slightly increased T $= 120$, Figure 12B shows some scatter in the recovered initial data. But a further increase in T renders the recovery of initial data all but impossible; as attested to by the widely scattered data points of Figure 12C. Because of the periodicity $\omega_i^\mu = [\omega_i^\mu, \bmod(1)]$, Figure 12C gives only a partial account of recovery failure in that recovered initial data points may

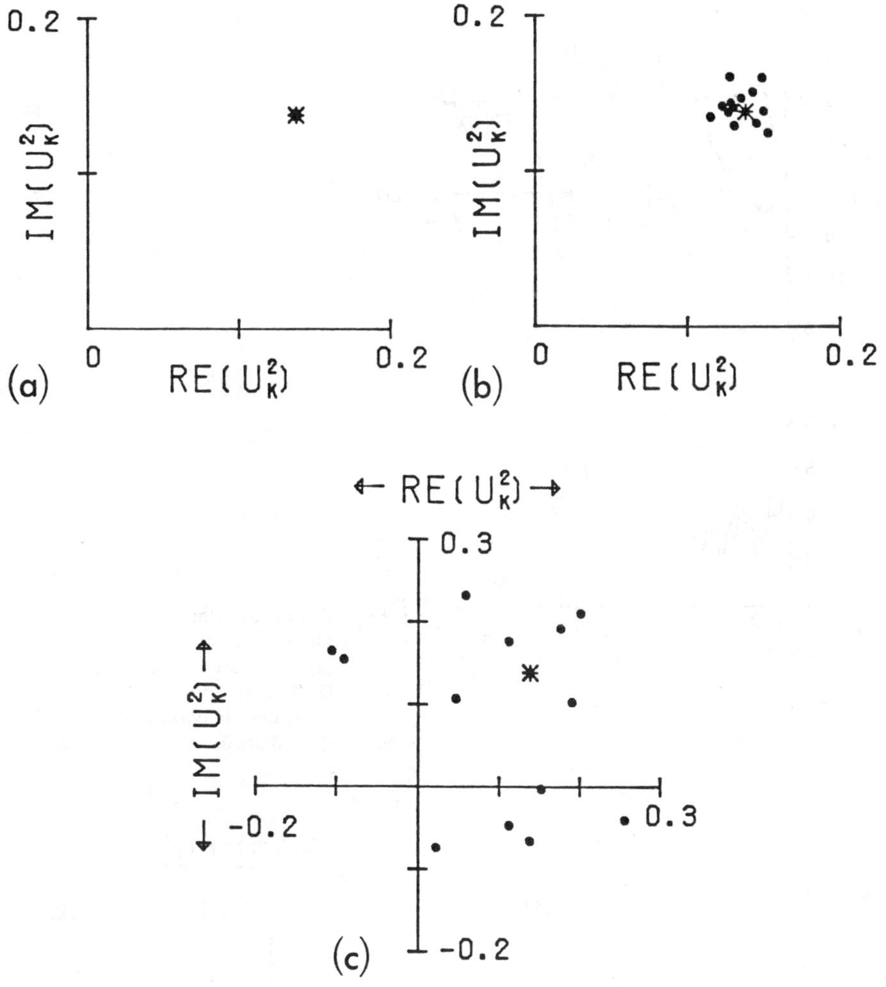

Figure 12. Forward-backward time integration of D(13) under $\xi = 3\pi/8$ and ABSERR $= 10^{-13}$. $*$, initial condition I(13); o, recovered initial data. Evolution time T = 110 (a); T = 120 (b); T = 130 (c).

not all lie on the same Riemann sheet. To examine this, we have presented the time histories of ω_3^1 and ω_4^2 during the forward time integration in Figure 13a, and during the backward time integration in Figure 13b. Since the recovered ω_4^2 falls below the initial value 1/8 more than unity, we now know that the recovered u_4^2 of Figure 12c lies in a Riemann sheet different from that of I(13).

In conclusion, based on the $\pm 0.75\%$ maximum deviation, the largest T has been found to be 110 under ABSERR $= 10^{-13}$. Although T varies with the choice of I and ξ, we shall take this T as a typical effective evolution time of D(13), beyond which evolution would give a pseudo-orbit. Now, to show strong dependence on the error tolerance, the forward-backward time integration of Figure 12a has been repeated, but under less stringent error controls. Under ABSERR $= 10^{-12}$ the

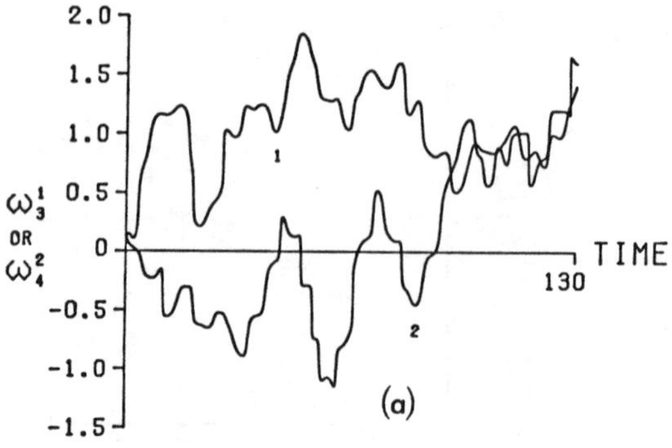

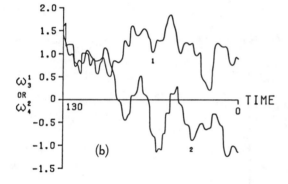

Figure 13. Time histories of ω_3^1 and ω_4^2. "1" = ω_3^1; "2" = ω_4^2. (a) Forward time integration D(13, 1(13), 3π/8, 0, 130) (b) Backward time integration D(13, D(13, I(13), 3π/8, 0, 130), 3π/8, 130, 0).

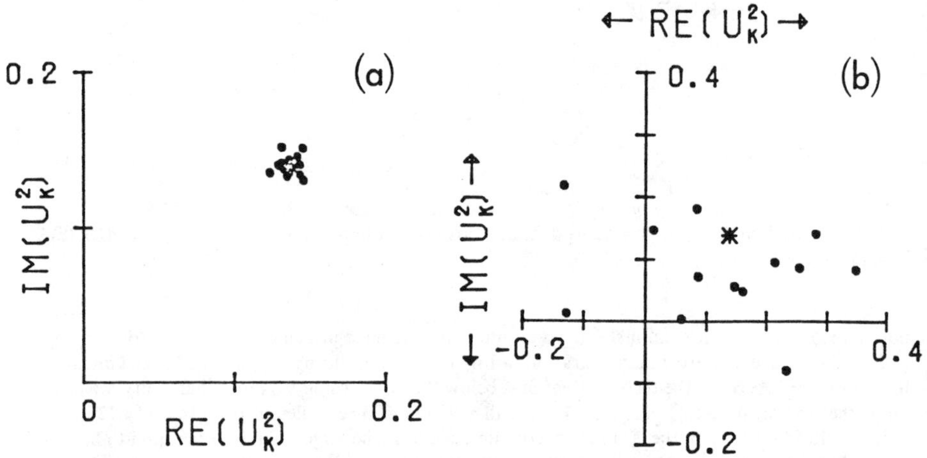

Figure 14. Forward-backward time integration of D(13) under $\xi = 3\pi/8$ over time T = 110. *, initial condition 1(13); o, recovered initial data. Error tolerance ABSERR $= 10^{-12}$ (a); $= 10^{-11}$ (b).

Table 2
Comparison of the Number of Derivative Evaluations

$\log_{10}$(ABSERR)	Total number of derivative evaluations for D(13, I(13), $3\pi/8$, 0, 110)	Reference
−13	8,559	Figure 12(a)
−12	5,429	Figure 14(a)
−11	3,457	Figure 14(b)

recovered initial data of Figure 14a show modest spread about I(13), whereas Figure 14b clearly indicates that T = 110 is too large when the error tolerance is ABSERR = 10^{-11}.

Numerically, energy and helicity are conserved only approximately. We have observed that deviations in energy and helicity conservations are less than ABSERR by at least one order of magnitude within the effective evolution time. This is why monitoring the energy and helicity conservations cannot forewarn of numerical catastrophe; a similar view was expressed by Glaz [7] for the two-dimensional turbulence.

Departing for now from the numerical accuracy, the set of computations underlying Figures 12a and 14a, b can provide a comparison of computing costs, which escalate rapidly with the more stringent error tolerance. We have summarized in Table 2 the total number of derivative evaluations incurred during the evolution of D(13, I(13), $3\pi/8$, 0, 110) under three ABSERR's.

Assuming that computing cost is directly proportional to the number of derivative evaluations, this being the case for a large degree-of-freedom system such as D(13), one can infer from the table that tightening ABSERR by two orders of magnitude has resulted in more than doubling the computing cost.

THE LOWEST ORDER SYSTEM

We shall present here a detailed numerical study of the lowest-order system D(13) of spherical truncation, following the theoretical guidelines presented earlier. Due to the limited evolution time, it is not possible to provide a direct test of ergodicity (Equation 23) and mixing (Equation 24), which necessitates an equilibrium trajectory. According to earlier theoretical discussion, however, the limited evolution time is a sufficient numerical testimony of trajectory instability that is also possessed by K-systems and Anosov systems. As a matter of fact, what we have shown is something stronger than ergodicity and mixing; the D(13) has a positive Kolmogorov entropy, and hence is a K-system.

Earlier we used the initial condition

$$I(13) = \{R_i^\mu = \sqrt{\tfrac{1}{26}}, \omega_i^\mu = \tfrac{1}{8}\,(\mu = 1, 2; i = 1 - 13)\}$$

which represents zero helicity (H = 0), for $\omega_i^1 = \omega_i^2$ at the initial time (c.f., the subset S′ in). By reversing the sign of ω_i^2 in I(13), we have

$$I'(13) = \{R_i^\mu = \sqrt{\tfrac{1}{26}}, \omega_i^1 = -\omega_i^2 = \tfrac{1}{8}\,(\mu = 1, 2; i = 1 - 13)\}$$

which, according to Equation 22, represents the maximum helicity, since $|\sin 2\pi(\omega_i^1 - \omega_i^2)| = 1$. An intermediate helicity may be represented for instance by

$$I''(13) = \{R_i^\mu = \sqrt{\tfrac{1}{26}}, \omega_i^1 = \tfrac{1}{8}, \omega_i^2 = 0\,(\mu = 1, 2; i = 1 - 13)\}\ .$$

Our contention is that I, I′, I″ can generate typical trajectories of zero helicity, maximum helicity, and an intermediate helicity, respectively (as will be discussed in the following section, "Sensitive

Dependence on the Initial Conditions." Unlike energy and ensophy of two-dimensional turbulence, which are interdependent, one may assign helicity quite arbitrarily (but between the zero and maximum values), independent of energy. For instance the zero helicity corresponds to isotropic turbulence; hence most of the numerical experiments to be reported below have been evolved from I(13). In an attempt to detect possible separation of the phase space into random and ordered regions (e.g., Hénon and Heiles [33]), we have tested other initial data besides I(13), I′(13), and I″(13). However, no such separations have been observed. This conclusion is, at best, tentative in that there are (infinitely) many other initial data yet to be tested.

A last comment before presenting numerical results is that overall (macroscopic) dynamics are not sensitive to the choice of ξ, except for $\xi = \pi/4$. As already seen by Equation 37, $\xi = \pi/4$ imposes extraneous symmetries on the coupling coefficients, and hence will be excluded from further consideration. In particular, we shall use one of the four values $\xi = 0, 3\pi/8, 2\pi/3$, and 0.87654π, more or less arbitrarily for the numerical results to be presented here.

Sensitive Dependence on the Initial Conditions

Let us consider another initial condition

$$\mathrm{I}'''(13) = \{\mathrm{R}_i^\mu = \sqrt{\tfrac{1}{26}}, \omega_i^\mu = \tfrac{1}{8} + 0.005\delta_{i,1}^{\mu,1} \ (\mu = 1, 2; \mathrm{i} = 1 - 13)\}$$

which differs slightly only in the phase of u_1^1; otherwise, it is completely identical with I(13). In Figure 15 we have presented the simultaneous trajectories of D(13, I(13), $3\pi/8$, 0, 25) and D(13, I‴(13), $3\pi/8$, 0, 25), which break away from each other in such a manner that, after a short time T = 25, they do not show any sign of the initial proximity in phase space. In other words, the divergence of initially nearby trajectories is completely unpredictable; hence, the precise details of trajectories are quite sensitive to the initial conditions. However, when certain time averages are considered, random behavior of the trajectory gets averaged out. This therefore permits us to use a single initial condition, such as I(13), to generate a generic trajectory, from which typical statistics may be computed, independently of the choice of initial data.

Random Trajectory

As was alluded to by the phase plots of Figure 15, D(13) develops apparently random trajectories. Figure 16 contains some typical phase plots of D(13, I(13), $3\pi/8$, 0, 130). The choice of T = 130 was made to match the evolution time of Figure 21 (described later), which calls for an even longer evolution time for autocorrelation computations. Two observations are relevant here. First of all, the phase plots show no inaccessible annular region such as found in Figures 8(d–f); hence, the trajectory flow is apparently unconstrained in the phase space. Secondly, the phase plots in general tend to cover all four quadrants. In particular, the phase plot of Figure 16a traverses the four quadrants more uniformly and randomly than the remaining Figures 16(b–d). When $\mathrm{Re}(u_i^\mu)$ and $\mathrm{Im}(u_i^\mu)$ are considered as phase functions, uniformly and randomly traversing phase plots will satisfy the zero-mean condition (Equation 25). In this respect, we may say the trajectory of Figure 16A obeys the zero-mean condition more faithfully than those of Figures 16(b–d).

Ergodic Behavior of Modal Energies

We shall present here a limited test for ergodicity (Equation 23) by restricting the phase functions to be the modal energies J_i^μ. The reason for this choice of phase functions is that their phase averages have already been computed, based on the canonical distribution, which approximates the microcanonical distribution quite satisfactorily for the present system of 26 degrees of freedom. Recall the following result: energy equipartition (Equation 30) prevails under zero helicity, whereas in nonzero helicity case the phase-averaged modal energies $\langle J_i^1 \rangle$ and $\langle J_i^2 \rangle$ have the same distribution

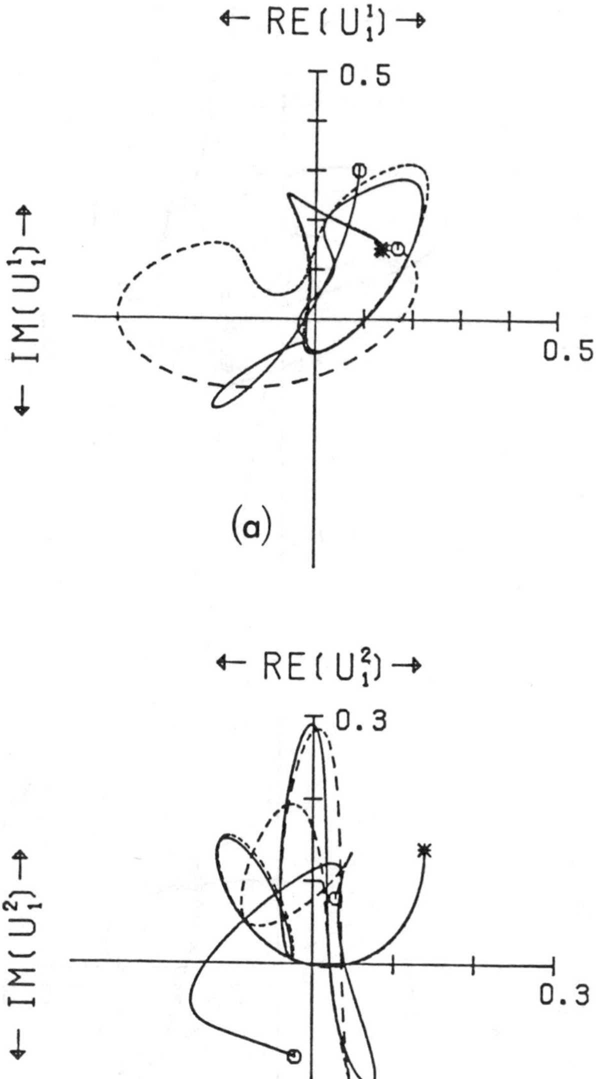

Figure 15. Exponential orbit separation of the (solid-line) trajectory of D(13, I(13), $3\pi/8$, 0, 25) from the (dotted-line) trajectory of D(13, I‴(13), $3\pi/8$, 0, 25). $*$, initial point; o, final point. Phase plots of u_1^1 (a); u_1^2 (b); u_5^1 (c); u_5^2 (d).

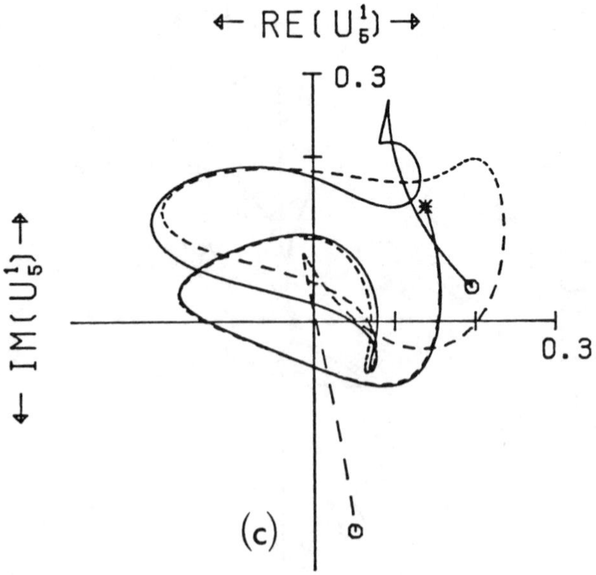

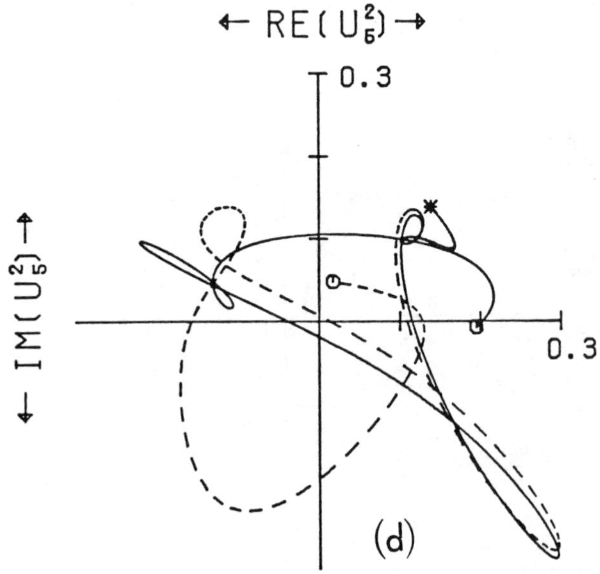

Figure 15. (Continued)

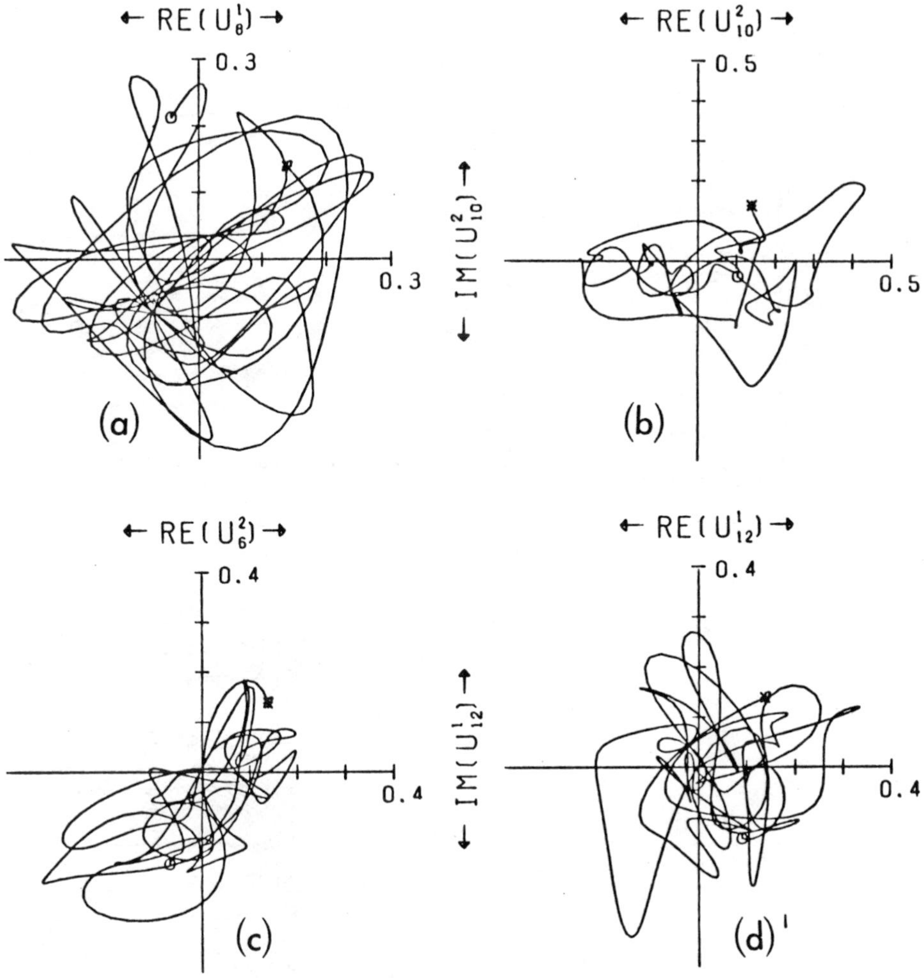

Figure 16. Random trajectory of D(13, I(13), 3π/8, 0, 130). ∗, initial point; o, final point. Phase plots of u_8^1 (a); u_{10}^2 (b); u_6^2 (c); u_{12}^1 (d).

(Equation 29a). We have presented in Figures 17 and 18 the time-averaged actions under the initial conditions I(13) and I′(13), respectively. For zero helicity, Figure 17 indicates a tendency towards energy equipartition. On the other hand, Figure 18 depicts the equality of modal energy distributions under a maximum helicity. It is noted that Figures 17 and 18 are intended for qualitative inference only. To be quantitative, we need to extend the evolution time much beyond T = 120, which we shall not do because the mixing property stronger than ergodicity will be demonstrated later on.

As mentioned in the beginning of this chapter, although extreme, the choice of zero and maximum helicity is not at all exceptional (non-generic). Theoretically, this is evident from the equilibrium distribution (Equation 29), which depends continuously on C_1 and C_2. To lend further numerical

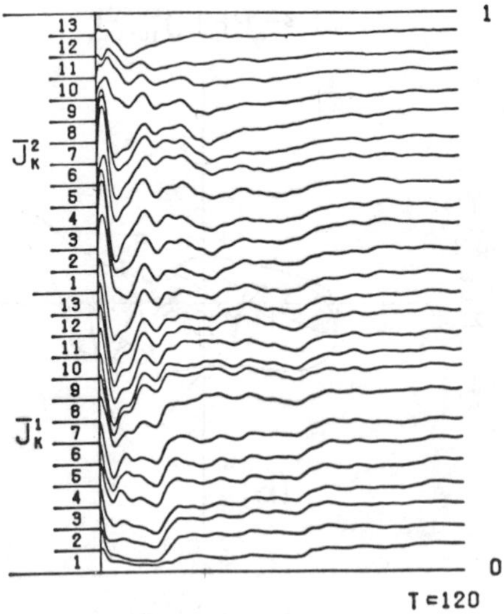

Figure 17. Time-averaged action of D(13, I(13), 0.87654π, o, 120) representing zero helicity.

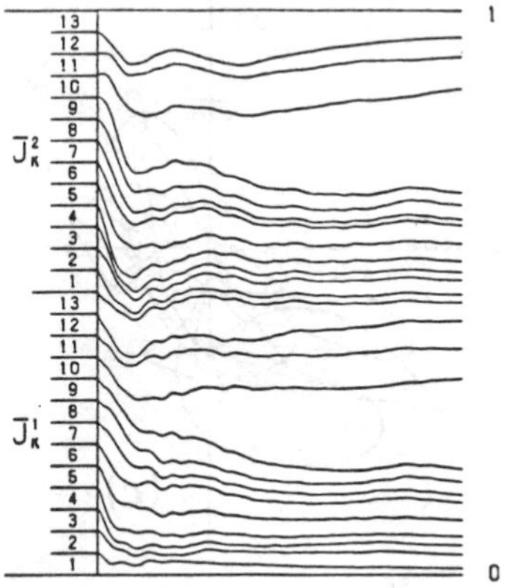

Figure 18. Time-averaged action of D(13, I′(13), 0.87654π, 0, 120) representing the maximum helicity.

support, we have presented in Figure 19 the time-averaged actions under an intermediate helicity. Note that qualitatively the long-time energy distribution lies somewhere in-between Figures 17 and 18.

Decay of U_4

Under energy equipartition, isotropy of the covariance spectral tensor calls for the disappearance of the real part of the reflexional asymmetry U_4, given by Equation 34. The phase average $\langle U_4 \rangle$ based on the canonical distribution is identically zero. We shall now compute its time average from the trajectory of D(13, I(13), 0, 0, 120). As shown in Figure 20, the asymptotic decay of $\bar{U}_4$ has been brought about by random oscillations of the time history of U_4.

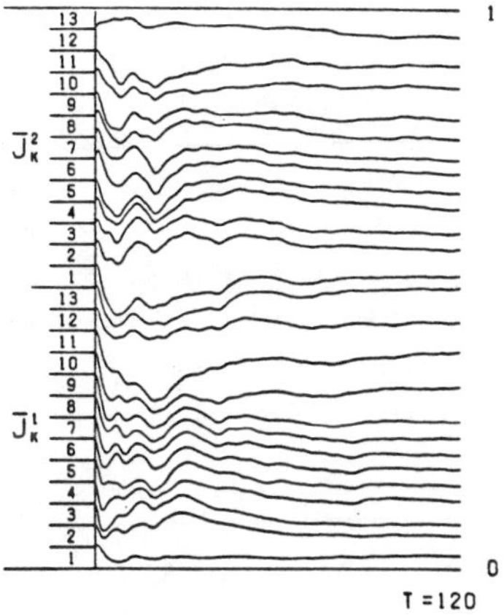

Figure 19. Time-averaged action of D(13, 1″(13), 0.87654π, 0, 120) representing an intermediate helicity.

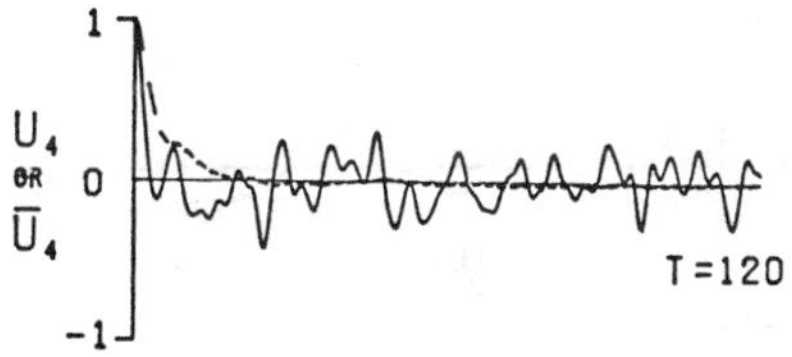

Figure 20. Evolution of the real part of the reflexional asymmetry of D(13, I(13), 0, 0, 120). ————, U_4; - - - - - - -, $\bar{U}_4$.

Decay of Autocorrelation Function

Under the ergodic assumption, let us compute the autocorrelation $\rho^\mu_i(T, \tau)$ for $Re(u^\mu_i)$ as defined by Equation 27. (The correlations for $Im(u^\mu_i)$ will not be discussed here, for they exhibit more or less the same decay behavior as $\rho^\mu_i(T, \tau)$.) Figure 21 presents autocorrelation functions computed from the same trajectory as in Figure 16. Figure 21a typifies the majority of ρ^μ_i, whereby correlation falls off rapidly and decays with damped oscillations. However, it would be misleading not to mention some anomalous case, such as those depicted by Figures 21(b–d). First of all, the correlation

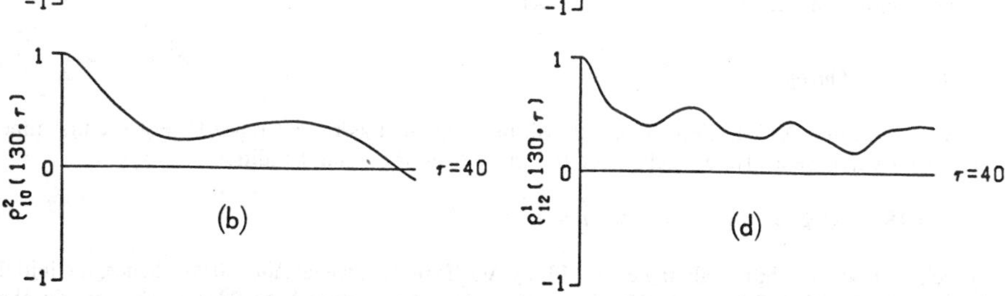

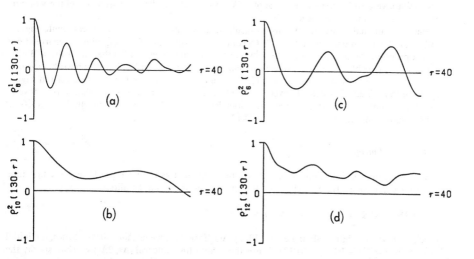

Figure 21. Autocorrelation functions generated from the trajectory of D(13, I(13), 3π/8, 0, 130). Correlations of $Re(u^1_8)$ (a); $Re(u^2_{10})$ (b); $Re(u^2_6)$ (c); $Re(u^1_{12})$ (d).

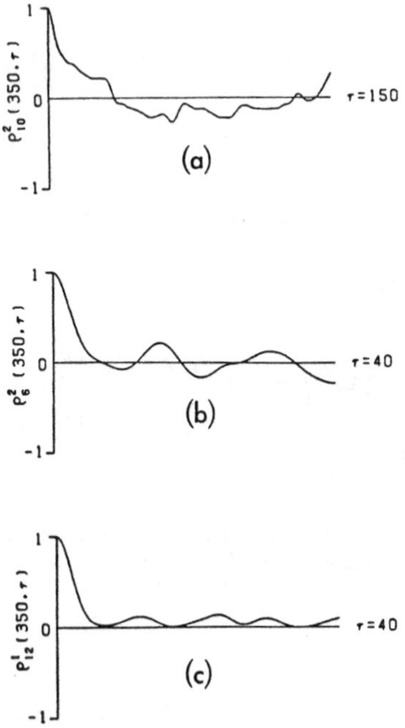

Figure 22. Autocorrelation functions generated from the trajectory of D(13, I(13), $3\pi/8$, 0, 350). Correlations of $\mathrm{Re}(u_{10}^2)$ (a); $\mathrm{Re}(u_6^2)$ (b); $\mathrm{Re}(u_{12}^1)$ (c).

of Figure 21b decays, but very gradually in the time range that ρ_8^1 has undergone several oscillations. That is, ρ_{10}^2 has a much longer correlation time than ρ_8^1. Secondly, the correlation of Figure 21c falls off initially, but remains at a higher level even toward the end of τ. This is clearly a reflection of the somewhat milder erratic trajectory of Figure 16c. Lastly, the correlation of Figure 21d lies well above zero for the entire range of τ. We suspect this to be a consequence of the violation of zero-mean condition (Equation 25).

Since the anomalies are due mainly to insufficient evolution time, we shall recompute Figures 21(b–d), but with a much longer T = 350. Of course, the justification for computing autocorrelations from a pseudo-orbit is derived from the observation of Benettin et al. [36] that the Anosov-Bowen theorem is applicable not only to Anosov systems, as originally intended, but to some other systems (e.g. the Hénon-Heiles model) exhibiting chaos. The recomputed correlations are shown in Figure 22. Note that Figure 22a extends over $\tau = 150$ to permit development of correlation over several correlation times. To a large degree, the anomalies of $\rho_i^\mu(130, \tau)$ have disappeared in $\rho_i^\mu(350, \tau)$, thereby indicating the mixing property of D(13).

Kolmogorov Entropy

The trajectory D(13, I(13), $3\pi/8$, 0, 300) will be designated as the reference U. For the first time interval $(0, \Delta\tau)$, we evolve a neighboring trajectory from the initial condition

$$i_0 = \{R_i^\mu = \sqrt{\tfrac{1}{26}}, \omega_i^1 = \omega_i^2 = \tfrac{1}{8} + \Delta\omega \ (\mu = 1, 2; i = 1 - 13)\}$$

which represents a shift of all angles of I(13) by $\Delta\omega$. Typically, we let $\Delta\omega = 0.002$; hence, the initial distance is $d = \|i_0 - I(13)\| \simeq 0.012566$. For the second time interval $(\Delta\tau, 2\Delta\tau)$ we shall specify the initial condition i_1 as follows. First, note that each mode of i_0 is displaced from I(13) by the same distance $[(1 - \cos 2\pi\,\Delta\omega)/13]^{1/2}$. Consider U at $T = \Delta\tau$, and denote its real and imaginary components by $\tilde{v}_i^\mu$ and $\tilde{w}_i^\mu$, respectively. Then, define each mode of i_1 by requiring that it has the same magnitude $[\tilde{v}_i^{\mu})^2 + (\tilde{w}_i^{\mu})^2]^{1/2}$ as U at $T = \Delta\tau$, but its angle is shifted by

$$\frac{1}{2\pi}\arccos\left(1 - \frac{1 - \cos 2\pi\,\Delta\omega}{26[(\tilde{v}_i^{\mu})^2 + (\tilde{w}_i^{\mu})^2]}\right)$$

to maintain the same distance $[(1 - \cos 2\pi\,\Delta\omega)/13]^{1/2}$. Then the distance of i_1 from U at $T = \Delta\tau$ would be exactly $d \simeq 0.012566$. A similar procedure can be repeated for the initial condition i_2, and so on. Although there is nothing unique about our choice of initial conditions $i_1, i_2, \ldots$, the only justification is that the computed $k_n(\Delta\tau, U, d)$ are insensitive to $\Delta\omega$ (and hence d), as called for by condition (iii) of the section "kolmogorov Entropy". For instance, either by halving $\Delta\omega$ (=0.001) or reversing the direction of angle shift, the maximum deviation in $k_n(\Delta\tau, U, d)$ has been found to be less than 0.1%.

Next, to check conditions (i) and (ii) we have computed $k_n(\Delta\tau, U, d)$ under several values of $\Delta\tau$. One finds from Figure 23 that $k_n(\Delta\tau, U, d)$ settle down to constant values for all $\Delta\tau$. Hence, condition (i) is apparently satisfied. Yet, the limiting values $k(\Delta\tau, U, d)$ are not the same, and thereby violate condition (ii). However, this is inevitable because the entropy computation is possible only for certain values of $\Delta\tau$. That is, there are upper and lower limits for $\Delta\tau$. Note that the smallest time interval $\Delta\tau = 3$ in Figure 23 is about half the typical correlation time estimated from Figure 21(a). For $\Delta\tau < 3$, it has been found that some of the terms $\ln(d_j/d)$ become negative, so that $k_n(\Delta\tau, U, d)$ decreases steadily with increasing n. That is, nearby trajectories are actually being drawn in together, rather than breaking away from each other, in some time intervals. Since the decreasing $k_n(\Delta\tau, U, d)$ imply a non-random trajectory [32], we may identify $\Delta\tau = 3$ as a threshold time for the apparent trajectory instability. Now, as $\Delta\tau$ (>3) increases, so does $k(\Delta\tau, U, d)$, but at a gradually decelerating rate. For $\Delta\tau > 18$, however, one finds that a decreasing trend sets in, as indicated by k(21, U, d) in Figure 23. This is because the Euclidean metric (Equation 35) cannot discern different Riemann sheets, so that the trajectories with a shrinking distance may indeed be diverging from each other when the angle variation is correctly taken into account.

In conclusion, the violation of condition (ii) is inevitable. It must, however, be noted that the computed $k(\Delta\tau, U, d)$ also depends on the reference trajectory U. Since the spread of $k(\Delta\tau, U, d)$ in Figure 23 is well within its variation for different reference trajectories, we shall accept Figure 23 as numerical evidence for positive Kolmogorov entropy. Hence, D(13) may be said to be a K-system.

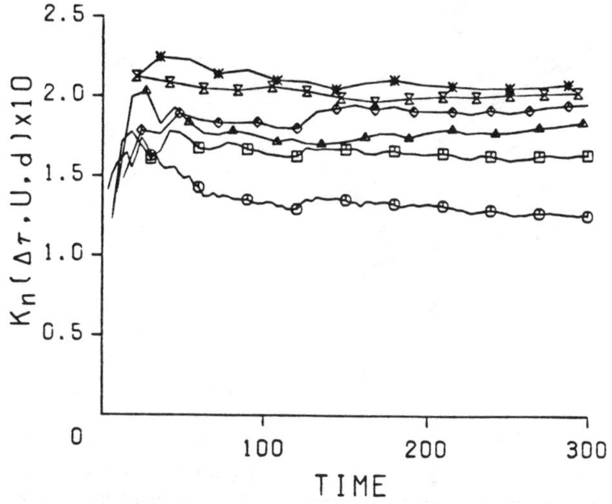

Figure 23. Kolmogorov entropy based on the reference trajectory D(13, I(13), 3π/8, 0, 300). ○, $\Delta\tau = 3$; ⊡, 6; △, 9; ◇, 12; ✱, 18; ⧖, 21.

TOWARDS MIXING AS THE TRUNCATION ORDER INCREASES

Because of trajectory instability, it was not possible to evolve the true trajectory of D(13) for $T \gg 110$ to provide a direct test for ergodicity and mixing. Yet, the limited trajectory evolution was a sufficient manifestation of mixing in that trajectory instability has further led to positive Kolmogorov entropy, which is a stronger dynamical characterization than mixing. Earlier the effective evolution time was adopted as a practical measure of limited evolution time. This has been motivated by the consideration that the effective evolution time is practically infinite in the absence of trajectory instability, whereas it is in the order of a typical characteristic correlation time when the system is mixing. However, it is well known [45] that in fully developed turbulence the characteristic time scale of a typical eddy decreases with decreasing eddy size (increasing wavenumber). One therefore expects that the effective evolution time should decrease with increasing N, if the truncated system D(N) exhibits enhanced mixing, as more and more triad interactions are included. To test this numerically we have carried out the forward-backward time integration of D(N) for five values of N beyond N = 13 to determine effective evolution time based on the $\pm 0.75\%$ recovery criterion set forth earlier. Figure 24 summarizes the results of forward-backward time integration under the same parameter $\xi = 3\pi/8$ and initial condition I(N) as in Figure 12. The effective evolution time under ABSERR $= 10^{-13}$ is denoted by o in Figure 24. The decreasing trend of the figure is a macroscopic indication of tendency towards mixing as increasingly many triad interactions are included in D(N). In the limit as $N \to \infty$, the effective evolution time will be in the order of the eddy-circulation time of the smallest eddy, which however decreases without bound in the inviscid case. Hence, the inviscid flow becomes virtually uncomputable. This has been demonstrated explicitly in a Burger's model [8].

Also included in Figure 24 are the effective evolution times of D(N) under less stringent error tolerance ABSERR $= 10^{-11}, 10^{-9}, 10^{-7}$. Two things are noteworthy. First, the effective evolution time falls off with increasing N under all ABSERRs. Secondly, for a fixed order of truncation, the effective evolution time decreases significantly with increasing ABSERR. Based on the estimate of Table 2, the computing cost can be more than halved by relaxing the ABSERR by two orders of magnitude. Therefore, although tempting to use a larger ABSERR for the sake of computing economy, one should be aware of the price paid for in terms of a smaller effective evolution time.

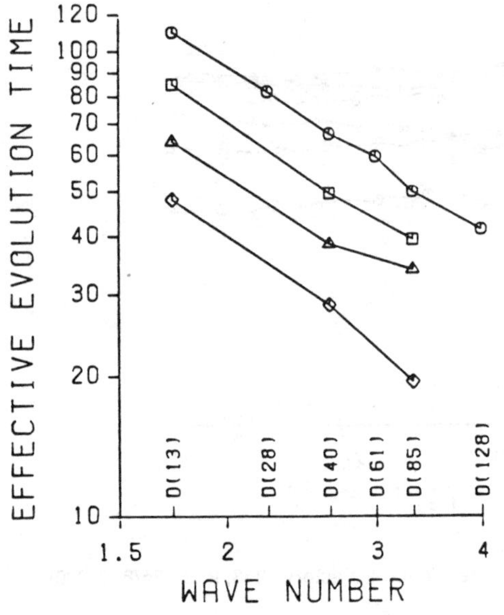

Figure 24. Effective evolution time determined from the forward-backward time integration of D(N) under $\xi = 3\pi/8$, and I(N). ○, ABSERR $= 10^{-13}$; □, 10^{-11}; △, 10^{-9}; ◇, 10^{-7}.

In conclusion, the effective evolution time decreases monotonically with increasing order of spherical truncation. Hence, the Navier-Stokes system develops mixing on the constant energy-helicity surface, as we have speculated from the consideration of isolating constants of motion.

NOTATION

E	energy	S	invariant sets
H	helicity	T	period
J_i^μ	action	t	time
k	wave vector	U	velocity
K	wave vector	ω_i^μ	angle
R_i^μ	amplitude	x, y, z	coordinate axes
ν	kinematic viscosity	ϵ^μ	polarization vector
$\rho(t)$	correlation factor	ξ	parameter in polarization vector
ϕ_{ij}	coupling coefficients	τ	time
ϕ_k^μ	coupling coefficients		

REFERENCES

1. Lee, J., *J. Math. Phys.*, 16, 1367, 1975.
2. Orszag, S. A., in *Fluid Dynamics*, (ed. R. Balian and J. L. Peube), p. 235, Gordon & Breach, 1977.
3. Birkhoff, G., and Fisher, J., *Rend. Circ. Mat.*, Palermo, 8, 77, 1959.
4. Robinson, G. D., *Quart. J. R. Met. Soc.*, 43, 409, 1967.
5. Lorenz, E. N., *Tellus*, 21, 289, 1969.
6. McLaughlin, J., *J. Stat. Phys.*, 15., 307, 1976.
7. Glaz, H. M., *Statistical Study of Approximations to Two Dimensional Inviscid Turbulence*, Lawrence Berkeley Lab. Rept. LBL-6708, 1977.
8. Lee, J., in *Nonlinear Phenomena in Mathematical Sciences*, (ed. V. Lakshmikantham), p. 629, Academic, 1982.
9. Liepmann, H. W., *Am. Scientist*, 67, 221, 1979.
10. Balescu, R., and Senatorski, A., *Ann. Phys.*, New York, 58, 587, 1970.
11. Herring, J. R., *Phys. Fluids*, 17, 859, 1974.
12. Lee, J., *J. Math. Phys.*, 16, 1359, 1975.
13. Moses, H. E., *SIAM J. Appl. Math.*, 21, 114, 1971.
14. Lee, J., *J. Fluid Mech.*, 120, 155, 1982.
15. Orszag, S. A., and Patterson, G. S., *Phys. Rev. Lett.*, 28, 76, 1972.
16. Landford III, O.E., *Ann. Rev. Fluid Mech.*, 14, 347, 1982.
17. Brissaud, A., Frisch, U., Leorat. J., Lesieur, M., and Mazure, A., *Phys. Fluids*, 16, 1366, 1973.
18. Kraichnan, R. H., *J. Fluid Mech.*, 59, 745, 1973.
19. Lee, T. D., *Quart. Appl. Math.*, 10, 69, 1952.
20. Nemytskii, V. V., and Stepanov, V. V., *Qualitative Theory of Differential Equations*, Princeton University Press, 1960.
21. Hemmer, P. C., *Dynamic and Stochastic Types of Motion in the Linear Chain*, Thesis, Trondheim, Norway, 1959.
22. Birkhoff, G. D., *Dynamical Systems*, Am. Math. Soc. Coll. Publ., Vol. 9., 1927.
23. Landford III, O.E., *Qualitative and Statistical Theory of Dissipative Systems*, Lecture Notes of 1976 CIME School of Statistical Mechanics, 1976.
24. Khinchin, A. I., *Mathematical Foundations of Statistical Mechanics*, Dover, 1949.
25. Kells, L. C., and Orszag, S. A., *Phys. Fluids*, 21, 162, 1978.
26. Arnold, V. I., and Avez, A., *Ergodic Problems of Classical Mechanics*, Benjamin, 1968.
27. Lebowitz, J. L., in *Statistical Mechanics: New Concepts, New Problems, New Applications*, (ed. S. A. Rice, K. F. Freed and J. C. Light), p. 41, University of Chicago Press, 1972.

28. Glaz, H. M., *Statistical Study of Approximations to Two-Dimensional Inviscid Turbulence*, Lawrence Berkeley Lab. Rept. LBL-6708, 1977.
29. Lee, J., *Phys. Fluids*, 25, 1480, 1982.
30. Moffatt, H. K., *J. Fluid Mech.*, 41, 435, 1970.
31. Sinai, Ya. G., in *The Boltzmann Equation—Theory and Application*, (Acta Phys. Austriaca, Suppl. X)., (ed. E. G. D. Cohen and W. Thirring), p. 575, Springer, 1973.
32. Benettin, G., Galgani, L., and Strelcyn, J. M., *Phys. Rev.*, A14, 2338, 1976.
33. Henon, M., and Heiles, C., *Astron. J.*, 69, 73, 1964.
34. Leith, C. E., *Ann. Rev. Fluid Mech.*, 10, 107, 1978.
35. Ruelle, D., *Ann. N.Y. Acad. Sci.*, 316, 408, 1979.
36. Benettin, G., Casartelli, M., Galgani, L., Giorgilli, A., and Strelcyn, J. M., *Nuovo Cim.*, 44B, 183, 1978.
37. Prigogine, I., *Astrophys. Space Sci.*, 65, 371, 1979.
38. Lee, J., *Phys. Fluids.*, 20, 1250, 1977.
39. Jackson, E. A., *J. Math. Phys.*, 4, 686, 1963.
40. Bellman, R., *Stability Theory of Differential Equations*, McGraw-Hill, 1953.
41. Hald, O. H., *Phys. Fluids*, 19, 914, 1976.
42. Lamb, H., *Higher Mechanics*, Cambridge University Press, 1943.
43. Lee, J., *Phys. Fluids*, 22, 40, 1979.
44. Shampine, L. F., and Gordon, M. K., *Computer Solution of Ordinary Differential Equations*, Freeman, 1975.
45. Monin, A. S., and Yaglom, A. M., *Statistical Fluid Mechanics*, Vol. 2, MIT Press, 1975.

CHAPTER 6

PRINCIPLES OF MICROMIXING

J. Baldyga

Instytut Inżynierii Chemicznej Politechniki Warszawskiej
PL 00-645, Warszawa, Polska

and

J. R. Bourne
Technisch-chemisches Laboratorium ETH
CH-8092 Zuerich, Switzerland

CONTENTS

NOTION AND DEFINITION OF MICROMIXING: INTRODUCTORY CONCEPTS

This chapter treats the effects of turbulent mixing of incompressible fluids in single phase isothermal systems on the course of homogeneous chemical reactions. Chemical reaction is essentially a molecular level process, and only mixing on the molecular scale can directly influence its course. For that reason any incorporation of chemical kinetic equations into mixing models can only be done locally, i.e., in those fine scale regions where homogeneity may be assumed.

Micromixing theory is concerned with all those features of mixing that cause the attainment of homogeneity on the molecular level. Let us consider the mixing of at least two streams of miscible fluids, which differ in some way. If one observed a complex mixing process, one could distinguish some simpler stages of mixing [1–4]:

- The process of spatially distributing blobs or lumps of one fluid through the other. The process is directed towards coarse scale homogenisation (attainment of average composition in all regions, without, however, local mixing)).
- Reduction of the scale of unmixed lumps by breakage and deformation.
- Mixing by molecular diffusion, whereby on the molecular scale a random (or so-called homogeneous) mixture results.

The particular events presented in this sequence can occur one after another or simultaneously, for instance, wandering and breaking apart of fluid elements in the process of turbulent dispersion, or viscous deformation with molecular diffusion in the viscous-diffusive subrange of the concentration fluctuation spectrum; but in any case molecular diffusion is the final step.

In this chapter, we are interested in turbulent mixing; the exact interpretation of such a kind of mixing from the point of view of fluid mechanics will be given subsequently.

It should also be pointed out that the perspective from which the mixing process is discussed (Eulerian or Lagrangian) is of the highest importance. In the classical, statistical theory of turbulent mixing, as presented in various review papers [2–8], mixing is discussed from the Eulerian or fixed frame perspective. The concepts of residence time distribution and micromixing, fundamental in the field of chemical reaction engineering, have been introduced [9–12] and developed, using the Larangian or material frame approach.

Unfortunately, neither of these descriptions of turbulent mixing is complete; their advantages and imperfections will be briefly discussed.

In attempts to understand and model the process of small scale mixing, presented subsequently in this chapter, the Lagrangian approach is used. The Lagrangian frame perspective is very convenient; following the history of a fluid element, we can identify and describe the elementary processes forming the overall micromixing process: engulfing, deformation, molecular diffusion, etc. Using residence time distribution functions, we are able to make predictions for a certain fluid element among all others in the system, or at least to find limits for various environments in the system. However, such an approach does not permit prediction of spatial distributions in the system, and that is really a disadvantage of the Largrangian formulation. If the Eulerian approach is applied, the situation is quite opposite. The averaging procedure, typical for turbulence descriptions, causes a loss of information about the local mixing and the age distribution. To solve the set of differential equations describing the process, additional information is necessary; this is called the closure problem. For systems without reaction, the turbulent diffusion concept is used and the spatial distribution can be calculated. Much more difficult is the closure approximation for the reaction term. There are many concepts of closure described [2–8]. The most important are: probability formulations (pdf) and moment closure (a good example for pdf closure is Toor's approach for 2-order, instantaneous, non-stoichiometric reaction—for moment closure Toor's invariance hypothesis), but none of them is general. Generally, it is impossible to predict theoretically the closure scheme, or even to use one scheme for another system (e.g., to model complex reactions when neither of them is slow based on the method for a single, fast reaction). Patterson has used the spiked distribution, obtained from a so-called interdiffusion model (similar to Miyawaki et al., [58]), to model moments and to use them later in the system of differential equations to model the real process; the modeling by Donaldson is very similar [7]. Quite primitive models are currently used for closure, thus more sophisticated models of local mixing should be applicable to formulate closure approximations. It is hoped the Lagrangian interpretation of the micromixing mechanism presented in this chapter will be useful in this context.

A basic concept in the Lagrangian approach to turbulent mixing is that of division of the whole mixing process into two subprocesses, namely macromixing and micromixing. The process of macromixing refers to those large-scale flow processes that cause the realization of large-scale distributions like the residence time distribution (RTD), or distribution of mean concentration in the Eulerian frame. Macromixing is clearly the first stage of mixing in the already presented sequence. Because the residence time distribution is identified as macromixing, the two other stages of mixing, as previously presented, form the micromixing process.

Micromixing is therefore a complex process of reduction of the size of large, unmixed eddies (i.e., erosive mixing [4, 15] or dissipation of the completely segregated zone [16–18]), as well as of molecular diffusion in deforming fluid elements [19–21]. In micromixing theory, the residence time distribution is treated as independent of the micromixing process; this means that molecular diffusion is very slow in comparison with the convective process. This assumption can be untrue in the case of mixing in gaseous systems, where molecular diffusion is much faster than in liquids. However, the opposite influence of RTD on molecular mixing is very strong and is considered as one of the most important branches of micromixing theory. RTD defines the environment surrounding the fluid elements during their residence time in the reactor [9, 10], or at least defines limitations for such environments and, as a consequence, the limits of molecular mixing.

The process of mixing with chemical reaction is usually considered for two kinds of single reactions: single-species reactions of the type A $\rightarrow$ products, and two-species reactions of the type A + bB $\rightarrow$ products, and for multiple reactions which are combinations of simple reactions. These reactions can occur in two kinds of systems: premixed feed systems and unpremixed feed systems. In premixed feed systems a single stream of reactants (uniform solution of A or uniform mixture of A and B for single-species and two-species reactions, respectively) is introduced into the reactor. The process of molecular mass exchange in such systems occurs between fluid elements of different age and thus different conversion and concentration of reactants. Such a process is called often self-mixing or back-mixing. When, however, two streams of different reactants are introduced into the reactor (unpremixed feed case), the process of mixing of different species occurs as well as back-mixing; mixing in this case is particularly important because mixing must precede chemical reaction. It should be noticed that also "in-between systems" are possible, if feeding streams are partially mixed, i.e., the system with premixed reactants diluted by a stream of non-reactive solvent or system fed by two streams with different stoichiometric ratios of premixed reactants.

Chemical reactions in premixed and unpremixed systems are very often classified from the viewpoint of competition between reaction and mixing. Due to this classification, the reaction can belong to one of the three groups:

- Instantaneous reactions [22], also called very rapid [6, 23] or fast [24]
- Fast reactions [22, 23], also known as in-between reactions [24], intermediate rate reactions [6], moderate rate reactions [8]
- Slow reactions [22–24], sometimes called very slow [6]

Although, there are some terminological complications (i.e., the first or second group of reactions can be called "fast"), the idea is clear. Because the interpretation of this classification is somewhat different for unpremixed feed and premixed feed systems, let us present them separately.

In unpremixed feed systems, the reaction is called instantaneous if it is so fast that it is totally controlled by mixing; slow reaction depends only on reaction kinetics and forms a second limiting case that is opposite to instantaneous reaction. The course of the fast reaction is dependent on both: mixing history and the nature of reactions. In the case of premixed feed systems, the reaction is considered to be instantaneous if it progresses so fast that the local change of concentration is caused only by local reactions and not by molecular mixing; such a system can be regarded as completely segregated. For slow reactions, there should be negligible progress of reaction during the time required to mix any fluid element on molecular scale with its environment. The rate of reaction becomes a function of mean concentration and perfect molecular mixing can be assumed. The reaction is fast, if the local changes of concentrations resulting from molecular mixing and reaction are of the same order of magnitude; the system is partially segregated in this case.

The interpretation for the "in-between systems" (feed streams partially mixed) is similar, although both methods of interpretation are often necessary. A good example of such a situation is the case of instantaneous reactions; in regions where two reactants are initially mixed on the molecular scale, so that reaction starts without molecular mixing the complete segregation may be assumed, and afterwards the system becomes diffusion controlled.

This classification of reactions in their mixing conditions has important consequences because the methods of reactor calculation are usually different for different groups. Numerical criteria are necessary. They are formulated by using particular micromixing models and will be given after the model description.

Finishing this introductory part of the chapter, we should touch upon the question of the importance of the subject and can look from two points of view: practical and theoretical. In practice, micromixing affects the course of several types of reactions. It is obvious from the definitions that the instantaneous and fast reactions are dependent upon mixing. For single reactions there is an influence on the time of reaction, conversion or reactor size; for multiple reactions there can be also a much more important influence on selectivity [25–29]. Mixing can change product properties and hence its quality. Such a case is present in precipitation process, where the size distribution of precipitating particles is strongly dependent on mixing [17, 30]. Also, the molecular weight distribution of polymer molecules, which determines the physical properties of polymer, is often controlled by mixing [31–35]. Further examples of processes dependent on mixing can be given: combustion of gaseous fuel, freezing or quenching of reaction composition, nitric oxide formation in combustors, reactions in highly viscous liquids, processes in chemical lasers and biological flow reactors, etc. From the theoretical point of view, micromixing investigations can supply new information about the mechanism of mixing, i.e., can be used to verify the proposed mechanisms of mixing. A good example is subsequently given, where calculations for three compared mechanisms of diffusion and reaction in shrinking laminae are confronted with experimental results [22].

MICROMIXING CRITERIA

To describe the degree of mixing, Danckwerts [11] introduced the concepts of a "point," "concentration at a point," and "age of the fluid at a point." Accordingly, the concentration and the age are averaged over a region small in comparison with the whole system, even small compared to the microscales of turbulence, but large enough to be independent of statistical fluctuations related to the molecular structure of the fluid, and thus much larger than any molecular length scale. Evidently, this definition is identical with the concepts of phenomenological limit and phenomenological point, i.e., with treating a fluid as a continuum. Using these concepts, Danckwerts [11] and Zwietering [10] were able to define two limiting cases of mixing: complete segregation when any mass exchange among points is excluded, and maximum mixedness when mass exchange is instantaneous among those points which, consistent with the limitations of the RTD, can exchange with each other. The completely segregated fluid is often called a "macrofluid" and the one in the state of maximum mixedness a "microfluid." Of course, in real mixing conditions, the intermediate state of partial segregation can often be expected. Now, we are interested in how existing criteria for mixing describe (or at least are related to) these states of micromixing.

Degree of Segregation Defined by "Age of the Fluid at a Point" [10, 11]

$$J = \frac{\text{var}\ \alpha_P}{\text{var}\ \alpha} = \frac{\overline{(\alpha_P - \bar{\alpha})^2}}{\overline{(\alpha - \bar{\alpha})^2}} \tag{1}$$

where α is the age of a molecule, $\bar{\alpha}$ is the mean age of all molecules that are at some particular moment in the system, and α_P is the mean age of the molecules at a point.

The variances are obtained by averaging over all molecules and all points in the system, respectively. The value of J is a measure of back-mixing in a continuous-flow system and may lie between J = 1 for completely segregated systems and a certain minimum value J_{min} for systems in a state of maximum mixedness. This minimum is only a function of RTD and equals 1 for a plug flow system and zero for ideal, exponential RTD. In practice, the J-value is not a very useful measure of the level of partial segregation, because it cannot be directly measured but must be computed using a micromixing model. The J_{min}-value can be used as a parameter characterizing RTD from the point of view of the possibilities for mixing molecules having different ages, or as one parameter characterizing back-mixing. It can be computed from the RTD function as follows:

$$J_{min} = (\text{var } \alpha_P)_{min}/\text{var } \alpha \tag{2}$$

$$(\text{var } \alpha_P)_{min} = \int_{\lambda_P=0}^{\infty} \left[\frac{1}{1 - F(\lambda_P)} \int_{\lambda_P}^{\infty} \{1 - F(S)\}\, dS - \bar{\alpha}\right]^2 \psi(\lambda_P)\, d\lambda_P \tag{3}$$

where $\psi(\lambda)$ is the life expectation frequency and equals:

$$\psi(\lambda) = \{1 - F(\lambda)\}/\tau \tag{4}$$

and the average age and the variance of molecules in the system are:

$$\bar{\alpha} = \int_0^{\infty} \alpha\phi(\alpha)\, d\alpha \tag{5}$$

$$\text{var } \alpha = \int_0^{\infty} (\alpha - \bar{\alpha})^2 \phi(\alpha)\, d\alpha \tag{6}$$

$\phi(\alpha)$ is the age distribution frequency:

$$\phi(\alpha) = \{1 - F(\alpha)\}/\tau \tag{7}$$

and F(t) is the cumulative RTD.

Scales of Segregation [36, 37]

Some information about the spatial structure of the concentration field is contained in the double Eulerian spatial correlation:

$$Q_{c,c}(r, t) = \overline{(c_i - \bar{c}_i)_\alpha (c_i - \bar{c}_i)_\beta} \tag{8}$$

where $c_{i,\alpha}$ and $c_{i,\beta}$ are the local, instantaneous concentrations of any substance i at the positions α and β, separated by a distance r from each other. Isotropic, homogeneous turbulence is considered for simplicity. The correlation coefficient is defined as follows:

$$R_{c,c}(r, t) = Q_{c,c}(r, t)/\overline{(c_i')^2} \tag{9}$$

where $\overline{(c_i')^2} = \overline{(c_i - \bar{c}_i)^2}$ (10)

is the variance of the local concentration c_i. The square root of the variance is called the root-mean-square concentration fluctuation:

$$\sqrt{\overline{(c_i')^2}} = \sqrt{\overline{(c_i - \bar{c}_i)^2}} \tag{11}$$

The correlation coefficient is a measurable function; using it we can find two scales:

Integral Scale for the Concentration Fluctuations

$$\Lambda_c = \int_0^\infty R_{c,c}(r, t)\, dr \tag{12}$$

This length is a measure of the largest concentration eddies and thus of the large scale break-up process.

Microsale for the Concentration Fluctuations

$$\left\{\frac{\partial^2 R_{c,c}(r, t)}{\partial r^2}\right\}_{r=0} = -(2/\lambda_S^2) \tag{13}$$

λ_S is a measure of the dimension of eddies which, at a given concentration variance, produce the same rate of dissipation of this variance as the considered turbulence. It should be cleared up that the last value is not the maximum dissipation scale as is stated sometimes in literature [4]. The dissipative eddy size for the concentration field in liquids (Sc ≫ 1) is given by Batchelor's microscale:

$$\lambda_B = (D^2 \nu/\epsilon)^{1/4} \tag{14}$$

and for gases, with Sc ≪ 1, by Corrsin's microscale:

$$\lambda_c = (D^3/\epsilon)^{1/4} \tag{15}$$

The microscales λ_B and λ_C are measures of the sizes of eddies in which most of the molecular mixing occurs.

Intensity of Segregation Defined by Concentration Variance

The concentration fractions $C_i = c_i/c_{i0}$ and $C_i = \overline{c_i}/\overline{c_{i0}}$ are used to define I_S [38]. This definition is valid for non-reacting A and B (also "in-between systems"):

$$I_S = \overline{(C'_A)^2}/(\overline{C_{A0}}\,\overline{C_{B0}}) = \overline{(C'_B)^2}/(\overline{C_{A0}}\,\overline{C_{B0}}) \tag{16}$$

If unpremixed feed system is considered (A and B initially unmixed):

$$I_S = (C'_A)^2/\{\overline{C_{A0}}(1 - \overline{C_{A0}})\} = \overline{(C'_B)^2}/\{\overline{C_{B0}}(1 - \overline{C_{B0}}')\} \tag{17}$$

More general definition, valid for both reacting and non-reacting systems, is cited by Hill [6]:

$$I \equiv (\overline{C'_A C'_B})/(\overline{C_A}\,\overline{C_B}) = 1 - (\overline{C_A C_B})/(\overline{C_A}\,\overline{C_B}) \tag{18}$$

I becomes I_S for non-reacting systems. The intensity of segregation I_S (or I) is the best measure of the degree of mixing at the molecular level. The value of I_S may lie between 1 for completely segregated fluid (A and B do not coexist anywhere) and zero for systems ideally mixed on molecular scale. Generally, I_S can vary in time and space. In such cases, the bar in Equations 16–18 means ensemble averaging; for a system at steady state time-averaged variables at a point can be used.

PROBLEMS OF MIXING ENVIRONMENT AND MIXING EARLINESS

Micromixing evidently consists of molecular diffusion within and among stretching fluid elements. Before trying to describe the mechanisms of this process, some important questions should be answered, namely, What is the environment of such fluid elements? What kind of fluid elements meet in a reactor and when?

Premixed Feed Systems

This problem was approached by Danckwerts [11] and Zwietering [10] 25 years ago, and is traditionally formulated using the RTD concept. To illustrate the problem let us consider a simple, numerical example. A second order, single-species reaction, having a rate given by:

$$R(c) = k_2c^2 \tag{19}$$

is occurring with premixed feed and with a RTD frequency function:

$$f(t) = \tfrac{1}{2}\delta\{t - \tfrac{2}{3}\tau\} + \tfrac{1}{2}\delta\{t - \tfrac{4}{3}\tau\} \tag{20}$$

where τ is the mean residence time, and $\delta(t)$ is the delta function (Figure 1A, B). All three reactors in Figure 1C–E reproduce the given RTD. There is piston flow in the reactors, perfect molecular mix-

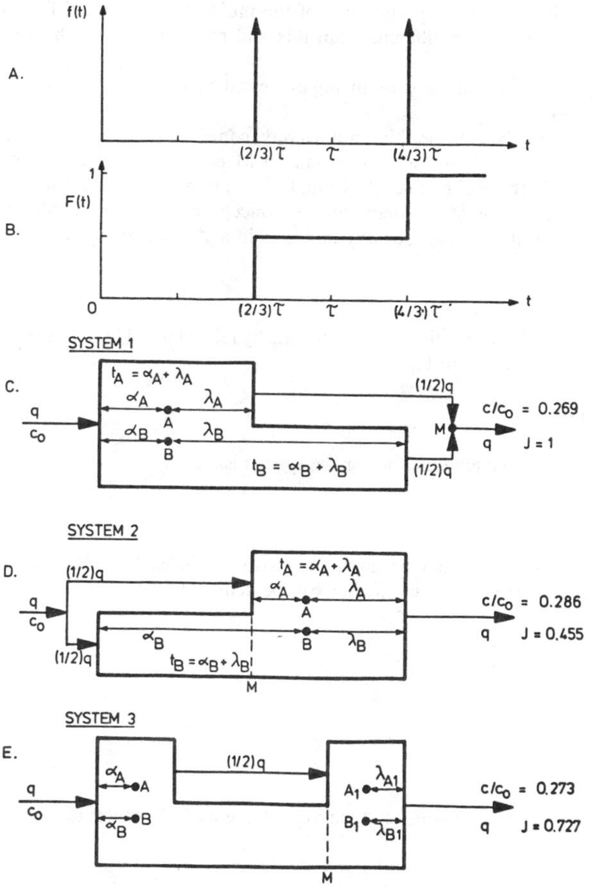

Figure 1. Example of non-ideal flow in reactors having premixed feed: (A) RTD frequency function; (B) RTD cumulative function; and (C–E) examples of reactors with this RTD.

ing in a vertical direction among elements from each horizontal layer is assumed, and the tubes outside have negligible volume. It is easy to imagine an infinite number of systems of this kind, reproducing a given RTD.

Now, let the concentration at the entrance to each reactor be c_0, and let $k_2c_0\tau = 3$. The calculated exit concentrations and the corresponding degrees of segregation J (Equation 1) are given in Figure 1. The conversions for these three systems are different and the higher conversion corresponds to the higher degree of segregation, J. It is clear from this example that the RTD by itself does not determine the state of mixing and that the state of mixing for a given RTD is somehow related to the degree of segregation J. What really distinguishes these systems is the direct environment of reacting molecules. In the first system, molecule A is surrounded by molecules that entered the system at the same instant. Because the age α of a molecule is defined as the time that has elapsed since it entered, one can say that any molecule in the first system is surrounded by others of the same age. Thus, two molecules A and B having different residence times t_A and t_B can meet only if they enter at the same moment ($\alpha_A = \alpha_B$) (Figure 1C).

In the second system, the environment of any molecule is composed of molecules that will leave the system at the same instant. The time a molecule will spend in the system from a specified time until it leaves is called the life expectation λ of the molecule (or residual lifetime). In this case, any A and B molecules having different t can mix and react only if they have equal λ ($\lambda_A = \lambda_B$) (Figure 1D).

In the third system, where the level of mixing expressed by J is moderate, both environments are possible (Figure 1E).

These differences may be expressed in another way, namely that the mixing of molecules leaving together is as late as possible in the first system (in M point) and as early as possible for given RTD in the second (on the broken line M). Again the situation in the third system is in-between (see the position of the broken line M). To describe these concepts in a general way, Zwietering considered α and λ distribution in the system. For any molecule in a steady state system:

$$\alpha + \lambda = t \tag{21}$$

and the age and residual lifetime distribution are directly related to RTD. The age frequency function for the system $\phi(\alpha)$ was found to be:

$$\phi(\alpha) = \{1 - F(\alpha)\}/\tau \tag{7}$$

and residual lifetime frequency function for the system is:

$$\psi(\lambda) = \{1 - F(\lambda)\}/\tau \tag{4}$$

The age and residual lifetime can be distributed also within each point. For each point P, the average age and the average residual lifetime can be defined by:

$$\alpha_P = \int_0^\infty \alpha\phi_P(\alpha)\,d\alpha \tag{22}$$

and:

$$\lambda_P = \int_0^\infty \lambda\psi_P(\lambda)\,d\lambda \tag{23}$$

when $\phi_P(\alpha)$ and $\psi_P(\lambda)$ are the frequency functions of age and life expectation distributions within points.

Using the functions above, it is possible to define precisely the limiting mixing states. The steady state system is completely segregated, if within each point the ages of molecules are all equal, and therefore equal to their mean value α_P, or:

$$\phi_P(\alpha) = \delta(\alpha - \alpha_P) \tag{24}$$

In the steady state premixed feed, completely segregated system, mixing among points of the same age is allowed; certainly it can change neither local concentration nor age distribution in a point. The variance between points var α_P is equal to the total variance in this case, and J = 1.

The system is said to be in a state of maximum mixedness (10), if:

1. All the molecules within each point have the same residual lifetime

$$\psi_P(\lambda) = \delta(\lambda - \lambda_P) \tag{25}$$

2. Points with equal mean residual lifetime λ_P are mixed or have identical age distributions:

$$\phi_P(\alpha) = \phi(\lambda_P, \alpha) \tag{26}$$

Zwietering has proved that these two conditions are sufficient to minimize the degree of segregation J. He modeled reactors in states of complete segregation and maximum mixedness using simple, piston flow systems. Figure 2 shows these models. Piston flow is assumed within a long tube with a large number of side exits (complete segregation) or side entrances (maximum mixedness). These side tubes are placed at small intervals and their volume is negligible. The flow through these tubes is controlled in such a way that a given RTD is reproduced. The exit concentrations for both systems can be obtained from material balances on both systems in Figure 2. The average exit concentration from the completely segregated system can be found from:

$$\overline{c_{out}} = \int_0^\infty c_{batch}(t)f(t)\,dt \tag{27}$$

where $c_{batch}(t)$ is the concentration in a completely segregated point in the reactor at time t, which is equal to that in a batch reactor after time t. For the maximum mixedness reactor, Zwietering [10] obtained the differential equation relating c with λ:

$$dc/d\lambda = R(c) + \frac{f(\lambda)}{1 - F(\lambda)}(c - c_0) \tag{28}$$

where R(c) is the reaction velocity as a λ function $\{$for $dc/dt = -R(c)\}$ we have $dc/d\lambda = R(c)\}$. The exit concentration is the c-value for $\lambda = 0$. The boundary condition for $\lambda \to \infty$ can be found from the condition:

$$\lim_{\lambda \to \infty} (dc/d\lambda) = 0 \tag{29}$$

The condition of Equation 29 results from the fact that c is bounded and positive. For reactors having a maximum finite residence time, such that $f(t) = 0$ for $t > t_{max}$, like in the reactor in Figure 1,

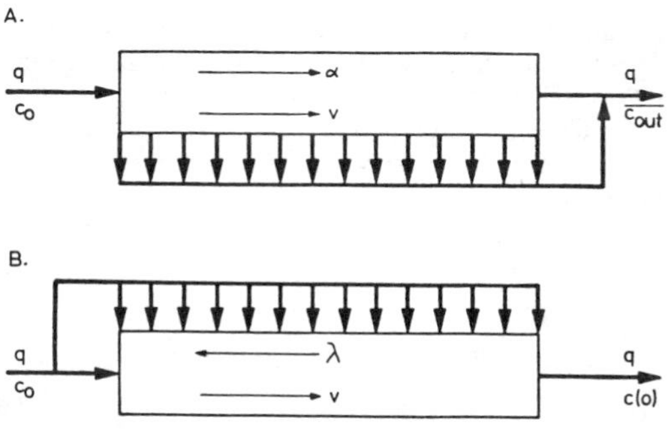

Figure 2. Limiting states for mixing [10]: (A) completely segregated system, (B) maximum mixedness system.

the boundary condition becomes:

$$c(\lambda = t_{max}) = c_0 \tag{30}$$

Conversions, thus calculated corresponding to micromixing limits, determine the upper and lower conversion bounds for simple, single-species reactions in premixed feed systems. This generalization was suggested by Danckwerts [11] and Zwietering [10], and proved afterwards by Chauhan, Bell, and Adler [39]. Let us consider the n-order reaction whose rate is described by a kinetic equation of the form:

$$R(c) = k_n c^n \tag{31}$$

For $n < 1$ micromixing accelerates reaction and complete segregation and maximum mixedness form for given RTD the lower and upper bounds on conversion, respectively. Differences in the level of molecular mixing do not influence the conversion of first order reaction ($n = 1$). For $n > 1$ complete segregation forms the upper, while maximum mixedness the lower bound of conversion. For a single reaction with a premixed feed, these limiting conversions are usually close to each other [10, 40–42]; thus precise modeling of intermediate mixing states is hardly necessary [13].

However, even in such simple systems the time needed to reduce the reactant concentration to a very small value (almost complete consumption) can be very different for these two limits. Generally, these differences become smaller as J_{min} (Equation 2) approaches 1. The influence of micromixing on complex reactions (selectivity, product quality etc.) can be much more important even in premixed feed systems [43–45]. However, the largest influence of micromixing on the course of reactions is observed in unpremixed feed systems.

Unpremixed Feed Systems

The two-species reaction $A + bB \rightarrow$ products is first considered. The concept of limiting states of mixing formulated by Danckwerts and Zwietering (Equations 24–26) can be easily adapted to unpremixed systems. However, there no longer exists any general proof that the limiting mixing states are related to conversion limits. The state of complete segregation may be interpreted in agreement with Danckwerts's [11] definition that any mass exchange among points is excluded (minimum species mixedness or complete species segregation). The conversion is then zero, and this is certainly the lower limit of yield. However, there is also a second state of mixing, defined by Treleaven and Tobgy [46], which fulfills Equation 24 and gives $J = 1$. This is complete segregation with respect to ages, but complete mixing with respect to species. Figure 3 illustrates these two concepts. Two feeds are introduced continuously to each reactor at flow rates q_1 for A and q_2 for B. Three RTD functions can then be considered: the overall RTD and one each for the A and B streams. They are related as follows:

$$(q_1 + q_2)f(t) = q_1 f_1(t) + q_2 f_2(t) \tag{32}$$

where f(t) is overall residence time function and $f_1(t)$ and $f_2(t)$ for A and B streams, respectively.

In the first system any contact of A with B is forbidden, in the second molecules of the same age are mixed. These are supposed to bound conversion [13], although this is still only a suggestion. The physical interpretation of system 3b is not clear; Treleaven and Tobgy [46] have given an example where the products of reaction are considerably more viscous than the reactants. The average exit concentration can be computed from Equation 33:

$$c_{out,A} = \int_0^\infty c_{batch,A}\{\overline{c_{A0}(t)}, \overline{c_{B0}(t)}, t\} f(t)\, dt \tag{33}$$

where: $$\overline{c_{A0}(t)} = \{q_1 f_1(t) c_{A0}\}/\{(q_1 + q_2) f(t)\} \tag{34}$$

and: $$\overline{c_{B0}(t)} = \{q_2 f_2(t) c_{B0}\}/\{(q_1 + q_2) f(t)\} \tag{35}$$

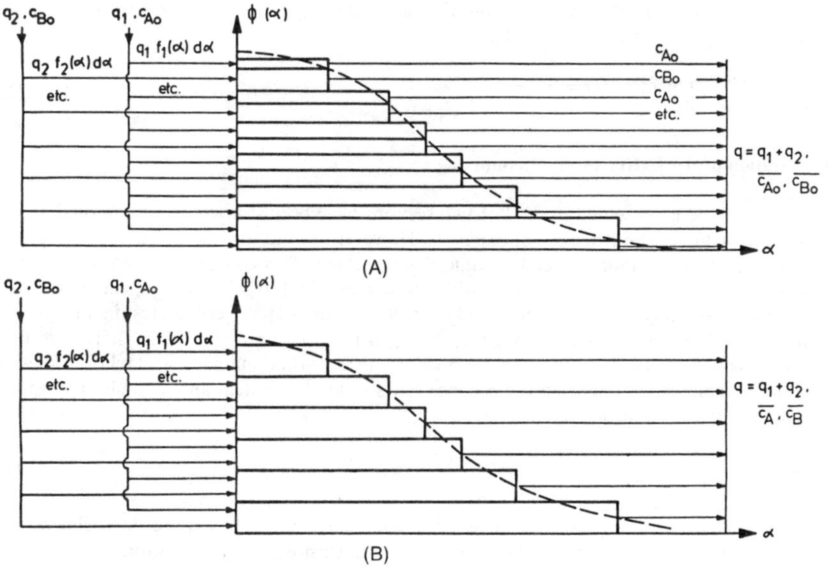

Figure 3. Complete segregation in unpremixed feed systems: (A) Complete segregation with respect to species, (B) Complete segregation with respect to ages, but complete mixing with respect to species.

$c_{batch,A}\{\overline{c_{A0}(t)}, \overline{c_{B0}(t)}, t\}$ is A concentration in a batch reactor with initial concentrations $\overline{c_{A0}(t)}$ and $\overline{c_{B0}(t)}$ after batch time t. The maximum species and age mixedness minimizes the degree of segregation J for overall RTD. In this case, all molecules of the same residual lifetime mix together as early as possible. The model is similar to that of Zwietering, with the appropriate modifications for the separate residence time distributions of both streams (Figure 4). The flow through side entrances is distributed in such a way that both RTD are modeled. The differential equations relating c_A and c_B with λ are similar to Equation 28:

$$dc_A(\lambda)/d\lambda = R_A + \frac{f(\lambda)c_A(\lambda) - f_1(\lambda)\overline{c_{A0}}}{1 - F(\lambda)} \tag{36}$$

$$dc_B(\lambda)/d\lambda = R_B + \frac{f(\lambda)c_B(\lambda) - f_2(\lambda)\overline{c_{B0}}}{1 - F(\lambda)} \tag{37}$$

where: $\overline{c_{A0}} = \{q_1/(q_1 + q_2)\}c_{A0}$ (38)

and: $\overline{c_{B0}} = \{q_2/(q_1 + q_2)\}c_{B0}$ (39)

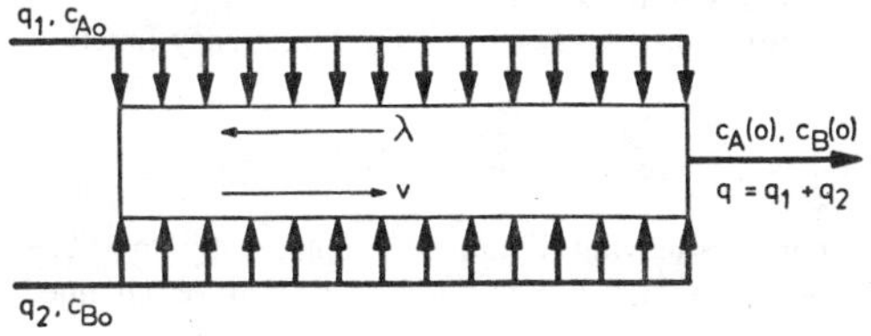

Figure 4. Maximum age and species mixedness in unpremixed feed systems.

$f(\lambda)$, $f_1(\lambda)$ and $f_2(\lambda)$ are related by Equation 32. The c_A and c_B concentrations must be of course bounded and positive, and thus normally:

$$\lim_{\lambda \to \infty} (dc_A/d\lambda) = \lim_{\lambda \to \infty} (dc_B/d\lambda) = 0 \tag{40}$$

Generalization of the Extremes of Micromixing

The considerations presented in the last two sections have been restricted to isothermal, steady state systems with constant density of reactants. However, during the last 15 years many papers have appeared, where a more general treatment is presented. This concerns micromixing in non-steady state systems [47–50] and systems with variable density [50]. The attainments in this field have been reviewed recently by Nauman [13] and Villermaux [4]. A very interesting and general treatment of the problem has been given by McCord [51]. Using a stochastic analysis of non-steady state, continuous flow systems, he has obtained an important mathematical relation between a conditional temporal distribution of residence times inside the system and a conditional spatial distribution of particles throughout the system:

$$f_{T|a}(t|s) = -\partial/\partial t \int_S f_{X|a}(s + t, x|s)\, dx \tag{41}$$

t is a fixed and T a random residence time; s is a fixed entrance time; S is the three-dimensional interior representing the system; x means a fixed three-dimensional position vector; dx denotes the product $dx_1\, dx_2\, dx_3$. X is the random three-dimensional position vector; $f_{T|a}(t|s)$ is the entrance residence time frequency function, and $f_{T|a}(t|s)\, dt$ equals the fraction of particles entering S during (s, s + ds) that will reside in S for a time between t and t + dt; $f_{X|a}(s + t, x|s)$ is the entrance spatial frequency function, and $f_{X|a}(s + t, x|s)\, dx$ the fraction of particles entering S during (s, s + ds) that are in (x, x + dx) at time t + s, t > 0.

The important conclusion drawn from Equation 41 is that for any spatial distribution $f_{X|a}$ only one residence time distribution function $f_{T|a}$ can be found. However, neither Equation 41 nor any other relation can be used to find the spatial distribution only from the knowledge of RTD. In fact, there is usually an infinite number of solutions of an integral equation; certain solutions can be interpreted as extremes of mixing.

McCord has found the concept of maximum mixedness to be unambiguous, but there are ambiguities associated with the concept of complete segregation. Therefore, McCord introduced two new concepts: minimum mixedness and maximum segregation. These limiting mixing states have been defined as follows:

Maximum Mixedness

For every u (u is fixed exit time), all particles leaving S during (u, u + du) that were in S at an earlier time t′ have remained throughout the time interval (t′, u) completely mixed together and completely segregated from all other particles. The mixing process in the state of maximum mixedness can be represented for each value of u by the following semibatch process (Figure 5a):

1. Process starts at $s = u - \lambda = -\infty$.
2. Particles are added at the rate $f_{T|b}(\lambda|u)$ at time $s = u - \lambda$, with material having intensive properties identical with feeding stream at time $s = u - \lambda$.
3. These particles are completely mixed. $f_{T|b}(t|u)$ is the exit residence time frequency function.

Minimum Mixedness

In the definition for maximum mixedness "particles leaving S" must be replaced by "particles entering S." All particles entering S during (s, s + ds), that are in S at a later time t′, have remained

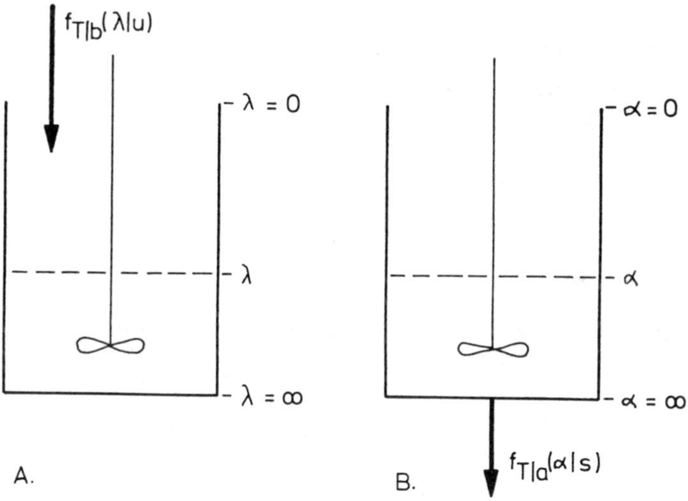

Figure 5. Limiting states for mixing [51]: (A) model for maximum mixedness, and (B) model for minimum mixedness.

throughout the time interval (s, t′) completely mixed together and completely segregated from all other particles. Minimum mixedness for each S can be represented by continual removing of particles at the rate $f_{T|a}(\alpha/s)$ at time $u = s + \alpha$ from the tank. The tank is full of material entering S during (s, s + ds) and during the process the particles are completely mixed (Figure 5b).

Maximum Segregation

Maximum segregation required no mixing within the system of particles entering the system either at different times or at different points on the entrance surface. A good representation of this model is streamline flow or a bundle of parallel tubes. In Figure 3, system A is in the state of maximum segregation, B in minimum mixedness.

The conversion for maximum mixedness can be computed from the system of differential equations, written for species i:

$$d(vc_i)/d\lambda = vR_i + \frac{f_{T|b}(\lambda|u)}{1 - F_{T|b}(\lambda|u)}\{vc_i - (v_a c_{ia})(u - \lambda)\} \tag{42}$$

where v is the specific volume, and $v(\lambda|u)$ and $c_i(\lambda|u)$ are unknown functions of λ for fixed u. The $v_a c_{ai}$ must have an average value specified for feed streams at the time $u - \lambda$. $F_{T|b}(\lambda|u)$ is a cumulative distribution:

$$F_{T|b}(\lambda|u) = 1 - \int_{\lambda}^{\infty} f_{T|b}(\lambda'|u)\, d\lambda' \tag{43}$$

The exit composition is given by $c_i(\lambda|u)$ at $\lambda = 0$; the boundary condition follows from:

$$\lim_{\lambda \to \infty} (d/d\lambda)(vc_i) = 0 \tag{44}$$

For minimum mixedness, a material balance for the system in Figure 5B should be made. The composition $c_i(\alpha|s)$ can be found from the system of equations:

$$(d/d\alpha)(vc_i) = -vR_i; \qquad c_i(0|s) = c_{0i}(s) \tag{45}$$

The average composition at time u can be found by integration of $c_i(\alpha, s)$:

$$(v_b c_{bi} g_b)(u) = \int_0^\infty (vc_i)(\alpha|u-\alpha) f_{T|a}(\alpha|u-\alpha) g_a(u-\alpha)\, d\alpha \tag{46}$$

where g_a and g_b are reactor entrance and exit rates, respectively.

Two-Environment Models

The idea of two-environment models consists in division of the reactor interior into two (or more) parts (three [57] or even four environment models for unpremixed feed reactors are possible [4]), being in the states of maximum mixedness and complete segregation. These zones can be placed in parallel [40, 52, 53] or in series [52, 54, 55] (Figure 6B and 6A, respectively). Usually the

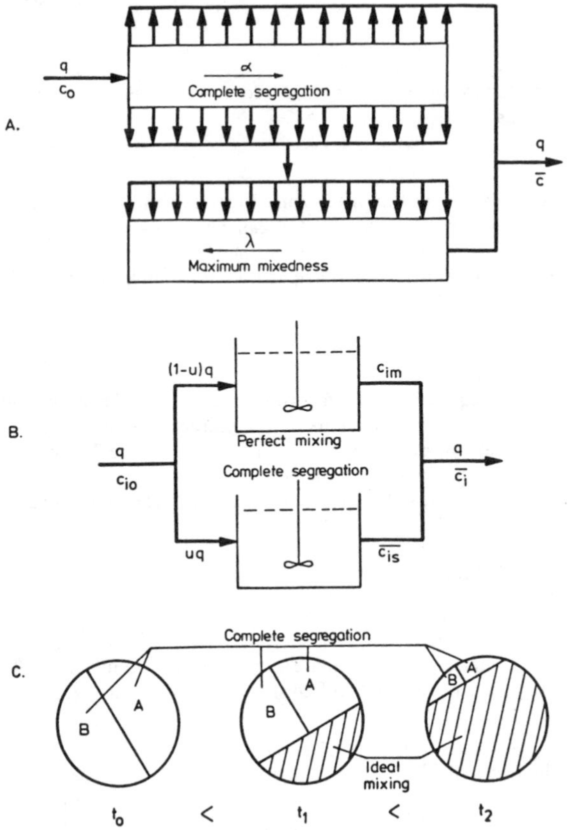

Figure 6. Examples of two-environment models: (A) Ng and Rippin [54], (B) Methot and Roy [40], and (C) Miyawaki, Tsuikawa, and Uraguchi [58].

environments are in series, with the region of maximum mixedness situated after a completely segregated region, forming entering and leaving environments. One parameter is used to describe the division of flow in parallel model, or mass transfer from entering to leaving environment in the case of consecutive composition. Such a parameter is usually identified experimentally. The limiting values of this parameter correspond to the situation when only one environment exists in the system.

For instance, in the case of the Ng and Rippin [54] model, the transfer from the entering to the leaving environment is described by:

$$-(dm/d\alpha) = mR_N \tag{47}$$

where m is an amount of fluid in the entering environment. For $R_N = 0$ there is complete segregation, and, for $R_N \to \infty$, the maximum mixedness (Figure 6A).

In the second example, for $u = 0$, the system is well mixed on molecular scale and for $u = 1$ completely segregated.

It should be noticed that u is identical with an empirical degree of segregation, which can easily be estimated when conversion is measured [40, 56, 4] (Figure 6B):

$$u = (\overline{c_i} - c_{im})/(\overline{c_{is}} - c_{im}) \tag{48}$$

where $\overline{c_i}$ is an exit value of concentration (measured), c_{im} the concentration in the system when well-mixed on molecular scale, and $\overline{c_{is}}$ is the concentration in the system when well-mixed on macro-scale but completely segregated.

In Figure 6c an example is presented where the decay of the completely segregated region is treated locally, as a process occurring in a small lump of fluid blending in the system [58].

As we already know from the discussion of Equation 41, an infinite number of such models can be proposed for any given RTD (except the plug flow reactor of course). The concept of two-environments was formally generalized and discussed by Nishimura and Matsubara [55]. They introduced a segregation function $s(\alpha, \lambda)$ which is the fraction of fluid of age α and residual lifetime λ in the completely segregated environment. $s(\alpha, \lambda)$ determines non-ambiguously the composition of two-environment models and mass transfer between enviroments. Each function $s(\alpha, \lambda)$ can be the segregation function, if:

$$0 \le \alpha < \infty; \qquad 0 \le \lambda < \infty$$

$$0 \le s(\alpha, \lambda) \le 1$$

$$s(0, \lambda) = 1$$

$$s(\alpha + \Delta t, \lambda - \Delta t) \le s(\alpha, \lambda) \qquad \text{for} \quad 0 < \Delta t \le \lambda$$

As we can see from Figure 6B, the third condition is not always fulfilled.

The concept of two-environments is the basis of two groups of models:

1. In the first group perfect molecular mixing is assumed among molecules of the same residual lifetime in the leaving environment. Thus, any lump of fluid is losing its identity immediately, at the moment of arrival in the environment [52–56]. The name "two-environment model" is usually used for this group.
2. In the second group of models, it is assumed that any lump of fluid is losing its identity slowly due to a certain mechanism of mixing. The environment of such a lump of fluid at any instant is determined by the two-environment model. In the entering environment any lump of fluid exchanges mass with the others of the same age and in the leaving environment with lumps of fluid of the same residual lifetime. The process of local mass exchange is usually simulated with the use of the coalescence-redispersion model [59–63]. However, any other mechanism of local mass exchange can be adopted; e.g. Villermaux [4] has proposed the use of the IEM

model (Interaction by Exchange with the Mean). Of course, any diffusional model, where mixing is modeled as molecular diffusion within certain fluid aggregates or lumps of fluid, can be used as well.

MICROMIXING IN THE LIGHT OF TURBULENCE THEORY

The aim here is to identify the key physical processes contributing to mixing on the molecular scale using information about turbulent flows available in the fluid mechanics literature. Such information will be used subsequently to interpret existing mechanistic models and to construct an improved mathematical model of micromixing.

Liquid mixtures are considered, where the diffusivity D is much smaller than the kinetic viscosity (i.e. $Sc = \nu/D \gg 1$). The system is assumed to be isothermal and very dilute solutions are considered, so that physical properties like density, viscosity, or diffusivity as well as chemical rate constant are not influenced by reaction. (These assumptions are not essential, but they simplify matters). At the small scales where micromixing occurs, turbulence will be considered, following Kolmogoroff, to be locally isotropic and homogeneous.

Concentration Spectra

The variance of the local instantaneous concentration c_i of any substance i, defined by Equation 10, characterizes the level of the unmixedness on the molecular scale. The double correlation defined by Equation 8 contains some information about the spatial structure of the concentration field; this information is more legible when the Fourier transformation of Q is used:

$$G(k, t) = 2/\pi \int_0^\infty Q_{c,c}(r, t) kr \sin(kr)\, dr \tag{49}$$

The Fourier transformation expresses the three-dimensional spectrum of the concentration fluctuations c_i' among eddies of various sizes. The spectral density G(k, t) is expressed as a function of the wave number k, which is proportional to the reciprocal of eddy size. The variance $\overline{(c_i')^2}$ is related to G through:

$$\overline{(c_i')^2} = \int_0^\infty G(k, t)\, dk \tag{50}$$

and thus contains contributions to unmixedness from eddies of all sizes. The spectral density of the concentration fluctuation distribution can, for liquid mixtures, be divided into three regions; the inertial-convective, the viscous-convective and the viscous-diffusive subranges [36, 37, 64].

The inertial-convective subrange of wave numbers characterizes eddies, whose size varies approximately from the largest scale encountered (k_{0c}^{-1}) to the Kolmogoroff scale $(k_K^{-1}) = \lambda_K = (\nu^3/\epsilon)^{1/4}$. The small scale flows are inertial and the spectral mass transfer, leading to mixing, is convective, while molecular diffusion plays no role. The spectral density is given by:

$$G(k, t) = Ba\epsilon_c(t)\epsilon(t)^{-1/3}k^{-5/3} \tag{51}$$

where $\epsilon(t)$ is the rate of dissipation of kinetic energy per unit mass, $\epsilon_c(t)$ is the rate of dissipation of concentration variance, and Ba (Batchelor constant) has been measured and is roughly 0.4 [36, 64].

Kinetic energy and concentration variance are neither created nor destroyed in this subrange, but in the course of time concentrated blobs of fluid are deformed and broken apart by fluid motions, so that their scale of segregation is reduced without simultaneous molecular mixing (without reduction of I_s). This is demonstrated in Figure 7 where, for $Sc \gg 1$, the spectra of the variance and its dissipation are shown. The variance in this subrange is estimated by setting the upper limit of Equation 50 equal to $\sim k_K$. As seen in Figure 7, most of the total variance is contained within

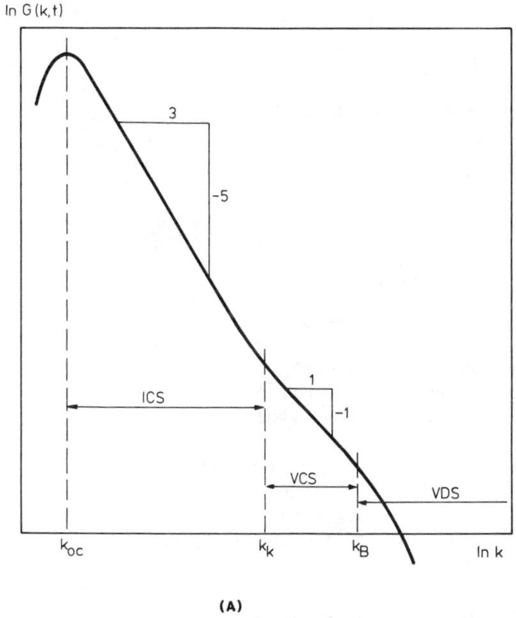

Figure 7A. Concentration spectrum for liquid mixtures having Sc ≫ 1. ICS = inertial-convective subrange; VCS = viscous-convective subrange; and VDS = viscous-diffusive subrange.

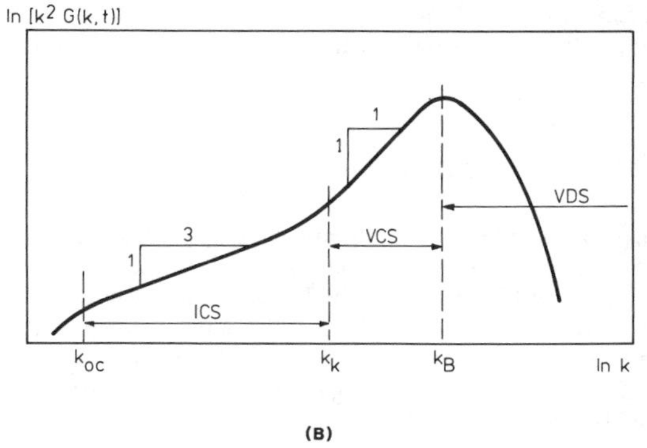

Figure 7B. Spectrum of dissipation of concentration variance. ICS = inertial-convective subrange, VCS = viscous-convective subrange, and VDS = viscous-diffusive subrange.

the inertial-convective subrange. Thus, in the language of micromixing theory, the eddies in this subrange are practically completely segregated.

The viscous-convective subrange contains smaller eddies and extends approximately from k_K up to k_B, where k_B (Batchelor wave number) is given by:

$$k_B = \lambda_B^{-1} = (\epsilon/\nu D^2)^{1/4} \tag{52}$$

Eddies are subjected to laminar strain, which depends upon viscosity, so that their scale is further reduced by viscous deformation, whilst in principle at least molecular diffusion is not yet active (it becomes the controlling mechanism when $k > k_B$).

The viscous-diffusive subrange ($k > k_B$) starts when laminar strain and diffusion become of equal importance for spectral transfer (i.e. when $k \simeq k_B$), whilst for still higher wave numbers molecular diffusion rapidly dissipates concentration variance (Figure 7). Equation 53, due to Batchelor [65]:

$$G(k, t) = C\epsilon_c(t)\{\nu/\epsilon(t)\}^{1/2}k^{-1}\exp(-Ck^2/k_B^2) \tag{53}$$

applies to the viscous-convective ($k < k_B$) as well as to the viscous-diffusive ($k > k_B$) subranges. The kinematic viscosity ν also appears and influences k_B. The diffusivity also enters k_B via Equation 52. The constant C in Equation 53 relates the time-average rate of strain γ to ϵ and ν through:

$$\gamma = -C^{-1}(\epsilon/\nu)^{1/2} \tag{54}$$

It has been determined experimentally and falls in the range 1 to 5, whilst the value 2 is often given [36, 37, 64].

The implications of this spectral information for micromixing now follow:

1. Micromixing, which proceeds by molecular diffusion, does not occur in the inertial-convective subrange, where only the scale of segregation is reduced. In any mixing process, time and mixer volume are required to reduce the scale of segregation as eddies pass through this subrange. Attempts to predict the completely segregated volume in a reactor are discussed subsequently.
2. Mixing by molecular diffusion dominates the viscous-diffusive subrange.
3. The viscous-convective subrange (Equation 53) is insensitive to diffusion when $k \simeq k_K$, whilst when $k \simeq k_B$ molecular diffusion is important. Laminar deformation concentrates material so that steep concentration gradients arise where a fluid element is thin and generates mixing by diffusion.

Any model of micromixing should include the viscous-convective and viscous-diffusive subranges. The size below which diffusion becomes important is on the order of λ_K, but any error in estimating this size by λ_K should be small since in Equation 53 diffusion is less important than convection at this scale. The error in initial size estimation will influence only the time necessary to start molecular diffusion.

Deformation Kinematics

Fluid deformation in the viscous-convective subrange has been modelled by specific kinematic schemes, namely elongation [19, 66–68] and one-dimensional shear [20], as well as by a non-mechanistic, statistical relationship, derived from the theory of initial, relative turbulent diffusion [69]. Generally, if we want to follow the history of a lump of fluid in a turbulent flow (lump of fluid consists a large number of particles), we may study separately the movement of the centers of mass of the lump and the relative movement of the fluid particles forming the lump [66]. The first is certainly macromixing, and is usually described in fluid mechanics using the equation of turbulent diffusion in the Eulerian frame.

The relative movement of particles resulting from the deformation and changing the shape of these lumps requires a Lagrangian analysis. The particular aspect of the relative motion needed here is how this diffusion proceeds in time for two points initially separated by a distance on the order of Kolmogoroff's microscale. For such a separation distance, one usually assumes that the rate of strain at all positions between the two points is the same, and deformation is viscous. This assumption is supported by interpretation of the spectra of kinetic energy and the rate of dissipation of kinetic energy (Figure 8). Both spectral functions become small in the region near Kolmogoroff's wave number. This means that the velocity fluctuations are not very active in this region {small E(k)} and the flow in this region can be interpreted as laminar. Because the scales of concentration fluctuations of our interest are smaller than those of velocity fluctuations, the velocity fluctuations will simply convect concentration eddies without any direct influence on deformation. The implication of this remark for micromixing modeling is very important; the processes of viscous deformation, molecular diffusion and chemical reaction in the Lagrangian frame can be described in a deterministic way without introduction of fluctuating values.

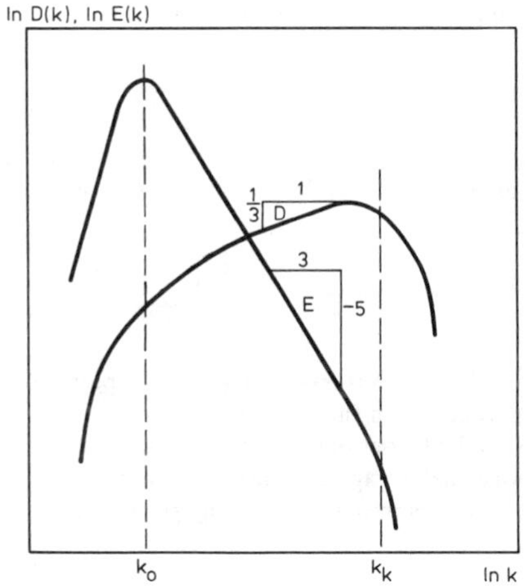

Figure 8. The kinetic energy E and dissipation of kinetic energy D spectra (schematic).

It has been determined that in Cartesian coordinates, which rotate and translate with the fluid, intensive elongation occurs in the direction of the local vortex line with rapid thinning in the second direction (denoted by x) and only a moderate increase of dimension in the third direction (Figure 9) [66, 70, 71]. As the thickness decreases, the concentration gradient in the x-direction greatly exceeds the gradients in the other two directions. Thus, mixing by molecular diffusion rapidly becomes one-dimensional. When x is measured from the center line of the thinning fluid, the velocity u at any distance x is directed towards the center line (i.e. the fluid is thinning), and is proportional to x [36, 71, 66], thus:

$$u = (x/\delta)\, d\delta/dt = -\Psi x \tag{55}$$

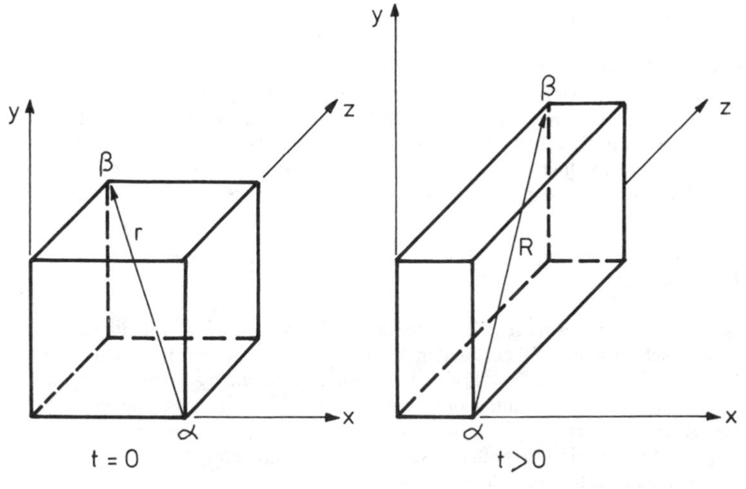

Figure 9. Initial deformation of fluid element in viscous subrange.

The Ψ-value is dependent on the specific kinematic scheme. A limiting solution for a sufficiently long time:

$$t \gg (\nu/\epsilon)^{1/2} \tag{56}$$

assuming pure elongation-compression kinematics, has been obtained by Batchelor [66], giving:

$$\delta/\delta_0 = \exp\{-0.55(\epsilon/\nu)^{1/2}t\} \tag{57}$$

and thus: $\Psi = 0.55(\epsilon/\nu)^{1/2}$ (58)

However, if the process of molecular mixing with chemical reaction is modeled, the initial period of deformation when the reaction in the unpremixed feed system is in the diffusional regime, is much more important. Thus, the fluid deformation derived from the theory of initial relative turbulent diffusion has been applied [36, 69]. According to this theory, the average distance between two points R, whose initial separation r was small relative to λ_K, increases initially according to [36, 69]:

$$R^2 = r^2 + \epsilon r^2 t^2/3\nu \tag{59}$$

Using this equation, the thinning of the slab is described by [69]:

$$\delta/\delta_0 = \sqrt{1 + 0.5(\epsilon/\nu)t^2 - \sqrt{\{1 + 0.5(\epsilon/\nu)t^2\}^2 - 1}} \tag{60}$$

and:

$$\Psi = (\epsilon/\nu)^{1/2}/(4 + \epsilon t^2/\nu)^{1/2} \tag{61}$$

Also, the shearing formulation, sometimes used to describe micromixing [20, 72], can be put in terms of ϵ, ν and t to give:

$$\delta/\delta_0 = \{1 + (0.5\{\epsilon/\nu\}^{1/2}t)^2\}^{-1/2} \tag{62}$$

and:

$$\Psi = 0.25(\epsilon/\nu)t/\{1 + 0.25(\epsilon/\nu)t^2\} \tag{63}$$

However, such a description of kinematics, assuming only pure shearing (without elongation), is not very much related to known mechanisms of turbulent mixing. Figure 10 compares the thinning of fluid elements according to these three formulations, whereby Equation 57 is invalid for short times. The range of validity of Equation 59 and thus of Equation 61 is unknown, although it is clear that $(\epsilon/\nu)^{1/2}t$ should not be too high. Fortunately, t does not need to be too long in micromixing modeling; in accordance with subsequent discussion the limiting values equals approximately $12(\nu/\epsilon)^{1/2}$ (Equation 68).

Zones of Micromixing

Consider a thin layer of a reagent B located in infinite volume of surrounding turbulent A-rich fluid. The sandwich structure of B enclosed in A will be deformed, so that B will tend to be convected inwards towards the center line x = 1, whilst simultaneously tending to diffuse outwards into the A-solution. Of course, A will simultaneously convect and diffuse inwards. The net movement of B, represented by its concentration profile, is of interest in relation to the description of the penetration zone of reagents; this description is often based on arbitrary assumptions. The concentration profile of B has been calculated for the following case:

$$A + B \xrightarrow{k_1} R \qquad R + B \xrightarrow{k_2} S$$

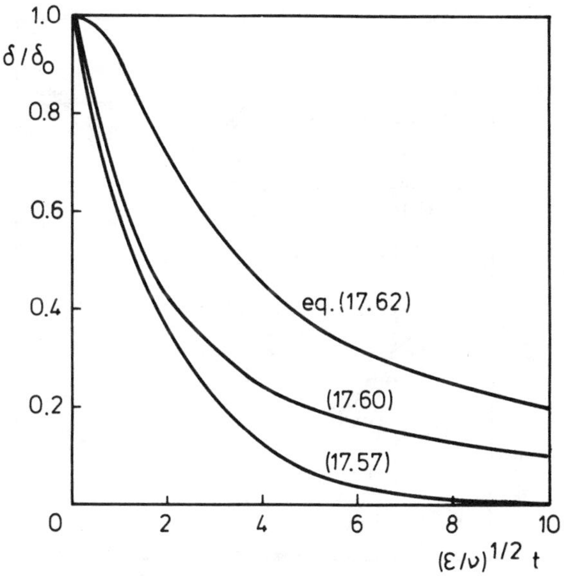

Figure 10. The change of the thickness of small fluid elements as predicted by three models of fluid deformation.

Batch System

Initial concentrations: $c_{B0}/c_{A0} = 100/1.05$

Initial thickness of B-rich fluid layer: $2\delta_0 = \lambda_K$

Volume of surrounding A-rich fluid: Infinite

Model

Diffusion-convection-reaction formulation based on Equation 64:

$$(\partial c_i/\partial t) + u(\partial c_i/\partial x) = D_i(\partial^2 c_i/\partial x^2) + R_i \qquad (64)$$

where u is given by Equations 55 and 61. Kinetics are second order, irreversible.

Conditions

No restrictions for molecular diffusion. Mixing modulus $M = k_2 c_{B0}\delta_0^2/D = 100$ $k_1 \gg k_2$—first reaction instantaneous. Results were obtained by numerical solution of Equation 64 for i = A, B and R.

Figure 11A gives concentration profiles of B at various times ($T = Dt/\delta_0^2$), where the abscissa is normalized relative to the current half-thickness δ. Figure 11B uses normalization relative to the initial half-thickness δ_0. Figure 11A shows that when T = 0.03 only 4.4% of B has reacted and yet over 50% of B has diffused out of the initial layer. Figure 11B indicates that in the viscous convective

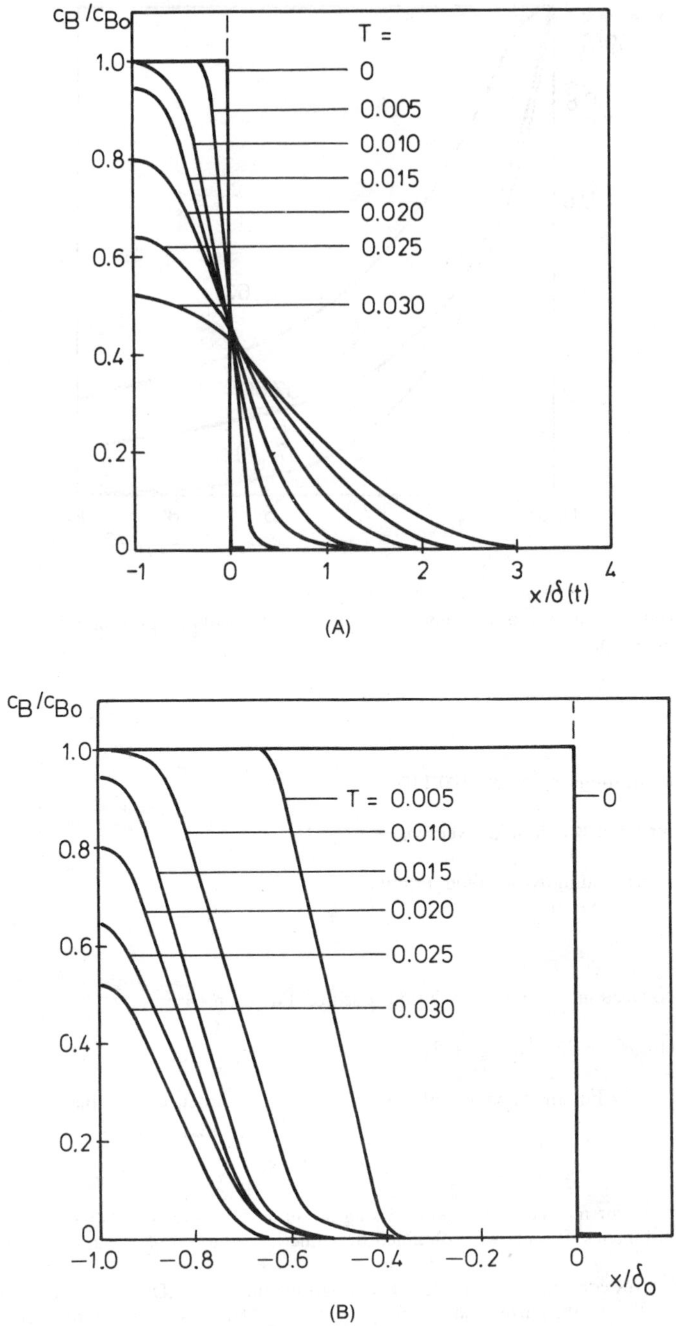

Figure 11. Concentration profiles of B normalised with respect to: (A) current half-thickness of laminae, and (B) initial half-thickness of laminae.

subrange convection (thinning) initially dominates, while later diffusion nearly balances convection. After a long time, the layer forms a long, elongated ribbon with reagent B escaping from the original B-rich element. The ribbon can be divided into smaller parts, which can be dispersed over the whole system, but because of the straining action the concentration spots keep their identity.

There is no evidence for the assumption that B remains within the initial layer [20, 67, 72], although it only penetrates a short distance into neighboring A-solution. Thus, small fluid elements initially mix by mass exchange with their immediate surroundings, not with the average content of their whole surroundings. These findings for a reactive system agree with turbulence theory, particularly with the concept of Batchelor's microscale of turbulence λ_B.

The Role of Vorticity

Turbulent flow is rotational and dissipative, and is characterized by high levels of fluctuating vorticity. (Vorticity is the curl or rotation of the velocity vector). Fluid deformation causes vortices to stretch and vorticity and kinetic energy to be transported from larger to smaller eddies [37]. The nature of the energetically active, small scale motions in a fluid influences the regions within and between which micromixing occurs; they are therefore considered here.

Vorticity is distributed over eddies of various sizes, whereby in the inertial subrange there is a steady increase with decreasing eddy size. As, however, the Kolmogoroff scale λ_K is approached, viscous forces rapidly become significant and cause a large vorticity decrease. Tennekes and Lumley [37] have demonstrated that the vorticity dissipation scale is proportional to Kolmogoroff's microscale λ_K.

The mean-square vorticity fluctuations would decay exponentially, if a mechanism were to exist for stopping the uptake of energy by strain from the rest of the fluid and if the production terms were to be removed from the vorticity budget. It can be shown that the time constant in this process is proportional to the Kolmogoroff time constant $(\nu/\epsilon)^{1/2}$ [73]. For a steady-state system production equals dissipation; therefore, the characteristic frequency of vorticity production is proportional to $(\epsilon/\nu)^{1/2}$.

These considerations are related to mean-square vorticity fluctuations. However, we are interested not in the whole spectrum of eddies, but in the most active eddies with highest vorticity; these eddies are responsible for a small scale engulfing process, which is important in mixing on the molecular scale. The characteristic time constant for an eddy of given size τ_W is known as the time scale of return to an isotropic state. It would then wait in this state until again being set in rotation by the straining action of its surroundings. (Such small scale motions have an intermittent character and vorticity is not homogeneously distributed but shows a spottiness—this is a very important feature of the structure of a turbulent field). The distribution of τ_W for a given eddy size does not seem to be known. In the steady state, where the level of vorticity is constant and energy transfer randomly initiates vortices, which then gradually deform and decay, in agreement with the remark for the mean-square vorticity fluctuations, the mean lifetime for a given eddy size should be close to τ_W. (Lifetime here means the period over which a vortex loses its hydrodynamical identity and returns to isotropy. The implication with respect to mass transfer and mixing will be discussed in the next section). τ_W as a function of eddy size [37] is given by:

$$\tau_W = 2\pi/w = 2\pi/\{k^3E(k)\}^{1/2} \tag{65}$$

where E(k) is the kinetic energy spectral density. The spectrum derived independently by Corrsin and Pao is:

$$E(k) = \alpha\epsilon^{2/3}k^{-5/3}\exp\{-1.5\alpha(k\lambda_K)^{4/3}\} \tag{66}$$

and describes the inertial subrange ($k < \lambda_K^{-1}$) and probably also at least the initial part of the viscous subrange ($k > \lambda_K^{-1}$). The constant α, obtained experimentally, is about 1.5 [36, 37, 64].

Substituting for E(k) from Equation 66 into Equation 65 and differentiating with respect to k, the minimum occurs when $k = 0.544\lambda_K^{-1}$. This corresponds to an eddy size $l = 2\pi/k$ of $11.5\lambda_K$. (The position of minimum τ_W was given on p. 278 of [37] as $k = \lambda_K^{-1}$, but this result appears to be in error). Thus, the size of the most active eddy l' and its mean lifetime are given by:

$$l' \simeq 12\lambda_K \simeq 12(\nu^3/\epsilon)^{1/4} \tag{67}$$

$$\tau' \simeq 12.7(\nu/\epsilon)^{1/2} \tag{68}$$

The variation of mean lifetime with eddy size, calculated from Equations 65 and 66 for small eddies, is shown in Figure 12. The vorticity of eddies larger than $12\lambda_K$ is increased by the straining action and energy transfer from still larger eddies, whilst the vorticity of eddies smaller than $12\lambda_K$ is decreased by viscosity.

The estimate given in Equation 67 for the size of the vortex, which is active in energy dissipation, is supported by some experimental results. Burst of vorticity caused spotty dissipation in structures not unlike cylinders, whose mean diameter in a particular experiment was 0.2 in., whilst at the same time λ_K was 0.0134 in. [70]. The ratio of these lengths is 15. When following the diffusion of heat spots in turbulent air, the regions over which laminar strain and dissipation occurred were as large as $15\lambda_K$ [71]. It seems that the ratio 12 in Equation 67 is an acceptable estimate.

Eddies of size $12\lambda_K$ form an important part of the fine, dissipating structure. They can engulf fluid, so that partially segregated laminae are formed within vortices (Figure 13). Furthermore because $D \ll \nu$ (i.e., $Sc \gg 1$), the fine concentration structure persists and deformation, diffusion and reaction at scales $\ll \lambda_K$ can continue after the vortex has died out. After a time of order τ', a vortex structure will, however, be rebuilt, incorporating fluid from the local environment, and diffusion and reaction will again follow. The process of reconstituting eddies with a mean frequency τ'^{-1} continues as a cascade and, after sufficient repetitions, diffusion will have completely homogenized the system.

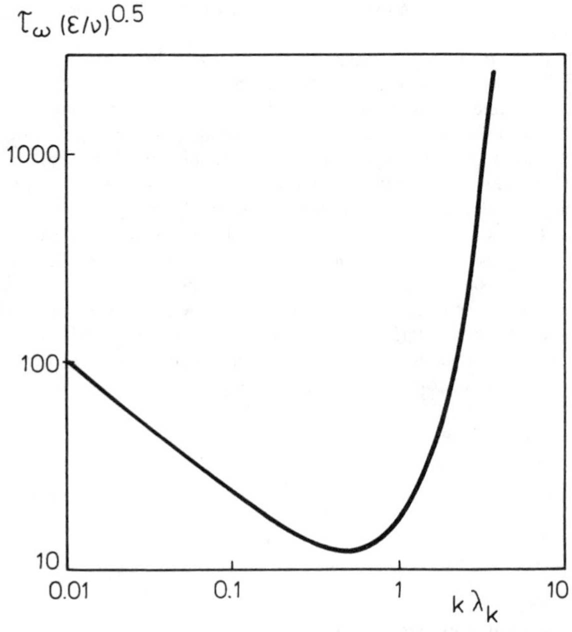

Figure 12. Lifetime of vortices as a function of their wave number.

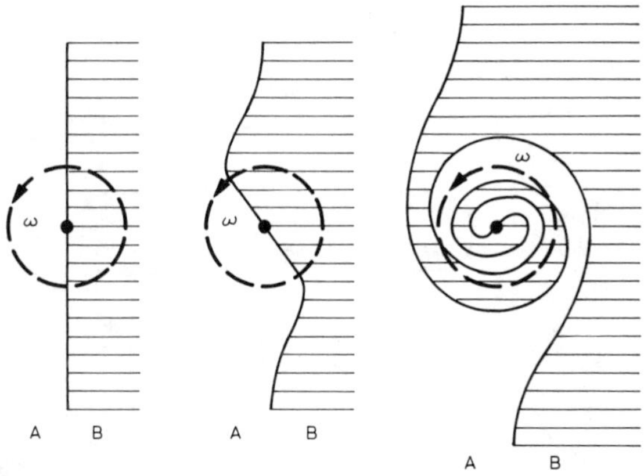

Figure 13. Formation of laminated structure through action of vorticity.

Description of Micromixing

Figure 14 represents schematically (without being drawn to scale) various steps in mixing. Figure 14A refers to the relative movement of two points α and β, situated sufficiently far apart that their separation distance falls in the inertial subrange (left-hand sketch). This relative movement results in deformation (middle sketch), so that on average the scale of the elements formed is reduced and the resulting elements move independently of each other (right-hand sketch). Figure 14B sketches further deformations, whereby the boundary of a fluid element becomes less regular. These reflect the action of inertial velocity fluctuations at scales smaller than in Figure 14A (left). Figures 14A and 14B refer to the inertial convective subrange, where no molecular mixing occurs, but where the structures, which will subsequently participate in molecular mixing, are formed.

Figure 14C indicates finer scale, laminar deformations occurring within the viscous subrange, whilst Figure 14D concentrates on a small region and follows the development of the smallest, most energetic, dissipative vortices. At time t_1, the tongue of B-rich fluid, which was recently formed by a velocity fluctuation, is set in rotation by a local burst of vorticity. At time t_2 the laminated structure, similar to that at Figure 13 and containing approximately equal volumes of A and B solutions, is forming, which at time t_3 contains more turns and has been stretched out along the vortex line. Thus, the diffusional mixing shown in Figure 15 proceeds in laminae which thin under the action of vortex stretching. At time t_4, the initial vortex has been retarded by viscous drag and, because approximately twice the initial volume is available, two new and independent vortices can now act. Whereas at t_2 the two solutions forming the vortex contained only A and B respectively, at t_5A is drawn into the vortex from its immediate environment, whilst the material from the first vortex at t_4, which forms the second layer at t_5, contains A, B and any of their reaction products. The next vortex is then also stretched and at time t_6 waits for a third burst of vorticity to form a third family of vortices, etc. The number of vortices follows a geometric progression with ratio 2. This mechanism of small scale mixing is general, and is not limited to the mixing of two pure species. In fact, the mechanism describes the behavior of any fluid element B in the environment denoted as A.

Let us imagine, for example, that the fluid element of pure substrate B has been introduced to the system, and let us try to answer the question: What is A, the fluid surrounding B, which is subsequently engulfed in an eddy? The answer depends on the conditions of mixing. In the case of unpremixed feed system, without backmixing and for a large volume ratio of substrates, pure second substrate can be engulfed. If there is backmixing in the system, A can represent the homogeneous mixture of the substrates and products, and if the volumes of reacting liquids are not very different, the non-homogeneous mixture can be engulfed. In the case of premixed feed systems, the

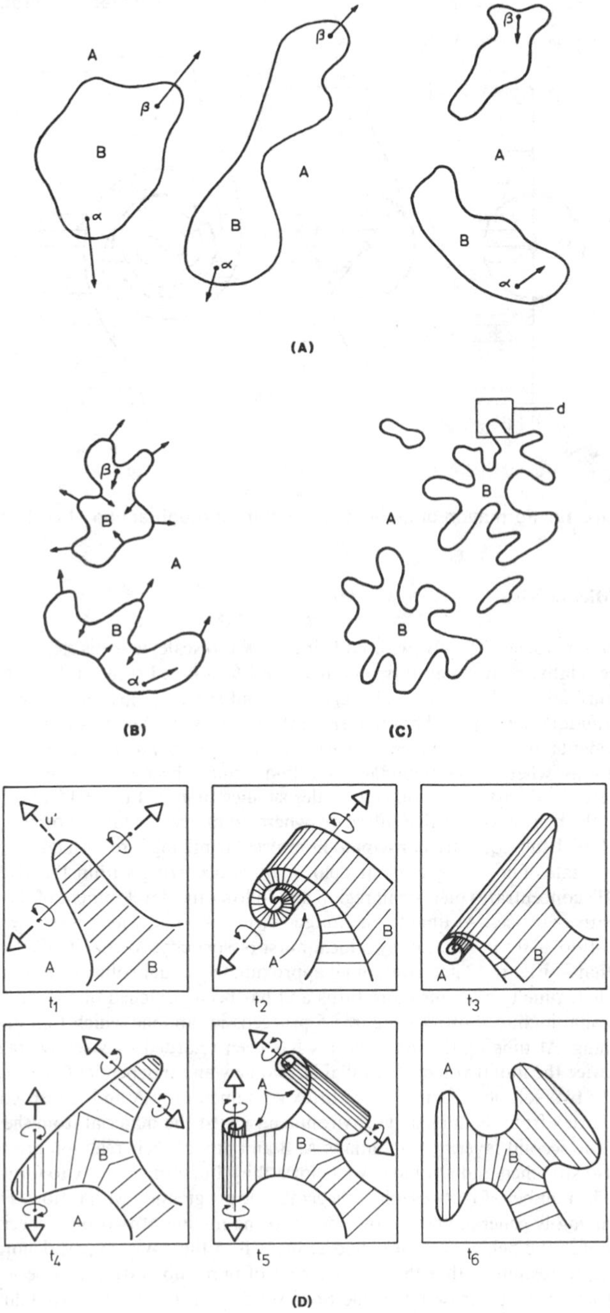

Figure 14. Fluid deformation and vorticity: (A/B) large- and small-scale deformations within the inertial subrange; (C) fine-scale, laminar deformation in viscous subrange; and (D) action of vorticity acting on fluid elements whose initial thickness is on the order of λ_K.

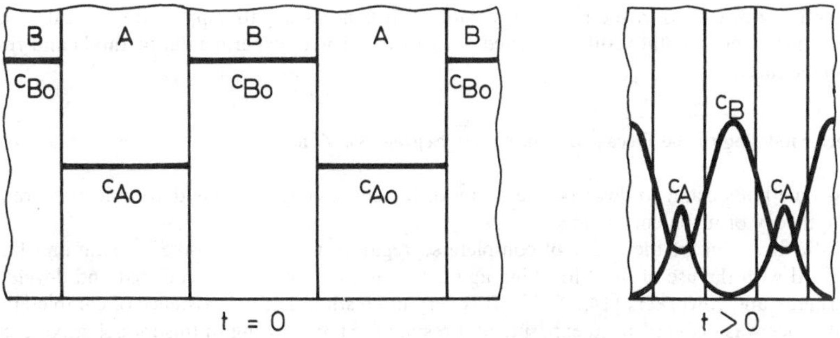

Figure 15. Concentration profiles during reaction and diffusion of substances A and B in progressively thinning laminae.

older and thus more converted material can be incorporated to the elements of fresh, concentrated substrate, etc. . . .

Thus, the micromixing of solutions, originally containing only A and B respectively, proceeds as follows:

1. Initially the scale of segregation is reduced to a value on the order of λ_K without molecular mixing. This process requires time and mixer volume—time of complete segregation and completely segregated volume in the system. The value of scale, above which the liquid is completely segregated and below which micromixing takes place, does not need to be specified accurately since:
   a. as seen in Figure 7, most of the concentration variance and thus most of the completely segregated volume is contained in large concentration eddies (which have small wave numbers) and any error in estimating the smallest, completely segregated eddies should not influence much the estimated value of the completely segregated volume;
   b. the concentration spectrum for the viscous-convective and the viscous-diffusive subranges Equation 53 is insensitive to molecular transport when $k \simeq k_K$, and
   c. fully in agreement with b) and as seen in Figure 11, convection predominates near k_K, while molecular diffusion becomes important at higher wave numbers (i.e. at smaller scales).
2. Vortices incorporate A and B fluid elements, having an initial thickness of order λ_K. The diameter of such vortices is approximately $l' = 12\lambda_K$ (Equation 67). Deformation of these elements (compare Figure 9) further reduces the scale of segregation, according to equations like Equation 60, so that molecular diffusion can effect micromixing (Figure 15). Unsteady state diffusion occurs in shrinking laminae as described by Equations 64, 61, and 55, which are isolated from the fluid outside the vortex, for a period of τ'. After this time, the vortex diameter has decreased from 12 to $0.94\lambda_K$ (calculated from Equation 60, which is practically equal to λ_K. This fluid has lost its vorticity, but still possesses internal concentration gradients of A, B, etc. (Diffusion is slower than momentum transport when $Sc \gg 1$). It is set in rotation again by a new burst of vorticity and brought together with a fresh layer of solution having the same thickness from its environment (Figure 13). Thus, Equation 64 has again to be solved for a closed region (i.e., for batch reaction within a stretching vortex) over the period τ'. This sequence stops only when the limiting reagent has been consumed by chemical reaction or, if no reaction occurs, when the composition of the A and B layers entering a new vortex are equal.

MODELING AND SIMULATION OF LOCAL MIXING

The aim in this section is to present briefly typical methods of modeling used to describe small scale, local mixing processes. These methods are subsequently confronted with the mechanism of micromixing discussed in the previous section. If it is necessary to support theoretical considerations, a few experimental results are cited—a wider review of experimental methods and results is presented later.

Direct Modeling of the Decay of Completely Segregated Zone

Two methods claim to describe the decay of the completely segregated volume and are related to the theory of turbulent mixing.

In the first concept the decay of complete segregation is called "erosive" mixing and has been described with the use of the "shrinking aggregate" model. This was introduced and developed by Villermaux and coworkers [14, 15, 74]. It is very interesting that the existence of completely segregated zones was deduced from experimental results [15]. According to this model, mixing consists in peeling off smaller fragments from the segregated lumps of fluid, by turbulent action at their external surface. In the model the peeling-off process is characterized by a mass transfer coefficient h, the same as in the case of mass transfer between small solid particles, dispersed in turbulent media, and the surrounding liquid. The Calderbank-Moo Young correlation was applied:

$$hl/D = 0.13l(\epsilon/\nu^3)^{1/4}Sc^{1/3} \tag{69}$$

where l is the diameter of shrinking aggregate. A spherical shape was assumed, giving the linear decrease of aggregates size with age:

$$l = l_0(1 - \alpha/t_e) \tag{70}$$

l_0 is the initial size of fluid elements, and t_e is the characteristic erosion time constant.

$$t_e = \{(l_0\lambda_K)/(0.26D)\}Sc^{-1/3} \sim l_0\nu^{5/12}D^{-2/3}\epsilon^{-1/4} \tag{71}$$

The authors [4, 15, 14, 74] noted some problems arising when t_e has to be computed from Equation 71. The first problem is the estimation of l_0 and especially the dependence of l_0 on the initial velocity of the feeding jet; the authors suggested that $l_0 \sim u_0$. The second problem is the energetic efficiency of mixing; the energy inputs from jet and impeller have different influences on mixing and thus two different efficiency coefficients were proposed. Erosive mixing may be identified as mixing in the inertial-convective subrange of wave numbers. However, if this identification is correct, the peeling-off process has inertial character and the rate of this process is independent of molecular properties like viscosity or diffusivity (the spectral function in the inertial- convective subrange, Equation 51, is independent of ν and D). Disintegration of the large fluid eddies in the inertial-convective subrange is fast enough to make molecular diffusion in these eddies ineffective. Evidently, the fluid aggregates do not have and can not have properties of solid particles, and the Calderbank-Moo Young correlation should not be used.

The second concept is based on the spectral interpretation of the scale reduction as it has been described in the previous section [16–18]. The reduction in size of completely segregated elements is considered as the transfer through the inertial-convective subrange and the rate of decay of concentration variance in this subrange can be evaluated from Rosensweig's [75] equation as the rate of transfer down the spectrum:

$$-\frac{d\overline{(c_i')_s^2}}{dt} = \frac{2\Gamma(5/6)}{\sqrt{\pi}\,\Gamma(1/3)C_\gamma} k_{0c}^{2/3}\epsilon^{1/3}\overline{(c_i')_s^2} = \overline{(c_i')_s^2}/t_{ms} \tag{72}$$

where C_γ is a universal constant. $\overline{(c_i')^2}$ is the concentration variance in the inertial-convective subrange, and t_{ms} is a time constant for the rate of decrease of the completely segregated region:

$$t_{ms} = \frac{\sqrt{\pi}\,\Gamma(1/3)C_\gamma}{2\Gamma(5/6)} k_{0c}^{-2/3}\epsilon^{-1/3} \tag{73}$$

As the value of unmixedness and the volume of completely segregated fluid are directly related to each other, Equation 72 may be transformed into a form containing the volume of completely segregated fluid V_s:

$$-(dV_s/dt) = V_s/t_{ms} \tag{74}$$

Using this equation, it is very easy to estimate the completely segregated volume in the system, i.e.,

In batch system: $V_s = V_0 e^{-t/t_{ms}}$ (75)

where V_0 is the initial volume of fluid considered;

in the well-mixed on macroscale CSTR: $V_s = q_0\tau\{1/(1 + \tau/t_{ms})\}$ (76)

Generally, the reduction of the completely segregated zone is a first order process and the volume of the completely segregated zone can be evaluated from a corresponding balance. In flow systems, the completely segregated volume can be found from the age distribution, and the exit value from the residence time distribution. When this model (Equations 73 and 74) is compared with that of Plasari et al. [15], two important differences should be pointed out:

1. t_{ms} is not related to ν and D (Equation 73), while t_e is (Equation 71).
2. Decay of segregation in batch system is exponential in this model (Equation 75) and linear in the previous (Equation 70).

It is very interesting that in both models the existence of a molecular diffusion zone, placed nearby, is assumed. In the case of Klein et al. [74], it was deduced from experimental results again, in the case of the second model [16, 18] from the analysis of the spectral function.

The Coalescence and Redispersion Model

This model was introduced originally to describe the behavior of drops in a two-liquid phase chemical reactor by Rietema [76], Harada et al. [77], and Curl [78]. The concept was adapted later to simulate mixing processes in a single-phase system, for premixed and unpremixed feed systems, also for arbitrary RTD [61]. The content of single-phase systems is divided into a large number of fluid elements, which are now used instead of drops. They have equal volumes and can mix with one another by coalescence, perfect mixing on molecular scale and immediate redispersion. The model can be simulated with a Monte Carlo process [59–61, 79] or with probability functions [7, 61]. The volume of the fluid elements in this model does not have any physical meaning; these volumes should be chosen small enough not to influence the resulting conversion.

The key parameter characterizing micromixing is I, the average number of coalescences undergone by a fluid element during its stay in the sytem. This parameter was determined by Patterson [79]:

$$I = 1{,}333(\tau/N)(\epsilon/\Lambda_c^2)^{1/3} \tag{77}$$

where N is the number of fluid elements, and Λ_c is the integral scale for concentration fluctuations.

The equivalent formulation of Equation 77 was determined by Canon et al. [80]:

$$I = 1{,}333(\tau/N)(\epsilon/k) \tag{88}$$

where k is the turbulent kinetic energy per unit mass.

The equivalence results from the spectral interpretation of the turbulent kinetic energy per unit mass. The model was adopted to simulate both local mixing and large scale turbulent dispersion [81, 82].

The coalescence and redispersion model simulates in a certain way both regions of mixing, as identified from the spectrum. Evidently, before the first coalescence with the fluid of different concentration occurs, any fluid element is completely segregated. The sum of these elements forms a completely segregated zone in the system. The subsequent acts of coalescence occur in the region of molecular mixing. The process of engulfing fluid to form partially segregated laminae within vortices can be compared with the coalescence act and the time when the shrinking laminae are isolated from the fluid outside can be identified with the batch time between collisions in the coalescence redispersion model. The molecular diffusion occurring in shrinking laminae within the vortices in the coalescence-redispersion formulation is assumed to be instantaneous. However, it is not the only simplification. The micromixing parameter I characterizes both kinds of mixing having different mechanisms. When Equation 77 is compared with Equations 68 and 73 (and remembering that $k_{0c} \sim 1/\Lambda_c$), it becomes clear that the coalescence-redispersion model expresses inertial-convective mixing.

Interaction by Exchange with the Mean (IEM Model)

This model was proposed independently by Harada et al. [77], Villermaux and Devillon [83], and Costa and Trevissoi [84, 85]. According to this model:

$$-(dc_i/dt) = k_m(c_i - \bar{c}_i) + R_i \tag{79}$$

The change of concentration of a component i in a "point" is caused by chemical reaction proceeding with the rate R_i and mass exchange between the point and its environment (exchange coefficient k_m). The micromixing parameter k_m is related to Corrsin's [86] time constant [85]:

$$k_m = 1/(2t_m) \tag{80}$$

The last equation gives a possibility to relate k_m to the turbulent field parameters and consequently to the working conditions of the reactor. The average concentration in the environment $\bar{c}_i$ can be evaluated with the help of the two-environment concept, as described before, or using any other information about the composition of the fluid surrounding the point.

For instance, for the well-mixed on macroscale CSTR $\bar{c}_i$ is the average concentration in the tank (outlet concentration as well), and in the unpremixed piston-flow reactor $\bar{c}_i$ is the average concentration of the fluid of the same age.

It should be noted that there is a very close connection between the IEM and the coalescence-redispersion models. In fact, Harada et al. [77] introduced Equation 79 as describing the coalescence-redispersion process of reacting droplets, which are dispersed in an inert phase and proposed to use this equation to describe reactions proceeding in a homogeneous flow reactor, applying the concept of fluid elements. Recently, Villermaux [87] has applied the IEM model to coalescence-redispersion process in liquid–liquid suspension. Equation 79 expressing the model is very simple and the physical meaning of the model is not related to the form of Equation 79, but to the formulation of k_m in terms of turbulent field parameters.

Let us consider some examples. If k_m is assumed to be proportional to $\epsilon^{1/3}\Lambda_c^{-2/3}$, as in works by Klein et al. [74] and Villermaux and David [14], Equation 79 should express the inertial-convective decay of complete segregation, and not diffusive mixing as supposed in [14, 74]. The proportionality of k_m to $\epsilon^{1/3}\Lambda_c^{-2/3}$ is equivalent to Rosensweig's equation (Equation 73). When the full Corrsin estimation of t_m is applied in Equation 80, k_m is as follows:

$$k_m = (k_{0c}^{2/3}\epsilon^{1/3})/\{3 + k_{0c}^{2/3}\epsilon^{-1/6}\nu^{1/2}\ln(Sc)\} \tag{81}$$

This equation is valid for the inertial-convective and the viscous-convective subranges. However, the qualitative difference complete segregation and the molecular dissipation of concentration variance is neglected. The viscous-diffusive subrange is neglected as well. Equation 79 can be applied as well to describe molecular mixing in the viscous-convective and the viscous-diffusive subranges, provided that k_m is appropriately evaluated.

The characteristic time constant, determined only for these subranges [16–18], equals:

$$t_{md} = 3.1(\nu/\epsilon)^{1/2}\{\ln(Sc) - 1.27\} \quad (82)$$

and consequently the micromixing parameter in the zone of molecular mixing is:

$$k_{md} = 0.16(\epsilon/\nu)^{1/2}\{\ln(Sc) - 1.27\}^{-1} \quad (83)$$

This version (Equations 79 and 83) was verified for single, instantaneous, and fast reactions. Equation 79 is very simple, and the calculations with the use of this model are also not very difficult. This is a real advantage. However, by reason of this simplicity, some information about mixing can be lost. This problem is particularly important in the case of reacting systems, which are very sensitive to concentration gradients on the molecular scale. An example is complex, competitive-consecutive reactions having very different initial concentrations of substrates [88].

Diffusion Models

There are two kinds of diffusion models: those assuming molecular diffusion in rigid fluid aggregates [44, 88–95] and those assuming diffusion with reaction in deforming fluid lamellae [19–21, 68, 67, 72, 96–98]. Both kinds of models claim to describe the finishing stage when molecular mixing occurs. From the spectral interpretation we know that diffusion becomes important in the viscous-convective and the viscous-diffusive subranges. Because mixing on molecular scale requires molecular diffusion, these models may be regarded as the most exact simulation of micromixing. However, as follows from the interpretation of turbulent mixing, fluid mechanics offers no support to models of simultaneous diffusion and reaction in a region having time-independent shape and size. These models were constructed with the wrong assumption that turbulence reduces the size of fluid elements until they cannot be further reduced. Thus, for liquid elements below this limiting size, only molecular diffusion can mix the species. But in reality, such a limiting scale of eddies does not exist, and although velocity fluctuations are no longer active, somewhere near Kolmogoroff's scale, further reduction of the scale is caused by deformation of the fluid elements. The stretching process is a key mechanism in micromixing and its omission may lead to large errors. The characteristic time constant for diffusional mixing in rigid aggregated is defined as:

$$t_D = \text{constant}(\delta^2/D) \quad (84)$$

where δ is a characteristic dimension of fluid elements.

The second group of models, where diffusion with reaction in stretched fluid elements is assumed, is much more interesting and realistic. These models are closely related to Batchelor's structure of turbulent fluid, based on laminar deformations of small scale fluid sheets. The spectrum of turbulence expressed by Equation 53 was obtained using this model.

The adaptation of this idea to reacting systems requires:

1. The reaction kinetics be introduced into differential convection-diffusion equations.
2. The environment of stretched sheets of fluid be precisely determined; when reaction occurs, the age of molecules becomes a very important factor.

Let us consider some representative models of this kind.

Lamellar Structure Model

Ottino, Ranz et al. [19, 68] consider a microflow element S_x, within which velocity gradients are uniform and thus whose size will not be significantly greater than the Kolmogoroff velocity microscale. Such elements exhibit a lamellar structure (Figure 16), whose component striations having thickness δ_A and δ_B exchange matter by molecular diffusion. Statistically considered, each thickness

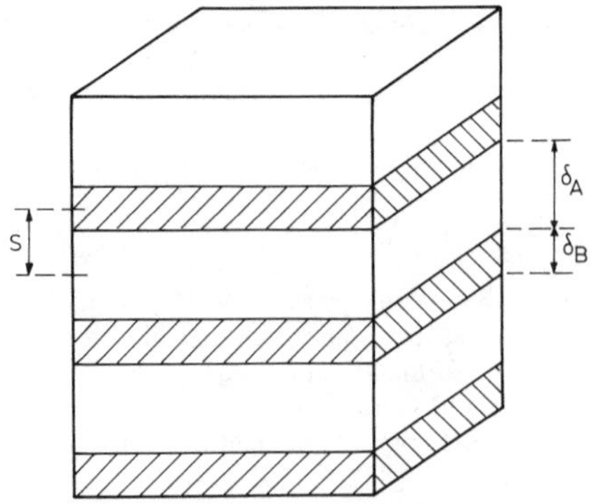

Figure 16. Lamellar structure consisting of layers of A- and B-rich solutions.

decreases with time, as a consequence of uniform fluid deformation, which accelerates the diffusion process. Within the frequently repeated region $S = (\delta_A + \delta_B)/2$ in Figure 16, convection (shrinkage), diffusion, and reaction are described by:

$$\partial c_i/\partial t + u(\partial c_i/\partial x) = D_i(\partial^2 c_i/\partial x^2) + R_i \tag{64}$$

where the coordinate system translates and rotates with the microflow element. This constitutes a model of micromixing. The exchange of material between the microflow elements S_x is excluded from Equation 64, as noted by Ottino [68]. This is not important in batch or plug flow reactors, where all microflow elements in the neighborhood of any fluid element have the same age and the same composition. When, however, their ages are distributed, macromixing should also be considered. Ottino [68, Equation 2.15] states that, provided the RTD {f(t)} measures the distribution of element ages, the mean outlet concentration from a flow reactor will be given by:

$$\overline{c_i} = \int_0^\infty \langle c_i \rangle f(t)\, dt \tag{85}$$

where $\langle c_i \rangle$ is the spatially averaged concentration of i over the region S_x within a microflow element having age t. This equation has also been used by other authors [44, 95] to model partial segregation in a continuous reactor operating with separate feed streams. Diffusion occurs within regions, which are segregated from each other. Evidently, the model is not general and exhibits maximum age and partial species segregation (Figure 17).

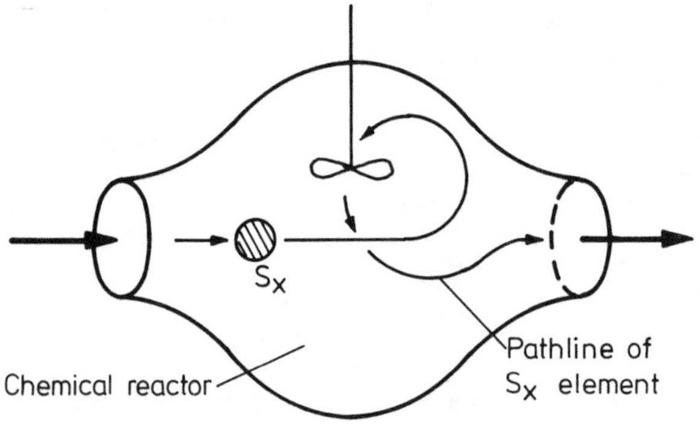

Figure 17. Combination of the lamellar mixing model with the RTD concept.

Some examples of the errors caused by the omission of exchange with the environment have been given [99].

A further problem is the choice of the thickness S at the moment when diffusion and therefore also reaction between unpremixed reagents begin. The microflow element S_x contains many striations S, so that S must be small compared to the Kolmogoroff scale. But, from the spectral interpretation, the initial value of S should be of order S_x. This, however, contradicts the idea that S_x contains many striations. The assignment of a reasonable initial value to S seems to be unclear. The lamellar structure, as proposed by Ottino [19, 68], raises another question, when the volumes to be mixed (V_A and V_B) are widely different. If $V_A = \alpha V_B$, where α might be large compared to unity, $\delta_A = \alpha\delta_B$ (Figure 16). When the densities and viscosities of the A and B solutions are similar, one might, however, expect that locally $\delta_A \simeq \delta_B$, since the deformation history leading to these fine striations would not depend upon their compositions.

Ottino et al. [19] have presented some interesting remarks about deformation kinematics. The Ψ value in Equation 55 has the highest, limiting value for the case when the internal kinetic energy is conserved in S_x:

$$\Psi_{max} = (1/\sqrt{2})(\epsilon/\nu)^{1/2} \tag{86}$$

It should be noticed that this is close to the value obtained by Batchelor [66]:

$$\Psi = 0.55(\epsilon/\nu)^{1/2} \tag{58}$$

for long deformation times. The local efficiency was defined as the ratio of real Ψ value in a certain mixing situation to Ψ_{max} [96]. However, one must remember that defined in this way efficiency refers to mixing in the viscous-convective and the viscous-diffusive subranges. An example of a wrong application of efficiency is given in [96]. Ottino estimated the efficiency of mixing from the observed rate of neutralization of hydrocholoric acid with sodium hydroxide solution in a tank. The solutions were initially separated in the tank by a rubber membrane, which broke when agitation began. Even at the highest impeller speed (1,198 rpm), not less than 2 s were needed for neutralization. Assuming that a lamellar structure existed during the whole time, its rate of shrinkage d ln S/dt was detemined by fitting Equation 64 to the conversion-time measurements. This rate was then compared with the maximum rate of deformation $(\epsilon/2\nu)^{1/2}$ to give an efficiency. The measured values of efficiency are very small (1.4 to 34%) and dependent on time and stirring rate. The circulation time in the tank was roughly 2 s [96], and thus it is impossible to conclude that a lamellar structure would be present after 0.25 s. Some 3 to 4 circulations are usually considered necessary to achieve macroscopic homogeneity. The observed neutralization times reflect much more macromixing and reduction of the size of segregated eddies than diffusion and reaction in lamellar structures, and the efficiencies are physically unrealistic. Evidently, the dependence of efficiency on mixing rate compensates the wrongly assumed mechanism of mixing. It is easy to check that the overall rate of process should be correlated to $\epsilon^{1/3}$ like in Equation 73 instead of $\epsilon^{1/2}$ in Equation 80.

Diffusion-Reaction Model

A diffusion-reaction model of micromixing was developed to apply to a particular type of experiment, when the volume ratio $\alpha = V_A/V_B$ greatly exceeds unity and for the competitive-consecutive reaction scheme [20, 67, 72]. In the semi-batch mode of operation, a small volume of highly concentrated, B-rich solution was pumped into a stirred reactor, containing a large volume of more dilute A-solution. The feed stream was supposed to be dispersed and the transient concentration gradients in the B-solution were described by Equation 64, while no reaction was supposed to occur in the A-solution [20]. A diffused into the dispersed B-solution, but B was assumed not to diffuse into the A-solution, which was furthermore considered to be homogeneous at the molecular level. These simplifications were originally introduced to save computer time (i.e., by ignoring concentration gradients in the A-solution), are, however, physically unrealistic and imply that the model is not general. Some examples of the significant errors caused by these simplifications have been given [99].

Micromixing in a Shrinking Laminated Structure Within a Small Energy Dissipating Vortex

The basic ideas for modeling micromixing, given earlier, have been applied to a pair of consecutive, competitive reactions conducted in semi-batch as well as in continuous stirred tank reactors [73]. The objections to the two last models no longer apply here, and we believe the present model to be the most realistic. Equation 64 is applied now to the region $-\delta < x < \delta$ in Figure 18:

$$\partial c_i/\partial t + u(\partial c_i/\partial x) = D_i(\partial^2 c_i/\partial x^2) + R_i \tag{64}$$

where the local, instantaneous rate of shrinkage is given by:

$$u = \{-(\epsilon/\nu)^{1/2}x\}/\{(4 + \epsilon t^2/\nu)^{1/2}\} \tag{87}$$

From symmetry, the boundary conditions are:

$$x = \pm\delta, \qquad \partial c_i/\partial x = 0$$

The initial thickness of each of the two fluid layers in Figure 18 is $2\delta_0$. The initial conditions for the first vortex are:

$$t = 0, \quad -\delta_0 < x < 0, \qquad c_i(x, 0) = c_{i1s}(x)$$

c_{i1s} is the initial distribution of B-rich solution in the left-hand slab (Figure 18).

$$t = 0, \quad 0 < x < \delta_0, \qquad c_i(x, 0) = c_{i1e}(x)$$

c_{i1e} is the initial distribution of A-rich solution coming from the environment to form the right-hand slab (Figure 18). After a time τ', which is the mean life-time of an energy dissipating vortex:

$$\tau' \simeq 12(\nu/\epsilon)^{1/2} \tag{68}$$

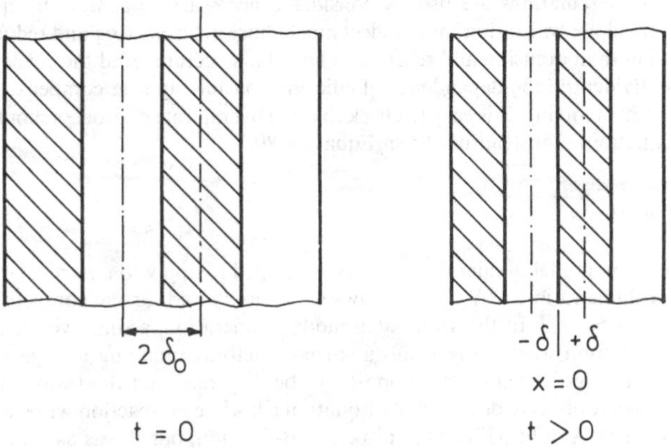

Figure 18. Diffusion-reaction equations solved over regions with thicknesses $2\delta_0$ when $t = 0$, and 2δ when $t > 0$.

two, second generation vortices are formed, so that their initial conditions are:

$$t = \tau' \quad \begin{matrix} -\delta_0 < x < 0 & c_i(x, 0) = c_{i2s}(x) \\ 0 < x < \delta_0 & c_i(x, 0) = c_{i2e}(x) \end{matrix}$$

c_{i2s} is the concentration distribution in the first vortex when $t = \tau'$. c_{i2e} is the concentration profile in the environment as it forms eddies of the second generation. (Its calculation is subsequently explained).

Similar considerations apply to later generations of vortices. After each eddy lifetime, the number of eddies doubles, so that the number of eddies forms a geometric progression with ratio 2. In dimensionless form:

$$\bar{X} = x/\delta \qquad T = tD/\delta_0^2 \qquad C_i = c_i/c_{j0}$$

Equation 64 becomes:

$$(\partial C_i/\partial T) = (\delta_0/\delta)^2(\partial^2 C_i/\partial \bar{X}^2) + (\delta_0^2 R_i/Dc_{j0}) \tag{88}$$

Each slab forming the vortex is assumed to have an initial thickness of λ_K, thus $2\delta_0 = \lambda_K = (\nu^3/\epsilon)^{1/4}$, and δ and δ_0 have been related using the theory of initial relative turbulent diffusion [69]. In dimensionless form:

$$\delta_0/\delta = \left\{1 + \frac{T^2Sc^2}{32} - \sqrt{\left(1 + \frac{T^2Sc^2}{32}\right)^2 - 1}\right\}^{-1/2} \tag{89}$$

The dimensionless eddy lifetime from Equation 68 is given by:

$$T_\omega = 48/Sc \tag{90}$$

Competitive, consecutive reactions:

$$A + B \xrightarrow{k_1} R \qquad R + B \xrightarrow{k_2} S$$

exhibit a mixing-sensitive product distribution, which has also been studied experimentally. Choosing B as reference material (j = B), the reaction terms in Equation 88 become:

$$i = A \qquad -(k_1/k_2)MC_AC_B \tag{91}$$

$$i = B \qquad -MC_B\{(k_1/k_2)C_A + C_R\} \tag{92}$$

$$i = R \qquad MC_B\{(k_1/k_2)C_A - C_R\} \tag{93}$$

where the mixing modulus M is defined as:

$$M = k_2c_{B0}\delta_0^2/D \tag{94}$$

The boundary conditions are:

$$\bar{X} = \pm 1 \qquad \partial C_i/\partial \bar{X} = 0$$

The initial conditions are related to the mode of reactor operation, as explained later.

The solution of the coupled, non-linear partial differential equations (Equation 88) is usually needed for volume-averaged concentrations (i.e., over the interval $-\delta < x < \delta$) and for long reaction times (t and $T \to \infty$), so that concentrations are obtained when the limiting reagent B has been consumed. Any concentration then depends upon:

$$c_i = f(k_1/k_2; N_{A0}/N_{B0}; V_A/V_B; M; Sc) \quad (95)$$

The product distribution will be expressed as:

$$X_S = 2c_S/(c_R + 2c_S) \quad (96)$$

representing the fraction of the limiting reagent B, which has reacted and is now present in the secondary product S. The computation has been carried out for $k_1/k_2 \gg 1$, which was the experimental condition (coupling of 1-naphthol with diazotized sulphanilic acid).

Semi-Batch Reactor

To secure a high yield of R, the whole number V_A of A-rich solution is charged to the reactor and agitated. The B-rich solution is slowly fed in. Modeling consists of supposing that the added volume V_B is divided into σ parts, each of them encounters a different environment. The solution of Equation 64 calls for double discretization, splitting the feed into σ parts, whereby each part reacts in N generations of eddies. After each part has been added to the reactor, diffusion and reaction run to completion in vortices, which finally mix with the solution in the reactor and change its composition.

After the first part of the feed has been added, the first generation of vortices is formed with equal volumes of A and B solutions, having initial concentrations, c_{A0} and c_{B0}. Diffusion, thinning, and reaction proceed in the laminated structure within this first generation of eddies for a time τ', when the vortex diameter has decreased from $\sim 12\lambda_K$ to $\sim \lambda_K$. This fluid together with an equal volume from its environment, here still of concentration c_{A0}, now form the second generation of vortices, doubling by engulfment the vortex volume, etc. After N generations, the B-concentration will have gone to zero, and the formation of further vortices homogenizes the solution. As the second part of the B-feed enters the reactor, it encounters the environment with concentration c_{A1} of A and concentrations of R and S, resulting from Equation 88 and the overall species balances. c_{A1} is valid for all the generations of vortices for the second part of B. After this second part of the feed has reacted, the concentrations A, R, and S are again updated to give c_{A2}, etc. This procedure is continued until the whole amount of B is consumed.

Results. The product distributions were computed for subsequent application to diazo coupling reactions, the parameters being correspondingly chosen. Thus, in all results the stoichiometric ratio (N_{A0}/N_{B0}) is 1.05, and the first reaction is instantaneous. Unless otherwise stated, Sc = 1,280, corresponding to reaction in aqueous solution at 295 K.

Initial thickness of reaction zone (*Table 1*). As seen earlier, little diffusive mixing and therefore little reaction is expected when $2\delta_0 \simeq \lambda_K$, but rather at smaller scales, i.e., the product distribution should be insensitive to the estimate of $2\delta_0$. This was checked here by doubling the estimate of $2\delta_0$ to $2\lambda_K$ and making the corresponding changes in the constants of Equations 89 and 90, as well as the value of M, defined in Equation 94. The X-values in the last column were calculated for the same λ_K (same

Table 1
Effect of Initial Thickness of Reaction Zone on X

Conditions:	σ	α	X when $2\delta_0 = \lambda_K$	X when $2\delta_0 = 2\lambda_K$
	10	10	0.3096 (M = 40)	0.3070
	10	50	0.2530 (M = 80)	0.2515
	10	100	0.2053 (M = 100)	0.2042

Table 2
Effect of α and M on Product Distribution X

α	M = 0.1	0.5	2	4	10	25	40	80	100
10	X = 0.001	0.004	0.028	0.053	0.118	0.233	0.310	—	—
50	—	—	0.012	0.023	0.054	0.116	0.164	0.253	—
100	—	—	0.007	0.015	0.034	0.076	0.110	0.179	0.205

ν and ϵ) as in the middle column. As Table 1 shows, X is insensitive to the estimate of $2\delta_0$. In all further calculations $2\delta_0 = \lambda_K$ was employed.

Density of discretization of feed addition. As also noticed earlier [72], X tends to a constant value as σ increases. A value of σ probably in excess of 100 – 1,000 represents the physical situation, but would be excessive in its demand on computer time. When $\alpha = 100$ and $M = 100$, increasing σ from 5 through 10 to 20 changed X from 0.2003 through 0.2053 to 0.2079. Since the analytical error in measuring X is around ± 0.005, $\sigma = 10$ gives a sufficiently accurate estimate of X from a semibatch reactor.

Mean life-time of energy dissipating vortex. Product distribution was sensitive to the value of the constant in Equation 68. When $\alpha = 10$, $\sigma = 10$ reducing this from 12 to 10 caused the following reductions in X: when $M = 2$ from 0.028 to 0.022, when $M = 10$ from 0.118 to 0.097, and when $M = 40$ from 0.310 to 0.269. In accordance with the derivation given earlier in "The Role of Vorticity" the value 12 was employed here. The resulting concentration profiles when $t = \tau'$ (i.e. when new vortices are just about to form) were rather flat. It made no difference when computing X for the next vortex generation whether mean values or the actual concentration profiles were used for the initial condition in the B-rich layer.

Volume ratio of reagent solutions and mixing modulus. For $N_{A0}/N_{B0} = 1.05$ and $k_1/k_2 = 7.3 \times 10^3/1.9 = 3{,}840$, reaction in the chemical regime would give $X_{min} = 0.001$. The results in Table 2 indicate that the chemical regime corresponds to $M < 0.1$ in this case. Figure 19 is based on Table 2.

Schmidt number. The Schmidt number can be changed for example by changing the reaction temperature and the viscosity of the solutions ($Sc = \nu/D$). It exerts a significant influence on product distribution, as indicated in Table 3. Figure 20 is based on Table 3.

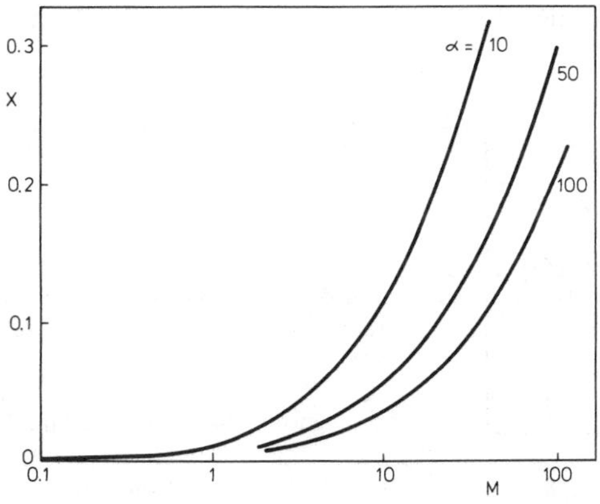

Figure 19. Effect of α and M on product distribution X for semibatch reactor.

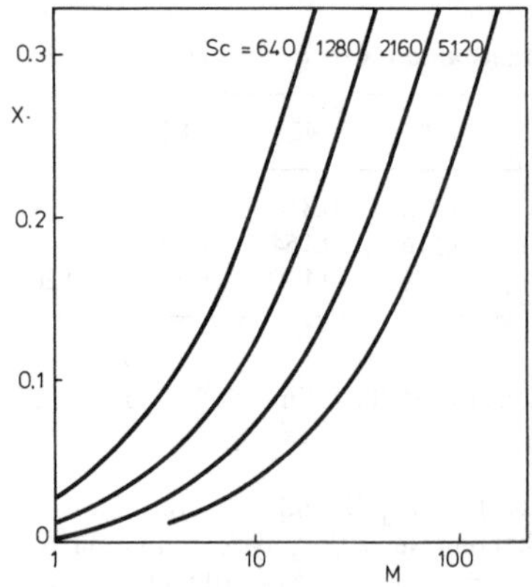

Figure 20. Effect of Schmidt number on product distribution X for semibatch reactor.

Table 3
Effect of Sc on X ($\alpha = 10$)

M	Sc = 640	1,280	2,560	5,120
1	0.028	—	—	—
2	0.053	0.028	0.011	0.007
4	0.098	0.053	0.028	0.014
10	0.201	0.118	0.065	0.034
25	0.350	0.233	0.141	0.079
40	—	0.310	0.199	0.117
80	—	—	0.309	0.199
150	—	—	—	0.296

Continuous Stirred Tank Reactor

The principle for modeling a CSTR is presented in Figure 21, where A-rich and B-rich reaction zones as well as non-reactive fluid are present. The zones are sketched for fast reactions, so that no unreacted limiting reagent leaves them. (If it does, the degree of reaction as a function of time needs to be convoluted with the RTD to determine the conversion of the CSTR). Since B is the limiting reagent ($\bar{c}_B \simeq 0$), computation showed that the A-rich zone serves simply to blend fresh A

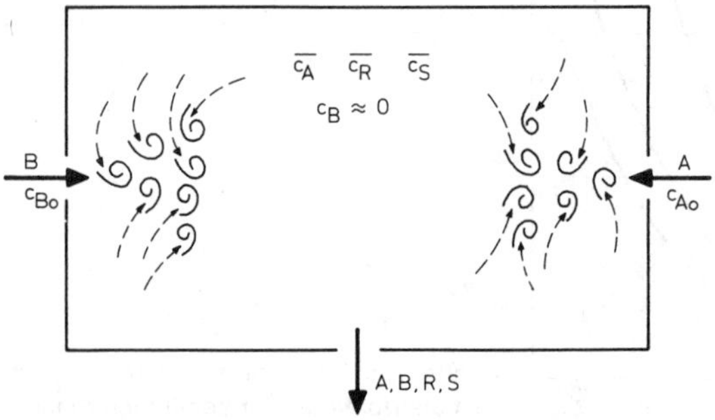

Figure 21. Schematic representation of CSTR with A- and B-rich reaction zones right and left, respectively.

$\mathcal{9}$ = vortex –→ = entrained fluid containing A, R and S).

Table 4
Effect of α and M on X (CSTR)

$\alpha = 10$	M = 0.5	2	4	10	25
	X = 0.017	0.056	0.102	0.188	0.328
$\alpha = 50$	M = 9.27	18.54	46.38	115.9	
	X = 0.079	0.132	0.245	0.393	
$\alpha = 100$	M = 4.59	18.4	91.8		
	X = 0.021	0.079	0.252		

with the reactor contents. The processes in the B-rich zone are in principle the same as for the semi-batch reactor. The first generation of vortices is formed from pure B and an environment now containing the concentrations $\bar{c}_A$, $\bar{c}_R$ and $\bar{c}_S$. The solution of Equation 64 over the period τ' yields the concentration distribution as the second generation of vortices is formed from the first generation together with again the environment, containing A, R, and S. This sequence repeats itself until B has been consumed, and the N-th generation of vortices then blends with the reator contents to yield the composition in the outlet stream.

Clearly, this sequence requires the composition of the environment to initiate the diffusion-reaction calculation for each generation of eddies. The iterative method [72] consists of guessing the composition of the environment (it is sufficient to guess X), then working through the sequence which itself yields this composition. Any failure of the calculated and guessed X-values to agree calls for further iteration. Only sufficient product distributions were computed to permit comparisons with experimental results for the diazo couplings. In the chemical regime, $X_{min} = 0.009$ (Table 4). These results, as well as those in Table 2, show that X decreases with increasing α when M is constant. This agrees with SS model formulation of micromixing [20]. For a given set of conditions (α, M, N_{A0}/N_{B0}, Sc), X is lower from a semi-batch reactor. Compared with diffusion-reaction descriptions of micromixing in shrinking B-rich laminae [20, 72], the present model predicts higher X_S-values and a more rapid increase in X as M rises. Figure 22 summarizes some of these results.

It should be noticed that the last model gives correct solution also in the case of single reactions [99]. The difficulties present in the other models, have been removed in this best formulation by the engulfing process caused by the bursts of vorticity. The limiting solution of Equation 64, if ϵ approaches infinity and thus the characteristic eddy life-time τ' approaches zero, is the same as for the homogeneous reactor.

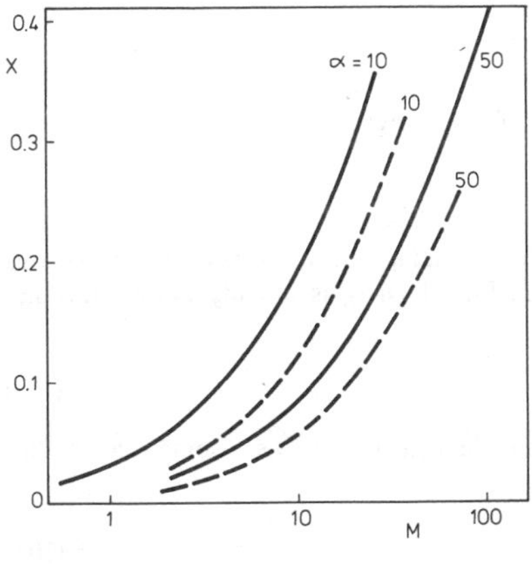

Figure 22. Effect of α and M on product distribution X for CSTR (———) and semibatch (-------) operations.

CLASSIFICATION OF REACTIONS FROM THE VIEWPOINT OF COMPETITION BETWEEN REACTION AND MIXING—NUMERICAL CRITERIA

Reactions may be classified by comparing the characteristic times for mixing and reaction, whereby the half-lifetimes are often used. For instance, for second-order kinetics and a stoichiometric ratio of order unity, the half-lifetime is:

$$t_R = (k\bar{c}_{A0})^{-1} \tag{97a}$$

Suppose that we are interested in whether the process of erosion in the inertial convective subrange influences the observed reaction rate (e.g., if reaction is faster than the reduction of the completely segregated volume). The half-lifetime in that case can be defined as the time needed to halve the completely segregated volume. From Equations 74 and 75, the mixing time t_M characterizing such a situation equals:

$$t_M = 0.69t_{ms} \tag{97b}$$

and the reaction is therefore:

instantaneous if $t_M \gg t_R$

fast for $t_M \simeq t_R$

and slow for $t_M \ll t_R$

An influence of scale reduction will be observed in the first two cases.

When mixing on the molecular scale is considered, the half-lifetime for molecular diffusion should be used. From many possible definitions let us choose the time needed to halve the concentration at the center of a slab whose original thickness is δ_0. Without deformation:

$$t_D = 0.38\delta_0^2/D \tag{98}$$

When deformation is elongational with constant shear rate, an analytical solution to the two-dimensional diffusion-convention problem (i.e., Equation 64 without a reaction term) exists. This considers the diffusional spreading of material originally concentrated at a point to give a diffuse spot and expresses the square of the average spot radius as proportional to $I_{11} + I_{22}$. Without deformation [37]:

$$(I_{11} + I_{22})_D = 4Dt \tag{99}$$

With strain [37]:

$$(I_{11} + I_{22})_{DS} = 4D\sinh(2st)/2s \tag{100}$$

where s is the rate of elongation. For the same degree of spreading:

$$t_D = \sinh(2st_{DS})/2s \tag{101}$$

relates the times needed by diffusion alone (t_D) and diffusion and shear (t_{DS}). It is assumed that this also applies to the penetration of a slab by diffusion, for which t_D is already known through Equation 98. Then:

$$t_{DS} = \text{arc}\sinh(0.76s\delta_0^2/D)/2s \tag{102}$$

The rate of strain applying to deformations of small fluid elements in vortices is somewhat smaller than $0.5(\epsilon/\nu)^{1/2}$ (Equation 54). Thus:

$$t_{DS} > (\nu/\epsilon)^{1/2}\,\text{arc}\sinh(0.1Sc) \tag{103}$$

Now, comparing the characteristic times for diffusion in stretched laminae and reaction:

$t_{DS} \gg t_R$ instantaneous reaction

$t_{DS} \simeq t_R$ fast reaction

$t_{DS} \ll t_R$ slow reaction

EXPERIMENTAL METHODS TO ASSESS THEORIES OF MICROMIXING

General

Micromixing refers to mixing on the molecular scale, whereby molecular diffusion operating at scales smaller than the Kolmogoroff scale is the ultimate mixing mechanism. In well-stirred water at room temperature (energy dissipation rate 1 W kg^{-1}), Kolmogoroff's velocity microscale is about 30 μm and concentrations at, say, intervals of 1 μm in moving, deforming fluid elements have not yet been measured. Thus, a direct comparison with the predicted time- and space-dependent concentration of any substance undergoing micromixing has not been attempted. A non-reactive (passive) substance would be suitable for such studies. All available results were obtained from the effects of inhomogeneity on mass transfer (100), but especially on chemical reactions. The application of chemical reactions to assess theories of micromixing is the main theme of this section.

When a single, irreversible reaction is to occur in a reactor and a single feed stream is used, it will contain all necessary reagents; these are, therefore, available in a premixed form. In a tubular, plug flow reactor, the composition of small fluid elements formed by dispersion of the feed stream will depend only upon the axial position. Since no mixing occurs between elements at different axial positions and the composition of the elements having a given axial position, but various radial positions, is uniform, micromixing will not influence composition distribution and reaction rate. This situation changes as soon as the fluid exhibits a residence time distribution. On the one hand this age distribution increases the differences in composition between the fluid elements, within which reaction occurs, whilst on the other hand mixing between these elements reduces such differences [42]. This mixing has no effect on reactions having first-order kinetics: reaction is accelerated in regions with reagent concentrations in excess of the mean, but is then retarded by exactly the same amount in regions, where concentrations fall below the mean. With non-linear kinetics, however, micromixing influences average reaction rate and conversion. Moreover this influence increases as the RTD progressively departs from plug flow. For a single, irreversible, second-order reaction conducted in a well-stirred, continuously operated tank reactor (with an exponential RTD), conversions have been calculated assuming (a) complete homogeneity and (b) complete segregation on the molecular scale [11]. (In the latter case the feed stream breaks up into small elements that are well-mixed among each other, but do not exchange mass between one another). For a given mean residence time in the reactor, the conversion is higher for case (b)—the square of the mean concentration, which represents the reaction rate in case (a), is less than the mean square concentration of case (b). Experiments have been carried out to detect this difference as well as to realize intermediate states of partial segregation, whose conversions fall between cases (a) and (b). The results proved to be rather inconclusive [54, 101], two significant difficulties being (a) insufficient sensitivity and (b) reactions too slow. Under (a), the analytical error in measuring conversion was of order $\pm 1\%$, whilst the maximum predicted difference (conversion in segregated minus conversion in perfectly mixed state) was only 7%. Under (b), only reactions, which are fast enough relative to the rates of diffusion of reagents, will cause inhomogeneity at molecular scale. Because of limited reactivity, agitation speeds were reduced to increase the size of the partially segregated elements. This, unfortunately, also changed the RTD with the result that the observed conversion could not uniquely be related to micromixing, but also reflected to some unknown extent macroscopic concentration gradients in the tank.

Some useful increase in sensitivity is obtained when the reaction is autocatalytic. The mean residence time for example can be chosen so that, when the mixture is chemically homogeneous (i.e., the feed mixes rapidly with the reactor contents containing the catalytically active product), reaction rate is a maximum. Although promising results are being obtained with such reactions, no complete and rigorous study has been published yet [15, 74, 102, 103]. Autothermal reactions, where the heat liberated works like a catalyst, also fall in this category.

Faster reactions, which create steeper concentration gradients and therefore partial segregation of reagents, require the reagent streams to enter the reactor separately, thus avoiding pre-reaction. This case is practically more important than that of a single feed, where little if any segregation occurs in low viscosity media, yet has received much less attention [23]. The theory of simultaneous diffusion and reaction in shrinking laminae, developed earlier, is applicable to both cases, although only the situation of separate feeds is considered here in detail. Greater sensitivity is provided by the wider conversion limits for a single second-order reaction—complete segregation gives zero conversion and complete homogeneity gives the highest conversion compatible with the available residence time and quantities of reagents. Multiple reactions give several products and in certain cases (e.g., series-parallel reactions, copolymerization, and parallel reactions having different orders) the distribution of the products from sufficiently fast reactions is strongly influenced by the degree of segregation. At a time of economic and especially ecological constraints it is necessary to recognise this influence, which determines the conversion efficiency of raw materials to useful products as well as the complexity and the energy demand of the separation operations following reaction.

Characteristics Required of Test Reactions

An outline will now be given of the characteristics, which reactions should exhibit in order to be suitable to detect partial segregation.

a. Reaction should occur to a substantial extent during the period when segregation is present and micromixing is still proceeding. Many reactions are sufficiently slow that their half-life times $t_{1/2R}$ greatly exceed the time required by diffusion to homogenize a system $t_{1/2D}$, whereas segregation with suitable reactions becomes detectable if $t_{1/2R} < t_{1/2D}$. For a single, second-order reaction using equal initial concentrations of reagents A and B, after the feed solutions have been mixed, namely $\bar{c}_{A0} = \bar{c}_{B0}$, then $t_{1/2R} = (k\bar{c}_{A0})^{-1}$ (Equation 96). The theory given earlier can, by dropping the reaction terms, also be employed to calculate $t_{1/2D}$, but simpler, cruder methods allow order of magnitude estimates: Setting $L = \lambda_K/4 = 0.25(\nu^3/\epsilon)^{1/4}$, $t_{1/2D} \simeq 0.4L^2/D$ (Equation 98). A better approximation considers the steady deformation of the region within which diffusion is occurring (Equation 103) $t_{1/2D} > (\nu/\epsilon)^{1/2}$ arc sinh(0.1Sc). When $\nu = 10^{-6}\,m^2\,s^{-1}$ (water at 293 K), $\epsilon = 1\ Wkg^{-1}$ and Sc = 1,280, then $t_{1/2D} > 6$ ms. Thus, any second-order reaction for which $(k\bar{c}_{Ao})^{-1} < 6$ ms would be strongly retarded by segregation.
b. If a single, irreversible reaction is employed, then the time required to achieve a given conversion will increase as the degree of segregation rises. The time defined in a micromixing model is, however, different from the residence time in, say, a tubular reactor, since the latter includes time required for equalization of concentration across the tube radius and for reduction of the scale of the lumps. Furthermore, any departure from plug flow complicates the specification of residence time. Since reaction should be rapid (see a), the residence time should be short, otherwise the fluid leaving the reactor will be fully converted. This condition is especially difficult to satisfy for stirred tank reactors and explains why rapid, single reactions have limited application to them.

   Rapid, irreversible, competitive reactions offer more potential, provided some products are formed when micromixing is rapid (concentration gradients are small), whilst others characterize slow micromixing. Irreversibility ensures that products formed during the initial, more segregated period are preserved as diffusion progressively reduces segregation. (They represent a sort of concentration/time integral photograph of the segregated state). Multiple reactions also avoid the difficulty of deciding during which part of the residence time micromixing is occurring, when all reactions occur and compete for the same period, even though it is not separable from macromixing and directly measurable.
c. The full reaction mechanism and the kinetics of each step should be known. This includes considerable chemical information and the effects of temperature, solvent, ionic strength, pH,

soluble catalysts and the concentrations of all species (partial reaction orders). Very few, if any of the reactions used to date have been characterized in this detail, which is necessary to exclude attributing observations to micromixing, when the real cause is a chemical artefact.

In addition to these three principal requirements, the following factors are also desirable:

d. Accurate quantitative analysis of all substances present. This permits checking mass balances. Rapid, on-line analysis would be welcome, permitting the temporal evolution of the species to be followed and thus giving a much fuller test of theories than when analyzing only after completion of reaction. This has, however, not yet been achieved.
e. Regeneration or safe disposability of solvent, reagents and products.
f. Suitability for application to large scale reactors, thereby providing information for industrial scale-up. This requires consideration of toxicity, fire hazard, light sensitivity, cost of reagents, corrosion etc.

Competitive, Consecutive Reactions

The pair of competitive, consecutive (also termed parallel, series) reactions:

$$A + B \xrightarrow{k_1} R \qquad R + B \xrightarrow{k_2} S \tag{104}$$

offer a valuable system of mixing-dependent reactions and were also employed earlier in the theoretical description of diffusion and reaction in shrinking laminae. The yield of the intermediate R responds to concentration gradients on the molecular scale as follows:

- Chemical regime ($t_{1/2R} \gg t_{1/2D}$) (slow reaction). The yield of R is a maximum, its value is determined below.
- Diffusional regime ($t_{1/2R} \ll t_{1/2D}$) (instantaneous reaction). The yield of R is always zero.
- Mixed regime ($t_{1/2D} \simeq t_{1/2D}$) (fast reaction). The yield of R is intermediate between its chemical and diffusional regime values. It is influenced by both chemical and physical factors.

The chemical regime serves as a convenient bench mark for experiments, since the measured yield of R will tend towards its value in the chemical regime as rates of mixing and reaction are increased and decreased respectively. This regime is the province of classical kinetics, so that the rates of formation of the various species may be written:

$$\begin{aligned} r_A &= -k_1 c_A c_B & r_B &= -k_1 c_A c_B - k_2 c_R c_B \\ r_R &= k_1 c_A c_B - k_2 c_R c_B & r_S &= k_2 c_R c_B \end{aligned} \tag{105}$$

These concentrations are related through mass balances on the resistant groups A and B, which for dilute, initially pure reagent solutions may be written:

$$\bar{c}_{A0} = c_A + c_R + c_S \qquad \bar{c}_{B0} = c_B + c_R + 2c_S \tag{106}$$

For a batch reactor and a tubular, plug flow reactor $r_i = dc_i/dt$, assuming that dilute solutions are used. The equations cannot be analytically integrated in time domain. Eliminating the time, e.g., by writing $r_R/r_A = dc_R/dc_A$, permits analytical integration relating two concentrations, e.g.:

$$c_R/\bar{c}_{A0} = 1/(1 - k_2/k_1)\{(c_A/\bar{c}_{A0})^{k_2/k_1} - (c_A/\bar{c}_{A0})\} \tag{107}$$

valid provided $c_{R0} = 0$ and $k_2 \neq k_1$. The constancy of k_2/k_1 implies either isothermal reaction (effective cooling, dilute solutions) or equal activation energies for the two reactions. This result relates c_R to c_A, whilst the two mass balances permit c_B and c_S to be calculated. Thus, the full composition is defined, although the time needed to obtain this mixture may only be found numerically.

By setting $dc_R/dc_A = 0$, the maximum yield of R may be calculated together with the corresponding conversions of A and B. To stop the reactions at this point, the reactor may be cooled or, if this is too slow relative to the rate of over-reaction of R to S, a limitation of the quantity of B added initially relative to that of A, namely N_{A0}/N_{B0}, may be used. In this case B is just completely consumed (i.e., N_B tends to zero) as the maximum in c_R is attained.

For a well-stirred, completely homogenized tank reactor, volume V, continuously operated with a volumetric throughput Q, r_i is given by $r_i = Q(c_i - c_{i0})/V$, where c_i and c_{i0} are the exit and inlet concentrations of i. Steady state is assumed. Combining this expression with the reaction kinetics, the mean residence time V/Q may easily be related to the various concentrations. If, however, V/Q is eliminated, c_R and c_A are related by:

$$c_R/c_{A0} = \frac{(c_A/\bar{c}_{A0})(1 - c_A/\bar{c}_{A0})}{(c_A/\bar{c}_{A0}) + (k_2/k_1)(1 - c_A/\bar{c}_{A0})} \tag{108}$$

With a given ratio k_2/k_1 and a given conversion of A, the yield of R from this mode of reactor operation is less than from a batch or tubular reactor. Otherwise reaction in well-mixed, continuous tanks is similar to that in batch reactors.

Batch operation presents a particular difficulty with fast reactions, namely reaction occurs during the filling of the reactor under hydrodynamically variable conditions. Semi-batch reaction largely avoids this problem and consists of slowly adding one of the reagent solutions to the other, which has initially been placed in the agitated tank. If B-rich solution is initially present, then, provided reaction occurs shortly after adding A, the concentration profiles are given in Figure 22. R, finding itself in an excess of B, is fully converted to S, even in the chemical regime. This mode is therefore unsuitable for studying micromixing. Figure 22 shows an appropriate mode, whereby A is initially in the tank and B is slowly added, whereby initially R and subsequently S are formed. These profiles correspond exactly to those of batch reaction, provided that the regime is chemical.

By writing the mass balances and kinetic expressions in dimensionless form, it is seen that in the chemical regime:

$$c_i(t)/\bar{c}_{A0} = f(k_1\bar{c}_{A0}t;\ k_1/k_2;\ \bar{c}_{B0}/\bar{c}_{A0} = N_{B0}/N_{A0};\ \text{operating mode}) \tag{109}$$

Clearly, physical variables (diffusivity, viscosity, energy dissipation rate etc.) are absent in this regime.

Other Mixing-Sensitive Reactions

A few potential alternatives to a pair of competitive, consecutive reactions are briefly listed here. although the list is not exhaustive. For a known or postulated kinetic scheme, the type of analysis given in the last section reveals the influence of the macroscopic flow pattern (e.g., plug flow and well-mixed represent extremes of no and complete backmixing) on yield. To investigate the corresponding influence of local, fine scale concentration gradients (segregation), the instantaneous local concentration is written as the spatially averaged value plus a spatially dependent deviation from this mean $c_i = \bar{c}_i + c_i'$ [23].

As a simple example, A isomerizes with first-order kinetics to a useful product P, whilst simultaneously forming dimers D and other oligomers. Thus, $r_p = k_1c_A$ whilst $r_D = k_2c_A^2$. Locally the selectivity is:

$$S = r_P/r_D = k_1c_A/k_2c_A^2 = k_1(\bar{c}_A + c_A')/k_2(\bar{c}_A + c_A')^2 \tag{110}$$

Experimentally, only the average selectivity from all such points can be measured:

$$\bar{S} = \bar{r}_p/\bar{r}_D = k_1\bar{c}_A/k_2(\bar{c}_A^2 + \overline{c_A'^2}) \tag{111}$$

Any segregation (mean square deviation exceeds zero) therefore influences product distribution and reduces selectivity relative to the homogeneous case. To quantify this result, the kinetics would be introduced into the diffusion-reaction model and the equations integrated numerically. This has not yet been done, because no suitable parallel reactions are currently available. Application of the simple concentration fluctuation method has, however, revealed a number of potentially valuable types of reactions [104] and subsequent exploitation may be expected.

Segregation influences include (a) the distribution of composition between the chains of a copolymer [31, 105, 106] and (b) the distribution of chain length in a homopolymer [31–34]. Despite some detailed analysis of copolymerization, little progress has been made on the corresponding experimentation and no well developed polymerization may currently be recommended for experimental work.

Mainly Qualitative Results from Mixing-Sensitive Reactions

Results will refer to product distributions from fast, single phase, multiple reactions. A 1926 study considered competitive reactions between various dissolved substances and added bromine water, and reported "difficulty arises of obtaining a homogeneous mixture of the reactants, since the reactions are so rapid that they take place at the point of entrance of the bromine water" [107]. It was also correctly recognized that such inhomogeneity reduces the differences (attributed to various chemical reactivities) in the rates of competition.

Nitration

The rate of nitration of toluene is some 21 to 27 times faster than that of benzene, as measured by the competitive method. The nitrating agent—the nitronium ion NO_2^+—has a concentration that is many orders of magnitude less than that of the nitric acid [108]. When the NO_2^+ concentration is greatly increased, by adding it as a soluble salt, toluene is nitrated only slightly faster than benzene. (Similar results have been obtained with heterogeneous nitration in a laminar jet [109]). However, this difference could be increased by faster stirring, lowering the amount of the NO_2^+-salt and substituting a magnetic stirrer by a more efficient agitator (Vibromixer) [110]. The author reported "reaction prior to homogenization in the layer adjacent to the entering nitronium salt solution," i.e., segregation.

Durene (1,2,4,5-tetramethylbenzene) should, when nitrated, form mono-(R) and di-nitrated (S) derivatives, yet little mononitrodurene has often been found [111]. This is surprising since $k_2/k_1 < 10^{-4}$ and thus in the chemical regime R should be highly favored over S. The resolution of this situation was the recognition that reaction was fast i.e. segregation promoted S over R. R could be obtained in significantly higher yield by lowering the NO_2^+ concentration (either by using more dilute solutions or adding water to convert NO_2^+ to nitric acid) or changing from a weakly agitated flask to an intensively agitated mixing head [111]. Further evidence of a falling R/S ratio as concentrations rise is available [112] and, together with the results of changing mixing intensity, is at least qualitatively consistent with the effect of the mixing modulus M on product distribution X.

A simplified model of simultaneous diffusion and reaction has been applied to the nitration of durene [113] as well as of other polymethylbenzenes [114]. Because the impeller and reaction vessel had not been independently characterized hydrodynamically and only one rate constant was known, a full comparison with diffusion-reaction theory was and is not worthwhile. Some semi-quantitative results e.g. ranking of k_2 values appear, however, to be quite reasonable. A critical review, written with the significance of masked reaction kinetics in mind, is available [108].

Iodination

A detailed study of the engineering factors influencing the yield of 3-mono-iodotyrosin (R) appeared in 1971 [25]. Semibatch iodination of 1-tyrosin (A) in tanks of two sizes using turbine impellers, having four diameters, was conducted at three concentration levels of iodine (c_{B0}) and therefore with three volume ratios ($\alpha = 5$, 10 and 20). The concentrated iodine solution was either added to the highly turbulent radial jet leaving a turbine or fed in slowly in the upper part of the tank. Three reaction temperatures were employed and in a few runs the viscosity was increased. Depending upon the conditions, the yield of R was up to 25% lower than in the chemical regime. The yield responded in the same direction as a diffusion-reaction model [115] to all experimental variables. The measured yield failed, however, to tend to the level calculated from the reaction kinetics as the stirrer speed was raised. Some reasons for this have been suggested [115] and it seems best to regard the agreement between theory and experiment as promising, but as yet unproven. Further kinetic meassurements would be worthwhile before adopting this reaction as a test system.

Mixing dependent product distributions during the iodination of p-cresol have been observed [116]. A full comparison of these with the predictions of a model does not seem to be available.

Bromination

The bromination of 1,3,5-trimethoxybenzene [27] was fast and gave only two products (mono- and dibromo-compounds). A clear effect of turbine speed was found. The solvent was, however, methanol, which presents hazards especially in larger than laboratory scale experiments, and moreover the kinetics have not been determined. However, some complications are likely, e.g., reaction of product Br^- with reagent Br_2, thus decreasing reaction rate.

The situation with the bromination of resorcin is similar [28]. Although the distribution of isomeric dibromoresorcins is sensitive to partial segregation, a full quantitative description would need some six rate constants. Complexity in modeling and experimentation make this reaction unattractive for routine work.

Alkaline Hydrolysis

The alkaline hydrolysis of glycol diacetate [117] has been characterized kinetically and, as might be anticipated, was chemically simple, with no side reactions. An influence of mixing on the yield of the intermediate (ethylene glycol monoacetate) has been reported [44, 116]. The reactions were, however, probably not fast enough to cause significant segregation, e.g., the half-lifetime of the first step was around 7 s and no difference in yield was observed when a premixed feed was replaced by separate feeds [116]. No analysis of the hydrodynamic regime was given [44, 116]. Substantially higher temperatures than those used (298–303 K) might be useful in shifting this reaction to the fast regime.

Diazo Coupling

The kinetics of the coupling of 1-naphthol-6-sulfonic acid (A) with the phenyl diazonium ion (B) in alkaline solution have been measured and the product distribution has been shown to be mixing-dependent [94]. The primary reaction between A and B is more complex than that postulated in the model ($A + B \rightarrow R$) and consists of two parallel steps ($A + B \rightarrow R$ and $A + B \rightarrow R'$). Although most of the primary coupling occurs in the 4 or p-position, some occurs simultaneously in the 2 or o-position. The secondary couplings ($R + B \rightarrow S$ and $R' + B \rightarrow S$) then follow in the o- and p-positions, respectively, whereby in both cases S is formed. From the point of view of checking models of micromixing, these additional reactions represent unnecessary complication. An experimental difficulty with this reaction is that the available analytical method is thin film chromatography, followed by photometry.

The mechanisms and the kinetics of diazo coupling reactions have been intensively studied [118] and, like the nitrations and halogenations already reffered to, are electrophilic substitutions. The couplings cited here take place between ions (simultaneous reactions between uncharged species are extremely slow), whose concentrations are related to the total concentrations of the respective substances (given by the masses added to the solution) through ionic equilibria and their pK values. The concentrations of the reactive species and their reaction rates thus depend strongly on pH. In electrophilic substitutions protons are liberated at the reaction site and in segregated mixtures the local pH can therefore sink below that in the bulk of the solution. In the bromination of resorcin, cited earlier, a local excess of protons shifted the phenolate (—O^-) form of resorcin to the phenol (—OH) form, accompanied by a dramatic fall in reactivity and changes in the reaction path and products [28]. Similarly, restricting the amount of buffer in diazo couplings (to cause the local pH at the reaction site to drop below that in the bulk) promoted changes in product distribution [94, 119]. A simplified diffusion-reaction model handled this situation well and helped to define the critical level of buffer concentration above which no pH shift can occur [120].

Quantitative Results from a Mixing-Sensitive Diazo Coupling

The coupling reactions between 1-naphthol (A) and diazotized sulfanilic acid (B) (O_3^-S-ϕ-N_2^+) near room temperature in dilute, alkaline buffered solutions are fast and may be represented as two competitive, consecutive reactions [26] (Equation 104). At pH = 10, the rates of each reaction attain their maximum values (this may be shown from the ionic pre-equilibria). When the concentration of the buffers ($NaHCO_3$ and Na_2CO_3) exceeds 50 to 100 times that of reagent A, the pH at the reaction site should also be buffered at 10. Since B is a zwitterion, no effect of ionic strength on reaction rate (primary salt effect) has been found. At 298 K, $k_1 = 7.3 \times 10^3$ $m^3\,mol^{-1}\,s^{-1}$, and $k_2 = 1.9$ $m^3\,mol^{-1}\,s^{-1}$. The activation energy of the second reaction is $E_2 = 37$ kJ mol^{-1}. The partial reaction orders are one.

Using 1-naphthol concentrations from 0.05 to 0.5 mol m^{-3}, the half-life times of the first reaction fall in the range 3 to 0.3 ms (Equation 96), which compared to a diffusive half-life time of order 20–80 ms (Equation 103) means that this reaction is certainly fast and might be instantaneous (A and B do not coexist). Integration of the kinetic equations for batch reaction shows that, when the initial A-concentration is 0.5 mol m^{-3}, B is completely consumed and the S-concentration is constant after 10 ms. This indicates how long reaction would need in the absence of segregation: since this time is shorter than a typical diffusional mixing time of at least 50 ms in turbulent water, the diazo coupling reactions occur in the fast regime.

Various possibilities exist to express the product distribution, e.g., yield of R (moles of R produced compared to moles of limiting reagent B consumed); ratio of quantities (moles) of R and S produced. In several studies the quantity of limiting reagent B converted to the secondary product S relative to the quantity of B reacting was chosen and denoted by X:

$$X_S = 2c_S/(c_R + 2c_S) \qquad (112)$$

Since B reacts fully, $c_R + 2c_S$ should finally be equal to $\bar{c}_{B0}$ (Equation 106), thus providing a mass balance check on the analysis. This was carried out on the product mixture, after reaction had been completed, using spectrophotometry. The absorption of the solution was measured at some 15 wavelengths between 400 and 650 nm. By assuming that the mono- and bis-azo dyestuffs absorb independently, a simple technique of linear regression then allows c_R and c_S to be determined. (As noted earlier, an on-line analytical method, giving the evolution of the various concentrations in time, would have been preferable, since it gives more information and permits a more detailed test of the model. It has not yet, however, been developed). The standard error of the X-values was around 0.005.

The solution of the partial differential equations, describing diffusion and reaction in a shrinking, laminated structure, has the following form for a given reactor operating mode:

$$c_i(x, t)/c_{A0} = f(x/\delta;\ k_1c_{A0}t;\ N_{A0}/N_{B0};\ k_1/k_2;\ \alpha = V_A/V_B;\ M = k_2c_{B0}\delta_0^2/D;\ Sc = \nu/D) \qquad (113)$$

The spectrophotometric analysis averages over x as $t \rightarrow \infty$. X then depends upon:

$$X = f(N_{A0}/N_{B0};\ k_1/k_2;\ \alpha;\ M;\ Sc;\ \text{operating mode}) \qquad (95)$$

This solution is sufficiently general to describe (a) the instantaneous, fully diffusion controlled regime ($M \rightarrow \infty$) where $X = 1$; (b) the chemical regime ($M \rightarrow 0$) where $X = X_{min} = f(N_{A0}/N_{B0};\ k_1/k_2)$; and

(c) the mixed regime with finite values of M where $X_{min} < X < 1$. For the coupling of 1-naphthol (A) with diazotized sulphanilic acid (B) at pH = 10 and near 298 K, $k_1 = 3{,}860k_2$ so that, with an initial stoichiometric ratio N_{A0}/N_{B0} of 1.05, $X_{min} = 0.001$ (batch and tubular reactors), and $X_{min} = 0.009$ (continuous, well-stirred tank reactor). (These X-values were calculated using Equations 107 and 108, respectively).

A summary of some experimental results obtained with this diazo coupling now follows.

Operating mode. Semi-batch operation involved feeding a relatively small volume of concentrated diazonium salt solution into a stirred, alkaline 1-naphthol solution. Calculation with the model showed a somewhat higher X-value from a semi-batch than from a batch reactor as soon as M was sufficiently high to reach the mixed regime [115, 121]. A macroscopically uniform, continuous stirred tank should give a still higher X-value, and this has been quantitatively confirmed [26, 72, 122]. No effect on X of the mean residence time (V/Q) in a CSTR and of the time of addition of B to a semi-batch reactor has been observed [121]. These times were much greater than the time needed for diffusion and reaction and therefore in all cases the limiting reagent (B) was essentially completely consumed, leading to the same product distribution.

Stoichiometric ratio. Increasing the ratio N_{A0}/N_{B0} reduces X as predicted [26, 114].

Volumetric ratio. When the volumetric ratio $\alpha = V_A/V_B$ is changed at constant stoichiometric ratio, nor change in X occurs in the chemical regime. Increasing α means, for example, that a given quantity of B will be added in a smaller volume, but at higher concentration to a given quantity of A. The model predicts steeper concentraion gradients (more segregation) and higher X-values, in quantitative agreement with experiments [26, 94, 121]. The model also predicts that at constant M, X decreases with increasing α, which has also been observed [25, 115]; an alternative model formulation, ignoring deformation of fluid elements, predicts an increase and may be rejected [123].

Mixing modulus. M has been varied by changing the feed concentration of the diazonium salt solution (c_{B0}), and the measured X-values were successfully correlated [22, 121]. The transition between the chemical regime ($X = X_{min}$ irrespective of M) and the mixed regime (X increases with rising M) is of course not sharp; nevertheless transition values of M seem to agree with the model [22, 121]. M depends upon the initial half-thickness of fluid layers $\delta_0 = 0.5\lambda_K = 0.5(\nu^3/\epsilon)^{1/4}$ which in turn is a function of the rate of energy dissipation and of the viscosity. These variables will be discussed separately.

Viscosity. An increase in viscosity causes essentially no change in ϵ in fully turbulent flow, an increase in λ_K, possibly a decrease in D and an increase in Sc. The resultant increase in M tends to increase X, although this is to some extent offset by the increase in Sc [22]. Although experimental results in viscous solutions were more scattered than when no viscosity-increasing additives were present, reflecting mainly increased analytical errors, there is no doubt that higher viscosities cause more segregation and higher M and X values [122]. Similar trends have been seen with other reactions [25, 115, 124]. Thus, although the Reynolds number remains sufficiently high for turbulent flow, micromixing is sensitive to viscosity (and diffusivity), as one would expect for processes in the viscous-convective and viscous-diffusive subranges.

Rate of energy dissipation. δ_0 is reduced by increasing ϵ, which in a fully turbulent, non-homogeneous flow may be written $\epsilon = \phi\bar{\epsilon}$, where $\bar{\epsilon}$ is the mean rate of energy dissipation, and ϕ is position-dependent. For a stirred tank, for instance, $\bar{\epsilon} = Po n^3 d^5/V$, where Po is the impeller power number (this is a form-dependent constant in fully turbulent flow), n and d are the rotational speed and diameter of the impeller, and V is the liquid volume. By varying impeller speed, size and type, it was shown that X depends only on ϵ [121], i.e., on the power transmitted to a tank. Experiments have also shown a very strong dependence of ϕ on position, i.e., upon where the feed stream of B enters the reactor [22, 121]. Using the attractive and simple, yet effective method first described in [125], the zone over which rapid reaction is occurring can be made visible and it was confirmed that micromixing was local [22], fluid often moving only a few cm during diffusion and fast reaction. In the slow reaction regime, X would have been 0.001 (batch, tubular and semi-batch reactors) or

0.009 (CSTR). Depending upon the experimental conditions [22, 121, 122] values of X as high as 0.5 have been measured and predicted by the model [22].

CONCLUDING REMARKS

Micromixing has long presented problems. What are the mechanisms by which it occurs? How should it be mathematically modeled? How secure is such modeling? Do unequivocal experiments agree well with the models, without having to fit various quantities? Although Danckwerts [11] in 1958 helped to structure and define micromixing, as late as 1977 and 1982, respectively, one could read "Matters are by no means so clear when it comes to micromixing. Many theoretical models have already been proposed, yet unfortunately their number greatly exceeds that of the available experimental results. Therefore, it would presently be too early to apply considerations of micromixing in industry. Certainly, its influence is only important in certain special cases, e.g., reactions in viscous media (for instance polymerization), or with fast reactions with on-line mixing of the reagents (for instance in flames)" [126]. "There is still a need for a unified theory allowing a priori predictions from the sole knowledge of physicochemical properties and operating parameters" [4]. The present authors have sought to meet these requirements by developing a description of micromixing, which is physically based. Thus results and ideas about the fine scale structure of turbulence were used to build up the descriptions. Its application to a pair of competitive, consecutive reactions was given and this was then compared with experimental results. The following procedure to treat any kind of single-phase reaction results.

1. *Physico-chemical information.* The following quantities should be known under the conditions pertaining during reaction: reaction scheme and mechanism, reaction kinetics, diffusivities of all reagents, kinematic viscosity of solution(s). Summarising: get k_j, D_i, and ν.
2. *Hydrodynamic information.* The rate of energy dissipation in the region, where diffusion and reaction occur, is the most critical quantity (the velocity distribution would be useful; it is not, however, essential). Often, the mean rate of energy dissipation, obtained from a friction factor (or power number) and characteristic velocities and lengths in the usual way, differs substantially from the local value, and information about the spatial distribution of ϵ is needed. Values resulting from computational fluid dynamics are now becoming available. Measurements in stirred tanks refer only to Rushton turbines and are not all mutually consistent [22]. Summarizing: get ϵ.
3. *Operating information.* Mode of reactor operation (semi-batch etc.), reactor volume, volumes (or volumetric flow rates) of reagent solutions and their initial concentrations. Summarizing get V, V_A, V_B, c_{A0}, c_{B0}, reactor type.
4. *Reaction regime.* As explained earlier $t_{1/2R}$ is determined from the reaction kinetics and $t_{1/2D}$ from Equation 103, which needs ν, ϵ and $Sc = \nu/D$. Suppose, after comparing these characteristic times, that reaction is "slow": then, kinetics and classical methods of reaction engineering suffice. If, however, reaction is "fast," the full modeling developed here is needed to find the product distribution, e.g., when the limiting reagent has been consumed or when micromixing has been operative for a known period of time.
5. *Application of full model.* The basic model equation is an extended Fick's law (Equation 64), describing unsteady state diffusion and reaction in a shrinking laminated structure (Figure 18). The specific reaction kinetics enter via the term R_i in Equation 88. The initial half-thickness $\delta_0 = 0.5\lambda_K = 0.5(\nu^3/\epsilon)^{1/4}$. The deformation kinematics are given by Equation 89. Numerical solution of the set of coupled Equation 88 is needed. (This involves solving partial differential equations and a simplification of the full model to, say, ordinary d.e. needs further study). The solution will have the same form as Equation 113 or, as the reaction time for fast reactions tends to infinity, as Equation 95. In this application of the full model, no parameters have to be fitted; all parameters have been related to information in the earlier groups 1–3 using fluid mechanical arguments.

Scope for further investigation of micromixing exists. Experimental goals include (a) extension of determinations of micromixing in viscous solutions, including Reynolds numbers in the transition region and non-Newtonian media; (b) studies in geometrically similar reactors of various scales and definition of scale-up criteria (the model predicts ϵ = constant); and (c) on-line analytical methods to follow the course of fast reactions in time. Objectives in the context of modeling include (a) improvement of parameter estimates in the model previously presented using new theoretical and especially experimental results from fluid mechanics; (b) more extensive and in particular more rigorous comparisons between models and experimental results for fast reactions, avoidance of "fitting"; (c) simplification of newest model and for instance reduction to ordinary differential equations; and (d) consideration of the influence on modeling of the statistical distributions of the various fluid mechanical parameters (in this text only their mean values have been employed).

The conversion of a second order reaction, when all reagents are present in a pre-mixed feed stream, is most sensitive to segregation when the residence time distribution corresponds to that of a macroscopically well-mixed tank. (There is no difference of conversion in plug flow reactors, irrespective of whether the reaction mixture exhibits total segregation or complete homogeneity at the molecular level). Considering steady-state operation of a continuous, stirred tank reactor, the maximum difference in conversion as the level of micromixing changes from total segregation to complete homogeneity is 7%. During unsteady-state operation of such a reactor, however, larger differences in conversion are predicted by McCord's analysis. The literature contains few experimental results obtained under unsteady conditions and such experiments and their evaluation are worth further investigation.

NOTATION

C	constant	M	mixing modulus
c_i	local concentration	$Q_{c,c}$	Eulerian spatial correlation
C_i	concentration factor	$R_{c,c}$	correlation coefficient
c_r	universal turbulence constant	R_i	reaction rate
D	diffusion coefficient	S	selectivity
k_1, k_2	reaction rate constants	Sc	Schmidt number
$G(k, t)$	spectral density	T_w	dimensionless eddy lifetime
$g_{a,b}$	reactor entrance and exit rates	t	time
h	mass transfer coefficient	t_D	time constant
J	degree of segregation	t_e	characteristic erosion time constant
k_B	Batchelor wave number	v	specific volume
l	diameter of shrinking aggregate	V_s	fluid volume
l_0	initial size of fluid elements	$\bar{X}$	dimensionless distance
m	fluid mass	X_s	product distribution

Greek Symbols

α	age of molecules	λ_s	dimension of eddies
γ	strain rate	ν	kinematic viscosity
δ	distance or thickness	τ	time
ϵ	void fraction	$\phi(\alpha)$	age distribution frequency
Λ_c	integral scale for concentration fluctuations	$\psi(\lambda)$	life expectation frequency

REFERENCES

1. Beek, J., Jr., and Miller, R. S. "Turbulent transport in chemical reactors." *Chem. Eng. Progr. Symp. Ser.*, vol. 55, no. 25, p. 23 (1959).
2. Brodkey, R. S. "Mixing in turbulent fields," Chapter 2, in *Turbulence in Mixing Operations.* R. S. Brodkey, ed. Academic Press, New York (1975).
3. Brodkey, R. S. "Fundamentals of turbulent motion, mixing, and kinetics." *Chem. Eng. Commun.*, vol. 8. p. 1 (1981).

4. Villermaux, J. "Mixing in chemical reactor," in *Chemical Reaction Engineering*. ACS Symp. Ser. no. 226, Washington (1983).
5. Patterson G. K. Chapter 5 in Handbook of Fluids in Motion, N. P. Cheremisinoff and R. Gupta editors, Ann Arbor Science Pub., Ann Arbor, Mich. (1983).
6. Hill, J. C. "Homogeneous turbulent mixing with chemical reaction." *Ann. Rev. Fluid Mech.*, vol. 8, p. 135 (1976).
7. Patterson, G. K. "Application of turbulence fundamentals to reactor modeling and scale-up." *Chem. Eng. Commun.*, vol. 8, p. 25 (1981).
8. Murthy, S. N. B. "Turbulent mixing in chemically reactive flows", in *Turbulence in Mixing Operations*. R. S. Brodkey, ed., Academic Press, New York (1975).
9. Danckwerts, P. V. "Continuous flow systems: Distribution of Residence Time." *Chem. Eng. Sci.*, vol. 2, p. 1 (1953).
10. Zwietering, T. N. "Degree of mixing in continuous systems." *Chem. Eng. Sci.*, vol. 11, p. 1 (1959).
11. Danckwerts, P. V. "Effect of incomplete mixing on homogeneous reaction." *Chem. Eng. Sci.*, vol. 8, p. 93 (1958).
12. Bosworth, R. C. L. "Distribution of reaction times for laminar flow in cylindrical reactors." *Phil. Mag.*, vol. 39, p. 847 (1948).
13. Nauman, E. B. "Residence time distributions, and micromixing." *Chem. Eng. Commun.*, vol. 8, p. 53 (1981).
14. Villermaux, J. "Recent advances in micromixing phenomena in stirred reactors." *Chem. Eng. Commun.*, vol. 21, p. 105 (1983).
15. Plasari, E., David, R., and Villermaux, J. "Micromixing phenomena in continuous stirred reactors using a Michaelis-Menten reaction in the liquid phase," chapter 11 in *Chemical Reaction Engineering* (*Houston*), ACS Symp. Ser. no. 65, Washington, D.C. (1978).
16. Baldyga, J. Doctoral Thesis, Politechnika Warszawska, Warsaw (1980).
17. Pohorecki, R. and Baldyga, J. "Use of a new model of micromixing for determination of crystal size in precipitation." *Chem. Eng. Sci.*, vol. 38, p. 79 (1983).
18. Pohorecki, R. and Baldyga, J. "New model of micromixing in chemical reactors: 1. General development and application to a tubular reactor. 2. Application to a stirred tank reactor." *Ind. Eng. Chem. Fund.*, vol. 22, p. 392, p. 398 (1983).
19. Ottino, J. M., Ranz, W. E., and Macosko, C. W. "A lamellar model for analysis of liquid–liquid mixing." *Chem. Eng. Sci.*, vol. 34, p. 877 (1979).
20. Angst, W., Bourne, J. R., and Sharma, R. N. "Dimensions of the reaction zone." *Chem. Eng. Sci.*, vol. 37, p. 585 (1982).
21. Spalding, B. D. "Chemical reactions in turbulent fluids." *Physicochem. Hydrodynamics, Advance Publications*, vol. 1, p. 321, Oxford (1978).
22. Baldyga, J. and Bourne. J. R. "Computational and experimental results for the new micromixing model." *Chem. Eng. Commun.* vol. 28, p. 259 (1984).
23. Toor, H. L. "Non-premixed reactions," chapter 3 in Ref. 2.
24. Brodkey, R. S. "Turbulent motion, mixing, and kinetics," p. 289 in Ref. 21.
25. Paul, E. L. and Treybal, R. E. "Mixing and product distribution for a liquid–phase, second-order, competitive, consecutive reaction." *AIChE. J.*, vol. 17, p. 718 (1971).
26. Bourne, J. R., Kozicki, F, and Rys, P. "Test reactions to determine segregation." *Chem. Eng. Sci.*, vol. 36, p. 1643 (1981).
27. Bourne, J. R. and Kozicki, F. "Mixing effects during the bromination of 1,3,5-trimethoxybenzene." *Chem. Eng. Sci.*, vol. 32, p. 1538 (1977).
28. Bourne, J. R., Rys, P., and Suter, K. "Mixing effects in the bromination of resorcin." *Chem. Eng. Sci.*, vol. 32, p. 711 (1977).
29. Ou, J. J. and Ranz, W. E. "Mixing and chemical reactions: Chemical selectivities." *Chem. Eng. Sci.*, vol. 38, p. 1015 (1983).
30. Barthole, J. P. et al. "Macroscopic kinetics of precipitation of barium sulphate in presence of EDTA" (in French). *J. de Chim. Phys.*, vol. 79, p. 719 (1982).
31. Nauman, E. B. "Mixing in polymer reactors." *J. Macromol. Sci.*, vol. C-10, p. 75 (1974).
32. Gerrens, H. "Polymerization reactors and polyreactions." *Proc. Symp. Chem. React. Eng.*, p. 585. DECHEMA, Frankfurt (1976).

33. Chatterjee, A., Park, W. S., and Graessley, W. W. "Free radical polymerization with long chain branching: Continuous polymerization of vinylacetate in t-butanol." *Chem. Eng. Sci.*, vol. 32, p. 167 (1977).
34. Baade, W. and Reichert, K. H. "Molecular weight distribution and branching of polyvinylacetate as functions of micro- and macro-mixing" (in German). *Chem. Ing. Tech.*, vol. 53, MS 910 (1981).
35. Ravindranath, K. and Mashelkar, R. A. "Modeling of poly(ethylene terephthalate) reactors: Final stages of polycondensation." *Poly. Eng. and Sci.*, vol. 22, p. 628 (1982).
36. Hinze, J. O. *Turbulence*, 2nd ed. McGraw-Hill, New York (1975).
37. Tennekes, J. H. and Lumley, I. L. *A First Course in Turbulence.* MIT-Press, Cambridge, Mass. (1972).
38. Danckwerts, P. V. "Definition and measurement of some characteristics of mixtures." *Appl. Sci. Res.*, vol. A3, p. 279 (1953).
39. Chauhan, S. P., Bell, J. P., and Adler, R. J. "On optimum mixing in continuous, homogeneous reactors". *Chem. Eng. Sci.*, vol. 27, p. 585 (1972).
40. Methot, J. C. and Roy, P. H. "Segregation effects on homogeneous second-order reactions." *Chem. Eng. Sci.*, vol. 26, p. 569 (1971).
41. Novosad, Z. and Thyn, S. "Absolute conversion limits for a series of backmix reactors". *Collect. Czech. Chem. Commun.*, vol. 31, p. 3710 (1966).
42. Levenspiel, O. *Chemical Reaction Engineering*, Chapter 10. John Wiley and Sons, New York (1972).
43. Tadmor, Z. and Biesenberger, J. A. "Influence of segregation on molecular weight distribution in continuous linear polymerizations." *IEC Fund.*, vol. 5, p. 336 (1966).
44. Truong, K. T. and Methot, J. C. "Segregation effects on consecutive competing reactions in a CSTR." *Can. J. Chem. Eng.*, vol. 54, p. 572 (1976).
45. Harada, M. et al. "Effect of micromixing on homogeneous polymerization of styrene." *J. Chem. Eng. Japan*, vol. 1, p. 148 (1968).
46. Treleaven, C. R. and Tobgy, A. H. "Conversion in reactors having separate reactant feed streams." *Chem. Eng. Sci.*, vol. 26, p. 1259 (1971).
47. Chen, M. S. K. "Theory of micromixing for unsteady state flow reactors." *Chem. Eng. Sci.*, vol. 26, p. 17 (1971).
48. Nauman, E. B. "Note on RTD in cyclic reactors." *Chem. Eng. Sci.*, vol. 28, p. 313 (1973).
49. Fan, L. T., Fan, L. S., and Nassar, R. F. "Stochastic model of unsteady state age distribution." *Chem. Eng. Sci.*, vol. 34, p. 1172 (1979).
50. Villermaux, J. "Génie de la Réaction Chimique—Conception et Fonctionnement des Réacteurs." *Technique et Documentation*, Paris (1982).
51. McCord, J. R., III. "Stochastic analysis of extremes of micromixing in non-steady state, continuous flow chemical reactor." *Chem. Eng. Sci.*, vol. 27, p. 1613 (1972).
52. Weinstein, H. and Adler, R. J. "Micromixing effects in continuous chemical reactors". *Chem. Eng. Sci.*, vol. 22, p. 65 (1967).
53. Villermaux, J. and Zoulalian, A. "State of mixedness of fluid in a continuous reactor" (in French). *Chem. Eng. Sci.*, vol. 24, p. 1513 (1969).
54. Ng, D. Y. C. and Rippin, D. W. T. "Effect of incomplete mixing on conversion in homogeneous reactions". *Proc. Symp. Chem. React. Eng.*, Amsterdam (1964). Pergamon Press, Oxford, p. 161 (1965).
55. Nishimura, Y. and Matsubara, M. "Micromixing theory via the two environment mode." *Chem. Eng. Sci.*, vol. 25, p. 1785 (1970).
56. Methot, J. C. and Roy, P. H. "Experimental evaluation of a model based degree of segregation in a CSTR." *Chem. Eng. Sci.*, vol. 28, p. 1961 (1973).
57. Ritchie, B. W. and Tobgy, A. H. "Three environment micromixing model for chemical reactors with arbitrary separate feed stream." *The Chem. Eng. J.*, vol. 17, p. 173 (1979).
58. Miyawaki, O., Tsujikawa, H., and Uraguchi, Y. "Chemical reactions under incomplete mixing." *J. Chem. Eng. Japan*, vol. 8, p. 63 (1975).
59. Spielman, L. A. and Levenspiel, O. "Monte Carlo treatment for reacting and coalescing dispersed phase systems." *Chem. Eng. Sci.*, vol. 20, p. 247 (1965).

60. Kattan A., and Adler, R. J. "Conceptual framework for mixing in continuous chemical reactors." Chem. Eng. Sci., vol. 27, p. 1013 (1972).
61. Treleaven, C. R. and Tobgy, A. H. "Residence times, micromixing and conversion in an unpremixed fixed reactor." *Chem. Eng. Sci.*, vol. 28. 413 (1973).
62. Goto, S. and Matsubara, M. "Generalized two environment model for micromixing in a continuous flow reactor." *Chem. Eng. Sci.*, vol. 30, p. 61 (1975).
63. Goto, H, Goto, S. and Matsubara, M. ibid., p. 71.
64. Leslie, D. L. *Developments in the Theory of Turbulence*, Chapter 8. Clarendon Press, Oxford (1973).
65. Batchelor, G. K. "Small scale variation of convected quantities like temperature in turbulent fluid." *J. Fluid Mech.*, vol. 5, p. 113 (1959).
66. Batchelor, G. K. and Townsend, A. A. "Turbulent diffusion", in *Surveys in Mechanics*, G. K. Batchelor and R. M. Davies, ed. Cambridge Univ. Press (1956).
67. Bolzern, O. and Bourne, J. R. "Extension of the reaction zone". *Chem. Eng. Sci.*, vol. 38, p. 999 (1983).
68. Ottino, J. M. "Lamellar mixing models for structured chemical reactions and their relationship to statistical models." *Chem. Eng. Sci.*, vol. 35, p. 1377 (1980).
69. Baldyga, J. and Bourne, J. R. "Initial deformation of material elements in isotropic, homogeneous turbulence." *Chem. Eng. Sci.*, vol. 39, p. 329 (1984).
70. Kuo, A. Y. and Corrsin, S. "Experiment on the geometry of the fine structure regions in fully turbulent fluid." *J. Fluid Mech.*, vol. 56, p. 447 (1972).
71. Townsend, A. A. "Diffusion of heat spots in isotropic turbulence." *Proc. Roy. Soc.*, vol. A-209, p. 418 (1951).
72. Bourne, J. R. and Rohani, S. "Deforming reaction zone model for the CSTR." *Chem. Eng. Sci.*, vol. 38, p. 911 (1983).
73. Baldyga, J. and Bourne, J. R. "Micromixing in the light of turbulence theory." *Chem. Eng. Commun.* vol. 28, p. 293 (1984).
74. Klein, J. P., David, R. and Villermaux, J. "Interpretation of experimental liquid phase micromixing phenomena in a CSTR with short residence times." *Ind. Eng. Chem. Fund.*, vol. 19, p. 373 (1980).
75. Rosensweig, R. F. "Idealized theory for turbulent mixing in vessels." *AIChE. J.*, vol. 10, p. 91 (1964).
76. Rietema, K. "Heterogeneous reactions in the liquid phase." *Chem. Eng. Sci.*, vol. 8, p. 103 (1958).
77. Harada, M. et al., *Memoirs Fac. Eng.*, Kyoto Univ., vol. 24, p. 431 (1962).
78. Curl, R. L. "Dispersed phase mixing." *AIChE. J.*, vol. 9, p. 175 (1963).
79. Patterson, G. K. "Simulating turbulent field mixers and reactors," Chapter 5 in Ref. 2.
80. Canon, R. M., Smith, K. W. and Patterson, G. K., "Turbulence level significance of the coalescence-dispersion rate parameter." *Chem. Eng. Sci.*, vol. 32, p. 1349 (1977).
81. Rao, D. P. and Dunn, I. J. "Monte Carlo coalescence model for reaction with dispersion in a tubular reactor." *Chem. Eng. Sci.*, vol. 25, p. 1275 (1970).
82. Patterson, G. K. "Modeling complex chemical reactions in flows with turbulent diffusive mixing." 70th Ann. AIChE Meeting, New York (1977).
83. Villermaux, J. and Devillon, J. C. "Representation of coalescence and redispersion of segregated fluid elements with a phenomenological model" (in French). *Proc. Symp. Chem. React. Eng.*, B-1-13. Elsevier, Amsterdam (1972).
84. Costa, P. and Trevissoi, C. "Some kinetic and thermodynamic features of reactions between partially segregated fluids." *Chem. Eng. Sci.*, vol. 27, p. 653 (1972).
85. Costa, P. and Trevissoi, C. "Reactions with non-linear kinetics in partially segregated fluids." *Chem. Eng. Sci.*, vol. 27, p. 2041 (1972).
86. Corrsin, S. "Isotropic turbulent mixer." *AIChE. J.*, vol. 10, p. 870 (1964).
87. Villermaux, J. "Drop break-up and coalescence. Micromixing effects in liquid–liquid reactors," in *Multiphase Chemical Reactors*, vol. I, A. E. Rodrigues, J. M. Cato, and N. H. Sweed, (eds.) Sijthoff Noordhoff, Alphen, Netherlands (1981).
88. Mao, K. W. and Toor, H. L. "Diffusion model for reactions with turbulent mixing." *AIChE. J.*, vol. 16, p. 49 (1970).

89. Mao, K. W. and Toor, H. L. "Second order chemical reactions with turbulent mixing." *IEC Fund.*, vol. 10, p. 192 (1971).
90. Rao, D. P. and Edwards, L. L. "Mixing effects in stirred tank reactors: Comparison of models." *Chem. Eng. Sci.*, vol. 28, p. 1179 (1973).
91. Nauman, E. B. "Droplet diffusion model for micromixing." *Chem. Eng. Sci.*, vol. 30, p. 1135 (1975).
92. Ott, R. J. and Rys, P. "Simple model of mixing-disguised reactions in solution." *Helv. Chim. Acta*, vol. 58, p. 2074 (1975).
93. Nabholz, F., Ott, R. J. and Rys, P. "Comparison of two versions of a simple model of mixing-disguised reactions in solution." *Helv. Chim. Acta*, vol. 60, p. 2926 (1977).
94. Bourne, J. R., Crivelli, E. and Rys, P. "Mixing-disguised azo coupling reaction." *Helv. Chim. Acta*, vol. 60, 2944 (1977).
95. Geurden, J. M. and Thoenes, D. "Influence of mixing on rate of conversion of chemical reactions in viscous liquids," in Ref. 83.
96. Ottino, J. M. "Efficiency of mixing from data on fast reactions in multi-jet reactors and stirred tanks." *AIChE. J.*, vol. 27, p. 184 (1981).
97. Ottino, J. M. "Description of mixing with diffusion and reaction in terms of the concept of material surfaces." *J. Fluid Mech.*, vol. 114, p. 83 (1982).
98. Ottino, J. M. and Macosco, C. W. "Efficiency parameter for batch mixing of viscous fluids." *Chem. Eng. Sci.*, vol. 35, p. 1454 (1980).
99. Baldyga, J. and Bourne, J. R. "Inadequacies of available methods." *Chem. Eng. Commun.* vol. 28, p. 231 (1984).
100. Batchelor, G. K. "Mass transfer from small particles suspended in turbulent fluid." *J. Fluid Mech.*, vol. 98, p. 609 (1980).
101. Worrell, G. R. and Eagleton, L. C. "Experimental study of mixing and segregation in a stirred tank reactor." *Can. J. Chem. Eng.*, vol. 42, p. 254 (1964).
102. Lintz, H. G. and Weber, W. "Study of mixing in a continuous, stirred tank reactor using an autocatalytic reaction." *Chem. Eng. Sci.*, vol. 35, p. 203 (1980).
103. Lintz, H. G. and Weber, W. "Experimental study of mixing in continuous reactors using an autocatalytic reaction." *Chem. Eng. Fund.*, vol. 1, p. 27 (1982/83).
104. Bourne, J. R. and Toor, H. L. "Simple criteria for mixing effects in complex reactions." *AIChE. J.*, vol. 23, p. 602 (1977).
105. Szabo, T. T. and Nauman, E. B. "Copolymerization and terpolymerization in continuous, non-ideal reactors." *AIChE. J.*, vol. 15, p. 575 (1969).
106. Mecklenburgh, J. C. "Influence of mixing on the distribution of copolymerization compositions." *Can. J. Chem. Eng.*, vol. 48, p. 279 (1970).
107. Francis, A. W. "Relative rates of certain ionic reactions." *J. Am. Chem. Soc.*, vol. 48, p. 655 (1926).
108. Schofield, K. *Aromatic Nitration.* Chapter 8. Cambridge Univ. Press (1980).
109. Hanson, C. and Ismail, H. A. M. "Macrokinetics of toluene and benzene nitration under laminar conditions." *Chem. Eng. Sci.*, vol. 32, p. 775 (1977).
110. Tolgyesi, W. S. "Relative reactivity of toluene-benzene in nitronium tetrafluoroborate nitration." *Can. J. Chem.*, vol. 43, p. 343 (1965).
111. Hanna, S. B. et al. "Problem of the mononitration of durene" (in German). *Helv. Chim. Acta*, vol. 52, p. 1537 (1969).
112. Hunziker, E., Penton, J. R. and Zollinger, H. "Nitration of durene and pentamethylbenzene with nitronium salts in nitromethane and acetonitrile." *Helv. Chim. Acta*, vol. 54, p. 2043 (1971).
113. Pfister, F., Rys, P. and Zollinger, H. "Mixing-disguised nitrations of aromatic compounds with nitronium salts." *Helv. Chim. Acta*, vol. 58, p. 2093 (1975).
114. Nabholz, F. and Rys, P. "Mixing-disguised nitrations of aromatic compounds with nitronium salts." *Helv. Chim. Acta*, vol. 60, p. 2937 (1977).
115. Bourne, J. R. aand Rohani, S. "Micromixing and the selective iodination of 1- tyrosine." *Chem. Eng. Res. Des.*, vol. 61, p. 297 (1983).
116. Zoulalian, A. and Villermaux, J. "Influence of chemical parameters on micromixing in a CSTR." *Adv. in Chem.* no. 133, p. 348. Am. Chem. Soc. (1974).

117. Aubry, C., Zoulalian, A., and Villermaux, J. "Kinetic study of alkaline hydrolysis of glycol diacetate" (in French). *Bull. Soc. Chim. France*, no. 7, p. 2483 (1971).
118. Zollinger, H. *Azo and Diazo Chemistry*. Interscience, New York (1961).
119. Belevi, H., Bourne, J. R., and Rys, P. "Influence of pH-gradients on the product distribution mixing-disguised azo coupling reactions." *Helv. Chim. Acta*, vol. 64, p. 1618 (1981).
120. Belevi, H., Bourne, J. R., and Rys, P. "Simple model of pH-dependence of product distribution in mixing-disguised azo coupling reactions." *Helv. Chim. Acta*, vol. 64, p. 1599 (1981).
121. Angst, W., Bourne, J. R., and Dell'Ava, P. "Comparison between models and experiments." *Chem. Eng. Sci.*, vol. 39, p. 335 (1984).
122. Bourne, J. R., Kozicki, F., Moergeli, U., and Rys, R. "Model-experiment comparisons." *Chem. Eng. Sci.*, vol. 36, p. 1655 (1981).
123. Angst, W., Bourne, J. R., and Sharma, R. N. "Influence of diffusion within the reaction zone on selectivity." *Chem. Eng. Sci.*, vol. 37, p. 1259 (1982).
124. Bourne, J. R., Schwarz, G., and Sharma, R. N. "New methods for studying diffusive mixing in a tubular reactor." Paper J3, 4th European Conf. on Mixing, BHRA Fluid Eng., Cranfield, U.K. (1982).
125. Rice, A. W., Toor, H. L., and Manning, F. S. "Scale of mixing in a stirred vessel." *AIChE. J.*, vol. 10, p. 125 (1964).
126. Cappelli, A. and Trambouze, P. "The significance of chemical reaction engineering for industry" (in German). *Chem. Ing. Tech.*, vol. 49, p. 5 (1977).

CHAPTER 7

RAYLEIGH-TAYLOR INSTABILITY

Louis Baker

Mission Research Corp.
Albuquerque, New Mexico

CONTENTS

INTRODUCTION

When a glass full of water is held upside down, the water cannot simply fall down; that would create a partial vacuum near the base of the glass. In fact, the inverted glass of water is in equilibrium, as may be demonstrated by covering the mouth of the glass with a piece of cardboard. Atmospheric pressure will hold the cardboard in place—nothing will fall. The equilibrium is unstable, however, and if the cardboard is removed, the water will leave the glass. The water-air interface will ripple; then sheets or "spikes" of water will descend while "bubbles" of air will rise into the glass. When a bubble extends continuously to the glass base, the water may then fall freely. This instability is the Rayleigh-Taylor instability.

The principal interest in this instability stems from circumstances in which acceleration mimics the effect of a gravitational field. The basis of Einstein's general theory of relativity is the principle of equivalence of gravitation and inertia, which states that a (uniform) gravitational field is indistinguishable from an acceleration. Rayleigh-Taylor instability can therefore arise not only when a dense material is above a less dense material, but when a dense material is pushed upon and accelerated by a lighter material. This can and does happen in a number of interesting circumstances.

One such circumstance occurs when explosives are used to implode the plutonium core of an atomic bomb. Concern about the deleterious effects of the instability on the bomb's performance lead to its study by the British meteorologist, G. I. Taylor [1], whose work was published several years later after it was declassified. It was then noted that Lord Rayleigh [2] had studied the problem many years earlier, whence the name Rayleigh-Taylor instability; one may find it referred to as Taylor instability in many papers dating from this period. Taylor instability has been observed in solids [3] subject to a shock wave from an explosive lens. Elastic-plastic effects play a role in this case.

The instability is of importance in many controlled fusion applications as well. If a magnetic field is used to accelerate material, we have an effectively massless fluid accelerating another, which can be unstable. If a laser or particle beam is being used in an inertial-confinement fusion reactor to compress some hydrogen fuel, we again have dense material being accelerated by the pressure in a low-density "blowoff" plasma. This produces a rocket effect which compresses the fuel, but it may be expected to be unstable. Many of these target designs have a complicated arrangement of shells, often with an innermost volume containing the hydrogen fuel surrounded by a dense metallic shell. As the target implodes, the pressure in the fuel increases along with the fuel temperature. The containing shell(s) eventually stagnate. At this point, they are being decelerated, i.e., accelerated outward, by the hydrogen and this interface is generally unstable at this time of shell "turn-around." Such instabilities are of great concern [4] since they can cause mixing of the fuel and the shell, degrading the burning of the fuel, or they can rupture the shell and prevent the successful implosion of the target.

A shock wave striking a rippled interface can cause a growth of the corrugations in an analogous fashion. The implosions mentioned are subject to such instablities, whether they are explosively-driven or driven by magnetic fields, lasers, or particle beams, as all will generally produce shock-waves in the target.

A kinetic sculpture called a "Color Window" created by Rachel bas-Cohain and marketed by Artex Co. of Concord, Mass., employs Rayleigh-Taylor instability and may be used to demonstrate its behavior.

More generally, there are many instabilities driven by buoyancy in which dense materials find themselves above lighter ones. This can occur when fluids are heated from below (Rayleigh-Benard instability, or "thermal convection"). It may even occur when certain microorganisms tend to swim upward. In the ocean, the combined effects of salinity and temperature gradients can produce stratifications that lead to complex motion patterns, resulting in a stepped or layered temperature structure, or patterns such as "salt fingers" descending into less dense layers below. In astrophysics, unstable density gradients due to compositional gradients can lead to "semiconvection," while those due to entropy gradients lead to convection. The Schwarzschild criterion specifies that convection will occur if a rising parcel of gas finds itself lighter than its surroundings, or if a sinking parcel finds itself heavier. This is the Rayleigh-Taylor instability modified by thermal transport and compressibility effects. Instead of the instability criterion being the density gradient opposite gravity, it is that the entropy gradient be in the direction of gravity. In the stable case, the analog of the Rayleigh-Taylor growth rate becomes the Brunt-Vaisala frequency.

LINEAR THEORY

The simplest case is that studied by Taylor and Rayleigh of two distinct, incompressible, inviscid fluids (as sketched in Figure 1). We may view gravity as acting downward, i.e., in the negative z direction, or, equivalently, assume the system is accelerating upward in the positive z direction. The amplitude of the surface perturbation then grows exponentially, with growth rate $\gamma = \sqrt{(gAk)}$. Here $k = 2\pi/\lambda$ is the wavenumber, λ is the wavelength of the perturbation, and $A = (\rho' - \rho)/(\rho' + \rho)$ is the Atwood number, where ρ' is the density of the upper fluid and ρ the density of the lower fluid.

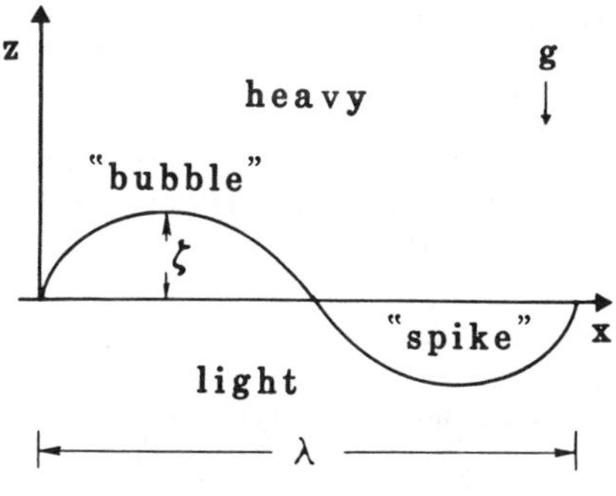

Figure 1. Sketch of the geometry and notation for the theoretical model of Rayleigh-Taylor instability.

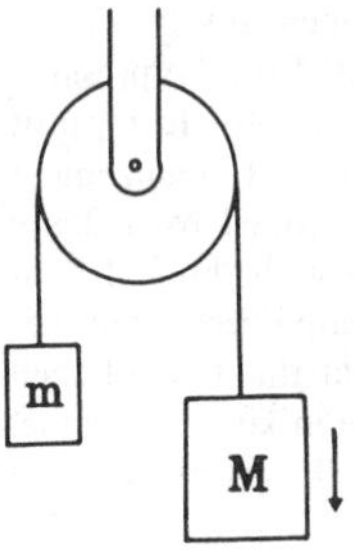

Figure 2. Atwood's machine.

Figure 2 shows Atwood's machine, a mechanism which "dilutes" gravity. The heavier weight M falls with acceleration a = Ag, where A = (M − m)/(M + m). The fluid case is analogous, giving rise to the terminology. Note that smaller wavelengths grow faster, with apparently no upper bound to growth rate. This is clearly unphysical and due to the idealizations of this simple model. In the linear regime, many effects, such as surface tension and viscosity, can limit the growth rate and have their largest effect on the small wavelengths (large k), resulting in a finite maximum growth rate. In addition, nonlinear terms become significant when the amplitude is of the same order as the wavelength.

These simple results can be derived as follows. Readers unfamiliar with hydrodynamic stability theory should consult the appendix "Rudiments of Stability Theory" at the end of this chapter or see Chandrasekhar's treatise[5] for the necessary background. (This reference also contains a chapter on the Rayleigh-Taylor instability.) Using the notation of Figure 1, we assume all perturbation quantities depend upon time t and the horizontal coordinate y through the factor $\exp\{iky + \gamma t\}$. The vertical velocity is w, the horizontal u, and the perturbation pressure p; denote the z-derivative by the operator D. We linearize and obtain in each fluid momentum conservation equations (Euler's equations):

$$\gamma u = -ikp/\rho$$

$$\gamma w = -Dp/\rho$$

The mass conservation equation for an incompressible fluid reduces to the condition that the divergence of the fluid velocities is zero, that is,

$$iku + Dw = 0$$

We may obtain from these the equation

$$[D^2 - k^2]w = 0$$

(and similar equations for p and u), showing that the general dependence upon the z-coordinate is A exp(kz) + B exp(−kz). In the upper fluid, we must take A = 0 and in the lower B = 0 in order to obtain a physically reasonable solution finite far from the interface. We will from this point on have to distinguish between the two fluids, and will use a prime to denote the upper fluid.

To complete the problem we must specify the interfacial conditions. Let the position of the interface be denoted by $\zeta = \eta \exp\{iky + \gamma t\}$. The basic kinematic conditions that the interface move with the fluid may be linearized [5] to give at the mean interface z = 0:

$$w = \gamma\eta = w'$$

The last condition serves to define the interface. There are several ways to obtain this condition. One is to impose continuity of pressure at the interface. The pressures in either fluid near the interface must be found using Bernoulli's theorem and taking into account the displacement of the interface.

The state of the fluid before the onset of instability was one of hydrostatic equilibrium, resulting in a pressure distribution of the form $P = P(z = 0) - g\rho z$. Taking this into account, we find at $z = 0$, $P' - P = g(\rho' - \rho)\eta$. Using $w = -kP/\gamma\rho$ and $w' = kP'/\gamma\rho'$ to substitute for P and P′ and eliminating w and w′ with the kinematic condition for the interface given, we then find that $\gamma^2 = gkA$, where A is as defined.

Surface tension may easily in incorporated in this analysis. Instead of the pressures being equal across the interface, they differ by $-Tk^2\eta$, where T is the surface tension constant (typical units: dynes/cm). One then finds:

$$\gamma^2 = gkA - Tk^2/(\rho + \rho')$$

Thus, wave numbers greater than $gA(\rho + \rho')/T$ (shorter wavelengths) do not grow; this is one way to remove the unboundedness problem previously discussed.

If the interface is not an ideal discontinuity, it may be shown that the growth rate cannot exceed $\sqrt{(gk)}$ for any wave number k, where $k = d\rho/\rho\, dz$ is a wave number based on the density gradient scale length [6, 7]. We can show this quite simply with a beautiful proof due to M. Frese (unpublished). To treat a continuous $\rho(z)$ variation, we use in place of the interfacial conditions employed above an equation for the change in local density $\delta\rho$ due to motion:

$$n\delta\rho = -wD\rho$$

obtained from the continuity equation by imposing the incompressibility condition [5]. We then find on using the momentum and continuity equations to eliminate p, u, and $\delta\rho$:

$$D(\rho Dw) = k^2[-g/n^2(D\rho)w + \rho w]$$

This equation may be used for the discontinuous interface by integrating over a small range about $z = 0$. Then the z-derivative D(F) of any field F becomes the "jump" in F across the interface, and we have:

$$\rho' Dw' - \rho Dw = k^2 g/n^2(\rho' w' - \rho w)$$

As $Dw' = -kw'$ and $Dw = kw$, and $w = w'$ at the interface, this reduces to $n^2 = kgA$.

Frese's proof is beautifully simple. The equation for the vertical velocity may be written:

$$D^2w + k^2[g/n^2(D\rho/\rho) - 1]w + (D\rho/\rho)Dw = 0$$

As we expect on physical grounds that w must decay to zero at large distances from the region of interest, w must have a local maximum somewhere if it is nonzero anywhere; here $D^2w < 0$ and $Dw = 0$ as well. Consequently, the factor in the square brackets must be positive; this yields $n^2 < g(D\rho/\rho)$.

Viscosity leads to a finite bound to the growth rate, but does not absolutely stabilize any mode. (See Chandrasekhar [5] and references cited therein for the details.) The analysis is complicated by the fact that the viscosity terms raise the order of the system, making the eigenvalues more complicated. If the kinematic viscosities of the fluids are equal, we can show using the results presented in [5] that the ratio of the growth rate to the inviscid growth rate $\sqrt{(gkA)}$ is a simple function of the Atwood number, and approximately 0.63 (0.618 for $A = 0.01$ and 0.6565 for $A = 1$). The wave number of maximum growth rate is proportional to $k = (\nu^2/g)^{1.3}$ and varies somewhat more, from 0.1134k for $A = 0.01$ to 0.4907k for $A = 1$. Here, ν is the kinematic viscosity of the fluids.

The theoretical discussion above was limited to incompressible fluids. Compressibility, interestingly enough, can either enhance or decrease the growth rate [13], depending upon the details of the fluid equation-of-state. For example, assume we have two isothermal layers in pressure equilibrium. For "adiabatic" (no flow of heat) perturbations, compressibility will enhance growth rates. For isothermal (constant temperature) disturbances, compressibility will enhance growth rates if the lower density material is also the more compressible, and decrease growth rates if not. A similar result holds for more general "polytropic" atmospheres. The effect decreases as Atwood number increases, and vanishes for Atwood number equal to one, i.e., when the lighter fluid is a vacuum. The effect is small in linear theory for perfect gases. In some cases, the effect can be large, if pressure changes are such that the speed of sound changes significantly in the materials involved and a large increase in the growth rate has been observed experimentally [8]. (The next chapter is by David L. Book on the subject of "Compressible Rayleigh-Taylor Instability.")

NONLINEAR BEHAVIOR AND SECONDARY INSTABILITIES

Beyond the linear regime, the acceleration slows. The dense material forms thin so-called "spikes" which fall with an acceleration of approximately g, while broad "bubbles" of the lighter material rise into the denser layer with approximately constant velocity. This has been demonstrated in the experiments of Lewis [9] and others. It is possible for the downward spike acceleration to briefly exceed g [10, 11]. Because the fluids are traveling in opposite directions on opposite sides of the interface, it is also possible that the shear induces Kelvin-Helmholtz instabilities which mix the fluids. This can cause turbulent mixing and inhibit further spike penetration into the lighter fluid. Belenkii and Fradkin [12] have considered this turbulence using a simple mixing length model and obtained the depth of the mixed region to be $\frac{1}{2}bgt^2$, a result that may also be obtained from dimensional reasoning. Youngs has found $b = 0.14A$ from computer simulations and experiments. Because as the Atwood number A increases the Taylor instability growth rate increases, while the Kelvin-Helmholtz instability growth rate decreases, one might expect turbulence to be of less importance and have later onset, for values of A near one. The spikes might in fact be in free fall (acceleration near g) before significant turbulence occurs.

A simple model for the nonlinear behavior of the instability is available [13]. It is assumed that only one wavelength is present initially in the perturbation. A second order, nonlinear ordinary differential equation is found for the spike and bubble amplitudes.

APPLICATIONS

The principal interest in Taylor instabilities to date is due to the deleterious effects such instabilities have in inertial confinement fusion (ICF) targets. High energies (from lasers or particle beams) are used to implode targets containing hydrogen fuel. The target shells must move at velocities of roughly 10 cm/microsecond. The fuel at the center of the target will be compressed, heated, and burn on a timescale of about 10 nanoseconds, during which time the shell is decelerated by the lower density fuel to zero velocity. This implies that the fuel-shell interface has an acceleration of approximately 10^{15} cm/sec^2. The Atwood number will be close to 1, so perturbations of wavelength 0.1 mm will have a growth rate on the order of one nanosecond, well below the implosion time of ten nanoseconds. Consequently, such instabilities must be taken seriously.

For applications such as ICF targets, one is concerned of the instability of shells of fluid. Hunt [14] has discussed the modifications that arise when the fluid layer is thin and sandwiched between other layers of different densities. One crucial result is that if there is a "free surface," i.e., an interface with unit Atwood number, there will always be a mode with growth rate $\sqrt{(gk)}$. Layering will introduce more weakly-growing modes, but will not eliminate this mode. Shock waves in such targets can cause perturbations on shell surfaces to grow. Richtmyer [15] first treated the problem theoretically, showing that the growth was not exponential in the linear regime but linear. This is because the acceleration "g" is zero except for a brief moment while the shock interacts with the interface. It may be shown [13] that the growth drastically slows once the amplitude is large enough to cause departure from the linear regime. Meshkov[16] has shown that the interface is unstable whatever the direction of shock passage, and has confirmed experimentally the instability's existence. A model [13] treats this case by setting $g = 0$ with an initial velocity and amplitude perturbation of the interface.

A popular rule-of-thumb used in estimating instability growth is that the spike penetration into the lighter fluid (and consequent mixing) cannot pass the "free-fall" line, i.e., the spike amplitude must be less than $\frac{1}{2}gt^2$, where g is the maximum acceleration. This can be false, especially if shock waves are involved. Another popular rule is that the "most damaging" wavelength for a shell (here we are concerned with bubble or light fluid penetration into the heavier fluid) is approximately equal to the shell thickness. Shorter wavelengths saturate early (i.e., depart from the linear regime of exponential growth soon and grow slowly), while longer wavelengths grow more slowly. This rule tends to be roughly true, although it requires assumptions about the spectra of the initial perturbation, for example.

Experimental studies of the instability employ principles similar to the apparatus shown in Figure 3, taken from [8]. The experiment was performed at Sandia National Laboratories, Albuquerque, N. M., by Dr. J. Asay and colleagues. A light gas gun is used to accelerate a projectile to a high velocity (a few km/sec). This projectile impacts on a material such as fused silica. This material "unshocks" and converts the inpulsive acceleration of its free surface by the projectile into a smoother, "ramped" pressure at its other surface. This provides an approximately constant acceleration of the materials to be studied. In the experiment in question, one of the materials (PMMA, polymethylmethacrylate or "Plexiglas") is solid under normal conditions and had its surface prefigured with a perturbation of known wavelength and amplitude. This was then silvered and a VISAR or laser velocity interferometer was used to obtain the time history of the velocity of the bubble and spike (VISAR is another technique developed at Sandia National Laboratories). The velocity could of course be integrated to find the position of the points, or differentiated to find the acceleration. The various experimental studies of Rayleigh-Taylor instability generally employ similar methods, with the diagnostics often being frames of motion picture films and the acceleration produced by high-pressure compressed air, and occasionally by laser or electron beams.

In some cases of interest, the gravitational acceleration varies in space or time. This can occur in astrophysical accretion disks and in magnetically accelerated plasmas that have finite resistivity and consequent magnetic field diffusion[17]. In such cases, the growth rates are even smaller than might be expected on the basis of the value of g at the velocity eigenfunction maximum [18]. When g is a function of time, a Mathieu equation gives the time dependence of the eigenfunctions [19]. This can stabilize ranges of wave numbers [20, 21]. Boris [22] has proposed using this effect in laser fusion targets by modulating the laser (and hence the acceleration) to stabilize the "most damaging" modes. Feedback [23] has been used in magnetic acceleration to stabilize modes as well.

The outer shell of an ICF target shell is accelerated by ablation, i.e., the "rocket" efect of heating the shell exterior to produce hot blowoff which drives the shell. Such ablation layers are subject to both Rayleigh-Taylor instabilities and convective modes, the Schwarzschild criterion governing the stability of the latter [24]. The problem is complicated by the fact that we are concerned with the growth of perturbations of the interface between the hot blowoff and the relatively cool shell rather than a well-defined interface between different materials. Heat and mass flow through this

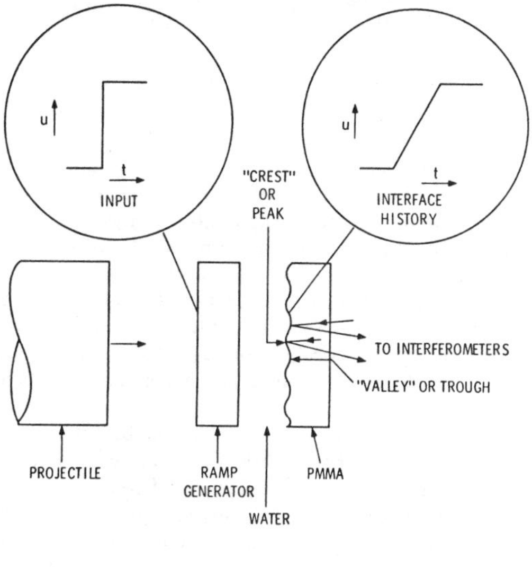

Figure 3. Sketch of the experimental geometry used in the study of compressible Rayleigh-Taylor instability.

interface, complicates the analysis. Theory [25, 26] shows these effects significantly smooth the interface and reduce growth rates, accounting for the fact that such instabilities have not been serious problems to date. They could catastrophically affect some "high-aspect-ratio" targets with thin shells, however.

CONCLUSION

Rayleigh-Taylor instability may be viewed as the result of pushing upon (accelerating or supporting against gravity) a dense fluid with a less dense one. The result is a tendency for narrow "spikes" of the latter to fall behind the mean interface. It is similar in effect to Taylor-Saffman instability. When a less viscous fluid is used to push on a more viscous fluid (as when water is injected into a well to force out oil), there is a tendency for "fingers" of the latter to be left behind.

The density differences may arise due to thermal gradients in a single fluid. Such an instability is referred to as "Rayleigh-Benard instability" or "thermal convection" [27]. Here thermal diffusion as well as momentum diffusion (viscosity) can play important roles. If the density differences arise from solutes "double diffusive convection" may arise [28]. The "salt fingers" of the ocean are due to such as instability. The descending "spikes" or "fingers" slow and an overturning occurs, mixing the layer and reducing the gradients that supply the energy for the motion. This may result in the "step" structure observed in the ocean thermocline, as layers are mixed and homogenized.

If a fluid configuration is unstable without diffusive effects, such effects generally reduce the growth rate of the instability when present. However, an inviscid, stable (or neutrally stable) configuration may be rendered unstable by diffusive effects. Historically, plane Poiseuille flow was the first example of this. It is a shear flow with a parabolic velocity profile that is neutrally stable (disturbances neither grow nor decay) for an inviscid fluid, but a slight viscosity renders it unstable. Welander[29] has shown that thermal diffusion can render other fluid configurations unstable. Typically, it produces a growing oscillation, often called "overstability." Without diffusion, a displaced fluid element is brought back to its equilibrium position by buoyancy forces. Diffusion introduces a phase lag, and the element overshoots the equilibrium point. If conditions are right, its oscillations about the equilibrium point grow is time.

Rayleigh-Taylor instability can be of importance in many cases of engineering, geophysical, or astrophysical interest. The linear behavior is well understood. Improved computational techniques [10, 11], as well as more powerful computers and renewed interest in the theory of the instability, should lead to much better predictive capability over the next few years.

Studies over many years have not revealed any ingenious way to circumvent the Rayleigh-Taylor instability, such as with special coatings or layers. Consequently, theory rightly concentrates on determining the effects of the instability in any configuration and guiding the designer in developing configurations that are least sensitive to the effects of the instability.

APPENDIX: RUDIMENTS OF STABILITY THEORY

This chapter is intended to briefly review the theory of fluid stability. It presupposes a familiarity with the basic equations of fluid motion. The references as well as useful background sources are in a special annotated bibliography. We concentrate here on reviewing the basic physics and the general character of instabilities, referring the more mathematically inclined readers to works cited.

A fluid motion is termed stable if a disturbance dies away and does not markedly alter the flow. Strictly speaking, there must be a well-defined equilibrium for which we ask, "Will this state be maintained?" In practice, it is often desirable to inquire into the stability of time-dependent flows. For example, will the surface of collapsing bubble, initially spherical, maintain its shape or will small perturbations result in a cavity that deviates markedly from a sphere at smaller radii? In such cases, we are concerned not with whether the amplitude of the disturbance grows, but whether the ratio of the amplitude of the disturbance to some other amplitude (as the mean bubble amplitude) grows. For the most part, stability theory is concerned with the stability of equilibrium configurations (or flows that are independent of time). Due to the mathematical difficulties of nonlinear problems, the stability with respect to infinitesimal disturbances is often the only study made of many flows, as this permits the solution of a linear system of equations and a consequent simplification of the problem. If the flow is unstable to infinitesimal disturbances ("linearly unstable") and if the growth rates are sufficiently high to be of concern on the time scales of interest, then a detailed (usually numerical) study must be made to see if nonlinear effects limit the effects of the instability on the mean flow (such nonlinear effects are often called "saturation," "mode coupling," etc. depending upon their nature).

Some of the terminology of stability theory is potentially confusing. A flow configuration is termed "neutrally stable" if an infinitesimal disturbance neither grows nor decays. The term "metastable" implies that the disturbance must be of the right character and have an amplitude exceeding some threshold to grow; infinitesimal disturbances are not unstable. The term "overstable" refers to an oscillation of increasing amplitude; the displacement of the system from equilibrium creates restoring forces that cause an overshoot of the equilibrium point, this overshoot increasing with each cycle of oscillation. One example is the onset of convection in a rotating fluid [5]. The term "superstable" refers to disturbances that oscillate and decay. Shock waves in an ideal gas are superstable. If, as some parameter is varied, the mean flow changes from stable to neutrally stable to unstable, and if the instability sets in as exponential growth (not overstability), then the "principle of exchange of stabilities" is said to hold.

If we assume infinitesimal perturbation amplitude, we can "linearize" the equations, i.e., retain only terms containing the first power of the perturbation amplitude. We can then expand the perturbation in a complete set of eigenfunctions that satisfy the boundary conditions of the problem. This is the classic approach to stability theory discussed in Chandrasekhar and the other references. If the equilibrium state is independent of time, the coefficients of the linear system will be independent of time. It will then suffice to assume that the eigenfunctions are proportional to $\exp(\gamma t)$, where t is the time and γ is the growth rate of the eigenfunction (or "linear mode"). If the mean flow whose stability is being studied is time-dependent, the situation is more complex. In the collapsing bubble problem mentioned earlier (see references that follow) the time dependence is in the form of a hypergeometric function, for example. Similarly, in many problems of plane-parallel flows, it is sufficient to treat the spatial dependence on some variable as $\exp(ikx)$; here k is referred to as the wave number, $k = 2\pi/\lambda$ where λ is the wavelength.

Instabilities generally convert energy available on a large scale in the mean flow into smaller scale motion. Rayleigh-Taylor instability is a good example of the conversion of gravitational potential energy into kinetic energy of smaller scale motions. Rayleigh-Benard instability or thermal convection is analogous for potential energy in a fluid with density stratification due to a temperature gradient. Note that, in general, not all of the potential energy is available for the instability; only that amount which exceeds the potential energy of a stable configuration. Note also that the mere fact that there is available potential energy does not mean that the equilibrium or mean flow is unstable; for thermal convection to occur; the unstable temperature gradient must be sufficiently large to overcome diffusion of heat and momentum. Horizontal temperature gradients can supply available potential energy as well. This gives rise to the baroclinic instability, which is important in geophysics.

Flows with velocity shear have kinetic energy that is available for conversion into kinetic energy of an instability (a uniform flow cannot do so and conserve energy and momentum). The simplest example is that of two adjacent streams of different velocities. This is called Kelvin-Helmholtz instability. The interface between the two fluids tends to wrinkle and form "cat's-eye" vortices. For inviscid fluids of different densities, we have instability no matter how small the velocity difference may be, or how great the density difference is. All wave numbers k above a certain minimum are unstable. Here the flow is assumed in the x direction and the perturbation is assumed to depend on x as $\exp(ikx)$.

Let us consider only two-dimensional flows in the x, y plane. Consider first inviscid, "ideal" fluids, incompressible, and let the flow be in the x direction ("parallel flow"). Lord Rayleigh showed that if the velocity is U(y), there must be a point of inflection ($U''(y) = 0$) for instability to occur. Hence, if we have a simple linear shear $U = ay + b$ ("plane Couette flow") or a quadratic velocity profile $U = ay^2 + by + c$ ("plane Poiseuille flow"), the flows are stable. While it was at one time generally believed that viscosity would only damp instabilities, it was shown by C. C. Lin that plane Poiseuille flow could indeed be unstable in a viscous fluid. This occurs if the Reynolds number, Ud/ν, where ν is the kinematic viscosity, U the mean velocity, and d the distance between boundaries, is large. Couette flow is typically studied between rotating cylinders and is not plane flow; many instabilities can form, depending upon the velocities of the inner and outer cylinders. For a discussion see Chandrasekhar. Pipe flow (Hagen-Poiseuille flow) in a viscous fluid has a quadratic dependence of velocity on radius. This flow appears to be metastable. Theoretically, it is found to be stable to infinitesimal disturbances, and experimentally it is stable up to quite large Reynolds' numbers if care is taken to minimize initial disturbances.

A stable density stratification (gravity in the negative y direction, shear flow U(y) in the x direction) can suppress the instability. Miles (see following references) proved that if the Richardson number R exceeds 1/4 everywhere, the flow is stable; $R = (N/U')^2$, where N is the Brunt-Vaisala frequency and is given by $N\gamma = -g(d\rho/dy)/\rho$.

One instability mechanism is wave-wave interactions. In a stably-stratified medium gravity waves may propagate. Instabilities are possible in a shear flow if gravity waves interact with a "critical level" in which the phase velocity of the waves equals that of the flow at that level. One of the proposed mechanisms for the generation of ocean waves by instabilities, proposed by J. Miles, involves such a resonant interaction between waves and a critical level. It is possible to "over-reflect" gravity waves at such a level, producing a stronger reflected wave than the incident wave. It might be noted here that the generation of ocean waves is still not understood with complete satisfaction. Furthermore, at a critical layer nonlinear effects rapidly become important, modifying the mean flow and affecting the wave-mean flow interaction. Consequently, nonlinear treatments are needed for a quantitative understanding of the phenomena involving critical layers.

We have sought to avoid mathematical detail in this overview of stability theory; readers who feel cheated can probably find more complicated and subtle mathematics than they want in some of the references below.

General References

Brodkey, R. S., *The Phenomena of Fluid Motions*, NY: Addison Wesley, 1967. This book briefly reviews the fundamentals, then provides a physical discussion of stability theory and experiment (Chapter 14). It is noteworthy in discussing phenomena of engineering interest (multiphase flow, non-Newtonian fluids, turbulence) often given short shrift.

Chandrasekhar, S., *Hydromagnetic and Hydrodynamic Stability*, Oxford: Clarendon Press, 1961. The bible of linear stability theory. Discusses Kelvin-Helmholtz, Rayleigh-Taylor, and Couette flow instabilities and thermal convection (Rayleigh-Benard instability) in detail, as well as the stability of jets, gravitationally-bound systems, and magnetohydrodynamic configurations.

Drazin, P. G., and Reid, W. H., *Hydrodynamic Stability*, Cambridge: University Press, 1981. A more recent review of stability theory than either Chandrasekhar's or Lin's work, this includes some discussion of nonlinear effects.

Swinney, H. L., and Gollub, J. P., ed., *Hydrodynamic Instabilities and the Transition to Turbulence*, Berlin: Springer, 1981. Many articles of interest, particularly strong on nonlinear effects and the relation between instability and the onset of turbulence.

Bubbles and Time-Dependent Mean Flows

Plesset, M. S., and Mitchell, T. P., "On the stability of the spherical shape of a vapor cavity in a liquid," *Quart. Applied Math.*, 13, 419–430, (1956). For this time-dependent mean flow, the perturbation time dependence is found to be a sum of hypergeometric functions. See also:

Birkhoff, G., "Stability of spherical bubbles," *Quart. Applied Math.*, 13, 451–3, (1956).

Bernstein, I. B., and Book, D. L., "Rayleigh-Taylor Instability of a self-similar spherical expansion," *Astrophys. J.*, 225, 633–640 (1978). In a series of papers (see also Ref. 24 of Chapter 12) these authors have treated instabilities of a non-equilibrium mean flow. In the case studied, hypergeometric functions again describe the perturbation time dependence.

Shear Flows

Betchov, R., and Criminale, W. Jr., *Stability of Parallel Flows*, NY: Academic Press, 1967. A good treatment of the basics and also of the stability of boundary layer flows.

Drazin, P. G., and Howard, L. N., "Hydrodynamic stability of parallel flow of inviscid fluid," *Advances in Applied Mathematics*, Kuerti, G. ed., Vol. 9 NY: Academic Press, 1966. A classic.
Lin, C. C., *Theory of Hydrodynamic Stability*, Cambridge: University Press, 1955. Another classic, devoted principally to Lin's work on plane Poiseuille flow (the Orr-Sommerfeld equation).
Miles, J., "On the stability of heterogeneous shear flows," *J. Fluid Mech.*, 13, 433–448 (1961). The theorem on the Richardson number discussed in the text is proved in this work.
Yih, C.-S., *Dynamics of Nonhomogeneous Fluids*, (NY: MacMillan Co., 1967. Chapter 4 is devoted to stability, particularly of density-stratified shear flows. Howard's beautiful "semicircle theorem" is discussed, along with Miles' theorem.

Oceans Waves

Miles, J., "On the generation of surface waves by shear flows," *J. Fluid Mech.*, 3, 185–204 (1957). The first paper by Miles on his theory for the wind generation of ocean waves.
Stern, M. E., *Ocean Circulation Physics*, NY: Academic Press, 1975. A good discussion contrasting Kelvin-Helmholtz instability and the Miles theory of ocean wave generation.
Barnett, T. P., and Kenyon, K. E., "Recent advances in the study of wind waves," *Reports Progress in Physics*, 38, 667–729 (1975). This review discusses the status of research in this area. Present theories underpredict the observed growth rates for wavelengths in excess of about 10 cm.

Geophysical and Astrophysical Applications

Pedlosky, J., *Geophysical Fluid Mechanics*, Berlin: Springer, 1979. Particularly strong on baroclinic instabilities.
Spiegel, E. A., "Convection in stars," *Ann. Rev. Astron. Astrophysics*, 9, 323, (1978) and 10, 261 (1979). A thorough review of thermal convection. The first installment treats the basic fluid mechanics, the second considers complications (rotation, compressibility, etc.)
Zahn, J. P., "Instability and Mixing Processes in Upper Main Sequence Stars," in Astrophysical Processes in Upper Main Sequence Stars, Thirteenth Advanced Course of the Swiss Society of astronomy and Astrophysics, Geneva Observatory, (1983).

NOTATION

A	Atwood number	m	mass
a	acceleration	p	pressure
D	vertical derivative	T	surface tension constant
g	acceleration due to gravity	u	velocity
k	wavenumber	w	vertical velocity
M	mass	x, y, z	coordinates

Greek Symbols

		λ	wavelength of perturbation
γ	growth rate	ρ	density
η	perturbation amplitude	ν	kinematic viscosity

REFERENCES

1. Taylor, G. I., "The instability of liquid surfaces when accelerated in a direction perpendicular to their planes. I", *Proc. Roy. Soc. London*, A201, p. 192–6, 1950.
2. Lord Rayleigh, Scientific Papers, Vol. 2, (N.Y.: Dover, 1945), pp. 200–207; from Proc. London Math. Soc., XIV, pp. 170–177, 1883.
3. Barnes, J. F., Blewett, P. J., McQueen, R. G., Myer, K. A., and Venable, D., "Taylor instability in solids," J. *Appl. Phys.*, 45, 727 (1974).

4. Roberts, P. D., Rose, S. J., Thompson, P. C., and Wright, R. J., "The stability of multiple-shell ICF targets," *J. Phys. D.*, 13, 1957 (1970).
5. Chandrasekhar, S., *Hydrodynamic and Hydromagnetic Stability*, Oxford: Clarendon Press, 1961).
6. Newcomb, W. A., "Convective instability induced by gravity in a plasma with a frozen-in magnetic field," *Phys. Fluids*, 4, 391 (1961).
7. Chakraborty, B. B., "Hydromagnetic Rayleigh-Taylor instability of a rotating stratified layer," *Phys. Fluids*, 24, 743 (1982).
8. Baker, L., "Compressible Rayleigh-Taylor instability," *Phys. Fluids*, 26, 950 (1983).
9. Lewis, D. J., "The instability of liquid surfaces when accelerated in a direction perpendicular to their planes. II," *Proc. Roy. Soc. London*, A202, p. 81–96, 1950.
10. Baker, G. R., Meiron, D. I., and Orszag, S. A., Vortex simulations of the Rayleigh-Taylor instability. *Phys. Fluids*, 23, 1485, 1980.
11. Menikoff, R., and Zemach, C., "Rayleigh-Taylor instability and the use of conformal maps for ideal fluid flow," *J. Comp. Phys.*, 51, 28, (1983).
12. Belen'kii, S. Z., and Fradkin, E. S., "The theory of turbulent mixing, *Ir. Fiz. Inst. Akad. Nauk SSSR*, 29, 207 (1965).
13. Baker, L., and Freeman, J. R., "Heuristic model of the nonlinear Rayleigh-Taylor instability," *J. Appl. Phys.*, 52, 655, (1981).
14. Hunt, J. N., "Taylor instability in a thin fluid layer," *Appl. Sci. Res.* A10, 45 (1961).
15. Richtmyer, R. D., "Taylor instability in shock acceleration of compressible fluids, *Comm. Pure Applied Math.* 13, 297 (1960).
16. Meshkov, E. E., "Instability of the interface of two gases accelerated by a shock wave," *Izv. AN SSSR Mekhanika Zhidkosti i Gaza* 4, 151, (1969).
17. Hussey, T. W., and Roderick, N. F., "Diffusion of magnetic field into an expanding plasma shell," *Phys. Fluids*, 24, 1384 (1981).
18. Baker, L., "Rayleigh-Taylor instability with spatially varying acceleration and illustration," *Phys. Fluids*, 26, 391 (1983).
19. Yih, C. S., *Dynamics of Nonhomogeneous Fluids*, (NY: McMillian), 1965.
20. Wolf, G. H., "The dynamic stabilization of the Rayleigh-Taylor instability and the corresponding dynamic equilibrium," *Z. Physik* 227, 291 (1969), and "Dynamic Stabilization of the Interchange Instability of a Liquid–Gas Interface," *Phys. Rev. Lett.*, 24, 444 (1970).
21. Troyon, F., and Gruber, R., "Theory of the dynamic stabilization of the Rayleigh-Taylor instability," *Phys. Fluids*, 14, 2069 (1972).
22. Boris, J. P., "Dynamic stabilization of the imploding-shell Rayleigh-Taylor instability," *Comments Plasma Phys. Controlled Fusion*, 3, 1 (1977).
23. Zrnic, D. S., and Hendricks, C. D., "Stabilization of the Rayleigh-Taylor instability with magnetic feedback," *Phys. Fluids*, 13, 618 (1970).
24. Book, D. L., and Bernstein, I. B., "Fluid instabilities of a uniformly imploding ablatively driven shell," *J. Plasma Phys.*, 23, 521 (1980).
25. Baker, L., "Analytic theory of ablation layer instability, " *Phys. Fluids*, 26, 627 (1983).
26. Baker, L., "Propagation and smoothing of nonuniform thermal fronts," *Phys. Rev.*, A26, 461 (1982).
27. Spiegel, E. A., "Convection in stars," *Annual Rev. Astron. Astrophys*, 9, 323 (1978) and 10, 261 (1979).
28. Turner, J. S., *Buoyancy Effects in Fluids*. (Cambridge, University Press, 1973).
29. Welander, P., "Instability due to heat diffusion in a stably stratified fluid," *J. Fluid Mech.*, 47, 51 (1971).

CHAPTER 8

RAYLEIGH-TAYLOR INSTABILITY IN COMPRESSIBLE MEDIA

David L. Book

Laboratory for Computational Physics
Naval Research Laboratory
Washington, DC 20375

CONTENTS

INTRODUCTION

The Rayleigh-Taylor instability [1–2] occurs when a fluid supports a denser fluid against gravity, whereupon the two tend to interchange positions. It is encountered frequently in nature and in the laboratory. For example, inertial confinement fusion experiments, in which an ablatively driven medium implodes, compressing the material ahead of the ablation front to high densities, can exhibit Rayleigh-Taylor instabilities in the ablative region, at the compression front, or (in the case of a layered target) at an interface between layers of different density.

When the time scale associated with the growth of the instability is short compared with the time $(kc_s)^{-1}$ for sound to traverse a wavelength $2\pi/k$, one should expect to have to include the finite compressibility of the fluid in calculating instability growth rates. It is not obvious *a priori* whether finite compressibility acts to increase or decrease growth rates. For example, compression absorbs some energy that might otherwise go into fluid motion. On the other hand, a compressible system exhibits more modes of propagation than an incompressible one, so the most unstable one might possibly have a faster growth rate than the most unstable mode in an incompressible medium. Moreover, Bernstein et al. [3] have shown that in a broad class of general compressible hydromagnetic systems, the unstable modes with lowest threshold are associated with incompressible perturbations. It is thus conceivable that compressible and incompressible systems might display the same Rayleigh-Taylor growth rate.

Theoretical research into the Rayleigh-Taylor instability can be divided into analytical and computational approaches. Some of the early analytical work done on Rayleigh-Taylor instability in compressible fluids was inconclusive or erroneous. Vandervoort [4] and Plesset and Hsieh [5] both analyzed the instability at the interface between two polytropic media. Recently Shivamoggi [6] pointed out that these two papers disagree with one another. Replying to his comment, Plesset and Prosperetti [7] attributed the contradiction to an error in Vandervoort's analysis, which invalidates the latter's treatment except when $\gamma = 1$. They went on to derive in a simple manner a general dispersion relation for an arbitrary equation of state. This derivation, however, itself makes use of an identity that is strictly true only for $\gamma = 1$, namely the statement that the ratio of the perturbed pressure to the perturbed density is equal to the square of the unperturbed sound speed.

Blake [8] wrote down without derivation a dispersion relation for Rayleigh-Taylor instability in compressible fluids that in fact is correct for the interface between two isothermal (uniform-temperature) fluids satisfying an isothermal ($\gamma = 1$) equation of state, and argued that compressibility effects are negligible except in the long-wavelength limit. Matthews and Blumenthal [9] derived the same "isothermal-isothermal" formula with the inclusion of volume radiation forces ($\propto \rho^2$). (The formally identical dispersion relation for waves propagating in a medium consisting of two *stably* stratified isothermal layers is well known to atmospheric scientists; e.g., Tolstoy derives it in his review article [10] for the case in which the fluids have an (identical) arbitrary γ and analyzes the different waves that arise.) McCrory et al. [11] made the physically plausible argument that pressure differences cannot be transmitted across the mode structure on a time scale shorter than $(kc_s)^{-1}$, so that compressibility effects must limit growth rates to values $\lesssim kc_s$. Takabe and Mima [12] wrote down a variational form for the growth rate in the presence of both compressibility and thermal conduction, but neglected to state clearly the assumptions they employed. Scannapieco [13] investigated the stability of a slab of ideal polytropic gas with an exponentially increasing or decreasing density profile confined between two rigid horizontal walls separated by a distance d and found that growth rates are enhanced by compressibility. However, in analyzing the limit in which the scale height H satisfies $H \gg d$, he allowed the density to vary while treating the sound speed as a constant, which is inconsistent unless the density decreases in the upward direction. Baker [14] solved the Blake [8] dispersion relation numerically for the ratio of the square of the calculated growth rate to the incompressible value as a function of Atwood number $A = (\rho_2 - \rho_1)/(\rho_2 + \rho_1)$, the ratio c_1^2/c_2^2 of the squares of the sound speeds, and the nondimensionalized wavelength $g/2kc_1^2$. Because of the assumptions implicit in the "isothermal-isothermal" model, however, the first two parameters are not independent: $c_1^2/c_2^2 = \rho_2/\rho_1 = (1 + A)/(1 - A)$. His conclusion that finite compressibility is sometimes stabilizing and sometimes destabilizing is therefore questionable.

On the whole, the problem with the analytical approach to studying compressibility effects on the Rayleigh-Taylor instability has most frequently failed to specify clearly the model being investigated. Many authors have attempted to derive model-independent dispersion relations, or at least formulas of wide applicability, which would not be restricted to a particular type of density profile. For a medium to behave compressibly with respect to a mode of wavenumber k, however, the dimensionless wavenumber must satisfy $kc_s^2/g = kH \lesssim 1$, where H is the equilibrium scale height. But h, the vertical scale of the perturbation, typically satisfies $h \sim k^{-1}$, so we must have $h \gtrsim H$. The mode samples the vertical variation of the equilibrium state and must therefore depend sensitively on it.

It is actually quite easy to show for ideal polytropic media that compressibility destabilizes the Rayleigh-Taylor instability [15]. The energy principle of Bernstein et al. [3] (an extension of the version given earlier by Chandrasekhar [16]) predicts that polytropic media with finite adiabatic index γ exhibit maximum growth rates which decrease with increasing γ. Incompressible fluids, which are obtained in the limit $\gamma \to \infty$, are thus more stable then compressible ones. This has been confirmed experimentally [14].

Most computational approaches to the problem strive so hard for realism that they treat too many physical processes simultaneously. When something happens in a calculation it is hard to say which process is responsible, especially when all the parameters in the code have been chosen so as to simulate a particular laboratory experiment. While such simulations are a major reason for computation, they are worthless unless every effect included in the model has been carefully validated against analytical theory or reliable measurements. Failure to do this properly vitiated some early simulations of Rayleigh-Taylor instability in imploding pellets. Another difficulty arose from the necessity of performing well-resolved multidimensional calculations in place of the one-dimensional ones used for studying unperturbed implosions.

A way around this difficulty was found using so-called "piggyback" codes [17]. The linearized equations of motion are analyzed into a superposition of angular (e.g., spherical) harmonics, and the equations for the radius-dependent amplitudes corresponding to one or more such modes are advanced in time together with the zeroth-order quantities. This technique, also successfully employed in connection with incompressible cylindrical liner implosions [18], eliminates the numerical resolution problem. The work of Shiau et al. [17] clearly showed for the first time that flow of plasma across an interface (a process that can only occur in compressible media), while stabilizing, is not by itself able to completely suppress the Rayleigh-Taylor instability. Subsequently, improvements in computational methods and techniques of code validation have enabled fully nonlinear two-dimensional calculations to be carried out which predict the linear and nonlinear evolution of the Rayleigh-Taylor instability with high accuracy [19, 20]. Stimulated partly by experiment and partly by code results, quite comprehensive theories have now been developed that consider such diverse effects as vortex shedding, compressibility, thermal conduction, and ablation [21]. The rest of this chapter will not further address the use of numerical simulations to study the Rayleigh-Taylor instability.

Another question not considered here involves the time required to establish a state as a result of an initial localized disturbance. This process, which is instantaneous in an incompressible fluid, lasts a time equal to that required for a sound wave to propagate a few times back and forth across the entire system. Instead, this chapter does consider linear eigenmodes, which by definition are initiated in "prepared" states involving the entire system. Finding the eigenmodes in compressible fluids presents enough analytical difficulty to dissuade one from seeking the solution of the general initial value problem. Relatively little has been written on this topic.

This chapter is organized as follows. A simple version of an argument originally employed by Schwarzschild (see, e.g., Landau and Lifshitz [22]) in discussing hydrodynamic interchange is used to derive threshold criteria for the Rayleigh-Taylor and convective instabilities in arbitrary stratified media. Then the energy principle is used to show that compressibility is always destabilizing, and Newcomb's extension of this result to higher eigenmodes of the system is presented [23]. Next, the exact dispersion relation for the Rayleigh-Taylor instability at the interface between two ideal polytropic fluids with different adiabatic indices, each fluid having uniform temperature, is derived following Bernstein and Book [24], and various limiting cases of this result are discussed. The extension to an arbitrary piecewise isothermal equilibrium is sketched, concluding the portion of the paper dealing with stability of static equilibria.

The only nonstationary fluid states in which the Rayleigh-Taylor instability can be treated analytically are self-similar expansions or contractions for which the velocity is proportional to the distance from the center of symmetry (uniform self-similar motions) [25]. Following a summary of the formalism used to discuss stability of such states, examples are given, first for implosions and then for expansions. A final section summarizes the main results and attempts to draw them together by pointing out the common themes that run through all of these examples.

PHYSICAL BASIS OF GRAVITATIONAL INTERCHANGE INSTABILITIES

Suppose a fluid with vertical density and pressure profiles $\rho(y)$, $p(y)$ is in hydrostatic equilibrium:

$$\frac{\partial p}{\partial y} + \rho g = 0, \tag{1}$$

where g is the constant gravitational acceleration. Note that p(y) must decrease monotonically as a function of y, but $\rho(y)$ need not. We assume that the adiabatic index (ratio of specific heats) γ is constant.

Now consider an element of fluid with differential volume ΔV at some arbitrary height y. It contains mass $\Delta m = \rho\, \Delta V$ and internal energy $p\, \Delta V/(\gamma - 1)$. Assume that it is displaced *adiabatically* to a new position y′, where it occupies a new volume $\Delta V'$ at a new density ρ' and pressure p′. By conservation of mass,

$$\rho' \Delta V' = \rho \Delta V = \Delta m \tag{2}$$

by adiabatic invariance (entropy conservation),

$$p'(\Delta V')^{\gamma} = p(\Delta V)^{\gamma} \tag{3}$$

To make room for the displaced parcel of fluid, a second differential volume $\overline{\Delta V}$ with initial density $\bar{\rho}$ and pressure $\bar{p}$ is displaced from location $\bar{y} = y'$ to the first location $\bar{y}' = y$. We assume that $y - \bar{y} = h > 0$, i.e., the first location is above the second (possibly by a finite distance). Evidently, the second parcel of fluid after displacement has density $\bar{\rho}'$ and pressure $\bar{p}'$ satisfying

$$\bar{\rho}' \overline{\Delta V}' = \bar{\rho} \overline{\Delta V} = \overline{\Delta m} \tag{4}$$

$$\bar{p}'(\overline{\Delta V}')^{\gamma} = \bar{p}(\overline{\Delta V}) \tag{5}$$

Since the displacements are adiabatic, the total change in internal energy is

$$\delta W_I = (p' \Delta V' - p \Delta V + \bar{p}' \overline{\Delta V}' - \bar{p} \overline{\Delta V})/(\gamma - 1) \tag{6}$$

This is accompanied by a total net change in gravitational energy given by

$$\delta W_G = \Delta m\, g(y' - y) + \overline{\Delta m}\, g(\bar{y}' - \bar{y}) = (\overline{\Delta m} - \Delta m)gh \tag{7}$$

If after the interchange the displaced fluids are in equilibrium with their surroundings but the total change in energy is *negative* (i.e., energy is reduced),

$$\delta W = \delta W_I + \delta W_G < 0 \tag{8}$$

the interchange is energetically favored, and the configuration is therefore *unstable.*

There are two conditions for a displaced differential volume to be in equilibrium with its surroundings: a kinematic condition, that it "fill the hole" left by its counterpart, and a dynamic condition, that it be in pressure balance. For the first parcel, these requirements imply

$$\Delta V' = \overline{\Delta V} \tag{9}$$

and

$$p' = \bar{p} \tag{10}$$

for the second, they imply

$$\overline{\Delta V}' = \Delta V \tag{11}$$

and

$$\bar{p}' = p \tag{12}$$

Equations 3 and 5 now both reduce to a relation connecting ΔV and $\overline{\Delta V}$:

$$\bar{p}(\overline{\Delta V})^{\gamma} = p(\Delta V)^{\gamma} \tag{13}$$

If we calculate the energy in each parcel of fluid as it undergoes displacement we see that the work done on the first one by the surrounding fluid exactly equals the expansion work done by the second. Thus

$$\delta W_I = 0 \tag{14}$$

and so

$$\delta W = \delta W_G = \rho\, \Delta V g h[\bar{\rho}/\rho)(p/\bar{p})^{1/\gamma} - 1] \tag{15}$$

We consider three cases of interest.

Case I: Discontinuous change in density. Here we can take y and $\bar{y}$ to be on opposite sides of the discontinuity, but contiguous, so that h is very small. Since the pressure is continuous, $\overline{\Delta V} = \Delta V$, and Equation 15 reduces to

$$\delta W = (\bar{\rho} - \rho)\, \Delta V g h \tag{16}$$

The system is thus unstable if $\rho > \bar{\rho}$. This is the usual criterion for the Rayleigh-Taylor instability.

Case II: Continuous density variation, $\rho > \bar{\rho}$. Now y and $\bar{y}$ need not be contiguous. Since $\bar{p} > p$ always holds, δW_G is again negative if $\rho > \bar{\rho}$. The limit as $h \to 0$ yields the criterion for Rayleigh-Taylor instability in a smoothly stratified medium

$$\nabla \rho \cdot \nabla p < 0 \tag{17}$$

Case III: Continuous density variation, $\rho < \bar{\rho}$. Even if the density decreases monotonically in the upward direction, it is still possible to satisfy $\delta W < 0$, provided that

$$\bar{p}/\bar{\rho}^{\gamma} > p/\rho^{\gamma} \tag{18}$$

This is the criterion for *convective* instability [22].

We thus see that in certain circumstances a fluid stratified under gravitational acceleration can be unstable to overturning, or interchange. When the instability is driven by a density inversion (dense fluid lying above less dense), the name "Rayleigh-Taylor instability" is used. By means of the energy principle, these ideas can be carried a step further to exhibit the manner in which the degree of compressibility affects stability.

ENERGY PRINCIPLE

The equations describing the evolution of a small perturbation about a specific equilibrium state are linear in the perturbed fluid variables. They can be reduced to a single homogeneous differential equation in terms of one dependent variable, e.g., the perturbed pressure or displacement. If we assume time dependence $\sim \exp(-i\omega t)$, where ω is the frequency, an ordinary differential equation, usually of second order, results for the spatial dependence. The quantity ω (or ω^2) appears as an eigenvalue, determined by solving the equation subject to appropriate boundary conditions. If this eigenvalue problem is of Sturm-Liouville form, a number of rigorous theorems apply. The most important of these says that ω can be found from a variational principle, i.e., by looking for the extremal (usually minimum) value of some functional over a class of those functions, which satisfy the boundary conditions and other physical constraints. In hydrodynamic stability problems, the variational principle has a natural interpretation in terms of energies.

Assuming an adiabatic equation of state, the potential energy W associated with a general small perturbation $\xi(x)$ about an allowed equilibrium of the ideal magnetohydrodynamic equations can be expressed in the form [3]

$$W = W_0 + \gamma \int d^3x p(\nabla \cdot \xi)^2 \tag{19}$$

where W_0 is a quadratic functional of ξ which is independent of γ. If the eigenvalues resulting from solution of the Sturm-Liouville problem determining ξ are ordered by magnitude according to

$$\omega_0^2 \leq \omega_1^2 \leq \omega_2^2 \leq \cdots \tag{20}$$

then the lowest (most unstable) eigenvalue is determined by a variational principle

$$\omega_0^2 = \min_{\xi} (W/K) \tag{21}$$

where K is a second (nonnegative) quadratic functional and the minimum is taken over all ξ satisfying the boundary conditions. By Equation 19, W/K is an increasing function of γ for any fixed ξ, so that ω_0^2 is also an increasing function of the adiabatic index γ. An incompressible medium ($\gamma \to \infty$) is thus more stable than any with finite γ.

Newcomb [23] has shown how this result can be extended to the higher modes of the system. If Σ_n is the subspace spanned by $\xi_0, \xi_1, \ldots, \xi_{n-1}$, the eigenvectors corresponding to $\omega_0^2, \omega_1^2, \ldots, \omega_{n-1}^2$, then the energy principle for the next eigenvalue takes the form

$$\omega_n^2 = \min_{\xi \epsilon C(\Sigma_n)} (W/K) \tag{22}$$

where $C(\Sigma)$ is the orthogonal complement of Σ. Let

$$F(\Sigma) = \min_{\xi \epsilon C(\Sigma)} (W/K) \tag{23}$$

so that $\omega_n^2 = F(\Sigma_n)$. The subspace Σ_n is distinguished from all others of dimension n as that which maximizes the function F (because it is spanned by the eigenvectors with the n *lowest* eigenvalues). Hence,

$$\omega_n^2 = \max_{\dim(\Sigma) = n} \; \min_{\xi \epsilon C(\Sigma)} (W/K) \tag{24}$$

from which it follows that ω_n^2 is an increasing function of γ for all $n = 1, 2, \ldots$.

We saw in the preceding section that the degree of compressibility plays no role in determining the *threshold* for instability. The present result implies that, of two otherwise identical unstably stratified fluid systems, the more compressible one has the larger growth rate. In the sequel we illustrate this conclusion by calculating growth rates in some situations where analytic solutions are possible.

RAYLEIGH-TAYLOR INSTABILITY AT THE INTERFACE BETWEEN TWO ISOTHERMAL LAYERS

An ideal polytropic fluid in a constant gravitational field evolves in time according to the system of equations

$$\frac{\partial \rho}{\partial t} + \nabla \cdot \rho \mathbf{v} = 0 \tag{25}$$

$$\rho\left(\frac{\partial \mathbf{v}}{\partial t} + \mathbf{v} \cdot \nabla \mathbf{v}\right) + \nabla p + \rho \mathbf{g} = 0 \tag{26}$$

$$\frac{\partial p}{\partial t} + \mathbf{v} \cdot \nabla p + \gamma p \nabla \cdot \mathbf{v} = 0 \tag{27}$$

The condition for a stationary equilibrium ($\mathbf{v} = 0$) is expressed by Equation 1. Let us leave p and ρ otherwise unspecified for a moment and suppose that this state is subjected to an infinitesimal perturbation defined by the local displacement $\xi(x, y, z, t)$ of an element of fluid. Using the subscript 1 to distinguish perturbed quantities, we have for the velocity

$$\mathbf{v}_1 = \frac{\partial \xi}{\partial t} \tag{28}$$

so that the perturbed density satisfies

$$\frac{\partial \rho_1}{\partial t} = -\nabla \cdot \rho \mathbf{v}_1 = -\frac{\partial}{\partial t} \nabla(\rho \xi) \tag{29}$$

whence ρ_1 is given by

$$\rho_1 = -\nabla \cdot (\rho \xi) \tag{30}$$

plus a time-independent quantity that can be set equal to zero. Likewise, the perturbed adiabatic law

$$\frac{\partial p_1}{\partial t} + \mathbf{v}_1 \cdot \nabla p + \gamma p \nabla \cdot \mathbf{v}_1 = 0 \tag{31}$$

has the solution

$$p_1 = -\gamma p \nabla \cdot \xi - \xi \cdot \nabla p \tag{32}$$

Substitution of Equations 30 and 32 in the perturbed momentum equation

$$\rho \frac{\partial^2 \xi}{\partial t^2} + \nabla p_1 + \rho_1 \mathbf{g} = 0 \tag{33}$$

yields

$$\rho \frac{\partial^2 \xi}{\partial t^2} - \nabla(\gamma p \nabla \cdot \xi \cdot \nabla p) - \mathbf{g} \nabla \cdot (\rho \xi) = 0 \tag{34}$$

Assuming that the perturbed quantities vary sinusoidally with frequency ω as functions of time, we can rewrite Equation 34 in the form

$$-\omega^2 \xi + (\gamma - 1)\mathbf{g}\sigma - \frac{\gamma p}{\rho} \nabla \sigma + \nabla(\mathbf{g} \cdot \xi) = 0 \tag{35}$$

where $\sigma = \nabla \cdot \xi$. It is this equation which must be solved, subject to boundary conditions, in order to determine the eigenvalues ω^2 and the spatial dependence of the eigenvectors ξ.

Evidently the simplest choice of the basic state is isothermal,

$$p/\rho \equiv c^2 = \text{const}, \tag{36}$$

which leads to an equation for ξ all of whose coefficients are constant. If g is directed downward, i.e., in the negative y direction, we then have

$$\rho(y) = \rho_0 \exp(-gy/c^2) \tag{37}$$

As will be seen, the specification (Equations 36–37) for the basic state results in eigenfunctions that are likewise exponential in y, and yields an algebraic dispersion relation. Any other choice gives rise to transcendental functions that must be evaluated at ordinary (nonsingular) points when the boundary conditions are imposed, greatly complicating the form of the dispersion relation.

For our present purposes it suffices to consider a piecewise isothermal state with just two regions (Figure 1). We take the interface separating them to be at y = 0, i.e., coinciding with the x-z plane. To distinguish between the regions we will label all quantities belonging to the lower one with bars. For the sake of generality we allow the adiabatic exponent to vary, $\bar{\gamma} \neq \gamma$.

With this choice, the coefficients of Equation 35 become constants. Assuming harmonic dependence in the transverse plane, with the x-axis chosen parallel to the wave vector k, we can look for solutions that are exponentials in y:

$$\xi(x, y, t) = (\mathbf{e}_x A + \mathbf{e}_y B) \exp[i(kx - \omega t) - \mu y] \tag{38}$$

A, B, and μ constant. Thus Equation 35 becomes

$$-\omega^2 A - ikc^2(ikA - \mu B) + ikgB = 0 \tag{39}$$

$$-\omega^2 B + (\gamma - 1)g(ikA - \mu B) - \mu c^2(ikA - \mu B) - \mu gB = 0 \tag{40}$$

Setting the determinant of Equations 39 and 40 equal to zero yields a quadratic for μ. The condition that the solution be well behaved as $y \to +\infty$ selects the root

$$\mu = \{-\gamma g + [\gamma^2 g^2 - 4\omega^2 \gamma c^2 + 4\gamma^2 k^2 c^4 - 4\gamma(\gamma - 1)g^2 k^2 c^2/\omega^2]^{1/2}\}(2\gamma c^2)^{-1} \tag{41}$$

Similarly, we obtain a quadratic for $\bar{\mu}$ in the lower half-plane; the condition that the eigenfunctions vanish at $y = -\infty$ yields

$$\bar{\mu} = \{-\bar{\gamma} g - [\bar{\gamma}^2 g^2 - 4\omega^2 \bar{\gamma} \bar{c}^2 + 4\bar{\gamma}^2 k^2 \bar{c}^4 - 4\bar{\gamma}(\bar{\gamma} - 1)g^2 k^2 \bar{c}^2/\omega^2]^{1/2}\}(2\bar{\gamma}\bar{c}^2)^{-1} \tag{42}$$

The kinematic boundary condition at the interface reduces to continuity of the vertical component of ξ, i.e.,

$$A = \bar{A} \tag{43}$$

The dynamic boundary condition requires that $p_1 + \xi \cdot \nabla p$ be continuous at y = 0, whence by Equation 32

$$\gamma p(0)\sigma = \bar{\gamma}\bar{p}(0)\bar{\sigma} \tag{44}$$

Since $p(0) = \bar{p}(0)$, we can combine Equations 43 and 44 as

$$\gamma(ikA - \mu B)/A = \bar{\gamma}(ik\bar{A} - \bar{\mu}\bar{B})/\bar{A} \tag{45}$$

Substituting B/A from Equation 39 or Equation 40 and writing μ in terms of ω^2 by means of Equation 41, and performing the analogous operations with the barred counterparts of these equations, we find from Equation 45 a relation determining ω^2. When $\gamma = \bar{\gamma}$ it becomes formally identical with the wave dispersion relation given by Tolstoy [10], and for the special case $\gamma = \bar{\gamma} = 1$ it reduces to the one given by Blake [8].

The treatment can, however, be carried a step further. Using the interactive symbolic manipulation system MACSYMA, we can transform this equation into an algebraic equation in $Z = \omega^2/kg$. This is done by squaring the equation twice to eliminate the square roots appearing in Equations 41 and 42 and factoring the result. The physical root is found [24] to satisfy the quartic

$$D'^2 Z^4 - 2DD'SZ^3 + (D^2 S^2 + 2DD' - D'^2)Z^2 - 2D^2(S - S')Z - D^4 = 0 \tag{46}$$

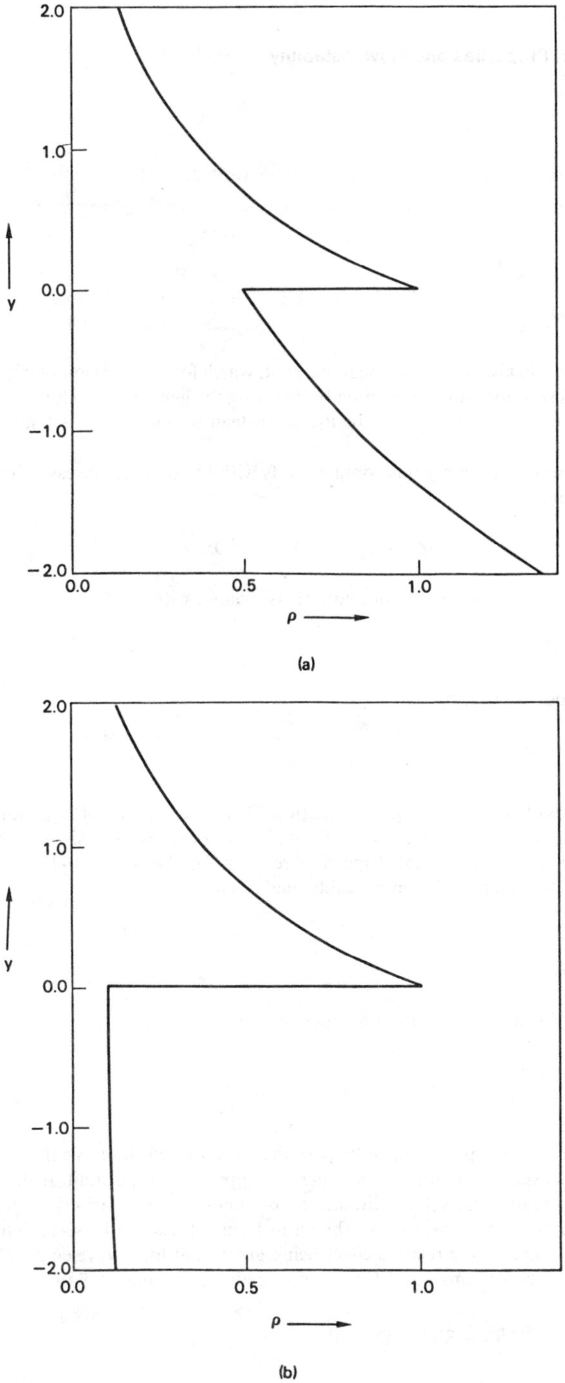

Figure 1. Density (horizontal axis) vs height for a system consisting of two constant-temperature media supported by pressure against gravity. The ratio of upper to lower density at the interface is (a) 2; (b) 10. The units are chosen to make the scale height in the upper region and the gravitational acceleration g both equal to unity.

where

$$D = k(c^2 - \bar{c}^2)g^{-1} \tag{47}$$

$$S = k(c^2 + \bar{c}^2)g^{-1} \tag{48}$$

$$D' = k(c^2/\gamma - \bar{c}^2/\gamma)g^{-1} \tag{49}$$

$$S' = k(c^2/\gamma - \bar{c}^2/\bar{\gamma})g^{-1} \tag{50}$$

Evidently Equation 46 always has one negative root, which for $D < 0$ is found numerically to satisfy Equation 45. This solution can be exhibited by applying the general procedure for solving a quartic, but the result is far too cumbersome to be useful. Instead we look at some limits and special cases of physical interest.

First let $\bar{\gamma} = \gamma$, so that both regions contain fluids with the same compressibility properties. Then Equation 46 becomes

$$Z^4 - 2\gamma SZ^3 + (\gamma^2S^2 + 2\gamma - 1)Z^2 - 2\gamma(\gamma - 1)SZ - \gamma^2D^2 = 0 \tag{51}$$

In the limit $\gamma \to \infty$, the solution of Equation 51 associated with the instability satisfies

$$S^2Z^2 - 2SZ - D^2 = 0 \tag{52}$$

whose negative root is given by

$$Z = [1 \pm (1 + D^2)^{1/2}]S^{-1} \tag{53}$$

For negative values of D the lower sign in Equation 53 yields a solution of Equation 45, as confirmed by numerical evaluation. When we take $kc^2 \gg g$, $k\bar{c}^2 \gg g$ (wavelength short compared with both scale heights), we recover the usual dispersion relation for the Rayleigh-Taylor instability at an interface between two uniform incompressible media, viz.,

$$-\frac{\omega^2}{kg} = \frac{\rho_0 - \bar{\rho}_0}{\rho_0 + \bar{\rho}_0} \tag{54}$$

At long wavelengths ($k \to 0$), Equation 53 goes over to

$$\omega^2 = -k^2\frac{(c^2 - \bar{c}^2)^2}{2(c^2 + \bar{c}^2)} \tag{55}$$

displaying the effect of the spatial dependence of the unperturbed state. We thus recover the incompressible result, as expected. Figure 2 illustrates the approach to this limit. It shows plots of ω^2/kg as functions of k obtained by solving Equation 46 numerically for various choices of γ between 1 and ∞, assuming the unperturbed states shown in Figure 1. As can be seen, finite compressibility increases the growth rates, the relative effect being greatest at long wavelengths. When $k \to 0$ then S, D, S′ and D′ all become small and Equation 46 reduces for finite γ, $\bar{\gamma}$ to

$$D'(2D - D')Z^2 - 2D^2(S - S')Z - D^4 = 0 \tag{56}$$

whence ω is proportional to k. For $\bar{\gamma} = \gamma$ Equation 56 yields

$$Z = \frac{\gamma}{2\gamma - 1}\{(\gamma - 1)S - [(\gamma - 1)^2 + (2\gamma - 1)D^2]^{1/2}\} \tag{57}$$

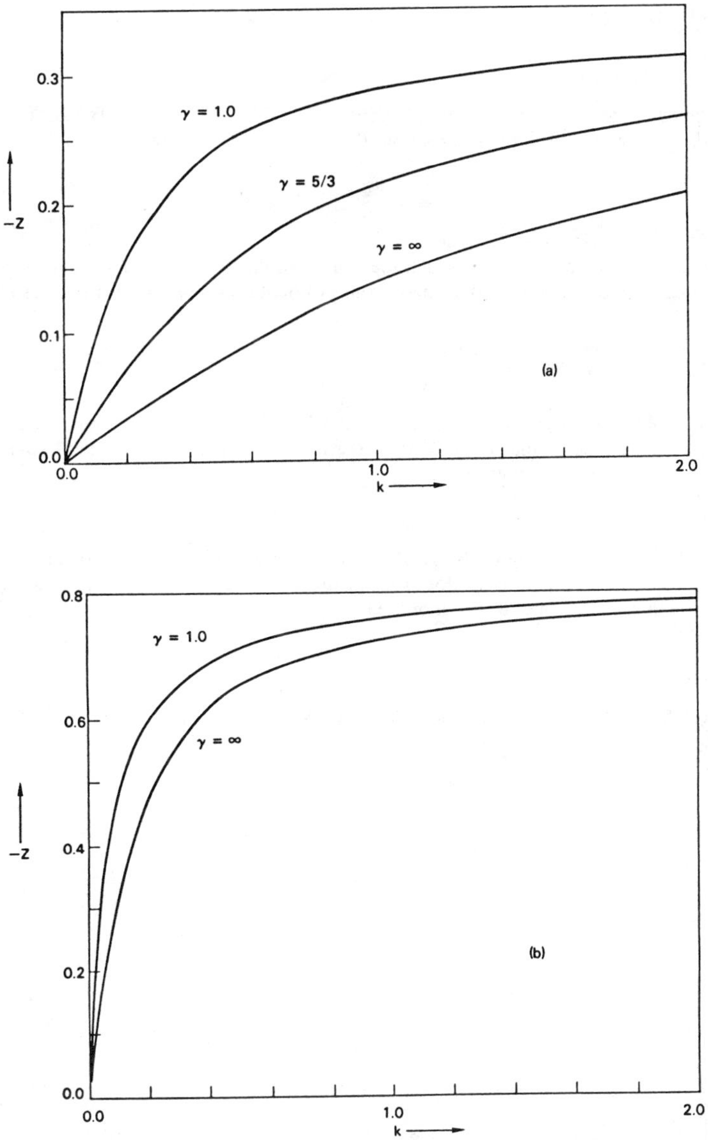

Figure 2. Dimensionless squared growth rate $-\omega^2$/kg vs wavenumber k for the two basic states shown in Figure 1, with the same choice of units. The adiabatic index γ in both regions is taken to be 1, 5/3 or ∞, as indicated by the label. Note that the curves asymptotically approach the value $(\rho - \bar{\rho})/(\rho + \bar{\rho})$, equal to 0.333 and 0.818, respectively.

When $\gamma = 1$ this becomes

$$\omega^2 = k^2(c^2 - \bar{c}^2) \tag{58}$$

This is to be compared with the corresponding incompressible result given in Equation 55. On the other hand, for short-wavelength perturbations ($k \to \infty$), Equation 46 reduces to

$$S^2Z^2 - D^2 = 0 \tag{59}$$

whose solution is identical with Equation 54.

Another interesting limit is that in which the density of the upper medium becomes infinite, so that $c = 0$. One of the extraneous roots factors out of Equation 46, which then reduces to a cubic.

$$Z^3 + (2\bar{\gamma} - 1)Z - \frac{k\bar{\gamma}\bar{c}^2}{g}(Z^2 - 1) = 0 \tag{60}$$

Equation 60 holds even if $\gamma \to \infty$ in such a way that γc^2 (the square of the sound speed) remains finite. If instead we assume that the density in the lower region vanishes (i.e., $\bar{c}^2 \to \infty$), then the dispersion relation is even simpler, becoming

$$Z = -1 \tag{61}$$

which coincides with the incompressible result. This is consistent with the behavior shown in Figure 2, which indicates that as $\bar{\rho}_0$ decreases (for fixed ρ_0), the difference between compressible and incompressible growth rates becomes less pronounced.

If we assume

$$\rho_0\gamma = \bar{\rho}_0\bar{\gamma} \tag{62}$$

then D' vanishes identically and Equation 46 simplifies to

$$S^2Z^2 - 2(S - S')Z - D^2 = 0 \tag{63}$$

whence

$$Z = \frac{S - S' - [(S - S')^2 + D^2S^2]^{1/2}}{2S^2} \tag{64}$$

Finally, if

$$\rho_0 = \bar{\rho}_0 \tag{65}$$

so that $c^2 = \bar{c}^2$ and $D = 0$, then even for $\bar{\gamma} \neq \gamma$

$$Z = 0 \tag{66}$$

i.e., the perturbations are marginally stable.

USE OF PIECEWISE ISOTHERMAL PROFILE TO APPROXIMATE AN ARBITRARY EQUILIBRIUM

More generally, we can specify an equilibrium state consisting of $N - 1$ slabs of finite thickness sandwiched between two semi-infinite regions, with density profiles in the various regions of the form

$$\rho_j(y) = \rho_0^j \exp(-gy/c_j^2) \tag{67}$$

pressure profiles given by

$$p_j(y) = \rho_j(y)c_j^2 \tag{68}$$

and adiabatic indices γ_j, $j = 0, 1, \ldots, N$. We take the interface separating layer j from layer j + 1 to be at $y = y_j$, $j = 0, 1, \ldots, N - 1$, with $y_0 = 0$. At each interface the changes discontinuously but the pressure is continuous, so that

$$\rho_0^j c_j^2 \exp(-gy_j/c_j^2) = \rho_0^{j+1} c_{j+1}^2 \exp(-gy_j/c_{j+1}^2) \tag{69}$$

$j = 0, 1, \ldots, N - 1$. Evidently, such a piecewise isothermal state can be made to approximate an arbitrary ideal hydrostatic equilibrium state as closely as desired if the number of interfaces N is allowed to increase without bound. It is thus analogous to the piecewise isopycnic (constant-density) model used by Mikaelian [26] to approximate an arbitrary incompressible equilibrium state.

Following the treatment employed in the previous section, we seek a solution for the perturbed displacement in the form

$$\xi_0 = (\mathbf{e}_x A_0^+ + \mathbf{e}_y B_0^+) \exp[i(kx - \omega t) - \mu_0^+ y] \tag{70}$$

$$\xi_j = \sum_{\pm} (\mathbf{e}_x A_j^{\pm} e^{-\mu_j^{\pm} y} + \mathbf{e}_y B_j^{\pm} e^{-\mu_j^{\pm} y}) \exp[i(kx - \omega t)] \tag{71}$$

$j = 1, 2, \ldots, N - 1$, and

$$\xi_N = (\mathbf{e}_x A_N^- + \mathbf{e}_y B_N^-) \exp[i(kx - \omega t) - \mu_N^- y] \tag{72}$$

where

$$\mu_j^{\pm} = \{-\gamma_j g \pm [\gamma_j^2 g^2 - 4\omega^2 \gamma_j c_j^2 + 4\gamma_j^2 k^2 c_j^4 - 4\gamma_j(\gamma_j - 1)g^2 k^2 c_j^2/\omega^2]^{1/2}\}(2\gamma_j c_j^2)^{-1} \tag{73}$$

Evidently, Equations 70–72 introduce 4N unknown quantites $A_j^{\pm}$, $B_j^{\pm}$. The boundary conditions

$$\sum_{\pm} A_j^{\pm} e^{-\mu_j^{\pm} y_j} = \sum_{\pm} A_{j+1}^{\pm} e^{-\mu_{j+1}^{\pm} y_j} \tag{74}$$

and

$$\gamma_j \sum_{\pm} (ikA_j^{\pm} - \mu_j^{\pm}) e^{-\mu_{j+1}^{\pm}} = \gamma_{j+1} \sum_{\pm} (ikA_{j+1}^{\pm} - \mu_{j+1}^{\pm} B_{j+1}^{\pm}) e^{-\mu_{j+1}^{\pm} y_j} \tag{75}$$

$j = 0, 1, \ldots, N - 1$, provide 2N linear relations among these. (Note that $A_0^- = B_0^- = A_N^+ = B_N^+ = 0$.) We can eliminate the $B_j^{\pm}$ in favor of the $A_j^{\pm}$ using the analog of Equation 39,

$$-\omega^2 A_j^{\pm} - ikc_j^2(ikA_j^{\pm} - \mu_j^{\pm} B_j^{\pm}) + ikgB_j^{\pm} = 0 \tag{76}$$

leaving 2N linear homogeneous equations in the 2N quantities $A_j^{\pm}$. The dispersion relation giving ω^2 as a function of k is then obtained by equating to zero the determinant of this system.

The resulting equation must be solved numerically. It is found that the number of roots associated with gravity modes equals N − 1, the number of interfaces. For short wavelengths ($k \to \infty$), one mode is localized at each interface position y_j and is stable or unstable according as $\rho_0^{j+1} < \rho_0^j$ or $\rho_0^{j+1} > \rho_0^j$. At longer wavelengths the identity of various modes becomes obscure, and the distinction between Rayleigh-Taylor and convective instability may be blurred. Many interesting limiting cases can be distinguished, e.g., $\gamma_j \to \infty$, $j = 0, 1, \ldots, N$ (incompressible fluid); $\rho_0^0 \to \infty$ (solid lower boundary); $\rho_0^N \to 0$ (free upper surface), etc. Of course, it might be argued that it is just as easy (in both incompressible and compressible cases) to approximate the differential equation for ξ by finite differences and obtain the spectrum using a standard eigenvalue-solving routine.

UNIFORM SELF-SIMILAR IMPLOSIONS AND EXPLOSIONS: FORMULATION

Many of the problems in which compressible fluids are subject to Rayleigh-Taylor instability involve nonsteady basic states (e.g., laser pellet implosions, the outward motion of gas following a stellar explosion). Usually in such problems even the unperturbed motion can only be treated by solving the fluid equations numerically using a code with one spatial variable. To analyze perturbations with angular dependence requires a two- or three-dimensional code. A compromise approach involves linearizing the perturbed equations, expanding them in cylindrical or spherical harmonics, and advancing the perturbed radius-dependent amplitude functions in parallel with the variables describing the basic state [17].

However, there exists a class of nontrivial ideal compressible flows that are sufficiently symmetric that both the unperturbed and perturbed equations can be solved analytically. These are the uniform self-similar solutions studied by Sedov [25]. They can readily be derived as follows [27].

We rewrite Equations 25–27 in the form

$$\dot{\rho} + \rho \nabla \cdot \hat{v} = 0 \tag{77}$$

$$\rho \dot{\hat{v}} + \nabla p = 0 \tag{78}$$

$$(p\rho^{-\gamma})^{\cdot} = 0 \tag{79}$$

where the raised dot ($^{\cdot}$) denotes a total time derivative. In a system with planar, cylindrical, or spherical symmetry ($\nu = 1, 2, 3$, respectively), Equations 77–78 become

$$\dot{\rho} + \rho R^{-\nu} \frac{\partial}{\partial R} (R^{\nu} v) \tag{80}$$

$$\rho \dot{v} + \frac{\partial p}{\partial R} = 0 \tag{81}$$

For uniform self-similar flow there is a function f(t) such that for an arbitary fluid element whose position at $t = 0$ was r, the position R at time t satisfies

$$R = rf(t) \tag{82}$$

with $f(0) = 1$ and $\dot{f}(0) = 0$. The Continuity Equation 80 then yields

$$\rho(r, t) = \rho_0(r) f^{-\nu} \tag{83}$$

and hence from the adiabatic law Equation (79)

$$p(r, t) = p_0(r) f^{-\nu\gamma} = s(r) \rho_0^{\gamma} f^{-\nu\gamma} \tag{84}$$

where we have introduced the entropy function $s(r) = p_0 \rho_0^{-\gamma}$. If we specify an initial density profile $\rho_0(r)$ on some interval $r_i \le r \le r_0$, then substitution in Equation 8 results in separation into a spatial part

$$\frac{dp_0}{dr} = \pm r \rho_0 \tau^{-2} \tag{85}$$

which determines the pressure p_0, and a time-dependent part

$$f^{\alpha+1} \ddot{f} = \mp \tau^{-2} \tag{86}$$

where $\alpha = \nu(\gamma - 1)$, determining f. In Equations 85 and 86 τ is a separation constant with units of time; the upper (lower) sign corresponds to implosion (expansion). If the pressures on the inner and outer surfaces, $p(r_i, t) = p_0(r_i)f^{-\nu\gamma}$ and $p(r_0, t) = p_0(r_0)^{f-\nu\gamma}$, are nonvanishing, they must be balanced by an equal pressure applied to the shell. It is difficult to imagine how this might be realized in practice, so we will assume $p_0(r_i) = 0$ for implosions and $p_0(r_0) = 0$ for expansions.

A quadrature can be performed on Equation 86, with the result

$$\tau^2\dot{f}^2 = \mp 2 \ln f \tag{87}$$

where $\gamma = 1$, and

$$\tau^2\dot{f}^2 = \mp\frac{2}{\alpha}(1 - f^{-\alpha}) \tag{88}$$

otherwise. When $\alpha = 2$ (corresponding to $\gamma = 3, 2, \frac{5}{3}$ for $\nu = 1, 2, 3$, respectively), Equation 88 can be integrated directly to give

$$f(t) = (1 \mp t^2/\tau^2)^{1/2} \tag{89}$$

For other values of γ, f is most conveniently found by numerical means. In the case of implosions, f vanishes at a time t_0 given by

$$t_0/\tau = \left(\frac{\alpha}{2}\right)^{1/2} \int_0^1 \frac{df}{(f^{-\alpha} - 1)^{1/2}} = \left(\frac{\pi}{2\alpha}\right)^{1/2} \frac{\Gamma(1/\alpha + 1/2)}{\Gamma(1/\alpha + 1)} \tag{90}$$

which decreases monotonically as a function of α.

To study the stability of uniform self-similar flows under small perturbations, we must solve the linearized form of Equation 78 for the first-order displacement ξ:

$$\rho(\ddot{\xi} - \xi \cdot \nabla_{\mathbf{R}}\ddot{\mathbf{R}}) - \ddot{\mathbf{R}}\nabla_{\mathbf{R}} \cdot \rho\xi - \nabla_{\mathbf{R}}(\gamma p \nabla_{\mathbf{R}} \cdot \xi + \xi \cdot \nabla_{\mathbf{R}} p) = 0 \tag{91}$$

Note that Equation 91 is identical to Equation 34, except that $\mathbf{g}$ is replaced by $-\ddot{\mathbf{R}}$. Like the unperturbed momentum equation, Equation 91 is separable in Lagrangian variables. Substituting

$$\xi(\mathbf{r}, t) = \xi(\mathbf{r})T(t) \tag{92}$$

we have, on writing $\nabla = \nabla_{\mathbf{r}}$,

$$(\lambda - 1)\xi + (\gamma - 1)\mathbf{r}\sigma \pm \tau^2(\gamma p_0/\rho_0)\nabla\sigma + \nabla(\mathbf{r} \cdot \xi) = 0 \tag{93}$$

(cf. Equation 35), and

$$\tau^2 f^{\alpha+2}\ddot{T} = \mp\lambda T \tag{94}$$

where λ is the new separation constant, obtained as an eigenvalue in connection with the solution of Equation 93. Once λ is known, Equation 94 can be converted into a hypergeometric equation in the new variable $x = 1 - f^{-\alpha}$ and solved to give

$$T(t) = T(0)F(a, b; 1/2; x) \mp \tau\dot{T}(0)(\mp 2x/\alpha)^{1/2}F(a + 1/2, b + 1/2; 3/2; x) \tag{95}$$

where F is the hypergeometric function [28]. Here,

$$a = (\alpha + 2 + \Delta)/4\alpha \tag{96a}$$

$$b = (\alpha + 2 - \Delta)/4\alpha \tag{96b}$$

with $\Delta = [(\alpha + 2)^2 - 8\alpha\lambda]^{1/2}$. When $\gamma \to 1$, Equation 95 goes over to

$$T(t) = T(0)\Phi(\lambda/2; 1/2; \ln f) \mp \tau\dot{T}(0)(\mp 2 \ln f)^{1/2}\Phi(\lambda/2 + \tfrac{1}{2}; \tfrac{3}{2}; \ln f) \tag{97}$$

where $\Phi(a; b; x)$ is the Kummer function [29].

Since T(t) is not exponential, we must decide what we mean by instability. A perturbation is defined to be unstable if the associated time-dependent factor satisfies

$$\lim_{t \to \infty} |T(t)|/f(t) = \infty \tag{98}$$

and stable if the limit of this ratio is finite. This is equivalent to saying that a perturbation is unstable if and only if its amplitude eventually becomes infinitely larger than the radius of the unperturbed state.

STABILITY OF UNIFORM SELF-SIMILAR IMPLOSIONS

We begin by considering imploding systems. As $f \to 0$, both terms in Equation 95 approach asymptotic forms containing terms proportional to $f^{(\alpha+2\pm\Delta)/4}$. With the lower sign this expression diverges whenever $\alpha + 2 < \Delta$, i.e., $\lambda < 0$. Since the condition for Δ to be real is that λ be no greater than $(\alpha + 2)^2/8\alpha$, whose minimum value as a function of α is unity, we see that $\lambda < 1$ is always sufficient to make T/f diverge. Thus, $\lambda > 1$ is the stability criterion for uniform self-similar implosions. If Δ is imaginary, T/f still diverges when $\alpha < 2$, i.e., $\gamma < 1 + 2/\nu$. Elsewhere [30] I have presented a simple argument involving conservation of wave action to show that this describes sound wave amplification as a consequence of the geometric properties of the implosion.

Taking the scalar product of Equation 93 with $\boldsymbol{\xi}$ and introducing notations for the transpose $\nabla\boldsymbol{\xi}^+$ and the curl $\boldsymbol{\omega} = \nabla \times \boldsymbol{\xi}$, we can multiply through by ρ_0 and integrate to obtain an energy principle [31]:

$$\tau^{-2}\lambda \int_V dV\, \rho_0\xi^2 = \int_V dV\, p_0[(\gamma - 1)\sigma^2 + \nabla\xi{:}\nabla\xi^+ - \omega^2] + \gamma \int_S dS\, p_0\sigma\mathbf{n}\cdot\xi \tag{99}$$

where V is the volume and S the surface of the shell, and n is the unit vector defined so as to point away from the shell on both inner and outer surfaces. The expression multiplying λ and the first two terms in the volume integral on the right-hand side are manifestly nonnegative. From this we see that relative instability ($\lambda < 1$) can only result if $\boldsymbol{\omega} \neq 0$ or if the surface integral is nonvanishing. The latter is the case whenever a perturbation exists at a point where the density changes discontinuously, e.g., at $r = r_0$, $r = r_i$, or an internal density jump. The external pressure, which enters the model as a boundary condition, produces an inward acceleration. Thus, there is an effective gravity in the outward direction. Hence, we anticipate that a Rayleigh-Taylor instability should occur localized at the outer surface. This case was first studied by Kidder [32], who assumed $\gamma = \frac{5}{3}$. It turns out we can solve for ξ provided the perturbation wavelengths also expand or contract self-similarly in time, i.e., whenever they do not introduce a definite length scale into the problem. This means that in spherical geometry ($\nu = 3$) there is no restriction on the form of the perturbation; in cylindrical geometry ($\nu = 2$), however, we must have $k_z = 0$; and no general solution is possible in planar geometry ($\nu = 1$).

Operating on Equation 93 with the divergence and with the curl, we get two equations:

$$[\lambda + \nu(\gamma - 1) + 1]\sigma \pm \tau^2\nabla\cdot\left(\frac{\gamma p_0}{\rho_0}\nabla\sigma\right) + \gamma\mathbf{r}\cdot\nabla\sigma = \mathbf{r}\cdot\nabla\times\omega \tag{100}$$

and

$$(\lambda - 1)\omega + \mathbf{r}\times\nabla\sigma \pm \tau^2\gamma p_0\,\nabla(\rho_0^{-1})\times\nabla\sigma = 0 \tag{101}$$

Let us suppose that $\omega = 0$. It follows from Equations 100 and 101 that σ also vanishes. If that happens,

$$\xi = \nabla\psi \tag{102}$$

where the potential ψ satisfies Laplace's equation

$$\nabla^2\psi = 0 \tag{103}$$

The general solution of Equation 103 is

$$\psi(r, \phi) = (\psi_+ r^l + \psi_- r^{-l})e^{il\phi} \tag{104}$$

in cylindrical coordinates (assuming $\partial/\partial z = 0$), and

$$\psi(r, \theta, \phi) = (\psi_+ r^1 + \psi_- r^{-\ell-1})Y_{lm}(\theta, \phi) \tag{105}$$

in spherical coordinates. The $\psi_\pm$ are constants, and the $Y_{lm}(\theta, \phi)$ are spherical harmonics. Evidently, the first term in Equations 104 and 105 corresponds to a solution localized at the outer surface of the shell, and the second to one localized at the inner surface, so we set $\psi_- = 0$. Substitution in Equation 93 then yields

$$(\lambda - 1)\xi + \nabla(\mathbf{r} \cdot \xi) = \nabla[(\lambda - 1)\psi + \mathbf{r} \cdot \nabla\psi] = \nabla[(\lambda - 1 + l)\psi_+ r^l] = 0 \tag{106}$$

whence

$$\lambda = -\ell + 1$$

For $\ell = 0, \lambda = 1$ and Equation 94 reduces to Equation 86, showing that the perturbations are marginally stable. For all $\ell > 0$, the limiting form of T/f diverges as

$$\frac{T}{f} \sim f^{\{\alpha - 2\lambda[(\alpha-2)^2 + 8\alpha l]^{1/2}\}} \tag{108}$$

when $f \rightarrow 0$. Evidently, the magnitude of the exponent in Equation 108 increases with both l and $\alpha = \nu(\gamma - 1)$. Thus, in contrast with the case of static equilibria considered previously, compressibility appears to be somewhat stabilizing. However, it must be noted from Equation 86 that as α increases, the motion becomes increasingly stiff and so the effective gravity also increases, rendering comparisons difficult.

The problem of perturbations with $\omega \neq 0$ is treated elsewhere [30–31]; it is completely analogous to that in the case of expanding solutions to be discussed shortly. Book and Bernstein [33] and Han and Suydam [34] have treated the stability of imploding cylindrical systems in detail.

STABILITY OF UNIFORM SELF-SIMILAR EXPANSIONS

Here, we consider expanding systems, in which $f \rightarrow \infty$. (Sedov distinguishes a third class of trajectories in which f varies between 0 and ∞ with no turning point; we do not treat them here.)

Bernstein and Book [35] and Han [36], who found the general solutions for the time dependence of arbitrary perturbations in spherical and in cylindrical geometry, respectively, both assumed that the unperturbed states were homentropic ($p\rho^{-\gamma}$ independent of radius). In both geometries the only instability was a Rayleigh-Taylor mode localized at the inner surface, where the driving pressure acts. Here, we analyze a class of states, parametrized by the adiabatic index and a shape parameter, which relax the requirement of uniform entropy distribution [37].

Suppose the initial density profile is given by:

$$\rho_0(r) = \hat{\rho}(1 - r^2/r_0^2)^\kappa \tag{109}$$

μ and $\hat{\rho}$ constant. Then from Equation 85,

$$p_0(r) = \hat{p}(1 - r^2/r_0^2)^{\kappa+1} \tag{110}$$

where

$$\hat{p} = \hat{\rho} r_0^2/2(\kappa + 1)\tau^2 \tag{111}$$

and

$$s(r) = p_0 \rho_0^{-\gamma} = (\hat{p}/\hat{\rho}^\gamma)(1 - r^2/r_0^2)^{\kappa+1-\kappa\gamma} \tag{112}$$

The pressure vanishes at the outer radius of the shell, as does the density provided $\kappa > 0$. At the inner radius the imposed (driving) pressure must have the form

$$p_i = \hat{p}(1 - r_i^2/r_0^2)^{\kappa+1} f^{-\nu\gamma} \tag{113}$$

unless $r_i = 0$. In this model the temperature $\theta = p/\rho$ always decreases quadratically as a function of radius:

$$\theta(r, t) = \hat{\theta}(1 - r^2/r_0^2) f^{-\nu(\gamma-1)} \tag{114}$$

where $\hat{\theta} = r_0^2/2(\kappa + 1)\tau^2$.

As in the implosion case, there is an incompressible irrotational perturbation mode satisfying Equations 102 and 103. The solution (which is independent of the shapes of the density and pressure profiles) is given again by Equation 104 or Equation 105. This time, though, it is the mode corresponding to ψ_- which is unstable, giving rise to eigenvalues $\lambda = \ell + 1$ (cylindrical geometry) or $\lambda = \ell + 2$ (spherical geometry). Using standard formulas [28] to evaluate hypergeometric functions of unit argument, we find from Equation 95 that as $f \to \infty$,

$$\frac{T}{f} \to \frac{\Gamma(1/2)\Gamma(1/\alpha)T(0)}{\Gamma(\frac{1}{4} + (2 + \Delta)/4\alpha)\Gamma[\frac{1}{4} + (2 - \Delta)/4\alpha]} + \frac{(2/\alpha)^{1/2}\Gamma(3/2)\Gamma(1/\alpha)\tau\dot{T}(0)}{\Gamma[\frac{3}{4} + (2 + \Delta)/4\alpha]\Gamma[\frac{3}{4} + (2 - \Delta)/4\alpha]} \tag{115}$$

Although this limit is finite, for large values of λ (large ℓ) the constants are found from Stirling's formula to grow exponentially:

$$\frac{T}{f} \sim \frac{\Gamma(1/\alpha)\exp[\pi(\lambda/2\alpha)^{1/2}]}{2\pi^{1/2}(\lambda/2\alpha)^{(2-\alpha)/4\alpha}} T(0) + \frac{\Gamma(1/\alpha)\exp[\pi(\lambda/2\alpha)^{1/2}]}{2(2\pi\alpha)^{1/2}(\lambda/2\alpha)^{(2+\alpha)/4\alpha}} \tau\dot{T}(0) \tag{116}$$

Figure 3 displays the late-time asymptotic behavior of these solutions as a function of the eigenvalue λ for $\alpha = 1, 2, 3$, and ∞.

To study the modes for which $\boldsymbol{\omega} \neq 0$, we use Equations 100 and 101 with ρ_0 and p_0 given by Equations 109 and 110. Using the lower sign, specializing to $\nu = 3$, and using r_0 to scale $\mathbf{r}$, we get

$$(\lambda + 3\gamma - 2)\sigma - \frac{\gamma}{2(\kappa + 1)}[\nabla \cdot (1 - r^2)\nabla\sigma] + \gamma\mathbf{r} \cdot \nabla\sigma = \mathbf{r} \cdot \nabla \times \boldsymbol{\omega} \tag{117}$$

and

$$(\lambda - 1)\boldsymbol{\omega} = \left(\frac{\gamma\kappa}{\kappa + 1} - 1\right)\mathbf{r} \times \nabla\sigma \tag{118}$$

Eliminating $\boldsymbol{\omega}$ between Equations 117 and 118 and assuming separation of the angular and radial dependence of σ by setting

$$\sigma(\mathbf{r}) = \sigma(r)Y_{lm}(\theta, \phi) \tag{119}$$

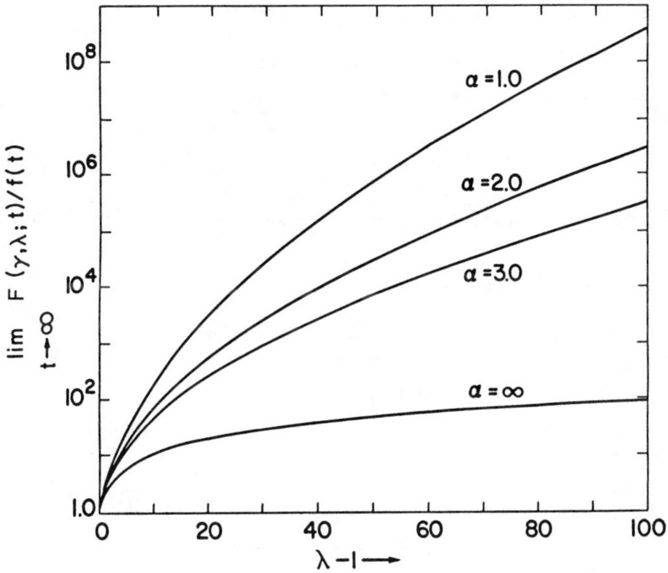

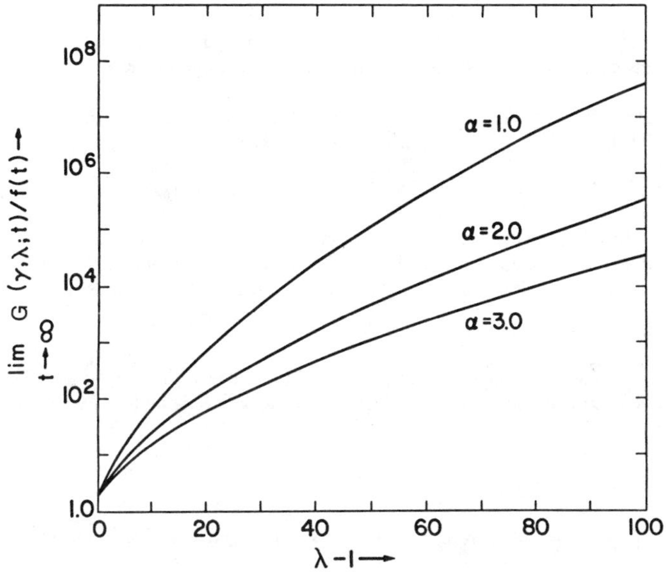

Figure 3. Limiting values (Equation 115) as $t \to \infty$ of (a) F/f, where $F(\gamma, \lambda; t) = T(t)$ defined by Equation 95 with $T(0) = 1$, $\dot{T}(0) = 0$, and (b) G/f, where $G(\gamma, \lambda; t) = T(t)$ with $T(0) = 0$, $\tau\dot{T}(0) = 1$, for $1 \leq \lambda \leq 100$. Shown are results for the cases $\alpha = 1, 2, 3$, and, in Figure 3a, the limiting value $\lim_{\gamma,t\to\infty} F(\gamma, \lambda; t)/f(t) = \lambda$. (The corresponding limit for G is infinite.)

we obtain a second-order equation for the radial factor $\sigma(\mathbf{r})$,

$$\frac{\gamma}{2(\kappa+1)}\left[\frac{1-r^2}{r^2}\frac{d}{dr}\left(r^2\frac{d\sigma}{dr}\right)-\frac{1-r^2}{r^2}\ell(\ell+1)\sigma-2r\frac{d\sigma}{dr}\right]$$

$$-\gamma r\frac{d\sigma}{dr}+\left[\frac{(\kappa+1-\kappa\gamma)\ell(\ell+1)}{(\gamma-1)(\kappa+1)}-3\gamma-\lambda+2\right]\sigma=0 \tag{120}$$

Rewriting this by means of the substitution $\sigma = r^{\ell}y$ and $x = r^2$, we obtain the hypergeometric equation

$$x(1-x)y''+[c-(a+b+1)x]y'-aby=0 \tag{121}$$

where

$$\left.\begin{matrix}a\\b\end{matrix}\right\}=\tfrac{1}{2}\{\kappa+1+\tfrac{5}{2}\pm[(\kappa+1+\tfrac{5}{2})^2-4K]^{1/2}\} \tag{122a,b}$$

$$c=\ell+\tfrac{3}{2} \tag{122c}$$

Here,

$$K=\frac{(\kappa+2)\ell}{2}+\frac{\kappa+1}{2\gamma}\left[\lambda-2+3\gamma-\frac{\ell(\ell+1)(\kappa+1-\kappa\gamma)}{(\lambda-1)(\kappa+1)}\right] \tag{123}$$

The general solution of Equation 121 is [28]

$$y=C_1F(a,b;c;x)+C_2x^{1-c}F(a-c+1,b-c+1;2-c;x) \tag{124}$$

where C_1, C_2 are constants. If $r_i \to 0$, the shell goes over to a gas sphere expanding under its own pressure. In this case only the first term in Equation 124 is finite at the origin and we must set $C_2 = 0$.

The boundary conditions are found from the requirement that the perturbed pressure vanish at both surfaces of the shell. At the inner surface this implies $\sigma = 0$ or

$$C_1F(a,b;c;r_i^2/r_0^2)+C_2(r_i/r_0)^{1-c}F(a-c+1,b-c+1;2-c;r_i^2/r_0^2)=0 \tag{125}$$

Since the unperturbed pressure vanishes at the outer surface, it is only necessary that y be finite at $x = 1$. The linear connection formulas [28] of both terms of Equation 124 contain a term that diverges as $(1-x)^{-(\kappa+1)}$ unless a or b is a nonpositive integer. Thus, we must have

$$\tfrac{1}{2}\{\kappa+\ell+\tfrac{5}{2}-[(\kappa+\ell+\tfrac{5}{2})^2-4K]^{1/2}\}=-n \tag{126}$$

$n = 0, 1, 2, \ldots$. From Equation 123 it follows that λ decreases with n. Hence, the fastest growth (largest positive λ) corresponds to $n = 0$, which implies $K = 0$. Solving for λ, we finally obtain the dispersion relation

$$\lambda-1=-\frac{\gamma\ell(\kappa+2)+(\kappa+1)(3\gamma-1)}{2(\kappa+1)}$$

$$\pm\frac{\{[\gamma\ell(\kappa+2)+(\kappa+1)(3\gamma-1)]^2+4\ell(\ell+1)(\kappa+1-\kappa\gamma)\}^{1/2}}{2(\kappa+1)} \tag{127}$$

For the upper branch, $\lambda > 1$ for all $\ell > 0$, provided that

$$\kappa<1/(\gamma-1) \tag{128}$$

From Equation 112, we see that this is just the condition that $s'(r) < 0$, i.e., that $s(r)$ decrease in the "upward" direction, that directed opposite to the effective gravitational acceleration $-\ddot{\mathbf{R}}$. This is identical with the Schwarzchild criterion (Equation 18) derived for the convective instability in static media. Indeed, an interchange argument along the lines of that used to obtain Equation 18 has been employed [37] to show that such a uniformly expanding fluid system should be convectively unstable whenever $s'(r) < 0$ holds.

For the $\gamma = 1$ case we can redo all the analysis in terms of confluent hypergeometric functions [29] instead of hypergeometric functions, or we can reach the same results by formally letting $\gamma \to 1$ in the previous equations. The solutions are qualitatively similar to those for $\gamma > 1$ except that now Equation 128 always holds.

Motivated by experiments investigating the use of imploding cylindrical liquid metal liners to compress and heat plasma [38], Book and Bernstein [33] studied the Rayleigh-Taylor instability on the inner surface of a liner during both the implosion and rebound phases, assuming $\gamma = 1$. Since both terms in Equation 97 diverge the same way at large t, it is useful to introduce in their place two new solutions that behave asymptotically like Ψ, the standard confluent hypergeometric function of the second kind [29]:

$$\left.\begin{matrix} P^{1}(t) \\ Q^{1}(t) \end{matrix}\right\} = \Phi[(1+\ell)/2; \tfrac{1}{2}; \ln f] \mp \frac{\Gamma(1+\ell/2)}{\Gamma[(1+\ell)/2]} \tau \dot{f} \Phi(1+\ell/2; \tfrac{3}{2}; \ln f) \qquad (129a,b)$$

For large arguments ($t \to +\infty$), we have

$$P^{l}(t) \sim (\ln f)^{-(l+1)/2} \qquad (130a)$$

$$Q^{l}(t) \sim f(\ln f)^{l/2} \qquad (130b)$$

Thus defined, $Q^{l}(t)$ has the property of increasing monotonically for all t, $-\infty < t < \infty$; at turnaround (the instant $t = 0$ when $f = 1$), $Q^{l}(0) = 1$ also. The only perturbations which are unstable both before and after turnaround are those whose time dependence is proportional to $Q^{\ell}(t)$. Figure 4 compares the behavior of $Q^{\ell}(t)$ for $\ell = 1$ and $\ell = 10$ with that of f.

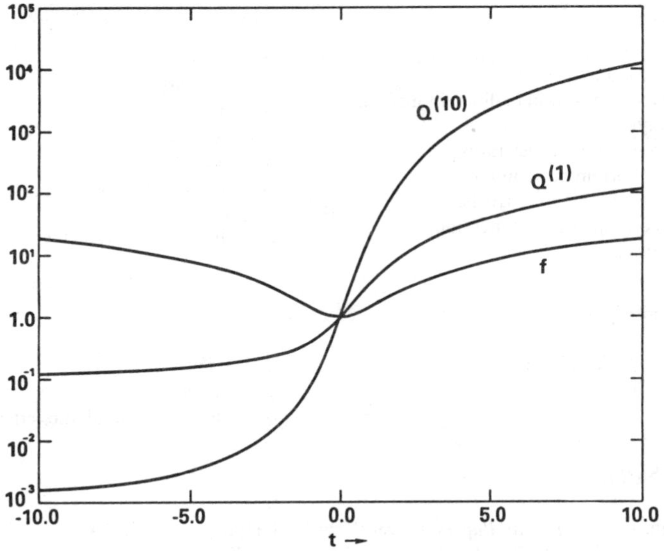

Figure 4. $Q^{\ell}(t)$ (Equation 129b) for $\ell = 1$ and $\ell = 10$, obtained by numerical solution of Equation 94 for $\alpha = 0$ and $\lambda = \ell + 1$, with f plotted for comparison.

CONCLUSIONS

In the theory of compressible fluids, the Rayleigh-Taylor instability at a density jump, the Rayleigh-Taylor instability in a continuously stratified medium, and the convective instability are close relatives. All are gravitational interchange modes. One can easily generate a series of examples that display a continuous transition from one mode to the next.

The energy principle applied to polytropic media shows that, by itself, compressibility increases instability. For only a handful of specific compressible states, however, is it possible to actually calculate growth rates in closed form. The only tractable equilibrium states involve contiguous isothermal layers of fluid satisfying the adiabatic law with constant γ. In the limit where the density of the lower layer vanishes, the growth rate reduces to the classical result found for incompressible fluids.

Closely related is the problem of the stability of uniformly imploding or expanding shells driven by pressure applied at a vacuum-material boundary. The unstable modes are incompressible and (if one allows for the nonexponential time dependence) the growth rates are given by the same classical incompressible fluid formula as in the static case.

Because the unstable eigenmodes are localized near a density jump within distances of order k^{-1}, one expects the growth rates not to change very much when the basic state is not isothermal, particularly at short wavelengths. Thus, dispersion relation (Equation 46) is at least qualitatively correct most of the time in static situations.

To determine growth rates with precision for any but the simplest piecewise isothermal unperturbed states or to study the effects of thermal conduction, flow through the interface, nonlinearity, etc., one must resort to computational means. Nevertheless, numerical solutions present their own difficulties. Validation against nontrivial analytical solutions such as those discussed in this review is indispensable in the development of any code.

Acknowledgment

This work was supported by the U.S. Department of Energy and the Office of Naval Research.

NOTATION

A	Atwood number
$C_{1,2}$	velocity of sound in different mediums
f	frequency
g	gravitational acceleration
H	equilibrium scale number
h	vertical scale of perturbation
K	second quadratic functional
k	wavelength
m	mass
p	pressure
t	time
V	volume
v	velocity
W_I	internal energy
x, y, z	coordinates
Z	dispersion parameter

Greek Symbols

γ	specific heat ratio
θ	temperature
μ	constant
ξ	eigenvector
ρ	density
ω_0	function of adiabatic index γ

REFERENCES

1. Rayleigh, Lord, *Scientific Papers*, Dover, New York (1964), vol. 2, p. 200.
2. Taylor, G. I., *Proc. R. Soc. London Ser.*, A201, 192 (1950).
3. Bernstein, I. B., Frieman, E. A., Kruskal, M. D., and Kulsrud, R. M., *Proc. R. Soc. London Ser.*, A244, 17 (1958).

4. Vandervoort, P. O., *Astrophys. J.*, 134, 699 (1961).
5. Plesset, M. S., and Hsieh, D.-Y., *Phys. Fluids*, 7, 1099 (1964).
6. Shivamoggi, B. K., *Phys. Fluids*, 25, 911 (1982).
7. Plesset, M. S., and Prosperetti, A., *Phys. Fluids*, 25, 911 (1982).
8. Blake, G. M., *Mon. Not. R. Astron. Soc.*, 156, 67 (1972).
9. Matthews, W. G., and Blumenthal, G. R., *Astrophys. J.*, 214, 10 (1977).
10. Tolstoy, I., *Rev. Mod. Phys.* 35, 207 (1963).
11. McCrory, R. L., Montierth, L., Morse, R. L., and Verdon, C. P., in *Laser Interaction and Related Plasma Phenomena*, ed. H. J., Schwarz, H., Hora, M. J., Lubin, and B., Yaakobi (Plenum, New York, 1981), vol. 5, p. 713.
12. Takabe, H., and Mima, K., *J. Phys. Soc.* Japan, 48, 1793 (1980).
13. Scannapieco, A. J., *Phys. Fluids*, 24, 1699 (1981).
14. Baker, L., *Phys. Fluids*, 26, 950 (1983).
15. Schmidt, G., *Physics of High Temperature Plasmas* (Academic, New York, 1965), p. 135.
16. Chrandrasekhar, S., *Hydrodynamic and Hydromagnetic Stability* (Oxford Univ. Press, Clarendon, 1961).
17. Shiau, J. N., Goldman, E. B., and Weng, C. I., *Phys. Rev. Lett.*, 32, 352 (1973).
18. Barcilon, A., Book, D. L., and Cooper, A. L., *Phys. Fluids*, 17, 1707 (1974)
19. Emery, M. H., Gardner, J. H., and Boris, J. P., *Phys. Rev. Lett.*, 48, 677 (1982).
20. Evans, R. G., Bennett, A. J., and Pert, G. J., *Phys. Rev. Lett.* 49, 1639 (1982).
21. Manheimer, W. M., and Colombant, D. G., *Phys. Fluids*, 27, 983 (1984).
22. Landau L. D., and Lifshitz, E. M., *Fluid Mechanics* (Addison-Wesley, Reading, MA, 1959), p. 9.
23. Newcomb, W. A., *Phys. Fluids*, 26, 3246 (1983).
24. Bernstein, I. B., and Book, D. L., *Phys. Fluids*, 26, 453 (1983).
25. Sedov, L. I., *Similarity and Dimensional Methods in Mechanics* (Academic, New York, 1959).
26. Mikaelian, K. O., *Phys. Rev.*, A26, 2140 (1982).
27. Keller, J. B., *Q. J. Appl. Math.*, 14, 171 (1956).
28. Oberhettinger, F., in *Handbook of Mathematical Functions*, ed. M., Abramowitz and I. A., Stegun (U.S. Govt. Printing Office, Washington, D.C., 1968), chap. 15.
29. Slater, L. J., in *Handbook of Mathematical Functions*, ed. M., Abramowitz and Stegun I. A., (U.S. Govt. Printing Office, Washington, D.C., 1968), chap. 13.
30. Book, D. L., *Phys. Rev. Lett.*, 41, 1552 (1978).
31. Book, D. L., and Bernstein, I. B., *J. Plasma Phys.*, 23, 521 (1980).
32. Kidder, R. E., *Nucl. Fusion*, 16, 3 (1976).
33. Book, D. L., and Bernstein, I. B., *Phys. Fluids*, 22, 79 (1979).
34. Han, S. J., and Suydam, B. R., *Phys. Rev.*, A26, 926 (1982).
35. Bernstein, I. B., and Book, D. L., *Astrophys. J.*, 225, 633 (1978).
36. Han, S. J., *Phys. Fluids* 25, 1723 (1982).
37. Book, D. L., *J. Fluid Mech.* 95, 779 (1979).
38. Book, D. L., Cooper, A. L., Ford, R., Gerber, K. A., Hammer, D. A., Jenkins, D. J., Robson, A. E., and Turchi, P. J., in *Plasma Physics and Controlled Nuclear Fusion Research* (IAEA, Vienna, 1977), vol. 3, p. 507.

CHAPTER 9

TAYLOR VORTICES AND INSTABILITIES IN CIRCULAR COUETTE FLOWS

Kunio Kataoka

Department of Chemical Engineering
Kobe University
Rokkodai, Kobe 657, Japan

CONTENTS

INTRODUCTION

The flow in the annular space between two concentric rotating circular cylinders is a very interesting shear flow without a pressure gradient in the direction of mean flow. In this chapter, we shall be concerned primarily with the case where the inner cylinder is rotated and the outer cylinder at rest. When the rotation speed of the inner cylinder is increased beyond a certain critical value, the basic laminar axisymmetric flow, known as Couette flow, becomes unstable. The instability leads to the transition to a laminar cellular vortex flow, referred to as Taylor vortex flow. In 1923, G. I. Taylor [1] made a brilliant contribution to the theory of hydrodynamic stability. He achieved the first success in solving the instability problem by using a linear stability analysis for the case of the small annular gap width compared to the mean radius of the cylinders. His prediction of the critical speed was in good agreement with his experimental observation. As the rotation speed is further increased, this flow system exhibits a sequence of distinct time-independent and time-dependent flow regimes until chaotic turbulence occurs. The instability of Taylor vortex flow leads to the transition to a wavy vortex flow which is established with traveling azimuthal waves superimposed on the Taylor vortex flow. The higher instabilities and the transition to turbulence have been investigated by experiment, using flow-visualization techniques and flow spectrum analyses. This hydrodynamic system is very attractive in connection with the transition-to-turbulence problem.

Following an overview of the problem, we shall discuss first the theoretical approach to the hydrodynamic instability and transition. This discussion is subdivided into three parts: (a) Taylor instability (the instability of circular Couette flow); (b) nonlinear theory of Taylor vortex flow, and (c) wave instability and wavy vortex flow. We next consider the experimental observations of the hydrodynamic behaviors. This section is subdivided into two parts: (a) sequential transitions and nonuniqueness by flow-visualization experiment; and (b) spectral analysis of time-dependent flows. This flow system is also of great importance not only in the design of rotating machinery, such as electric motors, multiple concentric drives, and turbine rotors, but also for the application to chemical equipment such as compact rotating heat exchangers and mixers. In the next section, we discuss the problem of torque and heat/mass transfer in this flow system. We often encounter in the engineering situation the axially developing flow in the annulus between concentric rotating cylinders. When the forced axial flow is weak compared to the tangential flow, the Taylor vortex structure occurs downstream after a certain axial length for development. In the last section we discuss briefly the problem of instability of such a flow and its practical application.

OVERVIEW

The viscous flow in the annulus between two concentric cylinders with the inner one rotating is of great importance not only in mechanical and chemical engineering but also in fluid physics because the flow may give a key to the better understanding of the transition-to-turbulence problem.

Figure 1 shows schematically the Taylor vortices and the coordinate system to be used. Let (r, θ, z) denote the usual cylindrical coordinates. The radii of the inner and outer cylinders are r_i and r_o, respectively. Both cylinders are assumed to be infinitely long. The gap width is given by $d = r_o - r_i$. The velocity components in the increasing r, θ, z directions are denoted by u, v, and w, respectively, and the pressure by p. The inner cylinder is rotated at an angular velocity Ω.

The basic flow, known as circular Couette flow, is stable when Ω is low enough. All streamlines form stationary axisymmetric circles. The Couette velocity distribution is given theoretically by the following hyperbolic function

$$u = w = 0 \quad \text{and} \quad v = Ar + \frac{B}{r} \tag{1}$$

where $A = -\dfrac{r_i^2}{r_o^2 - r_i^2}\Omega$ and $B = \dfrac{r_i^2 r_o^2}{r_o^2 - r_i^2}\Omega$

This equation indicates that the centrifugal force $\rho v^2/r$ is larger near the inner cylinder than near the outer cylinder. Therefore, when the speed of the inner cylinder exceeds a critical value, the

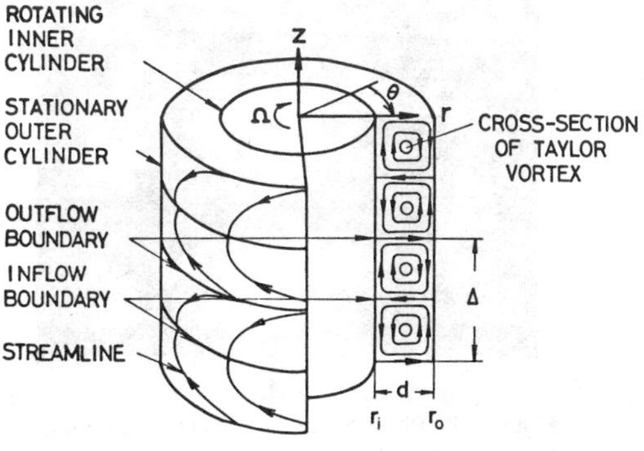

Figure 1. Schematic picture of Taylor vortices and the coordinate system to be used.

Couette flow becomes unstable and the instability leads to the transition to a laminar cellular vortex flow, which is called "Taylor vortex flow" in honor of the early excellent scientist.

Figure 2 is a photograph of Taylor vortex flow. The fluid behavior can be observed clearly as variations in the transmitted or reflected intensity from the scattered light of fine aluminum platelets aligned with the flow. The dark horizontal lines indicate the vortex boundary. Ring-shaped counter-rotating vortices are arrayed along the cylinder axis like stacked automobile tires. The net fluid motion is a helical superposition of the azimuthal Couette flow and the circular secondary flow around the axis of the vortices. The instability is also called "Taylor instability." The critical rotation speed to the onset of Taylor instability depends upon the radii of two cylinders and the kinematic viscosity of fluid. Therefore, the following dimensionless parameters are primarily used:

Reynolds number $$\mathrm{Re} = \frac{r_i \Omega d}{\nu}$$

Taylor number $$\mathrm{Ta} = \frac{r_i \Omega^2 d^3}{\nu^2}$$

It is sometimes convenient to use the square root of the Taylor number. This parameter is called in this chapter "the modified Reynolds number":

modified Reynolds number $$\mathrm{Rem} = \mathrm{Re}\left(\frac{d}{r_i}\right)^{1/2}$$

This instability problem for the case of small gap width was first investigated both theoretically and experimentally by G. I. Taylor in his famous paper [1]. He formulated the instability problem, and obtained an approximate formula for predicting the onset of Taylor instability by means of linear stability theory. According to his linear theory, for sufficiently small values of Ta, i.e. $\mathrm{Ta} < \mathrm{Ta}_c$, all initially infinitesimal disturbances are damped owing to the action of viscosity; for $\mathrm{Ta} > \mathrm{Ta}_c$, however, some of them are amplified with increasing time. The critical Taylor number Ta_c can be obtained as the parameter of the neutral stability. He confirmed the result of linear theory to be in

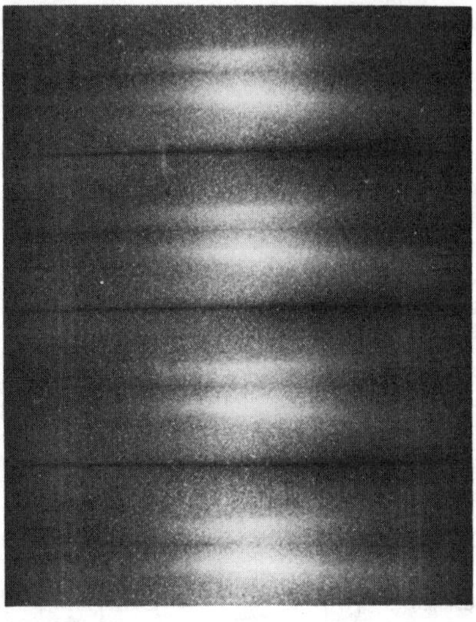

Figure 2. Photograph of axially symmetric Taylor vortex flow: $R^* = 4.47$, $\eta = 0.68$.

good agreement with his own experimental observation. For very narrow gap width $d/r_i \ll 1$, Ta_c approaches approximately 1,708. (Recent exact calculations [e.g., 7, 17, 24, 25] recommend 1,695 as the critical Taylor number.) The critical Taylor number tends to increase with increasing d/r_i (for example, $Ta_c = 2{,}453$ when $d/r_i = 0.33$).

The linear theory of hydrodynamic stability predicts correctly the critical Taylor number, but cannot predict the establishment of a new equilibrium flow (Taylor vortex flow) above the critical Taylor number. This time-independent vortex flow was calculated taking into account the nonlinear effect of supercritical disturbances by Stuart [2, 3], Watson [4] and Davey [5] for Taylor numbers slightly higher than Ta_c. Initially, time-dependent disturbances reach an equilibrium amplitude in the supercritical range. The mean motion (the radial distribution of azimuthal velocity averaged with respect to z) is distorted by nonlinear interaction of the equilibrium disturbances. As a result, the torque acting on the inner cylinder wall becomes much larger than that for purely laminar Couette flow.

As the rotation speed is increased further, the axisymmetric Taylor vortex flow becomes unstable and the instability leads to a new time-dependent vortex flow. As can be seen in the photograph of Figure 3, the vortex boundaries are S-shaped. This flow is established with azimuthally traveling waves superimposed on the Taylor vortices. This is known as wave instability. The second critical Taylor number Ta_w or Reynolds number Re_w depends upon the radius ratio of the two cylinders $\eta = r_i/r_o$.

Coles [6] investigated systematically the wavy vortex flow by applying a flow-visualization technique in an annulus with $\eta = 0.874$. He found that the wavy vortex flow shows many stable states at a given Reynolds number. This suggests nonuniqueness of the flow or the possibility of multiple solutions. Davey, DiPrima, and Stuart [7] studied theoretically the instability of the Taylor vortex flow for $\eta \simeq 1$. They showed that a wavy vortex flow with m waves in the azimuthal direction is stable for $Re > Re_w(m)$. This time-dependent flow has a single fundamental frequency and its harmonics. The fundamental frequency corresponds to the frequency of traveling azimuthal waves passing a point of observation. In this sense, this flow is called "singly periodic wavy vortex flow."

Higher transitions have been confirmed by experiment only [8, 9]. As the Reynolds number is raised further, the wavy vortex flow becomes more complex; the amplitude of the azimuthal waves varies periodically with time. This amplitude-modulated wavy vortex flow has two fundamental frequencies; the first fundamental frequency results from the basic wave motion and the second

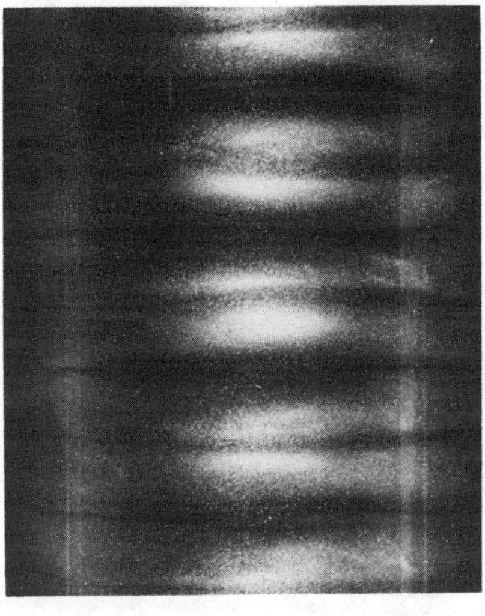

Figure 3. Photograph of singly periodic wavy vortex flow: $R^* = 11.13$, $\eta = 0.68$, $m = 3$.

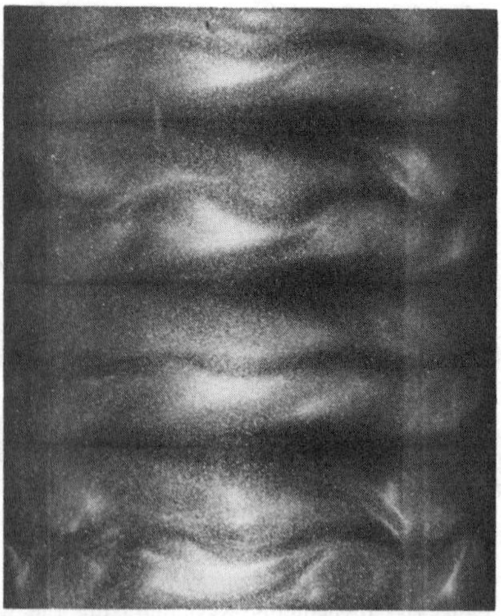

Figure 4. Photograph of doubly periodic wavy vortex flow: $R^* = 18.19$, $\eta = 0.68$, $m = 7$.

fundamental frequency from the periodicity of amplitude-modulation. In this sense, this flow is called "doubly periodic wavy vortex flow." A photograph of the visualized doubly-periodic flow is shown in Figure 4.

At further increased Reynolds numbers, the doubly periodic wavy vortex flow undergoes the transition to weakly turbulent wavy vortex flow. It can be seen from Figure 5 that chaotic turbulence is superimposed on the wavy vortex structure.

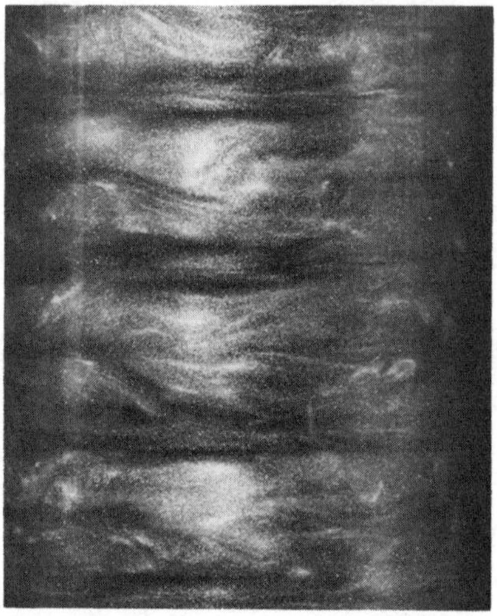

Figure 5. Photograph of weakly turbulent wavy vortex flow: $R^* = 35.28$, $\eta = 0.68$, $m = 6$.

As the Reynolds number goes beyond a critical value, the azimuthal waves disappear but the cellular vortex structure is preserved. This flow regime, called "turbulent vortex flow," is maintained up to a very high Reynolds number at which the vortex structure becomes indiscernible.

As mentioned, it is very interesting that this hydrodynamic system has several sequential transitions with the Reynolds number raised before chaotic turbulence occurs. An excellent review of the hydrodynamic stability and transition problems has been made by DiPrima and Swinney [10].

HYDRODYNAMIC INSTABILITIES AND TRANSITIONS

Taylor Instability—Linear Stability Theory

Taylor [1] achieved the first success in the calculation of the critical Reynolds number by using a linear stability theory. A number of investigators [e.g., 11–18] discussed the Taylor instability problem by improving the linearized theory. The linear stability theory is discussed in this section.

All quantities are made dimensionless by using d as the reference length, $r_i\Omega$ as the reference velocity, $\rho(r_i\Omega)^2$ as the reference pressure and d^2/ν as the reference time. It is known from experimental work that the instability of Couette flow leads to a time-independent axisymmetric cellular vortex flow with axial periodicity as time proceeds. It is therefore assumed that the flow is axisymmetrical, so that the dimensionless velocity components u, v, w are independent of θ. Each quantity can be divided into a mean value and a disturbance with respect to z (dimensionless).

$$
\begin{aligned}
u &= u'(r, z, t) \\
v &= \bar{v}(r, t) + v'(r, z, t) \\
w &= w'(r, z, t) \\
p &= \bar{p}(r, t) + p'(r, z, t)
\end{aligned} \tag{2}
$$

Substituting into the Navier-Stokes equations and averaging the resulting equations with respect to z, we obtain the following equations of mean motion (dimensionless), as given by Davey [5],

$$\frac{1}{r}\frac{\partial}{\partial r}(r\overline{u'^2}) - \frac{\bar{v}^2 + \overline{v'^2}}{r} = -\frac{\partial \bar{p}}{\partial r} \tag{3}$$

$$\frac{1}{Re}\frac{\partial \bar{v}}{\partial t} + \frac{1}{r^2}\frac{\partial}{\partial r}(r^2\overline{u'v'}) = \frac{1}{Re}\left(\nabla^2 - \frac{1}{r^2}\right)\bar{v} \tag{4}$$

where $\nabla^2 = \dfrac{\partial^2}{\partial r^2} + \dfrac{1}{r}\dfrac{\partial}{\partial r} + \dfrac{\partial^2}{\partial z^2}$

A bar above a quantity denotes a mean value with respect to z. By integrating Equation 4 for steady equilibrium flow ($\partial\bar{v}/\partial t = 0$), we obtain the mean velocity distribution (dimensionless) [19]:

$$\bar{v} = -A_0 r + \frac{B_0}{r} + Re\int_{r_i}^{r_o}\frac{\overline{u'v'}}{r}\,dr \times \left(-A_1 r + \frac{B_1}{r}\right) + Re\int_{r_i}^{r}\frac{\overline{u'v'}}{r}\,dr \tag{5}$$

where $A_0 = \dfrac{r_i}{r_o + r_i}$, $\quad B_0 = \dfrac{r_i r_o^2}{d^2(r_o + r_i)}$

$A_1 = \dfrac{r_o^2}{d(r_o + r_i)}$, $\quad B_1 = \dfrac{r_i^2 r_o^2}{d^3(r_o + r_i)}$

The first two terms describe the basic Couette flow. The remaining two terms imply that the velocity distribution is distorted by non-linear interaction of the equilibrium velocity disturbances.

The axisymmetric disturbance equations can be obtained by subtracting the mean motion equation from the Navier-Stokes equations (dimensionless), as given by Davey [5],

$$\frac{1}{\mathrm{Re}}\frac{\partial u'}{\partial t} - \frac{2\bar{v}v'}{r} + \frac{1}{r}\frac{\partial}{\partial r}(ru'^2 - \overline{ru'^2}) + \frac{\partial}{\partial z}(u'w') - \frac{v'^2 - \overline{v'^2}}{r} = -\frac{\partial p'}{\partial r} + \frac{1}{\mathrm{Re}}\left(\nabla^2 - \frac{1}{r^2}\right)u' \tag{6}$$

$$\frac{1}{\mathrm{Re}}\frac{\partial v'}{\partial t} + u'\left(\frac{\partial \bar{v}}{\partial r} + \frac{\bar{v}}{r}\right) + \frac{1}{r^2}\frac{\partial}{\partial r}(r^2u'v' - r^2\overline{u'v'}) + \frac{\partial}{\partial z}(v'w') = \frac{1}{\mathrm{Re}}\left(\nabla^2 - \frac{1}{r^2}\right)v' \tag{7}$$

$$\frac{1}{\mathrm{Re}}\frac{\partial w'}{\partial t} + u'\frac{\partial w'}{\partial r} + w'\frac{\partial w'}{\partial z} = -\frac{\partial p'}{\partial z} + \frac{1}{\mathrm{Re}}\nabla^2 w' \tag{8}$$

$$\frac{1}{r}\frac{\partial}{\partial r}(ru') + \frac{\partial w'}{\partial z} = 0 \tag{9}$$

If we assume the disturbance velocities u', v', w' to be small compared to the mean velocity $\bar{v}$, we can neglect their quadratic terms to linearize those equations. Eliminating p' by differentiating Equations 6 and 8 with respect to z and r, respectively, and eliminating w' by using the equation of continuity (Equation 9), we finally obtain two linearized partial differential equations for u' and v'.

The normal mode solutions to the equations are assumed to be of the form

$$\begin{aligned} u'(r, z, t) &= u_1(r)\exp[\sigma t + i\lambda z] \\ v'(r, z, t) &= v_1(r)\exp[\sigma t + i\lambda z] \end{aligned} \tag{10}$$

where the axial wave number λ is taken to be real and positive and the amplification rate σ is generally complex.

Substituting into the linearized equations, we obtain the following equations (dimensionless), as given by Davey [5],

$$\frac{1}{\lambda\,\mathrm{Re}}(DD^* - \lambda^2)(DD^* - \lambda^2 - \sigma)u_1 - 2\lambda\frac{\bar{v}}{r}v_1 = 0 \tag{11}$$

$$\frac{1}{\mathrm{Re}}(DD^* - \lambda^2 - \sigma)v_1 - (D^*\bar{v})u_1 = 0 \tag{12}$$

and the boundary conditions are

$$u_1 = Du_1 = v_1 = 0 \qquad \text{at } r = r_i \text{ and } r_o \tag{13}$$

Here, $D = \dfrac{d}{dr}$ and $D^* = \dfrac{d}{dr} + \dfrac{1}{r}$

The mean velocity $\bar{v}$ can be assumed to have the Couette velocity distribution just when the Taylor instability occurs:

$$\bar{v} = -A_0 r + \frac{B_0}{r} \tag{14}$$

If $\sigma = 0$, the disturbance velocities are neither attenuated nor amplified. We can determine Re_c as a function of λ at the marginal state. The eigenfunctions u_1, v_1 should be determined to within an arbitrary multiplicative factor. If the annular gap width d is small compared to the mean radius

$r_m = (r_0 + r_i)/2$, the critical modified Reynolds number approaches approximately $Rem_c = 41.2$ and the wavenumber λ of the critical disturbance is approximately π. (Recent exact calculation [e.g., 17, 24] recommends $Rem_c = 41.18$ and $\lambda_c = 3.127$ for $\eta \to 1$.)

As in many investigations of the linear stability theory [11–18], the eigenvalue problem can be simplified by approximation for the narrow gap geometry; for $d/r_m \ll 1$, $D^* \simeq D$. The angular velocity $\bar{v}/r$ for laminar Couette flow can be approximated by a linear profile of the form

$$\frac{\bar{v}}{r} = \Omega\left(\frac{1}{2} - x\right) \qquad -\frac{1}{2} \le x \le \frac{1}{2} \tag{15}$$

where x is the dimensionless radial distance from the mean radius; $r = r_m + xd$. At the neutral state ($\sigma = 0$; this condition, $Re\{\sigma\} = Im\{\sigma\} = 0$, is called the principle of exchange of stabilities.) the eigenvalue problem reduces to the following equations

$$(D^2 - \lambda^2)^2 u_1 = (\tfrac{1}{2} - x)v_1 \tag{16}$$

$$(D^2 - \lambda^2)v_1 = -2\lambda^2 \, Ta \, u_1 \tag{17}$$

and the boundary conditions are

$$u_1 = Du_1 = v_1 = 0 \qquad \text{at } x = \pm\tfrac{1}{2} \tag{18}$$

Here $D = \dfrac{d}{dx}$ and $Ta = -\dfrac{2A\Omega d^4}{\nu^2}$ is the Taylor number.

In the limiting case $\eta \to 1$, $Ta = Re^2 d/r_i$ because $A = -r_i^2\Omega/(r_o^2 - r_i^2)$.

The eigenvalue problem with $\sigma = 0$ can be solved usually by the Galerkin method. It should be kept in mind that a good approximation to the critical Taylor number can be obtained from the following formula given by Taylor [1]:

$$Ta_c = \frac{\pi^4(1 + d/2r_i)}{0.0571(1 - 0.652d/r_i) + 0.00056(1 - 0.652d/r_i)^{-1}} \tag{19}$$

Chandrasekhar [11] and DiPrima [14] give

$$Ta_c = 1{,}695(1 + d/2r_i) \qquad d \ll r_i \tag{20}$$

The critical Taylor and Reynolds numbers are a function of the radius ratio η for the case when d/r_i is finite. The critical modified Reynolds numbers and wavenumbers obtained theoretically are plotted against the radius ratio in Figure 6.

Taylor Vortex Flow—Nonlinear Theory

The nonlinear mechanism of supercritical disturbances was first studied by Stuart [2, 3] and Watson [4]. Davey [5] rigorously calculated the Taylor vortex flow by improving the Stuart's nonlinear theory. In this section, for simplicity and instructiveness, we shall discuss the nonlinear theory in our notation, following the Stuart's energy-balance method.

In the Taylor vortex flow, the disturbance velocity can be expressed as

$$u' = \sum_{n=1}^{\infty} u_n(r, t)\cos n\lambda z$$

$$v' = \sum_{n=1}^{\infty} v_n(r, t)\cos n\lambda z \tag{21}$$

$$w' = \sum_{n=1}^{\infty} w_n(r, t)\sin n\lambda z$$

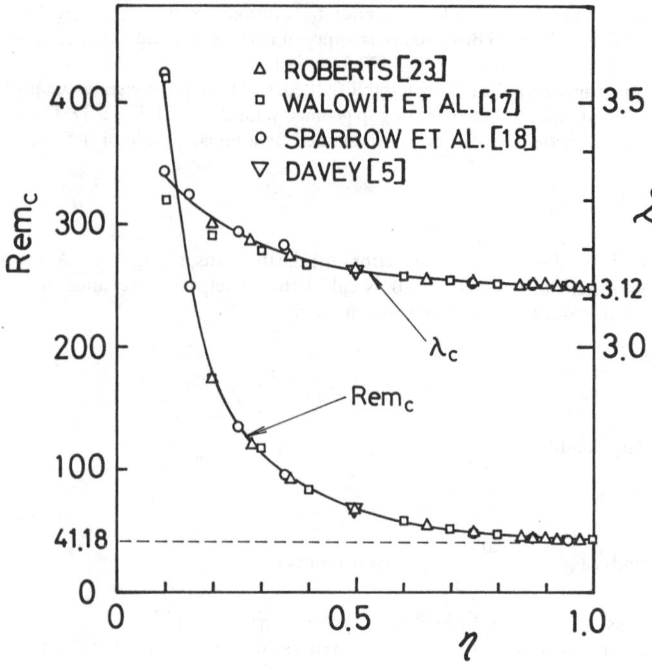

Figure 6. Critical modified Reynolds number and axial wavenumber predicted by linear stability analysis.

It is assumed that the radial profiles of the fundamental components $u_1(r, t)$, $v_1(r, t)$, $w_1(r, t)$ of the supercritical disturbance are similar in shape to those of the linear theory $u_1(r)$, $v_1(r)$, $w_1(r)$ and that the harmonic components of the fundamental are of negligible importance; then the fundamental components can be expressed as

$$u' \simeq Am(t)\, u_1(r) \cos \lambda z$$

$$v' \simeq Am(t)\, v_1(r) \cos \lambda z \qquad (22)$$

$$w' \simeq Am(t)\, w_1(r) \sin \lambda z$$

and the harmonic components can be neglected

$$u_n(r, t),\ v_n(r, t),\ w_n(r, t) \simeq 0 \qquad (n \geq 2)$$

Here Am(t) is a time-dependent amplitude. The wavenumber λ is also assumed to be equal to that predicted by linear theory: $\lambda = \lambda_c$.

If we multiply Equations 6, 7, and 8 by u′, v′, w′, respectively, combine the three resulting equations into an equation of disturbance energy, and integrate the equation over a space ($z = 0 \sim 2\pi/\lambda$ and $r_i \leq r \leq r_o$), we obtain the following energy-balance equation of an axisymmetric supercritical disturbance (dimensionless), as given by Stuart [2]:

$$\frac{1}{\mathrm{Re}}\frac{\partial}{\partial t}\iint\frac{1}{2}(u'^2 + v'^2 + w'^2)r\,dr\,dz = \iint(-u'v')\left(\frac{\partial\bar{v}}{\partial r} - \frac{\bar{v}}{r}\right)r\,dr\,dz - \frac{1}{\mathrm{Re}}\iint(\xi'^2 + \eta'^2 + \zeta'^2)r\,dr\,dz \tag{23}$$

where the vorticity components of the disturbance are given by

$$\xi' = -\frac{\partial v'}{\partial z}, \qquad \eta' = \frac{\partial u'}{\partial z} - \frac{\partial w'}{\partial r}, \qquad \zeta' = \frac{1}{r}\frac{\partial}{\partial r}(rv') \tag{24}$$

Equation 23 implies that the rate of increase in the disturbance energy is equal to the rate of transfer of kinetic energy from the mean flow to the disturbance, less the rate of viscous dissipation of kinetic energy. The mean velocity distribution for evaluating the term $(\partial\bar{v}/\partial r - \bar{v}/r)$ can be calculated substituting Equation 22 into Equation 5. Substituting Equation 22 into Equation 23, we obtain the following amplitude equation (dimensionless) [19]:

$$\frac{\Phi_4}{\mathrm{Re}}\frac{d}{dt}\mathrm{Am}^2 = \mathrm{Am}^2\left[\Phi_1\left(1 - \frac{\Phi_3/\Phi_1}{\mathrm{Re}}\right) - \mathrm{Am}^2\,\Phi_2\,\mathrm{Re}\right] \tag{25}$$

where $\Phi_1 = \dfrac{2\pi}{\lambda}B_o\displaystyle\int_{r_i}^{r_o}\frac{u_1 v_1}{r}\,dr$

$$\Phi_2 = \frac{2\pi}{\lambda}\left[\frac{1}{4}\int_{r_i}^{r_o} r(u_1 v_1)^2\,dr - \frac{B_1}{2}\left(\int_{r_i}^{r_o}\frac{u_1 v_1}{r}\,dr\right)^2\right]$$

$$\Phi_3 = \frac{\pi}{\lambda}\int_{r_i}^{r_o}\left[(\lambda v_1)^2 + \left(\lambda u_1 + \frac{dw_1}{dr}\right)^2 + \left(\frac{dv_1}{dr} + \frac{v_1}{r}\right)^2\right]r\,dr$$

$$\Phi_4 = \frac{\pi}{\lambda}\int_{r_i}^{r_o}\frac{1}{2}(u_1^2 + v_1^2 + w_1^2)r\,dr$$

The Φ's can be calculated using the linear theory solution of the velocity eigenfunctions u_1, v_1, w_1 and the wavenumber $\lambda = \lambda_c$. It should be kept in mind that these values depend upon the multiplicative factor used for calculation of the eigenvalue problem. In the condition for marginal stability, the Reynolds stress term (the last term of Equation 25) can be neglected, and the critical Reynolds number is given by

$$\mathrm{Re}_c = \frac{\Phi_3}{\Phi_1} \tag{26}$$

Since Am(t) behaves like exp(σt) as Am → 0, Equation 25 should be of the form

$$\frac{d}{dt}\mathrm{Am} = \sigma\,\mathrm{Am} + a_1\,\mathrm{Am}^3 \tag{27}$$

Hence, $\sigma = (\Phi_1/2\Phi_4)(\mathrm{Re} - \mathrm{Re}_c)$ and $a_1 = -\Phi_2\,\mathrm{Re}^2/2\Phi_4$

The equilibrium amplitude of the fundamental disturbance is given by [19] (dimensionless):

$$\mathrm{Am}_e^2 = \frac{\Phi_1}{\Phi_2}\frac{1}{\mathrm{Re}}\left(1 - \frac{\mathrm{Re}_c}{\mathrm{Re}}\right) \tag{28}$$

Using the eigenfunctions u_1, v_1, w_1 of the linear stability theory, the equilibrium amplitude can be calculated as a function of Re. For small-gap problem, Stuart [2] gives the equilibrium amplitude of the form

$$Am_e^2 = \frac{5.425 \times 10^4}{Re^2}\left(1 - \frac{Ta_c}{Ta}\right) \qquad \eta \to 1 \tag{29}$$

where $Ta_c = 1{,}708$.

For simplicity, this nonlinear theory does not take into account the effect of harmonics on the fundamental disturbance. As shown in Figure 7, the nonlinear interaction structure must be much more complex. Davey [5] made rigorously a numerical calculation of the equilibrium amplitude for both small and wide gap widths by extending Stuart's energy-balance method to the nonlinear interaction with the supercritical disturbances of the form (dimensionless)

$$\begin{aligned} u'(r, z, t) &= \sum_{n=1}^{\infty} Am(t)^n \left\{ u_n(r) + \sum_{m=1}^{\infty} Am(t)^{2m} u_{nm}(r) \right\} \cos n\lambda z \\ v'(r, z, t) &= \sum_{n=1}^{\infty} Am(t)^n \left\{ v_n(r) + \sum_{m=1}^{\infty} Am(t)^{2m} v_{nm}(r) \right\} \cos n\lambda z \end{aligned} \tag{30}$$

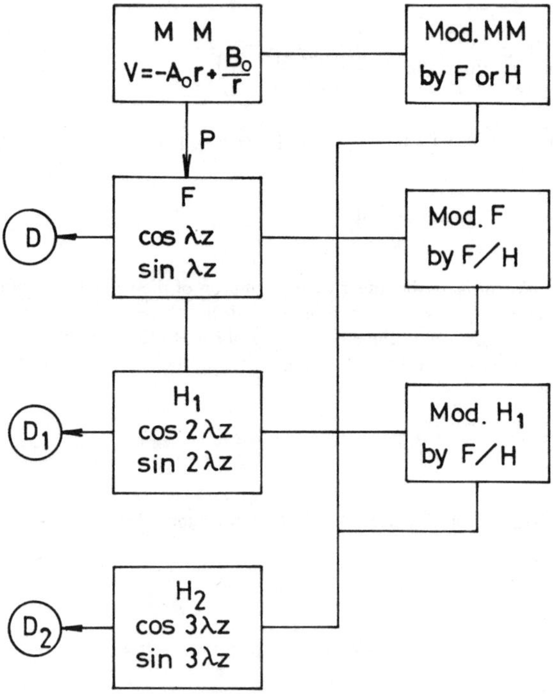

Figure 7. Structure of nonlinear interaction of supercritical disturbances: MM = mean motion, F = fundamental, H = harmonic, D = dissipation, P = production, Mod. = modification of radial dependence.

He took into consideration the transfer of kinetic energy from the fundamental mode to its harmonics and the radial distortion of the fundamental in the energy-balance equation. The normalizing condition $v_1(r_m) = 1.0$ was adopted as the multiplicative factor for calculation. As an example of the wide-gap problem, Davey obtained an equilibrium amplitude for $\eta = 1/2$ given by

$$Am_e^2 = 0.09017\left(1 - \frac{Ta_c}{Ta}\right) \tag{31}$$

For small-gap problem ($\eta \rightarrow 1$):

$$Am_e^2 = 0.3257\left(1 - \frac{Ta_c}{Ta}\right) \tag{32}$$

Once Am_e is known, the velocity distribution distorted by nonlinear interaction of supercritical disturbances can be calculated by the equation [19] (dimensionless):

$$\bar{v} = -A_0 r + \frac{B_0}{r} + \frac{Re}{2} Am_e^2\left(-A_1 r + \frac{B_1}{r}\right)\int_{r_i}^{r_o} \frac{u_1 v_1}{r}\,dr + \frac{Re}{2} Am_e^2 r \int_{r_i}^{r} \frac{u_1 v_1}{r}\,dr \tag{33}$$

Figure 8 shows the radial profiles of the fundamental component predicted in the supercritical range ($Re/Re_c = 1.22$) [19]. Mizushina et al. [20] measured the equilibrium velocity disturbances by taking a pair of stroboscopic photographs of a small number of fine resin particles suspended in a Taylor vortex flow. The experimental data are given by small open circles in the same figure.

Figure 9 shows the distorted radial profiles of the mean velocity predicted by the nonlinear theory [19].

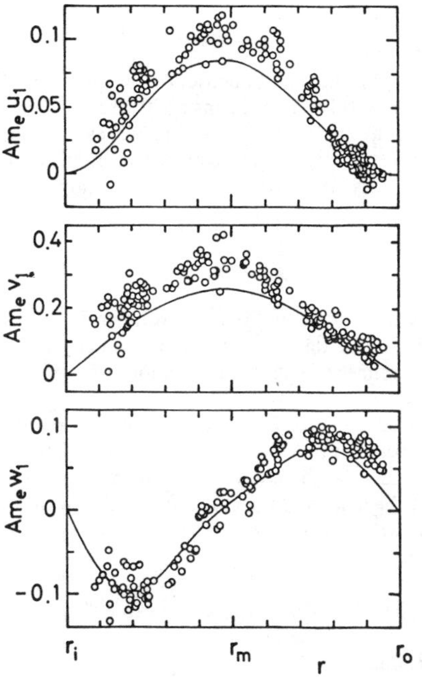

Figure 8. Radial profiles of the equilibrium velocity disturbance predicted in the supercritical range by nonlinear theory [19]: $R^* = 1.22$, $r_i\Omega = 0.101$ m/s, $d/r_i = 0.21$, open circle = experimental [20].

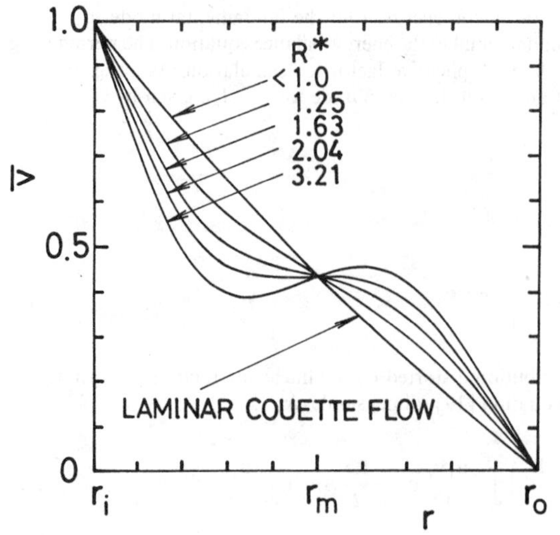

Figure 9. Radial profiles of the mean velocity predicted by nonlinear theory [19] ($d/r_1 = 0.5$).

As will be discussed later, the torque and heat transfer can be calculated from the gradients on the inner or outer cylinder wall of the distorted mean velocity and temperature. Meyer [21, 22] also made a numerical calculation of the Taylor vortex flow using a time-dependent finite-difference method. His predicted torque was in good agreement with the experimental data over a wider range than obtained by Davey [5].

Wave Instability and Wavy Vortex Flow

According to the experimental observation of Coles [6], the axisymmetric Taylor vortex flow becomes unstable at approximately $Re_w = 1.2Re_c$ for $d/r_i = 0.144$, and the instability leads to non-axisymmetric time-dependent Taylor vortex flow, which has azimuthally traveling waves.

The linear problem for the stability of Couette flow to non-axisymmetric disturbances has been studied by DiPrima [14], Roberts [23], and Krueger, Gross, and DiPrima [24]. For example, the azimuthal component of the disturbance is given by

$$v'(r, \theta, z, t) = v(r) \exp[\sigma t + i(\lambda z + k\theta)] \tag{34}$$

The eigenvalue σ is complex corresponding to the azimuthal wave motion. According to DiPrima, the critical Taylor number is a slowly increasing function of the azimuthal wave number k and the critical Taylor number for non-axisymmetric disturbances is higher than that for axisymmetric disturbances.

Davey, DiPrima, and Stuart [7] extended the nonlinear theory of Taylor vortex flow to the wave instability problem for small-gap case ($\eta \to 1$). They considered two disturbances; one is an axisymmetric Taylor vortex disturbance that is periodic in the axial direction only, and the other is a non-axisymmetric disturbance that is periodic in both the axial and the circumferential directions. Both disturbances were assumed to have the same axial wavelength $2\pi/\lambda$. The azimuthal component of velocity disturbance (dimensionless) is expressible as

$$v'(r, \theta, z, t) = v_{c10}(x, t) \cos \lambda z + v_{s10}(x, t) \sin \lambda z + [v_{c11}(x, t) \cos \lambda z + v_{s11}(x, t) \sin \lambda z]e^{ik\theta} \tag{35}$$

where k is the azimuthal wavenumber and $x = (r - r_m)/d$.

As in the nonlinear theory of the axisymmetric Taylor vortex flow (similarly to Equation 22), they considered the primary modes to be of the form (dimensionless):

$$\left.\begin{aligned} v_{c10}(x,t) &= A_c(t)f_0(x) + A_c(t)^3 f_1(x) + A_c(t)A_s(t)^2 f_2(x) + \cdots \\ v_{s10}(x,t) &= A_s(t)f_0(x) + A_s(t)^3 f_1(x) + A_s(t)A_c(t)^2 f_2(x) + \cdots \\ v_{c11}(x,t) &= B_c(t)h_0(x) + B_c(t)|B_c(t)|^2 h_1(x) + B_c(t)|B_s(t)|^2 h_2(x) + \cdots \\ v_{s11}(x,t) &= B_s(t)h_0(x) + B_s(t)|B_s(t)|^2 h_1(x) + B_s(t)|B_c(t)|^2 h_2(x) + \cdots \end{aligned}\right\} \tag{36}$$

Here, the f's and h's are solutions of the linear eigenvalue problems for axisymmetric and non-axisymmetric disturbances, respectively. The amplitude functions A_c, A_s are real while the amplitude functions B_c, B_s are complex-valued.

Nonlinear interaction (represented by quadratic terms in A_c, A_s, B_c, B_s) of the fundamentals gives rise to the generation of the first harmonics of the fundamentals and to the distortion of a mean motion. Then the interaction of the fundamentals with the distorted mean motion and the first harmonics leads to the generation of the second harmonics and to the distortion of the radial profile of the fundamentals. This interaction effect is represented by cubic terms in A_c, A_s, B_c, B_s.

The above expansion (Equation 36) should be consistent with the nonlinear disturbance equations. At cubic order, those four amplitude functions constitute a set of four coupled first-order nonlinear equations, as given by Davey et al. [7],

$$\frac{d}{dt}A_c = (\sigma + a_1A_c^2 + a_2A_s^2 + a_3|B_c|^2 + a_4|B_s|^2)A_c + (a_5B_c\tilde{B}_s + \tilde{a}_5\tilde{B}_cB_s)A_s$$

$$\frac{d}{dt}B_s = (\epsilon + b_1|B_s|^2 + b_2|B_c|^2 + b_3A_s^2 + b_4A_c^2)B_s + (b_3 - b_4)B_cA_sA_c + (b_1 - b_2)\tilde{B}_sB_c^2 \tag{37}$$

with similar equations for A_s and B_c, where a tilda is used to denote the complex conjugate. The parameter σ (real) and ϵ (complex) are the amplification rates for axisymmetric and non-axisymmetric disturbances, respectively. The a's and b's are definite numbers determined as functions of λ, k, Ta.

The experimental observations (e.g., Coles [6]) indicate the sequence of the successive transitions from Couette flow to Taylor vortex flow to wavy vortex flow as the Taylor number is raised. Therefore, the A_c and B_s equations, Equations 37, are necessary to be solved for this sequence. Depending upon the values of the a's and b's, these amplitude equations suggest the possibility of several flow states: laminar Couette flow $A_c = B_s = 0$ and $A_s = B_c = 0$; Taylor vortex flow $A_c \neq 0$, $B_c = B_s = 0$ but A_s a multiple of A_c; and wavy vortex flow $A_c \neq 0$, $B_s \neq 0$ and $A_s = B_c = 0$.

They successfully showed that for a fixed value $\lambda = \lambda_c$, a Taylor vortex flow exists in the supercritical range $Ta_c < Ta < Ta_w$ but becomes unstable at $Ta = Ta_w$ and that for $Ta > Ta_w$ a wavy vortex flow with k is attained. The second critical Taylor number Ta_w is a function of λ and k. If we assign the values of λ and k, we can obtain Ta_c by linear theory. For the case of the outer cylinder at rest, Ta_c occurs for $k = 0$ and $\lambda = \lambda_c$:

$$Ta_c \simeq 1695 \quad \text{and} \quad \lambda_c \simeq 3.13 \qquad (\eta \to 1) \tag{38}$$

For a fixed value $\lambda = \lambda_c$ and an assigned value $k \neq 0$, a Taylor vortex flow is stable when $a_1 < 0$, $\sigma > 0$ for $Ta > Ta_c$ but $Re\{\epsilon\} < 0$ up to $Ta_c(k) > Ta_c$. Here $Ta_c(k)$ is the critical Taylor number for non-axisymmetric disturbance with $k \neq 0$, which is calculated by Krueger et al. [24] and Roberts [23].

According to the Davey et al. [7], the equilibrium azimuthal velocity disturbance for Taylor vortex flow is given by

$$v'(x,z) = Am_e f_0(x)\cos\lambda z + Am_e^2 m_1(x)\cos 2\lambda z + \theta(Am_e^3) \tag{39}$$

where $(A_c^2 + A_s^2) \to Am_e^2 = -\sigma/a_1$ as $t \to \infty$.

The second term of Equation 39 indicates the generation of first harmonics due to the nonlinear interaction of the fundamentals.

The mean motion of Taylor vortex flow is given by

$$\bar{v}(x) = V(x) + Am_e^2 F_1(x) \tag{40}$$

where V is the radial distribution of Couette flow. The second term indicates the modification of the radial dependence due to the nonlinear interaction of the fundamentals.

The azimuthal velocity disturbance for wavy vortex flow is given by

$$\begin{aligned} v'(t, x, \theta, z) = {} & Am_e f_0(x) \cos \lambda z + \{Am_e^2 m_1(x) + \beta_e^2 m_3(x)\} \cos 2\lambda z \\ & + 2\, Re\{\beta_e h_0(x) \sin \lambda z \exp(i[k\theta + \omega(t - t_s)])\} \\ & + 2\beta_e^2\, Re\{[-p_1(x) \cos 2\lambda z + y_1(x)] \exp(i2[k\theta + \omega(t - t_s)])\} \\ & + 2\, Am_e \beta_e\, Re\{r_1(x) \sin 2\lambda z \exp(i[k\theta + \omega(t - t_s)])\} \end{aligned} \tag{41}$$

where $A_s = B_c = 0$

$$A_c^2 = Am_e^2 = \frac{\sigma\, Re\{b_1\} - a_4\, Re\{\epsilon\}}{a_4\, Re\{b_4\} - a_1\, Re\{b_1\}}$$

$$B_s = \beta_e \exp[i\omega(t - t_s)]$$

$$\beta_e^2 = \frac{a_1\, Re\{\epsilon\} - \sigma\, Re\{b_4\}}{a_4\, Re\{b_4\} - a_1\, Re\{b_1\}}$$

$$\omega = Im\{\epsilon\} + Im\{b_1\}\beta_e^2 + Im\{b_4\}Am_e^2$$

The m's, p's, and r's are the radial dependence of the first harmonics v_{c20}, v_{c22}, v_{c21}, respectively. The function y_1 is the radial dependence of the harmonic component independent of z, i.e., v_{02}. The last three terms of Equation 41 indicate the periodical wave motion that has the first fundamental frequency $f_1 = \omega/2\pi$. As will be discussed later, the first fundamental frequency was observed by Coles [6] by using a flow-visualization technique.

The mean motion of wavy vortex flow is given by

$$\bar{v}(x) = V(x) + Am_e^2 F_1(x) + \beta_e^2 F_3(x) \tag{42}$$

The third term indicates the distortion of the radial distribution of Couette flow due to the nonlinear interaction of non-axisymmetric disturbances.

Davey et al. [7] have successfully confirmed the sequence of transitions observed by experiment and found that the Taylor vortex flow for $\eta \to 1$ becomes unstable at $R^* = Re/Re_c \simeq 1.04$. Eagles [25] extended the Davey et al. theory to fifth order in the amplitudes for $\eta = 0.9512$ and calculated the critical Taylor number Ta_w for the wave instability. Eagles [26] also calculated the torque for wavy vortex flow with k = 1, 2, 3, and 4 for $\eta = 0.9512$ and compared with the torque measurements of Donnelly [27] and Debler, Fuener, and Schaaf [28].

There are some numerical analyses [29–35] of model systems for the investigation of the transition to chaos. The velocity and pressure are expanded in series of basic functions such as double Fourier series. Each term of the series indicates the corresponding mode. The velocity and pressure series suitably truncated are substituted into the Navier-Stokes equation to yield a set of coupled nonlinear ordinary differential equations for the time-dependent amplitudes of the different modes. The model equations can be numerically integrated to compare with the experimental spectral data.

Yahata [30–35] tried to calculate not only the wavy vortex flow but also the quasi-periodic vortex flow by using a numerical analysis of the model system. The velocity field was Fourier-analyzed in both the axial and azimuthal directions. Each coefficient in the double Fourier series was expanded in a complete set of radial distribution functions. The suitably truncated Fourier double series of the velocity and pressure were substituted into the Navier-Stokes equations. A set of time-dependent amplitude equations were derived using the Galerkin method. In spite of his severe truncation of the Fourier series, his power spectra calculated at four Reynolds numbers can describe qualitatively some aspects of the experimental data obtained by literature [8]. Yahata found in his model analysis four successive dynamical regimes: periodic, doubly periodic, triply periodic, and chaotic.

EXPERIMENTAL OBSERVATIONS OF HYDRODYNAMIC BEHAVIORS

Flow-Visualization Experiments—Sequential Transitions, and Nonuniqueness

In a classic study of the hydrodynamic stability, G. I. Taylor [1] observed the vortex flow pattern by using dye as a tracer, and found that when the rotation speed of inner cylinder (or the Reynolds number Re) exceeds a critical value Re_c, a laminar Couette flow changes into a time-independent toroidal vortex flow. When Re is increased beyond a second critical value Re_w, the Taylor vortex flow becomes unstable and the instability leads to the transition to a time-dependent cellular vortex flow that has azimuthally traveling waves. The time-dependent wavy vortex flow was observed successfully by Schultz-Grunow and Hein [36] and Coles [6], suspending very fine aluminum platelets in the fluid. Since the fine platelets are aligned with the flow, we can observe the fluid motion as variations in the transmitted or reflected intensity (see Figures 2 to 5).

As mentioned briefly earlier, Coles found that at a given Reynolds number there are several distinct stable flow states, depending upon the Reynolds number history, i.e., how to reach the final Reynolds number. He labeled a flow state of the wavy vortex flow by the number of Taylor vortices accommodated in the annular space and the number of azimuthal waves around the annulus, i.e. (N, m).

Figure 10 shows a three-dimensional representation of the state transitions he observed as Re varies slowly. As can be seen in the figure, he discovered the nonuniqueness and the associated hysteresis in the transitions between different flow states. Each transition is represented by a horizontal bridge or arrow from one column to another. As the Reynolds number is slowly increased from zero, 28 Taylor cells (N = 28) appear at $Re_c = 114$, followed by four waves (m = 4) at $Re_w = 143$. (i.e. $Re_w = 1.254\, Re_c$). The number of cells is approximately equal to the length/gap ratio (H/d = 27.9): $Rem_c = 43.3$ and $\lambda_c = 3.13$. He also measured the angular velocity ω of the azimuthally traveling waves, which corresponds to the frequency f_1 of the azimuthal waves passing a point of observation. He found that f_1/m is a function of Re, independent of the number of azimuthal waves m. The dimensionless frequency f_1/mf_r for his apparatus decreased gradually from 0.5 at the onset of wavy vortex flow to 0.34. In the large Reynolds number range ($8 < R^* < 23$), f_1/mf_r was kept at 0.34. Here $f_r = \Omega/2\pi$ is the frequency of the inner cylinder rotation.

Snyder [37] and Burkhalter and Koschmieder [38] found the nonuniqueness of axial state even in the time-independent Taylor vortex flow regime. Benjamin [39] and Cole [40] studied the effect of annulus height on the nonuniqueness and transitions. Koschmieder [41] studied turbulent vortex flow between long concentric cylinders to measure the axial wavelength of Taylor vortices by suspending aluminum powder in the fluid. He found that the wavelength becomes substantially larger than the critical wavelength of laminar Taylor vortices.

Gorman and Swinney [9] observed doubly periodic vortex flow by using flow-visualization and spectral techniques. They found that the second frequency f_2 characterizing the doubly periodic regime corresponds to a modulation of the azimuthal waves. They discovered many patterns of the amplitude modulation; for example, all waves may be flattened simultaneously and successive waves may be flattened in sequence. Their result will be discussed in the next section.

Spectral Analyses—Evolution of Time-Dependent Vortex Flows

Spectral analysis of time-dependent flows has been made by Gollub and Swinney [42]; Fenstermacher, Swinney, and Gollub [8]; Walden, and Donnelly [43]; Bouabdallah and Cognet [44]; Cognet, Bouabdallah, and Aider [45]; Gorman and Swinney [9]; and Krugljak et al. [46].

Fenstermacher et al. studied the transition-to-turbulence problem by applying a technique of laser-Doppler velocimetry in the fluid between concentric rotating cylinders ($\eta = 0.877$, H/d = 20.0). The power spectra of the radial component of velocity fluctuations were analyzed for different dynamical flow regimes. They found the critical Reynolds numbers to higher transitions: $Re_w \simeq 1.2\, Re_c$, $Re_q \simeq 10\, Re_c$, and $Re_t \simeq 22\, Re_c$. Here Re_q and Re_t are the critical Reynolds numbers for the quasi-

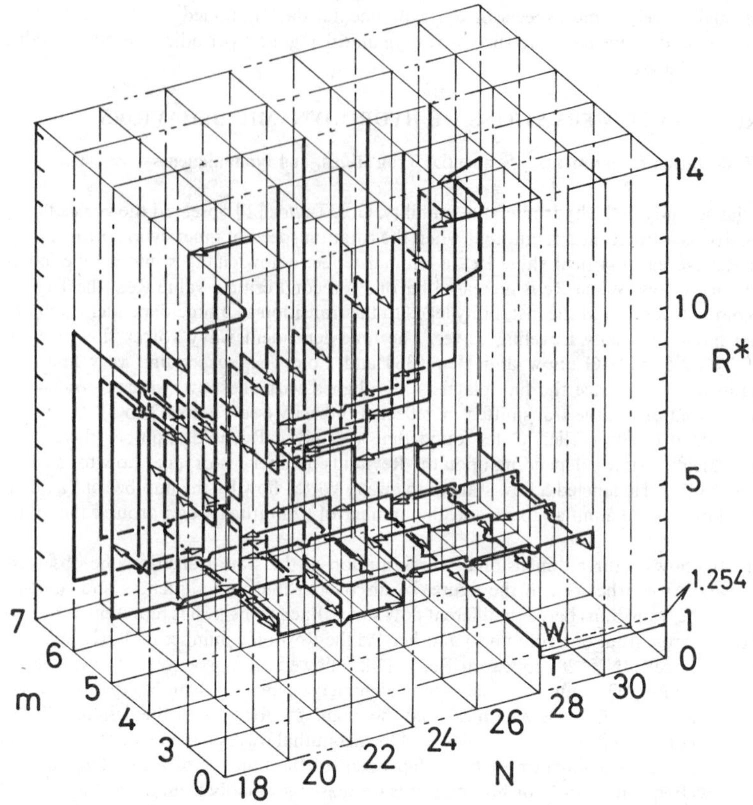

Figure 10. State transitions in wavy and quasi-periodic vortex flows observed by Coles [6]: $\eta = 0.874$, H/d = 27.9, T = Taylor boundary, W = wave boundary.

periodic flow and the turbulent vortex flow, respectively. They observed primarily the 17-vortex 4-wave state (i.e., N = 17, m = 4). Some of their spectra are shown in Figure 11.

As Gorman and Swinney pointed out, Fenstermacher et al. studied the transition from doubly periodic vortex flow to weakly turbulent vortex flow for the 4/1 state when the four waves were flattened successively in sequence. In the range of wavy vortex flow ($Re_w < Re < Re_q$), the power spectrum (Figure 11a) contains only a single frequency component f_1 (called the first fundamental frequency), which corresponds to the frequency of the azimuthal waves passing a point of observation.

In the range of quasi-periodic flow ($Re_q < Re < Re_t$), a second fundamental frequency f_2 appears in the power spectrum (Figure 11b). (The frequency f_2 was designated ω_3 in their original paper: they observed a transient component, designated ω_2 at lower R*.) The first fundamental component is usually identified as the intensest spectral component. It is, however, very difficult to identify which of the spectral components corresponds to the modulation frequency of the waves because the first fundamental component generates its harmonics and combinations intenser than the second fundamental component.

In the flow-visualization experiment by Gorman et al. [9], the second fundamental frequency was identified with the frequency of amplitude modulation of the first fundamental mode. When the Reynolds number is increased beyond $R^* \simeq 12$, there appears in the spectrum a weak broad component, labeled B in Figure 11c. This suggests a sign of the generation of chaotic turbulence. In the range of weakly turbulent flow ($Re > Re_t$), the power spectrum (Figure 11d) no longer con-

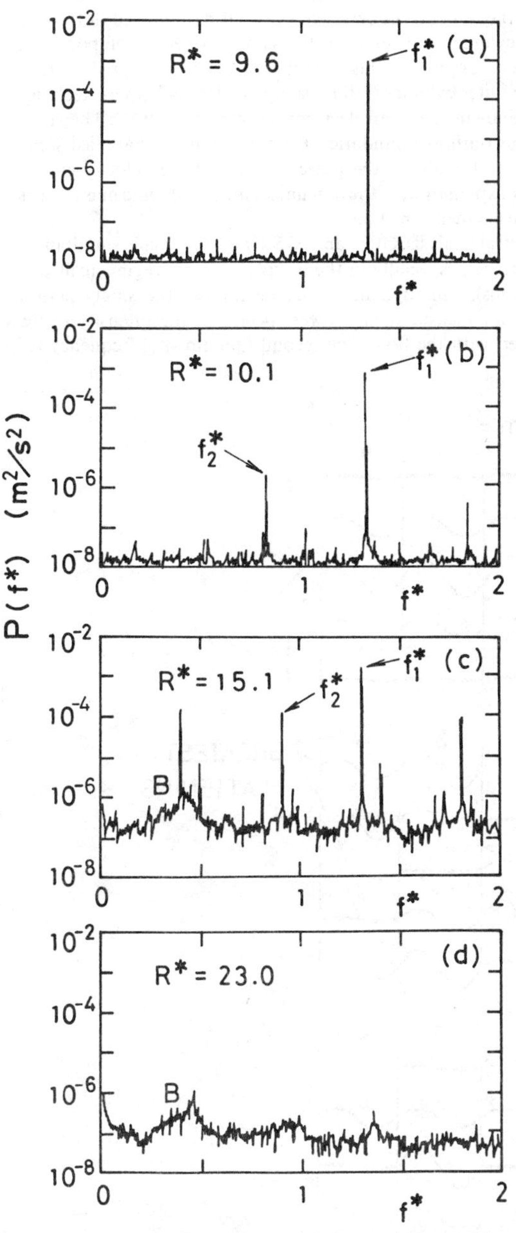

Figure 11. Power spectra of the fluctuations of radial velocity component measured by Fenstermacher et al. [8]: $\eta = 0.877$, B indicates a broad spectral component.

tains the two fundamental frequencies. This implies the disappearance of the azimuthal waves. As can be seen from the flow-visualization experiment, the flow still preserves a cellular vortex structure. Fenstermacher et al. found that chaotic turbulence occurs after a small number of sequential transitions: laminar Couette flow → axisymmetric Taylor vortex flow → singly periodic wavy vortex flow → quasi-periodic wavy vortex flow → weakly turbulent wavy vortex flow → turbulent vortex flow with no azimuthal waves.

Walden and Donnelly [43] measured the ion current between a collector embedded in the outer cylinder wall and the gold-plated inner cylinder and obtained the same frequency components f_1, f_2 and B as observed by Fenstermacher et al. Cognet and his coworkers [44, 45] measured the radial gradient of the azimuthal velocity on the outer cylinder wall by using an electrochemical technique and analyzed the velocity-gradient spectra by means of an electronic spectrum analyzer. They found that turbulence originated from the vortex outflow boundaries. Kataoka et al. [47] applied similar diffusion-controlled electrolytic reaction and analyzed the power spectra of the velocity-gradient fluctuations on the outer cylinder wall for explanation of the dynamical modes of ionic mass transfer measured on the outer cylinder wall. This is mentioned later.

Following the Fenstermacher et al. spectral analysis, Gorman and Swinney [9] made simultaneous spectral and flow-visualization measurements, especially in the doubly periodic regime (called the quasi-periodic regime in the early sections). This dynamical regime follows the singly periodic regime (i.e., wavy-vortex-flow regime). They measured the power spectra of the intensity of light scattered by the fine platelets that aligned with the flow. The second fundamental frequency f_2 in

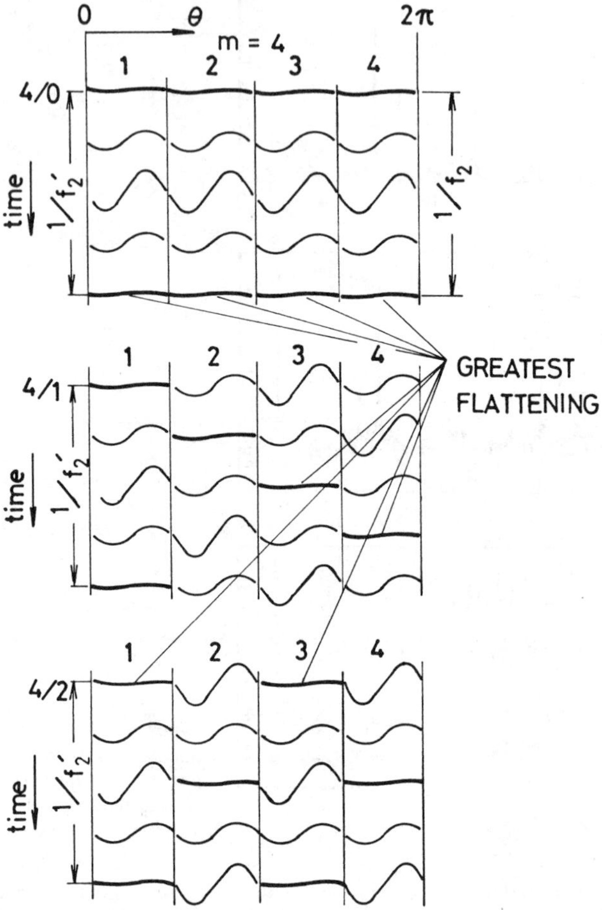

Figure 12. Modulation patterns for 4 wave states observed in a reference frame rotating with the waves, given by Gorman and Swinney [9]: $1/f_2'$ is one period of the modulation.

the spectrum was identified with the modulation frequency of the azimuthal waves from the simultaneous visualization measurements. The modulation can be observed as a periodic flattening of the wave motion. They discovered many distinct wave patterns in the doubly periodic flow. Each different wave pattern is characterized by two integers m, k*: m is the number of azimuthal waves and k* is related to the phase angle between the successively modulated azimuthal waves by $\Delta\theta = 2\pi k^*/m$.

As shown in Figure 12, the flow state with $k^* = 0$ indicates that the vortex outflow boundaries of all waves simultaneously flatten. For $k^* = 1$, successive waves are modulated in sequence. As aforementioned, Fenstermacher et al. [8] made a spectral analysis to the 4/1 state doubly periodic flow in which the modulation phase angle is $\Delta\theta = \pi/2$. For the 4/2 state the modulation angle is $\Delta\theta = \pi$; therefore, the third wave is in phase with the first. The second and fourth waves are modulated $\Delta\theta = \pi$ out of phase with the first and third waves.

Figure 13 shows comparison of the power spectra for the 4/0, 4/1 and 4/2 states obtained by Gorman and Swinney [9]. They found that the reduced frequency of the first fundamental

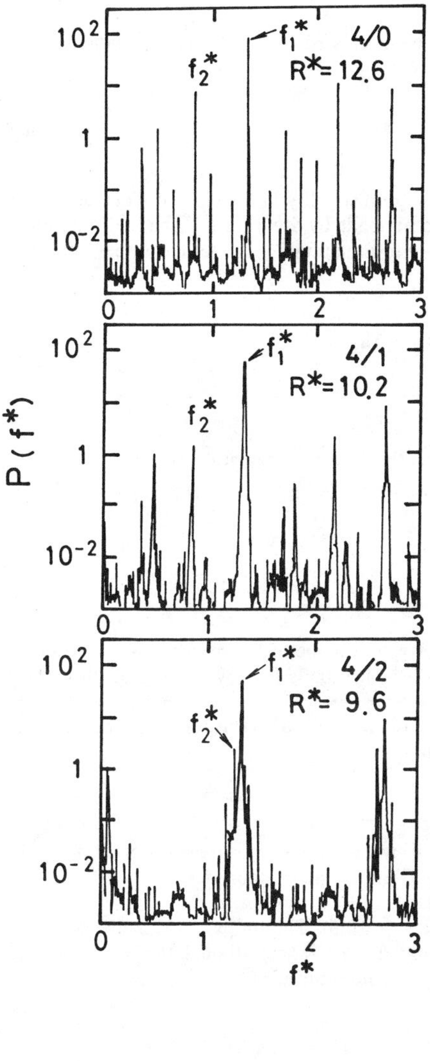

Figure 13. Power spectra of the intensity fluctuations of light scattered by fine aluminum platelets suspended in the doubly periodic wavy vortex flow, measured by Gorman and Swinney [9]: $\eta = 0.883$.

component is kept constant at $f_1/mf_r = 0.34 \pm 0.01$ for all Re, N, m, and k* in the doubly periodic flow regime.

TORQUE AND HEAT/MASS TRANSFER IN VORTEX FLOWS

Torque

The torque required for one of the two cylinders to rotate at a desired speed is a very important measurable factor not only for mechanical design of rotating machinery but also for confirmation of the validity of theories. We consider the case when the inner cylinder is rotated and the outer cylinder at rest.

The torque G acting on the inner cylinder of height H is written as

$$G = \tau_{wi} 2\pi r_i H \cdot r_i = 2\pi r_i^2 H \mu \left| r \frac{d(\bar{v}/r)}{dr} \right|_{r=r_i} \tag{43}$$

When the flow is purely laminar, the torque can be calculated by substituting Equation 1 into Equation 43:

$$G = 4\pi\mu H \frac{r_i^2 r_o^2}{r_o^2 - r_i^2} \Omega \qquad Re < Re_c \tag{44}$$

The validity of this relation is limited by the onset of instability in the laminar Couette flow. A calculation of the torque for values $Ta > Ta_c$ has been made by Stuart [2]. Using the equilibrium velocity of his finite amplitude theory (corresponding to Equation 29), the torque for axisymmetric Taylor vortex flow is given by

$$G = 2\pi r_i^2 H \mu \frac{r_m \Omega}{d} \left[1 + 1.4472 \left(1 - \frac{Ta_c}{Ta} \right) \right] \qquad (Ta > Ta_c) \tag{45}$$

where $Ta_c = 1{,}708$ for $\eta \rightarrow 1$.

In order to compare his theory with experiment, his theory can be rewritten as

$$G = a\Omega^{-1} + b\Omega \qquad Re > Re_c \tag{46}$$

where $a = -(2\pi r_i H \mu r_m \nu^2)(1.4472\,Ta_c)/d^4$

$b = (2\pi r_i^2 H \mu r_m)(1 + 1.4472)/d$

Stuart's theory is valid for the narrow gap geometry, and numerical agreement with experiment is limited to a very short range above Ta_c. However, his analysis provides a good theoretical background.

An empirical torque relation is given by Donnelly and Simon [48]. Their result has the following functional form as a modification of Stuart's relation

$$G = a\Omega^{-1} + b\Omega^{1.36} \tag{47}$$

The numerical constants a, b have been determined by the torque measurements of Taylor [49], Wendt [50], and Donnelly [51]. The relation by Donnelly and Simon describes well both the narrow and wide gap measurements over a wide range of Re above Re_c.

The exponent 1.36 is almost independent of the gap width. The torque measurements do not show any significant departure from Equation 47 until $Ta \geq 100\,Ta_c$, although the flow undergoes the successive dynamical transitions before the flow becomes turbulent.

We often need a torque relation at high speeds well above the critical for engineering calculations. We usually rely on dimensional analysis. One of the dimensionless relations is given by Wendt [50]:

$$\mathbf{f} = 0.46\left(\frac{r_o d}{r_i^2}\right)^{1/4} Re^{-0.5} \qquad 4 \times 10^2 < Re < 10^4 \tag{48a}$$

$$\mathbf{f} = 0.073\left(\frac{r_o d}{r_i^2}\right)^{1/4} Re^{-0.3} \qquad Re > 10^4 \tag{48b}$$

where **f** is the friction factor defined by

$$\mathbf{f} = \frac{\tau_{wi}}{\frac{1}{2}\rho(r_i\Omega)^2} = \frac{G}{\pi H \rho (r_i \Omega)^2 r_i^2} \tag{49}$$

Donnelly and Simon propose a similar dimensionless relation of the form

$$\frac{G}{\rho H (r_i \Omega)^2 r_i^2} \sim \left(\frac{d}{r_i}\right)^{0.31} Re^{-0.5} \tag{50}$$

When the flow is laminar, the friction factor is given by

$$\mathbf{f}_{lam} = \frac{4 r_o^2}{r_i (r_o + r_i)} \frac{1}{Re} \tag{51}$$

Using Equations 28, 33, and 51, Kataoka [19] gives a friction factor relation for an arbitrary gap width:

$$\mathbf{f}\,Re = (\mathbf{f}\,Re)_{lam}\left[1 + \frac{\lambda}{4\pi}\frac{(r_o + r_i)d}{r_o^2}\frac{\Phi_1^2}{\Phi_2}\left(1 - \frac{1}{R^*}\right)\right] \tag{52}$$

where Φ_1 and Φ_2 are the same as in Equation 25.

This does not hold for values of Re much above Re_c owing to the simplifying assumptions. An empirical equation, obtained by modification of Equation 52, is in much better agreement with experiment over a wide range up to $Re = 20\,Re_c$:

$$\mathbf{f}\,Re = (\mathbf{f}\,Re)_{lam}\left[1 + \frac{\lambda}{4\pi}\frac{(r_o + r_i)d}{r_o^2}\frac{\Phi_1^2}{\Phi_2}(\sqrt{R^*} - 1)\right] \tag{53}$$

Figure 14 shows comparison of these equations with experiment.

Bjorklund and Kays [52] give an empirical relation of the form

$$\frac{\mathbf{f}\,Rem}{(\mathbf{f}\,Rem)_{lam}} = 0.19\,Rem^{0.522} \qquad 90 < Rem < 10^4 \tag{54}$$

At very high values of the modified Reynolds number, Mizushina [53] and Kataoka [54] give a similar relation of the form

$$\frac{\mathbf{f}\,Rem}{(\mathbf{f}\,Rem)_{lam}} = 0.128\,Rem^{0.58} \qquad 10^3 < Rem < 10^6 \tag{55}$$

These two friction factors are convenient for the design calculation of the machinery to be rotated at high Reynolds numbers.

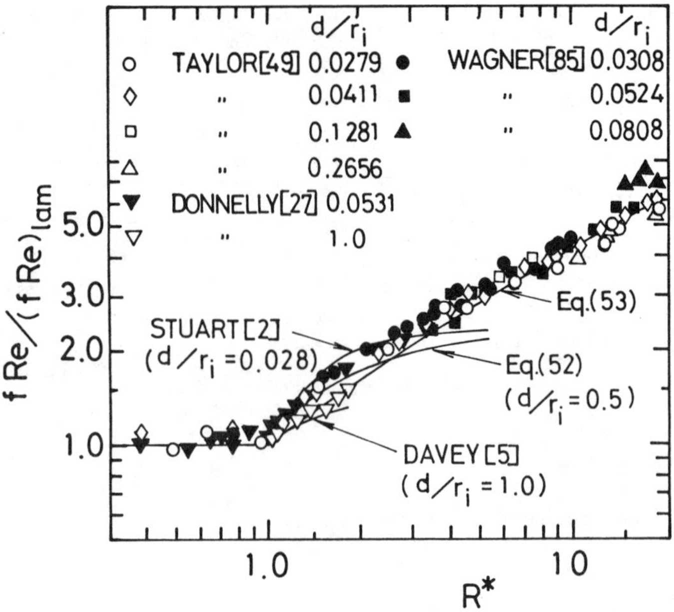

Figure 14. Comparison of friction factor between theory and experiment, given by Kataoka [19].

Heat Transfer

Heat transfer from a rotating cylinder to a concentric cylindrical shell is one of the most widely encountered problems in mechanical and chemical technology. This section deals with the heat transfer for the case of no axial flow of fluid in the annulus between the cylinders.

The overall coefficient of heat transfer from inner to outer cylinder is defined by

$$U = \frac{Q}{S(T_i - T_o)} \tag{56}$$

where Q is the heat transfer rate and the heat transfer area $S = 2\pi r_m H$.

Gazley [55], Bjorklund and Kays [52], Becker and Kaye [56], Tachibana, Fukui, and Mitsumura [57, 58], Ho, Nardacci, and Nissan [59] and Aoki, Nohira, and Arai [60] made measurements of the heat transfer from an inner rotating cylinder to a stationary outer cylinder.

As in the case of friction factor (Equation 54), Bjorklund and Kays proposed a heat transfer correlation of the form

$$\frac{Nu}{Nu_{lam}} = 0.175\ Rem^{1/2} \qquad 90 < Rem < 2{,}000 \tag{57}$$

where Nu_{lam} is the Nusselt number for pure conduction given by

$$Nu_{lam} = \frac{2d/r_i}{\ln(1 + d/r_i)} \tag{58}$$

Becker and Kaye proposed correlation equations given by

$$Nu = 0.128(Ta/Fg)^{0.367} \qquad 1{,}700 < Ta/Fg < 10^4 \tag{59}$$

$$Nu = 0.409(Ta/Fg)^{0.241} \qquad 10^4 < Ta/Fg < 10^7 \tag{60}$$

where Fg is a complicated radius-ratio geometric factor [60]:

$$Fg = (\pi^4/1{,}697)(1 - d/2r_m)^{-2}P^{-1}$$

$$P = 0.0571\left[1 - 0.652\left(\frac{d/r_m}{1 - d/2r_m}\right)\right] + 0.00056\left[1 - 0.652\left(\frac{d/r_m}{1 - d/2r_m}\right)\right]^{-1}$$

These correlations are valid for fluids whose Prandtl number is very close to that of air.

Ho et al. [59] and Aoki et al. [60] extended Stuart's small-gap finite amplitude theory to the heat transfer problem:

$$\text{Ho et al.:} \quad \frac{Nu}{Nu_{lam}} = 1 + 1.4472\left(1 - \frac{Ta_c}{Ta}\right) \tag{61}$$

$$\text{Aoki et al.:} \quad \frac{Nu}{Nu_{lam}} = 1 + 1.438\left(1 - \frac{Ta_c}{Ta}\right)Pr^{1/3} \tag{62}$$

These equations are also valid for the Prandtl numbers $Pr \simeq 1$ and in a very short range of Ta above Ta_c.

Aoki et al. proposed another empirical relation obtained over a wide range of Pr by experiment:

$$Nu = 0.44(Ta/Fg)^{1/4}Pr^{0.3} \qquad 5{,}000 < Ta/Fg < 2 \times 10^5, \quad 0.7 < Pr < 200 \tag{63}$$

Mizushina et al. [53] and Kataoka [54] proposed a similar correlation at high Reynolds numbers:

$$Nu = 0.44\,Rem^{1/2}\,Pr^{1/3} \qquad 10^2 < Rem < 10^5, \quad 0.7 < Pr < 160 \tag{64}$$

As in the case of friction factor, Kataoka [19] applied Stuart's finite amplitude theory to the heat transfer for wide gap geometry. He defined the heat transfer coefficient on the inner wall of the stationary outer cylinder:

$$\mathbf{h}_o = \frac{q_o}{T_b - T_o} \tag{65}$$

In this definition, the Nusselt number for laminar Couette flow is given by the equation (dimensionless)

$$(Nu_o)_{lam} = \frac{4}{r_o \ln(r_o/r_i)} \tag{66}$$

The linear stability equation for the axisymmetric temperature disturbance, similarly to Equations 11 and 12, is given by

$$(1/Re\,Pr)(DD^* - \lambda^2 - Pr\,\sigma)\Theta_1 - (D\bar{\Theta})u_1 = 0 \tag{67}$$

where $\Theta' = \Theta_1(r)\cos\lambda z\, e^{\sigma t}$ in the linear theory.

Similarly to Equation 22, the supercritical, axisymmetric temperature disturbance may be written as

$$\Theta' = Am(t)\,Pr\,\gamma_1(r)\cos\lambda z \tag{68}$$

Here the function $\gamma_1(r)$ can be assumed to be similar in shape to Θ_1. The equilibrium amplitude is given by Equation 28. The temperature distribution in the mean motion is distorted by the nonlinear interaction between velocity and temperature disturbances:

$$\bar{\Theta} = (Re\,Pr^2/2)\,Am(t)^2 \int_{r_i}^{r} \frac{1}{r}\left[\int_{r_i}^{r} r\left(u_1\frac{d\gamma_1}{dr} - \lambda w_1\gamma_1\right)dr\right]dr + C_1 \ln r + C_2 \tag{69}$$

where $C_1 = [1/\ln(r_o/r_i)]\left[1 - (Re\,Pr^2/2)\,Am(t)^2 \times \int_{r_i}^{r_o} \frac{1}{r}\left[\int_{r_i}^{r} r\left(u_1\frac{d\gamma_1}{dr} - \lambda w_1\gamma_1\right)dr\right]dr\right]$

$C_2 = -C_1 \ln r_i$

Figure 15 shows the distorted radial profiles of the mean temperature by the nonlinear theory. Differentiating Equation 69 with respect to r at $r = r_o$, the theoretical heat transfer, as given by Kataoka [19], becomes of the form

$$Nu_o = Nu_{o,lam}\left[1 + \frac{\Phi_1\Phi_5}{2\Phi_2}\left(1 - \frac{1}{R^*}\right)Pr^2\right] \tag{70}$$

where Φ_1 and Φ_2 are the same as in Equation 25 and

$$\Phi_5 = \int_{r_i}^{r_o} \frac{1}{r}\left[\int_{r_i}^{r} r\left(u_1\frac{d\gamma_1}{dr} - \lambda w_1\gamma_1\right)dr\right]dr - \ln(r_o/r_i)\int_{r_i}^{r_o} r\left(u_1\frac{d\gamma_1}{dr} - \lambda w_1\gamma_1\right)dr$$

This equation holds only in a very short range of Re above Re_c. An empirical equation, obtained by a modification of Equation 70, is in much better agreement with experiment over a wide range

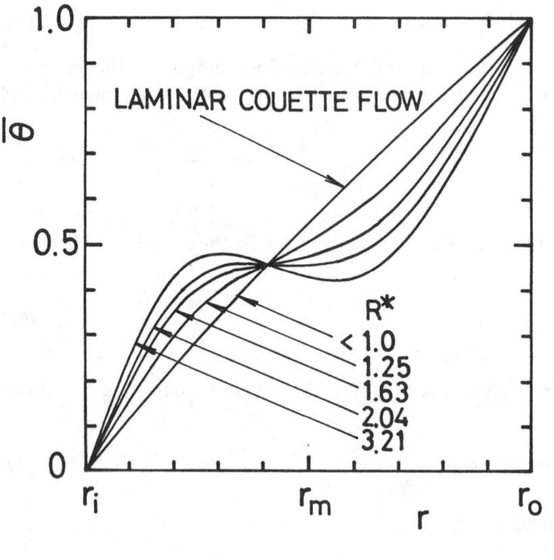

Figure 15. Radial profiles of the mean temperature predicted by nonlinear theory [19]: $d/r_1 = 0.5$, $Pr = 1$.

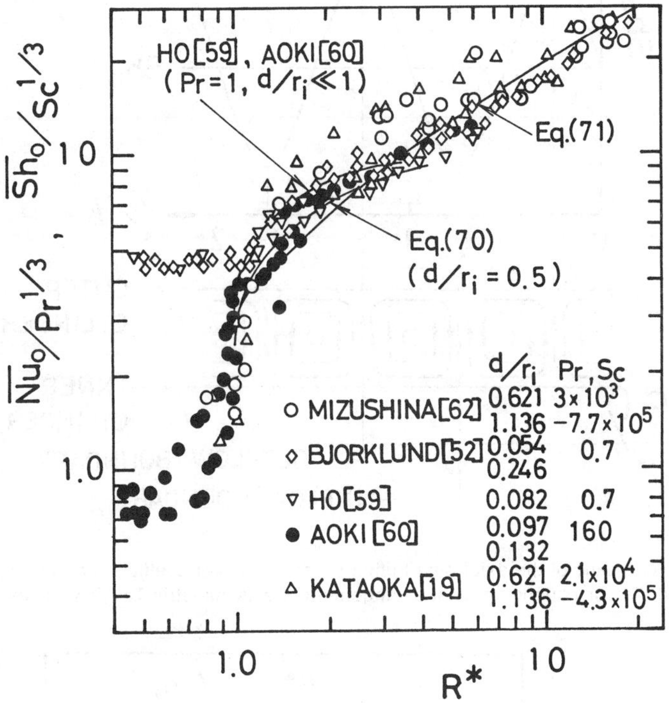

Figure 16. Comparison between theory and experiment of averaged Nusselt (Sherwood) number on the wall of outer cylinder, given by Kataoka [19].

up to $Re = 20\,Re_c$:

$$Nu_o = Nu_{o,lam}\left[1 + \frac{\Phi_1\Phi_5}{2\Phi_2}(\sqrt{R^*} - 1)\,Pr^{1/3}\right] \tag{71}$$

Figure 16 shows comparison of these equations with the heat transfer measurements. The experimental data are scattered in the range of laminar Couette flow ($R^* < 1$). This is attributable to the fact that $Nu_{o,lam}$ is independent of Pr (see Equation 66).

Mass Transfer

It is very difficult to measure local variation of the heat transfer in relation with the axial array of Taylor vortices. Mass transfer has been investigated under the assumption of analogy between heat and mass transfer, mainly for precise measurements of local time-dependent coefficient of mass transfer.

Eisenberg, Tobias, and Wilke [61] are perhaps the first to have investigated the mass transfer from an inner rotating cylinder by means of an electrochemical technique. In our notation, their correlation of the averaged mass transfer becomes

$$Sh_i = 0.225\,Rem^{0.65}\,Sc^{1/3} \tag{72}$$

They did not, however, measure local variation of the mass transfer.

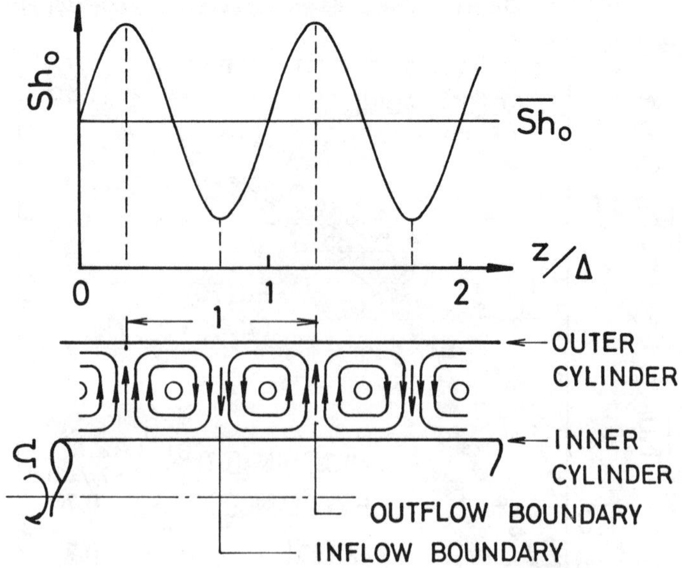

Figure 17. Schematic of the axial distribution of mass-transfer coefficient on the wall of outer cylinder and the corresponding axial arrangement of axisymmetric Taylor vortices.

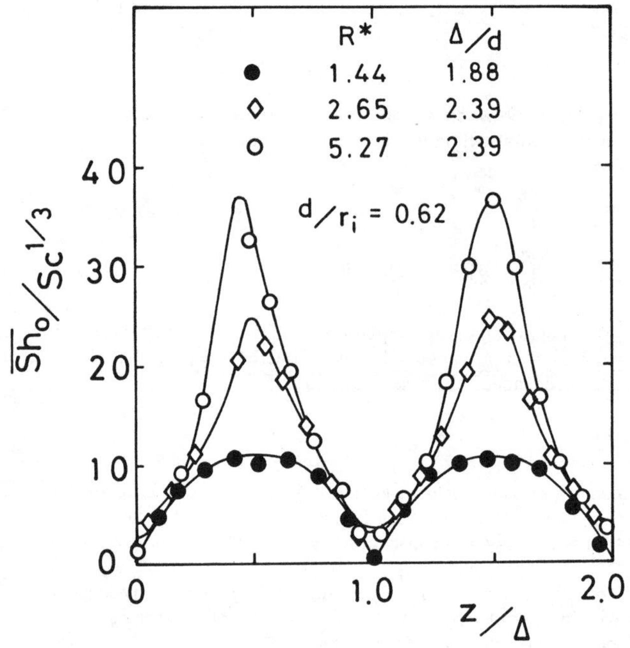

Figure 18. Axial variation of local mass-transfer coefficients on the wall of outer cylinder, measured by Kataoka et al. [63].

Mizushina et al. [62], Kataoka, Doi and Komai [63], and Kataoka et al. [47] measured local time-dependent mass transfer on the inner wall of a stationary outer cylinder by using a diffusion-controlled electrolytic reaction of copper ions. The mass-transfer coefficient is defined by

$$\mathbf{k}_o = \frac{N_{Ao}}{C_{Ab} - C_{Ao}} \tag{73}$$

The mass transfer measurements can be considered as those of heat transfer to a constant-temperature wall from a high Prandtl number fluid. The local mass flux or mass-transfer coefficient can be measured from the limiting current density on each isolated electrode embedded in the outer cylinder wall.

Figure 17 shows a schematic picture of the axial distribution of the mass-transfer coefficient on the outer cylinder and the corresponding axial arrangement of axisymmetric Taylor vortices. Figure 18 shows axial variation of the measured mass-transfer coefficient.

The mass-transfer coefficient shows clearly an axial periodicity; the maximum point corresponds to the vortex outflow boundary at which the secondary flow impinges on the outer cylinder wall and the minimum point corresponds to the vortex inflow boundary at which the secondary flow goes back toward the inner cylinder; the peak-to-peak distance is equal to the axial spacing of a couple of vortices.

Figure 19 shows axial distribution of the time-averaged mass transfer in turbulent vortex flow. The axially periodic distribution indicates the persistent presence of cellular structure. At the higher Reynolds number in the figure, the local mass-transfer coefficient shows a new secondary peak at

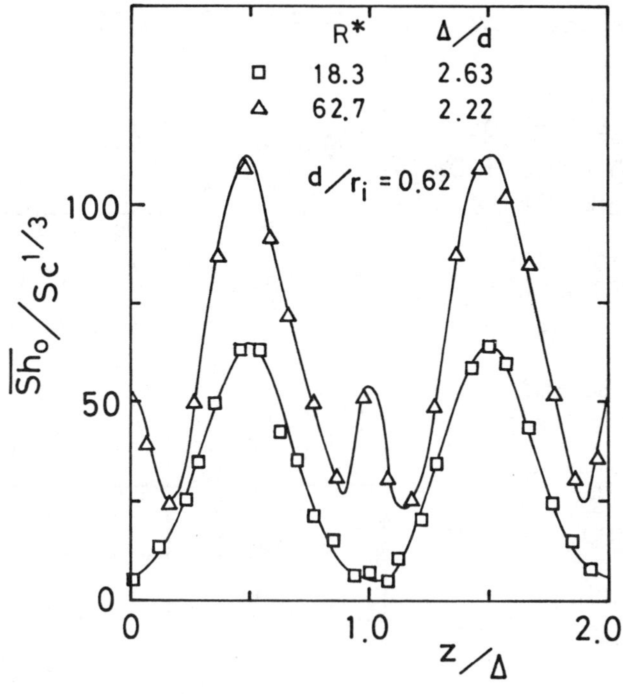

Figure 19. Axial variation of local mass-transfer coefficients on the wall of outer cylinder in turbulent vortex flow regime, measured by Kataoka et al. [63].

each vortex inflow boundary owing to the generation of turbulence at its position. It seems that the separation of boundary layer occurs at the outer cylinder wall corresponding to the vortex inflow boundary.

As mentioned, this flow system has several stable modes of flow. The modes of mass transfer can be classified by the time-dependency and axial periodicity in the mass-transfer coefficient, as shown in Figure 20.

It is striking that the wavy vortex flow does not show any time-dependency in mass transfer until $Re > Re_q$ although the velocity gradient on the outer cylinder wall has clear temporal periodicity corresponding to the azimuthally traveling waves.

In the wide range of Rem where several sequential transitions occur, the mass-transfer coefficient, averaged with respect to time and axial distance, can be well correlated with the Reynolds number

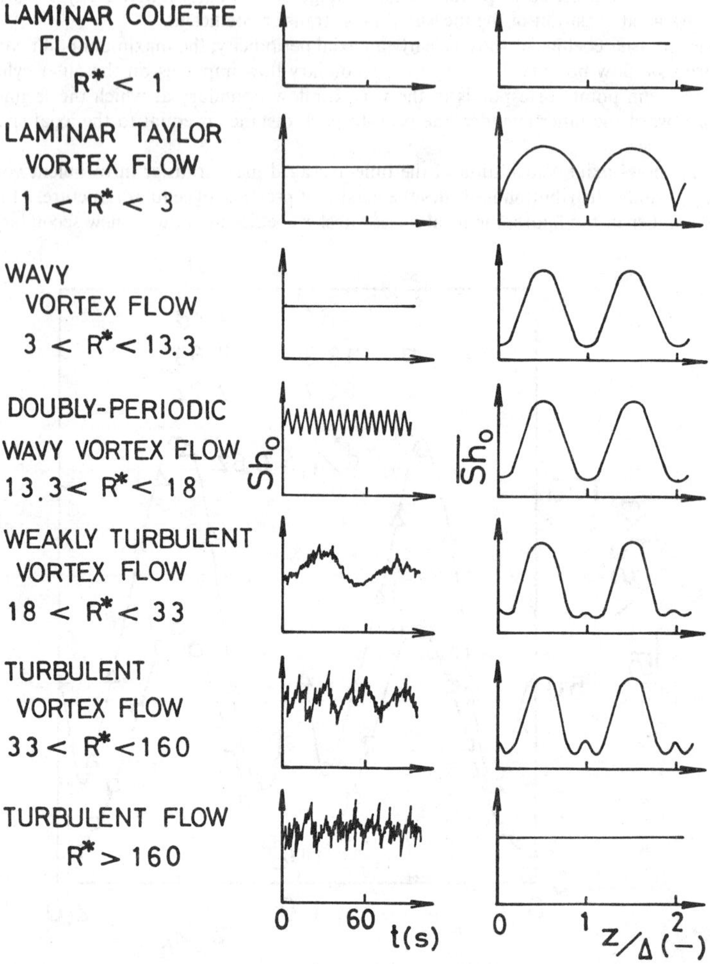

Figure 20. Modes of mass transfer classified by the time dependency and axial periodicity in the coefficient of mass transfer on the wall of outer cylinder, given by Kataoka et al. [47]: $d/r_i = 0.62$.

by the equation [47]:

$$\mathrm{Sh}_o = 6.04\,\mathrm{R}^{*1/2}\,\mathrm{Sc}^{1/3} \qquad 1 < \mathrm{R}^* < 160$$
$$3 \times 10^3 \le \mathrm{Sc} \le 7.7 \times 10^5 \qquad (74)$$

EFFECT OF AXIAL FLOW ON TAYLOR VORTEX FLOW

Instability and Development of Tangential Flow

As shown in Figure 21, when a uniform axial flow enters an annular space with the inner cylinder rotating, the tangential velocity profile is generated in a very thin fluid layer adjacent to the rotating inner cylinder while the axial velocity profile is generated near the stationary outer cylinder. The tangential boundary layer grows due to the outward diffusion of momentum and the tangential profile approaches the fully-developed state with distance downstream.

Taylor vortex flow in the presence of axial flow is of great importance in many problems of cooling the rotating machinery and this flow system suggests possible application to continuous chemical mixing machinery. In this flow system, the modes of flow can be characterized as a function of two independent parameters: the azimuthal Reynolds number $\mathrm{Re} = r_i\Omega d/\nu$ or $\mathrm{Rem} = \mathrm{Re}(d/r_i)^{1/2}$ and the axial Reynolds number $\mathrm{Rez} = 2\langle w\rangle d/\nu$. For small Rez, the instability of the basic flow leads to axisymmetric toroidal vortices propagating in the axial direction, and for slightly large Rez, the instability leads to a spiral vortex flow. According to the hot-wire measurements of Kaye and Elgar [64] and the flow-visualization experiment of Astill [65], this flow system has four regimes on a Rem—Rez map in the fully-developed region, shown schematically in Figure 22.

In the limit for $\mathrm{Rez} \to 0$ and $\eta \to 1$, the problem reduces to the instability of narrow-gap circular Couette flow which has a critical Taylor number Ta_c or Rem_c. In the opposite limit for $\mathrm{Rem} \to 0$ and $\eta \to 1$, the problem reduces to the instability of plane Poiseuille flow which has a critical Reynolds number Rez_c. The distance from the entrance to the point at which vortices occur increases with increasing Rez and decreases with increasing Rem.

The instability of axial forced flow between concentric cylinders with one or both of the cylinders rotating has been studied by many investigators [66–78]. Chandrasekhar [66] and DiPrima [67] applied linear stability theory to predict the critical Taylor number Ta_c in a narrow-gap annular flow for $\mathrm{Rez} < 2{,}000$ with both the cylinders rotating. The critical Taylor number predicted by DiPrima for an assumed averaged axial velocity distribution is, to some extent, supported by the measurements of Donnelly and Fultz [68] and Snyder [69] for $\mathrm{Rez} \le 200$ in an annulus with $\eta = 0.95$. The narrow-gap instability analysis was followed by Hughes and Reid [70] for $\mathrm{Rez} > 200$ with a uniform tangential velocity distribution and by Elliott [71] for $\mathrm{Rez} < 200$ with a linear tangential velocity distribution. For fully-developed laminar axial flow, Astill [72] and Astill, Ganley, and Martin [73] calculated the growth and displacement thickness of the tangential velocity profile by the momentum integral method and the method of separation of variables. The displacement

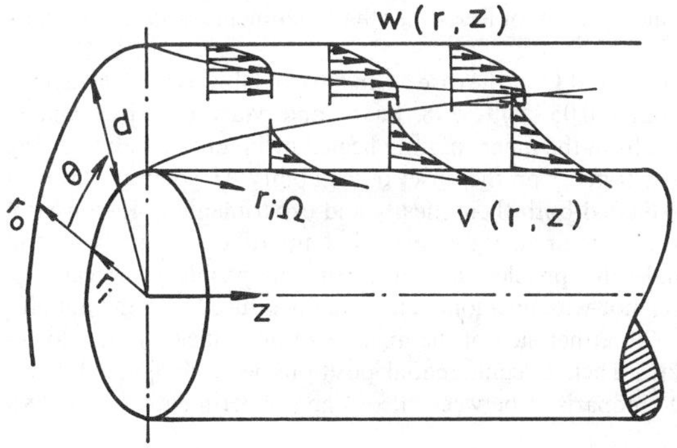

Figure 21. Schematic showing boundary-layer development in an annulus with the inner cylinder rotating.

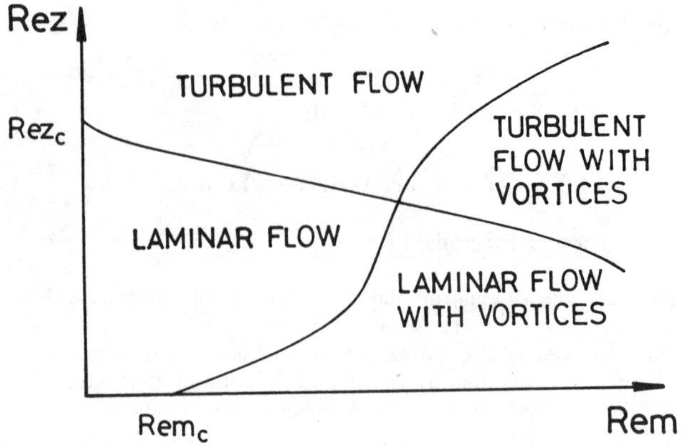

Figure 22. Schematic of modes of flow in an annulus with an axial flow.

thickness of the tangential velocity boundary layer can be defined as

$$\delta = \frac{1}{r_i\Omega}\int_{r_i}^{r_o} v\, dr \tag{75}$$

The latter analysis yielded acceptable predictions of the displacement thickness at high values of Rez. According to the flow-visualization experiment of the developing tangential flow by Astill [65], vortices originate as oscillating waves in the fluid layer adjacent to the inner rotating cylinder. The transition is characterized by a special form of Reynolds number Re_δ based upon the displacement thickness of tangential velocity boundary layer

$$Re_\delta = \frac{\Omega r_i \delta}{\nu}\left(\frac{\delta}{r_i}\right)^{1/2} \geq 24 \tag{76}$$

Martin and Payne [74] employed the numerical finite-difference method to predict the growth of the tangential velocity profile and the displacement thickness of the tangential velocity boundary layer. The dimensionless thickness $\delta^* = \delta/d$ can be considered to be a function of Rez, r_i/r_o and L, where L is the dimensionless axial length given by

$$L = \frac{2z/d}{Rez} \tag{77}$$

According to their analysis, δ^* is almost independent of Rez, other than that included in L, especially for $L > 0.015$.

Figure 23 shows their predicted relation δ^* and L for the case when the axial flow is fully developed at entry, for five values of η over the range $0.05 < \eta < 0.98$. Payne and Martin [75] calculated laminar heat transfer at uniform heat-flux from the inner rotating heated cylinder to axially flowing fluid in connection with the effect of heat transfer on the onset of instability. Hasoon and Martin [76] and Gravas and Martin [77] investigated both theoretically and experimentally the onset of instability in developed tangential flow. In their analysis, explicit finite-difference method and Galerkin method were used for an initial-value problem and an eigenvalue problem, respectively. In their experiment of Ta_c measurements, hot-wire anemometers were introduced into the annulus from the access holes at the top and at 90° either side of the axial position sufficiently far downstream to give developed tangential flow. Their circumferential positions were designated as 0°, −90°, and +90°. Figure 24 shows the comparison between theory and experiment for $\eta = 0.81$.

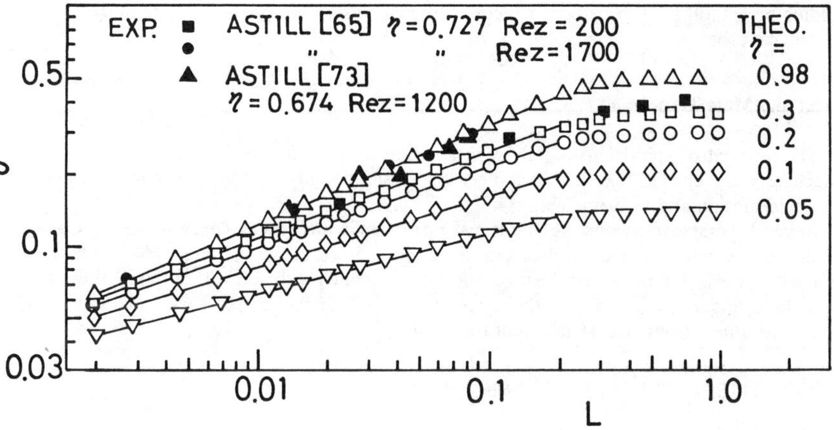

Figure 23. Development of displacement thickness predicted by Martin and Payne [74].

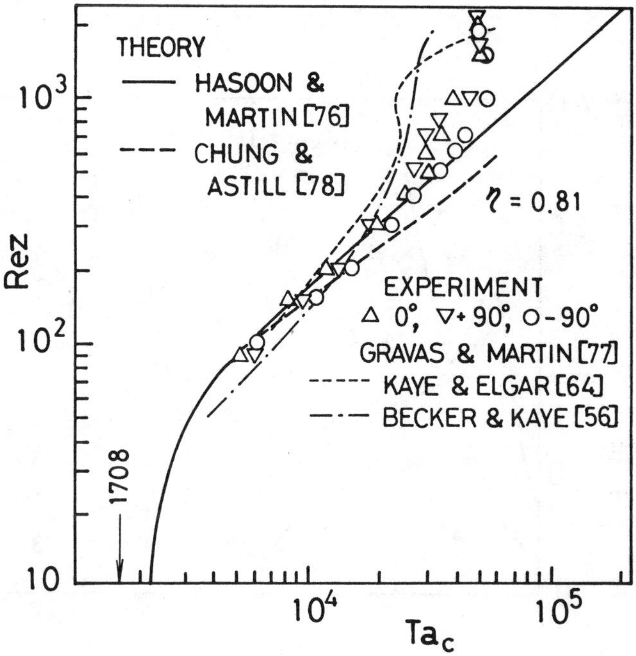

Figure 24. Comparison between theory and experiment of the critical Taylor number for $\eta = 0.81$, given by Gravas and Martin [77].

Chung and Astill [78] investigated theoretically the instability problem for initially non-axisymmetric disturbances.

Heat and Mass Transfer

The toroidal or spiral vortex flow with an axial flow is often encountered in rotating electric machinery and chemical mixing machinery. Successful design depends upon reliable knowledge of the fluid motion and heat transfer.

Luke [79] first investigated the heat transfer problem in this flow system, followed by Gazley [55], in connection with the cooling problem of electric motors. Kaye and Elgar [64] and Becker and Kaye [56] measured the rate of heat transfer in an annulus with and without axial flow. The inner rotating cylinder was heated and the outer stationary cylinder cooled. Tachibana and Fukui [80] proposed a heat transfer correlation of the form

$$\mathrm{Nu} = 0.015\left(1 + 2.3\frac{2d}{l}\right)(r_o/r_i)^{0.45}(\mathrm{Re})_{\mathrm{eff}}^{0.8}\,\mathrm{Pr}^{1/3} \tag{78}$$

where l is the axial length of the test section and $(\mathrm{Re})_{\mathrm{eff}}$ is the Reynolds number based upon Gazley's effective velocity. The defining equation is given by

$$v_{\mathrm{eff}} = [\langle w\rangle^2 + (r_i\Omega/2)^2]^{1/2} \tag{79}$$

Kuzay and Scott [81] made measurements of turbulent heat transfer from an electrically heated outer cylinder to axially flowing air in a very long vertical annular channel with an insulated inner

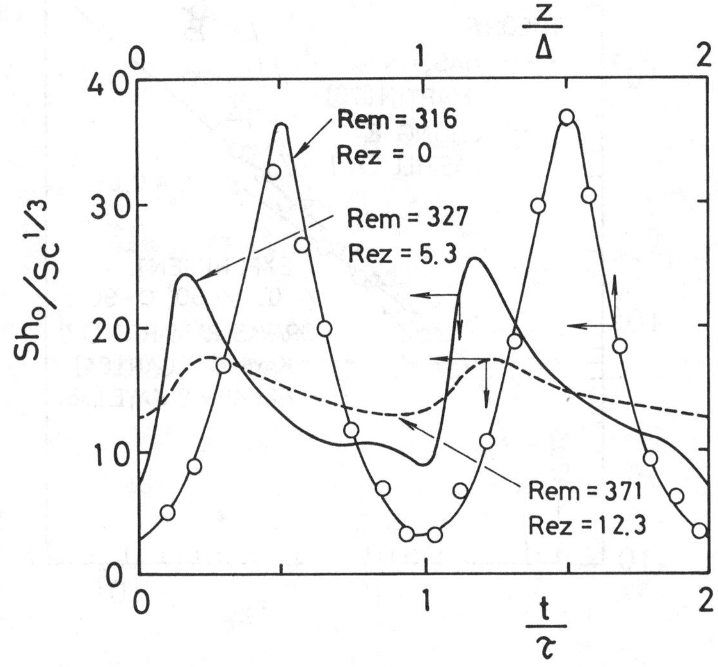

Figure 25. Comparison of the stationary axial distribution with the temporal variation of mass transfer, measured by Kataoka et al. [63]: $d/r_i = 0.62$.

rotating cylinder. Their heat transfer correlation is given by

$$Nu_o = Nu_{o,z}\left[1 + \frac{4}{\pi^2}\left(\frac{\Omega d}{\langle w \rangle}\right)^2\right]^{0.8714} \qquad 10^4 < Rez < 10^5 \quad \text{and} \quad 0 < \frac{r_i\Omega}{\langle w \rangle} < 2.8 \tag{80}$$

where $Nu_{o,z}$ is the Nusselt number for turbulent annular flow with no tangential flow given by

$$Nu_{o,z} = 0.022\,Rez^{0.8}\,Pr^{0.5} \qquad (Re = 0) \tag{81}$$

Kataoka, Doi, and Komai [63] made local measurements of mass transfer on the inner wall of outer stationary cylinder by means of an electrochemical technique. The axial Reynolds number and modified Reynolds number ranged, respectively, from Rez = 0 to 260 and from Rem = 35 to 9,200. The test section was far downstream from entry, so that Taylor vortex structure could be established in laminar axial flow. For a small forced axial flow, toroidal vortices march through in single file without breaking up.

Figure 25 shows comparison of the stationary axial distribution of mass transfer for zero axial flow rate with the temporal variation of mass transfer observed at a fixed measuring position for different axial flow rates at the same modified Reynolds number. Here, Δ refers to the axial spacing of a couple of Taylor vortices and τ is the periodic time required for a couple of vortices to axially pass the measuring point.

As the axial Reynolds number Rez is raised gradually, the regular sinusoidal variation of Sherwood number is not only distorted by the forced axial motion, but also its mean value and amplitude are reduced greatly. When Rem is relatively larger than Rez, the added axial motion tends to stabilize the circular Couette flow and to delay the initial formation of Taylor vortices. Figure 26 shows the time- and space-averaged Sherwood number plotted against the modified

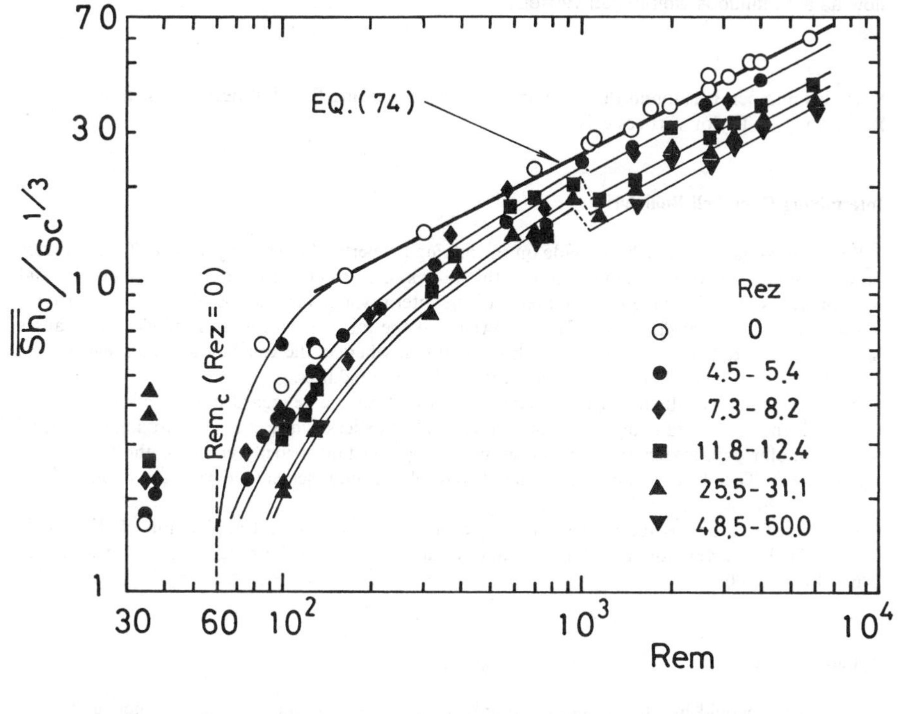

Figure 26. Effect of axial flow on the time- and space-averaged mass-transfer coefficient, given by Kataoka et al. [63]; $d/r_i = 0.62$.

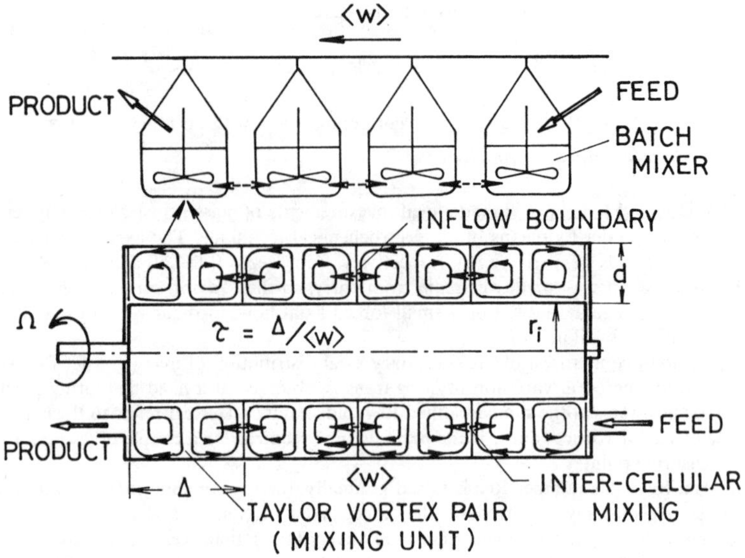

Figure 27. Schematic showing the possibility of application of the axially moving Taylor vortex flow as a continuous well-mixed vessel.

Reynolds number. It is found that Sh increases with increasing Rem but decreases with increasing Rez as long as Taylor vortices exist.

Intermixing Over Cell Boundaries

Some investigations have been made on the mixing property of the axially moving Taylor vortex flow in connection with the application of this flow system to chemical equipment. Kataoka and his coworkers [82, 83] made measurements of the intermixing over cell boundary between Taylor vortices for small axial flow rates. They showed that the toroidal motion of fluid elements causes highly effective radial mixing within cellular vortices, whereas the cell boundaries prevent fluid elements from being exchanged over the vortex inflow boundaries.

As shown schematically in Figure 27, each pair of vortices can be regarded as a well-mixed batch vessel, which moves axially at a constant velocity. The vortices, whose size is approximately equal to the annular gap, march through the annulus at a constant velocity equal to the mean axial velocity ⟨w⟩. Therefore, it can be considered that all the fluid elements leaving the annulus have the same residence time in the apparatus.

Kataoka, Ono, and Kubo [84] reported the possibility of practical application of the axially moving Taylor vortex flow to a liquid–liquid surface reaction in the range of $0 < \text{Rez} < 20$ and $246 < \text{Rem} < 500$.

Acknowledgment

The author would like to express his thanks to T. Mizusugi and H. Ueno for their assistance in the preparation of this chapter.

NOTATION

A	constant, $-r_i^2\Omega/(r_o^2 - r_i^2)$, (1/s)
A_c, A_s	amplitude function for axisymmetric disturbance, (—)
A_0	geometrical constant, $r_i/(r_o + r_i)$, (—)
A_1	geometrical constant, $r_o/[(r_o + r_i)d]$, (—)
Am	amplitude of fundamental component given by Equation 22, (—)
a_i	numerical coefficient of Equation 37, (—)
B	constant, $r_i^2 r_o^2\Omega/(r_o^2 - r_i^2)$, (m^2/s)
B_c, B_s	amplitude function for non-axisymmetric disturbance, (—)
B_0	geometrical constant, $r_i r_o^2/[(r_o + r_i)d^2]$, (—)
B_1	geometrical constant, $r_i^2 r_o^2/[(r_o + r_i)d^3]$, (—)
b_i	numerical coefficient of Equation 37, (—)
C_A	concentration, $(kmol/m^3)$
D	differential operator, d/dr, (1/m) or (—)
D*	differential operator, d/dr + 1/r, (1/m) or (—)
D_A	diffusion coefficient, (m^2/s)
d	annulus gap width, (m)
F_i	radial distribution function, (—)
f	frequency, (1/s)
f	friction factor, (—)
f*	dimensionless frequency, f/f_r, (—)
f_i	radial distribution function for axisymmetric disturbance, (—)
f_1	(first) fundamental frequency, (1/s)
f_2	second fundamental frequency (amplitude-modulation frequency), (1/s)
f'_2	modulation frequency in a reference frame rotating with waves, (1/s)
G	torque, (N·m)
H	annulus height, (m)
h	heat-transfer coefficient, $(W/m^2\ K)$
h_i	radial distribution function for non-axisymmetric disturbance, (—)
Im	imaginary part
k	azimuthal wavenumber, (—)
k	mass-transfer coefficient, (m/s)
k*	phase difference between successively modulated azimuthal waves, (—)
L	axial length parameter defined by Equation 77, (—)
l	axial length of heated section, (m)
m	number of azimuthal waves, (—)
m_i	radial distribution function, (—)
N	number of Taylor vortices accommodated in annulus, (—)
Nu	Nusselt number, $2hd/\kappa$, (—)
N_A	mass flux, $(kmol/m^2\ s)$
n	integer, (—)
P	power density function, (m^2/s^2) or (—)
Pr	Prandtl number, (—)
p	pressure, (Pa) or $p/\rho\ (r_i\Omega)^2$, (—)
p_i	radial distribution function, (—)
Q	heat transfer rate, (W)
q	heat flux, (W/m^2)
Re	Reynolds number, $r_i\Omega d/\nu$, (—)
Re	real part
$(Re)_{eff}$	Reynolds number based upon v_{eff}, $2v_{eff}d/\nu$, (—)
Rem	modified Reynolds number, $Re(d/r_i)^{1/2}$, (—)
Rez	axial Reynolds number, $2\langle w\rangle d/\nu$, (—)
R*	reduced Reynolds number, Re/Re_c, (—)
r	radial coordinate, (m) or r/d, (—)
r_i	inner cylinder radius, (m) or r_i/d, (—)
r_m	mean radius, $(r_o + r_i)/2$, (m) or (—)
r_o	outer cylinder radius, (m) or r_o/d, (—)
S	heat transfer area, (m^2)
Sc	Schmidt number, (—)
Sh	Sherwood number, $2kd/D_A$, (—)
T	temperature, (K)
Ta	Taylor number, $r_i\Omega^2 d^3/\nu^2$, (—)
t	time, (s) or $t/(d^2/\nu)$, (—)
U	overall heat-transfer coefficient, (W/m^2K)
u	radial component of velocity, (m/s) or $u/r_i\Omega$, (—)
u_1	eigenfunction given by Equation 10, (—)
V	azimuthal velocity of laminar Couette flow, (—)
v	azimuthal component of velocity, (m/s) or $v/r_i\Omega$, (—)
v_1	eigenfunction given by Equation 10, (—)
v_{eff}	effective velocity defined by Equation 79, (m/s)
w	axial component of velocity, (m/s) or $w/r_i\Omega$, (—)

x radial distance from mean radius, (—)

y_i radial distribution function, (—)

z axial coordinate or axial distance from entry, (m) or z/d, (—)

Greek Symbols

γ radial distribution function of temperature disturbance, (—)

Δ axial spacing of a couple of vortices, (m)

δ displacement thickness of tangential velocity boundary layer, (m)

δ^* dimensionless displacement thickness, δ/d, (—)

ϵ amplification rate of non-axisymmetric disturbance, (—)

η radius ratio, r_i/r_o, (—)

Θ dimensionless temperature, $(T_i - T)/(T_i - T_o)$, (—)

θ angular coordinate, (—)

κ thermal conductivity, (W/m K)

λ axial wave number, (—)

μ viscosity, (kg/m s)

ν kinematic viscosity, (m^2/s)

(ξ', η', ζ') vorticity components of velocity disturbance given by Equation 24, (—)

ρ fluid density, (kg/m^3)

σ amplification rate of axisymmetric disturbance, (—)

τ periodic time required for a couple of vortices to pass a point of observation, (s)

τ_w shear stress at wall, (N/m)

Ω angular velocity of inner cylinder, (1/s)

ω angular velocity of azimuthally traveling waves, (1/s)

Superscripts

disturbance with respect to z

* reduced or dimensionless

Subscripts

b bulk fluid

c critical or cosine

e equilibrium

i inner cylinder

lam laminar Couette flow

n integer or (n − 1)th harmonic

o outer cylinder

q quasi-periodic

r rotation

s sine or reference

w wave boundary (second critical) or at wall

t turbulent

1 fundamental

2 modulation (second fundamental) or first harmonic

δ displacement thickness

Overbars

$\overline{}$ averaged with respect to z or t

$\overline{\overline{}}$ averaged with respect to z and t

Brackets

$\langle\ \rangle$ averaged over a cross section of annular space

REFERENCES

1. Taylor, G. I., *Phil. Trans. Roy. Soc. London*, A223:289 (1923).
2. Stuart, J. T., *J. Fluid Mech.*, 4:1 (1958).
3. Stuart, J. T., *J. Fluid Mech.*, 9:353 (1960).
4. Watson, J., *J. Fluid Mech.*, 9:371 (1960).

5. Davey, A., *J. Fluid Mech.*, 14:336 (1962).
6. Coles, D., *J. Fluid Mech.*, 21:385 (1965).
7. Davey, A., DiPrima, R. C., and Stuart, J. T., *J. Fluid Mech.*, 31:17 (1968).
8. Fenstermacher, P. R., Swinney, H. L., and Gollub, J. P., *J. Fluid Mech.*, 94:103 (1979).
9. Gorman, M., and Swinney, H. L., *J. Fluid Mech.*, 117:123 (1982).
10. DiPrima, R. C., and Swinney, H. L., "Instabilities and Transition in Flow between Concentric Rotating Cylinders," In *Hydrodynamic Instabilities and the Transition to Turbulence* (ed. H. L. Swinney and J. P. Gollub), Chap. 6, p. 139 (Springer, 1981).
11. Chandrasekhar, S., *Mathematika*, 1:5 (1954).
12. DiPrima, R. C., *Quart. Appl. Math.*, 13:55 (1955).
13. Chandrasekhar, S. *Proc. Roy. Soc. London*, A246:301 (1958).
14. DiPrima, R. C., *Phys. Fluids*, 4:751 (1961).
15. Chandrasekhar, S., *Hydrodynamic and Hydromagnetic Stability*, Oxford: Clarendon Press, London (1961).
16. DiPrima, R. C., *J. Appl. Mech.* (*Trans. ASME*), 30:486 (1963).
17. Walowit, J., Tsao, S., and DiPrima, R. C., *J. Appl. Mech.* (*Trans. ASME*), 31:585 (1964).
18. Sparrow, E. M., Munro, W. D., and Jonsson, V. K., *J. Fluid Mech.*, 20:35 (1974).
19. Kataoka, K., *J. Chem. Eng. Japan*, 8:271 (1975).
20. Mizushina, T., Ito, R., Kataoka, K., Nakashima, Y., and Fukuda, A., *Kagaku-Kogaku* (*Chem. Eng. Japan*), 35:1116 (1971) (in Japanese).
21. Meyer, K. A., *Los Alamos Report LA-3497*, Los Alamos, New Mexico (1966).
22. Meyer, K. A., *Phys. Fluids*, 10:1874 (1967).
23. Roberts, P. H., *Proc. Roy. Soc. London*, A283:550 (1965).
24. Krueger, E. R., Gross, A., and DiPrima, R. C., *J. Fluid Mech.*, 24:521 (1966).
25. Eagles, P. M., *J. Fluid Mech.*, 49:529 (1971).
26. Eagles, P. M., *J. Fluid Mech.*, 62:1 (1974).
27. Donnelly, R. J., *Proc. Roy. Soc. London*, A246:312 (1958).
28. Debler, W., Fuener, E., and Schaaf, B., "Torque and Flow Patterns in Supercritical Circular Couette Flow," in *Proc. 12th Int. Congress Appl. Mech.* (ed. M. Hetényi and W. G. Vincenti), (Springer, 1969).
29. Sherman, J., and McLaughlin, J. B., *Commun. Math. Phys.*, 58:9 (1978).
30. Yahata, H., *Prog. Theor. Phys. Suppl.*, 64:176 (1978).
31. Yahata, H., *Prog. Theor. Phys.*, 61:791 (1979).
32. Yahata, H., *Prog. Theor. Phys. Suppl.*, 69:200 (1980).
33. Yahata, H., *Prog. Theor. Phys.*, 64:782 (1980).
34. Yahata, H., *Prog. Theor. Phys.*, 66:879 (1981).
35. Yahata, H., *Prog. Theor. Phys.*, 69:396 (1983).
36. Schultz-Grunow, F., and Hein, H., *Z. Flugwiss.*, 4:28 (1956).
37. Snyder, H. A., *J. Fluid Mech.*, 35:273 and 337 (1969).
38. Burkhalter, J. E., and Koschmieder, E. L., *J. Fluid Mech.*, 58:547 (1973).
39. Benjamin, T. B., *Proc. Roy. Soc. London*, A359:1 and 27 (1978).
40. Cole, J. A., *J. Fluid Mech.*, 75:1 (1976).
41. Koschmieder, E. L., *J. Fluid Mech.*, 93:515 (1979).
42. Gollub, J. P., and Swinney, H. L., *Phys. Rev. Lett.*, 35:927 (1975).
43. Walden, R. W., and Donnelly, R. J., *Phys. Rev. Lett.*, 42:301 (1979).
44. Bouabdallah, A., and Cognet, G., "Laminar-Turbulent Transition" *IUTAM Symposium*, Stuttgart (Sringer) p. 368 (1980).
45. Cognet, G., Bouabdallah, A., and Aider, A., "Stability in the Mechanics of Continua" *2nd Symposium*, Nümbrecht (Springer). p. 330 (1982).
46. Krugljak, Z. B., Kuznetsov, E. A., L'vov, V. S., Nesterikhin, Yu. E., Predtechensky, A. A., Sobolev, V. S., Utkin, E. N., and Zhuravel, F. A., "Laminar-Turbulent Transition" *IUTAM Symposium*, Stuttgart (Springer) p. 378 (1980).
47. Kataoka, K., Bitou, Y., Hashioka, K., Komai, T., and Doi, H., "Mass Transfer in the Annulus between Two Coaxial Rotating Cylinders" In *Heat and Mass Transfer in Rotating Machinery* (ed. D. E. Metzger and N. H. Afgan), p. 143 (Hemisphere, 1984).

48. Donnelly, R. J., and Simon, N. J., *J. Fluid Mech.*, 7:401 (1960).
49. Taylor, G. I., *Proc. Roy. Soc. London*, A157:546 (1936).
50. Wendt, F., *Ingen. Arch.*, 4:577 (1933).
51. Donnelly, R. J., *Proc. Roy. Soc. London*, A246:312 (1958).
52. Bjorklund, I. S., and Kays, W. M., *J. Heat Transfer* (*Trans. ASME*), 81:175 (1959).
53. Mizushina, T., Ito, R., Nakagawa, T., Kataoka, K., and Yokoyama, S., *Kagaku-Kogaku* (*Chem. Eng. Japan*, 31:974 (1976) (in Japanese).
54. Kataoka, K., *D. Thesis*, Kyoto Univ., Kyoto (1970) (in Japanese).
55. Gazley, C., Jr., *Trans. ASME*, 80:79 (1958).
56. Becker, K. M., and Kaye, J., *J. Heat Transfer* (*Trans. ASME*), 84:97 (1962).
57. Tachibana, F., Fukui, S., and Mitsumura, H., *Trans. JSME*, 25:788 (1959).
58. Tachibana, F., Fukui, S., and Mitsumura, H., *Trans. JSME*, 29:1366 (1963).
59. Ho, C. Y., Nardacci, J. L., and Nissan, A. H., *AIChE J.* 10:194 (1964).
60. Aoki, H., Nohira, H., and Arai, H., *Bull. JSME*, 10:523 (1967).
61. Eisenberg, E., Tobias, C. W., and Wilke, C. R., *Chem. Eng. Prog. Symp. Ser.*, 51:1 (1955).
62. Mizushina, T., Ito, R., Kataoka, K., Yokoyama, S., Nakashima, Y., and Fukuda, A., *Kagaku-Kogaku* (*Chem. Eng. Japan*), 32:795 (1968) (in Japanese).
63. Kataoka, K., Doi, H., and Komai, T., *Int. J. Heat Mass Transfer*, 20:57 (1977).
64. Kaye, J., and Elgar, E. G., *J. Heat Transfer* (*Trans. ASME*), 80:753 (1958).
65. Astill, K. N., *J. Heat Transfer* (*Trans. ASME*), 86:383 (1964).
66. Chandrasekhar, S., *Proc. Natn. Acad. Sci. Wash.* 46:137 and 141 (1960).
67. DiPrima, R. C., *J. Fluid Mech.*, 9:621 (1960).
68. Donnelly, R. J., and Fultz, D., *Proc. Natn. Acad. Sci. Wash.* 46:1150 (1960).
69. Snyder, H. A., *Proc. Roy. Soc. London*, A265:198 (1962).
70. Hughes, T. H., and Reid, W. H., *Phil. Trans. Roy. Soc.*, A263:57 (1968).
71. Elliott, L., *Phys. Fluids*, 16:577 (1973).
72. Astill, K. N., *Ph. D. Thesis*, Massachusetts Institute of Technology (1961).
73. Astill, K. N., Ganley, J. T., and Martin, B. W., *Proc. Roy. Soc. London*, A307:55 (1968).
74. Martin, B. W., and Payne, A., *Proc. Roy. Soc. London*, A328:123 (1972).
75. Payne, A., and Martin, B. W., *Proc. 5th Int. Heat Transfer Conf.*, Vol. 2:FC2.7, 80 (1974).
76. Hasoon, M. A., and Martin, B. W. *Proc. Roy. Soc. London*, A352:351 (1977).
77. Gravas, N., and Martin, B. W., *J. Fluid Mech.*, 86:385 (1978).
78. Chung, K. C., and Astill, K. N., *J. Fluid Mech.*, 81:641 (1977).
79. Luke, G. E., *Trans. ASME*, 42:646 (1923).
80. Tachibana, F., and Fukui, S., *Bull. JSME*, 7:No. 26 (1964).
81. Kuzay, T. M., and Scott, C. J., *J. Heat Transfer* (*Trans. ASME*), 99:12 (1977).
82. Kataoka, K., Doi, H., Hongo, T.. and Futagawa, M., *J. Chem. Eng. Japan*, 8:472 (1975).
83. Kataoka, K., and Takigawa, T., *AIChiE J.*, 27:504 (1981).
84. Kataoka, K., Ono, M., and Kubo, N., *Synopsis of Contributions, Taylor Vortex Flow Working Party*, Tufts Univ., Medford, MA, p. 49 (1981).
85. Wagner, E. M., *D. Thesis*, Stanford Univ., CA (1932).

SECTION II

FLOW DYNAMICS AND FRICTIONAL BEHAVIOR

CONTENTS

CHAPTER 10

PROPERTIES AND CONCEPTS OF SINGLE FLUID FLOWS

N. P. Cheremisinoff

Exxon Chemical Co.
Linden, NJ

CONTENTS

INTRODUCTION

Knowledge of fluid mechanics has expanded greatly with principles that are applied not only to fluid transport, but to an overwhelming number of process operations. It is simply not possible to rationally design a piece of equipment or process without an understanding of the physics of flow phenomena, regardless of the intent of design, whether simply to transport fluids; process them via mixing, for separating, drying or coating; transfer heat; remove certain constituents via adsorption or absorption; or promote chemical reactions.

This chapter presents an overview of fundamental principles and theorems important to single fluid flows. These principles underlie the analysis of virtually all problems of fluid mechanics and, in fact, are applied to understanding and modeling multicomponent flows, which are discussed in subsequent volumes. Specifically, this chapter reviews the general differential equations of conservation of mass and momentum, and the law of conservation of energy and relates to real or viscous fluids. The first is often referred to as the principle of continuity, which accounts for the mass of fluid flowing through any system. The last is the mechanical energy equation, which accounts for the energy interconversions that occur in the flowing fluid. Reviewed also in this chapter are the theories of turbulent flow, along with the semi-empirical models describing shear and velocity distribution for simple flow configurations. Discussions on compressible flows are also included.

GOVERNING EQUATIONS OF CONTINUITY, MOTION, AND ENERGY

The principle of *continuity* is the law of conservation of mass. Consider a cubical element of fluid flowing in space with respect to the coordinates shown in Figure 1. A simple material balance for this fluid element is as follows:

$$\begin{Bmatrix}\text{Mass}\\ \text{accumulation}\end{Bmatrix} = \begin{Bmatrix}\text{Mass into cubical}\\ \text{element}\end{Bmatrix} - \begin{Bmatrix}\text{Mass out of cubical}\\ \text{element}\end{Bmatrix} \tag{1}$$

The mass input into the x face of the element and the mass out at the $x + \Delta x$ face can be obtained over some differential time element, dt; denoting small increments as derivatives, the material balance becomes

$$\frac{\partial \rho}{\partial t}\, dx\, dy\, dz\, dt = \rho u\, dy\, dz\, dt - \left(\rho u + u\frac{\partial \rho}{\partial t} dx + \rho \frac{\partial u}{\partial x} dx\right) dy\, dz\, dt \tag{2}$$

Rearranging terms,

$$-u\frac{\partial \rho}{\partial x} - \rho\frac{\partial u}{\partial x} = \frac{\partial \rho}{\partial t} \tag{3a}$$

or

$$-\frac{\partial(\rho u)}{\partial x} = \frac{\partial \rho}{\partial t} \tag{3b}$$

where u refers to the fluid velocity in the x-direction.

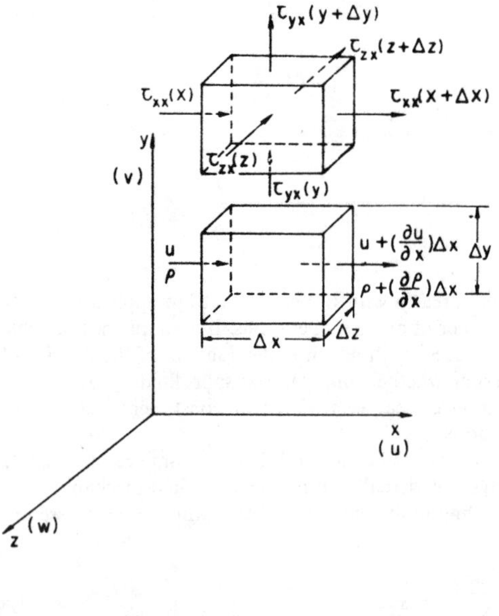

Figure 1. One-dimensional flow of a cubical element of fluid. The top element shows the directions in which the x-component of momentum is transported through the surfaces.

This differential equation is the law of conservation of mass, where both u and ρ are functions of x and t. However, describing flow in the x-direction the more general statement is

$$\frac{\partial \rho}{\partial t} = \left(\frac{\partial}{\partial x}\rho u + \frac{\partial}{\partial y}\rho v + \frac{\partial}{\partial z}\rho w\right) \tag{3c}$$

Table 1 summarizes the various forms of the continuity equation for both compressible and incompressible fluids under steady- and unsteady-state flow conditions. Table 2 gives the general statement of continuity in rectangular, cylindrical, and spherical coordinates.

The general statement of a momentum balance for laminar, rectilinear flow, following Figure 1, is

$$\begin{Bmatrix}\text{Rate of momentum}\\ \text{into a fluid element}\end{Bmatrix} - \begin{Bmatrix}\text{Rate of momentum}\\ \text{out of a fluid element}\end{Bmatrix} + \begin{Bmatrix}\text{Pressure forces and}\\ \text{gravity forces}\end{Bmatrix}$$
$$= \begin{Bmatrix}\text{Rate of momentum}\\ \text{accumulation}\end{Bmatrix} = 0 \text{ at steady state} \tag{4}$$

The arrows on the upper fluid element in Figure 1 indicate the direction in which the x-component of momentum is transported through the fluid volume surfaces. Momentum transport into and out of the fluid element occurs because of the bulk fluid motion (convection), and the velocity gradients (molecular transfer).

The unsteady-state momentum transport for the x-component is

$$\underbrace{\frac{\partial}{\partial t}\rho u_x}_{①} = \underbrace{-\left(\frac{\partial}{\partial x}\rho uu + \frac{\partial}{\partial z}\rho wu + \frac{\partial}{\partial y}\rho vu\right)}_{②} - \underbrace{\left(\frac{\partial}{\partial x}\tau_{xx} + \frac{\partial}{\partial z}\tau_{zx} + \frac{\partial}{\partial y}\tau_{yx}\right)}_{③} - \underbrace{\frac{\partial P}{\partial x} + \rho g_x}_{④} \tag{5}$$

where term (1) = rate of momentum accumulation
term (2) = rate of momentum in
term (3) = rate of momentum out
term (4) = sum of forces acting on fluid element

Equation 5 is the equation of motion for the x-component of momentum.

For the y- and z-components, the equations of motion are

$$\frac{\partial}{\partial t}\rho v = -\left(\frac{\partial}{\partial x}\rho uv + \frac{\partial}{\partial y}\rho vv + \frac{\partial}{\partial z}\rho wv\right) - \left(\frac{\partial}{\partial x}\tau_{xy} + \frac{\partial}{\partial y}\tau_{yy} + \frac{\partial}{\partial z}\tau_{zy}\right) - \frac{\partial P}{\partial y} + \rho g_y \tag{6}$$

$$\frac{\partial}{\partial t}\rho w = -\left(\frac{\partial}{\partial x}\rho uw + \frac{\partial}{\partial y}\rho vw + \frac{\partial}{\partial z}\rho ww\right) - \left(\frac{\partial}{\partial x}\tau_{xz} + \frac{\partial}{\partial y}\tau_{yx} + \frac{\partial}{\partial z}\tau_{zz}\right) - \frac{\partial P}{\partial z} + \rho g_z \tag{7}$$

Table 3 gives the general equations of motion in three different coordinate systems.

For incompressible, Newtonian fluids, the above equations are known as the Navier-Stokes equations, which in rectangular coordinates are as follows:

x-component:

$$\rho\left(\frac{\partial u}{\partial t} + u\frac{\partial u}{\partial x} + v\frac{\partial u}{\partial y} + w\frac{\partial u}{\partial z}\right) = -\frac{\partial P}{\partial x} + \mu\left(\frac{\partial^2 u}{\partial x^2} + \frac{\partial^2 u}{\partial y^2} + \frac{\partial^2 u}{\partial z^2}\right) + \rho g_x \tag{8}$$

y-component:

$$\rho\left(\frac{\partial v}{\partial t} + u\frac{\partial v}{\partial x} + v\frac{\partial v}{\partial y} + w\frac{\partial v}{\partial z}\right) = -\frac{\partial P}{\partial x} + \mu\left(\frac{\partial^2 v}{\partial x^2} + \frac{\partial^2 v}{\partial y^2} + \frac{\partial^2 v}{\partial z^2}\right) + \rho g_y \tag{9}$$

Table 1
Various Forms of the Continuity Equation*

	Compressible Fluids		Incompressible Fluids	
	Unsteady State	Steady State	Unsteady State	Steady State
One-Dimensional Flow (x-direction)	$\frac{\partial(\rho u)}{\partial x} = -\frac{\partial \rho}{\partial t}$	$\frac{\partial(\rho u)}{\partial x} = 0$	$\frac{\partial u}{\partial x} = 0$	$\frac{\partial u}{\partial x} = 0$
Two-Dimensional Flow (x and y direction)	$\frac{\partial(\rho u)}{\partial x} + \frac{\partial(\rho v)}{\partial y} = \frac{-\partial \rho}{\partial t}$	$\frac{\partial(\rho u)}{\partial x} + \frac{\partial(\rho v)}{\partial y} = 0$	$\frac{\partial u}{\partial x} + \frac{\partial v}{\partial y} = 0$	$\frac{\partial u}{\partial x} + \frac{\partial v}{\partial y} = 0$
Three-Dimensional Flow	$\frac{\partial(\rho u)}{\partial x} + \frac{\partial(\rho v)}{\partial y} + \frac{\partial(\rho w)}{\partial z} = -\frac{\partial \rho}{\partial t}$	$\frac{\partial(\rho u)}{\partial x} + \frac{\partial(\rho v)}{\partial y} + \frac{\partial(\rho w)}{\partial z} = 0$	$\frac{\partial u}{\partial x} + \frac{\partial v}{\partial y} + \frac{\partial w}{\partial z} = 0$	$\frac{\partial u}{\partial x} + \frac{\partial v}{\partial y} + \frac{\partial w}{\partial z} = 0$

* *v represents fluid velocity in the y-direction; w represents fluid velocity in the z-direction.*

Table 2
Equation of Continuity in Different Coordinates

Coordinate System	Axes	Equation
Rectangular	x-, y-, z-	$\frac{\partial \rho}{\partial t} + \frac{\partial}{\partial x}(\rho u) + \frac{\partial}{\partial y}(\rho v) + \frac{\partial}{\partial z}(\rho w) = 0$
Cylindrical	r-, θ-, y-	$\frac{\partial \rho}{\partial t} + \frac{1}{r}\frac{\partial}{\partial r}(\rho r u_r) + \frac{1}{r}\frac{\partial}{\partial \theta}(\rho u_\theta) + \frac{\partial}{\partial y}(\rho w) = 0$
Spherical	r-, θ-, ϕ-	$\frac{\partial \rho}{\partial t} + \frac{1}{r^2}\frac{\partial}{\partial r}(\rho r^2 u_r) + \frac{1}{r \sin\theta}\frac{\partial}{\partial \theta}(\rho u_\theta \sin\theta) + \frac{1}{r \sin\theta}\frac{\partial}{\partial \phi}(\rho u_\phi) = 0$

Table 3
General Equation of Motion in Different Coordinates

Coordinate System	Component	Equation
Rectangular (x, y, z)	x-	$\rho\left(\frac{\partial u}{\partial t} + u\frac{\partial u}{\partial x} + v\frac{\partial u}{\partial y} + w\frac{\partial u}{\partial z}\right) = -\frac{\partial P}{\partial x} - \left(\frac{\partial \tau_{xx}}{\partial x} + \frac{\partial \tau_{xx}}{\partial y} + \frac{\partial \tau_{zx}}{\partial z}\right) + \rho g_x$
	y-	$\rho\left(\frac{\partial v}{\partial t} + u\frac{\partial v}{\partial x} + v\frac{\partial v}{\partial y} + w\frac{\partial v}{\partial z}\right) = -\frac{\partial P}{\partial y} - \left(\frac{\partial \tau_{xy}}{\partial x} + \frac{\partial \tau_{yy}}{\partial_y} + \frac{\partial \tau_{zy}}{\partial z}\right) + \rho g_y$
	z-	$\rho\left(\frac{\partial w}{\partial t} + u\frac{\partial w}{\partial x} + v\frac{\partial w}{\partial y} + w\frac{\partial w}{\partial z}\right) = -\frac{\partial P}{\partial z} - \left(\frac{\partial \tau_{xz}}{\partial x} + \frac{\partial \tau_{xz}}{\partial y} + \frac{\partial \tau_{zz}}{\partial z}\right) + \rho g_z$
Cylindrical (r, θ, y)	r-	$\rho\left(\frac{\partial u_r}{\partial t} + u_r\frac{\partial u_r}{\partial r} + \frac{u_\theta}{r}\frac{\partial u_r}{\partial \theta} - \frac{u_\theta^2}{r} + u_y\frac{\partial u_r}{\partial y}\right) = -\frac{\partial P}{\partial r} - \left(\frac{1}{r}\frac{\partial}{\partial r}(r\tau_{rr}) + \frac{1}{r}\frac{\partial \tau_{r\theta}}{\partial \theta} - \frac{\tau_{\theta\theta}}{r} + \frac{\partial \tau_{ry}}{\partial y}\right) + \rho g_r$
	θ-	$\rho\left(\frac{\partial u_\theta}{\partial t} + u_r\frac{\partial u_\theta}{\partial r} + \frac{u_\theta}{r}\frac{\partial u_\theta}{\partial \theta} + \frac{u_r u_\theta}{r} + u_y\frac{\partial u_\theta}{\partial y}\right) = -\frac{1}{r}\frac{\partial P}{\partial \theta} - \left(\frac{1}{r^2}\frac{\partial}{\partial r}(r^2\tau_{r\theta}) + \frac{1}{r}\frac{\partial \tau_{\theta\theta}}{\partial \theta} + \frac{\partial \tau_{\theta y}}{\partial y}\right) + \rho g_\theta$
	y-	$\rho\left(\frac{\partial u_y}{\partial t} + u_r\frac{\partial u_y}{\partial r} + \frac{u_\theta}{r}\frac{\partial u_y}{\partial \theta} + u_y\frac{\partial u_y}{\partial y}\right) = -\frac{\partial P}{\partial y} - \left(\frac{1}{r}\frac{\partial}{\partial r}(r\tau_{ry}) + \frac{1}{r}\frac{\partial r_{\theta y}}{\partial \theta} + \frac{\partial \tau_{yy}}{\partial y}\right) + \rho g_y$
Spherical (r, θ, φ)	r-	$\rho\left(\frac{\partial u_r}{\partial t} + u_r\frac{\partial u_r}{\partial r} + \frac{u_\theta}{r}\frac{\partial u_r}{\partial \theta} + \frac{u_\phi}{r\sin\theta}\frac{\partial u_r}{\partial_\phi} - \frac{u_\theta^2 + u_\phi^2}{r}\right) = -\frac{\partial P}{\partial r} - \left(\frac{1}{r^2}\frac{\partial}{\partial r}(r^2\tau_{rr}) + \frac{1}{r\sin\theta}\frac{\partial}{\partial \theta}(\tau_{r\theta}\sin\theta) + \frac{1}{r\sin\theta}\frac{\partial \tau_{r\phi}}{\partial \phi} - \frac{\tau_{\theta\theta} + \tau_{\phi\phi}}{r}\right) + \rho g_r$
	θ-	$\rho\left(\frac{\partial u_\theta}{\partial t} + u_r\frac{\partial u_\theta}{\partial r} + \frac{u_\theta}{r}\frac{\partial u_\theta}{\partial \theta} + \frac{u_\phi}{r\sin\theta}\frac{\partial u_\theta}{\partial \phi} + \frac{u_r u_\theta}{r} - \frac{u_\phi^2\cot\theta}{r}\right)$ $= -\frac{1}{r}\frac{\partial P}{\partial \theta} - \left(\frac{1}{r^2}\frac{\partial}{\partial r}(r^2\tau_{r\theta}) + \frac{1}{r\sin\theta}\frac{\partial}{\partial \theta}(\tau_{\theta\theta}\sin\theta) + \frac{1}{r\sin\theta}\frac{\partial \tau_{\theta\phi}}{\partial \phi} + \frac{\tau_{r\theta}}{r} - \frac{\cot\theta}{r}\tau_{\phi\phi}\right) + \rho g_\theta$
	φ-	$\rho\left(\frac{\partial u_\phi}{\partial t} + u_r\frac{\partial u_\phi}{\partial r} + \frac{u_\theta}{r}\frac{\partial u_\phi}{\partial \theta} + \frac{u_\phi}{r\sin\theta}\frac{\partial u_\phi}{\partial \phi} + \frac{u_\phi u_r}{r} + \frac{u_\theta u_\phi}{r}\cot\theta\right)$ $= -\frac{1}{r\sin\theta}\frac{\partial P}{\partial \phi} - \left(\frac{1}{r^2}\frac{\partial}{\partial r}(r^2\tau_{r\phi}) + \frac{1}{r}\frac{\partial \tau_{\theta\phi}}{\partial \phi} + \frac{1}{r\sin\theta}\frac{\partial \tau_{\phi\phi}}{\partial \phi} + \frac{\tau_{r\phi}}{r} + \frac{2\cot\theta}{r}\tau_{\theta\phi}\right) + \rho g_\phi$

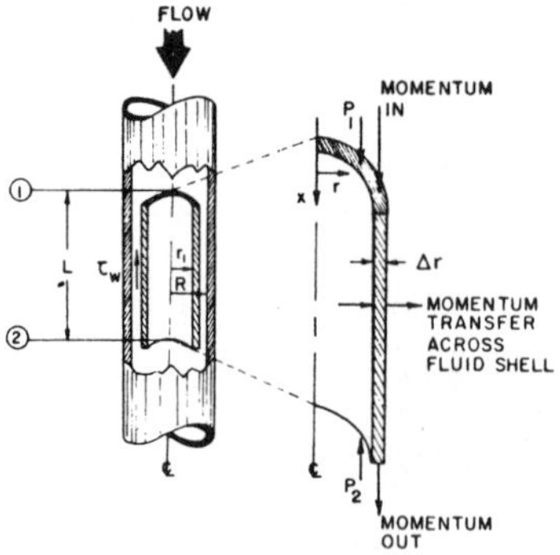

Figure 2. Incompressible fluid flowing downward in a vertical tube. A cylindrical shell of fluid can be imagined if one chooses to derive the velocity profile of the fluid from a momentum balance.

z-component:

$$\rho\left(\frac{\partial w}{\partial t} + u\frac{\partial w}{\partial x} + v\frac{\partial w}{\partial y} + w\frac{\partial w}{\partial z}\right) = -\frac{\partial P}{\partial z} + \mu\left(\frac{\partial^2 w}{\partial x^2} + \frac{\partial^2 w}{\partial y^2} + \frac{\partial^2 w}{\partial z^2}\right) + \rho g_z \tag{10}$$

By applying proper boundary conditions based on the flow system geometry and the physics of the flow phenomenon, the equation of motion can be solved for the appropriate flow parameter of interest. For example, consider the simple case of an incompressible fluid flowing down a vertical tube, as shown in Figure 2. To derive an expression for the velocity profile for laminar flow in cylindrical coordinates one may ignore end effects. Hence, by inspection of the equation in cylindrical coordinates in Table 3, noting that $u_\theta = u_r = 0$, the following is obtained:

$$\rho v\frac{\partial v}{\partial y} = -\frac{\partial P'}{\partial y} + \mu\left[\frac{1}{r}\frac{\partial}{\partial r}\left(r\frac{\partial v}{\partial r}\right) + \frac{\partial^2 v}{\partial y^2}\right] \tag{11}$$

The continuity equation reduces to

$$\frac{\partial v}{\partial y} = 0 \tag{12}$$

Also, $\partial^2 v/\partial y^2 = 0$. Hence, Equation 11 becomes

$$0 = -\frac{dP'}{dy} + \mu\frac{1}{r}\frac{d}{dr}\left(r\frac{dv}{dr}\right) \tag{13}$$

Applying a "no-slip" boundary condition: $v = 0$ at $r = R$, and integrating

$$v = \frac{\Delta P' R^2}{4\mu L}\left[1 - (r/R)^2\right] \tag{14}$$

Equation 14 is the expression for a parabolic velocity profile describing laminar flow in a tube.

Flow problems involving the determination of fluid velocity as a function of time and space require simultaneous solution of the momentum and continuity equations. For many situations the flow is such that these equations may be greatly simplified. In the one-dimensional flow problem treated above, the two equations in the y- and z-directions (or the θ- and r-directions in cylindrical coordinates) were eliminated. In two-dimensional flow problems, two momentum equations are required.

We now direct attention to the mechanical energy equation. The flow system shown in Figure 3 has fluid entering at point (1) and exiting at point (2), with the same mass flow rate, W. That is,

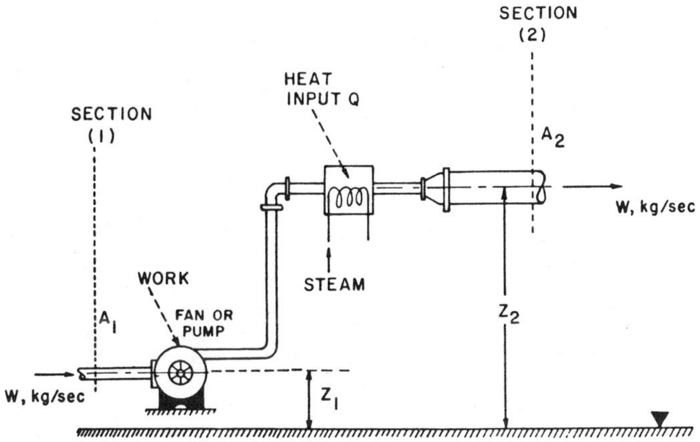

Figure 3. Example system for describing the total energy balance.

the flow system is at steady state and, hence, there is no progressive accumulation or depletion of fluid at any point.

The fluid entering at point (1) has a certain amount of energy associated with it. This energy exists in several forms. First, since the fluid has a velocity, U, it brings to the system *kinetic energy* defined as $U_1^2/2g$. Second, because the incoming flow is elevated at a distance, Z_1, above the specified horizontal datum plane, it possesses potential energy of Z_1-meters of fluid. It also brings *internal energy*, E_1, (thermal, chemical, or combinations of these). Finally, there is mechanical energy associated with the incoming fluid, defined as the product of the force exerted by the flowing fluid and the distance over which it acts. Hence, it is the fluid pressure per unit area times the cross-sectional area of flow, P_1A_1. If, for calculation purposes, the mass of fluid flowing per unit time past point (1) is 1 kg, then the distance through which the force acts is the specific volume of the fluid divided by the cross section, v_1/A_1. Furthermore, since the work performed is the product of force and distance, then the energy expended is equal to pressure times the specific volume:

$$W_M = P_1A_1(v_1/A_1) = P_1v_1 \tag{15}$$

Between points (1) and (2) of the system diagram occurs an exchange of energy between the surroundings and system. This interchange of energy is in the form of heat, Q, and work, W, which can be expressed in terms of the mass of flowing fluid. In Figure 3, both energies are added to the flow from the surroundings and, thus, are denoted by positive signs.

At point (2) the fluid departs from the system with energy in the same four forms that entered at point (1): kinetic, potential, internal, and mechanical energies. To maintain conservation of energy requires equality between the sum of all energies entering and the sum of those leaving.

$$z_1 = P_1v_1 + \frac{U_1^2}{2g} + E_1 + Q + W' = Z_2 + P_2v_2 + \frac{U_2^2}{2g} + E_2 \tag{16}$$

The kinetic energy term, $U^2/2g$ is based on a small stream of fluid having a local velocity of flow in which the velocity profile across the cross section is flat. In reality, there is a velocity gradient across the passage, as was shown by the derivation for laminar flow in a tube (Equation 14). For turbulent flow, the velocity profile is more flat, and it is customary to use an average velocity in the kinetic energy term. The error introduced by using $U^2/2g$ is generally not serious, but its magnitude depends on the geometry of the flow cross section. Also, since the $U^2/2g$ terms appear on each side of the energy balance equation, the error tends to cancel out. In the case of a parabolic velocity profile, the error introduced by using $U^2/2g$ can be significant. An exact expression for the kinetic energy term can be readily derived for laminar flow, by using Equation 14 to describe the velocity profile. The volumetric flow rate in a tube is then:

$$q = \int_0^R u 2\pi r \, dr = \frac{\pi R^4 \Delta P'}{8\mu L} \tag{17}$$

Defining the average velocity as the volumetric flow rate divided by the cross-sectional area of flow for a single cylindrical tube,

$$U = -\Delta P' R^2 / 8\mu L \tag{18}$$

The local kinetic energy through the cylindrical shell is

$$d(kE) = dw \, u^2/2g \tag{19}$$

where dw is the mass flow rate through an annulus, assuming the fluid flowing in a section of tube may be visualized as an infinite number of cylindrical shells of thickness r (refer to Figure 2). The mass of fluid flowing per unit time through the annulus is $2\pi \, dr \, u\rho$.

From Equation 14 it can be shown that

$$u = 2U(R^2 - r^2)/R^2$$

Then,

$$d(kE) = \frac{dw \, u^2}{2g} = \frac{(2\pi r \, dr \, u\rho)u^2}{2g} \tag{20}$$

The average kinetic energy per unit mass of fluid flowing is the integral of this expression divided by the fluid mass flowing through the entire cross section of pipe.

$$\begin{aligned} kE &= \frac{\int d(kE)}{\pi r_1^2 u\rho} = \frac{8U^2}{g r_1^8} \int_0^{r_1} (r_1^6 r \, dr - 3r_1^4 r^3 \, dr + 3r_1^2 r^5 \, dr - r^7 \, dr) \\ &= \frac{8U^2}{g r_1^8}\left(\frac{r_1^8}{2} - \frac{3r_1^8}{4} + \frac{3r_1^8}{6} - \frac{r_1^8}{8}\right) \\ &\simeq \frac{U^2}{g} \end{aligned} \tag{21}$$

Hence, for the parabolic velocity profile case, U^2/g provides a reasonable estimate of the kinetic energy term in the total energy equation.

Variations in point velocity over each section of flow in the system of Figure 3 must be evaluated wh^n applying the energy equation.

The total heat that is added to the system per unit mass of flowing fluid, Q, refers to heat that is supplied by the surroundings or some external source, such as steam to a heat exchanger. Excluded from this term is any heat generated by friction. This latter energy form is already dissipated in the flow and is accounted for in the internal energy term. The addition of heat is denoted by a positive sign on Q in Equation 16. Note that the positive addition of heat does not always involve a rise in temperature for the system and, in fact, it is not uncommon for a temperature drop to occur.

External work energy, W′, is most commonly supplied by a mechanical device such as a pump or compressor. Work energy also may be removed from the fluid system by such devices as engines, prime movers, etc. If no work is either added or removed from the system, the term W′ vanishes in Equation 16.

The internal energy term, E, accounts for thermodynamic properties. Note that both E and P are properties of state of the fluid that are uniquely determined by point conditions. The sum of these two terms is most readily treated as a single function, enthalpy:

$$H = E + Pv \tag{22}$$

Enthalpy itself is uniquely determined by the fluid's point conditions. Unfortunately, like E, the absolute value of H cannot be determined; however, differences in value above an arbitrary reference can be evaluated. Evaluation of thermodynamic properties is treated in the literature [1–5].

When treating problems involving steady flow of incompressible fluids under conditions in which friction is small and in the absence of external effects, the energy balance based on mechanical energy forms may be applied with confidence. For situations involving compressible flow, the total mechanical energy at the system discharge is often significantly greater than the upstream section. This is especially true when large pressure drops between sections occur. Any quantity of flowing fluid undergoes an expansion when it performs mechanical work. This work is then expended on the fluid immediately ahead of it. Realize, however, that the fluid element in question also acquires an equivalent quantity of mechanical energy from the fluid behind it. This results in an overall increase in the mechanical energy at the expense of the internal energy of the fluid or of externally added heat energy [6, 7].

To account for self-expansion work and for total friction due to fluid flow, the mechanical energy balance is modified:

$$Z_1 + P_1 v_1 + \frac{u_1^2}{2g} + \int_1^2 P\,dv = Z_2 + P\,dv = Z_2 + P_2 v_2 + \frac{U_2^2}{2g} + \sum F \tag{23}$$

In applying the correction term $\int_1^2 P\,dv$, it should be noted that the mechanical energy output is almost always less than the input. Unlike Pv, E and H, the integral is not a point function of the conditions. Instead, it is a function of the path of expansion or compression.

$\sum F$ represents the total friction generated by the flow. Fluid friction is basically the mechanical energy rendered nonavailable due to irreversibilities in the flow process [7]. For example, assume the pump between points (1) and (2) in Figure 3 delivers in reality W′ joules of mechanical work to the fluid. Note that W′ is less than the work term W′ in the total energy equation since it includes friction in the pump. That is,

$$W' = W' - F_p \tag{24}$$

where F_p is friction energy in the pump.

Hence, the most general form of Equation 23 is the Bernoulli equation:

$$Z_1 + P_1 v_1 + \frac{u_1^2}{2g} + {}_1^2\, P\,dv + W' = Z_z + P_2 v_2 + \frac{u_2^2}{2g} + \sum F \tag{25}$$

or

$$W' - \int_1^2 v\,dP = \Delta Z + \Delta u^2/2g + \sum F \tag{26}$$

where $\Delta u = u_2^2 - u_1^2$

and noting that

$$\int_1^2 P\,dv + P_1 v_1 - P_2 v_2 = -\int_1^2 v\,dP \tag{27}$$

Both the Bernoulli equation and the total energy equation represent the fundamental basis for solving problems in fluid mechanics. Note that application of Equation 27 requires not only the thermodynamic properties of the flowing system to be defined, but also the term $\sum F$.

REGIMES OF FLOW

Laminar Flow

Shearing stresses exist in the boundary layer of flowing fluid within a system of fixed boundaries (e.g., when flowing through a pipe or duct or over a surface). These stresses are exerted in the opposite direction to the flow and, therefore, may be thought of as the forces of resistance to flow. From a force balance over a length of tube, L, we obtain:

$$\pi R^2 P_1 - \pi R^2 P_2 + \pi R^2 L\rho g - 2\pi R L \tau_w = 0 \tag{28}$$

or

$$\frac{\Delta P}{\rho} + Lg = \frac{2L\tau_w}{\rho R} \tag{29}$$

Defining the sum of all the friction forces to be $\Delta P/\rho$, then

$$\sum F = \frac{\Delta P}{\rho} = \frac{P_1 - P_2}{\rho} - Lg \tag{30}$$

Solving Equation 29,

$$\tau_w = \frac{\Delta P}{\rho}\left(\frac{\rho R}{2L}\right) = \frac{R\,\Delta P}{2L} \tag{31}$$

Note that τ_w is actually τ_{rx} at $r = R$.

Hence, for any r, where $r < R$, the fluid shear stress is:

$$\tau_{rx} = \frac{r\,\Delta P}{2L} \tag{32}$$

Combining Equations 31 and 32, τ_{rx} is expressed in terms of the shearing force (shear stress) at the wall:

$$\tau_{rx} = \tau_w(r/R) \tag{33}$$

This last expression states that the shear stress distribution across the tube is linear and varies from zero at the tube centerline to τ_w at the tube wall. The shear stress is illustrated in Figure 4 along with the parabolic velocity profile for laminar flow.

Since shear stress is the force of resistance to flow, it is a measure of the friction losses. For laminar flow, a relatively simple relation can be obtained. By combining the expression for the velocity profile (Equation 14), and Newton's law of viscosity ($d_\tau = d(\mu A\, du/dr)$), an alternative expression for the wall shear is obtained:

$$\tau_w = \frac{4\mu U}{R} \tag{34}$$

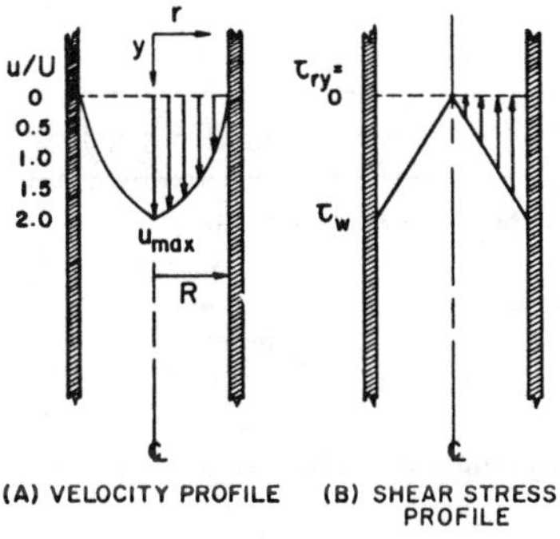

Figure 4. Velocity profile and shear stress distribution for laminar flow in a tube.

This expression defines wall shear stress in terms of fluid viscosity, average velocity, and the tube radius. Combining Equations 31 and 34 and expressing the tube radius, R, in terms of the inside diameter of the tube, the following expression written in differential form is obtained:

$$-\frac{dP}{dx}=\frac{32\mu U}{gD^2} \tag{35}$$

This equation is Poiseuille's law and defines the pressure loss due to friction in terms of fluid viscosity, average fluid velocity and tube diameter. The lost work due to friction is conventionally correlated by the parameter "friction factor." The friction factor is a dimensionless parameter, which enables a compact expression to be formulated to include tube diameter, fluid properties, fluid velocity, and the pressure gradient equivalent to the frictional resistance per unit length of tube.

Friction losses in turbulent flow are proportional to the kinetic energy of the fluid per unit volume and the area of the solid surface contacting the flow. The force per unit area, A_w, acting on a solid surface such as a pipe wall is:

$$F/A_w \propto \frac{\rho U^2}{2g_c}=f\frac{\rho U^2}{2g_c} \tag{36}$$

where f is a proportionality constant.

We note that F/A_w is really the wall shear, τ_w; hence, $\tau_w = f(\rho U^2/2g_c)$.

By substituting for τ_w in Equation 32, the proportionality constant, f, is solved for. Factor f, the friction factor, is referred to as the "Fanning friction factor" for pipe flow:

$$f=\frac{g_cR}{\rho U^2}\frac{\Delta P}{L} \tag{37a}$$

or

$$f=\frac{g_cD}{2\rho U^2}\frac{\Delta P}{L} \tag{37b}$$

where D is the pipe diameter.

For laminar flow in a tube, Poiseuille's equation may be used to derive the following relation:

$$f=\frac{8\mu}{RU\rho}=\frac{16}{Re} \tag{37c}$$

Re is the Reynolds number, ($Re = DU\rho/\mu$).

The friction factor definition in Equation 37c is good for Reynolds numbers up to 2,000.

Turbulent Flow

Laminar flow as previously described consists of a steady advance of fluid layers such that a streamline flow is maintained. If the fluid's flow rate is increased sufficiently, this pattern is no longer maintained and the flow becomes unsteady. This latter flow situation is characterized by chaotic movements of portions of the fluid in different directions superimposed on the main flow of the fluid; the phenomenon is referred to as turbulence. The complex movement of fluid elements within the flow can only be described in terms of time-averaged values.

Momentum transfers between adjacent regions of flowing fluid play a dominant role in establishing turbulence. These momentum transfers are inertial effects, which cause velocity and fluid density to take on important roles. By contrast, in laminar flow only purely viscous effects determine the nature of the flow. Turbulent flow dominates when inertial effects (denoted by ρu^2) become large in comparison to viscous forces (denoted by $\mu u/R$).

The transition from laminar to turbulent flow for a small portion of flowing fluid can be illustrated by visualizing the flow to consist of distinct parallel streamlines of fluid. By introducing a random disturbance, density and velocity are in part dumped out by viscous forces. As more momentum is transferred into the disturbance, inertial forces eventually become much greater than the viscous term, and eddy currents form. The Reynolds number best describes this flow transition since it is the ratio of inertial to viscous forces, i.e., $Re = (\rho u)/(\mu u/D)$.

Numerous investigators [8–14] have studied the flow of fluids in pipes, confirming the flow regime dependency on Reynolds number. For flow in smooth pipes, the flow is always laminar for Reynolds numbers up to about 2,000. Between Reynolds numbers of 2,000 and 4,000, the flow undergoes a transitional change from laminar to turbulent flow. Above Reynolds numbers of 4,000, the flow is generally fully established turbulent flow.

There are several situations that can lead to turbulent flow. One situation just described is that of rapid flow of a fluid past a solid surface. This situation leads to unstable, self-amplifying velocity fluctuations, which form in the fluid in the vicinity of the wall and spread outward into the main fluid stream. In a similar manner, turbulent eddies are formed from the velocity gradients established between a fast-moving fluid and a slower-moving fluid. A third general way in which turbulence is induced is by the relative movement of an object through the fluid streams. Examples of this last case are an impeller blade on an agitator and a sphere or cylinder falling through a fluid medium. These situations cause eddies to form in the wake, resulting in an increase in the resistance of the movement of the object (called "form drag").

In the case of stirred vessels, turbulence can be very intense near the tips of the rotor blades. Most of the turbulence throughout the vessel arises from velocity gradients, whereby portions of high-velocity fluid are thrown from impeller blades into slower-moving fluid. Some of this turbulence is attributed, however, to the high shearing over the blades themselves, which creates separation behind each blade or neighboring baffles.

Another mixing-related turbulence arises from submerged jets. In submerged jets or "free turbulent jets," the fluid is expelled from a nozzle into a mass of miscible fluid, which is essentially at rest with respect to the discharging stream. The free jet expands outward in the form of a cone, entraining large amounts of the surrounding fluid at the periphery of the jet. The sizes of eddies formed are relatively uniform across any given section of such a jet. When a submerged jet is introduced to an immiscible fluid, the stream is referred to as a *restrained turbulent jet*. Turbulent eddies protrude from the sides of the free jet but are restrained from breaking away by surface tension forces. This type of turbulence decays quickly, as fluctuations are damped by the elastic forces associated with surface tension effects.

Turbulence may be described quantitatively in terms of the instantaneous velocity fluctuations occurring in local patches of flowing fluid. The instantaneous velocity in the x-direction of flow is:

$$u_x = u_x \pm u'_x \tag{38}$$

where u_x is the time-averaged velocity at any point in the flowing fluid and u'_x is the instantaneous velocity fluctuation. Fluctuation velocities may be positive or negative at times, with their time average being zero. That is,

$$u'_x = 0 \tag{39}$$

The amplitude of fluctuation velocities in the x-direction can be expressed in terms of the mean of the squares of the fluctuating velocities, obviously making them positive. This is denoted by $(u'_x)^2$. A root mean square fluctuation velocity would be

$$u'_{x_{rms}} = [(u'_x)^2]^{1/2} \tag{40a}$$

where $u'_{x_{rms}}$ is always positive. $u_{x_{rms}}$ provides an indication of the *intensity of turbulence* in the x-direction.

$$u_y = u_y \pm u'_y$$
$$u'_y = 0 \tag{40b}$$

and

$$u'_{y_{rms}} = \sqrt{(u'_y)^2} \tag{40c}$$

The product $u'_x u'_y$ for any point in the fluid is not necessarily zero. The product of u'_x and u'_y has been time-averaged, and these fluctuations are correlated with a particular eddy. These fluctuating velocities are responsible for establishing the degree of dissipation of energy, mass transfer and heat transfer in turbulent flows.

Visualize the fluid as consisting of individual parcels. In turbulent flow there is a continuous interchange of these parcels between adjacent regions of the fluid. This exchange of mass involves both gains and losses of momentum. That is, each parcel of fluid leaving a region with some velocity takes momentum with it to the next fluid region, which, in turn, may be moving at a different mean velocity. Fluid parcels transposed laterally from a faster-moving region tend to accelerate the slower-moving fluid region; conversely, a slow-moving parcel transposed to a faster-moving region has a retarding effect.

The rate of momentum transfer per unit area is $\rho u'_y u'_x$ and can be expressed in terms of a stress, i.e., equivalent force per unit area:

$$\tau = -\rho u'_x u'_y \tag{41a}$$

where the time average of this product is the Reynolds stress:

$$\tau = -\rho u'_x u'_y \tag{41b}$$

Continuing with this model of parcel exchanges, assume that a particular parcel is carried a distance, y_1, from the wall to a new distance. Further assume that the parcel is moved from a slower-moving section to a faster-moving one. The instantaneous deficit in momentum occurring at distance y_2 may be expressed as follows:

$$\text{Momentum deficit} = -(\Delta y)d(\rho u_x)/dy \tag{42}$$

where $\Delta y = y_2 - y_1$.

If momentum is conserved, then

$$\rho u'_x = -(\Delta y)d(\rho u_x)/dy \tag{42a}$$

or

$$\tau = -\rho u'_x u'_y = u'_y(\Delta y)d(\rho u_x)/dy \tag{42b}$$

Alternately, this last expression may be written as follows:

$$\tau = -\rho u'_x u'_y = \rho l u'_{y_{rms}} \, du_x/dy \tag{43}$$

Equation 43 is obtained by representing the root mean squares of the displacement, Δy, and the velocity fluctuation, u'_y, by l and $u'_{y_{rms}}$, respectively. The parameter l is defined as the mean distance of travel of the parcels of fluid before they are mixed into the new fluid region. We can think of l as a measure of the "scale of turbulence eddies" (commonly referred to as the *mixing length or Prandtl eddy length*).

By analogy to Newton's law,

$$\epsilon = \tau/(du_x/dy) = \rho l u'_{y_{rms}} \tag{44}$$

where ϵ is the eddy viscosity. Equation 44 can be written in the alternative form:

$$\tau = \rho \epsilon_M \, du_x/dy \tag{45}$$

where ϵ_M is the eddy kinematic viscosity (or eddy viscosity of momentum), defined by $\epsilon_M = l u'_{y_{rms}}$.

Prandtl [15] hypothesized that ϵ_M is proportional to the velocity fluctuations and to their distances (this is analogous to the product of the average molecular velocity and the mean free path in the kinetic theory of gases).

Since $u'_{y_{rms}}$ is difficult to obtain directly, Prandtl assumed u'_y to be related simply to the velocity fluctuation ($u'_y \propto l \, du_x/dy$) and wrote the Reynolds stress in the following form:

$$\tau = \rho l^2 (du_x/dy)^2 \tag{46a}$$

or, alternately,

$$\tau = (\rho l^2 \, du_x/dy)(du_x/dy) \tag{46b}$$

Hence, Equation 44 may be expressed as:

$$\epsilon = \rho l^2 (du_x/dy) \tag{47}$$

With this definition, the eddy kinematic viscosity may be written as $\epsilon_M = l^2(du_x/dy)$.

Finally, we can account for both viscous and turbulent contributions with an equation analogous to Newton's law of viscosity:

$$\tau = (\nu + \epsilon_M) du_x/dy \tag{48}$$

In the last series of equations, only time-averaged values for the velocity components are necessary. From this point on we will drop the bar ($^-$), realizing that values are time-averaged. Equation 48 describes the shear distribution for turbulent flow in terms of the velocity gradient.

The parabolic velocity distribution characteristic of laminar flow is due to the effects of viscous forces between adjacent fluid layers. In turbulent flow, the shape of the velocity profile is very different. The majority of the velocity gradient occurs in the fluid region close to the wall where some streamline flow continues to persist; however, as we enter into the fluid core, the velocity profile is more blunt than in the laminar case and is nearly flat at the pipe centerline. The reason for this flatness is that in turbulent flow inertial forces are high and have influence over a large portion of the tube. Figure 5 compares the velocity profiles for laminar and turbulent flow in pipe. It is of interest that although the two profiles differ greatly, both flows have the same average velocity.

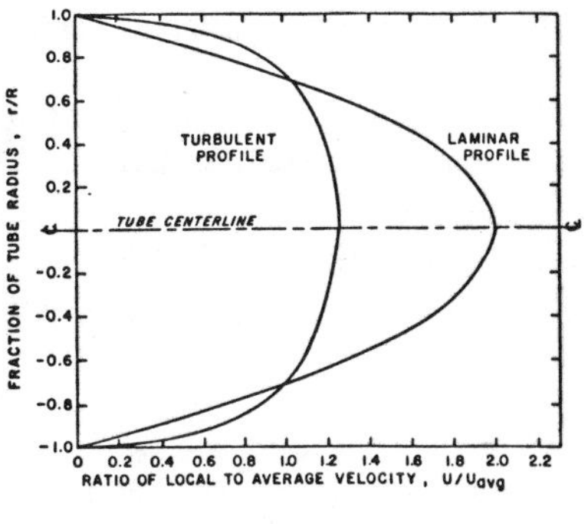

Figure 5. Comparison of the velocity profiles for turbulent and laminar flows in a tube. Note that the average fluid velocity is the same in each case.

Many early investigations were directed at studying the velocity profiles of turbulent flows. Nikuradse [9] extensively studied turbulent flow velocity distributions in smooth circular tubes, showing that the profile always takes on the same general shape shown in Figure 5. From these investigations, a number of attempts have been made to develop from first principles an equation for describing the velocity distributions in tubes. In the discussions to follow, we shall review the principal models developed for describing turbulent flow in pipes. Much of the early work between 1920–1955 forms the basis for our current understanding of turbulence and has enabled sound engineering principles to be established.

Turbulence Mixing Models

The most extensive study of turbulent velocity profiles was that of Prandtl [13, 15]. Prandtl's, as well as Nikuradse's [9], experiments showed evidence that the velocity of the fluid at the wall is zero for smooth pipes. Prandtl further postulated that there is a layer of fluid very close to the wall that is in laminar motion, and that only viscous forces contribute to shearing in this region. That is, at the wall $u \simeq 0$ for y close to 0. Hence, from τ_w the velocity profile in the laminar layer can be computed. The earliest attempt to describe the flow in the turbulent region resulted in a power law relation:

$$\frac{u}{u_{max}} = C(y/R)^{1/7} \tag{49}$$

where C is a constant, y is the distance from the pipe wall, and R is the tube radius. Equation 49 is the $\frac{1}{7}$th power law velocity distribution and is related to the Blasius [16] friction factor as follows:

$$\tau_w = \frac{f\rho U^2}{2g_c} \tag{50}$$

The Blasius friction factor equation has been determined experimentally to be valid for turbulent Reynolds numbers up to 100,000.

Prandtl further derived a logarithmic distribution from the mixing length theory (Equation 48):

$$\tau_w g_c = \underbrace{\mu \frac{du}{dy}}_{\text{viscous shear}} + \underbrace{\rho\left(l\frac{du}{dy}\right)^2}_{\text{turbulent shear}} \tag{51}$$

where y is defined as being the distance normal from the wall.

We know from laminar flow analysis that $\tau = \tau_w(r/R)$ and, noting the following approximation,

$$(r/R) \simeq 1 - y/R \tag{52}$$

The following relation is obtained:

$$\tau_w g_c(1 - y/R) = \rho\left(l\frac{du}{dy}\right)^2 \tag{53}$$

or

$$l\left(\frac{du}{dy}\right) = \left(\frac{\tau_w g_c}{\rho}\right)^{1/2}(1 - y/R)^{1/2} \tag{54}$$

The group of terms $(\tau_w g_c/\rho)^{1/2}$ is the *friction velocity* and is commonly denoted by u*:

$$u^* = \sqrt{\frac{\tau_w g_c}{\rho}} \tag{55}$$

For conditions at the wall, $1 - y/R$ reduces to 1 and

$$u^* = l\frac{du}{dy} \tag{56}$$

Prandtl hypothesized that the eddy mixing length is proportional to the distance from the wall by some constant that he hoped to be universal. That is,

$$l = ky \tag{57}$$

Equation 56 may be restated as

$$\frac{du}{dy} = u^*/ky \tag{58}$$

Or, on integration,

$$u = \frac{u^*}{k}\ln y + c \tag{59}$$

Using the eddy viscosity, Equation 47, which also may be written as $\epsilon = k^2y^2(du/dy)$ from the above considerations, and solving for the constant of integration, c, in Equation 59, Prandtl developed the following velocity distribution equation:

$$u = u_{max} + 2.5u^* \ln(y/R) \tag{60}$$

von Karman [17, 18] proposed the velocity distribution through similarity theory, developing the following series of equations:

$$\epsilon_M = \frac{k^2(du/dy)^3}{(d^2u/dy^2)^2} \tag{61}$$

$$l = \frac{k(du/dy)}{d^2u/dy^2} \tag{62}$$

$$\tau = \frac{\rho k^2(du/dy)^4}{g_c(d^2u/dy^2)^2} \tag{63}$$

Applying the following boundary conditions, for

$$y = 0, \qquad du/dy = \infty$$

and for

$$y = R, \qquad u = u_{max}$$

the velocity distribution expression was derived.

$$u = u_{max} + \frac{1}{k} u^*[\ln (1 - \sqrt{1 - y/R}) + \sqrt{1 + y/R}] \tag{64}$$

where the constant, k, must be evaluated empirically from the velocity profile data.

From Prandtl's theory and the data of Nikuradse [9], Reichardt [19] and others, von Karman was able to construct a universal velocity profile through three equations. Each equation describes a different region in the flow: a laminar boundary layer, a buffer-layer, and the turbulent fluid core.

From the friction velocity (Equation 55), the following expression was obtained:

$$\tau_w = \frac{u^*\rho}{g_c} = \frac{\mu}{g_c}\frac{u_{\delta_1}}{\delta_1} \tag{65}$$

Equation 65 defines conditions at the edge of the boundary layer, where $y = \delta_1$ (i.e., the boundary layer thickness) and du/dy is approximated by u_{δ_1}/δ_1. u_{δ_1} is the velocity of the fluid at the edge of the boundary layer.

For conditions at the edge of the boundary layer, Prandtl's velocity distribution equation (Equation 60) may be written as

$$\frac{u_{\delta_1}}{u^*} = \frac{u_{max}}{u^*} - 2.5 \ln(R/\delta_1) \tag{66}$$

The following manipulation can be made with the log term $R/\delta_1(u^*/u^*)$ and since $u_{\delta_1}/u^* =$ constant $= C_1$, then $u^* = C_1\nu/\delta_1$.

This term then becomes

$$\frac{Ru^*}{\delta_1\left(\frac{C_1\nu}{\delta_1}\right)}$$

and Equation 66 now becomes

$$\frac{u_{max}}{u^*} = \frac{u_{\delta_1}}{u^*} + 2.5\left[\ln\left(\frac{Ru^*}{\nu}\right) + \ln\left(\frac{1}{C_1}\right)\right] \tag{67}$$

The form of Equation 66 enables all the constants to be collected into one term:

$$\frac{u_{max}}{u^*} = C_2 + 2.5 \ln\left(\frac{Ru^*}{\nu}\right) \tag{68}$$

where $C_2 = \frac{u_{\delta_1}}{u^*} + 2.5 \ln(1/C_1)$

By substituting Equation 68 for u* back into Prandtl's equation (Equation 60), the following is obtained:

$$\frac{u}{u^*} = C_2 + 2.5 \ln\left(\frac{yu^*}{\nu}\right) \tag{69}$$

Equation 69 may be written in its most general form by introducing the following dimensionless terms:

$$u^+ = u/u^*, \text{ velocity parameter} \tag{70}$$

$$y^+ = yu^*/\nu, \text{ friction distance parameter} \tag{71}$$

where y^+ is the dimensionless distance from the wall and u* is the dimensionless velocity. Hence, the universal velocity distribution equation for turbulent flow in circular pipes is

$$u^+ = C_2 + 2.5 \ln y^+ \tag{72}$$

Further details of the derivation are given by Cheremisinoff [20].

von Karman evaluated the constant C_2 in terms of three equations that empirically fit different portions of the dimensionless velocity profile u^+ versus y^+:

Laminar sublayer .... $u^+ = y^+$, for $0 < y^+ < 5$ (73)

Buffer region .... $u^+ = -3.05 + 5.0 \ln y^+$, for $5 < y^+ < 30$ (74)

Turbulent core ...... $u^+ = 5.5 + 2.5 \ln y^+$, for $y^+ > 30$ (75)

The universal velocity profile has been used extensively in developing useful analogies between momentum and heat transfer. In later volumes, the above principles are applied to analyzing flows involving heat transfer and two-phase (gas–liquid) flows.

Deissler [11, 21–25] proposed a universal velocity profile model based on a description of the laminar and buffer layers in terms of the exponential decay of eddies penetrating these layers.

Using von Karman's model, $\tau = -(\mu + \epsilon\rho)(du/dy)$, in which the eddy viscosity could be defined for $y^+ > 30$ as

$$\epsilon = k \frac{(du/dy)^3}{(d^2u/dy^2)^2} \tag{76}$$

For the region $0 < y^+ < 30$, the following relationship was redeveloped:

$$\epsilon = n^2 uy(1 - e^{-\rho n^2 uy/\mu}) \tag{77}$$

where n is a constant equal to 0.10.

Equation 77 predicts that the eddy viscosity approaches zero as the wall is neared. In dimensionless form, the velocity profile is:

$$y^+ = \frac{1}{n}\frac{\sqrt{\frac{1}{2\pi}}\int_0^{nu^+} e^{-[(nu^+)^2/2]}\,d(nu^+)}{(1/\sqrt{2\pi})e^{-[(nu^+)^2/2]}} \tag{78}$$

This last expression applies to the laminar and buffer regions for $0 < y^+ < 26$. Equation 78 must be numerically integrated to obtain values for u^+.

For the turbulent core, Deissler obtained the following expression:

$$u^+ = 3.8 + 2.78 \ln y^+ \tag{79}$$

FRICTION FACTORS AND DRAG COEFFICIENTS

In smooth pipe flow, the viscous sublayer near the wall is very thin at high Reynolds numbers. The introduction of roughness elements limits the thinning of the viscous sublayer. Roughness elements are protrusions or interstices on the pipe walls. Standard pipe, even when new, has many imperfections and is considered to be quite rough. Drawn tubing, when new, and plastic pipe generally are smooth-walled. As fluid flows past these roughness elements, eddies are formed behind them, causing "form drag."

The height of a single unit of roughness is denoted by ϵ' and is referred to as the *roughness parameter*. From dimensional analysis, the friction factor, f, is both a function of the Reynolds and the *relative roughness*, ϵ'/D, where D is the pipe diameter. A different relationship between the friction factor, f, and Re is obtained for each magnitude of the relative roughness.

Studies by Nikuradse [26, 27], Rouse [28] and Moody [29] have led to the development of generalized friction factor charts.

Moody [29] has given a complete friction factor plot for smooth and rough pipes (a log-log plot summarizing experimental data on fricition factor as a function of Reynolds number and relative roughness), which is given in Figure 6. Moody also developed a chart giving the relative roughness of commercial pipe as a function of the diameter for different pipe materials. A portion of this plot is given in Figure 7.

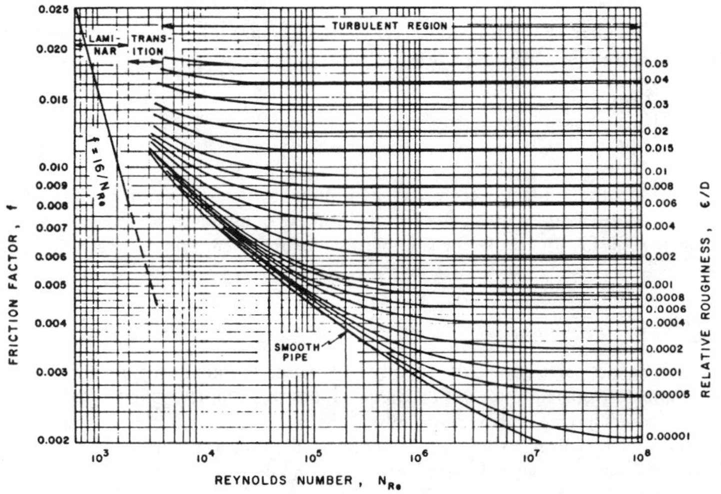

Figure 6. Moody chart for obtaining friction factors [29].

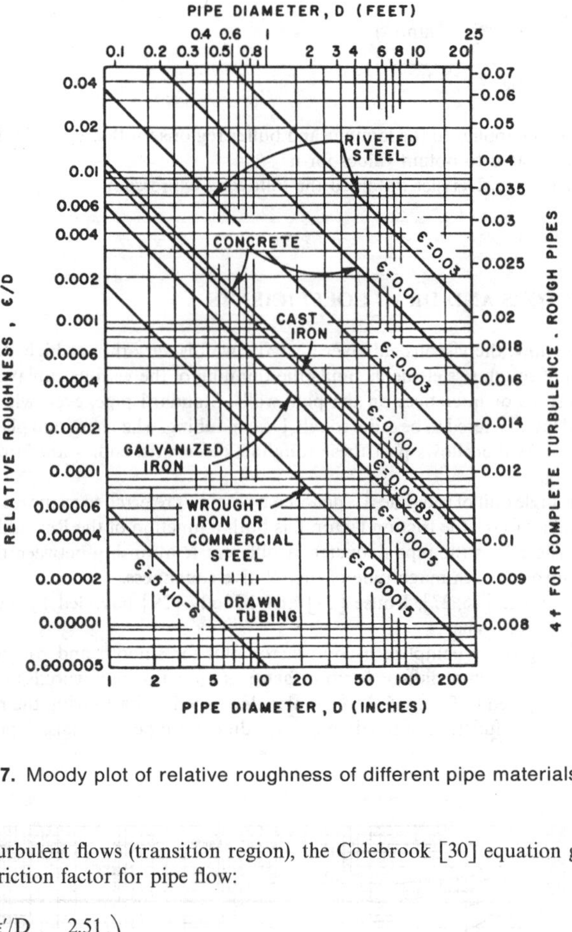

Figure 7. Moody plot of relative roughness of different pipe materials [29].

For partially turbulent flows (transition region), the Colebrook [30] equation gives reasonable estimates of the friction factor for pipe flow:

$$\frac{1}{\sqrt{f}} = 2 \log_{10}\left(\frac{\epsilon'/D}{3.7} + \frac{2.51}{Re\sqrt{f}}\right) \tag{80}$$

Table 4 lists various friction factor correlations for pipe flow and their range of applicability.

Few studies have been aimed at friction factors for flows in non-circular conduits. Knudsen et al. [1] note that the Blasius friction factor is applicable for turbulent flow for Reynolds numbers up to 100,000:

$$f_B = 0.079\,Re^{-1/4} \tag{81}$$

In lieu of more exact correlations, the Moody plot is used to estimate f from a Reynolds number based on a characteristic length term. For circular tubes this characteristic length is the diameter; however, for conduits having cross sections other than circular, an equivalent diameter is used. The equivalent diameter is defined as four times the hydraulic radius, which is the ratio of the cross-sectional area of flow to the wetted perimeter of the conduit:

$$D_E = 4R_H = 4\,\frac{\text{cross section of flow}}{\text{wetted perimeter}} \tag{82}$$

Table 4
Friction Factor Equations

Type of Flow	Equation Giving f	Range of Application
Laminar	$f = 64/N_{Re}$	$N_{Re} < 2{,}100$
Hydraulically Smooth or Turbulent Smooth	$f = 0.316/N_{Re}^{0.25}$	$4{,}000 < N_{Re} < 10^5$
	$\frac{1}{\sqrt{f}} = 2\log_{10}(N_{Re}\sqrt{f}) - 0.08$	$N_{Re} > 4{,}000$
Transition between Hydraulically Smooth and Wholly Rough	$\frac{1}{\sqrt{f}} = 2\log_{10}\left(\frac{\epsilon/D}{3.7} + \frac{2.52}{N_{Re}\sqrt{f}}\right) = 1.14 - 2\log_{10}\left(\frac{\epsilon}{D} + \frac{9.35}{N_{Re}\sqrt{f}}\right)$	$N_{Re} > 4{,}000$
Hydraulically Rough or Turbulent Rough	$\frac{1}{\sqrt{f}} = 1.14 - 2\log_{10}(\epsilon/D) = 1.14 + 2\log_{10}(D/\epsilon)$	$N_{Re} > 4{,}000$

This definition is suitable for conduits of constant cross section. There are, however, situations in which the conduit cross section varies periodically over the system. For these cases, Walker et al. [2] proposed the following definitions for the hydraulic radius:

$$(R_H)_{max} = \frac{\text{maximum cross section}}{\text{wetted perimeter}} \tag{83}$$

$$(R_H)_{min} = \frac{\text{minimum cross section}}{\text{wetted perimeter}} \tag{84}$$

$$(R_H)_v = \frac{\text{volume of free space per unit length}}{\text{area of wetted surface per unit length}} \tag{85}$$

$(R_H)_v$ is defined as the volumetric hydraulic radius, and the equivalent diameter is defined as before.

When the flow cross section changes, the fluid's velocity changes either in direction or magnitude. This results in additional friction over that formed in straight pipe flow. This additional friction includes form friction or drag, which produces vortices that develop during boundary layer separation. Figure 8 illustrates several situations in which this occurs.

Figure 8A shows that when the cross section of a conduit is enlarged suddenly, the fluid stream separates from the wall and flows into the enlarged section as a jet. This fluid jet expands, filling the entire cross section of the larger conduit. The region between the expanding jet and the wall is filled with fluid in vortex motion, which is a characteristic of boundary layer separation. There is considerable friction generated in this turbulent region.

Figure 8B shows the effect of suddenly reducing the conduit cross section. The fluid stream cannot flow around the sharp edge, causing it to break contact with the conduit wall. This situation also results in a jet that flows into the stagnant fluid in the smaller section. The jet first undergoes contraction and then expands, filling the smaller cross section. Downstream from the point of contraction, the normal velocity distribution is reestablished. The cross section of minimum flow area where the fluid jet changes from a contracted to an expanded stream is referred to as the *vena contracta*.

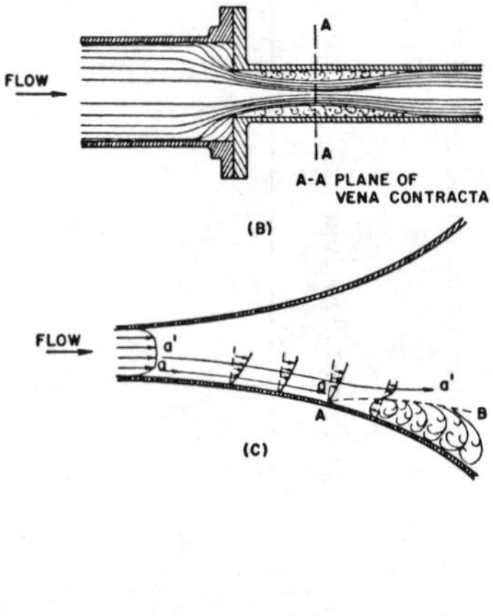

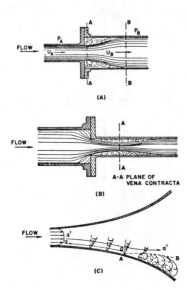

Figure 8. Illustrates flow situations resulting in boundary layer separation: (A) Sudden enlargement of conduit cross section; (B) sudden contraction of cross-section; (C) gradual enlargement of crosssection.

The last example shown in Figure 8 (sketch C) shows boundary separation in a gradually diverging channel. In this situation, because of the increase of cross section in the flow direction the velocity of the fluid decreases. In accordance with the Bernoulli equation, a decrease in fluid velocity must be accompanied by an increase in pressure. Visualize two fluid streamlines as shown in this sketch; a close to the wall, and a′ a′ at some short distance from the wall. The rise in pressure over a specified length of channel must be the same for both streamlines (the pressure across any flow area is uniform). This means that the loss in velocity head is the same for both filaments. The initial velocity head of a′ a′ is greater than that of aa, since aa is closer to the wall. At a specific point downstream, the velocity of streamline aa becomes zero, however, the velocities of other streamlines farther from the wall still remain positive. This point is denoted by A in Figure 8C. As shown, beyond point A the velocity nearest the wall will change sign, and a backflow situation is created between aa and the wall. In this region, we then see a separation of the boundary layer. Point A is referred to as the *separation point.* The dashed line AB denotes the edge of this fluid backflow region where the fluid velocity changes sign and is referred to as the *line of zero tangential velocity.* Again, the vortices formed between the wall and the separated fluid stream beyond the separation point result in large form-frictional losses.

Frictional losses from sudden expansions and contractions as illustrated in Figure 8 are not readily estimated and require the use of empirical correlations. Losses in sudden expansions are attributed mainly to turbulence generated by the impact of the high-velocity fluid stream into a region of slowly moving fluid downstream. This means that the energy dissipated depends on the difference in initial and final values of fluid velocity. Although difficult to justify from a theoretical basis, frictional pressure losses for sudden contractions and expansions have been shown to follow an equation of the following form:

$$\Delta P_f/\rho = \frac{KU^2}{2g_c} \tag{86}$$

Equation 86 is an approximation of the friction loss due to sudden enlargements and contractions (in English units $\Delta P_f/\rho$ is ft-lb_f/lb_m). Parameter K describes the fraction of the final kinetic energy content that is dissipated and is a function of the cross-section change. K, called the resistance coefficient, is shown plotted in Figure 9 as a function of the ratio of cross-section changes for sudden enlargements and contractions. If sections are noncircular in shape, equivalent diameters may be used in obtaining values of K. Note that the general relation given by Equation 86 is largely empirical

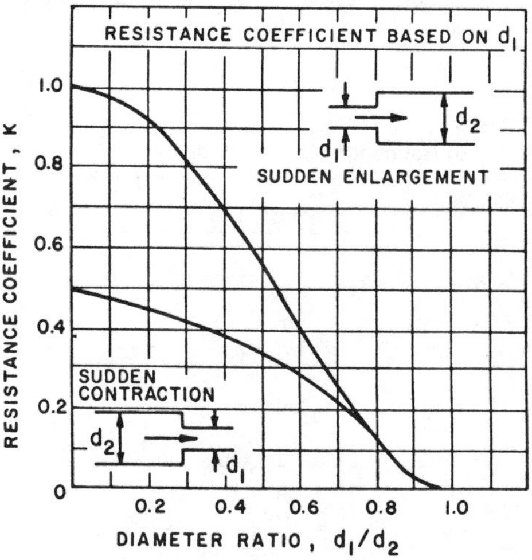

Figure 9. Resistance coefficients for cross-sectional changes [31].

and is known to be most reliable when conditions are decidedly turbulent. In actual use, Equation 86 is combined with the Bernoulli equation:

$$-\int_1^2 v\,dP = \frac{U_2^2}{2g} - \frac{U_1^2}{2g} + K\frac{U_2^2}{2g} \tag{87}$$

Flow Through Bends and Fittings

The effects of bends and fittings such as elbows, valves, tees, etc., are evaluated empirically through a fictitious *equivalent length* of straight pipe having the same diameter and that would develop the same frictional pressure loss. This equivalent length is added to the length of actual straight pipe whence the total friction loss can be computed from the Fanning pressure loss equation. That is,

$$L = \text{length of straight pipe} + L_e \tag{88}$$

where L_e is the equivalent length of all fittings whence the total frictional pressure loss may be computed from

$$(\Delta P)_f = \frac{1}{144}\left(\frac{4fL}{D_E}\right)\left(\frac{\rho U^2}{2g_c}\right) \tag{89}$$

where $(\Delta P)_f$ = frictional pressure loss (psi)
L = total pipe length (ft)
D_E = equivalent diameter (ft)

Typical values for the ratio of equivalent length to nominal diameter for standard fittings are given in Table 5.

Equivalent lengths also have been correlated in terms of resistance coefficients with the following formula:

$$L_e = \left(\frac{D_E}{4f}\right)\sum K \tag{90}$$

where $\sum K$ is the sum of the resistance coefficients of all fittings.

Typical values for K for bends, ells, and tees can be obtained from the plots given in Figure 10. The methods outlined in the subsection are best applied to turbulent flow conditions.

Table 5
Equivalent Length to Diameter (L_{eg}/D) Ratios for Standard Fittings

	L_{eq}/D
Globe Valves (in).	
1–2.5	45
3–6	60
7–10	75
Tees	
1–4	60
90° Elbows	
1–2.5	30
3–6	40
7–10	50

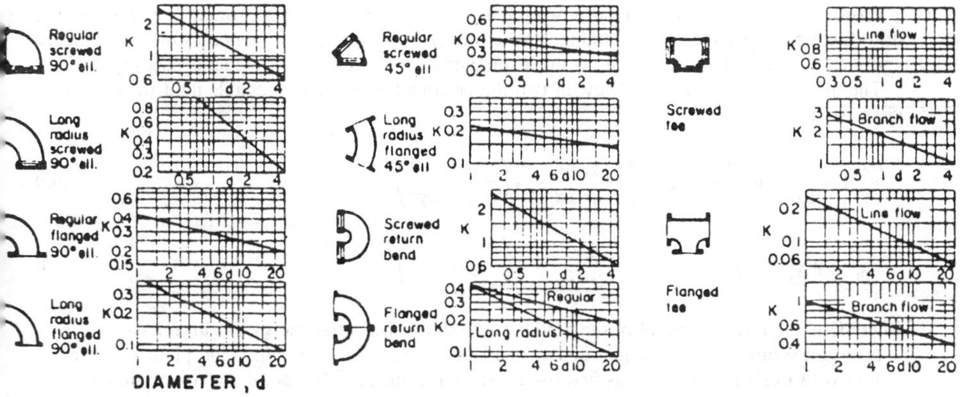

Figure 10. Resistance coefficients for bends, ells, and tees [32].

Flow Over Flat Surfaces

Flow around solid objects is of interest to a number of process operations. In tubular heat exchangers, for example, flow occurs parallel and perpendicular to tube bundles. Heat transfer from extended surfaces depends on the nature of the flow past these surfaces. A fluidized bed is another example in which the flow of fluid past solid particles governs both heat and mass transfer mechanisms. To understand the complex flow mechanisms for these situations, considerable analytical studies have been devoted to modeling and describing simple flow situations. Some of the conclusions arrived at in examining simple flow situations, such as flow over a flat plate, may be extended to more complex flow situations.

Drag is the resistance to movement of a solid in a fluid. It is caused by the shear stresses exerted in the boundary layer of the fluid adjacent to the solid surface. For a completely laminar boundary layer, shear forces are viscous only. If the boundary layer is turbulent, then the resistance results from velocity fluctuations of the fluid within the boundary layer. There are basically two types of drag. First, if the resistance is caused by stresses in the boundary layer alone, as in the case of flow over a flat surface, the resistance is referred to as surface drag. However, if the flow is around a two-dimensional object, such as a cylinder, separation of the boundary layer occurs and the fluid immediately behind the solid body becomes turbulent. This type of drag caused by the turbulent wake is called form drag and depends on the geometry of the body past which the fluid flows.

The drag coefficient is defined as follows:

$$F = \frac{f \rho U^2 A_w}{2 g_c} \tag{91}$$

or

$$F = \frac{f_D \rho U^2 A_p}{2 g_c} \tag{92}$$

F is the force of resistance exerted on the solid by the flowing fluid. In Equation 91, A_w is the wetted area of the immersed body. A_p in Equation 92 is the projected area of the body on the plane normal to the flow direction. Since the drag coefficient f in Equation 91 is based on the wetted area, it is mainly applicable for flat plates or thin streamlined struts. The coefficient f_D (Equation 92) is based on the projected area of the body and therefore applies to cylinders and bodies of revolution. Hence, when Equation 92 is used to compute the force, it includes both surface drag and form drag.

Consider an incompressible fluid flowing past a thin, flat plate, which is oriented parallel to the flow direction. The length of the laminar boundary layer will depend on the main stream turbulence and mean velocity, the distance from the leading edge, and the fluid's kinematic viscosity.

The momentum equations of flow in two dimensions for an incompressible fluid with constant viscosity are

$$\frac{\partial u}{\partial t} + u\frac{\partial u}{\partial x} + u\frac{\partial u}{\partial y} = -g_c\frac{\partial \Omega}{\partial x} - \frac{g_c}{\rho}\frac{\partial P}{\partial x} + v\left(\frac{\partial^2 u}{\partial x^2} + \frac{\partial^2 u}{\partial y^2}\right) \tag{93}$$

$$\frac{\partial v}{\partial t} + u\frac{\partial v}{\partial x} + v\frac{\partial v}{\partial y} = -g_c\frac{\partial \Omega}{\partial y} - \frac{g_c}{\rho}\frac{\partial P}{\partial y} + v\left(\frac{\partial^2 u}{\partial x^2} + \frac{\partial^2 v}{\partial y^2}\right) \tag{94}$$

where Ω is the force potential of the fluid, i.e., the fluid's potential energy per unit mass.

An exact solution to these equations for the laminar boundary layer on a flat plate immersed in a fluid with uniform velocity was first obtained by Blasius [33] by use of the stream function, Ψ:

$$\frac{\partial \Psi}{\partial y} = u \tag{95}$$

$$\frac{\partial \Psi}{\partial x} = -v \tag{96}$$

Equations 95 and 96 satisfy the two-dimensional continuity equation ($\partial u/\partial x + \partial v/\partial y = 0$).

Defining U as the velocity at the edge of the boundary layer and noting that $\partial U/\partial x = 0$, the momentum expression over the entire thickness of the boundary layer becomes

$$u\frac{\partial u}{\partial x} + v\frac{\partial u}{\partial y} = \nu\frac{\partial^2 u}{\partial y^2} \tag{97}$$

Blasius reduced Equation 97 to an ordinary differential equation by introducing the variables η and ϕ:

$$\eta = y/2\left(\frac{U\rho}{\mu x}\right)^{1/2} \tag{98}$$

$$\Psi = \left(\frac{\mu U x}{\rho}\right)^{1/2}\phi \tag{99}$$

The solution obtained by Blasius is in the form of a Taylor series solution and is expressed in terms of one of the constants of integration:

$$\phi = \frac{C_2\eta^2}{2!} - \frac{C_2^2\eta^5}{5!} + \frac{11C_2^3\eta^8}{8!} - \frac{375C_2^4\eta''}{11!} + \cdots \tag{100}$$

where $C_2 = 1.32824$. The derivation is given in detail by Cheremisinoff [20].

Howarth [34] computed values for ϕ, ϕ' and ϕ''' as function of η. Table 6 gives selected values computed by Howarth along with values of u/U. Using these values, flow conditions at any point in the x-y coordinate system may be obtained.

von Karman [35] analyzed the flow in the boundary layer using Newton's second law to develop an integral relationship for the velocity distribution. His analysis considered a two-dimensional region of fluid, which included the boundary layer of differential length, dx. It can be shown that in this region the total rate of momentum increase is equal in both magnitude and direction to the forces acting on the boundaries of the region.

Table 6
Computed Values for Blasius Parameters [34]

η	ϕ	ϕ'	ϕ''	u/U
0	0	0	1.32824	0
0.4	0.1061	0.5294	1.3096	0.2647
0.8	0.4203	1.0336	1.1867	0.5168
1.2	0.9223	1.4580	0.9124	0.7290
1.6	1.5691	1.7522	0.5565	0.8761
2.0	2.3058	1.9110	0.2570	0.9555
2.4	3.0853	1.9756	0.0875	0.9878
2.8	3.8803	1.9950	0.0217	0.9915
3.2	4.6794	1.9992	0.0039	0.9996
3.6	5.4793	2.0000	0.0005	1.0000
3.8	5.8792	2.0000	0.0002	1.0000

For the constant U, von Karman's integral momentum equation for the boundary layer on a flat plate is:

$$\frac{\tau_w g_c}{\rho} = \frac{\partial}{\partial x}\int_0^\delta u(U - u)\,dy \tag{101}$$

The boundary layer thickness, δ, is the distance normal from the wall where the point velocity is within 1% of the mainstream velocity. That is, δ is the y value where $u/U = 0.99$. Using the values given in Table 6, it can be shown at:

$$\frac{\delta}{x} = \frac{4.96}{\sqrt{Re_x}} \tag{102}$$

where Re_x is the local Reynolds number.

Pohlhausen [36] modified von Karman's integral equation (Equation 101) to obtain the following expressions for the laminar boundary layer thickness and velocity profile:

$$\delta = 4.64(\nu x/U)^{1/2} \tag{103}$$

$$\frac{u}{U} = \frac{1.5}{4.64}\frac{y}{\sqrt{\nu x/U}} - \frac{y^3}{199.8(\sqrt{\nu x/U})^3} \tag{104}$$

For laminar flow, the Blasius solution predicts the following relation for the total drag coefficient for a plate of length, L:

$$f = \frac{1.328}{\sqrt{Re}}, \text{ for } 2 \times 10^4 < Re_L < 5 \times 10^5 \tag{105}$$

where Re_L is the total Reynolds number for flow over the flat plate ($=UL/\nu$).

At low Reynolds numbers ($10 < Re < 3{,}000$), the following empirical equation may be used [37]:

$$f = 2.90(Re)^{-0.60} \tag{106}$$

Using the analogy for flow through circular tubes, von Karman [38] derived the form of the velocity-profile equation describing the turbulent portion of the boundary layer. For any point, x, on the plate, the relation between point velocity, u, and distance normal from the plate, y, is:

$$\frac{u}{U} = (y/\delta)^{1/7} \tag{107}$$

Equation 107 is the power law for the velocity distribution over flat plates.

The turbulent boundary layer thickness may be computed from the Blasius friction factor equation for circular tubes. Following von Karman [38], the friction factor equation is used in expressing the shear stress at the plate's boundary:

$$\tau_w = 0.0228 \frac{\rho U^2}{g_c}\left(\frac{\nu}{U\delta}\right)^{1/4} \tag{108}$$

Combining this expression with von Karman's integral equation (Equation 101),

$$\frac{\partial}{\partial x}\int_0^\delta u(U - u)\, dy = 0.0228U^2\left(\frac{\nu}{U\delta}\right)^{1/4} \tag{109}$$

Using the $\frac{1}{7}$th power law relation (Equation 107) for obtaining a value of u and substituting into Equation 109, integration can be performed:

$$\frac{\delta}{x} = 0.376(Re)^{-1/5}, \text{ for } 5 \times 10^5 < Re < 10^7 \tag{110}$$

For local Reynolds numbers above 6.5×10^5, the following relation may be used [41]:

$$\frac{\delta}{x} = 0.1285(Re)^{-1/7} \tag{111}$$

By substituting Equation 110 in Equation 108, it can also be shown that:

$$\frac{2\tau_w g_c}{\rho U^2} = 0.0585(Re)^{-1/5} \tag{112}$$

$$f' = 0.0585(Re)^{-1/5} \tag{113}$$

The last expression gives the local drag coefficient for the turbulent boundary layer. The total friction factor is obtained as outlined earlier:

$$f = 0.074(Re)^{-1/5}, \text{ for } 5 \times 10^5 < Re < 10^7 \tag{114}$$

Several other expressions for local and total drag coefficients have been derived for turbulent boundary layers on flat plates. These have been summarized by Cheremisinoff [20, 39, 40].

All drag coefficients given thus far in this chapter describe flow over a smooth plate. Surface roughness for flat plates is described in terms of the *admissible roughness*. This is the roughness above which an increase in the drag coefficient occurs over that for a smooth plate. The following expression can be used to determine admissible roughness:

$$\frac{U\epsilon_{ad}}{\nu} = 10^2 \tag{115}$$

where ϵ_{ad} is the admissible height of roughness projections.

For flat plates having a roughness below the admissible roughness, any of the smooth plate expressions are acceptable for estimating the drag coefficient. The following equation can be used for computing f for $U\epsilon_{ad}/\nu > 200$:

$$f = (1.89 + 1.62 \log_{10} L/\epsilon_{ad})^{-2.5} \tag{116a}$$

where $10^2 < L/\epsilon_{ad} < 10^6$.

The thickness of the turbulent boundary layer on the surface of a rough plate can be computed from the following expression [41–43].

$$\delta = 0.259\left(\frac{x}{\epsilon_{ad}}\right)^{2/3} \epsilon_{ad}(1 - 0.00059x) \tag{116b}$$

where quantities x, ϵ_{ad} and δ are in units of cm.

Flow Past Two- and Three-Dimensional Objects

Now consider fluid flowing past a two-dimensional object such as a cylinder. The fluid is accelerated as it passes over the front portion of the object and then decelerated after it passes the thickest portion of the object. This sequence of events causes a separation of the boundary layer.

The boundary layer undergoes separation at a point on the object where the pressure gradient is zero. The reason for this is as follows: The boundary layer thickness increases with distance in the flow direction. The portion of fluid in the mainstream accelerates because of its movement around the object. This acceleration is actually an increase in the fluid's kinetic energy, which must be accompanied by a decrease in the pressure, i.e., $\partial P/\partial x$ becomes negative. As the mainstream fluid travels past the object, the expanding cross section of flow requires a deceleration of the fluid and, hence, a corresponding increase in the pressure ($\partial P/\partial x$ becomes positive). Hence, the boundary layer actually flows against an adverse pressure gradient as it moves around the object, resulting in a distortion in the velocity profile in the boundary layer. To maintain flow in the direction of the adverse pressure gradient, the boundary layer must separate from the solid surface, and this point of separation is actually a stagnation point (that is, the velocity gradient $\partial u/\partial y$ at the surface becomes zero). Hence, at any point on the surface of the object prior to separation u = 0 and v = 0. Prandtl's [44] momentum equations for the boundary layer are:

$$u\frac{\partial u}{\partial x} + v\frac{\partial u}{\partial y} = -\frac{g_c}{\rho x}\frac{\partial P}{\partial x} + \nu\frac{\partial^2 u}{\partial y^2} \tag{117}$$

$$\frac{\partial u}{\partial x} + \frac{\partial v}{\partial y} = 0 \tag{118}$$

whence it follows that

$$\frac{\partial P}{\partial x} = \frac{\mu}{g_c}\frac{\partial^2 u}{\partial y^2} \quad \text{at } y = 0 \tag{119}$$

The significance of Equation 119 is that it predicts a change in sign of $\partial P/\partial x$, the term $\partial^2 u/\partial y^2$ changes sign; thus, the velocity-profile curve exhibits a point of inflection.

Defining δ as the boundary layer thickness for fluid flowing over a circular cylinder, and U and P as the velocity and pressure for the undisturbed portion of the flowing stream, respectively, we can develop relationships for the velocity and pressure distributions. The outer edge of the boundary layer represents the point at which the velocity gradient becomes zero. The velocity at the edge of the boundary layer, u_{max} differs from the mainstream velocity and is a function of positions on the surface of the cylinder in relation to the leading edge:

$$u_{max} = 2U \sin \theta \tag{120}$$

where θ represents the angle of departure of the boundary layer from the cylinder.

Bernoulli's steady-flow equation may be applied between the main-stream and the edge of the boundary layer (the implication being that the main fluid stream is irrotational); thus,

$$P_\theta + \frac{\rho U_{max}^2}{2g_c} = P + \frac{\rho U^2}{2g_c} \tag{121}$$

where P_θ is the static pressure in the boundary layer and is constant across thickness δ.

Rearranging to solve for u_{max},

$$u_{max} = U\sqrt{\frac{P - P_\theta}{\sqrt{\rho U^2/2g_c}} + 1} \tag{122}$$

Equation 122 can be used to compute the velocity at the edge of the boundary layer from the knowledge of the pressure distribution over the cylinder. If the effect of fluid viscosity is negligible, the pressure distribution over the surface of the cylinder can be estimated from

$$\frac{P_\theta - P}{\rho U^2/2g_c} = 1 - 4\sin^2\theta \tag{123}$$

When a two-dimensional cylinder is immersed in a flowing stream, it has a considerably greater drag on it than the case of a flat plate. This is due to the separation of the boundary layers on two-dimensional objects and the subsequent formation of a turbulent wake behind the body.

Several investigators [45–48] have measured total drag coefficients past cylinders. Equation 124 is just one expression for the drag coefficient for flow past circular cylinders when no boundary separation occurs [45].

$$f_D = \frac{8\pi}{Re_0(2.002 - \ln Re_0)} \tag{124}$$

Equation 124 applies only for low Reynolds numbers ($Re_0 < 0.5$). Note that for fluids flowing past immersed cylinders, Reynolds numbers are expressed in terms of the cylinders outside diameter, D_0. One of the most extensive studies on the drag on cylinders, including noncircular cylinders, is that of Delany and Sorenson [48]. Their investigation covered cylindrical, elliptical, square, rectangular, and triangular cylinders.

Bodies of revolution of interest are spheres, ellipsoids, and disks. We define the radius in the plane normal to the flow direction as r_0 and the Reynolds number for the flow as D_0U/ν, where $D_0 = 2r_0$. The projected area of the body is πr_0^2, and the overall length of the body in the direction of flow is L.

When flow occurs past any of these objects, the fluid motion near the object's surface is three-dimensional with respect to a rectangular coordinate system. The general boundary layer equations for flow over bodies of revolution are

$$u\frac{\partial u}{\partial x_1} + v\frac{\partial u}{\partial y} = -\frac{g_c}{\rho}\frac{\partial P}{\partial x_1} + \nu\left(\frac{\partial^2 u}{\partial y^2} + \frac{1}{r}\frac{\partial r}{\partial y}\frac{\partial u}{\partial y}\right) \tag{125}$$

$$\frac{\partial (ru)}{\partial x_1} + \frac{\partial (rv)}{\partial y} = 0 \tag{126}$$

Equations 125 and 126 are the momentum and continuity equations, respectively. The terms in these equations are as follows:

x_1 = distance from the leading edge of the body

y = distance normal to the body surface

r = radius of transverse cross section at point (x_1, y)

The definitions of distance terms used in Equations 125 and 126 for flow past spheres are given in Figure 11. The radius of curvature at any point is r_c, and at the sphere's surface (i.e., y = 0) the transverse cross section of the body has a radius r_t. These definitions assume $y < r_c$ and that δ is small in comparison to r. These assumptions tend to break down in the region of the leading edge as r goes to 0.

Solutions to Equations 125 and 126 provide expressions for the boundary layer velocity profile, the boundary thickness and local drag coefficients. The following three-dimensional integral momentum equations were developed from Equations 125 and 126 [49]:

$$\frac{\partial}{\partial x_1}\int_0^\delta u^2\,dy - u_{max}\frac{\partial}{\partial x_1}\int_0^\delta u\,dy + \frac{1}{r_t}\frac{\partial r_t}{\partial x_1}\left(\int_0^\delta u^2\,dy - u_{max}\int_0^\delta u\,dy\right) = \delta u_{max}\frac{\partial u_{max}}{\partial x_1} - \frac{\tau_w g_c}{\rho} \tag{127}$$

To solve this expression, information is needed on the pressure distribution over the immersed sphere. The pressure distribution is given by the following expression:

$$\frac{P_\theta - P}{\rho U^2/2g_c} = 1 - \frac{\delta}{4}\sin^2\theta \tag{128}$$

(Note the resemblance to Equation 123).

where θ = angle measured from the forward stagnation point of the sphere
P_θ = static pressure at angle θ
P = static pressure of the undisturbed stream

It can be shown from Bernoulli's equation that:

$$\frac{u_{max}}{U} = 1.5\sin\theta \tag{129}$$

A number of empirical correlations are available for estimating u_{max} (the velocity at the edge of the boundary layer).

A simple approach used in solving Equation 127, and one that also has been used in solving the two-dimensional equation, is to assume a velocity profile. A simplified approach is to use a velocity profile empirically evaluated to a polynomial fit:

$$u = c_1 y + c_2 y^2 + c_3 y^3 + c_4 y^4 \tag{130}$$

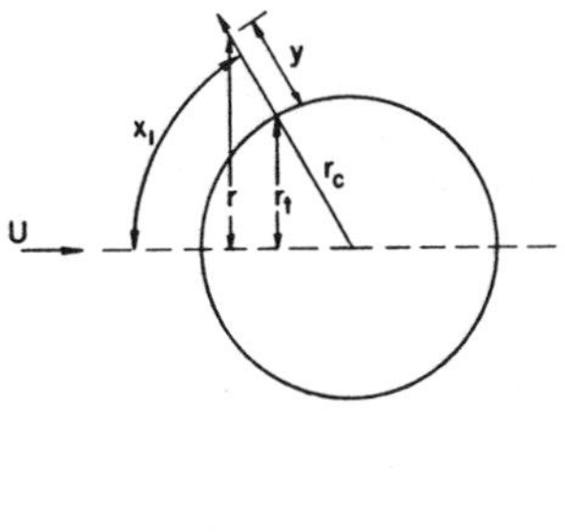

Figure 11. Defines distance terms for three-dimensional equations describing flow over a sphere.

Using this approach, Tomotika [50] obtained the following expression for the laminar boundary layer thickness:

$$\frac{U}{\nu r_0}\frac{\partial(\delta^2)}{\partial\theta} = \phi_1(\Delta)\frac{U}{u_{max}} - \phi_2(\Delta)\cot\theta\frac{U}{\partial u_{max}/\partial\theta} + \phi_3(\Delta)\frac{\delta^4}{U\nu^2}\frac{\partial^2 u_{max}}{\partial\theta^2} \tag{131}$$

where

$$\Delta = \frac{\delta^2}{r_0\nu}\frac{\partial u_{max}}{\partial\theta} \tag{132}$$

$$\phi_1(\Delta) = \frac{7{,}257.6 - 1{,}336.2\Delta + 37.92\Delta^2 + 0.8\Delta^3}{213.12 - 5.76\Delta - \Delta^2} \tag{133}$$

$$\phi_2(\Delta) = \frac{426.24\Delta - 3.84\Delta^2 - 0.4\Delta^3}{213.12 - 5.76\Delta - \Delta^2} \tag{134}$$

$$\phi_3(\Delta) = \frac{3.84 + 0.8\Delta}{213.12 - 5.76\Delta - \Delta^2} \tag{135}$$

A parabolic velocity profile also may be assumed:

$$\delta^2 = \frac{30\nu}{u_{max} 9 r_t^2}\int_0^{x_1} u_{max} 8 r_t^2\, dx_1 \tag{136}$$

To develop an expression for the drag coefficient of a sphere moving slowly through a viscous fluid, Stokes' law can be used to define the drag force:

$$F = \frac{3\pi D_0 \mu U}{g_c} \tag{137}$$

Equation 138 is based on the condition that the boundary layer does not separate from the sphere.

For the flow of a viscous fluid past a sphere, the following relation is applicable in computing the drag force:

$$F = \frac{3\pi D_0 \mu U}{g_c}\left(1 + \frac{3}{16}Re_0\right) \tag{138}$$

For $Re_0 < 1.0$, the drag coefficient from Stokes' law is:

$$f = \frac{24}{Re_0} \tag{139}$$

Combining the above relations, the total drag coefficient describing the motion of spheres (neglecting neighboring wall effects) is:

$$f_D = \frac{24}{Re_0}\left(1 + \frac{3}{16}Re_0\right) \tag{140}$$

Equation 140 is applicable in the range $Re < 2$. For drag coefficients for flow past a sphere contained in a cylinder of diameter D_w, use the following relation:

$$f_D = \frac{24}{Re_0}(1 + 2.4(D_0/D_w)) \tag{141}$$

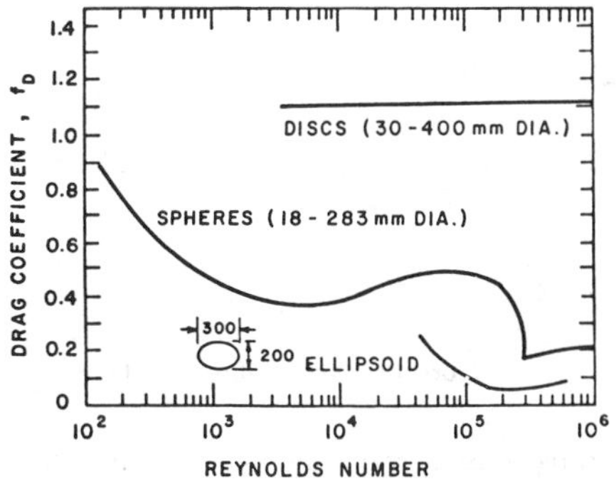

Figure 12. Plot of drag coefficient for flow past bodies of revolution at high Reynolds numbers (curves based on data of Wieselberger [46].

Equation 141 can be used for rough estimates of the total drag coefficient up to Reynolds numbers of about 100. To estimate drag coefficients at Reynolds numbers outside the range of Equations 140 and 141, the data of Wieselsberger [46] for spheres, disks and ellipsoids may be used. A summary of these data is given in Figure 12. It should be noted that there is a distinct effect of aspect ratio on the drag coefficient, where the aspect ratio is defined as the diameter: length ratio of the body. For ellipsoids, for example, the drag coefficient tends to decrease with values of the aspect ratio. As $2r_0/L$ decreases, the object becomes more streamlined and, hence, the resistance to the flow becomes less.

FLOW OF NON-NEWTONIAN FLUIDS

Definitions and Fluid Classification

Many materials exhibit both fluid- and solid-like properties. For example, certain materials will not deform continuously under the action of any non-zero shearing stress, but only if a certain "yield" stress is exceeded. Other materials exist that deform continuously, but when the stress is removed they will partially recover their original configuration. A fluid can be considered to be a form of matter that exhibits continuous deformation under some range of shearing stress and may partially recover its original configuration when the stress is removed.

Rheology is the study of the deformation and flow of materials [51–54]. Rheologists are concerned with both the measurement of the deformation of a fluid with stress and the formulation of relations between deformation rates and stress (τ). These relations, called rheological models, or "constitutive equations," can be used in the momentum balance equation to compute for a given laminar flow situation, velocity profiles, volumetric flow rates, etc. While many such equations have been formulated, their use often entails the evaluation of multiple parameters from data, which is difficult in practice. Further, the complexity of many of these equations precludes their use for engineering design. Unfortunately, parameters evaluated from measurements under steady shear often cannot represent data from, say, oscillatory shear. For these reasons, only very simple constitutive equations are in use for engineering purposes. These simple rheological models are limited in their ability to represent all aspects of fluid behavior and are suitable only under restricted conditions.

Fluid deformation is often described by the example shown in Figure 13, where a fluid held between two large parallel plates is separated by a small gap, Y. The lower plate is stationary while the upper plate is moved at a constant velocity, V, through the action of a force, F. A thin layer of fluid adjacent to each plate will move at the same velocity as the plate, (i.e., "no slip" at the solid boundary). Molecules in fluid layers between these two extremes will move at intermediate velocities. For example, a fluid layer, B, immediately below A will experience a force in the x-direction from layer A and a smaller retarding force from layer C. Layer B then will flow at a velocity lower than V. This progression will continue to the layer adjacent to the lower plate, whose

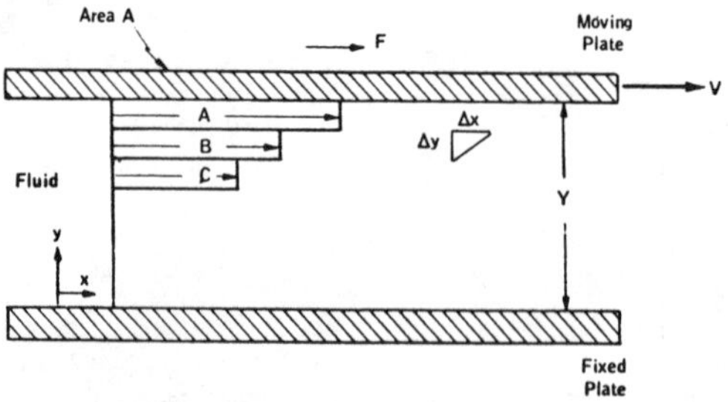

Figure 13. Deformation of a fluid.

velocity will be zero. Under steady-state conditions, the force, F, required to produce the motion becomes constant and is related to velocity

$$\frac{F}{A} = f(V/Y) \tag{142}$$

where A is the area of each plate, and f( ) is a function depending on the fluid in question and the temperature and pressure.

F/A = shear stress on the fluid at the upper plate, and V/Y = velocity gradient. At any point within a fluid, the expression is generalized as

$$\tau_{yx} = f\left(\frac{dv_x}{dy}\right) \tag{143}$$

where τ_{yx} = local shear stress (force per unit area) acting in the x-direction on a plane perpendicular to the y axis, v_x = velocity in the x-direction, and dv_x/dy = velocity gradient on a local basis. In graphic form Equation 143 is referred to as the flow curve, or rheogram for the fluid.

Note that the velocity gradient is

$$\frac{dv_x}{dy} = \frac{d}{dy}\left(\frac{dx}{dt}\right) = \frac{d}{dt}\left(\frac{dx}{dy}\right) \approx \frac{d}{dt}\left(\frac{\Delta x}{\Delta y}\right) \tag{144}$$

The term $\Delta x/\Delta y$ is the shear strain on the fluid, and therefore, dv_x/dy is the rate of shear strain, or simply shear rate. Thus, the shear stress is a function of the rate of shear strain, or shear rate. (For solids, on the other hand, the shear stress would be a function of strain rather than the rate of strain). The shear rate is sometimes called the rate of deformation.

For a homogeneous fluid containing small molecules, the constitutive equation (Equation 143) is usually simple, but for a multiphase mixture, a solution or a liquid containing large molecules, complex relations result. In the first case, the shear stress-shear rate relation is often linear through the origin, (i.e., fluids are Newtonian). For such fluids, the internal structure is unaffected by the magnitude of the imposed shear rate. For complex fluids, an imposed shear results in changes in the internal structure of the fluid so that the constitutive relation becomes more complex. For example, in a liquid containing large molecules (a polymer solution) at low shear rates, the molecules would remain randomly coiled, much as they are in the fluid at rest. The fluid structure would remain unchanged in this range, and the shear stress-shear rate relation would be linear, as for a Newtonian fluid. As the shear rate is increased, the randomly coiled molecules would tend to line up in the flow direction, changing the structure of the fluid and, hence, the nature of the constitutive relation. The progressive lining-up or disentangling of the molecules results in the fluid becoming less viscous, or "thinner," in its flow properties. Increasing the shear further would continue to cause more molecules to line up, so that the stress-shear rate relation would remain nonlinear, until a limiting shear is reached when all the molecules have relined. Further shear rate increases do not result in structural changes and the constitutive relation would return to a Newtonian relation, albeit a different one from that for low shear rates. Thus, the progression is from Newtonian to non-Newtonian and back to Newtonian as the shear rate is increased. Structural changes also occur for suspensions of solid particles in a liquid, or for liquid, or for liquid–liquid emulsions, resulting again in complex flow curves for such fluids.

Note that the system described in Figure 13 does not reflect time-dependent properties nor solid-like or elastic behavior. However, if the force moving the upper plate were removed, the upper plate would continue to move, but with decreasing velocity until it and the fluid eventually came to a stop. If the fluid were viscoelastic (that is, possessing both viscous and elastic properties), the plate and the fluid would first slow down as before, but after coming to a stop some motion in the negative x-direction would occur as the fluid sought to recover its original configuration. Only partial recovery would be attained.

Next, consider what happens if the shear rate on a fluid changed instantaneously. For many fluids, any resulting changes in internal structure occur very rapidly, and a new stress level corresponding to the new shear rate is reached instantaneously following Equation 143. Such fluids are *time-independent*. For *time-dependent* fluids structural changes are much slower, and the shear stress changes slowly until ultimately a steady value is reached corresponding to the new shear rate. Finally, some fluids do not deform until a certain "yield" stress value is exceeded. In our experiment, F/A would have to be greater than the yield stress for flow to occur for such a fluid.

It is convenient to divide fluids into three classes in terms of their flow behavior: time-independent purely viscous fluids, time-dependent purely viscous fluids, and viscoelastic fluids. Of these three classes, the time-independent purely viscous fluids are best understood. Time-dependent fluids are less common, and perhaps the least is known about these fluids.

We shall first discuss time-independent fluids. As noted, the simplest class of real fluids is comprised of Newtonians, where

$$\tau = \mu\delta \tag{145}$$

Equation 145 is Newton's law of viscosity. Viscosity, μ, is a fluid property and is a function of temperature and pressure only (for a single-phase, single-component system). The viscosity of liquids decreases, and that of gases increases, with temperature. An ideal or perfect fluid is one whose viscosity is zero, and thus can be sheared without the application of a shear stress.

The flow curve or rheogram of a Newtonian fluid is a straight line through the origin in rectangular coordinates. The slope of the line is the viscosity so that the entire class of Newtonian fluids can be represented by a family of straight lines through the origin. It is often convenient to plot the rheogram using logarithmic coordinates, as it allows a greater range of data and also permits easy comparison of Newtonian versus non-Newtonian behavior. On such coordinates, a Newtonian fluid has a rheogram that is a straight line of unit slope and whose intercept at a shear rate of unit is the viscosity. Figure 14 illustrates these alternative methods of representation. Most low-molecular-weight liquids and solutions, and all gases, display Newtonian behavior. Homogeneous slurries of small spherical particles in gases or liquids at low solids concentration are also frequently Newtonian. There is thus a large class of fluids for which a single property, the viscosity, characterizes the flow behavior.

For non-Newtonian fluids it is convenient to define an "apparent viscosity," η_a,

$$\eta_a = \tau\delta \tag{146}$$

The apparent viscosity is a function of δ for non-Newtonian materials and is analogous to the Newtonian viscosity, μ. However, whereas the Newtonian viscosity does not vary with shear rate, the non-Newtonian apparent viscosity does.

An important class of non-Newtonian fluids is the *pseudoplastic or shear-thinning* fluid. These have the property of a decreasing apparent viscosity with decreasing shear rate. In other words, the rate of increase of shear stress for such fluids decreases with increased shear rate. Increased shear breaks down the internal structure within the fluid very rapidly and reversibly, and no time-dependence is manifested. Examples of fluids that exhibit shear-thinning are polymer melts and solutions, mayonnaise, suspensions such as paint and paper pulp, and some dilute suspensions of

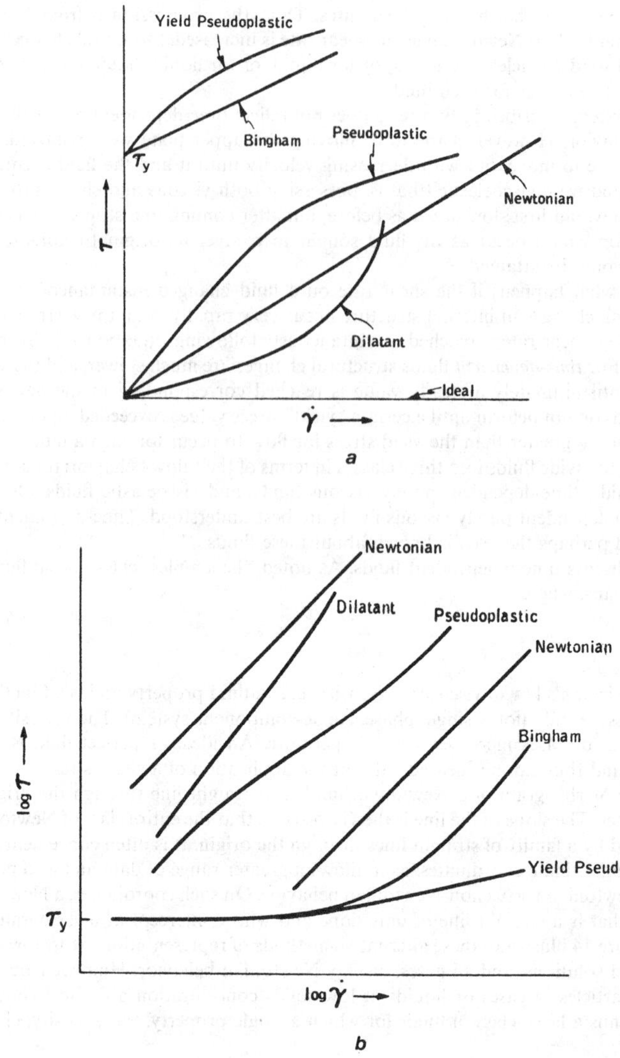

Figure 14. (A) Rheograms on arithmetic coordinates, (B) Rheograms on logarithms coordinates.

inert particles. Many of these fluids also exhibit other non-Newtonian characteristics, such as viscoelasticity in the case of polymer solutions and melts, and time-dependence in the case of paints. Thus, pseudoplasticity is but one important characteristic of such a non-Newtonian fluid and does not necessarily describe all of its features. Many pseudoplastics are shear-thinning at intermediate shear rates and are Newtonian at low and high shear rates. Rheograms for pseudoplastics are illustrated in Figure 14.

Many constitutive equations of varying complexity have been proposed for these fluids. Because of its simplicity, the power law is the most widely used, even though it does not describe the Newtonian extremes found for most pseudoplastics.

The power law, or Ostwald-de Waele equation is

$$\tau = K\delta^{n} \tag{147}$$

Parameter K is the *consistency index* and n is the *flow behavior index*. Both are functions of temperature and pressure, but K is more sensitive to temperature than n. Pressure dependence of these parameters has not been investigated.

It is clear that if $n = 1$, Equation 147 reduces to Newton's law. For $n < 1$, pseudoplastic or shear-thinning behavior is observed, whereas if $n > 1$, dilatant or shear-thickening behavior is shown. The apparent viscosity for a power law fluid is

$$\eta_a \equiv \tau/\delta = K\delta^{n-1} \tag{148a}$$

showing that if $n < 1$, η_a decreases with δ. The value of n is the slope of the rheogram on logarithmic coordinates.

The fit of the power law equation to experimental data is usually good over several orders of magnitude in δ. However, at low and high δ Newtonian behavior is obtained and agreement is poor. In spite of these limitations, use of the power law gives good predictions of pipeline pressure drop if the parameters are fitted over the same range of shear encountered in the wall region of the pipe.

Note that Equation 147 is valid only for $\delta > 0$. Depending on the coordinate system used to solve a flow problem, δ may be negative over some portion of the flow field, in which case the correct form of Equation 147 is [55–57]:

$$\tau = K|\delta|^{n-1}\delta \tag{148b}$$

Many other equations have been proposed for pseudoplastics, but they have not been used as widely as the power law. Also, they are generally more complex, and some involve three or more parameters, rather than only two. The principal advantage of these more complex models is that they can predict shear-thinning behavior as well as the tendency to Newtonian behavior at one or both extremes of shear rate. Since these models are not in wide use, we will not discuss them in detail, but rather refer to Table 7, where a few are listed along with their main features. Further details may be found in the literature [52, 57, 59].

Following the classes of fluids shown by the rheograms in Figure 14, we direct attention to *dilatant or shear-thickening fluids* and fluids with a yield value. With dilatant fluids apparent viscosity increases with shear rate. The rate of increase of shear stress for such fluids increases with shear rate. By using appropriate values of the parameters, equations developed for pseudoplastics can be applied to dilatant fluids. For example, the power law equation may be used with n greater than unity for dilatant fluids. Dilatancy is not as common as pseudoplasticity and generally is observed for fairly concentrated suspensions of irregular particles in liquids. A commonly accepted mechanism [60] for dilatancy is that at low shear rates the particles in such fluids are densely packed but still surrounded by liquid, which lubricates the motion of adjacent particles. At higher shear rates the dense packing breaks up progressively, forcing liquid out of more and more of the interstices between particles. There is now insufficient liquid to lubricate the motion of such adjacent "dried out" particles, and the shear stress increases with shear rate at a higher rate than before. Many dilatant fluids also exhibit volumetric dilatancy, that is, an increase in volume with shear rate, as well as viscous dilatancy, the increase in apparent viscosity with shear rate. Examples of dilatant fluids are aqueous suspensions of titaneium oxide, suspensions of starch and quicksand.

Certain fluids do not deform continuously unless a limiting or "yield" stress is exceeded. They are sometimes called plastic fluids or Bingham fluids, although the latter implies that they follow a particular constitutive equation. In the simplest case, such fluids behave exactly like Newtonian fluids once the yield stress is exceeded, in that their rheograms are Newtonian rheograms shifted upward. These are called Bingham fluids. Pure Bingham behavior is rare. In other cases, the flow curve is that of a pseudoplastic shifted upward, and such fluids are termed yield pseudoplastics.

Table 7
Some Constitutive Equations for Pseudoplastic (or Dilatant) Fluids

Model Name	Constitutive Equation	Apparent Viscosity	Limiting Newtonian Prediction		Remarks
			Low Shear	High Shear	
Power law (Ostwald-de Waele)	$\tau = k\delta^{n}$	$k\delta^{n-1}$	—	—	Two parameters: K, n Newtonian limits not predicted.
Prandtl-Eyring	$\tau = A^{-1}_{\sinh}(\delta/B)$	$A\delta^{-1}_{\sinh}(\delta/B)$	A/B	—	Two parameters: A, B Based on Eyring's kinetic theory of liquids
Ellis	$\delta = (\phi_0 + \phi_1\tau^{\alpha-1})\tau$	$(\phi_0 + \phi_1\tau^{\alpha-1})^{-1}$	ϕ_0^{-1} for $\alpha > 1$ only	ϕ_0^{-1} for $\alpha < 1$ only	Three parameters: ϕ_0, ϕ_1, α If $\phi_1 = 0$ gives Newtonian equation If $\phi_0 = 0$ gives power law equation
Reiner-Phillippoff	$\delta = \left[\mu_\infty + \frac{\mu_0 - \mu_\infty}{1 + (\tau/\tau_s)^2}\right]^{-1}\tau$	$\mu_\infty + \frac{\mu_0 - \mu_\infty}{1 + (\tau/\tau_s)^2}$	μ_0	μ_∞	Three parameters: $\mu_0, \mu_\infty, \tau_s$ μ_0 and μ_∞ are the limiting viscosities
Sisko	$\tau = a\delta = b\delta^{c}$	$a + b\delta^{c-1}$	a*	—	Three parameters: a, b, c. Combination of Newtonian and power law

* *All these equations predict shear-dependent viscosity.*

Yield dilatant behavior also may be encountered. The commonly accepted mechanism for the behavior of plastic fluids is that the fluid at rest contains a structure sufficiently rigid to resist shear stresses smaller than the yield stress, τ_y. When this stress is exceeded, the structure collapses and deformation is continuous, as for nonplastic fluids. Examples of fluids exhibiting plastic behavior are paint systems, suspensions of finely divided minerals such as chalk in water, and some asphalts. The yield values can be quite small for water suspensions and very large for materials such as asphalt. Plastic behavior is necessary in a paint to prevent flow when it is applied as a vertical film.

The Bingham equation is

$$\begin{aligned} \tau &= \tau_y + \eta\delta \qquad \tau > \tau_y \\ \delta &= 0 \qquad\qquad\;\; \tau < \tau_y \end{aligned} \tag{149}$$

where τ_y = yield stress, and η = "plastic viscosity" (or "coefficient of rigidity"). On logarithmic coordinates the curve is asymptotic to τ_y at low δ and approaches a slope of unity for high δ (see Figure 14). Note that for $\tau < \tau_y$, $\delta = 0$, implying that there is no deformation in this region. The apparent viscosity for Bingham fluids is

$$\eta_a \equiv \tau/\delta = \eta + \tau_y\delta^{-1} \tag{150}$$

showing that the apparent viscosity decreases with shear rate. For very high shear rates the effect of the yield stress becomes negligible, the apparent viscosity levels off to the plastic viscosity, η, and the fluid behaves as a Newtonian fluid.

Another useful expression is the yield-power model:

$$\begin{aligned} \tau &= \tau_y + K\delta^n \qquad \tau > \tau_y \\ \delta &= 0 \qquad\qquad\;\; \tau < \tau_y \end{aligned} \tag{151}$$

This model combines the Bingham and power law equations (refer to Figure 14). As before, pseudoplastic or dilatant behavior obtains, depending on whether n is less than or greater than unity. An example of yield-pseudoplastic liquids would be a clay-water suspension.

This discussions has been limited to time-independent behavior. There are, however, a large class of fluids that display time-dependency in their consistency index. For such fluids, the sudden application of a change in shear rate results in the shear stress changing slowly with time until a new equilibrium shear stress corresponding to the changed shear rate is reached. The postulated mechanism for time-dependence is that the time scale for structural changes within the fluid due to shear is large compared to the time scale of shear and the very short time scales for time-independent fluids. Since the flow behavior depends on the fluid structure, the shear stress responds slowly to an imposed change in shear rate. Thus, the shear stress becomes a function of shear rate and time until steady conditions are reached. Time-dependent fluids generally are classified as thixotropic or rheopectic.

Thixotropic fluids essentially break down under shear. At a given shear rate, the shear stress slowly decreases until an equilibrium state is reached. Such fluids behave as time-dependent pseudoplastics. An illustration of thixotropic behavior is shown in Figure 15, which shows the change in apparent viscosity of the fluid when a higher shear rate, δ_2, is suddenly imposed at time t_1. At times less than t_1 the fluid has been sheared at δ_1 for a long time, so that an equilibrium apparent viscosity, η_{a_1}, is manifested. For a time-independent pseudoplastic, the apparent viscosity will decay to η_{a_2} immediately. For a thixotropic fluid, there will be only a slow decay over some measurable time $(t_2 - t_1)$ until the equilibrium viscosity, η_{a_2}, is reached. If the new shear rate is a decrease over the original value, δ_1, the apparent viscosity will increase slowly with time. Such instantaneous changes in δ cannot be achieved in practice because of fluid and system inertia.

A more practical way to detect thixotropy is to subject the fluid to a programmed change in shear rate with time, increasing from zero shear rate to some peak value and then decreasing back

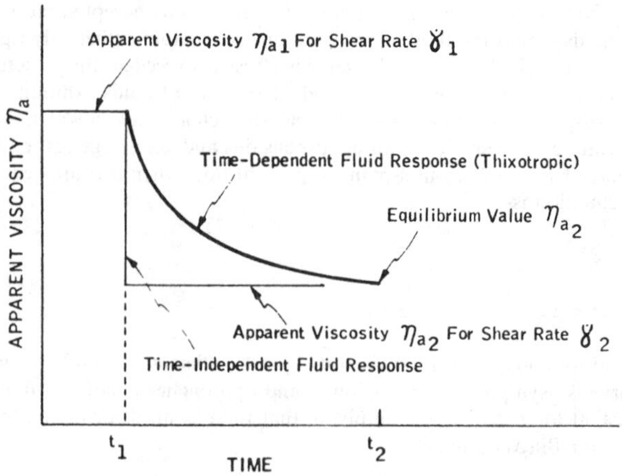

Figure 15. Response of time-dependent fluid to change in shear rate (thixotropic case).

to zero. A thixotropic fluid under such a program would produce a hysteresis loop for τ versus δ, as shown in Figure 16.

For a pseudoplastic, no hysteresis is observed, and the same curve would be traced out for increasing and decreasing shear. Note that the area within the loop depends both on the degree of thixotropy as well as the time scale of change of the shear rate. If the shear rate is changed slowly enough, even a highly thixotropic fluid would produce a curve with no hysteresis. Also to be noted is the position of the curve for increasing shear relative to that for decreasing shear, which follows from the fact that the fluid is basically pseudoplastic with an apparent viscosity that decreases with increasing shear rate. Examples of thixotropic fluids are paints, ketchup and food materials, oil well drilling muds, and some crude oils. Thixotropy is necessary in paint to allow it both to flow easily after the large shear imposed by brushing and then to recover a more viscous character after a short period of standing. Govier and Aziz [61] have summarized some of the rheological models proposed for thixotropic fluids. A review of the subject has been given by Bauer and Collins [62].

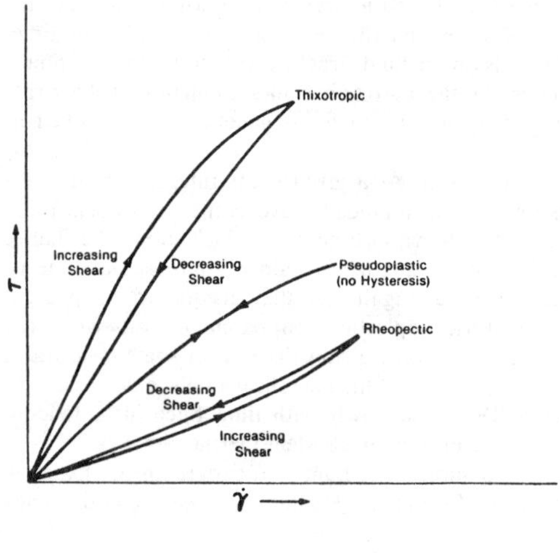

Figure 16. Hysteresis loops for time-dependent fluids (arrows show the chronology of the imposed shear rate).

With rheopectic fluids, the shear stress at a constant shear rate increases slowly with time until an equilibrium value is reached. Such fluids behave as time-dependent dilatant fluids. Under the programmed change in shear rate described above, the stress-shear rate curve forms a hysteresis loop, but of a different shape than a thixotropic fluid (Figure 16). As before, the location and shape of the loop depend on the shear rate program, as well as on the degree of rheopexy. Rheopectic fluids are rare, examples being gypsum suspensions and bentonite clay suspensions.

With the materials previously described, when the imposed shear stress is removed, deformation ceases, and there is no tendency for the fluid to recover its original undeformed state. Certain fluids do have the property of partially recovering their original state after the stress is removed. Such fluids thus have properties akin to elastic solids, as well as viscous liquids, and are termed viscoelastic. Examples of viscoelastic liquids are molten polymers and polymer solutions, egg white, dough, and bitumens.

The elastic property of such fluids leads to unusual behavior. The classic example is the phenomenon of rod-climbing, or the "Weissenberg effect." If a rotating cylinder or rod is immersed in a purely viscous liquid, the liquid surface is depressed near the rod because of centrifugal forces. In a viscoelastic liquid, on the other hand, liquid climbs up the rod because of normal stresses generated by its elastic properties. This can be observed during mixing of flour dough and in stirred polymerization reactors. Another phenomenon is the marked swelling in a jet of viscoelastic liquid issuing from a die. As a result, extrusion dies must be designed properly to produce the desired product cross section. These and other phenomena have been strikingly demonstrated by Markovitz [63].

For these materials, some of the work performed on the fluid in forcing it through a tube is stored as normal stresses as opposed to being dissipated into heat in the case of purely viscous liquids. This stored energy is released when the fluid emerges from the tube and results in swelling of the emerging fluid jet. The normal stresses generated within the tube relax when the fluid emerges from the tube, and such fluids are said to exhibit stress relaxation. Some liquid–liquid mixtures consisting of droplets of one liquid dispersed in the other also exhibit viscoelasticity. Elastic energy is stored when the spherical droplets are distorted by shear and is released through the action of interfacial tension when the shear is removed.

Elastic effects come into play mainly during the storage or release of elastic energy. Thus, such effects are important in the entrance and exit sections of tubes and during flow accelerations and decelerations caused by changes in cross section, or by imposed oscillations or by turbulence. For steady laminar flow in a tube or channel of constant cross section, elastic effects are not important except near the entrance and exit. On the other hand, for flow in fittings or in turbulent flow these effects may become important. Marked viscoelasticity is observed mainly in molten polymers and concentrated polymer solutions and generally is not considered important for pipeline flow of other non-Newtonians, such as slurries.

It is clear that for viscoelastic fluids the flow behavior cannot be represented as a relation between shear and shear rate alone. Rather, it will depend on the recent history of these quantities, as well as their current values. Constitutive equations for such fluids therefore must involve shear stress, shear rate and their local time derivatives. The time derivatives involved must be those applicable to a particular sample of fluid as it moves through the system, and not those from the viewpoint of, for example, a stationary observer. The latter has no direct impact on fluid behavior; the former represents time rates of change experienced by the fluid itself. A large number of constitutive equations of this general type have been proposed involving, among other features, various types of time derivatives. One example is the three-constant Oldroyd model, applicable for low shear rates:

$$\tau + \lambda_1 \frac{d\tau}{dt} = \mu\left(\delta + \lambda_2 \frac{d\delta}{dt}\right) \tag{152}$$

where λ_1, λ_2 = relaxation times, μ = viscosity. When λ_1 and λ_2 are both zero, the equation reverts to that for a Newtonian fluid. If only λ_2 is zero, the equation reduces to that for a Maxwell fluid. The Maxwell fluid is a simple model for a viscoelastic fluid based on a mechanical analogy. The fluid is represented as a spring and dashpot in series, the spring representing the elastic part and the dashpot the viscous part. Both the Maxwell fluid and the fluid represented by Equation 152 show stress relaxation. If flow is stopped,

$$\delta = \frac{\partial \delta}{\partial t} = 0$$

the stress decays or relaxes as e^{-t/λ_1}. If stress is removed, the shear rate in a Maxwell fluid becomes zero immediately, while for the fluid of Equation 152 the shear rate decays as e^{-t/λ_2}. This simple three-constant model has been shown to represent the behavior of certain viscoelastic fluids at low shear rates.

With the properties of non-Newtonians now described, we may now relate to practical flow situations. Following the outline covered in the first section of this chapter, attention is directed to the regimes of flow, namely, laminar and turbulent flows.

Laminar Flow of Non-Newtonians

Discussions of laminar flows of non-Newtonians will be restricted to power law fluids and Bingham-type or yield-power law fluids as these models are more widely accepted in engineering calculations.

The usual methodology for distributed parameter analysis of laminar flows may be divided into five steps [62–67]:

1. Simplify the velocity field by judicious assumptions.
2. Simplify the equations of conservation of mass and momentum based on (1).
3. Use the velocity field of (1) in the constitutive equation for the fluid and substitute for τ in the momentum conservation equation.
4. Apply the boundary conditions on velocity.
5. Solve for the velocity profile and other derived quantities such as flow rate, force on solid surfaces, etc.

This approach can be taken for several geometries such as tubes, annuli, falling films, and parallel plates. In all cases, only one velocity component is nonzero and, further, it depends on only one coordinate direction.

It is often simpler to take a more indirect route than the steps outlined above to arrive at the flow rate-pressure drop relation. An example of this is the Rabinowitsch-Mooney equation for flow in circular tube. This system is shown in Figure 17, where the velocity field is $v_z(r)$, and all other velocity components are zero. If the fluid is inelastic and time-independent, the only nonzero

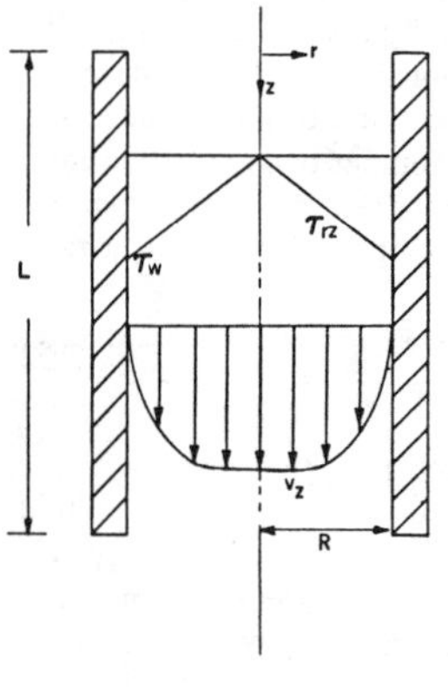

Figure 17. Flow in a tube.

component of stress is τ_{rz}, the shear stress in the z direction acting on a surface normal to the r direction, and τ_{rz} depends only on r. The equation of motion for the z direction is

$$\frac{1}{r}\frac{d}{dz}(r\tau_{rz}) = \frac{dP}{dz} \tag{153}$$

where gravity has been neglected, and dp/dz = pressure gradient. (Gravity can be included by defining a dynamic pressure $P = p - \zeta gz$; dp/dz may be shown to be a constant and is equal to $\Delta p/L$). Equation 153 can be obtained from a force balance on a differential volume within the tube as shown in the first section.

Applying the condition that $\tau_{rz} = 0$, at $r = 0$, we integrate to obtain

$$\tau_{rz} = \frac{r}{2}\frac{dp}{dz} \tag{154}$$

or

$$\frac{\tau_{rz}}{\tau_w} = \frac{r}{R} \tag{155}$$

where τ_w is the shear stress at the wall and, using Equation 154

$$\tau_w = \frac{R}{z}\frac{dp}{dz} = \frac{R}{z}\frac{\Delta P}{L} \tag{156}$$

The volumetric flow rate, Q, is

$$Q = \int_0^R v_2 2\pi r\, dr \tag{157}$$

Integrating by parts and applying the no-slip assumption we obtain the Rabinowitsch-Mooney equation:

$$Q = \frac{\pi R^3}{\tau_w^3}\int_0^{\tau_w} \tau^2 f(\tau)\, d\tau \tag{158}$$

where $\quad f(\tau_{rz}) = -\frac{dV_r}{dr}$

Equation 158 relates the volumetric flow rate to the pressure gradient (since τ_w depends on $\Delta p/L$ only). Equation 158 also can be used to measure the function, f, from capillary viscometer data.

The flow rate-pressure gradient relations for various non-Newtonian fluids flowing in a tube can be constructed from this equation by substituting the appropriate forms of f. Results are shown in Table 8. For yield-type fluids, $f(\tau)$ is discontinuous because for the shear stress is less than τ_y, the yield stress. That is, f is zero here, or that the fluid in this region moves in plug flow. The integral is readily evaluated, however. Note further that once the velocity gradient is known, the integral in Equation 158 can be evaluated. Hence, a rheological model is not required. If τ_w (or Δp) is known and Q is to be determined, an iterative procedure must be followed.

We now develop the procedure velocity distribution of a power law fluid. The following assumptions/steps are made:

1. Assume that the flow at steady-state and that the fluid has constant density. The simplified velocity field $v_z = v_z(r)$ with other components being zero.

Table 8
Flow Rate for Non-Newtonian Fluids in a Circular Tube (Laminar Flow)

Fluid	Constitutive Relation (Equation 158), f	Flowrate, Q
Newtonian	τ/μ	$\dfrac{\pi R^3 \tau_w}{4\mu}$
Power Law	$(\tau/k)^{1/n}$	$\dfrac{\pi R^3 n}{3n+1}\left(\dfrac{\tau_w}{k}\right)^{1/n}$
Bingham Plastic	$0 \leq r \leq r_c \quad f = 0$ $r_c \leq r \leq R \quad f = (\tau - \tau_y)/\eta$ $r_c = 2L\tau_y/\Delta\rho$	$\dfrac{\pi R^3 \tau_w}{4\eta}\left[1 - \dfrac{4}{3}\left(\dfrac{\tau_y}{\tau_w}\right) + \dfrac{1}{3}\left(\dfrac{\tau_y}{\tau_w}\right)^4\right]$
Yield-Pseudoplastic	$0 \leq r \leq r_c \quad f = 0$ $r_c \leq r \leq R \quad f = \left(\dfrac{\tau - \tau_y}{k}\right)^{1/n}$ $r_c = 2L\tau_y/\Delta\rho$	$\dfrac{\pi R^3 n(\tau_w - \tau_y)^{(n+1)/n}}{(K)^{1/n}\tau_w^3}\left[\dfrac{(\tau_w - \tau_y)^2}{1+3n} + \dfrac{2\tau_y(\tau_w - \tau_y)}{1+2n} + \dfrac{\tau_y{}^2}{1+n}\right]$

2. The continuity equation shows that if v_r and v_θ are both zero, then $\partial v_2/\partial z = 0$, confirming that $v_z = v_z(r)$ only. The components of the equations of motion (conservation of momentum) simplify as follows if we assume τ_{rz} is the only nonzero stress term and τ_{rz} is independent of z (fully developed flow).

r-equation: $$\frac{\partial p}{\partial r} = 0$$

θ-equation: $$\frac{1}{r}\frac{\partial p}{\partial \theta} = 0$$

z-equation: $$\frac{\partial p}{\partial z} = \frac{1}{r}\frac{d}{dr}(r\tau_{rz})$$

The first two expressions show that p = p(z) only and, therefore, that

$$\frac{dp}{dz} = \frac{\Delta p}{L}$$

We therefore obtain Equation 153 as before.

3. The constitutive equation for a power law fluid in our coordinate system reduces to:

$$\tau_{rz} = K\left(-\frac{dv_2}{dr}\right)^n \tag{159}$$

Substituting Equation 159 into Equation 153, we obtain:

$$\frac{1}{r}\frac{d}{dr}\left[rK\left(-\frac{dv_z}{dr}\right)^n\right] = \frac{dp}{dz} = \frac{\Delta p}{L} \tag{160}$$

which is a second-order equation for v_z.

4. The boundary conditions on velocity are:

$$r = R \qquad v_z = 0$$
$$r = 0 \qquad v_z = \text{finite}, \quad \text{or} \quad \frac{dv_z}{dr} = 0 \tag{161}$$

(The second condition is equivalent to the condition $r = 0$, $\tau_{rz} = 0$).

5. The solution of Equation 160 with boundary conditions (Equation 161) is

$$vz = \left[\frac{\Delta p}{2KL}\right]^{1/n} \frac{nR^{(n+1)/n}}{n+1}\left[1 - \left(\frac{r}{R}\right)^{(n+1)/n}\right] \tag{162}$$

or

$$\frac{v_z}{v} = \left(\frac{3n+1}{n+1}\right)\left[1 - \left(\frac{r}{R}\right)^{(n+1)/n}\right] \tag{163}$$

where v is the average velocity in the tube and is related to the flow rate, Q, by

$$Q = \pi R^2 v \tag{164}$$

Figure 18 shows various velocity profiles for a power law fluid flowing through a tube. The shapes of these profiles are shown to vary as a function of the flow index, n. For pseudoplastic fluids ($n < 1$) the profiles are flatter than for Newtonian fluids ($n = 1$), and, as n approaches zero, plug flow results. For dilatant fluids ($n > 1$) the profiles are less flat, and, as n becomes very large, a triangular profile is approached. This variation in the velocity profiles has a significant effect on heat transfer rates for such fluids.

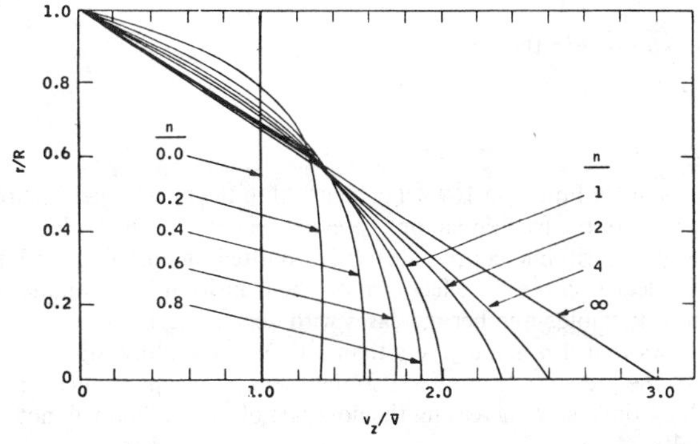

Figure 18. Velocity profiles for flow of a power law fluid in a tube.

The velocity gradient, or shear rate in the tube, may be obtained from Equation 162:

$$\delta = -\frac{dv_z}{dr} = \left(-\frac{\Delta p}{2KL}\right)^{1/n} r^{1/n} \tag{165}$$

At $r = 0$, δ is zero, and for small r, δ is also small.

The volumetric flow rate, Q, (see Table 8), can also be derived by inserting Equation 162 into Equation 156. The flow rate-pressure drop relation is useful in sizing pipelines and pumps and, for this purpose, is frequently represented in the form of the friction factor or drag coefficient.

$$E_v = 4f\frac{L}{D}\frac{v^2}{2} \tag{166}$$

where E_v represents the frictional "losses" per unit mass of fluid flowing. When gravity is neglected, the lumped parameter mechanical energy balance, or Bernoulli equation reduces to

$$\frac{\Delta p}{\zeta} = E_v$$

so that

$$\Delta p = 4f\frac{L}{D}\frac{v^2}{2} \tag{167}$$

The friction factor, f, for laminar flow of a power law fluid using Table 8 and Equations 164 and 165 is

$$\Delta p = \frac{2L\tau_w}{R} = 2KL\left(\frac{3n+1}{n}\right)^n \frac{v^n}{R^{n+1}} \tag{168}$$

Using Equation 167

$$f = 16\bigg/\left[\frac{D^n v^{2-n}\zeta}{K} 8\left(\frac{n}{6n+2}\right)^n\right] = 16/Re_{n,2} \tag{169}$$

where the power law-Reynolds number, $Re_{n,2}$ is defined as

$$Re_{n,2} = \frac{D^n v^{2-n}\zeta}{K} 8\left(\frac{n}{6n+2}\right)^n \tag{170}$$

This is sometimes expressed in the form:

$$Re_{n,1} = \frac{D^n v^{2-n}\zeta}{K} \tag{171}$$

Note that $Re_{n,2}$ results in Equation 169 being identical to the Newtonian friction factor equation, $f = 16/Re$. If $Re_{n,1}$ is used, the friction factor expression varies with n. A plot of f versus $Re_{n,2}$ is a single straight line on logarithmic coordinates. If f is plotted against $Re_{n,1}$, a line is produced for each value of n and clearly shows the effect of n on the transition from laminar to turbulent flow. In fact, the transition Reynolds number increases with decreasing n.

Equation 168 shows that if n is much less than 1 (highly pseudoplastic), Δp is insensitive to v, implying that use of the pressure drop over a straight length of pipe to measure flow rate will not give accurate results. Conversely, increasing the flow rate of such a fluid will not incur a substantial pressure drop penalty.

Velocity distributions for other non-Newtonian fluids flowing in tubes may be derived following the same procedure as previously described. A summary of these relations is given in Table 9. For

Table 9
Velocity Distributions and Friction Factors for Non-Newtonian Fluids in Circular Tubes (Laminar Flow)

Fluids	Velocity Distribution	Friction Factor
Newtonian	$\frac{\Delta\rho R^2}{4\mu L}\left[1-\left(\frac{r}{R}\right)^2\right]$	$f = 16/Re = 16/(Dv\zeta/\mu)$
Power Law	$\left(\frac{\Delta\rho}{2KL}\right)^{1/n}\frac{nR^{(n+1)/n}}{n+1}\left[1-\left(\frac{r}{R}\right)^{(n+1)/n}\right]$	$f = 16/Re_{n,2} = 16\Big/\left[\frac{D^n v^{2-n}\zeta}{K}8\left(\frac{n}{6n+2}\right)^n\right]$ $= \left(\frac{16}{Re_{n,1}}\right)\left(\frac{1}{8}\right)\left(\frac{2+6n}{n}\right)^n$
Bingham Plastic	$0 \le r \le r_c \quad v_z = \frac{\Delta\rho R^2}{4\eta L}\left(1-\frac{r_c}{R}\right)^2$	$\frac{1}{Re} = \frac{f}{16} - \frac{1}{6}\frac{He}{Re_B^2} + \frac{1}{3}\frac{He^4}{f^3 Re_B^8}$
	$r_c \le r \le R \quad v_z = \frac{\Delta\rho R^2}{4\eta L}\left[1-\left(\frac{r}{R}\right)^2\right] - \frac{\tau_y R}{\eta}\left[1-\left(\frac{r}{R}\right)\right]$	$He = \frac{\tau_y D^2\zeta}{\eta^2}, \quad Re_B = \frac{Dv\zeta}{\eta}$
	$r_c = 2L\tau_y/\Delta\rho$	
Yield-Power Law	$r_c \le r \le R \quad v_z = \frac{n}{n+1}\frac{2L}{\Delta\rho K^{1/n}}\left[\left(\frac{R\Delta\rho}{2L}-\tau_y\right)^{(n+1)/n} - \left(\frac{r\Delta\rho}{2L}-\tau_y\right)^{(n+1)/n}\right]$	$f = f\left(\frac{D^n v^{2-n}\zeta}{K}, \frac{\tau_y D^2\zeta}{K^2}, n\right)^*$
	$0 \le r \le r_c$ as above with $r = r_c$	
	$r_c = 2L\tau_y/\Delta\rho$	

* *Relation is complex; best approach is direct Q vs p relation.*

Bingham plastics, the velocity profile is a plug flow at the center of the tube surrounded by an annular region where the velocity decreases from the plug velocity to zero at the wall. The edge of the plug is r_c, and if $r_c = R$

$$\Delta p = 2L\tau_y/R \tag{172}$$

there will be no flow. Equation 172 gives the minimum pressure drop required for flow to develop. It also shows that if a Bingham fluid is held in a vertical tube, it may or may not flow out under the force of gravity, depending on the tube radius, R, and the yield stress, τ_y.

The friction factor for a Bingham fluid depends on two dimensionless groups. One form for the relation is shown in Table 9 and uses the Bingham Reynolds number $Re_B = Dv\zeta/\eta$, and the Hedstrom number, $He = \tau_y D^2\zeta/\eta^2$.

The Hedstrom number is independent of v and depends only on tube diameter and fluid properties, which is an advantage. The relation for f must be evaluated numerically and plots of f versus Re with He as a parameter may be constructed as in Figure 19. If $\tau_y = 0$ (He = 0), the usual Newtonian relation is obtained. The laminar-turbulent transition depends on He.

For *yield-power law* fluids, the velocity distribution is given in Table 9. There is a plug flow region in the central part of the tube and an annular region where the velocity varies. This is the most general velocity profile for the fluids considered in Table 9, and other profiles can be derived from this case. For n = 1, the profile reverts to that of a Bingham fluid, and if $\tau_y = 0$ the power law profile is obtained. The Newtonian case is obtained for $\tau_y = 0$ and n = 1. In dimensionless form, the friction factor depends on the power law-Reynolds number, the Hedstrom number and n.

A generalized scaleup and correlation procedure for laminar flow of time-independent non-Newtonians flowing in a tube is given by Metzner et al. [66], where the Rabinowitsch equation is rearranged as

$$\frac{Q}{\pi R^3} = \frac{1}{4}\left(\frac{8v}{D}\right) = \frac{1}{\tau_w^3}\int_0^{\tau_w} \tau^2 f(\tau)\,dt \tag{173}$$

where τ_w is a function of only (8v/D):

$$\tau_w = \frac{D}{4}\frac{\Delta p}{L} = \phi(8v/D) \tag{174}$$

This equation implies that we may scaleup the pressure drop—flow rate relation from a model to full scale by running the two at the same value of v/D using the same fluid. That is,

$$\tau_{w_{model}} = \tau_{w_{full\ scale}}$$

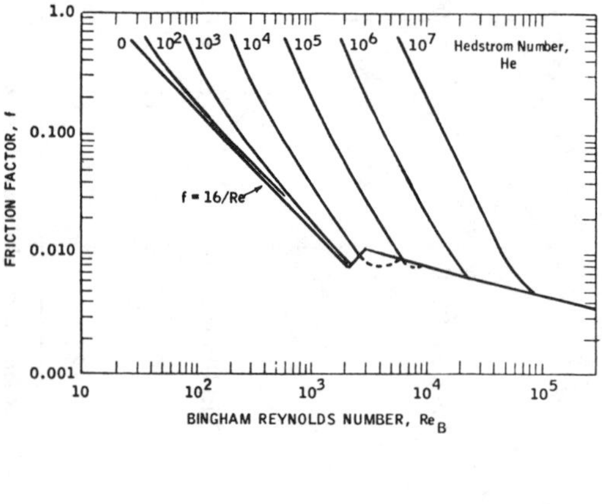

Figure 19. Friction factor-Reynolds number relation for Bingham plastics [61].

or

$$\frac{(\Delta p/L)_{\text{full scale}}}{(\Delta p/L)_{\text{model}}} = \frac{D_{\text{model}}}{D_{\text{full scale}}} \tag{175}$$

Thus, without knowing the rheology of the fluid, we can predict the fully developed laminar flow pressure drop in a full-scale unit using one data point from a laboratory scale unit at the same v/D.

Note that the function ϕ can be expressed in terms of a power law relation:

$$\tau_w = \frac{D}{4}\frac{\Delta p}{L} = K'(8v/D)^{n'} \tag{176}$$

Combining with Equation 167, an expression for the friction factor is obtained:

$$f = \frac{\left(\frac{D}{4}\frac{\Delta p}{L}\right)}{\zeta v^2/2} = \frac{K'(8v/D)^{n'}}{\zeta v^2/2} \tag{177}$$

or

$$f = 16\Bigg/\left(\frac{D^{n'}v^{2-n'}\zeta}{\delta}\right) \tag{178}$$

where

$$\delta = K'8^{n'-1}$$

Equation 178 is analogous to the friction factor relation for laminar Newtonian flow and indeed reverts to this relation when $n' = 1$ and $K' = 1$. Defining a generalized non-Newtonian Reynolds number

$$Re_{n'} = \frac{D^{n'}v^{2-n'}\zeta}{\delta} \tag{179}$$

The effective viscosity, μ_{eff}, of a non-Newtonian fluid may be used in the Reynolds number definition:

$$Re_{n'} = Re = Dv\zeta/\mu_{eff} \tag{180}$$

where

$$\mu_{eff} = K'8^{n'-1}v^{n'-1}D^{1-n'} \tag{181}$$

This viscosity that makes the non-Newtonian friction factor relation identical to the Newtonian relation.

The friction factor relationship is shown plotted in Figure 20. A plot of τ_w versus 8v/D must be constructed from data over the same range of 8v/D (or τ_w) as will be used in the design. Values of K' and n' are then calculated at the appropriate value of 8v/D, and the friction factor obtained from Figure 20. No particular form of constitutive equation is needed, thus avoiding problems such as the inadmissibility of the power law for low and high shear rates. On a double logarithmic plot of the D Δp/4L versus 8v = D, n' is the slope of the tangent to the curve at some 8v/D and K'

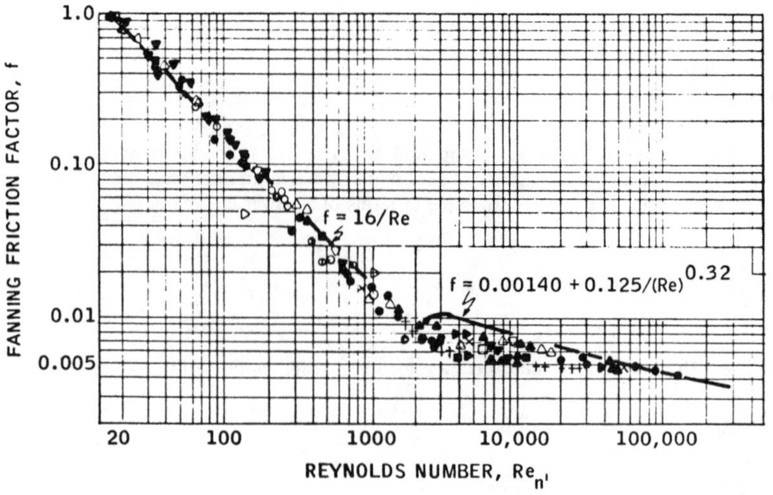

Figure 20. Friction factor vs. generalized Reynolds number: time-independent non-Newtonian flow in a tube (curves are Newtonian correlations [66]).

is the intercept made by the tangent at 8v/D equal to unity. For a general non-Newtonian fluid, n′ and K′ vary with 8v/D. Using the flow equations already developed for power law and Bingham fluids, the relations between n′ and K′ and the constants characterizing these fluids may be constructed as shown in Table 10. It may be seen that n′ and K′ are independent of 8v/D for power law fluids. $Re_{n'}$ for power law fluids is identical to $Re_{n,2}$.

Another important flow geometry is that of flow between two large stationary parallel plates. The practical case of this flow is flow in a rectangular duct of large aspect ratio or polymer flows through a slit die. The solution is valid away from the duct edges. Solutions that account for all the duct walls are available [68, 69] but do not provide much more physical insight and are necessarily quite complex.

The system is shown in Figure 21. The shear stress, τ_{yz}, acting in the z direction on a surface normal to the y direction is:

$$\tau_{yz} = y\,\Delta p/L \tag{182}$$

Table 10
The Metzner-Reed Constants, n′, K′ for Various Non-Newtonian Fluids 67

	n′	K′	δ
Newtonian	1	μ	μ
Power-Law	η	$K\left(\frac{1+3n}{4n}\right)^n$	$K\left(\frac{1+3n}{4n}\right)^n 8^{n-1}$
Bingham Plastic ($x = \tau_y/\tau_w$)	$\frac{1 - 4x/3 + x^4/3}{1 - x^4}$	$\tau_w\left[\frac{\eta}{\tau_w(1 - 4x/3 + x^4/3)}\right]^{n'}$	$8^{n'-1}\tau_w\left[\frac{\eta}{\tau_w(1 - 4x/3 + x^4/3)}\right]$

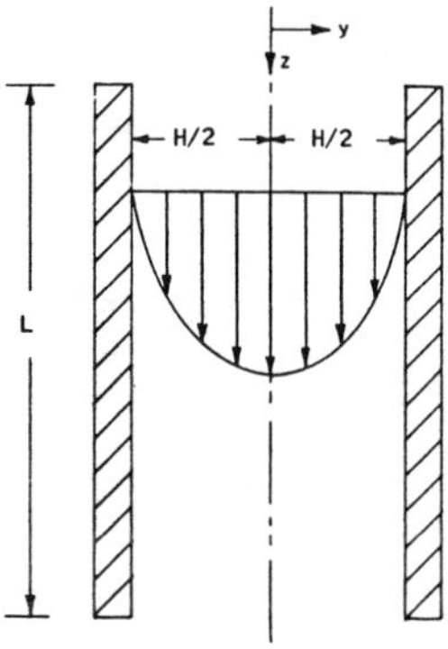

Figure 21. Flow between parallel plates.

and the shear stress at the wall, τ_w, is, therefore,

$$\tau_w = H\,\Delta p/2L \tag{183}$$

where H is the distance between the plates and $\Delta p/L$ is the pressure gradient.

An equation analogous to be the Rabinowitsch-Mooney equation for pipes can be derived as

$$Q = \frac{WH^2}{2\tau_w^2}\int_0^{\tau_w} \tau f(\tau)\,d\tau \tag{184}$$

where W is the width of the plates. Knowing the constitutive relation $f(\tau)$ the integral may be evaluated. Velocity profiles may be calculated as for flow in tubes and results for different models are given in Table 11.

Flow in other cross-section geometries are described in the literature [68–71]. Examples are flow in *concentric annuli* for Bingham plastics and power law fluids. For the Bingham plastic an annular plug flow develops as shown in Figure 22.

Charts are available for calculations [69, 70]. In an annulus the stress becomes zero at the point of maximum velocity. As noted earlier, the power law equation is a poor representation for fluid rheology at low stress levels and predicts infinite apparent viscosity at zero stress. Real non-Newtonians exhibit Newtonian behavior in this region. For flow in tubes where zero stress occurs at the centerline, the volumetric flow rate is not greatly affected because r is small near the centerline. In an annulus, appreciable error may result because r is large at the point of zero stress. Thus, flow rate predictions for power law fluids in annuli give erroneous results, as has been pointed out by Vaughn and Bergman [71] who outlines a method of using tube flow results to predict flow in annuli.

Solutions also may be developed for flow between *rotating cylinders* and for plane *Couette flow* or flow between flat plates, one of which is moving [57, 59]. Solutions of this type have application in rotational viscometers.

A variety of other problems have been considered, including flow in a falling film, in cone and plate viscometers, over submerged objects, radially between disks, etc. Some of these are discussed in the books cited in the references. Boundary layer flow of power law liquids has been treated by Acrivos et al. [72].

Turbulent Flow of Non-Newtonians

The critical Reynolds number marking the transition from laminar to turbulent pipe flow for Newtonian fluids is generally accepted as 2,100, although if external disturbances and vibrations are minimized and a very smooth entrance is provided, laminar flow may be observed for Reynolds

Table 11
Velocity Distributions and Flowrates for Non-Newtonian Flow Between Parallel Plates (Laminar Flow)

Fluid	Velocity Distribution, v	Flowrate, Q
Newtonian	$\frac{\tau_w H}{4\mu}\left[1-\left(\frac{2y}{H}\right)^2\right]$	$\frac{\tau_w{}^{H2}W}{6\mu}$
Power Law	$\left(\frac{2\tau_w}{HK}\right)^{1/n}\frac{n}{n+1}\left(\frac{H}{2}\right)^{(n+1)/n}\left[1-\left(\frac{2y}{H}\right)^{(n+1)/n}\right]$	$\frac{nWH^2}{2(2n+1)}\left(\frac{\tau_w}{k}\right)^{1/n}$
Bingham Plastic	$y_c \le y \le \frac{H}{2}$, $\frac{\tau_w H}{4\eta}\left[1-\left(\frac{2y}{H}\right)^2-2\frac{\tau_y}{\tau_w}\left(1-\frac{2y}{H}\right)\right]$	$\frac{WH^2\tau_w}{6\eta}\left[1-\frac{3}{2}\left(\frac{\tau_y}{\tau_w}\right)+\frac{1}{2}\left(\frac{\tau_y}{\tau_w}\right)^3\right]$
	$0 \le y \le y_c$ as above with $y = y_c = H\tau_y/2\tau_w$	
Yield-Power Law	$y_c \le y \le \frac{H}{2}$, $\frac{Hn}{2(n+1)}\left(\frac{\tau_w}{K}\right)^{1/n}\left[\left(1-\frac{\tau_y}{\tau_w}\right)^{(n+1)/n}-\left(\frac{2y}{H}-\frac{\tau_y}{\tau_w}\right)^{(n+1)/n}\right]$	$\frac{WH^2n}{2(n+1)}\left(\frac{\tau_w}{K}\right)^{1/n}\left(1-\frac{\tau_y}{\tau_w}\right)^{(n+1)/n}\left[1-\left(\frac{n}{2n+1}\right)\left(1-\frac{\tau_y}{\tau_w}\right)\right]$
	$0 \le y \le y_c$ as above with $y = y_c = H\tau_y/2\tau_w$	

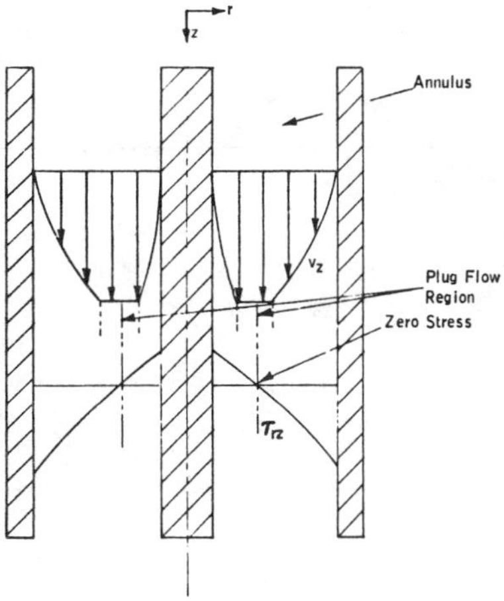

Figure 22. Typical velocity distribution for Bingham fluid in an annulus.

numbers as high as 40,000 [73]. For non-Newtonian fluids, the critical Reynolds number varies with the type of non-Newtonian and the degree of departure from Newtonian behavior, as measured by the flow index, n, or the yield stress, τ_y.

For power law fluids, the critical Reynolds number can be estimated from [74]:

$$(\mathrm{Re}_{n'})_c = (\mathrm{Re}_{n,2})_c = \frac{6{,}464n}{(1 + 3n)^2(2 + n)^{-(2+n)/(1+n)}} \tag{185}$$

Equation 185 reduces to the Newtonian critical Reynolds number of 2,100 when $n = 1$. As n decreases, $(\mathrm{Re}_{n,2})_c$ increases until $n = 0.4$, after which it decreases rapidly. This is at variance with data of friction factor versus generalized Reynolds number of Dodge and Metzner [75], which show that the critical Reynolds number increases monotonically as n decreases from unity. Further, values from Equation 185 are not in complete agreement with these data, even for $n > 0.4$. In any case, the largest critical Reynolds number reported for power law fluids is in the region of 3,000, which from the design point of view is not significantly different from the Newtonian value of 2,100. If the fluid is viscoelastic, the transition to turbulent flow is substantially affected, with $(\mathrm{Re}_{n,2})_c$ as large as 10,000 [76].

For Bingham Plastics, the critical Bingham Reynolds number, $(\mathrm{Re}_B)_c$ can be substantially different from 2,100, and can be estimated from [77]

$$(\mathrm{Re}_B)_c = \frac{\mathrm{He}}{8x_c}\left[1 - \frac{4}{3}x_c + \frac{1}{3}x_c^4\right] \tag{186}$$

where x_c is the critical value of x ($x = \tau_y/\tau_w$) given by

$$\frac{x_c}{(1 - x_c)^3} = \frac{\mathrm{He}}{16{,}800} \tag{187}$$

For Bingham fluids the critical Bingham Reynolds number depends on the degree of plastic behavior, as represented by the Hedstrom number. Figure 23 shows a plot of the above equation. When He becomes zero the Newtonian value applies, and as He increases, the critical Reynolds number increases monotonically. For $\mathrm{He} = 10^7$, transition to turbulent flow occurs at Reynolds numbers on the order of 50,000. Equation 186 predicts $(\mathrm{Re}_B)_c = 34{,}000$ when $\mathrm{He} = 10^7$, and underpredicts $(\mathrm{Re}_B)_c$ for He greater than 10^5.

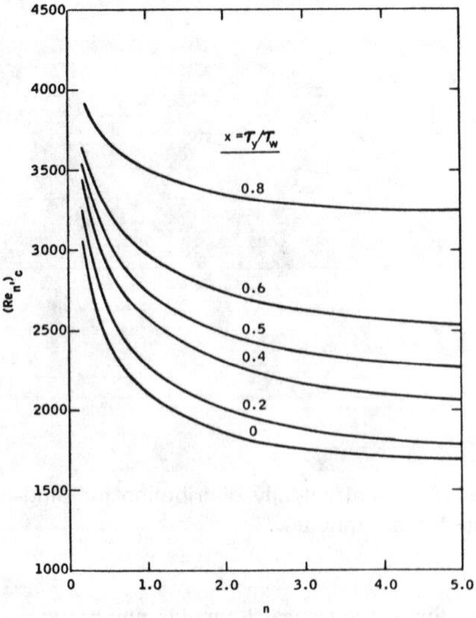

Figure 23. Critical Bingham-Reynolds number for flow in a pipe [77].

Foishteter et al. [78] give the following relation for the critical Reynolds number for a yield-power law fluid:

$$\left.\begin{aligned}
(Re_{n'})_c &= 2{,}100/C \\
C &= \frac{1}{w^3}\left[\frac{x^2}{2} + (1 - x^2)A + x(1 - x)B\right] \\
w &= x^2 + 2x(1 - x)\frac{m+1}{m+2} + (1 - x^2)\frac{m+1}{m+3} \\
A &= \frac{3(m+1)^3}{2(m+3)(m+2)(3m+5)} \\
B &= \frac{6(m+1)^3}{(m+2)(2m+3)(3m+4)} \\
m &= \frac{1}{n} \\
x &= \tau_y/\tau_w
\end{aligned}\right\} \tag{188}$$

Values of $(Re_{n'})_c$ from this equation are plotted as a function of n and x in Figure 24. It may be seen that for constant x, the critical Reynolds number decreases continuously as n increases (i.e., the fluid becomes less pseudoplastic) and levels off for n approximately equal to five. The critical Reynolds number increases as x increases (fluid becomes more plastic) for any value of n. For pseudoplastic fluids, Dodge and Metzner [75] found experimentally that $(Re_{n'})_c = 2{,}700$ for $n' = 0.726$, and 3,100 for $n' = 0.38$. Equation 188 predicts $(Re_{n'})_c$ of 2,253 and 2,636, respectively, showing only fair agreement.

Friction factors in turbulent flow can be estimated from a modified Fanning friction factor. Extending the laminar relation of Metzner and Reed [66] to turbulent flow, Dodge and Metzner [75]

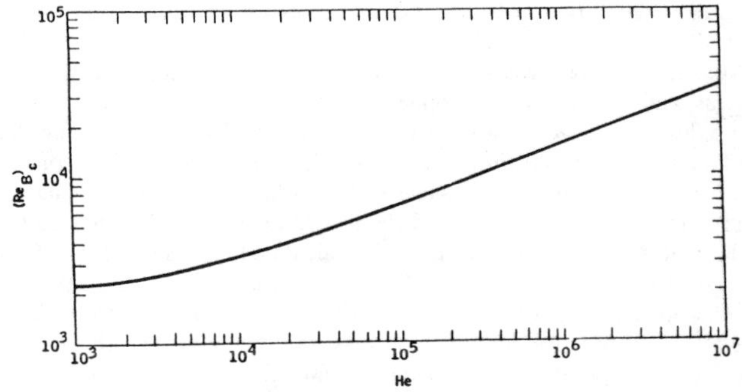

Figure 24. Critical Reynolds number for yield-power law fluid from Equation 188.

obtained the friction factor as a function of the generalized Reynolds number, $Re_{n'}$

$$\frac{1}{\sqrt{f}} = \frac{4.0}{(n')^{0.75}} \log_{10}[Re_{n'} f^{(1-n'/2)}] - \frac{0.40}{(n')^{1.2}} \tag{189}$$

This relation is implicit in f, and a chart of f versus $Re_{n'}$ is given in Figure 25. The chart shows the effect of the exponent n′ on the friction factor in turbulent flow. Values of n′ less than unity (pseudoplastic behavior) give lower friction factors (at the same $Re_{n'}$) than do Newtonian fluids, and for n′ greater than unity the friction factor becomes progressively larger. When $n' = 1$ (Newtonian fluid), Equation 189 reduces to the classical expression of Nikuradse given earlier. For highly dilatant fluids, $n' > 2$ Equation 189, and Figure 25 must be used with caution because certain assumptions underlying their development may no longer be valid. Further, no confirming data are available in this range, as is also the case for highly pseudoplastic liquids. (These regions are

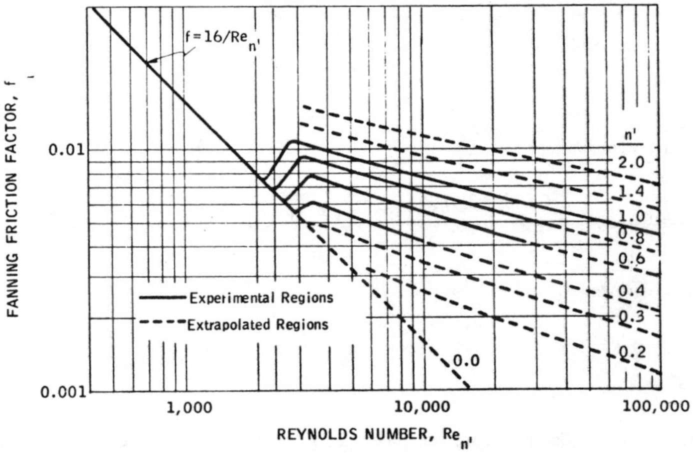

Figure 25. Friction factor-Reynolds number relation for non-Newtonian fluids [75].

represented by the dashed curves in Figure 25.) Extrapolation to $Re_{n'}$ greater than 10^5 is permissible but not for $n' > 2$.

Dodge and Metzner [75] have recommended the use of Figure 25 for Bingham plastics as well as power law fluids, and have verified its applicability with experimental data. For Bingham fluids, as with any other fluid, n′ and K′ must be evaluated at the wall shear stress prevailing in the flow. For power law fluids, this does not present any difficulty since n′ and K′ are constants. For other fluids, n′ and K′ vary with shear rate and shear stress, so that an iterative calculation is necessary. First the pressure drop must be assumed and the wall shear stress calculated from Equation 150. The values of K′ and n′ are then obtained at the calculated wall shear stress from the laminar flow curve τ_w versus 8v/D. These are used in Equation 189 to obtain f, and Δp is calculated using Equation 178. The procedure is repeated until agreement is obtained between the calculated and assumed values of Δp. In practice, for Bingham plastics flowing in smooth tubes for Re_B greater than $(Re_B)_c$ the friction factor depends only on Re_B for fully turbulent flow. The relation in this range is essentially that for a Newtonian fluid.

A particular problem with this method is that the parameters n′ and K′, which must be estimated at the wall shear stress prevailing in the turbulent flow, are generally evaluated from laminar data. The problem is that often these laminar data are not at high enough shear rate, usually because of the onset of turbulence in the experiments. Thus, n′ and K′ are estimated at too low a shear rate, which means that n′ is underestimated. This, in turn, leads to values of f from the Dodge and Metzner correlation that will be too low. Other approaches are outlined in the literature [61, 79].

Masuyama and Kawashima [79] provide design charts for friction factors for pseudoplastic liquids that are claimed to be more accurate than previous methods. Correlations for rough-walled pipes have been reviewed by Govier and Aziz [61]. In most cases, the Newtonian values for rough pipes are suitable for design for pseudoplastic fluids. Friction factor relations for annuli and for rectangular ducts are reported by Skelland [59]. For Newtonian fluids, turbulent pressure drop in noncircular cross-sectional ducts may be estimated using the hydraulic radius of the cross section. The literature does not report such a practice for non-Newtonian flows, and hence, this approach cannot be recommended.

For viscoelastic fluids, friction factors are found to be much lower than predicted by the Dodge and Metzner relation. Govier and Aziz [61] have reviewed some available correlations.

As in the case of Newtonian fluids, time-averaged velocity profiles of non-Newtonians in turbulent flow are much flatter than in laminar flow. Dodge and Metzner [75] have derived a profile for power law fluids analogous to the universal velocity distribution for Newtonian fluids. The Dodge-Metzner profile, as corrected by Skelland [59] is:

For the laminar sublayer:

$$u^+ = (y^+)^{1/n} \tag{190}$$

For the turbulent core:

$$u^+ = \frac{5.66}{n^{0.75}} \log_{10} y^+ - \frac{0.566}{n^{1.2}} + \frac{3.475}{n^{0.75}}\left[1.960 + 0.815n - 1.628n \log_{10}\left(3 + \frac{1}{n}\right)\right] \tag{191}$$

where $u^+ = u/u^*$

$y^+ = y^n u^{*2-n} \zeta / K$

$u^* = \sqrt{\tau_w / \zeta}$

and $y = R - r$ is the distance from the pipe wall into the fluid. By contrast to the Newtonian universal velocity distribution discussed earlier, the constants in the above were obtained from friction factor measurements rather than experimental velocity data. As a result, the thickness of the sublayer was not obtained.

Entrance Effects

In the discussions of flow in ducts of constant cross section (tubes, annuli, etc.), we assumed the flow to be fully developed. This implies that the axial velocity is independent of axial position, which is only true away from the entrance exit of the duct. Exit effects are much less significant than entrance effects, and not much is known about them. Entrance effects have been investigated to a limited extent for non-Newtonian fluids with the majority of studies limited to the laminar regime, to which initial discussions are now directed.

Figure 26 shows the velocity and pressure distribution at the entrance of a tube or duct. Fluid enters the duct from a large reservoir and the entering velocity distribution is flat, as shown. The velocity of the fluid adjacent to the wall becomes zero as soon as the fluid enters the conduit. As the fluid penetrates farther into the duct, adjacent layers are subjected to increasing drag from the wall layer and they, in turn, exert drag on fluid layers adjacent to them. A progressively thicker zone is set up where the velocity varies with distance from the wall. This zone is the boundary layer. Fluid in the boundary layer therefore flows at velocities lower than the entrance velocity. To satisfy continuity, the fluid in the core must have a higher velocity than the entrance velocity, so the core region accelerates. The boundary layer grows in thickness as we go farther into the duct and ultimately grows to fill the entire cross section. From this point on we have fully developed flow, and equations derived earlier apply as long as flow remains laminar. The entrance length, L_e, is usually defined as the distance from the entrance at which the centerline velocity becomes very close (within 98–99%) to the fully developed value.

Figure 26 also shows the pressure change with axial distance. The upper straight line shows the distribution if the flow were fully developed throughout the duct. The lower curve is the actual distribution. In the entrance region the pressure decreases more rapidly than in developed flow because of the acceleration of the core fluid (pressure energy converted to kinetic energy) and because of the greater velocity gradients in the wall boundary layer, which result in greater wall

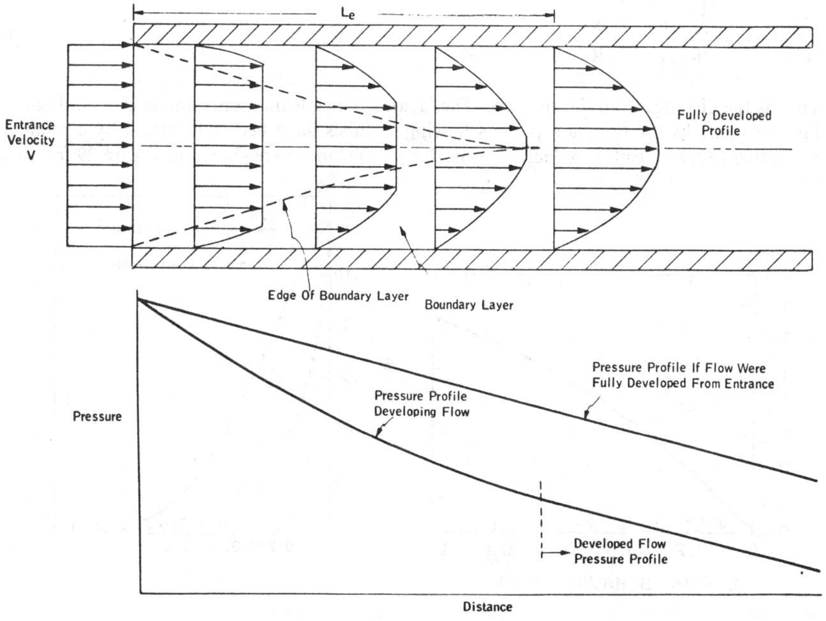

Figure 26. Velocity and pressure distributions in entrance region of a tube.

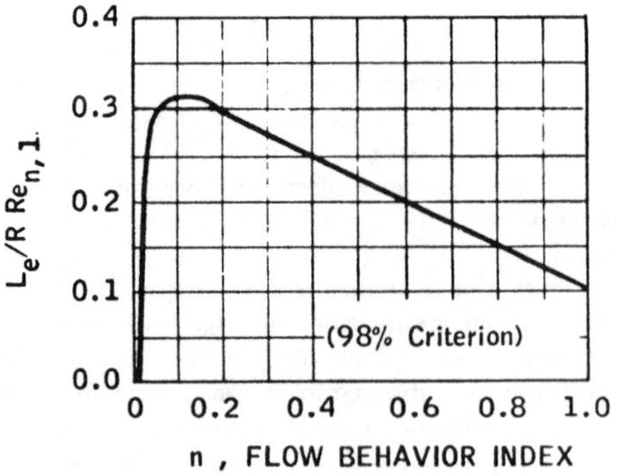

Figure 27. Entrance length for power flow in a tube [80].

shear stresses. The pressure profile is nonlinear in the developing region and becomes linear when flow is fully developed. The overall pressure gradient is greater if the developing region is included.

Collins and Schowalter [80, 81] developed entrance region solutions for power law fluids flowing in tubes and in channels. Their solutions utilize a boundary layer analysis in the near-entrance region and perturbation of the fully developed flow near the fully developed region, where boundary layer theory does not apply. Good agreement was obtained with previous results for Newtonian fluids and with data for both Newtonian and non-Newtonian fluids. Dimensionless entrance lengths, $L_e/R\ Re_{n.1}$ (98% of centerline velocity attained), as a function of the flow behavior index, n, are shown in Figure 27 from Collins and Schowalter [80]. It may be seen that for the same Reynolds number entrance lengths increase as n decreases up to n = 0.01. No results were provided for n > 1. The pressure drop in the entrance region was given by these authors as

$$\frac{\Delta p}{\zeta v^2/2} = \frac{2^{n+2}\left(\frac{1+3n}{n}\right)^n}{Re_{n,1}} \frac{x}{R} + C \qquad (192)$$

where x is the distance from the entrance. The factor c depends on n only and is given in Figure 28.

Entrance lengths for Bingham plastics flowing in tubes have been considered by a number of investigators [82, 83]. Entrance lengths from these two sources are shown in Figure 29 from Chen

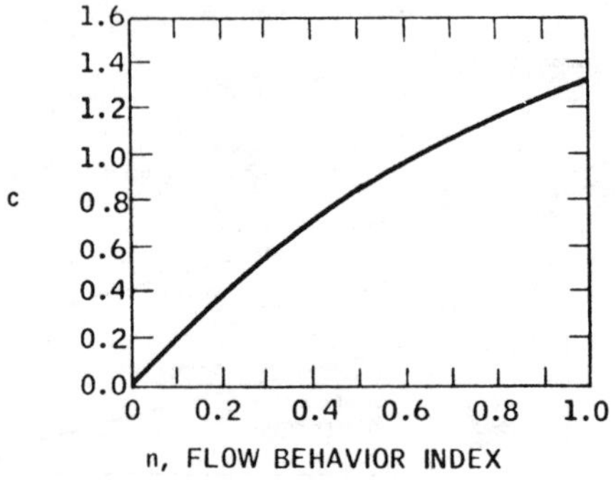

Figure 28. Entrance region pressure drop correction for power law flow in tubes [80].

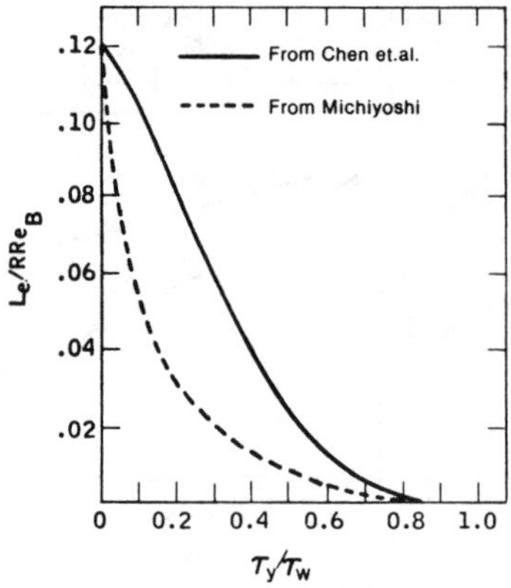

Figure 29. Entrance lengths for Bingham flow in a tube [83].

et al. [83], which give their results and those of Michiyoshi [82]. The dimensionless entrance length, $L_e/R\ Re_B$, is given as a function of τ_y/τ_w and shows that entrance lengths decrease as the yield stress, τ_y, increases. This agrees with physical intuition in that as τ_y increases, the plug flow core of the fully developed profile becomes larger and the profile becomes more and more similar to the flat profile at the tube entrance. The distance required to attain the developed profile therefore decreases. Entrance lengths are more conservative for design purposes. Pressure drops in the entrance region also have been given by these investigators.

When these non-Newtonian fluids approach Newtonian behavior ($n \rightarrow 1$, or $\tau_y \rightarrow 0$) the results cited above reduce approximately to the generally accepted entrance length relation for Newtonian fluids:

$$L_e = 0.125 R\ Re \tag{193}$$

No analytical treatment of flow through fittings is available. Weltmann and Keller [84] did experimental work on such flows for Bingham plastics and for pseudoplastic fluids. Pressure losses were found to be comparable to those for Newtonian fluids. Therefore, the conventional friction loss factors in fittings that are available for Newtonian fluids may be used for design for piping systems for non-Newtonian fluids. However, Heywood [85] has stated that at low Reynolds number, losses for non-Newtonian fluids may be as much as ten times greater than for Newtonian fluids. If the fluid is viscoelastic, greater losses will occur because of the elastic effects that come into play in such flows.

Few studies have been devoted to the entrance length necessary to achieve fully developed turbulent flow or on the pressure drop in this region. As in the case of laminar flow, the pressure gradient in the entrance region is greater than that in fully developed flow. Dodge and Metzner [75] found that

$$L_e/D < 15$$

for turbulent flow of non-Newtonians with $n' < 1$. This value is considerably less than that for a Newtonian fluid, for which some 50 diameters are usually needed to obtain a fully developed flow [86]. Limited data indicate that the pressure gradient in the entrance region is best estimated from Newtonian correlations.

Substantial deviations from Newtonian behavior occur for viscoelastic fluids. Much greater entry lengths are found in drag-reducing systems [61].

As for laminar flow, little information is available on frictional losses for turbulent flow in fittings. Current practice is to use the loss coefficients developed for Newtonian fluids as long as viscoelastic effects are negligible.

FLOW OF COMPRESSIBLE FLUIDS

General Principles and Flow Classifications

As noted in the first section, the assumption of constant fluid density greatly simplifies the equations of motion and thereby provides straightforward solutions. With minor exceptions, this assumption is justified for liquids and for most applications involving gases the use of a constant, mean value for the density provides a practical approximation.

However, there are systems that encounter gases at high velocities or induce large pressure ranges in which cases compressibility may be significant. Examples are depressuring operations, pressure relief systems, mechanical compressions, shock compressions, detonations, thermocouple measurements, pitot-tube measurements, flow through pipelines and flow through porous media.

We first direct attention to the case of steady, reversible (inviscid), quasi-one-dimensional flow. By neglecting viscous effects, most internal flows and many external flows can be approximated as quasi-one-dimensional in the local direction of flow. In steady flow, the path of a fluid particle follows a streamline. Streamlines passing through a closed curve form a stream tube, which represents the boundaries of a filament of fluid during changes in direction and velocity.

From Newton's second law, the rate of change of momentum is equal to the net sum of the applied forces. For the filament of fluid, the applied forces in the direction of flow are the pressure times the cross-sectional area, and the component of gravity in the direction of flow times the mass of fluid. The momentum equals the velocity times the mass rate of flow. The resulting balance on a differential length of fluid is:

$$-A\,dP - g\left(\frac{dz}{ds}\right)A\,ds = w\,du \tag{194}$$

where
A = cross-sectional area of filament
P = mean pressure across A
g = acceleration due to gravity
z = elevation (upward distance)
s = distance along filament
w = mass rate of flow in stream tube
u = mean velocity across A

Within the stream tube the mass rate of flow is constant and

$$w = u\rho A \tag{195}$$

where here, ρ = density. Replacing w in Equation 194 by $u\rho A$, and rearranging, we obtain

$$u\,du + g\,dz + \frac{dP}{\rho} = 0 \tag{196}$$

which is the differential form of Bernoulli's equation for compressible flow. Although it was obtained above from a force-momentum balance, this equation can also be interpreted as an energy balance since energy and momentum are not independent for reversible flows.

For all gas flows in which the effect of compressibility is significant, the gravitational term is negligible, and hence,

$$u\,du + \frac{dP}{\rho} = 0 \tag{197}$$

Differentiating Equation 195, dividing through by $u\rho A$ and rearranging terms gives

$$\frac{du}{u} + \frac{d\rho}{dP}\left(\frac{dP}{\rho}\right) + \frac{dA}{A} = 0 \tag{198}$$

The acoustic (sonic) velocity for any material is defined as

$$a = \left(\frac{dP}{d\rho}\right)^{1/2} \tag{199}$$

Substituting for dP/ρ from Equation 197 and for $dP/d\rho$ from Equation 199 permits re-expression of Equation 198 as

$$\frac{du}{u}(1 - M^2) = -\frac{dA}{A} \tag{200}$$

where $M = u/a$ = Mach number.

We note that the Mach number or, more specifically, the factor $(1 - M^2)$, characterizes the effect of compressibility.

For subsonic flows, $M < 1$, and the velocity decreases as the area increases, whereas for supersonic flows, $M > 1$, and the velocity increases.

Equation 197 can be integrated formally from any location 1 to any location 2 to obtain

$$\frac{u_2^2 - u_1^2}{2} = \int_{P_1}^{P_2} \frac{dP}{\rho} \tag{201}$$

This expression is an integral form of Bernoulli's equation for horizontal, compressible flow. Introducing Equation 195 yields

$$u_1^2 = \frac{2\int_{P_2}^{P_1} \frac{dP}{\rho}}{\left(\frac{P_1 A_1}{P_2 A_2}\right)^2 - 1} \tag{202}$$

Also,

$$u_2^2 = \frac{2\int_{P_2}^{P_1} \frac{dP}{\rho}}{1 - \left(\frac{\rho_2 A_2}{\rho_1 A_1}\right)^2} \tag{203}$$

and

$$w^2 = \frac{2(\rho_2 A_2)^2 \int_{P_2}^{P_1} \frac{dP}{\rho}}{1 - (\rho_2 A_2/\rho_1 A_1)^2} \tag{204}$$

For many expansions $(\rho_2 A_2/\rho_1 A_1)^2 \ll 1$, permitting the above corresponding simplifications. Equations 202–204 provide a basis for correlation of flow through venturi tubes, nozzles, and orifices. A relationship between P and ρ is required to evaluate the integral.

We now need to relate to certain state-properties of the fluid. The generalized equation of state for a gas may be stated as:

$$\rho = \phi\{P, T, y_i\} \tag{205}$$

where T = absolute temperature (°K)
y_i = mole fractions of components of gas

And for an ideal gas,

$$\rho = Pm/RT \tag{206}$$

where m = molar mass
R = universal gas law constant

Nonideal behavior generally can be represented by a compressibility factor Z:

$$\rho = Pm/RTZ \tag{207}$$

The compressibility factor for most gases is expressed in terms of the reduced pressure and temperature. Many other equations of state of varying complexity and range of applicability have been developed. An alternative approach is through the fugacity, defined as

$$\frac{dP}{\rho} = \frac{RT}{m} d\ln(f) \tag{208}$$

The fugacity coefficient $\eta = f/P$ also has been correlated in terms of the reduced pressure and temperature.

Correlations for Z and η, as well as a discussion of various equations of state, can be found in the literature [87, 88].

Churchill [89] notes that in general, some restraint (such as constant temperature) is necessary to permit evaluation of the integral in Equations 201–204 even when the thermodynamic behavior of the fluid is known. There are three special cases that are generally acceptable approximations in applications (mean value, isothermal, adiabatic). The *mean value*, in general, may be stated as:

$$\int_{P_2}^{P_1} \frac{dP}{\rho} = \frac{P_1 - P_2}{\rho_m} \tag{209}$$

If the variation of ρ is slight, the error in choosing a mean value cannot be great. If ρ varies monotonically with P, which is the usual case, ρ_m lies between ρ_1 and ρ_2, and the fractional error in using the arithmetic average cannot be greater than $(\rho_1 - \rho_2)/(\rho_1 + \rho_2)$.

For the *isothermal expansions* or compression of an ideal gas,

$$\int_{P_2}^{P_1} \frac{dP}{\rho} = \int_{P_2}^{P_1} \frac{RT}{Pm} dP = \frac{RT}{m} \ln\left(\frac{P_1}{P_2}\right)$$
$$= \frac{RT(P_1 - P_2)}{m}\left(\frac{\ln(P_1/P_2)}{P_1 - P_2}\right) = \frac{P_1 - P_2}{\rho_1 - \rho_2} \ln\left(\frac{P_1}{P_2}\right) = \frac{P_1 - P_2}{\rho_{lm}} \tag{210}$$

where ρ_{lm} = log-mean value of ρ.

For steady flow, the first law of thermodynamics can be stated in the integral form:

$$q - w = \Delta H + \frac{\Delta u^2}{2} + g\,\Delta z \tag{211}$$

where q = heat entering fluid stream from surroundings
w = work done by fluid stream on surroundings
H = specific enthalpy of fluid

For free flow, w = 0, and for an isothermal expansion of an ideal gas, $\Delta H = 0$ [89]. Hence, if the change in potential energy is neglected, the heat required to maintain a constant temperature is:

$$q_T = \frac{u_2^2 - u_1^2}{2} \tag{212}$$

From Equations 201 and 210 it follows that:

$$q_T = \frac{RT}{m} \ln(P_1/P_2) \tag{213}$$

Also,

$$dH = T\,dS + \frac{dP}{\rho} \tag{214}$$

$$q_T = T(S_2 - S_1) \tag{215}$$

where S = specific entropy of fluid. Thus, the required heat can be seen to be equal to the increase in the kinetic energy, also to (RT/m) ln (P_1/P_2) and $T(S_2 - S_1)$.

For free adiabatic reversible flow, w, q, and S are zero. For an ideal gas with a constant heat capacity, it can be shown

$$\frac{\rho}{\rho_1} = \left(\frac{T}{T_1}\right)^{1/(\gamma-1)} = \left(\frac{P}{P_1}\right)^{1/\gamma} \tag{216}$$

where $\gamma = C_p/C_v$ = heat capacity ratio
C_p = specific heat capacity at constant pressure
C_v = specific heat capacity at constant volume

Then,

$$\int_{P_2}^{P_1} \frac{dP}{\rho} = \int_{P_2}^{P_1} \left(\frac{P_1}{P}\right)^{1/\gamma} \frac{dP}{\rho_1} = \frac{\gamma P_1^{1/\gamma}}{(\gamma-1)\rho_1}\left(P_1^{(\gamma-1)/\gamma} - P_2^{(\gamma-1)/\gamma}\right) \tag{217}$$

$$= \left(\frac{\gamma}{\gamma-1}\right)\frac{P_1}{\rho_1}\left[1 - \left(\frac{P_2}{P_1}\right)^{(\gamma-1)/\gamma}\right] \tag{218}$$

$$= \frac{\gamma}{\gamma-1}\left(\frac{P_1}{\rho_1} - \frac{P_2}{\rho_2}\right) \tag{219}$$

$$= C_p(T_1 - T_2) \tag{220}$$

The latter expression follows from Equation 206, and

$$(C_p - C_v)m = R \tag{221}$$

For an adiabatic process without work and with a negligible change in potential energy, Equation 211 reduces to:

$$\frac{u_2^2 - u_1^2}{2} = H_1 - H_2 \tag{222}$$

It then follows that:

$$\frac{u_2^2 - u_1^2}{2} = C_p(T_1 - T_2) \tag{223}$$

Thus, in an adiabatic reversible (isentropic) expansion of an ideal gas, the temperature decreases in direct proportion to the increase in the kinetic energy.

The heat capacity ratio, γ, is relative invariant with pressure and varies only moderately with temperature.

Many, if not most, high-velocity processes are essentially adiabatic. The effect of heat exchange with the surroundings tends to shift the behavior toward that of an isothermal process. These two idealized conditions therefore generally indicate the limiting behavior of real processes. Assuming $\gamma = 1$ (at the appropriate point to avoid a singularity) in the previous expressions for adiabatic reversible processes the condition of isothermal behavior is described. Hence, a possible approximation for intermediate behavior is to replace γ with an empirical value, k, intermediate between 1 and 8.

With this background, we now direct attention to the practical problem of flow through nozzles. Combining Equation 218 with Equation 201 for negligible u_1 gives the following limiting expression for adiabatic flow from a large vessel:

$$u^2 = \left(\frac{2\gamma}{\gamma - 1}\right)\frac{P_1}{\rho_1}\left[1 - \left(\frac{P}{P_1}\right)^{(\gamma-1)/\gamma}\right] \tag{224}$$

where subscript 2 has been dropped to emphasize that this expression gives the velocity for any pressure. The corresponding expression for the mass rate of flow is

$$w^2 = \left(\frac{2\gamma}{\gamma - 1}\right)A^2P_1\rho_1\left[1 - \left(\frac{P}{P_1}\right)^{(\gamma-1)/\gamma}\right]\left(\frac{P}{P_1}\right)^{2/\gamma} \tag{225}$$

Since the mass rate of flow is invariant in a nozzle, this expression provides a relationship for the variation of the pressure with the cross-sectional area. Equation 225 indicates that the mass flux density, $G = w/A = u\rho$, approaches zero in the limits of $P = 0$ and $P = P_1$. Hence, a maximum possible mass flux density exists at some intermediate pressure. Equating the derivative dg/dP to zero gives the following critical (or throat) pressure corresponding to this maximum value of G:

$$\left(\frac{P_c}{P_1}\right) = \left(\frac{2}{\gamma + 1}\right)^{\gamma/(\gamma-1)} \tag{226}$$

It follows that

$$\frac{T_c}{T_1} = \left(\frac{\rho_c}{\rho_1}\right)^{\gamma-1} = \frac{2}{\gamma + 1} \tag{227}$$

$$G_c^2 = \gamma P_1\rho_1\left(\frac{2}{\gamma + 1}\right)^{(\gamma+1)/(\gamma-1)} \tag{228}$$

and

$$u_c^2 = \frac{2\gamma P_1}{(\gamma + 1)\rho_1} = \frac{\gamma P_c}{\rho_c} = \left(\frac{dP}{d\rho}\right)_{cs} \tag{229}$$

The latter expression indicates that the velocity corresponding to the critical conditions is sonic.

For air and other diatomic gases (with $\gamma \simeq 1.4$) we note that,

$$\frac{P_c}{P_1} \cong \left(\frac{2}{1.4 + 1}\right)^{1.4/(1.4-1)} = 0.528 \tag{230}$$

For converging nozzles, if $P_\infty > P_c$ (where P_∞ = external pressure) as given by Equation 226, substituton of the exit area in Equation 225 gives the flow rate. For $P_\infty < P_c$, the flow rate is given by Equation 228 (which is equivalent to Equation 225 in terms of P_c) and the throat area. The gas then expands irreversibly outside the nozzle from P_c to P_∞.

For converging-diverging nozzles the mass flow rate expression (Equation 225) has two real positive roots, i.e., two values of P for each A and w. The greater root corresponds to subsonic re-expansion in the diverging section, as indicated by curve 3 in Figure 30, whereas the lesser root corresponds to supersonic expansion in the diverging section (curve 5). If the external pressure is less than P_2, the flow is obtained from Equation 225 in terms of P_∞ and the exit area. If the external pressure is decreased below P_5, and indicated by P_6 in Figure 30, the flow within the nozzle will be unchanged, and an irreversible expansion from P_5 to P_∞ will occur outside the nozzle (called underexpansion). If the external pressure is between P_2 and P_5, as illustrated by P_4, completely reversible flow cannot be attained. An irreversible, shock recompression will occur within, or just outside, the nozzle (overexpansion). The same mass flux density, as given by Equation 228 and the area of the throat, is obtained for all external pressures less than P_2.

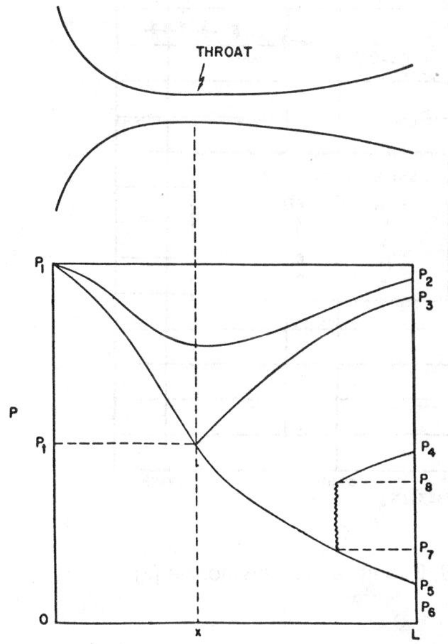

Figure 30. Pressure profiles in a nozzle [89].

Although $P_c/P_1 \simeq 0.5$, critical flow can be attained even with a ratio of external pressures, P_2/P_1, approaching unity. The minimum external pressure, P_2, which will produce sonic flow depends on the ratio of the area of the exit to the area of the throat, according to

$$\left(\frac{A_t}{A_e}\right)^2 = \frac{2}{\gamma - 1}\left(\frac{\gamma + 1}{2}\right)^{(\gamma+1)/(\gamma-1)}\left[1 - \left(\frac{P_2}{P_1}\right)^{(\gamma-1)/\gamma}\right]\left(\frac{P_2}{P_1}\right)^{\gamma/2}$$

Real Nozzles

The shape of a real nozzle, i.e., the relationship between area and length, is not determined by the above expression. The change in the area must be gradual enough to prevent separation. The converging section may be quite steep, but the total internal angle of the diverging section should not exceed 0.122 rad (7°). Deviations from ideality due to viscosity, two-dimensionality and separation are discussed by Emmons [87] and design guidelines can be found in [90]. Such deviations are sometimes correlated for a particular nozzle by multiplying the right side of Equations 225 and 228 by C_n^2 and then correlating the nozzle coefficient, C_n, with the Reynolds number at the throat. This is illustrated in Figure 31 for an ASME long-radius flow nozzle [91], with a converging section of length D_2 and a straight following section of length $0.6D_2$, where D_2 is the throat diameter.

Further discussions on nozzles, along with venturis and orifice plates are given in the measurement section of this volume. A few comments on the latter two devices is worthy at this point.

A venturi tube is a converging-diverging nozzle used to measure flow rate. Ordinarily, P_1 is measured upstream, and P_2 at the throat. Equation 203 is applicable, and the value of ρ_2 from Equation 216 should be used in the integral and in the denominator. However, a mean value of ρ is usually used as an approximation. The correlation coefficient for venturi tubes ranges from 0.9 to 0.99 and approaches the latter value for $Re_1 > 100{,}000$.

The fluid stream through the hole in an orifice plate necks down to a lesser diameter (vena contracta) somewhat downstream from the plate, and then re-expands. The flow is quite reversible down to this minimum jet diameter; hence, the above equations for isentropic expansions are applicable. However, the diameter of the orifice plate usually is used in the correlating equation, rather

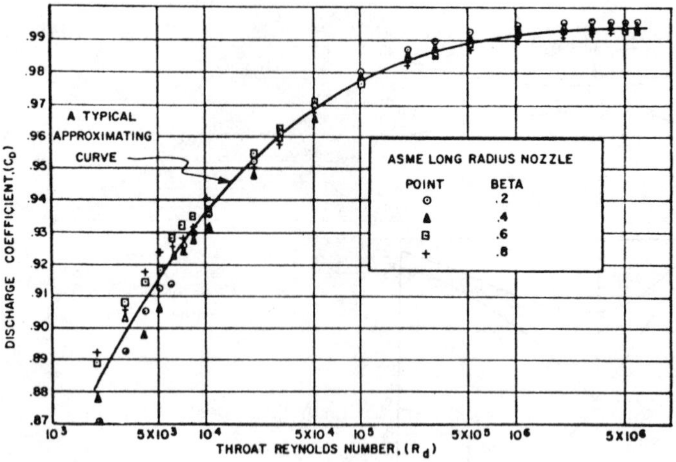

Figure 31. Discharge coefficient for AS, E long-radius flow-nozzle [9].

than the actual diameter at the downstream point of pressure measurement. As a result the coefficient of correction for an orifice differs significantly from unity. For subsonic flow, a mean density, rather than that given by Equation 216 usually is used as an approximation. The orifice coefficient for subsonic flow approaches the theoretical value of $\pi/(\pi + 2) \simeq 0.611$ for the vena contracta for larger Re [92]. For an orifice rounded off on the upstream side, a coefficient of 0.91 is recommended [93], and for rounding off on the downstream side, a coefficient of 0.82–0.86.

Flow Through Pipes

In the usual manner, we can write a force-momentum balance to represent flow through a constant cross-section, ignoring the net gravitational force.

$$\tau p\, dx + A\, dP + w\, du = 0 \tag{231}$$

where x = distance down channel
τ = mean shear stress over perimeter
p = perimeter of channel
A = cross-sectional area of channel
P = pressure
w = mass rate of flow
u = mean velocity

The wall shear stress can be expressed in terms of the Fanning friction factor,

$$f = \frac{2\tau}{\rho u^2} \tag{232}$$

For turbulent flow, this friction factor can be represented [94] by the correlating equation,

$$\frac{1}{\sqrt{f}} = -1.74 \ln\left\{\frac{e}{3.7D} + \frac{1.252}{\mathrm{Ref}^{1/2}}\right\}$$

where e = effective roughness (m)
$Re = Du\rho/\mu$ = Reynolds number for channel
μ = viscosity (Pa-s)

In a channel of constant cross-section, the mass flux density, G, is invariant and, hence, a more convenient variable than the linear velocity, u. Substituting for τ and u into the force balance gives

$$f\,dx = -\frac{D\rho\,dP}{2G^2} + \frac{D}{2}\frac{d\rho}{\rho} \tag{233}$$

Equation 233 holds for irreversible, as well as reversible, adiabatic flows. Substituting G/ρ for u, for C_p from Equation 221 and for ρ from the ideal gas law gives, on rearrangement,

$$\frac{4fL}{D} = \frac{\gamma+1}{\gamma}\ln\left(\frac{\rho_2}{\rho_1}\right) + \left[1 - \left(\frac{\rho_2}{\rho_1}\right)^2\right]\left(\frac{\gamma-1}{2\gamma} + \frac{P_1\rho_1}{G^2}\right) \tag{234}$$

Note that $G \to 0$ as $\rho_2 \to 1$ and, also, as $\rho_2 \to 0$. Hence, a maximum value exists for some intermediate, ρ_2, and can be found by equating $dG/d\rho_2$ to zero. If f again is to change negligibly with G, this gives

$$\frac{P_1\rho_1}{G_c^2} = \frac{\gamma+1}{2\gamma}\left(\frac{\rho_1}{\rho_{2c}}\right)^2 - \frac{\gamma-1}{2\gamma} \tag{235}$$

It can be shown further that

$$G_{2c}^2 = \gamma P_{2c}\rho_{2c} \tag{236}$$

Hence, that

$$u_{2c}^2 = \gamma P_{2c}/\rho_{2c} \tag{237}$$

Thus, despite the irreversibilities, the maximum flow in a channel occurs when the velocity at the exit becomes sonic; the irreversibilities simply increase the required external pressure ratio, P_1/P_2.

Eliminating ρ_1/ρ_2 produces

$$\frac{P_1\rho_1}{G_c^2} = \frac{4fL}{D} + \frac{1}{\gamma} + \frac{\gamma+1}{2\gamma}\ln\left\{\frac{\gamma-1}{\gamma+1} + \left(\frac{2\gamma}{\gamma+1}\right)\frac{P_1\rho_1}{G_c^2}\right\} \tag{238}$$

If the external pressure is less than P_{2c}, the flow rate is given by Equation 238 and ρ_1/ρ_{2c} then can be calculated from Equation 235 and P_{2c}/P_1 from Equation 237, thereby testing the postulate of $P_{2c} > P_\infty$. If the external pressure is greater than P_{2c}, the flow rate can be calculated from Equation 234 for a given ρ_2/ρ_1, and the corresponding P_2/P_1 from the ideal gas law. Trial and error is necessary to determine the appropriate ρ_2/ρ_1 for the specified P_∞. In either case, an additional, but rapidly converging, trial-and-error procedure is necessary to evaluate f from the friction factor correlation.

The above equations all can be reduced for the other limiting case of isothermal flow by setting $\gamma = 1$. Coulson and Richardson [95] assert that the adiabatic flow rate exceeds that for isothermal flow by less than 20%, for all practical cases, and by less than 5% for $L/D > 1{,}000$.

Wave Theory and Propagation

Compression waves are generated by a variety of causes. Examples are explosions, electrical discharges, and the movement of bodies at high velocities. Under some conditions these waves coalesce into a single strong wave traveling at supersonic velocity, thus constituting a shock wave. A shock wave exhibits a near discontinuity in pressure density and temperature. Such waves are important because of their potential for damage, but also because of their use to produce high temperatures in the laboratory.

The wave can be considered to create a finite discontinuity in pressure, temperature, density, and velocity in which case eddy and molecular transport are neglected. Ideal gas behavior and adiabatic conditions are postulated. Following Churchill [89], the equations of conservation are written with the velocities referring to the wave, in which case the mass and momentum force balances are

$$u_1'\rho_1 = u_2'\rho_2 \tag{239}$$

$$P_1 + u_1'^2\rho_1 = P_2 + u_2'^2\rho_2 \tag{240}$$

Equation 223 is applicable as an energy balance. From the ideal gas law and Equation 221:

$$\frac{u_1^2 - u_2^2}{2} = \frac{\gamma}{\gamma - 1}\left(\frac{P_2}{\rho_2} - \frac{P_1}{\rho_1}\right) \tag{241}$$

Combining Equations 239 and 240, the following relationships are developed:

$$\left.\begin{aligned}
&u_1'\rho_1 = u_2'\rho_2 = \left(\frac{P_2 - P_1}{\dfrac{1}{\rho_1} - \dfrac{1}{\rho_2}}\right)^{1/2}\\
&\frac{P_2}{P_1} = \frac{\beta(\rho_2/\rho_1) - 1}{\beta - \rho_2/\rho_1}\\
&\frac{u_1'}{u_2'} = \frac{\rho_2}{\rho_1} = \frac{1 + \beta(P_2/P_1)}{\beta + P_2/P_1}\\
&\frac{T_2}{T_1} = \frac{\beta + P_2/P_1}{\beta + P_1/P_2}\\
&M_1'^2 = \frac{\gamma - 1}{2\gamma}\left(1 + \beta\frac{P_2}{P_1}\right)\\
&M_2'^2 = \frac{\gamma - 1}{2\gamma}\left(1 + \beta\frac{P_1}{P_2}\right)\\
&M_1' = u_1'\left(\frac{\gamma P_1}{\rho_1}\right)^{1/2}\\
&M_2' \equiv u_2\left(\frac{\gamma P_2}{\rho_2}\right)^{1/2}
\end{aligned}\right\} \tag{242}$$

From Equation 214:

$$dS = \frac{dH}{T} + \frac{dP}{\rho T} = \frac{C_p\, dT}{T} - \frac{R}{m}\frac{dP}{P} \tag{243}$$

Integrating for constant C_p gives:

$$S_2 - S_1 = C_p \ln\left(\frac{T_2}{T_1}\right) - (C_p - C_v)\ln\left(\frac{P_2}{P_1}\right)$$

or,

$$\frac{S_2 - S_1}{C_p} = \ln\left\{1 - \frac{2}{\gamma + 1}\left(1 - \frac{1}{M_1'^2}\right)\right\} + \frac{1}{\gamma}\ln\left\{1 + \frac{2\gamma}{\gamma + 1}(M_1'^2 - 1)\right\} \tag{244}$$

where the following conditions arise:

$M_1' > 1 \quad S_2 - S_1 > 0$

$M_1' = 1 \quad S_2 - S_1 = 0$

$M_1' < 1 \quad S_2 - S_1 < 0$

The second law of thermodynamics shows that $M_1' > 1$. It follows that a shock wave is necessarily supersonic with respect to the gas ahead, and that P_2/P_1, ρ_2/ρ_1, and T_2/T_1 are all greater than unity. Thus, a shock wave is necessarily a compression wave; a shock expansion is excluded.

We now direct attention to reflected waves. Normal reflection off a solid body produces a further increase in pressure, temperature, and density. The velocity behind the reflected wave is zero. The frame of reference of the reflected wave produces a situation completely analogous to that of the initial wave in the standing frame of reference (which means that Equations 239–242 apply) with u_2'', u_3'', ρ_2, and ρ_3 replacing u_1', u_2', ρ, and ρ_2, respectively, etc.

This results in the following expressions:

$$\frac{P_3 - P_2}{P_2 - P_1} = \frac{\dfrac{\rho_1}{\rho_2} - 1}{1 - \rho_2/\rho_3} \tag{245}$$

$$\frac{P_3}{P_1} = \frac{(\beta + 2)\dfrac{P_2}{P_1} - 1}{1 + \beta(P_1/P_2)} \tag{246}$$

The other conditions behind the reflected wave are readily obtained by substituting P_3/P_2 for P_2/P_1 in the expressions for the initial wave.

A shock wave can be generated by a moving piston. In this case the velocity of the piston is represented by u_2.

$$\frac{P_2}{P_1} = 1 + \frac{\gamma(\gamma + 1)}{4}\left(\frac{u_2}{a_1}\right)^2\left\{1 + \left[1 + \left(\frac{4a_1}{(\gamma + 1)u_2}\right)^2\right]^{1/2}\right\} \tag{247}$$

where

$a_1 = (\gamma P_1/\rho_1)^{1/2}$

A shock wave can be generated in the laboratory by rupturing a diaphragm between the high- and low-pressure section of a shock tube, as illustrated in Figure 32. The generating equation is:

$$\frac{P_0}{P_1} = \frac{P_2}{P_1}\left[1 - \left(\frac{P_2}{P_2} - 1\right)\left(\frac{\beta - 1}{\beta_0 - 1}\right)\left(\frac{\gamma_1 T_1 m_0}{\gamma_0 T_0 m_1(\beta_1 + 1)\left(1 + \beta_1 \dfrac{P_2}{P_1}\right)}\right)^{1/2}\right]^{-(1+\beta_0)} \tag{248}$$

where the subscript 0 indicates the conditions in the driver section of the shock tube. This expression is based on the assumption that an isentropic rarefaction wave propagates upstream.

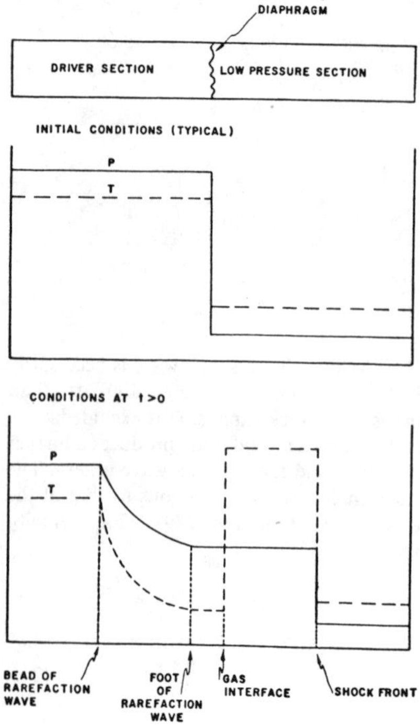

Figure 32. Shock tube behavior [92].

A detonation wave is a shock wave driven by a chemical reaction. Detonation waves have greater destructive power than explosions (isentropic expansions). Hence, their behavior is of direct interest in processing and storing explosive materials [96, 97].

Defining ΔH_{R1} as the enthalpy change due to reaction at T_1 and C_{P2} as mean heat capacity of reacted mixture between T_1 and T_2:

$$\frac{\Delta H_{R_1}}{C_{P_2} T_1} + \frac{T_2}{T_1} - 1 = \left(\frac{\gamma_2 - 1}{2\gamma_2}\right)\left(\frac{P_2}{P_1} - 1\right)\left(\frac{m_2}{m_1} + \frac{T_2 P_1}{T_1 P_2}\right) \tag{249}$$

where subscripts 1 and 2 designate the unreacted and reacted mixtures, respectively.

From Equations 222, 239, and 240:

$$H_2 - H_1 = \left(\frac{P_2 - P_1}{2}\right)\left(\frac{1}{\rho_2} + \frac{1}{\rho_1}\right) \tag{250}$$

Differentiating for fixed P_1, ρ_1, and H_1 gives

$$dH_2 = \left(\frac{1}{\rho_1} + \frac{1}{\rho_2}\right)\frac{dP_2}{2} - \left(\frac{P_2 - P_1}{2}\right)\frac{d\rho_2}{\rho_2^2} \tag{251}$$

From Equation 214 for condition 2 with the same restrictions,

$$dH_2 = T_2\, dS_2 + \frac{dP_2}{\rho_2}$$

Equating these two expressions for dH_2 and setting $dS_2 = 0$,

$$u'_2 = \left(\frac{dP_2}{d\rho_2}\right)_s^{1/2}$$

This expression indicates that the burned gas moves at acoustic velocity relative to the wave (Chapman-Jouguet condition).

Substitution of

$$u'_2 = (\gamma_2 P_2/\rho_2)^{1/2}$$

in Equation 242 and rearrangement leads to:

$$\frac{T_1}{T_2} = \frac{m_1}{\gamma_2 m_2}\left(\gamma_2 + 1 - \frac{P_1}{P_2}\right)\frac{P_1}{P_2}$$

Combining Equations 249 with this expression gives

$$\left(\frac{P_2}{P_1}\right)^2 + 2\left[\frac{m_1}{m_2}\left(\frac{\Delta H_{R_1} T_1}{C_{P_2}} - 1\right)\gamma_2 + \frac{\gamma_2 - 1}{\gamma_2 + 1}\left(\frac{3\gamma_2 + 1}{2}\right)\right]\frac{P_2}{P_1} - \frac{m_1}{m_2}\left(\frac{\Delta H_R}{C_{P_2} T_1} - 1\right)\frac{2\gamma_2}{\gamma_2 + 1}$$
$$- \gamma_2\left(\frac{\gamma_2 - 1}{\gamma_2 + 1}\right) = 0 \quad (252)$$

Computation details for evaluating P_2/P_1 as a function of m_2, ΔH_{R1}, C_{p2} and γ_2 are given by Churchill [92].

Adamson et al. [98] give the following approximate expression for the Mach number of a detonation wave:

$$M'_1 = \left[\left(\frac{\gamma + 1}{2}\right)\left(\frac{-\Delta H_R}{C_p T_1}\right)\right]^{1/2} + \left[\left(\frac{\gamma + 1}{2}\right)\left(\frac{-\Delta H_R}{C_p T_1}\right) + 1\right)\right]^{1/2} \quad (253)$$

Morrison [99] proposed that $(-\Delta H_R)$ be estimated as two-thirds of the value for complete reaction, and that γ and m be evaluated at the initial conditions.

Adamson and Morrison [98] further generalized the corresponding solution for the other characteristics for both detonation and shock waves:

$$\frac{P_2}{P_1} = 1 + \frac{\gamma F}{\gamma + 1}(M_1'^2 - 1) \quad (254)$$

$$\frac{u'_2}{u'_1} = \frac{\rho_1}{\rho_2} = 1 - \frac{F(M'_1 - 1)}{(\gamma + 1)M_1'^2} \quad (255)$$

$$\frac{T_2}{T_1} = \left[1 + \frac{\gamma F}{\gamma + 1}(M_1'^2 - 1)\right]\left[-1\frac{F(M_1'^2 - 1)}{(\gamma + 1)M_1'^2}\right] \quad (256)$$

and

$$M_2'^2 = \frac{(\gamma + 1 - F)(M_1'^2 - 1) + \gamma + 1}{F\gamma(M_1'^2 - 1) + \gamma + 1} \quad (257)$$

where F = 1 for detonation waves and 2 for shock waves.

If the further reaction on reflection is neglected, the following expression is applicable for P_3/P_2 for a detonation:

$$\frac{P_3}{P_2} = 1 + \frac{(\gamma + 1)\left(1 - \frac{P_1}{P_2}\right)^2}{4\gamma}\left[1 + \left(1 + \left[\frac{4\gamma}{\left(1 - \frac{P_1}{P_2}\right)(\gamma + 1)}\right]^2\right)^{1/2}\right] \tag{258}$$

See Churchill [92] for details.

Stagnation

Stagnation temperature is an important consideration in thermometry. When a gas stream is slowed down by flowing over or impacting on a surface, or by expanding into a large reservoir, the temperature increases due to the conversion of the kinetic energy to thermal energy. For free, adiabatic flow, Equation 223 applies, where $u_2 = 0$.

$$\frac{T_s}{T_1} = 1 + \frac{u_1^2}{2C_pT_1} \tag{259}$$

or

$$\frac{T_s}{T_1} = 1 + \left(\frac{\gamma - 1}{2}\right)M_1^2 \tag{260}$$

where T_s is the stagnation temperature.

It follows that T_s is invariant in any adiabatic process, whether reversible or not. Equation 260 is applicable, therefore, to flows with and without friction, e.g., to pipes as well as nozzles.

For adiabatic and reversible flows (such as approximated in nozzles) of an ideal gas with constant heat capacity, Equation 216 can be combined with Equation 260 to give

$$\frac{P_s}{P_1} = \left[1 + \left(\frac{\gamma - 1}{2}\right)M_1^2\right]^{\gamma/(\gamma - 1)} \tag{261}$$

or

$$\frac{\rho_s}{\rho_1} = \left[1 + \left(\frac{\gamma - 1}{2}\right)M_1^2\right]^{1/(\gamma - 1)} \tag{262}$$

Note that the stagnation pressure, P_s, and the stagnation density, ρ_s, also are invariant in such flows.

In series expansion form:

$$\frac{P_s}{P_1} = 1 + \frac{\gamma}{2}M_1^2\left(1 + \frac{M_1^2}{4} + \cdots\right) = 1 + \frac{u_1^2\rho_1}{2P_1}\left(1 + \frac{u_1^2\rho_1}{4\gamma P_1} + \cdots\right) \tag{263}$$

And for incompressible flow,

$$\frac{P_s}{P_1} = 1 + \frac{u_1^2\rho}{2P_1} \tag{264}$$

The second term inside the brackets of Equation 263 represents the first-order effect of compressibility.

Expanding Equation 262:

$$\frac{\rho_s}{\rho_1} = 1 + \frac{M_1^2}{2} + \frac{M_1^4}{4} + \cdots \tag{265}$$

The first-order effect of compressibility on the density, $(M_1^2/2)$, is seen to be twice that for the pressure, $(M_1^2/4)$.

From Equations 263 and 265 for incompressible, adiabatic flow,

$$\frac{T_s}{T_1} = 1 + \frac{u_1^2 \rho}{2P_1} \tag{266}$$

Comparison with Equation 260 indicates that the coefficient of u_1^2 for compressible flow is only $(\gamma - 1)/\gamma$ times that for incompressible flow, whereas ρ_s and P_s are greater.

The stagnation density for adiabatic flow of an ideal gas in a pipe can be determined by setting $u_2 = 0$ in Equation 223:

$$\frac{u_1^2}{2} = \frac{\gamma}{\gamma - 1}\left(\frac{P_s}{\rho_s} - \frac{P_1}{\rho_1}\right) \tag{267}$$

or

$$\frac{4fL}{D} = \ln\left(\frac{\rho_s}{\rho_1}\right)^2 + \frac{1}{\gamma}\left(\frac{1}{M_1^2} + \frac{\gamma - 1}{2}\right)\left[1 - \left(\frac{\rho_s}{\rho_1}\right)^2\right] \tag{268}$$

The stagnation pressure can, in turn, be calculated using the ideal gas law and T_s from Equation 260.

NOTATION

A	area; constant in Prandtl-Eyring model
a	acoustic velocity; constant in Sisko model
B	constant in Prandtl-Eyring model; constant in Froishteter-Vinogradov equation
b	constant
C	coefficient; constant in rheological model
C	specific heat capacity
c	constant
D	diameter
E	bulk modulus; internal energy
E_v	rate of irreversible conversion of mechanical to internal energy per unit fluid mass
e	effective roughness
F	force; wave coefficient (= 1 for detonation, = 2 for shocks)
f	friction factor; fugasity
G	mass flux density
g	gravitational acceleration
H	enthalpy; distance
He	Hedstrom number
K	power law consistency index; friction loss coefficient
Ke	kinetic energy
K′	factor in Metzner-Reed equation
k	empirical exponent; constant
L	length
L_e	equivalent length
l	Prandtl mixing length
M	Mach number
m	reciprocal of flow behavior index; molar mass
n	power law flow behavior index; constant
n′	factor in Metzner-Reed equation
P, p	pressure
p	perimeter
Q	heat; volumetric flow rate
q	volumetric flow rate; heat transferred to system
R	radius; universal gas constant
Re	Reynolds number
R_H	hydraulic radius
r	radius; radial coordinate
S	specific entropy
s	distance in direction of flow
T	temperature; torque
t	time
U	average velocity
u	velocity
u_0	superficial velocity
u_x, u'_y	instantaneous velocity components
u_{rms}	root mean square velocity
u^+	dimensionless velocity

u^*	friction velocity
v	velocity
W	mass flow rate; specific work done by system
W′	work
w	velocity component
x	coordinate; ratio of yield stress to wall stress for Bingham fluid
Y	yield number for Bingham fluid
y	coordinate
y^+	dimensionless distance
Z, z	elevation

Greek Symbols

α	kinetic energy correction factor
γ	specific heat ratio; shear rate
Δ	parameter
δ	wall thickness; boundary layer thickness; factor in Metzner-Reed equation
ϵ	roughness height
ϵ_M	eddy viscosity of momentum
η	fugacity coefficient; plastic viscosity for Bingham fluid
η_a	apparent viscosity
μ	viscosity
μ_{eff}	effective viscosity
ν	kinematic viscosity; specific volume
θ	coordinate
ρ	density
τ	shear stress
τ_y	yield stress
ϕ	coordinate
ϕ_0, ϕ_1	constants in Ellis model
ψ	stream function
Ω	angular velocity

REFERENCES

1. Reynolds, W. C., and Perkins, H. C., *Engineering Thermodynamics*, New York: McGraw-Hill Book Co., (1970).
2. Lee, J. F., Sears, F. W., and Turcotte, D. L., *Statistical Thermodynamics*, Reading, MA, Addison-Wesley Pub. Co., Inc., (1963).
3. Weber, H. C., *Thermodynamics for Chemical Engineers*, New York: John Wiley & Sons, Inc., (1939).
4. McAdams, W. H., *Heat Transmission*, New York: McGraw-Hill Book Co., 1942.
5. Badger, W. L., and McCabe, W. L., *Elements of Chemical Engineering*, New York: McGraw-Hill Book Co., 1936.
6. Knudsen, J. G., and Katz, D. L., *Fluid Dynamics and Heat Transfer*, New York: McGraw-Hill Book Co., 1936.
7. Walker, W. H., Lewis, W. K., McAdams, W. H., and Gilliland, E. R., *Principles of Chemical Engineering*, 3rd ed., New York: McGraw-Hill Book Co., (1937).
8. Stanton, T. E., Marshall, D., and Bryant, C. N., *Proc. Roy. Soc.* (London) 97A:413, (1920).
9. Nikuradse, J., VDI-Forschungsheft 356 (1932).
10. Reichardt, H., NACA TN 1047 (1943).
11. Deissler, R. G., NACA TN 2138 (1950).
12. Moody, L. F., *Trans. ASME* 66:671 (1944).
13. Prandtl, L. Z., *Ver. Deut. Ing.*, 77:105 (1933); NACA TN 720 (1933).
14. von Karman, T., NACA TN 611 (1931).
15. Prandtl, L. Z., *Angew. Math. Mech.*, 5:136 (1925); NACA TN 1231 (1949).
16. Blasius, H., Mitt. Forschungsarb., 131:1–40 (1913).
17. von Karman, T., NACA TN 611 (1931).
18. von Karman, T. J., *Aeronaut. Sci.*, 1:1 (1934).
19. Reichardt, H., NACA TN 1047 (1943).
20. Cheremisinoff, N. P., *Fluid Flow: Pumps, Pipes and Channels*, Ann Arbor, MI: Ann Arbor Science Publishers, Inc., (1981).
21. Deissler, R. G., NACA TN 2242 (1950).
22. Deissler, R. G., *Trans. ASME* 73:101 (1951).
23. Deissler, R. G., and Eian, C. S., NACA TN 2629 (1952).

24. Deissler, R. G., NACA TN 3145 (1954).
25. Deissler, R. G., and Tayloer, M. F., NACA TN 3451 (1955).
26. Nikuradse, J., *Proc. Int. Cong. Appl. Mech.*, (3rd Congress, Stockholm), 1:239 (1932).
27. Nikuradse, J., VDI Forschungsheft 361 (1933).
28. Rouse, H., *Elementary Mechanics of Fluids*, New York: John Wiley & Sons, Inc. (1948).
29. Moody, L. F., *Trans. ASME* 66:671 (1944).
30. Colebrook, C. F., and White, D., *J. Inst. Civil Eng.*, London 11: 133 (1938–1939).
31. Simpson, L. L., "*Process Piping: Functional Design*," Chem. Eng., 76(8):167–181, (1969).
32. *Chem. Eng.*, 75(13):198–199 (1968).
33. Blasius, H. Z., *Math. Phys.*, 56:1 (1908).
34. Howarth, L., *Proc. Roy. Soc.* (London), 164A:547 (1938).
35. von Karman, T. Z., *Angew. Math. Mech.* 1:233 (1921).
36. Polhausen, K. Z., *Angew. Math. Mech.* 1:252 (1921).
37. Janour, Z., NACA TN 1316 (1951).
38. von Karman, T., NACA TN 1092 (1946).
39. Cheremisinoff, N. P., and Azbel, D. S., *Fluid Mechanics and Unit Operations*, Ann Arbor Science Pub., Ann Arbor, MI (1983).
40. Cheremisinoff, N. P., and Gupta, R.,—eds., *Handbook of Fluids in Motion*, Ann Arbor Science Pub., Ann Arbor, MI (1983).
41. Falkner, V. M., *Aircraft Eng.*, 15:65 (1943).
42. Schlichting, H., NACA TN 1218 (1949).
43. von der Hegge-Zijnen, B. G., Verhandel. Koninkl. Akad. Wetenschap. Amsterdam Afdeel, Natuur, 31:499 (1928).
44. Prandtl. L., NACA TN 542 (1928).
45. Lamb. H., *Phil. Mag.*, 21:112 (1911).
46. Weisselberger, C., *Ergeb. Aerodyn. Versuchsantalt Gottingen* 2:22 (1923).
47. Weisselberger, C., *Ergeb. Aerodyn. Versuchsanstalt Gottingen* 1:120 (1923).
48. Delany, K., and Sorenson, N. E., NACA TN 3038 (1953).
49. Millikan, C. B., *Trans. ASME* 54: APM-3 (1932).
50. Tomotika, S., *Brit. Aeronaut. Res. Comm.*, R & M 1678 (1935).
51. Bird, R. B., Armstrong, R. C., and Hassager, O., *Dynamics of Polymeric Liquids*, Vols. I & II, New York: John Wiley & Sons, Inc. (1977).
52. Frederickson, A. G., *Principles and Applications of Rheology*, (Englewood Cliffs, NJ: Prentice-Hall, Inc., 1964).
53. Han, C. D., *Rheology in Polymer Processing*, New York: Academic Press, Inc. (1976).
54. McKelvey, J. M., *Polymer Processing*, New York: John Wiley & Sons, Inc. (1962).
55. Middleman, S., *The Flow of High Polymers: Continuum and Molecular Rheology*, New York: Interscience Publishers, (1968).
56. Schowalter, W. T., *Mechanics of Non-Newtonian Fluids*, Oxford, England: Pergamon Press, Inc., (1978).
57. Bird, R. B., Stewart, W. E., and Lightfoot, E. N., *Transport Phenomena*, New York: John Wiley & Sons, Inc. (1960).
58. Lodge, A. S., *Elastic Liquids*, New York: Academic Press, Inc. (1964).
59. Skelland, A. H. P., *Non-Newtonian Flow and Heat Transfer*, New York: John Wiley & Sons Inc., (1967).
60. Metzner, A. B., In *Advances in Chemical Engineering*, Vol. 1, T. B. Drew and J. W. Hoopes, Eds., (1956), pp. 79–150.
61. Govier, G. W., and Aziz, K. A., *The Flow of Complex Mixtures in Pipes*, New York: Krieger Publishing Co., (1977).
62. Bauer, W. H., and Collins, E. A., In *Rheology: Theory and Applications*, F. R. Eirich, Ed., Vol. 4, New York: Academic Press, Inc. (1967), pp. 423–459.
63. Markovitz, H., "*Rheological Behavior of Fluids, National Committee for Fluid Mechanics Films*," Encyclopedia Brittanica Educational Corp., Chicago, IL, (1972).
64. Van Wazer, J. R., Lyons, J. W., Kim, K. Y., and Colwell, R. E., *Viscosity and Flow Measurement*, New York: Interscience Publishers, (1963).

65. Cheremisinoff, N. P., *Fluid Flow: Pumps, Pipes, and Channels*, Ann Arbor, MI: Ann Arbor Science Publishers, Inc., (1981).
66. Metzner, A. B., and Reed, J. C., *AIChE J.*, 1:434 (1955).
67. Metzner, A. B., *Ind. Eng. Chem.*, 49:1429 (1957).
68. Laird, W. M., *Ind. Eng. Chem.*, 49:138 (1957).
69. Frederickson, A. G., and Bird, R. B., *Ind. Eng. Chem.*, 50:347 (1958).
70. Frederickson, A. G., and Bird, R. B., *Ind. Eng. Chem. Fund.*, 3:383 (1964).
71. Vaughn, R. D., and Bergman, P. D., *Ind. Eng. Chem. Proc. Dec. Dev.*, 5:44 (1966).
72. Acrivos, A., Shah, M. I., and Petersen, E. E., *AIChE J.*, 6:312 (1960).
73. Schlichting, H., *Boundary Layer Theory*, 6th Ed., New York: McGraw-Hill Book Co., (1968).
74. Ryan, N. W., and Johnson, M. M., *AIChE J.*, 5:433 (1959).
75. Dodge, D. W., and Metzner, A. B., *AIChE J.*, 5:189 (1959).
76. Metzner, A. B., and Park, M. G., *J. Fluid Mech.*, 20:291 (1964).
77. Hanks, R. W., *AIChE J.*, 9:306 (1963).
78. Froishteter, G. B., and Vinogradov, G. V., *Rheol. Acta.*, 16:620 (1977).
79. Masuyama, T., and Kawashima, T., *Bull. JSME*, 22:48 (1979).
80. Collins, M., and Schowalter, W. R., *AIChE J.*, 9:804 (1963).
81. Collins, M., and Schowalter, W. R., *AIChE J.*, 9:98 (1963).
82. Michiyoshi, L., Mizuno, K., and Hoshiai, Y., *Int. Chem. Eng.*, 6:373 (1966).
83. Chen, S. S., Fan, L. T., and Hwang, C. L., *AIChE J.*, 16:293 (1970).
84. Weltmann, R. N., and Keller, T. A., NACA TN 3389 (1957).
85. Heywood, N. I., *Chem. E. Symposium Series* No. 60:33 (1980).
86. Knudsen, J. G., and Katz, D. L., *Fluid Dynamics and Heat Transfer*, New York: McGraw-Hill Book Co., (1958).
87. Emmons, H. W., ed., *Fundamentals of Gas Dynamics.* Vol. III: *High Speed Aerodynamics and Jet Propulsion*, Princeton, NJ: Princeton University Press, (1958).
88. Balzhiser, R. E., and Samuels, M. R., *Engineering Thermodynamics*, Englewood Cliffs, NJ: Prentice-Hall, Inc., (1977).
89. Churchill, S. W., Chapter 8 in *Handbook of Fluids in Motion*, N. P. Cheremisinoff and R. Gupta (eds.), Ann Arbor Pub., Ann Arbor, MI, (1983).
90. Cheremisinoff, N. P., *Applied Fluid Flow Measurement*, Marcel Dekker Inc., New York, NY, (1979).
91. Benedict, R. P., "*Most Probable Discharge Coefficients for ASME Flow Nozzles*," J. Basic Eng. 880:734–744 (1966).
92. Churchill, S. W., *The Practical Use of Theory in Fluid Flow, Book* 1, *Inertial Flows*, Thornton, PA: Etaner Press, (1980).
93. Grace, H. P., and Lapple, C. E., "Discharge Coefficients of Small Diameter Orifices and Flow Nozzles," *Trans: ASME* 73:639–647 (1951).
94. Churchill, S. W. "Empirical Expressions for the Shear Stress in Turbulent Flow in Commercial Pipes," *AIChE J.*, 19:375–376 (1973).
95. Coulson, J. M., and Richardson, J. F., *Chemical Engineering*, Vol. I: *Fluid Flow Heat Transfer, and Mass Transfer*, (New York: McGraw-Hill Book Co. (1954), p. 57.
96. Ginsburgh, I., and Bulkley, W. L., "Hydrocarbon-Air Detonations-Industrial Aspects," *Chem. Eng. Prog.*, 59(2):82–86 (1963).
97. Randall, R. N., Bland, J. Dudley, W. M., and Jakobs, R. B., "The Effects of Gaseous Detonations Upon Vessels and Piping," *Chem. Eng. Prog.* 53(12):574–580 (1957).
98. Adamson, T. C., Jr., and Morrison, R. B., "On the Classification of Normal Detonation Waves," *Jet Propulsion*, 25:400, 403 (1955).
99. Morrison, R. B., "A Shock Tube Investigation of Detonative Combustion," Report AF33 (038) 12657, UMM-97, Willow Run Research Center, University of Michigan, Ann Arbor, MI (1952).

CHAPTER 11

THEORY OF MINIMUM ENERGY AND ENERGY DISSIPATION RATE

Chih Ted Yang

U.S. Department of the Interior
Bureau of Reclamation
Engineering and Research Center
Denver, CO 80225

and

Charles C. S. Song

St. Anthony Falls Hydraulic Laboratory
Department of Civil and Mineral Engineering
University of Minnesota
Minneapolis, MN 55414

CONTENTS

INTRODUCTION

The science of mechanics has been developed along two main lines, i.e., the vectorial and the variational approaches. The vectorial mechanics approach starts directly from Newton's laws of motion. Force and momentum are the basic concern of the vectorial approach. Thus, the motion of a particle can be uniquely determined by the known forces action on it at every instant. The variational approach to mechanics replaces the force by the work of the force or the work function, which is frequently replaceable by potential energy. At the same time, the momentum is replaced by kinetic energy. The variational approach is a scalar approach which the law of mechanics is stated as the minimization of the action integral. The mathematical tool used in the vectorial approach is the equation of motion, while the calculus of variation is used in solving minimization problems. Because the minimum value depends on the constraints that define the optimum path, the scalar approach is that of minimization with constraints instead of solving differential equations with boundary conditions.

Most of the recent developments in fluid mechanics and hydraulics are based on the vectorial approach. This approach is well suited to solving problems that can be described easily in a rectangular coordinate system, but becomes cumbersome for curvilinear coordinates. The process and resulting equations based on the scalar approach remain valid for an arbitrary choice of coordinates. For systems with constraints, the scalar approach is often simpler and more economical. Thus, the scalar or minimization approach is especially suitable for solving complicated problems of natural rivers. A more detailed discussion of the two approaches and their merits is given by Lanczos [1].

This chapter provides a brief summary of the history of development of the theory of minimum energy and energy dissipation rate and its special or simplified versions. The general theory of minimum energy and energy dissipation rate are explained and elaborated. Examples of application of the general theory and its simplified versions are used to demonstrate the flexibility and strength of the theory. Other related theories in minimization or maximization are reviewed and compared to resolve possible confusions on the theories and their applications.

HISTORY OF THEORY DEVELOPMENT

The magnitude and direction of forces must be given in the spatial and time domain before the vectorial approach can be applied to solving fluid mechanic problems. The equation of motion becomes an ideal and fundamental tool for rigid boundary fluid mechanics when and where the above two requirements are satisfied. In solving fluvial hydraulic problems, the description of a three-dimensional river of irregular shape by rectangular or curvilinear coordinates becomes cumberson. In many cases, hydraulic engineers find it difficult or impossible to solve river hydraulic problems analytically with the vectorial approach due to insufficient numbers of independent equations available. This difficulty stems from the fact that both the boundary and the roughness of a river are varibles. The lack of acceptable and reliable theories in predicting the variations of river boundary and roughness prompted engineers to seek other avenues for solution. The following history of theory development is evidence of this.

Minimum Variance

Based on U.S. Geological Survey river gaging station data, the hydraulic and channel geometry characteristics can be approximated as power functions of water discharge by the following equations:

$$V = kQ^m \tag{1}$$

$$D = cQ^f \tag{2}$$

$$B = aQ^b \tag{3}$$

$$S = iQ^j \tag{4}$$

where V = average flow velocity; Q = water discharge; D = average flow depth; B = channel width; S = channel slope; and k, m, c, f, a, b, i, j = coefficients. Leopold and Maddock [2] called these empirical relationships hydraulic geometry relationships. They also found that the ranges of variation of hydraulic exponents m, f, b, and j are relatively small compared with those of coefficients k, c, a, and i. Leopold and Maddock's work was later expanded by Stall and Yang [3] by relating water discharge to drainage area and flow frequency. The first attempt to provide a theoretical explanation of the variation of the hydraulic exponents was made by Langbein [4] based on minimum variance.

Langbein hypothesized that the changes in the variables shown in Equations 1 through 4 are such that the total effect, action, work, or adjustment is a minimum. At a cross section of a river, Langbein's hypothesis would suggest that all variables strive to resist any imposed changes, with the net result of having equal amount of adjustments among the dependent variables, subject to the external restrictions. The "variance" used by Langbein represents the square of hydraulic exponent. Thus, minimum variance means the sum of the squares of the hydraulic exponents is a minimum that is consistent with local restrictions. One of the basic problems of applying Langbein's hypothesis is the selection of correct combination of variables to be used in the minimization process. Different combinations of variables could lead to different answers. Although Williams [5] suggested a step-by-step computational procedure, the lack of sound theoretical basis for the hypothesis and the selection of variables are yet to be overcome before the hypothesis can be applied with consistent results. However, Langbein's contribution of introducing the concept of minimization to the study of river morphology prompted other researchers to devote their attentions to the advancement and application of minimization theories.

Entropy Analogy and Minimum Unit Stream Power

Leopold and Langbein [6] introduced the concept of entropy to the study of stream morphology. The entropy concept was further explained by Langbein [4]. Kennedy et al. [7] pointed out some mathematical inadequacy in Leopold and Langbein's derivation. Langbein [8] justified his use of entropy analogy because of the propriety of the analogy between temperature and elevation as shown by Scheidegger [9] from statistical mechanics. No further advancement on entropy analogy was made until 1971.

Yang [10] reintroduced the entropy concept to the study of natural stream systems. He considered the only useful energy per unit mass or weight of water in a stream system is its potential energy. He further assumed that the potential energy and elevation of a stream system are equivalent to thermal energy and absolute temperature, respectively, of a heat system. Based on this analogy and the direct application of entropy concept in thermodynamics, it can be shown that [10]

$$\frac{dY}{dt} = \text{a minimum} \tag{5}$$

where Y = potential energy per unit weight of water in a stream system; and t = time. Thus, Yang concluded that during the evolution toward its equilibrium condition, a natural stream chooses its course of flow in such a manner that the rate of potential energy dissipation per unit weight of water along its course is a minimum. The minimum value depends on the constraints applied to the stream. Unlike the minimum variance concept by Langbein, Equation 5 gives a clearly defined objective function and equation for minimization. Equation 5 can be rewritten as

$$\frac{dY}{dt} = \frac{dX}{dt}\frac{dY}{dX} = VS = \text{a minimum} \tag{6}$$

where X = distance along the course of flow; V = average flow velocity; and S = energy slope. Yang [11, 12] defined the VS product as the unit stream power. Thus, the theory of minimum unit stream power can be stated as [12]:

> "For flow in an alluvial channel of a given width, where the rate of energy dissipation due to sediment transport is negligible, the channel will adjust its velocity, slope, depth,

roughness in such a manner that a minimum amount of unit stream power is used to transport a given sediment and water discharge under equilibrium condition. The value of minimum unit stream power depends on the constraints applied to the channel. If the flow deviates from its equilibrium condition, it will adjust its velocity, slope, depth, and roughness until the unit stream power is minimized and regains equilibrium."

The minimum unit stream power theory was successfully applied by Yang [12] for the determination of velocity, slope, depth, and roughness of one-dimensional uniform subcritical flows. The average value of correlation coefficient between the measured and computed unit stream power for 449 sets of laboratory and river data is 0.997. In his solution, Yang [12] did not use any roughness equation because roughness varies with flow and sediment conditions. Encouraged by the theory of minimum unit stream power's ability to solving complicated alluvial hydraulic problems, further developments to expand the theory for more general applications have been made in the last few years.

Minimum Rate of Energy Dissipation and Equation of Motion

Some scientists and engineers remain skeptical of the application of minimization theories partially because they are more used to the vectorial approach of solving equations of motion. Another reason is the lack of more rigorous derivation or proof of the theory of minimum unit stream power, based on Yang's original assumption of entropy analogy. The following is a summary of theoretical derivation by Yang and Song [13] to show that the theory of minimum rate of energy dissipation can be derived from Navier-Stokes equations of motion for gradually varied flows without sediment when the inertia force due to time-averaged velocity distribution is small compared with the force due to gravity and shear.

For turbulent flows, the Reynolds' modification of the Navier-Stokes equations of motion for incompressible fluid is [14]

$$\rho\left(\frac{\partial \bar{u}_i}{\partial t} + \bar{u}_j \frac{\partial \bar{u}_i}{\partial x_j}\right) = -\frac{\partial \gamma \bar{h}}{\partial x_i} + \frac{\partial \sigma_{ij}}{\partial x_j}; \qquad i, j = 1, 2, 3 \tag{7}$$

where $\gamma\bar{h}$ = time-averaged gravitational potential; γ = specific weight of water; $\bar{u}_i$ = time-averaged velocity; and σ_{ij} = the Reynold's total stress tensor is

$$\sigma_{ij} = -\bar{P}\delta_{ij} + \bar{\tau}_{ij} - \rho\overline{u_i'u_j'} \tag{8}$$

where $\bar{P}$ = the average pressure; δ_{ij} = the unit second-order tensor; $\bar{\tau}_{ij}$ = the viscous stress tensor; and $-\rho\overline{u_i'u_j'}$ = the Reynold's turbulent stress tensor. According to Boussinques' concept, it is possible to define an eddy viscosity, ϵ, such that

$$\sigma_{ij} = -\bar{P}\delta_{ij} + \rho(\nu + \epsilon)\left(\frac{\partial \bar{u}_i}{\partial x_j} + \frac{\partial \bar{u}_j}{\partial x_i}\right) \tag{9}$$

and the equation of motion for the turbulent flow becomes similar to that of the laminar flows. The only difference is that, unlike the kinematic viscosity, the eddy viscosity is a flow-related variable.

The case of gradually varied steady open channel flows for which the Froude number is not too large will be considered herein. Under this condition, the left-hand side of Equation 7, the inertia terms, are negligible and the equation of motion reduces to

$$\frac{\partial}{\partial x_i}(\gamma\bar{h} + \bar{P}) = \rho\frac{\partial}{\partial x_j}\left[(\nu + \epsilon)\left(\frac{\partial \bar{u}_i}{\partial x_j} + \frac{\partial \bar{u}_j}{\partial x_i}\right)\right] \tag{10}$$

Analogous to the laminar flow case, we shall define the energy dissipation function, ϕ, or the rate of energy dissipation per unit volume of fluid, for the turbulent flows as follows [15]:

$$\phi = \frac{1}{2}\rho(\nu + \epsilon)\left(\frac{\partial \bar{u}_i}{\partial x_j} + \frac{\partial \bar{u}_j}{\partial x_i}\right)^2 \tag{11}$$

or, from Equation 9

$$\phi = \frac{1}{2}(\sigma_{ij} + \bar{P}\delta_{ij})\left(\frac{\partial \bar{u}_i}{\partial x_j} + \frac{\partial \bar{u}_j}{\partial x_i}\right) \tag{12}$$

By integrating ϕ over the total flow region of interest, the total rate of energy dissipation, E, is obtained, i.e.,

$$E = \iiint_V \phi \, dV \tag{13}$$

where V = the volume.

Let $\bar{u}_i$ be the velocity distribution that satisfies Equation 10, a given set of boundary conditions, and the equation of continuity

$$\frac{\partial \bar{u}_i}{\partial x_i} = 0 \tag{14}$$

Then consider any other hypothetical velocity distribution $(\bar{u}_i + \hat{u}_i)$ that does not satisfy Equation 10 but has the same eddy viscosity, ϵ, and satisfies the equation of continuity and the same boundary condition. According to Equation 11, the dissipation function of the alternate flow is

$$\phi(\bar{u}_i + \hat{u}_i) = \phi(\bar{u}_i) + \rho(\nu + \varepsilon)\left(\frac{\partial \bar{u}_i}{\partial x_j} + \frac{\partial \bar{u}_j}{\partial x_i}\right)\left(\frac{\partial \hat{u}_i}{\partial x_j} + \frac{\partial \hat{u}_j}{\partial x_i}\right) + \frac{1}{2}\rho(\nu + \varepsilon)\left(\frac{\partial \hat{u}_i}{\partial x_j} + \frac{\partial \hat{u}_j}{\partial x_i}\right)^2 \tag{15}$$

The total rate of energy dissipation of the alternate flow is the integral of Equation 15. It can be shown that the integral of the second term on the right-hand side of Equation 15 is zero. Therefore, by integrating Equation 15, we have

$$E(\bar{u}_i + \hat{u}_i) = E(\bar{u}_i) + E(\hat{u}_i) \tag{16}$$

Because E is nonnegative

$$E(\bar{u}_i) = E(\bar{u}_i + \hat{u}_i) - E(\hat{u}_i) \leq E(\bar{u}_i + \hat{u}_i) \tag{17}$$

for any $\hat{u}_i$.

It shows that the velocity distribution, $\bar{u}_i$, that satisfies the equation of motion, Equation 10, results in less rate of energy dissipation than any other velocity distribution that could not satisfy the equation of motion. That is, among all velocity distributions that satisfy the equation of continuity and a given set of boundary conditions, only the one that minimizes E satisfies the equation of motion. Formally, it can be said that the system, $\bar{u}_i$, which minimizes Equation 13 subject to Equation 14 and given boundary conditions, is the real flow solution. Thus, the minimization theory is equivalent to the equation of motion, and the flow problem reduces to a minimization problem.

The advantage of using the minimization approach as compared with the conventional approach of solving the equation of motion is in its flexibility of applications. Optimization can be carried out in terms of any number of variables or parameters and thus is adaptable to the purpose of finding approximate solutions. It can also be used to compare among many alternate approximate solutions and choose the best solution.

The foregoing analysis has been limited to steady flows. For unsteady creeping flows, without free surface, there is a theory by Korteweg [16] that states:

$$\frac{dE(u_i)}{dt} = -\rho \iiint_V \left[\left(\frac{du_1}{dt}\right)^2 + \left(\frac{du_2}{dt}\right)^2 + \left(\frac{du_3}{dt}\right)^2\right] dV \leq 0 \tag{18}$$

This means dE/dt is negative provided that the boundary is not accelerating (the boundary may move); it becomes zero only if du_i/dt are all zero. Equation 18 shows that, for unsteady flow, the total rate of energy dissipation is not at its minimum value. However, the rate of energy dissipation will continuously decrease until the motion becomes steady and the total rate of energy dissipation attains its minimum value. By using the analogy between laminar and turbulent flows shown in Equations 7 through 10, this theory can be extended to the case of turbulent flows provided that the convective accelerations $\bar{u}_i\,(\partial\bar{u}_i/\partial x_j)$ are negligible. It may be difficult to infer that the free surface will not be accelerating for an unsteady flow. However, the acceleration of free surface for most natural rivers may be small enough so Equation 18 can still be applied without much error.

The following derivation will show that the theory of minimum unit stream power is consistent with the above derivation and the equation of motion. First, we integrate Equation 13 by parts and follow the derivations shown in Equations 11 and 12. It can be shown that

$$E = \int_{S_b} \left[(\sigma_{ij} + P\delta^d{}_{ij})\bar{u}_i \right]_{x_{j1}}^{x_{j2}} \left(\frac{dx_1\, dx_2\, dx_3}{dx_j} \right) - \iiint_V \bar{u}_i \frac{\partial}{\partial x_j} (\sigma_{ij} + \bar{P}\delta_{ij})\, dV \qquad (19)$$

If, on the boundary S_b, either the velocity is zero (a stationary boundary) or the stress is zero (free surface boundary) then the double integral in Equation 19 is zero. According to the equation of motion, therefore

$$E = -\iiint_V \bar{u}_i \frac{\partial}{\partial x_i} (\gamma\bar{h} + \bar{P})\, dV \qquad (20)$$

in which

$$-\frac{\partial(\gamma\bar{h} + \bar{P})}{\partial x_i} = \gamma S_i \qquad (21)$$

is the rate of energy decrease in the direction of x_i. Thus, for a unidirectional flow

$$E(\bar{u}) = \iiint_V \gamma\bar{u}s\, dV = \text{total stream power of flow} \qquad (22)$$

Therefore, for a unidirectional flow, the total stream power of flow is also minimized. If the water discharge and energy slope vary mainly in the longitudinal direction, Equation 22 can be written as

$$E(\bar{u}_i) = \gamma \int Qs\, dx_i \qquad (23)$$

For a constant water discharge, Q, in a given reach, Equation 23 can be simplified to

$$E(\bar{u}_i) = \gamma Q \int s\, dx_i \qquad (24)$$

provided that the slope varies in the longitudinal direction only. If S = s in a given reach with reach length L and cross sectional area A, Equation 24 can further be simplified to

$$E(\bar{u}_i) = \gamma L(QS) \qquad (25)$$

in which QS is the stream power.

It is also possible to define an average stream power, $\psi(u_i)$, as

$$\psi(\bar{u}_i) = \frac{E(\bar{u}_i)}{\gamma V} = \frac{\iiint_V \bar{u}_i s\, dV}{V} = VS \qquad (26)$$

in which VS = unit stream power, and Equation 26 reduces to Equation 6 when the flow is one dimensional.

The preceding derivations also show that when the rate of energy dissipation due to sediment transport can be neglected, and when the velocity and slope does not vary across a channel or can be fairly well represented by the average velocity V and slope S, the theory of minimum unit stream power becomes a simplified and special case of the theory of minimum rate of energy dissipation. Where the local velocity and slope vary across a channel, the rate of energy dissipation has to be integrated, and the stream power QS is to be minimized. Thus, the theory of minimum stream power is a generalization of the theory of minimum unit stream power by replacing VS with QS. Needless to say, the theory of minimum stream power is also a special and simplified version of the theory of minimum rate of energy dissipation.

To further generalize the theory of minimum rate of energy dissipation, a variational formulation of linearized equation of motion suitable for irrotational flows, nonaccelerating, as well as the gradually varied flow was made by Song and Yang [17]. A summary of their derivation follows. For simplicity, the bar denoting time averaged value will be dropped.

Boussinesq assumed the existence of the phenomenological relation between the stress tensor and the rate of strain tensor:

$$\sigma_{ij} = \rho(\nu + \epsilon)\left(\frac{\partial u_i}{\partial x_j} + \frac{\partial u_j}{\partial x_i}\right); \qquad i, j = 1, 2, 3 \tag{27}$$

This assumption allows the equations of motion to be written in the same form as the original Navier-Stokes equations. For steady turbulent flows, the equations of motion may be written as

$$\frac{\partial(\gamma H)}{\partial x_i} - \epsilon_{ijk} u_j \omega_k = \frac{\partial}{\partial x_i}\left[\rho(\nu + \epsilon)\left(\frac{\partial u_i}{\partial x_j} + \frac{\partial u_j}{\partial x_i}\right)\right] \tag{28}$$

where γ = specific weight; ϵ_{ijk} = Kronecker delta; ω_k = the vorticity vector of the time averaged flow or the mean flow; and H = the total head, i.e.,

$$H = \frac{V^2}{2g} + \frac{P}{\gamma} + h \tag{29}$$

in which V = average velocity; P = pressure; h = elevation; and g = gravitational acceleration.

By virtue of its linearity, the time-averaged equation of continuity takes its original form

$$\frac{\partial u_i}{\partial x_i} = 0 \tag{30}$$

The second term in Equation 28 vanishes when the mean flow field is irrotational, nonaccelerating, or gradually varying.

Thus, for all these three classes of flow, the equation of motion takes the form

$$\frac{\partial(\gamma H)}{\partial x_i} = \frac{\partial}{\partial x_i}\left[\rho(\nu + \epsilon)\left(\frac{\partial u_i}{\partial x_j} + \frac{\partial u_j}{\partial x_i}\right)\right] \tag{31}$$

The theory of calculus of variation states that the necessary condition for

$$J = \iiint_V F(x_i, u_i, u_{i,j})\, dx_1\, dx_2\, dx_3 \tag{32}$$

to take a minimum value with respect to the variation of u_i, when the value of u_i on the boundary is specified, is for F to satisfy the Euler-Lagrange equation

$$\frac{\partial F}{\partial u_i} - \frac{\partial}{\partial x_j}\frac{\partial F}{\partial u_{i,j}} = 0 \tag{33}$$

in which $u_{i,j}$ = the partial derivative of u_i with respect to x_j. The sufficient condition for J to take a minimum value requires, in addition to Equation 33, a certain restriction concerning its second variation [18].

The variational principle stated herein applies whenever u_i is specified on the boundary. However, in order for the theory to be applicable for open channel flow problems, the free surface boundary condition has to be a natural boundary condition for the variational principle. For this reason, it is necessary to reexamine the formal derivation of the principle. Suppose that u_i is given a variation $\epsilon_i \eta_i(x_j)$ in which ϵ_i = the parameter of variation; and $\eta_i(x_j)$ = an arbitrary function of x_j. The first variation of J with respect to u_i is

$$\delta J = \epsilon_i \iiint_V \left(\frac{\partial F}{\partial u_i} \eta_i + \frac{\partial F}{\partial u_{i,j}} \frac{\partial \eta_i}{\partial x_j} \right) dx_1\, dx_2\, dx_3 \tag{34}$$

Equation 34 can be rewritten as

$$\delta J = \epsilon_i \iiint_V \eta_i \left(\frac{\partial F}{\partial u_i} - \frac{\partial}{\partial x_j} \frac{\partial F}{\partial u_{i,j}} \right) dx_1\, dx_2\, dx_3 + \epsilon_i \iiint_V \frac{\partial}{\partial x_j} \left(\eta_i \frac{\partial F}{\partial u_{i,j}} \right) dx_1\, dx_2\, dx_3 \tag{35}$$

By applying Green's theorem to the second integral in Equation 35, it follows that

$$\delta J = \epsilon_i \iiint_V \eta_i \left(\frac{\partial F}{\partial u_i} - \frac{\partial}{\partial x_j} \frac{\partial F}{\partial u_{i,j}} \right) dx_1\, dx_2\, dx_3 + \epsilon_i \iint_A \eta_i \frac{\partial F}{\partial u_{i,j}} n_j\, dA \tag{36}$$

in which n_j = the jth component of the outward unit vector normal to the area element dA. If u_i is specified on the boundary A, then η_i is zero on A and, therefore, the second integral vanishes. Because η_i is arbitrary in V, the stationary value of J is obtained only if Equation 33 is satisfied. If u_i is not specified over the complete boundary A, then the second integral of Equation 36 may vanish only under special cases. Obviously, one such case is when the other factor, the derivative of F with respect to $u_{i,j}$, vanishes.

Based on the solution obtained by Yang and Song [19], and after some trials, an appropriate function is found to be

$$F = \phi + u_i \frac{\partial}{\partial x_i} (\gamma H) \tag{37}$$

By differentiating Equation 37 with respect to $u_{i,j}$, it follows that

$$\frac{\partial F}{\partial u_{i,j}} = \rho(\nu + \epsilon) \left(\frac{\partial u_i}{\partial x_j} + \frac{\partial u_j}{\partial x_i} \right) = \sigma_{ij} \tag{38}$$

It is reasonable to assume that $\sigma_{ij} = 0$ on the free surface as well as on the two end sections cut normal to the streamlines. Therefore, because of Equation 38, the integrand of the surface integral in Equation 36 is zero on these surfaces. Because η_i is zero on the solid boundary where u_i is specified, the surface integral of Equation 38 is identical to zero. It follows that, for an open channel flow in which either u_i is specified or $\sigma_{ij} = 0$ on the boundary, the functional J, defined by Equations 32 and 37, takes a stationary value if Equation 33 is satisfied. Because F is essentially the rate of energy dissipation, which is positive definite, this stationary value should be a minimum. It can be easily checked that, by substituting Equation 37 into equation 33, the equation of motion, Equation 31, is obtained.

It has been shown so far that J is the functional from which the linearized equation of motion can be obtained. The next step is to show that, in the case of a long river, the minimization of J also minimizes E. When Equation 37 is substituted into Equation 32, the functional J can be written as

$$J = E + E_1 \tag{39}$$

such that E is defined by Equation 13 and

$$E_1 = \iiint_V u_i \frac{\partial}{\partial x_i}(\gamma H)\, dx_1\, dx_2\, dx_3 \tag{40}$$

Now consider the identity

$$\frac{\partial}{\partial x_i}(u_i \gamma H) = u_i \frac{\partial}{\partial x_i}(\gamma H) + \gamma H \frac{\partial u_i}{\partial x_i} = u_i \frac{\partial}{\partial x_i}(\gamma H) \tag{41}$$

Equation 41 holds because of the continuity relation given by Equation 30. It is now possible to apply the Green's theory and rewrite Equation 40 as

$$E_1 = \iint_A \gamma H(\vec{V} \cdot \vec{n})\, dA \tag{42}$$

where $\vec{n}$ = the outer normal to the boundary surface A, and $\vec{V} \cdot \vec{n}$ = the velocity component normal to the surface. Clearly, $\vec{V} \cdot \vec{n}$ is zero on the water surface and the channel boundary. The only places where it is not zero are the upstream and the downstream sections of a river reach. By denoting upstream and downstream sections by subscripts 1 and 2, respectively, Equation 42 reduces to

$$E_1 = \gamma Q(\alpha_2 H_2 - \alpha_1 H_1) \tag{43}$$

where H_1, H_2 = average heads; and α_1, α_2 = correction factors to account for variable velocity over the cross sections. For a gradually varied turbulent flow, α_1 and α_2 are known to be very close to unity. It should be observed that E_1 is a known constant if the discharge and the total head at the two end sections are known. Because E_1 is a constant, the minimization of J also means the minimization of E. Thus, when the rate of energy dissipation due to sediment transport can be neglected, the consistence between the theory of minimum rate of energy dissipation and the linearized equation of motion for irrotational, nonaccelerating, as well as gradually varied open channel flows can be proven theoretically.

Generalization of Minimization Theory

The entropy analogy is based on the assumption that there is an analogy between a river system and a thermal or heat system. The equation of motion approach is based on the assumption that rate of energy dissipation can be minimized by adjusting the velocity distribution. In order for the minimization theory to be applicable to both static and dynamic equilibrium systems without any of the foresaid assumptions and restrictions, further generalization of the theory was made by Song and Yang [20]. They applied the generalized theory to flow around bluff bodies, stability of falling bodies, dynamics of gas bubbles, and generation of ripples and dunes on alluvial bed. Due to the wide range of applications of the general theory, the theory is further developed and explained in the next section. The relationship between the general theory of minimum rate of energy dissipation and the simplified theory of minimum stream power and minimum unit stream will also be explained.

THEORY OF MINIMUM ENERGY AND RATE OF ENERGY DISSIPATION

Concept of Mechanical System

In classical mechanics, mass and energy are two fundamental properties that can be used to define a system. Energy is known to take one of two forms: the potential energy that depends on the position and the kinetic energy that depends on the speed of the particle. Energy also occurs in different scales. For example, the kinetic energy at molecular scale is known as heat energy, which is quite different from and independent of the macroscopic kinetic energy of fluid flow.

A mechanical system, which may contain solids as well as fluid, is defined as a system for which only the macroscopic energy is relevant. A mechanical system is called "closed" if there is no interchange of mass and macroscopic energy between the system and its surroundings. Otherwise, the system is "open." For example, a reach of river is an open system but two reservoirs joined by a channel is a closed system. It should be noted that the flow of smaller scale energy like heat energy is allowable in the definition of a closed mechanical system.

Theory of Minimum Energy

Consider a closed mechanical system whose total mechanical energy at any instant is E. In general, this total energy consists of the potential energy E_p and the kinetic energy E_k. It is then possible to write

$$E = E_p(r_i) + E_k(v_i) \geq 0 \tag{44}$$

In this equation, r_i = all the parameters necessary to define the system's configuration to which the potential energy depends; and v_i = all the velocity vectors which determine the system's kinetic energy.

If the system is dynamic, then not all v_i are zero and r_i and v_i may be functions of time. For an ideal system, E remains constant while E_p and E_k may vary with time. This section addresses a real closed system wherein the mechanical energy can decrease due to friction or viscosity but never increase. It follows that, for a real closed mechanical system,

$$\frac{dE}{dt} \leq 0 \tag{45}$$

In the end, when all motions cease, we have $v_i = 0$, $r_i = \bar{r}_i$, and

$$\frac{dE}{dt} = 0 \tag{46}$$

The static equilibrium condition represented by Equation 46 is the condition of minimum energy for the whole range of state that the system has taken during the evolution process. That is

$$E(\bar{r}_i) = E_p(\bar{r}_i) \leq E(r_i) \tag{47}$$

Now, consider a system at a state of static equilibrium $\bar{r}_i$. A small amount of energy is added and the system is disturbed to an arbitrary new dynamic state represented by

$$r_i = \bar{r}_i + \Delta r_i; \qquad v_i = \Delta v_i \tag{48}$$

The dissipation mechanism of the system will start to work immediately to reduce the energy until a new minimum condition is attained. The original equilibrium condition $\bar{r}_i$ is called a stable equilibrium if it is also equal to the new equilibrium for small but arbitrary disturbance. That is, Equation 47 holds true for all possible r_i sufficiently close to $\bar{r}_i$ if $\bar{r}_i$ represents a stable equilibrium condition.

The preceding argument has demonstrated the essential relationship between the state of stable equilibrium and the state of minimum energy. If the state variable r_i is unconstrained at $\bar{r}_i$, then

$$\frac{\partial E}{\partial r_i} = 0 \tag{49}$$

at the state of stable static equilibrium. Thus, the theory of minimum energy states that a stable static equilibrium condition of a closed mechanical system is equivalent to a state of local or global minimum energy over all the possible range of the state variables.

Relative Stability

In the preceding section, it was shown that Equation 49 can be used to find the state of stable static equilibrium. The concept of relative stability is introduced herein for the purpose of evaluating the degree of stability if Equation 49 leads to more than one set of solutions. For simplicity, consider a case when there is only one state variable r. Assume that $E_p(r)$ has two local minimums as illustrated by Figure 1. There are two local minimums A and B and a local maximum C located between A and B. Both A and B are clearly stable static equilibrium conditions. Point C is the point of unstable equilibrium because an infinitesimal amount of perturbation will bring the system to either A or B. Imagine that the state A is perturbed by introducing certain amount of energy from the external source. Let the energy input be

$$\Delta E = \Delta E_p + \Delta E_v \tag{50}$$

If ΔE is less than ΔE_1, then the system will surely return to the original state A. However, if ΔE is greater than ΔE_1, then it is possible to attain a state P with kinetic energy, ΔE_v, as shown in Figure 1. This would cause a shift in state to the other stable equilibrium, B. State A is stable to all perturbations less than ΔE_1 but state B is stable to all perturbations less than ΔE_2. Because ΔE_2 is larger than ΔE_1, state B is more stable than state A. The preceding argument leads to the concept of relative stability, which states that if there are two possible stable static equilibrium conditions, then the state with smaller energy is more stable than the one with greater energy.

Friction and Viscosity

In fluid mechanics, viscosity is responsible for the transfer of macroscopic energy to the heat energy. Because the energy dissipation due to viscosity is related to velocity gradient, dissipation and shear stress ceases when flow ceases. This property is the basis for the theory of minimum energy leading to a concept that the existence of energy dissipation is fundamental to the existence of a stable static equilibrium. Without energy dissipation, there will be no stable static equilibrium because any perturbation will result in perpetual motion.

For solid mechanics, there is another mechanism of energy dissipation and shear stress known as friction. In this case, energy dissipation is still related to relative motions but shear stress is not. In fact, according to the conventional concept, shear stress is not uniquely related to the system configuration and the theory of minimum energy fails to give a unique solution. We shall now reexamine the problem and try to resolve the paradox.

Consider first, a simple mass-spring system shown in Figure 2. A concentrated mass M hangs vertically on a spring with spring constant K in a gravity field g. Dotted line represents the neutral position of the spring and the solid line is the equilibrium position. The total potential energy consisting of the spring energy and the gravitational potential for a deflection x is

$$E_p(x) = \tfrac{1}{2}Kx^2 - Mgx \tag{51}$$

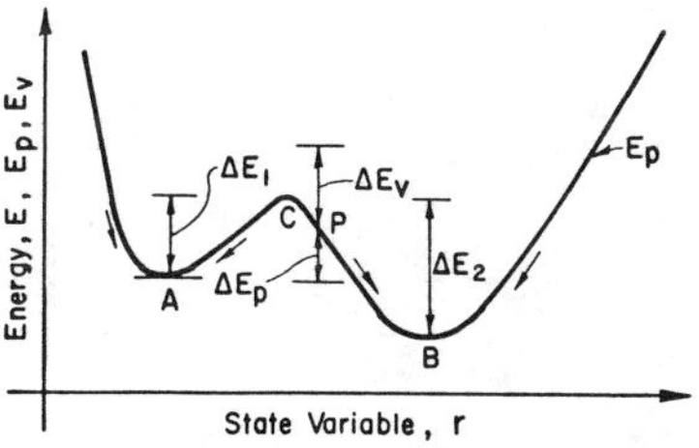

Figure 1. Schematic energy diagram.

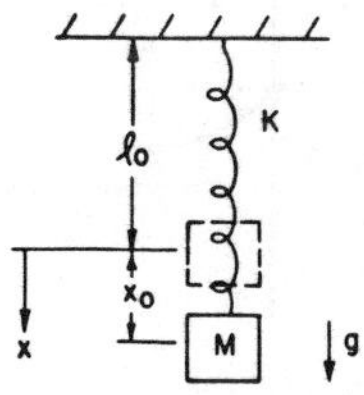

Figure 2. A mass-spring system.

By differential calculus, it is readily shown that the condition of minimum energy is

$$x_0 = Mg/K \tag{52}$$

Clearly, Equation 52 is the correct solution of the static equilibrium problem. In this case, the solution is unique and the equilibrium state is absolutely stable. That is, no amount of perturbation will alter the final equilibrium state. The energy diagram for the mass-spring system is shown in Figure 3.

Now, consider a similar mass-spring system placed on a horizontal surface as shown in Figure 4. Assume that an external force is used initially to place the center of the mass at a position x_0 to the left of the neutral position which is regarded as the origin of the coordinate. The resulting motion and the final equilibrium condition depend very much on the initial position x_0 and can be very complicated. For the purpose of illustration, only the first two simplest cases will be analyzed.

According to conventional mechanics, the mass will stay at the original position x_0 as long as

$$0 < x_0 < \frac{fMg}{K} \tag{53}$$

where f = the friction coefficient. When the initial position is in the following range,

$$\frac{fMg}{K} < x_0 < \frac{2fMg}{K} \tag{54}$$

Then, the mass will first slide to the right and stop at the position x, which will be calculated later. For larger x_0, the mass will oscillate before stopping at a certain position. In any event, the conventional minimum energy principle fails to give a unique solution.

To remedy the situation just described, it is proposed to interpret the friction phenomenon as an elastic phenomenon. As shown in Figure 4, the contact surfaces are visualized to contain microscopic roughness. The roughness of the two surfaces interlock each other when pressure is applied. When external shear force is applied, the interlocking roughness will perform elastic deformation to resist the shear force. If the shear force exceeds a limit, then the interlocking roughness fails and the body slides. A sliding body will not only lose mass but it will also dissipate energy due to viscous action between the broken particles of very large velocity gradient.

Let K_f be the elastic constant and ϵ_m be the maximum allowable deformation of the roughness, then we have

$$fMg = K_f \epsilon_m \tag{55}$$

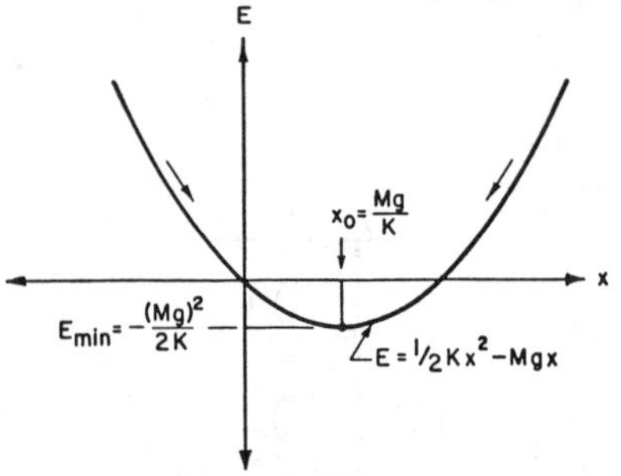

Figure 3. Energy diagram of a mass-spring system.

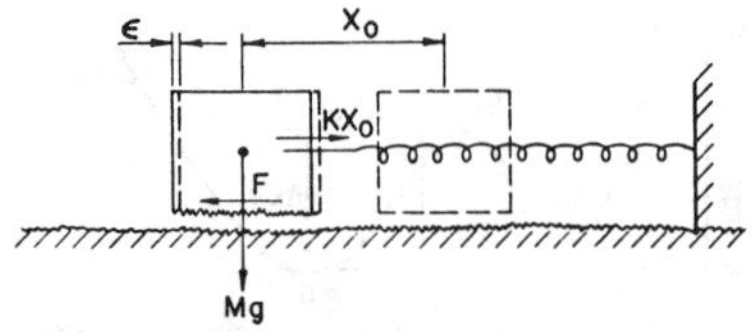

Figure 4. A mass-spring system with friction.

Thus, if the actual deformation of the roughness ϵ is less than ϵ_m, then the body will be stationary, otherwise the body will slide.

Now, consider the first case when the initial displacement x_0 satisfies Equation 53. After the removal of the external force, the roughness will deform by an amount ϵ causing the center of mass to move to $x_0 - \epsilon$. The total potential energy of the closed system is

$$E_p(\epsilon) = \tfrac{1}{2}K_f\epsilon^2 + \tfrac{1}{2}K(x_0 - \epsilon)^2 \quad (56)$$

By differential calculus, i.e., $dE_p/d\epsilon = 0$, the state of minimum energy is

$$\epsilon = \frac{Kx_0}{K_f + K} \quad (57)$$

It can be readily shown that Equation 57 agrees with the result that would be obtained by balancing the two forces. The minimum energy at the equilibrium condition is

$$(E_p)_{min} = \frac{1}{2}\frac{KK_f}{K + K_f}x_0^2 \quad (58)$$

and the energy dissipation is

$$\Delta E_p = \frac{1}{2}\frac{K^2}{K_f + K}x_0^2 = \frac{1}{2}\epsilon K x_0 \quad (59)$$

The present solution implies that, after the removal of the external force, the mass will oscillate until the viscous action dissipates energy in the amount given by Equation 59 and attains a static equilibrium condition. This equilibrium is a stable equilibrium provided that the perturbation is less than $\epsilon_m - \epsilon$.

Next, we consider the case when the initial displacement is within the range given by Equation 54. In this case, the body will first slide to the point x before it settles to the final equilibrium position $x - \epsilon$. The potential energy at position $x - \epsilon$ is

$$E_p(\epsilon) = \tfrac{1}{2}K_f\epsilon^2 + \tfrac{1}{2}K(x - \epsilon)^2 \quad (60)$$

Minimization of this energy with respect to ϵ leads to

$$\epsilon = \frac{Kx}{K_f + K} \quad (61)$$

The second condition is furnished by the law of conservation of energy. The energy lost during the slide is equated to the work done by the friction during slide. It can be shown that

$$x = \frac{2fMg}{K} - x_0 \quad (62)$$

The total amount of energy dissipated is

$$\Delta E_p = \frac{1}{2}Kx_0^2 - \frac{1}{2}\left(\frac{KK_f}{K + K_f}\right)x^2 \quad (63)$$

It can be shown that Equation 62 agrees with the result obtained by solving the equation of motion.

Theory of Minimum Rate of Energy Dissipation

It was stated in a previous section that a closed dynamic system will dissipate energy during its evolutional process to the final static equilibrium. Let Φ represent the rate of energy dissipation so that

$$\Phi = -\frac{dE}{dt} > 0 \tag{64}$$

Because static equilibrium is a state of minimum energy, a dynamic state that is not too far from the static state should have

$$\frac{d\Phi}{dt} = -\frac{d^2E}{dt^2} \leq 0 \tag{65}$$

That is, the rate of energy dissipation of all closed dynamic systems must eventually tend to decrease or stay the same after sufficient time. Analogous to the static equilibrium, a dynamic equilibrium condition is defined as the condition when the rate of energy dissipation is constant and

$$\Phi(\bar{r}_i, \bar{v}_i) = \text{constant} \tag{66}$$

Because of Equation 65, $\Phi(\bar{r}_i, \bar{v}_i)$ must be the minimum value the system has taken during the evolution process. If we apply an external energy to perturb this system, then the rate of energy dissipation $\Phi(r_i, v_i)$ will also change. A system is defined to be in a stable dynamic equilibrium state if the system will regain its original state after any small perturbation. That is, for all possible small perturbations, $\Phi(r_i, v_i) \rightarrow \Phi(\bar{r}_i, \bar{v}_i)$. According to this definition, the state of stable dynamic equilibrium is a state of minimum rate of energy dissipation. Under this condition, the state of stable dynamic equilibrium is equivalent to

$$\frac{\partial\Phi}{\partial r_i} = \frac{\partial\Phi}{\partial v_i} = 0 \tag{67}$$

and the variational method is applicable.

The restriction of not being too far from the state of static condition is not overly severe because it is always possible to imagine that the evolutional process has taken a long time starting from a much more dynamic condition than the condition under consideration. The argument used here for the theory of minimum rate of energy dissipation is essentially the same as that given by Song and Yang [20].

Relationship With Equations of Motion

About a century ago, Korteweg [16] showed that the velocity distribution that satisfies the equation of continuity and, at the same time, minimizes the rate of energy dissipation is equivalent to the solution of the Navier-Stokes equations with the acceleration terms neglected when the flow boundary consists of solid surfaces. This is the first indication that the theory of minimum rate of energy dissipation is applicable to Newtonian fluid flow at very low Reynolds number under certain type of boundary conditions. Korteweg also showed that the solution is unique and the flow is absolutely stable.

Some more recent researches have attempted to broaden the applicability of the theory of minimum rate of energy dissipation or other types of variational principles. Recently, Song and Yang [17] showed that Korteweg's theory is applicable to the case when the boundary consists of free surface as well as solid surface. As shown earlier they also showed that the theory is applicable to irrotational flows, non-accelerating flows, and gradually varied flows.

Closed or Open System

There has been some confusion among researchers from time to time on whether the energy dissipation rate should be minimized or maximized. Most variational principles of nondissipative systems require only a stationary condition that can be either maximum, minimum, or point of inflexion. For a "closed and dissipative system," the stationary condition has to be a minimum for a stable equilibrium. When the problem at hand is that of an open system, it is best to convert the problem into an equivalent closed system before the minimization process is carried out.

Consider, for an example, the flow in a straight pipe. Superficially, the system appears to be an open one and the minimization theory is not applicable. But, if we consider a more complete system of a long pipe attached to a very large reservoir at each end, then the system can be converted to a closed one as shown in Figure 5. For a given head difference H between the two reservoirs, the total energy dissipation rate Φ is

$$\Phi = \gamma QH \tag{68}$$

Because γ and H are given, minimization of Φ is equivalent to minimization of Q.

Several different flow regimes such as weir controlled, orifice controlled, friction controlled, etc., are known to occur in a pipe. In a later section, some examples will be given to show that the minimization of Q is a good criterion for determining the flow regime that will take place when head is considered given.

Based on the Darcy-Weisbach equation, the discharge and head may be related to the friction factor f as follows:

$$H = f\frac{L}{D}\frac{Q^2}{2gA^2} \tag{69}$$

where L, D, and A = the length, the diameter, and the cross-sectional area of the pipe, respectively, as shown in Figure 5. According to this equation, minimization of Q is equivalent to maximization of f because these are the only two variables. It will be shown later that the maximization of f is a good criterion for determining the transition between laminar and turbulent flows or determining the trend in the geometrical evolution of rivers under certain special conditions.

Take the same problem considered above but assume that the two reservoirs are small so that an equilibrium condition can only be maintained by adding an inflow to the upstream reservoir and an outflow from the downstream reservoir. In other words, Q is now assumed to be fixed by external means. The resulting system is an open one and the minimization theory is not applicable. Indeed, in this case, the energy dissipation rate must be maximized so that H is maximized according to Equation 68 and f is maximized according to Equation 69.

Using the language of systems engineering, H is the constraint and Q is the objective function for the closed system. For the open system, H is the objective function and Q is the constraint. The well known duality principle [21] states that for each minimization problem, there is a dual

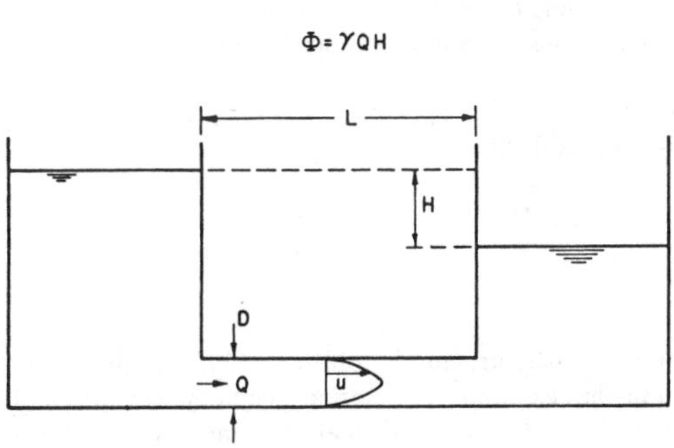

Figure 5. Closed dissipative hydrodynamic system.

maximization problem by interchanging the objective function and the constraint. Furthermore, the minimum of the original problem is equal to the maximum of the dual problem. For this reason, a hydrodynamic problem could be solved correctly by either minimizing or maximizing if the problem is properly set up.

Stream Power and Unit Stream Power

Chang and Hill [22] called the rate of energy dissipation represented by Equation 68 the stream power. If the cross-sectional area as well as the slope of a stream are constant, then Equation 68 may be written as

$$\Phi = \gamma LAVS \tag{70}$$

Yang [12] called VS the unit stream power. When γLA is assumed to be given, then minimization of Φ is equivalent to the minimization of the unit stream power. Clearly, the rate of energy dissipation, the stream power, and the unit stream power are referred to the same physical entity.

General Theory of Minimum Energy and Energy Dissipation Rate

The variational formulation and illustration stated in this section is applicable to a closed and dissipative system. Heat energy was not considered in the formulation. The general theory of minimum energy and energy dissipation rate thus obtained is based on physical argument rather than from the equations of motion or the analogy of entropy in thermodynamics. As a result, the general theory is independent of the constitutive relationship of the particular material concerned and should be applicable to multiphase mechanical problems as well as single-phase flow problems. The general theory of minimum energy can be stated as: *For a closed and dissipative system in a static stable equilibrium condition, its total energy is at its minimum value which is compatible with the constraints applied to the system.*

The general theory of minimum rate of energy dissipation can be stated as: *For a closed and dissipative system in a stable dynamic equilibrium condition, its total rate of energy dissipation is at its minimum value which is compatible with the constraints applied to the system.*

The common rules to be followed in the application of the general theory of minimum energy and energy dissipation rate are:

Rule 1—If a system is not at equilibrium, its energy or rate of energy dissipation is not at its minimum value. However, the system will adjust in such a manner that its energy or energy dissipation rate can be minimized to regain equilibrium.

Rule 2—If there is a unique minimum, the equilibrium state is absolutely stable.

Rule 3—When there are multiple equilibrium solutions, the solutions are only conditionally stable. The solution with smaller energy or rate of energy dissipation is more stable than the one with larger energy or energy dissipation rate.

Rule 4—The critical condition for stability occurs at the point of maximum energy or maximum rate of energy dissipation. This condition corresponds to unstable equilibrium.

Rule 5—An inflection point or a saddle point represents a neutrally stable equilibrium solution.

Examples will be used in the next section to illustrate the applications of the general theory or its simplified versions, in accordance with the foresaid rules.

APPLICATION OF THE THEORY

Static Equilibrium and Stability

Application of the theory of minimum energy to elastic structural problems, including equilibrium and stability, is widely known and needs no further elaboration. A spring-weight system with and without friction was used to explain the theory from the energy dissipation point of view. Only a few additional examples that are not likely to be seen in a typical structural mechanics book will be used to further clarify the concept and application of the theory.

Equilibrium and Stability of a Solid Block

Consider a two-dimensional rectangular block of height b and width a resting on a flat platform as shown in Figure 6. Assuming uniform density distribution so that the center of gravity coincides with the geometrical center. Obviously, there are two stable equilibrium positions, position 1 and position 2. Consider the block initially at position 1, which is given a perturbation by an external energy input, to a position shown by dotted lines. The potential energy of the block at this perturbed position is

$$E = MgR\cos(\theta_0 - \theta) \tag{71}$$

in which R = half the length of a diagonal; θ_0 = the angle between the diagonal AO and the vertical; and θ = the perturbation angle.

Because of the supporting surface constraint, the range of perturbation θ is between 0 and $\pi/2$ for a counterclockwise rotation. Within this region, Equation 71 gives one maximum at $\theta = \theta_0$ and two constrained minimums at $\theta = 0$ and $\theta = \pi/2$. The two minimums coincide with the two equilibrium conditions shown in Figure 6 and the corresponding minimums are

$$E_1 = MgR\cos\theta_0 \tag{72}$$

and

$$E_2 = MgR\cos\left(\frac{\pi}{2} - \theta_0\right) \tag{73}$$

The maximum potential energy is

$$E_m = MgR \tag{74}$$

A small perturbation $\theta < \theta_0$ causes the block to temporarily gain energy and produces the condition $E_m > E > E_1$. This would set the block in motion until the excess energy is dissipated and the original equilibrium condition ($\theta = 0$, $E = E_1$) is regained. If the perturbation exceeds θ_0, then the block will end up at the second equilibrium condition ($\theta = \pi/2$, $E = E_2$) rather than returning to the original position. A similar argument can also be made for the block originally at position 2 and find that it is stable for any clockwise perturbation of less than $\pi/2 - \theta_0$. As long as $b > a$, then $\theta < \pi/2 - \theta_0$ and $E_m > E_1 > E_2$. This means, if the block is subjected to a random perturbation field, then the block is more likely to stay at position 2 than position 1. The variation of energy as a function of the block's position is shown in Figure 7. In other words, the energy level of each equilibrium state may be used as a measure of its relative stability.

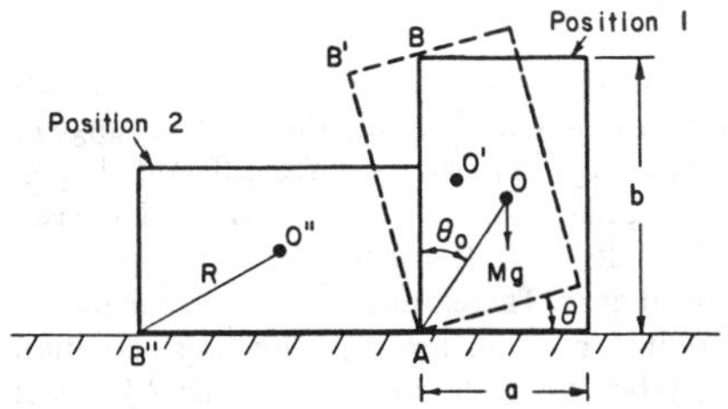

Figure 6. Equilibrium and stability of a two-dimensional block.

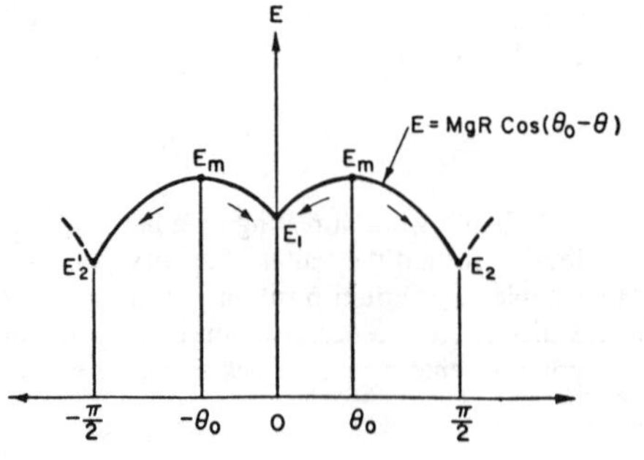

Figure 7. Energy diagram of a two-dimensional block.

Buckling of a Simple Column

A single degree of freedom buckling of an idealized column was analyzed in a book by Godden [23] based on the conventional force balance method. A similar analysis is repeated here using the minimum energy principle in order to further clarify the role of energy in mechanics. Consider an idealized column consisting of two rigid bars joined by an elastic hinge as shown in Figure 8a. The column is supported at its ends by nearly frictionless hinges A and C. The hinge C is fixed but the hinge A may slide along the vertical line as the column bend at the elastic hinge B. The structure is initially assumed to be perfectly straight and a vertical load W is applied at A.

To investigate the stability of this system, consider what happens when a disturbance causes the elastic hinge at B to rotate by θ and the point A falls by y. The disturbances θ and y are geometrically related through

$$y = 2h\left(1 - \cos\frac{\theta}{2}\right) \tag{75}$$

Taking the original condition as the reference, the total potential energy of the system at the disturbed state is

$$E = \frac{1}{2}m\theta^2 - 2Wh\left(1 - \cos\frac{\theta}{2}\right) \tag{76}$$

The first term on the right hand side of this equation represents the energy gained by the elastic hinge of coefficient m due to angular rotation θ. The second term is the loss in energy of mass W/g due to a decrease in elevation. It is more convenient to convert Equation 76 into a dimensionless form as

$$\frac{2E}{m} = \theta^2 - 4\frac{Wh}{m}\left(1 - \cos\frac{\theta}{2}\right) \tag{77}$$

which states that the dimensionless energy 2E/m is a function of two dimensionless variables θ and Wh/m.

The dimensionless energy is plotted as a function of θ for several fixed values of the dimensionless load Wh/m in Figure 8b. Because Equation 77 is an even function of θ, only the positive range of $0 \le \theta \le \pi$ is shown. It is interesting to observe that the energy curves take three different shapes. When Wh/m is less than 2, energy is minimum at the original position, $\theta = 0$. As Wh/m increases from 2 to π, the point of minimum energy gradually shifts from $\theta = 0$ to $\theta = \pi$. The point of minimum energy remains at $\theta = \pi$ for all Wh/m greater than π. Therefore, according to the theory of minimum energy, the structure is stable at its original position for small loading condition (Wh/m < 2) no matter how much the disturbance may be. When the loading exceeds 2 but less

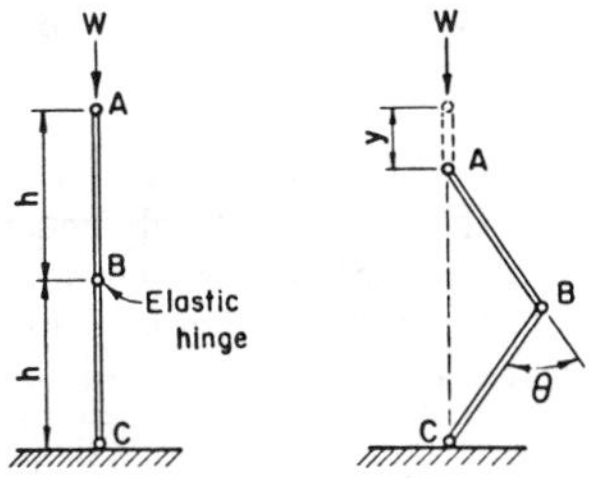

Figure 8a. Equilibrium and perturbed position of idealized column.

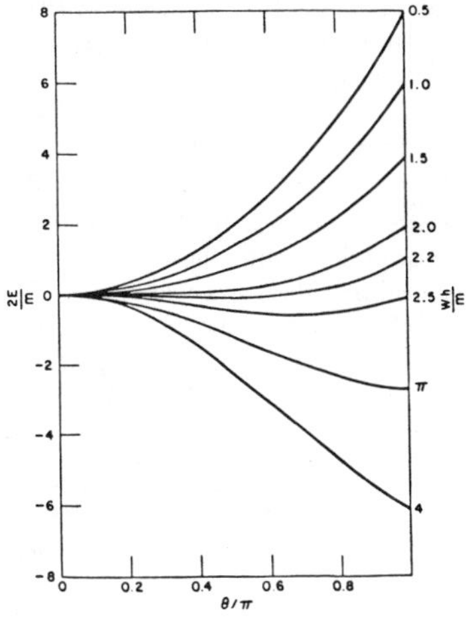

Figure 8b. Energy diagram for an idealized column under vertical load.

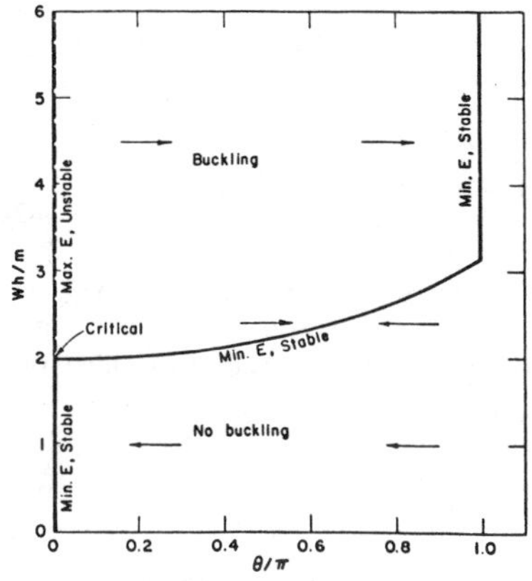

Figure 8c. The stability diagram for an idealized column under vertical load.

than π, the structure will deflect elastically to a stable position depending on the loading. Finally, for a large loading exceeding π, the structure will collapse to $\theta = \pi$ inelastically. By differentiating Equation 76 and setting the derivative to zero, the following equation is obtained:

$$\frac{\theta}{\sin\frac{\theta}{2}} = \frac{Wh}{m} \tag{78}$$

Equation 78 gives the stable equilibrium condition for the range $2 < Wh/m < \pi$, which is the unconstrained minimum. The constrained minimum at $\theta = 0$ and $\theta = \pi$ are not found without using the Lagrangian multiplier method.

The original position $\theta = 0$ is a point of maximum energy for $Wh/m > 2$. It is interesting to note that the condition is also equivalent to that of unstable equilibrium which may exist only under an ideal situation when the column is perfectly straight and no disturbance exists. The point with $\theta = 0$ and $Wh/m = 2$ is a very special point. It can be readily checked that the second derivative of E with respect to θ is equal to zero there. This is the point where a minimum and maximum meet and is called the critical point or point of neutral stability. That is, the structure is stable below this point, but buckling occurs above this point.

Figure 8c is the stability diagram that shows the lines of stable equilibrium (minimum energy) and the line of unstable equilibrium (maximum energy). The arrows shown in this figure indicate the direction of change when the structure is placed at unstable conditions.

Hydrostatic Water Surface

The problem of finding the water surface profile under hydrostatic condition by the conventional approach is very simple. Nevertheless, it serves a useful purpose of illustrating the working of the minimum energy principle. To simplify the analysis, it is assumed that there is a two-dimensional container of unit length in which water of volume **V** is to be placed. The shape of the container shown in Figure 9 can be expressed by

$$y = f_0(x) \tag{79}$$

The problem is to find the water surface profile

$$y = f_1(x) \tag{80}$$

which minimizes the total potential energy

$$E = \rho g \int y \, dx \, dy \tag{81}$$

subject to the condition that

$$\mathbf{V} = \int_{x_1}^{x_2} [f_1(x) - f_0(x)] \, dx \tag{82}$$

In Equation 82, x_1 and x_2 are the yet unknown points of intersection between the boundaries represented by Equations 79 and 80.

To solve this problem, Equation 81 is first integrated with respect to y between the two boundaries and yields

$$E = \tfrac{1}{2}\rho g \int_{x_1}^{x_2} [f_1^2(x) - f_0^2(x)] \, dx \tag{83}$$

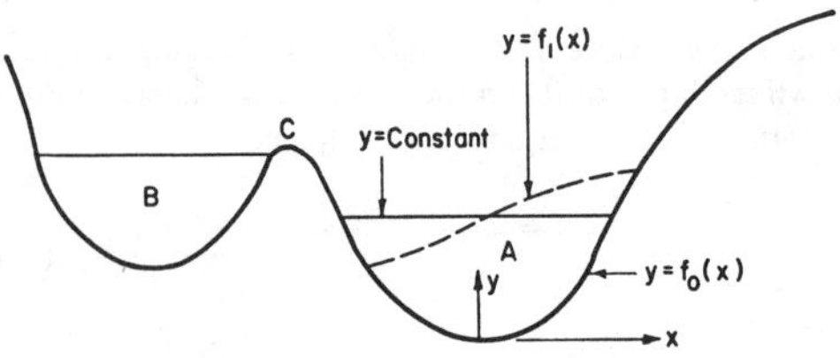

Figure 9. Water in hydrostatic equilibrium conditions.

The equivalent unconstrained problem is to minimize

$$L = \lambda V + \int_{x_1}^{x_2} [\tfrac{1}{2}\rho g(f_1^2 - f_0^2) - \lambda(f_1 - f_0)]\, dx \tag{84}$$

where λ = the Lagrangian multiplier. This is a classic variational problem having the associated Euler-Lagrange equation as follows:

$$\rho g f_1 - \lambda = 0 \tag{85}$$

The solution to Equation 85 is

$$f_1 = \frac{\lambda}{\rho g} = \text{constant} \tag{86}$$

stating that the line representing the water surface is a horizontal line. The actual values of f_1 and λ can be obtained by solving Equations 82 and 85.

Stability of Floating Block

Consider the equilibrium and stability of a floating two-dimensional block as sketched in Figure 10. In this case, a state is represented by two variables, the elevation of the center of gravity above the water surface y and the tilting angle θ. For convenience of calculation another variable, ℓ, is used to represent the submergence of the right side edge. There is a geometrical relationship

$$y = \left(\frac{a}{2} - \frac{b}{2}\tan\theta - \ell\right)\cos\theta \tag{87}$$

The potential energy of the block measured from the water surface E is

$$E_p = \rho_s abg\left(\frac{a}{2} - \frac{b}{2}\tan\theta - \ell\right)\cos\theta \tag{88}$$

in which ρ_s = the density of the block. To simplify the problem, the pool is assumed to consist of an infinitely large pool so that any amount of water displaced by the body will not cause the water surface to rise. Under this condition, it is only necessary to consider the change in potential energy due to the displaced water that may be considered to move up to the water surface. Therefore, the

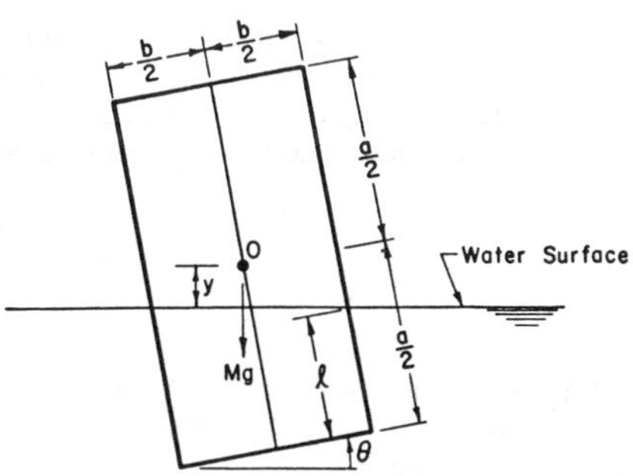

Figure 10. Equilibrium and stability of a two-dimensional block floating in water.

potential energy of the displaced water E_w is

$$E_w = \rho g b(\tfrac{1}{2}\ell^2 \cos\theta + \tfrac{1}{2}b\ell \sin\theta + \tfrac{2}{9}b^2 \tan\theta \sin\theta) \tag{89}$$

in which ρ = the density of water assumed to be greater than ρ_s. The total energy is

$$E = E_p + E_w \tag{90}$$

The stationary condition is obtained by differentiating E with respect to ℓ and θ, respectively, and set the derivatives to zero. From the ℓ derivative, the following equation is obtained

$$\ell + \frac{1}{2} b \tan\theta + \frac{\rho_s}{\rho} a = 0 \tag{91}$$

By eliminating ℓ between Equation 91 and the θ derivative, it follows that

$$\tan\theta\left[\frac{1}{2}\left(\frac{\rho_s}{\rho} a\right)^2 - \frac{1}{2}\frac{\rho_s}{\rho} a^2 + \frac{4}{9} b^2 + \frac{7}{71} b^2 \tan^2\theta\right] = 0 \tag{92}$$

This leads to either

$$\tan\theta = 0; \qquad \theta = 0 \tag{93}$$

or

$$\tan^2\theta = \frac{36}{7}\frac{\rho_s}{\rho}\left(\frac{a}{b}\right)^2\left(1 - \frac{\rho_s}{\rho}\right) - \frac{32}{7} \tag{94}$$

It can be readily checked that Equation 93 represents a minimum and Equation 94 gives a maximum. An equilibrium condition is, therefore, obtained by setting $\theta = 0$ in Equations 87, 90, and 91. The solution is

$$\ell_1 = \frac{\rho_s}{\rho} a \tag{95}$$

$$y_1 = \left(\frac{1}{2} - \frac{\rho_s}{\rho}\right) a \tag{96}$$

$$E_1 = \frac{1}{2} \rho_s g a^2 b\left(1 - \frac{\rho_s}{\rho}\right) \tag{97}$$

which can easily be verified by using a vector approach.

The solution is not absolutely stable because there is a maximum given by Equation 94. Denoting the principal solution to Equation 94, θ_m, the equilibrium state will be stable only to angular perturbations of magnitude less than θ_m. It is interesting to note that θ_m vanishes if the righthand side of Equation 94 vanishes. By setting the right-hand side of Equation 94 equal to zero, it follows that

$$\frac{a}{b} = \frac{\rho}{3\rho_s}\sqrt{\frac{8\rho_s}{\rho - \rho_s}} \tag{98}$$

When the geometry and the densities are such that Equation 98 is satisfied, then the stationary condition becomes the point of inflection and the solution becomes only neutrally stable. The energy diagram takes a complicated three-dimensional form and will not be shown here.

Dynamic Equilibrium

Many engineers have the painful experience of being unable to find sufficient number of independent equations for solving complicated fluid mechanics problems from vectorial approach. The variational formulation of minimum rate of energy dissipation is based on scalar approach and is completely independent of vectorial approach. Thus, the theory can provide additional independent equations. The general theory of minimum rate of energy dissipation or its simplified versions will be used in the following sections to solve some fluid mechanics problems that are difficult or impossible to solve analytically from vertorial approach alone. The theory will also be used to explain some interesting phenomena in fluid mechanics and river morphology.

Velocity Distribution in Open Channel Flows

For an open channel flow of fixed boundary with a given set of discharge, width, depth, slope, roughness, and average flow velocity, the only way for the flow to minimize its rate of energy dissipation is through the adjustment of its velocity distribution. Experimental evidence indicates that the velocity distribution of turbulent flow can be divided into three parts: the outer turbulent region, which follows parabolic distribution; the inner turbulent region, which follows the logarithmic distribution; and a laminar sublayer as shown in Figure 11. It was shown by Yang and Song [13] that under the condition of gradually varied subcritical flow, the rate of energy dissipation is a minimum when the velocity profile satisfies the equation of motion. The following derivations made by Song and Yang [19] show how the theory can be used to determine the velocity distribution under a given set of constraints.

The velocity distribution in a laminar sublayer can be approximated by the linear equation

$$\frac{u}{u_*} = \frac{u_* y}{\nu}; \qquad 0 \le y \le y_1 \tag{99}$$

where $u_* = \sqrt{gDS}$; u = velocity; D = water depth; S = slope; and ν = kinematic viscosity. The velocity distribution in the inner region of the turbulent flow can be approximated by the logarithmic profile

$$\frac{u}{u_*} = A_1 \ln \eta + A_2; \qquad \eta_1 \le \eta \le \eta_2 \tag{100}$$

where η = dimensionless depth = y/D; and A_1, A_2 = dimensionless coefficients to be determined. The outer region is assumed to be characterized by a parabolic velocity distribution and a constant eddy viscosity distribution, i.e., for the region $\eta_2 \le \eta \le 1$

$$\frac{u}{u_*} = A_3 + A_4\eta + A_5\eta^2 \tag{101}$$

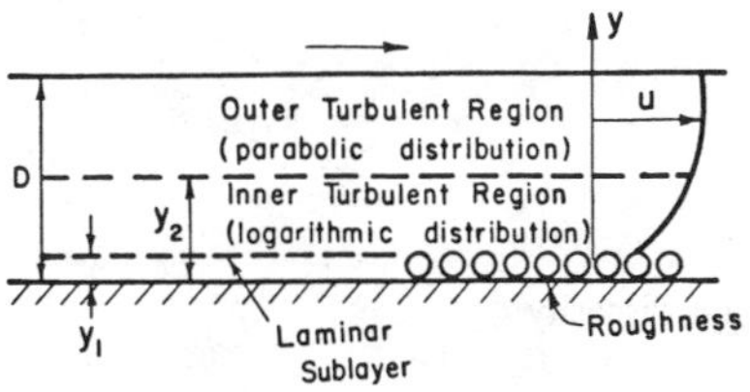

Figure 11. Schematics of turbulent flow velocity distribution.

and

$$\nu + \epsilon = \alpha u_* D \tag{102}$$

in which A_3, A_4, A_5, and α = dimensionless coefficients to be determined.

There are a total of seven constants (A_1, A_2, A_3 A_4, A_5, α, and η_2) to be determined. The other quantities like D, S, and η_1 are assumed to be given. The unknown constants will be determined by the minimization theory and an appropriate number of constraint equations. First, in order for the velocity distribution given by Equation 101 to be compatible with the eddy viscosity given by Equation 102, the following equation should hold:

$$\alpha\rho u_*^2\left(A_4 + \frac{2A_5 y}{D}\right) = \gamma S(D - y) \tag{103}$$

Because the equation must hold for any value of y between y_2 and D, it follows that

$$A_4 = \frac{1}{\alpha} \tag{104}$$

and

$$A_5 = -\frac{1}{2\alpha} \tag{105}$$

The boundary conditions or the constraints to be satisfied are: (1) the velocity is zero at the bottom of the channel, (2) the shear at the bottom of the channel is equal to γDS, (3) the shear is zero at the water surface, and (4) the velocity and shear are continuous at the junction, $\eta = \eta_2$.

The first two boundary conditions are automatically satisfied by Equation 99. Because of Equation 103, the third boundary condition is also automatically satisfied. The last boundary condition leads to two constraint equations. First, the velocity is continuous if

$$A_1 \ln \eta_2 + A_2 = A_3 + A_4\eta_2 + A_5\eta_2^2 \tag{106}$$

Because of Equation 103, the shear will be continuous if the eddy viscosity is continuous. Thus,

$$\frac{gSy_2(D - y_2)}{A_1 u_*} = \alpha u_* D \tag{107}$$

Equation 107 is equivalent to

$$\alpha = \frac{\eta_2(1 - \eta_2)}{A_1} \tag{108}$$

For uniform flows, only the rate of energy dissipation per unit length per unit width need to be considered i.e.,

$$\xi = \int_0^D \phi(y)\, dy = \int_0^D \rho(\nu + \epsilon)\left(\frac{du}{dy}\right)^2 dy \tag{109}$$

in which ϕ = rate of energy dissipation function; and μ = viscosity.

Neglecting the laminar sublayer, the rate of energy dissipation can be obtained by integrating Equation 109, i.e.,

$$\xi = \rho u_*^3 A_1 \left[\ln \eta_2 - \ln \eta_1 + \eta_1 - \eta_2 + \frac{(1-\eta_2)^2}{3\eta_2} \right] \tag{110}$$

which has the absolute minimum value at $\eta_2 = \frac{1}{2}$.

According to the theory of minimum rate of energy dissipation, $\eta_2 = \frac{1}{2}$ is the best choice. The fact that turbulent flow should consist of an inner layer and an outer layer of equal thickness was proven by Vanoni [24] in his laboratory tests. Vanoni's data were replotted by Song and Yang [25] as shown in Figure 12. Taking $\eta_2 = \frac{1}{2}$ as the best choice, the velocity distributions can be written as

$$\frac{u}{u_*} = \frac{1}{4\alpha} \ln \eta + A_2; \qquad \eta_1 \le \eta \le \frac{1}{2} \tag{111}$$

$$\frac{u}{u_*} = \frac{1}{\alpha} \eta \left(1 - \frac{\eta}{2}\right) + A_2 - \frac{1}{4\alpha} \ln 2 - \frac{3}{8\alpha}; \qquad \frac{1}{2} \le \eta \le 1 \tag{112}$$

Figure 13 shows a comparison between the measured velocity profile by Guy et al. [26] and the theoretical profile. The data shown in Figure 13 were collected with very low sediment concentration and the effect on rate of energy dissipation due to sediment transport and bed form can be neglected.

The minimum rate of energy dissipation theory can also be applied to determine the velocity distribution in a pipe. Let us consider the case of two very large reservoirs connected by a long pipe of length L and diameter D as shown in Figure 5. The head differential between the two reservoirs is H. Because the reservoir is assumed to be very large, it is possible to have a nearly equilibrium flow without adding water to the upstream reservoir. Thus, the system is closed, dissipative, and an equilibrium condition may exist. Since the pipe is assumed to be very long, the entrance and exit losses can be neglected. This problem is very similar to that of the flow in a wide open channel previously solved. Because of symmetry, the velocity profile of fully developed

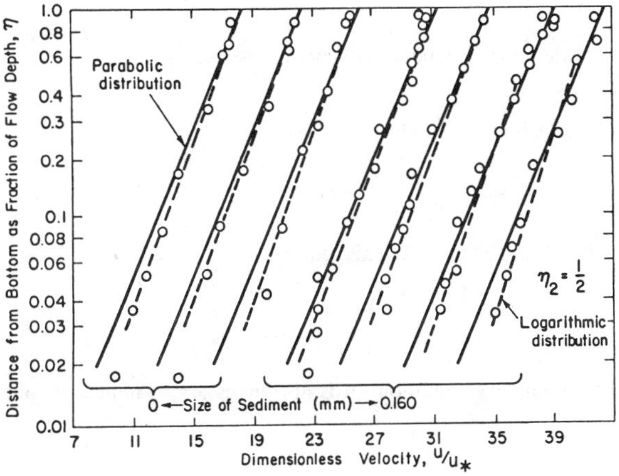

Figure 12. Dimensionless velocity profiles on center of flume (after Vanoni, 1946).

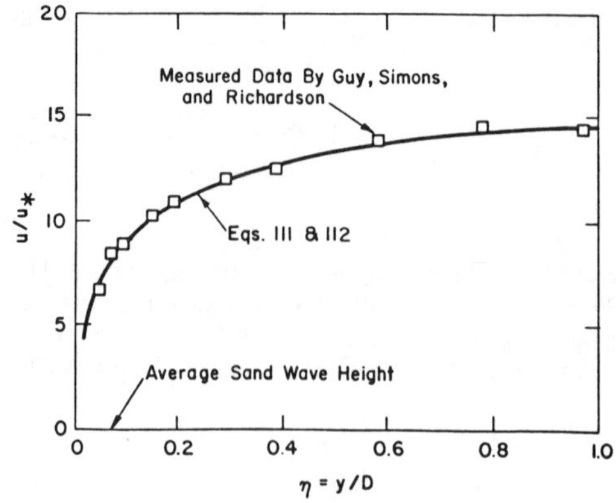

Figure 13. Comparison between measured and theoretical velocity profiles with low sediment concentration.

laminar flow may be represented by

$$u = \sum_{i=0}^{N} C_i r^i \tag{113}$$

where u = the velocity; r = the radial distance measured from the axis of the pipe; C_i = undetermined constants; and N = an integer as large as required. The dissipation function in a cylindrical coordinate system is given by Yuan [27]. For this case, it simplifies to

$$\phi = \mu \left(\frac{du}{dr}\right)^2 \tag{114}$$

Total rate of energy dissipation is,

$$\Phi = 2\pi\mu L \int_0^R \left(\frac{du}{dr}\right)^2 r\,dr \tag{115}$$

where R = radius of the pipe. The boundary conditions are:

1. By symmetry, $\frac{du}{dr} = 0$ at r = 0. This requires

$$C_1 = 0 \tag{116}$$

2. No slip condition, u = 0 at r = R. This requires

$$\sum_{i=0}^{N} C_i R^i = 0 \tag{117}$$

In addition to these boundary conditions, it is also necessary to impose a continuity constraint,

$$Q = 2\pi \int_0^R ur\,dr \tag{118}$$

Some readers may be curious as to why Equation 118 is required to serve as a constraint while the assumed velocity profile, Equation 113, automatically satisfies the differential equation of continuity. The answer to this question lies in the fundamental property of the variational principle.

As explained by Lanczos [1] and Serrin [28], because the action integral (and also Φ of this theory) contains no pressures term, the continuity constraint must always accompany the minimization process. The Lagrangian multiplier arising from the continuity constraint was shown to serve the role of the pressure gradient term in the equation of motion.

By substituting Equation 113 into Equation 118 and integrating, the continuity constraint becomes

$$Q = 2\pi \sum_{i=0}^{N} \frac{1}{i+2} C_i R^{i+2} \tag{119}$$

By eliminating C_0 from Equations 117 and 119 and solving for C_2, the following is obtained:

$$C_2 = -\left(\frac{2Q}{\pi R^4} + \frac{2}{R^4} \sum_{i=3}^{N} \frac{i}{i+2} C_i R^{i+2}\right) \tag{120}$$

Now, Equation 115 can be written as

$$\Phi = 2\pi L\mu \int_0^R \left(-\frac{4Qr}{\pi R^4} - 4r \sum_{i=3}^{N} \frac{i}{i+2} C_i R^{i-2} + \sum_{i=3}^{N} iC_i r^{i-1}\right)^2 r\,dr \tag{121}$$

The problem is now to determine C_i for $i \geq 3$ such that Φ given by Equation 121 takes a stationary value. By differentiating with respect to C_n for $n \geq 3$, performing the integration and setting the result equal to zero, the following equation is obtained, i.e.,

$$\sum_{i=3}^{N} iR^{i-1}\left(\frac{n+1}{n+i} - \frac{4}{i+2} R\right) C_i = 0; \qquad n = 3, 4, \ldots, N \tag{122}$$

Equation 122 constitutes a set of N-2 homogeneous linear equations with N-2 number of unknown C_i. Unless the determinent of the coefficient is zero, which is not possible, the solution to Equation 122 is

$$C_i = 0; \qquad \text{for all } i \geq 3 \tag{123}$$

Equation 121 can now be integrated to yield

$$\Phi = \frac{8L\mu Q^2}{\pi R^4} \tag{124}$$

The physical reality also required that

$$\Phi = \gamma QH \tag{125}$$

Combining all the results obtained thus far, the complete solution is obtained as

$$u = \frac{\gamma H}{4L\mu}(R^2 - r^2) \tag{126}$$

and

$$Q = \frac{\pi\gamma HR^4}{8L\mu} \tag{127}$$

The solution agrees with that obtained by the conventional approach.

Roughness

One of the most difficult problems in the study of fluvial mechanics of movable boundary is the prediction of variation of roughness. The theory of minimum unit stream power was applied by Yang [12] to the determination of roughness of alluvial channels with constant width where sediment loads are not high. The continuity equation of water can be written as

$$Q = BDV \tag{128}$$

where Q = water discharge; B = channel width; D = channel depth; and V = average flow velocity. The sediment concentration can be determined by Yang's [11] equation, i.e.,

$$\log C_t = 5.435 - 0.286 \log \frac{\omega d}{\nu} - 0.457 \log \frac{u_*}{\omega} + \left(1.799 - 0.409 \log \frac{\omega d}{\nu} - 0.314 \log \frac{u_*}{\omega}\right) \log\left(\frac{VS}{\omega} - \frac{V_{cr}S}{\omega}\right) \tag{129}$$

where C_t = total sediment concentration; in parts per million by weight; ω = terminal fall velocity; d = median sieve diameter of sediment particles, ν = kinematic viscosity; $u_* = \sqrt{gDS}$ = shear velocity; g = gravitational acceleration; VS = unit stream power, and $V_{cr}S$ = critical unit stream power required at incipient motion.

In addition to Equations 128 and 129, another equation is based on the theory of minimum unit stream power, i.e.,

$$VS = V_m S_m \tag{130}$$

The subscript m denotes the value determined from the theory of minimum unit stream power. The three unknowns V, D, and S, can be solved by Equations 128 through 130 for a given set of discharge Q and sediment concentration C_t without using any roughness equations. Figure 14

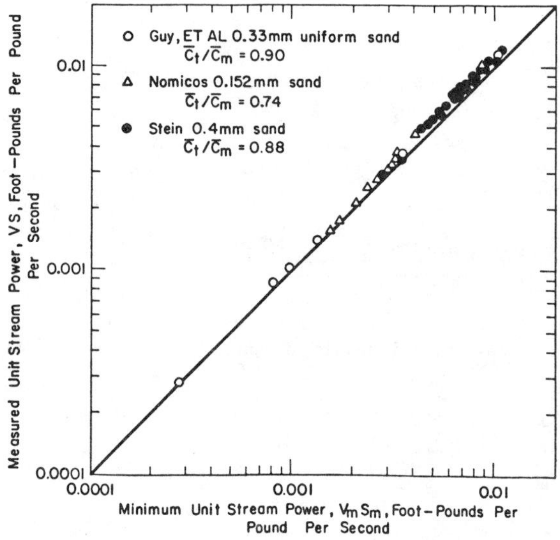

Figure 14. Relationship between measured and computed results from the theory of minimum unit stream power.

shows some examples of comparison between measured and computed unit stream power. Figure 15a shows the comparison between measured and computed results for the Rio Grande River. Once V, D, and S are determined, they can be substituted in Manning's equation to find the corresponding roughness coefficient n. An example of comparison between measured and computed n values given by Yang and Song [13] is shown in Figure 15b for the Rio Grande River. More

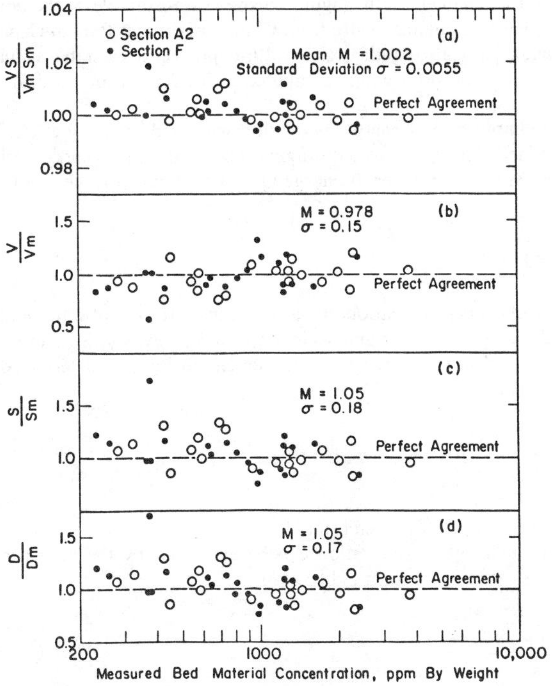

Figure 15a. Comparison between measured and computed results based on the theory of minimum unit stream power for the Rio Grande River near Bernalillo, New Mexico.

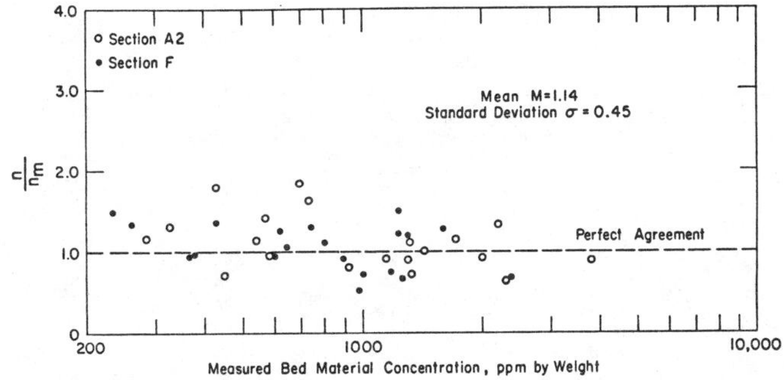

Figure 15b. Comparison between measured and computed Manning's roughness based on the theory of minimum unit stream power for the Rio Grande River near Bernalillo, New Mexico.

detailed comparisons between the roughness coefficient determined by other methods and by the application of minimum unit stream power theory were given by Parker [50].

Channel Geometry

Chang [29] applied the theory of minimum stream power to determine the stable channel geometry, especially the channel width. Figure 16 shows an example of the comparison between the computed and measured channel width from Chitale River and Punjab Canal. The agreements are very good. Chang [30] also studied the width-depth ratio of stable channel based on the theory of minimum stream power. Chang's computed width-depth ratio are also in good agreement with field data.

The theory of minimum rate of energy dissipation was used by Yang et al. [31] for the determination of the exponents in Equations 1 through 4. The total rate of energy dissipation Φ required to transport a given amount of water discharge Q and a sediment discharge Q_s through a reach of L is

$$\Phi = \Phi_w + \Phi_s = (Q\gamma + Q_s\gamma_s)LS \tag{131}$$

where Φ_w, Φ_s = rate of energy dissipation due to the transportation of water and sediment, respectively; γ, γ_s = specific weight of water and sediment, respectively; and S = slope.

Assume that water discharge in a rectangular open channel can be determined by the Manning-Strickler equation in metric units, i.e.,

$$Q = \frac{1}{n}\left(\frac{BD}{B + 2D}\right)^{2/3} S^{1/2}(BD) \tag{132}$$

in which n = Manning's roughness coefficient.

Also assume that sediment concentration can be expressed by the dimensionless unit stream power equation in its exponential form [11], i.e.,

$$\frac{Q_s}{Q} = I(P - P_c)^J; \qquad P > P_c = 0$$

$$= 0; \qquad P \le P_c = 0 \tag{133}$$

where

$$P = VS/\omega \tag{134}$$

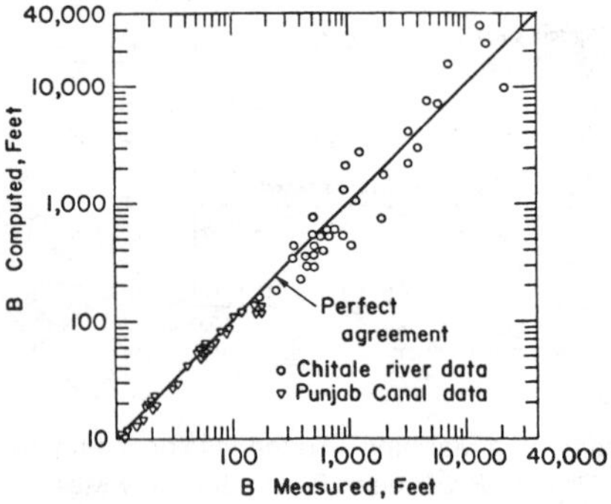

Figure 16. Comparison between computed channel width and measured channel width (from Chang, 1979).

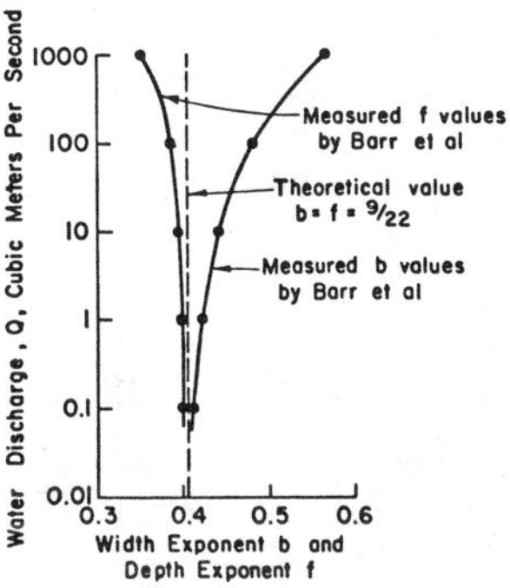

Figure 17. Comparison between the theoretical values of b and f based on the theory of minimum rate of energy dissipation and the values measured by Barr et al., 1980 (from Yang, Song, and Woldenberg, 1981).

is the dimensionless unit stream power; P_c = the critical dimensionless unit stream power required at incipient motion; and I and J = coefficients.

From Equations 132, 133, and 134

$$\frac{n^2Q^3(B + 2D)^{4/3}}{\omega(BD)^{13/3}} = P_c + \left(\frac{Q_s}{IQ}\right)^{1/J} \tag{135}$$

The problem is therefore to determine B and D so that the Φ value can be minimized subject to the constraint given by Equation 135.

The theoretical values of b, f, m, and j thus obtained are $\frac{9}{22}$, $\frac{9}{22}$, $\frac{2}{11}$, and $-\frac{2}{11}$, respectively. These values are very close to the observed mean values for natural rivers. The j value of $-\frac{2}{11}$ is also very close to the value of $-\frac{1}{6}$ used in Blench's regime equation [32]. The fact that the b and f should converge to $\frac{9}{22}$ has been recently verified by field and laboratory data collected by Barr, Alan, and Nishat [33] as shown in Figure 17.

Channel Pattern

The theory of minimum unit stream power was used by Yang [34] to provide a qualitative explanation of the formation of meandering rivers. Chang [30] applied the theory of minimum stream power to verify natural river patterns. He considered a river to be straight when the valley slope is equal to the channel slope. Sinuosity is defined as the ratio of valley slope to channel slope. Chang also considered a river to be meandering when the sinuosity is equal to or greater than 1.3. Based on the theory of minimum stream power, Chang found that there is only one solution for a straight channel that is equal to the valley slope. However, for meandering rivers, two minimum slopes can be obtained. One minimum value is larger while the other is smaller than the valley slope. Because it is physically impossible for the channel slope to be steeper than the valley slope, the one which is smaller than the valley slope must be the solution for meandering river. His results are shown in Figure 18. Thus, the change of channel patterns are also consistent with the theory of minimum rate of energy dissipation or the theory of minimum stream power.

Longitudinal Profile

River meandering and the formation of riffles and pools are two sides of the same coin in the process of minimizing a river's rate of energy dissipation. River meandering provides a way to minimize the rate of energy dissipation by lateral adjustment of a river's course. The formation of

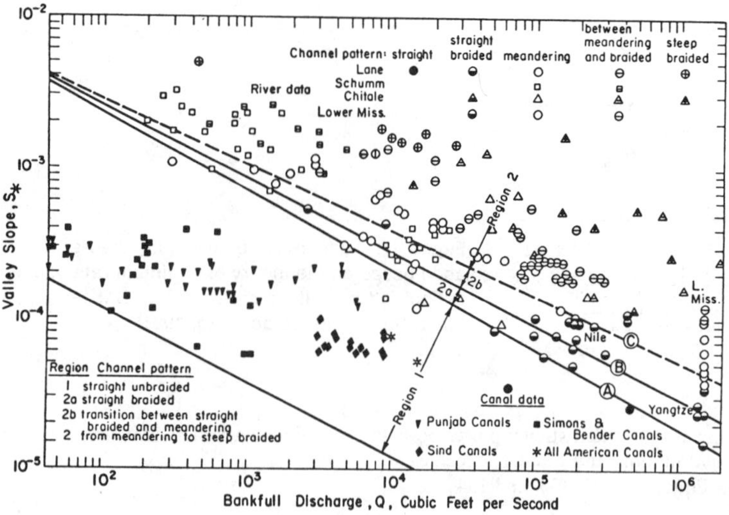

Figure 18. Channel pattern for sand streams (from Chang, 1980).

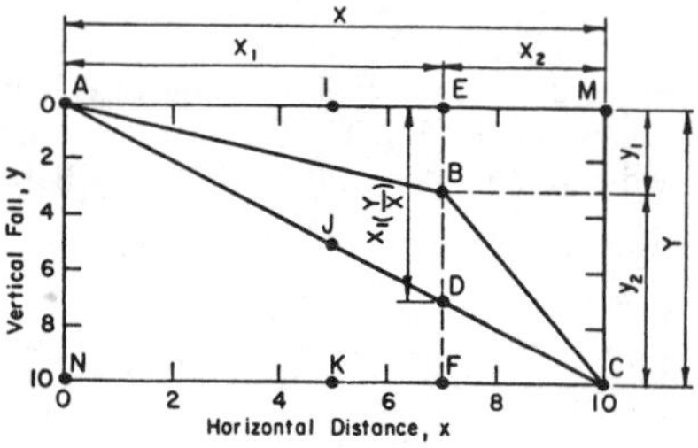

Figure 19. Alternative paths of a stream in the vertical plane along its thalweg (from Yang, 1971).

riffle and pool sequence creates an undulating longitudinal riverbed profile with shallows and deeps associated with steep and mild slopes. Consider the case of water flowing from point A to C in Figure 19. In the study reach AC, a river can choose either a straight longitudinal profile ADC or a pool and riffle sequence ABC. In either case, the total energy loss is the potential energy Y. Thus, minimization of rate of energy dissipation is equivalent to maximization of time of travel between A and C. The relation between time of travel and the position of point B was studied by Yang [35] as shown in Figure 20. It can be seen that the straight line profile ADC is the one with minimum time of travel among all possible profiles. According to the theory of minimum rate of energy dissipation, this straight profile is most unstable and should be avoided because it is the one with maximum rate of energy dissipation. Thus, the formation of riffles and pools is in compliance with the theory of minimum rate of energy dissipation.

The velocity distribution, roughness, sediment transport rate, channel geometry, channel pattern, and the pool-riffle sequence are different means available to a river for the minimization of its rate of energy dissipation at a given station or within a given reach. It is a commonly observed phenomenon that a river generally has a concaved overall longitudinal bed profile and the slope decreases in the downstream direction. This concaved longitudinal river profile can be obtained

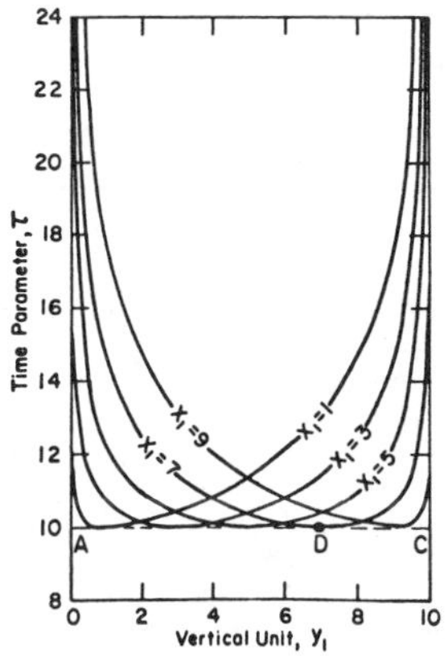

Figure 20. Travel time of a stream for different paths in the vertical plane along its thalweg (from Yang, 1971).

by minimizing its rate of energy dissipation along its course of flow [11], i.e.,

$$\frac{d}{dx}(QS) = S\frac{dQ}{dx} + Q\frac{dS}{dx} = 0 \tag{136}$$

where x = longitudinal distance along the course of flow.

Because water discharge increases in the downstream direction, dQ/dx in Equation 136 is positive. To satisfy Equation 136, dS/dx must be negative and the slope must decrease in the downstream direction. Thus, the overall longitudinal bed profile adjustment of a river is also consistent with the theory of minimum rate of energy dissipation.

It has been shown that a natural river can adjust its velocity distribution, roughness, channel geometry, channel pattern, and longitudinal bed profile, either individually or collectively, to minimize its rate of energy dissipation in accordance with the general theory of minimum rate of energy dissipation or its simplified versions. More detailed discussion on the application of this theory to the study of river morphology were given recently by Yang [11].

Laminar and Turbulent Flow Regime

Laminar flow and turbulent flow are regarded as two alternate equilibrium flow patterns possible for a given boundary condition. Actual selection of one pattern against the other by nature should depend on the relative stabilities of the two equilibrium conditions as represented by Φ and the initial condition, or the existence of perturbations.

For a closed system represented by Figure 5, the total rate of energy dissipation given by Equation 125 is applicable for both laminar and turbulent flows. To compare the two cases, it is convenient to use the Darcy-Weisbach equation expressed by Equation 69.

By eliminating Q between Equations 125 and 69, it follows that

$$\Phi = \gamma AH\sqrt{\frac{2gDH}{fL}} \tag{137}$$

This equation indicates that the rate of energy dissipation is decreased when the friction factor is increased. That is, the decrease in the rate of energy dissipation is achieved by the reduction in discharge caused by increased drag. For the laminar flow, the friction factor is

$$f = \frac{64}{Re} \tag{138}$$

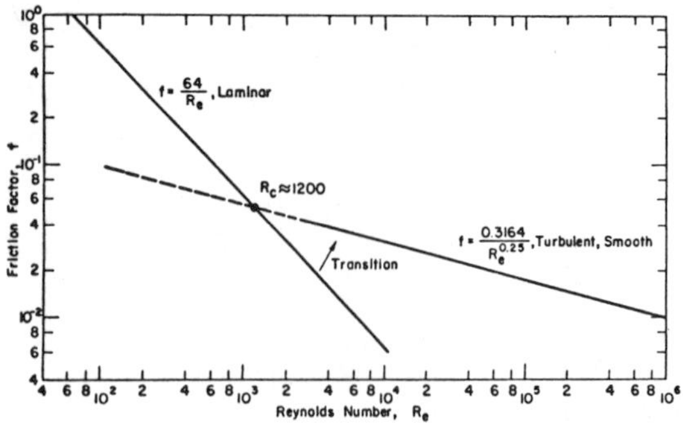

Figure 21. Friction factor for flow in a pipe.

where the Reynolds number Re = QD/νA. For the turbulent flow in a smooth pipe, there is a well-known Blasius' equation

$$f = \frac{0.3164}{Re^{0.25}} \tag{139}$$

The friction factor represented by Equations 138 and 139 are plotted in Figure 21. Two lines are seen to intersect roughly at Re = 1,200. According to the stability criterion, laminar flow should be preferred for Re < 1,200 and turbulent flow is preferred for Re > 1,200 if the existence of perturbation is not taken into consideration. In reality, the transition always occurs at a slightly larger Reynolds number as indicated by an arrow in Figure 21. That is, the laminar flow region extends somewhat beyond the relative critical Reynolds number.

The reason that nature appears to prefer laminar flow over turbulent flow near the relative critical Reynolds number of 1,200 is unknown. One possibility, simply due to the fluctuating nature of the turbulent flow, is that it naturally contains enough perturbation needed to push the flow back to the laminar condition once turbulent condition is produced. On the other hand, the laminar flow contains no internally generated perturbation and requires much greater amounts of externally induced perturbation to switch over to the turbulent condition. It is noteworthy that the predicted critical Reynolds number of 1,200 is much closer to any other theoretical values. For example, Davey and Drazin [36] and Salwen and Grosch [37] indicated that the critical Reynolds number for pipe flow obtained by a linear perturbation theory is infinity. Even the nonlinear extension of the classical stability theory has not been able to produce a conclusive result on the pipe flow transition [38]. The best available prediction is that of Joseph [39] who obtained the critical Reynolds number of 81.49 based on an energy method. Compared with these previous theoretical results, the present predicted critical value of 1,200 is much closer to the commonly accepted experimental value of 2,100.

Falling Particle

The stability of a falling particle in a large body of stationary ambient water was studied by Song and Yang [20]. The system approaches a dynamic equilibrium condition as the speed of fall particle approaches its terminal value. Under this condition, the rate of energy dissipation is equal to the rate at which the system loses its potential energy because the kinetic energy is constant. Therefore, the rate of energy dissipation is

$$\Phi = W_s \omega \tag{140}$$

and

$$W_s = C_D A \frac{\rho \omega^2}{2} \tag{141}$$

where W_s = submerged weight; ω = terminal fall velocity; C_D = drag coefficient; A = cross-sectional area; and ρ = density of water. For a steadily falling sphere or a circular cylinder, W_s and A are constants. Therefore, minimization of Φ is equivalent to minimization of ω and maximization of C_D.

The conclusion given above is borne out by the fact that the C_D − Re curve, as shown in Figure 22, concaves upward. Consider two points A and B on the curve; one corresponding to Stokes flow and the other representing the flow with the trailing eddies. The point B′ is then obtained by extending the straight portion of the graph to the point where the Reynolds number is equal to that of B. At this particular Reynolds number, the hypothetical flow without eddies, as represented by B′, would result in a smaller C_D than the actual flow represented by B. Because the flow with larger C_D results is less Φ, the nature selects B over B′, as expected. Similarly, nature favors point A without eddies over the alternate point A′ with eddies because A gives larger C_D than A′.

It was observed by Stringham et al. [40] that, depending on the value of Reynolds number, a disk may fall in one of the four patterns. They are steady-flat fall, regular oscillation, glide-tumble, and tumble. It was noted by Stringham et al. [40] that the first and fourth fall pattern were stable patterns in a sense that both the translational and the rotational speed were constants. In contrast, the second and the third patterns were unstable patterns in a sense that there was some randomness associated with the motions.

Song and Yang [20] applied the minimum rate of energy dissipation theory to explain the pattern and stability of a falling disk in water. The C_D − Re graph of falling disks shown in Figure 23 is strikingly similar to the Moody diagram for a circular pipe. First, there is a smooth descending curve traced by all the data points (circles) representing the steady-flat fall pattern. Secondly, it is possible to draw a nearly horizontal line fitting the data points (crosses) representing the stable tumble fall pattern. Most other data points representing the two unstable modes are seen to fall between the two lines just described. Now it is possible to postulate that there are two stable fall patterns corresponding to the two dynamic equilibrium states for a falling disk. These are the steady-flat fall and the tumble fall. According to the stability criterion based on the minimization theory, if there are two alternate equilibrium states, then the mode with less Φ (larger C_D in this case) is more likely to occur in practice. The relative position of the two lines in Figure 23 fit this requirement. The two unstable fall patterns can then be regarded as the state of transition (or the state of dynamic nonequilibrium) between the two equilibrium states.

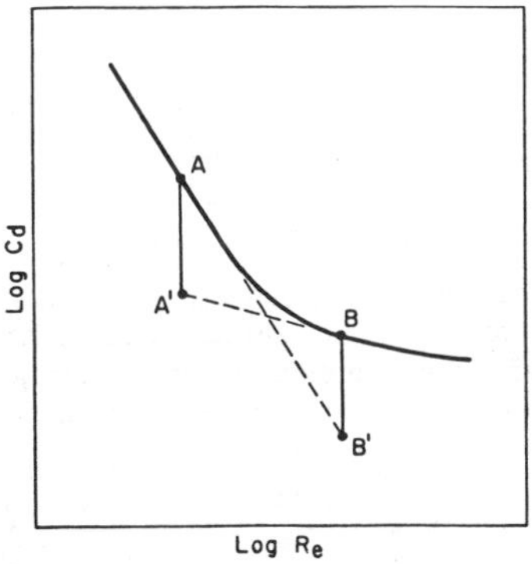

Figure 22. Drag coefficient and Reynolds number relationship for sphere on circular cylinder indicating minimization of energy dissipation rate and maximization of drag coefficient.

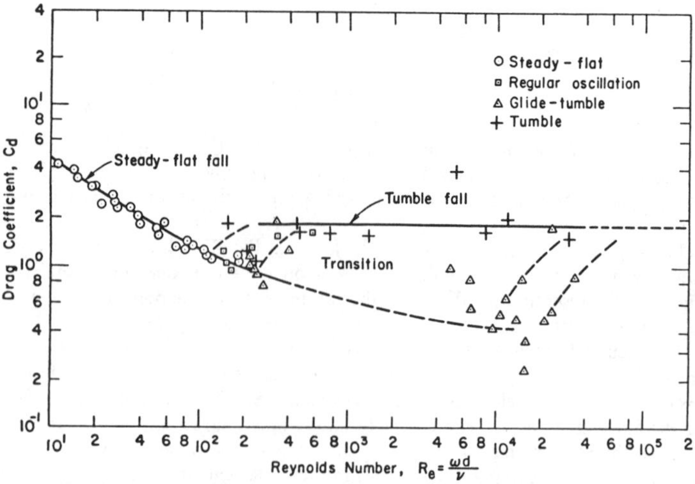

Figure 23. Drag coefficients of falling circular discs.

From Figure 23, it is quite obvious that the lower critical Reynolds number for a falling disk is about 100. It is interesting to point out that the drag coefficient for a disk in a tumbling mode is substantially greater than that of a flow around a fixed disk at the same Reynolds number. This is another indication that, given the chance by reduced constraint, the mechanical system will take the configuration that results in minimum energy or minimum rate of energy dissipation.

Air Bubble

The dynamics of a rising air bubble in a liquid is difficult to analyze because the shape of the bubble and the pattern of motion are not known beforehand. For this reason, an exact analytical solution of a rising bubble has not been found in spite of the wide interest and extensive investigations of the subject. Experiments show that a rising bubble may take one of the following three patterns:

Spherical bubbles. A small bubble may maintain a spherical shape because the surface tension energy dominates the kinetic energy. The motion is of Stokes' type and the drag coefficient is given by

$$C_D = \frac{16}{Re} \tag{142}$$

where Re = Reynolds number.

Oblate ellipsoidal bubble. As the bubble size grows and the rise velocity increases, the surface tension energy becomes less dominant and the bubble deforms. The bubble also orients itself in such a way that the maximum cross-sectional area faces the direction of motion similar to a falling body. The drag coefficient for this type of motion was shown to be closely approximated by the semi-empirical formula [41]:

$$C_D = \sqrt{\frac{288}{Re}} \tag{143}$$

Spherical-cap bubble. This is the mode for a typically large bubble shaped like an umbrella, and it rises steadily and rapidly. Although many investigators regard the flow as essentially nonviscous, flow visualization by Batchelor [42] clearly indicates the existence of a pair of large vortices trailing the bubble. The drag coefficient for this type of flow has been found to be a constant, i.e.,

$$C_D = 2.64 \tag{144}$$

Equations 142, 143, and 144 were plotted by Song and Yang [20] on a log-log paper and compared with some experimental data by Davidson et al. [41] as shown in Figure 24. As can be seen, the three equations plot as three straight lines of slope -1, $-\frac{1}{2}$, and 0, respectively. These lines intersect at the critical Reynolds number of 0.89 and 42.6. According to the theory of minimum rate of energy dissipation, a dynamic equilibrium condition requires maximum C_D. Therefore, a bubble will maximize the drag coefficient by changing its shape so the rate of energy dissipation can be minimized.

Bed Forms

One of the most difficult but fascinating problems in applied mechanics is the self-adjustment of alluvial channels. The self-adjustments of channel geometry, pattern, and longitudinal profile have been discussed in accordance with the theory of minimum rate of energy dissipation. The formation of ripples and dunes on an alluvial bed will be explained herein. Consider a closed system consisting of a large reservoir discharging water and bedload to a very long and straight channel with constant width. The channel length, L, the mild channel slope, S, the specific energy, E_s, and the mean sand diameter, d, are regarded as the given constants. The unknown variables to be determined are the friction coefficient, f, the flow depth, y, the water discharge per unit width, q, and bed load per unit width, q_b, etc. The total rate of energy dissipation per unit width of the system, including that of water and sediment is

$$\Phi = (\gamma q + \gamma_s q_b)SL \tag{145}$$

where γ, γ_s = specific weight of water and sediment, respectively. By neglecting the entrance and exit losses, the specific energy must be a constant and can be written as

$$E_s = \frac{q^2}{2gy^2} + y \tag{146}$$

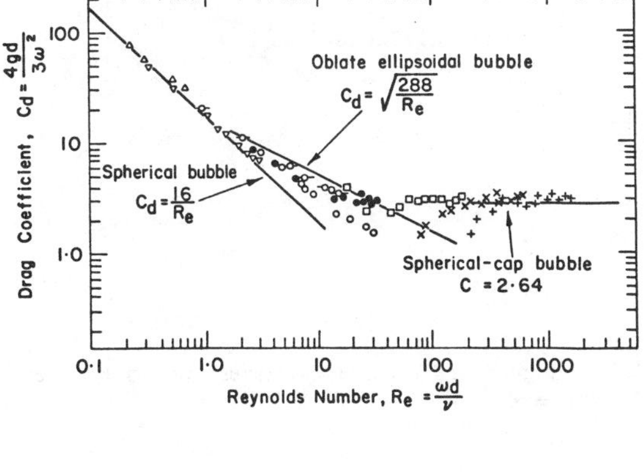

Figure 24. Drag coefficient as a function of Reynolds number for bubbles rising in liquids.

The resistance equation under an equilibrium condition may be written as

$$q = \sqrt{\frac{8g}{f}}\, y^{3/2} S^{1/2} \tag{147}$$

Without sediment transport, the friction factor in Equation 147 would be determined by the grain roughness and equal to a known constant f_d. Under this condition, the normal depth and the discharge may be determined by solving Equations 146 and 147 simultaneously. With sediment transport, f becomes a variable because the flow is capable of changing the roughness by generating bedforms. Under this condition, additional equations, including the bedload equation, are needed to determine the increased number of unknowns. Let us assume that the classical DuBoy's bedload equation can be used, i.e.,

$$q_b = \Psi \tau_0 (\tau_0 - \tau_c) \tag{148}$$

where Ψ = a function of sediment size; τ_0 = bed shear; and τ_c = critical shear stress. Under an equilibrium condition, τ_0 is related to y and S as

$$\tau_0 = \gamma y S \tag{149}$$

This is a classical indeterminate problem because there are six unknowns (Φ, q, q_b, τ_0, y, f) and five equations. The problem is posed as that of determining the six unknowns, which will minimize Φ subject to the constraints represented by Equations 146 through 149. Before solving the minimization problem, it is instructive to first eliminate four variables from the five equations and rewrite Equation 145 in a dimensionless form

$$\tilde{\phi} = \tilde{\phi}_w + \tilde{\phi}_b \tag{150}$$

The first term in Equation 150 is a dimensionless total rate of energy dissipation defined as

$$\tilde{\phi} = \Phi \sqrt{2g}\, \gamma S L E_s^{3/2} \tag{151}$$

The second term in Equation 150 is the rate of energy dissipation due to the flowing water which is found to be

$$\tilde{\phi}_w = 2\tilde{f}(1 + 2\tilde{f})^{-3/2} \tag{152}$$

in which

$$\tilde{f} = f/8S = Fr^{-2} \tag{153}$$

and Fr = the Froude number. The last term in Equation 150 is, for $\tilde{f} > \tilde{f}_c$

$$\tilde{\phi}_b = \tilde{\Psi}\left(\frac{2\tilde{f}}{1 + 2\tilde{f}} - \frac{2\tilde{f}_c}{1 + 2\tilde{f}_c}\right)\left(\frac{2\tilde{f}}{1 + 2\tilde{f}}\right) \tag{154}$$

in which

$$\tilde{\Psi} = \Psi \gamma \gamma_s S^2 \sqrt{E_s/2g} \tag{155}$$

and $\tilde{f}_c$ is the value of $\tilde{f}$ at incipient motion.

Equations 150, 152, and 154 are plotted on Figure 25. The curve $\tilde{\phi}(1)$ corresponds to $\tilde{\Psi} = 1$ and $\tilde{f}_c = 5$ while $\tilde{\phi}(2)$ corresponds to $\tilde{\Psi} = 2$ and $\tilde{f}_c = 5$.

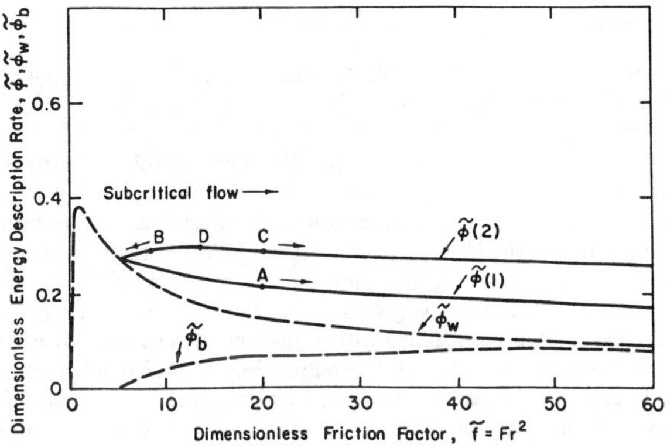

Figure 25. Relationship between dimensionless energy dissipation rate and dimensionless friction factor with bedload.

The following physical interpretations of Figure 25 were made by Song and Yang [20]:

1. If the given condition is such that no bedload is possible, $\tilde{f} < \tilde{f}_c$, then the bed is rigid and $\tilde{f}$ is a constant. In this case, an initially flat bed will remain flat.
2. Two different possibilities are indicated if bedload exists and $\tilde{f} \geq \tilde{f}_c$. The first case is when $\tilde{\phi}$ is monotonically decreasing, as indicated by the $\tilde{\phi}(1)$ curve in Figure 25. Regardless of the initial bedform, as long as $\tilde{f} > \tilde{f}_c$, there is a bedload and the bedform is capable of changing. This change can occur only in the direction of decreasing $\tilde{\phi}$ or increasing $\tilde{f}$, causing q to decrease, while q_b increases. This is believed to be the mechanism of ripple and dune formation based on the principle of variational mechanics. Further increase of sediment transport rate will cause a damping effect by suspended load. This effect was also explained by Song and Yang [20].
3. A second possibility is suggested by the shape of $\tilde{\phi}(2)$. Suppose that the initial condition is such that $\tilde{f} > \tilde{f}_c$ and represented by point B in Figure 25. Furthermore, if the initial bed is not flat, then the sediment movement should tend to reduce the roughness until the bed becomes flat or $\tilde{f}$ approaches $\tilde{f}_c$ and bedload stops. The former case indicates the mechanism of lower regime flat bed and the latter case corresponds to a subincipient flow condition. That is, bedload may be temporarily generated due to excess bed roughness, but the bedload stops as soon as the excess roughness is eliminated by the flow.
4. Now consider the case when the initial condition is given by point C which is located to the right of point D which is the maximum on curve $\tilde{\phi}(2)$. Like the case described under case (2), the tendency here is again to generate more roughness. However, because there is a point of maximum $\tilde{\phi}$ on the curve, a different interpretation must be given. There are two cases to be considered.
   a. If the initial bed is flat, then the flow will generate sand waves as in the case explained under case (2).
   b. This is the case when the grain roughness is so small that $\tilde{f}$ would be less than that of maximum $\tilde{\phi}$ if the bed were flat, but the initial roughness was created by disturbances imposed on the bed. That is, the instability is induced by an initial large amplitude disturbance.

   This type of instability is known in hydrodynamics as the subcritical instability.

It has been shown that many of the complicated phenomena in fluid mechanics can be explained by the application of theory of minimum rate of energy dissipation. The remaining examples will be devoted to the application of the theory to solve hydraulic engineering design problems.

Morning-Glory Spillway

One of the commonly used reservoir hydraulic structures is the morning-glory type of spillway. One design problem a hydraulic engineer has to face is the determination of different types of control that limit the discharge.

There are three types of controls, i.e, crest control, orifice control, and pipe control. The corresponding head discharge curves are shown by the solid lines in Figure 26. As the reservoir head is gradually increased from zero, the flow regime switches from crest control to orifice control and to pipe control. The reason that the head discharge curves followed these solid line segments rather than their extensions as indicated by dotted lines, can be explained by the theory of minimum rate of energy dissipation. Because head H is given, minimization of rate of energy dissipation means minimization of discharge Q in accordance with Equation 68. The solid lines will give smaller Q within the corresponding range of H than the dash lines. Figure 26 also indicates that there is a small range on either side of point h where the flow may be unstable. For a slight variation of H, flow may oscillate between orifice control and pipe control because both curves can give us about the same Q value. For a given H, equal Q means equal relative stability. In practice, we will try to avoid the regime near point h to avoid oscillation. Thus, the theory can be used to explain the change of controls and provide some guideline to improve drop inlet structure design.

Stable Channel Design

Chang [43] applied the theory of minimum stream power, i.e.,

$$QS = \text{a minimum} \qquad (156)$$

to stable channel design. The independent variables are water discharge and sediment load. He also assumed that sediment particle size and channel side slope are given. The dependent variables to be determined are flow velocity, channel width, depth, and slope. In addition to Equation 156, Chang also used the continuity equation of water discharge, a sediment transport equation, and a roughness equation. Thus, the four dependent variables can be solved by trial-and-error with the four equations just stated. An example of the design chart obtained by Chang is shown in Figure 27.

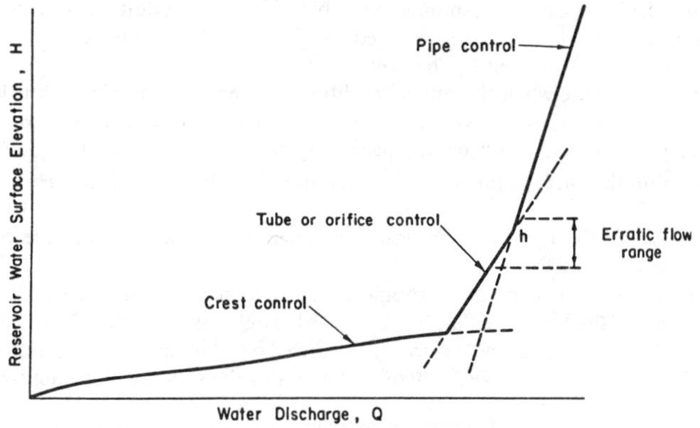

Figure 26. Nature of flow and discharge characteristics of a morning glory spillway.

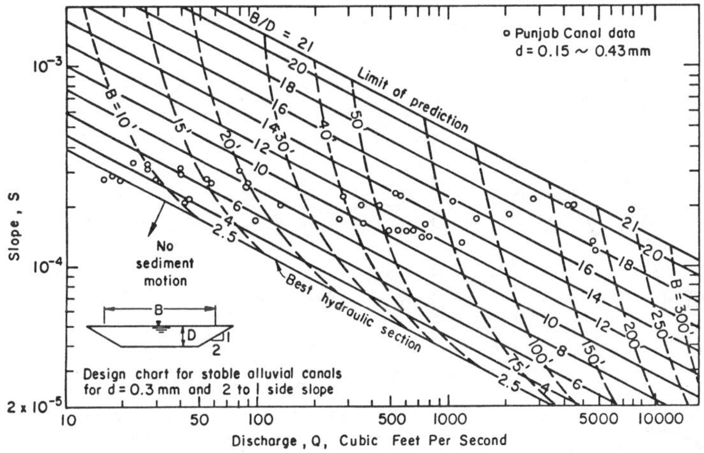

Figure 27. Design chart for stable alluvial canals for specified sediment size and side slope (after Chang, 1980).

Reservoir Sedimentation

Annandale and Rooseboom [44] applied the theory of minimum rate of energy dissipation to the study of reservoir sedimentation. The total stream power used by them is the integration of unit stream power through the total flow depth. The final reservoir bed equilibrium profile obtained by the application of minimum total stream power is

$$(DS)^{1.5} = \text{a constant} \tag{157}$$

Other equations used by them include the unit stream power equation for sediment transport, Equation 133, and the continuity equation for sediment along the reservoir bed. Favorable comparisons between the computed and measured reservoir bed profile due to sedimentation were made by them for six existing South African reservoirs.

Case Study of Natural Rivers

Yang and Song [19] studied the stability of the Mississippi River and other rivers in the United States. Historical data studied by them indicate that when the rate of energy dissipation was at its minimum value under the given constraints, the Mississippi River was relatively stable. Under similar constraints but with higher rate of energy dissipation, the Mississippi River was relatively unstable.

The theory was recently applied by Robbins and Simon [45] to study man-induced channel adjustments of Tennessee streams. They found that engineering construction could alter the equilibrium of natural streams. Field studies of Tennessee streams revealed that an originally stable stream can become unstable after engineering work. The rate of energy dissipation will increase first immediately after the disturbance and then gradually decrease as a stream readjusts itself to regain equilibrium.

Between 1964 and 1966, the U.S. Army Corps of Engineers enlarged and straightened the main-stem Forked Deer River from the confluence of the north and south forks to its confluence with the Obion River. Between 1968 and 1969, the Corps modified the lowest seven miles of the South Fork Deer River channel that had previously been left untouched. These activities made the river unstable and caused the river to readjust itself. Figure 28 shows that the unit stream power increases immediately after these man-induced activities and then decrease gradually as the river regains its

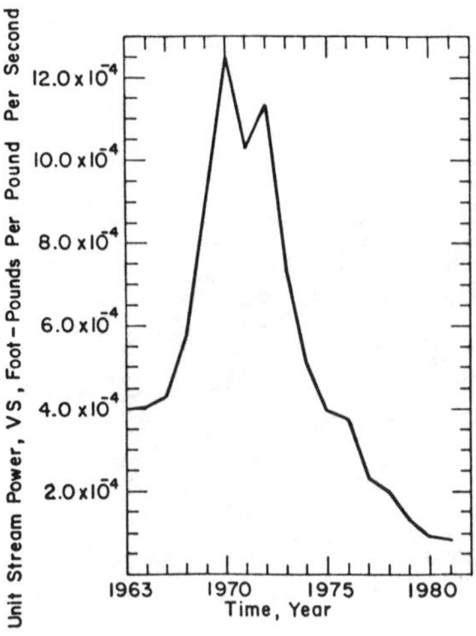

Figure 28. Unit stream power versus time for gaging station at Halls on South Fork Forked Deer River, Tennessee (after Robbins and Simon, 1983).

equilibrium. Robbins and Simon's field observation provides a direct confirmation of Rule 1 of the minimum rate of energy dissipation theory. This rule states that when a system is not at its equilibrium condition, its rate of energy dissipation is not at its minimum value. However, the system will adjust in such a manner that its rate of energy dissipation can be minimized to regain equilibrium.

OTHER RELATED THEORIES AND HYPOTHESES

The flexibility and potential of using minimization or maximization approach to solve complicated problems have gained increasing attention among researchers in recent years. Due to the difficulty of solving fluvial hydraulic problems from vectorial approach, most of the theories or hypotheses were developed for fluvial hydraulics of movable bed. A review and evaluation of these hypotheses will be made in the following sections.

Maximum Sediment Discharge

White, Bettess, and Paris [46] hypothesized that in the process of reaching an equilibrium condition for an alluvial channel, the rate of sediment transport will be maximized. They stated that although there is no physical justification to support their hypotheses, it could lead to acceptable predictions over a large range of flow and sediment transport conditions. They also found that minimization of stream power and maximization of sediment transport rate lead to the same answer as shown on Figure 29. Apparently, they consider sediment transport as a dependent variable instead of a given independent variable. When sediment transport rate and water discharge are considered as given constraints, the theory of minimum stream power requires an alluvial channel to adjust itself in such a manner that the given sediment load and water discharge can be transported with a minimum stream power. Conversely, for a given stream power, an alluvial channel will adjust itself so the sediment can be transported most efficiently, that is, maximum sediment transport rate. Thus, maximum sediment transport rate is only another way of stating minimum rate of energy dissipation theory under the special condition that sediment transport rate can be treated as a variable. Maximization of sediment transport rate is not valid when there is no sediment transport or where the rate of sediment transport is limited by supply or other constraints. It is also questionable whether the hypothesis of maximum sediment transport rate can be applied to explain the change of river pattern and longitudinal bed profile as stated earlier.

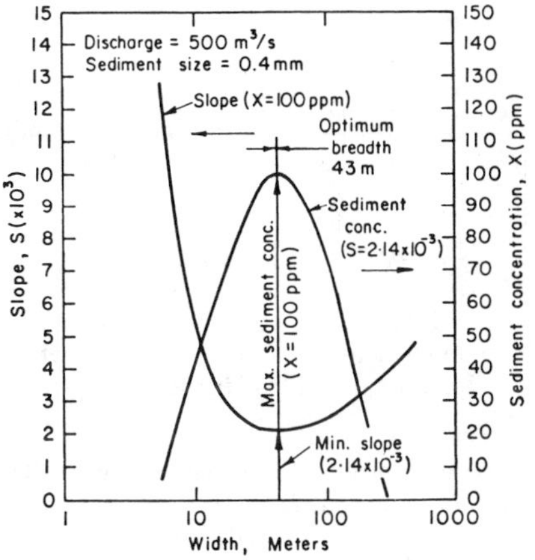

Figure 29. Slope and sediment concentration versus width (from White, Bettess, and Paris, 1982).

Maximum Friction Factor

Davies and Sutherland [47, 48] proposed an empirical hypothesis of maximum friction factor to explain the dynamic adjustments of an alluvial system. They stated that, "If the flow of a fluid past an originally plane boundary is able to deform the boundary to a nonplanar shape, it will do so in such a way that the friction factor increases. The deformation will cease when the shape of the boundary is that which gives rise to a local maximum of friction factor. Thus, the equilibrium shape of a nonplanar self-formed flow boundary or channel corresponds to a local maximum of friction factor." They also found in many cases that their hypotheses will give the same results as those from the theory of minimum stream power or minimum unit stream power. Figure 30 indicates that as water discharge increases, the friction factor increases initially and then decreases. However, the rate of energy dissipation can always be minimized along either course A or B, regardless whether the friction factor is increasing or decreasing as shown in Figure 31, by using the theory of minimum unit stream power and Yang's [11] sand transport equation.

Davies and Sutherland [48] stated that minimum unit stream power and maximum friction factor can lead to the same answer for alluvial channels under most conditions. The only exception is the case where water discharge and depth are independent variables. Under this condition, minimization of unit stream power would lead to minimization of channel slope and sediment discharge while maximization of friction factor would lead to maximization of slope and sediment discharge. They did not provide any computation or theoretical explanation to support the validity of their hypothesis under this condition. The data shown in Figure 14 were collected under the

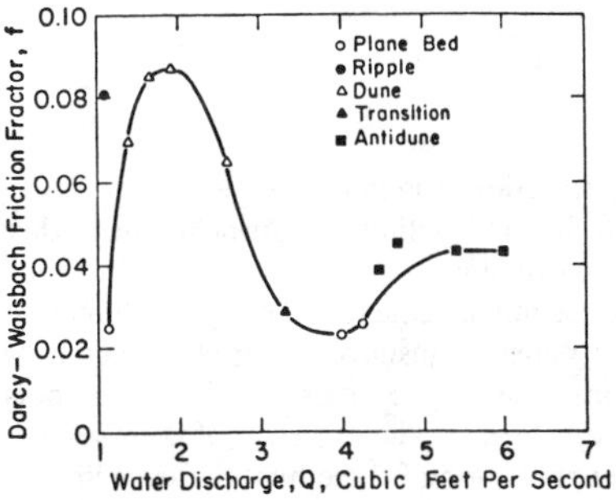

Figure 30. Relationship between friction factor and water discharge for the 0.33 mm uniform sand by Guy, Simons, and Richardson (1966).

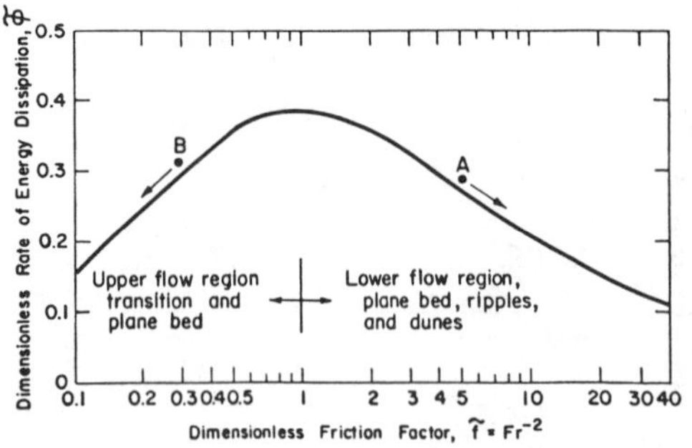

Figure 31. Relationship between dimensionless rate of energy dissipation and dimensionless friction factor.

The computed minimum unit stream powers are in good agreement with the measurements. Because Davies and Sutherland already stated that minimization of unit streaam power and maximization of friction factor under this condition would give opposite results, the only conclusion is that the hypothesis of maximum friction factor would not be valid in this case.

It should be pointed out that the hypothesis of maximum friction factor was based on limited observation of increase of bed roughness as bed surface changes from plane bed to ripple and dune. The friction factor actually decreases when the water and sediment discharge are further increased and bed form is changed from dune to plane bed or transition as shown on Figure 30. There are many different ways a system can minimize its rate of energy dissipation. Maximization of friction factor is only one way of adjustment to minimize the rate of energy dissipation under some special conditions where the friction factor is permitted to be maximized.

Maximum Sediment Discharge and Froude Number

According to Ramlett [49], the ideal morphological formation of a water course corresponds to the parameters that enable each of the following to be considered:

1. Flow at full volume of dominant discharge.
2. No sediment discharge or saturation of sediment discharge to ensure stability.
3. Maximization of Froude number so that the minimum amount of rate of energy dissipation is used for channel excavation.

Ramlett applied these hypotheses to stable channel design with reasonable results, although he made little theoretical justification for the use of them. His first statement is equivalent to the old regime concept that a channel is formed by a dominant water discharge. His second statement is equivalent to the hypothesis of maximum sediment discharge. The last statement is equivalent to the statement that minimization of energy dissipation rate can be achieved by maximization of Froude number under some special conditions. Since the minimum rate of energy dissipation theory does not prohibit these phenomena from happening, they may be interpreted as different means to satisfy the theory.

General Remarks

An alluvial channel system can adjust its velocity distribution, roughness or bed form, slope, channel geometry, channel pattern, longitudinal bed profile, and sediment transport rate, either individually or collectively, to minimize its rate of energy dissipation under certain constraints imposed on the system. The three hypotheses stated in this section indicate that the state of minimum rate of energy dissipation may be achieved by adjusting sediment transport rate, friction factor, or Froude number of the flow under some special conditions. These hypotheses describe some tools available to an alluvial system not the basic reason for adjustments. The objective of these adjustments is to minimize the rate of energy dissipation. The application of these hypotheses is limited to fluvial systems where those factors are permitted to adjust while the general theory of minimum rate of energy dissipation applies to all closed and dissipative systems. Energy and mass are the fundamental quantities of the universe while friction factor, sediment transport rate, and Froude number are derived quantities. Application of these hypotheses without recognizing the foresaid important points could lead to erroneous results and explanations of phenomena in fluvial hydraulics.

SUMMARY

The vectorial mechanics approach commonly used in fluid mechanics is based on Newton's laws of motion. Force and momentum are the basic concern and the equation motion is the basic mathematical tool used in vectorial mechanics. The limitations and difficulties of applying the equation of motion to some complicated problems have been briefly stated in this chapter. The variational approach replaces the force or the work of the force by energy. Calculus of variation or other minimization techniques become the mathematical tool for the variational approach. The history of development which leads to the formulation of the general theory of minimum energy and energy dissipation rate has been reviewed and further developed. The general theory of minimum energy and energy dissipation rate can be applied to a closed and dissipative system under static or dynamic equilibrium condition. Examples are given to illustrate how this general theory or its simplified versions can be applied. The limitations of other related theories or hypotheses and their relationships with the general theory have been reviewed to clarify possible confusions. The general theory and its simplified versions provide us a useful alternative to the vectorial mechanics approach in solving static and dynamic problems.

NOTATION

A	cross-sectional area
A_1, A_2	dimensionless coefficients
a, b, c	coefficients
B	channel width
C_1, C_2	constants
C_D	drag coefficient
D	average flow depth
E	total rate of energy dissipation
Fr	Froude number
f	friction coefficient
g	gravitational acceleration
H	total head
h	elevation
i, j	coefficients
k, k_f	spring constant and elastic constant
k	coefficient
L	length
l	length
M	mass
m	coefficient
P	pressure
Q	water discharge rate
q	water discharge per unit width
R	radius
Re	Reynolds number
r	radial distance
r_i	state variable
S	channel or energy slope
t	time
$\bar{u}_i$	time-averaged velocity
V	average flow velocity
W_s	submerged weight
X	distance
x, y, z	coordinates
Y	potential energy per unit weight of fluid

Greek Symbols

α	constant
$\alpha_{1,2}$	velocity correction factors
γ	specific weight
δ_{ij}	unit second-order tensor
ϵ	eddy viscosity or deformation
ϵ_{ijk}	Kronecka delta
η	dimensionless depth
μ	viscosity
ν	kinematic viscosity
ξ	energy dissipation rate per unit length per unit width
Ψ	function of sediment size
θ	angle
ρ	density
σ_{ij}	Reynold's total stress tensor
τ_c	critical shear stress
τ_0	bed shear
$\bar{\tau}_{ij}$	viscous stress tensor
ϕ	rate of energy dissipation per unit fluid volume
ω	terminal fall velocity
ω_k	vorticity vector

REFERENCES

1. Lanczos, C., *The Variational Principles of Mechanics*, 4th edition, University of Toronto Press, Toronto, Canada, 1974.
2. Leopold, L. B., and Maddock, T. Jr., "The Hydraulic Geometry of Stream Channels and Some Physiographic Implications," U.S. Geological Survey Professional Paper 252, 57 pp., 1953.
3. Stall, J. B., and Yang, C. T., "Hydraulic Geometry of 12 Selected Stream Systems in the United States," University of Illinois Water Resources Center, Research Report No. 32, 73 pp., 1970.
4. Langbein, W. B., "Geometry of River Channels," *American Society of Civil Engineers Proceedings*, Vol. 90, pp. 301–312, 1964.
5. Williams, G. P., "Hydraulic Geometry of River Cross Sections—Theory of Minimum Variance," U.S. Geological Survey Professional Paper 1029, 47 pp., 1978.
6. Leopold, L. B., and Langbein, W. B., "The Concept of Entropy in Landscape Evolution," U.S. Geological Survey Professional Paper 500-A, 1962.
7. Kennedy, J. F., Richardson, P. D., and Sutera, S. P., Discussion on "Geometry of River Channels," by W. B., Langbein, *Journal of the Hydraulics Division*, ASCE, Vol. 91, No. HY3, pp. 332–341, 1964.
8. Langbein, W. B., Closure on "Geometry of River Channels," *Journal of the Hydraulics Division*, ASCE, Vol. 91, No. HY3, pp. 297–313, 1965.
9. Scheidegger, A. E., "Some Implications of Statistical Mechanics in Geomorphology," *International Association of Scientific Hydrology Bulletin*, Vol. 9, No. 1, pp. 12–16, 1964.
10. Yang, C. T., "Potential Energy and Stream Morphology," American Geophysical Union, *Water Resources Research*, Vol. 7, No. 2, 311–322, 1971.
11. Yang, C. T., "Minimum Rate of Energy Dissipation and River Morphology," D. B. Simons Symposium on Erosion and Sedimentation, Colorado State University, Fort Collins, CO, pp. 3.2–3.19, July 1983.
12. Yang, C. T., "Minimum Unit Stream Power and Fluvial Hydraulics, "*Journal of the Hydraulics Division*, ASCE, Vol. 102, No. 4Y7, Proceeding Paper 12238, pp. 919–934, 1976.
13. Yang, C. T. and Song, C. C. S., "Theory of Minimum Rate of Energy Dissipation," *Journal of the Hydraulics Division*, ASCE, Vol. 105, No. HY7, Proceedings Paper 14677, pp. 769–784, 1979.
14. Hinze, J. O., *Turbulence, An Introduction to its Mechanism and Theory*, McGraw-Hill Book Co., Inc., New York, NY, pp. 17–23, 68–74, 1959.
15. Lamb, H., *Hydrodynamics*, 6th ed., Dover Publications, Inc., NY, pp. 43–45, 574–580, 617–619, 1932.
16. Korteweg, D. J., "On a General Theorem of the Stability of the Motion of a Viscous Fluid," *Philosophical Magazine*, London, England, Vol. XVI, No. 5, pp. 112–118, 1883.
17. Song, C. C. S., and Yang, C. T., "Minimum Stream Power: Theory," *Journal of the Hydraulics Division*, ASCE, Vol. 106, No. HY9, Proceedings Paper 15691, pp. 1477–1487, 1980.
18. Gelfand, I. M., and Fomin, S. V., *Calculus of Variations*, translated by R. A. Siberman, Prentice-Hall, Inc., Englewood Cliffs, NJ, pp. 23, 24, 125, 153, 1963.
19. Yang, C. T., and Song, C. C. S., "Dynamic Adjustments of Alluvial Channels," *Adjustments of the Fluvial System*, edited by D. D. Rhodes and G. P. Williams, Kendall/Hunt Publishing Company, pp. 55–67, 1979.
20. Song, C. C. S., and Yang, C. T., "Minimum Energy and Energy Dissipation Rate," *Journal of the Hydraulics Division*, ASCE, Vol. 108, No. HY5, Proceedings Paper 17063, pp. 690–706, 1982.
21. Hadley, G., *Nonlinear and Dynamic Programming*, Addison-Wesley Pub. Co., pp. 36, 204, 238, 467, 1964.
22. Chang, H. H., and Hill, J. C., "Minimum Stream Power for Rivers and Deltas," *Journal of the Hydraulics Division*, ASCE, Vol. 103, No. HY12, Proceeding Papaer 13394, pp. 1375–1389, 1977.
23. Godden, W. G., *Numerical Analysis of Beam and Column Structures*, Prentice-Hall, Inc., pp. 100–103, 1965.
24. Vanoni, V. A., "Transportation of Suspended Sediment by Water," *Transactions*, ASCE, Vol. 111, Paper No. 2267, pp. 67–102, 1946.

25. Song, C. C. S., and Yang, C. T., "Velocity Profiles and Minimum Stream Power," *Journal of the Hydraulic Division*, ASCE, Vol. 105, No. HY8, Proceeding paper 14780, pp. 981–998, 1979.
26. Guy, H. P., Simons, D. B., and Richardson, E. V., "Summary of Alluvial Channel Data from Flume Experiments," U. S. Geological Survey Professional Paper 462-I, 1966.
27. Yuan, S. W., *Foundations of Fluid Mechanics*, Prentice-Hall, Englewood Cliffs, NJ, pp. 121, 1967.
28. Serrin, J., Mathematical Principles of Classical Fluid Mechanics, *Handbook of Physics*, edited by S. Flugge and C. Truesdell, Springer-Verlag, Berlin, 1959.
29. Chang, H. H., "Geometry of Rivers in Regime," *Journal of the Hydraulics Div.*, ASCE, Vol. 105, No. HY6, Proceeding Paper 14640, pp. 691–706, 1979.
30. Chang, H. H., "Minimum Stream Power and River Channel Patterns," *Journal of Hydrology*, 41, pp. 303–327, 1979.
31. Yang, C. T., Song, C. C. S., and Woldenberg, M. J., "*Hydraulic Geometry and Minimum Rate of Energy Dissipation*, Water Resources Research, Vol. 17, No. 4, pp. 1014–1018, 1981.
32. Blench, T., *Regime Behavior of Canals and Rivers*, Butterworths, London, 1957.
33. Barr, D. I. H., Alan, M. K., and Nishat, A., "A Contribution to Regime Theory Relating Principally to Channel Geometry," *Proceedings of the Institute of Civil Engineers*, 69, pp. 651–670, 1980.
34. Yang, C. T., "On River Meanders," *Journal of Hydrology*, Vol. 13, pp. 231–253, 1971.
35. Yang, C. T., *Formation of Riffles and Pools*, Water Resources Research, Vol. 7, No. 6, pp. 1567–1574, 1971.
36. Davey, A., and Drazin, P. G., "The Stability of Poiseuille Flow in a Pipe," *Journal of Fluid Mechanics*, Vol. 36, pt. 2, pp. 209–218, 1969.
37. Salwen, H., and Grosch, E., "Stability of Poiseuille Flow in a Pipe of Circular Cross Section," *Journal of Fluid Mechanics*, Vol. 54, pt. 1, pp. 93, 1972.
38. Davey, A., "On Itoh's Finite Amplitude Stability Theory for Pipe Flow," *Journal of Fluid Mechanics*, Vol. 80, pt. 4, pp. 695–703, 1978.
39. Joseph, D. D., "Stability of Fluid Motions I," Springer Tracts in Natural Philosophy, Vol. 27, Springer-Verlag, Berlin, 1976.
40. Stringham, G. E., Simons, D. B., and Guy, H. P., "The Behavior of Large Particles Falling in Quiescent Liquids," U. S. Geological Survey Professional Paper 562-C, 1969.
41. Davidson, J. F., Harrison, D., and Guedes de Carvalho, "On the Liquidlike Behavior of Fluidized Beds," *Annual Review of Fluid Mechanics*, Vol. 9, pp. 55–86, 1977.
42. Batchelor, G. K., *An Introduction to Fluid Dynamics*, Cambridge University Press, pp. 229–263, also plates 1–15, 1967.
43. Chang, H. H., "Stable Alluvial Canal Design," *Journal of the Hydraulics Div.*, ASCE, Vol. 106, No. HY5, Proceeding paper 15420, pp. 873–891, 1980.
44. Annandale, G. W., and Rooseboom, A., "Reservoir Sedimentation and Stream Power," D. B. Simons Symposium on Erosion and Sedimentation, Colorado State University, Fort Collins, CO, pp. 7.2–7.32, July 1983.
45. Robbins, C. H., and Simon, A., "Man-Induced Channel Adjustment in Tennessee Streams," U. S. Geological Survey Water-Resources Investigations Report 82-4098, Nashville, TN, 1983.
46. White, W. R., Bettess, R., and Paris, E., "Analytical Approach to River Regime," *Journal of the Hydraulics Div.*, ASCE, Vol. 108, No. HY10, Proceedings paper 17399, pp. 1179–1193, 1982.
47. Davies, T. R. H., and Sutherland, A. J., "Resistance to Flow Past Deformable Boundaries," *Earth Surface Process*, 5, pp. 175–179, 1980.
48. Davies, T. R. H., and Sutherland, A. J., "Extremal Hypotheses for River Behavior," *Water Resources Research*, Vol. 19, No. 1, pp. 141–148, Feb. 1983.
49. Ramlette, M., "Guide to River Engineering," State Electricity Commission of Victoria, Electricite de France, Direction des Etudes et Recherches, Report No. HE/40/81.04, April 1981 (Translated from French by M. Dixon and G. H. Wheelhouse).
50. Parker, G., Discussion of "Minimum Unit Stream Power and Fluvial Hydraulics," by C. T. Yang, *Journal of the Hydraulics Div.*, ASCE, Vol. 103, No. HY7, pp. 811–816, 1977.

CHAPTER 12

PRINCIPLES OF DIMENSIONAL ANALYSIS

Thomas Z. Fahidy and Mohd S. Quaraishi

Department of Chemical Engineering
University of Waterloo
Waterloo, Ontario, Canada

CONTENTS

INTRODUCTION

Basic Notions

When a physical relationship is described by mathematical equations, each quantity can be represented by its appropriately chosen magnitudes. The search for the correct form of the relation connecting the magnitudes of employed quantities is known as dimensional analysis. An equation, whose form does not depend on the fundamental units of measurement is called dimensionally homogeneous; if the fundamental magnitudes have been fixed, quantities of the same kind are represented by magnitudes with the same dimensional relationships and any equation carrying these magnitudes must be dimensionally homogeneous. The relation between the magnitudes of quantities is *not* a complete solution of a problem: it only provides a specific combination of quantities, called dimensionless groups, which are expected to govern the physical system or phenomenon of interest. The exact relationship between the dimensionless groups must usually be found via experiment, although in a limited number of cases it may be derived from fundamental principles of the physical sciences (e.g., heat and mass transfer in natural convection at a vertical plate). The major value of dimensional analysis is in allowing significant simplifications in the number of the independent variables when treating complex physical phenomena, thereby offering important shortcuts towards a comprehensive analysis with the aid of dimensional consistency.

Dimensional analysis is closely related to the concept of physical (i.e., geometric, kinetic, dynamic, etc.) similarity and of scaling based on model-prototype similarity. Direct applications are found in the design of large-scale engineering systems.

The dimensional analysis approach suffers from two major limitations: the first, mentioned above, is the lack of a complete solution. The second is the skill (or "foresight") required to choose the right dimension base in order to obtain the correct number and type of dimensionless groups. Recent extensions of the classical techniques of dimensional analysis have been noticeably successful in minimizing the effect of the second limitation.

Historical

It is generally believed that dimensional analysis originates from Newton [1]. The indicial (exponent) method introduced by Lord Rayleigh [2, 3] was closely followed by Buckingham's method called the pi-theorem [4]; a variation of the latter by O'Rahilly [5] known as the measure-ratio method, has remained relatively unrecognized in spite of its rigorous mathematical structure [6]. The rationale behind the pi-theorem was demonstrated to be correct by Brand [7, 8] in terms of linear vector space theory and by Corrsin [9] in geometrical terms employing the concept of dimension-space. Directed graphs [10] and group theoretic methods [11–13] have been used to generalize the approach of dimensional analysis; the technique of Staicu [14] successfully reduces the extent of functional indeterminacy at the expense of stringent restrictions. A simplified method of dimensional analysis for the identification of all dimensionless groups governing the behavior of a system was recently introduced [15] and generalized [16] to all unit systems.

Dimensional analysis has been used extensively since its inception in various areas of science and engineering; the width of its applicability is illustrated in Table 1. Applications in fluid mechanics

Table 1
Major Areas of Application for Dimensional Analysis*

Application area	References
Cosmology and Cosmogony	[17–20]
Gas dynamics	[21–24]
General theory and methods (including use of computers) of dimensional analysis	[25–38]
Electrochemistry; electrochemical engineering; physical chemistry of electrolytes	[39–43]
Foam formation; detergency (soil-science)	[44–47]
Liquid–gas and liquid–liquid systems; mass and heat transfer; thermodynamics of phase and chemical equilibria; chemical reactors	[48–63]
Handling of solids and powders; milling operations; solid films; filtering	[64–69]
Structure of chemical compounds; polymers; magnetic properties	[70–72]

* *For fluid mechanics and heat transfer applications see Table 2.*

Table 2
Principal Sourcebooks For The Study of Dimensional Analysis

Author(s) and reference number	Emphasis and major subjects treated	Remarks
Bridgman [73]	Dimensional formulae, Pi-theorem, application to model experiments, engineering and theoretical physics.	A widely quoted introductory source on fundamentals.
Focken [74]	A thorough historical review of the early development of DA and the measure-ratio method. An overview of shortcomings and criticisms. Some applications to physics and engineering.	The historical polemic between Campbell, Buckingham, Ehrenfest-Afanassjewa, Bridgman, Dingle and O'Rahilly is carefully analyzed and commented.
Giles [75]	Principles of similarity; Rayleigh's method; Pi-theorem; applications to fluid mechanics.	Numerous examples; clear presentation of the recurring-variable principle in the Pi-theorem method.
Kurth [76]	Formal proof by Brand of the Pi-theorem; applications to astrophysics (e.g., stellar radiation and stellar structure).	Advanced reading.
Langhaar [77]	Algebraic theory of model testing; applications to strength of materials, fluid mechanics, heat transfer, electromagnetics, differential equations.	Discussion of a wide spectrum of applications. Formal derivation of the Pi-theorem.
Massey [78]	Rayleigh's method and the Pi-theorem, choice of dimension base; dimensional formulae; physical similarity.	A concise critical treatment. Numerous tables and a comprehensive list of frequently encountered dimensionless groups.
Pankhurst [79]	Fundamentals of dimensional analysis.	Thorough introduction for beginners.
Raghunath [80]	General theory; river channel models, wave action; hydraulic machinery models.	Strong emphasis on applications to hydraulics.
Schepatz [81]	Fundamentals of dimensional analysis and applications in the biomedical sciences.	Emphasis on the biomechanics of solids and liquids; surface phenomena; enzymes, metabolisms; allometry.
Sedov [82]	Fundamentals; similarity and modelling Applications to the motion of viscous fluids, turbulence, unsteady gas motion and astrophysics.	A thorough and comprehensive treatment.
de St. Q. Isaacson and de St. Q. Isaacson [83]	Systematic calculation of dimensionless numbers; geometric approach; directed graphs; extension of dimension sets; reduction of undetermined functions; model laws; cosmological aspects; limitations.	A comprehensive treatise of advanced material. Thorough in depth and width.
Taylor [84]	The choice and finding of dimensionless groups, geometric similarity. Application to mathematical formulations in the absence of a specific physical problem. Practical examples.	Strong emphasis on a variety of applications, e.g., eddy current brake, sleeve bearing, piston engines, sailing boats and airplanes (written with a sharp sense of humour).

and hydrology, one major area of the approach, have not been included in Table 1, inasmuch as they are amply discussed in principal texts and references assembled in Table 2.

Dimension Bases

Every physical quantity can be expressed in terms of fundamental dimensions; however, dimension bases can be chosen arbitrarily to include derived dimensions for the sake of efficiency. The international system of units (SI) defines seven fundamental dimensional quantities: length, mass, time, electric current, (thermodynamic) temperature, amount of substance, and luminous intensity. On the other hand, force may replace mass, and heat may be used as a separate (independent) quantity to great advantage in a dimension base if, for instance, the motion is unaccelerated or heat is not transformed into another form of energy. Similarly, the use of the electric charge instead of current is often more convenient in the dimensional analysis of electrical and magnetic systems. The mass-length-time (MLT) base and the force-length-time (FLT) base are the two most common dimension bases; Table 3 contains the dimensional representation of the most frequently occurring physical

Table 3
Dimensional Representation of Frequently Occurring Physical Quantities Via the MLT and the FLT Base

Quantity	Conventional symbol	Dimensional Representation MLT-base	FLT-base
Absolute viscosity	μ	$ML^{-1}T^{-1}$	FTL^{-2}
Acceleration	a, g	LT^{-2}	LT^{-2}
Angular velocity	ω	T^{-1}	T^{-1}
Area	A	L^2	L^2
Density	ρ	ML^{-3}	FT^2L^{-4}
Electric charge	Q	Q	Q
Electric current	I	QT^{-1}	QT^{-1}
Electric field strength	E	$MLT^{-2}Q^{-1}$	FQ^{-1}
Electric flux	ψ	Q	Q
Energy	E	ML^2T^{-2}	FL
Enthalpy	h, H	ML^2T^{-2}	FL
Entropy	s, S	$ML^2T^{-2}\theta^{-1}$	$FL\theta^{-1}$
Force	F	MLT^{-2}	F
Magnetic field strength	H	$L^{-1}T^{-1}Q$	$L^{-1}T^{-1}Q$
Magnetic flux	ϕ	$ML^2T^{-1}Q^{-1}$	$FLTQ^{-1}$
Magnetic permeability	μ	MLQ^{-2}	FT^2Q^{-2}
Mass	m, M	M	FT^2L^{-1}
Power	P	ML^2T^{-3}	FLT^{-1}
Pressure	p, P	$ML^{-1}T^{-2}$	FL^{-2}
Rate of flow	Q	L^3T^{-1}	L^3T^{-1}
Shear stress	τ	$ML^{-1}T^{-2}$	FL^{-2}
Specific conductance	σ	$M^{-1}L^{-3}TQ^2$	$FM^{-2}L^{-4}T^3Q^2$
Specific heat capacity	c	L^2T^{-2-1}	L^2T^{-2-1}
Surface tension	σ	MT^{-2}	FL^{-1}
Temperature	θ	θ	θ
Thermal conductivity	k	$MLT^{-3}Q^{-1}$	$FT^{-1}\theta^{-1}$
Thermal diffusivity	α	L^2T^{-1}	L^2T^{-1}
Torque	T	ML^2T^{-2}	FL
Unit weight	w	$ML^{-2}T^{-2}$	FL^{-3}
Velocity	v	LT^{-1}	LT^{-1}
Volume	V	L^3	L^3
Weight	W	MLT^{-2}	F

quantities in these two bases. Representation in other bases can be derived from the entries by a logical combination of dimensions; thus in the MLTH base heat appears as a fundamental dimension, whereas in the MLT base its dimension is ML^2T^{-2} and in the FLT base its dimension is FL. Force, as a vector may be split into its components *if* they are independent. Similarly, length can also be decomposed into its orthogonal components. This useful property is often combined with the symmetry principle, which allows assignment of equal weights to orthogonal components in the absence of a preferred orientation. Thus, the radius of a circle lying in the (xy) plane may be represented as $X^{1/2}Y^{1/2}$ and the z-component of surface tension (force per unit length exerted in a plane tangential to the surface, and in the direction perpendicular to the unit length of interest) is represented as $MX^{-1/2}Y^{-1/2}ZT^{-2}$, where $L = L(X, Y, Z)$.

Dimensionless Groups

In the study of physically similar systems, many dimensionless groups have been developed that are usually denoted by symbols abbreviating a certain scientist's or engineer's name associated with the particular group. In some instances, the same dimensionless group may have several names associated with it, or conversely, the same name may be associated with different groups. In Table 4 common dimensionless groups have been assembled; more groups can be found in Massey's monograph [78] and in the comprehensive compilation by Catchpole and Fulford [85]. Not all dimensionless groups obtained via dimensional analysis have names associated with them. Most of the important dimensionless groups, as shown in Table 4, represent ratios of forces, energies, resistances, and rates.

DIMENSIONAL ANALYSIS VIA RAYLEIGH'S INDICIAL METHOD

In the indicial (or exponent) method introduced by Lord Rayleigh, a dependent quantity is chosen and its dimensions are related to the dimensions of all independent quantities considered by means of a priori indeterminate indices or exponents assigned to each independent quantity. Then, the indices are determined by comparing every dimension appearing in the relationship. While the Rayleigh method is structurally simple and its application is straightforward, there is some arbitrariness in choosing the dependent variable, although it does not necessarily have to reflect a physical cause-effect relationship. The incorporation of the correct number and kind of quantities requires the right physical interpretation of the problem analyzed but this prerequisite equally applies to all approaches in dimensional analysis. Examples 1 and 2 illustrate Rayleigh's method.

Example 1—Gravity Waves Generated by Fluid Motion in Shallow Water

From physical considerations, the wavelength of the gravity waves λ is related to wave velocity v, acceleration due to gravity g, the density of water ρ and the depth of the channel d. Choosing the wavelength as the dependent quantity, $m_i \quad i = 1, 4$ as the dimensional indices, and the MLT base of dimensions, the governing relationship is written as

$$\lambda = f(v^{m_1}g^{m_2}\rho^{m_3}d^{m_4}) \qquad \text{or} \qquad \lambda = m_0 v^{m_1}g^{m_2}\rho^{m_3}d^{m_4} \tag{1a}$$

or, alternatively as

$$L = (LT^{-1})^{m_1}(LT^{-2})^{m_2}(ML^{-3})^{m_3}L^{m_4} \tag{1b}$$

The exponents are related according to the linear set of equations

$$L:\ 1 = m_1 + m_2 - 3m_3 + m_4$$

$$T:\ 0 = -m_1 - 2m_2$$

$$M:\ 0 = m_3$$

Table 4
List of Frequently Occurring Dimensionless Groups*

Symbol and name of dimensionless group	Definition	Notation	Physical significance
Ar; Archimedes number	$\frac{d_p^3 g \rho_f(\rho_s - \rho_f)}{\mu^2}$	ρ_s: solid density ρ_f: fluid density d_p: particle diameter g: weight/mass ratio μ: absolute viscosity	$\frac{\text{(Inertia force)(gravity force)}}{\text{(Viscous force)}^2}$
Bi; Biot number	hl/k_s	h: heat transfer coefficient l: characteristic length k_s: thermal conductivity of solid	$\frac{\text{Internal thermal resistance}}{\text{Surface thermal resistance}}$
Bi_m; mass transport Biot number	$k_m L/D_i$	k_m: mass transfer coefficient L: layer thickness D_i: interface diffusivity	$\frac{\text{Mass transport conductivity at solid/fluid interface}}{\text{Internal transport conductivity at solid wall of thickness L}}$
Bo; Bond number (also Eo, Eötvös number)	$d^2 g(\rho - \rho_f)/\gamma$	ρ: bubble or droplet density d: bubble or droplet diameter γ: surface tension	$\frac{\text{Gravity force}}{\text{Surface tension force}}$
Bq; Boussinesq number	$v/(2gm)^{1/2}$	m: mean hydraulic depth of open channel v: characteristic velocity	$\frac{\text{(Inertia force)}^{1/2}}{\text{(Gravity force)}^{1/2}}$
Dn; Dean number	$\left(\frac{d}{2R}\right)^{1/2} Re$	d: pipe diameter R: radius of curvature of channel centreline	Effect of centrifugal force on flow in a curved pipe
Ek; Ekman number	$(\nu/2\omega l)^{1/2}$	ω: angular velocity of fluid ν: kinematic velocity	$\frac{\text{(Viscous force)}^{1/2}}{\text{(Coriolis force)}^{1/2}}$

(Continued)

Table 4 (Continued)

Symbol and name of dimensionless group	Definition	Notation	Physical significance
Fo_f; Fourier flow number	$\nu t/l^2$	t: time	Used in undimensionalization.
Fo; Fourier number	$\alpha t/l^2$	α: thermal diffusivity	Indicates the extent of thermal penetration in unsteady state heat transport.
Fo_m; mass transport Fourier number	$k_m t/l$		Indicates the extent of substance penetration in unsteady state mass transport.
Fr; Froude number	$v/(gl)^{1/2}$		$\frac{(\text{Inertia force})^{1/2}}{(\text{Gravity force})^{1/2}}$
Ga; Galileo number	$l^3 g/\nu^2$		$\frac{(\text{Inertia force})(\text{Gravity force})}{(\text{Viscous force})^2}$
Gz; Graetz number	$\dot{m}cp/k_f l$	$\dot{m}$: mass flow rate C_p: specific heat capacity (constant pressure) k_f: thermal conductivity of fluid	$\frac{\text{Fluid thermal capacity}}{\text{Thermal energy transferred by conduction}}$
Gr; Grashof number	$l^3 g\, \Delta\rho/\rho\nu^2$	$\Delta\rho$: density driving force	$\frac{(\text{Inertia force})(\text{Bouyancy force})}{(\text{Viscous force})^{1/2}}$
Gr_n; mass transport Grashof number	$l^3 g\beta_c\, \Delta c/\nu^2$	β_c: volumetric expansion coefficient Δc: concentration driving force	
Ha; Hartmann number (M)	$lB(\sigma/\mu)^{1/2}$	B: magnetic flux density σ: electric conductivity	$\frac{(\text{Magnetically induced stress})^{1/2}}{(\text{Viscous shear stress})^{1/2}}$
Lu; Luikov number	$k_m l/\alpha$		$\frac{\text{Mass diffusivity}}{\text{Thermal diffusivity}}$

Ly; Lykoudis number			$\frac{(\text{Hartmann number})^2}{(\text{Grashof number})^{1/2}}$
Ma; Mach number	v/v_s	v_s: velocity of sound in fluid	$\frac{\text{Linear velocity}}{\text{Velocity of sound}}$
Ne; Newton number	$F/\rho v^2 l^2$	F: hydrodynamic drag force	$\frac{\text{Resistance force}}{\text{Inertia force}}$
Nu; Nusselt number	hl/k_f		$\frac{\text{Thermal energy transport in forced convection}}{\text{Thermal energy transport if it occurred by conduction}}$
Pe; Peclet number	lv/α		(Re)(Pr); $\frac{\text{Bulk thermal energy transport in forced convection}}{\text{Thermal energy transport by conduction}}$
Pe_m; mass transport Peclet number	lv/D		$\frac{\text{Bulk mass transport}}{\text{Diffusional mass transport}}$
Ps; Poiseuille number	$v\nu/gd_p^2(\rho_s - \rho_p)$		$\frac{\text{Viscous force}}{\text{Gravity force}}$
Pr; Prandtl number	ν/α		$\frac{\text{Momentum diffusivity}}{\text{Thermal diffusivity}}$
Ra; Rayleigh number			(Gr)(Pr)
Ra_m; mass transport Rayleigh number			$(Gr_m)(Sc)$
Re; Reynolds number	vl/ν		$\frac{\text{Inertia force}}{\text{Viscous force}}$
Re_R; rotational Reynolds number	D^2N/ν	N: rate of rotation (revolution per time)	

(Continued)

Table 4 (Continued)

Symbol and name of dimensionless group	Definition	Notation	Physical significance
Sc; Schmidt number	ν/D		$\frac{\text{Momentum diffusivity}}{\text{Molecular diffusivity}}$
Sh; Sherwood number	$k_m l/D$		$\frac{\text{Mass diffusivity}}{\text{Molecular diffusivity}}$
St; Stanton number			$\frac{\text{Nu}}{\text{Re Pr}}$; $\frac{\text{Thermal energy transferred}}{\text{Fluid thermal capacity}}$
St_m; mass transport Stanton number			$\frac{\text{Sh}}{\text{Re Sc}}$
Sk; Stokes number	$l\,\Delta p/\mu\nu$	Δp: pressure drop	$\frac{\text{Pressure force}}{\text{Viscous force}}$
Su; Suratnam number	$\rho l\gamma/\mu^2$		$\frac{\text{(Inertia force)(Surface tension force)}}{\text{(Viscous force)}^2}$
Ta; Taylor number	$\omega^2\bar{r}(r_0 - r_i)^3/\nu^2$	r_0: outer radius r_i: inner radius $\bar{r}$: mean radius	Criterion for Taylor vortex stability in rotating concentric cylinder systems
We; Weber number	$v(\rho l/\gamma)^{1/2}$		$\frac{\text{(Inertia force)}^{1/2}}{\text{(Surface tension force)}^{1/2}}$

* *Symbols are defined at the location of their first occurrence.*

Retaining m_1 as the characteristic index, $m_2 = -m_1/2$ and $m_4 = 1 - m_1/2$. Consequently, $\lambda = f(v^{m_1}g^{-m_1/2}d^{1-m_1/2})$, or $\lambda = m_0 v_1^{m_1} g^{-m_1/2} d^{1-m_1/2}$, or $\lambda/d = f[(v/\sqrt{gd})^{m_1}]$, or $\lambda = m_0 d(v/\sqrt{gd})^{m_1}$. Since m_1 remains numerically indeterminate, the simplest final form to adopt is

$$\lambda/d = f(v/\sqrt{gd}) = f(Fr) \qquad (2)$$

The exact quantitative relationship between λ/d and Fr has to be obtained by experiments or by an appropriate momentum balance [86]. Notice that ρ has turned out to be a redundant quantity.

Example 2—Drag Forces Exerted on a Solid Object in the Path of a Flowing Fluid in Steady State [75]

The physical quantities to be considered are the velocity, v, density, ρ, and the viscosity, μ, of the fluid and a characteristic length, L, of the solid object. In steady state, i.e., in the absence of acceleration, the force exerted may be retained as a fundamental quantity and the FLT dimension base may be chosen. The governing relationship is

$$F = f(v^{m_1}L^{m_2}\rho^{m_3}\mu^{m_4}) \quad \text{or} \quad F = m_0 v^{m_1}L^{m_2}\rho^{m_3}\mu^{m_4} \qquad (3a)$$

or, alternatively

$$F = (LT^{-1})^{m_1}L^{m_2}(FT^2L^{-4})^{m_3}(FTL^{-2})^{m_4} \qquad (3b)$$

The exponents are related as

$$F{:}\ 1 = m_3 + m_4$$

$$L{:}\ 0 = m_1 + m_2 - 4m_3 - 2m_4$$

$$T{:}\ 0 = -m_1 + 2m_3 + m_4$$

Retaining m_4 as the indeterminate index, $m_1 = 2 - m_4$, $m_2 = 2 - m_4$ and $m_3 = 1 - m_4$. It follows that $F = v^2L^2\rho f[(vL\rho/\mu)^{-m_4}]$ or $F = m_0 v^2 L^2 \rho [vL\rho/\mu]^{-m_4}$ and the relationship between the dimensionless groups may be written as

$$F/v^2L^2\rho = f(Re) \qquad (4)$$

The ordering of the independent quantities in Equation 3a and the retention of m_4 as the indeterminate index is arbitrary; however, in choosing the latter use was made of the dependence of the force on some power of the Reynolds number, i.e., of an a priori knowledge of physics.

In the foregoing examples the number of fundamental quantities is small enough to avoid encumbrance in solving for m_1. When the number of quantities is larger than five, the required manipulations become burdensome and the more structured pi-theorem is usually easier to employ.

THE PI-THEOREM AND ITS APPLICATION

Buckingham's pi-theorem yields a systematic procedure to establish the representative π_i, $i = 1, \ldots, n - r$ dimensionless groups if there are n quantities and r indicial equations associated with the n quantities and the dimensions of the problem. There are three major techniques to handle dimensional analysis via the pi-theorem: the indicial equation technique, the indicial (or dimensional) matrix technique and the historically first technique of recurring (or repeating) variables.

Theoretical Foundations

The following is a concise version of Brand's treatment of the pi-theorem [7, 8, 76]. If there is a functional relation between $Q_1, Q_2, \ldots, Q_n$ quantities, the task is to determine all $F(\lambda_1, \ldots, \lambda_n) = 0$

relationships where λ_i are the measures of Q_i. Let $\{\alpha_{ij}\}$, $i = 1, \ldots, n$, $j = 1, \ldots, N$ be the set of dimensions of the quantities, where N is the size of the units base; then α_{ij} are the elements of the $n \times N$ matrix of dimensions $\underline{A}$. Let $\{\tau_j\}$; $j = 1, \ldots, N$ be the set of coordinates of any unit transformation:

$$\bar{\lambda}_i = \tau_1^{-\alpha_{i1}} \tau_2^{-\alpha_{i2}} \cdots \tau_N^{-\alpha_{iN}} \lambda_i \qquad i = 1, \ldots, n \tag{5}$$

Then, three vectors, $\underline{x} = (\log \lambda_i)_n$; $\bar{\underline{x}} = (\log \bar{\lambda}_i)_n$ and $\underline{y} = (\log \tau_i)_N$ may be defined such that the transformation is expressed as $\bar{\underline{x}} = \underline{x} - \underline{A}\underline{y}$ for $\psi(\underline{x}) = F(\lambda_1, \ldots, \lambda_n)$. The problem can be stated at this point in the following manner: determine all equations $\psi(\underline{x}) = 0$ which, for an arbitrary vector $\underline{y}$ and transformation $\bar{\underline{x}} = \underline{x} - \underline{A}\underline{y}$, imply that $\psi(\bar{\underline{x}}) = 0$.

Brand's solution of the problem may be summarized by first assuming that the rank r of matrix $\underline{A}$ is less than n; if $r = n$ there is no functional relationship between $Q_1, Q_2, \ldots, Q_n$. Further, matrix $\underline{A}$ is partitioned into submatrices $\underline{A}_{11}, \underline{A}_{12}, \underline{A}_{21}, \underline{A}_{22}$ such that $\underline{A}_{11}$ is an $(r \times r)$ matrix and nonsingular ($\det \underline{A}_{11} \neq 0$). Vector $\underline{x}$ is also partitioned into $\underline{x}' = (\log \lambda_1 \cdots \log \lambda_r)$ and $\underline{x}'' = \log \lambda_{r+1} \cdots \log \lambda_n$). Then, the relationship between $Q_1, Q_2, \ldots, Q_n$ may be expressed as

$$\phi(\underline{x}'' - \underline{A}_{21}\underline{A}_{11}^{-1}\underline{x}') = 0 \tag{6}$$

where ϕ is a non-constant real function defined on the $(n - r)$ dimensional vector space R^{n-r}. Any real function defined on R^{n-r} is compatible with the above postulates. For a detailed proof of Brand's solution Sections 4.4–4.7 in Reference 76 are to be consulted.

The Indicial Equation Technique

This technique assigns an a priori indeterminate index to each quantity considered and represents each dimension via appropriate linear combinations of these indices, called the indicial equations. Their solution yields the numerical values of the indices and the proper form of the pi-groups. The technique slightly resembles Rayleigh's method. Since the number of indicial equations may well be less than the number of indices, there is some indeterminacy in assigning numerical values to outstanding indices in order to arrive at a numerical solution of the indicial equations. However, pi-groups obtained for the same problem via different numerical values for the indices are interconvertible on the basis of the pi-theorem stating that all real functions defined on the R^{n-r} space are compatible; in other words, multiplication and exponentiation of pi-groups yields equivalent bonafide pi-groups. Examples 3 and 4 illustrate the technique.

Example 3—The Gravity Wave Problem of Example 1 [83]

There are five quantities, hence five a priori unknown indices are assigned as shown in brackets: $\lambda(n_1)$, $v(n_2)$, $g(n_3)$, $\rho(n_4)$ and $d(n_5)$. Using the MLT base of dimensions, an auxiliary array is constructed from which the indicial equations may be written down by inspection:

	λ		v		g		ρ		d	
M							n_4			(=0)
L	n_1	+	n_2	+	n_3	−	$3n_4$	+	n_5	(=0)
T		−	n_2	−	$2n_3$					(=0)

There are five unknowns and three independent equations; any arbitrary numerical set of two indices will yield a numerical set of the remaining three indices. If, for instance, $n_1 = 1$; $n_2 = -2$ and $n_1 = -1$; $n_2 = 0$ are chosen, the remaining index values are $n_3 = 1$; $n_4 = 0$; $n_5 = 0$ and $n_3 = 0$; $n_4 = 0$;

$n_5 = 1$, respectively. Assembling these values into the auxiliary array:

	n_1	n_2	n_3	n_4	n_5
π_1	1	−2	1	0	0
π_2	−1	0	0	0	1

the two dimensionless groups are recognized immediately as $\pi_1 = \lambda v^{-2} g = g\lambda/v^2$ and $\pi_2 = \lambda^{-1} d = d/\lambda$. This can also be interpreted as the functional relationship $g\lambda/v^2 = f(d/\lambda)$.

Suppose that the two "starting" indices are taken arbitrarily as $n_1 = 1; n_2 = 0$ and $n_1 = 0; n_2 = 1$. Then, the remaining index values are $n_3 = 0; n_4 = 0; n_5 = -1$ and $n_1 = 0; n_3 = -\frac{1}{2}; n_4 = 0; n_5 = -\frac{1}{2}$, respectively. From the auxiliary array:

	n_1	n_2	n_3	n_4	n_5
π'_1	1	0	0	0	−1
π'_2	0	1	$-\frac{1}{2}$	0	$-\frac{1}{2}$

$\pi'_1 = \lambda/d$ and $\pi'_2 = v/\sqrt{gd}$ are immediately constructed. These pi-groups, which yield the result identical to the Rayleigh method illustrated in Example 1, are interconvertible with π_1 and π_2, inasmuch as $\pi'_1 = \pi_2^{-1}$ and $\pi'_2 = (\pi_1 \cdot \pi_2)^{-1/2}$. There are, in principle, an infinite number of variations to this procedure.

Example 4—The Drag Force Problem of Example 2

The quantities and their associated a priori unknown indices, shown in parentheses are: $F(n_1)$, $v(n_2)$, $L(n_3)$, $\rho(n_4)$ and $\mu(n_5)$. Using the FLT base of dimensions, the auxiliary array is constructed as shown:

	F	v	L	ρ	μ	
F	n_1		$+n_3$	$+n_4$	$+n_5$	(=0)
L		n_2	$+n_3$	$-4n_4$	$-2n_5$	(=0)
T		$-n_2$		$+2n_4$	$+n_5$	(=0)

As in Example 3, there are five unknowns and three independent equations; hence, two starting indices must be chosen arbitrarily to establish each pi-group. Choosing first $n_1 = 1; n_2 = -2$ and then $n_1 = 0; n_2 = 1$ the remaining indices are determined as $n_3 = -2; n_4 = -1; n_5 = 0$ and $n_3 = 1; n_4 = 1; n_5 = -1$, respectively. From the auxiliary array:

	n_1	n_2	n_3	n_4	n_5
π_1	1	−2	−2	−1	0
π_2	0	1	1	1	−1

the pertinent pi-groups $\pi_1 = F/\rho v^2 L^2$ and $\pi_2 = vL\rho/\mu = Re$ agree with the result of Example 2, since the $F/\rho v^2 L^2 = f(Re)$ relation follows directly. Pi-groups obtained as a result of different choices of the starting indices are interconvertible and will yield the above result upon appropriate multiplication and exponentiation, as indicated in Example 3.

The Indicial (or Dimensional) Matrix Technique

In the indicial matrix technique the matrix elements are the exponents of the row entries representing dimensions, which pertain to each column entry, i.e., to each quantity. A priori indeterminate indices are then assigned to each quantity, i.e., to each column. Since in each row the algebraic sum of the matrix element/index products is zero, there are as many linear equations obtained as the number of dimensions or rows. If the rank of the indicial matrix is r (r is the order of the largest non-zero determinant contained in the matrix) and n is the number of quantities, then the number of independent pi-groups is (n − r). As in the case of the indicial equation technique, the choice of the starting indices is arbitrary and, in principle, an infinite number of interconvertible pi-groups can be obtained for each set of starting indices. Examples 5 and 6 illustrate the technique.

Example 5—The Gravity Wave Problem of Example 1 and Example 3

The indicial matrix is shown in the array below.

	λ (1)	v (2)	g (3)	ρ (4)	d (5)
M	0	0	0	1	0
L	1	1	1	−3	1
T	0	−1	−2	0	0

The associated linear equations are

$$n_4 = 0$$

$$n_1 + n_2 + n_3 - 3n_4 + n_5 = 0$$

$$-n_2 - 2n_3 = 0$$

By inspection of the matrix, n = 5 and r = 3, hence there are two independent pi-groups. Choosing $n_1 = 1; n_2 = -2$ and $n_1 = -1; n_2 = 0$ as starting indices, the remaining indices are determined as $n_3 = 1; n_4 = 0; n_5 = 1$, respectively. The final result: $\pi_1 = g\lambda/v^2$ and $\pi_2 = d/\lambda$, which agrees with the outcome of Example 1 and Example 3, follows directly from the auxiliary array identical to the one in Example 3.

Example 6—The Drag Force Problem of Example 2 and Example 4

From the indicial matrix:

	F (1)	v (2)	L (3)	ρ (4)	μ (5)
F	1	0	0	1	1
L	0	1	1	−4	−2
T	0	−1	0	2	1

The associated linear equations are written down by inspection as

$$n_1 + n_4 + n_5 = 0$$

$$n_2 + n_3 - 4n_4 - 2n_5 = 0$$

$$-n_2 + 2n_4 + n_5 = 0$$

If the starting indices are chosen as $n_1 = 1; n_5 = 0$ and $n_1 = 0; n_4 = 1$ the remaining index values are computed as $n_2 = -2; n_3 = -2; n_4 = -1$ and $n_2 = 1; n_3 = 1; n_5 = -1$, respectively. The auxiliary array

	n_1 (F)	n_2 (v)	n_3 (L)	n_4 (ρ)	n_5 (μ)
π_1	1	−2	−2	−1	0
π_5	0	1	1	1	−1

yields upon inspection $\pi_1 = F/\rho v^2L^2$ and $\pi_2 = vL\rho/\mu = Re$, as in Example 4.

The Technique of Recurring (or Repeating) Variables

In the chronological order of its development, recurring variables were used first in employing the pi-theorem. This is a purely algebraic technique without recourse to geometric concepts, such as a matrix or an index array. Each pi-group is found by a systematic process where the result of each step is one of the pi-groups. First, out of the n quantities stated in the problem, k quantities are selected (k is the number of fundamental dimensions), whose product is raised to an unknown index (exponent) and one of the remaining quantities is raised to a specific index, usually unity. The only restriction for the k quantities is that they all have to have nonidentical dimensions. They are now retained as the recurring variables while each remaining variable is successively raised to a specific index (unity) to establish each successive pi-group. The technique offers certain numerical advantages in problems where the indicial matrix is large and the finding of the matrix rank would require involved calculations. On the other hand, certain combinations of the recurring variables may lead to nonsensical numerical relationships between the indices and a new combination has to be found. Example 7 illustrates the use of the technique.

Example 7—The Drag Force Problem in Examples 2, 4, and 6

The quantities are F, v, L, ρ and μ; hence $n = 5$. Since the FLT system is employed, $k = 3$. Let the arbitrary choices for the recurring variables be L, v and ρ. Then,

$$\pi_1 = L^{m_1}(LT^{-1})^{m_2}(FT^2L^{-4})^{m_3}F$$

if F is chosen as the first variable of the remaining quantities (F and μ) to be raised to the exponent of unity. The indices m_1, m_2, and m_3 are related by the three algebraic equations

F: $\quad m_3 + 1 = 0$

L: $\quad m_1 + m_2 - 4m_3 = 0$

T: $\quad -m_2 + 2m_3 = 0$

yielding $m_1 = -2$, $m_2 = -2$, and $m_3 = -1$. Hence, $\pi_1 = L^{-2}v^{-2}\rho^{-1}F$ or $F/\rho v^2L^2$.

The next and last remaining quantity is μ, hence

$$\pi_2 = L^{n_1}(LT^{-1})^{n_2}(FT^2L^{-4})^{n_3}(FTL^{-2})$$

The indices are related as

$$\text{F:} \quad n_3 + 1 = 0$$

$$\text{L:} \quad n_1 + n_2 - 4n_3 - 2 = 0$$

$$\text{T:} \quad -n_2 + 2n_3 + 1 = 0$$

yielding $n_1 = -1$, $n_2 = -1$, $n_3 = -1$. Hence, $\pi_2 = L^{-1}v^{-1}\rho^{-1}\mu$ or $\mu/Lv\rho$. Applying the principle of interconvertibility to π_2, it is replaced by its reciprocal and $\pi_2 = Lv\rho/\mu = Re$ is finally obtained. Once again, the $F/\rho v^2L^2 = f(Re)$ relationship is found.

Suppose that F, ρ and μ are taken to be the recurring variables and L is the first remaining quantity. In this instance

$$\pi_1 = F^{m_1}(FT^2L^{-4})^{m_2}(FTL^{-2})^{m_3}L$$

Consequently,

$$\text{F:} \quad m_1 + m_2 + m_3 = 0$$

$$\text{L:} \quad -4m_2 - 2m_3 + 1 = 0$$

$$\text{T:} \quad 2m_2 + m_3 = 0$$

Substituting $m_3 = -2m_2$ into the L-equation, $-4m_2 + 4m_2 + 1 = 0$ is obtained, which is nonsense. The procedure cannot be continued.

Similarly, if v is taken as the first remaining quantity, i.e.,

$$\pi_1 = F^{n_1}(FT^2L^{-4})^{n_2}(FTL^{-2})^{n_3}(LT^{-1})$$

is set, then

$$\text{F:} \quad n_1 + n_2 + n_3 = 0$$

$$\text{L:} \quad -4n_2 - 2n_3 + 1 = 0$$

$$\text{T:} \quad 2n_2 + n_3 - 1 = 0$$

and the last two equations yield the nonsensical statement that $0 = -1$.

ADVANCED CONCEPTS IN CLASSICAL DIMENSIONAL ANALYSIS

The Rayleigh method and the various techniques based on Buckingham's pi-theorem form the core of classical dimensional analysis. In this section some advanced concepts introduced in the recent past which are contributory to the power of the classical approaches are briefly described.

Directed Graphs: A Geometric Approach

Directed graphs are a geometric means of representing the development of dimensionless groups [9, 10, 83]. They may be regarded as essentially signal flow graphs [87] with the nodes representing physical quantities and the transmittances being positive or negative integers related to powers of dimensions. Since dimensionless groups have no dimension, they are represented by a vector

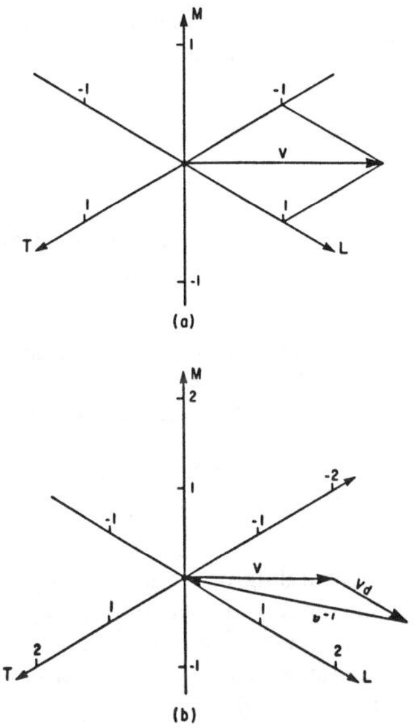

Figure 1. Directed graph for the representation of (a) velocity, (b) Reynolds number in MLT dimension base.

sequence which starts and terminates at the origin of the graph. Figure 1 illustrates the representation of velocity, as a dimensional quantity, and the Reynolds number as a dimensionless group via directed graphs whose principal axes are associated with the power of mass, length, and time (MLT base); similarly, if the FLT base is used, the principal axes are associated with the power of force, length and time. There are three major rules for constructing directed graphs: (1) multiplication of two or more quantities yields a vector sum where each vector component represents an individual quantity; (2) the nth power of any quantity is represented as a vector of that quantity multiplied by the scalar n; and (3) a dimensionless product forms a *cycle* whose individual vector elements add up algebraically to zero. Let a cycle consist of points $P_1, P_2, \ldots, P_n$ corresponding to vector tips (i.e., OP_1 is the first vector, P_1P_2 is the second, etc., and P_nO is the last vector in the cylce). Then at any arbitrary point P_j, the graph defines two identical quantities: $OP_1 + P_1P_2 + P_2P_3 + \cdots + P_{j-1}P_j$ and $P_jP_{j+1} + P_{j+1}P_{j+2} + \cdots + P_nO$, since these quantities correspond to vectors OP_j and P_jO. Extension to an n-dimensional space where $n > 3$ requires that the representative vectors of the n quantities must not lie in any space of $(n - 1)$ geometric dimensions, in order to guarantee independence of the n quantities considered.

Scale Change

Scale change is not a priori inherent in dimensional analysis, but it becomes necessary when numerical values of magnitudes are nonhomogeneous (e.g., SI, MKS, Imperial, etc., units are mixed together). In such an instance dimensional analysis automatically generates a multiplier to yield a dimensionally homogeneous dimensionless number. Transformation from one unit system to another also involves scale changes, as illustrated by the following example [83]. Assume that in a certain physical problem time, mass, area, and density are the quantities with a corresponding indicial matrix:

	t	m	A	ρ
M	0	1	0	1
L	0	0	2	−3
T	1	0	0	0

and that fundamental dimension units of SI have been chosen. Then, transformation to another (arbitrary) system of units may be affected by the three-parameter trannsformation group

$$t^* = (M^0L^0T^{-1})t \tag{7a}$$

$$m^* = (M^1L^0T^0)m \tag{7b}$$

$$A^* = (M^0L^2T^0)A \tag{7c}$$

$$\rho^* = (M^1L^{-3}T^0)\rho \tag{7d}$$

If the units in the system denoted by asterisks are foot, pound, and second, the transformation carries conversion factors M = 2.205 kg → lb, L = 3.281 m → ft and T = 1 s → s. Generalization of this relatively simple procedure in terms of group-theoretic aspects has been discussed in the literature [11–13].

Extension of the Reference Dimension Sets

The rational use of dimensional analysis requires a good understanding of the dimension set used in solving a particular problem. There is, in principle, a maximum number of dimensions beyond which analysis would run into logical inconsistencies. Earlier, a brief reference was made to the use of certain derived quantities such as force and heat as additional elements of a dimension base. In this section the subject of such an extension is treated in more detail.

As a first example consider the amount of heat built up (q) in time (t) in a gas ring, into which a gas of density ρ and caloric value c is flowing at a volumetric flow rate of Q [83]. From the indicial matrix

	q	t	ρ	c	Q
M	1	0	1	0	0
L	2	0	−3	2	3
T	−2	1	0	−2	−1

constructed in an MLT base of dimensions the indices are found to obey the equation set

$$n_1 + n_3 = 0 \tag{8a}$$

$$2n_1 - 3n_3 + 2n_4 + 3n_5 = 0 \tag{8b}$$

$$-2n_1 + n_2 - 2n_4 - n_5 = 0 \tag{8c}$$

If starting indices $n_1 = 1; n_2 = -1$ and $n_1 = -1; n_2 = 1$ are chosen, the remaining index sets are established as $n_3 = -1; n_4 = -1; n_5 = -1$ and $n_3 = 1; n_4 = 1; n_5 = 1$, respectively. The resulting dimensionless groups, $\pi_1 = q/t\rho cQ$ and $\pi_2 = t\rho cQ/q$ are obviously not independent and the solution is incomplete. While other index sets may be more successful (e.g., $n_1 = 0; n_2 = 4$ and $n_1 = -4; n_2 = 0$ yield independent pi-groups Q^2/t^4c^3 and ρ^4cQ^6/q^4) if chosen by chance, this success is random and masks the possibility of constructing redundant pi-groups. It is much better to include heat as an independent entity in the base of dimensions, for a straightforward and unambiguous result. In so doing, the indicial matrix:

	q	t	ρ	c	Q
M	0	0	1	−1	1
L	0	0	−3	0	3
T	0	1	0	0	−1
H	1	0	0	1	0

acquires the rank of $r = 4$ and since $n = 5$, there is obviously one dimensionless group only to be found. From the index relations

$$n_3 - n_4 = 0 \tag{9a}$$

$$-3n_3 + 3n_5 = 0 \tag{9b}$$

$$n_2 - n_5 = 0 \tag{9c}$$

$$n_1 + n_4 = 0 \tag{9d}$$

and choosing $n_1 = 1$ as starting index, $n_2 = n_3 = n_4 = n_5 = -1$ is found, hence $\pi = q/t\rho cQ$. The alternative form $q = f(t\rho cQ) = k \cdot t\rho cQ$, k a constant, is well known in heat transfer theory.

The enlargement of a dimension base does not always guarantee success, as demonstrated by the problem of the thermal energy h generated when a body of mass m falls to the ground from height s without bouncing [83]. Using the MLT dimension base, the index relationships

$$n_1 + n_4 = 0 \tag{10a}$$

$$2n_1 + n_2 + n_3 = 0 \tag{10b}$$

$$-2n_1 - 2n_3 = 0 \tag{10c}$$

may be solved, e.g., to yield $n_1 = 1, n_2 = -1, n_3 = -1$ and $n_4 = -1$; the n_i indices are associated with h, s, g and m, respectively (g is acceleration due to gravity). Consequently, $\pi = h/mgs$ or $h = f(mgs) = k \cdot mgs$, k is a constant. This is again a well known result in physics. It is now shown that the inclusion of heat into the dimension base would be useless since the thermal energy is simply converted mechanical energy and as such, it does *not* represent an independent quantity. The indicial matrix:

	h	s	g	m
M	0	0	0	1
L	0	1	1	0
T	0	0	−2	0
H	1	0	0	0

has the same rank as the number of quantities ($n = r = 4$) indicating the lack of a functional relationship between these quantities. Introducing an "ad-hoc" dimensional constant, which may be called the "mechanical equivalent of heat," J, would increase the number of quantities at the same rank of $r = 4$, by adding a J-column with elements $(-1, -2, 2, 1)$ to the above array. The index relationships can be solved, e.g., as $n_1 = 1$; $n_2 = n_3 = n_4 = 1$; $n_5 = -1$; hence, $\pi = sgmJ/h$ or $h = f(Jmgs)$, which yields nothing new; in fact, J has the same role as k in the MLT-base analysis. In SI, $J = k = 1$, yielding the simplest possible relationship.

Minimization of the Number of Pi-Groups

The relationship between Q_i, $i = 1, \ldots, n$ variables in a problem of dimensional analysis involves, in general, N dimensionless groups through a function

$$\Phi(\pi_1, \pi_2, \ldots, \pi_N) = 0 \tag{11}$$

It is of obvious importance to minimize the number of pi-groups entering into Φ in order to minimize the extent of functional indeterminacy. At first glance there are two basic means of achieving this goal: by increasing the dimensionality of the reference set and by restricting the selected list of variables, provided the physics of the problem is perfectly understood. There are three major methods whereby such minimization may be effected in a systematic manner.

Method I. If a subset of the variables of interest is connected by a dimensionally homogeneous equation, e.g., $Q_j = \phi(Q_{j+1}, Q_{j+2}, \ldots)$, then ϕ is employed to replace Q_j whenever the latter quantity occurs.

Method II. If $Q_j, Q_{j+1}, Q_{j+2}, \ldots$ are related *only* by the specific relationship ϕ, only Q_j is retained in the analysis.

Method III. The method of "empirical proportionality": let Q_j and Q_{j+1} be proportionally related, i.e., $Q_j = kQ_{j+1}$ or $Q_j = k/Q_{j+1}$. The pi-groups are arranged in such a manner that Q_j occurs only in, say, π_1 and Q_{j+1} occurs only in π_2. Then π_1 and π_2 can be condensed into a single pi-group. The application of the three methods is illustrated in Example 8.

Example 8—Illustration of the Three Major Methods of Minimizing the Number of Pi-Groups

Method I. In the analysis of the motion of a conical pendulum [83] the period T, string length L, cone height h, acceleration due to gravity g, angular velocity ω and inclination angle β are taken as the quantity set. In the LT dimension base four pi-groups are established, unless use is made of the trigonometric relationship $h = L \cos \beta$ and the definition of the period: $\omega T = 2\pi$; these are the associated ϕ-relationships. Using them, the number of pi-groups is reduced to two: T^2g/l and β.

On the other hand, in the problem of forced convection of water in a tube such an approach would not be advantageous. Straightforward dimensional analysis yields Nu = f(Re, Pr) and for water, the empirical relationship [88]

$$h = q/\Delta\theta = \alpha_1(1 + \alpha_2\theta)u^{0.8}/d^{0.2} \tag{12}$$

applies. The associated a priori quantities are heat flux q, velocity u, tube diameter d, thermal conductivity k, temperature driving force $\Delta\theta$, specific heat capacity c, density ρ, and dynamic viscosity μ. Introducing Equation 12 would result in the disappearance of the Reynolds number, which is a fundamental pi-group in fluid mechanics and heat transport theory; there is nothing to be gained in this instance from a reduction method.

Method II. In the pendulum problem above, L, β, and ω can individually be deleted from the list of quantities via the relationships $h = L \cos \beta$ and $\omega = 2\pi/T$ resulting in the final equation $T = f(\sqrt{h/g}) = k\sqrt{h/g}$.

Method III. The force f between the plates of a charged condenser [83] is related to the separation distance d, plate area A, voltage drop between plates V, and the medium permittivity ϵ. Dimensional analysis (carried out conveniently in the MLTϵ dimension base, although it is only one possible choice) yields $\pi_1 = f/v^2\epsilon$ and $\pi_2 = A/d^2$. Since the force must be proportional to the charged plate area, $f = K \cdot A$ where K is the proportionality factor. Thus, the ratio of π_1 and π_2 must be formed to satisfy this proportionality:

$$\pi_1/\pi_2 = (f/v^2\epsilon)/(A/d^2) = (f/A)d^2/v^2\epsilon = k$$

Consequently,

$$f = KA = kv^2\epsilon A/d^2$$

is the relationships sought; k is, of course, an a priori indeterminate constant.

The examples of this section represent relatively straightforward instances chosen for the sake of illustration; minimization of the dimensionless groups is not always easy and often requires a thorough physical understanding of the problem, as well as experience.

Negligible Quantities in Dimensional Analysis

If a quantity has no physical importance in a problem but it is included (erroneously) in the quantity set, its effect may in general be one of two alternatives:

1. It will drop out as the procedure progresses, i.e., it will appear in a dimensionless group raised to the zeroth power.
2. It will result in the appearance of an additional dimensionless group raised to the zeroth power. This event could cause confusion.

In certain problems a quantity may be negligible or unimportant within a certain range of an independent quantity or pi-group, and may serve for the establishment of the correct functional relationship. To illustrate this point the drag force problem discussed in Examples 2, 4, 6, and 7 is taken up again where $\pi_1 = F/\rho v^2 L^2$ and $\pi_2 = Re$ have been constructed. The functional form, a priori still indeterminate, may be written as

$$F = \rho v^2 L^2 \phi(Re) \tag{13}$$

If the flow rate is sufficiently low, drag is due essentially to viscosity since the influence of inertia is negligible and the drag force is proporational to the viscosity, velocity, and the characteristic length of the body. In other words, $F \propto 1/Re$. At the other extreme, when the flow rate is sufficiently high but still not high enough to bring about transition to a turbulent boundary layer, viscous effects are negligible and inertia effects dominate. Hence, the Reynolds number is expected to have no effect on the drag force and $F \propto \rho v^2 L^2$. Since the exponent of Re varies from -1 to 0 as v is increased, it is logical to suppose that the general functional relationship may be written as

$$F = k\rho v^2 L^2 \, Re^n \tag{14}$$

where $-1 \leq n \leq 0$ and k is a proportionality factor. In engineering practice drag is usually considered in terms of a drag coefficient C_D:

$$C_D = 2k \, Re^n \tag{15}$$

and its numerical dependence on Re has been well documented in the literature [e.g., 89].

The Problem of Undetermined Functions

It has been stated at the outset that dimensional analysis usually ends at the stage of finding an undetermined functional relationship between pi-groups. This limitation was phrased aptly by Bridgman [73], "... it is never possible to obtain factual information about any concrete physical situation by pure ratiocination ...." If fundamental mathematical/physical modeling is regarded as "ratiociation" (i.e, reasoning or a process of reasoning) then Bridgman's statement is not absolutely correct. In modeling natural convection at vertical solid surfaces in terms of fundamental convective diffusion theory and hydrodynamics, Levich [90] established on a purely theoretical basis the *quantitative* dimensionless relationship

$$Sh = 0.933 \, Ra^{1/4} \tag{16}$$

The proportionality factor has been corrected to about 0.67 via experimental measurements as well as by a more sophisticated *theory* of boundary layers [91]. It is true, of course, that dimensional analysis can only yield the undetermined functional relationships of Nu = f(Sc, Gr). Similarly, dimensional analysis of condensation in a vertical pipe [77] terminates at finding the pi-groups $\pi_1 =$ Nu; $\pi_2 = k\mu\,\Delta\theta/\lambda g\rho^3L^3$ (λ is the latent heat of vaporization, k the thermal conductivity of the condensate, $\Delta\theta$ the temperature driving force, μ the viscosity of the condensate, ρ the density of the condensate) whereas the fundamental ("ratiocinative") analysis of Nusselt [92] yields the quantitative relationship

$$\pi_1 = 0.943/\pi_2^{1/4} \tag{17}$$

In certain, perhaps privileged, instances dimensional analysis combined with appropriate mathematical manipulations *is* capable of yielding a quantitative functional relationship, as illustrated by the Poiseuille problem of steady laminar flow [83]. Dimensional analysis employing the MLT dimension base yields the functional relationship

$$v = r^2(dp/dx)/\mu \cdot \phi(a/r) \tag{18}$$

where r is the radial distance measured from the tube axis, a the radius of the tube, dp/dx the pressure drop along the tube and μ the fluid viscosity. Writing ϕ as a series expansion

$$\phi = k_0 + k_1(a/r) + k_2(a/r)^2 + \cdots \tag{19}$$

and substituting into the r-derivative of Equation 18, the expression

$$dv/dr = (dp/dx)/\mu \cdot [2k_0r + 2k_1a + \cdots - k_1a \cdots]$$

is obtained upon simplification. Knowing that dv/dr does not depend on a, $k_1 = 0$ and $k_3 = k_4 = \cdots = 0$ must be set. The non-slip condition $\phi(1) = 0$ imposes the relationship $k_0 + k_1 + \cdots = 0$, hence $k_0 = -k_2$ and the final quantitative relationship

$$v = r^2(dp/dx)/\mu \cdot k_0[1 - (a/r)^2] \tag{20}$$

immediately follows (Equation 20 may alternatively be obtained from the Navier-Stokes equations of hydrodynamics). Note that the series expansion approach requires that the series be convergent; moreover its validity is often restricted to a narrow range of the independent variable(s).

The Use and Abuse of Classical Dimensional Analysis

In its years of spiritual infancy, dimensional analysis was regarded by numerous scientists as a fundamental breakthrough in discovering ultimate natural laws governing the universe. The history of dimensional analysis [e.g., 73, 74, 83], especially its applications to cosmology and cosmogony, witness the various pitfalls of enthusiastic but uncautious treatments often leading to absurdities. On the other hand, there resulted many positive findings where correct physical insight (even intuition) coupled with dimensional analysis yielded successful new discoveries. Among the positive results is Einstein's [93] proposed relationship between interatomic forces determining elasticity of a solid and forces related to characteristic infrared frequencies, proved finally by Debye's studies. Another such result is Dirac's [94] noticing the unexpectedly frequent occurrence of the force constant $e^2/Gm_pm_e = 2.275 \times 10^{39}$ (e = electric charge of the electron; G = fundamental gravitational constant, m_e = electron mass, m_p = proton mass) and that the ratio of the present age of the universe (about 7×10^9 years) and the time required for light to cross the radius of an electron (about 10^{-23} second) is about the same number (about 2.4×10^{39}), suggested that this occurrence indicates a fundamental link between cosmology and atomic theory. Unfortunately, negative or at best, highly debatable results also abound sufficiently to devote discussion to a few selected cases.

Example 9—Relationship Between Gravitational Constant and Electron Charge/Mass Ratio

A "fundamental" relationship between the gravitational constant ($G = 6.658 \times 10^{-8}\ g^{-1}\ cm^2\ s^{-2}$) and the electron charge/mass ratio ($e/m = 5.3 \times 10^{17}$ statcoul/g) has been suggested by some physicists [73] in the form of $G = k(e/m)^2$. The numerical value of $k = 2.35 \times 10^{-43}$ is of an *extremely* small order of magnitude, well below the magnitudes encountered in various constants of atomic physics. The significance of this relationship is very doubtful.

Example 10—A Universal System of Units

There was a tendency in the past [e.g., 95] to regard the velocity of light ($c = 3 \times 10^{10}$ cm/s), the gravitational constant (G), the Planck constant ($h = 6.55 \times 10^{-27}\ g \cdot cm^2/s$) and the Boltzmann constant ($k = 1.38 \times 10^{-16}\ g \cdot cm^2/s \cdot K$) as fundamental constants of nature. It has been proposed that a new unit system could be constructed where each of these constants would have the value of unity; such a unit system could presumably be called a "universal (absolute) system of units". In the MLT dimension base the proposed transformation to unity would yield as a solution to the set of equations

$$[L][T]^{-1} = [C] = 3 \times 10^{10} \tag{21a}$$

$$[M]^{-1}[L]^3[T]^{-2} = [G] = 6.658 \times 10^{-8} \tag{21b}$$

$$[M][L]^2[T]^{-1} = [h] = 6.55 \times 10^{-27} \tag{21c}$$

the values of $[M] = 5.43 \times 10^{-5}$ g, $[L] = 4.02 \times 10^{-33}$ cm and $[T] = 1.34 \times 10^{-43}$ s. To find a new "absolute" unit of temperature, the equation

$$[M][L]^2[T]^{-2}[\theta]^{-1} = [k] = 1.38 \times 10^{-16} \tag{21d}$$

is to be solved using the numerical values of the mass, length, and time units in the new system. The computed value of $[\theta] = 3.54 \times 10^{32}$ is absurdly high and it follows that such a "universal" unit system would have no sense.

As a variation on this theme the suggestion of Lewis and Adams [96] that ". . . any set of absolute units will be found to bear a simple numerical relation to any other possible set of absolute units . . ." may be examined. As proposed by Bridgman [73] if the units of G, c, k, and e were arbitrarily set to unity, then the numerical values $[M] = 1.849 \times 10^{-6}$ g, $[L] = 1.368 \times 10^{-34}$ cm, $[T] = 4.56 \times 10^{-45}$ s and $[\theta] = 1.20 \times 10^{31}$ °C would be obtained. Earlier, a different set of "absolute" units was established and one can readily compute the ratios of units of the same kind; thus for the mass, e.g., $1.849 \times 10^{-6}\ g/5.43 \times 10^{-5}\ g = 1/29.367$. In fact, the other three ratios also yield 1/29.367, which is neither simple, nor an integer, nor the ratio of two integers, etc. The results of this example beg the question whether there is any possibility, indeed, of rationalizing fundamental units by speculation based on purely dimensional considerations.

The application of dimensional analysis to mathematical formulations without reference to a specific physical problem [84] also requires caution and circumspection. The underlying requirement that each element in a formulation has to be dimensionally identical does not necessarily identify all dimensions. Thus, in the expression $y = a + x$ one can state the $[y] = [x]$ dimensional equivalence only if a is independent of the unit system in which the variables x and y are measured. In the formulations $y = \ln(x/z)$; $y = \tan(x/z)$; $y = \epsilon^{(x/z)}$, etc., the dimensions [x] and [z] are indeterminate and only $[y] = 1$ can be stated with certainty. In the differential equation $y'' + a_1 y' + a_0 y = 0$, dimensional consistency requires that the condition $[a_1^2/a_0] = 1$ be satisfied, indicating again (partial) indeterminacy. More power may be seen in the approach of undimensionalizing equations without the identification of individual dimensions; as shown by Taylor [84], the dimensional differential equation of the linear vibrating system

$$m\frac{d^2x}{dt^2} + c\frac{dx}{dt} + kx = f\cos\omega t \tag{22}$$

may be undimensionalized to

$$(\omega^*)^2\frac{d^2x^*}{dt^{*2}} + 2\omega^*c^*\frac{dx^*}{dt^*} + x^* = \cos t^* \tag{23}$$

where the dimensionless quantities $x^* = x/[f/k]$; $t^* = [\omega t]$; $c^* = c/2\sqrt{km}$ and $\omega^* = \omega/\sqrt{k/m}$ do not depend on the units of the variables and parameters in Equation 22. Since physical equations are often closely related to a particular unit system, the scope of this approach is rather modest in spite of its intellectual merits and it has found so far limited acceptance in the scientific and engineering literature.

Less Known Applications of Dimensional Analysis

It is apparent from the study of pertinent literature that fluid mechanics, hydrology, heat transport theory, and astrophysics (with related areas) have been the major domains of scientific and engineering endeavour where dimensional analysis has made a very serious impact and contribution. Nevertheless, as it is demonstrated in Tables 1 and 2, the technique has proved successful in various areas of lesser general importance, or in specific applications not necessarily of *wide* interest to the technical community. Such applications have led to the creation of numerous dimensionless groups of specific (and somewhat limited) significance and yielded better tools of design and understanding of specific machinery and/or equipment. One good example here is dimensional analysis related to chemical transformations where a number of pi-groups were introduced in earlier approaches [e.g., 97–99] to link chemical reactor parameters to the kinetic characteristics of chemical reactions. In a thorough treatment of the subject matter [63] two fundamental pi-groups, the quasistationary group Qu, and the contact number K_0 were shown to represent in a dimensionless form the generalized reversible chemical reaction

$$\alpha_1 A_1 + \alpha_2 A_2 + \cdots = \beta_1 B_1 + \beta_2 B_2 + \cdots \tag{24}$$

with kinetics

$$-\frac{1}{\alpha_i}\frac{dc_i}{dt} = \frac{1}{\alpha_i}(k_1 C_{A_1}^{\alpha_1} C_{A_2}^{\alpha_2}\cdots - k_2 C_{B_1}^{\beta_2} C_{B_2}^{\beta_2}\cdots) \tag{25}$$

The pi-groups are defined as

$$Qu \equiv \frac{k_2 C_{B_1}^{\beta_1} C_{B_2}^{\beta_2}\cdots}{k_1 C_{A_1}^{\alpha_1} C_{A_2}^{\alpha_2}\cdots} \tag{26}$$

and

$$\begin{aligned} K_0 &\equiv k_1 C_{A_1}^{\Sigma\alpha_i - 1} t \\ &\equiv k_2 C_{B_1}^{\Sigma\beta_i - 1} t \qquad \text{(alternative definition)} \end{aligned} \tag{27}$$

In a steady-state process K_0 is approximately the contact time/decomposition time ratio and the zone width (i.e., the length required in a chemical reactor to effect a certain conversion at fluid velocity v) is $vK_0/k_1 C_{A_1}^{\Sigma\alpha - 1}$. The specific relationship between Qu and K_0 is the dimensionless equation relating conversion of reactant to product in a chemical reactor. The kinetics of ammonia synthesis serves as an illustrative example.

Example 11—Dimensional Analysis of Ammonia Synthesis (Summary)

The Temkin-Reuter reaction mechanism [100]

$$N_2 \rightleftharpoons N_2\ (ads.) \overset{+H_2}{\rightleftharpoons} 2NH\ (ads.) \overset{+H_2}{\rightleftharpoons} 2NH_3 \tag{28}$$

yields the kinetic equation (in terms of partial pressures) as

$$dp_{NH_3}/dt = k_1 p_N p_{H_2}^{3/2}/p_{NH_3} - k_2 p_{NH_3}/p_{H_2}^{3/2} \tag{29}$$

Then, using the second definition of K_0 in Equation 27 and the fundamental theory of chemical reaction kinetics, the pertinent pi-groups

$$Qu = (z/z^*)^2[(1 - z^*)/(1 - z)]^4 \tag{30}$$

$$K_0 = 273ak_2/\gamma^{3/2}p^{1/2}Tv \tag{31}$$

are derived; z is the mole fraction of ammonia in the reaction mixture, a the steric factor, γ the mole fraction of H_2 in the reaction mixture, p the total pressure in the reactor, T the thermodynamic temperature of the reactor (Kelvin), and v the volumetric flow rate through the reactor. The asterisk denotes conditions at thermodynamic equilibrium. On the basis of independent data from different catalytic converters the ammonia reactor model

$$1/\sqrt{Qu} = 1 + 1/K_0 \tag{32}$$

is obtained using the dimensional analysis approach [63] at good statistical accuracy: least-squares analysis [101] of the 58 data sets contained in Tables 9 and 10 of Reference 63 yields 0.9691 for the intercept and 1.182 for the slope of the $(1/K_0; 1\sqrt{Qu})$ regression line with an associated coefficient of determination of 0.996.

Dimensional analysis has been likewise successful in modeling the oxidation of the SO_2 process [63] and other chemical and physical processes described in the literature quoted in Tables 1 and 2.

PRINCIPLES OF PHYSICAL SIMILARITY AND MODEL LAWS

In the engineering design and construction of large-scale equipment, experiments conducted on small *models* serve for the testing of phenomena expected to occur in the full-size structure, or *prototype*. Typical examples are aircraft, sea vessels, large bodies of water (lakes, rivers and estuaries, dams, and reservoirs), bridges and in general, heavy machinery, and devices. Testing of behavior on a small-scale model is dictated not only by economic common sense but also by safety considerations. Experiments conducted on models can often be paired to advantage with computer-oriented mathematical simulation before design of a prototype is undertaken. The difference between prototype and a model is not necessarily size (in fact a model may be as large as its prototype or even larger); it can be different working material, fluid properties, temperature, fluid velocity, and so forth. Observations obtained on the model of a full-size structure, machine, or device can be applied to the design and the construction of the latter, if the conditions of functioning of model and prototype are physically similar. This is the essence of physical similarity (or physical similitude).

The four major types of physical similarity are geometric, kinematic, dynamic, and transport-oriented; the latter includes problems dealing with the transport of heat, mass, and substance. In this respect thermal, chemical, electrical, and electromagnetic similarities can be considered as

Table 5
The Representation of Physical Similarity Via Characteristic Quantity Ratios*

Type of physical similarity	Nature of the similarity	Characteristic Quantity Ratio(s)
Geometric	Shape	Scale factor, L_m/L_p; A_m/A_p
Kinematic	Motion	Velocity ratio, v_m/v_p Acceleration ratio, a_m/a_p Discharge (or flow) ratio, Q_m/Q_p
Dynamic	Force(s)	Total force ratio,** $\sum_i F_{i,m} \Big/ \sum_i F_{i,p}$

* *Notation: L = length; A = area; v = velocity, a = acceleration; Q = flow rate, F = force, Subscripts: m = model; p = prototype.*

** *Includes normally viscous, pressure, gravity, surface-oriented and elasticity-oriented forces.*

distinct groups. Similarities are mathematically represented in terms of certain quantity ratios as shown in Table 5; in a more general sense, dimensionless groups may be regarded as similarity representations. Similarity between a model and its full-scale prototype is true if the characteristic ratios are strictly sufficient for scale-up; otherwise similarity is distorted. An obvious example of distorted similarity is a flow system where a laboratory size model would have to have a vertical length so small that surface tension effects, negligible in the prototype (say a river, or lake), would seriously interfere with the hydrodynamic behavior of the model. The relationship between the principles of similarity and model laws may be phrased in the following manner: let the result of dimensional analysis of a physical problem be $\pi_1 = \phi(\pi_2, \cdots, \pi_n)$ in terms of n pi-groups. If, for model and prototype the set of identities

$$\pi_{j,p} = \pi_{j,m} \qquad j = 2, 3, \ldots, n \tag{33}$$

is true, then the identity

$$\pi_{1,p} = \pi_{1,m} \tag{34}$$

is also true. Thus a variable in the $\pi_{1,p}$ group may be estimated numerically via quantitative observations obtained on its model.

It is important to note that conditions in a model/prototype set may result in unrealistic quantity ratios by applying Equation 32 without sound physical judgment, as illustrated in Example 12.

Example 12—Physical Similarity Between Drag Force Acting on a Prototype and a Laboratory-Scale Model (83)

The relationship between associated pi-groups may be expressed in the form of Equation 33 as $Fr_m = Fr_p$ and $Re_m = Re_p$. It follows (see Table 4) that the characteristic quantity ratios must satisfy the equations

$$\left(\frac{v_m}{v_p}\right)^2 = \frac{l_m}{l_p} = \left(\frac{\nu_m}{\nu_p}\right)^2 \left(\frac{l_p}{l_m}\right)^2$$

hence,

$$\frac{\nu_m}{\nu_p} = \left(\frac{l_m}{l_p}\right)^{3/2} \tag{35}$$

Equation 35 indicates clearly that drag experiments in a laboratory would have to meet nearly impossible conditions to satisfy strictly physical similarity: for a scale factor as high as about $\frac{1}{2}$, the laboratory-scale experiment would require n-pentane as the fluid (assuming that the prototype fluid is water) since its kinematic viscosity to water viscosity ratio is $\frac{2}{3} = 0.373^{2/3} = 0.518$. A more realistic scale factor of $\frac{1}{10}$ would require a fluid whose kinematic viscosity is about 3.2% of the kinematic viscosity of water! In this kind of experiment complete similarity cannot be achieved.

RECENT DEVELOPMENTS IN DIMENSIONAL ANALYSIS

Dimensional analysis has gained further impetus in recent years by efforts to reduce the extent of functional indeterminacy in the classical methods and to develop a generalized approach to the treatment of physical problems, which may be applied without any modifications necessitated by the use of a particular system of units. The general dimensional analysis proposed by Staicu [14], although powerful and fairly simple, requires specific a priori conditions, among which the necessity of knowing precisely and beforehand the set of quantities, and the manner whereby an increase in the magnitude of one quantity generates an increase or decrease in the magnitude of another quantity, are rather restrictive. Major shortcomings of the technique [83] are the following: redundant variables may have to be introduced to arrive at the correct result; indical equation(s) may remain indeterminate; and there are, in principle, an infinite number of possible indices occurring during analysis. The technique, however, offers the "economy of effort" when used in a cautious and circumspect manner.

A Generalized Method of Dimensional Analysis: Standard Procedure

A generalized method of dimensional analysis for the identification of all pi-groups governing system behavior offers a systematic procedure to construct the minimum number of dimensionless groups [16]; in its initial form [15] the method was conceived specifically for manipulations in SI, but it is by no means linked to any unit system. The method is based on a quantity-unit equivalence structure, which yields quickly a first set of pi-groups that can be transformed into a final set usually by simple inspection of a table of dimensionless groups. The quantity-unit equivalences can be established conveniently by means of a *constitutive matrix* whose size is reduced systematically until all equivalences have been set. Any system of units and their combinations may be employed, moreover the number of units in the initial constitutive matrix is not constrained by any conventional (e.g., MLT and FLT) dimension base. The method minimizes the occurrence of redundant pi-groups and requires relatively simple mathematical manipulations.

The constitutive matrix has its minimum size when SI is employed since conversion factors relating fundamental units are always unity. In non-rational unit systems the number of conversion factors required to relate basic and derived units depends on the total number of units N and the number of basic units n; its numerical value is $N - n$. If M groups of units can be found such that they would appear as a single entry in the list of units needed, then the number of required conversion factors is $(N - n - M)$ if $M < (N - n)$, and $(q - N + M)$ if $M > (N - n)$; q is the number of quantities present. In the constitutive matrix quantities are listed in columns, units in rows and the elements are powers of a unit in a particular quantity. At least one column will have a single non-zero element; such a "pivotal" element defines a quantity-unit equivalence. If more than one pivotal element is found for a given unit, one with the most fundamental meaning is chosen for establishing equivalence. Removal of the rows that contain the pivotal elements reduces the original matrix to systematically reduced submatrices and finally to a single row, while each equivalence is established (the final equivalence might be written upon inspection). Once the equivalence set has been found, the first set of pi-groups is obtained by forming quantity/unit ratios and substituting into them appropriate equivalence relationships. Then, the number of pi-groups is reduced by a proper sorting of the pi-groups in the first set in order to eliminate redundant elements of the set. The entire procedure is illustrated in Example 13 for thermal free convection at a vertical plate. This straightforward application is followed by the handling of advanced problems.

Example 13—Thermal Free Convection at a Vertical Plate

There are nine quantities to be considered: film heat transfer coefficient h, length of the plate L, thermal conductivity of the fluid k, specific heat capacity of the fluid (at constant pressure) c_p, dynamic viscosity of the fluid μ, density of the fluid ρ, volumetric expansion coefficient of the fluid β, temperature driving force $\Delta\theta$, and acceleration due to gravity g. The unit system chosen is SI where the fundamental units of mass (kg), length (m), time (s), temperature (K) and energy (J) are used. The constitutive matrix is shown below where each column and row has been identified by a number for easy reference. There are three columns (2, 7, 8) containing a single non-zero entry

		(1) h	(2) L	(3) k	(4) c_p	(5) μ	(6) ρ	(7) β	(8) $\Delta\theta$	(9) g
(1)	kg	0	0	0	−1	1	1	0	0	0
(2)	m	−2	1	−1	0	−1	−3	0	0	1
(3)	s	−1	0	−1	0	−1	0	0	0	−2
(4)	K	−1	0	−1	−1	0	0	−1	1	0
(5)	J	1	0	1	1	0	0	0	0	0

that yield equivalences m $\doteqdot$ L, K $\doteqdot \beta^{-1}$ and K $\doteqdot \Delta\theta$; since the third equivalence is more fundamental than the second, the K $\doteqdot \beta^{-1}$ equivalence is disregarded. The second and fourth row are removed from the matrix since the pivotal points for the m $\doteqdot$ L; K $\doteqdot \Delta\theta$ equivalence pairs (1 and 1) occur in these rows. The resulting submatrix:

		(1) h	(3) k	(4) c_p	(5) μ	(6) ρ	(7) β	(9) g
(1)	kg	0	0	−1	1	1	0	0
(3)	s	−1	−1	0	−1	0	0	−2
(5)	J	1	1	1	0	0	0	0

contains two single non-zero entries, the pivotal point positions being 1(row 1, column 6) and −2 (row 3, column 9). The corresponding equivalences are kg = ρL^3 and $s^{-2} = gL^{-1}$, rewritten as $s = g^{-1/2}L^{1/2}$. Removal of row 1 and row 3 leaves only row 5 with column positions 1, 3, 4, 5 and 7. The unity elements in columns 1, 3 and 4 indicate that the last equivalence can be expressed in terms of h, k or Cp. The step-by-step constructions are

(a) $J \doteqdot h\,K\,s\,m^2 = h\,\Delta\theta(g^{-1/2}L^{1/2})L^2 = hL^{5/2}g^{-1/2}\,\Delta\theta$

(b) $J \doteqdot k\,K\,s\,m = k\,\Delta\theta(g^{-1/2}L^{1/2})L = kL^{3/2}g^{-1/2}\,\Delta\theta$

(c) $J \doteqdot c_p\,kg\,K = c_p(pL^3)\,\Delta\theta = c_pL^3\rho\,\Delta\theta$

Thus, in summary, degeneration of the constitutive matrix yields the equivalences m $\doteqdot$ D; K $\doteqdot \Delta\theta$; kg $\doteqdot \rho D^3$; s $\doteqdot g^{-1/2}L^{1/2}$; J $\doteqdot hL^{5/2}g^{-1/2}\,\Delta\theta$ (for the unit of energy equivalence (a) was arbitrarily chosen). The construction of the associated pi-groups is demonstrated in Table 6, where $\pi_i^{(0)}$ are the preliminary pi-groups and $\pi_i^{(1)}$ the final pi-groups (minimal set). The procedure is explained by considering the third entry (i = 3) whose associated quantity k(column 2) has the units $J\,s^{-1}\,m^{-1}\,K^{-1}$ in SI (column 3). The fourth column carries the ratio of k and $J\,s^{-1}\,m^{-1}\,K^{-1}$ which is ksmK/J.

Table 6
Construction of Pi-Groups in Example 13

Entry number, i	Quantity	Units (SI base)	Quantity / Units	$\pi_i^{(0)}$	Transformation	$\pi_i^{(1)}$
1	h	$J\,s^{-1}\,m^{-2}\,K^{-1}$	$h\,s\,m^2\,K/J$	1	—	1
2	L	m	L/m	1	—	1
3	k	$J\,s^{-1}\,m^{-1}\,K^{-1}$	$k\,s\,m\,K/J$	k/hL	$1/\pi_3^0$	hL/k = Nu
4	c_p	$J\,kg^{-1}\,K^{-1}$	$c_p\,kg\,K/J$	$c_p\rho L^{1/2}g^{1/2}/h$	$\pi_4^0\pi_5^0\pi_3^{(1)}$	$c_p\mu/k = Pr$
5	μ	$kg\,s^{-1}\,m^{-1}$	$\mu\,s\,m/kg$	$\mu/\rho g^{1/2}L^{3/2}$	$\pi_7^0/(\pi_5^0)^2$	$\rho^3 g L^3 \beta\,\Delta\theta/\mu^2 = Gr$
6	ρ	$kg\,m^{-3}$	$\rho\,m^3/kg$	1	—	1
7	β	K^{-1}	$\beta\,K$	$\beta\,\Delta\theta$	π_7	$\beta\,\Delta\theta$
8	$\Delta\theta$	K	$\Delta\theta/K$	1	—	1
9	g	$m\,s^{-2}$	$g\,s^2/m$	1	—	1

Substituting for each unit its quantity equivalence, π_3^0 is obtained:

$$\mathrm{k\,s\,m\,K/J} \rightarrow \mathrm{k(g^{-1/2}L^{1/2})(L)(\Delta\theta)/hL^{5/2}g^{-1/2}\,\Delta\theta} \rightarrow \mathrm{k/hL}$$

This entity is shown in column 5. Since this is the reciprocal of the well-known Nusselt number, the required transformation is the taking of the reciprocal of $\pi_3^0 = \mathrm{k/hL}$, yielding the final associated pi-group: $\pi_3 = \mathrm{Nu} = \mathrm{hL/k}$. The transformation step may involve several pi-groups, or none; in any transformation step, the most recent definition of π_i is to be used. Pi-groups represented by unity have, of course, no significance. Had the unit system been chosen differently, Table 6 would carry an additional entry related to an appropriate conversion factor; in, e.g., the Imperial system of units the conversion factor is $K_c = \mathrm{ft^2\,lb/Btu{\cdot}hr^2}$ and the corresponding pi-group is $\pi_{10} = K_c k\,\Delta\theta/\mu g L$; details of the analysis using Imperial units are given in Reference 16.

This scheme does *not* require that all units be fundamental, but if derived units are also used, appropriate conversion factors have to be supplied. In SI all conversion factors are (positive or negative) integer power of 10; in non-rational systems conversion is more complicated. Let, for instance, a dimensional analysis involve fundamental units m, kg, s and A, and derived units W and V. Then, it is convenient to augment the pi-construction table by two additional entries: using unity for the quantity column and conversion-equivalences, $\mathrm{kg\,m^2\,W^{-1}\,s^{-3}}(=1)$ and $\mathrm{W\,V^{-1}\,A^{-1}}$ $(=1)$, respectively are entered in the units column. The rest of the analysis is normal.

Deviation from the Standard Procedure:
Premature Disappearance of Columns With a Single Non-Zero Entry

If, in the course of generating the constitutive matrix, no column is found to possess a single non-zero entry, the technique of "unit-pairing" is employed in order to successfully complete the degeneration procedure. The technique is best explained by a specific example: heat transport in a fluid moving inside a tube. There are two major variants to this problem, depending on the choice of linear velocity or mass flux as the quantity representing fluid flow. The choice has some bearing on the specific unit-pairing step, hence both constructions are shown.

Example 14—Heat Transport in a Fluid Moving Inside a Tube

The constitutive matrix below carries both an Imperial and SI column for the sake of generality. The case of linear velocity as representative of flow is taken first.

	Imperial	SI	(1) h	(2) D	(3) v	(4) ρ	(5) c_p	(6) μ	(7) k
(1)	lb	kg	0	0	0	1	−1	1	0
(2)	ft	m	−2	1	1	−3	0	−1	−1
(3)	hr	s	−1	0	−1	0	0	−1	−1
(4)	°F	K	−1	0	0	0	−1	0	−1
(5)	Btu	J	1	0	0	0	1	0	1

The pivotal points for quantity-unit equivalences being the matrix elements (2, 2), (3, 3) and (1, 4) yielded by the first three degeneration steps, ft = D; $\mathrm{s = Dv^{-1}}$; $\mathrm{lb} = \rho D^3$ (Imperial) and m = D; $\mathrm{s = Dv^{-1}}$ $\mathrm{kg} = \rho D^3$ (SI) are the first three equivalences. However, this leaves the degenerate matrix

	Imperial	SI	(1) h	(2) D	(3) v	(4) ρ	(5) c_p	(6) μ	(7) k
(4)	°F	K	−1	0	0	0	−1	0	−1
(5)	Btu	J	1	0	0	0	1	0	1

with columns of full-zero entries or non-zero entries. The latter are completely symmetrical with unity entries of changing sign; this property allows the pairing of energy and temperature units in the following manner in Imperial units:

$$\text{Btu} = \text{h}^\circ\text{F s ft}^2 \rightarrow \text{Btu}/^\circ\text{F} = \text{hr ft}^2 = \text{h}(\text{Dv}^{-1})(\text{D}^2) = \text{hD}^3\text{v}^{-1}$$

$$\text{Btu} = \text{c}_\text{p}^\circ\text{F lb} \rightarrow \text{Btu}/^\circ\text{F} = \text{c}_\text{p}(\rho\text{D}^3) = \text{c}_\text{p}\rho\text{D}^3$$

$$\text{Btu} = \text{k}^\circ\text{F hr ft} \rightarrow \text{Btu}/^\circ\text{F} = \text{k}(\text{Dv}^{-1})(\text{D}) = \text{kD}^2\text{v}^{-1}$$

and in SI units,

$$\text{J} = \text{h K s m}^2 \rightarrow \text{J/K} = \text{h s m}^2 = \text{h}(\text{Dv}^{-1})(\text{D}^2) = \text{hD}^3\text{v}^{-1}$$

$$\text{J} = \text{c}_\text{p}\text{ K kg} \rightarrow \text{J/K} = \text{c}_\text{p}(\rho\text{D}^3) = \text{c}_\text{p}\rho\text{D}^3$$

$$\text{J} = \text{k K s m} \rightarrow \text{J/K} = \text{k}(\text{Dv}^{-1})\text{D} = \text{kD}^2\text{v}^{-1}$$

From here on, the construction of the final pi-groups Nu, Re and Pr is straightforward as shown elsewhere [16] in the instance of SI units.

A somewhat different unit pairing exercise is required when the mass flux G is considered as the representative of fluid flow. The constitutive matrix carries an extra entry for the c_p column between the third and the fourth row, which simply expresses the identity $1 - 1 = 0$; thus the (s, c_p) zero element is split into a -1 and a $+1$ entry:

	Imperial	SI	(1) h	(2) G	(3) D	(4) c_p	(5) μ	(6) k
(1)	lb	kg	0	1	0	−1	1	0
(2)	ft	m	−2	−2	1	0	−1	−1
(3)	hr	s	−1	−1	0	0	−1	−1
						1 − 1		
(4)	°F	K	−1	0	0	−1	0	−1
(5)	Btu	J	1	0	0	0	0	+1

If this splitting were not done, the normal degenerating procedure would have to stop after the establishment of the ft = D or m = D equivalence and subsequent removal of the second row. Consequently, c_p could not be related to Btu and °F, or J and K; similarly, lb and hr or kg and s could not be related to c_p (although h and k can be related to these units, as before, without any difficulty. With the $1 - 1 = 0$ splitting the degeneration of the constitutive matrix and the pi-group construction may be completed in the usual manner yielding again, of course, Nu, Re and Pr as the final pi-groups [16]. Notice that the use of hr as time unit in the Imperial system precludes a conversion factor (hence an additional pi-group) inasmuch as pairing of Btu and °F does not require conversion. The fact that velocity in the tube implies motion along the axial length, and not the diameter, is immaterial for the analysis in view of the purely formal sense of the [L] = D equivalence.

CONCLUDING REMARKS

Dimensional analysis offers essentially three major benefits to its users: first, it defines what physical quantities in which configuration are related to one another, although the exact quantitative relationship has to be found by means beyond the scope of dimensional analysis. Second, in its employment of dimensional homogeneity, it provides a reliable means of detecting errors and as such it can serve as a safeguard of consistency when other (e.g., algebraic) methods are used for the analysis of a physical problem. Third, it may direct the experimenter to minimize rationally the number of experiments required to find the numerical relationship between quantities, although statistical methods of planning experiments (e.g., factorial design, randomized-block design, latin square design, etc.) should always be considered along with deterministic methods of minimization. Applied within its obvious limitations, dimensional analysis is a powerful tool in the hands of the scientist and the engineer both in the conceptualization and the practical solution of physical problems.

NOTATION

A	area
a	tube radius
C_D	drag coefficient
c	specific heat capacity
c_i	concentration
D	diffusivity
d	diameter; depth
e	electric charge of electron
F	force; variable
Fr	Froude number
G	fundamental gravitational constant
g	acceleration due to gravity
h	heat transfer coefficient
K	proportionality constant
K_0	contact number
k	proportionality constant; thermal conductivity
L	length
l	length
M	mass
m_e	electron mas
m_p	proton mass
Nu	Nusselt number
p	pressure
Pr	Prandtl number
q	heat flux
Ra	Rayleigh number
Re	Reynolds number
r	radial distance
Sh	Sherwood number
T	temperature; time
t	time
u	velocity
V	voltage drop
v	wave velocity

Greek Symbols

ϵ	permittivity
θ	temperature
λ	wavelength
μ	viscosity
ν	kinematic viscosity
ρ	density
ω	frequency

REFERENCES

1. Newton, I. *Philosophiae Naturalis Principia Mathematica* (1686); English transl. (1729); revised Book II, Sec. 7, Prop. 32, Univ. Calif. Press, 1962.
2. Rayleigh, Lord., *Proc. Roy. Soc.*, LXVI:68 (1899/1900).
3. Rayleigh, Lord., *Nature*, 95:66. 202. 644 (1915).
4. Buckingham, E., *Phys. Rev.*, IV(4): 345 (1914).
5. O'Rahilly, A., *Electromagnetics: A Discussion of Fundamentals*. Longmans Green and Co., 1938.
6. Picken, D. K., *Math. Gazette*, 26:269:87 (1942).
7. Brand, L., *Arch. Rational Mech. Anal.*, 1:35 (1957).
8. Brand, L., *Math. Gazette*, 49:207 (1965).
9. Corrsin, S., *Am. J. Phys.*, 17:180 (1951).
10. Happ, W. W., *J. Franklin Inst.*, 292(6):527 (1971).
11. Hainzl, J., *J. Franklin Inst.*, 292(6):463 (1971).
12. Moran, M. J., *J. Franklin Inst.*, 292(6):423 (1971).
13. Na, T. Y., and Hansen, A. G., *J. Franklin Inst.*, 292(6):471 (1971).

14. Staicu, C. I., *J. Franklin Inst.*, 292(6):433 (1971).
15. Quraishi, M. S., and Fahidy, T. Z., *Can. J. Chem. Engrg.*, 59:563 (1981).
16. Quraishi, M. S., and Fahidy, T. Z., *Can. J. Chem. Engrg.*, 61:116 (1983).
17. Jordan, P., *Die Herkunft der Sterne*, Wissensch. Verlagsges. Stuttgart, 1947.
18. Schürer, M., "Das Alter der Welt," in ref. 19.
19. Kurth, R., and Schürer, M., *Zum Weltbild der Astronomie*, Stämpfli, Bern, 1954 (1st edn.); 1957 (2nd edn.).
20. Bondi, H., *Cosmology*, Cambridge Univ. Press, 1961.
21. Patel, R., *Glastech. Ber. Sonderband*, 32:30 (1959).
22. Rustamov, M. I., and Aliev, V. S., *Azerb. Khim. Fiz.*, No. 4:7 (1964).
23. Rustamov, M. I., and Aliev, V. S., *Azerb. Khim. Fiz.*, No. 2:85 (1965).
24. Sunavala, P. D., *Indian J. Technol.*, 3:372 (1965).
25. Ehrenfest-Afanassjeva, T., *Phil. Mag.*, 7:257 (1926).
26. Ciborowski, J., *Przemyst Chem.*, 28:206 (1949).
27. Pierce, J. E., *Chem. Eng.*, 61(4):185 (1954).
28. Silberberg, I. H., and McKetta, J. J., Jr., *Petroleum Refiner*, 32(4):179 (1953).
29. Silberberg, I. H., and McKetta, J. J., Jr., *Petroleum Refiner*, 32(5):147 (1953).
30. Silberberg, I. H., and McKetta, J. J., Jr., *Petroleum Refiner*, 32(6):101 (1953).
31. Benfey, O. T., *J. Chem. Educ.*, 34:286 (1957).
32. Standart, G., *Chem. Listy*, 51:1971 (1957).
33. Mosulin, B., *J. Chem. Educ.*, 41:622 (1964).
34. Szarawara, J., *Zesz. Nauk. Politech. Slask., Chem.*, No. 30:45 (1966).
35. Kerber, R., *Chem. Ing. Techn.*, 38:1133 (1966).
36. Henry, A., J. *Educ. Chem.*, 4:81 (1967).
37. Happ, W. W., *J. Appl. Phys.*, 38:3918 (1967).
38. Nakon, R., *Chemical Problem Solving Using Dimensional Analysis*, Prentice Hall, 1978.
39. Coull, J., and Roger Shields, J., *J. Eng. Educ.*, 31:473 (1941).
40. Ibl, N., *Electrochim. Acta*, 1:117 (1959).
41. Regner, A., and Roušar, I., Collection Czechoslov. Chem. Commun., 25:1132 (1960).
42. Halpern, O., *Acta Physica Austriaca* (1968).
43. Sadkowski, A., *J. Electroanal. Chem. Interfacial Electrochem.*, 105:1 (1979).
44. Finzi-Contini, B., *Chim. Industria*, 36:452 (1954).
45. Mankowich, A. M., *J. Am. Oil Chem. Soc.*, 42:629 (1965).
46. Mankowich, A. M., *NASA Accession*, No. N65-33404, Rept. No. AD 468289 (1965).
47. Mankowich, A. M., *J. Amer. Oil Chem. Soc.*, 45:152 (1968).
48. McNelly, M. J., *J. Imp. Coll. Chem. Eng. Soc.*, 7:18 (1953).
49. Ibl, N., *Chimia*, 9:135 (1955).
50. Luft, N. W., *Ind. Chemist*, 32:302 (1956).
51. Crico, A., *Génie* Chim., 73:57 (1955).
52. Filippov, L. P., *Zhurn. Fiz. Khim.*, 31:1999 (1957).
53. Kattwinkel, W., *Atomkernenergie*, 4:356 (1959).
54. Hakala, R. W., *J. Chem. Educ.*, 41:380 (1964).
55. Azizov, B. M., and Nikolaev, A. M., *Tr. Kazansk. Khim-Tekhnol. Inst.*, No. 32:103 (1965).
56. Degaleesan, T. E., and Laddha, G. S., *Chem. Eng. Sci.*, 21:199 (1966).
57. Aleksandrov, I. A., Gorechenkov, V. G., Korolenko, T. P., Sorokin, V. I., and Khalif, A. L., *Protsessy Khim. Tekhnol. Gidrodinam., Teplo i Massoperedacha, Akad Nauk SSSR, Otd. Obsch. i Tekhn. Khim* Sb. Statei, 44 (1965).
58. Tobilevich, N. Yu., Sagan, I. I., and Gordienko A. P., *Izv. Vyssh. Ucheb. Zaved., Pishch. Tekhnol.*, No. 5, 150 (1966).
59. Leland, T. W., Jr., *AIChE J.*, 12:1227 (1966).
60. Tairov, B. D., *Izv. Akad. Nauk Turkm. SSSR Ser. Fiz.-Tekhn., Khim. Geol. Nauk*, No 6, 115 (1966).
61. Arenas, A., and Herranz, A., *An. Fis.*, 73:112 (1977).
62. Dolan, W. J., Dimon, C. A., and Dranoff, J. S., *AIChE J.*, 11:1000 (1965).

63. Dyakonov, G. K., *Voprosi Teorii Podobia V Oblasti Fiziko-Khimicheskich Protzessov*, Izd. Akad. Nauk CCCP, Moscow, 1956.
64. Novosad, J., and Standart, G., *Collection Czechoslov. Chem. Commun.* 30:3247 (1965).
65. Mitra, P., Bhowmik, T. K., Krishnamurty, V. A., and Basu, A. N., *Indian J. Technol.*, 3:182 (1965).
66. Cerofolini, G., *Thin Solid Films*, 55:293 (1978).
67. Tokita, N., *Rubber Chem. Technol.*, 52:387 (1979).
68. Urun, H., Treat. "Disposal Liq. Solid Ind. Wastes," *Proc. Turk.-Ger. Environ. Eng. Symp.* (3rd) 89 (1979). Pergamon (ed. C. Kriton).
69. Matejicek, A., Kasha, J., Eichler, J., and Horky, J., *Farbe Lack.*, 89:9 (1983).
70. Ilyushin, A. A., and Ogibalov, P. M., *Mekh. Polim.*, No. 6:828 (1966).
71. Abeles, T. P., and Bos, W. G., *J. Chem. Educ.*, 44:438 (1967).
72. Strouf, O., and Fusek, J., *Collection Czechoslov. Chem. Commun.*, 44:1370 (1979).
73. Bridgman, P. W., *Dimensional Analysis*, Yale Univ. Press, 1956.
74. Focken, C. M., *Dimensional Methods and Their Applications*, E. Arnold Co. (London), 1953.
75. Giles, R. V., *Fluid Mechanics and Hydraulics, 2nd ed (SI edition)*, Schaum's outline series, McGraw-Hill, 1962.
76. Kurth, R., *Dimensional Analysis and Group Theory in Astrophysics*, Pergamon, 1972.
77. Langhaar, H. L., *Dimensional Analysis and Theory of Models*, R. E. Krieger, Huntington, N.Y., 1980 (Wiley, 1951).
78. Massey, B. S., *Units, Dimensional Analysis and Physical Similarity*, Van Nostrand Reinhold, 1971.
79. Pankhurst, R. C., *Dimensional Analysis and Scale Factors*, Chapman and Hall, 1964.
80. Ragunath, H. M., *Dimensional Analysis and Hydraulic Model Testing*, Asia Publ. House (London), 1967.
81. Schepatz, B., *Dimensional Analysis in the Biomedical Sciences*, C. C. Thomas (Springfield, Ill.), 1980.
82. Sedov, L. I., *Similarity and Dimensional Methods in Mechanics*, Acad. Press, 1959.
83. de St. Q. Isaacson, E., and de St. Q. Isaacson, M., *Dimensional Methods In Engineering and Physics*, E. Arnold (London), 1975.
84. Taylor, E. S., *Dimensional Analysis For Engineers*, Clarendon Press (Oxford), 1974.
85. Catchpole, J. P., and Fulford, G. D., *Dimensionless Groups*, Tables 9-44, 9-45, *Handbook of Tables for Applied Engineering Science*, first edition (eds. R. E. Bolz and G. L. Tuve); Tables 10-51, 10-52, ibid; second edition, Chemical Rubber Company, 1973.
86. Massey, B. S., *Mechanics of Fluids, 2nd ed.*, Section 11.8, Van Nostrand Reinhold, 1972.
87. Kuo, B. C., *Automatic Control Systems*, 2nd ed., Section 2.6, Prentice Hall, 1962.
88. Coulson, J. M., and Richardson, J. F., *Chemical Engineering*, Vol. I (SI edition), Section 7.4.3, 2nd edn., Pergamon Press, 1977.
89. *Handbook of Tables for Applied Engineering Science*, Figs. 5-20, Tables 5-21, 514–515, second edition (eds. R. E. Bolz and G. L. Tuve), Chemical Rubber Company, 1973.
90. Levich, V. I., *Physicochemical Hydrodynamics*, Section 23, Prentice Hall, 1962.
91. Fahidy, T. Z., and Mohanta, S., *Mass Transport in Electrochemical Systems*, Chapter 3 in *Advances in Transport Processes*, Vol. I (ed. A. S. Mujumdar), Sections 3.3, 3.5, Wiley-Eastern, 1980.
92. Nusselt, W., *Z. deut. Ing.*, 60:541, 569 (1916).
93. Einstein, A., *Ann. der Phys.*, 35:686 (1911).
94. Dirac, P. A. M., *Nature*, 139:323 (20 Feb. 1937).
95. Eddington, A. S., *Report on Gravitation*, London Phys. Soc., 91 (1918).
96. Lewis, G. N., and Adams, E. Q., *Phys. Rev.*, 3:92 (1914).
97. Damköhler, G., *Uspekhi Khimii*, VII:5 (1938).
98. Damköhler, G., *Chem. Fabrik*, 469–477 (1939).
99. Edgeworth-Johnston, R., *Trans. Inst. Chem. Eng.*, 17:129 (1939).
100. Temkin, M. I., and Reuter, V. A., *Zhurn. Fiz. Khim.*, 13:851 (1939).
101. Fahidy, T. Z., priv. comm. (1983).

CHAPTER 13

DRAG IN SINGLE FLUID FLOWS

Slobodan Koncar-Djurdjevic
Fedor Zdanski

Faculty of Technology and Metallurgy
University of Beograd,

Beogard, Yugoslavia

and

Aleksandar Dudukovic

Faculty of Technology,
University of Novi Sad,
Novi Sad, Yugoslavia

CONTENTS

INTRODUCTION

In the case of relative motion of fluids along flat surfaces, around submerged bodies, or through ducts, the drag force is induced. In general, frictional resistance and the form drag are involved. Sometimes, the total drag is referred to as only the frictional resistance in straight tubes and channels or along flat surfaces, *or* only the form drag in the case of the flow around a disk. Usually, both components appear simultaneously, as in the case of flow around a sphere, a cylinder, or any other geometrically complicated objects. Total drag force, F, is related to the relative motion, w, using a coefficient of drag, C_D, defined by the equation:

$$F = C_D A \tfrac{1}{2}\rho w^2 \tag{1}$$

where ρ is the fluid density and A the characteristic area. Drag coefficient is the function of Reynolds number and the geometry of the system. In the systems with free surfaces or in the case of compressible fluids, drag coefficient is the function of other dimensionless numbers, too (Froude number, Mach number, etc.).

If the fluid flow is parallel to the boundary surface, only frictional resistance is present, whereas the drag force and the corresponding pressure drop are usually calculated using the friction factor:

$$f = \frac{C_f}{4}\frac{d}{L} = \frac{\tau_w}{\rho w^2/2}$$

where C_f is the drag coefficient, commonly called the coefficient of friction when only tangential stress is involved. Sometimes, particularly in Europe, the alternative definition of the friction factor is used:

$$\lambda = C_D \frac{d}{L} = 4f$$

FRICTIONAL RESISTANCE

Frictional Resistance in Pipes

Experimental observations on the flow of water in cylindrical pipes led Darcy and Weisbach to express the head loss by

$$h_f = \frac{\Delta p}{\gamma} = 4f \frac{L}{d}\frac{w^2}{2g}$$

For laminar flows for which Newton's well understood law of fluid shear applies, it is possible to derive the relationship:

$$f = \frac{16}{Re} \tag{2}$$

This result is in good agreement with the experimental results of Nikuradse [1, 2] for Re < 2,100.

Values of the friction factors for turbulent flow can be obtained only by experiments, whereas the friction factor is not only a function of the Reynolds number but also of the relative roughness e/d. According to experiments, Blausius [3] has derived the expression for hydraulically smooth pipes:

$$f = 0.0791\,Re^{-1/4} \tag{3}$$

the limits of which are generally taken to be:

$$4{,}000 \le Re \le 10^5$$

Numerous investigations have been devoted to this problem [4–8], but the most systematic and mostly used were the experiments carried by Nikuradse [1, 2], who used a measurable roughness produced by uniform sand grains of diameter e. When his test results are plotted in a log-log coordinate system with the Reynolds number as abscissa and the friction factor as the ordinate, the following conclusions can be derived:

1. In the range of laminar flow ($Re < 2{,}100$) the friction factor is not dependent on relative roughness, but only on the Reynolds number. Results for this range are in agreement with the Equation 2.
2. In the turbulent range ($Re > 2{,}100$) the friction factor depends very much upon the surface roughness; $f = f(Re, e/d)$. If the roughness is small enough to remain within the laminar sublayer, the surface is called hydraulically smooth, whereas the friction factor is the function of the Reynolds number only.
3. At high Reynolds numbers and high relative roughness, the friction factor becomes only a function of the relative roughness and independent of the Reynolds number (a wholly rough surface).

Comparing the equation given by Blasius [3] and the results obtained by Nikuradse, it has been shown that they are in good agreement up to Re value of 10^5. For the wider range of Reynolds number, the equation given by Prandtl-Karman [9] is applied:

$$\frac{1}{\sqrt{f}} = 4.0 \log(Re\sqrt{f}) - 0.4 \tag{4}$$

the limits of which are usually taken as

$$4{,}000 \le Re \le 3 \times 10^6$$

Drew and Generaux [10] have also presented an equation very similar to Equation 4:

$$\frac{1}{\sqrt{f}} = 3.2 \log(Re\sqrt{f}) + 2.16 \tag{5}$$

In the case of Blasius' equation the upper limit of Reynolds number places severe restrictions on it, and in the cases of Equations 4 and 5 the fact that they are not written explicitly in f makes them awkward to use.

Colebrook [11] correlated the results of many laboratory tests on commercial pipes over a wide range of a Reynolds numbers and roughnesses and showed that these could be characterized by the equation:

$$\frac{1}{\sqrt{f}} = -4.0 \log\left(\frac{2.51}{2\sqrt{f}\,Re} + \frac{e}{3.7d}\right) \tag{6}$$

The equation is difficult to use as f appears on both sides. However, it can be easily solved using a computer. It may be noted in passing that Equation 6 reduces to Equation 4 when $e/d = 0$.

For practical purposes, the "Moody diagram" (Figure 1) proved to be convinient. Moody [12] plotted Equation 6 on the log–log diagram with Reynolds number as abscissa and the friction factor as the ordinate, and extended the work of Colebrook to other materials and roughnesses.

Round [13] proposed the term in which friction factor is explicitely expressed as the function of Re and e/d:

$$\frac{1}{\sqrt{f}} = 3.6 \log \frac{Re}{0.135\,Re\dfrac{e}{d} + 6.5}$$

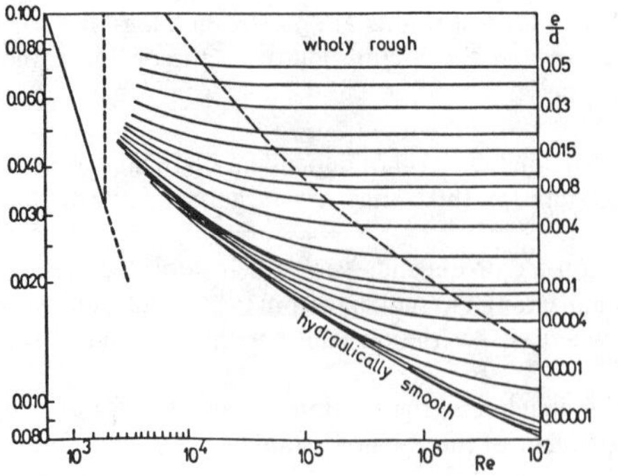

Figure 1. Moody diagram (friction factor as a function of Reynolds number and relative roughness).

For the wholly rough zone (high Reynolds number and high relative roughness) the friction factor is a function of relative roughness only and the following equation is valid [14]:

$$\frac{1}{\sqrt{f}} = 0.57 + \log\frac{d}{e}$$

what is in agreement with the Karman's expression [15]:

$$4f = \frac{1}{2\left(\log\frac{d}{2e} + 1.74\right)^2}$$

Entrance Lengths

The reviewed formulae for the calculation of the friction factors in smooth and rough pipes are referred to as drag in pipes, sufficiently long to eliminate the effect of the entrance lengths. At the pipe entrance the velocity profile is gradually formed involving the changes of the resistance law at the entrance part. For the laminar flow, the entrance length is

$$L_e = (0.03 \sim 0.05)\,\mathrm{Re}\,d$$

while for the turbulent flow it amounts 25 to 40 diameters according to the experiments carried out by Nikuradse [1].

Pipes with Noncircular Cross Sections

In the case of pipes with noncircular cross sections the same equations can be applied as for the circular cross section if instead of the pipe diameter the hydraulic diameter is used. It is defined as follows:

$$d_h = \frac{4A}{P}$$

where A is the cross-sectional area and P the wetted perimeter. For the circular cross section the hydraulic diameter is equal to the diameter. This approach is applicable for compressible fluids up to the Mach number Ma = 1 [16].

Friction Factor for Non-Newtonian Fluids

For non-Newtonian fluids whose behavior can be described by the Ostwald-de Waele model (power-law), it is convinient to define the generalized Reynolds number [17, 18]:

$$Re' = 2^{3-n}\left(\frac{n}{3n+1}\right)\frac{d^n \rho w^{2-n}}{K} \tag{7}$$

Using the generalized Reynolds number, the friction factor can be calculated as follows: For the laminar flow:

$$f = \frac{a_n}{(Re')}$$

For the turbulent flow:

$$\frac{1}{\sqrt{f}} = \frac{4.0}{n^{0.75}} \log(Re' f^{1-0.5n}) - \frac{0.4}{n^{1.2}}$$

Friction factor dependence on Re′ and n can be given in an explicit form too, similar to the Blasius equation:

$$f = \frac{a_n}{(Re')^{b_n}}$$

where a_n and b_n are the functions of n [17].

Frictional Resistance in Drag Reducing Fluids

Polymer drag reduction, an effect that shows itself in reduced flow friction when a trace of high-polymer substances are present in the flow, has been extensively studied in the literature [19–22]. It was found that the friction factor in turbulent pipe flow of dilute polymer solutions is bounded between two universal asymptotes, namely, the Prandtl-Karman law (Equation 4) and a maximum drag reduction asymptote:

$$\frac{1}{\sqrt{f}} = 19.0 \log(Re\sqrt{f}) = 32.4$$

Between these limits there is a polymeric regime in which the observed friction factor relations are approximately linear on Prandtl-Karman coordinates and can be characterized by two parameters: the wall shear stress at the onset of drag reduction and the slope increment by which the polymer solution slope exceeds Newtonian:

$$\frac{1}{\sqrt{f}} = (4.0 + \alpha) \log(Re\sqrt{f}) - 0.4 - \alpha \log(\sqrt{2}\,\alpha\beta)$$

where α and β are polymer solution parameters [21].

Frictional Resistance on a Plane Surface

On a plane surface parallel to the direction of the undisturbed flow, the flow is decelerated near the surface because of the fluid friction. The resulting velocity gradients are due only to that friction.

The boundary layer commences at the leading edge of the surface and grows thicker in a downstream direction. Within the first part of the boundary layer the flow is entirely laminar. After some distance the fluid motion becomes unsteady, and this short regime is called the transition boundary layer. At all places downstream of the transition area the flow in the boundary layer is turbulent. It has been observed that the transition between laminar and turbulent regime occurs for Reynolds numbers [23]:

$$Re_x = \frac{xU}{\nu} = 5 \times 10^5 \sim 2 \times 10^6$$

As the thickness of the boundary layer, both the laminar and the turbulent one, increases with the distance from the leading edge, the local coefficient of friction C_f', defined similar to the mean one, according to the Equation 1:

$$C_f' = \frac{\tau_w}{\frac{1}{2}\rho U^2}$$

will be the function of distance x.

For the laminar boundary layer the local friction coefficient can be calculated from

$$C_f' = \frac{0.332}{\sqrt{Re_x}}$$

and the mean coefficient of friction for the entire surface of length L

$$C_f = \frac{1.328}{\sqrt{UL/\nu}} = \frac{1.328}{\sqrt{Re_L}}$$

It can be observed that the drag coefficient changes with the root of the distance. This means that the initial parts of the boundary layer, where it is thinnest, proved to have the greatest effect on the drag. The above mentioned equations are valid up to Re_x, i.e., Re_L number of about $5 \times 10^5 \sim 2 \times 10^6$. For greater distances or higher velocities, the drag coefficient is calculated from the equations derived for turbulent boundary layer. For drag coefficients the following equations are usually used:

$$C_f' = 0.0592\, Re_x^{-1/5}$$

$$C_f = 0.074\, Re_L^{-1/5}$$

These equations are derived under the presumption of the same velocity profile as in the flow through pipes, and after correction of the coefficients according to the experiments; they can be used up to $Re = 10^7$ [24]. In a wider range of Reynolds number it is possible to use [25]:

$$C_f' = 0.370(\log Re_x)^{-2.584}$$

$$C_f = 0.427(\log Re_L - 0.407)^{-2.64}$$

or in a very close approximation [23]:

$$C_f = 0.455(\log Re_L)^{-2.58}$$

Rough Surfaces

Similary to the description given for the frictional resistance in pipes, surface roughness affects friction coefficients on the flat plate. Figure 2 shows the friction coefficient C_f in the function of

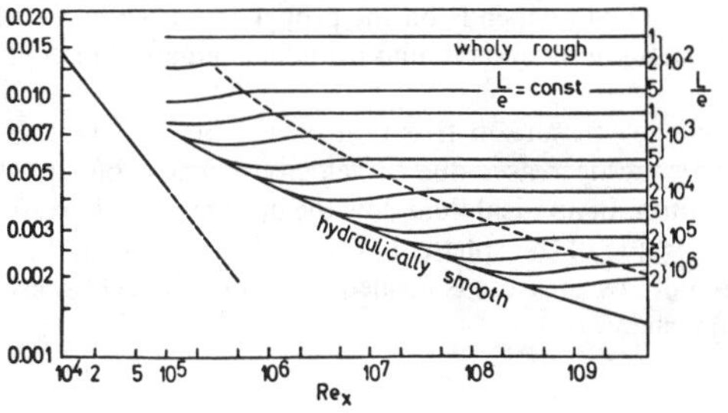

Figure 2. Frictional coefficient for a flat plate as a function of Reynolds number and relative roughness.

Re_L-number of different values of dimensionless roughness L/e. This diagram resembles the Moody diagram for pipes (Figure 1), and similarly to that the local friction coefficient C_f' can be given as a function of Re_x and x/e. For smooth surfaces C_f is controlled by the viscosity of the fluid, through its influence on the thickness of the laminar sublayer. In the case of the rough boundary the roughness size alone controls C_f, because it determines the size of the eddies thrown off in the wake of each individual roughness element. Between the true smooth and rough boundary cases there is an intermediate stage, when the laminar sublayer is about the same thickness as the height of the roughness; C_f depends on both Re_L and L/e.

For wholly rough surfaces the following equations are valid:

$$C_f' = \left(2.87 + 1.58 \log \frac{x}{e}\right)^{-2.5}$$

$$C_f = \left(1.89 + 1.62 \log \frac{L}{e}\right)^{-2.5}$$

End Effects

On the plates of finite widths the boundary layer formed is three-dimensional while on the sides there appear secondary flows, which remarkably increases the values of the local drag coefficients closer to the side edges. The above carried analysis can be applied in good approximation only when the width of the plate is sufficiently large to enable the end effects to be neglected.

Effects of drag increase, appear also on the bottom of rectangular channel and in some other cases with two or more plates parallel to the flow stream but at a particular angle to one another.

RESISTANCE OF BODIES MOVING THROUGH A FLUID

Form Drag and Skin Friction

In the case of fluid flow along the flat plate or any other solid surface, the boundary layer is formed in the way already described. The thickness of this layer increases with the distance. If along the surface there is the adverse pressure gradient, the phenomenon called boundary-layer separation will appear. Fluid layers close to the solid boundary have extremely small speeds, whereas their energy and momentum (in contrast to the layers far from the boundary which are less decelerated) may be insufficient to force their way through the system for very long, against an adverse pressure gradient. They are then brought to rest in the point that is called the separation point. Farther on, a slow back flow in the vicinity of the wall in the direction of the pressure gradient may set in. At the separation point boundary layer leaves the wall and makes its appearance in the external flow. Separation of flow can never occur in the region of negative pressure gradient (accelerated flow), but only in the region of positive pressure gradient (decelerated flow). Whether the separation will occur or not, in a region of positive pressure gradient, depends on the properties of the particular flow. It occurs primarily for blunt bodies, such as spheres and circular cylinders, but also in a highly divergent ducts.

Separation of the boundary layer changes the pressure distribution on the body surface. For example, on the front side of the sphere, pressure decreases due to velocity increase, but on the back side it would increase due to the deceleration. In an ideal fluid flow the distribution of pressure would be the same both on the front and back side of the sphere. However, in a real fluid, due to the separation of the boundary layer, the area of low pressure is formed behind the object, whereas form drag is the result of these pressure differences:

$$F_p = \oiint p \cos \psi \, dA$$

in which the integration is carried out over the entire surface, and ψ is the angle between the direction of flow and the normal to the surface element.

The total drag F is the sum of the form drag and frictional resistance termed also as the skin friction F_f:

$$F_f = \oiint \tau_w \sin \psi \, dA$$

The participation of these two drags in the total drag, depend on the flow conditions and the geometry of the system. For a flat plate and streamlined bodies separation, if it occurs at all, will be very near the rear of the body, and the form drag is very small. For bluff obstacles the skin friction is small compared to the form drag, except for very small Reynolds numbers when the separation does not occur. If the position of separation does not change very much, the form drag will be nearly constant. This is also the case with sharp edge bodies, such as a disk perpendicular to the flow direction, where separation must take place at the sharp edge. On the contrary, significant change of the position of the point at which separation of the boundary layer takes place, involves the change of the area of low pressure which in turn remarkably changes the value of form drag.

Drag of a Sphere

The interest in the behavior of a sphere moving through a fluid goes back many years, and has been the subject of numerous experimental and theoretical studies. Many diverse techniques have been used to make these measurements: freely falling spheres in liquid and air, atmospheric and variable-density wind tunnels, towed spheres in water and air, a sting-mounted sphere on an aircraft and the aeroballistic range. Many analyses and reviews of these data have been made [26, 27] and Hoerner [28] derived a "standard drag curve" for subsonic speeds (Figure 3). Recently, Bailey [27] suggested modifications to the generally accepted standard drag curve on the basis of the analysis of the experimental factors that affected the value of drag coefficient. This modification is based on the results published by several authors [29–36] which, contrary to the other results, were not in-

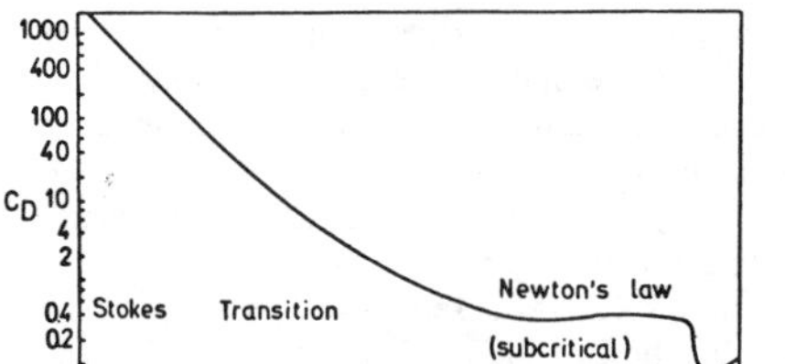

Figure 3. Standard drag curve for sphere.

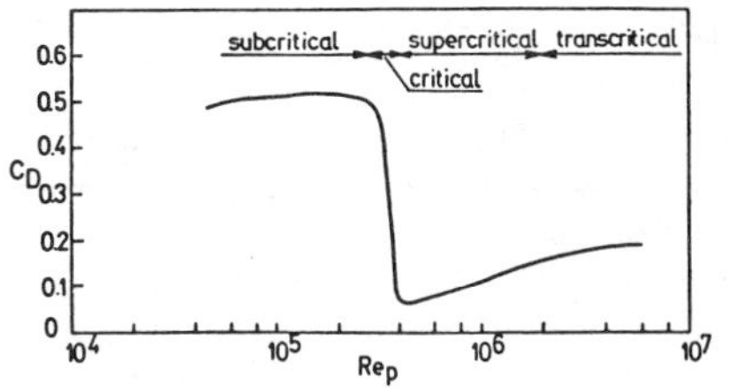

Figure 4. Drag curve for sphere at high Reynolds numbers.

fluenced by factors such as translational oscillations of the sphere, flow turbulence in the fluid, acceleration effects, model mounting technique, nonlinear trajectory of a sphere, etc.

Changes of drag coefficient defined by Equation 1 proved to be different for different ranges of Reynolds numbers, as shown in Figures 3 and 4, being the result of different flow regimes. (Here, the Reynolds number is defined in relation to the sphere diameter and the speed of undisturbed flow.)

At Reynolds numbers $Re_p < 0.1$ (Stokes law region) the drag coefficient falls linearly on a log–log plot with Reynolds number. The flow in this region has up- and down-stream symmetry. On the assumption that the inertia terms in the equation of motion of a viscous fluid can be neglected in comparison with the terms involving the viscosity, Stokes (1851) obtained the following expression for the drag coefficient:

$$C_D = \frac{24}{Re_p}$$

which corresponds exactly to experience up to a Reynolds number of 0.1.

Neglection of the inertia terms becomes increasingly important at Reynolds number above 0.1, and attempts have been made to deduce a more general relation for the drag coefficient of a sphere at small Reynolds numbers. Oseen (1910) pointed out that the missing inertia terms are important only in the region of the field approaching uniform flow, and suggested a new approximation in the linear form [26]. In 1929, Goldstein succeeded in providing a complete solution for the Oseen approximation, and the following drag coefficient expression resulted [37]:

$$C_D = \frac{24}{Re}\left(1 + \frac{3}{16} Re_p - \frac{19}{1{,}280} Re_p^2 + \frac{71}{20{,}480} Re_p^3 - \frac{30{,}179}{34{,}406{,}400} Re_p^4 + \frac{122{,}519}{550{,}502{,}400} Re_p^5 - \cdots\right)$$

The first term in the equation is the value obtained by Stokes, while the first two terms correspond to Oseen approximation.

An increase of the Reynolds number initates the separation process, deforms the streamline pattern and there appears the stationary vortex ring imbedded in the boundary layer at the rear of the sphere. The ring grows in size as the Reynolds number is increased, but all the vorticity generated is balanced by that which diffuses downstream from it.

As the Reynolds number further increases, the fixed ring vortex grows in size and decreases in stability and at Reynolds number of about 130 its downstream portion beings to oscillate. The oscillations become more severe until at Reynolds number of approximately 500 it detaches from the fluid system adhering to the body of the sphere and streams away at a velocity that is less than the fluid relative velocity [38].

A detailed review of the experimental and theoretical work dealing with the wakes of spheres is given by Torobin and Gauvin [38], and we will review as abridged as possible the most important differences in various Reynolds number regions related to the changes of drag coefficient values.

There were a few attempts to describe, by means of approximate equations, the dependence of drag coefficient on Reynolds number, beyond the range of creeping flow. Kaskas has derived an expression for Reynolds numbers smaller than the critical one [39]:

$$C_D = \frac{24}{Re_p} + \frac{4}{Re_p^{1/2}} + 0.4$$

Ihme et al. [40] have given the similar but slightly more precise dependance:

$$C_D = \frac{24}{Re_p} + \frac{5.48}{Re_p^{0.573}} + 0.36$$

This dependance has been proved by theoretical calculations for the range $0 \leq Re_p \leq 80$ and is in accordance with experiments up to $Re_p = 10^4$. Theoretical solutions are obtained by solving the differential equations, taking into account inertia terms. The field of streamlines, in the range $Re_p >$ 24, shows vortices behind the sphere.

In the range of Reynolds number from about 10^3 up to 2×10^5 (Figure 3) the value of the C_D coefficient is hardly changed with the Reynolds number. In the mentioned range Newton's law indicates its close validity, which was derived according to experiments in 1719 and which presumed the constant value of drag coefficient. It amounts 0.44 for this area.

In the vicinity of $Re_p = 2 \times 10^5$ the boundary layer attached to the body becomes unstable and enters into a transitional state tending toward turbulence. At the critical Reynolds number of about 3×10^5 a transition to turbulent flow occurs, what results with delayed separation. The wake is smaller, and the form drag is reduced. The drag due to skin friction is increased, but its magnitude is too small compared with the form drag, and there is a sharp drop of C_D from about 0.50 to approximately 0.1 [27, 41].

The relative contribution of skin friction and form drag varies with the flow regime. The skin friction accounts for two-thirds of the total drag in Stokes region and its contribution steadily diminishes to a neglible value as the Reynolds number is increased to higher values.

At the critical Reynolds number, the separation point moves suddenly to the rear, from an angle of about 82° to nearly 120° [41, 42].

In the subcritical region the vortex separation occurs at a point and that point of vortex release rotates around the sphere with the vortex shedding frequency. Above the critical Reynolds number, no frequency could be detected in the range $3 \times 10^5 < Re_p < 5 \times 10^6$ [43].

According to these results and those obtained in experiments for local mass transfer coefficients and drag in the flow around a sphere in a coaxial cylindrical tube, Duduković [44] concluded that in that region random separation in great number of points starts, forming the wake of many small eddies.

With further increase of the Reynolds number beyond the critical range, C_D slowly increases again (supercritical flow) up to the second nearly constant value (transcritical flow regime). The transition from supercritical to transcritical flow is rather floating (about $Re_p = 1.5 \times 10^6$). The transcritical flow is characterized by the shifting of the transition point in the direction of the front stagnation point, and by the immediate transition from a laminar to a turbulent boundary layer in the front part of the sphere.

Effect of Mach Number

The standard drag curve (Figure 3) corresponds to incompressible fluid or compressible fluid with the Mach number smaller than 0.3, the limiting value up to which it can be assumed that compressibility has no remarkable effect on the law of drag. At higher Mach numbers the effect of compressibility becomes significant and the drag coefficient becomes the function of Re_p and Ma numbers.

Analyses of experimental results at higher Mach numbers [45–47] indicate the increase of C_D values with an increase of Mach number up to $Ma = 1.6 \sim 1.8$ where C_D reaches the maximum. The marked dip on the curve that corresponds to the dependance of C_D on Re number becomes smaller with the increase of Mach number and levels out as sonic velocity is approached. A complete summary of drag coefficient as a function of Reynolds and Mach number is given in the paper by Miller and Bailey [46] and contains the analyses of contemporary results and also the results of the 18th and 19th century cannon firings.

Effects of Fluid Turbulence

The curve in Figure 3 (standard drag curve) is intended to be applied only to the smooth non-rotating spheres, moving at a constant relative velocity in an infinite fluid free from any disturbances. When any of the mentioned disturbances proved to be present, the value of C_D coefficient is changed, whereas this change is sometimes drastic.

The turbulence intensity in the fluid affects drag, first provoking premature turbulence in the boundary layer, which in turn moves to the critical range toward lower Reynolds numbers. This phenomenon and the fact that the model mounting technique may cause premature turbulence in the boundary layer, i.e., an earlier decrease of C_D value, have been noticed in an array of experimental investigations [35, 41, 48–50].

Torobin and Gauvin [51, 52] have shown that fluid turbulence indicates even more remarkable effects on the particles moving with the fluid stream. They have shown that this effect is referred to the relative turbulence intensity $\sqrt{u'^2}/U_R$, where U_R is the relative velocity, as the relatively small fluctuation velocities in relation to the mean fluid velocity could be important if related to the relative one. Figure 5 shows the results for fixed and moving spheres for various turbulence intensities. For a moving sphere the critical range is transferred toward the lower Reynolds number even for two potencies (about hundred times).

Effects of Acceleration, Rotation, and Roughness

Experiments for spheres moving in water or in air have shown that acceleration affects the increase of drag coefficient [53]. It is however, necessary to distinguish the relative acceleration of particles in relation to the fluid, which is decisive, from the acceleration in relation to the reference system.

There may appear two types of rotation of a sphere moving through the fluid: "screw motion"—the rotation with the axis parallel to the flow; and the "top spin"—with the axis perpendicular to the flow direction. In general, rotation appears to have little effect on the drag coefficient at low Reynolds numbers, but influences the linearity of the motion, thus affecting the accuracy of experiments, or the particle entrainment in conveyed solid–gas systems [54]. In the critical region the top spin causes an earlier laminar to turbulent transition, mainly by increasing the effective relative velocity between the surface of the sphere and the free stream.

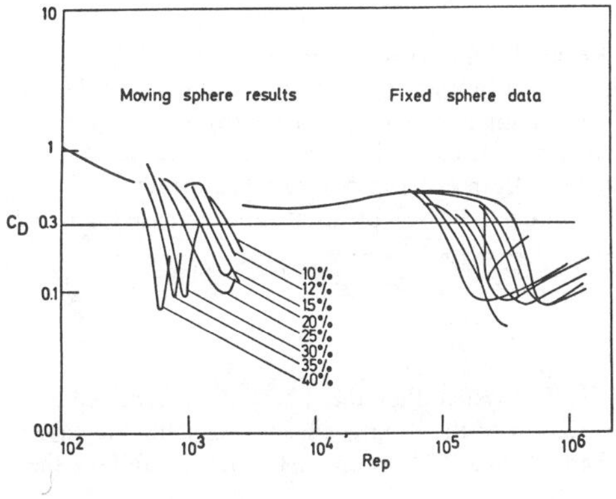

Figure 5. The effect of relative turbulence intensity on the drag coefficients of spheres [51].

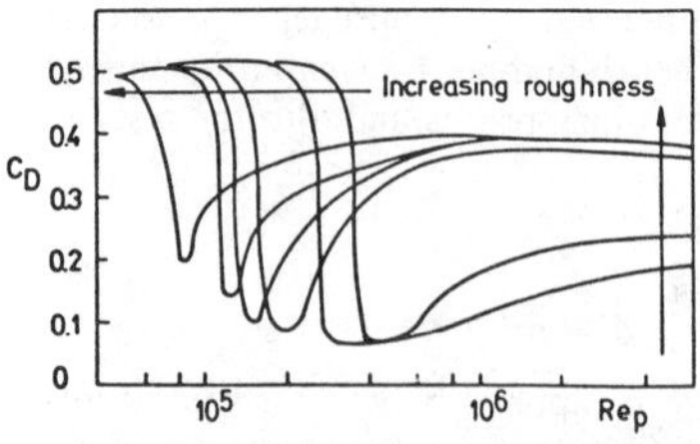

Figure 6. The effect of surface roughness on the drag coefficients of spheres [51].

Screw motion is of importance only with long slender bodies. The influence of this type of rotation diminishes as the form drag becomes predominant, so that for the sphere it causes only a slight increase of C_D at large values of the ratio of peripheral and mean velocity relative to a fluid.

Experiments with spheres of various surface roughness [55] have shown that a roughness increase will decrease the critical Reynolds number. At the same time the transcritical drag coefficient rises, having a maximum value of 0.4 as shown in Figure 6.

Effect of Blockage Ratio

In the Stokes range it is possible to conduct theoretical analyses of the effects of the wall of a coaxial tube, of a smooth surface, parallel or perpendicular to the moving direction of the sphere, or of another sphere moving close to the first one, on the drag coefficient for the sphere or spheroid [56–58].

For technical purposes the influence of the container wall vicinity is of special interest. The velocity of a free falling sphere in a circular tube has been investigated by many authors. After Ladenburg [59] and Faxen [60] the drag coefficient for a sphere falling in a tube whose diameter is d is given by:

$$C_D = \frac{24}{Re_p}\left(1 + 2.1\frac{dp}{d}\right)$$

In the range for which Newton's law is closely valid, i.e., where the drag coefficient is not dependant on Reynolds number, Duduković and Končar-Djurdjević [61] have correlated the results for spheres and discs (perpendicular to the flow direction) coaxially located in the tube, by means of a common equation that is supposed to be valid for the range of Re_p number between 2×10^3 and 2×10^4 for objects of various sphericity:

$$\frac{C_D}{C_{D\infty}} = (1 - d_p/d)^{-1.78}$$

where $C_{D\infty}$ is the drag coefficient at the same Reynolds number in an infinite fluid, and d_p/d represents the blockage ratio, i.e., the ratio of diameter of particle cross section perpendicular to the flow direction and tube diameter. These results differ at smaller blockage ratios from those given by Achenbach [55], probably due to different experimental conditions, i.e., fully developed velocity profile [61], and non-developed velocity profile [55]. Both experimental results indicate a relatively small increase of the drag coefficient up to $d_p/d \approx 0.5$, followed by an abrupt increase for higher d_p/d values.

The Drag of a Circular Cylinder

Since the cylinder represents a considerably simpler system than the sphere, it has received more theoretical and experimental attention. The variation of the drag coefficient with the increase of Reynolds number is similar to that of the sphere [49, 62, 63]. At small Reynolds numbers there isn't any separation of the boundary layer, whereas $C_D = f(Re_p)$ proved to be closely linear in the log–log coordinate system. If the Re_p number lies between about 2 and 40, the separation of the boundary layer takes place on each side of the cylinder, with two simultaneous eddies in the wake, which however, do not detach themselves periodically and do not create turbulence in the stream.

For Reynolds number greater than 40 separation occurs alternately on each side of the cylinder and the Karman's "vortex street" is formed. Although the cylinder is a two-dimensional analogue of the sphere, there is no simple extensions of the Karman's vortex street to three dimensions.

Up to the Reynolds number of about 150 (stable range) there is no laminar-turbulent instability present in the entire system. From $Re_p = 150 \sim 300$ (transition range) there is a laminar-turbulence instability, but this occurs in the separated boundary layer. Above 300 (irregular range) transition occurs at the separation point and the velocity fluctuations have both periodic and truly turbulent components [38]. In this range of Reynolds number C_D is almost constant. More precisely, the curve $C_D = f(Re_p)$ indicates the shallow minimum at about $Re = 1{,}800$ where C_D proved to be 0.95, and then increases up to about 1.2 for $Re_p \approx 3 \times 10^4 \sim 10^5$. At Reynolds number of about 5×10^5 the remarkable fall from 1.2 to about 0.36 has been registered. This is due to the appearance of turbulence in the boundary layer whereas the separation of the boundary layer appears at an angle of about 140° [64], resulting in a much narrower wake.

In the supercritical region the drag coefficient grows again. The transcritical range is reached when C_D is almost constant, and the angle of boundary layer separation is about 110° [64, 65].

The rhythmical and, therefore, non-turbulent component of the cyclic vortex phenomenon is usually described by the Strouhal number. The relationship between Strouhal and Reynolds number is much better established for cylinders than for spheres. In the subscritical regime the Strouhal number is almost constant ($Sr = 0.205$). In the critical flow regime quasi-regular flow fluctuations occur, and the Strouhal number abruptly increases up to about 0.5. The supercritical flow range ends at about $Re_p = 2 \times 10^6$, and the Strouhal number decreases step-wise down to a value of $Sr = 0.25$.

All of the complication factors described for the sphere that affect the drag, appear also in the case of a cylinder. Besides, there is also the appearance of end effects [64, 66], i.e., the effect of the relation of cylinder length and diameter, and for free-falling cylinders the problem of orientation.

As in the case of a sphere, an increase of the turbulence intensity in the fluid or an increase of roughness will transfer critical range toward the lower Reynolds numbers. In the subcritical range, where the boundary layer is laminar throughout, roughness has no influence on the drag. The value of the critical Reynolds number decreases with increasing roughness and at the same time the critical Reynolds number range is shrinking. In the transcritical flow range, C_D increases with increasing relative roughness, as the consequence of the decreased angle of boundary separation from 110° to 90°.

Drag Coefficient of Particles with Different Shapes

Many investigators examined the drag of differently shaped bodies. Data of the drag on discs are given by Schmiedel [67], Wieselsberger [49], and others. In the range up to $Re_p = 0.5$ the drag is

$$C_D = \frac{64/\pi}{Re_p}$$

For higher Re_p values, around hundred, the disc begins to oscillate (at falling experiments) and the C_D values rises, reaching the maximum at about $Re_p = 280$. After this it sinks again, reaching the constant value for $Re_p > 2{,}000$.

In experiments with other geometrical and irregularly shaped objects, it has been shown that the shape and orientation have a strong influence on the drag coefficient [54, 68–70].

In experimental studies the settling velocities of isometric (tetrahedrons, cubes, cuboctahedrons, octahedrons, and spheres), nonisometric (cylinders, square prisms, double cones, and spheroids of different aspect ratios and settling orientations), and irregularly shaped particles were measured and the drag coefficients and Reynolds numbers were calculated. However, the details of the complex interactions between fluid dynamics and particle shape remain largely unquantified.

A correlation that can accurately predict the settling velocity according to the known fluid and shape parameters for both regularly and irregularly shaped particles has not yet been found. One of the major difficulties in obtaining empirical correlations for the drag coefficients has been the inability to systematically and quantitatively describe the shape itself. At least 13 different shape parameters have been used in analyzing the effect of shape on settling velocity. These parameters include sphericity, circularity, flatness factor, roundness factor, etc. The use of these parameters has given some empirical equations with the limited applicability. Recently, the new morphological descriptors have been applied for the estimation of drag coefficients of regularly shaped particles in slow flow, with the probable possibility of widening on other conditions too [71].

ANALYSIS AND DETERMINATION OF DRAG BY THE ANALOGY WITH MASS TRANSFER AND VICE VERSA

The analogies between momentum, heat, and mass transfer enables, according to the value of one of the three mentioned characteristics measured in the given system and under defined conditions, the calculation of the other two characteristic values under the same conditions. Analogies had been evaluated, first of all due to fundamental reasons but also due to the necessity to calculate heat and mass transfer coefficients from more easy measurable friction coefficients. Sometimes, however, particularly in the case of local transfer coefficients, an opposite procedure is used.

Direct analogies between momentum, heat, and mass transfer are tied to the friction resistence in the flow over surfaces straight (or slightly curved) toward the flow direction (flat plates, channels, tubes, etc.). However, in a wider sense, momentum and heat or mass transfer could be in a certain way connected even at more complex geometrical conditions. Heat and mass transfer are in close relation to the thickness of a diffusion sublayer, and over the corresponding Schmidt and turbulent Schmidt number (for mass transfer) or the molecular and turbulent Prandtl number (for heat transfer) they are connected to the thickness of the hydraulic boundary layer. The knowledge of the local coefficients of mass transfer along the surface may give the clear picture of flow around the body and the zones of boundary layer separation. As the drag, in the case of flow around the bodies, is related to the point of boundary layer separation, distribution of the local coefficients of mass transfer also gives the idea of drag value and the flow regime.

Experimental Methods for the Determination of Mass Transfer Coefficients

Several methods have been used for determinating the mass transfer coefficients between solid surfaces and a fluid stream. In the case of gases the sublimation method and the absorption or evaporation from the thin liquid films on the solid surfaces have been used. Experiments for liquids have been done by applying the method of solid surface segment dissolution, as well as the electrochemical and adsorption method. Most of these methods evaluate the mean mass transfer coefficients. If electrochemical method is used with small electrodes placed in the main one, and isolated from it, it is possible to obtain local coefficients, but only on the places where the accessory electrodes are placed [72]. However, the optical method for the measurement of liquid evaporation rates [73] and the adsorption method [74–76] relatively simply enable the determination of the local mass transfer coefficients in a gas or a liquid, respectively.

The adsorption method is based on the experimentally determined fact that under particular conditions and at short adsorption periods, mass transfer with adsorption can be treated as the stationary process governed by mass transfer only. Silica gel is usually used as the adsorbent, deposited in thin layers on aluminium foils, which are then used to cover particular places of interest, and methylene blue is used as the adsorbing species. The obtained colorations on silica gel are known as adsorption spectra and visually indicate, depending on the degree of coloration, the thickness of the boundary layer and the point of its separation. Measuring of the reflected light intensity from individual points of this surface gives the local coefficients of mass transfer using an appropriate mathematical relation. The benefit of this method is that by only a single experiment it is possible to obtain a spectrum of the mass transfer coefficients on the entire surface in a qualitative sense, and by measuring the reflected light at any point of interest even the corresponding quantitative values can be obtained.

Analogies Between Momentum and Mass (or Heat) Transfer

The Reynolds analogy, where the identical mechanism for transfer of momentum and heat (or mass) is postulated, is of historical importance, and valid only for $Pr = Sc = 1$.

Prandtl divided the boundary layer into two parts: laminar layer where transfer is carried only by the molecular mechanism, and the turbulent one where only the turbulent mechanism is involved. Karman extended this concept by inserting the transitional layer between them, where transfer by molecular and turbulent mechanism are of the same order of magnitude. Finally, Levich inserted the diffusion sublayer within the laminar one, due to the fact that in liquid systems coefficient of molecular diffusion proved to be a thousand times smaller then the viscosity coefficient. Therefore, the turbulent pulsations penetrating to the laminar layer, due to their small intensities, contribute diminishingly to the momentum transfer; but for mass transfer the turbulent mechanism prevails up to the boundary of the diffusion sublayer [77]. All the mentioned concepts have been followed by the corresponding mathematical equations representing the analogies between momentum, heat, and mass transfer.

The Reichardt analogy [78] could be applied to mass transfer as follows:

$$\mathrm{Sh} = \frac{\mathrm{Re}\dfrac{f}{2}\dfrac{\mathrm{Sc}}{\mathrm{Sc_t}}}{\phi_m + \dfrac{\mathrm{Sc}}{\mathrm{Sc_t}}\left(\dfrac{f}{2}\right)^{1/2}\displaystyle\int_0^{w/w^*} \frac{dw^+}{1 + \dfrac{\mathrm{Sc}}{\mathrm{Sc_t}}\dfrac{\epsilon}{\nu}}}$$

From this equation several analogies for Newtonian, non-Newtonian, and drag reducing fluids were derived. Friend and Metzner [79] have experimentally obtained the value of the integral as $11.8\,\mathrm{Sc}^{-1/3}$ and for $\phi_m = 1.2$.

All of the mentioned analogies, as well as those we haven't mentioned, may be under particular conditions used for the calculation of mass (or heat) transfer coefficients from friction factors data and vice versa. In the case of mean coefficient determinations the frist procedure is usually applied. However, in the case of three-dimensional boundary layers [80] and any determination of local transfer coefficients, the experimental determination of local mass transfer coefficients, and by means of them the local drag coefficient, may represent a more convenient way.

Drag Analyses According to the Distribution of Local Mass Transfer Coefficients

The distribution of values of local mass transfer coefficients along the solid surface gives a clear idea of the boundary layer thickness and, particularly, of the places of its separation. When the position of boundary layer separation area is changed, it is obvious that the flow regime is changed, involving significant changes of drag coefficient, as already described. Such an approach to the qualitative investigation of the complex stream fields and drag proved to be extremely convinient for the determination of mutual effects between two objects, wall and object or some other even more complicated situations [76, 81, 82]. Applying the adsorption method, for example, according to the position of white areas (separation zones) it is easily possible and quite efficient to investigate what kind of changes in the geometry of the system ought to be carried out in order to reduce the drag and gain the effects of energy savings. Examples of such adsorption spectra are given in Figures 7 and 8 for cylinder and cone in a cross flow.

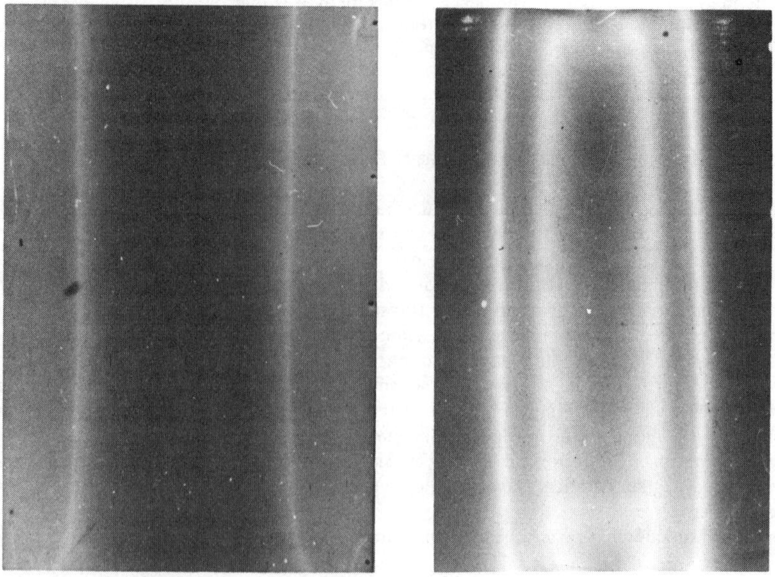

Figure 7. Adsorption spectra on the cover of a cylinder in cross flow: (a) laminar flow (front stagnation line is in the middle), (b) turbulent flow (front stagnation line is at the edges). White areas are the zones of the boundary layer separation [83].

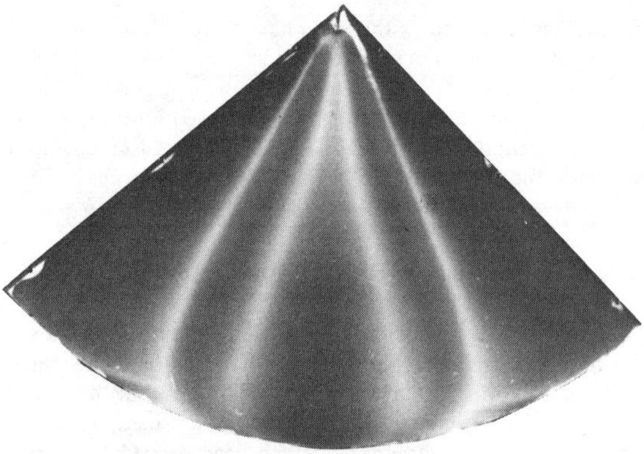

Figure 8. Adsorption spectra on the cover of a cone in turbulent flow perpendicular to the axis. White areas are the zones of the boundary layer separation (front stagnation line is at the edges) [83].

NOTATION

A	area, cross-sectional area	Pr	Prandtl number ($= \nu/\alpha$)
C_D	coefficient of drag	Re	Reynolds number ($= wd/\nu$)
C_f	coefficient of friction	Re′	generalized Reynolds number, Equation 7
C_f'	local coefficient of friction	Re_L	mean Reynolds number of a plate of length L ($= UL/\nu$)
c	velocity of sound	Re_p	particle Reynolds number ($= Ud_p/\nu$)
D	molecular diffusion coefficient	Re_x	local Reynolds number on a flat plate at a distance x from a leading edge ($= Ux/\nu$)
d	tube diameter	Sc	Schmidt number ($= \nu/D$)
d_h	hydraulic diameter	Sc_t	turbulent Schmidt number ($= \epsilon/\epsilon_m$)
d_p	diameter of sphere or cylinder	Sr	Strouhal number ($= Nd_p/U$)
e	roughness	U	velocity of undisturbed flow
F	drag force	U_R	mean fluid velocity relative to the particle
f	friction factor	u′	fluctuating component of velocity in the flow direction
h_f	head loss	w	velocity, mean velocity
K	flow consistency index	w*	friction velocity ($= \tau_w/\rho$)
L	length, of a plate	w^+	dimensionless velocity ($= w/w^*$)
L_e	entrance length	x	distance in the flow direction
Ma	Mach number ($= w/c$)		
N	shedding frequency of wake		
n	flow behavior index		
P	wetted perimeter		

Greek Symbols

α	thermal diffusivity	λ	alternative friction factor ($= 4f$)
ϵ	eddy viscosity	ν	kinematic viscosity
ϵ_m	eddy mass diffusivity	ρ	fluid density
ψ	angle between the direction of flow and the normal to the surface element	τ_w	shear stress at the wall

REFERENCES

1. Nikuradse, J., "Gesetzmässigkeit der turbulent Strömung in glatten Rohren," *Forschungsheft* Arb. Ing.-W es. No. 356 (1932).
2. Nikuradse, J., "Strömungsgesetze in rauhen Rohren," *Forschungsheft Arb. Ing.-Wes.*, No. 361 (1933).
3. Blasius, H., "Das Ahnlichkeitsgesetz bei Reibungsvorgängen in Flüssigkeiten, *Forschungsheft Arb. Ing.-West.*," No. 131 (1913).
4. Saph, V., and Schoder, E. H., "An Experimental Study of the Resistance to the Flow of Water in Pipes," *Trans. Amer. Soc. Civ. Engr.*, 51, 944 (1903).
5. Nusselt, W., "Wärmeübergang in Rohrleitungen," *Forschungsheft Arb. Ing.-Wes.*, No. 89 (1910).
6. Ombeck, H., "Druckverlust strömender Luft in geraden zylindrischen Rohrleitungen," *Forschungsheft Arb. Ing.-Wes.*, No. 158/159 (1914).
7. Stanton, T. E., "The Mechanical Viscosity of Fluids," *Proc. Roy. Soc. London*, A85, 366 (1911).
8. Stanton, T. E., and Pannel, J. R., Similarity of Motion in Relation of the Surface Friction of Fluids," *Phil. Trans. Roy. Soc.*, A214, 199 (1914).
9. von Karman, T., "Turbulence and Skin Friction," *J. Aeronaut. Sci.*, 1, 1 (1936).
10. Drew, T. B., and Genereaux, R. P., *Trans. Amer. Chem. Engrs.*, 32, 17 (1936).
11. Colebrook, C. F., "Turbulent Flow in Pipes with Particular Reference to the Transition Region Between the Smooth and Rough Pipe Laws," *J. Inst. Civil Engrs.* (London), 11 133 (1939).
12. Moody, L. F., "Friction Factors for Pipe Flow," *Trans. ASME*, 66, 671 (1944).
13. Round, G. F., "An Explicit Approximation for the Friction Factor—Reynolds Number Relation for Rough and Smooth Pipes," *Can. J. Chem. Eng.*, 58, 122 (1980).

14. Streeter, V. L., (ed.), *Handbook of Fluid Dynamics*, first edition, McGraw-Hill, New York (1961), p. 3–16.
15. von Karman, T., *Mechanische Ähnlichkeit und Turbulenz*, Nachr. Ges Wiss. Götingen, Math.-Physik. Klasse 58 (1930).
16. Schlichting, H., *Grenzschichttheorie*, Braun, Karlsruhe, Russian translation by Volpert, T. A., *Nauka*, Moscow (1974), pp. 551–553.
17. Dodge, D. W. and Metzner, A. B., "Turbulent Flow of Non-Newtonian Systems," *AIChE J.*, 5, 189 (1959).
18. McCabe, W. L., and Smith, J. C., *Unit Operations of Chemical Engineering*, 2nd edition, McGraw-Hill, New York (1967), pp. 106–108.
19. Hoyt, J. W., "Recent Progress in Polymer Drag Reduction," *Polymeres et Lubrification*, No. 233 (1974).
20. Little, R. C., Hansen, R. J., Hunston, D. L., Kim, O. K., Patterson, R. L., and Ting, R. Y., "The Drag Reduction Phenomenon: Observed Characteristics, Improved Agents, and Proposed Mechanisms," *Ind. Eng. Chem. Fundamentals*, 14 (4), 283 (1975).
21. Virk, P. S., "Drag Reduction Fundamentals," *AIChE J.*, 21, 625 (1975).
22. Kumor, S. M., and Sylvester, N. D., "Effects of a Drag-Reducing Polymer on the Turbulent Boundary Layer," *AIChE Symp. Ser.*, 69, No. 130 (1973).
23. Francis, J. R. D., *Fluid Mechanics for Engineering Students*," 4th edition, Edward Arnold, London (1976), pp. 174–208.
24. Schlichting, *Grenzschichttheorie*, pp. 571–600.
25. Schultz-Grunow, F. "Neues Widerstandsgesetz für glatte Platten," *Luftfahrtforschung*, 17, 239 (1940).
26. Torobin, L. B., and Gauvin, W. H., "Fundamental Aspects of Solid-Gas Flow, Part I: Introductory Concepts and Idealized Sphere Motion in Viscous Regime," *Can. J. Chem. Eng.*, 37, 129 (1959).
27. Bailey, A. B., "Sphere Drag Coefficient for Subsonic Speeds in Continuum and Free-Molecule Flows," *J. Fluid Mech.*, 65 (2), 401 (1974).
28. Hoerner, S. F., *Fluid-Dynamic Drag*, S. Hoerner, Midland Parka, New Jersey (1958), pp. 3–8.
29. Arnold, H. D., "Limitations Imposed by Slip and Inertia Terms Upon Stokes Law for the Motion of Spheres Through Liquids," *Phil. Mag.*, 22, 755 (1911).
30. Liebster, M., "Über den Widerstand von Kugeln," *Ann. Phys.*, 83, 541 (1927).
31. Schmiedel, J., "Experimentelle Unterschungen über die Fallbewegung von Kugeln und Scheiben in reibenden Flüssigkeiten," *Phys. Z.*, 29, 593 (1928).
32. Maxworthy, T., "Accurate Measurements of Sphere Drag at Low Reynolds Numbers," *J. Fluid Mech.*, 23, 369 (1965).
33. Shakespear, G. A., "Experiments on the Resistance of the Air to Falling Spheres," *Phil. Mag.*, 28 (6), 728 (1914).
34. Vlajinac, M., and Covert, E. E., "Sting-Free Measurements of Sphere Drag in Laminar Flow," *J. Fluid Mech.*, 54, 385 (1972).
35. Bacon, D. L., and Reid, E. B., "The Resistance of Spheres in Wind Tunnels and in Air," *NACA Rep.* No. 185 (1924).
36. Bailey, A. B., and Hiatt, J., "Free-flight Measurements of Sphere Drag at Subsonic, Transonic, Supersonic, and Hypersonic Speeds for Continuum, Transition and Near-Free-Molecular Flow Conditions," Arnold Engng. Develop. Center Rep. AEDC-TR-70-291 (1971).
37. Goldstein, S., (ed.) *Modern Developments in Fluid Dynamics*, Dover publications, New York (1965), pp. 491–505.
38. Torobin, L. B., and Gauvin, W. H., "Fundamental Aspects of Solid–Gas Flow. Part II: The Sphere Wake in Steady Laminar Fluids," *Can. J. Chem. Eng.*, 37, 167 (1959).
39. Brauer, H., *Grundlagen der Einsphasen- und Merphasenströmungen*, Sauerländer A. G., Aarau (Switzerland) (1971), p. 200.
40. Ihme, F., Schmidt-Traub, H., and Brauer, H., "Theoretische Untersuchung ůber die Umstrǒmung und den Stoffübergang an Kugeln," *Chemie-Ing.-Techn.*, 44 (5), 306 (1972).
41. Achenbach, E., "Experiments on the Flow Past Spheres at Very High Reynolds Numbers," *J. Fluid Mech.*, 51, 565 (1972).

42. Raithby, G. D., and Eckert, E. R. G., "The Effect of Support Position and Turbulence Intensity on the Flow Near the Surface of a Sphere." *Wärme- und Stoffübertragung*, 1, 87 (1968).
43. Achenbach, E., "Vortex Shedding from Spheres," *J. Fluid Mech.*, 62, 209 (1974).
44. Dudukovic, A. P., "The Effect of Dissolved Polymers on Mass Transfer and Fluid Flow," D. Sc. Thesis, Faculty of Technology and Metallurgy, University of Beograd, Beograd (1983).
45. Bailey, A. B., and Starr, R. F., "Sphere Drag at Transonic Speeds and High Reynolds Number," *A.I.A.A.J.*, 14, 1631 (1976).
46. Miller, D. G., and Bailey, A. B., "Sphere Drag at Mach Numbers from 0.3 to 2.0 at Reynolds numbers Approaching 10^7," *J. Fluid Mech.*, 93 (3), 449 (1979).
47. Miller, D. G., "Ballistic Tables for Spheres 7.5 to 25 mm (0.3 to 1 in) in Diameter," UCLR Lawrence Livermore Lab. Rep. UCLR-52755, Livermore (1979).
48. Lunnon, R. G., "Fluid Resistance to Moving Spheres," *Proc. Roy. Soc.*, A110, 302 (1926).
49. Wieselsberger, C., "Weitere Feststellungen über die Gesetze des Flüssigkeits- und Luftwiderstandes," *Physik. Z.*, 23, 219 (1922).
50. Flachsbart, O., "Neue Untersuchungen über den Luftwiderstand von Kugeln," *Physik.* Z., 28, 461 (1927).
51. Torobin, L. B., and Gauvin, W. H., "The Drag Coefficients of Single Spheres Moving in Steady and Accelerated Motion in a Turbulent Fluid," *AIChE J.*, 7, 615 (1961).
52. Torobin, L. B., and Gauvin, W. H., "Fundamental Aspects of Solid–Gas Flow. Part V: The Effects of Fluid Turbulence on the Particle Drag Coefficient," *Can. J. Chem. Eng.*, 38, 189 (1960).
53. Torobin, L. B., and Gauvin, W. H., "Fundamental Aspects of Solid–Gas Flow, Part III: Accelerated Motion of a Particle in a Fluid," *Can. J. Chem. Eng.*, 37, 224 (1959).
54. Torobin, L. B., and Gauvin, W. H., "Fundamental Aspects of Solid–Gas Flow. Part IV: The Effects of Particle Rotation, Roughness, and Shape," *Can. J. Chem. Eng.*, 38, 142 (1960).
55. Achenbach, E., "The Effects of Surface Roughness and Tunnel Blockage on the Flow Past Spheres," *J. Fluid Mech.*, 65 (1), 113 (1974).
56. Happel, J., and Brenner, H., *Low Reynolds Number Hydrodynamics*, Prentice-Hall (1965), Russian translation by B. Berman and B. Markov, *Mir*, Moscow (1976), pp. 329–409.
57. Cox, R. G., and Brenner, H., "Effect of Finite Boundaries on the Stokes Resistance of an Aribitrary Particle, Part 3. Translation and Rotation," *J. Fluid Mech.*, 28, 391 (1967).
58. Majumdar, S. R., and O'Neill, M. E., "On the Stokes Resistance of Two Equal Spheres in Contact in a Linear Shear Field," *Chem. Eng. Sci.*, 27, 2017 (1972).
59. Ladenburg, R., "Über den Einflus von Wänden auf die Bewegung einer Kugel in einer reibenden Flussigkeit," *Ann. Physik* 4, E23, 447 (1907).
60. Faxen, H., "Der Widerstand gegen die Bewegung einer starren Kugel in einer zähen Flüssigkeit, die zwischen zwei parallelen ebenen Wänden eingeschlossen ist," *Ann. Physik* 4, F68, 89 (1922).
61. Duduković, A. P., and Končar-Djurdjević, S., "The Effect of Tube Walls on Drag Coefficients of Coaxially Placed Objects," *AIChE J.*, 27 (5), 837 (1981).
62. Finn, R. K., "Determination of the Drag on a Cylinder at Low Reynolds Numbers," *J. Appl. Physics*, 24, 771 (1953).
63. Roshko, A., "Experiments on the Flow Past a Circular Cylinder at Very High Reynolds Number," *J. Fluid Mech.*, 10, 245 (1961).
64. Achenbach, E., and Heinecke, E., "On Vortex Shedding from Smooth and Rough Cylinders in the Range of Reynolds Numbers 6×10^3 to 5×10^6," *J. Fluid Mech.*, 109, 239 (1981).
65. Achenbach, E., "Influence of Surface Roughness on the Cross-Flow Around a Circular Cylinder," *J. Fluid Mech.*, 46, 321 (1971).
66. Slaouti, A., and Gerrard, J. H., "An Experimental Investigation of the End Effects on the Wake of a Circular Cylinder Towed Through Water at Low Reynolds Numbers," *J. Fluid Mech.*, 112, 297 (1981).
67. Schmiedel, J., "Experimentelle Untersuchung über die Fallbewegung von Kugeln und Scheben in reibenden Flüssigkeiten," *Physik. Z.*, 29, 593 (1928).
68. Pettyjohn, E. S., and Christiansen, E. B., "Effect of Particle Shape on Free Settling Rates of Isometric Particles," *Chem. Eng. Progr.*, 44, 157 (1948).
69. Christiansen, E. B., and Barker, D. H., "The Effect of Shape and Density on the Free Settling of Particles at High Reynolds Numbers," *AIChE J.*, 11, 145 (1965).

70. Becker, H. A., "The Effects of Shape and Reynolds Number on Drag in the Motion of a freely Oriented Body in an Infinite Fluid," *Can. J. Chem. Eng.*, 37, 85 (1959).
71. Carmichael, G. R., "Estimation of the Drag Coefficient of Regularly Shaped Particles in Slow Flows from Morphological Descriptors," *Ind. Eng. Chem. Process Des. Dev.*, 21 (3), 401 (1982).
72. Van Shaw, P., and Hanratty, T. J., "Fluctuations in the Local Rate of Turbulent Mass Transfer to a Pipe Wall," *AIChE J.*, 10, 475 (1964).
73. Utton, D. B., and Sheppard, M.A., "The Determination of Convective Heat Transfer by the Measurement of Mass Transfer Using a New Optical Technique," XVICHMT Symposium, Heat and Mass Transfer Measurement Techniques, Dubrovnik (1983).
74. Končar-Djurdjević, S., "Application of a New Adsorption Method in the Study of Flow of Fluids," *Nature*, 172, 858 (1953).
75. Končar-Djurdjeviće S., "Eine neue allgemeine Method zur Untersuchung der Wirkungsweise chemisher Apparaturen," *Dechema-Monographien*, 26, 139 (1956).
76. Končar-Djurdjević, S., and Duduković, A. P., "The Effect of Single Stationary Objects Placed in the Fluid Stream on Mass Transfer Rates to the Tube Walls," *AIChE J.*, 23, 125 (1977).
77. Levich, V. G., *Fiziko-himicheskaya gidrodynamica* (Physicochemical Hydrodynamics), Fiz. Mat. Giz., Moscow (1959), pp. 143–178.
78. Reichardt, H., "Der Einfluss der wandnahen Strömung auf den turbulenten Wärmeübergang," *Mitt, Max-Planck-Institut für Strömungfochung*, 3, 1 (1950).
79. Friend, W. L., and Metzner, A. B., "Turbulent Heat Transfer Inside Tubes and the Analogy Among Heat, Mass, and Momentum Transfer," *AIChE J.*, 4 393 (1958).
80. Vuković, O., Končar-Djurdjević, S., and Mitrović, M., "The Use of the Adsorption Method in the Study of Mass Transfer Through Three Dimensional Boundary Layers on Flat Plates," *Bull. Soc. Chim. Beograd*, 37, 297 (1972).
81. Duduković, A. P., and Končar-Djurdjević, S., "The Effect of the Array of Disks on Mass Transfer Rates to the Tube Walls," *AIChE J.*, 25, 895 (1979).
82. Duduković, and Končar-Djurdjević, S., "Effect of an Array of Objects on Mass Transfer Rates to the Tube Wall: An additional Note," *AIChE J.*, 26, 299 (1980).
83. Končar-Djurdjević, S., et al., Unpublished results.

CHAPTER 14

FREE STREAM TURBULENCE AND BLUFF BODY DRAG

W. H. Bell

Institute of Ocean Sciences
Sidney, B. C., Canada

CONTENTS

INTRODUCTION

The use of the Reynolds number as a correlating factor for drag force measurements on bluff bodies in relative motion with viscous fluids is well known, and much use is made of "standard" drag curves giving mean drag coefficient values, C_D, *vs* Reynolds number, R. Bluff bodies, simply stated, are objects that are not "streamlined." The use of the term "bluff" also implies that the Reynolds number is of sufficient magnitude that the boundary layer flow becomes detached, or separated, from the body at some point on its surface. The Reynolds number is a measure of the importance of viscous effects, compared to inertia effects, on the forces experienced by a body, when referred to a dimension and velocity appropriate to the phenomenon at hand. However, other features of the flow, in particular the turbulence characteristics of the free stream, may also have a substantial influence in determining the magnitude of the fluid-dynamic forces. (We are here examining only the effect of turbulence in modifying boundary layer and wake mean flows, and are not considering short-period fluctuating forces or buffeting, although these are of concern in some applications.)

The dependence of bluff body drag on the Reynolds number (based on a body dimension and the mean flow speed) is well documented for objects such as spheres and cylinders with circular cross-sections for steady, incompressible, non-turbulent flow. In addition, the effect of roughness in promoting an early transition from laminar to turbulent flow in the boundary layer, with a consequent change in drag, has been amply demonstrated. It is also well-known that free-stream turlulence can cause the same phenomenon. However, this action influences only a relatively small portion, termed the critical regime, of the total Reynolds number range of interest. Especially below this critical region, the effects of turbulence have not been determined with certainty. The picture is, perhaps, somewhat more clear in the case of sharp-edged objects, with their fixed separation points, than it is for bodies such as circular cylinders where a complex interaction exists between the boundary-layer separation points and the near wake, or base, pressures. Indeed, there has not yet been put forth any suggestion for a physical mechanism enabling the presence of free-stream turbulence to cause major changes in circular cylinder drag for Reynolds numbers below the critical regime.

The practical applications of a knowledge of the influence of free-stream turbulence on flow regimes are many, including the design of any device or structure exposed to turbulent fluids, such as oceanographic research instruments, heat exchangers, turbomachines, vehicles, buildings, bridges, etc. Also, test results from wind tunnels, tow tanks, and similar facilities can be highly dependent on the turbulence level. In fact, discrepancies noted in comparisons of early wind tunnel results gave the first clue to the importance of free-stream turbulence, the cause being traced to different turbulence levels in various tunnels [1, 2]. Subsequent investigations into the effects of turbulence, in the late 1920s and early 1930s, were generally restricted to relatively low turbulence intensities, consistent with the levels existing in early test facilities, that is, intensities not exceeding a few percent. Turbulence intensity, I, is defined as the root-mean-square value of the fluctuating velocities, normalized by the mean velocity. Any value of intensity greater than perhaps 0.1% must be of concern with respect to the modification of experimental results (and even less if boundary layer stability phenomena are being examined). Use of the term "high intensity" generally infers values of several percent or greater. The possible influence of the size, or scale, of turbulent eddies was not considered in the early experiments. Several statistical measures of size are available, the most often used in the present context being the integral of the autocorrelation function in the flow direction. This integral length scale, L_x, is representative of the larger eddies present, the so-called energy-containing eddies. When L_x is normalized with some body dimension, the "relative" scale is obtained. (Sometimes the length scale L_y, for the lateral direction, is used instead of L_x; in homogeneous turbulence, $L_x = 2L_y$).

The usual "standard" drag curves for spheres and circular cylinders (Figures 1 and 2) have been confirmed in test facilities having turbulence intensities as low as 0.02% [3], essentially non-turbulent in the present context. The effect on drag of variations in turbulence intensity and scale, with the exception of the critical regime, appears to have received little attention over the years. The probable reason for this is that the research centered around aeronautical developments, and the atmosphere, away from boundaries, is relatively free of turbulence scales of importance to the thin boundary layers associated with aircraft components. However, more down-to-earth situations involving high intensities are not uncommon close to boundaries in geophysical flows; for example, shear flows over rough ground, around buildings, in rivers and in tidal channels, in the wave zone in lakes and oceans where a body is continually re-immersed in its own turbulent wake by the oscillatory motion, or flow in the presence of breaking waves. The longitudinal turbulence intensity of natural winds over rough ground can reach 30% or more near ground level [4], and values approaching 20% have been reported for ocean currents close to the bottom of tidal channels [5].

The fact that the fluid-dynamic forces on a body may vary, depending on whether the free stream is laminar or turbulent, is not surprising when one considers that a turbulent fluid is characterized by high, fluctuating levels of vorticity at various scales. The shear layers generated in bluff body flow produce their own share of vortical motions. Interactions occur between the vorticity and mean velocity fields, affecting the internal stresses and strains in the fluid, the transport of mass and momentum in shear layers, the pressure gradients, the distribution and dissipation of energy, etc. One might then expect that the presence of high turbulence levels in the free stream could be responsible for interactions between the free stream and the boundary-layer flow about a body, and between the free stream and the wake flow, thus influencing the forces on the body.

In passing, it should be mentioned that, in addition to the steady streamwise drag force that is considered here, bluff bodies also experience unsteady lift (cross-stream) and drag forces related to the time-dependent shedding of eddies. These fluctuating forces, which are influenced in some measure by free-stream turbulence, can lead to vibrational instabilities if the body, or its support, is insufficiently stiff.

SMOOTH BODIES

Critical Regime

A critical Reynolds number regime occurs with bluff bodies when the shear layer separation point is not fixed by some surface discontinuity, such as a sharp edge or corner, for example, spheres, cylinders with circular or oval cross-sections, and air foils at some angle of attack. Such bodies exhibit a drag coefficient that may vary substantially with changes in the flow Reynolds number. The critical regime refers to the occurrence of a transition, in the boundary layer, from a laminar flow state to one of turbulent flow, i.e. from a well-organized, smooth flow to one of relative disorder. The transition doesn't occur instantaneously but, rather, results from the somewhat gradual, and selective, growth of certain natural (or deliberately-introduced) instabilities or disturbances, initially leading to random bursts of turbulence, but culminating in a final stage where the whole flow field is turbulent. The mechanism for the amplification of the disturbances in an initially-laminar flow is a linear process called the Tollmien-Schlichting (T-S) instability mechanism [3]. The disturbances may arise naturally from an increasing flow Reynolds number, or from surface roughness, acoustical excitation, surface vibration, and the presence of low-intensity turbulence in the free stream.

If the roughness elements are very large or the free-stream turbulence intensity is greater than perhaps 1%, then the linear T-S mechanism is bypassed and a non-linear process takes over. A physical explanation for the latter has been proposed by Taylor [6]. Basically, it is attributed to small pressure gradient variations associated with the turbulence or roughness causing momentary flow separations which trigger transition in the boundary layer, since a separated layer is less stable than an attached one. Development of the theory, supported by some empirical data, led Taylor to the result that the critical Reynolds number for spheres should be related to the parameter $I(L_y/D)^{-1/5}$, where D is the sphere diameter, I is the turbulence intensity, and L_y is the length scale. This Taylor parameter has been used with some success by various researchers for correlating results obtained with circular cylinders and spheres, generally at turbulence intensities of less than 5%. However, yet other experimenters have proposed different parameters for relating their data to the critical Reynolds number, as mentioned below.

Figure 1 shows a plot of the drag coefficient, C_D, for a smooth circular cylinder *vs* the flow Reynolds number in the vicinity of the critical regime, for various values of free-stream turbulence intensity. It is seen that, over some small range of Reynolds number, a sudden decrease in drag occurs. This

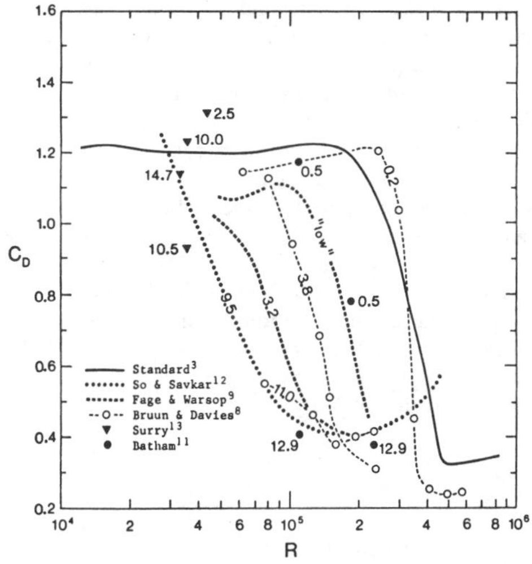

Figure 1. Steady drag coefficient, C_D, of a smooth circular cylinder *vs.* Reynolds number, R, near the critical region. Numbers on curves are percent turbulence intensity.

is due to the advent of transition in the boundary layer, either before separation or just after separation of the laminar shear layer, with the formation of a separation "bubble" and subsequent reattachment as a turbulent boundary layer [7]. The higher kinetic energy associated with the turbulent boundary layer permits greater penetration into the adverse pressure gradient on the rear half of the cylinder so that final separation is delayed and the wake width is reduced, resulting in the drastic reduction of the drag coefficient. (Surface stress is actually greater for the turbulent boundary layer, but this "skin friction" component of total drag is very small compared to the pressure component, for a bluff body.) The effect of increased free-stream turbulence levels is to shift the drag curves to a lower Reynolds number, as compared to the "standard" turbulence-free case. This is shown, for example, by the data of Brunn and Davies [8], where there is a substantial shift as the intensity goes from a relatively low value of 0.2%, nearly coincident with the standard curve, to the higher value of 3.8%. They also made two sets of measurements at an intensity of about 11% (only one is shown in the figure), with the relative scale values being less than unity, but differing by a factor of 1.6. There is very little difference in the results, which is, as one might expect, in the critical regime with the turbulence intensity presumably causing transition directly via the non-linear mechanism.

Among the earliest investigators of the effects of turbulence were Fage and Warsop [9], who used a rope net as a turbulence generator. They did not measure the actual intensity values but, as reported by Bearman [10], it has been possible to estimate these from the details given of the experimental arrangement. The intensity of their "standard stream," shown as "low" in Figure 1, is not given, but it may be close to 0.5%, as suggested by the proximity of this curve to the Batham [11] data. Also given in Figure 1 are two points by Batham for an intensity of 12.9%, with a relative scale of 0.5, which lie close to the similar Bruun and Davies data.

One of the more recent investigations was undertaken by So and Savkar [12], one of their curves for an intensity of 9.5% being presented in Figure 1. The data from which this curve was derived has not been corrected for the substantial wind tunnel blockage value of 16%, so the drag coefficient values are higher than they would be for the zero-blockage case. (Blockage is discussed in greater detail in a following section.) The curve shown is for a relative scale of the order of 0.5. Additional So and Savkar data suggest that, for a relative scale just over unity, there is a slight increase in the drag coefficient with the increase in scale. Several data points lying just on the lower fringe of the critical regime have been provided by Surry [13]. Two of these also suggest an increase in drag coefficient with an increase in scale, namely those for an intensity of about 10%. The relative scale for the uppermost of the two points in Figure 1 was 4.3, whereas it was only 0.36 for the lower point.

Additional confirmation of the reduction in drag coefficient for circular cylinders is provided by some Strouhal number data obtained in the vicinity of the critical regime, reviewed by Bell [14]. The Strouhal number is a non-dimensional representation of the eddy-shedding frequency of a body, given by the product of the frequency with a characteristic length and the inverse of a characteristic velocity. It has been shown to be inversely proportional to the drag coefficient for circular cylinders, which is not unexpected, since the eddy-shedding frequency increases as the wake width of a body decreases. The data show a dramatic increase in Strouhal number, in the presence of free-stream turbulence, more or less coincident with the great decrease in drag coefficient.

For spheres, a curve [15] of drag coefficient *vs* Reynolds number is found to be of similar form to that for a circular cylinder in the critical regime, including the marked change due to boundary layer transition. A number of experimenters have examined the influence of free-stream turbulence on sphere drag and critical Reynolds number, including: Bacon and Reid [16], who presented a review of the problem in 1924; Taylor [6], who first paid serious attention to turbulence scale and transition theory and proposed the previously-mentioned correlating parameter; Dryden et al. [17], who reported that sphere drag can vary "by a factor of 4 in air streams of different turbulence"; and Neve and Jaafar [18], with the most recent look at the subject.

Subcritical Regime

Very little work seems to address the effect of turbulence on the flow about bluff bodies below the critical regime, i.e., at pre-transitional Reynolds numbers. We are again restricting comment in this section to those bodies without fixed separation points. Perhaps the reason for this apparent neglect of an experimental opportunity can be attributed to the lack of any well-publicized or obvious physical mechanism comparable to the dramatic transition phenomenon of the critical regime. The data that has appeared in the literature is extremely inconsistent, one set with another, as will be seen from an examination of Figure 2. Various possible reasons exist for the discrepancies in this research, and these will be examined at some length in a later section.

To begin with, Figure 2 shows isolated data points from the work of Hegge Zijnen [19] and So and Savkar [12], both near the same Reynolds number and using turbulence with an intensity of about 9.5% and a relative scale of about 0.6. By comparing the corresponding drag coefficient values with those obtained in each respective low-turbulence case (i.e., no turbulence-generating apparatus present in the tunnel), we see that they disagree on the basic trend. So and Savkar provide additional data (not shown here) at increasing Reynolds numbers, which suggests that theirs is another case of the onset of boundary layer transition, so the decrease in drag coefficient in the presence of free-stream turbulence is as expected. In fact, pressure measurements around the periphery of their circular cylinder indicate that the reduction in drag is due to a combination of a more negative pressure minimum on the upstream half of the cylinder, an increased base pressure (i.e., the pressure in the near-wake), and a reduced wake width due to a downstream shift of the separation point location. Hegge Zijnen's pressure measurements, on the other hand, while showing a slight decrease in wake width, indicate a continuing decrease in base pressure with increasing turbulence intensity.

Another recent set of data, due to Kiya et al. [20], in the same Reynolds number region, tend to confirm the trend shown by the So and Savkar data, i.e. a reduction in drag with an increase in free-stream turbulence intensity. The relative scale in this case is about 0.8 and blockage corrections have been carried out. Kiya et al. have also examined other data, along with their own, and suggest that a good correlation is obtained by using the product of Taylor's parameter and the Reynolds number raised to the power 1.34.

Two quite extensive, and self-consistent, data sets are also shown in Figure 2, due to Ko and Graf [21] and Dyban et al. [22, 23]. While the two sets cover different Reynolds number ranges, they do have some overlap wherein it is obvious that they are in conflict as to the trend of the drag coefficient with increasing free-stream turbulence levels. The data from Dyban et al., for an intensity of 0.3% are of the same form as the standard curve, but the coefficient values at particular Reynolds numbers are substantially lower. This suggests a higher base pressure than one would normally anticipate for a low-turbulence situation, perhaps due to some factor associated with

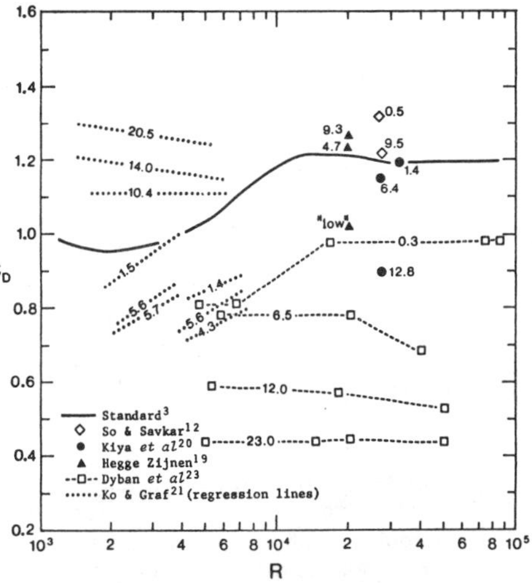

Figure 2. Steady drag coefficient, C_D, of a smooth circular cylinder *vs.* Reynolds number, R, in the subcritical regime. Numbers on curves are percent turbulence intensity.

their experimental arrangements (about which, more later). The pressure measurements, from which the drag coefficients were derived, show that the separation point moved downstream and that both the base pressure and the pressure at the point of maximum velocity around the cylinder increased substantially as the turbulence level was increased up to 23% at a constant Reynolds number of 1.8×10^4. Thus, the effect of turbulence in the experiments of Dyban et al. appears to be one of improving the pressure recovery at all points around the periphery of the cylinder. Turbulence scale varied with intensity in these experiments, but Dyban et al. state that they found no effect due to scale, within the limits of error of their measurements.

Ko and Graf, unfortunately, did not make any pressure measurements, obtaining drag values directly from a strain-gauge dynamometer. Each of their 12 data sets, each set containing about 12 individual measurements, is represented by a regression line in Figure 2. The break in the lines at intensity values of 5.7% and less is due to a change in relative scale by a factor of 2. This same change also occurred at the higher intensities, but the break was less pronounced, so two data sets were merged into one regression line for the intensities of about 10, 14, and 20%. Ko and Graf showed that their data correlated quite well with Taylor's parameter divided by the logarithm of the Reynolds number, producing a curve having a drag coefficient minimum at a free-stream turbulence intensity of about 5%. A similar curve is obtained by using intensity alone for the correlating factor, as shown in Figure 3. The length of the vertical bars is indicative of the combination of scale effect and Reynolds number dependence remaining in the data at the given value of intensity. An "eye-fit" dotted line is drawn through the bars to emphasize the trend.

The results of Dyban et al. are also given in Figure 3. Now the trend shown by the eye-fit dashed line shows the opposite tendency to the Ko and Graf data above an intensity value of about 5%. Which curve is acceptable? The answer must be "neither," for the present, although the weight of the experimental evidence does seem to favor a decrease in drag with an increase in turbulence intensity. Neither of the last two data sets has been independently verified, to the author's knowledge. Thus, until someone does a definitive and irrefutable experiment on circular cylinders over a range of Reynolds numbers in the subcritical regime, it is only possible to speculate on the true influence of free-stream turbulence in this region.

There is an experiment in the literature, involving direct measurements in the laminar boundary layer on a flat plate, which may assist in this speculation. It, also, is due to Dyban et al. [24], using a different flow facility than previously, and it appears to be quite straightforward with little possibility for error. They took "special," but unstated, measures to ensure a zero longitudinal pressure gradient in their experiment, thus avoiding undesirable flow accelerations. The measure of their success is proven by the very close comparison of their low-turbulence-intensity (0.3%) result for

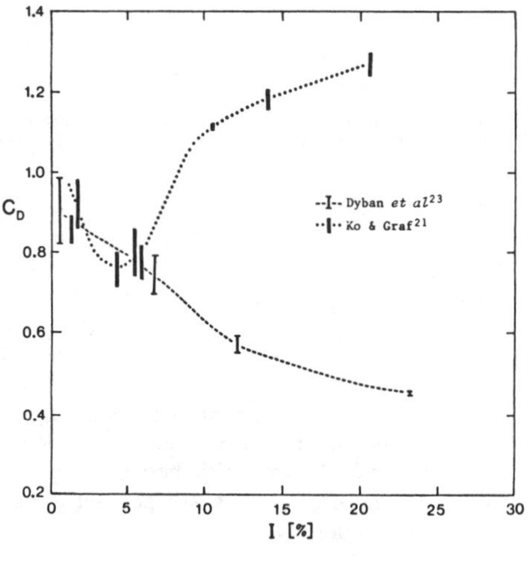

Figure 3. Steady drag coefficient, C_D, of a smooth circular cylinder *vs.* free-stream turbulence intensity, I. Dashed and dotted lines indicate data trends only.

the mean boundary layer velocity profile with the theoretical Blasius distribution [3]. They varied the free-stream turbulence intensity over the range of 0.3% to 25.2% at several subcritical Reynolds numbers (R_x, based on distance from the leading edge), and used a hot-wire anemometer to determine turbulent velocities in the flat plate boundary layer.

Dyban et al. found that the free-stream turbulent fluctuations penetrated the boundary layer and caused disturbances there. These included a distortion of the velocity distribution, resulting in an increase in velocity gradient near the plate and an overall thickening of the boundary layer, as normally occurs after transition. An interesting result is the finding that the largest relative change in the boundary layer behavior occurred for a free-stream turbulence intensity of about 4%. This is observed by plotting the turbulence intensity in the boundary layer against distance from the plate, at a given Reynolds number.

Distance from the plate is represented by the nondimensional number, η, obtained from the product of the actual distance normal to the plate and the Reynolds number, divided by the distance along the plate. From such a plot given by Dyban et al. for a Reynolds number of 6×10^3, it is seen that a free-stream turbulence intensity of 4.4% resulted in an increase in boundary-layer turbulence intensity, from a very small value near the plate to a maximum of about 9% at $\eta = 2.5$, thence decreasing to the free-stream value at η equal to 8 or 10. A free-stream intensity of 18.4% resulted in a maximum boundary layer intensity of about 20%, again at $\eta = 2.5$. The depth of penetration of the free-stream disturbances is thus apparently independent of the intensity. However, a direct dependence of penetration depth on Reynolds number was observed, the value of η being about 2.0 for the occurrence of the maximum boundary-layer intensities at a Reynolds number of 2×10^4.

The relative perturbations are shown in Figure 4, which is a plot of the ratio between boundary-layer and free-stream turbulence intensities *vs* free-stream intensity for two values of Reynolds number. The maximum relative changes are clearly shown to occur for free-stream turbulence intensities near 4%. Furthermore, the increasing peak amplitudes confirm that the boundary layer becomes more sensitive to external disturbances as the Reynolds number increases. Also, it is apparent that the relative energy in the boundary layer is substantially dependent on Reynolds number for free-stream intensities below 6 or 7%, and considerably less so for intensities above 10%. The effect of turbulence scale was not examined in these experiments.

The reason for some of the preceding results is not entirely certain. That there is an amplification of disturbances, or perhaps even a generation of new larger perturbations, in the boundary layer is clearly apparent in Figure 4 for free-stream turbulence intensities ranging up to 15%. To what mechanism the amplification should be attributed is another question, since the T-S theory suggests that damping of disturbances should occur for any Reynolds number less than about 9×10^4 [25]. (In fact, probably only if such damping is characteristic of a boundary layer, should it be called truly laminar.) Criminale [26] has presented a theory for the response of a boundary layer to external disturbances, which differs from the T-S theory in that it involves a selective impedance function, rather than an eigenfunction, approach. Criminale's theory permits a free stream of a certain turbulence intensity to generate disturbances of larger intensity in a laminar boundary layer. He

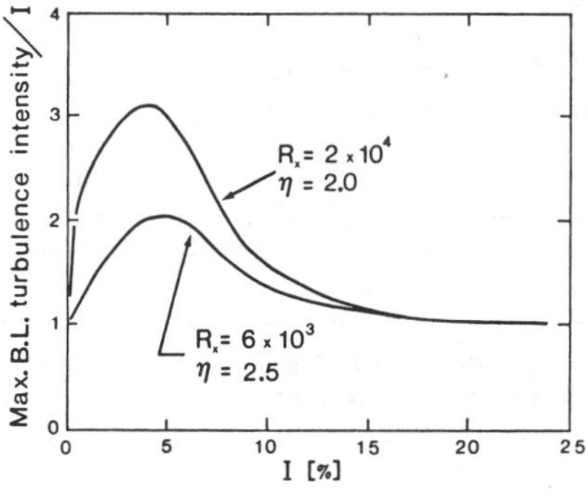

Figure 4. Relative boundary layer disturbance *vs.* free-stream turbulence intensity, I, for laminar flow over a flat plate [24].

attributes the stimulation to pressure, rather than vorticity, fluctuations, which is reminiscent of Taylor's suggestion [6] for a transition mechanism. Some of the observed effects, such as the change in shape of the mean velocity profiles and the increased layer thickness, can be caused by the very active mixing associated with a turbulent flow, which results in the insinuation of some of the free-stream turbulent energy into the laminar boundary layer. Perhaps "pseudo-turbulent" would be a more appropriate designation for such a layer, because it imitates the behavior of a turbulent boundary layer that has undergone transition in the normal way.

While the foregoing experiment of Dyban et al. [24] was performed under conditions of zero pressure gradient and zero surface curvature, the conclusions drawn from their data are generally applicable to other types of flow. Basically, then, a principal effect of free-stream turbulence is to substantially increase the kinetic energy in a laminar boundary layer enabling it to penetrate adverse pressure gradients to a greater distance before separation. In addition, of particular importance in regions bounded by separated shear layers, such a boundary layer exhibits enhanced entrainment or mixing properties compared to a purely laminar one.

For the flow of a turbulent free stream around a bluff body such as a circular cylinder, at pre-transition Reynolds numbers, one might thus expect the behavior to be somewhat analogous to that in the critical regime. Delayed separation of the more energetic boundary layer would result in a narrower wake and, therefore, a reduced drag if no change occurred in the pressure in the separated region, i.e., the base pressure. However, this pressure is affected by entrainment into the separated shear layers and may be either reduced or increased, as witnessed by the previously-mentioned measurements of Hegge Zijnen [19] and So and Savkar [12]. In addition, the interaction between base pressure and separation point location adds further uncertainty to the net effect. Thus, the situation with regard to the bluff body drag coefficient values that might be expected in this case remains open, with the answer awaiting a definitive experiment.

SHARP-EDGED BODIES

The term sharp-edged bodies is here applied to those bluff objects having an angularity of some sort that disrupts the flow sufficiently to cause shear layer separation. Thus, the location of the initial separation (in some cases there may be reattachment with a subsequent separation) is fixed in space and time. A consequence of this is that such bodies show, in general, little or no Reynolds number effects. Examples include rectangular-sectioned columns, flat plates normal to the flow direction, flanged beams and, in three dimensions, discs and blocks. The interest in turbulence-related phenomena with these shapes stems principally from architectural and engineering problems relating to the design of buildings and other structures, including bridges.

Much of the research effort has been concentrated on a study of two-dimensional rectangular cylinders. In the course of experiments with these forms, it became apparent that the depth of the afterbody was critical to the results. (The afterbody is that volume of a bluff body that extends downstream of the shear layer separation points. The depth and shape of the afterbody influence the flow in the base region which, in turn, may interact with the separating boundary layers.) In fact, there is a remarkable increase in the drag coefficient of a two-dimensional rectangular-sectioned cylinder, at zero angle of attack, as a particular depth-to-height ratio (d/h) is approached, followed by a subsequent reduction as the depth is further increased. This result, for a low-turbulence stream, is shown in Figure 5, after Bearman and Trueman [27] who confirmed and extended the work of previous investigators (in particular, Nakaguchi et al. [28]). The drag coefficient approaches a maximum value of 3 for $d/h = 0.62$. Bearman and Trueman attribute the phenomenon to entrainment from the base region, principally into the growing vortices that result from roll-up of the unstable shear layers separating from the front corners of the body. The base pressure is reduced accordingly, causing an increase in drag, and the shear layer curvature increases, thereby bringing the vortex formation region closer to the rear face of the cylinder. The corollary is that anything that produces an increased length of vortex formation region or, equivalently, reduced shear layer curvature, will cause a reduction in drag. Bearman and Trueman have verified this by inserting a "splitter" plate in the base region along the stream centerline, thus inhibiting interaction between the shear layers.

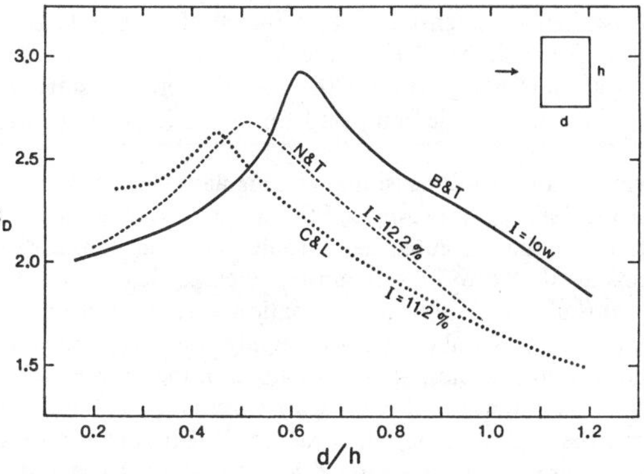

Figure 5. Steady drag coefficient C_D, of a rectangular cylinder *vs.* the ratio of cylinder depth to height, d/h, in a free stream of turbulence intensity, I [27, 30, 31].

The drag coefficient maximum shown in Figure 5 corresponds to an afterbody depth where, according to Bearman and Trueman, the downstream corners of the rectangular section just begin to interfere with the highly-curved separated shear layers. The interference is such that, for d/h values increasing beyond the critical value of 0.62, the shear layer curvature is progressively reduced, thus increasing the base pressure. No definite mechanisms was suggested to account for this, the probable one of reattachment being presumed not to occur except at a much higher value of d/h. (Bostock and Mair [29] suggest that reattachment does not happen for d/h < 2.8, approximately, for the two-dimensional case.) It does seem likely that some intermittent reattachment was taking place, especially in view of the highly curved nature of the shear layer at this point. The result would be an increased mean base pressure, even though permanent reattachment might not be achieved until after a further substantial increase in afterbody depth. Concurrently, entrainment would be thickening the shear layers and reducing their vorticity concentration so that they would react less strongly in the base region, forming vortices with lower core suctions and so also leading to a higher base pressure.

Of additional interest, the Bearman and Trueman [27] pressure data show that the curve for centerline base pressure coefficients *vs* d/h has the same form as the drag coefficient curve, while the front face pressures are essential independent of d/h, thus confirming that the drag coefficient is primarily controlled by the base pressure. Also, for d/h = 0.2 and 1.0, the base pressure coefficients are nearly uniform across the base region (and close to the same value), whereas, for d/h = 0.6, there is a large reduction in centerline pressure when compared to the near-corner pressures.

The previous discussion refers to smooth (low-turbulence) flow only, so we now examine the changes that occur in the presence of free-stream turbulence. The effects have been reported by various investigators, in particular, Nakamura and Tomonari [30], Courchesne and Laneville [31], and others mentioned in the following discussion. Nakamura and Tomonari [30] obtained drag coefficient information for a range of d/h values, using two-dimensional rectangular cylinders, with several values of turbulence intensity and scale. Their low-turbulence (0.1%) result confirms that of Bearman and Trueman [27] (Figure 5). For a free-stream turbulence intensity of 12.2% (and a relative scale of 0.74), they found that the effect was one of increasing the drag coefficient below the drag peak and decreasing it above the peak, as also shown in Figure 5. Note that the critical value of d/h is reduced, so the result of adding turbulence to the flow might be considered as an apparent increase in afterbody depth, fluid-dynamically speaking. The concensus seems to be that the effect is due to the enhanced entrainment from the base region with a resultant reduced radius of curvature (increased curvature) of the shear layers. This would account for the increase in drag below the critical value, but leaves one in doubt about the mechanism for providing a drag decrease at higher values. In this case, it is likely that the turbulence results in improved pressure recovery along the sides of the section when the intermittent reattachment begins. Little effect of turbulence scale is noted except, perhaps, in the vicinity of the drag maximum where a smaller scale may have caused a slight drag increase. However, comparative data is lacking.

Courchesne and Laneville [31] have also had an extensive look at the effect of turbulence on the relationship between drag and cross-sectional shape. Their curve for an intensity of 11.2% is presented in Figure 5, as well. Relative scale was not given, but the investigators did state that their work showed scale as playing only a secondary role compared to intensity. The data were subjected to blockage corrections.

Gartshore [32] conducted some experiments on two-dimensional rectangular cylinders in which he dispensed with the usual grid-generated turbulence, using instead the wake from small diameter ($\frac{1}{8}''$ and $\frac{1}{4}''$) circular rods. His contention was that, if enhanced turbulent mixing in the shear layers was the primary cause of the effects attributed to free-stream turbulence, then all that was necessary was the addition of small scale turbulence close to the stagnation streamline ahead of a bluff body in an otherwise smooth stream. The rod wake supplies roughly isotropic turbulence, with the scale depending on rod size and location. A plot, after Gartshore, of the results for two values of d/h is given in Figure 6. It is seen there that no obvious scale effect is present and that, for the cylinder having d/h = 1 (i.e. above the critical value), the drag coefficient is continuously reduced as the turbulence intensity increases. For the cylinder having d/h = 0.5, which is near-critical in the presence of turbulence, the drag at first increases with turbulence intensity until the apparent afterbody depth becomes critical, whence the drag is reduced with further increases in intensity. This is in agreement with previous considerations about the influence of turbulence.

We have previously examined researches that have not found any appreciable effect of turbulence scale on the flow around rectangular sections. Now, we consider some work showing a decided effect of scale. McLaren et al. [33], using values of turbulence intensity from about 3% to 10% and relative scales from less than 0.5 to about 5, with a square-section cylinder (d/h = 1), found a definite maximum in the drag coefficient for a scale in the vicinity of 1.5. It is interesting to note that this is the same value for which Hegge-Zijnen [19] reported a maximum in heat transfer rate when turbulence was introduced into the free stream. However, Hegge-Zijnen's family of curves at different intensities show convergence at higher scale values, as one might expect, whereas the family of curves due to McLaren et al. show a divergence. This may be due to experimental scatter and the small number of data points.

Lee [34] also worked with square cylinders, over approximately the same range of turbulence scales and intensities as did McLaren et al. He, too, found scale-dependent drag coefficient maxima, but for values near unity. In addition, he also observed minima occurring at relative scales near 1.5 (Figure 7). The magnitudes of the maxima and minima were intensity-dependent, with the whole drag *vs* scale curve shifting to lower drag values at higher intensities. Thus, in this respect, Lee's work is in agreement with that of other investigators. The suggested existence of minima and

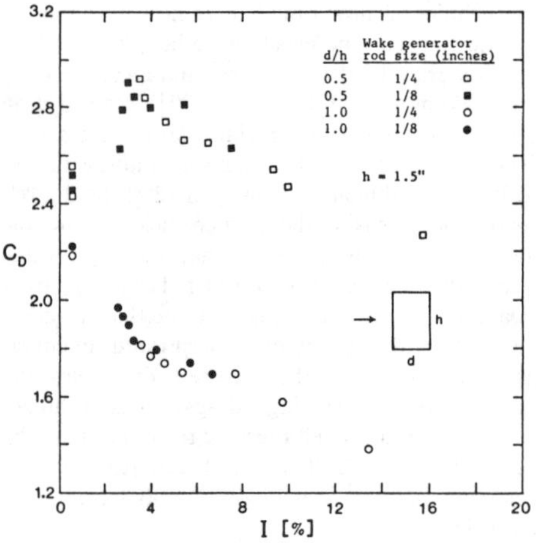

Figure 6. Steady drag coefficient, C_D, of a rectangular cylinder *vs*. turbulence intensity, I, for two values of depth/height ratio, d/h [32].

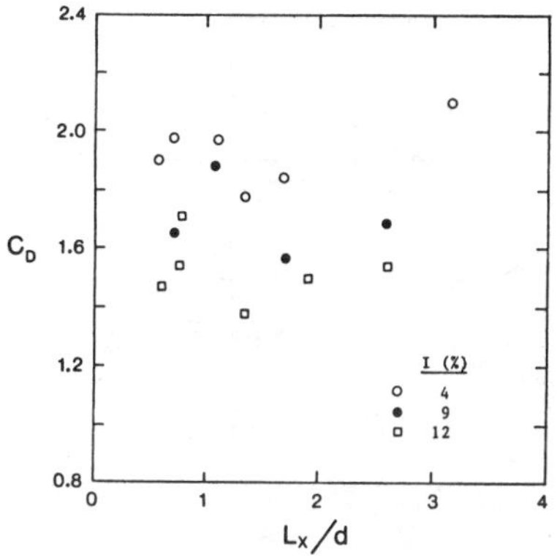

Figure 7. Steady drag coefficient, C_D, of a square cylinder *vs.* relative scale, L_x/d, for several values of turbulence intensity, I [34].

maxima offers a clue to the disagreement about a scale effect, since experiments might encounter the sampling problem known as "aliasing" if the scale values used are inappropriately selected.

We turn now to a look at the effect of turbulence on discs, blocks, and other three-dimensional sharp-edged objects normal to the flow. It is necessary here to note that aspect ratio (i.e. length/width ratio) is an important, and complicating, variable. For bodies of low aspect ratio, the flow in the base region is three-dimensional and the base pressure is relatively high. As aspect ratio increases, the flow over the central portion of the body tends to become more two-dimensional in character, with a concomitant reduction in base pressure and increase in drag. Much of the work done with blocks has been intended to simulate buildings and, consequently, involves low-aspect-ratio bodies in the presence of a ground plane.

Roberson et al. [35] have examined the flow about a variety of bluff body shapes, including plates, cubes, rods, and spools. They concluded that bodies that have shapes such that flow reattachment is not a factor (e.g. plates or discs) experience an increase in drag in the presence of a turbulent free stream. Bodies for which reattachment could be a factor may experience either an increase or decrease in drag with increasing turbulence intensity, depending upon the shape of the body. Bearman [36] has measured the mean base pressure for several sizes of square and circular plates in smooth and turbulent flows. He found that turbulence resulted in a lower base pressure and, therefore, a higher drag, with some dependence on both scale and intensity. The data correlated very well with the parameter $I(L_x^2/A)$, where A is the plate area, L_x is the longitudinal scale and I is the intensity. Humphries and Vincent [37], having conducted a similar series of experiments, preferred $I(L_x/D)$ as a correlating factor, where D is the plate diameter or similar dimension. Castro and Robins [38] examined the effects of turbulent shear flows around a surface-mounted cube, the velocity shear being included to represent the atmospheric boundary layer. They found that the size of the base recirculation region, i.e., the near-wake cavity, was substantially reduced as compared to smooth flow, and that reattachment to the cube sides was taking place. They also noted that velocity deficits in the wake were rapidly reduced by the turbulence.

Nakamura and Ohya [39] have investigated the mean flow about square rods (with face dimension h) of various lengths (d). The rods, with the square face normal to the stream, were suspended by means of a "sting" centered in the base region and piano wire at the downstream corners. The size ratio d/h of the rods ranged from 0.1 (essentially a flat plate) to 2.0. Turbulence intensity varied from 3.5 to 13% and relative scales (L_x/h) from 0.08 to 14. The scale was changed mainly by using rods with different values of h. Base pressure coefficients (i.e., the base pressures normalized with the dynamic pressure) were corrected for blockage. It was found (see Figure 8) that the base pressure is reduced, by turbulence, for short rods ($d/h < 0.6$), increased for long rods up to the point of reattachment, and then reduced again for still longer rods. Reattachment occurs at about $d/h = 1.6$ in smooth flows and $d/h = 1.1$—1.2 in turbulent flows. A definite scale effect is apparent for short rods, with $L_x/h = 2.4$ seemingly much more effective than $L_x/h = 0.24$ in causing pressure changes.

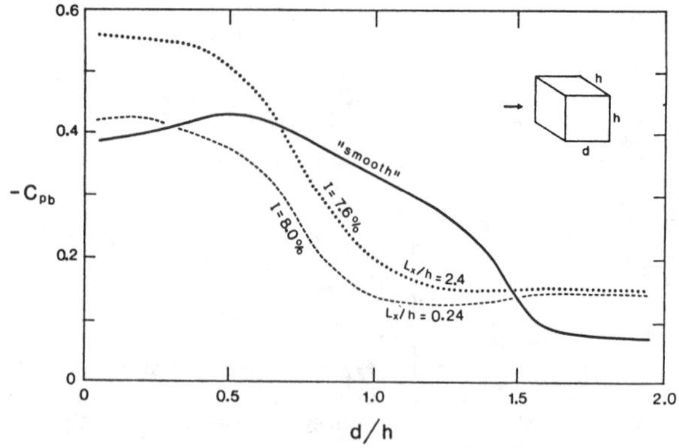

Figure 8. Base pressure coefficient, C_{pb}, of a three-dimensional square rod *vs.* depth/height ratio, d/h, for a free stream having turbulence intensity, I, and relative scale, L_x/h [39].

Side-face pressures and pressure recovery characteristics are shown by the Nakamura and Ohya data to be very much a function of both intensity and scale, and of d/h. Especially, for some cases, it appears that a large presssure recovery can occur over a very small distance in the immediate vicinity of the downstream edge. Altogether, the results suggest that no universal correlating parameter can be devised for sharp-edged bodies to account for the effects of turbulence. Nakamura and Ohya conclude with the observation that small-scale free-stream turbulence (i.e., that in which the size of the energy-containing eddies is of the order of the shear layer thickness) increases the growth rate of the shear layers, while large-scale turbulence (i.e., the energy-containing eddies and the near-wake cavity are comparable in size) enhances roll-up of the shear layers, and that the consequences of these depend upon the shape of the bluff body. This observation, in large measure, offers an explanation for results obtained in the various experiments with sharp-edged objects.

Thus far, we have concerned ourselves only with sharp-edged bluff bodies with an orientation normal to the flow. In practical applications, e.g., structures in natural winds, this will seldom be the case. Thus, a knowledge of the way in which forces are influenced by a stream impinging on a body at various angles of attack may be important. Certainly, separation and reattachment phenomena will be affected in some degree, with the distance to reattachment on the side closest to the direction of the oncoming stream becoming increasingly smaller as the angle of attack becomes larger. This results in asymmetrical cross-stream and torsional forces that, in turn, can lead to instabilities in structures with insufficient damping. A substantial body of literature exists in this regard. See, for example, the review by Scanlan [40] which, while emphasizing suspension bridge problems, gives an excellent list of references of widespread applicability. The variation in mean base pressure coefficient with change in angle of attack of both a smooth and a turbulent stream for a rigid, two-dimensional, square-sectioned cylinder is presented in Figure 9, after Vickery [41].

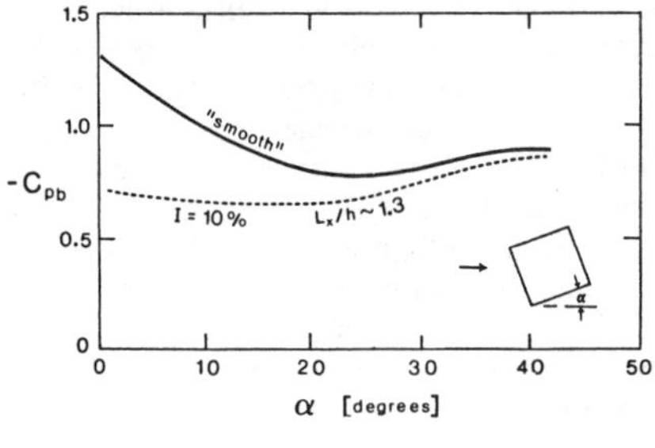

Figure 9. Base pressure coefficient, C_{pb}, of a square cylinder *vs.* angle of attack, α, for a "smooth" free stream and for one having turbulence intensity, I, and relative scale, L_x/h [41].

EXPERIMENTAL CONSIDERATIONS

A problem sometimes exists in trying to assess the quality of empirical data reported in the literature because of a lack of completeness in providing details of the experimental procedures. This has often been the case with experiments relating free-stream turbulence effects to bluff body drag forces, where the information has been incomplete for important details such as model mounting arrangements, tunnel speed measurement method and location, pressure measuring arrangements, blockage correction procedures (if any), turbulence generating apparatus, turbulence characteristics, and so on. In this section, we will examine the possibility that lack of adequate attention to such details may be responsible for some of the contradictions and differences to be found in the empirical data sets considered thus far.

Acoustical Limitations

It is certainly true that acoustical disturbances in the test facility, arising from some flow phenomenon such as separation from the tunnel wall or vibration of some component, can couple with the T-S mechanism to cause early transition in the boundary layer and, thereby, a change in the drag force [42]. However, this is only a problem for investigations, not considered here, seeking to determine the effect of free-stream turbulence at very low levels (less than 1%), for Reynolds numbers approaching the critical value. Otherwise, at higher levels, the turbulent fluctuations overwhelm the acoustic signal and induce transition directly by the transitory separation mechanism.

There was a suggestion that acoustical disturbances might contaminate the hot-wire anemometer signal (used to measure turbulence intensities), giving a spurious indication of higher levels of vorticity than were actually being produced by the turbulence generators. However, Bell [43], looked at this problem approximately, by applying the plane wave acoustic propagation equation. While the sonic velocity fluctuations from a strong acoustic wavefront would be detectable with a hot-wire anemometer, it would take a disturbance of great intensity to be of any significance in connection with experiments of the kind being considered here. In fact, using the plane wave equation in air at a "unit" Reynolds number (velocity $\div$ kinematic viscosity) of 10^4, a sound level intensity (referred to 10^{-12} W/m^2) of 130 dB is required to generate fluctuating velocities equivalent to a turbulence intensity of 1%. This level of sound is in the vicinity of the threshold of pain for the human ear. It seems unlikely that anything approaching it would be tolerated in the present context, so large differences in data sets cannot be attributed to acoustic problems.

Measurement Limitations

The accuracy of hot-wire anemometer readings at turbulence intensities in the neighborhood of 20% has been questioned [44, 45]. Substantial errors could result because of the lack of a cosine response pattern to the flow and because of non-linearities in the thermal response of this instrument. However, at least for intensities not exceeding 10–12%, the use of linearizing circuitry provides acceptable accuracy.

Of more concern is the measurement of pressures around the surface of a body, particularly when this is done with remote sensors connected to the measurement site by a network of tubes. Tubing, connectors, valves, and sensor support structures all have a certain resistance to the flow of fluid that occurs in response to pressure changes, as well as a storage volume, or capacitance. A system comprised of any of these components is analogous to an electrical filter circuit and, as such, has the usual characteristics of a filter, including a frequency-dependent modification of input signal amplitude and phase. Especially, long lengths of small-diameter tubing, such as is often used in wind-tunnel work, can cause amplification of the pressure signal near system resonance, or attenuation otherwise, plus a substantial phase shift between input and output. It is possible to correct these signal distortions in some measure [46] but, if any concern has been felt about this problem in the experiments under discussion here, it has seldom been mentioned. In particular, an error in pressure data due to this source could cause drag coefficient errors that would appear as shifts away from the standard drag curve.

Pressure measurements, correctly made, are probably the preferred method for obtaining drag coefficients for two-dimensional bodies. The reason is that the pressures are normally obtained at center span where any three-dimensional flow effects (see below) usually have the least influence. Away from center, base pressures may be elevated to a greater extent than at center by three-dimensional flow irregularities, resulting in a reduced drag in such regions. Thus, coefficient measurements made with balances or dynamometers tend to give values less than the true two-dimensional value because they integrate forces over the whole body.

Blockage

When tunnel walls confine the flow about an object, as in the tests considered here, corrections to the data may be required. This is because of the acceleration of the flow, in the vicinity of the object, as it speeds up through the reduced cross-section to satisfy continuity requirements, and because of energy losses in the wake and in the tunnel wall boundary layer. Experimenters are induced to use large object diameters and, hence, high blockage factors in order to obtain high Reynolds numbers. The lack of suitable corrections in such cases may be responsible for some of the differences in the data under consideration.

The acceleration effect is usually termed "solid blockage." It results in an increase of the axial velocity experienced by an object, and thereby causes an increase in drag force. This means that the tunnel speed, i.e., the calibrated speed applicable to the empty test section, usually determined by upstream measurements, must be corrected before it is used to calculate force coefficients and Reynolds number. Energy losses in wakes and in boundary layers give rise to similar phenomena, often called "wake blockage" and "horizontal buoyancy", respectively. Basically, the total pressure head is lower in wakes and boundary layers than in the flow outside these regions because some of the energy has been converted to heat by friction and turbulence, i.e. by "head losses." However, the static pressure head is practically constant across a given tunnel section (at least, once removed somewhat downstream of the test object). Therefore, the reduction in total head in the wake or boundary layer appears as a decrease in the local dynamic pressure. In other words, there is a velocity deficit or defect. This means that the mass flow rate in these regions is reduced below that of the flow upstream of the object. Once again, continuity requirements must be satisfied, so the flow outside of the wake or boundary layer compensates by gradually increasing its speed as it flows past the object and along the wake. This not only gives rise to a velocity gradient, but a pressure gradient as well, neither of which would exist in an unconfined stream. (Some test facilities attempt to counteract the influence of the wall boundary layer by designing in a gradual expansion.) Again, a correction is required to the tunnel speed U_∞.

The corrections are sometimes conveniently combined in terms of a single factor ϵ, which depends partly on test object and test section shapes, but principally on the ratio of object and cross-sectional areas, a ratio called the blockage factor B. Then, $U_F = U_\infty(1 + \epsilon)$ where U_F is the velocity experienced by the object. U_F is also the velocity, which would apply in unconfined flow, when $U_F = U_\infty$. It may be measured directly, rather than relying on the tunnel velocity calibrations, and some researchers have done this. Achenbach [47], in examining the effect of tunnel blockage on the flow about a sphere, reported using the mean equatorial velocity, i.e., the mean velocity at the minimum cross-section when the object is installed in the tunnel, as the appropriate reference velocity U_F. He then found that a blockage factor as high as 0.35 resulted in an error of only about 1% in the drag coefficient value C_D, as compared to the value C_{D_∞} for the unconfined case. Unfortunately, Achenbach does not give sufficient information to determine the error that would have arisen in the above case by using the tunnel speed as the reference velocity. He does mention, for the improbable case of B = 0.9, that using the mean equatorial velocity gives $C_D/C_{D_\infty} = 2$, whereas $C_D/C_{D_\infty} = 56$ if the tunnel speed is used. This is for the three-dimensional case of spheres, though, and the results for two-dimensional bluff bodies may be somewhat different.

Various techniques for correcting the effects of wall interference are presented in the literature, with most of the early ones being concerned primarily with streamlined bodies. A review of these has been given by Farell et al. [48]. Because of the greater expansion (in the absence of walls) of bluff body flow lines as compared to those of streamlined objects, and therefore the more substantial wall confinement of these, and because of the strong interaction between a bluff body boundary layer and its wake, it seems unlikely that techniques developed for streamlined objects can be suitable for correcting bluff body data, although such usage has been attempted occasionally. It is preferable to try, from the outset, to develop a correction scheme which relates directly to bluff body flow. Maskell [49] has done this with considerable success for the case of a flat plate normal to the stream, using a free-streamline wake model and momentum concepts. He suggests that his theory should be suitable for most bluff bodies having a fixed separation point location. A different free-streamline representation, and a brief review of similar models, is given by Fackrell [50]. A more recent study of the blockage problem has been undertaken by West and Apelt [51], who point out that none of the usual correction schemes are entirely suitable for the circular cylinder case, because the form of the pressure distribution changes with blockage factor. However, they also indicate that the effects of blockage are small if the blockage factor doesn't exceed about 6%. For rectangular cylinders, Courchesne and Laneville [52] have given a comparison of several correction procedures.

Other effects of high blockage ratios, besides the accelerations and pressure reductions outlined above, include changes in the wake geometry because of the restraint on the divergence of flow streamlines, so that dynamic similarity no longer exists in full measure. Also, because of the pressure gradient changes, increased blockage may promote early transition in the object boundary layer, change the separation point location and alter the vortex shedding frequency. These are just the kind of effects for which free-stream turbulence is also a suspect.

It is apparent, from the foregoing, that substantial errors can result if some attention is not paid to the blockage problem. Normally, at high blockage factors, the errors would be expected to give apparent values of C_{D_∞} which are too large, at apparent Reynolds numbers that are too low. Examining some of the data from Figure 2 in this light we find only So and Savkar [12] showing higher than normal drag coefficient values at low values of free-stream turbulence. They, in fact, state that no corrections to their data have been made to compensate for their blockage factor of 16%, so their results are at least consistent with the anticipated effect. None of the remaining investigators referred to in Figure 2 mention blockage corrections. Hegge Zijnen [19] used a blockage factor of 6%. The drag coefficient calculated by him from pressure measurements on a circular cylinder in an "empty" (no turbulence generators present) tunnel was $C_D = 1.08$, which is substantially lower than the standard value at the same Reynolds number and is opposite to the supposed effect of blockage. Ko and Graf [21] used blockage factors up to 13%, but measured drag forces directly with a dynamometer and measured free-stream velocities at the minimum cross-section, thus probably avoiding the need for blockage corrections. Dyban et al. [22, 23] apparently used blockage factors as high as 25%, yet show very low drag coefficient values, and make no mention of corrections or the location of the tunnel speed measurement.

It seems reasonable to state, then, that the large differences between the various data sets of concern here are not due to blockage problems, so we must look elsewhere. The most likely culprit on which to blame the discrepancies is the existence of three-dimensional flow in what were intended to be two-dimensional experiments. This is examined in the next section.

Three-Dimensional Flow

The provision of truly two-dimensional flow appears to be one of the more difficult problems of fluid dynamics. Unwanted transverse flow components can arise for a number of particular reasons but the underlying cause is the establishment, in some way, of small cross-stream variations in the pressure field. Fluid close to a body surface has lost much of its momentum due to the action of viscosity and it readily flows in the direction of any pressure gradient it may meet, rather than in the stream direction. This tendency is especially apparent near separation, where the velocity profile is cusp-like. By this means, small spanwise pressure variations can result in relatively large crossflows in the boundary layer. Some of the causes can be difficult to detect and most are difficult to rectify.

Deviations in test object shape can cause perturbations in the flow geometry, introducing a three-dimensionality to the flow that may cause large errors in measured or calculated drag values. Such things as free ends, gaps, steps at segment joints, non-uniformities in pressure tap holes, wall penetration seals, support interference, and the like can all be troublesome in this regard. Free ends and gaps along the length of a bluff body are obvious sources of error since they permit flow directly from regions of high pressure to those of lower pressure, changing the pressure forces on the object being tested. In particular, the base pressure will be increased under such circumstances. However, this may not result in a reduction in drag because the separation point locations, wake width, and eddy-shedding phenomena may also be modified. As an example of this type of problem, Ko and Graf [21] had small gaps (size not given) and steps of "less than $\frac{1}{16}$ in." at the ends of the central portion of their test cylinders of 0.25 and 0.50 in. diameters. The steps were due to bending of the support strut linking the central segment to the dynamometer. Any effects of three-dimensional flow were ignored, but they may have been substantial in view of the relative size of step to cylinder diameter.

Some transverse flows can originate in the test facility itself, when they are usually termed "secondary" flows. It is noted that most of the investigations considered above have involved tunnels with square or rectangular cross-sections. Townsend [53] has stated that turbulent flow of a viscous fluid in a straight length of pipe having any section other than circular is always accompanied by a cross-axis secondary flow. He attributes this to a pressure anomaly, arising from the cross-axis fluctuating velocity components, in the wall boundary layer that tends to propel fluid from the corners of rectangular or square sections toward the midpoint of the walls and thence toward the center of the section. The result, as seen in section, is eight separate roughly-triangular eddies that bring fluid of relatively high momentum into the corners and prevent stagnation there. Gessner [54] disagrees with the suggested cause, proposing instead an explanation based on turbulent shear stress gradients. In any case, the end result is the same, with the secondary flow attaining values as high as a few percent of the main flow speed. Undoubtedly, there is some influence on the three-dimensionality of the flow about test objects in the tunnel, which may cause discrepancies in the data. To alleviate the problem when test section shapes other than circular are used, they should be made to approach this desirable optimum, if possible. In the case of square or rectangular sections, large triangular fillets can be used to fill out the corners.

Another possibility for transverse flow problems was reported by Kline et al. [55]. They found discrepancies in some boundary layer measurements, which they traced to secondary flow arising from suction applied to the tunnel roof. It had been overlooked initially because its effect on the mean flow was very small, but it had a substantial effect on their test plate boundary layer flow. Pressure leaks in the tunnel walls might cause a similar difficulty.

Perhaps the most troublesome source of three-dimensional flow irregularities arises at the junction of a two-dimensional body with the tunnel wall. The wall boundary layer pressure field influences, or "interferes" with, the pressure field around the body, resulting in modifications to the base pressure and other parameters associated with the object. To assist in reducing the interference, it is helpful to make the fraction of body length that is immersed in the wall boundary layer as small as possible. Herein lies the occasionally-noted [51] dependency of measured pressures and drag coefficients on body fineness (aspect) ratio. An additional precaution that should always be considered is the use of suitable end plates properly located to assist in isolating pressures on the major portion of the body from those in the boundary layer. There is no concensus on exactly what constitutes a "suitable" end plate since, as its area increases, it undoubtedly provides greater isolation between wall boundary layer and test object but, at the same time, it develops its own boundary layer to compound the problem. However, a little judicious experimentation in a given case should suggest an optimum end plate arrangement. Some guidelines for the case of square-sectioned cylinders have been provided by Obasaju [56]. As an example of the magnitude of the error involved, Bearman and Wadcock [57] reported that three-dimensional effects from the wall boundary layer, in the absence of end plates, were felt over the whole span of a circular cylinder of aspect ratio near 30. The value of the base pressure coefficient (as shown in their Figure 1) varied from about 0.85 at the ends to 1.22 in the middle. When suitable end plates were used, the coefficient value became nearly uniform across the span, at about 1.29. Thus, even with only a small fraction of the total length of the cylinder actually immersed in the boundary layer, the response was severe.

In the work under discussion, the previously mentioned precautions were, apparently, not observed by any of the investigators dealing with circular cylinders, and by only a few of those working with other bodies, so this may be the major common flaw. The results of Dyban et al. [22, 23] might come into particular question here because of the very small aspect ratios used by them. Their data points for a turbulence intensity of 0.3% (Figure 2) follow the same general trend as the standard drag curve but the drag coefficient values are substantially lower, as previously mentioned. This is consistent with an increased base pressure due to three-dimensional flow effects.

EFFECTS OF TURBULENCE

Some possible reasons for the variance in empirical results have been examined above, but perhaps some blame can be assigned to the implicit assumption that the influence of turbulence manifests itself through simple overall dependencies on intensity and scale. In fact, the problem may be far more complex, probably requiring, at least, some consideration of the spectral distribution of free-stream turbulent energy. In the following, the role of various turbulence characteristics is reviewed, with the possible influences on a laminar boundary layer considered first, since the subcritical regime is of particular concern here. Then a brief examination is made of the effect of turbulence on an already turbulent boundary layer.

As already mentioned, one result of the energetic random motions in free-stream turbulence is to increase the entrainment of mass or momentum into the shear layers at the surface of an object. This can influence skin friction, pressure drag, and vortex shedding. In addition to the importance of the intensity level, one would expect some effect due to the length scale of the turbulence. If most of the turbulent energy occurs at scales (wavelengths) substantially larger than the body size, the flow should appear as quasi-steady to the body, with slight additional variations of velocity superimposed on the mean flow. Scales of the same order as the object dimensions might be expected to have some influence on the wake structure, and scales that are of the same order as the shear layer thickness should be relatively much more effective in modifying the boundary layer characteristics, as some investigators have suggested. Also, these smaller scales are subject to distortion by the mean flow field, about which more later. Thus, for entrainment phenomena, certain high-frequency bands are likely to be of more concern than others of lower frequency, and a knowledge of the turbulence spectrum could provide more appropriate information than does a measurement of relative scale or intensity alone.

The entrainment of momentum into the laminar boundary layer of an object results in a "filling out" of a velocity profile near the surface, with a concomitant increase in skin friction. However, the major drag contribution is through form drag and one might anticipate some reduction in this component because of the entrainment, as follows. At relatively low values of free-stream turbulence intensity, entrainment increases the kinetic energy of the boundary layer, permitting greater penetration into regions of adverse pressure gradients (assuming that transition has not yet occurred through lack of the appropriate excitation or of a sufficiently high Reynolds number). For the case of smooth cylinders, this can result in a downstream displacement of the separation points and a narrower wake which might, of itself, be reason for expecting a drag reduction. However, the form drag is a function of both the wake width and the base pressure, so the latter must also be considered.

Immediately downstream of a bluff body, in the near wake, lies the vortex formation region, where the separated shear layers begin to interact. The time-averaged flow in this region presents the picture of a "recirculation bubble," or "wake bubble," attached to the afterbody. The bubble size is stabilized by a feedback mechanism caused by entrainment of fluid from within the bubble. If the bubble lengthens, the surface for entrainment increases. The increased entrainment volume results in a decreased base pressure, increasing the curvature of the bounding shear layers and drawing the bubble termination back to its original location. Additional fluid enters the bubble at its downstream terminus to replace the loss due to entrainment. It has been suggested by various experiments [36, 37] that the penetration of free-stream turbulence, of a scale such that the energy-containing eddies are less than the body size, into the shear layers increases the rate of entrainment from the bubble, reducing the base pressure and the bubble length.

A short bubble implies a short vortex formation region, which is consistent with a higher average drag because the shear layers are "rolled-up" into vortices before much diffusion of vorticity has occurred, resulting in large vortex strengths and lower core pressures. However, another effect of the increased entrainment of mass into the shear layers is to thicken them and enhance the diffusion of the organized vorticity there. This could delay roll-up and reduce the strength of the vortices formed, thereby increasing the base pressure and leading to a possible reduction in average drag. Indeed, according to Bearman and Graham [58], one paper presented at the Euromech 119 conference described how turbulence influences vortex shedding by increasing diffusion and cancellation of vorticity, and reported on an experiment with a two-dimensional bluff body in which it was found that the addition of free-stream turbulence increased the base pressure and reduced the drag. (Perhaps, for a circular cylinder, the latter process holds sway at low values of turbulence intensity with the former becoming more effective at higher intensities, thus producing a drag minimum followed by a gradual drag increase.)

A similar situation prevails in the case of separation bubbles. A shear layer, having separated in a laminar fashion, finds itself in very unstable circumstances and quickly undergoes transition to turbulence. This transition may be encouraged by the presence of free-stream turbulence, but occurs rapidly in any case. Then, entrainment into the shear layer from the body base region increases substantially, the pressure between the layer and the afterbody decreases, and shear layer curvature increases. If the afterbody has sufficient length as, for example, in the case of the rectangular cylinders previously mentioned, then reattachment takes place, so that the shear layer now encloses a low-pressure separation bubble. For circular cylinders, the formation of such a bubble is part of the critical regime phenomenon, where the end result is a decrease in wake width and in drag. One effect of free-stream turbulence is to cause reattachment at a lower Reynolds number than would otherwise be the case.

The wake proper, downstream of the wake bubble, is also subject to the influence of free-stream turbulence. There appears to be little information on this aspect of wakes in the literature, most investigators having been more concerned with the vortex formation region. Except at Reynolds numbers less than about 400, the wake is inherently turbulent, either because the shear layers are turbulent when they separate or because they become so as they roll-up to form turbulent-cored vortices. Thus, it seems likely that the principal result of a turbulent stream will be to change wake width and internal pressure through the mechanism of entrainment, with the usual proviso that the most effective eddy sizes are of the same order as the body size, or less. Komoda [59] reported that for the wake of a circular cylinder exposed to turbulence having an intensity of about 14% the velocity defect at the wake centerline was reduced somewhat, the wake width increased substantially, and the wake turbulence intensity also increased (as compared to the "undisturbed" free-stream, with an intensity of less than 0.2%). These effects were observed for downstream distances equivalent to several hundred cylinder diameters.

A further complicating factor in the role played by turbulence arises from the distortion of the eddy motion as it progresses past a body. As is well known, angular momentum is conserved in vortical motions so that, if an axis of vorticity should be stretched (lengthened) for some reason, continuity requires a reduction in the diameter of the associated eddy and the linear velocity must increase. This increases the rms intensity of any such distorted turbulent motion, and reduces its scale. In addition, an axis of vorticity may be bent by the advective acceleration terms in the equation of motion so that the vorticity becomes re-oriented. Thus, lateral vorticity vectors approaching a bluff body become longitudinal vectors in the shear layers, with increased intensity relative to the mean tunnel speed (but perhaps not with respect to the local speed). This kind of distortion can be responsible for three-dimensional effects even in a low-turbulence test facility, where the vorticity generated by the no-slip condition at the wall distorts to form a "horseshoe" vortex around any object extending from the wall.

Since distortion influences the interactions between stream vorticity and shear layer phenomena, a good understanding of the process is required. Several proposals have been advanced to account for the effects. Sutera et al. [60] suggested, as an explanation of the enhanced heat transfer from a circular cylinder in the presence of free-stream turbulence, a theory wherein vorticity amplification occurs in the boundary layer through the stretching of vortex filaments in the diverging flow at the stagnation point. In their theory, the cross-stream, cross-axis vorticity is strongly amplified provided that the wavelengths (eddy sizes) exceed a certain neutral value. For a circular cylinder, for example, the neutral value is about 2.6 times the boundary layer thickness (based on 0.99U). Thus, eddies smaller than this value should dissipate through viscous damping before distortion effects become important, and larger eddies with a suitable orientation should undergo intensity amplification and reorientation. One consequence of this process is the production of some organized structure in the boundary layer and a reduction in the degree of isotropy of the turbulence as it flows past an object. Sutera et al. found that the amplified turbulence in the boundary layer had a much greater effect on heat transfer than on skin friction. The theory was extended by Sadeh et al. [61] to the flow field at some distance from the boundary layer. Subsequently, Sadeh and Brauer [62, 63] conducted flow-visualization studies, the results of which supported the previous proposals.

Hunt [64] presented another approach to the problem of distortion, building on the "rapid-distortion" theory advanced by several earlier investigators [65]. His theory is applicable to situations involving periods shorter than the time required for the turbulence field to change under the action of its own viscous and inertial forces, i.e., it assumes that distortions due to the mean velocity about an object occur before eddies of a particular wavelength can exchange energy nonlinearly with other eddy wavelengths, hence the designation "rapid." The distortion field is complicated, the exact effect depending on the distance away from the object surface, the alongstream position, and the orientation and size of the component eddies of the turbulence. The essential predictions of the theory are that cross-stream, cross-axis components of the turbulence will be amplified by vortex filament stretching when the relative scale is much less than unity and will undergo attenuation when the relative scale is much greater than unity. For relative scale of the order of one, a combination of these effects will occur, with amplification at high wave numbers and attenuation at low wave numbers. Several investigators [66, 67, 68] have conducted experiments whose results are in reasonable agreement with theoretical predictions, indicating the approach is valid. More recently, Durbin and Hunt [69] have extended the theory to calculate the surface pressure fluctuations on a bluff body subjected to a turbulent flow. One prediction of the new work is that the fluctuations are influenced only by turbulent eddies on the stagnation streamline, which lends support to Gartshore's use of a single small rod for a turbulence generator, as previously mentioned [32].

Another effect of free-stream turbulence, previously referred to in connection with the critical regime for smooth cylinders, is the triggering of the Tollmien-Schlichting instability mechanism, which leads to the transition from laminar to turbulent flow in the boundary layer. (More properly, one should say that the T-S mechanism leads to nonlinear disturbances, which then cause transition due to the fluctuating pressure gradients.) Much of the research in this connection has been done on flat plates with no pressure gradient, but the results should be qualitatively correct for other cases. Here, the distribution of turbulent energy is important because the T-S modes respond only to certain preferred wavelengths, especially to those that are a multiple of the boundary layer thickness. Hence, the turbulence intensity and the integral length scale, which are actually average values of the pertinent variables, are not the most significant parameters with regard to the excitation leading to transition. The minimum wavelength for instability in a flat plate boundary layer is remarkably long, being equivalent to about six boundary layer thicknesses [25].

Once a certain value of the Reynolds number (the "critical" value based on streamwise distance in the boundary layer for a flat plate, or momentum thickness in the case of flow with a pressure gradient) is exceeded, the preferred wavelength disturbances are amplified, whereas those at smaller or larger wavelengths are damped. The selective linear amplification of the T-S waves continues to the point of distortion, when there is a sudden generation of irregular high-frequency motions termed turbulent bursts or spots. As the flow proceeds downstream, the appearance of spots becomes more frequent, their turbulent wakes spreading out and merging with one another, finally culminating in a united front of turbulence. Thus, transition is complete only after the flow has traversed some small distance from the critical point where disturbances first start to amplify. This distance, the transition region, decreases in extent as the intensity of the turbulence at the appropriate wavelengths is increased because a lesser amount of amplification suffices to force the disturbance to the point of burst generation. Finally, when the turbulence intensity reaches a value near 1%, the points of instability inception and final transition are in near coincidence.

If the turbulence intensity exceeds the above-mentioned value, then the T-S phenomenon is bypassed because the nonlinear disturbances already exist in the form of fluctuating pressure gradients produced by the random vortical motions. If the change in local gradient is large enough, in the adverse sense, it can produce a condition of separation for a short period of time, which is conducive to the production of turbulence, as proposed by Taylor [6]. It appears, then, that intensity is of prime consideration in the case of pressure-gradient-induced transition, with scale playing a relatively minor role provided it is of the right order (i.e., not so large as to give quasi-steady flow).

Turning now to the case of a boundary layer that is already turbulent, by virtue of high Reynolds number flow or of a sufficiently-rough surface, one wonders what additional influence the imposition of a turbulent free-stream may have. Kline et al. [55] state that the self-generated turbulence intensity in a turbulent boundary layer (on a flat plate exposed to a low-turbulence stream) is about 5–10% near the wall. When the free-stream turbulence intensity equals or exceeds the self-generated values, substantial changes occur in the boundary layer characteristics. For example, at a Reynolds number of 2×10^6, Kline et al. found that only small changes occurred in the boundary layer velocity profile until after the free-stream intensity exceeded 4%. Beyond that, as the intensity was increased to about 20%, the profile became much more full (as in the laminar case) and there was a moderate increase in the wall shear stress. The boundary layer thickness increased substantially and the intensity distribution changed quite dramatically, remaining high at a distance from the wall. As might have been anticipated, the turbulence characteristics in the boundary layer were no longer wall-dominated. Also, a "universal" velocity profile correlation, which one expects for the case of self-generated turbulence in a non-turbulent stream [3], did not exist at high levels of free-stream turbulence.

More recently, Charnoy et al. [70] have carried out a similar investigation to that mentioned in the previous paragraph, except their maximum value of free-stream turbulence intensity was just over 5%. Even so, they still observed similar changes, including a strong increase in boundary layer thickness, a slight but systematic increase in skin friction, a high turbulence intensity away from the wall, and a lack of self-preservation in the velocity profiles. In addition to varying the free-stream intensity, they also used integral length scales ranging from 0.6 to 2.2 cm. Little effect of this changing scale on boundary layer characteristics was noted, except that it controlled the integral scale size of the outer edge of the boundary layer. Thus, their conclusion was that turbulence intensity of the free-stream was a much more important parameter than scale for changes in a turbulent boundary layer, at least for the range of scales investigated. This seems reasonable in view of the fact that the observed changes are primarily due to entrainment phenomena, i.e., to the advection of higher energy fluid from the main flow.

CONCLUSION

It is apparent that the presence of turbulence in the free stream can modify bluff body flow phenomena in various ways through interactions with the boundary layer and near wake. It is also apparent that the state of knowledge concerning such interactions and modifications is far from complete. In some cases, the effects of turbulence are uncertain, or even controversial, and physical mechanisms are unclear. Also, the turbulence parameters of importance would seem to vary with the phenomenon considered. Properties averaged over the entire turbulent spectrum appear not to be wholly satisfactory for correlating the various effects; most often it seems that the intensity over a particular bandwidth is of primary importance. Free-stream turbulence increases entrainment into the shear layer at the surface of a body, or in its wake, affecting the energy distribution and altering the locations of transition, separation and reattachment. There is an enhanced diffusion of organised vorticity, presumably affecting eddy-shedding phenomena and base pressures. Distortion of the vorticity field, with resultant changes in intensity, occur in the flow about an object. All of these factors influence the fluid-dynamic forces on bluff bodies in ways that have not yet been fully or satisfactorily explained.

Some of the difficulties in determining the influence of turbulence may be attributed to problems with experimental techniques. We have examined, above, some possible sources of errors or discrepancies in empirical data. No definitive statements can be made regarding the validity of certain data sets, but it is quite clear that future investigations should pay particular attention to experimental detail, including such problems as three-dimensional flow, blockage, measurement accuracies, and an adequate description of the turbulence characteristics. Then we may improve our understanding of the complex dynamics involving bluff bodies immersed in turbulent fluids.

Acknowledgment

The permission of Pergamon Press Ltd. to use some material that originally appeared in their serial publication *Ocean Engineering* is gratefully acknowledged.

NOTATION

A	area
C_D	drag coefficient
C_{pb}	pressure coefficient
D	diameter
h	height
I	turbulence intensity
$L_{x,y}$	length scale
R	Reynolds number
V	velocity

Greek Symbols

α	angle of attack
ϵ	correction factor
η	nondimensional number

REFERENCES

1. Dryden, H. L., "Turbulence and the Boundary Layer," *J. Aero. Sciences*, Vol. 6, No. 3, Jan. 1939, pp. 85–105.
2. Riabouchinsky, D. P., "Wind-Tunnel Turbulence Effects," *J. Aero. Sciences*, Vol. 14, Mar. 1947, pp. 190–192.
3. Schlichting, H., *Boundary Layer Theory*, 7th ed. New York: McGraw Hill, 1979.
4. Templin, R. J., "Aerodynamics Low and Slow," *Can. Aero. Space J.*, Vol. 16, 1970, pp. 318–328.
5. Bowden, K. F., and Proudman, J., "Observations on the Turbulent Fluctuations of a Tidal Current," *Proc. Roy. Soc. A*, Vol. 199, 1949, pp. 311–327.
6. Taylor, G., "Statistical Theory of Turbulence. Part V. Effect of Turbulence on Boundary Layer," *Proc. Roy. Soc. A*, Vol. 156, 1936, pp. 307–317.
7. Schewe, G., "On the Force Fluctuations Acting on a Circular Cylinder in Cross Flow from Subcritical up to Transcritical Reynolds Numbers," *J. Fluid Mech.*, Vol. 133, 1983, pp. 265–285.
8. Brunn, H. H., and Davies, P. O. A. L., "An Experimental Investigation of the Unsteady Pressure Forces on a Circular Cylinder in a Turbulent Cross Flow," *J. Sound Vibration*, Vol. 40, 1975, pp. 535–559.
9. Fage, A., and Warsop, J., "The Effects of Turbulence and Surface Roughness on the Drag of a Circular Cylinder," *Aero. Res. Comm. Lond.*, R. & M. No. 1283, 1929.
10. Bearman, P. W., "Some Effects of Turbulence on the Flow Around Bluff Bodies," *Nat. Phys. Lab.*, NPL Aero Rept. 1264, 1968.
11. Batham, J. P., "Pressure Distributions on Circular Cylinders at Critical Reynolds Numbers," *J. Fluid Mech.*, Vol. 57, 1973, pp. 209–228.
12. So, R. M. C., and Savkar, S. D., "Buffeting Forces on Rigid Circular Cylinders in Cross Flows," *J. Fluid Mech.*, Vol. 105, 1981, pp. 397–425.
13. Surry, D., "Some Effects of Intense Turbulence on the Aerodynamics of a Circular Cylinder at Subcritical Reynolds Number," *J. Fluid Mech.*, Vol. 52, 1972, pp. 543–563.
14. Bell, W. H., "The Influence of Turbulence on Drag," *Ocean Engng.*, Vol. 6, 1979, pp. 329–340.
15. Hoerner, S. F., *Fluid Dynamic Drag*. Published by the Author. 1965.
16. Bacon, D. L., and Reid, E. G., "The Resistance of Spheres in Wind Tunnels and in Air," *Nat. Advisory Comm. Aero.*, Rept. No. 185, 1924.
17. Dryden, H. L., et al. "Measurements of Intensity and Scale of Wind Tunnel Turbulence and Their Relation to the Critical Reynolds Number of Spheres," *Nat. Advisory Comm. Aero.*, Rept. No. 581, 1937.
18. Neve, R. S., and Jaafar, F. B., "The Effects of Turbulence and Surface Roughness on the Drag of Spheres in Thin Jets," *Aeronautical J.*, Vol. 86, Nov. 1982, pp. 331–336.
19. Hegge Zijnen, B. G. van der., "Heat Transfer from Horizontal Cylinders to a Turbulent Air flow," *Appl. Sci. Res.*, Vol. A7, 1958, pp. 205–223.
20. Kiya, M., et al., "A Contribution to the Free-Stream Turbulence Effect on the Flow Past a Circular Cylinder," *J. Fluid Mech.*, Vol. 115, 1982, pp. 151–164.

21. Ko, S. C., and Graf, W. H., "Drag Coefficients of Cylinders in Turbulent Flow," *J. Hydraulics Div., Proc. ASCE*, Vol. 98, No. HY5, 1972, pp. 897–912.
22. Dyban, Ye. P., Epik, E. Ya., and Kozlova, L. G., "Effect of Free-Stream Turbulence on the Flow Past a Circular Cylinder," *Fluid Mechanics—Soviet Research*, Vol. 3, No. 5, Sept.–Oct. 1974, pp. 75–78.
23. Dyban, Ye. P., Epik, E. Ya., and Kozlova, L. G., "Combined Influence of Turbulence Intensity and Longitudinal Scale and Air Flow Acceleration on Heat Transfer of Circular Cylinder," *Proc. Vth Int. Heat Transfer Conf.*, Tokyo, 1974, pp. 310–314.
24. Dyban, Ye. P., Epik, E. Ya., and Suprun, T. T., "Characteristics of the Laminar Boundary Layer in the Presence of Elevated Free-Stream Turbulence," *Fluid Mechanics—Soviet Research*, Vol. 5, No. 4, July–Aug. 1976, pp. 30–36.
25. White, F. M., *Viscous Fluid Flow.* New York: McGraw Hill, 1974.
26. Criminale, W. O., "Interaction of the Laminar Boundary Layer with Free-Stream Turbulence," *Phys. of Fluids Supp.*, Vol. 10, No. 9, 1967, pp. S101–S107.
27. Bearman, P. W., and Trueman, D. M., "An Investigation of the Flow Around Rectangular Cylinders," *Aero. Quart.*, Vol. 23, Aug. 1972, pp. 229–237.
28. Nakaguchi, H., Hashimoto, K., and Muto, S., "An Experimental Study on Aerodynamic Drag of Rectangular Cylinders," *J. Japan Soc. Aero. Space Sci.*, Vol. 16, 1968, pp. 1–5.
29. Bostock, B. R., and Mair, W. A., "Pressure Distributions and Forces on Rectangular and D-shaped Cylinders." *Aero. Quart.*, Vol. 23, Feb. 1972, pp. 1–6.
30. Nakamura, Y., and Tomonari, Y., "The Effect of Turbulence on the Drags of Rectangular Prisms." *Trans. Japan Soc. Aero. Space Sci.*, Vol. 19, No. 44, June 1976, pp. 81–86.
31. Courchesne, J., and Laneville, A., "An Experimental Evaluation of Drag Coefficient for Rectangular Cylinders Exposed to Grid Turbulence." *J. Fluids Engng., Trans. ASME*, Vol. 104, Dec. 1982, pp. 523–528.
32. Gartshore, I. S., "The Effects of Free Stream Turbulence on the Drag of Rectangular Two-Dimensional Prisms." *Univ. of Western Ontario Boundary Layer Wind Tunnel Laboratory*, BLWT-4-73, Oct. 1973.
33. McLaren, F. G., Sherratt, A. F. C., and Morton, A. S., "Effect of Free-Stream Turbulence on the Drag Coefficient of Bluff Sharp-Edged Cylinders," *Nature*, Vol. 224, No. 29, 1969, pp. 908–909.
34. Lee, B. E., "Some Effects of Turbulence Scale on the Mean Forces on a Bluff Body," *J. Industrial Aero.*, Vol. 1, 1975/1976. pp. 361–370.
35. Roberson, J. A., et al., "Turbulence Effects on Drag of Sharp-Edged Bodies," *J. Hydraulics Div., Proc. ASCE*, Vol. 98, No. HY7, July 1972, pp. 1187–1203.
36. Bearman, P. W., "An Investigation of the Forces on Flat Plates Normal to a Turbulent Flow," *J. Fluid Mech.*, Vol. 46, 1971, pp. 177–198.
37. Humphries, W., and Vincent, J. H., "Near Wake Properties of Axisymmetric Bluff Body Flows," *Appl. Sci. Res.*, Vol. 32, Dec. 1976, pp. 649–669.
38. Castro, I. P., and Robins, A. G., "The Flow Around a Surface-Mounted Cube in Uniform and Turbulent Streams," *J. Fluid Mech.*, Vol. 79, 1977, pp. 307–335.
39. Nakamura, Y., and Ohya, Y., "The Effects of Turbulence on the Mean Flow Past Square Rods," *J. Fluid Mech.*, Vol. 137, 1983, pp. 331–345.
40. Scanlan, R. H., "Developments in Low-Speed Aeroelasticity in the Civil Engineering Field," *AIAA Journal*, Vol. 20, No. 6, June 1982, pp. 839–844.
41. Vickery, B. J., "Fluctuating Lift and Drag on a Long Cylinder of Square Cross-Section in a Smooth and in a Turbulent Stream," *J. Fluid Mech.*, Vol. 25, 1966, pp. 481–494.
42. Shapiro, P. J., "The Influence of Sound Upon Laminar Boundary Layer Stability," *MIT*, Cambridge MA, Acoustics, and Vibration Lab. Rept. No. 83458-83560-1, 1977.
43. Bell, W. H., "Turbulence *vs* Drag—Some Further Considerations," *Ocean Engng.*, Vol. 10, No. 1, 1983, pp. 47–63.
44. Reynolds, A. J., *Turbulent Flows in Engineering.* New York: John Wiley & Sons, 1974.
45. Hinze, J. O., *Turbulence*, 2nd ed. New York: McGraw-Hill, 1975.
46. Irwin, H. P. A. H., Cooper, K. R., and Girard, R., "Correction of Distortion Effects Caused by Tubing Systems in Measurements of Fluctuating Pressures," *J. Industrial Aero.*, Vol. 5, 1979, pp. 93–107.

47. Achenbach, E., "The Effects of Surface Roughness and Tunnel Blockage on the Flow Past Spheres," *J. Fluid Mech.*, Vol. 65, 1974, pp. 113–125.
48. Farell, C., et al., "Effect of Wind Tunnel Walls on the Flow Past Circular Cylinders and Cooling Tower Models," *J. Fluid Engng., Trans. ASME*, Series I, Vol. 99, 1977, pp. 470–479.
49. Maskell, E. C., "A Theory of the Blockage Effects on Bluff Bodies and Stalled Wings in a Closed Wind Tunnel," *Aero. Res. Comm. Lond.*, R & M No. 3400, 1965.
50. Fackrell, J. E., "Blockage Effects on Two-Dimensional Bluff body Flow," *Aero. Quart.*, Vol. 26, 1975, pp. 243–253.
51. West, G. S., and Apelt, C. J., "The Effects of Tunnel Blockage and Aspect Ratio on the Mean Flow Past a Circular Cylinder with Reynolds Numbers Between 10^4 and 10^5," *J. Fluid. Mech.*, Vol. 114, 1982, pp. 361–377.
52. Courchesne, J., and Laneville, A., "A Comparison of Correction Methods used in the Evaluation of Drag Coefficient Measurements for Two-dimensional Rectangular Cylinders," *J. Fluids Engng., Trans ASME*, Vol. 101, Dec. 1979, pp. 506–510.
53. Townsend, A. A., "Turbulence," in *Handbook of Fluid Dynamics*, V. L. Streeter (editor-in-chief), New York: McGraw-Hill, 1961, p. 10–29.
54. Gessner, F. B., "The Origin of Secondary Flow in Turbulent Flow Along a Corner," *J. Fluid Mech.*, Vol. 58, 1973, pp. 1–25.
55. Kline, S. J., Lisin, A. V., and Waitman, B. A., "Preliminary Experimental Investigation of Effect of Free-Stream Turbulence on Turbulent Boundary-Layer Growth," *NASA*, Tech. Note D-368, 1960.
56. Obasaju, E. D., "On the Effects of End Plates on the Mean Forces on Square Sectioned Cylinders," *J. Industrial Aero.*, Vol. 5, 1979, pp. 179–186.
57. Bearman, P. W., and Wadcock, A. J., "The Interaction Between a Pair of Circular Cylinders Normal to a Stream," *J. Fluid Mech.*, Vol. 61, 1973, pp. 499–511.
58. Bearman, P. W., and Graham, J. M. R., "Vortex Shedding from Bluff Bodies in Oscillatory Flow: A Report on Euromech 119," *J. Fluid Mech.*, Vol. 99, 1980, pp. 225–245.
59. Komoda, H., "On the Effect of Free-Stream Turbulence on the Structure of Turbulent Wake," *J. Jap. Soc. Aeronaut. Engng.*, Vol. 5, 1957, pp. 274–279.
60. Sutera, S. P., Maeder, P. F., and Kestin, J., "On the Sensitivity of Heat Transfer in the Stagnation-Point Boundary Layer to Free-Stream Vorticity," *J. Fluid Mech.*, Vol. 16, 1963, pp. 497–520.
61. Sadeh, W. Z., Sutera, S. P., and Maeder, P. F., "Analysis of Vorticity Amplification in the Flow Approaching a Two-Dimensional Stagnation Point," *Z. angew. Math. Phys.*, Vol. 21, 1970, pp. 699–716.
62. Sadeh, W. Z., and Brauer, H. J., "A Visual Investigation of Turbulence in Stagnation Flow about a Circular Cylinder," *J. Fluid Mech.*, Vol. 99, 1980, pp. 53–64.
63. Sadeh, W. Z., and Brauer, H. J., "Coherent Substructure of Turbulence Near the Stagnation Zone of a Bluff Body," *J. Wind Engng. & Industrial Aero.*, Vol. 8, 1981, pp. 59–72.
64. Hunt, J. C. R., "A Theory of Turbulent Flow Round Two-Dimensional Bluff Bodies," *J. Fluid Mech.*, Vol. 61, 1973, pp. 625–706.
65. Batchelor, G. K., and Proudman, I., "The Effect of Rapid Distortion of a Fluid in Turbulent Motion," *Quart J. Mech. & Applied Math.*, Vol. 7, 1954, pp. 83–103.
66. Bearman, P. W., "Some Measurements of the Distortion of Turbulence Approaching a Two-Dimensional Bluff Body," *J. Fluid Mech.*, Vol. 53, 1972, pp. 451–467.
67. Petty, D. G., "The Distortion of Turbulence by a Circular Cylinder," *Symposium on External Flows*, University of Bristol, 1972, pp. d.1–d.7.
68. Britter, R. E., Hunt, J. C. R., and Mumford, J. C., "The Distortion of Turbulence by a Circular Cylinder," *J. Fluid Mech.*, Vol. 92, 1979, pp. 269–301.
69. Durbin, P. A., and Hunt, J. C. R., "Fluctuating Surface Pressures on Bluff Structures in Turbulent Winds: Further Theory and Comparison with Experiment," *Wind Engineering, Proc. of 5th Int. Conf.*, Fort Collins, CO, July 1979. Oxford: Pergamon Press, 1980, pp. 491–507.
70. Charnoy, G., Comte-Bellot, G., and Mathieu, J., "Development of a Turbulent Boundary Layer on a Flat Plate in an External Turbulent Flow," *Turbulent Shear Flows*, AGARD CP-93, 1971, pp. 27.1–27.10.

CHAPTER 15

STABILITY OF FLAT PLATES AND CYLINDRICAL SHELLS EXPOSED TO FLOWS

Yuji Matsuzaki

Department of Aeronautical Engineering
Nagoya University
Chikusaku, Nagoya, Japan

CONTENTS

INTRODUCTION

It is a common observation that flags flutter in a strong wind. Fluttering represents loss of stability. When an elastic flat plate is exposed to subsonic or supersonic flow, it buckles or self-excitedly oscillates beyond certain critical flow speeds; so do flexible straight cylindrical shells placed in flows, or straight tubes conveying internal fluids. Buckling is referred to as static instability, and flutter or self-excited oscillation as dynamic instability. Such instabilities [1] are caused by aero- or fluid-elastic interaction that occurs between deformations of structural components and fluid pressures acting on them.

In engineering there are many structural components that must be designed so as to avoid this kind of instability. Among them are airplane wings [1], outer skins of supersonic flight vehicles [1], liquid-fuel and oil pipelines [2a, 2b], and chemical and atomic power plant piping [2c]. Unfavorable interactions between flow and wall structure deformation sometime imperil normal operation of the whole system. Instability may also cause structural damage, that is, almost immediate destruction or vibration-induced fatigue which shortens the life span of the structures.

In medical and biological areas, similar instability phenomena occur due to aero/fluid-elastic interaction between biological tubes and fluids, such as between blood vessels and blood, airways and air. Biomechanical investigations aim mainly at understanding and interpreting mechanisms of occurrence of phenomena such as limitation of expiratory flow from the lung [3a], production of Korotokoff sound in arterial blood vessels [3b], and violent oscillation of large veins observed during open heart surgery [3c]. These biological phenomena are much more complicated than those encountered in engineering. Airways and veins are very flexible and are even completely collapsed. The airway consists of a large number of subairways. Therefore, fluid flow characteristics are complex and often not well defined.

Apart from biological complexities, there is a simple but unsolved problem, that is, dynamic instability of a single tube conveying fluid at subsonic speeds, *viz*, "Under what condition(s) will a tube flutter?" This question has been found to be much more difficult to answer than expected. Similar problems exist for flexible plates exposed to subsonic flow. On the other hand, extensive theoretical and experimental studies were made in the late 1950s and '60s for instability in the supersonic flow range. Excellent review articles, text books, and monographs have been published on this subject [4–7].

In order for this chapter to be as self-contained as possible, the fluid forces in the stability analysis are rederived. Details of the fundamental analytical procedure are given in each of the first problems presented for fluid force and stability analyses. Unified analytical approaches are taken throughout this chapter so as to be able to compare the assumptions used and the results obtained. We exclusively treat two-dimensional plates and cylindrical shells (or tubes) of finite length with the leading and trailing edges supported. (A cantilevered pipe or a plate supported only at the leading edge has different stability characteristics [8].) As it is important to gain a deeper insight into the nature of complicated instability phenomena, simplified models are introduced in the analysis. A two-dimensional model cannot take into account the effect of circumferential geometry, which plays a very important role in nonlinear vibration of the cylindrical shell. Therefore, some details are inevitably given on limit cycle oscillation (flutter) of the shell, too.

As for the supersonic flow problem, the fundamental aspects of instability are presented from the same viewpoint as that for subsonic flow analysis, so that comparison between the two problems illustrates the instability characteristics pertinent to each flow. In addition, the comparison may suggest future directions of research for the subsonic case, where much remains to be resolved, especially for the tube-flow problem.

We first discuss unsteady potential flow theories, and then treat stability analyses. Next, we present a brief comparison between experiments and theories, and finally summarize major results obtained in this articles and unsolved problems. Controversial points are emphasized and critically reviewed. Related topics are briefly mentioned.

UNSTEADY POTENTIAL FLOW THEORIES

Basic to fluid-elastic stability analyses is the determination of unsteady fluid dynamic forces. Exact solutions to the potential flow equations are usually too complicated to apply to analytical examinations of stability. It is, therefore, very important to introduce appropriate approximations into evaluation of fluid forces in order to derive rather simple but still meaningful expressions. With such useful tools the very essence of stability can directly be spotlighted.

For subsonic flows, Ishii [9, 10], Weaver and Unny [11], Ellen [12], Kornecki et al [13], etc., evaluated fluid pressures on a harmonically oscillating two-dimensional plate of finite chord. However, some of the earlier investigations were claimed [12] to have errors. Hence, the low subsonic and incompressible fluid pressures were reexamined by Matsuzaki and Ueda [14] with the aid of operational method. They cast doubt on validity of the incompressible flow assumption itself for the case of oscillating plates, and also found that most previous analyses on the unsteady incompressible flow were not necessarily correct. In addition, using the operational approach, Matsuzaki and Fung [15–17] analyzed the unsteady fluid forces on a cylindrical shell and a two-dimensional channel with internal flows. As for supersonic flow analyses, stability of the plates and cylindrical shells are most often examined by using two-dimensional aerodynamic forces such as piston theory [18] and the two-dimensional quasi-steady force.

The first problem will outline the fundamental analytical approach to the evaluation of unsteady potential fluid forces. The flow is assumed to be inviscid and irrotational. Deflection of plates and cylindrical shells is sufficiently small compared with their length along the flow direction, so that linearized potential flow approximations can be used and the principle of superposition is applicable. In addition, it is assumed that the plates and the shells oscillate harmonically.

Two-Dimensional Plate

An inviscid fluid with free-stream velocity, U, flows past a two-dimensional flexible plate in the X direction as shown in Figure 1. A panel is connected to two semi-infinitely long, flat rigid walls at the leading and trailing edges, i.e., X = 0 and L. No static pressure differential across the plate is assumed to exist.

The velocity potential $\phi(X, Z, t)$ is described by

$$\partial^2\phi/\partial X^2 + \partial^2\phi/\partial Z^2 - (1/a_0)^2(\partial/\partial t + U\partial/\partial X)^2\phi = 0 \qquad \text{for } Z \geq 0 \tag{1}$$

where a_0 is the velocity of sound in the fluid. The velocity potential is related to the panel deflection w(X, t) by

$$(\partial\phi/\partial Z)_{Z=0} = (\partial/\partial t + U\partial/\partial X)w \tag{2}$$

The potential must satisfy a radiation condition

$$\lim \phi(X, Z, t) = 0 \qquad \text{as} \qquad (X^2 + Z^2)^{1/2} \to \infty \tag{3}$$

The perturbation pressure p(X, t) on the panel is given by Bernoulli's formula:

$$p = -\{\rho(\partial/\partial t + U\partial/\partial X)\phi\}_{Z=0} \tag{4}$$

where ρ is the density of the free stream. Assuming the plate deflection in the form of a standing wave, we set

$$w(X, t) = W(X)\exp(i\omega t) \tag{5}$$

with $\quad W(0) = W(L) = 0$

Consistent with this plate motion, the potential and the pressure are given by

$$\phi(X, Z, t) = \Phi(X, Z)\exp(i\omega t), \qquad p(X, t) = P(X)\exp(i\omega t) \tag{6}$$

Then, Equations 1, 2, and 4 can be rewritten as

$$\partial^2\Phi/\partial x^2 + \partial^2\Phi/\partial z^2 - M^2(\partial/\partial x + ik)^2\Phi = 0 \tag{7}$$

$$(\partial\Phi/\partial z)_{z=0} = U(d/dx + ik)W \tag{8}$$

$$P(x) = -\{(\rho U/L)(\partial/\partial x + ik)\Phi\}_{z=0} \tag{9}$$

where $\quad x = X/L,\ z = Z/L,\ k = \omega L/U,\ M = U/a_0 \tag{10}$

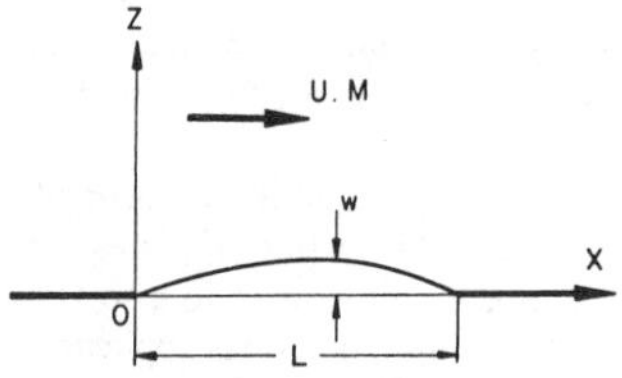

Figure 1. The geometry of a two-dimensional plate and the coordinate system.

Subsonic Flow (M < 1) [14]

First, we will evaluate subsonic fluid pressures. Applying the Fourier transformation, with respect to x, to Φ and W, i.e.,

$$\Phi^*(\xi, z) = (1/\sqrt{2\pi}) \int_{-\infty}^{\infty} \Phi(x, z) \exp(i\xi x)\, dx \tag{11a}$$

$$W^*(\xi) = (1/\sqrt{2\pi}) \int_{0}^{1} W(x) \exp(i\xi x)\, dx \tag{11b}$$

we obtain from Equations 7 and 8

$$d^2\Phi^*/dz^2 - \zeta^2\Phi^* = 0 \tag{12}$$

$$(d\Phi^*/dz)_{z=0} = iU(-\xi + k)W^* \tag{13}$$

where $\zeta = \{\xi^2 - M^2(\xi - k)^2\}^{1/2}$ (14)

The solution to Equation 12 is given by

$$\Phi^*(\xi, z) = A \exp(\zeta z) + B \exp(-\zeta z) \qquad \text{for } z \geq 0 \tag{15}$$

Satisfying Equation 13 and the radiation requirement, Equation 15 becomes

$$\Phi^*(\xi, z) = \{iU(\xi - k)/\zeta\}W^* \exp(-\zeta z) \tag{16}$$

The inverse Fourier transform of Equation 16 can analytically be evaluated:

$$\Phi(x, z) = \{iU/2(1 - M^2)^{1/2}\} \int_0^1 e^{iM\tau(x-\xi)} H_0^{(2)}[\tau\{(x - \xi)^2 + (1 - M^2)z^2\}^{1/2}] \times (d/d\xi + ik)W(\xi)\, d\xi \tag{17}$$

where $H_0^{(2)}$ is the Bessel function of the third kind and

$$\tau = Mk/(1 - M^2) \tag{18}$$

Equation 17 is equivalent to Equation 3 of Reference 9, where Ishii derived it with the aid of the source function. $\Phi(x, z)$ always vanishes as $\sqrt{(x^2 + z^2)}$ tends to infinity as long as the fluid is compressible.

Low Mach number approximation (M ≪ 1). Now, let us assume that

$$M \ll 1 \tag{19}$$

and that the order of k is of unity at most, i.e.,

$$0(k) \lesssim 1 \tag{20}$$

Now, Equation 17 is reduced, for $0 \leq x \leq 1$ and $z = 0$, to

$$\Phi(x, 0) = (U/\pi) \int_0^1 (\ln|x - \eta| + c)(d/d\eta + ik)W(\eta)\, d\eta \tag{21}$$

where $c = \ln(Mk/2) + \gamma^* + i\pi/2$ (22)

γ^* is Euler's constant and c may have a large negative value in its real part since $Mk \ll 1$. Substituting Equation 21 into Equation 9, we have

$$P(x) = (\rho U^2/\pi L)\left[\int_0^1 \{1/(x-\eta)^2 - 2ik/(x-\eta) - (ik)^2 \ln|x-\eta|\}W(\eta)\,d\eta - c(ik)^2 \int_0^1 W(\eta)\,d\eta\right] \tag{23}$$

Now, letting

$$W(x) = \sum_{m=1}^{N} \alpha_m W_m(x)$$

we will evaluate the generalized fluid forces defined by

$$Q_{mn} = L \int_0^1 P_m(x) W_n(x)\,dx \Big/ (\rho U^2/2) \tag{24}$$

where $P_m(x)$ is the pressure due to the deflection $W_m(x)$. It can be shown that the relations

$$Q_{mn} = (-1)^{m+n} Q_{nm} \tag{25}$$

hold [19], if the deflecction modes W_m and W_n are orthogonal and either symmetric or anti-symmetric:

$$\int_0^1 W_m(x) W_n(x)\,dx = 0 \qquad \text{if } m \neq n \tag{26}$$

$$W_m(x) = (-1)^{m+1} W_m(1-x) \tag{27}$$

Equation 25 shows that coupling between the odd and even modes is nonconservative [19]. When the panel is simply supported at the both edges, we may put

$$W_m(x) = \sin m\pi x \tag{28}$$

This is the m-th natural mode of the plate in vacuum. Substituting Equations 23 and 28 into Equation 24 gives

$$Q_{mn} = -Q_{mn}^{(0)} + (ik)^2 Q_{mn}^{(2)} \qquad \text{for } m+n = \text{even} \tag{29a}$$

$$Q_{mn} = ikQ_{mn}^{(1)} \qquad \text{for } m+n = \text{odd} \tag{29b}$$

where

$$Q_{nn}^{(0)} = 2n[\mathrm{Si}(n\pi) - \{1-(-1)^n\}/n\pi] \tag{30a}$$

$$Q_{nn}^{(2)} = (1/n\pi)^2[Q_{nn}^{(0)} - (4/\pi)[\mathrm{Ci}(n\pi) - \gamma^* - \ln(n\pi) + \{1-(-1)^n\}c]] \tag{30b}$$

$$Q_{mn}^{(0)} = 4mn\{\mathrm{Ci}(m\pi) - \mathrm{Ci}(n\pi) + \ln(n/m)\}/\{\pi(m^2-n^2)\} \qquad \text{for } m \neq n \tag{30c}$$

$$Q_{mn}^{(1)} = (8/\pi^2)\{n\mathrm{Si}(m\pi) + m\mathrm{Si}(n\pi)\}/(n^2-m^2) \qquad \text{for } m \neq n \tag{30d}$$

$$Q_{mn}^{(2)} = 4[n^2\{\mathrm{Ci}(m\pi) - \ln(m\pi) - \gamma^*\} - m^2\{\mathrm{Ci}(n\pi) - \ln(n\pi) - \gamma^*\}]/\{mn(m^2-n^2)\pi^3\} - (c/mn\pi^2)\{1-(-1)^m\}\{1-(-1)^n\} \qquad \text{for } m \neq n \tag{30e}$$

Si and Ci are the sine and cosine integral functions, respectively. The generalized forces given by Equations 29a and b are equivalent to those obtained for a low frequency limit $\omega L/a_0 \rightarrow 0$ in Reference 12.

Because of c in the last term of Equation 30b, the real part of $Q_{jj}^{(2)}$, where j is an odd integer, may be positive and large for a nearly incompressible flow. As discussed later under "Stability Analyses," it respresents a coefficient of virtual mass induced by the oscillating fluid. The effective mass consists of the generalized mass of the plate and the virtual mass. Therefore, no rapid oscillation in a natural symmetric mode is expected. If M tends to zero, then we may predict from Equations 22 and 30b that an infinite amount of virtual mass is induced and prevents the plate from oscillating in the symmetric mode.

However, it is clear from Equation 23 that a rapid oscillation in symmetric modes may occur if

$$\int_0^1 W(x)\,dx = 0 \tag{31}$$

since the term associated with c disappears. Let us consideer an arbitrary control surface enclosing the elastic plate. Then, we see that Equation 31 represents no change in fluid volume contained in the control volume. A simple symmetric mode that satisfies Equation 31 is given, for example, by

$$W_s(x) = (1/\sqrt{10})(\sin \pi x - 3 \sin 3\pi x) \tag{32}$$

Incompressible flow approximation ($M = 0$). When $M = 0$, we need to return to Equation 16. Because of $\zeta = |\xi|$, the inverse transform of Equation 16 is, for $z = 0$, written as

$$\Phi(x, 0) = -\{U/\sqrt{(2\pi)}\} \int_{-\infty}^{\infty} G(x - \eta)(d/d\eta + ik)W(\eta)\,d\eta \tag{33}$$

where

$$G(\eta) = \{1/\sqrt{(2\pi)}\} \int_{-\infty}^{\infty} \exp(-i\eta\xi)/|\xi|\,d\xi \tag{34}$$

The inverse Fourier integral, Equation 33, cannot be evaluated in an ordinary manner [20] because of a singularity at $\xi = 0$. However, if we resort to the theory of generalized functions [21], then Equation 34 becomes

$$G(\eta) = \{-2/\sqrt{(2\pi)}\}(\ln|\eta| + d) \tag{35}$$

where d is an arbitrary constant. Substitution of Equation 35 into Equation 33 yields

$$\Phi(x, 0) = (U/\pi) \int_0^1 (\ln|x - \eta| + d)(d/d\eta + ik)W(\eta)\,d\eta \tag{36}$$

Comparison of Equation 36 with Equation 21 shows that the disturbance velocity potential for $M = 0$ has the same form as that for $M \ll 1$. Therefore, the generalized forces for the incompressible flow can be determined from Equations 29 and 30 by substituting c for d. It should be noted, however, that Equation 21 is applicable only to the small range of x about the origin [14], whereas Equation 36 is valid for any value of x. When $|x| \gg 1$, Equation 36 is reduced to

$$\Phi(x, 0) = (ikU/\pi)(\ln|x| + d) \int_0^1 W(\eta)\,d\eta \tag{37}$$

Let us assume $k \neq 0$. As long as Equation 31 holds, the radiation condition is satisfied. If this is not the case, that is, if the fluid volume in the control surface is not constant during oscillation, then the radiation condition is violated unless d is set to a negative infinite number, such that $\ln|x| + d \to 0$ as $x \to \pm\infty$. In other words, if the deflection of the panel does not satisfy Equation 31, then no oscillation in such a mode can occur because of an infinitely large virtual mass induced. This agrees with the remark concerning $M \to 0$ given above.

Discussion

Let us begin with comparing Weaver and Unny's analysis [11]. There are two points to be mentioned. First, they partly introduced an assumption of a nearly incompressible flow to avoid the boundlessness of the pressure integral for M = 0. As is seen in Equation 34, we have the same problem. However, this has been solved by using the theory of generalized functions. Second, by arguing that the velocity potential does not attenuate along the z direction for $0 < \xi < \omega/a_0$, they changed the lower limit of the integral equation (18) of Reference 11 corresponding to the inverse Fourier transform of Equation (17) of this section. If we examine the equation (22) of Reference 14, we find the same situation there. However, without changing the limit, this transformation could be performed analytically to obtained $\Phi(x, t)$ defined by Equation 17 here. Because of an undue change of the integration limit, Weaver and Unny's velocity potential neglects a significant contribution, as pointed out by Ellen [12].

Ellen [12] asserted that Ishii [9] had errors in the generalized forces for subsonic flow, being suspicious of calculations based on the results for M = 0. However, it is considered that Ishii's numerical calculations for finite Mach numbers [9, 10b] were independent of the results for M = 0, since the generalized forces were evaluated by approximating the Hankel function in Equation 17 with an asymptotic expansion series with respect to its argument.

As for Ishii's incompressible flow analysis, however, his velocity potential, i.e., the equation (2) of Reference 9 or the equation (7) of Reference 10a, is not correct for those modes that do not satisfy Equation 31. Because the integrand of his integral velocity potential does not contain the constant d. As previously examined, when Equation 31 is dissatisfied, d must be negative infinite in order to fulfill the radiation requirement. In other words, Ishii's potential does not satisfy the radiation condition for a natural symmetric mode. The incompressible flow analyses of References 13, 22, etc. have the same defect.

In Table 1, the numerical results for the incompressible flow are presented in comparison with those of Reference 10a. When m and n are odd integers, that is, Equation 31 does not hold, $Q^{(2)}_{mn}$ calculated by the present analysis are infinite, whereas those of Reference 10a are finite. As previously mentioned, the radiation condition is violated in the latter. Table 2 shows $Q^{(2)}_{mn}$, (m, n:odd), at small Mach numbers with k = 1.0. They are complex numbers. Their real part grows as M decreases, and they become infinite at M = 0. The generalized forces associated with the symmetric mode described by Equation (32) are given in Table 3. It is noted that $Q^{(2)}_{ss}$ is smaller than the real part of $Q^{(2)}_{11}$ or $Q^{(2)}_{33}$ in Table 2. We may expect that the plate will oscillate more rapidly in the mode

Table 1
Generalized Forces[a] Q_{mn} for M = 0, Associated with Natural Modes [14]

n		m = 1	m = 2	m = 3
1	b	$-2.4304 + \infty(ik)^2$	$-1.3838ik$	$0.55461 + \infty(ik)^2$
	c	$-2.4247 + 0.45296(ik)^2$	$-1.3824ik$	$0.57312 + 0.06265(ik)^2$
2	b	d	$-5.6720 + 0.22230(ik)^2$	$-1.2327ik$
	c		$-5.6753 + 0.24173(ik)^2$	$-1.2320ik$
3	b	d	d	$-8.7759 + \infty(ik)^2$
	c			$-8.7755 + 0.13904(ik)^2$

[a] $Q_{mn} = -Q^{(0)}_{mn} + ikQ^{(1)}_{mn} + (ik)^2Q^{(2)}_{mn}$
[b] *Present analysis*
[c] *From Ishii [10a]*
[d] $Q_{mn} = (-1)^{m+n}Q_{mn}$

Table 2
$Q^{(2)}_{mn}$ at Small Mach Numbers with k = 1.0 for m and n Being Odd Numbers [14]

M	Real Part 0.01	0.001	0.00001	0	Imag. Part[a]
$Q^{(2)}_{11}$	1.6770	2.2711	3.4593	∞	−0.40529
$Q^{(2)}_{31}$ [b]	0.47067	0.66870	1.0648	∞	−0.13510
$Q^{(2)}_{33}$	0.27442	0.34043	0.47245	∞	−0.04503

[a] *Imaginary parts are independent of M.*
[b] $Q^{(2)}_{13} = Q^{(2)}_{31}$

Table 3
Generalized Forces for the Symmetric Mode Defined by Equation 32 [14]

Q_{s1} [a]	$-1.2947 + 0.083795(ik)^2$
Q_{s2}	$1.6070ik$
Q_{ss}	$-8.4741 + 0.13228(ik)^2$

[a] $Q_{ms} = (-1)^{m+1} Q_{sm}$

defined by Equation 32 than either in the first mode or third natural mode, since the virtual mass of the former is smaller than that of the latter. Therefore, it is reasonable that such a symmetric mode is considered together with the second natural mode in analyzing a coupled-mode flutter of the plate exposed to the incompressible flow, as discussed in "Stability Analyses."

Supersonic Flow ($M > 1$)

Now, we proceed to evaluating supersonic aerodynamic pressures acting on harmonically oscillating plates. Equations 7–9 are the equations from which the analysis restarts. Because no disturbance propagates upstream, supersonic flows can be treated with the aid of a Laplace transformation with respect to x:

$$\Phi^*_L(s, z) = \int_0^\infty \Phi(x, z) \exp(-sx)\, dx$$
$$W^*_L(s) = \int_0^\infty W(x) \exp(-sx)\, dx \qquad (38)$$

In the following, the subscript L is omitted because no confusion occurs between the Laplace and Fourier transforms, Φ^*_L and Φ^*, etc. Applying the Laplace transformation to Equations 7 and 8, we obtain

$$d^2\Phi^*/dz^2 - \zeta_L^2 \Phi^* = 0 \qquad (39)$$

$$(d\Phi^*/dz)_{z=0} = U(s + ik)W^* \qquad (40)$$

where $\partial\Phi/\partial x = \partial^2\Phi/\partial x^2 = 0$ at $x = 0$ have been used and

$$\zeta_L = \{M^2(s + ik)^2 - s^2\}^{1/2} \qquad (41)$$

The solution to Equation 39 is given by

$$\Phi^*(s, z) = -U(s + ik)W^* \exp(-\zeta_L z)/\zeta_L \tag{42}$$

which satisfies Equation 40 and the radiation condition, Equation 3. Application of the convolution theorem to Equation 42 yields

$$\Phi(x, 0) = -(U/\beta)\int_0^x \exp\{-ikM^2(x-\eta)/\beta^2\}J_0[kM(x-\eta)/\beta^2](\partial/\partial\eta + ik)W(\eta)\,d\eta \tag{43}$$

where $\beta = (M^2 - 1)^{1/2}$ (44)

Using Bernoulli's equation (Equation 9) and manipulating the resulting equation, we have

$$P(x) = (\rho U^2/L\beta)\Big[\{\partial/\partial x + ik(M^2-2)/(M^2-1)\}W(x) + \int_0^x W(\eta)(-\partial/\partial\eta + ik)\exp\{-ikM^2(x-\eta)/\beta^2\}\{(-ik/\beta^2)J_0[kM(x-\eta)/\beta^2] + (kM/\beta^2)J_1[kM(x-\eta)/\beta^2]\}\,d\eta\Big] \tag{45}$$

When $k^2 \ll 1$, we may obtain from Equation 45

$$p(x, t) = (\rho U^2/L\beta)[d/dx + \{(M^2-2)/(M^2-1)\}(L/U)d/dt]w(x, t) \tag{46}$$

for $M > \sqrt{2}$ [23]. Convergence of Equation 45 to Equation 46 is more rapid with increasing M. If M is sufficiently large, Equation 46 becomes

$$p(x, t) = (\rho U^2/LM)\{d/dx + (L/U)d/dt\}w(x, t) \tag{47}$$

Equation 46 is referred to as the two-dimensional quasi-static aerodynamic pressure and Equation 47 as piston theory [18]. When the plate motion is given by Equation 28, the generalized forces based on Equation 46 are:

$$Q_{mm} = ikQ_{mm}^{(1)} = (ik/\beta)(M^2-2)/(M^2-1) \tag{48a}$$

$$Q_{mn} = Q_{mn}^{(0)} = 4mn/\{\beta(n^2-m^2)\} \quad \text{for } m+n = \text{odd}$$
$$= 0 \quad \text{for } m+n = \text{even}, m \neq n \tag{48b}$$

Two-Dimensional Channel

We will restrict ourselves to internal subsonic flows with a free-stream velocity, U. The elastic upper and lower panels of a two-dimensional channel are connected to two semi-infinitely long rigid channels of the same width 2B. Three channels form a single infinitely long channel. As shown in Figures 2a and 2b, the deflection of the panels, w, is assumed in two different manners; either to be symmetric with respect to the X axis, or to be parallel in the Z direction.

The velocity potential, ϕ, and fluid pressure, p, are described by the same equations as Equations 1 and 4 with modification about the boundaries $Z = \pm B$. Instead of the radiation condition, it is required that ϕ remains finite at any X along the channel.

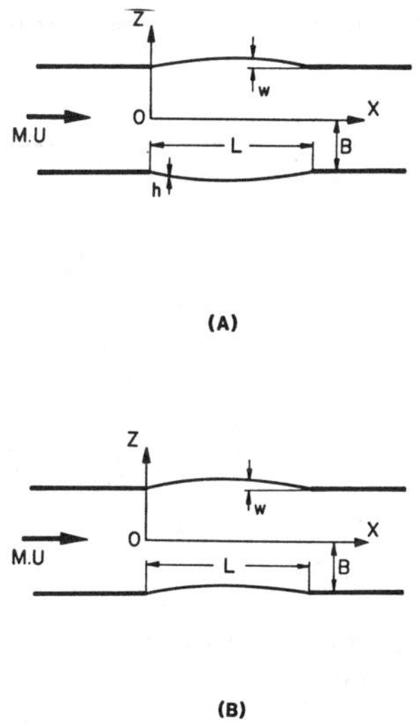

Figure 2. The geometry of a two-dimensional channel and the coordinate system: (A) symmetric deformation, (B) parallel deformation.

Symmetric deformation [17]. Owing to the symmetric wall motion, the flow pattern induced has symmetry with respect to the X axis. Therefore, only an upper half of the flow and channel is sufficient for analytical treatment. If the same analytical procedure is taken as in the single plate analysis, the Fourier transform of $\Phi(x, z)$, satisfying the finiteness condition, is written for $0 \le z \le b$ as

$$\Phi^*(\xi, z) = \{iU(-\xi + k)/\zeta\}(\cosh \zeta z/\sinh \zeta b)W^*(\xi) \tag{49}$$

where

$$b = B/L \tag{50}$$

If $M = 0$, then Equation 49 becomes

$$\Phi^*(\xi, z) = \{iU(-\xi + k)/|\xi|\}(\cosh \xi z/\sinh \xi b)W^*(\xi) \tag{51}$$

When k = 0, because of a singularity at $\xi = 0$, the inverse Fourier transformation of Equation cannot be carried out in an ordinary manner like that of the velocity potential for the single plate exposed to an incompressible flow. Therefore, we assume here $M \neq 0$. Then, the inverse transform of Equation 49 can analytically be evaluated [17]. When $M \ll 1$ and $W(x) = \sin m\pi x$, the fluid pressure for $z = b$ and $0 \le x \le 1$ is given by

$$\begin{aligned} P_m(x) = -(\rho U^2/m\pi L)\Big[& \coth bm\pi[\{-(m\pi)^2 + (ik)^2\} \sin m\pi x + 2ikm\pi \cos m\pi x] \\ & + b^2m^2 \sum_{s=1}^{\infty} [(-s\pi/b + ik)^2 \exp(-s\pi x/b) - (-1)^m(s\pi/b + ik)^2 \exp\{-(1 - x)s\pi/b\}] \\ & /\{s\pi(b^2m^2 + s^2)\} - (ik/2b)[(i/M)\{\sin kMx - (-1)^m \sin(1 - x)kM\} \\ & + 2\{\cos kMx + (-1)^m \cos(1 - x)kM\}]\Big] \end{aligned} \tag{52}$$

If M is so small that both cos kMx and sin kMx/kMx can be approximated by unity, the generalized forces Q_{mn} for m, n = 1, 2, are written as

$$Q_{mm} = -Q^{(0)}_{mm} + (ik)^2 Q^{(2)}_{mm} \tag{53}$$

$$Q_{mn} = ikQ^{(1)}_{mn} \qquad \text{for } m \neq n$$

where $Q^{(0)}_{mm} = m\pi \coth m\pi b - \bar{K}^{(0)}(m, m)$ (54a)

$Q^{(2)}_{mm} = \coth m\pi b/(m\pi) + (-1)^m 2^{3-2m}/(b\pi^2) + \bar{K}^{(2)}(m, m)$ (54b)

$Q^{(1)}_{mn} = 8mn \coth m\pi b/\{m\pi(n^2 - m^2)\} - 2\bar{K}^{(1)}(m, n) + 4(n - 2)/(b\pi^2)$ (54c)

$$\bar{K}^{(j)}(m, n) = (4mn\pi^2/b) \sum_{s=1}^{\infty} \{1 - (-1)^n \exp(-s\pi/b)\}(s\pi/b)^{-j+1} \times [\{(m\pi)^2 + (s\pi/b)^2\}\{(n\pi)^2 + (s\pi/b)^2\}]^{-1} \qquad \text{for } j = 0, 1, 2 \tag{55}$$

Parallel deformation [16]. When the deflections of the upper and lower panels are the same in the Z direction, the Fourier transform of Φ corresponding to Equation 51 for M = 0 is

$$\Phi^*(\xi, z) = \{iU(-\xi + k)/\xi\}(\sinh \xi z/\cosh \xi b)W^*(\xi) \tag{56}$$

Unlike Equation 51, Equation 56 is still inversely transformable. The generalized forces are evaluated in Reference 16.

For M = 0, the ordinary inverse transformation technique was applicable to the parallel deformation of both plates while it was inapplicable to the symmetric deformation because of the singularity. Any parallel motion demands no change in the volume of a control surface enclosing the flexible plates. However, this is not necessarily the case for the symmetric oscillation.

Numerical results. According to numerical evaluations, convergence of the infinite series of $\bar{K}^{(j)}$ (m, n) defined by Equation 55 and the similar infinite series for the parallel deformation is satisfactory with the first 20 terms being contained. Calculations for a wide range of parameters showed that $Q^{(0)}_{mm}$ and $Q^{(2)}_{mm}$ are positive for symmetric and parallel motions.

Circular Cylindrical Shell

An elastic circular cylindrical shell of length L is coaxially connected to two semi-infinitely long rigid tube with the same radius R_0 as shown in Figure 3. An inviscid fluid flows along the X axis either on the inside or outside of this cylinder. Symbols associated with the internal or external flow are distinguished by subscript 1 or 2, respectively.

The linearized equations for three-dimensional unsteady potential flows are

$$\left\{\frac{\partial^2}{\partial X^2} + \frac{\partial^2}{\partial R^2} + \left(\frac{1}{R}\right)\frac{\partial}{\partial R} + \left(\frac{1}{R^2}\right)\frac{\partial^2}{\partial \theta^2}\right\}\phi_j - \left(\frac{1}{a_{0j}^2}\right)\left(\frac{\partial}{\partial t} + U_j \frac{\partial}{\partial X}\right)^2 \phi_j = 0 \qquad \text{for } j = 1, 2 \tag{57}$$

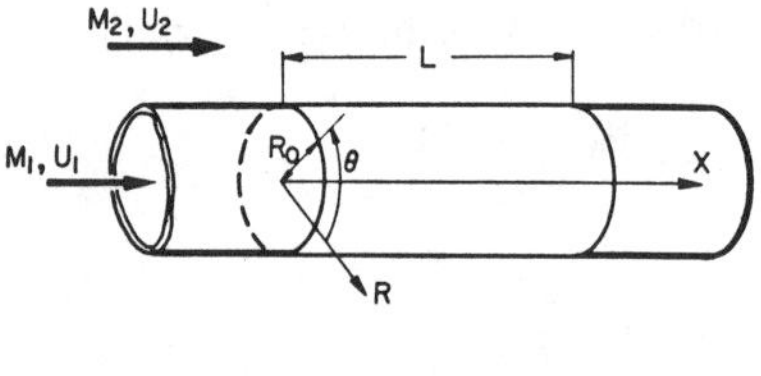

Figure 3. The shell geometry and the coordinate system.

and the relationship between the potential, ϕ_j, and the deflection of the tube, w, is described by

$$(\partial\phi_j/\partial R)_{R=R} = (\partial/\partial t + U_j\partial/\partial X)w \qquad \text{for } j = 1, 2 \tag{58}$$

where $R = R_j$ means $R = R_0-$ or $R = R_0+$ for the internal or external flow, respectively.

The fluid pressure on the internal or external surface is given by Bernoulli's equation:

$$p_j = -\rho_j(\partial/\partial t + U_j\partial/\partial X)_{R=R} \qquad \text{for } j = 1, 2 \tag{59}$$

The velocity potential of the internal flow, ϕ_1, is required to satisfy the finiteness condition and that of the external flow, ϕ_2, the radiation condition, respectively.

In the following, we shall consider two cases: an incompressible fluid pressure on the internal surface and a supersonic aeroforce on the external surface.

Incompressible Internal Flow with No External Fluid, i.e., $M_1 = 0$ and $\rho_2 = 0$ [15]

Let us omit the subscript 1 because no confusion would occur. Like in the preceding analyses, assuming w, ϕ and p in the standing wave forms as

$$\begin{aligned} w(x, \theta, t) &= W(x)\cos n\theta \exp(i\omega t) \\ \phi(x, r, \theta, t) &= \Phi_n(x, r)\cos n\theta \exp(i\omega t) \\ p(x, \theta, t) &= P(x)\cos n\theta \exp(i\omega t) \end{aligned} \tag{60}$$

and applying the Fourier transformation, we obtain from Equations 57 and 58:

$$d^2\Phi_n^*/dr^2 + (1/r)d\Phi_n^*/dr - [(\xi/\alpha)^2 + (n/r)^2]\Phi_n^* = 0 \qquad \text{for } 0 \le r \le 1 \tag{61}$$

$$(d\Phi_n^*/dr)_{r=1} = (iU/\alpha)(-\xi + k)W^* \tag{62}$$

where Φ_n^* and W^* are the Fourier transforms of Φ and W, respectively, and

$$x = X/L, \qquad r = R/R_0, \qquad \alpha = L/R_0 \tag{63}$$

Equation 61 is satisfied by the modified Bessel functions of order n. The solution, to Equation 61, that is finite and is subjected to Equation 62 is given by

$$\{\Phi_n^*(\xi, r)\}_{r=1} = iUV^*(\xi)W^*(\xi) \tag{64}$$

where $$V^*(\xi) = \{(-\xi + k)/\xi\}I_n(\xi/\alpha)/I_n'(\xi/\alpha) \tag{65}$$

with $I_n'(\xi)$ being the derivative of the modified Bessel function of the first kind $I_n(\xi)$.

When $n = 0$, the inverse transformation of $V^*(\eta)$ can be performed and

$$V^{(\pm)}(\eta) = i\sqrt{(2\pi)}\sum_{s=1}^{\infty}\{(\mp\alpha\sigma_{n\ell}' + ik)/\sigma_{n\ell}'\}\{J_n(\sigma_{n\ell}')/J''(\sigma_{n\ell}')\}\exp(\mp\alpha\sigma_{n\ell}'\eta) \qquad \text{for } \eta \gtrless 0 \tag{66}$$

J_n is the Bessel function of the first kind, and $\sigma_{n\ell}'$ is the ℓ-th root of $J_n'(\sigma) = 0$. Application of the convolution theorem to Equation 64 yields an analytical expression of $\Phi_n(x, r)$ on the cylindrical surface without difficulty like in the preceding section.

The generalized forces are given for a mode sin mπx cos nθ as

$$(Q_{mm})_n = -(Q^{(0)}_{mm})_n + (ik)^2(Q^{(2)}_{mm})_n \tag{67a}$$

$$(Q_{m\ell})_n = (Q^{(0)}_{m\ell})_n + ik(Q^{(1)}_{m\ell})_n + (ik)^2(Q^{(2)}_{m\ell})_n \qquad \text{for } m \neq \ell \tag{67b}$$

where

$$\begin{aligned}
(Q^{(0)}_{mm})_n &= \{(m\pi)^2/\alpha\}\bar{H}_n(m\pi/\alpha) + \bar{K}^{(0)}_n(m, m)\\
(Q^{(2)}_{mm})_n &= (1/\alpha)\bar{H}_n(m\pi/\alpha) - \bar{K}^{(2)}_n(m, m)\\
(Q^{(0)}_{m\ell})_n &= -\{1 + (-1)^{m+\ell}\}K^{(0)}_n(m, n)/2\\
(Q^{(1)}_{m\ell})_n &= \{1 - (-1)^{m+\ell}\}[\{8m\ell/\alpha(\ell^2 - m^2)\}\bar{H}_n(m\pi/\alpha) + 2\bar{K}^{(1)}_n(m, \ell)]/2\\
(Q^{(2)}_{m\ell})_n &= -\{1 + (-1)^{m+\ell}\}\bar{K}^{(2)}_n(m, \ell)/2
\end{aligned} \tag{68}$$

$$\bar{H}_n(m\pi/\alpha) = (\alpha/m\pi)I_n(m\pi/\alpha)/I'_n(m\pi/\alpha) \tag{69}$$

$$\bar{K}^{(j)}_n(m, \ell) = (4m\ell\pi^2/\alpha^4) \times \sum_{s=1}^{\infty} (\alpha\sigma_s)^{1-j}J_n(\sigma_s)\{1 - (-1)^{\ell}\exp(-\alpha\sigma_s)\} /[\sigma_s J''_n(\sigma_s)\{(m\pi/\alpha)^2 + \sigma_s^2\}\{(\ell\pi/\alpha)^2 + \sigma_s^2\}] \tag{70}$$

In Equation 70, σ'_{ns} is replaced by σ_s for the sake of simplicity.

Supersonic External Flow Without Internal Fluid, i.e., $M_2 > 1$ and $\rho_1 = 0$ [24]

Application of the Laplace transformation to Equations 57 and 58 yields

$$d^2\Phi^*_n/dr^2 + (1/r)\, d\Phi^*_n/dr - \{(n/r)^2 + \zeta^2\}\Phi^*_n = 0 \qquad \text{for } r \geq 1 \tag{71}$$

$$(d\Phi^*_n/dr)_{r=1} = (U/\alpha)(s + ik)W^* \tag{72}$$

where $\zeta^2 = (1/\alpha)^2\{M^2(s + ik)^2 - s^2\}$ (73)

The solution to Equation 71, subjected to Equation 72 and satisfying the radiation condition, can be written as

$$\Phi^*_n(s, 1) = (U/\alpha)[K_n(\zeta)/\zeta\, dK_n(\zeta)/d\zeta](s + ik)W^*(s) \tag{74}$$

where K_n is the modified Bessel function of the second kind. In order to evaluate the inverse transform defined above, it is necessary to take a numerical approach, as shown in Reference 24.

Assume $n \gg 1$ [25]. Now, let us assume that the circumferential wave number n is sufficiently large and that u_n/u_{n-1} may be approximated by unity where u_n is defined by

$$u_n = K_n(\eta)/K_{n-1}(\eta) \tag{75}$$

Since [26]

$$-\eta K'_n(\eta)/K_n(\eta) = n + \eta K_{n-1}(\eta)/K_n(\eta) = -n + \eta K_{n+1}(\eta)/K_n(\eta) \tag{76}$$

When $n \gg 1$, substitution of Equation 75 into Equation 76 yields a quadratic equation with respect to u_n. The solution to the quadratic equation is

$$u_n = \{n + (n^2 + \eta^2)^{1/2}\}/\eta \quad (77)$$

because $u_n > 0$. Substituting Equations 75 and 77 into Equation 74, we obtain

$$\Phi_n^*(s, 1) = -(U/\alpha)(n^2 + \zeta^2)^{-1/2}(s + ik)W^* \quad (78)$$

Application of the convolution and inverse transform theorems to Equation 78 and use of Bernoulli's equation with some manipulation give

$$P(x) = (\rho U^2/\beta L)\left[\{\partial/\partial x + ik(M^2 - 2)/(M^2 - 1)\}W(x) + (1/2)\int_0^1 W(\eta)I(x - \eta)\, d\eta\right] \quad (79)$$

where

$$I(\eta) = \{-(\bar{v}^2 + 2k^2/\beta^4)J_0(\bar{v}\eta) + (4ik\bar{v}/\beta^2)J_1(\bar{v}\eta) + v^2 J_2(\bar{v}\eta)\}\exp(-ikM^2\eta/\beta^2) \quad (80)$$

$$\bar{v} = \{(kM/\beta)^2 + (n\alpha)^2\}^{1/2}/\beta \quad (81)$$

If $k^2 \ll 1$ and $(n\alpha/\beta)^2 \ll 1$, then the last term, i.e., the integral in Equation 79 can be neglected and Equation 79 is reduced to Equation 46.

Slender body approximation ($L \gg R_0$) [24]. When α tends to infinity, it follows from Equation 73 that $\zeta \to 0$, and thus $K_n(\zeta)/\{\zeta K_n'(\zeta)\} \to -1/n$. The inverse transform of Equation 74 is, then, easily calculated and the pressure is expressed by

$$P(x) = (\rho U^2 R/nL^2)(d/dx + ik)^2 W(x) \quad (82)$$

Numerical results. Comparisons between the exact potential theory [24] based on Equation 74 and the approximated force given by Equation 79 were presented in References 25 and 27. According to Parthan and John's remark [27] on the approximation for $n = 10$ to 25, the agreement with the exact solutions is excellent unless the real or imaginary part of the generalized forces is very small. In this case, however, the percentage error becomes large and a change in sign may take place. In this regard, introduction of some minor modification to evaluate the approximated forces is necessary when they are used in stability analysis of the cylinder.

Subsonic and supersonic generalized forces on a harmonically oscillating cylindrical shell were numerically evaluated by Dowell and Widnall [28, 29]. For transient motions of cylindrical shells and plates, Dowell [30, 31] also performed numerical evaluations of supersonic generalized forces.

STABILITY ANALYSES

Flutter or divergence (buckling) of flexible plates and cylindrical shells does not necessarily represent an immediate failure because the deflection of these structural components is usually limited by edge constraints. Hence, it is of significance to know not only the critical speeds but also instability characteristics at flow speeds above the critical ones. Investigation of the structural behaviors with finite deflection needs plate and shell equations in which the effect of geometrical nonlinearity is considered.

For subsonic flows, it is commonly recognized that at a certain critical flow speed a flat plate or a straight cylinder loses static stability and starts to buckle with further increase in the flow speed. The postbuckling behavior is, however, not well understood yet, although many analytical investigations have been performed [9–17, 32–38]. Postbuckling flutter, especially, has often been a subject of controversy (e.g., discussion of [11]). Occurrence of the flutter was predicted by several investigators using linearized plate and shell theories and potential flow assumption. It will be discussed that such prediction is not appropriate because stability analyses based on nonlinear plate equations show no occurrence of postbuckling flutter.

On the other hand, supersonic panel flutter of both plates and cylindrical shells has fully been examined by theoretical and experimental studies, as mentioned previously. But only fundamental characteristics are discussed here so that comparison with subsonic flow cases be easily made.

In the preceding sections, the unsteady fluid forces have been derived under the assumption of harmonic oscillation. But it is assumed now that they are applicable to a slightly divergent or convergent oscillation. In other words, we assume that

$$(ik)^j Q_{mn}^{(j)} w(x, t) = Q_{mn}^{(j)} (L/U)^j \, d^j w(x, t)/dt^j \qquad \text{for } j = 1 \text{ and } 2 \tag{83}$$

for a vibration, w(x, t), whose characteristics is slowly changing. A basic analytical approach of stability will be outlined in the first example problem. We shall begin with stability analysis of a two-dimensional channel rather than that of a single plate because the former is more straightforward than the latter.

Two-Dimensional Channel [16, 17]

The geometry of the channel of width 2B and the coordinate system are shown in Figure 2. It is assumed that, in the initial no load condition, the flexible plates are assumed to be perfectly flat and their length is L. The upper and lower plates are assumed to deform parallelly or symmetrically with respect to the X axis when the channel conveys incompressible or low subsonic flow, respectively. In either case, the plates are able to vibrate in a natural symmetric mode, $\sin\{(2j + 1)\pi X/L\}$. It is sufficient to examine stability of either the upper plate or the lower plate alone. Von Karman's nonlinear equation of equilibrium of the upper plate undergoing cylindrical bending is given [39] by

$$\{Eh^3/12(1 - \nu^2)\} \, \partial^4 w/\partial X^4 + N_x \, \partial^2 w/\partial X^2 - p + c \, \partial w/\partial t + \rho_s h \, \partial^2 w/\partial t^2 = 0 \qquad \text{at } Z = B \tag{84}$$

where N_x is an induced in-plane force defined by

$$N_x = -\{Eh/2L(1 - \nu^2)\} \int_0^L (\partial w/\partial X)^2 \, dX \tag{85}$$

The boundary conditions for simply supported edges are

$$w = \partial^2 w/\partial X^2 = 0 \qquad \text{at } X = 0, L \tag{86}$$

Symbols w, h, E, ν, ρ_s, c and p are, respectively, the upward deflection of the middle plane of the upper plate, the plate thickness, Young's modulus, Poisson's ratio, density of the plate, coefficient of viscous structural damping and the fluid pressure.

Two-Mode Approximation

The plate motion is assumed to be in the form of

$$w(x, t) = \sum_{m=1}^{2} w_m(t) \sin m\pi x \qquad \text{for } 0 \le x \le 1 \tag{87}$$

This simple two-mode approximation enables us to analytically examine the stability to a larger extent. Equation 84 can be solved with the aid of Galerkin's method [40], since Equation 87 satisfies the boundary conditions (Equation 86). Substituting Equation 87 into Equation 84, multiplying the resulting equation by $\sin j\pi x$, $j = 1, 2$, and integrating from $x = 0$ to 1, we obtain

$$(a_1/\omega_0^2)\, d^2W_1/dt^2 + (\gamma/\omega_0)\, dW_1/dt + \{1 - Q_{11}^{(0)}Q + 3(W_1^2 + 4W_2^2)\}W_1 + (\mu Q)^{1/2}Q_{21}^{(1)}\, dW_2/dt = 0 \tag{88a}$$

$$(\mu Q)^{1/2}Q_{12}^{(0)}\, dW_1/dt + (a_2/\omega_0^2)\, d^2W_2/dt^2 + (\gamma/\omega_0)\, dW_1/dt + \{16 - Q_{22}^{(0)}Q + 12(W_1^2 + 4W_2^2)\}W_2 = 0 \tag{88b}$$

where

$$a_m = 1 + \mu Q_{mm}^{(0)} \tag{89}$$

$$W_m(t) = w_m(t)/h \tag{90}$$

$$Q = \{12(1 - \nu^2)/\pi^4\}(\rho U^2/E)(L/h)^3 \tag{91}$$

$$\mu = (\rho L/\rho_s h) \tag{92}$$

$$\gamma = (c/\pi^2)(L/h)^2\{12(1 - \nu^2)/E\rho_s\}^{1/2} \tag{93}$$

$$\omega_0 = (\pi/L)^2\{Eh^2/12(1 - \nu^2)\rho_s\}^{1/2} \tag{94}$$

The symbol W_m is redefined by Equation 90 and is now a function of t; in the preceding sections it was a function of X as defined by Equation 5. The coefficients of the generalized forces, $Q_{mn}^{(j)}$, are calculated by Equation 53 for symmetric motion of the upper and lower plates, or by the Equation 17 of Reference 16 for parallel motion. In Equations 88a and b, a_m/ω_0^2 effectively represents the mass associated with the m-th mode, $\sin m\pi x$. a_m consists of the generalized mass term of unity and the generalized force term, $\mu Q_{mm}^{(2)}$. Since $Q_{mm}^{(2)}$ is positive as mentioned previously, the effective mass is increased by the virtual mass induced by the oscillating fluid. If the virtual mass becomes infinite as in the case of the plate exposed to the incompressible flow, no motion can take place.

Static Equilibriums

Let W_{10} and W_{20} be the static equilibrium configurations of the first and second modes, respectively. Putting the time derivatives to zero in Equations 88a and b, we get three types of static equilibriums.

Flat Equilibrium:

$$W_{10} = W_{20} = 0 \qquad \text{for } Q > 0 \tag{95}$$

First-mode deflection:

$$W_{10} = \pm\{Q_{11}^{(0)}(Q - Q_1)/3\}^{1/2}, \qquad W_{20} = 0 \qquad \text{for } Q > Q_1 \tag{96}$$

Second-mode deflection:

$$W_{20} = \pm\{Q_{22}^{(0)}(Q - Q_2)/48\}^{1/2}, \qquad W_{10} = 0 \qquad \text{for } Q > Q_2 \tag{97}$$

where $\quad Q_m = m^4/Q_{mm}^{(0)} \qquad \text{for } m = 1, 2 \qquad (98)$

The first and second modes in equlibrium are always uncoupled, so that $W_{10}W_{20} = 0$ for any Q.

Stability of Static Configurations Against Infinitesimal Disturbances

Let us consider a disturbed motion slightly deviated from the equilibrium:

$$W_m(t) = W_{m0} + \alpha_m^* \exp(\epsilon\omega_0 t) \qquad \text{for } m = 1, 2 \tag{99}$$

On substituting Equation 99 into Equations 88a and b and neglecting the higher-order terms of α_1^* and α_2^*, we obtain a characteristic equation for nontrivial solutions of α_1^* and α_2^*:

$$C_4\epsilon^4 + C_3\epsilon^3 + C_2\epsilon^2 + C_1\epsilon + C_0 = 0 \tag{100}$$

where

$$C_4 = a_1 a_2, \qquad C_3 = \gamma(a_1 + a_2)$$

$$C_2 = \gamma^2 + a_1 I_2 + a_2 I_1 + \mu(Q_{12}^{(1)})^2 Q \tag{101}$$

$$C_1 = \gamma(I_1 + I_2), \qquad C_0 = I_1 I_2$$

$$I_1 = Q_{11}^{(0)}(Q_1 - Q) + 3(3W_{10}^2 + 4W_{20}^2) \tag{102a}$$

$$I_2 = Q_{22}^{(0)}(Q_2 - Q) + 12(W_{10}^2 + 12W_{20}^2) \tag{102b}$$

According to the Routh-Hurwitz criteria [41], the stability condition for the small disturbance to decrease with time is

$$C_4 > 0, \qquad C_3 > 0, \qquad C_2 > 0, \qquad C_1 > 0, \qquad C_0 > 0 \quad \text{and} \quad R^* > 0 \tag{103}$$

where

$$\begin{aligned} R^* &= C_1C_2C_3 - C_0C_3^2 - C_1^2C_4 \\ &= \gamma^2[(a_1 + a_2)(I_1 + I_2)\{\gamma^2 + \mu Q(Q_{12}^{(1)})^2\} + (a_1I_2 - a_2I_1)^2] \end{aligned} \tag{104}$$

It is well known that $R^* = 0$ and $C_0 = 0$ give the critical conditions for dynamic (periodic motion) and static (divergent motion) instabilites, respectively.

Flat equilibrium ($W_{10} = W_{20} = 0$). Below $Q = Q_1$, the stability condition (103) is always satisfied. But there is no stable flat configuration above $Q = Q_1$ because $C_0 < 0$ for $Q_1 < Q < Q_2$ and $C_0 > 0$ but $C_1 < 0$ for $Q > Q_2$.

First-mode deflection ($W_{10} \neq 0$). Substituton of Equation 96 into Equation 102a and b yields

$$I_1 = 2Q_{11}^{(0)}(Q - Q_1) = I_{11}, \qquad I_2 = 4Q_{11}^{(0)}(Q - Q_1) - Q_{22}^{(0)}(Q - Q_2) = I_{12} \tag{105}$$

Second-mode deflection ($W_{20} \neq 0$). Similarly, we have

$$I_1 = -I_{12}/4, \qquad I_2 = 2Q_{22}^{(0)}(Q - Q_2) \tag{106}$$

The first-mode buckling is stable for $Q_1 < Q < Q_2$, since Equations 103 are satisfied because $I_{11} > 0$ and $I_{12} > 0$. Generally speaking, in order to discuss the stability for $Q > Q_2$, it is necessary to evaluate Equations 103 numerically. However, in a range of Q above and close to Q_2 where $Q - Q_1$ is positive and finite and $Q - Q_2$ is positive but small, it can still be proved analytically that the first-mode deflection satisfies the stability condition whereas the second-mode deflection violates $C_0 > 0$. In this range, therefore, the first-mode deflection is still stable, but the second-mode buckling is unstable. It is evident that either the first- or the second-mode configuration must be unstable above $Q = Q_2$ because I_1 of the second mode and I_2 of the first mode have opposite signs.

Harmonic Balance Method

Next, we will examine the possibility of periodic flutter oscillation of finite amplitude about the flat or buckled configuration. Let us assume that the limit cycle solution of Equations 88a and b can be represented by

$$W_m(t) = A_{m0} + A_{ml} \cos(\omega t + \psi_m) \qquad \text{for } m = 1, 2 \tag{107}$$

Substituting Equation 107 into Equations 88a and b, and balancing the terms of the constants and first harmonics, we obtain six equations involving A_{mo}, A_{ml}, ω, and $\psi_2 - \psi_1$ as unknowns. For nontrivial solutions of A_{11} and A_{21}, we have

$$(\gamma\Omega)^2 + \{3(8A_{10}A_{20} \sin \Delta + A_{11}A_{12} \sin 2\Delta) - (\mu Q)^{1/2}Q_{12}^{(1)}\Omega \cos \Delta\}^2 = 0 \tag{108}$$

$$\text{where} \quad \Omega = \omega/\omega_0, \qquad \Delta = \psi_2 - \psi_1 \tag{109}$$

Since $\gamma \neq 0$, it follows from Equation 108 that

$$\Omega = 0, \quad \text{i.e.,} \quad \omega = 0 \tag{110}$$

and consequently that no limit cycle oscillation exists.

Numerical Integration Method

Equations 88a and b may numerically be solved by the Runge-Kutta-Gill method to examine the validity of analytical results, stability of static equilibriums against finite disturbances and possibility of flutter oscillations other than that defined by Equation 107.

Results and Discussions

Figure 4 illustrates W_{10} and W_{20} against the dynamic pressure for symmetric deformation of an air-conveying rubber passage. The parameters used in the calculation were $L/h = 100$, $b = 0.5$, $v = 0.5$, $\mu = 0.143$, and $\gamma = 0.01\gamma_c$ where γ_c is a critical damping coefficient of the first natural mode. A symbol $\bar{Q}$ used in Figure 4 is defined by

$$\bar{Q} = \{\pi^4/12(1 - v^2)\}(h/L)^3Q \tag{111}$$

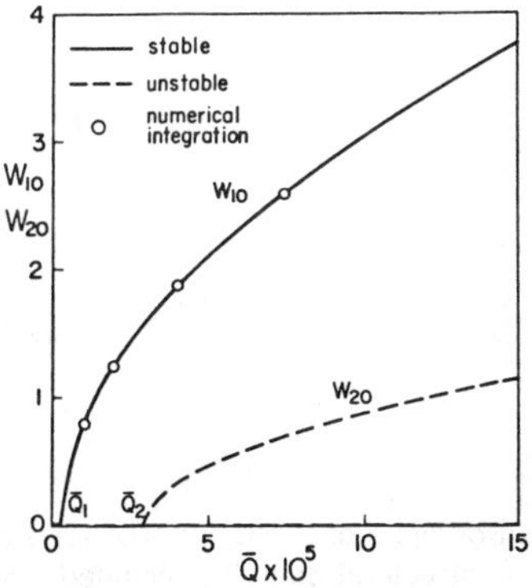

Figure 4. Amplitudes of the first and second mode bucklings and their stability versus dynamic pressure [17].

The horizontal coordinate represents the flat configuration also. Stable equilibrium is shown by a solid line, unstable equilibrium by a broken line. The flat configuration is stable up to $\bar{Q} = \bar{Q}_1$, that is, $Q = Q_1$ and the first-mode deflection appearing above $\bar{Q} = \bar{Q}_1$ is stable but the second mode is always unstable. Some of numerical results calculated by the Runge-Kutta-Gill method are also presented in Figure 4. Open circles indicate the values to which $W_1(t)$ converged. In most cases, the disturbed motions damped out and converged to the flat or buckled configurations, which were predicted to be stable by the stability analysis using the Routh-Hurwitz criteria, and no flutter was observed. However, if the disturbance was not small, then the oscillation of $W_2(t)$ was found to have extremely large amplitudes, such that the assumptions on which the flow and plate theories stood were clearly violated. At least as long as the disturbances remained comparatively small, the results obtained by the numerical integration method confirmed the stability characteristics predicted by the analysis based on the Routh-Hurwitz criteria.

Effect of Damping

If the damping coefficient γ is set to zero, then the characteristic equation (Equation 100) is reduced to a biquadratic equation

$$C_4\epsilon^4 + C_2\epsilon^2 + C_0 = 0 \tag{112}$$

Since C_4, C_2 and C_0 are real, if a set of $a' \pm ib'$ are a pair of complex conjugate roots, so are $-a' \pm ib'$, where a', b' are real. Hence, if Equation 112 has a complex root, then the small disturbances described by Equation 107 will increase oscillatorily and the static configuration is unstable. This means that, for stability, ϵ must be pure imaginary, i.e., ϵ^2 must be real and negative. Because $C_4 > 0$, the condition for stability is

$$C_2 > 0, \quad C_0 > 0, \quad \text{and} \quad D^* > 0 \tag{113}$$

where $$D^* = C_2^2 - 4C_4C_0 \tag{114}$$

Let Q_{C2} denote a pressure that satisfies $C_2 = 0$, and Q_{D1} and Q_{D2} be the pressures where $D^* = 0$. We will limit our discussion to the flat configuration ($W_{10} = W_{20} = 0$) and a case of $Q_{C2} > Q_2$. Then, D^* is positive at $Q = Q_2$ because $C_0 = 0$, but negative at $Q = Q_{C2}$ because $C_2 = 0$ and $C_0 > 0$ and, hence, $Q_2 < Q_{D1} < Q_{C2}$. Signs of C_2, C_0 and D^* can be easily evaluated as shown in Figure 5. It is seen that the flat configuration regains stability between Q_2 and Q_{D1}. However, as discussed previously, the flat configuration is always unstable above $Q = Q_1$ when $\gamma \neq 0$. Therefore,

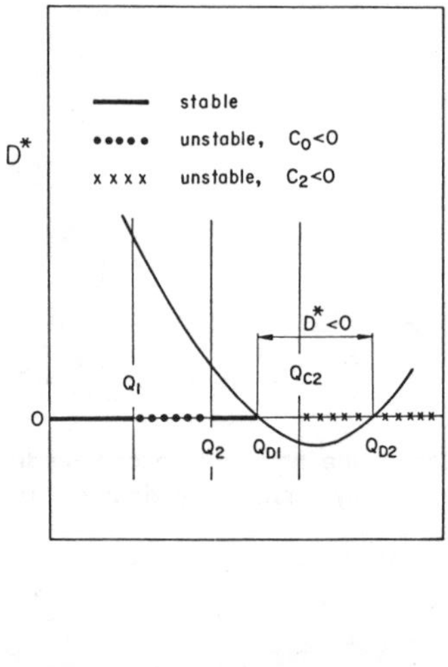

Figure 5. Evaluation of signs of C_2, C_0 and D^* for stability criteria with no structural damping included [15].

it is very important to take into account the damping that always exists in reality. This was first pointed out in a cylindrical shell problem [15]. In addition, the stabilized flat state becomes dynamically unstable (flutter) at $Q = Q_{D1}$ because $D^* = 0$ (and $C_0 > 0$).

There are many parameters that are influential on the buckling speed and postbuckling behaviors. Matsuzaki and Fung [16, 17] examined the effects of an initial inplane force applied to the plates, pressure differential across the channel plates and a distributed spring supporting the plate on the outside.

It should be noted in passing that, although the generalized forces are functions of kM for the case of symmetric deformation, they were approximately expressed by Equations 53 to be constant because M was assumed to be sufficiently small. If this is not the case, Equations 88a and b no longer represent a set of simultaneous ordinary differential equations with constant coefficients, and the stability cannot analytically be treated as shown here.

Two-Dimensional Plate

As shown in Figure 1, a two-dimensional plate of thickness h is simply supported at both edges with fluid flowing over at main stream velocity U. The flexible plate is assumed to be flat without flow. The equation of equilibrium of the plate at $z = 0$ and the edge conditions to use in the analyses are the same as Equations 84 and 86 for the channel plates.

Incompressible Flow [42]

When the fluid is incompressible, no motion of the plate in a symmetric natural mode such as $\sin(2m + 1)\pi x$ is possible as discussed in the derivation of the unsteady fluid force.

Three-mode approximation. In order to permit examination of instability due to dynamic coupling between modes that are symmetric and antisymmetric with respect to the midpoint of the chord, let the deflection of the plate be

$$w(x, t) = W_1 \sin \pi x + w_2(t) \sin 2\pi x + w_s(t)(\sin \pi x - 3 \sin 3\pi x) \tag{115}$$

The last term associated with $w_s(t)$ is a symmetric mode defined by Equation 32. This mode can oscillate even with exposition to the incompressible fluid because its generalized force is finite as shown in Table 3.

Substituting Equation 115 into the equation of equilibrium and applying Galerkin's method, we obtain a set of three ordinary differential equations involving W_1, W_2 and W_s. The time derivative of W_1 always vanishes. The mathematical structure regarding W_2 and W_s are similar to that of W_1 and W_2 in Equations 88a and b.

Stability analysis. Like in the two-dimensional channel case, the stability of static equilibrium configurations against small and finite disturbances can be analyzed by using the Routh-Hurwitz criteria and the Runge-Kutta-Gill method, respectively. The harmonic balance method is also applicable to examine flutter oscillation of finite amplitude. Figure 6 shows two sets of static deflections of the symmetric coupled-mode ($W_{10} \neq 0$, $W_{s0} \neq 0$, $W_{20} = 0$) and that of the second mode ($W_{20} \neq 0$, $W_{10} \neq W_{s0} = 0$) as well as their stability against the small disturbances. Solid and thin lines represent stability and instability, respectively. The stability of the flat configuration ($W_{10} = W_{20} = W_{s0} = 0$) is also indicated on the horizontal coordinate. It is stable up to $\bar{Q} = \bar{Q}_2$, that is, $Q = Q_2$. There are two symmetric coupled-mode configurations. (Actually, four configurations, two of which have a different sign but the same absolute value as the remaining two.) One exists above $\bar{Q} = \bar{Q}_{s-}$ and the other above $\bar{Q} = \bar{Q}_{s+}$. The former is stable but the latter is unstable. The second-mode deflection indicated by a chain curve above $\bar{Q} = \bar{Q}_2$ is also stable.

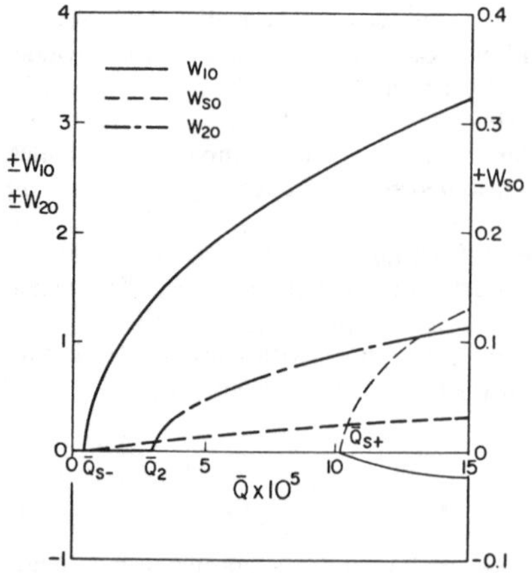

Figure 6. Amplitudes of symmetric and anti-symmetric mode bucklings and their stability (solid line: stable, thin line: unstable) [42].

Comparison with the channel conveying compressible flow. Let us compare Figure 6 with Figure 4 for the two-dimensional channel conveying a compressible, but nearly incompressible flow. The main difference between both results is the number of the stable static equilibrium states. At any dynamic pressure there always exists only a single stable pattern for the channel plate. On the other hand, there may be two patterns for the single plate. The addition of one more stable pattern is considered to be caused by the limitation on modal oscillation due to the assumption of incompressible flow. If the compressibility were accounted for, then, for instance, the flat configuration which is stable in the range between $\bar{Q}_{s-}$ and $\bar{Q}_2$ would not remain stable since the disturbance in the first natural mode might increase with time. This comparison suggests that the assumption of incompressibility of the flow may induce an excess stable static state. Therefore, it is very necessary to exercise a great care when the incompressible flow theories are used for stability problems.

Subsonic Flow

The generalized forces for subsonic flows are functions of k and M. Therefore, the governing equations of modal deflection are no longer ordinary differential equations with constant coefficients like Equations 88a and b as discussed in the stability analysis of the channel plates. It is, therefore, inevitable to investigate stability with the aid of numerical approach from the very beginning. Ishii [9, 10b] calculated the stability boundary using a linearized plate equation, which is derived from Equation 84 by setting $N_x = 0$. The fluid pressure was determined from the exact potential flow solution, i.e., Equation 17. The integral was numerically evaluated by approximating the Hankel function in the integrand with an asymptotic expansion series with respect to the argument of the function. The flat plate became unstable always due to buckling. As for a square flat plate, using von Karman's nonlinear plate theory, Dowell [43] showed that buckling occurred with no postbuckling flutter.

Supersonic Flow [23, 44]

The equation of equilibrium of the plate and the boundary conditions are again described by Equations 84 and 86. The aerodynamic pressure is evaluated by the two-dimensional quasi-steady expression, Equation 46.

Two-mode approximation. The two-mode approximation defined by Equation 87 is used to represent the plate motion. Then, the governing equations with respect to the modal amplitudes can be derived with the aid of Galerkin's method as

$$(1/\omega_0^2)\, d^2W_1/dt^2 + (\bar{\gamma}/\omega_0)\, dW_1/dt + \{1 + 3(W_1^2 + 4W_2^2)\}W_1 + Q_{21}^{(0)}QW_2 = 0 \quad (116a)$$

$$(1/\omega_0^2)\, d^2W_2/dt^2 + (\bar{\gamma}/\omega_0)\, dW_2/dt + 4\{4 + 3(W_1^2 + 4W_2^2)\}W_2 - Q_{21}^{(0)}QW_1 = 0 \quad (116b)$$

where $\bar{\gamma} = \gamma + \gamma_a$ (117)

$$\gamma_a = Q_{mm}^{(1)}(\mu Q)^{1/2} \quad (118)$$

The symbols W_m, Q, μ, γ, ω_0 are defined by Equations 90–94, and $Q_{21}^{(0)}$ is calculated from Equation 48b. Without initial compressive in-plane forces, the possible static equilibrium state is flat at any Q.

Stability analysis. The characteristic equation with respect to small disturbances about the flat static equilibrium is defined by the same equation as Equation 100 where the coefficients are given by

$$C_4 = 1, \quad C_3 = 2\bar{\gamma}, \quad C_2 = \bar{\gamma}^2 + 17, \quad C_1 = 17\bar{\gamma}, \quad C_0 = 16 + (8Q/3\beta)^2 \quad (119)$$

where β is given by Equation 44.

All the coefficients C_j's are always positive. According to the Routh-Hurwitz criteria defined by Equation 103, therefore, the plate becomes unstable (dynamically) only when the last equation of the criteria, $R^* > 0$, is violated. Because $\bar{\gamma}^2 \ll 1$, R^* is reduced to

$$R^* = (16\bar{\gamma}/3)^2(Q_f^2 - Q^2)/\beta^2 \quad (120)$$

where $Q_f = 45\beta/16$ (121)

For $0 < Q < Q_f$, $R^* > 0$ and, therefore, the flat configuration is stable. But if $Q > Q_f$, then $R^* < 0$. Hence, small disturbances will increase in an oscillatory way about the flat configuration which is always stable from a static view point because $C_0 > 0$ is satisfied for any Q.

Harmonic balance method. The next step is to examine whether or not the limit cycle oscillation exists especially in a region where the flat equilibrium is predicted to be dynamically unstable. Assuming the first and second axial modes of a periodic oscillation in the same form as Equation 107 with $A_{m0} = 0$ and applying the harmonic balance method, we obtain

$$\bar{\gamma}\Omega\delta^* = (16Q/3\beta)\delta \sin \Delta \quad (122)$$

$$(4Q/3\beta)(1 - \delta^2) = -A_{11}^2\delta\delta^* \cos \Delta \quad (123)$$

$$\delta[15 + A_{11}^2\{\tfrac{5}{4} + \cos 2\Delta + \delta^2(10 - \cos 2\Delta)\}] + (8Q/3\beta)\delta^* \cos \Delta = 0 \quad (124)$$

$$\Omega^2\delta^* = 1 + A_{11}^2\{\tfrac{3}{4} + \delta^2(2 + \cos 2\Delta)\} + \delta^2[16 + A_{11}^2\{(2 + \cos 2\Delta) + 12\delta^2\}] \quad (125)$$

where Δ and Ω are defined by Equation 109 and

$$\delta = A_{21}/A_{11}, \delta^* = 1 + \delta^2 \quad (126)$$

Elimination of A_{11} from Equations 123 and 124 yields

$$Q = (45\beta/\cos \Delta)\delta\delta^*/(1 + 11\delta^2 - 44\delta^4) \quad (127)$$

Because $\bar{\gamma}$ is very small compared with unity, Equation 122 reduces to

$$\sin \Delta = 0, \text{ that is, } \Delta = 0 \text{ or } \pi \quad (128)$$

It follows from Equation 123 that

$$\begin{aligned} &\delta > 1 && \text{for } \Delta = 0 \\ &0 < \delta < 1 && \text{for } \Delta = \pi \end{aligned} \tag{129}$$

Taking into account Equations 127 and 129, we determine

$$\Delta = \pi \quad \text{and} \quad \delta_m < \delta < 1 \tag{130}$$

where δ_m ($\doteqdot 0.5664$) is a solution of $1 + 11\delta^2 - 44\delta^4 = 0$

The upper limit $\delta = 1$ corresponds to the critical boundary $Q = Q_f$ predicted by the stability analysis based on small disturbances. It follows from Equation 127 that Q increases from Q_f with decreasing δ from 1 and tends to infinity as δ approaches δ_m. For any Q above Q_f, there always exists a limit cycle solution. Its amplitude ratio δ, amplitudes A_{11} and A_{21}, and frequency ω are calculated by Equations 127, 123, 126, and 125. The numerical results obtained by the harmonic balance method were confirmed by solving Equations 116a and b directly with the aid of an analogue computer [44] and the numerical integration method [23].

Circular Cylindrical Shell

The shell geometry and a small curved element of the shell are shown in Figure 7. Symbols u, v, w denote axial, circumferential and radial displacements of the middle plane of the shell, and N_x, N_θ, $N_{x\theta}$ are in-plane normal and shear forces, respectively. The flexible shell of thickness h is simply supported at both edges and coaxially connected to two rigid tubes with the same radius R_0. (See also Figure 3.) Donnell's nonlinear equations of equilibrium and compatibility of a thin cylindrical shell [45] are given by

$$\begin{aligned} &\{Eh^3/12(1-\nu^2)\}\nabla^4 w + N_\theta/R - N_x\, \partial^2 w/\partial X^2 - (N_\theta/R_0^2)\, \partial^2 w/\partial\theta^2 \\ &- (2N_{x\theta}/R_0)\, \partial^2 w/\partial X\, \partial\theta + \rho_s h\, \partial^2 w/\partial t^2 + c\, \partial w/\partial t + p = 0 \end{aligned} \tag{131}$$

$$\nabla^4 F = (E/R_0^2)\{(\partial^2 w/\partial X\, \partial\theta)^2 - \partial^2 w/\partial X^2\, \partial^2 w/\partial\theta^2 + R_0\, \partial^2 w/\partial X^2\} \tag{132}$$

where $\nabla^2 = \partial^2/\partial X^2 + (1/R_0^2)\partial^2/\partial\theta^2$ (133)

F is a stress function defined by

$$N_x = (h/R_0^2)\, \partial^2 F/\partial\theta^2, \qquad N_\theta = h\, \partial^2 F/\partial X^2, \qquad N_{x\theta} = -(h/R_0)\, \partial^2 F/\partial X\, \partial\theta \tag{134}$$

Internal Incompressible Flow ($M_1 = 0$, $\rho_2 = 0$) [15]

Let us consider here stability of the flexible cylinder within the framework of Donnell's linear shell theory. When the shell has no structural damping and is subjected to no fluid pressure, that

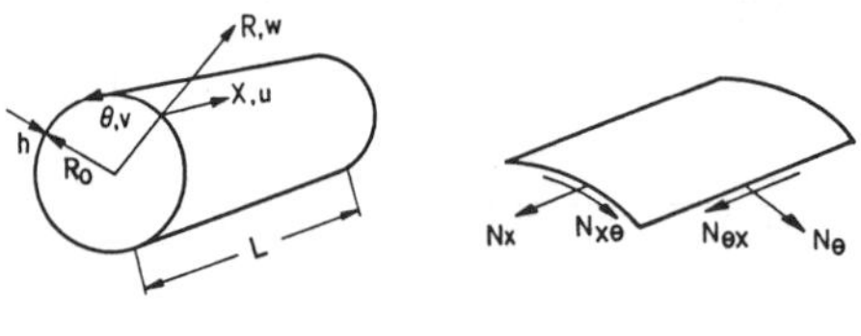

Figure 7. The shell geometry and a curved small element.

is, p = c = 0, the linearized versions of Equations 131 and 132 have the solutions

$$w = w_{mn} \sin(m\pi X/L) \cos n\theta \cos \omega_{mn} t \tag{135a}$$

$$F = F_{mn} \sin(m\pi X/L) \cos n\theta \cos \omega_{mn} t \tag{135b}$$

with being subjected to Equation 86. Equations 135a and b represent the radial deflection and in-plane stress state for a natural vibration. In addition to the simply supported conditions (Equation 86), Equations 135a and b satisfy the in-plane boundary conditions given by

$$N_x = v = 0 \qquad \text{at } X = 0 \text{ and } L \tag{136}$$

Eliminating F from the linearized versions of Equations 131 and 132, we have

$$\{Eh^3/12(1-\nu^2)\}\nabla^4 w + (Eh/R_0^2)\nabla^{-4}\,\partial^4 w/\partial X^4 + \rho_s h\,\partial^2 w/\partial t^2 + c\,\partial w/\partial t - p = 0 \tag{137}$$

The radial deflection is again assumed to consist of the first two axial modes:

$$w = \sum_{m=1}^{2} w_m(t) \sin m\pi x \cos n\theta \tag{138}$$

$$\text{where} \quad x = X/L \tag{10a}$$

Application of Galerkin's method to Equation 137 yields a set of two linear ordinary differential equations with respect to w_1 and w_2. Mathematical structures of the two equations are essentially the same as those of the Equations 88a and b with the nonlinear terms being deleted. According to the linear stability analysis [15], when the circumferential mode is fixed at n, the straight tube is stable below $Q = Q_{cr}$ where Q_{cr} is the critical boundary, that is, the smaller of $(Q_1)_n$ and $(Q_2)_n$ and

$$(Q_m)_n = L_{m0}/(Q^{(0)}_{mm})_n \qquad \text{for } m = 1, 2 \tag{139}$$

$$L_{m0} = \{m^2 + (n\alpha/\pi)^2\}^2 + \{12(1-\nu^2)m^4/\pi^4\}(\alpha L/h)^2/\{m^2 + (n\alpha/\pi)^2\}^2 \tag{140}$$

$$\alpha = L/R_0 \tag{63}$$

$(Q^{(0)}_{mm})_n$ is calculated by Equation 68.

Numerical results. In Figure 8 buckling boundaries $(Q_1)_n$ and $(Q_2)_n$ of the first and second axial modes for n = 2 to 4 are plotted against L/R_0 where $\bar{Q} = \{\pi^4/12(1-\upsilon^2)\}(h/L)^2Q$. As Donnell's shell equation is not accurate for small n [45], the actual numerical calculation was performed by using

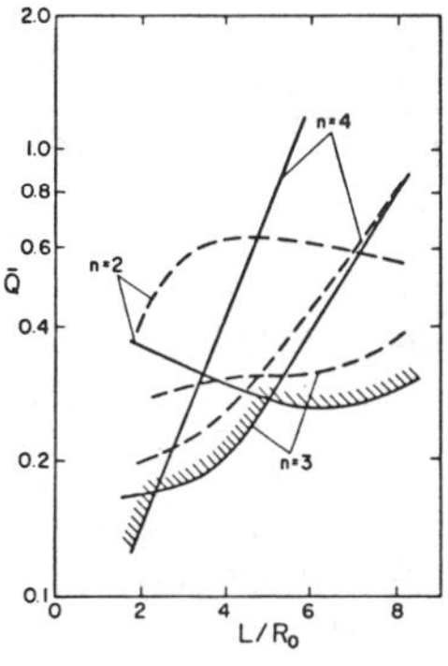

Figure 8. Divergence boundaries of a cylindrical shell versus length-to-radius ratio [15].

Morley's modification to Donnell's equation [15, 45]. This means that the first term of the right-hand side of Equation 140, $\{m^2 + (n\alpha/\pi)^2\}^2$, is replaced by $\{m^2 + (n^2 + 1)(\alpha/\pi)^2\}^2$. We need to search for the circumferential wave number n that gives the lowest boundary. The critical boundary is indicated by hatch marks. The circumferential wave number of the critical mode decreases as L/R_0 increases. When damping of the shell is accounted for, the undeformed, straight state is always unstable above the critical boundary like for the flat plate. In order to examine the static and dynamic behaviors at pressures above the critical boundary, it is necessary to consider the goemetrical nonlinearity in the analysis as shown in the plate and channel problems.

External Supersonic Flow ($M_2 > 1$, $\rho_1 = 0$) [46, 50]

To begin with, technical complexity of nonlinear vibration analysis of a cylindrical shell shall briefly be mentioned. Assumption of the radial deflection defined by Equation 135a is no more appropriate in the nonlinear analysis [47], because Equations 135a and b do not satisfy the circumferential continuity condition

$$\oint \partial v/\partial\theta \, d\theta = R_0 \int_0^{2\pi} [(1/E)\{\partial^2F/\partial X^2 - (\upsilon/R_0^2)\, \partial^2F/\partial\theta^2\} - w/R_0 - (1/2)(\partial w/R_0\, \partial\theta)^2]\, d\theta = 0 \tag{141}$$

This is a requirement that the circumferential displacement be continuous and single-valued around the circumference.

Two-mode approximation. Consistent with Equation 141, the deflection may be given as

$$w(x, \theta, t) = \{w_1(t) \sin \pi x + w_2(t) \sin 2\pi x\} \cos n\theta - (n^2/4R_0)\{w_1(t) \sin \pi x + w_2(t) \sin 2\pi x\}^2 \tag{142}$$

Because of additional nonlinear terms, however, Equation 142 dissatisfies one of the simply supported conditions, i.e., $\partial^2w/\partial x^2 = 0$ at $x = 0$ and 1. In addition, the in-plane boundary conditions, Equation 136 is satisfied only in an average sense around the circumference:

$$\oint N_x \, d\theta = \oint v \, d\theta = 0 \qquad \text{at } x = 0 \text{ and } 1 \tag{143}$$

In addition, we assume that the two-dimensional quasi-steady approximation defined by Equation 46 can substitute for the pressure, p, acting on the cylinder. Application of the generalized Galerkin method [40] to Equation 131 yields the governing equations with respect to the dimensionless amplitudes $W_1(t)$ and $W_2(t)$ as

$$(1/\omega_0^2)(d^2W_1/dt^2 + \lambda V_{11}/8) + (\gamma/\omega_0)(dW_1/dt + \lambda V_{12}/8) + Q_{21}^{(0)}Q(W_2 + \lambda V_{13}/5) + W_1\{L_{10} + L_{11}W_1^2 + L_{12}W_2^2 + L_{13}W_1^4 + L_{14}W_1^2W_2^2 + L_{15}W_2^4\} = 0 \tag{144a}$$

$$(1/\omega_0^2)(d^2W_2/dt^2 + \lambda V_{21}/8) + (\gamma/\omega_0)(dW_2/dt + \lambda V_{22}/8) - Q_{21}^{(0)}Q(W_1 + \lambda V_{23}/5) + W_2\{L_{20} + L_{21}W_1^2 + L_{22}W_2^2 + L_{23}W_1^4 + L_{24}W_1^2W_2^2 + L_{25}W_2^4\} = 0 \tag{144b}$$

where

$$V_{11} = (3W_1^2 + 2W_2^2)\, dW_1^2/dt^2 + 4W_1W_2\, d^2W_2/dt^2 + W_1\{3(dW_1/dt)^2 + 2(dW_2/dt)^2\} + 4W_2\, dW_1/dt\, dW_2/dt, \text{ etc.,} \tag{145}$$

$$\lambda = (n^2h/R_0)^2 \tag{146}$$

Q, ω_0 and γ are defined by Equations 91, 94, and 117, respectively. L_{mj} for j = 1 to 5 are functions of n and geometrical parameters of the shell like L_{m0} defined by Equation 140. Generalized forces are calculated by Equations 48a and b like in the case of the plate exposed to the supersonic flow. If R_0 tends to infinity, then Equations 142 and 144 are ultimately reduced to Equations 87 and 116 for the flat plate, respectively.

With no initial in-plane forces and no external pressure applied, the static equilibrium configuration is given as the undeformed state ($W_{10} = W_{20} = 0$) and becomes unstable due to dynamic coupling at a certain dynamic pressure Q_f when the disturbances are assumed to be small in the stability analysis. This instability characteristics is the same as that of the plate exposed to the supersonic flow. It has been shown that the plate has a single stable limit cycle at a pressure above $Q = Q_f$. But, as will be seen, the cylinder may have multiple limit cycle solutions. Stability characteristics of limit cycles of the cylinder is much more complicated. The limit cycle analysis will be given along the line of Matsuzaki and Kobayashi [46a].

Method of averaging and numerical integration. Instead of the harmonic balance method, Equations 144a and b are solved by using the method of averaging [48], which can also examine the stability of limit cycles. The limit cycle solutions are assumed to be in the same form as Equation 107. Like the nonlinear analyses discussed previously, the limit cycle solutions are also investigated by numerically integrating the modal Equations 144a and b with the aid of the Runge-Kutta-Gill method.

Numerical example. In Figure 9, limit cycle solutions for the n = 14 mode are plotted against Q/β. $\bar{A}_{10}$, etc, represent average values of A_{10}, etc, over one cycle. f denotes flutter frequency, $\omega/2\pi$. On the horizontal coordinate, the terminal point of the amplitude curves corresponds to the flutter boundary Q_f predicted by a linearized analysis [46B] with the circumferential mode being fixed at n = 14. Limit cycle solutions were found to exist at dynamic pressures, which were not only above but also below Q_f. At the pressures below Q_f there were two solutions. Only for the smaller amplitude do the amplitude curves approach the flutter boundary Q_f as Q increases. The limit cycles at Q above Q_f are associated with the larger amplitude oscillations.

Figure 10 illustrates the amplitude $\bar{A}_{10}$ for the circumferential modes n = 12 to 24. Stability and instability of the limit cycles against small disturbances examined by the method of averaging are also indicated by open circles and crosses, respectively. The critical, i.e., lowest flutter boundary was $(Q_f)_{cr}$ = about 280β for the mode n = 20. Many limit cycle solutions are seen to exist at any dynamic pressure except near Q = 0. However, the number of stable solutions is very limited. There was none at Q near $(Q_f)_{cr}$. But the numerical integration method showed that there were undamped limit cycle oscillations in the region where the limit cycle solutions had been predicted to be *unstable* by the method of averaging. Such undamped vibrations were of nonsinusoid and beating. In Figure 11, the amplitudes $\bar{A}_{10}$ and $\bar{A}_{20}$ for the flutter oscillations for n = 14 are presented by squares and triangles, respectively. The amplitude curves are redrawn from Figure 9. For the beating vibrations, the average values of the amplitudes are given by open symbols. Numerical integrations produced no flutter that was considered to correspond to the limit cycles with the smaller amplitudes. Deviation of the vibration from a sinusoidal motion grew with decreasing Q in the *unstable* region. In addition, it was necessary to give sufficiently large disturbances in order to establish the (nonsinusoidal) limit cycles at the pressures below $Q = Q_f$. With further decrease in

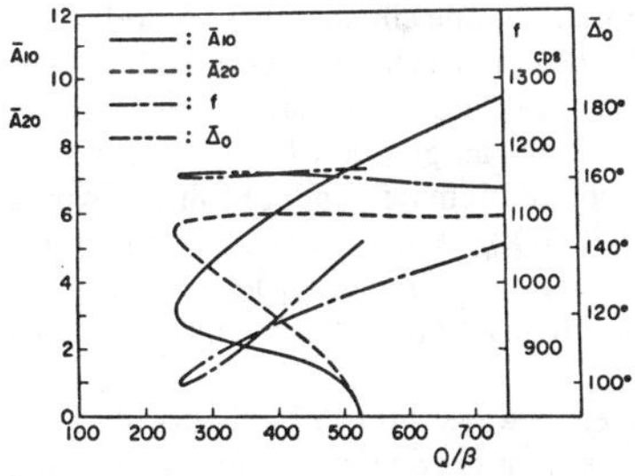

Figure 9. Limit cycle solutions of the n = 14 mode versus dynamic pressure [46a].

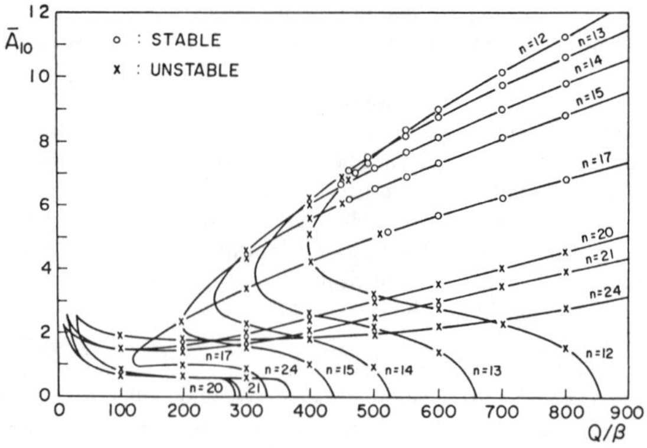

Figure 10. Amplitudes of the first axial mode for the n = 12 to 24 modes versus dynamic pressure [46a].

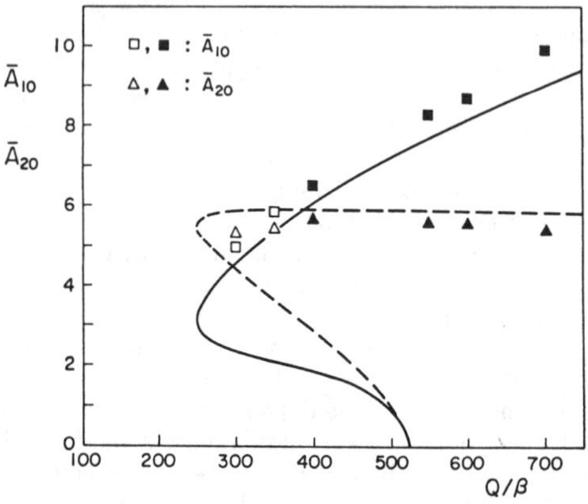

Figure 11. Comparison between amplitudes of the n = 14 mode obtained by the methods of numerical integration and averaging [46a].

Q, it became increasingly difficult to find undamped oscillations even though disturbed motions started from the limit cycle solutions, which had been calculated by the method of averaging. For example, all the disturbed motions examined by the numerical integration damped out below $Q = 280\beta$ for n = 14. For the mode n = 17, the lowest pressure at which a sustained vibration existed was Q = about 250β. This was lower than the critical flutter boundary $(Q_f)_{cr}$.

Every numerical integration mentioned above was carried out using Equations 144a and b for a single circumferential mode n. No mutual interaction of more than one circumferential modes was investigated. However, if a finite amplitude vibration is sustained in one circumferential mode, then it would act as a finite disturbance and excite other modal vibrations so that a limit cycle of another circumferential mode may develop and take over the finite amplitude vibration. Therefore, when Q increases and reaches $(Q_f)_{cr}$, the circumferential mode of ultimately induced flutter would not necessarily coincide with the mode directly associated with $(Q_f)_{cr}$. Furthermore, if disturbances are sufficiently large, then a limit cycle would be established even at a pressure below and close to $(Q_f)_{cr}$. This would be also true if the pressure decreases gradually from a value which is higher than the critical flutter boundary, $(Q_f)_{cr}$, and penetrates it.

Evensen and Olson [50] analyzed circumferentially traveling wave flutter of cylindrical shells. The complex feature of stability mentioned above has its origin in softening type nonlinear vibration characteristics of the cylindrical shell: the resonance amplitude decreases with increasing frequency [61, 62]. (The flat plate has an opposite relationship between the resonance amplitude and frequency.) An excellent review on nonlinear vibration of the cylindrical shell was written by Evensen [47].

COMPARISON BETWEEN THEORIES AND EXPERIMENTS

Because of the space limitation, the preceding sections have treated mainly stability of the simplest cases, that is, the initially unloaded, two-dimensional flat plates and straight cylinders exposed to or conveying uniform potential flows. In fact, there are many influential parameters on stability, including the edge restraint conditions, the initial plate and shell imperfections, in-plane loads, pressure differential across the walls, damping of the walls and the flows, flow conditions, etc. For meaningful comparisons between experimental results and theoretical analyses, we need to take into account all these experimental conditions in the analyses. Details of experimental parameters are not necessarily described in the literature. This is partly because it is very difficult to control all the experimental conditions and define them quantitatively.

For *supersonic flow ranges*, a large number of experiments were conducted to provide a vast amount of data. Generally speaking, agreement between theory and experiment is good for flutter boundaries of flat plates under a variety of in-plane applied loads and over a considerable range of Mach number. Agreement on flutter frequency is also good. As for cylindrical shells, experimental data and analyses are scarce. Direct comparison between theory and experiment is very limited. Olson and his coworkers [49, 50] and Matsuzaki and Kobayashi [46a] performed both experiments and nonlinear flutter analyses.

According to one of the experimental results at $M = 2$ [46a], which corresponds to the numerical calculations shown in Figures 9 to 11, an unpressurized cylinder made of steel alloy onset flutter with the circumferential wave number of $n = 14$ and frequency of about $f = 500$ cps. The amplitude of flutter was about five times the panel thickness. On the other hand, the critical dynamic pressure and frequency that were predicted by using the linearized shell equation were 90% of the experimental value and about 1,800 cps at the mode $n = 22$. Although the flutter boundaries were considered to agree, the modes and frequencies did not coincide at all. However, as seen in Figures 9 and 11, the nonlinear flutter analysis on an equivalent shell model indicated that the flutter oscillation in the $n = 14$ mode was possible in the range around the critical flutter boundary and that the corresponding flutter frequency was about 850 cps. This frequency was much closer to the experimental value than that obtained from linear theory. It should be noted, however, that the reason why the flutter of the $n = 14$ mode was induced at the critical boundary remains unanswered. Carter and Stearman [51] reported that even no agreement between theoretical and experimental flutter boundaries was obtained when the shell was subjected to the pressure differential. As for an unpressurized truncated conical shell, Ueda, Kobayashi, and Kihira [52, 53] conducted the experiment and nonlinear flutter analysis and showed that flutter characteristics were similar to that of the unpressurized cylinders.

Experimental data in *subsonic flow ranges* are very meager and the results are inconclusive. Using a two-dimensional aluminum panel resting on a continuous elastic foundation, Dugundji, Dowell, and Perkin [22] observed flutter of traveling-wave type at low subsonic speeds. Ishii [9, 10b] conducted experiments on two-dimensional aluminum or steel plates in both low speed and transonic wind tunnels. In *low subsonic* tests, divergence (or buckling) occurred in the first axial mode, but no flutter was observed. Experimental and theoretical divergence boundaries agreed well with each other. According to the results at *high subsonic* speeds, there were two types of postbuckling flutter, that is, a small amplitude oscillation with high frequency and a large amplitude oscillation with low frequency. Gislason [54] reported that no flutter of a square aluminum plate was found although the dynamic pressure was increased to twice the divergence boundary.

Unless structural damping is neglected, no flutter of the flat or buckled plates is predicted, as discussed earlier in the stability analyses based on the low subsonic potential flow assumption and von Karman's nonlinear plate equation. It appears that this may correspond to Ishii's and Gislason's experimental results in the low subsonic range. Mention should be made of Ishii's interesting finding that a positive static pressure gradient along the flow direction has a destabilizing effect and may cause postbuckling flutter of traveling-wave type. It is necessary, therefore, to examine analytically the effect of the pressure gradient on stability. The potential flow theories derived in this chapter are not considered to be useful for that purpose. Recently, Grotberg and Reiss [55] introduced Darcy's law in the stability analysis of a two-dimensional infinite channel in order to model the frictional resistance of the fluid and showed that the channel plates onset flutter without buckling.

As for tubes conveying water, occurrence of self-excited oscillation was reported by Rodbard and Saiki [56] in 1954 who used a thin rubber tube as a model of blood vessels. Conrad [57] and Katz, Chen and Moreno [58] performed extensive studies on water-conveying rubber tubes subjected to external pressure. In a certain range of flow rate and external pressure, the collapsed tube showed limit cycle oscillations accompanied by oscillatory change in the flow rate. More recently, Weaver and Paidoussis [37] observed two distinct modes of flutter, that is, a flapping oscillation with low frequency and a shell mode ($n = 2$) oscillation with high frequency. When the tube was initially flat or flattened, for instance, by pinching, flutter of the first type occurred at low flow speeds. The other type of flutter was induced if the tube was sufficiently circular in the initial state or almost rounded out by the internal pressure at higher flow speeds. Like in the case of the flat plates, flutter of the soft tubes is closely associated with either slight or full collapsing. Flow through the collapsed tube has much more complex characteristics than those predicted by a linearized potential flow theory.

CONCLUDING REMARKS

One of the main objectives of the present article is to present a fundamental approach to fluid-elastic stability problems from a unified point of view. Subsonic and supersonic fluid forces acting on a harmonically oscillating two-dimensional plate, channel wall or cylindrical shell of finite length have been rederived with the aid of the operational methods. Fluid/aeroelastic stability of the plate or shell is examined by using the Routh-Hurwitz criteria and the method of limit cycle analysis. As the unified approach has been taken, we may clarify the differences of the stability characteristics associated with the plate and shell in subsonic and supersonic flow ranges. Major results described in this chapter as well as unsolved problems may be summarized as follows:

1. The incompressible flow approximation may lead us to unrealistic results in the analysis of unsteady fluid dynamic forces and stability of the plates (and shells), because the incompressible flow is an imaginary product. Therefore, great care must be taken in evaluation of the analytical results. A seemingly reasonable solution is not necessarily a correct solution to the problem under consideration. For instance, when the two-dimensional plate exposed to the flow on one side is supported (or clamped) at the leading and trailing edges, no motion in a symmetric natural mode is allowable. It is recommended that the previous flow and stability analyses based on the incompressible approximation be carefully examined. Some of them are not necessarily correct; the radiation condition of the flow at infinity is dissatisfied.
2. No limit cycle flutter is predicted to occur in the stability of flat plates or channel plates at subsonic flow ranges, as long as the potential flow is assumed and the structural damping is accounted for. Holmes [59, 60] reached the same conclusion. However, many linearized analyses (e.g., 11, 22, 32–34, 36, 37) were performed to show occurrence of post-buckling flutter of the plate or cylinder after recovery from buckling. Such prediction is possible only when we discard the structural damping that always exists in reality.
3. In the stability analysis of cylindrical shells, it is important to use nonlinear shell equations. The analysis based on linearized equation is not so sufficient. For a specific circumferencial mode, the nonlinear shell analysis shows that limit cycle (flutter) solutions exist not only above but also below the *flutter boundary*, which is predicted by the linear analysis. This is illustrated by the numerical example of supersonic panel flutter given in Figures 9 and 11. (The nonlinear plate equation has flutter solutions only in the range demarcated by the *flutter boundary*.)
4. The predominant circumferential mode must be determined in order to make a definitive comparison with the phenomena observed in the experiment. A difficulty of the cylindrical

shell problem is that, unlike the plates, the critical mode may not be associated with the lowest circumferential (or cross-stream) mode, as shown in Figures 8 and 10. As consequence, together with item 3, there are many circumferential modes which have the limit cycle solutions above, at and below the most critical boundary, as shown in Figure 10. It is necessary to search for the circumferential mode which ultimately predominates over other modes.

5. The least-solved problem in the fluid-elasticity is nonlinearity of flow, particularly, the effect of flow viscosity. Bioengineers have made extensive studies on the collapsible tube by using one-dimensional models. Among others are Katz, Chen, and Moreno [58], Griffith [63], and Bertran and Pedley [64] who accounted for the pressure loss through the collapsed tube and showed limit cycle oscillations by numerical simulations, etc. One-dimensional analysis has, of course, weakness, caused through oversimplification of the three-dimensional objects. As mentioned in the text, limit cycle oscillation of the tube is closely associated with state of collapsing. The collapsed tube has convergent and divergent segments upstream and downstream of the severest constriction. Matsuzaki and Fung [65] reported that flow in a two-dimensional divergent channel may easily be separated at moderate Reynolds numbers when the angle of divergence is not small. Therefore, flow separation is expected to occur in the downstream of the severest constriction of the collapsed tube. It is well known in aeronautical engineering [1] that separation of flow over a wing of high-angle-of-attack plays a decisive role in onset of stall flutter of the wing. Matsuzaki [66] used an empirical relationship among the flow rate, pressure and divergence angle [65] and made a qualitative discussion on occurrence of flutter of the collapsed tube as an analogy to the stall flutter of the wing.
6. Although it is easy to observe flutter of collapsed tubes in the laboratory, quantitative measurement of the state of the collapsed tube and flow is extremely difficult. Yet, this is necessary to obtain a firm basis for the theoretical analysis and to fill the gap between experiment and theory. It is important to use well manufactured and preserved specimens in order to obtain useful experimental results. Most commerical soft tubes are initially collapsed and there are few that can be used as test specimens to compare with theoretical prediction of buckling.

This chapter is not intended to be a thorough review of the subject. For further information, the reader should consult the survey papers [4, 5] and books [1b, 6, 7, 64b]. In addition, Kornecki [67] recently presented an excellent review article on stability of infinitely long plates in subsonic and supersonic flows. Mention was also made of papers treating finite plates and cylinders. Chen [70] published an extensive survey paper on the cylinder-flow problems. Fluid-elasticity and determination of fluid forces in the electromagnetic field were reviewed by Librescue [68, 69]. A review of one-dimensional steady flows in collapsed tubes was presented by Shapiro [71] from a biomedical view point.

Acknowledgment

The author is grateful to Dr. S. Yamamoto of the U.S. Office of Naval Research, Tokyo, for his assistance.

NOTATION

a_0	velocity of sound	M	Mach number
c	coefficient of viscous structural damping	n	wave number
		p	pressure
E	Young's modulus	P_m	pressure due to deflection
H_0	Bessel function of third kind	t	time
h	thickness	U	velocity
J_n	Bessel function of first kind	W_m, W_n	deflection modes
K_n	modified Bessel function of second kind	w	deflection
		X, Y, Z	coordinates
L	length		

Greek Symbols

γ^*	Euler's constant
ν	Poisson's ratio
ρ	density
ϕ	velocity potential
ω	frequency

REFERENCES

1a. Fung, Y. C., *An Introduction to the Theory of Aeroelasticity*. New York: John Wiley and Sons, 1955.

1b. Bisplinghoff, R. L., and Ashley, H., *Principle of Aeroelasticity*. New York: John Wiley and Sons, 1962.

2a. Dodds, H. L., Jr., and Runyan, H. L., "Effect of High-Velocity Fluid Flow on the Bending Vibrations and Static Divergence of a Simply Supported Pipe." NASA TN D-2870, June 1965.

2b. Ashley, H., and Haviland, G., "Bending Vibrations of a Pipe Line Containing Flowing Fluid." *J. of Applied Mechanics*, Vol. 17, Sept. 1950, pp. 229–232.

2c. Paidoussis, M. P., "Flow-Induced Vibrations in Nuclear Reactors and Heat Exchangers: Practical Experiences and State of Knowledge." In *Practical Experiences with Flow-Induced Vibrations*, E. Naudascher and D. Rockwell Eds. Berlin: Springer-Verlag, 1980, pp. 1–81.

3a. Hyatt, R. E., Schilder, D. P., and Fry, D. L., "Relationship Between Maximum Expiratory Flow and Degree of Lung Inflation." *J. of Applied Physiology*, Vol. 13, 1958, pp. 331–336.

3b. McCutcheon, E. P., et al. "Korotkoff Sounds: An Experimental Critique." Circulation Res., Vol. 20, No. 2, Feb. 1967, pp. 149–161.

3c. Tsuji, T., et al. "Study on Hemodynamics During Cardiopulmonary Bypass (in Japanese)." *Artificial Organs*, Vol. 7, No. 2, 1978, pp. 435–438.

4. Fung, Y. C., "Some Recent Contributions to Panel Flutter Research." *AIAA J.*, Vol. 1, No. 4, April 1963, pp. 898–909.

5. Dowell, E. H., "Panel Flutter: A Review of the Aeroelastic Stability of Plates and Shells." *AIAA J.*, Vol. 8, No. 3, March 1970, pp. 385–399.

6. Dowell, E. H., *Aeroelasticity of Plates and Shells*. Leyden: Noordhoff International Publishing, 1975.

7. Librescu, L., *Elastostatics and Kinetics of Anisotropic and Heterogeneous Shell-Type Structures*. Leyden: Noordhoff International Publishing, 1975.

8. Benjamin, T. B., "Dynamics of a System of Articulated Pipes Conveying Fluid. I. Theory and II. Experiments." *Proceed. the Royal Society* (*London*) A, Vol. 261, 1961, pp. 457–486 and pp. 487–499.

9. Ishii, T., "Aeroelastic Instabilities of Simply Supported Panels in Subsonic Flow." AIAA Paper No. 65–772, Nov. 1965.

10a. Ishii, T., "Theoretical Study of Two-Dimensional Panel Flutter and Panel Divergence in Subsonic Flow, (I): Incompressible Flow Case (in Japanese)." NAL TR-83, National Aerospace Laboratory, Tokyo, Feb. 1965.

10b. Ishii, T., "Theoretical Study of Two-Dimensional Panel Flutter and Panel Divergence in Subsonic Flow, (II): Compressible Flow Case and (III): Experimental Studies of Two-Dimensional and Three-Dimensional Panels in Low Speed Wind Tunnel (in Japanese)." NAL TR-87, National Aerospace Laboratory, Tokyo, June 1965.

11. Weaver, D. S., and Unny, T. E., "The Hydroelastic Stability of a Flat Plate." *J. of Applied Mechanics*, Vol. 37, No. 3, Sept. 1970, pp. 823–827; Discussion: *J. of Applied Mechanics*, Vol. 38, No. 2, June 1971, p. 565.

12. Ellen, C. H., "The Stability of Simply Supported Rectangular Surfaces in Uniform Subsonic Flow." *J. of Applied Mechanics*, Vol. 40, No. 1, March 1973, pp. 68–72.

13. Kornecki, A., Dowell, E. H., and O'Brien, J., "On the Aeroelastic Instability of Two-Dimensional Panels in Uniform Incompressible Flow," *J. of Sound and Vibration*, Vol. 47, No. 2, 1976, pp. 163–178.

14. Matsuzaki, Y., and Ueda, T., "Reexamination of Unsteady Fluid Dynamic Forces on a Two-Dimensional Finite Plate at Small Mach Numbers." *J. of Applied Mechanics*, Vol. 47, No. 4, Dec. 1980, pp. 720–724.
15. Matsuzaki, Y., and Fung, Y. C., "Unsteady Fluid Dynamic Forces on a Simply Supported Circular Cylinder of Finite Length Conveying a Flow, With Applications to Stability Analysis." *J. of Sound and Vibration*, Vol. 54, No. 3, 1977, pp. 317–330.
16. Matsuzaki, Y., and Fung, Y. C., "Stability Analysis of Straight and Buckled Two-Dimensional Channels Conveying an Incompressible Flow." *J. of Applied Mechanics*, Vol. 44, No. 4, Dec. 1977, pp. 548–552.
17. Matsuzaki, Y., and Fung, Y. C., "Nonlinear Stability Analysis of a Two-Dimensional Model of an Elastic Tube Conveying a Compressible Flow." *J. of Applied Mechanics*, Vol. 46, No. 1, March 1979, pp. 31–36.
18. Ashley, H., and Zartarian, G., "Piston Theory—A New Aerodynamic Tool for the Aero-elastician." *J. of the Aeronautical Sciences*, Vol. 23, No. 12, Dec. 1956, pp. 1109–1118.
19. Miles, J. W., "On a Reciprocity Condition for Supersonic Flutter." *J. of the Aeronautical Sciences*, Vol. 24, Dec. 1957, p. 920.
20. Churchill, R. V., *Operational Mathematics*. New York: McGraw-Hill, 1944.
21. Lighthill, M. J., *Introduction to Fourier Analysis and Generalized Functions*. Cambridge: Cambridge Univ. Press, 1958, p. 43.
22. Dugundji, J., Dowell, E. H., and Perkin, B., "Subsonic Flutter of Panels on Continuous Elastic Foundations." *AIAA J.*, Vol. 1, No. 5, May 1963, pp. 1146–1154.
23. Fung, Y. C., "On Two-Dimensional Panel Flutter." *J. of the Aeronautical Sciences*, Vol. 25, No. 3, March 1958, pp. 145–160.
24. Dowell, E. H., and Widnall, S. E., "Generalized Aerodynamic Forces on an Oscillating Cylindrical Shell." *Quarterly of Applied Mathematics*, Vol. 24, No. 1, April 1966, pp. 1–17.
25. Matsuzaki, Y., and Kobayashi, S., "Unsteady Supersonic Aerodynamic Forces on an Oscillating Circular Cylindrical Shell." *AIAA J.*, Vol. 9, No. 12, Dec. 1971, pp. 2358–2362.
26. Watson G. N., *A Treatise on the Theory of Bessel Functions*, second ed. Cambridge: Cambridge Univ. Press, 1966, pp. 82–83.
27. Parthan, S., and Johns, D. J., "Aerodynamic Generalized Forces for Supersonic Shell Flutter." *AIAA J.*, Vol. 10, No. 10, Oct. 1972, pp. 1369–1371.
28. Dowell, E. H., and Widnall, S. E., "Generalized Aerodynamic Forces on an Oscillating Cylindrical Shell: Subsonic and Supersonic Flow." *AIAA J.*, Vol. 4, No. 4, April 1966, pp. 607–610.
29. Widnall, S. E., and Dowell, E. H., "Aerodynamic Forces on an Oscillating Cylindrical Duct With an Internal Flow." *J. of Sound and Vibration*, Vol. 6, No. 1, 1967, pp. 71–85.
30. Dowell, E. H., "Generalized Aerodynamic Forces on a Flexible Cylindrical Shell Undergoing Transient Motion." *Quarterly of Applied Mathematics*, Vol. 26, No. 3, 1968, pp. 343–353.
31. Dowell, E. H., "Generalized Aerodynamic Forces on a Flexible Plate Undergoing Transient Motion." *Quarterly of Applied Mathematics*, Vol. 24, No. 4, 1967, pp. 331–338.
32. Paidoussis, M. P., and Denise, J. P., "Flutter of Thin Cylindrical Shells Conveying Fluid." *J. of Sound and Vibration*, Vol. 20, No. 1, 1972, pp. 9–26.
33. Weaver, D. S., and Unny, T. E., "On the Dynamic Stability of Fluid-Conveying Pipes." *J. of Applied Mechanics*, Vol. 40, No. 1, March 1973, pp. 48–52.
34. Weaver, D. S., and Myklatun, B., "On the Stability of Thin Pipes With an Internal Flow." *J. of Sound and Vibration*, Vol. 31, No. 4, 1973, pp. 399–410.
35. Kornecki, A., "Static and Dynamic Instability of Panels and Cylindrical Shells in Subsonic Potential Flow." *J. of Sound and Vibration*, Vol. 32, No. 2, 1974, pp. 251–263.
36. Paidoussis, M. P., and Issid, N. T., "Dynamic Stability of Pipes Conveying Fluid." *J. of Sound and Vibration*, Vol. 33, No. 3, 1974, pp. 267–294.
37. Weaver. D. S., and Paidoussis, M. P., "On Collapse and Flutter Phenomena in Thin Tubes Conveying Fluid." *J. of Sound and Vibration*, Vol. 50, No. 1, 1977, pp. 117–132.
38. Ellen, C. H., "The Nonlinear Stability of Panels in Incompressible Flow." *J. of Sound and Vibration*, Vol. 54, No. 1, 1977, pp. 117–121.

39. Timoshenko, S. P., and Woinowsky-Krieger, S., *Theory of Plates and Shells*, second ed. New York: McGraw-Hill, 1959, pp. 415–428.
40. Washizu, K., *Variational Methods in Elasticity and Plasticity*. Oxford: Pergamon Press, 1968, p. 15.
41. Gantmacher, F. R., *The Theory of Matrics*. New York: Chelsea Publishing, 1959, Chapt. 15.
42. Matsuzaki, Y., "Reexamination of Stability of a Two-Dimensional Finite Panel Exposed to an Incompressible Flow." *J. of Applied Mechanics*, Vol. 48, No. 3, Sept. 1981, pp. 472–478.
43. Dowell, E. H., "Nonlinear Oscillations of a Fluttering Plate. II." *AIAA J.*, Vol. 5, No. 10, Oct. 1967, pp. 1856–1862.
44. Kobayashi, S., "Two-Dimensional Panel Flutter, I. Simply Supported Panel." *Transactions of the Japan Society for Aeronautical and Space Sciences*, Vol. 5, No. 8, 1962, pp. 90–102.
45. Kraus, H., *Thin Elastic Shells*. New York: John Wiley, 1967.
46a. Matsuzaki, Y., and Kobayashi, S., "A Theoretical and Experimental Study of Supersonic Panel Flutter of Circular Cylindrical Shells." Proceed. Eighth Int. Symp. Space Tech. Sci., Tokyo, Aug. 1969, pp. 281–290.
46b. Kobayashi, S., "Supersonic Panel Flutter of Unstiffened Circular Cylindrical Shells Having Simply Supported Ends." *Trans. Japan Soc. for Aero. Space Sci.*, Vol. 6, No. 9, 1963, pp. 27–35.
47. Evensen, D. A., "Nonlinear Vibrations of Circular Cylindrical Shells." In *Thin-Shell Structures: Theory, Experiment, and Design*, Y. C. Fung and E. E. Sechler Eds. New Jersey: Prentice-Hall, 1974, pp. 133–155.
48. Minorsky, N., *Nonlinear Oscillations*. Princeton: D. Van Nostrand, 1962, pp. 273–281.
49. Olson, M. D., and Fung, Y. C., "Supersonic Flutter of Circular Cylindrical Shells Subjected to Internal Pressure and Axial Compression." *AIAA J.*, Vol. 4, No. 5, May 1966, pp. 858–864.
50. Evensen, D. A., and Olson, M. D., "Circumferentially Traveling Wave Flutter of a Circular Cylindrical Shell." *AIAA J.*, Vol. 6, No. 8, Aug. 1968, pp. 1522–1527.
51. Carter, L. L., and Stearman, R. O., "Some Aspect of Cylindrical Shell Panel Flutter." *AIAA J.*, Vol. 6, No. 1, Jan. 1968, pp. 37–43.
52. Ueda, T., Kobayashi, S., and Kihira, M., "Supersonic Flutter of Truncated Conical Shells." *Trans. Japan Society for Aero. Space Sci.*, Vol. 20, No. 47, 1977, pp. 13–30.
53. Ueda, T., "Nonlinear Analysis of the Supersonic Flutter of a Truncated Conical Shell Using the Finite Element Method." *Trans. Japan Society for Aero. Space Sci.*, Vol. 20, No. 50, 1978, pp. 225–240.
54. Gislason, T., Jr., "Experimental Investigation of Panel Divergence at Subsonic Speeds." *AIAA J.*, Vol. 9, No. 11, Nov. 1971, pp. 2252–2258.
55. Grotberg, J. B., and Reiss, E. L., "A Subsonic Flutter Anomaly." *J. of Sound and Vibration*, Vol. 80, No. 3, 1982, pp. 444–446.
56. Rodbard, S., and Saiki, H., "Flow Through Collapsible Tubes." *American Heart J.*, Vol. 46, 1953, pp. 715–725.
57. Conrad, W. A., "Pressure-Flow Relationships in Collapsible Tubes." *IEEE Trans. Bio-Medical Eng.*, Vol. BME-16, No. 1, Jan. 1969, pp. 284–295.
58. Katz, A. I., Chen, Y., and Moreno, A. H., "Flow Through a Collapsible Tube. Experimental Analysis and Mathematical Model." *Biophysical J.*, Vol. 9, 1969, pp. 1261–1279.
59. Holmes, P. J., "Bifurcations to Divergence and Flutter in Flow-Induced Oscillations: A Finite Dimensional Analysis." *J. of Sound and Vibration*, Vol. 53, No. 4, 1977, pp. 471–503.
60. Holmes, P. J., "Pipes Supported at Both Ends Cannot Flutter." *J. Applied Mechanics*, Vol. 45, No. 3, Sept. 1978, pp. 619–622.
61. Evensen, D. A., "Nonlinear Flexural Vibrations of Thin-Walled Circular Cylinders." NASA TN D-4090, 1967.
62. Matsuzaki, Y., and Kobayashi, S., "A Theoretical and Experimental Study of the Nonlinear Flexural Vibration of Thin Circular Cylindrical Shells with Clamped Ends." *Trans. Japan Society Aero. Space Sci.*, Vol. 12, No. 21, 1970, pp. 55–62.
63. Griffiths, D. J., "Oscillations in the Outflow from a Collapsible Tube." *Med. & Biol. Eng. & Comput.*, Vol. 15, July 1977, pp. 357–362.
64a. Bertram, C. D., and Pedley, T. J., "A Mathematical Model of Unsteady Collapsible Tube Behavior." *J. of Biomechanics*, Vol. 15, No. 1, pp. 39–50.

64b. Pedley, T. J., *The Fluid Mechanics of Large Blood Vessels*. Cambridge: Cambridge Univ. Press, 1980, Chapt. 6.

65. Matsuzaki, Y., and Fung, Y. C., "On Separation of a Divergent Flow at Moderate Reynolds Numbers." *J. of Applied Mechanics*, Vol. 43, No. 2, June 1976, pp. 227–231.

66. Matsuzaki, Y., "Mechanism of Self-Excited Oscillation of a Collapsed Tube Conveying a Flow." Proceed. Second Int. Conf. on Mechanics in Med. & Biol., Osaka, June 1980, pp. 74–75.

67. Kornecki, A., "Aeroelastic and Hydroelastic Instabilities of Infinitely Long Plates. Parts I & II." *SM Archives*, Vol. 3, Nov. 1978, pp. 381–440, & Vol. 4, Nov. 1979, pp. 241–345.

68. Librescu, L., "Recent Contributions Concerning the Flutter Problem of Elastic Thin Bodies in an Electrically Conducting Gas Flow, a Magnetic Field Being Present." *SM Archives*, Vol. 2, Feb. 1977, pp. 1–108.

69. Librescu, L., "Developments in the Determination of Pressure Loads on Elastic Thin Panels Undergoing Arbitrary Motions in a Supersonic Ionized Gas Flow." *SM Archives*, Vol. 5, May 1980, pp. 91–130.

70. Chen, S. S., "Flow-Induced Vibrations of Circular Cylindrical Structures. Part I: Stationary Fluids and Parallel Flow." *The Shock and Vibration Digest*, Vol. 9, No. 10, Oct. 1977, pp. 25–38.

71. Shapiro, A. H., "Steady Flow in Collapsible Tubes." *J. of Biomechanical Engineering*, Vol. 99, No. 3, Aug. 1977, pp. 126–147.

CHAPTER 16

STEADY TRANSONIC FLOW

E. Dick

State University of Ghent
Department of Machinery
Ghent, Belgium

CONTENTS

INTRODUCTION

Small disturbances superimposed on a steady gas flow propagate with a velocity that is known as the velocity of sound. Physically, this propagation originates from collisions of individual molecules in the motion relative to the mean flow. Since the intensity of the molecular collisions is expressed by the pressure, propagating disturbances are called pressure waves. Due to the thermal agitation of the molecules, continuously small pressure waves are generated in each point of a flow and due to the propagation of these waves, the generating point spreads its influence into its surroundings. In the point itself, the properties of the mean flow are defined by taking mean values on a small volume surrounding the point, called a particle. When the particle velocity in a point is lower, equal, or higher than the local velocity of sound, the flow is called respectively subsonic, sonic, or supersonic in this point.

In this chapter, steady flows will be studied in which some regions of the flow field are subsonic and other regions are supersonic. These regions are separated by sonic regions. A flow in which these different regions occur simultaneously is called a transonic flow. Transonic flows usually are characterized by the occurrence of extreme steep gradients in flow properties when passing in the flow direction from supersonic regions to subsonic regions. Usually, these steep gradients mathematically are described by discontinuities called shock waves. The occurrence of shock waves often makes the calculation of a transonic flow extremely complex.

BASIC EQUATIONS FOR STEADY ADIABATIC FLOW OF AN INVISCID FLUID

In this section some basic equations necessary in the analysis of steady transonic flows are derived. However, the complete description of transonic flow by mathematical equations is postponed until later in this chapter.

γ-Gas Laws

In the sequel, it will be assumed that the perfect gas law applies:

$$p = \rho RT \tag{1}$$

in which p is pressure (Pa), ρ is density (kg/m^3), T is temperature (K) and R is the gas constant (air: R = 287 J/kg K).

It will also be assumed that the gas has constant specific heats:

$$\begin{aligned} e &= c_v T \\ h &= c_p T \end{aligned} \tag{2}$$

in which e is specific internal energy (J/kg), h is specific enthalpy (J/kg), c_v is specific heat for constant volume (air: $c_v = 717.5$ J/kg K) and c_p is specific heat for constant pressure (air: $c_p = 1004.5$ J/kg K).

By definition:

$$h = e + p/\rho \tag{3}$$

Combining Equations 1, 2, and 3 gives

$$c_p = c_v + R \tag{4}$$

The ratio of specific heats is denoted by

$$\gamma = c_p/c_v \tag{5}$$

For air: $\gamma = 1.4$

Gases that fulfill the laws (Equations 1–5) are called γ-gases. Most gases like air can be represented very well by γ-laws in their usual domain of application.

The entropy change ds during a transformation is given by

$$T\,ds = dh - \rho^{-1}\,dp \tag{6}$$

Using Equations 1 and 2, this gives

$$ds = c_p\,dT/T - R\,dp/p = c_v\,dp/p - c_p\,d\rho/\rho \tag{7}$$

From Equation 7, it follows that an isentropic transformation obeys:

$$p \sim \rho^{\gamma} \tag{8}$$

For this reason, the ratio of specific heats usually is called the isentropic exponent.

Conservation Laws

Figure 1 shows a control volume V with surface S and outward normal $\vec{n}$, fixed with respect to the coordinate system in which we measure the flow properties. For a flow that is steady with respect to this coordinate system, the law of conservation of mass reads

$$\int_S \rho \vec{v} \cdot \vec{n}\, dS = 0 \tag{9}$$

Equation 9 expresses that the net outflux of mass from a closed surface in a steady flow is zero.

When all flow properties are continuous functions of space, by Gauss' theorem, Equation 9 gives

$$\nabla \cdot (\vec{\rho} v) = 0 \tag{10}$$

Similarly, the law of conservation of momentum gives

$$\int_S \rho \vec{v}\vec{v} \cdot \vec{n}\, dS = \int_S \rho \vec{f}\, dV + \int_S \vec{T}\, dS \tag{11}$$

in which $\vec{f}$ is the force per unit mass acting in a point of the volume V and $\vec{T}$ is the force per unit area acting in a point of the surface S. Equation 11 expresses that the net outflux per unit time of momentum from a closed surface is equal to the sum of the forces acting on the surface and the volume enclosed by it.

The mass force $\vec{f}$ is generated by external force fields like gravitation. In the sequel we will only consider conservative mass forces. These are forces that can be derived from a potential energy U by

$$\vec{f} = -\nabla U \tag{12}$$

This means that we will not consider cases in which due to the motion of the coordinate system centrifugal forces or Coriolis forces are to be included.

For the analysis of transonic flows, viscosity will be neglected. In this case, the surface force is only due to pressure:

$$\vec{T} = -p\vec{n} \tag{13}$$

The effects of viscosity will be discussed later.

By Equations 12 and 13, assuming that flow properties are continuous, Equation 11 gives

$$\nabla \cdot (\rho \vec{v}\vec{v}) + \rho\, \nabla U + \nabla p = 0 \tag{14}$$

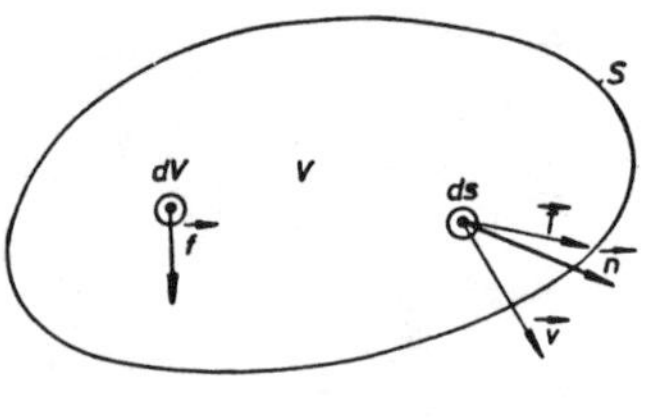

Figure 1. Control volume V with surface S, outward normal $\vec{n}$, mass force $\vec{f}$, surface force $\vec{T}$ and velocity $\vec{v}$.

Combining Equations 10 and 14 gives

$$\rho\vec{v} \cdot \nabla\vec{v} + \rho\, \nabla U + \nabla p = 0$$

or

$$\nabla(\tfrac{1}{2}\vec{v} \cdot \vec{v}) - \vec{v}x(\nabla x\vec{v}) + \nabla U + \rho^{-1}\, \nabla p = 0 \tag{15}$$

Projection of Equation 15 on the direction of the velocity gives

$$\vec{1}_v \cdot (\nabla\tfrac{1}{2}\vec{v} \cdot \vec{v} + \nabla U + \rho^{-1}\, \nabla p) = 0 \tag{16}$$

Equation 16 is called the equation of Bernoulli. It is to be recalled that this equation is only valid for continuous flow properties.

The law of conservation of energy gives

$$\int_S \rho E\vec{v} \cdot \vec{n}\, dS = \int_V \rho\vec{f} \cdot \vec{v}\, dV + \int_S \vec{T} \cdot \vec{v}\, dS \tag{17}$$

in which E is the total mechanical energy:

$$E = e + \tfrac{1}{2}\vec{v} \cdot \vec{v} \tag{18}$$

Equation 17 expresses that, in absence of heat exchange, the net outflux of energy per unit time from a closed surface is equal to the work per unit time of the forces acting on the surface and the volume enclosed by it.

With Equations 12 and 13, Equation 17 gives:

$$\int_S \rho E\vec{v} \cdot \vec{n}\, dS = -\int_V \rho\vec{v} \cdot \nabla U\, dV - \int_S p\vec{v} \cdot \vec{n}\, dS$$

and combining with Equation 9 gives

$$\int_S \rho H\vec{v} \cdot \vec{n}\, dS = 0 \tag{19}$$

in which the total enthalpy H is defined by

$$H = e + \tfrac{1}{2}\vec{v} \cdot \vec{v} + U + p/\rho$$

For continuous flow properties Equation 19 gives in combination with Equation 10:

$$\vec{1}_v \cdot \nabla H = 0 \tag{20}$$

Equation 20 expresses that the total enthalpy is constant on a streamline for an adiabatic flow. (A streamline is a line that is tangent to the local velocity in all its points.) It is to be noted that due to the integral law (Equation 19) the constancy of total enthalpy remains valid even if flow properties do not vary continuously.

For continuously varying flow properties, Equations 16 and 20 can be combined into

$$\vec{1}_v \cdot (\nabla h - \rho^{-1}\, \nabla p) = 0$$

and by using the definition of entropy (Equation 6):

$$\vec{1}_v \cdot \nabla s = 0 \tag{21}$$

Equation 21 expresses that for an adiabatic flow of an inviscid fluid with continuously varying flow properties, the entropy is constant on a streamline.

Velocity of Sound

In order to calculate the velocity of sound in a stationary fluid, we consider the propagation of a plane pressure wave in a long tube as sketched in Figure 2.

Since the disturbance is infinitesimally small, effects of viscosity and heat transfer can be neglected. Hence, from the previous section it follows that the changes caused by the sound wave are to be considered as being isentropic. To facilitate the calculation, we consider a control volume moving with the wave. For constant speed of the wave, this does not introduce any forces on the control volume. For a wave moving with speed c from left to right, relative to the control volume, a steady flow enters at the right side with velocity c, pressure p, and density ρ. A steady flow leaves at the left side with velocity $c - dc$, pressure $p + dp$, and density $\rho + d\rho$.

The conservation equations are

mass: $(\rho + d\rho)(c - dc) - \rho c = 0$

or, neglecting higher order terms:

$$c\,d\rho - \rho\,dc = 0 \tag{22}$$

momentum: $\rho c(c - dc) - \rho cc = p - (p + dp)$

or
$$\rho c\,dc = dp \tag{23}$$

Combination of Equations 22 and 23 gives

$$c^2 = dp/d\rho$$

For isentropic transformation j, Equation (8) gives:

$$dp/p = \gamma\,d\rho/\rho$$

such that $c^2 = \gamma p/\rho = \gamma RT$ (24)

For a constant property gas, the velocity of sound is only dependent on temperature. For air at $T = 290$ K: $c = 341.35$ m/s.

Mach Number

Figure 3a shows the spherical sound waves generated by a stationary point disturbance at time Δt, $2\Delta t$, $3\Delta t$.

Figure 3b shows similar sound waves for a point disturbance moving in a stationary gas with a subsonic velocity $v(v < c)$. It is obvious that the wave front moves ahead of the disturbance. The

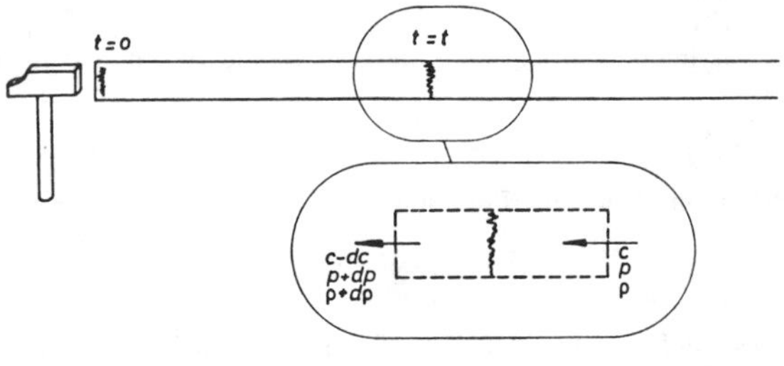

Figure 2. Propagation of a plane pressure wave in a tube.

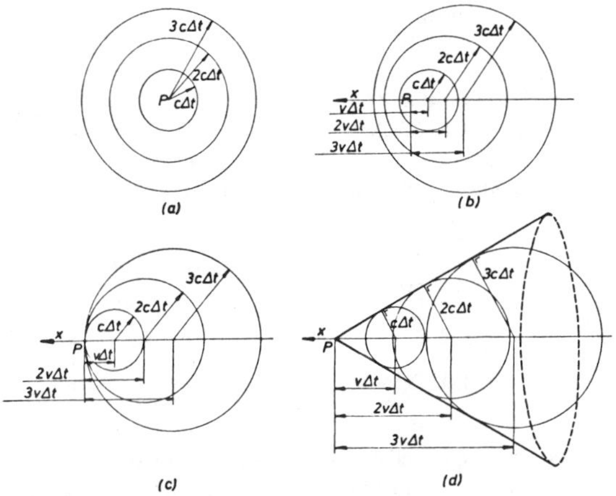

Figure 3. Sound waves generated by a point disturbance: (a) stationary point, (b) point moving with subsonic velocity, (c) point moving with sonic velocity, and (d) point moving with supersonic velocity.

ratio of the velocity of the point disturbance to the velocity of sound is called the Mach number:

$$M = v/c \tag{25}$$

In Figure 3b, clearly $M < 1$.

Figure 3c shows the sound waves for sonic conditions $M = 1$ or $v = c$. The sound waves coincide at the moving point. Finally, Figure 3d shows the wave front for a point moving with supersonic velocity. The envelope of the spherical sound waves is a conical surface called the Mach cone. The cone angle μ is given by

$$\sin \mu = c/v = 1/M \tag{26}$$

The effects of the disturbance are entirely contained by the Mach cone.

The effects of the sound waves shown in Figure 3 remain unchanged if a steady flow with velocity v from left to right is superimposed. In particular this means that in subsonic flow sound waves emitted from a point can reach all points in the neighborhood of the point itself. At the same time any point is influenced by sound waves coming from all surrounding points. In supersonic flow, on the contrary, a particular point only can influence points that are within the Mach cone. At the same time any point only can be influenced by points that are within a complementary cone.

From this preliminary analysis, it is already clear that subsonic and supersonic steady flows are completely different in nature.

Streamline Equations

Earlier, it was shown that for steady adiabatic flow of an inviscid fluid, under conservative mass forces, total enthalpy is constant on a streamline. If, additionally, all flow properties vary continuously, entropy is also constant on a streamline. For such a flow it is useful to define stagnation properties. Stagnation enthalpy, stagnation pressure, and stagnation temperature are the enthalpy, pressure, and temperature one would observe in a point if in this point the flow would be stopped under reversible adiabatic, hence isentropic, conditions.

By this definition, stagnation enthalpy h_0 is given by

$$H = h + v^2/2 + U = h_0 + U$$

or $\quad h_0 = h + v^2/2 \quad (27)$

Stagnation temperature is found from

$$h_0 = c_p T_0 = c_p T + v^2/2$$

$$\text{or} \quad T_0/T = 1 + v^2/(2cpT) = 1 + (\gamma - 1)M^2/2 \tag{28}$$

Stagnation pressure and stagnation density are found from the isentropic transformation law (Equation 8):

$$\frac{p_0}{p} = \left(\frac{T_0}{T}\right)^{\gamma/(\gamma-1)} = (1 + (\gamma - 1)M^2/2)^{\gamma/(\gamma-1)}$$

$$\frac{\rho_0}{\rho} = \left(\frac{T_0}{T}\right)^{1/(\gamma-1)} = (1 + (\gamma - 1)M^2/2)^{1/(\gamma-1)} \tag{29}$$

In most practical flows of gases, the change in potential energy on a streamline can be neglected and from Equation 27 it follows that in this case, isentropic flow on a streamline simply can be expressed be saying that the stagnation properties are constant. This means that all flow quantities can be expressed in function of one parameter—the Mach number.

When the flow is adiabatic but non-isentropic, when flow properties do not vary continuously, stagnation temperature remains constant by virtue of the constancy of total enthalpy, but due to the absence of constancy of entropy, stagnation pressure is not constant. In fact, there exists a direct relationship between stagnation pressure and entropy. From Equation 7 it follows, through integration between a point 1 and a point 2 on a streamline:

$$s_2 - s_1 = c_p \ln(T_2/T_1) - R \ln(p_2/p_1) = -R\left(\ln(p_2/p_1) - \frac{\gamma}{\gamma - 1}\ln(T_2/T_1)\right) \tag{30}$$

The relation between stagnation pressure and stagnation temperature gives

$$\frac{p_{02}}{p_2} = \left(\frac{T_{02}}{T_2}\right)^{\gamma/(\gamma-1)} \quad \frac{p_{01}}{p_1} = \left(\frac{T_{01}}{T_1}\right)^{\gamma/(\gamma-1)} \quad T_{02} = T_{01}$$

$$\text{Hence,} \quad \frac{p_{02}}{p_{01}} = \frac{p_2}{p_1}\left(\frac{T_1}{T_2}\right)^{\gamma/(\gamma-1)}$$

such that Equation 30 gives

$$(s_2 - s_1)/R = \ln(p_{01}/p_{02}) \tag{31}$$

When, due to irreversibility, entropy increases, stagnation pressure decreases.

ONE-DIMENSIONAL TRANSONIC FLOW

In this section, in order to gain more insight in the features of transonic flow, some one-dimensional examples are treated.

Generation of Shock Waves

We consider the propagation in positive x-direction of a plane compression wave of finite amplitude in a stationary gas as depicted in Figure 4a.

This compression wave can, for instance, be generated by the motion to the right of a piston in a tube in which initially gas is at rest at pressure p_1. The piston causes a compression from pressure

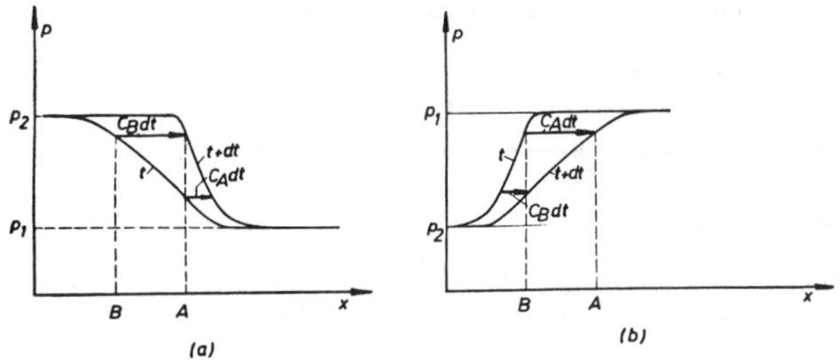

Figure 4. Propagation of plane waves in a stationary gas: (a) compression wave and (b) expansion wave.

p_1 to p_2. The wave front shown in Figure 4a at time t, in general, will vary with time. Since the wave front of finite amplitude can be considered as the sum of an infinite number of infinitesimal compression waves, it is clear that each point of the wave front moves with the local velocity of sound. In absence of heat exchange, a compression causes an increase in temperature. Hence, the temperature at the point A is lower than at point B. By Equation 24 it follows that the velocity of sound in the point A is lower than the velocity of sound in the point B. Therefore, the wave front of the compression wave becomes progressively steeper while the wave is traveling. Clearly, this steepening continues until a vertical wave front, called a shock wave, is formed.

The same analysis can be done on the expansion wave moving in positive x-direction shown at time t in Figure 4b. This expansion wave can, for instance, be generated by the motion of a piston to the left in gas initially at rest, causing an expansion behind the piston. In absence of heat exchange, we now see that the velocity of sound in point A is larger than the velocity of sound in point B. As a consequence, the wave front of the expansion wave becomes progressively flatter while the wave is traveling. This means that an expansion wave never can form a shock.

In this section we will not go further into the analysis of moving waves. This is the subject of a subsequent chapter. We only note that, once a steep moving shock wave is formed, in principle, a stationary shock wave can be obtained by superimposing a steady flow with the same velocity in opposite sense.

Analysis of Stationary Plane Normal Shock Waves

In the previous section, it was shown that a stationary compression shock can exist in a steady flow. We now analyze the relations between flow properties on both sides of the shock wave.

Figure 5 shows a control volume containing a plane stationary shock wave perpendicular to a uniform flow.

The flow direction is from left to right. We denote by index 1 the flow parameters in front of the shock and by index 2 the flow parameters behind it. From the previous section, we know that only compression shocks can exist ($p_2 > p_1$).

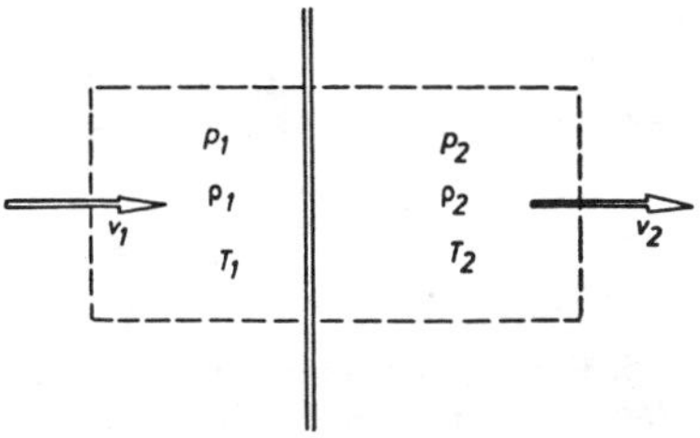

Figure 5. Stationary shock wave perpendicular to a uniform flow.

The conservation equations applied to the control volume are

mass: $$\rho_2 v_2 - \rho_1 v_1 = 0 \tag{32}$$

momentum: $$\rho_2 v_2 v_2 - \rho_1 v_1 v_1 = p_1 - p_2 \tag{33}$$

energy: $$\rho_2 E_2 v_2 - \rho_1 E_1 v_1 = p_1 v_1 - p_2 v_2 \tag{34}$$

Combining Equations 32 and 34 gives,

$$H_2 = H_1$$

and using the gas law,

$$v_2^2/2 + (\gamma/(\gamma - 1))p_2/\rho_2 = v_1^2/2 + (\gamma/(\gamma - 1))p_1/\rho_1 \tag{35}$$

Hence, as already remarked earlier the stagnation enthalpy is constant on a streamline through a discontinuity like a shock wave. As already mentioned, it is not to be expected that the transformation across the shock is isentropic and as a consequence stagnation pressure is not expected to be constant.

For adiabatic, but non-isentropic, flows it is useful to define a critical velocity as the velocity that would be obtained by an adiabatic transformation to a flow in which the local velocity is equal to the velocity of sound.

The critical velocity a is given by

$$h_0 = h + v^2/2 = (c_p/\gamma R)a^2 + a^2/2 = ((\gamma + 1)/2(\gamma - 1))a^2 \tag{36}$$

Hence, $$a^2 = (2(\gamma - 1)/(\gamma + 1))h_0 = (2\gamma/(\gamma + 1))RT_0 \tag{37}$$

The Laval number is then defined by

$$L = v/a \tag{38}$$

The relation with the Mach number is obtained by dividing Equation 36 by v^2:

$$(1/(\gamma - 1))c^2/v^2 + \tfrac{1}{2} = ((\gamma + 1)/2(\gamma - 1))a^2/v^2$$

or $$L^2 = (\gamma + 1)M^2/(2 + (\gamma - 1)M^2) \tag{39}$$

The Laval number and the Mach number take simultaneously the value 0 and 1. The relationship (Equation 39) is shown in Figure 6.

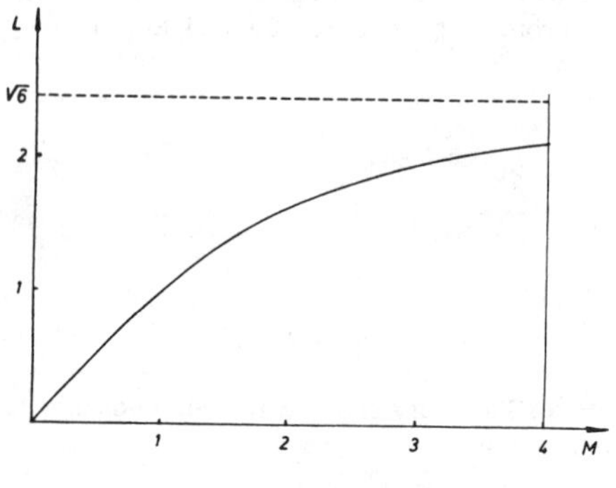

Figure 6. Relation between Laval number and Mach number for air: $\gamma = 1.4$.

By combining Equations 32 and 33, one obtains

$$p_2/(\rho_2 v_2) + v_2 = p_1/(\rho_1 v_1) + v_1 \tag{40}$$

Through multiplication by $(v_1 + v_2)$ and using Equation 32, Equation 40 gives

$$v_2^2 + p_2/\rho_1 + p_2/\rho_2 = v_1^2 + p_1/\rho_1 + p_1/\rho_2 \tag{41}$$

Combining Equations 35 and 41 gives

$$\frac{p_2}{\rho_1} - \frac{\gamma + 1}{\gamma - 1}\frac{p_2}{\rho_2} = \frac{p_1}{\rho_2} - \frac{\gamma + 1}{\gamma - 1}\frac{p_1}{\rho_1}$$

or $$\frac{p_2 - p_1}{p_2 + p_1} = \gamma \frac{\rho_2 - \rho_1}{\rho_2 + \rho_1} \tag{42}$$

Equation 42, which takes the form of an entropy equation, is known as the equation of Hugoniot. It also can be written:

$$\frac{\rho_2}{\rho_1} = \frac{\frac{\gamma + 1}{\gamma - 1}\frac{p_2}{p_1} + 1}{\frac{\gamma + 1}{\gamma - 1} + \frac{p_2}{p_1}} \tag{43}$$

whence, $$\frac{T_2}{T_1} = \frac{\frac{p_2}{p_1}}{\frac{\rho_2}{\rho_1}} = \frac{\frac{\gamma + 1}{\gamma - 1} + \frac{p_2}{p_1}}{\frac{\gamma + 1}{\gamma - 1} + \frac{p_1}{p_2}} \tag{44}$$

Equations 43 and 44 are known as the Hugoniot-Rankine equations.

Using the definition of critical velocity, Equation 35 also can be written as

$$v_1^2 + (2\gamma/(\gamma - 1))\frac{p_1}{\rho_1} = ((\gamma + 1)/(\gamma - 1))a^2$$

$$v_2^2 + (2\gamma/(\gamma - 1))\frac{p_2}{\rho_2} = ((\gamma + 1)/(\gamma - 1))a^2 \tag{45}$$

Substitution of the expressions of $\frac{p_1}{\rho_1}$ and $\frac{p_2}{\rho_2}$ from Equation 45 into Equation 40 gives

$$(a^2/(v_1 v_2))(v_1 - v_2) = v_1 - v_2$$

This equation has two possible solutions:

$$v_1 = v_2$$

which corresponds to the case of no shock and

$$v_1 v_2 = a^2 \tag{46}$$

or $$L_1 L_2 = 1$$

Equation 46 is known as the equation of Prandtl.

In combination with Equation 39 it gives

$$\gamma M_1^2 M_2^2 = 1 + (\gamma - 1)(M_1^2 + M_2^2)/2 \tag{47}$$

The Hugoniot-Rankine equations and the equation of Prandtl can be connected by remarking that

$$\frac{\rho_2}{\rho_1} = \frac{v_1}{v_2} = \frac{v_1^2}{(v_1 v_2)} = \frac{v_1 v_2}{v_2^2} = \frac{v_1^2}{a^2} = \frac{a^2}{v_2^2} = L_1^2 = 1/L_2^2$$

This gives

$$\frac{p_2}{p_1} = (2\gamma/(\gamma + 1))M_1^2 - ((\gamma - 1)/(\gamma + 1))$$

and

$$\frac{p_1}{p_2} = (2\gamma/(\gamma + 1))M_2^2 - ((\gamma - 1)/(\gamma + 1)) \tag{48}$$

Finally, the relationship between the stagnation pressures on both sides of the shock can be obtained by

$$\frac{p_{o2}}{p_{o1}} = (p_{o2}/p_2)(p_2/p_1)(p_1/p_{o1})$$

Inserting Equations 29 and 48, using Equation 47 one obtains

$$\frac{p_{o2}}{p_{o1}} = \left(\frac{\gamma + 1}{2\gamma M_1^2 - (\gamma - 1)}\right)^{1/(\gamma - 1)} \left(\frac{(\gamma + 1)M_1^2}{2 + (\gamma - 1)M_1^2}\right)^{\gamma/(\gamma - 1)} \tag{49}$$

Figure 7 shows the relations (Equations 47, 48, and 49).

One-Dimensional Adiabatic Flow in a Convergent-Divergent Nozzle

We consider the flow in a convergent-divergent duct with slowly varying section S such that the flow can be considered to be uniform in each section. Under these assumptions, the calculation of the flow through the duct can be done as if it were one-dimensional. Previously, it was shown that in a supersonic flow a shock perpendicular to the flow direction can occur resulting in a non-continuous transformation from supersonic to subsonic flow. When such discontinuities do not occur, we know that under adiabatic conditions the flow is isentropic. In this case the calculation is rather straight forward. The flow is then described by constancy of mass flux, $\rho v S$, together with

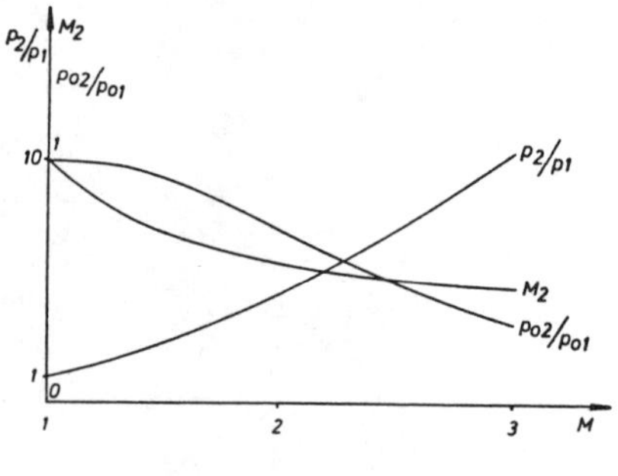

Figure 7 Mach number behind the shock, pressure ratio and stagnation pressure ratio for a stationary normal shock.

constancy of stagnation properties. This gives constancy of

$$\left(\frac{\rho}{\rho_0}\right) \cdot \frac{v}{\sqrt{\gamma R T}} \cdot \left(\frac{T}{T_0}\right)^{1/2} \cdot S$$

and with Equation 29:

$$\frac{MS}{\left(1 + \frac{\gamma - 1}{2} M^2\right)^{1/2 + 1/(\gamma - 1)}} \text{ is constant} \tag{50}$$

In case the convergent-divergent duct takes the fluid from a large reservoir, stagnation pressure and stagnation temperature are exactly pressure and temperature in the reservoir. In this case, the flow is governed by the pressure at the outlet of the duct, the so-called backpressure. For given backpressure, in principle, with Equation 29 the Mach number at outlet can be calculated, and with Equation 50 the Mach number, and hence all other variables, can be calculated in each section. However, this simple procedure does not always give a solution. To see this, an infinitesimal analysis of Equation 50 is first needed. Logarithmic differentiation of Equation 50 gives

$$\frac{dM}{M} + \frac{dS}{S} - \frac{\gamma + 1}{2} \frac{M\,dM}{1 + \frac{\gamma - 1}{2} M^2} = 0$$

or $$\frac{1 - M^2}{1 + \frac{\gamma - 1}{2} M^2} \frac{dM}{M} + \frac{dS}{S} = 0 \tag{51}$$

In the convergent part of the nozzle ($dS/dx < 0$) it follows from Equation 51 that in subsonic flow the Mach number increases in the flow direction, while in supersonic flow the Mach number decreases in the flow direction. In the divergent part of the nozzle ($dS/dx > 0$), in subsonic flow, the Mach number decreases in flow direction while in supersonic flow, it increases in flow direction. From this observation, it is clear that when the flow starts subsonic in the convergent part of the nozzle, the maximum Mach number in the throat is exactly unity. Supplementary, Equation 51 shows that in the throat section ($dS/dx = 0$) either the Mach number is equal to one, or it obtains an extremum ($dM/dx = 0$).

With the results of this analysis in mind, we can now see how the flow properties vary in the nozzle when backpressure exists. When the backpressure is equal to the reservoir pressure, clearly there is no flow. This is shown by curve 1 in Figure 8.

For a backpressure that is somewhat lower, the flow is subsonic in the whole nozzle. The Mach number increases in the convergent part, decreases in the divergent part, and has a maximum in the throat section. The pressure has a minimum in the throat section (curve 2). The limiting case of this subsonic flow is reached when the Mach number is exactly unity in the throat section (curve 3). The corresponding backpressure is denoted by p_{b3}.

For backpressures lower than p_{b3}, the calculation procedure based on the Equations 29 and 50 does not give a solution except for the special value of the backpressure denoted by p_{b4} in Figure 8. This solution can be found starting from sonic conditions in the throat and taking the supersonic solution of Equation 50 in each section of the divergent part of the nozzle. According to the previous infinitesimal analysis, the Mach number uniformally increases in the flow direction (curve 4).

For a backpressure lower than p_{b4}, the solution still is given by curve 4 since the maximum Mach number in the throat is unity. In this case there is a supplementary expansion from p_{b4} to the actual backpressure outside the nozzle. We will not further discuss this so-called Prandtl-Meyer expansion since it is a transformation involving only supersonic flow. Its analysis is not necessary for the further description of transonic flow.

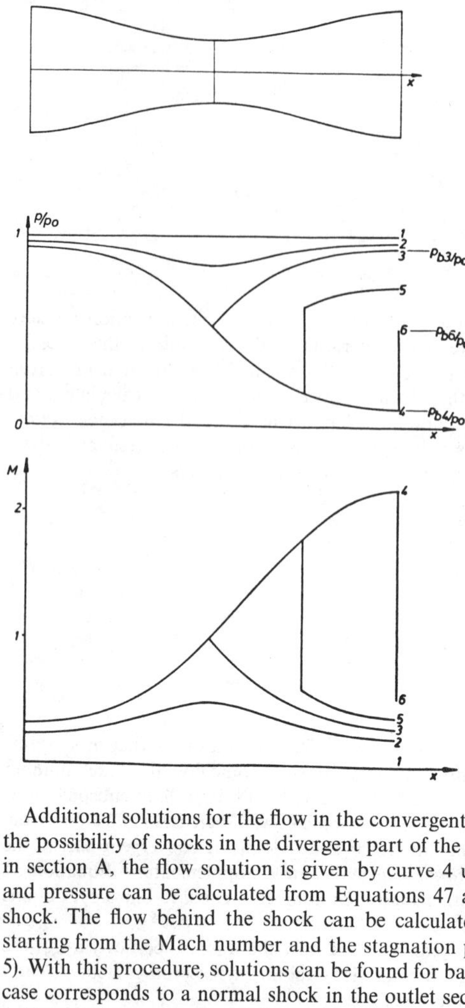

Figure 8. Pressure and Mach number variation in a convergent-divergent nozzle: (1) no flow, (2) subsonic flow, (3) subsonic flow with chocking, (4) smooth subsonic-supersonic flow, (5) transonic flow with shock within the nozzle, and (6) transonic flow with shock at exit section.

Additional solutions for the flow in the convergent-divergent nozzle can be found by considering the possibility of shocks in the divergent part of the nozzle. When, for instance, a shock is present in section A, the flow solution is given by curve 4 until this section. The jump in Mach number and pressure can be calculated from Equations 47 and 48, resulting in subsonic flow behind the shock. The flow behind the shock can be calculated with the isentropic Equations 29 and 50 starting from the Mach number and the stagnation pressure behind the shock in section A (curve 5). With this procedure, solutions can be found for backpressures between p_{b3} and p_{b6}. The limiting case corresponds to a normal shock in the outlet section.

No solution can be found for backpressures between p_{b6} and p_{b4}. For these backpressures the flow in the nozzle follows curve 4 and an additional compression between p_{b4} and the actual backpressure takes place outside the nozzle by a system of oblique shocks. The oblique shock is a supersonic flow phenomenon. As for the Prandtl-Meyer expansion, its discussion is not needed for the further analysis of transonic flows.

For all backpressures lower than p_{b3}, the flow parameters in the throat section are exactly equal. In particular this means that the mass flow is not depending on the backpressure for values lower than p_{b3}. This phenomenon is called choking.

TWO-DIMENSIONAL TRANSONIC FLOW

General Appearance

In the previous section, the transonic flow was studied in a convergent-divergent duct with slowly varying cross section. This allows a one-dimensional description of the flow and leads to analytical solutions. When the height of the duct is not small with respect to the length, the one-dimensional approximation is not valid. In this case, Equations 10, 14, and 19 must be solved when the effects of viscosity are neglected. In usual cases, the gradient of potential energy is negligible and the set of equations can be written as:

$$\begin{aligned} \nabla \cdot (\rho \vec{v}) &= 0 \\ \nabla \cdot (\rho \vec{v}\vec{v}) + \nabla p &= 0 \\ \nabla \cdot (\rho H \vec{v}) &= 0 \end{aligned} \tag{52}$$

More explicitly, in two dimensions, this is:

$$\begin{aligned} \frac{\partial}{\partial x}(\rho v_x) + \frac{\partial}{\partial y}(\rho v_y) &= 0 \\ \frac{\partial}{\partial x}(\rho v_x v_x + p) + \frac{\partial}{\partial y}(\rho v_x v_y) &= 0 \\ \frac{\partial}{\partial x}(\rho v_x v_y) + \frac{\partial}{\partial y}(\rho v_y v_y + p) &= 0 \\ \frac{\partial}{\partial x}(\rho H v_x) + \frac{\partial}{\partial y}(\rho H v_y) &= 0 \end{aligned} \tag{53}$$

Either set of Equations 52 or 53 is a set of Euler equations. In general, for transonic flow, this set cannot have an analytical solution and numerical techniques must be used to solve it. The discussion of the numerical techniques is treated in the following chapter. However, in this chapter, numerical results are used in order to clarify transonic flow patterns.

A second possibility to study transonic flows is to use results of wind tunnel experiments. Usually, velocity fields are measured by laser-doppler anemometry. In qualitative studies visualization of shock patterns by Schlieren techniques or interference-techniques are used. The description of transonic wind tunnel instrumentation is out of the scope of this chapter but wind tunnel results are used in order to clarify transonic flow patterns.

Figure 9 shows the iso-machlines in a channel obtained by the author through numerical integration of the Euler equations (Equation 52) on a 97 × 21 grid. The flow at inlet is a uniform flow

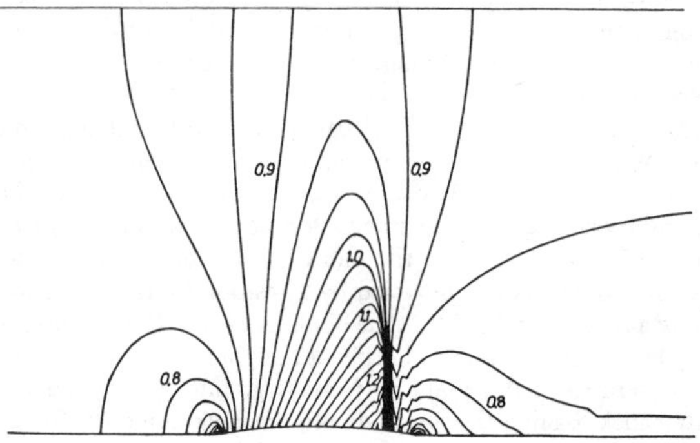

Figure 9: Isomachlines for unchocked transonic flow in a two-dimensional channel.

with Mach number 0.85. Figure 9 shows that the flow accelerates due to the contraction by the bump. Sonic velocity first is obtained in a point in the middle of the bump. Downstream of this sonic point there is a supersonic pocket terminated by a shock.

Summarizing the results of experiments and calculations, one can conclude that the following flow patterns are possible in a channel with a narrowing:

1. When the Mach number at inlet M_∞ is sufficiently low, the flow is subsonic in the complete flow field.
2. When the Mach number at inlet gradually is increased, for a value called the critical Mach number $M_{\infty,c}$ (< 1), just one sonic point is found in the flow field.
3. When the inlet Mach number is increased above the critical Mach number, a supersonic pocket is formed terminated by a shock, as shown in Figure 9.
4. When the Mach number further is increased up to a value called the choking Mach number $M_{\infty,ch}$ (< 1), the sonic line and the shock just span the whole section of the channel.
5. When the Mach number is increased above the shocking Mach number but still is lower than unity, the supersonic region increases while the shock moves further downstream. This flow pattern holds as long as the inlet Mach number is lower than unity.
6. For inlet Mach numbers equal to unity or larger, the flow is supersonic in the whole channel.

The Effect of Viscosity

In transonic flow, viscosity mainly has two effects. First, due to viscosity, shocks do not appear as sharp discontinuities but are spread over a finite distance. However, this distance is only of the order of 10 μm. As a consequence, from a practical point of view, one never is interested in determining the flow quantities in the shock region itself and since velocity gradients outside the shock can be considered to be small, the inviscid relations obtained in the previous sections remain valid across the shocks.

The second effect of viscosity is the formation of boundary layers along walls. Since at the wall, velocity is zero, within the boundary layer there is always a subsonic sublayer, even when the mainstream is supersonic. As a consequence, the qualitative description of transonic flow strictly cannot be valid close to the wall. However, if boundary layers remain attached, their physical extent is rather limited and the qualitative description of transonic flow, given in the previous section, remains globally valid.

However, since shock waves form a strong compression, there is always a danger of boundary layer separation at the point of impingement of the shock. In such a case, the inviscid flow pattern is seriously changed.

From the description in the previous section, it follows that shocks in transonic flow always are nearly normal to the flow. Hence, they are of the strong type. This means that the flow velocity passes across the shock from a supersonic to a subsonic value. For the further study of transonic flows, it therefore is not necessary to discuss the interaction of oblique weak shocks with boundary layers as it occurs in supersonic flow.

Experiments indicate that a near-normal shock can interact with a laminar boundary layer mainly in three different ways. When the shock is rather weak—this occurs when the upstream Mach number is just larger than unity—the shock only causes a gradual thickening of the boundary layer. The thickening of the boundary layer generates compression waves ahead of the shock that join the main shock. This is shown in Figure 10a. When the shock is somewhat stronger the laminar boundary layer may separate. This usually is accompanied by transition from laminar to turbulent flow and reattachment, forming a separation bubble in the boundary layer. If the bubble has a non-negligible extent, the boundary layer thickens rather sharply at the point of laminar separation such that the compression waves generated at this point may converge into an oblique shock. This generates the so-called lambda shock. Behind the shock, due to the deviation of the boundary layer into the direction of the wall, a reexpansion usually occurs forming a so-called supersonic tongue. In this supersonic tongue, the Mach number is only slightly higher than unity. The lambda shock is shown in Figure 10b. With still greater local supersonic Mach number ahead of the shock, the pressure

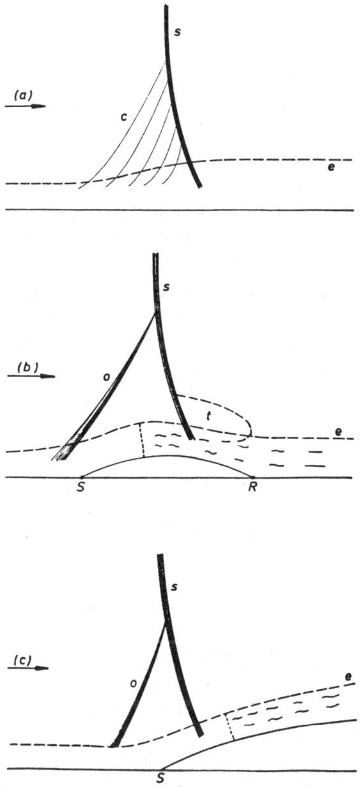

Figure 10. Laminar shock-boundary layer interaction: (a) gradual thickening of the boundary layer, (b) lambda-shock due to bubble generated by laminar boundary, layer separation with transition to turbulence and reattachment, (c) lambda-shock due to laminar boundary layer separation without reattachment, (c) compression waves, (e) edge of boundary layer, (s) impinging shock, (o) oblique shock, (S) separation point, and (R) reattachment point.

rise may be sufficient to cause separation of the laminar boundary layer with transition into turbulence but without reattachment. As shown in Figure 10c, in this case the shock also has a lambda form.

The interaction of a near-normal shock with a turbulent boundary layer is much more simple than with a laminar boundary layer. Due to the larger momentum in the boundary layer, the turbulent boundary layer is not likely to separate for mainstream Mach numbers ahead of the shock less than about 1.3. When no separation occurs, there is only thickening of the boundary layer generating compression waves ahead of the shock. The lambda profile does not appear. If separation of the boundary layer occurs, a lambda shock develops similar to the laminar case.

In this chapter, the discussion of the computation of shock-boundary layer interaction will not be treated. It is clear that such a computation involves an adequate modeling of the turbulence in the interaction region. To date this still is a largely unsolved problem.

From the previous discussion, it follows that, in absence of boundary layer separation, the effects of viscosity on transonic flow remain localized to the interaction region, so that in most cases, the analysis of transonic flow can be done neglecting viscosity.

Transonic Flow Past an Airfoil

When an airfoil is placed in a uniform flow, some of the transonic flow patterns are similar to these in transonic channel flow.

For sufficiently low Mach number of the oncoming flow M_∞, the flow is subsonic everywhere.

With gradually increasing upstream Mach number, a critical value $M_{\infty,c}$ is reached for which just one sonic point is found on the suction side of the airfoil.

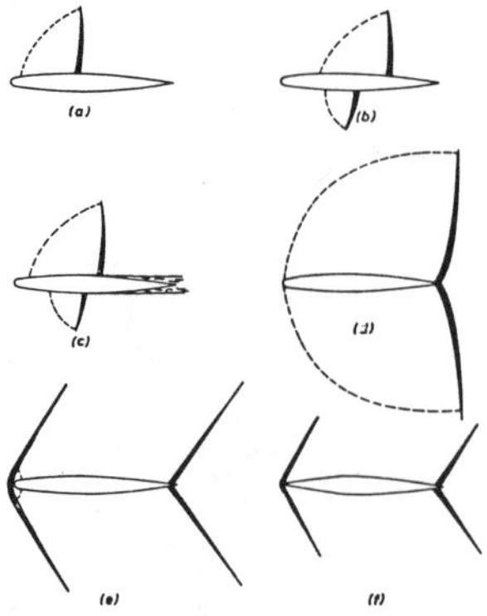

Figure 11. Flow past an airfoil: (a) supersonic pocket on suction side, (b) supersonic pockets on both suction- and pressure sides, (c) shock stall, (d) flow with upstream Mach number just below unity, (e) separated bow shock due to supersonic upstream flow and rounded leading edge, and (f) attached bow shock due to supersonic upstream flow and sharp leading edge.

When the Mach number is further increased, a supersonic pocket terminated by a near-normal shock is formed on the suction side. This flow pattern is shown in Figure 11a.

With increasing Mach number, a second supersonic pocket is formed on the pressure side of the profile. With increasing Mach number, the supersonic regions spread and the shocks move toward the trailing edge. A typical flow pattern of this kind is shown in Figure 11b.

For increasing upstream Mach number, the shock strengths increase. This may lead to boundary layer separation, with a consequent loss of lift and increase of drag. This so-called shock stall is shown in Figure 11c.

With increasing Mach number, the shocks reach the trailing edge for an upstream Mach number just below unity. In this case, there is no separation anymore. This is shown in Figure 11d.

As soon as M_∞ exceeds unity, a shock called bow shock forms upstream of the leading edge, if this leading edge is rounded. In this case, the flow is supersonic everywhere, except for a small region just behind the bow shock. The shocks at the trailing edge now are oblique shocks. This is shown in Figure 11e. If the leading edge is sufficiently sharp, the bow shock is attached consisting of two oblique shocks. In this case, shown in Figure 11f, the flow is entirely supersonic.

Figure 12 shows, qualitatively, the lift and drag coefficients for an airfoil in function of upstream Mach number. The lift coefficient C_L and the drag coefficient C_D are defined by

$$C_L = L/(\rho_\infty c w_\infty^2/2) \qquad C_D = D/(\rho_\infty c w_\infty^2/2)$$

in which L and D are the lift- and drag force per unit length of the profile, c is the chord length, ρ_∞ and w_∞ are density and velocity of the oncoming flow.

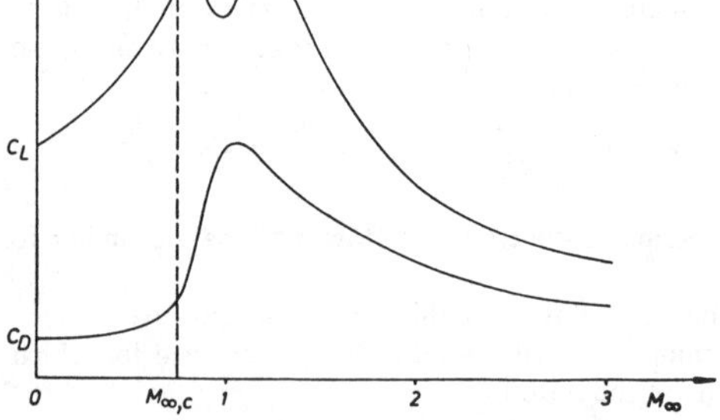

Figure 12. Lift- and drag coefficients for an airfoil in function of upstream Mach number.

For Mach numbers lower than the critical value, this means in complete subsonic or subcritical flow, the lift coefficient is approximately proportional to $1/\sqrt{1 - M_\infty^2}$. This is the so-called Prandtl-Glauert rule originating from subsonic flow theory. The typical dip in the lift coefficient in the transonic region is due to shock stall. In complete supersonic flow, the lift coefficient is approximately proportional to $1/\sqrt{M_\infty^2 - 1}$. This proportionality in supersonic flow is called the Ackeret rule.

The drag coefficient is roughly independent of Mach number in complete subsonic flow. There is a sharp increase in the transonic region due to the high losses associated with strong shocks and boundary layer separation. In supersonic flow, the shock strengths weaken with increasing upstream Mach number causing a decrease of the drag coefficient proportional with $1/\sqrt{M_\infty^2 - 1}$.

Transonic Flow in Cascades

Figure 13 shows the transonic flow in a typical convergent-divergent accelerating turbine cascade. In design conditions the incoming subsonic flow accelerates to completely supersonic flow showing oblique shock waves at the trailing edge (Figure 13a). When the backpressure is increased, the shock strengths increase (Figure 13b) until finally normal shocks develop leading to subsonic outflow (Figure 13c). From Figure 13 it is clear that the flow in a turbine cascade has similarity with both transonic nozzle flows and flows past airfoils.

Figure 14 shows the flow patterns in typical transonic decelerating compressor cascades. Figure 14a corresponds to subsonic inflow and subsonic outflow. A supersonic pocket is formed at the

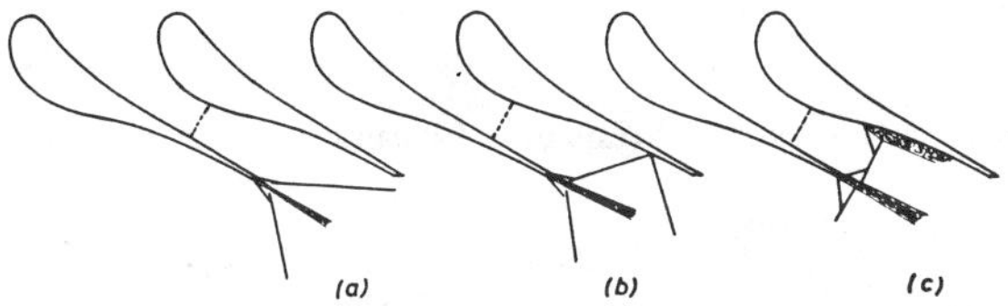

Figure 13. Flow patterns in a transonic turbine cascade: (a) design conditions, (b) increased backpressure causing stronger oblique shocks, and (c) high backpressure causing near normal shocks.

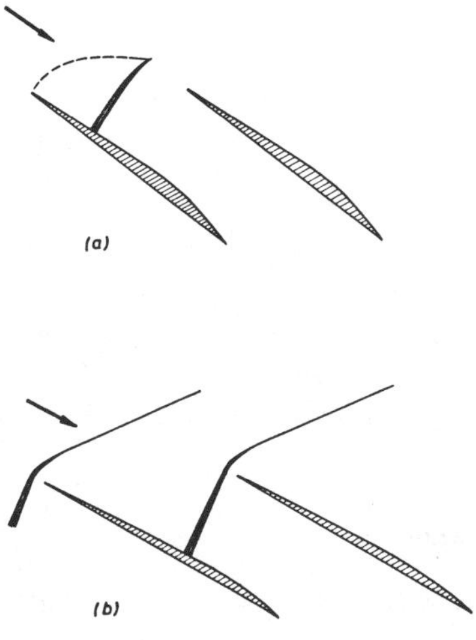

Figure 14. Flow patterns in a transonic compressor cascade: (a) subsonic inflow with supersonic pocket and (b) supersonic inflow with bow shock.

leading edge, terminated by a near-normal shock. Figure 14b corresponds to supersonic inlet and subsonic outlet. At the leading edge, a bow shock is formed consisting of an oblique shock outside the cascade and a near-normal shock within the blade passage.

Transonic Shockless Flow

It has been found both by experiments and by calculations that it is possible to design airfoils and cascades for supercritical conditions without any compression shock. These profiles are called supercritical profiles. Complete shockless transonic flow never can be reached for all flow conditions. However, it seems possible to design airfoils and cascades such that the flow is shock free in design conditions and such that shock losses are minimized in off-design conditions. To date, the problem of designing supercritical profiles is still largely unsolved. Further progress in this field is to be expected in the near future with promising applications in airplane and compressor design.

EQUATIONS DESCRIBING TRANSONIC FLOW

Neglecting viscosity, transonic flows can be described by Euler equations. The Euler equations form a rather complex set. Therefore, in many cases, one tries to describe transonic flow by simplified equations. In this section, the mathematical character of the Euler equations is analyzed and the potential flow simplification is described.

Euler Equations

The equations in the conservative set (Equation 52) can be combined to form a quasi-linear set of equations.

The mass equation can be written as

$$\vec{v} \cdot \nabla \rho + \rho \nabla \cdot \vec{v} = 0 \tag{54}$$

Combination of mass- and momentum equation gives

$$\rho \vec{v} \cdot \nabla \vec{v} + \nabla p = 0 \tag{55}$$

Using: $\nabla(\vec{v} \cdot \vec{v}/2) = \vec{v} \cdot \nabla \vec{v} + \vec{v} x (\nabla x \vec{v})$

Equation 55 can be written as

$$\nabla(\vec{v} \cdot \vec{v}/2) - \vec{v} x \vec{\omega} + \rho^{-1} \nabla p = 0 \tag{56}$$

with $\vec{\omega} = \nabla x \vec{v}$

Scalar multiplication of Equation 56 by $\vec{v}$ gives

$$\vec{v} \cdot \nabla(\vec{v} \cdot \vec{v}/2) + \rho^{-1} \vec{v} \cdot \nabla p = 0 \tag{57}$$

Remarking that

$$\rho H = \rho h + \rho v^2/2 = \gamma p/(\gamma - 1) + \rho v^2/2,$$

and using the mass equation, the energy equation can be written as

$$\nabla \cdot (\rho H \vec{v}) = (\gamma/(\gamma - 1)) p \nabla \cdot \vec{v} + (\gamma/(\gamma - 1)) \vec{v} \cdot \nabla p + \rho \vec{v} \cdot \nabla(\vec{v} \cdot \vec{v}/2) = 0 \tag{58}$$

Combining Equations 57 and 58 gives

$$\gamma p \nabla \cdot \vec{v} + \vec{v} \cdot \nabla p = 0 \tag{59}$$

Equations 54, 55 and 59 can be assembled in the system:

$$A_x \frac{\partial \xi}{\partial x} + A_y \frac{\partial \xi}{\partial y} + A_z \frac{\partial \xi}{\partial z} = 0 \tag{60}$$

with:

$$\xi = \begin{pmatrix} \rho \\ v_x \\ v_y \\ v_z \\ p \end{pmatrix} \qquad A_x = \begin{pmatrix} v_x & \rho & 0 & 0 & 0 \\ 0 & v_x & 0 & 0 & \rho^{-1} \\ 0 & 0 & v_x & 0 & 0 \\ 0 & 0 & 0 & v_x & 0 \\ 0 & \gamma p & 0 & 0 & v_x \end{pmatrix}$$

$$A_y = \begin{pmatrix} v_y & 0 & \rho & 0 & 0 \\ 0 & v_y & 0 & 0 & 0 \\ 0 & 0 & v_y & 0 & \rho^{-1} \\ 0 & 0 & 0 & v_y & 0 \\ 0 & 0 & \gamma p & 0 & v_y \end{pmatrix} \qquad A_z = \begin{pmatrix} v_z & 0 & 0 & \rho & 0 \\ 0 & v_z & 0 & 0 & 0 \\ 0 & 0 & v_z & 0 & 0 \\ 0 & 0 & 0 & v_z & \rho^{-1} \\ 0 & 0 & 0 & \gamma p & v_z \end{pmatrix}$$

The characteristic matrix associated to Equation 60 is

$$K = k_x A_x + k_y A_y + k_z A_z \tag{61}$$

with $\quad k_x^2 + k_y^2 + k_z^2 = 1$

The characteristic matrix has real eigenvalues:

$$\begin{aligned} \lambda_1 = \lambda_2 = \lambda_3 &= k_x v_x + k_y v_y + k_z v_z = \vec{k} \cdot \vec{v} \\ \lambda_4 &= \vec{k} \cdot \vec{v} + c \\ \lambda_5 &= \vec{k} \cdot \vec{v} - c \end{aligned} \tag{62}$$

Some of the eigenvalues of Equation 62 vanish for a non-zero set of parameters k_x, k_y, k_z. Hence, the set of equations (Equation 60) possesses characteristic surfaces and the system is not elliptic. The eigenvalues, however, are different in nature. For any $\vec{v}$, there is always a non-zero set of parameters k_x, k_y, k_z such that the eigenvalues $\lambda_1 = \lambda_2 = \lambda_3$ vanish. Hence, there is always a characteristic surface associated to $\lambda_1 = \lambda_2 = \lambda_3$. As a consequence, hyperbolic properties are associated to these eigenvalues. The characteristic surface described by a unit normal that is perpendicular to $\vec{v}$ degenerates to a line—the streamline. The eigenvalues λ_4 and λ_5 have different character. Indeed, in subsonic flow, $|v| < c$, it is not possible to find a non-zero set of parameters k_x, k_y, k_z such that λ_4 and λ_5 vanish. Hence, in subsonic flow, no characteristic surfaces, and as a consequence elliptic properties, are associated to λ_4 and λ_5. In supersonic flow, $|v| > c$, there exists always a set of non-zero parameters k_x, k_y, k_z such that λ_4 and λ_5 vanish. Clearly, the characteristic surface with unit normal $\vec{k}$ such that λ_4 and λ_5 vanish, is the Mach cone.

From the preceding characteristic analysis, it is concluded that Euler equations simultaneously have elliptic and hyperbolic features in subsonic flow. In supersonic flow they possess pure hyperbolic properties.

Due to the hybrid character in subsonic flow, it is hard to devise solution techniques for Euler equations in subsonic and transonic flows. The only property Euler equations have, to construct solution techniques on, is that the eigenvalues of the characteristic matrix always are real.

Of course, it is to be remarked that the quasi-linear Euler equations (Equation 60) only are valid for continuous flow properties. Equation 60 is used to illustrate the mathematical character of the equations but for transonic calculations, an integral form of Equation 52 is to be used, not requiring the continuity of flow properties.

The Transonic Potential Equation

Since solutions techniques for transonic Euler equations are difficult to construct, it is useful to derive simplifications of the Euler equations, which are easier to solve. In this section it is shown how this can be done in an approximate way.

In case all properties vary continuously, the Euler equations lead to the streamline equations derived earlier:

$$\vec{1}_v \cdot \nabla H = 0 \tag{63}$$

$$\vec{1}_v \cdot \nabla s = 0 \tag{64}$$

These equations express that total enthalpy and entropy are constant on a streamline. Hence, when at the inlet boundary of a flow field, total enthalpy and entropy have constant values on each streamline, total enthalpy and entropy are constant in the whole field. Such a flow is called homenthalpic and homentropic. The calculation of a flow with these properties is considerably simplified since the stagnation properties hold for the whole flow field.

The temperature can be calculated from

$$H = C_p T_0 = C_p T + \vec{v} \cdot \vec{v}/2$$

whence, $$T_0/T = H/(H - \vec{v} \cdot \vec{v}/2) \tag{65}$$

Density and pressure follow from

$$\rho_0/\rho = (T_0/T)^{1/(\gamma-1)} \qquad p_0/p = (T_0/T)^{\gamma/(\gamma-1)} \tag{66}$$

A further simplification even is possible. Using the definition of total enthalpy and entropy, Equation 56 can be written as

$$\nabla H - T\,\nabla s = \vec{v} \cdot \vec{\omega} \tag{67}$$

Equation 67 is called the equation of Crocco. It shows that for homentropic and homentalpic flow, the rotation vector always is parallel to the velocity vector and hence a relation holds of the form

$$\vec{\omega} = \chi \rho \vec{v} \tag{68}$$

Hence, $$\nabla \cdot \vec{\omega} = \chi \nabla \cdot (\rho \vec{v}) + \rho \vec{v} \cdot \nabla \chi = 0 \quad \text{and} \quad \vec{1}_v \cdot \nabla \chi = 0 \tag{69}$$

This means that for a homentropic and homentalpic flow, there is a third scalar that is constant on the streamline—the ratio of the rotation to the momentum. As a consequence, when the flow is irrotational at the inlet boundary, rotation vanishes in the whole flow field.

In this case, an extreme simplification of the Euler equations is possible. Indeed, Equations 63, 64, and 69 express that H, s, and χ are characteristic variables associated to the eigenvalues $\lambda_1 = \lambda_2 = \lambda_3$ of the characteristic matrix (Equation 61). These characteristic variables are known a priori and the corresponding equations can be eliminated from the Euler equations.

Logarithmic differentiation of Equation 66 gives

$$\rho^{-1}\nabla\rho = -\nabla(\vec{v}\cdot\vec{v}/2)/((\gamma-1)(H-\vec{v}\cdot\vec{v}/2)) \tag{70}$$

Remarking that $(\gamma-1)(H-\vec{v}\cdot\vec{v}/2) = (\gamma-1)h = c^2$

Equation 70 can be written as

$$\rho^{-1}\nabla\rho = -\nabla(\vec{v}\cdot\vec{v}/2)/c^2 \tag{71}$$

Combining Equation 71 with the mass Equation 54 gives

$$c^2\,\nabla\cdot\vec{v} - \vec{v}\cdot\nabla(\vec{v}\cdot\vec{v}/2) = 0 \tag{72}$$

Homenthalpic, homentropic, irrotational flow is then described by Equation 72 in combination with the condition of irrotationality

$$\vec{\omega} = 0 \tag{73}$$

Equation 73 can be satisfied by the introduction of a velocity potential

$$\vec{v} = \nabla\phi \tag{74}$$

Substitution of Equation 74 into Equation 72 gives the potential equation. In two dimensions, it is

$$(c^2 - v_x^2)\frac{\partial^2\phi}{\partial x^2} - 2v_xv_y\frac{\partial^2\phi}{\partial x\,\partial y} + (c^2 - v_y^2)\frac{\partial^2\phi}{\partial y^2} = 0 \tag{75}$$

The characteristic form associated to Equation 75 is:

$$(c^2 - v_x^2)k_x^2 - 2v_xv_yk_xk_y + (c^2 - v_y^2)k_y^2 = 0 \tag{76}$$

Equation 76 allows a non-zero solution for k_x and k_y if the discriminant is positive, this is if

$$c^2(c^2 - v_x^2 - v_y^2) \geq 0$$

Hence, Equation 75 is hyperbolic in supersonic flow and is elliptic in subsonic flow.

In comparison with Euler equations, the potential equation has the enormous simplification, that is not hybrid anymore. It still is mixed, which means that it can have either elliptic or hyperbolic properties, but it cannot have them simultaneously. This makes the construction of solution techniques for the potential equation much easier than for the Euler equations.

Although the potential formulation is only valid for flows with continuously varying flow properties, it is tempting to try to use it for transonic flows involving shocks. To make calculations possible across discontinuities, an integral formulation based on conservative equations is then necessary. The equation to be solved is then the mass equation:

$$\nabla\cdot(\rho\vec{v}) = 0$$

together with Equations 65, 66, and 74.

The approximation involved is then that the entropy increase across shocks and the associated loss of stagnation pressure are neglected. This approximation is not dramatic as can be seen from Figure 7. For a Mach number 1.3, the loss in stagnation pressure is about 2% for air. Hence, if one limits calculations to flows with relative weak shocks, the use of the potential equation only introduces errors in the order of a few percent. For most engineering applications, this accuracy is sufficient.

NOTATION

a	critical velocity
C_D	drag coefficient
C_L	lift coefficient
c	velocity of sound
c_p	specific heat for constant pressure
c_v	specific heat for constant volume
D	dragforce per unit length
E	total mechanical energy
e	specific internal energy
$\vec{f}$	mass force
H	enthalpy
h	specific enthalpy
h_0	stagnation enthalpy
L	lift per unit length; Laval number
M	Mach number
p	pressure
R	gas law constant
S	surface
s	entropy
T	temperature
$\vec{T}$	force per unit area acting in a point of the surface S
U	potential energy
V	volume
v	velocity

Greek Symbols

γ	ratio of specific heats
λ_i	eigenvalues
μ	cone angle
ρ	density
$\vec{\omega}$	rotational velocity vector

REFERENCES

1. Shapiro, A. H., *Dynamics and Thermodynamics of Compressible Fluid Flow.* Ronald Press, New York, 1953.
2. Liepmann, H. W., and Roshko, A., *Elements of Gasdynamics.* John Wiley, New York, 1957.
3. Von Mises, R., *Mathematical Theory of Compressible Fluid Flow.* Academic Press, New York, 1958.
4. Ferrari, C., and Tricomi, F., *Transonic Aerodynamics.* Academic Press, New York, 1968.
5. Adamson, T. C., and Platzer, M. F., ed., *Transonic Flow Problems in Turbomachinery.* Hemisphere Publishing Corporation, Washington, 1977.
6. Shreier, S., *Compressible Flow.* John Wiley, New York, 1982.

CHAPTER 17

NUMERICAL MODELING OF TRANSONIC FLOWS OVER AIRFOILS AND CASCADES

Xin-Hai Zhou
Song-Ling Liu
Fei-Da Fan

Northwestern Polytechnical University, Xi'an, China

CONTENTS

INTRODUCTION

Transonic airfoils have been widely adopted in modern aircraft. Transonic stages of compressors and turbines often appear in advanced gas turbine engines. Aerodynamic performance of these devices depends very much on the behavior of transonic flow passing through them. Therefore, it is very important for the designer to predict correctly the characteristics of transonic flow. As shown in Figures 1 and 2, the flow patterns of transonic flow around an airfoil, and especially those in cascades, are very complicated. In the past, it seemed that only through wind tunnel tests could it be studied. In recent years considerable attention has been given to the numerical modeling of the transonic flow and much work on it has been done. In the aerodynamic design of airfoils or cascades, the computational analysis not only replaces some experiments so as to cut costs, but also eliminates some defects inherent in experiments, such as the influence of measuring instruments on the flow, wind tunnel wall interference, and incorrect Reynolds-number scaling. In addition,

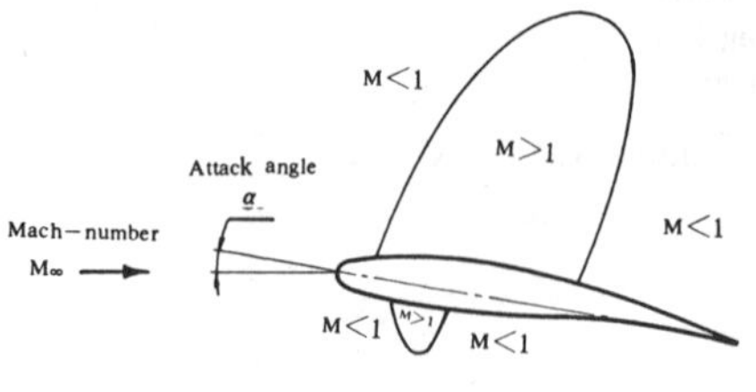

Figure 1. Transonic flow over an airfoil.

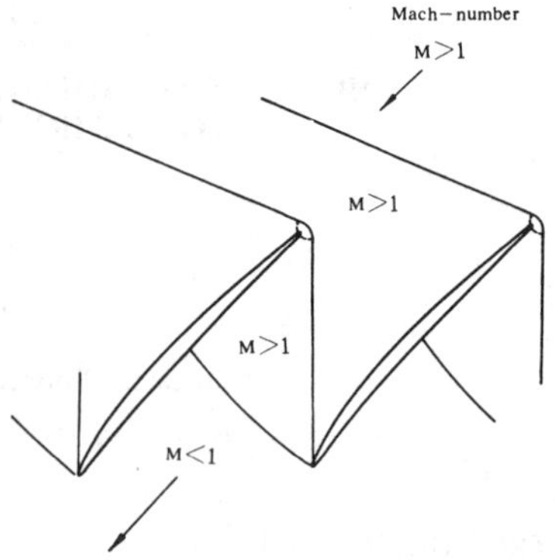

Figure 2. Transonic flow over a cascade.

modeling of an exact two-dimensional flow can be achieved only in computations. So the solution of numerical modeling is an indispensable complement to test data. As a result of the rapid increase in capacities and speeds of computers and advances in computational fluid mechanics, the techniques of numerical modeling of transonic flow have been significantly improved in the past decade.

The difficulty encountered in transonic flow computations arises from the fact that both subsonic and supersonic flows are always co-existing in the flowfield. Physically, the characteristics of perturbation propagation in these two zones are quite different. In subsonic flow, perturbation propagates in all directions, while in supersonic flow only within a limited downstream region. Mathematically, these two flows are governed by two different types of partial differential equations—elliptic and hyperbolic, respectively. Therefore, different computational schemes have to be used for different zones in numerical modeling. Unfortunately, the boundary between these two zones is not known in advance. Moreover, shock wave may appear in transonic flowfield and discontinuities in flow parameters result. The wave structure is rather complicated, particularly in transonic cascade. All these bring more difficulties to computations.

Finite difference methods play the most important role in computation of steady transonic flows. These methods are usually divided into two types: time-dependent and relaxation. For the time-dependent method, the steady-state solution is regarded as an asymptotic one of the unsteady equations as time marches forward. Thus, the governing equations, for either subsonic or supersonic flow, are always of hyperbolic type, and the same numerical scheme can be used over the entire field. Furthermore, shock wave can be calculated as a steep gradient part of the solution, and no special treatment is necessary.

For the transonic relaxation method, type-dependent finite difference schemes have to be used to solve velocity potential equation; upwind and central difference schemes are employed for supersonic and subsonic zones separately, then successive relaxation technique is applied to solve the finite difference equations. The major advantages of relaxation method are its comparatively high efficiency and accuracy. These features often make it the choice for computing transonic flows, especially for external flow problems. The shortcoming of this method is the potential flow assumption, which causes the calculated pressure rise to be in serious error as shock wave strength increases. This discrepancy arises from both the use of the isentropic shock model and incorrect modeling of vortex phenomena in the whole field. The latter may be even more serious [1].

In the computations of fully three-dimensional flow in turbomachinery, even subsonic, the irrotational flow assumption is not valid in general [2]. On the other hand the time-dependent method for solving the Euler equations involves no such restrictions. Furthermore, it is able to capture stronger shock and easier to program. Owing to these attractive aspects, the time-dependent method has been widely applied to the computation of transonic flows in cascades [3–25], especially in diffusion cascades involving strong shock. It seems that no other method can match it in this respect. This method was also successfully used for the computation of transonic flows around airfoils [26], but the transonic relaxation method was employed more often for such cases [27–32]. It should be pointed out that, it is very important in transonic flow analysis to consider the effects of viscosity. The ultimate methods accounting for the viscous effects is to solve the Navier-Stokes equations. Here, the relatively high cost in computer time may make it unacceptable in engineering design and analysis. Another method for solving this problem is the iteration calculation of inviscid flow and boundary layer. This seems to be more practical than the former. In this procedure, the inviscid flow analysis is an important part in the iteration calculation.

This chapter mainly deals with the numerical modeling of inviscid transonic flows around airfoils and in cascades. Major subjects include basic governing equations, computational mesh, discretization techniques for governing equations, boundary conditions and methods of accelerating convergence, etc. Finally, some numerical results are presented to illustrate the accuracy and the limitations of the numerical modeling.

This chapter concentrates on time-dependent method for solving the Euler equations. Naturally, cascade flow computations are discussed in some detail. For the sake of brevity and clarity, two-dimensional flows are taken as an example throughout the chapter. Of course, extension to three-dimensional flows is straightforward in many cases. Besides, we refrain from presenting various numerical methods in detail. The discussion emphasizes the most popular and typical methods for numerical modeling.

GOVERNING EQUATIONS

The governing equations of inviscid, no heat transfer, compressible flow consist of the conservation relations of mass, momentum, and energy. These equations are called Euler equations, and can be written in various forms. For solving transonic flow problems by finite difference methods, the equations in conservation form (or divergence form) should be employed. This is because all conservative finite difference schemes applied to the conservation equations satisfy the Rankine-Hugoniot relation so as to produce the correct jump condition across a shock, and this ensures the solution of finite difference equations to approach the proper weak solutions of the governing equations. So the use of the equations in conservation form is necessary for shock problems. In Cartesian coordinates, the governing equations in conservation form are expressed as

$$\partial U/\partial t + \partial F/\partial x + \partial G/\partial y = 0 \tag{1}$$

where

$$U = \begin{bmatrix} \rho \\ \rho u \\ \rho v \\ e \end{bmatrix}, \quad F = \begin{bmatrix} \rho u \\ \rho u^2 + p \\ \rho uv \\ (e + p)u \end{bmatrix}, \quad G = \begin{bmatrix} \rho v \\ \rho uv \\ \rho v^2 + p \\ (e + p)v \end{bmatrix}.$$

ρ is the density, u and v are the velocity components in the x and y directions respectively, e is the total internal energy per unit volume. The pressure is related to e and ρ by an equation of state

$$p = (k - 1)[e - \rho(u^2 + v^2)/2] \tag{2}$$

where k is ratio of specific heats. From Equation 1 the isentropic equation

$$dS/dt = 0 \tag{3}$$

can be derived, where S is the entropy. If the entropy is uniform on the upstream boundary then

$$p/\rho^k = \text{const} \tag{4}$$

holds in the whole field. By replacing the energy equation in Equation 1 with Equation 4, another set of equations, equivalent to Equation 1, is obtained [20]. This set of equations is applicable only to the shock free flow. In the stronger shock flow, serious error occurs due to the variation of the entropy through the shock. In order to reduce the number of equations and unknowns to be solved, it is better to replace the energy equation by a statement of constant stagnation enthalpy i* along each particle path [13], namely

$$di^*/dt = 0 \tag{5}$$

Equation 5 is exact for adiabatic steady flow. The replacement has no effect on the asymptotic solution when the flow becomes steady and changes nothing but the intermediate states of the solution. If the stagnation enthalpy at the upstream boundary is uniform and does not vary with time, we have

$$kp/[(k - 1)\rho] + (u^2 + v^2)/2 = \text{const} \tag{6}$$

Equation 6 and the first three equations of Equation 1 constitute a set of equations often used.

According to the forms of governing equations to be solved, the time-dependent methods can be classified as differential and integral. In the latter case, the difference equations are established for each finite volume (or area for two-dimensional problems) in the flowfield—the so-called finite-volume method. For the analysis of the flow in a rotating cascade, the effects of rotation, of the variations of radius and thickness of stream surface should be taken into account. Thus, it is preferable to choose an axisymmetric orthogonal coordinate system, which rotates with the same angular velocity as the machine, and to write the governing equations in a corresponding form. In such a coordinate system, the stream surface of revolution, the normal surface of it and the meridional plane constitute the orthogonal curved surfaces of coordinate (see Figure 3a). And the governing equations in integral form are expressed as following:

$$\frac{\partial}{\partial t}\int_v U\,dv + \int_S \vec{H}\cdot\vec{\xi}\,dS + \int_v K\,dv = 0 \tag{7}$$

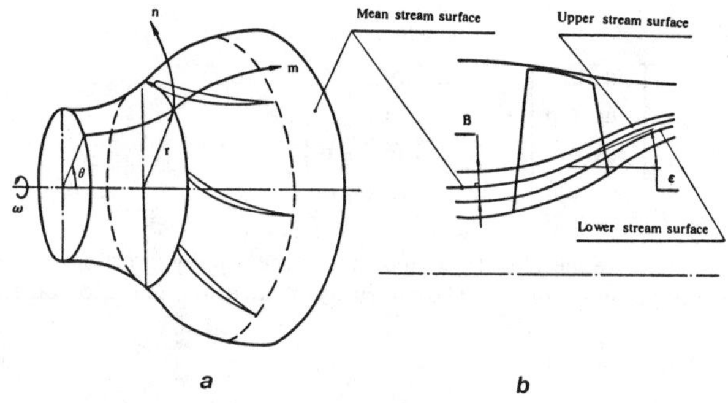

Figure 3. Orthogonal system of coordinates with axial symmetry on arbitrary surface of revolution: (a) stream surface of revolution and (b) meridional plane.

where

$$U = \begin{bmatrix} \rho \\ \rho w_m \\ \rho w_\theta \\ \rho I - p \end{bmatrix}, \qquad \vec{H} = \begin{bmatrix} \rho w_m \vec{i}_m + \rho w_\theta \vec{i}_\theta \\ (\rho w_m^2 + p)\vec{i}_m + \rho w_m w_\theta \vec{i}_\theta \\ \rho w_m w_\theta \vec{i}_m + (\rho w_\theta^2 + p)\vec{i}_\theta \\ \rho I w_m \vec{i}_m + \rho I w_\theta \vec{i}_\theta \end{bmatrix},$$

$$K = \begin{bmatrix} 0 \\ -p(d \ln B/dm) - p \sin \epsilon/r - \rho \sin \epsilon (w_\theta + \omega r)^2/r \\ \rho w_m \sin \epsilon (2\omega + w_\theta/r) \\ 0 \end{bmatrix}$$

$\vec{\xi}$ is the unit vector in the direction of outward normal to the surface S, which are the boundaries of the finite volume; w_m and w_θ are the relative velocity components and $\vec{i}_m$ and $\vec{i}_\theta$ are unit vectors in the m and θ direction respectively; r is radius, ω is angular velocity, B is the normal thickness of stream surface, ϵ is the angle between the meridional stream-line and the axis of the machine (see Figure 3b). I is named rothalpy and defined as

$$I = C_p T + w_m^2/2 + w_\theta^2/2 - \omega^2 r^2/2$$

where C_p and T are the specific heat at constant pressure and temperature respectively. Similarly, Equation 7 can be replaced by the relation of constant rothalpy along each particle path. If the upstream relative stagnation parameters are uniform, then I remains constant over the whole field, namely

$$I = kp/[(k-1)\rho] + (w_m^2 + w_\theta^2 - \omega^2 r^2)/2 = \text{const} \tag{8}$$

the pressure is computed by this equation. If setting $\epsilon = 0$, $\omega = 0$, and $B = \text{const}$, then Equation 7 becomes the integral form of Equation 1.

COMPUTATIONAL MESH

The basic principle involved in grid design in the computational domain is that for computational efficiency the number of mesh points must be kept to the minimum required to resolve spatially all significant features of the flow. Therefore, the computational mesh can be arranged in such a manner that the finer mesh is distributed in regions where the changing rates of the flow properties are great or where more accurate results are desirable, and that the coarser mesh prevails everywhere else. Typical meshes for computing the flows over airfoils and cascades are shown in Figures 4 and 5.

It can be seen from Figure 5 that the axial width of the mesh is gradually decreased from upstream boundary to leading edge of the profile and from downstream boundary to trailing edge, so that too many mesh points can be avoided. The spacing in the y direction is also nonuniform. It decreases gradually from the middle of cascade channel to the blade surface, in order to reduce the number of mesh points needed for adequate accuracy of flow parameters around blade surface. The meshes shown in Figures 4 and 5 are constructed by a simple geometric method. This procedure, although the simplest, usually leads to severe mesh distortion and coordinate singularities in the

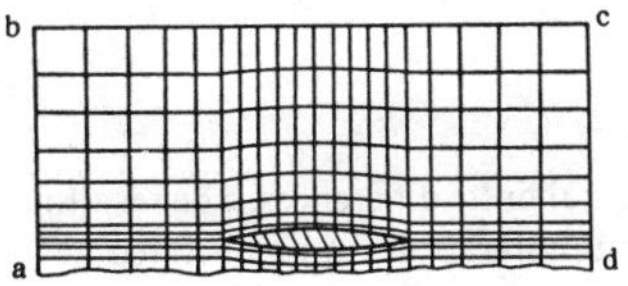

Figure 4. Computational mesh (airfoil).

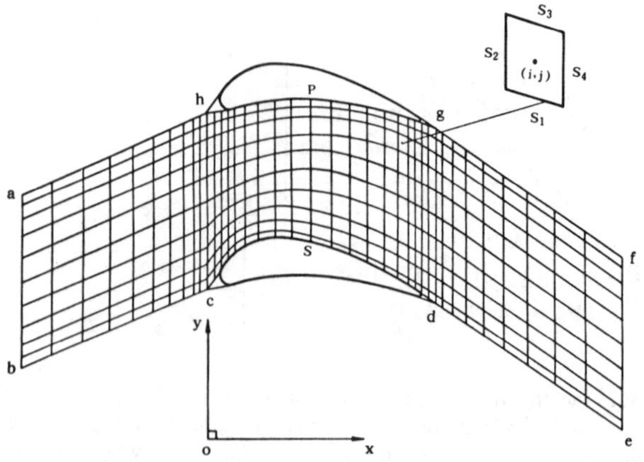

Figure 5. Computational mesh (cascade).

critical leading and trailing edge regions, especially for cascade problems. Several automatic mesh generation techniques have been developed in recent years. They involve two main types, namely, methods based on complex variables [33], and solutions to elliptical differential equations [34–35]. The former has limited control over the mesh point distribution and is also obviously restricted to two-dimensional problems. However, the latter developed by Thompson et al. is considerably more flexible. It is applicable to the general computational domain containing any number of arbitrarily shaped bodies, and there are no restrictions on the shape of the boundaries. Therefore, in many computations for transonic flows over airfoils and cascades it was employed to generate automatically the body-fitted curvilinear coordinate system [18, 36]. Moreover three-dimensional body-fitted coordinates in turbomachinery have been generated by the same approach [37]. The basic principle of this method is that making the general transformation from the physical plane (x, y) to a rectangular plane (ξ, η) is performed with the curvilinear coordinates generated by solving an elliptic system with Dirichlet boundary conditions. The system is

$$\begin{aligned} \xi_{xx} + \xi_{yy} &= P(\xi, \eta) \\ \eta_{xx} + \eta_{yy} &= Q(\xi, \eta) \end{aligned} \tag{9}$$

where the source functions P and Q are introduced to cluster mesh points to desired regions. Interchanging the dependent and independent variables in Equation 9, one can get the equations for computing the mesh coordinates in the transformed plane:

$$\begin{aligned} \alpha x_{\xi\xi} - 2\beta x_{\xi\eta} + \gamma x_{\eta\eta} + J^2(Px_\xi + Qx_\eta) &= 0 \\ \alpha y_{\xi\xi} - 2\beta y_{\xi\eta} + \gamma y_{\eta\eta} + J^2(Py_\xi + Qy_\eta) &= 0 \end{aligned} \tag{10}$$

where

$$\alpha = x_\eta^2 + y_\eta^2, \qquad \beta = x_\xi x_\eta + y_\xi y_\eta$$

$$\gamma = x_\xi^2 + y_\xi^2, \qquad J = x_\xi y_\eta - x_\eta y_\xi$$

All derivatives in Equation 10 are approximated by central finite differences. The resulting difference equations then can be solved by successive overrelaxation iterations. An associated key problem is how to make the mesh points cluster to the places where they are needed. In order to make the ξ = const lines approach the $\xi = \xi_i$ lines and (ξ_j, η_j) points; and meanwhile the η = const lines to the $\eta = \eta_i$ lines and (ξ_j, η_j) points, the source functions P and Q can be chosen to be [35]

$$P(\xi, \eta) = -\sum_{i=1}^{n} a_i[(\xi - \xi_i)/|\xi - \xi_i|]e^{-C_i|\xi-\xi_i|} - \sum_{j=1}^{m} b_j[(\xi - \xi_j)/|\xi - \xi_j|]e^{-d_j\sqrt{[(\xi-\xi_j)^2+(\eta-\eta_j)^2]}} \quad (11a)$$

$$Q(\xi, \eta) = -\sum_{i=1}^{n} a_i[(\eta - \eta_i)/|\eta - \eta_i|]e^{-C_i|\eta-\eta_i|} - \sum_{j=1}^{m} b_j[(\eta - \eta_j)/|\eta - \eta_j|]e^{-d_j\sqrt{[(\xi-\xi_j)^2+(\eta-\eta_j)^2]}} \quad (11b)$$

The difficulty with the source functions in the form previously mentioned is that no automatic way has been available to choose the coefficients of P and Q for a desired clustering. Much work has been done [38–40] in order to ensure that the mesh points automatically cluster to the required regions and to improve the orthogonality of the mesh. The technique developed by Steger [38] makes the mesh points automatically cluster to the inner boundary ($\eta = \eta_1$) and the mesh lines intersect the boundary in a nearly normal fashion. In this approach the source functions are taken as

$$\begin{aligned} P &= J^{-1}(y_\eta R_1 - x_\eta R_2)\Big|_{\eta=\eta_1} \cdot e^{-a(\eta-\eta_1)} \\ Q &= J^{-1}(-y_\xi R_1 + x_\xi R_2)\Big|_{\eta=\eta_1} \cdot e^{-b(\eta-\eta_1)} \end{aligned} \quad (12)$$

where

$$R_1 = -J^{-2}(\alpha x_{\xi\xi} - 2\beta x_{\xi\eta} + \gamma x_{\eta\eta})\Big|_{\eta=\eta_1}$$

$$R_2 = -J^{-2}(\alpha y_{\xi\xi} - 2\beta y_{\xi\eta} + \gamma y_{\eta\eta})\Big|_{\eta=\eta_1}$$

For an airfoil and a cascade, body-fitted meshes generated by Thompson's method are illustrated in Figures 6 and 7, respectively.

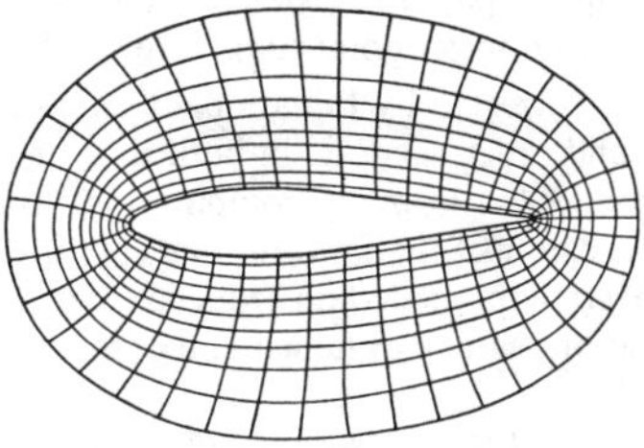

Figure 6. Curvilinear body-fitted mesh for an airfoil.

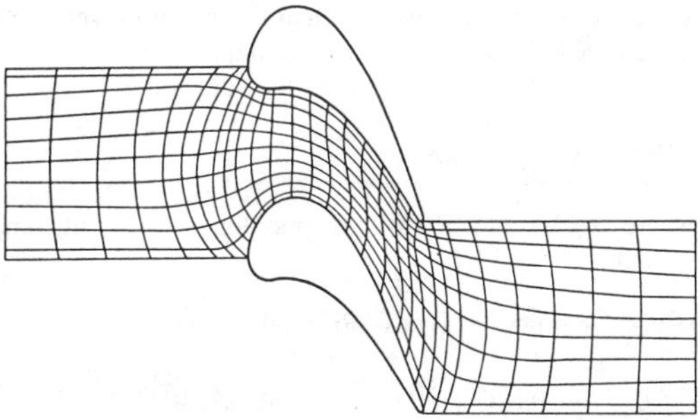

Figure 7. Curvilinear body-fitted mesh for a cascade.

In addition to the methods previously mentioned, some others, such as algebraic methods [41, 42], have been used to generate the computational mesh for transonic flow analysis. Recently, a hybrid technique was applied [43]. This procedure, briefly, is that the mesh point distributions along the airfoil surface and outer boundaries are determined first by using an electrostatic analog technique [44] and the inner-node coordinate are then determined by solving the Laplace equations. With this approach an orthogonal mesh for cascade flow analysis is generated and the mesh points are automatically clustered in regions of high surface curvature.

FINITE DIFFERENCE METHODS

Difference Scheme

To date, the explicit difference methods are still the most popular techniques for solving the time-dependent Euler equations. These methods are easily implemented, but the computation time needed to reach the convergence state is comparatively long because time step is restricted by the stability requirement. Recently, implicit difference methods were used to compute the flows in cascades [36]. These methods are unconditionally stable regardless of the size of the time step, so that fewer time steps are required to obtain the steady solutions. Therefore, the convergence of the implicit methods may be much faster than that of the explicit methods. Unfortunately, this advantage has not yet been fully taken in computing transonic flows in cascades.

The most widely used explicit method is MacCormack's scheme [45]. For solving numerically the basic equations in differential form, the MacCormack explicit two-step difference equations can be written as follows:

Predictor step

$$U_{i,j}^{\overline{n+1}} = U_{i,j}^{n} - (\Delta t/\Delta x)(F_{i,j}^{n} - F_{i-1,j}^{n}) - (\Delta t/\Delta y)(G_{i,j}^{n} - G_{i,j-1}^{n}) \tag{13a}$$

Corrector step

$$U_{i,j}^{n+1} = 0.5[U_{i,j}^{n} + U_{i,j}^{\overline{n+1}} - (\Delta t/\Delta x)(F_{i+1,j}^{\overline{n+1}} - F_{i,j}^{\overline{n+1}}) - (\Delta t/\Delta y)(G_{i,j+1}^{\overline{n+1}} - G_{i,j}^{\overline{n+1}})] \tag{13b}$$

Equation 13 shows that one-side difference is used to approximate spatial first derivative in each step. From the known solution, $U_{i,j}^{n}$, at time $t = n\,\Delta t$ and at each mesh point (i, j), a new value $U_{i,j}^{\overline{n+1}}$ is predicted by Equation 13a; then the predicted value is corrected by Equation 13b to find the solution, $U_{i,j}^{n+1}$, at time $t = (n + 1)\,\Delta t$. This scheme is second order accurate in both time and space, and has good shock fitting properties, so it has been adopted in many computations for transonic flows [3–12]. In addition, the MacCormack scheme can take the form of time splitting. Then, the computation of multi-dimensional flow may be converted into a certain sequence of one-dimensional flows. By this means, a simplification in computations can be obtained and computational efficiency improved.

For solving numerically the basic equations in differential form, it is necessary to transform the physical plane onto a rectangular computational domain. When using the finite-volume method to solve the basic equations in integral form, however, the coordinate transformation of the domain is not needed because arbitrarily shaped mesh in the physical plane can be used. The difference equations of the MacCormack time-split finite-volume method are [46]

$$U_{i,j}^{n+1} = \mathscr{L}_y(\Delta t)\mathscr{L}_x(\Delta t)U_{i,j}^{n} \tag{14}$$

where $\mathscr{L}_y(\Delta t)$ and $\mathscr{L}_x(\Delta t)$ are one-dimensional difference operators. The effect of the operator $\mathscr{L}_y(\Delta t)$ is defined by

$$U_{i,j}^{\overline{n+1/2}} = U_{i,j}^{n} - (\Delta t/\Delta A_{i,j})(\vec{H}_{i,j}^{n} \cdot \vec{S}_3 + \vec{H}_{i,j-1}^{n} \cdot \vec{S}_1) \tag{15a}$$

$$U_{i,j}^{n+1/2} = 0.5[U_{i,j}^{n} + U_{i,j}^{\overline{n+1/2}} - (\Delta t/\Delta A_{i,j})(\vec{H}_{i,j+1}^{\overline{n+1/2}} \cdot \vec{S}_3 + \vec{H}_{i,j}^{\overline{n+1/2}} \cdot \vec{S}_1)] \tag{15b}$$

The operator $\mathscr{L}_x(\Delta t)$ is defined similarly by

$$U_{i,j}^{\overline{n+1}} = U_{i,j}^{n+1/2} - (\Delta t/\Delta A_{i,j})(\vec{H}_{i,j}^{n+1/2} \cdot \vec{S}_4 + \vec{H}_{i-1,j}^{n+1/2} \cdot \vec{S}_2) \tag{16a}$$

$$U_{i,j}^{n+1} = 0.5[U_{i,j}^{n+1/2} + U_{i,j}^{\overline{n+1}} - (\Delta t/\Delta A_{i,j})(\vec{H}_{i+1,j}^{\overline{n+1}} \cdot \vec{S}_4 + \vec{H}_{i,j}^{\overline{n+1}} \cdot \vec{S}_2)] \tag{16b}$$

where $\vec{S}_i = \vec{\xi}_i S_i$ in which $\vec{\xi}_i$ is the outward unit vector normal to cell side S_i for i = 1, 2, 3, 4 (see Figure 5). It is not difficult to prove that the computation according to Equation 14 results in first order accuracy only. However, if the operators are arranged in the following symmetric sequence

$$U_{i,j}^{n+2} = \mathscr{L}_y(\Delta t)\mathscr{L}_x(\Delta t)\mathscr{L}_x(\Delta t)\mathscr{L}_y(\Delta t)U_{i,j}^n \tag{17}$$

then, advancing the computation from $U_{i,j}^n$ to $U_{i,j}^{n+2}$ has second order accuracy.

There exist several variants in the MacCormack two-step scheme. It can be seen from the difference equations given above that simple backward differences in the predictor steps (Equations 13a, 15a, and 16a) and forward differences in the corrector steps (Equations 13b, 15b, and 16b) are used to approximate spatial first derivatives. Similarly, forward differences in the predictor steps and backward differences in the corrector steps can also be used. We refer to the former as scheme (BF) and the latter as (FB). For illustrative purpose the operator of (FB) is presented here:

$$U_{i,j}^{\overline{n+1/2}} = U_{i,j}^n - (\Delta t/\Delta A_{i,j})(\vec{H}_{i,j+1}^n \cdot \vec{S}_3 + \vec{H}_{i,j}^n \cdot \vec{S}_1) \tag{18a}$$

$$U_{i,j}^{n+1/2} = 0.5[U_{i,j}^n + U_{i,j}^{\overline{n+1/2}} - (\Delta t/\Delta A_{i,j})(\vec{H}_{i,j}^{\overline{n+1/2}} \cdot \vec{S}_3 + \vec{H}_{i,j-1}^{\overline{n+1/2}} \cdot \vec{S}_1)] \tag{18b}$$

In addition, there exist some other variants in the time-split difference scheme. Obviously, for linear problems all these schemes are identical. For nonlinear problems, however, the results obtained by using various schemes may be different for the same problem. This is because the accumulation of their truncation errors is different [47]. For a transonic turbine cascade without shock waves, the distributions of the total pressure p* on the suction surface, computed by using scheme (BF) and (FB), are shown in Figure 8. In this case, the exact value of p* is equal to the total pressure at the upstream boundary p_1^*. Figure 8 shows that the difference between the computed and the exact value by (BF) scheme is less than that by (FB) scheme [48].

In addition to the MacCormack scheme, Denton's scheme [13], which is characterized by high rate of convergence, is well-known and has found wide use in the computations of two- and three-dimensional transonic flows in cascades [15–18]. This is a method for solving the governing equations in integral form. In order to ensure stability it uses up-wind differencing in the streamwise direction for fluxes of mass and momentum while downwind differencing for pressure. Therefore, this scheme has been termed the opposed difference technique.

Recently, the hopscotch method combining the features of both the explicit and the implicit schemes has been employed to compute two-dimensional transonic flows in cascades [43, 49]. It is an explicit difference method and has second order accuracy in both time and space. Owing to the introduction of a simple linear extrapolation formula in the scheme, the computation time of this method is much shorter than that of other explicit methods.

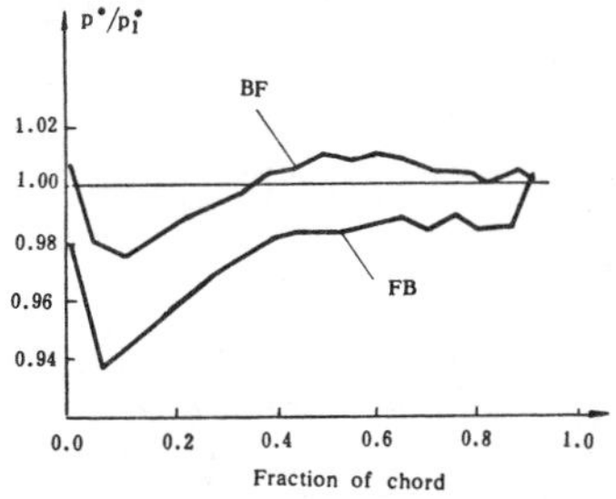

Figure 8. Total pressure distribution on suction surface of a turbine blade.

Stability of Difference Equations

The explicit difference schemes are conditionally stable. To obtain the stability conditions for the difference equations and to discuss the boundary conditions later, we will briefly present the concept of characteristics in two-dimensional flow here. For a general hyperbolic system in three independent variables (x, y, t), one can find only one linear combination system of the original partial differential equations if it is required that the dependent quantities in combined equations are differentiated in directions that are coplanar [50]. Such equations are said to be in characteristic form for the two-dimensional flow, and the corresponding plane is a characteristic plane. In order to obtain the equations in characteristic form, first of all, e is written in terms of p and ρ, then Equation 1 in nonconservative form can be written as

$$\partial\rho/\partial t + u(\partial\rho/\partial x) + v(\partial\rho/\partial y) + \rho(\partial u/\partial x) + \rho(\partial v/\partial y) = 0 \tag{19a}$$

$$\partial u/\partial t + u(\partial u/\partial x) + v(\partial u/\partial y) + (\partial p/\partial x)/\rho = 0 \tag{19b}$$

$$\partial v/\partial t + u(\partial v/\partial x) + v(\partial v/\partial y) + (\partial p/\partial y)/\rho = 0 \tag{19c}$$

$$\partial p/\partial t + u(\partial p/\partial x) + v(\partial p/\partial y) + a^2\rho(\partial u/\partial x + \partial v/\partial y) = 0 \tag{19d}$$

where a is the velocity of sound. Let α_i (i = 1, 2, 3, 4) denote the coefficient of linear combinations in making the linear combinations for the four equations above, then we have

$$\vec{\eta}_1 \cdot \nabla\rho + \vec{\eta}_2 \cdot \nabla u + \vec{\eta}_3 \cdot \nabla v + \vec{\eta}_4 \cdot \nabla p = 0 \tag{20}$$

where

$$\vec{\eta}_1 = (\alpha_1, u\alpha_1, v\alpha_1),$$

$$\vec{\eta}_2 = (\alpha_2, \rho\alpha_1 + u\alpha_2 + a^2\rho\alpha_4, v\alpha_2)$$

$$\vec{\eta}_3 = (\alpha_3, u\alpha_3, \rho\alpha_1 + v\alpha_3 + a^2\rho\alpha_4)$$

$$\vec{\eta}_4 = (\alpha_4, (\alpha_2/\rho) + u\alpha_4, (\alpha_3/\rho) + v\alpha_4)$$

The requirement of $\vec{\eta}_i$ (i = 1, 2, 3, 4) being coplanar demands that there exists a non-zero vector $\vec{N}$ such that

$$\vec{N} \cdot \vec{\eta}_i = 0 \qquad (i = 1, 2, 3, 4) \tag{21}$$

where $\vec{N} = (N_t, N_x, N_y)$ is the vector normal to the characteristic plane, with arbitrary length. Equation 21 is a set of linear equations in the four unknowns α_i. The necessary and sufficient condition of having a non-trivial solution for the linear system (Equation 21) is that determinant of the system has to be zero, i.e.

$$\begin{vmatrix} \ell & 0 & 0 & 0 \\ \rho N_x & \ell & 0 & a^2\rho N_x \\ \rho N_y & 0 & \ell & a^2\rho N_y \\ 0 & N_x/\rho & N_y/\rho & \ell \end{vmatrix} = 0$$

namely,

$$\ell^2[\ell^2 - a^2(N_x^2 + N_y^2)] = 0$$

where

$$\ell = N_t + uN_x + vN_y$$

Then, we find

$$\ell = 0 \qquad \text{(double root)} \tag{22a}$$

and

$$\ell = \pm a\sqrt{(N_x^2 + N_y^2)} \tag{22b}$$

Determining α_i by Equations 21 and 22, one can get the equations in characteristic form. Besides, the orientations of a characteristic plane can be found by Equation 22, so that the boundaries of domain of propagating perturbation follow. For the explicit difference equations of hyperbolic system, the necessary condition for stability is that the physical domain of dependence of the partial differential equations is contained completely in the numerical domain of dependence of the difference equations. This is the so-called Courant-Friedrichs-Lewy condition [51]. Accordingly, the stability condition for the difference equations (Equation 13) can be derived in terms of Equation 22, i.e.,

$$\Delta t \le [|u|/\Delta x + |v|/\Delta y + a\sqrt{(1/\Delta x^2 + 1/\Delta y^2)}]^{-1} \tag{23}$$

Replacing the energy equation by a simplified relation of constant stagnation enthalpy (Equation 5) is equivalent to assuming that during time marching a heat transfer, which keeps the stagnation enthalpy constant, will affect the speed of perturbation propagation. In order to find the stability condition of the difference equations for this case, first, Equation 5 is rewritten as

$$k(\partial p/\partial t) + u(\partial p/\partial x) + v(\partial p/\partial y) + a^2\rho(\partial u/\partial x + \partial v/\partial y) = 0 \tag{24}$$

Equation 19d is replaced by Equation 24, so that we yield

$$\ell = 0 \tag{25a}$$

and

$$\ell = 0.5\{-(k-1)N_t \pm \sqrt{[(k-1)^2N_t^2 + 4a^2(N_x^2 + N_y^2)]}\} \tag{25b}$$

From Equation 25, then, the stability condition for the difference system of the Euler equations with constant stagnation enthalpy can be found, i.e.,

$$\Delta t \le [|u|/\Delta x + |v|/\Delta y + \psi a\sqrt{(1/\Delta x^2 + 1/\Delta y^2)}]^{-1} \tag{26}$$

where

$$\psi = \sqrt{\{[(k-1)/2k]^2(M_x\,\Delta y + M_y\,\Delta x)^2/(\Delta x^2 + \Delta y^2) + 1/k\}}$$
$$- [(k-1)/2k](|M_x|\Delta y + |M_y|\Delta x)/\sqrt{(\Delta x^2 + \Delta y^2)} \qquad M_x = u/a \qquad M_y = v/a$$

It is easy to verify that $0 < \psi < 1/\sqrt{k}$ when $0 < M_x, M_y < \infty$. Obviously, the time step given by Equation 26 is larger than that given by Equation 23.

The stability condition for the time-split scheme is one-dimensional CFL condition. For the system with constant stagnation enthalpy, the stability conditions for the difference Equations 15 and 16 are

$$\Delta t_y \leq \Delta A_{i,j}/(|\vec{q}\cdot\vec{S}_1| + \phi(|M_y|)a|\vec{S}_1|)_{i,j} \tag{27a}$$

$$\Delta t_x \leq \Delta A_{i,j}/(|\vec{q}\cdot\vec{S}_2| + \phi(|M_x|)a|\vec{S}_2|)_{i,j} \tag{27b}$$

for the operators $\mathscr{L}_y(\Delta t)$ and $\mathscr{L}_x(\Delta t)$ respectively, where $\vec{q} = u\vec{i}_x + v\vec{i}_y$

$$\phi(|x|) = \sqrt{\{[(k-1)/2k]^2x^2 + 1/k\}} - [(k-1)/2k]|x|$$

and x represents M_x or M_y. The comparison between Equations 26 and 27 shows that the time-split scheme allows a larger time step to be employed.

INITIAL AND BOUNDARY CONDITIONS

Mathematically, the problems of time-dependent fluid dynamics are of the mixed initial and boundary value type. In order to ensure that the problem is properly posed, attention must be paid to the specification of initial conditions. Although steady-state solution of such a problem is not related to initial conditions, if it differs extremely from the steady solution, numerical unstability may occur in the early stage of the computing. Provided it is physically compatible with the problem, the initial conditions are rather arbitrary in the computation of transonic flows over airfoils and cascade. A linear variation of the flow variables from upstream to downstream boundary is usually assumed in specifying an initial field, and the tangency condition has to be satisfied on solid body surface. For cascade flows, initial flow field in cascade channel can also be obtained by using one-dimensional flow equations. It should be noticed that it is not worthwhile spending too much computational time using more complicated methods in improving these initial conditions. Because the increase in accuracy of initial flow field is not proportional to the reduction in the number of time steps required to reach convergence; the convergence rate mainly depends upon the nature of finite difference scheme adopted in the computation.

For mixed type problems, in order to obtain correct steady solution certain conditions must be specified on the entire boundary of computational domain for all times. For cascade flow, as shown in Figure 5, there are four different kinds of boundaries: upstream (ab), downstream (ef), body surface (csd and hpg) and periodic (ah, bc, gf, de). For an airfoil, there is a far field boundary bc (see Figure 4), and periodic boundary disappears. The main problems in the treatment of boundary conditions are how many flow variables must be specified and how to compute the remaining variables on the boundaries.

For two-dimensional flow boundary, the number of specified flow variables depends on the orientation of characteristic plane in the flow. Since the orientation of characteristic plane is also the propagating direction of perturbations in the field, the number of specified flow variables is related to that of characteristic planes coming from outside the computational domain to the boundary [4, 52]. For example, if we take the characteristic planes parallel to the upstream boundary and if $u < a$, then there are two characteristic planes corresponding to $\ell = 0, -a\sqrt{(N_x^2 + N_y^2)}$. These planes are tangent to the particle path and sonic cone, and come from outside the domain to the upstream boundary (see Figure 9). In this case, the corresponding three characteristic equations will be meaningless, so that three flow variables must be given on the upstream boundary. For body surface boundary, we take the characteristic planes parallel to the local surface, then one of those two characteristic planes corresponding to $\ell = \pm a\sqrt{(N_x^2 + N_y^2)}$ comes from outside the domain to the surface (see Figure 10), so only one flow variable has to be specified on this boundary. In Table 1 the number of specified flow variables for various boundaries in various cases is given. For the computation of airfoil and cascade flow, the total temperature, total pressure and flow angle are usually specified on upstream boundary. For compressor cascade, if $q > a$ while $u < a$, upstream Mach number should be specified instead of flow angle according to the concept of unique incidence angle of supersonic cascade flow. Usually, the static pressure is given on the downstream boundary and flow tangency condition imposed on body surface.

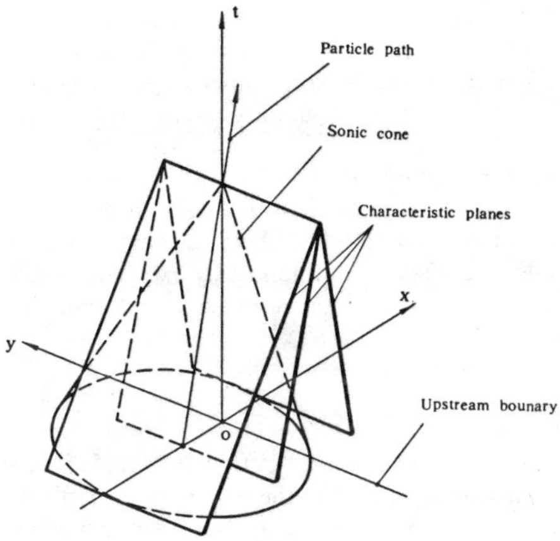

Figure 9. Characteristic planes and sonic cone—upstream boundary.

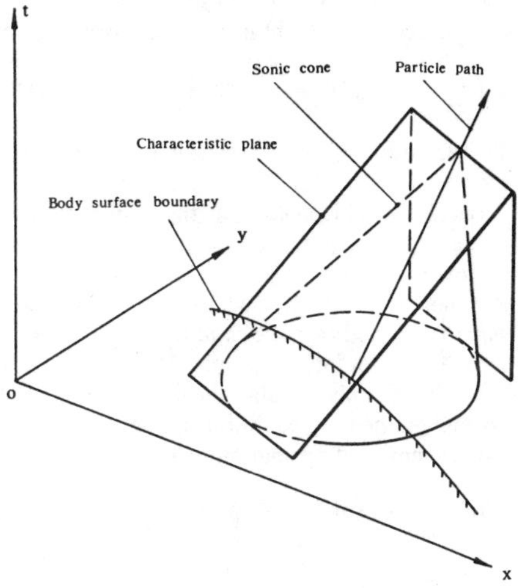

Figure 10. Characteristic planes and sonic cone—body surface boundary.

Table 1
Number of Specified Conditions on Boundaries

	Upstream boundary	Downstream boundary	Body surface boundary
$u < a$	3	1	1
$u > a$	4	0	1

On each boundary, except that specified, the remaining variables must be computed as time-dependent parameters. Usually characteristic equations, corresponding to the characteristic planes that come from the inside of computational domain to its boundary, can be used for this computation [4, 53]; and boundary flow variables of next time level are obtained from those at interior points of present time level. As an example, on the upstream boundary, if $u < a$, only one variable remains to be computed and the characteristic equation derived from $\ell = a\sqrt{(N_x^2 + N_y^2)}$ is the only equation to be used. Usually, numerical scheme for characteristic equation is the same as that for interior mesh points. Theoretically, it is rigorous to obtain boundary flow parameters through characteristic equations, but it is very cumbersome in practice. There are some other ways to compute boundary flow parameters. One of them uses the so-called "additional boundary conditions." Various techniques of such methods have been developed. All of them are easy to use with fairly good results. It should be pointed out that, numerical methods for treating the boundary condition have an important effect on the stability and accuracy of the whole computation. For various additional boundary conditions the stability and accuracy of the computation have been analyzed [54–56]. As an additional boundary condition, flow variables are often been obtained by extrapolation. If upstream and downstream boundary are far away from the body, even the zeroth order extrapolation such as

$$U_B^{n+1} = U_I^{n+1} \tag{28}$$

can give flow variable values on the boundary with satisfactory accuracy. In Equation 28, U is flow variable, subscripts B and I denote boundary and interior point next to the boundary respectively. Denton et al. [13] used Equation 28 to compute cascade flow to obtain upstream static pressure (for turbine cascade) and downstream velocity. Both stability and accuracy are fairly good. Another very simple approach is to use the one-side difference scheme on the boundary [46]. For example, the corrector step (16b) in the MacCormack time-split operation $\mathscr{L}_x(\Delta t)$ can be changed as

$$U_{i,j}^{n+1} = 0.5[U_{i,j}^{n+1/2} + U_{i,j}^{\overline{n+1}} - (\Delta t/\Delta A_{i,j})(\vec{H}_{i,j}^{\overline{n+1}} \cdot \vec{S}_4 + \vec{H}_{i-1,j}^{\overline{n+1}} \cdot \vec{S}_2)]$$

This equation provides a relation to obtain flow variables on exit boundary, but the accuracy of this scheme is lowered at first order.

It is more difficult to treat the boundary conditions on body surface, nevertheless, the accuracy of the whole computation depends largely upon the accuracy of surface flow variables. If the variation of flow parameters is smooth, extrapolation may be suitable. If flow parameters change abruptly, extrapolation may cause a large error in surface variables. One-side difference scheme offers a proper way to treat body surface conditions, but it is liable to make computation unstable for some difference scheme used for interior points. At body surface, because of the tangency condition (zero flux through the wall), the only variable needed for solving difference equations is the pressure on the surface. The normal momentum equation can be adopted to this effect [46, 57]. As a first order approximation, two-dimensional normal momentum equation can be written as:

$$p_B = p_I - (\Delta\eta)\rho_I(u_I^2 + v_I^2)/R_B \tag{29}$$

where subscripts B and I denote two points on the normal coordinate η (see Figure 11), R_B is the radius of curvature of the surface. The flow properties on point I are obtained by interpolation of those of mesh points in its vicinity. Reference 58 presents the general form of normal momentum equation appropriate for wall pressure computation. It is a finite difference equation is nonorthogonal coordinate system. The equation can be used to relate the variables on point B and those on mesh points with no interpolation needed, and the computation becomes simpler.

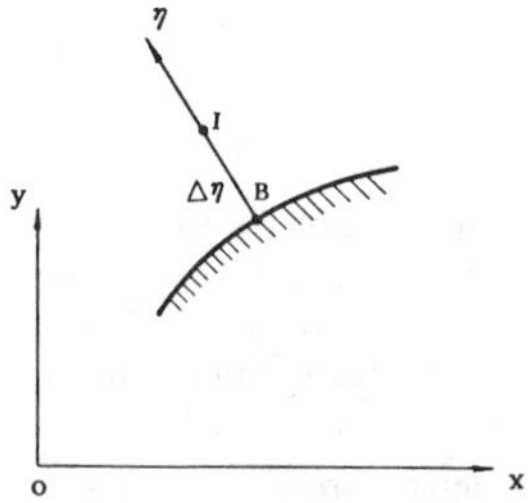

Figure 11. Body surface boundary.

Now we touch on periodic boundary conditions in cascade. Flow parameters on these boundaries are easy to compute by making use of the periodicity properties of the flow in upstream and downstream regions.

ARTIFICIAL DISSIPATION

It is found in practice that as shocks appear in the flowfield, the computed variables distribution has a sharp peak immediately in front of the shock and oscillates immediately behind. This is a numerical oscillation that can lead to computation breakdown. To reduce this oscillation one can artificially introduce a dissipation term (or so-called viscosity term) in the difference equations. Usually, such an artificial dissipation technique is applied to compute the transonic flows over airfoils and cascades. Moreover, when applying some scheme (such as MacCormack scheme) to the computation of transonic flow one will find that it is hard to reach convergence without an artificial dissipative term even if there is no shock in the flowfield. This is illustrated by an example of one-dimensional flow, to be discussed briefly. In this case, we have

$$\partial w/\partial t + A(\partial w/\partial x) = 0 \qquad (30)$$

where

$$w = \begin{bmatrix} \rho \\ u \\ p \end{bmatrix}, \qquad A = \begin{bmatrix} u & \rho & 0 \\ 0 & u & 1/\rho \\ 0 & \rho a^2 & u \end{bmatrix}$$

If A is assumed to be a constant matrix, then the MacCormack scheme applied to foregoing set of equations yields

$$w_i^{n+1} = w_i^n - (A\,\Delta t/\Delta x)(w_{i+1}^n - w_{i-1}^n)/2 + (A\,\Delta t/\Delta x)^2(w_{i+1}^n - 2w_i^n + w_{i-1}^n)/2 \qquad (31)$$

The amplification matrix G of Equation 31 can be shown easily to be

$$G = I - i(\Delta t/\Delta x)A \sin\alpha - (A\,\Delta t/\Delta x)^2(1 - \cos\alpha)$$

where I is the unit diagonal matrix, and $\alpha = \ell\Delta x$, ℓ being an arbitrary real number. The corresponding eigenvalue of G is

$$g = 1 - i(\Delta t/\Delta x)\lambda \sin\alpha - (\lambda\,\Delta t/\Delta x)^2(1 - \cos\alpha)$$

where λ is an eigenvalue of A, and has value u and $u \pm a$. At stagnation point or sonic point in the transonic flowfield, λ is zero. Then the corresponding eigenvalue g of the amplification matrix G is equal to 1 for all α rather than <1. Therefore, the difference system is not dissipative and in such case for quasi-linear gas dynamic equation, nonlinearities can destroy the stability.

If a dissipative term dependent on $\nu(\delta^2 w/\partial x^2)$($\nu > 0$, is artificial viscosity factor) is added with ν assumed constant, then the difference system becomes

$$w_i^{n+1} = \tilde{w}_i^{n+1} + \nu(\tilde{w}_{i+1}^{n+1} - 2\tilde{w}_i^{n+1} + \tilde{w}_{i-1}^{n+1}) \qquad (32)$$

where $\tilde{w}$ represents the magnitude of flow variables obtained without dissipative term in the system. The corresponding eigenvalue g of the amplification matrix G is

$$g = (1 - 2\nu)[1 - i(\Delta t/\Delta x)\lambda \sin\alpha + (\lambda\,\Delta t/\Delta x)^2(\cos\alpha - 1)] + \nu[2\cos\alpha - i(\Delta t/\Delta x)\lambda \sin 2\alpha + (\lambda\,\Delta t/\Delta x)^2(1 + \cos 2\alpha - 2\cos\alpha)]$$

It follows that, when λ is zero,

$$g = 1 - 2\nu(1 - \cos\alpha)$$

Thus, the introduction of artificial dissipation prevents the nonlinear instability in transonic flow problems.

There are various forms of artificial dissipation, and they can be divided into two types, linear and nonlinear.

Linear form [3, 59]

A simple linear dissipation form is usually employed in the computations. For two-dimensional problems, it is

$$\bar{U}_{i,j} = (U_{i+1,j} + U_{i-1,j} + U_{i,j+1} + U_{i,j-1} + \bar{\varphi}U_{i,j})/(\bar{\varphi} + 4) \tag{33}$$

where $\bar{U}$ is the modified flow variables, and $\bar{\varphi}$ is a damping factor to be chosen. As factor $\bar{\varphi}$ is a constant, this is a linear dissipation form, in which the numerical damping is introduced by weighted average of the flow parameters at point (i, j) and its four surrounding points. Nevertheless the introduction of artificial dissipation will bring some errors to the computations. By means of a Taylor expansion about $x = i\,\Delta x$, $y = j\,\Delta y$ one can get from Equation 33:

$$\bar{U} = U + [(\partial^2U/\partial x^2)(\Delta x)^2 + (\partial^2U/\partial y^2)(\Delta y)^2]/(\bar{\varphi} + 4) + O_1[(\Delta x)^4] + O_2[(\Delta y)^4]$$

It shows that when $\bar{\varphi} \geq \max\{1/\Delta x, 1/\Delta y\}$ the three-step scheme consisting of the combination of Equations 13 and 33 still has second order accuracy. For the time-split scheme (Equation 14), this form can be written as

$$\bar{U}_{i,j} = U_{i,j} + \varphi_y(U_{i,j+1} - 2U_{i,j} + U_{i,j-1}) \tag{34a}$$

$$\bar{U}_{i,j} = U_{i,j} + \varphi_x(U_{i+1,j} - 2U_{i,j} + U_{i-1,j}) \tag{34b}$$

for operators $\mathscr{L}_y(\Delta t)$ and $\mathscr{L}_x(\Delta t)$ respectively.

Nonlinear Form [60, 61]

In order to make the effect of artificial dissipation greater in regions where the flow variables oscillate sharply, and smaller in smooth regions, an automatically switched function is employed to control the dissipation. This approach is more reasonable than the linear dissipation form. Usually, such dissipation is introduced in the difference system by means of "fourth order damping." The procedure is to add a correction term $\Delta U_{i,j}$ to the operators $\mathscr{L}_y(\Delta t)$ and $\mathscr{L}_x(\Delta t)$. Here are the forms

$$(\Delta U_{i,j})_y = C_y[|p_{i,j+1} - 2p_{i,j} + p_{i,j-1}/(p_{i,j+1} + 2p_{i,j} + p_{i,j-1})] \times (U_{i,j+1} - 2U_{i,j} + U_{i,j-1})(a_{i,j} + |v_{i,j}|)(\Delta t/\Delta y) \tag{35a}$$

$$(\Delta U_{i,j})_x = C_x[|p_{i+1,j} - 2p_{i,j} + p_{i-1,j}|/(p_{i+1,j} + 2p_{i,j} + p_{i-1,j})] \times (U_{i+1,j} - 2U_{i,j} + U_{i-1,j})(a_{i,j} + |u_{i,j}|)(\Delta t/\Delta x) \tag{35b}$$

for $\mathscr{L}_y(\Delta t)$ and $\mathscr{L}_x(\Delta t)$ respectively, where C_y and C_x are dimensionless constants usually lying between 0 and 0.5. It is seen from Equation 35 that the coefficient of the dissipative term is proportional to the second derivative of pressure and is of significant magnitude in regions of pressure oscillation. Therefore, this is a nonlinear dissipation form. From Equation 35b we have

$$\bar{U} = U + C_x[(a + |u|)/(4p)]|\partial^2 p/\partial x^2|(\partial^2 U/\partial x^2)(\Delta x)^3 \, \Delta t + \ldots,$$

so that the error caused by this dissipative term is of magnitude to the fourth-order.

METHODS OF ACCELERATING CONVERGENCE

Owing to the introduction of the time variable, the computation time required by the time-dependent methods is longer than that by a direct algorithm for a steady problem. It is desirable to make the computation as efficient as possible. For an explicit difference scheme, the maximum allowable time step is determined by the CFL condition, and it depends on the size of mesh and local flow parameters that may be different for different points. If an uniform time step is used for the whole flowfield at every iteration, it may occur that at only one point the time step is equal to the maximum allowable and at all other points it is smaller than allowed. Obviously, it will slow down the convergence. In order to accelerate the convergence of computation, a variable time step, as large as is locally allowable to sustain stability, can be used at each mesh point. This does not affect the final asymptotic solution provided that the problem to be solved is properly posed. When we take as large a time step as possible at every point, the number of time steps required to approach the steady state is greatly reduced. In the case of a computation of flow around an airfoil, the effect of using a locally allowable time step on the convergence rate is given in Table 2, and it is seen that the convergence speed has double [11].

Generally, a finer grid is needed for ensuring accuracy. However, a reduction of mesh spacing results in not only an increasing of the number of mesh points, but also a decreasing of the allowable time step. The successive mesh refinement approach is one of the effective techniques for overcoming this difficulty. In this procedure, an improved initial field is obtained first by computing with a coarse mesh at a rapid convergence rate, then with interpolations, the computation proceeds with a refined mesh, so that the total computation time will be shortened. The effectiveness of this technique in the computation of a transonic flow through a turbine cascade is given in Table 3. Application of both the largest locally allowable time step and the successive mesh refinement approach (a 40 × 7 mesh is refined to 40 × 14), reduces computation time for the steady solution to 52% of what it was. It is to be noted that the number of mesh points needed to obtain adequate accuracy of the flow parameters on the body surface can be reduced by using a mesh with nonuniform spacing in the y direction (the spacing decreases toward the body surface). When such a mesh is used in conjunction with locally allowable time step, the finer mesh near the body surface does not cause a reduction in the time step in the whole field. It is seen from Table 3 that, in this case 57% reduction in overall computation time is obtained [11].

In addition to the techniques mentioned above, the computation time can be greatly saved by using the so-called multigrid method developed in recent years. This technique was first proposed by Fedorenko [62] and subsequently analyzed by Bakhavlov [63]. In 1979, the multigrid technique was successfully used in computing transonic potential flow by Jameson [64]—a great speed-up of the convergence. Differing entirely from the mesh refinement approach, the multigrid method utilizes simultaneously coarse and fine meshes. When applied to the Euler equations in integral form, the basic idea of this procedure is that the fluxes for the coarse control volume are found from the enclosed fine control volume, then the time changes of flow parameters on the coarse mesh are then interpolated into the fine mesh, thus updating the parameters on it. Essentially, the

Table 2
Relative Computation Times (Airfoil)

	Uniform time step	Local time step
Number of time steps to convergence	1,820	840
Relative computation time	100%	46%

Table 3
Relative Computation Times (Cascade)

	Uniform time step, uniform spacing in y direction, 40 × 14 mesh	Local time step, uniform spacing in y direction, 40 × 14 mesh	Local time step, mesh refinement 40 × 7 → 40 × 14	Local time step, nonuniform spacing in y direction, 40 × 9 mesh
Number of time steps to convergence	1,260	840	580 + 360	840
Relative computation time	100%	67%	52%	43%

fast perturbation propagation is assured by the coarse mesh so as to get quicker convergence while the accuracy is assured by the fine mesh. Denton [2], using this method to solve the Euler equations reported the convergence to be about four times faster than that of using single grid approach, the computer time required for computing a transonic cascade flow being only one third of the time required for single grid approach. Ni [65] used an algorithm with four levels of grid, and the saving in computation time is about 80% for a transonic channel flow.

COMPUTED RESULTS

In this section we present some computed results of transonic flows over airfoils and cascades to illustrate the accuracy of the numerical modeling. Some problems in the computations are discussed. The MacCormack time-split finite-volume method is adopted exclusively for all the examples, and all the results obtained are compared with experimental data.

Circular-Arc Airfoil

The flowfields around circular-arc airfoils described in Reference 66 are computed with a 32 × 22 mesh. The damping factors in Equation 34 for artificial dissipation are taken to be $\phi_x \leq 0.01$ and $\varphi_y = 0$. With angle of attack $\alpha_\infty = 0°$, the upstream Mach-number $M_\infty = 1.057$, 0.92, and 0.858, the computed pressure coefficients $C_p(=2(p - p_\infty)/(\rho_\infty q_\infty^2))$ are shown in Figures 12, 13 and 14, respectively.The experimental results are also shown. The agreement is quite satisfactory. Figure 13 shows that at $M = 0.92$ the measured shock appears at 70% chord length and the computed shock occurs a little beyond. When the damping factor φ_x has a small value, the pressure coefficient in front of the shock appears too low. Increasing the damping should have no effect on the smooth

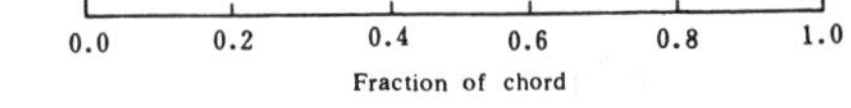

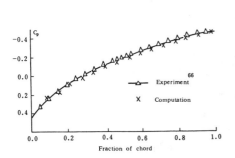

Figure 12. Pressure distribution around a circular-arc airfoil, $M_\infty = 1.057$.

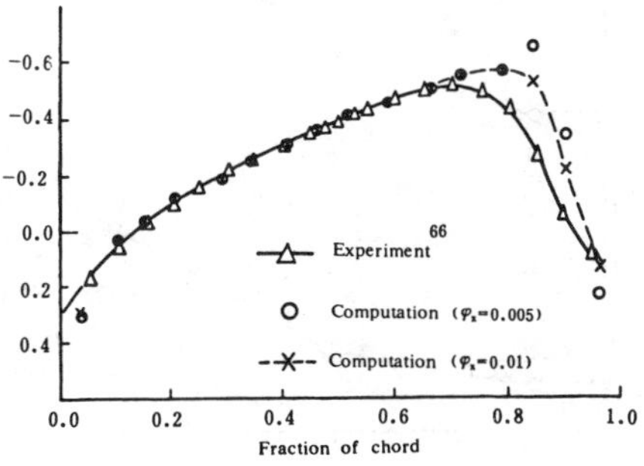

Figure 13. Pressure distribution around a circular-arc airfoil, $M_\infty = 0.92$.

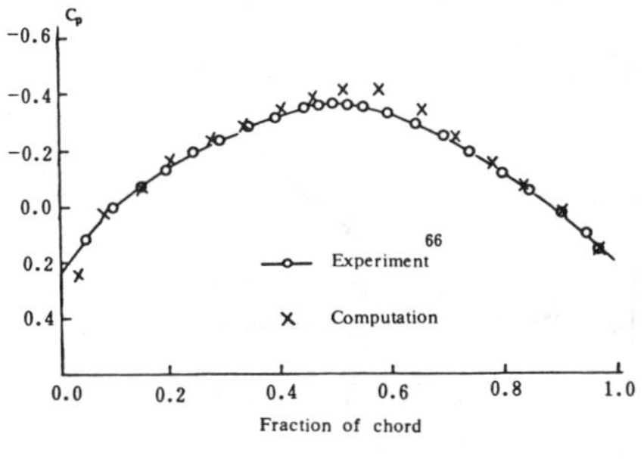

Figure 14. Pressure distribution around a circular-arc airfoil, $M_\infty = 0.858$.

solutions, but the pre-shock peak of the pressure coefficient diminishes, thus improving the computation of shock wave.

Transonic Turbine Cascade

The numerical results of several NACA turbine cascades [67] and a high turning turbine cascade [68] are now to be given. The meshes used are 40 × 14 (for NACA cascades) and 44 × 9 (nonuniform spacing in the y direction, for high turning cascade), respectively.

NACA Near-Sonic Turbine Cascade with Turning Angle 80°

Figure 15 shows the computed and experimental pressure coefficients $\bar{S}$ $(= 2(p_1^* - p)/(\rho_1, q_1^2)$, subscript 1 indicates upstream). The computed value on the suction surface is a little bit lower than the experimental values while on the pressure side they are in good agreement. The flow at 60% chord length on the suction surface is nearly sonic.

NACA Transonic Turbine Cascade with Turning Angle 95°

The computed pressure coefficients on the blade surface are in fairly good agreement with experiment data as shown in Figure 16. For this cascade the flow is supersonic at certain region on the suction surface of blade.

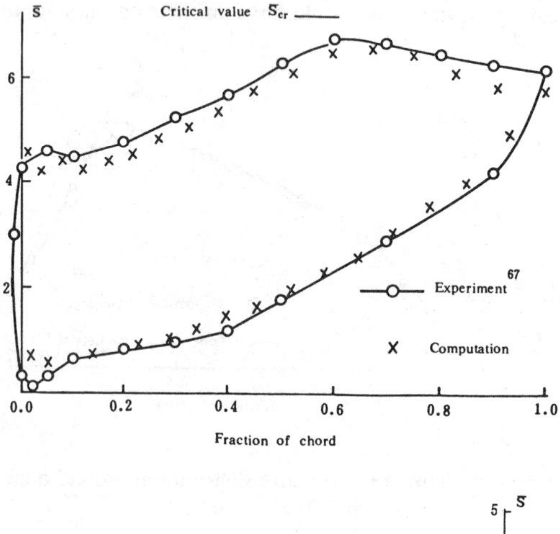

Figure 15. NACA near-sonic turbine cascade.

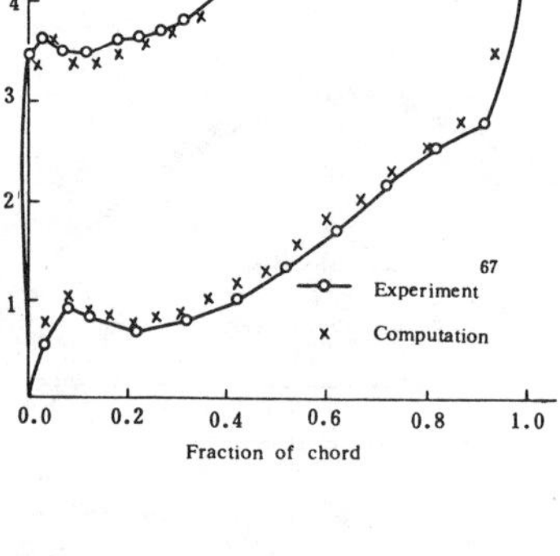

Figure 16. NACA transonic turbine cascade.

NACA Transonic Turbine Cascade with Turning Angle 95°, with Shock Waves

Reference 67 reports the results of a high-speed test for this cascade. It demonstrates that there is a shock at 20% chord length on the suction surface, which causes an abrupt fall in the pressure coefficient. Figure 17 compares the computed and experimental pressure coefficients. The position of the computed shock agrees approximately with that based on experiment, whereas the computed pressure coefficient is a little higher.

High-Turning Transonic Turbine Cascade

Reference 68 reports two different turbine profiles with turning angle 128°, called RA and RB. These profiles were designed to investigate the effect of shock-boundary-layer interaction on the performance of transonic turbine profiles. Figure 18 shows the computed and experimental blade surface M_{it} (isentropic Mach-number) distributions for the profile RA of design exit Mach-number $M_2 = 1.15$ and inlet angle $\alpha_1 = 60°$. It can be seen that the agreement of the experimental and computed M_{it} distributions is good on the pressure surface, that the computed M_{it} peak on the suction surface is a little lower than the experimental value, and that the computed position of shock is correct.

Transonic Compressor Cascade

Detailed information on design and experiment for a high-speed fan stage is available [69, 70]. The relative flow through the tip section of this fan is computed. To account for the effect on the solution of rotating and radial distance and thickness of stream surface, the solutions are obtained by using Equation 7. Figure 19 shows the computed and experimental pressure contours in the cascade channel. It can be seen that a strong shock at the trailing edge is correctly predicted by the computation, but the smearing of the oblique shock at leading edge is rather too much.

It is seen from the numerical results given that the computations of inviscid flow agree well with the test data in shock-free flows. As shocks appear, however, the computed pressure distributions on the airfoil surface present considerable discrepancies from experimental values. This is due to the effects of the shock-boundary-layer interaction neglected in the computation. In many cases, shocks in the flowfield of transonic airfoils and turbine cascades are not too strong and the discrepancy between computation and experiment is acceptable in general. However, the viscous

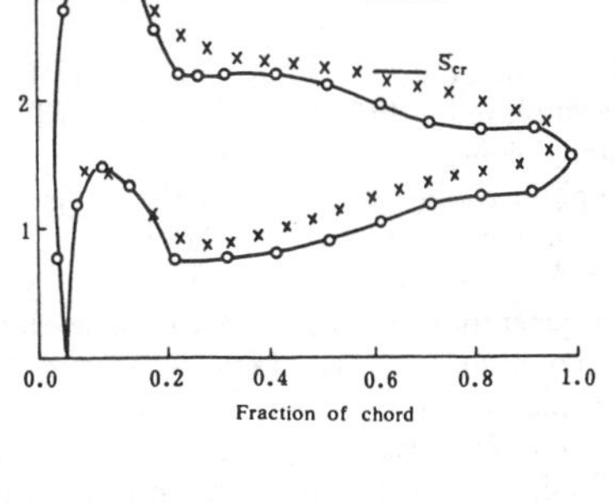

Figure 17. NACA transonic turbine cascade with shock wave.

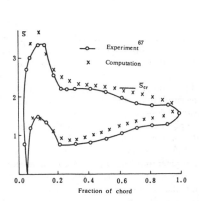

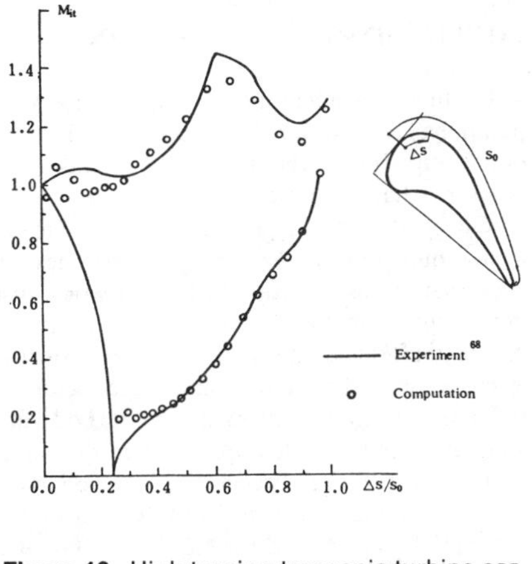

Figure 18. High turning transonic turbine cascade.

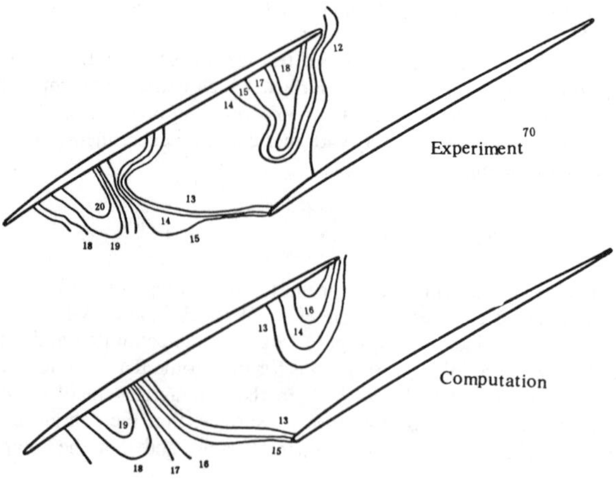

Figure 19. Computed and experimental static pressure contours for a high speed fan.

effects and, in particular, the shock-boundary-layer interaction becomes more important in the transonic compressor cascades. As shown by the experimental pressure contours in Figure 19, a λ-type shock takes place at the rear of the blade pressure surface because of the boundary layer separation. Moreover, the viscous blockage effects in the actual flow narrow the flow passage. All these phenomena, impossible to be correctly modeled by inviscid computation, result in the discrepancies between the computed and experimental data of both pressure distributions and shock locations. To consider viscous phenomena in cascades an inviscid-viscous flow iteration technique [71–73] has been developed to predict the effects of viscosity, including the effects of wake and shock-boundary-layer interaction. This procedure matches the inviscid flow with the blade surface boundary layer, to reach a mutually compatible solution. In the analysis of transonic flows in cascades, generally, the time-dependent technique is employed to compute the inviscid flow, and the integral method is used to compute the boundary layer.

CONCLUSIONS AND PROSPECTS

The time-dependent method for solving the Euler equations is an effective technique in the computation of the rotational transonic flow. This method plays an important role in numerical modeling of transonic flows over airfoils and cascades, especially involving strong shock in diffusion cascade. It seems that at present there does not exist a more powerful method to replace it. In the time-dependent techniques, the finite-volume method for solving the governing equations in integral form is quite flexible. Arbitrarily shaped meshes can be employed to fit irregular boundary without coordinate transformation in the numerical discretization. Thus, it is most suitable for computing flow around complicated body geometry. It should be pointed out that in using either differential or integral form equations, severe mesh distortion and coordinate singularities in the critical regions (such as the leading and trailing edge regions) must be avoided, else the computing accuracy will suffer. In recent years several automatic body-fitted mesh generation techniques have been developed and have received wide application in the computations of airfoil and cascade flows. Among these techniques, the mesh generation method based upon the solution of nonlinear elliptic partial differential equations is the most popular one. Efficient, automatic clustering of mesh points in the appropriate regions and the required orthogonality of mesh are two main problems in this method.

To date, though some research work on implicit finite difference methods has been done, explicit methods are still more prevalent. The MacCormack explicit two-step scheme is second order accuracy in both time and space and has good shock fitting properties, so it is the most widely used scheme at present. Denton's "opposed finite difference" scheme has higher rate of convergence and has been widely used for transonic cascade flow computations.

It is found in practice that other than the difference scheme the treatment of boundary conditions is another significant factor, which holds the key to the accuracy and stability of the whole computation. Among them body surface boundary condition is the most difficult to handle in airfoil and cascade flows. It affects the accuracy most and deserves special careful treatment. Simplified additional boundary conditions are often used to compute boundary flow variables. These methods are simpler to use and fairly good results can often be expected. Unfortunately, a general rule to choose numerical boundary conditions does not exist up to now.

The main disadvantage of time-dependent methods is their relatively high cost in computer time. In order to accelerate convergence, variable time step and successive mesh refinement approach can be used for this purpose. However, the multigrid technique may have greater potentiality. The multigrid technique has not been widely used for solving the Euler equations yet and not so much success as that with the relaxation method. It may, however, be anticipated that this technique will become a very powerful tool for finding steady solutions of the Euler and Navier-Stokes equations in the near future [74].

The numerical modeling of inviscid flow over airfoil and cascade free of strong shock can be done quite satisfactorily. If strong shock is present, then viscous effect including shock wave-boundary-layer interaction must be considered, or else a large discrepancy between numerical results and experimental data of actual flow will emerge. For cascade flow, inviscid-viscous iteration calculation may be more practical for the modeling of viscous effects. The advancement of this method depends upon the correct prediction of inviscid flow and boundary layer.

The numerical modeling techniques for transonic flow continue to make steady progress. The methods for solving the Euler equations are still far from perfect and their improvements involve mainly two areas: convergence rate and reliability. It is noteworthy that recently Jameson et al. developed a Runge-Kutta time stepping technique [75, 76] in the time-dependent finite-volume method. This technique significantly improves convergence and accuracy. Besides, attempts to solve the steady Euler equations directly for transonic flows have been made [77, 78]. It may be expected that, along with improved accuracy and efficiency in numerical modeling, more and more difficult problems, which could only be investigated by experiments, will become solvable by numerical modeling.

NOTATION

a	velocity of sound	q	magnitude of flow speed
B	normal thickness of stream surface	R_B	radius of curvature
C_p	specific heat at constant pressure	r	radius
e	total internal energy per unit volume	S	entropy
F	flux in x-direction	T	temperature
G	flux in y-direction	t	time
I	rothalpy	U	flow variable
i*	stagnation enthalpy	u, v	velocity
k	ratio of specific heats	w_m, w_θ	relative velocity components
M	Mach number	x, y, z	coordinates
P	pressure		

Greek Symbols

α_i	coefficient	ρ	density
ϵ	angle between meridional streamline and machine axis	$\bar{\psi}$	damping factor
ξ, η	normal coordinate	ω	angular velocity

REFERENCES

1. Rizzi, A., "Damped Euler—Equation Method to Compute Transonic Flow Around Wing—Body Combinations." *AIAA Journal*, Vol. 20, No. 10, Oct. 1982, pp. 1321–1328.
2. Denton, J. D., "An Improved Time Marching Method for Turbomachinery Flow Calculation." *ASME Paper* 82-GT-239, 1982.
3. Gopalakrishnan, S., and Bozzola, R., "Computation of Shocked Flows in Compressor Cascades." *ASME Paper* 72-GT-31, 1972.
4. Kurzrock, J. W., and Novick, A. S., "Transonic Flow around Rotor Blade Elements." *Journal of Fluids Engineering*, Transactions of the ASME, Vol. 97, No. 4, Dec. 1975, pp. 598–607.
5. Oliver, D. A., and Sparis, P., "Computational Aspects of the Prediction of Multidimensional Transonic Flows in Turbomachinery." *NASA SP*-347, 1975, pp. 567–585.
6. Palulon, P. J., and Veuillot, J. P., "Étude Theorique et Experimentale Ďune Grille Ďaubes Supersonique Annulaire Mobile." *ICAS Proc.*, 10th Confress of the ICAS, Ottawa, Ontario Canada, 1976, pp. 416–425.
7. Thompkins, W. T., Jr., and Epstein, A. H., "A Comparison of the Computed and Experimental Three Dimensional Flow in a Transonic Compressor Rotor." *AIAA Paper* 76-368, 1976.
8. Veuillot, J. P., "Calculation of Quasi Three-Dimensional Flow in a Turbomachine Blade Row." *Journal of Engineering for Power*, Transactions of the ASME, Vol. 99, No. 1, Jan. 1977, pp. 53–62.
9. Haymann-Haber, G., and Thompkins, W. T., Jr., "Comparison of Experimental and Computational Shock Structure in a Transonic Compressor Rotor." *ASME Paper* 80-GT-81, 1980.
10. Zhou, X. H., et al., "Application of the Time-Dependent Finite-Volume Method to the Calculation of Transonic Cascade Flow Field." *Journal of Engineering Thermophysics*, Vol. 2, No. 2, May 1981, pp. 121–129 (in Chinese).
11. Zhou, X. H., and Zhu, F. Y., "Numerical Computation of Transonic Flows over Airfoils and Cascades." *Computer Methods in Applied Mechanics and Engineering*, Vol. 37, No. 3, May 1983, pp. 277–288.
12. Lu, W. Q., and Chen, X. M., "The Time-Dependent Numerical Solution of Transonic Flow on S_1 Surface of Revolution Employing Non-Orthogonal Curvilinear Coordinates and Non-Orthogonal Velocity Components." *Journal of Engineering Thermophysics*, Vol. 4, No. 2, May 1983, pp. 119–126 (in Chinese).
13. Denton, J. D., "A Time Marching Method for Two-and Three-Dimensional Blade-to-Blade Flows." Aeronautical Research Council British R&M No. 3775, 1974.
14. Denton, J. D., and Singh, U. K., "Time Marching Methods for Turbomachinery Flow Calculation." VKI Lecture Series 1979-7, Application of Numerical Methods to Flow Calculations in Turbomachines, 1979.
15. Zhang, Y. K., et al., "Numerical Tests of Three-Dimensional Flow in Transonic Turbine." *Acta Aeronautica et Astronautica Sinica*, Vol. 2, No. 3, Sep. 1981, pp. 67–76 (China).
16. Zhang, Y. K., et al., "Numerical Test of Transonic Flow Passing Cascade of Arbitrary Airfoils on General Surface of Revolution." *Journal on Numerical Methods and Computer Applications.* Vol. 1, No. 4, Dec. 1980, pp. 243–252 (in Chinese).
17. Sarathy, K. P., "Computation of Three-Dimensional Flow Fields Through Rotating Blade Rows and Comparison with Experiment." *ASME Paper* 81-GT-121, 1981.
18. Younis, M. E., and Camarero, R., "Finite Volume Method for Blade-to-Blade Flows Using a Body-Fitted Mesh." *AIAA Journal*, Vol. 19, No. 11, Nov. 1981, pp. 1500–1502.
19. Gopalakrishnan, S., and Bozzla, R. A., "Numerical Technique for the Calculation of Turbomachinery Cascades." *ASME Paper* 71-GT-42, 1971.
20. McDonald, P. W., "The Computation of Transonic Flow Through Two-Dimensional Gas Turbine Cascades." *ASME Paper* 71-GT-89, 1971.
21. Delaney, R. A., and Kavanagh, P., "Transonic Flow Analysis in Axial Flow Turbomachinery Cascades by a Time-Dependent Method of Characteristics." *Journal of Engineering for Power*, Transactions of the ASME, Vol. 98, No. 3, July 1976, pp. 356–364.
22. Shuster, A. R., "Calculation of Transonic and Supersonic Flows in Plane Turbomachinery Cascades," Teplo Energetika, 3, March 1976, pp. 41–43 (In Russian).

23. Couston, M., "Time Marching Finite Area Method." VKI Lecture Series 84, Transonic Flows in Axial Turbomachinery, 1976.
24. Van Hove, W., "Time Marching Methods for Turbomachinery Flow Calculations: Methods of Improving Convergence." VKI Lecture Series 1979-7, Application of Numerical Methods to Flow Calculations in Turbomachines, 1979.
25. McDonald, P. W., et al., "A Comparison between Measured and Computed Flow Fields in a Transonic Compressor Rotor." *ASME Paper* 80-GT-7, 1980.
26. Magnus, R., and Yoshihara, H., "Inviscid Transonic Flow over Airfoils." *AIAA Journal*, Vol. 8, No. 12, Dec. 1970, pp. 2157–2162.
27. Murman, E. M., and Cole, J. D., "Calculation of Plane Steady Transonic Flows." *AIAA Journal*, Vol. 9, No. 1, Jan. 1971, pp. 114–121.
28. Jameson, A., "Iterative Solution of Transonic Flows over Airfoils and Wings, Including Flow at Mach 1." *Communications on Pure and Applied Mathematics*, Vol. 27, No. 3, May 1974, pp. 283–309.
29. Jameson, A., "Transonic Potential Flow Calculations in Conservation Form." *Proc. of the AIAA 2nd Computational Fluid Dynamics Conference*, Hartford, 1975, pp. 148–161.
30. Jameson, A., and Caughey, D. A., "A Finite-Volume Method for Transonic Potential Flow Calculations." *Proc. of the AIAA 3rd Computational Fluid Dynamics Conference*, Albuquerque, 1977, pp. 35–54.
31. Holst, T. L., and Ballhaus, W. F., "Fast, Conservative Schemes for the Full Potential Equation Applied to Transonic Flows." *AIAA Journal*, Vol. 17, No. 2, Feb. 1979, pp. 145–152.
32. Hafez, M., et al., "Artificial Compressibility Methods for Numerical Solution of Transonic Full Potential Equation." *AIAA Journal*, Vol. 17, No. 8, Aug. 1979, pp. 834–844.
33. Ives, D. C., and Liutermoza, J. F., "Analysis of Transonic Cascade Flow Using Conformal Mapping and Relaxation Techniques." *AIAA Journal*, Vol. 15, No. 5, May 1977, pp. 647–652.
34. Thompson, J. F., et al., "Automatic Numerical Generation of Body-Fitted Curvilinear Coordinates System for Field Containing Any Number of Arbitrary Two-Dimensional Bodies." *Journal of Computational Physics*, Vol. 15, No. 3, July 1974, pp. 299–319.
35. Thompson, J. F., et al., "Boundary-Fitted Curvilinear Coordinate System for Solution of Partial Differential Equations on Fields Containing Any Number of Arbitrary Two-Dimensional Bodies." *NASA CR*-2729, 1976.
36. Steger, J. L., et al., "An Implicit Finite-Difference Code for Inviscid and Viscous Cascade Flow." *AIAA Paper* 80-1427, 1980.
37. Camarero, R., and Reggio, M., "A Multigrid Scheme for Three-Dimensional Body-Fitted Coordinates in Turbomachine Applications." *Journal of Fluids Engineering*, Transactions of the ASME, Vol. 105, No. 1, Mar. 1983, pp. 76–82.
38. Steger, J. L., and Sorenson, R. L., "Automatic Mesh-Point Clustering Near a Boundary in Grid Generation with Elliptic Partial Differential Equations." *Journal of Computational Physics*, Vol. 33, No. 3, Dec. 1979, pp. 405–410.
39. Visbal, M., and Knight, D., "Generation of Orthogonal and Nearly Orthogonal Coordinates with Grid Control near Boundaries." *AIAA Journal*, Vol. 20, No. 3, May 1982, pp. 305–306.
40. Haussling, H. J., and Coleman, R. M., "A Method for Generation of Orthogonal and Nearly Orthogonal Boundary-Fitted Coordinate Systems," *Journal of Computational Physics*, Vol. 43, No. 2, Oct. 1981, pp. 373–381.
41. Eiseman, P. R., "A Multi-Surface Method of Coordinate Generation." *Journal of Computational Physics*, Vol. 33, No. 1, Oct. 1979, pp. 118–150.
42. Eiseman, P. R., "A Coordinate System for a Viscous Transonic Cascade Analysis." *Journal of Computational Physics*, Vol. 26, No. 3, Mar. 1978, pp. 307–338.
43. Delaney, R. A., "Time-Marching Analysis of Steady Transonic Flow in Turbomachinery Cascades Using the Hopscotch Method." *Journal of Engineering for Power*, Transactions of the ASME, Vol. 105, No. 2, Apr. 1983, pp. 272–279.
44. Farrell, C., and Adamczyk, J., "Full Potential Solution of Transonic Quasi-Three-Dimensional Flow Through a Cascade Using Artificial Compressibility." *Journal of Engineering for Power*, Transactions of the ASME, Vol. 104, No. 1, Jan. 1982, pp. 143–153.

45. MacCormack, R. W., "The Effect of Viscosity in Hypervelocity Impact Cratering." *AIAA Paper* 69-354, 1969.
46. MacCormack, R. W., and Paullay, A. J., "Computational Efficiency Achieved by Time Splitting of Finite Difference Operators." *AIAA Paper* 72-154, 1972.
47. MacCormack, R. W., and Baldwin, B. S., "A Numerical Method for Solving the Navier-Stokes Equations with Application to Shock-Boundary Layer Interactions." *AIAA Paper* 75-1, 1975.
48. Zhou, X. H., and Zhu, F. Y., "Calculation of Shocked Flow Along a Stream surface of Revolution in Transonic Compressor Cascade." *Journal of Engineering Thermophysics*, Vol. 4, No. 4, Nov. 1983, pp. 358–363 (in Chinese).
49. Liu, S. L., and Wu, C. Y., "The Hopscotch Finite Volume Method and Its Application to the Computation of Transonic Flow." *Selected Proceedings of the 1st China National Computational Fluid Mechanics Conference*, Mar. 1983, pp. 143–147.
50. Richtmyer, R. D., and Morton, K. W., *Difference Methods for Initial-Value Problems.* second ed. New York: Interscience Publishers, 1967, pp. 375–377.
51. Courant, R., et al., "Über die Partiellen Differenzengleichungen der Mathematischen Physik." *Mathematischen Annalen*, Vol. 100, 1928, pp. 32–74.
52. Gopalakrishnan, S., and Bozzla, R., "Numerical Representation of Inlet and Exit Boundary Conditions in Transient Cascade Flows." *ASME Paper* 73-GT-55, 1973.
53. Studerus, C. J., "Aerodynamic Effects of Surface Cooling-Flow Injection on Turbine Transonic Flow Fields." *AIAA Paper* 79-0048, 1979.
54. Gottlieb, D., "Boundary Conditions for Multistep Finite-Difference Methods for Time-Dependent Equations." *Journal of Computational Physics*, Vol. 26, No. 2, Feb. 1978, pp. 181–196.
55. Abarbanel, S. S., and Murman, E. M., "Stability of Two-Dimensional Hyperbolic Initial Boundary Value Problems for Explicit and Implicit Schemes." *Journal of Computational Physics*, Vol. 48, No. 2, Nov. 1982, pp. 160–167.
56. Blottner, F. G., "Influence of Boundary Approximations and Conditions on Finite-Difference Solutions." *Journal of Computational Physics*, Vol. 48, No. 2, Nov. 1982, pp. 246–269.
57. Rizzi, A. W., and Inouye, M., "Time Split Finite-Volume Method for Three-Dimensional Blunt-Body Flow" *AIAA Journal*, Vol. 11, No. 11, Nov. 1973, pp. 1478–1485.
58. Rizzi, A. W., "Numerical Implement of Solid-Body Boundary Conditions for the Euler Equations." *Zeitschrift für Angewandte Mathematik und Mechanik*, Bd. 58, Heft 7, Juli 1978, T301-304
59. Vliegenthart, A. C., "The Shuman Filtering Operator and the Numerical Computation of Shock Waves." *Journal of Engineering Mathematics*, Vol. 4, No. 4, Oct. 1970, pp. 341–348.
60. Harten, A., and Zwas, G., "Switched Numerical Shuman Filters for Shock Calculations." *Journal of Engineering Mathematics*, Vol. 6, No. 2, Apr. 1972, pp. 207–216.
61. Hung, C. M., and MacCormack, R. W., "Numerical Solutions of Supersonic and Hypersonic Laminar Compression Corner Flows." *AIAA Journal*, Vol. 14, No. 4, Apr. 1976, pp. 475–481.
62. Fedorenko, R. P., The Speed of Convergence of One Iterative Process, Journal of Computational Mathematics and Mathematical Physics, Vol. 4, No. 3, 1964, pp. 559–564 (in Russian).
63. Bakhvalov, N. S., On the Convergence of a Relaxation Method with Natural Constraints on the Elliptic Operator, Journal of Computational Mathematics and Mathematical Phys. Vol. 6, No. 5, 1966, pp. 861–883 (in Russian).
64. Jameson. A., "Acceleration of Transonic Potential Flow Calculations on Arbitrary Meshes by Multigrid Method." *AIAA Paper* 79-1458, 1979.
65. Ni, R. H., "A Multiple-Grid Scheme for solving the Euler Equations." *AIAA Journal*, Vol. 20, No. 11, Nov. 1982, pp. 1565–1571.
66. Knechtel, E. D., "Experimental Investigation at Transonic Speed of Pressure Distributions over Wedge and Circular-Arc Airfoil Sections and Evaluation of Perforated-Wall Interference." *NASA TN D*-15, 1959.
67. Dunavant, J. C., and Erwin, J. R., "Investigation of a Related Series of Turbine-Blade Profiles in Cascade." *NACA TN*-3802, 1956.
68. Graham, C. G., and Kost, F. H., "Shock Boundary Layer Interaction on High Turning Transonic Turbine Cascades." *ASME Paper* 79-GT-37, 1979.
69. Doyle, V. L., "Evaluation of Range and Distortion Tolerance for High Mach Number Transonic Stages." *NASA CR*-72720, 1970.

70. Bilwakesh, K. R., et al., "Evaluation of Range and Distortion Tolerance for High Mach Number Transonic Fan Stages." *NASA CR*-72880, 1972.
71. Singh, U. K., "Computation of Transonic Flow in Cascade with Shock and Boundary Layer Interaction." *Proc. 1st International Conference on Numerical Methods in Laminar and Turbulent Flow*, Swansea, 1978, pp. 697–707.
72. Gliebe, P. R., "Couple Inviscid/Boundary-Layer Flow-Field Predictions for Transonic Turbomachinery Cascades." *Proc. of the Workshop on Transonic Flow Problems in Turbomachinery*, Naval Postgraduate School, Monterey, California, Feb. 1976, pp. 434–453.
73. Calvert, W. J., and Herbert, M. V., "An Inviscid-Viscous Interaction Method to Predict the Blade-to-Blade Performance of Axial Compressors." *The Aeronautical Quarterly*, Vol. 31, Part 3, Aug. 1981, pp. 173–196.
74. Lomax, H., "Some Prospects for the Future of Computational Fluid Dynamics." *AIAA Journal*, Vol. 20, No. 8, Aug. 1982, pp. 1033–1043.
75. Jameson, A., et al., "Numerical Solutions of the Euler Equations by Finite Volume Methods Using Runge-Kutta Time-Stepping Scheme." *AIAA Paper* 81-1259, 1981.
76. Jameson, A., and Pimothy, J. B., "Solution of the Euler Equations for Complex Configurations." *AIAA Paper* 83-1929, 1983.
77. Hafez, M., and Lovell, D., "Numerical Solution of Transonic Stream Function Equation." *AIAA Journal*, Vol. 21, No. 3, Mar. 1983, pp. 327–335.
78. Atkins, H. L., and Hassan, H. A., "Transonic Flow Calculations Using the Euler Equations." *AIAA Journal*, Vol. 21, No. 6, June 1983, pp. 842–847.

CHAPTER 18

APPROACHES TO ANALYZING UNSTEADY LAMINAR FLOW IN LONG PIPES

Mario F. Letelier S.

Department of Mechanical Engineering
University of Santiago, Chile

CONTENTS

INTRODUCTION

This chapter is mainly concerned with the study of parallel, laminar unsteady flow of isothermal, incompressible Newtonian fluids in circular pipes. Parallel flow exists in long, rigid straight pipes of constant diameter where end effects are negligible. These are restrictions for which rigorous and rather ample methods of solution are available and applicable to a wide variety of problems.

Compressibility effects, non-parallel flows, and non-circular geometry are considered to some extent at the end of the chapter.

The presentation and discussion of the analytical methods are guided by a flow classification based upon the physics and mathematics of the phenomenon.

EQUATIONS OF MOTION

Laminar axisymmetric flow, subject to the restrictions previously stated, is governed by the linear forms of the Navier-Stokes and continuity equations, viz., in cylindrical coordinates (Figure 1),

$$\frac{\partial u}{\partial t} = -\frac{1}{\rho}\frac{\partial P}{\partial x} + \nu\left(\frac{\partial^2 u}{\partial r^2} + \frac{1}{r}\frac{\partial u}{\partial r}\right) \tag{1}$$

$$\frac{\partial P}{\partial \theta} = \frac{\partial P}{\partial r} = 0 \tag{2}$$

$$\frac{\partial u}{\partial x} = 0 \tag{3}$$

in which $P = p + \rho gh$ denotes the piezometric pressure; p = fluid pressure; u = local velocity; x, r, h = coordinate in, respectively, the axial, radial and vertical directions; θ = tangential coordinate; t = time; ρ = density; $\nu = \mu/\rho$ = kinematic viscosity; μ = dynamic viscosity; and g = acceleration due to gravity.

The continuity equation, Equation 3, states that, in parallel flow, u can only be a function of r and t. Accordingly, the momentum equations, Equations 1 and 2, suggest that $\partial P/\partial x$ can be a function neither of x nor r. Therefore,

$$-\frac{1}{\rho}\frac{\partial P}{\partial x} = \phi(t) \tag{4}$$

where ϕ, usually referred to as the forcing function of the system, is a function only of time. Equation 4 can be integrated with respect to x, yielding $-\Delta P = \rho L\phi$. $\Delta P = P_2 - P_1$ is the difference in piezometric pressure between two stations 1 and 2 a distance $L = x_2 - x_1$ apart. This shows that, at any instant, the piezometric pressure varies linearly with x. In terms of ϕ, the momentum Equation 1 becomes

$$\frac{\partial u}{\partial t} - \nu\left(\frac{\partial^2 u}{\partial r^2} + \frac{1}{r}\frac{\partial u}{\partial r}\right) = \phi \tag{5}$$

Its boundary condition is, by the no-slip hypothesis, $u(a, t) = 0$, where a is the radius of the pipe. Similarly, an initial condition can be imposed such as $u(r, o) = u_i(r)$, where u_i is the local velocity

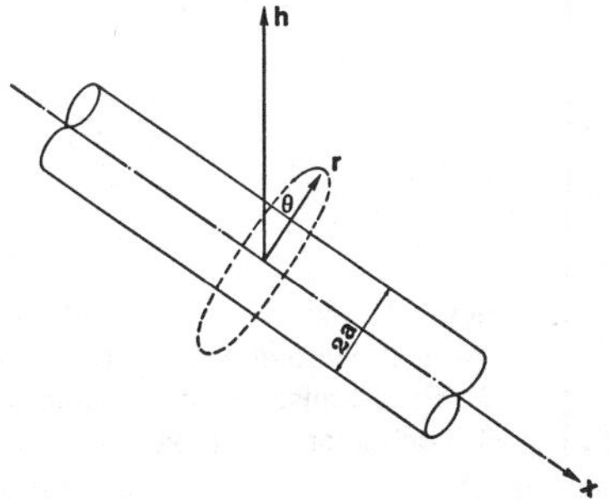

Figure 1. Coordinate System.

at t = 0. The instantaneous average velocity U over the cross-section of the pipe is defined as

$$U(t) = \frac{2}{a^2}\int_0^a ur\,dr = I(u) \tag{6}$$

in which I(u) is the associated integral operator.

The equivalent one-dimensional form of Equation 5, preferred for engineering calculations, is obtained by averaging this equation over the cross-section of the pipe, with the aid of the operator (Equation 6). This process yields

$$\frac{dU}{dt} + \frac{2\tau_w}{\rho a} = \phi \tag{7}$$

where

$$\tau_w = -\mu\left(\frac{\partial u}{\partial r}\right)_{r=a} = \frac{f}{4}\rho\frac{U^2}{2} \tag{8}$$

denotes the instantaneous wall shear stress, which has also been expressed in terms of the familiar Darcy-Weisbach friction coefficient f. The apparent simplicity of Equation 7 as compared with Equation 5, is solely in form but not in substance. It belies the fact that, for real utility of the one-dimensional approach, τ_w must still be suitably expressed in terms of the average velocity.

CLASSIFICATION OF FLOWS

A survey of the physical systems that can be grouped around the mathematical model described in Equations 1–4 reveals that, for the sake of convenience, three principal classes can be distinguished. These will, henceforth, referred to as, respectively, Class A, Class B and Class C [1]

Class A. This type of flow occurs when the piezometric pressures at two fixed stations in a pipe are explicitly known as a function of time. Some examples are portrayed schematically in Figure 2, which also gives the applicable forms of the forcing function, and information on previously

CASE	FLOW SYSTEM	FORCING FUNCTION	SOLUTION	PROBLEM CLASS
1	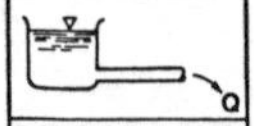 FLOW ESTABLISHMENT	constant	F.J. Szymanski [2] 1932	A $\phi=\phi(t)$
2	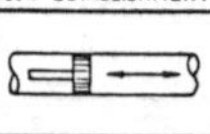 FORCED OSCILL. FLOW	$K_0 \cos \omega t$	T. Sexl [19] 1930	
3	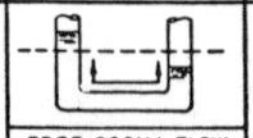 FREE OSCILL. FLOW	$\bar{\alpha}_1+\bar{\alpha}_2\iint urdrdt$	C. Clarion and R. Pelissier [4] 1975	B $\phi=\phi[U(t),t]$
4	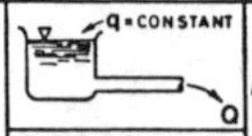 FL. EST. WITH INFLOW	$\bar{\beta}_1+\bar{\beta}_2 t+\bar{\beta}_3\iint urdrdt$	M.F. Letelier S. and H.J. Leutheusser [5] 1983	
5	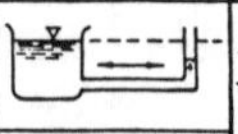 ONE-ARMED U-TUBE	$\dfrac{\bar{\gamma}_1+\bar{\gamma}_2\iint urdrdt}{\bar{\gamma}_3+\bar{\gamma}_4\iint urdrdt}$		

Figure 2. Examples of pipe-flow transients. (From M. F. Letelier S. and H. J. Leutheusser [5]. By permission of the American Society of Mechanical Engineers).

achieved solutions, if any. The forcing function can then be written simply as

$$\phi(t) = -\frac{P_2(t) - P_1(t)}{\rho L} \tag{9}$$

The mathematical consequences of this kind of forcing function are two-fold. First, Equation 5 is linear and has a general homogeneous solution. Second, there exist several methods of analysis of general applicability.

Class B. It is frequently found that the piezometric pressure at one or two stations of a pipe is affected by the fluid motion itself; typical examples are depicted in Figure 2. In these situations the functional dependence of the piezometric pressure on the time coordinate is not known explicitly but only through the medium of the average velocity.

Thus, in some cases, the elevation (relative to a horizontal datum plane) of a pipe station is fixed but the prevailing pressure may be changed by the motion. This occurs, for instance, at the inlet of a pipe connected to a supply tank where the head depends on the time history of the pipe discharge. In other instances, the pressure is known at a given station but the latter's elevation changes with the motion. This situation is exemplified by the conditions prevailing at the free ends of a liquid column where the applied pressure may be constant. Moreover, here the length of the inertially active liquid column (of radius a) may change with time. It follows that the forcing function of Class B problems must be assumed to have the general form

$$\phi = -\frac{P_2(U, t) - P_1(U, t)}{\rho L(U)} \tag{10}$$

which clearly includes, as a special case, the forcing function (Equation 9) for Class A. In most instances, the dependence on U of both the numerator and denominator of Equation 10 will be through the integral

$$\int_0^t U\, dt' = \frac{2}{a^2}\int_0^t\int_0^a ur\, dr\, dt' \tag{11}$$

which has the physical meaning of a length. The applicable expressions for P_1 and P_2 in terms of U may assume a great variety of forms, depending on the end conditions of the pipe system, i.e., the shape of supply and receiver tanks, compressibility of possibly enclosed gaseous volumes, etc. Indeed, it is very difficult, if not impossible, to enumerate in a general way all of the physical circumstances that may impose a dependence of the forcing function on the instantaneous average velocity. Nevertheless, it would appear that the presence of a moving free surface somewhere in the system will likely render ϕ functionally dependent on U.

The appearance of Equation 11 in the forcing function changes the mathematical structure of Equation 5. This may now be linear or non-linear. An obvious cause of non-linearity is a fluid column of varying length; but there are also other possible reasons.

Class C. There exists a wide range of technically important flow situations where the forcing function is not a priori known. Typical instances include biological flow system; pipes starting from, or ending in junction points; and flow systems where only the kinemamatics of the driving agent (i.e, such as a piston) is known. In these cases some alternative data may be given, such as the time history of the average velocity. The problem then becomes to determine the missing information, including the time history of the pressure at some point in the system and the instantaneous distribution of the flow velocity over the cross-section of the pipe. Clearly, this is the inverse of problems considered in Classes A and B, where the forcing function is, at least, implicitly known initially, and sought is information on the velocity field.

SOME METHODS OF ANALYSIS FOR AXISYMMETRIC PARALLEL FLOWS

The following presentation of analytical methods is by no means exhaustive. It includes those methods which, in opinion of the present author, have proved to be particularly useful.

Non-Dimensional Equations

In the following it is convenient to work in terms of non-dimensional variables. Thus, defining $u = U_0u^*$; $t = T_0t^*$; $r = ar^*$; where U_0 and T_0 are constant reference values of velocity and time, respectively, and asterisks denote dimensionless quantities, Equation 5 can be transformed into

$$\Omega\frac{\partial u^*}{\partial t^*} - \left(\frac{\partial^2 u^*}{\partial r^{*2}} + \frac{1}{r^*}\frac{\partial u^*}{\partial r^*}\right) = \phi^*(t^*) \tag{12}$$

Here $\phi^* = (a^2/\nu U_0)\phi$, and the boundary condition is $u^*(1, t^*) = 0$. The parameter Ω is given by

$$\Omega = \frac{a^2}{\nu T_0} \tag{13}$$

The physical meaning of Ω is always a measure of the unsteadiness of the motion. For this reason, Ω will henceforth be referred to as the "unsteadiness number" of the flow. The pipe-fluid system can be characterized by a diffusion time, or response time a^2/ν; on the other hand, T_0 is a characteristic time of the forcing function and, hence, may be termed "forcing time." The unsteadiness number is, thus, seen to represent the ratio of diffusion to forcing time; equivalently, the value of Ω indicates how important the acceleration is in a given flow situation.

The Concept of Quasi-Steady Flow

The steady flow, or Poiseuille counterpart of Equation 12 is

$$-\left(\frac{\partial^2 u_s}{\partial r^{*2}} + \frac{1}{r^*}\frac{\partial u_s^*}{\partial r^*}\right) = \phi_s^* \tag{14}$$

where subscript s denotes steady condition, and ϕ_s^* is always a constant. For convenience of notation, asterisks will henceforth be eliminated. Equation 14 states that in Poiseuille flow the piezometric pressure force per unit mass is exactly balanced by the shear force per unit mass. The well-known solution of Equation 14 yields, respectively, for local velocity,

$$u_s = \frac{\phi_s}{4}(1 - r^2) \tag{15}$$

average velocity

$$U_s = \frac{\phi_s}{8} \tag{16}$$

wall shear stress,

$$\tau_{ws} = -\frac{\partial u_s}{\partial r}(1) = \frac{\phi_s}{2} = 4U_s \tag{17}$$

and friction coefficient,

$$f_s = \frac{64}{\mathbf{R}_s} \tag{18}$$

where $\mathbf{R}_s$ is the Reynolds number of the steady flow.

It is possible to associate to any unsteady flow a "quasi-steady" velocity μ_{qs} defined as

$$u_{qs}(r, t) = \frac{\phi(t)}{4}(1 - r^2) \tag{19}$$

In this, $\phi(t)$ is the actual forcing function of the unsteady flow. Equation 19 is an exact solution of Equation 12 only when $\Omega \to 0$, i.e., when Equation 12 reduces to Equation 14 and, thus, Equation 19 becomes Equation 17.

On the other hand, Equation 19 is an approximate solution of Equation 12 when the following condition is satisfied, viz.

$$\left|\Omega \frac{\partial u}{\partial t}\right| \ll \left|\frac{\partial^2 u}{\partial r^2} + \frac{1}{r}\frac{\partial u}{\partial r}\right| \tag{20}$$

Thus, Equation 20 is the analytical condition for quasi-steady flow. When this condition is operative, the time-dependent inertia force is very small compared with the shear force. The latter, in turn, is either slightly larger or slightly smaller than the pressure force. As a consequence, the fluid slowly accelerates or, as the case may be, decelerates. Yet, at any instant of time conditions prevail that are essentially those of a steady flow.

It should be noted that Equation 19 is a mathematical definition that models accurately the physical concept of quasi-steady flow only when the flow acceleration is, indeed, very small. Otherwise u_{qs} is simply the "quasi-steady component" of the complete velocity which, when the acceleration is not negligible, may not necessarily be the most important component of the velocity. A quasi-steady wall shear stress τ_{wqs} can be derived from Equation 19 by means of Equation 8.

Fourier Analysis

Flows where the forcing function is explicit in time (i.e., Class A flows) admit a homogeneous solution of Equation 12 obtained by separation of variables, and given by [2],

$$u_h = \sum_{n=1}^{\infty} C_n J_0(b_n r) \exp(-b_n^2\, t/\Omega) \tag{21}$$

where J_0 is the Bessel function of the first kind and order zero, $J_0(b_n) = 0$, and the C_n's are arbitrary constant coefficients. There is not a general homogeneous solution for Class B flows. The particular solution of Equation 12, u_p, can be deduced through various techniques. The first of them to be considered here is Fourier analysis.

If the forcing function and its derivatives are piecewise continuous in $0 \leqq t \leqq T_0$, where T_0 is any finite physical time (Dirichlet's conditions), then ϕ can be expressed by a Fourier series, viz.

$$\phi = B_0 + \sum_{n=1}^{\infty} B_{cn} \cos nt + \sum_{n=1}^{\infty} B_{sn} \sin nt$$

or, in complex form,

$$\phi = B_0 + R_e\left\{\sum_{n=1}^{\infty} B_n e^{int}\right\} \tag{22}$$

in which B_0, B_{cn}, and B_{sn} are constant coefficients; $B_n = B_{cn} - iB_{sn}$; R_e denotes real part of its argument; $i = \sqrt{-1}$; and $\Omega = \pi a^2/\nu T_0$. The particular solution of Equation 12, when ϕ is given by Equation 22, takes the form [2, 6, 7]

$$u_p = \frac{B_0}{4}(1 - r^2) - R_e\left\{\sum_{n=1}^{\infty} \frac{iB_n}{n\Omega}\left[1 - \frac{J_0(r\sqrt{-in\Omega})}{J_0(\sqrt{-in\Omega})}\right]e^{int}\right\} \tag{23}$$

Equation 23 can be written in real notation by means of the Kelvin functions, through the relation (Equation 24), viz.

$$J_0(r\sqrt{-in\Omega}) = \mathrm{ber}_0(r\sqrt{n\Omega}) + i\,\mathrm{bei}_0(r\sqrt{n\Omega}) \tag{24}$$

The coefficients B_n are explicitly known for Class A flows, otherwise it is necessary to determine them from Equation 12 for other classes of flows. This method is especially useful for oscillatory or pulsating flows. In the cases where the forcing function has some finite asymptotic value for $t \to \infty$, Fourier integrals are applicable [2].

Eigenfunction Expansions

The presence of the one-dimensional Laplacian operator in Equation 12 makes it possible to consider the following relationship, derived from Bessel's equation, i.e.,

$$\frac{d^2J_0(b_nr)}{dr^2} + \frac{1}{r}\frac{dJ_0(b_nr)}{dr} = -b_n^2J_0(b_nr) \tag{25}$$

Since $J_0(b_nr)$ satisfies the boundary condition on the pipe wall, the velocity can be expressed as

$$u_p = \sum_{n=1}^{\infty} T_n(t)J_0(b_nr) \tag{26}$$

where the T_n's are functions of time to be determined a posteriori. The forcing function is next written in the form of a Fourier-Bessel series, viz.

$$\phi = 2\phi \sum_{n=1}^{\infty} \frac{J_0(b_nr)}{b_nJ_1(b_n)} \tag{27}$$

Substitution of Equations 26 and 27 into Equation 12, aided by Equation 25, leads to

$$\Omega\frac{dT_n}{dt} + b_n^2T_n = \frac{2\phi}{b_nJ_1(b_n)} \tag{28}$$

The homogeneous solution of Equation 28 reproduces Equation 21. The particular solution of Equation 28, T_{np}, is obtainable only if ϕ is explicitly given. In this case the methods of variation of parameters or Laplace transforms, among others, may yield suitable expressions for T_{np}. The first of these leads to

$$T_{np} = \frac{2\exp(-b_n^2t/\Omega)}{\Omega b_nJ_1(b_n)}\int \phi \exp(b_n^2t/\Omega)\,dt$$

This method is a powerful tool for Class A flows.

Laplace Transforms

Taking the Laplace transform of both sides of Equation 12 with zero initial condition, i.e., $u(r, 0) = 0$, it is found [9, 10]

$$\Omega s\tilde{u} - \left(\frac{d^2\tilde{u}}{dr^2} + \frac{1}{r}\frac{d\tilde{u}}{dr}\right) = \tilde{\phi} \tag{29}$$

in which

$$\tilde{u}(r, s) = \int_0^{\infty} u(r, t)\, e^{-st}\, dt$$

is the Laplace transform of the velocity, and the same definition holds for the transform of ϕ, viz. $\tilde{\phi}$. In this, ϕ is assumed to be piecewise continuous. The solution of Equation 29 is [9, 10]

$$\tilde{u}(r, s) = \frac{\tilde{\phi}}{\Omega s}\left[1 - \frac{J_0(ir\sqrt{\Omega s})}{J_0(i\sqrt{\Omega s})}\right] \tag{30}$$

The velocity is finally obtained by finding the inverse Laplace transform of Equation 30. The success of this operation depends on the complexity of $\tilde{\phi}(s)$. The method may become exceedingly difficult for Class B flows.

Method of Power Series

The structure of Equation 12 admits a series representation for u of the form [5]

$$u = A(1 - r^2) + \frac{\Omega}{4^2}\frac{dA}{dt}(1 - r^4) + \frac{\Omega^2}{4^2 6^2}\frac{d^2A}{dt^2}(1 - r^6)\cdots$$

$$= A(1 - r^2) + \sum_{n=1}^{N} \frac{\Omega^n}{4^2 6^2 \cdots (2n + 2)^2}\frac{d^nA}{dt^n}[1 - r^{2(n+1)}] \tag{31}$$

where A is a function of time only, and is related to the forcing function through the following ordinary, Nth order differential equation, viz.

$$A + \frac{\Omega}{2^2}\frac{dA}{dt} + \frac{\Omega^2}{2^2 4^2}\frac{d^2A}{dt^2} + \cdots = A + \sum_{n=1}^{N} \frac{\Omega^n}{2^2 4^2 \cdots (2n)^2}\frac{d^nA}{dt^n} = \frac{\phi}{4} \tag{32}$$

The preceding equations apply to the three classes of flows. If ϕ is a function of time through the average velocity (Equations 10 and 11), i.e., it describes Class B flows, then ϕ in Equation 32 is $\phi = \phi(A, t)$, where $A = A(t)$. The integer N depends on the number of nonzero derivatives of ϕ.

Equation 31 meets the boundary condition and reveals that the whole time dependence of the flow can be expressed by means of a single function of time, namely A. When $\Omega \to 0$, then $A \to \phi/4$ and Equation 15 is recovered. Equations 31 and 32 converge for any value of Ω provided ϕ is piecewise continuous [1]. For given forcing function it is first necessary to solve Equation 32; the velocity field is then found by substituting the result into Equation 31 and by applying the initial condition. Integration of Equation 31 with the aid of the operator of Equation 6 yields a series expansion of the average velocity, viz.

$$U = \sum_{n=0}^{N} \beta_n \Omega^n \frac{d^nA}{dt^n} \tag{33}$$

where $\beta_0 = 1/2$, and, for $n \geq 1$,

$$\beta_n = \frac{1}{2^2 4^2 \cdots (2n)^2 (n+1)(n+2)} \tag{34}$$

By virtue of Equation 31, the nondimensional form of Equation 8 can be written as

$$\tau_w = \sum_{n=0}^{N} \gamma_n \Omega^n \frac{d^n A}{dt^n} \tag{35}$$

where $\gamma_0 = 2$, and, for $n \geqq 1$,

$$\gamma_n = \frac{1}{2\{4^2 6^2 \cdots (2n)^2\}(n+1)} \tag{36}$$

Some simple explicit solutions for the velocity, average velocity and wall shear stress can be obtained from the set of Equations 31 and 32 for Class A flows. A particular solution of Equation 32, developed by inversion of Equation 32, is

$$A_p = \frac{\phi}{4} - \frac{\Omega}{16}\frac{d\phi}{dt} + \frac{3\Omega^2}{256}\frac{d^2\phi}{dt^2} - + \cdots$$

wherefrom, and combining Equations 21 and 31, it is found

$$u = \sum_{n=1}^{\infty} C_n J_0(b_n r) \exp(-b_n^2 t/\Omega) + \frac{\phi}{4}(1 - r^2) + \frac{\Omega}{64}\frac{d\phi}{dt}(-3 + 4r^2 - r^4)$$
$$+ \frac{\Omega^2}{2{,}304}\frac{d^2\phi}{dt^2}(19 - 27r^2 + 9r^4 - r^6) + \cdots \tag{37}$$

The corresponding average velocity is

$$U = \sum_{n=1}^{\infty} \frac{2C_n}{b_n} J_1(b_n) \exp(-b_n^2 t/\Omega) + \frac{\phi}{8} - \frac{\Omega}{48}\frac{d\phi}{dt} + \frac{11}{3{,}072}\Omega^2 \frac{d^2\phi}{dt^2} - + \cdots \tag{38}$$

Finally, the wall shear stress is given by

$$\tau_w = \sum_{n=1}^{\infty} C_n b_n J_1(b_n) \exp(-b_n^2 t/\Omega) + \frac{\phi}{2} - \frac{\Omega}{16}\frac{d\phi}{dt} + \frac{\Omega^2}{96}\frac{d^2\phi}{dt^2} - + \cdots \tag{39}$$

Equations 33 and 35 converge under very general conditions. Equations 37, 38 and 39 are, however, subject to greater restrictions. If ϕ and its derivatives are of the same order of magnitude, these series converge for $\Omega < b_1^2 = 5.7832$. In cases where the number of nonzero derivatives of ϕ is finite, Equations 38 and 39 become exact forms of, respectively, the velocity field, average velocity, and wall shear stress.

One-Dimensional Analysis

The basic idea behind the one-dimensional approach to the solution of unsteady pipe flow problems is to obtain directly the average velocity from the equation of motion. This avoids part of the mathematical complexity which is entailed by solving for the local velocity, i.e., its radial distribution. The dimensionless form of the one-dimensional momentum Equation 7 is

$$\Omega \frac{dU}{dt} + 2\tau_w = \phi \tag{40}$$

where

$$\tau_w = -\frac{\partial u}{\partial r}(1, t) \tag{41}$$

Equation 40 is of practical interest when τ_w is expressed in terms of the average velocity. This then yields an equation for U, which does not involve the local velocity u and, thus, has a simpler mathematical structure. However, the direct solution for U entails a degree of approximation, since the initial condition for U is not necessarily equivalent to the initial condition for u.

An approximate form of Equation 41, widely used over the years, is the quasi-steady form, derived from Equation 19. However, it should be evident, by inspection of Equation 39, that this approximation may lead to considerable error in flows where Ω is not small (i.e., when there is non-negligible acceleration) or when the initial transients (expressed through the homogeneous part of u) are important. For an assessment of the applicability of the quasi-steady approximation see references [1, 11, 12, 13]

Zielke [10] derived an analytical expression for τ_w from the full momentum Equation 12, by means of Laplace transforms. It contains the quasi-steady component of the shear plus an integral term, which reflects the history in time of the average velocity, viz., in dimensionless variables,

$$\tau_w = 4U + 2\int_0^t \frac{dU}{dt}(s)W(t - s)\, ds \tag{42}$$

in which s is a dummy variable and W is given by

$$W(t) = 0.28209\left(\frac{t}{\Omega}\right)^{-1/2} - 1.250000 + 1.057855\left(\frac{t}{\Omega}\right)^{1/2}$$
$$+0.937500\frac{t}{\Omega} + 0.396696\left(\frac{t}{\Omega}\right)^{3/2} - 0.351563\left(\frac{t}{\Omega}\right)^{2} + \cdots \qquad \frac{t}{\Omega} < 0.02$$

$$W(t) = \exp(-26.3744t/\Omega) + \exp(-70.8493t/\Omega) + \exp(-135.0198t/\Omega)$$
$$+ \exp(-218.9216t/\Omega) + \exp(-322.5544t/\Omega) + \cdots \qquad \frac{t}{\Omega} > 0.02 \tag{43}$$

Use of Equation 42 in Equation 40 is not simple, in general. On the other hand, Equation 42 may prove to be a convenient way to determine τ_w when the average velocity is known a priori and one is interested in evaluating the shear stress.

A different approach can be developed through the method of power series. Equations 32, 33, and 35 are series for, respectively, the forcing function, average velocity, and wall shear stress in terms of a unique function of time, A, and its derivatives. The function A can be eliminated and convenient relationships among ϕ, U and τ_w are thus deduced. In effect, the following relations hold [5]

$$\alpha_0 U + \alpha_1\Omega\frac{dU}{dt} + \alpha_2\Omega^2\frac{d^2U}{dt^2} + \cdots = \beta_0\phi + \beta_1\Omega\frac{d\phi}{dt} + \beta_2\Omega^2\frac{d^2\phi}{dt^2} + \cdots \tag{44}$$

$$\gamma_0 U + \gamma_1\Omega\frac{dU}{dt} + \gamma_2\Omega^2\frac{d^2U}{dt^2} + \cdots = \beta_0\tau_w + \beta_1\Omega\frac{d\tau_w}{dt} + \beta_2\Omega^2\frac{d^2\tau_w}{dt^2} + \cdots \tag{45}$$

$$\alpha_0\tau_w + \alpha_1\Omega\frac{d\tau_w}{dt} + \alpha_2\Omega^2\frac{d^2\tau_w}{dt^2} + \cdots = \gamma_0\phi + \gamma_1\Omega\frac{d\phi}{dt} + \gamma_2\Omega^2\frac{d^2\phi}{dt^2} + \cdots \tag{46}$$

The coefficients α_n, β_n, and γ_n are defined in, respectively, Equations 32, 34, and 35. Equations 44 through 46 converge for any value of Ω, provided ϕ is piecewise continuous.

Equation 44 is a general form of the one-dimensional momentum equation, equivalent to 40, but with τ_w eliminated. Curiously, Equation 44 can be deduced directly, without determining previously the wall shear stress. Equation 45 is a general form of the law of friction, and Equation 46 relates the wall shear stress to the forcing function. Equation 35 was also obtained by Achard and Lespinard [14]. In many phenomena the unsteadiness is wholly or in part described by exponential functions of time, of either real or imaginary argument. The following expressions relate the coefficients in Equation 31, α_n, β_n, and γ_n to the Bessel functions of the first kind, after putting $A(t) = \exp(-\lambda t)$ in, respectively, Equations 31, 32, 33, and 35, viz.

$$u = \frac{4}{\Omega\lambda}\left[J_0(\sqrt{\Omega\lambda}r) - J_0(\sqrt{\Omega\lambda})\right]e^{-\lambda t} \tag{47}$$

$$\phi = 4J_0(\sqrt{\Omega\lambda})\,e^{-\lambda t} \tag{48}$$

$$U = \frac{8}{\Omega\lambda}\left[\frac{1}{\sqrt{\Omega\lambda}}J_1(\sqrt{\Omega\lambda}) - \frac{1}{2}J_0(\sqrt{\Omega\lambda})\right]e^{-\lambda t} \tag{49}$$

$$\tau_w = \frac{4}{\sqrt{\Omega\lambda}}J_1(\sqrt{\Omega\lambda})\,e^{-\lambda t} \tag{50}$$

The preceding equations need to be transformed in terms of Kelvin functions (Equation 24) whenever λ is imaginary, i.e., when there exist periodic components of the unsteadiness. A comparison of Equation 44 with 40 reveals that an explicit law of friction (in contrast with the implicit form, (Equation 45) is

$$\tau_w = 4U + \frac{\Omega}{2}\frac{dU}{dt} + \alpha_2\Omega^2\frac{d^2U}{dt^2} + \alpha_3\Omega^3\frac{d^3U}{dt^3} + \cdots - \left(\beta_1\Omega\frac{d\phi}{dt} + \beta_2\Omega^2\frac{d^2\phi}{dt^2} + \cdots\right) \tag{51}$$

where it is seen that a general, explicit form for τ_w involves the sum total of the derivatives of U and of ϕ. Simpler versions of the momentum Equation 44 and of the law of friction (Equation 45), although of more restricted applicability, are generated by solving for ϕ in Equation 44 and for τ_w in Equation 45, wherefrom

$$\alpha'_0U + \alpha'_1\Omega\frac{dU}{dt} + \alpha'_2\Omega^2\frac{d^2U}{dt^2} + \cdots = \phi \tag{52}$$

and

$$\tau_w = \beta'_0U + \beta'_1\Omega\frac{dU}{dt} + \beta'_2\Omega^2\frac{d^2U}{dt^2} + \cdots \tag{53}$$

Equations 52 and 53 do not always converge for all valuses of Ω; when U and its derivatives are of order unity, Equation 52 converges for $\Omega < 26.3746$. The first values of α'_n and β'_n are given in Table 1. An expression for the friction coefficient follows from Equations 8 and 53, viz.

$$f = \frac{1}{\mathbf{R}_0}\left[\frac{64}{U} + \frac{8}{3}\frac{\Omega}{U^2}\frac{dU}{dt} - \frac{1}{18}\frac{\Omega^2}{U^2}\frac{d^2U}{dt^2} + \cdots\right]$$

in which $\mathbf{R}_0 = 2aU_0/\nu$.

The analytical methods already presented are next applied to some instances of parallel unsteady flow, following the flow classification previously defined.

Table 1
Values of α'_n in Momentum Equation 52 and of β'_n in Law of Friction (Equation 53)

n	0	1	2	3	4	5
α'_n	8	$\frac{4}{3}$	$-\frac{1}{144}$	$\frac{1}{2^5 \cdot 3^3 \cdot 5}$	$-\frac{7}{2^{11} \cdot 3^4 \cdot 5}$	$\frac{11}{3^5 \cdot 4^6 \cdot 5 \cdot 7}$
β'_n	0	$\frac{1}{6}$	$-\frac{1}{288}$	$\frac{1}{2^6 \cdot 3^3 \cdot 5}$	$-\frac{7}{2^{12} \cdot 3^4 \cdot 5}$	$\frac{11}{2^{13} \cdot 3^5 \cdot 5 \cdot 7}$

FLOWS OF CLASS A

Here the conditions at the ends of the pipe are explicitly defined in time so that the forcing function is, consequently, an explicit function of time (Equation 9). Three types of Class A flows are discussed herein, namely flow establishment under constant head, flow with forcing function linear in time, and oscillatory flow.

Flow Establishment Under Constant Head

The simplest case will be considered first, i.e., the problem where a steady flow is established in a pipe under a constant pressure gradient, starting from rest. A convenient physical model of the phenomenon is described in Figure 3. A valve at the end of a long cylindrical pipe, attached to a reservoir providing a constant head H_0, is suddenly completely opened at $t = 0$. Sought is the time history of the resulting volume rate of pipe efflux, given by U(t) in dimensionless variables. The physical forcing function is gH_0/L; its dimensionless counterpart is

$$\phi = \frac{gH_0}{L}\frac{a^2}{\nu U_0}$$

If the ultimate (i.e., $t \to \infty$) steady-state velocity of Poiseuille flow is chosen as a reference velocity, i.e., $U_0 = ga^2H_0/8\nu L$, then $\phi = 8$. A suitable choice for the reference time is $T_0 = 2\,[aL/gH_0]^{1/2}$, whence, by Equation 13,

$$\Omega = \frac{a}{2\nu}\sqrt{\frac{gaH_0}{L}} = \sqrt{\mathbf{R}_0}$$

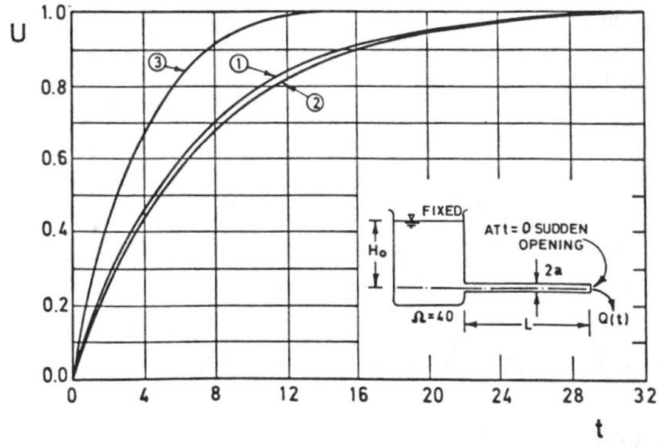

Figure 3. Flow establishment under constant pressure gradient. Curve 1: exact solution (Eq. 57); curve 2: approximation, (Equation 60); curve 3: approximation, (Equation 62). (From M. F. Letelier S. and H. J. Leutheusser [5]. By permission of the American Society of Mechanical Engineers).

in which $\mathbf{R}_0 = 2aU_0/\nu$. Thus, in this case, Ω is the square root of the ultimate steady-state Reynolds number $\mathbf{R}_0$. As such, Ω must not exceed a value of approximately $\sqrt{2000} \cong 45$ in order for laminar flow to exist.

This problem was first solved by Szymanski [2] by the method of separation of variables. The velocity field, with reference to Equation 21 and to the conditions for $t \to \infty$, is given by

$$u = \sum_{n=1}^{\infty} C_n J_0(b_n r) \exp(-b_n^2 t/\Omega) + 2(1 - r^2) \tag{54}$$

This is subject to the initial condition $u(r, 0) = 0$ with which the Fourier-Bessel coefficients C_n can be found. The final result is

$$u = 2\left\{1 - r^2 - \sum_{n=1}^{\infty} \frac{8J_0(b_n r)}{b_n^3 J_1(b_n)} \exp(-b_n^2 t/\Omega)\right\} \tag{55}$$

The unsteadiness of this flow is entirely due to the initial condition, since ϕ = constant; therefore, the particular solution is independent of time and becomes the full solution once the initial transients, described by the homogeneous part, have died out.

The method of eigenfunction expansions yields also simple results in this instance of unsteady flow. Equation 28 admits a solution of the form

$$T_n = C_n \exp(-b_n^2 t/\Omega) + \frac{16}{b_n^3 J_1(b_n)}$$

which, upon substitution into Equation 26, leads to

$$u = 16 \sum_{n=1}^{\infty} \frac{J_0(b_n r)}{b_n^3 J_1(b_n)} [1 - \exp(-b_n^2 t/\Omega)] \tag{56}$$

Equation 56 is equivalent to Equation 55, is of simpler structure than Equation 55, and, additionally, in its deduction was not necessary to find the coefficients C_n by integration, which is the standard procedure for Fourier-Bessel series. Equation 56 was also obtained by Gerbes [15], who solved the inverse problem, i.e., a decelerating flow under zero head, as well.

The problem under consideration can alternatively be solved by Laplace transforms. To this end, the transform of $\phi = 8$, i.e., $\tilde{\phi} = 8/s$, is substituted into Equation 30, wherefrom follows

$$\tilde{u} = \frac{8}{\Omega s^2}\left[1 - \frac{J_0(ir\sqrt{\Omega s})}{J_0(i\sqrt{\Omega s})}\right]$$

The inverse transform of $\tilde{u}$ is exactly given by Equation 54.

The method of power series, the last exact method to be discussed herein, proceeds first by solving for A in Equation 32. The corresponding equation for A is

$$A + \frac{\Omega}{2^2}\frac{dA}{dt} + \frac{\Omega^2}{2^2 4^2}\frac{d^2A}{dt^2} + \cdots = 2$$

where the full solution is given by

$$A = \sum C_n \exp(-b_n^2 t/\Omega) + 2$$

The function A is next substituted into Equation 31, which yields (Equation 47)

$$u = 2(1 - r^2) + \sum_{n=1}^{\infty} \frac{4C_n}{b_n^2} J_0(b_n r) \exp(-b_n^2 t/\Omega)$$

The final result is achieved by putting $u(r, 0) = 0$ with which the Fourier-Bessel coefficients C_n can be found, thus reproducing Equation 55.

From an engineering viewpoint, it is specially important that volume rate of flow, which in dimensionless form, is obtained by integrating Equation 55, i.e. [16],

$$U = 1 - \sum_{n=1}^{\infty} \frac{32}{b_n^4} \exp(-b_n^2 t/\Omega) \tag{57}$$

Equation 57 is plotted in Figure 3 for $\Omega = 40$.

The time of flow establishment, t_e, customarily defined as the time for which, dimensionally, $U = 0.99\, U_0$ (where $U(\infty) = U_0$), results directly from Equation 57 by putting $U = 0.99$, $t = t_e$, viz.

$$t_e = 0.7884\, \Omega \tag{58}$$

or, in dimensional variables,

$$t_e = 0.7884\, a^2/\nu \tag{59}$$

Equation 59 shows that the time necessary for the flow to become established depends only on the pipe-fluid time a^2/ν and is independent of the driving force. All other conditions invariant, high-viscosity fluids reach the steady state velocity faster than low-viscosity fluids. This rather nonintuitive result is explained considering that high-viscosity fluids reach, on the other hand, relatively small terminal velocities. Experiments support Equation 59 [11, 17].

The wall shear stress can be computed from Equation 55 and substituted into 8 in order to evaluate the friction coefficient f for this flow. A plot of f as a function of the Reynolds number is given in Figure 4. The Reynolds number was calculated from Equation 57. Figure 4 reveals that skin friction, does not follow in general the quasi-steady line $64/\mathbf{R}$, except for $t \sim t_e$, but is higher than quasi-steady skin friction. This is due to the fact that, except for $t \sim t_e$, the velocity profiles are not parabolic. The central portion of the fluid moves initially unaffected by the wall shear and a relatively steep velocity gradient develops close to the wall.

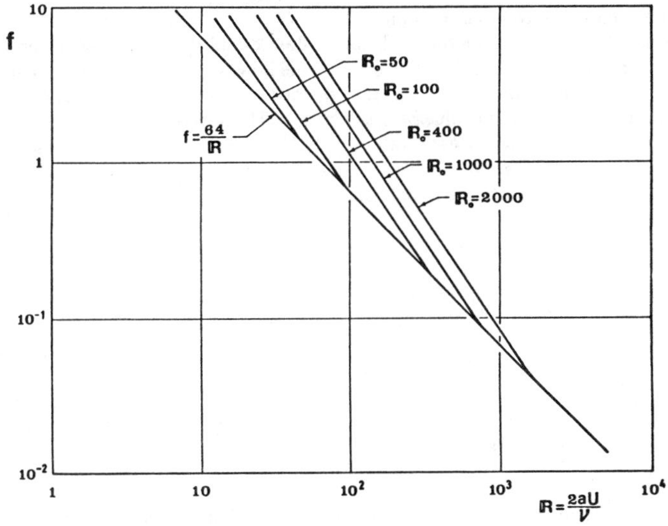

Figure 4. Friction coefficient in flow establishment under constant pressure gradient. (From M. F. Letelier S. and H. J. Leutheusser [11]. By permission of the American Society of Civil Engineers).

All the exact methods of solution discussed lead to expressions for u in terms of infinite series of Bessel functions. The spatially averaged velocity U is then obtained by integration of u over the cross-section of the pipe. A simpler method of solution, albeit approximate, is the one-dimensional approach, which yields directly U from the one-dimensional momentum. Equations 44 and 52 both apply to this problem. The solution of Equation 52 is of the form

$$U = 1 - \exp(-\lambda t) \tag{60}$$

which, once substituted into 52, yields

$$8 - \frac{4}{3}\Omega\lambda - \frac{(\Omega\lambda)^2}{144} - \cdots = 0 \tag{61}$$

The polynomial (Equation 61) can have only one positive root. It follows that the more terms considered, the better will be the accuracy of the solution. Thus, for two terms in Equation 61, the solution is $\Omega\lambda = 6$, for three terms, $\Omega\lambda = 5.82$, and for four terms, $\Omega\lambda = 5.79$. The exact solution is $\Omega\lambda = b_1^2 = 5.7832$. It may be noted that, in this problem, Equation 52 converges for any value of Ω. A plot of Equation 60 is included as curve 2 in Figure 3. The error entailed by using Equation 60 instead of the exact solution, Equation 57, amounts to less than 5% for $t > 8$, if $\Omega = 40$. The flow reaches $U = 0.99$ for $t = t_e = 31.536$.

The preceding results are obtainable as well from Equation 44. Since $\phi = 8 =$ constant, Equation 44 has a solution of form Equation 60, where λ is given by $J_0(\sqrt{\Omega\lambda}) = 0 = J_0(b_1)$.

It may be of interest to compare the present results with the prediction of a typical engineering approximation. A very common artifice is the assumption of a constant friction coefficient. Thus, with $f = 64/\mathbf{R}_0$ corresponding to the ultimate, steady Poiseuille flow, there follows

$$U = \tanh(8t/\Omega) \tag{62}$$

which is plotted as curve 3 in Figure 3. The presentation leaves little doubt about the error entailed by this arbitrary model of transient friction.

A more general formulation of the problem of flow establishment under a constant forcing function is to consider that, dimensionally, for $t < 0$ the flow is steady with nonzero initial average velocity U_i and constant initial forcing function ϕ_i. At $t = 0$ there is a step change of the forcing function, which takes the constant value $\phi_f \lesseqgtr \phi_i$ for $t \geq 0$. The flow is then accelerated or decelerated until it achieves an ultimate steady state with velocity U_f. Adopting U_f as the reference velocity U_0, the differential equation for this generalized version of the problem is the same as for Szymanski's problem. The solution, subject to the initial condition $u(r, 0) = u_i = 2(1 - r)U_i/U_f$ in dimensionless variables [17], is

$$u = 2(1 - r) + \frac{16(1 - \lambda')}{\lambda'} \sum_{n=1}^{\infty} \frac{J_0(b_n r)}{b_n^3 J_1(b_n)} \exp(-b_n^2 t/\Omega)$$

where $\lambda' = U_f/U_i$. The average velocity is given by

$$U = 1 + \frac{32(1 - \lambda')}{\lambda'} \sum_{n=1}^{\infty} \frac{1}{b_n^4} \exp(-b_n^2 t/\Omega) \tag{63}$$

The time of flow establishment t_e can be computed from Equation 63, wherefrom

$$t_e = 0.7884\Omega + 0.1729\Omega \ln \frac{|1 - \lambda'|}{\lambda'} \tag{64}$$

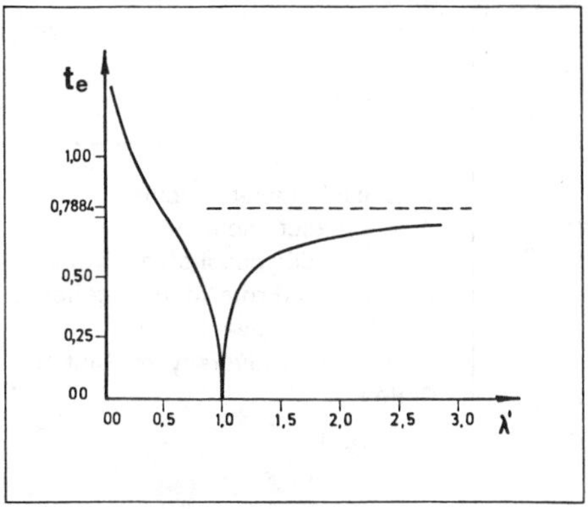

Figure 5. Time of flow establishment as a function of $\lambda' = U_f/U_i$, (Equation 64). (From M. F. Letelier S. and R. Hudson R. [17]. By permission of the University of Santiago, Chile).

A plot of Equation 64 is shown in Figure 5. An interesting feature of the phenomenon is that the predictions for t_e are quite different for accelerated flow as compared with decelerated flow. In the first case there is an asymptotic value of t_e when $\lambda' \to \infty$, i.e., for Szymanski's flow, namely $t_e = 0.7884\ \Omega$ (Equation 58). On the other hand, $t_e/\Omega \to \infty$ when $\lambda' \to 0$, that is, the phenomenon of flow deceleration may take a much longer time than flow acceleration.

Flow with $\phi = 1 + t$

Consider flow case 4 in Figure 2. A long pipe is attached to a finite-size tank with constant cross-sectional area. Initially, steady flow prevails with tank inflow q equaling pipe efflux Q_0 under a constant reservoir head. At time $t = 0$ the volume rate of tank inflow is suddenly increased to a new constant value $q > Q_0$, thereby causing the level in the tank to rise with time. Assume further that for some time interval $0 \leqq t \leqq T$, $Q_0 \ll q$, then, for $t \leqq T$, the head in the tank will increase linearly in time, and, in dimensionless form, the flow is governed by

$$\phi = 1 + t \tag{65}$$

with initial condition $u(r, 0) = 2(1 - r^2)$ (for details concerning nondimensional variables see the section "Flow Establishment with Inflow"). The velocity follows directly from Equation 37 in simple form, since ϕ has only one nonzero derivative. Once the initial condition is applied, the result is [18]

$$u = 2\Omega \sum_{n=1}^{\infty} \frac{J_0(b_n r)}{b_n^5 J_1(b_n)} \exp(-b^2 t/\Omega) + \frac{1}{4}(1 + t)(1 - r^2) + \frac{\Omega}{64}(-3 + 4r^2 - r^4) \tag{66}$$

An expression equivalent to Equation 66 is deduced from Equation 28 in terms of eigenfunction expansions, viz.

$$u = 2\Omega \sum_{n=1}^{\infty} \frac{J_0(b_n r)}{b_n^5 J_1(b_n)} [\exp(-b_n^2 t/\Omega) - 1] + 2(1 + t) \sum_{n=1}^{\infty} \frac{J_0(b_n r)}{b_n^3 J_1(b_n)}$$

Equation 66 shows that very soon be quasi-steady term $(1 + t)(1 - r^2)/4$ becomes dominant. This is an instance of flow where the acceleration is constant once the initial transients have died out, and nevertheless the flow is basically quasi-steady. A plot of some velocity profiles is given in Figure 6 for $\Omega = 5$; in it are depicted both the exact velocity (Equation 66) and the quasi-steady term $(1 + t)(1 - r^2)/4$.

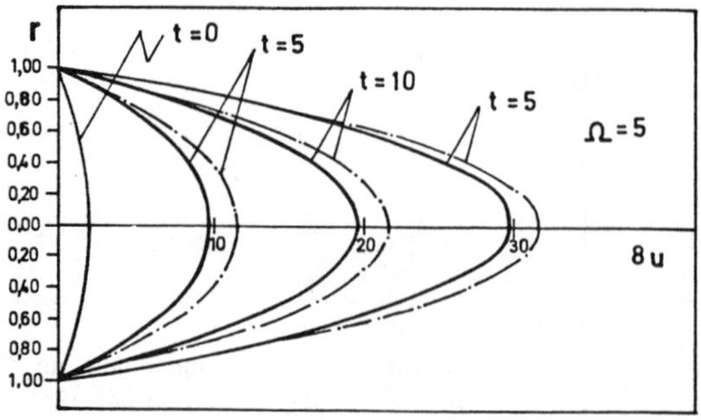

Figure 6. Velocity profiles for $\phi = 1 + t$; exact solution: ——— (Equation 66); quasi-steady velocity: -------. (From M. F. Letelier S. and E. Casanova [18]. By permission of the University of Santiago, Chile.)

Oscillatory Flow

In this third example of Class A, the driving force is oscillating periodically in time. Physically such oscillations may be imposed on a fluid contained in a pipe by the cyclic motion of a piston (see case 2 in Figure 2). It is well recognized [19] that this type of motion is one in which the assumption of quasi-steady friction fails to describe the phenomenon with sufficient accuracy. In particular, it is unable to account for the phase shift between average velocity, or wall shear stress, and the pressure oscillations. Because of their importance and diversity in both industrial applications and biomechanics, oscillatory flows have been studied more widely than any other type of unsteady fluid motion.

The simplest form the forcing function can take in this context is, dimensionally,

$$\phi = K_0 \cos \omega t$$

where ω is the angular frequency, and the amplitude K_0 is a constant. The dimensionless form of ϕ is

$$\phi = \cos t = R_e(e^{it}) \tag{67}$$

In this $T_0 = 1/\omega$; $U_0 = K_0 a^2/\nu$; and $\Omega = \omega a^2/\nu$; the particular form of Ω in this problem is usually referred to as the Womersley number. The following analysis pertains to steady oscillatory motion and, hence, no actual initial conditions need to be considered. The solution of Equations 12 and 67 was first given by Sexl [3] in a successful attempt to explain on theoretical grounds E. G. Richardson's "annular effect," i.e., some experimental findings of Richardson and Tyler [20], according to which, the maximum velocity in an oscillating flow, under certain conditions, appears close to the pipe wall and not in the pipe axis. By separating variables, he found

$$u = R_e\left\{\frac{ie^{it}}{\Omega}\left[\frac{J_0(r\sqrt{-i\Omega})}{J_0(\sqrt{-i\Omega})} - 1\right]\right\} \tag{68}$$

The corresponding average velocity is

$$U = R_e\left\{\frac{\sqrt{-i}\,e^{it}}{\Omega^{3/2}J_0(\sqrt{-i\Omega})}\left[2J_1(\sqrt{-i\Omega}) - \sqrt{-i\Omega}\,J_0(\sqrt{-i\Omega})\right]\right\} \tag{69}$$

It may be noted that Equation 68 is a special case of the particular solution for u when ϕ is expressed as a Fourier series (see Equation 22), where $B_0 = 0$, $n = 1$. Equations 68 and 69 need to be expressed in real form by means of Equation 24 and its equivalent for J_1. A plot of Equation 69 is shown in Figure 7 for $\Omega = 64$.

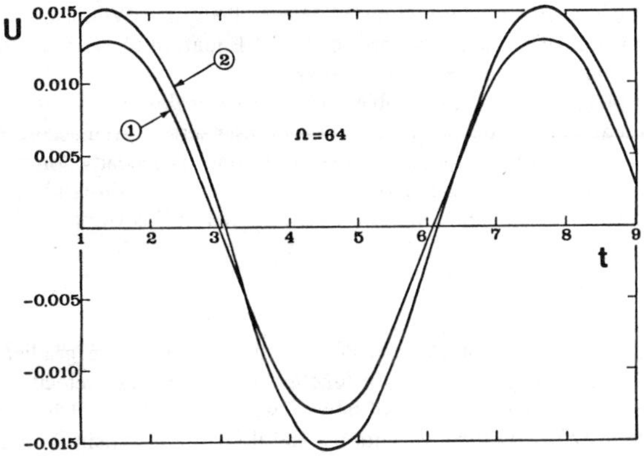

Figure 7. Oscillatory flow. Curve 1: exact solution (Equation 69); curve 2: approximation (Equation 71).

An alternative approach is the method of eigenfunction expansions. The solution of Equation 28 is

$$T_n = \frac{2}{(\Omega^2 + b_n^4)J_1(b_n)}\left[b_n \cos t + \frac{\Omega}{b_n}\sin t\right]$$

which, upon substitution into Equation 26, gives the series for u, i.e.,

$$u = 2\sum_{n=1}^{\infty} \frac{J_0(b_n r)}{(\Omega^2 + b_n^4)J_1(b_n)}\left[b_n \cos t + \frac{\Omega}{b_n}\sin t\right]$$

The instantaneous average velocity U can be obtained also from the one-dimensional momentum Equation 44 combined with Equation 49. Once the forcing function, in complex form, is substituted in the right hand side of Equation 44, the resulting differential equation for U is

$$U + \frac{\Omega}{2^2}\frac{dU}{dt} + \frac{\Omega^2}{2^2 4^2}\frac{d^2U}{dt^2} + \cdots = R_e\left\{\frac{e^{it}}{(-i\Omega)^{3/2}}\left[2J_1(\sqrt{-i\Omega}) - \sqrt{-i\Omega}\,J_0(\sqrt{-i\Omega})\right]\right\} \tag{70}$$

in which use was made of Equation 49. The solution is achieved by putting $u \sim R_e[\exp(it)]$ in the left-hand side of Equation 70. The resulting expression for U is exactly Equation 69. In this case the one-dimensional approach yields an exact result because initial transients, i.e., the homogeneous solution of u, are not taken into account. Equation 69, i.e., the exact solution for U, has a simpler counterpart in the solution that follows by substituting the quasi-steady wall shear stress, Equation 17, into the momentum equation 40; thus, the solution to Equations 40 and 17 is

$$U = \frac{1}{64 + \Omega^2}(8 \cos t + \Omega \sin t) \tag{71}$$

Equation 71 is plotted in Figure 7 for $\Omega = 64$, where it is seen that even for a relatively small value of the unsteadiness number the quasi-steady shear assumption leads to substantial error.

The velocity field, as given by Equation 68 may have a rather complex structure for high values of Ω. Asymptotic forms of Equation 68 for $\Omega \to \infty$ were developed by Sexl [3]. An asymptotic form of u, valid for $\Omega < b_1^2$, results from Equation 37. Harris, Peev, and Wilkinson [21], among others, realized an experimental study where the predictions of Equation 68 were tested, finding excellent agreement between theoretical and experimental values.

A more general presentation of the problem of oscillatory flow was made by Uchida [23]. In this, the forcing function is now defined as a general Fourier series, given according to Equation 22, where oscillating terms are superimposed on a constant one. The steady component of the flow is represented by the constant term B_0. The velocity u is then given by Equation 23. Velocity profiles, wall shear stress, and energy dissipation are also considered in this work.

FLOWS OF CLASS B

The identifying characteristic of Class B flows is that the pressure gradient depends on the motion, and, thus, the forcing function is a function of the average velocity. No general, exact methods of analysis exist for this variety of flows. Two cases of Class B flows will be presented, namely, free oscillatory flow (U-tube oscillations) and flow establishment with inflow (Figure 2).

Free Oscillatory Flow

Consider the U-shaped tube depicted in Figure 8. For $t < 0$, the ends of the liquid column inside the tube are displaced from the level $z = 0$ by an amount z_0. At $t = 0$ the pressure at both ends of the column is equalized and the column moves until a static equilibrium is reached with both ends at the level $z = 0$. According to the balance between inertia and viscous forces, the ensuing motion may be overdamped (nonscillatory) or oscillatory. It is desired to describe the motion of one end of the column, i.e., to find the function $z = z(t)$. Due to its importance in manometers and instrument lines, this problem has been investigated for many years, but only recently has an exact solution of the problem been found [24] within the limits of parallel flow theory. The linear equation of motion (Equation 12) may be considered to hold in this instance, when the tube is gently curved, and the column is long enough so that end effects are insignificant. The pressure gradient is proportional to 2z, where z depends on the motion itself, which makes this a typical instance of Class B flows.

The physical forcing function is [16]

$$\phi = \frac{2gz}{L} = \frac{2g}{L}\left(z_0 - \frac{2}{a^2}\int_0^a \int_0^t ur \, dr \, dt'\right) \tag{72}$$

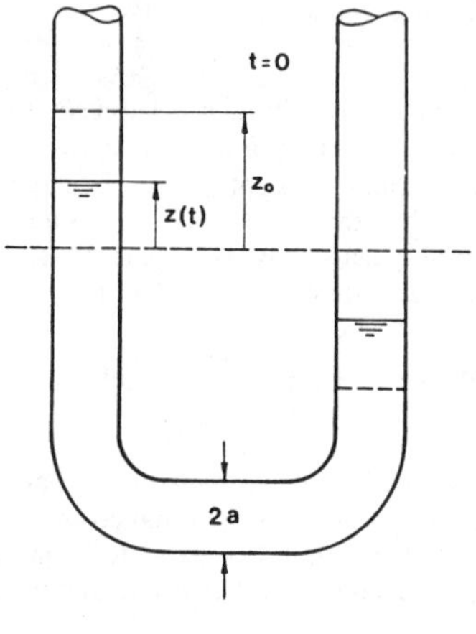

Figure 8. Definition diagram for U-tube motion.

A convenient dimensionless form of Equation 72 is obtained by defining $U_0 = 2a^2gz_0/\nu L$; $T_0 = \nu L/4ga^2$, wherefrom $\Omega = 4ga^4/\nu^2 L$. The resulting equation of motion is

$$\Omega\frac{\partial u}{\partial t} - \left(\frac{\partial^2 u}{\partial r^2} + \frac{1}{r}\frac{\partial u}{\partial r}\right) = 1 = -\int_0^1 \int_0^t ur\, dr\, dt' \tag{73}$$

subject to the initial condition $u(r, 0) = 0$. The solution of Equation 73 is rather involved and is not presented here. The reader is referred to the work of Clarion and Pelissier [4] for a study of an exact solution of it. The structure of Equation 73 makes all analytical methods outlined in this chapter very difficult to apply, except the one-dimensional approach, which yields simple and accurate results. The remaining discussion of U-tube motion is mainly concerned with the use of the one-dimensional momentum Equation 44. If the dimensionless form of z is defined by z/z_0, there holds, in dimensionless variables, $\phi = z$, and $U = -2\, dz/dt$. Once these expressions are substituted into Equation 44, there follows

$$-2\frac{d}{dt}\left(\alpha_0 z + \alpha_1\Omega\frac{dz}{dt} + \cdots\right) = \beta_0 z + \beta_1\Omega\frac{dz}{dt} + \cdots \tag{74}$$

which is an ordinary differential equation for z. The solution of Equation 73 is of the form

$$z = C\exp(-\lambda t) \tag{75}$$

where C is a constant and λ may be a complex number. Substitution of Equation 79 into Equation 74, aided by Equations 48 and 49, yields the transcendental Equation 76, viz.

$$\epsilon_n J_0(\epsilon_n)\left(\epsilon_n^4 + \frac{\Omega}{2}\right) - \Omega J_1(\epsilon_n) = 0 \tag{76}$$

in which $\epsilon_n = \sqrt{\Omega\lambda_n}$. Equation 75 defines infinite values of λ_n in terms of its infinite roots ϵ_n. For $\Omega > 23.6$ there are two complex conjugate roots of Equation 76 plus infinite real roots. Only real roots exist for $\Omega \leq 23.6$. Hence, the column motion is oscillatory for $\Omega > 23.6$, while it is overdamped for $\Omega \leq 23.6$. The final forms of Equation 75, subject to $z = 1$, $dz/dt = 0$ at $t = 0$, are, for oscillatory flow ($\Omega > 23.6$)

$$z = \left(\cos\lambda_i t + \frac{\lambda_r}{\lambda_i}\sin\lambda_i t\right)\exp(-\lambda_r t)$$

$$\epsilon_1 = \epsilon_r + i\epsilon_i = \text{complex root of Equation 76}$$

$$\lambda_1 = \frac{1}{\Omega}(\epsilon_r^2 - \epsilon_i^2); \qquad \lambda_i = \frac{2}{\Omega}\epsilon_r\epsilon_i$$

and, for overdamped motion ($\Omega \leq 23.6$)

$$z = \frac{1}{\lambda_1 - \lambda_2}\{\lambda_1\exp(-\lambda_2 t) - \lambda_2\exp(-\lambda_1 t)\}$$

where λ_1 and λ_2 are the first two roots of Equation 76. Experimental and additional information concerning U-tube motion can be found in references [4, 11, 22, 23, 24, 25, 26].

Flow Establishment with Inflow

The flow to be discussed now is basically the same physical system described in the section devoted to flow with forcing function linear in time (Figures 2 and 9). For $t < 0$, the tank inflow q_0 = pipe efflux Q_0, and, hence, the flow in the pipe is steady under a constant tank head H_0. At

$t = 0$, the tank inflow increases to a value $q > Q_0$, with which the tank level $H(t)$ rises until it reaches a final value H_∞ when both inflow and outflow are equal. It is desired to find the function $H = H(t)$.

Taking into account the initial head H_0, and the changes in head due to the constant tank inflow q and time-varying pipe efflux $Q(t)$, the dimensionless forcing function becomes

$$\phi = 1 + t - \eta \int_0^1 \int_0^t ur\,dr\,dt' \tag{77}$$

where

$$\eta = \frac{2\pi g H_0 a^4}{qL\nu} = 16\frac{H_0}{H_\infty}$$

$$U_0 = \frac{ga^2H_0}{L\nu}; \qquad T_0 = \frac{H_0A_T}{q}; \qquad \Omega = \frac{a^2q}{H_0A_T\nu}$$

The initial condition is $u(r, 0) = (1 - r^2)/4$. Equation 77 depends on the average velocity through the integral term, which in this problem accounts for pipe efflux. Mathematically speaking, Equation 77 is a rather general forcing function; special cases of Equation 77 have already been associated to flow with $\phi = 1 + t$ (negligible outflow) and U-tube motion (zero inflow).

The exact solution of Equations 12 and 77 is involved. Only the final result will be shown here, deduced from the system of Equations 31 and 32 of the method of power series. More details can be found in Letelier [1]. Thus, the exact solution of Equations 31 and 32 is [1, 5]

$$u = \frac{4}{\eta}(1 - r^2) + \sum_{n=1}^{\infty} C_n[J_0(\epsilon_n r) - J_0(\epsilon_n)]\exp(-\lambda_n t) \tag{78}$$

in which

$$C_n = \frac{V^2(\eta - 16)}{2\epsilon_n^2\eta J_0(\epsilon_n)[12V\epsilon_n^2 - (\epsilon_n^4 + V)^2]} \tag{79}$$

and $V = \eta\Omega/2$. The characteristic values ϵ_n are the real roots of

$$2VJ_1(\epsilon_n) - \epsilon_n(\epsilon_n^4 + V)J_0(\epsilon_n) \tag{80}$$

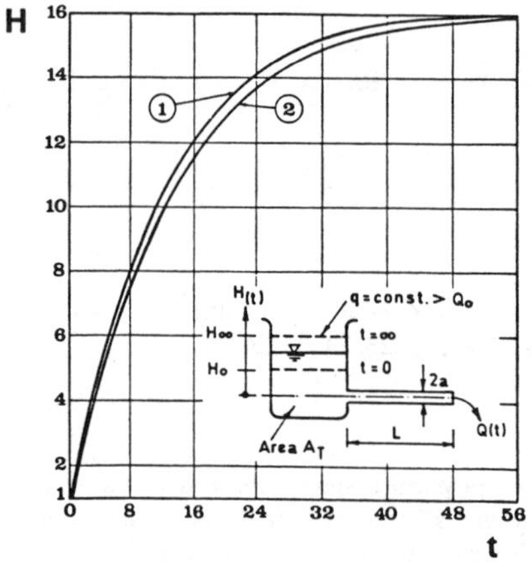

Figure 9. Flow establishment with inflow ($\Omega = 23.4$, $\eta = 1$). Curve 1: exact solution, (Equation 83); curve 2: approximation (Equation 86). (From M. F. Letelier S. and H. J. Leutheusser [5]. By permission of the American Society of Mechanical Engineers).

and the λ_n are given by $\lambda_n = \epsilon_n^2/\Omega$. Equation 80 is basically Equation 76; therefore, Equation 80 yields a physically meaningful solution only for $V \leqq 11.8$. This is valid also for the average velocity, viz.

$$U = \frac{2}{\eta} + \frac{1}{V}\sum_{n=1}^{\infty} \epsilon_n^4 J_0(\epsilon_n) C_n \exp(-\lambda_n t) \tag{81}$$

The continuity equation is

$$U = \frac{2}{\eta}\left(1 - \frac{dH}{dt}\right) \tag{82}$$

whence

$$H = 1 - \sum_{n=1}^{\infty} \epsilon_n^2 J_0(\epsilon_n) C_n [1 - \exp(-\lambda_n t)] \tag{83}$$

A plot of Equation 83 is included in Figure 3 for $\Omega = 23.4$; $\eta = 1$.

A much simpler method of solution, albeit approximate, may again be deduced from the one-dimensional equations of motion, Equations 44 and 52. Upon substitution of Equation 82 therein, these become differential equations for the nondimensional reservoir head H(t), i.e.,

$$H + \left(\frac{16}{\eta} + \frac{\Omega}{12}\right)\frac{dH}{dt} + \left(\frac{4}{\eta} + \frac{\Omega}{384}\right)\frac{d^2H}{dt^2} + \cdots = \frac{16}{\eta} \tag{84}$$

and

$$H + \frac{16}{\eta}\frac{dH}{dt} + \frac{8}{3}\frac{\Omega}{\eta}\frac{d^2H}{dt^2} + \cdots = \frac{16}{\eta} \tag{85}$$

with initial conditions $H = 1$; $dH/dt = (16 - \eta)/16$ at $t = 0$. Both Equations 84 and 85 are infinite-order ordinary differential equations for H. Equation 84 reproduces 80 when rearranged in terms of Equations 48 and 49. Since the flow is nonoscillatory, i.e., $\Omega\eta \leqq 23.6$, Equation 85 is especially appropriate because it is an asymptotic form of Equation 84 for relatively small values of Ω. The solution of both Equations 84 and 85 is

$$H = \frac{16}{\eta} + K_1 \exp(-\lambda_1 t) + K_2 \exp(-\lambda_2 t) \tag{86}$$

The coefficients K_1 and K_2 are determined with the aid of the initial conditions, λ_1 and λ_2 are the dominant roots of the characteristic polynomial obtained by substituting Equation 86 into Equation 85. Equation 86 is an approximation of Equation 83 based upon the two leading terms in its infinite series. A plot of Equation 86 appears in Figure 9. The small error entailed by the one-dimensional analysis is, as in the previous problems, due to lack of sufficient initial conditions that are necessary to determine the exact form of H.

FLOWS OF CLASS C

Flows where the mathematical structure of the forcing function is not known a priori have been grouped in this third and last class of flows. Such problems basically arise when it is not practically possible to predict the time variation of the pressure at one or two ends of a pipe. Usually, the information that is available is the time-history of the average velocity or centerline velocity some instantaneous velocity profiles, etc. In order to determine the missing information, i.e., the velocity field for all time, the wall shear stress, and the forcing function, it is necessary to employ special analytical expressions that are conveniently gathered in this section. It is understood that once ϕ is determined, the flow will likely show either Class A or Class B structure. The following discussion concerns the three kinds of initial data already mentioned, namely time-history of the average velocity, time-history of the velocity at a given value of the radius, and instantaneous velocity profiles.

Initial Data: U = U(t)

The results to be stated herein are mainly deduced from the method of power series, which shows adequate flexibility for these purposes. The forcing function can be determined in this case by means of Equation 44, which relates in exact form the average velocity to the forcing function. If the acceleration is not big, Equation 52 is a convenient alternative way for determining ϕ. Once ϕ is known, the basic momentum Equation 12 can be used to obtain u(r, t) and $\tau_w(t)$. However, it is also possible to determine u and τ directly from U(t). To this end, Equations 31 and 33 can be combined to yield

$$\beta_0 u + \beta_1 \Omega \frac{\partial u}{\partial t} + \cdots = U(1 - r^2) + \frac{\Omega}{4^2}\frac{dU}{dt}(1 - r^2) + \cdots \tag{87}$$

which is a differential equation that relates u(r, t) to U(t). Solution of Equation 87 gives the required velocity field. The wall shear stress follows from Equation 45 in similar form; Equations 42 and 43 provide an alternative method of achieving the same result.

Biological conduits, particularly blood vessels, are typical instances of pipe flow where the forcing function, for instance, of a given arterial branch, is not readily known. Consider the time-history of U given in Figure 10. This curve is a mathematical idealization of Figure 4 in reference [27], where measurements of the flow in the ascending aorta of a dog are presented. The average velocity in Figure 10 is represented by

$$U = \frac{D_0}{2} + \sum_{n=1}^{\infty} \{D_n \cos nt + E_n \sin nt\} \tag{88}$$

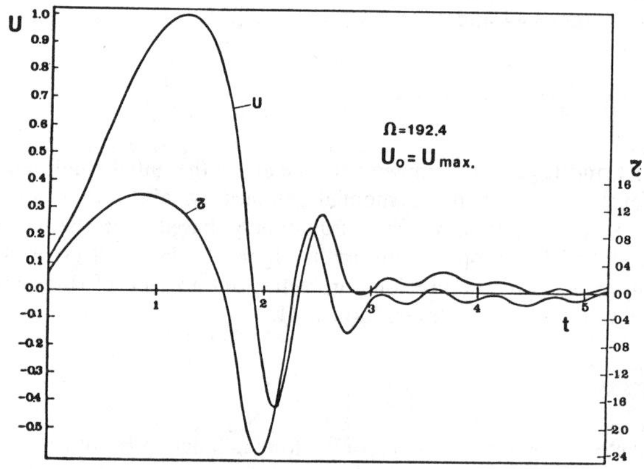

Figure 10. Prediction of wall shear stress from time-history of average velocity U (Equation 88). τ (Equation 89).

wherefrom, and using Equations 45, 49, and 50 [28, 29],

$$\tau_w = 2D_0 + \sum_{n=1}^{\infty} \frac{\xi^2/2}{\mathrm{ber}_2^2(\xi) + \mathrm{bei}_2^2(\xi)} \{[S_n \,\mathrm{bei}(\xi) + V_n \,\mathrm{ber}_2(\xi) - E_n\xi^2] \cos nt$$
$$+ [S_n \,\mathrm{ber}_2(\xi) + V_n \,\mathrm{bei}_2(\xi) + D_n\xi^2] \sin nt\} \tag{89}$$

where

$$S_n = D_n \,\mathrm{ber}_0(\xi) + E_n \,\mathrm{bei}_0(\xi)$$

$$V_n = E_n \,\mathrm{ber}_0(\xi) - D_n \,\mathrm{bei}_0(\xi)$$

$$\xi = \sqrt{n\Omega}; \qquad \Omega = \frac{\pi a^2}{\nu T_0}$$

In this T_0 is half the period. Plots of Equations 88 and 89 are included in Figure 10. Similar results were obtained by Tarbell, Chang and Hollis [30], and Letelier and Leutheusser [31].

Initial Data: $u(r_0, t)$

The data, according to Equation 31, can be written as

$$u_0(t) = u(r_0, t) = A(1 - r_0^2) + \frac{\Omega}{4^2}\frac{dA}{dt}(1 - r_0^4) + \cdots \tag{90}$$

which equation, after combined with 31, gives

$$u(1 - r_0^2) + \frac{\Omega}{4^2}\frac{\partial u}{\partial t}(1 - r_0^4) + \cdots = u_0(1 - r^2) + \frac{\Omega}{4^2}\frac{\partial u_0}{\partial t}(1 - r^4) \tag{91}$$

Equation 91 is a differential equation for $u(r, t)$ in terms of $u(r_0, t)$. In this, $0 \leqq r_0 \leqq 1$. Appropriate expressions for the forcing function and wall shear stress can also be obtained from, respectively, Equations 32, 90, and 35, viz.

$$\phi(1 - r_0^2) + \frac{\Omega}{4^2}\frac{d\phi}{dt}(1 - r_0^4) + \cdots = \alpha_0 u_0 + \alpha_1 \Omega \frac{du_0}{dt} + \cdots$$

$$\tau_w(1 - r_0^2) + \frac{\Omega}{4^2}\frac{d\tau_w}{dt}(1 - r_0^4) + \cdots = \gamma_0 u_0 + \gamma_1 \Omega \frac{du_0}{dt} + \cdots$$

Initial Data: $u(r, t_0)$

It can be shown [1] that the velocity field, and, hence, the forcing function and wall shear stress can be deduced from one single instantaneous velocity profile, provided this is not parabolic. Defining $u(r, t_0) = v(r)$, where $v(r)$ is the velocity distribution at time $t = t_0$, then

$$u(r, t) = v(r) + [f_1(r) - f_1(1)]t + [f_2(r) - f_2(1)]t^2 + \cdots \tag{92}$$

where

$$f_n = \frac{1}{n!\Omega^n} B_r^n v; \qquad B_r = \frac{\partial^2}{\partial r^2} + \frac{1}{r}\frac{\partial}{\partial r}$$

Once u is known from Equation 92, ϕ can be obtained from Equation 12 and τ_w from Equation 41.

COMPRESSIBILITY EFFECTS AND NON-PARALLEL FLOW

In the foregoing it was assumed that the flow was incompressible, i.e., pressure variation at the ends of the pipe are supposedly felt instantaneously over the whole pipe. However, in some applications, such as fluid control lines, the compressibility and flexibility of the pipe cannot be ignored in order to account for the response time involved in the transmission of pressure fluctuations along the pipe axis. Assuming flow of a slightly compressible fluid in a rigid tube, the equations of momentum and continuity are, respectively, in dimensional variables [9],

$$\frac{\partial u'}{\partial t} = -\frac{\partial P'}{\partial x} + \mu\left(\frac{\partial^2 u'}{\partial r^2} + \frac{1}{r}\frac{\partial u'}{\partial r}\right) \tag{93}$$

and

$$\frac{1}{K}\frac{\partial P'}{\partial t} + \frac{\partial v'}{\partial r} + \frac{v'}{r} + \frac{\partial u'}{\partial x} = 0 \tag{94}$$

in which u′ and v′ are the deviations of the velocities in the x and r directions from their steady-state values; P′ is the deviation of the piezometric pressure; and K is the fluid bulk modulus. In this it is assumed $u' \gg v'$. The average deviation of the velocity in the axial direction is

$$U' = \frac{2}{a^2}\int_0^a u' r\, dr$$

Assuming initial conditions $u'(x, r, 0) = 0$ and $P'(x, 0) = 0$, then it is found [9]

$$\frac{\tilde{U}'_1}{\tilde{P}'_1} = \frac{\dfrac{\tilde{U}'_2}{\tilde{P}'_2}\cos\dfrac{s\beta L}{c} + \dfrac{1}{\beta\rho c}\sin\dfrac{s\beta L}{c}}{\cos\dfrac{s\beta L}{c} - \dfrac{\tilde{U}'_2}{\tilde{P}'_2}\beta\rho c\sin\dfrac{s\beta L}{c}} \tag{95}$$

where $\tilde{P}'_1(s)$ and $\tilde{U}'_1(s)$ are the Laplace transforms of the pressure and velocity at the upstream section where $x = 0$; and $\tilde{P}'_2$, $\tilde{U}'_2$ are the corresponding Laplace transforms at the downstream end where $x = L$; the velocity of sound in the fluid is $c = (K/\rho)^{1/2}$ and β is defined by

$$\beta = \left\{\frac{2}{ia\sqrt{\dfrac{s}{\nu}}}\,\frac{J_1\left[ia\sqrt{\dfrac{s}{\nu}}\right]}{J_0\left[ia\sqrt{\dfrac{s}{\nu}}\right]} - 1\right\}^{-1/2} = \beta_r + i\beta_i \tag{96}$$

For the especially important class of oscillatory pressure fluctuations, the frequency-response equations are obtained by putting $s = i\omega$ (ω = frequency) in Equations 95 and 96. For complementary information on the subject see References 32–34.

Unsteady flow in pipes where the radius varies with x, i.e., non-parallel laminar flow, is much more difficult to deal with. Here $a = a(x)$ and the linear momentum Equation 12 no longer accurately models the phenomenon. Nevertheless, if da/dx is small, the quasi-parallel flow assumption may be acceptable, and this makes it possible to formulate an approximate mathematical model in terms of the average velocity, which now is $U = U(x, t)$. Assuming incompressible flow, the one-dimensional momentum equation is [10, 35]

$$\frac{\partial U}{\partial t} + U\frac{\partial U}{\partial x} + 2\frac{\tau_w}{a\rho} = -\frac{1}{\rho}\frac{\partial P}{\partial x} \tag{97}$$

with the continuity condition

$$\pi a^2(x)U(x, t) = Q(t)$$

If the flow is quasi-parallel, the parallel-flow law of friction can be used, in some of its versions, in Equation 97, e.g., Equations 42, 51, or 53. An application of Equation 97 and 42 for compressible flow by means of the method of characteristics is found in Zielke [10]. A solution of the problem of flow establishment in a pipe with $a = a_0\ (1 + \epsilon \cos 2\pi x/1)$ by perturbation analysis is given in Letelier and Pizarro [36].

NON-CIRCULAR PIPES

Parallel unsteady flow in non-circular ducts is governed by a linear form of the momentum equation, viz.

$$\frac{\partial u}{\partial t} - \nu \nabla^2 u = -\frac{1}{\rho}\frac{\partial P}{\partial x} \tag{98}$$

in which the Laplacian operator may be expressed in cylindrical coordinates, wherefrom Equation 12 follows, or in other coordinate systems. Flow establishment under constant pressure gradient and oscillatory flow in rectangular pipes was studied by Fan and Chao [37], where $\nabla^2 u$ is written in Cartesian coordinates. Flow establishment under constant head in annular ducts was analyzed by Yih [38] and Müller [39]. Hepworty and Rice [40] gave solutions, based upon Duhamel's theorem, to the problem of unsteady flow with arbitrary pressure gradient in conduits of circular-sector and circular-annular sector cross-sections. In this case $\nabla^2 u$ includes both the radial and tangential variations of u.

Some results, concerning the extension of the method of power series and of the one-dimensional approach to rectangular and annular sections are given in Letelier [41].

NOTATION

a	pipe radius
A	function of time in method of power series
A_T	area of tank
ber_n, bei_n	Kelvin functions
b_n	roots of J_0
B_0, B_{cn}, B_{sn}, B_n	Fourier Coefficients
$B_r(\ )$	differential operator in Equation 92
c	speed of sound
C, C_n	constant coefficients
D_0, D_n	Fourier coefficients
e	2.7183
$\exp(\ )$	exponential function
E_n	Fourier coefficients
f	friction coefficient
f_n	functions of r in Equation 92
g	acceleration due to gravity
h	vertical coordinate
H, H_0, H_∞	tank head
i	$\sqrt{-1}$
$I(\ )$	integral operator
J_0, J_1	Bessel's functions of the first kind
K_0	amplitude of oscillations, constant
K	fluid modulus of elasticity
K_1, K_2	constants in Equation 86
L	pipe length
P	fluid pressure
P, P_1, P_2	piezometric pressure
P'	deviation of P'
q, q_0	rate of volumetric inflow
Q, Q_0	rate of volumetric outflow
r, r_0	radial coordinate
$R_e(\ )$	real part
$\mathbf{R}_1, \mathbf{R}_0$	Reynolds number
s	Laplace variable
S_n	constant coefficients
t, t_0	time
t_e	time of flow establishment
t'	dummy variable

T_0	reference time
T_n	time-dependent coefficients in Equation 26
u, u_0	axial velocity
u'	deviation of axial velocity
U, U_1, U_2	average velocity
U_0	reference velocity
v	axial velocity at $t = t_0$
v'	deviation of radial velocity
V	$\eta\Omega/2$
V_n	constant coefficients
W	weight function in Equation 43
x	axial coordinate
z, z_0	displacement of U-tube column. Figure 8

Greek Symbols

α_n	constants defined by Equation 32
α'_n	constant coefficients defined in Equation 52
$\bar{\alpha}_1, \bar{\alpha}_2$	parameters in U-tube forcing function, Figure 2
β	complex constant defined in Equation 96
β_n	constants defined by Equation 34
β'_n	constant coefficients defined in Equation 53
$\bar{\beta}_1, \bar{\beta}_2, \bar{\beta}_3$	parameters in forcing function of case 4 in Figure 2
γ_n	constants defined in Equation 36
$\bar{\gamma}_1 \cdots \bar{\gamma}_4$	parameters in forcing function of case 5 in Figure 2
ϵ_n	roots of Equation 76
ξ	$\sqrt{n\Omega}$
η	dimensionless parameter in Equation 77
θ	tangential coordinate
λ, λ_n	constant factors in exponential arguments
λ'	U_f/U_i
μ	dynamic viscosity
ν	kinematic viscosity
ρ	density
τ_w	wall shear stress
ϕ	forcing function
ω	angular frequency
Ω	unsteadiness number

Subscripts

f	final state
h	homogeneous solution
i	initial state; imaginary part
p	particular solution
qs	quasi-steady
r	real part
s	steady state

Superscripts

*	dimensionless variables
~	Laplace transform

REFERENCES

1. Letelier S., M. F., "*An Approach to the Analysis of Unsteady Parallel Flow in Circular Pipes.*" Ph.D. thesis presented to the University of Toronto at Toronto, Ontario, Canada, 1979.
2. Szymanski, F. J., "Quelques Solutions Exactes des Équation de l'Hydrodynamique de Fluide Visqueux dans le Cas d'un Tube Cylindrique." *Journal des Mathématiques Pures et Appliquées*, Paris, France, Vol. 11, Series 9, 1932, pp. 67–107.
3. Sexl, T., "Ueber den von E. G. Richardson Endeckten 'Annulareffekt'." *Zeitschrift fuer Physik*, Vol. 61, 1930, pp. 349–362.
4. Clarion, C., and Pelissier, T., "A Theoretical and Experimental Study of the Velocity Distribution and Transition to Turbulence in Free Oscillatory Flow." *Journal of Fluid Mechanics*, Vol. 70, 1975, pp. 59–79.

5. Letelier S., M. F., and Leutheusser, H. J., "Unified Approach to the Solution of Problems of Unsteady Laminar Flow in Long Pipes." *ASME Journal of Applied Mechanics*, Vol. 50, No. 1, 1983, pp. 8–12.
6. Uchida, S., "The Pulsating Viscous Flow Superposed on Steady Laminar Motion of Incompressible Fluid in a Circular Pipe," *Zeitschrift für angewandte Mathematik und Physik*, Vol. 7, 1956, pp. 413–422.
7. Womersley, J. R., "Method for the Calculation of Velocity, Rate of Flow and Viscous Drag in Arteries When the Pressure Gradient is Known." *Journal of Physiology*, Vol. 127, 1955, p. 553.
8. Boyce, W. E., and DiPrima, R. C., *Elementary Differential Equations and Boundary Value Problems*, Third Edition. New York: John Wiley & Sons, 1977, pp. 544–547.
9. D'Souza, A. F., and Oldenburger, R., "Dynamic Response of Fluid Lines." *ASME Journal of Basic Engineering*, Vol. 86, No. 3, 1964, pp. 589–598.
10. Zielke, W., "Frequency-Dependent Friction in Transient Pipe Flow." *ASME Journal of Basic Engineering*, Vol. 90, No. 1, 1968, pp. 109–115.
11. Letelier S., M. F., and Leutheusser, H. J., "Skin Friction in Unsteady Laminar Pipe Flow." *ASCE Journal of the Hydraulics Division*, Vol. 102, No. HY1, 1976, pp. 41–56.
12. Stavitsky, D., and Macagno, E., "Approximate Analysis of Unsteady Laminar Flow." *ASCE Journal of the Hydraulics Division*, Vol. 106, No. HY12, 1980, pp. 1973–1980.
13. Letelier S., M. F., and Leutheusser, H. J., "Approximate Analysis of Unsteady Laminar Flow" (Discussion). *ASCE Journal of the Hydraulics Division*, Vol. 107, No. HY10, 1981, pp. 1278–1280.
14. Achard, J. L., and Lespinard, G. M., "Structure of Transient Wall-friction Law in One-dimensional Models of Laminar Pipe Flows." *Journal of Fluid Mechanics*, Vol. 113, 1981, pp. 283–298.
15. Gerbes, W., "Zur instationären, laminaren Strömung einer inkompressiblen zähen Flüssigkeit in kreiszylindrischen Rohren." *Zeitschrift für angewandte Mathematik un Physik*, Vol. 3, 1951, pp. 267–271.
16. Moraga B. N., Esteffan M. F., and Letelier S., M. F., "Pérdida de Energía en Flujo Laminar Transiente." *Contribuciones*, Santiago, Chile, No. 20, 1976, pp. 3–46.
17. Letelier S., M. F., and Hudson R., R. "Transicion Entre dos Flujos de Poiseuille por Cambio Instantáneo del Gradiente de Presiones." *Contribuciones*, Santiago, Chile, No. 27, 1977, pp. 5–11.
18. Letelier S., M. F., and Casanova S., E. "Flujo Laminar en Tuberías con Gradiente de Presiones Linealmente Dependiente del Tiempo." *Contribuciones*, Santiago, Chile, No. 27, 1977, pp. 13–20.
19. Brown, F. T., Margolis, D. L., and Shah, R. P., "Small-Amplitude Frequency Behavior of Fluid Lines with Turbulent Flow." *ASME Journal of Basic Engineering*, Vol. 19, No. 4, 1969, pp. 678–693.
20. Richardson, E. G., and Tyler, E., "The Transverse Velocity Gradient Near the Mouths of Pipes in Which an Alternating or Continuous Flow of Air Is Established." *Proceedings of the Physical Society*, Vol. 42, 1929, pp. 1–15.
21. Harris, J., Peev, G., and Wilkinson, W. L., "Velocity Profiles in Laminar Oscillatory Flow in Tubes,"*Journal of Scientific Instruments*. Vol. 2, 1969, pp. 913–916.
22. Biery, J. C., "Numerical and Experimental Study of Damped Oscillating Manometers: I. Newtonian Fluids." *American Institute of Chemical Engineers Journal*, Vol. No. 5, 1963, pp. 606–614.
23. Biery, J. C., "Numerical and Experimental Study of Damped Oscillating Manometers: II. Non-Newtonian Fluids." *American Institute of Chemical Engineers Journal*, Vol. 10, No. 4, 1964, pp. 551–557.
24. Biery, J. C., "The Oscillating Manometer: A Review of Experimental, Theoretical, and Computational Results." *American Institute of Chemical Engineers Journal*, Vol. 15, No. 4, 1969, pp. 631–634.
25. Letelier S., M. F., "*Skin Friction in Unsteady Laminar Flow.*" M. S. thesis presented to the University of Toronto. Canada, 1972.
26. Safwat, H. H., and Polder, J. v. d., "Friction-Frequency Dependence for Oscillatory Flows in Circular Pipe." *ASCE Journal of the Hydraulics Division*, Vol. 99, No. HY11, 1973, pp. 1933–1945.

27. Clark, C., and Schultz, D. L., "Velocity Distribution in Aortic Flow." *Cardiovascular Research*, Vol. 7, 1973, pp. 691–613.
28. Gutiérrez S., A. "*Determinación de la Función Forzante y Esfuerzo de Corte a Partir de Medidas de Velocidad Media en Arterias*." Engineering thesis, Department of Mechanical Engineering, University of Santiago of Chile, Santiago, Chile, 1981.
29. Vidal E., L. H., *Estudio de Aspectos Dinámicos del Flujo en Arterias a Partir de Mediciones Experimentales de la Velocidad Media Arterial.* Engineering Thesis, Department of Mechanical Engineering, University of Santiago of Chile, Santiago, Chile, 1982.
30. Tarbell, J. M., Chang, L. J., and Hollis, T. M., "A Note on Wall Shear Stress in the Aorta." *ASME Journal of Biomechanical Engineering*, Vol. 104, No. 4, 1982, pp. 343–345.
31. Letelier S., M. F., and Leutheusser, H. J., "Analytical Deduction of the Instantaneous Velocity Distribucion, Wall Shear Stress and Pressure Gradiente from Transcutaneous Measurements of the Time-varying Rate of Blood Flow." *Medical & Biological Engineering and Computing*, Vol. 19, 1981, pp. 433–436.
32. Iberall, A. S., "Attenuation of Oscillatory Pressure in Instrumental Lines," *Journal of Research*, National Bureau of Standards, Vol. 45, 1950, pp. 85–108.
33. Brown, F. T., "The Transient Response of Fluid Lines." *Journal of Basic Engineering*. Vol. 84. No. 3, 1962, pp. 550–555.
34. Goodson, R. E., and Leonard, R. G., "A Survey of Modeling Techniques for Fluid Line Transients." *ASME Journal of Basic Engineering*, Vol. 94, 1972, pp. 474–487.
35. Wylie, E. B., and Streeter, V. L., *Fluid Transients*, New York: McGraw-Hill, 1978.
36. Letelier S., M. F., and Pizarro O., L. "Establecimiento en el Tiempo de Flujo en Tuberías de Sección Variable." *Proceedings of the Second Latinoamerican Symposium of Applied Mathematics*, Río de Janeiro, Brasil, 1983.
37. Fan, C., and Chao, B. T., "Unsteady, Laminar, Incompressible Flow Through Rectangular Ducts." *Zeitschrift für angewandte Mathematik und Physik*, Vol. 16, 1965, pp. 351–360.
38. Yih, C. S., *Fluid Mechanics*, first ed. New York: McGraw-Hill, 1969, pp. 318–320.
39. Müller, W., "Zum Problem der Anlaufströmung Flüssigkeit im geraden Rohr mit Kreisring- und Kreisquerschnitt." *Zeitschrift für angewandte Mathematik und Mechanik*, Vol. 16, 1936, pp. 227–238.
40. Hepworth, H. K., and Rice, W., "Laminar Two-dimensional Flow in Conduits with Arbitrary Time-varying Pressure Gradient." *ASME Journal of Applied Mechanics*, September 1970, pp. 861–864.
41. Letelier S., M. F., "Laminar Unsteady Flow in Long Pipes: One-Dimensional Relations for Circular, Annular, Rectangular, and Parallel-plates Ducts." *Proceedings of the Ninth Canadian Congress of Applied Mechanics*, Saskatoon, Canada, 1983, pp. 499–500.

CHAPTER 19

STABILITY OF SPATIALLY PERIODIC FLOWS

K. GOTOH and **M. YAMADA**†

Research Institute for Mathematical Sciences
and
†Department of Physics, Faculty of Science
Kyoto University, Kyoto, Japan

CONTENTS

INTRODUCTION

For the two decades since Eisler published the Project Michigan Report [1] in 1962, there has been remarkable progress in the instability theory, which belongs to rather modern territory of hydrodynamics. Of all theories, those that concern the present subject are the instability of free flows, various computation techniques, the theory of nonlinear stage of transition, and the theory of spatially periodic flows. In this chapter we review the progress made in the theory of spatially periodic flows.

Instability of a laminar flow depends on the velocity distribution of the flow. A spatially periodic flow is certainly a kind of unbounded flow, or boundary-free flow, while it is different from other typical free flows such as jets, wakes, and free shear layers. In contrast to the uniform velocity distribution of free flows in a far field, the periodic flow repeats periodically its velocity profile eternally; in other words, there is no far field in this flow. The periodic flow within the fundamental period resembles a channel flow, but it does not satisfy the viscous boundary condition. The periodic flow is, therefore, expected to be different in its stability from either the other free flows or the channel flow. The two-dimensional free shear layer is known as the most unstable flow, for which it has been established that the critical Reynolds number is zero. In the case of channel flow, the value of the critical Reynolds number is of the order of several thousands. The critical Reynolds number of the two-dimensional parallel periodic flow is expected to be very small, but it would not vanish because the examples give the finite values of the critical Reynolds number.

Mathematical theory of the instability of laminar flow has been mostly established for the two-dimensional parallel flows. In this case the problem can be reduced to an eigenvalue problem of an ordinary differential equation. Extension to slightly three-dimensional or non-parallel flows has been studied, and the contribution of these effects has been evaluated, but they give only modification of the stability characteristics and the magnitude of the critical Reynolds number. Except for the spatially periodic flows, an essentially non-parallel flow can be dealt with by making full use of its periodic structure. This problem was first dealt with by Eisler, and is dealt again in this chapter by our own method.

The disturbance equation for periodic flow is a Floquet system. The Floquet system allows a set of solutions that contain a continuous parameter. This parameter leads to the continuous mode in this problem. In the development of the instability theory of laminar flows several epochs have been made, and the recent one is the discovery of the continuous mode in the unbounded flows [2–6]. It was verified earlier that any disturbance in the channel flow (i.e., bounded flows) could be given by a superposition of the discrete normal modes of disturbance [7]. It means that the eigenfunctions of the discrete modes compose a complete set for any disturbance in the bounded flow. In the case of unbounded flows, in general, the complete set of eigenfunctions consists of the discrete and the continuous modes. They are characterized by discrete and continuous eigenvalues of the stability characteristic and also by the behaviors in far field; the discrete mode vanishes at infinity, while the continuous one oscillates in finite amplitude there. In the case of the periodic flows no discrete modes can exist, and therefore all modes are continuous.

An orthodox approach to the Floquet system is to seek the solution in the form of Fourier series. In this case one must treat an infinite dimensional determinant to solve the eigenvalue problem. Instability of the sinusoidal basic flow has been investigated in this method by Eisler [1], Green [8] and Beaumont [9]. A different approach to this problem has been developed recently to seek the instability for general periodic flows [10]. Although the unbounded periodic flow is naturally concerned with phenomena in geophysics or astrophysics, it also arises in an idealization of the flow behind any periodic structure, such as a regularly spaced grid. The problems so far solved and being the subject of this paper are restricted to some fundamentally simple periodic flows, but the method will be applicable to more complicated and practical problems in applied physics and engineering.

In this chapter the problem is formulated in a general form and some properties of the Floquet system are covered. Later sections concern two-dimensional parallel flows. A sufficient condition for stability is derived using theorems for the Rayleigh equation which are extended, and the inviscid instability to the long wave disturbance and the estimation of the neutral wavenumber are presented. The critical Reynolds number of the viscous long wave disturbance is presented and results of the numerical approach are shown. An approach to the non-parallel periodic flow is also given along with comments about applying the nonlinear theory.

For the general theory of hydrodynamic stability with applications other than periodic flows see References 11, 12, and 13.

FORMULATION OF THE PROBLEM

Let us consider a three-dimensional viscous incompressible flow. The velocity $u = u(x, t)$ is governed by the Navier-Stokes equation from which, by eliminating the pressure, we have the vorticity equation:

$$\frac{\partial \omega}{\partial t} + (u\nabla)\omega = (\omega\nabla)u + \nu\,\Delta\omega \tag{1a}$$

where $\omega = \nabla \times u$ is the vorticity, x is the space variables in a Cartesian coordinate system (x, y, z), t the time, ρ is the constant density of the fluid, and ν is the kinematic viscosity. Use has been made of the continuity equation,

$$\nabla u = 0 \tag{1b}$$

A steady solution that is periodic in space is denoted by

$$u = U(x) \tag{2}$$

where

$$U(x) = U(x + a) = U(x + b) = U(x + c) \tag{3}$$

in which a, b, and c are constant vectors denoting the periods. If a, b, and c are linearly independent and if neither of them is a null vector, then the flow u = U(x) is three-dimensionally periodic. If one of them is a null vector, the flow is two-dimensionally periodic. If only one of them does not vanish, the flow is periodic only in one direction specified by the non-zero vector.

The evolution of a small disturbance $\hat{u}(x, t)$ superimposed upon the main flow (Equation 2) is governed by the linearized Equation (1),

$$\frac{\partial \hat{\omega}}{\partial t} + (U\nabla)\hat{\omega} + (\hat{u}\nabla)\bar{W} = (\bar{W}\nabla)\hat{u} + (\bar{\omega}\nabla)U + \nu\, \Delta\bar{\omega} \tag{4a}$$

$$\nabla\hat{u} = 0, \qquad \hat{\omega} = \nabla \times \hat{u}, \qquad \bar{W} = \nabla \times U \tag{4b}$$

If the time-dependence of the small disturbance is assumed to be exponential type, then

$$\hat{u}(x, t) = e^{\sigma t}\tilde{u}(x), \tag{5}$$

and Equation 4 is reduced to

$$\sigma\tilde{\omega} + (U\nabla)\tilde{\omega} + (\tilde{u}\nabla)\bar{W} = (\bar{W}\nabla)\tilde{u} + (\tilde{\omega}\nabla)U + \nu\, \Delta\tilde{\omega} \tag{6a}$$

$$\nabla\tilde{u} = 0, \qquad \tilde{\omega} = \nabla \times \tilde{u} \tag{6b}$$

where σ is to be determined as an eigenvalue of the problem. If Re(σ) (the real part of σ) is positive, the disturbance increases in time and the main flow is unstable. If Re(σ) is negative, the disturbance decreases in time and the main flow is stable. If Re(σ) vanishes, the main flow is neutrally stable.

For the periodic main flow U(x), we can find the boundary condition by the following inspection. Equation 6 is a set of linear differential equations whose coefficients are periodic functions with the periods a, b, and c. The theory of differential equations permits us to assume that each of the fundamental solutions of Equation 6 is obtainable in the form of a product of an exponential function and a periodic function with the same periods as those of coefficient functions, namely,

$$\tilde{u}(x) = e^{i\beta x}f(x) \tag{7}$$

$$f(x) = f(x + a) = f(x + b) = f(x + c) \tag{8}$$

where β is a constant vector. This fact has been known as the Floquet theorem or the Bloch theorem, though we do not give the proof to it here [14]. The form of the solution to Equation (7) gives us an idea to set the boundary condition on $\tilde{u}(x)$. If β is not real, the solution $\tilde{u}(x)$ diverges to infinity when $|x| \to \infty$ in some direction in [a, b, c], where [a, b, c] is the linear space spanned by the vectors a, b, and c. Therefore, we should adopt only the solutions with real β. Even for real β, it is emphasized that the solution $\tilde{u}(x)$ never vanishes but oscillates in space as $|x| \to \infty$ in [a, b, c]. When some of the periods a, b, and c are null, we must impose other boundary conditions in the directions normal to [a, b, c]. In the problems so far investigated the main flow is uniform along these directions, and in this case, $\tilde{u}(x)$ is required to vanish as $|x| \to \infty$ in these directions. This boundary condition will be used in the following sections. In non-uniform cases the same boundary condition seems to be plausible though we have had no example.

Substituting Equation 7 into Equation 6, we obtain the following equation for periodic function f(x),

$$\sigma g + [U(\nabla + i\beta)]g + (f \cdot \nabla)\bar{W} = [\bar{W} \cdot (\nabla + i\beta)]f + (g \cdot \nabla)U + \nu(\nabla + i\beta)^2 g \tag{9a}$$

$$(\nabla + 2\beta)f = 0, \qquad g = (\nabla + i\beta) \times f. \tag{9b}$$

Therefore, the solution (σ, f(x)) of these equations is a function of a real continuous parameter β. It can be easily seen from Equations 7 and 8 that the parameter β specifies the period of the eigenfunction $\tilde{u}(x)$. To see it for a, b, c $\neq 0$, it is convenient to make use of the following expression of β,

$$\beta = \frac{2\pi}{\{abc\}}[(b \times c)C_a + (c \times a)C_b + (a \times b)C_c] \tag{10}$$

where $\{abc\} = a(b \times c)$ and C_a, C_b and C_c are real constants. If $(C_a, C_b, C_c) = (l'/l, m'/m, n'/n)$, where l′, l, m′, m, n′, and n are integers, and l′/l, m′/m, and n′/n are simple fractions, then the eigenfunctions $\tilde{u}(x)$ has the periods la, mb, and nc. In other cases, if, for example, C_a is not a rational number, $\tilde{u}(x)$ is not periodic along the direction of the vector a. It is also seen from Equations 7, 8, and 10 that it is sufficient to consider the constants, C_a, C_b, C_c, within an interval [0, 1). When one of the periods, c, say, is zero, a similar argument holds by replacing c by $a \times b$ and putting $C_c = 0$. When two of the periods, b and c, say, are zero, we can obtain a similar result by putting $\beta = 2\pi C_a a/|a|^2$.

In many stability problems in fluid mechanics, eigenfunctions satisfy the same boundary conditions as a disturbance itself. (By "disturbance" we mean a localized perturbation of the main flow.) For example, in Rayleigh-Benard problem, Rayleigh-Taylor problem or the stability problems of parallel flow in a channel, the disturbance and the eigenfunctions must be subject to the same boundary conditions that they vanish at the walls. In contrast with these cases, the eigenfunctions in the stability problem of unbounded flows do not satisfy the same boundary condition as a disturbance, the disturbance is localized in space, but the eigenfunctions are not necessarily localized but remain finite as the space variables tend to infinity. The eigenfunctions that vanish at infinity are associated with the discrete eigenvalues and are called discrete modes. On the other hand, the eigenfunctions that do not vanish at infinity are associated with the continuous eigenvalues and are called continuous modes.

In the stability problem of spatially periodic flows, which we are concerned with, all the eigenfunctions have the form of Equation 7 and thus they do not vanish at infinity. In other words, they are all continuous modes. The fact that the eigenvalues are continuously distributed with a continuous parameter β has a simple but important physical meaning. Suppose that a critical mode is given by an eigenfunction with $\beta = \beta_0$, say, and that the flow is in a supercritical state, then there are unstable modes with the values of β close to β_0. Remembering that an irrational number of β gives an aperiodic eigenfunction $\tilde{u}(x)$, we can conclude that in a supercritical state of spatially periodic flow, there are growing disturbance with an arbitrarily long period, or with no period.

EQUATION OF TWO-DIMENSIONAL FLOW

We have mentioned the stability properties of general periodic flows. They are the direct consequences of the Floquet theorem (Equations 7 and 8), and are independent of details of velocity profiles. For stating of details of the stability properties, however, we must restrict ourselves to lower dimensionally periodic flows in the existing stability theory. In this and the following sections, we deal with the simplest unidirectional periodic flows that are periodic only in one direction normal to the stream and are uniform in other two perpendicular directions.

Take the x-axis of Cartesian coordinates along the stream and the y-axis in the direction where the main flow is periodic. The main flow and the period vector in Equations 3 and 8 are then expressed as

$$\mathbf{U}(\mathbf{x}) = (U(y), 0, 0), \qquad \mathbf{a} = \mathbf{c} = 0, \qquad \mathbf{b} = (0, b, 0) \tag{11}$$

In a strict argument, some force distribution is needed to maintain the steady periodic parallel flow, but the disturbance equation (9) is not altered in the presence of this force, and the theory has been applied to idealized unbounded flows given by extending the local periodic (or nearly periodic) velocity distribution to the entire space by taking no notice of its motive force. Nondimensionalization is made so as to give $b = 2\pi$, i.e., $U(y + 2\pi) = U(y)$. The real parameter β then becomes

$$\boldsymbol{\beta} = (0, \beta, 0) \qquad \text{where } 0 \leq \beta < 1 \tag{12}$$

The velocity is also nondimensionalized with an appropriate velocity, and the Reynolds number R is defined with these representative length and velocity and the kinematic viscosity ν.

Uniformity of the main flow in x- and z-directions permits us to assume the following form of the solution,

$$\mathbf{f}(\mathbf{x}) = (f_1(y), f_2(y), f_3(y)) \exp[i(\alpha_x x + \alpha_z z)] \tag{13}$$

Substituting Equation 13 into Equation 9 and operating $(\nabla + i\boldsymbol{\beta})$ to it, we can reduce the problem to a simpler one of $f_2(y)$:

$$(U - c)(\mathbf{D}^2 - (\alpha_x^2 + \alpha_z^2))f_z - U''f_z = \frac{1}{i\alpha_x R}(\mathbf{D}^2 - (\alpha_x^2 + \alpha_z^2))^2 f_2, \tag{14a}$$

$$f_2(y + 2\pi) = f_2(y) \tag{14b}$$

where $\mathbf{D} = d/dy + i\beta$, $U'' = d^2U/dy^2$ and $c = \sigma/i\alpha_x$. It can easily be seen from Equations 5, 7, and 13 that the value of c_r (= Re(c)) is the phase velocity along the x-axis and that if c_i (= Im(c)) > 0 or <0 for $\alpha_x > 0$, the main flow is unstable or stable, respectively. It should be noted that Equation 14a is just the Orr-Sommerfeld equation if $\mathbf{D}$ is replaced by $D = d/dy$.

In the case of the Orr-Sommerfeld equation, we can always assume $\alpha_x > 0$ without loss of generality. Care should be taken of this assumption in Equation 14 because $\mathbf{D}^* \neq \mathbf{D}$, where the asterisk denotes a complex conjugate. In the present case if $(c, \alpha_x, \alpha_z, \beta, f_2)$ is a solution of Equation 14, then $(c^*, -\alpha_x, \alpha_z, -\beta, f_2^*)$ is also a solution. Therefore, we can assume $\alpha_x > 0$ in Equation 14 so long as the equation is considered for every value of β in $[-\frac{1}{2}, \frac{1}{2}]$. Notice that this interval can be shifted arbitrarily, and reduced to $[0, \frac{1}{2}]$ when U(y) is even function of y. With this assumption and the inspection of Equation 14a, we have

$$(\alpha_x^2 + \alpha_z^2)^{1/2} R_{2c} = \alpha_x R_{3c}, \qquad \text{so that } R_{2c} < R_{3c}, \tag{15}$$

where R_{2c} is the critical Reynolds number for the two-dimensional mode ($\alpha_z = 0$), and R_{3c} is that for the three-dimensional mode ($\alpha_z \neq 0$). This means that Squire's theorem (see Reference 11, p. 155) has been extended to periodic parallel flows, and in the following we consider only the two-dimensional disturbances. Denoting f_2 and α_x by f and α respectively in Equation 14, we have

$$(U - c)(\mathbf{D}^2 - \alpha^2)f - U''f = \frac{1}{i\alpha R}(\mathbf{D}^2 - \alpha^2)^2 f \tag{16a}$$

$$f(y + 2\pi) = f(y) \tag{16b}$$

Without affecting the stability, an arbitrary real constant can be added to U(y), and this constant is chosen throughout in this paper by the condition,

$$\int_0^{2\pi} U(y)\, dy = 0 \tag{16c}$$

The eigenvalue c of Equation 16a is continuously distributed with a parameter β in **D** as stated in the previous section. Similarly the so-called neutral curve $R = R(\alpha, \beta)$ obtained by solving the equation $c_i = c_i(\alpha, R, \beta) = 0$ forms some two-dimensional area on the (R, α)-plane, which will be referred to as "neutral domain." All the neutral curves should be found within the neutral domain. Also inside the domain some growing modes exist throughout, and the critical Reynolds number is determined by the neutral curve on the border of this domain.

For channel flows between rigid walls, a universal sufficient condition for stability is well known as the Synge's theorem [15]. As in the proof of Synge's theorem, an integration of Equation 16a over one period after multiplying it by f* gives an inequality (see Reference 11, p. 161),

$$c_i \le \left\{ MI_0I_1 - \frac{1}{\alpha R}(I_2^2 + 2\alpha^2 I_1^2 + \alpha^4 I_0^2) \right\} \Big/ (I_1^2 + \alpha^2 I_0^2) \tag{17}$$

where

$$\left. \begin{aligned} I_0^2 &= \int_0^{2\pi} |f(y)|^2 \, dy, & I_1^2 &= \int_0^{2\pi} |\mathbf{D}f(y)|^2 \, dy \\ I_2^2 &= \int_0^{2\pi} |\mathbf{D}^2 f(y)|^2 dy, & M &= \max_{0 \le y < 2\pi} |U'(y)| \end{aligned} \right\} \tag{18}$$

and use has been made of the equality for arbitrary differentiable function F(y) and G(y),

$$\int_0^{2\pi} (\mathbf{D}F(y))^* G(y) \, dy = [F^*(y)G(y)]_0^{2\pi} - \int_0^{2\pi} F^*(y)\mathbf{D}G(y) \, dy. \tag{19}$$

If at least one of I_k's $(k = 0, 1, 2)$ remains finite as $R \to 0$, the inequality of Equation 17 assures the stability of the main flow in this limit.

Making use of an inequality valid for any real values of ξ and η,

$$\eta^2 I_2^2 \ge (\xi\eta + 2\eta - 4\pi^2\xi^2) I_1^2 + (\xi - 1) I_0^2 - 4\pi\xi(|f(0)|^2 + \eta|\mathbf{D}f(0)|^2) \tag{20}$$

which is derived from

$$\int_0^{2\pi} |f + \xi y \mathbf{D}f + \eta \mathbf{D}^2 f|^2 \, dy \ge 0 \tag{21}$$

we can eliminate I_2^2 from Equation 17 and have

$$c_i \le -\frac{1}{\alpha R \eta^2 (I_1^2 + \alpha^2 I_0^2)} [A_1 I_0^2 + A_2 I_0 I_1 + A_3 I_1^2 - 4\pi\xi(|f(0)|^2 + \eta|\mathbf{D}f(0)|^2)], \tag{22}$$

where

$$\left. \begin{aligned} A_1 &= \alpha^4\eta^2 - 1 + \xi \\ A_2 &= -\alpha R \eta^2 M \\ A_3 &= 2\eta + \xi\eta - 4\pi^2\xi^2 + 2\alpha^2\eta^2 \end{aligned} \right\} \tag{23}$$

A sufficient condition for stability, $c_i < 0$, is thus given by

$$\left. \begin{aligned} & A_1 > 0, \qquad A_3 > 0, \qquad 4A_1A_3 - A_2^2 > 0 \\ & \xi(|f(0)|^2 + \eta|\mathbf{D}f(0)|^2) \le 0 \end{aligned} \right\} \tag{24}$$

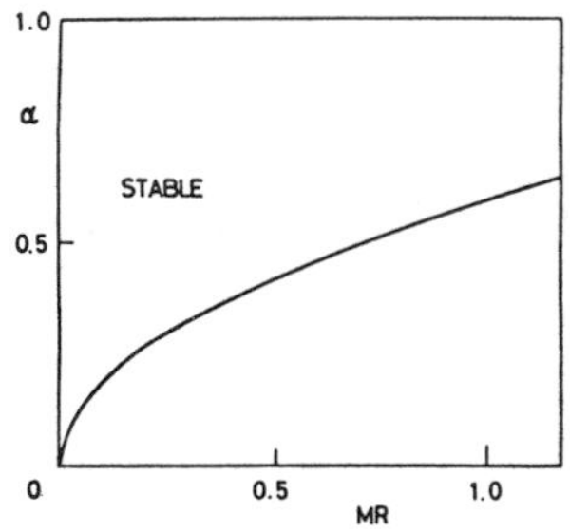

Figure 1. Sufficient condition for stability, Equation 26.

These inequalities are satisfied by

$$\xi = 0, \qquad \eta = A/\alpha^2 \tag{25a}$$

$$MR \leq \frac{2(A+1)\sqrt{2A(A-1)}}{A^2}\,\alpha^2 \tag{25b}$$

where A is an arbitrary constant larger than unity. Taking the limit of $A \to \infty$ in Equation 25b, we obtain the simplified condition for stability,

$$MR \leq 2\sqrt{2}\,\alpha^2 \tag{26}$$

This condition ensures the stability to the modes of large wavenumbers, but gives no information about the critical Reynolds number because R must go to zero as $\alpha \to 0$ to satisfy Equation 26 (Figure 1). A lower bound of the critical Reynolds number may be obtained by putting $\alpha = \beta = 0$. Optimum choice of ξ and η may yield some more marginal condition on MR, but there is not much merit in it because a formula for evaluating R_c for $\alpha = \beta = 0$ has been obtained for an arbitrary periodic flow.

INVISCID INSTABILITY

In the inviscid case ($R = \infty$), the eigenvalue problem (Equation 16) is reduced to a simpler one as

$$(U - c)(\mathbf{D}^2 - \alpha^2)f - U''f = 0 \tag{27}$$

It is easily seen that if (c, α, β, f) is a solution of Equation 27, so are $(c, -\alpha, \beta, f)$ and $(c^*, \pm\alpha, -\beta, f^*)$. This means that unstable modes exist for $\pm\beta$ simultaneously if $c_i = 0$, and that we can reduce the interval of β to $[0, \frac{1}{2}]$ even if U(y) is not an even function.

Equation 27 is just the Rayleigh equation if **D** is replaced by D. The Rayleigh equation has been investigated for about a century and many theorems have been proved. Proof of these theorems is often owing to anti-Hermitian property of $D = d/dy$ in partial integration. (We define the inner product of periodic differentiable functions F(y) and G(y) by $\int_0^{2\pi} F^*(y)G(y)\,dy$, and therefore, for example, the statement "D is anti-Hermitian" means $\int_0^{2\pi} (DF(y))^*G(y)\,dy = -\int_0^{2\pi} F^*(y)DG(y)\,dy$.) Noticing that $\mathbf{D}$ $(= d/dy + i\beta)$ with real β also has the anti-Hermitian property, we can expect that the theorems may be extended to Equation 27.

An example of this extension is given on Howard's semicircle theorem: The adjoint equation of Equation 27 is found to be

$$(\mathbf{D}^2 - \alpha^2)(U - c)f^\dagger - U''f^\dagger = 0 \tag{28}$$

or equivalently,

$$\mathbf{D}[(U - c)^2 Df^\dagger] - \alpha^2 (U - c)^2 f^\dagger = 0 \tag{29}$$

Comparison of Equation 28 with Equation 27 shows that $f^\dagger = f/(U - c)$. It is assumed that $c_i > 0$ so that $f^\dagger$ is nonsingular on real y-axis. Multiplying Equation 29 by $f^{\dagger *}$ and then integrating it over one period, we obtain

$$\int_0^{2\pi} (U - c)^2 Q \, dy = 0 \qquad \text{where } Q = |Df^\dagger|^2 + \alpha^2 |f^\dagger|^2 \tag{30}$$

The real and the imaginary parts of this equation give

$$\int_0^{2\pi} \{(U - c_r)^2 - c_i^2\} Q \, dy = 0 \qquad \text{and} \quad \int_0^{2\pi} (U - c_r) Q \, dy = 0 \tag{31}$$

Using these relations we obtain

$$\begin{aligned} 0 &\geq \int_0^{2\pi} (U - U_{min})(U - U_{max}) Q \, dy \\ &= \int_0^{2\pi} \{c_r^2 + c_i^2 - (U_{min} + U_{max}) c_r + U_{min} U_{max}\} Q \, dy \end{aligned} \tag{32}$$

and therefore

$$c_r^2 + c_i^2 - (U_{min} + U_{max}) c_r + U_{min} U_{max} \leq 0 \tag{33}$$

Rewriting this inequality as follows,

$$\{c_r - \tfrac{1}{2}(U_{max} + U_{min})\}^2 + c_i^2 \leq \{\tfrac{1}{2}(U_{max} - U_{min})\}^2 \tag{34}$$

we obtain the Howard's semicircle theorem extended to periodic velocity profiles; for unstable mode, c must lie in the semicircle (Equation 34) with $c_i > 0$.

It should be noted, however, that some of the extended theorems lose their proper meaning. A good example is the necessary condition for instability that the velocity profile should have an inflexion point (Rayleigh's inflexion-point theorem), or that $U''(U - U_s) < 0$ somewhere in the field of flow, where U_s is the value of U at which $U'' = 0$ (Fjørtoft's theorem). These theorems can be proved for periodic velocity profiles, but they are satisfied by all periodic velocity profiles, and therefore lose their meaning as a selection condition.

The stability characteristics of the long wave disturbance in Equation 27 are obtained by α-expansion of the solution. We treat here a rather special case of the expansion, in which α and β take small values of the same order, to show an existence of the unstable mode.

Let us assume

$$\beta = \beta_1 \alpha \tag{35}$$

where β_1 is a parameter independent of α, and expand f and c into power series of α:

$$f = f_0 + \alpha f_1 + \alpha^2 f_2 + \cdots \tag{36a}$$

$$c = c_0 + \alpha c_1 + \alpha^2 c_2 + \cdots \tag{36b}$$

where all the f_n's are periodic functions with the period 2π. Substituting Equations 35 and 36 into Equation 27, we get the successive equation for f_n,

$$(U - c_0) f_n'' - U'' f_n = F_n \qquad (n = 0, 1, 2, \ldots), \tag{37}$$

where the prime denotes the differentiation with respect to y and

$$F_n = \sum_{\substack{j+k=n \\ k\neq n}} c_j f_k'' + 2i\beta_1 \sum_{j+k=n-1} c_j f_k'$$
$$-2i\beta_1 U f_{n-1}' + (1+\beta_1^2)\left(U f_{n-2} - \sum_{j+k=n-2} c_j f_k \right), \tag{38}$$

in which the f_k and c_k with negative k should be put equal to zero. This rule is used throughout this chapter. Some ambiguities have been left in the f_n as for the component parallel to f_0, for removal of which a condition

$$\int_0^{2\pi} f_n^* f_0 \, dy = 2\pi \, \delta_{0n}, \tag{39}$$

is imposed on f_n without loss of generality, where δ_{0n} is Kronecker's delta, zero for $n \neq 0$. The solution is obtained as

$$f_n = (U - c_0)\left\{ \int_0^y (U - c_0)^{-2} \left[\int_0^y F_n \, dy + K_{n1} \right] dy + K_{n2} \right\} \tag{40}$$

where K_{n1} is determined by the periodic condition for f_n as

$$K_{n1} = -\int_0^{2\pi} \left[(U - c_0)^{-2} \int_0^y F_n \, dy \right] dy \Big/ \int_0^{2\pi} (U - c_0)^{-2} \, dy \tag{41}$$

The other constant K_{n2} is determined by the orthogonality condition of Equation 39, but its explicit form is not necessary in the lowest order approximation. The value of c_n is determined by the solvability condition of Equation 37,

$$2i\beta_1 \int_0^{2\pi} U f_{n-1}' \, dy - (1+\beta_1^2)\left[\int_0^{2\pi} U f_{n-2} \, dy - \sum_{j+k=n-2} c_j \int_0^{2\pi} f_k \, dy \right] = 0 \tag{42}$$

The fundamental solutions of equation 37 for $n = 0$ are

$$U - c_0 \quad \text{and} \quad (U - c_0) \int_0^y (U - c_0)^{-2} \, dy \tag{43}$$

The former solution satisfies the periodic condition, but the latter not necessarily, so that we utilize the former as f_0. The solutions f_n and c_n are then obtained successively without difficulty. The solvability condition (Equation 42) for $n = 1$ is automatically satisfied, and that for $n = 2$ becomes

$$-4\pi^2 \beta_1^2 = \int_0^{2\pi} (U - c_0)^2 \, dy \int_0^{2\pi} (U - c_0)^{-2} \, dy \tag{44}$$

which gives c_0 for given value of β_1. Especially when $\beta_1 = 0$, Equation 44 becomes

$$\int_0^{2\pi} (U - c_0)^2 \, dy = 0, \quad \text{or} \quad \int_0^{2\pi} (U - c_0)^{-2} \, dy = 0. \tag{45}$$

The former equation gives

$$c_0 = \pm\left(\frac{1}{2\pi} \int_0^{2\pi} U^2 \, dy \right)^{1/2} i = \pm\sqrt{\langle U^2 \rangle} i, \tag{46}$$

where the bracket $\langle \ \rangle$ means the average of the content over one period. Equation (46) means that the unstable mode exists for small values of α and β for an arbitrary periodic flow in the inviscid case. The latter in Equation 45 is nothing but the periodic condition for the second solution in Equation 43, and therefore is the condition for existence of another unstable mode. Equations 44, 45, and 46 were obtained by Beaumont [9]. His analysis was similar to that of Floquet's theory, while the above derivation is based on the solvability condition of Equation 37.

The instability to the long wave (Equation 46) suggests the neutral mode with some moderate wavenumber α_s, say, unstable for $\alpha < \alpha_s$ and stable for $\alpha > \alpha_s$. It is rather easy to find α_s numerically, for example, by Fourier spectral method. For approximate evaluation of α_s in some cases, however, it is advantageous to make use of a resemblance between Equation 27 and the Schroedinger equation. For a real value of c, Equation 27 is just the Schroedinger equation with the periodic potential $U''/(U - c)$. In the latter equation β represents the wavenumber of the wave function and $-\alpha_s^2$ the energy eigenvalue. The value of α_s depends continuously on β and the range of $-d_s^2$ is called the energy band, which is a fundamental concept in solid-state physics.

We consider only the neutral mode. Assume the existence of a real value of c such that $U''/(U - c)$ has no singularity and keeps its sign negative in the fundamental period of y, and fix c by this condition.

Since the energy eigenvalue for $\beta = 0$ is evaluated by

$$\int_0^{2\pi} \left\{ |f'|^2 + \frac{U''}{U - c} |f|^2 \right\} dy, \tag{47}$$

if we substitute U(y) for f(y), the integral takes a definite negative value:

$$c^2 \int_0^{2\pi} \frac{U''}{U - c} dy \qquad (<0) \tag{48}$$

and at least one negative eigenvalue is guaranteed by the variational principle. In the case of symmetric U(y), we expand the potential into an even power series of y,

$$\frac{U''}{U - c} = B_0 + B_1 y^2 + B_2 y^4 + \cdots \tag{49}$$

In the simplest approximation in which the first two terms of the expansion (Equation 49) are taken into consideration, the solution of the eigenvalue problem is the same as that of harmonic oscillator in quantum mechanics. This solution is improved by taking the quartic term in Equation 49 into account as a perturbation [10], namely,

$$\alpha_{sn}^2 = B_0 + (2n + 1)B_1^{1/2} + \frac{3B_2}{4B_1}(2n^2 + 2n + 1) \qquad (n = 0, 1, 2, \ldots, n_0), \tag{50}$$

where only a finite number of the solutions are physically permissible because α_{sn}^2 must be positive. This approximation is appropriate when the profile of the main flow is a superposition of widely separated flows, because in this case the eigenfunction takes significant values only in the region where the approximation (Equation 49) is valid.

In the case of $\beta \neq 0$, the neutral wavenumber varies continuously with β (so that for some time we denote $\alpha_s = \alpha_s(\beta)$). The same situation is encountered in band theory in solid state physics. There it has been proved [16] that the energy $-\alpha_s^2$ is a monotone function of β on $[0, \frac{1}{2}]$. In other words, the range of α where a neutral mode exists for some β is between the neutral wavenumbers $\alpha_s(0)$ and $\alpha_s(\frac{1}{2})$.

The dependence of the neutral wavenumber on β is studied in detail by β-expansion of the solution. Substitution of the following expansions:

$$f = f_0 + \beta f_1 + \beta^2 f_2 + \cdots \tag{51a}$$

$$\alpha_s^2 = A_0 + \beta A_1 + \beta^2 A_2 + \cdots \tag{51b}$$

into Equation 27 leads to the equation for f_n,

$$L(f_n) \equiv f_n'' - A_0 f_n - \frac{U''}{U - c} f_n = F_n \qquad (n = 0, 1, 2, \ldots) \tag{52}$$

where

$$F_n = -2if_{n-1}' + f_{n-2} + \sum_{\substack{j+k=n \\ k \neq n}} A_j f_k \tag{53}$$

and f_n's are as before periodic with the period 2π and the normalization (Equation 39) is again adopted. Noticing that the homogeneous equation $L(f_0) = 0$ is the eigenvalue problem for $\beta = 0$, we choose a real periodic solution f_{01}, say. Since the operator L is self-adjoint, the solvability condition of Equation 52 is

$$\int_0^{2\pi} f_{01} F_n \, dy = 0 \tag{54}$$

Under this condition, the solution of Equation 52 is successively obtained as

$$f_n = f_{02}\left[\int_0^y F_n f_{01} \, dy + k_{n2}\right] - f_1\left[\int_0^y F_n f_{02} \, dy + K_{n1}\right] \tag{55}$$

where f_{02} is the solution of $L(f_0) = 0$, which is linearly independent of f_{01}. The solutions f_{01} and f_{02} have been normalized so as to make their Wronskian unity, and K_{n1} and K_{n2} are integration constants to be determined by the periodic condition and the normalization condition for f_n. The coefficients A_n are determined from the solvability condition of Equation 54:

$$A_1 = 0, \qquad A_2 = -f_{01}(0)/[\delta f_{02}\langle f_{01}^2\rangle] \tag{56}$$

where

$$\delta f_{02} = f_{02}(2\pi) - f_{02}(0) \tag{57}$$

Thus the expression of the neutral wavenumber for small value of β is given by

$$\alpha_s(\beta) = \alpha_s(0) + \frac{A_2}{2\alpha_s(0)} \beta^2 \tag{58}$$

where the value of β^2 is at most $\frac{1}{4}$.

For the largest value of $\alpha_s(0)$, it has been known that the eigenfunction $f_{10}(y)$ does not vanish in the interval $[0, \frac{1}{2}]$ (e.g., see [17]), which enables us to put

$$f_{02} = f_{01} \int_0^y f_{01}^{-2} \, dy \tag{59}$$

and therefore

$$A_2 = -1/[\langle f_{01}^2\rangle\langle f_{01}^{-2}\rangle] \tag{60}$$

Thus, remembering that $\alpha_s(\beta)$ is a monotone function of β, we can conclude that $\alpha_s(0) > \alpha_s(\frac{1}{2})$. The eigenfunction for the second-largest value of α_s, however, vanishes somewhere on $[0, 2\pi]$, and Equation 60 is inapplicable. In such a case, we must go back to Equation 58 for the evaluation of A_2.

INSTABILITY TO VISCOUS LONG WAVE MODE OF DISTURBANCE

In this section, we continue to discuss the Orr-Sommerfeld type equation (Equation 16). The asymptotic behavior of the neutral curves in the long wave limit, i.e., $\alpha \to 0$, depends on the main flow. For parallel flows with rigid walls, R goes to infinity as $\alpha \to 0$ [15], and to zero for free shear layers whose velocity distribution is uniform asymptotically [18]. A different behavior has been found for the parallel periodic flow: R converges to a finite value R_0 as $\alpha \to 0$ on a neutral curve with $\beta = 0$ [10]. This is disclosed by α-expansion method as in the following.

Expand the quantities in Equation 16 into power series in the wavenumber,

$$f = f_0 + \epsilon f_1 + \epsilon^2 f_2 + \cdots, \tag{61a}$$

$$R = R_0 + \epsilon R_1 + \epsilon^2 R_2 + \cdots, \tag{61b}$$

$$c = c_0 + \epsilon c_1 + \epsilon^2 c_2 + \cdots, \tag{61c}$$

where $\epsilon = i\alpha$, all the f_n's are periodic functions with the period 2π. When the expansion (Equation 61) is considered along the neutral curves, the coefficients R_n and c_n are assumed to be real for even n and pure imaginary for odd n. Rewriting Equation 16a in the form,

$$f^{iv} = \epsilon R[(U - c)f'' - U''f] - 2\epsilon^2 f'' + \epsilon^3 R(U - c)f - \epsilon^4 f \tag{62}$$

and substituting Equation 61 into Equation 62, we get successive equations for f_n,

$$\begin{aligned} f_n^{iv} = & \sum_{j+k=n-1} R_j\left[Uf_k'' - U''f_k - \sum_{p+q=k} c_p f_q''\right] \\ & - 2f_{n-2}'' + \sum_{j+k=n-3} R_j\left[Uf_k - \sum_{p+q=k} c_p f_q\right] - f_{n-4} \qquad (n = 0, 1, 2, \ldots) \end{aligned} \tag{63}$$

where the f_k, R_k and c_k with negative k should be put equal to zero. The solutions f_n are normalized by Equation 39.

It is easily found that $f_0 = 1$. Noticing that it is also the solution of the adjoint homogeneous equation (63), we get by the familiar procedure the solvability condition of Equation 63,

$$\sum_{j+k=n-3} R_j\left[\int_0^{2\pi} Uf_k\, dy - 2\pi c_k\right] = 2\pi\delta_{n4} \tag{64}$$

From Equations 63 and 64, the f_n, R_n and c_n are successively obtained, for example,

$$f_0 = 1 \tag{65a}$$

$$f_1 = -R_0 J^2(U) \tag{65b}$$

$$f_2 = R_0^2 J^4(U''J^2(U) - U^2) \tag{65c}$$

$$R_0 = 1/\sqrt{\langle (J(U))^2 \rangle} \tag{66a}$$

$$R_1 = 0, \tag{66b}$$

$$R_2 = -\frac{R_0^3}{2}[\langle (J^2(U))^2 \rangle + \langle J^2(U) J^2(Uf_2'' - U''f_2) \rangle] \tag{66c}$$

$$c_0 = 0 \tag{67a}$$

$$c_1 = 0 \tag{67b}$$

$$c_2 = \langle Uf_2 \rangle = -R_0^2 \langle J^3(U) J(U'' J^2(U) - U^2) \rangle \tag{67c}$$

where J(F) denotes a function of y that is defined for arbitrary continuous function F(y) by

$$J(F)(y) = \int_0^y F(y)\,dy - \left\langle \int_0^y F(y)\,dy \right\rangle \tag{68}$$

and the bracket $\langle\ \rangle$ means as before an average of the content over one period, and

$$J^0(F) = F, \qquad J^{n+1}(F) = J(J^n(F)) \qquad (n = 0, 1, 2, \ldots) \tag{69}$$

It can be proved by mathematical induction that each f_n is an even function if U(y) is so.

This is all of the proof to the theorem that for periodic velocity profile, R converges to a finite value R_0 as $\alpha \to 0$ on the neutral curve with $\beta = 0$. From this theorem we can conclude that the periodic parallel flow is unstable at least for the Reynolds number larger than R_0. Equation 66a is used for evaluating the limiting Reynolds number R_0. From Equation 66a and the inequalities

$$\langle (J(U))^2 \rangle \leqq \langle (J'(U))^2 \rangle = \langle U^2 \rangle \leqq \langle U'^2 \rangle \tag{70}$$

we have

$$MR_0 \geq 1, \tag{71}$$

which supplements the sufficient condition for stability (Equation 26).

It should be noted that this theorem does not deny the existence of the neutral curves with $\beta = 0$ on which $R \to \infty$ as $\alpha \to 0$, because the expansion (Equation 61) implicitly assumes the existence of a finite value of R_0. Numerical calculation in fact shows that there are two types of the neutral curves.

For the neutral curves with $\beta \neq 0$, general analysis has not yet been established. We show below an asymptotic analysis for a periodic flow for which U(y) is an even function of y. In contrast with the case of $\beta = 0$, the neutral curves with small but nonzero value of β have the asymptotic behaviors similar to those of bounded flows: $R \to \infty$ on the lower branch as $\alpha \to 0$.

Here, to evaluate the growth rates of the long wave disturbance with a small value of β, define anew a small parameter $\epsilon = \alpha R$ and put

$$\alpha = \alpha_1 \epsilon, \qquad \beta = \beta_1 \epsilon \tag{72}$$

where α_1 and β_1 are new parameters independent of αR. Substituting Equation 72 and the expansions:

$$f = f_0 + \epsilon f_1 + \epsilon^2 f_2 + \cdots, \tag{73}$$

$$c = c_0 + \epsilon c_1 + \epsilon^2 c_2 + \cdots, \tag{74}$$

into Equation 16, we obtain successive equations for f_n,

$$\begin{aligned} f_n^{iv} = {} & -i \sum_{j+k=n-1} c_j f_k'' + 2\beta_1 \sum_{j+k=n-2} c_j f_k' + i(\alpha_1^2 + \beta_1^2) \sum_{j+k=n-3} c_j f_k - 4i\beta_1 f_{n-1}''' \\ & + i(Uf_{n-1}'' - U'' f_{n-1}) + 2(\alpha_1^2 + 3\beta_1^2) f_{n-2}'' - 2\beta_1 U f_{n-2}' - i(\alpha_1^2 + \beta_1^2) U f_{n-3} \\ & + 4i\beta_1(\alpha_1^2 + \beta_1^2) f_{n-3}' - (\alpha_1^2 + \beta_1^2)^2 f_{n-4} \end{aligned} \tag{75}$$

The same condition as Equation 39 is adopted to normalize the solution as before. When U(y) is an even function of y, there is an even solution f(y) for $\beta = 0$ as shown in Equation 65. Therefore, each of the f_n is put in the form

$$f_n = g_n + \beta_1 h_n \tag{76}$$

where g_n is even and h_n odd. According to the following two limiting cases.

$$\begin{aligned} &\text{case (i)} \quad \beta_1 \to 0 \quad \text{then } \alpha_1 \to 0 \\ &\text{case (ii)} \quad \alpha_1 \to 0 \quad \text{then } \beta_1 \to 0 \end{aligned} \tag{77}$$

the solvability condition becomes

$$\text{case (i)} \quad 2\pi c_{n-3} - \int_0^{2\pi} U g_{n-3}\, dy = 0 \tag{78}$$

$$\text{case (ii)} \quad 2\pi c_{n-3} - \int_0^{2\pi} U g_{n-3}\, dy + 2i \int_0^{2\pi} U h'_{n-2}\, dy = 0 \tag{79}$$

while in both cases the equations for g_n and h_n become

$$g_n^{iv} = -i \sum_{j+k=n-1} c_j g_k'' + i(U g_{n-1}'' - U'' g_{n-1}) \tag{80}$$

$$h_n^{iv} = -i \sum_{j+k=n-1} c_j h_k'' + i(U h_{n-1}'' - U'' h_{n-1}) + 2 \sum_{j+k=n-2} c_j g_k' - 4i g_{n-1}''' - 2U g_{n-2}'. \tag{81}$$

The solution of Equations 78-81 is successively obtained as

$$g_0 = 1 \tag{82a}$$

$$g_1 = -iJ^2(U) \tag{82b}$$

$$g_2 = J^4(U^2 - U''J^2(U)) \tag{82c}$$

$$h_0 = 1 \tag{83a}$$

$$h_1 = 0 \tag{83b}$$

$$h_2 = -4J^3(U) \tag{83c}$$

$$c_0 = 0 \tag{84a}$$

$$c_1 = \begin{cases} i/R_0^2 & \text{(case (i))} \\ -7i/R_0^2 & \text{(case (ii))} \end{cases} \tag{84b}$$

where R_0 is defined by Equation 66a [10]. Referring to Equation 84b the eigenvalue c_1 is different according to the order of taking the limits: the mode in case (i) is a growing mode, while that in case (ii) is a decaying mode. The former is consistent with Equation 66a. The latter shows that $R(\alpha, \beta)$ is larger than $1/\alpha$ for small values of α and β which satisfy the inequality $\beta \gg \alpha$. In other words, the neutral curve goes to infinity as $\alpha \to 0$.

EXAMPLES OF PARALLEL PERIODIC FLOWS

The analytical approach mentioned in the preceding sections is based on the parameter expansion, and cannot give information on stability characteristics for moderate values of the parameter. In this section, we present the overall structure of neutral curves (or neutral domains) for simple periodic velocity profiles to handle the eigenvalue problems.

We first show three examples in the inviscid case, two broken-line velocity profiles and a sinusoidal one [9], and then discuss the viscous problem for one-parameter family of velocity profiles that include the sinusoidal flow as a limiting case [10].

Triangular Velocity Profile

Consider the velocity profile;

$$U(y) = \begin{cases} 2y/\pi & (0 < y < \pi/2) \\ 2 - 2y/\pi & (\pi/2 < y < 3\pi/2) \\ 2y/\pi - 4 & (3\pi/2 < y < 2\pi) \end{cases} \tag{85}$$

with $U(y + 2\pi) = U(y)$. Because Equation 85 has discontinuities in its first derivative, continuity conditions should be imposed on the presssure and the normal velocity:

$$\left[\!\left[(U - c)\left(\frac{\partial}{\partial y} + i\beta\right)f - U'f \right]\!\right] = [\![f/(U - c)]\!] = 0 \tag{86}$$

where the square brackets denote the difference across the discontinuity of its content. The eigenvalue problem, Equations 27, 85, and 86, can be solved analytically and the eigenvalue of c is obtained as

$$c^2 = 1 + \frac{4(1 - \cosh 2\pi\alpha + \pi\alpha \sinh 2\pi\alpha)}{(\pi\alpha)^2(\cos 2\pi\beta - \cosh 2\pi\alpha)}. \tag{87}$$

The stability curves obtained from Equation 87 are depicted in Figure 2. For a fixed value of α, the disturbance of smaller value of β is more unstable, and the disturbance with the same period as the main flow ($\beta = 0$) is unstable for $\alpha < 0.76$.

Square Velocity Profile

Take a flow where U(y) is given by

$$U(y) = \begin{cases} 1 & (0 < y < \pi/2) \\ -1 & (\pi/2 < y < 3\pi/2) \\ 1 & (3\pi/2 < y < 2\pi) \end{cases} \tag{88}$$

with $U(y + 2\pi) = U(y)$. The eigenvalue of c is then given by

$$c_r = \pm\left\{\frac{1 - \cos 2\pi\beta}{\cosh 2\pi\alpha - \cos 2\pi\beta}\right\}^{1/2}, \qquad c_i = \pm\left\{\frac{\cosh 2\pi\alpha - 1}{\cosh 2\pi\alpha - \cos 2\pi\beta}\right\}^{1/2} \tag{89}$$

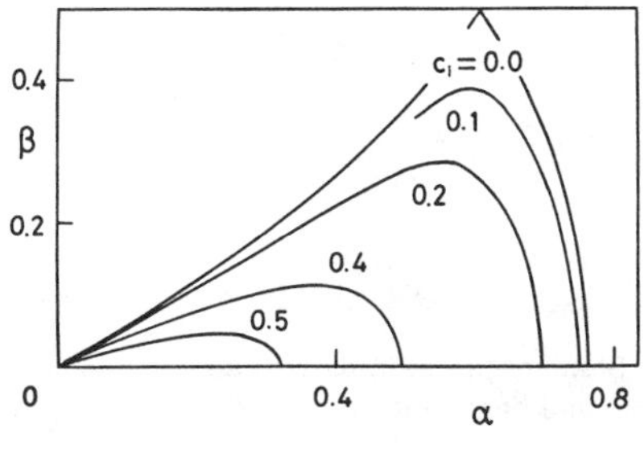

Figure 2. Stability curves for triangular velocity profile, Equation 85. Part of the $c_i = 0.1$ curve has been omitted for clarity [9].

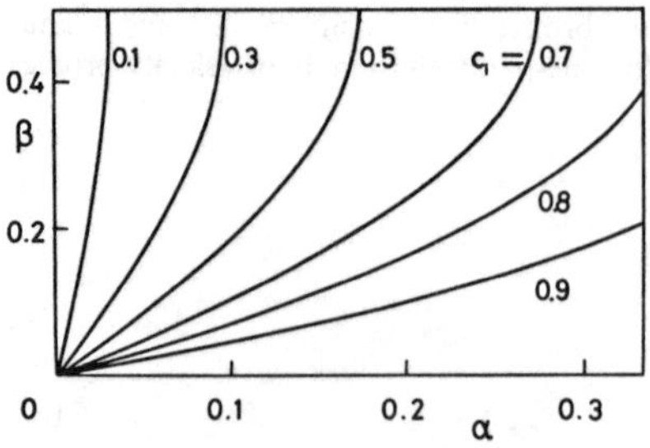

Figure 3. Stability curves for rectangular velocity profile, Equation 88 [9].

For a fixed value of α, the most unstable mode is given by the disturbance with $\beta = 0$ as before. But in contrast to triangular velocity profiles described earlier, unstable modes are possible for all wavenumbers, and $c_i \to \pm i$ as $\beta \to 0$ for all α. It should be remarked that $c \to \pm i$ as $\alpha \to \infty$, which are the eigenvalues for Helmholz instability of a vortex sheet at each discontinuity of Equation 88.

Sinusoidal Velocity Profile

The stability of the sinusoidal flow $U(y) = \sin y$ has been treated by several authors. It is clear that a solution of the inviscid eigenvalue problem is given by

$$f = 1, \qquad c = 0, \qquad \alpha^2 + \beta^2 = 1 \tag{90}$$

Numerical calculation (Figure 4) shows that this simple solution gives the marginal mode. For a fixed value of α, the most unstable mode is given, as well as in the previous cases, by the disturbance with $\beta = 0$.

One-Parameter Family of Velocity Profiles in the Viscous Case

The effect of various profiles has been investigated by using the one-parameter family of velocity distributions:

$$\begin{aligned} U(y) &= C(\gamma)\left\{1 - \frac{\gamma}{\sqrt{\pi}} \sum_{n=-\infty}^{\infty} \exp\left[-\left\{\frac{\gamma}{2\pi}(y - 2n\pi)\right\}^2\right]\right\} \\ &= C(\gamma)\left\{1 - \upsilon_3\left(\frac{y}{2}\middle| i\frac{\pi}{\gamma^2}\right)\right\} \\ &= -2C(\gamma)\sum_{n=1}^{\infty} \exp\left[-\left(\frac{\pi n}{\gamma}\right)^2\right]\cos ny \end{aligned} \tag{91}$$

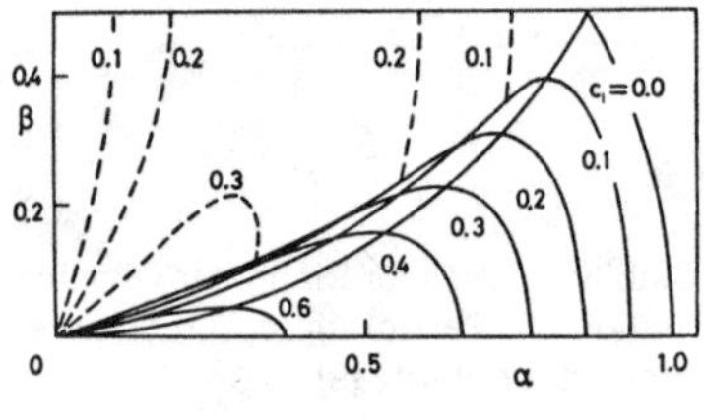

Figure 4. Stability curves for $U(y) = \sin y$. Solid lines have $c_r = 0$ and dashed lines have $c_r \neq 0$. [9]

where γ is a positive parameter that varies from zero to infinity, c is a constant depending on γ, and ϑ_3 is the theta function of the third type [19]. The constant c is fixed by the normalization condition $U_{max} - U_{min} = 1$ as

$$C(\gamma) = \frac{1}{4\sum_{n=0}^{\infty} \exp\left[-\frac{\pi^2}{\gamma^2}(2n+1)^2\right]} \tag{92}$$

This velocity profile has the asymptotic forms with respect to γ:

$$U(y) \sim \begin{cases} \frac{\sqrt{\pi}}{\gamma} - \exp\left[-\left(\frac{\gamma}{2\pi}y\right)^2\right] & (\gamma \to \infty, y \sim 0) \quad (93a) \\ -\frac{1}{2}\cos y \quad (\gamma \to 0) & \quad (93b) \end{cases}$$

which express a sparsely spaced wake (Equation 93a) and a sinusoidal flow (Equation 93b).

Examples of the neutral curves (neutral domains) are shown in Figure 5 for $\gamma = \pi$ and in Figure 6 for $\gamma = 2\pi$. In each case, there are two domains in the (R-α)-plane. The left neutral domain (called the first domain hereafter) in Figures 5 and 6 consists of the neutral curves with all $\beta \in [0, \frac{1}{2}]$, where the curve with $\beta = 0$ is on the left border of the domain, while that with $\beta = \frac{1}{2}$ is near the right border. The right neutral domain (the second domain) also consists of the neutral curves with all $\beta \in [0, \frac{1}{2}]$, the distribution order of which is, however, converse to that of the first domain. Both neutral domains rapidly become bigger as γ varies from 2π to π, that is, as the effects of neighboring wakes are intensified. Growing modes are found everywhere to the right of the neutral curves. The neutral curves in the second domain indicate a distinct family of neutral modes. On each of the neutral curves in both domains, α approaches some finite non-zero value at one end but zero at the other end.

As shown in the previous section, Reynolds number converges to a non-zero finite value in the long wave limit on the neutral curve with $\beta = 0$ in the first domain. The limiting Reynolds numbers are 2.828 for $\gamma = \pi$ and 3.129 for $\gamma = 2\pi$, which agree perfectly with the values analytically obtained from Equation 66a (Figure 7):

$$R_0 = \sqrt{8}\frac{\sum_{n=0}^{\infty} \exp\left[-\frac{\pi^2}{\gamma^2}(2n+1)^2\right]}{\left[\sum_{n=1}^{\infty} \frac{1}{n^2}\exp\left[-\frac{2\pi^2}{\gamma^2}n^2\right]\right]^{1/2}} \tag{94}$$

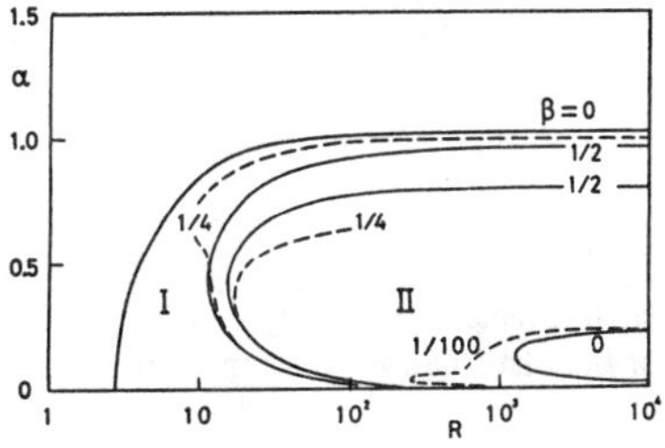

Figure 5. Neutral stability curves for U(y) in Equation 91 with $\gamma = \pi$. I: the first neutral domain, II: the second neutral domain [10].

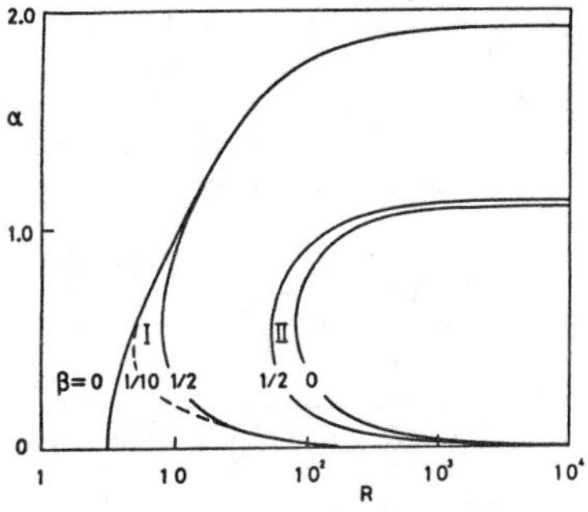

Figure 6. Neutral stability curves for U(y) in Equation 91 with $\gamma = 2\pi$. I: the first neutral domain, II: the second neutral domain [10].

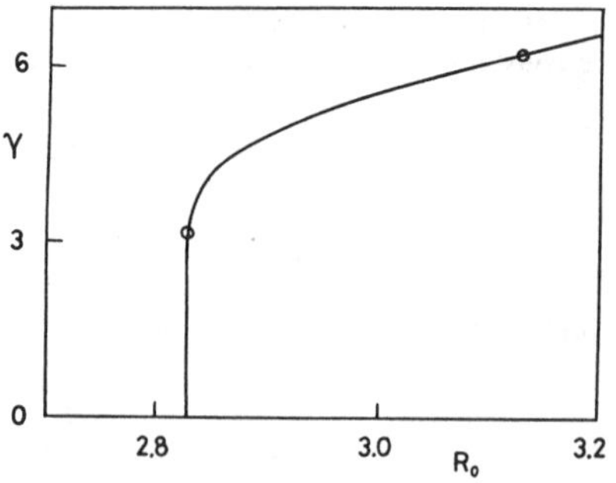

Figure 7. Limiting Reynolds number R_0 for U(y) in Equation 91. Two circles are from numerical calculation in Figures 5 and 6 [10].

Equation 94 is reduced asymptotically to

$$R_0 \sim \begin{cases} \sqrt{8}\left(1 - \frac{1}{8}\exp\left(-\frac{6\pi^2}{\gamma^2}\right) + \exp\left(-\frac{8\pi^2}{\gamma^2}\right) + \cdots\right) \quad (\gamma \to 0) & \text{(95a)} \\ 3\gamma^2/\pi^3 \quad (\gamma \to \infty) & \text{(96b)} \end{cases}$$

Thus, in the limit of small values of γ, R_0 approaches $\sqrt{8}$. This value for sinusoidal velocity profile was obtained earlier by Eisler [1] and Green [8] ($\sqrt{2}$ in their papers owing to different scaling) by making use of a special relation between the Fourier components.

In every case calculated, the critical Reynolds number $R_c = \min_{\alpha,\beta} R(\alpha, \beta)$ is given for $\alpha = \beta = 0$. This is consistent with Green's comment that the most unstable mode would have the same period as the main flow [8]. The critical mode was antisymmetric ($f(-y) = f(y)$), which is consistent with the perturbation expansion (Equation 65).

In the cases where $\beta \neq 0$, the asymptotes of the lower branch of the neutral curves go to infinity. This means that, in the long wave limit, the unstable modes are only those with the same period as the main flow.

NON-PARALLEL PERIODIC FLOW

In contrast with the case of parallel flows, the theory for non-parallel flows is complicated, because the disturbance equation is not reduced to an ordinary differential equation. Therefore, the part depending on computer may be bigger in the approach to the problem. If some Reynolds number of the neutral mode is obtained analytically as in the case of parallel periodic flow, it will work as a key in numerical approach.

Although the Squire's theorem has not (cannot, maybe) been extended to this case, if we restrict, for simplicity, our interest to the two-dimensional flow, we can introduce the stream function, and the disturbance equation is obtained as

$$\left(\frac{\partial}{\partial t} - \frac{1}{R}\Delta\right)\Delta\psi + \frac{\partial(\Delta\bar{\psi}, \psi)}{\partial(x, y)} + \frac{\partial(\Delta\psi, \bar{\psi})}{\partial(x, y)} = 0 \tag{96}$$

where $\bar{\psi}(x, y)$ is the stream function of the main flow, which satisfies $\bar{\psi}(x + 2\pi, y) = \bar{\psi}(x, y + 2\pi) = \bar{\psi}(x, y)$, and $\Delta = \partial^2/\partial x^2 + \partial^2/\partial y^2$. The stream function of the disturbance is assumed to be

$$\left.\begin{aligned} &\psi = e^{\sigma t} e^{i(\beta x + \beta' y)} f(x, y) \\ &f(x + 2\pi, y) = f(x, y + 2\pi) = f(x, y) \end{aligned}\right\} \tag{97}$$

as cited earlier. Taking the earlier result into account, we seek a solution for $\beta' = 0$. Substitution of the following expansion in terms of $\epsilon = i\beta$:

$$f = f_0 + \epsilon f_1 + \epsilon^2 f_2 + \cdots \tag{98a}$$

$$R = R_0 + \epsilon R_1 + \epsilon^2 R_2 + \cdots \tag{98b}$$

$$\sigma = \sigma_0 + \epsilon\sigma_1 + \epsilon^2\sigma_2 + \cdots \tag{98c}$$

reduces Equation 96 to

$$\begin{aligned}
&[\Delta^2 - \sigma_0 R_0 \Delta - R_0\{(\bar{\psi}_y \partial_x - \bar{\psi}_x \partial_y)\Delta - ((\Delta\bar{\psi}_y)\partial_x - (\Delta\bar{\psi}_x)\partial_y)\}]f_n \\
&\quad = -4\Delta\partial_x f_{n-1} - 2(\Delta + 2\partial_x^2)f_{n-2} - 4\partial_x f_{n-3} - f_{n-4} \\
&\qquad + \sum_{\substack{j+k+m=n \\ m\neq n}} \sigma_j R_k \Delta f_m + 2\sum_{j+k+m=n-1} \sigma_j R_k \partial_x f_m + \sum_{j+k+m=n-2} \sigma_j R_k f_m \\
&\qquad + (\bar{\psi}_y \partial_x - \bar{\psi}_x \partial_y)\left[\sum_{\substack{j+k=n \\ k\neq n}} R_j \Delta f_k + 2\sum_{j+k=n-1} R_j \partial_x f_k + \sum_{j+k=n-2} R_j f_k\right] \\
&\qquad - ((\Delta\bar{\psi}_y)\partial_x - (\Delta\bar{\psi}_x)\partial_y)\sum_{\substack{j+k=n \\ k\neq n}} R_j f_k + \bar{\psi}_y\left[\sum_{j+k=n-1} R_j \Delta f_k + 2\sum_{j+k=n-2} R_j \partial_x f_k\right. \\
&\qquad \left. + \sum_{j+k=n-3} R_j f_k\right] - (\Delta\bar{\psi}_y)\sum_{j+k=n-1} R_j f_k
\end{aligned} \tag{99}$$

where $\partial_x = \partial/\partial x$, $\partial_y = \partial/\partial y$, $\bar{\psi}_x = \partial_x\bar{\psi}$, and $\bar{\psi}_y = \partial_y\bar{\psi}$, and the summation is done for all integers k, m, and j subject to the condition specified under $\sum$. As in Equation 39, the normalization of f_n is given by

$$\int_0^{2\pi}\int_0^{2\pi} f_0^* f_n \, dx\, dy = 4\pi^2 \delta_{0n} \tag{100}$$

Inspecting that 1 is the solution of Equations 99 and 100 for n = 0, and also its adjoint problem, we obtain the solvability condition of Equation 99

$$\begin{aligned}
&-\delta_{n-4,0} + \sum_{j+k=n-2} \sigma_j R_k + \frac{1}{2\pi^2}\sum_{j+k=n-2} R_j \int_0^{2\pi}\int_0^{2\pi} \bar{\psi}_y \partial_x f_k \, dx\, dy \\
&+ \frac{1}{4\pi^2}\sum_{j+k=n-3} R_j \int_0^{2\pi}\int_0^{2\pi} \bar{\psi}_y f_k \, dx\, dy = 0
\end{aligned} \tag{101}$$

Equation 101 for first few values of n, we obtain on a neutral curve,

$$\sigma_0 = 0, \tag{102a}$$

$$\sigma_1 = -\frac{1}{\pi}\int_0^{2\pi}\int_0^{2\pi} \bar{\psi}_y \partial_x f_1 \, dx\, dy, \tag{102b}$$

$$\sigma_2 = 0, \tag{102c}$$

$$R_0 = 2\pi \Big/ \int_0^{2\pi}\int_0^{2\pi} \bar{\psi}_y (f_1 + 2\partial_x f_2) \, dx\, dy, \tag{103a}$$

$$R_1 = 0 \tag{103b}$$

As in the previous case, if there is a finite and nonzero limiting Reynolds number for $\beta' = 0$ and $\beta \to 0$, it can be evaluated by Equation 103a. For example, in the case treated by Eisler, where

$$\bar{\psi} = \tfrac{1}{2}(\cos x + \cos y) \tag{104}$$

we have [20]

$$f_1 = -\frac{R_0}{2}\sin y \tag{105a}$$

$$f_2 = 0 \tag{105b}$$

$$R_0 = \sqrt{8} \tag{105c}$$

R_0 in Equation 105c is verified to be the critical Reynolds number to the disturbance with small values of Floquet exponents β and β' [20]. The overall structure of the neutral domain may be obtained by numerical calculation, and it is still an open problem whether Equation 105c gives the critical Reynolds number to any disturbance or not.

CONCLUSION AND COMMENTS

As for all examples cited, the critical disturbance in the spatially periodic parallel flow is the long wave mode with the same period as the basic flow ($\alpha = \beta = 0$). The value of the critical Reynolds number is evaluated by Equation 66a. This conclusion will probably be extended to almost every periodic flows. It is not difficult to give an exceptional example, namely a row of almost uninteracted wakes for which the critical disturbance is the mode of moderate wavenumber as discussed in Reference 10. This example is quite alien from the usual conception of periodic flow, but even in this case, Equation 66a certainly works as a key to seek the neutral curve. In this sense, it may be said that the fundamental approach by the linear theory has been completed. Now the interest is in the nonlinear stage of the evolution of disturbance.

For the evolution of discrete modes in a slightly supercritical flow, the weakly nonlinear theory has been established [21–23]. On the basis of this theory, the behavior of the disturbance is governed by the Landau equation:

$$\frac{dA}{dt} = \alpha c A + \ell A|A|^2, \tag{106}$$

where A(t) is the amplitude of the fundamental mode of disturbance and ℓ is the Landau constant to be determined depending on the basic flow. Under the assumption of the most unstable mode of $\beta = 0$ to be fundamental, the application of the theory gives the Landau constant analytically in the long wave limit ($\alpha \to 0$). Especially for an arbitrary symmetric velocity profile U(y), the Landau equation becomes

$$\frac{dA}{dt} = \frac{2\alpha^2(R - R_0)}{R_0^2}A - 2\alpha^4 R_0^3\langle (J^2(U))^2\rangle A|A|^2, \tag{107}$$

where R_0 and J are the same as before (Equations 66a and 68) [24]. Equation 107 shows that the amplitude grows up to the equilibrium value,

$$A_e = \left\{\frac{R - R_0}{\alpha^2 R_0^5\langle (J^2(U))^2\rangle}\right\}^{1/2} \tag{108}$$

in the slightly supercritical state in which A_e is limited in small magnitude. It is remarked that the contribution of the second harmonics decreases the Landau constant in this case, while not in other cases of unbounded free flows.

As shown earlier, in a supercritical state of the periodic flow the growing mode exists continuously depending on β even for a fixed wavenumber. This is an essential aspect of the periodic flow, and the theory should be extended to this system. It will reveal the priority of some continuous mode in its evolution.

The approach of parameter expansion seems to be useful in the non-parallel periodic flow as well as in the parallel flow, because the disturbance having the same period as the basic flow seems to be essential. Nevertheless, no universal properties have been found yet. It is the primal subject to be solved in near future.

NOTATION

A(t)	amplitude	ℓ	Landau constant
A_n	coefficients	R	Reynolds number
a, b, c	constant vectors	t	time
$C_{a,b,c}$	real constants	u	velocity
f_n	periodicity		

Greek Symbols

α_s	wave number	ρ	density
β	constant vector	$\bar{\psi}$	stream function
ν	kinematic viscosity	ω	vorticity

REFERENCES

1. Eisler, T. J., "Hydrodynamic Stability of Some Spatially Periodic Flows." Rep. Proj. Michigan, 2900-327-T, Univ. Michigan, 1962.
2. Mack, L. M., "A Numerical Study of the Temporal Eigenvalue Spectrum of the Blasius Boundary Layer." *J. Fluid Mech.*, Vol. 73, 1976, pp. 497–520.
3. Grosch, C. E., and Salwen, H., "The Continuous Spectrum of the Orr-Sommerfeld Equation. Part 1: The Spectrum and the Eigenfunctions." *J. Fluid Mech.*, Vol. 87, 1978, pp. 33–54.
4. Gustavsson, L. H., "Initial-Value Problem for Boundary Layer Flows." *Phys. Fluids*, Vol. 22, 1979, pp. 1602–1605.
5. Salwen, H., and Grosch, C. E., "The Continuous Spectrum of the Orr-Sommerfeld Equation. Part 2: Eigenfunction Expansions." *J. Fluid Mech.*, Vol. 104, 1981, pp. 445–465.
6. Yamada, M., and Gotoh, K., "Continuous Mode in the Theory of the Barotropic Instability of Zonal flows." *J. Phys. Soc.* Japan, Vol. 52, 1983, pp. 3039–3046.
7. DiPrima, R. C., and Habetler, G. J., "A Completeness Theorem for Nonself-Adjoint Eigenvalue Problems in Hydrodynamic Stability." *Arch. Rat. Mech. Anal.*, Vol. 34, 1969, pp. 218–227.
8. Green, J. S. A., "Two-Dimensional Turbulence Near the Viscous Limit." *J. Fluid Mech.*, Vol. 62, 1974, pp. 273–287.
9. Beaumont, D. N., "The Stability of Spatially Periodic Flows." *J. Fluid Mech.*, Vol. 108, 1981, pp. 461–474.
10. Gotoh, K. et al., "The Theory of Stability of Spatially Periodic Parallel Flows." *J. Fluid Mech.*, Vol. 127, 1983, pp. 45–58.
11. Drazin, P. G., and Reid, W. H., *Hydrodynamic Stability*. Cambridge Univ. Press, 1981.
12. Betchov, R., and Criminale, W. O., *Stability of Parallel Flows*. Academic Press, 1967.
13. Davey, A., "A Difficult Numerical Calculation Concerning the Stability of the Blasius Boundary Layer." In *Stability in the Mechanics of Continua*, F. H. Schroeder Ed. Springer Verlag, 1982, pp. 365–372.

14. Tinkham, M., *Group Theory and Quantum Mechanics.* McGraw-Hill, 1974, p. 267.
15. Synge, J. L., "Hydrodynamic Stability." *Semi-centenn. Publ. Amer. Math. Soc.*, Vol. 2, 1938, pp. 227–269.
16. Jones, H., *The Theory of Brillouin Zones and Electronic States in Crystals.* 2nd ed. North-Holland, 1975, pp. 9, 24.
17. Ince, E. L., *Ordinary Differential Equations.* Dover, 1956, p. 247.
18. Tatsumi, T., and Gotoh, K., "The Stability of Free Boundary Layers Between Two Uniform Streams." *J. Fluid Mech.*, Vol. 7, 1960, pp. 433–441.
19. Whittaker, E. T., and Watson, G. N., *A Course of Modern Analysis.* 4th ed. Cambridge Univ. Press, 1973, p. 464.
20. Gotoh, K., and Yamada, M., "Instability of a Cellular Flow," *J. Phys. Soc.*, Japan, Vol. 53, 1984, pp. 3395–3398.
21. Stuart, J. T., "On the Nonlinear Mechanics of Wave Disturbances in Stable and Unstable Parallel Flows. Part 1: The Basic Behavior in Plane Poiseuille Flow." J. Fluid Mech., Vol. 9, 1960, pp. 353–370.
22. Watson, J., "On the Nonlinear Mechanics of Wave Disturbances in Stable and Unstable Parallel Flows. Part 2: The development of a Solution for Plane Poiseuille Flow and for Plane Couette Flow," *J. Fluid Mech.*, Vol. 9, 1960, pp. 371–389.
23. Stewartson, K., and Stuart, J. T., "A Nonlinear Stability Theory for a Wave System in Plane Poiseuille Flow." *J. Fluid Mech.*, Vol. 48, 1971, pp. 529–545.
24. Yamada, M., "Nonlinear Stability Theory of Parallel Periodic Flows" to be published in *J. Phys. Soc.*, Japan.

CHAPTER 20

TURBULENT SWIRLING PIPE FLOW

Chiaki Kuroda and **Kohei Ogawa**

Department of Chemical Engineering
Tokyo Institute of Technology
Tokyo, Japan

CONTENTS

INTRODUCTION

A vortex with an axial velocity component is generally called swirling flow, and it is one of the well-recognized configurations of flow that occurs not only, as is to be expected, in cyclone separators, hydrocyclone separators, swirling spray dryers, swirling furnaces, and vortex tubes used for thermal separation, but also in agitators, the piping associated with fluid turbo-equipments, etc.

The swirling flow condition in such equipment is mostly regarded as turbulent flow, where fluid fluctuations are three-dimensionally occurring in all radial, tangential, and axial directions. In contrast with general non-swirling pipe flow, many differences are found also in the turbulent condition, needless to say in the time-mean flow condition, e.g., mean velocity distributions and mean pressure distributions.

In this chapter, we will make a synthesis of various characteristic facts about axially symmetrical swirling flow in a circular pipe, which represents common characteristics of many kinds of practical swirling flow. We are convinced that this type of investigation is widely applicable for turbulent swirling flow.

GENERATION OF SWIRL

Many kinds of methods have been devised to generate axially symmetrical steady swirling flow in a circular pipe. They are classified into the following three types, which are summarily illustrated in Figure 1, according to their structure.

1. A fluid is poured into the pipe through plural inlets that are tangentially set on the wall of the pipe.
2. A fluid is let flow in the pipe through plural vanes that are set inside the pipe with a certain angle.
3. A fluid is let flow into the pipe through rotating bodies, such as a rotating pipe and an impeller.

The first method is adopted in the present study because of the simplicity of the structure and the facility for controlling the operation. The experimental apparatus used is shown in Figure 2. The test pipe consists of a vertical smooth acrylic-resin cylinder (inner diameter D = 160 mm), and water as the test fluid is tangentially poured into the test pipe through two inlet pipes (d = 16 mm) with the equal flow rate respectively.

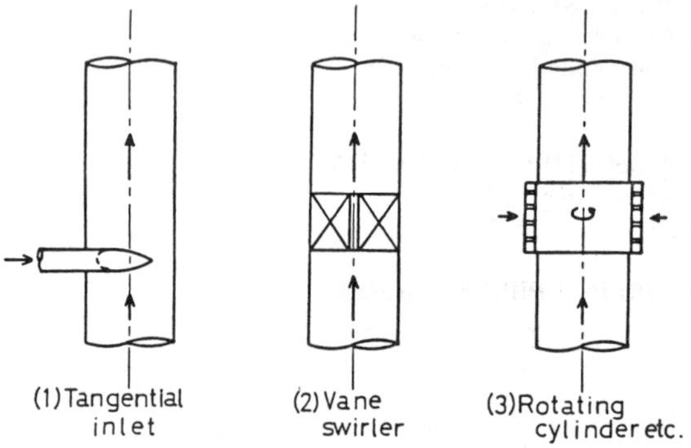

Figure 1. Generation of swirl.

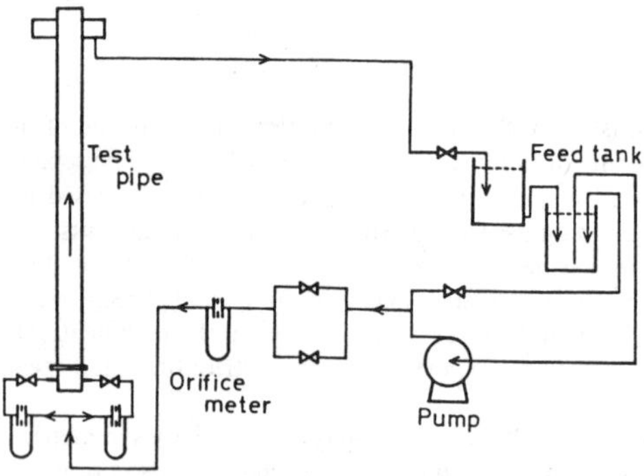

Figure 2. Schematic diagram of experimental apparatus.

THREE-DIMENSIONAL SWIRLING FLOW

The velocity component that most strongly dominates the state of swirling pipe flow is the swirling velocity, i.e., the tangential velocity component. On the basis of the investigation about the tangential velocity and the intensity of swirl, the three-dimensional mechanism of swirling pipe flow is explained as a whole in this section.

Tangential Velocity and Intensity of Swirl

The solid line in Figure 3 shows the practical radial distribution of the tangential mean velocity $\bar{U}_\theta$, and it is said that the radial region can be divided into two regions, i.e., a region of forced rotational flow in the center of the pipe, surrounded by a region of so-called quasi-free rotational flow. As a basic model of vortexes to express such a distribution curve, there is the Rankine's compound vortex [1], which is expressed by the combination of the forced vortex and the free vortex. These ideal vortexes are shown by the broken line in Figure 3.

However, $\bar{U}_\theta$ distributions are practically approximated by many kinds of empirical expressions, such as following typical ones:

Transformation [2, 3] of Rankine's compound vortex

$$\bar{U}_\theta = \omega_s r^n \qquad \text{for a forced vortex} \tag{1}$$

$$\bar{U}_\theta = \frac{k_1}{r^m} \qquad \text{for a quasi-free vortex} \tag{2}$$

($k_1 = \omega_s r_b^{n+m}$, r_b: boundary position)

Transformation [3, 4, 5] of Burgers' vortex [6]

$$\bar{U}_\theta = \frac{k_2}{r}\{1 - \exp(-k_3 r^2)\} \tag{3}$$

All of coefficients ω_s, k_1, k_2, k_3, exponents n, m, and r_b must be experimentally determined. Considering the values of them depend on the local flow condition, it is impossible to determine their respective constant values from operational conditions. On the other hand, some theoretical or numerical analyses [7–10] are tried on the basis of the momentum equation, but they are restricted to the case of the comparatively weak intensity of swirl or the laminar state. Therefore, it is important as an engineering problem to establish how to estimate $\bar{U}_\theta$ distributions for an arbitrary operational condition, and this is discussed in the following.

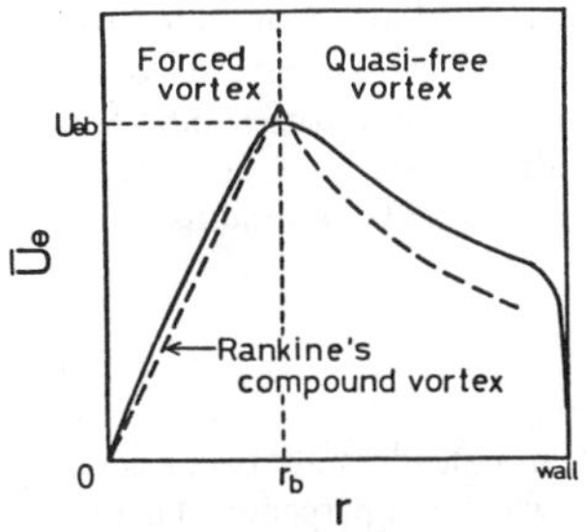

Figure 3. Tangential mean velocity $\bar{U}_\theta$ distribution.

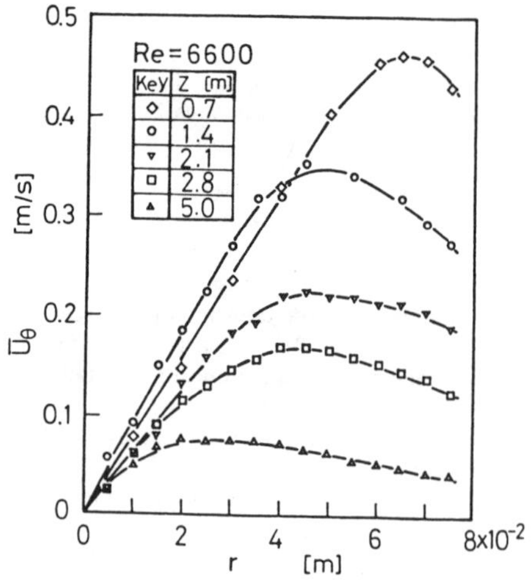

Figure 4. $\bar{U}_\theta$ distributions in the present apparatus.

Some examples of $\bar{U}_\theta$ distributions that were experimentally obtained by using our apparatus are shown in Figure 4, where the axial coordinate z is measured from the position of the inlet pipes and Reynolds number Re is based on the cross-sectional average velocity U_{av} and the inner diameter D of the test pipe. The value of $\bar{U}_\theta$ decreases with the increase of z, and it implies the decay of swirl.

One of the representative quantities that expresses the intensity of swirl is Swirl number, Sw [11], and it has been treated as an important similarity parameter of swirling flow. Sw is defined:

$$\mathrm{Sw} \equiv \frac{\text{(axial flux of angular momentum)}}{\mathrm{R} \cdot \text{(axial flux of linear momentum)}}$$

$$= \frac{\int_0^R \rho(\bar{U}_\theta r)\bar{U}_z 2\pi r\, dr}{R \int_0^R \rho \bar{U}_z^2 2\pi r\, dr + R \int_0^R \bar{P} 2\pi r\, dr} \tag{4}$$

and it was applied also to the investigation of swirling pipe flow [5, 7, 12, 13]. There have been, however, many problems, such as the difficulty in the practical computation of Sw and in the estimation of the effect of the turbulence on Sw.

Intensity of Swirl and Its Decay

In this study [14, 15], the authors will use the circulation for expression of the intensity of swirl in swirling pipe flow, considering the swirling pipe flow is one of many kinds of vortexes. Needless to say, the circulation is usually used to express the intensity of a vortex tube and a vortex filament, and it is one of the indispensable quantities to study a vortex [1, 3, 16, 17].

Let $\underset{\sim}{\Omega}$ be the vorticity vector of swirling flow, where the velocity vector is $\underset{\sim}{U}$ at a radius r. Then, the following general equation can be written:

$$(\underset{\sim}{\nabla} \cdot \underset{\sim}{\Omega}) = (\underset{\sim}{\nabla} \cdot [\underset{\sim}{\nabla} \times \underset{\sim}{U}]) = 0 \tag{5}$$

The axially symmetrical swirling flow of an incompressible fluid in a steady state is considered. If V is a cylindrical space enclosed between two planes S_1 and S_2, which are perpendicular to the

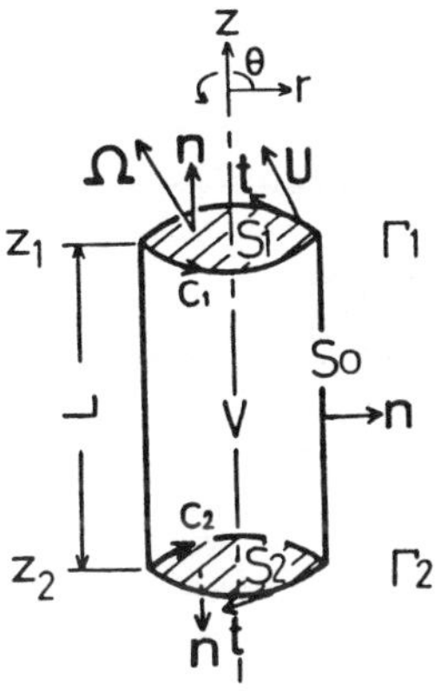

Figure 5. Integral space in swirling flow.

z axis as shown in Figure 5, Equation 5 can be integrated with respect to the volume, and can be transformed by using the Gauss divergence theorem as follows,

$$\iiint_V (\underset{\sim}{\nabla} \cdot \underset{\sim}{\Omega})\, dV = \iint_S (\underset{\sim}{\Omega} \cdot \underset{\sim}{n})\, dS = 0 \tag{6}$$

where $\underset{\sim}{n}$ is the unit normal vector directed away from the space bounded by the planes S (= S_0 + S_1 + S_2, S_0: the area of the curved surface around V). If C_1 and C_2 represent the closed curves surrounding S_1 and S_2, respectively, the Stokes' theorem can be used to derive the following:

$$\begin{aligned} \iint_S (\underset{\sim}{\Omega} \cdot \underset{\sim}{n})\, dS &= \oint_{C_1} (\underset{\sim}{U} \cdot \underset{\sim}{t})\, dC + \oint_{C_2} (\underset{\sim}{U} \cdot \underset{\sim}{t})\, dC + \iint_{S_0} (\underset{\sim}{\Omega} \cdot \underset{\sim}{n})\, dS \\ &= 0 \end{aligned} \tag{7}$$

where $\underset{\sim}{t}$ is the unit tangential vector along each of these closed curves.

If $U_{\theta 1}$ and $U_{\theta 2}$ are the tangential velocities along the curves C_1 and C_2, respectively, and Ω_z is the axial component of the vorticity,

$$\begin{aligned} \iint_{S_1} (\underset{\sim}{\Omega} \cdot \underset{\sim}{n})\, dS &= \iint_{S_1} \Omega_z\, dS \\ &= \oint_{C_1} (\underset{\sim}{U} \cdot \underset{\sim}{t})\, dC = 2\pi r U_{\theta 1} \\ &\equiv \Gamma_1 \end{aligned} \tag{8}$$

$$\begin{aligned} \iint_{S_2} (\underset{\sim}{\Omega} \cdot \underset{\sim}{n})\, dS &= \iint_{S_2} -\Omega_z\, dS \\ &= \oint_{C_2} (\underset{\sim}{U} \cdot \underset{\sim}{t})\, dC = -2\pi r U_{\theta 2} \\ &\equiv -\Gamma_2 \end{aligned} \tag{9}$$

Γ_1 and Γ_2 are the circulations along C_1 and C_2, respectively, and are defined as the intensities of swirl in each cross section. Thus, the decay in the intensity of swirl between z_1 and z_2 will be given by the difference $\Delta\Gamma$ between Γ_1 and Γ_2. From Equations 7, 8, and 9,

$$\Delta\Gamma = \Gamma_2 - \Gamma_1$$
$$= \iint_{S_0} (\underset{\sim}{\Omega} \cdot \underset{\sim}{n})\, dS = \iint_{S_0} \Omega_r\, dS \tag{10}$$

where Ω_r is the radial component of the vorticity.

The axial vorticity Ω_z, which is closely related to the intensity of swirl, is used to derive the following equation for Ω_r from the z-component of the vorticity equation,

$$\Omega_r = \frac{U_z \dfrac{\partial \Omega_z}{\partial z}}{\dfrac{\partial U_z}{\partial r}} - \frac{\nu \left\{ \dfrac{1}{r} \dfrac{\partial}{\partial r} \left(r \dfrac{\partial \Omega_z}{\partial r} \right) \right\}}{\dfrac{\partial U_z}{\partial r}} \tag{11}$$

where small terms, such as the term containing the radial velocity U_r, are ignored. (The aspect that U_r is very small in swirling flow in a constant diameter pipe will be treated later.)

The following dimensionless quantities will now be defined in terms of the characteristic parameters of the flow field, viz., the inner radius R of the test pipe, the initial swirling velocity U_i of the inflowing fluid at the inlet where the swirl is generated, the angular velocity ω_i ($= U_i/R$) at the inlet, and the distance L between S_1 and S_2:

$$\Gamma^* \equiv \frac{\Gamma}{2\pi R U_i} = \frac{\Gamma}{2\pi R^2 \omega_i}$$
$$r^* \equiv \frac{r}{R}$$
$$z^* \equiv \frac{z}{L}$$
$$S^* \equiv \frac{S}{2\pi R L} \tag{12}$$
$$\Omega_z^* \equiv \frac{\Omega_z}{\omega_i}$$
$$U_z^* \equiv \frac{U_z}{U_i} = \frac{U_z}{R\omega_i}$$

By using these dimensionless quantities and Equation 11, Equation 10 can be rewritten as the following dimensionless form, thus:

$$\Delta\Gamma^* = \Gamma_2^* - \Gamma_1^*$$
$$= \iint_{S_0^*} \frac{U_z^* \dfrac{\partial \Omega_z^*}{\partial z^*}}{\dfrac{\partial U_z^*}{\partial r^*}}\, dS^* - \frac{L}{R} \cdot \frac{\nu}{R^2 \omega_i} \iint_{S_0^*} \frac{\dfrac{1}{r^*} \dfrac{\partial}{\partial r^*} \left(r^* \dfrac{\partial \Omega_z^*}{\partial r^*} \right)}{\dfrac{\partial U_z^*}{\partial r^*}}\, dS^* \tag{13}$$

This equation means that, when the product of the dimensionless distance L/R and the reciprocal Reynolds number $\nu/R^2\omega_i$ based on the swirling motion, viz., $(L/R) \cdot (\nu/R^2\omega_i)$, is fixed, the dimensionless difference $\Delta\Gamma^*$ of the intensity of swirl will be fixedly given by Equation 13. That is to say, assuming the similarity of flow on the basis of the above-mentioned characteristic parameters, $\Delta\Gamma^*$ can be expressed as a function of $(L/R) \cdot (\nu/R^2\omega_i)$.

By applying this result to the case of swirling flow in a circular pipe, the whole intensity of swirl Γ_w over the cross section, which is defined to be Γ obtained from extrapolating to $r \rightarrow R$ the distribution curve, is indispensable. If S_2 is considered to be the surface at the inlet where the swirl is generated, Γ_2^* becomes unity because $U_{\theta 2} = U_i$. And replacing L by z, the following relational equation is obtained:

$$\Gamma_w^*(= \Gamma_1^* = 1 - \Delta\Gamma^*) = F\left(\frac{z}{R} \cdot \frac{\nu}{R^2\omega_i}\right) \tag{14}$$

In this equation, F is the function which should be experimentally determined.

Figure 6 shows the relationship between the experimental value of Γ_w^* and $(z/R) \cdot (\nu/R^2\omega_i)$. For a certain distance downstream from the inlet, there are approximately exponential relationships, and Γ_w^* is given by

$$\Gamma_w^* = \gamma \exp\left\{-\kappa\left(\frac{z}{R} \cdot \frac{\nu}{R^2\omega_i}\right)\right\} \tag{15}$$

where γ is a characteristic constant for the apparatus depending on the geometrical shape and the method of generating swirl, and κ is a coefficient that depends on the operational condition. If all relationships in Figure 6 are extrapolated to $(z/R) \cdot (\nu/R^2\omega_i) \rightarrow 0$, Γ_w^* will take an approximately constant value of 0.26 independently of Re, and therefore the value of γ becomes 0.26 for the present apparatus.

Figure 7 shows the relationship between κ and Re, and it quantitatively expresses the variation of the flow condition with Re. The value of κ changes from a constant value in the range of Re > 5,000 to another constant value in the range of Re < 2,000, and the flow condition is expected to be different for each of these ranges of Re. The results corresponding to Re within 2,000 < Re < 5,000 are considered to be in the intermediate range, and this range corresponds almost exactly to the range over which the flow condition changes from laminar to turbulent in the case of non-swirling pipe flow. Thus, according to the non-swirling downstream state of flow which is laminar or turbulent, the state of the upstream swirling flow will differ with regard to the mechanism of decay of the intensity of swirl.

Considering these results, it can be said that two kinds of Reynolds number, viz., the general Reynolds number Re $(= DU_{av}/\nu)$ and $R^2\omega_i/\nu$ based on the swirling motion, are important to explain swirling pipe flow.

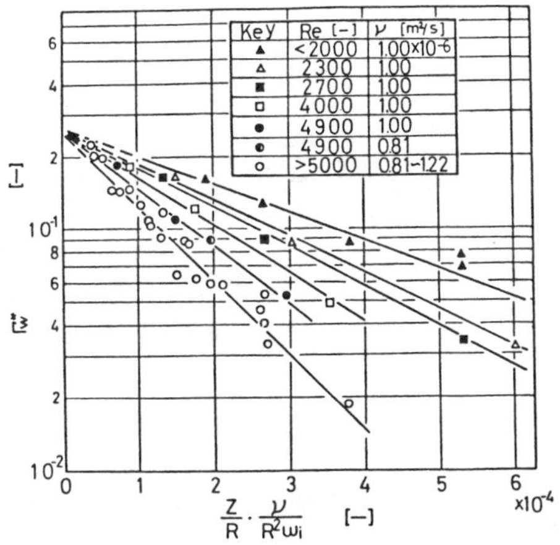

Figure 6. Decay of dimensionless intensity of swirl Γ_w^*.

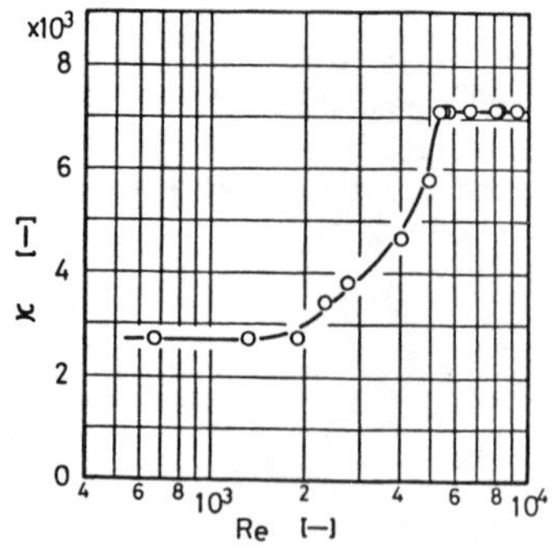

Figure 7. Relationship between a coefficient of swirl decay κ and Reynolds number Re.

Tangential Velocity Distributions

Even if Equation 1, 2, or 3 is used to approximate $\bar{U}_\theta$ distributions, there will remain a difficult problem, that is how to estimate the values of empirical constants in those equations from the local flow condition.

In most cases, swirling flow is in the turbulent state and this seems to be much dependent on the generation of the strong vorticity by the swirling motion. In this study [18], the swirling mechanism in turbulent swirling pipe flow is explained through the consideration on the radial distribution of Γ, which means the surface integral of the axial vorticity and the method of estimating $\bar{U}_\theta$ distributions is discussed by combining the results as described earlier.

Some examples of the radial distributions of $\bar{U}_\theta$ and $r\bar{U}_\theta(=\Gamma/2\pi)$ are shown in Figure 8. Since the boundary position between the forced vortex zone and the quasi-free vortex zone is expected to be the characteristic position also from the viewpoint of the turbulent mechanism, let the radial position where $\bar{U}_\theta$ takes the maximum value $U_{\theta b}$ be the representative boundary position r_b.

As mentioned later in this chapter, it is experimentally confirmed [19] that the distribution form of turbulent statistical quantities in the forced vortex zone is much different from that in the quasi-free vortex zone. In particular, the Reynolds stress $-\rho\overline{u_r u_\theta}$, which has influence on the swirling

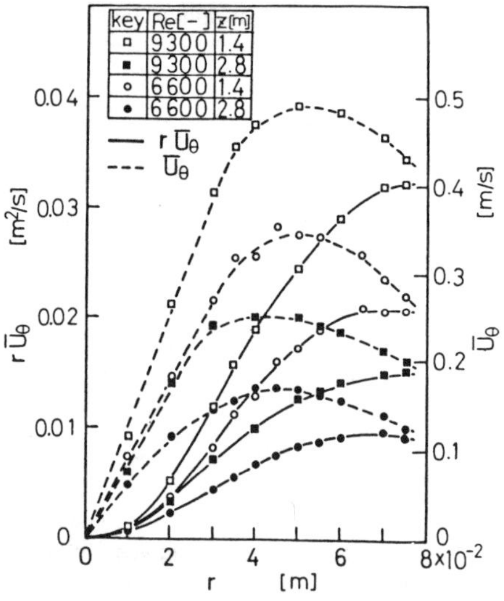

Figure 8. Distributions of $\bar{U}_\theta$ and $r\bar{U}_\theta$ $(=\Gamma/2\pi)$.

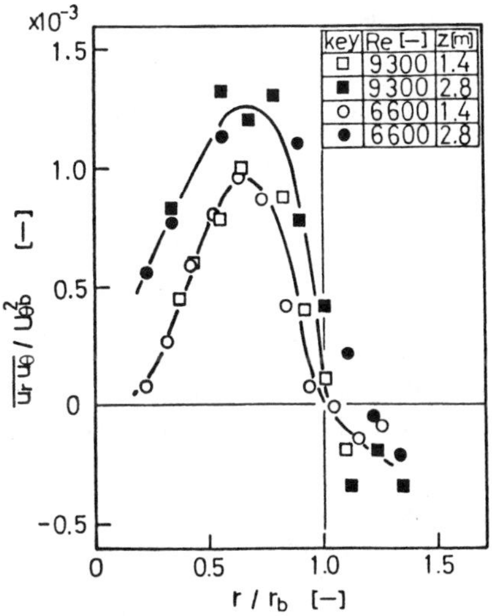

Figure 9. Reynolds stress $\overline{u_r u_\theta}$ distributions.

motion, takes zero value in the vicinity of the boundary position r_b as shown in Figure 9. Thus, a hypothetical cylindrical wall is assumed to be situated at r_b, where the molecular viscous stress $\tau_{r\theta,b}$ is dominant, so that the concept of the law of the wall in the turbulent boundary layer can be used.

The $r\theta$-component of the shear stress of Newtonian fluids in axially symmetrical turbulent flow is expressed as follows:

$$\tau_{r\theta} = \mu\left(\frac{\partial \bar{U}_\theta}{\partial r} - \frac{\bar{U}_\theta}{r}\right) - \rho\overline{u_r u_\theta} \tag{16}$$

The value of $\tau_{r\theta}$ at r_b is shown as the following form,

$$\tau_{r\theta,b} = -\mu\left(\frac{U_{\theta b}}{r_b}\right) \tag{17}$$

because both values of $\partial\bar{U}_\theta/\partial r$ and $\overline{u_r u_\theta}$ are zero at r_b.

The following friction velocity for swirling flow is defined by using the above shear stress on the hypothetical wall:

$$U_{\theta f} \equiv \left(\frac{|\tau_{r\theta,b}|}{\rho}\right)^{1/2} = \left(\frac{\nu U_{\theta b}}{r_b}\right)^{1/2} \tag{18}$$

And $U_{\theta f}$ is applied to the transformation of the radius r and the intensity of swirl Γ into dimensionless quantities r^+ and Γ^+ respectively in order to explain the radial distribution of Γ as follows:

$$r^+ \equiv \frac{U_{\theta f}}{\nu} r = \left(\frac{U_{\theta b}}{\nu r_b}\right)^{1/2} r \tag{19}$$

$$\Gamma^+ \equiv \frac{\Gamma}{2\pi r_b U_{\theta f}} = \frac{r\bar{U}_\theta}{(\nu r_b U_{\theta b})^{1/2}} \tag{20}$$

By making use of r_b^+ and Γ_b^+ at the boundary position, the respective relationships between $r_b^+ - r^+$ and $\Gamma_b^+ - \Gamma^+$ in the forced vortex zone and between $r^+ - r_b^+$ and $\Gamma^+ - \Gamma_b^+$ in the quasi-free vortex zone are experimentally investigated for both sides of the hypothetical wall as shown in Figure 10. Good correlations are obtained in each zone, and the relationships are approximated as follows for the fully turbulent condition:

$$\Gamma_b^+ - \Gamma^+ = (r_b^+ - r^+)^{1.06}$$
in the forced vortex zone (21)

$$\Gamma^+ - \Gamma_b^+ = (r^+ - r_b^+)^{0.92}$$
in the quasi-free vortex zone (22)

In $r^+ - r_b^+ = 40 \sim 70$ of Figure 10, however, some measured points in the vicinity of the pipe wall deviate slightly from the above relations. This is considered to be caused by the boundary layer on the pipe wall.

Considering from the above-mentioned results, if the values of r_b and $U_{\theta f}$ (or $U_{\theta b}$) can be obtained, the distributions of either Γ or $\bar{U}_\theta$ can be estimated for any position and any operational condition.

First, the decay process of the intensity of swirl Γ_b ($= 2\pi r_b \bar{U}_{\theta b}$) within the forced vortex zone will be discussed as follows. Here, as expressed by Equations 8 and 9, Γ is equivalent to the integrated value of the axial vorticity component Ω_z over the cross-sectional circular area. Since the value of Ω_z is much smaller in the quasi-free vortex zone than in the forced vortex zone (Ω_z is zero in the ideal free vortex), the most part of Γ_w is considered to be dependent on Γ_b. Consequently, it is expected that the decay process of Γ_b will be similar to that of Γ_w, and that there will be a characteristic relationship between $\Gamma_b^* (= \Gamma_b / 2\pi R^2 \omega_i)$ and $(z/R) \cdot (\nu / R^2 \omega_i)$ in the same manner as in the case of Γ_w^*.

The correlation between Γ_b^* and $(z/R) \cdot (\nu / R^2 \omega_i)$ is shown in Figure 11 and is expressed as follows,

$$\Gamma_b^* = \gamma' \exp\left\{ -\kappa' \left(\frac{z}{R} \cdot \frac{\nu}{R^2 \omega_i} \right) \right\} \tag{23}$$

where the value of γ' is 0.22 for the present apparatus and the value of κ' is 8,000 in the fully turbulent condition (about Re > 5,000).

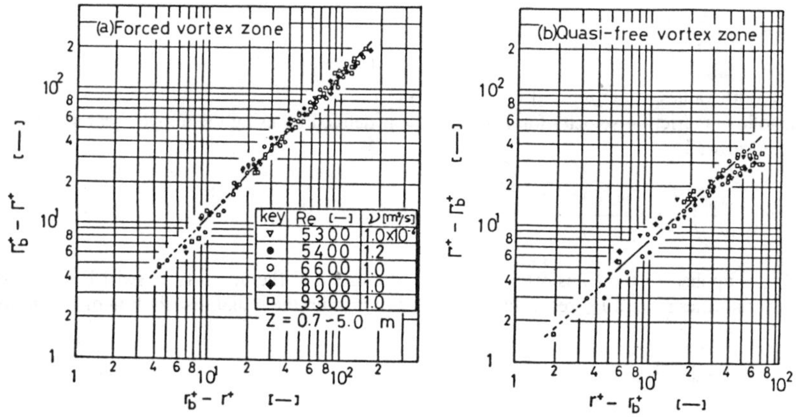

Figure 10. Relationship between dimensionless radius r^+ and dimensionless circulation Γ^+.

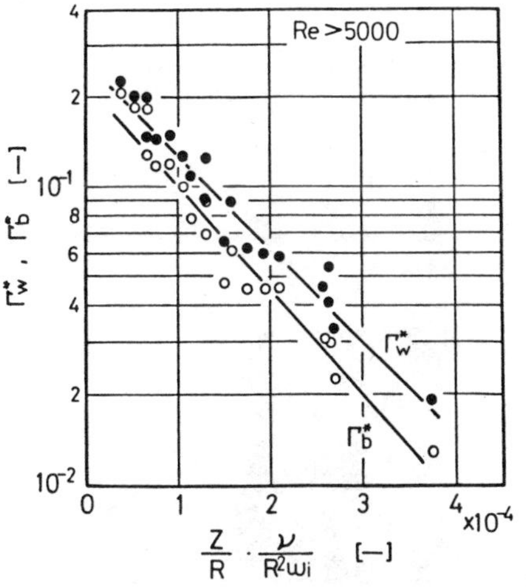

Figure 11. Decay of dimensionless intensity of swirl within the forced vortex zone Γ_b^*

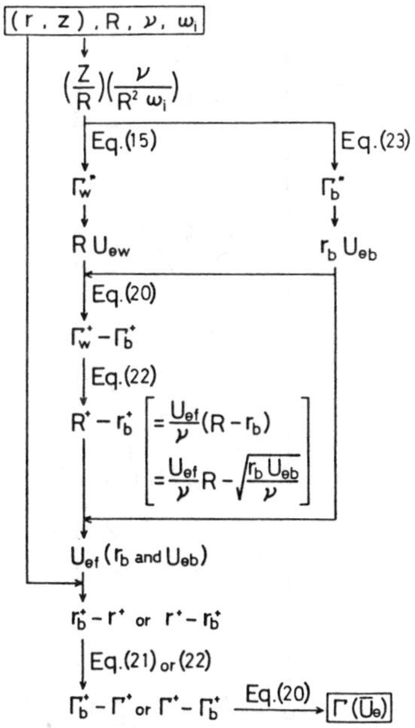

Figure 12. Calculation procedure for Γ ($\bar{U}_\theta$).

By using these results in the decay process of Γ_w^*, Γ_b^* and the relations between r^+ and Γ^+, either Γ or $\bar{U}_\theta$ for an arbitrary position and operational condition can be estimated by following the procedure of calculation shown in Figure 12.

In this section, it is made clear that the flow condition at the boundary between the forced vortex zone and the quasi-free vortex zone is important to know the mechanism of the swirling motion. And by using the newly defined friction velocity for turbulent swirling flow, the characteristic of the boundary and the distribution of Γ can be expressed and explained, respectively.

Axial Velocity and Reverse Flow

The radial distributions of the axial mean velocity $\bar{U}_z$ are very different from those in non-swirling pipe flow and have various shapes according to the degree of the swirl intensity. It is generally known [20, 21] that the $\bar{U}_z$ distributions are divided into the following three types, which are shown in Figure 13:

1. The axial flow goes in the direction of the average main flow over the whole cross section of the test pipe, notwithstanding the decrease of $\bar{U}_z$ in the range around the center axis.
2. The axial flow goes in the reverse direction of the average main flow in the range around the center axis and goes in the direction of the average main flow in the range near the pipe wall.
3. The axial flow goes in the direction of the average main flow in the both ranges of the vicinity of the center axis and the neighborhood of the pipe wall, and goes in the reverse direction of the average main flow in the range between the above two ranges.

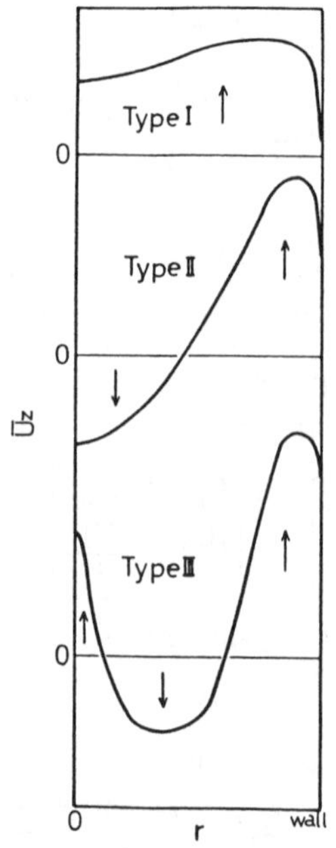

Figure 13. Three types of axial mean velocity $\bar{U}_z$ distributions.

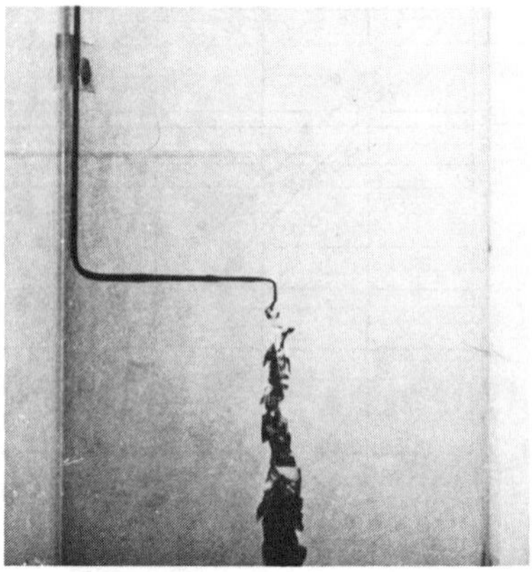

Figure 14. Mixing process of dye in reverse flow.

The axial flow that goes in the reverse direction of the average main flow, that is to say, the flow of $\bar{U}_z < 0$ is called reverse flow or recirculating flow. This is one of the important characteristic states of swirling flow and has been utilized positively in various practical apparatuses, for example swirling spray dryers in order to control the residence time of particles and swirling furnaces in order to stabilize flames. The unstable transitional flow between the types of (1) and (2) has been sometimes studied as vortex breakdown [1, 22–25]. Such phenomena are caused by the complicated pressure distributions in swirling flow, which is explained in a later section.

In usual operational conditions, the flow type of (1) or (2) is mostly generated and the flow type of (3) is sometimes found under the condition that the swirl is particularly strong, for example in the region just near the inlet pipe. If the general Reynolds number Re is constant, as the intensity of swirl becomes stronger, the flow type of (2) or (3) is more easily generated. This fact was clearly shown by Binnie [20] with visual experiments. Figure 14 shows the mixing process of dye that was injected into the vicinity of the center axis under the condition of the type (2). Most of the dye goes in the reverse direction of the average main flow (downwards in Figure 14) by the above reverse flow. Thus, it is clear that the axial flow has an important effect on the mixing process in swirling flow.

The distributions of $\bar{U}_z$ are considered to approach that of non-swirling pipe flow through the above flow types (1), (2), or (3) with the progress of decay of the intensity of swirl. Some empirical methods to estimate the change of the $\bar{U}_z$ distributions were explained in previous reports [5, 12, 13, 26], but there seems to be some insufficiency in wide usefulness. The theoretical or numerical analysis may be possible in the case of the relatively weak swirl [8], because its $\bar{U}_z$ distribution is not much different from that of non-swirling pipe flow. It is, however, very difficult to treat theoretically the above complicated flow condition.

In this section, referring to the treatments of $\bar{U}_z$ by Talbot [10] and Kreith et al. [8], an approximate equation that can be applied to calculate the $\bar{U}_z$ distributions of turbulent swirling pipe flow is explained [27].

The axial velocity $\bar{U}_z(r, z)$ at an arbitrary position is assumed to be divided into the fully-developed non-swirling turbulent pipe flow velocity $\langle U_z(r)\rangle$ and the velocity difference $\Delta\bar{U}_z(r, z)$ from $\langle U_z(r)\rangle$, of which the absolute value decreases toward zero with the increase of z. The distribution of $\langle U_z(r)\rangle$ is approximated by the one-seventh power law distribution in the fully turbulent condition. Then, the following relationship is obtained by using the maximum velocity $\langle U_z(r)\rangle_m$ of $\langle U_z(r)\rangle$ at the center axis in order to transfer the above $\bar{U}_z$, $\langle U_z\rangle$, $\Delta\bar{U}_z$ into the dimensionless form $\bar{U}_z^*$, $\langle U_z\rangle^*$, $\Delta\bar{U}_z^*$,

$$\begin{aligned}\bar{U}_z^* &= \langle U_z\rangle^* + \Delta\bar{U}_z^* \\ &= (1 - r^*)^{1/7} + \Delta\bar{U}_z^*\end{aligned} \tag{24}$$

where r* is the dimensionless radius divided by R.

Some examples of $\Delta\bar{U}_z^*$ distributions are shown in Figure 15. It is anticipated that all curves fall on an unique curve that is expressed by a function of only r* by changing suitably the scale of the vertical axis for each curve. Therefore, the following relationship can be assumed,

$$\Delta\bar{U}_z^* = kf(r^*) \tag{25}$$

where k is the experimental factor that depends on the axial position and the operational condition and decreases toward zero with decay of the intensity of swirl.

The function f(r*) should satisfy the required condition f(1) = 0 at the pipe wall. Therefore, it may be suitable to introduce the term $(1 - r^*)^m$ into the empirical function f(r*). Consequently, f(r*) is assumed to be approximated by using a polynomial expression as follows:

$$f(r^*) = (1 - r^*)^m(c_0 + c_1 r^* + \cdots + c_n r^{*n}) \tag{26}$$

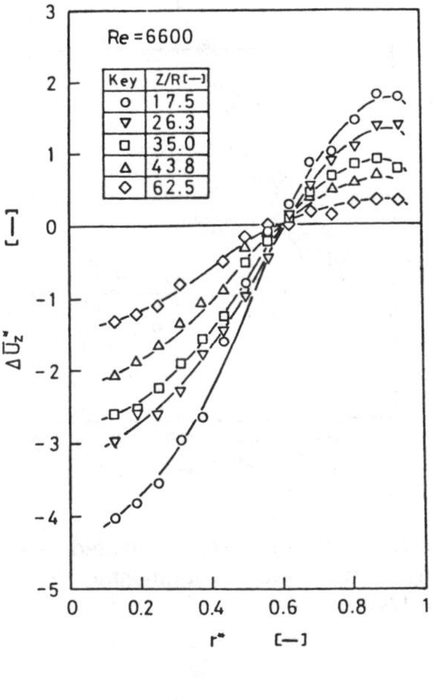

Figure 15. Distributions of dimensionless axial velocity difference $\Delta\bar{U}_z^*$.

Here, the following relation, which is expressed by using Beta function $B(\alpha, \beta)$, should hold for empirical constants c_i, m, n, considering the conservation of the amount of flow throughout the cross section:

$$\int_0^1 f(r^*)r^*\,dr^* = \int_0^1 (1 - r^*)^m(c_0r^* + c_1r^{*2} + \cdots + c_nr^{*n+1})\,dr^*$$
$$= c_0B(2, m + 1) + c_1B(3, m + 1) + \cdots + c_nB(n + 2, m + 1)$$
$$= 0 \tag{27}$$

As a trial, the values of empirical constants c_i, m, n, and the factor k were determined for our experimental results by using the trial-and-error method. And the following approximate equation is considered to be suitable for the present study:

$$\bar{U}_z^* = (1 - r^*)^{1/7} + k(1 - r^*)^{1/2}\left(-\frac{B(4, 1.5)}{B(2, 1.5)} + r^{*2}\right) \tag{28}$$

The comparisons of the experimental results with the calculated results of Equation 28 are shown in Figures 16 and 17. The region in the close vicinity of the center axis is excluded from the application of Equation 28.

Here, considering that the velocity difference $\Delta\bar{U}_z$ is caused by the swirling motion and that the factor k depends on the axial position and the operational condition, it seems to be adequate that the change of k is expressed by the dimensionless quantity $(z/R) \cdot (\nu/R^2\omega_i)$, which is used for the discussion on the decay process of the swirling motion in the preceding section. The relation between k and $(z/R) \cdot (\nu/R^2\omega_i)$ is shown in Figure 18, and it can be expressed by an exponential

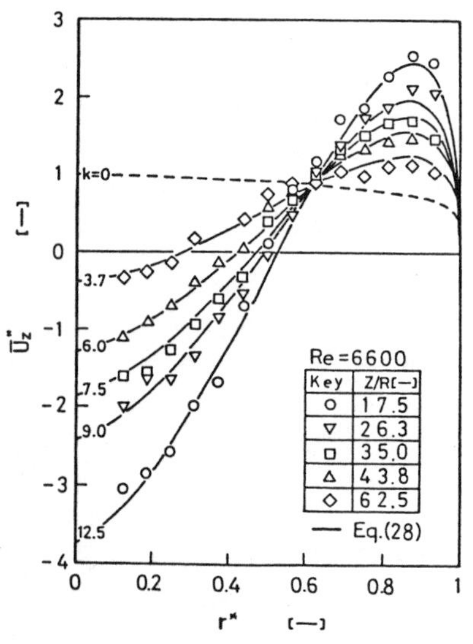

Figure 16. Comparison of calculated axial velocity curves with experimental results (Re = 6600)

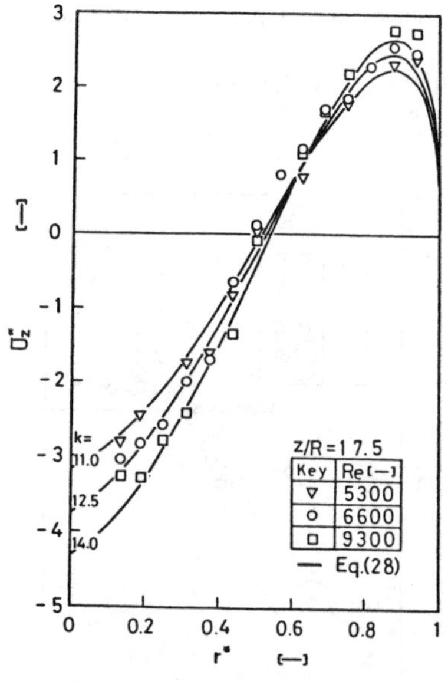

Figure 17. Comparison of calculated axial velocity curves with experimental results (z/R = 17.5)

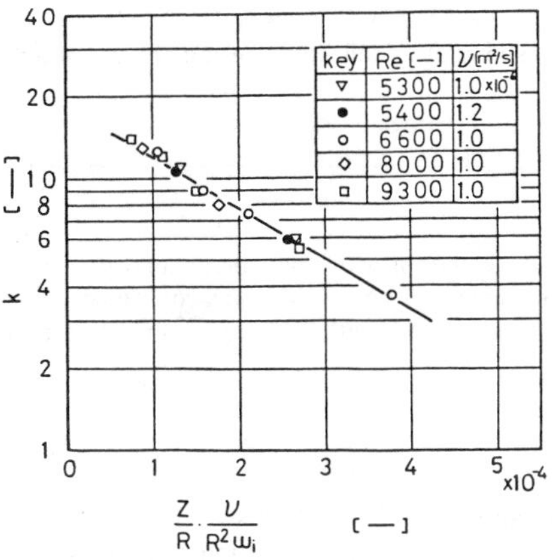

Figure 18. Relationship between experimental factor k and (z/R) · ($\nu/R^2\omega_i$).

function in the same manner as in the case of the intensity of swirl:

$$k = 18.0 \exp\left\{-4300\left(\frac{z}{R} \cdot \frac{\nu}{R^2\omega_i}\right)\right\} \tag{29}$$

The above results indicate that both axial and tangential velocities are changing in connection with each other as the flow goes downstream.

Radial Velocity

The effect of the inlet structure on the flow condition becomes negligibly small for several times as long as the inner diameter downstream from the inlet of the pipe [5, 28], and the radial mean velocity $\bar{U}_r$ becomes much smaller than either $\bar{U}_\theta$ or $\bar{U}_z$, so much so that it can be ignored [12, 28] except in the investigation about the flow just near the inlet.

Figure 19 shows examples of $\bar{U}_r$ distributions under condition that reverse flow is arising. Negative $\bar{U}_r$, namely inward flow, increases the reverse flow rate with the decrease of z, and the recirculated fluids are directed outwards in the vicinity of the inlet where the swirling motion is generated.

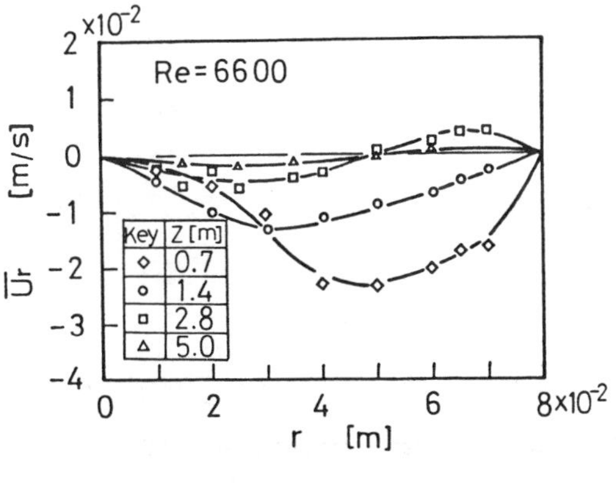

Figure 19. Radial mean velocity $\bar{U}_r$ distributions in the present apparatus.

When the $\bar{U}_z$ distribution change in the axial direction has been already known for axially symmetrical swirling flow, $\bar{U}_r$ distributions can be estimated as follows by using the continuity equation:

$$\bar{U}_r = -\frac{1}{r}\int_0^r r\frac{\partial \bar{U}_z}{\partial z}\,dr \tag{30}$$

The comparison of the calculated values by the above equation with the experimental ones was performed by Murakami et al. [12], and the good agreement of both values was confirmed.

In any case, the value of $\bar{U}_r$ is much smaller than the value of either $\bar{U}_\theta$ or $\bar{U}_z$ in the case of the constant diameter pipe.

STATIC PRESSURE IN SWIRLING FLOW

The generation of the radial distribution of the mean static pressure $\bar{P}$ in swirling pipe flow is considered to be caused mainly by the centrifugal force arising from the rotational motion of the fluid. Such $\bar{P}$ distributions give valuable information about not only the cause of the reverse flow but also the mechanical energy loss. Examples of the radial distributions of $\bar{P}$ relative to the hydrostatic pressure are shown in Figure 20. $\bar{P}$ seems to take the minimum value at the center axis and the maximum value on the pipe wall, and the difference between them decreases in the downstream direction with the decay of swirl.

In this study [29, 30], we examine the static pressure distributions in turbulent swirling pipe flow in connection with the decay process of the intensity of swirl, which is shown in the preceding section.

Static Pressure Drop Along the Pipe Wall

The mean static pressure $\bar{P}_w$ on the wall, relative to the hydrostatic pressure, decreases as the flow goes in the downstream direction as shown in Figure 21. Nissan et al. [2] calculated the dimensionless quantity that corresponded to the so-called pipe friction factor from the above static pressure drop $\Delta\bar{P}_w$ along the wall, and showed the quantity reached a value that is from several times to several tens of times higher than that of general non-swirling pipe flow. However, considering that $\bar{P}$ changes in the radial direction and that $\Delta\bar{P}_w$ in swirling pipe flow does not correspond to the total mechanical energy loss by friction, it seems unreasonable that only $\Delta\bar{P}_w$ is treated as the friction loss. Therefore, $\Delta\bar{P}_w$ is no more than a representative quantity of the overall static pressure drop. However, the decay process of $\bar{P}_w$ must be important to clarify the mechanism of swirling pipe flow.

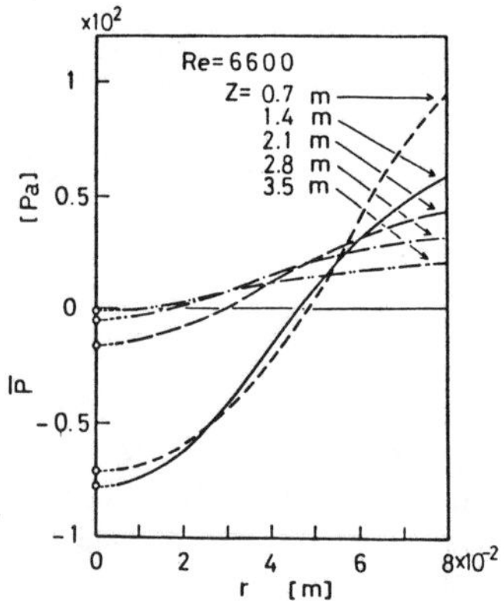

Figure 20. Mean static pressure $\bar{P}$ distributions.

Key	Re [−]	U_{av} [m/s]	ν [m²/s]
▲	1900	1.16×10⁻¹	
▲	4000	2.49	
▽	5300	3.32	1.00×10⁻⁶
○	6600	4.14	
◇	8000	4.97	
□	9300	5.80	
●	8200	4.14	
◆	9800	4.97	0.81×10⁻⁶
■	11500	5.80	

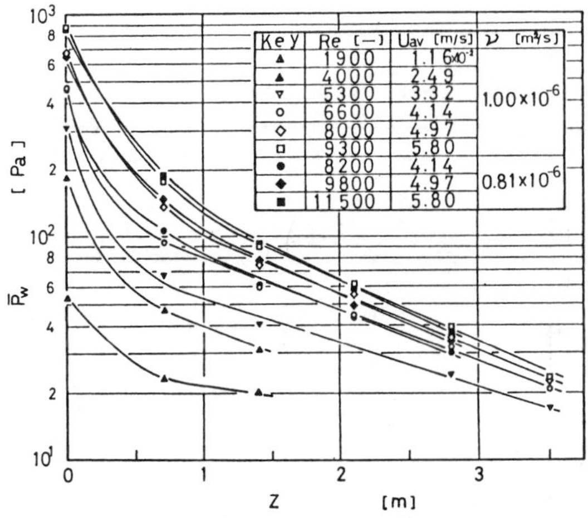

Figure 21. Decay of mean static pressure on the pipe wall $\bar{P}_w$.

$\Delta \bar{P}_w$ is expressed by the z-component of the Navier-Stokes equation as follows:

$$\Delta \bar{P}_w = \bar{P}_{w2} - \bar{P}_{w1} \qquad (z_1 > z_2)$$

$$= -\mu \int_{z_2}^{z_1} \left\{ \frac{1}{r} \frac{\partial}{\partial r} \left(r \frac{\partial \bar{U}_z}{\partial r} \right) \right\}_{r=R} dz \tag{31}$$

The inner radius R, the characteristic velocity U_i ($= R\omega_i$) of the inflowing fluid from the inlet pipe and the distance L between z_1 and z_2 are used to define the following dimensionless quantities, as was done in the case of the intensity of swirl:

$$\left.\begin{aligned}
\Delta \bar{P}_w^* &\equiv \frac{\Delta \bar{P}_w}{\rho U_i^2} \\[1ex]
\bar{U}_z^* &\equiv \frac{\bar{U}_z}{U_i} \\[1ex]
r^* &\equiv \frac{r}{R} \\[1ex]
z^* &\equiv \frac{z}{L}
\end{aligned}\right\} \tag{32}$$

By substituting these dimensionless quantities into Equation 31, the following equation is given:

$$\Delta \bar{P}_w^* = -\frac{L}{R} \cdot \frac{\nu}{R U_i} \int_{z_2^*}^{z_1^*} \left\{ \frac{1}{r^*} \frac{\partial}{\partial r^*} \left(r^* \frac{\partial \bar{U}_z^*}{\partial r^*} \right) \right\}_{r^*=1} dz^* \tag{33}$$

When the dimensionless quantity $(L/R) \cdot (\nu/R U_i)$ is fixed, the dimensionless difference $\Delta \bar{P}_w^*$ will be fixedly given by Equation 33. That is to say, assuming the similarity of flow on the basis of the above-mentioned characteristic parameters, $\Delta \bar{P}_w^*$ is considered to be expressed as a function of $(L/R) \cdot (\nu/R U_i)$. The dimensionless quantity $(L/R) \cdot (\nu/R U_i)$ corresponds to the previously used dimensionless quantity $(L/R) \cdot (\nu/R^2 \omega_i)$, which was useful for the discussion about the intensity of swirl, etc.

Our objective is to get this functional relationship from the empirical viewpoint. Let us consider the inlet position $(z = 0)$ as z_2 and look at $\Delta \bar{P}_w$ between each measuring position and the inlet

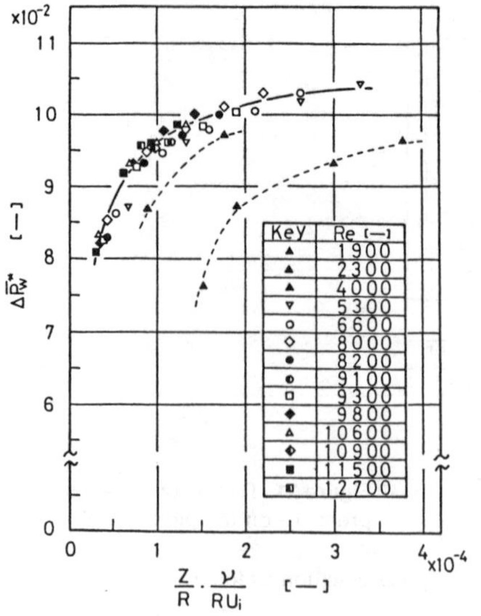

Figure 22. Relationship between dimensionless static pressure drop along the pipe wall $\Delta\bar{P}^*_\omega$ and $(z/R)\cdot(\nu/RU_i)$.

position. Figure 22 shows the relationship between $\Delta\bar{P}^*_w$ and $(z/R)\cdot(\nu/RU_i)$, where L is replaced by z. In the range of Re > 5,000, an almost unique correlation is obtained, but such an interrelation is not observed in the range of 2,000 < Re < 5,000. This range is considered to be the transitional one to fully-developed turbulent swirling flow and is the same as that observed for the decay process of the intensity of swirl in the preceding section.

Radial Distributions of Static Pressure

Radial distributions of $\bar{P}$ in swirling pipe flow have been often obtained either by direct measurement with static pressure tubes or by integrating the approximate relationship between $\bar{P}$ and $\bar{U}_\theta$, namely $\partial\bar{P}/\partial r = \rho\bar{U}_\theta^2/r$. However, in the case of turbulent swirling flow where the direction of flow is changing three-dimensionally, the above direct measurement necessitates considerable manipulative skill in the positioning of the static pressure tube, and thus leads easily to error. Furthermore, when the Navier-Stokes equation is applied to turbulent flow to determine the time-mean radial static pressure distributions, it is impossible to ignore entirely the terms of velocity fluctuations [28, 31].

In steady axially symmetrical swirling flow of an incompressible fluid, the terms containing the small quantity $\bar{U}_r$ and the small terms of the Reynolds stress in the r-component of the Navier-Stokes equation are neglected by comparison with the other terms, so that the following equation will be obtained:

$$\frac{\partial\bar{P}}{\partial r} = \rho\frac{\bar{U}_\theta^2}{r} - \rho\frac{1}{r}\frac{\partial}{\partial r}(r\bar{u}_r^2) + \rho\frac{\overline{u_\theta^2}}{r} \tag{34}$$

Integration of the above equation with respect to r between a fixed radial position r = r and the wall of the pipe r = R gives

$$\bar{P} = \bar{P}_w - \rho\int_r^R\left(\frac{\bar{U}_\theta^2 + \overline{u_\theta^2} - \overline{u_r^2}}{r}\right)dr - \rho\bar{u}_r^2 \tag{35}$$

When the value of $\bar{P}_w$ and the distributions of $\bar{U}_\theta$, $\overline{u_\theta^2}$ and $\overline{u_r^2}$ are known, the radial distribution of the static pressure $\bar{P}$ can be numerically obtained by using the above Equation 35.

Lines in Figure 20, which was shown at the beginning of this section, and Figure 23 show some examples of $\bar{P}$ distributions as calculated from Equation 35. Through these calculations, it is clear that the terms of velocity fluctuations have a several-percent effect on the value of $\bar{P}$.

The positive pressure gradient in the center region of the pipe reduces the axial velocity component, and this decelerating effect is considered to be a cause of generating the above-mentioned reverse flow. Furthermore, as found in the distributions at z = 0.7 m and z = 1.4 m, the negative pressure gradient just near the center axis is sometimes generated in the vicinity of the inlet when the intensity of swirl is especially strong. Such negative pressure gradient seems to be a cause of generating the axial flow type of (3) as shown in the preceding section.

Mechanical Energy Loss

The dissipation of mechanical energy of swirling pipe flow is discussed from the macroscopic mechanical energy balance for isothermal flow [32].

If a cylindrical space V shown in Figure 24 is set in the steady axially symmetrical swirling pipe flow, the mechanical energy loss F_μ per a unit time in V by friction is expressed by the following equation when $\bar{U}_r$ is negligibly small:

$$F_\mu = \iint_{S_2} \left(\frac{\rho}{2} \bar{U}^2 \bar{U}_z + \bar{P}\bar{U}_z - \tau_{z\theta}\bar{U}_\theta \right) dS - \iint_{S_1} \left(\frac{\rho}{2} \bar{U}^2 \bar{U}_z + \bar{P}\bar{U}_z - \tau_{z\theta}\bar{U}_\theta \right) dS \qquad (36)$$

By numerical integration of this equation based on experimental data, the value of F_μ can be obtained considering that the term of $\tau_{z\theta}\bar{U}_\theta$ is negligibly small in comparison with other two terms.

According to the general consideration that the quantity ΔE (= F_μ/Q) is the equivalent quantity to the pressure drop in non-swirling pipe flow, the so-called pressure loss in swirling pipe flow may

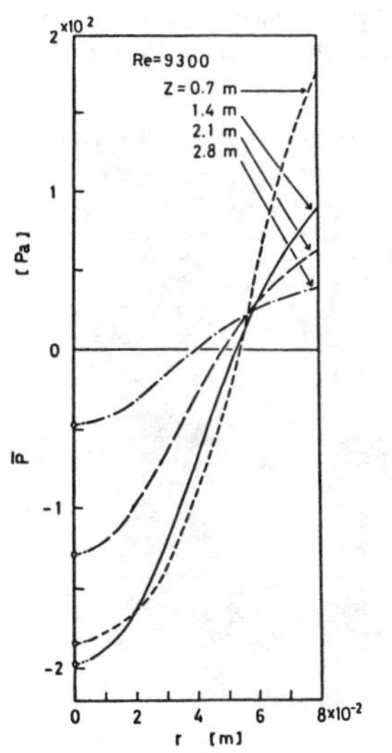

Figure 23. Calculated static pressure curves.

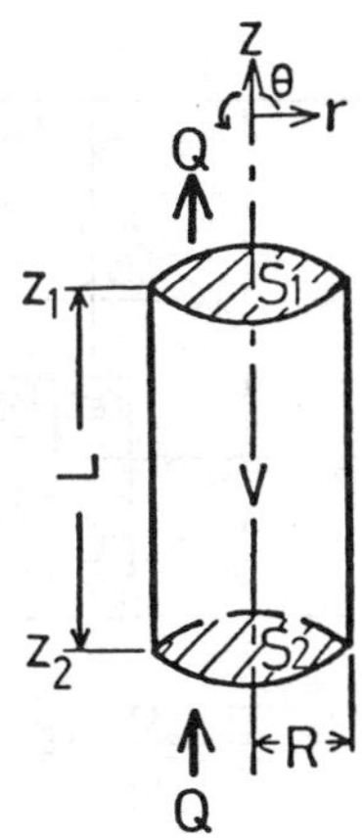

Figure 24. Integral space for mechanical energy loss.

be represented by ΔE. This ΔE can be transformed into the dimensionless form ΔE* by using ρU_i^2 in the same manner as the pressure drop $\Delta\bar{P}_w$, and the relation between ΔE* and $(z/R) \cdot (\nu/RU_i)$ is discussed. As shown in Figure 25, it was made clear that a good correlation was obtained in the fully turbulent range.

STATISTICAL QUANTITIES OF FLUCTUATIONS

In general, most swirling flow is in the turbulent state, and the effect of fluctuations on the flow condition should not be overlooked. However, there have been few experimental reports [19] about statistical quantities of fluctuations in turbulent swirling pipe flow. In this section, the turbulent characteristics will be made clear by using the experimental data.

The turbulent fluctuation is much promoted by the addition of the swirling motion, and this phenomenon seems to be caused chiefly by generation of the vorticity based on the swirling motion. The forced vortex holds the large vorticity, and the especially strong turbulence is generated in that zone. As a proof of it, the dye that is inserted into the vicinity of the center axis of the pipe is rapidly diffused in the positive radial direction, and the dyed cylindrical space is formed in the almost same region as the forced vortex zone as shown in Figure 26. This phenomenon possibly indicates the active turbulent diffusivity in the forced vortex zone, and also is related to the distributions of the Reynolds stress $-\rho\overline{u_r u_\theta}$ as mentioned in the preceding section.

Intensity of Turbulence

Some examples of the radial distributions of turbulence intensity, $(\overline{u^2})^{1/2}$, $(\overline{u_r^2})^{1/2}$, $(\overline{u_\theta^2})^{1/2}$, $(\overline{u_z^2})^{1/2}$, are shown in Figure 27, where the boundary radius r_b between two vortex zones is shown by a dotted line. Though the values of turbulence intensity are small near the center axis because of the small mean flow velocity, the turbulence is more powerful in the forced vortex zone than that in the quasi-free vortex zone on the whole. The radial distributions of the relative intensity $(\overline{u^2})^{1/2}/\bar{U}$ are shown in Figure 28. As known from this figure, the relatively powerful turbulence is generated in the forced vortex zone.

Considering that the vorticity has many effects on the generation of turbulent fluctuations, it is reasonable to suppose that the intensity of swirl Γ, which is the surface integral of the axial vorticity Ω_z, is closely related to the average intensity of turbulence over the whole cross section. Figure 29

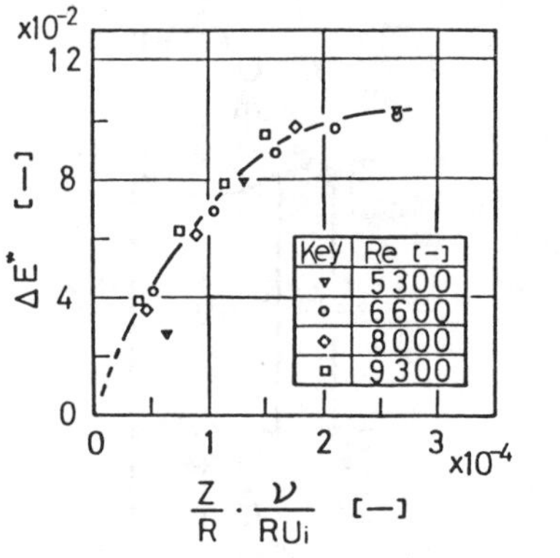

Figure 25. Relationship between dimensionless mechanical energy loss ΔE* and $(z/R) \cdot (\nu/RU_i)$.

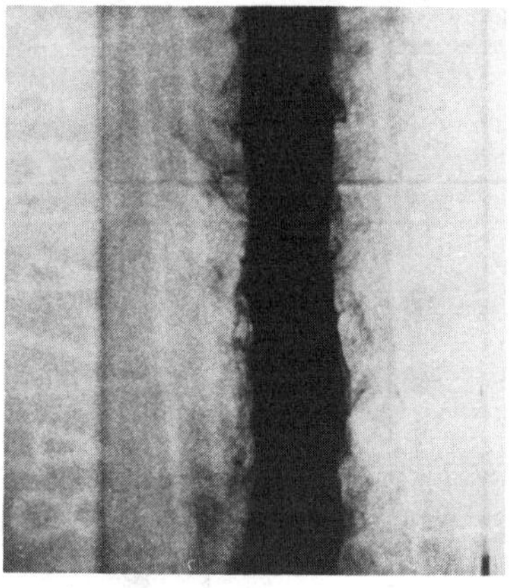

Figure 26. Diffusion of dye in the forced vortex zone.

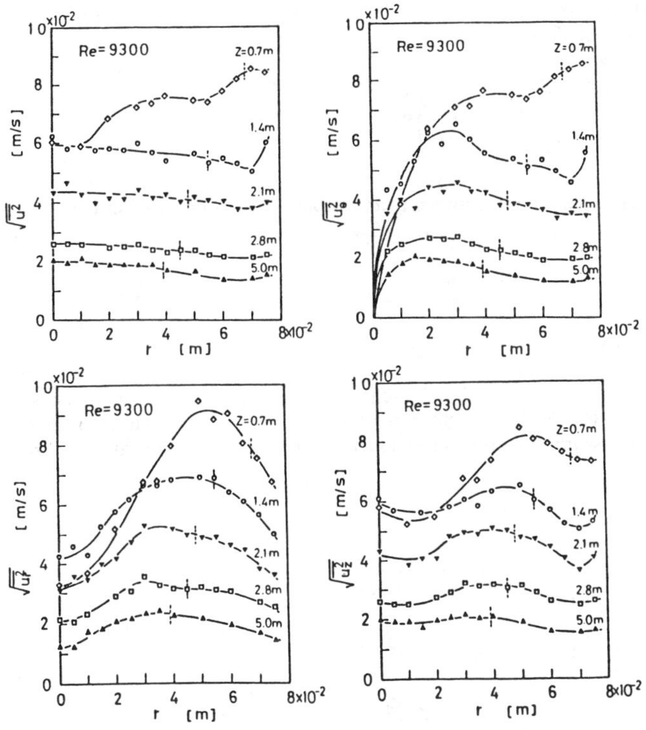

Figure 27. Distributions of turbulence intensity components.

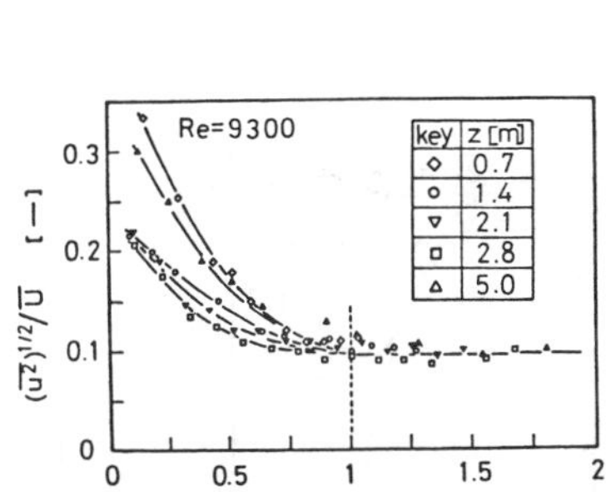

Figure 28. Distributions of relative turbulence intensity.

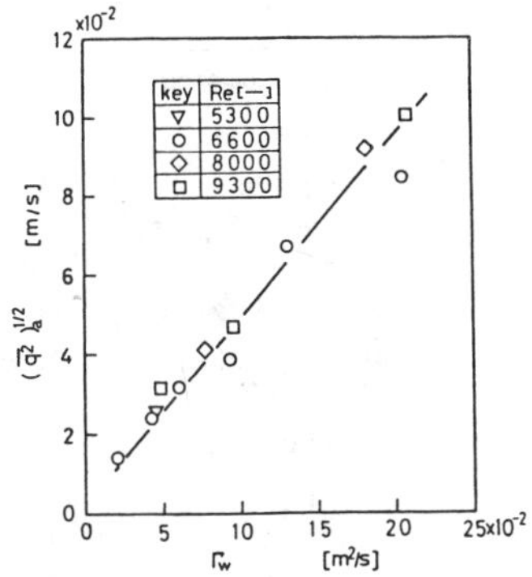

Figure 29. Relationship between swirl intensity Γ_w and cross-sectional average turbulence intensity $(\overline{q^2})_a^{1/2}$.

shows the relationship between Γ_w and the cross-sectional average root-mean-square value of the twice energy of turbulence $(\overline{q^2})_a^{1/2}$:

$$(\overline{q^2})_a^{1/2} = \left(\frac{\int_0^R (\overline{u_r^2} + \overline{u_\theta^2} + \overline{u_z^2}) 2\pi r \, dr}{\pi R^2} \right)^{1/2} \tag{37}$$

A good proportional relation between them is obtained as is expected in the case of the relatively strong swirl:

$$(\overline{q^2})_a^{1/2} \propto \Gamma_w \tag{38}$$

Turbulence Stress

Some examples of the radial distributions of turbulence stress (Reynolds stress) $\overline{u_r u_\theta}$, $\overline{u_r u_z}$, $\overline{u_\theta u_z}$ are shown in Figure 30. In particular, the $r\theta$-component $\overline{u_r u_\theta}$, which is an important factor for investigating the swirling motion, takes much larger value in the forced vortex zone than that in

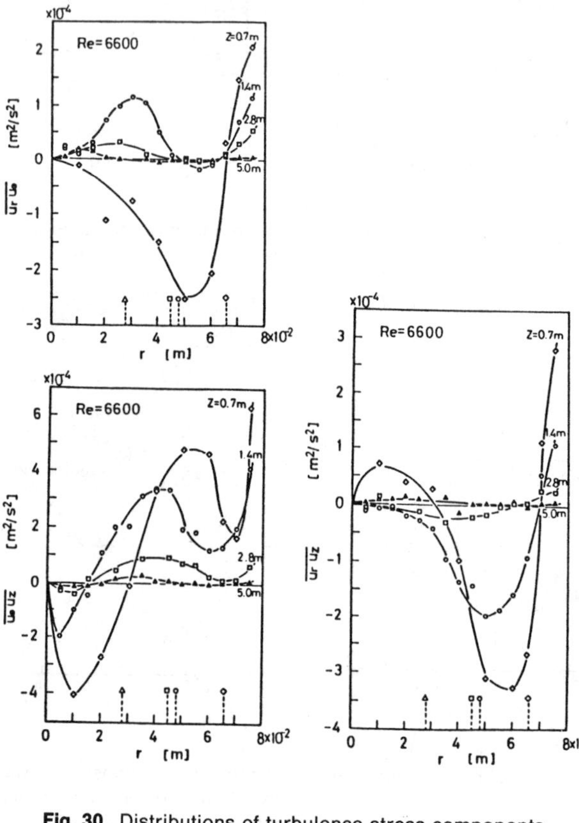

Fig. 30. Distributions of turbulence stress components.

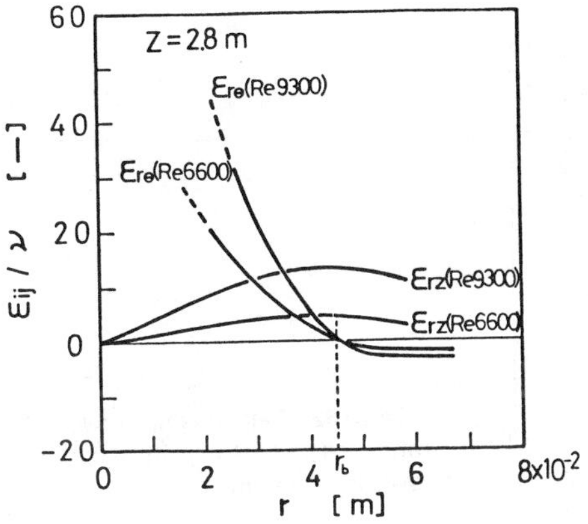

Figure 31. Distributions of two components $\epsilon_{r\theta}$, ϵ_{rz} of turbulence viscosity.

the quasi-free vortex zone. As it has been already mentioned in the preceding section, $\overline{u_r u_\theta}$ takes zero value just near the boundary position r_b.

Two components $\epsilon_{r\theta}$, ϵ_{rz} of turbulence viscosity [33] are calculated by using the following equations based on the previously mentioned experimental data,

$$-\overline{u_r u_\theta} = \epsilon_{r\theta}\left(\frac{\partial \bar{U}_\theta}{\partial r} - \frac{\bar{U}_\theta}{r}\right) \tag{39}$$

$$-\overline{u_r u_z} = \epsilon_{rz}\left(\frac{\partial \bar{U}_r}{\partial z} + \frac{\partial \bar{U}_z}{\partial r}\right) \tag{40}$$

and the relationship between each of them and r is discussed. As shown in Figure 31, $\epsilon_{r\theta}$ takes the positive large value especially in the forced vortex zone, and this is proved by the rapid diffusion of dye within that zone. The swirling motion greatly changes the turbulent mechanism and has many effects on the turbulent diffusion.

As is obvious from the above results, the value of the turbulence viscosity is changeable according to the position, the direction, and the operational condition. Some kinds of methods to estimate the turbulence viscosity in swirling flow have been reported [8, 16, 34–38], however, their application is restricted to each limited condition. At present, it has been left unsolved to complete the appropriate estimating method.

EXPANDING SWIRLING PIPE FLOW

Where the diameter of a pipe rapidly increases, swirling flow spreads in the positive radial direction. This expanding swirling flow is often called swirling jet flow. From the viewpoint of engineering, especially combustion engineering, it is a very useful configuration of flow [11, 36, 39, 40].

In this section, taking up such expanding swirling pipe flow as an example of more complicated swirling pipe flow, we will show some characteristics of it with reference to the above-mentioned way of treatments.

The experimental apparatus used consists of two pipes (their inner diameters are 0.1 m and 0.29 m), which are connected by adjusting each center axis. The steady fully-developed swirling flow in the small diameter pipe is let spread into the other large diameter pipe just like jet flow. Such expanding swirling flow in the large diameter pipe is the object of attention in this section.

Figure 32 shows the relationship between the dimensionless intensity of swirl Γ_w^* and $(z/R) \cdot (\nu/R^2\omega_i)$, where the inlet position is regarded as the joining place of two pipes. Also in this

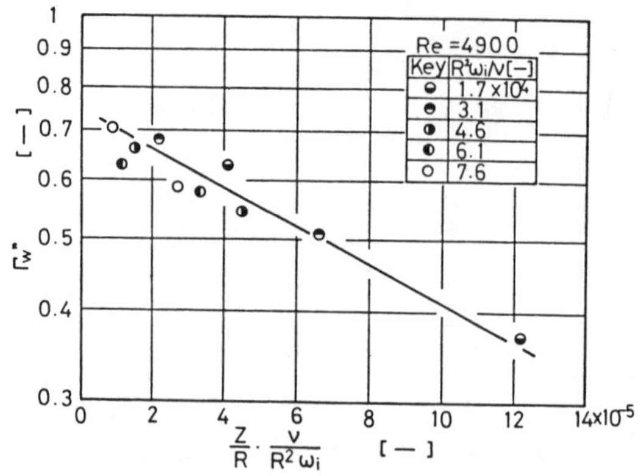

Figure 32. Relationship between Γ_w^* and $(z/R) \cdot (\nu/R^2\omega_i)$ in expanding swirling pipe flow.

case, an exponential relationship is found as

$$\Gamma_w^* = 0.75 \exp\left\{-6000\left(\frac{z}{R} \cdot \frac{\nu}{R^2\omega_i}\right)\right\} \tag{41}$$

where R is the inner radius of the large pipe. The value 6,000 in Equation 41 almost corresponds to the value of κ at Re = 4,900 in Figure 7.

Figure 33 shows the variation of the axial flow type which is explained in the preceding section. Three kinds of axial flow types are observed within the expanding swirling flow according to the operational conditions, which are expressed by the general Reynolds number Re $(= DU_{av}/\nu)$ and another Reynolds number $R^2\omega_i/\nu$ based on the swirling motion. In the same manner, as the swirl becomes stronger, the flow type of (2) or (3) is more easily generated.

Figure 34 shows the relationship between Γ_w and $(\overline{q^2})_a^{1/2}$. A good proportional relation is obtained also in this case.

These results may indicate the broad applicability of the way of treatments for simple turbulent swirling flow to more complicated swirling flow in practical apparatuses.

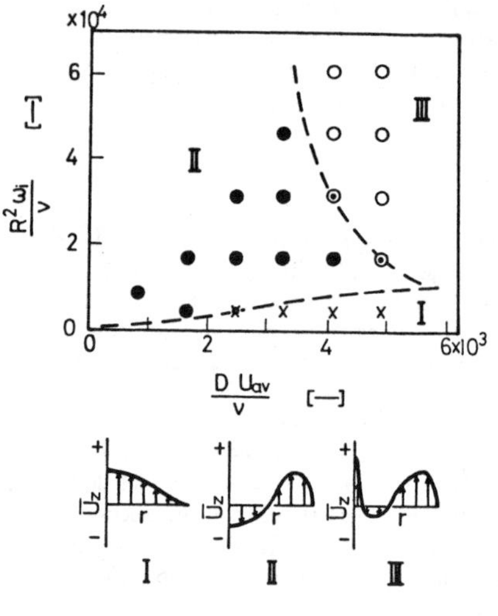

Figure 33. Three types of axial flow and operational conditions in expandings swirling pipe flow.

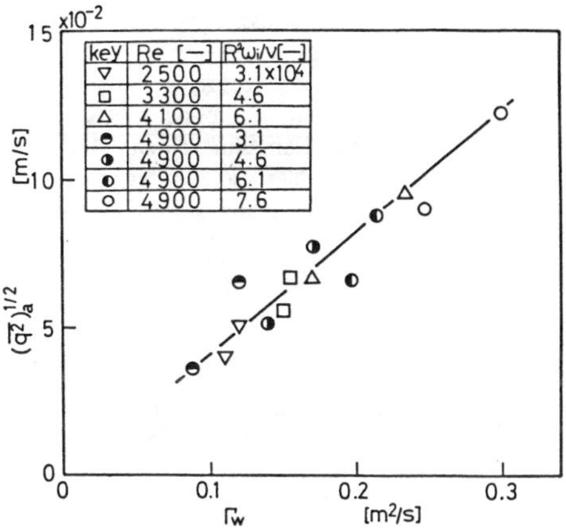

Figure 34. Relationship between Γ_w and $(\overline{q^2})_a^{1/2}$ in expanding swirling pipe flow.

The reader may refer to some reports [41, 42] in regard to swirling flow in a pipe of gradually variable cross section such as a conical pipe.

NOTATION

C	closed curve surrounding S, m
c_i	empirical constants of Equation 26, –
D	inner diameter of pipe, m
d	inner diameter
ΔE	energy dissipation rate/flow rate, kg/m s^2
ΔE^*	$\Delta E/\rho U_i^2$, –
F, f	empirical function,
F_μ	energy dissipation rate, kg m^2/s^3
k	experimental factor of Equation 25, –
k_1, k_2, k_3	empirical constants of Equations 2 and 3
L	axial distance between z_1 and z_2, m
m, n	empirical constants, –
P	static pressure, Pa
P_w	wall static pressure, Pa
P_w^*	$P_w/\rho U_i^2$, –
Q	flow rate, m^3/s
q^2	$u_r^2 + u_\theta^2 + u_z^2$, m^2/s^2
R	inner radius of pipe, m
Re	Reynolds number ($= DU_{av}/\nu$), –
r	radial coordinate, m
r_b	radial position of maximum U_θ, m
r^+	$rU_{\theta f}/\nu$, –
r^*	r/R, –
S	area of surface, m^2
Sw	Swirl number
$\underset{\sim}{U}$	velocity vector, m/s
U	velocity, m/s
U_{av}	cross-sectional average velocity, m/s
U_i	initial swirling velocity, m/s
U_r, U_θ, U_z	velocity components, m/s
$U_{\theta b}$	maximum value of U_θ, m/s
$U_{\theta f}$	friction velocity for swirling flow, m/s
U_z^*	$U_z/\langle U_z \rangle_m$, –
$\langle U_z \rangle$	general pipe flow velocity, m/s
$\langle U_z \rangle_m$	maximum $\langle U_z \rangle$ at center, m/s
u	velocity fluctuation, m/s
u_r, u_θ, u_z	velocity fluctuation components, m/s
V	closed space surrounded by S, m^3
z	axial coordinate, m

Greek Symbols

$B(\alpha, \beta)$	Beta function, –
Γ	circulation (intensity of swirl), m^2/s
Γ_b	$2\pi r_b U_{\theta b}$, m^2/s
Γ_w	extrapolating value of Γ to wall, m^2/s

Γ^+	$\Gamma/2\pi r_b U_{\theta f}$, −	τ	shear stress, Pa
Γ^*	$\Gamma/2\pi R U_i$, −	$\tau_{r\theta,b}$	$-\mu U_{\theta b}/r_b$, Pa
γ, γ'	coefficients of swirl decay, −	$\tilde{\Omega}$	vorticity vector, 1/s
ϵ	eddy viscosity, m^2/s	$\Omega_r, \Omega_\theta, \Omega_z$	vorticity components, 1/s
κ, κ'	coefficients of swirl decay, −	ω_i	initial angular velocity, 1/s
μ	molecular viscosity, kg/m s	ω_s	characteristic angular velocity in forced vortex zone, 1/s
ν	kinematic viscosity, m^2/s		
ρ	density, kg/m^3		

Subscripts

b	boundary position	w	wall position
r	radial direction	θ	tangential direction
z	axial direction	~	vector

Superscripts

$^+, ^*$	dimensionless quantity	$^-$	time-mean value

REFERENCES

1. Lugt, H. J., *Vortex Flow in Nature and Technology*, John Wiley & Sons, Inc., New York, 1983.
2. Nissan, A. H., and Bresan, V. P., *AIChE J.*, 7:543 (1961).
3. Ogawa, A., *Uzu Gaku*, Sankaido, Inc., Tokyo, 1981.
4. Rott, N., *Z. Angew. Math. Phys.*, 9 B:543 (1958).
5. Senoo, Y., and Nagata, T., *Trans. Soc. Mech. Eng. Japan B*, 38:759 (1972).
6. Burgers, J. M., *Advances in Appl. Mech.*, 1:171 (1948).
7. Kiya, M., Fukusako, S., and Arie, M., *Trans. Soc. Mech. Eng. Japan B*, 36:1865 (1970).
8. Kreith F., and Sonju, O. K., *J. Fluid Mech.*, 22:257 (1965).
9. Laban, Z., Nielsen, H., and Fejer, A. A., *Phys, Fluids*, 12:1747 (1969).
10. Talbot, L., *J. Appl. Mech.*, 21:1 (1954).
11. Syred, N., and Beer, J. M., *Combustion and Flame*, 23:143 (1974).
12. Murakami, M., Kito, O., Katayama, Y., and Iida, Y., *Trans. Soc. Mech. Eng. Japan B*, 41:1793 (1975).
13. Yajnik, K. S., and Sabbaiah, M. V., *J. Fluid Mech.*, 60:665 (1973).
14. Ito, S., Ogawa, K., and Kuroda, C., *Kagaku Kogaku Ronbunshu*, 4:247 (1978).
15. Ito, S., Ogawa, K., and Kuroda, C., *International Chem. Eng.*, 19:600 (1979).
16. Hoffmann, E. R., and Joubert, P. N., *J. Fluid Mech.*, 16:395 (1963).
17. Uberoi, M. S., *J. Fluid Mech.*, 90:241 (1979).
18. Ito, S., Ogawa, K., and Kuroda, C., *J. Chem. Eng. Japan*, 13:6 (1980).
19. Ito, S., Ogawa, K., and Kuroda, C., *Kagaku Kogaku Ronbunshu*, 1:121 (1975).
20. Binnie, A. M., *Quart. J. Mech. and Appl. Math.*, 10:276 (1957).
21. Nuttal, J. B., *Nature*, 172:582 (1953).
22. Bellamy-Knights, P. G., *Trans. ASME I, J. Fluids Eng.*, 98:322 (1976).
23. Bossel, H. H., *Phys. Fluids*, 12:498 (1969).
24. Harvey, J. K., *J. Fluid Mech.*, 14:585 (1962).
25. Suematsu, Y., Ito, T., Hayase, T., and Hori, F., *Trans. Soc. Mech. Eng. Japan B*, 47:1736 (1981).
26. Iinoya, K., *Trans. Soc. Mech. Eng. Japan*, 18:90 (1952).
27. Kuroda, C., Ogawa, K., and Inoue, I., *J. Chem. Eng. Japan*, 14:158 (1981).
28. King, M. K., Rothfus, R. R., and Kermode, R. I., *AIChE J.*, 15:837 (1969).
29. Ito, S., Ogawa, K., and Kuroda, C., *Kagaku Kogaku Ronbunshu*, 6:334 (1980).
30. Ito, S., Ogawa, K., and Kuroda, C., *International Chem. Eng.*, 22:295 (1982).
31. Reynolds, A., *ZAMP*, 12:149 (1961).

32. Bird, R. B., *Chem. Eng. Sci.*, 6:123 (1957).
33. Hinze, J. O., *Turbulence 2nd.*, McGraw-Hill, Inc., New York, 1975.
34. Kinney, R. B., *Trans. ASME E, J. Appl. Mech.*, 34:437 (1967).
35. Kubo, I., and Gouldin, F. C., *Trans. ASME I, J. Fluids Eng.*, 97:310 (1975).
36. Lilley, D. G., and Chigier, N. A., *Int. J. Heat Mass Trans.*, 14:573 (1971).
37. Rochino, A., and Lavan, Z., *Trans. ASME E, J. Appl. Mech.*, 36:151 (1969).
38. Uberoi, M. S., *Phys. Fluids*, 20:719 (1977).
39. Beltagui, S. A., and Maccallum, N. R. L., *J. Inst. Fuel.*, 49:183 (1976).
40. Beltagui, S. A., and Maccallum, N. R. L., *J. Inst. Fuel.*, 49:193 (1976).
41. Senoo, Y., Takesue, N., and Nagata, T., *Trans. Soc. Mech. Eng. Japan B*, 43:1803 (1977).
42. Wilks, G., *J. Fluid Mech.*, 34:575 (1968).
43. Deissler, R. G., and Perlmutter, M., *Int. J. Heat Mass Transfer*, 1:173 (1960).
44. Reynolds, A. J., *ZAMP*, 12:343 (1961).
45. Sibulkin, M., *J. Fluid Mech.*, 12:269 (1962).
46. Takahama, H., *Trans. Soc. Mech. Eng. Japan B*, 30:1419 (1964).
47. Takahama, H., and Soga, N., *Trans. Soc. Mech. Eng. Japan B*, 31:787 (1965).
48. Takahama, H., *Trans. Soc. Mech. Eng. Japan B*, 32:503 (1966).

CHAPTER 21

TURBULENT FLOWS WITHIN STRAIGHT DUCTS

A. Nakayama

Department of Mechanical Engineering
Shizuoka University
Johoku, Japan

and

W. L. Chow

Department of Mechanical and Industrial Engineering
University of Illinois at Urbana-Champaign
Urbana, IL

CONTENTS

INTRODUCTION

In a turbulent flow within a straight duct of non-circular cross-section, a transverse mean flow exists even when the flow is fully developed. This transverse flow, commonly known as the secondary flow, brings the fluids into complex lateral spiral motions as superimposed upon the axial mean flow. Thus, the turbulent flow in a non-circular duct is of a three-dimensional nature even for the mean velocity field. Nikuradse [1] was the first to report the details of the axial mean flow field in various non-circular ducts, and subsequently confirmed the existence of the secondary flow through the flow visualization technique [2]. The secondary flow of this type, termed by Prandtl [3, 4a, b] as the secondary flow of the second kind, exhibits the tendency for the core fluid to flow toward corner narrow regions, then parallel to the walls, and finally back into the core through the wider regions as indicated in Figure 1. Although the magnitude of the secondary flow scarcely amounts to a few percent of the bulk velocity and is usually negligible if the motion is also induced by a cross-sectional pressure gradient, its presence displaces the lines of constant axial mean velocity (i.e. isovels) considerably toward the narrow regions, yielding a comparatively high velocity field there.

The first explanation of the phenomenon was given by Prandtl [4] who suggested that the anisotropy of the cross planar (transverse) normal stresses is responsible for the secondary flow generation. Prandtl postulated that the velocity fluctuations tangential to an isovel exceed those perpendicular to it so that a centrifugal acceleration is induced in the region of isovel curvature, driving fluid radially outwards. Since the discovery of this phenomenon, not much effort was further directed to its study before sensitive instruments for more accurate observations were developed and improved.

Experimental investigations on fully developed flows in non-circular ducts were carried out by many workers such as Hoagland [5], Rodet [6], Leutheusser [7], Tracy [8], Gessner [9], and Launder and Ying [10]. The hot wire technique developed by Hoagland [5] for the secondary flow measurements was later followed by several workers with successive improvements in measurement accuracy (interested readers may consult Gessner [9]). A thorough description of the phenomenon was given first by Brundrett and Baines [11] through the hot wire measurements on all six Reynolds stress components in a square duct. It was shown that the gradients in the cross planar Reynolds stresses generate the streamwise vorticity and are fully responsible for the secondary motions. Gessner and Jones [12] carried out the hot wire measurements within a square duct by following a particular secondary flow streamline and showed that the principal axes of Reynolds stress field are not normal or tangent to the isovels. Similar evidence was reported by Perkins [13] who directly measured the orientation of the Reynolds stress principle planes in various ducts of non-circular cross sections.

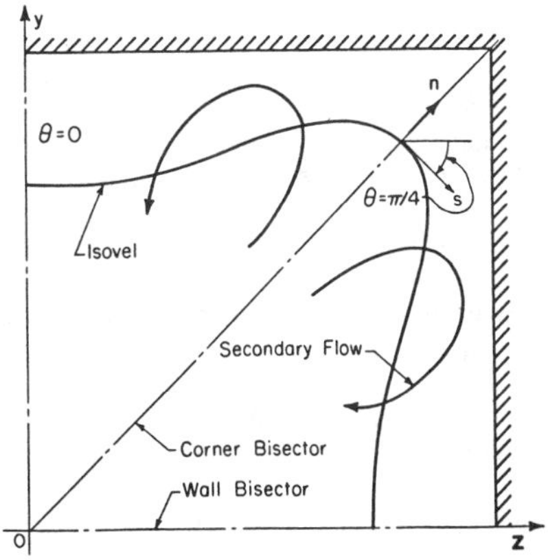

Figure 1. Typical isovel pattern.

For the developing turbulent flows in non-circular ducts on the other hand, experimental data are scarce especially on the development of the turbulent stress fields. In fact, only two sets of turbulence data are known to the authors at this time, namely those of Gessner, Po and Emery [14] and those of Melling and Whitelaw [15]. Gessner and his co-workers made a thorough experimental investigation on the three-dimensional developing turbulent flow in a square duct using a hot wire anemometer while exhaustive laser Doppler measurements were performed by Melling and Whitelaw at two axial stations of the square duct.

Some theoretical studies were carried out to analyze the turbulent flows in non-circular ducts. Diessler and Taylor [16] neglected the secondary flow motions and extended the law of the wall to the non-circular ducts. Their proposal was followed by Buleev [17], Ślykov and Carevskij-Djakin [18], Krajewski [19] and many others to determine the turbulent velocity fields in non-circular ducts. All these analyses were based on the assumption that the empirical and semi-empirical results available on the circular duct are also valid for other non-circular ducts when an isovel contour in the cross section is conformally transformed into a circle. These analyses all presume that the effective diffusion coefficient for momentum transfer is isotropic. This isotropic viscosity formulation cannot predict the secondary flow motions driven by the Reynolds stresses since the principal planes of the stress field and mean strain field under such a formulation are co-aligned everywhere. These analyses gave considerable deviations from the experimental results. Thus, suppression of the secondary flow in the simulation of turbulence only leads to an unrealistic prediction of the phenomenon.

Some workers did attempt to account for the secondary flow effects in the non-circular ducts. Gerard [20], Meyder [21], Ram and Johannsen [22] and Ibragimov, et al., [23] prescribed the directional dependency of the effective viscosity while Gerard [24] specified the secondary flow velocity field itself for the calculation of axial velocity distribution. The results obtained from these studies indicated correct trends of the secondary flow and gave improved agreement with experiments. These methods, however, too heavily rely on the direct experimental input. It is impossible to generalize these methods for these ducts of arbitrary cross sections.

Different Stress Models

A successful attempt to predict (rather than prescribe) the secondary flow motions in a non-circular duct was made by Launder and Ying [25] using the algebraic stress model reduced by themselves from the Reynolds stress transport equation with a number of approximations and assumptions. The prediction was made on a square duct by solving the vorticity transport equation, the axial momentum equation, the energy equation, and the turbulent kinetic energy transport equation with the aid of Buleev's proposal for the turbulence length scale [17]. The predicted mean velocity field was found to be in good agreement with the experiment. This innovative work on the secondary flow prediction was followed by many workers to calculate fully developed turbulent flows in other non-circular ducts. Aly, Trupp and Gerrard [26] and Gosman and Rapley [27] carried out the calculations on equilateral triangular ducts while the prediction was made on a trapezoidal duct by Nakayama, Chow, and Sharma [28]. Triangular rod bundle arrays were treated first by Crajilescov and Todreas [29] and later by Trupp and Aly [30] without neglecting the vorticity production term due to secondary shear stress. Gosman and Rapley [31] made extensive calculations and reported the results for the fully developed flows in various geometrical configurations including elliptical ducts and tube assemblies having variable pitch-to-diameter ratios.

Several workers have extended the above Launder-Ying algebraic stress model for the prediction of the three-dimensional turbulent flows. Gessner and Emery [32] developed a length scale model to prescribe the length scale in the three-dimensional space and coupled it to the Launder-Ying stress model for the calculations of the developing flow in a square duct [33]. Also for the developing flows, Tatchell [34] replaced Buleev's length scale employed in the initial work of Launder and Ying [25] by the transport equation for the dissipation rate of the turbulent kinetic energy. An identical scheme was applied by Neti [35] for the calculation of the developing flow in a square duct. Furthermore, Nakayama, Chow, and Sharma [36] carried out the three-dimensional fully elliptic calculations for the developing turbulent flow in a square duct, and substantiated the validity of the boundary layer approximation adopted in the parabolic marching calculations.

Instead of using the algebraic formulation, Reece [37] chose to solve the Reynolds stress transport equations using the differential transport model developed by Launder, Reece, and Rodi [38]. Naot, et al., [39] proposed a modified version of the differential transport model and performed fully developed flow calculations for a square duct. Despite the considerable complexity and increase in computational time, the results so obtained do not seem to exhibit any substantial improvement over the Launder-Ying algebraic stress model.

In this chapter we will examine the numerical solution of the secondary flows of the second kind within ducts of non-circular ducts. In view of the available stress models, the Launder-Ying algebraic stress model, which is representative of all previously mentioned alternatives, is selected to incorporate with the k-ϵ (turbulent kinetic energy and its dissipation rate) model for simulation of turbulence; thereby, its capability for describing the secondary flow effects can be explored. The following sections are devoted to the mathematical description of turbulence, the modeling procedure of the secondary flow, the detailed description of the numerical procedure, and finally the selective comparisons of the results of predictions with the available experimental data.

MATHEMATICAL DESCRIPTION

Reynolds Averaged Navier-Stokes Equation

For the mathematical description of turbulent viscous flows, one starts with the Navier-Stokes equation. A usual time averaging process on the mass and momentum conservation equations in a steady state leads to the time averaged continuity equation and the Reynolds-averaged Navier-Stokes equation which are given, in Cartesian tensor notations as

$$\frac{\partial \bar{u}_j}{\partial x_j} = 0 \tag{1a}$$

$$\frac{\partial}{\partial x_j} \bar{u}_i \bar{u}_j = -\frac{1}{\rho}\frac{\partial \bar{p}}{\partial x_i} + \frac{\partial}{\partial x_j}(2\nu \bar{s}_{ij} - \overline{u_i' u_j'}) \tag{1b}$$

where

$$\bar{s}_{ij} = \frac{1}{2}\left(\frac{\partial \bar{u}_i}{\partial x_j} + \frac{\partial \bar{u}_j}{\partial x_i}\right) \tag{1c}$$

In these equations, the bar and prime indicate the time averaged mean and the fluctuation from its mean value, respectively. Thus, the instantaneous velocity u_i, for example, is decomposed as $u_i = \bar{u}_i + u_i'$. The set of equations can be solved just as the laminar case of flow if the Reynolds stresses $-\overline{u_i' u_j'}$ are related to other mean flow quantities. The effective viscosity formulation, which is a direct extension of the laminar deformation law, is given by

$$-\overline{u_i' u_j'} = 2\nu_t \bar{s}_{ij} - \tfrac{2}{3} k\, \delta_{ij} \tag{2a}$$

where

$$k \equiv \tfrac{1}{2}\overline{u_i' u_i'} \tag{2b}$$

In Equation 2, k is the turbulent kinetic energy, and ν_t denotes turbulent kinematic viscosity which, unlike its laminar counterpart, varies spatially, and is not a property of the fluid.

Due to the reason already inferred, this isotropic viscosity formulation cannot predict the Reynolds stress driven secondary flow of the present concern. Nevertheless, it may be worthwhile to mention that the isotropic viscosity formulation in conjunction of the k-ϵ two equation turbulence model, has been successfully adopted to solve a number of complex turbulent flow problems.

k-ϵ Turbulence Model

The introduction of ν_t requires the task of expressing ν_t itself in terms of mean flow quantities. In order to accomplish this, many hypotheses have been proposed. Among them, Prandtl's mixing length hypothesis [3] appears to be one of the most popular proposals since it is simple and yet quite effective for simple flow cases. Many attempts were made to achieve universality in the turbulence modeling, and its applicability to more complex flow situations. Some more elaborate turbulence modelings are available at the present time. These models have been thoroughly reviewed by several investigators (Mellow and Herring [40] and Launder and Spalding [41]), and some of these models have been tested and compared with various experimental data in the establishment of their capabilities (Launder, et al., [42]).

The two equation model, known as k-ϵ model, was proposed by Harlow and Nakayama [43]. This model introduces the transport equations for the two turbulence quantities, namely, the turbulence kinetic energy k and its rate of dissipation ϵ. The following exact form of the transport equation for k may readily be extracted from the Navier-Stokes equation through the time averaging process:

$$\frac{\partial}{\partial x_j} k\bar{u}_j = \frac{\partial}{\partial x_j}\left[2\nu\overline{u_i's_{ij}'} - \overline{u_j'\left(\frac{1}{2}u_i'u_i' + \frac{p}{\rho}\right)}\right] - \overline{u_i'u_j'}\,\bar{s}_{ij} - 2\nu\overline{s_{ij}'s_{ij}'} \tag{3}$$

Upon performing an order of magnitude analysis of Equation 3, it can be shown that the first term of diffusion $\partial/\partial x_j\, 2\nu\overline{u_i's_{ij}'}$ becomes negligibly small compared with other terms when Reynolds number is sufficiently high. The following Prandtl's proposal is applied for the rate of diffusion:

$$\begin{aligned} 2\nu\overline{u_i's_{ij}'} - \overline{u_j'\left(\frac{1}{2}u_i'u_i' + \frac{p'}{\rho}\right)} &\simeq -\overline{u_j'\left(\frac{1}{2}u_i'u_i' + \frac{p}{\rho}\right)} \\ &= \left(\nu + \frac{\nu_t}{\sigma_k}\right)\frac{\partial k}{\partial x_j} \end{aligned} \tag{4}$$

where σ_k is the effective Prandtl number for k. Substitution of Equation 4 into Equation 3 yields

$$\frac{\partial}{\partial x_j} k\bar{u}_j = \frac{\partial}{\partial x_j}\left(\nu + \frac{\nu_t}{\sigma_k}\right)\frac{\partial k}{\partial x_j} + P - \epsilon \tag{5a}$$

where

$$P = -\overline{u_i'u_j'}\,\bar{s}_{ij} \quad \text{and} \quad \epsilon = 2\nu\overline{s_{ij}'s_{ij}'} \tag{5b, c}$$

P and ϵ are the rates of kinetic energy production and dissipation per unit mass, respectively. The transport equation for the kinetic energy dissipation proposed by Hanjalic [44] is given in a form similar to Equation 5a as

$$\frac{\partial}{\partial x_j} \epsilon\bar{u}_j = \frac{\partial}{\partial x_j}\left(\nu + \frac{\nu_t}{\sigma_\epsilon}\right)\frac{\partial \epsilon}{\partial x_j} + (c_1P - c_2\epsilon)\frac{\epsilon}{k} \tag{6}$$

where c_1 and c_2 are empirical constants, and σ_ϵ is the effective Prandtl number for the dissipation. The correlation, based on an isotropic viscosity assumption for the turbulent kinematic viscosity in terms of k and ϵ, as given by

$$\nu_t = c_D k^2/\epsilon \tag{7}$$

finally closes the system of Equations 1a, b, 2a, 5a, and 6, where c_D is another empirical constant.

Launder and Spalding [45] have demonstrated the wide range of applicability of this particular two-equation turbulence model by comparing its results with the available experimental data.

Reynolds Stress Transport Equation

Since the secondary flow of the second kind is caused by the anisotropy of Reynolds Stress, one cannot employ the isotropic viscosity formulation (Equation 2a) for the secondary flow prediction. To derive the stress model appropriate to this secondary flow simulation, one should start out with the Reynolds stress transport equation as given by

$$\frac{\partial}{\partial x_l}\left[\bar{u}_l\overline{u_i'u_j'} + \left\{\overline{u_l'u_i'u_j'} - \nu\frac{\partial}{\partial x_l}\overline{u_i'u_j'} + \overline{\frac{p'}{\rho}(\delta_{lj}u_i' + \delta_{li}u_j')}\right\}\right]$$

convection diffusion

$$= -\left(\overline{u_i'u_l'}\frac{\partial \bar{u}_j}{\partial x_l} + \overline{u_j'u_l'}\frac{\partial \bar{u}_i}{\partial x_l}\right) - 2\nu\overline{\frac{\partial u_i'}{\partial x_l}\frac{\partial u_j'}{\partial x_l}} + \frac{1}{\rho}\overline{2p's_{ij}'} \tag{8}$$

$+P_{ij}$: production $-\epsilon_{ij}$: dissipation $+R_{ij}$: redistribution.

The diagonal trace of this Reynolds stress transport equation, as it should, reduces to the k-equation (Equation 3). The terms on the left-hand side of Equation 8 represent the spatial transport of Reynolds stresses. As the divergence theorem indicates, these terms can influence the overall aspect of the Reynolds stresses only through the events occurring on the boundaries which are not subjected to the no-slip condition. P_{ij} on the right-hand side of the transport equation represents the rate of generation of $\overline{u_i'u_j'}$ by the mean rate of strain while ϵ_{ij} represents the rate of dissipation of $\overline{u_i'u_j'}$. Due to the continuity principle, the trace of the redistribution term R_{ij} vanishes. Therefore, it does not appear in the k-equation (Equation 3). This indicates that the term works only to redistribute the energy among the fluctuating velocity components by the action of the fluctuating pressure.

Although Equation 8 gives a logical set of partial differential equations for all six Reynolds stress components, it is quite formidable to deal with these Reynolds stress components directly through these differential equations. Even under the assumption that the dissipation of turbulent kinetic energy is isotropic, one must solve at least eleven partial differential equations to simulate a particular three-dimensional flow situation. These partial differential equations are: one equation for continuity, three equations for mean momentum, six equations for Reynolds stresses, and one equation for the dissipation rate of turbulent kinetic energy or an equivalent transport equation (the k-equation is implicit in Equation 8). All together, eleven partial differential equations must be solved simultaneously.

As an alternative to the above differential stress model, the algebraic stress models have been widely adopted for various turbulent flow situations (Launder [46] Rodi [47]). If the derivatives of $\overline{u_i'u_j'}$, which appear only on the left-hand side of Equation 8 are approximated by the expressions containing mean quantities such as $\bar{s}_{ij}$ and k, Equation 8 reduces from a system of the differential equations to an algebraic set of equations among $\overline{u_i'u_j'}$, k, ϵ, and $\bar{s}_{ij}$. Consequently, only two partial differential equations for the modeling of k and ϵ must be solved to close the system of equations. The algebraic stress model developed by Launder and Ying for the secondary flow of the second kind is discussed in the following section.

TURBULENCE MODEL FOR SECONDARY FLOW SIMULATION

Derivation of Launder-Ying Algebraic Stress Model

The study on a square duct by Brundrett and Baines [11] of the Reynolds stresses indicates that the effects due to the convection and diffusion of the Reynolds stresses are insignificant near the wall where the production of secondary flow vorticity is largest. Hence, as a first approximation, one may neglect the spatial transport terms on the left-hand side of Equation 8. Then, the stress transport equation can be written simply as

$$P_{ij} - \epsilon_{ij} + R_{ij} = 0 \tag{9}$$

The assumption that the small scales of motion responsible for ϵ_{ij} are isotropic leads to

$$\epsilon_{ij} = \tfrac{2}{3}\epsilon\delta_{ij} \tag{10}$$

The local equilibrium condition is implicitly included in this procedure since the trace of Equation 9 under the assumption described by Equation 10 yields

$$P \equiv -\overline{u_i'u_j'}\bar{s}_{ij} = \epsilon \tag{11}$$

Although Equation 11 may not hold for the region away from the wall, it may still serve as a good approximation there. For a straight duct flow, the coordinate x is chosen in the axial direction while y and z are taken in vertical and horizontal directions over the cross section. Accordingly, the primary (axial) velocity is denoted by $\bar{u}$ and the secondary flow velocities by $\bar{v}$ and $\bar{w}$. Upon noting that the secondary flow magnitude is small compared with the primary flow, a usual order of magnitude estimation on the basis of the continuity principle leads to

$$\frac{\partial\bar{v}}{\partial x}, \frac{\partial\bar{w}}{\partial x} \ll \frac{\partial\bar{u}}{\partial x}, \frac{\partial\bar{v}}{\partial y}, \frac{\partial\bar{w}}{\partial y}, \frac{\partial\bar{v}}{\partial z}, \frac{\partial\bar{w}}{\partial z} \ll \frac{\partial\bar{u}}{\partial y}, \frac{\partial\bar{u}}{\partial z} \tag{12}$$

Since the secondary flow velocity magnitude is usually found to be at most a few percent of the bulk velocity, the secondary flow velocity gradients within the cross-sectional plane, $\frac{\partial\bar{v}}{\partial y}, \frac{\partial\bar{w}}{\partial y}, \frac{\partial\bar{v}}{\partial z}$, and $\frac{\partial\bar{w}}{\partial z}$ are on the average two orders of magnitude smaller than the primary flow velocity gradients $\frac{\partial\bar{u}}{\partial y}$ and $\frac{\partial\bar{u}}{\partial z}$. Thus, from the definition, the production rate of the Reynolds stresses can be written approximately as

$$P_{ij} \approx \begin{Bmatrix} 2\epsilon - \left(\overline{v'^2}\frac{\partial\bar{u}}{\partial y} + \overline{v'w'}\frac{\partial\bar{u}}{\partial z}\right) - \left(\overline{w'^2}\frac{\partial\bar{u}}{\partial z} + \overline{v'w'}\frac{\partial\bar{u}}{\partial y}\right) \\ \text{--} \quad 0 \quad 0 \\ \text{--} \quad \text{--} \quad 0 \end{Bmatrix} \tag{13}$$

In this manipulation, the turbulent kinetic energy production rate has been replaced by its dissipation rate according to Equation 11 as

$$\epsilon = P \equiv -\overline{u_i'u_j'}\bar{s}_{ij} \approx -\left(\overline{u'v'}\frac{\partial\bar{u}}{\partial y} + \overline{u'w'}\frac{\partial\bar{u}}{\partial z}\right) \tag{14}$$

and the symbol "--" has been introduced to simplify the writing. Now, an exact expression for the pressure-strain correlation (Chou, [48]) suggests that the term can be decomposed depending on two types of physical processes, namely,

$$R_{ij} = (R_{ij})_1 + (R_{ij})_2$$

where

$(R_{ij})_1$: for the turbulent-turbulent interaction, (15)

and

$(R_{ij})_2$: for the turbulent-mean strain interaction.

Rotta [49] used the concept of "return to isotropy" for the turbulent-turbulent interaction as

$$(R_{ij})_1 = -c_{\phi 1}\frac{\epsilon}{k}\left(\overline{u_i'u_j'} - \frac{2}{3}k\delta_{ij}\right) \tag{16}$$

This model works in such a way that the flow decays into the isotropic state in an exponential manner when inserted into the Reynolds stress transport equation for the case of the anisotropic homogeneous flow with small mean shear.

For the turbulent-mean strain interaction, the following form suggested by Hanjalic and Launder [50] may be used:

$$(R_{ij})_2 = (a_{lj}^{mi} + a_{li}^{mj})\frac{\partial \bar{u}_l}{\partial x_m} \tag{17a}$$

with

$$\begin{aligned} a_{lj}^{mi} = {} & \alpha\overline{u_m'u_i'}\delta_{lj} \\ & + \beta(\overline{u_m'u_l'}\delta_{ij} + \overline{u_m'u_j'}\delta_{il} + \overline{u_i'u_j'}\delta_{ml} + \overline{u_i'u_l'}\delta_{mj}) \\ & + [\gamma\delta_{mi}\delta_{lj} + \eta(\delta_{ml}\delta_{ij} + \delta_{mj}\delta_{il})]k \\ & + c_{\phi 2}(\overline{u_m'u_i'}\cdot\overline{u_l'u_j'} - \overline{u_m'u_j'}\cdot\overline{u_i'u_l'} - \overline{u_m'u_l'}\cdot\overline{u_i'u_j'})/k \end{aligned} \tag{17b}$$

Tensor symmetry and mass conservation give the relations for the coefficients, α, β, γ, and η in terms of c_{ϕ_2} as

$$\alpha = -\tfrac{2}{11}(4c_{\phi 2} - 5), \qquad \beta = \tfrac{2}{11}(3c_{\phi 2} - 1) \tag{18a, b}$$

$$\gamma = \tfrac{4}{55}(3c_{\phi 2} - 1), \qquad \eta = -\tfrac{6}{55}(3c_{\phi 2} - 1) \tag{18c, d}$$

Under the assumption of the relation (Equation 12) for small secondary flow magnitude, Equation 17 may readily be simplified as

$$(R_{ij})_2 \simeq (av_{ij} + av_{ji})\frac{\partial \bar{u}}{\partial y} + (aw_{ij} + aw_{ji})\frac{\partial \bar{u}}{\partial z} \tag{19a}$$

with

$$\begin{aligned} av_{ij} = {} & \alpha\overline{v'u_i'}\delta_{j1} \\ & + \beta(\overline{u'v'}\delta_{ij} + \overline{v'u_j'}\delta_{i1} + \overline{u'u_i'}\delta_{j2}) \\ & + (\gamma\delta_{2i}\delta_{j1} + \eta\delta_{2j}\delta_{i1})k \\ & + c_{\phi 2}(\overline{v'u_i'}\cdot\overline{u_j'u'} - \overline{v'u_j'}\cdot\overline{u_i'u'} - \overline{u'v'}\cdot\overline{u_i'u_j'})/k \end{aligned} \tag{19b}$$

$$\begin{aligned} aw_{ij} = {} & \alpha\overline{w'u_i'}\delta_{j1} \\ & + \beta(\overline{u'w'}\delta_{ij} + \overline{w'u_j'}\delta_{i1} + \overline{u'u_i'}\delta_{j3}) \\ & + (\gamma\delta_{3i}\delta_{j1} + \eta\delta_{3j}\delta_{i1})k \\ & + c_{\phi 2}(\overline{w'u_i'}\cdot\overline{u_j'u'} - \overline{w'u_j'}\cdot\overline{u_i'u'} - \overline{u'w'}\cdot\overline{u_i'u_j'})/k \end{aligned} \tag{19c}$$

The sub-indices 1, 2, and 3 correspond to x, y, and z coordinates. Equation 19 along with Equation 14 determine the turbulent-mean strain interaction $(R_{ij})_2$ as

$$(R_{uu})_2 = 2c_{\phi 2}\epsilon\overline{u'^2}/k - 2(\alpha + 2\beta)\epsilon \tag{20a}$$

$$(R_{vv})_2 = 2c_{\phi 2}\epsilon\overline{v'^2}/k + 2\beta\left(\frac{\partial\bar{u}}{\partial y}\cdot\overline{u'v'} - \epsilon\right) \tag{20b}$$

$$(R_{ww})_2 = 2c_{\phi 2}\epsilon\overline{w'^2}/k + 2\beta\left(\frac{\partial\bar{u}}{\partial z}\cdot\overline{u'w'} - \epsilon\right) \tag{20c}$$

$$(R_{vw})_2 = 2c_{\phi 2}\epsilon\overline{v'w'}/k + \beta\left(\frac{\partial\bar{u}}{\partial z}\cdot\overline{u'v'} + \frac{\partial\bar{u}}{\partial y}\cdot\overline{u'w'}\right) \tag{20d}$$

$$(R_{uv})_2 = 2c_{\phi 2}\epsilon\overline{u'v'}/k + \frac{\partial\bar{u}}{\partial y}\left[(\alpha+\beta)\overline{v'^2} + \beta\overline{u'^2} + (\gamma+\eta)k\right] + (\alpha+\beta)\frac{\partial\bar{u}}{\partial z}\cdot\overline{v'w'} \tag{20e}$$

$$(R_{uw})_2 = 2c_{\phi 2}\epsilon\overline{u'w'}/k + \frac{\partial\bar{u}}{\partial z}\left[(\alpha+\beta)\overline{w'^2} + \beta\overline{u'^2} + (\gamma+\eta)k\right] + (\alpha+\beta)\frac{\partial\bar{u}}{\partial y}\cdot\overline{v'w'} \tag{20f}$$

where

$$(R_{ij})_2 = \begin{bmatrix} (R_{uu})_2 & (R_{uv})_2 & (R_{uw})_2 \\ --- & (R_{vv})_2 & (R_{vw})_2 \\ --- & --- & (R_{ww})_2 \end{bmatrix} \tag{20g}$$

The substitution of Equations 10, 13, 16, and 20 into Equation 9 gives a set of six algebraic equations for Reynolds stress components, namely,

$$\overline{u'^2} = c'_{ko} \tag{21a}$$

$$\overline{v'^2} = c'\frac{\partial\bar{u}}{\partial y}\cdot\overline{u'v'} + c'_k \tag{21b}$$

$$\overline{w'^2} = c'\frac{\partial\bar{u}}{\partial z}\cdot\overline{u'w'} + c'_k \tag{21c}$$

$$\overline{v'w'} = \frac{1}{2}c'\left(\frac{\partial\bar{u}}{\partial z}\cdot\overline{u'v'} + \frac{\partial\bar{u}}{\partial y}\cdot\overline{u'w'}\right) \tag{21d}$$

$$(c_{\phi 1} - 2c_{\phi 2})\overline{u'v'} - (\alpha+\beta-1)\left(\frac{\partial\bar{u}}{\partial y}\overline{v'^2} + \frac{\partial\bar{u}}{\partial z}\overline{v'w'}\right) - (\alpha+\eta+\beta\overline{u'^2})\frac{\partial\bar{u}}{\partial y} = 0 \tag{21e}$$

$$(c_{\phi 1} - 2c_{\phi 2})\overline{u'w'} - (\alpha+\beta-1)\left(\frac{\partial\bar{u}}{\partial y}\overline{w'^2} + \frac{\partial\bar{u}}{\partial y}\overline{v'w'}\right) - (\alpha+\eta+\beta\overline{u'^2})\frac{\partial\bar{u}}{\partial z} = 0 \tag{21f}$$

with

$$c'_{ko} = \tfrac{2}{3}\left[c_{\phi 1} - 3(\alpha+2\beta) + 2\right]/(c_{\phi 1} - 2c_{\phi 2}) \tag{21g}$$

$$c'_k = \tfrac{2}{3}(c_{\phi 1} - 3\beta - 1)/(c_{\phi 1} - 2c_{\phi 2}) \tag{21h}$$

$$c' = 2\beta/(c_{\phi 1} - 2c_{\phi 2}) \tag{21i}$$

It is understood that the Reynolds stresses and mean shears in these equations are temporarily normalized by k and ϵ/k for a concise presentation. The substitution of Equations 21a through 21d into 21e, f yields two equations containing only two Reynolds stress components $\overline{u'v'}$ and $\overline{u'w'}$. The subtraction between these two equations leaves the simple relation as

$$\overline{u'v'}\Big/\frac{\partial \bar{u}}{\partial y} = \overline{u'w'}\Big/\frac{\partial \bar{u}}{\partial z} \equiv -c_D^* \tag{22}$$

Equation 22, when combined with Equation 14, leads to

$$\left(\frac{\partial \bar{u}}{\partial y}\right)^2 + \left(\frac{\partial \bar{u}}{\partial z}\right)^2 = \frac{1}{c_D^*} \tag{23a}$$

In dimensional form, it is

$$\epsilon^2 = c_D^* k^2\left(\left(\frac{\partial \bar{u}}{\partial y}\right)^2 + \left(\frac{\partial \bar{u}}{\partial z}\right)^2\right) \tag{23b}$$

The summation between the same two equations for $\overline{u'v'}$ and $\overline{u'w'}$, along with the aid of Equations 22 and 23a reveals the constancy of c_D^* namely,

$$c_D^* = [^{I}(\alpha + \beta - 1)(c' - c'_k) - (\alpha + \eta + \beta c'_{k0})]/(c_{\phi 1} - 2c_{\phi 2}) \tag{24}$$

The explicit expressions for all the Reynolds stress components finally can be obtained from Equations 21a through 21d, 21g, through 21i, 22, and 24. The set of final expressions in dimensional form is given as follows:

$$-\overline{u'^2} = -c'_{k0}k \tag{25a}$$

$$-\overline{u'v'} = (c_D^* k^2/\epsilon)\frac{\partial \bar{u}}{\partial y} \tag{25b}$$

$$-u'w' = (c_D^* k^2/\epsilon)\frac{\partial \bar{u}}{\partial z} \tag{25c}$$

$$-\overline{v'^2} = (c' c_D^* k^3/\epsilon^2)\left(\frac{\partial \bar{u}}{\partial y}\right)^2 - c'_k k \tag{25d}$$

$$-\overline{w'^2} = (c' c_D^* k^3/\epsilon^2)\left(\frac{\partial \bar{u}}{\partial z}\right)^2 - c'_k k \tag{25e}$$

$$-\overline{v'w'} = (c' c_D^* k^3/\epsilon^2)\left(\frac{\partial \bar{u}}{\partial y}\right)\left(\frac{\partial \bar{u}}{\partial z}\right) \tag{25f}$$

where

$$c'_{k0} = \tfrac{2}{33}(11c_{\phi 1} - 12c_{\phi 2} + 4)/(c_{\phi 1} - 2c_{\phi 2}) \tag{25g}$$

$$c'_k = \tfrac{2}{33}(11c_{\phi 1} - 18c_{\phi 2} - 5)/(c_{\phi 1} - 2c_{\phi 2}) \tag{25h}$$

$$c' = \tfrac{4}{11}(3c_{\phi 2} - 1)/(c_{\phi 1} - 2c_{\phi 2}) \tag{25i}$$

$$c_D^* = \tfrac{2}{165}(22c_{\phi 1} - 64c_{\phi 2} - 11c_{\phi 1}c_{\phi 2} - 18c_{\phi 2}^2 + 5)/(c_{\phi 1} - 2c_{\phi 2})^2 \tag{25j}$$

These results satisfy an obvious relationship among normal stress and k, namely,

$$(\overline{u'^2} + \overline{v'^2} + \overline{w'^2})/k = c'_{ko} + 2c'_k - c' = 2 \tag{26}$$

It is most interesting to observe that the cross-planar stress components $\overline{v'^2}$, $\overline{w'^2}$, and $\overline{v'w'}$ responsible for the secondary flow generation, are not related to the mean strains in the cross-sectional plane as in the effective viscosity formulation (Equation 2a), but to the mean strains normal to it, namely, $\partial\bar{u}/\partial y$ and $\partial\bar{u}/\partial z$. The expressions for the axial stress components $\overline{u'v'}$ and $\overline{u'w'}$, on the other hand, happen to be identical to those in the effective viscosity formulation. Launder in his original work [46] did not give the explicit relations for $\overline{u'v'}$ and $\overline{u'w'}$, but suggested the validity of the effective viscosity concept for the axial stress components in his innovative work on the square duct with Ying [25]. Gessner and Emery [51] appear to be the first to show that $\overline{u'v'}$ and $\overline{u'w'}$ expressions identical to the effective viscosity formulation can be derived by directly solving all components of the Reynolds stress transport equation.

Choice of Empirical Constants

It is quite natural to expect that the constant stress layers exist near the wall even in the presence of the secondary flow. Since the secondary flow velocity is very small compared with the primary flow velocity, the Couette flow analysis may well be valid near the wall. Thus, under one-dimensional consideration, Equations 25b, c for the axial shear stresses should become compatible with Equation 2a. Naturally, the constant c_D^* must equal c_D already introduced in Equation 7, which is usually determined empirically using the near-wall relation based on the constant stress layer assumptions, namely,

$$c_D = (-\overline{u'v'}/k)^2 \tag{27}$$

Since the universality of the near-wall constant c_D (i.e., c_D^*) has been proven to some extent under various flow situations, there is essentially only one independent constant to be specified empirically if one is to be strictly consistent with the procedure taken so far. The $c_{\phi 1}$ for returning to isotropy may be chosen to be this last empirical constant, which can be determined from the observation on the decay of anisotropic homogeneous flow without mean shear. Upon setting $c_{\phi 1} = 2.6$ in consideration of the work by Rotta [52] and the well accepted value for the near-wall constant $c_D = 0.09$, one can solve the quadratic equation (25j) and obtain $c_{\phi 2} = 0.366$. This value happens to be almost identical to the value 0.365 determined semi-empirically by Launder and Ying [25]. This consideration may justify the final choice of these constant as

$$c_{\phi 1} = 2.6, \qquad c_{\phi 2} = 0.365, \qquad c_D = 0.09 \tag{28}$$

A comparison among sets of empirical constants used by previous workers is shown in Figure 2. Figure 2a shows the family of curves for c′ and $(c_{\phi 2} - 1/3)$ as function of $c_{\phi 1}$ and c_D (based on Equations 21g, h, i) along with the values of constants employed by various workers. The constant $c_{\phi 1}$ for returning to isotropy used by these workers lies between 2.6 and 2.8, which is consistent with the observation made by Rotta [52]. However, it is interesting to note the diversity in resulting levels of anisotropy coefficient c′ which, in fact, strongly governs the secondary flow generation as will be described shortly in consideration of the vorticity transport equation. Hanjalic and Launder [50] used the value $c' = 0.067$ for the calculation of the asymmetric flow, and the same set of constant are used by Gessner and Emery [33] except c_D. Aly et al., [26], on the other hand, set c′ about seven times smaller than that of Hanjalic and Launder [50] for the calculation of the equilateral triangular duct. The present choice of constant, with the exception of c_D, is essentially based on those employed by Launder and Ying [25]. As already indicated, the near wall constant c_D, however, is set to the well-accepted value 0.09, which consistently leads to a perfect agreement with the curves based on Equations 21g, h, i.

The variations of c'_k and c'_{ko} are shown in Figure 2b. The curves based on Equations 21g, h, i indicate that these constants are fairly insensitive to $c_{\phi 1}$ and all the values used by previous wokers lie on the same level.

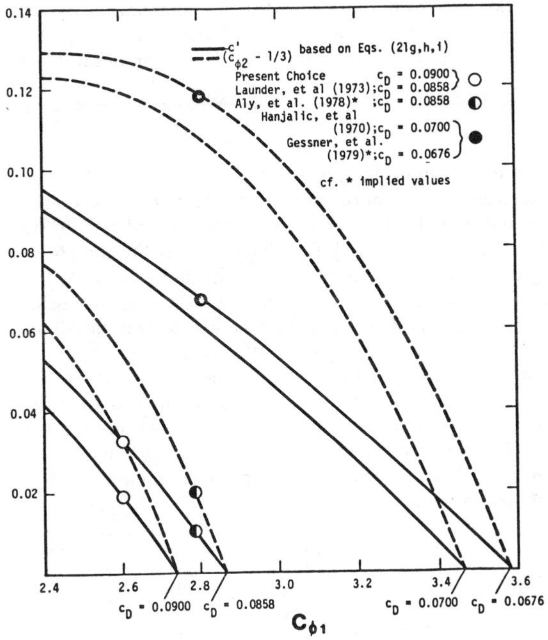

Figure 2a. c′ and $c_{\phi 2}$ variations with respect to $c_{\phi 1}$.

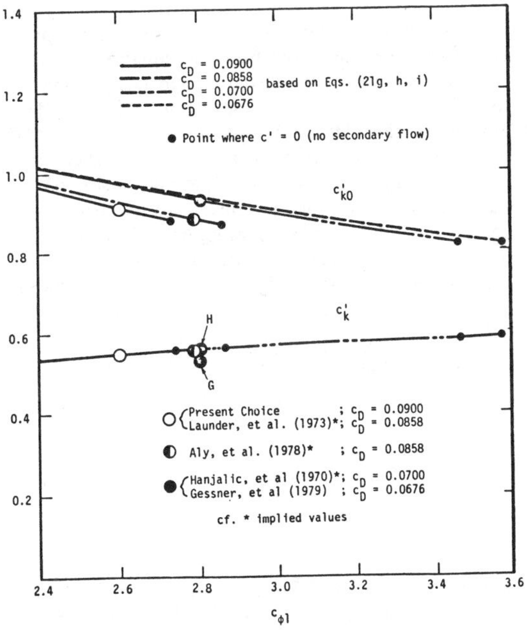

Figure 2b. c'_k and c'_{k0} variations with respect to $c_{\phi 1}$.

Having established the explicit algebraic relations for all Reynolds stress components, the local equilibrium condition given by Equation 11 may now be discarded. Instead, the transport Equations 5a and 6 for k and ϵ may be solved directly to acquire generality in the simulation of turbulence.

Vorticity Transport Equation

For the explanation of the secondary flow motion, Prandtl [4] postulated that the tangential fluctuations along a curved isovel are much greater than normal velocity fluctuations. This condition was vertified by many workers (Gessner and Jones [12], Perkins [13]). Alternate explanations were provided by several workers (Eichelbrenner and Toan [62], Eichelbrenner and Preston [63]) in consideration of the Reynolds averaged momentum equations. Brundrett and Baines [11] and Perkins [13] considered the vorticity transport equation and suggested that the streamwise vorticity generated in the proximity of the corner is due primarily to the secondary flow generation. All these hypotheses assume that the anisotropy of the cross-planar normal stresses is responsible for the secondary flow motion. Gessner [53], on the other hand, suggested that the gradients of the shear stress opposing to the primary flow in the corner region generate the secondary flow motions. Furthermore, instead of dealing with the momentum or vorticity transport, Hinze [54] considered the turbulent kinetic energy transport to explain the secondary flow generation mechanism.

Although there may be many alternative ways, the secondary flow production may best be explained in consideration of the vorticity transport equation as given by

$$\rho \frac{\partial}{\partial y}\left(\bar{v}\Omega - \nu \frac{\partial \Omega}{\partial y}\right) + \rho \frac{\partial}{\partial z}\left(\bar{w}\Omega - \nu \frac{\partial \Omega}{\partial z}\right) = \frac{\partial^2}{\partial y\, \partial z}(\tau_{zz} - \tau_{yy}) + \left(\frac{\partial^2}{\partial y^2} - \frac{\partial^2}{\partial z^2}\right)\tau_{yz} \tag{29a}$$

where

$$\Omega = \frac{\partial \bar{w}}{\partial y} - \frac{\partial \bar{v}}{\partial z} \tag{29b}$$

Reynolds stress components $-\rho\overline{v'^2}$, $-\rho\overline{w'^2}$, and $-\rho\overline{v'w'}$ are alternatively denoted by τ_{yy}, τ_{zz}, and τ_{zy}, respectively. This above transport equation is exact for a fully developed flow and may even be be regarded as a crude approximation to a developing flow. The terms on the left-hand side of Equation 29a may be identified as convection of the secondary flow vorticity by the secondary flow itself and the diffusion of the vorticity by the laminar viscosity. The terms on the right-hand side altogether may be referred to as the production of the vorticity by the action of Reynolds stresses and will be identified as P_Ω for future reference. As may be seen on the right-hand side of Equation 29a, the cross-planar shear stress τ_{yz} and the anisotropy of the normal stresses $(\tau_{zz} - \tau_{yy})$ are directly related to the generation of the secondary flow motions. Upon employing a coordinate transformation procedure (e.g. Nakayama [55]), the vorticity production terms on the right-hand side of Equation 29a can be transformed into the following form in the natural coordinates based on an isovel (s) and its normal (n) as depicted in Figure 1:

$$JP_\Omega = \frac{\partial^2}{\partial n\, \partial s}(\tau_{ss} - \tau_{nn}) + \frac{\partial}{\partial s}\frac{K}{J}(\tau_{ss} - \tau_{nn}) + \frac{\partial}{\partial n} J \frac{\partial \tau_{ns}}{\partial n} - \frac{\partial}{\partial s}\frac{1}{J}\frac{\partial \tau_{ns}}{\partial s} + 2K \frac{\partial \tau_{ns}}{\partial n} \tag{30a}$$

where

$$K \equiv \frac{d\theta}{ds} \quad \text{and} \quad J = 1 + Kn \tag{30b, c}$$

K and J are curvature of the isovel and the Jacobian transformation, respectively. For the case of a circular duct, the isovels are co-aligned with concentric circles. Thus, noting $dn = dr$, $ds = r\,d\theta$,

$K = \partial J/\partial n = 1/r$ and $J = 1$, one readily obtains P_Ω expressed in the cylindrical polar coordinates (r, θ) as

$$P_\Omega = \left(\frac{\partial^2}{r\,\partial\theta\,\partial r} + \frac{\partial}{r^2\,\partial\theta}\right)(\tau_{\theta\theta} - \tau_{rr}) + \left(\frac{\partial^2}{\partial r^2} - \frac{\partial^2}{r^2\,\partial\theta^2} + \frac{3}{r}\frac{\partial}{\partial r}\right)\tau_{r\theta} \tag{31}$$

Naturally, this expression becomes identical to the one employed by Trupp and Aly [30] for the calculation of secondary flows in rod bundles using the cylindrical polar coordinate system. Since $\partial\tau_{\theta\theta}/\partial\theta = \partial\tau_{rr}/\partial\theta = \tau_{r\theta} = 0$ over an entire cross section of the circular duct, P_Ω obviously vanishes, no secondary flow motions are generated in the circular duct.

For another example, the planes of symmetry in general may be considered. Since the symmetry conditions give

$$\frac{\partial\tau_{ss}}{\partial s} = \frac{\partial\tau_{nn}}{\partial s} = \tau_{ns} = \frac{\partial\tau_{ns}}{\partial n} = \frac{\partial^2\tau_{ns}}{\partial s^2} = 0 \quad \text{and} \quad \frac{\partial K}{\partial s} = \frac{\partial J}{\partial s} = 0$$

Equation 30a readily yields $P_\Omega = 0$. Therefore, the secondary flow vorticity production must take place inside the planes of symmetry.

It is interesting to evaluate P_Ω according to the algebraic stress model just derived. Upon transforming the algebraic expression (Equations 25d, e, f) into the coordinates tangent and normal to isovels under the assumption of local equilibrium as given by Equation 23b, one obtains remarkably simple expressions as follows:

$$\tau_{nn} = -(c'_k - c')k \tag{32a}$$

$$\tau_{ss} = -c'_k k \tag{32b}$$

$$\tau_{ns} = 0 \tag{32c}$$

Thus, the algebraic stress model of Launder and Ying implicitly assumes the principal planes of stress field ($\tau_{ns} = 0$ planes) to be everywhere normal and tangent to an isovel. Moreover, the model describes the anisotropy of cross-planar stress as $\tau_{nn} - \tau_{ss} = c'k$, essentially following the Prandtl's postulation that the level of the tangential velocity fluctuation $(-\tau_{ss})$ exceeds that of normal fluctuations $(-\tau_{nn})$.

The vorticity production P_Ω can be evaluated by substituting Equation 32 to Equation 30a as

$$JP_\Omega = -c'\left(\frac{\partial}{\partial s}\frac{Kk}{J} + \frac{\partial^2 k}{\partial n\,\partial s}\right) \tag{33}$$

All terms associated with the shear stress τ_{ns} vanish. Further assuming that the contour of the constant kinetic energy is nearly parallel to the isovel, a rough estimation may be made from Equation 33 as

$$JP_\Omega \simeq -c'\frac{k}{J^2}\frac{\partial K}{\partial s} \tag{34}$$

Equation 34 indicates that the sign of P_Ω depends solely on the derivative of the isovel curvature $\partial K/\partial s$. The result may be appreciated by referring to Figure 1. For the triangular sector $0 \le \theta \le \pi/4$, $\partial K/\partial s$ is positive; hence, the negative vortex (i.e., anti-clockwise vortex) is expected, and for the next sector, $\pi/4 \le \theta \le \pi/2$, the positive vortex is to be generated since $\partial K/\partial s$ here is negative. Thus, as suggested by Prandtl, the fluid flows away from the core through the region of convex isovels and then returns to the core through the region of concave isovels.

NUMERICAL METHOD

Governing Equations

Many numerical schemes exist that can be employed to solve the present problem. Among them, the primitive variable method based on the pressure correction procedure (Patanker and Spalding [56]) appears to be particularly suited for the solutions of the turbulent viscous flow problems of this kind.

For the numerical analysis, the governing equations appropriate for the three-dimensional developing turbulent flow in a straight duct are written below in a single common form as

$$\underline{\frac{\partial}{\partial x}\left(u\phi - \Gamma\frac{\partial \phi}{\partial x}\right)} + \frac{\partial}{\partial y}\left(v\phi - \Gamma\frac{\partial \phi}{\partial y}\right) + \frac{\partial}{\partial z}\left(w\phi - \Gamma\frac{\partial \phi}{\partial z}\right) = s_0 \tag{35}$$

where the general scalar ϕ stands for any one of the time-averaged dependent variables under consideration (the bars for the time averaged quantities are to be omitted hereafter). The diffusion coefficient Γ and source term s_0 in Cartesian form are listed below for each governing equation:

$$\phi = 1, \quad \Gamma = 0, \quad s_0 = 0 \text{ (continuity equation)} \tag{36a}$$

$$\phi = u, \quad \Gamma = \nu + \nu_t,$$

$$s_0 = -\frac{\partial}{\partial x}\frac{p}{\rho} + \underline{\frac{\partial}{\partial x}\left(\frac{\tau_{xx}}{\rho} - \nu_t\frac{\partial u}{\partial x}\right) + \frac{\partial}{\partial x}\left(\frac{\tau_{xy}}{\rho} - \nu_t\frac{\partial u}{\partial y}\right) + \frac{\partial}{\partial z}\left(\frac{\tau_{zx}}{\rho} - \nu_t\frac{\partial u}{\partial z}\right)} \tag{36b}$$

$$\phi = v, \quad \Gamma = \nu, \quad s_0 = -\frac{\partial}{\partial y}\frac{p}{\rho} + \underline{\frac{\partial}{\partial x}\frac{\tau_{xy}}{\rho}} + \frac{\partial}{\partial y}\frac{\tau_{yy}}{\rho} + \frac{\partial}{\partial z}\frac{\tau_{yz}}{\rho} \tag{36c}$$

$$\phi = w, \quad \Gamma = \nu, \quad s_0 = -\frac{\partial}{\partial z}\frac{p}{\rho} + \underline{\frac{\partial}{\partial x}\frac{\tau_{zx}}{\rho}} + \frac{\partial}{\partial y}\frac{\tau_{yz}}{\rho} + \frac{\partial}{\partial z}\frac{\tau_{zz}}{\rho} \tag{36d}$$

$$\phi = k, \quad \Gamma = \nu + \nu_t/\sigma_k, \quad s_0 = P - \epsilon \tag{36e}$$

$$\phi = \epsilon, \quad \Gamma = \nu + \nu_t/\sigma_\epsilon, \quad s_0 = (c_1 P - c_2\epsilon)\epsilon/k \tag{36f}$$

where

$$\nu_t = c_D\frac{k^2}{\epsilon} \tag{36g}$$

and

$$\rho P = \underline{\tau_{xx}\frac{\partial u}{\partial x} + \tau_{xy}\frac{\partial v}{\partial x} + \tau_{zx}\frac{\partial w}{\partial x}} + \tau_{xy}\frac{\partial u}{\partial y} + \tau_{zx}\frac{\partial u}{\partial z} + \tau_{yy}\frac{\partial v}{\partial y} + \tau_{zz}\frac{\partial w}{\partial z} + \tau_{yz}\left(\frac{\partial v}{\partial z} + \frac{\partial w}{\partial y}\right) \tag{36h}$$

For the case of a fully developed flow, the terms underlined in Equations 35 and 36 (the derivatives of the axial coordinate x) are dropped, and the equations becomes much simplified. The algebraic equations for Reynolds stresses, given by Equations 25a through 25f close the system of equations.

Coordinate Transformation

Even when dealing with complicated non-circular duct geometries, it is still possible to retain the Cartesian or cylindrical coordinate system. When using such systems, however, one must do extensive interpolative calculations in order to satisfy the required boundary conditions. Moreover, a considerable number of mesh points will be wasted since the points external to the flow field do not participate in the calculations in any meaningful manner. To overcome this problem, a coordinate transformation is introduced so that all mesh points lie within the flow field.

The general conservation Equation 35 common to all governing equations can be transformed into the following conservative form [55, 57] in an arbitrary coordinate system:

$$\frac{\partial}{\partial x^i} J\left[(\vec{g}^i \cdot \vec{u})\phi - \Gamma(\vec{g}^i \cdot \vec{g}^j)\frac{\partial \phi}{\partial x^j}\right] = J s_0 \tag{37a}$$

where

$$\vec{r} = x\vec{i} + y\vec{j} + z\vec{k} \tag{37b}$$

$$\vec{u} = u\vec{i} + v\vec{j} + w\vec{k} \tag{37c}$$

$$\vec{g}^i \cdot \left(\frac{\partial \vec{r}}{\partial x^j}\right) = \delta^i_j \tag{37d}$$

and

$$J = (\vec{g}^1 \cdot \vec{g}^2 \times \vec{g}^3)^{-1} \tag{37e}$$

The arbitrary generalized coordinates are given by $x^i (i = 1, 2, 3)$ while the Jacobian of transformation is denoted by J. Equation 37d defines the contravariant base vector $\vec{g}^i$. It should be noted that the velocity vector $\vec{u}$ of the mean flow is to take u, v, and w in Cartesian components even in the transformed coordinates ($\vec{i}, \vec{j}$, and $\vec{k}$ are the Cartesian unit base vectors). Naturally, each of these velocity components remain as the dependent variable ϕ in the corresponding momentum equation. Thus, the terms corresponding to the Christoffel symbols in a general tensor notation do not appear in the present momentum equations.

Now the general conservation Equation 37a may be rewritten as

$$\frac{\partial}{\partial x^i} J\left[(\vec{g}^i \cdot \vec{u})\phi - \Gamma|\vec{g}i|^2 \frac{\partial \phi}{\partial x^i}\right] = s_0^* \tag{38a}$$

where

$$s_0^* = J s_0 + \frac{\partial}{\partial x^i} J\Gamma[(\vec{g}^i \cdot \vec{g}^j) - |\vec{g}^i|^2 \delta^i_j]\frac{\partial \phi}{\partial x^j} \tag{38b}$$

Only the diagonal components of the metric tensor $(\vec{g}^i \cdot \vec{g}^j)$ are retained on the left-hand side of Equation 38a. Its counterpart, namely, the second term in s_0^*, accounts for the non-orthogonality since it obviously vanishes for the orthogonal set of $\vec{g}i$ (note, the summation rule does not apply to $|\vec{g}^i|^2$, hence, $|\vec{g}i|^2 = \vec{g}^i \cdot \vec{g}^i$ for a particular i).

The transformation for a straight duct of arbitrary cross section [28] may be specified in general as

$$x = x^1, \qquad y = y(x^2, x^3), \qquad z = z(x^2, x^3) \tag{39a}$$

Consequently,

$$\vec{g}^1 = \vec{i}, \qquad \vec{g}^2 = \frac{1}{J}\frac{\partial}{\partial x^3}(z\vec{j} - y\vec{k}), \qquad \vec{g}^3 = \frac{1}{J}\frac{\partial}{\partial x^2}(y\vec{k} - z\vec{j}) \tag{39b}$$

and

$$J = \left(\frac{\partial y}{\partial x^2}\right)\left(\frac{\partial z}{\partial x^3}\right) - \left(\frac{\partial y}{\partial x^3}\right)\left(\frac{\partial z}{\partial x^2}\right) \tag{39c}$$

Since both $\vec{g}^2$ and $\vec{g}^3$ in the cross-sectional plane are normal to $\vec{g}^1$ in the axial direction, s_0^* reduces to

$$s_0^* = Js_0 + \frac{\partial}{\partial x^2} J\Gamma(\vec{g}^2 \cdot \vec{g}^3)\frac{\partial \phi}{\partial x^3} + \frac{\partial}{\partial x^3} J\Gamma(\vec{g}^2 \cdot \vec{g}^3)\frac{\partial \phi}{\partial x^2} \tag{39d}$$

These equations are valid for the general three-dimensional developing flows in ducts of arbitrary cross section. For the case of fully developed flows, all the x^1 derivatives within the convection and diffusion terms on the left-hand side of Equation 38a vanish (with the exception of $\partial p/\partial x^1$).

A general finite difference form may be written once for all governing equations by discretizing Equation 38a. The resulting finite difference expressions are so universal that they can be used for any coordinate system simply by specifying $\vec{g}^i$ or equivalently the functions in Equation 39a appropriate to a given duct geometry.

Discretization

Upon following the procedure similar to the one employed by Patankar and Spalding [56], the general form of the difference equations is obtained by integrating Equation 38a over an elementary volume (Vol) as

$$\iint_{\mathrm{Vol}/\Delta x^i} J\left[(\vec{g}^i \cdot \vec{u})\phi - \Gamma|\vec{g}^i|^2 \frac{\partial \phi}{\partial x^i}\right] dS_i\Bigg|_{x^{i-}}^{x^{i+}} = \iiint_{\mathrm{Vol}} s_0^*\, dx^1\, dx^2\, dx^3 \tag{40a}$$

where

$$dS_i \equiv dx^1\, dx^2\, dx^3/dx^i \quad \text{and} \quad \mathrm{Vol} \equiv \Delta x^1\, \Delta x^2\, \Delta x^3 \tag{40b, c}$$

Referring to the grid of the transformed coordinate x^i as shown in Figure 3, integrations in Equation 40a may be approximated to yield the following finite difference equation:

$$(A_i^+ + A_i^-)\phi_i^0 = A_i^+\phi_i^{++} + A_i^-\phi_i^{--} + S_0 \tag{41a}$$

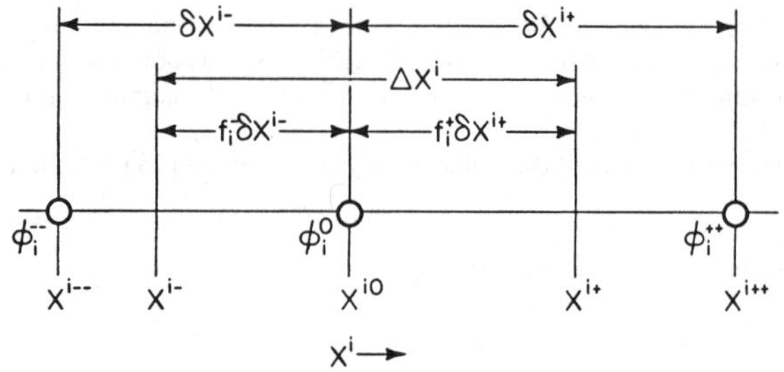

Figure 3. Grid nomenclature for discretization.

where

$$S_0 = s_0^*|_{x^{i0}}\mathrm{Vol} \tag{41b}$$

$$A_i^+ = -f_i^+ CO_i^+ + DI_i^+ \tag{41c}$$

$$A_i^- = f_i^- CO_i^- + DI_i^- \tag{41d}$$

$$CO_i^\pm = J(\vec{g}^i \cdot \vec{u})(\mathrm{Vol}/\Delta x^i)|_{x^{i\pm}} \tag{41e}$$

and

$$DI_i^\pm = (J\Gamma|\vec{g}^i|^2/\delta x^i)(\mathrm{Vol}/\Delta x^i)|_{x^{i\pm}} \tag{41f}$$

The locations of $(\phi_i^{++}, \phi_i^{--}, \phi_i^0)$, the interpolation factors (f_i^+, f_i^-) and the internodal distances $(\Delta x^i, \delta x^{i+}, \delta x^{i-})$ are shown in Figure 3. In Equation 41 the subscript i (such as in ϕ_i and A_i) is associated with a particular coordinate x^i, while the superscripts (such as 0, +, and −) refer to the relative locations along its coordinate. It is noted that the summation rule is effective in the finite difference Equation 41a while the rule does not apply in Equations 41c, d, e, f.

The resulting finite difference equation gives an algebraic equation for the scalar $\phi^0(\equiv \phi_1^0 = \phi_2^0 = \phi_3^0)$ in terms of the values at six neighboring nodes. Thus, if the pressure field is given, Equation 41a can be written for each variable at each node. This procedure then yields a closed set of algebraic equations, which can be solved by an iterative numerical algorithm. However, the velocity field thus obtained, usually does not satisfy the continuity equation, since the pressure field assumed beforehand does not represent the true pressure field. Therefore, these pressure and velocity fields must be subsequently adjusted to satisfy the continuity principle.

Pressure Correction Equation

For the correction on the pressure and velocity fields, Patankar and Spalding [56] formulated the "pressure correction equation" equivalent to a Poisson's equation for the pressure field by substituting an abbreviated momentum balance relationship into the continuity Equation 36a. This pressure correction equation can be solved most effectively using the so-called "staggered" grid system in which the velocities are defined midway between the pressure nodes. The details of this pressure correction procedure have been reported for a three-dimensional parabolic flow by Patankar and Spalding [56]. Its extension to the present system of generalized coordinates is straightforward. The final expression of the pressure correction equation (Nakayama [55]) takes the same form as the general finite difference Equation 41a.

The pressure correction procedure is essentially based on the following relations:

$$\nabla \cdot \vec{u}_{cor} = -\nabla \cdot \vec{u}_{est} \tag{42a}$$

and

$$\vec{u}_{cor} \propto -\nabla p_{cor} \tag{42b}$$

where the subscripts "est" and "cor" refer to the values based on the estimated pressure field (est) and its correction (cor). Equation 42a is obviously the continuity equation while the relation (42b) may be obtained from the discretized momentum equations. The substitution of the relation (42b) into Equation 42a leads to the following pressure correction relation (the actual correction equation differs somewhat from it):

$$\nabla^2 p_{cor} \propto \nabla \cdot \vec{u}_{est} \tag{43}$$

For an interior point in a volume element, the relation (43) indicates that the artificial accumulation of mass (i.e., $\nabla \cdot \vec{u}_{est} < 0$) tends to make p_{cor} at the interior point larger than that on the enveloping control surface. The resulting p_{cor} field, in return, produces the vector $\vec{u}_{cor}$ directed outward through the control surface by virtue of the relation (42b), hence, relaxing the mass accumulation within the control volume. The repetition of this self-correcting procedure eventually makes $\vec{u}_{cor}$ and p_{cor} identically zero within the control volume.

Solution Procedure

The resulting discretized equations can be solved by the successive use of tri-diagonal matrix algorithm. In principle, the iteration starts with solving three momentum equations for u, v, and w sequentially using the estimated pressure and turbulence fields currently available. Then the resulting velocity field is corrected by solving the pressure correction equation to bring the velocities into mass balance. Subsequently, the turbulence quantities k and ϵ are solved. The sequence is repeated until convergence is achieved.

This fully elliptic calculation scheme was employed for the three-dimensional developing flows. Some of these results will be presented in the following section. Other details on this elliptic scheme (such as the hybrid differencing, near-wall treatment and boundary conditions) may be found elsewhere [36, 55, 58].

Patankar and Spalding [56] and many others suggested that the developing flow in a straight duct can be treated as a three-dimensional parabolic flow. Thus, the axial diffusion terms may be neglected, and the economical downstream marching procedure may be employed.

For the case of fully developed flows, only the two-dimensional storage is required. The pressure gradient term, which is the only derivative associated with the x^1 (i.e., x) coordinate, may be evaluated by spatially averaging the u momentum Equation 36b as

$$-\frac{\partial p}{\partial x} \simeq 4\tau_{av}/D_h \simeq 4\rho c_D^{1/2} k_{av}/D_h \tag{44}$$

where τ_{av} and k_{av} are the wall shear stress and the near-wall turbulent kinetic energy averaged over the wall periphery, and are related to each other through Townsend's approximation. D_h is the hydraulic diameter. Since all x derivatives vanish due to the full development of the flow, the u momentum Equation 36b can be treated simply as another scalar conservation equation, such as k and ϵ equations. Thus, the three-dimensional calculation storage is no longer necessary. The fully developed calculation procedure using two-dimensional storage has already been described in detail by some workers. Interested readers may consult Gosman and Rapley [31] and Nakayama, Chow, and Sharma [28].

APPLICATIONS AND EXAMINATION OF STRESS MODEL

Preliminary Calculations

The laminar flow cases provide a useful preliminary check for the accuracy of the numerical procedure without the presence of ambiguity due to imperfection of turbulence models.

Laminar calculations were carried out for the developing flows in square and triangular ducts using the fully elliptic three-dimensional code with the downstream exit subjected to the zero gradient condition. Grid lay-outs on the cross-sectional planes of square (11 × 9 × 7) and triangular (7 × 7 × 13) shapes are shown in Figure 4. Both grid systems were intentionally made highly non-uniform in all three directions in order to see grid lay-out effects on the solutions. Reynolds numbers based on the hydraulic diameter (D_h) and bulk velocity (u_B) were set to 1 and 2/3 for the square and triangular ducts, respectively. The flows, in these cases, become nearly fully-developed after a short distance from the inlet. The resulting fully developed velocity profiles along the square duct wall bisectors are shown in Figure 5. The asymmetry of the grid lay-out with respect to y and z directions show no influence on the solution. Almost perfect symmetry has been attained as indicated by the result even with only seven grid nodes in the z direction. The numerical solution appears to be very close to the exact solution as indicated by the solid lines in the figure.

The fully developed profiles obtained for the triangular duct are shown in Figures 6a, b. The figures again indicate excellent agreement with the corresponding exact solution.

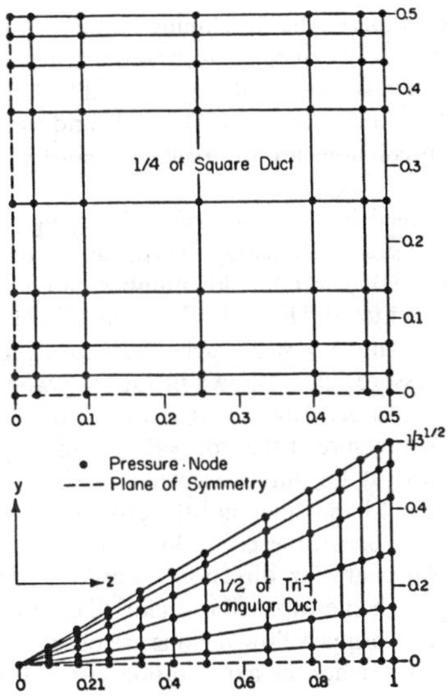

Figure 4. Grid layout.

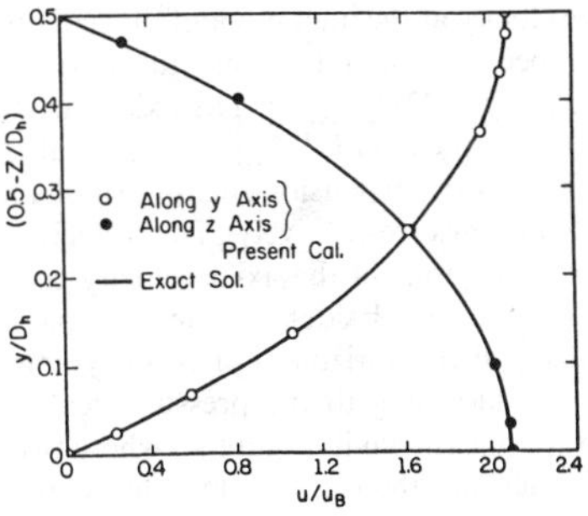

Figure 5. Velocity profile along a wall bisector of a square duct; fully developed, Re = 1.

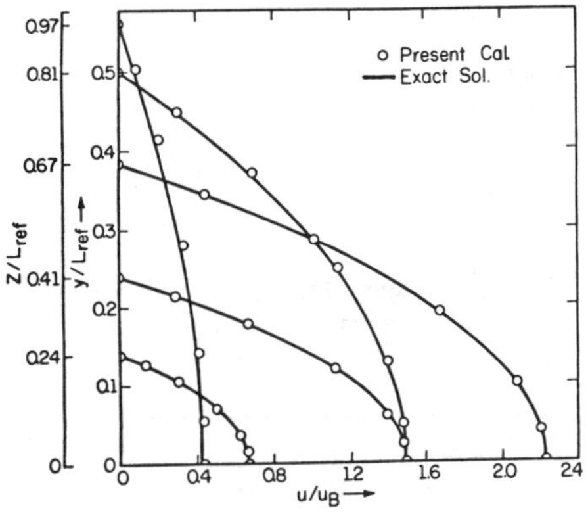

Figure 6a. Velocity profile in equilateral triangular duct; fully developed, Re = 2/3.

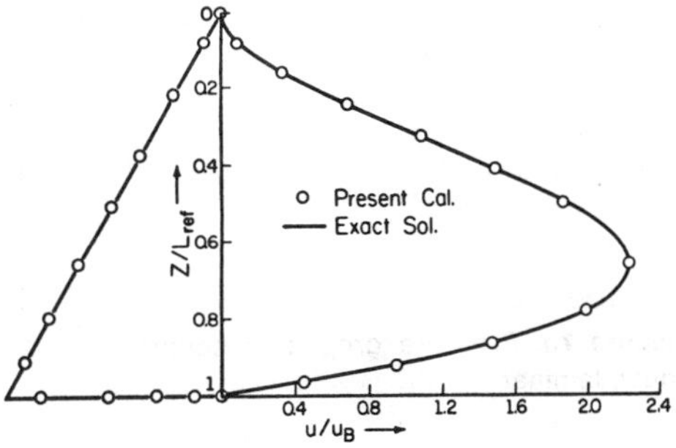

Figure 6b. Velocity profile in equilaternal triangular duct; fully developed, Re = 2/3, y = 0.

Reynolds number was increased to 100 for the calculation of the developing laminar flow in a square duct (15 × 9 × 7). The streamwise step was gradually expanded to cover the distance of $15D_h$. The results on the centerline velocity and pressure distribution along the axial distance are plotted in Figures 7a, b, which also present the available experimental data [59, 64] and the parabolic calculation results (11 × 11) by Neti [35]. Both predictions are in excellent accord with the experimental data.

For illustrative purpose, the calculation results obtained for the converging-diverging rectangular duct (30 × 9 × 9) by the present elliptic scheme are presented below (the transformation employed here is somewhat different from Equations 39 [55]). The Reynolds number based on the bulk velocity at the inlet u_{ref} and the half duct width was set to 50. The vertical distance Y_{tb} between the upper wall boundary and the horizontal plane of symmetry varies according to $Y_{tb}/L_{ref} = 1 - 0.5 \exp[-(2x^1/L_{ref})^2]$. The axial variation of the pressure field is shown in Figure 8 where the pressure at the inlet ($x^1/L_{ref} = -5$) is taken as the reference pressure p_{ref}. It is seen that the pressure field in the expansion region exhibits high- and low-pressure at the side wall (p_s) and top wall (p_t) centerlines, respectively. The velocity field is plotted at the three selected axial stations in Figure 9 in terms of the axial velocity contours (on the left-hand side) and the cross flow vectors (on the right-hand side). As the flow passes through the throat, the cross flow vectors pointing originally at the horizontal plane of symmetry (Figure 9a), change its direction toward the top wall center, conforming to the pressure field prevailing through the expansion region. The on-set of separation is shown in Figure 9b where two distinct reverse primary flow regions are observed near the corner and the top wall middle point. When the flow reaches the axial station at $x = 0.41L_{ref}$

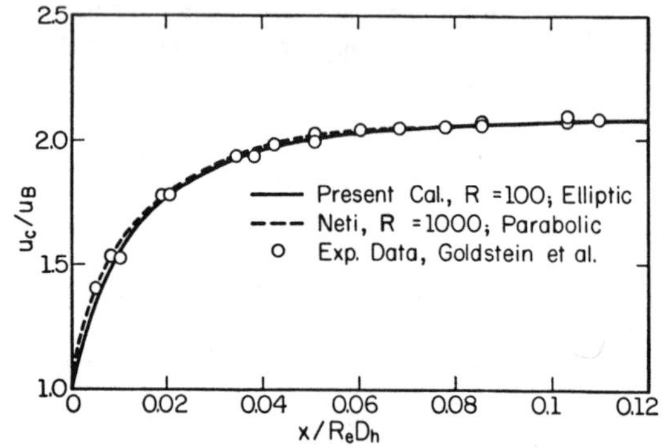

Figure 7a. Centerline velocity development in a square duct; laminar.

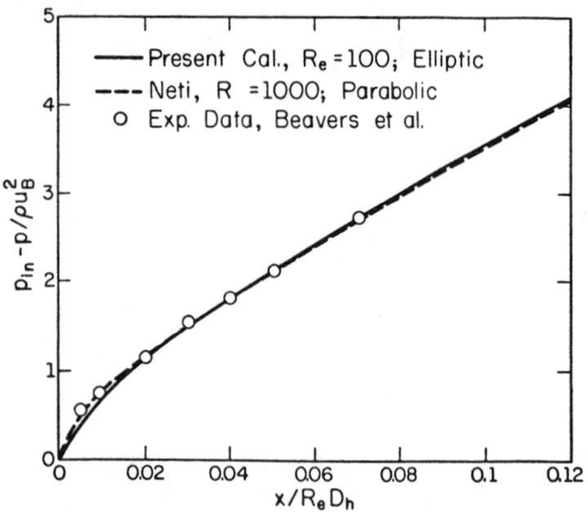

Figure 7b. Pressure drop in a square duct; laminar.

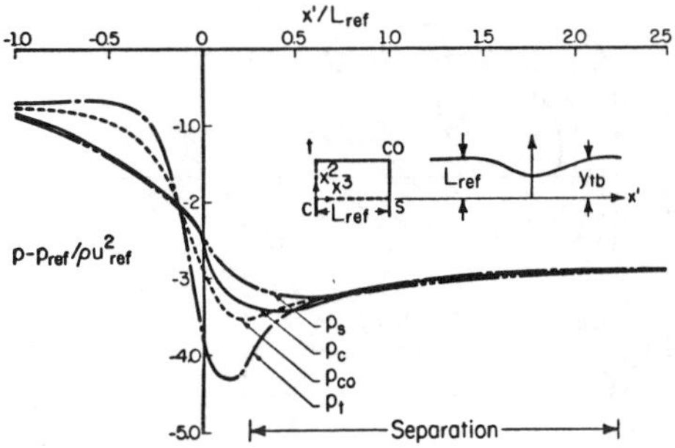

Figure 8. Axial variation of pressure.

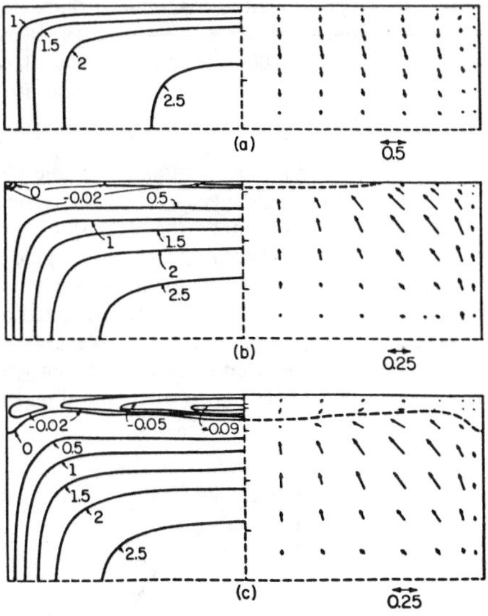

Figure 9. Development of axial and cross-flow velocity components (all values normalized by u_{ref}): (a) $x^1/_{ref} = -0.05$; (b) $x^1/L_{ref} = 0.29$; (c) $x^1/L_{ref} = 0.41$.

(Figure 9c), the separated flow on the center top wall expands towards the corner and these two regions coalesce to form one big reverse flow region with more fluid drawn into the secondary flow currents. The main flow finally reattaches around $x = 2.2L_{ref}$, onto the top wall.

Fully Developed Turbulent Flow in Non-Circular Ducts

Square Duct

For the study of turbulent duct flows, it is expedient to start with a simple geometry. The two-dimensional calculations were performed for the quadrant of a square (15 × 15) at Reynolds number (based on D_h and u_B) Re = 83,000, for which extensive measurements were carried out by Leutheusser [7] and later reproduced by Brundrett and Baines [11] using the same setup.

The isovels (u/u_B) in a square duct are shown in Figure 10a where the predicted velocity levels are in good agreement with the experimental data, despite the fact that the distortions of contours are overestimated and the velocity near to the corner is underestimated. The predicted secondary flow velocity vectors and streamlines (normalized by D_h and the centerline velocity u_c) are indicated in Figures 10b, c along with the streamlines obtained experimentally by Gessner and Jones [12]

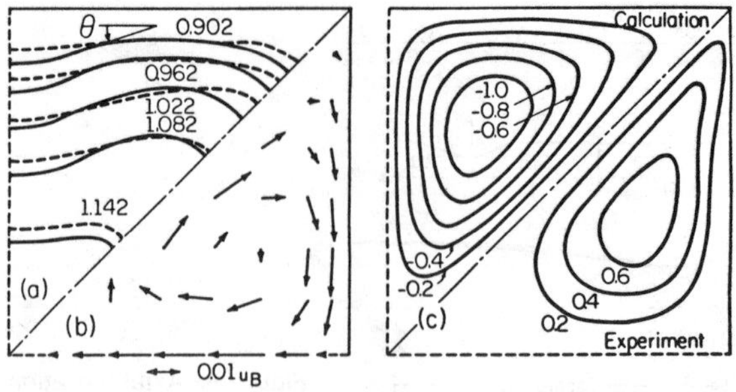

Figure 10. Mean velocity field, square duct. (a) isovels, u/u_B, Re = 83,000, ------- experiment (Leutheusser 1963), ——— present calculation; (b) secondary flow vectors, present calculation, Re = 83,000; (c) secondary flow streamlines, $(\psi/u_cD_h) \times 10^3$ upper triangle, present calculation, Re = 83,000; lower triangle, experiment (Gessner and Jones, 1965), Re = 150,000.

for Re = 150,000 (no experimental streamline plots are available for Re = 83,000). Even though there is experimental evidence [12] that the secondary flow magnitude (when normalized) increases for a lower Reynolds number, the secondary flow rate predicted here is considerably higher than that of the experimental data. Moreover, the prediction shows the underestimation of the secondary flow magnitude along the diagonal of the square (directed toward the corner, Figure 11b) and overestimation along the wall bisector (directed toward the core, Figure 11a).

The contour map of turbulent kinetic energy is shown in Figure 12a along with Brundrett and Baines' data. The predicted level of kinetic energy was found to be higher than that of the experiment. The contrast between prediction and experiment is especially evident near the corner. Yet,

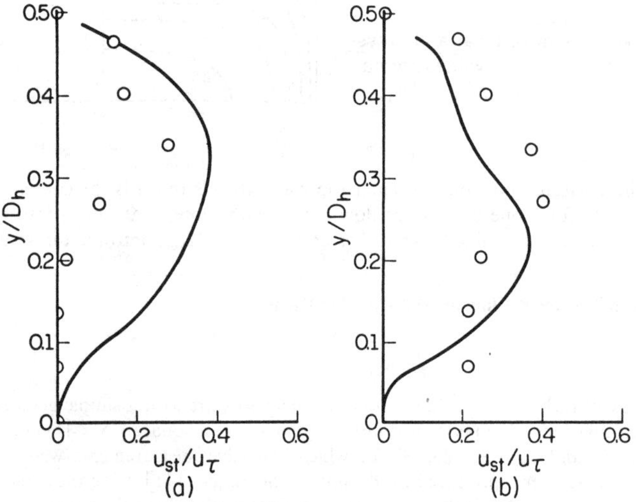

Figure 11. Secondary flow velocity mangitude, u_{st}/u_τ, square duct Re = 83,000; "o," experiment (Brunderett and Baines, 1964); ———, present calculation: (a) wall bisector, (b) corner bisector.

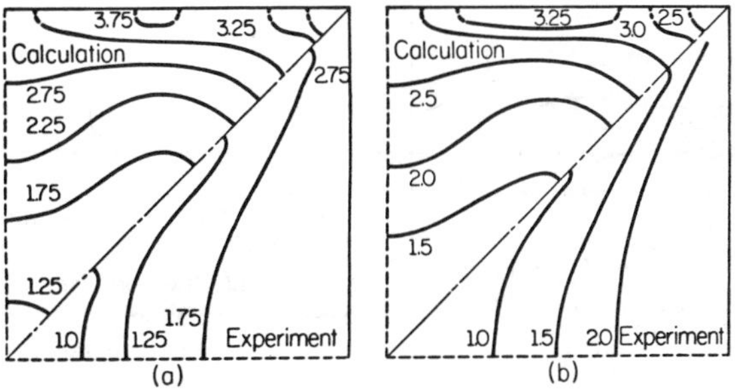

Figure 12. Kinetic energy and normal stress, τ_{xx}, square duct Re = 83,000: (a) k/u_τ^2, upper triangle, present calculation; lower triangle, experiment (Brundrett and Baines, 1964); (b) $-\tau_{xx}/\tau_{av}$, upper triangle, present calculation; lower triangle, experiment (Brundrett and Baines, 1964).

this is consistent with the underprediction of the secondary flow velocity along the diagonal since the dilution process near the corner due to the secondary flow current that carries the fluid of low kinetic energy at the core toward the corner becomes less efficient for a lower secondary flow velocity. The prediction and the experiment, however, share the fact that distortions of kinetic energy contour are more pronounced than those of the isovels. Essentially, the same comments for the kinetic energy contours may be made for the contour map of normal stress in the axial direction τ_{xx}, as shown in Figure 12b. The shear stress opposing the primary flow τ_{xy} is shown in Figure 13 along with the experimental data obtained at Re = 42,000 by Melling and Whitelaw [15]. It should be noted that Reynolds stresses when normalized by the averaged wall shear stress τ_{av} become fairly insensitive to Reynolds number. The predicted contours appear to be in good agreement with the experiment.

Rectangular Duct

Calculations were carried out with the grid system (14 × 24) for a quadrant of the rectangular duct of aspect ratio 1:3, for which experimental data of Re = 56,000 by Leutheusser [7] and Re = 30,000 by Hoagland [5] were available.

Isovel normalized by the centerline velocity u_c are shown for Re = 56,000 in Figure 14a, where Leutheusser's experimental data are also plotted. The prediction shows a good agreement with the experiment near the vertical wall bisector, yet overestimates the distortions of isovels toward the corner. The secondary flow velocity vector is indicated in Figure 14b. The secondary flow streamline pattern obtained experimentally by Haogland [5], which happens to be the only available information for details of the streamlines in the rectangular duct at this time, is shown in the lower

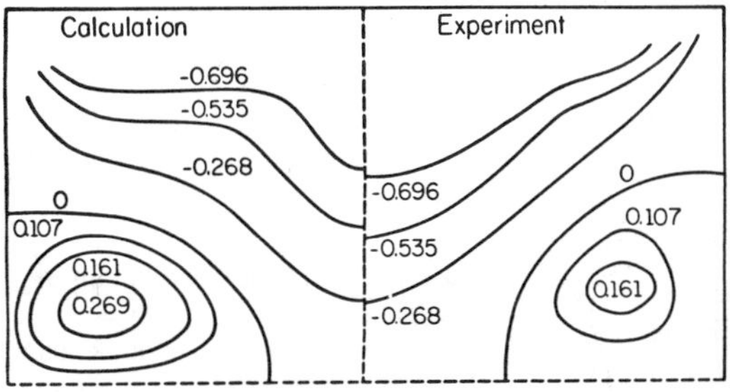

Figure 13. Shear stress τ_{xy}/τ_{av}, square duct, left half, present calculation, Re = 83,000; right half, experiment (Melling and Whitelaw, 1976), Re = 42,000.

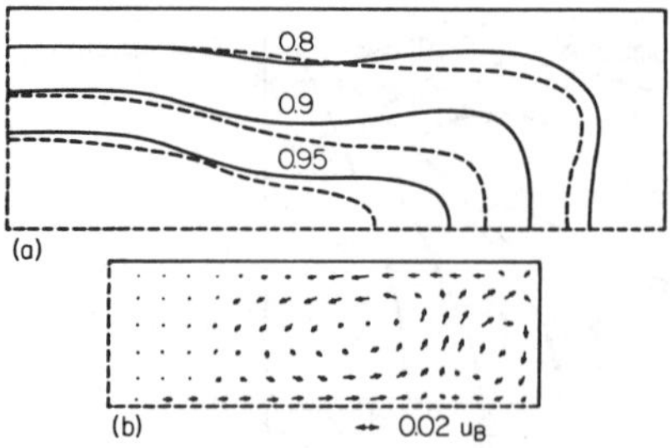

Figure 14. Mean velocity field, rectangular duct (1:3), Re = 56,000. (a) isovels, u/u_c, -------, experiment (Leutheusser, 1963); ———, present calculation; (b) secondary flow velocity vectors: present calculation.

portion of Figure 15. Although the intensity of the smaller vortex predicted is higher than that of the experiment, the predicted streamline pattern shown in the upper portion of the figure appears to be in good accord with the experiment. The anisotropy of cross-planar normal stresses is indicated in Figure 16 along with the experimental data of Hoagland [5]. The predicted pattern shows similar contours with the experimental data. The level of the predicted contours, however, is one order less than that of the experiment. This discrepancy seems to be inherent to this particular algebraic stress model as will be discussed later. Bulges on the predicted contour lines in the upper trapezoidal sector are found near the locations of the inflexion points along the isovels already shown in Figure 14a. The curvature change of this kind will be found in all predicted $(\tau_{zz} - \tau_{yy})$ contours of non-circular ducts, and again seems to be inherent to the model. The predicted wall shear distribution normalized by its average value over the periphery is shown in Figure 17 along

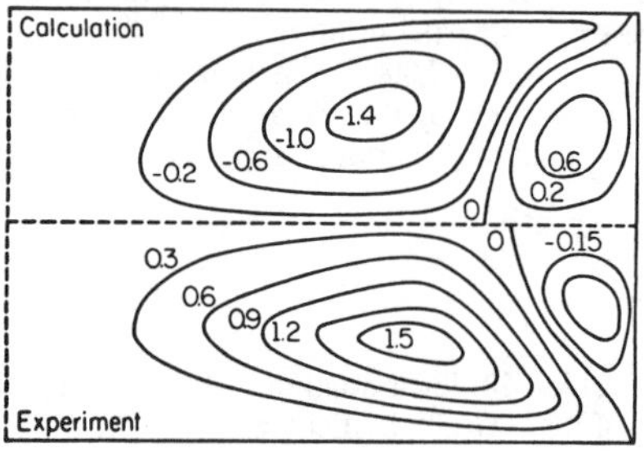

Figure 15. Secondary flow streamlines $[\psi/(u_B D_h)] \times 10^3$, rectangular duct (1:3): upper half, present calculation, Re = 56,000; lower half, experiment (Hoagland, 1960), Re = 30,000.

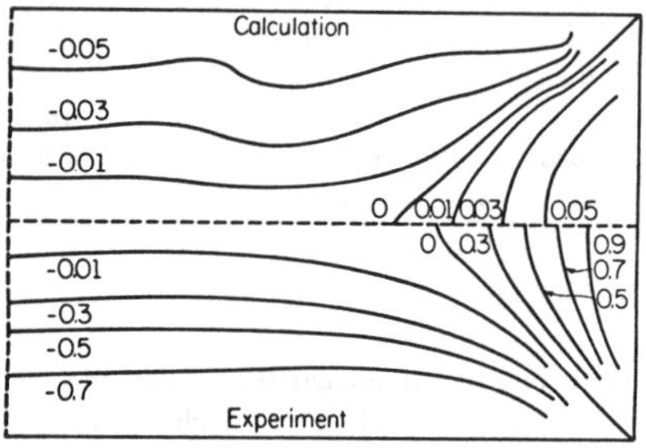

Figure 16. Anisotropy of normal stresses $(\tau_{zz} - \tau_{yy})/\tau_{av}$; rectangular duct (1:3). Upper half, present calculation, Re = 56,000; lower half, experiment (Hoagland, 1960), Re = 30,000.

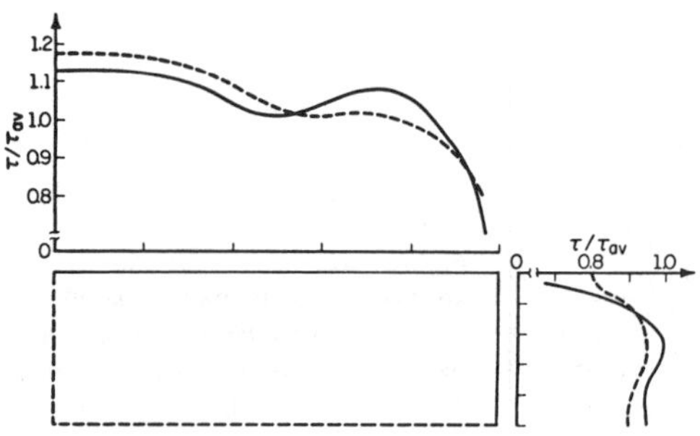

Figure 17. Wall shear distribution τ/τ_{av}; rectangular duct (1:3); Re = 56,000: -------, experiment (Leutheusser, 1963); ———, present calculation.

with the experimental data of Leutheusser [7]. Along the upper wall, the prediction underestimates the wall shear near the wall centerline and overestimates it near the corner. The relative difference between the upper and side wall shears, however, is in fairly good accord with that of the experiment.

Trapezoidal Duct

An attempt was also made to calculate the fully developed turbulent flow in a trapezoidal duct. A non-orthogonal coordinate system with a (14 × 27) grid nodes was employed for one-half the trapezoidal duct with the corner angles of 75 and 105 degrees at Re = 240,000. The corresponding coefficients in the finite difference equation may readily be evaluated according to Equations 39 and 41. The predicted isovels are plotted along with the experimental data [6] in Figure 18a. The prediction shows good agreement with the experiment except at the region near the corner with an acute angle where the prediction underestimates the velocity level.

The predicted secondary flow vectors are shown in Figure 18b. The figure indicates a big vortex surrounded by three smaller vortices. The vortex size may be estimated directly from the isovel pattern in Figure 18a according to Prandtl's [4] suggestion. Both experimental and predicted isovels in Figure 18a indicate a secondary flow pattern similar to the one actually observed in Figure 18b.

The antisotropy of normal stresses is shown along with the experimental data in Figure 18c. The anisotropy level is again underestimated to one order of magnitude less than that of the experiment. The distortion of contour lines near the upper wall appears at the middle of the contour line.

The wall shear distributions are indicated in Figure 19. The damping effect of the duct corner on the wall shear distribution becomes more effective for the corner with an acute angle. This may be explained by the fact that interferences of the walls to the cross-planar normal stresses are naturally more significant for the walls intersecting at a smaller angle.

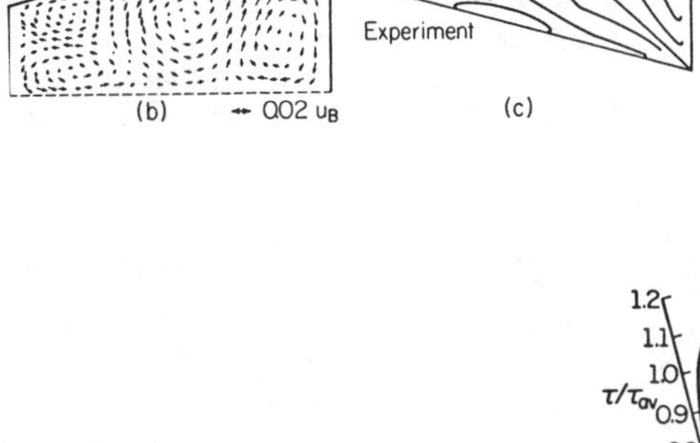

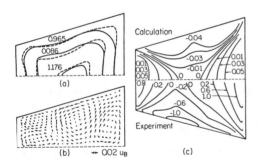

Figure 18. Mean velocity field and anisotropy of normal stresses, trapezoidal duct (corner angles 75° and 105°), Re = 240,000. (a) isovels, u/u_B, -------, experiment (Rodet, 1960); ———, present calculation; (b) secondary flow velocity vectors, present calculation; (c) $(\tau_{zz} - \tau_{yy})/\tau_{av}$, upper half, present calculation, lower half, experiment (Rodet, 1960).

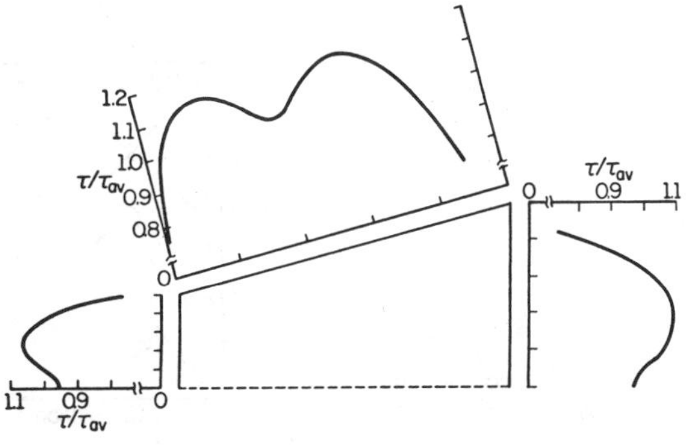

Figure 19. Wall shear distribution τ/τ_{av}, trapezoidal duct (corner angles 75° and 105°), present calculation, Re = 240,000.

Friction Coefficients

The predicted friction coefficients $C_f = 2\tau_{av}/\rho u_B^2$ of the square, rectangular, and trapezoidal ducts are plotted together in Figure 20. The square-duct prediction by Launder and Ying [25] based on a one-equation model and their results obtained by suppressing the secondary flow motions are also shown along with experimental data obtained by various workers [7, 10, 60]. The present prediction gives better agreement with most of the experiment data. Clearly, suppression of the secondary flow motions leads to gross errors from underestimating the friction coefficients.

Developing Turbulent Flows in a Square Duct

Extensive experimental study on the developing flow in a square duct with water as the working fluid was conducted by Melling and Whitelaw [15] using laser-Doppler anemometer for Re = 42,000. Exhaustive measurements were made at two axial stations, namely at $x = 5.6D_h$ and $36.8D_h$. Melling and Whitelaw are the first to report the turbulence structure measurements on an entire quadrant of the square cross-section. All the preceding experimental works have focused only on an octant of the square duct (one triangular sector) under the implicit assumption of symmetry about the bisectors (although the symmetry conditions have never been actually confirmed in these experiments). The experimental data obtained by Melling and Whitelaw, which are believed to be the most reliable among all in the sense of the usage of LDA to avoid the flow interference and the field covering an entire quadrant, however, manifested and extremely difficult nature of the secondary flow measurements, even though the experiment has confirmed that the axial velocity field has acquired a high degree of symmetry in all four quadrants. Such an evidence of uncertainty in the experiment may be observed in Figure 21, which shows the secondary flow velocity vectors obtained in their experiment with reference to the present prediction. The lack of the symmetry

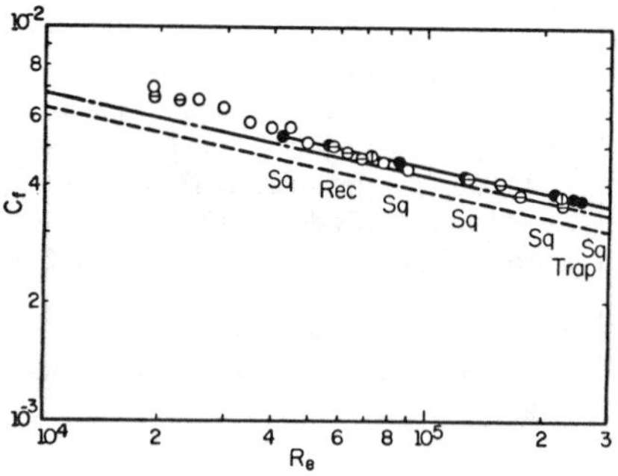

Figure 20. Friction coefficients C_f in non-circular ducts: "○," experiment (Leutheusser, 1963); ⦶, experiment (Launder and Ying, 1972); ⊖, experiment (Harnett, et al. (1962); ———, calculation without secondary flow (Launder and Ying, 1973); ————, calculation with secondary flow (Launder and Ying, 1973); —●—, present calculation (square, rectangle, trapezoid).

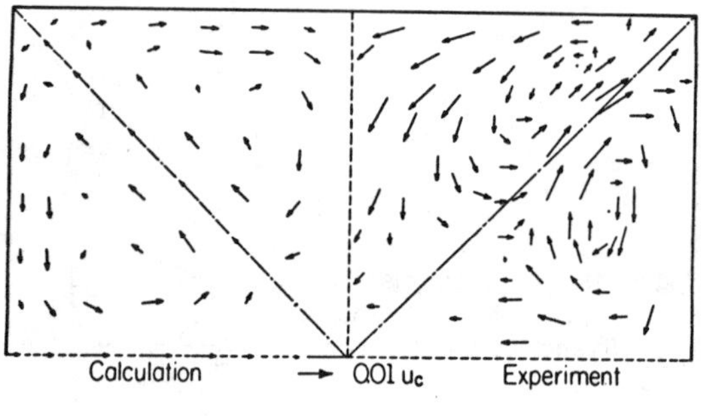

Figure 21. Secondary flow velocity vectors.

about the corner and wall bisectors appear to be considerable in the experimental data. All the experimental data in this section are those obtained by Melling and Whitelaw. However, as indicated above, the experimental data do not necessarily satisfy the required symmetry conditions as the predictions do.

Development of Center-Line Velocity and Turbulence Intensity

The development of the center-line velocity is shown in Figure 22a. The figure clealry shows the peak around $x = 25D_h$, which is obviously associated with the boundary layer merging process at the duct core. The present prediction overestimates the diffusion rate of the boundary layer. Thus, the predicted peak appears upstream of that of the experiment. It is possible to delay the diffusion process by increasing the turbulent Prandtl numbers σ_k and σ_ϵ. Figure 22b shows the axial variations of the kinetic energy k_c and turbulence intensity $(-\tau_{xx}/\rho)^{1/2}$ along the center line. For the turbulence intensity, the available experimental data are also plotted in the figure. The sharp increase observed in both curves around $x = 25D_h$ is again related to the boundary layer merging process. As the boundary layer diffuses toward the duct center, the kinetic energy distribution at the core exhibits a strong positive curvature, $(\nabla^2 k)$, and the kinetic energy gain process by diffusion attains its highest rate. Thus, the level of the kinetic energy increases sharply (note: $P \simeq 0$ since $\partial u/\partial x \simeq \partial u/\partial y \simeq 0$ at the duct core region). The strong amplification of the kinetic energy leads to the enhancement of the momentum exchange activities within the fluid, and naturally brings down the center line velocity level as already observed in Figure 22a. In the prediction, k_c has its maximum around $x = 45D_h$ where the center line velocity is nearly at its fully developed level. It is, however, the region downstream of this peak where the magnitude of the secondary flow is amplified and the bulges in the contour maps become appreciable. Unfortunately, the measurements of Melling and Whitelaw did not extend to this stage of the flow development. The falling-off behavior of the kinetic energy level near the exit may indicate that the axial distance of $84D_h$ as suggested by Gessner and Emery [33] for the establishment of the mean flow structure is not long enough for the turbulence structure to attain its fully developed stage in a strict sense.

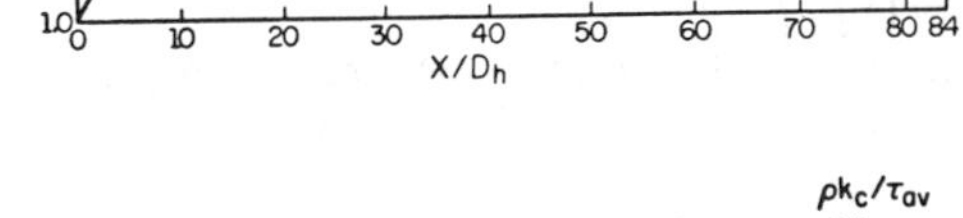

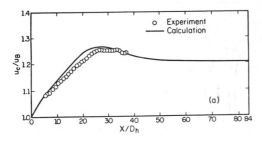

Figure 22a. Development of the center-line velocity.

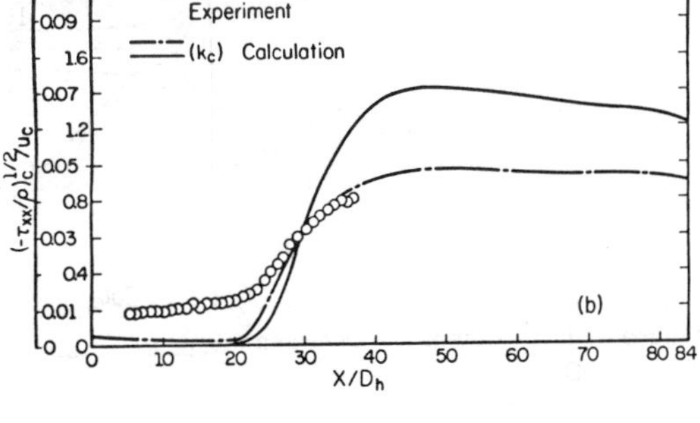

Figure 22b. Development of the turbulent kinetic energy.

However, no such falling-off behavior has been reflected in the axial velocity field. The discrepancy in the levels of $(-\tau_{xx}/\rho)^{1/2}$ is appreciable prior to the boundary layer merging as seen in Figure 22b. This may have been caused by the underestimation of the turbulence level at the inlet. Since no information on the details of the turbulence structure is available at the inlet, the calculations have been performed assuming arbitrarily small values, namely, $k/u_B^2 = 10^{-5}$ and $\epsilon D_h/u_B^3 = 2 \times 10^{-7}$.

Local Flow Structure at $x = 5.6\ D_h$

The calculation results (cal) at $x = 5.6D_h$ are compared with the experiment data (exp) in Figure 23. The predicted isovels plotted in the triangular sector above the diagonal of a square in Figure 23a show good agreement with the experiment in the other half of the square. The predicted

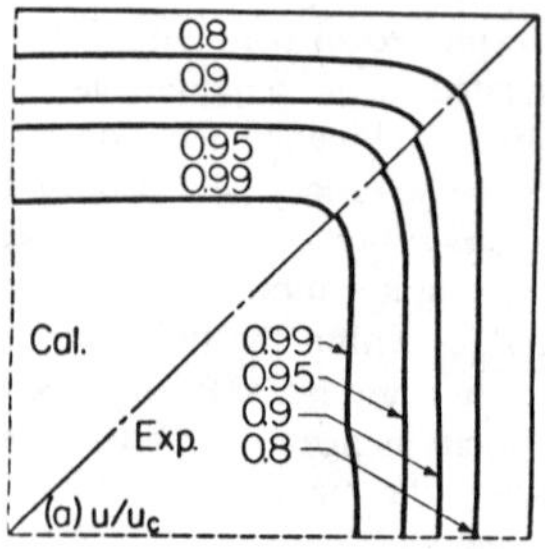

Figure 23a. Isovels of the developing flow at $x = 5.6D_h$.

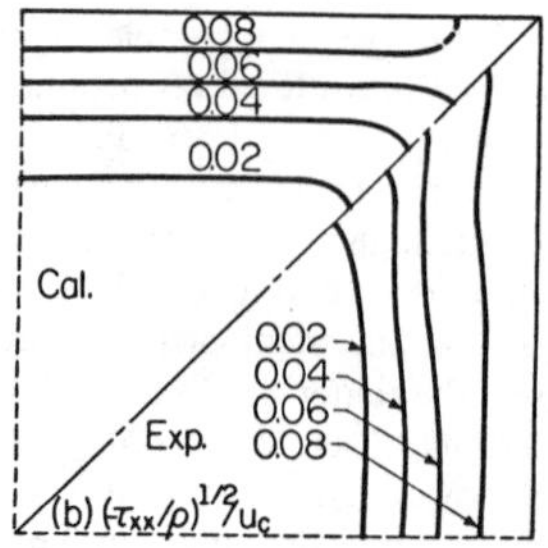

Figure 23b. Axial normal stress intensity at $x = 5.6D_h$.

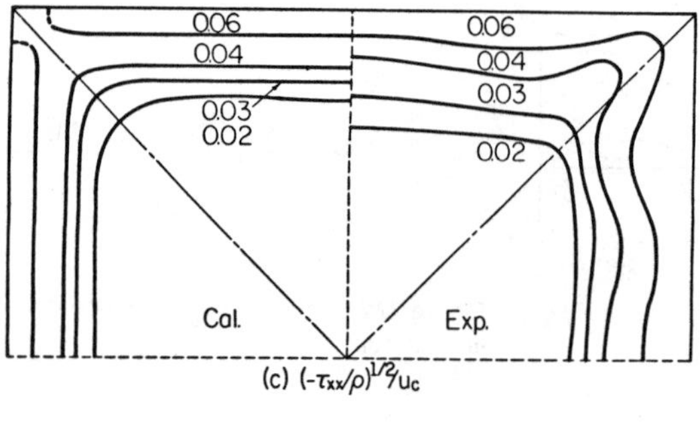

Figure 23c. Transverse normal stress intensity at $x = 5.6D_h$.

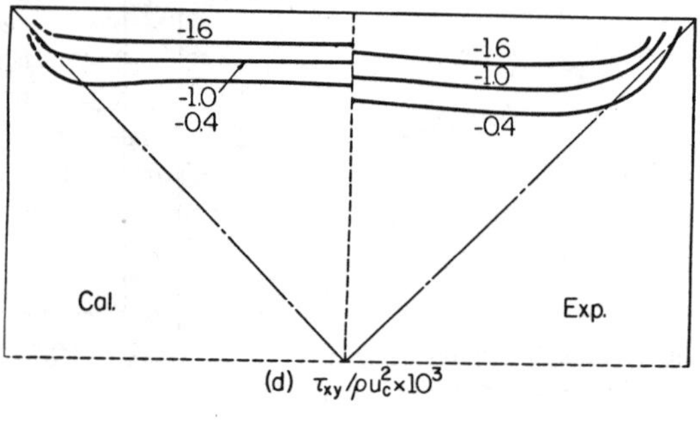

Figure 23d. Turbulent shear stress at $x = 5.6D_h$.

secondary flow vectors (which are not shown here) are found to direct from the wall toward the duct center, indicating the initial stage of the boundary layer build-up process.

The axial normal stress intensity $(-\tau_{xx}/\rho)^{1/2}$ is shown in Figure 23b where the predicted levels are generally in good accord with those of the experiment. The normal stress intensity $(-\tau_{yy}/\rho)^{1/2}$ is also indicated in Figure 23c. Because of the geometrical symmetry, the figure may also be regarded as $(-\tau_{zz}/\rho)^{1/2}$ contour map when it is rotated by $\pi/2$. Therefore, the anisotropy of normal stresses $(\tau_{zz} - \tau_{yy})$, the gradients of which are the major causes for the secondary flow generation, can be observed from the figure through its asymmetry about the diagonal of the square cross section. While the asymmetry is appreciable in the experiment, the prediction indicates an almost symmetrical distribution about the diagonal. Thus, the degree of the anisotropy of the cross-planar normal stresses is estimated too low by the prediction.

The shear stress τ_{xy} working against the primary flow is plotted in Figure 23d where neither the experiment nor prediction has yet indicated the existence of the positive shear region characteristic of the fully developed flow. The predicted level appears to be in good agreement with the experiment.

Local Flow Structure at $x = 36.8D_h$

The flow structures at $x = 36.8D_h$ are presented in Figure 24. As observed from the development of u_c in Figure 22a, this axial station is located just after the boundary layer merging point. The predicted levels of the isovels in Figure 24a are in fairly good agreement with those of the experiment. The distortion of the isovels observed in the experiment, however, is more significant than that of the prediction. The predicted contours for the w velocity component of the secondary flow are presented along with the experimental data in Figure 24b (the corresponding vector plot is already shown in Figure 21). The similarity in the pattern may be observed between experiment and prediction.

The kinetic energy and axial normal stress intensity $(-\tau_{xx}/\rho)^{1/2}$ are presented in Figures 24c, d. Both figures indicate that the degree of bulging toward the corner in the experiment is considerably

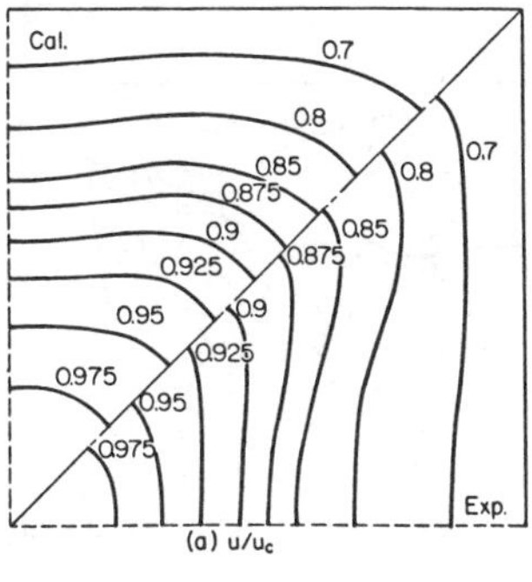

Figure 24a. Isovels of the developing flow at $x = 36.8D_h$.

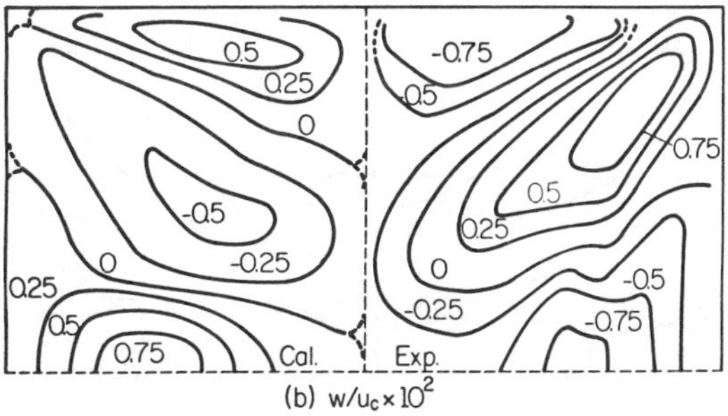

Figure 24b. Secondary flow contour at $x = 36.8D^h$.

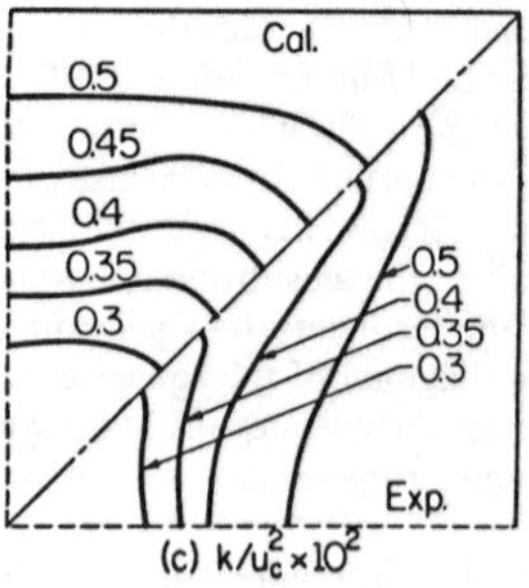

Figure 24c. Turbulence kinetic energy contour at $x = 36.8D_h$.

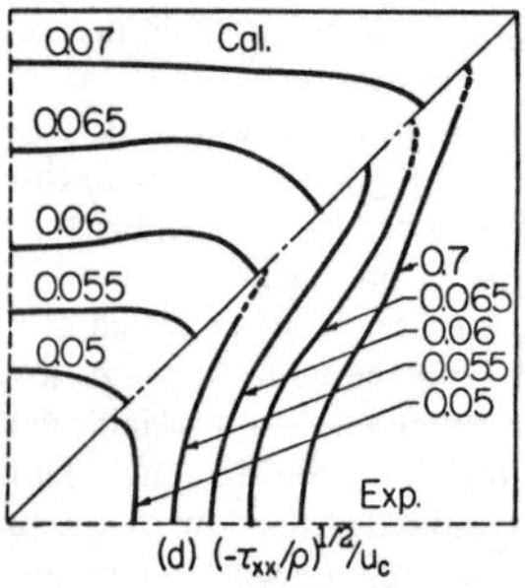

Figure 24d. Axial normal stress intensity at $x = 36.8D_h$.

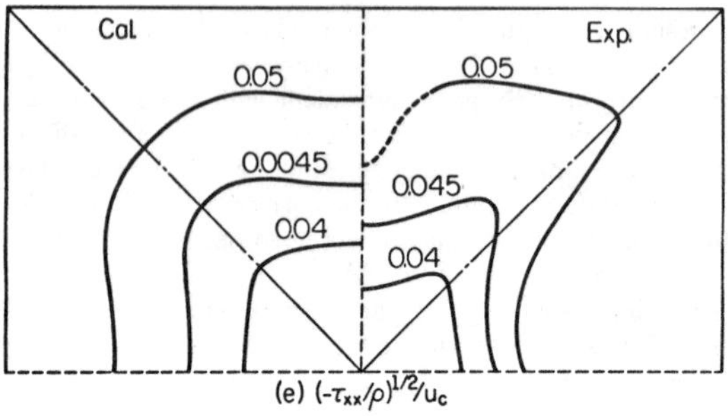

Figure 24e. Transverse normal stress intensity at $x = 36.8D_h$.

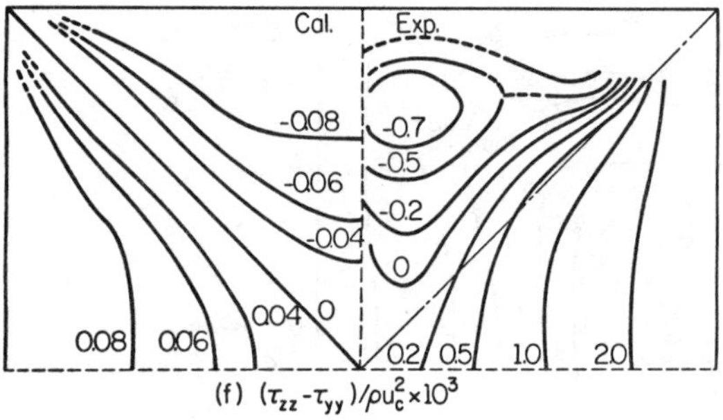

Figure 24f. Difference of the transverse normal stresses at $x = 36.8D_h$.

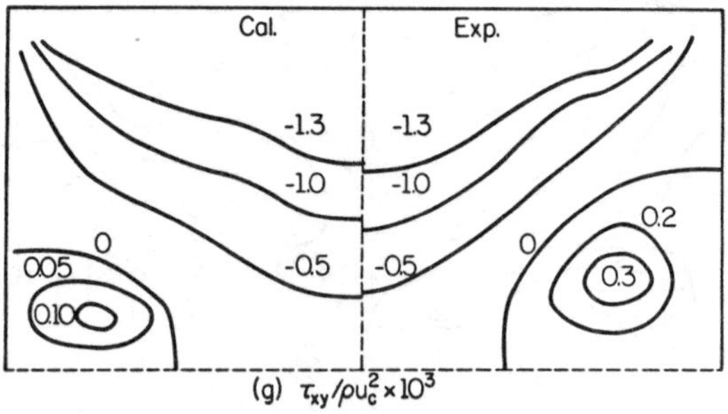

Figure 24g. Turbulent shear stress

higher than that of the prediction. The normal stress intensity $(-\tau_{yy}/\rho)^{1/2}$ is shown in Figure 24e. The degree of the anisotropy, namely, the asymmetry about the diagonal, is again much higher in the experiment than in the prediction. This aspect may well be appreciated from Figure 24f. which presents the contours of $(\tau_{zz} - \tau_{yy})$. The predicted level of the anisotropy is one order of magnitude less than that of the experiment. Moreover, the experiment considerably lacks in the skew-symmetric nature of the map pattern with respect to the diagonal while it is perfectly maintained in the prediction. In contrast to Figure 23d for $x = 5.6D_h$, both experiment and prediction at this axial station shown in Figure 24g already manifest the characteristic τ_{xy} pattern of the fully developed flow, namely, the region of the positive τ_{xy} around the side wall bisector points. The agreement of experiment and prediction in the τ_{xy} map pattern implies the validity of the strain-stress formula (Equations 25b, c) consistent with the effective velocity formulation. For the nearly fully developed flow, Equation 25b indicates that τ_{xy} changes its sign in accord with $\partial u/\partial y$. The experimental data shown in Figures 24a, g substantiate the relation between the axial velocity and shear stress.

Examination of Stress Model

As previously observed, the algebraic stress model developed by Launder and Ying yields a satisfying prediction on the developement of the mean flow field. Yet, some short-comings of the stress model have become evident especially on the cross planar stresses responsible for the secondary flow production. In the following, the algebraic stress model is re-examined in conjuction with the primary velocity field.

The degree of the anisotropy of normal stresses $(\tau_{zz} - \tau_{yy})$ and the cross planar stress τ_{yz} are indicated in Figures 25a, b for the fully developed flow in a square duct with Re = 83,000. The level of the anisotropy is again underestimated by a factor of 1/10 as in the developing flow. Moreover, the curvature change of the predicted contours observed in the middle of the upper triangular sector as shown in Figure 25a is more significant than that of the experiment [11]. The predicted τ_{yz} field is indicated along with the experimental data in Figure 25b where the consistent difference in the order of magnitude is seen. The shear stress of opposite sign, although small in magnitude, appears within a triangular sector. It is well known that the measurements of the cross-planar shear stress are extremely difficult. The only available experimental data on τ_{yz} contours reported [11], however, do not indicate the existence of such a region within a lower triangular sector as shown in Figure 25b. A possible explanation for the disagreement in the distributions of the cross-planar stress components follows.

Upon considering the natural coordinate system along an isovel as employed earlier, it can readily be shown that

$$-\frac{\partial u}{\partial y}\Big/|\nabla u| = \cos\theta, \qquad -\frac{\partial u}{\partial z}\Big/|\nabla u| = \sin\theta \tag{45a}$$

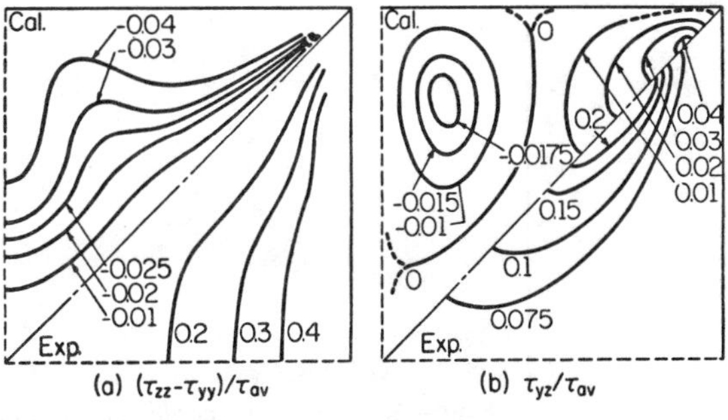

Figure 25. Stress fields in fully developed flow in a square duct.

where

$$|\nabla u| = \left\{ \left(\frac{\partial u}{\partial y} \right)^2 + \left(\frac{\partial u}{\partial z} \right)^2 \right\}^{1/2} \tag{45b}$$

where θ is the local angle between the tangent of the isovel and the z axial as indicated in Figure 1. The substitution of Equations 45 along with Equation 23b into Equations 25d, e, f readily gives

$$(\tau_{zz} - \tau_{yy})/\rho k \simeq -c' \cos 2\theta \tag{46a}$$

and

$$\tau_{yz}/\rho k \simeq \tfrac{1}{2} c' \sin 2\theta \tag{46b}$$

which are valid throughout the duct except at the region close to the corner or the duct center. Along the wall bisectors (namely, $y = 0$ or $z = 0$), Equation 46a reduces to

$$|\tau_{zz} - \tau_{yy}|/\rho k \simeq c' \tag{47}$$

The values corresponding to the anisotropy coefficient c' have been evaluated according to Equation 47 using the experimental data of Brundrett and Baines [11] and Melling and Whitelaw [15], and are plotted in Figure 26. The level of c' obtained on the basis of the experimental data is about 0.2, which is one order of magnitude higher than the value 0.0185 suggested by Launder and Ying and used throughout this chapter. This difference in c' results in the disagreement of the levels of τ_{yz} and $(\tau_{zz} - \tau_{yy})$ between prediction and experiment. As may be expected from the source term of Equation 33, a larger c' leads to higher and more realistic levels of τ_{yz} and $(\tau_{zz} - \tau_{yy})$, but at the same time leads to the amplification of the secondary flow motion and consequently to distorted isovels.

As indicated in Equation 46a, the relative magnitude of the anisotropy along a particular isovel is essentially determined by the local angle through $-\cos 2\theta$ since the kinetic energy level does not change significantly along the isovel.

Where the corresponding isovel nearest to the upper wall shown in Figure 10a is traced clockwise from a point on the wall bisector ($z = 0, \theta = 0$) to that on the diagonal ($y = z, \theta = \pi/4$), $-\cos 2\theta$ may be found along the isovel to vary according to

$$-1 \rightarrow -\cos 2\theta_i \, (\simeq -1 + 2\theta_i^2) \rightarrow -1 \rightarrow 0 \tag{48}$$

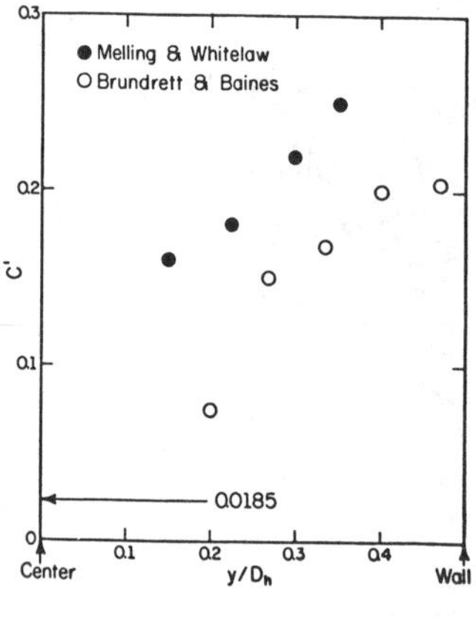

Figure 26. Anisotropy coefficient on the basis of experimental data.

where θ_i (<0) is the angle at the inflexion point of the isovel. The dip resulting from $2\theta_i^2$ reflected on the contour map of the anisotropy can be appreciated from the fact that the bulge on an anisotropy contour takes place around the inflexion point on the isovel as can be seen when Figure 25a is superimposed onto Figure 10a. The predicted contour map of τ_{yz} may also be explained in a similar fashion. According to Equation 46b, the cross-planar shear stress τ_{yz} is essentially governed by sin 2θ, which varies along the isovel in the clockwise direction as

$$0 \rightarrow \sin 2\theta_i\ (\simeq 2\theta_i) \rightarrow 0 \rightarrow 1 \tag{49}$$

Therefore, the region of opposite shear stress is to be centered around the inflexion point as may be seen by superimposing Figure 25b onto Figure 10a.

Even though the mean velocity field is predicted well by the algebraic stress model of Launder and Ying, the model fails to give a satisfactory agreement between prediction and experiment for the turbulence quantities associated with the secondary flow generation. It is not an easy task to analyze the behavior of the stress model and eventually to improve it since the model is based on many proposals and assumptions. It has, however, become clear through the preceding discussions that the relation between the cross planar stress and the axial velocity field is not as simple as that described by Equations 25d, e, f. As seen in the vorticity transport equation, the magnitude of the secondary flow motion strongly depends on the anisotropy coefficient c′ according to the algebraic stress model. The value of c′, which gives a reasonable mean stress field, however, turns out be only about 1/10th of the level that the Reynolds stress data suggest. Furthermore, Figure 26 indicates that c′ may not be taken as a constant over the duct cross section.

Some defects in the present stress model seem to stem directly from the assumption on the stress-strain relationship originally implied by Prandtl, namely, the principal planes of the stress field are normal and tangent to the isovel.

This assumption on the principal planes of the stress field is valid near the planes of symmetry. However, it becomes progressively worse away from the planes of symmetry where the major part of the secondary flow vorticity production takes place. Such evidence may be found in the experimental data obtained by Gessner and Jones [12] and Perkins [13]. Both sets of experimental data on the Reynolds stresses clearly indicate that the principal planes of the stress field do not lie perpendicular to the isovels inside a triangular sector of the square cross section. In order to account for such experiment evidence in the turbulence modeling, certain effort should be made by way of relaxing the strong requirements imposed on the principal planes of the stress field in conjunction with the isovels. The relations (48, 49) indicate that effort of this kind may eliminate somewhat unrealistic curvature change observed in $(\tau_{zz} - \tau_{yy})$ and the sign change observed in τ_{yz} contours.

CLOSURE

As surveyed in this chapter, the advances made in numerical schemes and turbulence modelings in recent decades have been quite significant. The turbulence models, especially the algebraic stress model of Launder and Ying and its derivatives, have greatly enhanced understanding of the hydrodynamic characteristics of the fully developed flow in straight, noncircular ducts. Nevertheless, much work remains to fully understand the complex turbulent viscous flow of this kind. The following suggestions, for future research conclude this chapter.

Improvement of Numerical Scheme

- A standard line relaxation procedure may be replaced by more efficient and economical procedure when such an algorithm becomes available.

Improvement of Stress Model

- The expression for the redistribution term should be modified so as to include the turbulence structure in the proximity of the duct corner [61].

- The strong requirement implicitly included in the algebraic stress model for the principal plane of the stress field to be normal to the isovel should be relaxed at the secondary flow vorticity production region between the symmetry planes.

Needs for Experimental Investigations

- The secondary flow measurement techniques are still at the stage of development. Continuous efforts are needed for the improvement of the techniques.
- Experimental data on non-circular ducts in general, especially on the cross-planar stress fields, are urgently needed for full exploration of the turbulent flow phenomena of this class.

NOTATION

c_1, c_2	empirical constants	p	pressure
C_D	empirical coefficient	R_{ij}	Reynolds stresses
C_f	friction coefficient	$\vec{r}$	position vector
c'	anisotropy coefficient	u_i	instantaneous velocity
D_h	hydraulic diameter	$\bar{u}, \bar{v}, \bar{w}$	flow velocity components
k	turbulent kinetic energy	x, y, z	coordinates
k_{av}	near-wall turbulent kinetic energy		

Greek Symbols

		η	coefficient
α, β, γ	coefficients	ν_t	turbulent kinematic viscosity
Γ	effective diffusion coefficient	σ_k	effective Prandtl number
ϵ	rate of energy dissipation	τ	Reynolds stress
θ	angle	Ω	vorticity

REFERENCES

1. Nikuradse, J., "Untersuchunguber die Geschwindingkeitsverteilung in Turbulenten Stromungen," *Diss. Gottingen, VDI-Forschungsheft*, 281 (1926).
2. Nikuradse, J., "Turbulente Stromung in nicht Kreisformigen Rohren," *Ing.-Arch.*, 1 (1930).
3. Prandtl, L., "Bericht uber Untersuchungen zur Ausgebildeten Turbulenz," *Z. Angew. Math. Mech.*, Vol. 5 (1925).
4. Prandtl, L., "Uber die Ausgebildete Turbulenz," *Verh. 2nd Int. Kong. fur Tech. Mech.*, Zurich (Trans. NACA Tech. Memo. No. 435) (1925).

4a. Prandtl, L., *Essentials of Fluid Dynamics*, London, Blackie (1952).

5. Hoagland, L. C., "Fully Developed Turbulent Flow in Straight Rectangular Ducts: Secondary Flow, Its Cause and Effect on the Primary Flow," Ph.D. Thesis, MIT (1960).
6. Rodet, E., "Etude de 1′ Ecoulement dun Fluide dans un Tunnel Prismatique de Section Trapezoidal," *Publications Scientifiques et Techniques du Ministere de L′ Air*, No. 369 (1960).
7. Leutheusser, H. J., "Turbulent Flow in Rectangular Ducts," *J. the Hydraulics Division*, HY3 (1963).
8. Tracy, H. J., "Turbulent Flow in a Three-Dimensional Channel," *J. the Hydraulics Division*, HY6 (1965).
9. Gessner, F. B., "Turbulence and Mean Flow Characteristics of Fully Developed Flow in Rectangular Channels," Ph.D. Thesis, Purdue University (1964).
10. Launder, B. E., and Ying, W. M., "Secondary Flows in Ducts of Square Cross-Section," *J. Fluid Mech.*, Vol. 54 (1972).
11. Brundrett, E., and Baines, W. D., "Production and Diffusion of Vorticity in Duct Flow," *J. Fluid Mech.*, Vol. 19 (1964).

12. Gessner, F. B., and Jones, J. B., "On Some Aspect of Fully Developed Turbulent Flow in Rectangular Channels," *J. Fluid Mech.*, Vol. 23 (1965).
13. Perkins, H. J., "The Formation of Streamwise Vorticity in Turbulent Flow," *J. Fluid Mech.*, Vol. 44 (1970).
14. Gessner, F. B., Po, J. K., and Emery, A. F., "Measurements of Developing Turbulent Flow in a Square Duct," 1st. Int. Symp. on Turbulent Shear Flow. Penn. State University (1977).
15. Melling, A., and Whitelaw, J. H., "Turbulent Flow in a Rectangular Duct," J. Fluid Mech., Vol. 78, Part 2 (1976).
16. Diessler, R. G., and Taylor, M. F., "Analysis of Turbulent Flow and Heat Transfer in Non-Circular Ducts," NACA Tech. Note 4384 (1958).
17. Buleev, N. E., "Theoretical Model of the Mechanism of Turbulent Exchange in Fluid Flow," *AERE Trans.* 957. Atomic Energy Res. Establish., Harwell, England (1963).
18. Ślykov, Jn., and Carevskij-Djakin., S. N., "Turbulentnoe tecenje i tepleobmen v gladkich prjamdinejnych kanalach proizvilnogo secenja," *Teploenergetika*, No. 12 (1966).
19. Krajewski, B., "Determination of Turbulent Velocity Field in a Rectangular Duct with Non-circular Cross-Section," *Int. J. Heat Mass Transfer*, Vol. 13 (1970).
20. Gerard, R., "Secondary Flow in Non-Circular Conduits," *ASCE J. The Hydraulics Division*, HY5 May (1978).
21. Meyder, R., "Turbulent Velocity and Temperature Distribution in the Central Sub-Channel of Rod Bundles," *Nuclear Eng. Design*, Vol. 35 (1975).
22. Ram, H., and Johannsen, K., "A Phenomenological Turbulence Model and Its Application to Heat Transport in Infinite Rod Arrays with Axial Turbulent Flow," *J. Heat Transfer*, Vol. 97 (1975).
23. Ibragimov, M. H., Petrishchev, V. S., and Sabelev, G. I., "Calculation of Heat Transfer in Turbulent Flow with Allowance for Secondary Flow," *Int. J. Heat Mass Transfer*, Vol. 14 (1971).
24. Gerard, R., "Finite Element Solution for Flow in Noncircular Conduits," *ASCE J. the Hydraulic Div.*, HY3, March (1974).
25. Launder, B. E., and Ying, W. M., "Prediction of Flow and Heat Transfer in Ducts of Square Cross-Section," *Heat and Fluid Flow*, Vol. 3, No. 2 (1973).
26. Aly, A. M. M., Trupp, A. C., and Gerrard, A. D., "Measurements and Prediction of Fully Developed Turbulent Flow in an Equilateral Triangular Duct," *J. Fluid Mech.*, Vol. 85, Part I (1978).
27. Gosman, A. D., and Rapley, C. W., "A Prediction Method for Fully Developed Flow through Noncircular Passages," *Numerical Methods in Laminar and Turbulent Flow, Proc. of the 1st Int. Conf.*, University College, Swansea (1978).
28. Nakayama, A., Chow, W. L., and Sharma, D., "Calculation of Fully Developed Turbulent Flows in Ducts of Arbitrary Cross-Section," *J. Fluid Mech.*, Vol. 128 (1983).
29. Crajilescov, P., and Todreas, N. E., "Experimental and Analytical Study of Axial Turbulent Flows in an Interior Subchannel of a Bare Rod Bundle," *J. Heat Transfer*, Vol. 98 (1975).
30. Trupp, A. C., and Aly, M. M., "Predicted Secondary Flows in Triangular Array Rod Bundles," *J. of Fluids Eng.*, Vol. 101 (1979).
31. Gosman, A. D., and and Rapley, C. W., "Fully Developed Flow in Passages of Arbitrary Cross-Section," *Recent Advances in Numerical Methods in Fluids*, Pineridge Press Ltd (1980).
32. Gessner, F. B., and Emery, A. F., "A Length-Scale Model for Developing Turbulent Flow in a Rectangular Duct," *J. of Fluids*, June (1977).
33. Gessner, F. B., and Emery, A. F., "The Numerical Prediction of Developing Turbulent Flow in Rectangular Ducts," *2nd Symp. on Turbulent Shear Flows*, Imperial College, London (1979).
34. Tatchell, D. G., "Convection Processes in Confined Three-Dimensional Boundary Layers," Ph.D. Thesis, University of London (1975).
35. Neti, S., "Measurement and Analysis of Flow in Ducts and Rod Bundles," Ph.D. Thesis, University of Kentucky. (1977).
36. Nakayama, A., Chow, W. L., and Sharma, D., "Calculation of the Secondary Flow of the Second Kind through a Fully Elliptic Procedure," *Proc. 1980–81 AFOSR-HTTM Stanford Conf. Complex Turb. Flows*, Vol. III (1982).
37. Reece, G. J., "A Generalized Reynolds Stress Model of Turbulence," Ph.D. Thesis, University of London (1976).

38. Launder, B. E., Reece, G. J., and Rodi, W., "Progress in the Development of a Reynolds Stress Turbulence Closure," *J. Fluid Mech.*, Vol. 68 (1975).
39. Naot, D., Shavit, A., and Wofshtein, M., "Numerical Calculation of Reynolds Stresses in a Square Duct with Secondary Flow," *Warme-und-Stoffubertgragung*, Vol. 7 (1974).
40. Mellow, G., and Herring, H. J., "A Survey of the Mean Turbulent Field Closure Methods," *AIAA J.*, Vol. 11 (1973).
41. Launder, B. E., and Spalding, D. B., *Mathematical Models of Turbulence*, Academic Press (1972).
42. Launder, B. E., Morse, A., Rodi, W., and Spalding, D. B., "The Prediction of Free Shear Flows," *Proc. NASA Conf. on Free Shear Flows*, Langley (1972).
43. Harlow, F. H., and Nakayama, P., "Transport of Turbulence Energy Decay Rate," Los Alamos Science Lab., LA-3854 (1968).
44. Hanjalic, K., "Two-Dimensional Asymmetric Turbulent Flow in Ducts," Ph.D. Thesis, University of London (1970).
45. Launder, B. E., and Spalding, D. B., "The Numerical Calculation of Turbulent Flows," *Comp. Methods in Appl. Mech. and Eng.*, Vol. 3 (1974).
46. Launder, B. E., "An Improved Algebraic Stress Model of Turbulence," Imperial College, Mech. Eng. Dept. Rep. TM/TN/A/9 (1971).
47. Rodi, W., "The Prediction of Free Turbulent Bondary Layers by Use of a Two-Equation Model of Turbulence," Ph.D. Thesis, University of London (1972).
48. Chou, P. Y., "On Velocity Correlations and the Solution of the Equations of Turbulent Fluctuation," *Quart. Appl. Math.*, Vol. 3 (1945).
49. Rotta, J., "Statistische Theorie Nichthomogener Tubulenz," *Zeitschrift fur Physik*, Vol. 129 (1951).
50. Hanjalic, K., and Launder, B. E., "A Reynolds Stress Model of Turbulence and Its Application to Thin Shear Flow," *J. Fluid Mech.*, Vol 52, Part 4.
51. Gessner, F. B., and Emery, A. F., "A Reynolds Stress Model for Turbulent Corner Flow, Part I," *J. Fluids Eng.*, June (1976).
52. Rotta, J., "Turbulent Boundary Layers in Incompressible Flow," *Progress in Aeronautical Sciences*, Vol. 2 (1962).
53. Gessner, F. B., "The Origin of Secondary Flow in Turbulent Flow Along a Corner," *J. Fluid Mech.*, Vol. 58, Part I (1973).
54. Hinze, J. O., "Experimental Investigation on Secondary Currents in the Turbulent Flow Through a Straight Conduit," *Appl. Sci. Res.*, Vol. 6–28 (1973).
55. Nakayama, A., "Three-Dimensional Flow Within Conduits of Arbitrary Geometrical Configurations," Ph.D Thesis, University of Illinois at Urbana-Champaign (1981).
56. Patankar, S. V., and Spalding, D. B., "A Calculation Procedure for Heat, Mass and Momentum Transfer in Three-Dimensional Parabolic Flows," *Int. J. Heat Mass Transfer*, Vol. 15 (1972).
57. Nakayama, A., "A Finite Difference Calculation Procedure for Three-Dimensional Turbulent Separated Flows," *Int. J. Num. Meth. Eng.*, June (1984).
58. Nakayama, A., Chow, W. L., and Sharma, D., "Three-Dimensional Developing Turbulent Flow in a Square Duct," *Bulletin of Japan Soc. Mech. Engrs.*, Vol. 27, No. 229, July (1984).
59. Beavers, G. S., Sparrow, E. M., and Magnuson, R. A., "Experiments on Hydrodynamically Developing Flow in Rectangular Ducts of Arbitrary Aspect Ratio," *Int. J. Heat Mass Transfer*, Vol. 13 (1970).
60. Hartnett, J. P., Koh, J. C., and McComas, S., "A Comparison of Predicted and Measured Friction Factors for Flow Through Rectangular Ducts," *J. Heat Transfer*, Vol. 84 (1962).
61. Launder, B. E., Dicussion to Paper (Gessner and Emery 1976), *J. Fluids Eng.*, Vol. 76 (1976).
62. Eichelbrenner, E. A., and Toan, N. K., "A Propos des vitesses secondaries dans la couche limit turbulente a linterrierur dun diedre," Comtes Rendus, Vol. 269 (1969).
63. Eichelbrenner, E. A., and Preston, J. H., "On the Role of Secondary Flow in Turbulent Boundary Layers in Corners (and Salients)," *J. Mecanique*, Vol. 10 (1971).
64. Goldstein, R. J., and Kreid, D. K., "Measurement of Laminar Flow Development in a Square Duct Using a Laser Doppler Flow Meter," *J. Appl. Mech.*, December (1967).

CHAPTER 22

RESIDENCE TIME DISTRIBUTION IN STRAIGHT AND CURVED TUBES

K. D. P. Nigam

Department of Chemical Engineering
Indian Institute of Technology, Delhi
Hauz Khas, New Delhi, India

and

Alok K. Saxena

Indian Institute of Petroleum
Projects Division
C.S.I.R. Complex, PUSA
New Delhi, India

CONTENTS

INTRODUCTION

Continuous flow systems are encountered in various engineering and physiological applications. For engineers these may be in the form of continuous flow reactors, hydraulic or pneumetic conveyors, flow through packed beds, heat exchangers or bubble columns, etc., while biologists may find it in the form of physiological flows. The characterization of all such flow systems requires the information regarding the time spent by different fluid elements in the system and mixing amongst the fluid elements of different ages. It is this aspect that is dealt by residence time theory. Here, the fluid elements may be atoms, molecules, brownian particles, or any other conserved entity of fluid, and the system is composed of a specified finite three-dimensional volume. The age of a fluid element is zero when it enters the system and acquires age at a rate equal to time spent in it. The age of the fluid element at the time of its exit from the system is called the residence time.

General Description and Scope

Ideal Flow Systems

The entire residence time theory is bounded by two extreme, ideal flow situations—the ideal plug flow and the complete back-mix flow. In the case of plug flow, generally relevant to tubular equipment, the fluid is assumed to travel through the system at constant velocity with the result that every small element of material gets processed for the same length of time. For this flow situation the rate of change of any attribute with axial distance, can be written as

$$\bar{u}\frac{dA}{dx} = (r_A) \tag{1}$$

where A = the attribute under consideration, e.g., composition, particle size, temperature etc. x = distance along the flow direction $\bar{u}$ = velocity of the fluid element in the direction of flow; and r_A = rate of change of the attribute.

In case of complete back mix-flow the assumption is that the value of the attribute is the same through out the system. For such systems one can write

$$QA_0 - QA = (r_A)V \tag{2}$$

where A_0 = value of the attribute in the feed
A = value of the attribute at the outlet and in the system itself
Q = flow rate of feed
V = volume of the system

Solutions of Equations 1 or 2 are quite straightforward and the performance predictions based on these will match the observations provided the system satisfies the assumptions made in idealizing it and there are not other complications due to the rate data, changes in temperature and other transport or thermodynamic properties.

From the point of view of a development engineer concerned with scale-up of the laboratory, pilot plant and full-scale equipment, all satisfy the same assumption, either of plug flow or back mix, a scale-up problem is not likely to arise. Also, when the flow changes from one type to the other on scale-up, rational scale-up is possible.

Such systems do exist where the assumptions of either plug flow or complete mix flow holds true. Turbulent flow of single fluid in straight tube, turbulent flow in shell and tube heat exchanger, one phase or multiphase flow in packed beds (provided length to diameter ratio is large enough), etc. can be approximated as plug flow, while moderate-sized plate columns for distillations, continuous stirred tank reactors, liquid phase in moderately sized bubble columns can be well represented as complete back mix flow systems.

On the other hand, situations are encountered in practice where neither plug flow nor back mix assumption can be expected to hold, e.g., laminar flow in pipes and coils, large plate towers, fluidized bed reactors, continuous polymerization tanks or tubes, rotating disc contactors, rotary kilns, etc. In such cases, the performance of equipment can not be evaluated on the basis of Equations 1 and 2. This necessitates knowing more about the mixing within the system. The question therefore arises as to what procedure can then be evolved for designing such equipment.

Non-Ideal Flow Systems

When the ideal plug flow or complete back mix assumption is not satisfied the continuous flow system is sometimes referred to as non-ideal system. It should be noted that there is nothing non-ideal with the flow system except that the flow of the fluid elements is such that it does not conform to the two idealized situations. How this deviation of actual flow systems from the ideal ones affect the flow performance, can be better explained by considering the system as a reactor and the attribute as an extent of reaction (or conversion); however, the principles discussed with suitable modifications can be used for other operations also.

Specifying the flow pattern implies a knowledge of the time (times) spent by the reactant in the reaction vessel and the environment through which the reactant passes. Thus, in the case of a plug flow reactor all reactant spends the same length of time before leaving the reactor and changes take place only along the flow direction. An element of fluid is always in contact with those of its own type. On the other hand, in a back mix reactor the composition is the same throughout the reactor; the possibility (or probability) of part of the feed leaving the reactor is the same irrespective of whether it has entered just now or it has been there for a long time. This implies that different elements of feed entering the reactor spend different lengths of time before they leave the reactor.

The performance of the reactor can be expected to depend on both these factors, i.e., how long the reactant stays in the vessel and what kind of environment it encounters during its stay in the vessel. Generally, plug flow reactor performance is superior to that of a back mix reactor. This is understood in two ways: (a) the reactor volume required for a given performance is less or (b) for the same reactor volume conversion is more in plug-flow reactor.

Exceptions do arise, e.g., a back mix reactor is superior for certain autocatalytic reactions, for some cases of adiabatic reactors, and for certain types of polymerization reactions. In the case of multiple reactions where selectivity, i.e., the amount of desired product formed per unit amount of feed is more important, in some cases plug flow reactor is superior. Calculations can be easily carried out (using Equations 1 and 2) in such cases to ascertain the basic superiority of the reactor type.

In view of this it is obvious that deviations from the two basic types are likely to influence the performance of the equipment in terms of the size of the equipment or the selectivity obtained. It is, therefore, of interest to examine whether in a given situation non-ideality can be expected to be appreciable, and, if so how to take this into account.

A clear-cut answer to the first question cannot be given for all situations and depends on the desired accuracy of prediction and the level of expected performance (i.e., conversion in case of reactors). Thus, if very high conversion is required, say above 99 percent as when recovering trace impurities like catalyst poisons, a small deviation from plug flow can have considerable significance. On the other hand, if there are other uncertainties, e.g., catalyst activity, kinetic data, thermal transport properties, etc., the small deviations from flow pattern acquire secondary significance. However, if reactor stability is important, as in control systems, for the more difficult cases, deviation from plug flow may yet have to be accounted for. For single reactions an acceptable expedient often followed is to increase the size of the equipment. However, for multiple reaction this can bring more harm than good. Again, the size of equipment is important; certain small equipment performs according to the idealized situation, whereas larger equipment with the same features may deviate considerably from idealized flow patterns. It is this aspect that has to be guarded against. Situations are known where the deviation becomes obvious only after the larger scale equipment has been built and found to perform unsatisfactorily. At that stage the question arises whether modifications can be carried-out that would restore the desired performance and it is here that the concept of residence time distribution (RTD) can be very helpful. In fact, RTD quantifies the exact status of the actual system in between the two ideal limiting cases of plug flow and complete back mix flow and provides a great deal of information about the state of mixing within the system. For equipment like fluidized bed reactors consideration of scale-up from the point of view of changes in flow pattern are of particular importance. The same is true for reactors where viscosity changes appreciably during the course of reaction, such as polymerization reactors or where two-phase operations are involved, e.g., gas-liquid reactors.

Age Distribution

The complete evaluation of the performance of a non-ideal flow reactor requires the knowledge of how each molecule travels through the reactor and what environmental changes it is subjected to. This complete information is not achievable and one must settle for something less. This is provided by a knowledge of time spent by different elements of feed entering at the same time. This is sufficient for predicting the performance in a large number of situations, even in other cases its use can be of considerable help. Here are some relevant concepts to understand the methods of obtaining information on residence time:

Age—The age of an element means the time spent by it in the system after its entry to the vessel. Two types of ages are to be distinguished.
Exit age—The time spent by the fluid since its entry up the time it leaves the system.
Internal age—The time spent by the fluid element since its entry while it is still in the system.

It should be noted that internal age is not necessarily equal to the exit age. Furthermore, for a given system there may not necessarily be a single value of exit age and either one or both may be distributions. For example, plug flow reactor exit age is a finite value and internal age is a uniform distribution. For back mix reactor both exit age and internal age distributions are identical. For any non-ideal system these distributions can be quantitatively specified using the functions defined in the following section.

Age Distribution Functions

The exit age distribution function E(t) is such that

$$\int_{t_1}^{t_2} E(t)\,dt \tag{3}$$

represents the fraction of fluid at the outlet that has spent the time between t_1 and t_2, by the time it leaves the reactor. With this representation the fraction of exit stream of age between t and t + dt is

$$E\,dt$$

the fraction younger than age t_1 is

$$\int_0^{t_1} E\,dt \tag{4}$$

where as the fraction of material older than t_1 is

$$\int_{t_1}^{\infty} E\,dt \tag{5}$$

Obviously,

$$\int_0^{\infty} E(t)\,dt = 1 \tag{6}$$

since it would represent all the elements of fluid entered in the feed to the system.

All these aspects of exit age distribution are graphically represented in Figure 1.

Similarly, we can define internal age distribution function I(t) such that

$$\int_{t_1}^{t_2} I(t)\,dt \tag{7}$$

represents the fraction of the fluid in the reactor that has spent time between t_1 and t_2 since its entry. No account is taken of how much more time it will spend. Once again I(t) must be evaluated in a way such that

$$\int_0^{\infty} I(t)\,dt = 1 \tag{8}$$

Knowing I(t) and E(t), we only must know about the ages of reactant. In fact, E curve is the distribution needed to account for non-ideal flow.

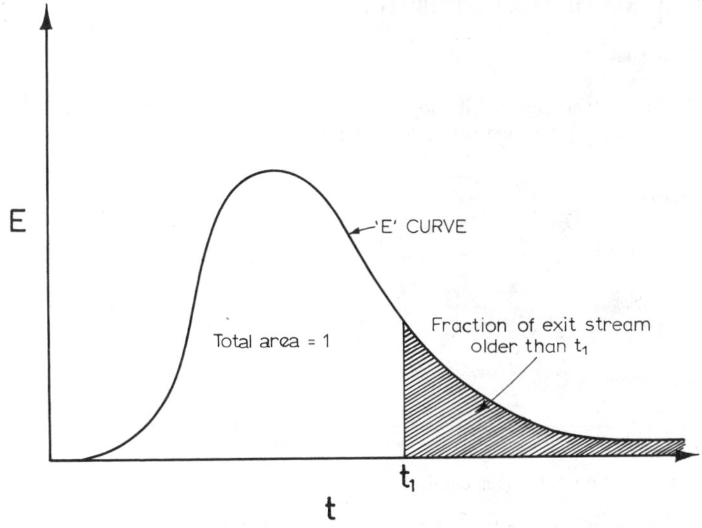

Figure 1. The exit age distribution curve E.

It is also useful to define the function F(t), the cumulative age distribution function, as the fraction of fluid elements in the outlet stream whose, in system or internal age is younger than t, i.e.

$$F(t) = \int_0^t E(t)\,dt \tag{9}$$

or

$$E(t) = \frac{dF(t)}{dt} \tag{10}$$

In engineering practice it is often convenient to present the information in dimensionless form. For this purpose, the time scale (t) is normalized with respect to holding time ($\bar{t}$), i.e.

$$\theta = \frac{t}{\bar{t}} \tag{11}$$

where

$$\bar{t} = \frac{V}{w} \tag{12}$$

In the RTD literature "holding time" is also referred as mean residence time or space time and the dimensionless time (θ) as reduced time. A simple algebric jugglery [1] can show that

$$F(\theta) = F(t) \tag{13}$$

$$E(\theta) = \bar{t}E(t) \tag{14}$$

$$I(\theta) = \bar{t}I(t) \tag{15}$$

DETERMINATION OF AGE DISTRIBUTIONS

Theoretical Methods

For ideal reactors these age distributions can be calculated from first principles. For example, in the cases of ideal plug flow and complete back mix systems, these functions can be written as

Plug Flow System

Internal age distribution—

$$\begin{aligned} I(\theta) &= 1 \qquad \theta < 1 \\ &= 0 \qquad \theta > 1 \end{aligned} \tag{16}$$

Exit age distribution—

$$E(\theta) = \delta(\theta - 1) \tag{17}$$

where Dirac's δ function $\delta(\theta - 1)$ is defined as

$$\begin{aligned} \delta(\theta - 1) = \lim_{h \to 0} P(h, \theta - 1) &= 0 \qquad \text{when } \theta \neq 1 \\ &= \infty \qquad \text{when } \theta = 1 \end{aligned}$$

and infinite pulse function $P(h, \theta - 1)$ as

$$P(h, \theta - 1) = \frac{1}{h} \qquad \text{for } 1 < \theta < 1 + h$$
$$= 0 \qquad \text{for } \theta < 1;\ \theta > 1 + h$$

and Cumulative age distribution—

$$F(\theta) = 0 \qquad \text{for } \theta < 1$$
$$= 1 \qquad \text{for } \theta > 1 \tag{18}$$

Complete Back Mix System

Internal age distribution—

$$I(\theta) = e^{-\theta} \tag{19}$$

Exit age distribution—

$$E(\theta) = e^{-\theta} \tag{20}$$

and Cumulative age distribution—

$$F(\theta) = 1 - e^{-\theta} \tag{21}$$

For non-ideal systems one would like to either derive the age distributions from first principle or obtain them from some suitably compiled information or correlation. Knowledge of velocity distribution in any system may provide sufficient information for computing these distribution functions. In the later part of this chapter we show how the velocity profiles are used to compute RTDs for diffusion free flow in straight and curved tubes.

However, it should be emphasized that in most cases the information on velocity distribution is not available. In such cases one has to perform experiments to determine the age distributions, although it limits the application of RTD studies for design purposes since at the design stage equipment is not available for experiments.

Experimental Determination of Age Distribution

Experimental determination of RTD is based on stimulus response technique. This consists in introducing a tracer, under nonreacting conditions, at the inlet in a known manner and measuring the tracer concentration or any other concentration dependent property at the outlet as a function of time.

Various aspects that require special attention to obtain reliable experimental data on RTD are—

1. Properties of tracer
2. Method of tracer introduction
3. Measurement of tracer concentration at the outlet

Properties of Tracer

An ideal tracer should have the following properties:

Identical physical properties. The tracer must be identical to the flowing fluid in all respects except that it should be distinguishable from the normally flowing fluid. It is particularly important that its density and viscosity be equal to those of normally flowing fluid. Even a density differences of the order of $\sim 10^{-4}$ can drastically effect the RTD [2–4].

Accurate detectability. Response to any tracer input is to be detected by measuring its concentration as a function of time at the outlet. The concentration can be determined by either direct chemical analysis [5, 6] or by measuring any concentration dependent property such as—electrical conductivity [6, 7], light adsorption [8–14], thermal conductivity [10, 15–19], radioactivity [20–22], and dielectric constant [9, 22].

It requires that the tracer's concentration or other concentration dependent property be detectable with a high degree of accuracy.

Linear dependence. If the concentration of the tracer at the outlet is to be determined by detecting any concentration dependent property, it is always desirable that the detected property should have a linear relationship with the tracer concentration over its expected range. It facilitates direct use of the measured property to obtain dimensionless concentrations (fraction of fluid in the outlet).

Chemical inertness. Another important property of an ideal tracer is its chemical inertness with normal flowing fluid. Also, it should not be decayed, absorbed or extracted by the solvent.

Method of Tracer Introduction

Ideally, a tracer should be injected so that is does not disturb the fully developed flow pattern of fluid. Commonly employed methods of tracer introduction are step input, pulse input, sinusoidal input, and arbitrary periodic functions.

Step input. At a specified time the feed is changed from normally flowing fluid to the tracer fluid. The two flow rates are the same and the tracer flow is continued for the duration of the experiment, i.e., until the outlet concentraction (c) becomes equal to the inlet concentration (c_0). This method is often more convenient than the pulse input technique, since the disturbance during tracer injection is minimized, but for large-scale equipment this technique requires unduly large amounts of tracer.

The ratio of c (observed at various instants, θ) and c_0 is the fractional change of concentration (c) in the outlet stream due to the change in feed, i.e., fraction of outlet stream whose age is younger than θ, and the time elapsed after the switching is made. Thus, the step response curve results in a cumulative distribution function $F(\theta)$.

Pulse input. A slug of concentrated tracer is introduced in a very short time and is preceded and followed by the normal fluid, or the normal fluid may continue during the injection time also. Pulse input (infinitesimal time of injection) is extremely difficult to achieve, however, if the injection is completed within one fiftieth of the mean holding time satisfactory results are usually obtained.

The normalized response curve to the pulse input is called C-curve. The normalization is performed by dividing the measured concentration at the outlet of system by the area under the concentration-time curve. Thus, the normalization yields

$$\int_0^\infty C\,dt = \int_0^\infty \frac{c}{Q}\,dt = 1 \tag{22}$$

where

$$Q = \int_0^\infty c\,dt$$

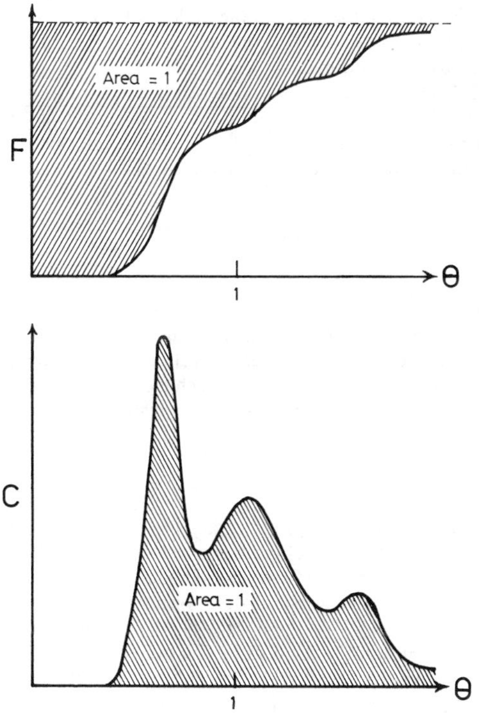

Figure 2. Typical C and F curves for a flow system.

It can be shown that this C-curve is the same as the exit age distribution curve i.e.

$$C(t) = E(t) \tag{23}$$

$$C(\theta) = \bar{t}C(t) = \bar{t}\,E(t) \tag{24}$$

Figure 2 shows the typical response curves for the step and pulse inputs.

Sinusoidal and periodic inputs. These methods involve change in inlet concentration sinusoidally or in any arbitrary, though specified, periodic functions. Such techniques are of particular advantage where the mean residence time is short (from fraction of a second to a few seconds). On the other hand, frequency response experiments can be carried out for a few cycles. These periodic inputs require recording of the tracer concentration at two positions down-stream and are tedious. Moreover in the case of these inputs relatively more sophisticated equipment is necessary and the experimental data require more involved analysis to be of any value.

Measurement of Tracer Concentration at Outlet

The measurement of tracer concentration at the outlet must be consistent with the concept of RTD. This assumes special significance under the conditions of low flow rates where radial variation in velocity can be appreciable. The tracer concentration obtained, at the outlet may be either cross-sectional area mean concentration or bulk mean (mixing cup) concentration. Ananthakrisnan et al. [2] stressed the need of distinction between these two while Ferrel and Himmelblau [23] proposed the conditions where the difference between these two can be ignored. Levenspiel and

Turner [24] pointed out that bulk mean concentration of the tracer at the outlet (with step intput) gives the true RTD. Nigam and Vasudeva [25] have also shown that the bulk mean concentration is of more relevance from the point of view of reactor performance.

Thus, the experimental determination of RTD requires a careful selection of a tracer, appropriate for the specific system at hand. Also, the method of tracer introduction and the measurement of its concentration at the outlet play important roles and should be given due attention. Information regarding various tracers used in liquid, gas and gas liquid systems has been complied by Wen and Fan [26].

Factors Affecting RTD

Defining a system geometry does not completely specify the distribution of residence times of different fluid elements. There are many other factors that significantly influencee RTD. These factors mainly comprise all the intrinsic properties of the flowing fluid, which governs the mass transfer of fluid within the system in one way or other. Following are some such factors that should be specified or checked, before reporting a RTD or using it for a given system, respectively.

Extent of Molecular Diffusion

Mass transfer due to concentration differences within the system comes under this heading. Superimposition of this mode of transfer over the normal flow of fluid (convective transfer) drastically influences the RTD. It is shown later in this chapter that the extent of its influence on RTD depends upon the molecular diffusion coefficient of diffusing substance (D_m) (generally tracer), the mean holding time (t), and the dimensions of the system across the normal flow (diameter in case of tubes). For the case where diffusional effects are negligible, the distribution in residence time is only due to the velocity distribution. In such cases knowledge of velocity profile provides the complete information for computing the RTD. The phenomenon of residence time distribution due to nonuniform flow pattern is often referred as convective axial dispersion.

Rheology of Flowing Fluid

Steady state flow pattern in a system is governed by the momentum balance. Since the rate of momentum transfer amongst the layers of fluid, moving at different velocities, depends on the viscosity, the shear dependent viscosity of non-Newtonian fluids brings about a change in the flow pattern and affects the RTD. Thus, the rheology of flowing fluid has a significant influence on RTD and needs to be specified.

Various models have been proposed for characterizing the non-Newtonian behavior of fluid [27, 28] some of which are reported in Table 1.

Buoyancy Effects

Difference in densities within the system results in a natural convection, super-imposition of which on normal flow effects the RTD. The density differences may arise either due to temperature gradients or due to the difference in chemical species. This aspect assumes special significance in view of the fact that density differences of the order of 10^{-4} may result in drastic changes in RTDs [2].

Extent of Chemical Reaction

This significantly influences the dispersion of different fluid elements in a flow reactor. Simultaneous diffusion and reaction in straight and curved tubes is discussed later in this chapter.

The discussion on residence time distribution in straight and curved tubes will be centered around these aspects.

Table 1
Models Relating Shear Stress and Shear Rates

Sl. No.	Model	Form	Empirical Constants
1.	Power law or Ostawalde Waele	$\tau_{yx} = \frac{K}{g_c}\left(\frac{du}{dy}\right)^n$	K-lb-mass sec^{n-2} ft^{-1} n-dimensionless
2.	Ellis	$\tau_{yx} = \frac{1}{A + B\tau_{yx}^{\alpha-1}}\left(\frac{du}{dy}\right)$	A ft^2 sec^{-1} lb $force^{-1}$ B $ft^{2\alpha}$ sec^{-1} lb $force^{-\alpha}$ α dimensionless
3.	Prandtl-Eyring	$\tau_{yx} = A \sinh^{-1}\left[\frac{1}{B}\left(\frac{du}{dy}\right)\right]$	A lb force/ft^2 B sec^{-1}
4.	De Haven	$\tau_{yx} = \frac{\mu_0/g_c}{1 + C\tau_{yx}^n}\left(\frac{du}{dy}\right)$	μ_0 lb mass/ft sec C $(lb\ f/ft^2)^{-n}$ n dimensionless
5.	Powell-Eyring	$\tau_{yx} = C\left(\frac{du}{dy}\right) + \frac{1}{B}\sinh^{-1}\left[\frac{1}{A}\left(\frac{du}{dy}\right)\right]$	A sec^{-1} B ft^2/lb force C lb_f/ft^2
6.	Sisko	$\tau_{yx} = A\left(\frac{du}{dy}\right) + B\left(\frac{du}{dy}\right)^n$	A lb f sec/ft^2 B lb f sec^m/ft^2 n dimensionless

RTD IN STRAIGHT TUBES

Non-Ideal Tubular Flow and Mixing Concepts

The non-ideal behavior of straight tubes is due to the incomplete mixing in the radial direction and some mixing in axial direction. Mixing in the axial direction is generally produced by vortex formation and eddies produced due to sudden variation in cross-sectional area of tube or because of other disturbances in the steady flow pattern. Such disturbances are commonly encountered at the inlet or outlet of the tube. Under the conditions of laminar flow in straight tubes, velocity distribution in radial direction assumes a parabolic distribution. Fluid elements near the wall, being at lower axial velocity have longer residence times while the elements entering at the center of the tube appear at the outlet at the earliest. In this case the mixing in a radial direction is possible due to the molecular diffusion, provided there exists a radial concentration gradient. Since the diffusion process is relatively slow, the annular elements of fluid flow through the reactor only slightly mixed in the radial direction. Moreover, in case of isothermal flow of single component fluid, where the question of radial concentration gradient does not arise, the deviation from ideal behavior is only due to nonuniform (parabolic) velocity distribution. On the other hand, a uniformly packed bed can be considered as a system in which the requirement of complete radial mixing is satisfied.

All deviations from ideal performance fall into two classifications—segregated flow and mixed flow.

In segregated flow, elements of fluid do not mix with in the system but follow separate paths through the system. The fluid elements can be assumed to flow in the form of discrete packets that never interact with each other. Each element of fluid within the packet has the identical residence time. Completely segregated reactors may be imagined to have no mixing between packets of different ages but this mixing takes place only at the system exit where the various packets merge to form the outlet.

In case of mixed flow the adjacent packets of fluid elements do mix. The effects of these deviations on the performance of the system can be evaluated provided we know the distribution of residence times in the fluid leaving the system and the extent of mixing. Such complete information is seldom available. However for well defined cases of "extent of mixing" the effect of residence time distribution on system performance can be evaluated. Zwietering [29] quantitatively analyzed this aspect and showed the amount of molecular level mixing that affects the performance of a reactor and the time when it occurs.

Convective Axial Dispersion

In case of unidimensional flow in straight circular tubes the velocity profile is a function of radial position only, i.e.

$$U = U(r)$$

For laminar flow it is considered to be a good approximation of segregated flow, which is exact if there is no diffusional effects. In this case, since the flow is segregated and the velocity profile is known, the RTD can be computed [30]. As the velocity distribution is monotonic in r, the fraction of fluid flowing through the radius r and r + dr can be given as

$$C(r) = U(r) 2\pi r \, dr/Q \tag{25}$$

and cumulative distribution function can be calculated as

$$F(r) = \frac{\pi}{Q} \int_0^r U(r') 2r' \, dr' \tag{26}$$

For laminar flow in straight tube

$$U(r) = \frac{2Q}{\pi R^2}\left(1 - \left(\frac{r}{R}\right)^2\right) \tag{27}$$

where R is the tube radius and Q is volumetric flow rate. Since axial velocity of fluid elements depend on r, the residence time also varies with r. If the length of the tube is L, the residence time (t) for fluid elements flowing at r is

$$t = L/u(r) = \frac{R^2}{2Q} \frac{\pi L}{(1 - (r/R)^2)} \tag{28}$$

Since $\pi R^2 L = V$, the volume of reactor, Equation 28 becomes

$$t = \frac{V/Q}{2(1 - (r/R)^2)} = \frac{\bar{t}}{2(1 - (r/R)^2)} \tag{29}$$

where $\bar{t}$ is the mean holding time. Substituting Equation 27 into Equation 26 and eliminating r, using Equation 29 we have

$$F(t) = 1 - \frac{(\bar{t})^2}{4t^2} \tag{30}$$

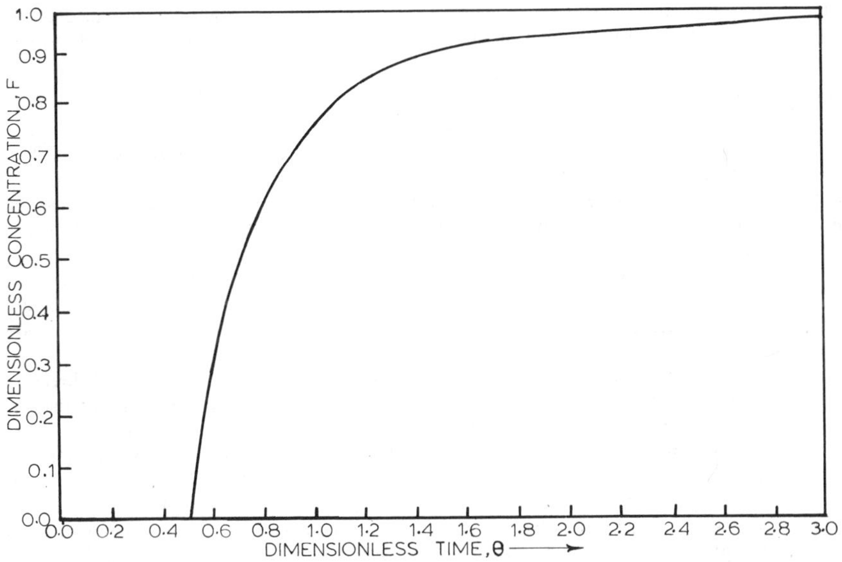

Figure 3. RTD for diffusion-free laminar flow through straight circular tubes.

or, in dimensionless form, Equation 30 can be written as

$$F(\theta) = 1 - \frac{1}{4\theta^2} \tag{31}$$

Equation 31 is the desired residence time distribution function for diffusion free laminar flow in straight circular tube and is shown in Figure 3. Of course, this RTD is subjected to certain simplifying assumptions. In the first place, entrance effects are ignored (as in the derivation of Poiseuille's formula) and the fluid is assumed to be in unaccelerated laminar flow. If the length of the tube is sufficiently large ($\geq 0.06\ N_{Re}$) so that the steady flow pattern (parabolic) is established, the entrance effect becomes unimportant. Secondly, the molecular diffusion is absent or negligible. Even for those cases where radial concentration gradients exist, Equation 31 is the RTD function [31] if the following conditions are satisfied:

$$R > 13\sqrt{D_m L/\bar{u}} \tag{32a}$$

$$L > 6.5 \times 10^4\ D_m/\bar{u} \tag{32b}$$

where D_m is the molecular diffusivity. The third assumption is the viscosity of the fluid is the same everywhere, and the fluid behaves as a Newtonian fluid.

Dispersion Under Significant Influence of Molecular Diffusion

Griffiths [32] appears to be the first to have studied the movement of material injected in to a flowing fluid. Using a colored tracer he made the observation that the tracer spreads out in a symmetrical manner about a plane in the cross-section, which moves with a mean speed of the fluid. Forty years later Taylor [33, 34] called this a rather startling result and provided a mathematical analysis of this situation, which has now become a cornerstone of the theory of dispersion.

Physical Description of Convective Diffusion

The qualitative aspects of movement of injected material can be examined by referring to Figure 4. Under laminar flow conditions the velocity profile is parabolic, i.e., the fluid at the tube center moves at twice the mean velocity and at the wall is stagnant. When a tracer is injected as a finite pulse (Figure 4A), say without appreciably disturbing the flow field, the tracer can be expected to spread out as a paraboloid (Figure 4B). Arising out of the concentration gradient at the periphery of this paraboloid, the tracer would diffuse in radial and axial directions. After sufficient time the process of diffusion tends to eliminate concentration gradient in the radial direction resulting in a uniform radial concentration (Figure 4C). By the time this is achieved (particularly in the case of liquids, i.e., low-diffusion coefficient) diffusion in the axial direction is negligible, but the tracer has been somewhat spread out in the axial direction. This spreading out in the axial direction is referred to as dispersion. The physical picture contains the essential components of the whole subject, i.e., the tracer tends to spread out due to velocity profile (convection). This spreading is reduced by diffusion in a direction perpendicular to the flow and enhanced by diffusion in the direction of flow.

An understanding of dispersion under the combined influence of convection and diffusion in the simple case of a straight circular tube or in other situations is of considerable value in many diverse fields. In the present context the results of such studies may be of use in measuring flow rates in physiological systems, studying the interaction between injected material and body fluids, and understanding and improving the design of artificial kidneys and heart pumps. To permit this we need a quantitative analysis of the situation that requires a mathematical description of the system along with the necessary solution and a priori estimate of any parameters in the mathematical model. An experimental confirmation of the model is highly desirable since it lends creditability to the model and can at times throw new light on the phenomenon.

Mathematical Description of Convective Diffusion

The exact convective diffusion equation governing an isothermal, incompressible, and steady flow of fluids for which both convective dispersion and molecular diffusion are considered to be

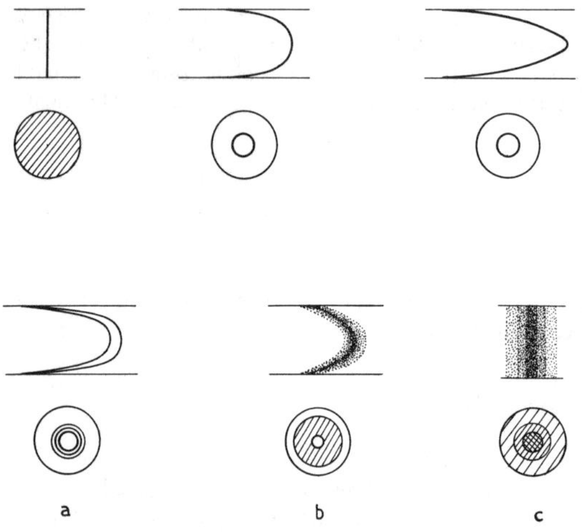

Figure 4. Development of tracer distribution.

equally important can be written as

$$\frac{\partial C}{\partial \theta} = \left(\frac{R}{L}\right)^2 \frac{D_L \bar{t}}{R^2} \frac{\partial^2 C}{\partial X^2} + \frac{D_R \bar{t}}{R^2}\left(\frac{\partial^2 C}{\partial Z^2} + \frac{1}{Z}\frac{\partial C}{\partial Z}\right) - U(Z)\frac{\partial C}{\partial X} \tag{33}$$

where D_L and D_R are the longitudinal and the radial molecular dispersion coefficients, respectively. Since in the case of Newtonian laminar flow

$$U(Z) = 2(1 - Z^2) \tag{34}$$

and

$$D_L = D_R \simeq D_m \tag{35}$$

Equation 33 can be written as

$$\frac{\partial C}{\partial \theta} = \left(\frac{R}{L}\right)^2 \frac{D_m \bar{t}}{R^2} \frac{\partial^2 C}{\partial X^2} + \frac{D_m \bar{t}}{R^2}\left(\frac{\partial^2 C}{\partial Z^2} + \frac{1}{Z}\frac{\partial C}{\partial Z}\right) - 2(1 - Z^2)\frac{\partial C}{\partial X} \tag{33a}$$

for which the appropriate boundary conditions are as follows:

1. At the time of tracer input ($\theta = 0$), there is no solute present at any radial (Z) and axial position (X) downstream ($X > 0$), i.e.,

   $$C(X, Z, 0) = 0 \qquad X > 0$$

2. If the tracer is introduced as a step input, the dimensionless concentration at the inlet ($X = 0$) is unity at all times (θ) and radial points (Z), i.e.,

   $$C(0, Z, \theta) = 1 \qquad \theta \geq 0 \tag{36}$$

3. If the tube wall is impermeable,

   $$\frac{\partial C}{\partial Z} = 0 \qquad \text{for } Z = 0 \text{ and } Z = 1 \tag{37}$$

4. If the tube is sufficiently long to eliminate any axial concentration gradiant at the outlet,

   $$\frac{\partial C}{\partial X} = 0 \qquad \text{at } X = 1 \tag{38}$$

Simplified Model for Laminar Dispersion

In most of the situations the axial molecular diffusion is negligible as compared to radial molecular diffusion, i.e.,

$$\frac{\partial^2 C}{\partial X^2} \ll \frac{\partial^2 C}{\partial Z^2} + \frac{1}{Z}\frac{\partial C}{\partial Z} \tag{39}$$

and the concentration gradient $\partial C/\partial X$ is independent of radial position (Z). Using these simplifying assumptions Taylor [33] showed that for

$$L/u_0 \gg R^2/3.8^2 D_m \tag{40}$$

the cross-sectional area mean concentration, C_m can be represented by

$$\frac{\partial C_m}{\partial \theta} = \frac{D}{\bar{u}L}\frac{\partial^2 C_m}{\partial X^2} - \frac{\partial C_m}{\partial X} \tag{41}$$

where

$$C_m = \frac{2}{R^2}\int_0^R C\, r dr \tag{42}$$

Physically, Equation 41 means that the dispersion phenomenon can at least under certain conditions be considered as equivalent to one where fluid moving at a radially uniform velocity is spread out by the phenomenon of diffusion with effective diffusion coefficient D. Taylor [33] obtained the expression for the effective dispersion coefficient as

$$D = \frac{\bar{u}^2 d_t^2}{192 D_m} \tag{43}$$

Aris [35] included the effect of axial molecular diffusion and showed that Equation 41 holds, provided Taylor's effective diffusion coefficient D is given by

$$D = D_m + \frac{\bar{u}^2 d_t^2}{192 D_m} \tag{44}$$

In practical terms Equations 43 and 44 give the same result for liquid phase system because for liquids

$$D_m \ll \frac{\bar{u}^2 d_t^2}{192 D_m} \tag{45}$$

Initial estimates of conditions under which the dispersion models defined by Equations 41 and 42 should be valid have been further refined [36] to show by experiment and numerical computations that it is a good approximation if the time after injection exceeds about $0.5R^2/D_m$. Physically, this means that the distance of diffusion during that time is comparable with the tube diameter, so that diffusion across the tube acts to average out the velocity of different particles of the tracer in input. This aspect is discussed in detail later in this chapter. After establishing the conditions for the applicability of simplified model, i.e., Equation 41, the next problem is to find its solution. To obtain a particular solution for Equation 41 three boundary conditions, relevant to physical situation, are required. The following four sets of boundary conditions have been employed in the literature:

1. Closed-Closed

$$\begin{aligned} &C_m = 1 && \text{at } X = 0 && \theta > 0 \\ &\frac{\partial C_m}{\partial X} = 0 && \text{at } X = 1 && \theta > 0 \\ &\text{and } C_m = 0 && \text{at } \theta = 0 && X > 0 \end{aligned} \tag{46}$$

2. Closed-Semi-infinite

$$C_m = 1 \quad \text{at } X = 0 \quad \theta > 0$$
$$C_m \to 0 \quad \text{at } X \to \infty \quad \theta > 0 \tag{47}$$
$$\text{and } C_m = 0 \quad \text{at } \theta = 0 \quad X > 0$$

3. Open-Closed

$$\frac{\partial C_m}{\partial X} = \frac{\bar{u}L}{D}(C_m - 1) \quad \text{at } X = 0 \quad \theta > 0$$
$$\frac{\partial C_m}{\partial X} = 0 \quad \text{at } X = 1 \quad \theta > 0 \tag{48}$$
$$\text{and } C_m = 0 \quad \text{at } \theta = 0 \quad X > 0$$

4. Doubly Infinite

$$C_m = 1 \quad \text{at } X \to -\infty \quad \theta > 0$$
$$C_m = 0 \quad \text{at } X \to +\infty \quad \theta > 0 \tag{49}$$
$$\text{and } C_m = 0 \quad \text{at } \theta = 0 \quad X > 0$$

The solutions for Equation 41 for different boundary conditions are reported in literature [37–39], most of which are quite difficult to use. It turns at that when $(D/\bar{u}L) < 0.01$, the choice of boundary conditions is unimportant and essentially the same result is obtained, in all cases. The solution for boundary conditions for doubly infinite is the simplest to use and is given by

$$F = \frac{1}{2}\left(1 - \text{erf}\left(\frac{1-\theta}{2\sqrt{\frac{D}{\bar{u}L}\theta}}\right)\right) \tag{50}$$

and

$$C = \frac{1}{4}\sqrt{\frac{N_{Pe}}{\theta}}\left(\frac{1+\theta}{\theta}\right)\exp\left(-\frac{(1-\theta)^2 N_{Pe}}{4\theta}\right) \tag{51}$$

The model and its solution have only one parameter namely $D/\bar{u}L$. This is called dispersion number and its inverse (or sometimes $\bar{u}$ dt/D) is called Peclet number (N_{Pe}). To avoid the confusion the term Peclet number may be reserved for $\bar{u}L/D$ and $\bar{u}d_t/D$ may be referred as effective Bodenstein number (N'_{Bo}) while the Bodenstein number (N_{Bo}) is ($\bar{u}$ dt/D_m).

Applicability of Dispersion Model

Gill et al. [2, 3, 40] obtained numerical solutions for Equation 33 and showed that Taylor's [33] analysis is valid for large $\bar{u}\, d_t/D_m$ (i.e., for liquids) provided $\tau > 0.8$. For the region $0.02 < \tau < 0.8$ no simple solution is available for the dispersion model. Gill et al. [41–43] have re-examined

Equation 34 and derived from first principle a generalized dispersion equation

$$\frac{\partial C_m}{\partial \hat{\tau}} = \sum_{i=1}^{\infty} K_i(\hat{\tau}) \frac{\partial^i C_m}{\partial X^i} \tag{52}$$

which represent the area mean concentration even at small values of τ. The expression for the coefficients $K_i(\hat{\tau})$ were obtained from first principles and found to be time dependent even when flow is steady. The series on the R.H.S. of Equation 52 can be truncated after the first two terms without any loss of accuracy. It is found that the values of the first two coefficients approach Taylor's constant values when $\tau \geq 0.5$.

Lighthill [44] has also analyzed the phenomenon of dispersion in liquids for $\tau < 0.1$. He examined the behavior of the front of the disturbance when tracer initially spread uniformly over a finite section of tube, which is convected by Poiseuille flow and subject to radial diffusion. A simple solution has been reported that is exact in the whole region where the tracer not yet interacted with the tube wall (i.e. $\tau < 0.1$). In fact, this result does not appear to have received the attention that it deserves and appears to have gone unchecked so far.

Yu [45, 46] developed a more generalized approach to analytically solve the diffusion convective equation. His analysis confirmed Gill's [41, 42, 47, 48] findings under certain conditions. For different values of Bodenstein numbers he compared predictions of the area mean concentration of a dispersed solute with those of [33, 41, 44, 49].

Regions for the applicability of various approximate solutions for dispersion of an injected solute in laminar flow of Newtonian fluid have been investigated by Anathkrishnan et al. [2]. Five approximate solutions are exaimined and compared with numerically solved results of exact convective diffusion equations. The regions in which these solutions fall are:

1. Region of pure convection or high velocity and completely segregated flow.
2. Region of axial dispersion model (Equation 41).
3. Region of Taylor's solution (Equation 43).
4. Region of Taylor-Aris solution (Equation 44).
5. Region of pure diffusion defined as

$$D_m \frac{\partial^2 C}{\partial X^2} = \frac{\partial C}{\partial t} \tag{53}$$

The regions of applicability of these approximate solutions are graphically shown in Figure 5. Since the regions 3, 4, and 5 are special cases of axial dispersion model, its region of applicability comprises that for the three.

Comments on the Use of Dispersion Model

The symmetry of Equation 51 belongs to the Gaussian nature of axial dispersion model, but such symmetry will not always exist in the physical process being characterized by the model. The possibility of reverse motion in the case of a diffusion-like process of mass transfer bring about the symmetry in exit age distribution; however, when overall dispersion coefficient results mainly from convection, reverse flow is not physically possible. This is the case with laminar flow, if the molecular diffusion is small the solution of convection diffusion equation is skewed [49]. It is also true when axial dispersion model is used to approximate packed bed reactors [50–51]. Sundarsan et al. [52] pointed out that it is not possible to approximate a flow system with skewed exit distribution by axial dispersion model with any simple modification. However, because of the simplicity of the model, many attempts have been made [50, 53, 54] by modifying the boundary conditions so that the model can be retained for such systems where D is mainly the result of convection.

Fitzgerald and Levenspiel [55] have warned against the use of a dispersion model as a black box for correlating the convective dispersion data. They have shown that even in the case of Gaussian distribution of residence time, the characterization of the system using dispersion model can be

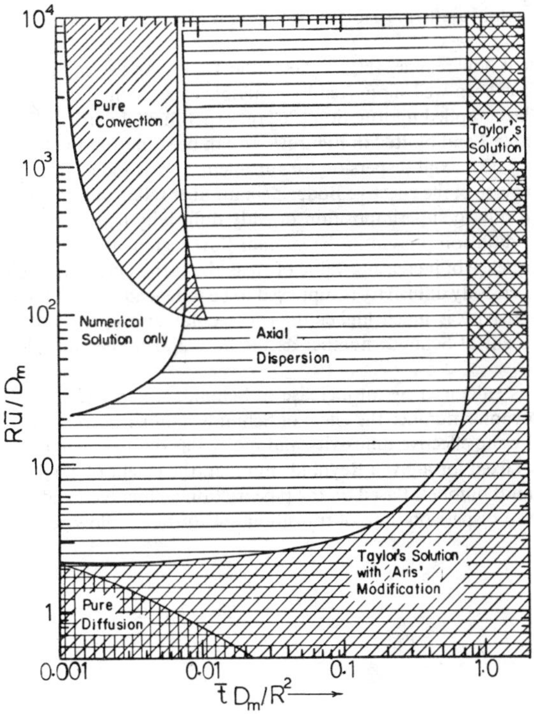

Figure 5. Applicable regions of various approximate solutions for Newtonian flow.

misleading. A dispersion model can characterize only those flow situations where variance of the exit age distribution curve (σ^2) is proportional to the flow length. It is true for the cases where axial dispersion is dominated by diffusion, such as in Brownian motion, molecular diffusion, and some forms of eddy motion. On the other hand, where axial dispersion is governed by convective transfer as in two-phase contactors, spray columns, bubble columns, and trickle beds, this condition is not expected to hold and use of dispersion model may be misleading and requires special attention [56].

Experimental Correlations for Dispersion Data in Straight Tubes

Having obtained [33] a simplified one-dimensional model and conditions of its applicability [2], much work has been conducted to establish experimental correlations of the dispersion coefficients. In case of laminar flow through pipes, the Taylor-Aris solution for axial dispersion coefficient is

$$D = D_m + \frac{\bar{u}^2 dt^2}{192 D_m} \tag{54a}$$

or

$$\frac{1}{N_{Bo}} = \frac{1}{N_{Re} N_{sc}} + \frac{N_{Re} \cdot N_{sc}}{192} \tag{54b}$$

This equation holds for $1.0 < N_{Re} < 2{,}000$ and $0.23 < N_{sc} < 1{,}000$. Wen and Fan [26] have compiled the available experimental information on axial dispersion in pipes, which confirms the accuracy of Equation 54. The reported experimental results are plotted in Figure 6 along with Equation 54 (reproduced from Reference 26).

In the cases of transition and turbulent flow regimes ($N_{Re} > 2{,}000$) the Taylor-Aris solution, considering only longitudinal dispersion, has been well supported by available experimental data over most of the part of the flow regions. The compiled information [26] on experimental work in these flow regions has been reproduced in Table 2 and plotted in Figure 7. As can be seen from the figure, unlike the laminar flow regime ($N_{Re} < 2{,}000$) axial dispersion for turbulent flow is independent of Schmidt number and is correlated by inverse of the effective Bodenstein number (N'_{Bo}) and Reynolds number. This phenomenon is expected due to superimposition of eddy diffusivity (ϵ_D) of mass in turbulent flow over the only molecular diffusion present in the laminar flow region. The reduction in axial dispersion is evidently a consequence of increasing value of ϵ_D as the turbulence being generated. Since for a given Reynolds number, eddy diffusivity is supposed to be independent of Schmidt number, and magnitude-wise eddy diffusivity is much higher than molecular diffusion, the axial dispersion obviously becomes independent of Schmidt number as the flow approaches turbulent region.

In transition regime the dispersion is enhanced by an increase in Schmidt number (Figure 7). This is because with the increase in Reynolds number the effect of Schmidt number becomes considerably less in transition regime where the effective diffusivity and molecular diffusion are of the same order. Since the eddy diffusivity is fixed for a given Reynolds number, the extent of axial dispersion reduces with a decrease in Schmidt number. Based on the pipe friction factor, Wen and Fan [26] have empircally correlated the axial dispersion data of transition and turbulent flow regimes as

$$\frac{1}{N_{Bo}} = \frac{3.0 \times 10^7}{(N_{Re})^{2.1}} + \frac{1.35}{(N_{Re})^{1/8}} \tag{55}$$

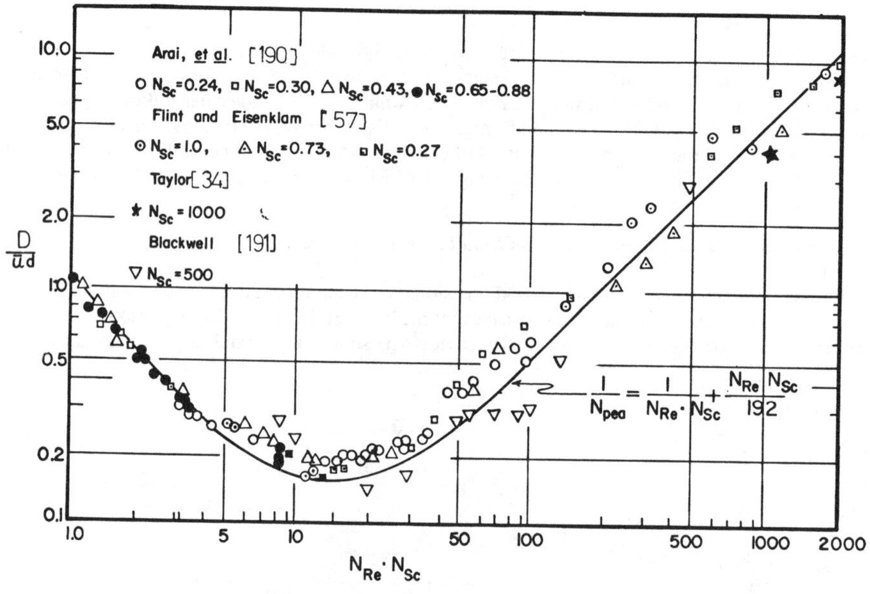

Figure 6. Correlation of axial dispersion coefficient for laminar flow through straight tubes.

Table 2
Summary of Experiments on Axial Dispersion in Pipes

Investigators	System Fluid/Tracer	Input	Diameter (in.)	Length (ft)	N_{Re}	N_{Pe_a}	Remarks
Taylor [33, 34]	Brine	Pulse	0.375	53.5	1.0×10^4 2.0×10^4	1.29–3.5	Smooth and rough pipes
Fowler and Brown [5]	Water/NaCl	Step	7.98	600,000	2.1×10^3 2.0×10^4	0.143–4.55	
Hull and Kent [20]	Oil (radioactive tracer)	Pulse	8.0 10.0 20.0	8,500 17,500 300,000	2.4×10^4 3.0×10^5	2.17–5.0	Data from pipe line gas distribution
Smith and Schulze [194]	Gasoline-Kerosene Kerosene-fuel oil	Step	6.0 12.0	299,060 2,279,583	2.4×10^4 6.0×10^5	2.04–4.0 0.66–6.6	Data from pipe line gas distribution
Keyes [19]	Air-CO_2		0.6	8.03	3.7×10^3 5.1×10^4	1.198–4.92	$N_{Sc} = 0.82$
Hays [192]	Water/NaCl	Pulse	2.0	70	6.0×10^3 $x.25 \times 10^5$	0.22–4.3	$N_{Sc} = 800$
Argo, et al. (196)	N_2/Water NaCl or KCl or tracer		1.84 3.75 17.65	27	8.0×10^4 1.5×10^6	0.60–1.0	Gas-liquid sparge
Sittel, et al. (193)	Water/NaCl	Pulse	0.5 1.0 2.0	70	6.0×10^3 2.0×10^5	1.62–4.96	$N_{Sc} = 800$
Flint and Eisenklam (57)	N_2/C_2H_4 N_2/A N_2/He	Pulse	1.25	9.5	1.0×10^3 1.0×10^4	0.22–2.08 0.33–2.7 0.625–3.13	$N_{Sc} = 0.27, 0.73, 1.0$
Allen and Taylor (195)	Water/NaCl		40.0	355	1.5×10^5	0.2	Only one reading
Davidson, et al. (197)	Refinery gas hydrogen as tracer		3.0	9,000	1.68×10^5 3.26×10^5	2.25–4.8	
Aria, et al. (190)	C_3H_8–N_2 C_4H_{10}–N_2 CH_4–N_2 CH_4–He		0.08 0.175	65.2 750	$1.0 \sim 50$	0.7–5.5	$N_{Sc} = 0.22 \sim 1.0$

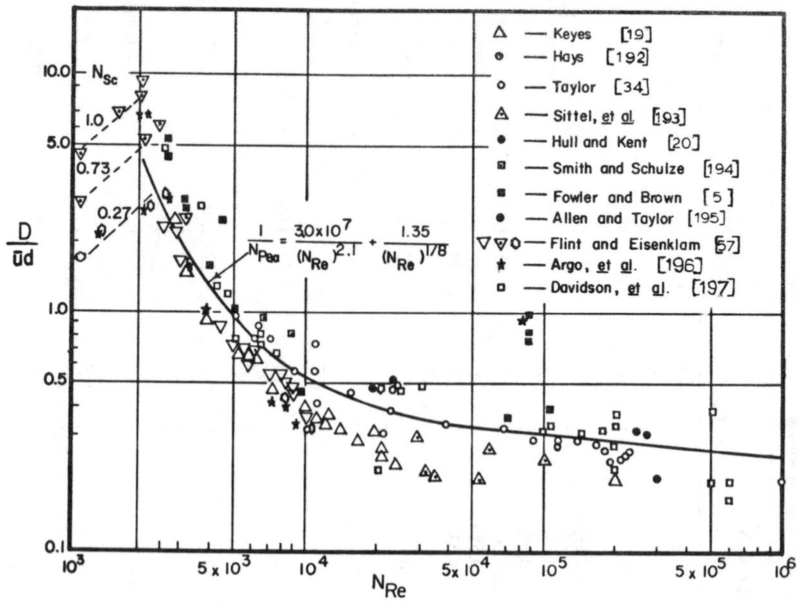

Figure 7. Correlation of axial dispersion coefficient for turbulent flow through straight tubes.

Equation 55 is also shown in Figure 7, indicating the accuracy of correlations. Other such correlation is presented by Flint and Eisenklam [57].

Recently, Wang and Stewart [58] have proposed a new approach to solve dispersion problems. They analyzed the convective dispersion of a solute in steady flow through a tube using convolution and collection methods. They found that orthogonal collection is a very good way to solve such problems and obtained the concentration profile for only Bodenstein number as a convolution of the profile for infinite Bodenstein number.

Dispersion of a Buoyant Solute

Taylor [33] showed that even small density differences of the two liquids flowing in horizontal tubes results in an increase in axial dispersion. Reejhsinghani et al. [3] observed that for low flow rates in horizontal tube, even extremely small density differences ($\Delta\rho/\rho \sim 10^4$) are sufficient to cause measurable effects under laminar flow conditions that are found to be pronounced for large diameter capillary (5 mm). The net effect of buoyancy may increase or decrease axial dispersion because radial mixing caused by cross-sectional circulation tends to decrease axial dispersion, while axial pressure gradients due to density gradients in axial direction enhance axial mixing and hence axial dispersion.

Erdogan and Chatwin [59] have also pointed out that in horizontal tube the effects of natural convection due to density difference between the dispersing solute and solvent are axial pressure gradient resulting in a change in the velocity distribution which may increase the dispersion; and secondary flow owing to density gradient in radial direction, which is expected to decrease dispersion. They [59] derived an expression to replace the diffusion equation if the buoyancy effects are small.

Reejhsinghani et al. [4] have quantitatively investigated the effect of buoyancy on axial dispersion. They have reported a theoretical analysis on the effect of combined natural and forced convection on dispersion in vertical straight circular tubes. Experimental data confirmed the validity of their analysis over the range $26.6 < N_{Bo} < 14{,}000$ and $4 < N_{Gr} < 2{,}510$. These authors have compared the experimental results for dispersion in vertical tubes with numerically computed results based on the absence of natural convection and found that for $N_{Gr} = 4$ theory and experiments agree well, indicating that natural convection at $N_{Gr} = 4$ is essentially negligible, whereas for large values of N_{Gr} ($N_{Gr} \geq 800$) the experimental results differ significantly from theoretical ones. For the case when light fluid being replaced by a heavy fluid at the bottom (the flow being upward), the mixing length over which concentration changes appreciably decreases as N_{Gr} increases for fixed values of τ and N_{Bo} hence results in a decreased dispersion. For the light fluid (on the bottom) replacing heavy fluid, velocity is enhanced in the center and retarded near the wall resulting in a elongated velocity profile and hence increased dispersion. However, they have explained that at small values of τ, pure laminar convection causes a parabolic finger of displacing fluid to grow into the displaced fluid. Thus, the central core is displacing into the displaced fluid by convection. Subsequently, transverse molecular diffusion reduces radial concentration gradients. Since the highest concentration of displacing (or bottom) fluid exists in central core, gravity will accelerate this core if the lighter fluid is on the bottom and retard the core if the heavier fluid is on the bottom. The former case results in an elongated velocity profile, while in the latter case (heavier fluid on bottom) obviously the dispersion coefficient is decreased.

Barton [60] analyzed the phenomenon of buoyant solute in straight horizontal pipe of circular cross-section where dispersion is affected by molecular diffusion, the laminar flow along the pipe, and density currents. He derived an approximate expression for cross-sectional mean concentration using Erdogan and Chatwin's [56] equation and obtained an asymtotic form for the second moment of distributions of buoyant solutes. This led to an important conclusion that the dispersion induced by the buoyancy effect at short and intermediate times is of greater order than the dispersion at asymptotically long tubes.

Recently Mazumdar [61] have theoretically analyzed the dispersion of a solute in natural convection through a vertical channel walls with a linear axial temperature variations. They found the dependence of effective diffusion coefficient on Rayleigh number ($= g\beta' N b^4/\alpha' \nu$). For small values of Rayleigh number, the Taylor diffusion coefficient is oscillatory in nature and increases gradually for large values of Rayleigh number.

Dispersion of Non-Newtonian Fluid in Laminar Flow

Non-Newtonian fluid. So far we have considered the axial dispersion in case of Newtonian flow in which the viscosity of the fluid is independent of shear rates (velocity gradient, $\partial u/\partial r$). Situations do occur in actual practice where this condition does not hold. The shear dependent viscosity brings about a change in velocity distribution, hence effects the extent of axial dispersion.

The non-Newtonian fluids are commonly divided into two broad groups.

1. Time independent fluids, for which the rate of shear at a given point is only dependent on shear stress at that point.
2. Time dependent fluids, which exhibit the dependence of shear rate on both magnitude and the duration of shear.

The time independent fluids can be further classified into those that exhibit yield stresses (τ_y) and those that do not. The yield stress is the minimum stress required to overcome the forces due to internal structure of fluid, which is capable of preventing its movement. Any stress less than τ_y does not yield any deformation (or flow) of the fluid.

In the present section we are mainly concerned with time independent non-Newtonian fluids with zero yield stress. Many models have been proposed to characterize such flow situation and reported in Table 1. The power law, or the Ostwald-de Waele model, is widely used in dispersion literature and given as

$$\tau_{rx} = -K(du/dr)^n \tag{56}$$

where K is the consistency index and τ_{rn} is the axial stress in plane r. The flow behavior index (n) describes the nature of the fluid. For Newtonian fluids n is one, while for $n > 1$ and $n < 1$ the fluid becomes dilatent and pseudoplastic, respectively.

Laminar velocity profile. Laminar flow velocity profile for a non-Newtonian fluid, characterized by Equation 56, flowing through a straight circular tube, without slip at the tube wall may be given as [28].

$$u(r) = u_0\left(1 - \left(\frac{r}{R}\right)^{(n+1)/n}\right) \tag{57}$$

$$u_0 = \left(\frac{3n+1}{n+1}\right)\bar{u} \tag{58}$$

where u(r) is the velocity at a point (x, r), u_0 is the maximum velocity at the axis of the tube and $\bar{u}$ is the mean velocity.

RTD for diffusion free flow. Novasad and Ulbrecht [62] have reported analytical expressions for responses to step and pulse forcing functions in a power-law fluid flowing in straight circular tube under diffusion free conditions. They derived these functions by using the velocity distribution given by Equation 57 and adopting the approach suggested by Danckwerts [30] for Newtonian laminar flow in straight tubes. The expressions can be written as:

Respond to step input (cumulative residence time distribution, F curve)

$$F = 0, \text{ for } \theta < \frac{n+1}{3n+1} \tag{59a}$$

$$F = \frac{3n+1}{n+1}\left(1 - \frac{n+1}{3n+1}\frac{1}{\theta}\right)^{2n/(n+1)} - \frac{2n}{n+1}\left(1 - \frac{n+1}{3n+1}\frac{1}{\theta}\right)^{(3n+1)/(n+1)} \quad \text{for } \theta \geq \frac{n+1}{3n+1} \tag{59b}$$

Response to pulse input (exit age distribution, E or C curve)

$$E(\theta) = 0 \qquad \text{for } \theta \leq \frac{n+1}{3n+1} \tag{60a}$$

$$E(\theta) = \left(\frac{2n}{n+1}\right)\frac{1}{\theta^2}\left[\left(1 - \frac{n+1}{3n+1}\frac{1}{\theta}\right)^{n-1/n+1} - \left(1 - \frac{n+1}{3n+1}\frac{1}{\theta}\right)^{2n/n+1}\right] \quad \text{for } \theta \geq \frac{n+1}{3n+1} \tag{60b}$$

These RTD functions are shown in Figures 8 and 9, which reveal a narrowing of distribution with decrease in power law index (n). Wen and Fan [26] have reported expressions for cross-sectional mean concentration in response to impulse input, step input, and a finite step or pulse input.

The conditions under which the RTD functions given by Equations 59 and 60 hold have not been firmly established, however, by implication one can expect the condition defined by Equation 32 should hold for non-Newtonian flow also. Moreover, numerical solution of the exact convective diffusion equation (Equation 34) reported by Ananthkrishnan et al. [2] indicates that when $D_m\bar{t}/R^2 < 0.012$ and Peclet number (N_{Re}) is large (say > 1,000), both axial and radial diffusion can be safely neglected and flow may be regarded as diffusion free. Since the diffusivity of non-Newtonian liquids is quite low and lies in the range (10^{-7} to 10^{-11} cm^2/sec), one can safely expect these conditions to meet within systems of practical sizes.

Dispersion with significant molecular diffusion. Under the combined influence of convection and molecular diffusion, the extent of dispersion for laminar flow of power law fluids in cylindrical tube is defined by the convective diffusion equation (Equation 33) provided the convective term in the equation is replaced by the appropriate velocity profile (Equation 57). Thus, the governing equa-

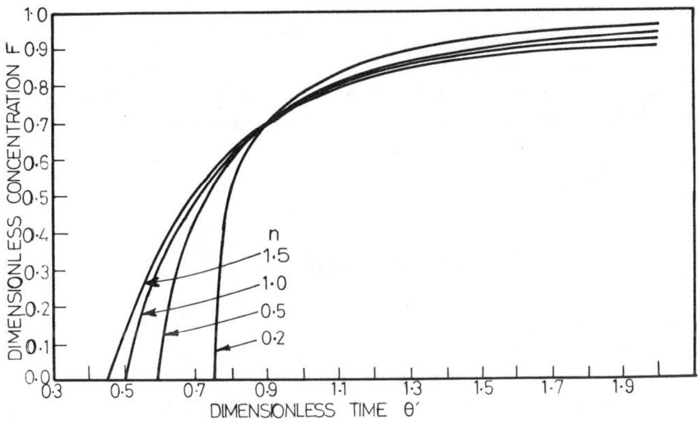

Figure 8. RTD for laminar flow of power law fluids through straight tubes.

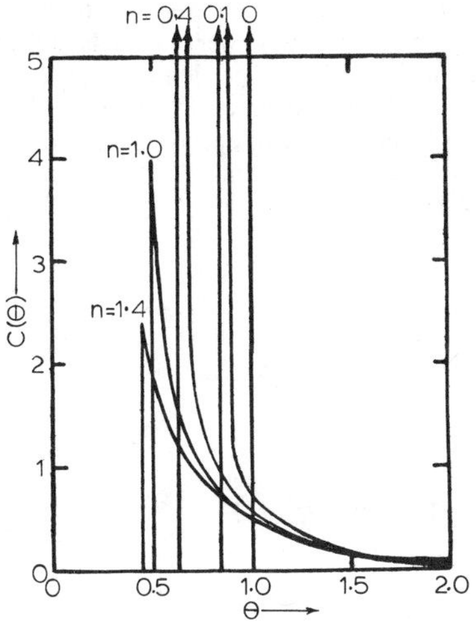

Figure 9. C curves for power law flow through straight tubes.

tion for the flow of isothermal non-Newtonian (power law) fluid becomes

$$\frac{\partial C}{\partial \theta} = \frac{D_m \bar{t}}{R^2}\left(\frac{\partial^2 C}{\partial Z^2} + \frac{1}{Z}\frac{\partial C}{\partial Z}\right) - \frac{3n+1}{n+1}(1 - Z^{(1+n)/n})\frac{\partial C}{\partial X} \tag{61}$$

and the relevant boundary conditions, are

$$\begin{aligned} &C(0, X, Z) = 0 \\ &C(\theta, 0, Z) = 1 \\ &C(\theta, \infty, Z) = 0 \\ &\frac{\partial C}{\partial Z}(\theta, X, 0) = \frac{\partial C}{\partial Z}(\theta, X, 1) = 0 \end{aligned} \tag{62}$$

Fan and Hwang [63] have solved Equation 61 imposing the limiting condition proposed by Taylor [33], i.e., the time necessary for appreciable effects to appear, owing to convective transport, is long compared with the "time of decay" during which radial variations of concentration are reduced to a fraction of their initial value through the action of molecular diffusion. They showed that this is true for laminar flow of power law fluids if the condition

$$\tau > 0.0682\left(\frac{3n+1}{n+1}\right) \tag{63}$$

is satisfied and expression for the effective dispersion coëfficient can be given as

$$\frac{\bar{u}L}{D} = \frac{2(3n+1)(5n+1)}{n^2}\tau \tag{64}$$

With this definition of effective diffusion coefficients the distribution curves are the same as given by Equations 50 and 51 and correlations given by Equations 54 and 55 are applicable for power law fluids also. All the assumptions involved in Taylor's analysis [33] are implicits in Fan and Hwang's solution.

Booras and Krantz [64] have numerically solved Equation 61 using the approach of Gill and Sankarasubramanian [41–43] and obtained the solution for dimensionless concentration in the form of eigenvalues. Their general unsteady state dispersion analysis for power law fluids indicated that in all cases it is sufficient to retain only one term in the generalized dispersion model, proposed by Gill et al. Furthermore, the time dependent behavior of the dispersion coefficient $K_2(\tau')$ for all power law fluids, including Newtonian flow, is similar as in all the cases it approaches steady state value at approximately the same rate. In view of these, the solution of Equation 61 for the boundary conditions defined by Equation 62 can be written as

$$C = \frac{1}{2}\left\{\mathrm{erf}\left[\frac{\frac{1}{2}X_s - X_1}{2\sqrt{\xi^*}}\right] + \mathrm{erf}\left[\frac{\frac{1}{2}X_s + X_1}{2\sqrt{\xi^*}}\right]\right\} \tag{65}$$

where

$$\xi^* = \left[\left(\frac{Ru_0}{D_m}\right)^{-2} + \frac{n^2(n+1)^2}{2(3n+1)^3(5n+1)}\right]\tau' - 4\sum_{m=1}^{\infty}\frac{B_m}{\lambda_m^2}(1-\exp(-\lambda_m^2\cdot\tau')) \tag{66}$$

$$B_m = -\frac{1}{2}\cdot A_m\int_0^1 y^{(2n+1)/n}J_0(\lambda_m Z)\,dz \tag{67}$$

$$A_m = \frac{\dfrac{n}{(3n+1)\lambda_m^2}J_2(\lambda_m) + \left(\dfrac{n}{3n+1}\right)^2\displaystyle\int_0^1 Z^{(4n+1)/n}J_0(\lambda_m Z)\,dz}{\dfrac{1}{2}J_0^2(\lambda_m)} \tag{68}$$

$J_i(y)$ is the ith order Bessel function having argument y and the eigenvalues λ_m must satisfy

$$J_1(\lambda_m) = 0 \tag{69}$$

The expression for the time dependent dispersion coefficient is reported as

$$K_2(\tau') = \left(\frac{Ru_0}{D_m}\right)^{-2} + \frac{n^2(n+1)^2}{2(3n+1)^3(5n+1)} - 4\sum_{m=1}^{\infty}B_m\exp(-\lambda_m^2\tau') \tag{70}$$

They [64] have shown that the dispersion coefficient defined by Equation 70 is not monotonic with n. Its value is $D_m^2/R^2u_0^2$ for n = 0, attains a maximum at n = 0.729 with the value $5.28 \times 10^{-3} + D_m^2/R^2u_0^2$ and asymtotically approaches $3.7 \times 10^{-3} + D_m^2/R^2u_0^2$ as n tends to infinity. This behavior is unexpected in view of narrowing of RTD due to flattening of velocity profile with decrease in n. In view of this, another dispersion coefficient $K_2^*(\tau)$ was defined [64] as

$$K_2^* = \left(\frac{3n+1}{n+1}\right)^2 K_2 \tag{71}$$

which is a monotonically increasing function of n. In fact the value of $K_2 - D_m^2/R^2u_0^2$ varies only $\pm 2\%$ over the range of flow behavior index $0.5 \leq n \leq 1.2$. This includes a broad range of both pseudoplastic and dilatant behavior for which the Newtonian value of the dispersion coefficient $K_2(\infty)$ can be used with little error.

Another interesting result obtained by Booras and Krantz [64] is that unlike the Newtonian fluid, purely convective dispersion in case of power low fluids yield a asymetrical mean concentration distribution. The distribution for pseudoplastic fluids is skewed down-stream due to larger velocities near the axis of the tube while dilatant fluids exhibit the opposite behavior.

Their analysis revealed that in the case of Newtonian fluid, Taylor-Aris dispersion theory applies if

$$\tau \gg \frac{0.000345}{(N_{Bo}^{-2} + 1/192)} \tag{72}$$

and for large Bodenstein number flow this condition reduces to

$$\tau \geq 0.6 \tag{73}$$

Equation 72 is believed to provide a good criterion for the applicability of the Taylor-Aris dispersion theory for moderate and large values of the Bodenstein number. It is evident from the fact that Equation 72 yields the value of τ as 0.0662 for $N_{Bo} = \infty$, while for $N_{Bo} = 50$ the value of τ is 0.0615. It covers a wide range of practical interest and Bodenstein number dependence is inconsequential at larger values of N_{Bo}.

While the phenomenon of axial dispersion in straight circular tubes discussed in this section has been confined to those fluids that are characterized by the power law model, Fan and Wang [65] have reported similar studies for Bihgham plastic and Ellis fluids. Many investigators [66–69] have also analyzed the phenomenon of solute dispersion in non-Newtonian fluids characterized by the Prandtl-Eyring model.

Determination of Molecular Diffusion Coefficient

Taylor [33] experimentally verified his theoretical analysis on dispersion of a solute in laminar flow through straight tubes by measuring the molecular diffusion coefficient of $KMnO_4$ in water. Since then, the technique has been extensively employed to determine molecular diffusion coefficients of different solutes in Newtonian fluids [49, 70, 71]. Non-Newtonian fluids are often encountered in polymer processing industries, but diffusivity data are not extensive and the methods of prediction are not well known. The different methods for measurement of diffusion coefficient in non-Newtonian fluids have been discussed in [72–76]. The technique of determining molecular diffusion coefficient using the extent of axial dispersion in non-Newtonian fluids [63] has been of recent origin [77–81].

The method essentially involves the experimental determination of the response to either step or pulse input under the condition of Taylor-Aris dispersion. The methods of determination of response curves are discussed previously (see page 681) and conditions for the applicability of dispersion model are given by Equations 63, 72 and 73. Once the response curve is obtained, the Peclet number can be estimated using any of the following methods.

Probability plot method. For small values of dispersion number ($\sim$0.001) the theoretical F curves for all boundary conditions yield straight lines on probability paper [1]. The adequacy of fit of the dispersion model is represented by the straightness of the line. The Peclet number can then be calculated using the expression

$$\frac{D}{\bar{u}L} = \frac{\sigma^2}{2} \tag{74}$$

where σ is equal to half the dimensionless time required for the value of F to change from 0.16 to 0.84 from the plot of F versus θ on a probability paper.

Method of moments. This method also uses Equation 74 to compute $D/\bar{u}L$. The difference lies in the method of interpretation of σ^2 for the experimental curve. It is determined by numerically computed second moment of the experimentally obtained C curve. The drawback of this method is that it does not provide any check for the applicability of dispersion model (hence, of Equation 74).

Method of matching experimental and theoretical F curves. This method involves the matching of experimental F curve with the theoretical F curve defined by Equation 50 for different assumed values of $D/\bar{u}L$. The criteria for best fit have been discussed by Trivedi and Vasudeva [82] in detail. This method not only provides the adequacy of fit but also permits a value to be assigned for the acceptability of the model. This method of evaluating of dispersion number for an experimental F curve, although computationally more involved, is more accurate as compared to first two methods and is conveniently carried out using a digital computer.

Knowing the value of $\bar{u}L/D$ the diffusion coefficient is calculated as

$$D_m = \frac{n^2}{2(3n + 1)(5n + 1)} \frac{\bar{u}L}{D} \frac{R^2}{\bar{t}} \tag{75}$$

This method of measuring diffusion coefficient is essentially based on the extent of axial dispersion in straight tubes. However, investigators [83–86] have used helically coiled tubes, for reasons of convenience, to measure molecular diffusion coefficients assuming coiling does not influence dispersion. In some of the situations the assumption may not hold. This aspect has been discussed in detail by Trivedi and Vasudeva [82]. Further, Dutta and Mashelkar [87] have pointed out that the effect of slip at the tube wall may influence the value of molecular diffusion coefficient and should be taken into account.

RTD in Noncircular Straight Tubes

In addition to tubes of circular cross sections, the flow through straight ducts of different cross-sectional geometries are also encountered in practice. The literature on RTD in non-circular ducts is very limited.

Residence time distribution function under diffusion-free conditions can be obtained for any flow system, if the information regarding flow distribution is available. It essentially involves similar methodology as used for obtaining RTD in straight circular tube. In the case of complex velocity distributions, where analytical derivation of RTD is not possible, numerical techniques can be employed. Nauman and Buffham [88] have outlined the basic features of such an approach, which is used by Saxena and Nigam [89] to compute the RTD in straight tube of square cross sections.

Here we present the available information on dispersion is non-circular straight ducts.

Parallel Plates

In a rectangular duct, if the width is very large compared to its height, the flow is approximately one dimensional and can be given as

$$u(y) = u_0(1 - (y/b)^2) = \frac{3\bar{u}}{2}\left(1 - \left(\frac{y}{b}\right)^2\right) \tag{76}$$

where 2b is the height of the channel and y is rectangular coordinate with origin as the center of the channel.

Flow in a parallel plate separators can be approximated by Equation 76. Zeevalkink and Brunsamaun [90] have shown that the separation efficiency of such separators depends on the RTD. They have quantitatively analyzed this aspect and reported the following equation for diffusion free RTD in parallel plates

$$F(\theta) = 0 \qquad \text{for } \theta \leq 2/3 \tag{77a}$$

$$F(\theta) = \frac{3}{2}\left(1 - \frac{2}{3}\frac{1}{\theta}\right)^{1/2} - \frac{1}{2}\left(1 - \frac{2}{3}\frac{1}{\theta}\right)^{3/2} \qquad \text{for } \theta > \frac{2}{3} \tag{77b}$$

It should be noted that this distribution of residence time is closer to ideal plug flow than that in straight circular tube. Interestingly, the same distribution can be achieved in a circular pipe by using a pseudoplastic fluid with $n = 1/3$ [88].

Another type of flow, possible between two parallel plates, is due to the relative motion of one with respect to other. In this case, for the Newtonian fluid, the velocity distribution is parabolic, hence, the RTD is the same as in case of straight circular tube (Equation 31).

Under the combined influence of natural and forced convection, the dispersion of a solute in the laminar flow through parallel plates, subjected to uniform axial temperature variation along the channel walls, has been considered by Mazumdar [91]. They observed an increase in the effective diffusion coefficient with the increase in Grashof number, which is the same for both heating and cooling of plates.

Dispersion of a solute in an electrically conducting fluid flowing between two parallel plates in the presence of a transverse magnetic field has been analyzed by Gupta and Chatterjee [92]. They concluded that the effective diffusion coefficient decreases with the increase in magnetic field.

Ducts of Square Cross Section

The velocity distribution in case of Newtonian laminar flow in striaght tube of rectangular cross-section can be written as [93]

$$u(x, y) = \frac{u_0}{0.571}\sum_{k=1,3,5-\infty}\frac{1}{k^3}(-1)^{(k-1)/2}\left[1 - \frac{\cosh\dfrac{k\pi y}{2a}}{\cosh\dfrac{k\pi b}{2a}}\right]\cos\frac{k\pi x}{2a} \tag{78}$$

where a = b for square cross sections. The overall average velocity can obtained by integrating Equation 78 over the entire cross section and dividing it by the cross-sectional area:

$$\bar{u} = 0.477u_0 \tag{79}$$

Velocity distribution being a series solution, it is not possible to derive an analytical expression for RTD. Saxena and Nigam [89] have numerically computed the RTD using the velocity profile given by Equation 78. The numerically computed RTD functions were fitted to Nauman's model

[94] for diffusion-free RTD as

$$F = 0 \text{ for } \quad \theta < 0.477 \qquad \text{for } \theta < 0.477$$
$$F = 1 - \frac{0.2316}{\theta^{1.908}} - \frac{0.0111}{\theta^2} \qquad \text{for } \theta \geq 0.477 \tag{80}$$

Comparison with Dankwert's [30] RTD for straight circular tube reveals that the diffusion-free RTD in case of a straight tube of square cross section is slightly broader than in a tube of circular cross section. In case of square duct, the dimensionless time at which the first trace of fluid element appears at the outlet (θ_{min}) is 0.477, as compared to 0.5 for circular tube. The probable explanation for the broader RTD may be that for the same cross-sectional area, the wetted periphery of a square cross section is 1.128 times higher than that of a circular cross section. This will cause a higher fraction of the fluid to be at lower axial velocity. Therefore, for the same cross-sectional area and volumetric flow rate, the fluid elements flowing at the center of the tube will move at a faster rate to compensate the higher fraction of fluid at lower axial velocity. This reduces the value of θ_{min} and results in a broader RTD.

Further reduction in the value of θ_{min} has been reported for ducts of equilateral-triangular cross section and star-shaped ducts [95, 96]. With equilateral triangle [95], θ_{min} is 0.45, while for star-shaped ducts [96], it is >0 depending upon the exact geometry.

Annulus Flow

Nigam and Vasudeva [97] have derived the RTD for Newtonian laminar flow in an empty annulus. Their analytical approach was similar to that used by Dankwerts [30] for a straight circular tube.

For an annulus the point velocity u(z) can be given as

$$u(z) = \frac{(\Delta P)R^2}{4\mu L}(1 - Z^2 + 2\bar{\beta}^2 \ln Z) \tag{81}$$

and the volumetric flow rate is

$$Q = \frac{(\Delta P)R^4}{8\mu L}(1 - \xi^4 - 2\bar{\beta}^2(1 - \xi^2)) \tag{82}$$

where

$$\bar{\beta} = \left(\frac{1 - \xi^2}{2\ln(1/\xi)}\right)^{1/2} \tag{83}$$

ξ is the ratio of radius of outer tube to radius of inner tube and $\bar{\beta}$ is the dimensionless radial distance (Z) at which the momentum flux is zero (i.e., the radius of the circle made of fluid elements moving at the fastest velocity). Since the annulus flow velocity function (Equation 81) is not monotonic in Z, there are two radial positions, at which the fluids flows at identical velocity. These positions Z and $\tilde{Z}$ such that $1/\xi < Z < \bar{\beta}$ and $\bar{\beta} < \tilde{Z} < 1$ can be related as

$$\tilde{Z}^2 - 2\bar{\beta}^2 \ln \tilde{Z} = Z^2 - 2\bar{\beta}^2 \ln Z \tag{84}$$

All fluid within the annulus, $Z < Z' < \tilde{Z}$, moves faster than the fluid outside this region. Therefore, RTD function can be derived as

$$F = \frac{1}{Q}\int_Z^{\tilde{Z}} 2\pi Z' U(Z')\, dZ' \tag{85}$$

Substituting Equations 81 and 82 in Equation 85 we have

$$F = \frac{4}{(1 - \xi^4 - 2\bar{\beta}(1 - \xi^2))}\left(\frac{\tilde{Z}^2 - Z^2}{2} - \frac{\tilde{Z}^4 - Z^4}{2}\right) + 2\bar{\beta}\left(\frac{\tilde{Z}}{4}(2 \ln \tilde{Z} - 1) - \frac{Z^2}{4}(2 \ln Z - 1)\right) \quad (86)$$

The corresponding value of dimensionless time can be obtained as

$$\theta = \frac{1}{2}\left(\frac{1 + \xi^2 - 2\bar{\beta}^2}{1 - Z^2 + 2\bar{\beta}^2 \ln Z}\right) \quad (87)$$

and

$$\theta_{min} = \frac{1}{2}\left(\frac{1 + \xi^2 - 2\bar{\beta}Z^2}{1 - \bar{\beta}^2(1 - 2 \ln \bar{\beta})}\right) \quad (88)$$

For an annulus of given ξ, using Equations 83, 88, 84, and 86 the value of F can be computed for any $\theta \geq \theta_{min}$. Typical F curves for annuli of different ξ values are shown in Figure 10 along with that for circular tube. The figure shows that even a very thin wire in a large diameter tube (e.g. $\xi = 10^{-9}$), appreciably decreases the ratio of the maximum to the mean velocity but has little influence on most of the F curve. With an increase in ξ the RTD is progressively narrowed. Experimental tubular reactors are often provided with axial thermowells and it is interesting to note that such thermowells would bring the reactor closer to plug flow.

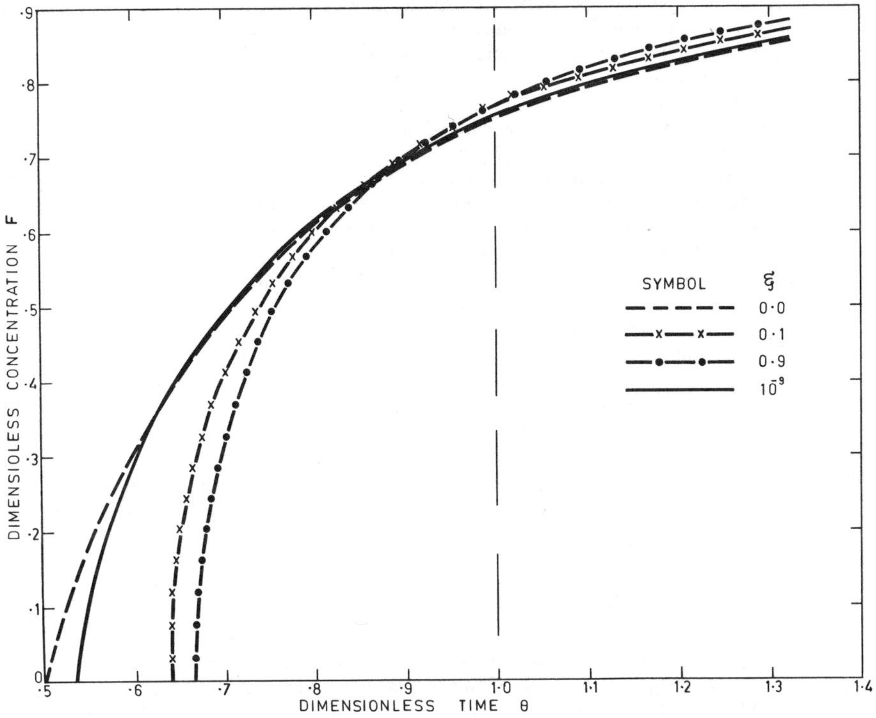

Figure 10. Theoretical step response curves for annulus.

Lin [98] and Pechoc [99] have obtained residence time distribution functions for the diffusion free laminar flow of non-Newtonian fluids in a annulus. Lin [98] used the approximate velocity distribution function reported by Mishra and Mishra [100] while Pechoc [99] used the exact velocity profile obtained by Fredrickson and Bird [101]. The approach in both the cases was similar to that used by Nigam and Vasudeva for Newtonian laminar flow [97]. Both the studies [98, 99] have reported the narrowing RTD with decrease in the value of power level index, which is obvious in view of the straight tube results [62].

Motionless Mixers

Motionless mixers are static devices that use the energy of the flowing fluid to promote mixing throughout the cross section of the duct. A typical device divides and recombines the flow stream. A series combination of such elements can approach homogeneity on a molecular scale [102, 103].

The complexity of flow pattern in such a geometry precludes the possibility of deriving the RTD theoretically from first principle. Therefore, one must depend on experimental information that can be helpful in the development of suitable models.

Nigam and Vasudeva [104] reported the experimentally obtained RTD in a 1.2 cm ID tube fitted with forty elements of Kenics type mixers. Under the negligible molecular diffusion, they observed a unique RTD (for low Reynolds numbers), which was narrower than that for empty straight tubes as well as for low Dean number flow in helical coils. On the other hand, in case of significant molecular diffusion a gradual narrowing of RTD with increase in Reynold's number was observed.

Recently, Nauman [105] presented a simplified theory to treat mixing process as a series of local disturbances in an otherwise open tube. He assumed mixing action to confine to a few isolated planes. At these planes, fluid is redistributed according to a mathematical transformation. If the motionless mixers provides the complete mixing, it can be considered a randomization process with respect to radial positions. In between the two planes of complete mixing, the fluid moves with parabolic velocity profiles for which RTD can be given by Equation 59. Complete mixing causes the sub-systems to be statistically independent and the overall distribution function could be found by simulating the flow of a particle through a flow system as a discreate system. With this approach Nauman [105] computed diffusion free washout functions, i.e., $(1 - F(\theta))$ for motionless mixers of kenics type. Numerical results indicate the narrowing of RTD with the increase in number of motionless elements. Recently, Nigam and Nauman [106] have provided the experimental support to this theory [105]. They considered the flow of moderately non-Newtonian fluids in a 3 cm. ID tube fitted with 20 elements of kenics type mixer. Figure 11 shows the experimental RTD data for aqueous solutions of sodium carboxymethyl-cellulose with polymer concentrations of 0.5, 1.0, and 2.0 g/l, giving power-law constants (n) in the range of 0.607 to 0.725. The arithmetic average for n was 0.647. The solid line in Figure 11 represents the theoretical curve for $n = 0.65$ and 5 planes of complete radial mixing (M). This curve is seen to provide a good fit to the experimental data. Figure 12 shows the computed curves for 4 and 6 planes of complete radial mixing (i.e., $M = 4$ and 6) at a constant value for the power law constant $n = 0.65$. The curve for $M = 6$ provides the best fit for the early part of the experimental tracer data while $M = 4$ gives a better fit for the tail of the distribution. Thus, $M = 5$ provides a good fit to the experimental data as a whole and the empirical result [105] that four kenics elements correspond to one plane of complete mixing is verified [106].

Channels of Varying Cross Section

Smith [107] has obtained a general expression for the longitudinal dispersion coefficient for a tracer in a varying channel. Weaver and Ultman [108] have experimentally measured the tracer gas dispersion for laminar flow through different constrictions. They found that each constriction generates a confined jet that significantly enhances axial dispersion at intermediate Reynolds numbers ranging from 100 to 1,000.

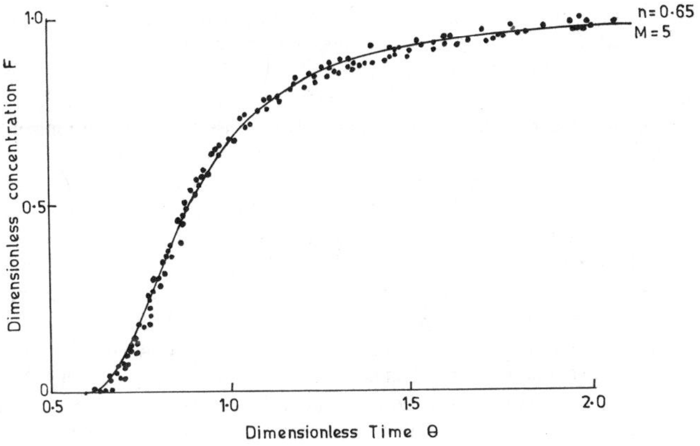

Figure 11. RTD of power law fluids in motionless mixers.

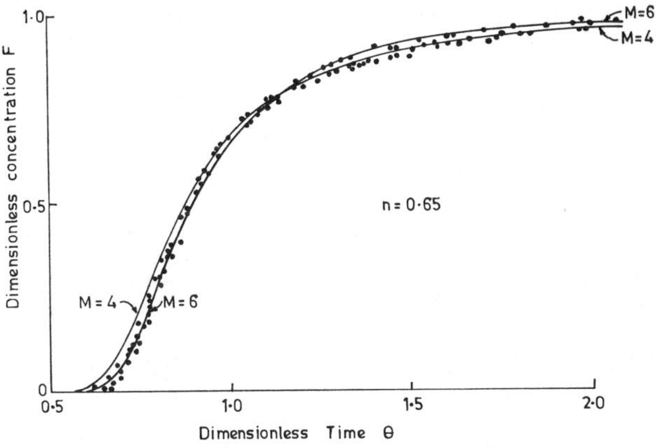

Figure 12. RTD of power law fluids in motionless mixers.

Information on the flow distributions for non-Newtonian flow in tapered ducts [109, 110] that can be used to compute the diffusion-free RTD in an annulus of non-uniform gap [111], in converging and diverging channels [112], and through tubes with sinusoidal axial variation in diameter [113] is available in literature.

RTD IN CURVED TUBES

Fluid flow in curved tubes such as helical and spiral coils occur in many engineering processes, particularly those involving viscometry or convective heat transfer. Blood flow in human arterial system is of particular interest to biologists. The largest vessel in this system, the aorta, is highly curved. The use of helical coils has been widely advocated for variety of applications including chemical reactors and other mass transfer equipment. The reasons for such a wide use of curved tubes are numerous. As against straight-tubes or tubes with bends, coiled tubes make it possible to house process equipment such as heat exchangers in very small space. More importantly they offer the advantages of higher heat and mass transfer coefficients and distribution of residence time reduced over straight tube. This difference in the performace is a consequence of secondary flow in cross-sectional plane induced due to the difference in centrifugal force experienced by different elements of fluid being at different axial velocities. Other important effects of secondary flow are high axial pressure gradient, a diametral pressure gradient, significant peripheral distributions of transfer rates, and higher critical Reynold's number for transition to turbulent flow.

Although systematic studies on axial dispersion in curved tubes are of fairly recent origin, the complexity of the flow has long been known. Thomson [114, 115] appears to be the first to have observed this fact in open curved channel flow. The effect of curvature on the flow through curved pipes was observed by Grindley and Gibson [116] while doing experiments on the vicosity of air. The existance of secondary flow in coiled pipe, as such was demonstrated by Eustice [117, 118], by injecting ink into water flowing through a coiled pipe, while Williams et al. [119] observed that the maximum in the axial velocity is shifted toward the outer wall of a curved pipe. The theoretical analysis of dispersion of an injected solute in a solvent flowing in tubes of any geometry is essentially based on combining the diffusional effects with the velocity distributions. It requires solving a relevant convective diffusion equation along with the equation of motion. Experimental information on the extent of axial dispersion in tubes can be obtained by employing step or pulse response experiments, which can be used to validate the theoretical analysis. The progress made with the two approaches to the problem in the case of curved tubes or/and helical coils is summarized here. First we address what is known about the flow in curved tubes that governs the convective axial dispersion and is the basic information required for solving convective diffusional equation.

Flow in Curved Tubes

Coiled Tubes of Circular Cross-Section

Much work has been done on flow through helical coils since the poineering work of Dean [120, 121]. Navier-Stokes equations for the fully developed steady flow of an incompressible Newtonian fluid in a curved tube of circular cross-section and for the coordinate system shown in Figure 13 are

$$u\frac{\partial u}{\partial r} + \frac{v}{r}\frac{\partial u}{\partial \phi} - \frac{v^2}{r} - \frac{w^2 \sin\phi}{R + r\sin\phi} = -\frac{\partial}{\partial r}\frac{P}{\rho} - \nu\left(\left(\frac{1}{r}\frac{\partial}{\partial \phi} + \frac{\cos\phi}{R + r\sin\phi}\right)\left(\frac{\partial v}{\partial r} + \frac{v}{r} - \frac{1}{r}\frac{\partial u}{\partial \phi}\right)\right) \tag{89}$$

$$u\frac{\partial u}{\partial r} + \frac{v}{r}\frac{\partial v}{\partial \phi} - \frac{uv}{r} - \frac{w^2 \cos\phi}{R + r\sin\phi} = -\frac{1}{r}\frac{\partial}{\partial \phi}\left(\frac{P}{\rho}\right) + \left(\left(\frac{\partial}{\partial r} + \frac{\sin\phi}{R + \sin\phi}\right)\left(\frac{\partial v}{\partial r} + \frac{v}{r} - \frac{1}{r}\frac{\partial u}{\partial \phi}\right)\right) \tag{90}$$

$$\begin{aligned} &u\frac{\partial w}{\partial r} + \frac{v}{r}\frac{\partial w}{\partial \phi} + \frac{uw\sin\phi x}{R + r\sin\phi} + \frac{vwx\cos\phi}{R + r\sin\phi} \\ &= -\frac{1}{R + r\sin\phi}\frac{\partial}{\partial \alpha}\left(\frac{P}{\rho}\right) + \nu\left[\left(\frac{\partial}{\partial r} + \frac{1}{r}\right)\left(\frac{\partial w}{\partial r} + \frac{w\sin\phi}{R + r\sin\phi}\right)\right. \\ &\left. + \frac{1}{\gamma}\frac{\partial}{\partial \phi}\left(\frac{1}{r}\frac{\partial w}{\partial \phi} + \frac{w\cos\phi}{R + r\sin\phi}\right)\right] \end{aligned} \tag{91}$$

and the equation of continuity is

$$\frac{\partial u}{\partial r} + \frac{u}{r} + \frac{u\sin\phi}{R + r\sin\phi} + \frac{1}{r}\frac{\partial v}{\partial \phi} + \frac{v\cos\phi}{R + r\sin\phi} = 0 \tag{92}$$

Under the assumptions of large coil to tube diameter ratio Dean [120] simplified these exact equations and solved for the velocity distribution using the perturbation analysis. He obtained the fol-

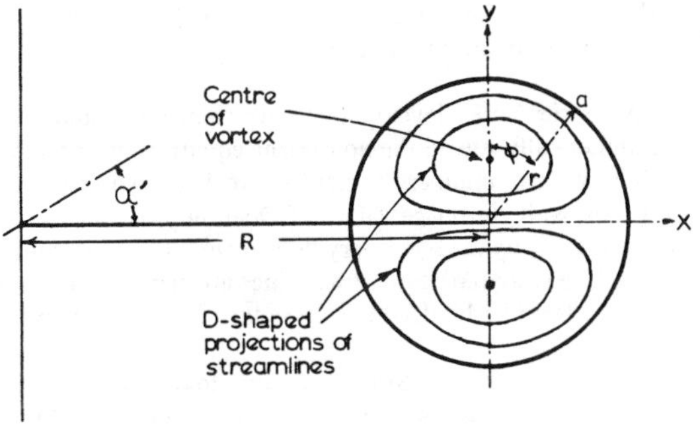

Figure 13. Coordinate system.

lowing expressions for the three components of velocity in r, ϕ, and axial directions, respectively:

$$\frac{u}{W_0} = \frac{N_{Re}}{288} \lambda^{-1} \sin \phi(1 - Z^2)(4 - Z^2) \tag{93}$$

$$\frac{v}{W_0} = \frac{N_{Re}\lambda^{-1}}{288} \cos \phi(1 - Z^2)(4 - 23Z^2 + 7Z^4) \tag{94}$$

$$\frac{w}{W_0} = (1 - Z^2)\left\{1 - \frac{3\lambda^{-1}}{4} \sin \phi + \frac{2N_{Re}^2}{11520} \sin \phi(19 - 21Z^2 + 9Z^4 - Z^6)\right\} \tag{95}$$

where $Z = r/a$

The equation for the projection of secondary flow stream lines on the cross-sectional plane was derived as

$$4 = KZ \cos \phi(1 - Z^2)^2(4 - z)^2 \tag{96}$$

As a result of his analysis Dean [119] concluded:

1. The secondary flow circulation is induced in cross-sectional plane due to the difference in centrifugal force experienced by different elements of fluid being at different axial velocities.
2. The fluid flowing in the central part of the curved tube is thrown toward the outer wall by centrifugal force and fluid near the wall flows along the surface to the inner wall.
3. Two symmetrical circulation patterns (vortices) are established in the plane perpendicular to the main flow. This is shown in Figure 13.
4. The apparent axial velocity is the maximum at two points where $Z = 0.429$ and $\phi = 0$ and π. The secondary flow components, u and v, are zero at these two points.
5. Flow in curved tubes does not show any definite velocity at which the pattern of fluid motion changes.
6. The above analysis is valid for large λ and $N_{De} < 17$, where N_{De} is the Dean number that characterizes the flow in curved tubes. The Dean number is equal to the ratio of the square root of the product of the inertia and centrifugal forces to the viscous force, i.e.,

$$N_{De} = \frac{2R\bar{W}\rho}{\mu\sqrt{\lambda}} \tag{97}$$

With the view to relax the condition of large λ implicit in Dean's velocity profile, Topakoglu [122] obtained an approximate solution for the laminar flow of an incompressible viscous fluid in curved tube by introducing a stream function and using series expressions for the axial velocity component and stream functions. Further attempts in this direction were of Larrain and Brilla [123] and Van Dyke [124], who after a certain amount of series manipulation and restructuring, claimed to find a solution valid for all Dean numbers.

McConalogue and Srivastava [125] used a Fourier series development with respect to the polar angle in the plane of tube cross section and numerically solved the non-linear equations of motion over the range $16.6 < N_{De} < 77$. For low values of Dean number the results revealed a symmetrical secondary flow pattern with respect to the y axis, while for large values of Dean number the axial momentum peak was found to be convected well away by the secondary flow, from the center of the cross section toward the outer wall of the bend. The secondary flow streamlines are accordingly not symmetrical between the inside and outside of the bend and indeed the secondary flow components take much greater values on the outside of the bend.

Truesdell and Adler [126] have numerically solved the Navier-Stokes equation to obtain the axial and secondary velocities in helical coils of finite pitch ($h_c/2\pi R_c < 0.2$) and of both circular and elliptical cross sections. They have reported twelve solutions covering the Dean number range 1 to 280. Their results have good accuracy up to $N_{De} = 200$. Austine and Seader [127] used a torodial coordinate system and employed an over-relaxation technique to solve governing Navier-Stokes equations in streamline and vorticity form. They reported the results for the ranges $5 < \lambda < 500$ and $1 < N_{De} < 1{,}000$. The important conclusion drawn by the authors is that the effect of curvature of tube is accounted for mainly in Dean number, which is the principal parameter to characterize the secondary flow in a curved tube. For $N_{De} = 1$ and $\lambda = 100$ the axial velocity profile is essentially parabolic and unaltered from the fully developed straight tube flow. However, for $N_{De} \geq 10$ the maximum velocity at the diameter of full circle was found to shift towards the outer wall of the curved tube. Tarbell and Samuels [128] have analyzed the flow in curved tubes by solving Navier-Stokes equations over the ranges of $20 < N_{De} < 580$ and $3 < \lambda < 30$. They employed an alternating direction implicit technique and reported separate influence of N_{Re} and λ to be more realistic than N_{De} alone to characterize the secondary flow in the curved tubes.

Smith [129–131] has extended the theoretical description of flow field in curved tubes for large values of Dean number ($N_{Re} \sim 1$) and laid down a formal basis for deriving the fundamental attributes for many physical situations arising from pulsation frequency, cross-sectional geometry, etc. He [130] applied his analysis to triangular and rectangular cross sections. In the latter case Smith found difficulties with the solution near the flat inner bend and speculated for the possibility of the separated solutions in this region. Other numerical studies on flow description in curved tubes are available [132–138]. Recently, Berger et al. [139] have reported a more comprehensive description of previous work on fluid flow in curved tubes. However, the problem of the large Dean number flow through a circular cross-section curve tube is still an unresolved matter.

Effect of Coil Pitch

Recently, there have been a few studies discussing the effect of coil pitch on the flow fields. Manlapaz and Churchill [140] obtained the solution for steady fully developed laminar flow of an incompressible Newtonian fluid through a helically coiled tube of finite pitch using a finite difference technique. They found that the behavior for helical coils is intermediate between that of a straight tube and a torus (coil of zero pitch) and for a moderate degree of pitch, differs only slightly from that for a torus. They developed a correlation for friction factor as a function of Reynolds number, coil and tube radius, valid for laminar flow conditions.

Wang [141] raise a point regarding the applicability of velocity profiles obtained for curved tubes (zero pitch) for a helical coil, which involves both curvature and torsion (i.e. twist due to non-zero pitch). By solving Navier-Stokes equations in non-orthogonal helical coordinate system he showed, torsion along with curvature in helical coils has a non-negligible effect on secondary flow pattern. The parameter $T(=h_c/2\pi R_c N_{Re})$ was found to characterize the effect of torsion on velocity distribution. The velocity profiles are reported later in this chapter. Germano [142] introduced an orthogonal coordinate system along a generic spatial curve and solved the Navier-Stokes equation. His results revealed that for a low Reynolds number the effect of curvature on flow field is more pronounced than that of pitch.

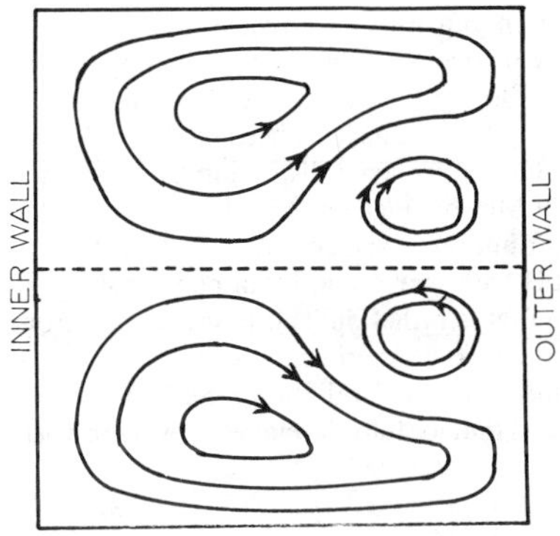

Figure 14. Secondary flow pattern in a coiled tube of square cross-section ($N_{De} = 107$, $\lambda = 100$) [145].

Flow Through Coiled Tubes of Non-Circular Cross-Section

The flow through coiled tubes of noncircular cross section is much less intensely analyzed. Cuming [143] applied Dean's approach [120] to analyze the phenomenon of secondary flow in coiled rectangular channels. He explicitly gave an approximate solution for the particular case of square cross section and showed that secondary flow is more intense is coiled tubes of square cross-section than that in circular cross section tubes; however, his analytical solution was limited to very low flow rates. Recently, some investigators [144–150] have shown interest in flow problems related to helical flow of non-circular geometries principally because of their practical significance, for example, in jacketted kettels and annular spacers between two concentric cylinders containing spiral spacer as in case of extruders. Specifically in the analysis of flow in nuclear reactor system, non-circular geometry in important components is the rule rather than the exception [151].

Joseph et al. [144] have numerically solved the Navier-Stokes equation using the finite difference method in right-handed orthogonal coordinate system. They reported velocity components and stream functions for Dean number ranging from 0.8 to 307.8. They observed that at a Dean number of 100 an abrupt transition occurs from twin counter-rotating vortex secondary flow to a new secondary flow pattern of four vortices, which was experimentally supported by their pressure drop study. Figure 14 shows the four counter-rotating vortex at $N_{De} = 107$ and $\lambda = 100$.

Non-Newtonian Flow Through Coiled Tubes

Very little has been reported on the flow of non-Newtonian fluids in curved tubes. Some investigators have reported [152–156] the pressure drop studies for non-Newtonian fluid flow in circular coiled tubes. Mashelkar and Devarajan [152, 153] have solved the equations of motion for laminar non-Newtonian flow in coiled tubes using a boundary layer approximation. They have reported a correlation for friction factor as a function of Dean number, curvature ratio, and power-law index, which was verified by their experimental data on pressure drop.

Clegg and Power [157] theoretically analyzed the flow of an incompressible Bingham fluid in a slightly curved pipe of circular cross section. Rathna [158] has solved the equations of motion for incompressible non-Newtonian fluids in circular curved tubes. Using Dean's approach, she obtained the expressions for velocity components in three dimensions, which are reported later in this chapter. By implication these velocity profiles are supposed to hold under the conditions reported by Dean, i.e., $N_{De} \leq 17$ and $\lambda \gg 1$.

Convective Axial Dispersion in Curved Tubes

RTD for Newtonian Flow in Circular Curved Tubes

Presence of three-dimensional flow in case of coiled tubes makes it quite complicated to evaluate residence time distribution. In the absence of molecular diffusion, the information about velocity distribution is sufficient to predict age distribution function, but it is not possible to propose any generalized approach (as in case of unidimensional flow in straight tubes) for computing RTDs. Although many investigators have computed velocity profiles for laminar flow of Newtonian fluid in curved tubes to cover higher values of Dean number, very few results have been used to calculate RTDs. McConalogue [159] has computed the laminar convective axial dispersion of an injected solute in the helical flow by using the velocity profiles of McConalogue and Srivastava [125]. He found that secondary flow produces a more uniform distribution of velocity over the fluid than in Poiseuille flow. He reported a gradual narrowing of RTD with increase in Dean number over the range $16.6 < N_{De} < 77$. This result is available in a form that is rather difficult to use at least from an engineering viewpoint, and we believe it has not been checked experimentally.

Ruthven [160] computed the residence time distribution functions for diffusion free laminar flow through coiled tubes using Dean's velocity profiles. He found that for fully developed flow there exists a unique RTD that can be given by the expression

$$F = 0 \qquad \theta < 0.613 \tag{98a}$$

$$F = 1 - \frac{1}{4\theta^{2.81}} \qquad \theta \geq 0.613 \tag{98b}$$

It should be noted that the RTD is independent of both curvature ratio (λ) and Reynolds number (N_{Re}) and is narrower than that for diffusion free laminar flow through straight tube as shown in Figure 15. Ruthven [160] estimated the condition for its validity as

$$\frac{N_{Re} \cdot N_{Sc}}{650} > L/a > N_{Re} \tag{99}$$

This restriction arises from two considerations—one that the effect of molecular diffusion should be negligible and the other that the fluid should undergo at least one complete secondary flow circulation. For liquids $N_{Sc} \approx 1{,}000$ so that the region of validity is

$$1.5N_{Re} > L/a > N_{Re} \tag{100}$$

This makes the analysis extremely restrictive although it may be added that this constraint is approximate. The additional constraints that $N_{De} < 17.7$ and $\lambda \gg 1$ make it impractical to check this analysis for fluid of $N_{Sc} \sim 1{,}000$. Trivedi and Vasudeva [161] have experimentally measured the RTD for diffusion free laminar flow in coiled tubes. To get around the restricted limits of applicability they used diethylen-glycol as the flowing fluid. They experimentally checked the condition of negligible diffusion and covered ranges of parameters of $10 < \lambda < 278$ and $4.7 < N_{Re} < 93.5$. Their results established the essential correctness of Ruthven's [160] analysis although the upper limit of N_{De} appeared to be 6 rather than 16, and the curvature ratio by itself is perhaps not very important. For higher values of N_{De} they have fitted the model:

$$\begin{aligned} F &= 0 \qquad \theta < \theta_{min} \\ F &= 1 - p/\theta^{q} \qquad \theta \geq \theta_{min} \end{aligned} \tag{101}$$

to the experimentally obtained RTDs and empirically correlated model constants (p, q) with curvature ratio (λ) as

$$p = 3.48/\lambda^{0.0294} \tag{102}$$

$$q = 2.03/\lambda^{0.049} \tag{103}$$

where

$$\theta_{min} = 0.613$$

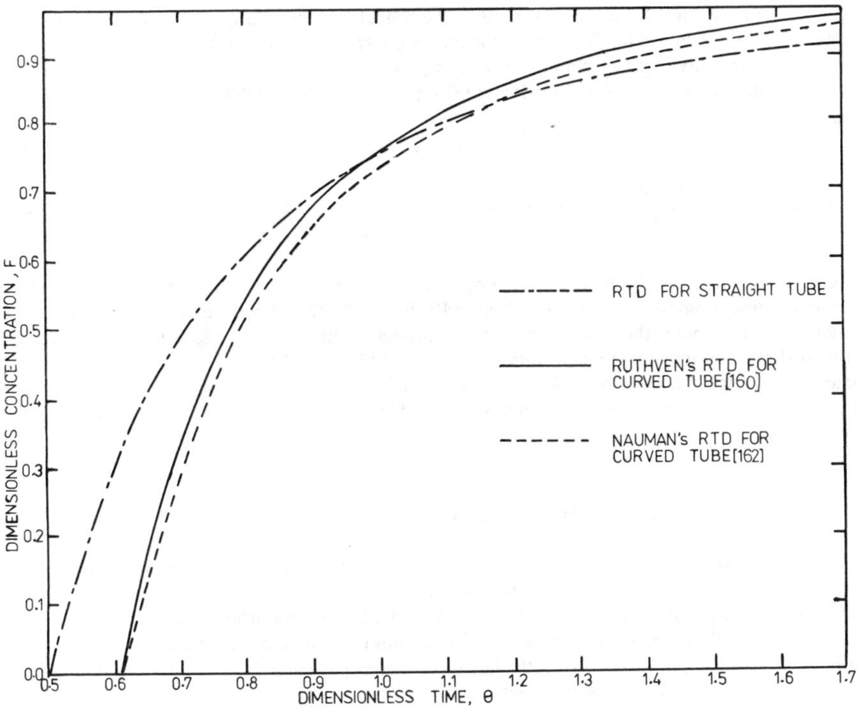

Figure 15. RTD's for diffusion-free laminar flow through coiled and straight tubes.

Later, Nauman [162] pointed out that the RTD reported by Ruthven does not satisfy the constraints imposed on first and second moments of RTD function, i.e.,

$$\int_0^{\infty} f(\theta)\, d\theta = \int_0^{\infty} \theta f(\theta)\, d\theta = 1 \tag{104}$$

and the experimental confirmation by Trivedi and Vasudeva [161] is rather in a semiqualitative sense. The reason for this discrepency is the failure to recognize that "fixation of one model constant in Equation 101 (p, q, or θ_{min}) will fix the RTD function if the condition given by Equation 104 is to be satisfied." Therefore, it requires a more flexible model to be fitted to either numerically computed or experimentally obtained RTDs, if Equation 104 is to be satisfied and inherences of a given system are to be retained in its RTD. In view of this, Nauman [162] proposed the following model for the RTD:

$$\begin{aligned} F(\theta) &= 0 && \theta < \theta_{min} \\ F(\theta) &= 1 - A_1\theta^{B+1} - A_2/\theta^2 && \theta \geq \theta_{min} \end{aligned} \tag{105}$$

The condition given by Equation 104 is satisfied if

$$A_1 = \frac{(B+2)(2\theta_{min} - 1)}{\theta_{min}^{B+2}(B+3)} \tag{106}$$

$$A_2 = \theta_{min}^2(1 - A_1\theta_{min}^{B+1}) \tag{107}$$

It leaves only to find the value of B so that the model best fits the experimental or numerically obtained F and θ data. The θ^{-2} term in the model governs the tail of distribution and results from an asymptotic analysis [162]. Nauman [162] employed an improved numerical scheme to compute RTD in coiled tubes using Dean's velocity profile and fitted the model given by Equation 105, as

$$\begin{aligned} F &= 0 && \theta < 0.6129 \\ F &= 1 - \frac{0.201}{\theta^{2.84}} - \frac{0.0744}{\theta^2} && \theta \geq 0.6129 \end{aligned} \tag{108}$$

Figure 15 also includes the RTD reported by Nauman. It can be seen from the figure that at higher values of dimensionless time error involved in Ruthven's RTD is quite substantial, though from an engineering view point the correction is only of academic interest [163]. Saxena and Nigam [164] provided an experimental support to the Nauman's RTD for helical coils and observed that Nauman's unique RTD is valid over the narrow range of $1.5 < N_{De} < 6.0$. For the low values of Dean number ($N_{De} < 1.5$) they found a gradual shift of RTD from that of straight tube to that for helical coils.

RTD for Non-Newtonian Flow in Curved Tubes

Shear dependent viscosity of non-Newtonian fluids brings about a change of the flow pattern and therefore of the residence time distribution within the flow system. Knowledge of the residence time distribution in helical flow of non-Newtonian fluids is of importance in the biomedical field, in the design of flow reactors for continuous fermentation or polymerization reactions, and in many other industrial applications. In particular, the thermal pasteurization of non-Newtonian liquid foods, where the death rate of micro-organisms is proportional to population density of micro-organisms is a process that could be modeled as a convective diffusion with a first-order chemical reaction. The RTDs for isothermal laminar flow of non-Newtonian fluids through coiled tubes have been numerically computed [165, 166]. Two independently carried out but similar studies by Ranade and Ulbrecht [165] and Saxena et al. [166] together provide a complete solution to the problem. In both the studies the velocity profile reported by Rathna [158] and the approach similar to that used by Nauman [162] are employed to compute RTDs. The velocity profiles for laminar flow of power-law fluid [158] can be written as

$$u(Z, \phi) = (A' + B'Z^{s-1} + C'Z^{(2n+2)/n} - D'Z^{(3n+3)/n}) \sin \phi \tag{109}$$

$$v(Z, \phi) = \left(A' + sB'Z^{s-1} + \frac{3n+2}{n} C'Z^{(2n+2)/n} - \frac{4n+3}{n} D'Z^{(3n+3)/n}\right) \cos \phi \tag{110}$$

$$w(Z, \phi) = (1 - Z^{1+1/n}) + \frac{w_1}{\lambda} \tag{111}$$

where

$$\begin{aligned} w_1 = \Bigg[\frac{N_{Re}^2 \lambda^2}{120\, n} Z^{1/n} \Big\{ & 30A'(1 - Z^{(n+1)/n}) + sC'(1 - Z^{(3n+3)/n}) \\ & - 3D'(1 - Z^{(4n+4)/n}) + \frac{60B'(n+1)^2}{(ns+1)(ns+2+n)} (1 - Z^{(s+1)/n}) \Big\} \\ & + Z - \frac{3n^2 - 5n - 4}{2n(3n+1)} Z^{(2+1)/n} - \frac{3n^2 + 7n + 4}{2n(3n+1)} Z^{1/n} \Bigg] \times \sin \phi \end{aligned} \tag{112}$$

The constants in Equations 109 to 112 are defined as:

$$s = \frac{n+1}{2n} + \frac{1}{2n}\sqrt{17n^2 - 2n + 1} \tag{113}$$

$$A' = \frac{n^3\{ns(21n^3 + 53n^2 + 38n + 8) - (60n^4 + 185n^3 + 200n^2 + 92n + 15)\}}{12(n+1)(1-s)(2n+1)(3n+1)(4n^2+9n+3)(n^2+4n+1)} \tag{114}$$

$$B' = \frac{n^3(13n^3 + 31n^2 + 23n + 5)}{4(1-s)(2n+1)(3n+1)(4n^2+9n+3)(n^2+4n+1)} \tag{115}$$

$$C' = \frac{n^4}{4(n+1)(3n+1)(n^2+4n+1)} \tag{116}$$

$$D' = \frac{n^4}{12(n+1)(2n+1)(4n^2+9n+3n)} \tag{117}$$

and the stream function can be represented as

$$\bar{K} = (A'Z + B'Z^s + C'Z^{(3n+2)/n} - DZ^{(4n+3)/n}) \cos \phi \tag{118}$$

Equation 118 represents two sets of D-shaped projection of stream lines on the cross-sectional plane of the coiled tube, which are symmetrical with respect to x and y axes. At the center of these two sets of D-shaped curves, the secondary velocity components u and v are zero and fluid elements falling on these points flow parallel to the axis of the coiled tube. As the distance to be covered by these elements is the least, their apparent axial velocity is the maximum and have minimum residence time in the coiled tube. Symmetry of stream function with respect to y axis implies that the center of vortex will fall on the line $\phi = 0$. Substituting $v(Z, \phi) = 0$ and $\phi = 0$ in Equation 110, the radial coordinate of the center of vortex (z_c) can be obtained by solving the equation:

$$A' + sB'Z^{s-1} + \frac{3n+2}{n} C'Z^{(2n+2)/n} - \frac{4n+3}{n} D'Z^{(3n+3)/n} = 0 \tag{119}$$

Symmetry of flow with respect to x axis facilitated all the computations to be carried out only upper half plane.

Numerical evaluation of RTD using Rathna's velocity profile has been based on the assumptions that secondary flow is fully developed and all the elements falling on one streamline (ψ) have the same apparent axial velocity.

The approach for computing RTD was similar to that used by Nauman [162], involving following steps—

1. Radial coordinate of the center of vortex can be computed by numerically solving Equation 119.
2. Dimensionless residence time for fluid element falling on streamline (ψ) can be given as

$$\theta(\psi) = \frac{W_{av}}{W(\psi)} = \frac{W_{av}\int ds/V_s}{\int W(Z, \phi)/V_s\, ds} \tag{120}$$

where W_{av} is the overall average velocity of the fluid and V_s and ds are velocity and incremental distance along a streamline, i.e.,

$$W_{av} = 1/\pi \int_0^1 2\pi W(Z, \phi) Z\, dz = \frac{n+1}{3n+1} W_0 \tag{121}$$

W_0 being the velocity at the center of tube

$$v_s = u^2 + v^2 \tag{122}$$

$$d_s = dx^2 + dy^2 \tag{123}$$

The integrals in Equation 120 can be numerically computed for different values of ψ. The net contribution of the asymmetric term involving sin ϕ in Equation 112 is zero if the integral occurring in the denominator of Equation 120 is evaluated over the entire upper half plane. Therefore, Equation 120 reduces to

$$\theta(\psi) = \frac{W_{av}}{W(\psi)} = \frac{W_{av} \int_{y_{min}}^{y_{max}} 1/v_s(y)\, ds(y)}{\int_{y_{min}}^{y_{max}} \frac{W_0(1 - Z^{1+1/n}(y))}{v_s(y)}\, ds(y)} \tag{124}$$

Once the streamlines are evaluated at various equidistant points on the line joining the center of the vortex and the center of the tube, lower limit of integrals (y_{min}) is known. The upper limit "y_{max}" for each streamline (ψ) can be obtained by solving Equation 118 over the interval $Z_c < z < 1$. The integrals of Equation 124 can be numerically evaluated.

The fraction of fluid associated with each $\theta(\psi)$ can be obtained as

$$F(\psi) = \frac{4}{\pi R^2 W_{av}} \int_{A^*}\int W(z, \phi)\, da \tag{125}$$

where A* is the domain of integration, i.e., first quadrant bounded by y axis and streamline (ψ). The antisymmetric term involved in $w(z, \phi)$ can again be dropped as discussed earlier, thus Equation 125 can be written as

$$F(\psi) = \frac{4}{\pi R^2 W_{av}} \int_{y_{min}}^{y_{max}}\int W_0(1 - (X^2 + Y^2)^{(n+1)/2n})\, dx\, dy \tag{126}$$

Ranade and Ulbrecht [165] have numerically integrated the double integral of Equation 126 while Saxena [167] has analytically solved the inner indefinate integral for integer values of 1/n. Despite this difference, both the studies [165, 166] have reported identical RTD's, which provides a truely definitive solution to the problem.

Numerically computed RTDs for various values of power-law index are shown in Figure 16. It can be seen from the figure that as the value of power-law index decreases RTD approaches closer to that for plug flow. Nauman's model (Equation 105) was fitted to these numerically computed RTDs for various values of power-law index ($0.2 < n < 2.0$) and the model parameters are correlated as [166].

$$-B = 13.7325 - 42.0n + 76.4527n^2 - 68.9796n^3 + 29.4561n^4 - 4.7625n^5 \tag{127}$$

$$\theta_{min} = 0.9484 - 0.9983n + 1.3439n^2 - 1.007n^3 + 0.3846n^4 - 0.0581n^5 \tag{128}$$

Model parameters A_1 & A_2 are related to B and θ_{min} by Equations 106 and 107.

Saxena et al. [166] have carried out step response experiments to measure the RTD under the conditions of negligible molecular diffusion with the view to verify numerically computed RTDs. They used the sodium carboxy methyl solutions of power-law indicies 0.6 and 0.725. The experimentally obtained RTDs are shown in Figures 17 and 18. The excellent agreement between numerically computed and experimental RTDs can be seen from the figures, which also provide an indirect sensitive test for the velocity profiles reported by Rathna [158].

None of the studies have investigated the conditions of applicability of these RTDs; however, by implication one can expect the necessary conditions are the same as stipulated in [160, 161, 164]. Higher Dean number flow of non-Newtonian fluid in curved tubes is still unreported.

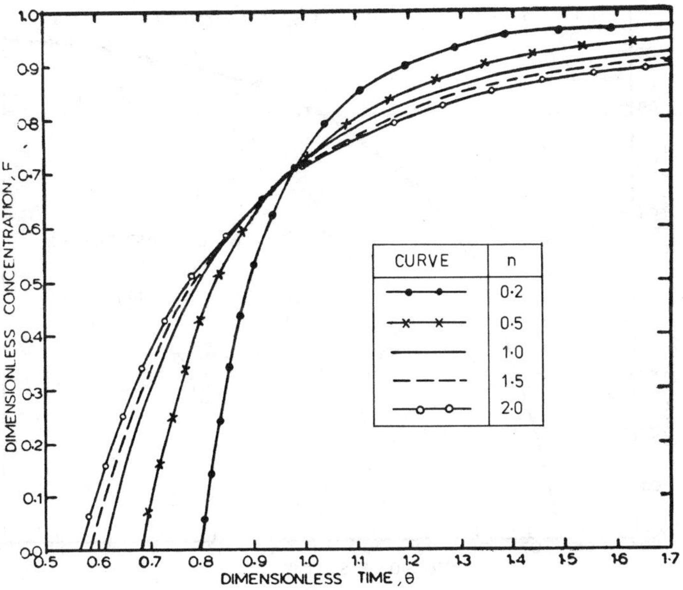

Figure 16. Effect of power law index on diffusion-free RTD in helical coils.

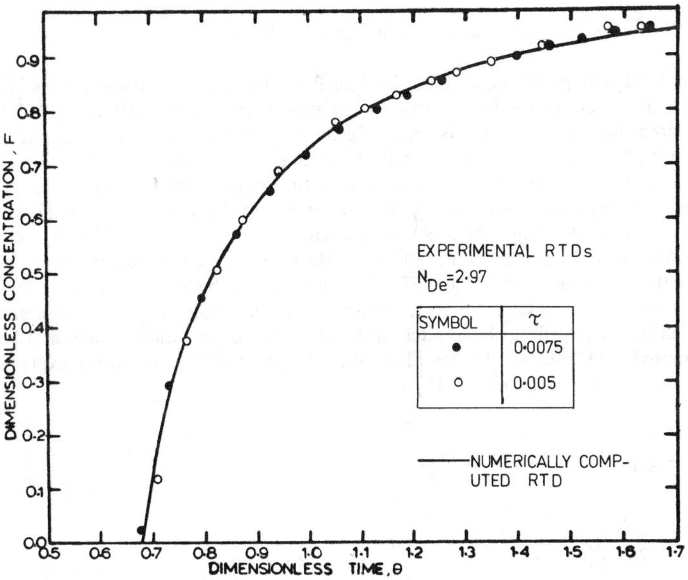

Figure 17. Experimental verification of numerically computed RTD for n = 0.6.

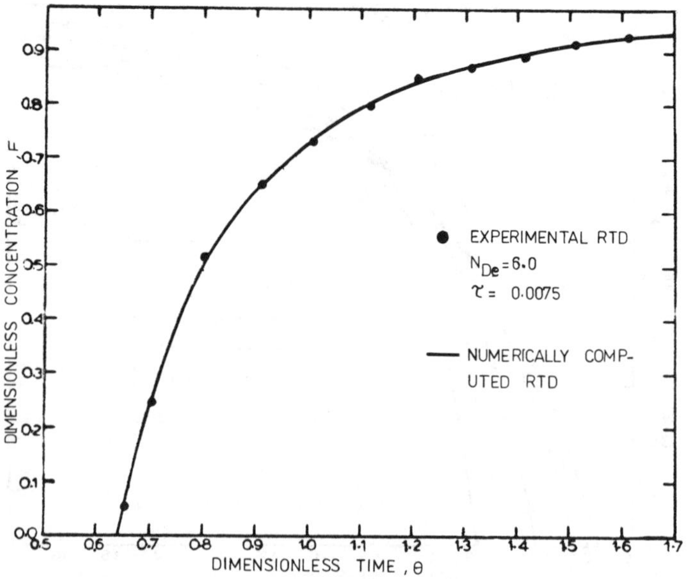

Figure 18. Experimental verification of numerically computed RTD in coiled tube for n = 0.725.

RTD for Newtonian Flow in Helical Coils of Finite Pitch

Nauman [162] computed the RTD for laminar flow in coiled tubes using Dean's [120] velocity profiles, which are essentially for a torus (coil of zero pitch). Recently, Wang [141] asked, "Can the results from the analysis of a tube curved in the form of a circle (torus) be applied to a helical coil?" He introduced a non-orthogonal helical coordinate system to study the effect of curvature and torsion on flow in helical pipe and found that both curvature and torsion induce non-negligible effects when the Reynolds number is less than about 40. He showed that major effect of curvature is on flow rate while the major effect of torsion (twist due to non-zero pitch) is on secondary flow. In view of this, the RTD predicted using Dean's velocity profile should not be applicable to helical coils. Since the experimentally measured RTDs in helical coils by Trivedi and Vasudeva [161] and Saxena and Nigam [164] have supported Ruthven's or Neuman's RTD's it requires a quantitative examination of the effect of coil pitch on RTD. With this intention Sexana and Nigam [168] have computed RTD's using Wang's [141] velocity profiles. The analytical expression for the velocity profiles can be written as [141]

$$u(z, \phi) = \frac{N_{Re}}{288}(z^6 - 6z^4 + 9z^2 - 4)\cos\phi \tag{129}$$

$$v(z, \phi) = \frac{-N_{Re}}{288}(7z^6 - 30z^4 + 27z^2 - 4)\sin\phi + \frac{h}{2\pi R_c}(z^3 - z) \tag{130}$$

$$W(z, \phi) = W_0(1 - z^2) - \epsilon\cos\phi\left(\frac{3}{4}(z^3 - z) + \frac{N_{Re}^2}{11{,}520}(z^9 - 10z^7 + 30z^5 - 40z^3 + 19z)\right) \tag{131}$$

where

$$\epsilon = 4\pi^2 R_c R/(n^2 + 4\pi^2 R^2) \tag{132}$$

and stream function was defined as

$$\frac{\Psi}{N_{Re}} = \frac{\sin\phi}{288}(z^7 - 6z^5 + 9z^3 - 4z) - \frac{1}{4}\frac{h}{2R_c N_{Re}}(z^4 - 2z^2 + 1) \tag{133}$$

Comparison of Equation 133 with the expression of stream function, obtained by Dean for torus (Equation 96), reveal that Wang's [141] analysis results in an additional term involving $h/2\pi R_c N_{Re}(\equiv T)$, which describe the effect of torsion on secondary flow. For $T = 0$ (i.e., zero pitch) the secondary flow pattern is symmetrical to the x axis (i.e., Dean's [120] velocity profile) and for $T > 0$ there exists two asymmetric counter-rotating vortices. The effect of torsion on secondary flow is so dominant that the two counter-rotating vortices become one vortex when $T \geq 1/24$ as shown in Figure 19.

Saxena and Nigam [168] have used Wang's [141] velocity profile to compute RTD in coiled tubes of non-zero pitch, for the range $0.0 < T < 1.0$. Asymmetric secondary flow pattern with respect to x axis is a consequence of the effect of torsion, therefore they carried out the computations over the entire left half plane of the tube cross section. The computational scheme was essentially the same as employed by Nauman [162].

Numerically computed RTD's for various values of torsion parameter are shown in Figure 20. It can be seen from the figure that for $T = 0.0$, RTD is identical to that reported by Nauman [162] and as T increase, RTD approaches straight tube RTD. This behavior is expected on physical grounds.

It is interesting to note that for all practical purposes, RTD in helically coiled tubes over the range $0 < T < 0.001$ (i.e., $h/R_c < N_{Re}/500$) can be approximated by Nauman's RTD. It covers a wide range of practical interest over which helical coils are employed in industrial practice. Most of the experimental studies reported in litrature [161, 164, 166] fall in this range only.

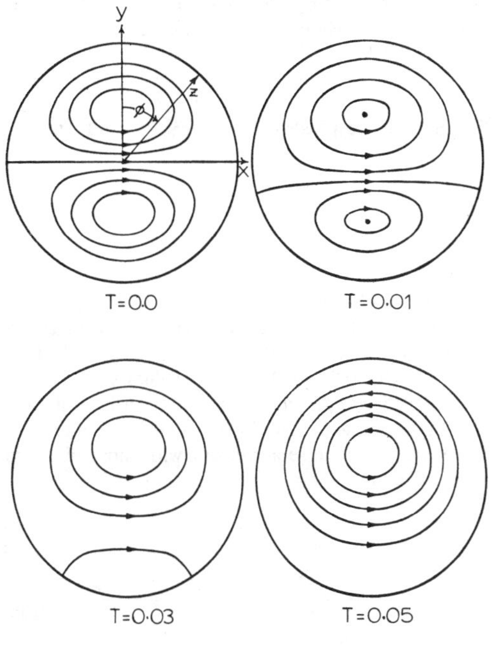

Figure 19. Effect of coil pitch on secondary flow pattern.

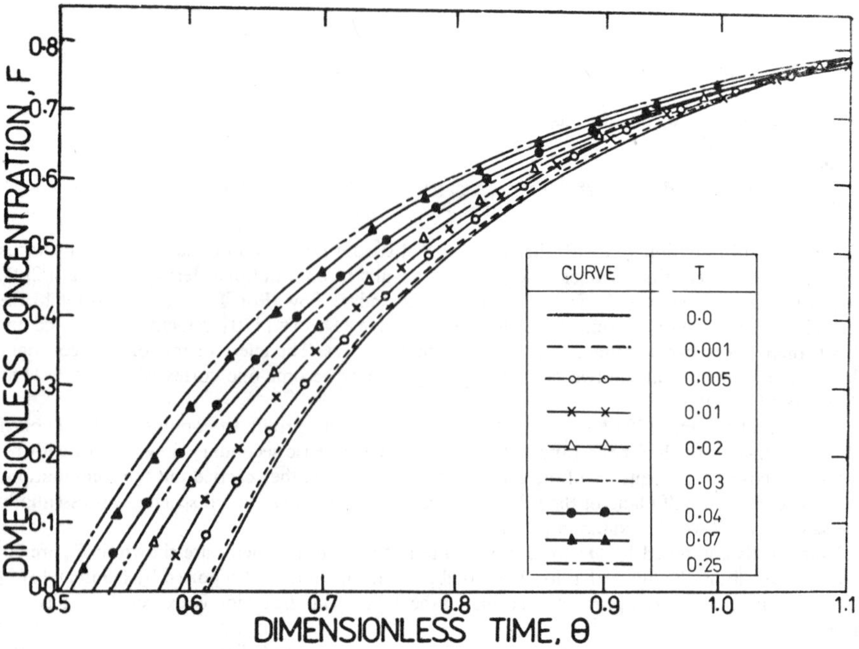

Figure 20. Effect of torsion on diffusion-free RTD in coiled tubes.

Nauman's [162] model (Equation 105) was fitted to the numerically computed RTD's. Variation of θ_{min} with torsion parameter T is shown in Figure 21, which can be correlated as

$$\theta_{min} = 0.60763 - 2.9604T + 29.9981T^2 - 124.3018T^3 + 178.163T^4 \quad (134)$$

Over the range $0.0 < T < 0.015$, the best-fit values of $-B$ are also plotted against T in Figure 21, which can be correlated as

$$-B = 3.882 - 1.544\beta + 1.4793\beta^2 - 0.575\beta^3 \quad (135)$$

where $\beta = T \times 10^2$ (136)

The model constants A_1 and A_2 can be computed using Equations 106 and 107.

For the values of Torsion parameter (T) greater than 0.015 no definite trend of best fit values of $-B$ with T could be observed. This is because of the fact that RTD approaches straight tube RTD as the value of T increase. Therefore, for the cases $T > 0.015$ best-fit values of B did not observe any fixed trend. In view of this, for $T > 0.015$ two-parameter model (given by Equation 101) was fitted to the numerically computed RTD's. It was observed [168] that best-values of p and q obtained using the least square fit procedure are in close agreement with those obtained by satisfying the constraints defined by Equation 104 as

$$q = -\frac{1}{1 - \theta_{min}(T)} \quad (137)$$

$$p = \theta_{min}^{q} \quad (138)$$

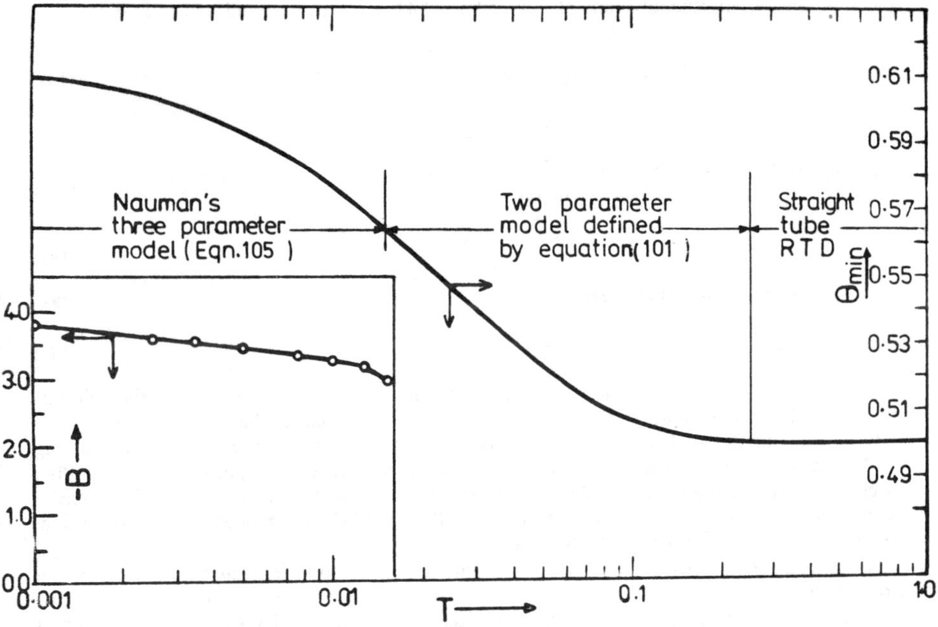

Figure 21. Variation of θ_{min} and B with T and applicability of different models.

The value of θ_{min} can be obtained using Equation 134 or can be read from Figure 21. The maximum value of q (for $T = 0.015$) is 2.29, for which deviation from the asymtotic behavior of RTD in straight tube is not very serious and for practical purposes p and q for the RTD model given by Equations 101 can be obtained using Equations 137 and 138, over the range $0.015 < T < 0.25$. Another interesting inference that could be drawn from the numerically computed RTDs is that for $T > 0.25$ (i.e., $h/R_c > N_{Re}/2$) RTD in helically coiled tubes can be approximated by straight tube RTD or coil behaves like a straight tube.

RTD for Newtonian Flow in Helical Coils of Non-Circular Cross Section

Coiled tube of elliptical cross section. Cuming [143] employed Dean's [120] approach to analyze the phenomenon of secondary flow in coiled tubes of elliptical cross section. Using perturbation analysis he solved the Navier-Stoke's equations in right-handed orthogonal system of axes (Figure 13) to obtain following velocity profiles:

$$u = \frac{N_{Re}^2}{R_c\bar{\alpha}^2}(1 - x^2 - y^2)((1 - x^2 - y^2)(B_1 + B_2x^2 + 3B_3y^2) - 4y^2(B_1 + B_2x^2 + B_3y^2)) \tag{139}$$

$$v = \frac{2N_{Re}^2}{R_c\bar{\alpha}}(1 - x^2 - y^2)(2(B_1 + B_2x^2 + B_3y^2) - B_2(1 - x^2 - y^2))xy \tag{140}$$

$$w = \frac{1}{2}N_{Re}(1 - x^2 - y^2) + \frac{a}{R_c}N_{Re}^2 w_1 + \frac{a}{R_c}w_2 \tag{141}$$

where w_1 and w_2 are the functions of x, y and $\bar{\alpha}$, which are even in y and odd in x. Expression for stream function was given as

$$\psi = (1 - x^2 - y^2)^2(B_1 + B_2x^2 + B_3y^2)y \tag{142}$$

where

$$B_1 = -\frac{\bar{\alpha}^4(375 + 820\bar{\alpha}^2 + 1{,}114\bar{\alpha}^4 + 212\bar{\alpha}^6 + 39\bar{\alpha}^8)}{360(5 + 2\bar{\alpha}^2 + \bar{\alpha}^4)G(\bar{\alpha})} \tag{143}$$

$$B_2 = -\bar{\alpha}^4(75 + 2\bar{\alpha}^2 + 3\bar{\alpha}^4)/360G(\bar{\alpha}) \tag{144}$$

$$B_3 = -\bar{\alpha}^4(15 + 26\bar{\alpha}^2 + 39\bar{\alpha}^4)/360G(\bar{\alpha}) \tag{145}$$

$$G(\bar{\alpha}) = 35 + 84\bar{\alpha}^2 + 114\bar{\alpha}^4 + 20\bar{\alpha}^6 + 3\bar{\alpha}^8 \tag{146}$$

where $\bar{\alpha}$ is the aspect ratio defined as the axis of elliptic cross section parallel to the coil axis divided by the axis of elliptic cross section perpendicular to the axis of the coil.

By considering the total vorticity as the measure of the intensity of the secondary flow, Cuming [143] showed that the secondary flow is more intense in coiled tubes of elliptical cross section than in circular cross sections of $\bar{\alpha} < 2.2$. Further increase in $\bar{\alpha}$ was found to diminish the intensity of the secondary flow. In view of this, aspect ratio ($\bar{\alpha}$) should influence residence time distribution, which is quantitatively examined by Saxena and Nigam [168].

Numerically computed RTD's for different values of $\bar{\alpha}$ are shown in Figure 22. It can be seen from the figure that reduction in the value of aspect ratio ($\bar{\alpha}$) narrows the RTD by reducing the fraction of fluid that is moving with higher or equal to average axial velocity (i.e. appearing at the outlet at $\theta = 1$); however, the value of dimensionless residence time at which the first element tracer appears at the outlet is practically unchanged. The parameters in Nauman's model [162] for diffusion free RTD can be correlated to aspect ratio ($\bar{\alpha}$) as

$$-B = 2.263 - 0.5677\bar{\alpha} + 5.714\bar{\alpha}^2 - 4.858\bar{\alpha}^3 + 1.3388\bar{\alpha}^4 \tag{146}$$

and

$$\theta_{min} = 0.61756 - 0.0046\bar{\alpha} \tag{147}$$

RTD for Laminar Newtonian Flow Through Coiled Tubes of Square Cross-Section

Although the developing and developed flow in curved tubes of square/rectangular cross sections has been analyzed by many investigators [143–150], none of the analyses appears to be used for computing RTD functions. Recently, Saxena and Nigam [89] reported an experimental study on diffusion-free RTD in coiled tubes of square cross section.

Since the flow in non-circular tubes is not axi-symmetric, the orientation of cross-sectional plane with respect to the direction of centrifugal force, in case of helical flow, brings about a change in secondary flow pattern, which in turn affects the RTD. This aspect has been recognized by Saxena and Nigam [89]. They obtained RTD's in coiled tubes of square cross section of two types:

Type A—Direction of centrifugal force makes an angle of 45° with the diagonals of square cross section, as shown in Figure 23.
Type B—Centrifugal force acts parallel to one of the diagonals, Figure 23.

Figure 24 shows the experimentally obtained RTD's in helix of Type A over the Dean number range 0.2 to 8.0. It can be inferred from the figure:

1. There exists a unique RTD (i.e., insensitive to Dean number) for laminar diffusion-free flow in coiled tube of square cross section over the range $1.8 < N_{De} < 5.0$.

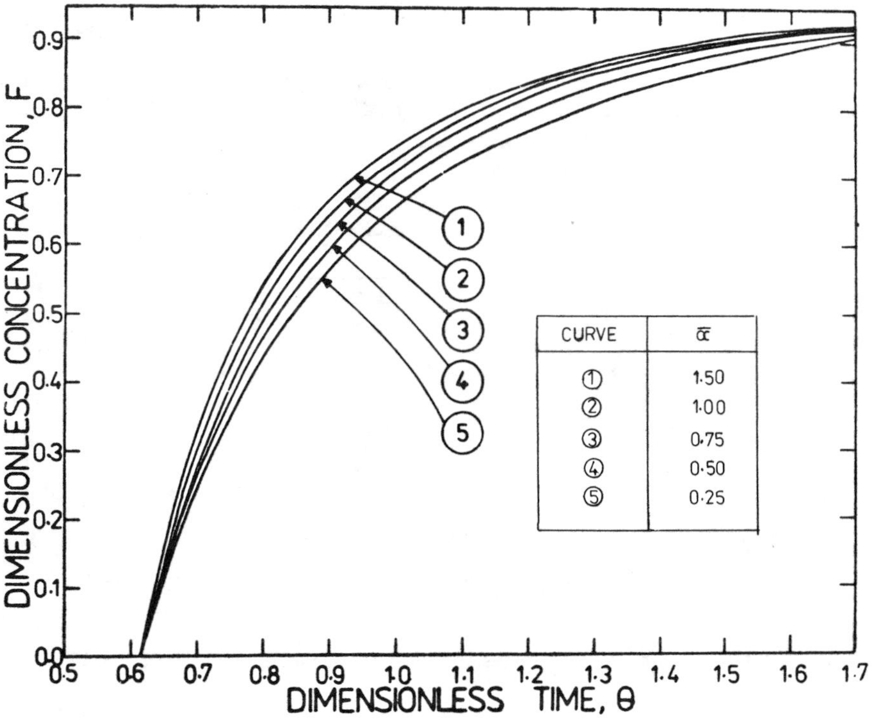

Figure 22. Effect of cross-sectional ellipticity on diffusion-free RTD in helical coils.

2. For $N_{De} < 1.8$, a gradual shift in RTD from that for straight tube to that for a coiled tube occurs with the increase in Dean number. This trend is similar to that reported by Saxena and Nigam [164] for coiled circular tubes. Similar observations were reported for coils where the centrifugal force acts parallel to one of the diagonals (i.e., Type B). Nauman's model [162] for diffusion-free RTD (Equation 105) was fitted to the experimental RTD data, which

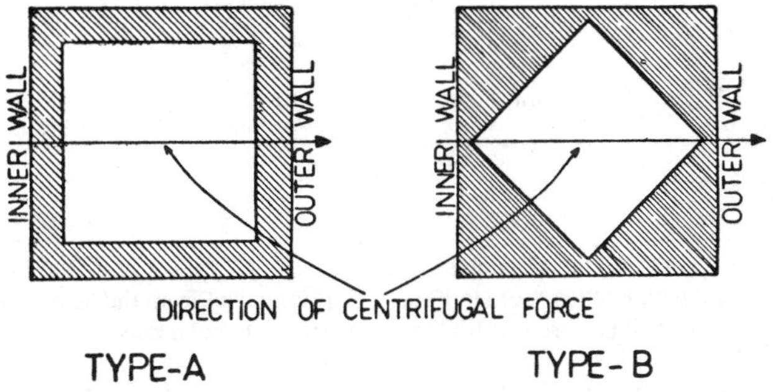

Figure 23. Two extreme cases of cross-sectional orientation with the direction of centrifugal force.

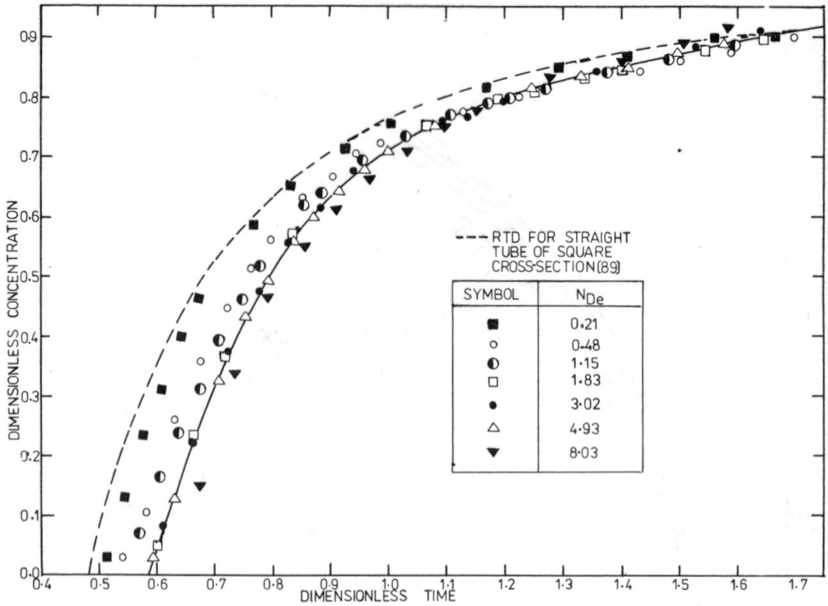

Figure 24. Effect of Dean number of diffusion-free RTD in coiled tube of square cross-section.

can be written as

$$F = 0 \qquad \text{for } \theta < 0.587$$
$$F = 1 + \frac{0.2782}{\theta^{1.6973}} - \frac{0.5819}{\theta^2} \qquad \text{for } \theta \geq 0.587 \tag{148}$$

for coils of Type A and

$$F = 0 \qquad \text{for } \theta < 0.572$$
$$F = 1 + \frac{0.5222}{\theta^{1.853}} - \frac{0.8087}{\theta^2} \qquad \text{for } \theta \geq 0.572 \tag{149}$$

for coils of Type B.

The unique but distinct RTD's obtained in helical coils of Types A and B are compared with Nauman's RTD for circular coiled tubes in Figure 25. It is evident from the figure that coiled tubes of square cross section result in a broader RTD as compared to circular cross section. The orientation of the centrifugal force with the diagonals of the cross-sectional plane has an appreciable effect on RTD. The broadest RTD in a helix of square cross section is observed when centrifugal force acts parallel to on of the diagonals (Type B). This result suggests that for the flow in helically coiled tubes of non-circular cross section the orientation of the cross-sectional plane with the direction of centrifugal force has non-negligible effect on velocity distribution.

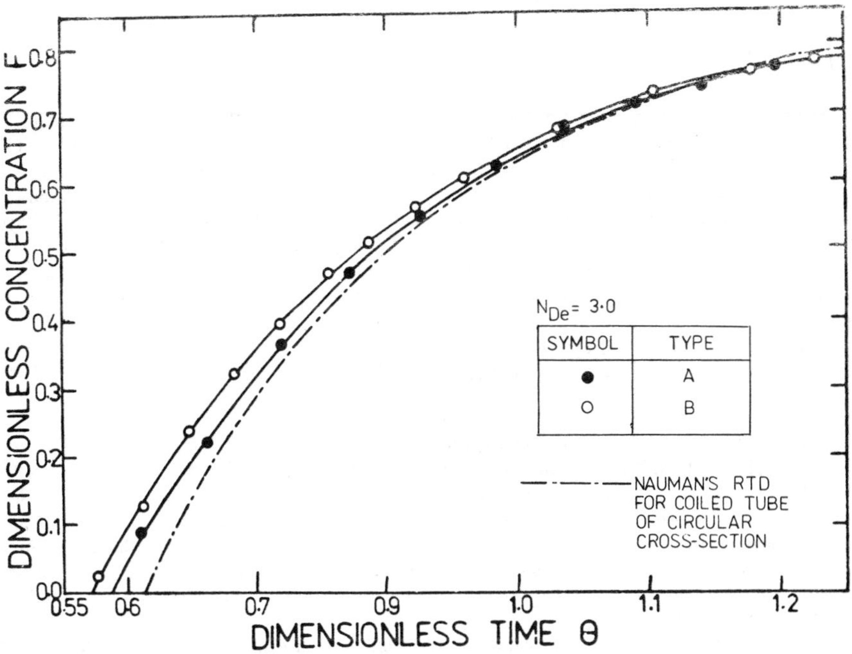

Figure 25. Effect of cross-sectional orientation on diffusion-free RTD in coiled tubes of square cross-section.

It should be noted that RTD's for diffusion-free flow in coils of different types are applicable to a very narrow range of Dean numbers due to the restrictions on velocity profiles used to compute these, i.e. $N_{De} \leq 17$ and $\lambda \gg 1$.The experimental findings of Trivedi and Vasudeva [161] and Saxena and Nigam [164] further contract the region of their applicability. The experimentally obtained unique RTD in coils of square cross section are also expected to hold under similar conditions in view of their almost, identical trend with increase in Dean number. Although a comparatively large number of studies have been reported on the analysis of higher Dean number flow in curved tubes, none of these are used (except McConalogue [159]) for computing RTD's. It leaves a wide scope for providing this information.

Dispersion in Coiled Tubes Under Significant Influence of Molecular Diffusion

Laminar dispersion in helical coils has been a subject of many papers in recent years. Several theoretical and experimental studies have been directed towards the representation of dispersion under the conditions of appreciable effect of diffusion in the complex three-dimensional flow field by axial dispersion model. The information is of importance not only engineering applications but also finds a considerable use in physiological systems. Chang and Mockros [169] have theoretically studied the rates of hemoglobin saturation and carbon dioxide reduction in blood in a curved channel membrane. Gilrog et al. [170] have experimentally assessed the theory on the convective dispersion of blood gases by secondary flow in curved channels. The dispersion of iodine in plasma and whole pig's blood has been studied by Patel and Sirs [171].

Dispersion in Newtonian Flow Through Coiled Circular Tubes

Theoretical Work

The theoretical analysis of laminar dispersion in helical coils, or in the simpler case of a curved tube, consists of solving the convective diffusion equation along with the equations of motion. However, due to the complexity of the equations involved, various investigators have obtained solutions only under restricted conditions using various approximations. Erdogen and Chatwin [59] incorporating the diffusional effect with Dean's [120] velocity profile provided analytical solution for effective diffusion coefficient within the framework of Taylor's [33] dispersion theory. Their expression for effective diffusion coefficient in coiled tubes can (by incorporating Aris [35] modification to account for axial molecular diffusion) be written as

$$\frac{DD_m}{\bar{u}^2 d_t^2} = \left(\frac{1}{N_{Bo}^2} + \frac{1}{192}\right) + \frac{4N_{Re}^4\lambda^{-2}}{576^2 \times 160}\left(-\frac{2{,}569}{15{,}840}N_{Sc}^2 + \frac{109}{43{,}200}\right) \quad (150)$$

This analysis is expected to hold when curvature is sufficiently large and $N_{De} \leq 17$, i.e., the conditions implicit in Dean's velocity profile. A comparison with Taylor-Aris expression (Equation 44) for effective diffusion coefficient in straight tubes reveal that terms involving $N_{Re}^4\lambda^{-2}$ (i.e., N_{De}^4) represent the influence of coiling on dispersion. The expression indicates that the coiling results in a reduced dispersion to that in straight tube if $N_{Sc} > 0.124$, which is the case with all common liquids and most gases. They have also reported that Taylor's [33] argument regarding symmetrical distribution of solute with respect to a plane moving with average axial velocity is also valid for the flow in curved tubes.

Nunge et al. [172] analytically treated the problem of axial dispersion in curved tubes using Topokoglu's [122] velocity distribution expressions and derived the following expression for effective diffusion coefficient:

$$\begin{aligned}\frac{DD_m}{\bar{u}^2 d_t^2} &= K_c \\ &= \left(\frac{1}{N_{Bo}} + \frac{1}{192}\right) + \left\{\frac{4N_{Re}^4}{576^2 \times 160}\left(-\frac{2{,}569}{15{,}840}N_{Sc}^2 + \frac{109}{43{,}200}\right)\right. \\ &\quad \left. + \frac{2N_{Re}^2}{576 \times 144}\left(\frac{31}{60}N_{Sc} - \frac{25{,}497}{13{,}440}\right) + \left(\frac{419}{8 \times 15 \times 96} + 0.25\left(\frac{1}{N_{Bo}^2} + \frac{1}{192}\right)\right)\right\}\lambda^{-2}\end{aligned} \quad (151)$$

This expression is subjected to rather relaxed conditions on curvature and Dean number, implicit in Dean's [120] velocity profile and carried over to Erdogan and Chatwin's analysis. The comparison of Equations 150 and 151 reveals that the expression derived by Nunge et al. [172] comprises additional terms involving N_{Re}^2 and N_{Re}^0, which describe the effect of coiling on axial dispersion. The extent of reduction being determined by the parameters N_{Re}, N_{Sc}, and λ. At low values of Reynolds number this behavior, contrary to that observed by Erdogan and Chatwin [59], was explained as an, "increase in axial dispersion due to coiling is a consequence of two mechanisms of dispersion in curved tubes: (a) secondary flow causes a mixing in cross-sectional plane as a result of which dispersion decreases; and (b) the curvature increases the variation in residence time across the flow as compared to that in straight tubes." Under the conditions when a second effect dominates, an increased dipersion could be obtained.

The requirement of non-negative effective diffusion coefficient imposes severe limitations on the range of N_{Re} up to which analysis can be expected to hold. This has been ascribed by non-availability of higher order terms in the velocity profile expressions. The condition for the applicability of dispersion model has been estimated to be

$$\tau_{min} > 101K_c^{0.9} \quad (152)$$

Janssen [173] theoretically analyzed the phenomenon of axial dispersion in helically coiled tubes for conditions under which molecular diffusion play a dominating role. He numerically solved the convective diffusion equation, obtained by combining diffusional effects with Dean's [120] velocity profiles. His analysis revealed $N_{De}^2N_{Sc}$ is the parameter that characterizes the axial dispersion in coiled tubes. For $N_{De}^2N_{Sc} < 100$ no significant reduction in axial dispersion as compared to straight tube was reported while for $100 < N_{De}^2 \cdot N_{Sc} < 5{,}000$ axial dispersion coefficient decreases more than three fold. He also pointed out that analytical approximate solutions of Erdogan and Chatwin [59] and Nunge et al. [172] immediately deviate from numerical results when the effect of secondary flow becomes significant, which is due to the poor convergence of power series in their analyses.

Influence of Coil Geometry and N_{Re} on Dispersion

Koutskey and Adler [174] have experimentally studied the axial dispersion in coiled tubes of finite pitch using pulse response techniques. They covered the range of Reynolds number from 300 to 4,000 and observed that secondary flow present in helical coils results in markedly reduced axial dispersion over that in straight tubes, especially at higher flow rates. The parameters they observed to be effective in reducing axial dispersion (for $300 < N_{Re} < 3{,}000$) are Reynolds number, curvature ratio, and cross-sectional ellipticity. In coils transition from laminar to turbulent flow is reported to be more gradual and at higher values of N_{Re} than that in straight tubes. Axial dispersion over the range $7{,}000 < N_{Re} < 40{,}000$ decreases greatly as compared to laminar flow conditions. They have graphically correlated the dispersion number to Reynolds number, curvature ratio, cross-sectional ellipticity, and the tube length. They have also reported that for equal power consumption helical coils facilitate less axial dispersion than in straight tubes.

van Andel et al. [175] employed step response technique to investigate the extent of axial dispersion in helical flow of gas (air-butane) and liquid (water-electrolyte). The experimentally obtained F curves using water as flowing media over the range $100 < N_{Re} < 2{,}000$ in two different coils of $\lambda = 30$ and 50 at $d_t = 0.6$ cm, did not correspond to that for the dispersion model; however, they used the variances of these curves to compute effective Bodenstein number as,

$$N'_{Bo} = \frac{\bar{u}d_t}{D} = \frac{2d_t}{\sigma^2 L} \tag{153}$$

They observed that in case of liquids, the calculated values of Bodenstein number have no definite trend with any of the operating parameters; however, the scatter was confined to the range $0.015 < N_{Bo} < 0.05$. The step response curves for gas phase experiments in five coils of $d_t = 0.6$, $\lambda = 30.55$ and 104; $d_t = 0.85$ cm, $\lambda = 39$ and 74 over the range $15 < N_{Re} < 8{,}000$ were found to correspond to those for dispersion model. The calculated values of the Bodenstein number for the gas phase system were found to vary between 0.5 to 2.0. While the reduction in dispersion was observed for both systems, a clear-cut dependence of dispersion on the system parameters could not be established.

Trivedi and Vasudeva [82] have reported an experiment study on dispersion of solute in helical coils under laminar Newtonian flow conditions over the ranges of variables $10 < \lambda < 280$, $10 < N_{Re} < 1{,}600$ and $1.5 \times 10^3 < N_{Sc} < 8.7 \times 10^3$. They computed the values of dispersion number for the experimentally obtained F curves by matching it with the solution of one dimensional Taylor's [33] dispersion model given by Equation 50. The criterion for the best fit value of $D/\bar{u}L$ was minimum $\sum |F_{exp} - F_{theo}|$. They have reported a scatter in data when the values of effective Bodenstein number were plotted against N_{Re} or λ; however, the order of magnitude values obtained were comparable to those reported by van Andel et al. [175] for liquids. In view of the analysis by Erdogan and Chatwin [59] and Nunge et al. [172] the experimentally obtained values of $DD_m/\bar{u}^2 d_t^2$ $(=K_c)$ were plotted against Reynolds numbers, and it was observed that by using K_c the data for axial dispersion in coiled tubes are well correlated with N_{Re} and λ. They reported 1.5 to 500-fold reduction in the values of K_c depending upon the values of N_{Re} and λ. Their results indicate that actual reduction in axial dispersion due to coiling is much less than that predicted by theoretical analyses of Erdogan and Chatwin [59] and Nunge et al. [172]. They have correlated

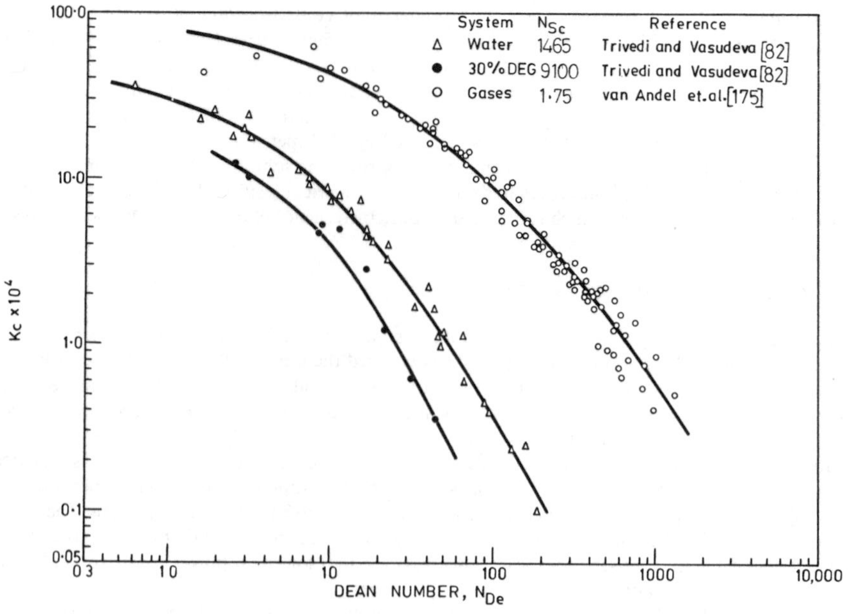

Figure 26. Effect of Dean and Schmidt numbers on axial dispersion in helical coils.

the experimental data for the dispersion of tracer in water against N_{De} (Figure 26) with maximum error in K_c being 49% and standard deviation of 19.6%. These authors have also found that K_c vs. N_{De} plots can be represented by series of curves according to Schmidt number and decrease in Schmidt number increases axial dispersion. Over the range of parameters studied they have reported the condition of the applicability of Taylor's dispersion model in helical flow as

$$\tau > 6.0/N_{Re} \tag{154}$$

which is less severe than those reported for straight tubes [41].

Nigam and Vasudeva [176] carried out an experimental study to investigate the conditions under which the effect of coiling on axial dispersion is negligible and the extent of axial dispersion in helical coils can be used to predict molecular diffusion coefficient using Taylor-Aris analysis. Their experimental results showed that for nearly 80-fold variations in curvature ratio, the effect of coiling on K_c is negligible if $N_{De} < 0.15$ and analysis for straight tubes is applicable in coils. Their experimental findings also revealed that the theoretical analysis of Erdogan and Chatwin [59] is quite accurate in predicting the minimum Dean number at which the effect of curvature on K_c is significant but is of little value in predicting the extent of reduction. Later on Anderson and Berglin [71] in a theoretical and experimental study supported this criterion. Moulijn et al. [177] using gaseous systems in coiled tubes of $9 < \lambda < 47$, observed that Taylor-Aris model applies up to $N_{De}(N_{Sc})^{0.5} \approx 10$. The criterion proposed by Nigam and Vasudeva [176], $N_{De} < 0.15$, in fact correspond to $N_{De}(N_{Sc})^{0.5} < 6.0$, which is well in agreement.

Singh and Singh [178] have measured the extent of axial dispersion in gases flowing in helical columns under laminar flow conditions. The ranges of variables covered were $26.6 < \lambda < 98$, $10 < N_{Re} < 100$, and $0.176 < N_{Sc} < 1.359$. They observed that the parameter $N_{De}(N_{Sc})^{0.5}$ correlate the data well. Their results reveal that for $N_{De}(N_{Sc})^{0.5} < 6.5$ (Figure 3 of Singh and Singh [178]), the effect of coiling on axial dispersion is not significant, which again supports the finding of Nigam and Vasudeva [176].

Influence of Schmidt Number on Dispersion

The influence of the Schmidt number on dispersion in helical coils was investigated by Trivedi and Vasudeva [82]. Their experimental results using water and 29% DEG solution in water revealed that about sixfold increase in the Schmidt number reduces the value of $K_c(DD_m/\bar{u}^2 d_t^2)$ by a factor of 2–4 depending on the Dean number (see Figure 26). They could also correlate the experimental data of van-Andel et al. [175] in terms of K_c vs N_{De}, which brought out the effect of the Schmidt number on dispersion over the range of N_{Sc} from 1.75 to 8.7×10^3 and revealed that extent of reduction in dispersion by coiling in the case of gases is much less than in the case of liquids.

In view of the strong influence of Schmidt number on axial dispersion [82, 173], Shetty and Vasudeva [179] have also examined this aspect in coiled tubes. Their experimental results for N_{Sc} of 7,020 and 55,600 are shown in Figure 27. The figure also includes the data of different workers [59, 82, 173, 175].

The suitability of the single parameter $N_{De}N_{Sc}^{1/2}$ to correlate the reduction in dispersion due to coiling is clearly established in Figure 27. Over the wide ranges of $1.75 < N_{Sc} < 55{,}600$, $0.02 < N_{De} < 50$ and $10.3 < \lambda < 785$ the agreement within the experimental results of different group of investigators appears well within the experimental errors. The limiting value of $N_{De}N_{Sc}^{1/2}$, up to which the validity of straight tube analysis is predicted by the two theoretical analyses [59, 173], is well supported by the experimental results and can be written as

$$N_{De}N_{Sc}^{1/2} < 7 \tag{155}$$

Shetty and Vasudeva [179] empirically correlated the experimental results as

$$\frac{D_c - D_m}{D_s - D_m} = 1.52 - 0.275 \ln(N_{De}N_{Sc}^{1/2}) \qquad \text{for } N_{De}N_{Sc}^{1/2} > 7 \tag{156}$$

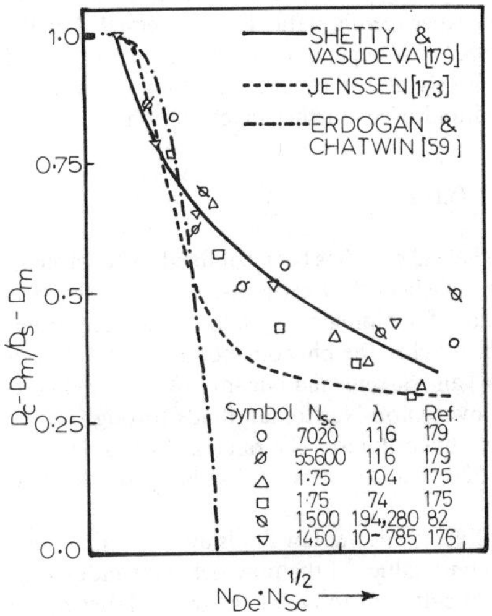

Figure 27. Correlation of axial dispersion coefficient for flow through helical coils [179].

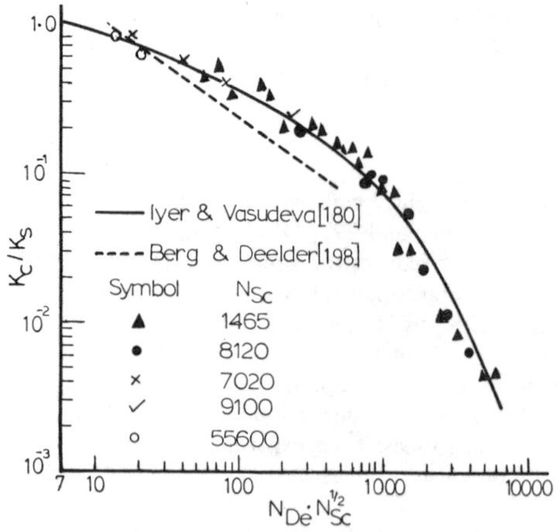

Figure 28. Correlation of axial dispersion data for flow through coiled tubes [180].

More recently, Iyer and Vasudeva [180] have further modified this correlation to cover a wider range of system parameters. Their correlation is shown in Figure 28 and can be written as

$$\frac{D_c - D_m}{D_s - D_m} = 1, \text{ for } N_{De}N_{Sc}^{1/2} < 7 \tag{157}$$

$$\frac{D_c - D_m}{D_s - D_m} = 1.65 - 0.863 \log(N_{De}N_{Sc}^{1/2}) + 0.113 \log(N_{De}N_{Sc}^{1/2})^2, \text{ for } N_{De}N^{1/2} > 7 \tag{158}$$

Influence of Pulsatile Flow on Laminar Dispersion

Pulsatile flow appears to be of interest in increasing heat and mass transfer rates [181, 182] and in physiological applications [139, 183]. The recent literature on unsteady flow in curved tubes has been compiled by Berger et al. [139]. Practically no information is available in literature on the influence of pulsations on laminar dispersion in straight and curved tubes except the work of Nigam and Vasudeva [176], which is also confined to limited experimental data. They observed that the pulsations may result in an increase or decrease of dispersion, depending upon the operating and system parameters. Though the results of the study show a significant effect of pulsations on laminar dispersion, the phenomenon is not yet well understood and needs further investigation.

Non-Newtonian Laminar Dispersion in Circular Coiled Tubes

Most of the available literature on axial dispersion in helical coils has been confined to Newtonian flow only. The knowledge of dispersion phenomenon in helical flow on non-Newtonian fluids is of importance in biomedical field, design of flow reactors for biological systems, and many other industrial applications. Despite its importance in different fields, the phenomenon has not yet been theoretically analyzed. Recently, Singh and Nigam [184] and Saxena and Nigam [185] have experimentally investigated the axial dispersion in laminar flow of non-Newtonian fluids through coiled tubes using the aqueous solutions of sodium carboxy methyl cellulose. The ranges of variables covered by Saxena and Nigam [185] were $10.5 < \lambda < 220$, $0.6 < n < 1.0$, $0.1 < N_{Regn} < 140$, and $0.04 < \tau < 2.2$.

Their experimental data for laminar dispersion with different power law fluids are plotted against N_{Regn} in Figure 29. The different sets of curves for different values of the flow behavior index suggests that the use of generalized Reynolds number is not sufficient to characterize axial dispersion

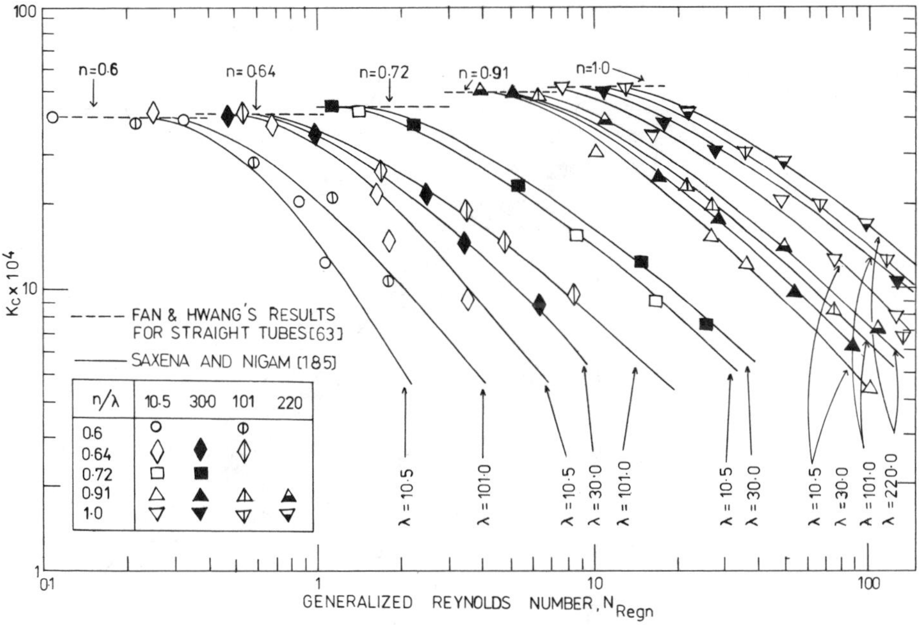

Figure 29. Effect of N_{Regn} and λ on axial dispersion in non-Newtonian helical flow.

in helical flow of non-Newtonian fluids. The use of the dimensionless parameter M instead of generalized Reynold's number, as suggested by Mujawar and Raja Rao [186] also led to the similar conclusion.

Since the product of N_{Regn} and N_{Sc} nullifies the deviation in N_{Regn} due to large deviation in apparent viscosity the use of the parameter N_{Regn}-N_{Sc} may characterize the dispersion in coils. With this intution Saxena and Nigam [185] plotted K_c vs $N_{Regn}N_{Sc}$ as shown in Figure 30. The Schmidt number was evaluated as

$$N_{Sc} = \mu_a/\rho D_m \tag{159}$$

where μ_a is the apparent viscosity and can be calculated as

$$\mu_a = K(8\bar{u}/d_t)^{n-1} \tag{160}$$

Equation 160 is essentially for straight tubes. Mashelkar and Devrajan [152] have shown that the shear rates in radial direction are negligible as compared to axial shear rates therefore the apparent viscosity can be computed using Equation 160.

It can be concluded from the Figure 30 that

1. Axial dispersion decreases with increase in the value of $N_{Regn}N_{Sc}$ for the helical flow of non-Newtonian fluids.
2. For a fixed value of $N_{Regn}N_{Sc}$, axial dispersion increases with increase in curvature ratio (λ).
3. Lower values of $N_{Regn}N_{Sc}$ where the values of K_c are closer to that in straight tubes (i.e., there is no effect of curvature on axial dispersion) suggests that for $N_{Regn}N_{Sc} < 20{,}000$ fluid elements are blind to curvature and coils behave like a straight tube.

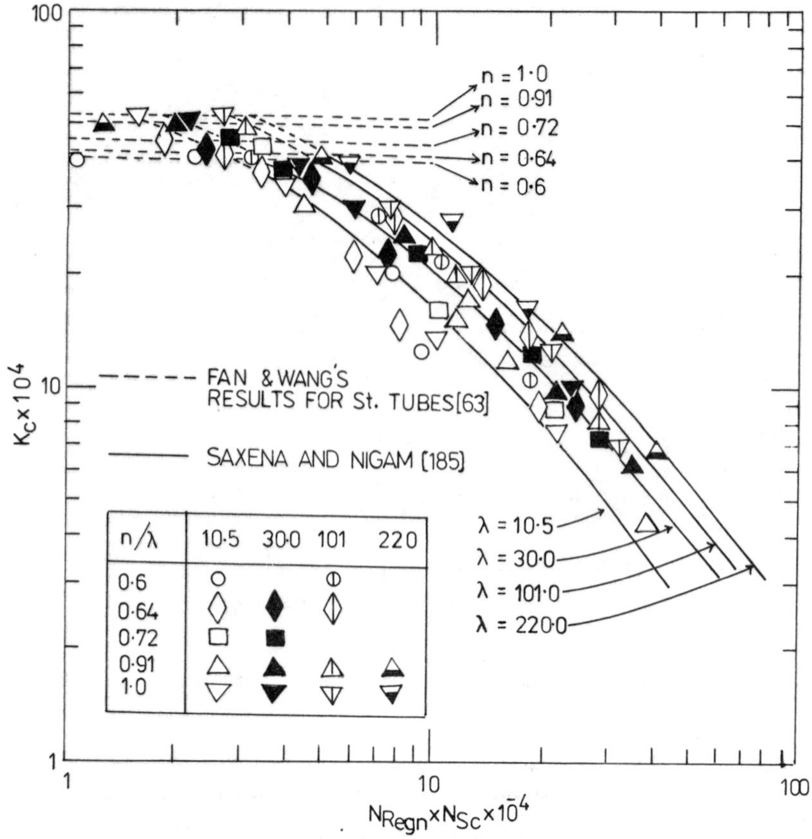

Figure 30. Effect of $N_{Regn} \cdot N_{Sc}$ on axial dispersion in non-Newtonian helical flow.

4. About twelve-fold reduction in the value of K_c due to coiling was observed depending upon the value of curvature ratio and $N_{Regn}N_{Sc}$.

The condition for the validity of the dispersion model for the helical flow of non-Newtonian fluids, over the range of variables studied by Saxena and Nigam [185] was reported as

$$\tau > 0.22N_{De}^{-0.6} \tag{161}$$

The results presented in this section on axial dispersion is non-Newtonian flow through curved tube are restricted to the ranges of variables covered by the investigators and more or less can be considered as semiqualitative. The area is still open for theoretical study so that the phenomenon of non-Newtonian axial dispersion in curved tubes could be well understood.

Dispersion in Coiled Tubes of Square Cross Section

The phenomenon of axial dispersion in curved tubes of noncircular cross-section has received very little attention. Sakra et al. [187] have experimentally studied axial dispersion in a coiled tube of rectangular cross section. They carried out pulse response experiments in a single coil of curvature ratio 4.41 using water ($N_{Sc} = 2{,}720$) as the flowing media. The ranges of parameters covered were $106 < N_{Re} < 640$, $50.5 < N_{De} < 305$, and $0.006 < \tau < 0.04$. They correlated their experimental values of dispersion number ($D/\bar{u}L$) with Reynolds number and Schmidt number as

$$D/\bar{u}L = 7.5917 \times 10^5/(N_{Re}N_{Sc})^{1.416} \tag{162}$$

Recently, Saxena and Nigam [89] have reported an experimental study on laminar dispersion of Newtonian fluids in coiled tubes of square cross section. They recognized that the orientation of non-circular cross-sectional plane with the direction of centrifugal force is an important parameter and carried out step response experiments in coiled tubes of the two types as shown in Figure 23. The ranges of parameters covered were Dean numbers from 2 to 100, Reynolds number from 10 to 300, dimensionless characteristic times (τ) from 0.04 to 1.5, and curvature ratio λ from 10 to 235. They analyzed their dispersion data within the framework of Taylor's dispersion theory [33] and results are shown in Figure 31.

It can be seen from the figure that in both types of coils axial dispersion decreases with an increase in Reynolds number and decrease in curvature ratio. The extent of reduction in axial dispersion is larger in coils of Type A (in which the direction of centrifugal force makes an angle of 45° with the diagonals of square cross section) as compared to Type B (in which centrifugal force acts along one of the diagonals). The effect of orientation of the cross-sectional plane with the direction of

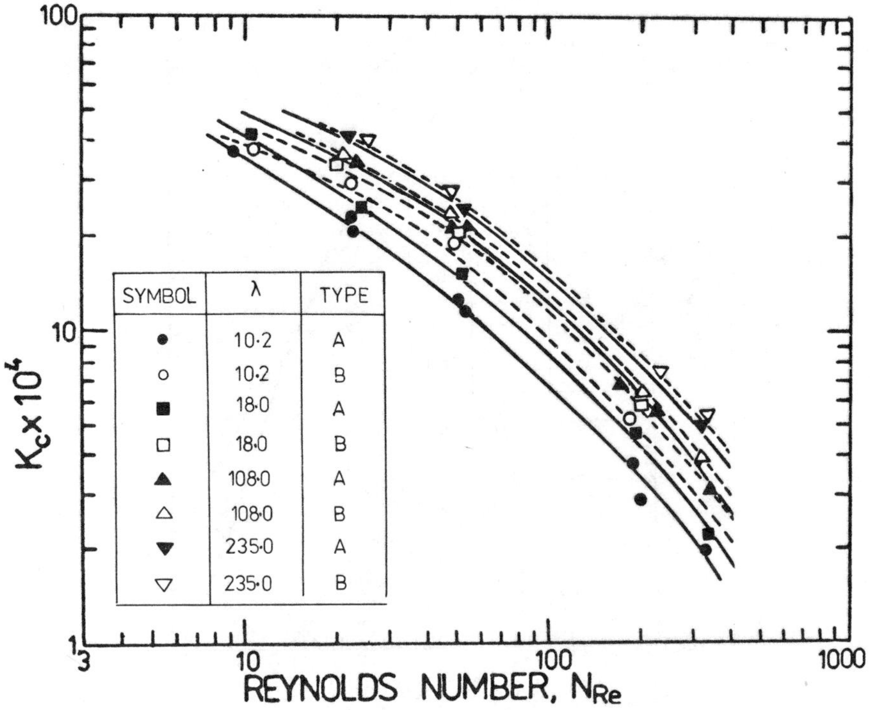

Figure 31. Effect of N_{Re}, λ, and cross-sectional orientation on axial dispersion in coiled tubes of square cross-section.

cetrifugal force diminishes as the curvature ratio (λ) increases. This can be seen from Figure 31, where the data for coils of Types A and B are very close to each other for $\lambda = 230$. This behavior may be explained on physical grounds. As the value of λ approaches infinity, the coil behaves like a straight tube for which orientation of cross-sectional plane carries no meaning.

The experimental data are also plotted against Dean number in Figure 32. It can be seen from the figure that the dispersion data can be represented by single but distinct curves for the coils of both the types. It is interesting to mention that the axial dispersion is more in coils where the centrifugal force acts parallel to one of the diagonals. Figure 32 also includes the dispersion data for coiled circular tubes [82] and rectangular coiled tubes [187]. It is evident from the figure that the axial dispersion is higher in coiled tubes of non-circular geometries. However, the experimental data of Sakra et al. [187] on a single rectangular coiled channel shows a reduction in axial dispersion as compared to a coiled circular tube. Although it is difficult to pointout the exact reason for this discripancy, it should be noted that the ratio of wetted periphery to cross-sectional area is 3.02 times higher in the coil used by Sakra et al. [187]. Since the increase in this ratio results in a broader RTD, the reported reduction in a rectangular coiled channel is rather surprising.

Saxena and Nigam [89] have also pointed out that the orientation of cross-sectional plane with the direction of centrifugal force has a negligible effect on the validity of dispersion model in a

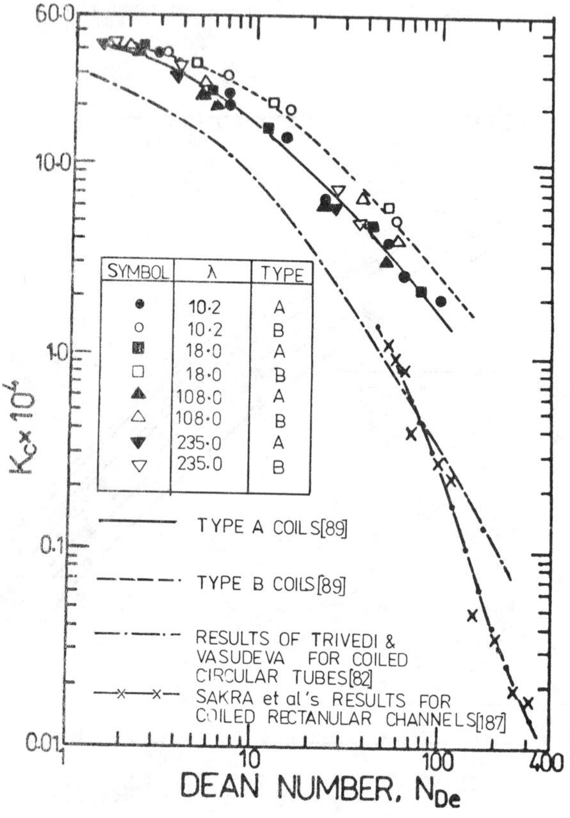

Figure 32. Effect of N_{De} and cross-sectional orientation on axial dispersion in coiled tubes of square cross-section.

coiled tube of square cross section. They reported the condition of applicability of the dispersion model as

$$\tau > 9.5/N_{Re} \tag{163}$$

which is about 1.5 times more rigid than that reported by Trivedi and Vasudeva [82] for coiled circular tube.

COILED CONFIGURATION FOR FLOW INVERSION

It is always required to achieve uniform reaction conditions and weaker temperature gradients within the fluid to improve the performance of flow reactors and heat exchangers. Commercial motionless mixers and flow inverters are some available mechanical devices used in industry to enhance heat transfer coefficient and provide a more uniform thermal and compositional environment. Such devices are usually effective in eliminating severe temperature and composition gradients, but have very high capital costs and high pumping costs as compared to open duct [188]. The experimental data of Nigam and Vasudeva [97] show that the improvement caused by motionless mixers is not as significant as may be intuitively expected and building a reactor of this complexity does not appear practical. Nauman [188] has introduced a comparatively econominal alternate to motionless mixers, called flow inverters, which may be installed midway or at more locations and are separated by relatively long lengths of open pipe. His analysis shows about 25 to 30% improvement in Nusselt number even with a single inverter installed midway in a heat exchanger for Graetz parameter above about 10.

The experimental studies reported so far on coiled tube reveal that very high Dean numbers are required in order to have enough mixing in cross-sectional planes and in case of motionless mixers the pumping cost is very high as compared to narrowing the RTD. In coils Dean number being the only parameter, it is practically difficult to narrow the RTD beyond a certain limit. This is because for a coil of fixed curvature ratio (i), as the Dean number is increased, volume of the helical coil should be more in order to maintain certain residence time, which increases initial cost; and (ii) to maintain higher Dean numbers, flow rate should be more, which tends to increase the operating cost.

To overcome these problems a very simple and economical alternative "bending of helical coils" is introduced by Saxena and Nigam [189], which is very efficient in inverting the flow and improving the mixing in cross-sectional plane.

Concept Behind Bending

Using perturbation analysis, Dean [120] obtained the velocity profiles for laminar flow in helix (Equations 93–95) and the equation of the projection of streamlines on the cross-sectional plane can be given as

$$4 = KZ(1 - Z^2)^2(4 - Z^2)\cos\phi \tag{164}$$

or in rectangular coordinates

$$4 = Ky(1 - x^2 - y^2)^2(4 - x^2 - y^2) \tag{165}$$

where K is a parameter that defines a particular streamline and is always greater than or equal to 3.67. $K = 3.67$ is the center of vortices and $K = \infty$ corresponds to a curve defined by the line $y = 0$ and the tube wall. These D-shaped streamlines are shown in Figure 33A along with coordinate system involved.

Though the axial velocity component is almost parabolic with respect to Z (Equation 95) and independent of K, the apparent axial velocity distribution, incorporating the effects of the other two velocity components u and v can be defined as a function of a single parameter K. At $K = 3.67$,

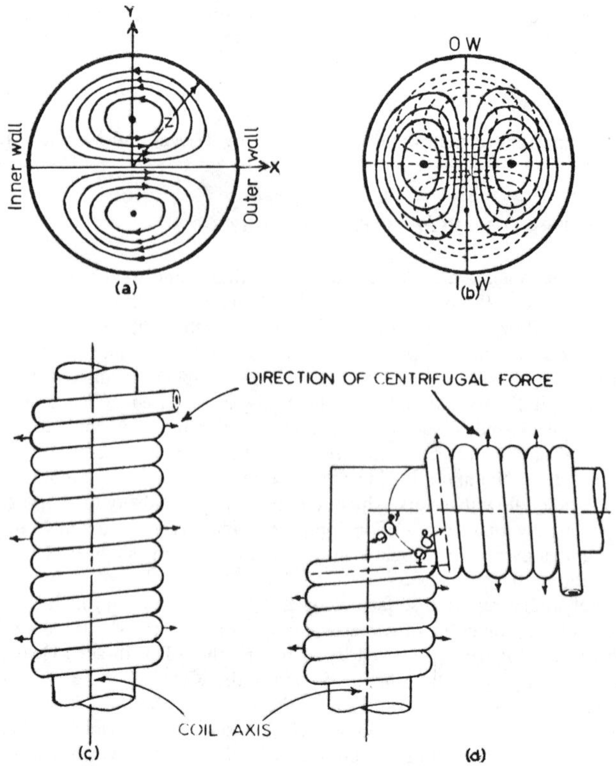

Figure 33. Inversion of flow due to bending of coils.

for which projection of streamline on cross-sectional plane is a single point, apparent axial velocity is the maximum and as K tends to infinity apparent axial velocity vanishes, i.e., the appearance of fluid elements falling on a particular streamline, K, will be delayed at the outlet as the value of K increases.

Now at any stage in helical flow if we change the direction of centrifugal force by any angle, the plane of vortex formation also rotates with the same angle. If this rotation is by an angle of 90°, the readjustment of streamlines will be as shown in Figure 33B. It can be seen from the figure that the points at which apparent axial velocity was maximum ($K = 3.67$) before changing the direction of centrifugal force are now lying on the streamline $K = \infty$, which corresponds to the least axial velocity and new points of maximum velocity are induced on the streamline, which was at the lowest axial velocity before. Thus, in helical flow a 90° rotation in the direction of centrifugal force induces a complete flow inversion.

The direction of centrifugal force is always perpendicular to the axis of the coil as shown in Figure 33C. Hence, it can be changed by any angle just by bending the axis of the helical coil with the same angle. Figure 33D shows a 90° shift in the direction of centrifugal force.

In view of the inversion of flow induced by a sharp band of 90°, it was also of interest to observe the effect if, instead of a sudden shift in the direction of centrifugal force, the plane of vortex formation is gradually rotated. This gradual rotation in the direction of centrifugal force can be obtained simply by coiling a helical coil over a cylindrical base. The effect of these two aspects on mixing in cross-sectional plane was investigated by Saxena and Nigam [189] by measuring residence

Figure 34. Typical coiled inverter.

Figure 35. Typical coiled coil configuration.

time distribution in bent coils and coiled coils. Figure 34 shows a typical coil of sixteen bends and a coiled coil of four turns is shown in Figure 35.

Parameters Affecting RTD in Bent Coils

Saxena and Nigam [189] identified the following parameters that may affect the performance of bent coils:

1. Number of bends (N).
2. Volume ratio of each arm to the total volume of bent coil.
3. Angle between different arms of helix (δ).
4. Dean number (N_{De}).

Effect of these parameters on step response curve was studied [189] under the conditions of negligible molecular diffusion and significant molecular diffusion.

Increased complexity of coils geometry, with the increase in number of bends of different angles, enforced the necessity to develop some symbolic representation of bent coils geometry, which is defined on page 753. The effects of these parameters on RTD are discussed under following heads.

Effect of Number of Bends

The experimental results with single bent coils of different angle of bend (0 to 180°) and different arm proportions revealed that a 90° bend and equal arm lengths before and after the bend provide the maximum narrowing of RTD. This is because a 90° bend induces a complete flow inversion and an odd number of equispaced bends provide an equal opportunity for the centrifugal force to act in two perpendicular directions. Figure 36 shows the effect of equispaced 90° bends on diffusion free RTD in bent coils. It can be seen from the figure that for more than 3 bends an increase in number of bends drastically narrows that RTD and for the coil of 57 bends the dimensionless time, at which the first element of the fluid appears at the outlet is as high as 0.85.

Effect of Dean Number of RTD in Bent Coils

Figure 37 shows the effect of Dean number on diffusion-free RTD in a typical coiled flow inverter (bent coil) having 15 bends of 90° each (C′4). Diffusion-free RTD's for straight helix and straight tube are also shown in Figure 37. The gradual narrowing of RTD with increase in Dean number is evident from the figure. It is worth mentioning that in the case of bent coils the unique RTD is

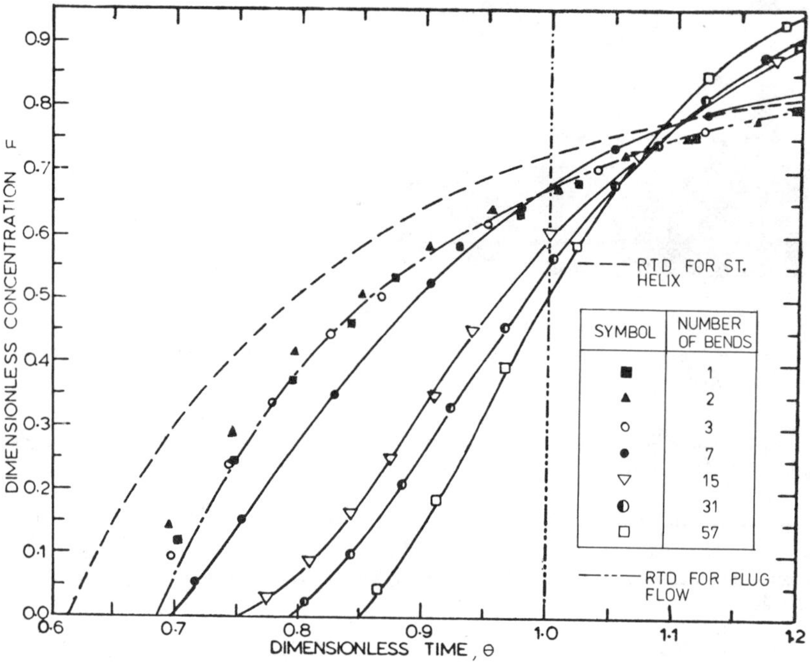

Figure 36. Effect of number of bends on diffusion-free RTD.

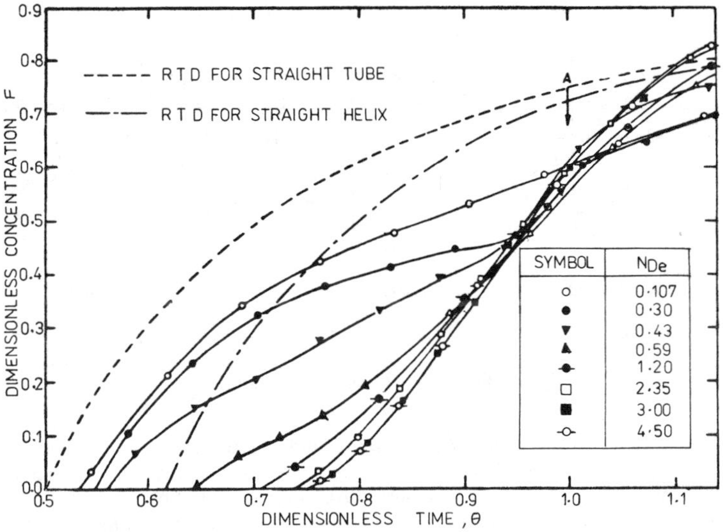

Figure 37. Effect of Dean number on diffusion-free RTD in coiled flow inverter C'4.

obtained at $N_{De} \simeq 3$, which is higher as compared to $N_{De} = 1.5$ reported for straight helix [164]. This may be because of the fact that in bent coils the narrowing of RTD is caused by two mechanisms: development of secondary flow in each arm of the helix that is fully developed for $N_{De} \geq 1.5$; and interchange of velocities among the fluid elements of different ages due to the shift in the direction of centrifugal force. Therefore, for the increase in Dean number above 1.5 the first mechanism is not affecting the RTD, but the second effect is causing more efficient shifting at the bends to further narrow the RTD and is growing up to $N_{De} \simeq 3$. This argument is supported by the experimental results shown in Figure 37, which reveal that for the range of Dean number $0.1 < N_{De} < 1.2$ where both the mechanisms are likely to grow with increase in Dean number, the rate of narrowing is much faster as compared to that in the range $1.2 < N_{De} < 3$, where only a second mechanism is active.

Another interesting point worth noting in Figure 37 is that unlike the straight helix, the narrowing of RTD with increase in Dean number (>0.1) starts from a point A which corresponds to $\theta = 1$, i.e., fluid elements flowing with approximately average axial velocity. This is probably because the narrowing of RTD is caused by mixing of fluid elements of different ages at each bend. At very low Dean number (~ 0.1) where the secondary flow is very weakly developed, mixing will take place only among those fluid elements falling on such streamlines that have enough secondary momentum, before as well as after the bend, to induce mixing. For very weakly developed secondary flow these streamlines cannot be close to $K = 3.67$ as well as $K = \infty$, naturally these will be for which $\theta = 1$.

Order of Inversion and Its Effect on RTD

In case of diffusion-free flow in a coiled flow inverter, having fixed number of bends, the narrowest RTD could be obtained when bends are equispaced and each bend is 90°. This is because of complete flow inversion caused by 90° bends and equal opportunity for centrifugal force to act in two perpendicular directions. This is referred as the inversion of first order. It may be argued on physical grounds that after the action of centrifugal force in two perpendicular directions in the

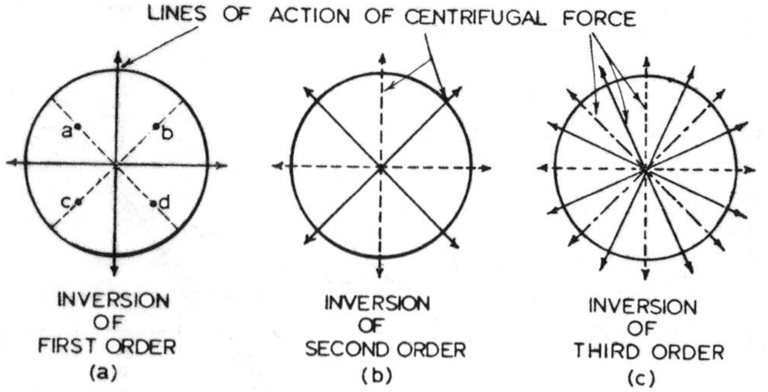

Figure 38. Order of inversion.

tube cross section (in one cycle of set of cycles) the fluid elements that are to appear at the outlet at $\theta = \theta_{min}$ should fall somewhere on diameters that make the angle of 45° with the direction of centrifugal force (points a, b, c, and d in Figure 38A). In view of this if two cycles (or two equal sets of cycles) are connected to each other so that the angle between last arm of the first cycle and first arm of second cycle is 45° (i.e., the coiled configuration shown in Figure 49E of this chapter's Appendix,) it should induce another flow inversion which we will call inversion of second order. It is owing to the angle of 45° between the lines of action of centrifugal force in both the cycles (or set of cycles as shown in Figure 38B. A similar concept can be extended to inversion of third order, which may be achieved by connecting two equal sets of such cycles (in which inversion of second order is already attained) at an angle of 22.5° (see Figure 49F in Appendix). Figure 38C shows the lines of action of centrifugal force in the cross-sectional plane of the coiled tube.

This aspect has been experimentally examined in bent coils having inversions of first, second, and third orders. It was observed that for the fully developed secondary flow ($N_{De} \sim 3$) the inversions of higher order have very marginal effect on the narrowing of RTD while in case of weakly developed secondary flow ($N_{De} \sim 1$) its effect is quite substantial [167].

Effect of Coiling a Coil

So far the effect of sudden shift in direction of centrifugal force on the mixing of fluid elements of different age groups has been discussed. Saxena and Nigam [189] have also studied the effect of gradual change in the direction of centrifugal force in two coiled coils of different λ_{cc} and identical λ_c. They concluded that the sudden shift in the direction of centrifugal force (Bent coils) is more effective in narrowing the RTD than the gradual change.

Laminar Dispersion in Bent Coils

Under the condition of significant molecular diffusion the effect of different parameters on axial dispersion was investigated [167, 189]. The range of Reynolds number was varied from 10 to 200, which corresponds to a Dean number of 30 to 60. The number of equispaced bends were changed from 0 to 57. They [189] analyzed the experimental data using Taylor's [33] dispersion model. The observed values of $D/\bar{u}L$ are plotted against the Dean number in Figure 39. It is evident from the figure that in coiled flow inverters the dispersion number is independent of Dean number. It is interesting to mention that about twenty-fold reduction in dispersion number as compared to a straight helix can be obtained in a coil having 57 equispaced bends of 90° each. The experimentally obtained values of dispersion number were correlated to the design parameter R_A as shown in Figure 40 and can be written as

$$\frac{D}{\bar{u}L} = 0.016\, R_A^{0.58} \qquad \text{for } 3 \le N_{De} \le 60 \text{ and } R_A \le 0.5 \tag{166}$$

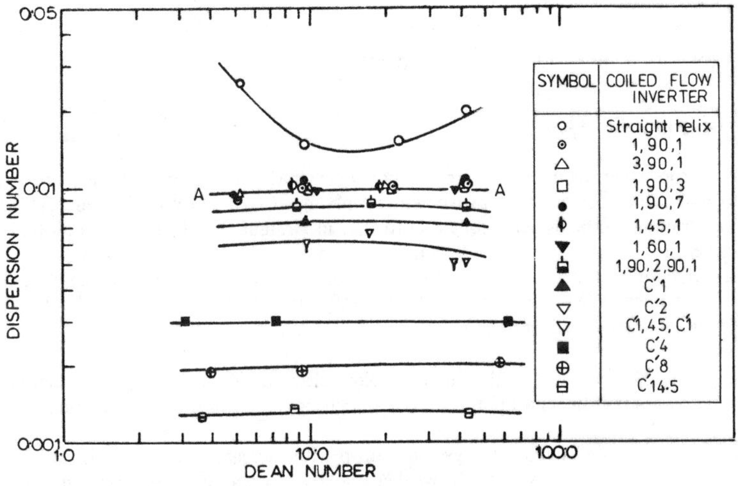

Figure 39. Effect of Dean number on dispersion number in coiled flow inverters.

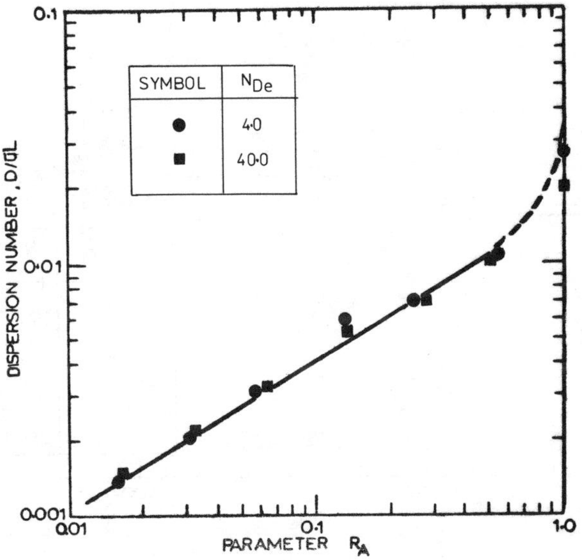

Figure 40. Dispersion number vs. R_A.

where

$$R_A = 1/(N + 1) \qquad \text{(for odd number equispace bends)} \tag{167}$$

Pressure Drop in Coiled Flow Inverters

Pressure drop experiments were carried out by Saxena and Nigam [189] in coiled flow inverters to assess the cost of the improvement in mixing in terms of pumping energy. The ratio of observed friction factor in bend coils (f_{cb}) to that in straight coils (f_{cs}) [155] is plotted against Dean number in Figure 41 which reveals that as the number of bends (N) are increased, surprisingly there is a reduction in friction factor up to two bends and then it starts increasing. The probable reason for this unexpected behavior may be the influence of two factors that should affect the pressure drop in coiled flow inverters:

1. Dissipation of energy due to the mixing in fluid elements of different ages at different bends.
2. Viscous forces that depend upon the axial velocity gradient in tube cross section.

The first factor should increase the pressure drop with increase in number of bends while the second one tends to reduce it owing to the weaker velocity gradients caused by interchange of velocities at the bends. When the number of bends is less ($N \leq 3$), the first factor is less effective, but the second one is showing its substantial effect due to significant narrowing of RTD even with single 90° bend, causing a reduction in pressure drop. As the number of bends is increased ($N > 3$), the influence of first factor becomes dominating, which enhances the pressure drop. It can be seen from Figure 41, that maximum enhancement in friction factor due to bending of coils (with 57

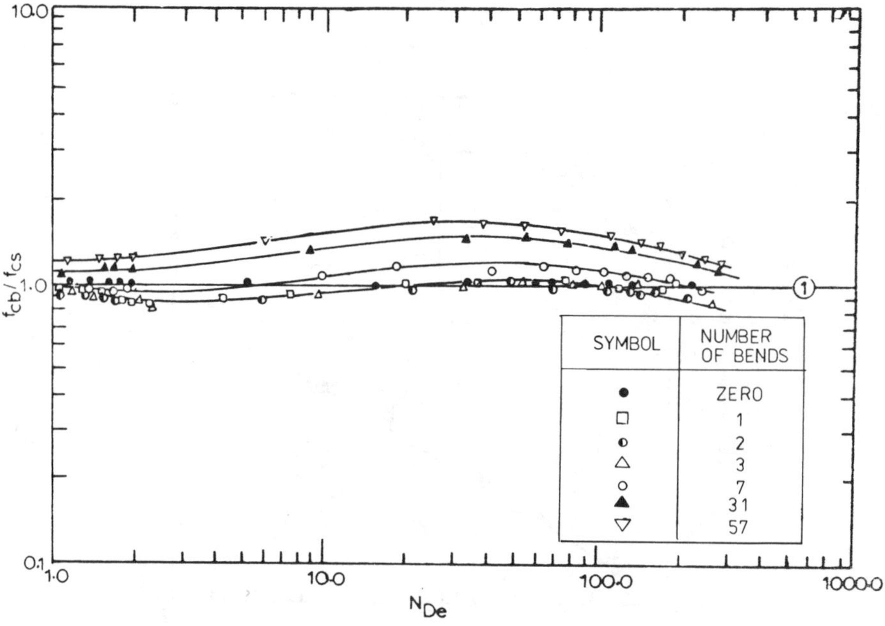

Figure 41. Effect of number of bends on friction factor in coiled flow inverters.

bends) is about 1.7 fold at $N_{De} = 35$. The reduction in axial dispersion is equally significant even for $N_{De} = 3$ for which friction factor is only about 1.3 times higher.

Thus, low pressure drop, compactness, easy fabrication, and narrower RTD in case of coiled flow inverter establish its superiority over any other mechanical device known in literature for inducing mixing in cross-sectional plane and making flow closer to plug flow.

SIMULTANEOUS DIFFUSION AND REACTION IN TUBULAR REACTORS

The dispersion of a solute injected into a flowing fluid formed the subject of our earlier discussion. However, the situation where the injected material undergoes physical or chemical changes is of even greater importance. The changes one can think of include chemical transformation of the solute by itself or by its interaction with the normally flowing fluid, adsorption of the solute on the walls of the containing vessel, diffusion of the solute through the walls, and so on. Such situations can be of interest in many fields, e.g., performance of chemical reactors, gas chromatography (particularly in the case of capillary columns), fluids injected into human body for treatment purposes, and so on. One can perhaps intuitively feel that the possibility of such changes in the solute can substantially alter the dispersion phenomenon. The implications of this can be perhaps best appreciated from an example of a study in the early 1930s. When heavy water was discovered, Forkas drank some to measure its residence time in the human body. By measuring the concentration of heavy water in urine he determined the half-life of water in the body was 13 days. Later, this was found to be erroneous, the error being due to the exchange of deuterium with the tissue. The occurrence of this process was not appreciated at that time.

In the present discussion we will examine the laminar dispersion of solute in straight and coiled tubes under the conditions where it can by itself undergo homogeneous reaction. We will also assume that the rate at which the solute decays, or reacts is directly proportional to its concentration, i.e., linear first-order process.

Steady State Laminar Dispersion with Reaction

The requirement of steady state implies that conditions do not change with time. In the absence of reaction, a steady state situation is of no interest as it leads to trivial solution for all types of inputs. However, in the presence of reaction, steady state behavior, although of no interest for pulse input, is of interest for many other inputs, such as step and sinusoidal inputs. The physical situation involves the laminar flow of feed of a given constant concentration at constant flow rate through a circular tube. Under the assumption of constant temperature, diffusivity and viscosity the governing equation can be written as

$$\left(\frac{R}{L}\right)^2 \tau \frac{\partial^2 C}{\partial X^2} + \tau\left(\frac{\partial^2 C}{\partial Z^2} + \frac{1}{Z}\frac{\partial C}{\partial Z}\right) - 2(1 - Z^2)\frac{\partial C}{\partial X} - \beta C = 0 \quad (168)$$

The relevant boundary conditions are

$$C = 1 \quad \text{at } X = 0 \quad (169a)$$

$$\frac{\partial C}{\partial X} = 0 \quad \text{at } X = 1 \quad (169b)$$

$$\frac{\partial C}{\partial Z} = 0 \quad \text{at } Z = 0 \quad (169c)$$

$$\frac{\partial C}{\partial Z} = 0 \quad \text{at } Z = 1 \quad (169d)$$

If the heterogeneous reaction is also occurring at the wall of tubular reactor, the boundary condition of Equation 169d will be modified as

$$\frac{\partial C}{\partial Z} = -f\,C \qquad \text{at } Z = 1 \tag{170}$$

and is known as third-kind boundary conditions [199].

Solution of Two-Dimensional Convective Diffusion Equations

The solution of Equation 168 depends on two parameters, α and β. α represents the ratio of characteristic time for reaction to characteristic time for diffusion, whereas β represents the ratio of characteristic time for convection to that for reaction. Thus, large value of α means that diffusion is "faster" than reaction and a large value of $k\bar{\tau}$ means time for flow is large enough to "complete" the reaction.

Lauwerier [200] investigated the homogeneous reaction without axial diffusion in a tube. Katz [201] analyzed the catalytic wall reaction in a tube. Kaufman [202] obtained a solution for the extreme ideal case that the rate of reaction was equal to the convective motion carried by the reactant, although he pointed out the importance of radial and axial diffusion, along with the wall reaction in the system. Cleland and Wilhelm [203] obtained experimental results and a numerical solution for convective diffusion with the first-order reaction but negligible axial diffusion. Dranoff [204, 205] obtained eigenfunction solutions for convective diffusion of the reactant with catalytic wall reactions in a tubular or annular reactor. Ogren [206] obtained a solution for laminar tube flow with radial diffusion and first-order homogeneous and tube wall reactions. Walker [207] obtained an asymptotic solution downstream of the tube for chemical reactions with axial and radial diffusion in a catalytic tubular reactor. In addition to the asymptotic nature of the solutions, Walker [207] did not consider the effect of axial diffusion at the inlet of the tube. Dang and Steinberg [208] presented an exact analysis of the same problem of Walker [207] by using the method of nonorthogonal series expansion. Recently, Nigam et al. [209] presented a simple closed-form analytical solution for convective diffusion equation with first-order simultaneous homogeneous and heterogeneous reactions in a tubular reactor by using the Galerkin technique [210, 211] in the Laplace-transform domain. The solution is very simple to use and saves significant numerical effort in comparison with that required in the case of the conventional variable separable method. A review of the convective diffusion problem till 1975 was given by Hoyermann [212], the more recent work is due to Dang [213, 214] and Apellelat [215].

One-Dimensional Model—Steady State Case

Although Taylor [33] originally offered his analysis for unsteady state dispersion under non-reacting conditions the possibility of using Taylor dispersion model for predicting steady state reactor performance under laminar flow conditions has received considerable attention. Houghton [216] using two approximate methods found that for the case of laminar flow reactor chemical reaction has no influence on axial dispersion and Taylor diffusivity can be used irrespective of the order. The condition for this result to hold was obtained as

$$3.8^2\tau + \beta \gg 2 \tag{171}$$

Horn and Parish [217], however, concluded that Taylor dispersion model would hold provided N_{Pe} is large. For the Taylor dispersion model this implies a large values of τ. Bischoff [218] observed that the applicability of Taylor dispersion model depends on the value of α. For slow reactions ($\alpha > 0.1$) the agreement with the exact solution was found to be satisfactory, but for fast reactions ($\sim$0.01) large deviations were observed. Using a perturbation method Wissler [53] obtained a modified inlet boundary condition and, using a solution for the semi-infinite reactor, a better agreement with the exact solution was claimed. Wan and Ziegler [219], using Danckwerts [30] boundary conditions, concluded that at constant value of β the agreement between the exact model and the Taylor dispersion model improves with increasing value of α until both approach the plug flow model. For most values of β and for $\alpha < 0.01$ the segregated flow model is claimed to be a good representation and for $\alpha > 0.5$ regardless of the value of N_{Pe} or β Taylor dispersion model was reported to hold. Woodhead and Yesberg [220] following Taylor's approach showed that for first-order reaction in laminar flow reactor effective dispersion coefficient is given by

$$D = \frac{0.333\bar{u}^2R^2/D_m}{16 + \dfrac{1}{\alpha}} \tag{172}$$

The predictions of their model were compared with the experimental data of Cleland and Wilhelm [203]. Due to some error in plotting the data they erroneously concluded that there was no definite trend in the improvement offered by their analysis. Kulkarni and Vasudeva [221] have re-examined this problem and have provided the criterion for the validity of this model along with the estimate of the error involved. The Taylor model is reported to hold when τ (and not α) is large ($\tau > 0.25$ for error to be $<0.5\%$). For many realistic situations likely to be encountered in the case of liquid phase reactions the practical values of τ are often much less. Under these conditions dispersion coefficient was argued to be dependent on the reaction rate constant and the following empirical expressions were obtained by Nigam and Vasudeva [25]:

$$N_{Pe} = 3.1 + 39.4\tau + 0.4\beta - 4\alpha, \text{ for } \alpha \geq 0.032 \tag{173a}$$

$$N_{Pe} = 1.4 + 1.03\beta + 105\alpha, \text{ for } \alpha < 0.032 \tag{173b}$$

The use of a dispersion model with Equation 173 gives the bulk mean concentration over a wide range of values of system parameters [221]. Recently, Nigam et al. [222] have obtained closed-form analytical solution for the effective dispersion coefficient using the Galerkin technique.

One-Dimensional Model—Unsteady State Case

The axial dispersion model for the unsteady state reacting system may be written as

$$\frac{\partial C}{\partial \theta} = \frac{D}{\bar{u}L}\frac{\partial^2 C}{\partial X^2} - \frac{\partial C}{\partial X} - \beta C \tag{174}$$

Before considering the applicability of dispersion model for unsteady state tubular reactor, it is necessary to examine the choice of suitable boundary conditions. The different kind of boundary conditions and solutions generally used in the literature [30, 223–227] are given in Table 3. In studies of the steady state behavior this aspect has received considerable attention and it was concluded that use of Danckwerts [30] boundary conditions gives physically realistic answers for all values of Peclet number. Despite this result, somewhat indiscriminate use of boundary conditions is still being made, e.g., Levenspiel [1] in his recent book states that the solution for the dispersion model with a first-order reaction term is the same for all types of inlet and outlet conditions. Shankar Subramanian et al. [228] also have not made use of this result and have instead used the boundary conditions for the closed-semi-infinite case despite the fact that for low values of α and β the Taylor model predicts low values of Peclet number. They [228] concluded that Taylor model predictions show large deviations from the exact result. However, using the correct boundary conditions, Nigam and Vasudeva [25] have shown that the disagreement is considerably reduced.

Table 3
Different Set of Boundary Conditions and Solutions for Steady State Reactor Performance [225]

Physical Description	Boundary Conditions	Solutions
Closed-closed	$C = 1$ at $X = 0$ $\frac{dC}{dX} = 0$ at $X = 1$	$C = 2a\exp(N_{Pe}/2) \Big/ \left\{(1+a)\exp\left(\left(\frac{1+a}{2}\right)N_{Pe}\right) - (1-a)\exp\left(\left(\frac{1+a}{2}\right)N_{Pe}\right)\right\}$
Closed-Semi-infinite	$C = 1$ at $X = 0$ $C \to 0$ at $X \to \infty$	$C = \exp\left(N_{Pe}\left(\frac{1-a}{2}\right)\right)$
Open-closed	$\frac{dC}{dX} = \frac{\bar{u}L}{D}(C-1)$ at $X = 0$ $\frac{dC}{dX} = 0$ at $X = 1$	$C = 4a\exp(N_{Pe}/2) \Big/ \left\{(1+a)^2\exp\left(N_{Pe}\left(\frac{1+a}{2}\right)\right) - (1-a)^2\exp\left(\left(\frac{1-a}{2}\right)N_{Pe}\right)\right\}$

where $a = \sqrt{1 + \frac{4K\bar{t}}{N_{Pe}}}$

While the steady state problem has received considerable attention the unsteady state problem in the reacting system has received very limited attention [25, 228, 229]. Shankar Subramanian et al. [30] using the generalized dispersion model approach with time dependent dispersion coefficients have obtained the exact area mean concentration. Although the method is very elegant, the solution is rather difficult to use. Further, in reactor applications one is more interested in the bulk mean concentration rather than the area mean concentration [25].

The influence of choice of boundary conditions on the predictions for the unsteady state behavior was examined by Nigam and Vasudeva [25]. They found that the discrepancy between the solutions for the two sets of boundary conditions, open-closed, and closed-semi-infinite, is very large at low values of β and N_{Pe}. However, for $N_{Pe} > 10$ and $\beta > 1$ the two solutions were found to be essentially identical for all times. Typical results are shown in Figures 42 and 43.

Nigam and Vasudeva [25] have also experimentally investigated the unsteady state behavior of a laminar flow reactor with acetic anhydride hydrolysis as the reacting system. The system parameters were chosen such that the Taylor model does not hold. The results are summarized in Figures 44 and 45, where the experimental results are compared with axial dispersion model predictions for both Taylor diffusivity (Equation 43) and reaction dependent dispersion coefficient (RDDC) (Equation 173). In both the cases open-closed boundary conditions were used and the concentration interpreted as bulk mean concentration. In addition the predictions of two other models, segregated flow model (i.e., the effect of molecular diffusion neglected) and plug flow model (effect of molecular diffusion as well as that of velocity variation neglected) are also shown. The superiority of reaction dependent dispersion coefficient (RDDC) to model the experimental data is evident from Figures 44 and 45. The conjecture of Shankar Subramanian et al. [228] that dispersion coefficient is independent of reaction rate constant does not hold. At large values of τ, the reaction-independent Taylor dispersion model holds but at low values of α and τ axial dispersion model with RDDC (Equation

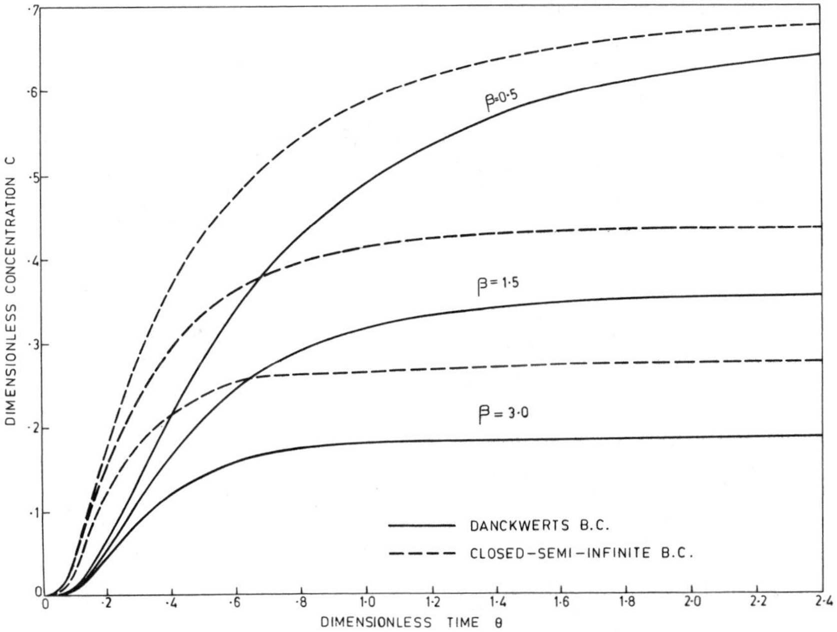

Figure 42. Influence of B.C. on dispersion model results, $N_{Pe} = 1.0$.

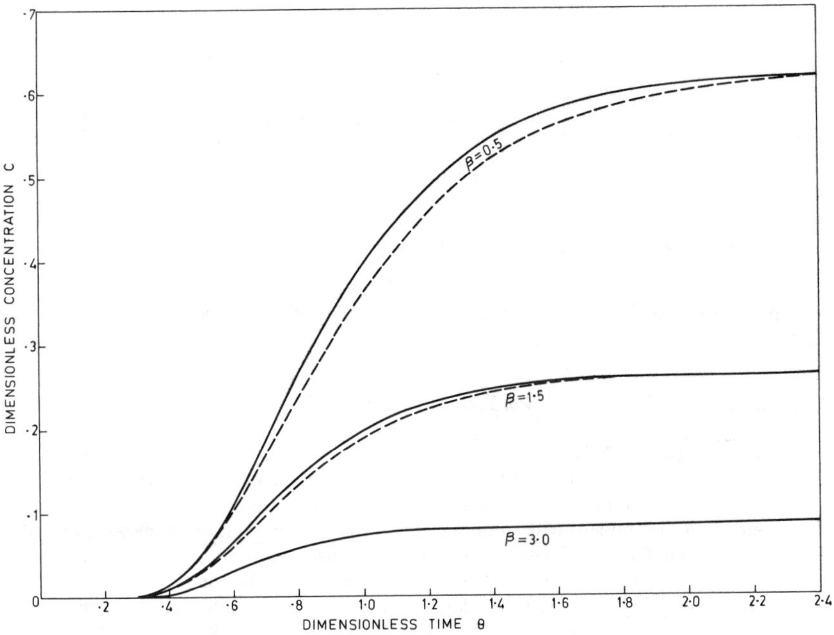

Figure 43. Influence of B.C. on dispersion model results, $N_{Pe} = 10.0$.

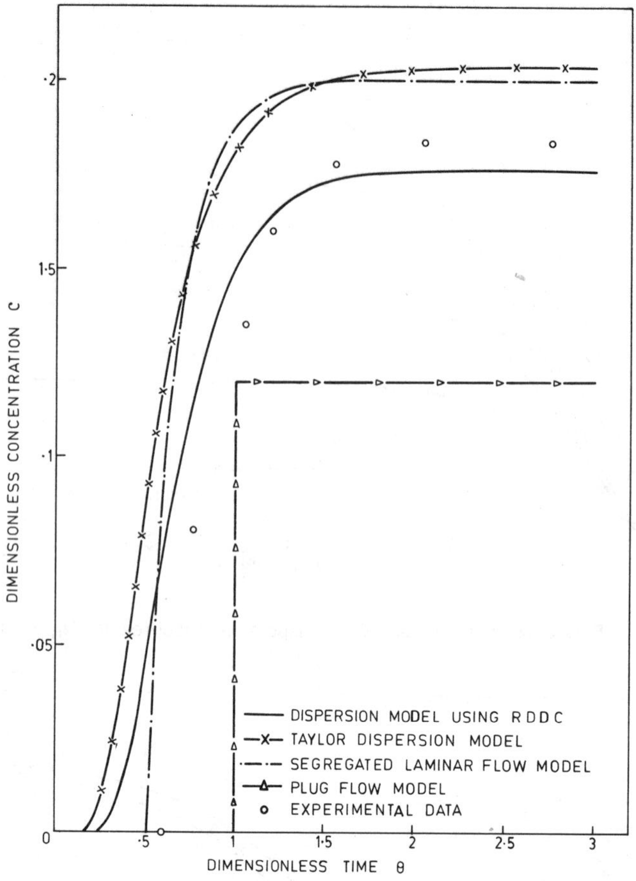

Figure 44. Transient behavior of laminar flow reactor $\alpha = 0.037$ and $\beta = 1.59$.

173) predicts the results quite satisfactorly. The use of RDDC to model the multiple reactions in a laminar flow reactor has also been examined [230].

Laminar Dispersion with Reaction in Power Law Fluids

While the problem of diffusion and reaction under conditions of laminar flow of Newtonian fluids has received considerable attention the analogous problem for non-Newtonian fluid has received much less attention. Homsy and Strohman [231] have solved the two-dimensional convective diffusion equation and obtained the results in the form of eigenvalues and eigenvectors. Shanker Subramanian and Gill [232] used their generalized dispersion model approach to obtain the solution. Both these studies [231, 232] are not entirely suitable for comparison or generalization. Recently, Nigam [233] has solved the two-dimensional convective diffusion equation:

$$\frac{\partial^2 C}{\partial Z^2} + \frac{1}{Z}\frac{\partial C}{\partial Z} - \frac{1}{\alpha}C - \frac{n' + 3}{2(n' + 1)}(1 - Z^{n'+1})\frac{\partial C}{\partial X} = 0 \tag{175}$$

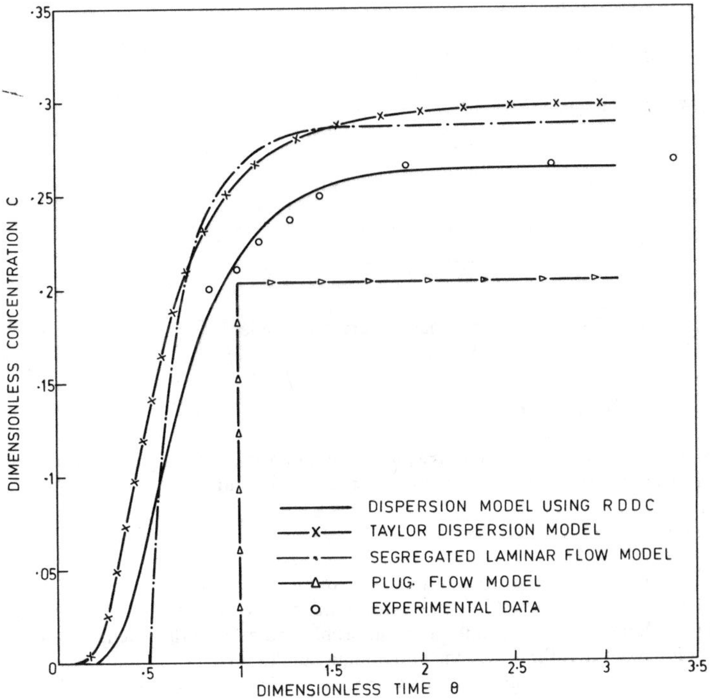

Figure 45. Transient behavior of laminar flow reactor $\alpha = 0.037$ and $\beta = 2.12$.

With boundary conditions

$$C = 1 \qquad \text{at } X = 0 \tag{176a}$$

$$\frac{\partial C}{\partial Z} = 0 \qquad \text{at } Z = 0 \tag{176b}$$

$$\frac{\partial C}{\partial Z} + f\,C = 0 \qquad \text{at } Z = 1 \tag{176c}$$

He has used the Galerkin method in space and Laplace transforms in time to obtain the closed-form analytical solution. The solution for first approximation can be written as

$$C' = \frac{3}{2}(n' + 7)\frac{f(n' + 7) + 4(n' + 5)}{\phi_1 f^2 + \phi_2 f + \phi_3}(f + 2)\left(1 - \frac{f Z^2}{f + 2}\right)\exp(-pX) \tag{177}$$

$$C_m = \frac{3}{4}\frac{n' + 7}{n' + 5}\frac{[f(n' + 7) + 4(n' + 5)]^2}{\phi_1 f^2 + \phi_2 f + \phi_3}\exp(-pX) \tag{178}$$

$$N_{Sh} = 4(n' + 5)/(n' + 7) \tag{179}$$

where

$$\phi_1 = n'^2 + 14n' + 57 \tag{180}$$

$$\phi_2 = 6n'^2 + 84n' + 294 \tag{181}$$

$$\phi_3 = 12(n' + 5)(n' + 7) \tag{182}$$

$$p = 2(n' + 5)(n' + 7)\frac{6f(f + 4) + \frac{1}{\alpha}(f^2\ 6f + 12)}{\phi_1 f^2 + \phi_2 f + \phi_3} \tag{183}$$

The solutions for higher order approximations are also available [233].

Simple Models for Non-Newtonian Systems

For unsteady state dispersion in non-reacting system Fan and Hwang [63], following the method of Taylor [33], obtained the expression for dispersion coefficient as

$$N_{Pe} = 2(n' + 3)(n' + 5)\tau \tag{184}$$

The suitability of Fan and Hwang's analysis to predict the reactor performance was examined by [234, 235]. Nigam and Vasudeva [235] have shown that reaction rate constant is important in determining the development of radial concentration profile and as such the extent of dispersion. They have also pointed out that the development of radially uniform concentration, a tenet of dispersion model under nonreacting conditions, is of less importance in the reacting system. However, from a practical point of view the Fan and Hwang analysis holds in the case of reacting systems provided $\tau > 0.15$ and $\beta > 1.0$, the error being less than 1%. Nigam and Vasudeva [235] have also developed empirical correlations for the effective dispersion coefficient, which were applicable over a wide range of parameter values, i.e.,

$$10^{-3} \le \tau \le 10,\ 0.4 \le n \le 2 \qquad \text{for } \beta > 1$$

with the maximum error being 0.5%. The expression for the effective dispersion coefficient can be written as follows [235]: For dilatent fluids $n > 1$

$$N_{Pe} = 2.7 + 1.64(n' + 3)(n' + 5)\tau + 1.56n' \qquad \text{for } \alpha \ge 0.032 \tag{185a}$$

$$N_{Pe} = 2.46 + 4.37(n' + 3)(n' + 5)\alpha - 1.07n' + (0.26 + 0.8n')\beta \qquad \text{for } \alpha < 0.32 \tag{185b}$$

For pseudoplastic fluids $n < 1$

$$N_{Pe} = 4.0 - 0.81n' + 1.64(n' + 3)(n' + 5)\tau - (1 + n')^2 + (0.76n' - 0.45)\beta \qquad \text{for } \alpha \ge 0.032 \tag{186a}$$

$$N_{Pe} = 1.76 - 0.15n' + 4.37(n' + 3)(n' + 5) + (0.56 + 0.43n') \qquad \text{for } \alpha < 0.032 \tag{186b}$$

Recently, Nigam and Nigam [236, 237] have studied the problem of diffusion with a homogeneous first-order reaction in the bulk and a heterogeneous reaction at the reactor wall in a non-Newtonian laminar flow tubular reactor. The have obtained an analytical solution for the effective diffusion coefficient that was dependent on reaction rate constant. It was found that for the same mean velocity of the flow, the effective dispersion coefficient decreases with increase in the chemical reaction rate constants.

SIMULTANEOUS DIFFUSION AND REACTION IN COILED REACTORS

The influence of secondary flow on the extent of dispersion in helical coils has been discussed. The influence of such secondary flow on a process of convective diffusion with reaction in coiled tubes has received very little attention. Seader and Southwick [238] have experimentally demonstrated the advantages of secondary flow in enhancing the rates of reaction. More recently, Mashelkar and Venkatasubramanian [239] have analyzed the problem of the influence of secondary flow on spatial concentration distribution as well as the bulk average concentration in a reacting system. They have considered the first-order reaction in non-Newtonian fluids under low Dean number flow through coiled tubes.

Assuming the concentration independent molecular diffusion coefficient and neglecting the contribution of axial molecular diffusion (i.e., a $W_{av}/D_m > 1{,}000$ [2]) the governing equation in the coordinate system showing Figure 13 can be written as

$$\frac{1}{r}\frac{\partial(rUc')}{\partial r} + \frac{1}{r}\frac{\partial(Vc')}{\partial \phi} + \frac{\partial(Wc')}{\partial L} = D_m\left(\frac{\partial^2 c'}{\partial r^2} + \frac{1}{r}\frac{\partial c'}{\partial r} + \frac{1}{r^2}\frac{\partial^2 c'}{\partial \phi^2}\right) - kC \tag{187}$$

Equation 187 can be non-dimensionalized as

$$C' = c'/c_0, \qquad z = r/a, \qquad L' = L\bigg/\left(a^2\frac{W_{av}}{\nu}\right), \qquad U' = \frac{aU}{\nu}$$

$$V' = \frac{aV}{\nu}, \qquad W' = W/W_{av}, \qquad \alpha = ka^2/D_m$$

Substituting these into Equation 187 reduces to

$$\frac{1}{z}\frac{\partial(U'C')}{\partial z} + \frac{1}{z}\frac{\partial(V'C')}{\partial \phi} + \frac{\partial(W'C')}{\partial L'} = \frac{1}{N_{Sc}}\left(\frac{\partial^2 C'}{\partial z^2} + \frac{1}{z}\frac{\partial C'}{\partial z} + \frac{1}{z^2}\frac{\partial^2 C'}{\partial \phi^2}\right) - \frac{\alpha C'}{N_{Sc}} \tag{188}$$

Boundary conditions are—

$$\frac{\partial C'}{\partial \phi} = 0 \qquad \text{at } \phi = \pm\pi/2 \tag{189a}$$

$$\frac{\partial C'}{\partial z} = 0 \qquad \text{at } z = 1 \tag{189b}$$

$$C' = C_0 \qquad \text{at the inlet} \tag{189c}$$

Mashelkar and Venkatasubramanian [239] have numerically solved Equation 188 along with the velocity profiles for low Dean number flow through coiled tubes reported by Raju and Rathna [240]. They employed Brain's [241] modification of the Peaceman and Rachford [242] alternating direction implicit technique to obtain detailed spatial concentration distributions and bulk mean concentration at the outlet.

Since the flow in coiled tubes in not axisymmetric, the local concentration profiles at a given axial distance (i.e., $C' = C'(r, \phi)$), will be functions of both angular and radial coordinates. This point is more clearly brought out in Figure 46 in which the local concentration along the cross-sectional diameter, parallel and perpendicular to coil axis, at two different axial distance $\xi\,(=L'/N_{Sc})$ are shown for $\alpha = 0.1$. The figure also includes the local concentration profiles for a laminar tubular reactor and a plug flow reactor. A considerable effect of axial dispersion on the radial concentration distribution, in case of a laminar tubular reactor, is evident from the figure. In contrast, in coiled-tube reactors there is remarkable narrowing of concentration distributions. The concentration profile is almost flat along the cross-sectional diameter perpendicular to the coil

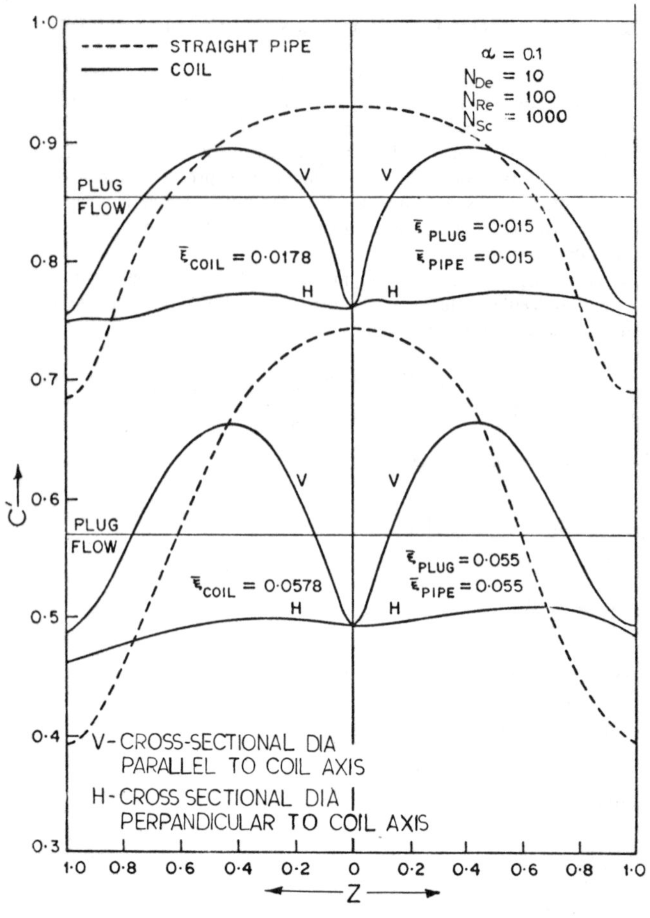

Figure 46. Local concentration profiles for laminar tubular, helically coiled, and plug flow reactors.

axis while has two maximas along the diameter parallel to the axis of helical coil. These maxima correspond to the centers of the two counter-rotating vortices (Figure 13) at which the apparent axial velocity of fluid elements is the maximum. A similar trend of concentration profiles has been reported for higher values of reaction rates (lower value of $\alpha = D_m/Ka^2$); however, it was observed that the flatness of the concentration profile, which exists in case of higher values of α, disappears with increase in the rate of reactions.

Figure 47 shows the bulk average concentration c_{av} as a function of tube length. As expected from the previous discussion, the performance obtainable in a coiled reactor is in between the plug flow reactor and laminar tubular reactor. The reduction in the reactor length to achieve a fixed level of conversion $(1 - c_{av})$ can be considered as the measure of improvement in performance of a flow reactor. The numerical results of Mashelkar and Venkatasubramanian [239] suggest that for a conversion level of 80% the improvement in using a coiled tube instead of laminar tubular reactor is 10% and 20% for α = 0.10 and 0.01, respectively; however, as the magnitude of α decreases,

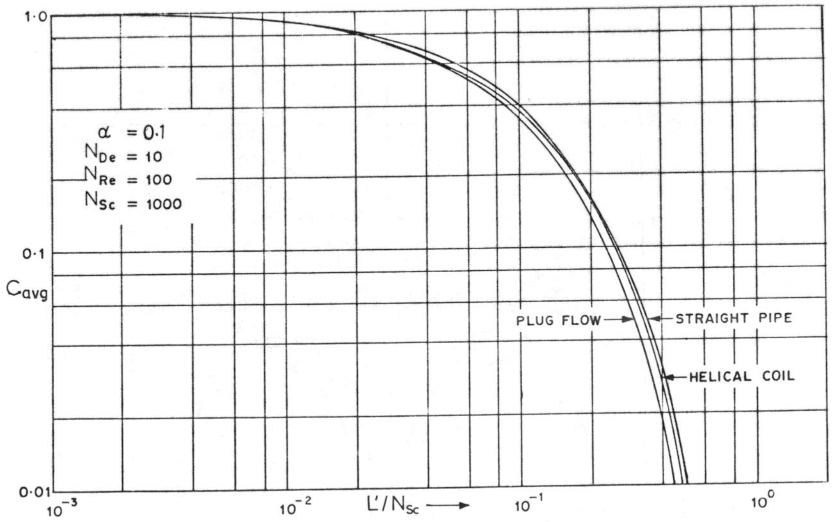

Figure 47. Bulk average concentrations for laminar tubular, helically coiled, and plug flow reactors.

the improvement over a straight tube reactor appears to diminish. This phenomenon could be physically explained by considering the interaction of kinetic, diffusive, and secondary convective time scales. If the reaction rate is very high, the difference in residence time of different elements of reactant will hardly have an influence on the conversion level. In such cases the use of coiled tubes instead of laminar tubular reactor will not have much influence on the reactor performance.

Spital concentration profiles for $\alpha = 0.1$ and power-law indices (n) of 0.75 and 0.5 are shown in Figure 48 [239]. As the extent of pseudoplasticity increases (i.e., n decreases) the local minima on the tube diameter parallel to coil axis tend to become smaller while its trend along the diameter perpendicular to coil axis is somewhat peculiar toward the inner half of the tube, the local concentration increases with an increase in pseudoplasticity, whereas for the elements near the outer half, this value decreases.

They [239] observed that whether the fluid is Newtonian or non-Newtonian, there is always an improvement when one switches over from tubular reactor to a coiled tube reactor; however, the extent of improvement in the performance reduces marginally with increase in pseudoplasticity. This is because of the stronger flattening effect of pseudoplasticity on the axial velocity profile that dominates the weaker secondary flow effect in the low Dean number range. The influence of strong secondary flow (i.e., $N_{De} > 17$) on diffusion with chemical reaction in coiled tubes has not been studied.

APPENDIX—NOTATION USED TO DEFINE THE GEOMETRY OF COILED FLOW INVERTERS

Up to three bends the geometry of coiled flow inverters is defined as

$$H_1, \delta_1, H_2, \delta_2, H_3, \delta_3, H_4$$

where H_i (i = 1, 2, 3, 4) is the arm proportion of ith arm and δ_j (i = 1, 2, 3 and j = i − 1) is angle between jth and j + 1th arm.

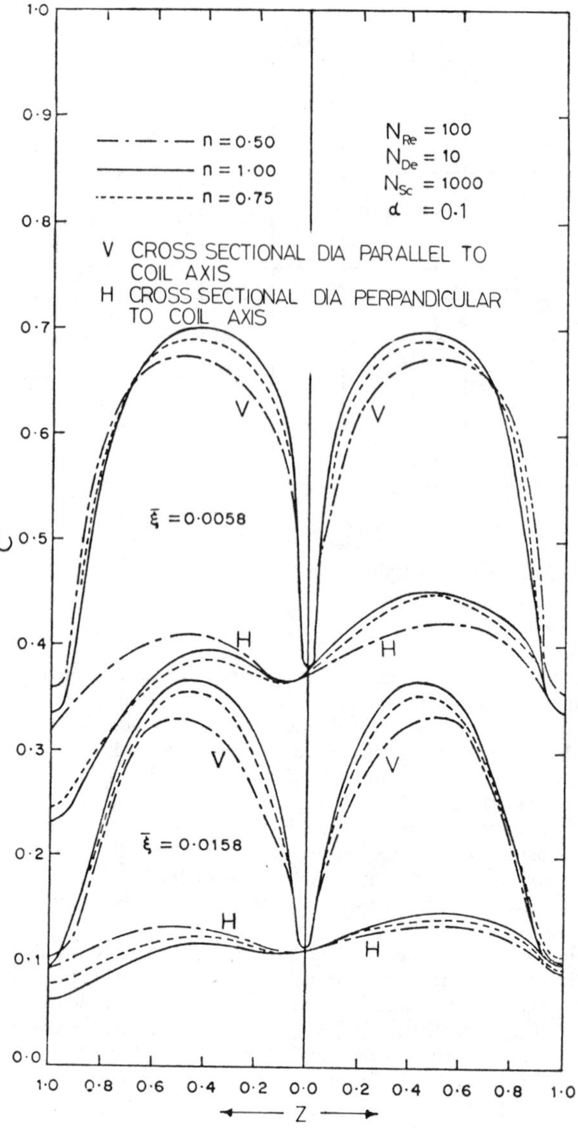

Figure 48. Influence of pseudoplasticity on local concentration profiles.

Examples

- Coiled flow inverter 1, 90, 2 implies a coil having two arms attached to each other at an angle of 90°. Proportion in the lengths (or volume) of first and second arm is 1:2. This configuration is shown in Figure 49a.
- Coiled flow inverter represented by 1, 90, 1, 45, 1, 90, 1 is shown in Figure 49b which has three bends and four equal arms. The angle between first and second arm is 90°. The third arm is attached to the second arm at an angle of 45° and the fourth arm to the third arm at an angle of 90°.

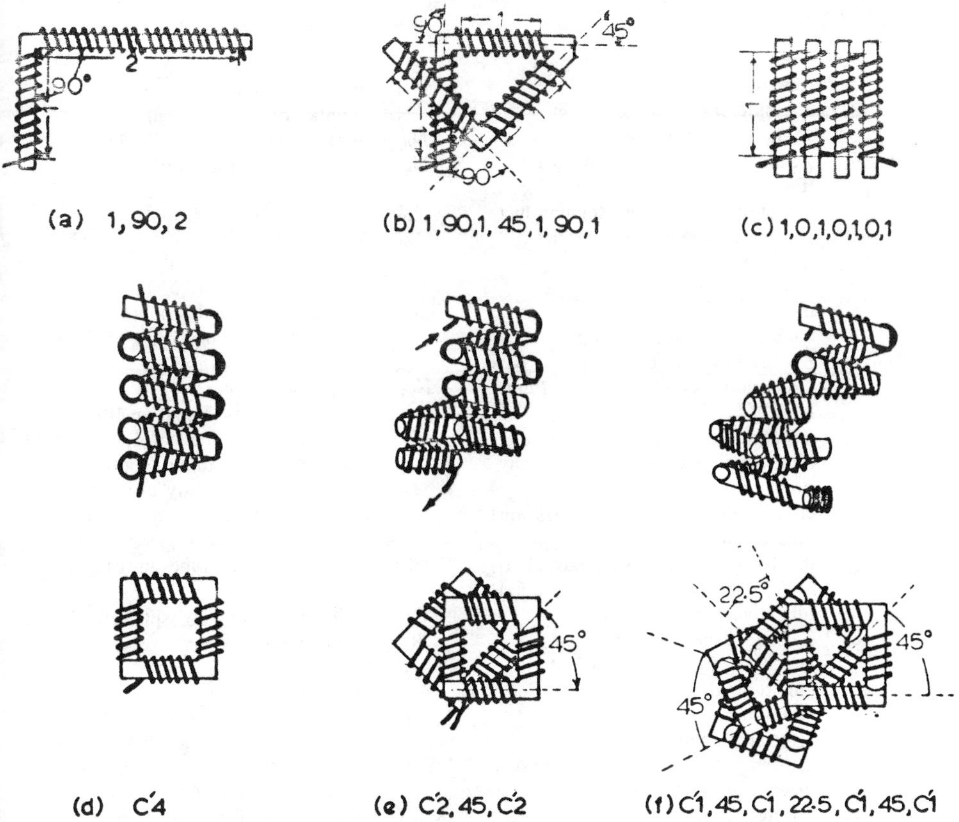

Figure 49. Typical examples of coiled flow inverters.

- Coiled flow inverter, 1, 0, 1, 0, 1, 0, 1 represents four parallel equal arms as shown in Figure 49c. For coiled flow inverters having more than three bends, the geometry is defined in terms of cycle. By one cycle (C′1) it means a square formed by helix arms, i.e.,

$$C'1 = 1, 90, 1, 90, 1, 90, 1$$

If such m cycles are attached to each other such that angles between last and first arms of two consecutive cycles are 90° and planes of all the cycles are parallel, the geometry is represented by C′m.

The term $C'm_1, \delta_1, C'm_2$ means there are two sets of m_1 and m_2 cycles that are attached to each other such that the angle between the last arm of the first set and the first arm of the second set is δ_1 degree. Also

$$C'm_1, \delta_1, C'm_2 = C'(m_1 + m_2)$$

If $\delta_1 = 90°$. In general it can be written as $C'm_1, \delta_1, C'm_2, \delta_2 \ldots C'm_{i-1}, \delta_{i-1}\ C'm_i$ which represents i sets of $m_1, m_2, \ldots, m_i$ cycles attached to each other such so that the angles between the arms connecting two consecutive cycles are $\delta_1, \delta_2, \ldots, d_{i-1}$ degrees, respectively.

For example, C′4 is shown in Figure 49d; C′2, 45, C′2 is shown in Figure 49e; and C′1, 45, C′1, 22.5, C′1, 45, C′1 is shown in Figure 49f.

NOTATION

- a tube radius
- A attribute under consideration
- A_0 value of A at the inlet
- A_1, A_2 parameter in Nauman's Model (Equation 105)
- b half of the height of rectangular duct
- $+B$ parameter in Nauman's Model (Equation 105)
- c concentration at outlet
- c_0 inlet concentration
- c' concentration at any point
- C curve response to ideal pulse input
- C dimensionless concentration at outlet
- C_0 dimensionless concentration at inlet
- C' dimensionless concentration at any point
- C_m cross-sectional mean concentration
- C_{av} bulk mean concentration
- C'_m m cycles of bent coils attached to each other at 90° (See Appendix)
- d_t tube diameter
- D effective dispersion coefficient
- D_m molecular diffusion coefficient
- D_L longitudinal dispersion coefficient
- D_R radial dispersion coefficient
- D_c dispersion coefficient in coils
- D_s dispersion coefficient in straight tubes
- $+E$ exit age distribution curve
- f curve derivative of F curve
- f_{cb} friction factor in bent coils
- f_{cs} friction factor in straight coils
- F response to step input
- F_{exp} experimental value of F
- F_{theo} value of F predicted by dispersion model
- g acceleration due to gravity
- h time interval
- h_c coil pitch
- H_i length proportion of ith arm in a bent coil
- I curve internal age distribution curve
- k reaction rate constant
- $\bar{K}$ consistency index parameter defining a streamline (Equation 96)
- K_c dispersion coefficient ($=DD_m/\bar{u}\, d_t^2$)
- K_s dispersion coefficient for straight tube
- L tube length
- L' dimensionless axial distance, $L/a^2 w_{\bar{av}}/\nu$
- n power-law index
- N number of bends in bent coil
- N_{Bo} Bodenstein number ($=\bar{u}d_t/D_m$)
- N'_{Bo} effective Bodenstein number ($=\bar{u}\, d_t/D$)
- N_{De} Dean number ($=N_{Re}/\sqrt{\lambda}$)
- N_{Gr} Grashof number ($=d_t^3 g\, \Delta\rho/\rho\nu^2$)
- N_{Pe} Peclet number ($\bar{u}L/D$)
- N_{Re} Reynolds number ($d_t\bar{u}/\nu$)
- N_{Regn} generalized Reynolds number $= (4n/3n + 1)^n(1.0/8^{n-1})(d_t^n\bar{u}^{2-n}\rho/\bar{K})$
- N_{Sc} Schmidt number ($\mu/\rho D_m$)
- N_{Sh} Sherwood number
- p, q model constants (Equation 101)
- Q volumetric flow rate
- r radial distance, radial coordinate
- r_A rate of change of any attribute A
- R tube radius, outer tube radius in case of annulus
- R' radius of inner tube in case of annulus
- R_A parameter defined by equation 167
- R_c radius of coil
- t time
- $\bar{t}$ mean holding time
- T torsion parameter ($=h_c/2\pi R_c N_{Re}$)
- u axial velocity in straight tube; radial velocity component in coiled tubes
- $\bar{u}$ average axial velocity
- u_0 axial velocity at the tube centre
- U dimensionless axial velocity
- U' dimensionless radial velocity component (au/ν)
- v tangential velocity component in coiled tubes
- v_s velocity along the stream line
- V system volume
- V' dimensionless tangential velocity component (av/ν)
- w axial velocity component in coiled tubes
- $\bar{W}$ average axial velocity in coiled tubes
- W_0 axial velocity at the center in coiled tube
- W_{av} average axial velocity in coiled tubes
- x axial distance, x coordinate
- X dimensionless axial distance
- y y coordinate
- Z dimensionless radial distance (r/R)
- $Z, \tilde{Z}$ dimensionless radial distances of equal velocity in case of annulus.

Greek Symbols

α	D_m/kR^2
α'	$1/\alpha$
$\bar{\alpha}$	aspect ratio
β	$k\bar{t}$
$\bar{\beta}$	reddi of zero momentum flux
β'	coefficient of thermal expansion
δ'	direction delta function
δ_i	angle of ith bend
θ	dimensionless time $(t/\bar{t})$
τ	dimensionless characteristic time $(\bar{t}D_m/a^2)$
τ	tD_m/a^2
τ_{rn}	shear stress
γ	variance
ξ	R/R'
ν	kinematic viscosity
λ	curvature ratio (R/R_c)
ψ	streamline
ϕ	tangential coordinate (Figure 13)
ϵ_D	eddy diffusivity
ρ	density

REFERENCES

1. Levenspiel, O., *Chemical Reaction Engineering*, Second Edition, Wiley Eastern Private Ltd., New Delhi, (1974).
2. Ananthkrishnan, V., Gill, W. N., and Barduhn, A. J., *A.I.Ch.E. J.*, vol. 11, pp. 1063 (1965).
3. Reejhsinghani, N. S., Gill, W. N., and Barduhn, A. J., *A.I.Ch.E. J.*, vol. 12, pp. 916, (1966).
4. Ibid, vol. 14, pp. 100 (1968).
5. Flower, F. C., and Brown, G. G., *Trans.* A.I.Ch.E, vol. 39, pp. 491 (1943).
6. Asbjørnsen, O. A., *Chem. Eng. Sci.*, vol. 14, pp. 211 (1961).
7. McHenry, K. W., and Wilhelm, R. H., *A.I.Ch.E. J.*, vol. 3, pp. 83 (1957).
8. van Andel, E., Kramers, H., and Devooged, A., *Chem. Eng. Sci.*, vol. 19, pp. 77 (1964).
9. Lapdius, L., *Ind. Eng. Chem.*, vol. 49, pp. 1000 (1957).
10. Carberry, J. J., and Bretton, R. H., *A.I.Ch.E. J.*, vol. 4, pp. 367 (1958).
11. Harrison, D., Lane, M., and Walne, D., *Trans. Inst. Chem. Eng.*, vol. 40, pp. 214 (1962).
12. Ebach, E. A., and White, R. R., *A.I.Ch.E. J.*, vol. 4, pp. 161 (1958).
13. Liles, A. W., and Geankoplis, C. J., *A.I.Ch.E. J.*, vol. 6, pp. 591 (1960).
14. Strang, D. A., and Geankoplis, C. J., *Ind. Eng. Chem.*, vol. 50, pp. 1305 (1958).
15. Gilliland, E. R., Mason, E. A., and Oliver, R. C., *Ind. Eng. Chem.*, vol. 45, pp. 1177 (1953).
16. Gilliland, E. R., and Mason, E. A., *Ind. Eng. Chem.*, vol. 44, pp. 218 (1952).
17. Davidson, J. F., Farguharson, D. C., Picken, J. Q., and Taylor, D. C., *Chem. Eng. Sci.*, vol. 4, pp. 201 (1955).
18. Overcashier, R. H., Todd, D. B., and Olney, R. B., *A.I.Ch.E. J.*, vol. 5, pp. 54 (1959).
19. Keys, J. J., *A.I.Ch.E. J.*, vol. 1, pp. 305 (1955).
20. Hull, D. E., and Kent, J. W., *Ind. Eng. Chem.*, vol. 44, pp. 2745 (1952).
21. Bartok, W., Heath, C. E., and Weises, M. A., *A.I.Ch.E. J.*, vol. 6, pp. 625 (1960).
22. Turner, G. A., *Chem. Eng. Sci.*, vol. 10, pp. 14 (1959).
23. Ferrel, R. J., and Himmelblau, D. M., *A.I.Ch.E. J.*, vol. 13, pp. 702 (1967).
24. Levenspiel, O., and Turner, J. C. R., *Chem. Eng. Sci.*, vol. 25, pp. 1605 (1970).
25. Nigam, K. D. P., and Vasudeva, K., *Cand. J. Chem. Eng.*, vol. 54, pp. 203 (1976).
26. Wen, C. Y., and Fan, L. T., *Models for Flow Systems and Chemical Reactors*, Marcel Dekker, Inc. New York (1975).
27. Skelland, A. H., *Non-Newtonian Flow and Heat Transfer*, John Wiley Publication, New York, (1967).
28. Bird, R. B., Dai, G. C., and Yarusse, B. J., *Reviews Chem. Eng.*, vol. 1, pp. 1 (1982).
29. Zwietering, Th. N., *Chem. Eng. Sci.*, vol. 11, pp. 1 (1959).
30. Danckwerts, P. V., *Chem. Eng. Sci.*, vol. 2, pp. 1 (1953).
31. Bosworth, R. C. L., *Phil. Mag. Ser.*, vol. 39, pp. 7 (1948).
32. Griffith, A., *Proc. Phys. Soc.*, London, vol. 23, pp. 190 (1911).
33. Taylor, G. I., *Proc. Roy. Soc.*, vol. A219, pp. 186, (1953).

34. Ibid, vol. A223, pp. 446 (1954).
35. Aris, R., *Proc. Roy. Soc.*, vol. A235, pp. 67 (1956).
36. Batley, H. R., and Gogarty, W. B., *Proc. Roy. Soc.*, vol. A228, pp. 473 (1962).
37. Smith, J. M., *Chemical Engineering Kinetics*, 2nd edition, McGraw-Hill Publication, New York, (1971).
38. Brenner, H., *Chem. Eng. Sci.*, vol. 17, pp. 229 (1962).
39. Drew, T. B., Hoopes, J. W., and Vermeulen, T., *Advances in Chemical Enginnering*, Academic Press, New York, vol. 4, pp. 118 (1963).
40. Gill, W. N., and Ananthkrishnan, V., *A.I.Ch.E. J.*, vol. 12, pp. 906 (1966).
41. Gill, W. N., and Sankarasubramanian, R., *Proc. Roy. Soc.*, vol. A316, pp. 341 (1970).
42. Ibid, vol. A322, pp. 100 (1971).
43. Ibid, vol. A327, pp. 191 (1972).
44. Lighthill, M. J., *J. Inst. Applics*, vol. 2, pp. 97 (1966).
45. Yu, J. S., Trans. ASME, *J. Appl. Mech.*, vol. 43, pp. 537 (1976).
46. Ibid, vol. 46, pp. 750 (1979).
47. Gill, W. N., *Proc. Roy. Soc.* (London), vol. 295, pp. 335 (1966).
48. Gill, W. N., and Ananthkrishnan, V, *A.I.Ch.E. J.*, vol. 13, pp. 801 (1967).
49. Hunt, B., *Int. Jl. Heat Mass Trans.*, vol. 20, pp. 393 (1977).
50. Wicke, E., *Chem. Ingr., Tech.*, vol. 47, pp. 547 (1975).
51. Froment, G. F., and Bishoff, K. B., *Chemical Reactor Analysis and Design*, Wiley-New York, (1979).
52. Sunderesan, S., Amundson, N. R., and Aris, R., *A.I.Ch.E. J.*, vol. 26, pp. 529 (1980).
53. Wissler, E. H., *Chem. Eng. Sci.*, vol. 24, 527 (1969).
54. Gill, W. N., *Chem. Eng. Sci.*, vol. 30, pp. 1123 (1975).
55. Levenspiel, O., and Fitzgerald, T. J., *Chem. Eng. Sci.*, vol. 38, pp. 489 (1983).
56. Deckwer, W. D., Nauyen-Tien, K., Kelkar, B. G., and Shah, Y. T., *A.I.Ch.E. J.*, vol. 29, pp. 915 (1983).
57. Flint, L. F., and Eisenklam, P., *Cand. J. of Chem. Eng.*, vol. 47, pp. 101 (1969).
58. Wang, J. C., and Stewart, W. E., *A.I.Ch.E. J.*, vol. 29, pp. 493 (1983).
59. Erdogan, M. E., and Chatwin, P. C., *J. Fluid Mech.*, vol. 29, pp. 465 (1967).
60. Barton, N. G., *J. Fluid Mech.*, vol. 74, pp. 81, (1976).
61. Mazumdar, B. S., *Int. J. Eng. Sci.*, vol. 99, pp. 771 (1981).
62. Novasad, Z., and Ulbrecht, J., *Chem. Eng. Sci.*, vol. 21, pp. 405 (1966).
63. Fan L. T., and Hwang, W. S., *Proc. Roy. Soc.* (London), vol. A282, pp. 576 (1965).
64. Booras, G. S., and Krantz, W. B., *I. & EC (Fund.)*, vol. 15, pp. 249 (1976).
65. Fan, L. T., and Wang, C. B., *Proc. Roy. Soc.*, vol. 292, pp. 203 (1966).
66. Ghoshal, S., *Chem. Eng. Sci.*, vol. 26, pp. 185 (1971).
67. Cox, K. E., and Shah, S. N., *Chem. Eng. Sci.*, vol. 29, pp. 1051 (1974).
68. Sawinsky, J., and Simandi, B., *Trans. Inst. Chem. Eng.*, vol. 60, pp. 188 (1982).
69. Fran, K. T., Osborne, *Chem. Eng. Sci.*, vol. 30, pp. 151 (1975).
70. Chen, H. R., and Blanchard, L. P., *Cand. J. Chem. Eng.*, vol. 53, pp. 476 (1975).
71. Anderson, B., and Berglin, T., *Proc. Roy. Soc.*, vol. A377, pp. 251 (1981).
72. Nishijima, Y, and Oster, G., *J. Appl. Poly. Sci.*, vol. 19, pp. 337 (1956).
73. Clough, S. B., Read, H. E., Metzner, A. B., and Behn, V. C., *A.I.Ch.E. J.* vol. 8, pp. 346 (1962).
74. Astarita, G., *I & EC (Fund.)*, vol. 5, pp. 14 (1966).
75. Honsford, G. S., and Litt, M., *Chem. Eng. Sci.* vol. 23, pp. 869 (1968).
76. Astarita, G., and Mashelkar, R. A., *The Chem. Engr.*, pp. 101 (Feb., 1977).
77. Shah, S. N., and Cox, K. E., *Chem. Eng. Sci.*, vol. 31, pp. 241 (1976).
78. Deo, P. V., and Vasudeva, K., *Chem. Eng. Sci.*, vol. 32, pp. 328 (1977).
79. Singh, D., and Nigam, K. D. P., *J. Appl. Poly. Sci.*, vol. 23, pp. 3021, (1979).
80. Baldauf, W. and Knapp, H., *Chem. Eng. Sci.*, vol. 38, pp. 1031 (1983).
81. Hancell, V., and Vacek, V., *I & EC (Fund.)*, vol. 22, pp. 269 (1983).
82. Trivedi, R. N., and Vasudeva, K., *Chem. Eng. Sci.*, vol. 30, pp. 317 (1975).
83. Ouana, A. C., *I & EC (Fund.)*, vol. 11, pp. 268 (1972).
84. Pratt, K. C., Slater, D. H., and Wakeham, W. A., *Chem. Eng. Sci.*, vol. 28, pp. 1901 (1973).

85. Pratt, K. C., and Wakeham, W. A., *Proc. Roy. Soc.*, vol. A338, pp. 393 (1974)
86. Bohemen, J., and Purnell, J. H., *J. Chem. Soc.*, vol. 1, pp. 360 (1961).
87. Dutta, A., and Mashelkar, R. A., *J. Appl. Poly. Sci.*, vol. 27, pp. 2739, (1982).
88. Nauman, E. B., and Bufftham, B. A., *Mixing in Continuous Flow Systems.*, John Wiley and Sons, New York, (1983).
89. Saxena, A. K., and Nigam, K. D. P., *Cand. J. Chem. Eng.*, vol. 61, pp. 53 (1983).
90. Zeevalkink, J. A., and Brunsmann, J. J., *Water Res.*, vol. 17, pp. 365, (1983).
91. Mazumdar, B. S., *Int. J. Eng. Sci.*, vol. 32, pp. 211 (1979).
92. Gupta, A. S., and Chatterjee, A. S., *Proc. Comb. Phil. Soc.*, vol. 64, pp. 1209 (1968).
93. Holmes, D. B., and Vermeulen, J. R., *Chem. Eng. Sci.*, vol. 22, pp. 717 (1968).
94. Nauman, E. B., *Chem. Eng. Science*, vol. 32, pp. 287 (1977).
95. Berker, R., "Integration des equations du movement dun fluid visqueux incompressib," *Handbuch der Physick*", vol. 8, pp. 71 (1963).
96. Nauman, E. B., *Chem. Eng. Sci.*, vol. 8, pp. 53 (1981).
97. Nigam, K. D. P., and Vasudeva, K., *I & EC* (*Pro. Des. Dev.*), vol. 15, pp. 473 (1976).
98. Lin, S. H., *Chem. Eng. Sci.*, vol. 35, pp. 617 (1976).
99. Pechoc, V., *Chem. Eng. Sci.*, vol. 38, pp. 1341 (1983).
100. Mishra, P., and Mishra, I., *A.I.Ch.E. J.*, vol. 22, pp. 617 (1976).
101. Fredrickson, A. G., and Bird, R. B., *Ind. Eng. Chem.*, vol. 50, pp. 357 (1958).
102. Lehtola S., and Kuoppamaki, R., *Chem. Eng. Sci.*, vol. 37, pp. 185 (1982).
103. Ottino, J. M., *A.I.Ch.E. J.*, vol. 29, pp. 159 (1983).
104. Nigam, K. D. P., and Vasudeva, K., *Cand. J. Chem. Eng.*, vol. 58, pp. 543 (1980).
105. Nauman, E. B., *Cand. J. Chem. Eng.*, vol. 60, pp. 136 (1982).
106. Nigam, K. D. P., and Nauman, E. B., *Cand. J. Chem. Engrs.* vol. 63, pp. 519 (1985).
107. Smith, R., *J. Fluid. Mech.*, vol. 130, p. 299 (1983).
108. Weaver, D. W., and Ultman, J. S., *A.I.Ch.E. J.*, vol. 26, pp. 9 (1980).
109. Jarzebski, A. B., and Thullie, J., *Chem. Eng. J.*, vol. 22, pp. 243 (1981).
110. Jerzebski, A. B., and Wilkinson, W. L., *J. Non-Newtonian Fluid Mech.*, vol. 8, pp. 239 (1981).
111. Markatos, N. C. G., Sala, R., and Spadling, O. B., *Trans. Inst. Chem. Eng.* vol. 56, pp. 28 (1978).
112. Hooper, A., Dutty, B. R., and Moffatt, H. K., *J. Fluid Mech.*, vol. 117, pp. 283 (1982).
113. Deiber, J. A., and Schowalter, W. R., *A.I.Ch.E. J.* vol. 25, pp. 138 (1979).
114. Thomson, J., *Proc. Roy. Soc.*, London Ser., vol. A25, pp. 5 (1876).
115. Thomson, J., *Proc. Roy. Soc.*, London Ser., vol. A26, pp. 356 (1877).
116. Grindley, J. H., and Gibson, A. H., *Proc. Roy. Soc.*, vol. A80, pp. 114 (1908).
117. Eustice, J., *Proc. Roy. Soc.*, London Ser., A84, pp. 107 (1910).
118. Ibid, vol. A85, pp. 119 (1911).
119. Williams, G. S., Hubell, C. W., and Finkell, G. H., *Trans. ASCE*, vol. 47, pp. 7 (1902).
120. Dean, W. R., *Phil. Mag.*, vol. 28, pp. 208 (1927).
121. Ibid, vol. 30, pp. 673 (1928).
122. Topakoglu, *J. Math. Mech.*, vol. 16, pp. 1321 (1967).
123. Larrain, J., and Brilla, C. F., *Trans. Soc. Rheol.*, vol. 14, pp. 135 (1970).
124. Van Dyke, M., *J. Fluid Mech.*, vol. 86, pp. 129 (1978).
125. McConalogue, D. J., and Srivastava, R. S., *Proc. Roy. Soc.*, London Ser., vol. A307, pp. 37 (1968).
126. Truesdell, L. C., Jr., and Adler, R. J., *A.I.Ch.E. J.*, vol. 16, pp. 1010 (1970).
127. Austin, L. R., and Seader, J. D., *A.I.Ch.E. J.*, vol. 19, pp. 85 (1973).
128. Tarbell, J. M., and Samuels, M. R., *Chem. Eng. J.*, vol. 5, pp. 117 (1973).
129. Smith, F. T., *J. Fluid Mech.*, vol. 71, pp. 15 (1975).
130. Smith, F. T., *Proc. Roy. Soc.*, London Ser., vol. A347, pp. 345 (1976).
131. Ibid, vol. A351, pp. 71 (1976).
132. Akiyama, M., and Cheng, K. C., *Int. J. Heat Mass Trans.*, vol. 14, pp. 1659 (1971).
133. Greenspan, A. D., *J. Fluid Mech.*, vol. 57, pp. 167 (1973).
134. Patankar, S. V., Pratap, V. S., and Spalding, D. B., *J. Fluid Mech.*, vol 62, pp. 539 (1974).
135. Zapryanov, Z., and Christov, Ch., *Theo. Appl. Mech.*, vol. 8, pp. 11 (1977).

136. Collins, W. M., and Dennis, S. C. R., *Q. J. Mech. Appl. Math.*, vol. 28, pp. 133 (1975).
137. Dennis, S. C. R., and Ng., M., *Q. J. Mech. Appl. Math.*, vol. 35, pp. 305 (1982).
138. Dennis, S. C. R., *J. Fluid Mech.*, vol. 99, pp. 449 (1980).
139. Berger, S. A., Talbot, L., and Yao, L. S., Ann. Rev., *Fluid Mech.*, vol. 15, 461 (1983).
140. Manlopaz, R. L., and Churchill, S. W., *Chem. Eng. Commun.*, vol. 7, p. 57 (1980).
141. Wang, C. Y., *J. Fluid Mech.*, vol. 108, pp. 185 (1981).
142. Germano, M., *J. Fluid Mech.*, vol. 125, pp. 1 (1982).
143. Cuming, H. G., *Aeronaut. Res. Conc. Rep.*, Mem. No. 2880 (1952).
144. Joseph, B., Smith, E. P., and Adler, R. J., *A.I.Ch.E. J.*, vol. 21, pp. 965 (1975).
145. Joseph, B., and Adler, R. J., *A.I.Ch.E. J.*, vol. 21, pp. 974 (1975).
146. Humphrey, J. A. C., Taylor, A. M. K., and Whitlaw, J. H., *J. Fluid Mech.*, vol. 83, pp. 509 (1977).
147. Humphrey, J. A. C., Whitlaw, H. H., and Yee, G., *J. Fluid Mech.*, vol. 103, pp. 443 (1981).
148. Yee, G., Chilukuri, R., and Humphrey, J. A. C., *J. Heat. Trans.*, vol. 102, pp. 285 (1980).
149. De Vriend, J. H., *J. Fluid Mech.*, vol. 107, pp. 423 (1981).
150. Abdallah, S., and Hamed, A., *A.I.A.A. J.*, vol. 19, pp. 993 (1981).
151. Wisman, J., *Elements of Nuclear Reactor Design*, pp. 192, Elseiver Scientific Publishing Co., Amesterdam, (1977).
152. Mashelkar, R. A., and Devarajan, G. V., *Trans. Inst. Chem. Eng.*, vol. 54, pp. 100 (1976).
153. Ibid, vol. 54, pp. 108 (1976).
154. Rajasekaren, S., Kubair, V. G., and Kuloor, N. R., *Ind. J. Techn.*, vol. 8, pp. 391 (1970).
155. Mishra, P., and Gupta, S. N., *I & EC* (*Proc. Des. Dev.*), vol. 18, pp. 130 (1979).
156. Mujawar, B. A., and Raja Rao, M., *I & EC* (*Proc. Des. Dev.*), vol. 17, pp. 22 (1978).
157. Clegg, D. B., and Power, G., *Appl. Sci. Res.*, Section A, vol. 12, pp. 119 (1962).
158. Rathna, S. L., *Proc.* of Conf. on Fluid Mech., Indian Institute of Science, Bangalore, pp. 378 (1967).
159. McConalogue, D. J., *Proc. Roy. Soc.*, vol. A315, pp. 99 (1970).
160. Ruthven, D. M., *Chem. Eng. Sci.*, vol. 26, pp. 1113 (1971).
161. Trivedi, R. N., and Vasudeva, K., *Chem. Eng. Sci.*, vol. 29, 2291 (1974).
162. Nauman, E. B., *Chem. Eng. Sci.*, vol. 32, pp. 287 (1977).
163. Ruthven, D. M., *Chem. Eng. Sci.*, vol. 83, pp. 629 (1978).
164. Saxena, A. K., and Nigam, K. D. P., *Chem. Eng. Sci.*, vol. 34, pp. 425 (1979).
165. Ranade, V. R., and Ulbrecht, J. J., *Chem. Eng. Commun.*, vol. 8, pp. 165 (1981).
166. Saxena, A. K., Nigam, K. M., and Nigam, K. D. P., *Cand. J. Chem. Eng.*, vol. 61, pp. 50 (1983).
167. Saxena, A. K., Ph.D. Thesis, Indian Institute of Technology, Delhi (1982).
168. Saxena, A. K., and Nigam, K. D. P., *Chem. Eng. Commun.*, vol. 23, pp. 277 (1983).
169. Chang, H. K., and Mackros, L. F., *A.I.Ch.E. J.*, vol. 17, pp. 541 (1971).
170. Gilroy, K., Brighton, E., and Gaylor, J. D. S., *A.I.Ch.E. J.*, vol. 17, pp. 107 (1977).
171. Patel, I. C., and Sirs, J. A., *Med. and Biol. Eng. and Comput.*, vol. 21, pp. 113 (1983).
172. Nunge, R. J., Lin, T. S., and Gill, W. N., *J. Fluid Mech.*, vol. 51, pp. 363 (1972).
173. Janseen, L. A. M., *Chem. Eng. Sci.*, vol. 31, pp. 215 (1976).
174. Koutsky, J. A., and Adler, R. J., *Cand. J. Chem. Eng.*, vol. 42, pp. 239 (1964).
175. van Andel, E., Kramers, H., and DeVooged, A., *Chem. Eng. Sci.*, vol. 19, pp. 77 (1964).
176. Nigam, K. D. P., and Vasudeva, K., *Chem. Eng. Sci.*, vol. 31, pp. 835 (1976).
177. Moulijn, J. A., Spijker, R., and Kolk, J. F. M., *J. Chromatog*, vol. 142, pp. 155 (1977).
178. Singh, P. C., and Singh, S., *Cand. J. Chem. Eng.*, vol. 61, pp. 254 (1983).
179. Shetty, V. D., and Vasudeva, K., *Chem. Eng. Sci.*, vol. 32, pp. 782 (1977).
180. Iyer, R. N., and Vasudeva, K., *Chem. Eng. Sci.*, vol. 36, pp. 1103 (1981).
181. Patel, R. D., McFeeley, J. J., and Jolls, K. R., *A.I.Ch.E. J.*, vol. 21, pp. 259 (1975).
182. Gupta, S. K., Patel, R. D., and Ackeberg, R. C., *Chem. Eng. Sci.*, vol. 37, pp. 1727 (1982).
183. Wood, B. N., Seed, S. W., and Neven, M. R., *J. Fluid Mech.*, vol. 52, pp. 137 (1972).
184. Singh, D., and Nigam, K. D. P., *J. App. Poly. Sci.*, vol. 26, pp. 784 (1981).
185. Saxena, A. K., and Nigam, K. D. P., *J. Appl. Poly. Sci.*, vol. 26, pp. 3475 (1981).
186. Mujawar, B. A., and Raja Rao, M., *I & EC* (*Proc. Des. Dev*), vol. 17, p. 22 (1978).
187. Sakra, T., Lesek, F., and Carmenkova, H., *Collect. Czeck. Chem. Commun.*, vol. 36, pp. 3543 (1971).

188. Nauman, E. B., *A.I.Ch.E. J.*, vol. 25, pp. 246 (1979).
189. Saxena, A. K., and Nigam, K. D. P., *A.I.Ch.E. J.*, vol. 30, pp. 363 (1984).
190. Arai, K., Saita, S., and Maeda, S., Kagaku Kogaku (*Chem. Eng.* (Japan)) vol. 31, pp. 25 (1967).
191. Blackwell, R. J., Paper presented at Local section of A.I.Ch.E. meeting Galveston, Texas, October, (1957).
192. Hays, J. R., Thesis, M. S., Vanderbilt Univ. (1964).
193. Sittel, C. N., Jr., Threadgill, K. B., and Schnelle, K. B., Jr., *Ind. Eng. Chem.*, vol. 7, pp. 39 (1968).
194. Smith, S. S., and Schultz, R. K., *Petrol Engr.*, vol. 19, pp. 94 (1948).
195. Allen, C. M., and Taylor, E. A., *Trans. Am. Soc. Mech. Engr.*, vol. 45, pp. 285 (1923).
196. Argo, W. B., and Cora, D. R., *I & EC* (*Proc. Des. Dev.*), vol. 4, pp. 352 (1965).
197. Davidson, J. F., Farguharson, D. C., Picken, J. O., and Taylor, D. C., *Chem. Eng. Sci.*, vol. 4, pp. 201 (1955).
198. van den Berg, J. H. M., and Deelder, R. H., *Chem. Eng. Sci.*, vol. 34, pp. 1345 (1979).
199. Tamir, A., and Taitel, Y., *Chem. Eng. Sci.*, vol. 28, pp. 1921 (1973).
200. Lauwerier, H. A., *Appl. Sci. Res.*, vol. A8, pp. 366 (1959).
201. Katz, S., *Chem. Eng. Sci.*, vol. 10, pp. 202 (1959).
202. Kaufman, F., "Reactions of Oxygen Atoms in Progress" in *Reaction Kinetics*, Porter, G., Ed., Pergamon Press, New York, vol. 1. (1961).
203. Cleland, F. A., and Wilhelm, R. H., *A.I.Ch.E. J.*, vol. 2, pp. 489, (1956).
204. Dranoff, J. S., *Mathematics of Computation*, vol. 15, pp. 403 (1962).
205. Lupa, A. J., and Dranoff, J. S., *Chem. Eng. Sci.*, vol. 21, pp. 861 (1966).
206. Ogren, P. J., *J. Phys. Chem.*, vol. 79, pp. 1749 (1975).
207. Walker, R. E., *Phys. Fluids*, vol. 4, pp. 1211 (1961).
208. Dang, V. D., and Steinberg, M., *J. Phys. Chem.*, vol. 84, pp. 214 (1980).
209. Nigam, K. M., Srivastava, V. K., and Nigam K. D. P., *Chem. Eng. J.*, vol. 25, pp. 147 (1982).
210. Kantoyovich, L. V., and Krylov, V. I., *Approximation Methods of Higher Analysis*, Noordhoff, Groningen (1958).
211. Finlayson, B. A., *Method of Weighted Residuals and Variational Principles*, Academic Press, New York (1972).
212. Hoyermann, K. H., *Physical Chemistry an Advanced Treatise*, Jost, W. Ed., Academic Press, New York, vol. VIB, pp. 931 (1975).
213. Dang, V. D., *Chem. Eng. Sci.*, vol. 33, pp. 1179 (1978).
214. Dang, V. D., *A.I.Ch.E. J.*, vol. 29, pp. 19 (1983).
215. Apelblat, A., *Chem. Eng. J.*, vol. 23, pp. 193 (1982).
216. Houghton, G., *Cand. J. Chem. Eng.*, vol. 40, pp. 188 (1962).
217. Horn, F. J. M., and Parish, T. D., *Chem. Eng. Sci.*, vol. 22, pp. 1549 (1967).
218. Bishoff, K. B., *A.I.Ch.E. J.*, vol. 14, pp. 820 (1968).
219. Wan, C. G., and Ziegler, E. N., *Chem. Eng. Sci.*, vol. 25, pp. 723 (1970).
220. Woodhead, J. R., and Yesberg, D., *Chemica 70*, Butterworths and Co., Sydney (1970).
221. Kulkarni, S. N., and Vasudeva, K., *Cand. J. Chem. Eng.*, vol. 54, pp. 238 (1976).
222. Nigam, K. M., Nigam, K. D. P., and Srivastava, V. K., *Int. Commun. Heat & Mass Transf.*, vol. 10, pp. 341 (1983).
223. Hulburt, H. M., *Ind. Eng. Chem.*, vol. 36, pp. 1012 (1944).
224. Wehner, J. F., and Wilhelm, R. H., *Chem. Eng. Sci.*, vol. 6, pp. 89 (1956).
225. Fan, L. T., and Ahn, Y. K., *Ind. Eng. Chem. Proc. Des. Develop*, vol. 1, pp. 190 (1962).
226. Raines, G. E., and Corrigan, T. E., *C. E. P. Symp.*, Ser. No. 72, vol. 63, pp. 90 (1967).
227. Standart, G., *Chem. Eng. Sci.*, vol. 23, pp. 645 (1968).
228. Shankar Subramanian, R., Gill, W. N., and Marra, R. A., *Cand. J. Chem. Eng.*, vol. 52, pp. 563 (1974).
229. Rahman, M., *J. de Mecanique*, vol. 14, pp. 339 (1975).
230. Nigam, K. D. P., and Vasudeva, K., *Chem. Eng. Sci.*, vol. 32, pp. 1119 (1977).
231. Homsy, R. V., and Strohman, R. D., *A.I.Ch.E. J.*, vol. 17, pp. 215 (1971).
232. Shankar Subramanian, and Gill, W. N., *Cand. J. Chem. Eng*, vol. 54, pp. 121 (1976).
233. Nigam, K. D. P., *J Non-Newtonian Fluid Mech.* (to be communicated 1985).
234. Mashelkar, R. A., *Cand. J. Chem. Eng.*, vol. 51, pp. 613 (1973).

235. Nigam, K. D. P., and Vasudeva, K., *Chem. Eng. Sci.*, vol. 32, pp. 673 (1977).
236. Nigam, K. M., and Nigam, K. D. P., *Chem. Eng. Sci.*, vol. 35, pp. 2358 (1980).
237. Nigam, K. M., and Nigam, K. D. P., *J. Appl. Poly. Sci.*, vol. 28, pp. 887 (1983).
238. Seader, J. D., and Southwick, L. M., *Chem. Eng. Commun.*, vol. 9, pp. 175 (1981).
239. Mashelkar, R. A., and Venkatasubramanian, C. V., *A.I.Ch.E. J.*, vol. 31, pp. 440 (1985).
240. Raju, K. K., and Rathna, S. L., *J Indian Inst. Sci.*, vol. 52, pp. 34 (1970).
241. Brain, P. L. T., *A.I.Ch.E. J.*, vol. 7, pp. 367 (1961).
242. Peaceman, D. W., and Rachfold, H. H., *J. Soc. Indust. Appl. Math.*, vol. 3, pp. 28 (1955).

CHAPTER 23

FLOW ACROSS TUBE BANKS

Tatsuo Nishimura

Hiroshima University
Japan

CONTENTS

INTRODUCTION

This chapter is devoted to the flow across tube banks with respect to heat exchangers, etc. The previous investigations can be broadly grouped into three categories:

1. Hydraulic resistance of tubes
2. Fluid-elastic vibration of tubes
3. Localized flow characteristics, such as the surface pressure, shear stress, and velocity distributions around a tube in a bank.

The studies of 1 and 2 have been already reported and also several reviews have been presented [1, 2, 3, 4]. Although the study of 3 is important to the understanding of transport phenomena from tubes, this problem has not been considered enough. This chapter provides a review of the localized flow characteristics in tube banks and indicates future work.

The flow pattern around a tube in a bank is influenced by the following parameters: the arrangement and geometrical parameters of the bank and Reynolds number. Banks of staggered and in-line arrangement of tubes are most common as shown in Figure 1, and they are usually defined by the transverse pitch ratio, $P_t = T/D$ and longitudinal pitch ratio, $P_l = L/D$. The Reynolds number based on mean velocity in the minimum cross section between adjacent tubes, Re_{max}, is usually used in the studies of tube banks, since the velocity in the free cross section in a bank increases rapidly with a decrease of transverse pitch ratio. In this chapter, another Reynolds number based on an approaching velocity far upstream from tube banks, Re_a, is also employed to make clear the effect of pitch ratio on the flow characteristics.

Figure 2 shows variations of Nusselt number with Reynolds number for staggered and in-line banks with various pitch ratios [1]. At low Reynolds numbers the Nusselt numbers are proportional to about 1/3 power of the Reynolds number for both staggered and in-line banks. With increasing

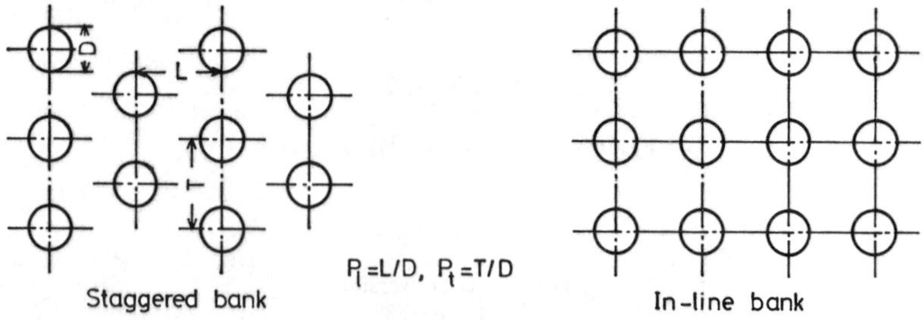

Figure 1. Arrangements of tube banks.

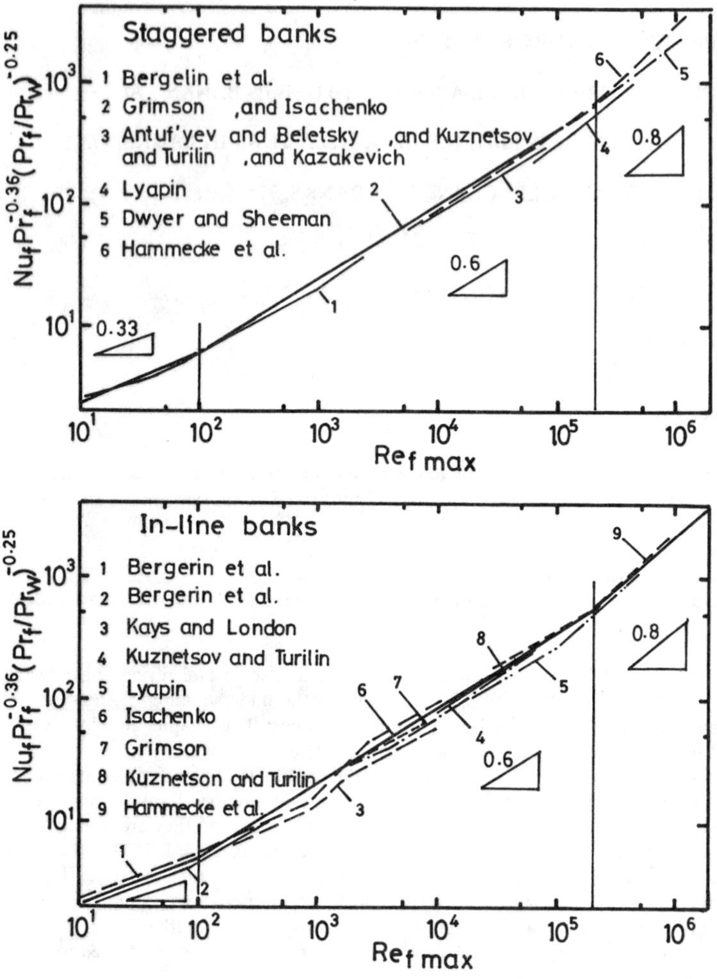

Figure 2. Variation of Nusselt number with Reynolds number for staggered and in-line banks [1].

the Reynolds number, the power undergoes considerable variations. In the range of $100 < Re_{max} < 2 \times 10^5$ the power has about 0.6. At higher Reynolds numbers the Nusselt numbers are proportional to about 0.8. This power variation would correspond to the flow variation. Thus, we distinguish three flow regimes in tube banks with respect to the Reynolds number, Re_{max}, as well as Zukauskas [1]; a laminar flow regime at $Re_{max} < 100$, a mixed flow regime at $10^2 < Re_{max} < 2 \times 10^5$, and a turbulent flow regime at $Re_{max} > 2 \times 10^5$.

We discuss the flow characteristics for these flow regimes and treat the fully developed flow region only. The problem of the developing flow region refers to Nishimura and Aiba's reports [5, 6].

This chapter is divided into several parts, and the following part, Part 1, gives the flow pattern to aid the understanding of the pressure and shear stress profiles around a tube in a bank. Surface pressure and shear stress profiles are described in Parts 2 and 3, respectively. The final part, Part 4, presents the velocity profile and turbulence in a bank.

FLOW PATTERN IN TUBE BANKS

A knowledge of the flow pattern is necessary to understand the surface pressure and shear stress profiles around a tube in a bank. There have been several investigations of the flow visualizations that have been performed by means of the aluminium dust method, Schlieren technique, and hydrogen bubbles method [5, 6, 7]. Besides this experimental approach, an alternative approach is direct numerical analysis of the Navier-Stokes equations using finite difference method or finite element method [9, 10, 11]. The numerical analysis would be limited to the laminar flow regime only, at the present time, because the flow pattern shows an unsteady motion according to the vortex shedding for mixed and turbulent flow regimes.

Laminar Flow Regime

Nishimura et al. [9] reported the numerical analysis by the finite element method. Figure 3 shows streamlines for the staggered banks. As shown in Figures 3a, b, and c, the circulation vortex

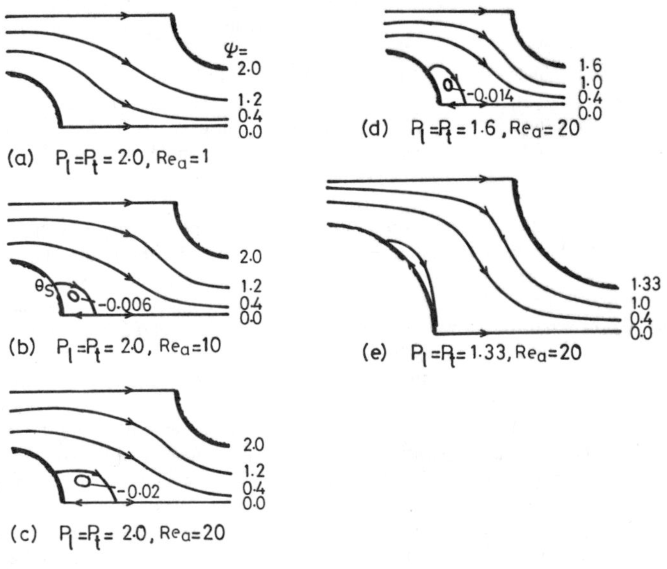

Figure 3. Streamlines for staggered banks in the laminar flow regime [9].

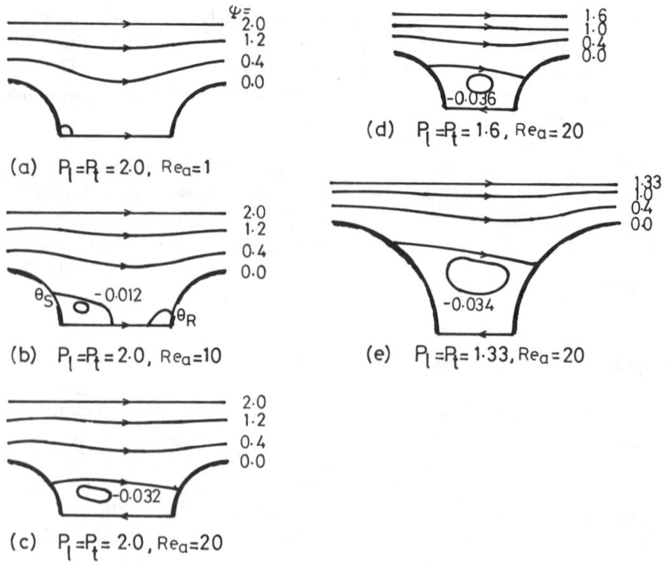

Figure 4. Streamlines for in-line banks in the laminar flow regime [9].

is formed at the rear part of the cylinder and grows on a large scale, with increasing the Reynolds number. The effect of pitch ratio is shown in Figure 3c, d, and e. As the pitch ratio decreases, the streamlines begin to lie along the cylinders and thus the length of the circulation vortex measured along the axis $\theta = 180°$ becomes smaller, but the reverse flow region along the cylinder surface shows no significant differences.

Figure 4 shows streamlines for the in-line banks. As shown in Figure 4a, b, and c, the circulation vortex is formed at the rear part of the cylinder for $Re_a = 1$, and its vortex becomes larger and another one is formed at the front part of the cylinder of the following row for $Re_a = 10$; furthermore, the two vortices grow and merge into one with increasing the Reynolds number. The effect of pitch ratio is shown in Figures 4C, D, and E. The large circulation vortex fills the gap between adjacent cylinders in the flow direction, at any pitch ratio.

Figure 5 shows a variation of separation and reattachment points with Reynolds number for $P_1 = P_t = 2.0$, as an example. For the staggered bank, no reattachment point exists. The zone near the cylinder surface from the front stagnation point, $\theta = 0°$ to the separation point, $\theta = \theta_S$ indicates the forward flow region, and the zone at the downstream side from $\theta = \theta_S$ is the reverse flow region. The onset of the flow separation is about $Re_{max} = 4$ and then the separation point gradually shifts upstream from $\theta = 180°$ with increasing the Reynolds number.

For the in-line bank, the zone near the cylinder surface from the reattachment point, $\theta = \theta_R$ to the separation point, $\theta = \theta_S$ indicates the forward flow region, and two zones at the upstream side from the reattachment point and the downstream side from the separation point are the reverse flow regions. The behavior of the separation point is similar to the case of the staggered bank, but the onset of the flow separation is earlier. The reattachment point shifts downstream from $\theta = 0°$ with increasing Reynolds number. For other pitch ratios, these behaviors are similar to the case of $P_1 = P_t = 2.0$, but the onset of the flow separation significantly differs for each pitch ratio [9].

Thus, it is characteristic for the flow pattern in the laminar flow regime that the regions of the forward flow and the reverse flow along the cylinder surface strongly depend not only on the pitch ratio, but also on the Reynolds number.

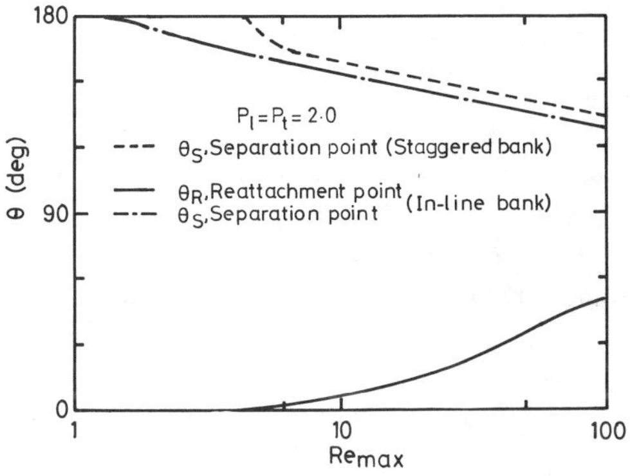

Figure 5. Variation of separation and reattachment points with Reynolds number in the range of laminar flow [18].

Mixed Flow Regime

Figure 6 shows photographs of a cylinder's wake for the staggered and in-line banks with $P_1 = P_t = 2.0$. This flow visualization was performed by the hydrogen bubbles method [7]. There are vortex sheddings for both staggered and in-line banks. Accordingly, when the flow pattern changes from the laminar flow to the mixed one, the circulation vortex formed at the rear part of the cylinder produces periodic oscillations. These vortex shedding phenomena disturb not only the flow around the cylinder of its own, but also the flow around other cylinders. Of course, the degree of flow disturbance depends on the pitch ratio and the Reynolds number.

Figure 7 shows a variation of separation and reattachment points with Reynolds number for $P_1 = P_t = 2.0$, as an example. The separation and reattachment points scarcely depend on the Reynolds number for both staggered and in-line banks, in contrast to the case of the laminar flow regime. For other pitch ratios ($P_1 = P_t = 1.6$, 1.33 and 1.2), the positions of the separation and reattachment points are not so much different from those for $P_1 = P_t = 2.0$ [5].

Ishigai et al. [8] found out various wake patterns in tube banks by the Schieren technique. Figure 8 shows typical wake patterns for the in-line and staggered banks. For the in-line banks, Pattern A indicates the case of vortex shedding as shown in Figure 5. This vortex shedding behavior is influenced by both P_1 and P_t. Pattern A changes into Pattern C when P_1 decreases, and its pattern is the case of no vortex shedding. The transition point from Pattern A to Pattern C is about $P_1 = 1.5$. Pattern B shows that an accelerated flow like a two-dimensional jet passing through between adjacent cylinders perpendicular to the flow direction deflects to one side in each row and also oscillates vertically according to the vortex shedding.

For the staggered banks, Pattern A indicates the case of vortex shedding as shown in Figure 5. This vortex shedding behavior is influenced by P_t only. Pattern B occurs when P_1 is larger than P_t, and it is similar to the case of Pattern B for the in-line banks. Pattern C takes place when P_1 is smaller than P_t, and it is similar to the case of Pattern A for the in-line banks. When P_1 decreases further, Pattern C resembles Pattern C for the in-line banks.

Schematic diagram of wake patterns as previously mentioned is shown in Figure 9. The boundary line between two different patterns is not always fixed but would be dependent on the Reynolds number, etc. Symbols A, B, and C in this diagram correspond to the patterns shown in Figure 8, respectively. Symbol D indicates the area where the flow deflection occurs and its direction does not oscillate according to the vortex shedding. Symbol E is the area where the wake pattern is similar to that of a single cylinder. Symbol C′ for the staggered banks shows the area where the wake pattern resembles Pattern C for the in-line banks as described in the explanation of Figure 8.

The flow deflection in a bank as previously mentioned induces tube vibrations and a significant non-uniformity in heat transfer from tubes and, thus, may damage the heat exchangers. However, this problem is not considered serious enough at the present time. Further investigations of its mechanism and control are needed.

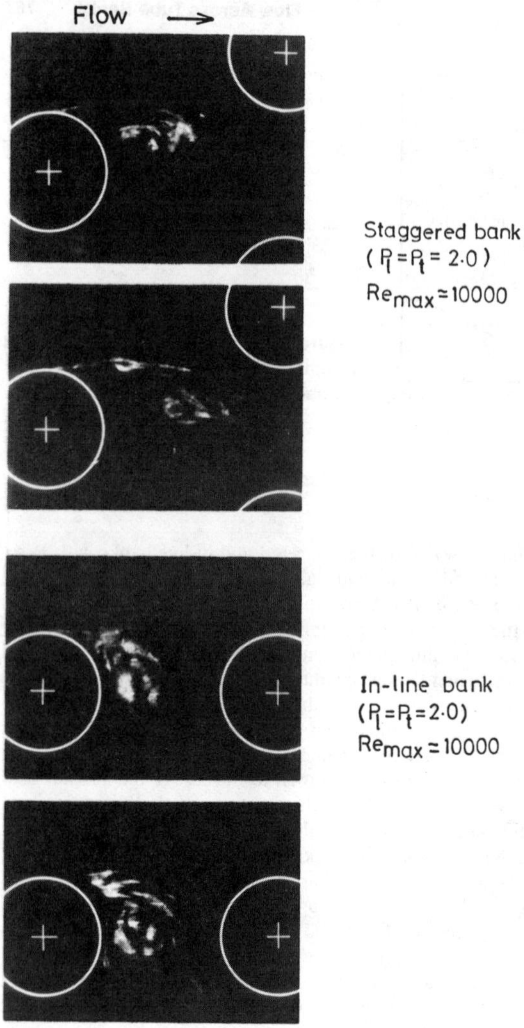

Figure 6. Typical flow patterns for staggered and in-line banks in the mixed flow regime [7].

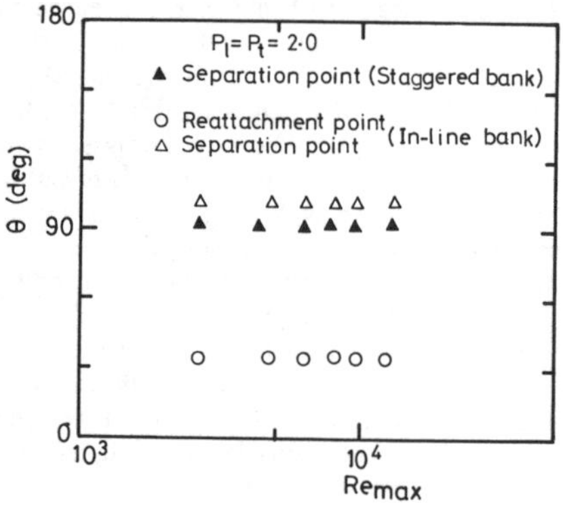

Figure 7. Variation of separation and reattachment points with Reynolds number in the range of mixed flow [18].

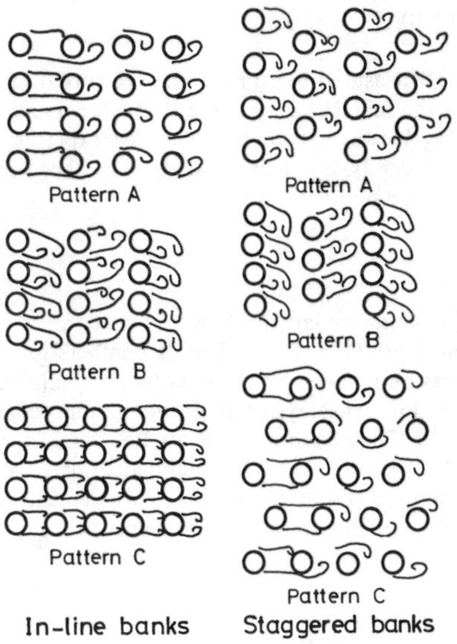

Figure 8. Various wake patterns for staggered and in-line banks in the mixed flow regime [8].

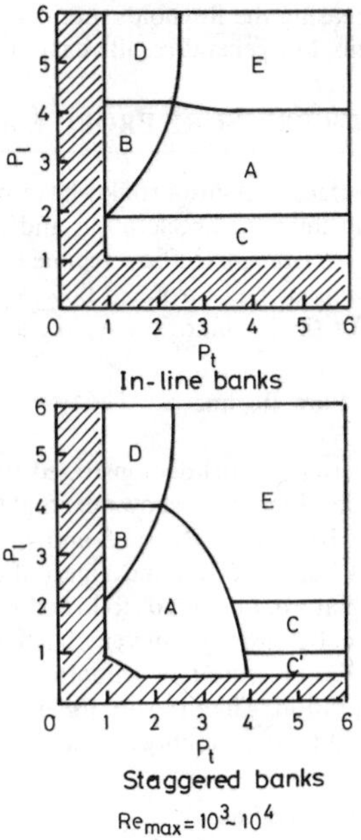

Figure 9. Classes of wake pattern for staggered and in-line banks in the mixed flow regime [8].

Turbulent Flow Regime

There have been a few investigations in the turbulent flow regime. Zukauskas et al. [12, 13, 14] have given useful information. Figure 10 shows a variation of separation and transition points with Reynolds number for the staggered bank with $P_l = 1.4$ and $P_t = 1.5$. The separation point shifts far downstream ($\theta_S \simeq 150°$), as compared with the case of the mixed flow regime ($\theta_S \simeq 95°$). The turbulent boundary layer first appears just before the separation point and its area expands

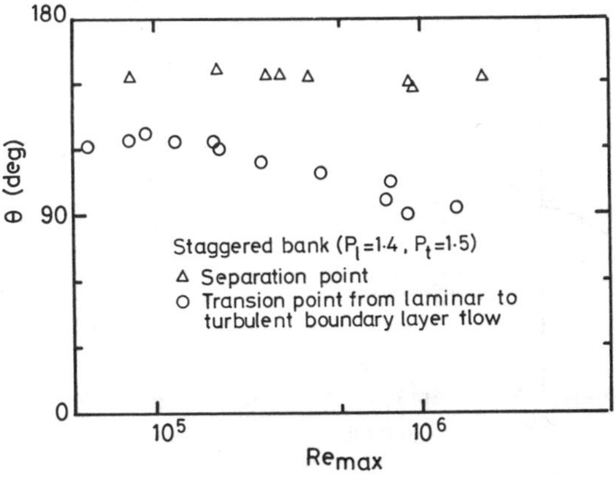

Figure 10. Variation of separation and transition points with Reynolds number in the range of turbulent flow [12].

with increasing the Reynolds number. These behaviors are influenced by the arrangement and the pitch ratio, but general results have not yet been presented.

SURFACE PRESSURE PROFILE AROUND A TUBE IN BANKS

The surface pressure profiles have been numerously measured to estimate the pressure drag of tubes and infer the acceleration and deceleration of the mainstream passing through a bank, in particular for the mixed flow regime [15, 16, 17, 18, 19]. However, there have been few experiments for the laminar and turbulent flow regimes [20, 21]. Recently, a numerical analysis has been performed for the laminar flow regime and has given the pressure profiles [10].

Laminar Flow Regime

Launder et al. [10] demonstrated the numerical prediction of laminar flow and heat transfer in tube banks. Figure 11 shows examples of the pressure profile at two Reynolds numbers for the staggered bank with $P_l = 1.5$ and $P_t = 3.0$. The pressure is given by the non-dimensional form, $C_{p\,max}$. At $Re_{max} = 50$ a typical inertial-dominated variation is displayed with the minimum pressure occurring at about $\theta = 90°$ followed by a slow recovery. The pressure profile for $Re_{max} = 3$ is, not unexpectedly, entirely different from the case of $Re_{max} = 50$. Due to the dominance of viscous stresses, the pressure falls smoothly from the front part to the rear one of the cylinder with no flow separation. The pressure difference between the maximum and minimum values of $C_{p\,max}$ at $Re_{max} = 50$ is considerably smaller than that at $Re_{max} = 3$.

Mixed Flow Regime

Figure 12 shows the pressure profile for the staggered bank with $P_l = P_t = 2.0$. The profile is qualitatively similar to that at inertia-dominated Reynolds numbers in the laminar flow regime as

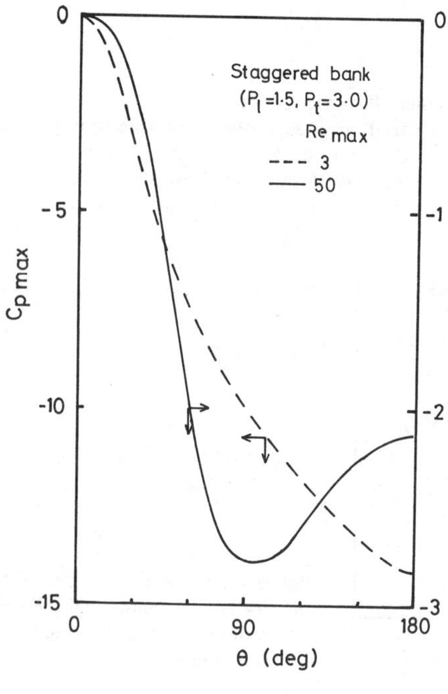

Figure 11. Surface pressure profile of cylinder in staggered bank in the laminar flow regime [10].

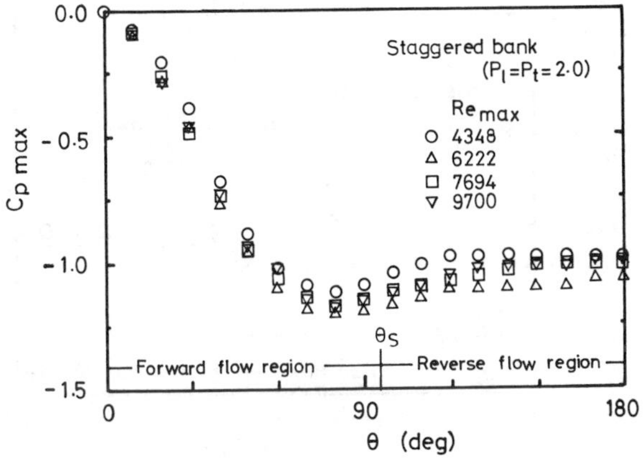

Figure 12. Effect of Reynolds number on surface pressure profile for staggered bank in the mixed flow regime [18].

shown in Figure 11. The pressure fall and recovery zones correspond to the forward and reverse flow regions as described earlier. The values of $C_{p\,max}$ are almost constant with respect to the Reynolds number and thus the influence of the Reynolds number is not significant, in contrast to the case of the laminar flow regime.

The effect of pitch ratio is shown in Figure 13. In this figure the pressure is given by C_p, instead of $C_{p\,max}$. The profile for the large pitch ratio resembles that of a single cylinder shown by a dotted

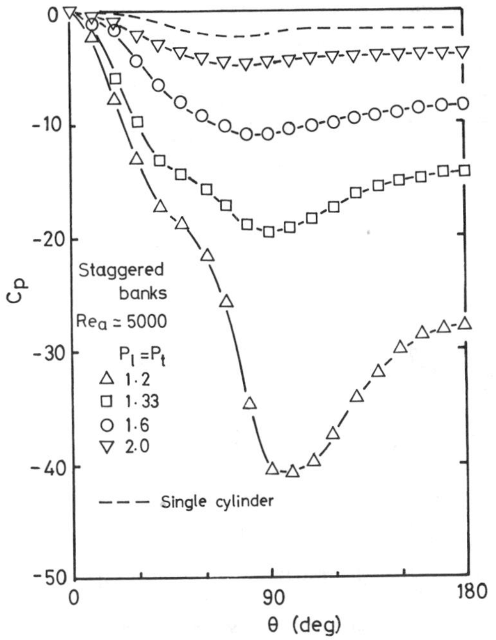

Figure 13. Effect of pitch ratio on surface pressure profile for staggered bank in the mixed flow regime [18].

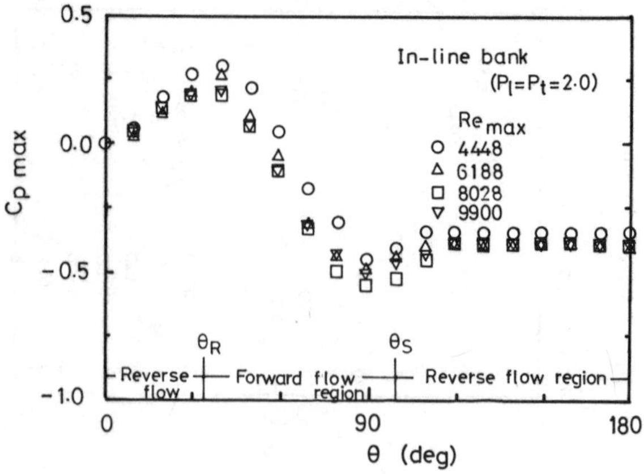

Figure 14. Effect of Reynolds number on surface pressure profile for in-line bank in the mixed flow regime [18].

line. However, as the pitch ratio decreases the pressure fall becomes remarkably, and two zones of the pressure fall exist at the front part of the cylinder ($\theta = 0° \sim 30°$ and $50° \sim 90°$) for $P_1 = P_t =$ 1.33 and 1.2. The former is due to the impingement of the accelerated flow like a two-dimensional jet formed at the minimum cross section between adjacent cylinders of the preceding row, and the latter is due to the accelerating flow passing through the convergence cross section between adjacent cylinders of the aimed row.

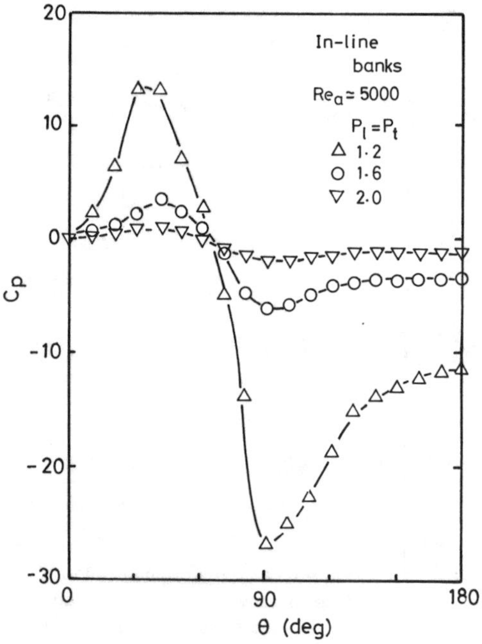

Figure 15. Effect of pitch ratio on surface pressure profile for in-line bank in the mixed flow regime [18].

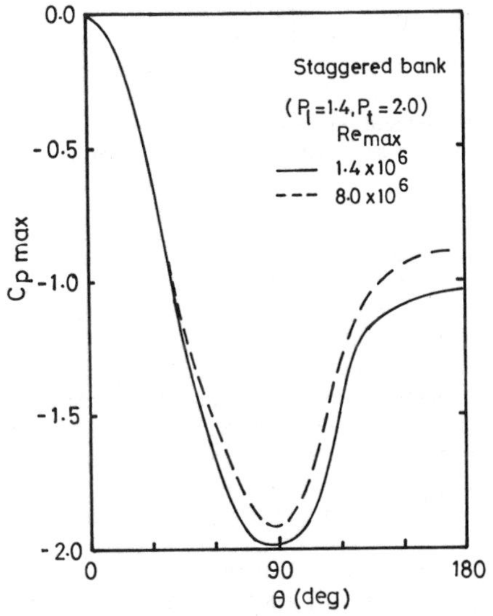

Figure 16. Surface pressure profile for staggered bank in the turbulent flow regime [20].

Figure 14 shows the pressure profile for the in-line bank with $P_l = P_t = 2.0$. The profile is different from the case of the staggered bank and the maximum pressure appears near the flow reattachment point ($\theta_R = 35°$).

The effect of pitch ratio is shown in Figure 15. The maximum and minimum values of C_p considerably increases with decreasing the pitch ratio.

In these cases, the wake pattern belongs to Pattern A at any pitch ratio, for both staggered and in-line banks as shown in Figure 9. The pressure profile for other wake patterns refers to the report of Nishikawa et al. [17].

Turbulent Flow Regime

There have been few investigations in this flow regime [20, 21]. Figure 16 shows the pressure profile for the staggered bank with $P_l = 1.4$ and $P_t = 2.0$, which was measured by Achenbach [20]. The pressure recovery is considerably larger than that for the mixed flow regime. This behavior is similar to the case of a single cylinder in the critical flow regime. The influence of Reynolds number on the pressure appears except for the front part of the cylinder, in contrast to the case of the mixed flow regime.

SURFACE SHEAR STRESS PROFILE AROUND A TUBE IN BANKS

The surface shear stress is an important factor to deduce the mechanism of heat transfer from tubes in a bank. However, measurements of surface shear stress have been reported few, due to the problem in the experimental techniques mentioned as follows.

There have been the experimental techniques such as thermal meter [22], electrochemical probe [23], and sublayer fence [24] for shear stress measurements. The principles of thermal meter and electrochemical probe are based on the measurements of heat and mass transfer rates at a small portion of the cylinder surface respectively. The calibration is necessary for the former, but is not for the latter. The principle of sublayer fence is based on the measurement of a pressure difference at a small edge, which vertically projects into the viscous sublayer from the cylinder surface. The advantage of this probe compared with other ones is the change of the sign in the reading out, when the flow changes direction. This probe also requires the calibration.

All these techniques have a high reliability for a steady boundary layer flow zone, but have a low reliability for a unsteady separation flow zone. This creates the problem in measurement on tube banks, because the unsteady separation flow zone occupies widely on the surface of a tube in a bank, in the mixed and turbulent flow regimes. However, the shear stress measurement by these techniques would provide at least a qualitative information in the separation flow zone.

For the laminar flow regime, the shear stress can be also obtained by a numerical analysis.

Laminar Flow Regime

Nishimura et al. [9] presented the surface shear stress profiles by the use of the finite element method. Figure 17 shows examples of the surface shear stress profile at three Reynolds numbers for the staggered bank with $P_l = P_t = 2.0$. The shear stress is given by the non-dimensional form, $C_{f\,max}$. At $Re_{max} = 2$ the profile has a simple shape with the maximum shear stress occurring at about $\theta = 85°$. With increasing Reynolds number, the maximum point shifts up-stream side considerably; the maximum shear stress for $Re_{max} = 40$ is located at about $\theta = 30°$. For $Re_{max} = 20$ and 40, the shear stress at the rear part of the cylinder has a small negative value due to the flow separation as shown in Figures 3b and c. The value of $C_{f\,max}$ decreases throughout the cylinder surface with increasing the Reynolds number, which shows a remarkable influence of the Reynolds number, as well as the case of $C_{p\,max}$ as shown in Figure 11.

The effect of pitch ratio is shown in Figure 18. The shear stress is given by C_f, instead of $C_{f\,max}$, in this figure. As the pitch ratio decreases, the value of C_f significantly increases and also the profile undergoes considerable variations. The two maximum shear stresses occur at the front part of the cylinder ($\theta = 20°$, $\theta = 85°$) for $P_l = P_t = 1.33$. The first peak is due to the impingement of the accelerated flow like a two-dimensional jet formed at the minimum cross section between adjacent cylinders of the preceding row, and the second peak is due to the accelerating flow passing through the convergence cross section between adjacent cylinders of the aimed row.

Figure 19 shows the surface shear stress profile for the in-line bank with $P_l = P_t = 2.0$. The profile at each Reynolds number has a simple shape with the maximum shear stress occurring at about $\theta = 80°$. For $Re_{max} = 20$ and 40, the shear stress near the front and rear stagnation points has a small negative value due to the flow separation as shown in Figures 4b and c.

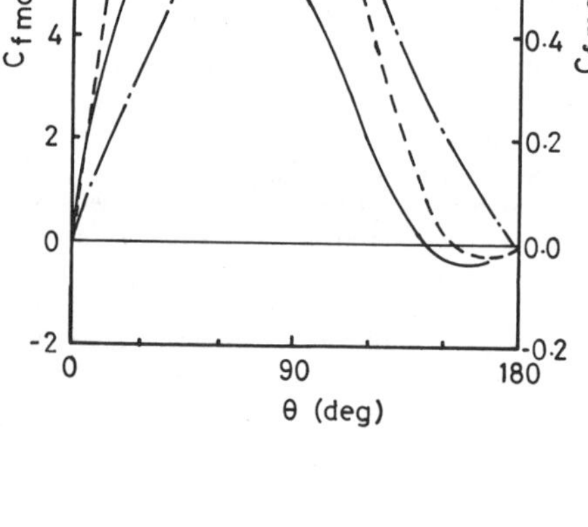

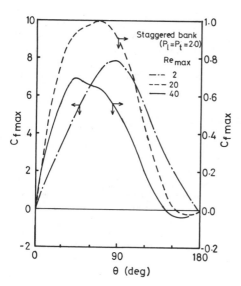

Figure 17. Effect of Reynolds number on surface shear stress profile for staggered bank in the laminar flow regime [9].

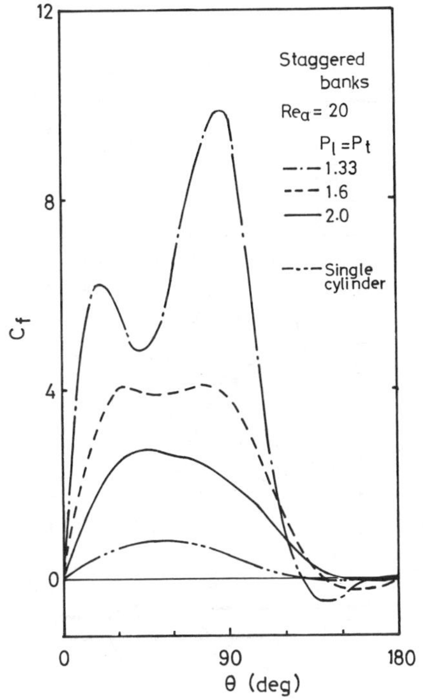

Figure 18. Effect of pitch ratio on surface shear stress profile for staggered bank in the laminar flow regime [9].

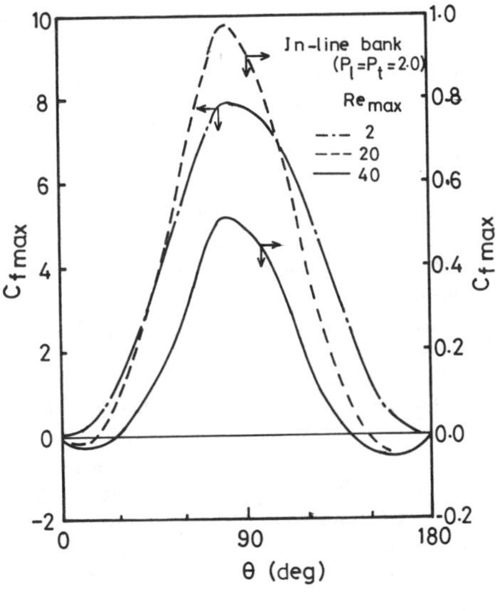

Figure 19. Effect of Reynolds number on surface shear stress profile for in-line bank in the laminar flow regime [9].

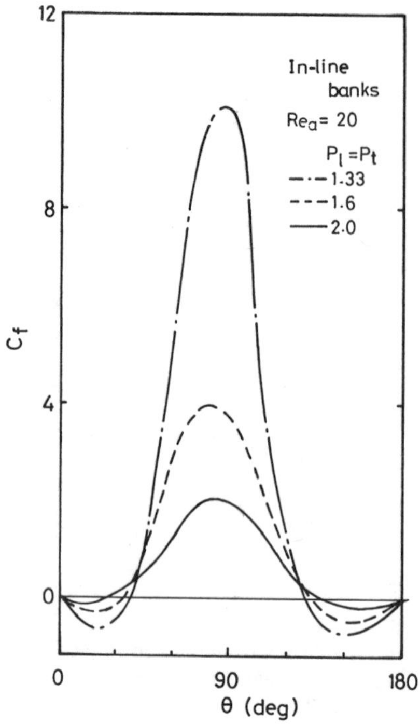

Figure 20. Effect of pitch ratio on surface shear stress profile for in-line bank in the laminar flow regime [9].

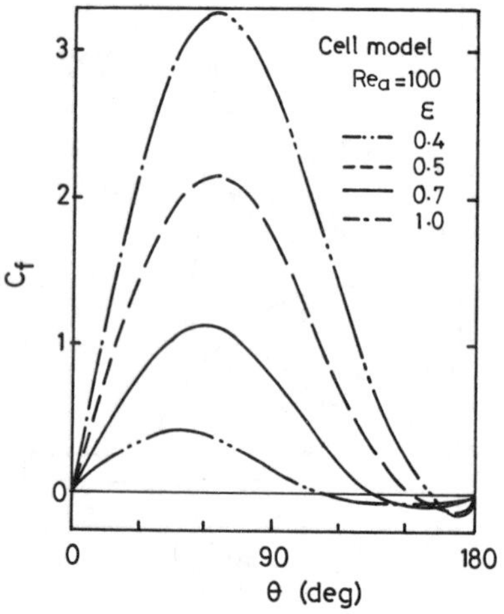

Figure 21. Surface shear stress profile of cylinder in bank obtained by cell model [27].

The effect of pitch ratio is shown in Figure 20. The profile is not sensitive to the pitch ratio, in contrast to the case of the staggered banks. The maximum shear stress significantly increases with decreasing the pitch ratio.

Before the numerical analysis in which the geometry of tube banks is considered exactly can be performed easily, a simple calculation technique called a cell model was employed to predict the flow characteristics in tube banks. This model was first developed for the creeping flow without inertia effects by Happel [25] and Kuwabara [26]. A cell model represents an assemblage of uniform cylinders equally spaced in the radial and longitudinal directions, all experiencing the same flow field. Each cylinder with a circular envelope of fluid, whose outer boundary depends on the void fraction, constitutes a cell for which the Navier-Stokes equations are solved. The interaction between adjacent cylinders is accounted for in the boundary conditions specified on the outer surface of the cell. After that, LeClair et al. [27] obtained the pressure and shear stress profiles at inertia-dominated Reynolds numbers by the use of the cell model. Figure 21 shows examples of the shear stress profile reported by them. As the void fraction corresponding to the pitch ratio decreases, the shear stress considerably increases, but the profile does not change with the void fraction. As compared with the real profiles as shown in Figures 18 and 20, there are important differences in the profile for both staggered and in-line banks, in particular with a small void fraction or pitch ratio. However, it is worth mentioning that the frictional drag coefficient, C_{Df} obtained by the integration of C_f shows an approximate agreement between cell model and real analysis, as can be seen from Figure 22. This agreement can only be explained by the existence of compensating effects in the calculation of C_{Df}.

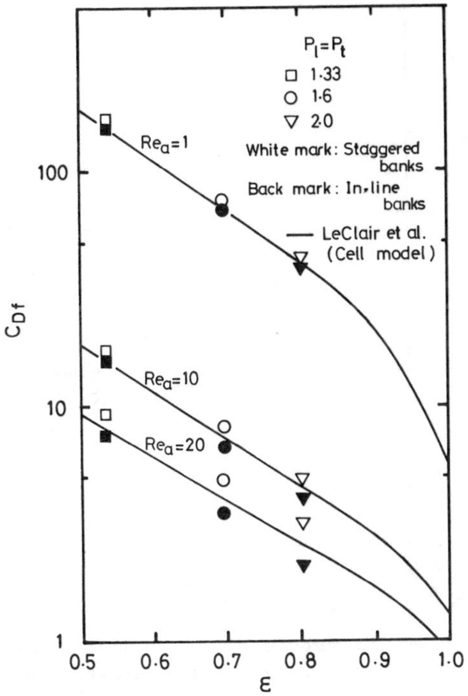

Figure 22. Comparison of frictional drag coefficient for tube banks with cell model's solution [18].

Mixed Flow Regime

Nishimura et al. [7] measured the surface shear stress profiles by the use of the electrochemical probe. Figure 23 shows examples of the shear stress profile for the staggered bank with $P_l = P_t = 2.0$. The profile is qualitatively similar to that at inertia-dominated Reynolds numbers in the laminar flow regime as shown in Figure 17. In the forward flow region ($\theta = 0° \sim \theta_S$), the value of $C_{f\,max}$ is proportional to $Re_{max}^{-1/2}$. This is, the flow behavior is regarded as characteristic of a laminar boundary layer flow. In the reverse flow region ($\theta = \theta_S \sim 180°$), this relationship does not hold due to the flow separation accompanied by the vortex shedding.

The effect of pitch ratio is shown in Figure 24. As the pitch ratio decreases the profile undergoes considerable variations, as well as the case of the laminar flow regime as shown in Figure 18. Two maximum shear stresses appear at the front part of the cylinder for $P_l = P_t = 1.2$. The occurrence of these peaks is the same reason as that described in the laminar flow regime. The shear stress at the rear part of the cylinder is smaller than that at the front part, at any pitch ratio, due to the flow separation.

Figure 25 shows the shear stress profile for the in-line bank with $P_l = P_t = 2.0$. The shear stress sharply increases from the flow reattachment point, θ_R, which is different from the case of the staggered bank. In the forward flow region ($\theta = \theta_R \sim \theta_S$), the value of $C_{f\,max}$ is nearly proportional to $Re_{max}^{-1/2}$, as well as that of the staggered bank. In the reverse flow regions, the shear stress at the up-stream side from the reattachment point is larger than that at the down-stream side from the separation point. This result leads to the following inference. The flow in the gap between adjacent cylinders becomes vortical with a higher degree of turbulence according to the vortex shedding. Although the shear stress at the front part of the cylinder is strongly influenced by the vortical flow, the effect at the rear part is smaller.

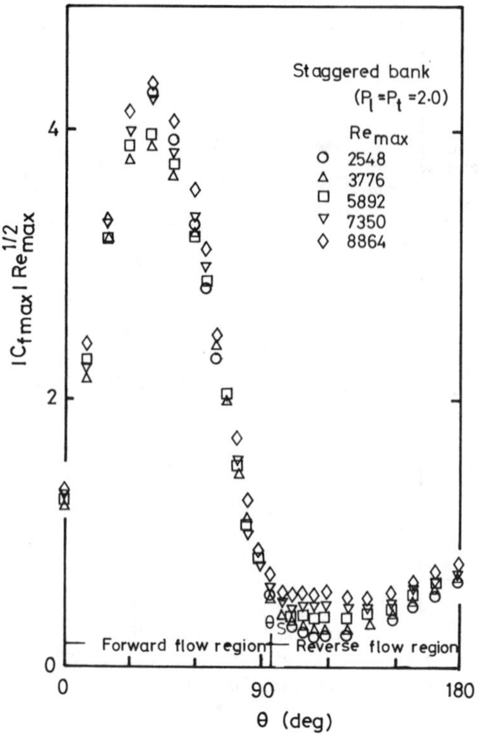

Figure 23. Effect of Reynolds number on surface shear stress profile for staggered bank in the mixed flow regime [7].

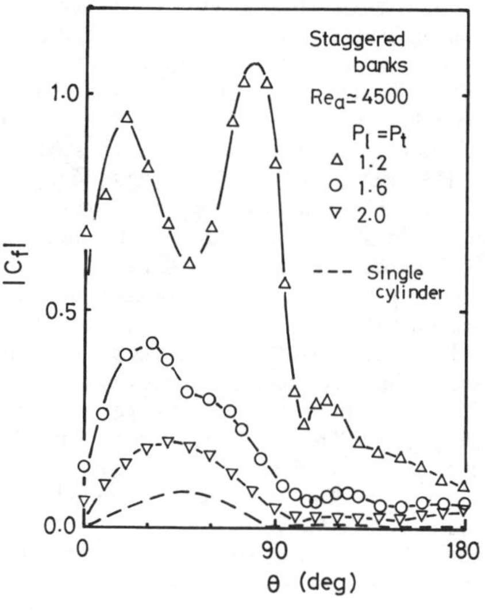

Figure 24. Effect of pitch ratio on surface shear stress profile for staggered bank in the mixed flow regime [7].

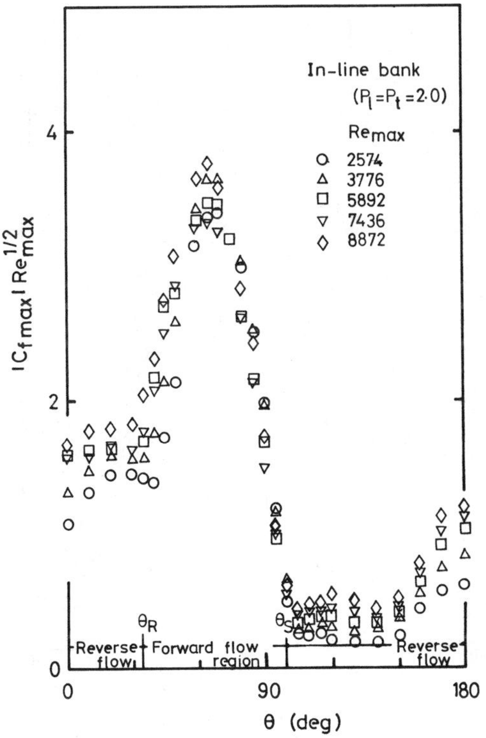

Figure 25. Effect of Reynolds number on surface shear stress profile for in-line bank in the mixed flow regime [7].

The effect of pitch ratio is shown in Figure 26. The profile is almost identical, at any pitch ratio, but the shear stress increases with decreasing pitch ratio, in particular at the front part of the cylinder.

Turbulent Flow Regime

Achenbach [20] measured the shear stress profile for the staggered bank with $P_l = 1.4$ and $P_t = 2.0$ by the use of the sublayer fence. Figure 27 shows examples of the shear stress profile at two Reynolds numbers. The influence of the Reynolds number is most evident in the region ($\theta = 40° \sim 100°$), which is different from the case of the mixed flow regime as shown in Figure 23. This result deduces that the laminar boundary layer is displaced by the turbulent one in the forward flow region, except for the region near the front stagnation point.

VELOCITY AND TURBULENCE IN TUBE BANKS

The velocity and turbulence provide useful information to infer the mechanism of heat transfer from tubes or make a turbulence model* in a bank for the purpose of computing the flow in the mixed and turbulent flow regimes. However, there have been few investigations of these regimes,

* Massey et al. [28] have recently presented a prediction for mixed and turbulent flow regimes by the use of a two-equation model of turbulence, solving equations for the kinetic of turbulence, and its dissipation.

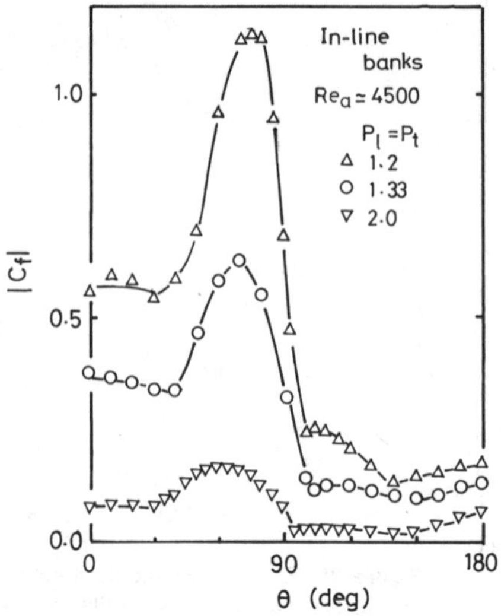

Figure 26. Effect of pitch ratio on surface shear stress profile for in-line bank in the mixed flow regime [7].

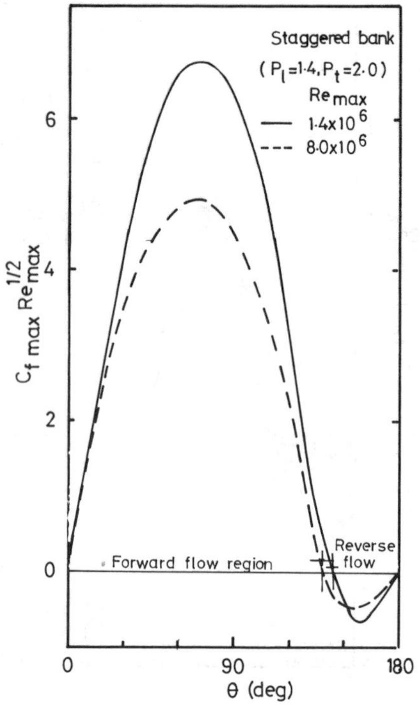

Figure 27. Surface shear stress profile for staggered bank in the turbulent flow regime [20].

as well as the case of surface shear stress. This reason would be that the measurements are not easy in a complicated vortical and high turbulent flow through a bank. In the previous studies, measurements of velocity and turbulence have been performed by a hot wire anemometer and a electrochemical probe in the mixed flow regime. The accuracies may be not sufficient, at the present time, but these results would suggest a qualitative characteristic of turbulence in a bank.

Mixed Flow Regime

Turner et al. [29] first determined flow patterns using a single hot wire probe and anemometer. The procedure involves rotating the wire to three discrete positions and solving the three equations thus obtained by an iterative process. Figure 28 shows the time mean velocity of air flow across a three by three tube bank with in-line arrangement of $P_1 = P_t = 2.0$ at $Re_a = 67{,}000$. The vectors are drawn in the direction of flow with the length of the vector proportional to velocity. The general trend of circulation flow between the first and second cylinders is clearly of a different nature to that between the second and third cylinders: velocity is generally higher and the angle of entrainment is more severe in the latter.

Aiba et al. [15, 16] measured the time mean and turbulent fluctuating velocities of air flow with a hot wire anemometer. Figure 29 shows the results for the in-line bank with $P_1 = P_t = 1.6$ at about $Re_a = 15{,}400$. The measuring section is the midway between the fourth and fifth rows as shown in this figure. The shear layer separated from the fourth cylinder has a very large velocity gradient and

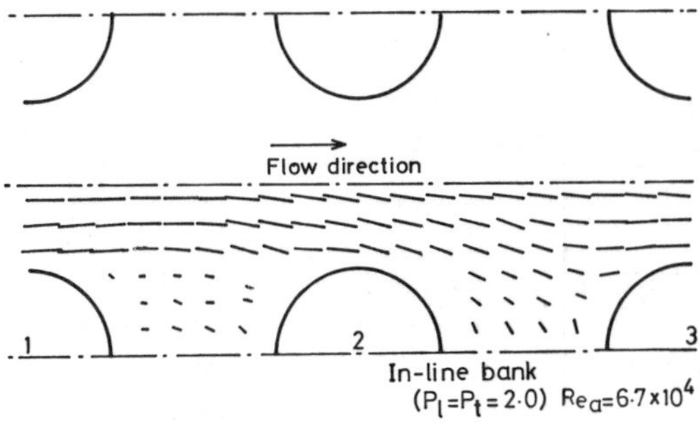

Figure 28. Velocity vectors for in-line bank [29].

a high turbulence intensity, and its maximum value is about 50%. The velocity on the central axis between adjacent cylinders reaches a maximum and the turbulence intensity is slightly smaller, as compared with the maximum value.

Figure 30 describes the results for the staggered bank with $P_l = P_t = 1.2$ at about $Re_a = 5{,}000$. The measuring section is the midway between the fifth and sixth rows as shown in this figure. The general trend is similar to the case of Figure 29, but the velocity on the central axis has a minimum value and also the maximum turbulence intensity is nearly 100%. The reason for the defect of this

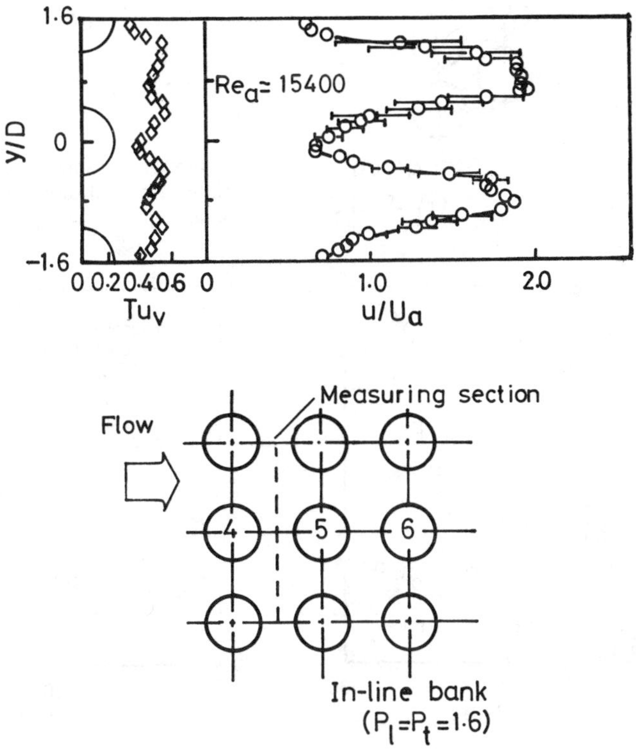

Figure 29. Streamwise velocity and turbulence intensity profiles for in-line bank [15].

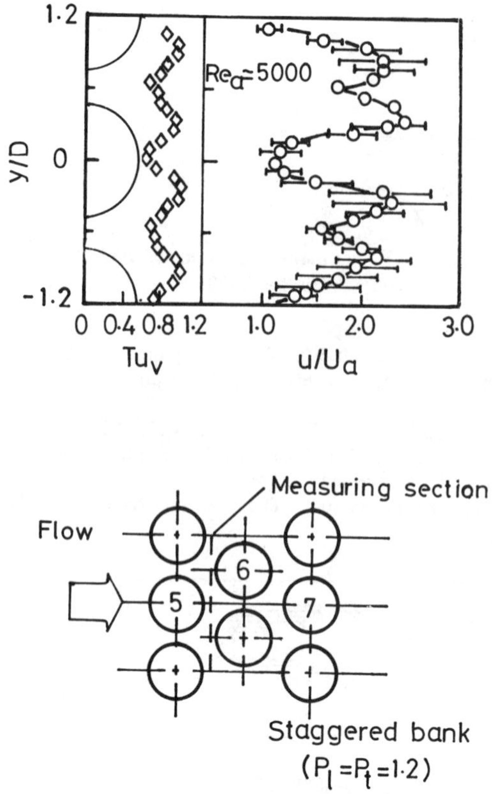

Figure 30. Streamwise velocity and turbulence intensity profiles for staggered bank [16].

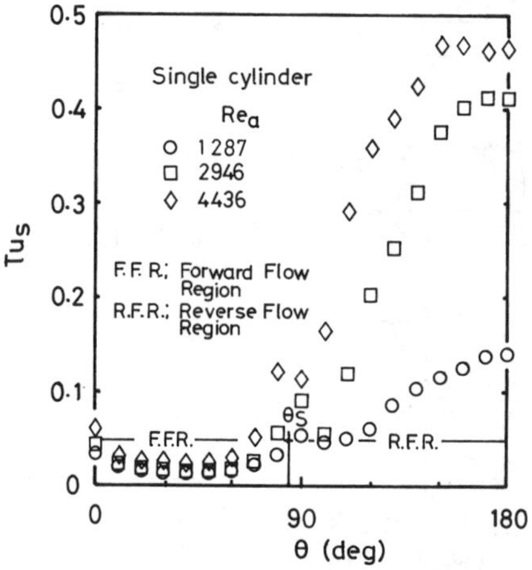

Figure 31. Turbulence intensity profile of surface shear stress for a single cylinder [18].

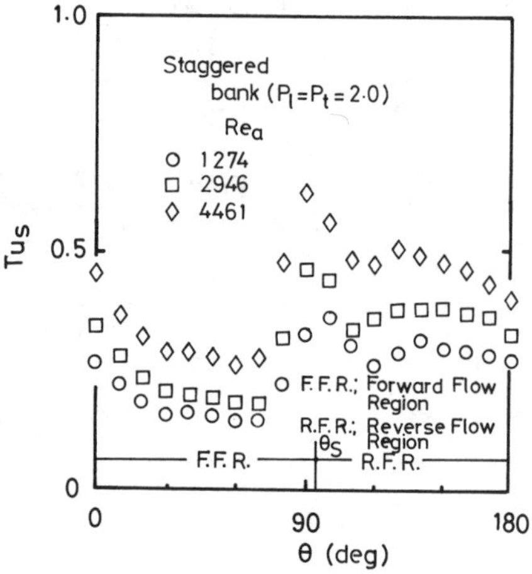

Figure 32. Turbulence intensity profile of surface shear stress for staggered bank [18].

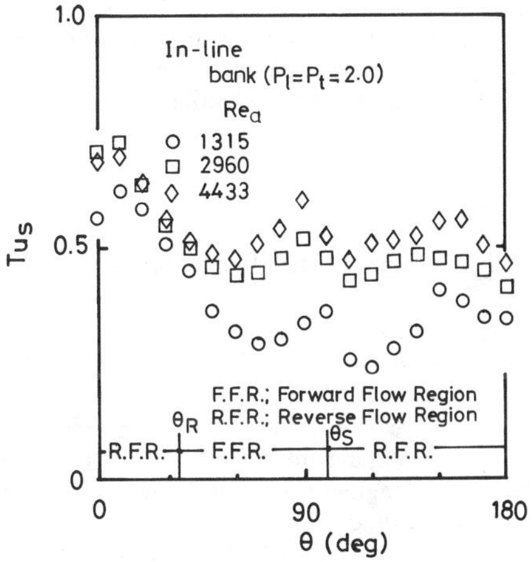

Figure 33. Turbulence intensity profile of surface shear stress for in-line bank [18].

velocity is that the accelerated flow like a two-dimensional jet formed at the minimum cross section in the fifth row impinges on the front stagnation point of the cylinder of the sixth row.

Besides these turbulence characteristics, it is also important to know of the existence of turbulence near the cylinder surface with relation to heat transfer from tubes. This measurement can be performed by the electrochemical probe as described earlier. Tournier et al. [30] investigated the behavior of naturally oscillating three-dimensional flow around a single cylinder using this probe. Also, Nishimura et al. [18] obtained the turbulence characteristics in tube banks by the same technique.

Figure 31 shows turbulence intensities of the surface shear stress of a single cylinder. In the forward flow region, where laminar boundary layer is formed, the turbulence intensity is small and independent of the Reynolds number. However, in the reverse flow region, where the flow separation occurs, the turbulence intensity becomes larger and depends on the Reynolds number.

Figure 32 shows the turbulence intensity profile for the staggered bank with $P_1 = P_t = 2.0$. The maximum turbulence intensity occurs near the separation point and the turbulence intensity is influenced by the Reynolds number in both forward and reverse flow regions, which is different from that of a single cylinder as shown in Figure 31.

Figure 33 shows the result of the in-line banks with $P_1 = P_t = 2.0$. The maximum turbulence intensity occurs near the front stagnation point and its value is over 50%. In the forward flow region, where a laminar boundary layer is formed, the turbulence intensity has the same order as that in the reverse flow region ($\theta = \theta_S \sim 180°$), which is different from that of the staggered bank as shown in Figure 32.

Thus, these experimental results indicate that although a laminar boundary layer persists on the cylinder surface in the forward flow region, the flow is under the influence of an intensive vortical flow according to the vortex shedding of the surrounding cylinders, which is quite different from that of a single cylinder.

CONCLUSION

The work presented here indicates the extremely complex nature of the flow across tube banks. In the laminar flow regime, the theoretical analysis gives a great deal of useful flow information. On the other hand, in the mixed and turbulent flow regimes, the flow pattern becomes unsteady vortical motion with a high turbulence according to the vortex shedding from each tube in a bank, and the flow disturbance depends strongly on the Reynolds number and the arrangement of tubes. So, theoretical analysis is difficult at the present time. However, modeling of turbulence will be developed by the appearance of a super computer in the near future, as well as the theoretical analysis in the laminar flow regime. But in order to establish turbulent models, a detailed set of experimental data for these flows are also needed in testing and perhaps refining turbulence models. Therefore, the need for further experimental research for the turbulence is identified.

Acknowledgment

The author expresses his appreciation to Prof. Yuji Kawamura of Hiroshima University for his kind interest and encouragement.

NOTATION

Re_a	Reynolds number ($=U_aD/\nu$)
Re_{max}	Reynolds number ($=U_{max}D/\nu$)
C_p	pressure coefficient ($=2(p-p_{\theta=0})/\rho U_a^2$)
$C_{p\,max}$	pressure coefficient ($=2(p-p_{\theta=0})/\rho U_{max}^2$)
C_f	frictional coefficient ($=2\tau_s/\rho U_a^2$)
$C_{f\,max}$	frictional coefficient ($=2\tau_s/\rho U_{max}^2$)
C_{Df}	frictional drag coefficient
P_l	longitudinal pitch ratio ($=L/D$)
P_t	transverse pitch ratio ($=T/D$)
D	cylinder diameter
U_a	approaching velocity
U_{max}	velocity through minimum cross section in bank ($=U_aP_t/(P_t-1)$)
L	longitudinal length between adjacent cylinders
T	transverse length between adjacent cylinders
Nu	Nusselt number
Pr	Prandtl number
Tu_v	turbulence intensity of velocity ($=\sqrt{\overline{U_a'^2}}/\lvert U_a\rvert$)
Tu_s	turbulence intensity of surface shear stress ($=\sqrt{\overline{\tau_s'^2}}/\lvert\tau_s\rvert$)
p	surface pressure

Greek Symbols

ρ	density
ν	kinematic viscosity
τ_s	surface shear stress
ψ	stream function
θ	angle in cylindrical coordinate

REFERENCES

1. Zukauskas, A., "Heat Transfer from Tubes in Cross-flow." *Advances in Heat Transfer*, Vol. 8, Irvine, T. F., and Hartnett, J. P., Eds. New York: Academic Press, 1972, pp. 93–160.
2. Zukauskas, A., and Ulinskas, R., "Some Aspects of the Heat Transfer for Banks of Tubes in Cross-Flow for the Low Range of Reynolds Numbers." *Proceedings of the 6th International Heat Transfer Conference*, 1978, HX-10.
3. Chen, Y. N., "Excitation Sources of The Flow-induced Vibration and Noise in Tube Bank Heat-exchangers." *Symp. Noise Fluids Eng.*, 1977, pp. 239–246.
4. Weaver, D. S., and Grover, L. K., "Cross Flow Induced Vibrations in A Tube Bank." *ASME Paper* No. 78-PVP-30, 1978, 7p.
5. Nishimura, T., "Transport Phenomena in Closely-spaced Tube Banks." Ph.D. Thesis, Hiroshima Univ., Japan, 1981.
6. Aiba, S., "Heat Transfer over Closely Spaced Tubes in Cross Flow of Air." Ph.D. Thesis, Hokkaido Univ., Japan, 1981.
7. Nishimura, T., and Kawamura, Y., "Flow Pattern in Tube Banks in the Fully Developed Region." *Kagaku Kogaku Ronbunshu*, Vol. 7, No. 3, 1981, pp. 222–227.
8. Ishigai, S., et al., "Structure of Gas Flow and Vibration in Tube Banks with Tube Axes Normal to Flow." *Journal of the MESJ*, Vol. 10, No. 1, January 1975, pp. 66–73.
9. Nishimura, T., and Kawamura, Y., "Analysis of Flow Across Tube Banks in Low Reynolds number region." *Journal of Chemical Engineering of Japan*, Vol. 14, No. 4, 1981, pp. 267–272.

10. Launder, B. E., and Massey, T. H., "The Numerical Prediction of Viscous Flow and Heat Transfer in Tube Banks." *Transactions of the ASME, Journal of Heat Tranfer*, Vol. 100, November 1978, pp. 565–571.
11. Gordon, D., "Numerical Calculations on Viscous Flow Fields through Cylinder Arrays." *Computers and Fluids*, Vol. 6, 1978, pp. 1–13.
12. Zukauskas, A., and Ulinskas, R. V., "Analysis of Lateral Streamline Flow of Water Through Tube Bundles During the Process of Heat Transfer at Critical Reynolds Numbers. "*International Chemical Engineering*, Vol. 17, No. 4, October 1977, pp. 673–676.
13. Poshkas, P. S., et al., "Local Heat Transfer Rate at a Tube in a Corridor Bundle Lateral to an Air Stream at Large Reynolds Numbers." *International Chemical Engineering*, Vol. 18, No. 2, April 1978, pp. 337–343.
14. Poshkas, P. S., et al., "Drag and Flow Patterns for a Tube in Various Rows of Bundles with Lateral Cross Flow of Air at High Reynolds Numbers." *International Chemical Engineering*, Vol. 20, No. 3, July 1980, pp. 503–512.
15. Aiba, S., et al., "Heat Transfer Around Tubes in In-line Tube Banks." *Bulletin of the JSME*, Vol. 25, No. 204, June 1982, pp. 919–926.
16. Aiba, S., et al., "Heat Transfer Around Tubes in Staggered Tube Banks." *Bulletin of the JSME*, Vol. 25, No. 204, June 1982, pp. 927–933.
17. Nishikawa, E., and Ishigai, S., "Structure of Gas Flow and Its Pressure Loss in Tube Banks with Tubes Axes Normal to Flow." *Trans. JSME*, Vol. 43, No. 373, 1978, pp. 3310–3319.
18. Nishimura, T., and Kawamura, Y. Unpublished works.
19. Kitaura, Y., et al. "Fluid Flows and Pressure Drop in Tube Banks." *Kagaku Kogaku*, Vol. 32, No. 9, 1968, pp. 883–889.
20. Achenbach, E., "Investigation on the Flow Through a Staggered Tube Bundle at Reynolds Numbers up to $Re = 10^7$." *Wärme und Stoffübertragung*, Vol. 2, 1969, pp. 47–52.
21. Zukauskas, A., "Influence of the Geometry of the Bundle on the Local Heat Transfer Rate in the Critical Region of Streamline Flow." *International Chemical Engineering*, Vol. 17, No. 4, October 1977, pp. 744–751.
22. Bellhouse, B. J., and Schltz, D. L., "The Determination of Mean and Dynamic Skin Friction, Separation and Transition in Low-speed Flow with a Thin Film Heated Element." *J. Fluid Mech.*, Vol. 24, 1966, pp. 379–400.
23. Mizushina, T., "The Electrochemical Method in Transport Phenomena." *Advances in Heat Transfer*, Vol. 7, Irvine, T. F., and Hartnett, T. P., Eds. New York: Academic Press, 1971, pp. 87–161.
24. Achenbach, E., "Distribution of Local Pressure and Skin Friction Around a Circular Cylinder in Cross-flow up to $Re = 5 \times 10^6$." *J. Fluid Mech.*, Vol. 34, 1968, pp. 625–639.
25. Happel, J., "Viscous Flow Relative to Arrays of Cylinders." *A.I.Ch.E. J.*, Vol. 5, 1959, pp. 174–177.
26. Kuwabàra, S., "The Forces Experienced by Randomly Distributed Parallel Circular Cylinder or Spheres in a Viscous Fluid at Small Reynolds Numbers." *J. Phys. Soc. Japan*, Vol. 14, 1959, pp. 527–532.
27. LeClair, B. P., and Hamielec, A. E., "Viscous Flow Through Particle Assemblages at Intermediate Reynolds Numbers." *Ind. Eng. Chem. Fundam.*, Vol. 9, No. 4, 1970, pp. 608–613.
28. Massey, T. H., et al., "A Prediction Procedure for Fully Developed Turbulent Flow and Heat Transfer in Tube Banks." *Proceedings of the 6th International Heat Transfer Conference*, 1978, HX-13.
29. Turner, J. R., and Eastop, T. D., "A Hot Wire Anemometry Method for the Flow Patterns in an Array of Heat Exchanger Tubes." *Trans. I Chem E*, Vol. 57, 1979, pp. 139–142.
30. Tournier, C., and Py, R., "The Behavior of Naturally Oscillating Three-dimensional Flow Around a Cylinder." *J. Fluid Mech.*, Vol. 85, 1978, pp. 161–186.

CHAPTER 24

FLOW IN RECTANGULAR CHANNELS

W. R. C. Myers

and

J. F. Lyness

Department of Civil Engineering
University of Ulster
Newtownabbey, County Antrim
Northern Ireland

CONTENTS

INTRODUCTION

Open channel or free surface flow presents one of the commonest and most challenging problems in fluid mechanics. It may be defined as flow of a liquid with a free surface that is generally, though not necessarily, subject to atmospheric pressure. Open channel flow is encountered in natural streams, rivers, estuaries, and oceans as well as in all manner of artificial aqueducts. This chapter is concerned with rectangular channels, and hence attention will be confined to artificial channels, although some natural water courses do approximate to a rectangular cross section. Hence, only fixed bed channels will be considered, sediment transport being outside the scope of this chapter. Furthermore, behavior of rectangular channel flow will be the primary concern although measurement of flow will also be considered.

Definitions

Open channel flow may be classified in various ways. Classifications according to parameter variability leads to the following definitions:

Steady uniform flow—no variation of any parameter with time or distance.
Steady non-uniform flow—variation of parameters with distance but not with time.
Unsteady non-uniform flow—variation of parameters with time and distance.

Steady uniform flow does not exist in a real fluid in that because of frictional effects there will be variation of velocity between channel center-line and channel boundary. Furthermore, in turbulent flow fluid velocity is constantly varying with time. However, if the average values of the relevant parameters do not vary with time or distance, then flow is classified as steady uniform. Even within this definition steady uniform flow is rare outside the laboratory.

Steady non-uniform flow is very common and often termed varied flow. This class is subdivided into rapidly varied and gradually varied, depending on whether the changes in parameter values occur rapidly or gradually. Generally, in the former case, frictional effects are very small, while in the latter they are significant.

Unsteady non-uniform flow, usually abbreviated to unsteady flow, is the most complex to analyze, but is also the most commonly occurring, especially in the natural realm.

The fourth possibility, namely unsteady uniform flow does not occur in open channels because of the freedom of the free surface to rise and fall.

Further classifications of channel flow lead to the following definitions:

Laminar flow—fluid particles move in discrete paths and there is no mixing of adjacent fluid paths; laminar channel flow is very rare.
Turbulent flow—mixing of fluid paths and continual change of direction and velocity of fluid particles.
Rapid flow—flow where the Froude number ($=v/\sqrt{gd}$) is greater than unity; usually as a result of steep channel gradients.
Tranquil flow—flow characterized by a Froude number of less than unity.

Critical flow, defined as when Froude number is equal to unity, divides rapid flow from tranquil. The nature of critical, rapid, and tranquil flow is further described in the following section.

BASIC CONCEPTS

Many rectangular channel problems may be analyzed using the basic equations of continuity, energy and momentum, which are based on the fundamental principles of conservation of mass energy and momentum.

Continuity

For steady flow the continuity equation states that any two points on a rectangular channel may be related by the expression

$$V_1A_1 = V_2A_2 = Q \tag{1}$$

where V_1V_2 = average cross-sectional velocities at sections 1 and 2
A_1A_2 = cross-sectional areas at sections 1 and 2
Q = discharge defined as volume per unit time

But for a rectangular channel

$$A_1 = b_1d_1 \quad \text{and} \quad A_2 = b_2d_2 \tag{2}$$

where b_1b_2 = channel widths at sections 1 and 2
d_1d_2 = channel depths at sections 1 and 2

Therefore, Equation 1 becomes

$$V_1b_1d_1 = V_2b_2d_2 = Q \tag{3}$$

This may be written as

$$V_1d_1 = \frac{Q}{b_1} = q_1 \quad \text{and} \quad V_2d_2 = \frac{Q}{b_2} = q_2 \tag{4}$$

where q_1q_2 = discharges per unit width at sections 1 and 2.

The discharge per unit width concept, which in applicable only to rectangular channels, is a useful parameter often leading to simplification of equations describing rectangular channel flow.

Energy

Energy considerations are encompassed in the Bernoulli equation as applied to open channel flow. Figure 1 shows a definition sketch of the various terms that are related by the following expression:

$$d + \frac{V^2}{2g} + Z = H_T \tag{5}$$

where d = depth of flow

V = average cross-sectional velocity = $\frac{Q}{A}$

g = gravitational acceleration
Z = height of channel bed above datum
H_T = total energy or total head

Equation 5 identifies the various forms of energy present in a rectangular open channel, namely potential energy (Z), kinetic energy ($V^2/2g$), and energy due to pressure (d). In open channel flow these terms are expressed in length units or "head" for convenience, and the summation of the three parts is termed total energy or total head. Proof of the energy equation may be found in Francis [1].

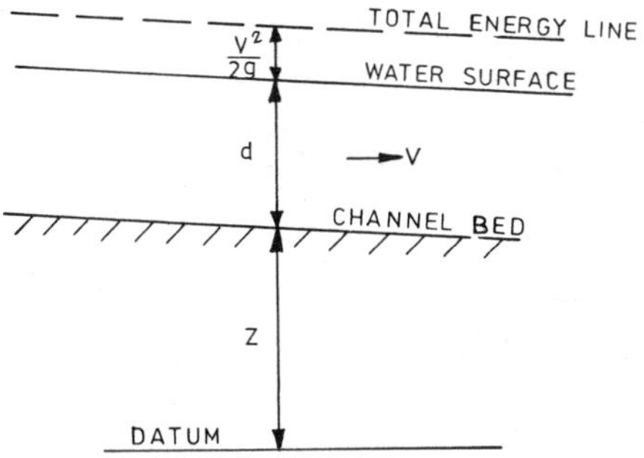

Figure 1. Definition sketch for energy equation.

Kinetic energy must be modified by multiplication by a kinetic energy factor (α), because of non-uniformity of velocity pattern as a result of boundary friction. α can vary between 1.03 and 1.6, although it is generally between 1.03 and 1.1. The kinetic energy factor is often neglected as being small with respect to other terms in Equation 5.

In an ideal fluid, where frictional resistance is zero, total energy (H_T) remains constant from one section of a channel to the next. Hence, any two or more points on a channel could be connected by the equation:

$$d_1 + \frac{V_1^2}{2g} + Z_1 = d + \frac{V_1^2}{2g} + Z_2 \qquad (6)$$

Such an assumption may be valid over short lengths of channel where frictional resistance is negligible when compared with other terms. Thus, in certain rapidly varied flow problems the energy equation may be used as shown in Equation 6. When channel length and geometry is such as to produce significant energy loss, then Equation 6 may be rewritten:

$$d_1 + \frac{V_1^2}{2g} + Z_1 = d_2 + \frac{V_2^2}{2g} + Z_2 + h_L \qquad (7)$$

where h_L = head loss due to friction and form drag.

Specific energy. It is sometimes convenient to express energy as that in relation to the channel bed. Thus, Equation 5 becomes

$$E = d + \frac{V^2}{2g} \qquad (8)$$

where E = specific energy. For a rectangular channel:

$$V = \frac{Q}{bd} = \frac{q}{d} \qquad (9)$$

Therefore, Equation 8 becomes

$$E = d + \frac{1}{2g}\left(\frac{q^2}{d^2}\right) \qquad (10)$$

If discharge per unit width (q) is constant, then Figure 2 shows the variation of specific energy with depth. Conversely, if specific energy is constant the variation of discharge per unit width of

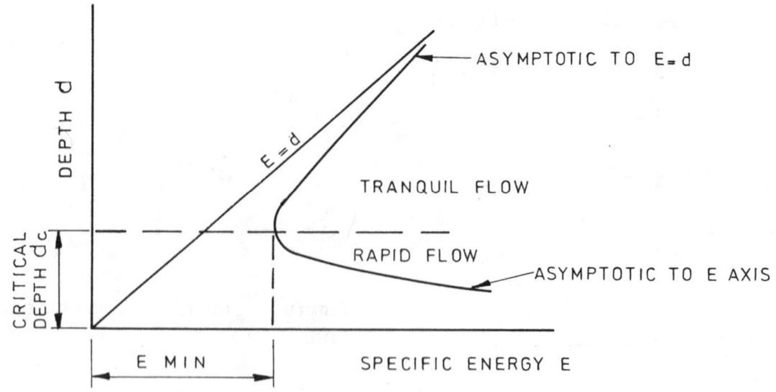

Figure 2. Variation of specific energy with depth.

depth is as shown on Figure 3. On both figures a depth (d_c) can be identified, called the critical depth, where specific energy is minimum for a given discharge per unit width and conversely where discharge per unit width is maximum for a given specific energy.

Critical conditions may be specific by differentiating Equation 8 and putting the derivative equal to zero. This yields the expression for critical depth

$$d_c = \left(\frac{q^2}{g}\right)^{1/3} \tag{11}$$

Substituting for this value of q into Equation 8 gives

$$E_{min} = d_c + \frac{gd_c^3}{2gd_c^2} \tag{12}$$

which reduces to

$$E_{min} = \frac{3}{2} d_c \tag{13}$$

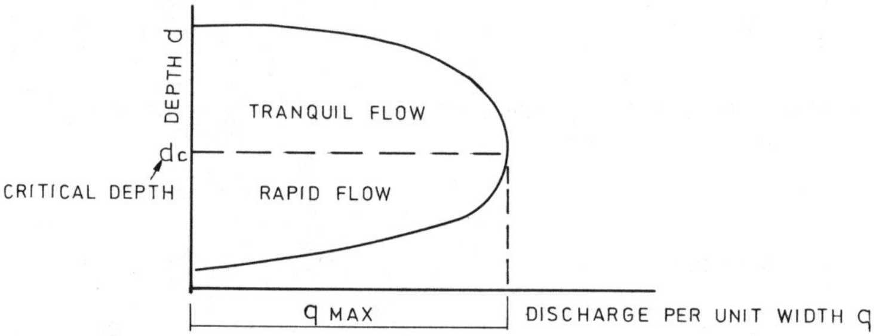

Figure 3. Variation of discharge per unit width with depth for constant specific energy.

Critical velocity may be found by noting

$$V_c = \frac{q}{d_c} = (gd_c)^{1/2} \tag{14}$$

The critical depth and velocity divide rectangular flow into two regions, tranquil flow where depth is greater than critical, and rapid flow where depth is less than critical. At the critical point, Equation 14 yields

$$\frac{V_c}{\sqrt{gd_c}} = 1 \tag{15}$$

Since the Froude number is defined as

$$Fr = \frac{V}{\sqrt{gd}} \tag{16}$$

Hence, the Froude number is unity at the critical point, greater than unity for rapid flow and less than unity for tranquil flow. These definitions were referred to earlier.

Additionally, it may be noted that the velocity of a small surface wave relative velocity is given by

$$C = \sqrt{gd} \tag{17}$$

where C = celerity of wave.

Equations 14 and 17 yield the conclusion that in rapid flow any small surface disturbances will be unable to travel upstream, since the flow velocity is greater than the celerity of the wave. Conversely, if flow is tranquil and velocity is less than the critical value, small surface disturbances will be capable of traveling upstream. Hence the occurrence of critical conditions acts as a channel control, effectively isolating the upstream reach from any changes in water level downstream of the critical section. This phenomenon is often used to advantage in the design of open channels and hydraulic structures.

Momentum

The momentum equation concerns forces on fluids and the reactives forces by fluids in motion on hydraulic structures, bridge piers, energy dissipating obstructions, etc. For steady flow, assuming hydrostatic pressure distributions, negligible frictional forces and weight forces, the momentum equation for a rectangular channel may be expressed as

$$F = \left[\frac{\rho g d_1^2}{2} + \frac{\rho q^2}{d_1}\right] - \left[\frac{\rho g d_2^2}{2} + \frac{\rho q^2}{d_2}\right] \tag{18}$$

where F = force on a control volume of the fluid by the obstruction
ρ = fluid density
g = gravitational acceleration
q = discharge per unit width of channel
$d_1 d_2$ = flow depths before and after the obstruction

Figure 4 is a definition sketch, illustrating a typical situation. The indeterminate obstruction to flow may be a bridge pier, or a concrete block for energy dissipation purposes or any type of obstruction encountered by a moving fluid. Equation 18 enables analysis of any such situation.

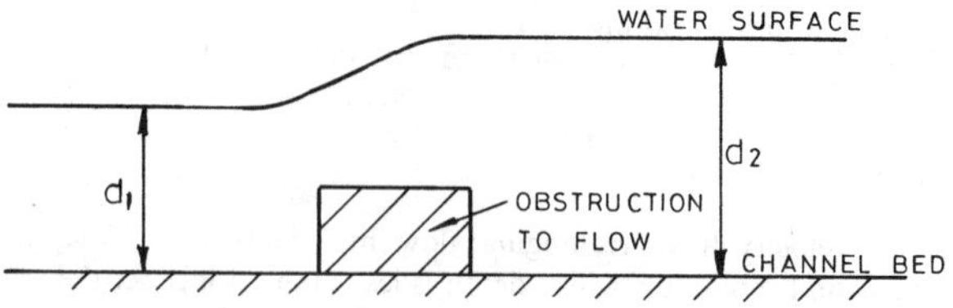

Figure 4. Definition sketch for momentum equation.

Equation 18 may be rewritten

$$F = M_1 - M_2 \tag{19}$$

where M_1, M_2 = the momentum functions or flow forces of the stream before and after the obstruction.

In an analogous manner to specific energy, the momentum function may be plotted against flow depth for constant discharge per unit width or alternatively depth may be plotted against discharge per unit width for constant momentum function. Figures 5 and 6 show these relationships, which

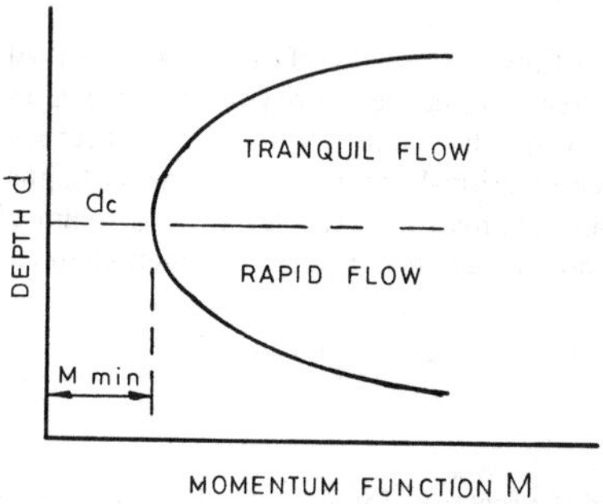

Figure 5. Variation of momentum function with depth.

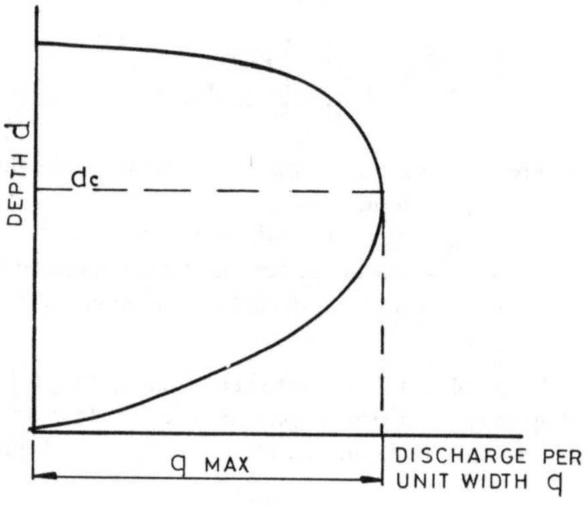

Figure 6. Variation of discharge per unit width with depth for constant momentum function.

are comparable to Figures 2 and 3. Once again critical conditions emerge, giving alternative definitions of critical depth as that at which momentum function is a minimum for a given discharge per unit width or that at which discharge per unit width is a maximum for a given momentum function.

Expressions for critical depth and velocity can be derived from Equation 18 by differentiating with respect to depth, and putting the derivative equal to zero to yield the minimum condition. This exercise yields

$$d_c = \left(\frac{q^2}{g}\right)^{1/3} \tag{20}$$

$$V_c = (gd_c)^{1/2} \tag{21}$$

The Equations 20 and 21 are exactly similar to Equations 11 and 14.

It is often convenient to render parameters such as specific energy and specific force in a dimensionless form. This allows generalization and, hence, comparison of data regardless of units and dimensions. Dividing Equation 10 by d_c yields

$$\frac{E}{d_c} = \frac{d}{d_c} + \frac{q^2}{2gd^2d_c} \tag{22}$$

and, hence,

$$\frac{E}{d_c} = \frac{d}{d_c} + \frac{1}{2}\left(\frac{d_c}{d}\right)^2 \tag{23}$$

The expression for momentum function from Equation 18 is

$$M = \frac{\rho g d^2}{2} + \frac{\rho q^2}{d} \tag{24}$$

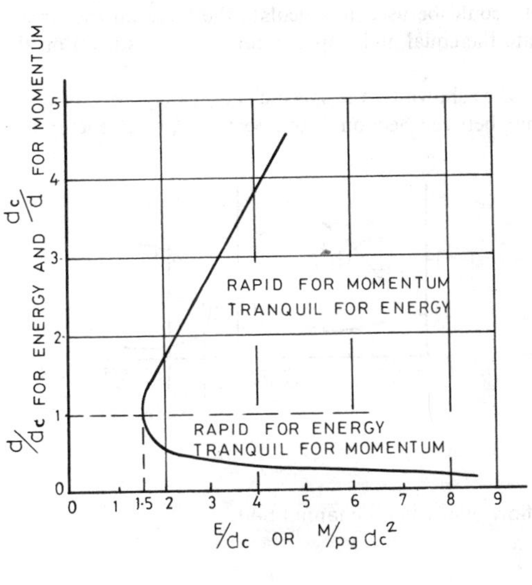

Figure 7. Generalized energy and momentum functions for rectangular channels.

Dividing through by $\rho g d_c^2$ gives

$$\frac{M}{\rho g d_c^2} = \frac{1}{2}\frac{\rho g d^2}{\rho g d_c} + \frac{\rho q^2}{\rho g d_c^2 d} \tag{25}$$

which may be simplified to

$$\frac{M}{\rho g d_c^2} = \frac{1}{2}\left(\frac{d}{d_c}\right)^2 + \frac{d_c}{d} \tag{26}$$

It is possible to plot Equations 23 and 24 on the same axis. They lie on the same curve provided d/d_c is used for one and d_c/d is used for the other (Figure 7).

APPLICATIONS OF BASIC EQUATIONS

The equations of continuity, energy, and momentum may be used to analyze a wide variety of rectangular channel situations provided that the assumptions on which the equations are based are not violated. These include a stipulation of steady flow and also that frictional resistance is negligibly small. Hence, application of these equations is confined to rapidly varied flow, where parameters are constant with time but change over relatively short channel lengths. Even where practical situations diverge somewhat from the theoretical conditions governing the equations, allowance may be made by the inclusion of experimentally determined coefficients. Some of the commoner rapidly varied flow problems are dealt with below.

Raised Channel Bed

Frequently channel flows pass over obstructions placed on the channel bed, usually for flow measurement or energy dissipation purposes. If the obstruction is smooth, causing minimal energy loss, both energy and momentum equations apply. If there is significant energy loss, only the momentum equation is applicable.

Figure 8 shows tranquil flow approaching a raised portion of channel bed. The obstruction exerts a restraining force on the fluid thereby decreasing its momentum function M. The specific force diagram predicts, therefore, that there will be a decrease in water level from Section 1 to Section 2. If necessary, the momentum Equation 18 could be used to calculate the force on the stream by the obstruction. This would also evaluate the equal and opposite force by the stream on the obstruction.

There is also a decrease in specific energy E as shown by the specific energy diagram. However, total energy H (Equation 5) does not change between Section 1 and Section 2, since energy loss

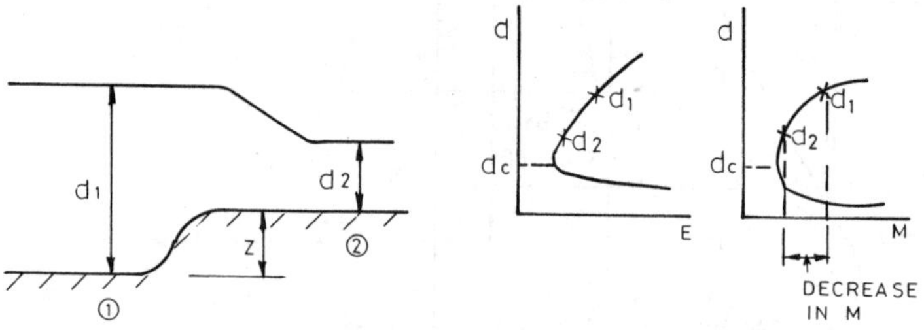

Figure 8. Tranquil flow over raised channel bed.

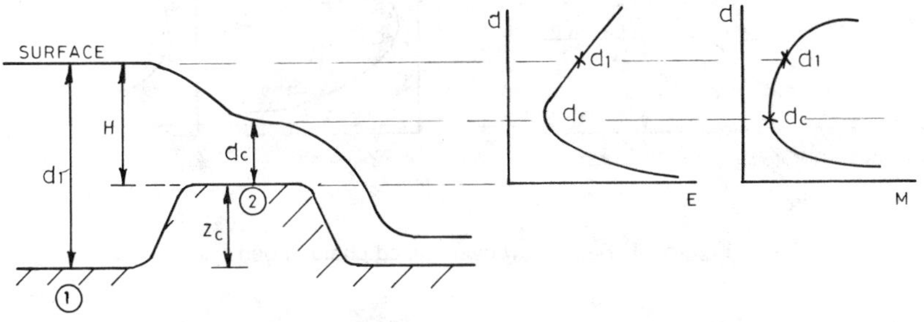

Figure 9. Tranquil flow over broad crested weir.

is negligibly small. Hence, the energy equation may be written to link Sections 1 and 2. If depth and velocity are known at Section 1, use may be made of continuity and energy equations to determine the corresponding paramaters at Section 2 or on the obstruction itself.

If the height of the obstruction Z is further raised, critical flow will be produced as shown on Figure 9. The height of obstruction required to produce critical flow is known as the critical height (Zc). Downstream of such an obstruction the specific energy rises as bed level drops, and flow moves into the rapid region. If an obstruction of height greater than Zc is encountered by a tranquil stream, the depth on the obstruction remains at d, but the upstream depth increases so as to produce the necessary decreases in specific energy and momentum produced by the obstruction.

Recalling Equation 11 it will be noted that there is a relationship between critical depth and discharge per unit width. Hence, if critical depth could be measured, the discharge could be calculated from Equation 11. This principle underlies the operation of a family of open channel flow measuring devices known collectively as critical depth devices, of which a raised channel bed, known as a broad crested weir is one. Full details of the range of such devices may be found in Ackers et al [2]. It is not convenient to measure the actual critical depth on the broad crested weir, partly because of the difficulty of identifying it and partly because critical conditions produce surface disturbances that make depth measurement difficult. Hence, depth upstream is measured and related to the critical depth by the energy equation.

Combining Equations 11 and 13 yields the following expression

$$q = 1.705E_{min} \tag{27}$$

Assuming that specific energy, with respect to the weir crest does not change between Section 1 upstream and Section 2 on the crest (Figure 9) then, for S.I. units

$$q = 1.705\left[H + \frac{V_1^2}{2g}\right] \tag{28}$$

and

$$Q = B1.705\left[H + \frac{V_1^2}{2g}\right] \tag{29}$$

where B = width of weir crest.

Upstream velocity head is small and is often incorporated using a coefficient of discharge, which also allows for any frictional effects and the curvature of the streamlines which combine to reduce

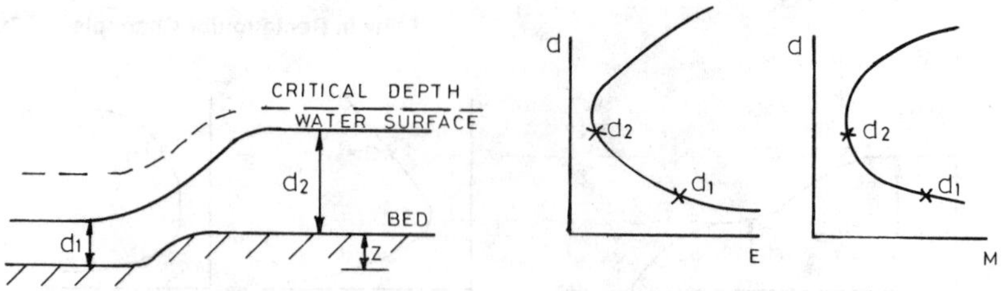

Figure 10. Rapid flow over raised channel bed.

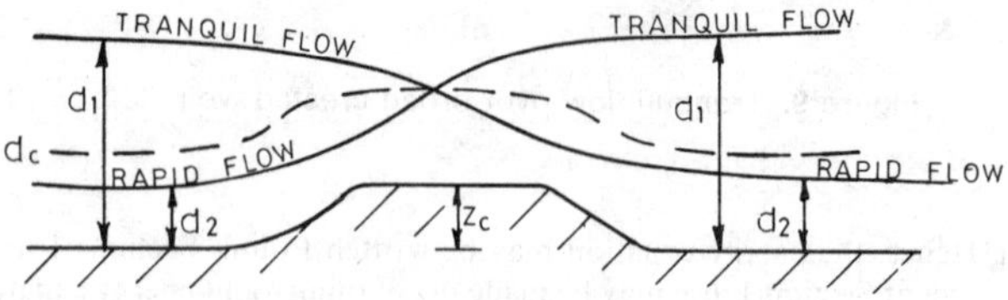

Figure 11. Rapid and tranquil flow over raised bed of critical height.

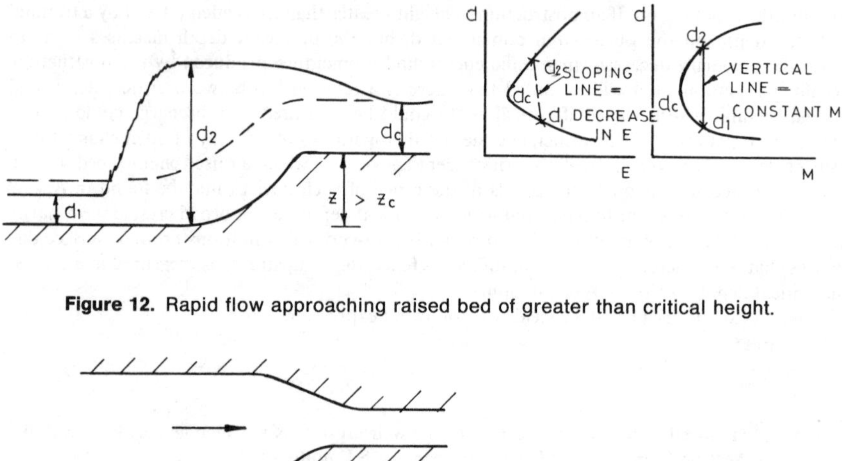

Figure 12. Rapid flow approaching raised bed of greater than critical height.

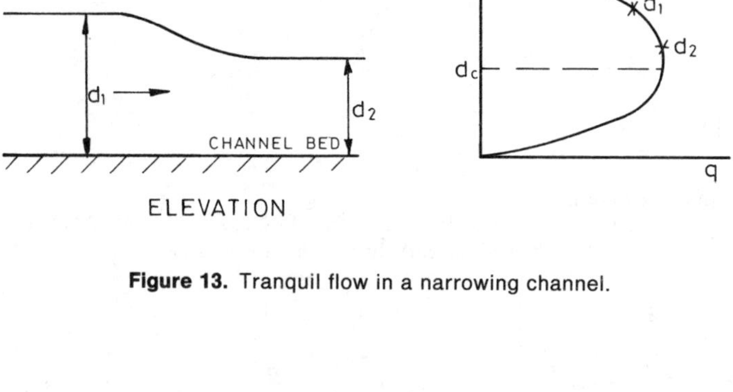

Figure 13. Tranquil flow in a narrowing channel.

discharge. Hence, the broad crested weir equation becomes:

$$Q = C_D B 1.705 H \tag{30}$$

The coefficient of discharge C_D can be determined only by experiment and is generally in the range 0.96 to 0.99.

Figure 10 shows rapid flow approaching an obstruction on the channel bed. As both momentum function and specific energy decrease, flow depth increases. As in the tranquil flow case, a critical height Z_c is reached when critical flow will occur. This leads to the general conclusion shown in Figure 11 where both tranquil and rapid flow pass through the critical depth when passing over an obstruction of the critical height Z_c. Either tranquil or rapid flow will be produced downstream of such a raised bed depending on downstream conditions.

If the bed hump is greater than Z_c for any specific approaching rapid flow, this will give rise to a hydraulic jump as shown in Figure 12. This will produce tranquil flow, which will adjust itself so as to pass through the critical depth at the hump.

Narrowing Channel

Figure 13 shows tranquil flow in narrowing rectangular channel, which produces an increase in discharge per unit width q. From the graph of depth versus discharge per unit width, it is obvious that this leads to a decrease in flow depth. If the narrowing is sufficient, critical conditions will be produced and further narrowing will result in an increase in upstream depth.

In an analogous manner to the broad crested weir, this phenomenon may be used as a flow measuring device, and is called the Venturi flume. By similar reasoning to that applied to the broad crested weir, an equation for S.I. units may be developed as follows:

$$Q = C_D 1.705 T H \tag{31}$$

where C_D = a coefficient of discharge in range 0.96–0.99
T = width of throat
H = head of upstream water level above bed at throat

As before, the coefficient of discharge C_D takes account of upstream velocity head, frictional energy loss, and losses due to the curvature of the streamlines. The Venturi flume is used extensively as an open channel flow measuring device. On occasions, a narrowing of the channel is combined with raising of the bed, so as to ensure critical conditions at all discharges. The channel is usually widened to its normal width downstream of the throat, though at a small angle to minimise energy loss.

Rapid flow approaching a narrowing channel generally undergoes a rise in water level as shown in Figure 14. If the channel narrows sufficiently, critical conditions would be produced. If the channel is further narrowed a hydraulic jump will occur in a similar manner to that depicted on Figure 12.

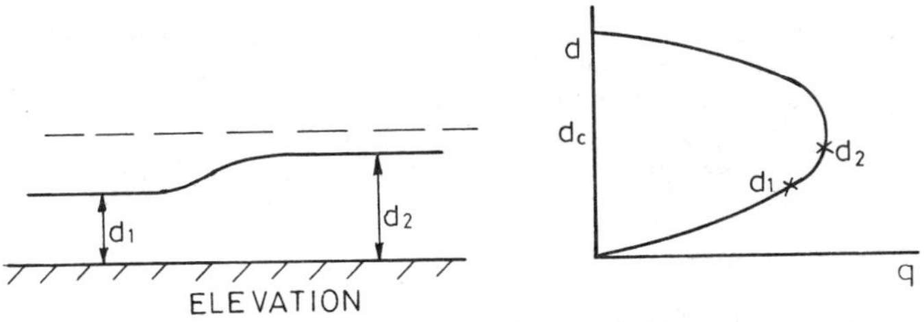

Figure 14. Rapid flow in a narrowing channel.

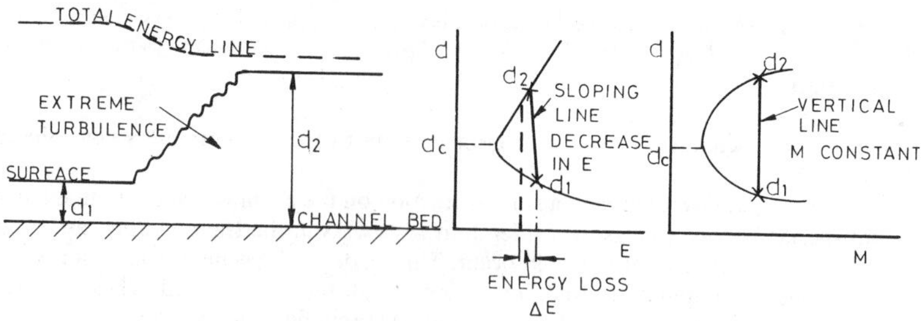

Figure 15. Hydraulic jump.

Hydraulic Jump

This is a phenomenon whereby rapid flow changes to tranquil with an associated decrease in energy due to considerable turbulence (Figure 15). The hydraulic jump occurs where rapid flow, produced by a steep slope or other means, cannot be sustained. An example would be flow at the base of a steep rectangular spillway that flows onto a horizontal or gently sloping apron. Because of the shallowness of the slope, there is insufficient energy input to sustain rapid flow. The specific energy of the flow decreases but cannot pass smoothly through the critical depth because there is no source of energy to push the flow up the tranquil limb of the specific energy diagram. Instead the depth increases rapidly, with much turbulence and consequent energy loss.

Because of the significant energy loss, or strictly transformation into heat energy, the energy equation cannot be used to analyze the hydraulic jump. However, momentum is conserved so the momentum equation is applicable. Neglecting frictional and weight forces, this may we written as

$$\frac{\rho g d_1^2}{2} + \frac{\rho q^2}{d_1} = \frac{\rho g d_2^2}{2} + \frac{\rho q^2}{d_2} \tag{31}$$

Equation 31 effectively equates momentum functions before and after the jump. This results in a quadratic equation in either d_1 or d_2 with a positive root:

$$d_2 = \left[\frac{d_1^2}{4} + \frac{2q^2}{g d_1}\right]^{1/2} - \frac{d_1}{2} \tag{32}$$

d_1, d_2 are known as initial and sequent depths and either may be evaluated if the other and the discharge are known. Once d_1 and d_2 have been evaluated, the energy loss may be calculated from

$$\Delta_E = \frac{(d_2 - d_1)^3}{4 d_1 d_2} \tag{33}$$

The hydraulic jump is used to transform rapid flow that may cause damage to channel bed and banks into slower flowing tranquil flow. Thus, steps are taken to arrange hydraulic jumps at the base of spillways and similar locations where rapid flow occurs. The slower flowing tranquil stream may then be returned safely to the river downstream. Another use of the hydraulic jump is as an energy dissipator, perhaps to settle out suspended matter or similar function.

STEADY UNIFORM FLOW

The foregoing has been based on the assumption that frictional effects are negligibly small. This is generally true over short channel lengths but not when longer channels are considered. In this case frictional effects cannot be ignored.

The simplest type of real fluid flow is steady uniform, defined as constant flow parameters in both time and space. Such flow is impossible to achieve due to variations of velocity with a cross section as a result of boundary friction and also due to turbulent fluctuations of velocity in all three dimensions. Steady uniform flow is therefore redefined as that in which there is no change in average flow parameters with time or distance. Even thus defined, uniform flow occurs very rarely outside the laboratory. However, being easy to analyze, uniform flow is often assumed as an approximation to the truth and, furthermore, uniform flow equations are incorporated in more complex expressions that describe gradually varied and unsteady flow situations.

It was stated earlier that channel flow could either be laminar or turbulent. While this is true in theory, in practice laminar flow is very rare, since it requires very low depths and very small channel velocities. Hence, attention will be confined to turbulent flow in rectangular channels.

Uniform Flow Equations

Chow [4] gives a comprehensive review of the development of uniform flow equations and their modifications. The first usable formula was developed empirically by Antoine de Chezy, though it is possible now to derive his equation mathematically; the proof is given by Massey [3]. Chezy's equation may be stated as follows:

$$V = C\sqrt{Ri} \tag{34}$$

where
- V = average cross-sectional velocity = Q/A
- Q = channel discharge
- A = cross-sectional area
- R = hydraulic radius = A/P
- P = wetted perimeter
- i = longitudinal bed slope
- C = Chezy's roughness coefficient

Equation 34 may be rewritten to give channel discharge

$$Q = AC\sqrt{Ri} \tag{35}$$

The equation may be used to calculate channel discharge provided the geometrical properties are known and provided an accurate value of Chezy's roughness coefficient may be estimated. The roughness coefficient considers such factors as bed and bank roughness, channel geometry, and channel bends; in fact, all factors contributing to flow resistance.

Not surprisingly, the roughness coefficient is the most difficult parameter to quantify in Equation 35, and many ingenious formulae have been suggested to evaluate C, as reported by Chow [4]. The attempt that has stood the test of time is that attributable to Manning, which states

$$C = \frac{R^{1/6}}{n} \tag{36}$$

where n = Manning's roughness coefficient. This gives the full Manning equation as

$$V = \frac{1}{n} R^{2/3} i^{1/2} \tag{37}$$

Equation 37 applied to S.I. units. The corresponding equation in imperial units is

$$V = \frac{1.486}{n} R^{2/3} i^{1/2} \tag{38}$$

Table 1
Values of Manning's Roughness Coefficient

Bed and Bank Material	Manning's n
Glass, plastic, machined metal	0.010
Planed timber, joints flush	0.011
Concrete, steel trowelled	0.012
Cast iron, brickwork	0.013
Concrete, timber forms	0.014
Vitrified clay, asphalt	0.015
Untreated gunite	0.015–0.017
Rubble set in cement	0.017
Firm gravel	0.020
Smooth earth, no weeds	0.025
Earth with some stones and weeds	0.025
Natural channels: clean and straight	0.025–0.030
Natural channels: winding	0.033–0.040
Natural channels: overgrown	0.075–0.150

The Manning equation is the most widely used uniform flow equation. As with the Chezy roughness coefficient, so Manning's n attempts to consider all factors contributing to flow resistance. Equation 36 indicates that Manning's n is assumed as a constant depending on the bed and bank material. Roughness coefficient behavior is more complex than this as will be shown below, but for many engineering purposes it may be assumed constant. Chow [4] gives a very comprehensive presentation of Manning's n values using both descriptions and photographs to aid the correct choice of value in any situation. A selection of values is given in Table 1.

The Darcy-Weisbach formula, which was developed to analyze steady flow in pipes, is also applicable to open channel flow. The formula is generally expressed as:

$$h_L = f\frac{L}{D}\frac{V^2}{2g} \tag{39}$$

where h_L = head loss due to friction
f = Darcy-Weisbach friction factor
L = pipe length
D = pipe diameter
V = average cross-sectional velocity
g = gravitational acceleration

To render Equation 39 applicable to channel flow, it is necessary to substitute the hydraulic radius R for the pipe diameter D according to

$$D = 4R \tag{40}$$

Substituting Equation 40 into Equation 39 and rearranging gives

$$V = \sqrt{\frac{8gRi}{f}} \tag{41}$$

Equation 41 is the form of the Darcy-Weisbach formula application to channel flow, and may be compared with Equation 37 and Equation 34, yielding the conclusion that:

$$C = \frac{R^{1/6}}{n} = \sqrt{\frac{8g}{f}} \tag{42}$$

The American Society of Civil Engineers' Task Force on Friction Factors in Open Channels [5] suggests that the Darcy-Weisbach equation and friction factor should be more widely applied to open channels, especially those of simple cross-sectional shape, such as rectangular. A major advantage is that the friction factor f is dimensionless, allowing comparison of results regardless of units used. Additionally, it also allows investigators in open channels to take full advantage of the significant advances in the knowledge of friction factor behavior in pipes in this century. For these reasons many recent investigations into open channel behavior have adopted the use of the Darcy-Weisbach friction factor.

It will be noted that Chezy's C and Manning's n are referred to as roughness coefficients, whereas the Darcy-Weisbach f is called a friction factor. This is due to the different backgrounds from which the terms have come. Open channel flow resistance is made up of many factors such as bed and bank material, sinuosity, and channel geometry, whereas resistance to flow in pipes takes the form predominantly of boundary shear. However, it is convenient to consider these terms as synonymous whether applied to channels or pipes. For convenience, the term friction factor will generally be used to include the effects of all flow resistance from whatever sources.

Flow Resistance

The determination of flow resistance represents one of the most difficult problems in open channel hydraulics. Sources of resistance to flow in rectangular channels would include boundary shear stress, secondary circulation, effects of bends and channel irregularities, changes of channel depth and width, and types of bank and bed material, amongst others, Not surprisingly, therefore, as Henderson states [6]: "While the behavior of f in circular pipe flow has been thoroughly explored . . . a similarly complete investigation of the behavior of C has never been made . . . "

Choice of a value of friction factor or roughness coefficient for use in uniform flow equations has been made on empirical or experiential grounds without the assistance of any universal theory of resistance such as is available for pipe flow. A significant body of data is available in this form, as noted in the preceding section, and many channel designs have been thus adequately executed. However, more recently, attempts have been made to evolve universal resistance relationships similar to those available for pipe flow and based on similar reasoning.

One complicating feature of flow even in straight uniform channels is non-uniformity of shear stress patterns around the periphery. Many investigators including Myers and Elsawy [7], Cruff [8], Rajaratnam and Muralidhar [9] have measured boundary shear in rectangular channels. Figure 16 shows a typical boundary shear pattern for a rectangular channel. Typically, maximum bed shear occurs at the centerline if the channel is symmetrical, and maximum wall shear occurs somewhat below the free surface. The latter point reflects the isovel pattern typically shown on Figure 16, where the filament of maximum velocity occurs below the free surface due to secondary circulation.

If the rectangular channel is wide, having a width to depth ratio of ten or more, there is a region of constant bed shear around the channel centerline as shown in Figure 17. This indicates the

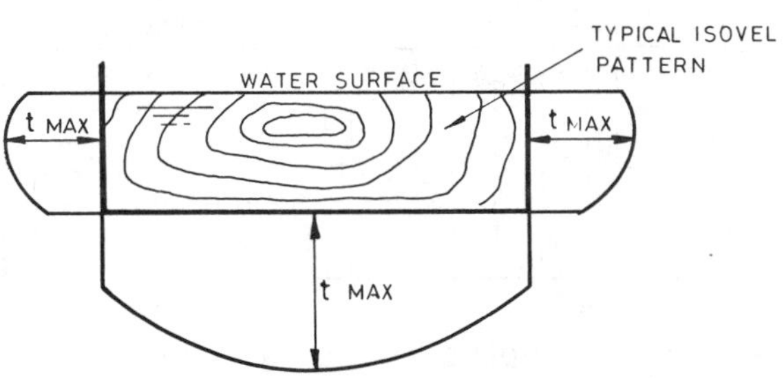

Figure 16. Typical boundary shear stress and isovel patterns in rectangular channel.

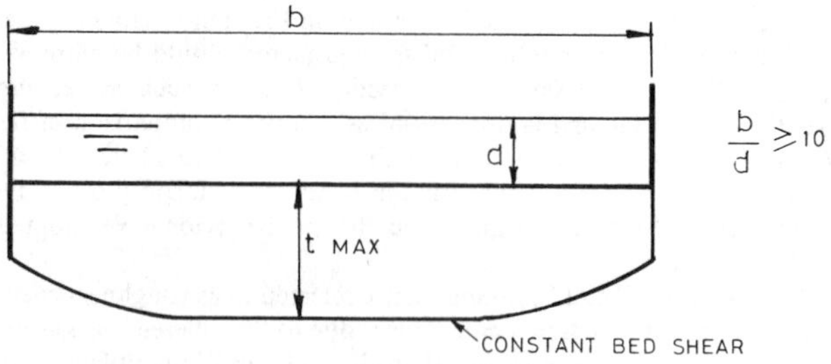

Figure 17. Bed shear stress pattern in wide rectangular channel.

presence of two-dimensional flow where there is no lateral momentum transfer between adjacent blocks of fluid [8].

If flow is uniform, then by definition the water surface is parallel to the bed. Hence, the weight force component in the flow direction is equal in value to the frictional force in the opposing direction. The frictional force is the integration of the boundary shear pattern in Figures 16 and 17. In the central region of a wide channel (Figure 17) the weight component and the bed shear force are equal for each unit of width. Hence, it is possible to derive a formula for τ_{max} on Figure 17 as follows:

$$\tau_{max} = \rho gdi \tag{43}$$

where τ_{max} = 2-dimensional shear stress value
ρ = fluid density
g = gravitational acceleration
d = depth
i = longitudinal bed slope

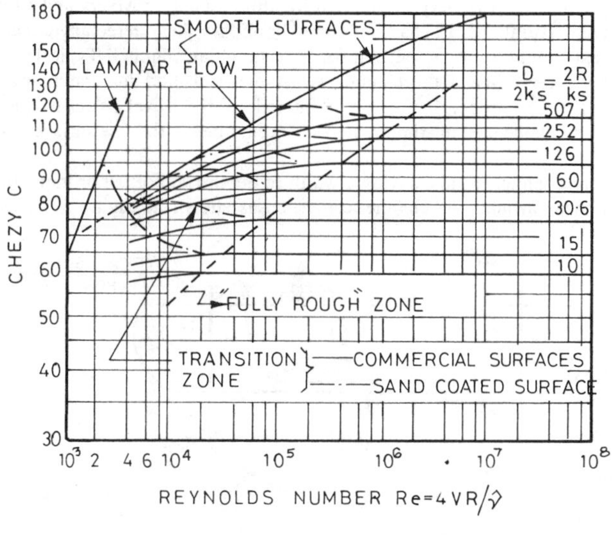

Figure 18. Modified Moody diagram showing behavior of Chezy C (After Henderson [6]).

The understanding of friction factor behavior in open channels has been greatly enchanced by developments in pipe flow. Based on earlier theoretical work by Prandtl and Von Karman, the experimental investigations by Nikuradse [10], Colebrook [11], friction factor behavior in pipes is now substantially understood. Ackers [14] and Chow [4] give fuller details of this background. The results have been presented graphically by Moody [13], while tables and charts [14, 15] have been prepared at the Hydraulics Research Station in England.

Using the relationship shown as Equation 42, it is possible to modify the Moody diagram to show the behavior of Chezy's C. This is shown on Figure 18, which is presented by Henderson [6], and where it may be seen that such a procedure assumes that Chezy's C, like Darcy's f, is a function of Reynolds number and boundary roughness. Hence, no allowance is made in this approach for cross-sectional shape or the presence of the free surface.

Referring to Figure 18, three regions of turbulent flow may be identified as *smooth*, *transitional*, and *rough*. In smooth turbulence, the roughness projections on the channel boundary are submerged in the laminar sub-layer. In this region the friction factor would be described by

$$\frac{1}{\sqrt{f}} = 2.0 \log_{10}(R_e\sqrt{f}) - 0.8 \tag{44}$$

or

$$C = 4\sqrt{2g}\log_{10}\left(\frac{R_e\sqrt{8g}}{2.51C}\right) \tag{45}$$

where f = Darcy-Weisbach friction factor
C = Chezy roughness coefficient
R_e = Reynolds number = $4VR/\nu$
R = hydraulic radius
V = average cross-sectional velocity
ν = coefficient of kinematic viscosity = μ/ρ
g = gravitational acceleration
μ = coefficient of absolute viscosity

More recent studies have confirmed the validity of the general form of Equation 44 in rectangular channels, but with minor modifications to the numerical constants. Keulegan [15] derived the following expression for two-dimensional channel flow:

$$\frac{1}{\sqrt{f}} = 2.03 \log(R_e\sqrt{f}) - 1.08 \tag{46}$$

Several experimental investigations have been carried out in smooth rectangular channels confirming that the general form of friction factor behavior is

$$\frac{1}{\sqrt{f}} = A \log(R_e\sqrt{f}) - B \tag{47}$$

Numerical constants A and B have been found as shown on Table 2. The comparatively minor differences in the numerical constants are probably due to experimental error. The differences between the numerical constants in Equation 44 and those in Table 2 are due to shape effects in rectangular channels.

Referring again to Figure 18, at high Reynolds number, friction factor behavior depends on boundary roughness in the form of the ratio of height of roughness elements to hydraulic radius,

Table 2
Values of Numerical Constants in Friction Factor Relationship for Rectangular Channels

Investigator	Value of A	Value of B
Keulegan [15]	2.03	1.08
Reinus [17]	2.00	1.06
Tracy & Lester [18]	2.03	1.30
Rao [19]	2.12	1.83
Myers [20]	2.10	1.56
Kazemipour & Apelt [21]	2.00	1.13
Kazemipour & Apelt [22]	2.00	1.62

known as relative roughness. For pipe flow this relationship is (Figure 18)

$$\frac{1}{\sqrt{f}} = \frac{C}{\sqrt{8g}} = 2 \log_{10}\left(\frac{14.84R}{K_s}\right) \tag{48}$$

where K_s = Nikuradse's equivalent sand roughness.

Relatively fewer studies have been undertaken in rough walled open channels when compared with the smooth case. Physically, flow resistance is governed entirely by the size of roughness projections since the laminar sub-layer has reduced in size, thereby allowing the roughness projections to project fully into the flow.

Henderson [6] suggests a modification of Equation 48 to suit open channels as follows:

$$\frac{1}{\sqrt{f}} = \frac{C}{\sqrt{8g}} = 2 \log_{10}\left(\frac{12R}{K_s}\right) \tag{49}$$

The only change is to the value of the numerical constant. Henderson presents a table of K_s values for concrete and masonry surfaces as shown on Table 3. Considerably more data is necessary, particularly from field measurements to establish Equation 49 for general use. This is especially so since natural and artificial surface roughnesses do not bear any resemblance to Nikuradse's sand roughness.

Table 3
Values of Equivalent Sand Roughness (K_s) for Concrete and Masonry Surfaces

Surface	K_s in feet	K_s in mm
Concrete—very high quality finish	0.0005	0.15
Concrete—very smooth finish	0.001	0.30
Concrete—smooth finish	0.0016	0.49
Planed timber	0.002	0.61
Glazed brickwork, smooth gunite	0.005	1.52
Small diameter concrete pipe	0.008	2.44
Concrete—rough finish	0.01	3.05
Roughly made concrete conduits	0.014	4.27
Rubble masonry	0.02	6.10
Untreated gunite	0.01–0.03	3.05–9.14

Between rough turbulence and smooth turbulence lies a transitional region where friction factor behavior depends on both Reynolds number and relative roughness (Figure 18). Friction factor behavior in this region for pipe flow is described by the Colebrook-White equation:

$$\frac{1}{\sqrt{f}} = \frac{C}{\sqrt{8g}} = -2\log_{10}\left(\frac{K_s}{14.83R} + \frac{2.52}{R_e\sqrt{f}}\right) \tag{50}$$

Henderson suggests a modification as follows for small channels of uniform cross-section, such as rectangular channels:

$$\frac{C}{\sqrt{8g}} = -2\log_{10}\left(\frac{K_s}{12R} + \frac{2.5}{R_e\sqrt{f}}\right) \tag{51}$$

As with Equation 49 much more data is necessary to establish the validity of Equation 51 for a wide range of open channel cases.

The three types of flow—smooth, transitional and rough—are distinguished from each other by the size of the dimensionless parameter K_sV^*/ν, where V^* is called the shear velocity and is defined

$$V^* = \sqrt{\frac{\tau_0}{\rho}} = \sqrt{gRS_f} \tag{52}$$

where τ_0 = boundary shear stress

S_f = friction slope $= \dfrac{V^2}{C^2R}$

Shear velocity, cannot be identified with any physically real velocity, although it has the dimensions of velocity. It may, however, be related to average cross-sectional velocity by combining Equations 35 and 52 to give

$$\frac{V}{V^*} = \frac{C}{\sqrt{g}} = \sqrt{\frac{8}{f}} \tag{53}$$

The transitional region is defined approximately by the limits

$$4 < \frac{V^*K_s}{\nu} < 100 \tag{54}$$

It must be re-emphasized that equations based on pipe flow experience, or variations thereof, are applicable only to small rectangular channels made up of fairly smooth commercial materials. For large natural channels having large scale roughness, recourse must be had to empirical data such as the Manning's equation, together with values of roughness coefficient as shown on Table 1. Henderson has shown that the Manning equation in particular is suitable for the fully rough turbulent region, for which in the context of natural channels it was originally derived. However, it should also be noted that the empirical equations of Manning and Chezy with their usual combinations with constant roughness coefficients are not suitable for application in the smooth and transitional turbulence regions.

Economical Cross Section

A typical aim in the design of rectangular channels for uniform flow is minimizing the cross-sectional area for a given value of discharge, or alternatively maximizing discharge for a given value of cross-sectional area. Reference to the Chezy Equation 34 shows that for a given channel bed slope and Chezy C, velocity and hence discharge are maximum if hydraulic radius R is a maximum. This in turn implies that for a given value of cross-sectional area, wetted perimeter P must be a minimum. By differentiating the wetted perimeter expression with respect to depth and putting equal to zero, it is found that for minimum P

$$b = 2d \tag{55}$$

This corresponds to a rectangular section inside which a semi-circle may be drawn, and represents the channel giving the maximum discharge for a given cross section. However, in channel design many other factors enter in such as flow resistance, sediment transport, and ease of construction, so that it is not always true to say that Equation 58 describes the optimum design in any given situation.

GRADUALLY VARIED FLOW

Steady non-uniform flow was defined as characterized by variation of fluid parameters with distance but not with time. Steady non-uniform flow, as previously stated, may also be called varied flow, and may be rapidly varied if the distances over which changes take place are short and gradually varied if such distances are long. The essential difference is whether or not frictional forces may be neglected. Rapidly varied flow has already been dealt with as it may be analyzed using the basic concepts of energy and momentum previously explained. Gradually varied flow involves significant frictional forces and hence must include terms to account for flow resistance based on ideas introduced earlier. Gradually varied flow is a common phenomenon in both natural and artificial channels, and hence the following analysis is applicable to a wide range of problems in rectangular channels.

Varied Flow Equation

Figure 19 is a definition sketch showing a gradually varied flow water surface profile with relevant parameters identified. Several assumptions underlie the derivation of a useful equation, namely hydrostatic pressure distributions and a kinetic energy factor of unity.

By equating energy at Sections 1 and 2 as shown on Figure 19, and applying the continuity equation, it is possible to derive an equation for change of depth with distance, which for a rectangular channel is

$$\frac{dd}{dL} = \frac{i - S_f}{1 - \left(\frac{d_c}{d}\right)^3} \tag{56}$$

where d = depth
L = channel length between adjacent sections
i = channel bed slope
S_f = slope of total energy line
d_c = critical depth

A full derivation of Equation 56 may be found in Francis [1]. Equation 56 describes depth and hence velocity changes with distance in a rectangular channel under gradually varied flow. If bed slope and total energy line slope are equal, i.e. $i = S_f$, flow is uniform and change of depth with distance is zero. As depth approaches the critical depth, the denominator of the right-hand side of Equation 56 approaches zero, indicating infinite depth changes. Hence, the equation does not apply

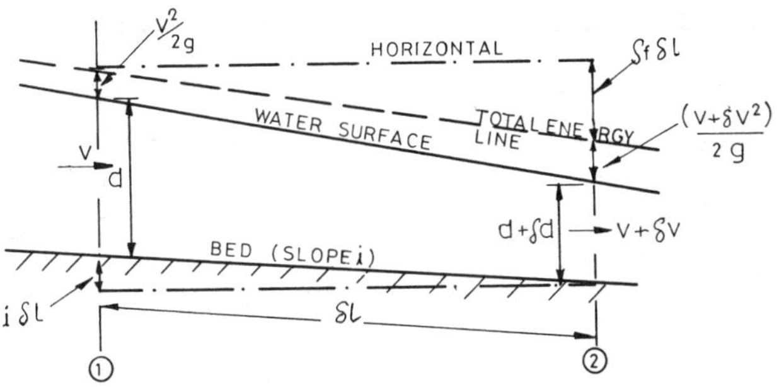

Figure 19. Definition sketch for gradually varied flow equation.

at or near the critical depth. This point however explains why critical conditions are accompanied by small surface waves, since the change of depth with distance is continually changing sign.

To integrate Equation 56, an expression must be found for the slope of the total energy line S_f, known as the friction slope. This is done by assuming that the friction slope has the same value under gradually varied flow, as for the same depth and velocity under uniform flow. Hence, friction slope may be expressed using the Chezy, Manning, or Darcy-Weisbach equations. Usually, Chezy or Manning are adopted giving:

$$S_f = \frac{V^2}{C^2R^2} \tag{57}$$

or

$$S_f = \left(\frac{Vn}{R^{2/3}}\right)^2 \tag{58}$$

Either Equation 57 or 58 is substituted into Equation 56, which is then integrated numerically by dividing the operation into discrete steps of depth. The change of depth with length dd/dL is evaluated for each discrete step, and the summation gives a total length for total depth change over the reach in question.

For a very wide rectangular channel, usually defined as having a width to depth ratio of at least 10, Equation 56 may be further simplified by considering any unit width of channel. This analysis, when the Chezy Equation is used, gives:

$$\frac{dd}{dL} = \frac{i\left(1 - \left(\frac{d_0}{d}\right)^3\right)}{1 - \left(\frac{d_c}{d}\right)^3} \tag{59}$$

where d_0 is the normal depth.

If the Manning uniform flow equation is used, Equation 59 becomes

$$\frac{dd}{dL} = \frac{i\left(1 - \left(\frac{d_0}{d}\right)^{10/3}\right)}{1 - \left(\frac{d_c}{d}\right)^3} \tag{60}$$

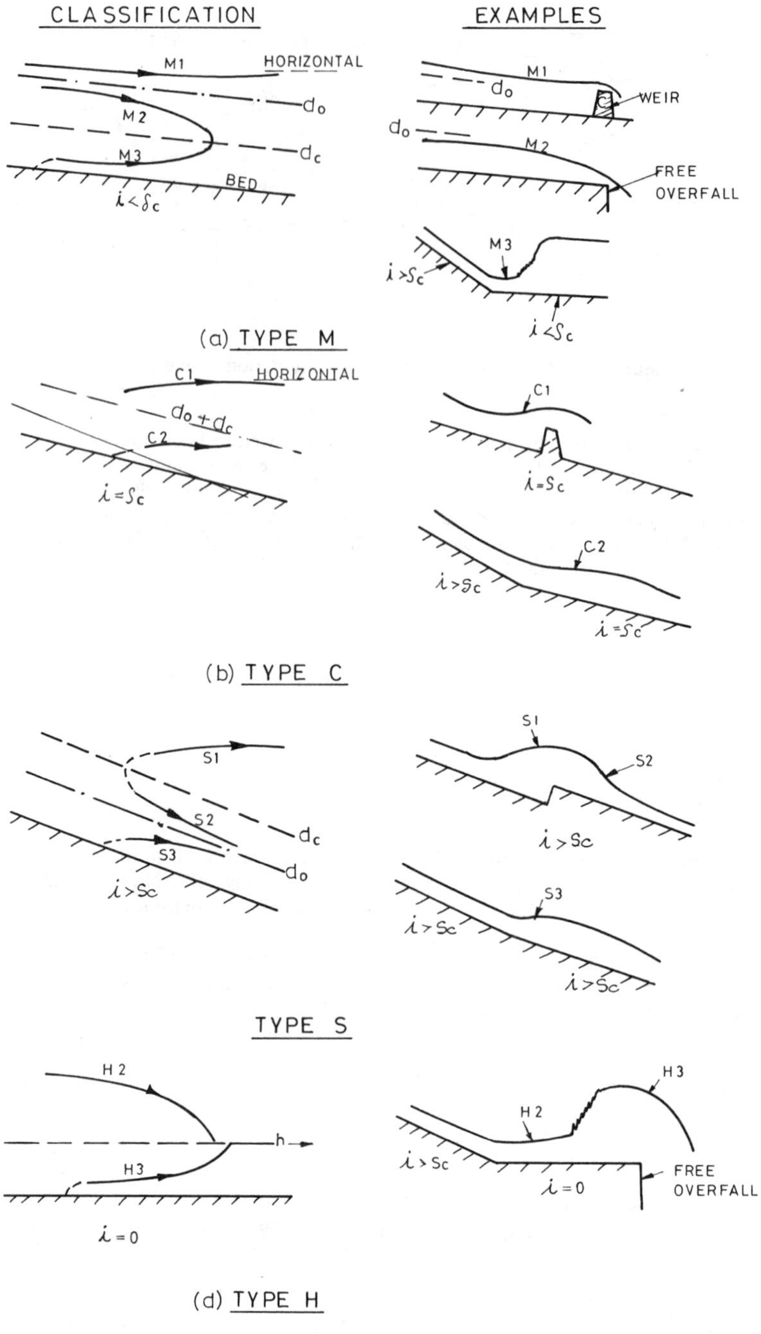

Figure 20. Gradually varied flow profiles and examples.

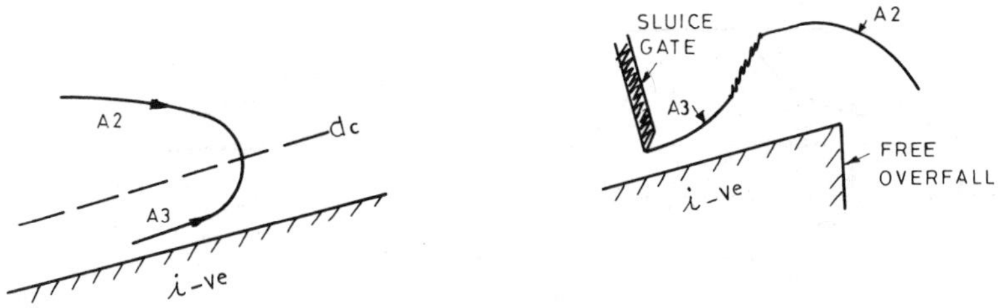

(e) TYPE A

Figure 20. (Continued)

Classification of Surface Profiles

Gradually varied flow surface profiles are classified according to bed slope and depth of flow. Bed slopes are classified as follows:

Mild:	(M)	$i < S_c$
Critical:	(C)	$i = S_c$
Steep:	(S)	$i > S_c$
Horizontal:	(H)	$i = 0$
Adverse:	(A)	i is negative

where S_c is the slope at which normal and critical depths coincide.

Because critical depth is a function of discharge, so too is slope classification. Hence, a slope may be mild for one flow but steep for another.

Depths may be in one of three regions:

Region 1: $d > d_c$ and $d > d_0$

Region 2: d between d_0 and d_c

Region 3: $d < d_c$ and $d < d_0$

The letters indicating slope are combined with numbers indicating depth to give a symbol for each possible profile, consisting of a letter and a number. Since there are five letters and three numbers, this would seem to imply fifteen possible combinations. However, since there can be no normal depth on horizontal or adverse slopes, profiles A1 and H1 do not exist. Also, since at the critical slope, normal and critical depths coincide, there is no C2 profile. This leaves twelve possible gradually varied flow profiles, and examples of each are shown in Figure 20.

The classification as shown in Figure 20 may be used to break down a gradually varied flow situation into its component profiles, each of which may be analyzed using the gradually varied flow equations previously developed.

UNSTEADY FLOW

Unsteady flow is characterized by variation of flow parameters with time. The full term is unsteady non-uniform flow indicating variation with both time and distance. Unsteady uniform flow is not possible in a channel because the free surface is free to move in the vertical plane.

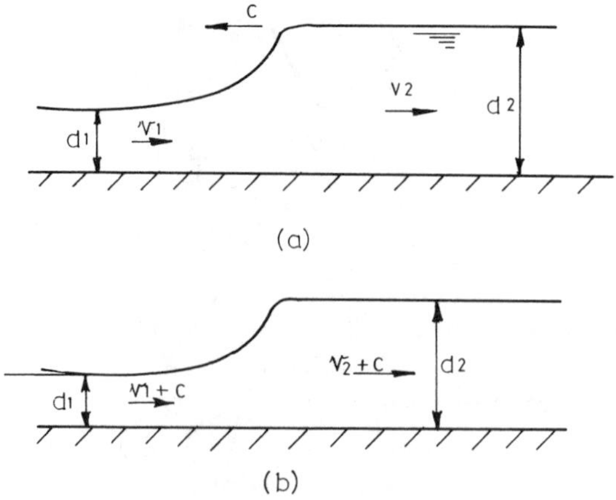

Figure 21. Upstream positive surge.

Unsteady non-uniform flow may be divided into two types—rapidly varied unsteady flow, where changes take place over relatively short distances; and gradually varied unsteady flow, where distances over which changes take place are relatively long. The most commonly occurring example of the former is the surge, while gradually varied unsteady flow will be dealt with under the heading of flood routing.

Surges

A surge is produced by a rapid change in the rate of flow, for example by the rapid opening or closing of a control gate in a channel. The phenomenon may be produced naturally in a tidal reach, by rapid tidal fluctuations that give rise to a surge, which is then termed a "bore."

If an increase in depth results from the passage of the surge, it is termed a positive surge, whereas a negative surge produces a decrease in depth. Surges may travel upstream or downstream.

Surges, although unsteady in nature, may be analyzed using the steady flow momentum Equation 18. This is possible by changing the reference point from that of a stationary observer to that of an observer moving in the direction of and with the celerity (velocity) of the surge. This transforms the problem from unsteady to steady, allowing application of Equation 18. The resulting equations for rectangular channels are presented in the following section.

The positive surge moving upstream with a celerity C is shown on Figure 21a, while Figure 21b shows the transformation of this to a steady flow situation by moving the reference point upstream with a speed equal to the surge celerity. Application of the momentum Equation 18 and continuity Equation 1 leads to the following expression of the celerity of an upstream positive surge in a rectangular channel [23]:

$$C = \left[\frac{gd_2}{2} \cdot \frac{d_2 + d_1}{d_1}\right]^{1/2} - v_1 \tag{61}$$

The hydraulic jump is a special case of the surge where celerity C is zero. Putting $C = 0$ in Equation 61 yields the hydraulic jump equation as given in Equation 32. Another special case is that of small surges in deep water where

$$d_1 = d_2 = d \tag{62}$$

Substituting d for d_1 and d_2 in Equation 61 yields

$$C = \sqrt{gd} - v_1 \tag{63}$$

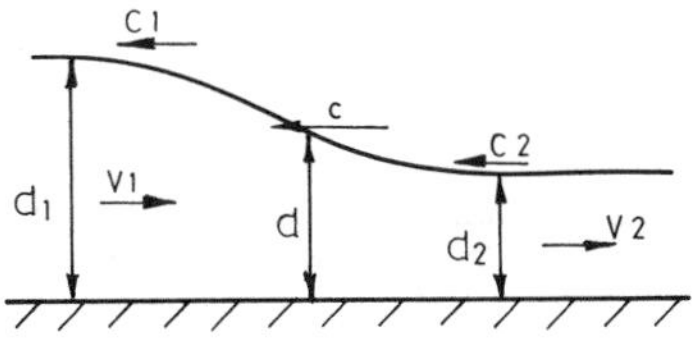

Figure 22. Upstream negative surge.

In still water where $v_1 = 0$, the celerity of a surge of small amplitude is

$$C = \sqrt{gd} \tag{64}$$

Note that Equation 64 is exactly equivalent to Equation 17 and Equation 14, showing the reasoning behind the assertion that when critical conditions occur, small surface waves cannot be transmitted upstream.

By similar reasoning to that which led to Equation 61, the celerity of a positive surge moving downstream may be expressed by

$$C = \left[\frac{gd_1}{2d_2}(d_1 + d_2)\right]^{1/2} + v_2 \tag{65}$$

The negative surge that appears to the observer as a lowering of the liquid surface, occurs downstream of a control gate when the opening has been suddenly reduced or in the upstream channel as the gate is suddenly opened. The wave front may be considered to be composed of a series of small waves superimposed on each other. Since the uppermost wave has the greatest depth it travels faster than those beneath; the retreating wave front therefore becomes flatter.

The average celerity of an upstream negative surge (Figure 22) is given by [23]

$$C = 3\sqrt{gd} - 2\sqrt{gd_1} - v_1 \tag{66}$$

The celerities at the crest and trough are given respectively by

$$C_1 = \sqrt{gd_1} - v_1 \tag{67}$$

$$C_2 = 3\sqrt{gd_2} - 2\sqrt{gd_1} - v_1 \tag{68}$$

The corresponding equations for a negative surge moving downstream are

$$C = 3\sqrt{gd} - 2\sqrt{gd_2} + v_2 \tag{69}$$

$$C_1 = 3\sqrt{gd_1} - 2\sqrt{gd_2} + v_2 \tag{70}$$

$$C_2 = \sqrt{gd_2} + v_2 \tag{71}$$

Remember that these equations for surge analysis, based on the application of the momentum equation, apply only to frictionless channels. They may not therefore be used to predict accurately the celerity of a surge at considerable distance from its inception. Generally, frictional effects will tend to attenuate the surge, which will eventually disappear completely if it travels sufficiently far upstream.

Flood Routing

This is a mathematical procedure whereby the progress of a flood discharge is charted as it moves downstream. In other words it is a method of analysis for unsteady flow in channels and rivers. Such

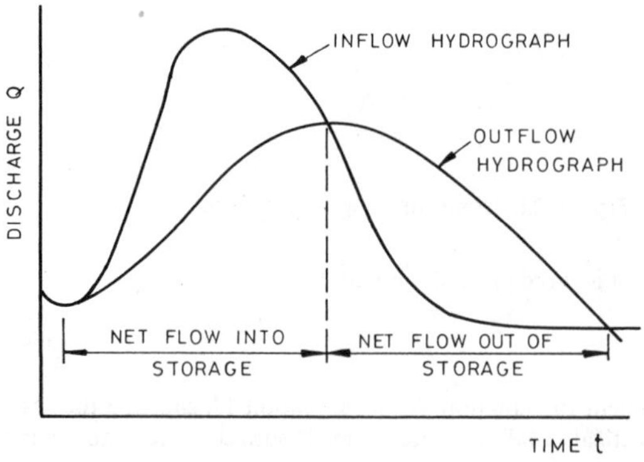

Figure 23. Observed inflow and outflow hydrographs for channel reach.

a method is very important in that many cases of artificial and natural channel flow fall into the unsteady classification in such a way that they cannot be analyzed by simplifying procedures such as those previously outlined.

Flood routing methods may be divided into storage methods, called hydrological routing, and dynamic methods, called hydraulic routing.

Hydrological routing or storage routing involves the solution of the storage equation, which for a channel reach may be written as

$$I - D = \frac{dS}{dt} \tag{72}$$

where I = instantaneous inflow
D = instantaneous outflow
S = storage volume
t = time

The problem is complicated by the fact that for a channel, storage is a function of both outflow and inflow. The most widely used hydrological routing method is the Muskingum technique, which suggests the following relationship:

$$S = K[xI + (1 - x)D] \tag{73}$$

where K and x are parameters that must be determined by hydrograph observation.

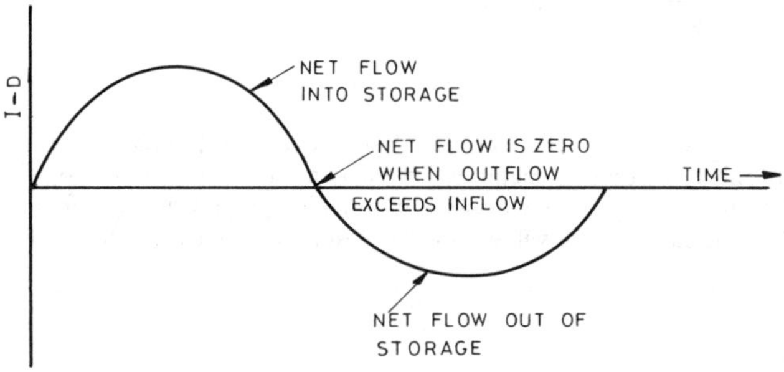

Figure 24. Difference diagram.

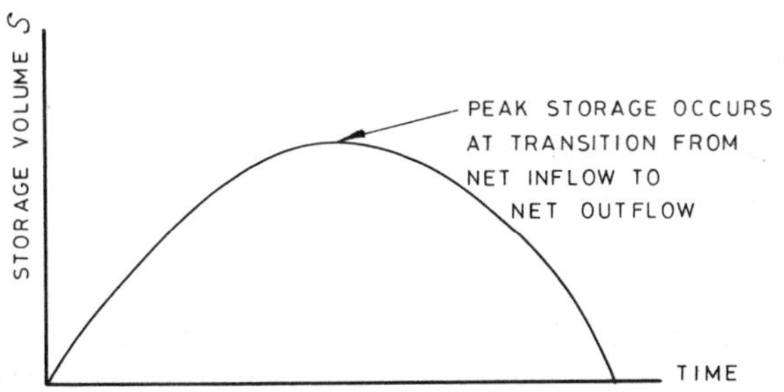

Figure 25. Cumulative graph of storage volume.

Figure 23 shows observed inflow and outflow hydrographs for a channel reach. These may be used to produce a difference diagram (Figure 24), which represents net flow into and out of storage. Hence, a cumulative graph of storage volume may be produced (Figure 25) by multiplying average net flow into or out of storage by time for each unit period for the entire storm.

Recalling Equation 73 it will be noted that there are two unknowns, x and K and only one equation. The method of solution therefore involves estimating a value for x, which usually lies between 0.1 and 0.5, and hence evaluating the right-hand side of Equation 73 for several equally spaced time intervals throughout the entire storm. Since Figure 25 yields values of storage S for these same time periods, it is possible to plot a graph of S versus $[xI + (1 - x)D]$. If the correct value of x has been estimated, the relationship will plot as a straight line (Figure 26) and the slope may be used to evaluate K, which has units of time. If the wrong value of x has been chosen, the resulting plot will be in the form of a storage loop (Figure 26) necessitating repetition of the process with a new value of x.

Once x and K have been evaluated, they may be inserted into Equation 73. Considering a short time interval t, known as the routing period, then Equations 72 and 73 may be rewritten as

$$\frac{I_1 + I_2}{2} t - \frac{D_1 + D_2}{2} t = S_2 - S_1 \tag{74}$$

and

$$S_2 - S_1 = K[x(I_1 - I_2) + (1 - x)(D_1 - D_2)] \tag{75}$$

Combining Equations 74 and 75 gives a routing equation for a channel reach which is:

$$D_2 = C^0 I_2 + C^1 I_1 + C^2 D_1 \tag{76}$$

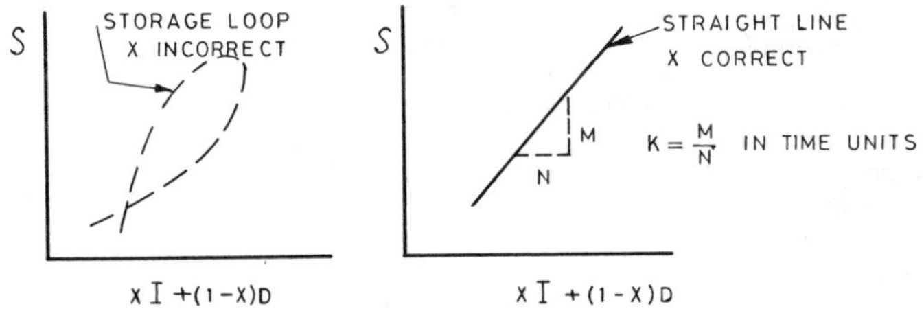

Figure 26. Determination of K by plotting storage loops.

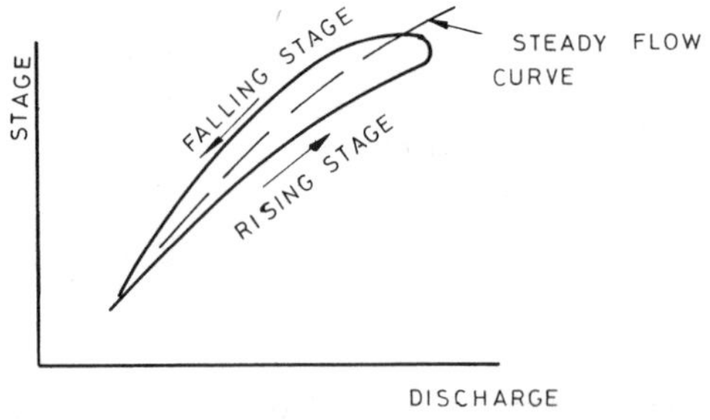

Figure 27. Typical looped stage-discharge curve for unsteady flow.

where $$C^0 = -\frac{Kx - 0.5t}{K - Kx - 0.5t} \tag{77}$$

$$C^1 = \frac{Kx + 0.5t}{K - Kx + 0.5t} \tag{78}$$

$$C^2 = \frac{K - Kx - 0.5t}{K - Kx + 0.5t} \tag{79}$$

For the routing procedure to be completed, the inflow hydrograph must be known, as also must the value of outflow at the commencement of the storm. Typically, inflow and outflow have the same value at the commencement of the flood. Information obtained from hydrological routing is in the form of outflow discharge only, which must be converted to depth using a stage discharge relationship and hence to average velocity by dividing discharge by cross-sectional area.

Herein lies one of the disadvantages of hydrological routing, since the stage discharge relationship under unsteady flow conditions is not unique but assumes a looped form as shown in Figure 27. Furthermore, the method is not very precise near the crest of the hydrograph, so it is difficult to be sure whether the flood wave is or is not subsiding as it moves downstream. For these and other reasons [4, 6, 24], if greater accuracy is required, recourse must be made to hydraulic methods.

Hydraulic flood routing involves solution of the equations of motion (St. Venant equations) for unsteady flow. A simple case for a rectangular channel is given below.

The equation of motion describing flow in a rectangular channel shown in Figure 28 is:

$$S_f = \underbrace{\underbrace{\underbrace{i}_{\text{steady uniform flow}} - \frac{\partial d}{\partial x} - \frac{v}{g}\frac{\partial v}{\partial x}}_{\text{steady non-uniform flow}} - \frac{1}{g}\frac{\partial v}{\partial t}}_{\text{unsteady non-uniform flow}} = \frac{v^2}{C^2 R} \tag{80}$$

where i = bed slope
S_f = friction slope

The equation of continuity is

$$\frac{\partial q}{\partial x} + \frac{\partial d}{\partial t} = 0 \tag{81}$$

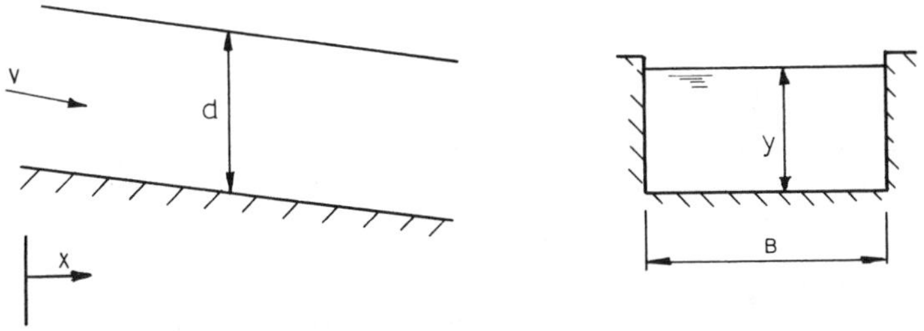

Figure 28. Flow in rectangular channel.

Let c = speed of long low wave in water of depth d. Inserting $q = vy$ and $c^2 = gd$ into Equations 80 and 81 and taking the sum and the difference of the modified equations gives two simultaneous partial differential equations:

$$\frac{\partial v}{\partial t} + (v \pm c)\frac{\partial v}{\partial x} \pm 2\frac{\partial c}{\partial t} \pm 2(v \pm c)\frac{\partial c}{\partial x} = g(i - S_f) \qquad (82)$$

The independent variables in Equation 82 are distance x and time t. The dependent variables are average velocity v and wave velocity c, which are the unknowns to be found. Once wave velocity, c, is known, water depth d can be calculated.

These equations may be solved by the method of characteristics for the simple wave problem, which describes the propagation of a disturbance in a parallel-sided horizontal channel, neglecting frictional effects, with flow initially steady and uniform. The method of characteristics provides a solution for the simultaneous hyperbolic partial differential Equation 82. The "characteristic directions" in the x − t plane are directions along which ordinary differential equations apply. Since v and c are functions of both x and t the "equations of variation" for v and c can be obtained using the rules of partial differentiation.

$$\left.\begin{aligned} dv &= \frac{\partial v}{\partial x}dx + \frac{\partial v}{\partial t}dt \\ dc &= \frac{\partial c}{\partial x}dx + \frac{\partial c}{\partial t}dt \end{aligned}\right\} \qquad (83)$$

Listing Equations 82 and 83 in matrix form for the simple wave problem gives,

$$\begin{bmatrix} (v + c) & 1 & 2(v + c) & 2 \\ (v - c) & 1 & -2(v - c) & -2 \\ dx & dt & 0 & 0 \\ 0 & 0 & dx & dt \end{bmatrix} \begin{Bmatrix} \dfrac{\partial v}{\partial x} \\ \dfrac{\partial v}{\partial t} \\ \dfrac{\partial c}{\partial x} \\ \dfrac{\partial c}{\partial t} \end{Bmatrix} = \begin{Bmatrix} 0 \\ 0 \\ dv \\ dc \end{Bmatrix} \qquad (84)$$

Taking the determinant of the 4 × 4 matrix and setting it equal to zero gives the case when the derivatives of the dependent functions are indeterminate.

$$\begin{vmatrix} (v+c) & 1 & 2(v+c) & 2 \\ (v-c) & 1 & -2(v-c) & -2 \\ dx & dt & 0 & 0 \\ 0 & 0 & dx & dt \end{vmatrix} = 0 \tag{85}$$

$$\therefore \frac{dx}{dt} = v \pm c \tag{86}$$

The two real roots for dx/dt show that the system (82) is hyperbolic with characteristic directions that are functions of v and c. To find the equations that apply along the characteristic directions, the matrix Equation 84 is rearranged and the determinant of the 4×4 matrix is set equal to zero.

$$\begin{vmatrix} (v+c) & 1 & 0 & 2 \\ (v-c) & 1 & 0 & -2 \\ \frac{dx}{dt} & 1 & \frac{dv}{dt} & 0 \\ 0 & 0 & \frac{dc}{dt} & 1 \end{vmatrix} = 0 \tag{87}$$

Integration of Matrix 87 gives $(v \pm 2c)$ = constants that apply along the characteristic directions $dx/dt = v \pm c$.

Let $\frac{dx}{dt} = v + c$ be known as the C_1 family of characteristics

Let $\frac{dx}{dt} = v - c$ be known as the C_2 family of characteristics

For initially undisturbed flow the C_1 family are straight lines and C_2 family are curved, as shown in Figure 29. From the relationship along the characteristic directions,

$$\left.\begin{aligned} v_A + 2c_A &= v_B + 2c_B = \text{constant along } C_1 \\ v_A - 2c_A &= v_c - 2c_C = \text{constant along } C_2 \end{aligned}\right\} \tag{88}$$

$$v_A = v_B \text{ and } c_A = c_B \text{ along the straight } C_1 \text{ characteristics} \tag{89}$$

The characteristic network and the property of C_1 characteristics as straight lines means that water velocity and wave velocity can be calculated at points in the x-t plane providing boundary conditions are known along the t axis.

Either average water velocity v(t) or water depth and hence wave velocity c(t) are specified as functions of time. If water velocity v(t) is specified along the t axis then, using the properties of the C_1 and C_2 characteristics, c(t) can also be found, providing the initial conditions are known. Given the water depth a distance x_p from the origin in x-t plane, the time t_p taken for this depth to occur can be found from the following steps:

1. From the C_2 characteristic passing through the point 1 on the t-axis and the first C_1 characteristic as shown in Figure 30,

   $v_0 - 2c_0 = \text{constant}$

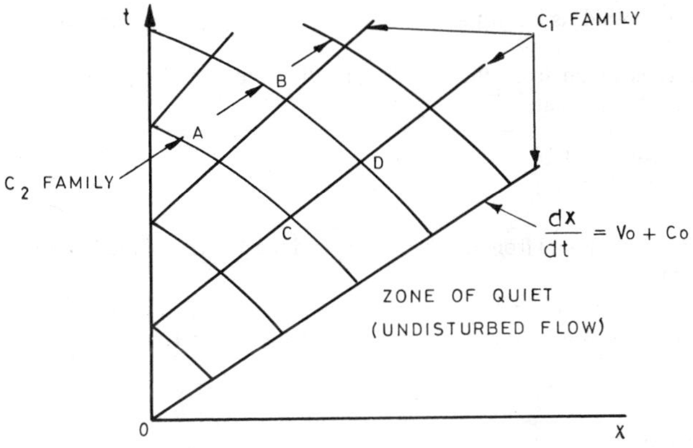

Figure 29. Characteristic network for simple wave problem.

2. If v(t) is specified along the t-axis then at point 1

$v(t_1) - 2c(t_1) = v_0 - 2c_0 =$ constant along c_2 characteristic,

$$\therefore c(t_1) = \frac{v(t_1)}{2} - \frac{v_0}{2} + c_0$$

3. Slope of c_1 characteristic through P which crosses t-axis at 1

$$\frac{dt}{dx} = \frac{1}{v(t_1) + c(t_1)}$$

$$= \frac{1}{\left(\frac{3}{2}v(t_1) - \frac{v_0}{2} + c_0\right)}$$

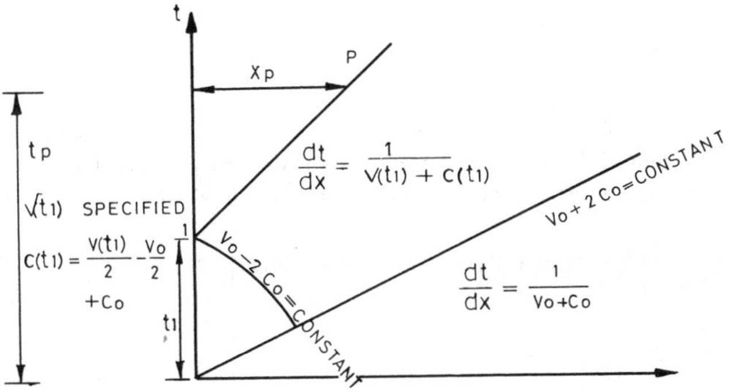

Figure 30. Use of characteristic network.

4. If the water depth d_p is known at P then $c(t_1) = c_p = \sqrt{gd_p}$ along C_1 characteristic through P and point 1.
5. The time t_1 is found from the specified relationship v(t) along the t-axis, and the initial conditions knowing that,

$$v(t_1) = v_0 - 2c_0 + 2c(t_1)$$
$$= v_0 - 2c_0 + 2\sqrt{gd_p}$$

6. The time $t_p - t_1$ is found from the distance x_p and the gradient of the C_1 characteristic through P and point 1.

$$\frac{dt}{dx} = \frac{t_p - t_1}{x_p}$$

$$= \frac{1}{v(t_1) + c(t_1)}$$

$$\therefore t_p - t_1 = \frac{x_p}{\left(\frac{3}{2} v(t_1) - \frac{v_0}{2} + c_0\right)}$$

The total time t_p can therefore be calculated knowing the initial conditions v_0, c_0 and the boundary condition v(t).

The finite difference scheme shown in Figure 31 is based on the finite difference form of the equations of motion and continuity. If Δx is constant throughout the computations, then Δt must be chosen so that the C_1 and C_2 characteristics passing through all the staggered points P, in time stage 2 enclose the appropriate points L and R in time stage 1, that is the domain of dependence of point P must include points L and R.

$$\left.\begin{aligned} \left(\frac{\partial v}{\partial x}\right)_M &\simeq \frac{v_R - v_L}{\Delta x} \\ \left(\frac{\partial c}{\partial x}\right)_M &\simeq \frac{c_R - c_L}{\Delta x} \\ \left(\frac{\partial v}{\partial t}\right)_M &\simeq \frac{v_P - v_M}{\Delta t} \\ \left(\frac{\partial c}{\partial t}\right)_M &\simeq \frac{c_P - c_M}{\Delta t} \end{aligned}\right\} \quad (90)$$

Substituting the finite difference approximations (90) into the equations of motion and continuity enables water velocity and wave speed to be calculated at time stage 2.

$$\left.\begin{aligned} v_P &= v_M + \frac{\Delta t}{\Delta x}\left[2c_M(c_L - c_R) + v_M(v_L - v_R) + g\,\Delta x(i - S_f)_M\right] \\ c_P &= c_M + \frac{1}{2}\frac{\Delta t}{\Delta x}\left[2v_M(c_L - c_R) + c_M(v_L - v_R)\right] \end{aligned}\right\} \quad (91)$$

Using these formulae an explicit finite difference time marching scheme produces average water velocities and wave speeds from specified initial and boundary conditions. To ensure that Δt is

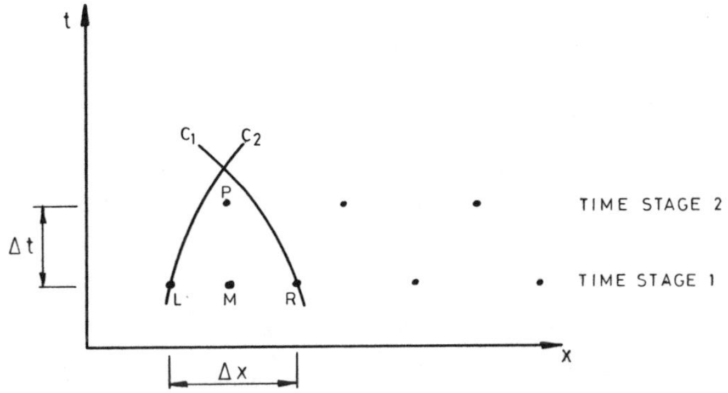

Figure 31. Finite difference scheme.

chosen so that the domain of dependence criterion is satisfied the ratio

$$\frac{\Delta x}{\Delta t} > 2(v + c)$$

This solution applies to a simple wave in a rectangular channel. More complex unsteady flow problems may be solved in a similar manner using finite differences and fuller details may be obtained [6, 24].

FLOW MEASUREMENT

It is frequently necessary in the laboratory or in the field to measure velocity and/or discharge in open channels. Methods of flow measurement were briefly touched on previously when dealing with applications of the energy equation to the broad crested weir and the venturi flume. These and other methods are reviewed in the following sections.

Weirs and Flumes

For small channels, thin plate weirs may be used for discharge measurement. These devices, also known at notches, consist of thin plates in which a rectangular or triangular opening has been made to allow the flow to pass downstream. Figure 32 shows a rectangular notch while Figure 33 dipicts a triangular or V-notch.

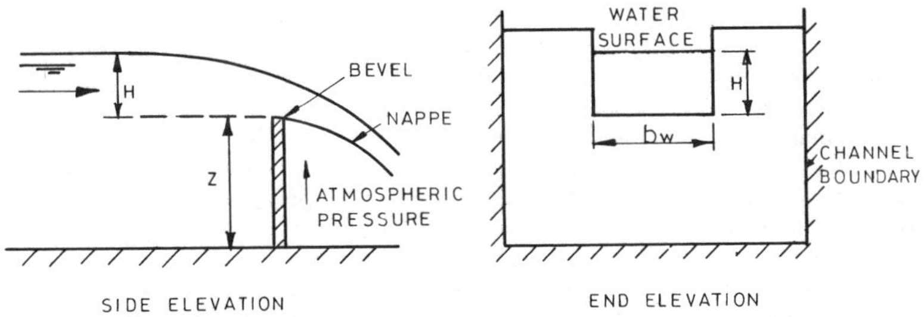

Figure 32. Rectangular thin plate weir.

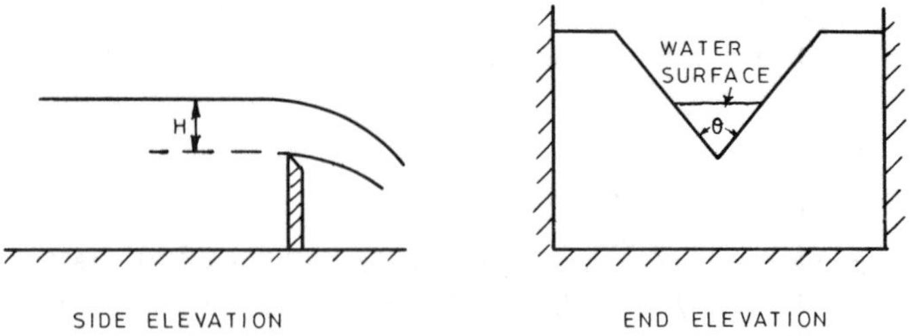

Figure 33. Triangular thin plate weir.

The analysis of flow over notches may be carried out using the energy equation (Bernoulli equation) or dimensional analysis. Using the former method yields an equation for the rectangular notch as follows:

$$Q = C_D \cdot \frac{2}{3} \cdot b_w \sqrt{2g}\, H^{3/2} \tag{92}$$

where Q = discharge in vol. per unit time
b_w = width of rectangular notch
g = gravitational acceleration
H = head over weir defined as depth from upstream water surface to sill of weir
C_D = coefficient of discharge taking account of simplifications necessary for energy equation to be applied.

The coefficient of discharge, typically in the range 0.6 to 0.65, takes account of such factors as contraction of the nappe, viscous effects, non-hydrostatic pressure distributions upstream, and variations from atmospheric pressure in the nappe, all of which are assumed in the theory. Other factors that affects C_D are depth Z (Figure 32) from the weir sill to the channel bed, and the ratio of the width of the weir b to the width of the channel. Generally, the more the nappe must contract the lower the value of discharge coefficient.

Subject to the same simplifying assumptions, it is possible to derive an equation for the V-notch as

$$Q = C_D \cdot \frac{8}{15} \tan\frac{\theta}{2} \sqrt{2g}\, H^{5/2} \tag{93}$$

where θ = included angle of the notch, which is typically in the range 30° to 90°
C_D = coefficient of discharge, which is typically 0.59

Long base weirs have already been referred to when discussing the broad crested weir as an application of the energy equation above. More modern weirs have been developed, notably the Crump weir (see Herschy, White and Whitehead [25]) which is of triangular profile and has the advantage of minimizing rise in upstream water level known as afflux. However, the principle of operation is the same in that critical depth is produced at or near the crest, and the general form of equation is the same as Equation 30. The venturi flume as previously discussed is also widely used as a flow measuring device. A common type of flume is that incorporating both a rise in channel bed and a narrowing of the channel width—a combination of the broad crested weir and the venturi flume. Again, the principle of operation involves the arrangement of critical conditions thereby invoking the relationship between discharge and critical depth. The form of equation used is as given in Equation 31. Each of these equations incorporates a coefficient of discharge that must be determined experimentally either by laboratory model or by field calibration. Full details of theory and operation of these and less commonly used weirs and flumes are given in Ackers, White, Perkins and Harrison [2].

Velocity Area Method

Where weirs and flumes are not available, or where information on point velocities is required, recourse must be made to the velocity area method. This involves the measurement of velocities at a number of points in the channel cross-section as shown in Figure 34. The cross section is divided into rectangles using verticals and horizontals and velocity measurements taken at the nodes to form a grid of values. The discharge through the channel may then be determined either by plotting isovels (contours of equal velocity) and multiplying the area between adjacent isovels by the appropriate velocity, or by calculating the average velocity for each vertical and multiplying by the area of flow represented by the vertical. The latter is shown in Figure 34 as the area bounded by vertical lines on either side of the vertical, which bisect the distance to the two adjacent verticals. For economy of time, a good estimate of average velocity for any vertical may be obtained by averaging the velocities at 20% and 80% of the depth. Or, if there is time for only one measurement per vertical, that at 60% of the depth is very close to the average value.

Velocity measurement is usually by means of a propeller meter, whereby the speed of rotation of a propeller by the flow is measured electronically. The speed of rotation is proportional to water velocity and conversion is usually by means of the manufacturer's chart. Other more modern methods of velocity measurement are now available, including hot wire and hot film anemometry, which operates on the principle that fluid velocity is proportional to heat transfer from a heated element (wire or film) held in the stream. More recently, laser doppler anemometry has become popular whereby velocity is measured by means of the interference caused by the moving fluid as it passes the intersection of two light beams in the flow. This method requires the use of channels made of transparent material but has the great advantage of zero interference of the flow pattern by the measuring device.

Modern Techniques

More modern techniques of flow measurement include the electromagnetic method whereby an electromotive force (e.m.f) is induced in the water by an electric current passed through a large coil buried beneath the channel bed. The e.m.f. is directly proportional to the average velocity through the cross section and is recorded by electron probes at each side of the river. The method is costly but shows promise.

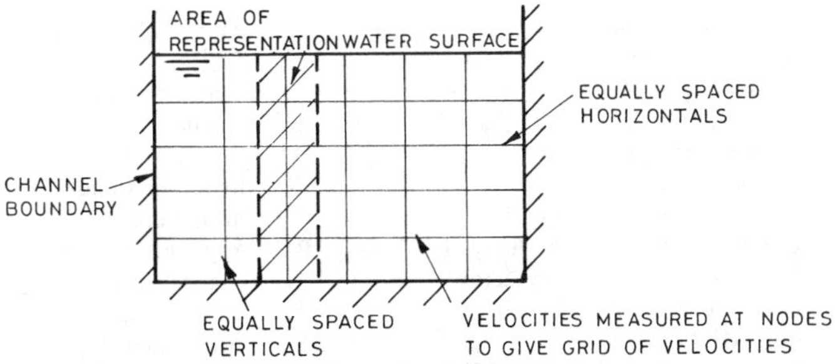

Figure 34. Velocity distribution measurements for velocity area method.

The ultrasonic method utilizes sound pulses to measure mean velocity at a specific depth in the channel. Sound pulses are beamed and received across the channel diagonally, to and from both banks. Because the speed of the pulses is increased by water velocity in one direction and decreased in the opposite direction, there is a difference in pulse velocity that is proportional to the water velocity at that depth. This velocity must be related to average cross-section velocity for discharge calculation. This method is now well proven and gives good accuracy.

Further information on these and other more unusual methods is given in Herschy [26] and Sargent [27].

CONCLUSION

Interest in channel flow has a long history stretching back to early engineering achievements in the ancient Near East and the Roman Empire. The subject has evolved by both experimental and mathematical means, and the state of knowledge today owes much to a close blend of the two approaches.

There are still many aspects of channel flow that are poorly understood and require further investigation. But perhaps the greatest developments in the most recent past and for the forseeable future are in using computer methods to analyze channel flow problems. This has enabled the solution of hitherto intractable equations describing complex flow patterns, so that a wider range of real open channel problems can now be handled more easily than in the past, when simplification and abstraction was necessary. This has led to a new subject in the realm of open channel flow, namely computational hydraulics, which is and continues to be one of the major growing edges of the discipline.

While life on earth exists, the supply of usable water will be conveyed by open channel means, where the motive force in the form of gravity is free. The study of channel flow therefore has an assured future while civilization exists, so that its governing principles are of abiding relevance and usefulness.

NOTATION

A	cross-sectional area, L^2
b	channel width, L
bw	width of thin plate weir, L
B	width of long base weir, L
c	wave celerity, LT^{-1}
C	Chezy's roughness coefficient, $L^{1/2}T^{-1}$
C_D	coefficient of discharge
C^0, C^1, C^2	coefficients in channel routing equation
C_1, C_2	families of characteristics
D	outflow discharge, L^3T^{-1}
d	channel depth, L
d_c	critical depth, L
d_0	normal depth, L
E	specific energy, L
E_{min}	minimum specific energy, L
ΔE	energy loss in hydraulic jump, L
f	Darcy-Weisbach friction factor
F	force on water flow by obstruction, MLT^{-2}
g	gravitational acceleration, LT^{-2}
H	head of upstream water surface above weir sill, L
H_T	total energy (=total head), L
h_L	head loss due to friction, L
i	longitudinal bed slope of channel
I	inflow discharge, L^3T^{-1}
K	storage factor in Muskingum routing equation, T
K_s	Nikuradse' equivalent sand roughness, L
L	channel length, L
M	momentum function, MLT^{-2}
n	Manning's roughness coefficient, $L^{1\ 3}T$
P	wetted perimeter, L
Q	channel discharge, L^3T^{-1}
q	discharge per unit width, L^3T^{-1}
R	hydraulic radius (=hydraulic mean depth), L
R_e	Reynolds number
S	channel storage, L^3
S_f	friction slope
S_c	critical bed slope
t	routing period, T
T	throat width in venturi flume, L

v average cross-sectional velocity, LT^{-1}
v_c critical velocity, LT^{-1}
V_* shear velocity, LT^{-1}
x dimensionless parameter in Muskingum equation
Z height above datum, L
Z_c critical height of bed hump, L

Greek Symbols

α kinetic energy correction factor
ρ fluid density, ML^{-3}
τ_0 boundary shear stress, $ML^{-1}T^{-2}$
τ_{max} maximum boundary shear stress, $ML^{-1}T^{-2}$
ν coefficient of kinematic viscosity, L^2T^{-1}
μ coefficient of absolute viscosity, $ML^{-1}T^{-1}$

REFERENCES

1. Francis, J. R. D., *Fluid Mechanics for Engineering Students*, 4th edition, Edward Arnold, London, 1976.
2. Ackers, P., White, W. R., Perkins, J. A., and Harrison, A. J. M., *Weirs and Flumes for Flow Measurement*, Wiley, Chichester, 1978.
3. Massey, B. S., *Mechanics of Fluids*, 5th edition, Van Nostrand Reinhold (UK), London, 1983.
4. Chow, V. T., *Open Channel Hydraulics*, McGraw-Hill, New York, 1959.
5. Carter, R. W., Einstein, H. A., Hinds, J., Powell, R. W., and Siberman, E., "Friction Factors in Open Channels." *Proc. Am. Soc. Civil Engrs.*, 89, Hy2, 1963, 97–143.
6. Henderson, F. M., *Open Channel Flow*, Macmillan, New York, 1966.
7. Myers, W. R. C., and Elsawy, E. M., "Boundary Shear in Channel with Floodplain," *Jour. Hyd. Div., Proc. Amer. Soc. Civ. Engrs.*, 101, Hy7, July 1975, 933–946.
8. Cruff, R. W., "Cross-Channel Transfer of Linear Momentum in Smooth Rectangular Channels," US Geological Survey, Water Supply Paper 1592-B, 1965.
9. Rajaratnam, N., and Muralidhar, D., *Boundary Shear Stress Distribution in Rectangular Open Channels*, La Houille Blanche, Grenoble, France, Vol. 6, 1969, 603–609.
10. Nikuradse, J., *Stronungsgesetze in Ranhen Rohen*, V. D. I., Forschungshaft, 1933, 361.
11. Colebrook, C. F., "Turbulent Flow in Pipes with Particular Reference to the Transition Region Between Smooth and Rough Pipe Laws," *Jour. Inst. Civ. Engrs.*, London, Vol. 11, 1939, 133.
12. Ackers, P., "Resistance of Fluids Flowing in Channels and Pipes," Hydraulics Research Paper No. 1, London, Her Majesty's Stationery Office, 1958.
13. Moody, L. F., "Friction Factors for Pipe Flow," *Trans. Am. Soc. Mech. Engrs.*, 66, 1944, 671–84.
14. Ackers, P., "Charts for the Hydraulic Design of Channels and Pipes," 3rd edition, Hydraulic Research Papers No. 2, London, Her Majesty's Stationery Office, 1969.
15. Ackers, P., "Tables for the Hydraulic Design of Storm-drains, Sewers, and Pipelines," 2nd edition, Hydraulic Research Paper No. 4, London, Her Majesty's Stationery Office, 1969.
16. Keulegan, G. H., "Laws of Turbulent Flow in Open Channels," *Jour. Research*, Nat. Bur. Stand, 21, 1938, 707.
17. Reinus, E., Steady Uniform Flow in Open Channels, *Trans. Royal Institute of Technology*, Stockholm, Sweden, No. 179, 1961.
18. Tracy, H. J., and Lester, C. M., "Resistance Coefficient and Velocity Distribution: Smooth Rectangular Channel," US Geol. Surv., Water Supply Paper 1592-A, Washington DC, 1961.
19. Rao, K. K., "Effect of Shape on the Mean Flow Characteristics of Turbulent Flow Through Smooth Rectangular Open Channel," Thesis presented to Univ. of Iowa, at Iowa City in partial fulfilment of requirements of Ph.D Degree, 1967.
20. Myers, W. R. C., "Flow Resistance in Wide Rectangular Channels," *Jour. Hyd. Div., Proc. Amer. Soc. Civ. Engrs.* Vol. 108, Hy4, April 1982, 471–482.
21. Kazemipour, A. K., and Apelt, C. J., "Shape Effects on Resistance to Flow in Smooth Semi-circular Channels," Univ. of Queensland Research Report No. CE18, Nov. 1980.

22. Kazemipour, A. K., and Apelt, C. J., "New Data on Shape Effects in Smooth Rectangular Channels," *Jour. Hyd. Res.* Vol. 20, No. 3, 1982, 225–233.
23. Featherstone, R. E., and Nalluri, C., *Civil Engineering Hydraulics*, Granada Publishing, London, 1982.
24. Price, R. K., *Flood Routing Studies*, Vol. III of Flood Studies Report, Natural Environment Research Council, London, 1975.
25. Herschy, R. W., White, W. R., and Whitehead, E., "The Design of Crump Weirs," Tech. Memo. No. 8, Department of Environment Water Data Unit, 1977.
26. Herschy, R. W., (Ed) *Hydrometry, Principles and Practices*, John Wiley, New York, 1978.
27. Sargent, D. M., "The Development of a Viable Method of Streamflow Measurement Using the Integrating Float Technique," *Proc. Inst. Civil Engrs.*, London, Vol. 17, No. 2, 1981, 1–15

CHAPTER 25

STRUCTURE OF TURBULENT CHANNEL FLOWS

Arne V. Johansson

and

P. Henrik Alfredsson

Department of Mechanics
The Royal Institute of Technology
Stockholm, Sweden

CONTENTS

INTRODUCTION

Turbulent phenomena are of importance in most technical, geophysical, biological, and a variety of other flow situations. Boundary layer turbulence is the major source of drag for both aircraft and ships, as well as for flows in channels, pipes, and conduits of various shapes used in technical applications. Existing theories are unable to give detailed and quantitative explanations of the mechanism of turbulence generation in shear flows near solid boundaries, and turbulence models rely to a large extent on empirical results. Experimental investigations of turbulence structure therefore play a key role in the development of knowledge in this field.

The present review concerns the state-of-the-art of the structure of turbulent channel flows, with emphasis on experimental findings. It will be restricted to flow of a Newtonian fluid in a straight, "two-dimensional" channel with smooth walls, at Reynolds numbers well above transition. (Other aspects of channel flow will, however, be treated elsewhere in this volume.) This restriction is not as limiting as it may seem. Many of the cited results are indeed applicable to, and some obtained from, other wall-bounded turbulent shear flows, such as pipe and boundary layer flows. It appears from the vast body of experimental studies that most characteristics of near-wall turbulence are in common for all wall bounded shear flows without strong pressure-gradient effects. Such effects are weak in the case of plane turbulent channel flow.

Turbulent flow in channels with a cross section of relatively large aspect ratio has, due to its geometrical simplicity, attracted attention from many investigators both from a theoretical and an experimental point of view. Aspect ratios for channels used in experimental studies of this type usually fall in the range 5–20. For the smaller values only the flow in a rather limited portion (say, two channel heights, where the height is assumed to be smaller than the width) of the channel can be considered as two-dimensional in the sense of being practically free from effects of secondary flows caused by side walls and corners.

Among the early experimental studies of turbulent channel flow may be mentioned those of Dönch [1], Nikuradse [2], Reichart [3], and Wattendorf [4]. In the first two cases only mean velocities were measured. The first turbulence measurements of good quality seem to be those of Reichart [5], who measured the fluctuation intensities, with hot-wire anemometry, both in the streamwise direction and in the direction normal to the wall. For both quantities the results compare well with modern measurements, even very close to the wall in the buffer region where the intensities are a maximum.

Laufer [6] and Comte-Bellot [7], using hot-wire anemometry, carried out extensive studies of turbulent channel flow in air (see also the pipe flow work of Laufer [8]). Comte-Bellot addressed the question of how long entrance region is needed to obtain a fully developed flow independent of downstream distance. She found that the higher order statistical moments of the streamwise fluctuations on the channel centerline are very sensitive to the stage of development, and that the fully developed regime is established at approximately 60 channel heights from the inlet. Both Laufer and Comte-Bellot measured all three components of velocity with single and X-wires, at different Reynolds numbers. The structure of turbulence was studied, in terms of statistical moments, one- and two-point correlation functions, spectra, etc.

The development of hot-film probes made it possible to do turbulence measurements also in liquids. Eckelmann and co-workers of the Göttingen group have used hot-film anemometry in a number of studies of turbulent channel flow with oil as the working fluid [9]. An advantage in this case, for the low Reynolds numbers used, is a thick (2–3 mm) viscous sublayer, enabling detailed investigations of the flow in this region. This is also possible, to some extent, in low-Reynolds-number flows of water [10], but is considerably more difficult (although not impossible) in air flows. Actually, the viscous lengthscale, and hence the thickness of the viscous sublayer, can readily be shown to be uniquely determined by e.g., the channel height and the Reynolds number, and for a fixed Reynolds number it is simply proportional to the channel height.

Among other hot-wire and hot-film studies of turbulent channel flow may be mentioned Clark [11], Hussain and Reynolds [12, 13], Kreplin [14], Wallace, Eckelmann and Brodkey [15], and Zarič [16]. Data from these and many other studies will be referred to in the following sections. Laser doppler velocimetry (LDV) has so far been applied to channel flow only in a few studies. Reischman and Tiederman [17] used LDV both in pure water and in water with drag-reducing polymers for measurement of mean velocity and streamwise fluctuations. Two-component measurements have been made by Willmarth and Velasquez-Saavedra [18].

A breakthrough in experimental turbulence research is marked by the discovery from visual studies [19–21] that a major part of the turbulence production takes place close to solid surfaces during short, intermittent periods of violent activity ("bursting periods"). Several methods have been devised to detect such events from velocity signals by means of computer algorithms of varying complexity [22]. The use of computers in the laboratory has opened up vast new possibilities to analyze data in the search for the mechanisms of turbulence production. This has resulted in many efforts in this direction during recent years, and are reflected in this chapter.

First, basic equations for turbulent channel flow and some definitions are presented. Next, some experimental results for mean flow characteristics are described, followed by a state-of-the-art review of conventional turbulence characteristics, such as probability density distributions of the fluctuations, intensity and Reynolds stress profiles, flow angles, and power spectra. Then, organized flow structures and their relation to turbulence producing mechanisms are discussed, and finally the chapter concludes with a review of numerical "experiments."

BASIC EQUATIONS AND DEFINITIONS FOR TURBULENT CHANNEL FLOW

The restrictions imposed by the presence of enclosing walls renders a homogeneity in the streamwise (x) direction, which is a characteristic feature of fully developed turbulent channel and pipe flows, but is in contrast to boundary layer flow. In the present work homogeneity is assumed also in the spanwise (z) direction, i.e., only "two-dimensional" channel flows are considered. The mean velocity, U, in the x-direction is, hence, a function of the coordinate normal to the wall (y) only. The equations for the mean flow can for this case be written as*

$$\frac{d\overline{uv}}{dy} = -\frac{1}{\rho}\frac{\partial P}{\partial x} + \nu\frac{d^2U}{dy^2}$$

$$\frac{d\overline{v^2}}{dy} = -\frac{1}{\rho}\frac{\partial P}{\partial y}$$

$$0 = \frac{\partial P}{\partial z}$$

where ρ is the density, ν is the kinematic viscosity and overbar denotes long-time average. These equations can be integrated to give

$$-\overline{uv} + \nu\frac{dU}{dy} = u_\tau^2(1 - y/b) \tag{1}$$

where b is the channel half-height (i.e. the centerline is at y/b = 1), and u_τ is the friction velocity, defined by

$$u_\tau = (\tau_w/\rho)^{1/2} = \left(\nu\left.\frac{dU}{dy}\right|_{y=0}\right)^{1/2}$$

where τ_w is the shear stress at the wall. The turbulent shear stress (or Reynolds stress) is defined as $-\rho\overline{uv}$. The left-hand side of Equation 1 is, hence, the sum of turbulent and viscous shear stresses (per unit mass). From Equation 1 it is also seen that the Reynolds stress distribution has a one-to-one relation with the mean velocity distribution in this case, and that the total shear stress varies linearly with the distance from the wall.

A Reynolds number can be defined from u_τ, b and ν. However, the velocity scale is often taken as the centerline velocity (U_{CL}), since this quantity can readily be obtained in an experiment. The Reynolds number that will be used in what follows is defined as

$$Re = 2bU_{CL}/\nu$$

Linear stability theory predicts a critical Reynolds number (Re) of about 11,500 for plane Poiseuille flow [23], whereas transition to turbulence in experimental cases usually occurs at much lower values. Patel and Head [24] made measurements of mean velocity profiles and skin friction at low Reynolds numbers. In their case the flow remained laminar for Reynolds numbers below about

* The fluctuating velocity components are denoted u, v and w in the x, y and z-directions, whereas P and p are the mean and fluctuating parts of the pressure.

2,000, whereas above Re ≈ 3,500 the flow was found to be fully turbulent. The results reviewed here fall in the Reynolds number range 5,600 [9] to about 500,000 [7].

As can be seen from Equation 1, the three quantities, u_τ, ν, and b, define the mean flow. Well away from solid boundaries, in what is usually called the outer region, the mean viscous stress is small compared to the Reynolds stress, and the flow is here practically independent of fluid viscosity. This property may be termed Reynolds number similarity, and implies that

$$(U_{CL} - U)/u_\tau = F(y/b) \tag{2a}$$

$$\overline{uv}/u_\tau^2 = G(y/b) \tag{2b}$$

where F and G are universal functions independent of Reynolds number. Equation 2a is called the velocity defect law. The distance from the wall is normalized with the channel half-height, here taken to be the "outer lengthscale." The relevant length (L) and time (t_0) scales for the most energetic velocity fluctuations in this region are

$$L = b, \qquad t_0 = b/U_{CL} \tag{3}$$

referred to as the outer scales.

Close to the walls ($y/b \ll 1$), in what is usually referred to as the inner region, the mean velocity and Reynolds stress profiles may be assumed to be independent of the outer lengthscale, and thereby uniquely defined by u_τ and ν. From these quantities one can form the inner length (ℓ_*) and time (t_*) scales

$$\ell_* = \nu/u_\tau, \qquad t_* = \nu/u_\tau^2 \tag{4}$$

whereas the inner velocity scale is u_τ. Denoting normalization with inner scales by superscript + we get

$$U^+ = f(y^+) \tag{5a}$$

$$\overline{uv}/u_\tau^2 = g(y^+) \tag{5b}$$

where f and g are functions independent of Reynolds number. Equations 5a and b are termed "law of the wall." Whether or not the inner scales also are relevant scales for the velocity fluctuations near the walls is not equally clear. This will be discussed later.

Estimates of the smallest scales of motion are usually taken to be the Kolmogorov scales determined from the viscous dissipation of turbulent energy, ϵ, which with the use of tensor notation (i = 1, 2, 3 corresponding to x, y, z), can be expressed as

$$\epsilon = \frac{\nu}{2}\overline{\left(\frac{\partial u_i}{\partial x_j} + \frac{\partial u_j}{\partial x_i}\right)^2}$$

The Kolmogorov length and time scales are defined as

$$\ell_K = (\nu^3/\epsilon)^{1/4}, \qquad t_K = (\nu/\epsilon)^{1/2}$$

If the characteristic size and velocity of the largest eddies are L and V, respectively, the turbulent dissipation may be qualitatively taken as the "Taylor estimate" V^3/L (see Townsend [25], p. 33). The timescale then varies with velocity as $V^{-3/2}$ (if L and ν are fixed). The total rate of work (per unit mass) exerted on the fluid by the wall shear stress is often used as another rough approximation for the turbulent dissipation. For channel flow it can be approximated by $U_b u_\tau^2/b$, where U_b is the bulk velocity. The corresponding Kolmogorov timescale then is essentially the geometric mean of inner and outer timescales

$$t_m = \left(\frac{\nu}{u_\tau^2}\frac{b}{U_{CL}}\right)^{1/2} = (t_* t_0)^{1/2} \tag{6}$$

It will be here referred to as the mixed timescale, and will be used in discussions of scaling of velocity fluctuations later in this chapter.

An expression for the mean velocity, which is assumed to be valid for all y-values may be written [26]

$$U^+ = f(y^+) + \frac{\Pi}{\chi} W(y/b) \tag{7}$$

where the second term is called the wake component, and Π is the wake strength parameter. W is a normalized function ($W(0) = 0$, $W(1) = 2$) usually referred to as the "law of the wake."

The two-layer approach requires an overlap region between the inner (wall) region and the outer region [27, 28]. The two descriptions F and f of the velocity profile must hence be valid simultaneously in the limit y/b approaching zero and y^+ approaching infinity. This is only possible for a functional dependence of a logarithmic type, i.e.

$$(U_{CL} - U)/u_\tau = \frac{-1}{\chi} \ln y/b + A \tag{8a}$$

$$U^+ = \frac{1}{\chi} \ln y^+ + B \tag{8b}$$

where χ is the von Karman constant, usually assigned the value 0.41, and A and B are constants. The universality of Equation 8b will be discussed in some detail later. For a more detailed description of matching procedures, etc., the reader is referred to, e.g., Tennekes and Lumley [29] and Townsend [25]. The two-layer approach has recently been questioned by Long and Chen [30] and Afzal [31]. However, the existence of log-laws (Equations 8a and b) seems to be experimentally well verified.

By adding Equations 8a and b we get

$$U_{CL}/u_\tau = \frac{1}{\chi} \ln \frac{u_\tau b}{\nu} + C \tag{9}$$

($C = A + B$) which is called the logarithmic friction law. Often used in this context is the skin friction coefficient (c_f), which is related to the U_{CL}/u_τ-ratio in the following manner.

$$c_f = 2(u_\tau/U_{CL})^2 \tag{10}$$

Very close to the wall, where the Reynolds stress is negligible compared to the viscous shear stress, we see from Equation 1 that

$$U^+ = y^+ \tag{11}$$

i.e., the variation of the velocity is linear in the immediate vicinity of the wall. This region is called the viscous (not laminar) sublayer, and extends to roughly $y^+ = 5$. Between this layer and the log-region is the buffer region, where turbulence intensities and production are a maximum.

So far nothing has been said explicitly about the similarity of turbulence intensity distributions. It may be assumed that they also conform to dependences of y/b and y^+ in the outer and inner regions, respectively. This will be discussed later. An equation of interest in this context is the turbulent energy equation, which for channel flow may be written

$$0 = -\overline{uv}\frac{dU}{dy} - \epsilon - \frac{d}{dy}\left[\frac{1}{\rho}\overline{pv} + \overline{qv} - \nu\frac{d}{dy}\overline{(q + v^2)}\right] \tag{12}$$

where q is the instantaneous turbulent kinetic energy per unit mass ($q = (u^2 + v^2 + w^2)/2$). The zero left-hand-side of Equation 12 signifies that there is no net change in turbulent energy for a fully developed channel flow. The first term is the production of turbulent energy, i.e., the transfer of energy from the mean flow to the turbulent fluctuations due to the action of Reynolds stress. The third term represents a redistribution of turbulent energy, constituted by the divergence of transport terms due to pressure, turbulent velocity fluctuations, and viscous stresses. Globally, however, production is exactly balanced by dissipation for the channel flow case. By multiplying Equation 1 by dU/dy and differentiating, one can show that the maximum turbulence production occurs at about where the slope of the mean velocity profile attains a value of half of that at the wall, i.e., $dU^+/dy^+ = 0.5$. This is at about $y^+ = 12$, independent of Reynolds number.

A characteristic feature of turbulent flow is the presence of strong vorticity fluctuations and vortices (or eddies) of various scales. The vorticity, ω_i, is defined as the curl of the velocity field,

$$\omega_x = \frac{\partial w}{\partial y} - \frac{\partial v}{\partial z}, \text{ etc.}$$

To measure vorticity fluctuations one must hence determine derivatives of the fluctuating velocity components. This is indeed a difficult task, and only a very limited amount of experimental data is available. The equation governing the fluctuating part of the vorticity comprises terms that can be interpreted as intensification of vorticity due to vortex stretching caused by the mean flow and the velocity fluctuations. It can readily be shown that these "production" terms vanish for two-dimensional turbulence, which illustrates the crucial role of three-dimensionality (of the fluctuations) for the persistence of turbulence (for a more complete discussion of vorticity dynamics refer to Tennekes and Lumley [29]).

MEAN FLOW

Results concerning the mean velocity distribution and frictional resistance are of basic importance in turbulence research, and in engineering problems, but are of interest also in the context of, e.g., interpretation of coherent structures and their scaling behavior. However, before we go to into the details of the experimental results it may be appropriate to give a brief description of the channels used in the studies discussed in this work. Some pertinent data are given in Table 1 for eight different channels. One reference is given in each case, although in several of the facilities described there have been a number of investigations carried out that will be cited later. Typical Reynolds numbers are also tabulated.

Table 1
Basic Data for Some Channel-Flow Facilities (all measured in m)

Reference	L	L/2b	Cross Section	Fluid	Re × 10^{-3}	Exp. Method
Bogard [32]	4.9	82	0.06 × 0.575	water	8.2	visualization, hot-film
Comte-Bellot [7]	12	67	0.18 × 2.4	air	126–514	hot-wire
Eckelmann [33]	7.0	32	0.22 × 0.9	oil	5.6–8.2	hot-film
Hussain and Reynolds [13]	14	219	0.064 × 1.14	air	21–64	hot-wire
Johansson and Alfredsson [10]	6.0	75	0.08 × 0.40	water	13–130	hot-film
Kastrinakis and Eckelmann [34]	9.5	53	0.18 × 1.40	air	25.2	hot-wire, vorticity meas.
Oldaker and Tiedermann [35]	2.5	67	0.038 × 0.45	water	8.9–16	visualization
Willmarth and Velasques-Saavedra [18]	2.6	98	0.027 × 0.30	water	29.4	LDV (2-comp.)

The channels in Table 1 represent a wide variety of design and size. The length of the channels ranges from 2.5 m to 14 m, and in terms of channel heights (2b) the range is from 32 to 219. The development length to attain a fully developed flow, independent of downstream distance, is normally found to be about 60 channel heights [7], but depends on the inlet conditions (and Reynolds number). The Göttingen oil channel is remarkably short, but has been shown [33] to give a fully developed flow. The aspect ratio of that channel is about 4, whereas the largest aspect ratio is almost 18 [13]. The working fluids used are air, water and oil, the latter with a viscosity of 6×10^{-6} m^2/s, and the studies represent both probe measurements and visualization studies. Bogard [32] used hydrogen bubbles, and dye visualization combined with hot-film anemometry. The study of Willmarth and Velasquez-Saavedra [18] represents the only laser doppler velocimetry measurements included in Table 1.

Laser doppler velocimetry is not yet as established a method as hot-wire and hot-film anemometry, but is of growing importance, and has, in some respects, superior qualities. Especially the fact that it is a non-intrusive method enables measurements that are not possible with other techniques. A problem usually encountered with the the LDV-technique is to obtain a reasonably high data rate without having more than one particle at a time in the measuring volume (when using counters). It is also essential to have particles of as uniform a size as possible in order to maintain good signal quality. An issue specific for LDV is the "signal-weighting" that has to be applied in order to avoid high-velocity bias, caused by the higher probability of finding a high-velocity particle in the measurement volume, as compared with particles having velocities lower than the mean. For a description of the LDV-technique the reader is referred to e.g., Buchhave, George and Lumley [36] and Somerscales [37].

Hot-wire and hot-film anemometry (for a description, see Blackwelder [38]) have been used for turbulence measurements over several decades, and has been proven to be a well-suited technique. There are, however, some problems that are important to be aware of, such as vortex shedding from the sensor at high velocities, heat conduction to prongs, and tangential cooling for X-wires. Problems with imperfect spatial resolution and heat conduction to the wall will be discussed below, in conjunction with the presentation of experimental data.

Expressions for the distribution of mean velocity in the viscous sublayer and the log-region were given earlier. Figure 1 shows mean velocity profiles, in law-of-the-wall coordinates, measured with hot-film anemometry in a water-channel flow. For the low-Reynolds-number case it is seen that the linear velocity distribution in the viscous sublayer is well reproduced, which means that heat conduction to the wall is negligible in this case. This problem is often severe in air flows, with highly disparate thermal conductivities of the air and the wall material. The wall then acts as a heat sink, causing too high velocity readings in the immediate vicinity of the wall. The first measurement point for the low Reynolds number is at $y^+ = 1.3$, and the linear law seems to be valid up to at least $y^+ = 6$ in this case.

The wall friction is usually obtained from a fit of the log-law to the measured velocity profile. For the low Reynolds number case in Figure 1, however, it could be determined directly from the

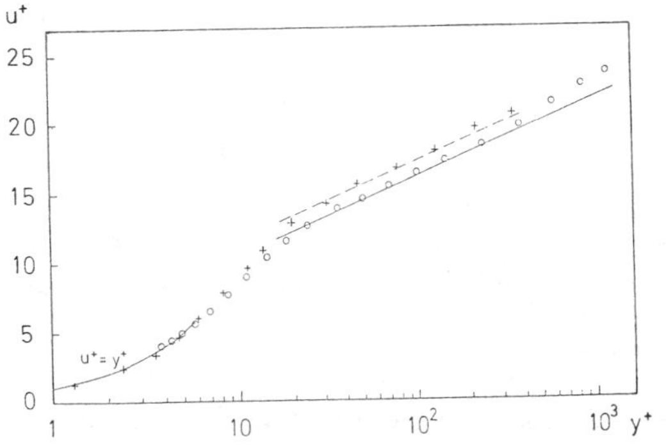

Figure 1. Mean velocity distributions for channel flow (logarithmic y^+-axis). +, Re = 14,300; ○, Re = 54,300. ——— and ------- represent log-laws with $\mathscr{K} = 0.41$ and $B = 5.0$ and 6.0, respectively [39, 40].

velocity gradient at the wall. This enables an independent check of log-law constants and of the additive constant in the logarithmic friction law. The logarithmic intercept, B, is for this case as high as 6.0, whereas for high Reynolds numbers (Figure 1) it seems to attain a value of 5.0, for channel flows as well as for pipe and boundary layer flows [41]. Also, the data from the Göttingen oil channel [33], the pipe flow data of Patel and Head [24], and the boundary layer data of Murlis, Tsai, and Bradshaw [42] show higher values of B at low Reynolds numbers (although in the latter case the authors ascribe the trend to possible measurement errors).

A characteristic feature of channel flow is the relatively weak wake component (Equation 7) of the velocity law in the outer region, as is clearly illustrated in Figure 1. The wake component can be seen as the deviation from the log-law in this region. Although the wake strength parameter, Π, appears to increase with increasing Reynolds number, it is in all cases considerably smaller than for boundary layer flows. This may be seen as a reflection of the different characters of the outer-region flow for the two cases. For instance, the intermittency found in boundary layers is for natural reasons not present in channel flow. For the low Reynolds number in Figure 1 the wake component is practically non-existing, whereas for the higher Reynolds number the wake strength parameter can be estimated to 0.27. A typical value for zero-pressure-gradient boundary layers is 0.55 [26].

The velocity defect law (Equation 2a) gives a universal representation of the mean velocity distribution in the outer region, as can be seen from Figure 2. Data for three different Reynolds numbers are shown in this case.

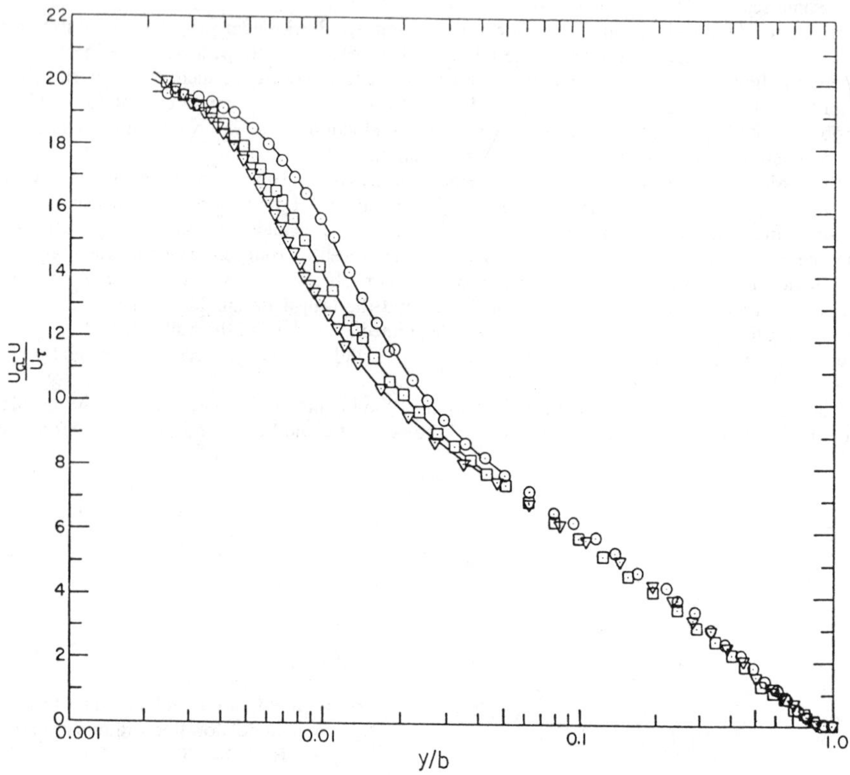

Figure 2. Velocity defect law. ○, Re = 27,600; □, Re = 46,400; ▽, Re = 64,600 [13].

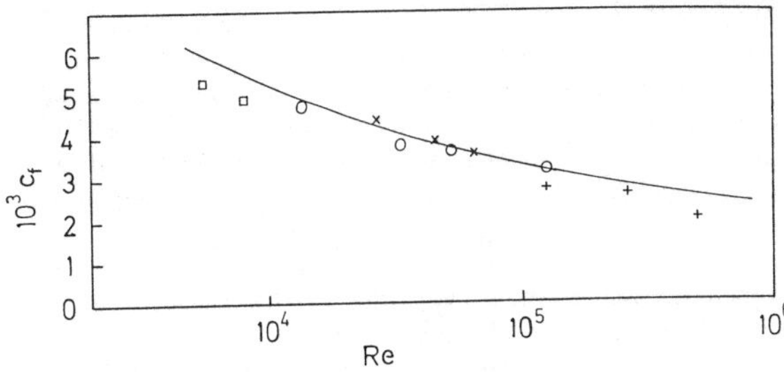

Figure 3. Skin friction coefficient as function of Reynolds number. Data from: +, Comte-Bellot [7]; □, Eckelmann [9]; ×, Hussain and Reynolds [12]; ○, Johansson and Alfredsson [10, 40]. Solid curve represents logarithmic friction law with additive constant equal to 6.0.

For the low-Reynolds-number case in Figure 1 the U_{CL}/u_τ-ratio is 20.6. Application of the logarithmic friction law (Equation 9), with an additive constant of 6.0 and $\chi = 0.41$, would for this case yield a value of 20.3, hence in good agreement with the experimental finding. The value 6.0 for the additive constant is suggested by Tennekes and Lumley [29] for pipe flow, but hence appears to be applicable also to channel flow. Also, for higher Reynolds numbers the logarithmic friction law gives reasonable values of the friction velocity, and thereby the wall friction. For the higher Reynolds number in Figure 1 the computed U_{CL}/u_τ-ratio is 23.2, whereas 23.5 was obtained from the experimental data.

The U_{CL}/u_τ-ratio has a one-to-one relation with the skin friction coefficient (Equation 10). Figure 3 shows c_f-data from some channel flow studies spanning a Reynolds number range of approximately two decades. The experimental data points are compared with a curve representing the logarithmic friction law (Equation 9).

TURBULENT FLUCTUATIONS

Turbulence in a wall-bounded shear flow, such as a turbulent channel flow, comprises structures that are coherent over sizeable regions in space and time, as well as more or less chaotic (predominantly small scale) motions. A review of research on ordered flow structures is given later. However, it is of interest to first try to establish "the state of the art" of knowledge about the conventional, long-time averaged, characteristics of turbulence. As mentioned earlier, there are considerable difficulties associated with turbulence measurements. Laser doppler velocimetry is a technique undergoing a rapid and strong development, and will probably soon be the standard method for turbulence measurements. However, to this date, most of the accurate and established results, especially for near-wall turbulence, have been obtained with hot-wire and hot-film anemometry, which will be reflected in the following presentation.

Velocity Fluctuations

Turbulence in a two-dimensional straight channel flow has, similarly with other turbulent shear flows, an anisotropic character with non-Gaussian probability density distributions of velocity and pressure fluctuations. This is illustrated in Figure 4, where the mean velocity profile is complemented with distributions of measured minimum and maximum velocities. The largest span of streamwise velocity fluctuations is found in the buffer region around $y^+ = 15$. The highest velocity at this position is about as high as the centerline mean velocity, and the lowest is about $2u_\tau$.

Probability density distributions (of u) are schematically shown in the same figure for three different positions, viz., one in the viscous sublayer, one in the buffer region and one in the outer part of the logarithmic region. In the viscous sublayer the probability density distribution is markedly

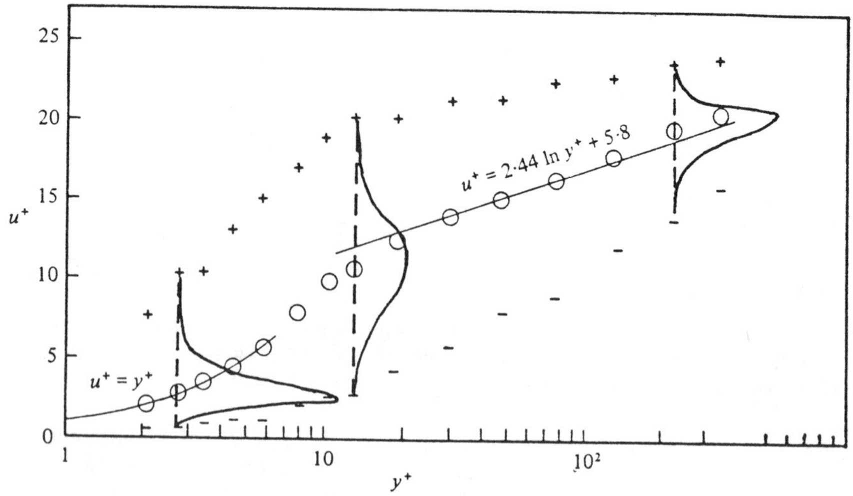

Figure 4. Distributions of measured maximum (+), minimum (−) and mean velocities (○) (Re = 13,800), complemented with probability density distributions of the streamwise velocity [10].

asymmetric due to the presence of the wall. A long tail is seen for velocities higher than the mean. If one thinks of the velocity fluctuations as being caused by displacements of fluid elements in the direction normal to the wall, the long tail implies that fluid from the buffer region quite frequently reaches positions very close to the wall in the viscous sublayer. In the buffer region the mean velocity is about half the centerline velocity, and the probability density distribution is approximately symmetric around the mean. It may again be tempting to interpret the probability density distribution in terms of fluid displacements (although this is of course not altogether correct). The large extreme values then indicate that fluid elements from the center of the channel (the outer region), as well as from the viscous sublayer, occasionally reach positions in the buffer region. The former suggests that motions on outer scales may be of influence even as close to the wall as $y^+ = 15$, or even closer. The distribution in the outer part of the log-region is skewed in the opposite direction as compared to that in the viscous sublayer, which is natural if the amplitude of vertical displacements in the outer region are equal in positive and negative directions.

The probability density distributions of the other velocity components (Figures 5a and b) show that the span of these fluctuations is smaller than for the streamwise component. The distributions for w are, for obvious reasons, symmetric around zero and extend to larger values than for v in the near-wall region.

The probability density distributions give a gross picture of some of the characteristics of turbulent fluctuations, but do not contain any time-related information or information about the frequency contents. Instead, they represent long-time averages of the fluctuating signals, from which quantities such as turbulence intensities, skewness (S) and flatness (F) factors may be determined as second, third and fourth moments, respectively. For the streamwise velocity they become

$$u_{rms}^2 = \int_{-\infty}^{\infty} u^2 P(u)\, du \tag{13a}$$

$$S(u) = \int_{-\infty}^{\infty} u^3 P(u)\, du / u_{rms}^3 \tag{13b}$$

$$F(u) = \int_{-\infty}^{\infty} u^4 P(u)\, du / u_{rms}^4 \tag{13c}$$

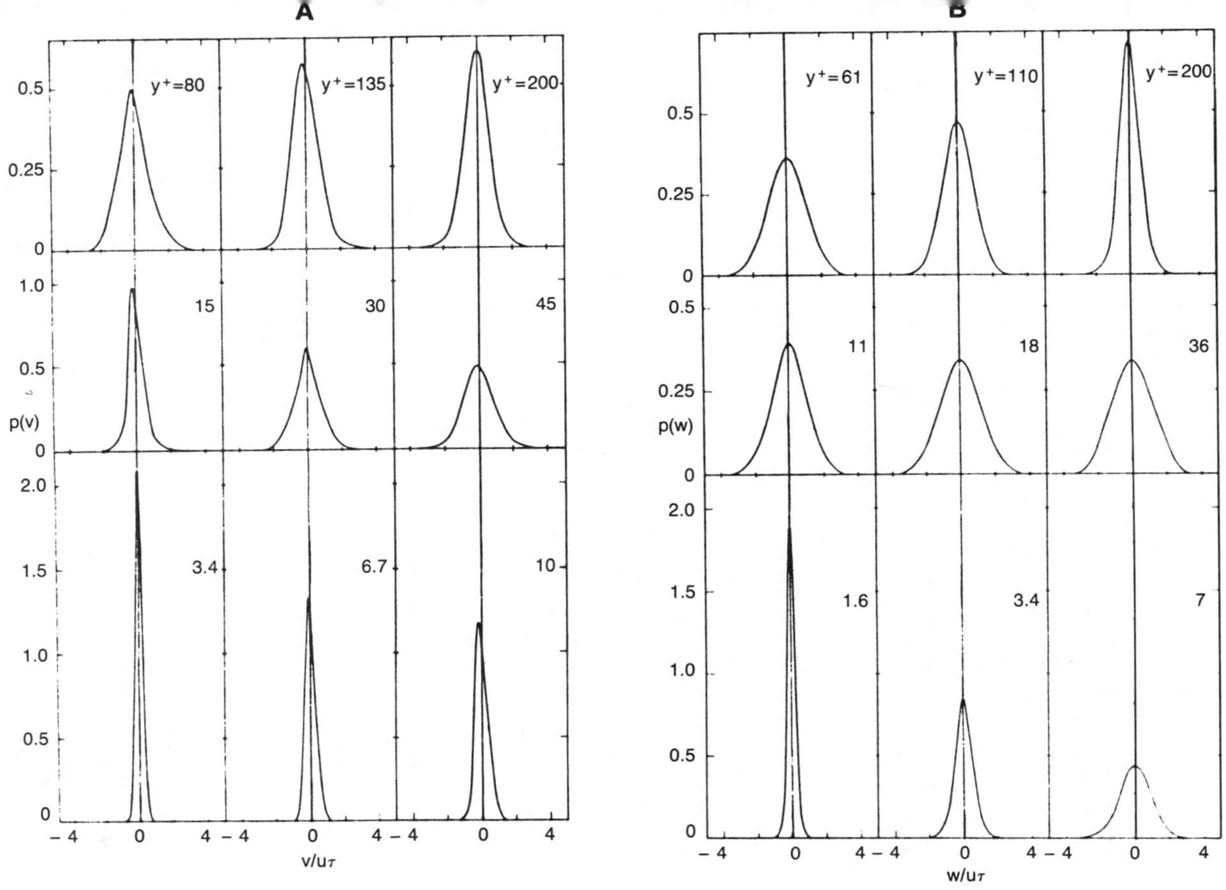

Figure 5. Probability density distributions of (a) v and (b) w at various distances from the wall [43].

where P(u) is the normalized probability density (similarly for v and w). The quantities S and F at the channel centerline may be used, as already mentioned, to judge the stage of development of the flow, but also to characterize the turbulence. For instance, a high value of the flatness factor is usually associated with an intermittent character of the corresponding velocity signal. Equations 13a–c also represent an efficient method to calculate turbulence quantities from digital data, instead of calculation from long-time integrals of u^n (n = 2, 3, 4).

Turbulence intensity distributions for the three components are shown in Figure 6 for a low-Reynolds-number channel flow. In analogy with the similarity considerations for the mean flow the friction velocity is used for normalization. The intensity of the v-component reaches a value of about 1.0, which is found considerably farther from the wall than the maxima for the other two components. For this low Reynolds number the v-maximum occurs at $y/b \approx 0.2$ ($y^+ \approx 40$). The maximum streamwise turbulence intensity attains a value of about 2.9. According to what is generally referred to as wall similarity this should then be a universal value independent of Reynolds number. When normalized with the local mean velocity the streamwise intensity is well above 30% in the viscous sublayer.

Large discrepances exist concerning measured values of the streamwise turbulence intensity in the near-wall region, as illustrated in Figure 7. These large differences cannot be solely attributed to Reynolds number effects. Since normally the length of a hot-wire or hot-film sensor cannot be considered small compared with the smallest scales of the turbulence, spanwise spatial averaging could be expected to influence the measurements. Effects of imperfect spatial resolution on measured turbulence intensity and (velocity) spectra have been studied by Uberoi and Kovasznay [45], Wyngaard [46], and recently, with emphasis on near-wall turbulence, by Johansson and Alfredsson [40]. A similar study of the effects of transducer size on measurements of wall pressure fluctuations has recently been reported by Schewe [47] (see also Willmarth [48]).

In the study of Johansson and Alfredsson [40] data on the maximum turbulence intensity from hot-wire (and hot-film) measurements of channel and boundary layer flows were compiled (Figure 8). There is a strong correlation between the measured value of the maximum turbulence intensity and the probe length. The maximum nondimensional intensity decreases from about 2.9 for

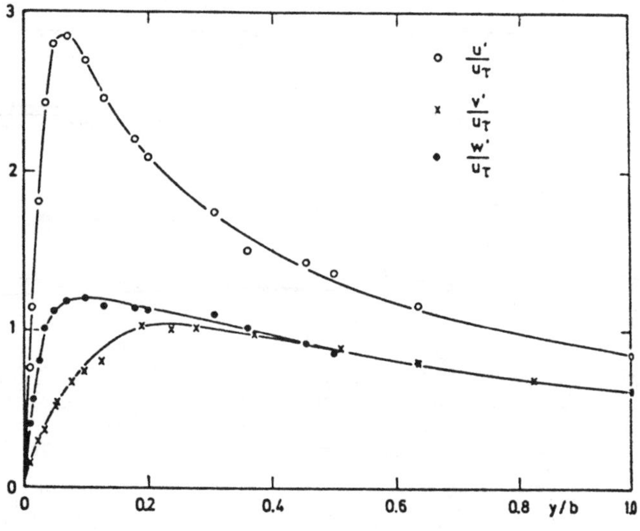

Figure 6. Distributions of the three turbulence intensities normalized with uτ [43].

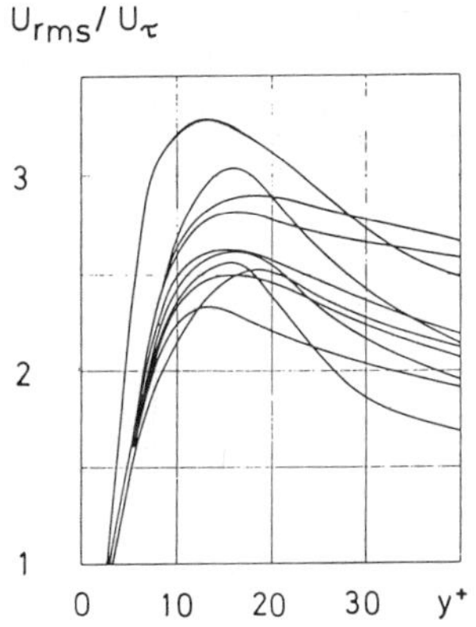

Figure 7. Streamwise turbulence intensity in the near-wall region. Results from various studies of channel, pipe and boundary layer flows [44].

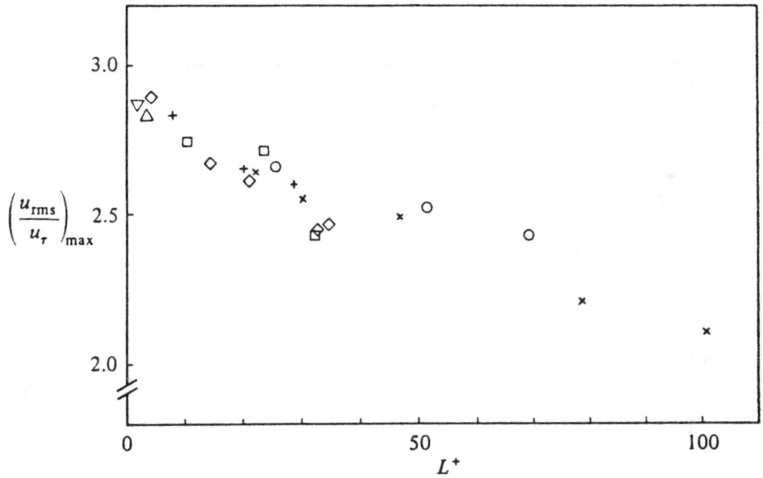

Figure 8. The maximum streamwise turbulence intensity as function of the probe length in viscous units (L^+). Data from: ▽, Echelmann [33]; ○, Hussain and Reynolds [12]; □, ◇, Johansson and Alfredsson [10, 40]; ×, Karlsson [49]; △, Kreplin [14]; +, Purtell, Klebanoff, and Buckley [50].

the smallest probes, to 2.1 for a probe 100 viscous units long. Hence, much of the reported discrepances concerning turbulence intensity in the near-wall region appears to be attributable to imperfect spatial resolution and not so much to Reynolds number effects. Only data obtained with constant-temperature anemometry and with linearization are shown in Figure 8. However, earlier "constant-current" data, such as those of Laufer [6, 8] and Klebanoff [51] are also in reasonable agreement with the presented results.

To further illustrate the validity of wall similarity for the streamwise turbulence intensity, measurements with two different probes are shown in Figure 9a. Considerable differences exist between the two sets of data with different probe lengths in viscous units ($L^+ = 14$ and 32) but with approximately the same Reynolds number. (The Kolmogorov length in this case corresponds to about 3

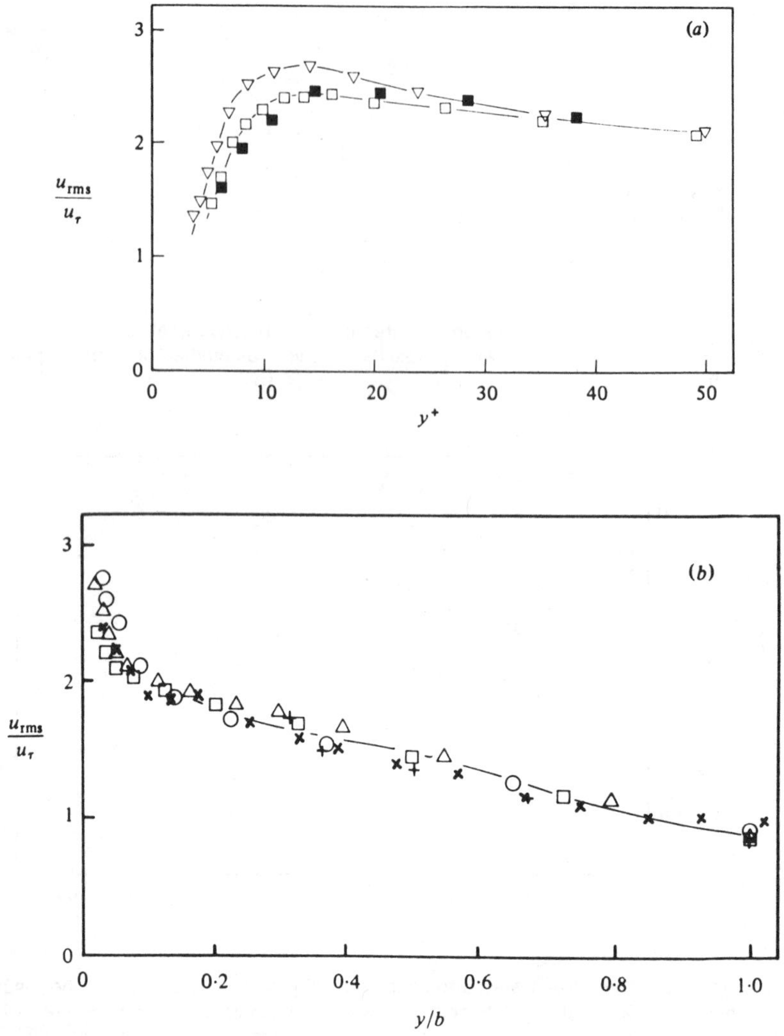

Figure 9. Distribution of streamwise turbulence intensity in: (a) the near-wall region: ▽ and □, Re = 50,000 (L^+ = 14 and 32, resp.), ■, Re = 129,000 (L^+ = 33) [40]; (b) the outer region: ○, Re = 13,800; △, Re = 34,600; □, Re = 48,900 [10]; +, Re = 7,700 [14]; and ×, Re = 29,400 [18].

viscous units.) On the other hand, there is good agreement between the cases with different Reynolds number (both with $L^+ \approx 30$). The conclusions to be drawn from Figures 8 and 9a are that wall similarity is valid for the streamwise turbulence intensity, i.e., u_{rms}/u_τ is a universal function of y^+ in the wall region, secondly that small probes are indeed needed if all turbulence scales are to be resolved, and thirdly that the spanwise scale of the energetic fluctuations increase with increasing distance from the wall. The first of these is by no means a new conclusion, but as discussed in conjunction with Figure 7, much of the data in the literature may seem to contradict this point. Wall similarity for the skewness and flatness factors may also be verified in the same manner as in Figure 9a [40]. The third conclusion is based on the fact that probe-size effects become less pronounced with increasing distance from the wall. As a rule of thumb, one may say that effects of imperfect spatial resolution in the near-wall region are small for wall distances larger than the probe length, i.e., for $y^+ > L^+$ (Figure 9A).

Away from the wall the energetic fluctuations become practically independent of fluid viscosity and the various moments (u_{rms}, S, F etc.) depend on y/b only, where b (the channel half-width) represents the relevant lengthscale for this case. For pipe and boundary layer flows b is replaced by the pipe radius and boundary layer thickness, respectively. This behavior, usually termed Reynolds-number similarity, is illustrated in Figure 9b for the streamwise turbulence intensity. The (universal) value on the channel centreline is about $0.9u_\tau$.

The bulk of the turbulence intensity data shown in Figures 8 and 9 was obtained with hot film or hot wire anemometry. However, a comparison with LDV-results for the turbulence intensity in the outer region [18] is included in Figure 9B. The agreement between the two methods of measurement is good.

Distributions of higher moments (S and F) for all three velocity components over half the channel width are shown in Figure 10. This set of data is unique in that results for v could be obtained with

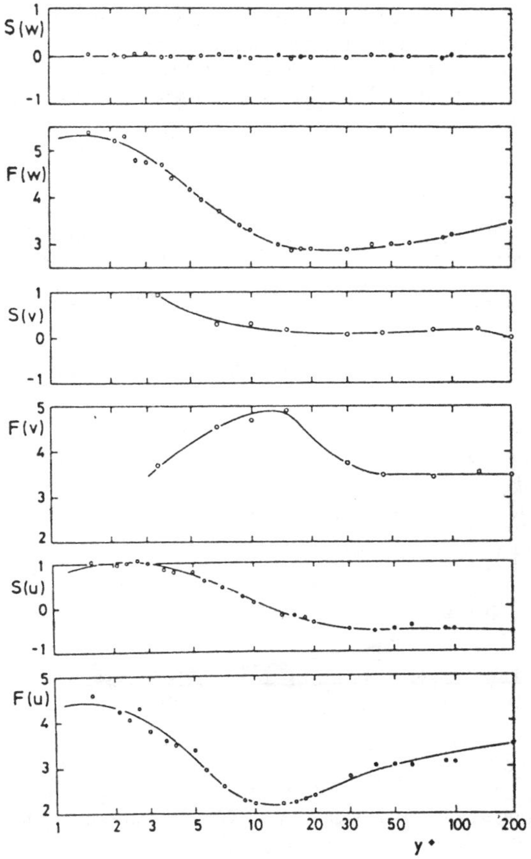

Figure 10. Distributions of skewness and flatness factors for all three velocity components over the channel half-width [43].

an X-probe in the viscous sublayer. The skewness of both u and v in this region is positive. The positive value of S(u) is probably caused by sweep motions from "outside," approaching the wall at a small angle, and hence with a relatively small negative value of the velocity component normal to the wall. The positive value of S(v) and associated high value of F(v), on the other hand, seem to be caused by intermittently occurring lift-up of low-speed fluid from positions very near the wall. Such events are closely coupled to turbulence production, and will be discussed in detail later.

The spanwise velocity component has a zero skewness factor, owing to symmetry, but a high flatness value in the viscous sublayer. The latter is associated with large deviations (from zero) of the flow angle in the u-w-plane. This was indeed observed already by Fage and Townend [52] in a flow visualization study. The extremely near-wall w-measurements presented in Figure 10 were accomplished by means of a specially designed V-type hot-film probe [43].

Reynolds Stress and Flow Angles

The long-time average of the product $-uv$, constitutes the only non-zero off-diagonal element of the Reynolds-stress tensor for plane turbulent channel flow. It will hereafter simply be referred to as the Reynolds stress (and be denoted by $-\overline{uv}$). To get the actual Reynolds stress, $-\overline{uv}$ should be multiplied by the fluid density. The probability density of uv (Figure 11) has quite a different character than that for single velocity components. The presence of long tails of the distribution (and a large flatness factor) signifies an intermittent character of the uv-signal. The measured distribution in Figure 11 was obtained in a turbulent boundary layer, but we have found similar results also in channel flow (unpublished results). The solid curve represents a distribution calculated under the assumption of joint Gaussian distributions for u and v with a correlation coefficient of -0.4, which is seen to give good agreement with measured values. This may seem somewhat surprising since P(u) and P(v) are both non-Gaussian.

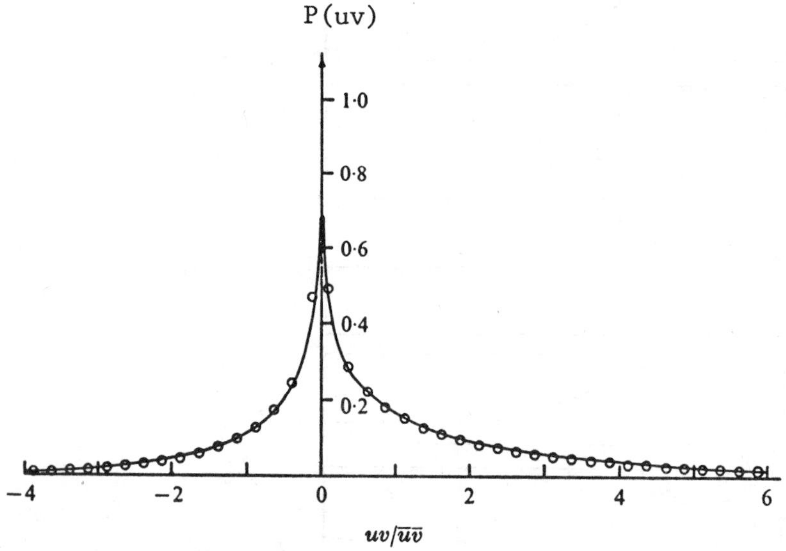

Figure 11. Probability density distribution of the uv-signal ($y^+ = 235$). ○, measurement: ————, computed [53].

The correlation coefficient of u and v is seen in Figure 12A to attain a constant value of -0.4 over a large portion of the channel. This is a result that has been verified in many studies for low as well as for high Reynolds numbers. As described earlier the Reynolds stress (Figure 12B) can be computed from the mean velocity profile. In the outer region (away from the wall) the distribution is well described by the straight line $1 - y/b$. The position of maximum Reynolds stress tends to get closer to wall, in terms of y/b, with increasing Reynolds number. The corresponding y^+-value, however, increases as the square root of the Reynolds number (based on u_τ).

The uv-signal is characterized by large intermittent peaks as illustrated in Figure 13A. The fractional contributions to the Reynolds stress from portions of the signal with an amplitude larger than $Hu_{rms}v_{rms}$, are shown in Figure 13B, for $y^+ = 50$. The four quadrants of the u-v-plane are analyzed separately, and it is seen that peaks larger than, say, $4u_{rms}v_{rms}$ ($\approx 10|\overline{uv}|$) are almost exclusively found in the second quadrant. Such peaks occur during only 1.5% of the total time, but represent a 20% contribution to the Reynolds stress. Second quadrant motions may be described as ejection of fluid elements away from the wall, and their fractional contribution to the total Reynolds stress ($H = 0$) is about 0.8. The corresponding values for the fourth, first, and third quadrants are 0.6, -0.2 and -0.2, respectively. Sweep motions towards the wall ($v < 0$) with a streamwise velocity higher than the mean ($u > 0$) also produce a considerable part of the total Reynolds stress. However, the associated uv-peaks are of smaller amplitude than for the ejections. The first and third quadrants are often called interaction quadrants, since, for instance, motions with $u < 0$ and $v < 0$ may be thought of as ejections being pushed back towards the wall.

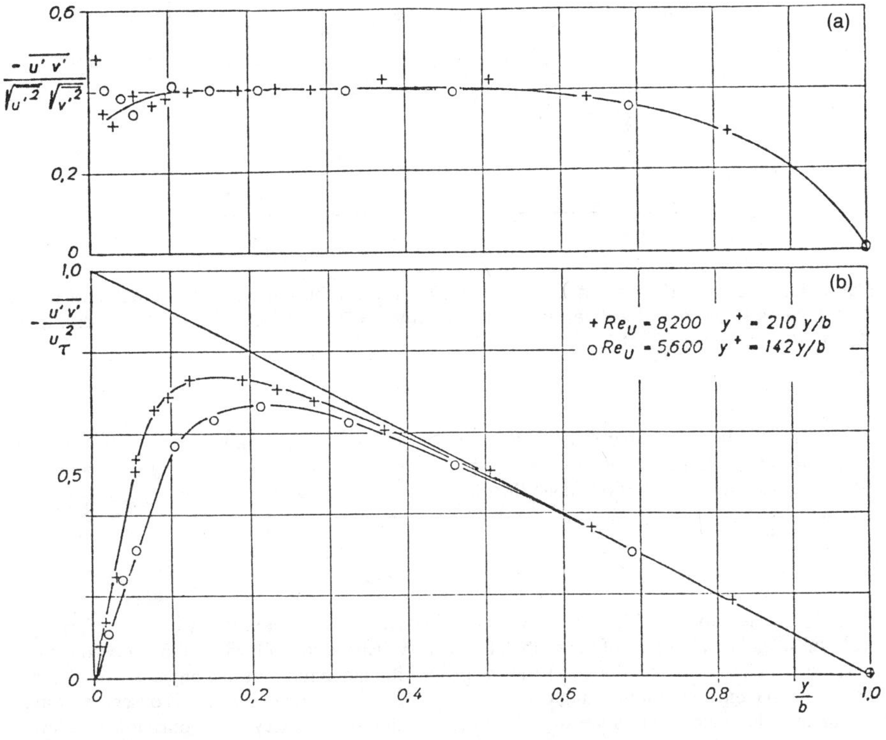

Figure 12. Distributions of (a) the correlation coefficient between u and v, and (b) the normalized Reynolds stress, over the channel half-width [9].

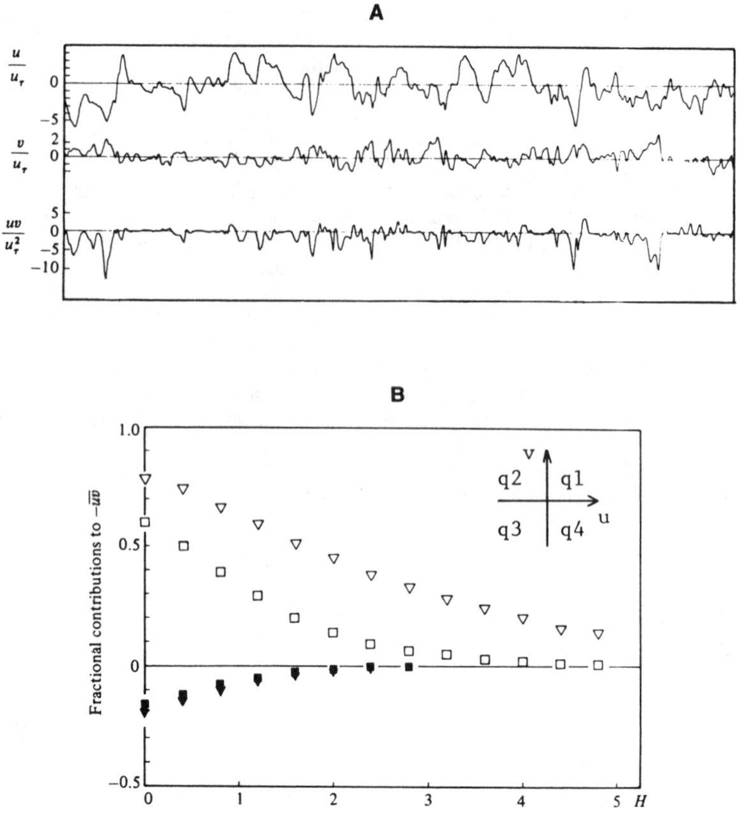

Figure 13. a, u, v, and uv-signals from $y^+ = 50$, showing the intermittent character of uv. b, contributions to uv from the different quadrants at $y^+ = 50$, as function of the threshold level: ■, q1; ▽, q2; ▼, q3; □, q4 [39].

It can readily be shown that large uv-peaks in the second quadrant imply large outflow angles. Consider such an ejection type of uv-peak with an amplitude of Cu_τ^2. The following relation for the associated instantaneous outflow angle then holds

$$\tan \beta > \frac{4C}{U^{+2}} \tag{14}$$

where U^+ is the normalized local mean velocity [39]. Hence, the intermittently occurring large uv-peaks must lead to a long tail of the probability density distribution of the instantaneous flow angle (in the u-v-plane). It is indeed seen in Figure 14 that the maximum outflow angle is considerably larger than the maximum inflow angle. At e.g., $y^+ = 37$ as large an angle as 21° is observed. It may appear that these large values are of little practical interest since they occur during a very small portion of the total time. However, the associated contribution to the Reynolds stress is significant.

Probability density distributions of the flow angle (α) in the u-w-plane have been measured by Kreplin and Eckelmann [55] at various y^+-positions. In the viscous sublayer, these distributions

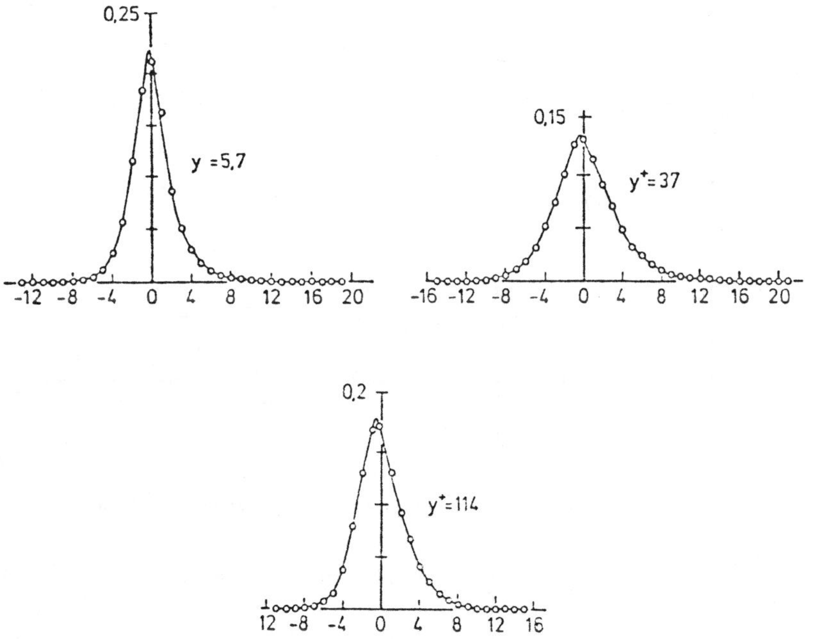

Figure 14. Probability density distributions of the instantaneous flow angle, β, in the u-v-plane [54].

are very broad as compared with that in Figure 14. This can be ascribed to the restriction imposed on the v-component by the presence of the wall for the latter case (Figure 5).

Vorticity Fluctuations

There are indications that streamwise vortices, occurring in counter-rotating pairs, are associated with lift-up of fluid elements away from the wall, giving rise to relatively large flow angles. This issue will be further discussed later, but the importance of vorticity dynamics for the persistence of turbulence motivates a discussion about vorticity fluctuations and their measurement at this point. Kovasznay [56] proposed and developed a four sensor hot-wire probe for the measurement of streamwise vorticity. The four wires are interconnected, forming a Wheatstone bridge, and oriented as two perpendicular X-wires. Both the streamwise velocity and vorticity could, in principle, be measured instantaneously with this probe, which is usually referred to as a Kovasznay-type probe, and has been used by Corrsin and Kistler [57] and Kastrinakis, Eckelmann, and Willmarth [58]. However, in the latter study it was shown that all three velocity components influence the measurement of streamwise vorticity, making accurate measurements in turbulent flows impossible with that probe. They suggested, therefore, that all three velocity components be measured with four independent wires (requiring eight prongs). Kastrinakis and Eckelmann [34] have recently reported successful measurements with such a probe, and showed that the intensity of streamwise vorticity fluctuations is a maximum in the buffer region (see Figure 15). The measured maximum value is about 16% of the mean velocity gradient at the wall.

An even more complex probe has been developed by Wallace [59] and co-workers. It consists of nine independent wires and measures all three components of velocity and vorticity. It requires

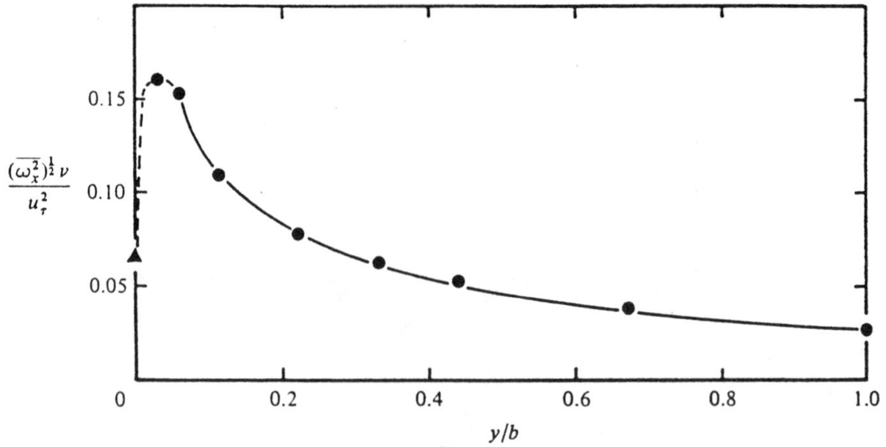

Figure 15. Distributions of the intensity of the streamwise vorticity fluctuations [34].

a rather extensive calibration procedure with more than 30 calibration constants to be determined from a set of non-linear algebraic equations.

Frequency Spectra

The results for the turbulent fluctuations described hitherto do not contain information on their temporal or spatial structure, although some indications on the spatial scale may be derived indirectly from Figure 8. In order to obtain detailed such information, frequency and wave-number spectra of the velocity components may be measured.

The frequency spectrum has a one-to-one relation with the autocorrelation function, in that the energy density (E) can be expressed as the Fourier transform of the latter. The frequency range of dominating energy content of a fluctuating signal can be clearly shown by plotting the product of E and the frequency (f) as function of the logarithm of f. Shown in Figure 16 are channel-flow spectra for two different Reynolds numbers from a position where the streamwise turbulence intensity is a maximum. In order to make the comparison easier the areas under the spectra have been normalized to unity. Three different scalings of the frequency axis are shown, viz. normalization with the inner timescale ($t_* = \nu/u_\tau^2$), the outer ($t_0 = b/U_{CL}$), and the mixed scale ($t_m = (t_* t_0)^{1/2}$). It is interesting to note that neither inner nor outer scaling collapses the data. The fact that mixed scaling does, indicates that the governing timescale in the buffer region is composed of a mixture of inner and outer variables. The maximum energy content is found at about 0.2 in mixed scaling, and there is practically no energy above 1.0.

Far away from the channel walls viscous effects are negligible for the energetic velocity fluctuations. Spectra from the outer region of the flow can hence be made to collapse for different Reynolds numbers by normalizing the frequency with the outer timescale [61]. This should be seen in conjunction with Reynolds number similarity for the turbulence intensity (Figure 9B) and higher moments.

The spatial structure of the turbulence may be said to be given by the three-component wave-number spectrum. In practice it is a difficult task to obtain even a one-component wave-number spectrum. Measurements must be carried out with (at least) two probes at several different separations to obtain the spatial correlation function, say $R_{uu}(x)$. The one-component spectrum $\phi(k_x)$ may then be computed as the Fourier transform of the correlation function. The streamwise wave-number, k_x, may be expressed as a ratio between the angular frequency, ω, and a convection velocity, c_x, which outside the near-wall region can be approximated by the local mean velocity. The convection velocity

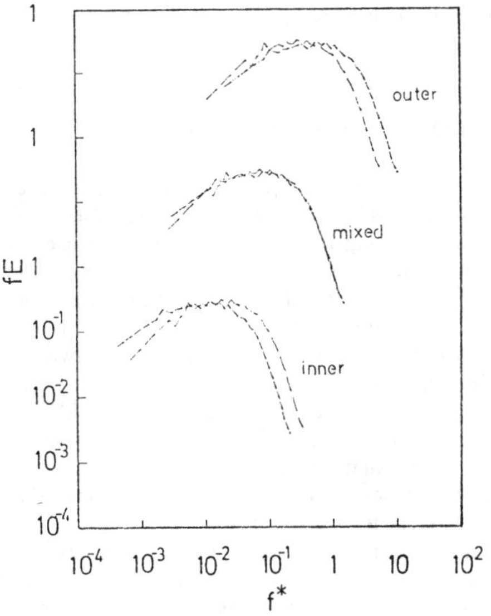

Figure 16. Power spectra of u, with outer, mixed and inner scaling of the frequency. f* denotes nondimensional frequency, i.e. ft_o for outer, ft_m for mixed, and ft_* for inner scaling. Note shifted ordinate scale. ————, Re = 13,800, $y^+ = 13$; -------, Re = 54,300, $y^+ = 14$ [60].

of the disturbances can be obtained by measuring the two-dimensional frequency-wave-number sepctrum. Such measurements have been reported by Morrison, Bullock and Kronauer [62], and a typical example from the viscous sublayer is shown in Figure 17. The spectral density is here multiplied by ωk_x, and the volume under the surface is normalized to unity. The distribution of energy is confined to a narrow band along an almost straight "ridge-line." This line defines the convection velocity (as ω/k_x). An interesting result is that the convection velocity in the near-wall region is considerably higher than the local mean velocity. For the case in Figure 17 c_x was estimated to $8u_\tau$, which should be compared with a mean velocity of about $3u_\tau$ at the position of measurement. Morrison et al. interpreted this in terms of wave-like disturbances.

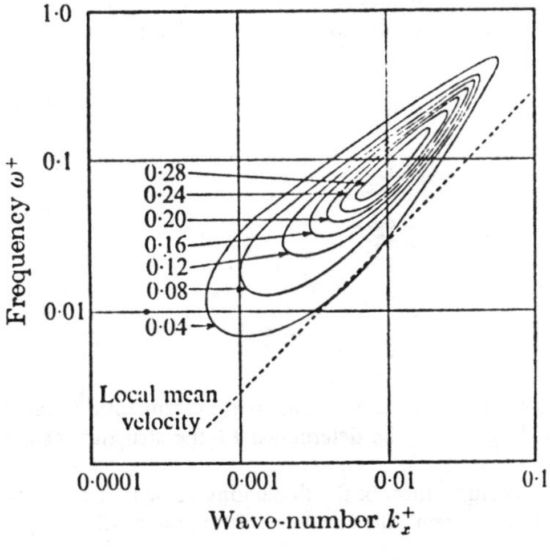

Figure 17. Two-dimensional frequency-wave-number spectrum from the viscous sublayer in pipe flow. Re = 34,200, $y^+ = 3$ [62].

Measurements by Kreplin [14] of the convection velocity from space-time correlations of the wall shear stress yielded a value of c_x as high as $12u_\tau$.

FLOW STRUCTURES AND TURBULENCE PRODUCING PROCESSES

Organized structures in turbulent flows have since the 1960's been a field of extensive research. However, already Reynolds [63] in his pioneering experiments on the transition to turbulence observed, by means of flow visualization with dye and flash lighting, large-scale structures that he called eddying motions. Such structures are a pronounced feature of many flows, especially in conjunction with transition, and are often denoted "coherent structures." For instance, in the transition region of jets and mixing layers they occur repetitively in the flow and are highly organized [64]. Flow visualization studies of the near-wall region of wall-bounded turbulent shear flows have shown that coherent structures are present also in such fully developed turbulent flows. However, in these cases they are obscured by random or chaotic small scale motions which here constitute a considerable part of the turbulent energy. In the following description of organized structures we will mainly consider the events associated with turbulence production that take place in the near-wall region. It is assumed that these structures are similar in boundary layers and channel flows, although their interaction with the outer flow may be somewhat different for the two cases.

It was first recognized in flow visualization studies that organized flow patterns in the near-wall region play a dominant role in the turbulence production process. These events occur randomly in space and time, and an Eulerian observer will see structures of different stage of development passing by, thereby seeing the flow as more chaotic than it really is. This may partly explain why these coherent events remained unrecognized during several decades of turbulence research. The main difference between probe measurements and flow visualization is that the former yields temporal information at a fixed position, whereas the latter shows the spatial structure and development of flow structures. There have, however, been certain developments of the methods for analysis of turbulent signals measured by fixed probes in the flow in order to sort out the interesting deterministic parts that are associated with what are believed to be dynamically important events.

The fluctuating velocity field at a fixed position, $u_i (i = 1, 2, 3)$, may be written as

$$u_i = u_{di} + u_{ri}$$

where subscripts d and r denote deterministic and random parts, respectively. Both parts contribute to the long-time statistics of the flow field. If we assume that they are uncorrelated and that the random part is Gaussian distributed, we obtain the following expressions for the rms-value, the skewness and flatness factors (for simplicity only the expressions for the streamwise velocity are given):

$$u_{rms} = (u_d)_{rms}(1 + \gamma)^{1/2} \quad (15a)$$

$$S(u) = S(u_d)/(1 + \gamma)^{3/2} \quad (15b)$$

$$F(u) = (F(u_d) + 6\gamma + 3\gamma^2)/(1 + \gamma)^2 \quad (15c)$$

where

$$\gamma = \bar{u}_r^2/\bar{u}_d^2$$

is the ratio between the energy content of the random and deterministic parts. If this ratio is small the long-time statistics of the flow will to a large extent be determined by the structure of the deterministic part.

Lumley [65] proposed another method of decomposition of the fluctuating velocity field. With the use of a "proper orthogonal decomposition theorem" the structure of "characteristic eddies" may be obtained from measured spatial correlation functions. Such a decomposition approach was applied by Bakewell and Lumley [66] to measured signals of the streamwise velocity. To obtain the velocity component normal to the wall they used simple mixing length arguments, whereas the spanwise velocity was obtained from the continuity equation. In this way they were able to compute streamlines in the yz-plane in the near-wall region, which resembled those of two counter-rotating streamwise vortices. Moin [67] has applied this technique to a database from numerical simulations of turbulent channel flow. Since he had access to all three velocity components it was possible to determine the dominant features of the three-dimensional structure of the "characteristic eddies" from the decomposition.

Flow Visualization Techniques

At this point it may be appropriate to describe briefly some of the visualization techniques used in the search for coherent structures. Visualization studies of the near wall region have almost exclusively been done in liquids; especially the hydrogen bubble method has been used extensively in water flows. It employs a thin wire (usually held stationary in the flow) through which an electric current is fed. Small hydrogen bubbles are then formed by electrolysis at the wire, and are swept away by the flow, thereby acting as markers. If the applied current is constant, a sheet of bubbles is produced. If instead the current is pulsed, time-lines of small bubbles are formed. The main advantage with this method is that a large region of the flow can be marked without too large a disturbance, and that the wire can be positioned in various ways, e.g., normal or parallel to the wall. Photographs of the flow will show the cumulative motion of the marked fluid elements over time since the bubbles were released from the wire. This makes it possible to infer information from still pictures [68].

Dye injection is in principle similar to the hydrogen bubble method, but is limited to either injection through a wall slot, thereby marking the fluid within the viscous sublayer, or by injecting it through one (or a few) syringe needles in the flow. The latter may be difficult to do without severely disturbing the flow. Dye injection and hydrogen-bubbles have in some studies been used simultaneously [32, 68].

In air flows smoke can be used as a fluid marker. However, it has mostly been used to study the outer intermittent region of turbulent boundary layers, either by injection from a slot, marking the whole boundary layer [69] or by the smoke-wire-method [70]. In the latter case a thin sheet of smoke is produced by heating a thin oil-covered wire (by an electrical current). Falco [71] have made studies also of the wall region with smoke injected through a wall slot.

Neutrally buoyant trace particles have been used in some studies of the wall region. Fage and Townend [52] used a trace particle technique, and were able to observe the flow in the wall region of turbulent pipe flow (of square cross section). They found strong spanwise velocity fluctuations very close to the wall. Trace particles together with a stereoscopic recording (film) system, may enable determination of the three-dimensional velocity field with an advanced analysis system. Tentative experiments with such a system were reported by Johnson et al. [72]. However, a fully developed system is not yet at hand. At Ohio State University a series of experiments have been carried out with a qualitative approach. By photographing only the near wall region of a pipe flow, with a moving film camera, Corino and Brodkey [21] were able to study the turbulence producing processes in some detail. Later stereoscopic photography studies also involved the outer region of a boundary layer flow. The trace-particle-moving-camera technique has the advantage over hydrogen bubbles in that it can follow an individual structure for a longer time than is possible with a stationary bubble wire. One disadvantage, however, is the difficulty, even with a stereoscopic technique, to observe streamwise vorticity in the flow.

The Streaky Structure of the Wall Region

The near wall region is of special interest in wall bounded shear flows, since a large part of the turbulence production takes place very close to the wall. It is now well established that the viscous sublayer has a fairly ordered three-dimensional structure that consists of alternating high and low speed regions. This structure seems to be present in all types of wall bounded turbulent shear flows and may be used as a diagnostic tool to determine whether a given flow is turbulent or not [73]. The existence of this streaky sturcture was first reported by Hama (see Corrsin [74]) who injected dye through a wall slot and found that the dye formed a streaky pattern. Later studies (for instance at Stanford and Lehigh Universities) have mainly used the hydrogen bubble technique to investigate these structures.

The tendency to evolve a streaky structure can easily be visualized in a low-speed water flow by introducing some particulate matter at a (horizontal) wall, as e.g., salt or fine grained sand, which then collects into regularly spaced streaks. Smith [75] showed that these streaks correspond to the low-velocity streaks by simultaneously visualizing the flow in the viscous sublayer with hydrogen bubbles.

Knowledge of the three-dimensional nature of the streaky structure is mainly obtained from flow visualization, since the very small dimensions of the viscous sublayer make probe measurements difficult in that region. Both hydrogen bubbles released from a wire parallel to the wall and dye injected from a wall slot collect into regions of low velocity. This indicates that there are coherent spanwise velocity fluctuations present near the wall, and that the fluid in the high speed regions comes from the unmarked flow outside the viscous sublayer. The low-speed regions have been found to be coherent over large streamwise distances of the order of $1{,}000\ell_*$ [76], but they do not extend straight down in the flow direction (Figure 18). This explains why a fixed probe located in the wall region does not see very long time-periods of low velocity. However, space-time correlations of the wall shear stress show that the flow is correlated over large streamwise distances (Figure 19). There is still a significant correlation at a separation of $1{,}280\ell_*$ between the probes, which for this case corresponds to almost three channel heights.

The statistical properties of the wall streaks (mainly their mean spacing) have been studied in several visualization experiments (Kline et al. [19] and Smith and Metzler [76], both using hydrogen bubbles, and Oldaker and Tiedermann [35], using dye slot injection). This has been done manually, i.e., the spacing between streaks has been measured from photographs. This involves some arbitrariness, but results from different studies are remarkably consistent. The spacing has been found

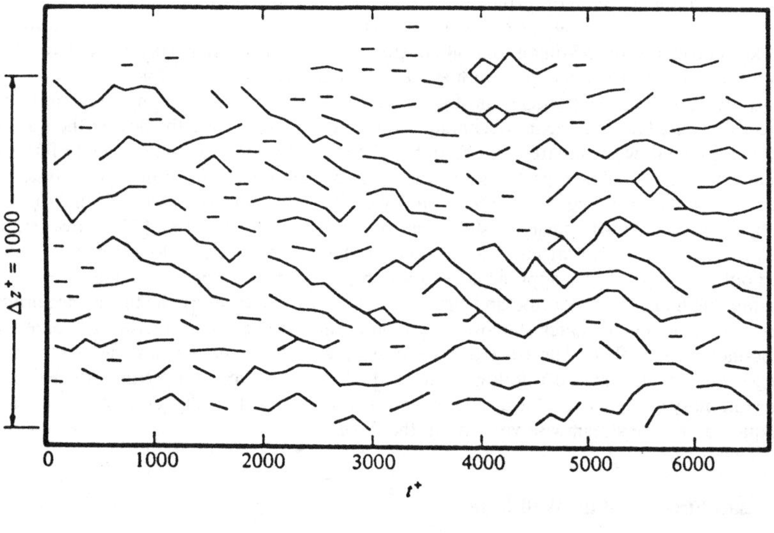

Figure 18. Low-speed streaks registered close to the hydrogen-bubble wire at $y^+ = 5$ as function of time [76].

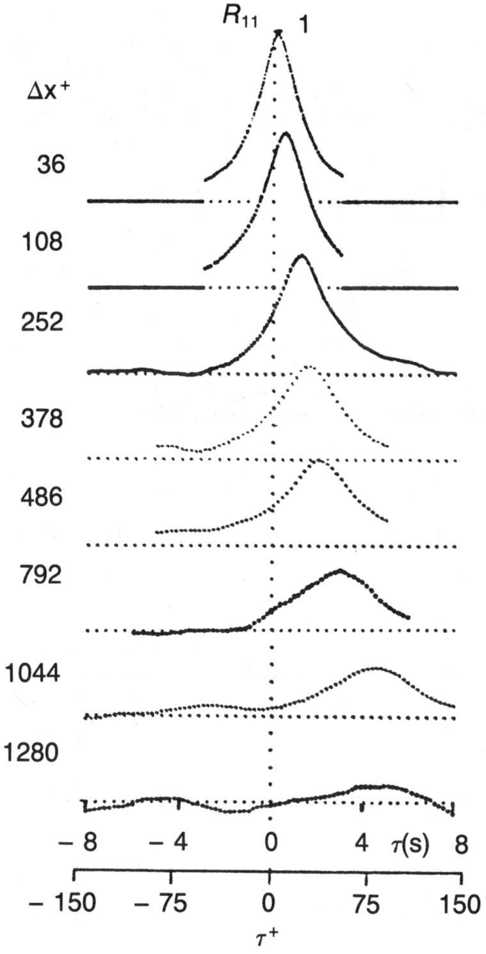

Figure 19. Space-time correlations of the streamwise wall shear stress as function of Δx^+ [77].

to be log-normally distributed (Figure 20), with a standard deviation of about 40% of the mean. The mean spacing in the viscous sublayer is close to $100\ell_*$ practically independent of Reynolds number (Figure 21). Of the studies in Figure 21 only that of Oldaker and Tiedermann [35] was carried out in a fully developed channel flow. Their data were taken over a rather limited Reynolds number range but seem to conform rather well to the data from the boundary layer studies.

The streaks have a width in the range $10\text{–}30\ell_*$, and thereby an aspect ratio of 30–100. They are most clearly seen within the viscous sublayer, but can also be found somewhat further from the wall.

There are also some probe measurements that have been made to study the streaky wall-region structure. Lee, Eckelman, and Hanratty [78] and Kreplin and Eckelmann [77] carried out correlation measurements of the transverse wall shear stress (Figure 22). These data show a minimum at a transverse separation of about $50\ell_*$, which is consistent with a streak spacing of $100\ell_*$. Also the correlation of the streamwise shear stress has a minimum at about the same spacing, which can be interpreted as a result of alternating high and low speed regions. The minimum in the

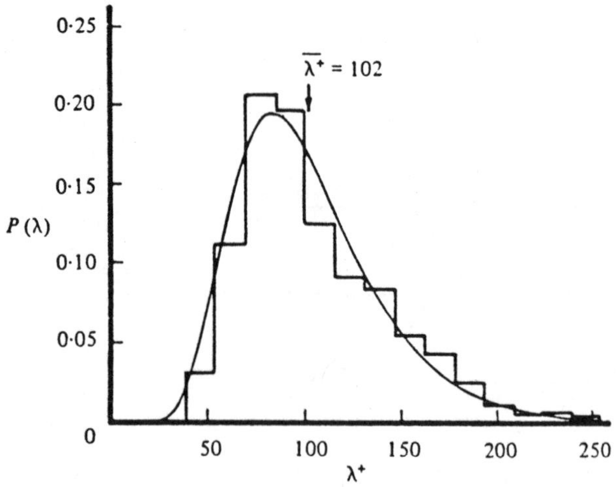

Figure 20. Probability density histogram of spacing between low-speed streaks at $y^+ = 5$ [76].

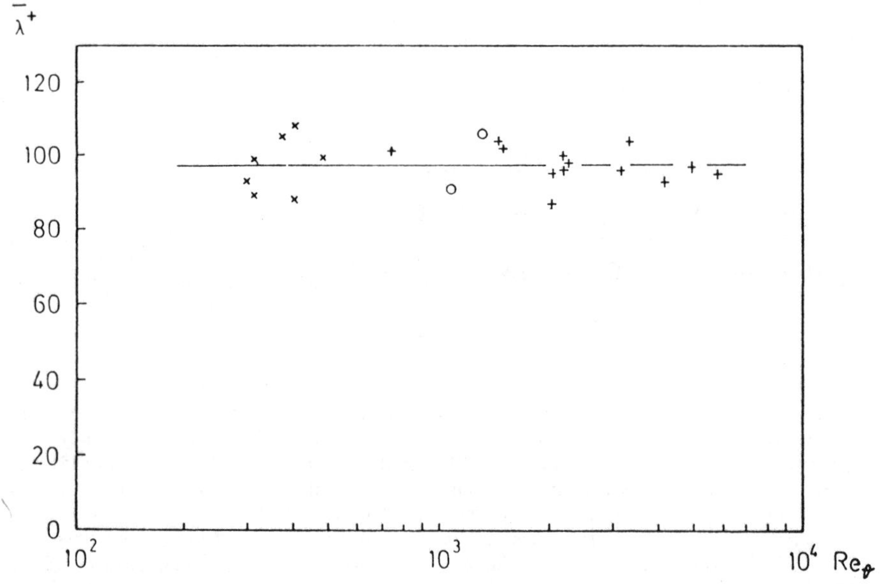

Figure 21. Mean non-dimensional streak spacing as a function of momentum thickness Reynolds number: ○ [19]; × [35]; + [76].

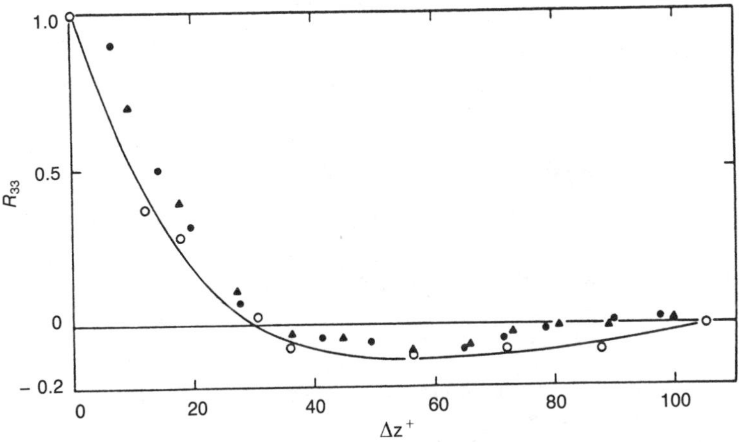

Figure 22. Spanwise correlation as function of Δz^+ for the spanwise shear stress (○). Data from Lee et al. [78] in pipe flow, ●: Re = 26,900; ▲: Re = 35,900 [77].

correlation for the transverse shear stress may be seen as a support the idea of streamwise counter-rotating vortices close to the wall.

Short-time correlations and spectral methods have been used by Gupta, Laufer and Kaplan [79] and Haritonidis [80] to determine the spacing between low-speed streaks from measurements of the streamwise velocity in the viscous sublayer. These studies have given higher values than $100\ell_*$ at high Reynolds numbers, although this may be associated with spatial resolution problems of the hot-wire arrays used.

The Bursting Process

The low-speed regions would not be very important if they were confined to the viscous sublayer and merely acted as passive flow structures. The interest in them stems from the early work at Stanford where the streaks were observed to be lifted from the wall into the buffer region and there usually begin an oscillating motion terminating in a chaotic break-up of the flow. The formation of a low-speed streak is hence one part of a quasi-cyclic scenario, during which a major part of the turbulence production occurs.

Dye visualizations have shown that when a low-speed streak has formed its outer edge slowly migrates out from the wall, possibly due to viscous effects. The rapid lift-up of the streak into the buffer region, which eventually follows may than be caused by some large-scale flow structure imposing an adverse pressure gradient at the wall (this idea was presented by Offen and Kline [81]). Another hypothesis regarding this behavior is that both the formation and the subsequent lift-up of the low-speed streaks is associated with a pair of long counter-rotating streamwise vortices [82]. Low-speed fluid may be pumped away from the wall in between these, which could explain the outward migration of the streaks, and possibly the subsequent, rapid lift-up of the streaks into the buffer region. In this context the theory of Landahl [83] should also be mentioned, which offers an alternative explanation for the lift-up and formation of shear layers. This theory will be further discussed later in this chapter.

After lift-up a localized region of low-speed fluid appears in the buffer region. The instantaneous velocity profile is hence inflexional both in the normal and in the spanwise directions. Such a flow field is unstable and undergoes oscillations followed by a rapid and violent mixing, a phase usually referred to as break-up. During this phase, fluid elements from the decelerated region are ejected

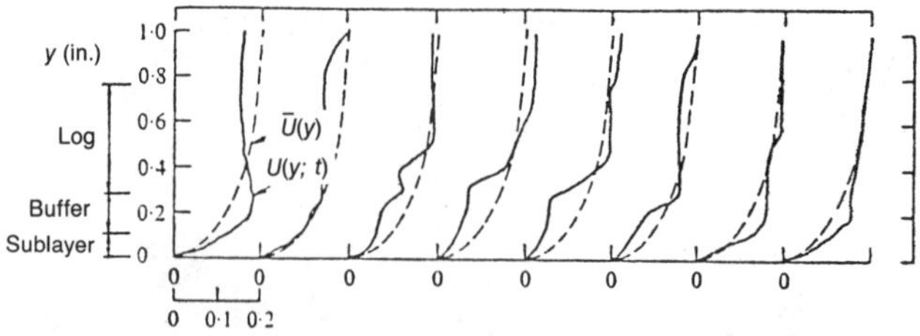

Figure 23. Instantaneous (———) velocity profiles over a typical bursting sequence obtained from hydrogen bubble visualization as compared to the mean velocity profile (-------) [20].

outwards, and an intense small scale mixing sets in. During this phase the affected region increases considerably in size, and the motions may be coherent over a region extending several hundred viscous length units in the direction normal to the wall. The ejected fluid has been observed to have rather steep trajectories out from the wall, with angles occasionally larger than 20° [21]. After this chaotic phase a large-scale, relatively quiescent, motion approaches the wall at a rather small angle. This is often called a sweep.

The sequence of events can be illustrated by the instantaneous velocity profiles during the process. Figure 23, taken from the work of Kim et al. [20], shows streamwise velocity profiles obtained from hydrogen-bubble flow visualization pictures during one bursting event. The different phases are clearly seen, viz. the lift-up causing the inflexional profile, the sweep motion of high speed fluid moving toward the wall, and the return of the flow to the mean profile. Kim et al. [20] as well as Corino and Brodkey [21] estimated that almost all of the turbulence production takes place during this sequence of events which was estimated to occur during 25% of the time.

Conditional Sampling of the Bursting Process

The aim of a conditional sampling method is usually twofold, namely to determine the frequency of occurrence of a specific type of event and to obtain an ensemble average of the events detected.

This process is illustrated in Figure 24, where the principal part is the detection algorithm, with which the specific events are sorted out from some signal(s) and a reference time for each event is set. From the series of reference times ($t_1, \ldots, t_N$) so obtained the frequency of occurrence can be computed and ensemble averages constructed over a suitable time window as

$$\langle u(\tau)\rangle = \frac{1}{N}\sum_{n=1}^{N} u(t_n + \tau)$$

where τ is a time relative to the reference time t_n.

In the case of an excited flow [84] the excitation signal itself can be used to give the phase information (i.e., to set the reference times), and in that case the detection algorithm is simple and straightforward. However, for naturally developed turbulence no such signal is available. Two approaches have been taken for fully developed turbulence—visual detection or detection with a scheme that uses time-series of some relevant measured quantity, e.g., the streamwise velocity or the uv-product.

The detection algorithm usually employs some parameter(s), such as a threshold level and/or an averaging time. It has been found for most methods that the number of events depends strongly on the threshold level, indicating that a detection scheme cannot give a well-defined number for the frequency of occurrence of the bursting process. One may then ask whether it is possible to obtain any relevant information at all from such a scheme. However, by comparing results obtained with only one method at the same setting of the parameters, it may at least be possible to obtain

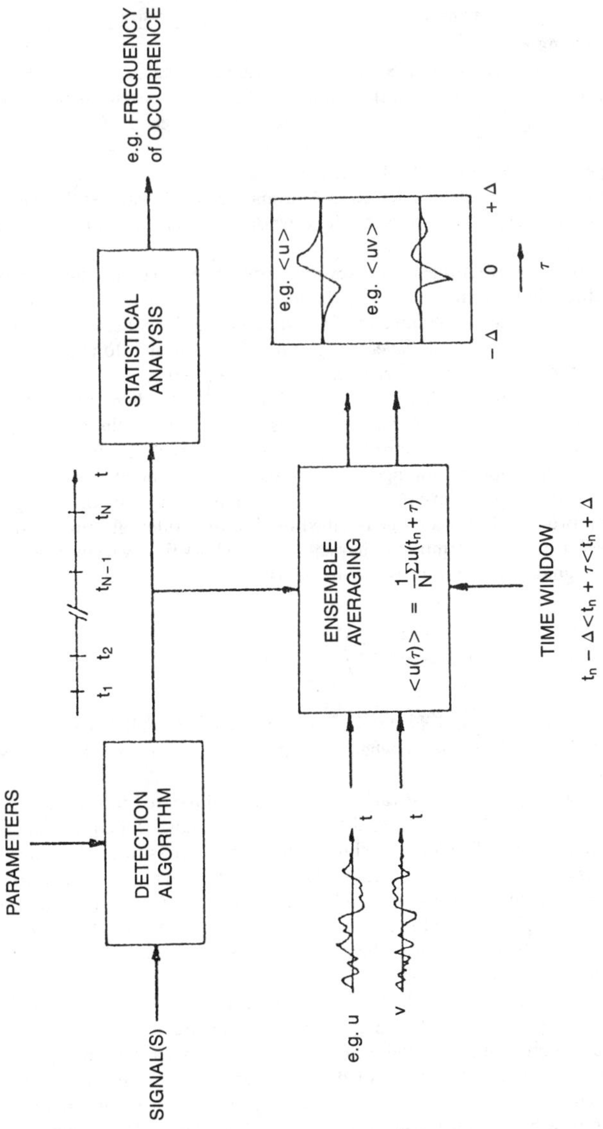

Figure 24. Schematic layout of typical detection scheme.

information about the scaling behavior of the events. For conditional averages it is usually found that the amplitude depends strongly on the threshold but that the qualitative features do not. In order to find a region of relevant thresholds it is helpful to construct ensemble averages of the uv-signal, in order to check that the events really are associated with large contributions to the Reynolds stress. This should in fact be a prerequisite for any detection scheme that is used to find the turbulence producing events.

For turbulent boundary layers the first type of conditional measurements were made in the intermittent region (i.e., outer flow region) to, among other things, determine the passage time between turbulent bulges [85]. In that case the demarcation between the free stream flow and the turbulent bulges is fairly clear. This is not the case for the bursting process in the near wall region because of the large turbulence intensity in that region. Several ideas for the detection have been explored, and for a review the reader is referred to Antonia [22]. Two features that will be discussed here at some length are large contributions to the Reynolds stress and shear layers created by the lift-up.

The detection of large contributions to the Reynolds stress in the second quadrant is an obvious method, as the ejections have been found to be the major contributor to the Reynolds stress. This method was introduced by Lu and Willmarth [53], and has later been used by Sabot and Comte-Bellot [86], Comte-Bellot, Sabot, and Saleh [87] and Alfredsson and Johansson [39]. A major problem with the method is that the physical size of X-probes makes measurements in the buffer region difficult, limiting the applicability of the method to low Reynolds numbers. The conditional averages of uv, obtained with this technique, may be used to determine the timescale of uv-peaks, i.e., typical size of the Reynolds stress producing events. However, the limited amount of data available precludes any attempts of finding the Reynolds number dependence of that quantity.

Blackwelder and Kaplan [88] developed a technique that focuses on the sharp shear layers formed by the lift-up process. The technique is called the Variable Interval Time Averaging (VITA) technique and is based on the intermittent character of the short-time variance of the fluctuating streamwise velocity signal. The variance is calculated as

$$\mathrm{var}(t; T) = \frac{1}{T}\int_{t-T/2}^{t+T/2} u^2\, dt' - \left(\frac{1}{T}\int_{t-T/2}^{t+T/2} u\, dt'\right)^2$$

When T becomes large the right-hand side tends to u_{rms}^2, whereas for small T the short-time variance tends to zero. An event is said to occur when the variance function exceeds ku_{rms}^2 where k is a threshold level.

The effect of the averaging time on the variance function is illustrated in Figure 25. The velocity signal shown is taken at $y^+ = 12$ in a channel flow and the variance function was calculated for three different averaging times. The VITA-technique has a band-pass-filter character that for the relatively slow deceleration at A gives the variance function a peak, which increases with increasing averaging time. However, this deceleration is followed by a rapid acceleration at B, which gives the variance function its highest value for the shortest averaging time. At C the acceleration of the flow has a timescale somewhere in between the other two, and the variance function has its highest value for the intermediate averaging time.

The band-pass-filter character of the VITA-technique means that the most probable timescale (duration) of this type of events can be found by determining the frequency of occurrence of events as function of the averaging time. In Figure 26 the accelerations have been separated from the decelerations. The most probable timescale, or duration, for the accelerations, in this case, corresponds to approximately 16 viscous time units. The large reduction in the number of detected events for increasing threshold is also demonstrated, as well as the predominance of accelerations for short averaging times.

The conditional averages of u in the buffer region obtained with the VITA-technique show a region of low velocity followed by a rapid acceleration of the flow. By employing a rake with 10 hot wires mounted at different heights (from $y^+ = 5$ to 100) Blackwelder and Kaplan [88] showed, by employing the detection at $y^+ = 15$, that the events (shear layers) detected are associated with a coherent motion occupying at least the distance spanned by their rake. They plotted their results

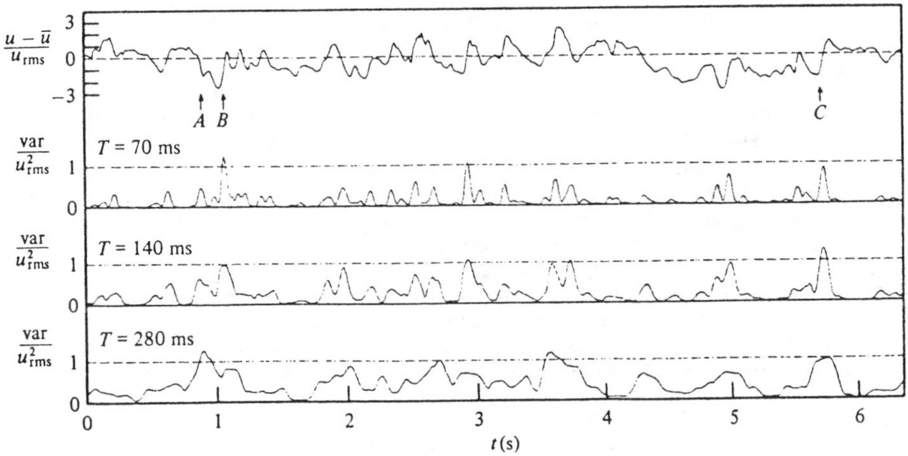

Figure 25. The streamwise velocity at $y^+ = 12$ and the short-time variance calculated for three different averaging times ($t_* = 14$ ms) [10].

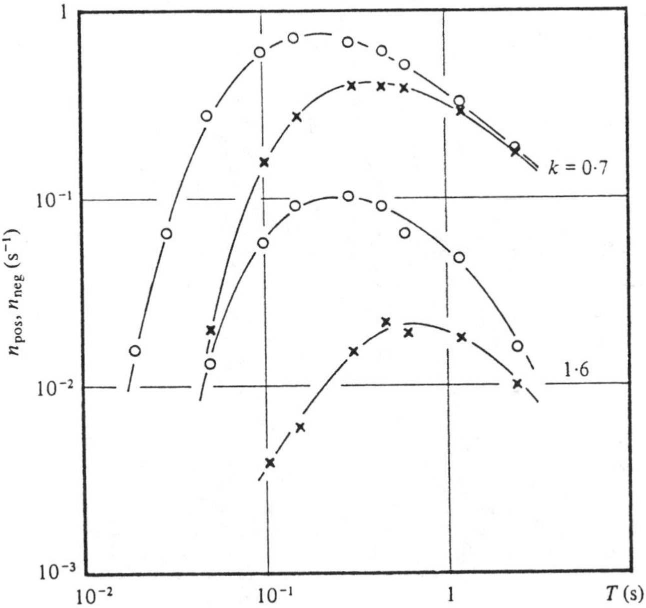

Figure 26. The number of detected events with the VITA technique as function of the averaging time for two threshold levels. The events have been sorted into accelerations (○) and retardations (×) [10].

in the form of conditionally averaged velocity profiles (Figure 27), which resemble the instantaneous profiles in Figure 23.

Johansson and Alfredsson [10] constructed ensemble averages at several different y^+-positions of the events detected by VITA at the respective position. Figure 28 shows typical examples at three different positions. The results for three different thresholds are plotted (k = 0.5, 1.0, and 2.0), and the ensemble average are normalized with the square root of k. With this scaling the resulting collapse for all thresholds, showing that the qualitative features of the detected events are independent of the threshold level. In the viscous sublayer the events are characterized by large positive excursions of the velocity, which give large contributions to the skewness and flatness factors (of u). As shown earlier (Equations 15A–C), there is a close correspondence between the deterministic part of the velocity signal and long-time statistics if the events detected contain a major part of the turbulent energy. This is clearly supported by the large values of the skewness and flatness factors at this position, viz. 1 and 4, respectively. At $y^+ = 12$ the skewness is approximately zero and hence the conditional average is symmetric around the mean, whereas it in the outer flow region is shifted toward the negative side corresponding to a negative value of the skewness factor. The same type of argument explains the variation of the flatness factor.

The conditionally sampled Reynolds stress, in a time window centered around the detection times obtained with VITA, was shown by Blackwelder and Kaplan to be an order of magnitude larger than the long time mean value. In Figure 29 results are shown for $y^+ = 50$ for the accelerations. The associated Reynolds stress production occurs mainly when fluid is moving outwards prior to the detection time, whereas the following sweep approaches the wall at a small angle and almost no Reynolds stress is produced during this phase.

Bogard [32] carried out a study where dye visualization and velocity measurements were carried out simultaneously, in order to compare the outcome of some detection schemes with what was

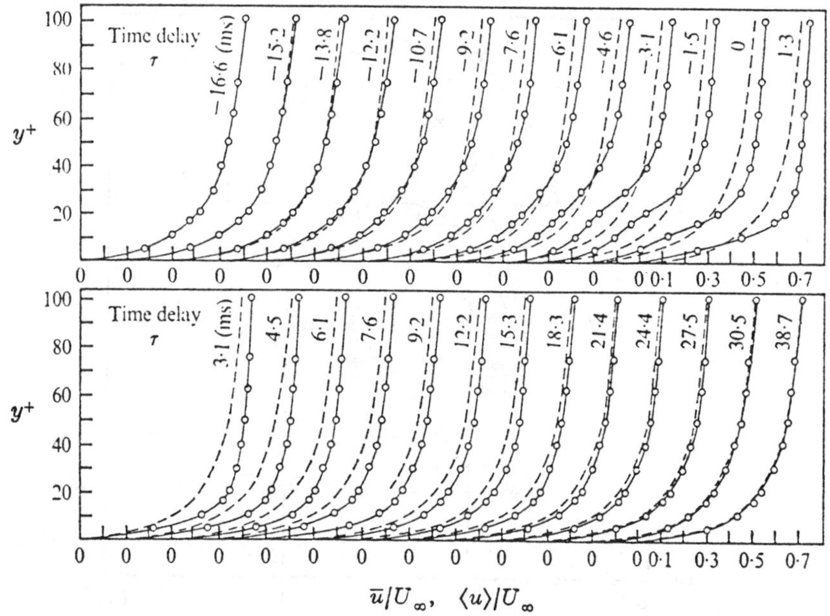

Figure 27. Conditionally averaged (–○–○–) and mean velocity (-------) profiles around the detection time obtained with the VITA-technique in a boundary layer. $T^+ = 10$, k = 1 detection at $y^+ = 15$. A viscous time unit corresponds to 0.6 ms [88].

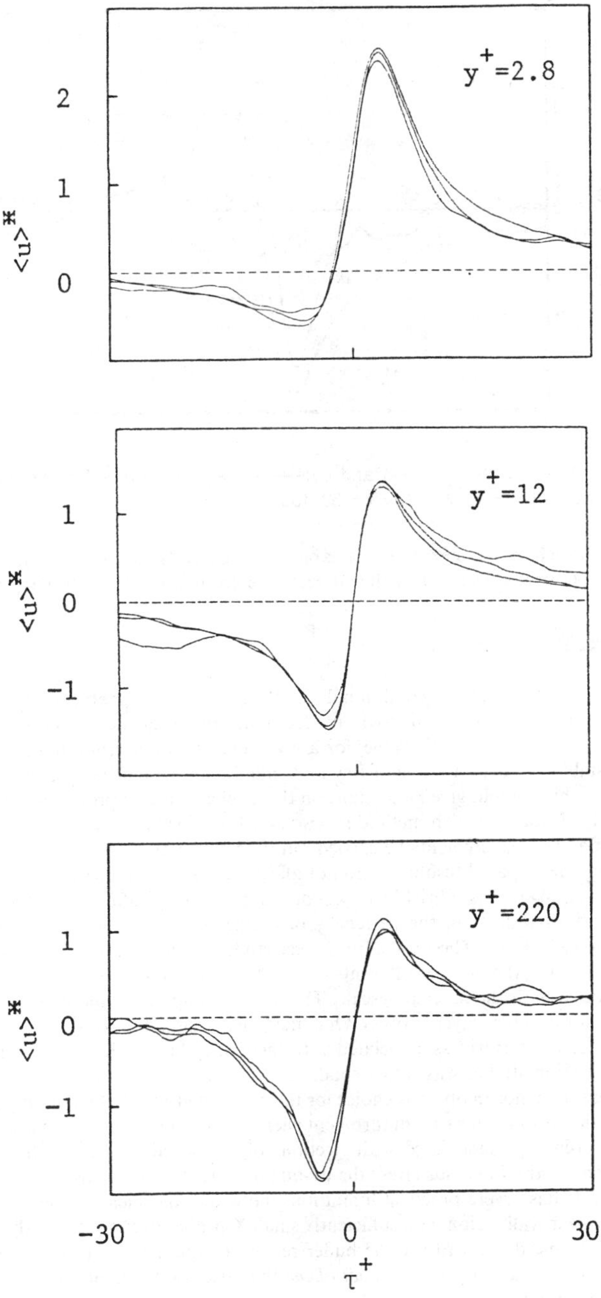

Figure 28. Conditionally averaged streamwise velocity signatures for events detected with the VITA-technique (with $T^+ = 10$) at three different positions in the boundary layer [10].

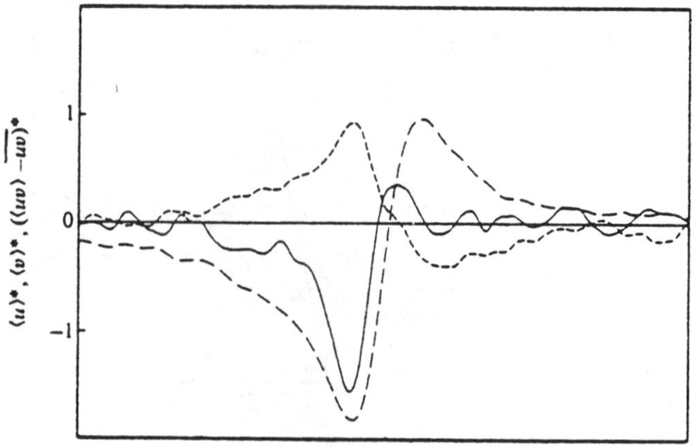

Figure 29. u(————), v(-------) and uv(————) signatures for events detected with the VITA-technique. ($T^+ = 10$, $k = 1$, $y^+ = 50$) [39].

visually observed. Such measurements are valuable, and the results indicated that the two techniques discussed above performed well, albeit they detect different phases of the bursting process.

Timescales for the Bursting Process

One of the motives for applying conditional sampling methods in analyzing the velocity field in the near wall region of turbulent flows is to determine the frequency of occurrence (n) of the turbulence generating events. If this value, for a given detection algorithm (and threshold), could be accurately established over a range of Reynolds numbers it would be possible to determine its scaling behavior, which would give indications on the mechanism that produces the events. Several questions arise: for instance, which method is best suited for this task, as well as at which position in the flow should the measurements be carried out. For probe measurements it is also important that effects of imperfect spatial resolution do not affect the results. This section deals with some of these problems, and also gives a brief historical account of work dealing with this subject.

The first quantitative data on the frequency of occurrence of the bursting events came from visualization studies [19, 20]. The results from these studies indicated that the bursting rate scales with inner variables, i.e., the number of events detected per unit time was independent of Reynolds number when scaled with the inner timescale. The inner scaling of the bursting phenomenon was a natural choice for an inner layer process. One must bear in mind, however, that visual methods inherently have some arbitrariness associated with them and that the Reynolds number range for these flow visualization studies was rather small.

For probe measurements an obvious choice for the position at which the bursting process should be detected is in the region of maximum turbulent energy and production, which is around $y^+ = 12$. As the bursting events are associated with ejections of low-speed fluid from the wall and hence large contributions to the Reynolds stress the uv-quadrant technique seems to be a natural choice for the detection. It has severe practical limitations, however, and measurements have as yet not been taken in the near wall region with sufficiently small X-probes over a sufficiently large Reynolds number interval. If one decides to use the buffer region for the detection one is hence limited to measurements of the streamwise velocity, *nota bene* that also for single probes effects of imperfect spatial resolution can be severe.

The first study employing the streamwise velocity signal for the detection was that of Rao, Narasimha, and Badri Narayanan [89]. They used an analog detection scheme, in which the hot-wire signal was fed into a band-pass filter, and from the output of that filter it was judged if an event had occurred. The results showed that the frequency of occurrence of these events scales with outer variables and this study had a great influence on the turbulence research community during most of the 1970s. The idea brought forward was that large scale structures (which supposedly are governed by the outer scales) trigger the events in the wall region. The number given by Rao et al. for the nondimensional frequency was 0.2, when n was scaled with the free stream velocity and boundary layer thickness.

Several other studies using other detection schemes, such as Lu and Willmarth [53], Blackwelder and Kaplan [88], and Chen and Blackwelder [90] obtained approximately the same number by giving the parameters involved in their detection schemes suitable values. However, as already mentioned, in order to make studies of scaling behavior only results obtained with the same method should be compared. These studies were only carried out at one or over a limited range of Reynolds numbers, which did not allow any definite conclusions about the scaling behavior to be drawn. Later several questions have arisen about the results of Rao et al. [89] and some of these indicate that the results may not be valid for the governing timescale in the wall region. For instance Blackwelder and Haritonidis [91] pointed out that the results may have been influenced by spatial resolution problems. Another more fundamental problem is that their detection method has not been proven to detect events associated with large contributions to the Reynolds stress.

Alfredsson and Johansson [60] carried out measurements of the streamwise velocity at $y^+ = 12$ in a fully developed turbulent channel flow and used the VITA-technique for data analysis. In order to avoid effects of imperfect spatial resolution, measurements were carried out with two probes of different size at two Reynolds numbers, chosen so that the larger probe at the low Re had the same size in viscous units as the smaller probe at the high Re. In this way it was possible to do comparisons over a Reynolds number range of almost one decade without any severe influence of probe-size effects on the results. Results from the VITA-analysis for two low Reynolds numbers from that study are shown in Figure 30. This figure is similar to Figure 26 but here only the result for the more numerous acceleration-type of events are shown. The data plotted are for one threshold

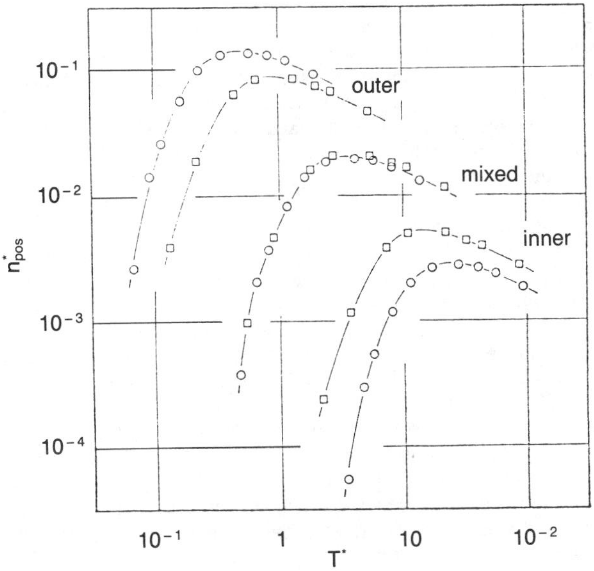

Figure 30. The number of detected events (accelerations) as function of the averaging time. The data are scaled with the outer, mixed and inner timescales ($y^+ = 13$, $L^+ = 12$). □: Re = 13,800, ○: Re = 54,300 [60].

($k = 1$) but are similar for other thresholds as well. Both the timescale of the events as well as their frequency of occurrence scale with the mixed timescale. This scaling behavior was also found for spectra (Figure 17). Also in this case, there is a clear correspondence between the long-term statistics and the conditional results, which is natural if the detected events are associated with a considerable part of the turbulent energy.

A complicating factor is that two recent studies that have been carried out in boundary layers using the VITA-technique have found that inner scaling of the bursting frequency is valid [91, 92]. The reason for this discrepancy is not clear, but may indicate that the interaction process between outer and inner regions is different for boundary layers as compared to channel flows (and probably also pipe flows). Further work is needed in order to give definite answers to the scaling problems.

Theoretical Modeling of Structures

Theoretical models or approaches towards an understanding of the structure of the turbulence producing processes may (somewhat arbitrarily) be divided into three different categories, namely

1. Models based on dimensional arguments.
2. Conceptual models, i.e., models in which the flow is described in terms of one or a few idealized flow structures.
3. Models directly based on the Navier-Stokes equations, but with input from experimental findings.

Dimensional reasoning was applied at an early stage of turbulence research, and has proven to be a successful approach to describe the qualitative features of long-time statistics of turbulent flows. For instance, Karman [27] and Millikan [28] demonstrated the existence of the overlap region between the outer and inner flow regions, and were able to obtain its logarithmic behavior from purely dimensional arguments. The works of Long and Chen [30] and Afzal [31] are also based on dimensional reasoning, although in their case a type of three-layer model is proposed instead of the classical two-layer approach. Another result that can be obtained from purely dimensional arguments is the behavior of frequency spectra in the so-called inertial subrange [93]. Although this theory is in a strict sense only valid for isotropic turbulence, the $k^{-5/3}$-behavior (where k is the wave-number) seems to be well established also for spectra of the streamwise velocity in, for instance, channel flows at high enough Reynolds numbers. This indicates that the small scale structure also in these flows is isotropic.

Among the conceptual models of turbulence, vortex structures of different types have been used to explain experimental observations of structures in the flow. The horseshoe vortex concept was introduced by Theodorsen [94], and has gained new interest, as structures of this type have been visually observed in several studies [69]. Kline et al. [19] hypothesized that the lift-up of a low-speed streak is associated with a structure resembling a horseshoe vortex (Figure 31). Later Offen and Kline [81] suggested that the different types of instabilities observed with the hydrogen bubble technique could be explained as observations of different parts of this structure. Perry and Chong [95] have published a theory in which the boundary layer is assumed to consist of a hierarchy of such eddies.

Another vortex structure that has been proposed to be the cause for the streaky wall structure consists of two counter-rotating streamwise vortices, with their center at $y^+ = 20 - 30$. This idea has been pursued by, among others, Blackwelder [82], and today there are several experimental findings seemingly in accordance with such a model (especially different types of correlation measurements, some of which were described earlier). Some researchers have proposed that this pair of eddies are the "legs" of a horseshoe vortex. Another closely related model has been proposed by Townsend [25], and consists of two counter-rotating growing conical wall eddies. The eddies are assumed to be born, grow, and breakdown in a regenerative process, and thus be associated with the quasi-cyclic bursting process.

Only a few modeling attempts have started from Navier-Stokes equations, which of course is due to the inherent difficulties of a non-linear coupled system of partial differential equations and

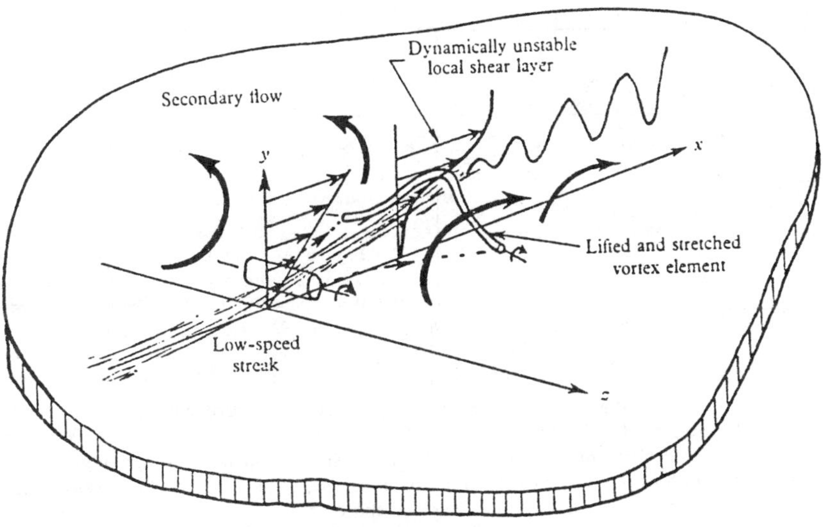

Figure 31. A conceptual model of lift-up of a low-speed streak, showing the formation of a vortex structure similar to the horseshoe vortex [19].

makes substantial approximations necessary. However, qualitative as well as quantitative agreement with experimental results has been obtained in some models.

Einstein and Li [96] assumed that the viscous sublayer is in a state of periodic growth and decay. They proposed that the thickness of the sublayer increases with time until it becomes unstable and breaks down. If this process takes place uniformly over an area with linear dimensions much larger than the thickness of the layer, the Navier-Stokes equations are essentially reduced to a one-dimensional diffusion equation. With the present knowledge of the three-dimensional structure of the viscous sublayer this is of course not a good approximation. However, the model predicts a linear velocity profile close to the wall and a maximum of the streamwise turbulence intensity around $y^+ = 10$. The timescale appropriate for the growth and break-down of the sublayer was found to be expressable as

$$T_{growth} = (U_0/u_\tau)^2 t_*$$

where U_0 is the velocity scale of the flow outside the sublayer. It should be noted that this timescale is not simply proportional to the inner timescale, but becomes increasingly larger with increasing Reynolds number.

The experimental results of Morrison et al. [62] suggested that the dominant streamwise velocity fluctuations have a wave-like character. This can be seen as a support for the type of wave-guide turbulence model as proposed by Landahl [97]. He showed that lightly damped waves may play an important role for near wall turbulence, and may dominate frequency wave-number spectra of, e.g., pressure fluctuations. He assumed the turbulent mean velocity profile to be known, and used that as an input to the Orr-Sommerfeld equation. The resulting convection velocity and decay rate of pressure fluctuations in a boundary layer were in fair agreement with experimental data. Bark [98] pursued this approach, using a non-homogeneous Orr-Sommerfeld equation with a driving term stemming from the Reynolds stresses produced by bursts. Indeed, the key assumption of the analysis was that nonlinear terms in the Navier-Stokes equations are of crucial importance only during bursting periods, i.e., during only a small fraction of the total time. He computed wave-number frequency spectra of the streamwise velocity in the wall region, and found dominant length and

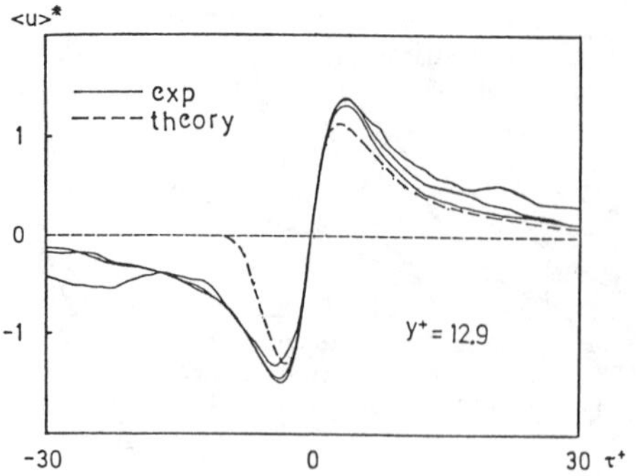

Figure 32. The VITA-educed streamwise velocity pattern (--------) obtained with theoretical model [83]. Experimental data (———) (same data as in Figure 28).

time scales deduced from these, which were in good agreement with experimental results of Morrison et al. [62]. However, the bandwidth of the computed spectra was much too small. This is probably mainly due to the idealized description of the Reynolds stress produced during bursting periods. In a real turbulent flow the Reynolds stress producing events span a wide range of scales in space and time, giving rise to a broad long-time averaged spectrum.

Landahl [83, 99] has recently developed a linearized three-dimensional theory in which a disturbance of large horizontal dimensions compared to its vertical extent is followed as it travels downstream [100]. The disturbance is assumed to be created initially by turbulent stresses produced by a previous burst. He showed that for a "flat" eddy the pressure gradient terms are negligible in the sense that they produce only a very small change in the perturbation velocities as the eddy travels a distance comparable to its own length. Similarly, viscous effects are negligible during this stage provided that the thickness of the eddy is not too small. Instead, the development of the disturbance is predominantly governed by horizontal inertia and continuity.

For the representation of the initial disturbance, Landahl [99] used a set of counter-rotating vortices, whereas the small scale random turbulent stress was described by a simple statistical model. He applied the VITA-technique to the resulting flow field, and was able to educe velocity signatures in good qualitative agreement with experimental findings (Figure 32). These signatures are associated with shear layers, which are created by vertical displacement of fluid elements. These were shown to be sharpened and stretched as they travel downstream, and an inflexional profile develops. The flow may become unstable due to the inflexional profile, which results in a new burst. This may in turn set up a new large-scale disturbance through its resulting Reynolds stress field. Hence, the model is consistent with the quasi-cyclic occurrence of bursts in form of a regenerative process where the large scales are instrumental for the momentum transport in a shear flow. It also offers an explanation for the lift-up process, which is contradictory to the "adverse-pressure-gradient" idea [81]. In this context it should also be mentioned that Thomas and Bull [101] concluded from an order-of-magnitude estimate that the adverse pressure gradient effects associated with typical pressure patterns in boundary layers are insufficient to initiate lift-up.

NUMERICAL "EXPERIMENTS"

Modern "super-computers" have opened up possibilities to supplement turbulence measurements with numerical simulations of turbulent shear flows. For such calculations turbulent channel flow has, due to its geometrical simplicity, attracted special attention. Direct simulation of the full time-dependent three-dimensional Navier-Stokes equations is not feasible today at typical Reynolds numbers described in previous sections. However, research in this area has reached the point where full turbulent simulations, without sub-grid-scale modeling can be carried out for very low Reynolds numbers [102].

Since the late sixties [103], successful numerical "experiments", i.e., simulations of turbulent channel flow, have been pursued with the large-eddy simulation (LES) technique. With LES the flow is divided into a large-scale field of motion, which is computed from three-dimensional time-dependent dynamical equations, and small scales that are modeled more or less crudely with, e.g., an eddy viscosity model. Since only the small scales are modeled, the LES-technique is qualitatively different from traditional turbulence modeling, and may be used to provide data in a way similar to physical experiments. This type of simulation approach has earlier mainly been used in meteorological applications.

In a recent paper Moin and Kim [104] describe extensive numerical simulations of channel flow with up to 516,096 grid points (64 × 63 × 128 in the x, y and z directions, respectively). They were able to compute the flow in the near-wall region, even in the viscous sublayer, at a Reynolds number of 27,600. An eddy-viscosity model was used for the sub-grid scales. From the computed data they obtained mean-velocity and Reynolds stress distributions in reasonable agreement with experiments, as well as many of the dominant features of near-wall turbulence. The demands for computer speed and capacity are quite extensive in connection with LES. The largest case run by Moin and Kim took more than 90 hours on an ILLIAC IV computer. For details of the simulation technique the reader is referred to the paper of Moin and Kim where also some of the earlier work is reviewed. A recent review of various aspects of LES is given by Rogallo and Moin [105].

Despite the large number of grid-points in the study of Moin and Kim, with e.g., a spanwise separation of $15.7\ell_*$, effects of imperfect spatial resolution were still rather severe in the wall-region. The computed streamwise turbulence intensity reaches a maximum value of about 2.3 u_τ (Figure 33) at $y^+ \approx 30$, which may be compared with the experimental value of 2.9 u_τ occurring at about $y^+ = 15$. This effect of imperfect spatial resolution is in some respects similar to the probe size effects described by Figure 8. Comparing the computed turbulence intensities with those in Figure 6, one also notices considerable attenuation of the v-component in the immediate vicinity of the wall. The computed v_{rms} at $y^+ = 20$ is roughly a factor of two smaller than the experimentally found value. This of course means that a significant part of the energy of the v-component resides in sub-grid-scale motions. A similar situation prevails also for other quantities, such as the streamwise vorticity, and suggests that in order to resolve all relevant scales of motion, one would have to increase the number of grid points enormously. However, a considerable amount of interesting information about turbulence and its structure may be found already from existing simulations, especially about the spatial structure. Figure 34 shows contour plots of the instantaneous streamwise velocity at the edge of the viscous sublayer ($y^+ = 6.3$). We clearly see elongated regions of low (and high) velocity, resembling the near-wall structure found in visualization experiments (Figure 18). This streaky structure is also reflected in the spanwise two-point correlation of the streamwise velocity near the wall. The mean streak-spacing inferred from computed correleations is about $250\ell_*$. This large value can again be ascribed to effects of imperfect spatial resolution.

Moin and Kim also made a computer motion picture simulating flow visualization experiments. They observed intermittent ejections of fluid elements from the buffer region, which occasionally penetrated out as far as to about $y^+ = 400$. In experiments such particle traces would be extremely difficult to follow. Many other interesting features can also be observed, and the resemblance with actual hydrogen-bubble flow visualization is striking.

Further analyses of the simulated channel-flow data are described by Kim [106] and Moin [67]. Kim applied the VITA-technique to the computed data, and obtained conditional averages of u

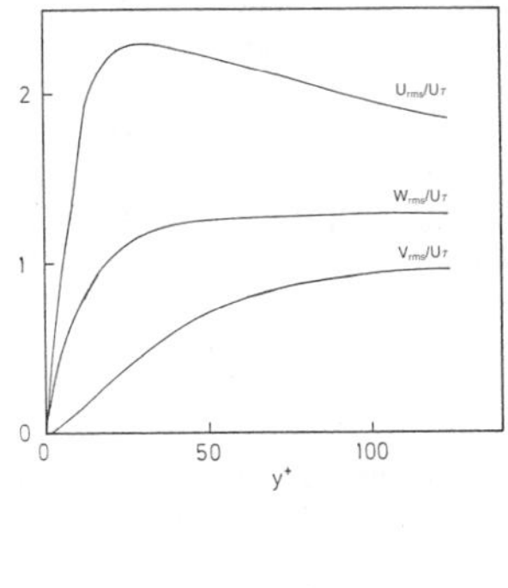

Figure 33. Turbulence intensities in the wall region from simulated channel-flow data (Re = 27,600) [104].

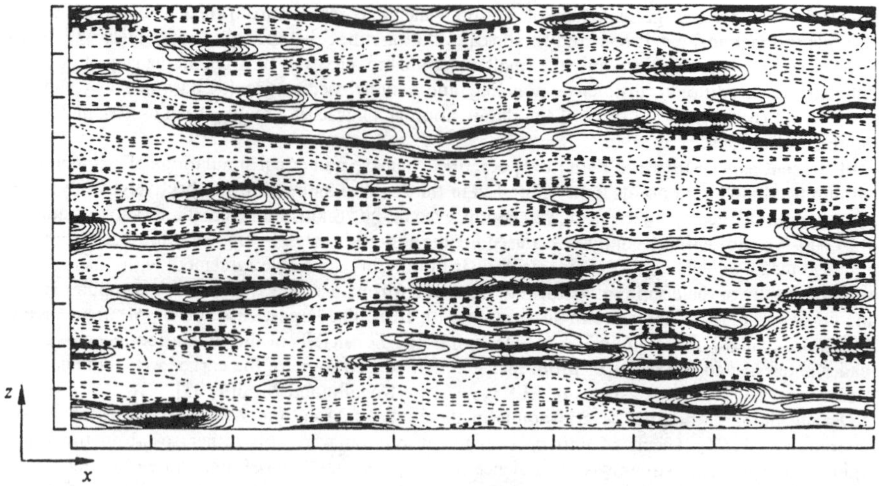

Figure 34. Contours of the fluctuating streamwise velocity in the x-z-plane at $y^+ = 6.26$, from simulated data. Streamwise and spanwise extents are $4{,}021\ell_*$ and $2{,}010\ell_*$ [104].

and v in qualitative agreement with those in Figure 28. Moin showed that the distribution of inclination angle (in the x-y-plane) of the vorticity vector attains a maximum at 45°. Also the existence of hairpin vortices was demonstrated. These results seem to be in good agreement with the boundary-layer results of Head and Bandyopadhyay [69], although the hairpin vortices were not found to be a dominating feature in the simulated data.

Numerical solution of the Navier-Stokes equations for turbulent (as well as transitional) channel flow without subgrid-scale modeling has been pursued by Orzag and coworkers. Computations at a Reynolds number of 10,000 are described by Orzag and Patera [107] and in a recent paper by Handler et al. [108]. In these cases the velocity is spectrally decomposed in terms of Fourier series in x and z and Chebyshev polynomials in y. Patera and Orzag obtained Reynolds stress profiles in fair agreement with experiments, whereas the computed w_{rms} was off by about a factor 2. Handler et al. analyzed the wall pressure field and obtained an intensity and convection velocities in reasonable agreement with experimental values.

CONCLUSIONS

As you certainly have noticed, there has been a tremendeous increase during the last decade in experimental, theoretical, and numerical efforts attacking the problem of wall-bounded turbulence. More than 50% of the cited references are from a date later than 1975. Computers and reliable anemometer equipment have meant an increased accuracy of measurements, and significant expansion of the output from the turbulence research laboratories. Especially the search for coherent structures in the flow has attracted much attention, since their discovery in flow visualization studies. The concept of a coherent structure is hitherto somewhat vaguely defined, and the relation between structures observed in visualization studies and patterns observed in the velocity signals is not altogether clear. More research is needed to formulate more objective definitions of coherent structures, and to elucidate the details of the regenerative process responsible for the generation of turbulence.

The experiments carried out in the laboratory are today, due to the development of super computers, being accompanied by numerical "experiments." In these simulations one has access to all three velocity components, as well as the pressure at all grid points. This is more than one can ever get in the laboratory. However, the small scales must be modeled, and the computations have so far given only qualitatively correct results. Simulations are as yet also limited to low Reynolds number flows, which hitherto precludes any conclusions to be drawn about scaling behavior. Anyway, it seems possible to use such simulations in the search for some aspects of the three-dimensional structure of coherent motions.

Although the experimental program devoted to turbulence research, on a global basis, is very large indeed, much of the experimental work is carried out at institutions with very limited experimental resources. This issue was discussed by Coles [109] already in 1968 in a paper entitled "On the Need for Better Experiments." Despite the importance of turbulent phenomena in technical, geophysical, biological, and a variety of other flow situations, there has never in turbulence research been a very large-scale joint experimental effort of the type seen in other areas of physics, such as astronomy or particle physics. For instance, a big channel-flow (or other type of shear-flow) facility with a large Reynolds number range, aspect ratio, and length would be an invaluable tool in obtaining data necessary to get closer to the resolution of the riddle of turbulence.

Acknowledgments

Part of this work has been sponsored by the Swedish Maritime Research Centre (SSPA). We gratefully acknowledge this support, and the many fruitful discussions with professors Mårten Landahl and Fritz Bark at the Royal Institute of Technology.

REFERENCES

1. Dönch, F., "Divergente und Konvergente Turbulente Strömungen mit Kleinen Öffnungswinkeln." Forschungsarbeiten auf dem Gebiete des Ingenieurwesens, Heft 282, 1926.
2. Nikuradse, J., "Untersuchungen über die Strömungen des Wassers in Konvergenten und Divergenten Kanälen." Forschungsarbeiten auf dem Gebiete des Ingenieurwesens, Heft 289, 1929.
3. Reichart, H., "Die Quadratischen Mittelwerte der Längsschwankungen in der Turbulenten Kanalströmung." *Z.A.M.M.*, vol. 13, no. 3, 1933, pp. 177–180.
4. Wattendorf, F. L., "Investigations of Velocity Fluctuations in a Turbulent Flow." *J. Aero. Sci.* vol. 3, no. 6, 1936, pp. 200–202.
5. Reichart, H., "Messungen Turbulenter Schwankungen" *Die Naturwissenschaften*, vol. 26, Heft 24/25, 1938, pp. 404–408.
6. Laufer, J., "Investigation of Turbulent Flow in a Two-dimensional Channel," NACA Rep. 1053, 1951.
7. Comte-Bellot, G., "Écoulement Turbulent Entre Deux Parois Parallèles." Publications Scientifiques et Techniques du Ministère de l'Air, No. 419, 1965.
8. Laufer, J., "The Structure of Turbulence in Fully Developed Pipe Flow," NACA Rep. 1174, 1954.
9. Eckelmann, H., "An Experimental Investigation of a Turbulent Channel Flow With a Thick Viscous Sublayer." Max-Planck Institut für Strömungsforschung, Göttingen, Bericht 101, 1973.
10. Johansson, A. V., and Alfredsson, P. H., "On the Structure of Turbulent Channel Flow." *J. Fluid Mech*, vol. 122, 1982, pp. 295–314.
11. Clark, J. A., "A Study of Incompressible Turbulent Boundary Layers in Channel Flow," *Trans. ASME D, J. Basic Engng*, vol. 90, 1968, pp. 455–468.
12. Hussain, A. K. M. F., and Reynolds, W. C., "The Mechanics of a Perturbation Wave in Turbulent Shear Flow." Dept, of Mech. Engineering, Stanford Univ., Rep. FM-6, 1970.
13. Hussain, A. K. M. F., and Reynolds, W. C., "Measurements in Fully Developed Turbulent Channel Flow." *Trans. ASME E, J. Fluids Engng*, vol. 97, 1975, pp. 568–580.
14. Kreplin, H.-P. "Experimentelle Untersuchungen der Längsschwankungen und der Wandparallelen Querschwankungen der Geschwindigkeit in Einer Turbulenten Kanalströmung."

Mitteilungen aus dem Max-Planck Institut für Strömungsforschung und der AVA, Göttingen, no. 63, 1976.
15. Wallace, J. M., Eckelmann, H., and Brodkey, R. S., "The Wall Region in Turbulent Shear Flow." *J. Fluid Mech.*, vol. 54, 1972, pp. 39–48.
16. Zarič, Z., "Wall Turbulence Structure and Convection Heat Transfer." *Int. J. Heat Mass Transfer*, vol. 18, 1975, pp. 831–842.
17. Reischman, M. M., and Tiederman, W. G., "Laser-Doppler Anemometer Measurements in Drag-Reducing Channel Flows." *J. Fluid Mech.*, vol. 70, 1975, pp. 369–392.
18. Willmarth, W. W., and Velasquez-Saavedra, I., "High Resolution Laser-Doppler Anemometer for Turbulence Measurements in a Channel." In *Proc. 8th Biennal Symposium on Turbulence*, Univ. of Missouri-Rolla, ed. J. L. Zakin and G. K. Patterson Sept. 26–28, 1983, pp. 18:1–18:12.
19. Kline, S. J., Reynolds, W. C., Schraub, F. A., and Runstadler, P. W., "The Structure of Turbulent Boundary Layers." *J. Fluid Mech.*, vol. 30, 1967, pp. 741–773.
20. Kim, H. T., Kline, S. J., and Reynolds, W. C., "The Production of Turbulence Near a Smooth Wall in a Turbulent Boundary Layer." *J. Fluid Mech.*, vol. 50, 1971, pp. 133–160.
21. Corino, E. R., and Brodkey, R. S., "A Visual Investigation of the Wall Region in Turbulent Flow." *J. Fluid Mech.*, vol. 37, 1969, pp. 1–30.
22. Antonia, R. A., "Conditional Sampling in Turbulence Measurement." *Ann. Rev. Fluid Mech.*, vol. 13, 1981, pp. 131–156.
23. Stuart, J. T., "Instability and Transition in Pipes and Channels." In *Proc. Symp. on Transition and Turbulence*, Univ. of Wisconsin-Madison, 1980, ed. R. E. Meyer, Academic Press, 1981, pp. 77–94.
24. Patel, V. C., and Head, M. R., "Some Observations on Skin Friction and Velocity Profiles in Fully Developed Pipe and Channel Flows." *J. Fluid Mech.*, vol. 38, 1969, pp. 181–201.
25. Townsend, A. A., *The Structure of Turbulent Shear Flow*, Cambridge Univ. Press, 1976.
26. Coles, D. E., "The Law of the Wake in the Turbulent Boundary Layer." *J. Fluid Mech.*, vol. 1, 1956, pp. 191–226.
27. Karman, T., von "Mechanische Ähnlichkeit und Turbulenz." Nachrichten der Akademie der Wissenschaften, Göttingen, Math.-Phys. Klasse, 58, 1930.
28. Millikan, C. B., "A Critical Discussion of Turbulent Flows in Channels and Circular Tubes." In *Proc. of the Fifth Int. Cong. on Appl. Mech.*, Cambridge, Mass. 1938, Wiley, New York, 1939, pp. 386–392.
29. Tennekes, H., and Lumley, J. L., *A First Course in Turbulence*, Mass. Inst. Tech. Press, 1972.
30. Long, R. R., and Chen, T. C., "Experimental Evidence for the Existence of the 'Mesolayer' in Turbulent Systems." *J. Fluid Mech.*, vol. 105, 1981, pp. 19–59.
31. Afzal, N., "A Sub-Boundary Layer Within a Two-dimensional Turbulent Boundary Layer: An Intermediate Layer." *J. de Mechanique Théorique et Appliquée*, vol. 1, no. 6, 1982, pp. 963–973.
32. Bogard, D. G., "Investigation of Burst Structures in Turbulent Channel Flows Through Simultaneous Flow Visualization and Velocity Measurements." Dept. Mechanical Engng., Purdue Univ. Doctoral Thesis, 1982.
33. Fckelmann, H., "The Structure of the Viscous Sublayer and the Adjacent Wall Region in a Turbulent Channel Flow." *J. Fluid Mech.*, vol. 65, 1974, pp. 439–459.
34. Kastrinakis, E. G., and Eckelmann, H., "Measurement of Streamwise Vorticity Fluctuations in a Turbulent Channel Flow." *J. Fluid Mech.*, vol. 137, 1983, pp. 165–186.
35. Oldaker, D. K., and Tiedermann, W. G., "Spatial Structure of the Viscous Sublayer in Drag-Reducing Flows." *Phys. Fluids*, vol. 20 Suppl., 1977, pp. 133–144.
36. Buchhave, P., George W. K., Jr, and Lumley, J. L., "The Measurement of Turbulence with the Laser-Doppler Anemometer." *Ann. Rev. Fluid Mech.*, vol. 11, 1979, pp. 443–503.
37. Somerscales, E. F. C., "Laser-Doppler Velocimeter" in *Methods of Experimental Physics*, vol. 18, part A, ed. R. J. Emrich, Academic Press, 1981, pp. 93–240.
38. Blackwelder, R. F., "Hot-Wire and Hot-Film Anemometers," in *Methods of Experimental Physics*, vol. 18, part A, ed. R. J. Emrich, Academic Press, 1981, pp. 259–315.
39. Alfredsson, P. H., and Johansson, A. V., "On the Detection of Turbulence Generating Events." *J. Fluid Mech.*, vol. 139, 1984, pp. 325–345.

40. Johansson, A. V., and Alfredsson, P. H., "Effects of Imperfect Spatial Resolution on Measurements of Wall-Bounded Turbulent Shear Flow." *J. Fluid Mech.*, vol. 137, 1983, pp. 411–423.
41. Coles, D. E., "The Young Person's Guide to the Data." In *Proc. AFOSR-IFP-Stanford Conf. on Computation of Turbulent Boundary Layers*, ed. D. E. Coles and E. A. Hirst vol. 2, 1968, pp. 1–46.
42. Murlis, J., Tsai, H. M., and Bradshaw, P., "The Structure of Turbulent Boundary Layers at Low Reynolds Numbers." *J. Fluid Mech.*, vol. 122, 1982, pp. 13–56.
43. Kreplin, H.-P., and Eckelmann, H., "Behavior of the Three Fluctuating Velocity Components in the The Wall Region of a Turbulent Channel Flow." *Phys. Fluids*, vol. 22, 1979, pp. 1233–1239.
44. Zarič, Z., "Wall Turbulence Studies." *Adv. Heat Transfer*, vol. 8, 1972, pp. 285–350.
45. Uberoi, M. S., and Kovasznay, L. S. G., "On Mapping and Measurement of Random Fields." *Q. Appl. Math.*, vol. 10, 1953, pp. 375–393.
46. Wyngaard, J. C., *J. Phys. E: Sci. Instrum.*, vol. 1, 1968, pp. 1105–1108.
47. Schewe, G., "On the Structure and Resolution of Wall-Pressure Fluctuations Associated With Turbulent Boundary-Layer Flow." *J. Fluid Mech.*, vol. 134, 1983, pp. 311–328.
48. Willmarth, W. W., "Pressure Fluctuations Beneath Turbulent Boundary Layers." *Ann. Rev. Fluid Mech.* vol. 7, 1975, pp. 13–38.
49. Karlsson, R., "Skin Friction in Turbulent Boundary Layers." Dept. of Appl. Thermo and Fluid Dynamics, Chalmers Univ. Tech. Gothenburg, Doctoral thesis, 1980.
50. Purtell, L. P., Klebanoff, P. S., and Buckley, F. T., "Turbulent Boundary Layer at Low Reynolds Number." *Phys. Fluids*, vol. 24, 1981, pp. 802–811.
51. Klebanoff, P. S., "Characteristics of Turbulence in a Boundary Layer With Zero Pressure Gradient." NACA Rep. 1247, 1955.
52. Fage, A., and Townend H. C. H., "An Examination of Turbulent Flow With an Ultramicroscope." *Proc. R. Soc. London*, Ser. A vol. 135, 1932, pp. 656–677.
53. Lu, S. S., and Willmarth, W. W., "Measurements of the Structure of the Reynolds Stress in a Turbulent Boundary Layer." *J. Fluid Mech.*, vol. 60, 1973, pp. 481–511.
54. Johnson, F. D., and Eckelmann, H., "A Variable-Angle Method of Calibration for X-probes Applied to Wall-Bounded Turbulent Shear Flow." Exp. in Fluids, vol. 2, 1984, pp. 121–130.
55. Kreplin, H.-P., and Eckelmann, H., "Instantaneous Direction of the Velocity Vector in a Fully Developed Turbulent Channel Flow." *Phys. Fluids*, vol. 22, 1979, pp. 1210–1211.
56. Kovasznay, L. S. G., *Physical Measurements in Gasdynamics and Combustion*, Princeton Univ. Press, 1954.
57. Corrsin, S., and Kistler, A. L., "The Free-Stream Boundaries of Turbulent Flow." NACA Rep. 1244, 1955.
58. Kastrinakis, E. G., Eckelmann, H., and Willmarth, W. W., "Influence of the Flow Velocity on a Kovasznay Type Vorticity Probe." *Rev. Sci. Instrum.*, vol. 50, 1979, pp. 759–767.
59. Wallace, J. M., "Measurement of Vorticity Using Hot-Wire Anemometry." *Bull. Am. Phys. Soc.*, vol. 28, Nov. 1983, p. 1401 (abstract).
60. Alfredsson, P. H., and Johansson, A. V., "Timescales in Turbulent Channel Flow." Phys. Fluids, vol. 27, 1984, pp. 1974–1981.
61. Perry, A. E., and Abell, C. J., "Scaling Laws for Pipe-Flow Turbulence." *J. Fluid Mech.*, vol. 67, 1975, pp. 257–271.
62. Morrison, W. R. B., Bullock, K. J., and Kronauer, R. E., "Experimental Evidence of Waves in the Sublayer." *J. Fluid Mech.*, vol. 47, 1971, pp. 639–656.
63. Reynolds, O., "An Experimental Investigation of the Circumstances Which Determine Whether the Motion of Water Shall Be Direct or Sinuous, and of the Law of Resistance in Parallel Channels." *Phil. Trans. Roy. Soc. London*, vol. 174, 1883, pp. 935–982.
64. Hussain, A. K. M. F., "Coherent Structures—Reality and Myth." *Phys. Fluids.* vol. 26, 1983, pp. 2816–2850.
65. Lumley, J. L., "The Structure of Inhomogeneous Turbulent Flows." In *Proc. Atmospheric Turbulence and Radio Wave Propagation*, Nauk, Moscow, 1967.
66. Bakewell, H. P., Jr and Lumley, J. L., "Viscous Sublayer and Adjacent Wall Region in Turbulent Pipe Flow." *Phys. Fluids*, vol. 10, 1967, pp. 1880–1889.

67. Moin, P., "Probing Turbulence via Large Eddy Simulation" *AIAA-Paper* 84-0174, presented at the AIAA 22nd Aerospace Sciences Meeting, Reno, USA, January 1984.
68. Offen, G. R., and Kline, S. J., "Combined Dye-Streak and Hydrogen-Bubble Visual Observations of a Turbulent Boundary Layer." *J. Fluid Mech.* vol. 62, 1974. pp. 223–239.
69. Head, M. R., and Bandyopadhyay, P., "New Aspects of Turbulent Boundary-Layer Structure." *J. Fluid Mech.* vol. 107, 1981, pp. 297–338.
70. Nagib, H. M., Guezennec, Y., and Corke, T. C., "Applications of a Smoke-Wire Visualization Technique to Turbulent Boundary Layers." Paper presented at the Workshop on Coherent Structures in Turbulent Boundary Layers, Lehigh Univ., 1978.
71. Falco, R. E., "New Results, a Review and Synthesis of the Mechanism of Turbulence Production in Boundary Layers and its Modification." *AIAA-paper*-83-0377, 1983.
72. Johnson, R. R., Elkins, R. E., Lindgren, E. R., and Yoo, J. K., "Experiments on the Structure of Turbulent Shear in Pipe Flows of Water." *Phys. Fluids,* vol. 19, 1976, pp. 1422–1423.
73. Kline, S. J., "The Role of Visualization in the Study of the Structure of the Turbulent Boundary Layer." Paper presented at the Workshop on Coherent Structures in Turbulent Boundary Layers, Lehigh Univ., 1978.
74. Corrsin, S., "Some Current Problems in Turbulent Shear Flows." In *Proc. 1st Symp. on Naval Hydrodyn., NAS-NRC Publ.* 515, 1957, pp. 373–407.
75. Smith, C. R., "A Synthesized Model of the Near-Wall Behavior in Turbulent Boundary Layers." Paper to appear in the *Proceedings of Eighth Symposium of Turbulence,* Univ. Missouri-Rolla (ed. G. K. Patterson and J. L. Zakin) 1983.
76. Smith, C. R., and Metzler, S. P., "The Characteristics of Low-Speed Streaks in the Near-Wall Region of a Turbulent Boundary Layer." *J. Fluid Mech.* vol. 129, 1983, pp. 27–54.
77. Kreplin, H.-P., and Eckelmann, H., "Propagation of Pertubations in the Viscous Sublayer and Adjacent Wall Region." *J. Fluid Mech.* vol. 95, 1979, pp. 305–322.
78. Lee, M. K., Eckelman, L. D., and Hanratty, T. J., "Identification of Turbulent Wall Eddies Through the Phase Relation of the Components of the Fluctuating Velocity Gradient." *J. Fluid Mech.* vol. 66, 1974, pp. 17–33.
79. Gupta, A. K., Laufer, J., and Kaplan, R. E., "Spatial Structure in the Viscous Sublayer." *J. Fluid Mech.,* vol. 50, 1971, pp. 493–512.
80. Haritonidis, J. H., "Some New Measurements of Streaks in a Turbulent Boundary Layer." *Bull. Am. Phys. Soc.,* vol. 24, no. 8, 1979, p. 1142.
81. Offen, G. R., and Kline, S. J., "A Proposed Model of the Bursting Process in Turbulent Boundary Layers." *J. Fluid Mech.,* vol. 70, 1975, pp. 209–228.
82. Blackwelder, R. F., "The Bursting Process in Turbulent Boundary Layers." Paper presented at the Workshop on Coherent Structures in Turbulent Boundary Layers, Lehigh Univ., 1978.
83. Landahl, M. T., "Theoretical Modeling of Coherent Structures in Wall Bounded Shear Flows." Paper to appear in the *Proceedings of the Eighth Symposium of Turbulence,* Univ. Missouri-Rolla (ed. G. K. Patterson and J. L. Zakin) 1983.
84. Hussain, A. K. M. F., and Thompson, C. A., "Controlled Symmetric Perturbation of the Plane Jet: An Experimental Study in the Initial Region." *J. Fluid Mech.,* vol. 100, 1980, pp. 397–431.
85. Kovasznay, L. S. G., Kibens, V., and Blackwelder, R. F., "Large-Scale Motion in the Intermittent Region of a Turbulent Boundary Layer." *J. Fluid Mech.,* vol. 41. 1970, pp. 283–325.
86. Sabot, J., and Comte-Bellot, G., "Intermittency of Coherent Structures in the Core Region of Fully Developed Turbulent Pipe Flow." *J. Fluid Mech.,* vol. 74, 1976, pp. 767–796.
87. Comte-Bellot, G., Sabot, J., and Saleh, I., "Detection of Intermittent Events Maintaining Reynolds Stress." In the Proceedings of the Dynamic Flow Conference, 1978, pp. 213–230.
88. Blackwelder, R. F., and Kaplan, R. E., "On the Wall Structure of the Turbulent Boundary Layer." *J. Fluid Mech.,* vol. 76, 1976, pp. 89–112.
89. Rao, K. N., Narasimha, R., and Badri Narayanan, M. A., "The 'Bursting' Phenomenon in a Turbulent Boundary Layer." *J. Fluid Mech.,* vol. 48, 1971, pp. 339–352.
90. Chen, K. K., and Blackwelder, R. F., "Large-Scale Motion in a Turbulent Boundary Layer: A Study Using Temperature Contamination." *J. Fluid Mech.,* vol. 89, 1978, pp. 1–31.
91. Blackwelder, R. F., and Haritonidis, J. H., "Scaling of the Bursting Frequency in Turbulent Boundary Layers." *J. Fluid. Mech.,* vol. 132, 1983, pp. 87–103.

92. Willmarth, W. W., and Sharma, L. K., "Study of Turbulent Structure with Hotwires Smaller than the Viscous Length." *J. Fluid Mech.*, vol. 142, 1984, pp. 121–149.
93. Monin, A. S., and Yaglom, A. M., *Statistical Fluid Mechanics*, The MIT Press, 1971, p. 15.
94. Theodorsen, T., "Mechanism of Turbulence," in *Proc. of the Second Midwestern Conf. on Fluid Mech.*, Ohio State Univ., 1952, pp. 1–18.
95. Perry, A. E., and Chong, M. S., "On the Mechanism of Wall Turbulence." *J. Fluid Mech.*, vol. 119, 1982, pp. 173–217.
96. Einstein, H. A., and Li, H., "The Viscous Sublayer Along a Smooth Boundary." in *Proc. Amer. Soc. Civil Eng.*, EM2, Paper 945, 1956.
97. Landahl, M. T., "A Wave-Guide Model for Turbulent Shear Flow." *J. Fluid Mech.*, vol. 29, 1967, pp. 441–459.
98. Bark, F. H., "On the Wave Structure of the Wall Region of a Turbulent Boundary Layer." *J. Fluid Mech.*, vol. 70, 1975, pp. 229–250.
99. Landahl, M. T., "On the Dynamics of Large Eddies in the Wall Region of a Turbulent Boundary Layer." Paper in the *Proceedings of the IUTAM Symposium of Turbulence and Chaotic Phenomena in Fluids*, Kyoto, Japan, Sept. 1983.
100. Russell, J. M., and Landahl, M. T., "The Evolution of a Flat Eddy Near a Wall in an Inviscid Shear Flow." *Phys. Fluids*, vol. 27, 1984, pp. 557–570.
101. Thomas, A. S. W., and Bull, M. K., "On the Role of Wall-Pressure Fluctuations in Deterministic Motions in the Turbulent Boundary Layer." *J. Fluid Mech.*, vol. 128, 1983, pp. 283–322.
102. Reynolds, W. C., "Progress, Insights, and Challenges of Large Eddy Simulation." *Bull. Am. Phys. Soc.*, vol. 28, 1983, p. 1369.
103. Deardorff, J. W., "A Numerical Study of Three-Dimensional Turbulent Channel Flow at Large Reynolds Numbers." *J. Fluid Mech.*, vol. 41, 1970, pp. 453–480.
104. Moin, P., and Kim, J., "Numerical Investigation of Turbulent Channel Flow.," *J. Fluid Mech.*, vol. 118, 1982, pp. 341–377.
105. Rogallo, R. S., and Moin, P., "Numerical Simulation of Turbulent Flows." *Ann. Rev. Fluid Mech.*, vol. 16, 1984, pp. 99–138.
106. Kim, J., "On the Structure of Wall-Bounded Turbulent Shear Flows," *Phys. Fluids*, vol. 26, 1983, pp. 2088–2097.
107. Orzag, S. A., and Patera, A. T., "Calculation of von Kármán's Constant for Turbulent Channel Flow." *Phys. Rev. Lett.* vol. 47, 1981, pp. 832–835.
108. Handler, R. A., Hansen, R. J., Sakell, L., Orzag, S. A., and Bullister, E., "Calculation of the Wall-Pressure Field in a Turbulent Channel Flow." *Phys. Fluids*, vol. 27, 1984, pp. 579–582.
109. Coles, D. E., "On the Need for Better Experiments." In Proc. AFOSR-IFP-Stanford Conf. on Computation of Turbulent Boundary Layers. 1968. vol. 1, pp. 434–436.

CHAPTER 26

PRESSURE LOSSES IN RECTANGULAR BENDS

Naohiro Shiragami*

Chemical Engineering Laboratory
Institute of Physical and Chemical Research
Wako, Japan

and

Ichiro Inoue

Department of Chemical Engineering
Tokyo Institute of Technology
Tokyo, Japan

CONTENTS

INTRODUCTION

A curved configuration of tube, as well as a straight tube, is frequently seen in fluid transport piping systems, heat exchangers, chemical reactors, and other apparatus, equipments, or devices. Considering the geometry that gives rise to the change in flow direction, these curved configurations

* Presently at Dept. of Chemical Engineering, Tokyo Inst. of Tech., Tokyo, Japan.

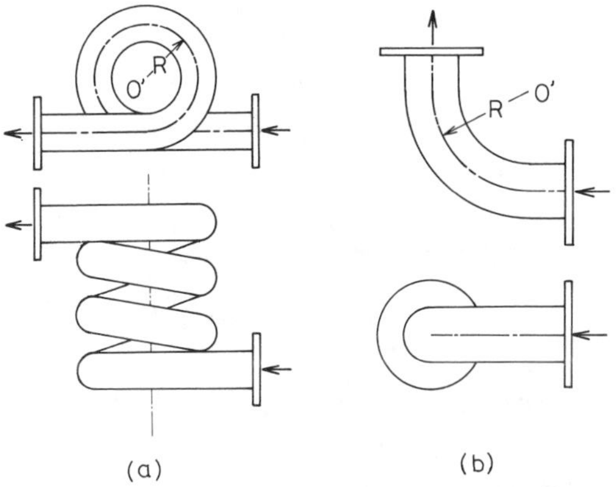

Figure 1. (a) Helically coiled tube, and (b) bend.

of tubes can be classified into the two types of curved tube as shown in Figure 1. One is a helically coiled tube, which is wound spirally at a constant curvature, R, and another is a bend. A helically coiled tube is widely used in heating and refrigeration in order to transfer heat from one fluid to another. Also, a bend is frequently used as a fitting in fluid transport piping systems.

The presence of curvature generates a centrifugal force that acts at right angles to the main flow direction. This effect causes the distortion of the flow field from the flow in a straight tube. Besides flow in a tube that has a curved configuration, a curved flow is commonly encountered in engineering practice. For example, it is often used plates or fins in chemical equipment such as tubular mixers and heat exchangers as shown in Figure 2. These flow fields are considered to be composed of curved flows. Such a curved flow has a considerable effect on the performance in a chemical equipment. Therefore, fluid flow in a curved tube is of importance in several engineering applications such as bends, cooling and heating coils, and in chemical equipment.

For laminar and turbulent flows in a straight tube, a large number of analytical investigations and supporting experimental studies have been reported, and much information about transport phenomena, including transport with simultaneous chemical reaction, have been presented. However, the problem of curved flow has received far less attention in the literature despite their frequent sight in industrial applications. Therefore, investigations are needed into the transport phenomena in a curved tube as well as those in a straight tube.

Since circular cross-section pipes are extensively used in industrial applications, it is important to study the flow field in a circular cross-section curved pipe. However, the flow in a square section curved tube has a significant meaning for considering the curved flow in heat exchangers, chemical reactors, and other apparatus, equipment, or devices. For example, the annular region between two cylinders often contains a spiraling spacer which, together with the cylinder walls, forms a

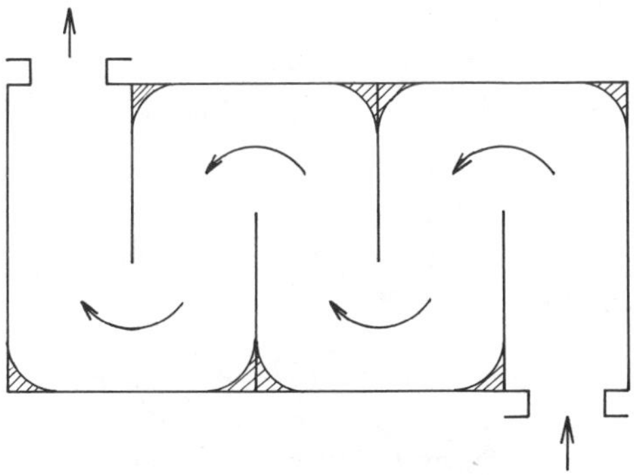

Figure 2. Flow pattern in a chemical equipment with fins.

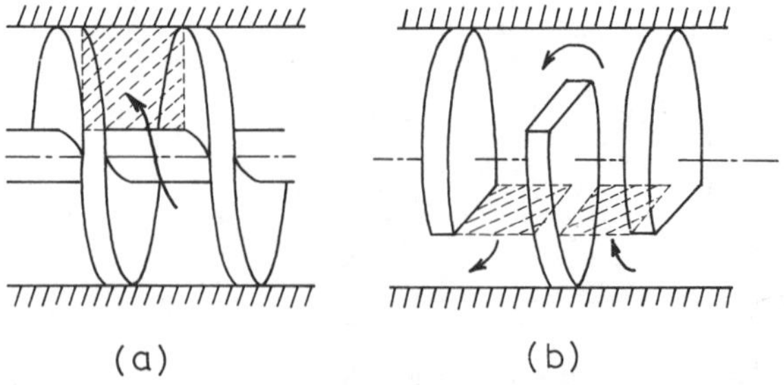

Figure 3. Flow pattern in a chemical equipment: (a) with spiraling spacer, (b) with fins.

curved tube with square cross section. Another example is that the flow in chemical equipment that has plates or fins may be considered to be the curved flow, which has a square cross-section as shown in Figure 3. The curved flow in chemical equipment, such as tubular mixers and heat exchangers, can decompose the flow field into several flows in a 90° square section bend. Therefore, the feature of the flow field in a 90° square section bend is of fundamental interest to the development of a complete understanding of the curved flow in chemical equipment. A 90° square section bend is considered to be an essential tube element on studying the curved flow in a chemical equipment.

PREVIOUS WORKS

If a fluid, flowing along an initially straight duct, encounters a change in direction, the fluid near the axis of the duct, having the highest velocity, is subjected to a greater centrifugal force than the slower-moving fluid in the neighborhood of the duct wall. Centrifugal force gives rise to the superposition on the primary axial flow of a transverse motion, known as secondary flow, in which the fluid in the central region of the duct moves away from the center of curvature, and the fluid near the duct wall flows toward the center of curvature. Thus, with respect to the plane of curvature, the secondary flow consists of a pair of helical vortices as shown in Figure 4. Dye injection experiments, [1, 2] were fundamental in demonstrating secondary flow.

It has been reported that the secondary flow causes a higher heat transfer coefficient [3–10], enhances cross-sectional mixing, and reduces axial dispersion [11–18]. The geometrical configuration of a duct wound into a curve is suggested as a convenient and efficient means of producing the secondary flow. Consequently, as compared with a straight duct, a curved duct has many advantages for heat exchangers and tubular mixers.

Pressure Drop

The problems of determining the pressure drop produced by a bend has a significant meaning, because of its considerable engineering importance in the design of fluid machinery and piping

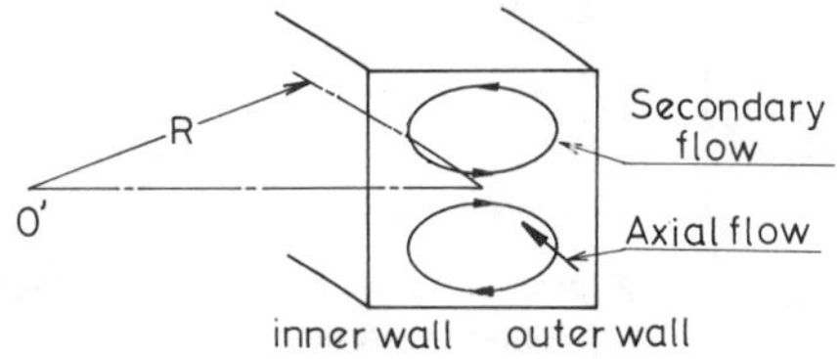

Figure 4. Secondary flow pattern.

systems. The relationship between the pressure drop and the flow rate is the key to the questions engineers most often ask: What is the pressure drop? What pumping power is required? How much fluid is delivered? For turbulent flow in a straight pipe, the relationship between the friction factor and the Reynolds number has been studied for a long time. A bend is frequently used in piping systems, so curved flow is also encountered in chemical process equipment. Therefore, it is expected that the pressure drop produced by a bend for both laminar and turbulent flows can be estimated as well as can those in a straight pipe.

Eustice [19] confirmed that the frictional resistance in a curved pipe increases considerably in comparison with that in a straight pipe. Experiments on the pressure drop for fully developed laminar flow in helically coiled pipes have been performed by several investigators [20–22]. Using the perturbation method, Dean [23, 24] solved analytically the Navier-Strokes equations for fully developed laminar flow in a curved pipe as a deviation from Poiseuille flow. He found that the dynamical similarity of the flow depends on a dimensionless parameter that is now known as the Dean number, K ($=Re\sqrt{a/R}$), where Re is the Reynolds number, a the radius of pipe, and R the radius of curvature. Dean assumed the perturbation expansion in power series of K and calculated up to the fourth power of K. Several investigators [25–28] tried to extend the series to higher order terms in the same manner as Dean's analysis. Ito [29] also calculated for the rectangular section case. McConalogue et al. [30] used a Fourier-series method to solve the equations of motion. However, these analyses were restricted to low Dean numbers, $K < 17$. It was confirmed that the friction factor of a helically coiled pipe increases markedly in comparison with that of a straight pipe. These results, together with the analytical and numerical work, were very well correlated by White's [22] empirical formula.

At high Dean numbers, the flow field for fully developed laminar flow in a curved tube can be assumed to be subdivided into a central core, in which the secondary flow has a nearly uniform velocity directed away from the center of curvature, together with a boundary layer in the immediate vicinity of the tube wall. The motion within the boundary layer is directed toward the center of curvature and hence the condition of continuity of mass within the secondary flow is achieved. By the boundary layer technique assuming the velocity profile within the boundary layer, many investigators [5, 8, 9, 31–34] have obtained the friction factor formulae for laminar and turbulent flows in a curved tube. For turbulent flow in circular section bends, the pressure drop was studied by Ito [35]. He also confirmed that the distorted flow condition in the downstream tangent away from the bend persists for some distance downstream, and most of the pressure drop produced by a bend takes place within 30 diameters of the downstream tangent for 90° circular section bends.

As previously mentioned, there are several problems concerning pressure drop, but the studies for a 90° square section bend have apparently not been published.

For turbulent flow in a bend, it was known that there is marked increase in pressure along the outer wall of a bend, accompanied by a corresponding decrease in pressure along the inner wall. Therefore, the difference between the pressure at outer wall and that at inner wall in the 45° cross section can be shown. From the friction factor, total pressure drop produced by a bend can be estimated, but the distribution of pressure through the cross section of bend cannot be obtained. This distribution of pressure is significant because the separation phenomena in a bend are closely related to the pressure gradient. In order to determine the distribution of pressure through the cross section of bend, one must estimate the difference between the pressure at outer wall and that at inner wall in the 45° cross section.

Separation Phenomena

For turbulent flow in a bend it was known that the pressure gradients in the axial direction take positive values in the neighborhood of both the inlet on the outer wall and the outlet on the inner wall [35]. Therefore, it is expected that the separation phenomena can be seen in these regions.

If a region of adverse pressure gradient exists in the flow, the pressure gradient will exert itself along the wall near which the fluid velocity is small. The momentum contained in the fluid layers that are adjacent to the wall will be insufficient to overcome the force exerted by the pressure gradient, so that a region of reversed flow will exist. The problems that separation phenomean pose for chemical engineers are the area of the reversed flow, the estimation of separation point and a calculation procedure for predicting the occurrence of separation.

The separation phenomena in a bend are significant. The process of separation is associated with large rates of dissipation of mechanical energy; therefore, the avoidance of separation is an important factor in the design of internal flow systems. These phenomena have a significant effect on the turbulent flow field in a bend. Ito [36] and Weske [37] have noted that the pressure drop produced by a bend increases markedly due to the separation phenomena. Using the flow visualization technique, Wirt [38] confirmed the separation phenomena for turbulent flow in a 90° square section bend. Futher, it was found [39] that separation occurs in 90° square section bends having a dimensionless curvature, a/R of about 0.3, where a and R are half the width of the square cross section and the radius of curvature, respectively.

However, the studies on the estimation of the separation point and a calculation procedure for predicting the occurrence of separation for turbulent flow in a bend are thought to be insufficient.

The pressure drop produced by a bend increases markedly due to the separation phenomena, and these phenomena significantly affect the turbulent flow field in a bend. Since the avoidance of separation is an important factor in the design of internal flow systems, it is expected that the occurrence of separation for turbulent flow can be estimated.

SCOPE

It is expected that the pressure drop produced by a 90° square section bend for both laminar and turbulent flows can be estimated, as well as can those in a circular section bend. Since a distorted flow condition affects the turbulent flow field in the downstream tangent away from the bend, the distorted flow length must be determined. Also, the pressure difference between the outer wall and the inner wall in the 45° cross section of bend must be estimated. Theoretical and experimental studies on the pressure drops for fully developed laminar and turbulent flows in helically coiled pipes have been performed by many investigators. For turbulent flow in circular section bends, the pressure drop was also studied. However, there is no empirical formula for determining the pressure drop in a 90° square section bend. Then, it is needed the friction factor formulae for both laminar and turbulent flows in a 90° square section bend. One approach to turbulence theory is to postulate a relation between stress and rate of strain that involves a turbulence-generated viscosity, which then supposedly plays a role similar to that of molecular viscosity in laminar flows. Phenomenological concepts like eddy viscosity were developed by several investigators. The apparent turbulent viscosity is newly defined so that the product of the friction factor for turbulent flow and the Reynolds number for turbulent flow has the same value as that for laminar flow in a square section straight duct. Using this apparent turbulent viscosity the friction factor for turbulent flow is correlated with the Reynolds number and the dimensionless curvature. Since the distorted flow condition affects the turbulent flow in downstream tangent away from the bend, the distorted flow length must be determined.

Separation phenomena in a bend have been reported in several studies. However, there is no report concerning the estimation of the separation point nor a calculation procedure for predicting the occurrence of separation in a 90° square section bend. Then, the study of separation phenomena in a bend is thought to be insufficient. When separation occurs in a bend, the axial velocity close to the wall becomes negative and so a region of reversed flow will be established. This reversed flow is considered to have a significant effect on the turbulent flow field in a bend. The boundary line between the reversed and main flows is also significant and the separation point should be determined. The problem, whether or not separation actually occurs in a bend, is solved, and then occurrence of separation for various bends is determined. The curve of onset condition is determined by experiments. Since the separation phenomena are closely related to the pressure gradient, the occurrence of separation should be discussed based on the axial distribution of pressure in a bend. Assuming the values of axial pressure gradients in a bend center plane took larger values than other positions, the occurrence of separation in a bend may be estimated on the basis of the axial distribution of pressure in a bend center plane. A calculation procedure for predicting the occurrence of separation will be proposed by considering the axial distribution of pressure. Finally, the calculation results are compared with experimental results, and it is shown that the proposed calculation is reasonable.

The axial distribution of pressure in a bend with the downstream tangent away from the bend are clarified, and the friction factor, the distorted flow length and the pressure difference in the 45° cross section are estimated. A calculation procedure for predicting the occurrence of separation in a 90° square section bend is also proposed.

EXPERIMENTAL APPARATUS AND PROCEDURE

The experimental apparatus for pressure drop is illustrated schematically in Figure 5. All square test ducts were made of brass castings and the length of each side was 14.0 mm. In order to measure the pressure along upstream tangent, bend and downstream tangent, the pressure taps were set in the each four wall of the duct. The diameter of each hole at the duct wall for measuring the pressure by manometers was 1.0 mm. A length of about 71 times the width of square cross-section was taken as the calming section before the first pressure tap. Pressure taps in a bend were located at three cross sections ($\theta = 0$, 45° and 90°). The axial distribution of pressure was measured for bends that have dimensionless curvatures, a/R, 0.0365, 0.0729, 0.146, and 0.292.

Separation phenomena were observed by the flow visualization technique. All square test ducts were made of transparent acrylic resin and the length of each side was 16.0 mm. The streamlines were measured by means of an aluminum power tracer suspended in the liquid. Intense plane-collimated beams of light were used to illuminate a horizontal 0.1 cm-wide slice of the bend center plane. The suspended tracer illuminated by the beam was photographed with a camera using an exposure time from 1/15 to 1/100 sec. The separation phenomenon was observed by the trajectories

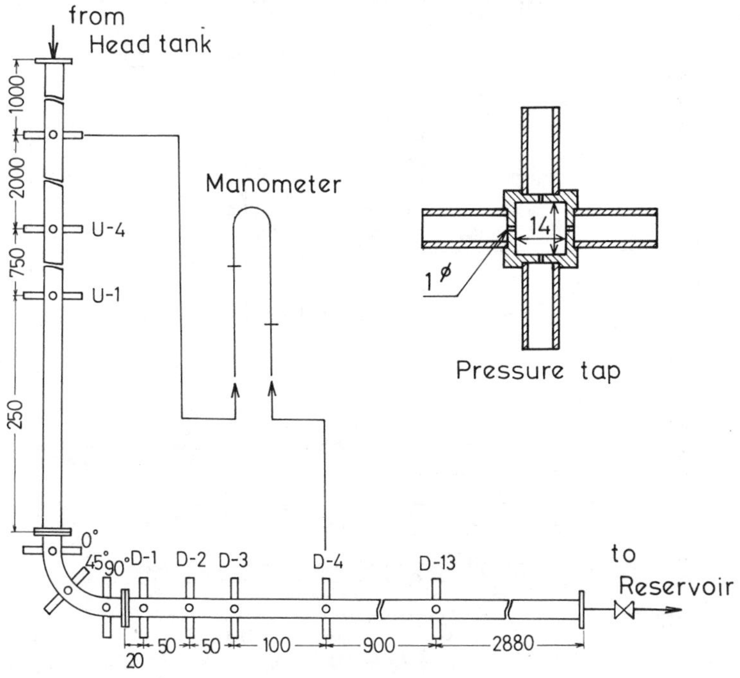

Figure 5. Schematic flow diagram of experimental apparatus.

of the tracer. Also, by injecting dye along bend walls, the dye tracer was photographed and the existence of separation was observed. The experiments were performed for six bends which has the dimensionless curvatures, a/R, from 0.125 to 0.50.

EFFECT OF CURVATURE ON FRICTION FACTOR

Typical data of the axial distribution of pressure are shown in Figure 6. There was a marked increase in pressure along the outer wall of a bend, accompanied by a corresponding decrease in pressure along the inner wall. Also, the distorted flow condition in the downstream tangent away from the bend was observed. Considering that the distorted flow condition persists for some distance downstream, the friction factor value is determined according to the following formula:

$$f_c = \Delta p_c a / \rho w_m^2 L_c \tag{1}$$

where Δp_c and L_c are shown in Figure 6 and f_c denotes the total pressure drop including the drop that produces in the downstream tangent.

The dependence of curvature on the friction factor is shown in Figure 7. The theoretical relationship between the friction factor and the Reynolds number for laminar flow in a square section straight duct was obtained [40] as follows:

$$f_s \, Re = 2/\alpha = 14.2 \tag{2}$$

where the constant, α, is calculated from the following infinite series:

$$\alpha = \frac{1}{3} - \frac{64}{\pi^5} \sum_{n=1}^{\infty} \frac{1}{(2n-1)^5} \tanh\left(\frac{2n-1}{2}\pi\right) \fallingdotseq 0.141 \tag{3}$$

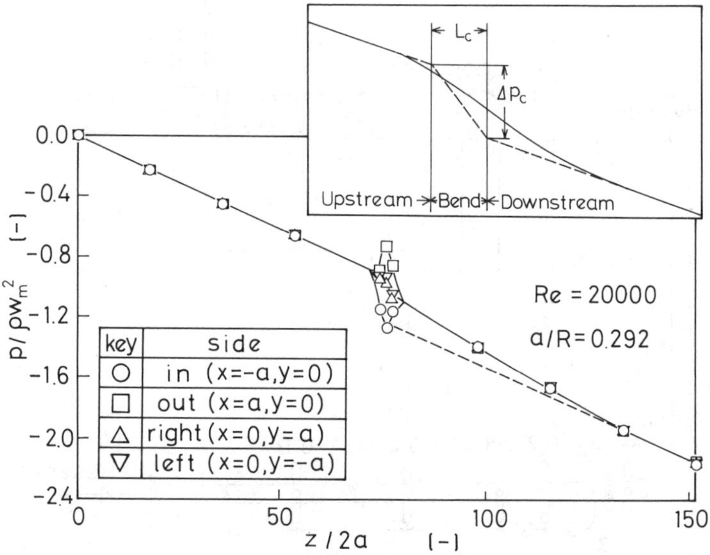

Figure 6. Axial distribution of pressure.

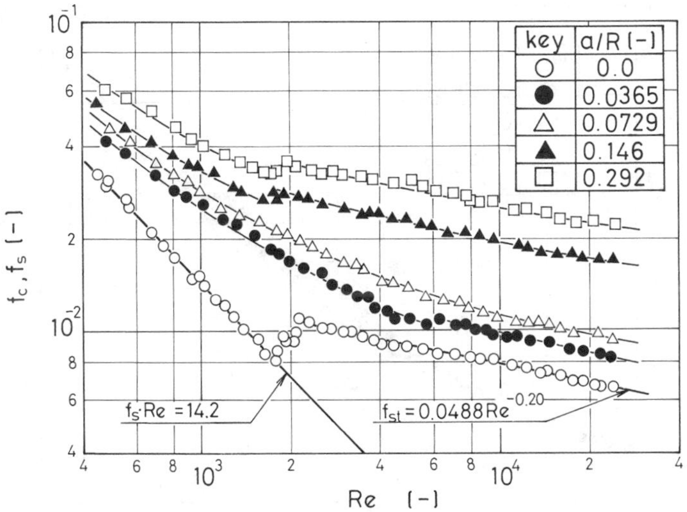

Figure 7. Friction factor for bend.

The friction factor for turbulent flow in a square section straight duct was written experimentally by

$$f_{st} = 0.0488\ Re^{-0.2} \tag{4}$$

The presssure drop produced by a bend for both laminar and turbulent flows were considerably increased in comparison with a straight duct. The pressure drop for a bend depended on the dimensionless curvature, and the pressure drop increased as the dimensionless curvature increase under the constant Reynolds number. The transition region from laminar to turbulent flows for a bend was not clear. On the basis of Adler's theoretical formula [31] and Prandtl's empirical formula [21] for a helically coiled pipe, the ratio of the friction factor for a square section bend, f_c, to that for a square section straight duct, f_s, is plotted against the Dean number, K. The correlation between the friction factor ratio and the Dean number is shown in Figure 8. The following empirical formula can be obtained for Dean number ranging from 100 to 400.

$$f_c/f_s = 0.291K^{0.35} \tag{5}$$

Similar to the helically coiled pipe, the friction factor can be correlated with the Dean number, and the laminar flow condition may be occurred up to the Dean number is 400. Using the previous formula, the friction factor for laminar flow in a bend can be estimated when the dimensionless curvature and the Reynolds number are given.

From Figure 8, when the Dean number is greater than about 400, the friction factor ratio does not correlate with the Dean number. Then the friction factor for $K \geqq 400$ must be related to the Reynolds number and the dimensionless curvature. When the Dean number is greater than about 400, the flow field is thought to be turbulent. Then, the friction factor may be correlated by the consideration based on the concept of the apparent turbulent viscosity. From Equation 5, the friction factor for laminar flow in a bend can be written as

$$f_c = 4.13\ Re^{-0.65}(a/R)^{0.175} \tag{6}$$

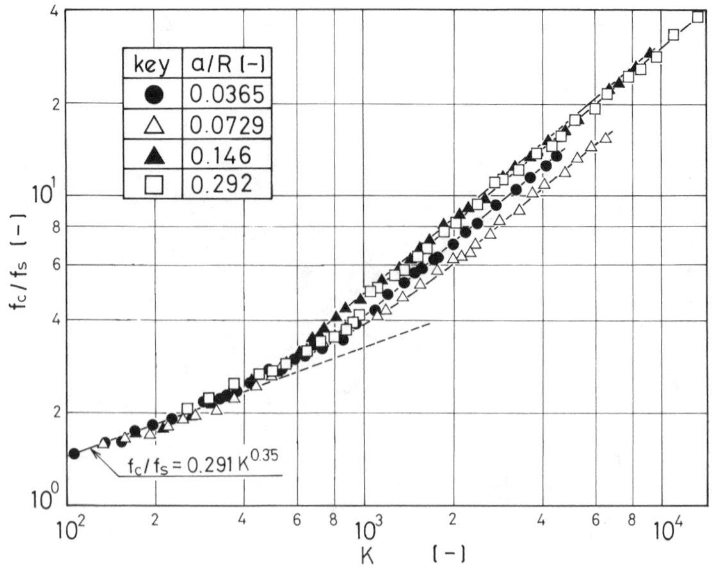

Figure 8. Correlation between friction factor ratio and Dean number.

The apparent turbulent viscosity [41], η, is newly defined so that the product of the friction factor for turbulent flow, f_{st}, and the Reynolds number for turbulent flow, Re_η $(=2\rho w_m a/\eta)$, has the same value as that for laminar flow in a square section straight duct as expressed in Equation 2.

$$f_{st}\, Re_\eta = 14.2 \tag{7}$$

Since the friction factor for turbulent flow in a square section straight duct can be written experimentally as Equation 4, the relationship between Re_η and Re can be written as

$$Re_\eta = 291\, Re^{0.20} \tag{8}$$

The friction factor for turbulent flow in a bend is assumed to be of the same form as Equation 6.

$$f_{ct} = C_1\, Re_\eta^{-0.65} (a/R)^{0.175} \tag{9}$$

where C_1 is experimental constant. From Equations 2, 8, and 9, the ratio of the friction factor for turbulent flow in a bend to that for laminar flow in a square section straight duct can be written:

$$f_{ct}/f_s = C_2 \{Re(a/R)^{0.20}\}^{0.87} \tag{10}$$

The friction factor ratio for turbulent flow is plotted against $Re(a/R)^{0.20}$, and the following empirical formula was obtained for the dimensionless curvature such as $a/R \leq 0.0729$ as shown in Figure 9.

$$f_{ct}/f_s = 0.0040 \{Re(a/R)^{0.20}\}^{0.87} \tag{11}$$

A good correlation between the friction factor ratio for turbulent flow and $Re(a/R)^{0.20}$ can be obtained by making use of newly defined apparent turbulent viscosity. In Figure 9 all of the solid

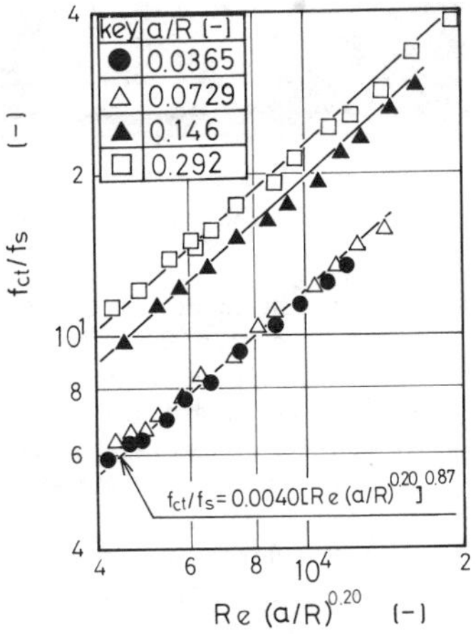

Figure 9. Correlation of friction factor ratio for turbulent flow.

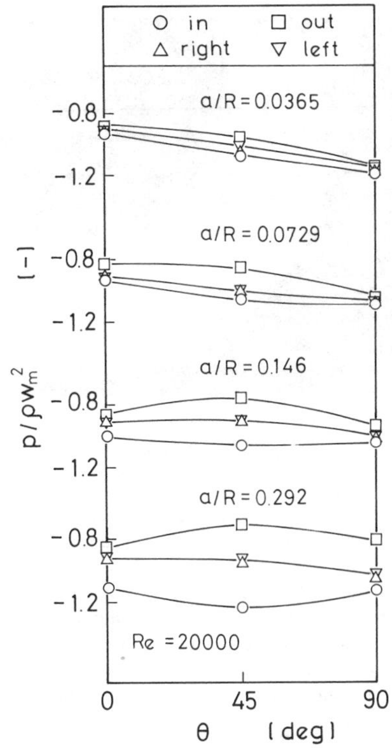

Figure 10. Axial distribution of pressure in a bend.

lines had almost the same slope, but in the cases of a/R = 0.146 and 0.292 the coefficients in Equation 11 had larger values than the others. In order to qualitatively explain these increases in friction factors, the following considerations are made.

Comparison of the axial distribution of pressure with dimensionless curvatures as parameter is shown in Figure 10. The pressure gradients for the dimensionless curvatures 0.146 and 0.292 took positive values in the neighborhood of both the inlet on the outer wall and the outlet on the inner wall. Therefore, it is expected that separation phenomena can be seen in these regions. By injecting dye in the neighborhood of both the inlet on the outer wall and of the outlet on the inner wall, the dye tracer was photographed. As shown in Figure 11, it was confirmed that separation occurred in those regions previously discussed. Therefore, it can be considered that for the dimensionless curvatures 0.146 and 0.292 separation phenomena affect the turbulent flow fields significantly.

When there is no separation in any region, the friction factor for ordinary turbulent flow in a bend can be estimated by Equation 11.

DISTORTED FLOW LENGTH IN DOWNSTREAM TANGENT

From Figure 6, the influence from bend on the pressure distribution was confirmed in the upstream and downstream tangents. The influence of bend in the upstream tangent was located close to the bend, as compared with that in the downstream tangent. The distorted flow condition in the downstream tangent is considered to have a significant effect on the pressure distribution.

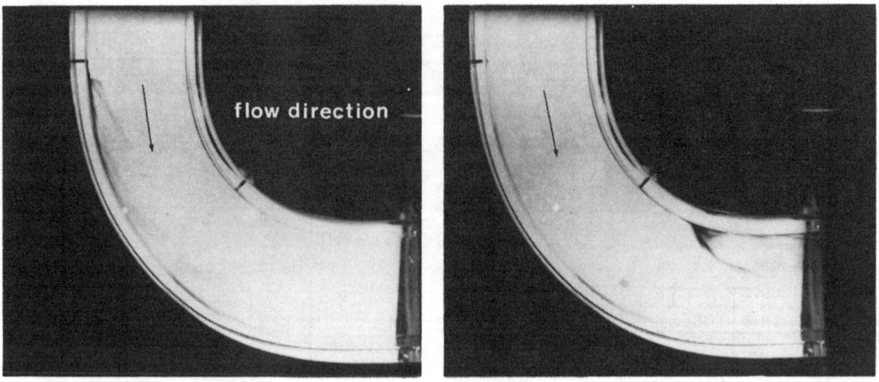

Figure 11. Separation phenomena in a bend (a/R = 0.25, Re = 3,100).

Therefore, in the following discussions the distorted flow length only in the downstream tangent is treated. Typical data of the axial distribution of pressure in the downstream tangent are shown in Figure 12, where z denotes the distance measured along the downstream tangent away from the bend. The influence of the bend on the pressure distribution existed a large distance downstream from the bend. In the bend exit, namely z = 0, the pressure at the outer wall was larger than that at the inner wall, and the pressure at inner and outer walls, except close to the bend had almost the same value. Also, the pressure decreased almost linearly with distance downstream, except close to the bend, and attained the value for fully developed straight duct flow. Assuming a constant pressure gradient in the region of distorted flow condition, the axial distribution of pressure in the

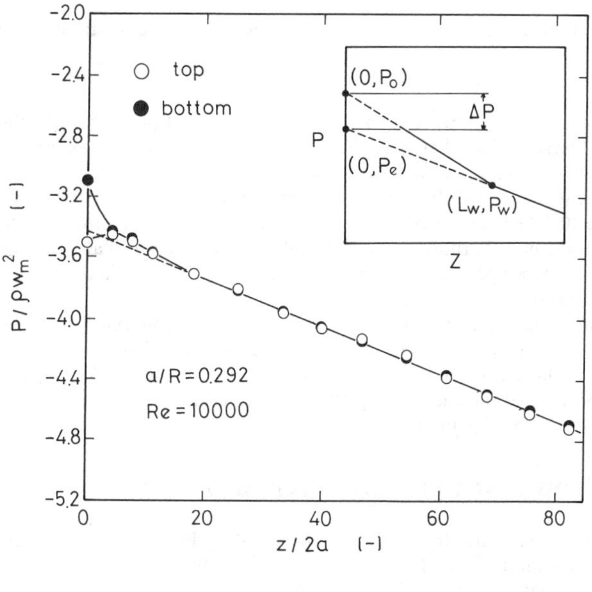

Figure 12. Axial distribution of pressure in downstream tangent.

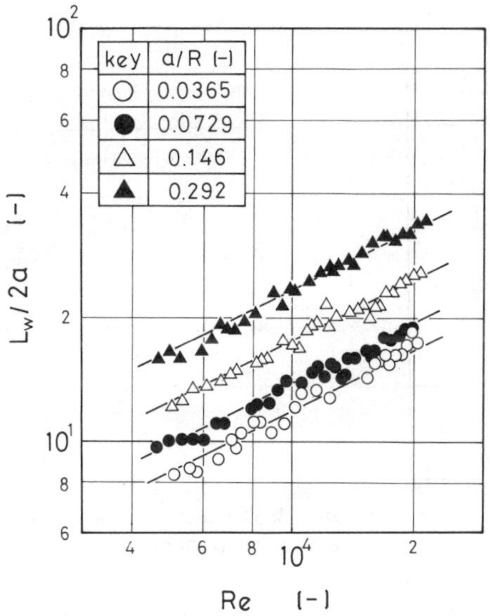

Figure 13. Distorted flow length.

downstream tangent is illustrated in Figure 12. L_w and P_0 are defined as the distorted flow length and the extrapolated value of pressure gradient line in the region of distorted flow condition, respectively. The dependence of curvature on distorted flow length is shown in Figure 13. It was obvious that the distorted flow length became greater with increasing value of dimensionless curvature, and the distorted flow length took from about 10 to 30 times the width of square cross section. In order to correlate the distorted flow length with the dimensionless curvature and the Reynolds number, the following considerations are made.

The mean apparent friction factor for distorted flow condition, f_w, and that for the fully developed straight duct flow, f_{st}, are defined as follows:

$$f_w = (P_0 - P_w)a/\rho w_m^2 L_w \tag{12}$$

$$f_{st} = (P_e - P_w)a/\rho w_m^2 L_w \tag{13}$$

where P_e, P_0, P_w and L_w are shown in Figure 12. Substracting Equation 13 from Equation 12, the distorted flow length becomes

$$L_w/2a = (\Delta P/\rho w_m^2)/2(f_w - f_{st}) \tag{14}$$

where $\Delta P = P_0 - P_e$.

The dependence of curvature on $\Delta P/\rho w_m^2$ and f_w are discussed in the following section. From the experimental results, P_0 showed almost the same value as the average pressure between the outer wall and the inner wall at $z = 0$, and ΔP is considered to be the average difference of pressure at the bend exit. The average difference of pressure increased with increasing value of the dimensionless curvature. A good correlation among the average difference of pressure, the dimensionless curvature and the Reynolds number is obtained as shown in Figure 14. Under the present experimental conditions, the average difference of pressure at the bend exit is estimated by the following empirical formula:

$$\Delta P/\rho w_m^2 = 0.0102\,(a/R)^{0.32} Re^{0.28} \tag{15}$$

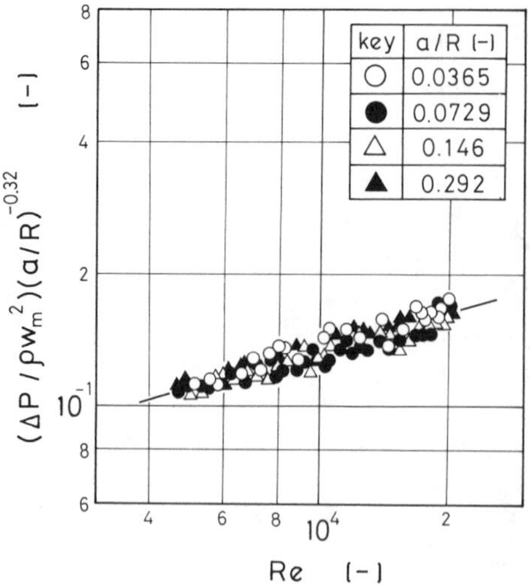

Figure 14. Correlation of average difference of pressure at bend exit.

The relation between the mean apparent friction factor for distorted flow, f_w, and the Reynolds number is shown in Figure 15. From the figure it was obvious that f_w was independent of the value of dimensionless curvature. Under the present experimental conditions, the mean apparent friction factor for distorted flow is estimated by the following empirical formula.

$$f_w = 0.0612\ Re^{-0.2} \tag{16}$$

The friction factor of a square section straight duct, f_{st}, was obtained as Equation 4 in the range of $4{,}000 \leqq Re \leqq 20{,}000$. Substituting Equations 15, 16, and 4 into Equation 14, the distorted flow length can be expressed as

$$L_w/2a = 0.411(a/R)^{0.32}Re^{0.48} \tag{17}$$

From Equation 17 a good correlation among the distorted flow length, the dimensionless curvature and the Reynolds number is obtained as shown in Figure 16. Using Equation 17, the distorted flow length can be estimated when the dimensionless curvature and the Reynolds number are given.

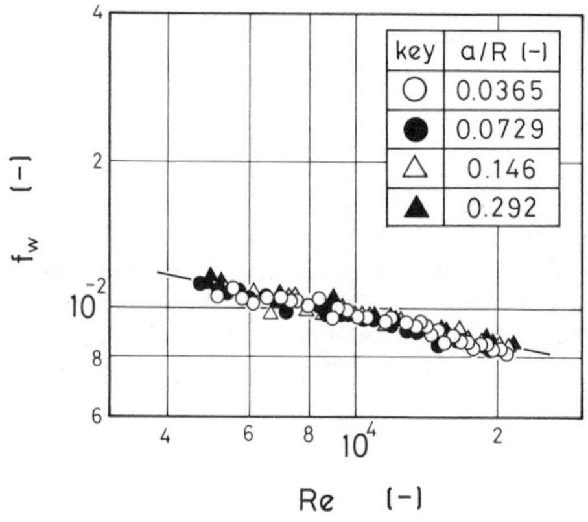

Figure 15. Mean apparent friction factor for distorted flow.

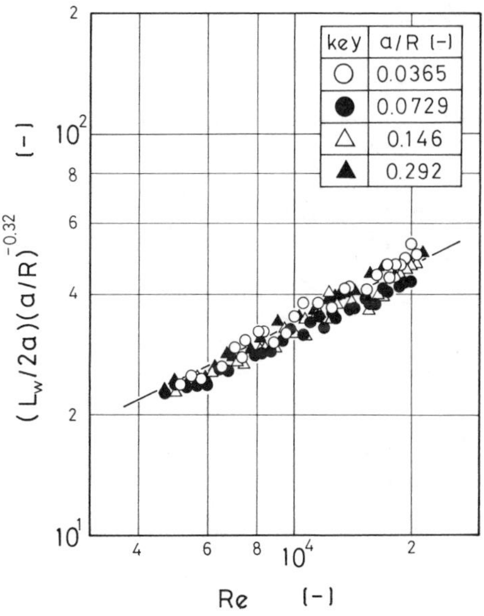

Figure 16. Correlation of distorted flow length.

The average difference of pressure at bend exit, the mean apparent friction factor for distorted flow and the distorted flow length can be estimated by Equations 15, 16 and 17, respectively.

PRESSURE DIFFERENCE IN 45° CROSS SECTION

The pressure at the outer wall was larger than that at the inner wall in the 45° cross section for any value of the dimensionless curvature, and the difference between the pressure at the outer wall and that at the inner wall increased with the degree of curvature as shown in Figure 10.

The difference between the pressure at the outer wall and that at the inner wall in the 45° cross section was often expressed in the term of discharge coefficient. The discharge coefficient, C_d, is defined [34, 42, 43] by:

$$C_d = \{2(P_{out} - P_{in})/\rho w_m^2\}^{-1/2} \tag{18}$$

where P_{out} and P_{in} are the pressure at the outer and inner walls, respectively. Also, ρ and w_m denote the density of the fluid and the cross-sectional mean flow rate, respectively.

The discharge coefficient is plotted against the Reynolds number as shown in Figure 17. Under the condition such as $Re \geqq 3{,}000$, the discharge coefficient was an almost constant value against the Reynolds number.

In the same manner as in the previous work [34, 43], the value of $C_d\sqrt{a/R}$ is plotted against the Reynolds number as shown in Figure 18. This relation was presented in the following formula:

$$C_d\sqrt{a/R} = 0.491 \tag{19}$$

Addison [44] reported the following formula by applying the free-vortex theory:

$$C_d\sqrt{a/R} = \frac{1-(a/R)^2}{4(a/R)} \ln \frac{1+(a/R)}{1-(a/R)} \tag{20}$$

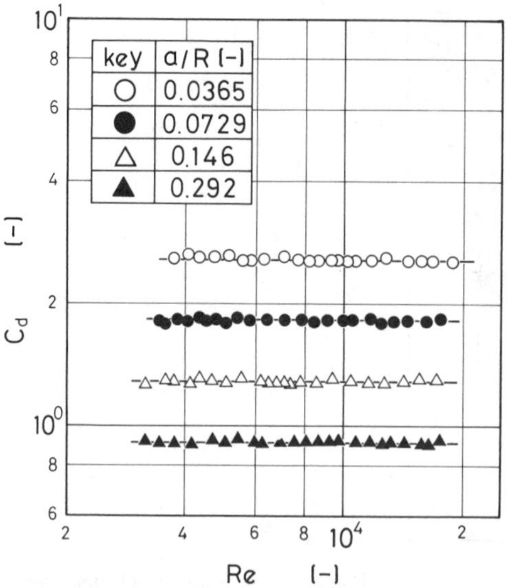

Figure 17. Discharge coefficient.

Tsuji et al. [45] also reported the formula for a circular section bend by applying the forced-vortex theory. In a similar manner, the formula for the square section bend is obtained as follows:

$$C_d\sqrt{a/R} = 0.5 \tag{21}$$

The empirical formula expressed as Equation 19 is compared with these theoretical forrmulae as shown in Figure 19. These theoretical formulae are different from the empirical formula, because the two theories have both neglected the problem of secondary flow in a bend.

The pressure difference in the 45° cross section can be estimated by Equation 19. Under the Reynolds number range of experimental conditions employed, the term on the right-hand side in Equation 19 was independent of the cross-sectional mean flow rate. The flow rate, Q, is easily determined by:

$$Q = C_d A\sqrt{2(P_{out} - P_{in})/\rho} \tag{22}$$

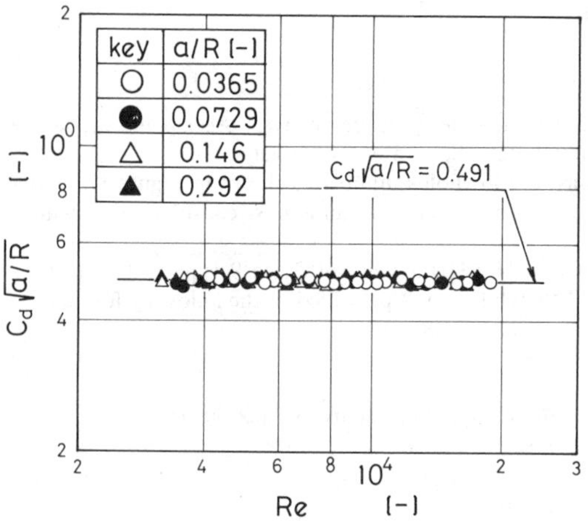

Figure 18. Correlation of discharge coefficient.

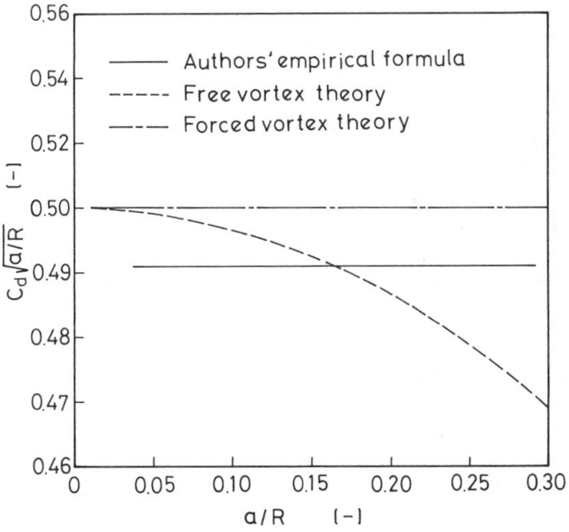

Figure 19. Comparison of experimental results with theoretical calculation.

where A is the cross-sectional area of the bend. It can be considered that the Reynolds number range of experimental conditions employed, such as $3{,}000 \leqq Re \leqq 18{,}000$, can be attained the admissible limit Reynolds number.

The pressure drop produced by a bend and the admissible Reynolds number are compared with those of a orifice flowmeter as shown in Figure 20. The pressure drop of the bend, $\zeta\ (=2\Delta p_c/\rho w_m^2)$, was calculated from the value of fraction factor, and that of the orifice flowmeter was chosen from the handbook.

Orifice flowmeter

m [-]	Re_T [-]	ζ [-]
0.1	30000	226
0.2	50000	47.8
0.3	80000	17.5
0.4	125000	7.8
0.5	170000	3.8

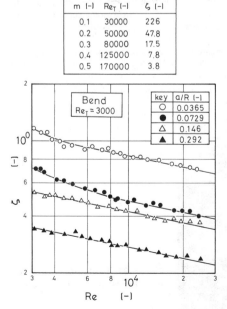

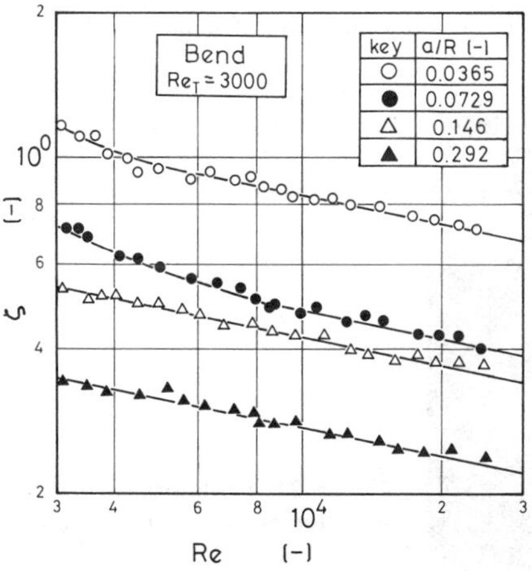

Figure 20. Pressure drop of bend and those of orifice flowmeter.

For Re = 30,000, the comparison is as follows:

	Bend				Orifice
Dimensionless curvature a/R [−]	0.0365	0.0729	0.146	0.292	
Pressure drop ξ [−]	0.67	0.39	0.34	0.22	226

It was found that the pressure drop of the bend was reduced considerably in comparison with that of the orifice flowmeter. Since the flow rate in a bend is easily determined by Equations 19 and 22 by measuring the pressure difference between the outer wall and the inner wall in the 45° cross section, the flow field in a bend can be estimated. Also, it is found that a bend is worth using as a flow rate measuring device.

SEPARATION POINT AND SEPARATED REGION

A typical flow pattern of separated region in a bend is shown in Figure 21. From the trajectories of the tracer, a separated flow was observed in the neighborhood of the outlet on the inner wall. Downstream of the separation point, the axial velocity close to the inner wall became negative and so a region of reversed flow was established.

Since the axial pressure gradient in the neighborhood of the inlet on the outer wall took positive value, it can be expected that the separation occurs not only on the inner wall but also on the outer wall. However, there was a marked decrease in pressure along the outer wall in the downstream tangent away from the 45° cross section of the bend. Thus, the separated flow region in the neighborhood of the inlet on the outer wall is considered to be located close to the wall, as compared with that in the neighborhood of the outlet on the inner wall. The separated flow region on the inner wall can be considered to have a significant effect on the pressure drop produced by a bend. Therefore, in the following discussions only the separation phenomena in the neighborhood of the outlet on the inner wall are treated.

From Figure 21 it was understood that the flow pattern consisted of two regions, i.e., a main flow and a separated flow region. Figure 22 shows a typical flow region of separation in a bend, where z_s denotes the separation point expressed as the distance from the bend inlet along the bend center line. From a number of photographs as shown in Figure 21, an average separation point was determined. The separation phenomena can be observed for the dimensionless curvature, such as $a/R \geqq 0.25$. The separation angle, θ_s, took the values from about 55° to 85° under the present Reynolds number range. The separation angle decreased with an increase in both the value of the dimensionless curvature and the Reynolds number. The boundary line between the two regions is obtained as shown in Figure 23. From the figure it was obvious that the boundary line can be expressed as an unique curve within a length about two times the width of square cross section

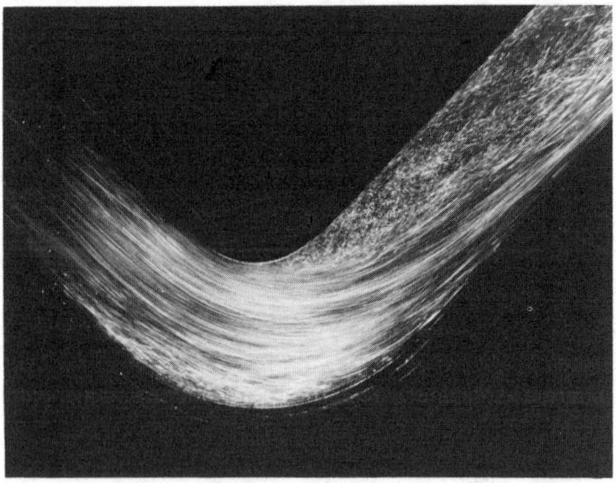

Figure 21. Photograph of tracer trajectories (Re = 10,900, a/R = 0.50).

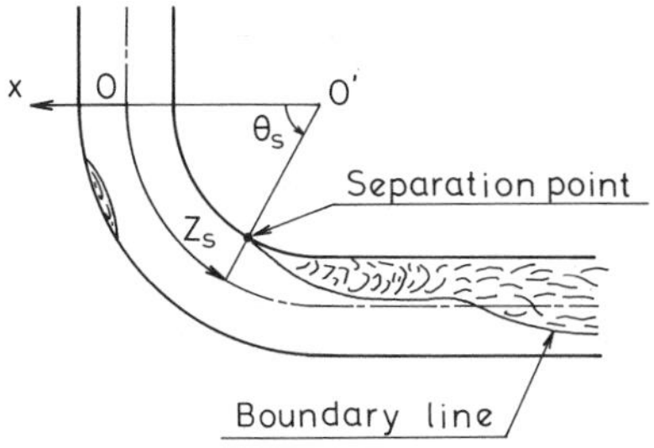

Figure 22. Typical flow region of separation in a 90° square section bend.

from the separation point. The separated flow region was located not only within a bend but also in the downstream tangent. The separation point. z_s, can be correlated with the dimensionless curvature and the Reynolds number as shown in Figure 24. This correlation was presented in Equation 23.

$$z_s/2a = 0.76\ \mathrm{Re}^{-0.06}(a/R)^{-1.24} \tag{23}$$

where

$4{,}000 \leqq \mathrm{Re} \leqq 20{,}000, \quad 0.25 \leqq a/R \leqq 0.50$

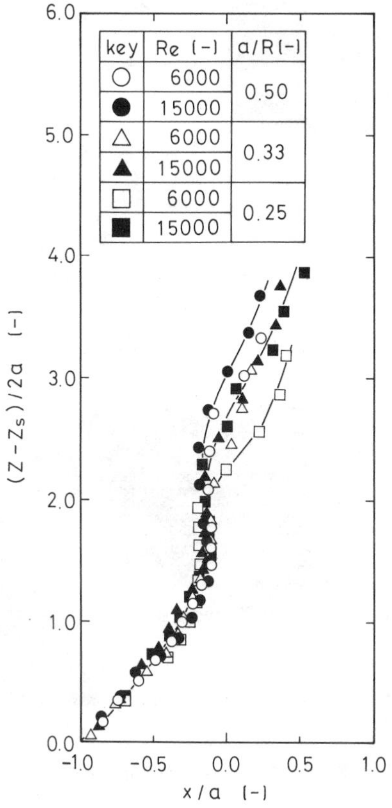

Figure 23. Boundary line between a main flow and a separated flow region.

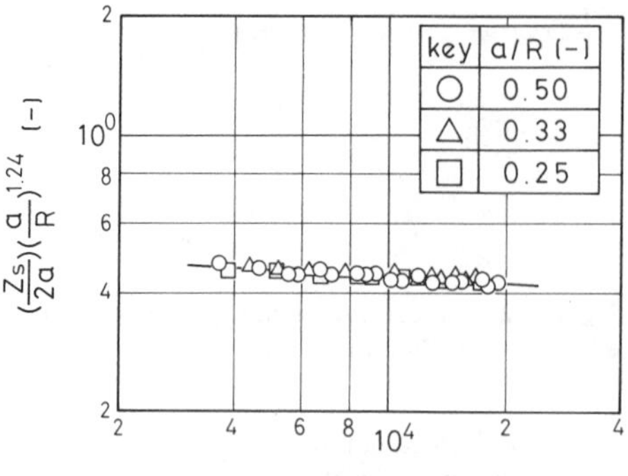

Figure 24. Correlation of separation point.

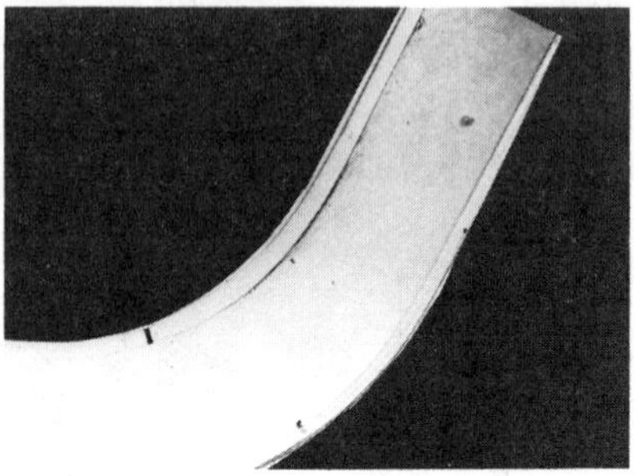

Figure 25. Photographs of dye tracer.

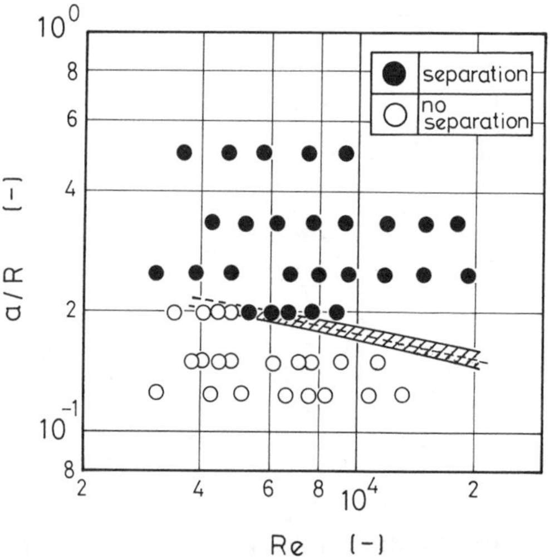

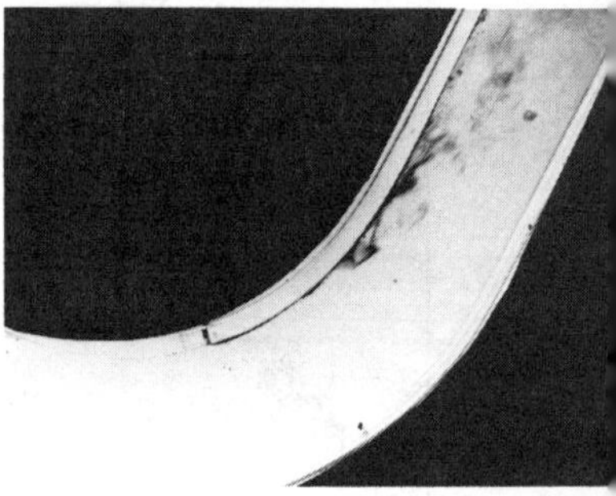

Figure 26. Existence of separation.

ONSET CONDITION OF SEPARATION

The existence of separation was observed by the dye tracer. For $a/R = 0.20$, the flow field in a bend without separation for $Re = 4{,}490$ and with separation for $Re = 6{,}070$ was observed as shown in Figure 25. Figure 26 was obtained from the experiments for the dimensionless curvatures from 0.125 to 0.5 under the Reynolds number ranging from about 4,000 to 20,000. For the value of dimensionless curvatures such as $a/R \geqq 0.25$, the separation occurred under the present Reynolds number range. Also, for the dimensionless curvature such as $a/R = 0.20$, the separation occurred only if the Reynolds number was larger than about 5,000. For $a/R = 0.15$ and 0.125 there was no separation under the present Reynolds number range. From Figure 26 it can be expected that the curve of onset condition of separation lies in the neighborhood of the dimensionless curvature such as $a/R = 0.20$.

Considering the separation phenomena are closely related to the pressure gradient, the occurrence of separation can be considered based on the axial distribution of pressure in a bend.

It was observed that there was marked increase in pressure along the outer wall of a bend, accompanied by a corresponding decrease in pressure along the inner wall. Assuming that there are constant pressure gradients in a bend and in the downstream tangent, the axial distribution of pressure is illustrated in Figure 27. P_{out} and P_{in} denote the pressures at the outer wall and at the inner wall in the 45° cross section, respectively. Also, P_e and P_0 denote the imaginary and the average pressures at the bend exit, respectively. The value of $P_e/\rho w_m^2$ and $P_0/\rho w_m^2$ can be obtained by the following relationship:

$$P_e/\rho w_m^2 = f_{ct} L_c/a \tag{24}$$

$$P_0/\rho w_m^2 = (\Delta P + P_e)/\rho w_m^2 \tag{25}$$

where f_{ct} and L_c are the friction factor for a bend and the axial length of a bend center line, respectively. ΔP denotes the difference between the average pressure and the imaginary one at the bend exit as shown in Figure 27. The value of f_{ct} and $\Delta P/\rho w_m^2$ for the dimensionless curvatures, such as $0.0365 \leqq a/R \leqq 0.292$, can be calculated from the value of friction factor. In the case of

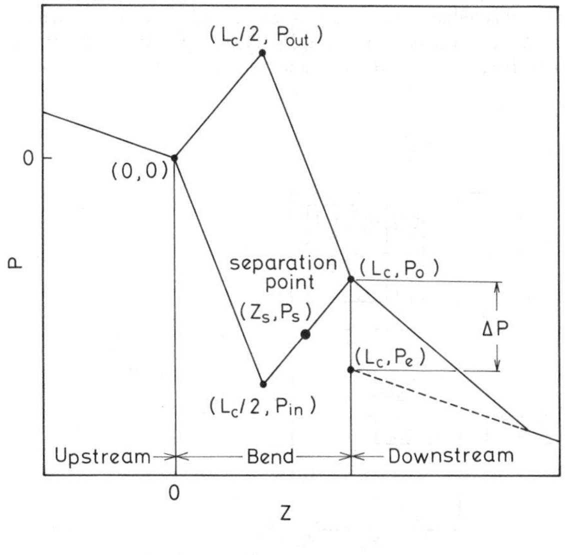

Figure 27. Axial distribution of pressure.

a/R = 0.5 the value of f_{ct} and $\Delta P/\rho w_m^2$ can be calculated from the following formulae which are obtained by the same experiments [46, 47]:

$$f_{et} = 0.454\ Re^{-0.16} \tag{26}$$

$$\Delta P/\rho w_m^2 = 0.00817\ Re^{0.28} \tag{27}$$

The relationship between P_{out} and P_{in} has been obtained, and this relation can be written as Figure 28. The further relationship expressed as Equation 29 is made on the assumption that the shape of the axial distribution of pressure in a bend builds up the parallelogram.

$$(P_{out} - P_{in})/\rho w_m^2 = 2(a/R) \tag{28}$$

$$(P_{out} + P_{in})/\rho w_m^2 = P_0/\rho w_m^2 \tag{29}$$

The values of $P_{out}/\rho w_m^2$ and $P_{in}/\rho w_m^2$ can be obtained by solving simultaneously Equations 28 and 29. The axial distribution of pressure in a bend can be determined from the previously described procedure.

Separation will occur when the momentum contained in the fluid, ρw_m^2, is insufficient to overcome the force exerted by the pressure gradient, $\{(P_0 - P_{in})/(L_c/2)\} \cdot (z_s - L_c/2)$. Then, the ratio of the force exerted by the pressure gradient to the momentum contained in the fluid may be the index for the occurrence of separation. The value of force exerted by the pressure gradient can be calculated from Equations 23–29. For the dimensionless curvatures, such as $0.25 \leqq a/R \leqq 0.50$, this ratio took the value within the hatched area as shown in Figure 28. It was obvious that this ratio was almost independent of not only the dimensionless curvature but also the Reynolds number; hence, the average value, the dotted line as shown in Figure 28, was presented as Equation 30.

$$\frac{\{(P_0 - P_{in})/(L_c/2)\} \cdot (z_s - L_c/2)}{\rho w_m^2} \fallingdotseq 0.125 \tag{30}$$

The ratio of the force exerted by the pressure gradient to the momentum contained in the fluid took the constant. Since this constant was independent of both the dimensionless curvature and the Reynolds number, this value is considered to be the index for the occurrence of separation.

Separation occurred within the bend, i.e., $z_s \leqq L_c$. Then, separation will occur when the following condition, which is derived from Equation 30 and $z_s \leqq L_c$ is satisfied:

$$(P_0 - P_{in})/\rho w_m^2 \geqq 0.125 \tag{31}$$

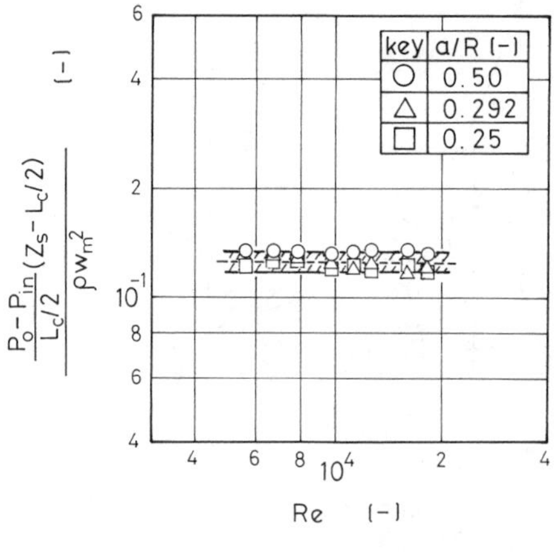

Figure 28. Ratio of force exerted by pressure gradient to momentum contained in fluid.

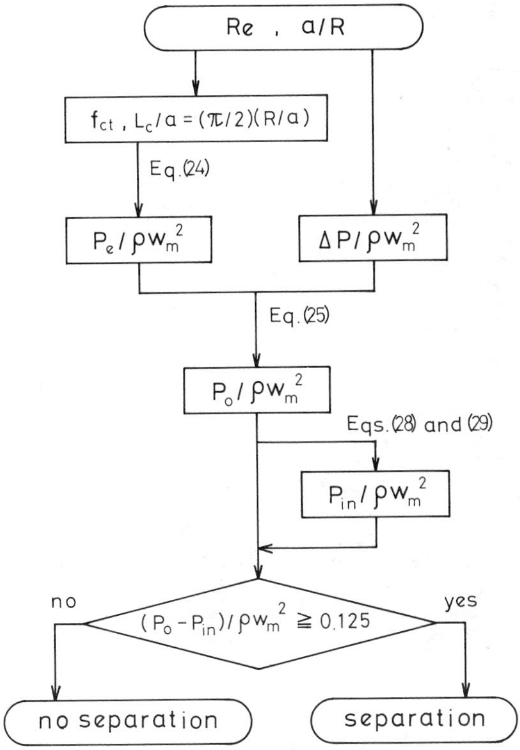

Figure 29. Calculation procedure for prediction of occurrence of separation in a 90° square section bend.

For the dimensionless curvatures such as $0.146 \leqq a/R \leqq 0.500$, one can predict the occurrence of separation in a bend in the following way as shown in Figure 29.

First, the value of $P_0/\rho w_m^2$ is calculated from the value of the dimensionless curvature and the Reynolds number by Equations 24 and 25. Second, the value of $P_{in}/\rho w_m^2$ can be calculated from solving simultaneously Equations 28 and 29. Finally, the value of $(P_0 - P_{in})/\rho w_m^2$ can be obtained.

This value is shown in Figure 30. If the value of $(P_0 - P_{in})/\rho w_m^2$ is larger than about 0.125, the separation phenomena was certainly observed.

The curve of onset condition of separation is calculated by the above procedure. This curve is shown in Figure 31.

Since there is good agreement between the calculation and the experiment, the index for separation is thought to be reasonable.

CONCLUSION

The axial distribution of pressure in a 90° square section bend within the downstream tangent away from the bend was obtained, and the following conclusions were made.

For laminar flow, the ratio of the friction factor for a 90° square section bend to that for a square section straight duct can be correlated with the Dean number. The friction factor for laminar flow in a 90° square section bend can be estimated by the following formula:

$$f_c/f_s = 0.291K^{0.35}$$

$$(100 \leqq K \leqq 400)$$

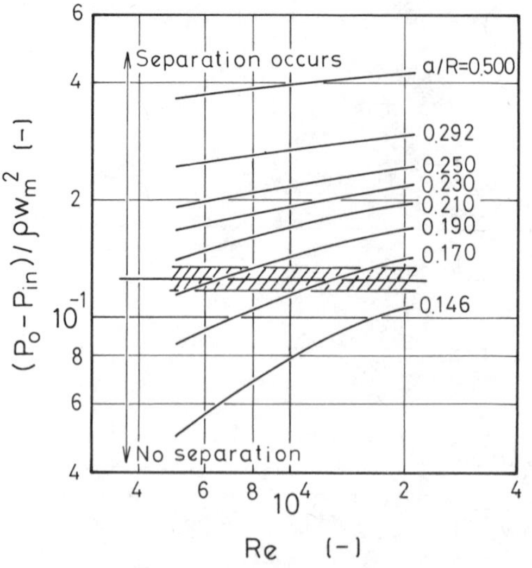

Figure 30. Difference between pressure at inner wall in 45° cross section and that at bend exit.

For turbulent flow, this friction factor ratio can be correlated with the dimensionless parameter, Re $(a/R)^{0.20}$, by making use of the newly defined apparent turbulent viscosity. The friction factor for turbulent flow in a 90° square section bend can be estimated by the following formula:

$$f_{ct}/f_s = 0.0040\{Re(a/R)^{0.20}\}^{0.87}$$

$$(K \geqq 400,\ a/R \leqq 0.0729)$$

The total pressure drop including the drop that produces in the downstream tangent can be estimated by the above formulae.

The distorted flow condition was confirmed in the downstream tangent away from the bend. It was seen that this distorted flow condition has a significant effect on the pressure distribution in the downstream tangent. The distorted flow length was determined experimentally from the axial distribution of pressure in the downstream tangent. The distorted flow length can be estimated by:

$$L_w/2a = 0.411(a/R)^{0.32}Re^{0.48}$$

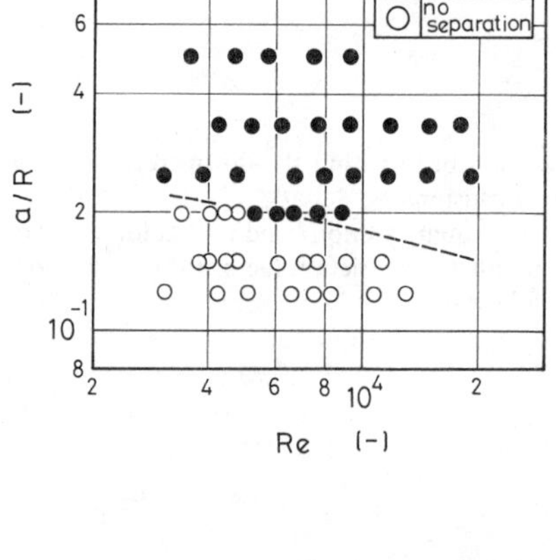

Figure 31. Curve of onset condition of separation.

The another outstanding features of pressure distribution in the downstream tangent were the average difference of pressure between the outer wall and the inner wall at the bend exit and the mean apparent friction factor for distorted flow. They can be estimated by:

$$\Delta P/w_m^2 = 0.0102(a/R)^{0.32}Re^{0.28}$$

$$f_w = 0.0612Re^{-0.2}$$

It was seen that the difference of pressure between the outer wall and the inner wall in the 45° cross section depended on the dimensionless curvature. The pressure difference expressed as the discharge coefficient can be related to the dimensionless curvature. The pressure difference can be estimated by:

$$C_d\sqrt{a/R} = 0.491$$

where

$$C_d = \{2(P_{out} - P_{in})/\rho w_m^2\}^{-1/2}$$

Consequently, the feature of pressure drop in the bend with upstream and downstream tangents can be made clear.

The separation phenomena in a 90° square section bend were studied by flow visualization technique, and the following conclusions were obtained.

The flow pattern of a separated region in a 90° square section bend were obtained by flow visualization technique. Reversed flows were confirmed in the neighborhood of both the inlet on the outer wall and the outlet on the inner wall. The separated flow region in the neighborhood of the inlet on the outer wall was located close to the wall, as compared with that in the neighborhood of the outlet on the inner wall. This phenomenon can be explained by the fact that there is marked decrease in pressure along the outer wall in the downstream tangent away from the 45° cross section of the bend. The boundary line between the reversed and main flows was obtained, and this line can be expressed as a unique curve within a length about two times the width of square cross section from the separation point. Also, it was seen that the separated flow region was located not only within a bend but also in the downstream tangent. The separation point can be correlated with the Reynolds number and the dimensionless curvature. The separation point can be estimated by:

$$z_s/2a = 0.76\,Re^{-0.06}(a/R)^{-1.24}$$

$$(4{,}000 \leq Re\ 20{,}000,\ 0.25 \leqq a/R \leqq 0.50)$$

The existence of separation was observed by the dye tracer. For the dimensionless curvatures, such as $a/R \geqq 0.25$, separation occurred under the present Reynolds number range. Also, for the dimensionless curvatures, such as $a/R = 0.20$, separation occurred only if the Reynolds number was greater than about 5,000. For $a/R = 0.15$ and 0.125 there was no separation under the present Reynolds number range. The curve of onset condition of separation was obtained experimentally.

The occurrence of separation was considered based on the axial distribution of pressure in a bend. It was shown that the ratio of the force exerted by the pressure gradient to the momentum contained in the fluid took the constant. This constant was independent of both the dimensionless curvature and the Reynolds number. By applying this value to the index for separation, a calculation procedure for predicting the occurrence of separation was proposed. There was good agreement between the calculation and the experiment. It can be predicted whether or not separation actually takes place in a 90° square section bend.

In the course of this study, the feature of the flow field for both laminar and turbulent flows in a 90° square section bend were made clear. These results may be introduced improvement of chemical equipment and more effective operations. Also, the intensive use of characteristics of curved flow is expected to produce new types of chemical equipment.

NOTATION

A	cross-sectional area	L_w	flow length
a	distance	P	pressure
C_1	experimental constant	P_0	extrapolated value of pressure gradient
C_D	discharge coefficient	Q	volumetric flowrate
f_c	laminar flow friction factor	R	radius
f_{st}	turbulent flow friction factor	Re	Reynolds number
K	Dean number	w_n	cross-sectional mean flowrate
L	length or characteristic dimension		

Greek Symbols

α	turbulence coefficient	η	apparent turbulent viscosity
ζ	pressure drop coefficient of a bend	ρ	density

REFERENCES

1. Eustice, J., *Proc. Roy. Soc.* London, A85, 119 (1911).
2. Taylor, G. I., *Proc. Roy. Soc.*, London A124, 243 (1929).
3. Akiyama, M., and Cheng, K. C., *Intl. J. Heat and Mass Trans.*, 14, 1659 (1971).
4. Cheng, K. C., and Akiyama, M., *Intl. J. Heat and Mass Trans.*, 13, 471 (1970).
5. Mori, Y., and Nakayama, W., *Intl. J. Heat and Mass Trans.*, 8, 67 (1965).
6. Mori, Y., and Nakayama, W., *Intl. J. Heat and Mass Trans.*, 10, 37 (1967).
7. Mori, Y., and Nakayama, W., *Intl. J. Heat and Mass Trans.*, 10, 681 (1967).
8. Mori, Y., and Uchida, Y., *Trans. J.S.M.E.*, 33, 1836 (1967).
9. Uchida, Y., and Koizumi, H., *Trans. J.S.M.E.*, 45, 1708 (1979).
10. Yee, G., Chilukuri, R., and Humphrey, J. A. C., *Trans. A.S.M.E.*, C102, 285 (1980).
11. Koutsky, J. A., and Adler, R. J., *Can. J. Chem. Eng.*, 42, 239 (1964).
12. McConalogue, D. J., *Proc. Roy. Soc.* London, A315, 99 (1970).
13. Nauman, E. B., *Chem. Eng. Sci.*, 32, 28 (1977).
14. Ruthven, D. M., *Chem. Eng. Sci.*, 26, 1113 (1971).
15. Sakra, T., Lesek, F., and Cermankova, H., *Collection Czechoslov. Chem. Commun.*, 36, 3543 (1971).
16. Saxena, A. K., and Nigam, K. D. P., *Chem. Eng. Sci.*, 34, 425 (1979).
17. Trivedi, R. N., and Vasudeva, K., *Chem. Eng. Sci.*, 29, 2291 (1974).
18. Trivedi, R. N., and Vasudeva, K., *Chem. Eng. Sci.*, 30, 317 (1975).
19. Eustice, J., *Proc. Roy. Soc.* London, A84, 107 (1910).
20. Mishra, P., and Gupta, S. N., *Ind. Eng. Chem., Process Des. Dev.*, 18, 130 (1979).
21. Prandtl, L., *Essentials of Fluid Dynamics*, Blackie and Son, London (1952).
22. White, C. M., *Proc. Roy. Soc.*, London, A123, 645 (1929).
23. Dean, W. R., *Phil. Mag.*, 4, 208 (1927).
24. Dean, W. R., *Phil. Mag.*, 5, 673 (1928).
25. Dyke, M. V., *J. Fluid Mech.*, 86, 129 (1978).
26. Larrain, J., and Bonilia, C. F., *Trans. Soc. Rheology*, 14, 135 (1970).
27. Smith, F. T., *Proc. Roy. Soc.* London, A347, 345 (1976).
28. Topakoglu, H. C., *J. Math. Mech.*, 16, 1321 (1967).
29. Ito, H., *Rep. Inst. High Sp. Mech.*, Tohoku Univ., 1, 1 (1950).
30. McConalogue, D. J., and Srivastava, R. S., *Proc. Roy. Soc.*, London, A307, 37 (1968).
31. Adler, M., *Z. Angew. Math. Mech.*, 14, 257 (1934).
32. Barua, S. N., *Quart. J. Mech. Appl. Math.*, 16, 61 (1963).
33. Ito, H., *Trans. A.S.M.E.*, D81, 123 (1959).
34. Ito, H., *Z. Angew. Math. Mech.*, 49, 653 (1969).

35. Ito, H., *Trans. A.S.M.E.*, D82, 131 (1960).
36. Ito, H., *Sokuken Hokoku*, 15, 1 (1960).
37. Weske, J. R., *J. Appl. Mech.*, 15, 344 (1948).
38. Wirt, L., *General Electric Review*, 30, 286 (1927).
39. Smith, A. J. W., *Internal Fluid Flow*, Clarendon Press, Oxford (1980).
40. Shiragami, N., and Inoue, I., *J. Chem. Eng. Japan*, 14, 173 (1981).
41. Ito, S., Ogawa, K., and Shiragami, N., *J. Chem. Eng. Japan*, 13, 1 (1980).
42. Ito, H., and Miyakawa, T., *Trans. J.S.M.E.*, 43, 4562 (1977).
43. Murdock, J. W., Foltz, C. J., and Gregory, C., *Trans. A.S.M.E.*, D86, 498 (1964).
44. Addison, H., *Engineering*, 145, 227 (1938).
45. Tsuji, S., and Kawashima, G., *Trans. J.S.M.E.*, 31, 1209 (1965).
46. Shiragami, N., and Inoue, I., *J. Chem. Eng. Japan*, 15, 394 (1982).
47. Shiragami, N., and Inoue, I., *J. Chem. Eng. Japan*, 16, 24 (1983).

CHAPTER 27

MODELING FLOWS IN RECTANGULAR ENCLOSURES AND CAVITIES

Asok K. Sen

Department of Mathematical Sciences
Purdue School of Science at Indianapolis
Indianapolis, IN, USA

and

Geo-Chem Research Associates, Inc.

Bloomington, IN, USA

CONTENTS

INTRODUCTION

Flows in rectangular enclosures and cavities are encountered in a wide variety of industrial and engineering applications. Because of their broad scope, such flows have been the subject of research for many decades. An attempt at a complete review of this subject is indeed a formidable task and is not intended here. Instead, a discussion of a few particular types of flows is presented reflecting the current interest of the author and their importance in technological processes.

The two major types of flows discussed here are buoyancy-driven flows and flows driven by surface tension gradients. The buoyancy and surface tension forces can arise from a number of causes. They can be induced, for example, by applying a temperature or a concentration gradient in the fluid. This review is restricted to flows driven by buoyancy and surface tension forces resulting solely from an imposed temperature gradient. Flows driven by concentration gradients also constitute an important class of problems. Unfortunately, a discussion of such flows is beyond the scope of this chapter.

There are many possible configurations in which the flows mentioned can be studied. Consider, for instance, a vertical rectangular enclosure with differentially heated sidewalls or heated from below, and an inclined rectangular cavity. Our discussion here will primarily deal with vertical rectangular enclosures with lateral heating although reference will be made of the other two configurations. First, we discuss buoyancy driven flows in tall and shallow rectangular enclosures. This is followed by a discussion of thermally induced surface tension flows in shallow cavities.

BUOYANCY-DRIVEN FLOWS IN ENCLOSURES

General Considerations

Buoyancy-driven flows, also known as free or natural convection flows are important in many diverse fields of application such as thermal insulation of buildings [1], solar energy storage [2], crystal growth [3], fluid filled cavities surrounding a nuclear reactor core [4], cooling of superconducting electric machinery [5], atmospheric dynamics [6], and pollutant disposal in estuaries [7]. Direct modeling of the physical processes in these systems is rather complex. As a result, natural convection flows in idealized configurations, for example, rectangular enclosures and horizontal circular cylinders have been examined by numerous researchers. An excellent review of the early work on natural convection in enclosures is given by Ostrach [8]. A review of the more recent literature has been presented by Rogers [9] and Catton [10].

A systematic study of natural convection in rectangular enclosures began with the pioneering work of Batchelor [1]. In order to estimate the insulation properties of gas-filled enclosures, he analyzed the air flow in a vertical rectangular cavity with isothermal sidewalls and perfectly conducting (or insulating) top and bottom surfaces. The configuration is shown in Figure 1. The sidewalls are of height d and distance ℓ apart; these are maintained at different temperatures T_H and T_C, with $T_H > T_C$. The temperature difference induces density variations in the fluid. The buoyancy forces resulting from these density variations will set the fluid in motion. Batchelor showed that convection will result no matter how small the temperature difference is. To describe the fluid motion and determine the rate of heat transfer through the sidewalls, he started with the conservation equations for mass, momentum, and energy of a viscous incompressible fluid subject to the so-called Boussinesq approximation. According to this approximation, the density variations can be ignored everywhere in the governing equations except in the term representing the buoyancy

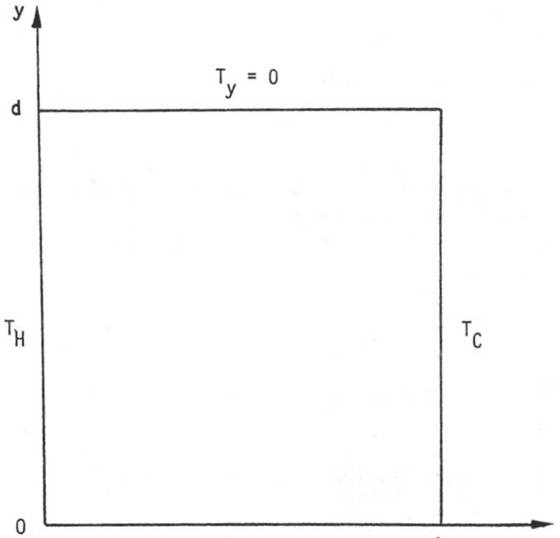

Figure 1. Schematic diagram of a rectangular enclosure.

force. Then, in the absence of viscous dissipation, the time dependent two-dimensional motion of the fluid is governed by the equations:

$$u_x + v_y = 0 \tag{1a}$$

$$u_t + uu_x + vu_y = \rho^{-1}p_x + \nu\nabla^2 u \tag{1b}$$

$$v_t + uv_x + vv_y = \rho^{-1}p_y + \nu\nabla^2 v + g\beta(T - T_C) \tag{1c}$$

$$T_t + uT_x + vT_y = \alpha\nabla^2 T \tag{1d}$$

and the boundary conditions

$$x = 0: \quad u = v = 0, \qquad T = T_H \tag{2a}$$

$$x = \ell: \quad u = v = 0, \qquad T = T_C \tag{2b}$$

$$y = 0, d: \quad u = v = 0, \qquad T_y = 0 \tag{2c}$$

Here u and v are the x- and y- components of velocity; p is the pressure and g is the acceleration due to gravity. The variables T and t represent the temperature and time. The quantities ρ, ν, α, β denote the density, kinematic viscosity, thermal diffusivity, and coefficient of thermal expansion of the fluid, respectively. The subscripts x, y, and t have been used to denote partial derivatives with respect to the corresponding variable. The Laplacian operator is defined by

$$\nabla^2 = \partial^2/\partial x^2 + \partial^2/\partial y^2 \tag{3}$$

Note that the thermal boundary conditions in Equation 2c apply for an enclosure whose top and bottom surfaces are insulated. The velocity field (u, v) is assumed to satisfy no-slip conditions on all four boundaries of the enclosure.

The problem can be made dimensionless as follows. Let

$$x = dx', \qquad y = dy', \qquad u = u_* u', \qquad v = u_* v',$$

$$t = (d/u_*)t', \qquad T = T_C + (T_H - T_C)T' \tag{4}$$

where $u_* = \alpha/d$ is a characteristic velocity, and introduce a streamfunction ψ such that

$$u' = \psi_{y'}, \qquad v' = -\psi_{x'} \tag{5}$$

By eliminating the pressure from the momentum Equations 1b, c and using the scalings of Equation 4, the Equations 1a-d can be written in the following dimensionless form:

$$Pr^{-1}[\omega_t + J(\omega, \psi)] = RaT_x + \nabla^2\omega \tag{6a}$$

$$T_t + J(T, \psi) = \nabla^2 T \tag{6b}$$

where

$$\omega = -\nabla^2\psi \tag{7}$$

is the dimensionless vorticity. In writing these equations, the primes on all variables have been dropped and the notation

$$J(F, G) = F_x G_y - F_y G_x \tag{8}$$

has been used. The Prandtl and Rayleigh numbers have the definitions

$$\mathrm{Pr} = \nu/\alpha, \qquad \mathrm{Ra} = (T_H - T_C) g\beta d^3/\nu\alpha \tag{9}$$

The Prandtl number is a measure of the relative rates of diffusion of vorticity and heat, and the Rayleigh number characterizes the relative importance of the buoyancy and the diffusive (or viscous) forces. Many authors use a different dimensionless parameter called the Grashof number, in place of the Rayleigh number. The Grashof number is defined by

$$\mathrm{Gr} = (T_H - T_C) g\beta d^3/\nu^2 \tag{10}$$

Clearly, the Rayleigh number is the product of the Prandtl and Grashof numbers. The boundary conditions of Equation 2b transform into

$$x = 0: \quad \psi = \psi_x = 0, \qquad T = 1; \tag{11a}$$

$$x = A^{-1}: \quad \psi = \psi_x = 0, \qquad T = 0; \tag{11b}$$

$$y = 0, 1: \quad \psi = \psi_y = 0, \qquad T_y = 0; \tag{11c}$$

Here A is the aspect ratio of the enclosure; $A = d/\ell$.

An estimate of the insulation property of the enclosure may be given in terms of the so called Nusselt number which measures the rate of heat transfer through the sidewall, say, at $x = 0$. The Nusselt number is defined as

$$\mathrm{Nu} = \int_0^1 T_x \Big|_{x=0} dy \tag{12}$$

and may, in general, depend on the parameters A, Pr, and Ra.

Enclosures with Large Aspect Ratio

Batchelor [1] analyzed the problem just described to determine the heat transfer rate through air gaps in buildings. He was primarily interested in tall enclosures with aspect ratios between 5 and 200 and examined three asymptotic limits of the parameters A, Pr, and Ra. For $\mathrm{Ra} \ll 1$ and arbitrary values of A and Pr, he found that the flow is conduction dominated, with convection playing only a subsidiary role. He also examined enclosures with large aspect ratio ($A \gg 1$) for arbitrary Rayleigh and Prandtl numbers. Here again, the primary mode of heat transfer is conduction. In the third limit, characterized by large Rayleigh numbers, the solutions were found to exhibit a boundary layer behavior. The flow pattern, in this limit, consists of an isothermal core of constant vorticity surrounded by boundary layers near the sidewalls. This is known as the boundary layer regime. Here, heat is transferred mainly by horizontal convection across the sidewalls. The flow near the sidewalls is quite complex and for this reason a complete analytical description of the boundary layer regime is lacking even today. Although Batchelor was unable to solve the boundary layer flows in detail, he derived two alternative expressions for the Nusselt number for this boundary layer regime in the form

$$\mathrm{Nu} = \delta \mathrm{Gr}^{0.25} A^{-0.75} \tag{13}$$

with $\delta = 0.38$ and $\delta = 0.48$. It turns out, however, that the value of Nu given by Equation 13 with $\delta = 0.3$ matches adequately with the experimental observations of Mull and Reiher [11]. The Nusselt number based on Jakob's correlation [12] of Mull and Reiher's experimental data is given by the expression

$$\mathrm{Nu} = 0.18 \mathrm{Gr}^{1/4} A^{1/9} \tag{14}$$

for $2 \times 10^4 \le Gr \le 2 \times 10^5$. Batchelor's analysis showed that a boundary layer regime would appear at $Ra = 1.37 \times 10^4$ for aspect ratios greater than 42 and at $Ra = 10^9/A^3$ for smaller values of A. Unfortunately, the boundary layer solution obtained by Batchelor was inconsistent with his core solution.

Poots [13] obtained a numerical solution for the model considered by Batchelor by expanding the temperature and the streamfunction in the form of two doubly infinite series of orthogonal polynomials. He presented results for a square cavity with a prescribed linear variation of temperature along the top and bottom surfaces for the range $500 < Ra < 10^3$. The numerical solution was in agreement with Batchelor's hypothesis of an isothermal core, although such a flow has never been experimentally observed.

The first set of careful experiments on natural convection in rectangular enclosures was done by Eckert and Carlson [14]. Using a Mach-Zehnder interferometer, they measured the temperature in an air layer between two vertical isothermal plates, for aspect ratios between 2.1 and 46.7. By increasing the Rayleigh number to about 2×10^5, they observed three different flow regimes: a conduction regime with linear temperature gradients for low Rayleigh numbers, a boundary layer regime with significant convection for high Rayleigh numbers and a transition regime for intermediate values of Ra. The transition regime is characterized by nonlinear temperature gradients in the core with weak convective effects. In the boundary layer regime, the core temperature is horizontally uniform, but increases almost linearly along the vertical. According to their measurements, the vertical temperature gradient in the core is inversely proportional to the aspect ratio. Clearly, these observations are in contradiction with Batchelor's model of an isothermal core. Nevertheless, there seems no reason to discard Batchelor's hypothesis on purely theoretical grounds.

Based on their experiments, Eckert and Carlson [14] arrived at the following empirical relation for the Nusselt number:

$$Nu = 0.119(Gr)^{0.3} A^{-0.1} \tag{15}$$

It should be mentioned that no velocity measurements were made by Eckert and Carlson.

A few years later, Elder [15] measured the temperature and velocity fields in a vertical rectangular slot that was heated from one side and cooled at the other. The slot was filled with a high Prandtl number fluid such as silicone oil or paraffin. In his experiments, Elder varied the Rayleigh number up to 10^8, the aspect ratio from 1 to 60 and used a value of $Pr = 1{,}000$. His principal findings can be summarized as follows. Below $Ra = 10^3$, the flow is dominated by conduction; but a weak unicellular motion is set up with the fluid rising near the hot sidewall and falling near the cold. For $10^3 < Ra < 10^5$, boundary layers develop near the sidewalls and the core flow becomes stably stratified. Near $Ra = 10^5$, secondary flows appear in the core region leading to a multicellular structure with streamlines resembling a "cats-eye" pattern. On increasing the Rayleigh number further, the flow becomes extremely complex. For Ra near 10^6, tertiary flows appear in the slot, although such flows are sometimes difficult to observe without careful experimentation. Similar observations have also been reported by Oshima [16].

From his measurements, Elder was able to obtain excellent results for the velocity and temperature profiles in the various flow regimes but he made no attempt to correlate the Nusselt number with the Rayleigh number, aspect ratio or other parameters affecting the flow. A particularly interesting interpretation of Elder's experimental results has been given by Ayyaswamy [17], which has also been reproduced in the review by Catton [10].

Like Eckert and Carlson, Elder found that in the boundary layer regime the centerline vertical temperature gradient in the core varies inversely as the aspect ratio. The proportionality constant has been estimated to be 0.60 by these authors. This was later confirmed by Grondin [18] via numerical computation.

In the same paper [15], Elder proposed a simple theory to explain some of his observations. However, this simple theory is incomplete in the sense that it only describes the flow in the stratified core interior to the boundary layers and not the form of the boundary layers themselves. The inadequacies of this theory have been largely removed by Gill [19] who confirmed the interior solutions of Elder and also presented a detailed analysis of the boundary layer flows.

Gill [19] developed a boundary layer theory by assuming that the volume flux in the sidewall boundary layers is large, as observed in Elder's experiments. This assumption is directly opposite to that of Batchelor [1] who postulated that the volume flux in the core is much larger. Using the fact that these boundary layers are much thinner than the slot height, Gill derived a set of boundary layer equations from the full system of conservation Equations 6. He then obtained an approximate solution of these equations in the large Prandtl number limit, using a modified Oseen approach. Unfortunately, this solution cannot simultaneously satisfy all the boundary conditions. More will be said about Gill's solution later. The main significance of this work is that it provided the first analytical description of the boundary layer regime without an isothermal core.

Motivated by the experiments of Elder, several numerical studies were undertaken, including one by Elder [20] himself. The purpose of these studies was to delineate the experimentally observed flow regimes by numerical integration of the governing differential equations. The first successful attempt at solving the two-dimensional time dependent Boussinesq equations is due to Wilkes and Churchill [21]. Using an implicit alternating direction finite difference approach, they obtained transient and steady state solutions for aspect ratios 1, 2, and 3 and Rayleigh numbers on the order of 10^5. The flow structure computed by these authors was consistent with the experimental results of Eckert and Carlson [14] and Elder [15], but the Nusselt numbers reported were up to 70 percent higher than those given by Jakob [12]. Moreover, the computations of Wilkes and Churchill could not be extended to high enough Rayleigh numbers, since at large values of Ra, their numerical scheme developed instabilities.

In his numerical study, Elder [20] used a finite difference algorithm based on an iterative method of solution of Poisson's equation. The numerical solution clearly demonstrates the gradual development of the boundary layer regime from the conduction regime as the Rayleigh number increases. In the boundary layer regime, the core flow is characterized by a strong stratification. On the basis of these computations, the Nusselt number can be correlated by the expression

$$\mathrm{Nu} = 0.231\ \mathrm{Gr}^{0.25} \tag{16}$$

for a square enclosure with Pr = 0.71 (the value for air). A similar study has been carried out by Han [22] who reports

$$\mathrm{Nu} = 0.782\ \mathrm{Gr}^{0.3594} \tag{17}$$

in place of Equation 16. Elder also claims to have numerically computed the secondary flows of the type observed in his experiments. Among others, Rubel and Landis [23] and Polezhaev [24] have found such secondary flows from their numerical simulations.

MacGregor and Emery [25, 26] made an experimental and numerical investigation with fluids of moderate and large Prandtl number. For numerical computations, they used an explicit finite difference method in conjunction with a Gauss Seidel scheme. The effects of Grashof number, Prandtl number, and aspect ratio on the flow structure are studied in detail both numerically and experimentally. As found in previous studies, the effect of increasing the Grashof number is to change the flow pattern from a conduction regime to a boundary layer regime. An increase in aspect ratio produces a temperature field that becomes progressively linear indicating the increased importance of conduction. Furthermore, with large aspect ratios, there is a significant decrease in the mean Nusselt number. For the boundary layer regime, the numerical calculations show that

$$\mathrm{Nu} = 0.38(\mathrm{Ra})^{0.35}\ \mathrm{A}^{-0.25} \tag{18}$$

when $\mathrm{Pr} > 1$ and $\mathrm{Ra} > 10^4$. The correlation based on their experimental data yields

$$\mathrm{Nu} = 0.42(\mathrm{Ra})^{0.25}\ \mathrm{Pr}^{0.012}\ \mathrm{A}^{-0.3} \tag{19}$$

under the same conditions. The influence of the Prandtl number (exclusive of its appearance in the Rayleigh number) is found to be minor when $\mathrm{Pr} > 1$. For $\mathrm{Pr} < 1$, however, the velocity field and heat transfer rates have a strong dependence on Pr. Similar results were obtained by Elder [20] for $\mathrm{Pr} > 1$, but for $\mathrm{Pr} < 1$, Elder's numerical scheme became unstable.

Newell and Schmidt [27] used a modified Crank-Nicholson method to integrate the time dependent equations. They obtained steady state solutions for aspect ratios 1, 2, 5, 10, and 20 in the range $4 \times 10^3 < Gr < 1.4 \times 10^5$. Their numerical results give the following expressions for the Nusselt number.

$$Nu = 0.547(Gr)^{0.397} \tag{20}$$

for a square enclosure and

$$Nu = 0.155(Gr)^{0.315} A^{-0.265} \tag{21}$$

for $2.5 \le A \le 20$. The main conclusion of this study is that two different regimes of heat transfer may exist depending on the aspect ratio. For $A > 2$, Nu decreases as A is increased. This is because most of the heat transfer into the fluid takes place near the bottom of the hot wall and so adding more height to the enclosure has little effect on increasing the total heat transfer. When $A < 2$, an increase in height will result in a more effective flow pattern to heat transfer and therefore the mean Nusselt number will increase. On the contrary, the numerical calculations of Thomas and De Vahl Davis [28] suggest that the heat transfer rate in the boundary layer regime is independent of the aspect ratio. A comparison of the finite difference methods used for the computation of natural convection flows has been given by Torrance [29].

The Nusselt number correlations given by the various investigators can be written in the form

$$Nu = a(Ra)^b A^c \tag{22}$$

these have been tabulated by Ostrach [8]. However, the controversy regarding the exact values of the constants a, b, and c still remains unresolved.

Quon [30] performed finite difference calculations for the boundary layer regime in a square cavity with a variety of dynamical boundary conditions. He demonstrated that the type of boundary conditions on the top and bottom surfaces has little effect on the interior or boundary layer solutions. Introduction of free vertical boundaries, on the other hand, may increase the vertical velocities in the boundary layers by a factor of two. In keeping with the results of MacGregor and Emery [25] and Elder [20], Quon showed that the flow in the boundary layer regime is insensitive to the magnitude of the Prandtl number for $7.14 < Pr < 900$.

It has been pointed out by Quon [30] that the vertical velocity field calculated by Gill [19] does not agree with Elder's [15] experiments as well as it appears. To be sure, the maximum vertical velocity predicted by Gill's solution is 25% too high. Gill's solution also underestimates the core temperature gradient observed by Elder. As mentioned earlier, the boundary layer solution of Gill cannot satisfy the conditions on all four boundaries. Gill expressed his solutions for the velocity and temperature fields in terms of a free unknown constant, which is determined by the use of appropriate boundary conditions. However, as Quon suggests, care should be taken in deciding which boundary conditions to use. Gill used the impermeability conditions, namely $\psi = 0$ at $y = 0, 1$ to determine the constant and thus obtained a self-contained solution without any reliance on experimental results. Unfortunately, this solution predicts infinite horizontal velocities and infinite vertical temperature gradients along the horizontal boundaries. Quon [31] improved on Gill's theory by determining this free constant via numerical computation and showed that this 'new' solution agreed more closely with experimental observations.

Blythe and Simpkins [32] reconsidered Gill's problem and analyzed the flow in the large Rayleigh number, large Prandtl number limit. Using integral techniques, they deduced analytical solutions for the velocity and temperature profiles. Other attempts to improve on Gill's solution are due to Bejan [33] and Graebel [34].

Bejan [33] argued that the net vertical energy flux should be zero near the top and bottom surfaces of the enclosure. Using this fact, he evaluated the free constant in Gill's boundary layer solution. The overall Nusselt number derived from Bejan's solution is given by the relation

$$Nu = 0.364(Ra/A)^{0.25} \tag{23}$$

as $RA^{1/7}A \rightarrow \infty$. This result is found to agree well with the experimental data of Eckert and Carlson [14], MacGregor and Emery [25], Seki et al. [35] and Yin et al. [36], and with the numerical simulations of de Vahl Davis [37], Newell and Schmidt [27] and Pepper and Harris [38], among others. Bejan also showed that the constants a, b, and c in the Nusselt number relationship (Equation 22) may depend on the Rayleigh number and the aspect ratio and presented a theoretical explanation for the disagreement regarding the dependence of the Nusselt number on these parameters. Using the method of Bejan, Graebel [34] generalized Gill's analytical solution for arbitrary Prandtl numbers so as to include experiments on liquid metals (which have low values of Pr on the order of 10^{-2}.)

In an attempt to study the effect of sidewall boundary conditions, Catton, Ayyaswamy, and Clever [39] performed numerical calculations in a vertical rectangular slot of aspect ratio unity and smaller, for $Ra = 3 \times 10^5$. Their computations reveal that the rate of heat transfer is lower for perfectly conducting sidewalls than adiabatic sidewalls. The difference becomes significant at lower aspect ratios. This is attributed to the more pronounced thermal interaction at the perfectly conducting walls. The heat transfer results obtained by Catton et al. agreed very well with the correlations of Berkovsky and Polevikov [40] who report that

$$Nu = 0.22A^{-1/4}\left\{\frac{Ra\ Pr}{0.2 + Pr}\right\}^{0.28} \tag{24a}$$

for $2 < A < 10$, $10^{-3} < Pr < 10^5$, $Ra < 10^{10}$ and

$$Nu = 0.18\left\{\frac{Ra\ Pr}{0.2 + Pr}\right\}^{0.29} \tag{24b}$$

for $1 < A < 2$, $10^{-3} < Pr < 10^5$, $10^3 < Ra\ Pr/(0.2 + Pr)$

These correlations are among the very few available that incorporate the effect of Prandtl number explicitly, especially when Pr is small. We note that the work of Catton et al. [39] is perhaps the first published work on rectangular cavities with aspect ratios smaller than one.

Over the past ten years, many advanced numerical techniques have been employed for solving the Boussinesq equations of natural convection in rectangular enclosures. Denny and Clever [41] used both ADI finite difference and Galerkin methods in a square cavity with perfectly conducting sidewalls. These authors stipulate that the Galerkin method may be superior to the finite difference methods for two-dimensional problems. Chu and Churchill [42] presented a systematic approach for the development of finite difference schemes, emphasizing the effect of grid size on the accuracy of the solution. Among others, Roux et al. [43] used a fourth-order Hermitian algorithm to transform the finite difference equations into a block tridiagonal form, to expedite convergence. Küblbeck et al. [44] utilized the concept of coordinate stretching by the use of nonuniform grids and obtained higher accuracy. In their computation, they also allowed one sidewall to be nonuniformly heated with the other held adiabatic. This situation may be considered a crude model for a room heated by a radiator on one side and cooled by a window on the other. The numerical simulation also describes the temporal evolution of the flow in detail. Lauriat [45] implemented four different higher order schemes for a cavity of aspect ratio 10. It should be pointed out that although these numerical calculations lead to better accuracy, they add little physical insight about the flow behavior.

The feasibility of using the finite element method to analyze natural convection flows in rectangular enclosures has been studied by several investigators [46–52]. Among them, Tobarrok and Lin [46] used the stream function-vorticity-temperature formulation while Gartling [47] adopted the pressure-velocity-temperature model. Heinrich et al. [48, 49] and Reddy and Mamidi [50] employed the so-called penalty function approach. Space limitation precludes a discussion of these methods here. The interested reader should refer to the work of Reddy and Satake [52] for further details. Suffice it to say that the results obtained by the finite element methods agree favorably with those of finite difference computations and experimental observations.

In a recent study, Schinkel et al. [53] reexamined natural convection in vertical enclosures with particular attention to the stratification in the core region. They tried to define precisely the emergence of the boundary layer regime from the conduction regime. Eckert and Carlson [14] characterize the starting of the boundary layer regime by the condition that the horizontal temperature gradient at the center of the enclosure becomes zero. Gill [19] on the other hand, uses the condition that the shear stress at the center vanishes. The numerical calculations of Schinkel et al with perfectly conducting sidewalls indicate that zero shear stress occurs at $Ra/A = 3 \times 10^4$ whereas the horizontal temperature gradient at the center vanishes when $Ra/A = 3.3 \times 10^3$ for $2 \leq A \leq 18$. Grondin [18] has shown that, with adiabatic sidewalls, zero horizontal temperature gradient occurs at $Ra/A = 2.5 \times 10^3$. The results obtained by Gilly et al. [54] support the fact that for a given aspect ratio, the boundary layer regime starts at a higher value of Ra with perfectly conducting than adiabatic sidewalls. This is due to heating of the fluid as it moves from the cold wall toward the hot wall in the lower parts of the sidewall region when the sidewalls are perfectly conducting.

In the boundary layer regime, Schinkel et al. [53] show that secondary flows will occur near the sidewalls at $Ra/A \geq 2.5 \times 10^5$ for $A \geq 2$, with perfectly conducting sidewalls. The secondary motion is the result of stratification in the sidewall regions and has a large influence on the stratification in the center of the enclosure. They find that the vertical temperature gradient (τ) at the center is, in general, not inversely proportional to the aspect ratio, in contradiction to the predictions of Eckert and Carlson [14] and Elder [15]. For $A > 4$, τA remains constant up to $Ra/A = 1.4 \times 10^4$, the constant value being 0.56, which is somewhat smaller than Grondin's [18] result for adiabatic sidewalls ($\tau A = 0.60$). When $Ra/A > 1.4 \times 10^4$, τA increases with Ra/A. When $A \leq 3$, A becomes constant again for large values of Ra/A; this constant increases with increasing aspect ratio. Secondary flows near the center are found for $Ra/A > 7 \times 10^4$ while at higher values of Ra/A tertiary motion has also been detected. These results are in good agreement with the flow visualization experiments of Linthorst and Schinkel [55].

Schinkel et al. [53] also made a careful study of the dependence of the overall Nusselt number on the Rayleigh number and the aspect ratio, for the case of perfectly conducting sidewalls. This is illustrated in the following table from their original paper [53]:

A	a	b
2	0.176	0.26
3	0.193	0.25
4	0.190	0.25
6	0.180	0.25
8	0.173	0.25
11	0.162	0.25
18	0.148	0.25

These results are valid in the range $10^4 \leq Ra \leq 10^6$ and can be approximated by the relation

$$Nu = 0.263(Ra)^{0.265} A^{-0.2} \tag{25}$$

for $8 \leq A \leq 18$.

From their numerical computations, Schinkel et al. [53] plotted the variation in the centerline vertical velocity with horizontal position. This is shown in Figure 2 where the results are also compared with Elder's [15] analytical solution. The figure is drawn for a cavity with perfectly conducting sidewalls. Here $A = 8$ and $Ra = 1.1 \times 10^5$. The effect of stratification can be represented in terms of a parameter $\gamma = (\tau Ra/4)^{0.25}$. For the values of Ra and A under consideration, $\gamma = 6.77$. We see from the figure that, for this value of γ, Elder's solution agrees well with that of Schinkel et al. only near the center of the cavity ($x = 0.5$). In the same figure, another velocity profile corresponding to $\gamma = 6.35$ is shown. This value of γ is chosen so as to match the maximum velocities of Elder and Schinkel et al. Now the agreement in the boundary layers is much better but near the center the agreement between the two solutions is not so good.

Korpela et al. [56] analyzed the convective flow of air through a double pane window and studied more fully the effect of aspect ratio. They examined aspect ratios between 10 and 20, using a value of $Pr = 0.71$ and $Gr = 6.4 \times 10^7$. At $A = 20$, the flow is found to be unicellular and conduction dominated. As A decreases, the flow becomes multicellular and then returns to a unicellular pattern near $A = 10$. Very little background stratification exists in the multicellular regime, for the value of

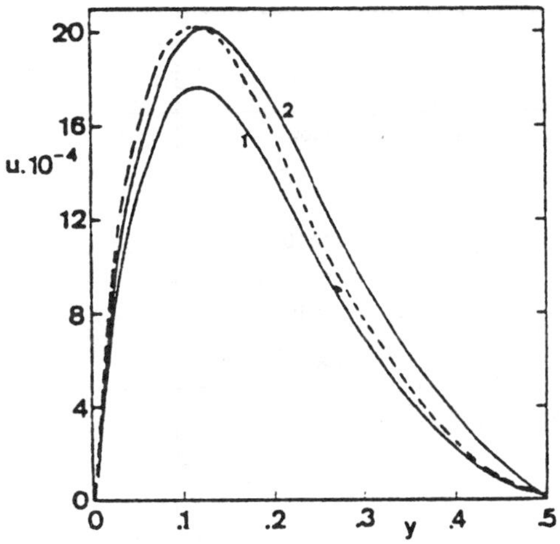

Figure 2. Variation of the vertical velocity at midheight with horizontal position. The dashed curve is from the calculations of Schinkel *et al.* [53]. The solid curves 1 and 2 correspond to Elder's [15] theory with $\gamma = 6.77$ and $\gamma = 6.35$, respectively [53].

Pr used. This is in contrast with the situation found in large Prandtl number flows (see, for example, de Vahl Davis and Mallinson [57] and Seki et al. [35]) where a stable vertical temperature gradient persists and distortion of the isotherms by the cellular flow structure perturbs this only mildly.

The main findings of Korpela et al. [56] are shown in Figure 3. Here, the overall Nusselt number is plotted against 1/A, with the Grashof number as a parameter. From the figure, one can easily see how the rate of heat transfer can be changed by varying the gap width, for a fixed height of the enclosure. For comparison, the correlations of Thomas and de Vahl Davis [28] and Elsherbiny et al. [58] are also drawn in the figure. Clearly, all these results are in close agreement with each other. It should be pointed out that the experimental data reported by Elsherbiny et al. [58] are perhaps the most accurate available today. An important feature of this work is that cavities with aspect ratios as high as 110 have been examined here experimentally. The figure also shows that the multicellular flow does not alter the heat transfer significantly. The reason for this is that most of the energy is transferred across the ends. Korpela et al. conjecture that only for cavities with aspect ratio on the order of 40, the multicellular convection would be significant. They also present a working formula to determine the gap width, which leads to an optimum insulating capacity. This corresponds to the situation when the flow is on the verge of undergoing a transition to multicellular structure. The "optimum" gap width in terms of the aspect ratio is given by

$$A^3 + 5A^2 = 1.25 \times 10^{-4}\,\mathrm{Gr} \tag{26}$$

For the value of $\mathrm{Gr} = 6.4 \times 10^7$ considered by Korpela et al., this last formula gives $A = 18.5$, approximately. The structure of the multicellular flows has been investigated in detail in a more recent paper by Lee and Korpela [59]. Their work will be discussed later in this section.

Apart from providing the motivation for many of the numerical studies discussed above, Elder's [15] experiments raised the question of stability of natural convection flows. It is now known that the multicellular secondary flows observed by Elder and others are the result of hydrodynamic instabilities. Gershuni [60] was the first to analyze the stability of these flows in the conduction regime. His analysis was limited to small values of Prandtl number. Rudakov [61] extended Gershuni's results to Prandtl numbers up to ten and found that the conduction dominated flow becomes unstable when the Grashof number reaches the value 7.7×10^3. This critical value of the Grashof number has been confirmed experimentally by Vest and Arpaci [62] who also studied the instabilities of the flow in the boundary layer regime. The stability of the conduction regime has also been examined by Birikh et al. [63], Korpela et al. [64] and Ruth [65]. Korpela et al. [64] report that when the Grashof number reaches the critical value, the conduction regime may become unstable in two ways. For $\mathrm{Pr} < 12.7$, the instability sets in as horizontal stationary cells, with the critical

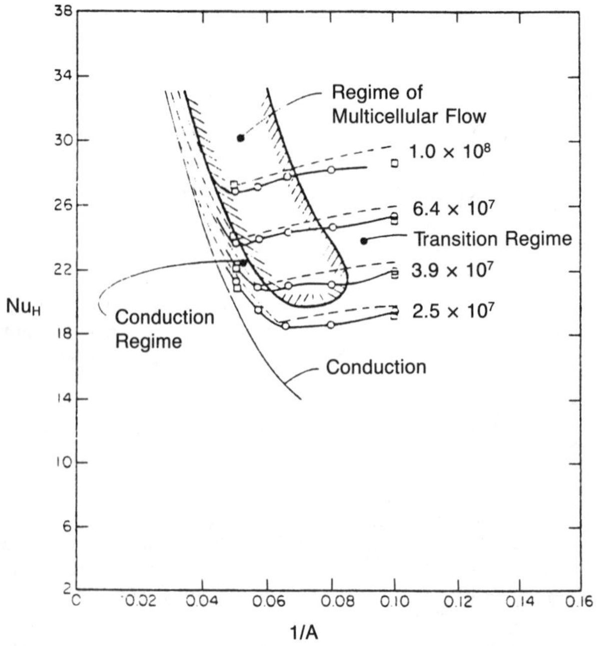

Figure 3. Average Nusselt number as a function of inverse aspect ratio for air [56]. The four curves correspond to four different values of Grashof number as shown in the figure: —○— [56], -----[28], —□—[58].

Grashof number nearly independent of Pr. When Pr > 12.7, the instability is manifested in the form of traveling waves.

In all of the mentioned stability analyses, the coefficient of thermal expansion of the fluid is assumed constant in the temperature range of interest. Very recently Shaaban and Ozisik [66] examined the effect of nonlinear density stratification on the stability of the conduction regime. A case in point is the flow of water in a temperature range that includes 4°C. Their numerical calculations reveal that the critical states of stability is independent of Prandtl number but depends on the combinations of the temperature of the hot and cold sidewalls. The instabilities lead to traveling waves moving against gravity. Stationary waves may also be found under certain conditions.

The first successful attempt for analyzing the stability of the boundary layer regime is due to Vest and Arpaci [62]. They considered a base flow that is essentially the same as Gill's [19] boundary layer solution. On the basis of a linear stability analysis, Vest and Arpaci predicted the Rayleigh number at which the boundary layer flow becomes unstable. They claim that multicellular secondary flow observed experimentally by Elder and themselves result from an instability of the basic flow. According to Vest and Arpaci [62], the flow undergoes transition to multicellular pattern at Gr = 8,000. This estimate has been verified more recently by Bergholz [67] who also contends that for fluids of low Pr, traveling wave solutions arise from the instability of the boundary layer regime whereas the instability sets in through stationary states for high Prandtl number fluids. Recently, Schinkel [68] observed these traveling waves in air with cavities of aspect ratios between 5 and 9. Elder [15] observed multicellular flows for an oil with Pr = 1,000 in a cavity of aspect ratio ranging from 12 to 60. We now turn to a discussion of the multicellular flows.

As mentioned before, multicellular natural convection in rectangular cavities arises from hydrodynamic instabilities. These flows were first calculated numerically by de Vahl Davis and Mallinson [57]. Their results agree quite well with Elder's [15] visual observations of secondary and tertiary cells. Computations by Pepper and Harris [38] for a high Prandtl number fluid in a cavity of aspect ratio 10 exhibit a three-cell structure in the center of the cavity. For air, Grondin and Roux [69] have demonstrated the existence of multicellular flows in a cavity of aspect ratio 16. These numerical calculations have been extended by Raithby and Wong [70] to aspect ratios as high as 80. Very recently Lee and Korpela [59] carried out extensive numerical computations of multicellular flows in cavities of aspect ratios up to 40 for a wide range of Prandtl numbers. They find that for low Prandtl number fluids (Pr $\ll$ 1) such as liquid metals, multicellular flows can develop when the aspect ratio is as low as six. For air (Pr = 0.71), the aspect ratio must be at least 12 before the flow becomes more complex as a result of instability. With A = 20, Lee and Korpela [59] found that transition to multicellular flows in air takes place for Gr in the interval 10^4 and 1.1×10^4. This range of Gr for transition appears to be in good agreement with the experiments of Hollands and Konicek [71] who observed instability to set in at Gr = 11,000 ± 500. By increasing the aspect ratio to 40, Lee and Korpela did not find a significant change in the qualitative flow behavior. The vertically averaged Nusselt number estimated by these authors are sketched in Figure 4 for aspect ratios from 5 to 40. The numerical results of Raithby and Wong [70] and the experimental correlations of Elsherbiny et al. [58]

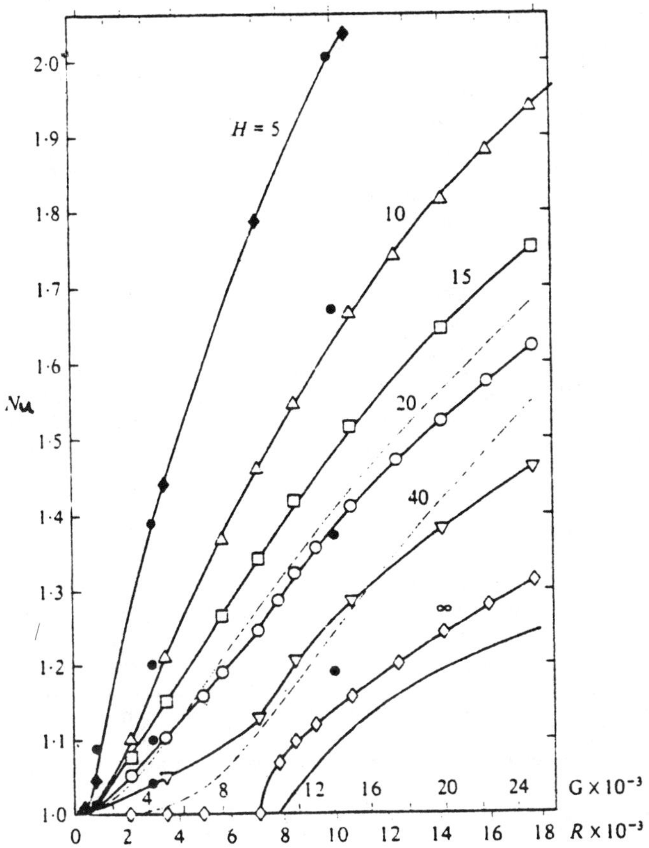

Figure 4. Average Nusselt number as a function of Grashof number for air flow in cavities of various aspect ratios [59]. The solid circles are calculations of Raithby and Wong [70]; the dashed lines are from the experiments of Elsherbiny *et al.* [58]; all other curves are from Lee and Korpela [59] except the lowest curve, which is due to Hollands and Konicek [71].

are also shown in the figure. The lowest curve in the figure is due to Hollands and Konicek [71] for a cavity with $A = 44$.

For high Prandtl number fluids, Lee and Korpela, like many others, find that transition to multicellular flow occurs at much smaller values of Gr for cavities with moderate aspect ratios. They examined a fluid with $Pr = 1{,}000$ in a cavity of aspect ratio 15 where weak cellular motion began near $Gr = 400$. Bergholz [67] predicts a value of $Gr = 240$ under similar conditions. For comparison, we note that in Elder's [15] experiments transition to multicellular flow occurs at $Gr = 330 \pm 10\%$ with $A = 19$ and it happens at $Gr = 411 \pm 10\%$ with $A = 20$ in the experiments of Vest and Arpaci [62].

Our discussion thus far has been confined to the study of two-dimensional motion in vertical rectangular cavities with differentially heated sidewalls. Three-dimensional calculations of natural convection in a box have also been carried out, beginning with the work of Aziz and Hellums [72]. Other attempts in this area of research include those of Mallinson and de Vahl Davis [73, 74] and Oze et al. [75] who computed the flow structure numerically using finite difference methods. These numerical computations indicate that the flow is truly three-dimensional although the heat transfer rate has been experimentally [76] found to be independent of cavity height, provided the height is not too small. The three dimensionality results presumably from the fact that the rotating stratified core induces an Ekman layer flow near the sidewalls of a cubical box.

Another configuration that has received a great deal of attention for many years is that of a rectangular cavity heated from below. If the imposed vertical temperature gradient is large enough, convection will occur. This process is usually known as Bénard convection. The literature on Bénard convection is very extensive and is beyond the scope of this review.

In recent years, there has been considerable interest in the study of natural convection in inclined rectangular cavities because of their importance in solar collector design. The interested reader should refer to the works of Elsherbiny et al. [77], Meyer et al. [78], and Linthorst and Schınkel [55] for more details.

In all of the previous discussion, the density variations in the fluid are assumed to be only due to an imposed temperature gradient. Density variations can be caused by yet another mechanism. If the fluid contains a solute substance, a variation in the solute concentration may lead to a density gradient even under isothermal conditions. Flows driven by buoyancy forces resulting from the combined effect of temperature and concentration gradients are commonly known as double diffusive convection. A study of these phenomena is essential to the understanding of the layered structures observed in the ocean and energy storage systems. Double diffusive convection is presently an active area of research. Some of the interesting problems in this field have been described by Rogers [9] and Turner [79].

Natural Convection in Shallow Cavities

To understand the transport phenomena in estuaries, Cormack et al. [80–82] and Imberger [83] made a number of investigations on buoyancy-driven flows in a shallow rectangular cavity. Natural convection flows in shallow cavities that are important in the study of crystal growth processes have been reviewed recently by Ostrach [84] and will be discussed later. In the first of a series of papers, Cormack et al. [80] examined the flow in a two-dimensional slot with differentially heated sidewalls, in the limit that the aspect ratio of the slot goes to zero. The configuration of Figure 1 still applies with $A = d/\ell \ll 1$. They nondimensionalize the equations of motion (Equations 1a–d) using the same variables as in Equation 4 except now the characteristic velocity is chosen to be

$$u_* = g\beta d^3(T_H - T_C)/\nu\ell \tag{27}$$

Then with the streamfunction ψ given by Equation 5, the dimensionless equations can be written in the form

$$Gr\, A^2[\omega_t + J(\omega, \psi)] = T_x + A\nabla^2\omega \tag{28a}$$

$$Gr\, Pr\, A[T_t + J(T, \psi)] = \nabla^2 T \tag{28b}$$

The Jacobian function J used in these equations is defined in Equation 8. The variables T and ψ must also satisfy the boundary conditions (Equation 11) for a cavity with all solid boundaries whose sidewalls are maintained at different temperatures T_H and T_C and the horizontal surfaces are insulated.

Cormack et al. [80] analyze the asymptotic problem for steady motion in which $A \to 0$ with Gr and Pr held fixed. In this limit the flow structure in the cavity consists of a parallel flow region in the center of the cavity, which is joined by a boundary layer region near each sidewall where the flow turns around and recirculates. The temperature field in the cavity is conduction dominated. However, the dominance of conduction in the present case, as Cormack et al. [80] points out, is quite different from that considered by Batchelor [1] and Gill [19] for A fixed (and finite) with $Ra \to \infty$. In the latter case, the flow is dominated by boundary layers at the sidewalls where almost all of the temperature drop takes place and the interior core flow is driven primarily by the entrainment-detrainment process associated with these boundary layers. The overall Nusselt number in this case is proportional to Ra^n ($n > 0$). On the other hand, the conduction dominated regime for $A \to 0$ results from the cumulative contribution of locally small viscous effects acting over a large distance. Here, the major part of the temperature drop occurs across the core and the Nusselt number becomes independent of Ra even when the Rayleigh number is large.

The velocity and temperature fields in the core region, as deduced by Cormack et al. [80] are relatively simple. Their core streamfunction may be written as

$$\psi = -K_1 y^2 (y - 1)^2 / 24 \tag{29}$$

The corresponding temperature distribution is given by

$$T = K_1(1 - Ax) + \frac{K_1^2}{720} \mathrm{Gr}\,\mathrm{Pr}\,A^2 y^3 (6y^2 - 15y + 10) + K_2 \tag{30}$$

with

$$K_1 = c_1 + c_2 A + c_3 A^2 + \ldots, \qquad K_2 = c_1^* + c_2^* A + c_3^* A^2 + \ldots \tag{31}$$

Clearly the result (Equation 29) shows that the flow in the core is horizontal to all orders in A and the temperature there is independent of height, to leading order. The coefficients c_i and c_i^* are determined by invoking the centro-symmetry of the problem and then matching the core solutions (Equations 29 and 30) with the appropriate solutions in the boundary layers near the sidewalls. Cormack et al. [80] were able to do this successfully using the method of matched asymptotic expansions. They also derived the detailed structure of the boundary layer flows, correct to $0(A^3)$. The requirement that the solutions for velocity and temperature are symmetric about the center of the cavity yields the relation

$$\frac{1}{2} K_1 + \frac{1}{1440} K_1^2 \mathrm{Gr}\,\mathrm{Pr}\,A^2 + K_2 = \frac{1}{2} \tag{32}$$

Then matching the core and boundary layer solutions to $0(A^3)$ leads to

$$K_1 = 1 - 3.48 \times 10^{-6}\,\mathrm{Gr}^2\,\mathrm{Pr}^2\,A^3 \tag{33}$$

The overall Nusselt number, to this order is given by the expression

$$\mathrm{Nu} = A + 2.86 \times 10^{-6}\,\mathrm{Gr}^2\,\mathrm{Pr}^2\,A^3 \tag{34}$$

Their analysis suggests that for these results to be useful, the aspect ratio should be limited by the condition

$$\mathrm{Ra}^2 A^3 \lesssim 10^5 \tag{35}$$

This inequality has been approximately verified by the numerical calculations of Cormack et al. [81] and the experimental observations of Imberger [83].

Cormack et al. [81] examined cavities of small but finite aspect ratio with arbitrary Grashof numbers. Their computations cover the parameter range $10 \leq Gr \leq 2 \times 10^4$ and $0.05 \leq A \leq 1$ with $Pr = 6.983$ (the value for water). The numerical method is based on a two-step ADI algorithm with non-uniform grid. These results verify the asymptotic solution obtained earlier by the authors [80] for $A \ll 1$. But more importantly, the numerical solution establishes the role of Grashof number in the development of the flow structure for small but finite aspect ratios. In particular, the solution clearly illustrates the transition from the parallel flow regime predicted by the asymptotic theory to an intermediate (and boundary layer) regime where the asymptotic theory becomes inappropriate. Typical flow patterns in a square cavity for air and water respectively are shown in Figures 1 and 2 of their paper. It is apparent from Figure 1B that large temperature gradients occur near the sidewalls. Since buoyancy forces are directly proportional to temperature gradients, it means that the flow is essentially driven by buoyancy forces in the sidewall boundary layers. Accordingly, the regions of largest vertical velocity are also confined to the sidewalls.

The effect of increasing Prandtl number from air to water is seen from their Figures 1 and 2. The thermal boundary layer becomes thinner as Pr increases. Another effect of increasing Pr is a decrease in the maximum value of the stream function within the cavity. Furthermore, a secondary flow develops as seen by the two streamfunction maxima. The secondary flow not only causes inflection in the isotherms but also convects the negative vorticity from the sidewalls into the bulk of the fluid. This picture is consistent with the numerical results of de Vahl Davis [37], Rubel and Landis [23], and Quon [30], and the experimental observations of Elder [15].

Streamlines, isotherms, and vorticity contours for a cavity of small aspect ratio ($A < 1$) are presented in Figures 3 to 8 of their paper [81]. A close examination of these figures reveals that the streamlines and vorticity lines become increasingly parallel as A decreases, consistent with the asymptotic theory. The figures also illustrate the gradual disappearance of the thermal boundary layers near the sidewalls, as A is decreased with Gr and Pr fixed. Consider, for definiteness, Figures 3–5 which are for water ($Pr = 6.983$) with $Gr = 2 \times 10^4$. In their Figure 3 where $A = 0.2$, the flow is driven primarily by the entrainment and detrainment of fluid in and out of the buoyancy driven boundary layers. In Figure 5, which corresponds to $A = 0.05$, these boundary layers have almost disappeared. Most of the temperature drop now occurs across the core and the flow is driven by the buoyancy

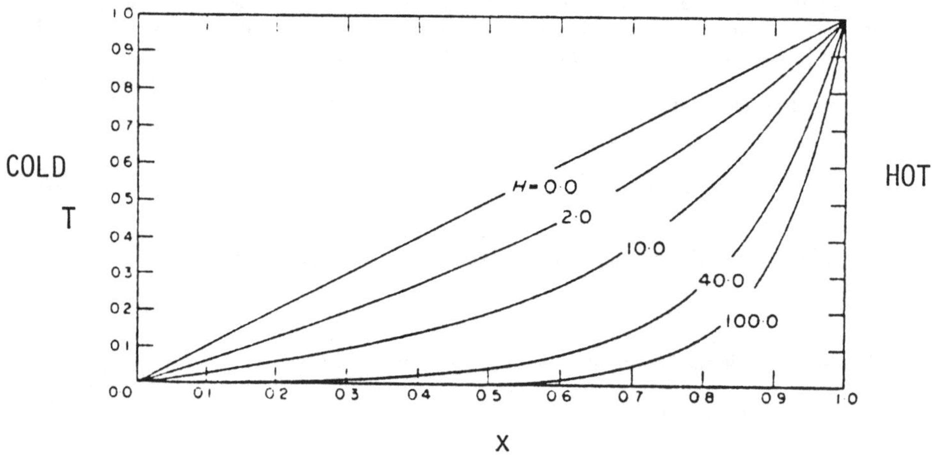

Figure 5. First order temperature profile for surface heat flux prescribed as a function of surface temperature [82].

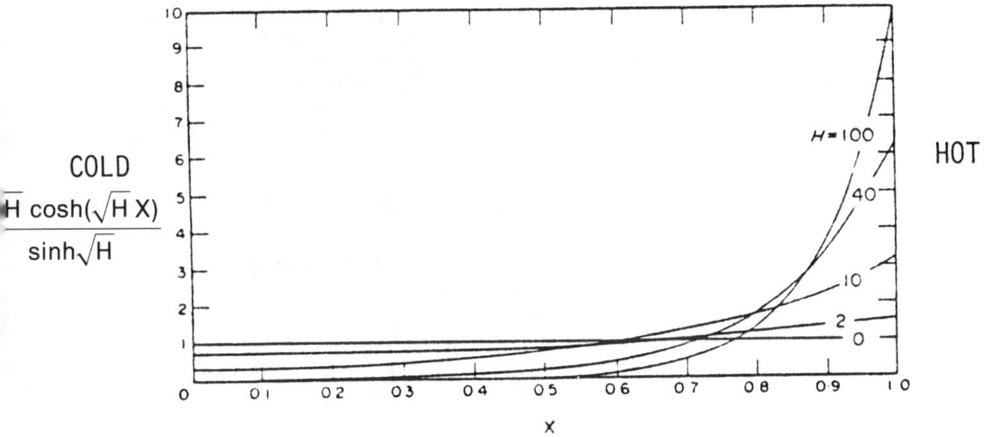

Figure 6. Magnitude of first order stream function for surface heat flux prescribed as a function of surface temperature [82].

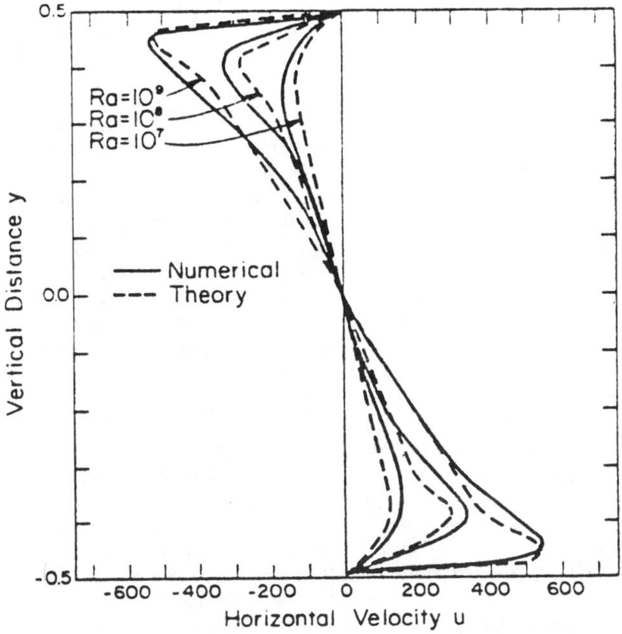

Figure 7. Horizontal velocity profiles in the core for A = 0.1, Pr = 1.0, and Ra = 10^7, 10^8, 10^9. The solid curves represent analytical solutions; the numerical results are shown by dashed lines [93].

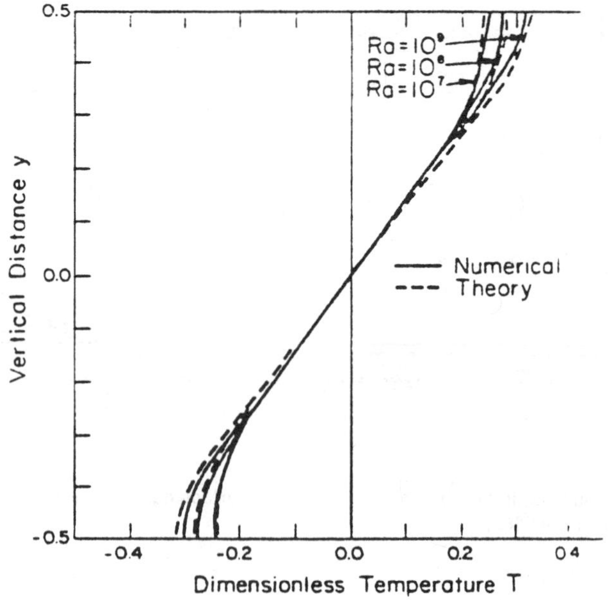

Figure 8. Core temperature profiles for $A = 0.1$, $Pr = 1.0$, and $Ra = 10^7$, 10^8, 10^9. The solid curves represent analytical solutions; the numerical results are shown by dashed lines [93].

forces resulting from the temperature gradients in the core region. Similar conclusions can be reached from a comparison of their Figures 6–8, which are for air.

In the experiments of Imberger [83], the aspect ratio of the cavity was varied between 0.01 and 0.019 with the value of Ra^2A^3 ranging from 10^6 to 10^{11}. These experiments agree quite well with the numerical solutions of Cormack et al. over the range in which they overlap. They also verify the asymptotic theory [80] as Ra^2A^3 decreases. Moreover, Imberger's results point out that as Ra^2A^3 increases, the sidewall regions begin enroaching into the core. In fact, for $A < 1$, a slow central circulation is found throughout the whole cavity at $Ra^2A^3 = 0(10^{11})$. When this happens, about half the fluid from the central part of the cavity turns around before it reaches the sidewall boundary layer. The situation is different for cavities that have a larger aspect ratio. For a square cavity with Ra^2A^3 on the order of 10^{10}, no circulation was found by Cormack et al. Quon [30] increased the value of Ra^2A^3 to about 10^{12} in a square cavity and reported no return flow. In Elder's experiments as well with $A = 0(10)$, the circulation was absent even at $Ra^2A^3 = 0(10^{20})$.

In a subsequent paper, Cormack et al. [82] studied the effect of upper surface conditions on the flow structure. They imposed the following conditions on the upper surface: uniform shear stress with zero heat flux, uniform heat flux with zero shear stress and zero shear stress with a heat flux varying linearly with temperature. A change in the upper surface conditions is shown to have a significant influence on the velocity and temperature fields in the cavity when $A \ll 1$. This is in contrast with the cavities with larger aspect ratios where no qualitative change in the flow pattern or temperature is brought about by a change of boundary conditions on the upper surface [30].

Consider, for example, that a uniform shear stress condition is used in Equation 11c instead of the condition $\psi_y = 0$. In particular, let

$$y = 1: \quad \psi_{yy} = B \tag{36}$$

B being a constant proportional to the imposed stress. Cormack et al. [82] finds that, when $B = 0(1)$, the core streamfunction is given by the expression

$$\psi = -\frac{K_1}{48} y^2(2y^2 - 5y + 3) + \frac{1}{4} By^2(y - 1), \tag{37}$$

where K_1 now depends on B also. The overall Nusselt number for this case is given by

$$Nu = A + Pr^2 Gr^2 A^3(13.10 \times 10^{-6} - 1.736 \times 10^{-4}B + 5.952 \times 10^{-4}B^2) + 0(A^4) \tag{38}$$

As expected, the result (Equation 37) with $B = 0$ shows that a free upper surface allows larger horizontal velocities than a rigid boundary. Even larger velocities may result when the upper surface is subjected to a shear stress. The free surface condition also leads to a reduction in the core temperature gradient and an enhancement in longitudinal heat transport. Under nonzero shear, the heat transfer rate is found to be minimum when $B = 0.1458$, with Ra^2A^3 fixed.

Let us now examine the effect of a nonzero heat flux prescribed on the upper surface. For definiteness, consider the condition

$$y = 1: \quad T_y = -A^2Q \tag{39}$$

in place of the no flux condition $T_y = 0$. The parameter Q represents a suitably scaled measure of the applied heat flux. If $Q > 0$, the surface cooling may destroy the slightly stable stratification that exists when the upper surface is insulated and may lead to a drastic change in the flow pattern. Under appropriate conditions, a strong cooling may result in a modified Bénard convection process. A strong surface heating ($Q < 0$), on the other hand, would enhance the stable stratification present in the case of an insulated surface and thus restrict free (vertical) movement of the fluid causing blockage near the sidewalls. The core solutions for prescribed heat flux are much more complex and the core velocity is no longer parallel except at the leading order. Furthermore, since there is heat exchange from the surface, the average Nusselt number at the sidewalls would differ by the amount QA. It can be shown that to first order in A, the amount of heat added through the upper surface is discharged equally by the two sidewalls. However, Cormack et al. [80] points out that for $Q \leq 1$, which is usually the case in estuaries, the first order velocity and temperature profiles in the core are quite similar to the insulated surface profiles with nearly parallel streamlines and almost uniform horizontal temperature gradient.

The picture is substantially different when the prescribed heat flux is a function of the surface temperature. Cormack et al. [82] consider the condition

$$y = 1: \quad T_y = -HA^2T \tag{40}$$

and show that in the limit $H \to 0$, the core solutions are identical to those for the insulated surface. As H increases, the temperature gradients increase at the hot end and decrease at the cold end. The increased temperature gradient in the hot end increases the driving force for the core flow; as a result the core streamfunction at this end increases with increasing H. The converse is true for the cold end. These trends can be seen from Figures 5 and 6 where the first order temperature and streamfunction are plotted for several values of H. It is clear from Figure 6 that as H becomes sufficiently large, there is a tendency of the streamfunction to vanish over a region near the cold end. As H increases, this region extends toward the hot end. Cormack et al. conjecture that in the limit $H \to \infty$, the flow would become unaware of the cold end of the cavity. In other words, the flow would behave as if the cavity was semi-infinite and would be independent of the aspect ratio. According to the numerical results shown in Figure 6, this transition to a semi-infinite cavity occurs when $H \sim 40$.

A few years later, Bejan and Tien [85] developed an approximate theory for natural convection in cavities with aspect ratios small but finite. They analyzed three flow regimes within the cavity: a core dominated regime for small Rayleigh numbers, an intermediate regime for $Ra = 0(1)$ and a boundary layer regime with $Ra \rightarrow \infty$. The primary goal of this work is to derive expressions for the Nusselt number for these various regimes. It is shown that as the Rayleigh number increases, the flow pattern changes substantially so that the asymptotic Nusselt number theory of Cormack et al. [80] no longer applies. When Ra is sufficiently large, the main resistance to heat transfer comes from the thin boundary layers near the sidewalls. Bejan and Tien [85] suggest that a core-dominated solution would exist as long as $RaA^2 < 72$ whereas a boundary layer would develop for $RaA^{5/3} > 4.4 \times 10^4$. From their analysis, the Nusselt number is found to be independent of the Prandtl number in all three regimes.

The expression for the Nusselt number is given by

$$Nu = A\left[1 + \left\{\left[\frac{(RaA)^2}{362{,}880}\right]^n + (0.623\, Ra^{0.2}/A)^n\right\}^{1/n}\right] \tag{41}$$

with $n = -0.386$. In another paper [86], these authors analyze the natural convection heat transfer through a long horizontal channel connecting two fluid reservoirs at different temperatures. This configuration may be important in the study of cooling of superconducting generators [5]. Not surprisingly, they find that for the same values of Ra and A, the Nusselt number for this case is considerably higher than that for a shallow cavity [85]. This is due to the absence of vertical side-walls in the present configuration and the boundary layer thermal resistance associated with them.

In many applications of cavity flows, for example, in river estuaries, there is uusually a net discharge through the cavity from one end to the other. Such a flow leads to forced convection since the horizontal temperature gradient is convected by the flow. Bejan and Imberger [87] studied the effect of this discharge on the natural convection flow set up in a parallel plate channel whose ends are maintained at different temperatures. The analysis is based on the same asymptotic limit as treated by Cormack et al. [80], namely the aspect ratio of the channel goes to zero. In this case, distinct from the natural convection flows analyzed by Cormack et al. [80], a net discharge results in a net transport of heat down the channel. When the walls of the channel are adiabatic, this heat cannot be balanced by vertical diffusion. It turns out that the temperature field compensates for the net flow by a horizontal variation strong enough so that the longitudinal diffusion can balance the forced convection. The flow becomes strictly nonparallel for this case and the Nusselt number, as expected, increases with a higher rate of discharge for Ra fixed.

Recently, Shiralkar and Tien [88] used a more efficient numerical approach for the problem of natural convection in shallow cavities with lateral heating and insulated top and bottom surfaces. Their algorithm is based on an unconditionally stable exponential difference scheme. It also uses the ADI method with underrelaxation. They examined Rayleigh numbers up to 10^6. For aspect ratios small but finite, the numerical results agree quite well with the approximate theory of Bejan and Tien [85] when Pr is moderately large. The agreement is further improved at high and low values of Ra than at intermediate values. They propose the following asymptotic relation for the rate of heat transfer:

$$Nu = 0.35\, Ra^{0.25}\, Pr^{0.25} \tag{42}$$

This formula is applicable for shallow and square cavities when $Pr > 1$. At lower Prandtl Numbers in the boundary layer regime, the numerical solution of Shiralkar and Tien [88] exhibits a clear dependence of the heat transfer rate on Pr. Note that this is in contrast with the predictions of Bejan and Tien [85] which, in fact, do not apply as well for low values of Pr as they do for higher values.

In another paper, Shiralkar et al. [89] investigated shallow cavity flows for very high Rayleigh numbers ($Ra \gg 10^6$). It is found that at such high values of Ra, the flow regime is characterized by boundary layers lining both the vertical sidewalls and the adiabatic horizontal surfaces. This is qualitatively different from the flows considered by Cormack et al. [80–82], Bejan and Tien [85], and Shiralkar and Tien [88] for lower Rayleigh numbers where the horizontal boundary layers are absent. This new flow structure results from the momentum imparted to the core by the fast moving boundary layers near the sidewalls. Flows of this kind have also been observed experimentally by Al-Hommoud and Bejan [90] in a cavity with $A = 0.0625$, $Ra = 1.6 \times 10^9$ and $Pr = 6.3$, and by Ostrach [91] with $A = 0.2$, $Ra = 1.8 \times 10^7$ and $Pr = 1.4 \times 10^3$.

Shiralkar et al. [89] developed a theoretical model in which the new core flow is matched with the boundary layer flows of Bejan and Tien [85]. They also computed the flow numerically using a finite difference scheme proposed by Patankar and Spalding [92] and found good agreement with the theory. The overall Nusselt number based on their numerical solution is given by

$$Nu = 2^{-3/2}\, Ra^{m} \tag{43a}$$

with

$$m = 0.25 - qRa^{-0.1917} \tag{43b}$$

Here q has the value 0.38 and 0.472 depending on $A = 0.2$ and $A = 0.1$, respectively. The Nusselt number reported by Al-Hommoud and Bejan [90] under similar conditions is

$$Nu = (0.0168 \pm 0.0002)Ra^{0.375} \tag{44}$$

while Imberger [83] gives the relationship

$$Nu = (0.098 \pm 0.008)Ra^{0.291} \tag{45}$$

Clearly the result (Equation 43) shows that the exponent m would, in general, depend on Ra and has the asymptotic value of 0.25 as $Ra \to \infty$.

Tichy and Gadgil [93] considered a similar problem [89] with particular attention to the detailed flow profiles. In the asymptotic limit $Gr \to \infty$, they deduced analytical solutions for the velocity and temperature distributions in the core as well as in the horizontal and vertical boundary layers. These calculations have been substantiated by numerical calculations in the same paper [93] and by experimental observations [94]. The horizontal velocity profiles predicted by these calculations are shown in Figure 7 and the corresponding temperature profiles are plotted in Figure 8 for three typically large values of Ra. Figure 7 clearly illustrates the development of the boundary layers near the horizontal boundaries as Ra is increased continuously through large values.

Buoyancy driven flows of low Prandtl number fluids in shallow cavities has been examined in detail in a very recent paper by Hart [95]. This paper describes the development of the unicellular flow in the cavity and the secondary instabilities associated with it at low Prandtl numbers. Such secondary flows are known to occur in typical crystal growth experiments [96]. Hart [95] points out that the previous analyses by Cormack et al. [80–82] and Bejan and Tien [85] use a roughly diffusive test function solution for the turning regions near the sidewalls and thus do not allow for secondary motions there or in the core. These secondary motions become especially important at low values of Pr and can only be studied using a model that retains the full effect of the inertial terms in the governing equations. Hart analyzed such a model with large Grashof numbers to show that for $Pr \leq 0.1$ and aspect ratios less than about 0.1, a parallel core flow would exist for Grashof numbers up to 8,000, beyond which secondary vortices would appear near the sidewalls. These secondary circulations propogate from the sidewalls into the core as Gr increases. We now turn to a more general discussion of modeling natural convection flows in cavitites that are applicable to crystal growth processes.

From the standpoint of technological applications, perhaps the most important property of a single crystal is its structural uniformity. It is now well known that the transport phenomena in the fluid phase in a crystal growth process have a pronounced effect on the structure and quality of the solid crystal. Natural convection may affect the process of crystal growth in two ways. It enhances the overall transport rate in the melt, which is desirable. But it also makes the mass flux of the growing crystal nonuniform and thus adversely affects the local growth conditions. The simplest geometry in which these effects can be studied is that of a shallow enclosure with lateral heating. Unfortunately, as Solan and Ostrach [97] point out, both theoretical and experimental work on natural convection in shallow enclosures is rather limited, inexact, and most of all in a parameter range inappropriate for crystal growth conditions. Also, in most of these investigations, the convective motion is caused by buoyancy forces resulting solely from an imposed temperature difference. A further complication in crystal growth problems arises from the presence of solute gradients in the melt fluid, which give rise to additional buoyancy forces. Only recently, due to the need for growing high quality crystals for semiconductor and other high technology applications, a considerable amount of research effort is directed to the study of the convective effects in crystal growth processes. A comprehensive review of this subject is given by Pimputkar and Ostrach [84, 98] and Hurle and Jakeman [99].

To study natural convection effects in crystal growth, Ostrach et al. [100] performed a series of experiments in differentially heated shallow rectangular cavities over an appropriate range of parameters, namely, $0.05 < A < 0.5$, $27.7 < Gr < 10^6$, and $0.72 < Pr < 1.38 \times 10^3$. The results of their observations can be summarized as follows. The overall flow in the cavity is predominantly unicellular. For $A \leq 0.1$, the flow in the interior is nearly horizontal, while for $A > 0.1$ it tends to be skewed or nonparallel and is inclined to the horizontal. The skewness of the streamlines is proportional to the Grashof number and the aspect ratio. Secondary cells are also formed near the sidewalls when $A = 0.2$ and $Gr \sim 10^{13}$.

Similar experiments have been carried out recently by Kamotani et al. [101] to show that the secondary cells that appear for $A > 0.1$ and higher Prandtl numbers, can significantly alter the temperature distribution near the sidewalls and therefore the rate of heat transfer through the cavity. It should be emphasized, however, that the Prandtl numbers used in these experiments were higher than those typical of crystal growth processes. In another recent paper, Jhaveri and Rosenberger [102] presented a numerical study of natural convection in shallow rectangular cavities with particular attention to the method of closed tube transport. The following parameter values have been used: $A = 0.1$, $Sc = 0.5$, $Gr = 3.3 \times 10^4$, and $Pr = 0.7$; here $Sc = \nu/D$ is the so-called Schmidt number, D being the mass diffusion coefficient. In their simulation, the mass flux from the source to the crystal is taken into account by prescribing a concentration difference and a mass average velocity at the ends. The mass average velocities are derived from Fick's law of diffusion. For the range of parameters considered, the numerical results exhibit a secondary cell near each end. The cells are formed due to the coupling between the mass flux at the ends and the transport in the interior. As discussed earlier, the numerical calculations of Hart [95] also indicate that secondary flows can occur in shallow cavities with differentially heated sidewalls, when the Prandtl number is sufficiently low ($Pr \sim 10^{-2}$).

We mentioned before that in crystal growth processes, the buoyancy forces are often induced by the combined action of temperature and concentration gradients. However, there exists very little work on natural convection flows in enclosures driven by both these effects. Ostrach [103] discussed some of the earlier work in this field and also outlined a few unsolved problems.

When both temperature and concentration gradients are present, a number of qualitatively different configurations can arise depending on the orientation of these gradients relative to each other and also with respect to the direction of gravity. For a cavity with lateral heating, there are four possible cases. These have been discussed in some detail in a recent review by Ostrach [84]. Of these cases, the configuration that has received the most attention is the one where a stable vertical concentration gradient exists in a rectangular cavity whose sidewalls are maintained at different temperatures [104]. Here, the flow becomes unstable leading to a layered cell structure. This problem has been studied more fully by Wirtz [105]. However, the range of parameter values used in these studies are more suited for geophysical applications than for crystal growth processes.

Wang [106] has experimentally studied two configurations in which the temperature and concentration gradients are both horizontal and can augment or oppose each other. He used a copper sulphate-acid solution in a rectangular cavity whose sidewalls are made into electrodes and are differentially heated. Whether or not the gradient fields are augmenting or opposing, the fluid in the cavity is layered in three horizontal cells, one near each of the top and bottom surfaces and one in the interior. When the two gradients augment each other, the circulation in each layer is in the same sense and the overall fluid flow is slower than that if the concentration gradient was absent. For opposing gradients, on the other hand, the boundary layer flows near the horizontal surfaces are found to have an opposite sense to that of the interior flow. Although Wang's work is mostly qualitative, the parameter values used here are relevant to crystal growth techniques.

Clearly more work is needed for a better understanding of natural convection flows in cavities driven by the combined action of temperature and concentration gradients.

FLOWS DRIVEN BY SURFACE TENSION GRADIENTS

General Considerations

When the surface tension along the interface between two fluids varies from point to point, motion results. This is commonly known as the Marangoni effect and is induced from a balance of the surface tension gradient along the interface and the jump in the bulk fluid shear stress across it. The interfacial motion is transmitted to the bulk fluids by viscous forces.

Surface tension variations can arise, for example, from a temperature or a concentration gradient. Flows driven by thermal variations in surface tension are called thermocapillary flows while those resulting from concentration gradients are referred to as diffusocapillary flows. This latter type of flows is encountered in applications such as spreading of oil and paint films, jet decay, and corrosion processes and has received more attention [107, 108] than thermocapillary flows. Yih [109, 110] and Adler and Sowerby [111] analyzed surface tension flows in shallow channels due to the presence of a contaminant. Thermocapillary flows occur in the problems of flame spreading over liquid fuels [112, 113], boiling heat transfer, and spot welding processes. Other examples include migration of bubbles in a thermal gradient field [114, 115].

Thermocapillary flows are also known to be important in the containerless processing of materials in space, e.g., aboard a spacecraft and other orbiting facilities [116]. There has been a great deal of interest in this area over the past few years. Under such microgravity conditions, the fluid flow in the melt is driven by thermal variations in surface tension along the liquid-gas interface. The nature of the crystal formed depends on the thermal, concentration and flow fields in the melt, and there is a strong coupling between the fluid dynamics and the growth dynamics of the crystal. Therefore, an understanding of thermocapillary flows is of fundamental importance to the study of feasibility of these processes. A discussion of the surface tension driven flows at reduced gravity has been presented recently by Ostrach [117].

Thermocapillary Flows in Cavities

Steady thermocapillary flows in thin unbounded layers have been examined by Levich [118], Birikh [119], and Adler [120], among others. In these early works the authors used lubrication approximation (explicitly or implicitly) to simplify the analyses. Thermocapillary flows in bounded regions have been treated by Babskiy et al. [121] who examined flows in a square cavity with a flat free surface. Ostrach [122] points out that several aspects of their analysis are questionable and also summarizes the earlier attempts on thermocapillary flows.

Recently, Sen and Davis [123] and Sen et al. [124] investigated steady thermocapillary flows in a two-dimensional slot. In the limit that the aspect ratio of the slot goes to zero, these authors derived analytical expressions for the flow field, thermal field and the shape of the interface, using the method of matched asymptotic expansions. The effect of surface deformation on the flow structure was also established. Very recently, Strani et al. [125] solved similar problems both analytically and numerically and extended the results of Sen and Davis [123] to higher orders of approximation. Strani et al. [125] found that if the surface deformation is small, it has a negligible influence on the flow characteristics. We now discuss the work of Sen and Davis in some detail.

The configuration used Sen and Davis [123] is shown schematically in Figure 9. They consider a rectangular cavity filled with an incompressible Newtonian liquid of density ρ, thermal diffusivity α, and kinematic viscosity $\nu = \mu/\rho$; μ is the dynamic viscosity. The cavity is differentially heated as shown in the figure. The lower surface of the cavity ($y = 0$) is insulated. The upper surface, described by $y = h(x)$ is a free surface bounded by a passive gas of negligible density and viscosity. This free surface is associated with a surface tension σ, which depends on the local temperature. The differential heating induces a temperature gradient along the liquid-gas interface; this in turn induces a

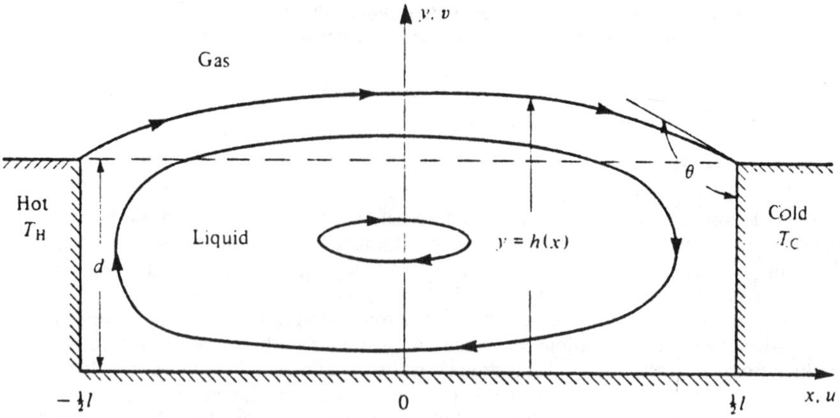

Figure 9. Schematic diagram of a two-dimensional slot with a free surface. The liquid sticks to a sharp edge at the sidewalls as shown for Case I [123].

surface tension gradient resulting in motion near the interface. Since the cavity is of finite length, the near-surface flow, to conserve mass, turns around at the sidwalls and recirculates below.

In the absence of gravity, the two dimensional motion of the liquid is governed by Equations 1a–d with $g = 0$. For the present configuration, these equations are subject to the following boundary conditions.

$$x = -\ell/2: \quad u = v = 0, \qquad T = T_H; \tag{46}$$

$$x = \ell/2: \quad u = v = 0, \qquad T = T_C; \tag{47}$$

$$y = 0: \quad u = v = 0, \qquad T_y = 0; \tag{48}$$

$$y = h(x): \quad v = uh_x, \qquad S_{ij}n_jn_i = \sigma K \tag{49a, b}$$

$$S_{ij}n_jt_i = \sigma_s, \qquad k_\ell T_n + k_g(T - T_g) = 0 \tag{49c, d}$$

Equation 49a is the kinematic boundary condition at the liquid-gas interface. The stress balances at the interface in the normal and tangential directions are given by Equations 49b and c. The jump in the normal stress across the interface is balanced by surface tension times curvature and the jump in the shear stress equals the surface-tension gradient along the interface. In these equations S_{ij} are the components of the stress tensor of the liquid defined by

$$S_{ij} = -p\delta_{ij} + 2\mu\epsilon_{ij}, \qquad \epsilon_{ij} = \frac{1}{2}(u_{i,j} + u_{j,i}) \tag{50}$$

u being the velocity vector and δ_{ij} the Kronecker delta. Note that in Equation 49c, σ_s denotes the tangential derivative of surface tension along the interface. The curvature K in Equation 49b has the definition

$$K(h) = h_{xx}/N^3 \tag{51}$$

The thermal boundary condition at the interface is given by Equation 49d in which k_ℓ is the thermal conductivity of the liquid, k_g is the heat transfer coefficient in the gas and T_g is the temperature in the gas phase. The outward unit normal vector **n** and the unit tangent vector **t** at the interface are defined as follows:

$$\mathbf{n} = (-h_x, 1)/N, \qquad \mathbf{t} = (1, h_x)/N \tag{52}$$

with

$$N = (1 + h_x^2)^{1/2} \tag{53}$$

Apart from the boundary conditions (Equations 46–49), the velocity field must also satisfy the condition

$$\int_0^{h(x)} u(x, y)\, dy = 0 \tag{54}$$

which follows from the fact that there exists no net mass flow into and out of the cavity. Furthemore, since the liquid is incompressible, its total volume must remain constant; thus the relation

$$\int_{-\ell/2}^{\ell/2} h(x)\, dx = V \tag{55}$$

must hold, where V is the total two-dimensional volume occupied by the liquid.

Finally, to close the problem, the type of contact made by the free surface at the sidewalls must be specified. Sen and Davis [123] consider the following two cases:

Case I. The liquid sticks to a sharp edge at the sidewalls with

$$h\left(\pm\frac{1}{2}\ell\right) = d \tag{56}$$

Case II. A contact angle θ is prescribed at each sidewall so that

$$h_x\left(\pm\frac{1}{2}\ell\right) = \mp\tan(\theta - \pi/2) \tag{57}$$

In order to determine the velocity and temperature fields in the liquid, the temperature distribution T_g in the overlying gas must be known, a priori. Sen and Davis [123] assume that the heated and cooled sidewalls of the cavity induce a sensibly conduction-dominated temperature in the gas so that a linear variation of temperature exists. This profile is given by

$$T_g = \frac{1}{2}(T_H + T_C) - (T_H - T_C)x/\ell \tag{58}$$

The temperature boundary condition (Equation 49d) reflects the existence of a thin convective thermal boundary layer in the gas.

Sen and Davis use a linear equation of state for surface tension. In particular, they take

$$\sigma(T) = \sigma_0 - \gamma_T\left[T - \frac{1}{2}(T_H + T_C)\right] \tag{59}$$

where σ_0 is the mean surface tension of the liquid and the constant γ_T is the negative of the derivative of surface tension with respect to temperature.

The problem is nondimensonalized in terms of lubrication-type variables as follows.

$$\begin{gathered} x = \ell x', \qquad y = dy', \qquad h = dh', \qquad u = u_* u', \qquad v = Au_* v', \\ p = (\mu u_* \ell/d^2)p', \qquad T - \frac{1}{2}(T_H + T_C) = (T_H - T_C)T', \qquad \sigma = \sigma_0\sigma'. \end{gathered} \tag{60}$$

with

$$u_* = \gamma_T A(T_H - T_C)/\mu \tag{61}$$

Recall that the aspect ratio has the definition $A = d/\ell$. The characteristic velocity u_* is derived from the so-called Marangoni effect, i.e., the jump in shear stress along the interface balances the surface tension gradient.

With these scales, the dimensionless equations (after dropping the primes) can be written:

$$u_x + v_y = 0 \tag{62a}$$

$$RA(uu_x + vu_y) = -p_x + A^2u_{xx} + u_{yy} \tag{62b}$$

$$RA^3(uv_x + vv_y) = -p_y + A^2(A^2v_{xx} + v_{yy}) \tag{62c}$$

$$MA(uT_x + vT_y) = A^2T_{xx} + T_{yy} \tag{62d}$$

Here, the Reynolds number R and the Marangoni number M are defined by

$$R = \gamma_T A(T_H - T_C)d/\mu\nu, \qquad M = \gamma_T A(T_H - T_C)/\mu\alpha \tag{63}$$

respectively. The dimensionless boundary conditions are

$$x = \pm\frac{1}{2}: \quad u = v = 0, \qquad T = \mp\frac{1}{2}; \tag{64}$$

$$y = 0: \quad u = v = 0, \qquad T_y = 0; \tag{65}$$

$$y = h(x): \quad v = uh_x, \tag{66a}$$

$$-p + 2A^2(1 + A^2h_x^2)^{-1}[(v_y - h_xu_y) + A^2h_x(-v_x + h_xu_x)] \\ = A^3C^{-1}h_{xx}(1 + A^2h_x^2)^{-3/2}(1 - A^{-1}CT) \tag{66b}$$

$$(1 - A^2h_x^2)(u_y + A^2v_x) + 2A^2h_x(v_y - u_x) = -(1 + A^2h_x^2)^{1/2}(T_x + h_xT_y) \tag{66c}$$

$$(1 + A^2h_x^2)^{-1/2}(T_y - A^2h_xT_x) + L(T + x) = 0 \tag{66d}$$

Note that the capillary number C and the Biot number L have the definitions

$$C = \gamma_T A(T_H - T_C)/\sigma_0, \qquad L = k_g d/k_\ell \tag{67}$$

The capillary number is a measure of the degree of deformation of the free surface and the Biot number measures the relative rates of heat transport between the liquid and the gas. The dimensionless forms of the no mass flux condition (Equation 54) and the incompressibility condition (Equation 55) are

$$\int_0^{h(x)} u(x, y)\,dy = 0, \quad \int_{-1/2}^{1/2} h(x)\,dx = 1 \tag{68}$$

Similarly, the contact angle conditions transform into

$$h\left(\pm\frac{1}{2}\right) = 1 \tag{69}$$

for Case I and

$$h_x\left(\pm\frac{1}{2}\right) = \mp A^{-1}\tan\left(\theta - \frac{\pi}{2}\right) \tag{70}$$

for Case II. This completes the problem description.

Sen and Davis [123] examined this problem in the limit $A \to 0$, with R and M fixed. In this limit the flow within the cavity can be divided into two regions: a core region where the flow is nearly parallel and a boundary layer region near each sidewall with turning flows. The complete flow structure in the cavity is obtained by solving the flows in the core and the boundary layers and joining them using the method of matched asymptotics. The calculations are analogous to those of Cormack et al. [80]. Sen and Davis neglected the convective transfer to momentum and energy to a first approximation by assuming that both the Reynolds and Marangoni numbers are small. In particular, they assumed that $R = 0(A)$ and $M = 0(A)$ although their analysis is valid even for higher values of these parameters. A further assumption in their analysis is that the capillary number C is small, i.e. the mean surface tension of the liquid is large. They considered the distinguished limit $C = 0(A^4)$.

The main result of Sen and Davis is that for the asymptotic values of the various parameters as considered here, the core flow is parallel to the bottom surface of the cavity, to a leading approximation. To this approximation, the interface remains flat and the temperature field in the cavity follows the prescribed gas temperature. The core velocity is given by the parabolic profile

$$u(y) = \frac{1}{4}y(3y - 2) \tag{71}$$

with $T = -x$. At this order the axial pressure gradient in the cavity is constant. They also obtained an explicit analytical solution for the surface deformation by writing

$$h(x) = 1 + Ah_1(x) + A^2h_2(x) + \dots \tag{72}$$

In the case when the liquid sticks to a sharp edge at the sidewalls so that the conditions (Equation 69) hold, the 0(A) perturbation of the free surface from a flat interface is given by

$$h_1(x) = -\frac{1}{4}\bar{C}x\left(x^2 - \frac{1}{4}\right) \tag{73}$$

where $\bar{C} = CA^{-4} = 0(1)$. Alternatively, if the contact angles at the sidewalls are prescribed by the conditions (Equation 70) with

$$\tan(\theta - \pi/2) = mA^2, \qquad m = 0(1) \tag{74}$$

the result is

$$h_1(x) = -\frac{1}{16}A\bar{C}\left\{x(4x^2 - 3) + \frac{4}{3}m(12x^2 - 1)\right\} \tag{75}$$

The boundary layer flows near the sidewalls are quite complex and are solved numerically using finite difference methods. It is in these sidewall regions where the contact line conditions have their effects. The composite flow structure is obtained by joining the core flows with the boundary layer flows. At the next order of approximation, the core flow becomes nonparallel, depending on the extent of surface deformation. The streamline pattern and the interface shape correct to $0(A^2)$ are shown in Figures 10, 11, and 12 for the two cases, with typical parameter values. The leading-order pressure gradient is directed from the hot end toward the cold end. Accordingly, the pressure is

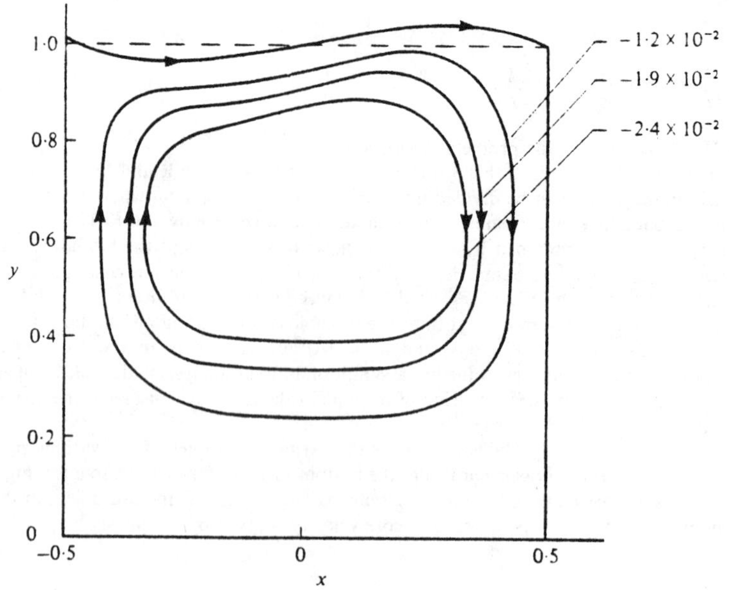

Figure 10. Streamlines and interface shape to order A^2 with L = 1 and A = 0.2 for Case I [123].

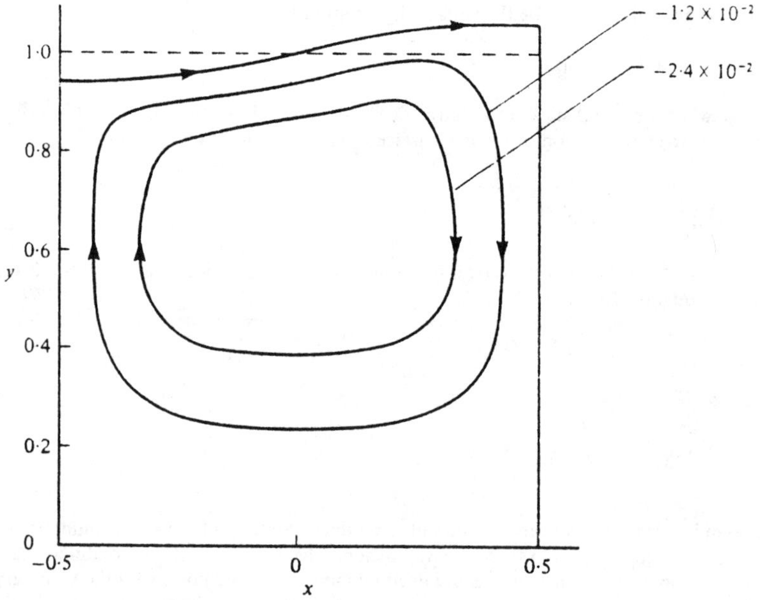

Figure 11. Streamlines and interface shape to order A^2 with L = 1, A = 0.2 and m = 0 for Case II [123].

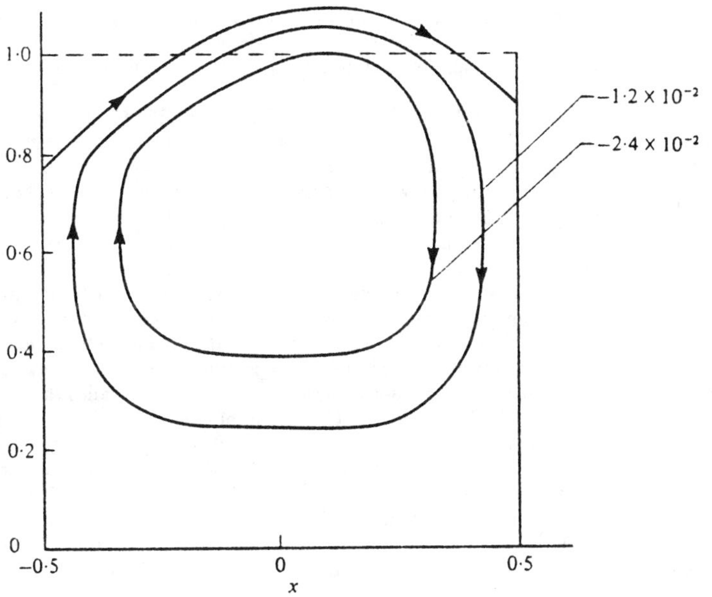

Figure 12. Streamlines and interface shape to order A^2 with $L = 1$, $A = 0.2$ and $m = 1$ for Case II [123].

higher at the cold sidewall. As a result, the interface bulges at the cold end and is constricted near the hot end, as shown in these figures. As mentioned earlier, Strani et al. [125] extended these results to higher approximations. They also considered moderate to large values of Reynolds' numbers and 0(1) aspect ratios and numerically computed the flow and temperature fields in the cavity.

Very recently, Homsy and Meiburg [126] examined the effect of free surface contamination on thermocapillary flows in cavities. They study the effect of an insoluble surfactant on the steady thermocapillary flow in a slot treated by Sen and Davis [123]. The surface tension here is assumed to be linear in both temperature and surfactant concentration. They analyze the problem in the limit of vanishingly small aspect ratio and low thermal Marangoni number. Under these conditions,the effect of the surfactant can be described by only two parameters: a surface Peclet number, Pe, and an elasticity parameter, E. These have the definitions

$$Pe = u_* \ell / D, \qquad E = \gamma_c C_0 / \gamma_T (T_H - T_C) \tag{76}$$

Here D is the diffusion coefficient of the surfactant and C_0 is its average concentration. The constants γ_T and γ_C are the rates of change of surface tension with temperature and concentration, respectively. Note that the elasticity parameter E is the ratio of compositional elasticity to the thermal variation of surface tension. Using the method of matched asymptotic expansions, they derive a nonlinear integral-algebraic equation for the core flow, which is then solved numerically and in the various asymptotic limits based on the values of Pe and E. Homsy and Meiburg [126] show that the general effect of surfactants is to retard the strength of the motion in the slot, but this retardation is not always uniform in space. Their main results can be summarized as follows. For small elasticity ($E \ll 1$), the surface velocity is nearly constant. Furthermore, if $Pe \ll 1$ (large surface diffusion), the surface concentration is also nearly uniform. For large Pe, on the other hand, with $E \ll 1$, a boundary layer may develop near the stagnation point at the cold sidewall and result in high local surface concentrations leading possibly to a surface phase change, monolayer collapse and/or resolubilization of the surface species. Homsy and Meiburg point out that, although these phenomena do not influence the main flow significantly, they may produce pronounced effects at the liquid-solid interface.

Consider next the situation when $Pe \ll 1$ but $E = 0(1)$ Here, the surface concentration remains nearly uniform with a uniform reduction in surface velocity and interfacial displacement. This behavior persists as Pe increases with a rapid decrease in the magnitude of the surface velocity. Sharp gradients in surface concentration will not form, in general, when $E = 0(1)$. They stipulate that this behavior is due to the strong coupling between the flow and the interfacial stress. In the limit of large Peclet number ($Pe \rightarrow \infty$), they arrive at the somewhat surprising result that the less mobile a surface species, the larger is its effect on surface speed.

Unsteady thermocapillary flows in thin liquid layers have been examined by Pimputkar and Ostrach [127], Lai, Ostrach, and Kamotani [128] and a few others. Pimputkar and Ostrach were the first to consider transient phenomena in such systems and identify the multiple time scale nature of these problems. For a family of imposed surface temperature distributions, they numerically computed the flow field and the interface shape at various times. Lai et al. studied the effect of surface deformation on the unsteady flow structure. Three cases of time-dependent, two-dimensional thermocapillary flows were analyzed. Cases I and II deal with flows in a thin infinite layer and a rectangular pool while in Case III oscillatory thermocapillary flows are examined. Their results show that in Cases I and II, the relative magnitudes of the various time scales are important in determining the flow field and that the flow field in the transient period may be strongly influenced by surface deformation. They also found that when the flow is oscillatory (Case III) there exists a time lag between the velocity and the temperature fields due to the effect of deformation of the free surface.

Smith and Davis [129, 130] analyzed the stability of thermocapillary flows in thin planar layers and shallow slots. They used the flow structure obtained by Sen and Davis [123] as the basic state and performed a linear stability analysis. According to their results, two types of instabilities can exist in these flows: a convective or thermal instability that results from a balance of heat conduction and heat convection at the free surface, and a surface wave instability that involves transfer of momentum from the basic state to the disturbances via Reynolds stresses. In the former type, the free surface deformation is relatively unimportant whereas in the latter the surface deformation plays a significant role. Their estimates show that surface waves are preferred in cavities of small aspect ratios over a considerable range of Prandtl numbers. For deeper slots however, both surface waves and hydrothermal waves can co-exist depending on the value of the Prandtl number.

CONCLUSION

We have presented a comprehensive review of natural convection and surface tension flows in rectangular enclosures and cavities. The configuration used in studying these flows is that of a vertical rectangular enclosure with lateral heating. The buoyancy and surface tension forces are considered to result primarily from the temperature difference applied across the vertical sidewalls. The natural convection flows have been discussed in enclosures with both large and small aspect ratios whereas the study of thermocapillary flows has been limited to shallow cavities.

It is apparent from our discussion that the subject of natural convection in large aspect-ratio enclosures has a rather long history. Although the elementary aspects of these flows are now well understood, more research is needed to fully interpret the flow behavior in the high Rayleigh number regime. In particular, the structure of the multicellular flows must be clearly determined since such flows are known to lead to turbulence and other complex phenomena.

Despite considerable progress in the study of thermally-driven natural convection flows in shallow cavities, further work is necessary to explore the various parameter ranges especially in relation to their application to crystal growth processes. Natural convection in cavities due to the combined action of temperature and concentration gradients is also a fertile area of research today.

Studies on thermocapillary flows in cavities are of relatively recent origin. Because of their importance in the materials processing experiments in space, a great deal of research on thermocapillary flows is currently in progress. Clearly, much work is still needed in order to completely understand the nature of these flows and their suitability for application to space processing.

NOTATION

A	aspect ratio	**n**	outward normal vector
B	constant proportional to imposed stress	p	pressure
C	capillary number	Pe	Peclet number
C_0	average concentration	Pr	Prandtl number
D	diffusivity	R	Reynolds number
d	diameter	Ra	Rayleigh number
E	elasticity parameter	Sc	Schmidt number
Gr	Grashof number	S_{ij}	stress tensor
J	Jacobian function	T	temperature
K_1	coefficient	t	time
ℓ	length	**t**	tangent vector
M	Marangoni number	u_*	characteristic velocity
Nu	Nusselt number	u, v	velocity components
		x, y, z	coordinates

Greek Symbols

α	thermal diffusivity	μ	viscosity
β	coefficient of thermal expansion	ν	kinematic viscosity
γ_T	mean surface tension	ρ	density
γ_T, γ_C	rates of change of surface tension with temperature and concentration, respectively	σ	surface tension
δ	boundary layer thickness	τ	stress
		ψ	streamfunction

REFERENCES

1. Batchelor, G. K., "Heat Transfer by Free Convection Across a Closed Cavity Between Vertical Boundaries at Different Temperatures." *Quart. Appl. Math.*, Vol. 12, No. 3, 1954, p. 209.
2. Buchberg, H., Catton, I., and Edwards, D. K., "Natural Convection in Enclosed Spaces—A Review of Applications to Solar Energy Collection." *J. Heat Transfer*, Vol. 98, 1976, p. 182.
3. Carruthers, J. P., "Crystal Growth From the Melt." *Treatise on Solid State Chemistry*, Vol. 5, Plenum Press, 1975, p. 325.
4. Petuklov, B. S., "Actual Problems of Heat Transfer in Nuclear Power Engineering." *International Seminar on Future Energy Production*, Hemisphere Publishing, Washington, D. C. 1976, p. 151.
5. Schwoerer, J. A., and Smith, J. L., Jr., "Transient Cooling of a Fault Worthy Superconducting Electric Generator." Paper presented at the *Cryogenic Engineering Conference*, Madison, Wisconsin, 1979.
6. Hart, J. E., "Stability of Thin Non-Rotating Hadley Circulation." *J. Atmos. Sci.*, Vol. 29, 1972, p. 687.
7. Fischer, H. B., "Mass Transport Mechanisms in Partially Stratified Estuaries." *J. Fluid Mech.*, Vol. 53, 1972, p. 671.
8. Ostrach, S., "Natural Convection in Enclosures." *Adv. Heat Transfer*, Vol. 8, 1972, p. 161.
9. Rogers, R. H., "Convection." *Rep. Prog. Phys.*, Vol. 39, 1976, p. 1.
10. Catton, I., "Natural Convection in Enclosures." *Proc. Sixth Int. Heat Transfer Conf.*, Toronto, Canada, Vol. 6, 1978, p. 13.

11. Mull, W., and Reiher, H., "Der Wärmeschultz von Luftschichten." *Beihefte Zum Gesundh–Ingenieur*, Reihe 1, Heft 28, 1930.
12. Jakob, M., *Heat Transfer*, Vol. 1, Wiley, New York, 1949.
13. Poots, G., "Heat Transfer by Laminar Free Convection in Enclosed Plane Gas Layers." *Quart. J. Mech. Appl. Math.*, Vol. 11, 1958, p. 257.
14. Eckert, E. R. G., and Carlson, W. O., "Natural Convection in an Air Layer Enclosed Between Two Vertical Plates with Different Temperatures." *Int. J. Heat Mass Transfer*, Vol. 3, 1961, p. 106.
15. Elder, J. W., "Laminar Free Convection in a Vertical Slot." *J. Fluid. Mech.*, Vol. 23, 1965, p. 77.
16. Oshima, Y., "Experimental Studies of Free Convection in a Rectangular Cavity." *J. Phys. Soc. Japan*, Vol. 30, 1971, p. 872.
17. Ayyaswamy, P. S., Ph.D. dissertation, Univ. California, Los Angeles, 1971.
18. Grondin, J. C., "Contribution à l'ètude de la Convection Naturelle Dans un Capteur Solaire Plan." These Institut du Mécanique des Fluides de Marseille, 1978.
19. Gill, A. E., "The Boundary Layer Regime for Free Convection in a Rectangular Cavity." *J. Fluid Mech.*, Vol. 26, 1966, p. 515.
20. Elder, J. W., "Numerical Experiments with Free Convection in a Vertical Slot." *J. Fluid Mech.*, Vol. 24, 1966, p. 823.
21. Wilkes, J. O., and Churchill, S. W., "The Finite Difference Computation of Natural Convection in a Rectangular Enclosure." *AIChE Journal*, Vol. 12, 1966, p. 161.
22. Han, J. T., "Numerical Solutions for an Isolated Vortex in a Slot and Free Convection Across a Square Cavity." M. S. thesis, Dept. Mech. Eng., Univ. Toronto, 1967.
23. Rubel, A., and Landis, F., "A Numerical Study of Natural Convection in a Vertical Rectangular Enclosure." *Phys. Fluids Suppl. II*, Vol. 12, 1969, p. 208.
24. Polezhaev, V. I., "Numerical Solution of a System of Two-Dimensional Unsteady Navier Stokes Equations for a Compressible Gas in a Closed Region." *Izvestia*, MZG, No. 2, 1967, p. 103.
25. MacGregor, R. K., and Emery, A. F., "Free Convection Through Vertical Plane Layers–Moderate and High Prandtl Number." *J. Heat Transfer*, Vol. 91, No. 3, 1969, p. 391.
26. MacGregor, R. K., and Emery, A. F., "Prandtl Number Effects on Natural Convection in an Enclosed Vertical Layer." *J. Heat Transfer*, Vol. 93, 1971, p. 253.
27. Newell, M. E., and Schmidt, F. W., "Heat Transfer by Laminar Natural Convection Within Rectangular Enclosures." *J. Heat Transfer*, Vol. 92, 1970, p. 159.
28. Thomas, R. W., and de Vahl Davis, G., "Natural Convection in Annular and Rectangular Cavities." *Proc. Fourth Int. Heat Transfer Conf.*, 1970.
29. Torrance, K. E., "Comparison of Finite Difference Computation of Natural Convection." *J. Research., Nat. Bur. Standards*, Vol. 72B, No. 4, 1968.
30. Quon, C., "High Rayleigh Number Convection in an Enclosure—A Numerical Study." *Phys. Fluids*, Vol. 15, 1972, p. 12.
31. Quon, C., "Free Convection in an Enclosure Revisited." *J. Heat Transfer*, Vol. 99, 1977, p. 340.
32. Blythe, P. A., and Simpkins, P. G., "Thermal Convection in a Rectangular Cavity." *Physico-chemical Hydrodynamics*, edited by D. B. Spalding, Advance, New York, 1977, p. 511.
33. Bejan, A., "Note on Gill's Solution for Free Convection in a Vertical Enclosure." *J. Fluid Mech.*, Vol. 90, 1979, p. 561.
34. Graebel, W. P., "The Influence of Prandtl Number on Free Convection in a Rectangular Cavity." *Int. J. Heat Mass Transfer*, Vol. 24, 1981, p. 125.
35. Seki, N., Fukusako, S., and Inaba, H., "Visual Observations of Natural Convection Flow in a Narrow Vertical Cavity." *J. Fluid Mech.*, Vol. 84, 1978, p. 695.
36. Yin, S. H., Wung, T. Y., and Chen, K., "Natural Convection in an Air Layer Enclosed Within Rectangular Cavities." *Int. J. Heat Mass Transfer*, Vol. 21, 1978, p. 307.
37. de Vahl Davis, G., "Laminar Natural Convection in an Enclosed Rectangular Cavity." *Int. J. Heat Transfer*, Vol. 11, 1968, p. 1675.
38. Pepper, D. W., and Harris, S. D., "Numerical Simulation of Natural Convection in Closed Containers by a Fully Implicit Method." *ASME J. Fluids Engng.*, Vol. 99, 1977, p. 649.

39. Catton, I., Ayyaswamy, P. S., and Clever, R. M., "Natural Convection Flow in a Finite Rectangular Slot, Arbitrarily Oriented With Respect to the Gravity Vector." *Int. J. Heat Mass Transfer*, Vol. 17, 1974, p. 173.
40. Berkovsky, B. M., and Polevikov, V. K., "Numerical Study of Problems on High-Intensive Free Convection." *Heat Transfer and Turbulent Buoyant Convection*, edited by D.B. Spalding and H. Afgan, Hemisphere Publ., Vol. 2, 1977, p. 443.
41. Denny, V. E., and Clever, R. M., "Comparison of Galerkin and Finite Difference Methods for Solving Highly Nonlinear Thermally Driven Flows," *J. Comput. Phys.*, Vol. 16, 1974, p. 271.
42. Chu, H. N. S., and Churchill, S. W., "The Development and Testing of a Numerical Method for Computation of Laminar Natural Convection in Enclosures." *Computers and Chemical Engineering*, Vol. 1, 1977, p. 103.
43. Roux, B., Grondin, J.C., Bontoux, P., and Gilly, B., "On a High Order Scheme for Natural Convection in a Vertical Square Cavity." *Numer. Heat Tfranfer*, Vol. 1, 1978, p. 331.
44. Kublbeck, K., Merker, G. P., and Straub, J., "Advanced Numerical Computation of Two Dimensional Time Dependent Free Convection in Cavities." *Int. J. Heat Mass Transfer*, Vol. 23, 1980, p. 203.
45. Lauriat, G., "Numerical Study of Natural Convection in a Narrow Vertical Cavity: An examination of High Order Schemes." *ASME–AIChE Heat Transfer Conf.*, Orlando, 1980, ASME Paper No. 80-HT-90.
46. Tobarrok, B., and Lin, R. C., "Finite Element Analysis of Free Convection Flows." *Int. J. Heat Mass Transfer*, Vol. 20, 1977, p. 945.
47. Gartling, D. K., "Convective Heat Transfer Analysis by the Finite Element Method." *Comput. Meth. Appl. Mech. Eng.*, Vol. 12, 1977, p. 305.
48. Heinrich, J. C., Marshall, R. S., and Zienkiewicz, O. C., "Penalty Function Solution of Coupled Convective and Conductive Heat Transfer." *Int. Conf. Numer. Meth. Laminar Turbulent Flow*, 1978.
49. Marshall, R. S., Heinrich, J. C., and Zienkiewicz, O. C., "Natural Convection in a Square Enclosure by a Finite Element Penalty Function Method Using Primitive Fluid Variables." *Numer. Heat Transfer*, Vol. 1, 1978, p. 315.
50. Reddy, J. N., and Mamidi, D. R., "Penalty Velocity - Stream Function Finite Element Models for Free Convection Heat Transfer Problems." *Recent Adv. Eng. Sci.*, edited by R. L. Sierakowski, Univ. Florida, Gainesville, 1978, p. 381.
51. Reddy, J. N., "Penalty Finite Element Methods for the Solution of Advection and Free Convection Flows." *Finite Element Methods in Engineering*, edited by A. P. Kabaila and V. A. Pulmano, The Univ. New South Wales, Sydney, Australia, 1979, p. 583.
52. Reddy, J. N., and Satake, A., "A Comparison of a Penalty Finite Element Model with the Stream Function–Vorticity Model of Natural Convection in Enclosures." *J. Heat Transfer*, Vol. 102, 1980, p. 659.
53. Schinkel, W. M. M., Linthorst, S. J. M., and Hoogendoorn, C. J., "The Stratification in Natural Convection in Vertical Enclosures." *J. Heat Transfer*, Vol. 105, 1983, p. 267.
54. Gilly, B., Bontoux, P., and Roux, B., "Influence Des Conditions Thermiques de Paroi sur la Convection Naturelle Dans Une Cavite Rectangulaire Verticale, Differentiellement Chauffée." *Int. J. Heat Mass Transfer*, Vol. 24, 1981, p. 289.
55. Linthorst, S. J. M., and Schinkel, W. M. M., "Flow Structure with Natural Convection in Inclined Air-Filled Enclosures." *J. Heat Transfer*, Vol. 103, 1981, p. 535.
56. Korpela, S. A., Lee, Y., and Drummond, J. E., "Heat Transfer Through a Double Pane Window." *J. Heat Transfer*, Vol. 104, 1982, p. 539.
57. de Vahl Davis, G., and Mallinson, G. D., "A Note on Natural Convection in a Vertical Slot." *J. Fluid Mech.*, Vol. 75, 1975, p. 87.
58. Elsherbiny, S. M., Raithby, G. D., and Hollands, K. G. T., "Heat Transfer by Natural Convection Across Vertical and Inclined Air Layers." *J. Heat Transfer*, Vol. 104, 1982, p. 96.
59. Lee, Y., and Korpela, S. A., "Multicellular Natural Convection in a Vertical Slot." *J. Fluid Mech.*, Vol. 126, 1983, p. 91.
60. Gershuni, G. Z., "Stability of Plane Convective Motion of a Liquid." *Zh. Tech. Fiz.*, Vol. 23, 1953, p. 1838.

61. Rudakov, R. N., "Spectrum of Perturbations and Stability of Convective Motion Between Vertical Planes." *PMM J. Appl. Math. Mech.*, Vol. 31, 1967, p. 349.
62. Vest, C. M., and Arpaci, V. S., "Stability of Natural Convection in a Vertical Slot." *J. Fluid Mech.*, Vol. 36, 1969, p. 1.
63. Birikh, R. V., Gershuni, G. Z., Zhukhovitskii, E. M., and Rudakov, R. N., "On Oscillatory Instability of Plane Parallel Convective Motion in a Vertical Channel." *PMM J. Appl. Math. Mech.*, Vol. 36, 1972, p. 7.
64. Korpela, S. A., Gözüm, D., and Baxi, C. B., "On the Instability of the Conduction Regime of Natural Convection in a Vertical Slot." *Int. J. Heat Mass Transfer*, Vol. 16, 1973, p. 1683.
65. Ruth, D. W., "On the Transition to Transverse Rolls in an Infinite Vertical Fluid Layer—A Power Series Solution." *Int. J. Heat Mass Transfer*, Vol. 22, 1979, p. 1199.
66. Shaaban, A. H., and Ozisik, M. N., "The Effect of Nonlinear Density Stratification on the Stability of a Vertical Water Layer in the Conduction Regime." *J. Heat Transfer*, Vol. 105, 1983, p. 130.
67. Bergholz, R. F., "Instability of Steady Natural Convection in a Vertical Fluid Layer." *J. Fluid Mech.*, Vol. 84, 1978, p. 743.
68. Schinkel, W. M. M., "Natural Convection in Inclined Air Filled Enclosures." Ph.D. thesis, Delft University, 1980.
69. Grondin, J. C., and Roux, B., "Recherche de Corrélations Simples Experiment les Pertes Convectives dan une Cavitié Bidimensionnelle Incline, Chauffée Differentiellement." *Revue Physique Appliquée*, Vol. 14, 1979, p. 49.
70. Raithby, G. D., and Wong, H. H., "Heat Transfer by Natural Convection Across Vertical Air Layers." *Numer. Heat Transfer*, Vol. 4, 1981, p. 447.
71. Hollands, K. G. T., and Konicek, L., "Experimental Study of the Stability of Differentially Heated Inclined Air Layers." *Int J. Heat Mass Transfer*, Vol. 16, 1973, p. 1467.
72. Aziz, K., and Hellums, J. D., "Numerical Solution of the Three Dimensional Equations of Motion for Laminar Natural Convection." *Phys. Fluids*, Vol. 10, 1967, p. 314.
73. Mallinson, G. D., and de Vahl Davis, G., "The Method of the False Transient for the Solution of Coupled Elliptic Equations." *J. Comput. Phys.*, Vol. 12, 1973, p. 435.
74. Mallinson, G. D., and de Vahl Davis, G., "Three Dimensional Natural Convection in a Box: A Numerical Study." *J. Fluid Mech.*, Vol. 83, 1977, p. 1.
75. Oze, H., Okamoto, T., Churchill, S. W., and Sayama, H., "Natural Convection in Doubly Inclined Rectangular Boxes." *Proc. Sixth Int. Heat Transfer Conf.*, Toronto, Canada, Part 2, 1978, p. 293.
76. Morrison, G. L., and Tran, V. Q., "Laminar Flow Structure in Vertical Free Convective Cavities," *Int. J. Heat Mass Transfer*, Vol. 21, 1978, p. 203.
77. Elsherbiny, S. M., Hollands, K. G. T., and Raithby, G. D., "Effects of Thermal Boundary Conditions on Natural Convection in Vertical and Inclined Air Layers." *J. Heat Transfer*, Vol. 104, 1982, p. 515.
78. Meyer, B. A., Mitchell, J. W., and El-Wakel, M. M., "The Effect of Thermal Wall Properties on Natural Convection in Inclined Rectangular Cells." *J. Heat Transfer*, Vol. 104, 1982, p. 111.
79. Turner, J. S., *Buoyancy Effects in Fluids*, Cambridge Univ. Press, 1979, p. 251.
80. Cormack, D. E., Leal, L. G., and Imberger, J., "Natural Convection in a Shallow Cavity with Differentially Heated End Walls Part I. Asymptotic Theory." *J. Fluid Mech.*, Vol. 65, 1974, p. 209.
81. Cormack, D. E., Leal, L. G., and Seinfeld, J. H., "Natural Convection in a Shallow Cavity with Differentially Heated End Walls Part II. Numerical Solutions." *J. Fluid Mech.*, Vol. 65, 1974, p. 231.
82. Cormack, D. E., Stone, G. P., and Leal, L. G., "The Effect of Upper Surface Conditions on Convection in a Shallow Cavity With Differentially Heated End Walls." *Int. J. Heat Mass Transfer*, Vol. 18, 1975, p. 635.
83. Imberger, J., "Natural Convection in a Shallow Cavity with Differentially Heated End Walls Part III. Experimental Results." *J. Fluid Mech.*, Vol. 65, 1974, p. 247.

84. Ostrach, S., "Fluid Mechanics in Crystal Growth—The 1982 Freeman Scholar Lecture." *ASME J. Fluids Engng.*, Vol. 105, 1983, p. 5.
85. Bejan, A., and Tien, C. L., "Laminar Natural Convection Heat Transfer in a Horizontal Cavity With Different End Temperatures," *J. Heat Transfer*, Vol. 100, 1978, p. 641.
86. Bejan, A., and Tien, C. L., "Laminar Free Convection Heat Transfer Through Horizontal Duct Connecting Two Fluid Reservoirs at Different Temperatures," *J. Heat Transfer*, Vol. 100, 1978, p. 725.
87. Bejan, A., and Imberger, J., "Heat Transfer by Forced and Free Convection in a Horizontal Channel with Differentially Heated Ends." *J. Heat Transfer*, Vol. 101, 1979, p. 417.
88. Shiralkar, G. S., and Tien, C. L., "A Numerical Study of Natural Convection in Shallow Cavities." *J. Heat Transfer*, Vol. 103, 1981, p. 226.
89. Shiralkar, G. S., Gadgil, A., and Tien, C. L., "High Rayleigh Number Convection in Shallow Enclosures with Different End Temperatures." *Int. J. Heat Mass Transfer*, Vol. 24, 1981, p. 1621.
90. Al-Hommoud, A. A., and Bejan, A., "Experimental Study of High Rayleigh Number Convection in a Horizontal Cavity with Different End Temperatures." Report CUMER-79-1; Dept. Mech. Eng., Univ. Colorado, Boulder, 1979.
91. Ostrach, S., "Natural Convection in Low Aspect-Ratio Rectangular Enclosures," *Proc. 19th Nat'l Heat Transfer Conf.*, Orlando, Florida, 1980, p. 1.
92. Patankar, S. V., and Spalding, D. B., "Numerical Prediction of Three dimensional Flows." Mech. Eng. Dept. Rep. No. EF/T-N/A 46, Imperial College, London, 1972.
93. Tichy, J., and Gadgil, A., "High Rayleigh Number Laminar Convection in Low Aspect Ratio Enclosures with Adiabatic Horizontal Walls and Differentially Heated Vertical Walls." *J. Heat Transfer*, Vol. 104, 1982, p. 103.
94. Bauman, F., Gadgil, A., Kammerud, R., and Greif, R., "Buoyancy Driven Convection in Rectangular Enclosures: Experimental Results and Numerical Calculations." ASME publication 80-HT-66; *paper presented at the 19th Nat'l. Heat Transfer Conf.*, Orlando, Florida, 1980.
95. Hart, J. E., "Low Prandtl Number Convection Between Differentially Heated End Walls." *Int. J. Heat Mass Transfer*, Vol. 26, 1983, p. 1069.
96. Hurle, D. T. J., "Temperature Oscillations in Molten Metals and Their Relationship to the Growth State in Melt Grown Crystals." *Phil. Mag.*, Vol. 13, 1966, p. 305.
97. Solan, A., and Ostrach, S., "Convection Effects in Crystal Growth by Closed Tube Chemical Vapor Transport." *Preparation and Properties of Solid State Materials*, edited by W. Wilcox, Marcel Dekker, Vol. 2, 1979, p. 63.
98. Pimputkar, S. M., and Ostrach, S., "Convective Effects in Crystals Grown From Melt." *J. Crystal Growth*, Vol. 55, 1981, p. 614.
99. Hurle, D. T. J., and Jakeman, E., (eds.). "The Role of Convection and Fluid Flow in Solidification and Crystal Growth." *Physico-chemical Hydrodynamics*, Vol. 2, 1981, p. 239.
100. Ostrach, S., Loka, R. R., and Kumar, A., "Natural Convection in Low Aspect Ratio Rectangular Enclosures." *Natural Convection in Enclosures*, edited by K. E. Torrance and I. Catton, HTD—Vol. 1, ASME, 1980.
101. Kamotani, Y., Wang, L. W., and Ostrach, S., "Experiments on Natural Convection Heat Transfer in Low Aspect Ratio Enclosures," *AIAA Journal*, Vol. 21, 1983, p. 290.
102. Jhaveri, B. S., and Rosenberger, F., "Expansive Convection in Vapor Transport Across Horizontal Rectangular Enclosures." *J. Crystal Growth*, Vol. 56, 1982, p. 57.
103. Ostrach, S., "Natural Convection with Combined Driving Forces." *Physicochemical Hydrodynamics*, Vol. 1, 1980, p. 233.
104. Chen, C. F., Briggs, D. G., and Wirtz, R. A., "Stability of Thermal Convection in a Salinity Gradient Due to Lateral Heating." *Int. J. Heat Mass Transfer*, Vol. 14, 1971, p. 56.
105. Wirtz, R. A., "The Effect of Solute Layering on Lateral Heat Transfer in an Enclosure." *Int. J. Heat Mass Transfer*, Vol. 20, 1979, p. 841.
106. Wang, L. W., "Experimental Study of Natural Convection in a Shallow Horizontal Cavity with Different End Temperatures and Concentrations." Ph.D. thesis, Dept. Mech. Eng., Case Western Reserve Univ., Cleveland, Ohio, 1982.

107. Kenning, D. B. R., "Two Phase Flow With Non-Uniform Surface Tension." *Appl. Mech. Rev.*, Vol. 21, 1968, p. 1101.
108. Levich, V. R., and Krylov, V. S., "Surface Tension Driven Phenomena." *Ann. Rev. Fluid Mech.*, Vol. 1, 1969, p. 293.
109. Yih, C. S., "Fluid Motion Induced by Surface Tension Variation." *Phys. Fluids*, Vol. 11, 1968, p. 477.
110. Yih, C. S., "Three Dimensional Motion of a Liquid Film Induced by Surface Tension Variation or Gravity." *Phys. Fluids*, Vol. 12, 1969, p. 1982.
111. Adler, J., and Sowerby, L., "Shallow Three Dimensional Flows with Variable Surface Tension." *J. Fluid Mech.*, Vol. 42, 1970, p. 549.
112. Sirignano, W. A., and Glassman, I., "Flame Spreading Over Liquid Fuels: Surface Tension Driven Flows." *Comb. Sci. Tech.*, Vol. 1, 1970, p. 307.
113. Torrance, K. E., "Subsurface Flows Preceding Flame Spreading Over a Liquid Fuel." *Comb. Sci. Tech.*, Vol. 3, 1971, p. 113.
114. Young, N. O., Goldstein, J. S., and Block, M. J., "The Motion of Bubbles in a Vertical Temperature Gradient." *J. Fluid Mech.*, Vol. 6, 1959, p. 350.
115. Subramanian, R. S., "Slow Migration of a Gas Bubble in a Thermal Gradient." *AIChE Journal*, Vol. 27, 1981, p. 646.
116. Ostrach, S., "Convection Phenomena of Importance for Materials Processing in Space." *Proc. COSPAR Symposium on Materials Sciences in Space*, Philadelphia, 1976, p. 3.
117. Ostrach, S., "Low Gravity Fluid Flows." *Ann. Rev. Fluid Mech.*, Vol. 15, 1982, p. 313.
118. Levich, V. G., *Physico Chemical Hydrodynamics*, Prentice Hall, 1962, p. 384.
119. Birikh, R. V., "Thermocapillary Convection in a Horizontal Layer of Liquid." *J. Appl. Mech. Tech. Phys.*, Vol. 7, 1966, p. 43.
120. Adler, J., "Fluid Mechanics of a Shallow Fuel Layer Near a Burning Wick." *Comb. Sci. Tech.*, Vol. 2, 1970, p. 105.
121. Babskiy, V. G., Skolvskaya, I. L., and Sklovskiy, Y. B., "Thermocapillary Convection in Weightless Conditions." *Space Studies in the Ukraine, No. 1: Space Materials Studies and Technology*, edited by G. S. Pisarenko, Nankova Domka, Kieve, 1973, p. 121.
122. Ostrach, S., "Motion Induced by Capillarity." *Physico Chemical Hydrodynamics*, edited by V. G. Levich, Vol. 2, 1977, p. 571.
123. Sen, A. K., and Davis, S. H., "Steady Thermocapillary Flows in Two-Dimensional Slots." *J. Fluid Mech.*, Vol. 121, 1982, p. 163.
124. Sen, A. K., Smith, M. K., Davis, S. H., and Homsy, G. M., "Thermocapillary Flows and Their Stability." *Materials Processing in the Reduced Gravity Environment of Space*, edited by Guy E. Rindone, Elsevier Publishing, 1982, p. 173.
125. Strani, M., Piva, R., and Graziani, G., "Thermocapillary Convection in a Rectangular Cavity: Asymptotic Theory and Numerical Simulation." *J. Fluid Mech.*, Vol. 130, 1983, p. 347.
126. Homsy, G. M., and Meiburg, E., "The Effect of Surface Contamination on Thermocapillary Flow in a Two-Dimensional Slot." *J. Fluid Mech.*, Vol. 39, 1984, p. 443.
127. Pimputkar, S. M., and Ostrach, S., "Transient Thermocapillary Flow in Thin Layers." *Phys. Fluids*, Vol. 23, 1980, p. 1281.
128. Lai, C. L., Ostrach, S., and Kamotani, Y., "Effects of Surface Deformation on Unsteady Thermocapillary Flows." *J. Fluid Mech.*, to appear.
129. Smith, M. K., and Davis, S. H., "Instabilities of Dynamic Thermocapillary Liquid Layers Part 1. Convective Instabilities." *J. Fluid Mech.*, Vol. 132, 1983, p. 119.
130. Smith, M. K., and Davis, S. H., "Instabilities of Dynamic Thermocapillary Liquid Layers Part 2. Surface Wave Instabilities." *J. Fluid Mech.*, Vol. 132, 1983, p. 145.

CHAPTER 28

MODELING WAVY FILM FLOWS

S. P. Lin

Department of Mechanical & Industrial Engineering
Clarkson University
Potsdam, NY

and

C. Y. Wang

Department of Mathematics
Michigan State University
East Lansing, MI

CONTENTS

INTRODUCTION

The flow of a thin layer of liquid can be observed when rain overflows the eaves trough, flows down a window pane, or when one hoses down a driveway. More important applications of such "film flow" are chemical engineering processes, including film cooling, gas absorption, reactors, condensers, evaporators, slot coating, and dip painting. In these cases the film is usually bounded by a solid boundary on one side with a free surface on the other side bounded by a gas. The solid boundary may be planar or curved. In order to simplify the discussion, we shall concentrate on the basically plane film flow on an incline. One may refer to Dukler and Wicks [1] for a description of film flow inside a tube.

Most experiments on film flow use water flowing down an inclined flat surface. If we define a Reynolds number as (average velocity) (average thickness)/(kinematic viscosity), experiments show the film appears to be smooth when the Reynolds number is very small, about 3 csc β, where β is the angle of inclination [2]. In this regime the flow is parallel with a parabolic velocity profile. At higher Reynolds numbers small wavy ripples appear on the surface. Generally, there seems to be a preferred wavelength and wave speed. At still higher Reynolds numbers, less regular larger "roll waves" appear. These roll waves may be laminar or turbulent with an amplitude as large as the film thickness. For Reynolds numbers above about 1,500 the surface is definitely stochastic and

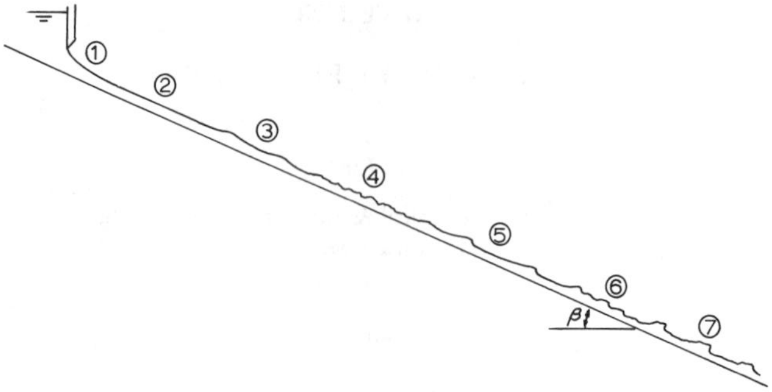

Figure 1. Representative experimental result for film flow over a long incline: (1) entrance region, (2) smooth parallel flow, (3) smooth sinusoidal waves, (4) zone of reorganization, (5) roll waves with laminar tail, (6) turbulent bursts, (7) turbulent roll waves.

turbulent. Excellent reviews of film flow regimes may be found in Levich [3], Fulford [2], Levich and Krylov [4], and Dukler [5].

The Reynolds number is not the only determinant of the condition of the film. The Froude number, a measure of gravity, can be shown to be related to the Reynolds number and the slope. Incipient wavy flows are observed when the Froude number is about unity [2]. The addition of surface active agents greatly suppresses the capillary waves but has little effect on the larger roll waves [6, 7]. The damping mechanism seems to involve not only surface tension represented by the Weber number, but also surface elasticity and surface viscosity.

Another important determinant of film condition is the entrance length. Using an inclined channel of 24 feet length Mayer [8] showed that, starting from the reservoir downstream, one encounters perfectly smooth uniform flow, not recordable long smooth "sinusoidal" waves, capillary ripples, zone of rearrangement and coalescence, roll waves separated by laminar film, turbulent bursts, and turbulent roll waves (slug flow) (Figure 1). This data leads us to suspect that the conclusions of earlier experimental results, where short channels of several feet in length were used, may not reveal the complete picture. Further evidences were presented by Takahama and Kato [9] who showed the onset of turbulence was highly dependent on entrance length, and Brauner and Maron [10] who showed the inception of waviness is also quite length dependent.

The wavy regime of film flow is of considerable interest since the waves enhance mass, momentum, and energy transfer across the free surface [5, 11]. Experimental results show mass transfer may be increased by several fold [6, 12]. The waviness of the surface does increase the transport surface area, but the increase is only several percent [13], which cannot explain the large observed increases. Most of the increase in transfer is probably due to convective mixing, both inside and outside the film [5, 14]. Thus, a knowledge of the flow is essential in the prediction of viable transfer rates.

In what follows, we shall present some current methods in the modeling of wavy film flows. As pointed out by Dukler [5], the literature in this area is vast, particularly in recent years. We apologize here for not being able to include all of the worthy papers in this short review.

BASIC EQUATIONS

Electromagnetic forces, mass and heat transfer, and phase changes will be excluded from consideration. Moreover only incompressible Newtonian fluids are considered. The governing differential equations are:

$$\rho(\partial_t \underline{V} + \underline{V} \cdot \underline{\nabla}\underline{V}) = -\underline{\nabla}P + \rho\underline{g} + \underline{\nabla} \cdot \underline{\underline{\tau}} \tag{1}$$

$$\underline{\nabla} \cdot \underline{V} = 0 \tag{2}$$

where ρ is the fluid density, t is time, $\underline{V}$ is the velocity, P is the pressure, g is the gravitational acceleration, and $\underline{\underline{\tau}}$ is the stress tensor given by

$$\underline{\underline{\tau}} = -P\underline{\underline{\delta}} + \rho\nu(\underline{\nabla}\underline{V} + (\underline{\nabla}\underline{V})^{\dagger})$$

in which $\underline{\underline{\delta}}$ is the unit diadic, ν the kinematic viscosity, and $(\underline{\nabla}\underline{V})^{\dagger}$ is the transpose of $(\underline{\nabla}\underline{V})$.

The boundary conditions at the fluid-fluid interface Y = H(X, Z, t) are [15, 16]

$$[\underline{\underline{\tau}}] = \underline{\nabla}'\sigma + \underline{n}\sigma(R_1^{-1} + R_2^{-1}) \tag{3}$$

$$[\underline{V}] = 0 \tag{4}$$

$$\dot{Y} = \partial_t H + \underline{V} \cdot \underline{\nabla}\underline{H} \tag{5}$$

where (X, Y, Z) denote Cartesian coordinates, [ ] denotes the change of the quantity it brackets in the direction of the unit normal $\underline{n}$, $\underline{\nabla}'$ is the surface gradient, σ is the surface tension, R_1 and R_2 are the principal radii of curvature defined as positive if directing from the centers of curvature in the direction of $\underline{n}$, and the upper dot denotes total time differentiation. Equation 3 is the consequence of modeling the interface as a massless geometric surface endowed with only one mechanical property σ, and the second law of Newton. Equations 4 and 5 result from the assumption that the interfacial velocity is the same as the fluid velocity at the interface. The boundary condition at the homogeneous solid-liquid interface is the no-slip condition

$$\underline{V} = 0 \tag{6}$$

WAVY FILM

Consider a layer of liquid running down a flat inclined plane under the action of gravity as shown in Figure 2. The wavy film flow on a wavy surface is discussed later. When the flow is almost parallel, the film can be approximated by an infinite sheet of liquid with a smooth free surface parallel to the incline. In addition, if the viscous and inertia effects of the overlying air can be neglected, then the exact solution of Equations 1–6 gives the well-known Nusselt's semi-parabolic parallel flow solution

$$\bar{U}(Y) = (gd^2 \sin\beta/2\nu)[2(Y/d) - (Y/d)^2] \tag{7}$$

$$\bar{P}(Y) = \rho g d \cos\beta[1 - (Y/d)] \tag{8}$$

where $\bar{U}$ is the velocity in the X-direction, Y the distance measured from the inclined plane into the liquid, d is the uniform film thickness, and β is the angle of inclination.

The basic flow described by Equations 7 and 8 easily becomes unstable. The literature on the onset of instability in a liquid film on a smooth surface is vast and has been recently surveyed by Solesio [17, 18] and Lin [19, 20]. The nonlinear evolution of the film flow subsequent to the onset

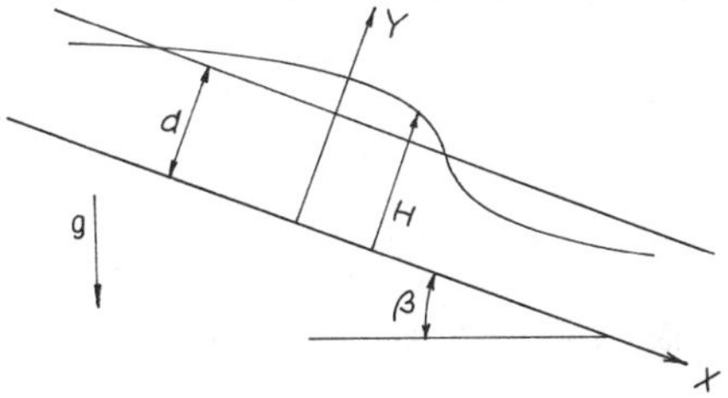

Figure 2. Definition sketch.

of instability has also been extensively, though as yet incompletely, studied. There have been mainly three different approaches to modeling wavy film flows. The first approach, initiated by Benney [21] and Mei [22], is to model waves of wavelength much greater than the film thickness with the long wave or shallow water expansion solution [23, 24] of the Navier-Stokes equations. The second approach, initiated by Kapitza [25, 26], is to seek the stationary wave solution of the Navier-Stokes equations with the boundary-layer approximation. Various refinements and extensions of Kapitza's original approach have appeared. The third approach, proposed by Dressler [27], is to use a shock condition to find a piecewise solution to an integral equation. These three approaches will be presently described.

Long Wave Expansion

For two dimensional wavy flows, Equations 1 and 2 can be written in a dimensionless form

$$p_x = \frac{1}{2}\alpha(u_{xx} + \alpha^{-2}u_{yy}) - \frac{1}{2}R(u_\tau + uu_x + vu_y) + \alpha^{-1} \tag{1'}$$

$$p_y = \frac{1}{2}\alpha\{(v_{yy} + \alpha^2 v_{xx}) - \alpha R(v_\tau + uv_x + vv_y)\} - \cot\beta \tag{2'}$$

where

$x = 2\pi X/\lambda, \qquad (y, h) = (Y, H)/d, \qquad u = U/U_m, \qquad v = V/\alpha U_m$

$p = P/\rho g d \sin\beta, \qquad \tau = t/(\lambda/2\pi U_m), \qquad U_m = U_{max} = gd^2 \sin\beta/2\nu$

$\alpha \equiv$ wave number $= 2\pi d/\lambda$

$\lambda \equiv$ wavelength

$R \equiv$ Reynolds number $= U_m d/\nu$

Equation 2 allows us to write

$$u = \bar{u} + \psi_y, \qquad v = -\psi_x$$

where ψ is the perturbation stream function, $\bar{u} = 2y - y^2$ and the subscripts denote partial differentiations. Introducing these velocity components into Equation 1′ we have after elimination of the pressure terms by cross-differentiation,

$$\begin{aligned}\psi_{yyyy} &= \alpha R\{\psi_{yy\tau} + (\bar{u} + \psi_y)\psi_{xyy} - (\bar{u}_{yy} + \psi_{yyy})\psi_x\} - 2\alpha^2\psi_{xxyy} \\ &\quad + \alpha^3 R\{\psi_{xx\tau} + (\bar{u} + \psi_y)\psi_{xxx} - \psi_x\psi_{xxy}\} - \alpha^4\psi_{xxxx}\end{aligned} \tag{1''}$$

Similarly, the dimensionless boundary conditions are the tangential and normal components of surface forces balance

$$(\bar{u}_y + \psi_{yy} - \alpha^2\psi_{xx})(1 - \alpha^2 h_x^2) - 4\alpha^2\psi_{xy}h_x = 0 \quad \text{at} \quad y = h \tag{3'}$$

$$\frac{W\alpha^2 h_{xx}}{(1 + \alpha^2 h_x^2)^{3/2}} + p + \alpha\psi_{xy}\frac{1 + \alpha^2 h_x^2}{1 - \alpha^2 h_x^2} \equiv N[x, h(\tau, x)] \equiv f(\tau, x) = 0 \quad \text{at} \quad y = h$$

where p is given by Equations 1′ and 2′. The kinematic condition:

$$h_\tau + (\bar{u} + \psi_y)h_x + \psi_x = 0 \quad \text{at} \quad y = h \tag{5'}$$

The no slip condition:

$$\psi_x = \psi_y = 0 \quad \text{at} \quad y = 0 \tag{6'}$$

where

$$W \equiv \text{Weber number} = \sigma/\rho g d^2 \sin\beta$$

To apply Equations 1′ and 2′ for p in Equation 3′, we use the relation

$$\partial_x f(\tau, x) = \partial_x N + \partial_h N \cdot h_x$$

Note that integration of Equation 2 from y = 0 to y = h gives, by virtue of Equation 5′, the following conservation equation:

$$h_\tau + \left[\int_0^h (\bar{u} + \psi_y)dy\right]_x = 0 \tag{5''}$$

The solution to this system has been expanded in terms of the small parameter α (wavelength long compared with film thickness) as follows [21, 22]:

$$\psi(x, y, \tau) = \sum_{n=0} \alpha^n \psi^{(n)}(x, y, \tau)$$

with

$$\psi^{(n)} = \sum_{m=0} A_m^{(n)}(x, \tau) y^m$$

The first two coefficients in the series are chosen to be zero so that the boundary conditions of Equation 6′ are satisfied. The rest of the coefficients are determined by demanding that the resulting series satisfies Equations 1″ and 3′ order by order. Substitution of the series solution, thus obtained up to $0(\alpha^2)$ into the condition of Equation 5″ then leads to

$$h_\tau + A(h)h_x + \alpha \frac{\partial}{\partial x}[B(h)h_x + C(h)h_{xxx}] + \alpha^2 \frac{\partial}{\partial x}[D(h)h_x^2 + E(h)h_{xx} + F(h)h_{xxxx}$$
$$+ G(h)h_x h_{xxx} + H(h)h_{xx}^2 + I(h)h_x^2 h_{xx}] + 0(\alpha^3) = 0 \tag{9}$$

where

$$A(h) = 2h^2, \qquad B(h) = \frac{8}{15}Rh^6 - \frac{2}{3}\cot\beta h^3, \qquad C(h) = \frac{2}{3}\alpha^2 W h^3$$

$$D(h) = \frac{1{,}016}{315}R^2h^9 + \frac{14}{3}h^3 - \frac{32}{15}R\cot\beta h^6$$

$$E(h) = \frac{32}{63}R^2h^{10} + 2h^4 - \frac{40}{63}R\cot\beta h^7$$

$$F(h) = \frac{40}{63}\alpha^2 RWh^7, \qquad G(h) = \frac{16}{3}\alpha^2 RWh^6$$

$$H(h) = \frac{16}{5}\alpha^2 RWh^6, \qquad I(h) = \frac{32}{5}\alpha^2 RWh^5$$

In arriving at Equation 9, it was assumed that $\alpha^2 W = 0(1)$ [28, 29]. This is the case in most experiments. When we put $\alpha^2 W = 0$, Equation 9 is reduced to Benney's [21] results for the case of $W = 0(1)$, except there were some minor algebraic errors. Benney's equation for $W = 0(1)$ was extended by Nakaya [30, 31] to include α^3—terms, and by Roskes [32] to the case of three-dimensional waves. Three-dimensional versions of Equation 9 up to $0(\alpha)$ and $0(\alpha^2)$ can be found respectively in the works of Atherton and Homsy [33, 34] and Krishna and Lin [35]. Equations similar to Equation 9 for cylindrical films can be found in Lin and Liu [36] and Atherton and Homsy [34].

Equation 9 can be linearized to give

$$f_\tau + 2f_x + \alpha(Bf_x + Cf_{xxx})_x + \alpha^2(Ef_{xx} + Ff_{xxxx})_x = 0 \tag{10}$$

where $f = h - 1$, and B, C, E, and F are the same functions given above but evaluated at $h = 1$. The eigenvalue of Equation 10 is

$$c = c_r + ic_i = 2 + \alpha^2(F - E) + i\alpha(B - C)$$

It follows that the infinitesimal waves will travel at wave speeds $c_r = 2 + \alpha^2(F - E)$ depending on the wave number α, and that the film is stable or unstable depending on whether $c_i < 0$ or $c_i > 0$. Thus, the neutral stability curve $c_i = 0$ is given by

$$\alpha[8R/15 - (2/3)(\cot\beta + \alpha^2 W)] = 0$$

Hence, there are two branches of neutral curves: the lower branch, $\alpha = 0$, and the upper branch $R = (5/4)(\cot\beta + \alpha^2 W)$. These two branches intersect at the critical Reynolds number $R_c = (5/4)\cot\beta$. These findings are consistent with the results of Benjamin [37] and Yih [38] who obtained the solution of the Orr-Sommerfeld equation for small α. The solution for finite α have been obtained by various authors [17, 18, 39–42] by use of different approximations with varied accuracy. No critical Reynolds number smaller than that given above was found up to $R = 0(100)$. Some sample results are given in Figures 3 and 4.

As is to be expected, experimental points near $\alpha = 0$ are very scarce. They require a long test section due to exceedingly long wavelengths and small amplification rates. This is the inherent

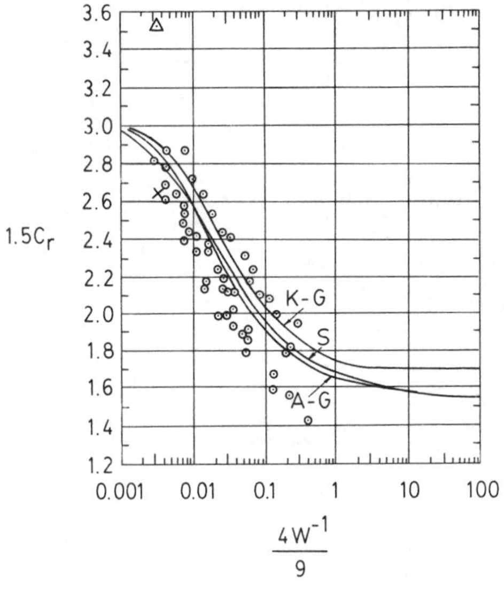

Figure 3. Variation of the wave speed c_r corresponding to the fastest growing wave with max (αc_i) as a function of W; $\beta = \pi/2$, $(\sigma/\rho)(3/g\nu^4)^{1/3} = 4587$. Experimental results: ○, Krantz and Goren (1971); Δ, Binnie (1957); x, Kapitza and Kapitza (1949). Theoretical results; A-G, Anshus and Goren (1966); S, Solesio (1977); K-G, Krantz and Goren (1971).

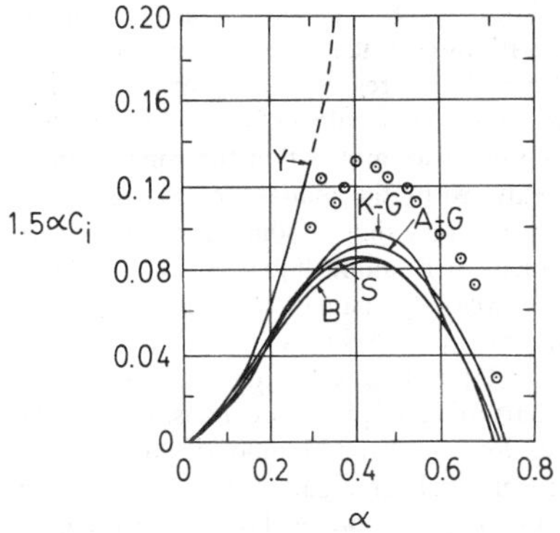

Figure 4. Variation of the amplification factor αc_i with the wave number α; $\beta = \pi/2$, R = 1.935, W = 0.844. Symbols as in Figure 2. B, Benjamin's theory (1957); Y, Yih's theory (1963).

difficulty with the experimental verification of the linear theory. The theoretical amplification rates in Figure 4 are smaller than the measured ones by as much as 50%, which is much larger than the claimed experimental accuracy. The difference cannot be due to the surface contamination, although a trace amount is known to have very large effects. This is because the surface contamination has been shown [38, 43–46] to stabilize the film, and thus if the surface could be made cleaner the difference would be even greater.

All waves speeds except that of Binnie [47] given in Figure 3 are smaller than twice the surface velocity, which is the maximum theoretical value. The scatter on wave speed data is not entirely due to experimental error. Different wave speeds for the same value of W correspond to different values of R. The larger wave speed observed by Binnie may be due to the larger R in his experiment. The larger R may correspond to larger wave amplitudes. No records of wave amplitude were given for all experimental points except that of Kapitza and Kapitza [25, 26] entered in Figure 3. Thus, the correlations given in these figures are incomplete. This is symptomatic of existing experimental records, including Binnie [47, 48], Jones and Whitaker [49], Kapitza and Kapitza, [26], Krantz and Goren [41], Stainthorp and Allen [50], Tailby and Portalski [51], and Fulford [2]. Complete records must include values of α, c_r, αc_i, β, R, W, and the entire time history of wave profile evolution. Note R and W are independent parameters of the flow. Various combinations of R and W used as single parameters in some of the existing works are the consequence of particular approximations used in the theories. These combined parameters are not appropriate for use in experimental correlations.

While all of the mentioned theories are for plane films, the measurements of Binnie and Kapitza and Kapitza were taken on annular films on the outer wall of a cylinder. The disturbances in such films indeed have higher amplification rates than those in plane films at the same flow parameters [36]. However, 50% increases in amplification rates do not occur until the film thickness and the cylinder radius become of the same order. The film thicknesses encountered in the experiments of Binnie and Kapitza and Kapitza were less than one-hundredth of the cylinder radii, and the curvature effects are not strong enough to account for the discrepancy. All evidences seem to suggest that the neglected nonlinear effects should not be overlooked.

Before looking into the nonlinear effects, we address the question of the relevance of the temporally varying disturbances to the onset of ripples [52, 53]. Experimental observations show clearly that ripples grow or decay as they travel downstream. Thus, they are spatially varying but not temporally varying everywhere with the same rate as is assumed in the temporal formulation. However, Gaster [54] proved that the temporal and spatial eigenvalues in plane parallel flows are simply related

$$c_r = \omega/\alpha_r \tag{11}$$

$$c_i = -(\alpha_i/\alpha_r)\partial(\alpha_r c_r)/\partial\alpha_r = -(\alpha_i/\alpha_r)c_g \tag{12}$$

where c_g is the group velocity of the temporally varying linear waves. The Gaster theorem implies that the neutral curves obtained from two different formulations coincide, since $\alpha_i = 0$ when $c_i = 0$. It should be pointed out that Gaster's theorem is valid only in the region $|\alpha_i(\alpha_r c_i) \max| \ll c_g$. This region turns out to be quite large in the present problem. Numerical results of Solesio [17] confirm Gaster's theorem near the neutral curve. The Gaster theorem also implies that the conclusion on stability resulting from temporal formulation will not be altered by the spatial formulation results except when the group velocity of temporal disturbances changes sign at some flow parameters. The first such example was given by the stability problem of a falling liquid curtain stretched between two verticle guide wires [55] in which the group velocity changes sign at $2\sigma/\rho QU = 1$, where Q is the volumetric discharge and U is the maximum curtain velocity.

A brief discussion on the nonlinear waves in a film on a flat surface is in order. Lin [44, 56, 57] solved the Navier-Stokes equation by extending the Stuart-Watson [58, 59] expansion formulas as reformulated by Reynolds and Potter [60] for parallel flows with rigid boundaries to that with a free surface. He showed that near the upper branch of the neutral curves, an unstable infinitesimal wave may evolve into a supercritically stable, small finite amplitude wave. These waves may be described adequately with the first three harmonics with its fundamental harmonic amplitude A evolving according to the Landau-Stuart equation

$$\frac{dA}{d\tau} = c_i A - a_2 |A^2| A \tag{13}$$

where a_2 is the second Landau coefficient depending on β, W, R and α. An exact solution of Equation 13 with the initial condition $A(-\infty) = 0$ is

$$A = |A_1| \exp[iB(\tau)\tau] \tag{14}$$

$$|A_1| = \left\{ \frac{c_i \exp[2c_i(\tau - \tau_0)]}{1 + a_{2r} \exp[2c_i(\tau - \tau_0)]} \right\}^{1/2}$$

$$B(\tau) = -\frac{a_{2i}}{\tau} \int_{\tau_0}^{\tau} |A_1|^2 \, d\tau$$

where τ_0 is an arbitrary time constant reflecting the arbitrary initial phase, and a_{2r} and a_{2i} stand respectively for the real and imaginary parts of a_2. It follows from Equation 14 that an equilibrium amplitude of c_i/a_{2r} and the corresponding reduction in wave speed of magnitude $-a_{2i}|A_1(\infty)|^2$ are reached as $\tau \to \infty$. The existence of supercritically stable nonlinear waves and the dependence of their wave speeds on wave amplitudes were also demonstrated by Gjevik [28] and Nakaya [30, 31] with different approaches. The former expanded the solution of Equation 9 without the $0(\alpha^2)$ terms into Fourier series with time dependent coefficients. A set of nonlinear first order differential equations of these coefficients was obtained by demanding the coefficient of each harmonic to vanish. The solution to this set of equations truncated at the third harmonics was obtained from the initial value problem. Equilibrated stationary waves were sought. Similar analysis was carried out by Nakaya who based his analysis on an equation similar to Equation 9, but for the case of weak surface tension such that $W = 0(1)$. Javdani and Goren [61] assumed that an equilibrated nonlinear wave travels with a constant velocity $c_r = 2$ and is the most amplified wave according to the linear theory. With this assumption, they sought periodic stationary solutions of Equation 9 without $0(\alpha^2)$ terms by expanding the solutions in Fourier series with constant coefficients. The major difficulty with this approach is that the dependence of the wave speed on the wave amplitude is denied from the outset. Table 1 gives some typical results of equilibrated monochromatic waves which can be constructed from the first several harmonics. In this table $r = (h_{max} - h_{min})/(h_{max} + h_{min})$. Some experimental results of Kapitza and Kapitza and Binnie are included for comparison.

Table 1
Supercritical Waves in Liquid Films, $\beta = 90°$

Data	Authors	c_r	r
Water, 15°C	Kapitza & Kapitza [26]	1.76	0.16
$c_i = 0.155$	Lin [56, 57]	1.78	0.225
$\alpha = 0.092$	Gjevik [28]	2.21	0.350
$R = 8.07$	Gjevik [29] (spatial)	1.71	0.260
$W = 463.3$	Lin [62]	1.81	0.174
	Agrawal [52] (spatial)	1.90	0.218
Alcohol, 15°C	Kapitza & Kapitza [26]	1.67	0.163
$c_i = 0.174$	Lin [56, 57]	1.73	0.221
$\alpha = 1.44$	Gjevik [28]	—	—
$R = 5.04$	Gjevik [29] (spatial)	1.66	0.270
$W = 107.2$	Lin [62]	1.80	0.178
	Agrawal [52] (spatial)	1.89	0.220
Water, 19°C	Binnie [48]	2.34	Not Observed
$c_i = 0.121$	Lin [56, 57]	2.01	—
$\alpha = 0.066$	Gjevik [28]	—	—
$R = 6.60$	Gjevik [29]	—	—
$W = 616.7$	Lin [62]	1.72	—
	Agrawal [52] (spatial)	2.34	—

It should be pointed out that in addition to the weakly nonlinear waves data shown in Table 1, Kapitza and Kapitza also obtained some experimental points for highly nonlinear waves the amplitudes of which are of the same order of magnitude as the film thickness. Theories of highly nonlinear waves will be mentioned shortly.

The equilibrated monochromatic waves have been shown [62] to be unstable with respect to side-band disturbances unless the side-band is sufficiently narrow. In the presence of the two-dimensional side-band disturbances the wave amplitude was shown to modulate according to the nonlinear Schrödinger equation

$$\frac{dA}{d\tau} - c_i A + a_1 A_{\xi\xi} + a_2 |A|^2 A = 0 \tag{15}$$

where ξ is the distance measured in the flow direction in a frame moving with the group velocity of the wave packet, and a_1 is a function of α, W, R and β. Equation 15 was derived from Equation 9 as the solubility condition for the amplitude perturbation solution of Equation 9. An equation more general than Equation 15 was obtained by Krishna and Lin [35] for the case of three-dimensional side-band disturbances. It was shown that the equilibrated monochromatic wave is also stable with respect to side-band disturbances of sufficiently narrow band-width. Stokes waves, on the contrary, are unstable with respect to narrow side-band disturbances [43]. The difference is due to the fact that in the Stokes waves there is no agent such as surface tension to store the energy generated by the side-band resonance.

Near the lower branch of the neutral stability curve, Equations 13 and 15 are inadequate to describe the nonlinear wave evolution. This is because the generated first several harmonics, unlike those near the upper branch of the neutral curve, still lie in the unstable region predicted by the linear theory (Figure 5), thus the nonlinear interaction is stronger, which requires more than the first several harmonics to describe. The evolution equation near the lower branch of the neutral curve was derived by Krishna and Lin [35] from the three-dimensional version of Equation 9. Only

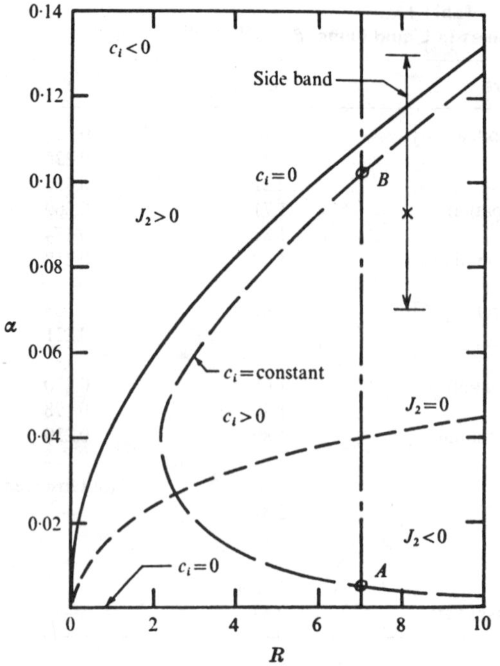

Figure 5. Supercritically stable waves; $\beta = \pi/2$, W = 463.3. X, Kapitza and Kapitza's experiments. B, weak modal interaction described by Equation 15. A, Strong modal interaction described by Equation 16.

the two-dimensional version of the evolution equation is given here,

$$f_\tau = -2f_x - 4ff_x - c_i f_{xx} - (2\alpha^3 W/3)f_{xxxx} - \alpha^2\{[2 + (32R/63)(R - 5\cot\beta/4)]f_{xxx} + (40\alpha^2 WR/63)f_{xxxxx}\} \quad (16)$$

where f = h − 1. The solution of this equation has not yet been reported.

Neglecting the α^2-terms in Equation 16, Atherton [33] described numerically the nonlinear time evolution of ripples until they reached a stationary periodic state. The equilibrated waves were reported to be insensitive to the initial data. Similar wave forms were found by Tougou [63], Nepomnyaschchii [64, 65], and Tsvelodub [66]. Sivashinsky and Michelson [67] found numerically that the nonlinear evolution based on the same equation led to chaotic irregular quasi-steady wave forms. They concluded that the deterministic equation is capable of exhibiting stochastic properties without explaining the discrepancy between their finding and the findings of existing works. However, it should be pointed out that before the free surface becomes chaotic, for whatever reason, the neglected stabilizing three-dimensional effects must become important. Thus, it is conjectured that if the z-dependence is included in Equation 16 [35] in computation, even for the flow condition that leads to two-dimensional chaos, one may still be able to find stable three-dimensional ripples with crests wavy in the cross-stream direction. Moreover, Lin and Krishna [35] showed that the final outcome of the wave evolution depends very sensitively on the initial condition. It should be pointed out that all of the calculated stationary wave forms based on Equation 16 without $0(\alpha^2)$ terms do not exhibit the characteristics of "single waves" observed by Kapitza and Kapitza. The profile of single waves have a steep front on which small wavelets ride. This wave front is connected by a smooth and gentle slope of the rear of the long single wave ahead of it. The wavelets on the steep crest suggest that the dispersion effects represented by the third and fifth derivatives terms in Equation 16 may not be neglected in describing the "single waves." It is anticipated that Equation 16 will be encountered in other weakly nonlinear phenomena in which nonlinear steepening, diffusion, dispersion, and restoring force (surface tension in the present problem) are of equal importance.

For large ripples whose amplitudes are of the same order as the film thickness, Equation 9 may still be used as long as the slope of the wave profile is everywhere much smaller than one. The possibility of associating the homoclinic trajectory of Equation 9 with the large amplitude solitary waves exists. Nonlinear studies of film waves of arbitrary wavelengths cannot be accomplished either by use of Equation 9 or by use of the boundary-layer approximation to be described next. A Galerkin's method of solving the full Navier-Stokes equation has been suggested by Levitch and Krylov [4] to model the stationary large amplitude film waves of arbitrary wavelength.

Boundary-Layer-Lubrication Theory

Earlier works based on this approach were reviewed by Dukler [5]. This approach, first adopted by Kapitza [26], was based on the boundary-layer approximation that neglects the pressure variation across the film thickness. Thus, the right side of Equation 2′ was completely neglected. Moreover, the velocity and pressure fields were made to satisfy Equation 1′ after it was integrated across the film thickness rather than the partial differential equation itself. Kapitza's original work has been refined and extended by many authors [68–76]. The approximate equations of motion for this approach are,

$$\frac{1}{2}R\int_0^h (u_\tau + uu_x + vu_y)\,dy = \int_0^h \left[\frac{1}{2}\alpha(u_{xx} + \alpha^{-2}u_{yy}) - p_x + \alpha^{-1}\right]dy$$

$$p_y = 0 \tag{17}$$

$$u_x + v_y = 0$$

For the dynamic boundary condition only the first order terms in Equation 3′ have been retained, i.e. at $y = h$

$$u_y = 0 \quad \text{and} \quad -p = \alpha^2 W h_{xx} \tag{18}$$

Note only the case of $\beta = \pi/2$ and $\alpha^2 W = 0(1)$ have been considered. For the free surface kinematic Equation (5″) is used:

$$h_\tau + \left[\int_0^h u\,dy\right]_x = 0 \tag{19}$$

The no-slip condition on the incline is

$$u = v = 0 \quad \text{at} \quad y = 0 \tag{20}$$

A stationary wave solution of the above system can be sought in the form

$$u(\xi) = \frac{3\bar{u}(\xi)}{h(\xi)}\left[y - \frac{y^2}{2h(\xi)}\right], \qquad h(\xi) = 1 + \phi(\xi) \tag{21}$$

where $\xi = x - c\tau$. c being a constant wave speed, and $\bar{u}(\xi)$ is the average film speed defined by

$$\bar{u}(\xi) = h^{-1}\int_0^h u\,dy$$

It can be shown that for the stationary wave solution, $\phi(\xi)$ must satisfy:

$$\alpha^3 W(1+\phi)^3\phi''' + \frac{1}{2}\alpha^2\left[\left(c-\frac{3}{2}u_0\right)+\frac{1}{2}(c-3u_0)\phi - c\frac{\phi^2}{2}\right]\phi''$$
$$+\frac{\alpha R}{2}\left[\left(c^2-\frac{12}{5}cu_0+\frac{6}{5}u_0^2\right)+\frac{1}{5}c^2\phi(\phi+2)+\frac{3\alpha}{R}(u_0-c)\phi'\right]\phi'$$
$$+\left[3-\frac{3}{2}c+3\phi+\phi^2\right]\phi+\left(1-\frac{3}{2}u_0\right)=0, \qquad (22)$$

where $u_0 = U_0/U_m$, U_0 being the average velocity in the basic flow. Equation 22 is a dimensionless form of the equation given by Webb [72], except there are some errors in his results. Different authors who used this approach neglected various terms in Equation 22. The linearized form of Equation 22 is

$$\alpha^3 W\phi''' + \frac{\alpha^2}{2}\left(c-\frac{3}{2}u_0\right)\phi'' + \frac{\alpha R}{2}\left(c^2-\frac{12}{5}u_0c+\frac{6}{5}u_0^2\right)\phi' + \left(3-\frac{3}{2}c\right)\phi + \left(1-\frac{3}{2}u_0\right)=0$$

For neutral periodic waves $\phi = e^{i\xi}$, c in $\xi = x - c\tau$ being real, this equation yields

$$1-\frac{3}{2}u_0 = 0$$

$$\frac{\alpha R}{2}\left(c^2-\frac{12}{5}u_0c+\frac{6}{5}u_0^2\right)-\alpha^3 W=0$$

$$\left(3-\frac{3}{2}c\right)-\frac{\alpha^2}{2}\left(c-\frac{3}{2}u_0\right)=0$$

It follows that

$$u_0=\frac{2}{3}, \qquad c=\frac{6-\alpha^2}{3-\alpha^2}=2+\frac{\alpha^2}{3}+\ldots, \qquad \alpha\left(\frac{2R}{3}-\alpha^2 W\right)=0$$

Thus, the average velocity is two third of the maximum velocity as it should be. The first term in the above wave speed agrees with that obtained from the solution of the Orr-Sommerfeld equation or from Equation 10. The neutral curves are $\alpha = 0$ and $R = 3\alpha^2 W/2$ at $\beta = \pi/2$. The second branch differs from $R = (5/4)(\cot\beta + \alpha^2 W)$ obtained with the linear stability theory. The differences arise from the fact that terms of higher order in α are not consistently included in Equations 17 and 18. The solution of Equation 22 is usually expanded in the Fourier series

$$\phi(\xi) = \bar{\beta}\sin\xi + \sum_{n=2}^{N}\bar{\beta}^n[A_n\sin(n\xi)+B_n\cos(n\xi)]$$

where $\bar{\beta}$ is a constant amplitude of the fundamental harmonic. Substituting this into Equation 22 or its reduced forms, and equating the coefficient of each harmonic to zero, one obtains a set of 2N nonlinear algebraic equations in 2N + 1 unknowns α, $\bar{\beta}$, c, A_n, B_n (n = 2 to n) for given parameters R, F and W. An additional equation is required to close the problem. Kapitza [25] suggested an equation obtained from a principle of minimum energy dissipation. Shkadov [69] suggested an equation based on minimization of the film thickness for a given film flow. An alternative is to treat c as a parameter. Hirshburg and Florschuetz [76] assumed that the small finite-amplitude wave most frequently observed corresponds to or is related to the most amplified wave predicted by linear stability theory. They retained six harmonics in their computation. The principles of minimum energy dissipation and minimum film thickness for a given flow lack rational bases. Moreover, the implied assumption that the equilibrated wave is unique for a given flow is inappropriate since analyses of Krishna and Lin [35, 54, 62] indicated that the equilibrated waves depend on the characteristic of the initial conditions.

In place of Equation 21, Ruckenstein and Berbente [70] expanded both of the velocity components into power series in $(1 - y)$ with the coefficient depending on ξ. The rest of the analysis was parallel to the Kapitza method, except that the closure problem was solved by correlating the amplitude of the fundamental harmonic with experimental data. It should be emphasized that the closure problem did not originate from any particular mathematical method but rather from the assumption that the equilibrated wave is unique regardless of initial conditions.

Telles and Dukler [77, 78] adopted the boundary-layer approximation i.e. $p_y = 0$ and solved the equation of motion corresponding to Equation 1″ with terms of orders higher than $0(\alpha^2)$ neglected. However, their equation was expressed in terms of the total stream function Ψ related to u and v by $u = \Psi_y$, $v = -\Psi_y$, and thus there was an additional term due to gravity, which was cancelled by viscous terms in mean flow of the long wave expansion approach. Moreover, they neglected the surface tension effect in the boundary conditions. The resulting differential system was solved iteratively utilizing the quasi-linearization method of Bellman [79] and Gram-Charlier series. The closure problem was solved by using the experimentally measured unperturbed part of the film thickness in the solution for long "single" waves. They reported that their description of wave profiles were in satisfactory agreement with experiments up to $R = 700$, while the previous results based on the Kapitza method were only satisfactory up to $R = 100$. However, various methods under the boundary-layer-lubrication approach described in this section neglect the acceleration effects represented by the right side of Equation 2′. For this reason alone, we cannot expect good agreements between theory based on the approach described here and experiments in the range of $R = 0(100)$.

Modeling Roll Waves Using Shocks

Roll waves are ridges of intense vorticity that occur in film flow at higher Reynolds numbers and/or larger distances from the entrance region. Since roll waves also occur in shallow water channel flow, they are of great interest to hydraulic engineers.

In general, roll waves have amplitudes comparable to the mean film thickness. Each wave has a steep front and a long, gently sloping tail. The mixing inside a roll wave is most likely turbulent while the long tail may be laminar or turbulent. Small secondary ripples are observed to superimpose on the roll waves. The distance between crests may remain constant as in a wave train but sometimes the waves may continuously overtake each other [8, 80, 81].

The mixing in roll waves seems to be the most important factor in the enhancement of mass transfer in wavy film flow. Oliver and Atherinos [82] showed roll waves increase mass transfer by 250% while regular sinusoidal waves (caused by artificial vibrations) have little effect. Simultaneous measurements of film thickness and transfer rate by Brauner and Maron [10, 83] also show small waves have negligible effect on transfer rates in comparison to roll waves. Thus, in modeling roll waves, the secondary ripples are ignored. As discussed before, surfactants may delay the formation of roll waves but have little effect on established roll waves. Thus, surface tension effects are also ignored in the modeling.

Due to the fact that roll waves have a steep front, the previously discussed modeling approaches, which assume small surface slopes, cannot be uniformly valid. A generally accepted method is to treat the sharp front as a mathematical shock or a hydraulic jump. Across the shock we require the conservation of mass and momentum but not energy. Dressler [27] first used this strategy on turbulent roll waves. He assumed the Chezy formula for water resistance and constructed a periodic permanent roll wave solution. Dressler's method was later extended to laminar tail by Ishihara, Iwagaki, and Iwasa [84] and to larger inclinations by Brock [81, 85].

Since Dressler's original work is quite complicated, we shall illustrate the modeling method with the simpler version of Ishihara et al. [84] as modified by Tamada and Tougou [86].

We normalize all lengths by a characteristic depth H (to be defined later), all velocities by the wave speed U, the time by H/U and pressure by ρU^2. If the long sloping tail is laminar, it can be described by the boundary layer equations

$$u_x + v_y = 0 \tag{23}$$

$$u_t + uu_x + vu_y = -p_x + k \sin \beta + \frac{1}{R} u_{yy} \tag{24}$$

$$0 = -p_y - k \cos \beta \tag{25}$$

Here (u, v) are velocity components in the (x, y) directions, R is the Reynolds number UH/ν, β is the inclination angle and $k \equiv gH/U^2$. The boundary conditions are that the velocities are zero at the bottom and that on the surface $y = h(x, t)$, shear and pressure are zero

$$u_y\big|_h = 0, \qquad p\big|_h = 0 \tag{26}$$

The kinematic surface condition gives

$$v\big|_h = h_t + uh_x \tag{27}$$

In the case of a turbulent tail, the last term in Equation 24 is substituted by the Chezy resistance formula

$$-\gamma^2 \frac{u|u|}{h} \tag{28}$$

where γ^2 is the roughness coefficient.

Using the momentum integral method, we integrate Equations 23 and 24 across the film. The result is

$$h_t + (\bar{u}h)_x = 0 \tag{29}$$

$$(\bar{u}h)_t + \frac{\partial}{\partial x}\int_0^h u^2\, dy = -k \cos \beta h h_x + k \sin \beta h - \frac{1}{R} u_y\big|_{y=0} \tag{30}$$

where $\bar{u}$ is the y average of u(x, y, t). Then we substitute for u the semi-parabolic profile of Nusselt:

$$u = 3\bar{u}\left(\frac{y}{h} - \frac{y^2}{2h^2}\right) \tag{31}$$

Equation 30 becomes

$$(\bar{u}h)_t + \frac{6}{5}\frac{\partial}{\partial x}(\bar{u}^2 h) = -k \cos \beta h h_x - k \sin \beta h - \frac{3\bar{u}}{Rh} \tag{32}$$

Now for a permanent wave train, where the flow will be steady to an observer moving with the wave, we set $\xi = x - t$ and Equation 29 can be integrated to obtain

$$H(\xi)[1 - \bar{u}(\xi)] = q \tag{33}$$

where q is the constant relative flow rate in the moving coordinates (ξ, y). Then, Equation 32 becomes

$$\frac{dh}{d\xi} = \frac{k \sin \beta h^3 + \frac{3}{R}(q - h)}{k \cos \beta h^3 + \frac{1}{5}h^2 - \frac{6}{5}q^2} \tag{34}$$

It can be shown, a posteriori, that the denominator in Equation 34 becomes zero somewhere in the tail. We define our characteristic depth H such that h = 1 at the singularity. Since the slope $dh/d\xi$ is finite, the numerator must be zero there also. Thus,

$$k \cos \beta + \frac{1}{5}(1 - 6q^2) = 0 \tag{35}$$

$$k \sin \beta + \frac{3}{R}(q - 1) = 0 \tag{36}$$

From Equations 35 and 36 we obtain the wave speed

$$U = \left\{\frac{75\, \nu g \cos^2\beta(1 - q)}{\sin \beta(6q^2 - 1)^2}\right\}^{\frac{1}{3}} \tag{37}$$

and the reference height

$$H = \frac{(6q^2 - 1)}{5g \cos \beta} U^2 \tag{38}$$

Across the shock from state 1 (supercritical) to state 2 (subcritical) mass and momentum are conserved.

$$\int_0^{h_1} (1 - u_1)\, dy = \int_0^{h_2} (1 - u_2)\, dy = q \tag{39}$$

$$\int_0^{h_1} (1 - u_1)^2\, dy + \int_0^{h_1} p_1\, dy = \int_0^{h_2} (1 - u_2)^2\, dy + \int_0^{h_2} p_2\, dy \tag{40}$$

Integrating Equation 40 with Equations 25, 31, and 35 gives the shock relation

$$h_2 = \frac{1}{2}\left\{\left(h_1 + \frac{2}{6q^2 - 1}\right)^2 + \frac{48q^2}{h_1(6q^2 - 1)}\right\}^{1/2} - \frac{1}{2}\left(h_1 + \frac{2}{6q^2 - 1}\right) \geq 1 \geq h_1 \tag{41}$$

Integration of Equations 34–36 yields the tail profile

$$\xi \tan \beta + \text{constant} = h + A \ln\left[h + \frac{1}{2} + \frac{1}{2}\sqrt{\left(\frac{1 + 3q}{1 - q}\right)}\right] + B \ln\left[h + \frac{1}{2} - \frac{1}{2}\sqrt{\left(\frac{1 + 3q}{1 - q}\right)}\right] \tag{42}$$

$$\begin{pmatrix} A \\ B \end{pmatrix} = \frac{1}{2(6q^2 - 1)}\left[1 \pm \frac{(1 - 3q)(1 + 4q)}{\sqrt{(1 - q)(1 + 3q)}}\right] \tag{43}$$

The wavelength is then

$$\xi(h_2) - \xi(h_1) \equiv \lambda(\alpha, q, h_1, h_2) \tag{44}$$

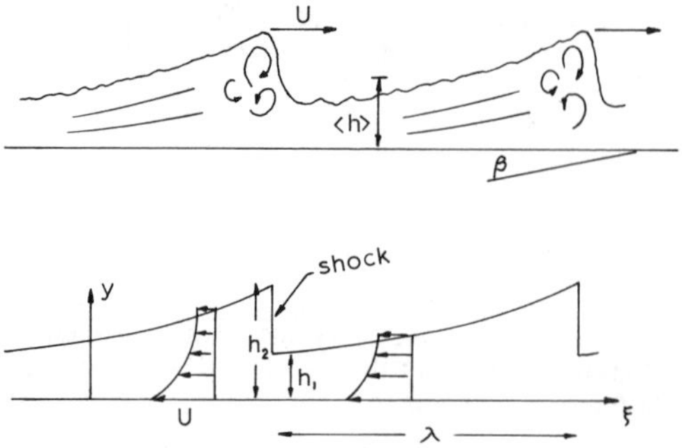

Figure 6. Top: Roll wave train profile reconstructed from bottom pressure measurements. The vertical scale has been exaggerated. Bottom: Typical modeled profile using moving coordinates (ξ, y).

The average film depth is

$$\langle h \rangle = \frac{1}{\lambda} \int_{\xi_1}^{\xi_2} h \, d\xi \tag{45}$$

and the average discharge, normalized by UH, is

$$Q = \langle h \rangle - q \tag{46}$$

The average film velocity is

$$\langle u \rangle = Q/\langle h \rangle \tag{47}$$

Figure 6 shows a typical modeled profile.

Thus, given the angle of inclination β and any two of the flow characteristics g, h_1, h_2, λ, $\langle h \rangle$, $\langle u \rangle$, U, Q one can theoretically predict all the other parameters. Using Q and $\langle h \rangle$, Ishihara et al. [84] found excellent agreement with experimental results for both laminar and turbulent tails. A more accessible reference of their results is in Ishihara et al. [87].

Of slightly different nature is the work of Brauner and Maron [83]. They modeled the roll wave by balancing y-averaged mass and momentum in three regions. The wave front region is a shock of finite width where the weight of the fluid is partially taken into account. The length of the wave back region is determined by a growing boundary-layer, and a thin substrate region where a plate withdrawal theory is applied.

As in the integral approach, the modeling of roll waves introduces one more unknown than relationships. Thus, prescribing only β and Q would lead to a family of solutions. One more characteristic need to be measured (e.g. $\langle h \rangle$ or frequency U/λ) or one more assumption needs to be made in order to make the problem closed. For roll waves Brauner and Maron [83] used Kapitza's [25] criteria of minimum $\langle h \rangle$ while Tamada and Tougou [86] studied the stability of established roll waves and used the most unstable wave.

Although we have some success in predicting the wave velocity and wave profile, the theory of roll waves still needs substantial improvement. This is because the velocity distribution inside the film has never been theoretically determined. The internal velocity is important in the viable calculation of transfer coefficients especially when the vertical motion under the crest has been shown to be a major factor in the increase in transfer rate.

DISCUSSION

The linear stability theory based on the parallel basic flow of Nusselt has produced predictions that agree qualitatively with experiments on the onset of surface waves. It should be pointed out that film flows may also become unstable due to shear waves at large R and small β [48, 88–90] as well as due to the as yet poorly understood mechanism of film rupture at extremely small R [91]. The works on the stability of multi-layered liquid films, which are of considerable importance for coating industries, were reviewed by Lin [20]. Among others, it was shown that a top heavy configuration may be stable due to interfacial tension [20, 92, 93]. The results based on nonlinear theories indicate that the equilibrated small finite amplitude waves are not unique but depend on the initial conditions. This appears to be the origin of the physically difficult closure problem encountered when the equilibrated waves are assumed to be unique. The boundary-layer-lubrication approach is mathematically simpler but less consistent than the long wave expansion approach in retaining successively smaller terms. For laminar wavy flow of amplitude comparable with the film thickness but with wavelength much larger than the amplitude, the long wave expansion approach will yield more accurate results if efficient synergestic analytical-computational methods can be found. If the stabilizing three-dimensional surface effects are included [35], and the initial conditions are considered [92], more useful results for large R may follow from the three-dimensional version of Equation 9. The modeling of laminar wavy flows of arbitrary amplitude and wave length still remains a virgin territory.

Since wavy film flow enhances heat and mass transfer, there exist experimental efforts to artificially promote wavy flows by altering the character of the bottom plate. It was found that random roughness of the plate does not seem to affect the laminar regime [2], while in turbulent flow roughness changes slightly roll wave characteristics [87]. Large-scale roughness or irregularities of the bottom plate, however, can substantially increase turbulent mixing. Davis and Warner [94] found that evenly spaced square ridges, of height comparable to film thickness, may increase mass transfer by 350%. Theoretically, Tougou [95] considered a wavy inclined plate where the wave length is large compared to the film thickness. He found, to first order, the small waviness of the plate does not affect film stability. Wang [96] also considered laminar film flow over a small amplitude wavy bottom although the wavelength is of same order as the film thickness. He determined the rippling on the surface but convective recirculation was not found. There exists no theoretical modeling for large-scale perturbances for either laminar or turbulent flow. For an estimation of turbulent mass transfer over ridges one may refer to a semi-empirical model of Brumfield et al. [97].

With the advent of large efficient computers, film flow may be determined by complete numerical integration of the Navier-Stokes equations. Using finite element methods, Orr and Scriven [98] studied film flow inside a rotating cylinder and Saito and Scriven [99] studied coating flows. So far, these numerical methods have been unable to obtain solutions of wavy film flows, whether the waviness is caused by instability or bottom corrugations.

NOTATION

a_2	Landau coefficient	P, p	pressure
c_g	group velocity	Q	average volumetric discharge
c_r	phase velocity	q	flow rate
d	film thickness	$R_{1,2}$	radii of curvature
g	gravitational acceleration	t	time
H	height	$\bar{U}$	average velocity
h	height	V	velocity
k	coefficient		

Greek Symbols

α	wave number	ξ	distance
β	angle of inclination	ρ	density
γ^2	roughness coefficient	σ	surface tension
$\underline{\underline{\delta}}$	unit diadic	τ	stress tensor
ν	kinematic viscosity	τ_0	time constant

REFERENCES

1. Dukler, A. E., and Wicks, M., "Gas-Liquid Flow in Conduits," in *Modern Chemical Engineering*, (A. Acrivos, ed.) Van Nostrand-Reinhold Press, New York (1963), 349.
2. Fulford, G. D., "The Flow of Liquids in Thin Films," *Advances in Chemical Engineering*, 5 (1964), 151.
3. Levich, V. E., *Physiochemical Hydrodynamics*, Prentice Hall, Chap. 12 (1962).
4. Levich, V. G., and Krylov, V. S., "Surface-Tension-Driven Phenomena," *Annual Review of Fluid Mech.* 1 (1969), 293.
5. Dukler, A. E., "Characterization, Effects on Modeling of the Wavy Gas-Liquid Interface," in *Progress in Heat and Mass Transfer* (G. Hetsroni, S. Sideman and J. P. Hartnet, eds.), Pergamon Press, New York, 6 (1972), 207.
6. Emmert, R. E., and Pigford, R. L., "A Study of Gas Absorptin in Falling Liquid Films," *Chem. Eng. Prog.* 50 (1954), 87.
7. Narasimhan, T. V., and Davis, E. J., "Surface Waves and Surfactant Effects in Horizontal Stratified Gas-Liquid Flow," *Ind. Eng. Chem. Fund.* 11, (1972), 490.
8. Mayer, P. G., "Roll Waves and Slug Flows in Inclined Open Channels," *Trans. ASCE*, 126 (1961), 505.
9. Takahama, H., and Kato, S., "Longitudinal Flow Characteristics of Vertically Falling Liquid Films Without Concurrent Gas flow, *Int. J. Multiphase Flow*, 6 (1980), 203.
10. Brauner, N., and Maron, D. M., "Characteristics of Inclined Thin Films," Waviness and the Associated Mass Transfer, *Int. J. Heat Mass. Trans.* 25 (1982), 99.
11. Chand., R., and Rosson, H. F., "Local Heat Flux to Water Film Flowing Down a Vertical Surface, *Ind. Eng. Chem. Fund.*, 4(1965), 356.
12. Sherwood, T. K., and Pigford, R. L., *Absorption and Extraction*, McGraw Hill, New York, 1952.
13. Portalski, S., and Clegg, A. J., "Interfacial Area Increase in Rippled Film Flow on Wetted Wall Columns, *Chem. Eng. Sci.* 26 (1971), 773.
14. Davis, J. T., and Rideal, E. K., *Interfacial Phenomena*, Academic Press, p. 266 (1961).
15. Aris, R., *Vectors, Tensors and Basic Equations of Fluid Mechanics*, Prentice Hall, Englewood Cliffs, N.J. (1962).
16. Joseph, D. D., *Stability of Fluid Motion*, Springer-Verlag, New York (1976).
17. Solesio, J. N., "Instabilities des films liquides isothermes," Rapport CEA-R-4835, Centere d'Etudes Nucleaires de Grenoble, 1977.
18. Solesio, J. N., "La methode de quadrature par differentiation appliquee a l'hydrodynamique des films liquides," Rapport CEA-R-4888, ibid, 1977.
19. Lin, S. P., "Effects of Surface Solidification on the Stability of Multilayered Liquid Films," *J. fluid Eng.* 105 (1983), 119.
20. Lin, S. P., "Film Waves," in *Waves on Fluid Interfaces* (R. E. Meyer ed.) Academic Press, New York (1983), 261.
21. Benney, D. J., "Long Wave in Liquid Film," *J. Math. Phys.* 45 (1966), 150.
22. Mei, C. C., "Nonlinear Gravity Waves in a Thin Sheet of Viscous Fluid," *J. Math. Phys.* 45 (1966), 266.
23. Whitham, G. B., *Linear and Nonlinear Waves*, Wiley-Interscience, New York, (1974).
24. Lighthill, J., *Waves in Fluids*, Cambridge University Press, New York (1978).
25. Kapitza, P. L., "Wave Flow of Thin Layers of a Viscous Fluid," *Zh, Eks. Teor. Fiz.* 18 (1948), 3. (see Collected papers of P. L. Kapitza, Pergamon Press, New York, 1965, 662).
26. Kapitza, P. L., and Kapitza, S. P., "Wave Flow of Thin Layers of a Viscous Fluid." *Zh. Eks. Teor. Fiz.* 19 (1949), 105.
27. Dressler, R. F., "Mathematical Solution of the Problem of Roll-Waves in Inclined Open Channels, *Comm. Pure Appl. Math.*, 2 (1949), 149.
28. Gjevik, B., "Occurrence of Finite Amplitude Surface Waves on Falling Liquid Films, *Phys. Fluids*, 13 (1970), 1918.
29. Gjevik, B., "Spatially Varying Finite-Amplitude Wave Trains on Falling Liquid Films, Acta Polytechnicia Scandinavica, Me 61, 1971.

30. Nakaya, C., "Long Waves on a Thin Fluid Layer Flowing Down an Inclined Plane," *Phys. Fluids*, 18 (1975), 1407.
31. Nakaya, C., "Waves of Large Amplitude on a Fluid Film Down a Vertical Wall," *J. Phys. Sci.* Jpn. 43 (1977), 1821.
32. Roskes, C. J., "Three-dimensional Long Waves on a Liquid Film, *Phys. Fluids*, 13 (1970), 1440.
33. Atherton, R. W., "Studies of the Hydrodynamics of a Viscous Liquid Film Flowing Down an Inclined Plane," Engineer's Thesis, Stanford University, 1972.
34. Atherton, R. W., and Homsy, G. M., "On the Evolution Equations for Interfacial Waves," *Chem. Eng. Comm.* 2 (1976), 57.
35. Krishna, M. V. G., and Lin, S. P., "Nonlinear Stability of a Viscous Film with Respect to Three-Dimensional Side-Band Disturbances," *Phys. Fluids*, 20 (1977), 1039.
36. Lin, S. P., and Liu, W. C., "Instability of Film Coating of Wires and Tubes, *A.I.Ch.E. J.* 21 (1975), 775.
37. Benjamin, T. B., "Wave Formation in Laminar Flow Down an Inclined Plane," *J. Fluid Mech.* 2 (1957), 554.
38. Yih, C. S., "Stability of Liquid Flow Down an Inclined Plane," *Phys. Fluids*, 6 (1963), 321.
39. Anshus, B. D., and Goren, S. L., "A Method of Getting Approximate Solution to the Orr-Sommerfeld Equation for Flow on a Vertical Wall," *AIChE J.* 12 (1966), 1004.
40. Graef, M., "Uber die Eigenschaften zwei-und dreidimensionaler Störungen in Riesel filmen an geneigten Wänden, Mitt." Max-Planck Inst. Strömungsforshung aeron. Versuchsanstalt 26, 1966.
41. Krantz, W. B., and Goren, S. L., "Stability of Thin Liquid Films Flowing Down a Plane, *Ind. Eng. Chem. Found.* 10 (1971), 91.
42. Whitaker, S., "Effects of Surface-Active Agents on the Stability of Falling Liquid Films," *Ind. Eng. Chem. Fundam.* 3 (1964), 132.
43. Benjamin, T. B. and Feir, J. E., "The Disintegration of Wave Trains on Deep Water," *J. Fluid Mech.* 31 (1967), 209.
44. Lin, S. P., "Roles of Surface Tension and Reynolds Stress on a Finite Amplitude Stability of a Parallel Flow with a Free Surface," *J. Fluid Mech.* 40 (1970), 307.
45. Lin, S. P., "Finite Amplitude Stability of a Contaminated Liquid Film," in *Progress in Heat and Mass Transfer* (G. Hetsroni, S. Sideman and J. P. Hartnet, eds.), Pergamon Press, New York, 6 (1972), 263.
46. Whitaker, S., and Jones, L. O., "Stability of Falling Liquid Films, Effect of Interface and Interfacial Mass Transport," *A.I.Ch.E. J.* 12 (1966), 421.
47. Binnie, A. M., "Experiments on the Onset of Wave Formation on a Film of Water Flowing Down a Vertical Plane," *J. Fluid Mech.* 2 (1967), 551.
48. Binnie, A. M., "Instability in a Slightly Inclined Water Channel," *J. Fluid Mech.* 5 (1959), 561.
49. Jones, L. O., and Whitaker, S., Experimental Studies of Falling Liquid Films," *AIChE J.* 12 (1966), 525.
50. Stainthorp, F. P., and Allen, J. M., "The Development of Ripples on the Surface of a Liquid Film Flowing Inside a Vertical Tube," *Trans. Inst. Chem. Eng.* (London) 43 (1964), 185.
51. Tailby, S. R., and Portalski, S., "Determination of the Wavelength on a Vertical Film of Liquid Flowing Down a Hydrodynamically Smooth Plate," *Trans. Inst. Chem. Eng.* (London) 40 (1962), 114.
52. Agrawal, S., "Spatially Growing Disturbances in a Film," Ph.D. thesis, Part I, Clarkson College of Technology, 1972.
53. Agrawal, S., and Lin, S. P., "Nonlinear Spatial Instability of a Film Coating on a Plate." *J. Appl. Mech.* 42 (1975), 580.
54. Gaster, M., "The Role of Spatially Growing Waves in the Theory of Hydrodynamic Stability," *Prog. Aero. Sci.*, 8 (1965), 251.
55. Lin, S. P., "Waves in Viscous Liquid Curtain," *J. Fluid Mech.* 112 (1981), 443.
56. Lin, S. P., "Finite Amplitude Stability of a parallel flow with a Free Surface," *J. Fluid Mech.* 36 (1969), 113.
57. Lin, S. P., Profiles and Speed of Finite Amplitude Waves in a Falling Liquid Layer, *Phys. Fluids*, 14 (1971), 263.

58. Stuart, J. T., "On the Nonlinear Mechanics of Wave Disturbances in Stable and Unstable Parallel Flows. Part I: The Basic Behavior in Plane Poiseuille Flow," *J. Fluid Mech.* 9 (1960), 353.
59. Watson, J., "On the Nonlinear Mechanics of Wave Disturbances in Stable and Unstable Parallel Flows, Part 2. The Development of a Solution for Plane Poiseuille Flow and for Couette Flow," *J. Fluid Mech.* 9 (1960), 371.
60. Reynold, W. C., and Potter, M. C., "Finite-Amplitude Instability of Parallel Shear Flows," *J. Fluid Mech.* 27 (1967), 465.
61. Javdani, K., and Goren, S. L., "Finite-Amplitude Wavy Flow of Thin Films," *Progress in Heat and Mass Transfer*, 6 (1971), 253.
62. Lin, S. P., Finite Amplitude Side-Band Stability of a Viscous Film, *J. Fluid Mech.* 63 (1974), 417.
63. Tougou, H., "Deformation of Supercritically Stable Waves on a Viscous Liquid Film Down and Inclined Plane Wall with the Decrease of Wave Number," *J. Physical Soci.* Japan, 50 (1981), 1017.
64. Nepomnyashchii, A. A., "Stability of Wave Regimes in a Film Flowing Down an Inclined Plane," *Izv. Akad. Nauk SSR, Mekh. Zhidk. Gaza*, No. 3 (1974).
65. Nepomnyashchii, A. A., "Stability of Wave Motions in a Layer of Viscous Liquid on an Inclined Plane," in *Nonlinear Wave Process in Two-Phase Media* [in Russian], Novosibirsk (1977).
66. Tsvelodub, O. Yu., Stationary Traveling Waves in a Film Flowing Down an Inclined Plane, *Fluid Dynamics* (1981) 591–594. Translated from Izvestiya Akademii Nank SSSR, Mekhanka Zhidkosti i Gaza, No. 4 (1980), 142.
67. Sivashinsky, G. I., and Michelson, D. M., "On Irregular Wavy Flow of a Liquid Film Down a Vertical Plane," *Prog. Theor. Phys.* 63 (1980), 2112.
68. Massot, C., Irani, F., and Lightfoot, E. N., "Modified Description of Wave Motion in a Falling Film," *A.I.Ch.E. J.* 12 (1966), 445.
69. Shkadov, V. Y., "Wave Flow Regimes of a Thin Layer of Viscous Fluid Subject to Gravity," *Fluid Dynamics*, 2 (1967), 29.
70. Ruckenstein, E., and Berbente, C., Mass Transfer to Falling Liquid Film at Low Reynolds Number, *Int. J. Heat and Mass Transfer*, 11 (1968), 743.
71. Rushton, E., and Davis, G., "Linear Analysis of Liquid Film Flow," *A.I.Ch.E. J.* 17 (1971), 670.
72. Webb, D. R., "A Note on Periodic Solution to Flow in a Liquid Film," *A.I.Ch.E. J.* 18 (1972), 1068.
73. Gollan, A., and Sideman, S., "On the Wave Characteristics of Falling Films," *AIChE Journal* 15 (1969), 301.
74. Lee, J., "Kapitza's Method of Film Flow Description, *Chem. Eng. Sci.*, 24 (1969), 1309.
75. Esmail, M. N., "Wave Profile on Inclined Falling Film, *Cana. J. of Chem. Eng.* 58 (1980), 156.
76. Hirshburg, R. I., and Florschuetz, "Laminar-Film Flow: Part 1, Hydrodynamic analysis," *J. of Heat Transfer* 104 (1982), 452.
77. Telles, A., "Liquid Film Characteristics in Vertical Two-Phase Flow," Ph.D. thesis, Univ. of Houston, 1968.
78. Telles, A., and Dukler, A. E., "Statistical Characteristics of Thin, Vertical, Wavy, Liquid Films," *Ind. Eng. Chem. Fund.* 9 (1970), 412.
79. Bellman, R. T., and Kalaba, *Quasilinearization and Nonlinear Boundary Value Problems*, Elsevier, New York, 1965.
80. Iwagaki, Y., and Iwasa, Y., "On the Hydraulic Characteristics of the Roll-Wave Trains, *J. Jap. Soc. Civil. Eng.*, 40, No. 1, (1955).
81. Brock, R. R., "Periodic Permanent Roll Waves," *J. Hydraulics Div.*, ASCE, 96 (1970), 2565.
82. Oliver, D. R., and Atherinos, T. E., "Mass Transfer to Liquid Film on an Inclined Plane, *Chem. Eng. Sci.*, 23 (1968), 525.
83. Brauner, N., and Maron, D. M., "Modeling of Wavy Flow in Inclined Thin Films," *Chem. Eng. Sci.* 38 (1983), 775.
84. Ishihara, T., Iwagaki, Y., and Iwasa, Y., "Theory of Roll-Wave Trains in Laminar Water Flow on a Steep Slope Surface," *Trans. Jap. Soc. Civil Eng.*, Vol. 19, No. 5 (1954).

85. Brock, R. R., "Development of Roll-wave Trains in Open Channels," *J. Hydraulics Div.*, ASCE, 95 (1969), 1401.
86. Tamada, K., and Tougou, H., "Stability of Roll-Waves on Thin Laminar Flow Down an Inclined Plane Wall," *J. Phys. Soc.* Jap. 47 (1979), 1992.
87. Ishihara, T., Iwagaki, Y., and Iwasa, Y., Comments to Mayer (1961), *Trans. ASCE*, 126 (1961), 548.
88. Lin, S. P., "Instability of Liquid Film Flowing Down an Inclined Plane," *Phys. Fluids*, 10 (1967), 308.
89. De Bruin, G. J., "Stability of a Layer of Liquid Flowing Down an Inclined Plane," *J. of Eng. Math.* 8 (1974), 259.
90. Chin, R. W., "Stability of Flows Down an Inclined Plane," Ph.D. thesis, Division of Applied Sciences, Harvard University, 1981.
91. Lin, S. P., and Brenner, H., "Tear Film Rupture," *J. Colloid and Interface Sci.* 89 (1982), 226.
92. Lin, S. P., and Krishna, M. V. G., "Stability of a Liquid Film with Respect to Initially Finite Three-Dimensional Disturbances," *Phys. Fluids*, 20 (1977), 2005.
93. Wang, C. K., Seaborg, J. J., and Lin, S. P., "Instability of Multi-Layered Liquid Films," *Phys. Fluids* 21 (1978), 1669.
94. Davis, J. T., and Warner, K.V., "The Effect of Large-Scale Roughness in Promoting Gas Absorption," *Chem. Eng. Sci.* 24 (1969), 231.
95. Tougou, H., "Long Waves on a Film Flow of a Viscous Fluid Down an Inclined Uneven Wall," *J. Phys. Soc.* Japan, 44 (1978), 1014.
96. Wang, C. Y., "Liquid Film Flowing Slowly Down a Wavy Incline," *A.I.Ch.E. J.* 27 (1981), 207.
97. Brumfield, L. K., Houze, R. N., and Theofanous, T. G., "Turbulent Mass Transfer at Free, Gas-Liquid Interfaces, With Applications to Film Flows," *Int. J. Heat Mass Transfer*, 18 (1975), 1077.
98. Orr, F. M., and Scriven, L. E., "Rimming Flow: Numerical Simulation of Steady, Viscous, Free Surface Flow with Surface Tension," *J. Fluid Mech.* 84 (1978), 145.
99. Saito, H., and Scriven, L. E., "Study of Coating Flow by the Finite Element Method," *J. Comp. Phys.* 42 (1981), 53.
100. Banerjee, S., Rhodes, E., and Scott, D. S., "Mass Transfer to Falling Wavy Liquid Films at Low Reynolds Numbers," *Chem. Eng. Sci.* 22 (1967), 43.

CHAPTER 29

HYDRODYNAMICS OF CREEPING FLOW

Joel Koplik

Schlumberger-Doll Research
Ridgefield, Connecticut, USA

CONTENTS

INTRODUCTION

Creeping flows are fluid motions for which inertial effects are negligible compared to viscous effects. Other names for this class of phenomena include slow viscous flow, Stokes flow, and low-Reynolds number flow. As the latter name suggests, a quantitative definition of this regime may be given in terms of the appropriate dimensionless group. Consider the Navier-Stokes equation [1] for an incompressible Newtonian fluid of velocity u, density ρ and viscosity μ, acted on by an external force per unit mass f,

$$\rho \frac{\partial u}{\partial t} + \rho u \cdot \nabla u = -\nabla p + \mu \nabla^2 u + \rho f \tag{1}$$

If the flow has a characteristic velocity U and spatial variation on a length scale L, then the inertial term $\rho u \cdot \nabla u \sim \rho U^2/L$, while the viscous term $\mu \nabla^2 u \sim \mu U/L^2$, and the ratio of these two estimates defines the Reynolds number

$$Re = \rho UL/\mu \tag{2}$$

Creeping flow is formally the asymptotic limit $Re \to 0$, and physically requires at least one of the following: small length scales, very viscous or very light fluids, or very low velocities. In this limit the non-linear term in the Navier-Stokes equation may be neglected, and some simplification results.

This chapter discusses the creeping flow of incompressible Newtonian fluids, and for the most part considers a single fluid in steady flow. After stating the equations of motion and boundary conditions of creeping flows, some general features of the subject (energy dissipation, uniqueness, bounds, potentials, and reversibility) are described. Then the chapter presents exact solutions for flows in channels and discusses entrance effects of the slowly-varying channel and lubrication approximations, and the properties of some simple rotary internal flows. Following some general aspects of two-dimensional flows—complex variable and orthogonal function expansions, the flow field near corners, and the presence of viscous eddies—this chapter addresses the problems of obstacles in prescribed external flows, giving some exact results, general expansions, and approximation methods, including the Oseen correction to Stokes flow and the resistance matrix formalism. The effects of interactions between particles and interactions with boundaries are then surveyed, with emphasis on bispherical coordinate expansions and the method of reflections. The chapter concludes with a discussion of several time-dependent problems.

Creeping flow is a discipline of long standing, and has been the subject of monographs, review articles on specialized topics, and is usually discussed to some degree in textbooks. In the preparation of this article I have drawn heavily on various references [1–7]. My principal aim is to survey a wide variety of analytic solutions, approximation methods and results, and, for the most part, detailed arguments are left to the references. One important subject that I barely mention is that of techniques for numerical calculation; good discussions are available elsewhere [8–10].

GENERAL FEATURES OF CREEPING FLOW

The steady creeping flow of an incompressible Newtonian fluid satisfies the "Stokes equations"

$$\nabla p = \mu \nabla^2 u + \rho f \tag{3}$$

and

$$\nabla \cdot u = 0 \tag{4}$$

Since the stress tensor of such a fluid is [1]

$$\sigma_{ij} = -p\delta_{ij} + \mu(\partial_i u_j + \partial_j u_i) \tag{5}$$

(using the notation $\partial_i \equiv \partial/\partial x_i$) the Stokes equations express the fact that in creeping flow the net force on a fluid element is zero. The linearity of the equations is of great practical value, as it simplifies dimensional scaling analyses, reduces the labor required to construct solutions, and allows superposition of solutions to fit boundary conditions. Note that the divergence of Equation 3 gives (for conservative external forces)

$$\nabla^2 p = 0 \tag{6}$$

The boundary conditions that may be applied to Equations 3 and 4 are of the following types.

Solid Boundaries

The component of velocity normal to the boundary must be zero, because the solid provides a material boundary the fluid cannot penetrate, and the experimental evidence strongly favors the "no-slip condition" that the tangential velocity is equal to the solid's velocity. One may think of the latter condition as the adherence resulting from intermolecular attraction between solid and fluid molecules. Recent work, particularly that concerned with moving contact lines [11], has revived the possibility of a non-zero microscopic slip velocity, but the characteristic length scale of such effects is that of asperities on the solid surface or of a thin film region and will not be considered here.

Fluid-Fluid Boundaries

The normal velocity must be continuous by the kinematic requirement that the boundary is a material surface, and the requirement of differentiability of u (i.e., finiteness of the stress tensor) requires that the tangential velocity be continuous as well. The discontinuity in the fluid stress across the boundary must be balanced by the force exerted by the interface itself, and with the usual assumption that the latter is a surface tension force normal to the interface we have

$$[\hat{n} \cdot \sigma \cdot \hat{t}] = 0, \qquad [\hat{n} \cdot \sigma \cdot \hat{n}] = 2\gamma\kappa \tag{7}$$

In the latter equation, $\hat{n}$ and $\hat{t}$ are unit normal and tangent vectors $[\cdot]$ denotes discontinuity across the interface, γ is the coefficient of surface tension, and κ is the mean curvature of the interface (the average of the two principal curvatures; for a spherical cap of radius R, the two are equal and $\kappa = 1/R$). Some workers [12] adopt more sophisticated rheological models of the interface, in which case further surface dependent terms appear on the right hand side of Equation 7. It should be noted that surface tension effects in real materials are notoriously subject to the effects of contaminants and chemical and temperature variation [13–14]. In case one of the fluids is a gas, one usually assumes that its pressure is constant and its viscosity is effectively zero.

Free Surfaces

If the boundary surface between the fluids is not known but is to be determined, an additional condition is needed. If the equation of the surface is $\phi(x, t) = 0$, then the kinematical condition that a fluid element at x(t) on the surface remain there is

$$\frac{d\phi}{dt} = \frac{\partial\phi}{\partial t} + u \cdot \nabla\phi = 0 \tag{8}$$

Asymptotic Conditions

For flow in unbounded regions one usually prescribes the flow at infinity in some simple way. For example, in channel flows (described later in this chapter) a total fluid flux and reference pressure may be given, and in flow around obstacles ("Two-Dimensional Flows" later in this chapter) an asymptotic uniform or extensional velocity field are often assumed. (Strictly speaking, these are "matching" rather than boundary conditions.)

We now consider the energy dissipation in a viscous fluid; this will provide a proof of uniqueness of solutions of the Stokes equations as well as some useful bounds. The net external force on a system of fluid in a volume V is the sum of the body force and the force exerted by the material on the boundaries, so that the rate of work done on the fluid is

$$\int_V d^3x\, \rho\, f \cdot u + \int_{\partial V} dS \cdot \sigma \cdot u = \int_V d^3x\, \partial_i u_j\, \sigma_{ij}$$

using the divergence theorem and the Stokes equation ($dS = \hat{n}dS$, where $\hat{n}$ is the unit normal on the boundary ∂V, oriented out of the fluid region). The rate of work done on the fluid is the same as the energy dissipation rate W (in creeping flow), and introducing the strain tensor $e_{ij} = \frac{1}{2}(\partial_i u_j + \partial_j u_i)$ we have

$$W[u] = 2\mu \int_V d^3x\; e{:}e \tag{9}$$

Suppose now that u, p, and e satisfy the Stokes equations with a prescribed velocity U on ∂V, and let $\tilde{u}$, $\tilde{p}$, and $\tilde{e}$ be any other incompressible flow fields with the same boundary condition but not necessarily satisfying the Stokes equations. Then,

$$\int_V d^3x\, \tilde{e}:\tilde{e} = \int_V d^3x[e:e + (e - \tilde{e}):(e - \tilde{e}) + 2e:(\tilde{e} - e)]$$

The third term on the right vanishes after some manipulation using Equations 3 and 4 and the boundary conditions, and since the second is positive definite,

$$W[\tilde{u}] \geq W[u] \tag{10}$$

Thus, Stokes flow minimizes the energy dissipation of all incompressible flows with the same boundary conditions. If $\tilde{u}$, $\tilde{p}$, and $\tilde{e}$ satisfy the Stokes equations as well, one can show similarly that

$$\int_V d^3x\, (e - \tilde{e}):(e - \tilde{e}) = 0$$

Hence $e = \tilde{e}$ and the difference between the two flows is a strainless motion, a combination of a rigid translation and rotation, which must be zero by the boundary conditions. Thus, Stokes flows with prescribed boundary conditions are unique.

We may also obtain a lower bound for creeping flow [15] by considering a flow field $\hat{u}$, $\hat{p}$, and $\hat{e}$ which *does* satisfy the Stokes equations but *not* the boundary conditions. One can then show by similar manipulations that

$$W[u] \geq W[\hat{u}] - 2\int_{\partial V} dS \cdot \hat{\sigma} \cdot (U - \hat{u}) \tag{11}$$

where U is the boundary velocity. Given a trial flow field ($\hat{u}$, etc.) the right-hand side can be evaluated and a lower bound on W obtained. For applications of the bounds given in Equations 10 and 11 see References 15–19.

A useful property related to energy dissipation is the reciprocity theorem, due to Lorentz [20]. If V is a closed volume of fluid and $(u_i, \sigma_i)i = 1, 2$, are two solutions of the Stokes equations within V, then

$$\int_{\partial V} dS \cdot \sigma_1 \cdot u_2 = \int_{\partial V} dS \cdot \sigma_2 \cdot u_1 \tag{12}$$

a relation easily proved by evaluating $\int d^3x e_1 : e_2$ using Equations 3 and 4 and the divergence theorem.

The linearity of the Stokes equations suggests the use of Greens function methods. The first step is to construct the flow field in the presence of a point force, called a "Stokeslet." In an unbounded fluid, the solution of the Stokes equations with a unit-strength point force at the origin oriented in the j-direction,

$$\partial_i U_{ij} = 0, \qquad \partial_i P_j = \mu \nabla^2 U_{ij} + \delta_{ij}\delta(x) \tag{13}$$

is

$$U_{ij}(x) = \frac{1}{8\pi\mu}\left[\frac{\delta_{ij}}{x} + \frac{x_i x_j}{x^3}\right], \qquad P_i(x) = \frac{x_i}{4\pi x^3} \tag{14}$$

This result is easily derived by Fourier transform [21] or other [22] methods. The analogous formulae in two dimensions are

$$U_{ij}(x) = \frac{1}{4\pi\mu}\left[-\delta_{ij}\log|x| + \frac{x_i x_j}{x^2}\right], \qquad P_i = \frac{x_i}{2\pi x^2} \tag{15}$$

Further singular solutions can be constructed by derivatives of this solution, or superpositions of Stokeslets at neighboring points; catalogs may be found in Hasimoto and Sano [22] and Chwang and Wu [23]. They are particularly useful at large distances from small objects (see later sections), in studies of aquatic animal locomotion [24] and in fitting boundary conditions in simple geometries using image-like methods [22].

We may now represent a general flow as an integral over Stokeslet contributions on the boundaries of the flow region. Using Green's theorem and some algebra one can show [21, 25]

$$u_i(x) = \int_V d^3y U_{ij}(x - y)\rho f_j(y) + \int_{\partial V} dS_j(y)[\sigma_{jk}(y)U_{ki}(x - y) - V_{ijk}(x - y)u_k(y)] \tag{16a}$$

where $V_{ijk}(x) = 3x_i x_j x_k/4\pi x^5$, and

$$p(x) = \int_V d^3y P_i(x - y)\rho f_i(y) + \int_{\partial V} dS_i(y)\left[\sigma_{ij}(y)P_j(x - y) + 2\mu u_j(y)\frac{\partial}{\partial x_i} P_j(x - y)\right] \tag{16b}$$

This representation is employed in practice by first using the known value of u on the boundary to solve for the stress σ there, and then regarding Equations 16a and b as integral equations for u and p in the interior of the fluid (see later section). Ladyzhenskaya [21] has also used the representation (Equation 16) to prove the existence and uniqueness of creeping flows.

Further Green's function solutions of the Stokes equations are available: for example, Garabedian [26] treats the case of the interior of a circle, Liron and Shahar [27] consider a Stokeslet in a pipe, Hasimoto [28] a periodic array of spheres, and Blake [29] a Stokeslet near a no-slip boundary. Flows involving variation in only two coordinate directions allow the use of scalar stream functions, which automatically enforce incompressibility. For flow in the plane, with $u(x) = (u(x, y), v(x, y), 0)$, one can write a general solution of Equation 4 as

$$u = \frac{\partial \psi}{\partial y}, \qquad v = -\frac{\partial \psi}{\partial x} \tag{17a}$$

and taking the curl of Equation 3, the stream function $\psi(x, y)$ satisfies the biharmonic equation:

$$\left[\frac{\partial^2}{\partial x^2} + \frac{\partial^2}{\partial y^2}\right]^2 \psi = 0 \tag{17b}$$

From the definition (Equation 17a), the velocity at a point is parallel to the contour of constant ψ passing through it. Since the stream function is undefined up to a constant, the solid wall boundary conditions may be taken to be that ψ and its normal derivative vanish. Once ψ is known, the pressure gradient may be found from Equation 3. In polar coordinates (r, θ), the velocity components are written

$$u_r = \frac{1}{r}\frac{\partial \psi}{\partial \theta}, \qquad u_\theta = -\frac{\partial \psi}{\partial r} \tag{18}$$

and in Equation 17b the square of the Laplacian operator is used.

Three-dimensional axisymmetric problems also permit the use of a stream function: in cylindrical coordinates (r, θ, z), if there is no motion along or dependence upon the azimuthal angle θ, one may solve Equation 4 by

$$u_z = \frac{1}{r}\frac{\partial \psi}{\partial r} \qquad u_r = -\frac{1}{r}\frac{\partial \psi}{\partial z} \tag{19a}$$

where ψ satisfies

$$\left[\frac{\partial^2}{\partial r^2} - \frac{1}{r}\frac{\partial}{\partial r} + \frac{\partial^2}{\partial z^2}\right]^2 \psi = 0 \tag{19b}$$

Alternatively, in spherical coordinates (r, θ, ϕ) with $u_\phi = \partial/\partial\phi = 0$, the incompressibility condition is satisfied by

$$u_r = \frac{-1}{r^2 \sin\theta}\frac{\partial\psi}{\partial\theta} \qquad u_\theta = \frac{1}{r\sin\theta}\frac{\partial\psi}{\partial r} \tag{20a}$$

and the Stokes equation leads to

$$\left[\frac{\partial^2}{\partial r^2} + r^{-2}\frac{\partial^2}{\partial\theta^2} - r^{-2}\cot\theta\frac{\partial}{\partial\theta}\right]^2 \psi = 0 \tag{20b}$$

Happel and Brenner [2] discuss stream functions in other orthogonal curvilinear coordinate systems.

The last general feature of steady Stokes flows to be discussed is reversibility. Since the equation of motion contains no time derivatives and is linear in u and p, if we reverse the forces (and directions of boundary motions, if any), we reverse the flow. Thus, if a straight line of visible particles is placed on a rotating dish of fluid, they will be carried into an irregular shape, but if the rotation is reversed they will be brought back into line [30]. Bacteria and small aquatic creatures swim at low Reynolds numbers by applying non-reversible (e.g., corkscrew) forces [24, 31], for otherwise the motion resulting from part of a cycle would be reversed on the remainder. Other consequences of reversibility will be mentioned later.

CHANNEL AND INTERNAL FLOWS

A class of creeping flows of relative simplicity and great practical importance concerns motion in channels and ducts. First, consider flow along the z-axis of an infinitely long cylinder of arbitrary cross-sectional shape (see Figure 1). We expect a z-independent velocity profile of the form $u = \hat{z}w(x, y)$, and substitution into Equation 3 shows that the pressure in a linear function of z alone, $p = p_0 - Gz$. The longitudinal velocity then satisfies

$$\left(\frac{\partial^2}{\partial x^2} + \frac{\partial^2}{\partial y^2}\right) w = \frac{-G}{\mu} \tag{21}$$

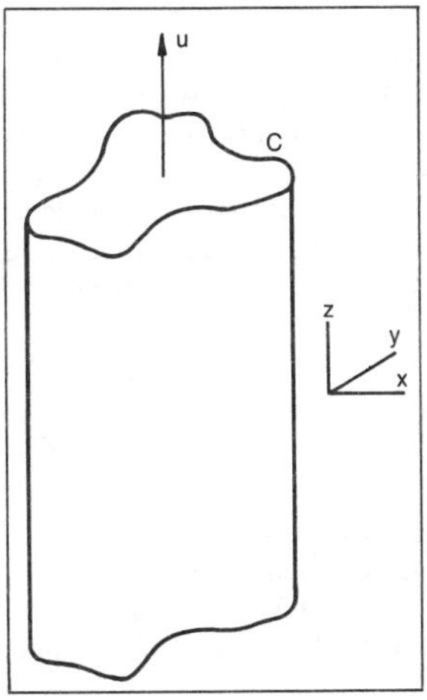

Figure 1. A cylinder of arbitrary cross section.

with w = 0 on C, the boundary curve of the cylinder. Closed-form solutions of Equation 21 exist for a two-dimensional channel of width h. (A two-dimensional flow field may be realized when the scale of variation in the third direction is much larger than that in the others in which the flow is imposed. A different sort of two-dimensional flow occurs in a Hele Shaw cell, a narrow gap between parallel plates [1].)

$$w = \frac{G}{2\mu} x(h - x) \tag{22a}$$

a cylinder of radius R

$$w = \frac{G}{8\mu} (R^2 - r^2) \tag{22b}$$

an ellipse of semi-axes a and b

$$w = \frac{G}{2\mu} \frac{a^2 b^2}{a^2 + b^2} \left(1 - \frac{x^2}{a^2} - \frac{y^2}{b^2}\right) \tag{22c}$$

and for an equilateral triangle of height h

$$w = \frac{G}{4\mu h} (y - h)(3x^2 - y^2) \tag{22d}$$

Note the famous Poiseuille parabolic velocity profile in Equations 22a and b. For a rectangular pipe a series solution is available, and further results for other shapes are given by Berker [5]. If the region can be mapped onto a circle by a conformal transformation, the problem can be reduced to quadratures [3]. Note that Equations 22a–d are all solutions of the full Navier-Stokes equations, because the velocity does not vary along its own direction and the non-linear term is identically zero.

For all three dimensional cases, the flux of fluid through the channel may be written

$$Q = \int dx\, dy\, w = \frac{\lambda A^2 G}{\mu} \tag{23}$$

where A is the cross-sectional area and the effects of shape are incorporated into λ. The maximum λ for a given A occurs for a circle, which has the least perimeter for a given area, and so to speak, maximizes the flux by minimizing the boundary on which the no-slip condition restrains the fluid.

Real pipes are of finite length and entry and exit effects must be allowed for. While this is a complicated subject in general [32], matters are simpler in creeping flow because the pipe radius R is the only natural length scale in the problem. The velocity profiles given above should then apply except for a distance O(R) from the ends, and corrections to Equations 22 and 23 are O(R/L) where L is the length of the pipe. For example, detailed calculations for a circular pipe connecting two infinite half spaces give the net pressure drop (between points at infinity to either side of the pipe), to 1% accuracy, as [33]

$$\Delta p = \frac{8\mu QL}{\pi R^4} \left(1 + \frac{3\pi R}{8L}\right) \tag{24}$$

Note that the entry and exit effects are identical, due to the reversibility of Stokes flow. The effects of different entrance profiles on the pressure drop has been examined by Brenner [34].

One-dimensional flows are particularly simple, but analytic solutions exist for some rotary flows as well. In Couette flow, for example, one has a fluid between two coaxially rotating infinite solid cylinders. If the radii of the cylinders are R_1 and R_2 and the angular velocities about the z-axis are Ω_1 and Ω_2, respectively, then the solution of the Stokes equations (and, in fact, the full Navier-Stokes equations as well) in cylindrical coordinates (r, θ, z) is

$$u_\theta(r) = \frac{\Omega_1 - \Omega_2}{R_1^{-2} - R_2^{-2}} \frac{1}{r} + \frac{\Omega_1 R_1^2 - \Omega_2 R_2^2}{R_1^2 - R_2^2} r \tag{25}$$

The torque per unit length exerted on the inner cylinder is then

$$N = -4\pi\mu \frac{\Omega_1 - \Omega_2}{R_1^{-2} - R_2^{-2}} \tag{26}$$

and its measurement can be used to determine the fluid's viscosity. The flow between rotating coaxial cylinders or concentric spheres rotating about a diameter can be found similarly, as discussed by Langlois [3]. These flows are somewhat prosaic, in that the velocity profile smoothly interpolates between the motions of the bounding surfaces. At higher Reynolds number [35] or with eccentric rotating cylinders [36] interesting secondary circulation patterns may develop. Similar closed form solutions also exist for the flow exterior to a rotating sphere [3] or other rotating axisymmetric bodies [37, 38].

Flow in a curved channel is of some theoretical and practical interest. The simplest such problem is the flow in a two-dimensional circular ring, of outer and inner radii $R \pm a$. In polar coordinates with a constant axial pressure gradient $\partial p/\partial\theta = -G$, one finds $u_r = 0$ and

$$u_\theta = \frac{G}{2\mu}\left(r \log r + \alpha r + \frac{\beta}{r}\right) \tag{27a}$$

with α and β adjusted to fit the no-slip boundary condition. The effects of curvature may be gauged by comparing the fluid flux to that in a straight channel of width 2a at the same pressure gradient. The ratio of the two fluxes is found to be

$$\frac{3}{4}\left[\xi^2 - \frac{1}{4}(\xi^2 - 1)^2 \log^2\left(\frac{\xi + 1}{\xi - 1}\right)\right] \tag{27b}$$

with $\xi = R/a$. For the "physical" range $\xi > 1$, this function increases monotonically from $\frac{3}{4}$ to 1. Thus, the curvature retards the flow, but the effect is rather weak.

Qualitatively speaking, in channel flows in the absence of inertial effects, the streamlines simply follow the boundary of the flow region as it bends. This phenomenon may be understood intuitively from Equation 16: the velocity and pressure at any point in the fluid can be thought of as a sum of contributions from all boundary surfaces, weighted by the functions U and P. Since the latter fall off as a power of the distance, the nearest parts of the boundary dominate the sum, and in the absence of rapid variation in shape the fluid effectively sees a locally straight boundary.

In three dimensions exact solutions are not available, and most approximate work [39] on flow in curved tubes considers the flow at non-negligible Reynolds number where again secondary circulation is found. The case of creeping flow in a helical pipe with weak curvature and torsion has been studied using series expansions [40–42] where it is found that curvature slightly enhances the flow while torsion is a higher order effect. The fact that bending of a pipe has little effect on the flow through it in the creeping limit can be easily verified experimentally, by allowing glycerine to flow down a flexible tube under gravity, and monitoring the flow rate as the tube is bent [43].

Exact solutions are available for flow in channels of hyperbolic cross section; in two dimensions if coordinates (ξ, η) are defined via $x + iy = c \cosh(\xi + i\eta)$, then curves of constant η are hyperbolas and the biharmonic equation may be solved using complex variable methods [44]. The three-dimensional analog of this geometry is a Venturi tube whose boundary is a hyperboloid of revolution, and the Stokes equations can be integrated here as well [2, 45].

A channel of arbitrarily varying cross section must be treated numerically, but if the variation is weak an expansion about a straight channel is possible. For simplicity, we discuss flow in a slowly varying two-dimensional straight channel, but the method extends easily to the other shapes previously discussed. Suppose fluid flows through a channel whose boundaries are specified by $x = \pm h(z)$, where h is weakly varying. An obvious first approximation is to assume that the flow field is locally Poiseuille in terms of the local channel radius, that is

$$w(z) = \frac{6Q}{h^3(z)} x(h(z) - x)$$

In this equation, we have rewritten Equation 22a in terms of the (constant) fluid flux Q. The local value of the pressure gradient can be found from Equation 3 to be $\partial p/\partial z = -12 Q\mu/h^3(z)$, plus terms of order $h'(z)$ and $h(z)h''(z)$. When the slope and curvature of the channel walls are small, these terms and the other components of velocity can be neglected, and the pressure drop between z_1 and z_2 is just

$$p_1 - p_2 = 12Q\mu \int_{z_1}^{z_2} \frac{dz}{h^3(z)} \tag{28}$$

If the slope is not so small, one can expand the velocity and pressure in a power series in $h'(z)$, while if only the curvature is small the channel may be approximated locally as flow in part of a wedge, for which the velocity field is known analytically (see "Two-Dimensional Flows"). The details of these procedures are given by Langlois [3]. Another method suitable for channels with a constriction of finite extent, involving an expansion in the ratio of the maximum constriction height to the unconstricted radius, is discussed by Bungay and O'Neill [46].

Note from Equation 28 that flow in a narrow channel can develop quite large pressure differences, and this fact is the basis of lubrication [47], in which a thin fluid layer acts to reduce the friction of solids sliding past each other by keeping them separated even under high stress. Consider the case shown in Figure 2, in which a fixed solid is bounded below by $z = h(x, y, t)$ while another solid bounded above by $z = 0$ translates past it with velocity $U\hat{x} + V\hat{y}$. If h is slowly varying and small compared to the transverse extent of the solids, we may assume as above that the velocity is

$$u = \frac{-1}{2\mu}\frac{\partial p}{\partial x} z(h - z) + U\left(1 - \frac{z}{h}\right)$$

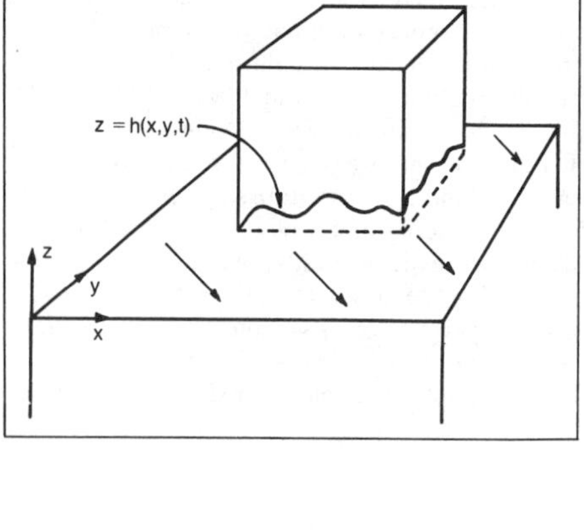

Figure 2. Example of lubrication flow.

similarly for v, and $w = O(\partial_i h)$. Integrating the incompressibility condition (Equation 4) across the gap with respect to z, and using the no-slip boundary condition, we have

$$0 = \int_0^{h(x,y,t)} dz \left(\frac{\partial u}{\partial x} + \frac{\partial v}{\partial y}\right) = \frac{\partial}{\partial x} \int_0^h dz\, u + \frac{\partial}{\partial y} \int_0^h dz\, v - U\frac{\partial h}{\partial x} - V\frac{\partial h}{\partial y}$$

Inserting u and v, and using the surface condition (Equation 7) to rewrite the last two terms, we derive the "Reynolds equation"

$$\frac{\partial h}{\partial t} = \frac{\partial}{\partial x}\left[\frac{\partial p}{\partial x}\frac{h^3}{12\mu} - \tfrac{1}{2}Uh\right] + \frac{\partial}{\partial y}\left[\frac{\partial p}{\partial y}\frac{h^3}{12\mu} - \tfrac{1}{2}Vh\right] \tag{29}$$

a fundamental result in lubrication.

In the simplest illustrative lubrication problem [1], the upper boundary is a fixed plane of length L inclined downward at an angle α and immersed in a fluid of ambient pressure p_0 while the lower boundary moves at constant velocity $U\hat{x}$. The pressure can be found by integration of Equation 29 to be

$$p(x) - p_0 = \frac{6\mu U}{\alpha}\frac{[h_1 - h(x)][h(x) - h_2]}{h^2(x)(h_1 + h_2)} \tag{30}$$

where $h_1 > h_2$ are the gap thicknesses at the ends. The flux Q is constant, and given by $Q = Uh_1h_2/(h_1 + h_2)$. The normal force exerted on the boundaries is essentially the integral of the pressure with respect to x; from Equation 30 this is $O(\mu U/\alpha^2)$ which can be quite large when α is small. Note that in order for the fluid layer to keep the solids apart the pressure difference must be positive, which requires that α be positive, i.e., that fluid be dragged from a wide region into a narrow one. A notable general feature of this type of flow is that because a typical horizontal length scale $L \gg h$, the ratio of inertial to viscous terms is

$$\frac{\rho U^2/L}{\mu U/h^2} = \frac{\rho U h}{\mu}\cdot\frac{h}{L}$$

As this ratio is much less than the ostensible Reynolds number, the lubrication approximation has a wider range of applicability than globally creeping flow alone.

Lubrication approximations may also be relevant in free boundary problems where a thin fluid film is present. Bretherton [48], for example, studied the slow motion of a gas bubble in a cylindrical tube when the bubble is too large to float freely. The bubble's boundary can be divided into spherical caps at its two ends in the interior of the tube, a cylinder of constant radius bounding a thin film of fluid at rest against the wall, and front and rear transition regions. In the latter, the diameter of the bubble varies slowly with longitudinal distance and one can derive a differential equation for the boundary shape by arguments analogous to those leading to Equation 29. The solution determines the film thickness, which in turn determines the velocity of the fluid U in terms of that of the bubble U_b, as

$$U \approx U_b[1 - 2.68\, Ca^{2/3}] \tag{31}$$

when the (dimensionless) capillary number,

$$Ca = \frac{\mu v}{\gamma} \tag{32}$$

is small. Some further discussion of lubrication follows, and extensive treatments, particularly for problems involving circular geometries and bearings, are available [49–51].

TWO-DIMENSIONAL FLOWS

This section presents some general methods for two-dimensional problems, along with a detailed discussion of several flow configurations of specific interest. A general approach to two-dimensional problems makes use of complex variable methods [3, 22, 52]. Introducing the complex coordinates $z = x + iy$, $\bar{z} = x - iy$, the biharmonic equation becomes

$$\left(\frac{\partial^2}{\partial z\, \partial \bar{z}}\right)^2 \psi(z, \bar{z}) = 0 \tag{33}$$

A general solution of this equation is a linear function of $\bar{z}$, so we may write the stream function in terms of two analytic functions α and β

$$\psi = -\mathrm{Re}[\bar{z}\alpha(z) + \beta(z)] \tag{34}$$

The velocity can be shown to be

$$\chi \equiv v - iu = \alpha(z) + z\overline{\alpha'(z)} + \overline{\beta'(z)} \tag{35}$$

and expressions for the pressure and stress tensor are given in the references. Some examples of this representation are as follows. For the uniform shear $u = -ey\hat{x}$, superimposed upon a rotation about the origin with angular velocity Ω, the complex "potentials" are

$$\alpha = \tfrac{1}{2}\Omega z + ez/4, \qquad \beta = -ez^2/4 \tag{36}$$

while for a Stokeslet at the origin of strength $4\pi\mu S$

$$\alpha = c(\log z - 1), \qquad \beta' = \bar{c} \log z \tag{37}$$

with $c = \frac{1}{2}(iS_x - S_y)$

Suppose the velocity is given on the boundaries of a plane region and ψ is sought. If the region is a circle, then using Equation 35 one can obtain α and β in terms of Cauchy integrals of χ and $\bar{\chi}$ around the circle (although for this case there are simpler integral methods available [53]). For arbitrary regions which can be mapped into a circle by a conformal transformation $z = m(s)$, Equation 35 transforms into

$$X(s) = A(s) + \frac{1}{\overline{m'(s)}}[m(s)\overline{A'(s)} + \overline{B'(s)}] \tag{38}$$

where $X(s) = \chi(m(s))$, and similarly for A and B. The Cauchy integrals of Equation 38 and its complex conjugate around the circle in the s-plane whose image under m is the original bounding region in z-plane provides two coupled integral equations for A and B. Further discussion and application of these results are presented in References 54–59.

A second (more popular) approach to solving the biharmonic equation employs separation of variables and expansion in eigenfunctions. Thus, in Cartesian coordinates, functions of the form

$$(a_m + b_m x + c_m y)e^{(mx + imy)} \tag{39}$$

with arbitrary complex values of m, solve the biharmonic equation, and one can attempt to fit the boundary conditions by choice of m or by superpositions of such functions. Analogously in polar coordinates, the functions

$$r^n[a_n e^{in\theta} + b_n e^{i(n-2)\theta}] \tag{40}$$

are solutions [60] for arbitrary n. The cases n = 0, 1, 2 are degenerate, and the appropriate solutions are instead

$$r^0[a_0 + b_0 e^{2i\theta} + c_0\theta] \quad \text{or} \quad r^0[A_0 + B_0\theta + C_0\theta^2 + D_0\theta^3] \tag{41a}$$

$$r^1[a_1 e^{i\theta} + b_1\theta e^{i\theta}] \quad \text{or} \quad r^1[A_1\theta + B_1 \log r]e^{i\theta} \tag{41b}$$

$$r^2[a_2 e^{2i\theta} + b_2\theta + c_2] \tag{41c}$$

We will use the functions (Equations 39–41) in some following examples. The biharmonic equation also occurs in elasticity theory, where such expansion methods are discussed further [61–62].

We now examine some general local features of two-dimensional flows, such as behavior near corners, the presence of eddies and stagnation points, and separation of flow. Typically in such problems the overall magnitude of the fluid velocity is undetermined; it must be fixed by matching the local flow to the "asymptotic flow" at large distance from the region of interest. The procedure used in this matching and some of the dangers of the method will be discussed briefly.

The simplest corner flow problem is that resulting from the scraping motion of solid boundaries [1]. In Figure 3 we consider the plane $\theta = 0$ moving to the right with constant velocity U past the plane $\theta = \alpha$ at rest. Dimensional analysis suggests that the stream function takes the form $\psi = Urf(\theta)$, so f must be as given in the first entry of Equation 41b. The no-slip boundary conditions become $f(0) = f(\alpha) = f'(\alpha) = 0$, $f'(0) = 1$, and one finds

$$\psi = \frac{Ur}{\alpha^2 - \sin^2\alpha}[\alpha(\alpha - \theta)\sin\theta - \theta\sin(\alpha - \theta)\sin\alpha] \tag{42}$$

Some streamlines of this flow are shown in Figure 3. The ratio of inertial to viscous stress can be calculated from Equation 42 to be proportional to r, so this solution is only valid near the corner itself. The stress tensor itself is proportional to r^{-1} so something is wrong just at the corner: the assumptions of perfect plane surfaces and linear rheology cannot be taken too far.

Another simple problem concerns fluid introduced with flux Q at the vertex of a fixed wedge whose sides are $\theta = \pm\alpha$, the Jeffery—Hamel [63] problem in the creeping flow limit. A suitable stream function is given by Equation 41a

$$\psi = \tfrac{1}{2}Q\frac{\sin 2\theta - 2\theta\cos 2\alpha}{\sin 2\alpha - 2\alpha\cos 2\alpha} \tag{43}$$

In this case the velocity is proportional to r^{-1}, and the solution is relevant not too near the vertex. It has, however, the embarrassing feature of a singularity for all θ when the denominator vanishes, at $2\alpha \approx 257°$. The source of the difficulty has been examined by Moffatt and Duffy [64], and traces from the attempt to describe the flow in a sub-region by a "similarity solution" independent of external scales (see also Barenblatt and Zeldovich [65]). When this problem is considered as part of a flow in a larger region, it is found that other terms in the stream function dominate Equation

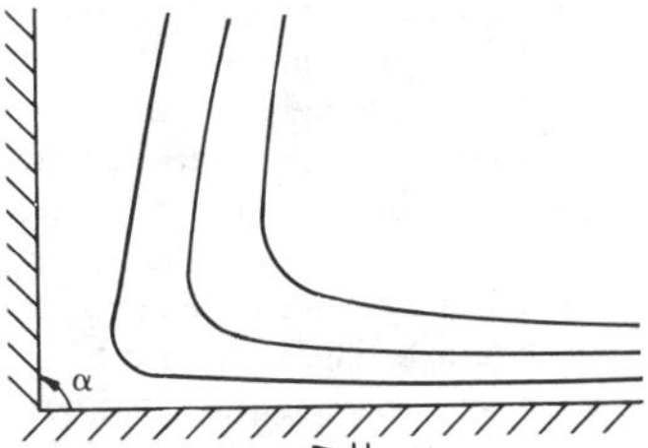

Figure 3. Scraping flow in a corner [1].

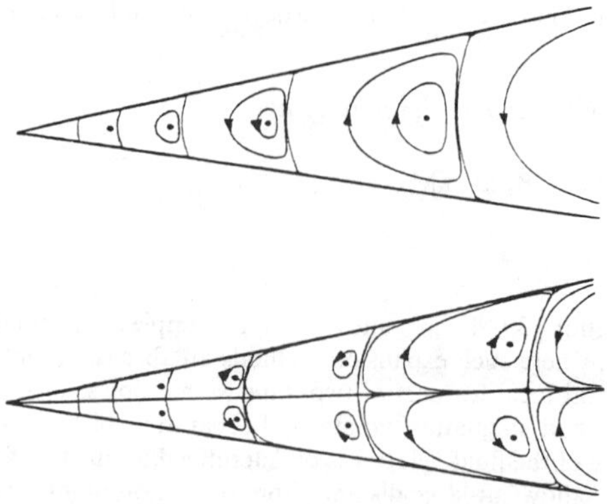

(A)

(B)

Figure 4. Flow in a wedge, stirred at a distance [73].

43 at or above the critical angle. In order for such a similarity solution to be relevant, it must be the case that other terms present outside the region of interest are negligible within it.

More interesting behavior in corner regions results when the boundaries are at rest and the flow is induced by, say, stirring at a far distance from the corner [66]. The solution of the biharmonic equation will be a sum of terms of the form seen in Equation 40, with the boundary conditions $\psi = \partial\psi/\partial\theta = 0$ on $\theta = \pm\alpha$, and we may write the stream function as

$$\psi \sim r^m[\cos m\theta \cos(m-2)\alpha - \cos(m-2)\theta \cos m\alpha] \tag{44a}$$

where m satisfies

$$m \cos(m-2)\alpha \sin m\alpha - (m-2)\sin(m-2)\alpha \cos m\alpha = 0 \tag{44b}$$

The solutions of Equation 44b can be found numerically; there is an infinite sequence of roots that can be ordered by their real parts $1 < \mathrm{Re}(m_1) < \mathrm{Re}(m_2) < \cdots$. Sufficiently close to the origin the solution of smallest Re(m) will dominate the flow. For opening angles $2\alpha \lesssim 146°$ it is found that all roots are complex, and the resulting flow is the sequence of eddies shown in Figure 4A, first discussed by Moffatt [67] in this situation. If the corner angle is large enough for the leading root to be real, it is found that the oscillatory behavior associated with the secondary roots is not strong enough to produce eddies. The qualitative features of the eddies can be seen algebraically, by computing the velocity on the center line of the wedge. Taking the real part of Equation 45 and setting $m_1 = 1 + p + iq$,

$$u_\theta(\theta = 0) \propto r^p \sin\left(q \log\frac{r}{r_0} + \phi_0\right)$$

As $r \to 0$ this expression oscillates in sign, crossing zero at a sequence of stagnation points $r_k = r_0 e^{-(\phi_0 + k\pi)/q}$, $k = 0, 1, 2, \ldots$. Thus, there is an infinite set of eddies, decreasing geometrically in size as the corner is approached. Similarly, the maximum velocity in each eddy can be shown to decrease geometrically with ratio $e^{-\pi p/q}$. The values of p and q, and therefore the characteristics of the eddy sequence, depend only on the corner angle α. The appearance of eddies can be understood by noting that in their absence the fluid steamlines would follow the solid boundary and be forced to bend abruptly near the corner, leading to very high strain there, which the minimum energy dissipation principle for the Stokes equations tends to discourage. The relevance of the corner solution (Equation 44), in the light of the previous remarks on similarity solutions, has been investigated as well by Moffatt [68], who considered a specific stirring mechanism and showed that it indeed emerges as the limiting form of the solution near the corner. The decisive verification of these ideas is experimental: part of the sequence of eddies in a corner has been observed by Taneda [69] in a flow visualization study.

In addition to the antisymmetric flow field just discussed, there is also a solution for a symmetric flow shown in Figure 4B. In this case it is found that eddies occur for $2\alpha \lesssim 159°$, and other aspects of the flow are qualitatively similar. The special case $\alpha = 0$ is also of interest, as it corresponds to the flow in a straight channel some distance away from a disturbance, or to the entry flow in a straight channel. This problem is most easily approached through the rectangular functions (Equation 39); if the disturbance is at the origin and the channel has sides $y = \pm a$, a suitable stream function is

$$\psi = e^{-k|x|}(A \cos ky + By \sin ky) \tag{46}$$

and the boundary conditions require $2ka + \sin 2ka = 0$. The resulting values of k are all complex, with that of smallest real part being $2ka \approx 4.2 + 2.3i$, and correspond to an infinite sequence of eddies of the same size and of geometrically decreasing amplitude. Pan and Acrivos [70] have presented detailed calculations showing that Equation 46 correctly describes the fluid behavior in a rectangular slot driven by a moving top plate, and part of this sequence of eddies has also been seen experimentally [69].

The flow fields shown in Figure 4 differ from those discussed previously by the presence of isolated stagnation points, where fluid in the interior is at rest, and separation, in which a streamline enters the fluid from the solid boundary. A streamline intersecting the boundary divides the fluid into two noncommunicating regions (except for molecular diffusion) and is of great value in understanding the path that the fluid and particles suspended therein actually take. At a separation point on a smooth, locally plane boundary the tangential stress vanishes, and a local solution for the stream function is [1]

$$\psi \sim r^3 \sin^2 \theta \sin(\theta - \theta_0) \tag{47}$$

Thus, separation at a smooth boundary can occur at any angle, which can be determined only by matching to a global solution. More generally, Michael and O'Neill [71] show that any number of streamlines can meet a smooth wall at arbitrary angles, and have also considered the possibilities for separation at cusps and corners.

One might suppose that the qualitative features of straight-corner flows are present in other wedge-shaped regions, and this has been verified in calculations of two-dimensional flow past a cylinder near a plane [72]. When the cylinder is tangent to the plane and the asymptotic flow is uniform shear, an infinite sequence of eddies is found. When there is a gap of any finite size between the two surfaces, it is found that fluid always moves through the gap and, while it is therefore not possible to have the infinite sequence of stagnant eddies show in Figure 4, there is an intricate pattern of eddies attached alternately to the wall and to the cylinder and the fluid paths (Figure 5) are quite tortuous. One can develop a qualitative understanding of this flow [73] by thinking of the region near the gap as a straight wedge with a line source of fluid in the corner, for which the appropriate stream function is just the sum of Equations 43 and 44. The source strength increases with gap width (see "Time-Dependent Creeping Flows") and has the effect of pushing the eddies outward from the corner. Realistically, there can only be a finite number of eddies, because away from the gap region the outer shearing flow controls the flow and the corner solution is inapplicable, so as the gap width increases the number of eddies decreases and at a gap width of about 1.7 times the cylinder radius they disappear entirely.

A similar problem [74] concerns the asymptotically uniform flow perpendicular to the axes of two parallel cylinders. When the cylinders touch, an infinite sequence of corner-like eddies is present, and as they move apart the situation is similar to the cylinder-plane problem except for the additional presence of "free eddies," not attached to the cylinder boundaries. Another type of situation in which eddies appear involves flow past an irregular wall. The earliest result is that of Dean [75] (the first prediction of eddies in Stokes flow, in fact) who studied shear flow past a projection hanging over a trough in a wall. Other authors have studied shear flow past a wall

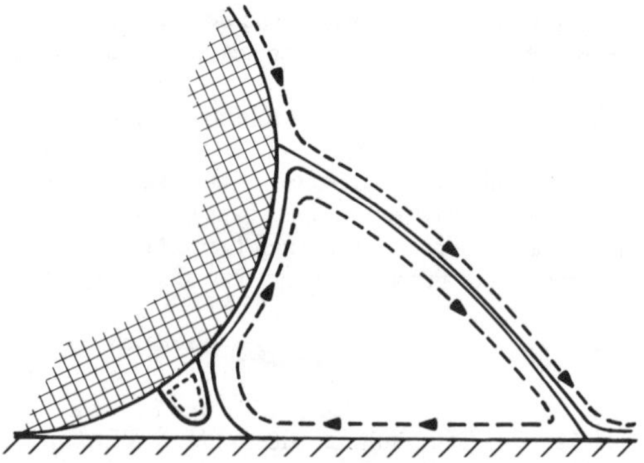

Figure 5. Shear flow past a cylinder near a plane [72].

with a circular ridge or trough, flow around a cylinder rotating near a plane and between eccentric rotating cylinders, and rotating cylinders in shear flow. Good reviews of these problems are given by O'Neill and Ranger [4] and Hasimoto and Sano [22].

Another interesting two-dimensional problem involves a cylinder at rest in an asymptotically uniform flow field, where the flow is at right angles to the cylinder axis. If the velocity at infinity is $U\hat{x}$, the asymptotic form of the stream function is $\psi \sim Ur \sin \theta$. We seek a solution in the form $\psi = f(r) \sin \theta$ with the no-slip boundary conditions $f(a) = f'(a) = 0$ at the cylinder radius $r = a$. The solution, unique up to an overall constant A, is

$$\psi = A\left[\frac{a}{r} - \frac{r}{a} + 2\frac{r}{a}\log\frac{r}{a}\right]\sin\theta \tag{48}$$

which has the wrong behavior at infinity. Other angular dependences are of no help and there is no solution of the Stokes equation for this situation [76]. The difficulty is known as Stokes' paradox, and its resolution is discussed in the next section.

FLOW PAST AN OBSTACLE

This section considers the flow past a single particle in an unbounded medium in three dimensions. It first discusses some general methods for solving the Stokes equations, then treats flow past a sphere in several situations, and finally introduces and discusses the resistance matrix formalism. Since the pressure satisfies a Laplace equation (Equation 6) we may write (in spherical coordinates)

$$p = \sum_{n=0}^{\infty} p_n, \qquad p_n = (A_n r^n + B_n r^{-n-1})Y_{nm}(\theta, \phi) \tag{49}$$

where the Y_{nm} are spherical harmonics. This leads to

$$u = \sum_{n=0}^{\infty}\left[\nabla \times (x\alpha_n) + \nabla\beta_n + \frac{(n+3)x^2\nabla - 2nx}{2\mu(n+1)(2n+3)}p_n\right] \tag{50}$$

where α_n and β_n are also harmonics of the form given in Equation 49. This expansion is due to Lamb [6] and further discussed by Brenner [77]. Similarly, in cylindrical coordinates (r, θ, z) Happel and Brenner [2] show that a general solution of the Stokes equation is

$$p = -2\mu\frac{\partial^2\Pi}{\partial z^2}, \qquad u = \left(r\frac{\partial}{\partial r}\nabla + \hat{z}\frac{\partial}{\partial z}\right)\Pi + \nabla\alpha + \nabla \times (\hat{z}\beta) \tag{51a}$$

where Π, α and β are harmonic, and may conveniently be taken to be functions of the form

$$\int_0^\infty d\lambda \sum_{n=-\infty}^{\infty} I_n(\lambda r)[A_n(\lambda)\cos n\theta + B_n(\lambda)\sin n\theta]\cos \lambda z \tag{51b}$$

where I_n is a modified Bessel function. More generally, one can write any Stokes flow in terms of one vector (A) and two scalar (α,β) harmonic functions as [78]

$$p = 2\mu\nabla\cdot A, \qquad u = \nabla\alpha + x \times \nabla\beta + \nabla(x\cdot A) - 2A \tag{52}$$

For problems with axial symmetry, one may introduce the stream function (Equation 20a) and solve Equation 20b to find (in spherical coordinates)

$$\psi(r,\theta) = \sum_{n=2}^{\infty}\left[A_n r^n + B_n r^{-n+1} + C_n r^{n+2} + D_n r^{-n+3}\right]\frac{P_{n-2}(\cos\theta) - P_n(\cos\theta)}{2n-1} \tag{53}$$

where the P_i are Legendre functions; explicit forms for u and p may be found in the references. In writing Equation 53 we omit terms singular along the z-axis. We note for later use that the force exerted on any sphere by the fluid outside it is $4\pi\mu D_2$, as found by surface integration of the stress tensor resulting from (53). For axisymmetric problems in other coordinate systems, a stream function may again be introduced and is found to satisfy an equation similar to Equation 20b, of the schematic form $(E^2)^2\psi = 0$, with an appropriate second order differential operator E^2. General solutions are found by first solving $E^2W = 0$, and then $E^2\psi = W$; each function in W has two constants of integration and thus ψ has the required four.

A general approach to axisymmetric problems involving shapes other than spheres was instigated by Payne and Pell [79], who use results of potential theory to write a solution of the Stokes equation in terms of solutions to the Laplace equation in various space dimensions. This method facilitates the use of unusual coordinate systems and has been employed to discuss asymptotically uniform flow past lens-shaped bodies, spherical caps, closed tori, spheroids, and other shapes. As in the two-dimensional examples discussed in the previous section, various separation phenomena, isolated stagnation points, and free eddies appear. For a review of these methods and results, see O'Neill and Ranger [4].

The classical problem of flow past an obstacle concerns a sphere at rest in a flow field that is uniform at infinity. The simplest procedure is to note that the stream function has the asymptotic behavior $\psi \sim \frac{1}{2}Ur^2\sin^2\theta$ (where the asymptotic flow is $-\hat{z}U$), and look for a solution of the form $\psi = U\sin^2\theta f(r)$. Using Equation 53 (or, more simply in this case, substitution into Equation 20b and some algebra to find the appropriate powers of r) along with the boundary conditions $\psi = \partial\psi/\partial r = 0$ at $r = a$, we find

$$\psi = \frac{1}{2}Ur^2\sin^2\theta\left[\frac{1}{2}\left(\frac{a}{r}\right)^3 - \frac{3a}{2r} + 1\right] \tag{54}$$

The streamlines for this flow are plotted in Figure 6 (note the symmetry about the plane $z = 0$, a consequence of the reversibility of Stokes flow), and explicit forms for the velocity and pressure

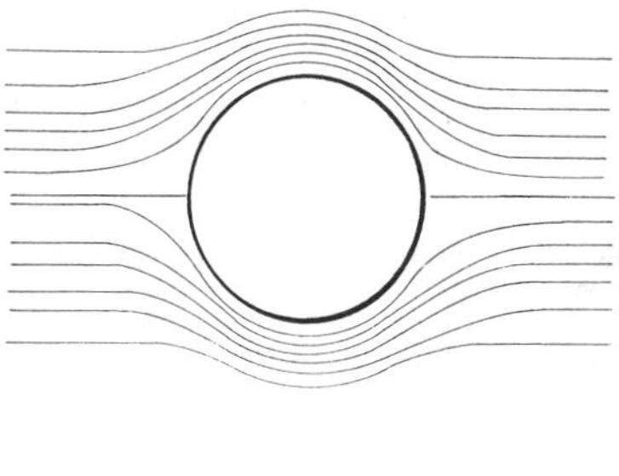

Figure 6. Stokes flow past a sphere at rest.

are given below. The force exerted on the sphere by the fluid is, according to the remark below (Equation 53),

$$F = 6\pi\mu aU \tag{55}$$

To solve the related problem of a sphere translating with uniform velocity $+U\hat{z}$ in a fluid at rest at infinity, one may simply superpose a uniform flow $\psi = -\frac{1}{2}Ur^2 \sin^2\theta$ on the stream function given in Equation 54 (and reverse the sign of the force in Equation 55 to account for the redefinition of U).

If we consider flow past a spherical drop of a different fluid of viscosity μ_{in}, we may obtain the stream function by similar reasoning. Suppose first that the surface tension coefficient is so large as to prevent distorsion of the drop away from a spherical shape. Using the expansion, Equation 53, the asymptotic condition of uniform flow suggests a solution in the outer fluid of the form

$$\psi_{out} = \tfrac{1}{2}U \sin^2\theta(r^2 + B/r + Dr) \tag{56a}$$

Continuity of velocity at the boundary implies $\psi_{in} = \psi_{out} = 0$ and $\partial\psi_{in}/\partial r = \partial\psi_{out}/\partial\psi_r$ at $r = a$. Thus, ψ_{in} must have the same angular dependence as Equation 56a and finiteness of u at the origin leads then to

$$\psi_{out} = \sin^2\theta(Ar^2 + Cr^4) \tag{56b}$$

The remaining boundary condition is continuity of tangential stress; there is no condition on normal stress because any imbalance is compensated by capillary pressure as in Equation 7. The constants above are then found to be ($\xi \equiv \mu_{out}/\mu_{in}$)

$$A = -C = -\frac{1}{4}U\frac{\xi}{1+\xi} \qquad B = \frac{1}{2}\frac{a^2}{1+\xi} \qquad D = -\frac{1}{2}a\frac{3+2\xi}{1+\xi} \tag{56c}$$

If the normal stress on the sphere is computed from Equation 56, one finds a constant independent of angle. Thus, as long as the Stokes approximation is valid, the external flow does not distort an initially spherical drop. A useful application is to the rise or fall of a drop under gravity. The drop will reach a steady velocity when the drag force $4\pi\mu_{out}D$ equals the buoyancy force associated with the density difference $(4/3)\pi a^3(\rho_{in} - \rho_{out})g$, and thus the terminal settling speed is

$$U = \frac{2a^2(\rho_{in} - \rho_{out})}{3\mu_{out}}\frac{\mu_{out} + \mu_{in}}{2\mu_{out} + 3\mu_{in}}g \tag{57}$$

An instructive alternative method in flow past a sphere is to use linearity, dimensional analysis, and the rotational invariance of the problem to argue that the solution must be of the form

$$u = Uf(w) + x\frac{U \cdot x}{x^2}g(w), \qquad p = \mu\frac{U \cdot x}{x^2}h(w) \tag{58a}$$

where $w = x/a$. Substitution into Equation 3 gives ordinary differential equations for f, g, and h, with solutions

$$f = 1 - \frac{3}{4w} - \frac{1}{4w^3}, \qquad g = \frac{3}{4}\left(\frac{1}{w^3} - \frac{1}{w}\right), \qquad h = -\frac{3}{2w} \tag{58b}$$

A simple application of this technique is to the problem of a sphere rotating with angular velocity Ω in a fluid at rest at infinity. Rotational invariance suggests

$$u = (\Omega \times x)\phi(w), \qquad p = 0 \tag{59}$$

(p must be a scalar, while quantities like $\Omega \cdot x$ are pseudoscalar), and the resulting ordinary differential equation yields $\phi = w^{-3}$. The net force on the sphere vanishes, and the torque is found to be

$$N = \int x \times (dS \cdot \sigma) = -8\pi\mu a^3 \Omega \tag{60}$$

Analogously, if a sphere is in a flow field which is asymptotically a uniform shear, $u \to E \cdot x$ as $x \to \infty$, with E a constant symmetric tensor, then rotational invariance requires

$$u = E \cdot x\, F(w) + x \frac{x \cdot E \cdot x}{x^2} G(w), \qquad p = \mu \frac{x \cdot E \cdot x}{x^2} H(w) \tag{61a}$$

and one finds The case $E = \text{diag}(1, 1, -2)$ could be solved easily by noting that the asymptotic stream function.)

$$F = 1 - \frac{1}{w^5}, \qquad G = \frac{5}{2}\left(\frac{1}{w^5} - \frac{1}{w^3}\right), \qquad H = -\frac{5}{w^3} \tag{61b}$$

More generally, problems involving a prescribed vector or tensor asymptotic flow can often be so attacked in this way by expressing the flow field in terms of unknown scalar functions multiplied by prefactors linear in the external quantity.

There are elegant general formulae by Faxèn [80] for the force and torque on a sphere in an arbitrary asymptotic flow. If the sphere translates with velocity U and rotates with angular velocity Ω in a flow field which far from the sphere is given by $U_\infty(x)$, then the force and torque on it are

$$F = 6\pi\mu a(U_\infty - U) + \pi a^3 \mu \nabla^2 U_\infty$$

$$N = 8\pi\mu a^3(\tfrac{1}{2}\nabla \times U_\infty - \Omega) \tag{62}$$

In Equation 62, U, Ω, and the derivatives of U_∞ are evaluated at the center of the sphere. These "Faxen laws" are most easily proven [5] by using the reciprocity relation (Equation 12) with a judicious choice of auxiliary flow. An alternate derivation and a generalization to flow around an ellipsoid is given by Brenner [81].

Note from Equation 54 that the presence of a spherical obstacle causes a long-range perturbation in the flow, the correction to the asymptotic velocity falling off only as r^{-1}. In consequence, other obstacles or rigid boundaries may cause significant effects even when much further away than the sphere size (see "Particle and Wall Interactions"). The long-range nature of the perturbation also has consequences for the validity of the Stokes approximation, for now estimates of the inertial and viscous terms in the Navier-Stokes equation are asymptotically $\rho U^2/r^2$ and $\mu U/r^3$, respectively. Thus, inertial effects are negligible only when $\rho U r/\mu \equiv \text{Re} \cdot r/a \ll 1$, and Equation 54 is not a solution of the Navier-Stokes equation at large enough r for any Reynolds number. This difficulty was first discussed systematically by Oseen [82], who proposed to remedy it by arguing that at large r, where the difficulty appears, the perturbation to the velocity is quite small and the inertial term may be approximated by keeping only the part linear in u. If the solution of the resulting "Oseen equation"

$$\rho U \cdot \nabla u = -\nabla p + \mu \nabla^2 u \tag{63}$$

has the property that $|u| = O(x^{-1})$ at large x, then the left-hand side will be a good approximation to the inertial term there. Similarly, if at small x the scale of variation of the velocity is the sphere size, the left-hand side will be negligible at sufficiently small Re. The approximate solution of Equation 63 is

$$\psi = \tfrac{1}{2}Ur^2 \sin^2\theta\left(1 + \frac{a^3}{2r^3}\right) - \frac{3U}{2\,\text{Re}}(1 + \cos\theta)\left(1 - \exp - \left[\frac{r\,\text{Re}}{2a}(1 - \cos\theta\right]\right) \tag{64}$$

Near the sphere, Equation 54 is recovered (and therefore the drag has been computed correctly), while at large distances the additional terms correspond to a point source at the position of the sphere fed by an influx along a narrow wake.

The original idea of Oseen was to provide a solution to the Navier-Stokes equations whose velocity and velocity derivatives were uniformly valid in all space, and use it as a starting point for a perturbation expansion in powers of Re. This procedure had been known to fail ("Whitehead's paradox") if Equation 54 were used as the starting point. As the complicated form of Equation 64 suggests, it becomes difficult to carry out the calculation to higher orders, and in particular to implement the no-slip condition at higher orders. The modern approach to these questions [83–85] is to consider two separate asymptotic expansions of the Navier-Stokes equations, one for the region near the sphere where the no-slip condition applies and the other for the far region where the uniform flow condition is imposed. Each expansion separately contains undetermined coefficients (reflecting the behavior in the other region), and these are fixed by requiring the two expansions to coincide in an intermediate region.

Returning to the problem of two-dimensional flow past a cylinder, discussed earlier, we find a similar situation. The ratio of inertial to viscous terms in Equation 48 is $O(\rho Ur/\mu)$ and the former cannot be neglected at large distances. Approximating the inertial terms as in Equation 63, a solution analogous to Equation 64 has been given by Lamb [86]. The large r behavior is somewhat complicated but at small r reduces to Equation 48 if the identification

$$A = \frac{\frac{1}{2}U}{\log\frac{4}{Re} - \gamma + \frac{1}{2}} \tag{65}$$

is made (γ is Euler's constant). A discussion in terms of matched asymptotic expansions and further results are in References 83–85.

Some general statements about the effects of a solid particle in an imposed flow field can be neatly described in terms of resistance matrices, as developed by Brenner [2, 87–88]. Suppose an arbitrary rigid object is slowly translating and rotating in a fluid at rest at infinity. Choosing the origin of coordinates at some point in the body, the fluid satisfies the boundary conditions $u = U + \Omega \times x$ on the surface of the object, where U and Ω are the velocity of the reference point and the angular velocity about it, and $|u| \to 0$ at infinity. The linearity of the Stokes equations imply that the fluid velocity, pressure, and stress, and therefore the force F and torque N exerted by the fluid on the object, are linear in U and Ω. We can thus write

$$\begin{pmatrix} F \\ N \end{pmatrix} = -\begin{pmatrix} A & B \\ C & D \end{pmatrix}\begin{pmatrix} U \\ \Omega \end{pmatrix} \equiv -RV \tag{66}$$

where the second-rank tensors A, B, C, and D must be proportional to the viscosity and will depend on the shape and orientation of the body and the choice of origin in the body. If the fluid at infinity is uniformly translating and rotating, then Equation 66 still holds with the asymptotic velocity and angular velocity subtracted from U and Ω, respectively. A uniform shear E can be accommodated by extra terms on the right. A further generalization arises in the study of suspensions where one is interested in the particle's "stresslet" S, the symmetric part of the force dipole exerted on it, which is its contribution to the bulk stress tensor [89]. Linearity implies that Equation 66 can be extended to a 3×3 linear system of vector equations [90] relating (F, N, S) to (U, Ω, E).

The reciprocity theorem can be used to show that the resistance matrix R in Equation 66 is symmetric: If we consider a second rigid motion of the object $(\tilde{U}, \tilde{\Omega})$ and integrate over its surface then

$$\begin{aligned}\int dS \cdot \sigma \cdot \tilde{u} &= \int dS \cdot \sigma \cdot \tilde{U} + \int x \times (dS \cdot \sigma) \cdot \tilde{\Omega} = F \cdot \tilde{U} + N \cdot \tilde{\Omega} \\ &= -U \cdot A \cdot \tilde{U} - U \cdot B \cdot \tilde{\Omega} - \Omega \cdot C \cdot \tilde{U} - \Omega \cdot D \cdot \tilde{\Omega}\end{aligned}$$

From Equation 12 however, this expression must be symmetric under the interchange $U \leftrightarrow \tilde{U}$, $\Omega \leftrightarrow \tilde{\Omega}$ which implies $A = A^T$, $B = C^T$ and $D = D^T$ or, in short, $R = R^T$. Thus, the force on a rotating particle is directly related to the torque on a translating one. The energy dissipation in the flow (U, Ω) is

$$W = U \cdot A \cdot U + 2U \cdot B \cdot \Omega + \Omega \cdot D \cdot \Omega = V^T \cdot R \cdot V \tag{67}$$

and since W is positive definite, the resistance matrix R (as well as A and D separately) are positive definite and invertible. If F and N are given, and R is known, the motion of the body can be found directly. This is one advantage of the resistance matrix formalism: all dependence on the shape of the object is contained in R, which can be computed once per object and applied to any flow or force-torque pair. A second advantage is that geometric symmetries of the body reflect themselves directly in R.

Suppose for example that the body is orthotropic, having three mutually perpendicular planes of symmetry, such as an ellipsoid or rectangular solid. If a coordinate system is chosen with origin at the body's center and axes along the symmetry axes, we have first that the "coupling" tensor $B = 0$ (because under reflection through a symmetry plane $U \to -U$ while $\Omega \to +\Omega$), and second that the "translation" and "rotation" tensors A and D are diagonal (because, e.g., translation along a symmetry axis requires a force along that axis). If all three symmetry axes are equivalent, as for a sphere or regular polyhedron, then the diagonal elements must be equal. Thus for a sphere, we have from Equations 55 and 60

$$A = 6\pi\mu a I, \qquad D = 8\pi\mu a^3 I \tag{68}$$

while for an ellipse of semi-axes a_i, $i = 1, 2, 3$, it can be shown [91] that

$$A_{11} = \frac{16\pi\mu}{L + a_1^2 K_1}, \qquad D_{11} = \frac{16\pi\mu(a_2^2 + a_3^2)}{3(a_2^2 K_2 + a_3^2 K_3)} \tag{69a}$$

while the other diagonal elements given by permutation, and all other elements of R equal to zero. The quantities L and K_i are defined by

$$L = \int_0^\infty \frac{d\lambda}{\Delta(\lambda)}, \qquad K_i = \int_0^\infty \frac{d\lambda}{\Delta(\lambda)(a_i^2 + \lambda)} \tag{69b}$$

where $\Delta^2 = \Pi(a_i^2 + \lambda)$. Special cases of Equation 69 are often of practical interest. We may approximate a flat disk as an ellipsoid with $a_1 \ll a_2 = a_3$ for which

$$F_1 = 16\mu a_2 U_1, \qquad F_{2,3} = \frac{32}{3}\mu a_2 U_{2,3}$$

$$N_i = \frac{8}{3}\mu a_2^3 \Omega_i \tag{70}$$

and similarly for a long needle with $a_1 \gg a_2 = a_3$

$$F_1 = \frac{4\pi\mu a_1}{\eta - \frac{1}{2}} U_1, \qquad F_{2,3} = \frac{8\pi\mu a_1}{\eta + \frac{1}{2}} U_{2,3}$$

$$N_1 = \frac{16\pi}{3}\mu a_1 a_2^2 \Omega_1, \qquad N_{2,3} = \frac{(8/3)\pi\mu a_1^3}{\eta - \frac{1}{2}} \Omega_{2,3} \tag{71}$$

where $\eta = \log(2a_1/a_2)$. Note that forces given in Equation 70 are numerically not far from those on a sphere of radius a_2, and that the logarithms and divergences as $a_1 \rightarrow \infty$ in Equation 71 are related to the Stokes paradox for the cylinder discussed earlier. More generally, suppose the body is symmetric under the coordinate rotation specified by the orthogonal matrix α [92]

$$x = \alpha \cdot x' \tag{72}$$

in that the shape of the body is the same in unprimed and primed coordinates. The boundary velocity U transforms the same way, while for the (pseudo-vector) angular velocity, $\Omega = \det(\alpha)\ \alpha \cdot \Omega'$. The energy dissipation is unaffected by this change of coordinates, which requires

$$A = \alpha \cdot A \cdot \alpha, \qquad B = \det(\alpha)\ \alpha \cdot B \cdot \alpha, \qquad D = \alpha \cdot D \cdot \alpha \tag{73}$$

The consequences of Equation 73 for various partial symmetries are discussed by Happel and Brenner [2].

An instructive application of the resistance formalism is to the settling speed under gravity of an arbitrary particle. If with respect to some interior reference point, the body translates with velocity U and rotates with angular velocity Ω, the conditions of equilibrium are zero net force,

$$A \cdot U + B \cdot \Omega + (m_f - m_b)g = 0 \tag{74}$$

and zero net torque

$$B^T \cdot U + D \cdot \Omega + (m_b x_b - m_f x_f) \times g = 0 \tag{75}$$

In these equations, m_b and m_f are the mass of the body and the displaced fluid, while x_b and x_f are the center of mass and the center of buoyancy (the latter two need not coincide for a non-uniform body). In order for the body to settle without rotation, both expressions above must hold at $\Omega = 0$, and eliminating U we find

$$B^T \cdot A^{-1}(m_f - m_b) \cdot g = (m_b x_b - m_f x_f) \times g \tag{76}$$

When the body is uniform and orthotropic, both sides of Equation 75 vanish identically, and such a body will fall without rotation. However, it need not fall vertically: unless it is initially oriented along a symmetry axis or unless it is spherically isotropic, we see from Equation 74 that U is not parallel to g. Thus, an ellipsoid will in general drift to the side as it settles. Some further discussion of this subject is given by Happel and Brenner [2, 88].

A general approximate method for flow past bodies that are long and thin is to regard the perturbation in the asymptotic flow produced by the body as due to a distribution of Stokeslets on its axis, chosen so as to approximately satisfy the no-slip condition on the surface. One may think of this as the "slender body" simplification of the fact that flow past *any* body can be represented by a distribution of singularities on the body's surface (Equation 16). For a leading order one has a uniform distribution of Stokeslets; indeed from Equation 14 such a superposition with strength S along the line $-d < z < +d$ gives a velocity field near the line of magnitude

$$u_i(x) = \frac{1}{4\pi\mu\epsilon}\left[S_i + \delta_{i3}S_3 + O(\epsilon)\right] \tag{77}$$

where $\epsilon^{-1} = \log(2d/R)$, R being a characteristic transverse size. A suitable choice of S can thus provide a constant velocity near the body's surface, appropriate for imposing a no-slip condition in a uniform asymptotic flow. The higher order corrections take the form of a series expansion in ϵ, determined by a matching procedure involving an inner region at a particular point on the body and an outer region consisting of the body as a whole in the imposed flow. The slender body approximation is discussed in detail by Batchelor [93], Cox [94], and Tilley [95].

PARTICLE AND WALL INTERACTIONS

I noted earlier that the presence of a rigid body causes a long-range perturbation in an asymptotically specified flow, and thus significant interaction effects are expected when other bodies or boundaries are present. This section first discusses the case of two or more particles translating in an unbounded medium.

If two particles whose boundary surfaces S_i, $i = 1, 2$, translate with velocities U_i in a fluid at rest at infinity, then the motion satisfies the Stokes equations with boundary conditions $u = U_i$ on S_i and $|u| \to 0$ at infinity. In general, boundary conditions can only be conveniently imposed when the boundary coincides with a surface of constant coordinate, and for almost all situations the appropriate coordinate system does not exist. A notable exception is the case of two spheres translating along the lines joining their centers, at fixed separation and at constant velocity U, for which bispherical coordinates may be employed. These are defined with respect to cylindrical coordinates by

$$r = \frac{c \sin \eta}{\cosh \xi - \cos \eta}, \qquad z = \frac{c \sinh \xi}{\cosh \xi - \cos \eta} \tag{78}$$

The range of variation is $0 \leq \xi < \infty$, $0 \leq \eta \leq 2\pi$, and a sphere of radius r_1 centered at $(0, 0, z_1)$ is specified by $\xi = \alpha$ with $r_1 = c \cosh \alpha$, $z_1 = c \coth \alpha$. The second sphere is specified by $\xi = \beta$ with $\alpha > 0 > \beta$. A stream function is introduced via Equation 20a, satisfying Equation 20b and no-slip conditions on the surface of each sphere. Transforming into bispherical coordinates and solving by separation of variables, Stimson and Jeffrey [96] found

$$\psi = (\cosh \xi - \cos \eta)^{-3/2} \cdot \sum_{n=0}^{\infty} U_n(\xi) C_{n+1/2}^{-1/2}(\cos \eta) \tag{79a}$$

where $C_{n+1/2}^{-1/2}$ is a Gegenbauer polynomial and

$$U_n(\xi) = a_n \cosh(n - \tfrac{1}{2})\xi + b_n \sinh(n - \tfrac{1}{2})\xi + c_n \cosh(n + \tfrac{1}{2})\xi + d_n \sinh(n + \tfrac{1}{2})\xi \tag{79b}$$

The four constants are fixed by the no-slip conditions on two velocity components on two spheres. The forces on the two spheres (α and β, respectively) required to maintain their motion are

$$F_z = \frac{2\pi\mu\sqrt{2}}{c} \sum_{n=1}^{\infty} (a_n \pm b_n + c_n \pm d_n) \tag{79c}$$

When the spheres are the same size, the forces deviate from the Stokes formula (Equation 55) by a numerical coefficient that increases from 0.645 to 1 as the gap between them increases from 0 to infinity; thus, the hydrodynamic resistance on a sphere is decreased by the presence of a second one. Extensive tabulations of the forces for unequal spheres are given by Cooley and O'Neill [97]. The same authors modify the solution previously given to discuss the motion of a sphere towards a second stationary sphere along the line joining their centers. While this is intrinsically a time-dependent problem, since the geometry is changing, the Stokes equations will still apply if the approach is sufficiently slow. The more general problem of asymmetric motion of two spheres has been solved, in semi-analytic form, by decomposing the motion into a superposition of linked relative rotations and translations. This work is reviewed by Ranger and O'Neill [4], with emphasis on the forces and torques exerted and on the limit of vanishing separation.

The solution of the two-sphere problem, or for that matter any n-particle problem, can be cast into the form of a generalization of the resistance matrix introduced above. The key idea [98] is to represent the force and torque on any one particle as a superposition of terms linear in the velocity and angular velocity of each of them, i.e.,

$$F_i = \sum_{j=1}^{n} (A_{ij} \cdot U_j + B_{ij} \cdot \Omega_j) \tag{80}$$

with a similar expression for N_i. As before, an asymptotic translation and rotation of the fluid can be subtracted off and an asymptotic shear would contribute an extra term. The coefficient matrices in Equation 80 satisfy the same positivity and symmetry properties as their one-particle counterparts. Unfortunately, the matrix elements are known explicitly only for the two-sphere system [4]. A discussion of the two-sphere interaction oriented towards the behavior of suspensions under shear is given by Batchelor and Green [99].

For interaction problems in which exact solutions are unknown, the best available approximation technique is the method of reflections [2, 100], related to the method of images in electrostatics [101], which is an iteration scheme where one attempts to implement the boundary conditions on one surface at a time. Suppose for example we have n spherically isotropic particles translating with velocities U_α, $\alpha = 1, \ldots, n$, in a fluid at rest at infinity. Define velocity and pressure fields u_1 and p_1 as the solution of the Stokes equations, which vanishes at infinity and satisfies the no-slip condition on the surface of particle α, $u_1(\alpha) = U_\alpha$. The boundary conditions on the other particles are in general not satisfied, and one can define the "reflection" of all the other particles on α as $u_2 = \sum_{\beta = \alpha} u_{2\beta}$, where the latter are the solutions of the Stokes equation vanishing at infinity with boundary conditions $u_{2\beta} = U_\beta - u_1(\beta)$ on particle β. The effect of the other particles on α is then approximately given by a correction u_3, the solution of the Stokes equation with boundary condition $u_3 = -u_2$ on α. At least when the particles are widely separated, one would expect u_3 to be suppressed compared to U_α and u_1 by powers of the ratio of particle size to particle separation, and to leading order the velocity field in the neighborhood of α equals $u_1 + u_3$. The force and torque on α can then be found to the same degree of approximation; note however that this flow field is *not* a good approximation near any other particle, and the calculation must be repeated for each of the others.

To find u_5, the flow field at α in next order, one first solves for $u_{4\beta}$, the Stokes solution for each $\beta \neq \alpha$ with the boundary condition $u_{4\beta} = -u_3$ on β. u_5 is then the Stokes solution satisfying the boundary condition $u_5 = -\sum_{\beta = \alpha} u_{4\beta}$ on α. The procedure can be iterated to give $u = \sum_{k=1}^{\infty} u_{2k+1}$, along with corresponding expansions for the pressure, force, and torque. There is no proof of the convergence of this expansion, but some confidence in it follows from its application to the two-sphere problem, where the results agree well with the exact solution. We wll illustrate the method of reflections by the simplest case, referring the reader to Happel and Brenner [2] for a detailed review.

Consider two widely separated spherically isotropic particles α and β translating in a plane with respect to fluid asymptotically at rest. The first approximation is a flow field vanishing at the surface of particle α; at large distances this should be the same as would be produced by a point force at the center of α [54] whose strength is the force required to maintain the translation of α. This force is given in terms of the resistance matrix by $F_{1\alpha} = +A_\alpha U_\alpha$ and the lowest order velocity field u_1 is given by Equation 14 dotted into this force, with x taken as the distance between particle centers. The reflection velocity can be calculated in the same way from the force on β; the latter is $F_{2\beta} = -A_\beta[U_\beta - u_1(\beta)]$, and considering the effect of β at the position of α to be a point force, the resulting velocity perturbation $u_{2\beta}(\alpha)$ is given just as above, with the replacement $F_{1\alpha} \to F_{2\beta}$.

Proceeding order-by-order, one finds a geometric series which can be summed to yield

$$F_\alpha = -A_\alpha\left[\hat{x}\,\frac{U_{\alpha x} - A_\beta U_{\beta x}/2\Delta}{1 - A_\alpha A_\beta/4\Delta^2} + \hat{z}\,\frac{U_{\alpha z} - A_\beta U_{\beta z}/\Delta}{1 - A_\alpha A_\beta/\Delta^2}\right] \tag{81}$$

where the particles are located at 0 and (0, 0, d), and $\Delta = 4\pi\mu d$. Note that the interaction corrections only fall off as d^{-1}, leading to interesting "renormalization" effects in calculations of the bulk rheological properties of suspensions [102]. If this formula is applied to particles settling under gravity, then one sees for example that (a) the resistance of either particle is reduced by the presence of the other; (b) a pair of identical particles will settle in parallel with the same velocity, although they will drift to the side unless they are placed one above the other or side by side; and (c) a large (i.e., of greater mass to resistance ratio) particle placed above a smaller one will tend to catch up to it, although they will never quite touch due to slow film thinning (see "Time-Dependent Creeping Flows").

A second class of interaction phenomena concerns the effects of rigid walls on the motion of particles. Again, there is only one known exact solution, corresponding to a sphere translating and rotating in the presence of an infinite plane boundary. Different particular cases of motion have been considered by various authors, including Jeffrey, Brenner, O'Neill and their collaborators, and a systematic review and citation of original papers is given in Reference 4. As in the two-sphere problem, the method used is separation of variables in bispherical coordinates, and typical results take the form of infinite series for the force and torque on the sphere. The limit in which the sphere approaches the plane is of particular interest, as it provides an opportunity to check the validity of lubrication appproximations; it is found that the latter indeed provides a good approximation *locally* in the region where the fluid gap is narrow, but to compute such global quantities as the force exerted on the sphere one must augment the local analysis by a matched asymptotic expansion to a global solution. The values of the force and torque diverge logarithmically as the gap tends to zero, reflecting the difficulty of forcing viscous fluid through a narrow region with a no-slip boundary condition.

When applied to a sphere settling parallel to a plane vertical wall, the results of these calculations are an increase in drag force compared to a freely-settling sphere (increasing linearly as particle size/separation, when this quantity is small) and a torque tending to rotate the sphere in the same direction as if it were rolling down the wall. There is no horizontal drift, a fact that can be demonstrated directly using the reversibility property of Stokes flow [103]. For a sphere translating perpendicular to a plane wall one also finds an increased drag but (by symmetry) no torque. Earlier, Lorentz [20] obtained the small-sphere limits of these results by the method of reflections, which operates in much the same manner as previously described. In view of its generality, we show how this goes for the case of pure translation [104].

Consider a particle translating with velocity U near a solid wall. To leading order we ignore the wall and obtain the force the fluid exerts on the particle (from its resistance matrix) as $F_1 = -A \cdot U$. The velocity field at the wall, provided the particle size a is much less than the distance to the wall d, is given by a Stokeslet (Equation 14) of opposite strength, whose value at any point S on the wall must be of the form $u_1(S) = \Lambda_S \cdot F_1$ where the tensor Λ_S is $(\mu d)^{-1}$ times a dimensionless function of the direction cosines of the line joining S to the particle center. We now require a Stokes solution u_2 with equal and opposite velocity on the wall, and this must be of the form $u_2(x) = \Lambda(x) \cdot F_1$ where Λ is $(\mu d)^{-1}$ times dimensionless function of x/d, reducing to $-\Lambda_S$ on S. The corrected velocity field near the particle in the presence of u_2 is then found, and exerts a force $F_3 = -A \cdot u_2(0)$ upon it, 0 referring to the center of the particle. Thus $F_3 = -A \cdot \Lambda(0) \cdot F_1$. As before, the higher order reflections sum geometrically, giving further powers of $A \cdot \Lambda(0)$, and one obtains

$$F = (\Lambda(0) - A^{-1})^{-1} \cdot U \tag{82}$$

The corrections due to the finite size of the body result in terms O(a/d) in the denominator. The real difficulty in this method is of course in the explicit calculation of the reflection solutions.

Considerable work has been done on the problem of a sphere translating through a cylindrical tube, being a comparatively simple problem of practical importance. Brenner and Happel [105] have used the reflection method in the form just described, and singular perturbation methods have been used for the limit where the sphere nearly touches the wall by Bungay and Brenner. [106]. For a small sphere settling in a quiescent fluid, the results are qualitatively similar to those found for a plane wall, in that the drag force increases as the sphere approaches the wall (i.e., the sphere is slowed by a neighboring wall), and a torque is induced. When the fluid in the cylinder is in motion, one finds that an additional pressure drop is required to maintain Poiseuille flow, whose value rises sharply as the sphere nears the wall. A non-spherically-isotropic particle will in general move radially as it falls in this situation.

Some work has been done on the motion of spheres between two plane walls, originally using the method of reflections [107] and, more recently, boundary collocation techniques [108]. Many related problems concerning particles of various shapes in confined geometries are surveyed by Happel and Brenner [2]. A still more difficult class of problems concerns particle motion in the vicinity of a fluid-fluid interface; some analytically-solvable cases are of symmetrical objects straddling or approaching an interface are reviewed by O'Neill and Ranger [4], while Leal [109] discusses solution methods for a moving particle deforming a nearby interface. An alterative form of the method of reflections has been given by Mazur and collaborators [110]. Reviews of various further aspects of interaction effects are given by Goldsmith and Mason [111], Caswell [112], and Leal [113].

TIME-DEPENDENT CREEPING FLOWS

Situations in which the flow is changing in time may be divided into two classes, depending on whether the time derivation term in the Navier-Stokes equation is small enough to be negligible. In the derivation of the Reynolds equation earlier for example, we assumed that the fluid velocity is determined by the instantaneous configuration of the fluid gap h. This assumption should be checked *a posteriori* by using the solution to compute $\rho\partial u/\partial t$ in terms of $\partial h/\partial t$. In oscillatory problems involving, say, a solid boundary motion of frequency ω with velocity scale U and length scale L, the time derivative term $\sim\rho\omega U$ while the viscous term $\sim\mu U/L^2$ and the former must be retained when the dimensionless group $\rho\omega L^2/\mu$ is of order unity. The remainder of this section discusses some illustrative examples of both types. Further discussion of time-dependent flows is given by Telionis [114].

First consider [6] a half-space of fluid in the region $y > 0$ bounded by a solid wall which at time 0 begins to move with constant velocity $U\hat{x}$. The fluid will move in the x-direction with a profile independent of x, so the pressure is constant and Equation 1 reduces to

$$\rho\frac{\partial u}{\partial t} = \mu\frac{\partial^2 u}{\partial y^2} \tag{83}$$

The time derivative term is required here because the wall is dragging an infinite mass of fluid and some acceleration is always present. The boundary conditions are $u = 0$ for all y at $t \leq 0$, $u = U$ at $y = 0$ for all $t > 0$, and $u \to 0$ as $y \to \infty$, and the solution is

$$u(y, t) = U\left(1 - \frac{2}{\sqrt{\pi}}\int_0^{y/2\sqrt{\nu t}} d\xi e^{-\xi^2}\right) \tag{84}$$

where $\nu = \mu/\rho$ is called the kinematic viscosity. Effectively, the disturbance due to the wall motion propagates diffusively into the fluid with a finite speed, and there is a moving fluid layer of thickness proportional to $t^{1/2}$. In a confined geometry, one would then expect a transient starting flow when a steady force is applied, asymptoting to a steady flow. Thus, if a constant pressure gradient G is applied along the axis of a fluid-filled cylindrical tube of radius a, an equation similar to Equation 83 applies, and one finds [1]

$$w(r, t) = \frac{G}{4\mu}(a^2 - r^2) - \sum_{n=1}^{\infty} c_n J_0(\lambda_n r/a) e^{-\lambda_n^2 \nu t/a^2} \tag{85}$$

where the λ_n satisfy $J_0(\lambda_n) = 0$ and the c_n are constants. The time-dependent effects decay exponentially with the obvious line scale a^2/ν and the asymptotic flow is Poiseuille (Equation 22b). Since problems of this type satisfy diffusion-like equations, solutions may often be found in that literature [115–116].

Persistent time-dependent effects result from oscillatory boundary motion, a simple half-space example having the plane $y = 0$ oscillating with frequency ω [6]. The differential equation (Equation 83) still applies, but now with the boundary condition $u(0, t) = U\cos(\omega t)$, and so

$$u(y, t) = Ue^{-ky}\cos(\omega t - ky) \tag{86}$$

where $k = \sqrt{\omega/2\nu}$. The fluid motion is then confined to a layer near the wall of constant thickness $\sim\sqrt{\omega/\nu}$, referred to as the viscous skin depth. The stress exerted on the wall by the fluid is

$$\sigma_{xy} = -\sqrt{\omega\mu\rho}\, U \cos(\omega t - \pi/4)$$

and oscillates out of phase with the wall motion. The fact that the viscosity of a fluid damps out oscillatory motion, an expected phenomenon since viscosity represents a frictional force, can be seen in a general qualitative way by considering the vorticity equation [117]. Taking the curl of Equation 1 and dropping the inertial term, one has

$$\frac{\partial}{\partial t} \nabla \times u = \nu\nabla^2(\nabla \times u) \tag{87}$$

An oscillatory motion $u \sim e^{i\omega t}$ then typically leads to an exponential decay in space $\sim e^{-k|x|}$, with k as above. Extensive discussions of oscillatory behavior in viscous fluids are given by Rayleigh [118], Lamb [6], and Whitham [119].

The slow drainage of thin films is ubiquitous phenomenon in creeping flow, and we illustrate this by the simple example of two infinitely long solid plates of width 2W, immersed in a viscous liquid at pressure p_0, pushed horizontally together with constant force per length of strength F on each plate. The flux of fluid leaving the gap between the plates equals the rate of change of volume, so if the gap h(t) is narrow compared to W and slowly varying in time, we may write (cf. Equation 29)

$$W\dot{h}(t) = Q(t) = \frac{h^3(t)}{12\mu}\left|\frac{\partial p}{\partial x}\right|$$

Therefore, the pressure gradient is independent of x, and since the total normal force on the upper plate is $\int_{-W}^{W} dx\,(p - p_0)$, we have

$$p(x, t) = p_0 + (W - |x|)\frac{F}{W^2}$$

The form of p just given incorrectly predicts a sharp discontinuity at the center of the plates and is also incorrect for $|x| \geq W$ because the fluid should spread out radially there. However, as in the earlier discussion of entry and exit effects in channel flows, the errors are confined to regions of size O(h), with consequent relative errors O(h/W), which we can neglect for thin films. Combining the latter two equations, we find

$$\frac{1}{h^2(t)} - \frac{1}{h_0^2} = \frac{Ft}{6\mu W^3} \tag{88}$$

and the film requires an infinite time to drain completely. Conversely, a quite large force is required to pull the plates apart rapidly, an effect that underlies the use of adhesive tape. Realistically, one cannot use continuum mechanics for films that are too thin; at distances O(10^2–10^3 Å) van der Waals forces must be included [120] and eventually of course one must consider individual molecules.

In problems involving a fluid-fluid interface it is necessary to ask whether the assumed or computed form of the interface is stable under small perturbations. With fixed boundaries the uniqueness theorem for Stokes flows obviates any question of instability, but when present a free interface effectively leads to non-linearity and the possibility of other boundary shapes. The first step is linear stability analysis: add a small perturbation to the assumed shape and use the Navier-Stokes equation to compute its time evolution. If the perturbation is found to grow, the interface is unstable and one may then study the evolution of finite disturbances. The stability of the interface to *finite* perturbations should be investigated as well, but this is a considerably more difficult task not always attempted. I will illustrate this class of problems by an example related to the household faucet: the instability of a stationary cylinder of viscous fluid in air, first considered by Rayleigh [121]. The essence of the situation is that surface tension tends to minimize the surface area of the fluid-air interface, and a small sinusoidal perturbation is easily shown to decrease the surface area of a cylinder when its wavelength is greater than the cylinder's circumference. Thus, the fluid cylinder is unstable; to calculate its evolution for small axisymmetric disturbances introduce the stream function (Equation 19a), which in a time-dependent situation satisfies

$$\left(E^2 - \frac{1}{\nu}\frac{\partial}{\partial t}\right)E^2\psi = 0 \tag{89}$$

where E^2 is the differential operator whose square appears in Equation 19b. One looks for a solution of the form $\psi = \epsilon e^{\sigma t + ikz}f(r)$, with $\epsilon \ll a$, the cylinder's radius, and the boundary conditions given in Equation 7 applied at the perturbed position of the interface, $r = a + \epsilon \cos kz$. The differential equation is solved by a sum of two Bessel functions, and after some algebra the growth rate of the unstable modes is found to be

$$\sigma \approx \frac{\gamma(1 - k^2a^2)}{6\mu a} \tag{90}$$

Thus, the most unstable modes are those of longest wavelength, and the cylinder tends to break at widely-separated places. When a thread of viscous fluid is surrounded by a second viscous fluid, it is found that the thread tends to break into evenly-spaced spherical drops [122–123], whose size depends on the viscosity ratio. This work has been extended to the breakup of a thread in an extensional flow [124], where the instability is seen to be inhibited. A further elaboration [125] concerns the evolution of a thread of fluid surrounded by a second fluid, both contained in a solid cylinder with straight or constricted sides. A related problem is the instability of fluid running down an inclined plane, either in the form of a layer [126] or a rivulet [127, 128]. A clear discussion of the other "classical" interface stability problems is contained in Chandrasekhar's book [129], and further developments may be found in References 14, 130–135.

The deformations of small drops of fluid in an external flow have been studied extensively recently because of their importance in the rheology of suspensions [13, 136–138]. The typical situation studied has a fluid drop of linear size a floating freely in a second fluid with a prescribed flow varying on a length scale $L \gg a$. The spatial dependence of the flow can be expanded in a Taylor series about the drop's center

$$U(x) = U(0) + x \cdot \nabla U(0) + O(a^2/L^2)$$

The drop will translate with velocity U(0), which can subsumed by choosing a coordinate system moving with the center of the drop. The problem is then to solve the Stokes equations with boundary conditions (Equation 7) and with an asymptotic flow given by a constant shear of magnitude $E = \nabla U(0)$. Theoretical analyses of the problem fall largely into three classes, depending on the degree of deformation of the drop shape away from spherical. For small deformations, one may use the spherical harmonic expansion (Equations 49–50), noting that an asymptotic shear only couples directly to the $n = 2$ mode, while other values of n decay under surface tension. Taylor [139] thus derived an evolution equation for the drop shape with terms corresponding to a rotation with the ambient angular velocity, a distorsion due to the imposed strain, and a restoring force due to surface tension. When the capillary number (Equation 32) is small, the drop arrives at a constant deformation representing a balance between the latter two effects, while at higher capillary number the deformation continues and the drop bursts. When the fluid in the drop is much more viscous than the outer fluid, the drop can simply rotate as a nearly-rigid body.

For drops of high elongation, on the other hand, the slender body approximation described earlier is useful. The normal stress boundary condition is then a balance between the external shear, a distribution of singularities along the drop axis, surface tension, and a nearly-Poiseuille flow in the drop's interior. For low viscosity drops at high shear and large capillary number there is a unique stable shape $R \propto 1 - (z/L)^2$, while drops with a constricted waist have no steady state solution. The detailed behavior at the ends of the drop is an unsettled question. These calculations lead to prediction for the capillary number at which the drop bursts, in reasonable agreement with experiment. At intermediate deformation numerical methods are required, the most popular [25] making use of the boundary integral formulae (Equation 16). These are first written separately for the fluids inside and outside the drop, then combined into one integral equation relating the velocity and surface tension stress on the drops surface to the asymptotic flow. The motion of the drop is found by a time-stepping procedure using the computed instantaneous velocity on the drop's surface. The advantage of the method is that it involves variables defined on the surface alone rather than in a three-dimensional volume. A recent systematic review of drop deformation, including the experimental situation, is given by Rallison [138].

NOTATION

A	cross-sectional area
a	radius
C	constant
Ca	capillary number
c_n, d_n	constants
E	extensional flow tensor
F	force
f	external force per unit mass
G	pressure gradient
h	channel width
K_i	parameter defined by Equation 69b
L	characteristic dimension or length
N	torque per unit length
$\hat{n}$	normal vector
P, p	pressure
R	radius
Re	Reynolds number
r	radius
S	surface area
$\hat{t}$	tangent vectors
U	boundary velocity
u, v	velocity
V	volume
W	energy dissipation rate
w	velocity
Y_{nm}	spherical harmonics
x, y, z	coordinates

Greek Symbols

α_n, β_n	harmonics
γ	Euler's constant; coefficient of surface tension
ϵ	parameter in Equation 77
θ	azimuthal angle
κ	mean curvature of the interface
μ	viscosity
ρ	density
σ	growth rate of unstable modes
σ_{ij}	stress tensor
χ	velocity parameter defined by Equation 35
ψ	stream function
Ω	angular velocity

REFERENCES*

1. Batchelor, G. K., *An Introduction to Fluid Dynamics*, Cambridge (1970).
2. Happel, J., and Brenner, H., *Low Reynolds Number Hydrodynamics*, Noordhoff (1973).
3. Langlois, W. E., *Slow Viscous Flow*, Macmillan (1964).
4. O'Neill, M. E., and Ranger, K., "Particle-Fluid Interactions," *in Handbook of Multiphase Systems*, G. Hetsroni, ed. McGraw-Hill (1982).
5. Berker, R., "Integration des equations du mouvement d'un fluide visqueux incompressible," *in Handbuch der Physik*, vol. VIII/2, S. Flügge, ed. Springer (1963).

* We have abbreviated the *Journal of Fluid Mechanics* as *JFM* and *Annual Reviews of Fluid Mechanics* as *ARFM*.

6. Lamb, H., *Hydrodynamics*, Dover (1945).
7. Hinch, E. J., unpublished lecture notes, Cambridge University (1983).
8. Peyret, R., and Taylor, T. D., *Computational Methods for Fluid Flow*, Springer (1982).
9. Kollmann, W., *Computational Fluid Dynamics*, 2 vol., McGraw-Hill (1978).
10. Chung, T. J., *Finite Element Analysis in Fluid Dynamics*, McGraw-Hill (1978).
11. Dussan, V., E. B., "On the Spreading of Liquids on Solid Surfaces: the Moving Contact Line," *ARFM* 11, 371 (1979).
12. Slattery, J. C., and Flumerfeldt, R. W., "Interfacial Phenomena," in *Handbook of Multiphase Systems*, G. Hetsroni, ed. McGraw-Hill (1982).
13. Levich, V. L., *Physiochemical Hydrodynamics*, Prentice-Hall (1962).
14. Sorensen, T. S., ed., *Dynamics and Instability of Fluid Interfaces*, Springer Lecture Notes in Physics, vol. 105 (1979).
15. Slattery, J. C., *Momentum, Heat and Mass Transfer in Continua*, McGraw-Hill (1972).
16. Howard, L. N., "Bounds on Flow Quantities," *ARFM* 4, 473 (1972).
17. Hill, R., and Power, G., *Quart. J. Mech. Appl. Math.* 9, 313 (1956).
18. Keller, J. B., Rubenfeld, L. A., and Molyneux, J. E., *JFM* 30, 97 (1967).
19. Skalak, R., *JFM* 42, 527 (1970).
20. Lorentz, H., *Abhandl. Theor. Phys.* 1, 23 (1906).
21. Ladyzhenskaya, O. A., *The Mathematical Theory of Viscous Incompressible Flow*, 2nd ed., Gordon & Breach (1969).
22. Hasimoto, H., and Sano, O., "Stokeslets and Eddies in Creeping Flow," *ARFM* 12, 335 (1980).
23. Chwang, A., and Wu, T. Y. T., *JFM* 67, 787 (1975).
24. Lighthill, M. J., *Mathematical Biofluiddynamics*, SIAM (1975).
25. Youngren G. K., and Acrivos, A., *JFM* 76, 433 (1976).
26. Garabedian, P., *Partial Differential Equations*, John Wiley (1964).
27. Liron N., and Shahar, R., *JFM* 86, 227 (1978).
28. Hasimoto, H., *JFM* 5, 317 (1959); see also Sangani, A. S., and Acrivos, A., *Int. J. Multiphase Flow* 8, 343 (1982) and Zick, A. A., and Homsy, G. M., *JFM* 115, 13 (1982).
29. Blake, J. R., *Proc. Cambr. Phil. Soc.* 70, 303 (1971).
30. Taylor, G. I., "Low Reynolds Number Flows," National Committee for Fluid Mechanics Films, distributed by Encyclopedia Britannica Educational Corp., and Illustrated Experiments in Fluid Mechanics, MIT Press (1972).
31. Purcell, E. M., *Am. J. Phys.* 45, 3 (1977).
32. Benedict, R. P., *Fundamentals of Pipe Flow*, John Wiley (1980).
33. Dagan, Z., Weinbaum, S., and Pfeffer, R., *JFM* 115, 505 (1982).
34. Brenner, H., *JFM* 6, 542 (1959).
35. DiPrima R. C., and Swinney, H. L., in *Hydrodynamic Instabilities and the Transition to Turbulence*, Swinney, H. L., and Gollub, J. P., eds. Springer, *Topics in Applied Physics*, vol. 45 (1981).
36. Wannier, G. H., *Quart. Appl. Math.* 8, 1 (1950).
37. Kanwal, R. P., *JFM* 10, 17 (1960).
38. Chwang, A., and Wu, T. Y. T., *JFM* 63, 607 (1974).
39. Berger, S. A., Talbot, L., and Yao, L. S., "Flow in Curved Pipes," *ARFM* 15, 461 (1983).
40. Topakoglu, H., *J. Math. Mech.* 16, 1322 (1967).
41. Larrain, J., and Bonilla, C. F., *Trans. Soc. Rheology* 14:2, 135 (1970).
42. Wang, C. Y., *JFM* 108, 185 (1981).
43. Hinch, E. J., private demonstration, Cambridge (1980).
44. Green, G., *Phil. Mag.* (7)35, 280 (1944).
45. Sampson, R. A., *Phil. Trans. Roy. Soc.* A182, 499 (1891).
46. Bungay, P. M., and O'Neill, M. E., *J. Coll. Int. Sci. 71*, 216 (1979).
47. Saibel, E. A., and Macken, N. A., "The Fluid Mechanics of Lubrication," *ARFM* 5, 185 (1973).
48. Bretherton, F. P., *JFM* 10, 166 (1961).
49. Mitchell, A. G. M., *Lubrication*, Blackie (1950).
50. Pinkus, P., and Sterlicht, B., *Theory of Hydrodynamic Lubrication*, McGraw-Hill (1966).
51. Cameron, A., *Principles of Lubrication*, Wiley (1966).

52. Milne-Thompson, L. M., *Theoretical Hydrodynamics*, 5th ed., Macmillan (1958).
53. Mills, R. D., *JFM* 79, 609 (1977).
54. Muskhelishvili, N. I., *Some Basic Problems in the Mathematical Theory of Elasticity*, Noordhoff (1958).
55. Kantorovich, L. V., and Krilov, V. I., *Approximate Methods of Higher Analysis*, Noordhoff (1958).
56. Dean, W. R., *Phil. Mag.* 15, 929 (1933); 21, 727 (1936); *Proc. Cambr. Phil. Soc.* 32, 598 (1936); 35, 27 (1939); 36, 300 (1940); 50, 623 (1954); *Mathematika* 5, 85 (1958).
57. Garabedian, P. R., *Comm. Pure Appl. Math.* 19, 421 (1966).
58. Richardson, S., *JFM* 33, 475 (1968); 58, 115 (1973).
59. Koplik, J., *JFM* 119, 219 (1982).
60. Mitchell, J. H., *Proc. Lond. Math. Soc.* 31, 100 (1899).
61. Love, A. E. H., *Theory of Elasticity*, Dover (1944).
62. Sokolnikoff, I. S., *Theory of Elasticity*, McGraw-Hill (1946).
63. Jeffrey, G. B., *Phil. Mag.* (6)29, 455 (1915); Hamel, G. *J. Deut. Math.-Vereingung* 25, 34 (1917).
64. Moffatt, K., and Duffy, B. R., *JFM* 96, 299 (1980).
65. Barenblatt, G. I., and Zeldovich, Ya. B., "Self-Similar Solutions as Intermediate Asymptotics," *ARFM* 4, 285 (1972).
66. Dean, W. R., and Montagnon, P. E., *Proc. Cambr. Phil. Soc.* 45, 389 (1949).
67. Moffatt, K., *JFM* 18, 1 (1964).
68. Moffatt, K., *Arch. Mech. Stosowanej.* 16, 365 (1964).
69. Taneda, S., *J. Phys. Soc. Japan* 46, 1935 (1979); see also Van Dyke, M., *An Album of Fluid Motion*, Parabolic Press (1982).
70. Pan, F., and Acrivos, A., *JFM* 28, 643 (1967).
71. Michael, D. H., and O'Neill, M. E., *JFM* 80, 785 (1977).
72. Davis, A. M. J., and O'Neill, M. E., *JFM* 81, 551 (1977).
73. Jeffrey, D. J., and Sherwood, J. D., *JFM* 96, 315 (1980).
74. Dorrepaal, J. M., and O'Neill, M. E., *Quart. J. Mech. Appl. Mech.* 32, 95 (1979).
75. Dean, W. R., *Proc. Cambr. Phil. Soc.* 40, 18 (1944).
76. Finn, R., and Noll, W., *Arch. Rat. Mech. Anal.* 1, 98 (1957).
77. Brenner, H., *Chem. Eng. Sci.* 19, 519 (1964).
78. Bretherton, F. P., *JFM* 14, 284 (1962).
79. Payne, L. E., and Pell, W. H., *JFM* 7, 529 (1960).
80. Faxen, H., *Arkiv. Mat. Astr. Fys.* 17, no. 1 (1922); 20, no. 8 (1927).
81. Brenner, H., *Chem. Eng. Sci.* 21, 97 (1966).
82. Oseen, C. W., *Arkiv. Mat. Astr. Fys.* 6, no. 29 (1922).
83. Lagerstrom, P. A., and Cole, J. D., *J. Rat. Mech. Anal.* 4, 817 (1955).
84. Proudman, I., and Pearson, J. R. A., *JFM* 2, 237 (1957).
85. Van Dyke, M., *Perturbation Methods in Fluid Mechanics*, annotated ed., Parabolic Press (1975).
86. Lamb, H., *Phil. Mag.* (6)21, 112 (1911).
87. Brenner, H., *Chem. Eng. Sci.* 18, 1 (1963); 19, 599, 631 (1964).
88. Brenner, H., *Adv. Chem. Eng.* 6, 287 (1966).
89. Batchelor, G. K., *JFM* 41, 544 (1970).
90. Hinch, E. J., *JFM* 54, 423 (1972).
91. Jeffrey, G. B., *Proc. Roy. Soc. A*102, 161 (1922).
92. Goldstein, H., *Classical Mechanics*, Addison-Wesley (1950).
93. Batchelor, G. K., *JFM* 44, 419 (1970).
94. Cox, R. G., *JFM* 44, 791 (1970); 45, 625 (1971).
95. Tillet, J. P. K., *JFM* 44, 401 (1970).
96. Stimson, M., and Jeffrey, G. B., *Proc. Roy. Soc. A*111, 110 (1926).
97. Cooley, M. D. A., and O'Neill, M. E., *Proc. Cambr. Phil. Soc.* 66, 407 (1969).
98. Brenner, H., and O'Neill, M. E., *Chem. Eng. Sci.* 27, 1421 (1972).
99. Batchelor, G. K., and Green, J. T., *JFM* 56, 375 (1972).
100. Smoluchowski, M., *Bull. Inter. Acad. Polon. Sci.* Lett. 1A, 28 (1911).
101. Stratton, J. A., *Electromagnetic Theory*, McGraw-Hill (1941).

102. Hinch, E. J., *JFM* 83, 695 (1977).
103. Saffman, P., *JFM* 1, 540 (1956).
104. Brenner, H., *JFM* 12, 35 (1962); 18, 144 (1964).
105. Brenner, H., and Happel, J., *JFM* 4, 195 (1958); see also Hirschfield, B. R., Brenner, H., and Falade, A., *Physiochem. Hydro.* (1984).
106. Bungay, P. M., and Brenner, H., *JFM* 60, 81 (1973); *Int. J. Multiphase Flow* 1, 75 (1973).
107. Faxen, H., *Arkiv. Mat. Astr. Fys.* 17, no. 27 (1923).
108. Ganatos, P., Pfeffer, R., and Weinbaum, S., *JFM* 99, 739, 755 (1980).
109. Leal, L. G., in *Waves on Fluid Interfaces*, ed. R. E. Meyer, Academic (1983); see also Berdan C. and Leal, L. G., *J. Coll. Int. Sci.* 87, 62 (1982) and Lee, S. H., and Leal, L. G., *ibid.* 87, 81 (1982).
110. Mazur, P., and van Saarloos, W., *Physica* 115A, 21 (1982), *ibid.* 120A 77 (1983); Weisenborn, A. J., and Mazur, P., *ibid.* 123A, 191, 209 (1983).
111. Goldsmith, H. L., and Mason, S. G., "The Microrheology of Dispersions," *Rheology* 4, 85 (1967).
112. Caswell, B., *Chem. Eng. Sci.* 27, 373 (1970).
113. Leal, L. G., "Particle Motions in a Viscous Fluid," *ARFM* 12, 435 (1980).
114. Telionis, D. P., *Unsteady Viscous Flows*, Springer (1983).
115. Carslaw, H. S., and Jaeger, C. J., *Conduction of Heat in Solids*, Oxford (1959).
116. Morse, P. M., and Feshbach, H., *Methods of Theoretical Physics*, McGraw-Hill (1953).
117. Landau, L. D., and Lifshitz, I. M., *Fluid Mechanics*, Pergamon (1959).
118. Rayleigh, Lord, *Theory of Sound*, Dover (1945).
119. Whitham, G. B., *Linear and Nonlinear Waves*, John Wiley (1974).
120. Sheludko, A.., "Thin Liquid Films," *Advances in Colloid and Interface Science 1*, 319 (1967).
121. Rayleigh, Lord, *Phil. Mag.* (5)34, 145 (1892).
122. Taylor, G. I., *Proc. Roy. Soc.* A146, 501 (1934).
123. Tomokita, S., *Proc. Roy. Soc.* A150, 322 (1935).
124. Mikami, T., Cox, R. G., and Mason, S. G., *Int. J. Multiphase Flow* 2, 113 (1975).
125. Hammond, P. S., *JFM* 137, 363 (1983) and Ph.D. thesis, Cambridge University (1982).
126. Yih, C.-S., *Phys. Fluids* 6, 321 (1963).
127. Towell, G. D., and Rothfeld, L. B., *AIChE J.* 12, 972 (1966).
128. Davis, S. H., *JFM* 98, 225 (1980); Weiland, R. H., and Davis, S. H., *JFM* 107, 261 (1981).
129. Chandrasekhar, S., *Hydrodynamic and Hydromagnetic Stability*, Dover (1981).
130. Mollo-Christensen, E. L., "Flow Instabilities," in Ref. 30.
131. Dussan V., E. B., *Arch. Rat. Mech. Anal.* 57, 363 (1975).
132. Joseph, D. D., *Stability of Fluid Motions*, 2 vol., Springer (1976).
133. Drazin, P., and Reid, W. H., *Hydrodynamic Stability*, Cambridge (1981).
134. Lin, C. C., *Theory of Hydrodynamic Stability*, Cambridge (1955).
135. Davis, S. H., *J. Appl. Mech.* 50, 977 (1983).
136. Harper, J. F., "Motion of Bubbles and Drops Through Liquids," *Adv. Appl. Mech.* 12, 59 (1972).
137. LaCroissette, D. H., ed., *Proceedings of the Second International Colloquium on Drops and Bubbles*, Jet Propulsion Laboratory (1982).
138. Rallison, J. M., "Deformation of Small Viscous Drops and Bubbles in Shear Flow," *ARFM* 16, 45 (1984).
139. Taylor, G. I., *Proc. Roy. Soc.* A138, 41 (1932); see also Cox, R. G., *JFM* 37, 601 (1969).

CHAPTER 30

STEADY NON-NEWTONIAN FLOW ABOUT A RIGID SPHERE

R. P. Chhabra†

Department of Chemical Engineering
University College of Swansea
Singleton Park, Swansea, U. K.

CONTENTS

INTRODUCTION

The motion of particles in a viscous liquid impinges on our everyday life to such an extent that it would be no exaggeration to say that the phenomenon is ubiquitous. The water we drink has suspended particles in it, some contributing to our well-being, while others have harmful effects. Blood, a vital element of life, immediately comes to mind with its red and white blood cells suspended in plasma. Thus, to understand the behavior of particles in a viscous medium has been a challenge throughout the history of man.

From the technological point of view, numerous operations in chemical and processing industries involve fluid-particle systems. Fluidization technology relies almost solely on fluid-particle interactions. Hydraulic and pneumatic pipeline transport involves the interaction between the carrier fluid and the material to be transported. Other examples include filtration, sedimentation and thickening of slurries, rheological behavior of suspensions, etc. Less appreciated applications of

† Present address: Department of Chemical Engineering, Indian Institute of Technology, KANPUR 208016, India

fluid-particle systems are the motion of red blood cells in capillary blood flow, separation of macromolecules according to size, etc. The mathematical complexity involved has hindered progress in the modeling of such systems. On the other hand, the study of the behavior of a single spherical particle in viscous medium has yielded valuable information, which has led to an understanding of multiparticle/fluid systems. Thus, the flow of a fluid past a single sphere represents an idealization of many industrially important processes.

The developments in this field began with the celebrated work of Stokes in 1851 [1] concerning the slow flow of particles in a viscous medium, in which he derived his famous expression for the drag force acting on a rigid spherical particle falling under the influence of gravity in a quiescent fluid of infinite extent. This initial work has subsequently led to extensions and developments in a variety of ways that consider such effects as the rotation of a particle, inertial effects, wall effects, shape of the particle, particle-particle and fluid-particle interactions in multiparticle systems, and magnetic and electrical fields. In recent years the knowledge of the detailed flow field facilitated by the modern instrumentation has revealed the complex nature of the flow structure not hitherto expected, with wakes and eddies forming in many instances.

All of us have been accustomed for three hundred years to accept the Newtonian fluid model as a standard fluid behavior. The resulting Navier-Stokes equations are well known for their outstanding complexity due to the nonlinear inertial terms. During the last thirty to forty years, the development of elegant numerical schemes has considerably alleviated the problem. Yet, it is a simple task to generate experiments that could never be explained (even qualitatively) by the Newtonian fluid model. Indeed, Newtonian fluid behavior seems to be the exception rather than the rule. During the last thirty years or so, there has been a recognition that many materials do not conform to the Newtonian fluid behavior, and are accordingly known as "non-Newtonian." The simplest non-Newtonian behavior is shear–thinning (or pseudoplasticity) whereby the apparent viscosity of a fluid decreases with increasing shear rate. Besides, some of these materials behave like a viscous fluid in long-time experiments, while the initial response to applied external stress is like that of elastic solids. Qualitatively, such behavior can be explained by hypothesizing that these viscoelastic fluids possess a memory for their past configuration, while purely viscous fluids have no such memory. A viscoelastic material further differs from a purely elastic solid, the latter only remembering its non-deformed state. On the other hand, viscoelastic material has a time dependent memory that is stronger for events in the recent past than those in the distant past.

In many applications of fluid-particles systems, the fluid phase exhibits complex non-Newtonian behavior. Therefore, the knowledge of the drag force experienced by a spherical particle moving in non-Newtonian media is necessary, and provides a starting point in understanding multiparticle systems. In the following we examine the influence of non-Newtonian behavior (particularly shear thinning and viscoelasticity) on the drag force and on the wall effects due to the presence of a cylindrical boundary experienced by a freely settling sphere. It is also appropriate to mention here that the non-Newtonian substances possessing yield stress (although equally important and common in occurrence) are not dealt with in this chapter.

In the mechanics of non-Newtonian fluids, it is customary to divide the fluid response into two components, viz. shear thinning and elasticity. Whereas such a distinction may not be strictly justifiable on scientific grounds, it does often provide a qualitative understanding of the flow problem. Therefore, this classification is adopted here for convenience.

In the following section we begin with the governing equations describing the behavior of a rigid spherical particle moving steadily in an incompressible liquid. This assists particularly in appreciating the complexity and the associated difficulties in seeking the consequent solutions.

GOVERNING EQUATIONS

The physical situation under consideration is that a rigid sphere of radius R is fixed in the unbounded fluid flowing at a constant velocity V. A polar spherical coordinate system (r, θ, ϕ) is elected with the origin at the center of the sphere. The configuration is schematically shown in Figure 1.

Noting symmetry, it is assumed that none of the physical quantities varies with the ϕ-coordinate

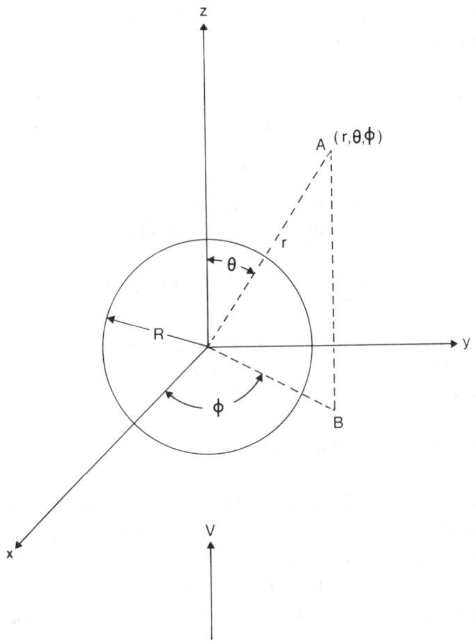

Figure 1. Coordinate system for the flow configuration.

and so their derivatives with respect to the variable ϕ vanish, hence

$$v_r = v_r(r, \theta), \qquad v_\theta = v_\theta(r, \theta), \qquad v_\phi = 0 \tag{1}$$

Under these conditions, the steady isothermal flow of an incompressible fluid in the absence of body forces is described by the r- and θ- components of the momentum equation as [2]

r-component,

$$\rho\left(v_r \frac{\partial v_r}{\partial r} + \frac{v_\theta}{r}\frac{\partial v_r}{\partial \theta} - \frac{v_\theta^2}{r}\right) = -\frac{\partial p}{\partial r} + \left[\frac{1}{r^2}\frac{\partial}{\partial r}(r^2\tau_{rr}) + \frac{1}{r \sin\theta}\frac{\partial}{\partial\theta}(\tau_{r\theta}\sin\theta) - \frac{\tau_{\theta\theta} + \tau_{\phi\phi}}{r}\right] \tag{2}$$

θ-component,

$$\rho\left(v_r \frac{\partial v_\theta}{\partial r} + \frac{v_\theta}{r}\frac{\partial v_\theta}{\partial \theta} + \frac{v_r v_\theta}{r}\right) = -\frac{1}{r}\frac{\partial p}{\partial \theta} + \left[\frac{1}{r^2}\frac{\partial}{\partial r}(r^2\tau_{r\theta}) + \frac{1}{r \sin\theta}\frac{\partial}{\partial\theta}(\tau_{\theta\theta}\sin\theta) + \frac{\tau_{r\theta} - \tau_{\phi\phi}\cot\theta}{r}\right] \tag{3}$$

and by the continuity equation,

$$\frac{1}{r^2}\frac{\partial}{\partial r}(r^2 v_r) + \frac{1}{r\sin\theta}\frac{\partial}{\partial\theta}(v_\theta \sin\theta) = 0 \tag{4}$$

By introducing the stream function ψ such that the velocity components (v_r, v_θ) are given by

$$v_r = \frac{1}{r^2 \sin\theta}\frac{\partial\psi}{\partial\theta}, \qquad v_\theta = -\frac{1}{r\sin\theta}\frac{\partial\psi}{\partial r} \tag{5}$$

the equation of continuity is automatically satisfied.

For uniform flow past a rigid sphere the velocity must vanish at the surface (no slip) and it must approach the streaming velocity V far away from the sphere, i.e.

$$v_r = 0 \quad \text{and} \quad v_\theta = 0 \qquad \text{at} \quad r = R \tag{6}$$

$$v_r = -V\cos\theta \quad \text{and} \quad v_\theta = V\sin\theta \qquad \text{as} \quad r \rightarrow \infty \tag{7}$$

For a given constitutive equation of the fluid behavior (relationship between shear stress and rate of strain), the solution of Equations 2–4 subjected to the boundary conditions given by Equations 6 and 7 yields expressions for all the physical quantities such as ψ, τ_{rr}, $\tau_{r\theta}$, $\tau_{\phi\phi}$, p etc. These in turn can be manipulated and integrated appropriately to obtain the drag force on the rigid sphere exerted by the fluid as:

$$F_D = \int_0^{2\pi}\int_0^{\pi}\left(p\Big|_{r=R}\cos\theta\right)R^2\sin\theta\,d\theta\,d\phi + \int_0^{2\pi}\int_0^{\pi}\left(\tau_{r\theta}\Big|_{r=R}\sin\theta\right)R^2\sin\theta\,d\theta\,d\phi \tag{8}$$

Usually the two terms in the right-hand side of Equation 8 are known as the "form drag" and the "friction drag," respectively.

The treatment and equations presented here are perfectly general and thus are applicable to all incompressible fluids including Newtonian and non-Newtonian liquids. The usefulness of these equations in yielding expressions for the drag force is illustrated for a wide range of fluid behavior in the following sections. It is instructive to start with the case of the simplest type of fluid behavior, namely Newtonian. This sets the stage for the subsequent treatment for non-Newtonian liquids.

NEWTONIAN FLUIDS

The Newtonian fluid represents the simplest realistic class of material and, as such, the behavior of rigid spherical particles in Newtonian media has received the greatest amount of attention. A Newtonian fluid is characterized by a constant viscosity, which depends upon temperature and pressure. The constitutive relationship connecting the components of the stress tensor to the components of rate of deformation tensor is written as

$$\tau_{ij} = 2\mu\epsilon_{ij} \qquad \text{where } i, j = r, \theta, \phi \tag{9}$$

ϵ_{ij} represents the components of the rate of deformation tensor, and are related to the two components of velocity (v_r, v_θ) in a spherical coordinate system [2] via the following equations:

$$\left.\begin{aligned}
\epsilon_{rr} &= \frac{\partial v_r}{\partial r}\\
\epsilon_{\theta\theta} &= \left(\frac{1}{r}\frac{\partial v_\theta}{\partial \theta} + \frac{v_r}{r}\right)\\
\epsilon_{\phi\phi} &= \left(\frac{v_r + v_\theta\cot\theta}{r}\right)\\
\epsilon_{r\theta} = \epsilon_{\theta r} &= \frac{1}{2}\left[r\frac{\partial}{\partial r}\left(\frac{v_\theta}{r}\right) + \frac{1}{r}\frac{\partial v_r}{\partial \theta}\right]\\
\epsilon_{\theta\phi} &= \epsilon_{\phi\theta} = 0\\
\epsilon_{r\phi} &= \epsilon_{\phi r} = 0
\end{aligned}\right\} \tag{10}$$

If required, Equation 10 can be expressed in terms of the stream function (ψ) using Equation 5 which, in turn, allows Equations 2 and 3 to be rewritten in a form that contains only ψ as the dependent variable:

$$E^4\psi^* = \frac{Re}{2}\left[\frac{\partial\psi^*}{\partial\theta}\frac{\partial}{\partial x}\left(\frac{E^2\psi^*}{x^2\sin^2\theta}\right) - \frac{\partial\psi^*}{\partial x}\frac{\partial}{\partial\theta}\left(\frac{E^2\psi^*}{x^2\sin^2\theta}\right)\right]\sin\theta \tag{11}$$

where $E^2 = \frac{\partial^2}{\partial x^2} + \frac{\sin\theta}{x}\left(\frac{1}{\sin\theta}\frac{\partial}{\partial\theta}\right)$

and ψ^* is a dimensionless stream function defined as (ψ/VR^2), x is a dimensionless radial coordinate defined as (r/R), Re is the celebrated Reynolds number (the ratio of the inertial forces to the viscous forces) and for the situation in hand is defined as $\rho V(2R)/\mu$. Now attention is turned to the progress made in obtaining the solution to Equation 11 eventually leading to expressions for the drag force F_D.

Drag Force

The highly nonlinear form of Equation 11 is evident and to date, determination of the exact and general solutions has proved to be a formidable task. Therefore, analyses only with varying degrees of approximations are available. For instance, for creeping flow or slow flows, the inertial terms on the right-hand side of Equation 11 are neglected in comparison to the viscous forces. Thus, Equation 11 simplifies to

$$E^4\psi^* = 0 \tag{12}$$

This equation was solved by Stokes almost a century and a half ago and the drag force experienced by a spherical particle is given by

$$F_D = 6\pi\mu RV \tag{13}$$

This total drag force draws a contribution of $4\pi\mu RV$ from the friction and the remainder is due to the form drag.

It is customary to introduce a dimensionless drag coefficient, C_D, defined as

$$C_D = \frac{F_D}{\frac{1}{2}\rho V^2\pi R^2} \tag{14}$$

which yields the more familiar form of Equation 13

$$\text{i.e.}\quad C_D = \frac{24}{Re} \tag{15}$$

Equations 13 and 15 are known as Stokes' law or equation. It is important to remember that Equation 15 or 13 is applicable in the absence of inertial effects. The slowness of flow is judged by the value of the Reynolds number (Re). Experiments have shown that Equation 15 is useful for Reynolds numbers up to 0.1 [2]. Stokes' equation finds use in the motion of colloidal particles under the influence of an electric field, in the theory of sedimentation and in the study of the movement of aerosol particles etc. Perhaps by far the most important application of Equation 15, from a technological point of view, has been the development of the falling ball viscometer for Newtonian liquids.

For values of Reynolds number >0.1, Equation 15 shows an increasing deviation from experimental data as the Reynolds number increases. Oseen [3] extended the range of applicability of

Stokes' law by partially taking into account the inertial forces and arrived at:

$$C_D = \frac{24}{Re}\left(1 + \frac{3}{16} Re\right) \tag{16}$$

Equation 16 predicts the drag coefficient for Reynolds numbers up to 1.0 with a maximum error of ±1%. In the intervening years, numerous expressions of varying complexity and accuracy have appeared in the literature. Most of these have been well summarized by Clift, Grace, and Weber in their classic treatise [4]. O'Neill [5] has written a comprehensive survey article on the theoretical developments in this area.

At high values of the Reynolds number (> 1), the inertial effects become increasingly important and the contribution of the terms on the right-hand side of Equation 11 is no longer negligible. Jenson [6] employed a finite difference scheme to obtain numerical solutions up to Re = 40. Other attempts [7, 8] to obtain numerical solutions have met with varying degrees of succcess. However, the numerical solutions have been only obtained for Reynolds number up to 400. All the work reported for Re > 400 has been empirical in nature. Consequently, there is no scarcity of experimental correlations relating drag coefficient with Reynolds number. These have been reviewed, again, by Clift et al. [4]. Their recommendations for calculating the value of drag coefficient at a given value of Reynolds number covering the entire range of the conventional standard drag curve are given in Table 1.

Wall Effects

All theoretical treatments and the predictive empirical correlations described in the preceding section assume the steady fall of a sphere in an infinite extent of fluid. This poses a substantial practical

Table 1
Recommended Drag Correlation: Standard drag curve [4]

Range	Correlation
$Re < 0.01$	$C_D = \frac{24}{Re}\left(1 + \frac{3}{16} Re\right)$
$0.01 < Re \le 20$	$C_D = \frac{24}{Re}(1 + 0.1315\, Re^{(0.82 - 0.05\omega)})$
$20 \le Re \le 260$	$C_D = \frac{24}{Re}(1 + 0.1935\, Re^{-0.6305})$
$260 \le Re \le 1{,}500$	$\log C_D = 1.6435 - 1.1242\omega + 0.1558\omega^2$
$1{,}500 \le Re \le 1.2 \times 10^4$	$\log C_D = -2.4571 + 2.5558\omega - 0.9295\omega^2 + 0.1049\omega^3$
$1.2 \times 10^4 < Re \le 4.4 \times 10^4$	$\log C_D = -1.9181 + 0.637\omega - 0.0636\omega^2$
$4.4 \times 10^4 < Re \le 3.38 \times 10^5$	$\log C_D = -4.339 + 1.5809\omega - 0.1546\omega^2$
$3.38 \times 10^5 < Re \le 4 \times 10^5$	$C_D = 29.78 - 5.3\omega$
$4 \times 10^5 < Re \le 10^6$	$C_D = 0.1\omega - 0.49$
$10^6 < Re$	$C_D = 0.19 - \frac{8 \times 10^4}{Re}$
	$\omega = \log Re$

difficulty. Usually, the experimental work is performed in cylindrical tubes and the walls of tube exert an extra retarding force on the freely settling sphere. Thus, the knowledge of this retardation effect (also known as wall effects) is important, particularly while using Stokes' law to calculate the viscosity of a Newtonian liquid [9].

Wall effects are usually quantified by defining a "wall factor." One of the simplest definitions of the wall factor is the ratio of the terminal fall velocity of a sphere in the bounded fluid to that in the unbounded fluid, i.e.

$$\text{wall factor} = f = \frac{\text{Terminal fall velocity in bounded fluid}}{\text{Terminal fall velocity in unbounded fluid}} = \frac{V_m}{V} \tag{17}$$

Evidently, f takes on values between zero and unity. Other definitions of the wall factor are used and have been adequately dealt with by Happel and Brenner [10] and by Clift et al. [4].

From a theoretical point of view, the effect of the containing walls is to change the boundary conditions for the equations of motion and continuity of the continuous phase. In place of the condition of uniform flow remote from the sphere the containing walls impose conditions that must be satisfied at definite boundaries.

Under creeping flow conditions, and for the particle-to-tube-diameter ratio <0.1, Ladenberg [11] and Faxen [12] have obtained approximate solutions to the pertinent equations. Within this range of conditions, the most widely accepted wall correction is that of Faxen [12]:

$$f_0 = 1 - 2.105(d/D) + 2.087(d/D)^3 - 0.95(d/D)^5 \tag{18}$$

Often, the last two terms are neglected (such as in Ladenberg [11]). This is only applicable for Re <0.1 and (d/D) <0.1. Haberman and Sayre [13] have given approximate correction factors for the values of diameter ratio up to 0.8; these are reproduced in Table 2 and are plotted in Figure 2. The values of wall factors given in Table 2 are valid up to Re = 2.0. Besides these analytical results, many empirical correlations are available, most of which have been listed by Iwaoka and Ishii [14]. Some of them are valid in the creeping flow regime for diameter ratios up to 0.97. One such equation that is simple to use is that of Francis [9]:

$$f_0 = \left[\frac{1 - (d/D)}{1 - 0.475(d/D)}\right]^4 \qquad \text{for } (d/D) \leq 0.97 \tag{19}$$

Equation 19 is also plotted in Figure 2. The results of Haberman and Sayre [13] and those calculated using Equation 19 are virtually indistinguishable from each other. Typical comparison with experimental data, as reported by Uhlherr [15] and Chhabra et al. [16], is also shown in the same figure, where excellent agreement is seen to exist.

Table 2
Values of Wall Correction Factor as a Function of (d/D) [13]

Diameter Ratio (d/D)	Wall Correction Factor f_0
0.00	1.00
0.10	0.792
0.20	0.595
0.30	0.422
0.40	0.278
0.50	0.168
0.60	0.0898
0.70	0.0401
0.80	0.0136

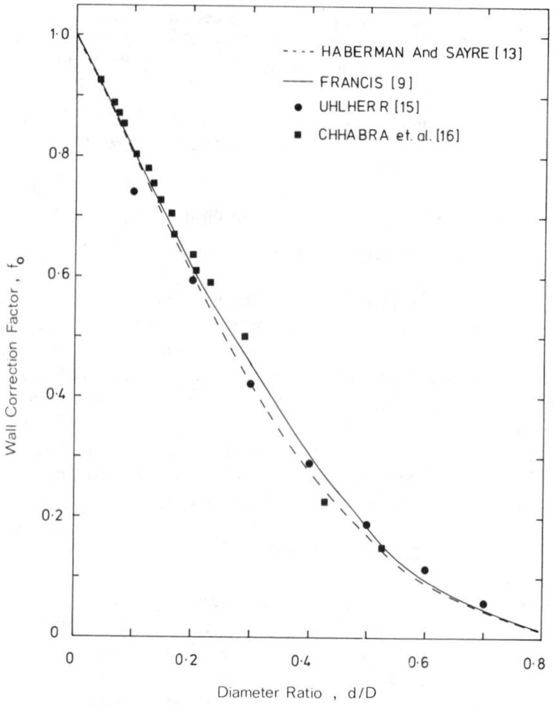

Figure 2. Comparison of Newtonian experimental wall factors with theory and empirical correlation in the low Reynolds number region.

The highly nonlinear form of the inertial terms in the governing equations precludes the possibility of analytical treatments at high values of Reynolds number; consequently, very little theoretical work has been done. Using a semi-empirical approach, Fayon and Happel [17] have provided an expression for wall factor valid up to Re = 40 and diameter ratios 0.3125. All work reported for Re > 40 is empirical in nature. In fact, there are only three sets of results available for wall effects at high Reynolds number [15, 18, 19]. Combined together, these three investigations cover diameter ratios up to 0.9 and Reynolds numbers up to 20,000. The wall factor is dependent only on diameter ratio—both in creeping flow regime ($Re_m < 0.1$) as well as in the region of $Re_m > 1{,}000$–$2{,}000$, while in the intermediate region it is a function of both the diameter ratio and the value of the Reynolds number. Munroe [18] suggested the following empirical equation for $Re_m > 1{,}000$, and $(d/D) \leq 0.8$:

$$f_\infty = 1 - (d/D)^{1.5} \tag{20}$$

Typical comparison between the experimental values of wall factor and those calculated using Equation 20 is shown in Figure 3.

For the intermediate region, $0.1 \leq Re_m \leq 1{,}000$, the functional dependence of f on (d/D) and Re is shown in Figure 4. The solid lines are based on the experimental data of Fidleris and Whitmore [19] and of Uhlherr [15]. Numerical values used in plotting the results shown in Figure 4 are reproduced in Table 3 for reference purposes. The use of Equations 19 and 20 is recommended to

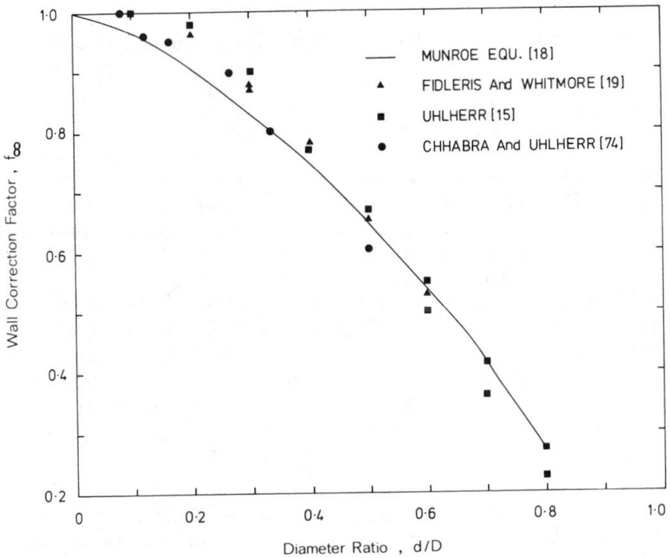

Figure 3. Dependence of Newtonian wall correction factor on diameter ratio at high Reynolds numbers.

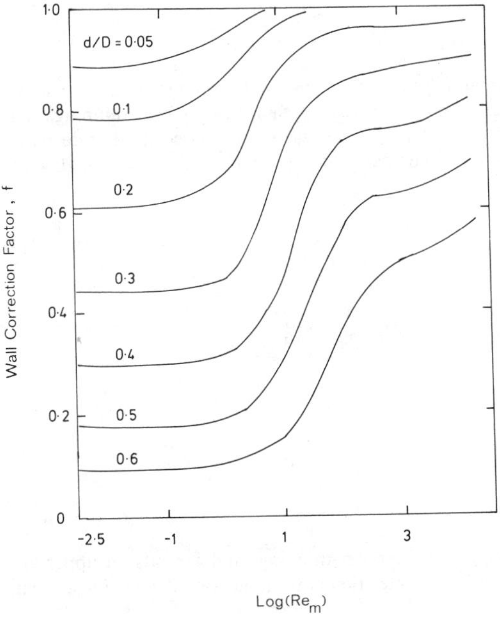

Figure 4. Effect of diameter ratio on the value of Newtonian wall correction factor in the intermediate Reynolds number region. (From Fidleris and Whitmore [19]).

Table 3
Numerical Values (f) Corresponding to Figure 4 (from the data of Uhlherr [15])

(d/D) \ Re_m	0.5	1.0	2	5	10	20	50	100	200	1,000	10,000
0.0	1.00	1.00	1.00	1.00	1.00	1.00	1.00	1.00	1.00	1.00	1.00
0.1	0.74	0.77	0.826	0.910	0.950	0.971	0.980	1.00	1.00	1.00	1.00
0.2	0.59	0.63	0.700	0.770	0.826	0.862	0.910	0.926	0.940	0.970	0.980
0.3	0.42	0.465	0.530	0.610	0.677	0.725	0.800	0.847	0.860	0.910	0.910
0.4	0.29	0.325	0.370	0.444	0.532	0.676	0.725	0.770	0.770	0.770	0.770
0.5	0.19	0.213	0.250	0.303	0.370	0.454	0.550	0.606	0.633	0.667	0.680
0.6	0.115	0.130	0.154	0.190	0.233	0.303	0.370	0.417	0.465	0.500	0.555
0.7	0.0625	0.0714	0.085	0.106	0.136	0.172	0.233	0.263	0.303	0.360	0.413
0.8	0.0308	0.0330	0.040	0.051	0.067	0.0855	0.116	0.146	0.182	0.222	0.272
0.9	0.0104	0.0120	0.0141	0.018	0.022	0.0308	0.0465	0.0625	0.078	0.108	0.152

calculate the values of wall factor in the creeping and the high Reynolds number regions respectively while one must refer to the graphical presentations given in Figure 4 for the intermediate region.

Thus, based on a combination of theoretical analyses applicable in the range of low Reynolds number and small values of the diameter ratio, and on the experimental work in the high Reynolds number region, a coherent picture of the motion of a rigid sphere in Newtonian fluids has emerged. Satisfactory means of calculating the values of drag coefficient and wall factor covering the entire range of practical interest are available. This now provides an appropriate perspective for non-Newtonian flow past a sphere. In the next section the influence of fluid shear-thinning behavior (in the absence of fluid elasticity) on drag coefficient and wall effects in low as well as high Reynolds number regions is examined.

INELASTIC SHEAR-THINNING FLUIDS

The most common inelastic non-Newtonian fluid behavior observed is pseudoplasticity, characterized by an apparent viscosity that decreases with increasing shear rate (shear-thinning). The pseudoplastic fluid displays a constant viscosity at very low shear rates, a viscosity that decreases with shear rate at intermediate shear rates, and a constant viscosity in the limit of very high shear rates, that is

$$\lim_{\gamma \to 0} [\tau/\dot{\gamma}] = \eta_0 \quad \text{(zero shear viscosity)} \tag{21}$$

and

$$\lim_{\gamma \to \infty} [\tau/\dot{\gamma}] = \eta_\infty \quad \text{(infinite shear viscosity)} \tag{22}$$

The apparent viscosity defined by

$$\eta = \frac{\tau}{\dot{\gamma}} \tag{23}$$

then decreases with shear rate from η_0 to η_∞ and the shear-thinning fluid exhibits an upper and lower region of Newtonian behavior. Figure 5 illustrates the entire range of behavior for a shear-thinning polymer solution [20].

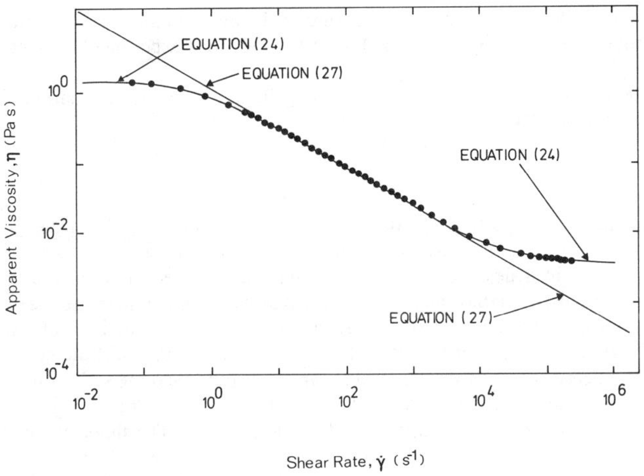

Figure 5. Apparent viscosity as a function of shear rate for 0.4% polyacrylamide aqueous solution displaying zero and infinite shear viscosities η_0 and η_∞ with shear thinning at intermediate shear rates. (From Tiu et al. [24] and based on the data of Boger [20]).

Expressions of varying complexity have been proposed and generalized to model pseudoplastic behavior [21, 22]. The Carreau model [23] contains η_0 and η_∞ and two other parameters, λ and n,

$$\frac{\eta - \eta_\infty}{\eta_0 - \eta_\infty} = [1 + (\lambda\dot{\gamma})^2]^{(n-1)/2} \tag{24}$$

In Equation 24, λ is considered to be a characteristic time of liquid, and n is a measure of the rate of change of viscosity with shear rate in the shear-thinning region. Usually, η_0 and η_∞ are evaluated from direct measurements, and the remaining two parameters, namely λ and n, are treated as the adjustable parameters giving the best fit of experimental data. Notice the excellent fit of the Carreau model to the experimental data shown in Figure 5. Rarely is such a complete set of data available and in flow fields shear rates larger than $10^4\ s^{-1}$ are not generally experienced. Hence, the use of such a refined model is not normally possible or warranted in the solution of hydrodynamic problem involving shear-thinning fluids.

The next simplest representation of the data shown in Figure 5 is the Ellis model,

$$\eta = \frac{\eta_0}{1 + (\tau/\tau_{1/2})^{\alpha-1}} \tag{25}$$

Equation 25 is a good approximation of the data shown in Figure 5 for low and intermediate shear rates but is not adequate at high shear rates. The parameter $\tau_{1/2}$ denotes the characteristic value of shear stress at which the apparent viscosity has dropped to half the value of the zero shear viscosity, and α is again a measure of rate of decrease of viscosity. The Carreau model (Equation 24) also reduces to a three-parameter equation if η_∞ is neglected. That is

$$\eta = \eta_0[1 + (\lambda\dot{\gamma})^2]^{(n-1)/2} \tag{26}$$

Even though the Ellis model and the truncated Carreau model (Equation 26) are good compromises between accurate representations of physical picture and simplicity, they have not often been used for the solution of flow problems, possibly because of the mathematical complications arising with a three-parameter model, but more likely due to the fact that experimentalists rarely measure a zero shear viscosity.

Many workers have characterized shear thinning fluids using the two-parameter Ostwald de Waele or power law model

$$\eta = K\dot{\gamma}^{n-1} \tag{27}$$

This model has the obvious advantage of simplicity in that deviations from Newtonian behavior are characterized by a single parameter, viz. the flow behavior index n. Thus, an investigation of shear-thinning effects in a flow field becomes a study of the effect of n on the flow field where $0 < n < 1$. Figure 5 illustrates the difficulties and the weaknesses inherent in the use of the power-law model. It does not describe the viscosity of a shear-thinning fluid in the limit of low or high shear rates. The deficiency of the power law at high shear rates is perhaps of little consequence but its inability to predict zero shear viscosity at low shear rates could be quite serious in non-viscometric flows. The problem is illustrated more clearly in Figure 6, taken from Boger [25], where the viscosity behavior of three hypothetical shear-thinning fluids is depicted. The fluids show identical power law behavior with $n = 0.4$ and $K = 2.83$ Pa $s^{0.4}$, yet their zero shear viscosities vary from 10 to 40 Pa s. A theoretical prediction based on power law representation of these three fluids would yield identical results even though their molecular weights would be vastly different. Another weakness of the power law model lies in the fact that unlike the Carreau viscosity equation and the Ellis model, it neither contains a time parameter nor can the two constants (n, K) be arranged to construct a characteristic time that is believed to be important for non-Newtonian liquids [26]. In view of these latter points, in what follows for flow past a sphere it will be important not to read too much into conclusions reached on the basis of the power law model.

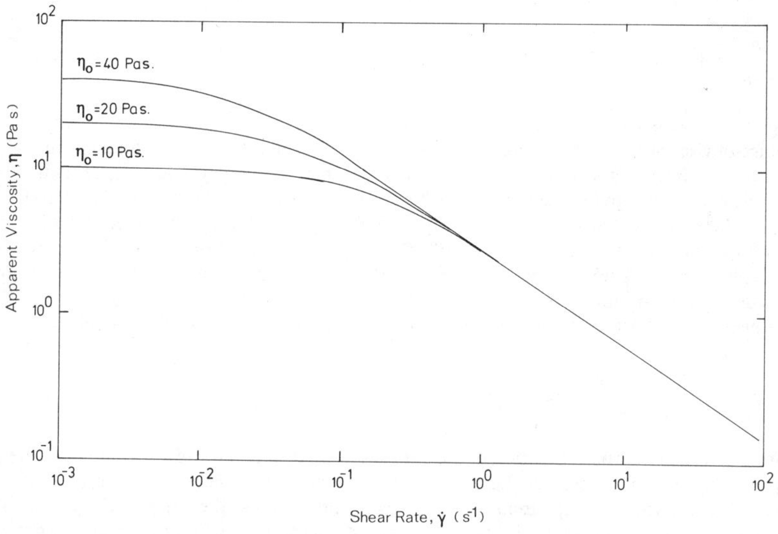

Figure 6. Apparent viscosity as a function of shear rate for three hypothetical fluids that could be characterized by the identical power law constants ($n = 0.4$ and $K = 2.83$ Nm^{-2} s^4), yet are vastly different and accurately characterized by the following sets of values for Ellis model parameters: (a) $\eta_0 = 10$ Pa s, $\alpha = 2.95$ and $\tau_{1/2} = 1.64$ Pa; (b) $\eta_0 = 20$ Pa s, $\alpha = 2.95$ and $\tau_{1/2} = 1.18$ Pa; (c) $\eta_0 = 40$ Pa s, $\alpha = 2.95$ and $\tau_{1/2} = 0.80$ Pa. (Modified after Boger [25]).

Likewise some care must be taken in the interpretation of experimental data where the power law has been used, not only because the power law model may not adequately describe the shear-thinning characteristics of the fluid in question, but also because highly shear-thinning fluids are almost invariably viscoelastic. It is not clear in many instances whether an investigator is observing shear-thinning effects, elastic effects, or a combination of both. Experimental work where the researchers attempt to study the influence of shear-thinning on a flow field and have not established the absence of significant elasticity of their test fluids must therefore be treated with extreme caution.

Drag Force

Creeping Flow

To obtain an equivalent of the Stokes law for inelastic shear-thinning liquids, the solution of governing equations (Equations 2–4) with the appropriate boundary conditions requires a suitable constitutive relation. Unlike the Newtonian case, one is immediately confronted with the question of selecting one constitutive equation from a wide variety available in the literature. It therefore seems unlikely that a unique solution that can rival the Stokes law will ever emerge. Instead, the form of drag expression strongly depends upon the choice of constitutive equation. In this section, the discussion is restricted essentially to the analyses that are based on the use of either the power law model or the Ellis fluid model or the Carreau viscosity equation. We start with the simplest of these models, that is, the power law.

Power law fluids. Due to the nonlinear relationship between the stress tensor and the rate of deformation tensor, an exact solution of the equation of motion and continuity is not possible even when the nonlinear inertial terms are neglected. In seeking approximate solutions to the governing equations, variational principles, originally developed by Johnson [27, 28], have been extensively used. This approach yields upper and lower bounds on the drag force experienced by a sphere in the creeping flow regime but does not provide any insight into the physics of the flow. Detailed descriptions concerning the use of variational principles are available elsewhere [29].

Although the deficiencies of the power law model are now well recognized, the creeping motion of a sphere in a power law fluid has received considerable attention. The results are usually expressed in the form of a correction factor X, which measures the deviation of the drag force from that given by the Stokes formula, that is,

$$X = \frac{C_D \, Re'}{24} \tag{28}$$

In the case of power law model fluids X simply becomes a function of the flow behavior index n and the corresponding Reynolds number Re′ is defined as

$$Re' = \frac{\rho V^{2-n} d^n}{K} \tag{29}$$

Most of the results reported in the literature [30–40] are plotted in Figure 7. There is an inherent arbitrariness in each solution as the application of variational principles requires a trial stream function that merely satisfies the boundary conditions. The different stream functions corresponding to each analysis shown in Figure 7 are given in a paper by Chhabra et al. [38]. Note that the correction factor X proposed by Acharya et al. [36] (and subsequently corrected by Kawase and Ulbrecht [37]) is not based on variational principles. Although Figure 7 seems to suggest that the drag is increased above its Newtonian value as a consequence of shear-thinning viscosity, there is very little quantitative agreement between different investigators.

Likewise, at the experimental end of this problem numerous workers have studied the influence of shear thinning on drag coefficients. In most instances aqueous polymer solutions were used as

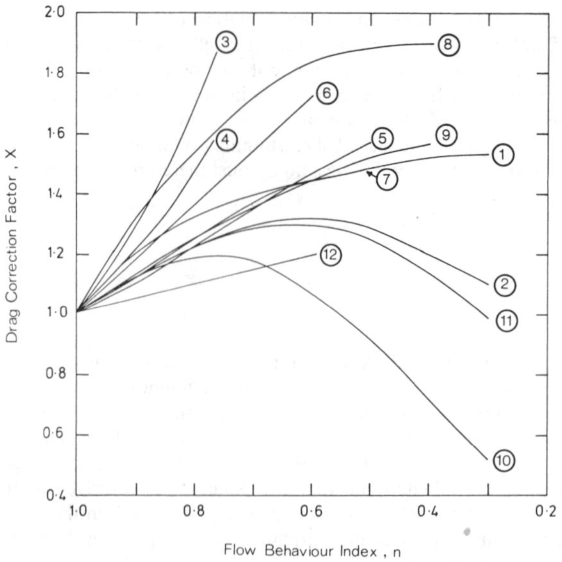

Figure 7. Comparison between various theoretical solutions based on different stream functions (modified after Chhabra et al. [38]). Curve numbers refer to : (1, 2) upper and lower bounds respectively due to Cho and Hartnett [39]; (3, 4) zeroth and first approximation respectively due to Slattery [32]; (5) from the calculations of Nakano and Tien [35]; (6) Acharya et al. [36, 37]; (7) approximation due to Chhabra et al. [38]; (8) Tomita's analysis [30, 31]; (9, 10) upper and lower bounds, respectively, due to Wasserman and Slattery [33]; (11) corrected lower bound [34, 40]; and (12) approximate upper bound based on the calculations of Mohan [34].

the experimental power law fluids. Results drawn from various sources [36, 38, 40–44] are plotted in Figure 8 where the theoretical analyses shown in Figure 7 are also included. Clearly, the situation is confused and there does not seem to be any consensus with regard to the influence of shear-thinning viscosity. The results of Chhabra et al. [38], Uhlherr et al. [44], and Slattery [41] seem to suggest that the drag force is increased above its Newtonian value due to fluid shear-thinning while the data of Acharya et al. [36], Kato et al. [43], Turian [42], and Yoshioka and Adachi [40], though exhibiting significant scatter, show no such increase in the drag force, and in fact the results of Kato et al. [43] and Turian [42] show a slight reduction in the drag force below the Newtonian value. In the past the wide scatter seen in Figure 8 has been attributed primarily to the unaccounted viscoelastic effects and to a lesser extent to the uncertainty regarding wall effects. Neither of these criticisms are applicable to the results of Uhlherr et al. [44] and of Chhabra et al. [38] as recently explained by Chhabra [45]. Thus, even in those instances where experiments are free from viscoelastic effects and uncertainty regarding the wall effects, the results lie almost outside the range of rigorous upper and lower bounds. One is therefore compelled to question the suitability of the power law model for this flow configuration.

The central problem in the use of the power law model is that it is not yet known over what proportion of the sphere surface the power law model breaks down. Presumably if the areas of the sphere surface about the front and rear stagnation points and the volume of fluid far from the sphere do not contribute significantly to the total drag of a sphere, then a power law model description of the flow may yield an acceptable result for the drag. Thus, Chhabra et al. [38] described an empirical scheme that concluded that the analysis of Tomita [30, 31] gives the best agreement

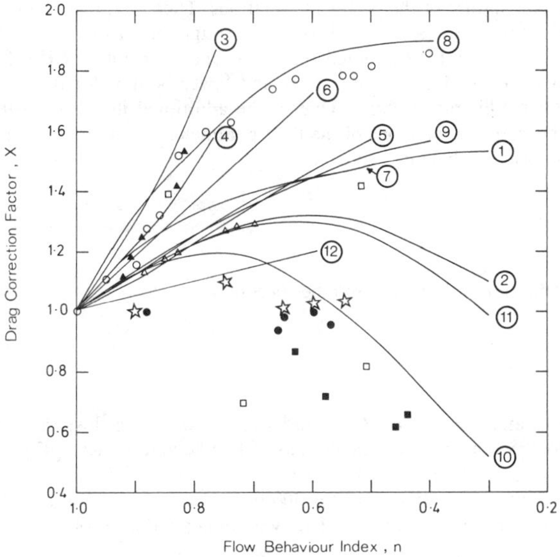

Figure 8. Comparison of theoretical solutions and experimental data from the literature for X as a function of n. The curve numbers have the same meaning as in Figure 7. Key to symbols: (▲) Uhlherr et al. [44]; (■) Turian [42]; (○) Chhabra et al. [38]; (△) Slattery [41]; (●) Kato et al. [43]; (☆) Yoshioka and Adachi [40]; (□) Acharya et al. [36].

with experimental results in the absence of measurable elastic effects. But this agreement was subsequently shown to be fortuitous when Chhabra and Uhlherr (46) investigated the influence of zero shear viscosity on the drag force on a sphere, while keeping the power law parameters (namely n and k) at constant values. Such an experiment, though exceedingly difficult in practice, can be simulated using the hypothetical fluids whose apparent viscosity-shear rate characteristics are shown in Figure 6. The findings of this study are quite revealing and can be summarized as follows:

1. The extent of observed disagreement (between theory and experiment) can be explained entirely in terms of zero shear viscosity—which is neglected in the power law analysis.
2. No particular power law analysis can be specified a priori as the best; agreement with experiment is fortuitous in all cases.
3. Drag predictions from a power law analysis should be avoided if possible; a theory based on a fluid model which is more appropriate for the description of the flow field is preferred. The Ellis model and the Carreau viscosity equation are two suitable models.

Thus, the inability of the power law model to predict zero-shear viscosity seems to be the main cause for the poor predictions of the drag force. Boger [25] has highlighted similar deficiencies in the use of the power law model for analyzing laminar entry flows in circular pipes. Now, we will consider some of the analyses based on fluid models containing zero shear viscosity available in the literature with particular emphasis on the Ellis model and the Carreau viscosity equation.

Ellis model fluids. The Ellis model is one of the simplest three-parameter generalized Newtonian fluid models. It has the advantage of being able to describe shear thinning behavior over a wide range of shear rate, including the zero shear viscosity region. But the presence of the third parameter further compounds the nature of the pertinent equations. Therefore, theoretical and experimental work on the creeping motion of a spherical particle through an Ellis model fluid has been sparse. In 1970, Hopke and Slattery [47] adopted the variational principles of Hill [48] and Hill and Power [49] to calculate the approximate upper and lower bounds for the drag coefficient of a sphere falling through Ellis model fluids. Due to the additional fluid parameter, the upper and lower bounds in this case are functions of the fluid parameter α and another dimensionless group called the Ellis number El, which is defined as

$$El = \frac{\sqrt{2}\eta_0 V}{d\tau_{1/2}} \tag{30}$$

and the relevant definition of Reynolds number becomes

$$Re_0 = \frac{\rho V d}{\eta_0} \tag{31}$$

Therefore, the upper and lower bounds, X_U and X_L, supposedly enclose the exact value of X as defined by Equation 28. The original calculations of Hopke and Slattery [47] covering the rather narrow ranges $1 \leq \alpha \leq 3$ and $El < 10$ have been recently extended [50] to encompass as high values of El as 150 and eventually up to El = 300 by Dolecek et al. [51]. For reference purposes, the results are plotted in Figures 9 and 10. The two bounds are non-coincident and show greater

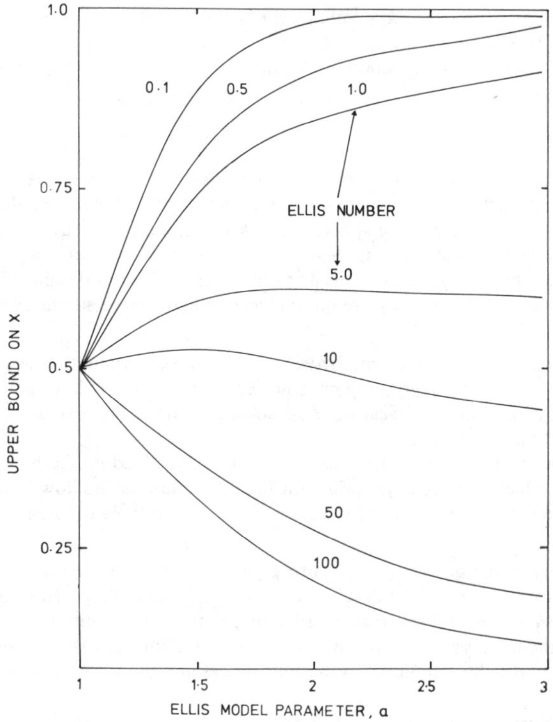

Figure 9. Upper bound on the drag correction factor X in an Ellis model fluid [47, 50, 51].

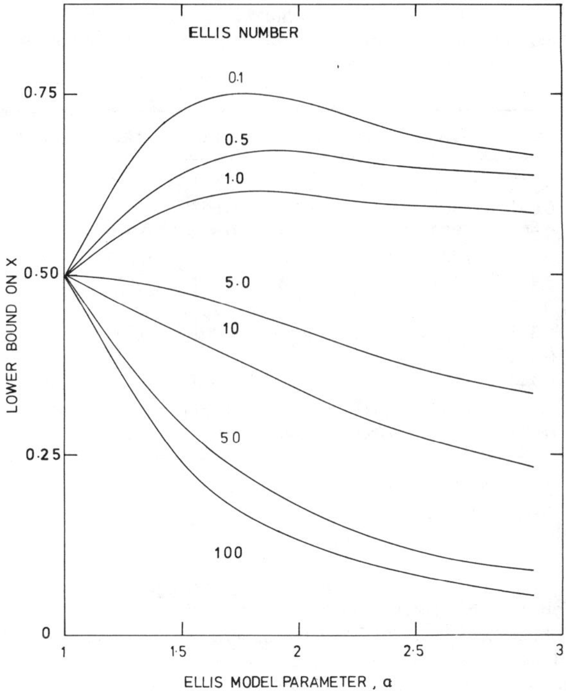

Figure 10. Lower bound on the drag correction factor X in an Ellis model fluid [47, 50, 51].

divergence for increasing values of the fluid parameter α. In the first instance Hopke and Slattery [47], in the absence of any definite criterion, suggested the use of the arithmetic average of the two bounds. It is interesting to note that for low values of Ellis number the values of X is close to unity and it increasingly deviates from unity with increasing values of Ellis number and α. The values of X smaller than unity suggest that the drag is reduced below its Newtonian value (in a liquid of the same viscosity as η_0). This behavior is contrary to the conclusions reached by all the analyses based on the use of the power law model, which predict an increase in drag due to fluid shear-thinning. In his exhaustive review on laminar entry flows, Boger [25] has noted similar apparent contradictions. The numerical values used to construct Figures 9 and 10 are also presented in Table 4 to facilitate their use in new applications.

The first comprehensive attempt to measure drag coefficients of spheres in test fluids whose steady shear behavior could be approximated by the Ellis fluid model was made by Chhabra, Tiu, and Uhlherr [52]. A wide variety of chemically different polymers were used to prepare the test fluids. Numerous precision spheres varying in size (1.59 mm–12.69 mm) and density (1,190 kg-m^{-3}–16,600 kg m^{-3}) were used. The measurements of terminal fall velocities were carried out in each of five to eight cylinders of different diameters. This permitted the elimination of wall effects without recourse to published wall factors of doubtful applicability. Test fluids were characterized by measuring the shear stress as a function of shear rate with an R-16 Weissenberg Rheogoniometer in the shear rate range of interest. The zero shear viscosity for the test fluids was measured directly and the remaining two model parameters ($\tau_{1/2}$ and α) were estimated using a nonlinear regression approach. The values of these parameters (η_0, $\tau_{1/2}$, α) along with the other physical properties are

Table 4
Numerical Values of Upper and Lower Bounds on X (shown in Figures 9 and 10) [50, 51]

α	1.5		2.0		2.5		3.0		3.5	
El	UB	LB	UB	LB	UB	LB	UB	LB	UB	LB
0.15	0.871	0.718	0.997	0.720	0.996	0.691	1.000	0.664	1.000	0.642
0.50	0.800	0.651	0.905	0.665	0.953	0.653	0.976	0.638	0.988	0.623
1.00	0.745	0.604	0.837	0.609	0.889	0.598	0.919	0.584	0.938	0.570
2.00	0.674	0.552	0.746	0.536	0.784	0.512	0.807	0.489	0.823	0.467
5.00	0.596	0.477	0.601	0.424	0.601	0.377	0.597	0.339	0.593	0.308
10.00	0.527	0.419	0.488	0.340	0.461	0.281	0.440	0.239	0.423	0.207
20.00	0.460	0.363	0.384	0.263	0.339	0.201	0.309	0.161	0.287	0.135
50.00	0.376	0.294	0.270	0.182	0.216	0.124	0.184	0.093	0.163	0.0732
100.00	0.320	0.247	0.202	0.134	0.150	0.0847	0.122	0.060	0.104	0.0455
200.00	0.256	0.206	0.149	0.098	0.103	0.0571	0.079	0.038	0.065	0.0280
300.00	0.230	0.184	0.124	0.0811	0.082	0.0452	0.061	0.0293	0.0495	0.0211

UB = Upper Bound; LB = Lower Bound

given in Table 5. The excellent fit of the Ellis model to the apparent viscosity data of two of the test fluids is exemplified in Figure 11. In addition to the shear-thinning behavior, all the test fluids employed also exhibited fluid elasticity. Thus, in this series of experiments, the combined effects of shear-thinning and viscoelasticity must have been observed.

The drag coefficient for steady motion of a sphere is obtained by balancing the drag, gravity and buoyancy forces in the usual way:

$$C_D = \frac{4}{3}\frac{gd}{V^2}\left(\frac{\rho_p - \rho}{\rho}\right) \tag{32}$$

Thus, the experimental value of the drag coefficient is obtained from a measurement of terminal velocity of each falling sphere and a knowledge of the relevant physical properties of the test fluids and spheres used. For each sphere drop test, the experimental value of X was calculated as a function of Ellis number El (Equation 30) and the fluid parameter α.

Typical comparisons of the experimental values of X with the theoretical values (Figures 9 and 10) are shown in Figures 12 and 13. Evidently, most of the points are seen to lie within the two bounds. There appears to be a weak tendency for the data points to be closer to the upper bound at low values of Ellis number, and shifting towards the lower bound with increasing values of Ellis number. Note that, under certain conditions, the drag can be as low as one tenth of the Newtonian value. This is predicted by the theory and is also borne out by the experimental work. In all there are about 140 data points extending over the following ranges of dimensionless variables:

$$4.13 \times 10^{-5} \le Re_0 \le 1.5 \times 10^{-2}, \quad 0.10 \le El \le 141$$

$$1.53 \le \alpha \le 3.13, \quad 0.1 \le X \le 1.00$$

The average error between theory (based on the arithmetic average of two bounds) and experiments is 13.6%.

Further examination of results is facilitated by plotting them in the form of X as a function of Ellis number El. This is shown in Figure 14, where the entire data population is included. Clearly, the fluid parameter α does not appear to be a significant variable and also the drag begins to deviate

Table 5

Values of the Ellis Model Parameters for the Test Fluids Used by Chhabra et al [52]

Test Fluid	Temp. (K)	ρ (kg m^{-3})	η_0 (Pa s)	$\tau_{1/2}$ (Pa)	α	Shear rate range (s^{-1})	(2V/d) range (s^{-1})	Symbols Used in Figure 14
0.85% Separan AP 30	292.0	1003	8.2	1.64	2.70	0.005–113	0.14–16.4	△
1.46 Methocel	295.5	1003	3.3	28.8	2.38	0.5–896	0.48–11.2	■
1.25% Separan AP 30	294.0	1005	19.9	2.39	2.93	0.003–283	0.09–15.4	□
2.00% Separan AP 30	289.5	1014	59.0	6.49	3.13	0.004–113	0.06–13.6	◒
0.75% Separan MG 500 + 1.25% CMC	292.0	1006	11.5	1.69	2.71	0.036–225	0.35–25.2	○
1.62% Separan AP 30	291.0	1010	28.8	3.24	2.94	0.005–283	0.12–22.5	◑
0.75% Separan AP 30 + 0.75% CMC	292.0	1003	10.5	1.47	2.18	0.011–89.5	0.13–16.6	◪
0.167% Separan AP 30 in 8.17% H_2O/91.83% corn syrup MCY41N	297.0	1300	13.0	14.3	1.53	0.07–5.4	0.16–12.5	▲
0.243% Separan AP 30 in 11.2% H_2O/88.8% corn syrup MCY41N	295.0	1310	16.9	10.4	1.66	0.07–6.8	0.24–6.6	●

Note: All solutions in water.

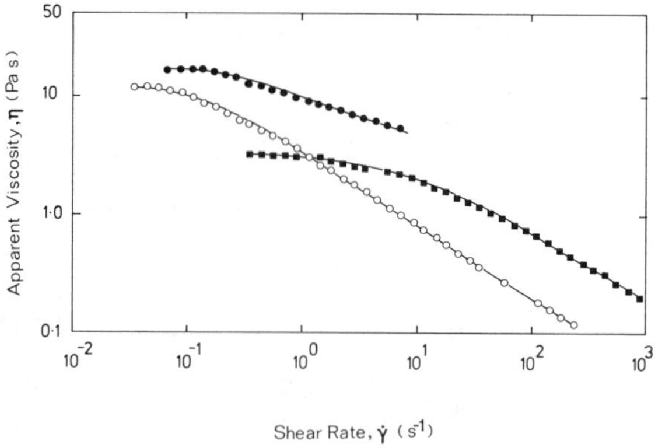

Figure 11. Apparent viscosity—shear rate data for three polymer solutions. The points are experimental, while the lines represent the fitted Ellis model. (■) 1.46% Methocel; (○) 0.75% separan MG500 + 1.25% carboxymethyl cellulose and (●) 0.243% separan AP 30 in 11.2% H_2O/88.8% corn syrup. (Adapted from Chhabra et al. [52]).

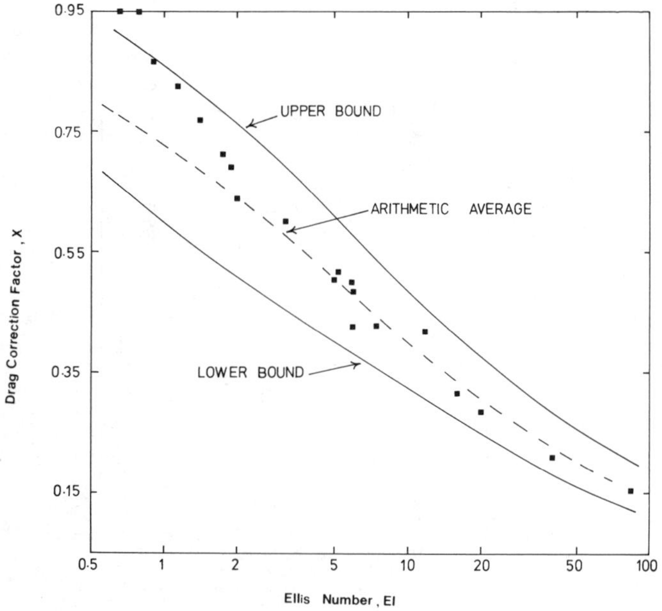

Figure 12. Comparison of theoretical upper and lower bounds with the experimental results in a (0.75% separan AP30 + 0.75% carboxymethyl cellulose) aqueous solution. (From Chhabra et al. [52]).

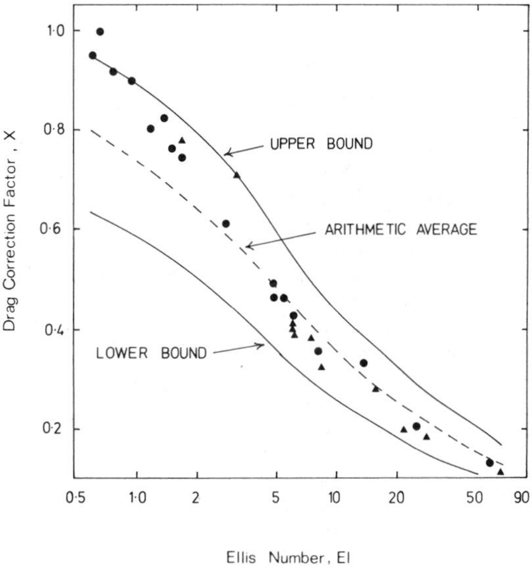

Figure 13. Comparison of theoretical upper and lower bounds with the experimental results of Chhabra et al. [52]. (●) 0.75% separan MG500 + 1.25% carboxymethyl cellulose in water; (▲) 0.85% separan AP30 in water.

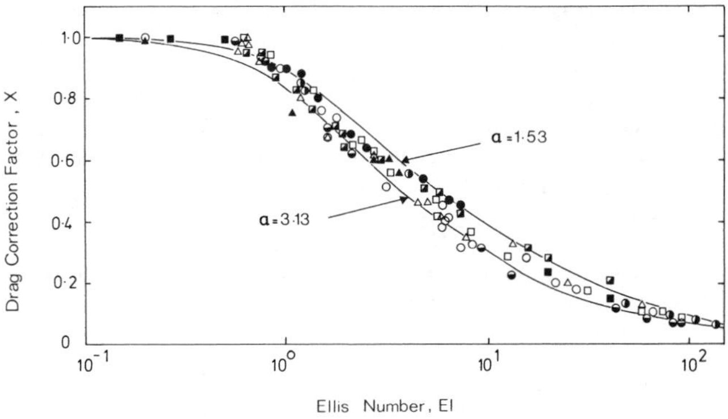

Figure 14. Drag correction factor X as a function of Ellis number El. The solid lines correspond to Equation 33 and the symbols have been identified in Table 5. (Replotted from [52]).

from its Newtonian value (i.e. $X = 1$) only when $El > 0.5$. It also proved possible to empirically correlate the results, shown in Figure 14, by the following [52]:

$$C_D = \frac{24}{Re_0}\left(1 + 0.50\ El^{1.65}(\alpha - 1)^{0.38}\right)^{-0.35} \tag{33}$$

The maximum deviation between Equation 33 and the experimental results is nearly 36% whilst the average deviation is just under 10%. Undoubtedly, Equation 33 is convenient to use.

The correlation proposed by Chhabra et al. [52] has received further support from the independent study of Dolecek et al. [51], who measured drag coefficients of a variety of spheres in 12 different polymer solutions covering the fluid properties in the ranges $0.85 \leq \eta_0 \leq 10.25$ Pa s, $1.32 \leq \tau_{1/2} \leq 111$ Pa, and $1.65 \leq \alpha \leq 3.22$. The value of Ellis number varied from 0.1 to about 100. Attention is drawn to the fact that these values of parameters are somewhat outside the range of conditions studied by Chhabra et al. [52]. A comparison between the predictions of Equation 33 and the results of Dolecek et al. [51] is shown in Figure 15. Included in the same figure are the arithmetic averages of upper and lower bounds corresponding to three representative values of the fluid parameter α. Evidently, in general terms, the results of Dolecek et al. [51] show a better agreement with Equation 33 than with the average of theoretical bounds. Although the deviations between the experiments and Equation 33 are well within its limits, there is, however, a weak trend that Equation 33 seems to overpredict the values of X.

Partial explanation of this behavior lies in the fact that Dolecek et al., while fitting the Ellis fluid model to shear-stress/shear-rate data made use of the extrapolation procedure of Turian [42] to obtain the values of zero shear viscosity. This method has been shown [53, 54] to over-estimate the value of η_0 and hence the value of X is underestimated. Therefore, the agreement seen in Figure 15 is considered to be acceptable. Other sets of data available in the literature [42, 55] (although somewhat sparse) are well correlated by Equation 33.

Carreau model fluids. It is now generally agreed that the Carreau viscosity equation is capable of describing shear-thinning behavior of a wide range of materials [56, 57] over the entire range of shear rate including the two asymptotic regions. Furthermore, Carreau et al. [57] demonstrated that in many instances this model provides a much better representation of data than the Ellis model fluid. Despite the advantages of the Carreau equation, it has been employed only recently to solve hydrodynamic problems [58, 59] and to interpret experimental data [60]. In 1980, Chhabra and Uhlherr [61] employed the variational principle, as formulated and expanded upon by Astarita [62], to obtain an approximation to the drag force on a sphere exerted by a Carreau model fluid. The mathematical details of the analysis are not reproduced here and may be found in the original paper [61]. It would suffice to say that the approximation obtained by these authors does not represent an upper bound or a lower bound and is only an approximation. The error involved in such an approach, however, is incalculable, and the final test of its usefulness therefore should lie in its ability to reproduce experimental measurements.

The drag correction factor X, defined by Equation 28, was obtained as a function of the flow behavior index (n), a dimensionless number $\Lambda(\lambda V/R)$, and the ratio of the two limiting viscosities (η_0/η_∞). In most cases, $\eta_\infty \lll \eta_0$ as demonstrated by Boger [20] and suggested by Abdel-Khalik et al. [56]. Here, therefore, the results are presented only for the case where η_∞ can be omitted from Equation 24 which then reduces to Equation 26.

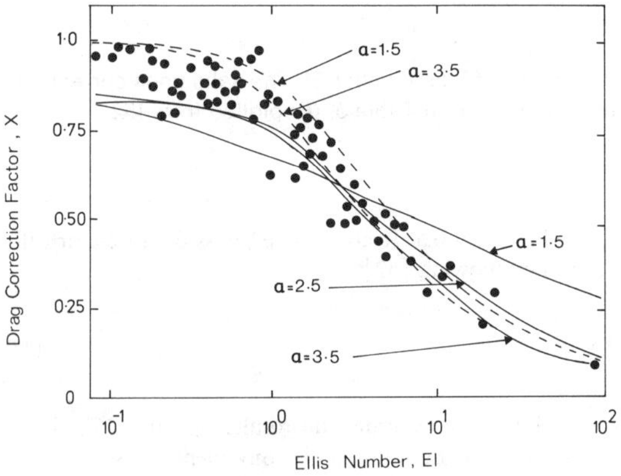

Figure 15. Drag correction factor X as a function of Ellis number El. The points are experimental results of Dolecek et al. [51] and the broken lines represent Equation 33 while the continuous curves are arithmetic average of upper and lower bounds for the different values of the fluid parameter α.

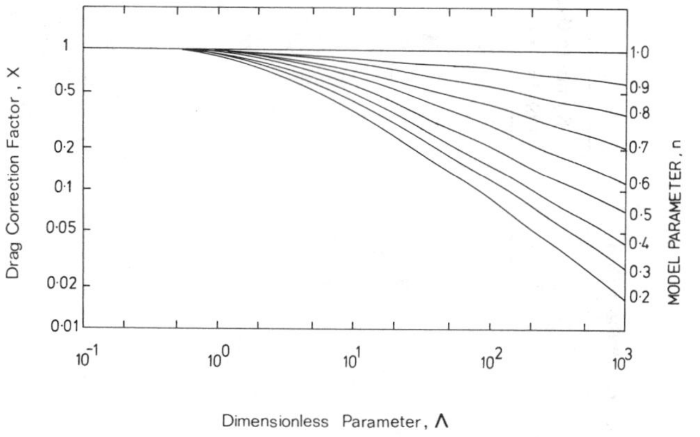

Figure 16. Theoretical values of drag correction factor X as a function of dimensionless parameter Λ and the Carreau model parameter n. (Replotted from [61]).

Approximation to X, obtained by Chhabra and Uhlherr [61], is portrayed in Figure 16. Clearly, the drag correction factor X is always smaller than unity, thereby meaning that the drag is reduced below its Newtonian value. This observation is consistent with the results obtained when the Ellis model fluid is used to represent the rheology of the test fluids. In fact, it is a general feature of all analyses based on non-Newtonian models containing zero shear viscosity as one of the parameters [63, 64, 68]. Furthermore, X does not seem to deviate from unity for values of $\Lambda < 1.0$, which is also similar to the behavior observed in the results in the Ellis model fluids presented in the preceding section. The value of X deviates from unity with increasing values of the dimensionless quantity Λ, which denotes the increasing importance of non-Newtonian effects on the drag of the sphere.

In the same paper, Chhabra and Uhlherr [61] reported the results of a systematic experimental study where the drag coefficients of a variety of spheres was measured as a function of n and Λ. The values of Reynolds number were, sufficiently low ($Re_0 \leq 0.015$) such that creeping flow requirements were satisfied [65].

All the test fluids exhibited shear-thinning behavior as well as other non-Newtonian effects—notably viscoelasticity. Shear-thinning behavior was modeled using the simplified version of the Carreau viscosity equation given by Equation 26. Values of relevant rheological parameters (namely n, λ and η_0) and other physical properties are presented in Table 6 where these parameters are seen to cover wide ranges and include the ranges of shear rates encountered by falling spheres in each case. The goodness of fit is obvious in Figure 17 where the experimental data are compared with the fitted curves for some of the test liquids.

From the knowledge of terminal falling velocity and other properties, the value of drag coefficient can be calculated using Equation 32 and the relevant definition of Reynolds number is given by Equation 31. These two quantities enable the experimental value of X to be determined using Equation 28. Thus X may be obtained as a function of Λ and n. A comparison between the predicted and the experimental values of X is given in Figure 18 for two of the test fluids. Included in this figure are the results from a preliminary study of Bush [66] who used a boundary integral method to solve the equations of motion and of continuity for the Carreau viscosity equation. Admittedly, the results of Chhabra and Uhlherr [61] are consistently lower (about 5%) than those of Bush [66]. But nonetheless, keeping in mind the simplicity and the approximate nature of the analysis of Chhabra and Uhlherr [61], the agreement between the two is regarded as satisfactory. Agreement between experiments and analyses is also seen to be excellent.

Table 6
Rheological Characteristics of Carreau Model Fluids (Modified after Chhabra and Uhlherr [61]

Test Fluid	Temp. (K)	ρ (kg m^{-3})	η_0 (Ns m^{-2})	n	λ(s)	Symbols Used in Figures 18 and 19
0.85% Separan AP 30	292.0	1003	8.20	0.49	11.00	◒
1.46% Methocel HG 90	295.5	1003	3.31	0.55	0.33	▲
1.25% Separan AP 30	294.0	1005	19.9	0.46	15.30	■
2.00% Separan AP 30	289.5	1014	59.0	0.42	19.07	☆
0.75% Separan MG 500 + 1.25% Carboxymethyl cellulose	292.0	1006	11.5	0.40	8.13	□
1.62% Separan AP 30	291.0	1010	28.8	0.47	15.85	◑
0.75% Separan AP 30 + 0.75% Carboxymethyl cellulose	292.0	1003	10.5	0.53	15.75	○
0.17% Separan AP 30/8.17% water in corn syrup	297.0	1300	13.0	0.75	6.26	●
0.24% Separan AP 30/11.2% water in corn syrup	295.0	1310	16.9	0.67	5.75	△
0.07% Separan AP 30/8% water in Maltose	295.0	1310	8.50	0.81	1.33	◪

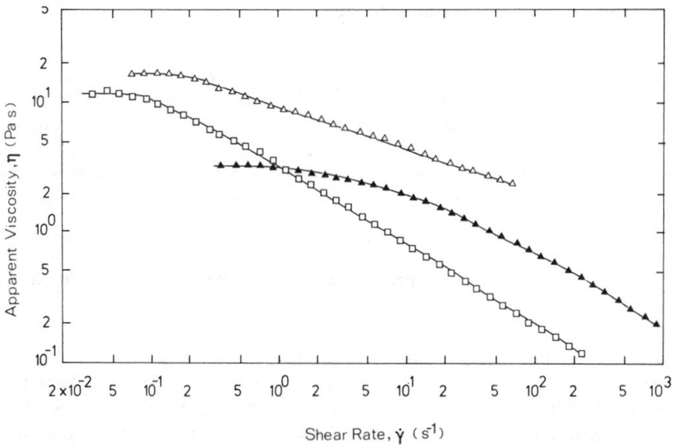

Figure 17. Apparent viscosity-shear rate data for three of the test fluids used by Chhabra and Uhlherr [61]. The points are experimental, while the lines represent the fitted Carreau viscosity equation. (△) 0.243% separan AP30/11.2% water/88.56% corn syrup; (□) 0.75% separan MG500 + 1.25% carboxymethyl cellulose in water; (▲) 1.46% methocel in water. (Replotted from [61]).

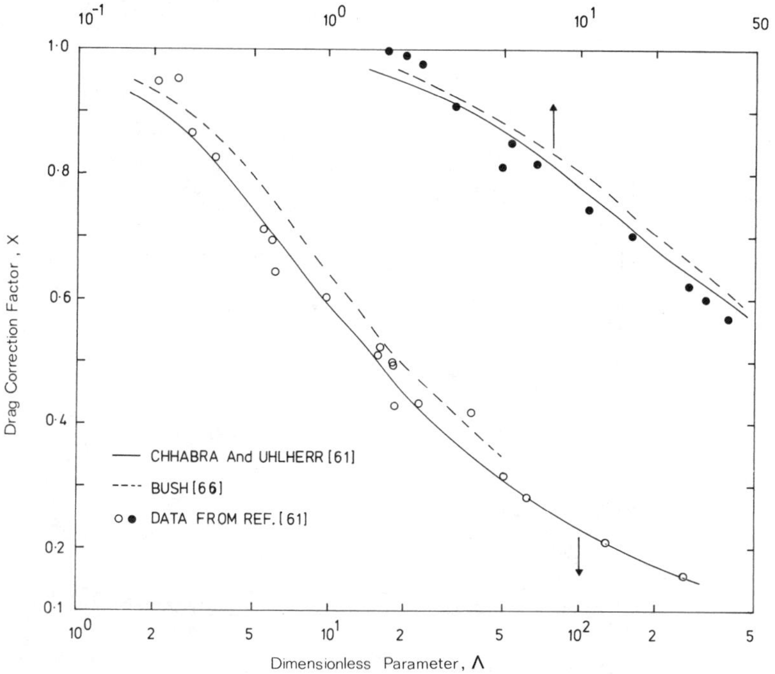

Figure 18. Effect of dimensionless parameter Λ on drag correction factor X. The symbols are identified in Table 6.

In order to compare all the results for all the test fluids the theoretical value of X is plotted against experimental value of X in Figure 19. Note that not all experimentals are shown owing to overlap. In fact, the average deviation between theoretical and experimental results is 8.4% for 150 data points, with 16 points displaying errors larger than 20% and with no discernible trends. It can be seen that X reaches as low a value as 0.1. The data embrace the following ranges of variables:

$$0.4 \leq n \leq 1.0 \quad \text{and} \quad 0.16 \leq \Lambda \leq 360$$

It can therefore be concluded that the approximate analysis based on the Carreau viscosity equation presented here is well supported by experimental data, and is also in agreement with the preliminary numerical study of Bush [66].

The foregoing discussion clearly demonstrates that the use of generalized Newtonian models containing zero shear viscosity yields much better results on the drag coefficient of spheres than the power law model. Since the mathematical technique employed still relies on the choice of an arbitrary stream function, the resulting improvement must be entirely attributable to the inclusion of zero shear viscosity. This conclusion is consistent with the study of Chhabra and Uhlherr [46] and the conclusions reached by others [40]. Once more the caveat: the use of the power law equation should be avoided (if possible) in calculating the drag force on a sphere.

Another interesting and significant observation is that even though the model fluids used by Chhabra et al. [52] and by Chhabra and Uhlherr [61] exhibited viscoelastic behavior, the results are in excellent agreement with the theoretical analyses based on the Ellis model fluid and the Carreau viscosity equation. This agreement is believed to be realistic and this aspect is covered in more detail later.

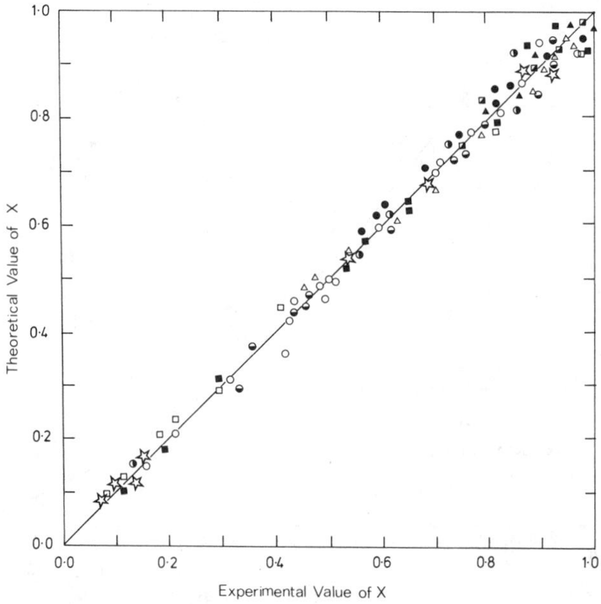

Figure 19. Comparison between theoretical and experimental values of drag correction factor X in the Carreau Model fluids. The symbols are identified in Table 6. (Replotted from [61]).

High Reynolds Number Flow

In contrast to the appreciable work carried out in the creeping flow regime, very little attention has been paid to the motion of spheres at high Reynolds number where the nonlinear inertial terms in the equations of motion are no longer negligible. Thus, from a theoretical point of view, even with the simplest non-Newtonian fluid model, namely the power law, the governing equations are hopelessly complex. Admittedly, some theoretical attempts [67–69] to analyze the high Reynolds number have been made but, unfortunately, virtually no real success has been achieved.

On the other hand, few investigators have studied the behavior of spheres falling through shear-thinning liquids at high Reynolds numbers. There are only three sets of data available in the literature [36, 55, 70]. Based on an arbitrary definition of effective viscosity, Dallon [55] presented a correlation giving a drag coefficient as a function of relevant dimensionless groups. Acharya et al. [36] reported extensive measurements of drag coefficients in power law model fluids. Recently, Chhabra and Uhlherr [70] examined the influence of rheological parameters on drag coefficient of spheres in the region of high Reynolds numbers. The Carreau viscosity equation was used to model the shear-thinning behavior of the test fluids and the corresponding values of η_0, n and λ are summarized in Table 7.

Using a dimensional analysis approach, Chhabra and Uhlherr [70] modified the expression of Schiller and Naumann [71] to extend its applicability to shear-thinning liquids. Based on their own data and those of Dallon [55], Chhabra and Uhlherr [70] have proposed the following expression

$$C_D = \frac{24}{Re_0}[1 + 0.15\,Re_0^{0.687}][1 + 0.65(n - 1)\Lambda^{0.20}] \tag{34}$$

In Equation 34 the terms within the first pair of brackets are Schiller-Naumann's formula for Newtonian liquids, and hence the multiplying term involving n and Λ is a measure of the importance

Table 7
Values of the Carreau Viscosity Equation Parameters for the Aqueous Polymer Solutions Used by Chhabra and Uhlherr [70] and by Dallon [55]

Test Fluid	Temp. (K)	η_0 (Nm^{-2} s)	n	λ(s)	ρ (kg m^{-3})	Shear Rate Range (s^{-1})
0.6% Polyox Coagulant*	292	0.125	0.69	0.0946	1000.0	3–1125
0.35% Methocel/0.45% Carboxymethyl cellulose	292	0.327	0.76	0.0679	1001.7	0.3–895
0.02% Carbopol	292	0.053	0.78	0.292	1000.0	1–895
0.75% Carboxymethyl cellulose/ 0.1% Separan MG500	292.8	0.103	0.80	0.232	1003.5	1.4–895
0.5% Methocel	292.8	0.163	0.79	0.0662	1000.9	1.4–1125
0.2% Methocel/0.02% Separan MG 500	293	0.193	0.61	1.41	1006.6	0.2–895
0.08% Separan MG 500	290	0.302	0.59	2.34	1001.3	0.28–895
1.1% Carboxymethyl cellulose	291	0.140	0.89	0.0515	1000.0	1–1125
Dallon's Test Fluids [55]						
1% Carboxymethyl cellulose	302.7	0.265	0.52	0.0176	998.1	17–1320
0.6% Carboxymethyl cellulose	303.1	0.108	0.58	0.0178	997.9	1.29–1183
1.5% Carboxymethyl cellulose	303.6	0.088	0.74	0.008	1002.1	3–1040
0.7% Hydroxyethyl cellulose	303.5	0.082	0.59	0.0174	996.5	11–1268

* *Exhibited measurable viscoelasticity in steady shear flow.*

of shear-thinning effects. Note that as $n \rightarrow 1$ or $\Lambda \rightarrow 0$ (or both), Equation 34 reduces to the proper Newtonian limit. The data used to establish the values of adjustable parameters in Equation 34 extend over the following ranges of variables:

$$0.032 \leq \Lambda \leq 630; \quad 0.032 \leq Re_0 \leq 400; \quad \text{and} \quad 0.52 \leq n \leq 1.0$$

The maximum deviation between Equation 34 and experiment is 14%. There is a paucity of data in this field. Unlike the standard drag curve for Newtonian fluids, the results obtained so far cover much narrower range of Reynolds number and the other dimensionless groups. This remains an area for future research activity.

Wall Effects

Among several investigators who have studied the motion of spheres in shear-thinning fluids, some have altogether ignored the wall effects [32, 36], while others have used the Newtonian wall correction factor [42, 55]. Generally, neither of these procedures is justified. Caswell [72] clearly demonstrated that, in creeping flow, the retarding effects on the fall velocity due to the presence of walls are less in non-Newtonian liquids than those in Newtonian liquids.

The experimental evaluation of wall effects is rather a cumbersome task as data must be obtained in fall tubes of different diameters. Partly due to this reason, wall effects have not received much attention. Chhabra and Uhlherr [73, 74] carried out a systematic study on the wall effects concerning sphere motion in inelastic shear-thinning polymer solutions. The experimental scheme entailed the measurement of terminal velocity for each sphere in a number of fall tubes containing test fluids whose rheological characteristics were measured using a Weissenberg Rheogoniometer. Thus, the terminal velocity V_m was obtained as a function of (d/D) and fluid rheology. In all, about nineteen polymer solutions were used. Since none of the test liquids exhibited measurable primary normal stress difference, all the polymer solutions can be treated as inelastic in nature. Complete details of the spheres and fall tubes used are available in the original publications [73, 74]. The diameter ratio (d/D) varied between 0.005 and 0.50. The maximum value of (d/D) was 0.5 and was chosen essentially to avoid radial migration and spinning of the sphere as noted by some earlier investigators [75].

Although the use of the power law model yields an entirely unacceptable description of the drag coefficient for spheres it was found to be quite adequate in analyzing the results for wall factors. Values of the power law model parameters along with the other physical properties of test liquids are summarized in Table 8.

The calculation of the wall factor f (Equation 17) requires the value of terminal velocity in unbounded medium V for each sphere used. The latter is obtained by plotting the measured terminal velocity V_m of a particular sphere against the diameter ratio (d/D) following Turian [42], Okuda [76], and Uhlherr et al. [44].

Typical plots of the variation of the measured terminal velocity with the diameter ratio of some of the spheres used are shown in Figures 20 and 21. As expected, the measured terminal velocity increases with decreasing values of (d/D) (i.e. the increasing value of fall tube diameter). An inspection of these figures, however, reveals two different types of characteristics:

1. In some cases (Figure 20), the dependence of the measured terminal velocity V_m on the diameter ratio (d/D) is linear and can easily be extrapolated to the zero value of (d/D) and the corresponding intercept on the ordinate is the value of terminal velocity in unbounded medium V. This observation is in line with the work of Sato et al. [77] and of others [42, 44]. This behavior corresponds to data obtained in the creeping flow regime ($Re'_m < \sim 0.5$).
2. In other cases (Figure 21), the measured terminal velocity increases in a nonlinear manner as the diameter ratio decreases and, in the vicinity of $(d/D) \sim 0.1$, V_m becomes almost independent of the diameter ratio and this limiting value is assumed to correspond to V. This type of behavior was observed at high values of Reynolds number and is again consistent with the previous work involving Newtonian fluids [15, 19].

Table 8
Rheological Properties of Inelastic Shear-Thinning Fluids Used by Chhabra et al. [73, 74]

Test Fluid	Temp. (K)	ρ (kg m^{-3})	n (–)	K (Nm^{-2} s^{-n})	Symbols Used in Figures 22 and 23
27% Aqueous Glycerine	293.0	1065.00	1.00	0.0022	○
0.35% Methocel + 0.45% CMC	292.0	1001.70	0.76	0.460	▲
0.02% Carbopol	292.0	1000.00	0.78	0.065	△
0.75% CMC + 0.1% Separan	292.8	1003.50	0.80	0.123	●
0.5% Methocel	292.8	1000.90	0.79	0.260	□
0.08% Separan	290.0	1001.30	0.59	0.225	◒
1.1% CMC	291.0	1000.00	0.89	0.150	■
0.5% CMC	290.8	1000.00	0.95	0.027	◑
0.5% CMC + 0.23% Separan	292.0	1000.00	0.67	0.340	◪
0.6% Polyox Coagulant*	292.0	1000.00	0.69	0.232	▽
0.77% CMC	294.0	1000.00	0.95	0.044	These liquids were used to study the wall effects in creeping flow regime
1.6% CMC	291.8	1000.00	0.86	0.300	
2.1% CMC	293.0	1000.00	0.83	0.485	
1% CMC + 0.15% Separan	293.0	1000.00	0.78	0.350	
0.5% Carbopol in 2% NaOH solution	291.0	1070.00	0.74	0.083	
1% CMC + 0.25% Separan	293.0	1000.00	0.63	1.08	
0.1% Separan MG 500	294.0	1000.00	0.55	0.205	
0.15% Separan MG 500	294.0	1000.00	0.53	0.237	
0.60% Carbopol in 1.5% NaOH solution	292.0	1020.00			
$4 < \dot{\gamma} < 112$			0.50	1.800	
$\dot{\gamma} < 4$			0.40	2.100	

*N.B.: * Exhibited measurable viscoelasticity in steady shear.*
CMC—Carboxymethyl cellulose.

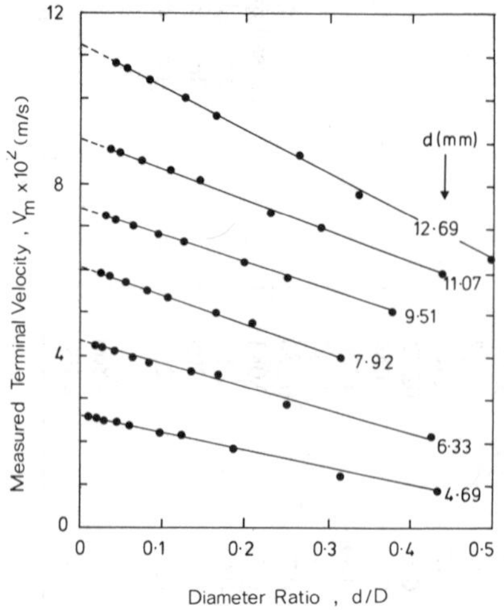

Figure 20. Effect of diameter ratio on terminal fall velocity of perspex spheres in a 0.5% aqueous Methocel solution. (From Chhabra et al. [73]).

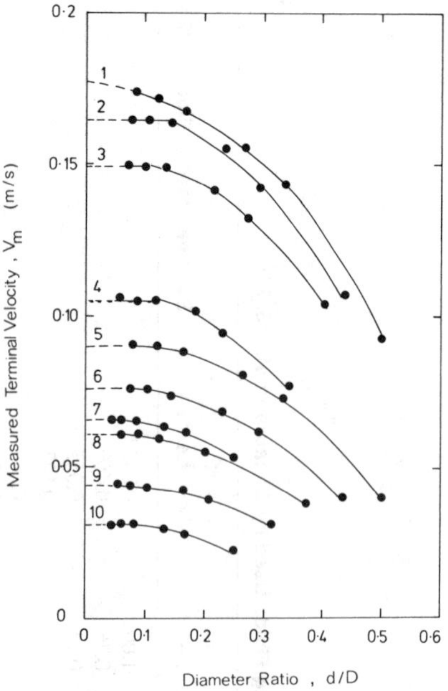

Figure 21. Effect of diameter ratio on terminal fall velocity of spheres in a 0.75% carboxymethyl cellulose +0.1% separan aqueous solution. The sphere diameters in mm corresponding to curves 1 to 10 are respectively as follows: 12.686, 11.115, 10.3, 8.749, 12.686, 11.065, 6.368, 9.51, 7.915, 6.33. Sphere materials are 1–4, 7 (PVC); 5, 6, 8–10 (perspex). (Adapted from [74]).

In both cases, once the value of terminal velocity is known, the value of the corresponding wall factor f can be computed. In the case of Newtonian liquids the wall factor depends on the diameter ratio and Reynolds number. Here it is expected to be a function of the non-Newtonian fluid parameters (K, n) in addition to (d/D) and Reynolds number Re'_m. For the power law model, the appropriate definition for the Reynolds number is:

$$Re'_m = \frac{\rho V_m^{2-n} d^n}{K} \tag{29}$$

In order to assess the relative roles of Reynolds number and diameter ratio, the measured terminal velocity was interpolated for the fixed values of (d/D) at 0.1, 0.2, 0.3, 0.4, and 0.5. Thus, the wall factor f was found for fixed values of (d/D) as a function of Reynolds number. The results so obtained are shown in Figures 22 and 23. Included in the same figures are the corresponding wall factor in Newtonian liquids [13, 19].

The prominent features of these two figures may be summarized as:

1. Wall effects in inelastic shear-thinning liquids are smaller than in Newtonian liquids over the entire range of Reynolds number. This is supported in creeping flow by the approximate analysis due to Caswell [72].

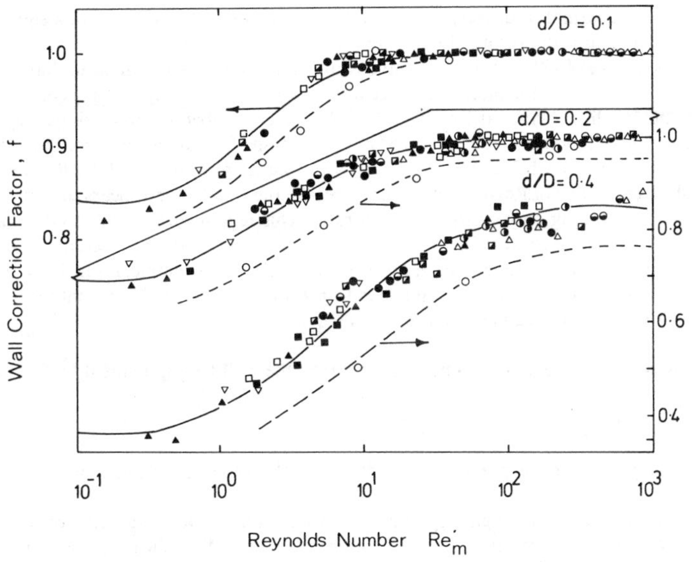

Figure 22. Dependence of wall correction factor on the Reynolds number based on measured terminal fall velocity for constant values of diameter ratio. The solid curves correspond to Equations 35 to 37 while the broken curves show the Newtonian results of Fidleris and Whitmore [19]. The symbols have been defined in Table 8.

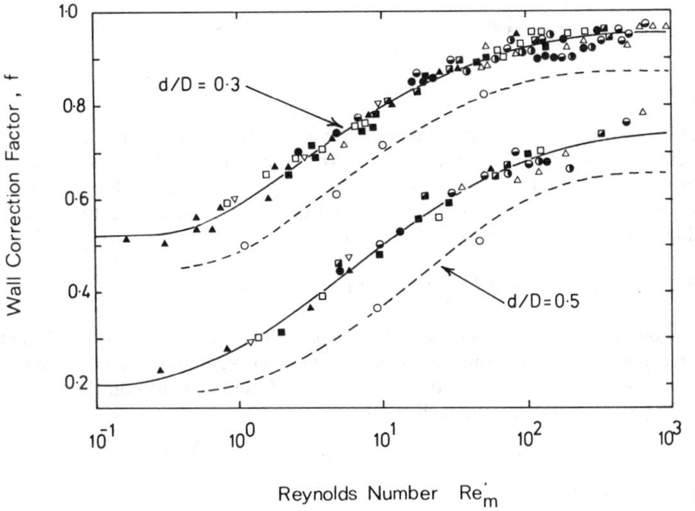

Figure 23. Dependence of wall correction factor on the Reynolds number based on measured terminal fall velocity for constant values of diameter ratio. The solid curves correspond to Equations 35 to 37 while the broken curves show the Newtonian results of Fidleris and Whitmore [19]. The symbols have been defined in Table 8.

2. Although one would intuitively expect the flow behavior index (n) to have a significant effect on the wall factor, the results do not appear to confirm this notion.
3. The dependence of wall factor on Reynolds number is qualitatively similar to that in Newtonian liquids (see Figure 4). Furthermore, three distinct regions are easily recognized. The chief characteristics of these being: at low values of Reynolds number (creeping flow, $Re'_m < 0.5$), f is independent of Reynolds number and is only a function of the diameter ratio. The second region is characterized by f being dependent on both diameter ratio as well as Reynolds number. Finally, as the value of Reynolds number increases, the wall factor appears to approach a limiting value depending only on the diameter ratio. This region corresponds to the plateau areas in Figure 21. The critical value of Reynolds number above which f becomes independent of Re'_m seems to depend strongly upon the diameter ratio. For example, for $(d/D) = 0.1$, this critical value is in the vicinity of 15 and rises to 600 for $(d/D) = 0.5$. Now analytical expressions for each of the three regimes are presented.

At the low values of Reynolds number ($Re'_m < 0.5$), the wall factor, denoted by f_0, is given by

$$f_0 = 1.0 - 1.6(d/D) \tag{35}$$

This empirically obtained expression also describes the other data available in the literature [73, 78].

In a similar fashion, in the high Reynolds number region, the limiting value of wall factor (f_∞) depends only upon the diameter ratio and is approximated by the following equation [74]:

$$f_\infty = 1 - 3(d/D)^{3.5} \tag{36}$$

Equation 36 reproduces the experimental values to within $\pm 2\%$.

In the intermediate region where f is a function of both (d/D) and Re'_m, Chhabra and Uhlherr [74] proposed the following predictive correlation:

$$\frac{\dfrac{1}{f} - \dfrac{1}{f_\infty}}{\dfrac{1}{f_0} - \dfrac{1}{f_\infty}} = (1 + 1.3\,Re'^{2}_{m})^{-0.33} \tag{37}$$

It is interesting to note that the contribution of diameter ratio (d/D) is embodied entirely in f_0 and f_∞. Furthermore, Equation 37 approaches the expression for f_0 with decreasing values of Re'_m (<0.5). Equations 35 to 37 are plotted as solid lines in Figures 22 and 23. Clearly, in a new application Equations 35–37 can readily be used to predict the value of the wall factor in inelastic shear-thinning fluids in the following ranges of conditions:

$$0.01 \le Re'_m \le 1{,}000, \qquad 0 < (d/D) < 0.5, \quad \text{and} \quad 0.53 < n < 0.95$$

In conclusion, it can be said that the influence of shear thinning on drag coefficient and wall effects in creeping as well as intermediate Reynolds number region is reasonably well understood.

Now attention will be focussed on the possible role of fluid elasticity on the two quantities of interest, namely drag coefficient and wall effects.

VISCOELASTIC FLUIDS

Most viscoelastic fluids exhibit shear-thinning, but in addition in steady shear also exhibit inequal normal stresses. They also show an extensional viscosity that differs from the Newtonian value. The behavior of a viscoelastic fluid at any time is dependent on its recent deformation history.

Elasticity in the viscoelastic fluid is often associated with the unequal and non-zero normal stresses present within the sheared fluid. Possibly the most well-known manifestation of fluid elasticity is the rod climbing effect (Weissenberg effect). Other phenomena include die swell and the open-channel syphon. An enlightening discussion on viscoelastic fluid flow phenomena can be found in the recent book by Bird et al. [22]. The studies of viscoelastic fluids have received considerable impetus from observations in the polymer processing industry where unusual and sometimes detrimental flow phenomena are observed, which are attributed to fluid elasticity.

It has been common practice to describe viscoelastic fluids by measuring the shear stress, τ and the first (primary) normal stress difference N_1, as a function of shear rate. Generally, a fluid characteristic time (or a spectrum of relaxation times) is defined to describe the elasticity of the fluid. There are several ways of defining this characteristic time by combining shear stress and primary normal stress difference in an arbitrary manner. The simplest definition of characteristic time is

$$\theta = \frac{N_1}{2\tau\dot{\gamma}} \tag{38}$$

This is also known as the Maxwellian relaxation time and is evidently shear-rate dependent. Bird and co-workers [22, 79, 80] have presented a number of other definitions that combine shear stress and primary normal stress difference. Typical of these is

$$\theta = \left(\frac{K'}{2K}\right)^{1/(n'-n)} \tag{39}$$

where n, K, and n,′ K,′ ($N_1 = K'\dot{\gamma}^{n'}$) are empirical power law constants for the shear stress-shear rate and the primary normal stress difference-shear rate data respectively. The factor of 2 in Equation 39 is quite arbitrary but is retained here to be consistent with others [16, 79]. In the limit of vanishingly low shear rates Equations 38 and 39 coincide.

An entirely different approach has been suggested by Bird [26]. He asserts that a characteristic time constant derived purely from shear-thinning viscosity data is also a measure of the fluid elasticity, e.g. the parameter λ in the Carreau viscosity equation and in the case of the Ellis model fluid the quantity $(\eta_0/\tau_{1/2})$ has the dimensions of time. In fact, Elbirli and Shaw [81] have demonstrated an interconnection between the fluid characteristic time derived from viscosity data and that calculated using the Rouse equation. Although a considerable body of largely experimental evidence [60, 61, 79, 80, 82, 83] supports this hypothesis, not all authors are in agreement on this issue [84, 85]. Other parameters including stress ratio, shear modulus etc. have also been used to quantify the importance of fluid elasticity [25].

In the mechanics of viscoelastic fluids at least one additional dimensionless group is needed to account for fluid elasticity. Many authors have recognized this group as the Weissenberg number [86, 87]:

$$We = \frac{\theta V}{R} \tag{40}$$

where V and R are a characteristic velocity and linear dimension of the flow system, respectively. Some authors [88] prefer an elasticity number

$$\text{Elasticity number} = \frac{\eta\theta}{\rho d^2} \tag{41}$$

which has been shown by Astarita and Marrucci [89] to be a measure of the ratio of elastic to inertial forces. Another popular dimensionless group is the Deborah number [87]

$$De = \frac{\text{characteristic fluid time}}{\text{time scale of the process}} \tag{42}$$

Metzner et al. [87] demonstrated that fluid response is expected to be fluid-like for small values of De and solid-like for large values of De. For steady flows Astarita and Marrucci [89] proposed

$$De = \frac{\theta V}{d} \tag{43}$$

It is now generally agreed that characterization of fluid elasticity from steady shear viscometric measurements and the use of parameters like those defined by Equations 38 to 43 may not be sufficient to make a fine distinction between viscoelastic fluids. In this context, the inclusion of extensional properties immediately comes to mind. Thus, a central issue remaining in non-Newtonian fluid mechanics is that put by Boger [25], viz, what are the continuum properties and what is the minimum number required to specify completely the behavior of a viscoelastic fluid? Linked to the continuing consideration and development of experimental techniques for measuring fundamental elastic properties is the continuing search for appropriate constitutive relations, which when used in conjunction with the governing equations should, in principle, explain the unusual flow phenomena observed for viscoelastic fluids. The search is far from complete as limited success has been achieved in making direct comparison between theoretical analyses and experimental observations in any flow field other than viscometric flows. When the constitutive equation is simple enough to obtain a solution to a non-viscometric flow problem, invariably solutions can only be obtained for weakly elastic fluids and large differences between Newtonian and viscoelastic fluid behavior are not predicted. In constrast is the experimentalist who deals with real viscoelastic materials that are not only highly elastic but are almost always shear-thinning. Hence, the theoretician predicts small effects on the flow field for idealized fluids (which can be related to fluid elasticity) while the experimentalist, although he may observe large effects on the flow, is not clear whether he is seeing elastic effects, shear-thinning effects, or a combination of both. Consequently, great confusion results and conflicting conclusions are often reached regarding the influence of fluid elasticity. Bridging this gap between theory and experiments is, perhaps, the major challenge in non-Newtonian fluid mechanics.

Drag Force

Low Reynolds Number Region

During the last two decades or so, many attempts have been made to elucidate the influence of fluid elasticity on the drag force of a sphere in the creeping flow regime. Most of the solutions available to date are of the perturbation type [90–94]. In this approach the nonlinear term contributions are embodied in a framework of a small parameter perturbation expansion about the creeping flow Newtonian kinematics. The analysis of Leslie and Tanner [90], based on the Oldroyd fluid model, is a typical example of this class. All such solutions are applicable only for Reynolds number and Weissenberg number very much less than unity. The restriction on Reynolds number limits the applicability to creeping flow while the more severe Weissenberg number limitation restricts the analysis to very low levels of fluid elasticity. Within the validity of these expansions, these analyses predict very little difference between the kinematics of flow around a sphere for Newtonian and viscoelastic fluids. Solutions of this class for flow around a sphere, all predict a small downstream shift in the streamlines and a small degree of reduction in drag below its Newtonian value. Hill [95] has questioned the validity of the perturbation approach for analyzing the flow around a sphere.

Contrasted with these perturbation analyses are the results of Ultman and Denn [96] and Sigli and Coutanceau [97]. Both used a generalized Maxwell fluid model. Ultman and Denn [96] employed the Oseen type of linearization in an attempt to include significant elastic effects in their approach. But very little effect of the fluid elasticity on the drag force is observed. Their approach, however, has come under some criticism [98–100].

More recently, with the development of sophisticated numerical schemes, and the advent of powerful computers, some researchers [101–103] have sought numerical solutions to the governing equations with a suitable constitutive equation. Crochet [101] reported an extensive and detailed study of the flow and stress field around a sphere moving slowly (creeping flow) in a Maxwell model fluid. His analysis predicted a small downstream shift in the streamlines, which is consistent with previous analyses [90–94] and the experiments of Zana et al. [98] and of Mena and Caswell [100]. Furthermore, Crochet [101] reported that the velocity field was virtually unaffected by the fluid elasticity (up to $We \leq 1.5$); the stress field, on the other hand, was found to be significantly different from its Newtonian counterpart. Despite these appreciable differences in the stress field, the drag force showed no significant deviation from the Stokes value. Crochet goes on to remark that some interesting phenomena may be present for $We = 2$ or larger, but unfortunately the numerical solutions do not converge under these conditions.

Almost simultaneously in an independent numerical study, Tiefenbruck and Leal [102] studied the behavior of a solid sphere in an Oldroyd type fluid. This model is capable of portraying both the non-equal normal stresses as well as the shear-thinning property. The conclusions reached in this study are consistent with those of Crochet [101] and others [103], and no significant reduction in drag is predicted. The numerical solution is not yet possible to include the realistic levels of shear-thinning and viscoelasticity. In view of the near equivalence of all fluid models to the second order fluid in the low shear rate region such as those generated by creeping sphere motion, it is perhaps appropriate to say that all the analyses cited here examine the influence of fluid elasticity in the absence of shear-thinning behavior. Although it is readily acknowledged [104, 105] that the idealized models such as the Maxwell or the Second Order fluid models are inadequate for a quantitative characterization of the behavior of real viscoelastic fluids, analyses for such models can still be qualitatively useful and can provide a launching pad for studies involving realistic models. The main conclusion reached so far is that the drag force on a sphere is slightly reduced below its Newtonian value as a consequence of the fluid elasticity, which is in accordance with the perturbation type studies. Leal [106, 107] has written informative accounts of the theoretical developments in this area.

On the other hand, many investigators have carried out experiments to demonstrate the influence of the fluid elasticity. In most of these experiments, polymer solutions that exhibited shear-thinning and viscoelasticity were used. It is not surprising, therefore, that the results are not in agreement. For example, some workers [97, 98, 100] concluded that the drag is decreased due to the fluid elasticity, while others [36, 43] have suggested that fluid elasticity has no influence on drag.

In 1977, Boger [108] reported a fluid that exhibited an almost constant viscosity but significant levels of elasticity in steady shearing flow. These fluids have since been widely used around the world for experimental work [109–111] and have become known as Boger fluids. Indeed, these idealized fluids provide a unique opportunity to investigate the influence of fluid elasticity in the near absence of shear-thinning on the drag of a sphere. This, in turn, will enable a direct comparison with the theoretical approaches detailed above.

The Boger fluids are constructed by dissolving very small amounts of polyacrylamide (Separan AP-30 or MG-500 from Dow Chemical) in commercially available corn syrup. Corn syrups are viscous Newtonian fluids available with viscosities up to 400 Pa s. Chhabra et al. [78, 112] carried out a detailed study to highlight the possible role of elasticity in the flow around a sphere by using Boger fluids. Weak to moderate elastic effects were also investigated by using three polymethylsiloxanes (Dow Corning, U.S.A.). These oils are known to be Newtonian in shear and to exhibit an extended region of second order behavior in the first normal stress difference [113]. Typical shear stress and first normal stress difference for one silicone oil and one Boger fluid are shown in Figures 24 and 25, respectively. Clearly, in the case of the silicone oil a line having a slope of unity can be drawn through the shear stress data and a line of slope 2 gives an excellent fit of the primary normal stress difference data. In other words, in steady shear these silicone oils can be described accurately by

$$\eta(\dot{\gamma}) = \eta_0 \tag{44}$$

and

$$N_1(\dot{\gamma}) = \alpha\dot{\gamma}^2 \tag{45}$$

Similarly, it should be noted that the Boger fluid is only very slightly shear-thinning (the slope of $\log \tau - \log \dot{\gamma}$ data is 0.98), is highly elastic (as indicated by the high normal stress difference to

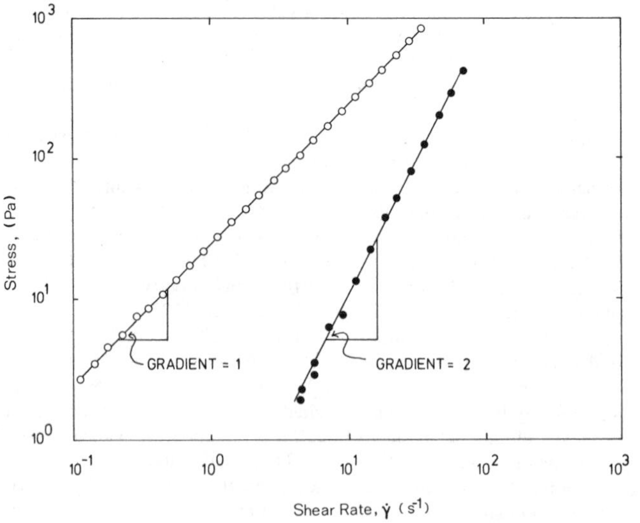

Figure 24. Typical shear stress (○) and first normal stress difference (●) for a silicone oil (adapted from Chhabra et al. [112]).

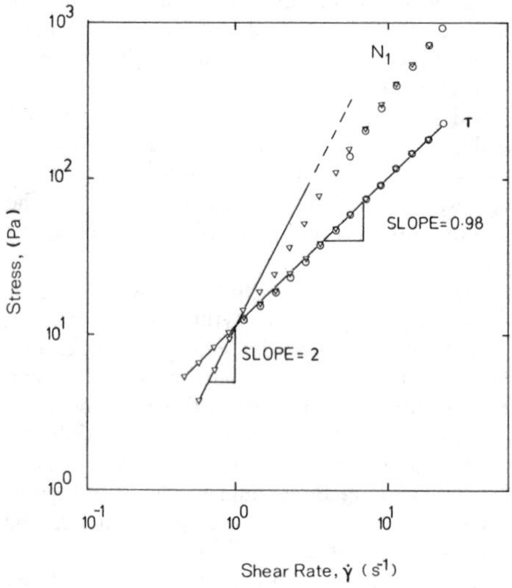

Figure 25. Typical shear stress and first normal stress difference for a Boger fluid: 0.038% separan, 4.6% water in glucose syrup solution (replotted from Chhabra et al. [112]).

shear stress ratio) and exhibits a quadratic relationship between the first normal stress difference and shear rate over a finite range at low shear rate values. Such normal stress difference behavior is expected of all viscoelastic fluids in the limit of zero shear rate but is rarely observed. Thus, these fluids can also be characterized by a constant shear viscosity and a constant value of relaxation time at low shear rates. Values of the zero shear viscosity and the maximum relaxation times calculated using Equation 38 are given in Table 9. Note that the value of relaxation time for the Boger fluids becomes shear-rate dependent outside the quadratic range.

Table 9
Characteristics of Boger Fluids and Silicone Oils (Modified after Chhabra et al. [112])

Test Fluid	Temp. (K)	η_0 (Nsm^{-2})	θ (s)	ρ ($kg\ m^{-3}$)	n	Symbols Used in Figure 26
Silicone Oil	293.0	1.15	1.21×10^{-4}	971.00	1.00	△
Silicone Oil	295.0	13.71	1.03×10^{-3}	971.00	1.00	○
Silicone Oil	292.8	24.11	1.97×10^{-3}	971.00	1.00	●
0.2% Separan MG 500/ 2.0% water in corn syrup	292.0	17.30	0.18	1414.00	1.00	□
0.033% Separan MG 500/ 4.6% water in corn syrup	292.0	10.73	0.46	1387.00	0.98	■
0.002% Separan AP 30/ 3.8% water in corn syrup	293.0	6.30	0.037	1400.00	0.98	▽
0.02% Separan AP 30/ 3.8% water in corn syrup	294.0	7.48	0.30	1395.00	0.98	◒
0.038% Separan AP 30/ 4.6% water in corn syrup	292.0	10.79	0.51	1391.00	0.98	▲
0.068% Separan AP 30/ 9.5% water in corn syrup	292.0	5.60	0.13	1390.00	1.00	◪
0.1% Separan AP 30/ 6.8% water in corn syrup	292.5	10.60	2.77	1395.00	0.97	◑

Note: In the case of Separan/water/corn syrup solutions, the values of θ refer to its maximum value in the low shear rate (quadratic behavior) range.

The drag coefficients of numerous spheres were measured as a function of Reynolds number and other variables. Since these fluids exhibit nearly constant shear viscosity, the appropriate definition of Reynolds number is simply

$$Re_0 = \frac{\rho V d}{\eta_0} \tag{31}$$

Wall effects were eliminated as described elsewhere [78]. The influence of fluid elasticity on the drag coefficient in creeping flow was expressed as the departure from Newtonian drag i.e.

$$X = \frac{C_D\, Re_0}{24} \tag{28}$$

In the absence of shear-thinning characteristics and the inertial effects ($1.69 \times 10^{-5} \leq Re_0 \leq 0.081$), X should be a unique function of Weissenberg number We, which in the case of a sphere is defined as

$$We = \frac{\theta V}{R} \tag{46}$$

Although Equation 38 defines the Maxwellian relaxation time, a difficulty arises with Boger fluids for conditions where the relaxation time becomes shear rate dependent. Chhabra et al. [112] arbitrarily assumed the Newtonian value of average shear rate, that is V/R, to be applicable in the present case. This approximation is partially justifiable as the fluids used exhibit constant shear viscosity and it is not appreciably influenced by the fluid elasticity [101]. Besides, the choice of shear rate expression does not influence the values of X but only the values of the Weissenberg number.

The observed values of X are shown in Figure 26 as a function of the Weissenberg number evaluated at the appropriate value of shear rate. The data cover almost a five-fold range of the values of Weissenberg number. For $0 \leq We \leq 0.1$, no significant deviation from the Newtonian (Stokes) drag is observed, and this is borne out by both the silicone oils as well as the Boger fluids. Significant reduction in drag is observed for $We \geq 0.1$. More important is the fact that reduction in Stokes drag is only observed when the surface average shear rate of the falling sphere is outside the region of second order behavior. This conclusion is consistent in a qualitative sense with the calculations of Crochet [101]. A substantial reduction in drag is observed in the range $0.1 \leq We \leq 0.7$–0.8 and

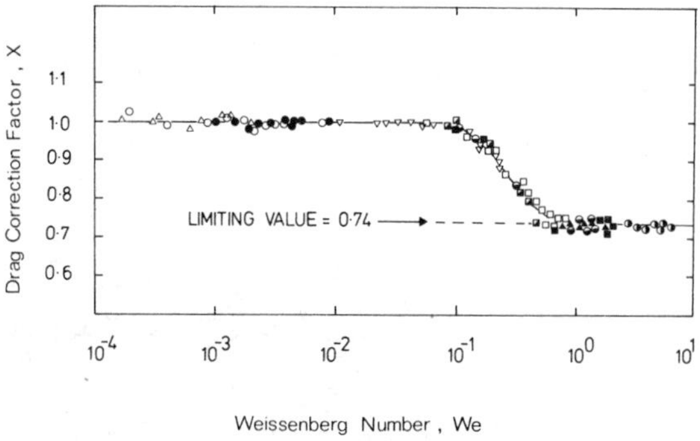

Figure 26. Deviation from Stokes drag X as a function of Weissenberg number We. The symbols are identified in Table 9. (Based on data of Chhabra [78]).

eventually X appears to reach an asymptotic value of approximately 0.74. No appreciable reduction in the Stokes drag in the quadratic region as demonstrated in this work is also the conclusion reached by all available theoretical analyses [90–94]. None, however, have successfully coped with a shear-rate-dependent shear modulus. From these experiments, it is abundantly clear that it is conceptually inadequate to apply second order or Maxwell thinking in the analysis of flow around a sphere.

Now attention is given to the case where both shear thinning and fluid elasticity are encountered. From the theoretical point of view, the mathematical complexity, as noted by Tiefenbruck and Leal [102], precludes the possibility of even an approximate solution. Some progress, however, has been made based on the hypothesis of Bird [26] and of others [57, 114] who have argued that a fluid characteristic time defined from shear-thinning data has some bearing on the importance of visco-elastic effects. Indeed, Abdel-Khalik et al. [56] and others [115–117] have reported some success in predicting the values of the primary normal stress difference from shear viscosity data of a variety of materials. Typical comparisons between the values of primary normal stress difference calculated from shear viscosity data using the method outlined by Abdel-Khalik et al. [56] and the experimental values are shown in Figures 27 and 28. Admittedly the method involves an arbitrary parameter but, nevertheless, it does yield good results for the primary normal stress difference. In view of this inter-connection between shear-thinning and fluid elasticity, an analysis of the flow around a sphere based on a constitutive equation, which either has a time constant (e.g. Equations 24 and 26) or whose parameters can be combined to construct a fluid characteristic time (e.g., Equation 25) should, in principle, also be applicable to viscoelastic fluids. Similar arguments have been advanced by numerous other workers who have rationalized the role of fluid elasticity in a variety of flow problems purely via considerations based on the shear-thinning behavior [60, 79, 80, 82, 83, 118]. Thus over the years, a significant volume of evidence (largely experimental) has accumulated which lends support to Bird's intuitive assertion [26].

Indeed, the results on drag coefficients reported by Chhabra et al. [52, 61] were obtained with polymer solutions that exhibited both shear-thinning effects as well as the viscoelasticity, yet they show excellent agreement with the approximate analyses based on the Ellis fluid model (wherein

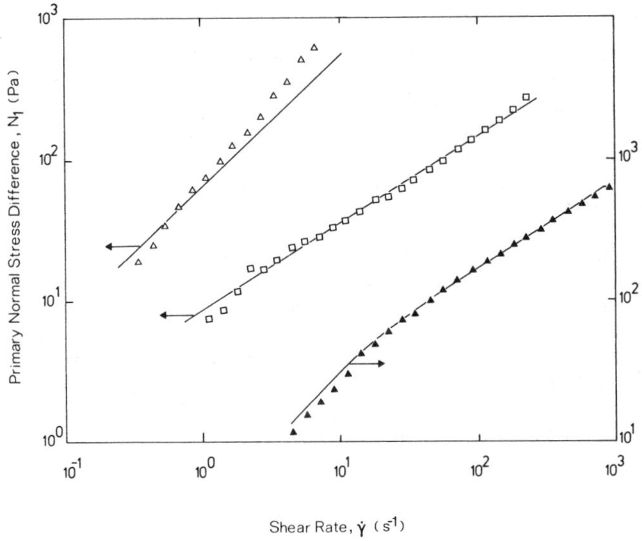

Figure 27. Primary normal stress difference as a function of shear rate for different polymer solutions. The points are experimental values reported by Chhabra and Uhlherr [61] and the lines are the predictions made using the method of Abdel-Khalik et al. [56].

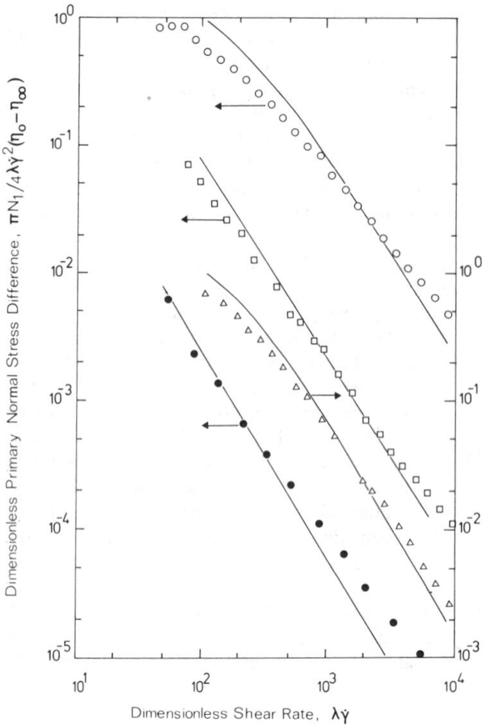

Figure 28. Dimensionless primary normal stress difference as a function of dimensionless shear rate for different polymer solutions. The points are experimental values and the solid lines are the predictions. (Replotted from Abdel-Khalik et al. [56]).

$\eta_0/\tau_{1/2}$ has the dimensions of time) and the Carreau viscosity equation (in this case λ has the dimensions of time). This agreement is realistic. Although one may have the temptation to conclude that either the shear-thinning effects completely overshadow the elastic effects or the fluid elasticity (in the presence of shear-thinning viscosity) exerts little or no influence on the drag force. But such a conclusion is premature and at best is conjecture. Definite conclusions regarding the role of fluid elasticity over a sufficiently wide range of dynamic and model parameters should await further experimental work, particularly using the ideal fluids developed by Boger [108] and recently by others [119].

Another interesting observation is that in the absence of fluid shear-thinning, the fluid elasticity causes a gradual reduction in the value of X, which asymptotically reaches a value of 0.74, but when both (shear-thinning and fluid elasticity) are encountered then X decreases to the much lower value of 0.1; thus suggesting strong interactions between the two. However, it is not evident to us how this reduction in the drag force may be explained on physical grounds. Whatever the nature of these interactions between shear-thinning and fluid elasticity may be, it is tentatively suggested that the analysis based on the Ellis fluid model or the Carreau viscosity equation should be used to calculate the drag coefficient of a sphere in viscoelastic fluids in creeping flow.

High Reynolds Number Region

Acharya et al. [36] experimentally examined the influence of fluid elasticity on the drag force of a sphere outside the creeping flow region. The rheological behavior in terms of both shear stress and primary normal stress difference was approximated using power law expressions. The expression for the drag coefficient for inelastic liquids was empirically modified by incorporating a Weissenberg number to account for viscoelastic effects. The fact that their original expression is in error, and has been corrected by Kawase and Ulbrecht [37], raises some doubt about the reliability of their correlation for viscoelastic liquids.

In the study of Chhabra and Uhlherr [70], one of the test fluids used was viscoelastic, yet the values of drag coefficients of spheres falling in this test fluid were in line with those obtained in inelastic liquids under identical values of Reynolds number. It is difficult to understand at this stage the possible influence of fluid elasticity. This situation is perhaps the most complex one as the results are influenced by the shear-thinning viscosity, by the viscoelastic and inertial effects. Until more definite information becomes available, Equation 34 may be used to calculate the value of drag coefficient for a given sphere under such circumstances.

Wall Effects

Even with the choice of the simplest constitutive relationship for viscoelastic liquids, namely the Maxwell model, the resulting set of equations describing the slow motion of a sphere in bounded fluids is not amenable to analysis. Therefore, not even qualitative inferences can be drawn from theory with respect to the possible roles of fluid elasticity on wall factor.

The preliminary experimental studies of Sigli and Coutanceau [97] and of Cho et al. [120] suggested that the presence of fluid elasticity suppressed the wall effects as compared to those in inelastic shear-thinning liquids. In 1981, Chhabra et al. [16, 78] reported an extensive experimental study to gauge the effect of fluid elasticity on the wall effects. The results correspond to the conditions when the fluid exhibits no shear-thinning, i.e. Boger fluids and silicone oils, as well as when the fluid possesses shear rate dependent viscosity.

The values of wall factor calculated from the experiments performed (exactly in the same manner as described in the preceding section for inelastic shear-thinning liquids) in Boger fluids and three silicone oils are particularly interesting. In the quadratic region of primary normal stress difference, the values of wall factor show excellent agreement with the Newtonian wall correction factors. The results plotted in this fashion are shown in Figure 29. Included in the same figure is the correlation of inelastic shear-thinning liquids in the limit of creeping flow (Equation 35). Evidently the wall effects are smaller than those in inelastic shear-thinning fluids. The fact that results are in agreement with the Newtonian case reaffirms that within the limits of quadratic behavior, the kinematics of flow is not significantly different from its Newtonian counterpart. On the other hand, when the sphere motion is such that it corresponds to the fluid behavior outside the quadratic range of Boger fluids, there are virtually no wall effects. Indeed the wall factor f_0 varies between only 0.97 and 1.00. Thus,

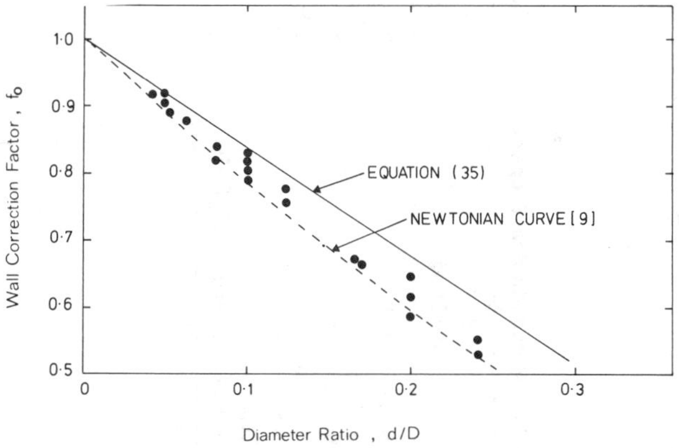

Figure 29. Wall correction factor in second-order silicone oils. (Based on data of Chhabra [78]).

Table 10
Rheological and Physical Properties of Polymer Solutions Used for Studying Wall Effects in Viscoelastic Fluids (Modified after Chhabra et al. [16])

Test Fluid	Temp. (K)	ρ (kg m^{-3})	η_0 (Pa s)	λ (s)	n	n′	K (Pa s^n)	K′ (Pa $s^{n'}$)	θ (s)
0.85% Separan AP 30	292.0	1003	8.20	11.00	0.49	0.62	2.70	6.10	2.61
1.46% Methocel	295.5	1003	3.31	0.33	0.55	1.00	3.00	0.20	5.22×10^{-4}
1.25% Separan AP 30	294.0	1005	19.9	15.30	0.46	0.67	5.20	8.20	0.32
2.00% Separan AP 30	289.5	1014	59.0	19.07	0.42	0.59	11.6	25.00	1.55
0.75% Separan MG 500 + 1.25% CMC	292.0	1006	11.5	8.13	0.40	0.60	3.15	7.10	1.10
1.62% Separan AP 30	291.0	1010	28.8	15.85	0.47	0.59	7.00	16.00	2.94
0.75% Separan AP 30 + 0.75% CMC	292.0	1003	10.5	15.75	0.53	0.60	3.08	6.80	3.87

there is no need to apply any correction due to the presence of walls in the following ranges of variables:

$$0 \leq (d/D) \leq 0.25, \qquad 1 \times 10^{-4} \leq We_m \leq 5.00, \qquad Re_{0,m} \leq 10^{-2}$$

Finally, we come to the case where the liquid exhibits both shear-thinning and elastic behavior. The fluids whose properties are summarized in Table 10 were used to investigate the influence of fluid elasticity on the wall effects. Typical variation of the measured terminal velocity with diameter ratio is shown in Figure 30 for one of the liquids used. Since the value of Reynolds number $Re_{0,m}$ (based on the measured velocity) never exceeded 0.01, creeping flow conditions were satisfied. Under these conditions, by analogy with Newtonian and inelastic shear-thinning fluids, the dependence of V_m on (d/D) is expected to be linear and indeed this linearity is borne out by the results shown in Figure 30 over the entire range of diameter ratios. Thus, the value of the terminal velocity in unbounded medium V is obtained simply by extrapolating these lines to (d/D) = 0.

In this manner the wall factor can be calculated for each sphere drop test as a function of diameter ratio, Reynolds number, and other dimensionless groups depending upon the choice of the constitutive relation. Due to its demonstrated suitability, the Carreau viscosity equation was elected to approximate the shear stress-shear rate behavior of the test fluids. Under these conditions,

$$f_0 = f(d/D, Re_{0,m}, \Lambda_m, n) \tag{47}$$

A statistical scrutiny of the entire data bank revealed that at 95% confidence level neither $Re_{0,m}$ nor n was a significant variable. This observation is in accordance with the work on inelastic shear-thinning fluids presented earlier. Therefore, Equation 47 reduces to

$$f_0 = f(d/D, \Lambda_m) \tag{48}$$

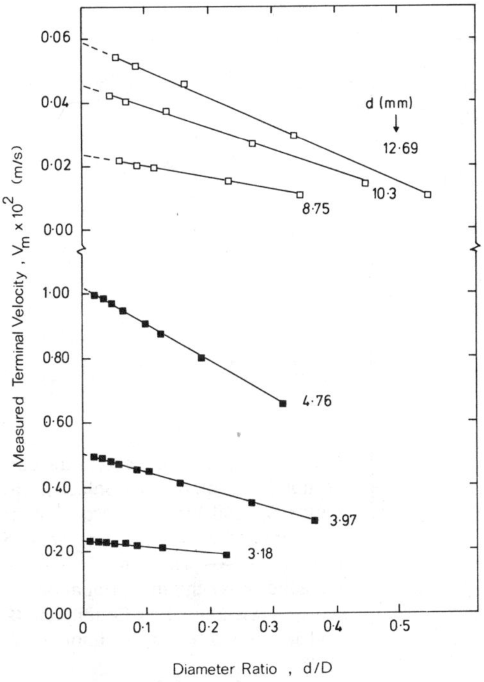

Figure 30. Effect of diameter ratio on terminal fall velocity of spheres in a 2% separan AP30 aqueous solution. Sphere materials are (□) PVC; (■) Stainless steel. (Replotted from [16]).

To isolate the effects of the dimensionless parameter Λ_m and the diameter ratio on the wall factor, f_0 is plotted against Λ_m for fixed values of diameter ratio. Such a plot is shown in Figure 31. Included in this figure are the Newtonian theory [13] and the results obtained with inelastic shear thinning liquids under creeping flow conditions (Equation 35). The wall effect for viscoelastic fluids is evidently smaller than for Newtonian and for inelastic shear-thinning liquids. Furthermore, the wall effect decreases as Λ_m increases. This is qualitatively similar to the behavior observed for purely elastic liquids if one views the dimensionless parameter Λ_m (as Bird suggests [26]) as being equivalent to the Weissenberg number. Furthermore, this observation is in accordance with the recent numerical study of Hassager and Bisgaard [103], who concluded that the wall effects decreased with the increasing value of Weissenberg number. Subsequently, Bisgaard [121] carried out a flow visualization study and confirmed that the fluid elasticity further suppressed the wall effects. A physical explanation of this behavior is not immediately obvious.

The results were correlated simply as [16]:

$$f = 1 - 1.3(d/D)^{0.94}\Lambda_m^{-0.077} \tag{49}$$

This equation predicts the wall factor in viscoelastic liquids with an average error of 4% and a maximum error of 8% in the following ranges of variables: $0.005 < (d/D) < 0.5$; $0.9 < \Lambda_m < 300$ and $0.4 < n < 0.55$.

In order to add weight to the claim that parameter Λ is indeed related to fluid elasticity, a Weissenberg number is defined based on the relaxation time of the fluid derived from the primary normal stress difference. The relaxation time given by Equation 39 is used to define a Weissenberg number as

$$We_m = \frac{\theta V_m}{R} \tag{50}$$

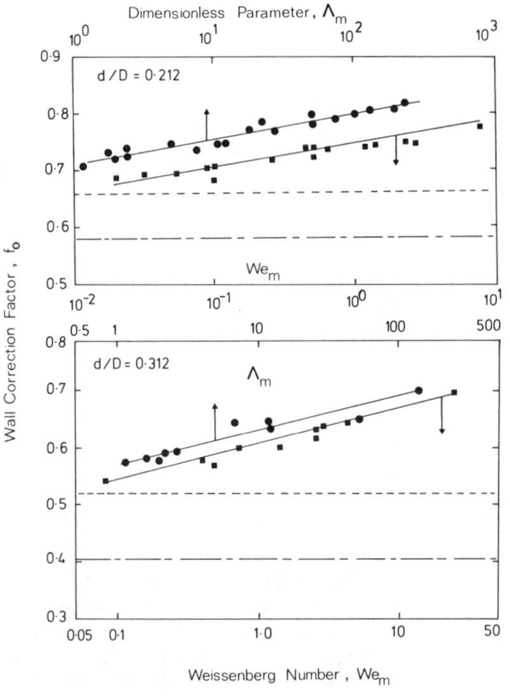

Figure 31. Dependence of wall correction factor on dimensionless parameter Λ_m and Weissenberg number We_m for fixed values of diameter ratio (d/D). —·—·— Newtonian; - - - - - inelastic shear-thinning, Equation 35; ——— viscoelastic, Equations 49 and 51. (Replotted from Chhabra et al. [16]).

Also shown in Figure 31 are the experimental data replotted as a function of the Weissenberg number defined by Equation 50. The similarity between the two is immediately obvious. Based on this comparison, perhaps one can argue that λ and θ are interrelated, though the relationship is not straightforward. Chhabra et al. [16] also proposed an equation for wall factor involving We_m as:

$$f = 1 - 0.94(d/D)^{0.80}(We_m)^{-0.073} \tag{51}$$

Attention is drawn to the similar values of indices and coefficients in Equations 49 and 51. Equation 51 is applicable for the following ranges of variables:

$$0.005 \leq d/D \leq 0.5, \qquad 0.02 \leq We_m \leq 11, \quad \text{and} \quad 0.4 \leq n \leq 0.55$$

Hence to recapitulate the state of affairs with regard to the wall effects in creeping flow, one can qualitatively state that—

$$f_{\substack{\text{Newtonian +}\\ \text{quadratic region}}} < f_{\substack{\text{inelastic}\\ \text{shear thining}}} < f_{\text{viscoelastic}} < f_{\text{pure elastic}}$$

Thus in creeping flow, our understanding of the wall effects related to sphere motion is reasonably satisfactory. One of the terrains not explored to date is the combined effects of elasticity, shear-thinning, and inertial forces at high Reynolds numbers. From the point of view of theoretical analysis, the prospects are less than bright, but this is a topic that can be easily exploited following the experimental path described in this chapter.

CONCLUDING REMARKS AND FUTURE NEEDS

The foregoing discussion clearly shows that one can estimate the values of drag coefficient and wall effects of spheres in Newtonian liquids over the entire range of interest. These macroscopic quantities, however, shed very little light on the structure of the flow field. In the future perhaps more effort should be diverted to gleaning such information. Numerical tools are also now available to extend the range of solutions of the governing equations to cover a significant part of the standard drag curve.

On the other hand, in the case of non-Newtonian fluids the resulting picture of the motion of a rigid sphere (particularly with reference to the drag force) is far from coherent. Though it is now well established and generally accepted that the use of the power law model is totally inadequate for the calculation of the drag coefficient, it does seem to yield satisfactory results for wall effects. The use of a generalized Newtonian fluid model containing zero shear viscosity results in a much better description of drag force in creeping as well as high Reynolds number regime. This has been convincingly demonstrated in the case of the Ellis fluid model and the Carreau viscosity equation. However, all the theoretical treatments are approximate and have an inherent weakness with regard to the choice of an arbitrary stream function (and/or stress profile). This problem can only be resolved by seeking the direct rigorous solutions to the relevant equations in conjunction with an appropriate generalized Newtonian fluid model. Such a study will not only remove the arbitrariness inherent in the use of variational principles, but will also provide insight into the flow field around a sphere. Both the Carreau viscosity equation and the Ellis fluid model are potential candidates for this exercise. Indeed, the preliminary indications [66] for the use of the Carreau viscosity equation are encouraging.

Our understanding of the role of fluid elasticity—with and without shear-thinning behavior—on drag coefficient and wall effect is still in an embryonic state. Admittedly the development of ideal elastic fluids (Boger fluids) has enabled a direct comparison between the viscoelastic theories (for sphere motion) and corresponding experiments. Beyond the quadratic region, the results obtained with Boger fluids reveal some interesting effects such as a significant reduction in drag below its Newtonian value. None of the existing theories so far has been successful in explaining this phenomenon in a quantitative manner. The chief difficulty in extending the range of available analyses to include significant levels of elasticity (high values of We) parameters has been of a numerical nature whereby the numerical schemes seem to fail to converge beyond a critical value of the Weissenberg number. The effort to develop effective numerical techniques for solutions at larger values of Weissenberg number (or equivalent parameters) is presently common to a number of research groups including those of Professor Leal (Caltech) and of Professor Crochet (Belgium). This eventually leads to the final problem where both the fluid elasticity and shear-thinning characteristics play important roles in determining the flow field around a sphere, and thus on drag coefficient. In this context Bird's approach merits serious consideration. If proven, it can provide a simple answer to an otherwise extremely difficult problem. The preliminary indications are affirmative.

Considerable scope exists for experimentalists as very little is known about the nature of flow field in well characterized non-Newtonian materials covering a wide range of dynamic and material parameters. Such measurements may hold the key to unravelling the mysterious roles played by the fluid elasticity and shear-thinning behavior. Needless to say that there is an equally important need to extend considerably the range of experiments on drag coefficients and wall effects.

Despite all the deficiencies previously mentioned, it is evident that considerable progress has been made in a relatively short period towards the understanding of the flow of non-Newtonian fluids around a sphere, especially when compared with its Newtonian analogue, which has been studied for almost a century and a half.

NOTATION

C_D	drag coefficient, Equation 14 or 32
d	diameter of sphere
D	diameter of cylinder
De	Deborah number, Equation 43
El	Ellis number, Equation 30
E^2	differential operator, Equation 11
F_D	drag force
f	wall correction factor
f_0	wall correction factor in the creeping region
f_∞	wall correction factor at high values of Reynolds number
g	acceleration due to gravity
K	consistency index in Power law model, Equation 27
K′	power law constant for primary normal stress difference
n	flow behavior index in Power law model, Equation 27
n′	index, $N_1 = K'(\dot{\gamma})^{n'}$
N_1	primary normal stress difference
p	pressure
r	radial coordinate
R	radius of sphere
Re′	Reynolds number based on Power law model parameters, Equation 29
Re_0	Reynolds number based on zero shear viscosity, Equation 31
v_r, v_θ	radial and angular components of velocity
V	free stream velocity or terminal fall velocity of a sphere
We	Weissenberg number, Equation 40
X	drag correction factor, Equation 28
x	dimensionless radial coordinate ($=r/R$)

Greek Symbols

α	parameter in the Ellis fluid model, Equation 25
$\dot{\gamma}$	shear rate
ϵ_{ij}	components of rate of deformation tensor, $i, j = r, \theta, \phi$
η	apparent viscosity ($=\tau/\dot{\gamma}$)
η_0	zero shear viscosity
η_∞	asymptotic value of apparent viscosity in the high shear rate region
θ	angular coordinate, or fluid characteristic time, Equation 38 or 39
λ	parameter in the Carreau viscosity model, Equation 24 or 26
Λ	dimensionless parameter ($=\lambda V/R$)
μ	viscosity of a Newtonian fluid
ρ	density of fluid
ρ_p	density of sphere material
τ_{ij}	components of stress tensor, i and j = r, θ, ϕ
τ	shear stress
$\tau_{1/2}$	parameter in the Ellis fluid model, Equation 25

Subscripts

m Refers to a quantity with wall effects

REFERENCES

1. Stokes, G. G., "On The Effect of the Internal Friction of Fluids on the Motion of Pendulums," *Trans. Cam. Phil. Soc.*, 9 (1851) 8–27.
2. Bird, R. B., Stewart, W. E., and Lightfoot, E. N., *Transport Phenomena*, Tokyo: Wiley International Edition (1960).
3. Oseen, C. W., "Neuere Methoden und Ergebnisse in der Hydrodynamik," Leipzig: Akademische Verlagsgesellschaft (1927).
4. Clift, R., Grace, J. R., and Weber, M. E., *Bubbles, Drops and Particles*, Academic Press, New York (1978).
5. O'Neill, M. E., "Small Particles in Viscous Media," *Sci. Prog.*, 67 (1981) 149–184.
6. Jenson, V. G., "Viscous Flow Around a Sphere at Low Reynolds Number (<40)," *Proc. Roy. Soc.*, 249A (1959) 346–366.
7. Clift, R., and Gauvin, W. H., "Motion of Entrained particles in Gas streams," *Proc. Powtech 71;* International Powder Technology and Bulk Granular Solids Conf. (1971) p. 47.
8. Le Clair, B. P., Hamielec, A. E., and Pruppacher, H. R., "A Numerical Study of the Drag on a Sphere at Low and Intermediate Reynolds Number," *J. Atmos. Sci.*, 27 (1970) 308–315.
9. Francis, A. W., "Wall Effect in Falling Ball Method for Viscosity," *Physics*, 4 (1933) 403–406.
10. Happel, J., and Brenner, H., *Low Reynolds Number Hydrodynamics*, Prentice-Hall, New Jersey (1965).
11. Ladenberg, R., "Uber den Einfluβ von Wanden auf die Bewegung einer Kugel in einer reibenden Flussigkeit." *Ann. Phys.* (Leipzig), 23 (1907) 447–458.
12. Faxen, H., "Die Bewegung einer Starren Kugel langs der Achse eines mit zaher Flussigkeit gefullten Rohres," *Ask. Mat., Arstron. Fys.*, 17 (1923) 1–28.
13. Haberman, W. L., and Sayre, R. M., "Motion of Rigid and Fluid Spheres in Stationary and Moving Liquids Inside Cylindrical Tubes," David Taylor Model Basin Report no. 1143, Oct. (1958).
14. Iwaoka, M., and Ishii, T., "Experimental Wall Correction Factors of Single Solid Spheres in Circular Cylinders," *J. Chem. Eng. Japan*, 12 (1979) 239–242.
15. Uhlherr, P. H. T., "Fluid Particle Systems," Ph.D. Dissertation, Monash University, Australia (1965).
16. Chhabra, R. P., Tiu, C., and Uhlherr, P. H. T., "A Study of Wall Effects on the Motion of a Sphere in Viscoelastic Fluids," *Can. J. Chem. Eng.*, 59 (1981) 771–775.
17. Fayon, A. M., and Happel, J., "Effect of a Cylindrical Boundary on a Fixed Rigid Sphere in a Moving Viscous Fluid," *A.I.Ch.E.J.*, 6 (1960) 55–58.
18. Munroe, H. S., "The English versus the Continental System of Jigging—Is Close Sizing Advantageous?" *Trans. A.I.M.M.E.*, 17 (1888/89) 637–659.
19. Fidleris, V., and Whitmore, R. L., "Experimental Determination of the Wall Effect for Spheres Falling Axially in Cylindrical Vessels," *Brit. J. Appl. Phys.*, 12 (1961) 490–494.
20. Boger, D. V., "Demonstration of Upper and Lower Newtonian Fluid Behavior in a Pseudoplastic Fluid," *Nature*, 265 (1977) 126–127.
21. Bird, R. B., "Useful non-Newtonian Models," *Ann. Rev. Fluid Mech.*, 8 (1976) 13–34.
22. Bird, R. B., Armstrong, R. C., and Hassager, O., *Dynamics of Polymeric Liquids*, Wiley (1977).
23. Carreau, P. J., "Rheological Equations from Molecular Network Theories," *Trans. Soc. Rheo.*, 16 (1972) 99–127.
24. Tiu, C., Uhlherr, P. H. T., and Halmos, A. L., "Rheology in Food Processing," *Chem. Eng. Aus.*, ChE7 (1982) 27–35.
25. Boger, D. V., "Circular Entry Flows of Inelastic and Viscoelastic Fluids," *Adv. Trans. Proc.*, 2 (1982) 43–104.

26. Bird, R. B., "Experimental Tests of Generalized Newtonian Models Containing a Zero Shear Viscosity and a Characteristic Time," *Can. J. Chem. Eng.*, 43 (1965) 161–168.
27. Johnson, M. W., "Some Variational Theorems for Non-Newtonian Flow," *Phys. Fluids*, 3 (1960) 871–878.
28. Johnson, M. W., "On Variational Principles for Non-Newtonian Fluids," *Trans. Soc. Rheo.*, 5 (1961) 9–21.
29. Slattery, J. C., *Momentum, Energy and Mass Transfer in Continua*, McGraw Hill (1972).
30. Tomita, Y., "On the Fundamental Formula of Non-Newtonian Flow," *Bull. J.S.M.E.*, 2 (1959) 469–474.
31. Wallick, G. C., Savins, J. G., and Arterburn, D. R., "Tomita Solution for the Motion of a Sphere in a Power Law Fluid," *Phys. Fluids*, 5 (1962) 367–368.
32. Slattery, J. C., "Approximations to the Drag Force on a Sphere Moving Slowly Through Either an Ostwald-De Waele or a Sisko Fluid," *A.I.Ch.E.J.*, 8 (1962) 663–667.
33. Wasserman, M. L., and Slattery, J. C., "Upper and Lower Bounds on the Drag Coefficient of a Sphere in a Power-Model Fluid," *A.I.Ch.E.J.*, 10 (1964) 383–388.
34. Mohan, V., "Fall of Liquid Drops in Non-Newtonian Media," Ph.D. Dissertation, I.I.T., Madras (1974).
35. Nakano, Y., and Tien, C., "Creeping Flow of Power Law Fluid Over Newtonian Fluid Sphere," *A.I.Ch.E.J.*, 14 (1968) 145–151.
36. Acharya, A., Mashelkar, R. A., and Ulbrecht, J., "Flow of Inelastic and Viscoelastic Fluids Past a Sphere," *Rheo. Acta*, 15 (1976) 454–478.
37. Kawase, K., and Ulbrecht, J., "Newtonian Fluid Sphere with Rigid or Mobile Interface in a Shear-Thinning Liquid: Drag and Mass Transfer," *Chem. Eng. Commun.*, 8 (1981) 213–231.
38. Chhabra, R. P., Tiu, C., and Uhlherr, P. H. T., "Shear-Thinning Effects in Creeping Flow about a Sphere," *Rheology*, Vol. II (1980) 9–16.
39. Cho, Y. I., and Hartnett, J. P., "Drag Coefficients of a Slowly Moving Sphere in Non-Newtonian Fluids," *J. Non-New. Fluid Mech.*, 12 (1983) 243–247.
40. Yoshioka, N., and Adachi, K., "Some Deductions from the Extremum Principles for Non-Newtonian Fluids," *J. Chem. Eng. Japan*, 6 (1973) 134–140.
41. Slattery, J. C., "Non-Newtonian Flow about a Sphere," Ph.D. Dissertation, Uni. Wis., U.S.A. (1959).
42. Turian, R. M., "An Experimental Investigation of the Flow of Aqueous Non-Newtonian Polymer Solutions Past a Sphere," *A.I.Ch.E.J.*, 13 (1967) 999–1006.
43. Kato, H., Tachibana, M., and Oikawa, K., "On the Drag of a Sphere in Polymer Solutions," *Bull. J.S.M.E.*, 15 (1972) 1556.
44. Uhlherr, P. H. T., Le, T. N., and Tiu, C., "Characterization of Inelastic Power Law Fluids Using Falling Sphere Data," *Can. J. Chem. Eng.*, 54 (1976) 497–502.
45. Chhabra, R. P., "Some Remarks on Drag Coefficients of a Slowly Moving Sphere in Non-Newtonian Fluids," *J. Non-Newt. Fluid Mech.*, 13 (1983) 225–227.
46. Chhabra, R. P., and Uhlherr, P. H. T., "Shortcomings of the Power Law in Describing Creeping Flow about a Sphere," *Proc. 2nd Nat. Conf. Rheo.*, Sydney (1981) pp. 89–92.
47. Hopke, S. W., and Slattery, J. C., "Upper and Lower Bounds on the Drag Coefficient of a Sphere in an Ellis Model Fluid," *A.I.Ch.E.J.*, 16 (1970) 224–229.
48. Hill, R., "New Horizons in the Mechanics of Solids," *J. Mech. Phys. Solids*, 5 (1956) 66–74.
49. Hill, R., and Power, G., "Extremum Principles for Slow Viscous Flow and the Approximate Calculation of Drag," *Quart. J. Mech. Appl. Math.*, 9 (1956) 313–319.
50. Chhabra, R. P., and Uhlherr, P. H. T., unpublished work.
51. Dolecek, P., et al., "Vypocet Padore Ry chlosti Kulovych Castee V. Ellisove Kapalim," Paper presented in CHISA, Prague (1983).
52. Chhabra, R. P., Tiu, C., and Uhlherr, P. H. T., "Creeping Motion of Spheres Through Ellis Model Fluids," *Rheo. Acta* 20 (1981) 346–351.
53. Chhabra, R. P., and Uhlherr, P. H. T., "Estimation of Zero Shear Viscosity of Polymer Solutions from Falling Sphere Data," *Rheo. Acta*, 18 (1979) 593–599.
54. Huilgol, R. R., *Continuum Mechanics of Viscoelastic Liquids*, Hindustan Publishing Co., New Delhi (1975).

55. Dallon, D. S., "A Drag Coefficient Correlation for Spheres Settling in Ellis Fluids," Ph.D. Dissertation, Uni. Utah, U.S.A. (1967).
56. Abdel-Khalik, S. I., Hassager, O., and Bird, R. B., "Prediction of Melt Elasticity from Viscosity Data," *Polym. Eng. Sci.*, 14 (1974) 859–867.
57. Carreau, P. J., Dekee, D., and Daroux, M., "An Analysis of the Viscous Behavior of Polymer Solutions, *Can. J. Chem. Eng.*, 57 (1979) 135–140.
58. Co, A., "Inelastic Flow from a Tube into a Radial Slit," *Ind. Eng. Chem. Fundam.*, 20 (1981) 340–346.
59. Co, A., and Stewart, W. E., "Viscoelastic Flow from a Tube into a Radial Slit," *A.I.Ch.E.J.*, 28 (1982) 644–655.
60. Kemblowski, Z., and Michniewicz, M., "Correlation of Data Concerning Resistance to Flow of Generalized Newtonian Fluids Through Granular Beds," *Rheo. Acta*, 20 (1981) 352–359.
61. Chhabra, R. P., and Uhlherr, P. H. T., "Creeping Motion of Spheres Through Shear-Thinning Elastic Fluids Described by the Carreau Viscosity Equation," *Rheo Acta*, 19 (1980) 187–195.
62. Astarita, G., "Variational Principles and Entropy Production in Creeping Flow of Non-Newtonian Fluids," *J. Non-Newt. Fluid Mech.*, 2 (1977) 343–351.
63. Ziegenhagen, A. J., "The Very Slow Flow of a Powell Eyring Fluid Around a Sphere," *App. Sci. Res.*, 14A (1964–65) 43–56.
64. Mitsuishi, N., Yamanaka, A., and Miyahara, F., "A Study on the Drag Coefficient of the Spheres Falling in the Non-Newtonian Fluids," *Tech. Rep.* Kyushu Uni. (Japan), 44 (1971) 192.
65. Hopke, S. W., and Slattery, J. C., "Note on the Drag Coefficient for a Sphere," *A.I.Ch.E.J.*, 16 (1970) 317–318.
66. Bush, M. G., Personal communication to Uhlherr, P. H. T., (1983).
67. Adachi, K., Yoshioka, N., and Yamamoto, K., "On Non-Newtonian Flow Past a Sphere," *Chem. Eng. Sci.*, 28 (1973) 2033–2043.
68. Adachi, K., Yoshioka, N., and Sakai, K., "An Investigation of Non-Newtonian Flow Past a Sphere," *J. Non-Newt. Fluid Mech.*, 3 (1977/78) 107–125.
69. Hua, T. N., and Ishii, T., "Momentum Transfer for Multi-Solid-Particle Power Law Fluid Systems at High Reynolds Number," *J. Non-Newt. Fluid Mech.*, 9 (1981) 310–319.
70. Chhabra, R. P., and Uhlherr, P. H. T., "Sphere Motion Through Non-Newtonian Fluids at High Reynolds Number," *Can. J. Chem. Eng.*, 58 (1980) 124–128.
71. Schiller, V. L., and Naumann, A., "Uber die grundlegenden Berechnungen bei der Schwerkraft aufbereitung," *Z.V.D.I.*, 77 (1933) 318–320.
72. Caswell, B., "The Effect of Finite Boundaries on the Motion of Particles in Non-Newtonian Fluids," *Chem. Eng. Sci.*, 25 (1970) 1167–1176.
73. Chhabra, R. P., Tiu, C., and Uhlherr, P. H. T., "Wall Effect for Sphere Motion in Inelastic Non-Newtonian Fluids," *Proc. 6th Australasian Hydraulics and Fluid Mech. Conf.*, Adelaide (1977) pp. 435–438.
74. Chhabra, R. P., and Uhlherr, P. H. T., "Wall Effect for High Reynolds Number Motion of Spheres in Shear-Thinning Fluids," *Chem. Eng. Commun.*, 5 (1980) 115–124.
75. Tanner, R. I., "Observations on the Use of Oldroyd Type Equations of State for Viscoelastic Liquids," *Chem. Eng. Sci.*, 19 (1964) 349–355.
76. Okuda, K., "Pipe Wall Effects on Suspension Velocities of Single Freely Suspended Spheres and on Terminal Velocities of Single Spheres in a Pipe," *Bull. J.S.M.E.*, 18 (1975) 1142–1150.
77. Sato, T., Taniyama, I., and Shimokawa, S., "Flow of Non-Newtonian Fluid—Drag Coefficient of a Sphere," *Kagaku-Kogaku*, 4 (1966) 215.
78. Chhabra, R. P., "Non-Newtonian Fluid Particle Systems—Sphere Drag," Ph.D. Dissertation, Monash Uni., Australia (1980).
79. Leider, P. J., "Squeezing Flow Between Parallel Disks," *Ind. Eng. Chem. Fundam.*, 13 (1974) 342–346.
80. Grimm, R. J., "Squeezing Flows of Polymeric Liquids," *A.I.Ch.E.J.*, 24 (1978) 427–439.
81. Elbirli, B., and Shaw, M. T., "Time Constants from Shear Viscosity Data," *Trans. Soc. Rheo.*, 22 (1978) 561–570.
82. Sadowski, T. J., and Bird, R. B., "Non-Newtonian Flow Through Porous Media," *Trans. Soc. Rheo.*, 9 (1965) 251–271.

83. Khillar, K. C., Weinberger, C. B., and Tallmadge, J. A., "Postwithdrawal Drainage of a Viscoelastic Separan Solution," *Ind. Eng. Chem. Fundam.*, 19 (1980) 266–275.
84. Astarita, G., "Comments on the Paper—Experimental Tests of Generalized Newtonian Models Containing a Zero Shear Viscosity and a Characteristic Time," *Can. J. Chem. Eng.*, 44 (1966) 59–60.
85. Slattery, J. C., "Dimensional Considerations in Viscoelastic Flows," *A.I.Ch.E.J.*, 14 (1968) 516–518.
86. White, J. L., "Dynamics of Viscoelastic Fluids, Melt Fracture, and the Rheology of Fibre Spinning," *J. Appl. Polym. Sci.*, 8 (1964) 2339.
87. Metzner, A. B., White, J. L., and Denn, M. M., "Constitutive Equation for Viscoelastic Fluid for Short Deformation Periods and for Rapidly Changing Flows: Significance of the Deborah Number," *A.I.Ch.E.J.*, 12 (1966) 863–866.
88. Broer, L. J. F., "On the Hydrodynamics of Viscoelastic Fluids," *App. Sci. Res.*, 6A (1957) 226–236.
89. Astarita, G., and Marrucci, G., *Principles of Non-Newtonian Fluid Mech.*, McGraw-Hill, London (1974).
90. Leslie, F. M., and Tanner, R. I., "The Slow Flow of a Viscoelastic Liquid Past a Sphere," *Quart. J. Mech. Appl. Maths.*, 14 (1961) 36–48.
91. Caswell, B., and Schwarz, W. H., "The Creeping Motion of a Non-Newtonian Fluid Past a Sphere," *J. Fluid Mech.*, 13 (1962) 417–426.
92. Clegg, D. B., and Power, G., "The Instantaneous Slow Flow of a Viscoelastic Fluid Between Two Concentric Spheres," *App. Sci. Res.*, 13A (1964) 423–431.
93. Verma, P. D., and Sacheti, N. C., "Low Reynolds Number Flow of a Second Order Fluid Past a Porous Sphere," *J. Appl. Phys.*, 46 (1975) 2065–2069.
94. Giesekus, H., "Die simultane translations und rotations bewegung einer kugel in einer elastovisken flussigkeit," Rheo Acta 3 (1963) 59–71.
95. Hill, C. T., "Viscoelastic Fluid Flow in the Disk and Cylinder System," Ph.D. Dissertation, Uni. Wis., U.S.A. (1969).
96. Ultman, J. S., and Denn, M. M., "Slow Viscoelastic Flow past Submerged Objects," *Chem. Eng. J.*, 2 (1971) 81–89.
97. Sigli, D., and Coutanceau, M., "Effect of Finite Boundaries on the Slow Laminar Isothermal Flow of a Viscoelastic Fluid around a Sphere," *J. Non Newt. Fluid Mech.*, 2 (1977) 1–21.
98. Zana, E., Tiefenbruck, G., and Leal, L. G., "A Note on the Creeping Motion of a Viscoelastic Fluid Past a Sphere," *Rheo. Acta*, 14 (1975) 891–898.
99. Broadbent, J. M., and Mena, B., "Slow Flow of an Elasticoviscous Fluid Past Cylinders and Spheres," *Chem. Eng. J.*, 8 (1974) 11–19.
100. Mena, B., and Caswell, B., "Slow Flow of an Elasticoviscous Fluid Past an Immersed Body," *Chem. Eng. J.*, 8 (1974) 125–134.
101. Crochet, M. J., "The Flow of a Maxwell Fluid Around a Sphere," *Finite Elements in Fluids*, 4 (1982) 573–598.
102. Tiefenbruck, G., and Leal, L. G., "A Numerical Study of the Motion of a Viscoelastic Fluid Past Rigid Spheres and Spherical Bubbles," *J. Non Newt. Fluid Mech.*, 10 (1982) 115–155.
103. Hassager, O., and Bisgaard, C., "A Lagrangian Finite Element Method for the Simulation of Flow of Non-Newtonian Liquids," *J. Non Newt. Fluid Mech.*, 12 (1983) 153–164.
104. Boger, D. V., "Separation of Shear-Thinning and Elastic Effects in Experimental Rheology," *Rheology*, Vol. I, (1980) 195–218.
105. Crochet, M. J., and Walters, K., "Numerical Methods in Non-Newtonian Fluid Mechanics," *Ann. Rev. Fluid Mech.*, 15 (1983) 241–260.
106. Leal, L. G., "The Motion of Small Particles in Non-Newtonian Fluids," *J. Non Newt. Fluid Mech.*, 5 (1979) 33–78.
107. Leal, L. G., "Particle Motions in a Viscous Fluid," *Ann. Rev. Fluid Mech.*, 12 (1980) 435–476.
108. Boger, D. V., "A Highly Elastic Constant Viscosity Fluid," *J. Non Newt. Fluid Mech.*, 3 (1977/78) 87–91.
109. Boger, D. V., and Binnington, R., "Separation of Elastic and Shear Thinning Effects in the Capillary Rheometer," *Trans. Soc. Rheo.*, 21 (1977) 515–534.

110. Boger, D. V., and Hang Nguyen, "A Model Viscoelastic Fluid," *Polym. Eng. Sci.*, 18 (1978) 1037–1043.
111. Cochrane, K., Walters, K., and Webster, M. F., "On Newtonian and Non-Newtonian Flow in Complex Geometries," *Phil. Trans. Roy. Soc.*, 301 (1981) 163–181.
112. Chhabra, R. P., Uhlherr, P. H. T., and Boger, D. V., "The Influence of Fluid Elasticity on the Drag Coefficient for Creeping Flow Around a Sphere," *J. Non Newt. Fluid Mech.*, 6 (1980) 187–199.
113. Giesekus, H., "Die Sekundarstromung in einer Kegel-Platt-Anordung: Abhangigkeit von der Rotationsgeschwindigkeit bei verschiedenen Polymersystemen," Rheo. Acta, 6 (1967) 339–353.
114. Cross, M. M., "Relation Between Viscoelasticity and Shear-Thinning Behavior in Liquids," *Rheo. Acta*, 18 (1979) 609–614.
115. Wagner, M. H., "Prediction of Primary Normal Stress Difference from Shear Viscosity Data Using a Single Integral Constitutive Equation," *Rheo. Acta*, 16 (1977) 43–50.
116. Shroff, R. N., and Shida, M., "Significance of Predicting the First Normal Stress Difference and the Dynamic Moduli from the Steady Flow Viscosity," *Trans. Soc. Rheo.*, 21 (1977) 327–335.
117. Stastna, J., and Dekee, D., "On the Prediction of the Primary Normal Stress Coefficient From Shear Viscosity," *J. Rheo.*, 26 (1982) 565–570.
118. Chhabra, R. P., and Raman, J. R., "Slow Non-Newtonian Flow Past an Assemblage of Rigid Spheres," *Chem. Eng. Commun.*, 27 (1984) 23–46.
119. Choplin, L., Carreau, P. J., and Aitkadi, A., "Highly Elastic, Constant Viscosity Fluids," *Polym. Eng. Sci.*, 23 (1983) 459–464.
120. Cho, Y. I., Hartnett, J. P., and Kwack, E. Y., "A Study of Wall Effect for Viscoelastic Fluids in the Falling Ball Viscometer," *Chem. Eng. Commun.*, 6 (1980) 141–149.
121. Bisgaard, C., "Velocity Fields Around Spheres and Bubbles Investigated by Laser Doppler Anemometry," *J. Non-Newt. Fluid Mech.*, 12 (1983) 283–302.

CHAPTER 31

POWER-LAW FLUID VELOCITY PROFILES IN TURBULENT PIPE FLOWS

A. V. Shenoy

Chemical Engineering Division
National Chemical Laboratory, India

CONTENTS

INTRODUCTION

Turbulent flow in non-Newtonian fluids has received much less attention than Newtonian fluids because it was believed that conditions of turbulence would theoretically never be reached in the case of non-Newtonian fluids due to their high consistencies. This was, however, later realized to be false and it is now generally accepted that turbulent flow of non-Newtonian fluids is not uncommon. The complexities of the rheological characteristics of the non-Newtonian fluids coupled with the chaotic random motions of the fluid particles when under turbulent conditions make the fluid mechanics rather difficult to handle theoretically. Hence, the approach to understanding turbulence in non-Newtonian fluids has been largely restricted to empirical correlations based on accumulated results of experimental observations of the time-averaged velocity profile and the pressure gradient, both of which are highly sensitive parameters.

This chapter deals with the velocity profiles in turbulent pipe flows of inelastic non-Newtonian fluids. It is restricted to only one type of fluid model to describe the rheological characteristics of the fluid, namely, the power-law model, as it is the simplest in form and yet describes the nature of several non-Newtonian fluids of industrial importance. In the case of turbulent flow of inelastic non-Newtonian fluids, this is the only model that has been used mainly because the added complexities of the other models lead to highly intricate mathematics. This chapter is limited to the fluid type but not to the flow situation, and considers turbulent flows through different geometrical pipe cross-sections, through annular ducts, and even curved tubes. Power-law fluid velocity profiles are considered in all cases for smooth internal surfaces. Only in the case of the straight tubes are the velocity profile changes due to the effect of roughness of the pipe considered. In all cases, the fully developed velocity profiles are studied but an approach to estimate the hydrodynamic entrance lengths in the case of power-law fluids is given.

FLUID MODEL

It is well known that liquids with complex structure, such as macromolecular fluids, soap solutions, and two-phase fluid systems, do not follow the usual simple Newtonian relationship between the shear stress and shear rate. In fact, the proportionality parameter relating shear stress and shear rate is not a constant for complex fluids but varies with shear-rate. The shear-rate dependency of this parameter, which is nothing but the viscosity of the system, is a very important property of the fluid to be considered during flow situations. Depending on whether the viscosity increases or decreases with shear rate, we have a dilatant (shear-thickening) fluid or pseudoplastic (shear-thinning) fluid. A typical rheogram for a shear-thinning fluid is shown in Figure 1. Note that the constant viscosity behavior of a Newtonian fluid is only shown for comparison. Most non-Newtonian fluids behave as Newtonian fluids in the two limiting cases of very low shear rates and very high shear rates. The Newtonian viscosities in these two limits, namely, η_0 and η_∞, are of course quite different in value and often by orders of magnitude. The very low shear rate limiting viscosity would be of importance during laminar flow conditions; however, it is quite unlikely that the very high shear rate viscosity η_∞ would be reached even under normal turbulent conditions. The dashed line in Figure 1 shows the portion of the rheogram where the viscosity varies linearly with shear rate on the log-log plot. This tilted straight line can be described by a power-law expression of the following

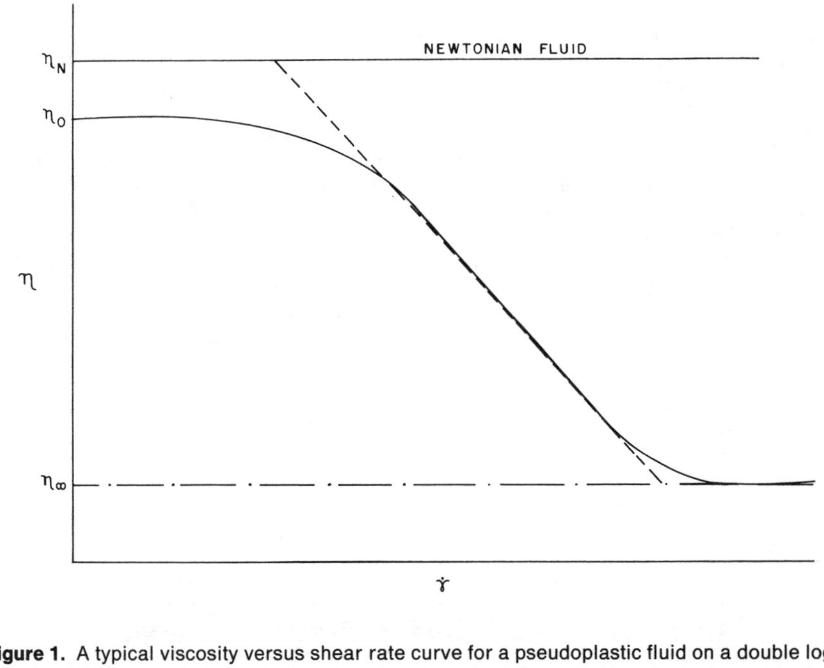

Figure 1. A typical viscosity versus shear rate curve for a pseudoplastic fluid on a double logarithmic plot. The constant Newtonian viscosity of a typical simple fluid is shown for comparison.

form:

$$\eta = K\dot{\gamma}^{n-1} \tag{1}$$

In Equation 1, K is often termed as the consistency index and n the power-law index. It is this linear portion of the viscosity versus shear rate curve described by Equation 1 that has assumed the most importance in the case of non-Newtonian fluid flow under turbulent conditions. The relationship between shear stress and shear rate for such fluids, whose rheogram is represented by Equation 1 in the linear portion is given by

$$\tau = K\dot{\gamma}^{n} \tag{2}$$

Equation 2 is conventionally known as the Ostwald-de Waele or power-law representation of the non-Newtonian fluids. Many polymer solutions or melts, biological fluids, suspensions, etc. can be described by this model. There are many useful models for describing the non-Newtonian characteristics of inelastic fluids [1], but the Ostwald-de Waele model is the most well known and widely used empiricism in many engineering problems. The flow model works excellently in all steady-state situations of pragmatic importance and describes the flow behavior of Newtonian ($n = 1$), shear-thinning ($n < 1$) and shear-thickening ($n > 1$) fluids. Values of n between 0.3 and 0.6 are quite common among most non-Newtonian fluids showing pseudoplastic behavior. It is possible to generate turbulence in pseudoplastic fluids but not in dilatant fluids because the increasing viscosity with shear rate in the latter case would dampen the turbulent motion.

VELOCITY PROFILES

The general methods for the development of expressions of power-law fluid velocity profiles and friction factors are on the same lines as those used in the case of Newtonian fluids, with of course the introduction of an additional degree of freedom by way of the flow behavior index n. Thus, in line with the traditional concepts that have been successfully used for Newtonian fluids, the turbulent flow of non-Newtonian fluids in smooth pipes, too, can be artificially divided into three zones as shown in Figure 2.

1. A laminar sublayer lying next to the pipe wall in which the effects of turbulence are negligible and whose thickness extending from the wall ($y = 0$) to $y = y_L$ is so thin that the shear stress across the layer can be assumed to be constant and equal to the wall shear stress.
2. A turbulent core comprising of the bulk of the fluid stream wherein the momentum transfer that accompanies the random velocity fluctuations characteristic of turbulent motion determines the velocity profile and the effects of viscosity are negligible.
3. A transition zone, between the laminar sublayer and the turbulent zone, in which the effects of turbulence and viscous shear are of comparable orders of magnitude.

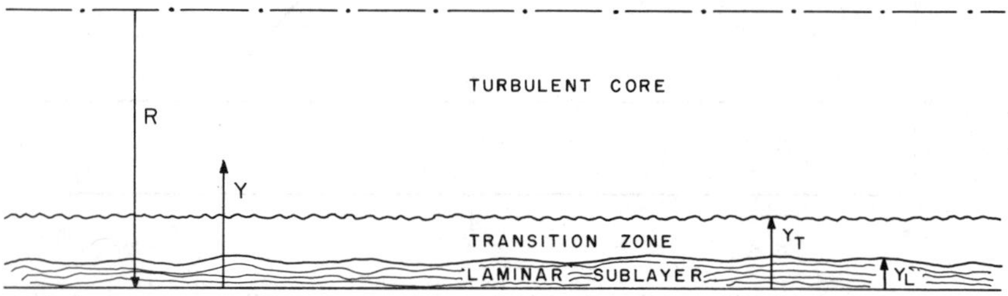

Figure 2. Schematic diagram of the highly idealized three-layer model for turbulent flow.

The thickness of the respective zones can be represented as follows:

Laminar sublayer: $0 < y < y_L$
Transition zone: $y_L < y < y_T$
Turbulent core: $y_T < y < R$

Smooth Straight Circular Pipes

For power-fluid turbulent flow in smooth straight circular pipes, the time-averaged velocity at a point would depend on six independent variables, namely, R, ρ, τ_w, K, n, y. Using conventional dimensional analysis as given by Langhaar [2], it can be shown that

$$\frac{u}{u^*} = \phi(R^+, y^*, n) \tag{3}$$

where u* is the friction or shear velocity defined as $\sqrt{\tau_w/\rho}$, R^+ is the Reynolds number based on friction velocity defined as $R^n u^{*2-n}\rho/K$, and y* is the dimensionless location parameter defined as y/R.

The theoretical development of the power-law fluid velocity profiles in turbulent flow through smooth straight circular pipes has been done by a number of investigators [3–6] and the typical nature of the predicted velocity profile is shown in Figure 3. The region of the transition is not well defined for power-law fluids. For Newtonian fluids, it has been established that the transition zone is given by $5 < y^+ < 30$, whereas the same is not the case for power-law fluids, and hence the point of change from the dashed line to the solid line in Figure 3 ought not to be taken seriously.

In each of the theoretical developments [3–6] there is unanimity in the expression for the velocity profile in the laminar sublayer. This is based on the highly idealized reasoning that there exists a viscous sublayer of thickness y_L in all turbulent flow situation where the velocity profile is linear and where the shear stress is constant and equal to the wall shear stress. For power-law fluids, the expression for the velocity profile in the viscous sublayer can be easily derived using Equation 2. Thus, for a velocity u at a distance y from the wall in the sublayer, we have

$$\tau_w = K\left(\frac{u}{y}\right)^n \tag{4}$$

Using the definition of friction velocity $u^* = \sqrt{\tau_w/\rho}$ as noted earlier, the expression for the velocity profile in the viscous sublayer can be written as

$$\frac{u}{u^*} = \left(\frac{\rho u^{*2-n} y^n}{K}\right)^{1/n} \tag{5}$$

or

$$u^+ = y^{+1/n} \quad \text{for} \quad y \leq y_L \tag{6}$$

where u^+ is a dimensionless velocity and y^+ is distance based Reynolds number written in terms of the friction velocity, distance from the wall and fluid properties.

Dodge and Metzner [3] used a semi-theoretical analysis to derive the friction factor-Reynolds number relationship and the velocity profile in the turbulent core of a flowing purely viscous non-Newtonian fluid of the power-law type. Their expression for the velocity profile in the turbulent core is as follows:

$$u^+ = \frac{5.66}{n^{0.75}} \log y^+ \frac{-0.40}{n^{1.2}} + \frac{2.458}{n^{0.75}}\left[1.960 + 1.255n - 1.628n \log\left(3 + \frac{1}{n}\right)\right] \tag{7}$$

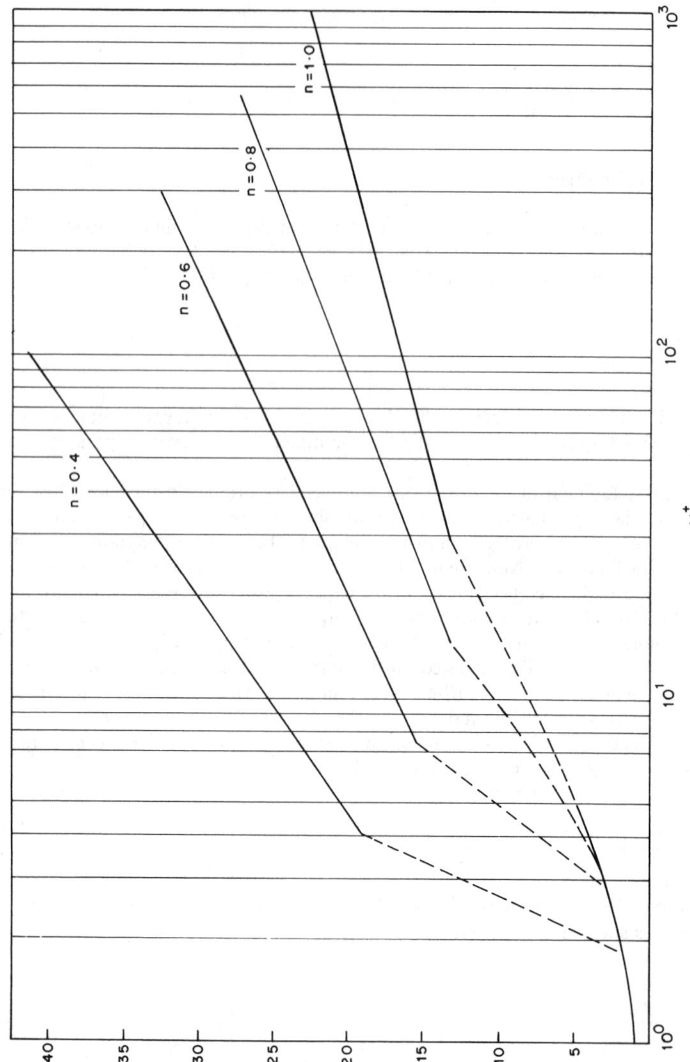

Figure 3. The typical shape of the velocity profile in the three-layer model for turbulent flow of power-law fluids.

An improvement in this expression was suggested by Bogue and Metzner [4], and a new expression of the following form was suggested for the velocity profile in the turbulent core:

$$u^+ = 5.57 \log y^{+1/n} + C(y^*, f) + I(n, Re_{gen}) \tag{8}$$

where C(y*, f) is a correction function defined as

$$C(y^*, f) = 0.05\sqrt{\frac{2}{f}} \exp \frac{-(y^* - 0.8)^2}{0.15} \tag{9}$$

with the expression for friction factor for power-law fluids given as

$$\frac{1}{\sqrt{f}} = \frac{4.0}{n^{0.75}} \log Re_{gen}\, f^{(2-n)/2} - \frac{0.4}{n^{1.2}} \tag{10}$$

The values of the function $I(n, Re_{gen})$ are tabulated in Table 1 as taken from Bogue and Metzner [4].

The expressions proposed by Dodge and Metzner [3] and Bogue and Metzner [4] presume a two-layer rather than the three-layer model. They have not explicitly provided an expression for the buffer layer (transition zone). Clapp [5], however, has followed the three-layer model and provided expressions for the velocity profile for power-law fluids in turbulent flow by using the Prandtl and von Karman approach for Newtonian fluids. The three expressions for the universal velocity profile as obtained by Clapp [5] are:

Laminar sublayer—

$$u^+ = y^{+1/n} \qquad (0 \le y^+ \le 5^n) \tag{11}$$

Buffer layer—

$$u^+ = 5.0 \ln y^{+1/n} - 3.05 \qquad (5^n \le y+ \le y_2^+) \tag{12}$$

Turbulent core—

$$u^+ = 2.78 \ln y^{+1/n} + \frac{3.8}{n} \qquad (y^+ > y_2^+) \tag{13}$$

The logarithmic expressions for the velocity profile in the turbulent core as given by Equations 7, 8, and 13 have an advantage to predict the velocity distributions beyond the range of experimentation due to their asymptotic nature. But each of the three expressions for the turbulent core have an incongruity implicitly hidden within them, namely that they fail to predict a zero velocity

Table 1
Values of the Function $I(n, Re_{gen})$ [4]

	Re_{gen}			
n	**5,000**	**10,000**	**50,000**	**100,000**
1.0	5.57	5.57	5.57	5.57
0.8	6.01	5.92	5.69	5.58
0.6	6.78	5.51	5.89	5.60
0.4	8.39	7.70	6.27	5.60

gradient at the center. However, the new velocity profile model as suggested by Shenoy and Saini [6] is devoid of this limitation and is, therefore, the correct theoretical velocity profile expression for turbulent core in the case of power-law fluids.

Shenoy and Saini [6] assumed a similar form of expression for the velocity profile as that of Bogue and Metzner [4], such that

$$u^+ = A(n)[\ln y^{+1/n} + C(y^*, n)] + B(n) \tag{14}$$

The correction function $C(y^*, n)$, however, was assumed to be of a different form unlike the one used earlier [4] as given by Equation 9. The following form was chosen:

$$C(y^*, n) = \sigma_1(n) \exp\left\{-\frac{1}{2}\left[\frac{y^* - 0.8}{\sigma_2(n)}\right]^2\right\} \tag{15}$$

This form for the correction function had two distinct advantages. Firstly, they could determine the expressions for $\sigma_1(n)$ and $\sigma_2(n)$ with a precondition such that it would allow the velocity gradient to go to zero at the centerline and secondly, because Equation 15 is independent of friction factor f, the correction function dependence is explicit on n as against the implicit dependence in earlier case of Equation 9.

Shenoy and Saini [6] derived the following expressions for $\sigma_1(n)$ and $\sigma_2(n)$:

$$\sigma_1(n) = 0.1944 - \frac{0.1313}{n} + \frac{0.3876}{n^2} - \frac{0.0109}{n^3} \tag{16}$$

$$\sigma_2(n) = \frac{0.254}{n} \tag{17}$$

From Equation 14, an implicit expression for friction factor can be easily derived and then compared with expression for friction factor of Dodge and Metzner [2] for power-law fluids as given by Equation 10 so that A(n) and B(n) can be determined. The final expression of Shenoy and Saini [6] for the velocity profile in the turbulent core is given as follows:

$$u^+ = 2.46n^{0.25}\left\{\ln y^{+1/n} + 0.1944 - \frac{0.1313}{n} + \frac{0.3876}{n^2} - \frac{0.0109}{n^3} \right.$$
$$\left. \exp\left[-\frac{n^2(y^* - 0.8)^2}{0.129}\right] + \frac{1.3676}{n} + \ln 2^{(2+n)/2n}\right\} - \frac{0.4\sqrt{2}}{n^{1-2}} \tag{18}$$

This velocity profile model was found to agree with the predictions of Bogue and Metzner [4] within a maximum deviation of $\pm 1.25\%$. It presently would thus stand as the most correct velocity profile for power-law fluids in the turbulent core based on its predictions and also its ability to attain a zero velocity gradient at the centerline for all values of n, which has been the main drawback in all earlier works.

The value of y^* in the Equation 18 may be obtained by readjusting the terms in the definition of y^+. Thus

$$y^* = \left(\frac{y^+ 2^{(2+n)/2}}{Re_{gen} f^{(2-n)/2}}\right)^{1/n} \tag{19}$$

It is not always easy to get quick estimates of f required in the above expression without the use of an iterative procedure as can be seen from Equation 10. The following expression was, however, found to give values of f by direct substitution and simple calculations within an error bound of

±2.4% for the entire range of n values between 1 and 0.3. Thus, the following equation may be used in preference to Equation 10, when friction factor values are to be estimated for power-law fluids for a range of generalized Reynolds number values between 4×10^3 and 10^6:

$$\frac{1}{\sqrt{f}} = 3.57 \log \frac{Re_{gen}^{n^{1/0.615}}}{6.5^{n^{1/(1+0.75n)}}} \tag{20}$$

where

$$Re_{gen} = \frac{D^n V^{2-n} \rho}{\gamma} \tag{21}$$

$$\gamma = K\left(\frac{1+3n}{4n}\right)^n 8^{n-1} \tag{22}$$

Though Equation 18 gives the correct description of the velocity profile for turbulent flow of power-law fluids, there is always a tendency to look for simpler expressions that could form a good engineering approximation to the exact one at least in a limited region. The one-seventh power-law expression for the velocity profile in the case of Newtonian fluids is an example of such an approach. The extension of the same for power-law fluids can be done very easily [7].

Dodge and Metzner [3] have provided the following Blasius type of approximate equation for friction factor $f = \phi(Re_{gen}, n)$ for turbulent flow of power-law fluids.

$$f = \frac{\alpha}{Re_{gen}^{\beta}} \qquad (5 \times 10^3 \le Re_{gen} \le 10^5) \tag{23}$$

The α and β values for varying n have been reported by them [3]. Noting that $\tau_w = (f/2)\rho V^2$, the following expression can be written using Equation 23:

$$\tau_w = \frac{\alpha}{2^{1+\beta n}} \rho V^{2-\beta(2-n)} \left(\frac{\gamma}{\rho}\right)^{\beta} R^{-\beta n} \tag{24}$$

By definition, $u^* = (\tau_w/\rho)^{1/2}$, and from Skelland [7] for turbulent flow of power-law fluids, V can be expressed in terms of the maximum velocity u_m:

$$V = \psi u_m \tag{25}$$

where

$$\psi = \frac{[2-\beta(2-n)][2-\beta(2-n)]}{[1-\beta(1-n)][4-\beta(4-3n)]} \tag{26}$$

Equation 24 can thus be transformed to give

$$\frac{u_m}{u^*} = \left[\frac{2^{1+\beta n}}{\alpha \psi^{2-\beta(2-n)}}\right]^{1/[2-\beta(2-n)]} \left[\frac{R^n u_m^{2-n} \rho}{\gamma}\right]^{\beta/[2-\beta(2-n)]} \tag{27}$$

The key assumption in order to get the simple velocity profile expression lies in an approximation that Equation 27 is valid at any wall distance y rather than only at R. Thus,

$$u^+ = A_1 y^{+B_1} \tag{28}$$

where

$$A_1 = \left[\frac{2^{1+\beta n}}{\alpha\psi^{2-\beta(2-n)}}\right]^{1/[2-\beta(2-n)]}\left[\left(\frac{4n}{1+3n}\right)8^{1-n}\right]^{\beta/[2-\beta(2-n)]} \quad (29)$$

$$B_1 = \frac{\beta}{[2-\beta(2-n)]} \quad (30)$$

Equation 28 has several limitations. Its range of validity is limited to Reynolds number between 5×10^3 and 10^5. Further, it cannot predict a zero velocity gradient at the centerline; and due to the simplifying assumptions made in its derivation, it does not give an accurate description of the velocity profile. However, with all its shortcomings, it still would be a useful expression to be used in engineering design such as for calculating the turbulent entrance lengths for power-law fluids as will be shown later.

Arbitrary Cross-Sectional Pipes

Non-Newtonian flow through non-circular pipes has been studied in considerable detail by Kozicki et al. [8] and Salem and Embaby [9]. The work of Kozicki et al. [8] deals with laminar flow of incompressible time-independent non-Newtonian fluids in ducts of arbitrary cross-section, while Salem and Embaby [9] have dealt with laminar as well as turbulent flow conditions. However, there is no expression for turbulent velocity profile derived by them for power-law fluids. Salem and Embaby [9], nevertheless, have given an expression for the friction factor in non-circular pipes based on the equivalent pipe diameter as was used by Kozicki et al. [8] in their studies. Their suggested expression is as follows:

$$\frac{1}{\sqrt{f}} = \frac{4}{n^{0.75}}\log_{10}[Re^*_{gen}\, f^{(2-n)/2}] - \frac{0.4}{n^{1.2}} + 4.0n^{0.25}\log_{10}\left[\frac{4(a+bn)}{1+3n}\right] \quad (31)$$

where Re^*_{gen} is the generalized Reynolds number for power-law fluids in non-circular pipes based on the hydraulic radius r_H and defined as follows:

$$Re^*_{gen} = \frac{r_H^n V^{2-n}\rho}{2^{2n-3}\left[\dfrac{a+bn}{n}\right]^n K} \quad (32)$$

The form of Equation 31 is the same as Equation 10 for friction factor in smooth straight pipes, except for an extra term that incorporates the geometrical correction factor due to the non-circular nature of the pipe. A similar approach can be adopted to obtain an expression for the velocity profile in non-circular pipes of arbitrary cross-section. Equation 18 can be modified to include the geometrical parameters for the cross-sectional shape and thus the following equation can be written:

$$u^+ = 2.46n^{0.25}\left\{\ln y^{+1/n} + \left(0.1944 - \frac{0.1313}{n} + \frac{0.3876}{n^2} - \frac{0.0109}{n^3}\right)\exp\left[\frac{-n^2(y^*-0.8)^2}{0.129}\right] + \frac{1.3676}{n} + \ln 2^{(2+n)/2n} + \ln\frac{4(a+bn)}{1+3n}\right\} - \frac{0.4\sqrt{2}}{n^{1.2}} \quad (33)$$

The definitions of u^+, y^+, y^* are the same as those in the smooth straight pipe case except for the difference in definition of the Reynolds number term (see Equation 32). The geometric parameters a and b for a wide variety of cross-sectional shapes of ducts have been determined by Kozicki et al. [8] and their expressions as well as the values for certain specific cases have been well tabulated.

Equations 18 and 33 are identical but for the following term:

$$T = 2.46n^{0.25} \ln \frac{4(a + bn)}{1 + 3n} \tag{34}$$

This term controls the change in the shape of the velocity profile depending on the geometry of the conduit. Hence, it is worth looking at the variations of the term T for different geometrical shapes and different values of n. The values of T for a wide variety of specific cases are evaluated and tabulated in Table 2. The cases include various types of cross-sectional geometries such as elliptical, rectangular, and isosceles triangular ducts for typical values of a and b taken from Kozicki et al. [8]. The parametric space covered, thus, includes a number of power-law fluid types flowing through ducts of different shapes and sizes. It should be noted that the term T would be identically equal to zero for all values of power-law index for a circular pipe. It can be seen from Table 2 that T takes values between −0.55 and +0.33 in all cases and hence does not form an important term in Equation 33. Thus, the velocity profile in effect could be predicted by Equation 18 given earlier for smooth straight circular pipes. This, however, does not imply that the velocity profiles are the same for all cross-sectional shapes. It must be noted that the dimensionless parameter now includes a modified Reynolds number based on the hydraulic radius factor, which would be different for various geometrical shapes. Thus, though the dimensionless velocity profiles are predicted as identical by Equation 33, the actual variation in the velocity shape does exist. There is presently no experimental work in the literature to substantiate these findings.

Annular Pipes

Flow of non-Newtonian fluids in annuli has been comprehensively studied by Tan and Tiu [10–13] and Bhattacharyya and Tiu [14 and 15]. However, they deal with only the laminar case and as far as turbulent flow of non-Newtonian fluids is concerned, there is a complete lack of information in the literature. Nevertheless, the fully developed turbulent velocity profile for flow of an inelastic non-Newtonian fluid can be easily derived based on the simple power-law type expression given by Equation 28 for smooth circular pipe.

In flow through an annulus, the flow model ought to be divided into two regions, namely, the outer region and the inner region as a consequence of the two wall boundaries. Each region would therefore have a separate velocity profile. It is now assumed that the effect of the inner cylinder on the velocity profile in the outer region is small enough to be neglected. It can thus be treated in a manner analogous to a smooth straight pipe so that Equation 28 would hold reasonably well. Thus,

$$u_o^+ = A_1 y_o^{+B_1} \tag{35}$$

where the subscript o represents the outer region of the annulus and A_1 and B_1 have the same forms as given by Equations 29 and 30, respectively.

It is now assumed that an expression of the similar form as Equation 35 would hold for the inner region as

$$u_i^+ = \bar{A}_1 y_i^{+B_1} \tag{36}$$

where the subscript i represents the inner region of the annulus, and $\bar{A}_1$ and $\bar{B}_1$ are to be determined for this region. In order to obtain expressions for $\bar{A}_1$ and $\bar{B}_1$, we would have to go through complicated mathematical procedures leading to expressions that would be too cumbersome for

Table 2
Values of T Given by Equation 34, for Different Non-Circular Pipe Shapes at Typical Values of n

Pipe shape	a	b	n	T
Elliptical $\frac{b'}{a'} = 0.3$	0.2796	0.8389	1.0	0.2645
			0.8	0.2507
			0.6	0.2341
			0.4	0.2127
$\frac{b'}{a'} = 0.5$	0.2629	0.7886	1.0	0.1235
			0.8	0.1168
			0.6	0.1088
			0.4	0.0983
$\frac{b'}{a'} = 0.7$	0.2538	0.7614	1.0	0.0371
			0.8	0.0351
			0.6	0.0327
			0.4	0.0295
Rectangular $\frac{b'}{a'} = 0.25$	0.3212	0.8182	1.0	0.3209
			0.8	0.3211
			0.6	0.3217
			0.4	0.3222
$\frac{b'}{a'} = 0.50$	0.2440	0.7276	1.0	−0.0709
			0.8	−0.0664
			0.6	−0.0610
			0.4	−0.0540
$\frac{b'}{a'} = 0.75$	0.2178	0.6866	1.0	−0.2472
			0.8	−0.2388
			0.6	−0.2289
			0.4	−0.2163
Isoceles Triangular Duct $2\alpha' = 20°$	0.1693	0.6332	1.0	−0.5413
			0.8	−0.5334
			0.6	−0.5252
			0.4	−0.5156
$2\alpha' = 40°$	0.1840	0.6422	1.0	−0.4697
			0.8	−0.4592
			0.6	−0.4474
			0.4	−0.4326
$2\alpha' = 60°$	0.1875	0.6462	1.0	−0.4474
			0.8	−0.4369
			0.6	−0.4251
			0.4	−0.4101

engineering applications. The situation, however, becomes greatly simplified if it is assumed that the slopes of the u_o^+ versus y_o^+ and the u_i^+ versus y_i^+ are the same, thus giving $\bar{B}_1 = B_1$.

The momentum balance for the inner and outer flow region of the annuli gives the following expressions for wall shear stresses τ_{wi} and τ_{wo}:

$$\tau_{w_i} = \frac{r_i}{2}\left[\frac{\lambda^2}{k^2} - 1\right]\frac{\Delta P}{L} \tag{37}$$

$$\tau_{w_o} = \frac{r_o}{2}[1 - \lambda^2]\frac{\Delta P}{L} \tag{38}$$

where

$$\lambda = \frac{r_m}{r_o}, \qquad k = \frac{r_i}{r_o} \tag{39}$$

In these equations, r_i, r_o, and r_m represent the radius of the inner cylinder, radius of the outer pipe and radius of maximum velocity, respectively.

Using the definition of $u^* = \sqrt{\tau_w/\rho}$ for the inner and outer region, the following expression can be easily derived. Thus,

$$\frac{u_o^*}{u_i^*} = \left[\frac{k(1 - \lambda^2)}{(\lambda^2 - k^2)}\right]^{1/2} \tag{40}$$

Using the fact that $u_i = u_o$ at $r = r_m$, the expression for $\bar{A}_1$ can be obtained as

$$\bar{A}_1 = A_1\left[\frac{k(1 - \lambda^2)}{(\lambda^2 - k^2)}\right]^{B'}\left[\frac{1 - \lambda}{\lambda - k}\right]^{B_1 n} \tag{41}$$

where

$$B' = \frac{(2 - n)B_1 + 1}{2} \tag{42}$$

Thus, the velocity profile for the inner region of the annulus can be written in the following form:

$$u_i^+ = A_1\left[\frac{k(1 - \lambda^2)}{(\lambda^2 - k^2)}\right]^{B'}\left[\frac{1 - \lambda}{\lambda - k}\right]^{B_1 n} y_i^{+B_1} \tag{43}$$

Equations 35 and 43 thus define the complete velocity profile for power-law fluids in an annulus. There is no experimental data in the literature to support these theoretical predictions but the expressions just given would provide the guidelines for the expected trend.

Curved Pipes

The study of transport processes in curved pipes is of considerable pragmatic importance because of the significant advantages that such a configuration offers when used for heat/mass transfer purposes or as chemical reactors. From a fluid mechanics point of view, curved pipes represent an interesting flow situation due to the effects of curvature. A secondary flow is set up by virtue of the centrifugal force that gets superimposed on the axial velocity flow field. The central fluid is actually driven towards the outer wall and is then pushed back along the wall towards the inner

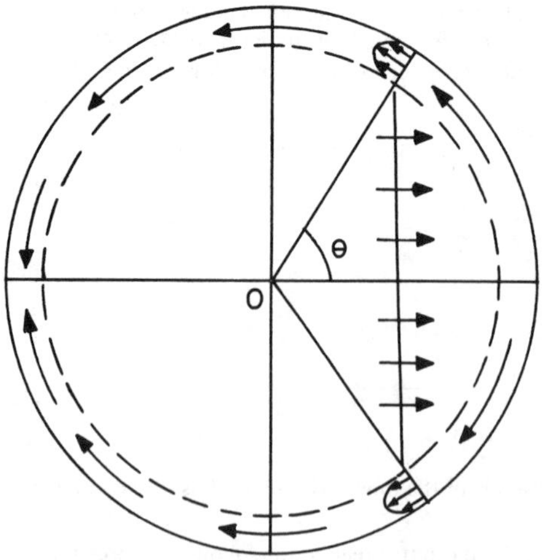

Figure 4. Schematic representation of the flow model for curved pipes showing the boundary layer, the inviscous core and the continuity of secondary flow. (From Mashelkar and Devarajan [16].)

side. A double vertical motion is thus set up and a flow situation of the kind shown schematically in Figure 4 results.

The only work on turbulent flow of power-law fluids in curved pipes available in the literature is that of Mashelkar and Devarajan [16]. Their theoretical approach considers the flow to be representable by a two region model—a central inviscous core and a thin boundary layer near the wall as shown in Figure 4. The momentum integral equations for the turbulent boundary layer flow were solved numerically by them to obtain the velocity distribution in the curved pipe. Mashelkar and Devarajan [16] assumed the following forms for the angular and axial velocity distributions in the boundary layer:

$$V = D_1\left(\frac{\xi}{\delta}\right)^{\beta n/[2-\beta(2-n)]}\left(1-\frac{\xi}{\delta}\right) \tag{44}$$

$$w = w_1\left(\frac{\xi}{\delta}\right)^{\beta n[2-\beta(2-n)]} \tag{45}$$

where $\zeta = \bar{a} - r$ with $\bar{a}$ as the radius of the pipe and r the radial position. δ is the boundary layer thickness.

Using the following expressions:

$$\delta = \delta_c \bar{a}\{[\mathrm{Re}(\bar{a}/\bar{R})^{1/2\beta}\}^{-\beta/(\beta n+1)} \tag{46}$$

$$D_1 = D_c(v_m\sqrt{\bar{a}/\bar{R}}) \tag{47}$$

$$w_1 = w_c(v_m) \tag{48}$$

the differential equations are non-dimensionalized and the values of δ_c, D_c, and w_c are obtained by expanding in the neighborhood of $\theta = o$ using

$$\delta_c = \delta_1(1 + \delta_2\theta^2 + \cdots) \tag{49}$$

$$D_c = D_2\theta(1 + D_3\theta^3 + \cdots) \tag{50}$$

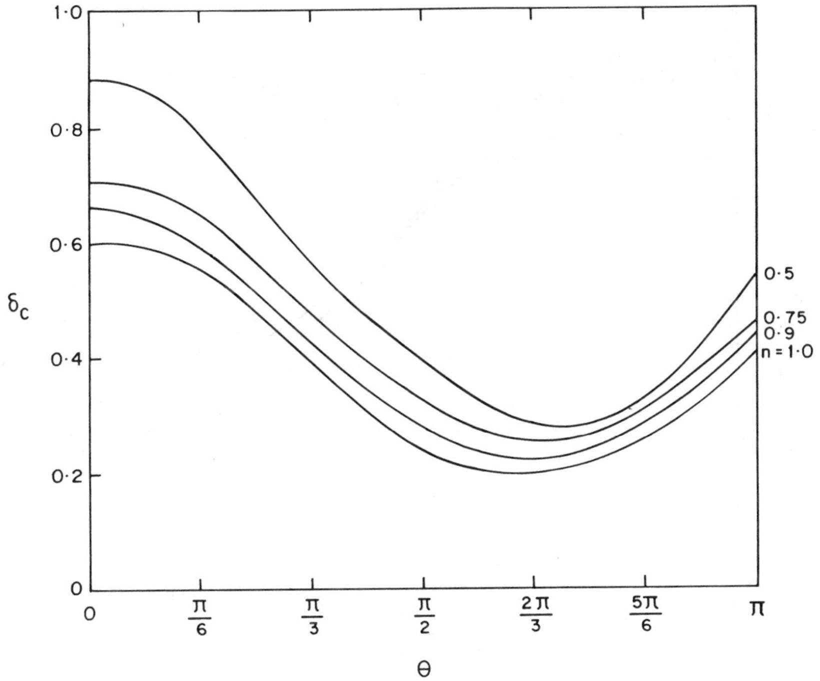

Figure 5. Variation of dimensionless turbulent boundary layer δ_0 with θ for varying pseudoplasticity index n. (From Mashelkar and Devarajan [16].)

and

$$w_c = 1 + \frac{\bar{B}\bar{a}}{V_m}\left(1 - \frac{\theta^2}{2!} + \frac{\theta^4}{4!} - \cdots\right) \tag{51}$$

In these equations $\bar{R}$ is the radius of curvature of the pipe, v_m is the mean axial velocity, and θ is the angular position in cylindrical co-ordinates. Using a Runge-Kutta-Merson technique, the integration of the differential equations yields the variation of the non-dimensionalized turbulent boundary layer thickness, δ_0, the non-dimensionalized axial velocity w_c at the outer edge of the boundary layer and the non-dimensionalized characteristic angular velocity D_c with θ. Figures 5–7 show the functions $\delta_c(\theta)$, $w_c(\theta)$, and $D_c(\theta)$ for power-law fluids with different pseudoplasticity indices. It can be seen that the dimensionless angular velocity decreases with increasing pseudoplasticity over the entire range of θ whereas the dimensionless axial velocity decreases with increasing n for $0 < \theta < \pi/2$ but increases with increasing n for $\pi/2 < \theta < \pi$ indicating the change in the steepness of the velocity profile for increasing pseudoplasticity. The information available from these figures along with Equations 44–48 describes the velocity profiles in the boundary layer during the turbulent flow of power-law fluids through curved pipes. The velocity distribution in the central core is given by the following expression through a straightforward derivation from the equation of motion:

$$w = v_m\left[1 + \left(\frac{\bar{B}\bar{a}}{v_m}\right)\frac{r}{\bar{a}}\cos\theta\right] \tag{52}$$

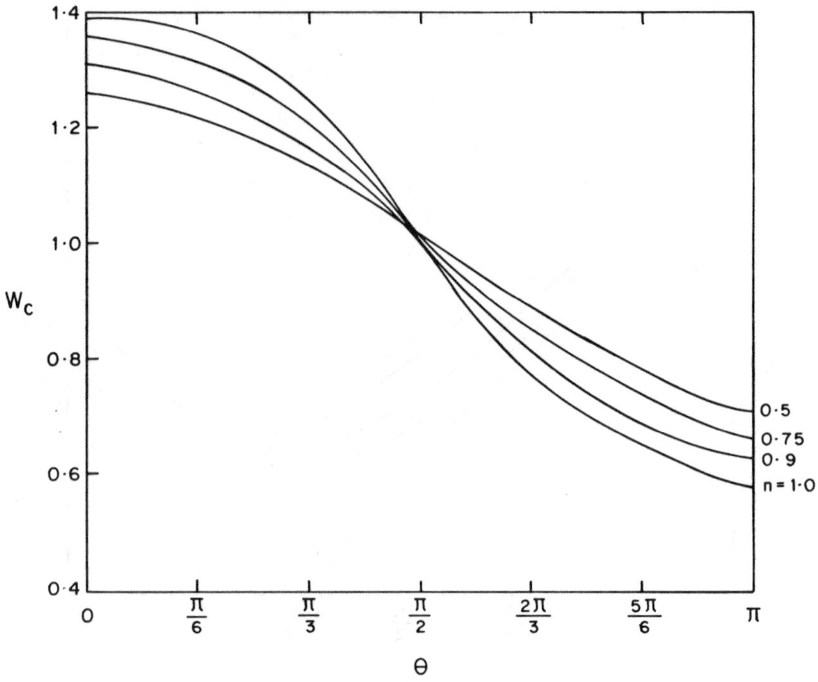

Figure 6. Variation of dimensionless axial velocity at the outer edge of the boundary layer w_0 with θ for varying pseudoplasticity index n. (From Mashelkar and Devarajan [16].)

The values of $\bar{B}\bar{a}/v_m$ for varying pseudoplasticity index are given in Table 3. The entire velocity distribution for turbulent flow of power-law fluids through curved pipes is thus made available. Mashelkar andd Devarajan [16] have provided experimental data on only friction factor and have not made any velocity profile measurements for verification of their theoretical predictions.

Rough Pipes

The developments of all the velocity profiles in the foregoing sections were based on the assumption that the inside wall of the pipe surface is smooth and free from defects. This, of course, is an idealized situation as in practice all pipes have a certain degree of roughness even when they are new. Those that have been in service for a while, invariably develop some surface defects. In order to give a complete description of the roughness, one would have to give detailed dimensions of the protrusions or indentations, which is impractical. However, it is understandable that the actual dimensions of the roughness should not have as great an effect as the relative roughness due to the size of the protrusions or indentations in comparison to the dimensions of the pipe. Figure 8 shows a schematic diagram of the roughness of a surface. The average height of the roughness projections is expressed as ϵ and thus the relative roughness of the tube is written by the factor ϵ/D. This would mean that for smooth tubes, the relative roughness factor is zero.

For laminar flow, there is no influence of the relative roughness on the velocity distribution. However, for turbulent flow, the nature of the flow is intimately related to surface roughness. The roughness will be effective as long as the laminar sublayer near the wall is thin compared to the ϵ/D and this would happen at increased Reynolds numbers. Thus, a rough surface does not always

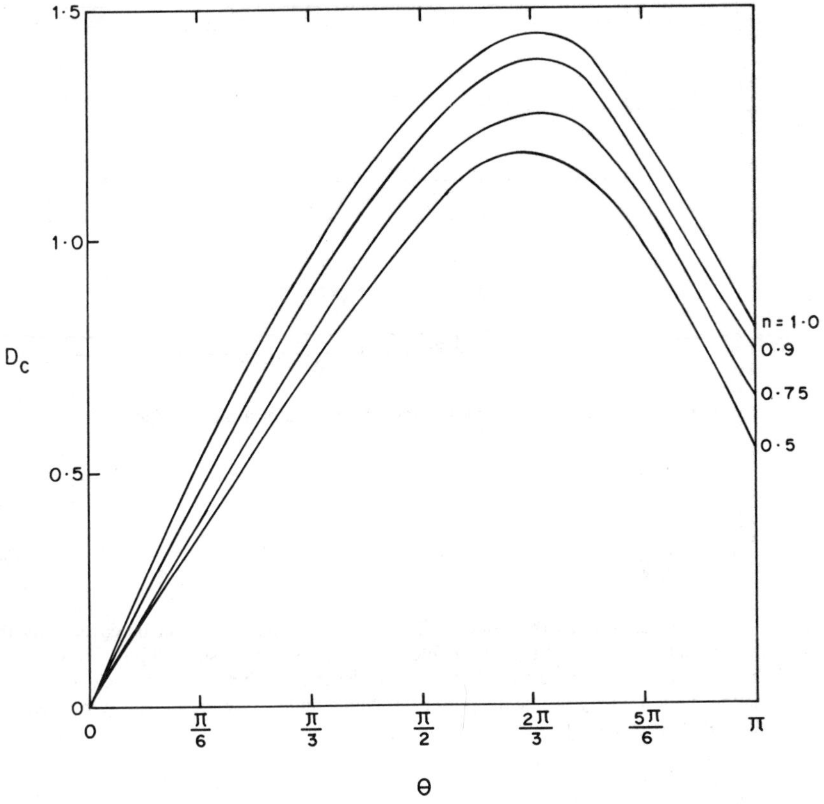

Figure 7. Variation of dimensionless characteristic angular velocity D_0 with θ for varying pseudoplasticity index n. (From Mashelkar and Devarajan [16].)

act rough but depends on the relative magnitude of the size of the surface roughness elements and the thickness of the viscous sublayer. A roughness Reynolds number may be defined for power-law fluids by substituting in the expression for y^+ given in Equation 5 the value of $y = \epsilon$ to give

$$Re_\epsilon = \frac{\rho u^{*2-n}\epsilon^n}{K} \tag{53}$$

Table 3
Values of $\bar{B}\bar{a}/V_m$ for Varying Pseudoplasticity Index [16]

n	$\frac{\bar{B}\bar{a}}{V_m}$
1.0	0.37803
0.9	0.36988
0.75	0.31824
0.5	0.26894

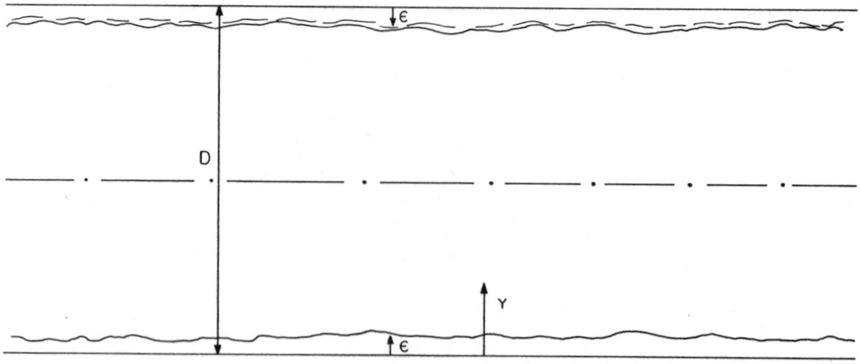

Figure 8. Schematic diagram of the roughness of a pipe surface.

or

$$\mathrm{Re}_\epsilon = \left(\frac{\epsilon}{D}\right)^n \mathrm{Re}_{gen}\left(\frac{f}{2}\right)^{(2-n)/2} \tag{54}$$

For power-law fluids, as long as the value of $\mathrm{Re}_\epsilon < 5^n$ the roughness would be covered by the laminar sublayer and then the resulting flow would take place as if in the smooth pipe. However, for $\mathrm{Re}_\epsilon > 5^n$ the roughness of the pipe would affect both the friction factor and the velocity profile. The expression for the velocity profile for power-law fluids under fully rough wall turbulence has been provided by Torrance [17] as

$$u^+ = 2.5 \ln\left(\frac{y}{\epsilon}\right)^{1/n} + 8.5 \tag{55}$$

Equation 55 again suffers from the limitation of not being able to predict the velocity gradient as zero at the centerline. A new expression similar to Equation 18 could certainly be derived by making some reasonable assumptions.

A modified form of Equation 18 can be written as

$$u^+ = 2.46n^{0.25}\left\{\ln\left(\frac{y}{\epsilon}\right) + \ln\left(\frac{\rho u^{*2-n}\epsilon^n}{K}\right)^{1/n} + \left(0.1944 - \frac{0.1313}{n} + \frac{0.3876}{n^2} - \frac{0.0109}{n^3}\right) \times \exp\left[-\frac{n^2(y^* - 0.8)}{0.129}\right] + \frac{1.3676}{n} + \ln 2^{(2+n)/2n}\right\} - \frac{0.4\sqrt{2}}{n^{1.2}} \tag{56}$$

The only extra term that comes in due to the roughness of the pipe is the term T_1 as defined below, which needs to be evaluated:

$$T_1 = 2.46n^{0.25} \ln\left(\frac{\rho u^{*2-n}\epsilon^n}{K}\right)^{1/n} \tag{57}$$

For the evaluation of this term we consider the velocity profile expression for Newtonian fluid for smooth and rough pipes and find the difference. Thus,

For smooth pipes—

$$u^+ = 2.5 \ln y^+ + 5.5 \tag{58}$$

For rough pipes—

$$u^+ = 2.5 \ln\left(\frac{y}{\epsilon}\right) + 8.5 \tag{59}$$

It can be concluded that the value of the term for Newtonian fluid is as follows:

$$2.5 \ln \frac{\rho u^* \epsilon}{K} = 3.0 \tag{60}$$

Thus,

$$\frac{\rho u^* \epsilon}{K} = 3.32 \tag{61}$$

The roughness Reynolds number is taken as 3.32 because, in fact, for Newtonian fluids [18] the roughness makes itself felt at $Re > 3$. In the case of non-Newtonian power-law fluids, we assume that the roughness would make itself felt at $Re > 3^n$. Thus, in an approximation we have

$$T_1 = 3.0n^{0.25} \tag{62}$$

Thus, the velocity profile for power-law fluids in rough pipes can be written as

$$u^+ = 2.46n^{0.25}\left\{\ln\frac{y}{\epsilon} + \left(0.1944 - \frac{0.1313}{n} + \frac{0.3876}{n^2} - \frac{0.0109}{n^3}\right)\exp\left[-\frac{n^2(y^* - 0.8)^2}{0.129}\right] + 1.2195 + \frac{1.3676}{n} + \ln 2^{(2+n)/2n}\right\} - \frac{0.4\sqrt{2}}{n^{1.2}} \tag{67}$$

A comparison between the predictions of Equations 55 and 63 shows reasonably close agreement.

ENTRANCE LENGTH ESTIMATION

Smooth Circular Pipes

The theoretical expressions for the velocity profiles given in foregoing sections of this chapter are all for the fully developed case. In order to get an idea of the fully developed region in any flow situation, one must have an estimate of the hydrodynamic entrance length. There is no information in the literature concerning turbulent entrance region flow of inelastic non-Newtonian fluids except for the work of Shenoy and Mashelkar [19] who have provided a method to get an engineering estimate of the hydrodynamic entrance lengths in non-Newtonian turbulent flow through smooth circular pipes. Their method involves an ordering technique to derive the turbulent hydrodynamic entrance lengths but they find that the predictions of their approximation approach give good agreement with existing data. They assumed that the simple velocity profile expression given by Equation 28 could be applied to the edge of the boundary layer (i.e. at $y = \delta$) where $u = u_o$, thus giving

$$\frac{u_o}{u^*} = \left[\frac{2^{1+\beta n}}{\alpha\psi^{2-\beta(2-n)}}\right]^{1/[2-\beta(2-n)]}\left[\frac{\delta^n u^{*2-n}\rho}{\gamma}\right]^{\beta/[2-\beta(2-n)]} \tag{64}$$

Rearranging Equation 64 and using $\tau_w = \rho u^{*2}$, gives the following

$$\tau_w = \frac{\alpha\psi^{2-\beta(2-n)}}{2^{1+\beta n}} \rho u_o^{2-\beta(2-n)} \left[\frac{\gamma}{\rho\delta^n}\right]^\beta \tag{65}$$

An alternative form for τ_w in the turbulent boundary layer has been provided by Skelland [7] as

$$\tau_w = \psi_1 \rho u_o^2 \frac{d\delta}{dx} \tag{66}$$

where

$$\psi_1 = \frac{[2-\beta(2-n)]}{[2-2\beta(1-n)]} - \frac{[2-\beta(2-n)]}{[2-\beta(2-3n)]} \tag{67}$$

Comparing Equations 65 and 66 and solving for δ gives

$$\delta = \left[\frac{(1+\beta n)}{\psi_1}\frac{\alpha\psi^{2-\beta(2-n)}}{2^{1+\beta n}}\right]^{1/(1+\beta n)} \left[\frac{\gamma}{\rho u_o^{2-n}}\right]^{\beta/(1+\beta n)} \left[\frac{1}{1+\beta n}\right] \tag{68}$$

In order to get an estimate of the turbulent entry length, x is set equal to x_e (the entrance length), δ is set equal to D/2 (the pipe centerline), and u_o is set equal to u_m (the maximum velocity at pipe centreline $= V/\psi$). Thus, by simplification the following expression for the turbulent entrance length for power-law fluids in smooth straight circular pipes is derived:

$$\frac{X_e}{D} = \left[\frac{\psi_1}{(1+\beta n)\alpha\psi^2}\right] Re_{gen}^\beta \tag{69}$$

Figure 9 shows a plot of the normalized dimensionless entrance lengths for turbulent flow through smooth straight circular pipes. The results indicate that increasing pseudoplasticity gives rise to larger entrance lengths. Dodge and Metzner [3] show that in a limited range $5 \times 10^3 < Re_{gen} < 10^4$, $0.6 < n < 1$, entrance length results for non-Newtonian fluids qualitatively agree with similar Newtonian measurements and are of a comparable magnitude or somewhat shorter. The results of Shenoy and Mashelkar [19] show a contrary trend and hence it might be erroneous to use Newtonian values for power-law fluids. In fact, making velocity measurements after an entrance length for Newtonian fluid could lead to errors as the flow is not fully developed until much greater lengths for pseudoplastic fluids. These findings would need experimental verification which are, unfortunately, not available in the existing literature.

Arbitrary Cross-Sectional Pipes

Presently, there is no clue whatsoever in the literature as to the magnitude of the turbulent entrance lengths in the case of power-law fluids flowing through conduits of various geometrical cross-sectional shapes. An exact solution of the hydrodynamic problem of turbulent non-Newtonian entrance region in non-circular pipes would, undoubtedly, be a more formidable task than in circular pipes. However, the design equation suggested by Shenoy and Mashelkar [19] for estimating the hydrodynamic turbulent entrance lengths for power-law fluids in circular pipes can be extended to include flow through conduits of arbitrary cross-sectional shape. Using an approach similar to that used earlier to obtain the velocity profile for arbitrary cross-sectional pipes, Equation 69 can be modified by appropriate manipulation of the terms to give

$$\frac{x_e}{2r_H} = \lambda_1 \left[\frac{\psi_1}{(1+\beta n)\alpha\psi^2}\right] Re_{gen}^{*\beta} \tag{70}$$

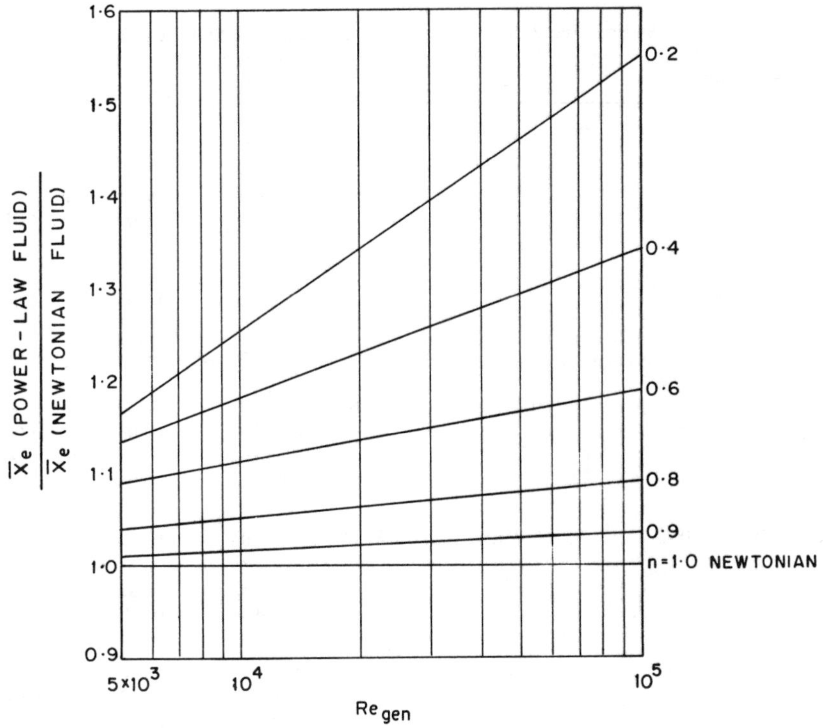

Figure 9. Estimated normalized turbulent entrance length ratios as a function of Reynolds number and the pseudoplasticity index n.

where

$$\lambda_1 = \left[\frac{4(a + bn)}{1 + 3n}\right]^{\beta n} \tag{71}$$

Defining the dimensionless entrance length $\bar{x}_c = x_e/D$ for circular pipes and $\bar{x}_{Nc} = x_e/2r_H$ for non-circular pipes, the following expression can be written when the Reynolds numbers in the circular and non-circular pipes are of the same magnitude. Thus,

$$\frac{\bar{x}_{Nc}}{\bar{x}_c} = \lambda_1 \tag{72}$$

The values of λ_1 for a wide variety of specific cases are evaluated and tabulated in Table 4. The cases include various types of cross-sectional geometries such as elliptical, rectangular, and isosceles triangular ducts for typical values of a and b taken from Kozicki et al. [8]. The values of λ_1 are seen to be not more than $\pm 5\%$ different from 1. Values of λ_1 less than one imply that the velocity profile develops earlier in these cases compared to the case of circular pipes. The fact that the values of λ_1 are not too different from one does not, however, imply that the entrance lengths

Table 4
Values of λ_1 Given by Equation 71 for Different Non-Circular Pipe Shapes at Typical Values

Pipe shape	a	b	n	β	λ_1
Elliptical $\frac{b'}{a'} = 0.3$	0.2796	0.8389	1.0	0.250	1.0284
			0.8	0.263	1.0238
			0.6	0.281	1.0191
			0.4	0.307	1.0138
$\frac{b'}{a'} = 0.5$	0.2629	0.7886	1.0	0.250	1.0126
			0.8	0.263	1.0106
			0.6	0.281	1.0085
			0.4	0.307	1.0062
$\frac{b'}{a'} = 0.7$	0.25338	0.7614	1.0	0.250	1.0038
			0.8	0.203	1.0032
			0.6	0.281	1.0025
			0.4	0.307	1.0019
Rectangular $\frac{b'}{a'} = 0.25$	0.3212	0.8182	1.0	0.250	1.0382
			0.8	0.263	1.0295
			0.6	0.281	1.0254
			0.4	0.307	1.0204
$\frac{b'}{a'} = 0.50$	0.2440	0.7276	1.0	0.250	0.9928
			0.8	0.263	0.9940
			0.6	0.281	0.9953
			0.4	0.307	0.9966
$\frac{b'}{a'} = 0.75$	0.2178	0.6866	1.0	0.250	0.9752
			0.8	0.263	0.9786
			0.6	0.281	0.9823
			0.4	0.307	0.9865
Isosceles Triangular Duct $2\alpha' = 20°$	0.1693	0.6332	1.0	0.250	0.9465
			0.8	0.263	0.9529
			0.6	0.281	0.9599
			0.4	0.307	0.9682
$2\alpha' = 40°$	0.1840	0.6422	1.0	0.250	0.9589
			0.8	0.263	0.9593
			0.6	0.281	0.9658
			0.4	0.307	0.9732
$2\alpha' = 60°$	0.1875	0.6462	1.0	0.250	0.9555
			0.8	0.263	0.9613
			0.6	0.281	0.9674
			0.4	0.307	0.9746

in pipes of different cross-sectional shapes are the same. It must be noted that the dimensionless entrance length is non-dimensionalized by twice the hydraulic radius, which changes with pipe cross-sectional shape. Thus, though the dimensionless entrance lengths for turbulent flow in pipes of all types of shapes are almost identical, the actual entrance length at the same Reynolds number is different by a factor equal to the hydraulic radius. As an engineering estimate, for turbulent flow in any non-circular pipe, the hydrodynamic entrance length for power-law fluids can be estimated by using a conservative value of $\lambda_1 = 1.05$ in Equation 70.

Annular Pipes

Entry region flow of non-Newtonian fluids in annular pipes has been treated for the laminar case by Tan and Tiu [10] and Bhattacharyya and Tiu [11, 12]. However, again there is no literature on similar treatise for turbulent flow. The procedure of Shenoy and Mashelkar [19] for estimating turbulent entrance lengths for power-law fluids could, nevertheless, be used for getting engineering estimates of entrance lengths in annular pipes.

As noted earlier, the shear stresses at the inner and outer wall of the annuli are different and therefore, it would be natural to expect two simultaneously developing boundary layer thicknesses—one starting from the inner wall, namely, δ_1 and the other from the outer wall, namely, δ_o, It is now assumed that the velocity profiles given by Equations 35 and 43 are valid at the edge of the corresponding boundary layers as a first approximation. Then using similar arguments as those used by Shenoy and Mashelkar [19] as detailed earlier, the expressions for the entrance lengths can be written:

$$\frac{x_{eo}}{F_o D_o} = \left[\frac{\psi_1}{(1+\beta n)\alpha\psi^2}\right] Re_o^{\beta} \tag{73}$$

$$\frac{x_{ei}}{F_i D_i} = \left[\frac{\psi_1}{(1+\beta n)\alpha\psi^2}\right] Re_i^{\beta} \tag{74}$$

where x_{eo} and x_{ei} are the entrance lengths for the outer and inner region, D_o and D_i are the outer and inner pipe diameters, Re_o and Re_i are the Reynolds numbers based on outer and inner equivalent diameters as defined below and F_o and F_i are geometric factors as shown in the following:

$$Re_o = \frac{D_{eo}^n V^{2-n} \rho}{\gamma}, \qquad Re_i = \frac{D_{ei}^n V^{2-n}}{\gamma} \tag{75}$$

$$D_{eo} = 2r_o[1-\lambda^2], \qquad De_i = 2r_i\left[\frac{\lambda^2}{k^2} - 1\right] \tag{76}$$

$$F_o = \frac{(1-\lambda)}{(1+\lambda)^{\beta n}}, \qquad F_i = \frac{(1-\lambda^2)}{(\lambda+k)}\left[\frac{k(1-\lambda)}{(\lambda^2-k^2}\right]^{\beta n} \tag{77}$$

The radius of maximum velocity is determined as discussed by Singh et al. [20] from the following:

$$\frac{\lambda - k}{1-\lambda} = k^{0.343} \tag{78}$$

This equation is general in nature as can be seen from its validity in the two limiting cases of the annular geometry, namely, a circular pipe ($\lambda = o$) and a parallel plate channel ($\lambda = 1$), irrespective of the type of the fluid and hence could be used for power-law fluids. Figure 10 shows a plot of x_{ei}/F_iD_i, x_{eo}/F_oD_o as functions of Reynolds number in the range of $10^4 \leq Re \leq 10^5$.

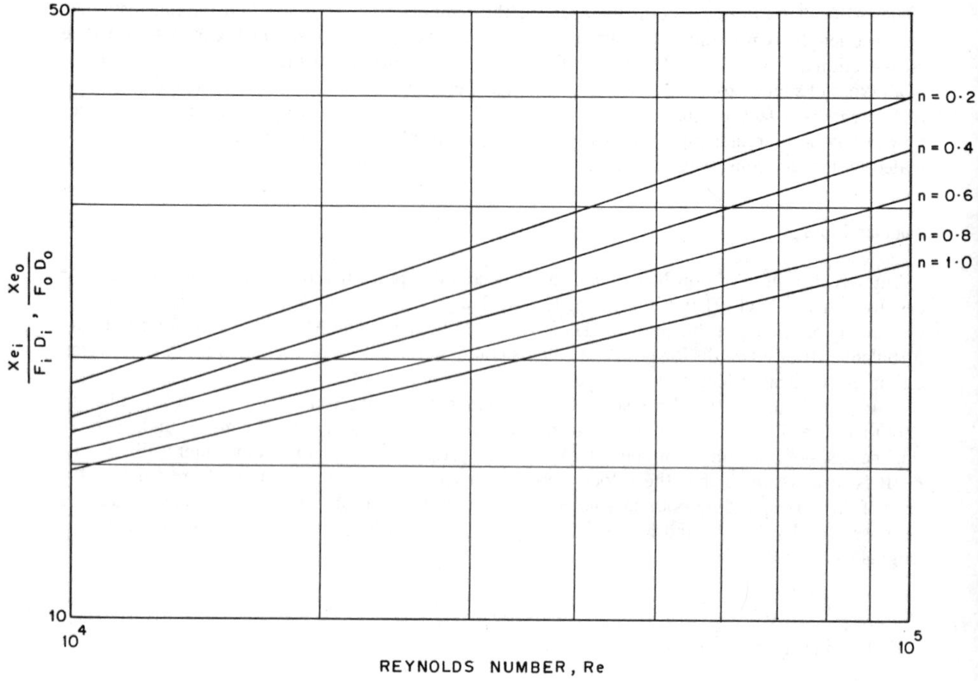

Figure 10. Entrance lengths for the inner and outer region of an annular pipe as a function of Reynolds number and pseudoplasticity index n.

CONCLUSION

This chapter deals with the nature of the time-averaged velocity profiles of turbulent pipe flow situations for inelastic non-Newtonian fluids in the absence of any anomalous effects, such as wall slip or drag reduction. The non-Newtonian characteristic of the fluid has been described by the power-law type Ostwald-de Waele fluid model. Several flow situations have been presented, such as those through circular and non-circular pipes as well as annular and curved pipes. In most cases, it is found that only a theoretical treatise has been done due to the dearth of experimental data in the literature. For smooth circular pipes alone experimental data is available and these have been reviewed by Edward and Smith [21]. The conclusions reached by them was that, if non-Newtonian velocity profiles were made dimensionless on a law of the wall basis with variables based on the apparent viscosity at the wall instead of Newtonian viscosity, the velocity profiles for non-Newtonian flows coincided with those for Newtonian flows.

Velocity profile measurements in turbulent flow with the use of heated or impact probes has had a variety of problems [22] due to the inherent flow disturbances caused by probes and their inaccurate sensitivity in non-Newtonian fluids. Presently, however, through the popular use of Laser-Doppler-Anemometry (LDA), it should be possible to determine velocity profiles in turbulent flow. In LDA measurement technique, the probes are photons, which are non-contact measuring elements. The frequency of the laser light is influenced by the moving particles of the fluid and the frequency shift is proportional to the particle velocity, thus making it an effective method for getting point to point data of the velocity field. One could thus expect that the theoretical expressions for power-law fluid velocity profiles in turbulent flow presented in this chapter would see experimental verification in the near future. However, it must be noted that all the velocity profile expressions given in the foregoing sections are valid only for inelastic non-Newtonian fluids and would not provide correct predictions for viscoelastic fluids as the elasticity of such fluids plays a major role in altering the flow field.

NOTATION

a	geometrical parameter for various cross-sectional shapes of ducts as tabulated in Tables 2 and 4
a′	dimensions of the major axis for various cross-sectional shapes of ducts as given in Tables 2 and 4
$\bar{a}$	radius of curved pipe
A	coefficient appearing in Equation 14 as a function of n
A_1	coefficient appearing in Equation 28 and defined by Equation 29
$\bar{A}_1$	coefficient appearing in Equation 36 and defined by Equation 41
b	geometrical parameter for various cross-sectional shapes of ducts as tabulated in Tables 2 and 4
b′	dimensions of the minor axis for various cross-sectional shapes of ducts as given in Tables 2 and 4
B	coefficient appearing in Equation 14 as a function of n
$\bar{B}$	coefficient appearing in Equation 52
B_1	exponent appearing in Equation 28 and defined by Equation 30
$\bar{B}_1$	exponent appearing in Equation 36 and which is taken to be equal to B_1
B′	exponent appearing in Equation 41 and defined by Equation 42
C	Correction factor defined by Equations 9 and 15 for respective cases
D	pipe diameter
D_1	coefficient appearing in Equation 44 and defined by Equation 47
D_i	diameter of the inner pipe of an annulus
D_o	diameter of the outer pipe of an annulus
D_{ei}, D_{eo}	equivalent diameters defined by Equation 76
f	friction factor
F_i, F_o	geometric factors defined by Equation 77 for an annulus
I	function appearing in Equation 8, values of which for varying Reynolds number and pseudoplasticity index are given in Table 1
k	geometric factor defined by Equation 39 for an annulus
K	consistency index for a power-law fluid as from Equation 2
L	length of pipe
n	pseudoplasticity index for a power-law fluid as from Equation 2
ΔP	pressure drop across the length of the pipe
r	radial position
r_H	hydraulic radius appearing in Equation 32
r_i	radius of the inner cylinder of an annulus
r_m	radius of maximum velocity inside an annulus
r_o	radius of the outer pipe of an annulus
R	pipe radius
$\bar{R}$	the radius of curvature of a curved pipe as appearing in Equations 46 and 47
R^+	Reynolds number based on pipe radius and friction velocity
Re	Reynolds number
Re_i, Re_o	Reynolds number for power-law fluids as given in Equation 75 based on equivalent inner and outer diameters of an annulus
Re_ϵ	roughness Reynolds number for power-law fluids as defined by Equations 53 and 54
Re_{gen}	generalized Reynolds number for power-law fluids as defined by Equation 21
Re^*_{gen}	generalized Reynolds number for power-law fluids for non-circular pipes as defined by Equation 32
T	controlling term in the velocity profile for non-circular pipes as defined by Equation 34
T_1	controlling term in the velocity profile for rough pipes as defined by Equation 57
u	time-averaged velocity at a point
u*	friction velocity defined as $\sqrt{\tau_w/\rho}$

u^+ dimensionless velocity defined as u/u^*
u_i^* friction velocity for inner region of the annulus
u_o^* friction velocity for outer region of the annulus
u_i^+ dimensionless velocity for inner region of the annulus
u_o^+ dimensionless velocity for outer region of the annulus
u_m maximum velocity as defined by Equation 25
v angular velocity for a curved pipe as defined by Equation 44
v_m average axial velocity for the curved pipe appearing in Equations 47 and 48
V mean velocity of power-law fluid in a smooth circular pipe as defined by Equation 25
w axial velocity for a curved pipe as defined by Equation 45
w_1 coefficient appearing in Equation 45
w_c dimensionless axial velocity for a curved pipe as defined by Equations 48 and 51
x_e entrance length for smooth circular pipe as defined by Equation 69
x_{ei} entrance length for the inner region of an annulus and defined by Equation 74
x_{eo} entrance length for the outer region of an annulus and defined by Equation 73
$\bar{x}_c$ dimensionless entrance length for a circular pipe in Equation 72
$\bar{x}_{Nc}$ dimensionless entrance length for a non-circular pipe in Equation 72
y distance from the wall as shown in Figure 2
y_L laminar sublayer thickness as shown in Figure 2
y_T transition zone thickness as shown in Figure 2
y^* dimensionless location parameter defined by y/R
y^+ distance based Reynolds number as defined by Equations 5 and 6
y_i^+ distance based Reynolds number for inner region of an annulus.
y_o^+ distance based Reynolds number for outer region of an annulus
y_2^+ transition zone thickness appearing in Equations 12 and 13

Greek Symbols

α coefficient appearing in Equation 23
α' half angle of an isosceles triangular duct appearing in Table 2
β exponent appearing in Equation 23
γ modified consistency index term defined by Equation 22
$\dot{\gamma}$ shear rate appearing in Equation 2
δ boundary layer thickness appearing in Equations 44 and 45
δ_c dimensionless boundary layer thickness as defined by Equation 46
ϵ average height of the roughness projections as shown in Figure 8.
η fluid viscosity
η_N Newtonian viscosity as shown in Figure 1
η_o zero-shear viscosity as shown in Figure 1
η_∞ infinite-shear viscosity as shown in Figure 1
λ dimensionless parameter for an annulus as defined by Equation 39
λ_1 dimensionless parameter for a non-circular pipe as defined by Equations 71 and 72
θ angular co-ordinate for a curved pipe as shown in Figure 4
ρ density of the fluid
σ_1 function of n defined by Equation 16
σ_2 function of n defined by Equation 17
τ shear stress apppearing in Equation 2
τ_w wall shear stress
τ_{wi} wall shear stress for inner region of an annulus as given by Equation 37
τ_{wo} wall shear stress for outer region of an annulus as given by Equation 38
ξ distance from the wall for curved pipe
ψ function of β and n as defined by Equation 26
ψ_1 function of β and n as defined by Equation 67

REFERENCES

1. Bird, R. B., Armstrong, R. C., and Hassager, O., *Dynamics of Polymeric Liquids: Volume I, Fluid Mechanics*, John Wiley and Sons, New York, 1977, pp. 205–212.
2. Langhaar, H. L., *Dimensional Analysis and Theory of Models*, John Wiley and Sons, New York, 1951.
3. Dodge, D. W., and Metzner, A. B., "Turbulent Flow of Non-Newtonian Systems," *AIChE J.*, Vol. 5. 1959, pp. 189–204.
4. Bogue, D. C., and Metzner, A. B., "Velocity Profiles in Turbulent Pipe Flow," *Ind. Eng. Chem. Fundam.*, Vol. 2, 1963, pp. 143–152.
5. Clapp, R. M., International Developments in Heat Transfer, Part III. 652–61; D-159; D–211-5, A.S.M.E., New York, 1961.
6. Shenoy, A. V., and Saini, D. R., "A New Velocity Profile Model for Turbulent Pipe-Flow of Power-Law Fluids," *Can. J. Chem. Eng.*, Vol. 60, 1982, pp. 694–696.
7. Skelland, A. H. P., *Non-Newtonian Flow and Heat Transfer*, John Wiley and Sons, New York, 1967, pp. 288–291.
8. Kozicki, W., Chou, C. H., and Tiu, C., "Non-Newtonian Flow in Ducts of Arbitrary Cross-Sectional Shape," *Chem. Eng. Sci.*, Vol. 21, 1966, pp. 665–679.
9. Salem, E., and Embaby, M. H., "Theoretical and Experimental Investigations of Non-Newtonian Fluid Flow Through Non-Circular Pipes," *Appl. Sci. Res.*, Vol. 33, 1977, pp. 119–139.
10. Tan, K. L., and Tiu, C., "Entry Flow Behavior of Viscoelastic Fluids in an Annulus," *J. Non-Newtonian Fluid Mechanics*, Vol. 3, 1977/1978, pp. 25–40.
11. Tan, K. L., and Tiu, C., "An Experimental Investigation of Upstream Flow Characteristics of Viscoelastic Fluids in an Annular Die Entry," *J. Non-Newtonian Fluid Mechanics*, Vol. 6, 1979, pp. 21–45.
12. Tan, P. K. L., and Tiu, C., "Velocity Profiles of Viscoelastic Fluids at the Inlet of an Annulus," *AIChE J.*, Vol. 26, No. 1, 1980, pp. 162–165.
13. Tiu, C., and Tan, K. L., "Boundary Layer Analysis of Viscoelastic Flow in Annuli," *Rheol. Acta*, Vol. 16, 1977, pp. 497–509.
14. Bhattacharyya, S., and Tiu, C., "Developing Pressure Profiles for Non-Newtonian Flow in an Annular Duct," *AIChE J.*, Vol. 20, No. 1, 1974, pp. 154–158.
15. Tiu, C., and Bhattacharyya, S., "Developing and Fully Developed Velocity Profiles for Inelastic Power-Law Fluids in an Annulus," *AIChE J.*, Vol. 20, No. 6, 1974, pp. 1140–1144.
16. Mashelkar, R. A., and Devarajan, G. V., "Secondary Flows of Non-Newtonian Fluids: Part III—Turbulent Flow of Visco-Inelastic Fluids in Coiled Tubes: A Theoretical Analysis and Experimental Verification," *Trans. Instn. Chem. Engrs.*, Vol. 55, 1977, pp. 29–37.
17. Torrance, B., McK., *South African Mechanical Engr.*, Vol. 13, 1963, p. 89.
18. Schlichting, H., *BoundaryKLayer Theory* , McGraw-Hill, New York, 1968.
19. Shenoy, A. V., and Mashelkar, R. A., "Engineering Estimate of Hydrodynamic Entrance Lengths in Non-Newtonian Turbulent Flow," *Ind. Eng. Chem. Process Des. Dev.*, Vol. 22, 1983, pp. 165–168.
20. Singh, R. P., Nigam, K. K., and Mishra, P., "Developing and Fully Developed Turbulent Flow in an Annular Duct," *J. Chem. Eng.* Japan, Vol. 13, 1980, pp. 349–352.
21. Edwards, M. F., and Smith, R., "The Turbulent Flow of Non-Newtonian Fluids in the Absence of Anomalous Wall Effects," *J. Non-Newtonian Fluid Mech.*, Vol. 7, 1980, pp. 77–90.
22. Seyer, F. A., and Metzner, A. B., "Turbulence in Viscoelastic Fluids," Sixth Symp. Naval Hydrodynamics, Sept. 28–Oct. 4, 1966, pp. 19–38.

CHAPTER 32

MOLECULAR INTERACTIONS IN DRAG REDUCTION IN PIPE FLOWS

Neil S. Berman

Department of Chemical and Bio-Engineering
Arizona State University
Tempe, Arizona, USA

CONTENTS

INTRODUCTION

When very small concentrations of large polymer molecules, fibers, or particles are present in a fluid, the frictional resistance in turbulent flow can be reduced in comparison to the turbulent friction for the pure fluid. This drag reduction, as the phenomenon is called, requires additional explanation because some properties of the pure fluid are changed when the additive is incorporated in solution. Although drag reduction effects are found in many flow geometries, most experiments relate to pipe flow and here we can give a precise definition. For small changes in fluid properties we can assume that the viscosity and density of the fluid are not changed. Then the percent drag reduction in turbulent pipe flow is defined as the pressure loss due to friction per unit length of pipe for the fluid alone, $\Delta P_S/L$, minus the pressure loss due to friction per unit length for the fluid plus additive, $\Delta P_A/L$, divided by $\Delta P_S/L$ and multiplied by 100. For the comparison, the flow rate and temperature must be constant. This definition would mean a positive value of percent drag reduction is a true savings in energy required to pump the fluid. Even if the fluid properties are changed, we can still use the definition to show energy savings but comparison of the two fluids to explain the phenomenon becomes difficult.

This chapter focuses on high molecular weight polymers dissolved in a liquid. The molecules can interact with the flow leading to changes in the molecular conformation as well as to changes in the flow. In addition, we must consider the behavior of groups of molecules and molecular interactions. Under some conditions other additives, such as particulates or fibers, will be similar to the polymer solutions.

Observations of decreased drag or increased flow rate due to contaminants in the fluid may date back to the 19th century. Reviews of the history of drag reduction by White [1] and Zakin and Hunston [2] credit early experiments to Hele-Shaw in 1898 [3] or Forest and Grierson in 1931 [4]. The modern work with polymeric additives stems from the studies by Toms [5] and Mysels [6] during World War II. Both have published short accounts of these initial studies [7, 8]. Mysels measured the pressure drop and flow rate for napalm (gasoline with aluminum disoaps) in several circular pipes. On June 27, 1945 he observed "The thickened gasoline flows faster than the unthickened." The results formed the basis for a patent issued in 1949 but publication in the scientific literature did not occur until 1954 [9]. Toms measured the pressure drop and flow rate of poly(methyl methacrylate) in monochlorobenzene. In his recollections he stated "When, however, I plotted the rate of flow at constant pressure against the polymer concentration, a remarkable anomaly appeared.... under turbulent conditions, a polymer solution clearly offered less resistance to flow, under constant pressure, than the solvent itself." Toms work was published in 1949 and he is often credited with the first observations of drag reduction by polymeric additives. These early investigators noted that the polymer solutions had a higher viscosity than the solvent when viscosity was measured by a standard method at a low rate of flow. In fact, the solutions were also shear-thinning and the viscosity varied with the shear rate. But the most interesting properties of the polymer solutions were the elastic and pituitous or stringiness nature.

Lower frictional resistance in dilute fiber suspensions and sediments were well known by the early 1950s, as can be seen from the references cited by Daily and Bugliarello [10]. This reference was often cited by early investigators, perhaps before the work of Toms was well known.

In 1959 Dodge and Metzner [11] showed that shear-thinning fluids had a lower frictional resistance than Newtonian fluids. Non-Newtonian fluids can appear to be drag reducing on this basis alone. In most cases the solution viscosity is much higher than for the solvent so there is no net drag reduction compared to the pure solvent based on the previous definition. Dodge and Metzner [11] found that 0.3% aqueous solutions of carboxymethyl cellulose (CMC) did not behave in the same manner as other non-Newtonian fluids. These solutions were truly drag reducing after the shear-thinning property was accounted for. Similar results were obtained by Shaver and Merrill [12]. Photographs of dye streaks in a 0.25% CMC solution showed that a dye streak in the center of the pipe 185 diameters downstream of injection was not mixed over the cross section as well as similar streaks in a Newtonian fluid. At the wall, dye streaks showed a thickened layer compared to a Newtonian liquid and a supression of the generation of turbulent eddies that form near the wall and move to the center.

At the Fourth International Congress on Rheology, August, 1963, Fabula showed that a one ppm solution of poly (ethylene oxide) with an average molecular weight over 10^6 could give a 20% friction reduction at a Reynolds number of 10^5 in a 1.02 cm I.D. pipe flow [13]. Similar results were also given by Hoyt and Fabula in 1964 [14]. This paper was cited by Gadd [15] as arousing interest in the subject of drag reduction. There has been a virtual explosion of experiments after 1963 and summaries of these experiments as well as theoretical studies have appeared in many reviews in the published literature. A selection of these reviews is listed chronologically in the references [15–25], and the reader is referred to these for additional background.

FRICTION FACTOR—REYNOLDS NUMBER STUDIES

Drag reduction is observed experimentally by comparing the pressure drop and flow rate in turbulent pipe flow for the polymer solution and pure solvent. Figure 1 shows a typical result for a 50 part per million by weight solution of a poly (ethylene oxide) in a solution of glycerine and water. Two alternate methods of plotting are shown in Figures 2 and 3. The following definitions

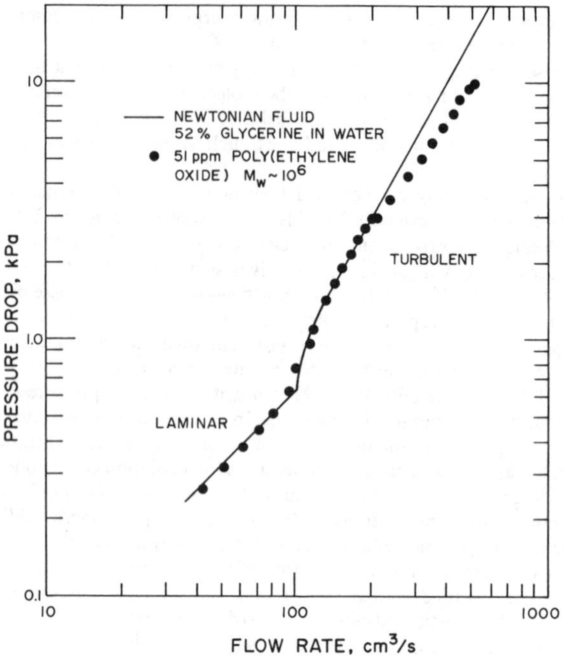

Figure 1. Pressure drop vs. flow rate for a drag reducing polymer solution compared to the pure solvent.

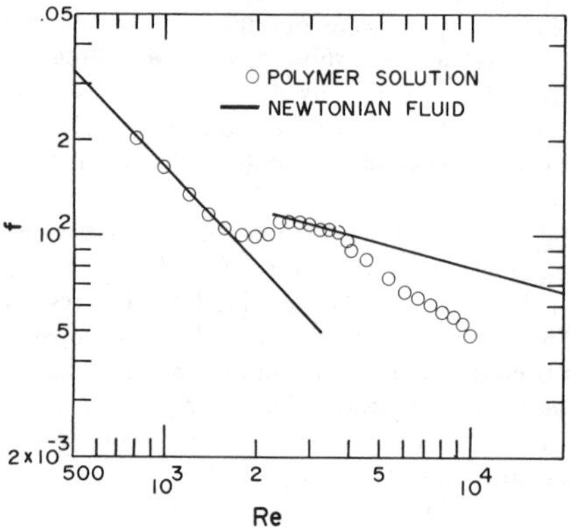

Figure 2. The data shown in Figure 1 replotted in the form Reynolds number vs. friction factor.

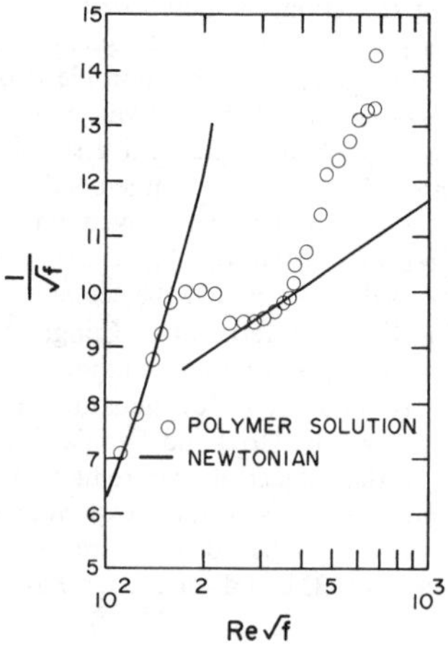

Figure 3. A third way to represent pipe flow data. The differences between polymer solution and solvent on this plot can be related to molecular parameters.

are used for the Reynolds number, Re, and friction factor, f:

$$Re = Dv_{ave}\rho/\mu$$

and

$$f = \frac{D(-\Delta P)}{2L\rho v^2_{ave}}$$

where D is the pipe diameter, ρ is the fluid density, μ is the fluid viscosity without polymers, v_{ave} is the average velocity, L is the length between pressure taps, and $-\Delta P$ is the pressure drop. The experiments plotted in Figures 1–3 show an early transition to turbulence and an onset of drag reduction. Other experiments using different pipe diameters or different polymers lead to a later transition to turbulence. The onset also may not be present and the polymer solution can be drag reducing for all turbulent Reynolds numbers.

The curves of Reynolds number vs. friction factor can be used to determine the importance of molecular properties of the polymers. Although a range of Reynolds numbers and pipe sizes is required to test interactions between molecules and turbulence scales, a coarse screening for the drag reducing ability of a polymer molecule can be made with a single Reynolds number and pipe size. The most extensive tabulation was done by Hoyt [26] where we find that almost any polymer with average molecular weight above 10^5 can give drag reduction when dissolved in an appropriate solvent.

Onset of Drag Reduction

Figures 1–3 show a well defined onset and other experiments show that the onset shear rate is a function of solution viscosity, temperature, and pipe diameter. Let us consider a non-ionic linear polymer dissolved in a solvent and experimentally measure flow rate and pressure drop in smooth tubes with diameters less than about 20 mm. Molecular models of polymers in dilute solutions show that the longest relaxation time of the molecule is

$$T_1 = \frac{b[\eta]\eta_s M}{RT} \tag{1}$$

where b is a constant dependent upon the model, $[\eta]$ is the intrinsic viscosity, η_s is the solvent viscosity, M is the polymer molecular weight, R is the gas constant, and T is the temperature. For tube diameters less than 20 mm there is ample experimental evidence to show that onset occurs when the characteristic time of the flow is about twice the relaxation time of the molecules. From Equation 1 we can see that changes in solvent viscosity or temperature can be used to test the time scale hypothesis. Although changes in molecular weight and intrinsic viscosity could also be used for such a test, polymer samples do not consist of a single molecular weight and interpretation of experiments can be difficult. Figure 4 shows the effect of adding glycerine to change the viscosity of a polymer solution. Additional experiments are reported in the literature [27–30]. The experiments show that the onset Reynolds number is reduced when the viscosity is increased or the temperature is decreased. The qualitative behavior is exactly as predicted by the time scale hypothesis.

Another possibility for relating the interaction between polymer molecules and turbulent flow is a length scale correlation. For a single molecule, the rms radius is proportional to the cube root of $M[\eta]$. Therefore, the experimental evidence of the effect of solvent viscosity and temperature are not consistent with the length scale correlation. In fact, no change in onset is predicted by the length scale hypothesis for the experiment shown in Figure 4. Virk [20] shows that the shear velocity is constant at onset, which does suggest a length scale interaction. The constant shear velocity, however, follows from the elongation of the molecules in the flow as a result of the time scale interaction. If the molecules become elongated, the viscosity in the center of the pipe will

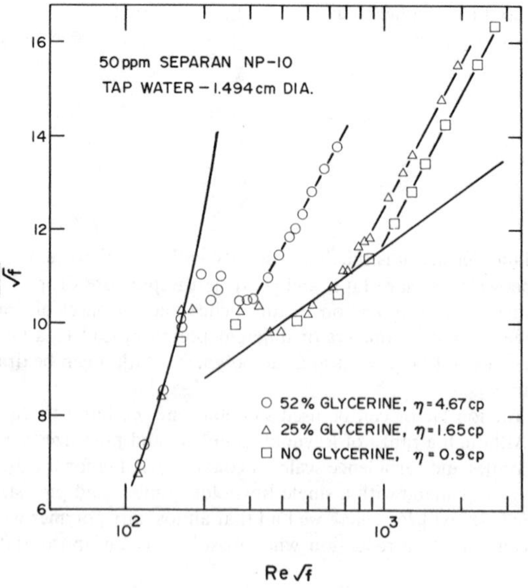

Figure 4. The effect of solution viscosity on onset of drag reduction. (From Berman [44].)

become larger and the velocity profile becomes

$$u_+ = 2.45 \ln y_+ + 5.66 + \Delta B \tag{2}$$

where $u_+ = u/u_*$, $y_+ = yu_*/\nu$, $u_* = \sqrt{\tau_w/\rho}$, and $\tau_w = (-\Delta P)D/4L$

Here, y is the distance from the wall and u is the local velocity. Equation 2 can be integrated to get the relationship between friction factor and Reynolds number for small amounts of drag reduction. When the friction factor of the solvent is subtracted we find that

$$(1/f)^{1/2} - (1/f_s)^{1/2} = (\text{constant})\,\Delta B \tag{3}$$

Now, if we fit a straight line to the drag reduction curves of Figure 3 or 4,

$$(1/f)^{1/2} = a \log(Ref^{1/2}) + b \tag{4}$$

Then, Equations 3 and 4 can be combined to give

$$\Delta B = \delta \log(u_*/u_{*o}) \tag{5}$$

where δ is a constant slope increment, u_{*o} is the onset shear velocity and the viscosities of polymer solution and solvent are assumed to be the same near the wall. Equation 5 implies a constant shear velocity at onset, but the ratio u_*/u_{*o} appears in the logarithm so the time scale may be related to the onset flow time:

$$T_1 = \frac{1}{2}\frac{\nu}{u_{*o}^2}$$

and

$$\Delta B = (\delta/2) \log(2u_*^2 T_1/\nu)$$

Therefore, the observed constant u_{*o} or the correlation of T_1 with u_{*o}^2/ν are not different. Only experiments that change the molecular scales independently can be used to study the relationship between polymer molecules and turbulent scales. The analysis is limited to small amounts of drag reduction for random coil polymers in solution with no interaction between molecules. Under these conditions the drag reduction trajectory is determined by two constants δ and the onset shear velocity.

Location of Polymer-Flow Interaction

When drag reduction experiments are conducted in larger pipes, it is found that higher concentrations are required to observe onset and the onset shear stress and the slope increment, δ are different from the small tube values. In order to find out why the small tubes are more effective, many experiments have been done to locate the region in the pipe where the polymer molecules interact with turbulence. The time scale relationship suggests that this interaction occurs in the region of highest strain rate, that is the molecules are stretched only when the product of the time averaged strain rate of the most energetic eddies and the molecular time scale exceeds 1/2. Wells and Spangler [31] injected polymer solutions at the pipe center and at the wall and observed drag reduction as a function of distance from the injection. The wall injection led to immediate drag reduction while the center injection was not effective until the polymer solution diffused to the wall. McComb and Rabie [32] measured the local concentration and the drag reduction at the wall and found that the polymer molecules were most effective in a thin ring near the wall but not in the viscous sublayer. This explains the experimental data for larger pipes because the cross-sectional area in the ring is smaller relative to the total cross section for larger pipes.

The evidence from onset and location in the flow where interaction of turbulence with the molecules occurs correlates with the distribution of strain rate in pipe flows as sketched in Figure 5. As discussed by Durst et. al. [32], the strain rate has a peak near the wall and exceeds $T_1/2$ in a ring. The polymer molecules become stretched in that ring and move away from the wall in an elongated state. Models for the elongation of polymer molecules in pure strain are discussed by Bird et al. [34] The stretched molecules have a large increase in elongational viscosity so the viscosity seen by the turbulence is greater than the viscosity in the viscous sublayer near the wall. On a dimensionless scale the large eddies scale with dimensionless distance from the wall and the smallest eddies scale with this distance from the wall to the one-quarter power. We know that turbulence cannot exist if the small eddies exceed the large eddies in size. Therefore, when the wall viscosity is used to scale the small eddies and the elongational viscosity is used to scale the large eddies, the viscous sublayer is moved toward the center of the pipe. As a result, the smallest *turbulent* eddies are larger

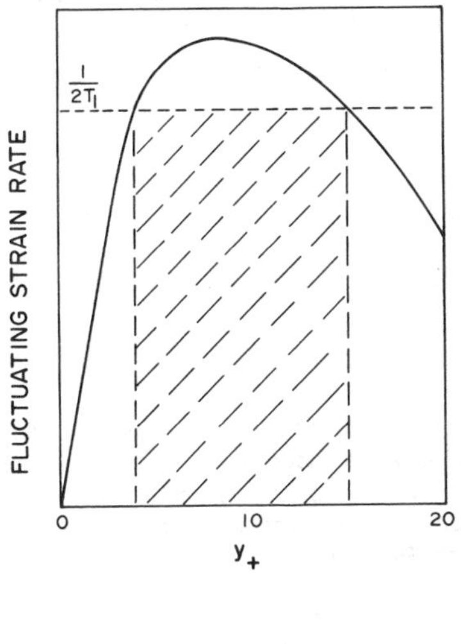

Figure 5. Average fluctuating strain rate on an arbitrary scale near the wall of a pipe. When the fluctuating strain rate exceeds $1/2T_1$, the polymer molecules are stretched out (shaded region).

in size compared to the solvent turbulent flow and the drag at the wall is reduced. Further discussion is given by Lumley [18].

The increase in size of the viscous sublayer does not imply that velocity fluctuations are not present in the expanded viscous sublayer region. The small fluctuations are not present, but large ones remain. Although some authors call the entire expanded viscous sublayer by the name viscous sublayer, the presence of large fluctuations at sufficient distances from the wall does not give an extension of a laminar velocity profile. It is more appropriate to call the extension of the viscous sublayer by the name elastic sublayer or buffer layer. Drag reduction reduces the size of the smallest eddies in this layer and dissipation occurs at a larger eddy size. In Newtonian fluids the dissipative or smallest eddies increase if Reynolds number is decreased. Thus, drag reduction changes the dynamics to that of a lower Reynolds numbers. We will examine more details of the effect of polymers on the turbulent structure later.

Intrinsic Drag Reduction

The good qualitative relation between turbulent and molecular time scales suggests that there should be an intrinsic drag reduction or drag reduction for an individual molecule. Oliver and Bakhtiyarov [35] found that drag reduction was proportional to concentration when the concentration of a high molecular weight polyacrylamide was below 0.05 ppm. There are difficulties in studying effects of concentration because of the polydispersity of the polymers and the problems of preparation of homogeneous solutions. To obtain homogeneous solutions Oliver and Bakhtivyarov mixed the solution for 15 hours. The effects of mixing have been noted by others but not well studied so both increased drag reduction after mixing and decreased drag reduction after mixing have been found. It appears that polyacrylamides can increase the effectiveness of drag reduction when the mixing time is increased and poly(ethylene oxide) solutions are more effective when freshly mixed.

In order to reduce the effect of the molecular weight distribution that is present in a commercial sample, studies on fractionated polymers have been made. Narrow fractions from the polymer molecular weight distribution were collected by passing polyethylene oxide solution through a gel permeation column [36]. When a 400-ppm solution of polyethylene oxide with a viscosity average molecular weight of 2×10^5 was fractionated, a 10.5-ppm solution containing the molecular weights greater than 10^6 gave the same drag reduction as the original 400-ppm solution. Additional experiments to verify that only the high molecular weight tail of the distribution is effective in drag reduction were reported by Berman and Yuen [37] and Hunston and Reischman [38]. A variety of concentrations and molecular weights were obtained in the fractionation experiments. One way to examine the data to find the concentration dependence of drag reduction is based on an analysis of the slope increment, δ, of Equation 5. Figure 6 shows the concentration dependence of $\delta/[\eta]$ where $[\eta]$ is taken to be the same as the specific viscosity divided by concentration measured in laminar tube flow. The results show that $\delta/[\eta]$ is proportional to $C^{2/3}$ for concentrations from

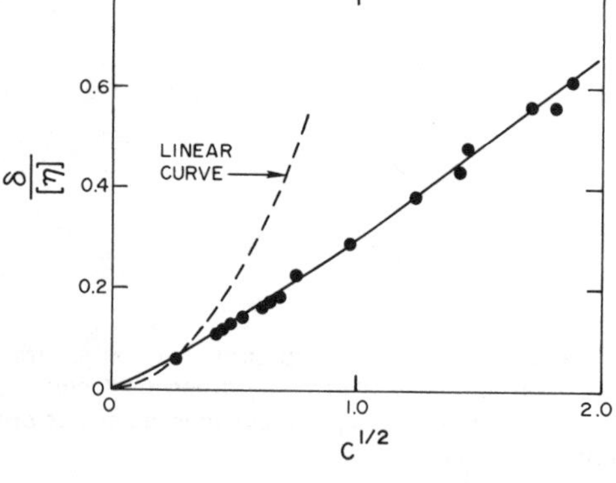

Figure 6. Slope increment, intrinsic viscosity and concentration relationship for fractionated poly(ethylene oxide) solutions at low concentrations. (Data are from Berman and Yuen [37].)

0.1 to 1.0 ppm and to $C^{1/2}$ beyond 1.0 ppm. Virk also finds that the concentration dependence is $C^{1/2}$. In Figure 6 a curve proportional to concentration passing through the origin and the lowest concentration experimental point is shown to illustrate the range corresponding to intrinsic drag reduction. It appears that when concentrations are larger than 0.05 ppm, drag reduction for an individual molecule is not observed.

Molecular Models

It is instructive to look at models of extensional viscosity to see if the experimental dependence on concentration is predicted. For a bead-rod model of a polymer, Hassager [39] finds that

$$\frac{\eta_T - 3\eta_o}{3\eta_o} = C[\eta]N[1 - 0.8/ST_1 - \cdots]$$

where η_o is the shear viscosity, η_T is the elongational viscosity, N is the number of backbone units in the molecule; and S is the strain rate. The strain rate in turbulent pipe flow is proportional to u_*^2/ν and the polymer relaxation time is related to u_{*o}^2/ν. Then

$$\frac{\eta_T - 3\eta_o}{3\eta_o} = Kc[\eta]N \ln(u_*/u_{*o}) \tag{6}$$

where K is a constant. Equation 6 becomes the same as Equation 5 if

$$\Delta B = \frac{\eta_T - 3\eta_o}{3\eta_o}$$

This is exactly what is obtained for small amounts of drag reduction when the viscosity in the pipe core is increased to η_T while the wall viscosity remains at η_o. Thus, the model of molecular stretching is consistent as long as the concentration is low. At higher concentrations we must examine molecular interactions but the concept of molecular extension can be retained. Batchelor [40] found that a suspension of rods becomes non-dilute when the volume of spheres circumscribing the rods equals the total volume. A large increase in extensional viscosity is predicted for such a suspension. The left-hand side of Equation 6 in non-dilute solutions would be proportional to $C^{2/3}$ at this point. If the concentration of rods is increased so that they overlap, the concentration dependence becomes $C^{1/2}$.

For random coil molecules in solution with molecular weights necessary to observe drag reduction the molecules do not have to be extended very much in order to reach the non-dilute state even for a concentration of 0.1 ppm. For example polyethylene oxide completely extended at this concentration and with a molecular weight of 10^6 has a length divided by distance between molecules of 100. Extension of 100 times less than complete uncoiling would lead to molecular interactions. Therefore, the molecular weight dependence for random coil molecules only relates to the ability of the molecules to extend in high strain rate regions of the turbulent flow. The number and size of backbone units is not a factor at least for low amounts of drag reduction. The flexibility of the molecule does appear to be important as indicated by Hoyt. Experiments with organic dyes that can bridge across sites within polyethylene molecules to stiffen the molecule show that drag reduction is reduced when small amounts of the dye are present [41]. Many dyes also degrade the polymers but congo-red does not degrade and can form intramolecular bridges as well as intermolecular bridges. Higher concentrations of congo-red led to greater drag reduction indicating that the intramolecular interactions also have some influence on drag reduction.

Extended Polyelectrolytes

Polymers are not always random coils in solution. Many polyelectrolytes have extended conformations in stagnant solutions of water that does not contain other ions. The charged groups on the polyelectrolyte molecule will tend to repel each other and expand the molecule in deionized water or very low concentration salt solutions. Drag reduction in these solutions of extended molecules has no onset. Similar drag reduction effects are found with fibers whose aspect ratio is greater than 10. Evidences of differences between the drag reduction trajectories for the fibers and polyelectrolytes are discussed by Virk [42] and Radin et al. [43].

It is possible to test molecules that are present in rigid rod conformation where the dimensions are known from light scattering. When this is done, we find that the volume occupied by the molecules in the sense of circumscribed spheres about each molecule must be larger than the total volume before drag reduction is observed. This means that in order for the rigid rods to remain aligned and to increase the viscosity in the pipe core the solution must be non-dilute. Experiments verifying this conclusion are discussed by Berman [44] and Berman et al. [45]. Some rigid rod solutions are more drag reducing in larger pipes compared to smaller ones in contrast to the random coil solutions. It is not clear if this effect is the result of an entrance disturbance or an interaction with the turbulent eddies that breaks up the alignment in the smaller tubes. This is a difficult problem because we are dealing with groups of molecules rather than single molecules that can be described by a molecular model. Rigid rods that can interact with the flow give drag reduction trajectories as shown in Figure 7b. At low Reynolds numbers the drag reduction is greater in the larger pipe. This is in contrast to the time scale behavior of random coiling polymers which gives the much different diameter effect shown in Figure 7a.

The DNA molecules are in a partially extended conformation at these low Reynolds numbers. As the Reynolds number is increased the flow can break-up the aligned groups of molecules or in some cases the flow can actually alter the conformation. For example, Xanthan gum is a polysaccharide with the side groups collapsed along the backbone in solutions at rest. This gives a rigid helical structure that is partially extended. In high shear it is possible that the side groups are moved out, the molecule becomes more flexible and the molecule collapses to a random coil. As the Reynolds number is increased beyond the laminar flow region, a Xanthan gum solution in deionized water will be highly drag reducing for low Reynolds numbers and turbulent flow. Drag reduction is reduced at first as Reynolds number is increased because the strain rate is not high enough to

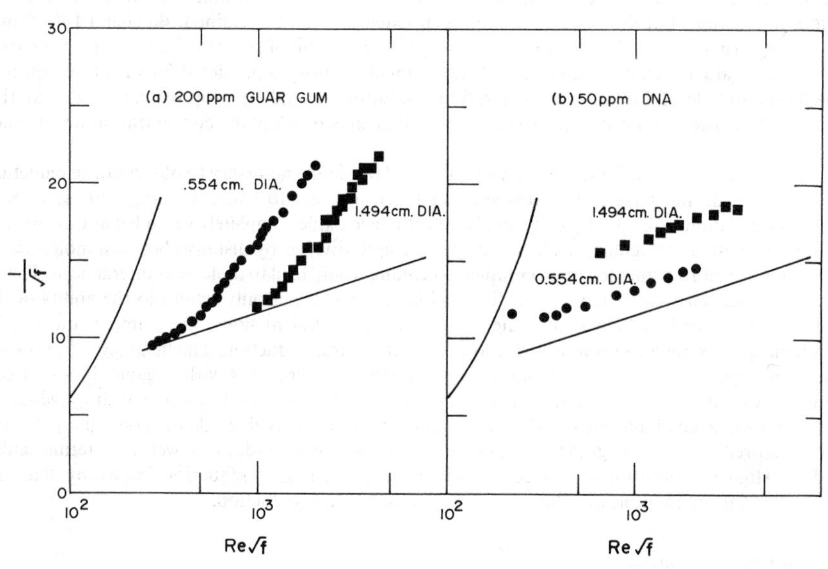

Figure 7. Drag reduction trajectories showing effects of pipe size for a non-ionic random coiling polymer (a) and for an ionic rod like polymer (b).

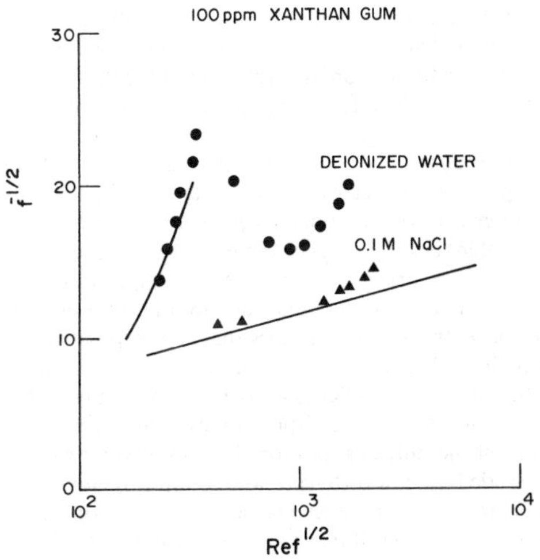

Figure 8. Drag reduction trajectories for a polyelectrolyte extended in deionized water and collapsed in salt solution.

stretch the molecule. Higher Reynolds numbers lead to the same onset behavior as is found for random coils. In fact, if we collapse the rods by changing the chemical environment (for Xanthan gum this is achieved in basic solution or by adding salt) we find the onset of drag reduction occurs at the same Reynolds number. Figure 8 shows the onset for Xanthan gum in two solutions corresponding to molecules collapsed by the flow and molecules collapsed by ions. The onset Reynolds numbers are seen to be the same.

It appears from these studies that there can be at least two types of drag reduction. For random coil molecules they must be elongated in the high strain field near the wall and for rigid rod molecules they must be aligned in the core of the pipe. Another difference appears at relatively high concentrations and high Reynolds numbers well beyond onset. The relatively low concentration solutions degrade at high Reynolds numbers and drag reduction is reduced. Degradation is an important problem in the use of polymer additives for drag reduction and a recent review of degradation is given by Hunston and Zakin [46].

Maximum Drag Reduction

When sufficient random coil polymer molecules are present, a "maximum drag reduction asymptote" appears as shown by Virk [20]. High concentrations of polyelectrolytes in extended conformations, fibers in suspension and combinations of fibers and polymers give drag reduction that exceeds this "maximum drag reduction asymptote." It appears that the "maximum drag reduction asymptote" results from a combination of factors including the increase in solution viscosity with polymer concentration and the break-up of groups of aligned molecules. When asbestos fiber suspensions are reused after passing through a turbulent pipe flow, the drag reduction drops. Similarly, when injected poly(ethylene oxide) polymer solutions are used to get drag reduction and then the mixed solution is recycled, the second pass has lower drag reduction. These experiments suggest that network formations in the solution may give higher drag reduction than solutions without such structures.

Non-Homogeneous Solutions

Non-homogeneous drag reduction may form a third class of drag reduction by polymer additives. Vleggaar and Tels [51] found that injection of a polymer solution that remained as a thread in the center of the pipe gave reduced pressure drop near the wall. This drag reduction experiment was repeated by Brewersdorf [52]. Berman and Sinha [53] found that the injected polymer could mix rapidly in the turbulent flow or not mix and remain as a wavy or fixed thread. Only the wavy thread led to drag reduction. This drag reduction by polymer threads comes from an interaction of an unstable jet with the large-scale turbulence and only depends on the properties of the thread. The polymer thread is held together by the polymer network, which can be broken up by the turbulence at high Reynolds numbers.

The high strain region in turbulent pipe flow is thought to be connected with the ejection of fluid from the wall region into the pipe center. The polymer molecules will elongate and align in this process. The wall region appears to contain counter-rotating eddy pairs, which are connected with the bursting process [19, 54]. In between the counter-rotating vortices there is a region of elongational flow where polymer molecules can undergo a high strain with little shear. The resulting high elongational viscosity of the fluid that moves into the core of the pipe leads to a large resistance to change. Small eddies are suppressed. As the concentration of molecules that can respond to the strain rate of the flow increases, the region near the wall that cannot sustain turbulence increases. The limit to this process occurs when the elastic sublayer or buffer layer reaches the center of the pipe. Then the only eddies that can exist are large and the characteristics of these eddies are determined by the properties of the internal structure of the polymer solution. It is not clear if we can assume the fluid is homogeneous, because the elastic liquid in turbulent flow contains regions with different histories. Therefore, now homogeneous interactions between elastic threads and turbulent eddies may exist in well mixed solutions as well as when such threads are purposefully introduced into the flow.

TURBULENT STRUCTURE

Although considerable information on the drag reduction process can be obtained from pressure drop vs. flow rate experiments, more detailed studies on the turbulent structure are needed to proceed further. Two types of experimental techniques are appropriate for such studies—laser Doppler velocimetry and flow visualization. Other probes may disturb the flow or give erroneous results in polymer solutions. Hot film probes or pitot tubes are usually small enough so that the flow around the probe can stretch the polymers. It is interesting that the onset of pitot tube error and the onset of excess pressure drop in porous media both correlate with the same time scale as the onset of drag reduction in small tubes [30, 55]. However, the elongational flow in a pipe entrance or constriction does not appear to have a time scale correlation [56]. Some of the difficulties with the probes and the explanation for the unusual experimental results may depend upon the presence or formation of molecular aggregates in the solution. There have been tests for errors in hot film velocity measurements that showed no difference between solvent and polymer solution for polymer molecules dissolved in organic solvents [57]. The organic solvents do not have a high degree of polymer-solvent interaction and would be expected to form homogeneous solutions with small amounts of polymer additives. Therefore, we shall include hot film studies in this discussion when the solutions can be considered very dilute and the amount of drag reduction is small, but unusual results in these cases could be questionable.

Mean Velocity Profile

First, let us look at the changes in the mean velocity profile in pipe flows of well mixed polymer solutions compared to Newtonian fluids. Figure 9 shows examples of experimental data [47, 57–60]. The profiles are a function of the amount of drag reduction, which depends upon the number of polymer molecules that are effective. In Figure 9 the lowest curve from Patterson et al. [57] is for Reynolds number 7,500, a concentration of 0.2%, and a drag reduction of 21%. The highest curve for asbestos fibers is for a Reynolds number of 9,000, a concentration of 300 ppm, and a drag reduction of 63%. The other high curve from Mizushina and Usui [59] is for a Reynolds number of 14,400, a concentration of 300 ppm, and a drag reduction of 72%. The three experiments that group in the middle all have drag reduction of about 35%. Other details will be given when we discuss Figure 10. For drag reduction less than about 30%, the displacement of the log region is ΔB of Equation 2. Very little if any change is seen for $y_+ < 5$ when the local velocity and y_+ are made dimensionless by the actual shear velocity and the solution viscosity measured at low shear rates.

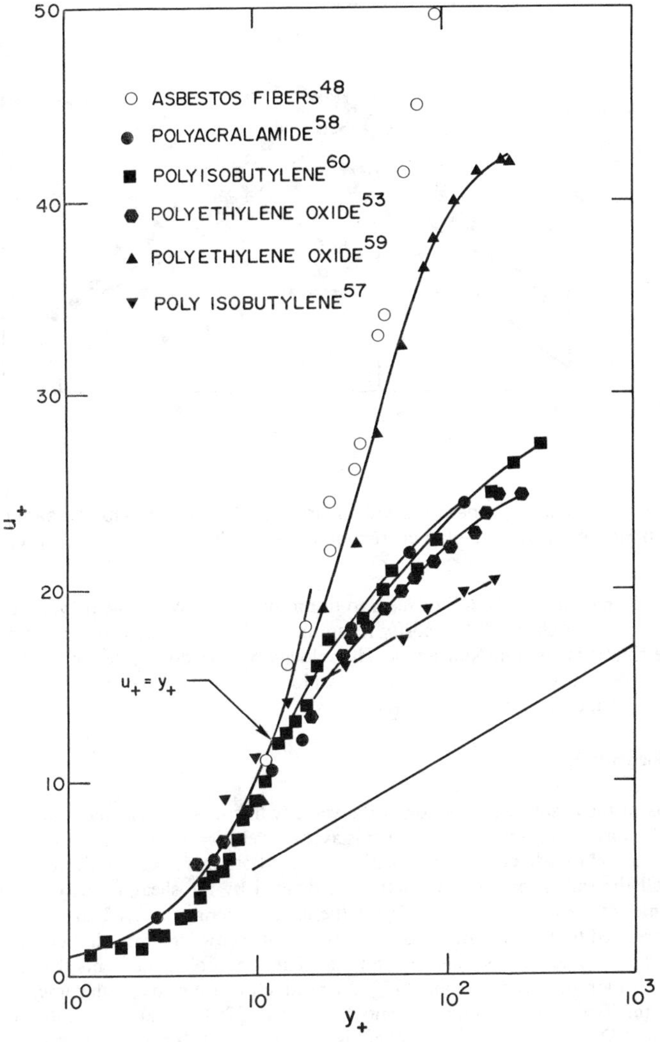

Figure 9. Velocity profiles in pipe flows of drag reducing agents. The solvent is water for all but poly-iso-butylene, which is dissolved in Newtonian oils.

The major effect of polymers on the dynamics of the flow is illustrated by the increase in size of the buffer layer. If we increase the number of effective molecules by increasing the concentration or the Reynolds number the buffer region grows and the log profile retains the same slope but moves upward. Finally, the log region disappears completely and a high enough concentration will produce a non-Newtonian laminar flow with superimposed large fluctuations. Virk's "maximum drag reduction asymptote" gives a velocity profile with the equation

$$u_+ = 11.7 \ln y_+ - 17.0 \tag{7}$$

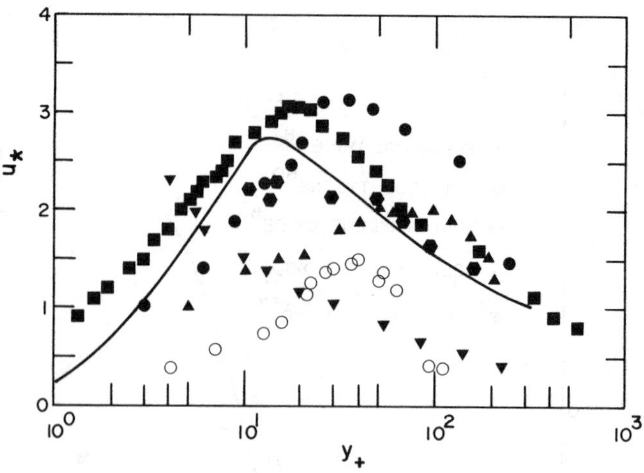

Figure 10. Turbulent intensity near the wall in pipe and channel flows for the same experiments as shown in Figure 9 (symbols also correspond). The solid line is a typical solvent curve.

Velocity profile measurements do not seem to fit on this curve. When the polymer drag reduction exceeds 30%, it is not clear what viscosity to use in the dimensionless distance from the wall, y_+, because the solutions are non-Newtonian. The velocity profiles do not follow the viscous relationship $u_+ = y_+$ beyond $y_+ = 5$, and we will have to look at details of the velocity fluctuations to see what happens in the expanded buffer layer.

Velocity Fluctuations

If we look at the absolute magnitude of the velocity fluctuations near the wall, we find that the polymer solutions usually lead to reduced rms average velocity fluctuations compared to pure solvent. When the velocity fluctuation are scaled with u_*, both increases and decreases are observed.

Let us call the rms axial velocity fluctuation divided by the shear velocity the dimensionless turbulent intensity. Experimental results for the same experiments used to obtain the data for Figure 9 were used to draw Figure 10. Some experiments show larger maximum values than Newtonian fluids and some show smaller magnitudes. The group of experiments that gave essentially the same velocity profile and the same 35% drag reduction are of particular interest. The data of Reischman and Tiederman [58] for a Reynolds number of 20,000 and a concentration of 100 ppm and the data of Durst et al. [60] for a Reynolds number of 22,500 and a concentration of 800 ppm show a higher maximum turbulence intensity compared to pure solvent. The location of the maximum is moved toward the pipe center, and the turbulent intensity at distances between the maximum and the pipe center is higher than in the solvent. At closer distances from the wall than the location of the maximum in turbulent intensity, the two experiments differ. The organic solvent leads to higher fluctuations for polymer solutions compared to pure solvent, and the aqueous solvent leads to lower fluctuations. Berman and Sinha [53] found a maximum turbulent intensity close to that of pure solvent and other characteristics in agreement qualitatively with the aqueous solvent result of Reischman and Tiederman. Berman and Sinha used a Reynolds number of 9,000 and a concentration of 100 ppm. The lower Reynolds number comparison for water is not shown but the curve is slightly lower than the solid line representing a Newtonian fluid at a Reynolds number of 20,000, which is shown in Figure 10.

The location of the peak with respect to distance from the wall varies from closer to the wall to farther from the wall for polymer solutions compared to Newtonian fluids in turbulent flow if we include the hot film experiment of Patterson et al. [57]. Only one consistent trend is clear. For high drag reduction the maximum turbulent intensity is decreased compared to Newtonian fluids and the peak is located further from the wall. In these cases the peak moves closer to the pipe center as drag reduction is increased. In addition, the velocity fluctuations in the very center of the pipe are comparable to the Newtonian values (i.e., the dimensionless turbulent intensity is about one). It is possible that the turbulent intensity would be reduced to zero if measurements were made in the non-Newtonian laminar flow, but there is some evidence that fluctuations due to large instabilities will still exist and the flow is not truly laminar.

Experimental measurements of turbulent intensity showing increases compared to solvent when drag reduction is not near the maximum are hard to explain. One key to what may be happening is found from the laser Doppler velocity studies in jets [61]. When flow visualization of a polymer solution jet shows elastic threads as pictured in Figure 11, laser Doppler measurements give increased axial turbulent intensities. The model of elongation of polymer molecules by the vortices in the buffer region could lead to the creation of similar jets within the turbulent flow. As the jet interacts with the surrounding fluid, instabilities are produced that break-up the jet into strings. These strings move in the flow producing large-scale motions that contribute to the turbulent intensity, but the motions in the axial and radial directions are uncorrelated so there is no contribution to the Reynolds stress. At high polymer concentrations the polymer solution jet looks like Figure 12 and only the largest scale instabilities are present. If similar jets are found in the turbulent boundary layer, lower amounts of drag reduction would lead to jets with string break-up and high

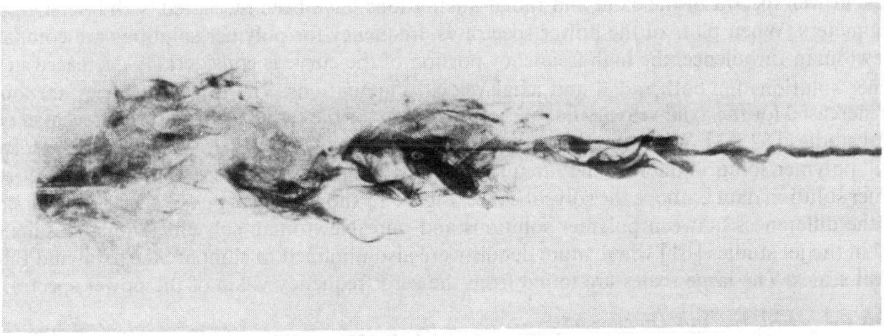

Figure 11. Photograph of a 2-mm jet of 100-ppm poly(ethylene oxide) M $\sim 2 \times 10^6$ flowing at about 5 m/s into the same stagnant fluid. (From Berman and Tan [61].)

Figure 12. Photograph of a 2-mm jet of 100-ppm polyacrylamide (M $\sim 10^7$) dissolved in deionized water.

observed turbulence intensities in the buffer layer. For high drag reduction with high polymer concentrations the jet would not break-up and lower measured turbulence intensities would be found. This is the observed behavior so the jet analogy leads to a qualitative explanation.

The normalized radial intensities also may be higher or lower than the Newtonian fluid in turbulent flow. Examples of experimental results are shown in Figure 13. The same explanation holds for the radial intensities as the axial intensities. Radial intensities are suppressed for high drag reduction and may be increased or decreased for low drag reduction. The largest decreases in radial fluctuations shown in Figure 13 are for a 1,000-ppm solution at a Reynolds number of 22,400 with 75% drag reduction.

The Reynolds stress or correlation between axial and radial fluctuations appears to drop near the wall in dilute polymer solutions [57, 60] compared to Newtonian fluids in turbulent flow. When the higher concentrations of polymer solutions are used, the Reynolds stress scaled with shear velocity is less than the results for a Newtonian fluid over the entire pipe [62]. The drop in Reynolds stress is consistent with the picture of large fluctuations in axial velocity that are uncorrelated with the radial velocity fluctuations shown. Examples of the measured Reynolds stress in turbulent pipe flows are shown in Figure 14.

The solutions with the large decrease in Reynolds stress across the entire pipe were 75% drag reducing at Reynolds numbers near 23,000. The experiment showing only a change close to the wall was for 21% drag reduction.

We have discussed the averaged values of the velocity fluctuations and we can also look at the details. Power spectra of the axial and radial fluctuations have been measured with laser Doppler anemometers. When plots of the power spectra vs. frequency for polymer solutions are compared to Newtonian turbulence, the high frequency portion of the curve is considerably decreased in the polymer solutions for both radial and axial velocity fluctuations. The low frequencies are found to be increased for the axial velocity fluctuations and also for the radial fluctuations in recent experimental studies [62, 63]. When the power spectra are scaled by the inner or small scales, the solvent and all polymer solution data for high frequencies will collapse to one curve but the low frequency polymer solution data is above the solvent curve. Figure 15 shows typical power spectra, which illustrate the differences between polymer solutions and pure Newtonian solvents. Similar results are found in the jet studies [61] where more details were also obtained to compare the axial and radial integral scales. The large scales are found from the zero frequency value of the power spectra. At

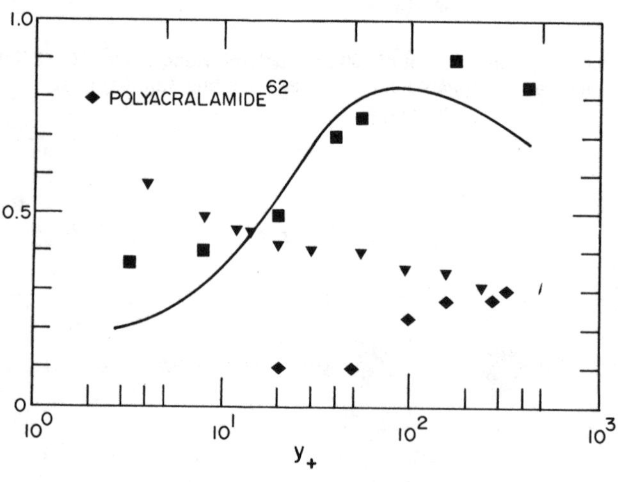

Figure 13. Radial velocity fluctuations near the wall in pipe flows of polymer solutions. Two experiments illustrated correspond to the experiments and symbols in Figures 9 and 10. The solid line is for pure solvent.

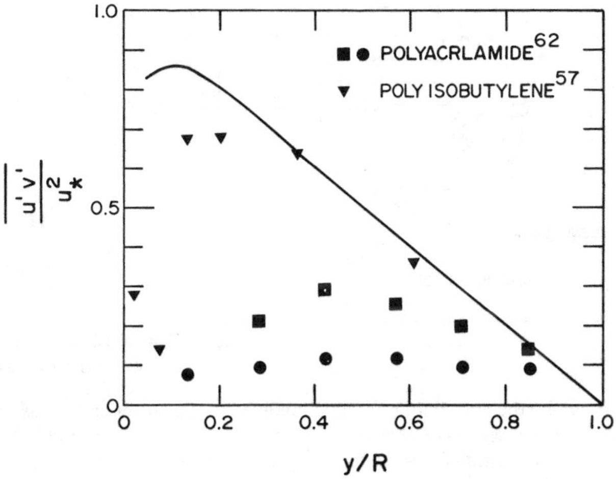

Figure 14. Reynolds stress in polymer solution flows. The solid line represents pure solvent.

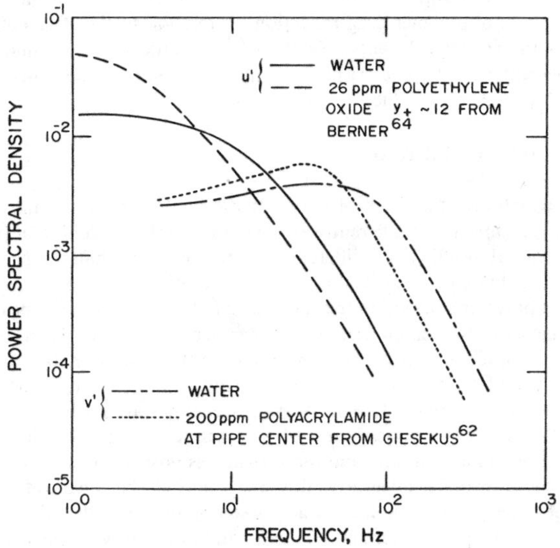

Figure 15. Power spectral density for axial and radial velocity fluctuations.

the jet centerline the ratio of axial to radial integral scales was 2.3 for water, 1.7 for a 100-ppm polyethylene oxide solution, and 3.8 for a 100-ppm polyacrylamide solution. These latter two solutions correspond to Figures 11 and 12, respectively. Therefore, the polyacrylamide solution clearly shows stretching of large eddies in the axial direction in power spectra and flow visualization. The polyethylene oxide solution appears to have stretched large eddies in the photographs but the break-up of the elastic strings distorts the low frequency power spectra to give the lower ratio of integral scales.

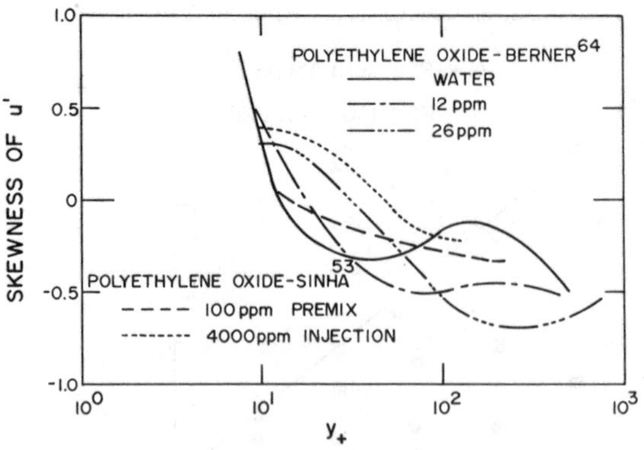

Figure 16. Skewness of the axial velocity fluctuations.

It is difficult to handle the vast amount of data involved in the power spectra as a function of distance which appear in the results of McComb and Rabie [63] or Berner and Scrivener [64]. Another informative plot that emphasizes the large eddies is the plot of skewness vs. y_+. The skewness or third moment near the wall was measured by Usui and Sano [65] and Berner [64] and is shown in Figure 16. Their results show that the skewness in the inertial sublayer is negative and lower for polymer solutions than the solvent. Very close to the wall, the skewness is the same for polymer solutions and solvent when the skewness is scaled with the rms velocity fluctuation and the amount of drag reduction is low. The implication is that more slow moving fluid from the wall is found in the pipe core in the case of polymer drag reduction. In relation to the stretching of the polymers in the vortices near the wall the skewness result means that the stretched fluid persists and does not break-up in the center of the pipe. At higher drag reduction the faster moving fluid very close to the wall is found at higher y_+ values for the polymer solutions.

Non-Homogeneous Structure Differences

When the drag reduction is the result of the fluctuating polymer threads in the pipe center, the turbulent structure is changed in a different way compared to the stretching of polymer molecules near the wall. The velocity profile has a different slope in the center than the Newtonian or homogeneous drag reducing fluids [52, 53]. The rms axial fluctuations behave in exactly the same way as the homogeneous polymer solution at the same drag reduction except the peak in the turbulent intensity is moved closer to the center. The major difference is shown in Figure 16 where the skewness becomes negative much further from the wall than for the homogeneous drag reduction. The non-homogeneous type of drag reduction is produced by an interaction of the fluctuating elastic thread with the large scale turbulence. Some of the large eddies are pushed to the wall leading to an increase in the average eddy size near the wall. As a result, the high velocity fluid from the pipe center is found more often near the wall and the skewness is positive. The presence of larger eddies in the dissipation region or buffer layer gives drag reduction. Perhaps we can also get this type of drag reduction when the polymer molecules are stretched in the wall region forming an elastic thread which persists in the pipe center and fluctuates and interacts with the large eddies. It is possible that both types of drag reduction exist when the drag reduction is a maximum. To exceed the maximum there must be some way to increase the size or persistence of the elastic threads.

Bursting Near the Wall

We return again to the action of polymers in the bursting process near the wall. A recent review by Cantwell [66] summarizes the turbulent dynamics near a wall for Newtonian fluids. Flow visualization studies in boundary layers and shear layers have led to a qualitative picuture of the interaction between sweeps coming from the outer region ($y_+ > 100$) and low-speed streaks in the inner region located between counter-rotating streamwise vortices. Quantitative measurements can be made of the distance between streaks in the spanwise direction, the length of streaks in the streamwise direction and the time between bursts. The averages appear to correlate with inner variables giving

$$\lambda_+ = \lambda u_*/\nu \cong 100$$

$$\Delta x_+ = \Delta x\, \mu_*/ \cong 1{,}000$$

$$T_{b+} = T_b u_*^2/\nu \cong 100$$

where λ is the average spanwise distance between streaks, Δx is the average length of streaks, and T_b is the average time between bursts. In polymer solutions all of the dimensionless quantities associated with bursting appear to increase [63, 67–69]. These results are consistent with the previous description of the inhibition of small eddies and the expansion of the inner layer in polymer solution turbulent flow. The numerical value of the dimensionless streak spacing, streak length, and bursting period depends on the amount of drag reduction. Some difficulty has been experienced in measuring bursting rates because the probe size and the method of analysis affect the result [68, 70, 71].

FRICTION FACTOR CORRELATIONS

In order to use these high molecular weight polymers in practical situations, scale-up equations are necessary. For small amounts of drag reduction and homogeneous solutions Equations 4 and 5 can be used as already indicated. Only two constants are required, the onset shear velocity, which is a function of the polymer time scale, and the slope increment, which is a function of concentration of the polymer. As concentration is increased the increased buffer layer must be included in the integration of the velocity profile to get the friction factor. No additional parameters are necessary if some universal relationship between u_+ and y_+ is assumed for the viscous sublayer ($y_+ < 5$) and the buffer layer ($5 < y_+ <$ the interaction with the displaced log profile).

Thickened Sublayer Models

If the viscous sublayer is extended to the intersection with the log profile, we get a thickened viscous sublayer model [72].

$$\sqrt{2/f} = A(1-\xi)^2 \ln Re\sqrt{f} + (1-\xi)^2[\Delta B - A \ln 2\sqrt{2}] - G$$

where $\xi = y/R$ is the intersection of the viscous sublayer with the log profile in dimensionless form, A is the slope of the log profile, and G is a function of ξ and Ru_*/ν. For high Reynolds numbers G is approximately 3. This model introduces a third parameter that is the viscosity of the polymer solution, ν. At the higher concentrations of polymer, the time scale becomes a function of concentration so the model really has only two parameters and we can reduce this to one when viscosity of the solvent and the polymer solution are taken to be equal.

Burger et al. [73] modified this equation slightly and were successful in correlating data from 1-, 2-, 14-, and 48-inch pipes. The results of the correlation tests were used to design a polymer drag reducing system for the Trans-Alaska Pipeline. The final equations that were used are as follows:

$$\sigma[1 + \psi(2^{3/2} \log \sigma + 1.454\gamma_o\theta_P\sigma - 0.8809]^2 = 1 \qquad (8)$$

$$\theta_P = a\phi^b\gamma_P^c \qquad (9)$$

$$\psi = (f_o/8)^{1/2}$$

$$\gamma_P = \gamma_o\sigma$$

where σ is $(1-\beta)$ and β is the fractional drag reduction. The subscript o refers to a Newtonian fluid, ϕ is the concentration in parts per million, and γ is the shear rate at the wall u_*^2/ν. The constants a, b, and c were found by linear regression from the test results. The procedure is to measure fractional drag reduction (flow rate and pressure drop) for several pipe sizes and concentrations. Then, θ_P can be calculated from Equation 8 so the experimental results a, b, and c are then calculated. For scale-up we choose pipe diameter and drag reduction required for a given flow rate. Then θ_P is found from Equation 8 and ϕ is calculated from Equation 9. Another possibility is to find the drag reduction given ϕ and Newtonian conditions. A transcendental equation for σ is obtained from Equations 8 and 9, and is solved by trial and error.

This type of model can be further modified by using Virk's maximum drag reduction profile for the buffer layer. The result becomes complicated for engineering applications.

Damping-Factor Models

When numerical computations are necessary, a modified Van Driest Model is useful. The velocity distribution is

$$u_+ = \int_0^{y_+} \frac{2(1 - y/R)\,dy_+}{1 + [1 + 4l_n^2(1 - y/R)]^{1/2}} \tag{10}$$

where l_n is Prandtls' mixing length given by

$$l_n = fnD_f \tag{11}$$

Mizushina and Usui [59] have used

$$fn = 0.4y_+ - 0.44y_+^2/R_+ + 0.24y_+^3/R_+^2 - 0.06y_+^4/R_+^3 \tag{12}$$

$$D_f = 1 - \exp[-(y_+/26)(-\alpha + \sqrt{\alpha^2 + 1})^{1/2}] \tag{13}$$

$$\alpha = 2T_1(u_*^2/\nu)/26^2 \tag{14}$$

The friction factor—Reynolds number expression is found by integrating Equation 10 across the pipe. The only parameter in this equation is the ratio of time scales in Equation 14, which appears in the damping factor expression of Equation 13. For Newtonian fluids, $\alpha = 0$. The equation for the polymer time scale here is

$$T_1 = 0.4\eta_s[\eta]^2Mc/RT \tag{15}$$

which applies at relatively high polymer concentrations where T_1 is concentration dependent. The maximum value of α was found to be 40.

Another damping factor model was used by Darby and Chang [74]. They relate the energy dissipation in an oscillating fluid to the ratio of polymer time scale to flow time scale or Deborah number, De.

$$f/f_s = 1/(1 + De^2)^{1/2} \tag{16}$$

$$De = 0.145T_1^*[Re_s\,\eta_s/(1 + T_1^{*2})\eta_o]^{0.285} \tag{17}$$

$$T_1^* = 8v_{ave}T_1/D \tag{18}$$

where η_o is the zero shear rate limiting viscosity and the subscript s refers to solvent. The two numerical constants in Equation 17 were found by regression analysis and may differ for polymer solutions other than those tested by Darby and Chang. This model contains two constants η_o and T_1 that are found from the rheology of the polymer solution in non-turbulent flows. It is not clear if this model can be used at low concentrations or with T_1 from Equation 1 or 15. Many other correlations can be found in the literature, but they usually contain more than two adjustable parameters.

The correlations provide another bit of evidence for a time-scale relationship between the molecular and turbulent motions. In terms of the friction factor vs. Reynolds number or the fractional drag reduction, the relation is complex because the entire velocity profile must be examined. The viscous sublayer and buffer layer may extend over a large fraction of the pipe radius. The correlations are useful when drag reduction exceeds 30%. The previous simpler analysis can be used for lower amounts of drag reduction with random coiling polymers. The drag reduction for partially extended electrolytes or nonhomogeneous drag reduction have not been successfully modeled.

CONCLUSION

Interactions between molecules and turbulent flow can occur in several forms when solutions of high molecular weight polymers are drag reducing. Polymer molecules can be stretched by high strain regions in the buffer layer near the wall. After the initial stretching the molecules interact with turbulence where shear layers separating high and low velocities exist. When concentrations are above 0.05 ppm, the stretched molecules interact with each other to produce networks in the flow that resist interaction with the turbulence. As a result, the smallest eddies that can exist in the polymer solution flow become larger than the smallest eddies in Newtonian flow at the same velocity.

Other types of interactions that occur in turbulent flow include degradation of polymer molecules, degradation or break-up of networks and the interaction of networks in elongated threads with large eddies. These are in addition to the stretching of single molecules or groups of molecules. The complex behavior of large molecules in turbulent flow can only be qualitatively explained at present. In this chapter experimental results have been discussed but the details of the interaction between a viscoelastic fluid and turbulent flow are not known. This is not surprising because the detailed behavior of Newtonian fluids in turbulent flows are also not well understood. A better understanding of turbulence and of the interaction of turbulence with molecules will continue to be of importance in theoretical fluid mechanics for some time in the future.

NOTATION

b	constant	R	universal gas constant
C	concentration	Re	Reynolds number
D	pipe diameter	S	strain rate
De	Deborah number	T	temperature
f	friction factor	T_1	relaxation time
K	constant	u_*	friction velocity
L	length	u_+	dimensionless velocity
l_n	Prandtl mixing length	u	local velocity
M	molecular weight	v_{ave}	average velocity
N	number of backbone units in molecule	y	distance from wall
P	pressure		

Greek Symbols

β	fraction drag reduction	δ	constant slope increment
γ	shear rate at wall	η	intrinsic velocity

REFERENCES

1. White, A., *Drag Reduction by Additives*. BHRA Fluid Engineering, 1976.
2. Zakin, J. L., and Hunston, D. L., "Introductory Remarks." *Polymer Engineering and Science*, Vol. 20, 1980, pp. 449–450.

3. Hele-Shaw, H. S., "Experiments on the Nature of the Surface Resistance in Pipes and on Ships." *Trans. Inst. Nat. Architects*, Vol. 39, 1897, pp. 145–156.
4. Forrest, F., and Grierson, G. A., *Paper Trade J.*, Vol. 92, No. 22, 1931, pp. 39–41.
5. Toms, B. A., "Some Observations on the Flow of Linear Polymer Solutions Through Straight Tubes at Large Reynolds Numbers," *Proc. 1st Int. Rheology Congress*, Part 2, North Holland, 1949, pp. 135–142.
6. Mysels, K. J., Flow of Thickened Fluids, U.S. Patent 2 492 173, December 27, 1949.
7. Mysels, K. J., "Early Experiences with Viscous Drag Reduction," *AIChE Chem. Engr. Prog.* Vol. 67. Symposium Series No. 111, 1971, 45–49.
8. Toms, B. A., "On the Early Experiments on Drag Reduction by Polymers." *Physics of Fluids*, Vol. 20, 1977, pp. 53–35.
9. Agoston, G. A., Harte, W. H., Hottel, H. C., Klemm, W. A., Mysels, K. J., Pomeroy, H. H., and Thompson., J. M., "Flow of Gasoline Thickened by Napalm," *Ind. Engr. Chem.*, Vol. 46, 1954, pp. 1017–1019.
10. Dailey, J. W., and Bugliarello., G., "Basic Data for Dilute Fiber Suspensions in Uniform Flow with Shear," *TAPPI*, Vol. 44 1961, pp. 497–512.
11. Dodge, D. W., and Metzner, A. B., "Turbulent Flow of Non-Newtonian Systems," *AIChE Journal*, Vol. 5, 1959, pp. 189–204.
12. Shaver, R. G., and Merrill, E. W., "Turbulent Flow of Pseudoplastic Polymer Solutions in Straight Cylindrical Tubes," *AIChE Journal*, Vol. 5, 1959, pp. 181–188.
13. Fabula, A. G., "The Toms Phenomenon in the Turbulent Flow of Very Dilute Polymer Solutions." in *Proc. Fourth Inst. Congress on Rheology, Part 3*, 1963, Interscience, New York, 1965, pp. 455–479.
14. Hoyt, J. W., and Fabula, A. G., "The Effect of Additives on Fluid Friction," in *Proc. Fifth Symposium on Naval Hydrodynamics*, 1964, Office of Naval Research, Washington, 1966, pp. 947.
15. Gadd, G. E., "Friction Reduction," in *Encyclopedia of Polymer Science and Technology*, Vol. 15, Interscience, 1971, 224–253.
16. Lumley, J. L., "Drag Reduction by Additives," *Ann. Rev. Fluid Mechanics*, Vol. 1, 1969, pp. 367–384.
17. Hoyt, J. W., "The Effect of Additives on Fluid Friction." *Transactions of the ASME Journal of Basic Engineering*, Vol. 94, 1972, 258–285.
18. Lumley, J. L., "Drag Reduction in Turbulent Flow by Polymer Additives," *J. Poly. Sci. Macromolecular Reviews*, Vol. 7, 1973, pp. 283–290.
19. Landahl, M. T., "Drag Reduction by Polymer Addition," In *Proc. Thirteenth Int. Congress Theoretical and Applied Mechanics* (Moscow), Berker, E., and Mikhaelov, G. K., eds., Berlin Springer-Verlag, 1973, pp. 177–199.
20. Virk, P. S. "Drag Reduction Fundamentals," *AIChE Journal*, Vol. 21, 1975, pp. 625–656.
21. Little, R. C., Hansen, R. J., Hunston, D. L., Kim, O. K., Patterson, R. L., and Ting, R. Y., "The Drag Reduction Phenomenon Observed Characteristics Improved Agents, and Proposed Mechanisms," *I&EC Fundamentals*, Vol. 14, 1975, pp. 283–296.
22. Lumley, J. L., "Two-Phase and Non-Newtonian Flows," in *Topics in Applied Physics*, Vol. 12 Turbulence, P. Bradshaw Ed., Berlin Springer-Verlag, 1976, pp. 290–324.
23. Berman, N. S., "Drag Reduction by Polymers," *Ann. Rev. Fluid Mechanics*, Vol. 10, 1978, pp. 47–64.
24. Sellin, R. H. I., Hoyt, J. W., and Scrivener, O., "The Effect of Drag Reducing Additives on Fluid Flows and Their Industrial Applications. Part I: Basic Aspects," *J. Hydraulic Res.*, Vol. 20, 1982, pp. 29–68.
25. Sellin, R. H. J., Hoyt, J. W., Pollet, J., and Scrivener, O., "The Effect of Drag Reducing Additives on Fluid Flows and Their Industrial Applications. Part 2: Present Applications and Future Proposals," *J. Hydraulic Research*, Vol. 20, 1982, pp. 235–292.
26. Hoyt, J. W., "Drag-Reduction Effectiveness of Polymer Solutions in the Turbulent-flow Rheometer: A Catalog," *J. Polymer Sci. B. Polymer Letters*, Vol. 9, 1971, pp. 851–862.
27. Berman, N. S., and George, W. K., "Onset of Drag Reduction in Dilute Polymer Solutions," *Phys. Fluids*, Vol. 17, 1974, pp. 250–251.
28. Cox, L. R., North, A. M., and Dunlap, E. H., "Evidence for a Time Scale Effect On Drag

Reduction In Solutions of Polystyrene in Toluene," in *Proc. Int. Conf. Drag Reduction*, St. John's Coll. Cambridge N. Coles Ed. pp. C217–20.
29. Berman, N. S., "Flow Time Scales and Drag Reduction," *Phys. Fluids*, Vol. 20, 1977, pp. S168–S174.
30. Durst, F., Haas, R., and Interthal, W., "Laminar and Turbulent Flows of Dilute Polymer Solutions: A Physical Model," *Rheol. Acta*, Vol. 21, 1982, pp. 572–577.
31. Wells, C. S., and Spangler, J. G., "Injection of Drag-Reducing Fluid into Turbulent Pipe Flow of a Newtonian Fluid," *Phys. Fluids*, Vol. 10, 1967, pp. 1890–1894.
32. McComb, W. D., and Rabie, L. H., "Development of Local Turbulent Drag Reduction Due to Nonuniform Polymer Concentration," *Phys. Fluids*, Vol. 22, 1979, pp. 183–185.
33. Durst, F., Haas, R., Interthal, W., and Keck, T., "Polymerwirkung in Stromungen—Mechanismen Und Praktische Anwendungen," *Chem. Ing. Tech.*, Vol. 54, 1982, pp. 213–221.
34. Bird, R. B., Hassager, O., Armstrong, R. C., and Curtiss, C. F., "Dynamics of Polymeric Liquids, Volume 2 Kinetic Theory," John Wiley, 1977.
35. Oliver, D. R., and Bakhtiyarov, S. I., "Drag Reduction in Exceptionally Dilute Polymer Solutions," *J. Non-Newtonian Fluid Mech.*, Vol. 12, 1983, pp. 113–118.
36. Berman, N. S., "Drag Reduction of the Highest Molecular Weight Fractions of Polyethylene Oxide." *Physics Fluids*, Vol. 20, 1977, pp. 715–718.
37. Berman, N. S., and Yuen, J., "The Study of Drag Reduction Using Narrow Fractions of Polyox," in *Proc. 2nd International Conference on Drag Reduction 1977 BHRA Fluid Engineering*, Cranfield England, pp. Cl-1–10.
38. Hunston, D. L., and Reischman, M. M., "The Role of Polydispersity in the Mechanism of Drag Reduction," *Phys. Fluids.*, Vol. 18, 1975, pp. 1626–30.
39. Hassager, O., "Kinetic Theory and Rheology of Bead Rod Models for Macromolecular Solutions," *J. Chem. Phys.*, Vol. 60, 1974, pp. 2111–2124.
40. Batchelor, G. K., "The Stress Generated in a Non-Dilute Suspension of Elongated Particles by Pure Straining Motion," *J. Fluid Mech.*, Vol. 46, 1971, pp. 813–829.
41. Berman, N. S., Berger, R. B., and Leis, J. R., "Drag Reduction of Well-Mixed Solutions of Poly(Ethylene Oxide) and Organic Dyes in Water," *J. Rheology*, Vol. 24, 1980, pp. 571–587.
42. Virk, P. S., "Drag Reduction of Collapsed and Extended Polyelectrolytes," *Nature*, Vol. 253, 1975, pp. 109–110.
43. Radin, I., Zakin, J. L., and Patterson, G. K., "Drag Reduction in Solid-Fluid Systems," *AIChE J.* Vol. 21, 1975, p. 358–371.
44. Berman, N. S., "Evidence for Molecular Interactions in Drag Reduction in Turbulent Pipe Flows," *Polymer Engineering and Science*, Vol. 20, 1980, pp. 451–455.
45. Berman, N. S., Griswold, S. T., Elihu, S., and Yuen, J., "An Observation of the Effect of Integral Scale on Drag Reduction," *AIChE Journal*, Vol. 24, 1978, pp. 124–130.
46. Hunston, D. L., and Zakin, J. L., "Flow-Assisted Degradation in Dilute Polystyrene Solutions," *J. Polymer Engineering and Science*, Vol. 20, 1980, pp. 517–523.
47. Rollin, A., and Seyer, F. A., "Velocity Measurements in Turbulent Flow of Viscoelastic Solutions," *Can. J. Chem. Eng.*, Vol. 50, 1972, pp. 714–718.
48. McComb, W. D., and Chan, K. T. J., "Drag Reduction in Fibre Suspensions: Transitional Behavior Due to Fibre Degradation," *Nature*, Vol. 280, 1979, pp. 45–46.
49. Kale, D. D., and Metzner, A. B., "Turbulent Drag Reduction in Dilute Fiber Suspensions: Mechanistic Considerations," *AIChE J.*, Vol. 22, 1976, pp. 669–674.
50. Lee, W. K., Vaseleski, R. C., and Metzner, A. B., "Turbulent Drag Reduction in Polymeric Solutions Containing Suspended Fibers," *AIChE J.*, Vol. 20, 1974, pp. 128–133.
51. Vleggaar, J., and Tels, M., "Drag Reduction by Polymer Threads," *Chem. Engr. Sci.*, Vol. 28, 1973, pp. 965–968.
52. Bewersdorff, H. W., "Effect of a Centrally Injected Polymer Thread on Drag in Pipe Flow," *Rheol. Acta*, Vol. 21, 1982, pp. 587–589.
53. Berman, N. S., and Sinha, P., "Drag Reduction in Pipe Flow for Non-Homogeneous Injection of Polymer Additives," to be published, Proc. of Drag Reduction, 84 Cong. Bristol, England. See also Sinha, P. K., "Drag Reduction by Centrally Injected Polymer Jet," Thesis Arizona State University, 1984.

54. Seyer, F. A., and Metzner, A. B., "Turbulence Phenomena in Drag Reducing Systems," *AIChE J.*, Vol. 15, 1969, pp. 426–434.
55. Berman, N. S., and George, W. K., "Time Scale and Molecular Weight Distribution Contributions of Dilute Polymer Solution Fluid Mechanics," *Proc. 1974 Heat Trans. Fluid Mech. Inst.* Davis, L. R., and Wilson, R. E., eds., Stanford Univ. Press, 1974, pp. 348–364.
56. Quibraham, A., "Excess Pressure Drop of Polymer Solutions in Laminar Capillary Tube Flows," *Physics Fluids*, Vol. 22, 1979, pp. 784–785.
57. Patterson, G. K., Chosnek, J., and Zakin, J. L., "Turbulence Structure in Drag Reducing Polymer Solutions," *Phys. Fluids*, Vol. 20, 1977, pp. 589–599.
58. Reischman, M. M., and Tiederman, W. G., "Laser-Doppler Anemometer Measurements in Drag-Reducing Channel Flows," *J. Fluid Mech.*, Vol. 70, 1975, pp. 369–392.
59. Mizushina, T., and Usui, H., "Reduction of Eddy Diffusion for Momentum and Heat in Vescolastic Fluid Flow in a Circular Tube," *Phys. Fluids*, Vol. 20, 1977, pp. 100–108.
60. Durst, F., Keck, T., and Kleine, R., "Turbulence Quantities and Reynolds Stress in Pipe Flow of Polymer Solutions Measured by Two-Channel Laser-Doppler Anemometry," in *Turbulence in Liquids*, Patterson, G. K., and Zakin, J. L., eds. U. Missouri Rolla, 1981, pp. 55–65.
61. Berman, N. S., and Tan, H., "Two-Component Laser Doppler Velocimeter Studies of Submerged Jets of Dilute Polymer Solutions," *AIChE Journal*, Vol. 31, 1985, pp. 208–215.
62. Giesekus, H., "Structure of Turbulence in Drag Reducing Flows," in Von Karman Institute for Fluid Dynamics Lecture Series 1981–6, Non-Newtonian Flows, May 4, 1981, pp. 12. See also Schummer, P. and Thielen, W. "Structure of Turbulence in Viscoelastic Fluids," *Chem. Eng. Commun.* Vol. 4, 1980, pp. 563–606.
63. McComb, W. D., and Rabie, L. H., "Local Drag Reduction Due to Injection of Polymer Solutions into Turbulent Flow in a Pipe. Part II: Laser Doppler Measurements of Turbulent Structure," *AIChE Journal*, Vol. 28, 1982, pp. 558–565.
64. Berner, C., and Scrivener, O., "Drag Reduction and Structure of Turbulence in Dilute Polymer Solutions," in "*Viscous Flow Drag Reduction.*" G. R. Hough ed. 1980, Amer. Inst. of Aeronautics and Astronautics, pp. 290–299. See also Berner, C., "Etude Statistique de la Structure de Turbulence Dans Les Fluides Non-Newtoniens," Thesis Louis Pasteur, University of Strasbourg, 1980.
65. Usui, H., and Sano, Y., "Turbulence Structure of Dilute Drag-Reducing Polymer Solutions in a Rectangular Open Channel Flow, *AIChE Journal*, Vol. 29, 1983, pp. 611–617.
66. Cantwell, B. J., "Organized Motion in Turbulent Flow," *Ann. Rev. Fluid Mech.*, Vol. 13, 1981, pp. 457–515.
67. Achia, B. U., and Thompson, D. W., "Structure of the Turbulent Boundary in Drag-Reducing Pipe Flow," *J. Fluid Mech.*, Vol. 81, 1977, pp. 439–464.
68. Tiederman, W. G., and Luchek, T. S., "Bursting Rates in Channel Flows and Drag-Reducing Channel Flows," *Proc. 8th Symposium on Turbulence*, Zakin J. L., and Patterson, G. K., eds., 1983, U. Missouri-Rolla, Paper No. 2.
69. Donohue, G. L., Tiederman, W. G., and Reischman, M. M., "Flow Visualization of the Near Wall Region in a Drag-Reducing Channel Flow," *J. Fluid Mech.*, Vol. 56, 1972, pp. 559–575.
70. Blackwelder, R. F., and Haritonedis, J. H., "Scaling of the Bursting Frequency in Turbulent Boundary Layers," *J. Fluid Mech.*, Vol. 132, 1983, pp. 87–103.
71. Berman, N. S., "The Effect of Sample Probe Size on Sublayer Period in Turbulent Boundary Layers," *Chem. Engineering Comm.* Vol. 5, 1980, pp. 337–345.
72. Savins, J. G., and Seyer, F. A., "Drag Reduction Scale-Up Criteria," *Phys. Fluids*, Vol. 20, 1977, pp. 578–584.
73. Burger, E. D., Chorn, L. G., and Perkins, T. K., "Studies of Drag Reduction Conducted over a Broad Range of Pipeline Condition when Flowing Prudhoe Bay Crude Oil," *J. Rheology*, Vol. 24, 1980, pp. 603–626.
74. Darby, R., and Chang, H. D., "A Generalized Correlation of Friction Loss in Drag Reducing Polymer Solutions," *AIChE Journal*, Vol. 30, 1984, pp. 274–280.

CHAPTER 33

A GENERALIZED CORRELATION FOR FRICTION LOSS IN DRAG REDUCING POLYMER SOLUTIONS

Ron Darby

Chemical Engineering Dept.,
Texas A&M University
College Station, Texas, USA

CONTENTS

INTRODUCTION

It is well known that the addition of small amounts (measured in parts per million) of high molecular weight polymeric additives to a liquid can reduce the friction loss in turbulent pipe flow by as much as 80% below that exhibited by the solvent alone at the same flow rate. This phenomenon has been studied extensively, and a number of review papers have been written on the subject [1–6].

Despite the large volume of work on this subject, a generalized expression or correlation relating the turbulent friction loss characteristics of these solutions to basic, readily measurable, rheological properties has not heretofore been obtained. Although it is generally agreed that turbulent drag reduction is a consequence of the viscoelastic nature of the polymer solutions, experimental characterization of these properties is very difficult because of the very low concentrations of the solutions.

Nonetheless, it has been shown that these solutions are definitely nonlinear viscoelastic fluids [7–10].

The mechanism by which drag reduction occurs has, likewise, not been firmly established. Various approaches to an explanation of drag reduction have been taken, such as reduced energy dissipation [11], modified transient shear response [12, 13, 14], boundary layer thickening [15, 16, 17] and resistance to extensional flow [18, 19]. The approaches range from purely hypothetical to essentially re-arrangement of turbulent flow data, although none of these studies have resulted in a method of quantitatively predicting the pressure drop for a given solution from fundamental measurable physical properties of these solutions.

However, a study by Darby and Chang [20] has shown that it is possible to relate the friction loss in turbulent drag reducing polymer solutions to measurable nonlinear viscoelastic properties of these solutions, in a generalized dimensionless form applicable to fresh and degraded solutions of various concentrations in a wide range of tube sizes. A rheological model that describes the non-Newtonian properties of the solutions is utilized, and the mechanism of drag reduction is explained by the influence of viscoelastic properties on the rate of energy dissipation in the locally oscillating dissipative turbulent eddies. These first order concepts are used to develop a simple model for drag reduction in terms of a generalized friction factor, which includes a dimensionless fluid time constant to account for the influence of elasticity in reducing the energy dissipation. The fluid time constant can be readily determined from measurements of the fluid apparent viscosity function, and the friction factor reduces to the usual Fanning friction factor for Newtonian non-drag reducing fluids. The details of this study will be presented, along with a discussion of the implications of results.

EXPERIMENTAL DATA

Concentrations of 100, 250, and 500 wppm of Separan AP-30, a partially hydrolyzed polyacrylamide manufactured by Dow Chemical Company, in distilled water were used in this study. This additive is a very popular drag reducer for aqueous turbulent flows and, according to Hoyt [21], exhibits maximum drag reduction at a concentration somewhat below 100 wppm. Variations in the drag reducing effectiveness of solutions made with the same commercial polymer may occur due to slight variations in molecular weight distribution from batch to batch, aging, method of formulating the solution, purity of the water especially with respect to the presence of certain ions, etc. According to the classification of Liaw et. al. [22], these solutions would be considered to be "concentrated" and would be classed as strong drag reducers.

Tube flow data, consisting of pressure gradient versus flow rate, were obtained for both fresh and shear degraded solutions of the above concentrations, in stainless steel tubes with diameters of 0.178, 0.216, 0.406, 0.467, 0.616, 0.704, and 1.021 cm. Two different flow arrangements were used. For diameters greater than 0.5 cm, horizontal tubes with piezometric pressure taps located a minimum of 150 diameters from the ends were used. For the smaller tubes, a vertical arrangement was used with the pressure drop being obtained from the measured pressure in an upstream reservoir. In the latter case, two tubes of different lengths for each diameter were used, and the pressure gradient was obtained from the difference in pressure drop and the difference in lengths. All tube arrangements were checked and calibrated with distilled water. Further equipment details and procedures for data reduction are given by Chang [23]. The range of data covered both laminar and turbulent flow, corresponding to solvent Reynolds numbers from about 10^2 to 10^5.

The solutions exhibit pronounced nonlinear viscoelastic properties, similar to those previously reported by Tsai and Darby [9], Darby [7], and others. Apparent viscosity and first normal stress difference versus shear rate were measured for all solutions, both fresh and shear degraded. Viscosity data were obtained over a shear rate range of 10^{-2} to 10^6 s^{-1} by using four different instruments. The details of these measurements and the equipment used are given by Chang and Darby [24].

Shear degradation of the solutions was achieved by repeatedly cycling them through a 0.460 cm I.D., 0.511 m long tube at a constant reservoir pressure of 515 kPa (60 psig). The flow rate was measured for each pass, and the fluid was recycled until the flow rate no longer changed (after 8 to 12 passes).

Turbulent Flow Behavior

A typical shear history is shown in Figure 1. As can be seen, the velocity increases with successive cycles at constant stress, leveling off after 10–12 cycles. This indicates that the degree of drag reduction increases with shear degradation. Although this is contrary to expectations, a similar effect has been previously noted by Ng and Hartnet [25].

The drag reducing properties of the solutions are shown in Figures 2–4 in terms of the Fanning friction factor versus solvent Reynolds number. The use of the Reynolds number based on the solvent viscosity provides a direct indication of the degree of drag reduction, which is defined as a reduction in pressure drop due to the addition of the polymer relative to that for the pure solvent at the same flow rate, i.e. the same solvent Reynolds number. Thus, a comparison of the friction factor for the solution with that of the (Newtonian) solvent at a given Reynolds number provides a direct measure of the degree of drag reduction. The solvent (water) data follow the familiar Hagen-Poiseuille law for laminar flow, and the classical Blasius equation for turbulent flow. As expected for such solutions, the friction factor is greater than that of the solvent in laminar flow, but is as much as 70%

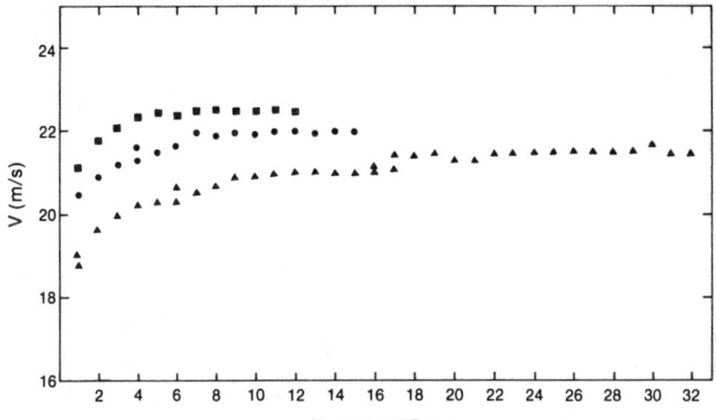

Fig. 1. Typical shear history during degradation: (▲) 500 wppm, (●) 250 wppm, (■) 100 wppm [20].

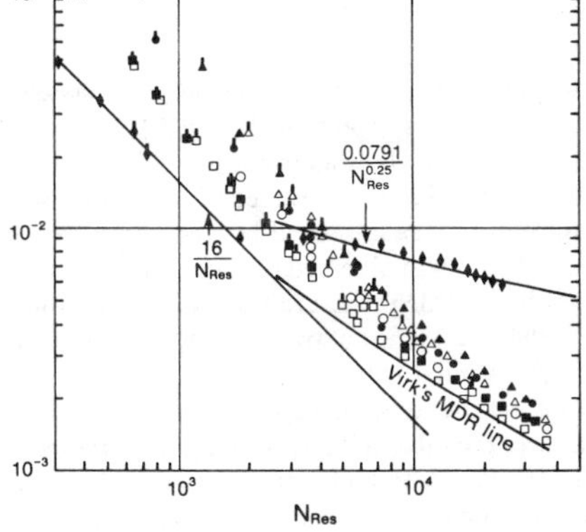

Figure 2. Fanning friction factor versus solvent Reynolds number: (▲) 500 wppm fresh, (●) 250 wppm fresh, (■) 100 wppm fresh, (△) 500 wppm sheared, (○) 250 wppm sheared, (□) 100 wppm sheared, (◆) water. Symbols without tic are for 0.178 cm I.D. tube; symbols with tic are for 0.216 cm I.D. tube [20].

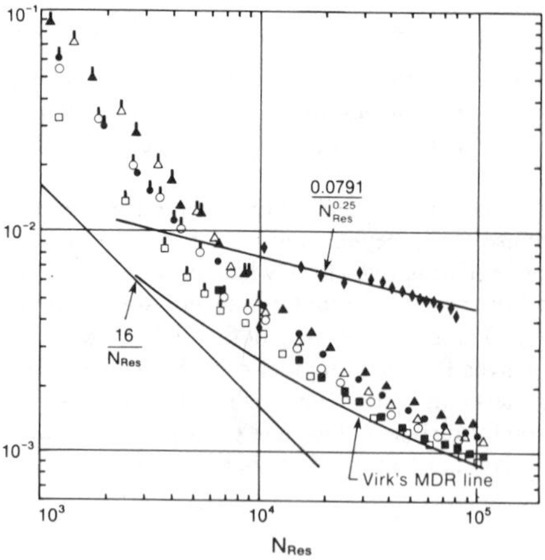

Figure 3. Fanning friction factor versus solvent Reynolds number: (▲) 500 wppm fresh, (●) 250 wppm fresh, (■) 100 wppm fresh, (△) 500 wppm sheared, (○) 250 wppm sheared, (□) 100 wppm sheared, (◆) water. Symbols without tic are for 0.406 cm I.D. tube; symbols with tic are for 0.467 cm I.D. tube [20].

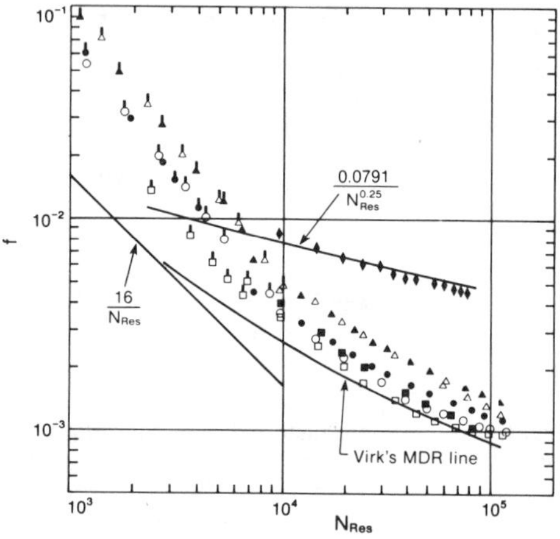

Figure 4. Fanning Friction Factor versus Solvent Reynolds Number: (▲) 500 wppm fresh, (●) 250 wppm fresh, (■) 100 wppm fresh, (△) 500 wppm sheared, (○) 250 wppm sheared, (□) 100 wppm sheared, (◆) water. Tube diameter 0.460 I.D. [20].

below that of the solvent in turbulent flow. The 100 wppm sheared solutions show the greatest degree of drag reduction, approaching the maximum drag reduction asymptote of Virk [26].

Although the polymer solution data in Figures 2–4 appear to show an extension of laminar flow behavior to higher Reynolds numbers, with no obvious transition to turbulence, a transition was indeed observed. For those runs in which the pressure drop was measured between pressure taps using a differential pressure transducer, the transducer output was observed on a chart recorder. For runs with pure water, the increase in fluctuations on the recorded pressure trace was noticeable at a Reynolds number of about 2,100, corresponding to the usual point of transition to turbulence for Newtonian fluids. For the runs with polymer solutions, a similar increase in the pressure drop fluctuations occurred at various points, which depended upon the solution, tube size and flow velocity. The results are shown in Table 1, which shows the value of the solvent Reynolds number corresponding to this transition point ($N_{Res,cr}$) for all six solutions, in two different tube sizes. Note that the critical Reynolds number increases with concentration and with tube diameter, and is higher for the shear degraded solutions than for the fresh, reaching a value above 10^4 for the 500 ppm sheared solution.

Table 1
Critical Reynolds Numbers for Separan AP-30 Solutions in Horizontal Tubes

Diameter (cm)	Solution	$N_{Res,cr}$
0.216	100 ppm Fresh	3,261
	250 ppm Fresh	3,569
	500 ppm Fresh	5,760
	100 ppm Sheared	6,066
	250 ppm Sheared	6,447
	500 ppm Sheared	9,860
0.467	100 ppm Fresh	—
	250 ppm Fresh	4,082
	500 ppm Fresh	6,455
	100 ppm Sheared	6,560
	250 ppm Sheared	8,680
	500 ppm Sheared	10,170

Note: $N_{Res,cr}$ is the critical Reynolds number based on the solvent (water) viscosity.

Rheological Properties

The nonlinear viscoelastic rheological properties of the solutions are shown in Figures 5 and 6 (from Chang and Darby [24]). The apparent viscosity data in Figure 5 show that the effect of shear degradation is to lower the zero-shear viscosity by about one order of magnitude, although the high shear limiting viscosity is virtually unaffected. Also, the shear rate at which transition from low shear Newtonian behavior to non-Newtonian behavior occurs is increased by about one order of magnitude for the degraded solutions. Figure 6 shows the first normal stress difference function for the solutions, which is commonly assumed to be indicative of elastic character. As can be seen, the effect of shear degradation is to reduce the normal stress difference at all shear rates by about the same factors for a given concentration.

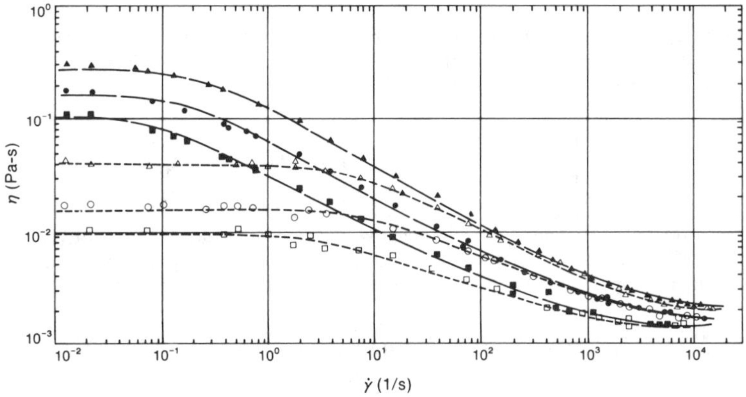

Figure 5. Apparent viscosity versus shear rate [24]. (▲) 500 wppm fresh, (●) 250 wppm fresh, (■) 100 wppm fresh, (△) 500 wppm sheared, (○) 250 wppm sheared, (□) 100 wppm sheared, (◆) water.

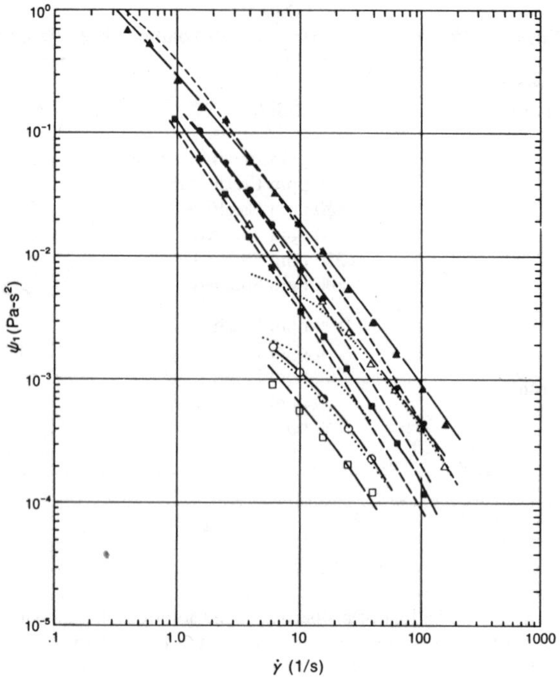

Figure 6. First normal stress function versus shear rate: long dashes, Equation 4; short dashes, Equation 5 [24].

RHEOLOGICAL MODEL

A constitutive equation has been proposed by Chang and Darby [24], which accurately represents the measured data. The general form of the model is given by the following integral or hereditary expression relating shear stress and shear rate:

$$\tau^{ij}(t) = \int_{-\infty}^{1} \phi[t - t', II'] \left[\left(1 + \frac{\epsilon}{2}\right) \frac{\partial x^i}{\partial z^m} \frac{\partial x^j}{\partial z^n} \Delta^{mn}(z, t') - \frac{\epsilon}{2} g^{ir} g^{js} \frac{\partial z^m}{\partial x^r} \frac{\partial z^n}{\partial x^s} \Delta_{mn}(z, t')\right] dt' \quad (1)$$

where Δ^{ij} is the rate of strain tensor, II is the negative of its second invariant, the x^i are current coordinates, the z^i are material coordinates, g^{ij} is the conjugate metric, and $\phi[\ ]$ is the nonlinear relaxation function given by

$$\phi[t, II] = \frac{(\eta_0 - \eta_\infty)}{\theta}(1 + \zeta^2 II)^\beta \exp[-(1 + \zeta^2 II)^{\Omega+\beta} t/\theta] + \eta_\infty \delta(t) \quad (2)$$

The apparent viscosity function predicted by this model is

$$\eta = \eta_\infty + \frac{\eta_0 - \eta_\infty}{(1 + \zeta^2 \dot{\gamma}^2)^\Omega} \quad (3)$$

Table 2
Values of Rheological Parameters for Separan AP-30 Solutions

Solutions	η_0 (Pa-s) X 10	ζ (s)	Ω	η_∞ (Pa-s) X 1,000	θ (s)	β
100 ppm (fresh)	1.113	11.89	0.266	1.30	21.6	0.213
250 ppm (fresh)	1.714	6.67	0.270	1.40	7.10	0.133
500 ppm (fresh)	3.017	3.53	0.300	1.70	2.43	0.04
100 ppm (sheared)	0.098	0.258	0.251	1.30	0.107	0.104
250 ppm (sheared)	0.169	0.106	0.270	1.40	0.0091	0.137
500 ppm (sheared)	0.397	0.125	0.295	1.70	0.187	0.139

and the first normal stress difference function is

$$\psi_1 = \frac{\tau_{11} - \tau_{22}}{\dot{\gamma}^2} = \frac{2(\eta_0 - \eta_\infty)\theta}{(1 + \zeta^2\dot{\gamma}^2)^{2\Omega+\beta}} \tag{4}$$

These functions involve six parameters, or material properties. Four of them (η_0, η_∞, ζ and Ω) serve to define the viscosity function, although all four of these properties also appear in the normal stress function, along with the remaining two (θ and β). The parameter ζ is a characteristic time constant of the fluid, and Ω is related to the flow index, n, in the power law region by $n = (1 - 2\Omega)$. The time constant ζ is the reciprocal of the shear rate at the point where the power law region of the viscosity curve intersects with η_0. The physical significance of the limiting viscosity parameters η_0 and η_∞ is evident. Values of the six material properties are shown in Table 2 for each of the fresh and sheared solutions. The lines on Figure 5 represent Equation 3, and the long dashed lines on Figure 6 represent Equation 4, with the parameter values from Table 2.

Although the viscous and elastic material properties are distinct, there is a strong interaction between the apparent viscosity and normal stress functions, as indicated by the number of parameters common to Equations 3 and 4. This relationship has been elaborated on by Abdel-Khalik et al. [27] who proposed the following semi-empirical formula relating the normal stress and viscosity functions:

$$\psi_1 = \frac{4K}{\pi} \int_0^\infty \frac{\eta(\dot{\gamma}) - \eta(\dot{\gamma}')}{\dot{\gamma}'^2 - \dot{\gamma}^2} \, d\dot{\gamma}' \tag{5}$$

where $K = 2$ for solutions and 3 for melts. The values predicted by Equation 5 are shown as the dotted lines on Figure 6. Although the agreement is not perfect, it is seen to be quite reasonable. The point to be noted is that the material parameters that influence elastic behavior also influence viscous behavior and vice versa, so that these effects are not entirely independent.

ENERGY DISSIPATION MODEL

In tube flow, wall drag, pressure drop, and friction loss are all synonymous with the rate of energy dissipation. Quantitatively, the relationships are:

$$\dot{e} = \frac{\tau_w(\pi DL)V}{\rho \underset{\sim}{V}} = \frac{-\Delta PQ}{\rho \underset{\sim}{V}} = \left[\frac{2fL\rho V^2}{D}\right]\left[\frac{\pi D^2 V}{4}\right]\left[\frac{4}{\pi D^2 L\rho}\right] = \frac{2fV^3}{D} \tag{6}$$

where $\dot{e}$ is the mean rate of energy dissipation per unit mass of fluid, and f is the Fanning friction factor. In time dependent deformations, energy is stored by elastic properties and dissipated by viscous properties. Since turbulence is locally time dependent, and polymeric fluids are viscoelastic, it is logical that energy stored by elastic properties would represent a reduction in the amount of energy that would otherwise be dissipated by viscous forces, i.e. drag reduction.

Linear Viscoelastic Fluid Response

The effect of elastic properties on energy dissipation can be illustrated by considering a linear viscoelastic fluid and the simplest possible model for a time dependent deformation, e.g. a local element undergoing a small amplitude oscillating shear deformation in a large body of fluid. If the frequency of oscillation of the element is ω and the amplitude of the oscillating velocity is v_0, a shear wave will be propagated by the oscillation. If the motion is the x direction, the wave will be propagated in the y direction. The equation of motion for the shear field is

$$\rho \frac{\partial v_x}{\partial t} = \frac{\partial \tau_{yx}}{\partial y} \tag{7}$$

with boundary conditions

$$\begin{aligned} v_x &= v_0 e^{j\omega t} \quad \text{at} \quad y = 0 \\ v_x &= 0 \qquad\quad \text{at} \quad y = \infty \end{aligned} \tag{8}$$

For any linear viscoelastic fluid in an oscillatory deformation, the local shear stress is related to the local velocity gradient by the complex viscosity of the fluid:

$$\tau(\omega t) = \eta^*(j\omega)\dot{\gamma}(\omega t) = \eta^*(j\omega)\frac{dv_x}{dy} \tag{9}$$

where $\eta^*(j\omega) = \eta'(\omega) - j\eta''(\omega)$ (10)

The real part of the complex viscosity, η', represents the fluid viscosity whereas the imaginary part, η'', represents the elasticity. The solution of Equations 7 and 9 for the velocity field is

$$v(\omega t, y) = v_0 e^{j\omega t} e^{-\alpha^*(j\omega)y} \tag{11}$$

where $\alpha^*(j\omega) = [j\omega\rho/\eta^*(j\omega)]^{1/2}$ (12)

From Equations 9 and 11, the local stress at the oscillating element (y = 0) is

$$\tau(y = 0) = \tau_0 = -v_0(j\omega\rho\eta^*)^{1/2} e^{j\omega t} \tag{13}$$

and the corresponding shear rate is

$$\dot{\gamma} = \tau_0/\eta^* = v_0(j\omega\rho/\eta^*)^{1/2} e^{j\omega t}$$

The local mean rate of energy dissipation per unit volume of fluid is given by

$$\overline{\dot{e}_v} = \overline{Re(\tau_0)\,Re(\dot{\gamma}_0)} = \tfrac{1}{2}[\overline{Re(\tau_0\dot{\gamma}_0)} + \overline{Re(\tau_0\dot{\gamma}_c)}] \tag{15}$$

where $\dot{\gamma}_c$ is the complex conjugate of $\dot{\gamma}_0$. Using Equations 13 and 14 to evaluate the right side of Equation 15, it can easily be shown that the first term in the bracket is zero. Evaluation of the second term gives

$$\dot{e} = \overline{\dot{e}_v}/\rho = \frac{v_0^2\omega\eta'}{2[(\eta')^2 + (\eta'')^2]^{1/2}} \tag{16}$$

The simplest possible model for a linear viscoelastic fluid is the Maxwell model:

$$\tau + \lambda \dot{\tau} = \eta_0 \dot{\gamma} \tag{17}$$

for which

$$\eta' = \frac{\eta_0}{1 + \omega^2 \lambda^2}, \qquad \eta'' = \frac{\omega \lambda \eta_0}{1 + \omega^2 \lambda^2} \tag{18}$$

For this case, Equation 16 becomes

$$\dot{e} = \frac{v_0^2 \omega}{2(1 + \omega^2 \lambda^2)^{1/2}} \tag{19}$$

Since the Maxwell fluid reduces to a purely viscous Newtonian fluid for a relaxation time (λ) of zero, it is evident from Equation 19 that elastic properties (represented by λ) serve to reduce the rate of energy dissipation relative to that of an inelastic fluid by a factor equal to the radical in the denominator.

Dissipation in Turbulent Eddies

Newtonian Fluid

In order to apply these results to turbulent flow, we consider turbulence to consist of a spectrum of oscillating eddies with amplitude v_0' and frequency ω, following Davies [28]. The local mean rate of energy dissipation by the eddies per unit mass in a viscous fluid (e.g. the solvent) is given by

$$\dot{e}_s = \frac{\overline{\tau_e^{\dagger} \dot{\gamma}_e^{\dagger}}}{\rho} \tag{20}$$

where the turbulent eddy (Reynolds) stress and equivalent turbulent shear rate are given by:

$$\tau_e^{\dagger} = \rho(\bar{v}')^2, \qquad \dot{\gamma}_e^{\dagger} = \bar{v}'/\ell_e = \omega \tag{21}$$

where $\bar{v}'$ is the rms eddy velocity, i.e. $\bar{v}' = v_0'/\sqrt{2}$. Equation 20 thus becomes

$$\dot{e}_s = \bar{v}'^2 \omega = \tfrac{1}{2}(v_0')^2 \omega \tag{22}$$

Now according to Davies [28], the maximum rate of energy dissipation by turbulent eddies in a viscous fluid occurs when the eddy Reynolds number has a value of about 9, i.e.

$$N_{Re}' = \frac{\ell_e \bar{v}'}{\nu} = \frac{(\bar{v}')^2}{\omega \nu} \simeq 9 \tag{23}$$

or

$$(\bar{v}')^2 \simeq 9 \omega \nu \tag{24}$$

Equating the total rate of energy dissipation from Equation 6 to the eddy dissipation from Equation 22, and using Equation 24 to eliminate $\bar{v}'$, provides a relation between the friction factor and the dissipative eddy frequency for a purely viscous fluid (solvent):

$$f_s = \frac{\dot{e} D}{2V^3} = \frac{9}{2}\left(\frac{D\omega}{V}\right)^2 \left(\frac{1}{N_{Res}}\right) \tag{25}$$

Re-arranging Equation 25 gives the following expression for the most dissipative eddy frequency:

$$\omega = (2f_s N_{Res}/9)^{1/2} V/D \tag{26}$$

For a Newtonian fluid in a smooth tube, the friction factor can be determined by the Blasius equation:

$$f_s = \frac{0.0791}{N_{Res}^{1/4}} \tag{27}$$

Substituting Equation 27 into Equation 25, we get

$$\omega = 0.0166\left(\frac{8V}{D}\right) N_{Res}^{3/8} \tag{28}$$

for the frequency of the dissipative eddies. For a Reynolds number of about 5×10^4, Equation 28 reduces to $\omega \simeq 8V/D$.

Linear Viscoelastic Fluid

For the viscoelastic polymer solution, the rate of energy dissipation by the eddies is modified by the elastic properties according to Equation 19. Using this expression instead of Equation 22 in Equation 25 results in the following expression for the polymer solution friction factor

$$f_p = \frac{f_s}{\sqrt{1 + \lambda^2 \omega_p^2}} = \frac{9}{2}\left(\frac{D\omega}{V}\right)^2 \frac{1}{N_{Res}\sqrt{1 + \lambda^2 \omega_p^2}} \tag{29}$$

If it is assumed that the dissipative eddy frequency for the polymer solution is related to the solution friction factor and Reynolds number by an expression analogous to that for the solvent, i.e. Equation 26:

$$\omega_p = (2f_p N_{Rep}/9)^{1/2} V/D \tag{30}$$

substitution of f_p from Equation 29 into Equation 30 gives the following relation between ω and ω_p:

$$\omega_p = \frac{\omega}{(1 + \lambda^2 \omega_p^2)^{1/4}}\left(\frac{N_{Rep}}{N_{Res}}\right)^{1/2} \tag{31}$$

where

$$N_{Rep} = \frac{DV\rho}{\eta_p}$$

Equation 31 shows that the spectrum of the dissipative eddies in the polymer solution should be shifted to slightly lower frequencies as a consequence of a finite relaxation time, as well as a consequence of a higher viscosity, since $N_{Rep} \leq N_{Res}$. This is consistent with previous experimental observations [29–32]. In general, the values of ω and ω_p are not greatly different.

Equation 29 relates the Newtonian (solvent) friction factor, f_s, to the corresponding friction factor for the polymer solution, f_p. This can be written

$$f_p = f_s/(1 + N_{De}^2)^{1/2} \tag{32}$$

where $N_{De} = \omega_p\lambda$ is the characteristic dimensionless eddy frequency (Deborah number) for the system.

Both ω_p and λ are dependent upon the rheological properties of a given polymer solution and can be evaluated as follows. The appropriate dissipative eddy frequency in the polymer solution is assumed to be given by Equation 31 in which the solvent frequency, ω, is given by Equation 28. However, for simplicity, the ω_p in the denominator of Equation 31 may be replaced by (8V/D), since we have seen that these are of the same magnitude (Equation 28), and ω_p does not differ greatly from ω. Also, the ratio $(N_{Rep}/N_{Res}) = \mu_s/\eta_p$ is the ratio of the solvent to polymer solution viscosity. However, since the solution is non-Newtonian, η_p is not a parameter, but a function whose maximum value is η_0. If it is assumed that the appropriate characteristic viscosity should be that associated with the force required to deform the polymer molecules from their equilibrium configuration, or conversely, the maximum force resisting the rate at which equilibrium configuration could be achieved, then the zero shear viscosity should be the appropriate parameter. With these approximations, then, combination of Equations 28 and 31 gives the following expression for N_{De}:

$$N_{De} = \omega_p\lambda = \frac{0.0166\, N_\lambda N_{Res}^{3/8}(\mu_s/\eta_0)^{1/2}}{[1 + N_\lambda^2]^{1/4}} \tag{33}$$

where

$$N_\lambda = 8V\lambda/D \tag{34}$$

Nonlinear Viscoelastic Fluid

The results to this point are based upon simple, first order concepts, and the assumption that the polymer solutions behave as a Maxwell fluid, i.e. a linear viscoelastic fluid with a single, constant relaxation time (λ). In reality, as we have seen, the solutions are markedly nonlinear viscoelastic fluids, with properties described by Equation 1. In principle, this equation could be solved along with the equation of motion for the oscillating element to yield an expression analogous to Equation 19, which would express the energy dissipation rate as a function of all appropriate model parameters. However, in practice the expression would be too complex to be of much practical value.

Instead, we take a simpler approach, in which we utilize the results derived for the Maxwell model, but replace the constant relaxation time (λ) by an equivalent quantity that is appropriate to the nonlinear properties of the fluid. The result will be a "relaxation time function," since real nonlinear viscoelastic fluids cannot be described by a single constant relaxation time, but can instead be characterized by a function of shear rate that has the same significance as the relaxation time for a Maxwell fluid. This relaxation time function relates to the constant Maxwell relaxation time in much the same way as an apparent viscosity for a non-Newtonian fluid relates to a constant Newtonian viscosity. Such a function can be determined as follows:

Since the relaxation time is defined by the linear Maxwell model, Equation 17, a generalized form of this model can be written that does predict nonlinear behavior:

$$\tau_{ij} + \lambda\frac{D\tau_{ij}}{Dt} = \eta_0\Delta_{ij} \tag{35}$$

where the time derivative has been replaced by the Jaumann or corotational derivative, which follows the translation and rotation of local fluid elements. The viscosity function predicted by this model is

$$\eta = \frac{\eta_0}{1 + \lambda^2\dot{\gamma}^2} \tag{36}$$

which exhibits qualitatively correct shear-thinning behavior. A slight modification to Equation 35 can be made to account for the observed high shear limiting viscosity, η_∞, as follows:

$$\tau_{ij} + \lambda \frac{D\tau_{ij}}{Dt} = \eta_0 \Delta_{ij} + \lambda \eta_\infty \frac{D\Delta_{ij}}{Dt} \tag{37}$$

This model is also sometimes known as the Jeffreys model, and predicts the following viscosity function:

$$\eta = \eta_\infty + \frac{\eta_0 - \eta_\infty}{(1 + \lambda^2 \dot{\gamma}^2)} \tag{38}$$

Comparison of Equations 3 and 38 shows that the Jeffreys model viscosity function is qualitatively similar to that predicted by the more general model for our polymer solutions. In the previous equations, the significance of λ is still the same, i.e. the characteristic relaxation time constant, which we have related through our simplified model to energy dissipation in turbulent eddies. In fact, by equating the viscosity functions in Equations 3 and 38, we can solve for λ:

$$\lambda = [(1 + \zeta^2 \dot{\gamma}^2)^\Omega - 1]^{1/2}/\dot{\gamma} \tag{39}$$

which is the equivalent "relaxation time function" for our solutions. It is seen to be a function of shear rate and the two material parameters ζ and Ω.

For application to turbulence, we have seen that the equivalent eddy shear rate is ω, which is of the order of (8V/D). Making this substitution in Equation 39 results in the following expression:

$$N_\lambda = \frac{8V\lambda}{D} = |(1 + N_\zeta^2)^\Omega - 1|^{1/2} \tag{40}$$

where

$$N_\zeta = 8V\zeta/D \tag{41}$$

If $N_\zeta^2 \gg 1$, which is often the case, this reduces to:

$$N_\zeta \cong N_\zeta^\Omega = N_\zeta^{1-n/2} \tag{42}$$

It is seen from Equations 3 and 38 that the nonlinearity in the viscosity function depends on the same fluid parameters that influences the elastic behavior, i.e. λ or ζ. This is also illustrated by Equation 4 for the normal stresses, and is the principle behind the development of Equation 5 relating the normal stress and viscosity functions. This result is really quite fortuitous, since it enables the determination of the elastic parameter (e.g. ζ) from appropriate characteristics of the viscosity function. In this case, ζ is the reciprocal of the value of $\dot{\gamma}$ where the power law region of the viscosity function intersects with η_0, the low shear Newtonian region.

DATA CORRELATION

Our model for turbulent drag reduction indicates that the friction factor for the polymer solution, f_p, should be related to the friction factor of the Newtonian solvent, f_s, by Equation 32, i.e.

$$f_p = f_s/(1 + N_{De}^2)^{1/2} \tag{32}$$

where N_{De} is given by Equation 33. Because of the simplifications and first order approximations involved in the derivation of Equation 33 it could not be expected to provide a precise quantitative representation of the data with the values of the numerical parameters as given. However, if the concepts behind the model are valid, it should be correct in form, and could be expected to provide a useful guide for correlation of data with suitable values of the numerical parameters. Hence, an equation of the form:

$$N_{De} = \frac{aN_\lambda N_{Res}^b}{(1 + N_\lambda^2)^c}\left(\frac{\mu_s}{\eta_0}\right)^d \quad (43)$$

was used to correlate the drag reduction data, in which the parameters a, b, c, and d were determined by regression analysis, and Equation 40 was used for N_λ. The resulting expression is

$$N_{De} = \frac{0.0867 N_\lambda N_{Res}^{0.34}}{(1 + N_\lambda^2)^{0.33}}\left(\frac{\mu_s}{\eta_0}\right)^{0.237} \quad (44)$$

If the correlation is valid, the product $f_p\ (1 + N_{De}^2)^{1/2}$ should be identical to f_s, i.e. the friction factor for a Newtonian fluid in smooth tubes, which is a well-known function of N_{Res}. All of the data, for three concentrations of both fresh and degraded solutions in six tube sizes were correlated by this expression, with a correlation coefficient (r^2) of 0.956. The resulting correlation is shown in Figure 7, in which the solid line on the plot represents the Colebrook equation [33] for Newtonian fluids smooth tubes:

$$f_s = 0.41[\ln(N_{Res}/7)]^{-2} \quad (45)$$

which is slightly more accurate than the Blasius equation. The fact that the values of the exponential parameters in Equation 44 are reasonable and in the same range as those of the approximate model, Equation 33, is encouraging and enhances the validity of the model.

Inasmuch as the values of all of the exponential parameters in Equation 44 are nearly the same, it was found that the following expression, involving only two regression parameters, provided almost an equally good representation of the data ($r^2 = 0.952$):

$$N_{De} = 0.145 N_\lambda\left(\frac{N_{Res}}{(1 + N_\lambda^2)}\left(\frac{\mu_s}{\eta_0}\right)\right)^{0.285} \quad (46)$$

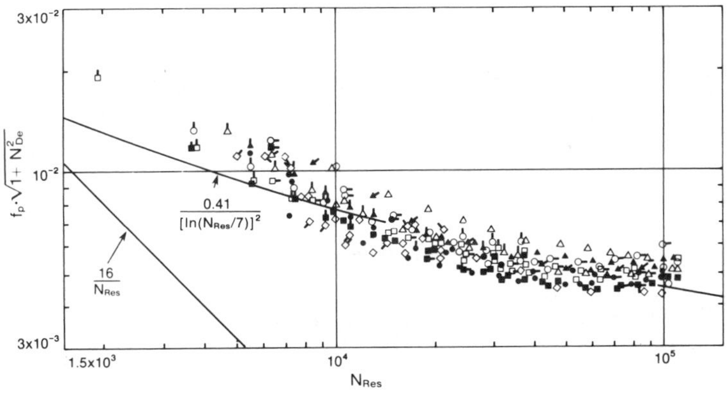

Figure 7. Generalized friction factor versus solvent Reynolds number. Points shown include entire range of data. For clarity, not all data points are shown. [20].

Application of the model to prediction of friction loss is straightforward. The rheological parameters η_0, ζ and Ω must be known, and can be determined directly from data for the viscosity function. For a given tube diameter and velocity, ζ is used to determine N_ζ from Equation 41. This value, along with Ω, is used in Equation 40 to determine N_λ, which, along with η_0 and the solvent properties, is used to evaluate N_{De} from Equation 44. From a knowledge of the solvent friction factor at the same value of N_{Res} (e.g. Equation 45) and N_{De}, the friction factor for the polymer solution is determined from Equation 32.

EFFECT OF CONCENTRATION

As previously noted, the maximum degree of drag reduction for the type of polymer used in this study has been reported to occur at a concentration near, or slightly below, 100 ppm [21]. In fact, significant drag reduction has been observed at concentrations as low as 10–20 ppm. Since the concentrations used in this study were all greater than the critical concentration for which maximum drag reduction is expected (which we will call C_c), they can be referred to as "concentrated" in this context. These concentrations were used in order to clearly determine the influence of rheological properties and degradation upon drag reduction. For these concentrations (above C_c), it is seen that the degree of drag reduction decreases with increasing concentration, whereas drag reduction increases with concentration below C_c. The relation between these trends and the concentration dependence of the rheological properties can be explained in light of the drag reduction model as follows.

The concentration dependence of the rheological parameters η_0, η_∞, Ω, and ζ, for both the fresh and degraded solutions, are shown in Figures 8–11. As seen, the parameters η_0, η_∞, and Ω increase with concentration, although the change in Ω is relatively small. However, the time constant ζ decreases with increasing concentration over the range shown. Now we know that ζ must approach

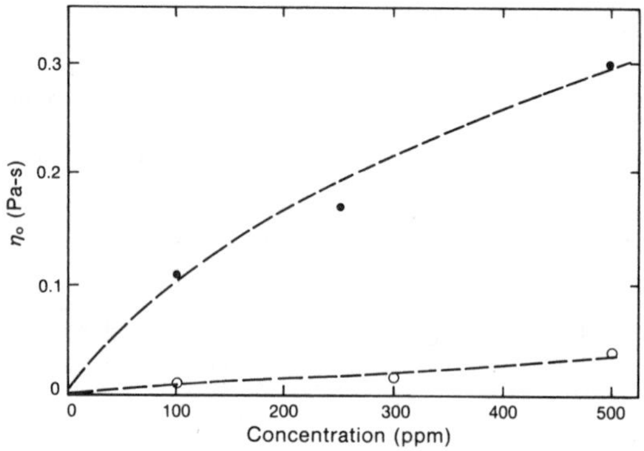

Figure 8. Concentration dependence of low shear limiting viscosity: (●) fresh solutions and (○) degraded solution [24].

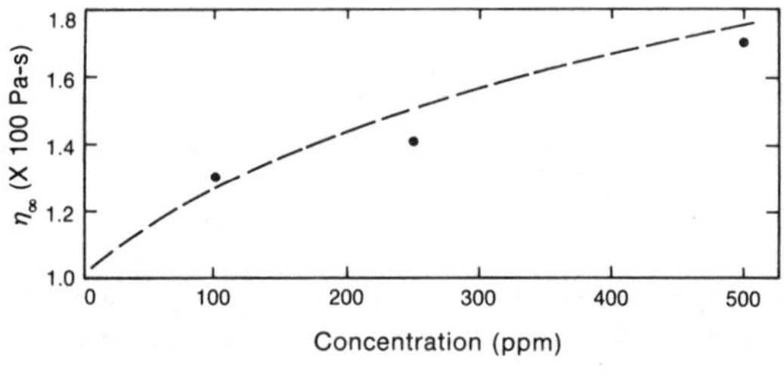

Figure 9. Concentration dependence of high shear limiting viscosity: (●) fresh solutions and (○) degraded solution [24].

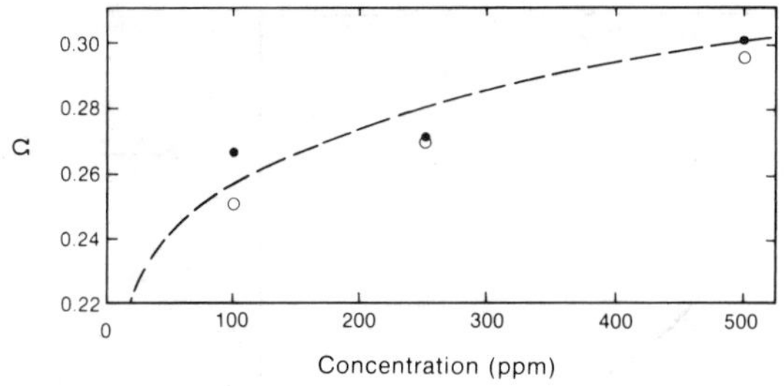

Figure 10. Concentration dependence of parameter Ω: (●) fresh solutions and (○) degraded solutions.

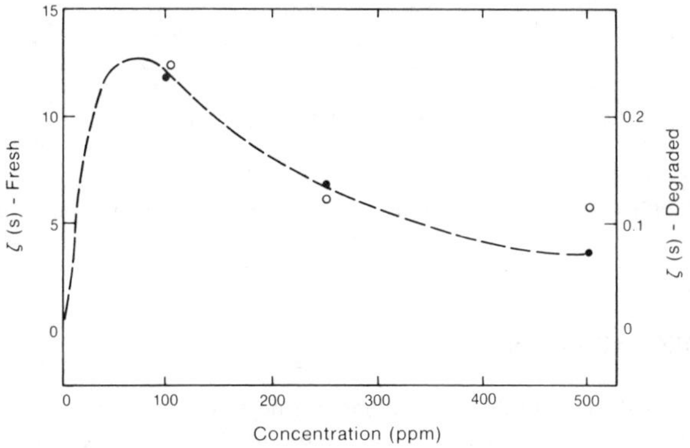

Figure 11. Concentration dependence of time constant ζ: (●) fresh solutions and (○) degraded solutions [24].

zero as the concentration approaches zero (i.e. the solvent properties), which means that it must exhibit a maximum at some concentration below 100 ppm. Since the degree of drag reduction, which is directly related to the value of ζ, has been reported to be maximum in the range of 75–100 ppm, we may assume that this also corresponds to the concentration at which the maximum in ζ occurs (i.e. C_c).

Now the energy dissipation model provides a quantitative relation between the degree of drag reduction, and the rheological properties ζ, η_0, and Ω. If $N_\zeta^2 \gg 1$ (which is the case for the data in this study), the relation is

$$N_{De} \sim \zeta^{x\Omega}/\eta_0 \tag{47}$$

where x = 1 from the simplified model (Equation 33) and x = 4/3 from the correlation equation (Equation 44). Since Ω varies only from about 0.25 to 0.3, Equation 47 can be written

$$N_{De} \sim \zeta^y/\eta_0 \quad \text{where} \quad 0.25 < y < 0.4 \tag{48}$$

Thus, it is seen that N_{De}, and hence the degree of drag reduction, is much more sensitive to changes in η_0 than to ζ. The corresponding relative dependence of ($\zeta^{1/3}/\eta_0 \sim N_{De}$) upon concentration is shown in Figure 12. It is clear that this is consistent with an increase in drag reduction with concentration below C_c, and a decrease in drag reduction with concentration above C_c, as observed experimentally.

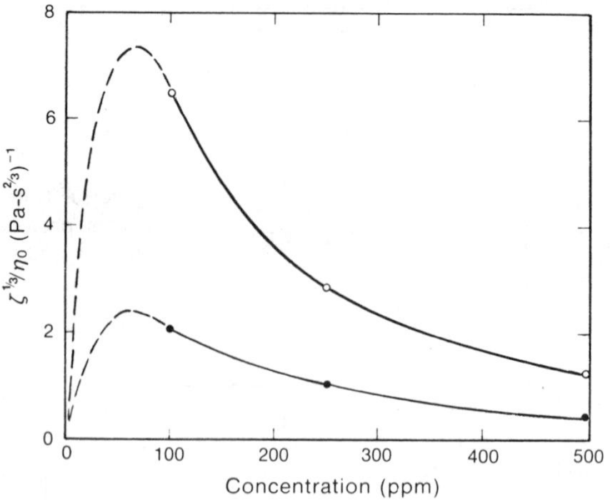

Figure 12. Concentration dependence of $\zeta^{1/3}/\eta_0$: (●) fresh solutions and (○) degraded solutions.

EFFECT OF DEGRADATION

The increase in the degree of drag reduction with degradation exhibited by these solutions is contrary to most results reported for dilute solutions, although there have been other reports of similar effects in the literature [25]. This result, as well as the difference between the effect of degradation on concentrated and dilute solutions, can also be explained in light of the energy dissipation model and the concentration dependence of rheological parameters.

As previously shown, the degree of drag reduction is approximately proportional to $\zeta^{1/3}/\eta_0$. For the concentrated solutions in this study, the effect of degradation is to reduce both η_0 and ζ by more than an order of magnitude (Table 2), with a greater effect on ζ by about a factor of 3. However, since drag reduction is less sensitive to changes in ζ than η_0, the net effect of degradation is to increase the value of $\zeta^{1/3}/\eta_0$, and hence drag reduction, for these solutions. This is clearly shown in Figure 12.

The situation is quite different for more dilute solutions, however. For such solutions, both ζ and η_0 are significantly smaller than for the more concentrated solutions. In fact, as $C \to 0$, $\zeta \to 0$ and $\eta_0 \to \mu_s$ (the solvent viscosity). Since the effect of degradation is to reduce both ζ and η_0, a significant decrease in ζ would be expected. However, since η_0 for very dilute solutions is of the same order of magnitude as μ_s, the degree to which it can be reduced by degradation is constrained by this limit. Hence the net expected effect of degradation would be to lower the value of $\zeta^{1/3}/\eta_0$ and hence to reduce the degree of drag reduction, as is normally observed for very dilute solutions.

EFFECT OF PIPE DIAMETER

It has been widely observed that the degree of drag reduction decreases as the pipe diameter increases [34–36], but no satisfactory explanation of this effect has been given. However, this is a natural consequence of the energy dissipation model, as can readily be seen. Since the predicted degree of drag reduction depends directly on the value of $N_\zeta = 8V\zeta/D$, as the diameter increases the value of N_ζ, and hence drag reduction, decreases for a given fluid at a given velocity. This is illustrated in Figures 13 and 14, which show the predicted friction factor versus N_{Res} for diameters from 2 in. to 30 in. for the 250 ppm fresh and degraded solutions, respectively.

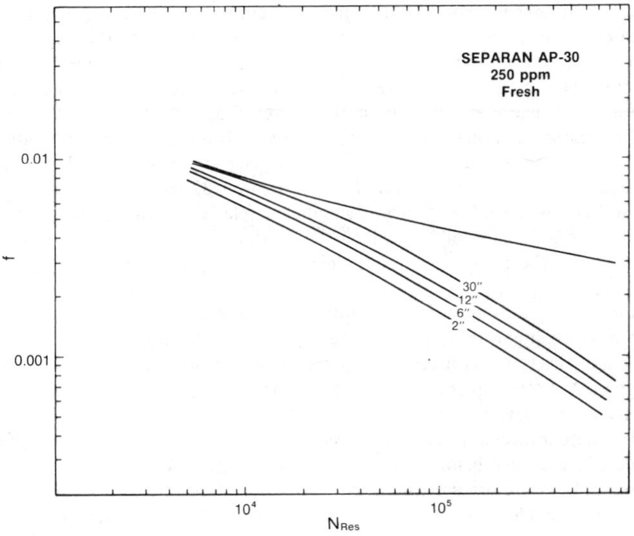

Figure 13. Predicted dependence of friction factor on pipe diameter, 250-ppm fresh solution.

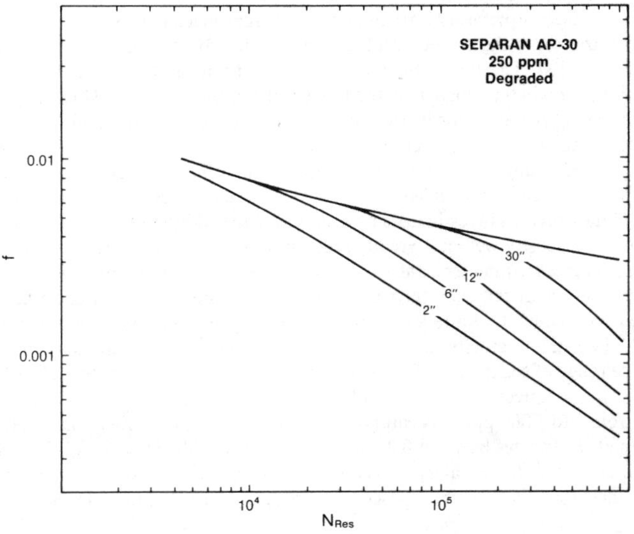

Figure 14. Predicted dependence of friction factor on pipe diameter, 250-ppm degraded solution.

There are several interesting effects illustrated by Figures 13 and 14. It is evident that drag reduction decreases as diameter increases for both solutions, and that the degree of decrease is greater for the degraded solution. This is because the value of ζ, and hence N_ζ, is lower for the degraded solution, which makes it relatively more sensitive to changes in diameter. It can also be seen that, although the degraded solutions shows greater drag reduction than the fresh solution in small tubes, the situation is reversed in the larger tubes. This is due to the combination of the relative changes in the magnitude of N_{De} (or N_ζ) with diameter for the two solutions and the changes in absolute value of N_{De}^2 relative to unity. From Equation 32 it is evident that significant drag reduction occurs only if the value of N_{De}^2 is significant relative to unity. The point at which this is satisfied corresponds to the point at which the friction factor curve for the polymer solution deviates from that for the Newtonian fluid. This is quite evident in Figure 14 for the degraded solution.

This behavior also explains another frequently observed phenomenon that has previously defied explanation. Many investigators have noted the difference in the drag reduction behavior of some systems, in which the friction factor relation appears to follow an extension of the laminar flow line such as in Figures 2–4, and others in which the friction factor follows that for Newtonian fluids in turbulent flow before deviating, as in Figure 14. These effects have variously been attributed to two completely different mechanisms for drag reduction [2, 36] or to the existence of a critical wall stress which must be exceeded before drag reduction can occur [34, 37]. However, as we have seen, all of these types of behavior are consistent with, and predictable by, the energy dissipation model, and are simply a consequence of the relative magnitudes of N_{De} compared to unity for the various systems.

CONCLUSION

The correlation of drag reduction data based upon the energy dissipation model permits the prediction of pipe friction loss for relatively concentrated polymer solutions from a knowledge of appropriate viscous and elastic rheological parameters, which can be obtained directly from data for the apparent viscosity function. The results are also consistent with reported observations on very dilute solutions and provide a simple explanation for the effects of concentration, degradation, and pipe diameter on drag reduction for any concentration, as well as differences in the characteristics of drag reduction previously attributed to different mechanisms.

The key to utilization of the correlation lies in being able to determine values of the rheological parameters η_0, ζ, and Ω. These may be determined directly from data for the complete apparent viscosity function, if available, since η_0 is the low shear limiting viscosity, Ω is equal to one half of the slope of the log η vs log $\dot{\gamma}$ plot in the linear (i.e. power law) region, and ζ is the value of the reciprocal of the shear rate at the intersection of the lines that define these two regions. For very dilute solutions, these values are difficult to obtain since the viscosity is of the same order of magnitude as that of the solvent and the non-Newtonian characteristics are not easy to quantify. In fact, many of the previous investigators have assumed that dilute solutions are Newtonian, since measured values of viscosity were not greatly different from that of the solvent. Of course, the magnitude of the rheological parameters, and hence the degree of non-Newtonian behavior, will depend greatly upon the particular polymer, the solvent, the relative ionic character of both, concentration, etc. However, even extremely dilute solutions normally assumed to be Newtonian may in reality exhibit a non-Newtonian viscosity function if sufficiently sensitive measurements are made over a sufficiently wide range of shear rates. This is illustrated in Figure 15, taken from Friebe [38], which shows the viscosity of aqueous polyacrylamide (Separan) solutions versus shear rate for concentrations ranging from 5 to 1,000 ppm. As can be seen, non-Newtonian character is evident even for the 5 ppm solution at shear rates between 0.2 and 1 s^{-1}. However, the viscosity of this solution is very low (only slightly greater than water) and is not easy to measure at such low shear rates. As the concentration increases, the low shear viscosity increases markedly, making measurements easier in this range.

In summary, we will review the procedure involved in order to use the relations presented herein in order to predict the pipe flow friction factor for a given polymer solution, in a given pipe at a

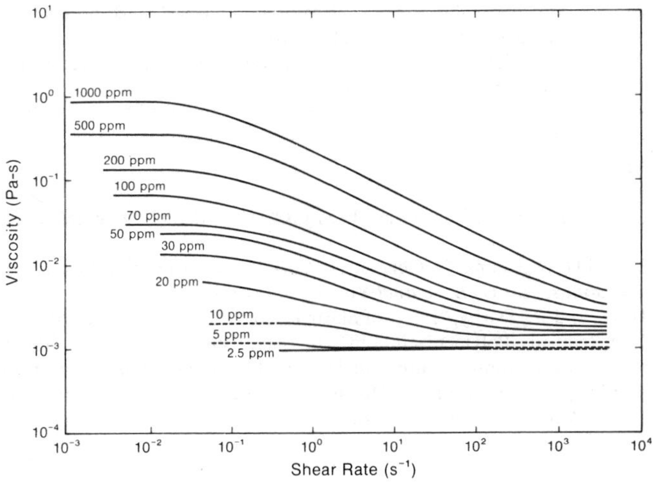

Figure 15. Concentration dependence of apparent viscosity functions of aqueous separan solutions from 5 to 1,000 ppm [38].

given flow rate:

1. The rheological parameters are determined from data for the apparent viscosity versus shear rate, such as shown in Figure 5. η_0 is the value of the low shear limiting Newtonian viscosity, Ω is equal to $(1 - n)/2$, where $(n - 1)$ is the slope of the linear intermediate shear rate (power law) range, and ζ is the value of the reciprocal of $\dot{\gamma}$ at the intersection of the lines defining these two ranges.
2. These values and the values of D and V are used to calculate N_ζ, N_λ, N_{Res}, N_{De}, f_s, and, finally, f_p as follows:

$$N_\zeta = \frac{8V\zeta}{D}$$

$$N_\lambda = [(1 + N_\zeta^2)^\Omega - 1]^{1/2}$$

$$N_{Res} = \frac{DV\rho}{\mu_s}$$

$$N_{De} = \frac{0.0867\, N_\lambda N_{Res}^{0.34}}{(1 + N_\lambda^2)^{0.33}} \left(\frac{\mu_s}{\eta_0}\right)^{0.237}$$

$$f_s = 0.41[\ln(N_{Res}/7)]^{-2}, \quad \text{or} \quad f_s = \frac{0.0791}{N_{Res}^{1/4}}$$

$$f_p = f_s/(1 + N_{De}^2)^{1/2}$$

This procedure should result in a reasonable prediction ($\pm 15\%$) for the Fanning friction factor for the polymer solution over the range $8 \times 10^3 < N_{Res} < 10^5$. Extrapolation beyond this range may give reasonable values, but caution is advised since no data outside of this range have been included in evaluation of the correlation, and the lower end of the range corresponds to the critical Reynolds number for transition to turbulence, which varies with fluid properties, diameter, etc.

Acknowledgment

Much of the material presented herein has been taken from the following references:

1. Darby, R., and H. D., Chang, "A Generalized Correlation for Friction Loss in Drag Reducing Polymer Solutions," *A.I.Ch.E. Journal* Vol. 30, 274 (1984).
2. Chang, H. D., and R., Darby, "Effect of Shear Degradation on the Rheological Properties of Dilute Drag Reducing Polymer Solutions," *Journal of Rheology*, 27(1), 77 (1983).
3. Chang, H. D., "Correlation of Turbulent Drag Reduction in Dilute Polymer Solutions with Rheological Properties by an Energy Dissipation Model," Ph.D. Dissertation, Chemical Engineering, Texas A & M University (1982).

NOTATION

C	polymer concentration, wppm
D	tube diameter, m
D/Dt	Jaumann or co-rotational time derivative, s^{-1}
$\dot{e}$	rate of energy dissipation per unit mass, w/kg
f	Fanning friction factor
ℓ	characteristic length, m
L	tube length, m
n	power law flow index
N_{De}	Deborah number $= \omega_p \lambda$
N_{Re}	Reynolds number
N_λ	dimensionless relaxation time
N_ζ	dimensionless fluid time constant, $8V\zeta/D$
ΔP	pressure change, Pa
Re	real part of complex quantity
Q	volumetric flow rate, m^3/s
$\bar{v}'$	root mean squared velocity fluctuation, m/s
v_0'	amplitude of velocity fluctuation, m/s
V	average tube velocity, m/s
v_0	amplitude of oscillating velocity, m/s
$\underset{\sim}{V}$	volume, m^3
β	parameter in Equation 2,
$\dot{\gamma}$	shear rate, s^{-1}
Δ_{ij}	rate of strain tensor, s^{-1}
ζ	time constant of polymer solution, s
ν	kinematic viscosity, m^2/s
ρ	density, kg/m^3
η	apparent viscosity, Pa-s
η^*	complex viscosity, Pa-s
η'	real part of complex viscosity, Pa-s
η''	imaginary part of complex viscosity, Pa-s
η_0	zero shear rate limiting viscosity, Pa-s
η_∞	high shear rate limiting viscosity, Pa-s
θ	parameter in Equation 2, s
λ	relaxation time, s
μ	Newtonian viscosity, Pa-s
τ	shear stress, Pa
τ_{11}	normal stress in direction of motion, Pa
τ_{22}	normal stress in direction of gradient, Pa
τ_{ij}	shear stress tensor, Pa
ψ_1	first normal stress difference function, $Pa\text{-}s^2$
ω	frequency of oscillation, s^{-1}
Ω	parameter in Equation 1

Subscripts

cr	cirtical value for transition to turbulence
e	eddy
p	polymer solution
s	solvent
w	wall value

Superscript prime—eddy property (except in Equation 5)

REFERENCES

1. Lumley, J. L., "Drag Reduction in Turbulent Flow by Polymer Additives," *J. Polym. Sci. Macromol. Rev* 7, 263 (1973).
2. Patterson, G. K., Zakin, J. L., and Rodriguez, J. M., "Drag Reduction," *I. & E. C*. 61, 22 (1963).
3. Hoyt, J. W., "The Effect of Additives on Fluid Friction," *J. Basic Eng.*, 94D, 258–285 (1972).
4. Berman, N. S., "Drag Reduction by Polymers," *Ann. Rev. Fluid Mech.*, 10, 47 (1978).
5. Sellin, R. H. J., Hoyt, J. W., and Scrivener, O., "The Effect of Drag Reducing Additives on Fluid Flows and Their Industrial Applications, Part 1: Basic Aspects," *J. Hydraulic Resh.*, 20(1), 29 (1982).
6. Sellin, R. H. J., et. al. "The Effect of Drag Reducing Additives on Fluid Flows and Their Industrial Applications, Part 2: Present Applications and Future Proposals," *J. Hydraulic Resh.*, 20(3), 235 (1983).
7. Darby, R., "Transient and Steady State Rheological Properties of Very Dilute Drag Reducing Polymer Solutions," *Trans. Soc. Rheol.*, 14, 185 (1970).
8. Bruce, C., and Schwarz, W. H., "Rheological Properties of Ionic and Nonionic Polyacrylamide Solutions," *J. Poly. Sci.*, A2, 7, 909 (1970).
9. Tsai, C. F., and Darby, R., "Nonlinear Viscoelastic Properties of Very Dilute Drag Reducing Polymer Solutions", *J. Rheol.*, 22, 219 (1978).
10. Argumedo, A., Tung, T. T., and Chang, K. I., "Rheological Property Measurements of Drag-Reducing Polyacrylamide Solutions," *J. Rheol.*, 22 (5), 449–470 (1978).
11. Patterson, G. K., and Zakin, J. L., "Prediction of Drag Reduction with a Viscoelastic Model," *AIChE J.*, 14, 434 (1968).
12. Hanson, R. J., "A Theoretical Study of Transient Shear Effects in Drag Reduction," in *Drag Reduction in Polymer Solutions*, Sylvester N. D. (ed.). AIChE Symp. Ser. 130, v. 69, 20 (1973).
13. Barnes, H. A., Townsend, P., and Walters, K. "On Pulsatile Flow of Non Newtonian Liquids," *Rheol. Acta*, 10 517–527 (1971).
14. Townsend, P., "Numerical Solutions of Some Unsteady Flows of Elasto-Viscous Liquids," *Rheol. Acta*, 12, 13 (1973).
15. Wang, C. B., "Correlation of the Friction Factor for Turbulent Pipe Flow of Dilute Polymer Solutions," *I&EC Fund.*, 11, 546 (1972).
16. Elata, C., Lehrer, J., and Kahanovitz, A., "Turbulent Shear Flow of Polymer Solutions," *Israel J. Tech.*, 4, 87 (1966).
17. Seyer, F. A., and Metzner, A. B., "Turbulence Phenomena in Drag Reducing Systems," *AIChE J.*, 15, 426 (1969).
18. Metzner, A. B., and Metzner, A. P., "Stress Levels in Rapid Extensional Flows of Polymeric Fluids," *Rheol Acta*, 9, 174 (1970).
19. Baid, K. M., and Metzner, A. B., "Rheological Properties of Dilute Polymer Solutions Determined in Extensional and in Shearing Experiments," *Trans. Soc. Rheol.*, 21:22, 237–260 (1977).
20. Darby, R., and Chang, H. D., "A Generalized Correlation for Friction Loss in Drag Reducing Polymer Solutions," *AIChE J.*, 30, 274 (1984).
21. Hoyt, J. W., "Drag-Reduction Effectiveness of Polymer Solutions in the Turbulent-Flow Rheometer: A Catalog," *Polymer Letters*, 9, 851–862 (1971).
22. Liaw, G-C., Zakin, J. L., and Patterson, G. K., "Effects of Molecular Characteristics of Polymers on Drag Reduction," *AIChE J.*, 17, 391 (1971).
23. Chang, H. D., "Correlation of Turbulent Drag Reduction in Dilute Polymer Solutions with Rheological Properties by an Energy Dissipation Model," Ph.D. Dissertation, Chemical Engineering, Texas A&M University, (1982).
24. Chang, H. D., and Darby, R., "Effect of Shear Degradation on the Rheological Properties of Dilute Drag Reducing Polymer Solutions," *J. Rheol.* 27(1), 77 (1983).
25. Ng, K. S., and Hartnett, J. P., "Effect of Mechanical Degradation on Pressure Drop and Heat Transfer Performance of Polyacrylamide Solutions in Turbulent Pipe Flow," in *Studies in Heat Transfer*, Edited by J. P. Hartnett, T. F. Irvine, Jr., E. Pfender and E. M. Sparrow, Hemisphere Publishing Corp., N.Y., McGraw-Hill Book Co., 297 (1979).

26. Virk, P. S., "Drag Reduction Fundamentals," *AIChE J.* 21, 505 (1975).
27. Abdel-Khalik, S. I., Hassager, O., and Bird, R. B., "Prediction of Melt Elasticity from Viscosity Data," *Polym. Eng. and Sci.*, 14, 859 (1974).
28. Davies, J. T., *Turbulence Phenomena*, 51–62, Academic Press N.Y. (1972).
29. Baker, S. J., "Laser-Doppler Measurements on a Round Turbulent Jet in Dilute Polymer Solutions," *J. Fluid Mech.*, 60, part 4, 721–731 (1973).
30. Hanratty, T. L., and Chorn, L. G., "Turbulence Properties in the Region of Maximum Drag Reduction," *Proced. of the Fifth Biennial Symp. on Turbulence*, Univ. of Missouri-Rolla, Edited by G. K. Patterson and J. L. Zakin, 169 (1977).
31. Fortuna, G., and T. J., Hanratty, "Use of Electrochemical Techniques to Study the Effect of Drag-Reducing Polymer on Flow in the Viscous Sublayer in Drag Reduction," J. G. Savins and P. S. Virk (eds.) *CEP Symp. Series* 111, 67, 90 (1971).
32. Ivanyuta, Yu F., Zheltukhin, I. D., Sergievskii, N. A., "Measurement of the Spectrum of the Longitudinal Component of the Pulsation Velocity in the Turbulent Boundary Layer of a Non-Newtonian Fluid," *Fluid Mech.-Soviet Res.*, 1, 3, 82–90 (1972).
33. Colebrook, C. E., "Turbulent Flow in Pipes with Particular Reference to the Transition Region Between Smooth and Rough Pipe Laws," *J. Inst. Civil Engrs.*, London, 11, 133 (1939).
34. Hansen, R. J., and Little, R. C., "Pipe Diameter, Molecular weight and Concentration Effects on the Onset of Drag Reduction," in *Drag Reduction*, J. G. Savins and P. S. Virk (eds.) CEP Symposium Series 111, 67, 93 (1971).
35. Savins, J. G., "Drag Reduction Characteristics of Solutions of Macromolecules in Turbulent Pipe Flow," *Soc. Petr. Eng. J.*, 203, Sept. (1963).
36. Hershey, H. C., and Zakin, J. L., "Existence of Two Types of Drag Reduction in Pipe Flow of Dilute Polymer Solutions," *I&EC Fund*, 6, 381 (1967).
37. Virk, P. S., Merrill, E. W., Mickley, H. S., Smith, K. A., and Mollo-Christensen, E. L., "The Toms Phenomenon: Turbulent Pipe Flow of Dilute Solutions," *J. Fluid Mech.*, 30, 305 (1967).
38. Friebe, H. W., "Das Stabilitätsverhalten verdünnter Lösungen sehr langkettiger Hochpolymerer in der Couette-Stromong," *Rheol. Acta*, 15, 329 (1976).

CHAPTER 34

ANALYSIS OF POWER LAW VISCOUS MATERIALS USING COMPLEX STREAM, POTENTIAL AND STRESS FUNCTIONS

Y. S. Lee

and

L. C. Smith

Westinghouse Electric Corporation
Pittsburgh, Pennsylvania, USA

CONTENTS

INTRODUCTION

Mooney and Black [1] mentioned that a non-Newtonian fluid described by the power law is equivalent to Nadai's three-dimensional law for a steady state creep material expressed in terms of the octahedral shearing stress and the rate of shear, which is simply the relation between the effective stress and strain rate. Thus there are analogies between the constitutive equation as well as equation of motion for non-Newtonian fluids and those for steady state creep materials. These analogies can be applied to investigation of the flow pattern for the metal forming process.

There are many articles regarding the flow behavior of non-Newtonian fluids available [2–6]. From a numerical analysis point of view, Kalipeda and Fenner [7] have investigated non-Newtonian channel flow with slow velocity by using finite element analysis. The behavior of non-Newtonian lubricants with various constitutive equations has been studied by numerous investigators [8–12]. Most, however, are limited to one-dimensional flow.

Presented in this chapter is the application of the stream function approach to the solution of one and two-dimensional non-Newtonian (non-linear viscous materials) fluid flow problems. In a previous paper [13], the governing equation is expressed in terms of a complex conjugate stream function gradient with respect to z and $\tilde{z}$ by modifying the equilibrium equation derived by Shabaik [14]. The general solution of the governing equation is given in [15]. Three different approaches for creep deformation solutions are presented in [15]. In the separation of variables case, both the summation and product forms of the conjugate flow functions are derived. When the variables cannot be separated, a mixed mode solution expressed in terms of an independent conjugate function is obtained.

The solution for non-Newtonian fluid problems in two-dimensional space may also be obtained by modifying the solution of rate sensitive viscous materials given in [15].

This chapter consists of four main sections. Following this introduction, channel flow and Couette flow behavior are investigated by reducing the two-dimensional equation to a one-dimensional equation. The results of these solutions are then compared to existing Newtonian and non-Newtonian fluid results. The purpose of this exercise is to illustrate the application of the complex stream function approach to non-Newtonian fluid problems. The second part of this section deals with two-dimensional flow behavior around a circular cylinder in an infinite non-Newtonian fluid medium flowing with a uniform horizontal approach velocity. The abstracts of these analyses were published in [16a] and the analysis in [16a] were modified. The next section of the chapter presents a discussion of two-dimensional flow behavior in a sudden contracted channel and demonstrates that the problem can be solved by using the Schwartz-Christoffel conformal transformation [39]. Since analogy between non-Newtonian fluids and steady state creep materials is valid, the velocity solution is applied to a metal extrusion problem. Analyses of the power law materials using pseudo-stress functions and simple boundary value problems are illustrated [34] in the last section.

ANALYSIS OF NON-NEWTONIAN FLUIDS USING COMPLEX STREAM FUNCTIONS

Analysis

Detailed derivations for the solutions of non-linear viscous materials under plane strain conditions are given in [15]. The governing equation results in a non-linear differential equation because of the nature of the constitutive equation. Consequently, it may be difficult to obtain a closed form solution. This difficulty, however, is circumvented by using complex variables. The solution determined in [15] was obtained by assuming the summation form of the complex stream function. Additionally, the solution was found by using the mixed mode solution expressed in terms of non-separable, independent conjugate complex variables. To obtain the solution for non-Newtonian fluid flow problems, it is beneficial to briefly describe the solution shown in [15].

The governing equation for non-Newtonian fluids described by the power law under plane strain and assuming an incompressible condition can be derived from the two equilibrium equations. The constitutive equation, $\sigma_{ij} = \dot{\lambda}(\sigma_{ij} + \sigma_p\delta_{ij})$, and $\tilde{\sigma} = \sigma_0\dot{\tilde{\epsilon}}^m$, the strain rate, $\dot{\epsilon}_{ij} = \frac{1}{2}(u_{i,j} + u_{j,i})$ and the incompressibility condition under a plane strain condition are substituted into the two equilibrium equations. Using the stream function definition, $u = \partial\phi/\partial y$ and $v = -\partial\phi/\partial x$, the equilibrium equations are expressed in terms of the hydrostatic pressure gradient and the higher derivatives of the stream function with respect to x and y. Eliminating the hydrostatic pressure gradient from the equilibrium equation gives

$$\frac{\partial^2}{\partial x\,\partial y}\left[\frac{2}{\dot{\lambda}}\frac{\partial^2\phi}{\partial x\,\partial y}\right]+\left(\frac{\partial^2}{\partial y^2}-\frac{\partial^2}{\partial x^2}\right)\left[\frac{1}{2\dot{\lambda}}\left(\frac{\partial^2\phi}{\partial y^2}-\frac{\partial^2\phi}{\partial x^2}\right)\right]=0 \tag{1}$$

On transformation of the stream function ϕ into the complex plane, Equation 1 becomes

$$\frac{\partial^2}{\partial z^2}\left[\left(\frac{\partial^2\phi}{\partial z^2}\right)^{(m-1)/2}\left(\frac{\partial^2\phi}{\partial \bar{z}^2}\right)^{(m+1)/2}\right]+\frac{\partial^2}{\partial \bar{z}^2}\left[\left(\frac{\partial^2\phi}{\partial z^2}\right)^{(m+1)/2}\left(\frac{\partial^2\phi}{\partial \bar{z}^2}\right)^{(m-1)/2}\right]=0 \tag{2}$$

where ϕ is the stream function and $\dot{\lambda}$ is the scalar multiplier that is a function of the second invariant of the strain rate and is given by

$$\dot{\lambda}=\frac{3}{2}\frac{\dot{\bar{\epsilon}}}{\bar{\sigma}}=\frac{3}{2\sigma_0}\left(4\sqrt{\frac{1}{3}}\right)^{1-m}\left[\left(\frac{\partial^2\phi}{\partial z^2}\frac{\partial^2\phi}{\partial \bar{z}^2}\right)^{(1-m)/2}\right] \tag{3}$$

$\bar{\sigma}$ and $\dot{\bar{\epsilon}}$ are the effective stress and strain rate.

The reciprocal of Equation 3 corresponds to the apparent viscosity. The solution of Equation 2 is obtained by assuming the summation form of the complex conjugate stream function, and it is given in [15] and

$$\phi(z,\bar{z})=\phi_1(z)+\phi_2(\bar{z})=2\,\mathrm{Re}\left\{\frac{(m-1)^2}{2m(m+1)A_0^2}(A_0 z+A_1)^{2m/(m-1)}+A_2 z+A_3\right\} \tag{4}$$

The detailed derivation of Equations 1–4 are given in Appendix A. The other solution of Equation 2 is the mixed mode solution expressed in terms of non-separable, independent complex conjugate variables,

$$\phi(z,\bar{z})=\left(\frac{K_0}{A_0}\right)^{1/m}\left\{\frac{1}{K_0A_0}\frac{m^2}{(1+m)(1+2m)}(A_0 z+K_0\bar{z})^{(1+2m)/m}+A_2 z\bar{z}+A_3 z+A_4\bar{z}\right\} \tag{5}$$

The detailed derivation of Equation 5 is given in Appendix B.

Using Equation 5, the velocity states are derived. The horizontal and vertical velocity components are:

$$\left.\begin{aligned} u&=\frac{2}{r_0}\frac{m}{1+m}\eta^{(1+m)/m}\sin\frac{1}{m}\{(1+m)\zeta-(2+m)\theta_0\}+A_1\\ v&=\frac{2}{r_0}\frac{m}{1+m}\eta^{(1+m)/m}\cos\frac{1}{m}\{(1+m)\zeta-(2+m)\theta_0\}+A_2\end{aligned}\right\} \tag{6}$$

The strain rate components are:

$$\dot{\epsilon}_x=-\dot{\epsilon}_y=2\eta^{1/m}\sin\frac{1}{m}\left\{\zeta-2(1+m)\theta_0\right\}$$

$$\dot{\epsilon}_{xy}=-2\eta^{1/m}\cos\frac{1}{m}\left\{\zeta-2(1+m)\theta_0\right\} \tag{7}$$

and the stress components are:

$$\left.\begin{aligned}\sigma_x &= 2c\,\mathrm{Im}\left\{\left(\frac{\partial^2\phi}{\partial z^2}\right)^{(m-1)/2}\left(\frac{\partial^2\phi}{\partial \tilde{z}^2}\right)^{(m+1)/2}\right\} - \sigma_p \\ \sigma_y &= -2c\,\mathrm{Im}\left\{\left(\frac{\partial^2\phi}{\partial z^2}\right)^{(m-1)/2}\left(\frac{\partial^2\phi}{\partial \tilde{z}^2}\right)^{(m+1)/2}\right\} - \sigma_p \\ \sigma_{xy} &= -2c\,\mathrm{Re}\left\{\left(\frac{\partial^2\phi}{\partial z^2}\right)^{(m-1)/2}\left(\frac{\partial^2\phi}{\partial \tilde{z}^2}\right)^{(m+1)/2}\right\}\end{aligned}\right\} \tag{8}$$

One Dimensional Flow

The application of the mixed mode solution is illustrated in this section. The two-dimensional mixed mode solution, shown in Equations 5–8, is reduced to the one-dimensional case and the problem associated with flow between two parallel flat plates as well as Couette and slit flow problems are addressed. The solutions obtained by using the mixed mode approach are compared with the existing solutions for non-Newtonian fluids. The purpose of this section is to illustrate the mixed mode method by comparing the solutions resulting from the application of the mixed mode approach to the solution of classical one-dimensional fluid flow problems.

Flow Between Parallel Flat Plates

The schematic diagram illustrating this problem is shown in Figure 1a. One-dimensional flow behavior can be obtained from the reduction of the two-dimensional solution. Since the non-vanishing stress component is the shear stress, Equation 5 is reduced to

$$\phi(z, \tilde{z}) = \left\{\frac{m^2}{(1+m)(1+2m)}\right\} a_0^{(1+2m)/m}\{(z+\tilde{z})^{(1+2m)/m}\} + A(z+\tilde{z}) \tag{9}$$

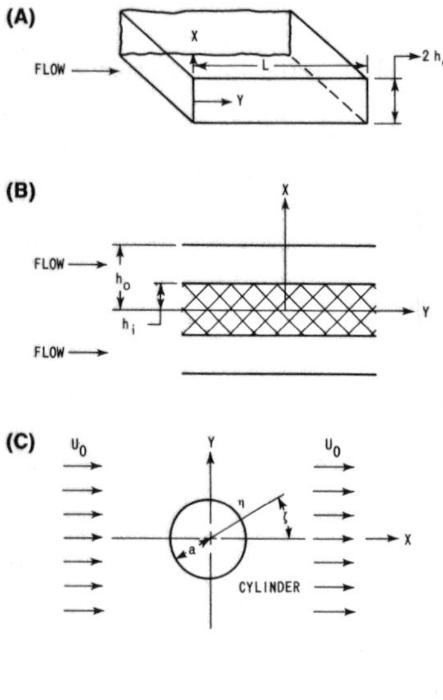

Figure 1. (A) Flow between parallel plates; (B) schematic diagram of slite flow; (C) schematic diagram of two-dimensional non-Newtonian flow.

and the velocity component in the axial direction is given by

$$V = -\frac{\partial \phi}{\partial x} = -(2)^{(1+2m)/m}\left(\frac{m}{1+m}\right)a_0^{(1+2m)/m}x^{(1+m)/m} + A \tag{10}$$

The constant A is obtained from the homogeneous boundary condition, $v = 0$ at $x = h_0$, thus

$$A = (2)^{(1+2m)/m}\frac{m}{1+m}(a_0)^{(1+2m)/m}(h_0)^{(1+m)/m} \tag{11}$$

If the feeding rate through the cross section is 2 in.^{3}/sec/unit width, the continuity equation implies that

$$\int_0^{h_0} V\,dx = 1 \tag{12}$$

The integral constant a_0 from Equation 11 is found to be

$$a_0^{(1+2m)/m} = \frac{(1+2m)}{(2)^{(1+2m)/m}m(h_0)^{(1+2m)/m}} \tag{13}$$

On substituting into Equation 10, the velocity profile is given by

$$V = \frac{1+2m}{h_0(1+m)}\left[1-\left(\frac{x}{h_0}\right)^{(1+m)/m}\right] \tag{14}$$

From Equation 5, $A_0 = K_0$. Hence, the only non-vanishing strain rate component is the shear rate and is given by

$$\dot{\epsilon}_{xy} = -(2)^{(1+m)/m}\left\{\frac{1+2m}{2^{(1+2m)/m}m(h_0)^{(1+2m)/m}}\right\}x^{1/m} \tag{15}$$

The stress components are given from the constitutive equation. The stress component in the x and y direction is

$$\sigma_x = \sigma_y = K \qquad \text{(hydrostatic pressure term)}$$

The shearing stress is given as

$$\sigma_{xy} = -4c\left[\frac{(1+2m)}{2^{(1+2m)/m}m(h_0)^{(1+2m)/m}}\right]^m x \tag{16}$$

If the material is rigid perfectly plastic, the velocity profile is obtained from Equation 14 as

$$V = \frac{1}{h_0} \qquad \text{since } x/h_0 \leq 1$$

In this case the shear rate vanishes through the cross section except at the corner of the opening which is a singular point. The force associated with the two parallel flat plates is obtained from the equilibrium condition,

$$h_0 p = \sigma_{xy}(x = h_0)L \tag{17}$$

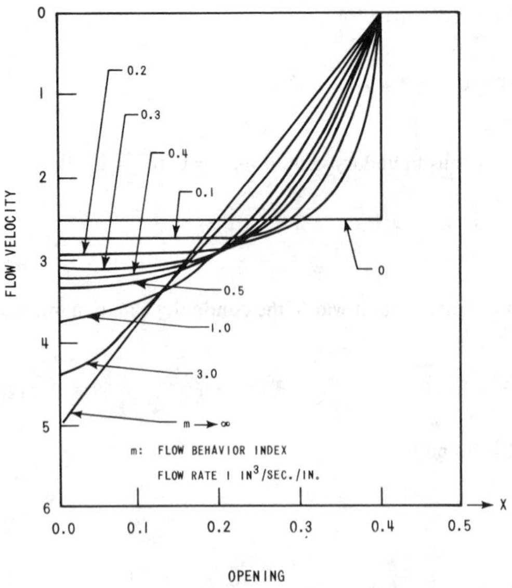

Figure 2. Velocity distribution of two flat plates for Non-Newtonian fluids.

The resulting expression for the pressure drop p is

$$\frac{p}{\sigma_0 L} = \left(\frac{1 + 2m}{m}\right)^m \left(\frac{1}{\sqrt{3}}\right)^{m+1} \frac{1}{h_0^{2m}} \tag{18}$$

where L and $2h_0$ are the length and opening size of the plate, respectively.

Figure 2 shows the velocity profiles for this solution. For very small values of m, except for the local discontinuity at the boundary $x = h_0 = 0.4$, the velocity throughout the section is approximately uniform. For increasing m values, the velocity gradient becomes progressively sharper and reaches the maximum peak at $x = 0$ as $m \to \infty$. For $m = 1$, that is, a Newtonian fluid, the velocity profile is a parabolic distribution which is in agreement with the Newtonian fluid solution [16]. The parabolic velocity distribution for $m = 1$ is confirmed from Figure 2, which shows the linear distribution of the shearing strain rate. The shearing strain rate shown in Figure 3 indicates a similar trend, that is, the strain rate gradients become progressively sharper and more localized near the boundary ($x = h_0 = 0.4$) as the flow behavior index decreases.

A Slit Flow

The flow behavior of a non-Newtonian liquid flowing through a slit (illustrated in Figure 1b) is studied neglecting entrance effects. The stream function for the mixed mode solution given in Equation 5 is used. The stream function for this problem is given as

$$\phi = \frac{m^2}{(1 + m)(1 + 2m)} a^{(1+2m)/m} x^{(1+2m)/m} + A_1 x^2 + A_2 x \tag{19}$$

The horizontal velocity V is given as

$$V = -\frac{\partial \phi}{\partial x} = -\left\{\frac{m}{1 + m} a_0^{(1+2m)/m} x^{(1+m)/m} + A_1 x + A_2\right\} \tag{20}$$

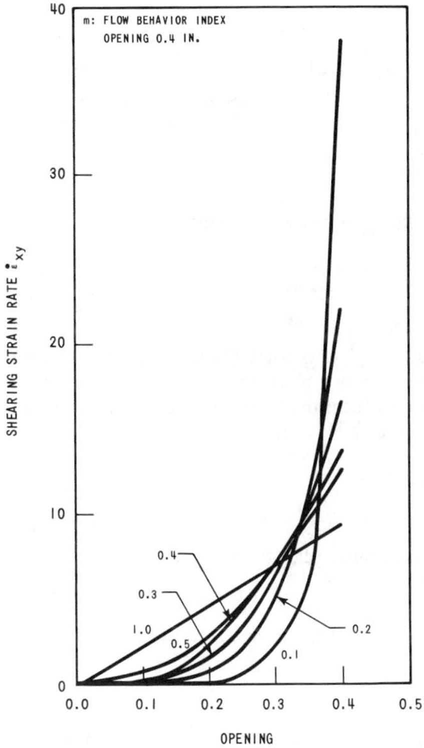

Figure 3. Shearing strain rate distribution for two flat plates for non-Newtonian fluid.

The integral constants a_0, A_1, and A_2 are obtained from the boundary conditions,

$$\left.\begin{array}{ll} V = 0 & \text{at } x = h_i \\ V = 0 & \text{at } x = h_0 \end{array}\right\} \tag{21}$$

where h_i and h_0 are the distances from the centerline of the block to the inner and outer plate, respectively.

The third boundary condition is a unit flow rate through a unit thickness,

$$\int_{h_i}^{h_0} v\, dx = 1 \tag{22}$$

Using the three boundary conditions, the three integral constants are determined and the resulting velocity profile is given by

$$V = \frac{m}{1+m} a_0^{(1+2m)/m} \left[x^{(1+m)/m} + \frac{1}{h_0 - h_i} \times \{(h_i^{(1+m)/m} - h_0^{(1+m)/m})x + (h_i h_0^{(1+m)/m} - h_0 h_i^{(1+m)/m})\} \right] \tag{23}$$

where

$$a_0^{(1+2m)/m} = \frac{1+m}{c}$$

$$c = m\left[\frac{m}{1+2m}(h_i^{(1+2m)/m} - h_0^{(1+2m)/m}) + \frac{1}{2(h_0 - h_i)}\right.$$

$$\left.\times \{(h_i^2 - h_0^2)(h_i^{(1+m)/m} - h_0^{(1+m)/m}) + 2(h_i - h_0)(h_i h_0^{(1+m)/m} - h_0 h_i^{(1+m)/m})\}\right]$$

Couette Flow

A Couette flow problem is one in which the bottom plate is fixed, the upper plate is moving with a velocity, (u_0), and the fluid is flowing between the two plates (1 in.^{3}/in.sec.). This problem can easily be solved by reducing Equation 5 to the one-dimensional problem. The one-dimensional stream function is given for the Couette flow problem as

$$\phi(z, \tilde{z}) = - \tag{24}$$

The vertical velocity, u, vanishes at y = 0 and the horizontal velocity, v is given as

$$u = i\left(\frac{\partial \phi}{\partial z} - \frac{\partial \phi}{\partial \tilde{z}}\right) = 0$$

$$v = -\left(\frac{\partial \phi}{\partial z} + \frac{\partial \phi}{\partial \tilde{z}}\right) = -\frac{m}{1+m}\, a_0^{(1+2m)/m} x^{(1+m)/m} - Ax \tag{25}$$

Using the boundary conditions

$$\left.\begin{array}{ll} v = u_0 & \text{at } x = h \\ v = 0 & \text{at } x = 0 \end{array}\right\}$$

The integral constant, A, is found as

$$A = -\frac{u_0}{h} - \frac{m}{1+m}\, a_0^{(1+2m)/m} h^{1/m} \tag{26}$$

The other boundary condition is

$$1 = \int_0^h v\, dx = \frac{m(h)^{(1+2m)/m}}{2(1+2m)(1+m)}\, a_0^{(1+2m)/m} + \frac{u_0 h}{2}$$

The integral constant a_0 is found from the previous equation,

$$a_0^{(1+2m)/m} = \frac{2}{m}\left(1 - \frac{u_0}{2} h\right)(1+m)(1+2m)\left(\frac{1}{h}\right)^{(1+2m)/m}$$

The velocity profile for Couette flow becomes

$$v = -\frac{2}{h}(1+2m)\left(1 - \frac{1}{2}u_0 h\right)\left\{\left(\frac{x}{h}\right)^{(1+m)/m} - \frac{x}{h}\right\} + \frac{u_0}{h}x \tag{27}$$

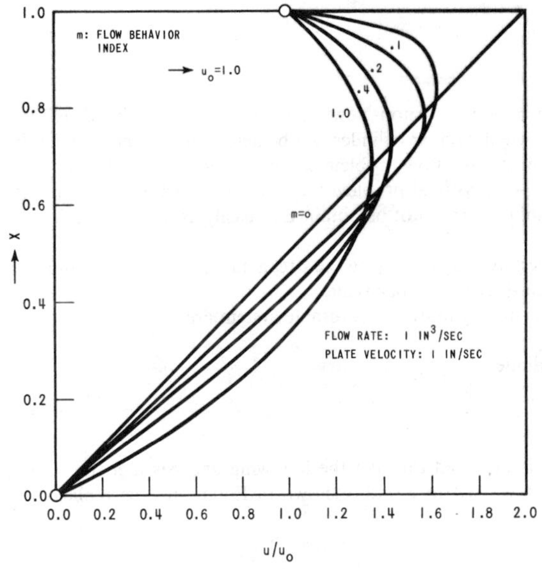

Figure 4. Velocity profile of Couette flow for non-Newtonian fluids

The shear strain rate is given as

$$\dot{\epsilon}_{xy} = \frac{-2}{h}(1 + 2m)\left(1 - \frac{u_0 h}{2}\right)\left\{\frac{1 + m}{m}\left(\frac{1}{h}\right)^{(1+m)/m} x^{1/m} - \frac{1}{h}\right\} + \frac{u_0}{h} \tag{28}$$

and the shear stress, σ_{xy} is given by

$$\begin{aligned} \sigma_{xy} = 2c&\left|\left\{\frac{2}{h^2}(1 + 2m)\left(1 - \frac{1}{2}u_0 h\right)\left\{1 - \frac{1 + m}{m}\left(\frac{x}{h}\right)^{1/m}\right\} + \frac{u_0}{h}\right\}\right|^{m-1} \\ &\times \left\{\frac{1}{h^2}(1 + 2m)\left(1 - \frac{1}{2}u_0 h\right)\left\{1 - \frac{1 + m}{m}\left(\frac{x}{h}\right)^{1/m}\right\} + \frac{u_0}{h}\right\} \end{aligned} \tag{29}$$

The velocity profile, from Equation 27, is illustrated in Figure 4, assuming $h = 1.0$ and the upper plate moves with a unit velocity. When Equation 27 is reduced to Newtonian flow, $m = 1.0$, the results shown in Figure 4 correspond to the velocity profile in Reference 17 for a unit flow rate and

$$P = 3.0 = \frac{1}{2\mu}\left(-\frac{dp}{dx}\right)$$

The velocity at the moving plate may be unstable when the flow behavior index (m) = 0.0 and the velocity gradient (dv/dx) = 2.0. When the flow behavior index approaches infinity the velocity profile is indeterminate as shown in Equation 27. The velocity profile, however, can be predicted by induction from Figure 4. The velocity profile as $m \to \infty$ is linear through the section which can be predicted from the velocity variations with respect to the behavior index.

Illustrated herein is the stream function approach to solve a one-dimensional non-Newtonian fluid problem. The next section deals with a similar two-dimensional problem.

Two Dimensional Flow

Analysis

The two-dimensional velocity distribution of a non-Newtonian fluid flowing with a uniform horizontal approach velocity around a rigid circular cylinder will be determined by employing the stream function approach. The one-dimensional flow problem can be solved readily without using the stream function approach. The two-dimensional problem requires a more complicated mathematical evaluation, consequently the solution may not be obtained as easily as the solution for the one-dimensional problem.

The two-dimensional problem can be solved by modifying the stream function shown in Equation 4 and by assuming a slip fluid with non-Newtonian behavior.

The considered boundary conditions in a cylindrical coordinate system are

$$\left.\begin{aligned} &U_\eta = 0 \quad \text{at} \quad \eta = a \text{ (radius of the cylinder)} \\ &U = U_\infty = \{U_\eta^2 + U_\xi^2\}^{1/2} \quad \text{at} \quad \eta \to \infty \end{aligned}\right\}$$

The modified solution of Equation 4 is obtained through the following analysis. Let the second derivatives of the stream function with respect to z and $\tilde{z}$ shown in Equation 2 be replaced by

$$\left.\begin{aligned} \left(\frac{\partial^2\phi}{\partial z^2}\right)^{(m-1)/2}\left(\frac{\partial^2\phi}{\partial \tilde{z}^2}\right)^{(m+1)/2} &= \frac{\partial^2\Phi}{\partial \tilde{z}^2} \\ \left(\frac{\partial^2\phi}{\partial z^2}\right)^{(m+1)/2}\left(\frac{\partial^2\phi}{\partial \tilde{z}^2}\right)^{(m-1)/2} &= \frac{\partial^2\Phi}{\partial z^2} \end{aligned}\right\} \tag{30}$$

Solving these two equations with respect to $\partial^2\phi/\partial z^2$ and $\partial^2\phi/\partial \tilde{z}^2$ gives

$$\left.\begin{aligned} \frac{\partial^2\phi}{\partial z^2} &= \left(\frac{\partial^2\Phi}{\partial z^2}\right)^{(1+m)/2m}\left(\frac{\partial^2\Phi}{\partial \tilde{z}^2}\right)^{(1-m)/2m} \\ \frac{\partial^2\phi}{\partial \tilde{z}^2} &= \left(\frac{\partial^2\Phi}{\partial z^2}\right)^{(1-m)/2m}\left(\frac{\partial^2\Phi}{\partial \tilde{z}^2}\right)^{(1+m)/2m} \end{aligned}\right\} \tag{31}$$

where $m \neq 0$.

If Equation 30 is substituted into Equation 2, the resulting equation will be the biharmonic equation. The solution of the biharmonic equation can be obtained as

$$\Phi = \tilde{z}\Omega(z) + \psi(z) + z\tilde{\Omega}(z) + \tilde{\psi}(z) \tag{32}$$

and

$$\begin{aligned} \frac{\partial^2\phi}{\partial z^2} &= \{\tilde{z}\Omega''(z) + \psi''(z)\}^{(1+m)/2m}\{z\tilde{\Omega}''(z) + \tilde{\psi}(z)\}^{(1-m)/2m} \\ \frac{\partial^2\phi}{\partial \tilde{z}^2} &= \{z\tilde{\Omega}''(z) + \tilde{\psi}''(z)\}^{(1+m)/2m}\{\tilde{z}\Omega''(z) + \psi''(z)\}^{(1-m)/2m} \end{aligned} \tag{33}$$

where $\Omega(z)$, $\psi(z)$, $\tilde{\Omega}(z)$, and $\tilde{\psi}(z)$ are assumed to be holomorphic functions of z and $\tilde{z}$ in the given domain. The solution obtained from Equation 33 is not complete since the uniqueness of ϕ is not apparent. The proof of uniqueness of ϕ can be easily shown. Let ϕ_1 and ϕ_2 be two solutions obtained from Equation 33. Since the stream function is a real valued function, the conjugate of the second equation in Equation 31 should be identical with the second expression in Equation 31.

Namely,

$$\left(\overline{\frac{\partial^2\phi_2}{\partial\tilde{z}^2}}\right) = (\tilde{z}\Omega'' + \psi'')^{(1+m)/2m}(z\tilde{\Omega}'' + \tilde{\psi}'')^{(1-m)/2m} = \frac{\partial^2\phi_2}{\partial z^2} \tag{34}$$

Equation 34 is identical with the first expression in Equation 33. Hence, $\phi_1 = \phi_2$ and the uniqueness of ϕ is demonstrated. This means the stream function obtained from the first expression in Equation 33 is identical with that obtained by using the second expression in Equation 33.

The hydrostatic pressure is needed to determine the stress solution. The solution for the hydrostatic solution is given in the next subsection.

Pressure in the Non-Newtonian Fluid

The fluid pressure is required to obtain the stress distribution in a Non-Newtonian medium. The pressure, σ_p, can be obtained from two equilibrium equations written in terms of the stream function gradients:

$$\left.\begin{aligned} -\frac{\partial\sigma_p}{\partial x} + \frac{\partial}{\partial x}\left\{\frac{1}{\lambda}\frac{\partial^2\phi}{\partial x\,\partial y}\right\} + \frac{\partial}{\partial y}\left\{\frac{1}{2\lambda}\left(\frac{\partial^2\phi}{\partial y^2} - \frac{\partial^2\phi}{\partial x^2}\right)\right\} = 0 \\ -\frac{\partial\sigma_p}{\partial y} - \frac{\partial}{\partial y}\left\{\frac{1}{\lambda}\frac{\partial^2\phi}{\partial x\,\partial y}\right\} + \frac{\partial}{\partial x}\left\{\frac{1}{2\lambda}\left(\frac{\partial^2\phi}{\partial y^2} - \frac{\partial^2\phi}{\partial x^2}\right)\right\} = 0 \end{aligned}\right\} \tag{35}$$

Differentiating the first equation with respect to y and the second with respect to x, and adding yields

$$-2\frac{\partial^2\sigma_p}{\partial x\,\partial y} + \left(\frac{\partial^2}{\partial x^2} + \frac{\partial^2}{\partial y^2}\right)\left\{\frac{1}{2\lambda}\left(\frac{\partial^2\phi}{\partial y^2} - \frac{\partial^2\phi}{\partial x^2}\right)\right\} = 0$$

In the complex plane this equation can be rewritten as

$$i\left(\frac{\partial^2\sigma_p}{\partial z^2} - \frac{\partial^2\sigma_p}{\partial\tilde{z}^2}\right) = -2C\frac{\partial^2}{\partial z\,\partial\tilde{z}}\left(p^{(m+1)/2}q^{(m-1)/2} + p^{(m-1)/2}q^{(m+1)/2}\right) \tag{36}$$

where $c = \dfrac{2\sigma_0}{3}\left(\dfrac{\sqrt{3}}{4}\right)^{1-m}$

$$p = \frac{\partial^2\phi}{\partial z^2}$$

$$q = \frac{\partial^2\phi}{\partial\tilde{z}^2}$$

p and q are obtained from Equations 30, 32, and 33. If $\Omega(z) = \tilde{\Omega}(z) = 0$ in Equation 32, then Equation 36 yields

$$\frac{\partial^2\sigma_p}{\partial z^2} - \frac{\partial^2\sigma_p}{\partial\tilde{z}^2} = 0$$

and the solution is

$$\sigma_p = f(z + \tilde{z}) + g(z - \tilde{z}) \tag{37}$$

Equation 37 is independent of p and q or ϕ. This is not realistic. Therefore, the pressure is obtained from the Bernoulli's equation. An analysis to obtain the two-dimensional fluid flow solution so far has been presented. The next subsection illustrates how to obtain the velocity field around the rigid circular cylinder with uniform approach velocity.

Solution for a Rigid Circular Cylinder with Uniform Approach Velocity

Considering a perfect Newtonian fluid, associated with the circular theorem, Φ (pseudo-stream function) in Equation 32,

$$\Phi = \psi(z) + \bar{\psi}(z)$$

Equation 31 results in

$$\frac{\partial^2 \phi}{\partial z^2} = \left(\frac{\partial^2 \psi}{\partial z^2}\right)^{(1+m)/2m} \left(\frac{\partial^2 \bar{\psi}}{\partial \bar{z}^2}\right)^{(1-m)/2m} \tag{38}$$

where $\Omega(z) = \bar{\Omega}(z) = 0$ in Equation 32

and

$$\frac{\partial^2 \psi}{\partial z^2} = K\left(\frac{a^3}{z^3}\right)$$
$$\frac{\partial^2 \bar{\psi}}{\partial \bar{z}^2} = \bar{K}\left(\frac{a^3}{\bar{z}^3}\right) \tag{39}$$

substituting Equation 39 into 38, the resulting equation is

$$\frac{\partial^2 \phi}{\partial z^2} = (R)^{(1-m)/m} K(z)^{-3(1+m)/2m} (\bar{z})^{-3(1-m)/2m} (a)^{3/m}$$
$$\frac{\partial^2 \phi}{\partial \bar{z}^2} = (R)^{1-m/m} \bar{K}(z)^{-3(1-m)/2m} (\bar{z})^{-3(1+m)/2m} (a)^{3/m} \tag{40}$$

where $R^2 = (K\bar{K})$

using the definition of the total differentiation

$$\frac{\partial \phi}{\partial z} = \int_{\Gamma 1} \frac{\partial^2 \phi}{\partial z^2} dz + \int_{\Gamma 1} \frac{\partial^2 \phi}{\partial z \, \partial \bar{z}} d\bar{z} \tag{41}$$

and

$$\frac{\partial^2 \phi}{\partial z \, \partial \bar{z}} = \int_{\Gamma 1} \frac{\partial}{\partial \bar{z}} \left(\frac{\partial^2 \phi}{\partial z^2}\right) dz + \int_{\Gamma 1} \frac{\partial}{\partial z} \left(\frac{\partial^2 \phi}{\partial \bar{z}^2}\right) d\bar{z}$$

where $\Gamma 1$ is the integration path described by $\zeta = 0$ to $\zeta = \zeta$ at $\eta \to \infty$ and $\eta \to \infty$ to $\eta = \eta$ at given ζ value.

$$\frac{\partial}{\partial \bar{z}} \left(\frac{\partial^2 \phi}{\partial z^2}\right) = \frac{3m-3}{2m} (R)^{(1-m)/m} K(z)^{-3(1+m)/2m} (\bar{z})^{(m-3)/2m} (a)^{3/m}$$
$$= \frac{3(m-1)}{2m} (R)^{(1-m)/m} (K)(a)^{3/m} \eta^{-(3+m)/m} e^{-2i\zeta}$$

$$\frac{\partial}{\partial z}\left(\frac{\partial^2\phi}{\partial \bar{z}^2}\right) = \frac{3m-3}{2m}(R)^{(1-m)/m}(\bar{K})(z)^{(m-3)/2m}(\bar{z})^{-3(1+m)/2m}(a)^{3/m}$$
$$= \frac{3(m-1)}{2m}(R)^{(1-m)/m}(\bar{K})(a)^{3/m}\eta^{-(3+m)/m}e^{2i\zeta}$$

Performing the integration of the second equation in Equation 41, the mixed derivative can be found as

$$\frac{\partial^2\phi}{\partial z\,\partial \bar{z}} = \frac{(1-m)}{2}(R)^{(1-m)/m}(a)^{3/m}\eta^{-3/m}(Ke^{-i\zeta} + \bar{K}e^{i\zeta})$$

The first equation in Equation 41 results in

$$\frac{\partial\phi}{\partial z} = (R)^{(1-m)/m}(a)^{3/m}\eta^{(m-3)/m}\left(\frac{m}{2}\right)\left\{-Ke^{-2i\zeta} + \frac{1-m}{m-3}\bar{K}\right\} + D$$

$$\frac{\partial\phi}{\partial \bar{z}} = (R)^{(1-m)/m}(a)^{3/m}\eta^{(m-3)/m}\left(\frac{m}{2}\right)\left\{-\bar{K}e^{2i\zeta} + \frac{1-m}{m-3}K\right\} + \bar{D}$$

since

$$\phi = \int_{\Gamma 1}\frac{\partial\phi}{\partial z}\,dz + \int_{\Gamma 1}\frac{\partial\phi}{\partial \bar{z}}\,d\bar{z} + Dz + \bar{D}\bar{z}$$
$$= \frac{m^2(2-m)}{(2m-3)(m-3)}(R)^{(1-m)/m}(a)^{3/m}\eta^{(2m-3)/m}(Ke^{-i\zeta} + \bar{K}e^{i\zeta}) + \eta(De^{i\zeta} + \bar{D}e^{i\zeta}) \tag{42}$$

Equation 42 can be rewritten as

$$\phi = \frac{m^2(2-m)}{(2m-3)(m-3)}(R)^{(1-m)/m}(a)^{3/m}Q\eta^{(2m-3)/m}\sin\zeta + B\eta\sin\zeta \tag{43}$$

where $K = -Qi$ and $\bar{K} = Qi$ and Q is real constant. $D = Bi$, $\bar{D} = -Bi$ and B is real constant. Equation 43 can be written as

$$\phi = \left(-A\left(\frac{a}{\eta}\right)^{(3-m)/m}\eta + B\eta\right)\sin\zeta \tag{44}$$

or

$$\phi = \left\{-A\left(\frac{a}{\eta}\right)^{(3-m)/m} + B\right\}\eta\sin\zeta$$

The radial and tangential velocity distribution is given as

$$\left.\begin{aligned} U_\eta &= \frac{1}{\eta}\left(\frac{\partial\phi}{\partial\zeta}\right) = \left\{B - A\left(\frac{a}{\eta}\right)^{(3-m)/m}\right\}\cos\zeta \\ U_\zeta &= -\frac{\partial\phi}{\partial\eta} = -\left\{B + A\frac{3-2m}{m}\left(\frac{a}{\eta}\right)^{(3-m)/m}\right\}\sin\zeta \end{aligned}\right\}$$

From the boundary condition, $U_\eta = 0$ at $\eta = a$, $A = B$ from the second boundary condition,

$$U_\infty = \sqrt{U_\eta^2 + U_\zeta^2} \qquad \text{as } \eta \to \infty$$

$$B = U_\infty$$

Thus, the velocity distribution around the rigid cylinder with uniform approach velocity is given as

$$\left.\begin{aligned} U_\eta &= U_\infty\left\{1 - \left(\frac{a}{\eta}\right)^{(3-m)/m}\right\}\cos\zeta \\ U_\zeta &= -U_\infty\left\{1 + \frac{3-2m}{m}\left(\frac{a}{\eta}\right)^{(3-m)/m}\right\}\sin\zeta \end{aligned}\right\} \tag{44a}$$

If the fluid is Newtonian, the velocity distribution reduces to

$$\begin{aligned} U_\eta &= U_\infty\left\{1 - \left(\frac{a}{\eta}\right)^2\right\}\cos\zeta \\ U_\zeta &= -U_\infty\left\{1 + \left(\frac{a}{\eta}\right)^2\right\}\sin\zeta \end{aligned} \tag{44b}$$

This agrees with the existing solution [16].

The hydrodynamic pressure is obtained by applying Bernoulli's theorem. The pressure distribution is obtained from

$$\sigma_p + \tfrac{1}{2}\rho U^2 = \P + \tfrac{1}{2}\rho U_\infty^2 \tag{45}$$

where σ_p is the hydrodynamic pressure, U is the absolute velocity, $U = (U_\eta^2 + U_\zeta^2)^{1/2}$ and $\P$ is the hydrostatic pressure, $\sigma_p = \P$ at $\eta \to \infty$. The pressure at any arbitrary point in the medium is given as

$$\begin{aligned} \sigma_p - \P = \frac{1}{2}\rho U_\infty^2 - \frac{1}{2}\rho U^2 = \frac{1}{2}\rho U_\infty^2\Bigg[1 - \Bigg\{1 + \left(\frac{2(3-2m)}{m}\sin^2\zeta - 2\cos^2\zeta\right)\left(\frac{a}{\eta}\right)^{(3-m)/m} \\ + \left(\frac{a}{\eta}\right)^{2(3-m)/m}\left(\cos^2\zeta + \frac{(3-2m)^2}{m^2}\sin^2\zeta\right)\Bigg\}\Bigg] \end{aligned}$$

where ρ is the fluid density. The pressure distribution on the cylinder surface, at $\eta = a$, is given as

$$\sigma_p - \P = \frac{1}{2}\rho U_\infty^2\left(1 - \frac{m^2 - 6m + 9}{m^2}\sin^2\zeta\right) \tag{46}$$

If Equation 46 is reduced by letting $m = 1$, the pressure distribution becomes

$$\sigma_p - \P = \frac{1}{2}\rho U_\infty^2(1 - 4\sin^2\zeta) \tag{47}$$

Equation 47 agrees with the existing solution for a Newtonian fluid associated with the potential flow regime [16].

Results and Discussion

This section illustrates the application of the stream function approach to the solution of non-Newtonian fluid problems neglecting the inertia term. The three solutions, the mixed mode solution expressed in terms of non-separable independent complex conjugate variables, the summation form of the solution and the pseudo-stream function solution, are given. The boundary value problems are illustrated by using the mixed mode solution and the pseudo-stream function solution. The summation form of the solution shown in Equation 4 suggests the possibility of instability for certain values of m. For example, the velocity obtained from Equation 4 shows divergence for large η values and for $m > 1$. However, as seen from Equation 5, the solution obtained using the mixed mode formulation does not have this instability associated with the flow behavior index provided the solution domain is finite when applying these solutions to real boundary value problems. These limitations must be considered. From the solution of simple one-dimensional problems, the application of the complex stream function technique is illustrated neglecting inertia. Some solutions can be reduced to existing Newtonian fluid solutions and some solutions are the same as existing solutions for non-Newtonian fluids.

Many one-dimensional non-Newtonian fluid problems can be solved without difficulties. However, when the problem is two-dimensional, the apparent viscosity in general will contain $\dot{\epsilon}x$, $\dot{\epsilon}y$, and $\dot{\epsilon}xy$. The mathematical manipulation of the second strain rate invariant, $\{(\dot{\epsilon}x - \dot{\epsilon}y)^2 + 4\dot{\epsilon}xy^2\}^{1/2}$, in Cartesian coordinates could be difficult. This difficulty can be circumvented by introducing the complex variable since the second strain rate invariant contained in the effective strain rate can be expressed in terms of the product form of the second derivatives of the stream function with respect to z and $\tilde{z}$ as shown in Equation 3. It appears that the major advantage of the complex variable approach is that the effective strain rate can be expressed by the product of the derivatives of the stream function with respect to z and $\tilde{z}$. As seen in Equation 32, Φ resembles the complex stream function of a Newtonian fluid and the functions Ω, ψ, $\tilde{\Omega}$ and ϕ can be *sometimes* found by referring to the expression for a Newtonian fluid in the potential flow regime. Thus $\tilde{\phi}$ may be referred to as the pseudo stream function of a non-Newtonian fluid.

Most of this section deals with the analysis of the two-dimensional velocity field associated with a rigid cylinder in an infinite non-Newtonian fluid. The velocity field is obtained by using the so-called pseudo stream function of the non-Newtonian fluid shown in Equation 32. The constant stream functions for $m = 1.0$ and $m = 0.5$ are shown in Figure 5. The variation of the stream function with the fluid behavior indices may not be significant.

The tangential velocity distribution on the solid wall for various fluid behavior indices is shown in Figure 6. The stagnation points are $\zeta = 0°$ and $180°$, which is the case for Newtonian fluids, and the peak velocity is obtained at $\zeta = \pi/2$ and $3\pi/2$ as expected. The magnitude of the tangential velocity decreases with increasing m values. However, the velocities decay more rapidly for small values of m than for larger m values as shown in Equation 43. The pressure distribution on the cylindrical surface is shown in Figure 7. The dimensionless quantity, $2(\sigma_p - \P)/\rho U_\infty^2$ decreases with decreasing m values and the maximum pressure is observed at the stagnation point. The zero

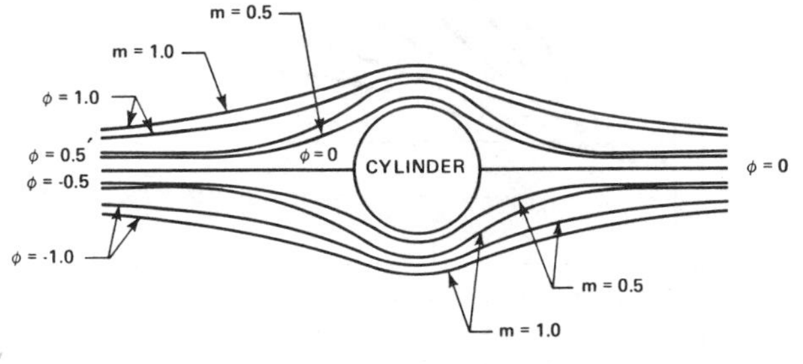

Figure 5. Constant stream function for m = 1.0 and m = 0.5.

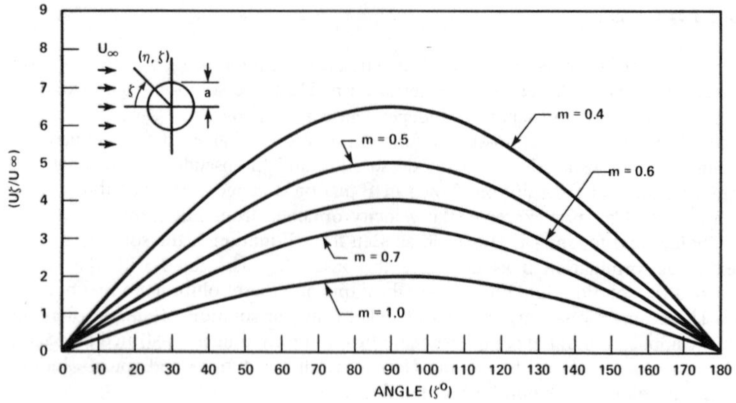

Figure 6. Tangential velocity distribution for various fluid index behavior on the surface $\eta = a$.

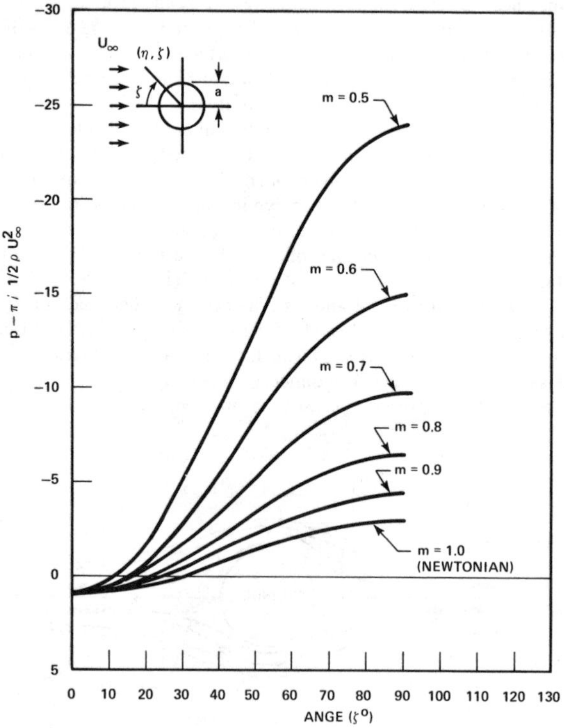

Figure 7. Pressure distribution on the surface for variour ζ values ($\eta = a$).

pressure (pressure at the free stream) is found between $O < \zeta \leq 30°$ for all m values. $\zeta = 30°$ for $m = 1.0$ and ζ decreases with decreasing m values.

This analysis demonstrates the application of complex variables associated with the stream function to non-Newtonian fluids. The expression shown in Equation 32 can be sometimes found by referring to the Newtonian fluid solution.

APPLICATION OF THE SCHWARTZ-CHRISTOFFEL CONFORMAL MAPPING AND THE COMPLEX POTENTIAL FUNCTION TO THE CORNERED CHANNEL EXTRUSION OF POWER LAW MATERIALS

As mentioned earlier, there are analogies between the constitutive equation and the governing equations for non-Newtonian fluids and those for steady state creep materials. These analogies may be applied to the investigation of the flow pattern for the metal forming process. Yang and Thomson [18] compared the results obtained from potential theory with results obtained for a flow pattern of a lead extrusion. The results showed that potential theory cannot be applied to metal flow except as a first approximation. Thomson [19] stated that the failure of potential theory to yield correct answers to metal flow problems does not preclude that other analogies systems may be found that are in closer agreement. He [19] devised a method of extruding a viscous oil simulating a metal extrusion. Whether or not these two systems would give the same flow patterns could not be predicted since the flow equations, as pointed out by Hill [23] are not the same. Thomson [19] compared the results obtained from oil (SAE 90) extrusion to those obtained by lead extrusion and showed the flow patterns are in close agreement except for a region near the cylinder wall. He [19] concluded from his limited experiments that it appears that viscous fluid flow studies may yield at least qualitative information about the behavior of metals in some forming process.

This section deals with the problem of viscous material processing. The constitutive equation of this material is the same expression as for a non-Newtonian fluid. The velocity solutions shown in this section may be equivalent to those for a non-Newtonian fluid assuming the fluid can slip on the wall boundary.

The knowledge of stress and strain rate associated with plastic flow during the metal forming process is of great practical importance from the viewpoint of optimally designing the die shape. The governing equations of plastic flow, however, are highly non-linear due to such common factors as, temperature, frictional effects, and non-linear material behavior. The boundary conditions are complex. A complete exact solution of the problem is therefore extremely difficult. Exact analytical solutions are virtually non-existent in the literature.

Generally the metal flow pattern and velocity field solutions are obtained by using simplifying assumptions. These are often the most significant factors needed for the design of the die shape.

Axially symmetric extrusion and drawing processes have been investigated analytically by many investigators [20–22]. Extrusion problems using a cornered die have been investigated by using a slip line theory without including the hardening effects as an upper bound solution [23, 24]. Three-dimensional extrusion problems have been performed by a few researchers [25–27]. Beyer [28] calculated the strain at the exit section by a conformal transformation of the exit section onto the entrance section without considering the plastic deformation in the intermediate passage. Goon and Pelukhim [29] proposed a variational method utilizing conformal transformation to find the flow and stresses when a very long section is extruded or rolled through slightly tapered dies. Nagpal and Altan [30] performed the analysis of a three-dimensional extrusion problem employing the dual stream function to obtain a kinematically admissible velocity field. Yang, et al. [31] performed the analysis of the extrusion problem having a generalized cross section using the conformal transformation in order to find a kinematically admissible velocity solution.

Two-dimensional analysis associated with rate dependent materials by using the complex stress and stream functions were proposed by Lee, et al. [15, 32–34]. Lee and Patel [13] performed the analysis of an extrusion problem of a perfect viscous material presuming a square cornered die. The superposition principle was applied [13] by taking advantage of the linear field expression. However, the analysis of strain rate hardening material extrusion problems using a cornered die may be very complicated due to the many boundary conditions and the nature of the non-linear field equation. In this section, an analysis of the extrusion of power law materials using a cornered die is performed by employing the complex potential function and the conformal transformation. This analysis was submitted for publication [39].

Analysis

The equilibrium equation written in terms of the stress gradient with respect to x and y is expressed in terms of the second derivative of the stream function using Cartesian coordinates. The equations are transformed on to the complex plane by eliminating the hydrostatic pressure gradient from the two equations. The transformation equation expressed in terms of the higher derivatives of the stream function with respect to the complex conjugate variables is solved by using a semi-inverse method. Similarly, the equilibrium equation expressed in terms of the derivatives of the velocity potential or complex potential function with respect to the complex conjugate variables is derived.

The analysis of power law materials extruded through cornered dies is performed by applying these solutions and using the Schwartz-Christoffel conformal transformation. Through the analysis, stream functions, velocity potential functions, horizontal and vertical velocities and strain rate components in the die are found. Additionally, stress components for various strain rate hardening exponents are found.

The assumptions made in the analysis are: (1) a plane strain condition, (2) an incompressibility condition, (3) an isothermal condition, and (4) a frictionless condition between the die and materials.

The field equation in the complex plane can be derived by assuming plane strain and incompressibility and by using the constitutive equation. Detailed derivations are given in [15]. The constitutive equation for power law materials is given by

$$\left.\begin{aligned} &\tilde{\sigma} = \sigma_0 \dot{\tilde{\epsilon}}^m \\ &\dot{\lambda}(\sigma_{ij} + \sigma_p \delta_{ij}) = \dot{\epsilon}_{ij} \end{aligned}\right\} \tag{48}$$

where $\tilde{\sigma}$ and $\dot{\epsilon}_{ij}$ are the effective stress and the strain rate, respectively, m is the strain rate hardening exponent, $\dot{\lambda} = \frac{3}{2}\dot{\tilde{\epsilon}}/\tilde{\sigma}$ and σ_p is the dilatational stress. The equilibrium equations and definitions of the strain rates and stream functions are

$$\left.\begin{aligned} &\sigma_{ij,j} = 0 \\ &\dot{\epsilon}_{ij} = \frac{1}{2}(u_{i,j} + u_{j,i}) \\ &u = -\frac{\partial \phi}{\partial y}, \qquad v = \frac{\partial \phi}{\partial x} \end{aligned}\right\} \tag{49}$$

where u and v are the horizontal and vertical velocities and ϕ is the stream function.

Using the constitutive equations, substituting the second and third expressions of Equation 49 into the equilibrium equations and eliminating the dilatational stress from the two equilibrium equations, results in the following equations:

$$\frac{\partial^2}{\partial x\,\partial y}\left[\frac{2}{\dot{\lambda}}\frac{\partial^2 \phi}{\partial x\,\partial y}\right] + \left(\frac{\partial^2}{\partial y^2} - \frac{\partial^2}{\partial x^2}\right)\left\{\frac{1}{2\dot{\lambda}}\left(\frac{\partial^2 \phi}{\partial y^2} - \frac{\partial^2 \phi}{\partial x^2}\right)\right\} = 0 \tag{50}$$

Transforming the stream function, ϕ, onto the complex plane, Equation 50 becomes

$$\frac{\partial^2}{\partial z^2}\left[\left(\frac{\partial^2 \phi}{\partial z^2}\right)^{(m-1)/2}\left(\frac{\partial^2 \phi}{\partial \bar{z}^2}\right)^{(1+m)/2}\right] + \frac{\partial^2}{\partial \bar{z}^2}\left[\left(\frac{\partial^2 \phi}{\partial z^2}\right)^{(1+m)/2}\left(\frac{\partial^2 \phi}{\partial \bar{z}^2}\right)^{(m-1)/2}\right] = 0 \tag{51}$$

$\dot{\lambda}$, the scalar multiplier, is given in the complex plane as

$$\dot{\lambda} = \frac{3}{2\sigma_0}\left(\frac{4}{\sqrt{3}}\right)^{1-m}\left(\frac{\partial^2\phi}{\partial z^2}\frac{\partial^2\phi}{\partial \bar{z}^2}\right)^{(1-m)/2} \tag{52}$$

Equation 51 is the equilibrium equation expressed in terms of the stream function and complex conjugate variables.

The derivation of the equilibrium equation in terms of the velocity potential function is required to analyze the extrusion problem. The definition of the velocity potential function is

$$\left.\begin{aligned} u &= -\frac{\partial\psi}{\partial x} = -\left(\frac{\partial}{\partial z} + \frac{\partial}{\partial \bar{z}}\right)\psi \\ v &= \frac{-\partial\psi}{\partial y} = -i\left(\frac{\partial}{\partial z} - \frac{\partial}{\partial \bar{z}}\right)\psi \end{aligned}\right\} \tag{53}$$

The strain rates in terms of the velocity potential function are

$$\left.\begin{aligned} \dot{\epsilon}_x &= \frac{\partial u}{\partial x} = -\frac{\partial^2\psi}{\partial x^2} = -\left(\frac{\partial}{\partial z} + \frac{\partial}{\partial \bar{z}}\right)^2\psi \\ \dot{\epsilon}_y &= \frac{\partial v}{\partial y} = -\frac{\partial^2\psi}{\partial y^2} = \left(\frac{\partial}{\partial z} - \frac{\partial}{\partial \bar{z}}\right)^2\psi \\ \dot{\epsilon}_{xy} &= \frac{1}{2}\left(\frac{\partial u}{\partial y} + \frac{\partial v}{\partial x}\right) = -\frac{\partial^2\psi}{\partial x\,\partial y} = -i\left(\frac{\partial^2\psi}{\partial z^2} - \frac{\partial^2\psi}{\partial \bar{z}^2}\right) \end{aligned}\right\} \tag{54}$$

The incompressibility condition, which limits the solution of the velocity potential function, is

$$\dot{\epsilon}_x + \dot{\epsilon}_y = 0 = 4\frac{\partial^2\psi}{\partial z\,\partial \bar{z}} \tag{55}$$

The solution of the velocity potential function should be in the form of

$$\psi = \psi(z) + \bar{\psi}(z) \tag{56}$$

where $\psi(z)$ and $\psi(\bar{z})$ are holomorphic functions of z and $\bar{z}$, respectively. The effective strain rate under plane strain presuming incompressibility is given by

$$\dot{\bar{\epsilon}} = \frac{2}{\sqrt{3}}(\dot{\epsilon}_x^2 + \dot{\epsilon}_{xy}^2)^{1/2} = \frac{4}{\sqrt{3}}\left(\frac{\partial^2\psi}{\partial z^2}\frac{\partial^2\psi}{\partial \bar{z}^2}\right)^{1/2} \tag{57}$$

Equation 57 is true if and only if $\partial^2\psi/\partial z\,\partial\bar{z} = 0$. The scalar multiplier, $\dot{\lambda}$, is found as

$$\dot{\lambda} = \frac{3}{2\sigma_0}\left(\frac{4}{\sqrt{3}}\right)^{1-m}\left(\frac{\partial^2\psi}{\partial z^2}\frac{\partial^2\psi}{\partial \bar{z}^2}\right)^{(1-m)/2} \tag{58}$$

Eliminating the dilatational stress gradient from the two equilibrium equations, the equilibrium equation, derived in terms of the velocity potential function, is given by

$$\frac{\partial^2}{\partial z^2}\left\{\left(\frac{\partial^2\psi}{\partial z^2}\right)^{(m-1)/2}\left(\frac{\partial^2\psi}{\partial \bar{z}^2}\right)^{(m+1)/2}\right\} - \frac{\partial^2}{\partial \bar{z}^2}\left\{\left(\frac{\partial^2\psi}{\partial z^2}\right)^{(m+1)/2}\left(\frac{\partial^2\psi}{\partial \bar{z}^2}\right)^{(m-1)/2}\right\} = 0 \tag{59}$$

Equation 59 is valid only for $\psi = \psi(z) + \bar{\psi}(z)$.

When Equations 51 and 59 are compared, it is seen that the expressions are identical except for the sign. The stream function solution given by Equation 51 does not have any restriction. However, Equation 59 does have the strict limitation for the velocity potential function solution because of the limitation imposed by the incompressibility condition.

The equilibrium equation expressed in terms of the complex potential function and complex conjugate variables will be derived. Defining the complex potential functions as

$$\left.\begin{aligned} W &= \psi + i\phi \\ \bar{W} &= \psi - i\phi \end{aligned}\right\} \tag{60}$$

Assuming W and $\bar{W}$ are holomorphic functions of z and $\bar{z}$, respectively, and using the form $\psi = \psi(z) + \bar{\psi}(z)$, the velocity potential function and the stream function can be expressed as:

$$\left.\begin{aligned} \psi &= \tfrac{1}{2}(W + \bar{W}) \\ \phi &= \frac{1}{2i}(W - \bar{W}) \end{aligned}\right\} \tag{61}$$

Substituting the first expression of Equation 61 into Equation 59, the equilibrium equation is expressed in terms of the complex potential function. The resulting equation is given by

$$\frac{\partial^2}{\partial z^2}\left\{\left(\frac{\partial^2 W}{\partial z^2}\right)^{(m-1)/2}\left(\frac{\partial^2 \bar{W}}{\partial \bar{z}^2}\right)^{(m+1)/2}\right\} - \frac{\partial^2}{\partial \bar{z}^2}\left\{\left(\frac{\partial^2 W}{\partial z^2}\right)^{(m+1)/2}\left(\frac{\partial^2 \bar{W}}{\partial \bar{z}^2}\right)^{(m-1)/2}\right\} = 0 \tag{62}$$

Similarly, substituting the second expression of Equation 61 into Equation 51 results in the identical relationship. Equations 51, 59, and 62 are equilibrium equations expressed in terms of the stream, velocity potential, and complex potential functions, respectively. The velocity potential and complex potential function solutions are limited to the form:

$$\psi = \psi(z) + \bar{\psi}(z) \quad \text{or} \quad W = W(z) + \bar{W}(z)$$

Assuming the solution of Equation 62 as

$$\left.\begin{aligned} W(z) &= A \ln z \\ \bar{W}(z) &= A \ln \bar{z} \end{aligned}\right\} \tag{63}$$

and

$$\left.\begin{aligned} \partial^2 W/\partial z^2 &= -A\frac{1}{z^2} \\ \partial^2 \bar{W}/\partial \bar{z}^2 &= -A\frac{1}{\bar{z}^2} \end{aligned}\right\} \tag{64}$$

and substituting Equation 64 into Equation 62 results in

$$(-A)^m\left\{\bar{z}^{-(1+m)}\frac{\partial^2}{\partial z^2}z^{1-m} - z^{-(1+m)}\frac{\partial^2}{\partial \bar{z}^2}\bar{z}^{1-m}\right\} = 0 \tag{65}$$

Equation 63 is therefore a solution of Equation 62. Using Equations 59 and 51, a further investigation of the velocity potential function and stream function will be made. The stream function is written in terms of the complex potential function (Equation 61) and by using Equation 62 it can be expressed as

$$\phi = \frac{A}{2i} \ln(z/\bar{z}) \tag{66}$$

Equation 66 satisfies Equation 51. Similarly, the velocity potential function is expressed by

$$\psi = \frac{A}{2} \ln z\bar{z} \tag{67}$$

which satisfies Equation 59.

The equilibrium equation expressed by the stream function, the velocity potential function and the complex potential function are derived. Equations 63, 66, and 67 illustrate solution sets for these functions. These solutions of the potential functions are independent of the rate hardening exponent. Furthermore, the velocity and strain rate field are independent of the strain rate hardening exponent, m.

Schwartz-Christoffel Transformation for a Cornered Channel

The cornered die can be transformed onto the upper half plane, which includes the real axis by using the Schwartz-Christoffel conformal transformation. The inside of the die can be mapped onto the upper half plane and each corner of the die can be mapped into the real axis in the transformed plane. This mapping process is illustrated in References 16 and 35. A schematic diagram of the cornered die and the transformed plane is presented in Figure 8. The extrusion processing of a power law material for a cornered die can be replaced by the sink problem in the

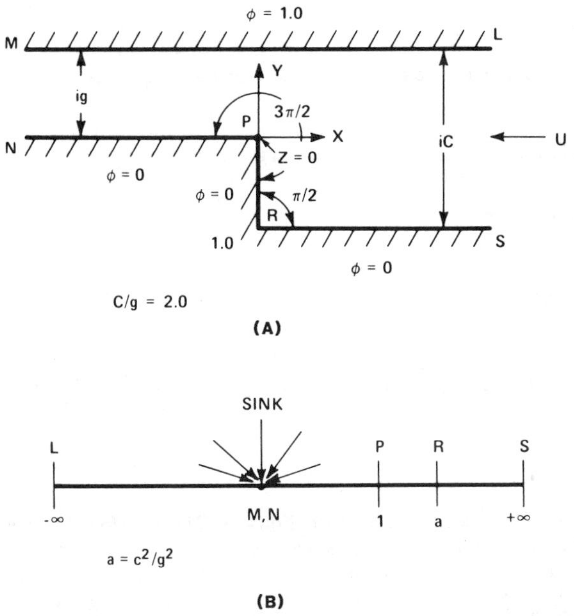

Figure 8. (A) Cornered die geometry (Z-plane); (B) conformerly transformed plane (t-plane).

transformed plane. The solution of the sink problem is available [16] for a perfect Newtonian fluid. The complex potential function, W, is given by

$$W = S \ln t \tag{68}$$

where S is the intensity of the sink. S for the present problem is given by

$$S = \frac{Uc}{\pi} \tag{69}$$

where U is the inlet velocity of the die and c is the opening dimension of the inlet. The mapping function from the z-plane (primitive plane) to the transformed plane (t-plane) is given [16, 35] as

$$\frac{dz}{dt} = K_1 \frac{1}{t}\sqrt{\frac{t-1}{t-a}} \tag{70}$$

where $a = c^2/g^2$, g is the outlet opening dimension of the cornered die and $K_1 = c/\pi$. Performing the integration of Equation 70 the resulting equation [35] is given as

$$Z = \frac{c}{\pi}\left\{\cosh^{-1}\frac{2t-(a+1)}{a-1} - \frac{1}{\sqrt{a}}\cosh^{-1}\frac{(a+1)t-2a}{(a-1)t}\right\} + D + iE \tag{71}$$

Referring to the transformed plane in Figure 8, constants D and E can be found from the mapping conditions,

$$t = 1, \quad W = 0, \quad \text{and} \quad z = 0$$

These constants are found to be $D = 0$ and $E = -c\left(1 - \dfrac{1}{\sqrt{a}}\right)$ (72)

The relation between the z-plane (cornered die) and the transformed plane (t-plane) can be found as

$$z = \frac{c}{\pi}\left\{\cosh^{-1}\frac{2t-(a+1)}{a-1} - \frac{1}{\sqrt{a}}\cosh^{-1}\frac{(a+1)t-2a}{(a-1)t} - \pi i\left(1 - \frac{1}{\sqrt{a}}\right)\right\} \tag{73}$$

or using Equation 68

$$z = \frac{c}{\pi}\left\{\cosh^{-1}\frac{2e^{KW}-(a+1)}{(a-1)} - \frac{1}{\sqrt{a}}\cosh^{-1}\frac{(a+1)e^{KW}-2a}{(a-1)e^{KW}} - \pi i\left(1 - \frac{1}{\sqrt{a}}\right)\right\} \tag{74}$$

where $K = \dfrac{\pi}{Uc}$

The position, $z = x + iy$ in the z-plane to satisfy Equation 74 can be found numerically for a given geometry. Once the velocity potential and stream function distribution in terms of x and y coordinates are found, the velocity or strain rate distribution in the cornered die can be obtained. The stress distribution can be found from the constitutive equation. However, the stress calculation requires the dilatational stress. The dilatational stress can be found from the equilibrium equation and from the known stream function. The field equation of the dilatational stress can be derived from the two equilibrium equations.

$$\left.\begin{aligned} -\frac{\partial \sigma_p}{\partial x} - \frac{\partial}{\partial x}\left\{\frac{1}{\lambda}\frac{\partial \phi}{\partial x\,\partial y}\right\} - \frac{\partial}{\partial y}\left\{\frac{1}{2\lambda}\left(\frac{\partial^2 \phi}{\partial y^2} - \frac{\partial^2 \phi}{\partial x^2}\right)\right\} = 0 \\ -\frac{\partial \sigma_p}{\partial y} + \frac{\partial}{\partial y}\left\{\frac{1}{\lambda}\frac{\partial^2 \phi}{\partial x\,\partial y}\right\} - \frac{\partial}{\partial x}\left\{\frac{1}{2\lambda}\left(\frac{\partial^2 \phi}{\partial y^2} - \frac{\partial^2 \phi}{\partial x^2}\right)\right\} = 0 \end{aligned}\right\} \tag{75}$$

Differentiating the first equation with respect to y and the second with respect to x and adding yields:

$$-2\frac{\partial^2 \sigma_p}{\partial x\,\partial y} - \left(\frac{\partial^2}{\partial x^2} + \frac{\partial^2}{\partial y^2}\right)\left\{\frac{1}{2\lambda}\left(\frac{\partial^2 \phi}{\partial y^2} - \frac{\partial^2 \phi}{\partial x^2}\right)\right\} = 0 \tag{76}$$

In the complex plane Equation 76 can be rewritten as

$$i\left(\frac{\partial^2 \sigma_p}{\partial z^2} - \frac{\partial^2 \sigma_p}{\partial \bar{z}^2}\right) = 2c_1 \frac{\partial^2}{\partial z\,\partial \bar{z}}\left\{\left(\frac{\partial^2 \phi}{\partial z^2}\right)^{(m+1)/2}\left(\frac{\partial^2 \phi}{\partial \bar{z}^2}\right)^{(m-1)/2} + \left(\frac{\partial^2 \phi}{\partial z^2}\right)^{(m-1)/2}\left(\frac{\partial^2 \phi}{\partial \bar{z}^2}\right)^{(m+1)/2}\right\} \tag{77}$$

where

$$c_1 = \frac{2\sigma_0}{3}\left(\frac{\sqrt{3}}{4}\right)^{1-m}$$

It should be noted that the stream function, is known in the transformed plane, and is given as

$$\phi = \frac{Uc}{\pi}\frac{1}{2i}(\ln t - \ln \bar{t}) \tag{78}$$

Substituting Equation 78 into 77, the resulting equation is

$$\frac{\partial^2 \sigma_p}{\partial t^2} - \frac{\partial^2 \sigma_p}{\partial \bar{t}^2} = -2c_1\left(\frac{Uc}{2\pi}\right)^m (\bar{t}^{-2-m} t^{-m} - t^{-2-m}\bar{t}^{-m})(1+m)(m-1) \tag{79}$$

The solution of Equation 79 is

$$\sigma_p = -\frac{(1-m)}{m}(2c_1)\left(\frac{Uc}{2\pi}\right)^m t^{-m}\bar{t}^{-m} + A \tag{80}$$

where $t = e^{KW}$
$\bar{t} = e^{K\bar{W}}$
$K = \pi/Uc$

or Equation 80 can be expressed in terms of the complex potential function or the velocity potential function,

$$\sigma_p = -\frac{1-m}{m}(2c_1)\left(\frac{1}{2K}\right)^m e^{-2mK\psi} + A \tag{81}$$

Since ψ can be expressed in terms of z in the z-plane, σ_p can be found in the z-plane if A can be determined. Since the process is extrusion, the horizontal stress in the exit should be zero and the unknown constant, A, can be found from this boundary condition. Equation 81 shows that the constant velocity potential functions correspond to the isobaric lines in the extrusion process.

The velocity potential and stream function distributions can be obtained from Equation 74. The velocity and strain components can be obtained from the first and the second derivatives of the complex potential function with respect to z (Equation 74). The deviatoric stress components and the dilatational stress are obtained from the constitutive equations and Equation 81.

Numerical Calculations

The numerical calculations are performed considering the following extrusion conditions:

Inlet opening dimension (c), 2 in.
Feed velocity of the material (U), 0.5 in./sec.
Outlet opening dimension (g), 1 in.

$$a = c^2/g^2 = 4.0$$

$$K = \frac{\pi}{Uc} = \pi$$

The reduction ratio, r, for this particular extrusion process problem is 0.5. Substituting these values into Equation 74, the complex variable, z, can be expressed by

$$z = \frac{2}{\pi}\left\{\cosh^{-1}\frac{2e^{\pi W} - 5}{3} - \frac{1}{2}\cosh^{-1}\frac{5 - 8e^{-\pi W}}{3} - \frac{\pi i}{2}\right\} \tag{82}$$

Equation 82 can be calculated using the relation given in [36]

$$\begin{aligned} \cosh^{-1}(u_1 \pm iv_1) &= \cosh^{-1}\tfrac{1}{2}\{\sqrt{(1 + u_1) + v_1^2} + \sqrt{(1 - u_1)^2 + v_1^2}\} \\ &\pm i\cos^{-1}\tfrac{1}{2}\{\sqrt{(1 + u_1)^2 + v_1^2} - \sqrt{(1 - u_1)^2 + v_1^2}\} \end{aligned} \tag{83}$$

and $W = \psi + i\phi$

In Equation 82

$$\left.\begin{aligned} &\tfrac{1}{3}(2e^{\pi W} - 5) = \tfrac{1}{3}\{(2e^{\pi\psi}\cos\pi\phi - 5) + 2ie^{\pi\psi}\sin\pi\phi\} \\ &\tfrac{1}{3}(5 - 8e^{-\pi W}) = \tfrac{1}{3}\{(5 - 8e^{-\pi\psi}\cos\pi\phi) + 8ie^{-\pi\psi}\sin\pi\phi\} \\ &u_1 = \tfrac{1}{3}(2e^{\pi\psi}\cos\pi\phi - 5), \qquad v_1 = \tfrac{2}{3}e^{\pi\psi}\sin\pi\phi \\ &u_2 = \tfrac{1}{3}(5 - 8e^{-\pi\psi}\cos\pi\phi), \qquad v_2 = \tfrac{8}{3}e^{-\pi\psi}\sin\pi\phi \end{aligned}\right\}$$

Substituting these values into Equation 82 and using Equation 83 for given ψ and ϕ values, z values are found. The result is shown in Figure 9.

The velocity components, u and v, in the die can be obtained by using the relation,

$$\frac{dW}{dz} = -u + iv \tag{84}$$

From Equation 82

$$\frac{dW}{dz} = \frac{1}{2}(e^{\pi W} - 4)^{1/2}(e^{\pi W} - 1)^{-1/2} = \frac{1}{2\eta_1^{1/2}}\{(e^{2\pi\psi} - 5e^{\pi\psi}\cos\pi\phi + 4) + 3i\sin\pi\phi e^{\pi\psi}\}^{1/2} \tag{84a}$$

where $\eta_1 = e^{2\pi\psi} - 2e^{\pi\psi}\cos\pi\phi + 1$

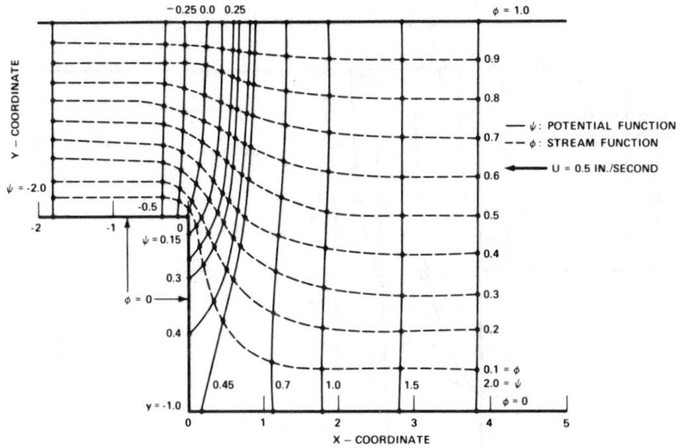

Figure 9. Stream and velocity potential function distribution in cornered die.

and $\rho_1 = \{(e^{2\pi\psi} - 5e^{\pi\psi}\cos\pi\phi + 4)^2 + 9e^{2\pi\psi}\sin^2\pi\phi\}^{1/2}$

$$\tan\Delta_1 = \frac{3e^{\pi\psi}\sin\pi\phi}{e^{2\pi\psi} - 5e^{\pi\psi}\cos\pi\phi + 4}$$

The velocity components, u and v are

$$-u + iv = \frac{\rho_1^{1/2}}{2\eta_1^{1/2}}\left(\cos\frac{\Delta_1}{2} + i\sin\frac{\Delta_1}{2}\right) \tag{85}$$

If $\phi = 0$ and $e^{2\pi\psi} - 5e^{\pi\psi}\cos\pi\phi + 4 < 0$, then

$$-u + iv = \frac{|\rho_1|^{1/2}}{2\eta_1^{1/2}}\left(-\sin\frac{\Delta_1}{2} + i\cos\frac{\Delta_1}{2}\right) \tag{86}$$

The strain rate components, $\dot{\epsilon}_x$, $\dot{\epsilon}_y$, and $\dot{\epsilon}_{xy}$ are found from the relations,

$$\frac{d^2W}{dz^2} = -\frac{\partial u}{\partial x} + i\frac{\partial v}{\partial x}$$
$$\frac{d^2W}{dz^2} = \frac{\partial v}{\partial y} + i\frac{\partial u}{\partial y} \tag{87}$$

From Equation 84a and Equation 87, the strain rates are given by

$$\dot{\epsilon}_x = -\dot{\epsilon}_y = -\frac{3}{4}\frac{1}{\eta_1^2}(e^{2\pi\psi}\cos 2\pi\phi - 2e^{\pi\psi}\cos\pi\phi + 1) \tag{88}$$

$$\dot{\epsilon}_{xy} = -\frac{1}{2\eta_1^2}(e^{\pi\psi}\cos\pi\phi - 1)e^{\pi\psi}\sin\pi\phi$$

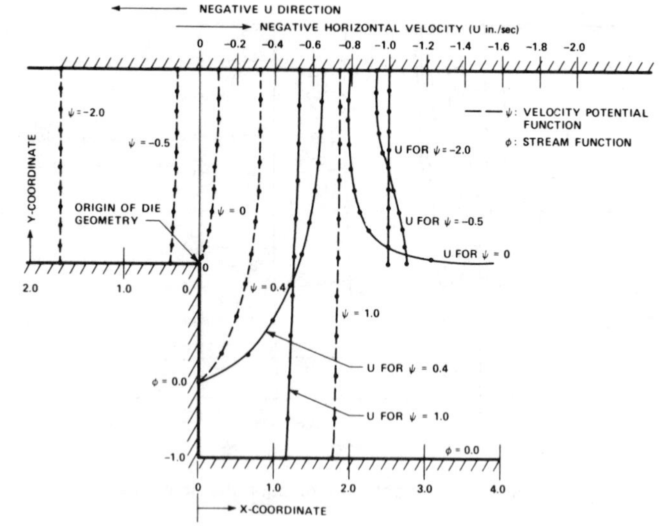

Figure 10. Horizontal velocity distributions corresponding to various velocity potential functions.

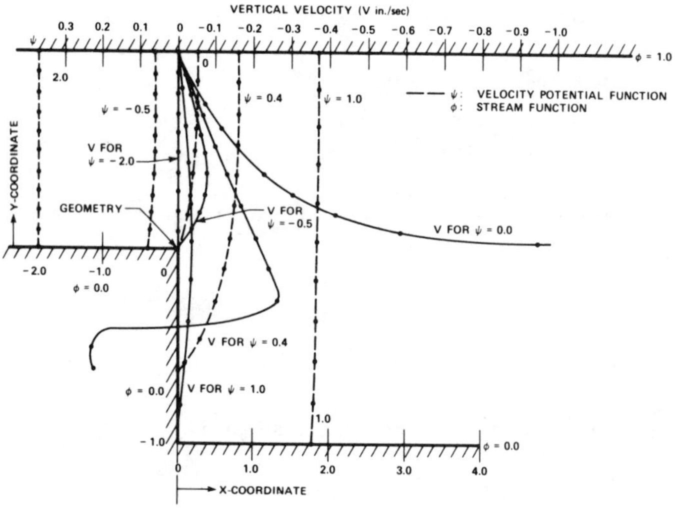

Figure 11. Vertical velocity distributions corresponding to various velocity potential functions in the die.

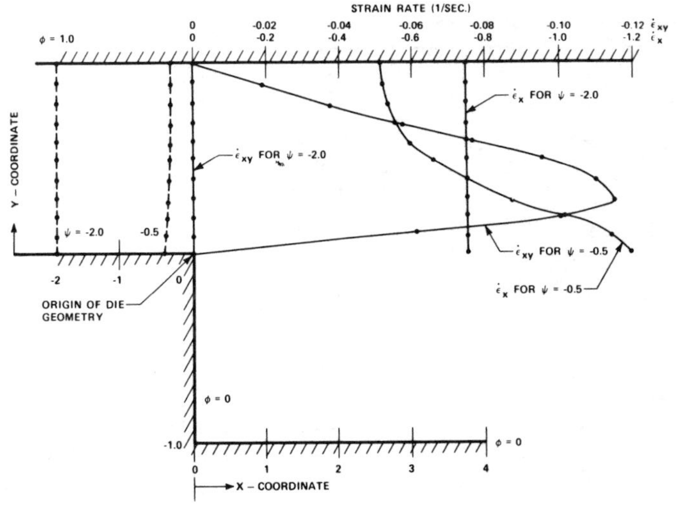

Figure 12. Strain rates corresponding to various velocity potential functions.

Velocity and strain components for various ψ and ϕ values are given in Figures 10–14. The effective strain rate, $\dot{\epsilon}$ is given by

$$\dot{\bar{\epsilon}} = \frac{2}{\sqrt{3}} (\dot{\epsilon}_x^2 + \dot{\epsilon}_{yx}^2)^{1/2} \tag{89}$$

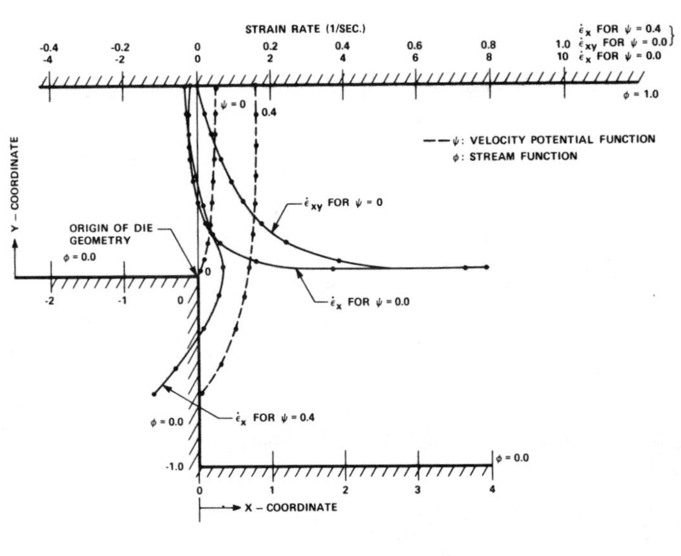

Figure 13. Strain rates corresponding to various velocity potential functions.

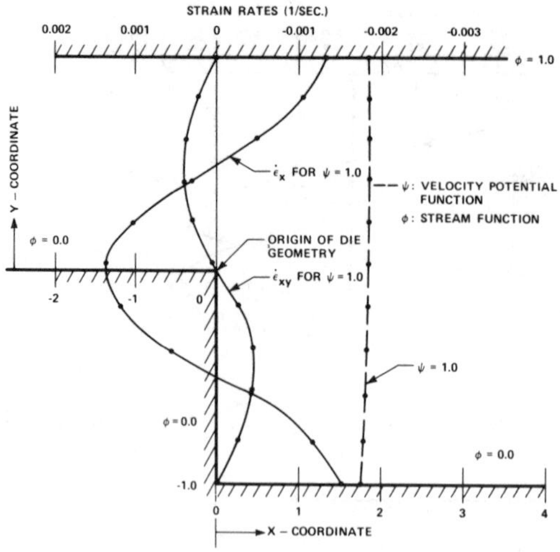

Figure 14. A strain rates corresponding to the velocity potential function, $\psi = 1.0$.

presuming plane strain and presuming incompressible conditions. This can be obtained from Equation 88 for this particular application.

The proportionality constant, $\dot{\lambda}$ is given by

$$\dot{\lambda} = \frac{3}{2}\frac{1}{\sigma_0}\dot{\bar{\epsilon}}^{1-m} \tag{90}$$

stress components are given by

$$\begin{aligned}
\sigma_x &= \frac{1}{\dot{\lambda}}\dot{\epsilon}_x - \sigma_p \\
\sigma_y &= \frac{1}{\dot{\lambda}}\dot{\epsilon}_y - \sigma_p \\
\sigma_{xy} &= \frac{1}{\dot{\lambda}}\sigma_{xy}
\end{aligned} \tag{91}$$

If the exit plane corresponds to $\psi = -2$ in Figure 9, then $\dot{\epsilon}_x/\dot{\lambda}$, at $\psi = -2.0$ is approximately given by,

$$\dot{\epsilon}_x/\dot{\lambda}\Big|_{\psi=-2.0} \cong -\frac{2}{3}\left(\frac{\sqrt{3}}{2}\right)^{m-1}\left(\frac{3\sigma_0}{4}\right) \tag{92}$$

Substituting Equation 92 into Equation 91 and using Equation 81, the constant A can be found using the condition

$$\sigma_x\Big|_{\psi=-2.0} \cong 0 = -\frac{\sigma_0}{\sqrt{3}}\left(\frac{\sqrt{3}}{2}\right)^m + \frac{1-m}{m}(2c_1)\left(\frac{1}{2K}\right)^m e^{4mK} - A$$

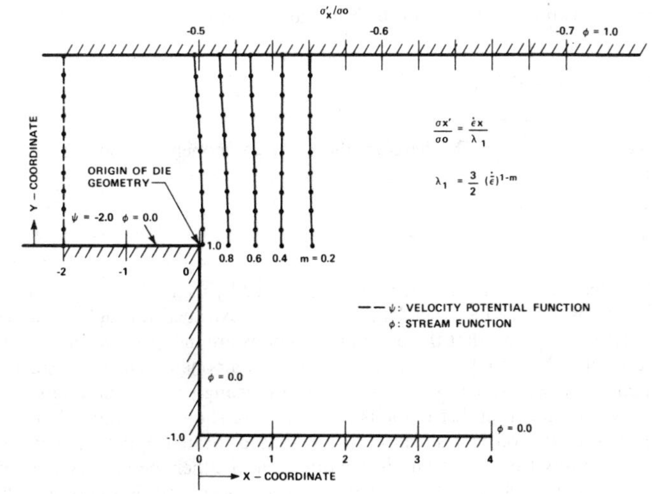

Figure 15. Deviatoric stress (σ_x'/σ_0) corresponding to $\psi = -2.0$ for various rate hardening exponents.

and $$A = -\frac{\sigma_0}{\sqrt{3}}\left(\frac{\sqrt{3}}{2}\right)^m + \frac{1-m}{m}(2c_1)\left(\frac{1}{2K}\right)^m e^{4mk}$$

$$\sigma_p = \frac{1-m}{m}(2c_1)\left(\frac{1}{2K}\right)^m (e^{4mK} - e^{-2mK\psi}) - \frac{\sigma_0}{\sqrt{3}}\left(\frac{\sqrt{3}}{2}\right)^m \tag{93}$$

Using Equations 91 and 93 the stress components are found. The deviatoric stress components are illustrated in Figures 15 and 18. If m = 1, and $\phi = 1/2i(\ln t - \ln \bar{t})$ is substituted into Equation 77 the right hand side of Equation 77 equals zero. Therefore Equation 77 should be considered separately. Equation 77 for m = 1.0 is given by

$$\frac{\partial^2 \sigma p}{\partial t^2} - \frac{\partial^2 \sigma p}{\partial \bar{t}^2} = 0 \tag{93a}$$

The solution of Equation 93a is $\sigma_p = Af_1(t + \bar{t}) + Bg_1(t - \bar{t})$. However, σ_p, for m = 1.0 cannot be an arbitrary function of $t + \bar{t}$ or $t - \bar{t}$. The dilatational stress for m = 1.0 corresponds to a constant velocity potential function as shown in Equation 81. Therefore, when the function $f_1(t + \bar{t})$ and $g_1(t - \bar{t})$ are chosen, this restriction should be imposed. The functions for the dilatational stress are

$$Af(t + \bar{t}) = \frac{c_1}{4k}(t + \bar{t})^2 = c_1 e^{2k\psi} \cos^2 \pi\phi/K$$

$$Bg(t - \bar{t}) = \frac{-c}{4k}(t - \bar{t})^2 = c_1 e^{2k\psi} \sin^2 \pi\phi/K$$

and $$\sigma_p = \frac{c_1}{K} e^{2K\psi} + F$$

where F is the unknown constant. From the boundary condition,

$$\sigma_x\Big|_{\psi=-2.0} \cong -\frac{\sigma_0}{2} - F = 0$$

therefore the constant, $F = -\sigma_0/2$. Therefore the stress components for $m = 1.0$ can be obtained from Equation 91.

Results and Discussion

The velocity potential and stream functions are obtained by using the Schwartz-Christoffel conformal transformation when rate dependent materials are extruded through frictionless cornered dies. From analysis it is found that the potential functions are independent of the rate hardening exponent. The results show that the velocity or strain components are independent of the rate hardening exponent. However, the stress components are strongly dependent on the rate hardening exponents. The stream function distributions in the die are shown in Figure 9. The stream lines have equal spacing in the vertical direction at the section located approximately two times the opening dimension away from the origin in the direction of either the inlet or the outlet portion. Therefore, the horizontal velocity at this section can be considered as uniform and the vertical velocity vanishes. The stream and velocity potential functions are significantly distorted in the region $-0.35 < \psi < 0.45$, therefore a large deformation may take place in this region. Because of the frictionless condition assumed in the analysis, the dead zone cannot be obtained. Since the friction factor between the material and the die is dependent on the materials, including frictional effects would alter the predicted shape of the potential functions. The horizontal and vertical velocity corresponding to various velocity potential functions are shown in Figures 10 and 11. The outlet velocity is twice the inlet velocity as expected and the horizontal velocity at $\psi = \pm 2.0$ is uniform which is consistent with the results shown in Figure 9. The horizontal velocity (u) corresponding to $\psi = 0.0$ is complicated and u at the origin of the die is indeterminate due to the geometrical singularity. The u in the region of $\psi < 0.4413$ and $\phi = 0.0$ is zero because of the boundary conditions.

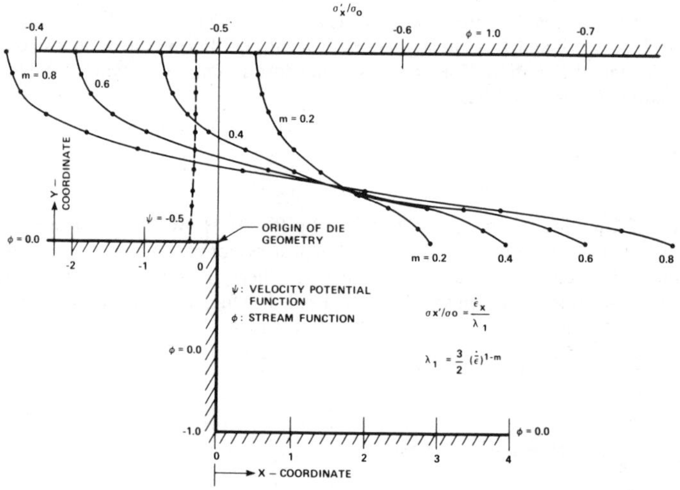

Figure 16. Deviatoric stress (σ_x'/σ_0) corresponding to the velocity potential function, $\psi = -0.5$ for various rate hardening exponents (m).

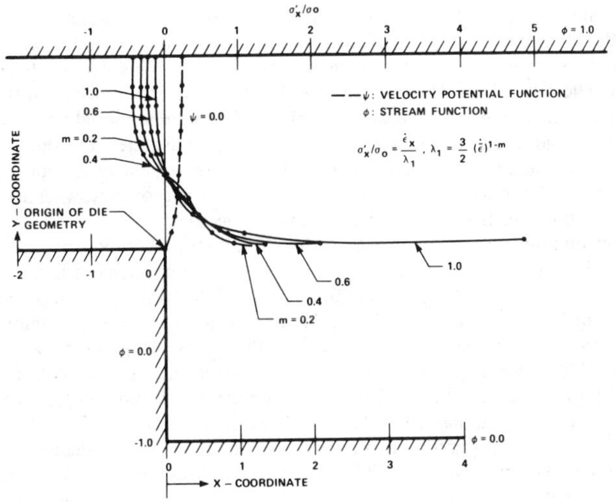

Figure 17. Deviatoric stress (σ_x'/σ_0) corresponding to rate hardening exponent velocity function, $\psi = 0.0$ for various rate hardening exponents.

The strain rate distributions corresponding to various velocity potential functions are shown in Figures 12, 13 and 14. As expected from the velocity profile, the maximum strain rate (σ_x) is obtained for $\psi = 0.0$ and the shearing and strain rates (σ_{xy}) are small compared with those of σ_x. The strain rates corresponding to the velocity potential function, $\psi > 0.0$ decrease rapidly and eventually diminish in the region $\psi > 2.0$.

The deviatoric stress distributions (σ_x'/σ_0), corresponding to various ψ values, are shown in Figures 15–18 for various rate hardening exponents. They are significantly larger than the shearing stress

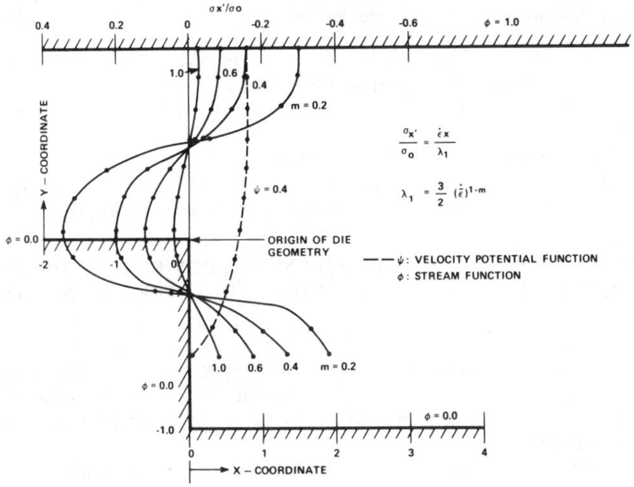

Figure 18. Deviatoric stress (σ_x'/σ_0') corresponding to the velocity potential function, $\psi =$ 0.4 for various rate hardening exponents (m).

for all m values and the stress distributions corresponding to all ψ values are strongly dependent on the rate hardening exponents. The maximum tensile stress is obtained in the region of the surface at $\psi = 0.0$. The dilatational stress is dependent on the rate hardening exponents and ψ as shown in Equation 93. The dilatational stress when $m < 1.0$ does not vary much with the length of the inlet portion of the die, if $2mk = 2m\pi\psi > 5.0$, $e^{-2m\pi\psi} \cong 0$. The dilatational stress when $m = 1.0$ is very sensitive to the variation of ψ as well as the length of the die inlet. The dilatational stress for $m = 1.0$ increases exponentially with increasing ψ values. Therefore, the magnitude of the hydrostatic pressure, when $m < 1.0$, is sensitive to the variation of the outlet geometry, whereas the hydrostatic pressure for $m = 1.0$ is sensitive to the variation in inlet geometry.

The important finding from analyses is the velocity and strain components are independent of the strain hardening exponent or the fluid behavior index. This statement is not true in general. When the solution of the complex potential function is of the form $w = A \ln z$, then the velocity field is independent of the fluid behavior index. No analytical or experimental information or data related to this problem is available to the best of author's knowledge except one [37]. Oh [37] analyzed the isothermal forging of a titanium alloy by using the rigid viscoplastic finite element method. He [37] obtained the solutions by using two materials, β and $\alpha + \beta$ microstructure characterized by Dadras and Thomas [38]. Dadras et al. found that the flow stress of the $\alpha + \beta$ microstructure Ti-6242-0.1Si was found to be independent of strain and dependent on strain rate at the temperature range investigated (816°C ~ 1,010°C). The flow stress of β microstructure Ti-6242 alloy is a function of the strain, the strain rate and the rate hardening exponent. The m values are different for two materials. When these two materials are forged under identical boundary conditions, he [37] compared the deformation pattern and found that the deformation patterns are not affected by the different flow stress behaviors ($\alpha + \beta$) and β microstructure. In fact, the grid distortions of the model are almost identical and he mentioned that this behavior may not be considered as a general case, but should be treated as a restricted response under his process condition and die geometry.

Even though present and Oh's boundary conditions are different, he [37] reached similar conclusions to the results and conclusions in the presented here. The analysis described in this section deals with steady state creep materials (power law materials). The velocity field and potential function distributions in the channel or the die obtained through the analysis are identical to those of non-Newtonian fluids.

From the analysis of rate hardening materials described by the power law extruded through cornered dies, the following conclusions are derived:

1. If friction between the materials and the die is neglected, the stream and velocity functions are independent of the rate hardening exponents.
2. A large amount of deformation occurs in the region of $-0.25 <$ velocity potential function value < 0.45.
3. The maximum deformation takes place at $\psi = 0.0$.
4. The hydrostatic pressure for $m < 1.0$ is sensitive to the variation of the outlet length of the die whereas for $m = 1.0$ it is sensitive to the variation of the inlet die length.

AN ANALYSIS OF THE VOLTERRA PROBLEM FOR POWER LAW VISCOUS MATERIALS USING COMPLEX STRESS AND PSEUDO-STRESS FUNCTIONS

Previous sections discussed applications of the stream function to non-linear viscous materials described by the power law. This section deals with the application of complex stress functions to non-linear viscous materials. This method may be applicable to solid media rather than fluid. This method can be used for certain boundary value problems but is not yet general. The following analysis was published in [34].

An increasingly large number of engineering structures are exposed to high temperature loadings often resulting in the materials exhibiting creep behavior. Many investigators have determined the material behavior and have performed stress analysis of structures subjected to creep behavior. Since the material behavior is complex, making it difficult to formulate the constitutive equation representative of the entire creep behavior region for the structure, the stress analysis of the material is often limited to a continuum subjected to simple boundary condition. The constitutive equation described by the power law accurately represents the steady state creep behavior at elevated temperatures. The steady creep behavior can be explained by the mechanism of the competition between strain hardening and the recovery process of a crystal solid [40]. Nadai [41, 42] mentioned the contradiction regarding the power law expression for the uni-axial stress field. That is, the stress should be an odd function of the rate of flow if the shearing direction is reversed.

Even though the constitutive equation describing the actual material behavior is found for the steady state creep region, the stress analysis did not show significant improvement because the governing field equation is a nonlinear equation. Lee, et al. [13, 32, 33] obtained the stress solutions by using the complex stress or stream function and illustrated a very simple boundary value problem under plane strain presuming incompressibility.

The method described in the previous paper [33] was based on the variable separation or semi-inverse method for solving the equilibrium equation. The present approach is simply an extension of the previously described method. A pseudo-stress function may be obtained from the linear elastic stress function. The dilatational stress is expressed in terms of the integration of the derivatives of the pseudo-stress function rather than the derivatives of the stress function. When the integrated function associated with the dilatational stress is a complicated expression, the stress may be obtained by numerical integration using the specified integration path. If the dilatational stress is obtained through the integration as shown in the section, the method developed may be applied to more complicated boundary value problems than those described in [33].

This section describes the solution methodology for two stress fields—a stress field containing σ_x, σ_y, and σ_{xy}, and a stress field that has only shearing stresses such as σ_{xz} and σ_{yz}. Using the two methods, boundary value problems are illustrated, namely the stress analysis of a steady state creep material subjected to a constant velocity in the radial direction and a constant velocity in the axial direction. If the constant velocity is considered to be a constant displacement and the material is linear elastic then the problem is a Volterra problem for which solutions are available [43, 44].

The solution of the Volterra problem for a linear elastic continuum is applied to a model for an edge or screw dislocation and is very useful in finding the self-stress or energy and interaction forces of dislocation by applying the Peach-Koehler equation. Dislocation interaction forces in the steady state creep material include the chemical force corresponding to the diffusion of vacancies and the mechanical interaction force. The total interaction force due to the diffusion of vacancies and the mechanical interaction force due to the interaction of two dislocations can be found from the generalized Peach-Koehler equation [43]. The solution presented herein may be applied to the modeling of an edge or screw dislocation associated with a power law creep material under low temperatures.

Analysis

This section deals with the two-dimensional stress analysis in the steady state creep behavior region of materials described by the power law.

The solutions obtained by using the complex stress function and the pseudo-stress function show behavior similar to the Airy stress function solutions. Plane strain, incompressibility, and isothermal conditions are assumed. The analysis consists of two parts: In the first part the stress field contains σ_x, and σ_y and σ_{xy}, the second part contains only the shearing stresses, σ_{xz} and σ_{yz}.

The governing equation is derived by using two equations,

$$\tilde{\sigma} = \sigma_0 \dot{\tilde{\epsilon}}^m \tag{95}$$

$$\dot{\lambda}(\sigma_{ij} + \sigma_p \delta_{ij}) = \dot{\epsilon}_{ij}$$

where $\tilde{\sigma}$ is the effective stress. $\dot{\tilde{\epsilon}}$ is the dimensionless effective strain rate and equals the effective strain rate divided by the material constant. The scalar multiplier, which is time dependent, is defined as

$$\dot{\lambda} = \frac{3}{2}\frac{\dot{\tilde{\epsilon}}}{\tilde{\sigma}}$$

The compatibility equation, expressed in terms of the second derivatives with respect to x and y of the strain rate, can be expressed in terms of the stress function gradient by using the constitutive equation shown in Equation 95 and by presuming a plane strain and an incompressibility condition. Substituting the stress function definitions, $\sigma_{xx} = U_{yy}$, $\sigma_{yy} = U_{xx}$ and $\sigma_{xy} = -U_{xy}$ into the compatibility equation, results, as derived in [33], in

$$\left(\frac{\partial^2}{\partial y^2} - \frac{\partial^2}{\partial x^2}\right)\left\{\frac{\dot{\lambda}}{2}\left(\frac{\partial^2 U}{\partial y^2} - \frac{\partial^2 U}{\partial x^2}\right)\right\} + \frac{\partial^2}{\partial x\,\partial y}\left\{2\dot{\lambda}\frac{\partial^2 U}{\partial x\,\partial y}\right\} = 0 \tag{96}$$

The scalar multiplier, $\dot{\lambda}$, is expressed in terms of the second derivatives with respect to the complex conjugate variables,

$$\dot{\lambda} = \frac{3\Delta}{2C_1}(4)^{(1-m)/m}\left\{\frac{\partial^2 U}{\partial z^2}\frac{\partial^2 U}{\partial \bar{z}^2}\right\}^{(1-m)/2m} \tag{97}$$

where $\Delta = (\frac{3}{2})^{(1-m)/m}$
$C_1 = (\sigma_0)^{1/m}$

transforming Equation 96 into the complex plane gives

$$\frac{\partial^2}{\partial z^2}\left\{\left(\frac{\partial^2 U}{\partial z^2}\right)^{(1-m)/2m}\left(\frac{\partial^2 U}{\partial \bar{z}^2}\right)^{(1+m)/2m}\right\} + \frac{\partial^2}{\partial \bar{z}^2}\left\{\left(\frac{\partial^2 U}{\partial z^2}\right)^{(1+m)/2m}\left(\frac{\partial^2 U}{\partial \bar{z}^2}\right)^{(1-m)/2m}\right\} = 0 \tag{98}$$

If the complex function in Equation 98 is assumed such that

$$\left.\begin{aligned}\left(\frac{\partial^2 U}{\partial z^2}\right)^{(1-m)/2m}\left(\frac{\partial^2 U}{\partial \bar{z}^2}\right)^{(1+m)/2m} &= \frac{\partial^2 \Lambda}{\partial \bar{z}^2}\\ \left(\frac{\partial^2 U}{\partial z^2}\right)^{(1+m)/2m}\left(\frac{\partial^2 U}{\partial \bar{z}^2}\right)^{(1-m)/2m} &= \left(\frac{\partial^2 \Lambda}{\partial z^2}\right)\end{aligned}\right\} \tag{99}$$

then Equation 98 can be written as a biharmonic equation. The solution of a biharmonic equation can be easily determined. Since Λ satisfies the biharmonic equation and it is similar to the stress function, Λ may be denoted as a pseudo-stress function for power law materials. Solving Equation 99 with respect to $\partial^2 U/\partial z^2$ and $\partial^2 U/\partial \bar{z}^2$, the results are

$$\left.\begin{aligned}\frac{\partial^2 U}{\partial z^2} &= \left(\frac{\partial^2 \Lambda}{\partial z^2}\right)^{(1+m)/2}\left(\frac{\partial^2 \Lambda}{\partial \bar{z}^2}\right)^{(m-1)/2}\\ \frac{\partial^2 U}{\partial \bar{z}^2} &= \left(\frac{\partial^2 \Lambda}{\partial z^2}\right)^{(m-1)/2}\left(\frac{\partial^2 \Lambda}{\partial \bar{z}^2}\right)^{(m+1)/2}\end{aligned}\right\} \tag{100}$$

Since Λ is known and is differentiable in the given domain, $\partial^2 U/\partial z^2$ and $\partial^2 U/\partial \bar{z}^2$ can be obtained. Equation 100 is not complete since the uniqueness of the stress function, U obtained from the two expression in Equation 100 is not apparent.

Let U_1 and U_2 be solutions obtained from $\partial^2 U_1/\partial z^2$ and $\partial^2 U_2/\partial \bar{z}^2$, respectively. Since U is a real valued function

$$\frac{\partial^2 U_2}{\partial z^2} = \overline{\frac{\partial^2 U_2}{\partial \bar{z}^2}} = \overline{\left\{\left(\frac{\partial^2 \Lambda}{\partial z^2}\right)^{(m-1)/2}\left(\frac{\partial^2 \Lambda}{\partial \bar{z}^2}\right)^{(m+1)/2}\right\}} \tag{101}$$

and since $\Lambda = \bar{z}\phi(z) + z\bar{\phi}(z) + \psi(z) + \bar{\psi}(z)$

$$\frac{\partial^2 \Lambda}{\partial z^2} = (\bar{z}\phi''(z) + \psi''(z))$$

$$\frac{\partial^2 \Lambda}{\partial \bar{z}^2} = (z\bar{\phi}''(z) + \bar{\psi}''(z))$$

where $\phi''(z) = \partial^2\phi/\partial z^2$
$\bar{\phi}''(z) = \partial^2\bar{\phi}/\partial \bar{z}^2$

then Equation 101 can be written as

$$\frac{\overline{\partial^2 U_2}}{\partial \bar{z}^2} = (\bar{z}\phi''(z) + \psi''(z))^{(1+m)/2}(z\bar{\phi}''(z) + \bar{\psi}''(z))^{(m-1)/2} \tag{102}$$

Equation 102 is identical to the first expression of Equation 100. Therefore, $U_1 = U_2$. This means the stress function obtained from the first equation in Equation 100 is identical to the one obtained using the second expression in Equation 100.

The next step is to obtain the mixed derivative of the stress function since it is necessary in order to obtain the dilatational stress. The mixed derivative is given by

$$2d\left(\frac{\partial^2 U}{\partial z\, \partial \bar{z}}\right) = \frac{\partial}{\partial z}\left(\frac{\partial^2 U}{\partial z\, \partial \bar{z}}\right) dz + \frac{\partial}{\partial \bar{z}}\left(\frac{\partial^2 U}{\partial z\, \partial \bar{z}}\right) d\bar{z} \tag{103}$$

Equation 103 is integrated by using Equation 100,

$$\frac{\partial^2 U}{\partial z\, \partial \bar{z}} = \frac{1}{2}\left\{\int \frac{\partial}{\partial \bar{z}}\left\{\frac{\partial^2 \Lambda}{\partial z^2}\left(\frac{\partial^2 \Lambda}{\partial z^2}\frac{\partial^2 \Lambda}{\partial \bar{z}^2}\right)^{(m-1)/2}\right\} dz + \int \frac{\partial}{\partial z}\left\{\frac{\partial^2 \Lambda}{\partial \bar{z}^2}\left(\frac{\partial^2 \Lambda}{\partial z^2}\frac{\partial^2 \Lambda}{\partial \bar{z}^2}\right)^{(m-1)/2}\right\} d\bar{z}\right\} \tag{104}$$

Performing the differential calculus of the inside integration in Equation 104 and simplifying results in

$$\frac{\partial^2 U}{\partial z\, \partial \bar{z}} = \int (PQ)^{(m-3)/2}\left[\frac{1+m}{2}(PQ)\left(\frac{\partial \phi}{\partial z} dz + \frac{\partial \bar{\phi}}{\partial \bar{z}} d\bar{z}\right) + \frac{m-1}{2}\left\{P^2\left(z\frac{\partial^2 \bar{\phi}}{\partial \bar{z}^2} + \frac{\partial \bar{\psi}}{\partial \bar{z}}\right) dz \right.\right.$$
$$\left.\left. + Q^2\left(\bar{z}\frac{\partial^2 \phi}{\partial z^2} + \frac{\partial \psi}{\partial z}\right) d\bar{z}\right\}\right] \tag{105}$$

If

$$P = \frac{\partial^2 \Lambda}{\partial z^2} = \bar{z}\phi' + \psi = \eta e^{i\Delta}$$

$$Q = \frac{\partial^2 \Lambda}{\partial \bar{z}^2} = z\bar{\phi}' + \bar{\psi} = \eta \bar{e}^{i\Delta}$$

$$\frac{\partial \phi}{\partial z} = \eta_1 e^{i\delta_1}, \quad \frac{\partial^2 \phi}{\partial z^2} = \eta_2 e^{i\delta_2}, \quad \frac{\partial \bar{\phi}}{\partial \bar{z}} = \eta_1{}^{-i\delta_1}$$

$$\frac{\partial^2 \phi}{\partial \bar{z}^2} = \eta_2 e^{-i\delta_2}, \quad \psi = \rho_0 e^{i\zeta_0}, \quad \bar{\psi} = \rho_0 e^{-i\zeta_0}$$

$$\partial\psi/\partial z = \rho_1 e^{i\zeta_1}, \quad \partial\bar{\psi}/\partial \bar{z} = \rho_1 e^{-i\zeta_1}, \quad z = re^{i\theta}, \text{ and } \bar{z} = r\bar{e}^{i\theta}$$

Equation 104 can be simplified as

$$\frac{\partial^2 U}{\partial z\,\partial \bar{z}} = \frac{1}{2}\Bigg[\int \{(1+m)\eta_1 \cos(\delta_1+\theta) + (m-1)\eta_2 r\cos(2\Delta+2\theta-\delta_2) + \rho_1(m-1)\cos(2\Delta+\theta-\zeta_1)\}\eta^{m-1}\,dr - \int\{(1+m)\eta_1 r\sin(\delta_1+\theta) + \eta_2 r^2(m-1)\sin(2\Delta+2\theta-\delta_2) + \rho_1(m-1)r\sin(2\Delta+\theta-\zeta_1)\}\eta^{m-1}\,d\theta\Bigg] \tag{106}$$

The stress components can be obtained by using Equations 100 and 106.

$$\sigma_x + \sigma_y = 4\frac{\partial^2 U}{\partial z\,\partial \bar{z}}$$

$$\sigma_y + \sigma_x + 2i\sigma_{xy} = 4\frac{\partial^2 U}{\partial z^2} = 4\left(\frac{\partial^2 \Lambda}{\partial z^2}\right)^{(1+m)/2}\left(\frac{\partial^2 \Lambda}{\partial \bar{z}^2}\right)^{(m-1)/2} \tag{107}$$

or $\quad \sigma_\theta - \sigma_r + 2i\sigma_{r\theta} = 4e^{2i\theta}\dfrac{\partial^2 U}{\partial z^2}$

If the stress field is associated with only the shearing stress, such as σ_{xt} and σ_{yt}, where t is the axis perpendicular to x-y plane. (z is used as a complex variable), the stress components have the following relationship with the strain rate components

$$\left.\begin{aligned}\dot{\lambda}\sigma_{xt} &= \dot{\epsilon}_{xt}\\ \dot{\lambda}\sigma_{yt} &= \dot{\epsilon}_{yt}\end{aligned}\right\} \tag{108}$$

where $\dot{\lambda}$ is the scalar multiplier defined by

$$\dot{\tilde{\epsilon}} = \frac{3}{2}\frac{\dot{\tilde{\epsilon}}}{\tilde{\sigma}}$$

and the effect strain rate in this analysis is given by

$$\dot{\tilde{\epsilon}} = \frac{2}{\sqrt{3}}(\dot{\epsilon}_{xt}^2 + \dot{\epsilon}_{yt}^2)^{1/2} \tag{109}$$

The definition of the shearing strain rates are

$$\left.\begin{aligned}\dot{\epsilon}_{xt} &= \frac{\partial W}{\partial x}\\ \dot{\epsilon}_{yt} &= \frac{\partial W}{\partial y}\end{aligned}\right\} \tag{110}$$

where W is the velocity in the t-direction. Introducing the complex conjugate variables,

$$z = x + iy$$

$$\bar{z} = x - iy$$

Equation 110 can be rewritten as

$$\left.\begin{aligned} \dot{\epsilon}_{xt} &= \frac{\partial W}{\partial z} + \frac{\partial W}{\partial \bar{z}} \\ \dot{\epsilon}_{yt} &= i\left(\frac{\partial W}{\partial z} - \frac{\partial W}{\partial \bar{z}}\right) \end{aligned}\right\} \tag{111}$$

Substituting Equation 111 into Equation 109, the effective strain rate is obtained as

$$\dot{\lambda} = \frac{4}{\sqrt{3}}\left(\frac{\partial W}{\partial z}\frac{\partial W}{\partial \bar{z}}\right)^{1/2} \tag{112}$$

and the scalar multiplier $\dot{\lambda}$ is found as

$$\begin{aligned} \dot{\lambda} &= \frac{3}{2\sigma_0}\dot{\bar{\epsilon}}^{1-m} \\ &= \frac{3}{2\sigma_0}\left(\frac{4}{\sqrt{3}}\right)^{1-m}\left(\frac{\partial W}{\partial z}\frac{\partial W}{\partial \bar{z}}\right)^{(1-m)/2} \end{aligned}$$

The stress components, σ_{xt} and σ_{yt} can now be written from Equation 108

$$\left.\begin{aligned} \sigma_{xt} &= \frac{2\sigma_0}{3}\left(\frac{\sqrt{3}}{4}\right)^{1-m}\left(\frac{\partial W}{\partial z}\frac{\partial W}{\partial \bar{z}}\right)^{(m-1)/2}\left(\frac{\partial W}{\partial z} + \frac{\partial W}{\partial \bar{z}}\right) \equiv c(pq)^{(m-1)/2}(p+q) \\ \sigma_{yt} &\equiv ci(pq)^{(m-1)/2}(p-q) \end{aligned}\right\} \tag{113}$$

where $c = \dfrac{2\sigma_0}{3}\left(\dfrac{\sqrt{3}}{4}\right)^{1-m}$

$$p = \frac{\partial W}{\partial z}$$

$$q = \frac{\partial W}{\partial \bar{z}}$$

The equilibrium equation in the t-direction is

$$\frac{\partial \sigma_{xt}}{\partial x} + \frac{\partial \sigma_{yt}}{\partial y} = 0 \tag{114}$$

Transforming on to the complex plane and substituting Equation 113 into Equation 114, results in the equation:

$$\frac{\partial}{\partial z}\left(p^{(m-1)/2}q^{(m+1)/2}\right) + \frac{\partial}{\partial \bar{z}}\left(p^{(m+1)/2}q^{(m-1)/2}\right) = 0 \tag{115}$$

Let

$$\left.\begin{aligned} \frac{\partial \phi}{\partial \bar{z}} &= p^{(m-1)/2}q^{(m+1)/2} \\ \frac{\partial \phi}{\partial z} &= p^{(m+1)/2}q^{(m-1)/2} \end{aligned}\right\} \tag{116}$$

Substituting Equation 116 into Equation 115, results in the harmonic equation for which a solution can be found. Solving Equation 116 with respect to $(\partial W/\partial z)$ and $(\partial W/\partial \bar{z})$, the following expressions are obtained

$$\left.\begin{aligned} \frac{\partial W}{\partial z} &= \left(\frac{\partial \phi}{\partial z}\right)^{(1+m)/2m}\left(\frac{\partial \phi}{\partial \bar{z}}\right)^{(1-m)/2m} \\ \frac{\partial W}{\partial \bar{z}} &= \left(\frac{\partial \phi}{\partial z}\right)^{(1-m)/2m}\left(\frac{\partial \phi}{\partial \bar{z}}\right)^{(1+m)/2m} \end{aligned}\right\} \tag{117}$$

If ϕ satisfies the harmonic equation, then the derivatives of the velocity in the t-direction with respect to z and $\bar{z}$ can be found by using Equation 117. However, Equation 117 is not complete since the uniqueness of the velocity obtained from the two expressions in Equation 117 is not apparent.

Let W_1 and W_2 be solutions obtained from $\partial W_1/\partial z$ and $\partial W_2/\partial \bar{z}$ in Equation 117. Since W is a real valued function

$$\frac{\partial W_2}{\partial \bar{z}} = \left(\overline{\frac{\partial W_2}{\partial \bar{z}}}\right) = \left(\overline{\frac{\partial \phi}{\partial z}}\right)^{(1-m)/2m}\left(\overline{\frac{\partial \phi}{\partial \bar{z}}}\right)^{(1+m)/2m} \tag{118}$$

Since ϕ is a solution of the harmonic equation, ϕ can be obtained as

$$\phi = f(z) + \bar{f}(z) \tag{119}$$

and $$\frac{\partial \phi}{\partial z} = \frac{df}{dz} = f'(z)$$

$$\frac{\partial \phi}{\partial \bar{z}} = \frac{df}{d\bar{z}} = \bar{f}'(z)$$

Substituting into Equation 118

$$\frac{\partial W_2}{\partial \bar{z}} = \left(\overline{\frac{\partial W_2}{\partial \bar{z}}}\right) = \{\bar{f}'(z)\}^{(1-m)/2m}\{f'(z)\}^{(1+m)/2m} \tag{120}$$

The first expression in Equation 117 becomes

$$\frac{\partial W_1}{\partial z} = (f'(z))^{(1+m)/2m}(\bar{f}'(z))^{(1-m)/2m} \tag{121}$$

When Equations 120 and 121 are compared, they are found to be identical. Hence, $W_1 = W_2$. This means the velocity obtained from the integration of the two expressions given in Equation 117 should be identical. Therefore, the velocity and the strain rate can be found from Equation 117 and the stress components can be found by using the constitutive equations.

Boundary Value Problems

Using the solutions obtained in the previous section, two boundary value problems are investigated. When a solid cylinder dislocates with a constant velocity in the radial or in the axial direction, the stress states are obtained by assuming steady creep behavior during the constant velocity phase. In the linear theory of elasticity, this kind of stress field is known as a Volterra problem. The Volterra solution can be used to find the self-energy created by an edge of screw dislocation of a solid crystal. As mentioned earlier, a Volterra problem associated with power law creeping materials can be applied to the modeling of an edge or a screw dislocation in the steady creep region if the vacancy diffusion mechanism is not significant.

Case 1—Constant Velocity in the Radial Direction

Figure 19 illustrates the edge dislocation problem. A material subjected to constant radial velocity is analyzed in this section for the steady creep continuum. The stress solution can be obtained by using the pseudo-stress function solution given in Equations 100 and 106.

Presuming the pseudo-stress function satisfies the biharmonic equation as

$$\Lambda = \bar{z}\phi_1(z) + z\bar{\phi}_1(z) + \psi_1(z) + \bar{\psi}_1(z) \tag{122}$$

where

$$\left.\begin{aligned} \phi_1(z) &= A_1 z \ln z + B_1 \ln z \\ \psi_1(z) &= C_1 z \ln z + D_1 \ln z \end{aligned}\right\} \tag{123}$$

$$A_1 = D_1 = 0$$

Equation 28 gives

$$\Lambda = (B_1 \bar{z} \ln z + C_1 z \ln z) + (B_1 z \ln \bar{z} + C_1 \bar{z} \ln \bar{z})$$

where $B_1 = -Bi$
$C_1 = Bi$

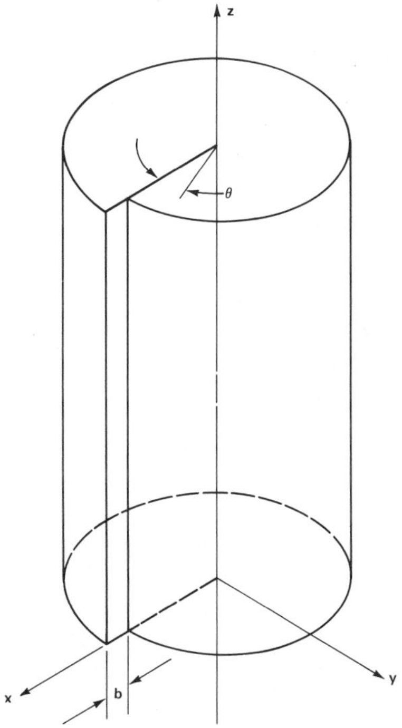

Figure 19. Schematic diagram when cylinder is dislocated in radial direction with constant velocity.

The second derivative of Λ with respect to z is obtained as

$$\frac{\partial^2\Lambda}{\partial z^2} = \frac{B\bar{z}i}{z^2} + \frac{Bi}{z} \equiv \bar{z}\frac{\partial\phi}{\partial z} + \psi(z) = (B/r)\{e^{-i(3\theta-\pi/2)} + e^{-i(\theta-\pi/2)}\} \equiv \eta e^{i\Delta} \tag{124}$$

where $$\eta = \frac{2B}{r}\cos\theta \tag{125}$$

$$\Delta = -2\theta + \pi/2$$

$$\left.\begin{aligned} \frac{\partial\phi}{\partial z} &= \frac{-\partial}{\partial z}\left(\frac{Bi}{z}\right) = \frac{B}{r^2}e^{-i(2\theta-\pi/2)} \equiv \eta_1 e^{i\delta_1} \\ \frac{\partial^2\phi}{\partial z^2} &= -\frac{2B}{r^3}e^{-i(3\theta-\pi/2)} \equiv \eta_2 e^{i\delta_2} \\ \frac{\partial\psi}{\partial z} &= -\frac{B}{r^2}e^{-(2\theta-\pi/2)i} \equiv \rho_1 e^{i\zeta_1} \end{aligned}\right\} \tag{126}$$

where $\eta_1 = B/r^2, \delta_1 = -2\theta + \dfrac{\pi}{2}$

$$\eta_2 = 2B/r^3, \delta_2 = -3\theta + \frac{3\pi}{2}$$

$$\rho_1 = B/r^2, \zeta_1 = -2\theta + \frac{3\pi}{2}$$

From Equation 106, the dilatational stress component is given by

$$\begin{aligned} \frac{\partial^2 U}{\partial z\,\partial\bar{z}} &= \frac{1}{2}\Bigg[\int\Bigg[(1+m)\cos\left(-\theta+\frac{\pi}{2}\right) + 2(m-1)\cos\left(\theta-\frac{\pi}{2}\right) \\ &\quad + (m-1)\cos\left(-\theta-\frac{\pi}{2}\right)\Bigg]\left(\frac{B}{r^2}\right)\left(\frac{2B}{r}\right)^{m-1}\cos^{(m-1)}\theta\, dr \\ &\quad - \int\Bigg[(1+m)\sin\left(\frac{\pi}{2}-\theta\right) + 2(m-1)\sin\left(\theta-\frac{\pi}{2}\right) \\ &\quad - (m-1)\sin\left(\theta+\frac{\pi}{2}\right)\Bigg]\left(\frac{B}{r}\right)\left(\frac{2B}{r}\right)^{m-1}\cos^{(m-1)}\theta\, d\theta\Bigg] \\ &= \frac{1}{2}\left[m\int(2B)^m r^{-(1+m)}\sin\theta\cos^{m-1}\theta\, dr - (2-m)\left(\frac{2B}{r}\right)^m\int\cos^m\theta\, d\theta\right] \\ &= -(2)^{m-1}\left(\frac{B}{r}\right)^m\left[\sin\theta\cos^{m-1}\theta + (2-m)\int\cos^m\theta\, d\theta\right] \end{aligned} \tag{127}$$

The hydrostatic pressure in the slip plane, $\theta = 0$, equals zero. Hence Equation 127 becomes

$$\frac{\partial^2 U}{\partial z\,\partial\bar{z}} = -(2)^{m-1}\left(\frac{B}{r}\right)^m\left[\sin\theta\cos^{m-1}\theta + (2-m)\int_0^\theta\cos^m\theta\, d\theta\right]$$

From Equation 13

$$\sigma_x + \sigma_y = 4\frac{\partial^2 U}{\partial z\,\partial \bar{z}} = -2\left(\frac{2B}{r}\right)^m\left[\sin\theta\cos^{m-1}\theta + (2-m)\int_0^\theta \cos^m\theta\,d\theta\right] \tag{128}$$

$$\sigma_y - \sigma_x + 2i\sigma_{xy} = 4\frac{\partial^2 U}{\partial z^2} = 4\left(\frac{\partial^2 \Lambda}{\partial z^2}\right)^{(1+m)/2}\left(\frac{\partial^2 \Lambda}{\partial \bar{z}^2}\right)^{(m-1)/2} = 4\left(\frac{2B}{r}\right)^m e^{i(\pi/2-2\theta)}\cos^m\theta \tag{129}$$

Separating the real part and imaginary part in Equation 129

$$\left.\begin{aligned} \sigma_y - \sigma_x &= 4\left(\frac{2B}{r}\right)^m \cos^m\theta\sin 2\theta \\ \sigma_{xy} &= 2\left(\frac{2B}{r}\right)^m \cos^m\theta\cos 2\theta \end{aligned}\right\} \tag{130}$$

Solving for σ_y and σ_x from Equations 128 and 130

$$\left.\begin{aligned} \sigma_y &= \left(\frac{2B}{r}\right)^m\left\{2\cos^m\theta\sin 2\theta - \sin\theta\cos^{m-1}\theta - (2-m)\int_0^\theta \cos^m\theta\,d\theta\right\} \\ \sigma_x &= -\left(\frac{2B}{r}\right)^m\left\{2\cos^m\theta\sin 2\theta + \sin\theta\cos^{m-1}\theta + (2-m)\int_0^\theta \cos^m\theta\,d\theta\right\} \\ \sigma_{xy} &= 2\left(\frac{2B}{r}\right)^m \cos^m\theta\cos 2\theta \\ \sigma_{zz} &= -\left(\frac{2B}{r}\right)^m\left\{\sin\theta\cos^{m-1}\theta + (2-m)\int_0^\theta \cos^m\theta\,d\theta\right\} \end{aligned}\right\} \tag{131}$$

if $m = 1.0$, then Equation 131 can be reduced to the elastic solution [41]. Equation 131 for $m = 1.0$ is

$$\left.\begin{aligned} \sigma_y &= \frac{4B}{r}\sin\theta\cos 2\theta = K\frac{y(x^2-y^2)}{(x^2+y^2)^2} \\ \sigma_x &= \frac{4B}{r}\sin\theta\,(2\cos^2\theta + 1) = -K\frac{y(3x^2+y^2)}{(x^2+y^2)^2} \\ \sigma_{xy} &= \frac{4B}{r}(\cos\theta)(\cos 2\theta) = K\frac{(x^2-y^2)x}{(x^2+y^2)^2} \\ \sigma_{zz} &= \frac{2B}{r}\sin\theta = -\frac{K}{2}\frac{y}{x^2+y^2} \end{aligned}\right\} \tag{132}$$

Equation 132 is identical to the solution given in [43]. Equation 132 is used to study the edge dislocation. Equation 131 or 132 is incomplete, as the constant B has not yet been determined. This constant can be found from the boundary condition,

$$b = \lim_{y\to -0}\int \dot{\epsilon}_x\,dx - \lim_{y\to +0}\int \dot{\epsilon}_x\,dx \tag{133}$$

at r = R, where b is the constant velocity in the radial direction at $\theta = 0$. Using Equations 129 and 97, the scalar multiplier, $\dot{\lambda}$ can be found

$$\dot{\lambda} = \frac{3}{2}\frac{1}{\sigma_0^{1/m}}(\tilde{\sigma})^{(1-m)/m}$$
$$= \frac{3\Delta}{2C_1}(4)^{(1-m)/m}\left(\frac{\partial^2 U}{\partial z^2}\frac{\partial^2 U}{\partial \bar{z}^2}\right)^{(1-m)/2m}$$
$$= \frac{3\Delta}{2C_1}(4)^{(1-m)/m}\eta^{(1-m)} = \frac{3\Delta}{2C_1}(4)^{(1-m)/m}\left(\frac{2B}{r}\right)^{(1-m)}(\cos\theta)^{(1-m)} \tag{134}$$

From the constitutive equation,

$$\left.\begin{aligned} &\dot{\lambda}(\sigma_y + \sigma_p) = \dot{\epsilon}_y = -\dot{\epsilon}_x \\ &\dot{\epsilon}_y = -\dot{\epsilon}_x = 2\Gamma\left(\frac{2B}{r}\right)\cos\theta\sin 2\theta \end{aligned}\right\} \tag{135}$$

where $\Gamma = \dfrac{3\Delta}{2C_1}(4)^{(1-m)/m}$

Since $x = r\cos\theta$, $dx = -\sin\theta r\, d\theta$ at r = R, substituting Equations 135 and 133

$$b = 2\Gamma\int_0^{2\pi}(\sin 2\theta)^2\, d\theta = 2B\pi\Gamma \quad \text{and} \quad B = \frac{b}{2\pi\Gamma} \tag{136}$$

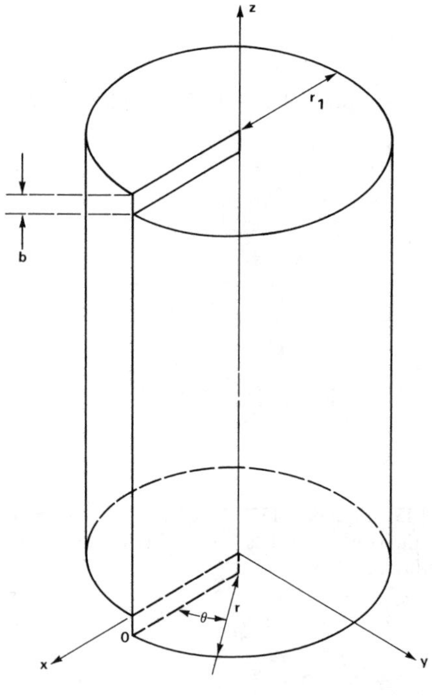

Figure 20. Schematic diagram when cylinder is dislocated in axial direction with constant velocity.

Substituting Equation 136 into Equations 131 and 133, the stress and strain components can be found where the steady creep material dislocates at constant velocity, b, in the radial direction at $\theta = 0$.

Case 2—Constant Velocity in the Axial Direction

This problem shown in Figure 20 can be solved easily since the velocity field can be defined as

$$W = b\theta/2\pi, \qquad \theta = \tan^{-1}\frac{y}{x} \tag{137}$$

where W is axial velocity.

The strain rates $\dot{\epsilon}_{xt}$ and $\dot{\epsilon}_{yt}$ are obtained directly and are given by

$$\left.\begin{aligned} \dot{\epsilon}_{xt} &= -\frac{b}{2\pi}\frac{y}{x^2 + y^2} \\ \dot{\epsilon}_{yt} &= \frac{b}{2\pi}\frac{x}{x^2 + y^2} \end{aligned}\right\} \tag{138}$$

The velocity and strain rates are independent of the strain rate hardening exponent. The stress components are found from the constitutive equation,

$$\left.\begin{aligned} \sigma_{xt} &= -\frac{2\sigma_0}{3}\left(\frac{\sqrt{3}}{2}\right)^{1-m}\left(\frac{b}{2\pi}\right)^m\left(\frac{1}{r}\right)^m \sin\theta \\ \sigma_{yt} &= \frac{2\sigma_0}{3}\left(\frac{\sqrt{3}}{2}\right)^{1-m}\left(\frac{b}{2\pi}\right)^m\left(\frac{1}{r}\right)^m \cos\theta \end{aligned}\right\} \tag{139}$$

and if $m = 1$

$$\left.\begin{aligned} \sigma_{xt} &= -\frac{2\sigma_0}{3}\frac{b}{2\pi}\frac{\sin\theta}{r} \\ \sigma_{yt} &= \frac{2\sigma_0}{3}\frac{b}{2\pi}\frac{\cos\theta}{r} \end{aligned}\right\} \tag{140}$$

Equation 140 is identical to the linear elastic solution [41].

Results and Discussion

The objectives of this section are to illustrate the applicability of the pseudo-stress function approach and to perform a stress analysis of a "Volterra Type" problem for a steady creep material described by the power law.

The pseudo-stress function approach is illustrated by solving a boundary value problem and comparing the solution to the existing linear elastic solution. The horizontal, vertical, shearing, and axial stress distribution with respect to the angular coordinate for various rate hardening exponents are shown in Figures 21–24. The horizontal and vertical stress distribution is shown to be skew-symmetric with respect to the slip plane ($\theta = 0$). The vertical stress σ_y, vanishes in the neighborhood of $\theta = \pi/4$ and $3\pi/4$ in addition to $\theta = 0$ and π. σ_y vanishes at $\theta = \pi/4$ and $3\pi/4$ for a

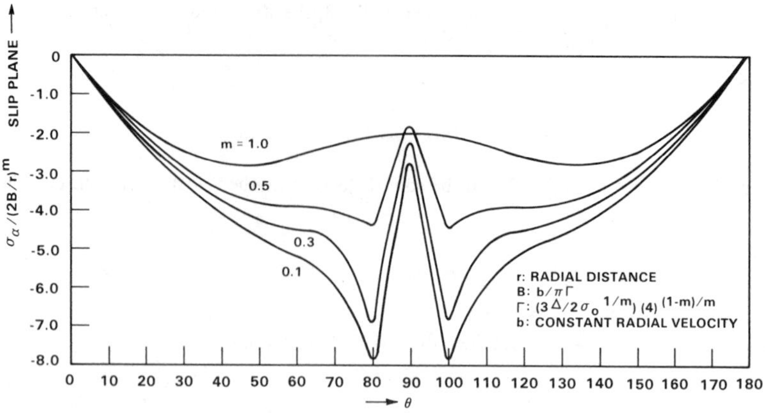

Figure 21. Horizontal stress distribution with respect to angular coordinate (θ) for various rate hardening exponents (constant velocity in radial direction).

perfectly viscous material (m = 1.0) and the angle that allows $\sigma_y = 0$ is shifted toward the slip plane when the rate hardening exponent, m, is decreased. The absolute value of the horizontal stress, σ_x, for $0 \leq \theta \leq \pi$ and the absolute value of the vertical stress $\pi/4 \leq \theta \leq 3\pi/4$ increases with increasing m values.

The stresses, σ_x, σ_y, and σ_z are unstable when θ approaches $\pi/2$ as shown in Figure 21-23. The tangent $\partial\sigma_{ij}/\partial\theta$ is indefinite at $\theta = \pi/2$ for m < 1.0, whereas $\partial\sigma_{ij}/\partial\theta$ = zero for m = 1.0. Equation 131 shows that a steady value is obtained for m = 1.0. The tangent of σ_{xy} with respect to θ gives a finite value when $0 < \theta < \pi$. The contribution to the unsteady state stress distribution with respect to θ in the neighborhood of $\theta = \pi/2$ is caused by the dilatational stress rather than the deviatoric stress. From Equation 131 the first term of σ_x and σ_y is seen to correspond to the deviatoric stress component. The second and third term correspond to the dilatational stress component which gives an indefinite value of $\partial\sigma_{ij}/\partial\theta$ at $\theta = \pi/2$. Figure 23 illustrates the axial stress

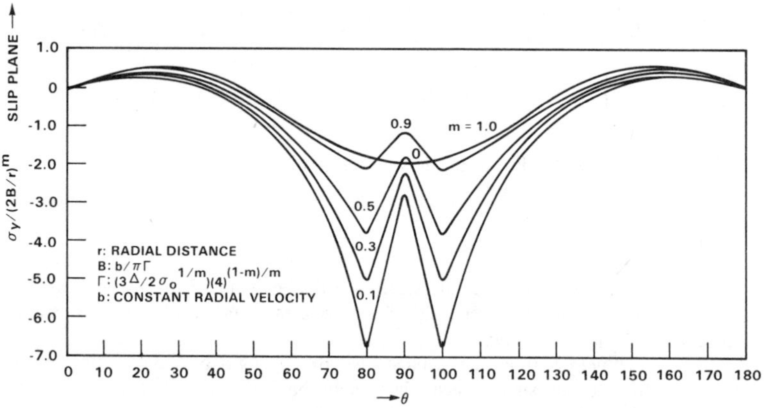

Figure 22. Vertical stress distribution with respect to angular coordinate (θ) for various rate hardening exponents (constant velocity in radial direction).

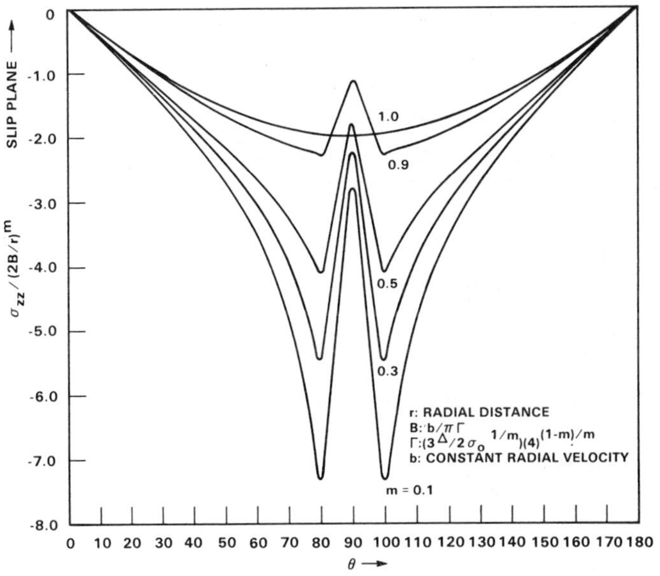

Figure 23. Axial stress distribution with respect to angular coordinate (θ) for various rate hardening exponents (constant velocity in radial direction).

corresponding to the dilatational stress under plane strain and presuming incompressibility. It shows that the variation of σ_{zz} near $\theta = \pi/2$ is more predominant than the variation of σ_x and σ_y.

The shearing stress distribution with respect to the angular coordinate, θ, in the x-y plane is shown in Figure 24. Stable solutions are obtainable for the entire domain. The shearing stress vanishes at $\theta = \pi/4$, $\pi/2$ and $3\pi/4$ for all m values. The variance in the magnitude of σ_{xy} for various m values in the region of $0 < \theta < \pi/4$ and $3\pi/4 < \theta < \pi$ is small whereas a significant

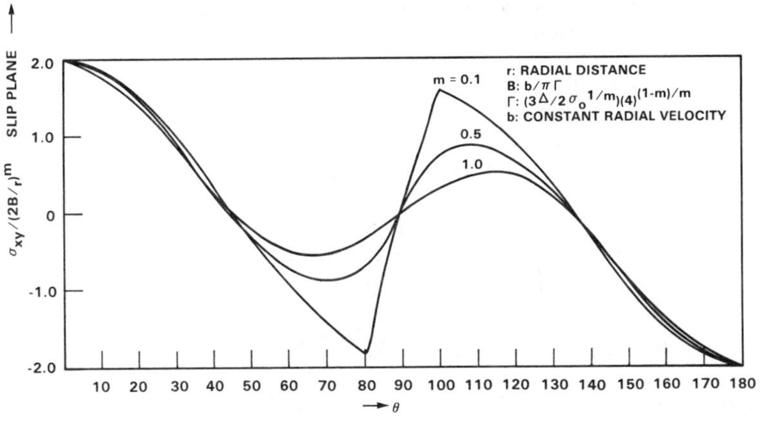

Figure 24. Shearing stress distribution with respect to angular coordinate (θ) for various rate hardening exponents (constant velocity in radial direction).

difference in the region $\pi/4 < \theta < 3\pi/4$ is noticed. The magnitude of σ_{xy} increases with decreasing m values.

APPENDIX A

This appendix deals with the detailed derivations of Equations 1–4. The equilibrium equation under a plane strain condition for an incompressible material is given as

$$\left.\begin{aligned} \frac{\partial \sigma_x}{\partial x} + \frac{\partial \sigma_{xy}}{\partial y} = 0 \\ \frac{\partial \sigma_y}{\partial y} + \frac{\partial \sigma_{xy}}{\partial x} = 0 \end{aligned}\right\} \text{(zero inertial)} \tag{141}$$

The constitutive equation for a non-Newtonian fluid is considered to be

$$\left.\begin{aligned} \tilde{\sigma} = \sigma_0(\dot{\tilde{\epsilon}})^m \\ (\sigma_{ij} + p\delta_{ij}) = \frac{1}{\dot{\lambda}}\dot{\epsilon}_{ij} \end{aligned}\right\} \tag{142}$$

where $\dot{\lambda} = \dfrac{3}{2}\dfrac{\dot{\tilde{\epsilon}}}{\tilde{\sigma}}$

Substituting Equation 142 into Equation 141

$$\left.\begin{aligned} -\frac{\partial p}{\partial x} + \frac{\partial}{\partial x}\left(\frac{1}{\dot{\lambda}}\dot{\epsilon}_x\right) + \frac{\partial}{\partial y}\left(\frac{1}{\dot{\lambda}}\dot{\epsilon}_{xy}\right) = 0 \\ -\frac{\partial p}{\partial y} + \frac{\partial}{\partial y}\left(\frac{1}{\dot{\lambda}}\dot{\epsilon}y\right) + \frac{\partial}{\partial x}\left(\frac{1}{\dot{\lambda}}\dot{\epsilon}_{xy}\right) = 0 \end{aligned}\right\} \tag{143}$$

Using the incompressibility condition, $\epsilon_x = -\epsilon_y$, the definitions of the stream function, $u = \partial\phi/\partial y$ and $v = -\partial\phi/\partial x$ and the linear strain rate, $\dot{\epsilon}_{ij} = (u_{i,j} + u_{j,i})/2$, Equation 143 can be rewritten as

$$\left.\begin{aligned} -\frac{\partial p}{\partial x} + \frac{\partial}{\partial x}\left(\frac{1}{\dot{\lambda}}\frac{\partial^2\phi}{\partial x\,\partial y}\right) + \frac{\partial}{\partial y}\left\{\frac{1}{\dot{\lambda}}\left(\frac{\partial^2\phi}{\partial y^2} - \frac{\partial^2\phi}{\partial x^2}\right)\right\} = 0 \\ -\frac{\partial p}{\partial y} - \frac{\partial}{\partial y}\left(\frac{1}{\dot{\lambda}}\frac{\partial^2\phi}{\partial x}\right) + \frac{\partial}{\partial x}\left\{\frac{1}{2\dot{\lambda}}\left(\frac{\partial^2\phi}{\partial y} - \frac{\partial^2\phi}{\partial x^2}\right)\right\} = 0 \end{aligned}\right\} \tag{144}$$

Eliminating hydrostatic pressure from Equation 144 as given

$$\frac{\partial^2}{\partial x\,\partial y}\left(\frac{2}{\dot{\lambda}}\frac{\partial^2\phi}{\partial x\,\partial y}\right) + \left(\frac{\partial^2}{\partial y^2} - \frac{\partial^2}{\partial x^2}\right)\left\{\frac{1}{2\dot{\lambda}}\left(\frac{\partial^2\phi}{\partial y^2} - \frac{\partial^2\phi}{\partial x^2}\right)\right\} = 0 \tag{145}$$

The second strain rate invariant for an incompressible material is given as

$$\dot{\tilde{\epsilon}} = \frac{2}{\sqrt{3}}(\dot{\epsilon}_x^2 + \dot{\epsilon}_{xy}^2)^{1/2} = \frac{2}{\sqrt{3}}\left\{\left(\frac{\partial^2\phi}{\partial x\,\partial y}\right)^2 + \frac{1}{4}\left(\frac{\partial^2\phi}{\partial y^2} - \frac{\partial^2\phi}{\partial x^2}\right)^2\right\}^{1/2} \tag{146}$$

since x = x + iy and z = x − iy

$$\frac{\partial}{\partial x} = \left(\frac{\partial}{\partial z} + \frac{\partial}{\partial \tilde{z}}\right) \quad \text{and} \quad \frac{\partial}{\partial y} = i\left(\frac{\partial}{\partial z} - \frac{\partial}{\partial \tilde{z}}\right) \tag{147}$$

$$\frac{\partial^2}{\partial x\,\partial y} = i\left(\frac{\partial^2}{\partial z^2} - \frac{\partial^2}{\partial \tilde{z}^2}\right)$$

Substituting Equation 147 into Equation 146 yields

$$\dot{\epsilon} = \frac{4}{\sqrt{3}}\left\{\left(\frac{\partial^2\phi}{\partial z^2}\right)\left(\frac{\partial^2\phi}{\partial \tilde{z}^2}\right)\right\}^{1/2} \tag{148}$$

Combining Equations 148 and 142, the scalar multiplier, is obtained as

$$\dot{\lambda} = \frac{3}{2\sigma_0}(\dot{\epsilon})^{1-m} = \frac{3}{2\sigma_0}\left(\frac{4}{\sqrt{3}}\right)^{1-m}\left(\frac{\partial^2\phi}{\partial z^2}\frac{\partial^2\phi}{\partial \tilde{z}^2}\right)^{(1-m)/2} \tag{149}$$

Substituting Equations 149 and 147 into 145 and simplifying the resulting equation gives

$$\frac{\partial^2}{\partial z^2}\left\{\left(\frac{\partial^2\phi}{\partial z^2}\right)^{(m-1)/2}\left(\frac{\partial^2\phi}{\partial \tilde{z}^2}\right)^{(m+1)/2}\right\} + \frac{\partial^2}{\partial z^2}\left\{\left(\frac{\partial^2\phi}{\partial z^2}\right)^{(m+1)/2}\left(\frac{\partial^2\phi}{\partial \tilde{z}^2}\right)^{(m-1)/2}\right\} = 0 \tag{3}$$

To satisfy Equation 3 the stream function is assumed to be of the summation form,

$$\phi = \phi_1(z) + \phi_2(\tilde{z}) \tag{150}$$

where $\phi_1(z)$ and $\phi_2(\tilde{z})$ is a holomophric function of z and $\tilde{z}$ respectively in the given domain. Substituting Equation 150 into Equation 3, $\partial^2\phi_1/\partial z^2$ and $\partial^2\phi_2/\partial\tilde{z}^2$ are found as

$$\left.\begin{aligned} \frac{\partial^2\phi_1}{\partial z^2} &= (A_0 z + A_1)^{2/(m-1)} \\ \frac{\partial^2\phi_2}{\partial \tilde{z}^2} &= (K_0 \tilde{z} + K_1)^{2/(m-1)} \end{aligned}\right\} \tag{151}$$

If A_i and K_i (i = 0, 1) are conjugate, then the stream function, ϕ can be found from Equation 151.

$$\phi = \phi_1(z) + \phi_2(\tilde{z}) = 2\,\mathrm{Re}\left\{\frac{(m-1)^2}{2m(m+1)}\frac{1}{A_0^2}(A_0 z + A_1)^{2m/(m-1)} + A_2 z + A_3\right\} \tag{4}$$

APPENDIX B

In this appendix, a method of solution is presented when the stream function cannot be separated into functions of $\phi_1(Z)$ and $\phi_2(\tilde{Z})$. The governing equation (Equation 3) is satisfied if the following general solutions are assumed

$$\left.\begin{aligned} \left(\frac{\partial^2\phi}{\partial Z^2}\right)^{(m-1)/2}\left(\frac{\partial^2\phi}{\partial \tilde{Z}^2}\right)^{(m+1)/2} &= \Lambda(A_0 Z + A_1 + K_0\tilde{Z} + K_1) \\ \left(\frac{\partial^2\phi}{\partial Z^2}\right)^{(m+1)/2}\left(\frac{\partial^2\phi}{\partial \tilde{Z}^2}\right)^{(m-1)/2} &= (A_0 Z + A_1 + K_0\tilde{Z} + K_1) \end{aligned}\right\} \tag{152}$$

and the constant Λ is chosen so as to yield a unique function $(Z, \tilde{Z})$ from either of the equations above. Solving Equation 152 with respect to $\partial^2\phi/\partial Z^2$ and $\partial^2\phi/\partial\tilde{Z}^2$, the solutions are given by

$$\left.\begin{aligned} \frac{\partial^2\phi}{\partial Z^2} &= \Lambda^{(1-m)/2m}\xi^{1/m} \\ \frac{\partial^2\phi}{\partial\tilde{Z}^2} &= \Lambda^{(1+m)/2m}\xi^{1/m} \end{aligned}\right\} \tag{153}$$

where $\xi = A_0Z + A_1 + K_0\tilde{Z} + K_1$

Integrating the first equation of 153 twice with respect to Z, the stream function is obtained to be

$$\phi = \Lambda^{(1-m)/2m}\frac{1}{A_0^2}\left\{\frac{m^2}{(1+m)(1+2m)}\right\}\xi^{(1+2m)/m} + A_2Zf_1(\tilde{Z}) + A_3f_2(\tilde{Z}) \tag{154}$$

similarly, from the second equation of 153

$$\phi = \Lambda^{(1+m)/2m}\frac{1}{K_0^2}\left\{\frac{m^2}{(1+m)(1+2m)}\right\}\xi^{(1+2m)/m} + K_2\tilde{Z}g_1(Z) + K_3g_2(Z) \tag{155}$$

If the unknown constant Λ is chosen as

$$\Lambda = \frac{K_0^2}{A_0^2}$$

Equations 154 and 155 become congruent and the stream function for the mixed mode solution is

$$\phi(Z, \tilde{Z}) = \left(\frac{K_0}{A_0}\right)^{1/m}\left\{\frac{1}{K_0A_0}\frac{m^2}{(1+m)(1+2m)}\right\}\xi^{(1+2m)/m} + A_2Z\tilde{Z} + A_3Z + A_4\tilde{Z} \tag{156}$$

Once the stream function is known, the velocity, strain rate and stress components can be obtained as shown previously in Equations 6–8.

NOTATION

a	radius of the rigid cylinder; c^2/g^2, where c is inlet opening of channel and g is outlet
A_i, K_i	integral constants, (i = 0, 1, 2)
c, c_1	$\frac{2\sigma_0}{3}\left(\frac{\sqrt{3}}{4}\right)^{1-m}$; also inlet opening of channel.
g	outlet opening of channel
h_i, h_0	distance between centerline and the inside and outside of the slit respectively
K	π/Uc
K_1	c/π
L	length of the plate
m	fluid behavior index; strain rate hardening exponent
r	reduction ratio
r_o	modulus of complex integral constant, A_o
s	sink intensity Uc/π
t, $\bar{t}$	complex conjugate variables in transformed plane
u_i, u, v	velocity tensor, horizontal, and vertical velocity respectively
u_0	free stream velocity
u_η, u_ζ	radial and tangential velocity
U	inlet velocity
u, v	horizontal and vertical velocities
W	complex potential function
$\bar{W}$	complex conjugate potential function
z, $\tilde{z}$	$z = x + iy$, $\tilde{z} = x - iy$

Greek Symbols

$\tilde{\sigma}, \dot{\tilde{\epsilon}}$	effective stress and strain rate	σ_{ij}	stress tensor
σ_{ij}	stress tensor	ρ	density
$\dot{\epsilon}_{ij}$	strain rate tensor	σ_p or p	hydrostatic pressure
$\dot{\tilde{\epsilon}}$	effective strain rate	ϕ	stream function
σ_0	fluid consistency index	$\dot{\lambda}$	scalar multiplier, $\dot{\lambda} = \frac{3}{2}\frac{\dot{\tilde{\epsilon}}}{\tilde{\sigma}}$
σ_p	dilatational stress	$\psi(z), \bar{\psi}(\bar{z})$	holomorphic functions of z and $\bar{z}$
η, ζ	modulus and argument of z	ψ	velocity potential function
θ_o	argument of A_o		

REFERENCES

1. Mooney, M., and Black, S. A., "A Generalized Fluidity Power Law and Law of Extrusion," *J. of Colloid Science*, p. 204–217, 1952.
2. Fredrickson, A. G., and Bird, R. B., "Non-Newtonian Flow in Annuli," *Industrial Engineering Chemistry*, Vol. 50, No. 3, p. 347–352, 1958.
3. Mishra, Padmaker and Mishira, Indramani, "Flow Behavior of Power Law in an Annulus," *AICHE J.*, Vol. 22, No. 3, p. 617–619, 1976.
4. Mohan, V., "Creeping Flow of a Power Law over Non-Newtonian Fluid Sphere," *AICHE J.* Vol. 20, No. 1, p. 180–182, 1974.
5. Miller, Chester, "Predicting Non-Newtonian Flow Behavior in Ducts of Unusual Cross Section," *Ind. E. Chem. Fundam.* Vol. 11, No. 4, p. 524–528, 1972.
6. Skvorova, V., "Non-Newtonian Flow in a Square Duct," Collection Czechoslav. Chem. Commun., Vol. 40, p. 2605–2610, 1975.
7. Palit Kalipada and Fenner, R. T., "Finite Element Analysis of Slow Non-Newtonian Channel Flow," *AICHE J.*, Vol. 18, No. 3, p. 628–633, 1972.
8. Swamy, S. T. N., "The Influence of Non-Newtonian Oil Film Short Journal Bearing on the Stability of a Rigid Rotor," *WEAR*, Vol. 43, 1977, p. 155–164.
9. Swamy, S. T. N., Prabhu, B. S., and Rao, B. V. A., "Steady State and Stability Characteristics of a Hydrodynamic Journal Bearing with a Non-Newtonian Lubricant," *WEAR*, Vol. 42, 1977, p. 229–244.
10. Hsu, Y. C., "Non-Newtonian Flow in Infinite Length Full Journal Bearing," *Trans. of the ASME, Journal of Lubrication Technology*, Vol. 89, 1967, p. 329.
11. Shukla, J. B., "Thermal Effects in Squeeze Films and Externally Pressurized Bearing with Power Law Lubricant," *WEAR*, Vol. 51, 1978, p. 237–251.
12. Ng, C. W., and Saibel, E., "Non-Linear Viscosity Effects in Slider Bearing Lubrication," *Trans. of the ASME, Journal of Basic Engineering*, Vol. 84, 1962, p. 192.
13. Lee, Y. S., and Patel, M. R., "An Analysis for Plane Strain Plastic Deformation in Metal Working Process," *J. of Eng. for Ind., Trans. ASME Series B*, Vol. 99, No. 3, p. 727–732, 1976.
14. Shabaik, A. H., and Thomson, E. G., "Theoretical Method for the Analysis of Metal-Working Problem," *J. of Eng. for Ind. Trans. ASME Series B*, Vol. 90, No. 2, p. 343–352, 1968.
15. Lee, Y. S., and Male, A. T., "Deformation Analysis of Strain Rate Hardening Materials Under Plane Strain Condition," *Int. J. of Mech. Sci.*, Vol. 25, No. 4, p. 251, 1983.
16. Milne-Thomson, L. M., *Theoretical Hydrodynamics*, The MacMillan Co., 1961, p. 154–156.

16A. Lee, Y. S., and Smith, L. C., "7th Canadian Congress of Applied Mechanics" (abstract) at University of Sherbrooke, Quebec, Canada, May 28th to June 1, 1979.

17. Schlichting, M., *Boundary Layer Theory*, McGraw-Hill Book Co., 1960, p. 67.
18. Yang, C. T., and Thomsen, E. G., "Plastic Flow in a Lead Extrusion," *Trans. ASME*, Vol. 75, p. 575, 1953.
19. Thomson, E. G., "Plane-Strain and Axially Symmetric Velocities and Pressure in Extrusions," Conference on the Properties of Materials at High Rates of Strain Sponsored by the Institution of Mechanical Engineers, Session 3, Paper 1, 1957.
20. Avitzur, B., *Metal Forming; Processes and Analysis*, McGraw Hill, N.Y. 1968.

21. Zimerman, Z., and Avitzur, B., *Journal of Engineering Industry*, Trans. ASME, 92B.
22. Johnson, W., and Kudo, H., *The Mechanics of Metal Extrusion*, Manchester University Press, Manchester, 1962.
23. Hill, R., *The Mathematical Theory of Plasticity*, Oxford University Press, 1950.
24. Johnson, W., Sowerby, R., and Haddow, J. B., *Plane-Strain Slip-line Fields: Theory and Bibliography*, American Elsevier Publishing Co., New York, 1970.
25. Juneja, B. L., and Prakash, R., *Int. J. Mach. Tool. Des. Res.* Vol. 15, 1, 1975.
26. Yang, D. Y., and Lange, K., "On the Equivalent Friction Factor in Hydrofilm Extrusion," *Int. J. Mech. Sci.*, Vol. 25, No. 4, p. 277, 1983.
27. Cho, N. S., and Yang, D. Y., "Analysis of Hydrofilm Extrusion of Elliptic Shapes Using Perturbation Method," *Int. J. Mech. Sci.* Vol. 25, No. 4, p. 283, 1983.
28. Beyer, K., *Neue Huette*, Vol. 16, p. 715, 1971.
29. Gron, G. Y., and Polukhin, *Izvest*, V.U.Z. Chernaya Met., Vol. 10, p. 63, 1971.
30. Nagpal, Y., and Altan, T., *Proc. 3rd North Am. Metal Working Conf*., p. 26, Pittsburgh, Pa., 1975.
31. Yang, D. Y., and Lee, C. H., "Analysis of Three Dimensional Extension of Sections Through Curved Dies by Conform Transformation," *Int. J. Mech. Sci.*, Vol. 20, No. 9, p. 541, 1978.
32. Lee, Y. S., and Patel, M. R., "Deformation Analysis of Plastic Strain Rate Hardening Material Using Stress Function Under Plane Strain Condition," *Int. J. Mech. Sci.*, Vol. 22, No. 6, p. 355, 1980.
33. Lee, Y. S., and Smith, L. C., "An Analysis of Power Law Viscous Materials Under Plane Strain Condition Using Complex Stream and Stress Function," *J. App. Mech. Trans. ASME.*, Vol. 48, p. 486, 1981.
34. Lee, Y. S., and Smith, L. C., "An Analysis of the Volterra Problem for the Steady State Creep Materials Using Complex Stress and Psuedo-Stress Function," *Acta Mechanica*, Vol. 49, p. 95–109, 1983.
35. Walker, M., *The Schwartz-Christoffel Transformation and Its Application—A Simple Exposition* (Formerly titled: *Conjugate Functions for Engineer*), Dover Publications, Inc., New York, 1964.
36. Kennelly, A. E., *Tables of Complex Hyperbolic and Circular Functions*, Cambridge Harvard University Press, 1921.
37. Oh, S. I., Park, J. J., Kobayashi, J., and Altan, T., "Application of FEM Modeling to Simulate Metal Flow in Forging a Titanium Alloy Engine Disk," *Trans. ASME, J. of Eng. for Ind.*, Vol. 105, p. 251, 1983.
38. Dadras, P., and Thomas, J. F., "Characterization of Modeling for Forging Deformation of Ti-6Al-2Sn-4Zr-2Mo-0.1 si," *Met., Trans. A*, 12A, p. 1867, 1981.
39. Lee, Y. S., and Smith, L. C., "An Application of Schwartz-Christoffel Conformal Mapping and Complex Variables to the Cornered Die Extrusion of Steady State Creep Materials," submitted to International J. of Mech. Science, vol. 27, no. 1, pp. 13–27, 1985.
40. Finnie lain, and William R. Heller, *Creep of Engineering Materials*, McGraw-Hill Book Co., Inc., 1959, Chap. 4.
41. Nadai, A., and McVetty, P. G., "Hyperbolic Sine Chart for Estimating Stresses of Alloys at Elevated Temperatures," ASTM, Vol. 43, p. 735, 1943.
42. Nadai, A., "On the Creep of Solid at Elevated Temperature," *J. App. Phy.*, Vol. 8, No. 6, p. 418, June 1937.
43. Hirth, J. P., and Lothe, J., *Theory and Dislocations*, McGraw-Hill Book Co., 1968.
44. Weertman, J., and Weertman, J. R., *Elementary Dislocation Theory*, MacMillian Co., New York, 1964.

CHAPTER 35

RHEOLOGICAL PROPERTIES OF THERMOPLASTICS

G. Akay

School of Industrial Science
Cranfield Institute of Technology
Cranfield, Bedford, U.K.

CONTENTS

INTRODUCTION

Rheology is not only used in characterizing material properties and understanding the structure of matter, but is also used extensively in various stages of the manufacture, handling, and processing of many materials, particularly polymers. The mechanical and physico-chemical properties of polymeric materials are strongly influenced by the process history, due to the presence of microstructure either in discrete (such as glass fibers in polymer melt) or semi-discrete (such as the cluster of macromolecules in the melt) form. In some cases, tne presence of fluid microstructure is only manifested during the flow, although it may persist for a long time after the cessation of flow. Owing to this long relaxation time, the microstructure may easily be frozen in the material at the end of processing. Such a process, in which flow induced microstructure is wholely or partially preserved, often presents unique advantages in the processing of polymeric materials. It also necessitates the understanding of flow microstructure interactions and their manifestations, which may sometimes prove detrimental in the final product.

Homogeneous polymer melts are often regarded as a structured fluid, in which the fluid microstructure may be characterized by the entanglement density. With more widespread use of copolymers, liquid crystalline polymers (LCP's), polymer blends, reinforced and filled polymers and foamed polymers, it is inevitable that flow induced phenomena should come under close scrutiny since most of these polymers are used as engineering materials. In this work, we are mainly concerned with the microstructural aspects of polymer melt rheology not only because of its significance in processing but also because of their implications in the basic concepts of rheology.

Steady Viscometric Flows

We briefly describe some common viscometric methods for determining the rheological properties of fluids. Detailed descriptions of these techniques and others may be found in some well established text-books and monograms [1–12]. In these techniques one essentially measures the "stress response" of the fluid as a result of imposed deformation (flow). The state of stress in the fluid is described by the stress tensor P_{ij} (here we use Cartesian tensor notation), which can be resolved into isotropic pressure p, and extra stress, P_{ij}

$$P_{ij} = -p\delta_{ij} + P_{ij} \qquad (\delta_{ij} = 0 \quad i \neq j; \quad \delta_{ij} = 1 \quad i = j) \tag{1}$$

In a steady simple shear flow which can be generated between two parallel plates (Figure 1) by moving one of the plates with constant velocity v_0 so that the components of the velocity are (referred to the Cartesian coordinates, x_i; i = 1, 2, 3)

$$v_1 = 0, \qquad v_2 = \dot{\gamma}x_1, \qquad v_3 = 0 \tag{2}$$

where $\dot{\gamma} = v_0/h$ is the shear rate and h is the plate seperation.

The corresponding stress distribution can be written as

$$P_{12} = P_{21} = \dot{\gamma}\eta(\dot{\gamma}) \tag{3}$$

$$P_{22} - P_{11} = N_1(\dot{\gamma}) \tag{4}$$

$$P_{11} - P_{33} = N_2(\dot{\gamma}) \tag{5}$$

where P_{12} is the shear stress, η is viscosity (function), N_1 and N_2 are the first and the second normal stress difference respectively. Since $N_1 \gg N_2$, we will exclude N_2 from our discussions.

Another important type of flow is called uni-axial elongational flow, in which the components of the velocity may be written as

$$v_1 = \dot{\epsilon}x_1, \qquad v_2 = -\tfrac{1}{2}\dot{\epsilon}x_2, \qquad v_3 = -\tfrac{1}{2}\dot{\epsilon}x_3 \tag{6}$$

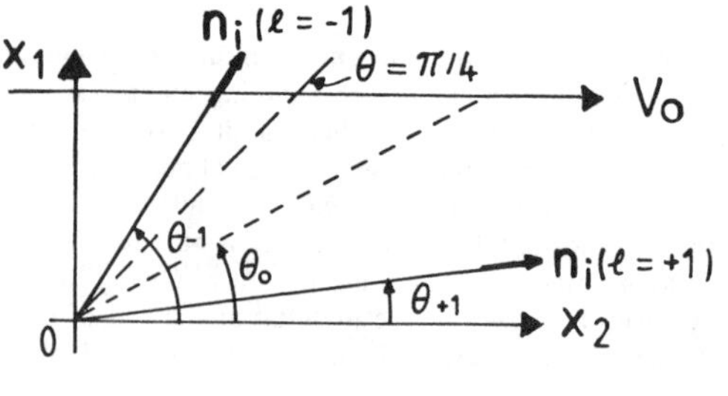

Figure 1. Simple shear flow between parallel plates and microstructure orientation.

where $\dot{\epsilon}$ is the elongation rate. In this type of flow, the diagonal components of P_{ij} are non-zero, but P_{11} is the primary stress that can be measured, while P_{22} and P_{33} are generally equal to the pressure of the environment. The elongational viscosity η_e is defined by

$$P_{11} = \eta_e(\dot{\epsilon})\dot{\epsilon} \quad (7)$$

Elongational flows are extremely important in fibre spinning and in entry flows.

These rheological functions can be evaluated experimentally using either "homogeneous" or "non-homogeneous" flow fields. In a homogenous flow field, the velocity gradient does not charge spatially over the domain of interest. Pressure-driven flow in a capillary is perhaps the simplest example of a non-homogeneous flow field in which shear rate varies from a maximum at the wall to zero along the centerline. For a Newtonian fluid, this variation is linear and, for a large number of non-Newtonian fluid models, analytical or numerical solutions of the velocity distribution are available [13, 14].

In a capillary flow, shear stress at the wall is given by

$$\tau_w = D\,\Delta P/4L \quad (8)$$

where D is the capillary diameter, ΔP is the pressure drop along the capillary length of L. The shear rate at the capillary is given by

$$\dot{\gamma}_w = \Gamma(3/4 + d(\ln \Gamma)/4d(\ln \tau_w)) \quad (9)$$

where

$$\Gamma = 32Q/\pi D^3 \quad (10)$$

is the wall shear rate for a Newtonian fluid, Q is the volumetric flow rate. From τ_w and Γ, $\dot{\gamma}_w$ and viscosity function $\eta(\dot{\gamma}_w)$ can be evaluated using Equations 3 and 9. The quantities τ_w and Γ are the directly measurable variables but Γ does not have the same significance for any general fluid as does $\dot{\gamma}_w$. However, it is sometimes more meaningful to use Γ rather than $\dot{\gamma}_w$, such as when there is instability in the flow so that $\dot{\gamma}_w \to \infty$ as $d \ln \tau_w \to 0$.

In capillary flow, ideally, pressures at two points far from the ends of the capillary should be measured. Due to the practical difficulties involved, the usual practice is to measure the pressure in a reservoir feeding the capillary and expose the other end to the atmosphere. However, this introduces a number of additional complications, such as capillary entrance and exit loss (known as ends-pressure loss) and finally die-swell as the fluid emerges from the capillary. Although these complications may not be desirable in viscometry, they, nevertheless, are useful phenomena in understanding fluid microstructure.

Total pressure loss ΔP_T in capillary flow may be written as

$$\Delta P_T = \Delta P + \Delta P_{ent} + \Delta P_{exit} \quad (11)$$

where ΔP_{ent} and ΔP_{exit} are the entrance and exit pressure losses that may be combined to yield the ends-pressure loss ΔP_e,

$$\Delta P_e = \Delta P_{ent} + \Delta P_{exit} \quad (12)$$

From Equations 8 and 12, we have

$$\Delta P_T = 4(L/D)\tau_w + \Delta P_e \quad (13)$$

The earliest technique of estimating ΔP_e is due to Bagley [15] and involves the plot of ΔP_T against capillary length-diameter ratio L/D for a series of dies with varying L and a constant D at a constant flow rate Q. From such a plot one may determine ΔP_e from the intercept on the ΔP_T

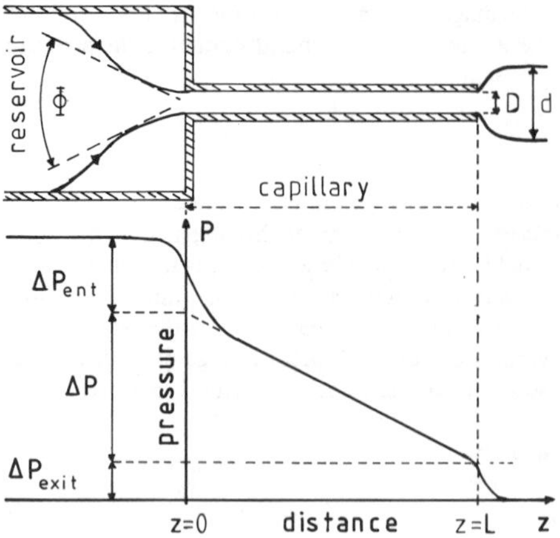

Figure 2. Flow in the reservoir and capillary: (a) definition of the flow cone angle Φ and extrudate die swell, $B = d/D$; (b) definition of entry pressure loss, ΔP_{ent}, exit pressure loss ΔP_{exit}, and capillary pressure loss ΔP.

axis according to Equation 13. There are two drawbacks to this technique; firstly, it does not allow [16] the separation of ΔP_{ent} and ΔP_{exit} and, although ΔP_{ent} is independent of L/D, ΔP_{exit} depends on the L/D ratio [2, 17]. Secondly, if the pressures involved are high, then a non-linear variation of ΔP_T with L/D is obtained [18, 19].

The technique developed by Han and Kim [20] can separate ΔP_{ent} and ΔP_{exit} and it involves the measurement of pressure along the length of capillary or slit dies and in the reservoir. It is found that ΔP_{ent} is about three times larger than ΔP_{exit} and in some cases is ignored. Han's method is illustrated in Figure 2.

The entrance pressure loss in capillary flow is closely related to the flow pattern in the reservoir and the type of microstructure of the fluid. Capillary entrance flow patterns of various polymers have been studied using flow visualization. It was found that low density polyethylene (LDPE), exhibited large vortices in the die entrance but high density polyethylene (HDPE) melts did not. Flow into the capillary is sustained from a cone with an apex angle of ϕ as shown in Figure 2(a). The entrance angle may decrease with increasing flow rate until the start of unstable flow, when the shape and size of the vortex become time dependent.

Fluid emerging from a capillary die often has a diameter (d) generally different from the diameter of the capillary (D) and the ratio of diameters B (= d/D) is called die-swell. The extent of die-swell depends on flow conditions (capillary length, die-entry geometry, flow rate and the steadiness of the flow) as well as on the material itself. In polymer melts and solutions $B > 1$ while in filled polymer melts and LCP's negative die-swell ($B < 1$) can be observed. Die-swell also depends on the type of measurement and the distance from the die-exit where the measurements are taken. The most common method is to measure the diameter of the extrudate after it has solidified. However, if the extrudate diameter is measured during flow under isothermal conditions or if the temperature of the solid extrudate is raised to its softening point before measuring its diameter, then larger die-swell values are recorded.

Time-Dependent Viscometric Flows

Time-dependent flows are important in the characterization and processing of polymeric fluids. Flow is generated by the exitation of the fluid and the response is detected as a function of time. We classify time-dependent flows as

1. Transient flows in which the exitation is imposed impulsively while the fluid is in a steady state and the response of the fluid is followed until a new steady state is reached; i.e. start-up flow and flow decay.

2. Programmed flows in which the time dependence of exitation is controlled and the response is followed. The best known example of this type of flow is small amplitude oscillatory shear flow.
3. Unstable flows in which although the exitation has no time dependence, the response is time periodic.

As a result of flow, fluid microstructure reaches an equilibrium state, i.e. a new equilibrium entanglement density or a new orientation distribution. Changes in the microstructural state of fluid can be described with respect to a reference (or rest) state, which is attained in the absence of any external forces. When a mechanical exitation is imposed on to a fluid, a new structural state is reached after a certain length of time. The material response reflects the difference between the initial and final structural states. When this exitation is removed, return to the equilibrium microstructural state is usually slow since mechanical forces accelerate the microstructural changes while the relaxation process may only involve molecular motions. Therefore, if a new exitation is applied to this material before full recovery, the response of the material will be different. Hence, transient and programmed flows can be used to evaluate the time dependence of the fluid microstructure.

In polymer melts, flow instability represents the temporary breakdown of the fluid microstructure in the flow fluid. There appears to be various mechanisms by which flow can be interrupted temporarily and we will discuss these mechanisms in detail.

Microstructure, Rheology, and Processing

The interaction of fluid microstructure and flow field can result in a number of phenomena (such as microstructure orientation and migration) influencing the macroscopic properties of the fluid. Rheological measurements, supplemented by the measurements of the structural changes in the fluid will be invaluable in understanding the flow properties of these fluids. Flow birefringence [21] does in fact provide one of the earliest examples of the simultaneous macroscopic and microscopic characterization of polymeric fluids.

The rheology of polymeric materials in the melt state is complicated by some practical difficulties. These are, high operating temperatures, extremely high pressures and non-transparency of the melt. When direct microstructure study is not possible, one can either use model systems (i.e., clear viscoelastic solutions at ambient temperature) or to freeze the flow-induced microstructure for subsequent morphological study. The former approach has been made more attractive by the discovery of so-called highly elastic constant viscosity Boger fluid [22]. It is possible to prepare model fluids with the desired elasticity and shear thinning properties. Laser-Doppler Anonometry is now widely used in studying the flow properties of such model systems. The latter approach has the advantage of being directly relevant to processing and is independent of the type of fluid, but suffers from the fact it is an off-line technique.

We have mainly adopted this latter approach in our studies of flow induced phenomena and the rheology of polymeric materials. Many mold inserts with well-defined geometries enable us to interpret observed rheological properties. These mold inserts are shown in Figure 3. In each of these inserts, the fluid emerging from the injection molding machine flows through a slowly diverging conical sprue before entering the mold cavity. In Figure 3a, a center gated disc mold (23-25) is illustrated. The mold inserts in Figures 3b and c have the same initial flow path. As the fluid emerges from the sprue, it is split into two, one stream feeding the mold cavity while the other is blocked. The mold insert in Figure 3b consists of twelve cylindrical reservoirs connected by capillaries [26]. In our microstructure studies we use the first three capillary reservoir systems so as to eliminate the transient nature of the injection molding, as each reservoir has to be filled before the flow can start in the next reservoir, even if jetting is present. The mold insert in Figure 3c simulates the flow between two flat plates and two diverging and converging flat plates. The angle of divergence or convergence is 45°. Before entering into the test section, the flow is allowed to develop and finally terminates in a rectangular shaped sink [27].

To a large extent, microstructure orientation or its distribution is dictated by the flow during mold filling although it may be influenced by packing and the non-isothermal and transient nature

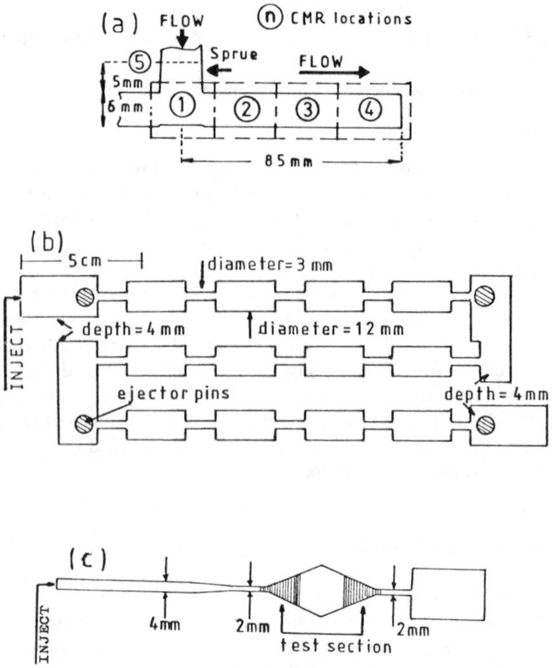

Figure 3. Mold inserts used in microstructure orientation studies: (a) center gated disc mold; (b) mold insert with cylindrical expansions and contractions; (c) mold insert with uniform, divergent, and convergent sections.

of the mold filling. The molding is carried out using a medium-sized injection molding machine, which is fitted with a closed-loop adaptive controller. Injection rate (input flow rate), temperature of the melt and mold, and the cooling time of the moldings can all be selected.

Microstructure orientation and distribution are determined by taking x-ray pictures (known as contact micro-radiography, CMR [28]) of thin slices cut from the moldings, along a plane of symmetry. In some cases, scanning electron microscopy and optical microscopy are also used.

Due to the extremely high pressures required in capillary flow of polymer melts at high shear rates, we have used the same injection molding machine for rheological measurement. In this case, a capillary assembly replaces the standard nozzle of the machine. Basically, we have three different capillary assemblies: (a) single capillary assembly in which the reservoir is the barrel of the machine, (b) two-stage capillary assembly in which a second reservoir and a second capillary are added [23, 25] and, (c) multi-stage assembly in which a number of capillaries separated by small reservoirs are connected in series [26]. The two-stage assembly is used to investigate the pressure dependence of viscosity [29, 30] flow instability [23, 25] while the multi-stage capillary assembly is used to investigate entrance pressure loss, and the transient and unstable flows of polymer melts [26]. A diagrammatic description of the two-stage capillary assembly is shown in Figure 4. Polymer melt in reservoir I is kept at temperature T_I, while the whole of the capillary assembly can be kept at this temperature. Pressures, P_I and P_{II} in reservoirs I and II together with the plunger (screw) position, z, are recorded using a UV-recorder. The input flow rate Q_0 (volumetric displacement of the plunger per unit time) is calculated from the slope of the plunger position.

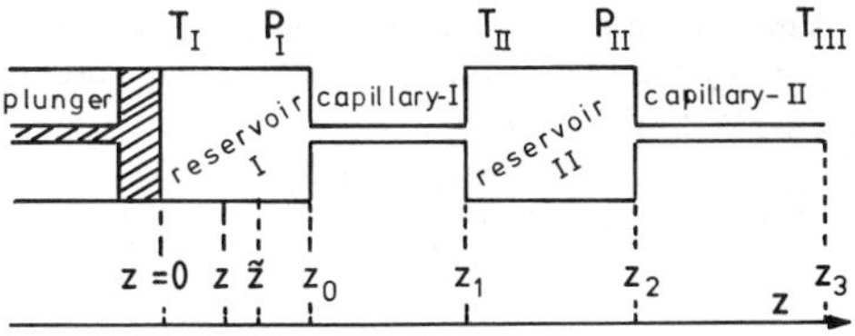

Figure 4. Two-stage rheometer. Here z̃ is the maximum plunger displacement.

FLOW INDUCED ORIENTATION

Theoretical

In order to highlight the significance of fluid microstructure in rheology we investigate the flow properties of polymer melts (single-phase polymers, polymer blends, filled and reinforced polymers and liquid crystalline polymers) under various flow-induced phenomena. Here, emphasis is on the fluids with orientable microstructure rather than single-phase polymer melts.

Flow-induced microstructure orientation in fluids, is not only important from a rheological point of view, but is also important in certain heat [31] and mass [32] transfer operations and in processing [23–28, 33–36] and in dictating the mechanical and physical properties of the finished product. The rheology of fibre reinforced thermoplastics, FRTP's [17, 19, 23, 25, 26, 37–42] and LCP's [43–51] indicates that, as a result of microstructure orientation there exists certain similarities between these two types of fluids. These similarities are: high capillary entrance pressure loss [17, 19, 26, 42, 45, 50] and low or negative die-swell [26, 40, 41, 45, 50]. However, LCP's show some anomalous flow phenomena that are not reported for FRTP melts. Depending on the melt temperature, LCP's may exhibit shear thickening [46, 49] transient [46] and steady negative primary normal stress differerence [48, 49, 51] while FRTP melts exhibit enhanced [38, 39, 41] or reduced [10] primary normal stress difference. Liquid crystalline polymer melts appear to have yield stress [43–46, 49, 51] while FRTP melts do not, although melts filled with sub-micron particles do exhibit yield stress [38]. These fluids may be considered to have orientable rigid microstructure dispersed in a viscoelastic continuous phase.

Folkes and Russell [33] have shown that the molecular orientation is affected by the orientation of glass fibers in FRTP's when the fiber concentration is high. Therefore, the presence of orientable rigid microstructure results in a high degree of anisotropy in the fluid during flow. Available experiments indicate that molecular orientation on a large scale is present in LCP's [27, 52, 53]. Molecular orientation in low molecular weight compounds that form liquid crystals have been studied extensively [54–56]. Due to the similarities between these three types of fluids, there have been various papers describing LCP's and FRTP's as an anisotropic fluid using the continuum model developed by Ericksen [57–59] and Leslie [56, 60, 61]. In so-called "Leslie-Ericksen" fluids [49], the orientable fluid microstructure is described by an objective vector, n_i, called the "director," which represents the direction of preferred orientation of the microstructure. This vector is regarded as an independent kinematic variable, in the sense that under certain circumstances it could vary independently of the other kinematic variables, although in general intimately linked to them through the equation of motion. This concept is commonly accepted in the classical anisotropic fluid theory, largely on the grounds that there are experiments in which the microstructure orientation can be changed with no detectable flow of the liquid crystals [54, 55]. Abhiraman and George [62] have relaxed the usually assumed restriction that the magnitude of the director is constant and allowed it to change with the shear rate. They were able to interpret some of the flow phenomena encountered in polymeric fluids.

Recently, Akay and Leslie [27] proposed an equation of state that can describe fluids with orientable rigid microstructure. The equation for the orientation vector n_i is

$$cn_i/ct = \xi_0 n_k n_i + \alpha_0 e_{ik} n_k + \alpha_1 e_{jk} n_j n_k n_i + \beta_0\, ce_{ik}/ct\, n_k + \beta_1\, ce_{jk}/ct n_j n_k n_i \qquad (14)$$

while the extra stress P_{ij} for the base material is modified in the presence of orientable microstructure so that for a base fluid that can be described by the Maxwell model, the stress equation of state is

$$P_{ij} + \lambda c P_{ij}/ct = 2\mu e_{ij} + (\mu_0 + \mu_1 e_{kr} n_k n_r) n_i n_j - 2\mu_2 (e_{ik} n_k n_j + e_{jk} n_k n_i) \tag{15}$$

where

$$cn_i/ct = \partial n_i/\partial t + v_k n_{i,k} - w_{ik} n_k \tag{16}$$

$$ce_{ij}/ct = \partial e_{ij}/\partial t + v_k e_{ij,k} - w_{ik} e_{kj} - w_{jk} e_{kj} \tag{17}$$

$$e_{ij} = \tfrac{1}{2}(v_{i,j} + v_{j,i}) \tag{18}$$

$$w_{ij} = \tfrac{1}{2}(v_{i,j} - v_{j,i}) \tag{19}$$

Here we use Cartesian tensor notation in which comma represents covariant differentiation and summation over the repeated suffices is assumed. We assume that coefficients in Equations 14 and 15 are material constants, rather than material functions of the appropriate invariants of the objective vectors and tensors. Material constants λ and μ are associated with the "base fluid" while the others $\mu_0, \mu_1, \mu_2, \xi_0, \alpha_0, \alpha_1, \beta_0, \beta_1$ are associated with the microstructure of the fluid. Although the stress tensor is symmetric and therefore such a fluid exhibits isotropic rheological properties, fluid microstructure is anisotropic. When $\beta_0 = \beta_1 = 0$; $\lambda = 0$ we have the constitutive equations that are often used to describe polymeric fluids with orientable microstructure [49, 62–64]. However, when $\beta_0 = \beta_1 = 0$ orientation in simple shear flow does not depend on shear rate.

Orientation in Steady Simple Shear Flow

In a simple shear flow (Figure 1) it can be shown that the orientation of the microstructure is given by

$$n_2^2/n_1^2 = (\alpha_0 + 1)(A + \beta_0 \dot{\gamma}^2)/(\alpha_0 - 1)(A - \beta_0 \dot{\gamma}^2) \tag{20}$$

where

$$A = \ell \dot{\gamma} (\alpha_0^2 - 1 + \beta_0^2 \dot{\gamma}^2)^{1/2} \qquad \ell = \pm 1 \tag{21}$$

$$E = 2\xi_0(\alpha_0 A + \beta_0 \dot{\gamma}^2) - \alpha_1(\alpha_0^2 - 1)\dot{\gamma}^2 - \beta_1(\alpha_0 \beta_0 \dot{\gamma}^2 + A)\dot{\gamma}^2 \tag{22}$$

$$N^2 = n_1^2 + n_2^2 = (\alpha_0 A + \beta_0 \dot{\gamma}^2)A/E \tag{23}$$

The nature of the microstructure orientation is best illustrated by considering two important limits as $\dot{\gamma} \to 0$ or $\dot{\gamma} \to \infty$. It can be show that

$$n_2^2/n_1^2 \to (\alpha_0 + 1)/(\alpha_0 - 1); \qquad N^2 \to 0 \quad \dot{\gamma} \to 0 \tag{24}$$

The limits of various variables as $\dot{\gamma} \to \infty$ are tabulated in Table 1. As seen from these limits, the orientations corresponding to $\ell = \pm 1$ are distinctly different. If θ_{+1} and θ_{-1} are the orientations corresponding to these two solutions, then the following relationship applies:

$$\cot \theta_{+1} \cot \theta_{-1} = \cot^2 \theta_0 = (\alpha_0 + 1)/(\alpha_0 - 1) \tag{25}$$

where θ_0 is the orientation angle when $\dot{\gamma} \to 0$ as shown in Figure 1. For large values of α_0 ($\alpha_0 > 1$)

Table 1
Limits of Various Functions

Function	$\ell = +1$	$\ell = -1$
$\lim_{\gamma\to\infty} \{n_1^2\dot{\gamma}^2\}$	$-(\alpha_0 - 1)^2/4\beta_0\beta_1$	$-\beta_0\dot{\gamma}^2/\beta_1$
$\lim_{\gamma\to\infty} \{n_2^2\dot{\gamma}^2\}$	$-\beta_0\dot{\gamma}^2/\beta_1$	$-(\alpha_0^2 - 1)/4\beta_0\beta_1$
$\lim_{\gamma\to\infty} \{n_1^2 n_2^2\dot{\gamma}^2\}$	$(\alpha_0 - 1)^2/4\beta_1^2$	$(\alpha_0^2 - 1)/4\beta_1^2$
$\lim_{\gamma\to\infty} \{n_1^2/n_2^2\}$	$(\alpha_0 - 1)^2/4\beta_0^2\dot{\gamma}^2$	$4\beta_0^2\dot{\gamma}^2/(\alpha_0^2 - 1)$

$n_1 \to n_2$ and therefore $\theta_0 \to \pi/4$ or $3\pi/4$ indicating that slow flows will result in a small bias of orientation at about an angle of $\pi/4$ or $3\pi/4$.

The possible orientations in steady simple shear flow are shown in Figure 1, where only the first quadrant is considered. Unlike the case when $\beta_0 = \beta_1 = 0$, or the case considered by Abhiraman and George [62], here, orientation is shear rate dependent, ranging from θ_0 when $\dot{\gamma} \to 0$ to θ_ℓ when $\gamma \to \infty$. The magnitude of the director $(n_i n_i)$ is unity in Leslie-Ericksen theory of anisotropic fluids, while in the formulation of Abhiraman and George; $n_i n_i \to 0$ when $\dot{\gamma} \to 0$ and $n_i n_i \to 1$ when $\dot{\gamma} \to \infty$. According to Ericksen [59], the variable magnitude of n_i represents the elasticity of suspended particles, while Kaloni [65] assumes that $(n_i n_i)$ is a measure of "stretch." Denn and Metzner [63] when applying the anisotropic fluid theory to polymeric liquids interpret the orientation vector as representing the tendency of long polymeric molecules to align under shear, while its magnitude is a measure of the degree of molecular uncoiling or molecular stretching. Gordon and Schowalter [66] assume that $(n_i n_i)$ is some measure of the "degree of orientation." As seen from this sample of interpretations, the physical significance of $(n_i n_i)$ is not clear. Experiments with short glass fiber reinforced polymer melts indicate that "overall orientations" in shear flow increases with shear rate [41], as shown in Figure 5. This is an important result since although this type of behavior has been known, no quantitative description has been previously available. We therefore take the view that $(n_i n_i)$ is some measure of the "degree of orientation" as interpreted by Gordon and Schowalter [66]

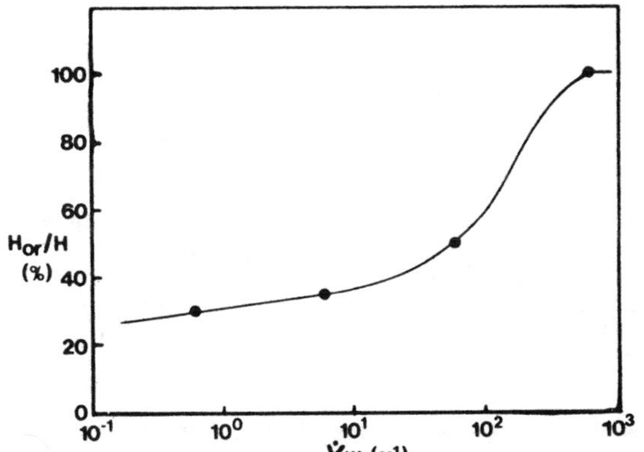

Figure 5. Variation of the orientation ratio (H_{or}/H) with wall shear rate $\dot{\gamma}_w$. Here, H_{or} is the number of fibers that are oriented parallel to the flow direction and H is the total number of fibers [41]. With permission from John Wiley and Sons Pub.

Viscosity and Primary Normal Stress Difference in Steady Simple Shear Flow

In the steady simple shear flow of a fluid described by Equations 14 and 15, shear stress and the primary normal stress difference are given by

$$P_{12} = P_{21} = \dot{\gamma}(\mu + \mu_2 N^2) + (\mu_0/\dot{\gamma} + \mu_1 n_1 n_2)(n_1 n_2 + \lambda(n_1^2 - n_2^2)\dot{\gamma}/2)/(1 + \lambda^2\dot{\gamma}^2) \tag{26}$$

$$P_{22} - P_{11} = 2\lambda\dot{\gamma}P_{12} + \mu_1 n_1 n_2(n_2^2 - n_1^2)\dot{\gamma} \tag{27}$$

It can be shown that the zero shear rate viscosity is

$$\eta_0 = \lim_{\dot{\gamma}\to 0} P_{12}/\dot{\gamma} = \mu + \mu_0(\alpha_0^2 - 1)/4\xi_0\alpha_0 \tag{28}$$

while the limiting viscosity as $\dot{\gamma} \to \infty$ is

$$\eta_\infty(\ell = +1) = (\mu - \mu_2\beta_0/\beta_1 + \beta_0\lambda(2\mu_0\beta_1 + \mu_1(\alpha_0 - 1))/4\beta_1^2)/\lambda^2\dot{\gamma}^2 \tag{29}$$

$$\eta_\infty(\ell = -1) = (\mu - \mu_2\beta_0/\beta_1 - \beta_0\lambda(2\mu_0\beta_1 + \mu_1(\alpha_0^2 - 1)^{1/2})/4\beta_1^2)\lambda^2\dot{\gamma} \tag{30}$$

Therefore, viscosity at high shear rates decays with $\dot{\gamma}^2$ (property of the base fluid) but the rate of decay depends on the type of orientation as described by $\ell = \pm 1$. When the base fluid is Newtonian ($\lambda = 0$), then

$$\eta_\infty(\ell = +1) = \eta_\infty(\ell = -1) = \mu - \mu_2\beta_0/\beta_1 \tag{31}$$

Therefore, a constant viscosity is obtained which is greater than the viscosity of the base fluid by an amount $-\mu_2\beta_0/\beta_1$. Equation 28 indicates that zero shear rate viscosity, η_0 can be very high, or in fact $\eta_0 \to \infty$ (i.e. yield stress) if, $\xi_0 \to 0$, unless $\mu_0 = 0$.

It can be seen from Equation 27 that even for a Newtonian base fluid ($\lambda = 0$), primary normal stress difference is present. Let us denote this extra normal stress difference (due to microstructure orientation) by ψ, i.e.,

$$\psi = \mu_1(n_2^2 - n_1^2)n_1 n_2\dot{\gamma} \tag{32}$$

In the limiting cases,

$$\lim_{\dot{\gamma}\to 0} \psi = 0 \tag{33}$$

$$\lim_{\dot{\gamma}\to\infty} \psi = -\mu_1(\alpha_0 - 1)\beta_0/2\beta_1^2 \qquad \ell = +1 \tag{34}$$

$$\lim_{\gamma\to\infty} \psi = \mu_1(\alpha_0^2 - 1)^{1/2}\beta_0/2\beta_1^2 \qquad \ell = -1 \tag{35}$$

which indicate the possibility of positive or negative normal stress difference. Since the "degree of orientation" is zero as $\dot{\gamma} \to 0$, then, there is no extra normal stress difference at low rates of shear.

When $\lambda > 0$, then we have (with $\beta_0/\beta_1 < 0$, and $\mu_0 > 0$)

$$\lim_{\dot{\gamma}\to\infty} (P_{22} - P_{11}) = +\ell\mu_0\beta_0/\beta_1 + 2(\mu - \mu_2\beta_0/\beta_1)/\lambda \qquad \ell = \pm 1 \tag{36}$$

Therefore, the sign of the normal stress difference depends on the relative magnitude of the two terms involving ℓ and λ. If λ is very large, then the sign of $P_{22} - P_{11}$ is dictated by ℓ. The significance of this result will be discussed later.

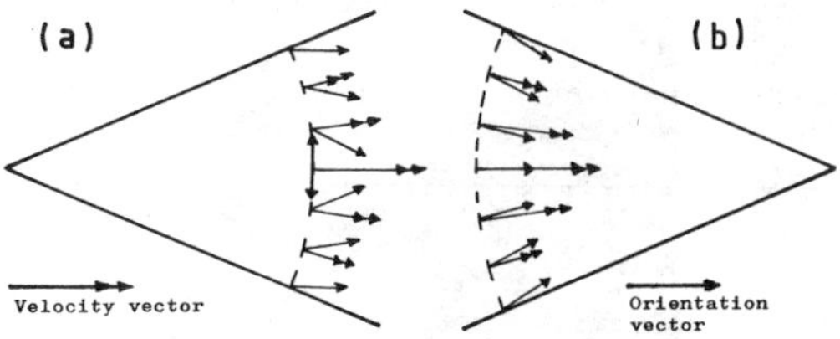

Figure 6. Stable microstructure orientation in divergent (a) and convergent (b) flow between two plates. [27].

Orientation in Flow Between Covergent and Divergent Channels

Leslie [67] has given an exact solution for the flow of a class of anisotropic fluids (described by Equations 14 and 15 with $\lambda = \xi_0 = \beta_0 = \beta_1 = 0$ and $\alpha_0 = -\alpha_1$) in covergent or divergent channels. The presence of two types of orientation is shown. These two types of orientation may be described as parallel (to the direction of flow or at a small angle to the velocity vector) or transverse orientation approximately perpendicular to the velocity vector. By applying the stability analysis of Ericksen [50], Leslie [67] has shown that the parallel orientation represents the stable solution in a convergent flow, and therefore the solution that yields transverse orientation may be disregarded. If, however, the direction of the flow is reversed (so that the flow is in fact divergent) then the transverse orientation represents the stable solution. The variation of the director and velocity across the channel is shown in Figure 6 for convergent and divergent flows. This is an important result since it gives two distinct microstructure orientations and therefore could be verified experimentally as has been in the next section. Any theory that claims to account for microstructure orientation should be checked in convergent as well as in divergent flows. Such flows may be used to determine the material constants associated with the microstructure of the fluid.

Experimental

Orientation in Capillary Flow

Fiber orientation in the extrudate of FRTP's through a capillary has been investigated by Crowson et al. [19] and Knutsson et al. [41] who also evaluated the flow rate dependence of the fibre fraction in which the fibres are oriented along the stream lines, as shown in Figure 5. They found that at low flow rates, the fraction of aligned fibres is low but that fibre orientation increases rapidly when the wall shear rate is above 100 sec^{-1}. Results of Crowson et al. [19] indicate that if the die length is short, fiber orientation is along the flow direction.

Fiber orientation in capillary flow of FRTP's has recently been investigated by Akay [23, 26, 42]. Fiber orientation at the entrance to the capillary is parallel to the direction of the flow, so that all the fibers are aligned parallel to the flow direction as shown in Figures 7a and c. The same fiber orientation is still maintained at the exit of the capillary as shown in Figure 7b. The presence of a skin/core structure in LCP's is illustrated in Figure 7d. There are two reasons for the parallel orientation of fibers in the capillary. Firstly, the capillary length (L/D = 4) is probably not long enough for the establishment of a core region where the fiber orientation is random or perpendicular to the flow direction. This explains why there are more randomly oriented fibers present when L/D = 100 in the experiments of Crowson et al. [19] compared with the case when L/D = 0.3 at the same flow rate. Secondly, if the fibre length is high compared with the capillary radius, a geometrical restriction will be present in fibre orientation. If the capillary diameter is large, transverse or random fibre orientation may be accommodated as shown in Figures 8a–d even if the initial fibre orientation is parallel to the flow direction. However, transverse fibre orientation in Figures 8a and b may be partly due to slight expansion of the capillary.

Such a geometrical restriction will not be present in the flow of LCP's and the development of transverse orientation will be dictated by the length of the flow path and the mobility of the molecules under the influence of the flow field. As seen in Figure 7d, a skin/core structure (in which

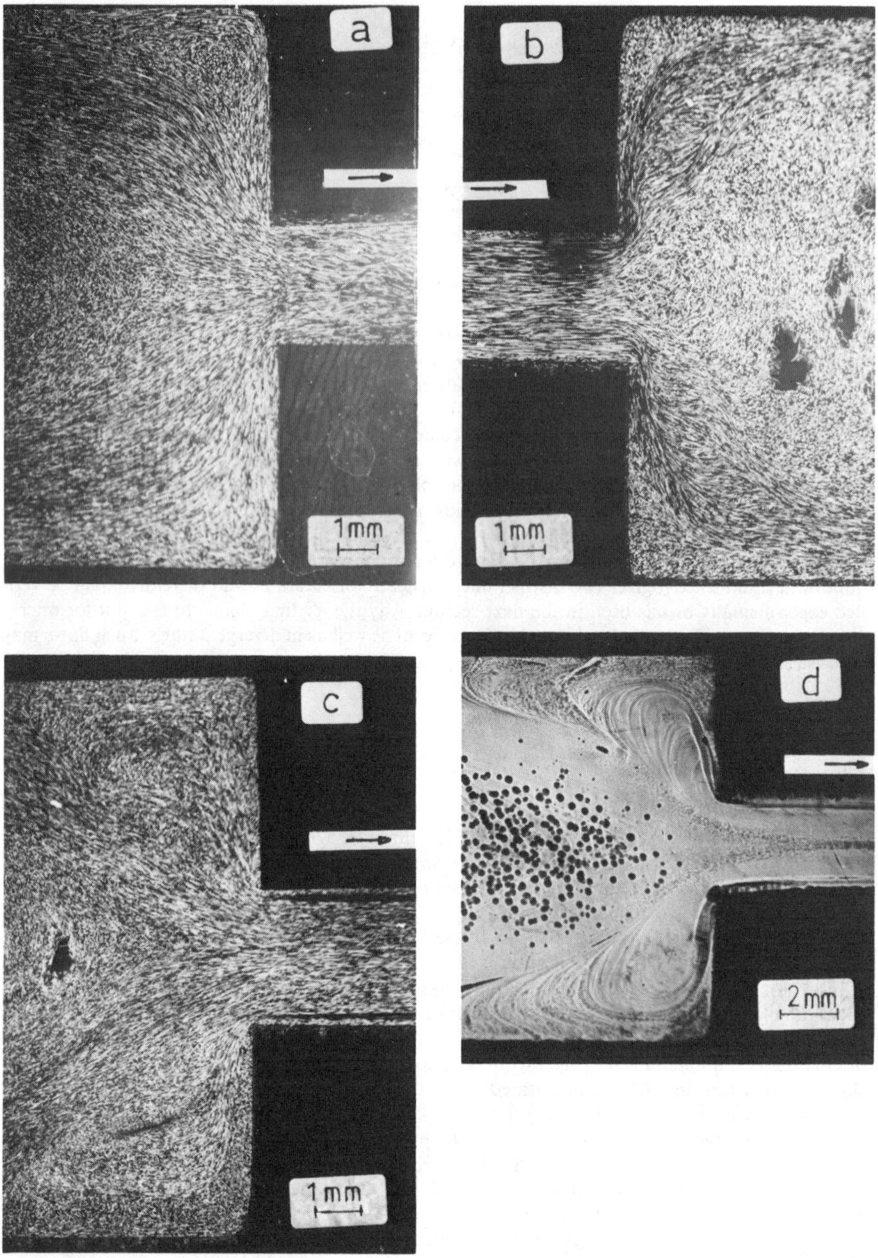

Figure 7. Microstructure orientation at the capillary entrance and exit: (a) capillary entry orientation in 30% glass-fiber-filled polypropylene when $Q_0 = 4$ cc/sec; (b) capillary exit orientation in 30% glass-fiber-filled polypropylene when $Q_0 = 4$ cc/sec; (c) capillary entry orientation in 32% glass-fiber-filled nylon 6.6. when $Q_0 = 20$ cc/sec; (d) capillary entry pattern and orientation in the capillary in liquid crystalline polymer (supplied by ICI) when $Q_0 = 4$ cc/sec [42]. With permission from Elsevier Applied Science Publishers.

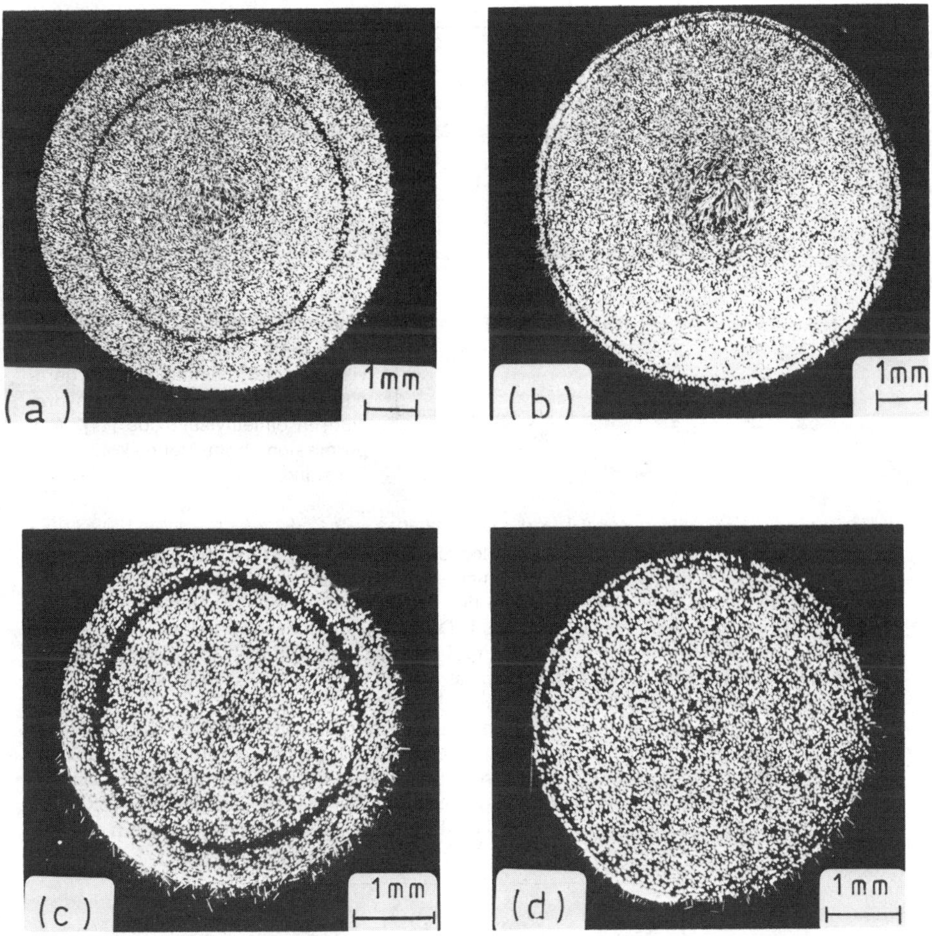

Figure 8. Contact microradiographs of sections cut from the sprue (perpendicular to the main flow direction) of short-glass-fiber-filled polypropylene (30% glass) moldings. Melt temperature at the nozzle is 210°C and mold temperature is 25°C: (a) $Q_0 = 4$ cc/sec, $L_s = 5$ mm; (b) $Q_0 = 160$ cc/sec, $L_s = 5$ mm; (c) $Q_0 = 4$ cc/sec, $L_s = 80$ mm; (d) $Q_0 = 160$ cc/sec, $L_s = 80$ mm. Here L_s is the distance from the gate along the sprue where the slice is cut for CMR. Fibers are aligned in the flow direction (seen as dots) initially ($L_s = 80$ mm) and as the flow develops in the sprue, fibers at the center assume transverse orientation ($L_s = 5$ mm) [23]. With permission from the Society of Plastics Engineers, Inc.

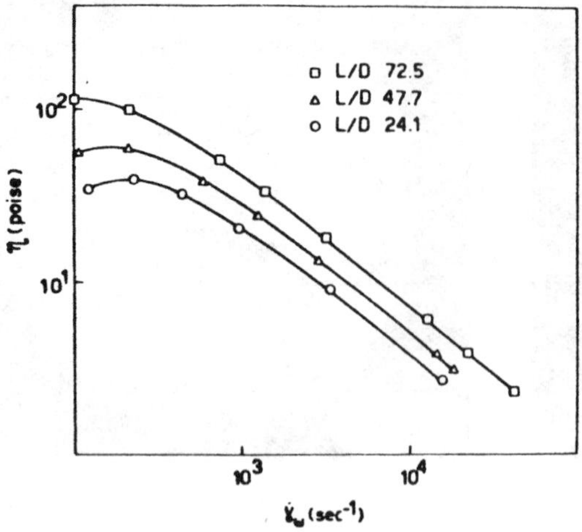

Figure 9. Length to diameter ratio (L/D) dependence of the capillary viscosity of 6% solution of poly(terephthalamide of p-aminobenhydrazide) in dimethylsulfoxide [44]. With permission from John Wiley and Sons, Inc.

the orientation in the skin is parallel to the flow direction and in the core transverse orientation is present) develops immediately. Scanning electron microscopy study of this type of skin/core structure is illustrated in the next section. Support for the developing skin/core structure is also provided by Baird et al. [44] who have shown that the apparent viscosity of solutions of aromatic polyamide—hydrazides were smaller when the L/D ratio is small, as shown in Figure 9. This type of behavior has not been reported for FRTP's, probably due to the fact that the measured viscosity is associated with full orientation of the fibers, as well as macromolecular orientation enhanced by parallel fiber orientation as described earlier [33].

In the example illustrated in Figure 9, the effective viscosity near the capillary wall is low due to molecular orientation, and the L/D ratio dependence of the apparent viscosity is due to development of the core. It is also possible to impose perpendicular orientation at the wall in nematic liquid crystals so that the effective viscosity at the wall is higher than the bulk viscosity [68]. As a result, viscosity increases with decreasing capillary diameter at the same capillary length L and flow

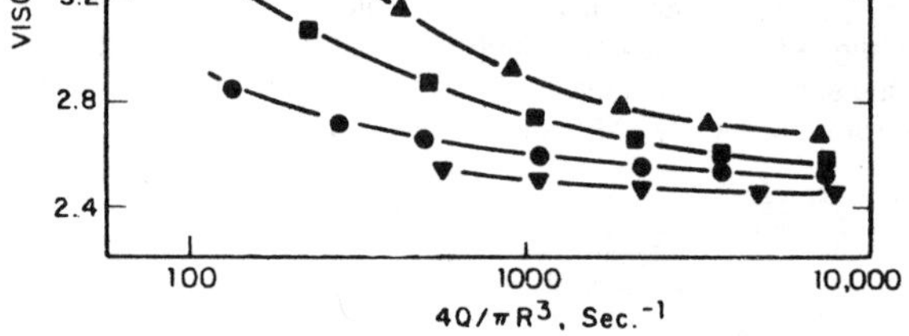

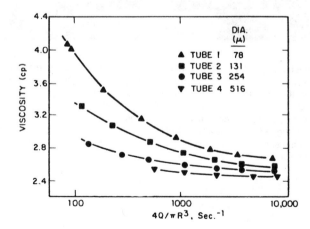

Figure 10. Viscosity behavior of nematic p-azoxyanisole in capillaries surface treated for perpendicular orientation at the boundary [68]. With permission from Gordon and Breach Science Publishers, Inc.

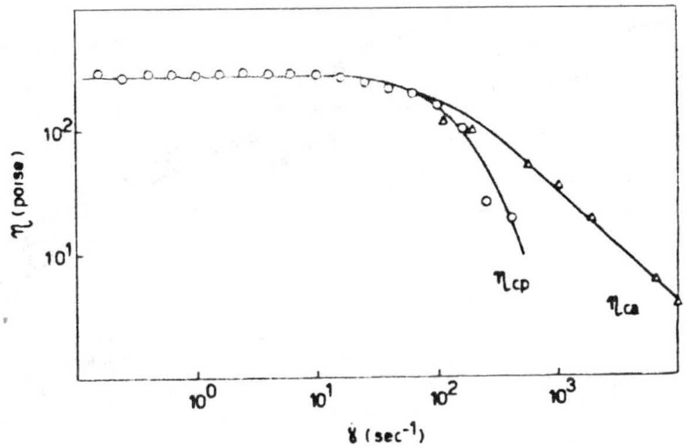

Figure 11. Comparison of the shear-rate dependence of the viscosity of 6% solution of poly(terephthalamide of p-aminobenhydrazide) in dimethylsulfoxide as measured by cone-and-plate (η_{cp}) and capillary (η_{ca}). In the latter case the shear rate is that at the wall (γ_w) [44]. With permission from John Wiley and Sons, Inc.

rate as illustrated in Figure 10. Clearly, enhanced viscosity is a result of small D, rather than large L/D.

Another technique whereby the presence of skin/core structure can be detected is by comparing the viscosities obtained from rotational (using for example cone-and-plate rheometer) and capillary viscometers. Such a study has been done by Baird et al. [44] whose results are reproduced in Figure 11. It is clearly shown that capillary viscometry yields higher viscosity, due to the formation of transverse orientation at the core region of the capillary. It is therefore possible that at high shear rates full orientation is obtained near the capillary wall but transverse orientation at the capillary centre, while due to homogeneous flow in cone-and-plate rheometer all the macromolecules undergo the same orientation. It is important to note that the formation of a reduced viscosity region near the wall is self-enhancing since such a flow will create a highly sheared region near the wall and a very low shear region in the core.

Unlike LCP's in solutions, the capillary viscosity function of FRTP's is consistent with that obtained using a cone-and-plate rheometer. The capillary viscosity function appears to be independent of capillary diameter as shown in Figure 12.

The presence of two solutions resulting in parallel or transverse orientation in steady shear flow may explain the shear thickening behavior of LCP's in simple shear flow. Wissbrun [46] has shown that the viscosity of some LCP melts shows shear thickening, a behavior strongly influenced by melt temperature, as shown in Figure 13. The state of orientation as described by $\ell = \pm 1$ is determined by the stable solution, which, in turn, is determined by the kinematics of the flow and the values of the constants in the equation of state. It is therefore possible that when the transverse orientation represents the stable solution, shear thickening is to be expected.

According to the fluid model described by Equations 14 and 15, negative normal stress difference is possible if transverse orientation represents the stable solution $\ell = -1$ and if the fluid is highly shear thinning and/or has yield stress and parallel orientation represents the stable solution $\ell = +1$. The possibility of transverse orientation causing negative primary normal stress difference has been proposed before [69, 70] and also demonstrated here. The alternative explanation has been summarized by Wissbrun [49], based on the collection of data obtained from different systems. Negative normal stress difference has been observed by Duke and Chapoy [71] in solutions of lecithin, by Wissbrun [46] and Prasadarao et al. [51] in LCP melts by Hutton [72] with lubricating greases and Huang [73] with block copolymers. These fluids had yield stresses or were extremely shear thinning, and in the case of the systems studied in [46, 71, 72] the negative normal stress was

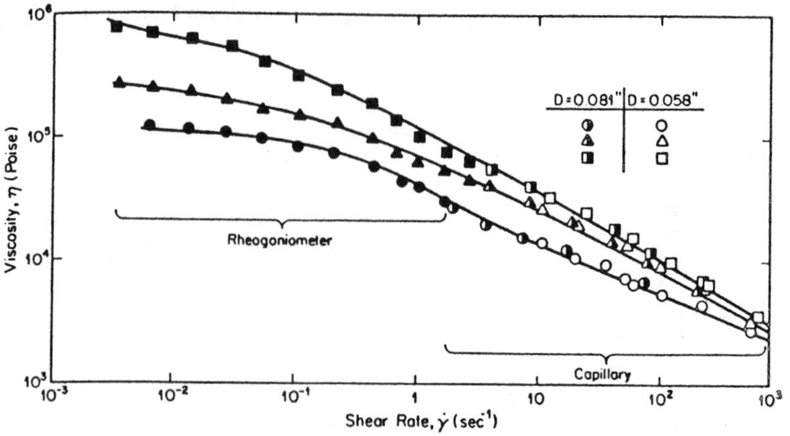

Figure 12. Steady shear viscosity η as a function of shear rate for the HDPE with 0 (circles), 20 (triangles), and 40 (squares) weight percent glass fillers when melt temperature is 180°C. Viscosity is obtained either using a cone-and-plate Weissenberg rheogoniometer (uniform shear rate $\dot{\gamma}$) or an Instron capillary rheometer (non-homogeneous shear rate, in which wall shear rate $\dot{\gamma}_w$ is used for $\dot{\gamma}$) with various L/D ratios (5, 15, and 40) and 0.081 or 0.058 inch capillary diameter [37]. With permission from the Society of Rheology, Inc.

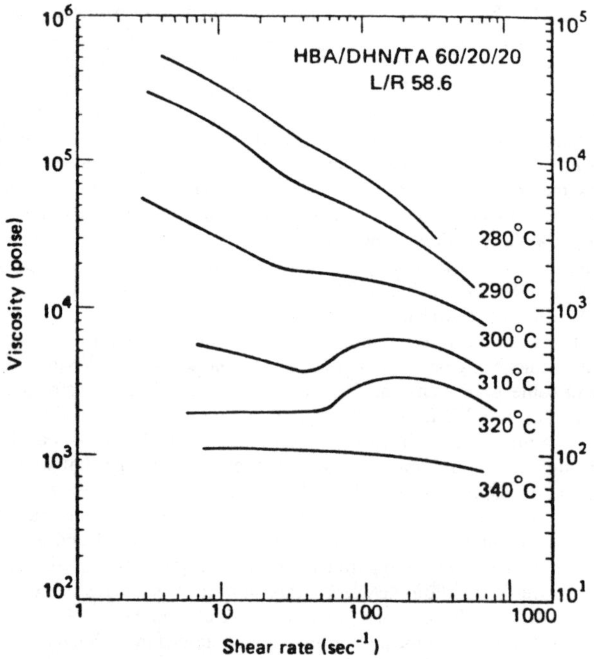

Figure 13. Temperature dependence of viscosity illustrating shear thickening in a copolymer of p-hydroxybenzoic acid, 2,6-dihydroxy-naphthalene, and terephthalic acid [46].

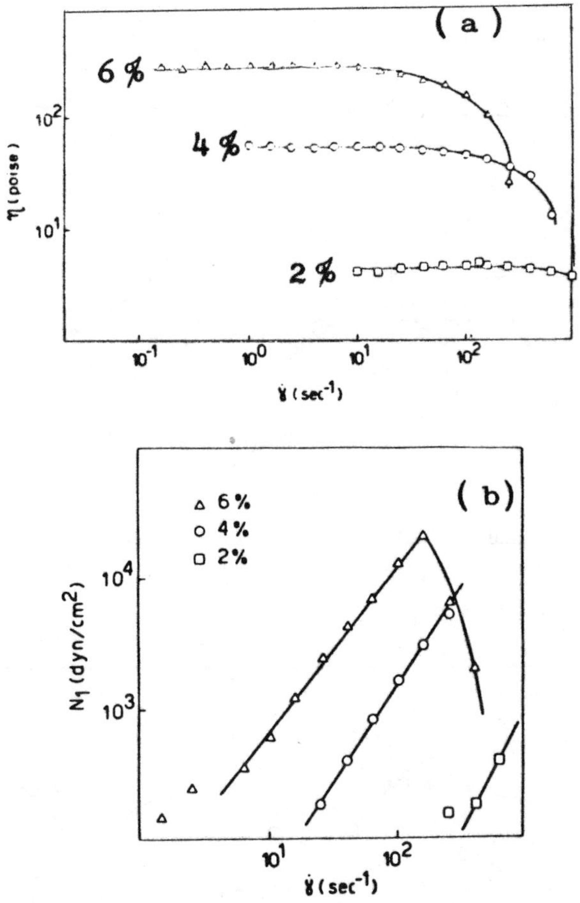

Figure 14. Viscosity (η) (a) and primary normal stress difference (N_1) and (b) of poly(terephthalamide of p-aminobenhydrazide) in dimethylsulfoxide at various concentrations. Note the rapid decrease in N_1 when the viscosity starts decreasing with shear rate [44].

associated with shear history, e.g. reversal of the direction of rotation during the measurements. We also note that primary normal stress difference starts decreasing (after reaching a maximum) when the viscosity decreases rapidly with shear rate [44, 51] in confirmation with the prediction made by Equation 36. These observations are best summarized by the results of Prasadarao et al. [51] and Baird, et al. [44] as illustrated in Figures 14 and 15.

Orientation During Flow Between Parallel Plates

Fiber orientation during flow between parallel plates has been studied extensively [19, 27, 42, 33, 74]. In all of these studies, a skin/core structure develops in which the fiber orientation in the skin is parallel to the flow direction and in the cone, transverse to the flow direction. In this flow

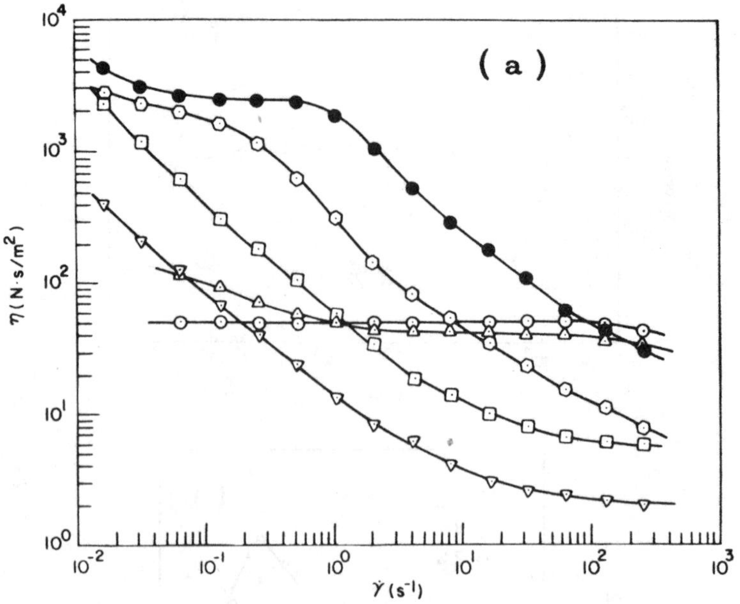

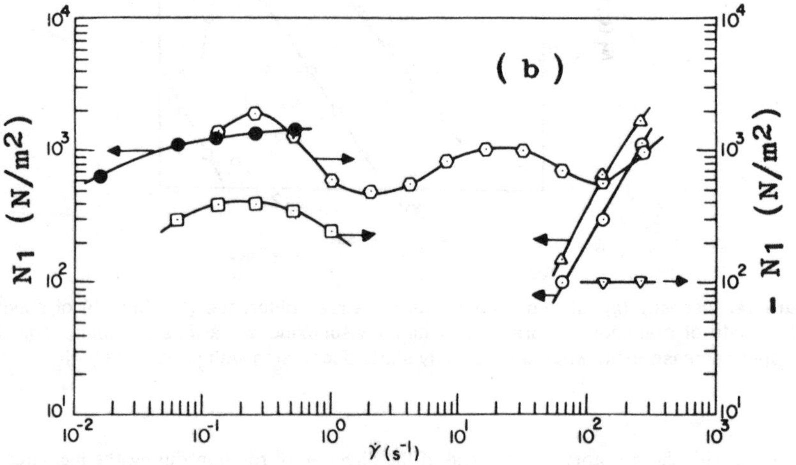

Figure 15. Viscosity (a) and the primary normal stress difference and (b) vs. shear rate for copolyesters of poly(ethylene terephthalete) (PET) with p-acetoxybenzoic acid (PAB), p-hydroquinone diacetate/tetramethylterephthalic acid (HQTA) at various compositions (molar ratio) when temperature is 260°C. (○) PET homopolymer; (△) PET/(PAB + HQTM) = 82/18; (□) PET/(PAB + HQTM) = 54/46; (▽) PET/(PAB + HQTM) = 43/57; (⬡) PET/(PAB + HQTM) = 33/67; (●) PET/(PAB + HQTM) = 25/75 [51].

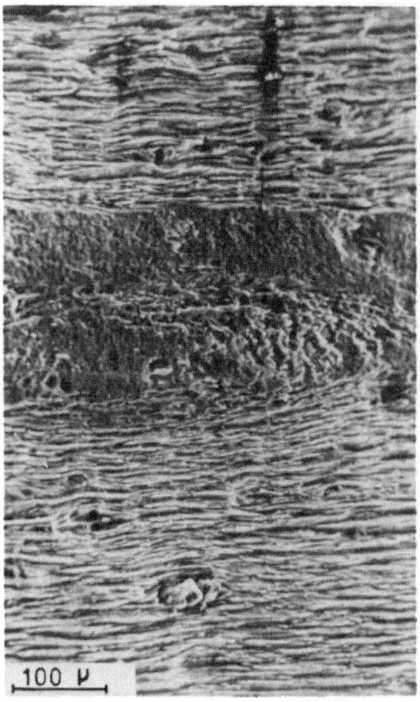

Figure 16. Scanning electron micrograph illustrating molecular orientation in liquid crystalline polymer (a copolyester of phydroxybenzoic acid and 2,6 hydroxynaphthalic acid, supplied by ICI) when melt temperature is 295°C and flow rate is 4 cc/sec [42].

geometry, there is no restriction for the fiber orientation if the width of the flow channel is large. Molecular orientation in LCP's is very similar to fiber orientation, as illustrated in Figure 16.

Orientation in Divergent and Convergent Flows

There have been many investigations of microstructure orientation in convergent flows [27, 75, 76]. It has been concluded that elongational flow is more dominant in orienting fibers compared with shear flows. Fiber orientation in divergent flows has also received some attention using a center-gated circular disc [23–25] as illustrated in Figure 3a, capillary expansion geometry [26, 42] as in Figure 3b and finally using divergent flat plates [27] as shown in Figure 3c. Unlike the microstructure orientation in convergent flow, orientation in divergent flows yields the familiar skin/core structure in which the size of the core can be very large.

Radial divergent flow between two circular discs from a source at the centre of one of the discs has been studied in some detail [23–25]. It is found that the size of the core (transverse orientation) increases with the elasticity of the base material; i.e. by decreasing melt temperature and increasing flow rate. If the melt elasticity is inherently low (i.e. compare nylon 6.6. (P.A.) with polyproplene (PP)) then the fiber orientation is mainly radial [25].

The divergent and convergent flow of anisotropic fluids between flat plates has been recently studied [27], using FRTP's and LCP and HDPE. It is found that HDPE does not yield any large scale molecular orientation, and any orientation is confined to a small region near the walls. Microstructure orientations in FRTP's and LCP are remarkably similar in all cases, microstructure orientation is parallel to the direction of flow in convergent flow while a skin (parallel orientation) and core (transverse orientation) are present in divergent flow. In Figures 17 and 18, microstructure orientation in FRTP (fiber reinforced nylon 6.6.) and LCP in convergent and divergent flows are shown. Typical micrmasctructure orientation in convergent and divergent flows of various FRTP's and LCP is shown diagrammatically in Figure 19. In the micrographs, flow is in the plane of the page, fibers oriented parallel to the flow direction appear as white streaks while those with transverse orientation (perpendicular to the plane of the page) appear as white dots. In the case of LCP, molecular orientation results in the usual fibrillar appearance as shown in Figures 16 and 18. Transverse fiber orientation in PP, HDPE, and PA melts is mainly perpendicular to the plane of the paper

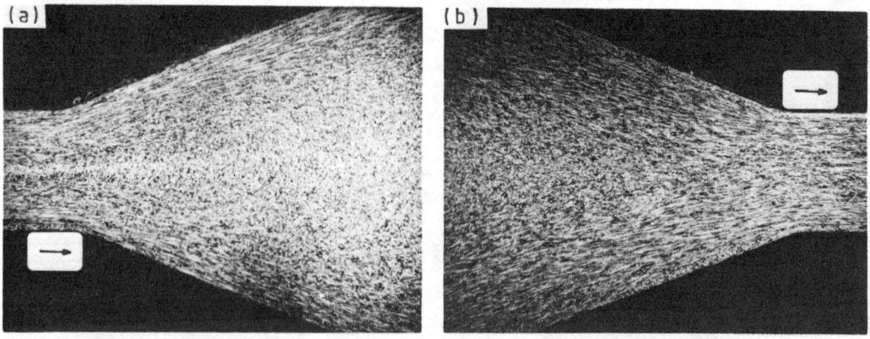

Figure 17. Microstructure orientation in divergent (a) and convergent (b) channel flow of fiber-reinforced nylon 6.6., when $T_N = 290°C$, $T_M = 25°C$ and $Q_0 = 160$ cc/sec [27].

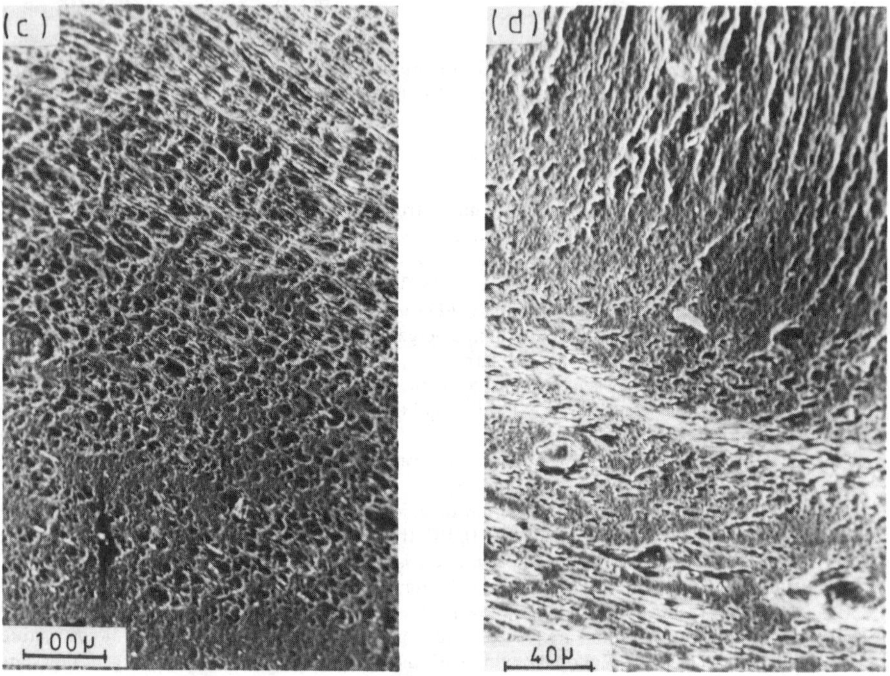

Figure 18. Molecular orientation in liquid crystalline polymer (a copolyester of phydroxybenzoic acid and 2,6 hydroxynaphthalic acid, supplied by ICI) when melt temperature is 295°C and flow rate 160 cc/sec: (a) Divergent flow; (b) Convergent flow; (c) SEM of the area marked as "B;" (d) SEM of the area marked as "C." [27]

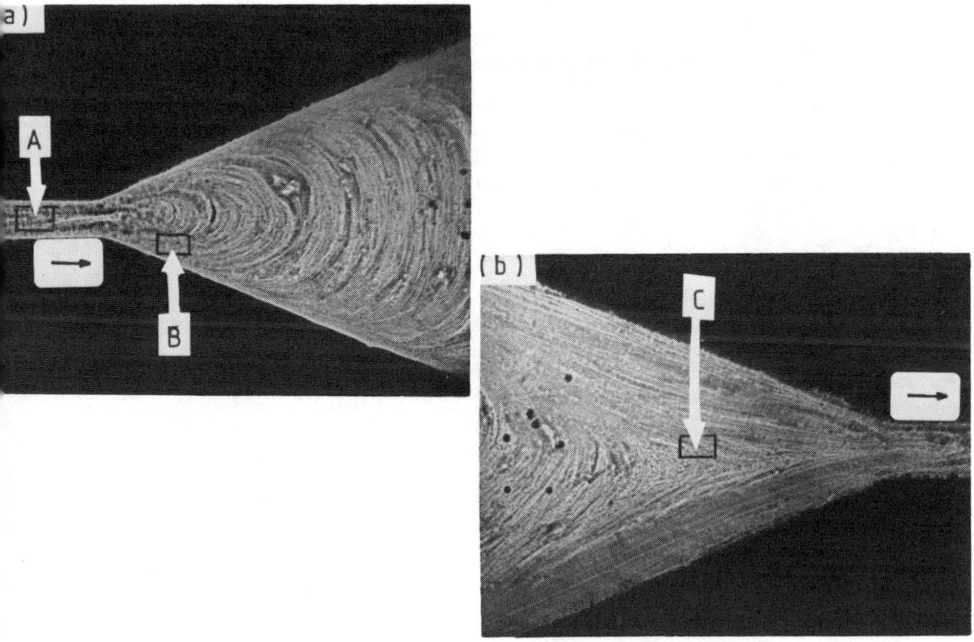

Figure 18. (Continued)

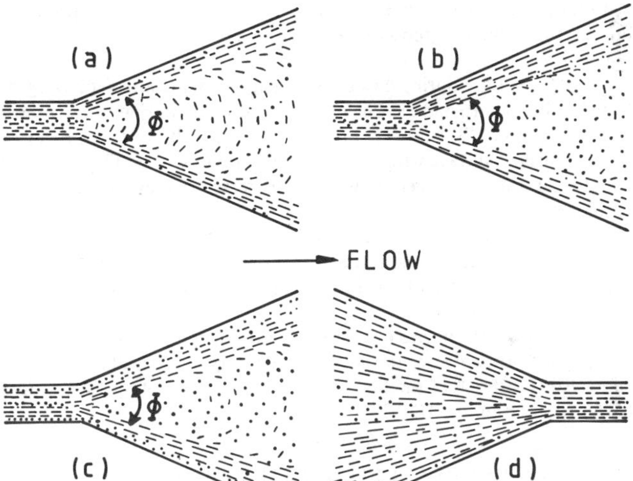

Figure 19. Diagram of the microstructure orientation in divergent and convergent channel flow and the definition of the core angle Φ: (a) divergent flow of fiber-reinforced polystyrene; (b) divergent flow of fiber reinforced polypropylene, nylon 6.6., and high density polyethylene or LCP melts in fast flows; (c) same as in (b) but slow flow; (d) convergent flow of all the materials studied above [27].

Table 2
Variation of Core Angle Φ for Various Materials

Material	T_N(°C)	Q_0(cc/s)	ϕ
30% glass-fiber-reinforced polypropylene	210	4	28°
	210	20	30°
	210	80	38°
	210	160	42°
	270	160	35°
20% glass-fiber-reinforced HDPE	210	4	28°
	210	160	42°
32% glass-fiber-reinforced nylon 6.6	290	4	24°
	290	160	38°
20% glass-fiber-reinforced polystyrene	260	4	28°
	260	160	42°
Liquid crystalline polymer (ICI)	295	20	25°
	295	80	40°
	295	160	42°

while in fiber reinforced PS, a large number of fibers are transversely oriented in the plane of the paper as illustrated diagrammatically in Figure 19a. In all cases, a skin core structure develops in the entry region (parallel plates) to the divergent section as shown in Figures 16, 17a, 18a and 19a, b and c. In the divergent section, the size of the core can be quantified by an angle as defined in Figure 19. The variation of ϕ with input flow rate, Q_0 for various systems is summarized in Table 2. It can be seen that ϕ is smaller for fiber reinforced PA when compared with fiber reinforced PP, HDPE and PS, a result similar to those obtained in divergent radial flow between circular discs, as discussed earlier [25].

Molecular orientation in LCP during flow between parallel plates and in convergent and divergent flow are illustrated in Figures 16 and 18. The region marked "A" in Figure 18a is enlarged in Figure 16 while the regions marked "B" and "C" in Figures 18a and b are enlarged in Figures 18c and d, after etching the polished surface with concentrated sulphuric acid. The development of parallel molecular orientation during convergent flow and transverse orientation in divergent flow are illustrated in these figures.

In convergent flow, complete microstructure orientation is achieved both in FRTP's and LCP. In the uniform section following the convergent section, a skin-core structure is gradually set-up finally resulting in an orientation pattern similar to the structure at the entrance to the divergent section.

Molecular orientation in HDPE has also been studied using the same molds. There is no large-scale orientation present probably due to the partial recovery of molecular coiling after the completion of flow. It is also possible that the flow field-microstructure interaction is not strong enough to effect molecular orientation in HDPE.

FLOW-INDUCED MICROSTRUCTURE REDISTRIBUTION

The so-called "Fahraeus-Lindquist" phenomenon in which the viscosity of blood decreases with decreasing tube diameter in capillary flow has been known for a long time; see for example Goldsmith and Skalak [78]. This effect is a result of red blood cell migration away from the capillary wall where the shear rate is maximum. Since then many examples of flow-induced microstructure "separation," "migration," or "diffusion" have been observed and studied in fluids with microstructure. Reviews of flow-induced microstructure redistribution and its effects on momentum, heat, and mass transfer are available [23, 25, 26, 32, 79–82]. As pointed out by Mashelkar and Dutta [81], the implications of flow-induced microstructure redistribution are enormous; not only a wide range of transport processes are affected [32, 81, 83] but more importantly perhaps, serious doubts are cast on a number of flow techniques used for testing theories that have been proposed to evaluate the apparent transport coefficients based on observation of transport rates in the non-homogeneous flows of structured fluids. We have already given an account of flow-induced microstructure orientation and how it effects the rheological properties of such fluids. In interpreting the rheological measurements of structured fluids it is probably equally important to consider the influence of flow induced microstructure redistribution.

Microstructure redistribution may manifest itself in a variety of ways. In some cases more than one mechanism may be responsible for its manifestation. Various primary mechanisms that are ultimately responsible for flow-induced microstructure redistribution may be best understood by considering the equation of continuity

$$\frac{\partial c}{\partial t} = -(J_{i,i} + v_i c, i) - Kc^{\alpha} \tag{37}$$

where c is the concentration of microstructure (or in fact any identifiable species such as small molecules, macromolecules or particles), K is a "mechanochemical rate constant" and α is the order of the mechanochemical reaction and J_i is the net diffusion flux, which may be written as [84]

$$J_i = J_i^* + J_i^+ \tag{38}$$

Here, J_i^* is the Fickean diffusion flux and J_i^+ is the "flow-induced diffusion flux,"

$$J_i^* = -D_0 c, i \tag{39}$$

$$J_i^+ = -(D_0 c/kT)F, i \tag{40}$$

where D_0 is the Fickean diffusivity, k is Boltzmann's constant, T is the absolute temperature, and F is an entropic potential field responsible for the stress-induced diffusion. As seen from Equation 37, concentration distribution is affected by a diffusion type of process in which the flux is increased as a result of flow and a mechanochemical process in which the reaction is thermally and/or mechanically activated. In the proceeding sections we examine these two processes.

Flow-Induced Diffusion: Theoretical

Tirrell and Malone [84] take the free energy of extension of a Gaussian chain in dilute polymer solutions for this potential field. Since F is of a conformational entropic origin, it will be proportional to the second invariant of the rate of deformation tensor, e_{ij} [80]

$$I = e_{ij}e_{ij} \tag{41}$$

The specific form of F assumed in [80] is

$$F = kTf_0 I^n \tag{42}$$

where f_0 and n are two constants. Janssen [85] argues that, since shear rate is inversely proportional to the local polymer concentration, then F can be written as [85]

$$F = kTf_0/c^2 \tag{43}$$

This type of formulation is valid for concentrated polymer solutions or more significantly for polymer melts. Another important feature of this potential field is that the flow-induced diffusion

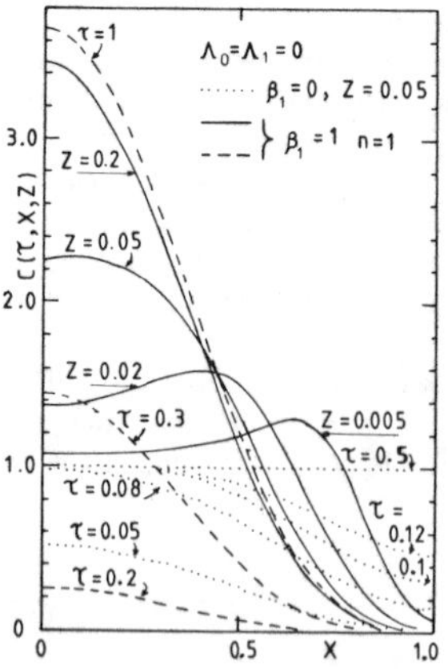

Figure 20. Development of polymer concentration profile during steady capillary flow of viscoelastic fluids in the absence of any reaction and when the flow is effected by the concentration of the macromolecules. The variation of the dimensionless concentration C $(=c/c_0)$ with dimensionless radial position X $(=r/R)$ at various dimensionless times τ $(= tD/R^2)$ and dimensionless axial distance Z $(= zD/\Delta PR^4)$. Dimensionless reaction rate parameters Λ_0 and Λ_1 are defined by $\Lambda_i = K_iR^2c_0^{\alpha-1}/D$ (i = 0, 1), and while β_1^* $(=f_0(v_0/R)^{2n})$ is the dimensionless stress induced diffusion parameter and β_2^* $(=g_0(4/v_0\,\Delta P)^m)$ is the dimensionless stress power parameter and v_0 $(=\Delta PR^2/4\mu)$ is the characteristic velocity [80].

process is self-enhancing. Based on the free energy expression obtained by Marrucci [86] for a dumbbell model, Metzner [87, 88] has shown that the concentration during flow is related to the concentration in the rest state by, c_0, by

$$c_0/c = \exp(P_{ii}/2ckT) \tag{44}$$

In the approach adopted by Tirrell and Malone [84] and Metzner [87] the reduction in the free energy of the fluid due to macromolecular extension and orientation is compensated by the concentration profile set-up in the flow field due to macromolecular migration from high-shear rate regions to low-shear-rate regions. In recent years, molecular descriptions of the polymeric fluids have been widely considered by Aubert et al. [89] and Sekhon et al. [90] have shown that flexible polymers can migrate across the streamlines. Continuum theories of structured fluids, such as those developed by Eringen [91] and Kline [92] also predict radial microstructure migration in nonhomogeneous flow fields.

Development of microstructure concentration profile in capillary flow of viscoelastic fluids in which the flow kinematics are not effected by microstructure concentration is shown in Figure 20. Initially, at the entrance to the capillary, concentration is constant but becomes a function of dimensionless axial (Z) and radial (X) coordinates. Maximum concentration is near the capillary wall when Z is small but gradually shifts towards the capillary center (X = 0) as $Z \to \infty$. This figure indicates that a concentration depleted "slip" layer is present near the capillary and the thickness of this layer increases with increasing axial distance. The significance of such a slip layer in rheology has recently been investigated by Dutta and Mashelkar [93].

Flow-Induced Diffusion: Experimental Results

The flow-induced diffusion process requires a certain flow length and time to develop. The driving force for microstructure diffusion depends on the linearity of polymer chains and the solvent power in polymer solutions while molecular weight distribution and chain type probably influence molecular migration in polymer melts. Although the consequences of this process are easily detected in dilute polymer solutions [93] any such process is probably obscured by thermal or mechanochemical degradation at the capillary wall. However, microstructure migration (glass fibres or filler particles) in polymer melts has been shown to be significant in certain cases. Kubat and Szalanczi [94] and Hegler and Menning [95] have shown that the so-called "Fahraeus effect" is significant in polymer melts filled with large spherical particles. Kubat and Szalanczi [94] used a spiral mold and axial concentration of filler is determined (Figure 21) from the moldings, indicating more pronounced radial migration with increasing filler size.

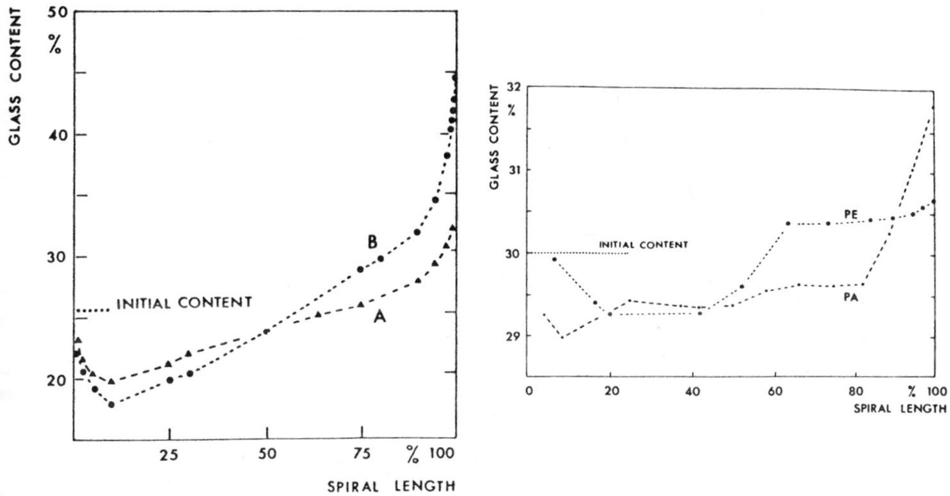

Figure 21. The variation of the filler concentration along the length of the spiral in spiral molding of various filled polymers: (a) low-density polyethylene filled with 25.7% large glass spheres (50–100 μm in diameter) for two spiral lengths; A:7.0 m, and B:1.6 m; (b) high-density polyethylene and nylon 6.6., filled with 30% glass fibers. PE: Polyethylene with spiral length of 0.48 m and PA: nylon 6.6., with spiral length of 1.22 m [94].

In the case of fiber-filled polymers, there are conflicting results on fiber migration, probably due to differences in the experimental conditions and the base fluid. Takono (as reported by Lee and George [96]) and Wu [97] observed significant fiber migration during flow through capillaries and converging channels. Wu [97] reports that, owing to normal stresses and/or the eccentric rotation of fibers, fiber migration is present and the form of it depends on the flow rate. The maximum fiber concentration is at the center of the extrudate, but the minimum fiber concentration is at a relative radial position of $r/R = 0.63$ where R is the capillary radius. This is opposite to the tubular pitch effect where the maximum particle concentration is at this position [98, 99]. Newman and Trementozzi [100] observed increasing inward radial migration with shear rate in glass-bead filled polyethylene melts. More recently, it has been found that (Figure 22) the melt elasticity increases radial migration during flow through channels with multiple expansions and contractions [42].

Microstructure migration at the capillary exit has been demonstrated in hemotology by Karino and Goldsmith [101] who observed that red blood cells migrate from the recirculatory vortex region into the main stream in circular expansion flows, both in steady and pulsatile flow. The size of the cell-free vortex increases with increasing Reynolds number. This type of migration was observed with platelets and with latex spheres of diameters <20 μm. In contrast, the larger latex spheres and aggregates of hardened human red cells with diameter >30 μm remain in the vortex at all Reynolds number. Thus, the vortex in steady or pulsitile flow acts as a region in which large particles and aggregates can accumulate while the smaller particles eventually rejoin the main stream of the suspension.

Capillary entry-flow-induced phase separation in the blends of two incompatible polymers with different viscosities is reported by Kanu and Shaw [102]. They observed that low viscosity component accumulates at the recirculating vortex at the capillary entrance and periodically feeds into the capillary along the wall, forming a skin around the high viscosity component thus causing slip and reducing the apparent viscosity. If the entry angle is reduced, apparent viscosity increases due to the absence of any segregation at the capillary entrance.

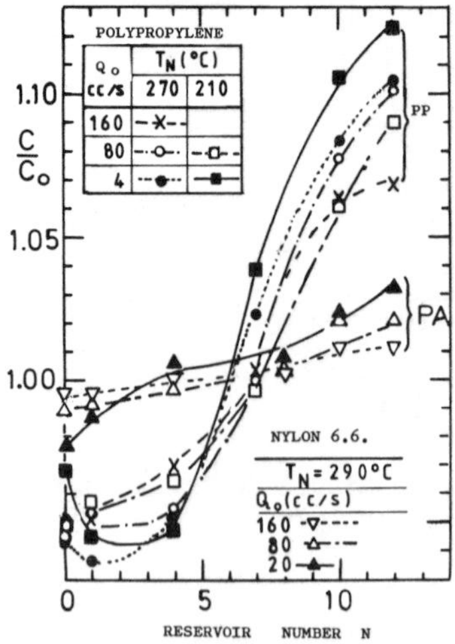

Figure 22. Variation of the dimensionless glass fiber concentration C/C_0 (C_0 is the mean bulk concentration) in the reservoirs in injection moldings of fiber-reinforced polypropylene and nylon 6.6 at various input flow rates (Q_0) [42].

Phase separation (or fiber migration) at the capillary entrance and exit have been recently observed [26, 42]. This type of phase separation appears to be enhanced with increasing melt temperature and shear rate, as a result of which fiber-free regions are created at the entrance (at the corners or at the reservoir) or at the exit reservoir. Since the formation of fiber-free regions create a slip layer, reduction in the capillary entry-exit pressure loss and flow instability may be expected as discussed later.

Mechanochemical Degradation

It is well known that stress distribution on a macromolecule is not uniform both in the solid and liquid state because of the presence of entanglements and macromolecular orientation. Stress concentration either directly causes scission, or thermally (or by radiation) activated chain scission may be accelerated by stress [103–106]. In the solid state, it is easy to separate the thermally and mechanically activated chemical process, while in polymer melts, these two processes occur, in most cases, simultaneously. However, in dilute polymer solutions, polymer degradation is mechanically activated. Mechanically activated macromolecular chain scission in flow is selective and macromolecules rupture mainly near the middle of the chain where the stress is maximum, while thermal degradation is random and therefore changes in the molecular weight, and molecular weight distributions can reveal the dominant type of degradation [103, 107–112]. In polyisobutene [108] stress-induced degradation is dominant while in polystyrene [109] degradation is thermally induced. It is proposed that the shear stress is the rate controlling factor of degradation and a definite relationship between stress and the limiting molecular weight (below which no chain scission occurs) is present [107]. Abbas [110] has shown that shear rate also increases the rate of degradation as a result of viscous heating. Merrill and Leopariat [113] have shown that transient elongational flows are effective in inducing chain scission.

Bestul (see Casale and Porter [103]) has combined the influence of shear stress and shear rate through the introduction of "stress power," which may be defined as volumetric power consumption during flow. The concept of stress power has been generalized [80] by assuming the "stress power" dependent reaction rate constant in Equation 37. The stress power is defined as

$$G = P_{ij}e_{ij} \tag{45}$$

and the reaction rate constant in Equation 37 is written as

$$K = K_0 + K_1 \exp\{-g_0 G^{-m}\} \tag{46}$$

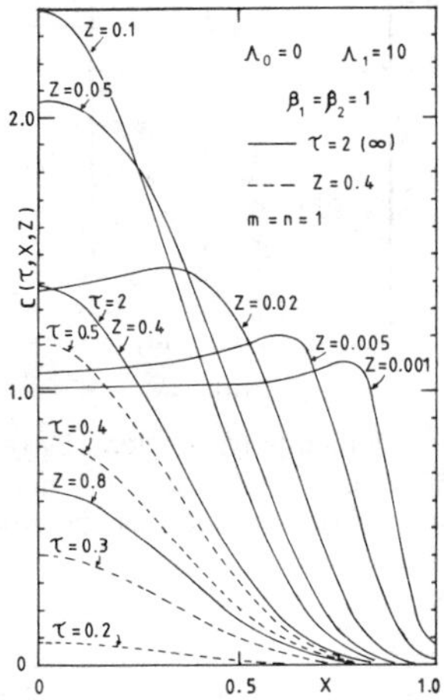

Figure 23. Dispersion with stress-induced first-order reaction ($\alpha = 1$, $\Lambda_0 = 0$, $\Lambda_1 = 10$, $\beta_1^* = \beta_2^* = 1$, $n = m = 1$) during the steady capillary flow of viscoelastic fluids in the presence of stress induced diffusion. The variation of the dimensionless concentration C with dimensionless radial distance X at various dimensionless axial distance Z and dimensionless time τ. See Figure 21 for the definitions of the various dimensionless quantities [80].

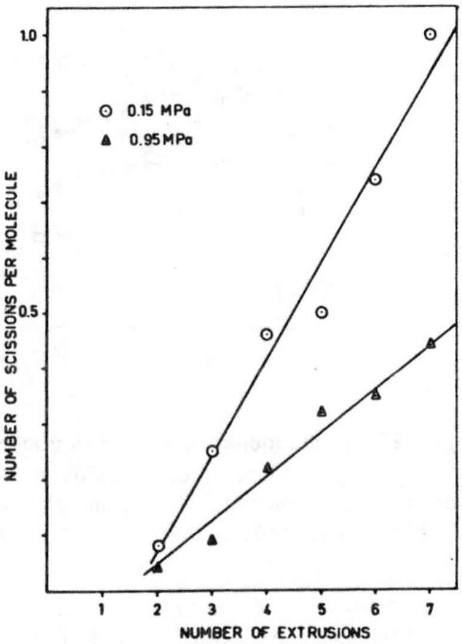

Figure 24. The variation of number of chain scissions per molecule as a function of the number of extrusions in polycarbonate at two constant wall shear stress, $\tau_w = 0.15$ MPa (⊙) and $\tau_w = 0.95$ MPa (△) when the melt temperature is 320°C [110].

where K_0, K_1, g_0, and m are constants. If $K_1 = 0$, then the reaction is thermally activated and K_0 has the usual Arrhenius type of temperature dependence. When $K_0 = 0$ then the process is mechanically activated. Equation 46 may also be interpreted as lowering the activation energy of the thermally activated process in the presence of stress.

Several techniques are available to induce mechanical degradation in macromolecular fluids. In high-shear rotational viscometers, all the fluid elements are subjected to the same stress power, but in capillary flow, degradation is mainly confined to wall regions, as a result of which a slip layer will develop. This slip layer thus creates a plug flow, preventing further degradation. If the macromolecules also undergo flow-induced diffusion, the development of a slip layer will be faster, thus reducing the rate of degradation. In Figure 23, the effect of flow-induced diffusion on the flow induced reaction is illustrated [80].

Here, we are concerned with degradation of polymer melts during repeated extrusion through capillaries. This type of degradation is important if polymer is recycled or if polymer is degraded deliberately in order to obtain certain desired rheological characteristics [112]. Macromolecular chain rupture during degradation may be characterized by the average number of polymer bonds scissioned per original polymer molecule [113] and is defined as $\bar{M}_n^0/\bar{M}_n - 1$ where $\bar{M}_n^0$ is the number average molecular weight of the original polymer and $\bar{M}_n$ is that of the degraded polymer. The effect of wall shear stress on the number of chain scissions per molecule is shown in Figure 24 for polycarbonate. Abbas [111] observed that the presence of filler increases the rate of degradation in polycarbonate due to extra viscous heating. As mentioned before [33], the presence of glass fibers influences the orientation of the macromolecules. It is therefore possible that the rate of degradation should be reduced in fiber-filled polymer melts provided that the polymer is thermally stable. The

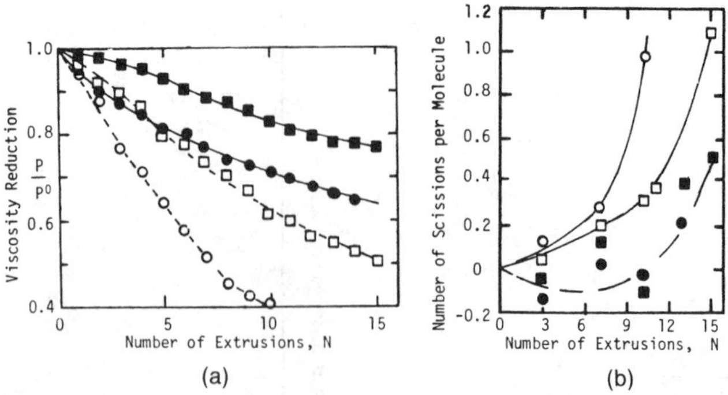

Figure 25. Mechanochemical degradation of glass fiber-reinforced polypropylene by repeated extrusions: (a) reduction of viscosity measured at 1,500 sec^{-1} shear rate; (b) number of chain scissions per molecule; (0) unstabilized base polymer; (□) stabilized base polymer; (●) 20% fiber-reinforced polymer; (■): 30% fiber-reinforced polymer [106].

mechanochemical degradation of glass-fiber-reinforced polypropylene is conducted by repeated extrusion through a long capillary and viscosity is measured using a short capillary as soon as the melt emerges from the first capillary. In Figure 25a the reduction in viscosity (as measured by the reservoir pressure ratio P/P^0, where P^0 is the reservoir pressure for the undegraded melt) is shown as a function of the number of extrusions. Highly filled PP degrades slower than the unfilled grades. The variation of the number of chain scissions per molecule with the number of extrusions indicates that at the initial stages of the process, chain addition, rather than chain scission, takes place, as shown in Figure 25b. This is attributed to the presence of additives, such as siloxane homopolymer and silane coupling agents.

Mechanochemical reactions are influenced by the presence of additives and by the chemical structure (presence of unsaturated end groups) of the polymer. The results illustrated in Figure 25 may in part be attributed to the presence of additives, mainly silane coupling agents used to coat the glass fibers. Riedal and Padget [114] have shown that the presence of reactive end groups may lead to extensive side-chain branching rather than chain scission during degradation in a mixer.

Flow-Induced Crystallization

Ultra-oriented high-strength fibers may be obtained directly from the melt by crystallization during flow through converging dies before any significant chain relaxation occurs. This process requires a certain flow regime, shear rate range, melt temperature, and temperature gradient [115–118]. An important aspect of flow-induced crystallization is that the crystallization induction time is reduced by several orders of magnitude compared with crystallization in the absence of stress fields as shown in Figure 26. Since the presence of fibres effects the molecular orientation, flow-induced crystallization may be enhanced in FRTP's.

In the injection moulding of reinforced polymers, a particle depleted or fiber-free region develops in the moldings. The presence of such regions was observed by Schmidt [119] and Bright et al. [120] and studied in some detail [23–25]. It is proposed that fiber-free (or particle-depleted) regions are the result of fiber (or particle) exclusion from those regions, where macromolecules are undergoing flow-induced orientation. Since this process can occur well above the normal melting point of the polymer, low-temperature rheological properties of the melt may be influenced by flow-induced crystallization and the formation of fiber-free regions that act as slip layers. The development of the fiber-free layers and the regions of different morphology are illustrated in Figures 27 and 28.

Another technique of producing ultra-high modulus fibers is by solid-state extrusion. Although these structures are highly ordered, they contain inter- and intra-fibrillar amorphous domains [104, 105, 121]. In these studies, it is suggested that highly ordered structures are capable of excluding small molecules or additives and pushing them into the amorphous regions.

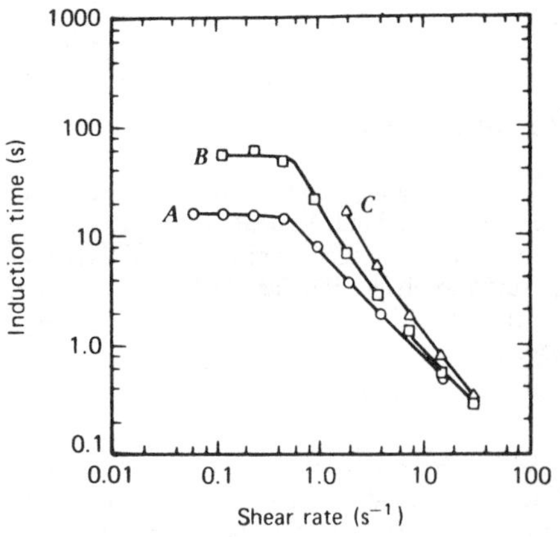

Figure 26. Induction time for flow-induced crystallization as a function of shear rate of high-density polyethylene at various temperatures: (A) 125°C, (B) 126°C, (C) 129°C [116].

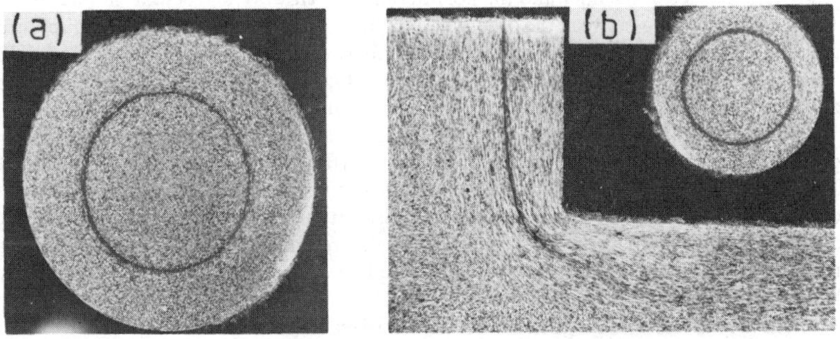

Figure 27. Fiber free regions and fiber orientation at the sprue and at the gate of a center gated disc mold in injection molded reinforced (32% glass) nylon 6.6, when $Q_0 = 4$ cc/sec, $T_M = 270°C$, $T_N = 25°C$: (a) Sprue, $L_s = 5$ mm, (b) Gate and the sprue, $L_s = 80$ mm [25].

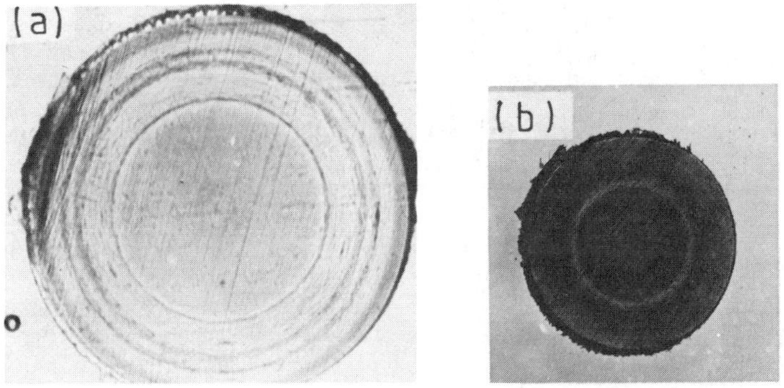

Figure 28. Transmission optical micrographs of the slices cut from the sprue, showing the morphologically different regions as concentric rings when $Q_0 = 4$ cc/sec: (a) unfilled high-density polyethylene when $T_N = 210°C$, $T_M = 25°C$ and $L_S = 5$ mm; (b) unfilled nylon 6.6. when $T_N = 270°C$, $T_M = 25°C$, and $L_s = 80$ mm [25].

INTERFACIAL EFFECTS IN POLYMER MELT RHEOLOGY

In structural fluids, the microstructure is defined by the interfaces that separate constituents of the bulk fluid. In some cases, these interfaces are well defined and independent of time. The nature of these interfaces are important since they can dominate the macroscopic properties of these materials. Here we confine our attention to inorganic surface-polymer interfaces, which also implies possible boundary effects in single-phase polymer melts.

Silane and Titanate Coupling Agents

Coupling agents used in the coating of fillers and reinforcements in polymers have two important functions. One is to improve the processibility of the melt by modifying its rheological properties, and the second is to promote adhesion between matrix and the filler thus facilitating the stress transfer across the phase interface in the final product. In general, coupling agents may be divided into two as "silane" and "titanate" based compounds whose chemical composition allows them to "react" with the surface of the filler and the polymer matrix. However, "reaction" implies the setting up of the primary bonding between the otherwise incompatible phases, although the coupling agent/filler and coupling agent/polymer interaction may only involve secondary bonding. The difference in the chemical structure of the silane and titanate coupling agents is reflected by the type of coating formed on the inorganic surface. In Figure 29, we have illustrated the type of coating obtained by these coupling agents [122].

The effect of coupling agents on the rheological properties of filled polymer melts may be found in [10, 38, 123–130]. In Figure 30, the variation of shear viscosity and the primary normal stress difference with shear stress are shown. In general, viscosity (η and η_e) decreases with the addition of the coupling agent while primary normal stress difference may be unaffected. In the case of calcium carbonate filled HDPE, the presence of coupling agents appears to lower the viscosity of the filled polymer below that of the unfilled material.

The function of the coupling agents can be summarized as follows:

Dispersion of the filler. Sub-micron spherical particles exhibit very strong tendencies towards agglomeration and network formation at high concentrations while particles with more complex geometry (fibers and flakes) tend to interact at lower concentrations [128, 129]. Coating of the

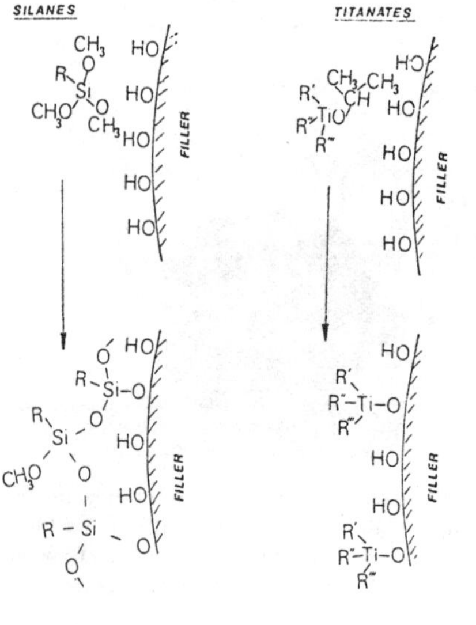

Figure 29. Reaction of silane coupling agent (represented by R-Si$(OCH_3)_3$ and titanate coupling agent (represented by R′R″R‴TiO(CHC_2H_3)) with the filler surface [122].

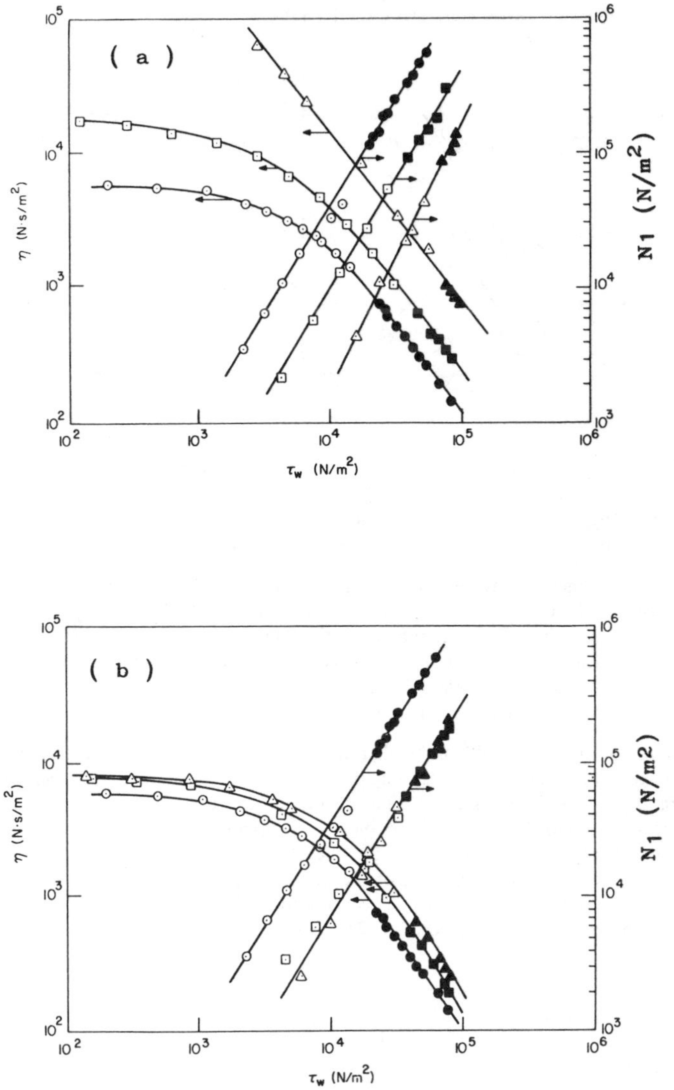

Figure 30. The effect of titanate coupling agent (Kenrich Petrochemicals Inc., TTS) on the viscosity (η) and primary normal stress difference (N_1) of pure or filled polypropylene (PP), and high-density polyethylene (HDPE) melts. Data obtained by a cone-and-plate rheometer (open symbols) and by a slit/capillary rheometer (closed symbols) are consistent with each other: (a)—(0, ●) pure PP; (△, ▲) PP/$CaCO_3$ = 50/50 (by weight), (□, ■) PP/$CaCO_3$ = 50/50 with 1 weight % TTS at 200°C; (b)—(0, ●) pure PP, (△, ▲) PP/glass fiber = 50/50, (□, ■) PP/glass fiber 50/50 with 1 weight % TTS at 200°C; (c)—(0) HDPE (pure), (□) HDPE/$CaCO_3$ = 30/70 with 1% TTS, (▽) HDPE/$CaCO_3$ = 45/55 with 1% TTS, (△) HDPE/$CaCO_3$ = 60/40 with 1% TTS at 240°C. Note that the viscosity of 40% $CaCO_3$ filled HDPE with 1% TTS (△) is lower than that of the base polymer and at high filler loadings melt exhibits yield stress [10].

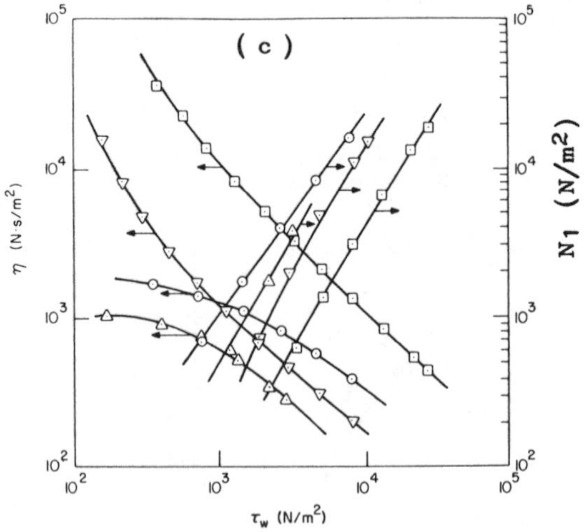

Figure 30 (Continued)

particles prevent agglomeration and results in more uniform dipersion of the particles in the continuous phase, thus reducing viscosity and the yield stress if it is present [128, 129].

Plastization. Presence of low-molecular-weight organic compounds (molecular, oligomeric, or polymeric) reduces the viscosity of the polymer melts below that of the base polymer. In addition to simple plastization, these additives may alter the flow mechanism of the base polymer, as proposed by Andrianova [131]. During processing (compounding stage of polymer, filler and coupling agents) some of the coating from the filler surface may be removed or some coupling agent may form oligomers and subsequently act as plastizer.

Compatability with polymer. When the filler has similar polarity with polymer then polymer may fully or partially wet the filler particles, thus provide a secondary bonding between them. In most applications this type of bonding is sufficient in providing stress transmission across the interface and also results in good dispersion and reduces bulk viscosity due to slip between polymer and filler. In non-polar polymers filled with polar fillers, the adhesion (or compatibility) between the phases is provided by surface treatment of the filler. Surface treatment may result in secondary bonding between the phases or the surface coating may also adhere the filler to the polymer, in which case an increase in viscosity occurs [128, 129].

Macromolecules at Interfaces

A macromolecule would be under a major conformational constraint near a surface. Brunn [132] and Aubert and Tirrell [133] have shown that such a conformational restriction near a wall results in concentration depletion. According to Silberberg [134], the presence of an interface can be seen as a means of producing a phase separation in a polymer system that would otherwise be stable. Such a phase separation may be at solid boundary or may be at filler particle-polymer interface as a result of "bound polymer" [135, 136]. Grubb and Keller [137] have shown that during the shearing of isotactic polystyrene melt between two glass plates, a stationary layer of 15 to 20 μm adhered to the glass. More recently, Burton et al. [138] have shown that torque developed during shearing of polystyrene between parallel plates was dependent on the plate separation, as a result of molecular chain conformation at a solid-melt interface. Gap separation dependence of the torque (and hence viscosity) is more pronounced for high molecular weight polymer as illustrated in Figure 31. This effect is explained by either assuming a stationery absorbed polymer layer at the solid interface, thereby reducing the amount of polymer being sheared, or by assuming molecular orientation at the wall, thus reducing the bulk viscosity by providing a slip layer. However, molecular orientation is present in either case, but whether it is immediately near the solid surface or adjacent

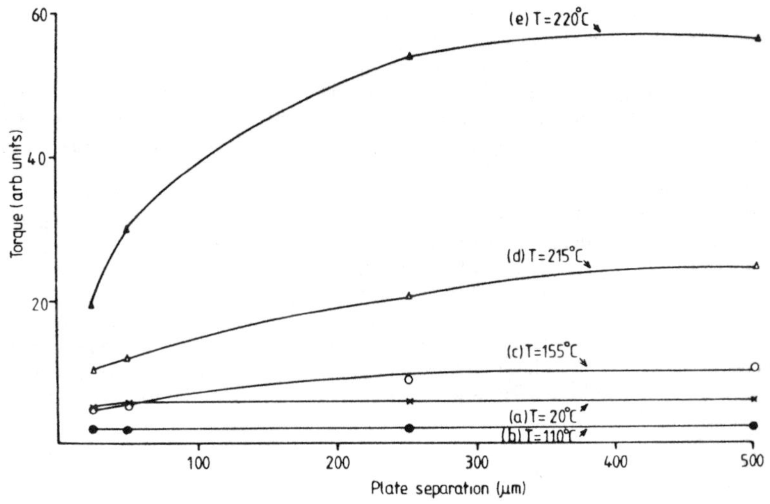

Figure 31. The gap effect on monodisperse polystyrene melts of varying molecular weights and a silicone fluid. Average torque against plate separation for ten minutes shearing. The speed of the rotating plate was adjusted to maintain a constant shear rate of 1 sec^{-1} for all plate separations: (a) silicone fluid; (b) atactic polystyrene (a-PS) with molecular weight ($\bar{M}_w$) = 1.5×10^3; (c) monodisperse a-PS with $M_w = 32 \times 10^3$; (d) monodisperse a-PS with $\bar{M}_w = 470 \times 10^3$; (e) monodisperse a-Ps with $\bar{M}_w = 970 \times 10^3$. [138].

to the immobilized polymer layer at the wall, is yet to be resolved. Nevertheless, it clearly indicates the importance of interaction between solid surfaces and macromolecules.

TIME-DEPENDENT FLOWS

Constitutive Equations and Time-Dependent Viscometric Flows

Polymer processing often involves large deformation rates and highly time-dependent deformation histories, which result in time-dependent rheological properties, thus adding to the complexity of the processing technology. Understanding and being able to predict the time dependence of the rheological properties can therefore be invaluable and may lead to new process routes.

With the advent of sophisticated instrumentation, and the necessity for more detailed characterization of the flow properties of structured fluids, time-dependent viscometric flows are likely to be more widely used. In recent years, some very important advances have been made in molecular modelling of polymeric fluids. Most of these models are successful in predicting steady flow properties and time-dependent flows can provide a more critical test for these constitutive equations. In this section we briefly summarize the recent advances in modelling of polymer solutions and simple phase polymer melts [139].

Until recently, the only molecular theories for polymer solutions and polymer melts were the "network theories" where the microstructure is characterized by the physical entanglements of the macromolecular chain. Recently, network theories have further advanced to account for time-dependent and non-linear viscoelastic properties of polymeric fluids. Soong and Shen [140, 141] have developed a kinetic network model in which flow-induced structure variation is controlled by the simultaneous existence of entanglement loss and regeneration. This dynamic equilibrium between the entanglement loss and generation results in non-linear visoelastic behavior in steady flow. The transient network concept is also used by Jongchaap [142] to develop a constitutive equation that reduces to those previously derived by Acierno et al. [143] and Phan-Thien and Tanner [144].

A totally different molecular modelling has been adopted by Doi and Edwards [145, 146] in their "tube model" for describing the dynamics of polymeric fluids. They assumed that entangled macromolecules rearrange their conformations by reptation; i.e., diffusion along their own contours, a concept introduced by de Gennes [147]. Adopting equations from the theory of rubber elasticity, they calculated the contribution of individual chains to the stress following a step strain and related the subsequent relaxation of stress to conformational rearrangement and they have subsequently extended the analysis to general deformations, arriving at constitutive equations of Kaye-BKZ type [148, 149]. An important aspect of the Doi-Edwards model is that it can predict the molecular weight dependence of the rheological properties. Curtiss and Bird [150] have generalized the Doi-Edwards model to cover the full polymer concentration range. These constitutive equations have been reviewed by Bird [151]. Graessley [152] has analyzed the Doi-Edwards model critically, and proposed a new model that better predicts the elasticity of entangled networks.

Currie [153] has shown that the model possesses a potential function that completely determines the strain-dependent memory fading, both in shear and extension, and is dependent on the first and second invariants of the Finger strain tensor. He concludes that differences between theory and experiments are due to differences in the relative importance of the two invariants between one material to another.

Transient Flows

Most of the transient flow experiments are conducted using a rotational type of rheometer in which the flow is homogeneous and therefore the shear history of all the fluid elements are the same. A large proportion of these transient experiments [154–159] deal with concentrated polymer solutions. The well known non-linear response of a viscoelastic fluid to a step increase in shear rate is illustrated in Figure 32, where the ratio of transient viscosity $\eta^+(t, \dot{\gamma})$ to steady shear viscosity $\eta(t \rightarrow \infty, \dot{\gamma})$ is plotted against the time t. In this figure, predictions based on the kinetic network model [140, 141] are also shown to be in excellent agreement with experiments [157]. The stress overshoot at high shear rates is a result of the inability of the polymer chains to respond (disentanglement) sufficiently fast to the imposed exitation. As discussed later, the amount of overshoot depends on the initial microstructural state; if the maximum entanglement density is present before the start of flow, then the maximum stress overshoot is obtained. Transient viscometric flow of unfilled and particle filled polymer melts has recently been investigated by Suetsugo and White [159].

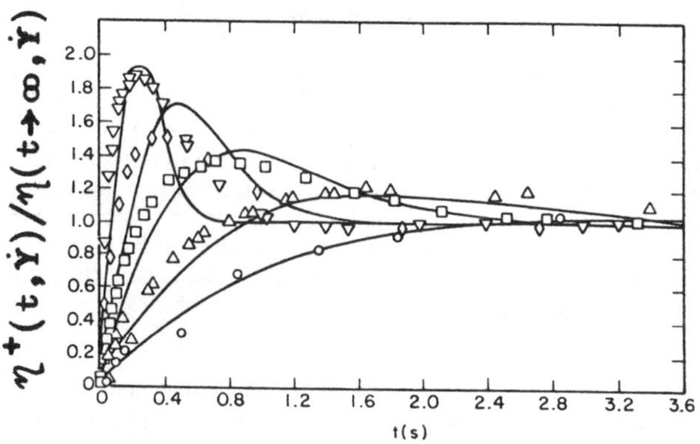

Figure 32. Decay of the transient viscosity η^+ (t, $\dot{\gamma}$) and steady viscosity $\eta(t \rightarrow \infty, \dot{\gamma})$ ratio with time for 4% solution of monodisperse polystyrene in aroclor at different shear rates (sec^{-1}): (0) 0.528; (△) 1.67; (□) 5.28; ( ) 16.7; (▽) 52.8. Solid curves are computed using the transient network model, and symbols are experimental data [157].

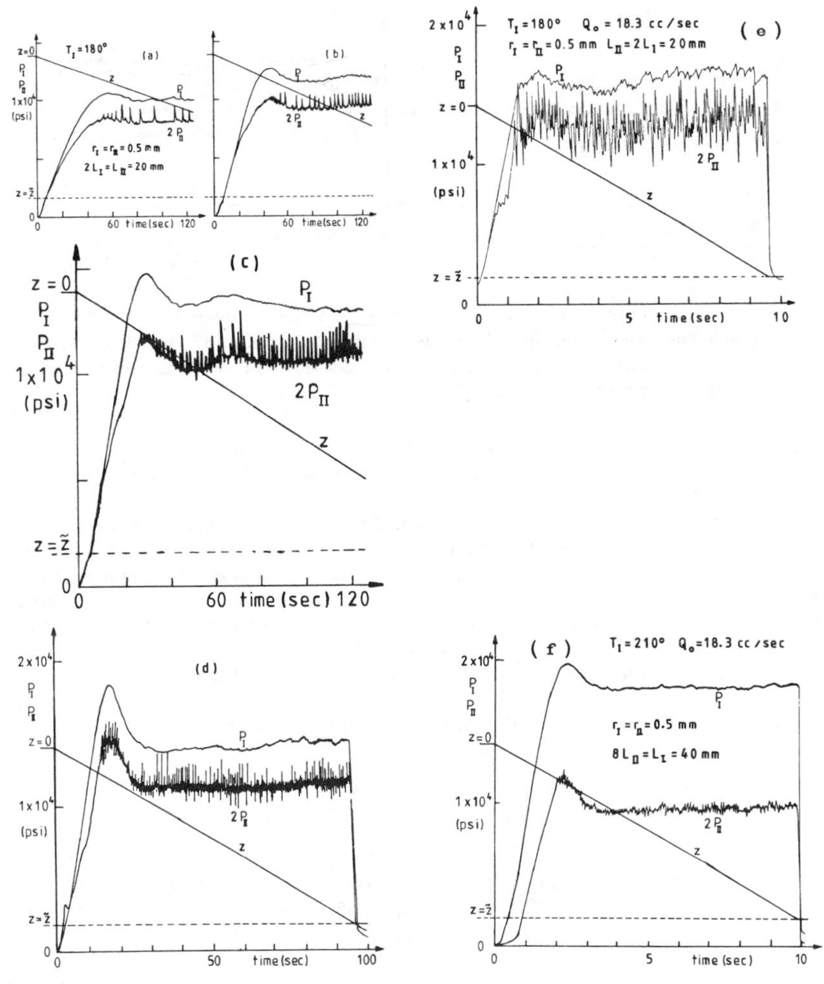

Figure 33. Development of the reservoir pressures P_I and P_{II} with time during steady injection capillary flow of unfilled and filled polypropylene (PP) melts through 1 mm-diameter capillaries in series. (a, b, c, d): 40% (by weight) calcium carbonate filled PP at 180°C. Input flow rates, Q_0, are (in cc/sec): (a) 0.40, (b) 0.66, (c) 1.15, (d) 1.80. (e) Pure PP at 180°C, $Q_0 = 18.3$ cc/sec. (f) 30% (by weight) glass-fiber-filled PP at 210°C when the resistance to flow is increased by increasing the capillary length [23].

The transient injection capillary flow of polymer melts has been studied recently [23, 25, 26, 42]. In these experiments, flow is induced by the displacement of a plunger and the pressure is recorded in the reservoirs(s), using either a single capillary, two capillaries (Figure 4) or five capillaries in series. It is found that unfilled PP and filled nylon 6.6. melts do not yield any pressure overshoot even at low melt temperatures, while glass fibre or calcium carbonate filled PP or unfilled nylon 6.6. yield reservoir pressure overshoot, at high injection rates (input flow rate) Q_0. Figures 33–35 illustrate the development of the pressures for various melts. In all cases, similar to the

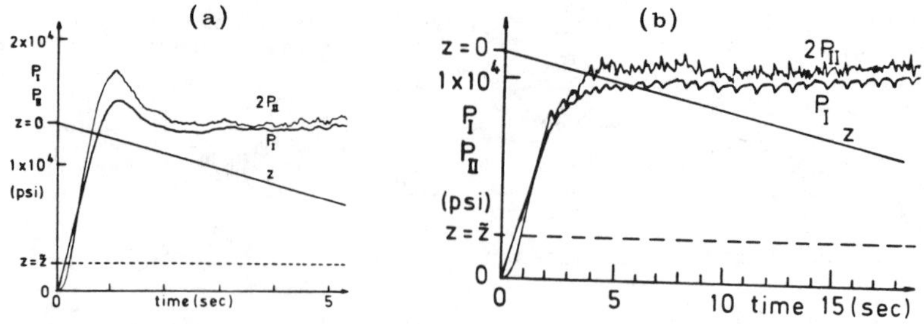

Figure 34. Development of the reservoir pressures P_I and P_{II} during steady injection capillary flow of: (a) unfilled nylon 6.6., when input flow rate (Q_0) is 18.3 cc/sec, and $T_I = 270°C$; and (b) 32% (by weight) glass-fiber-filled nylon 6.6., when $Q_0 = 3.85$ cc/sec and $T_I = 300°C$. Two capillaries in series are used as in the previous figure [25].

development of shear stress in rotational rheometers, reservoir pressure increases monotonically at small input flow rates. At large flow rates, pressure overshoot may occur depending on the fluid and flow conditions.

Except in unfilled nylon 6.6., pressure overshoot disappears with increasing temperature and decreasing input flow rate. In nylon 6.6. pressure overshoot is confined to a narrow range of input flow rate for a given capillary assembly. Pressure overshoot becomes more pronounced with increasing resistance to flow, which can be divided into three regions; entrance, capillary, and exit. Since the fluid in the reservoir is at rest initially, a stress build up takes place before the start of flow and before the flow rate in the capillary is equal to Q_0. This is particularly true for fluids with yield stress, which is probably why calcium carbonate filled PP yield the most marked pressure overshoot, as seen in Figure 33d. Since the yield stress decreases rapidly with increasing temperature, very small pressure overshoot is present if the melt temperature is increased from 180°C to 210°C.

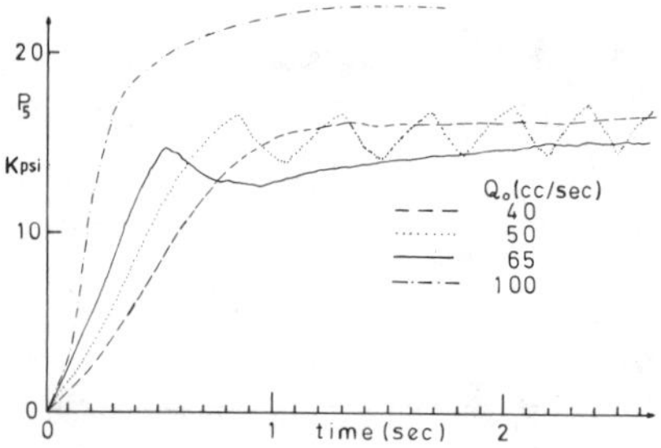

Figure 35. Development of the reservoir pressure (P_5) with input flow rate (Q_0) during transient and steady state flow of 30% glass-fiber-filled polypropylene through 1 mm-diameter five capillaries in series when the temperature in the first reservoir is $T_I = 300°C$ [26].

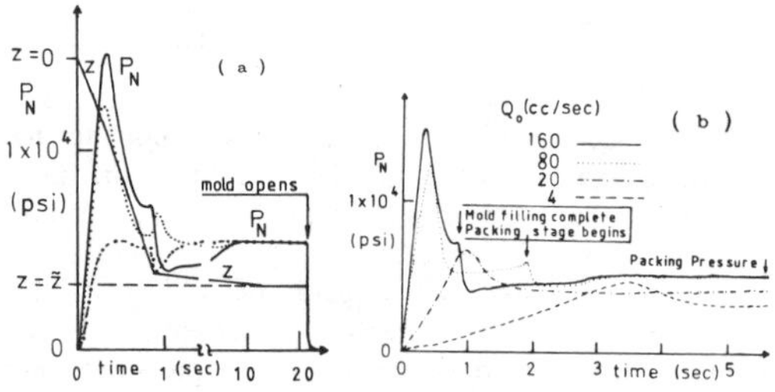

Figure 36. Development of the nozzle pressure P_N during mold filling for various polypropylene (PP) melts at different input flow rates (Q_0) when the melt temperature at the nozzle (T_N) is 210°C and the mold temperature (T_M) is 25°C: (a) the effect of filler when $Q_0 = 160$ cc/sec, (—·—) unfilled PP, (······) 30% glass-fiber-filled PP, (———) 60% calcium carbonate filled PP: (b) the effect of input flow rate when the material is 60% calcium carbonate filled PP [23].

Flow-induced degradation is probably responsible for the unusual behavior of unfilled nylon 6.6., in which pressure overshoot is confined to a narrow shear rate range. Molecular degradation results in lowering the resistance to the flow.

An important consequence of pressure overshoot is encountered in injection molding. At high injection rates, low melt and mold temperatures, a large reservoir (in this case nozzle) pressure overshoot is observed as shown in Figure 36. The amount of overshoot is reduced with low injection rate, increased melt and mold temperatures and with decreasing filler concentration.

Programmed Flows

Small and large amplitude oscillatory shear flows of polymeric fluids are well known. In recent years, another type of programmed flow has been widely used to characterize the time-dependent microstructure parameters of the fluids [23, 155–159]. In these studies, impulsively imposed shear rate is kept for a predetermined period and then lowered (or removed) for a certain length of time and finally re-imposed and the response of the fluid being monitored throughout. These experiments are useful in determining the characteristic structure break-down (i.e. disentanglement) and re-structuring (i.e. entanglement) times. In structured fluids, re-structuring time can be very long when compared with stress relaxation time. This implies that the isotropy of stress is not a sufficient condition for the fluid to be in its equilibrium state [156]. This is an important conclusion as most constitutive models are built on the assumption that the isotropy of the stress implies that the material is in its rest state. In the anisotropic fluid model with isotropic stress tensor, the rest state is characterized by the initial conditions on the orientation vector and the stress can be zero independent of the state of orientation.

The interrupted shear experiments of Dealy and Tsang [155, 156] are shown in Figure 37. Initially, the rest period t_r is infinity and maximum shear stress to a step change in shear rate is $\tau_m(\infty)$. After a fixed period of shearing, the fluid is allowed to rest and then shear flow is restarted. The variation of maximum shear stress $\tau_m(t_r)$ is shown in Figure 37b. Dealy and Tsang [155] have also given estimates of the relaxation time for shear stress and primary normal stress difference and found that the re-entanglement time was the longest followed by normal stress relaxation time. These results may also explain transient negative normal stress difference observed in LCP's [46] if the rates of relaxation of normal stresses are also different. Dealy and Tsang [155, 156] found that the model proposed by Acierno et al. [143] and Phan-Thien and Tanner [144] both predicted characteristic re-entanglement times much smaller than observed values.

The programmed injection capillary flow of PP based filled and unfilled melts was studied using a two-stage capillary assembly [23]. In this case, input flow rate Q_0 jumps from a minimum flow

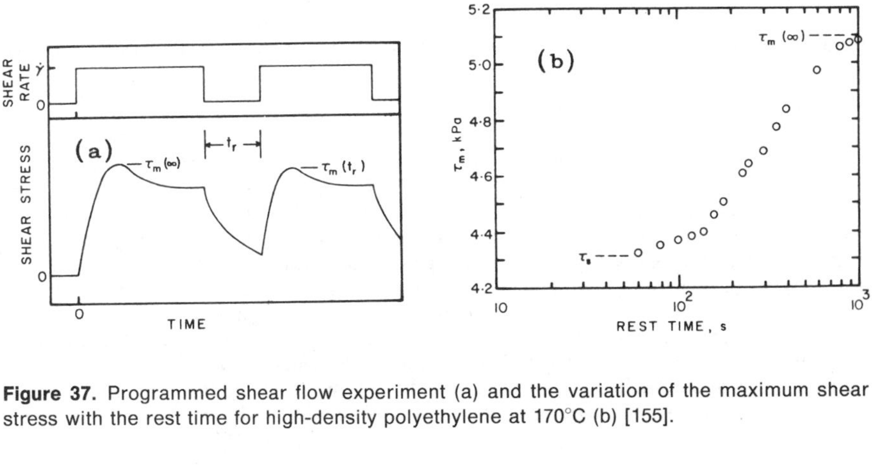

Figure 37. Programmed shear flow experiment (a) and the variation of the maximum shear stress with the rest time for high-density polyethylene at 170°C (b) [155].

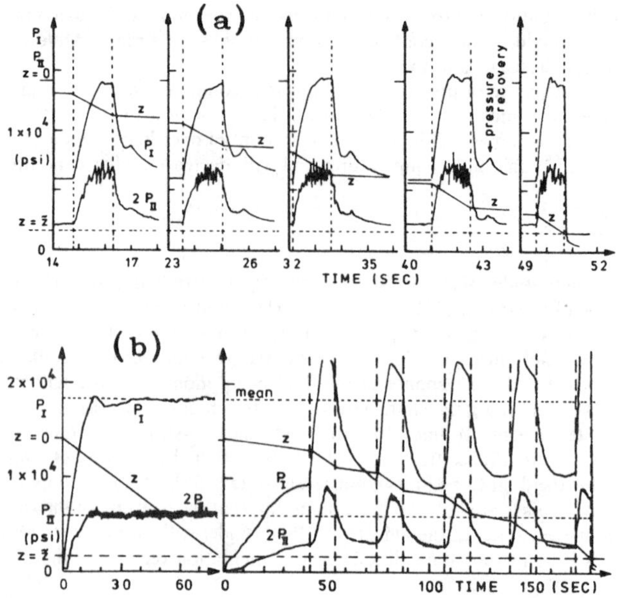

Figure 38. Programmed injection capillary flow of polypropylene (PP) melts at 210°C, through 1 mm-diameter capillaries in series in which the reservoir pressures P_I and P_{II} are recorded when the input flow rate changes periodically between Q_0(min) and Q_0(max): (a) Unfilled PP when Q_0(min) = 2.29 cc/sec and Q_0(max) = 18.3 cc/sec. (b) 60 percent calcium carbonate filled PP when Q_0(min) = 0.40 cc/sec and Q_0(max) = 2.29 cc/sec. Also shown in this figure is the development of the reservoir pressures when there is no flow initially [23].

rate Q_0 (min) to a maximum flow rate Q_0 (max) periodically. Results shown in Figure 38 indicate that there is a recovery of pressure after the reduction of flow rate from Q_0 (max) to Q_0 (min) in the unfilled polymer while this recovery is small in fiber-filled or particle-filled melts. The presence of an initial flow (Q_0 (min)) does not effect the maximum pressure reached in the reservoir except in calcium carbonate filled melt. Pressure recovery in unfilled PP is due to the relaxation of polymer chains following stretching and orientation in the capillary. As a result of particle-polymer chain interaction, such a recovery is not present (or is a long process) in filled polymer melts.

Unstable Flows

Instability during a capillary flow of unfilled polymer melts has been studied extensively and excellent periodical reviews are available [2, 160–167]. During the flow of polymeric fluids from a cylindrical reservoir into a capillary, a characteristic flow pattern at the capillary entrance develops. In general, this flow pattern [164] and also the type of instability [168, 169] for various polymers may be broadly classified as either typical of low density polyethylene or typical of HDPE. During flat entry capillary flow at low flow rates, the streamlines move radially into the orifice in a manner similar to Newtonian fluids in laminar flow. At high flow rates, the entry flow pattern changes gradually until finally flow become unstable, as manifested by several phenomena that depend on the type of flow exitation and the material. Recently, a new type of flow instability has been observed for FRTP's in which the flow through five capillaries in series, separated by small reservoirs [26]. The important aspect of this instability is that its manifestations are also shown by a number of polymeric fluids under various flow conditions. We therefore use the characteristics of this instability to illustrate the unstable flows in polymeric systems.

Capillary Entry Flow Pattern

During the flat entry capillary flow of some polymers such as LDPE, a "flow cone" develops at the capillary entrance before the start of unstable flow. The flow cone is characterized by an entry angle ϕ as shown in Figure 2, and is surrounded by a circulating stagnant region, which becomes unstable with increasing flow rate and eventually fractures periodically as the flow rate is increased even further. During fracture of the flow cone, flow is sustained from the stagnant region and "melt fracture" occurs. We refer to this type of flow as "primary instability." Unlike LDPE, primary instability in HDPE is not accompanied by a marked region of circulating flow, although there is some evidence of periodic alternating flow as a result of slip between two flow regions in the reservoir [162, 164, 170, 171].

A mechanism by which vortices are formed is proposed by White and Kondo [166]. From a critical review of the experiments, it is concluded that the vortices are induced by the viscoelastic characteristics of the converging fluid; if fluid can develop large elongational viscosity (LDPE and PS) the balance of forces makes radial flow impossible and secondary flows appear as means of stress relief. The correlation between vortex size and the elongational viscosity has been observed by Cogswell [172] who noted that polymer melts exhibiting large and increasing elongational viscosities exhibit large vortices and those with constant or decreasing elongational viscosity do not.

Ballenger and White [173, 174] have proposed that the die entry angle is determined by the Weissenber Number, We, a ratio of viscoelastic to viscous forces defined as We = $N_1/2\tau_\omega$ evaluated at the capillary wall. Since, for single phase polymer melts the primary normal stress difference, N_1, is related to the entrance pressure loss [166], then Ballenger and White [173] were able to correlate the capillary entry angle ϕ with the dimensionless capillary entrance pressure loss $\Delta P_{ent}/\tau_\omega$. Their correlation is shown in Figure 39. It can be seen from this figure that the entry angle ϕ for PP and HDPE is large but appears to remain constant.

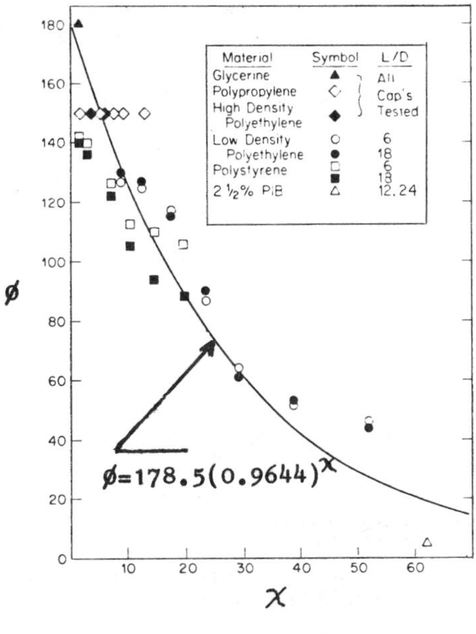

Figure 39. Ballenger-White correlation of capillary entrance angle ϕ with the dimensionless capillary entrance pressure loss, χ [173].

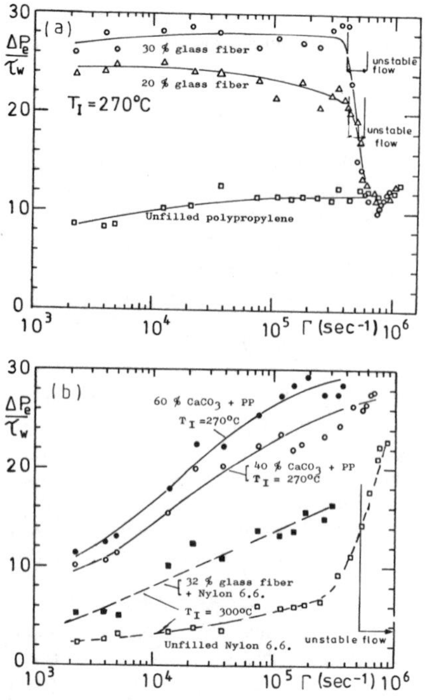

Figure 40. Variation of the dimensionless ends pressure loss ($\Delta P_e/\tau_w$) with apparent shear rate ($\Gamma = 32Q_0/\pi D^3$) during capillary flow of unfilled, fiber-or calcium-carbonate filled polypropylene (PP) and nylon 6.6. (PA): (a) Unfilled and fiber-filled PP at 270°C; (b) Calcium carbonate filled PP (circles) at 270°C and unfilled and fiber-filled PA at 300°C (squares) [42].

Since entry pressure loss is related to the normal stress difference, it may in principle be used to estimate melt elasticity in single-phase polymer melts or solutions. This argument is not valid for filled polymer melts [17], although in some cases (foamed polymers) it is a good measure of fluid elasticity [175]. In filled polymer melts it is easy to see that $\Delta P_e/\tau_\omega$ (or $\Delta P_{ent}/\tau_\omega$) is not a measure of melt elasticity. In general, $\Delta P_e/\tau_\omega$ for these systems is greater than the unfilled base material irrespective of the filler type while the normal stress is enhanced in some glass-fiber-filled melts and the opposite is true of particle-filled polymers and also in some fiber-filled polymers. The variation of dimensionless ends pressure loss $\chi = \Delta P_e/\tau_\omega$ with Γ ($=32Q_0/\pi D^3$) is shown in Figure 40 for various filled and unfilled polymer melts. The ends pressure loss is evaluated by using the multiple capillary (five capillaries in series) rheometer in which the total pressure loss δP_5 and the input flow rate Q_0 are measured when total L/D ratio is 20. Multiple capillary assembly is then replaced by a single capillary of the same L/D ratio and reservoir pressure δP_1 is measured at the same flow rate. From δP_5 and δP_1, $\Delta P_e/\tau_\omega$ is evaluated using [42]

$$\chi^* = \frac{\Delta P_e}{\tau_\omega} = 4\frac{L}{D}\frac{\Delta P_e}{\Delta P_1} \tag{47}$$

$$\Delta P_e/\Delta P_1 = (\Delta P_e/\delta P_1)/(1 - \Delta P_e/\delta P_1) \tag{48}$$

$$\Delta P_e/\delta P_1 = [\delta P/\delta P_1 + 1 - \exp(\epsilon\delta P)]/M - \exp(\epsilon\delta P)],\ M = 5 \tag{49}$$

$$\delta P = \delta P_5 - \delta P_1 \tag{50}$$

ϵ and γ are pressure and temperature coefficients of viscosity, which are evaluated separately. In the absence of such corrections, the Equation 49 is simplified further.

As seen in Figure 40, PP based melts yield higher dimensionless entrance pressure loss compared with PA based melts. This is probably due to low elasticity of PA base compared with PP. In fibre-

Table 3
Variation of Capillary Entry Angle (Φ) with Input Flow Rate (Q_0) and Initial Melt Temperature (T_N)

Material	T_N(°C)	Q_0(cc/s)	Φ
30% glass-fiber-reinforced polypropylene	210	4	130°
	210	80	120°
	270	4	145°
	270	80	135°
	270	160	125°
32% glass-fiber-reinforced nylon 6.6	290	20	75°
	290	80	70°
	290	160	65°
Liquid crystalline polymer (ICI)	295	4	70°

filled melts the capillary entrance pressure loss is related to the elasticity of the melt and the state of the fiber orientation. However, since the orientation is also dictated by the elasticity of the base fluid [23, 25, 26], then the base fluid elasticity dictates the entrance pressure loss for a given filler and filler concentration.

By using the injection moulded parts employing the mold insert illustrated in Figure 3b the flow entry angle ϕ for these two materials is measured, and the results are tabulated in Table 3. The entry angle for fiber-filled PP and PA appear to be similar to those reported by Ballenger and White [173, 174] for corresponding unfilled melts; although the shear rates are widely different in the present experiments. This suggests that the size of the vortex is mainly determined by the base fluid.

Reservoir Pressure Oscillations

The unstable flow of polymer melts may be accompanied by pressure oscillations in the reservoir, if the flow is brought about by the steady displacement of a plunger. A notable example is HDPE where the periodic oscillations in the reservoir pressure may have frequency as high as 1 cps [2, 176, 177]. These oscillations are found to create surface irregularities on the emerging extrudate [176]. The oscillations in pressure can be divided into two regions; compression (pressure increasing) region with period T_c and decompression (pressure increasing) with period T_d [177].The development of pressure oscillations in HDPE has been investigated by Okubo and Hori [177] and by Akay [26] for FRTP melts. As seen in Figure 35, the first oscillation cycle appears as a pressure overshoot, if the input flow rate is above a critical value, Q_0^*,the initial oscillation is repeated and the flow is unsteady. When the flow rate is increased to an upper critical value, Q_0^+, stable flow is restored, and the first pressure cycle does not repeat itself thus appearing as a pressure overshoot. However, unlike the pressure overshoot encountered in Figure 33, this overshoot does not grow with increasing flow rate. The development and the decay of the pressure oscillations are shown in Figure 41 for glass fiber reinforced PP. As seen in this figure, the intensity of the oscillations increases with increasing temperature; a situation opposite to that encountered in single-phase polymer melts through a single capillary.

Discontinuity in the Flow Curve

During the extrusion of HDPE through a capillary at constant pressure, and at a critical flow rate in addition to extrudate distortion, there is a discontinuity in the plot of output flow rate versus applied pressure (flow curve); the output flow rate becomes a double-valued function of pressure

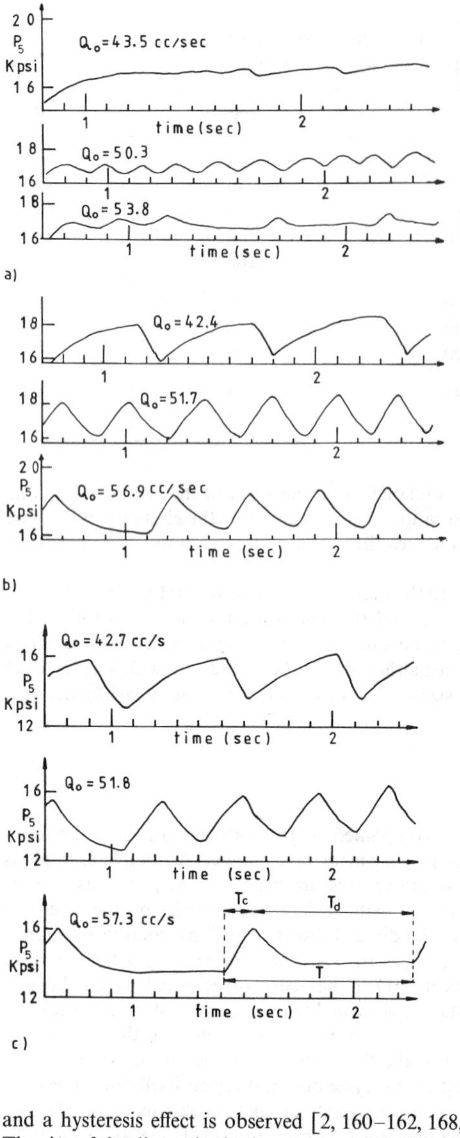

Figure 41. Development of reservoir pressure oscillations (P_5) with input flow rate Q_0 during the unstable flow of short glass fiber-reinforced polypropylene (PP) through five capillaries in series at various initial melt temperature (T_I), i.e., the temperature of the first reservoir: (a) 20% fiber-reinforced PP, $T_I = 270°C$; (b) 30% fiber-reinforced PP, $T_I = 270°C$; (c) 30% fiber-reinforced PP, $T_I = 300°C$. Capillary diameter is 1 mm [26].

and a hysteresis effect is observed [2, 160–162, 168, 169, 178, 179], as seen in Figures 42 and 43. The size of the discontinuity increases with L/D, while the temperature decrease has the same effect [179]. The critical shear stress at the discontinuity increases with temperature [160]. The flow curve below the discontinuity is sensitive to molecular weight, but above the discontinuity there appears to be no molecular weight dependence [181]. These observations are compatible with the explanation that unsteady flow is a result of flow-induced migration of the high molecular weight fraction of the macromolecules resulting in a slip layer at the wall.

Similar to single-phase polymer melts, instability in fiber-filled PP through multiple capillaries [26] yields discontinuity in the flow curve, however in this case the size of discontinuity increases with increasing temperature while the critical shear rate decreases with temperature. It is found that during unstable flow, the ends pressure loss of the filled PP is reduced by a factor of 2 to 3 and

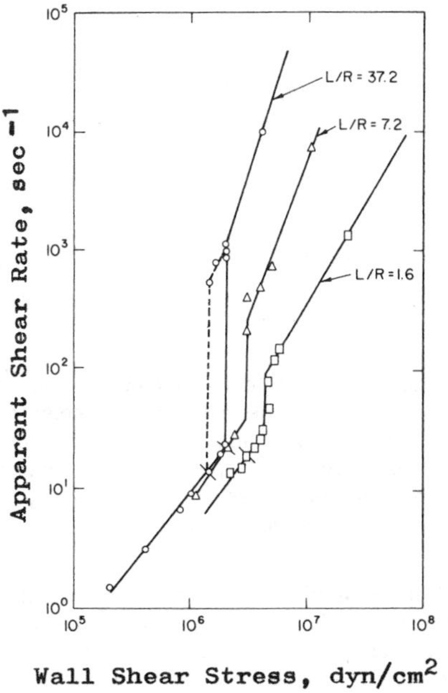

Figure 42. Flow data for tetrafluoroethylene hexafluoropropylene copolymer at 380°C, illustrating the discontinuity and hysteresis in the flow curve. Slash indicates stress at the onset of ripples on the extrudate; continuous line indicates flow for increasing stress and dashed line indicates flow as stress is decreased from those on the upper branch of the curve (hysteresis). Discontinuity increases with increasing length to radius ratio (L/R) of the capillary [160].

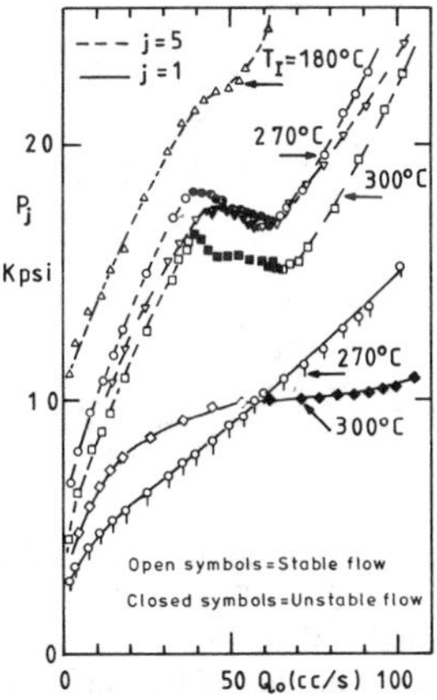

Figure 43. Variation of reservoir pressure (P_j) with input flow rate (Q_0) during the capillary flow of unfilled and glass fiber-reinforced polypropylene (PP) and nylon 6.6. (PA) melts through single capillary ($j = 1$, ———) or multiple capillaries in series ($j = 5$, ------) at various initial melt temperatures (T_I). (◆, ◇) unfilled PA, $T_I = 300$°C, $j = 1$; (▼, ▽) 20% fiber-reinforced PP, $T_I = 270$°C, $j = 5$; 30% fiber-filled PP; (△) $T_I = 180$°C, $j = 5$; (●, ○) $T_I = 270$°C, $j = 5$; (■, □) $T_I = 300$°C, $j = 5$. (○) $T_I = 270$°C, $j = 1$. Stable flow is represented by open symbols and unstable flow is represented by filled symbols. Capillary diameter is 1 mm [26].

becomes equal to that of the unfilled material as seen in Figure 40a. This behavior indicates that instability starts at the entrance and propagates into both the capillary and exit reservoir resulting in fiber-free layer near the wall and causes slip. However, similar to finding of Kanu and Shaw [102] these fiber free regions are not present at all times. Decompression in the reservoir occurs after the formation of the fiber free regions and melt causes slip. Compression in the reservoir starts when the fiber free region is eliminated due to feeding through the capillaries and fiber diffusion.

Time-Dependent Die-Swell

In general, polymers in solutions appear to have the same unstable flow characteristics as melts although polymeric solutions show a number of unstable flow characteristics not reported for polymeric melts. Using dilute polyacrylamide solutions in a plunger type of rheometer, Ramamurthy [182] observed a longitudinally oscillating exit stream, which resulted in oscillating flow rate when

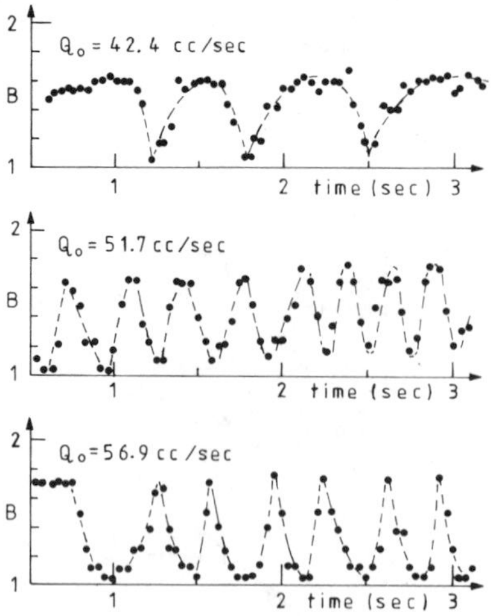

Figure 44. Variation of dimensionless extrudate diameter B (ratio of the extrudate diameter to capillary diameter) with time during unstable flow of glass-fiber-reinforced polypropylene melts through multiple capillaries in series when $T_1 = 270°C$. Capillary diameter is 1 mm [26].

the shear increased above a critical value. However, further increase in shear rate restored the stable flow and the flow in the reservoir was confined to a cylindrical volume with a diameter almost equal to that of the capillary itself. Although the exit stream was steady, the diameter of the jet was less than that of the capillary, indicating negative die-swell, as illustrated by Lenk [183]. Tomita et al. [183, 184] using aqueous polymer solutions observed that when the shear exceeded the critical shear rate (for the primary unstable flow) by a factor of ten, or so, the extrudate did not show any die-swell at the capillary exit, but the die-swell was observed at some distance from the exit and this distance changed periodically with time.

During the unstable flow of fiber-filled PP, the emerging extrudate appears to have time periodic expanding-contracting behavior as illustrated in Figure 44, where the dimensionless extrudate diameter B (B is the ratio of the extrudate diameter to capillary diameter) is plotted against time. Before the start of the unstable flow when $Q_0 = 34$ cc/s, B = 1.63 ($Q_0 < Q_0^*$) while after the restoration of the stable flow when $Q_0 = 60$ cc/s (i.e., $Q_0 > Q_0^+$) B = 1.07 for this fluid.

Restoration of Stable Flow

In some cases, if the flow rate is increased further, after the establishment of the primary unstable flow, a second stable flow regime may be encountered in which the extrudate is again smooth [160, 161]. Amongst the polymers that yield this second stable flow regime are HDPE, polytetrafluorethylene, PTFE, [160, 161, 186], and PP [186]. Ui et al. [186] remark that this second stable flow regime is followed by a new second region of unstable flow referred to as "secondary unstable flow." Tordella [160] indicates that the second stable flow regime for PTFE requires a combination of high temperature and high pressure.

As remarked earlier, the unstable multi-capillary flow of fiber-filled PP also yields a second stable flow regime. The characteristics of this flow are similar to that of the unfilled base material. This behavior is probably due to the permanent establishment of fiber-free layers. Therefore, the rate of formation of fiber-free layers is greater than the rate of disappearance. The reduction of die-swell is probably the result of extremely high molecular orientation in the fiber-free layer.

Mechanism of Unstable Flow

Various mechanisms for the unstable capillary flow of polymer melts and solutions have been discussed by Boudreaux and Cuculo [164]. These mechanisms can be summarized as follows:

Microstructure orientation. According to Spencer and Dillon [187], the extrudate distortion is due to differential flow-induced molecular orientation between the extrudate skin (highly oriented molecules) and the core where molecular orientation is not significant or even perpendicular to the flow direction. As the extrudate emerges from the die, it buckles due to this differential orientation.

As we have already elaborated, microstructure orientation develops with increasing capillary length to diameter ratio and therefore with higher L/D should increase the intensity of the instability as is shown in Figure 42. According to this mechanism the site of instability is the capillary wall and flow instability in HDPE can be explained by this mechanism.

Melt elasticity. According to Tordella [180], extrudate distortion is a result of polymer molecules reaching their elastic limit of storing energy thus causing fracture as a means of stress relief either at the capillary wall or at the capillary entrance.

After these two earlier theories, a number of other mechanisms have been suggested. Schreiber et al. [188] have combined the melt fracture theory of Tordella [180] and the differential flow orientation and extrudate buckling aspects of Spencer and Dillon's theory [187]. Benbow et al. [189] introduced a slip-stick mechanism to explain flow instability and extrudate distortion. Above a critical stress, polymer melt experiences intermittent slipping due to a lack of adhesion between the melt and the die wall, in order to relieve excessive deformation energy absorbed during flow. Since then, the slip-stick mechanism has been studied further [23, 26, 85, 176–179, 190–192], both experimentally and theoretically. In all cases, slip is the result of viscosity difference between two adjacent regions. Balmer [193], Hlavacek et al. [194], and Akay [23], have given thermodynamic arguments for flow instability. According to these arguments melt fracture or flow instability can initiate anywhere in the flow field when reduction in the fluid entropy due to molecular orientation reaches a critical value beyond which the second law of thermodynamics is violated, flow instability starts [193].

Orientation induced entropy reduction argument can be generalized to account for flow instability in polymeric fluids in general. As discussed earlier, the creation of differential molecular orientation (i.e., formation of skin-core structure in capillary flow and circulating vortex in the capillary entrance or exit) essentially causes phase separation as a result of conformational entropy differences in these regions and thus polymer molecules at the boundary of these thermodynamically different regions do not transmit stress, hence causing slip, which, in turn, as an entropic homogenization mechanism. During the recovery period, differences between the conformational states are reduced and phase adhesion is restored so that polymer phase appears to stick to the wall, although any sticking is in fact at the phase boundary created before slip. However, this adhesion does not last long and phase separation occurs as a result of restoration of the differential entropic states. During melt fracture some stress is transmitted across the phase boundary through tie-molecules or entanglements, but their numbers are large enough to support the applied stress, therefore these molecules probably undergo irreversible chain scission. The same mechanism has been shown to be valid for solid state deformation of polymers [135, 195].

It is known that [23, 161], the presence of fillers and fibers results in stable flow by reducing the conformational differences between high and low deformation regions in the flow field. Fillers, by restricting the macromolecular conformations in the melt or infact by promoting orientation, act as an "entropic homogenizer," delaying melt fracture and unstable flow until another mechanism causes phase separation, leading to unstable flow with different flow characteristics [23, 26]. The unstable flow of nylon 6.6. reported in [26] is attributed to phase separation caused by degradation at the capillary wall. Arguments based on phase separation can also explain flow geometry dependence of unstable flow and its characteristics [23, 26].

It appears that unstable flow in polymeric fluids can be explained by phase separation caused by the fluid microstructure-flow field interaction. Phase separation may be due to orientation, degradation, flow induced cyrstallization, flow-induced diffusion, and macromolecular conformation at the solid boundary. Phase separation can occur at the capillary entrance or exit or in the capillary. If it occurs at the capillary entrance, unstable flow propagates into the capillary where it may be suppressed or enhanced. If the capillary entry phase separation is not significant enough to cause unstable flow (such as in HDPE when the convergent flow field results in total orientation) then phase separation can occur in the capillary, causing flow instability.

CONCLUSION

In this chapter, the microstructural aspects of polymer melt rheology are covered. The emphasis has been given to fluids with orientable microstructure and their rheological properties are compared and contrasted with single phase polymer melts. White and his colleagues [38, 159] have proposed an equation of state to describe particle-filled polymer melts and shown that it can predict the steady and transient viscometric behavior of these fluids. The other important classes of polymeric fluids are polymer blends and rubber toughened polymers. The rheology of these materials can be found in works by Bucknall [196], Plochocki [197], Munstedt [198], Carley and Crossan [199], Utracki and Kamal [200], and Utracki [201].

The application of rheology to the processing of polymers is also treated. The realistic modelling of polymer processing operations often requires extensive computation. A comprehensive survey of numerical methods in viscoelastic fluid flow has recently given by Crochet and Walters [13]. Most of the available techniques deal with the differential type of constitutive equations using finite difference or finite element techniques in two space dimensions. The next stage of development is likely to be non-isothermal and time dependent flows.

NOTATION

A	function defined by Equation 21
B	die-swell
c	concentration
D	diameter
e_{ij}	rate of deformation tensor
f_0	material constant describing entropic potential field
F	entropic potential field
G	stress power
g_0	constant
H	number of fibers
h	plate separation
I	invariant of deformation rate tensor
J_i^+	flow-induced diffussion flux
j_i^*	diffusion flux
K	rate constant
K_0, K_1	constants
k	Boltzmann's constant
L	length
M	number of capillaries connected through reservoirs in multi-capillary rheometer
M_w	molecular weight
m	constant
N	normal stress
n	constant
p	pressure
Pe	Peclet number
P_{ij}	stress tensor
Q	volumetric flow rate
r	radical coordinate
T	absolute temperature
t	time
v_0	velocity
We	Weissenberg Number
W_{ij}	vorticity tensor

Greek Symbols

α	reaction order
Γ	apparent shear rate
$\dot{\gamma}$	shear rate
γ	temperature coefficient of viscosity
ε	material constant
$\dot{\varepsilon}$	elongation rate
η	viscosity
η_e	elongational viscosity
θ	angle
λ	material constant
μ	material constant
τ	shear stress or dimensionless time
τ_w	wall shear stress
ϕ	angle
χ	dimensionless ends pressure loss
ψ	normal stress differential

REFERENCES

1. Walters, K., *Rheometry*, Chapman and Hall, London, 1975.
2. Han, C. D., *Rheology in Polymer Processing.* Academic Press, New York, 1976
3. Middleman, S., *Fundamentals of Polymer Processing.* McGraw-Hill, New York, 1977.
4. Nielsen, L. E., *Polymer Rheology*, Dekker, New York, 1977.

5. Bird, R. B., Armstrong, R. C., and Hassager, O., *Dynamics of Polymeric Liquids*, Vol. 1, Wiley, New York, 1977.
6. Bird, R. B., et al., *Dynamics of Polymeric Liquids*, Vol. 2, Wiley, New York, 1977.
7. Lenk, R. S., *Polymer Rheology*, Applied Science Publishers, London, 1978.
8. Tadmor, Z., and Gogos, C. G., *Principles of Polymer Processing*, Wiley, New York, 1979.
9. White, J. L., in *Rheometry: Industrial Applications*, K. Walters, (ed.), Research Studies Press, Chichester, England, 1980.
10. Han, C. D., *Multiphase Flow in Polymer Processing*. Academic Press, New York, 1981.
11. Cogswell, F. N., Polymer Melt Rheology, George Godwin, London, 1981.
12. Walters, K., in *Rheometry: Industrial Applications*, K. Walters, (ed.), Research Studies Press, Chichester, England, 1980.
13. Crochect, J. M., and Walters, K., *Ann. Review Fluid Mech.*, Vol. 15, p. 241, (1983).
14. Akay, G., *Rheol. Acta*, Vol. 18, p. 256, (1979).
15. Bagley, E. B., *J. Appl. Phys.*, Vol. 28, p. 624 (1957).
16. Han, C. D., *Trans. Soc. Rheol.*, Vol. 17, p. 375 (1973).
17. Han, C. D., *Polym. Eng. Reviews*, Vol. 1, p. 363 (1981).
18. Penwell, R. C., Porter, R. S., and Middleman, S., *J. Polym. Sci.*, A2, Vol. 9, p. 731 (1971).
19. Crowson, R. J., Folkes, M. J., and Bright, P. F., *Polym. Eng. Sci.*, Vol. 20, p. 925 (1980).
20. Han, C. D., and Kim, K. U., *Rheol. Acta*, Vol. 11, p. 323 (1972).
21. Lodge, A. S., *Elastic Liquids*, Academic Press, New York, 1964.
22. D. V., Boger, *J. Non-Newtonian Fluid Mech.*, Vol. 3, p. 87, (1977).
23. Akay, G., *Polym. Eng. Sci.*, Vol. 22, p. 1027 (1982).
24. Akay, G., *Polym. Composites*, Vol. 4. p. 256 (1983).
25. Akay, G., in *Interrelations Between Processing, Structure and Properties of Polymeric Materials*, J. C., Seferis and P. S., Theocaris (eds.), Elsevier, The Netherlands, (1984).
26. Akay, G., *J. Non-Newtonian Fluid Mech.*, Vol. 13, p. 309 (1983).
27. Akay, G., and Leslie, F. M., (to appear).
28. Darlington, M. W., and McGinley, P. L., *J. Mater. Sci.*, Vol. 10, p. 906 (1975).
29. Crowson, R. J., Scott, A. J., and Saunders, D. W., *Polym. Eng. Sci.*, Vol. 21, p. 748 (1981).
30. Akay, G., *Rheology and Injection Moulding of Fibre Reinforced* Thermoplastics-II. Cranfield Institute of Technology Report, November 1981, Cranfield, England.
31. White, J. L., and Knutsson, B. A. *Polym. Eng. Reviews*, Vol. 2, p. 71, (1982).
32. Akay, G., in "European Federation of Chemical Engineering Symposium Series", No-27, p. 33.1, (1983).
33. Folkes, M. J., and Russell, D. A. M., *Polymer*, Vol. 21, p. 1252, (1980).
34. White, J. L., and Agrawal, A., *Polym. Eng. Reviews*, Vol. 1, p. 267 (1981).
35. Katti, S. S., and Schultz, *J. M.*, *Polym. Eng. Sci.*, Vol. 22, p. 1001, (1982).
36. Kamal, M. R., and Moy, F. H., *Polym. Eng. Reviews*, Vol. 2, p. 381, (1982).
37. Chan, Y., White, J. L., and Oyanagi, Y., *J. Rheol.*, Vol. 22, p. 507, (1978).
38. White, J. L., Czarnecki, L., and Tanaka, H., *Rubber Chem. Technol.*, Vol. 53, p. 823, (1980).
39. Czarnecki, L., and White, J. L., *J. Appl. Polym. Sci.*, Vol. 25, p. 1217 (1980).
40. Crowson, R. J., and Folkes, M. J., *Polym. Eng. Sci.*, Vol. 20, p. 934 (1980).
41. Knutsson, B. A., White, J. L., and Abbas, K. B., *J. Appl. Poly. Sci.*, Vol. 26, p. 2397 (1981).
42. Akay, G., Paper presented at the Conference on Engineering Rheology, Imperial College of Science and Technology, London, September 1983.
43. Baird, D. G., in *Liquid Crystalline Order in Polymers*, A. Blumstein, (ed.), Academic Press, New York, 1978.
44. Baird, D. G., et al., *J. Polym. Sci., Polym. Phys. Ed.*, Vol. 17, p. 1649 (1979).
45. Baird, D. G., in *Rheology*-3, G. Astarita, G. Marrucci, and L. Nicolais (eds.), Plenum Press, New York (1980).
46. Wissbrun, K. F., *Brit. Polym. J.*, Vol. 12, p. 163 (1980).
47. Cogswell, F. N., *Brit. Polym. J.*, Vol. 12, p. 170 (1980).
48. Kiss, G. and Porter, R. S., *J. Polym. Sci., Polym. Phys. Ed.*, Vol. 18, p. 361 (1980).
49. Wissbrun, K. F., *J. Rheol.*, Vol. 25, p. 619 (1981).
50. Jerman, R. E., and Baird, D. G., *J. Rheol.*, Vol. 25, p. 275 (1981).

51. Prasadarao, M., Pearce, E. M., and Han, C. D., *J. Appl. Polym. Sci.*, Vol. 27, p. 1343 (1982).
52. Ide, Y., and Ophir, Z., *Polym. Eng. Sci.*, Vol. 23, p. 261 (1983).
53. Thapar, H., and Bevis, M., *J. Mater. Sci., Letters*, Vol. 2, p. 733, (1983).
54. de Gennes, P. G., *The Physics of Liquid Crystals*, Clarendon Press, Oxford, 1974.
55. Chandrasekhar, S., *Liquid Crystals*, Cambridge University Press, Cambridge, 1977.
56. Leslie, F. M., *Advances in Liquid Crystals*, Vol. 4, p. 1 (1979).
57. Ericksen, J. L., *Advances in Liquid Crystals*, Vol. 2, p. 233 (1976).
58. Ericksen, J. L., *Arch. Rational Mech. Anal.*, Vol. 4, p. 231 (1960).
59. Ericksen, J. L., *Kolloid-Z*, Vol. 173, p. 117 (1960).
60. Leslie, F. M., *Quart. J. Appl. Math.*, Vol. 19, p. 357 (1966).
61. Leslie, F. M., *Arch. Rational Mech. Anal.*, Vol. 28, p. 265 (1968).
62. Abhiraman, A. S., and George, W., *J. Polym. Sci., Polym. Phys. Ed.*, Vol. 18, p. 127 (1980).
63. Denn, M. M., and Metzner, A. B., *Trans. Soc. Rheol.*, Vol. 10, p. 215 (1966).
64. Goettler, L. A., Leib, R. I., and Lambright, *A. J., Rubber Chem. Technol.*, Vol. 52, p. 838 (1979).
65. Kaloni, P. N., *Int. J. Engng. Sci.*, Vol. 3, p. 515 (1965).
66. Gordon, R. J., and Schowalter, W. R., *Trans. Soc. Rheol.*, Vol. 16, p. 79 (1972).
67. Leslie, F. M., *J. Fluid Mech.*, Vol. 18, p. 595 (1964).
68. Fisher, J., and Fredrickson, A. G., *Mol. Cryst. Liq. Cryst.*, Vol. 8, p. 267 (1969).
69. Okagawa, A., Cox, R. G., and Mason, S. G., *J. Colloid and Interface Sci.*, Vol. 45, p. 303 (1973).
70. Currie, P. K., *Mol. Cryst. Liq. Cryst.* Vol. 73, p. 1 (1981).
71. Duke, R. W., and Chapoy, L. L., *Rheol. Acta*, Vol. 15, p. 548 (1976).
72. Hutton, J. F., *Rheol. Acta*, Vol. 14, p. 979 (1975).
73. Huang, T. A., Thesis, Ph.D., University of Wisconsin, 1979.
74. McNally, D., *Polym.-Plast. Technol. Eng.*, Vol. 8, p. 101 (1977).
75. Lee, W. K., and George, H. H., *Polym. Eng. Sci.*, Vol. 18, p. 146 (1978).
76. Harris, J. B., and Pittman, J. F. T., *Trans. Instn. Chem. Engrs.*, Vol. 54, p. 73 (1976).
77. Bright, P. F., and Darlington, M. W., *Plast. Rubber Proc. Appl.*, Vol. 1, p. 139 (1981).
78. Goldsmith, H. L., and Skalak, R., *Ann. Rev. Fluid Mech.*, Vol. 7, p. 213 (1975).
79. Meldon, J. H., and Lederman, M. S., AIChE Symposium Series, Biomedical Engineering, No. 227, Vol. 79, p. 21 (1983).
80. Akay, G., *Polym. Eng. Sci.*, Vol. 22, p. 798 (1982).
81. Mashelkar, R. A., and Dutta, A., *Chem. Eng. Sci.*, Vol. 37, p. 969 (1982).
82. Bedford, A., and Drumheller, D. S., *Int. J. Engng. Sci.*, Vol. 21, p. 863 (1983).
83. Akay, G., in *Rheology*-3, G. Astarita, G. Marrucci, and L. Nicolais (eds.), Plenum Press, New York 1980.
84. Tirrell, M., and Malone, M. F., *J. Polym. Sci.*, Polym. Phys. Ed., Vol. 15, p. 1569.
85. Jonssen, L. P. B. M., *Rheol. Acta*, Vol. 19. p. 32 (1980).
86. Marrucci, G., *Trans. Soc. Rheol.*, Vol. 16, p. 321 (1972).
87. Metzner, A. B., in *Improved Oil Recovery by Surfactant and Polymer Flooding*, D. O. Shah, (ed.), Academic Press, New York, 1977.
88. Metzner, A. B., Cohen, Y., and Rangel-Nafaile, C., J. *Non-Newtonian Fluid Mech.*, Vol. 5, p. 449 (1979).
89. Aubert, J. H., and Tirrell, M., *J. Chem. Phys.*, Vol. 72, p. 2694, (1980).
90. Sekhon, G., Armstrong, R. C., and Jhon, M. S., *J. Polym. Sci., Polym. Phys. Ed.*, Vol. 20, p. 947 (1982).
91. Kang, C. K., and Eringen, A. C., *Bull. Math. Biol.*, Vol. 38, p. 135 (1976).
92. Kline, K. A., *Acta Mechanica*, Vol. 27, p. 239 (1977).
93. Dutta, A., and Mashelkar, R. A., *Rheol, Acta* Vol. 22, p. 455 (1983).
94. Kubat, J., and Szalanczi, A., *Polym. Eng. Sci.*, Vol. 14, 873 (1974).
95. Hegler, R. P., and Mennig, G., Paper presented at the Conference on Engineering Rheology, Imperial College of Science and Technology, London, September 1983.
96. Lee, W. K., and George, H. H., *Polym. Eng. Sci.*, Vol. 18, p. 146 (1978).
97. Wu, S., *Polym. Eng. Sci.*, Vol. 19, p. 638 (1979).
98. Segre, G., and Silberberg, S., *J. Fluid Mech.*, Vol. 14, p. 115 (1962).
99. Segre, G., and Silberberg, S., *J. Fluid Mech.*, Vol. 14, p. 136 (1962).
100. Karino, T., and Goldsmith, H. L., *Phil. Trans. R. Soc. London*, Vol. B279, p. 413 (1977).
101. Newman, S., and Trementozzi, Q. A., *J. Appl. Polym. Sci.*, Vol. 9, p. 3071 (1965).

102. Kanu, R. C., and Shaw, M. T., *Polym. Eng. Sci.*, Vol. 22, p. 507 (1982).
103. Casale, A., and Porter, R. S., *Polymer Stress Reactions*, Academic Press, New York, Vol. 1 (1978); and Vol. 2 (1979).
104. Akay, G., and Tincer, T., *Polym. Eng. Sci.*, Vol. 21, p. 8 (1981).
105. Tincer, T., and Akay, G., *Polym. Eng. Sci.*, Vol. 22, p. 410 (1982).
106. Akay, G., Saunders, D. W., Tincer, T., and Cimen, F., in *Polymer Processing and Properties*, G. Astarita and L. Nicolais (eds.), Plenum Press, New York (1984).
107. Basedow, A. M., Ebert, K. H., and Hunger, H., *Makromol. Chem.*, Vol. 180, p. 411 (1979).
108. Abbas, K. B. and Porter, R. S., *J. Appl. Polym. Sci.*, Vol. 20, p. 1289 (1976).
109. Abbas, K. B. and Porter, R. S., *J. Appl. Polym. Sci.*, Vol. 20, p. 1301 (1976).
110. Abbas, K. B., *Polym. Eng. Sci.*, Vol. 20, p. 703 (1980).
111. Abbas, K. B., *Polymer*. Vol. 22, p. 836 (1981).
112. Yamane, H., and White, J. L., *Polym. Eng. Reviews*, Vol. 2, p. 167 (1982).
113. Adams, J. H., *J. Polym. Sci.*, A-1, Vol. 8, p. 1077 (1970).
114. Riedal, G. R., and Padget, J. C., *J. Polym. Sci.*, Polym. Symposium Ed., Vol. 57, p. 1 (1976).
115. Spruiell, J. E., and White, J. L., *Polym. Eng. Sci.*, Vol. 15, p. 660 (1975).
116. Lagasse, R. R., and Maxwell, B., *Polym. Eng. Sci.*, Vol. 16, p. 189 (1976).
117. Tan, V., and Gogos, C. G., *Polym. Eng. Sci.*, Vol. 16, p. 512 (1976).
118. Carter, D. H., Cuculo, J. A., and Boudreaux, Jr., E., *Polym. Eng. Sci.*, Vol. 20, 324 (1980).
119. Schmidt, L. R., *Polym. Eng. Sci.*, Vol. 17, p. 666 (1978).
120. Bright, P. F., Crowson, R. J., and Folkes, M. J., *J. Mater. Sci.*, Vol. 13, p. 2497 (1978).
121. Prevorsek, D. C., Kwon, Y. D., and Sharma, R. K., in *Science and Technology of Polymer Processing*, N. P. Suh and N. H. Sung (eds.) The MIT Press, Cambridge, Mass., 1979.
122. Mascia, L., *Thermoplastics: Materials Engineering*, Applied Science Publishers, London, 1982.
123. Plueddeman, E. P., in *Additives for Plastics* Vol. 1, R. B. Seymour (ed.), Academic Press, New York, 1978.
124. Morrell, S. H., *Plastics and Rubber Proc. Appl.* Vol. 1, p. 179 (1981).
125. Han, C. D., et al., *Polym. Eng. Sci.*, Vol. 21, p. 196 (1981).
126. Monte, S. J., and Sugerman, G., SPE NATEC Paper, October 1982.
127. Utracki, L. A., and Fisa, B., *Polym. Composites*, Vol. 3, p. 193 (1982).
128. Bigg, D. M., *Polym. Eng. Sci.*, Vol. 22, p. 512 (1982).
129. Bigg, D. M., *Polym. Eng. Sci.*, Vol. 23, p. 206 (1983).
130. Luo, H. L., Han, C. D., and Mijovic, J., *J. Appl. Polym. Sci.*, Vol. 28, p. 3387 (1983).
131. Andrianova, G. P., *J. Polym. Sci., Polym. Phys. Ed.*, Vol. 13, p. 95 (1975).
132. Brunn, P., *Rheol. Acta*, Vol. 15, p. 23 (1976).
133. Aubert, J. H., and Tirrell, M., *Polym. Preprints*, Vol. 33, p. 82 (1981).
134. Silberberg, A., *J. Macromol. Sci., Phys.*, Vol. B18, p. 677 (1980).
135. Kendall, K., and Sherliker, F. R., *Brit. Polym. J.*, Vol. 12, p. 111 (1980).
136. Meissner, B., *J. Appl. Polym. Sci.*, Vol. 18, p. 2483 (1974).
137. Grubb, D. T., and Keller, A., *J. Polym. Sci., Polym. Lett. Ed.*, Vol. 12, p. 419 (1974).
138. Burton, R. H., et al., *J. Mater. Sci.*, Vol. 18, p. 315 (1983).
139. Akay, G., and Bucknall, C. B., in *Specialist Periodical Reports on Marcromolecular Chemistry*, Vol. 3, Royal Society of Chemistry London, (1984).
140. Soong, D. S., and Shen, M., *Polym. Eng. Sci.*, Vol. 20, p. 1177 (1980).
141. Soong, D. S., and Shen, M., *J. Rheol.*, Vol. 25, p. 259 (1981).
142. Jongchaap, R. J. J., *J. Non-Newtonian Fluid Mech.*, Vol. 8, p. 183 (1981).
143. Acierno, D., et al., *J. Non-Newtonian Fluid, Mech.*, Vol. 1, p. 125 (1976).
144. Phan-Thien, N., and Tanner, R. I., J. *Non-Newtonian Fluid Mech.*, Vol. 2, p. 353 (1977).
145. Doi, M., and Edwards, S. F., *J. Chem. Soc., Faraday Trans. II*, Vol. 74, p. 1789; p. 1802; p 1818; (1978).
146. Doi, M., and Edwards, S. F., *J. Chem. Soc., Faraday Trans. II*, Vol. 75, p. 38 (1979).
147. de Gennes, P. G., *J. Chem. Phys.*, Vol. 55, p. 572 (1971).
148. Kaye, A., College of Aeronautics (Cranfield Institute of Technology) Report, No. 134, (1962); *Brit. J. Appl. Phys.*, Vol. 17, p. 803 (1966).
149. Bernstein, B., Kearsley, E. A., and Zapas, L. J., *Trans. Soc. Rheol.*, Vol. 7, p. 391 (1963).
150. Curtiss, C. F., and Bird, R. B., *J. Chem. Phys.*, Vol. 74, p. 2016, and p. 2026 (1981).
151. Bird, R. B., *J. Rheol.*, Vol. 26, p. 277 (1982).

152. Graessley, W. W., *Advances in Polymer Science*, Vol. 47, p. 67 (1982).
153. Currie, P. K., *J. Non-Newtonian Fluid Mech.*, Vol. 11, p. 53 (1982).
154. Wagner, M. H., and Meissner, J., *Makromol. Chemie*, Vol. 181, p. 1533 (1980).
155. Dealy, J. M., and Tsang, W. K.-W., *J. Appl. Polym. Sci.*, Vol. 26, p. 1149 (1981).
156. Tsang, W. K.-W., and Dealy, J. M., *J. Non-Newtonian Fluid Mech.*, Vol. 9, p. 203 (1981).
157. Liu, T. Y., Soong, D. S., and Williams, M. C., *Polym. Eng. Sci.*, Vol. 21, p. 675 (1981).
158. Menezes, E. V., and Graessley, W. W., *J. Polym. Sci., Polym. Phys. Ed.*, Vol. 20, p. 1817 (1982).
159. Suetsugo, Y., and White, J. L., *J. Non-Newtonian Fluid Mech.*, Vol. 14, p. 121 (1984).
160. Tordella, in *Rheology, Vol.* 5, F. R. Eirich (ed.) Academic Press, New York, 1969.
161. Bagley, E. B., and Schreiber, H. P., in *Rheology*, Vol. 5, F. R. Eirich (ed.), Academic Press, New York, 1969.
162. White, J. L., *Appl. Polym, Symp.*, Vol. 20, p. 155 (1973).
163. Cogswell, F. N., *Appl. Polym. Symp.*, Vol. 27, p. 1 (1975).
164. Bourdreaux, Jr., E., and Cuculo, J. A., *J. Macromol. Sci., Rev. Macromol. Chem.*, Vol. C16, p. 39 (1977).
165. White, J. L., in *Science and Technology of Polymer Processing*, N. P. Suh and N. H. Sung (eds.), The MIT Press, Cambridge, Mass. 1979.
166. White, J. L., and A. J. Kondu, *J. Appl. Polym. Sci.*, Vol. 21, p. 2284 (1977)
167. Boger, D. V., in *Advances in Transport Processes*, A. S. Mujumdar and R. A. Mashelkar (eds.), Vol. 2, Wiley International, New Delhi, 1982.
168. den Otter, J. L., *Plast. Polym.*, Vol. 38, p. 155 (1970).
169. den Otter, J. L., *Rheol. Acta*, Vol. 10, p. 200 (1971).
170. Oyanagi, Y., *Appl. Polym. Symp.*, Vol. 20, p. 123 (1973).
171. Southern, J. H., and Paul, D. R., *Polym. Eng. Sci.*, Vol. 14, p. 560 (1974).
172. Cogswell, F. N., *Polym. Eng. Sci.*, Vol. 12, p. 64 (1972).
173. Ballenger, T. F., and White, J. L., *J. Appl. Polym. Sci.*, Vol. 15, p. 1949 (1971).
174. Ballenger, T. F., and White, J. L., *Chem. Eng. Sci.*, Vol. 25, p. 1191 (1970).
175. Han, C. D., and Ma, C.-Y., *J. Appl. Polym. Sci.*, Vol. 28, p. 831 (1983).
176. Weill, A., *Rheol. Acta*, Vol. 19, p. 623 (1980).
177. Okubo, S., and Hori, Y., *J. Rheol.*, Vol. 24, p. 253 (1980).
178. Uhland, E., *Rheol, Acta.* Vol. 18, p. 1 (1979).
179. Weill, A., *J. Non-Newtonian Fluid Mech.*, Vol. 7, p. 303 (1980).
180. Blyler, Jr., L. L., and Hart, A. C., *Polym. Eng. Sci.*, Vol. 10, p. 193 (1970).
181. Ramamurthy, A. V., *Trans. Soc. Rheol.*, Vol. 18, p. 431 (1974).
182. Lenk, R. S., *J. Appl. Polym. Sci.*, Vol. 22, p. 1781 (1978).
183. Tomita, Y., and Shimbo, T., *Appl. Polm. Symp.* Vol. 20, p. 137 (1973).
184. Tomita, Y., Shimbo, T., and Ishibashi, Y., *J. Non-Newtonian Fluid Mech.*, Vol. 5, p. 497 (1979).
185. Ui, J., et al., *SPE, Trans.*, Vol. 4 p. 295 (1964).
186. Spencer, R. S., and Dillon, R. E., *J. Colloid Sci.*, Vol. 4, p. 241 (1949).
187. Tordella, J. P., *J. Appl. Phys.*, Vol. 27, p. 454 (1956).
188. Schreiber, H. P., Bagley, E. B., and Birks, A. M., *J. Appl. Polym. Sci.*, Vol. 4, p. 362 (1960).
189. Benbow, J. J., Charley, R. V., and Lamb, P., *Nature*, Vol. 223 (1961).
190. Uhland, E., *Rheol, Acta*, Vol. 15, p. 30 (1976).
191. Bersted, B. H., *J. Appl. Polym. Sci.*, Vol. 28, p. 2777 (1983).
192. Ruckenstein, E., and Rajora, P., *J. Colloid and Interface Sci.*, Vol. 96, p. 488 (1983).
193. Balmer, R. T., *J. Appl. Polym. Sci.*, Vol. 18, p. 3127 (1974).
194. Hlavacek, B., Carreau, P., and Schreiber, H. P., in *Science and Technology of Polymer Processing*, N. P. Suh and N. H. Sung (eds.), MIT Press, Cambridge, Mass. 1979.
195. Akay, G., Cimen, F., and Tincer, T., in *Proceedings of the European Meeting on Polymer Processing and Properties*, Plenum Press, New York (1984).
196. Bucknall, C. B., *Toughened Plastics*, Applied Science Publishers, London, 1977.
197. Plochocki, A., in *Polymer Blends*, Vol. 2, D. R. Paul and S. Newman, (eds.), Academic Press, New York, 1978.
198. Munstedt, H., *Polymer Eng. Sci.*, Vol. 21, p. 259 (1981).
199. Carley, J. F., and Crossman, S. C., *Polym. Eng. Sci.*, Vol. 21, p. 249 (1981).
200. Utracki, L. A., and Kamal, M. R., *Polym. Eng. Sci.*, Vol. 22, p. 96 (1982).
201. Utracki, L. A., *Polym. Eng. Sci.*, Vol. 23, p. 602 (1983).

SECTION III

FLOW AND TURBULENCE MEASUREMENT

CONTENTS

CHAPTER 36

TECHNIQUES FOR TURBULENCE MEASUREMENT

Stavros Tavoularis

Department of Mechanical Engineering
University of Ottawa, Canada

CONTENTS

INTRODUCTION

Despite the intensive efforts of basic and applied researchers, the understanding and prediction of turbulent flows remains limited. An exact, self-consistent theoretical solution appears hopeless, at least within the existing mathematical formulation and physical comprehension of the problem, because of the non-linearity and instability of the governing equations. Considering the space-time variability of the instantaneous turbulent fields, it appears necessary to devise some kind of statistical presentation of the results [1]. Again, an exact (theoretical or experimental) statistical description of turbulence is impossible, since it would require the specification of all possible statistical properties of all turbulent (random) variables. An alternative formulation of the problem requires the, equally impossible, specification of the characteristic functional of the velocity. Thus, by necessity and, often, also by scope limitations of a particular application, statistical information on turbulence is restricted to lower-order statistical quantities, such as averages, variances and one-point correlations, and, depending on the sophistication of the study, also two-point correlations, frequency spectra, probability densities, etc. Nearly always, the results are presented under the assumption that the turbulent random processes are statistically stationary (i.e. independent of shifting the time origin when averaging) or nearly so.

Recent computational advances [2, 3, 4] based on semi-empirical modeling and/or direct simulation of the equations of motion appear promising, at least for relatively simple flow configurations. However, it is generally admitted that computational techniques are unlikely to replace completely the experimental studies, if only because they require experimental results for their verification. Even slight computational extrapolations beyond successfully represented experimental ranges have occasionally proved risky. At this moment, it appears that unless a major theoretical breakthrough is achieved, engineers must base their predictions of turbulent flows on an insightful interpretation of experimental and computational results.

Historically, the first scientific studies of turbulence have been experimental. Sawdust was used by Hagen [5] and dye by Reynolds [6] for the visualization of transition in pipe flow. Both correlated their observations with mean pressure drop measured through wall taps. Experimentation on turbulent flows was constrained to qualitative visualization and mean velocity and pressure measurement until the temporal response of a suitable instrument, the hot-wire anemometer, was improved [7, 8] to a degree that it could resolve the turbulent velocity fluctuations. Subsequent developments in hot-wire and hot-film anemometry permitted the collection of a wealth of turbulence data, often beyond the contemporary level of theoretical understanding of the problem. Following a rapid evolution during the past two decades, laser-Doppler velocimetry and, to a lesser degree, chronophotography have joined hot-wire anemometry as the prevailing, reliable turbulence tools. Fast-response instruments for measuring pressure, temperature and composition fluctuations are also available.

Turbulence Transducer Requirements

The measurement of turbulence is a challenging task imposing the following severe requirements on any measurement system.

Spatial Requirements

Since turbulent energy is spread continuously over a wide range of wavelengths, i.e. typical eddy sizes, the transducer must resolve large-, medium- and small-scale motions. In a typical, large Reynolds number laboratory experiment, this might cover the orders of magnitude between one meter and a hundredth of a millimeter. No current technique can resolve the entire range. However, both hot-wire and laser-Doppler anemometers can measure accurately local velocity within control volumes of the order of 10^{-4} to 10^{-6} mm^3. Traversing a single probe through the flow region can provide the spatial variation of turbulence statistics, although not simultaneously. Simultaneous measurement can, in principle, be achieved by using multi-probe setups, combined with high-speed data acquisition systems. Multi-probe hot-wire studies have been performed successfully, but they are subject to flow obstruction and probe cross-talk. Multi-point laser-Doppler systems appear feasible but they are currently prohibitive due to high cost. The best technique for resolving large-scale structure is still flow visualization. In combination with chronophotography, this technique can resolve all except, perhaps, the finest scales of turbulence. However, manual processing of visualization records is extremely tedious, so in practice the technique must be combined with digital image processing.

Temporal Requirements

Turbulent events occur over a wide temporal range, exhibiting typical durations between tens of seconds and tens of microseconds. State-of-the-art turbulence instrumentation is generally adequate for measuring lower-order statistics such as mean squares and, with proper care, can also measure fast events, such as energy dissipation. Commercially available hot-wire and laser-Doppler anemometers present a frequency response from DC to over 100 kHz. A different requirement is related to long-time response. A good measuring system should have negligible drift or other performance contamination during a time sufficiently long for a meaningful statistical average to be obtained and, in some cases, during the interval between two successive calibration checks. Since modern electronic circuitry has good stability, long-time errors are more likely to be introduced by the transducer itself. Hot-wire performance is known to deteriorate due to wire aging and to accumulation of dirt contained in the flow. On the other hand, laser-Doppler systems do not require calibration and, therefore, appear insensitive to this problem. The long-time averaging of wide-band signals is relatively simple with the use of high quality FM tape recorders and of digital data acquisition systems.

Amplitude Requirements

Ideally, a turbulent transducer should resolve all velocities in the order-of-magnitude range between the mean velocity and the Kolmogoroff velocity $(\nu\epsilon)^{1/4}$ (ν is the kinematic viscosity and ϵ is the turbulent kinetic energy dissipation rate). For large Reynolds number turbulence, this requires a high sensitivity and an enormous dynamic range. In this respect, hot-wire anemometers appear to be limited mainly by the thermal noise level of the circuitry, which can be reduced by improved circuitry design combined with signal filtering and statistical corrections. On the other hand, laser-Doppler systems have an inherently limited dynamic range, which can be extended only by using special techniques.

Linearity Requirements

Since turbulence measurements nearly always involve averaging, it is essential that the various averages of the velocity or other flow quantities can be calculated from corresponding averages of the transducer outputs. Transducer nonlinearities can distort significantly the measurement of turbulence statistics. For example, the hot-wire anemometer output must be linearized (e.g. with an analog linearizer) or converted to instantaneous flow velocity by applying a non-linear calibration relation on digital records of the signals; non-linearity errors are negligible when the flow velocity fluctuations are extremely small, typically less than 1% of the mean velocity.

Flow Distortion

In every experiment, it must be ascertained that the flow properties are not substantially altered by the presence or operation of the measuring system; since some flow distortion is inevitable when a probe is inserted in the flow, its effects must be minimized by proper design of the probe and the insertion mechanism and by further application of empirical or theoretical corrections. Optical techniques do not usually interfere with the flow, except near the tracer source and when the tracer concentration and/or size become significant.

Objectives and Scope of the Present Study

Measuring techniques in turbulent flows attracted significant academic and industrial interest, so it appears paradoxical that general, comprehensive, up-to-date reviews of the field are limited in number. Among the best known earlier reviews, are the monograph by Bradshaw [9], the article by Corrsin [10] and the special chapters in the textbooks by Reynolds [11] and Hinze [12]. Specialized articles covering particular measuring techniques have been contributed by several expert authors in the recent collective works edited by Frost and Moulden [13], Richards [14], Emrich [15], and Goldstein [16].

The present review attempts to update the informed layman, such as a starting graduate student or a non-specialist research engineer, on the currently prevailing techniques for the measurement of turbulence. Space limitations do not permit a thorough examination of all topics, however, specialized references are included for the person who wishes to engage in turbulence research. Only the most common techniques are discussed, while techniques which are only historically important or not entirely developed have been omitted.

The presently discussed techniques are applicable to fluids which can be assumed "continuous" in the classical fluid mechanical sense, (i.e. with continuously distributed properties) and essentially single-phase. Compressibility effects are occasionally included, but, most techniques apply to incompressible flows. Finally, a certain amount of accessibility and control of the experimenter over the flow has been assumed as would be the case in laboratory experiments but not necessarily in industrial or environmental applications.

FLOW VISUALIZATION TECHNIQUES

General Comments

Air, water, and most other common fluids are optically homogeneous and isotropic; thus, relative motions within the fluid volume would be indistinguishable to human eye and to light recording instruments, except by the introduction of localized optical disturbances. Occasionally, optical disturbances occur "naturally," or, rather, unintentionally: river turbulence becomes visible by means of transported air-bubbles, leaves, and other debris; vortex ring motion and instability by means of smoke particles exhaled by the skilled smoker; natural convection of air above a residential radiator or automobile engine by means of refractive index changes associated with density differences. Naturally or artificially visualized fluid motion must, undoubtedly, have fascinated countless generations of observers, but it also appears that particle tracing has been the first scientific experimental technique for the measurement of flow velocity. Leonardo da Vinci's (1452–1519) sketches of turbulent eddies and wakes, shown in Figure 1, combine artistic inspiration and modern concepts of turbulence structure. Da Vinci also proposed the observation of internal fluid motion by means of suspended particles while, in 1689, van Leeuwenhoek measured blood flow velocity by timing blood cell displacement under the microscope [17]. Recent developments in illumination, photography, and computer-aided image processing have made flow visualization indispensable to all areas of experimental fluid mechanics and, particularly, to experimental turbulence research.

Figure 1. Sketches of turbulent flows by Leonardo da Vinci (about 1510).

The main advantages and specific tasks of visualization vis à vis transducer measurement techniques in turbulent flows can be summarized as follows:

1. They usually describe relative fluid motions in an extensive flow region; under certain conditions, localized information is also provided. Mean flow and turbulence characteristics can be determined by statistical analysis of visualization records.
2. They normally introduce negligible physical disturbance of the flow.
3. They can demonstrate the existence and defines the boundaries of "irregularities", such as reverse flow regions and organized, "coherent" structures.
4. They permit the observation and recording of "frozen" as well as evolving flow patterns as functions of both space and time.
5. They define the boundaries between flow regimes, such as the interface between turbulent and "irrotational" regions in free shear layers and the transition region from laminar to turbulent flow.
6. In combination with transducer techniques (e.g., smoke visualization combined with hot-wire anemometry [18, 19], they facilitate the interpretation of possible patterns in the transducer signals.

Most visualization techniques can be classified in one of the two main groups: (a) visualization using moving particles, including surface flow pattern techniques, and (b) visualization by means of density differences.

Thorough reviews of flow visualization and related techniques, including extensive bibliographies, can be found in the monograph by Merzkirch [20] and in Chapters 1, 2 and 8 of Reference 15. Latest developments in this rapidly advancing field are likely to be found in the proceedings of specialized conferences [21].

Visualization Using Moving Particles

The essence of this technique is to infer fluid motion from the observed motion of visible, alien particles embedded in the fluid. For a correct interpretation of the particle motion, it is necessary to understand:

1. The physical procedure for generating or introducing such particles in the flow.
2. The relationship between particle and fluid motion, especially in a turbulent flow.
3. The physical significance of each particle motion record.

It is possible to distinguish the following patterns produced by moving particles:

1. *The instantaneous particle position*, recorded on a film using a time exposure that is short compared to the time that it takes for the particle to move by a distance equal to its size.

2. *A pathline*, namely the locus of consecutive positions of a particle during a given time interval, recorded by a long-time exposure of the film.
3. *A streakline*, namely the locus of all particles that have coincided with a certain fixed point in the flow; streaklines are obtained by short-time exposure of a continuous point source of particles.
4. *A time-line*, namely the position of all particles that at fixed time in the past formed a continuous line in the fluid; for a time-line, short-time exposure of an instantaneous line source of particles is required.
5. *Time-streakline* combinations, which can be obtained by time-pulsing a line source of particles.

It must be emphasized that *streamlines* (i.e., fluid lines that are everywhere tangent to the local velocity vector) are produced only if the flow is steady and if the particle's relative motion with respect to the fluid is negligible. Since turbulent velocity is time dependent, "averaged" pathlines correspond to "mean streamlines," only when the flow is statistically stationary.

Time-dependence as well as the presence of a wide range of length, velocity and time scales in every turbulent flow impose rather severe requirements on suitability of flow tracing particles. The motion of small spherical rigid particles in an infinite, incompressible flow is described by the BBO (Basset-Boussinesq-Oseen) equation [22], which, however, is too complicated to permit an analytical solution. A simplified model, assuming heavy particles, negligible pressure gradient and body forces, results in the so-called type III approximation

$$\tau \frac{du_p}{dt} + u_p = u_F \tag{1}$$

where u_p is the particle velocity, u_F is the local fluid velocity and the time constant is

$$\tau = \frac{\rho_p d_p^2}{18\mu_F} \tag{2}$$

where ρ_p is the particle density, d_p is the particle diameter, and μ_F is the fluid viscosity. Thus, accurate response to flow changes will be produced by particles with sufficiently small time constant. A somewhat more accurate analytical study [23] introduces the particle transfer function $H(i\omega)$, which has been tabulated for several particle fluid systems as a function of the density ratio ρ_p/ρ_F and the Stokes number $N_s = (\nu_F/\omega d_p^2)^{1/2}$ (ν_F is the fluid kinematic viscosity and ω is the frequency of a sinusoidal fluid motion). Then, the statistical properties of particle velocity can be expressed in terms of $H(i\omega)$ and the frequency spectrum $E(\omega)$ of turbulence; for example, the mean square velocity ratio becomes

$$\frac{\overline{u_p^2}}{\overline{u_F^2}} = \frac{\int_{-\infty}^{\infty} |H(i\omega)|^2 E(\omega)\, d\omega}{\int_{-\infty}^{\infty} F(\omega)\, d\omega} \tag{3}$$

Since $E(\omega)$ is usually unknown in advance, Equation 3, combined with a rough spectral estimate, can only serve as a guide for the selection of particles. Another criterion is that the velocity due to possible body forces, such as gravitational, electric, magnetic, and aerodynamic forces must be negligible with respect to the r.m.s. turbulent velocity. For very small spherical particles, Stokes' law provides an asymptotic estimate for the terminal sink or rise velocity due to gravity as

$$v_s = \frac{|\rho_p - \rho_F| g d_p^2}{18\mu_F} \tag{4}$$

Selection of tracing particle size for turbulent flows is a compromise between two conflicting criteria: good dynamic response requires small particle size and small particle-fluid density differ-

Figure 2. Flow visualization of a mixing layer using dye injection [24].

ences, while good visibility of the particle requires large size and large refractive index ratio. The size requirement is very strict; particles with submicron sizes are often necessary for accurate turbulence measurement, thus imposing extremely demanding specifications on the illumination and light recording systems. For example, 5 μm diameter polystyrene beads in air have a 3 dB cutoff frequency about 1 KHz, while 0.2 mm diameter hydrogen bubbles in water a mere 20 Hz.

The type of tracing particles that are suitable for turbulent flow visualization depends on the flow geometry, the mean speed range and turbulence intensity, the fluid phase (i.e., whether liquid or gas), as well as the fluid physical, chemical and electrical properties. Among the most commonly used particles are the following:

1. *Tufts*, usually short (5–20 mm) pieces of thread fastened on the surface of immersed bodies or on thin wires stretched across the flow. This procedure is suitable for aerodynamic investigations, especially for detecting separation regions, transition points and vortex shedding patterns.
2. *Natural tracers*, such as lint, dust, micro-organisms, etc. that might be suspended in unfiltered air and water flows. Although, at times, such particles have been used for visualization as well as in laser-Doppler velocimetry, their use is not recommended since their concentration and size distribution cannot be easily controlled.
3. *Dyes*, such as food coloring, milk, ink, and various chemical dyes, including fluorescent (e.g., fluorescene) and reacting dyes (e.g., acid and pH indicator). A main disadvantage of dyes is that they diffuse rapidly in turbulent flows and, thus, they cannot be traced accurately away from their source. However, the same property of large turbulent diffusivity can be utilized for demonstrating regions of increased turbulent activity, as shown in Figure 2. Dyes can be introduced in a liquid flow through surface orifices and protruding hypodermic ejectors or generated locally by using electrolytic and photolytic techniques [25].
4. *Smoke*. This technique is similar to dye injection in liquids, but is used in air and other gas flows. Smoke is usually generated by temperature-controlled vaporization of mineral oils, such as kerosene, or by burning of solid substances, such as wood or tobacco. Smoke can be ejected at a point, producing streaklines, or in bulk, in which case the visualized "smoke patterns" provide a rough indication of the turbulence structure. The "smoke-wire" technique [26, 27] (Figure 3) produces smoke streaklines by electrically heating a thin wire stretched across the flow and thus, burning oil droplets driven along the wire. Electric pulses, illumination, and light recording are synchronized to minimize smoke contamination of the flow.
5. *Hydrogen bubbles*, produced by electrolysis of the water and released on the immersed cathode of an electric field. The cathode is usually a thin platinum or stainless steel wire and the bubble size is comparable to the wire diameter (0.01 to 0.2 mm). Unlike dyes and smoke, hydrogen bubbles remain distinct and can be traced for long distances; their disadvantages are that they cannot be produced in extremely small sizes and that they can be only used over limited range of flow speeds (a lower speed limit is imposed by buoyancy effects while a higher speed limit is imposed by the decreasing bubble production rate). Pulsing of the cathode produces time-lines whose evolution can be traced in the flow (Figure 4 [28]), while use of kinked

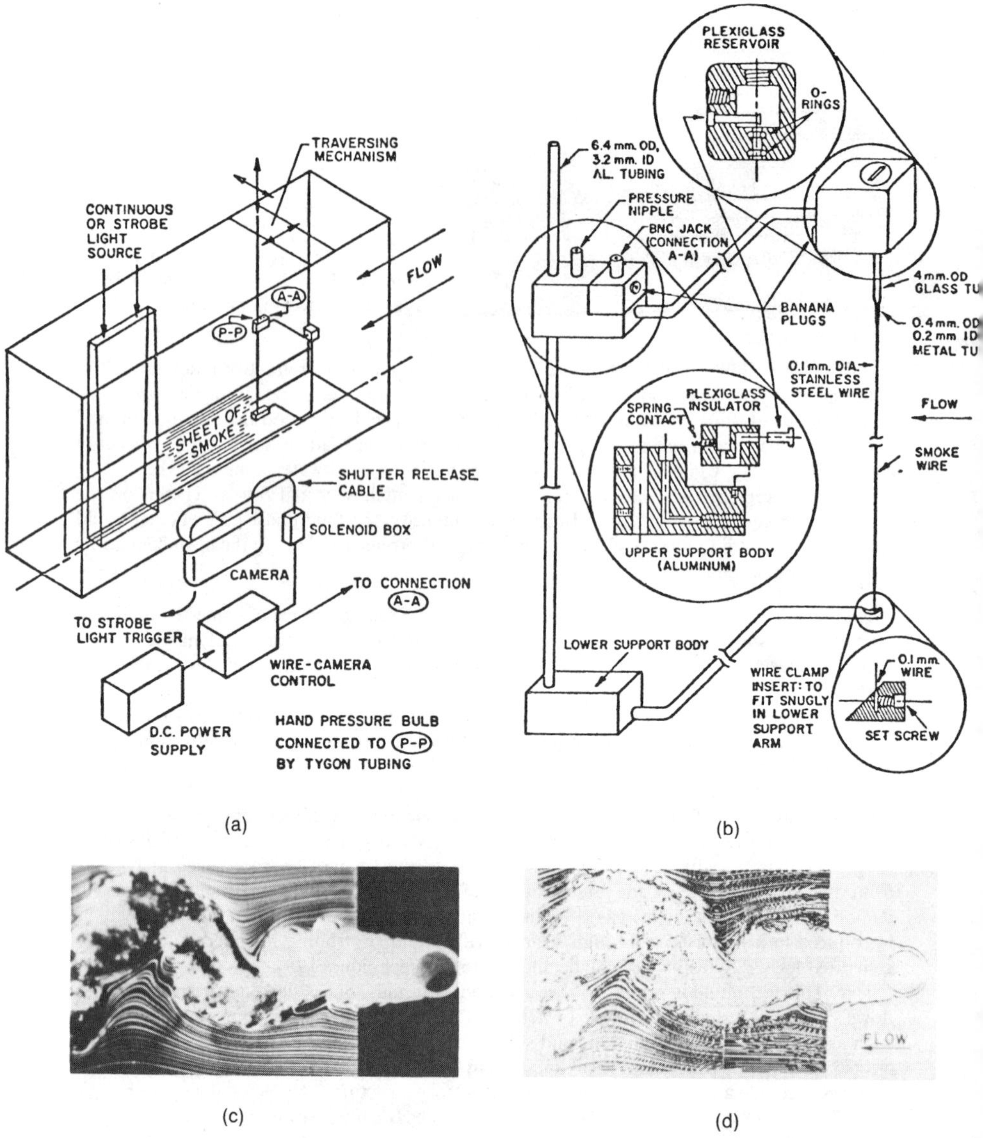

Figure 3. Smoke-wire visualization technique: (a) sketch of the facility [26]; (b) smoke-wire probe assembly [26]; (c) typical application of the technique for flow past a cylinder [27]; (d) digital plot of the same flow obtained through image processing [27].

or partially insulated wires produces streaklines. Depending on the liquid conductivity, the distance between electrodes and the flow speed, a DC voltage between 20 and 1,000 volts with a current between 0.02 and 1A is required. Practical details about bubble production, response, and illumination can be found in the general references and in the reviews by Schraub et al. [29] and Thompson [30].

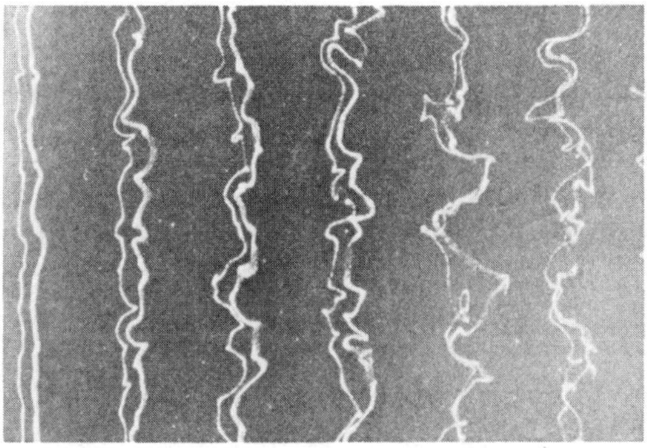

Figure 4. Hydrogen-bubble visualization of grid turbulence [28].

Figure 5. Surface visualization of flow past an airfoil (Prandtl, 1930).

6. Besides the above, a variety of other particles, including gas bubbles, liquid droplets, solid spheres and powders are also suitable for general flow visualization as well as for velocity measurement by particle tracing. Aluminum, lycopodium, and cosmetic powders are suitable for visualization of free liquid surface flow (Figure 5). Air bubbles, aluminium flakes and glass polystyrene beads have been used widely in liquid flows, while oil droplets, soap bubbles, marble dust and other powdered solids are common in gas flows. Table 1 contains the values of density and refractive index for common tracer materials. Small density difference between particle and flow medium is desirable for a good dynamic response while a large refractive index difference is desirable for a good particle visibility.

Visualization Based on Density Differences

The refractive index, n, of fluids is generally a function of their density, ρ. For example, classical electrodynamic analysis [20, 31] predicts that for non-ionized gases and for light frequencies

Table 1
Properties of Common Tracing Particles

Material	Air, Hydrogen	Water	Alumina	Polysterene	Peanut Oil	Silicon Carbide	Glass
Density [kg/m^3]	<1	1,000	3,900	1,050	910	3,300	2.6
Refractive Index	1	1.33	1.76	1.6	1.5	2.6	1.5

away from the molecular resonant frequencies of the gas, n and ρ are related by the Gladstone-Dale formula

$$n - 1 = K\rho \quad (5)$$

where the Gladstone-Dale constant K for each gas is a weak function of temperature and of light wave length. Density differences in a flow can be produced in three ways:

1. By external heating or cooling of the fluid.
2. By mixing two fluids of different densities.
3. For compressible gas flows, internally, as a result of conversion of kinetic energy to potential energy near flow obstructions.

There are three main variable density visualization techniques:

1. *The Shadowgraph Technique* (Figure 6a), in which density differences are recorded as regions of variable light intensity. In fact, it can be shown that light intensity on the recording plane

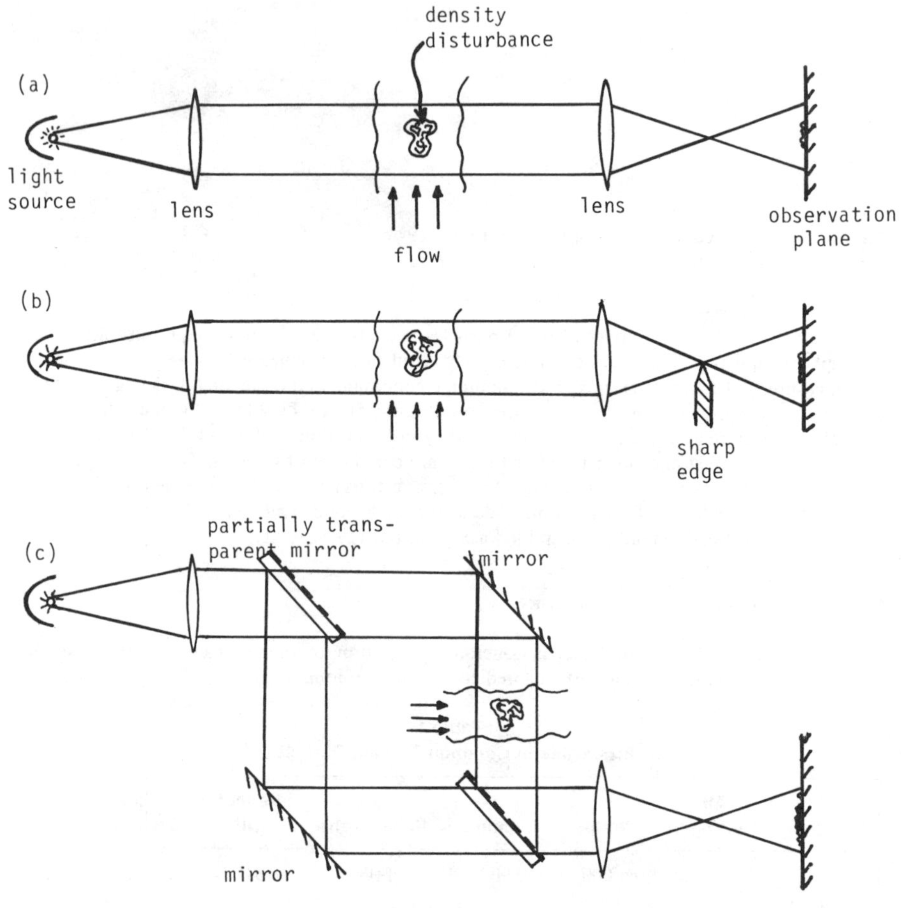

Figure 6. Sketch of apparatus for flow visualization in gases using density differences: (a) shadowgraph technique; (b) Schlieren technique; (c) interferometry.

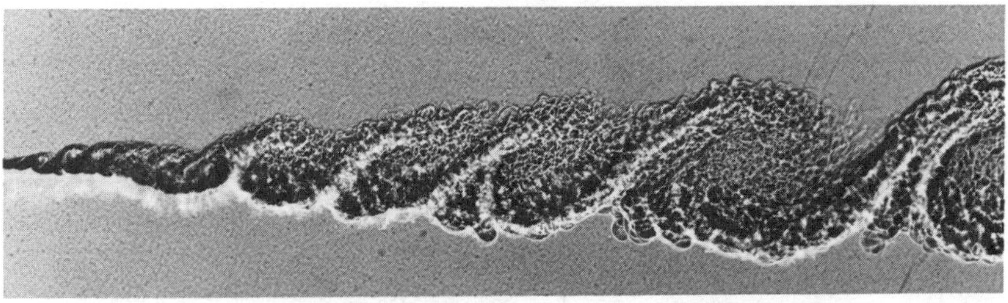

Figure 7. Spark shadowgraph of a mixing layer between helium (upper part) and nitrogen (lower part), visualized through density differences between the two streams (C. L. Brown and A. J. Roshko, *J. Fluid Mech.*, vol. 64, pp. 775–816, 1974).

is proportional to the second derivative or refractive index (and by G-D formula, also of density) transverse to the light path. Thus, shadowgraphs emphasize small-scale density fluctuations at the expense of large-scale ones and can give a deceiving image of turbulence structure. Examples of shadowgraphs are shown in Figures 7 and 8.

2. *The Schlieren Technique* (Figure 6b), in which the collected light intensity is distorted by a sharp blade partially covering the focal area of the collector lens. This technique is more complicated than the shadowgraph, but it provides better resolution, since the light intensity is proportional to the first derivative of density. If properly calibrated, this technique can also be used for quantitative studies of density fields.
3. *Interferometry* (Figure 6c), in which phase differences between two light beams traveling through an optically inhomogeneous medium create fringe patterns on the recording plane. Lines of constant density are recorded as continuous fringes, so that a quantitative analysis of the density field is straightforward.

Illumination, Light Recording, and Image Processing Techniques

Special light sources and illumination techniques are often necessary for the successful visualization of turbulent flows. In particular, light sources of high intensity and/or short duration are required for resolving the smallest scales of turbulence. Among the commonly used devices, one could distinguish two general groups:

1. *Thermal sources*, including continuous mercury and tungsten filament lamps, flash-lamps, spark sources and exploding wire sources.
2. *Lasers*, which produce coherent monochromatic light and are suitable for general illumination as well as for interferometric and holographic studies.

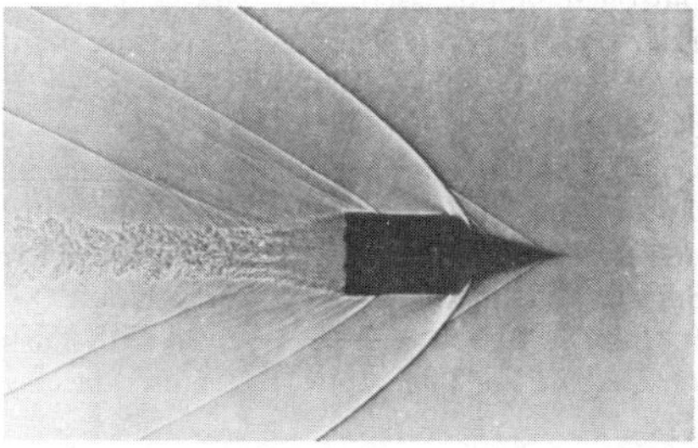

Figure 8. Shadowgraph of the flow around a supersonic pellet (photograph by Ballistic Research Laboratory, Aberdeen Proving Ground).

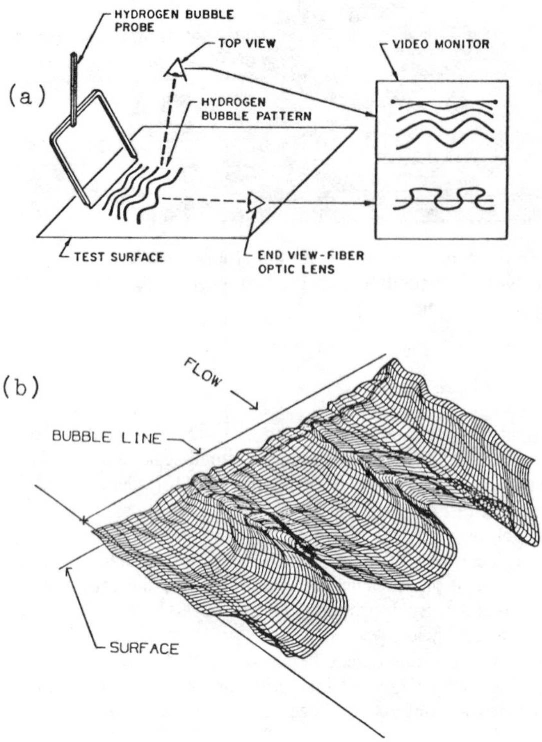

Figure 9. Three-dimensional, hyrogen-bubble visualization technique applied in turbulent boundary layer studies: (a) sketch of the system (b) typical computer plot of the bubble sheet pattern [32].

Since turbulent flows are three-dimensional, uniform illumination of a fluid volume with evenly distributed particles would produce an overlap of particle path projections on the recording plane and would be unable to resolve the particles' distance from that plane. This problem is resolved either by introducing the particles at an isolated point or along a fixed line in the flow (e.g. hydrogen-bubble wire) or by providing illumination in the form of a thin planar sheet. The latter technique is quite common, since a light sheet can be easily produced by passing a laser beam or other columnated beam of light through a cylindrical lens. Special optical arrangements permitting the simultaneous view and recording of the flow from two orthogonal directions have been developed in order to resolve the three-dimensional particle motion (Figure 9) [32].

The intensity of scattered light and, thus, particle visibility, depend strongly on the angle between incident light and observation plane. Mie's scattering theory [33, 34] provides a means for estimating the intensity of scattered light for various angles, particle refractive indices, and light wavelengths. Trial-and-error-type adjustments of the optical system would further improve the quality of visualization records. Forward scattering is preferable to backward scattering when the flow facility permits it. For small particles and well-controlled lighting conditions, dark field illumination (in which particles appear as bright spots in a dark background) gives higher contrast than bright field illumination; the latter is preferable when there is a significant amount of stray light combined with the presence of undesirable, small, dirt particles.

Quality of light recording appears to be limited only by cost, since high quality equipment is available. Although video recording systems find increasing use in flow visualization studies with relatively low speed and resolution requirements, conventional photography and cinematography are still the most reliable light recording methods. Special films are available with resolution exceeding several thousand lines per mm. Exposure times above 0.25 ms can be achieved with conventional cameras, while a variety of short duration (e.g. spark) or interrupted (e.g. plused lasers,

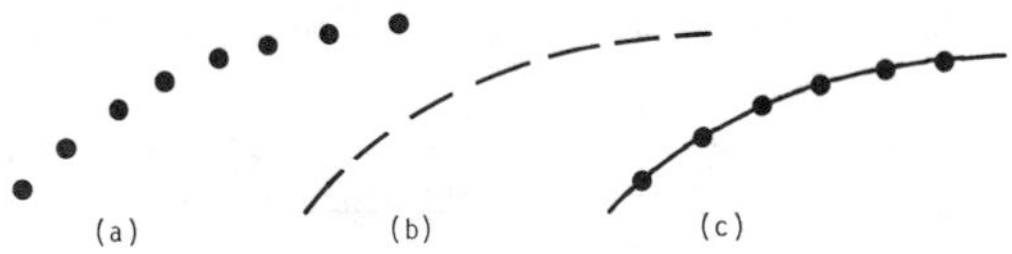

Figure 10. Illumination techniques for velocity measurement using particle tracing: (a) "dot" illumination; (b) "dash" illumination; (c) multiple illumination.

flash lamps, mechanical interrupters) light sources permit the recording of events of less than 1 μs duration, which is more than adequate for most turbulence requirements.

Particular care must be taken when pathline records are used for the mapping of the local velocity vector field. In this case, light recording must be synchronized with accurate time recording (chronophotography). Velocity direction is estimated from the tangent to the pathline, while velocity magnitude from the distance between the particle positions at two known times. When the particle concentration is so low that the trajectories of different particles would not be confused, "dot" illumination (Figure 10a), i.e. multiple exposure to short duration light pulses, is preferable. Otherwise, "dash" illumination (Figure 10b), i.e. multiple exposure to somewhat longer duration light pulses separated by short dark intervals, can be used, but at some expense of spatial resolution. "Multiple" illumination (Figure 10c), i.e. combination of weak continuous and strong interrupted illumination can be used to provide both precise particle positioning and unambiguous determination of trajectory.

Manual processing and statistical analysis of visualization records is an extremely tedious procedure and it is also subject to human bias. This task is significantly reduced with the use of automated discretization techniques and further digital image processing in digital computers [27, 32] (see also Figures 3 and 9). The accuracy of such techniques is mainly determined by the sensitivity of the discretizing instrument. Future developments in discretizing hardware and software are likely to establish digital image processing as one of the main techniques in experimental fluid mechanics.

PRESSURE MEASUREMENT

Pressure Transducers

Following the classical fluid mechanical approach, pressure is defined as the average normal stress at a point in the fluid (by convention, compressive pressure is positive):

$$P = -\tfrac{1}{3}(\sigma_{11} + \sigma_{22} + \sigma_{33}) \tag{6}$$

Since the trace of a second-rank tensor is an invariant, pressure is a scalar quantity independent of the orientation of the coordinate system. Through the previous definition, pressure is introduced into the equations of motion and, therefore, it is coupled with the velocity field. For Newtonian, incompressible fluids, Newton's second law results in the Navier-Stokes equations (where the summation convention, $a_k b_k = a_1 b_1 + a_2 b_2 + a_3 b_3$, applies)

$$\frac{\partial U_i}{\partial t} + U_k \frac{\partial U_i}{\partial x_k} = -\frac{1}{\rho}\frac{\partial P}{\partial x_i} + \nu \frac{\partial^2 U_i}{\partial x_k \partial x_k}, \qquad i = 1, 2, 3 \tag{7}$$

Most pressure transducers utilize the above definition (Equation 6) and measure pressure from a stress, that is as a force per unit area. Other instruments utilize the somewhat different definition of pressure in the kinetic theory of gases as

$$P = NRT \tag{8}$$

where N is the number of molecules per unit volume, R is a gas constant and T is the absolute temperature. In thermodynamics, pressure is defined by the rate of work δW produced during a change of state of a gas per unit volume change, δV (frictional losses are assumed negligible)

$$P = \frac{\delta W}{\delta V} \tag{9}$$

Finally, pressure can be computed through its relation to other measurable flow properties such as gas viscosity and thermal conductivity. A great variety of manometers, micromanometers, mechanical transducers, and other devices have long been used and provide accurate measurement of constant or slowly varying pressure [35, 36, 37]. However, the interest of turbulence researchers usually involves the measurement of pressure fluctuations as well. This restricts significantly the number of suitable pressure transducers, since a frequency response range of several kHz is usually required. Among the common fluctuating pressure transducers are the following:

1. *Piezoelectric Transducers.* They contain crystals of quartz, Rochelle salt, barium titanate or lead-zirconium-titanate, which generate a potential difference between two surfaces when subject to stress along certain directions. Piezoelectric elements are made in cylindrical, disc or beam shapes and are loaded in a normal or shearing mode.
2. *Variable Capacitance Transducers.* In a common configuration, the displacement of a metallic diaphragm or a membrane due to pressure fluctuations causes changes in the air gap between two electrically charged surfaces. In this version, a polarizing DC voltage of a few hundred volts is required, however, the so-called "electret" transducers require no such voltage, since they contain permanent electric charges embedded within a polymer diaphragm.
3. *Variable Resistance Transducers.* The most common type contains strain gages, bonded on an elastic link, which is deformed by the applied pressure fluctuations.
4. *Variable Reluctance Transducers.* In these, the motion of a metallic diaphragm due to pressure differences on its two sides modifies the magnetic flux linkage of two electromagnets. Such instruments have extremely high sensitivity but only moderate frequency response.
5. *Linear Variable Differential Transformers.* In these, displacement of a magnetic core attached to an elastic diaphragm causes an imbalance between the voltages of two, symmetrically placed, secondary coils of a transformer.

High-quality pressure transducers are readily available from specializing manufacturers (e.g., Figure 11), at a moderate cost and, unlike other sensors (e.g., hot-wires), they are rarely home-made by the researchers. The sensitivity and frequency response of several transducer types is generally adequate for most turbulence applications. Pressure transducers with sensing elements of diameters equal to a fraction of a mm are available, but sensitivity generally decreases with size. Other parameters that should influence the selection of a pressure transducer are linearity, temperature and humidity sensitivity and ease and repeatability of calibration. Careful study and comparisons between various manufacturers' specifications before the purchase of a transducer are recommended. Besides commercial catalogs, general information on pressure transducers and fluctuating

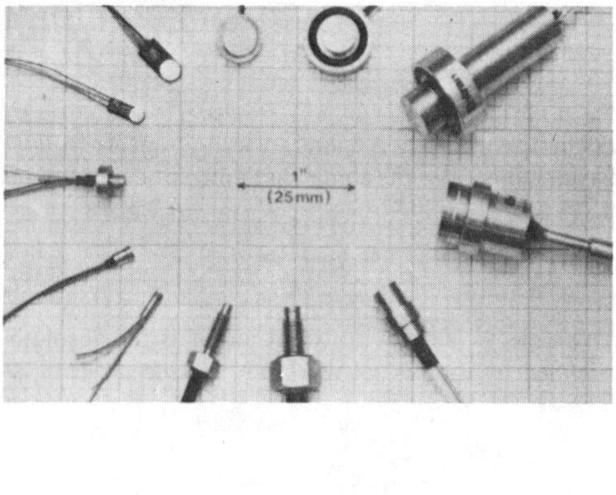

Figure 11. Miniature pressure transducers (courtesy of Entran Devices Inc.).

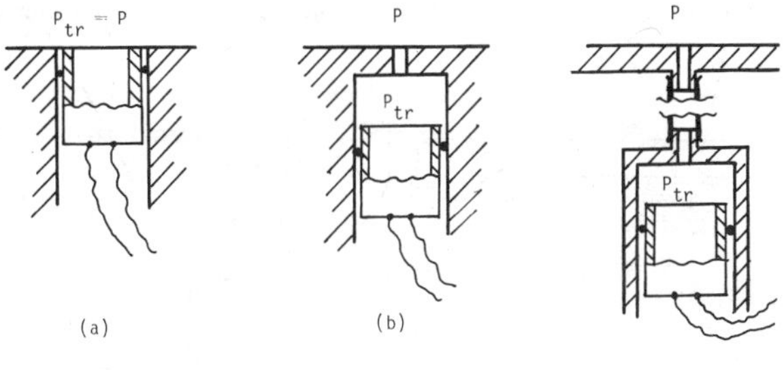

Figure 12. Typical transducer mounting configurations for wall pressure measurement: (a) flush mounting; (b) cavity mounting; (c) remote mounting.

pressure measurement techniques can be found in the articles by Bernstein [38], Willmarth [39], Bynum et al. [40], Corcos [41], and Saloukhin et al. [42].

Wall Pressure Measurement

The measurement of wall pressure fluctuations is, in principle, straightforward, provided that the wall material and shape permit the mounting of a transducer. Whenever possible, a flat diaphragm transducer should be flush-mounted (Figure 12a) on the wall surface in order to eliminate damping and other transmission line effects. If flush installation is impractical for a given apparatus or would result in space averaging of pressure fluctuations due to transducer size, pressure can be transmitted from the wall surface to the transducer through a pressure tap and/or tube (Figures 12b, c). Depending on the size, shape and machining quality of the tap and on the roughness of the wall, an error of several percent points might be introduced on the mean pressure measurement. More significant might be the measuring errors for pressure fluctuations, which depend on the dynamic response of the transducer-transmission system. A review of dynamic models for such systems, including specialized bibliography, can be found in Reference 36. Other techniques for measuring wall pressure, for example the "orifice-hot-wire" technique [43], have at times been proposed but their use remains limited.

In-Stream Pressure Measurement

In-stream pressure is by far more difficult to measure than wall pressure is, in part due to practical difficulties in inserting a pressure probe in the flow, but mainly due to the distortion of the pressure field by the probe itself. Mean pressure in a stream is usually measured with the use of static-tubes, aligned with the flow, which consist of a hollow tube with its upstream end sealed and on which a number of small holes are drilled circumferentially. Inaccuracies in statis-tube response are introduced by improper shaping of the tube nose and positioning of the holes, misalignment with the mean flow direction, flow turbulence, proximity of wall, and compressibility of the fluid.

The measurement of in-stream pressure fluctuations and pressure-velocity correlations remains a challenging task. Attempts to equip static pressure tubes with a variable capacitance microphone [44, 45, 46] or a piezoelectric transducer [47, 48, 49] have been only moderately successful. A promising technique [50, 51] estimates pressure fluctuations from laminar flow rate fluctuations, measured with a hot-film in a special bleed-type tube (Figure 13). This transducer appears insensitive to flow velocity fluctuations and, with electronic compensation, has a flat frequency response up to at least 10 kHz. Non-intruding, optical techniques, measuring pressure indirectly through its relation to other flow variables (for example, the electron beam technique, which provides gas pressure from measurements of flow temperature and density) are currently under development and might find wider application in the future.

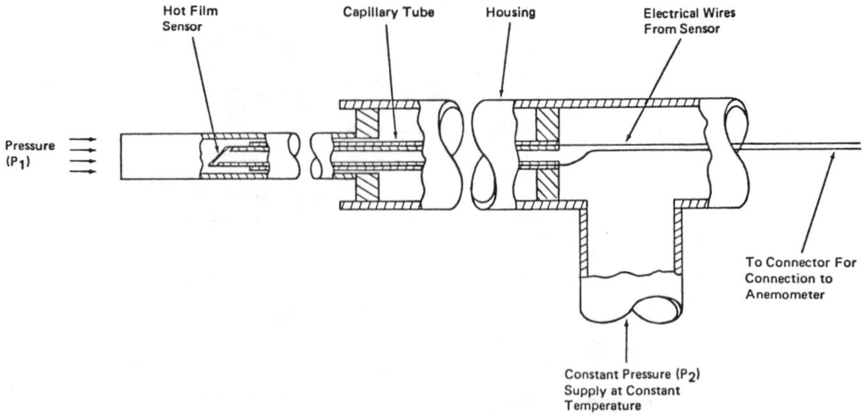

Figure 13. Bleed-type pressure probe. (Courtesy of Thermosystems Inc.).

FLOW VELOCITY MEASUREMENT

Definitions and Overview of Measuring Principles

The velocity of a fluid particle, which (within the range of validity of the continuum hypothesis) occupies an initial position $\vec{X}_0$ at time t_0 and a position $\vec{X}(\vec{X}_0,t)$ at time t, is defined as

$$\vec{V}(\vec{X}_0, t) = \lim_{\Delta t \to 0} \frac{\vec{X}(\vec{X}_0, t + \Delta t) - \vec{X}(\vec{X}_0, t)}{\Delta t} \tag{10}$$

This definition conforms with the material, or Lagrangian, description of fluid motion, which specifies the velocity field in terms of initial position of fluid particle $\vec{X}_0$ and current time t. On the other hand, the spatial, or Eulerian, description of fluid motion specifies velocity $\vec{U}(\vec{x}, t)$ in terms of current space coordinates $\vec{x}$ and time t. Obviously, at all positions and times,

$$\vec{V}(\vec{X}_0, t) = \vec{U}(\vec{x}, t) \tag{11}$$

Lagrangian velocities are obtained when the velocity transducer follows the particle position, as in optical tracing of specially marked fluid particles. In most cases, however, the transducer remains fixed in space and moves along a predetermined path, unrelated to the particle trajectories; the output of such transducers is related to the Eulerian velocity.

A great variety of techniques for the measurement of flow velocity have been developed, based on several different physical principles that can be summarized as follows:

1. *Particle Tracing.* Besides the addition of foreign substances, such as solid spheres, gas bubbles and liquid droplets, fluid elements can also be marked with heat injection, chemical concentration differences and vortex streets shed by upstream objects. Optical tracing of particle motion is the most common procedure, but transducer techniques, sensitive to variations of the marking property have also been used, such as the sensing of heated fluid passage by thermal sensors [52, 53, 54].
2. *Pressure Differences.* Under certain conditions, the use of the equations of motion permits the computation of flow velocity from pressure differences. Mean velocity is routinely estimated from pressure drop measurements in pipe flow or from the dynamic pressure provided by pitot-static tubes. However, since the measurement of in-stream pressure fluctuations remains a difficult task, it is unlikely that the above techniques will be extended to provide the turbulent velocity as well.

3. *Mechanical Effects.* Instruments of this type have stationary or moving elements permitting the measurement of flow velocity through its theoretical or empirical relationship to a measurable force, moment, deflection, or rotational speed. Cup, vane, and drag anemometers with sufficient sensitivity and temporal resolution to resolve turbulence characteristics have found wide application in atmospheric measurements [55]. The turbine flow meter [56, 57], the fiber anemometer [58], the airfoil anemometer [59], and various other lift and/or drag sensors [60] have also found occasional application in turbulence measurements.
4. *Thermal Effects.* In forced convective heat transfer, the heat flux from a heated body (or to a cooled body) is a function of flow velocity. This category includes the most important turbulence transducers, the hot-wire and hot-film anemometers, which will be examined separately.
5. *From Frequency Shifting by Moving Particles.* Two classes of instruments belong to this category, the laser Doppler anemometer (see later section) and the sonic Doppler anemometer [61, 62]. They both utilize the Doppler-Fizeau phenomenon, namely the frequency shift of an incident sound or light wave scattered by a moving particle.
6. *From Wave Propagation.* The most important instruments in this category are sonic [55, 63, 64] and ultrasonic anemometers [65] used respectively in atmospheric and industrial flows. They infer flow velocity from time delays in the propagation of acoustic waves transmitted through a fixed distance in the medium.
7. *From Electric Discharges.* Gas flow velocity can be related through a proper calibration to voltage and current of an electric discharge taking place between two electrodes. Glow discharge or corona discharge anemometers have been used successfully in high speed, turbulent gas flows [66, 67, 68].

Following sections will describe in some detail the two most common turbulent velocity transducers, the hot-wire/hot-film and the laser Doppler anemometers. A listing of bulk-velocity measuring techniques (as a reminder of the need for flow rate monitoring in research and industrial flow facilities) as well as a brief discussion on pressure tubes (reflecting their extensive use as calibration references) are also included.

Bulk Velocity Measurement

Various techniques for the measurement of bulk velocity, i.e. the space-averaged velocity through the cross-section of a pipe or duct, are described in other chapters of the present volume and can also be found in standard references [69, 70, 71]. In brief, one could list the following:

1. Differential pressure meters, including orifice plates, Venturi-tubes, Dall tubes, nozzle flow meters, laminar (or capillary) flow meters and averaging pitot tube meters.
2. Float meters (or rotameters).
3. Positive displacement meters.
4. Turbine meters.
5. Acoustic flow meters.
6. Electromagnetic flow meters.
7. Drag flow meters.
8. Vortex street flow meters.

Pressure Tubes

Integration of the momentum equation along a streamline in an ideal (i.e. inviscid, frictionless), incompressible, weightless, steady flow stream yields the *steady Bernoulli equation*, which, applied between a point in the flow and a stagnation point on the same streamline, becomes (P_0 is the "total" or "stagnation" pressure)

$$P_0 = P + \frac{\rho}{2} U^2 \tag{12}$$

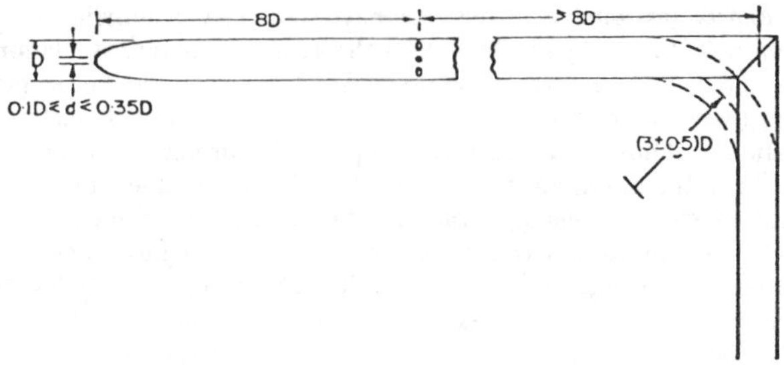

Figure 14. Standard, modified ellipsoidal nose, pitot-static tube (National Physical Laboratory, England).

Pressure tubes are thin hollow tubes, inserted in the flow, which measure the total and/or static pressure and thus permit the calculation of velocity as

$$U = \sqrt{2(P_0 - P)/\rho} \tag{13}$$

The size and shape of pressure tubes has been optimized for different flow regimes. A typical recommended geometry is shown in Figure 14 and more details can be found in the references [72, 73]. The response of pressure tubes in turbulent shear flows has been studied by several investigators but it is not yet completely understood. The main phenomena one has to account for are the following:

1. *Misalignment Effects.* Total pressure tubes are generally insensitive to misalignment with the mean flow up to a few degrees. The least sensitive are square-ended, thin-walled tubes (up to 20°). Special designs (shielded tubes) are available for increasing the insensitive range to over 40°. Particular care should be taken when tubes are used in highly turbulent flows and near separation regions.
2. *Shear Effects.* In shear flows, pitot tubes generally indicate a velocity U_{ind} higher than the actual one. This error can be expressed as the displacement δ of the actual to the effective pressure center of the tube along the y axis:

$$\delta = (U_{ind} - U)/(dU/dy) \tag{14}$$

A collection of several experimental results [74] with tubes having internal to external diameter ratio d/D about 0.6, can be represented by the semi-empirical relation

$$\frac{\delta}{D} \approx 0.19 \qquad K > 0.3 \tag{15a}$$

$$\frac{\delta}{D} \approx 1.025K - 4.05K^3 \qquad K < 0.3 \tag{15b}$$

where the shear parameter is defined as

$$K = D(dU/dy)/(2U) \tag{16}$$

3. *Turbulence Effects.* Our present understanding of these effects is incomplete. It appears that turbulence intensity effects are nonlinearly coupled with probe shape effects, shear effects, yaw effects due to large transverse turbulence fluctuations and effects of size of tube relatively

to the integral length scale of turbulence. As a guide for the order-of-magnitude estimation of turbulence errors in the mean velocity measurement by pitot-static tubes, one can use the expression

$$U = U_{ind}(1 + kq^2/U^2)^{-1/2} \tag{17}$$

where q^2 is the mean squared turbulent kinetic energy and the empirical coefficient k is between 0.3 (for very small turbulence scale) and 2 (for very large turbulence scale).

4. *Wall Effects.* Interference of tube with the flow becomes significant when the tube axis is closer to the wall than about three external diameters (or tube heights for flat tubes). The net effect is that the indicated velocity is lower than the actual one in the absence of the tube.
5. *Viscous Effects.* For extremely small tubes positioned near walls or in highly viscous flows, the assumption of inviscid flow is no longer valid. An estimation of viscous effects can be made based on the empirical relation

$$\frac{P_0 - P}{\frac{1}{2}\rho U^2} = \begin{cases} \dfrac{4.1}{Re_d} & Re_d < 0.7 \\ 1 + 2.8(Re_d)^{-1.6} & Re_d > 0.7 \end{cases} \tag{18}$$

where $Re_d = Ud/\nu$ is the Reynolds number based on the internal tube diameter.

6. *Compressibility Effects.* In subsonic gas flow, the pressure coefficient becomes a function of Mach number, M. A useful expression, assuming isentropic change of state from the undisturbed upstream flow to the stagnation conditions, is

$$\frac{P_0 - P}{\frac{1}{2}\rho U^2} = \left[\left(1 + \frac{\gamma - 1}{2} M^2\right)^{\gamma/(\gamma - 1)} - 1\right]\frac{2}{\gamma M^2} \tag{19}$$

where γ is the ratio of specific heats of the gas. In supersonic flow, a shock wave forms upstream of or in contact with the tube and the correct velocity can only be determined by careful calibration.

7. *Vibration Effects.* A crude correction for the reading of a tube vibrating with frequency f and amplitude A in the direction of the flow is

$$U = U_{ind}\left[1 + \left(\frac{2\pi Af}{U}\right)^2\right]^{-1/2} \tag{20}$$

These corrections should only be considered as order-of-magnitude estimates and should only be applied, when the corresponding causes cannot be eliminated. Considering that the velocity response of pressure tubes is non-linear one should avoid the superposition of corrections for various effects.

Thermal Anemometers

This class includes the most important turbulence transducers, the hot-wire and hot-film anemometers, which have been studied extensively as evidenced by the voluminous bibliography in the field [75]. Excellent reviews of the state-of-the-art for these instruments have been compiled by researchers [10, 76, 77, 78] as well as by specialized manufacturers. Hot-wires were the first to develop historically, these are better understood, less expensive and should be preferred whenever possible. However, hot-wires cannot be used in electrically conductive fluids and dirty or abrasive flow. Hot-films were developed particularly for water and atmospheric flow measurements, but are now used in many other applications. The main aspects of thermal anemometry are summarized in the following.

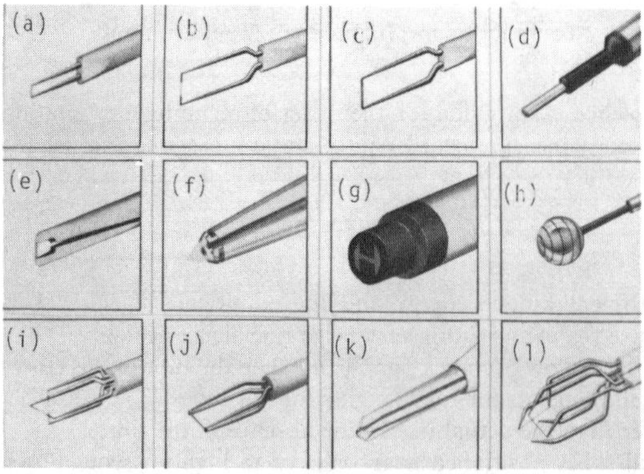

Figure 15. Examples of commercial hot-wire and hot-film probes: (a) platinum-plated, tungsten hot-wire, (b) partially gold-plated hot-wire; (c) cylindrical hot-film; (d) steel clad sensor; (e) wedge-shaped film probe; (f) conical probe; (g) flush-mounted probe; (h) omnidirectional probe; (i) cross-wire; (j) split-fiber probe; (k) V-film probe; (l) triple-wire probe. (Courtesy of DANTEC).

Probe Types, Materials, and Manufacturing Procedures

Hot-wires and, to a lesser extent, hot-films can be manufactured fairly easily and inexpensively from raw materials. Reliable instruments in great variety of shapes and sizes are also available commercially (see Figure 15). Materials that are commonly used as sensing elements are tungsten, platinum, and platinum alloys. Tungsten has good mechanical properties, but can only be used in moderate temperatures, since it oxidizes above 350°C. Platinum has a high coefficient of resistivity, does not oxidize, but has a low mechanical strength. Platinum alloys with 10% Rh or 20% Ir alloys platinum have somewhat inferior thermal qualities, but significantly improved strength and are commonly used as sensors. Hot-wires are soldered, spot-welded or, rarely, mechanically attached on slender prongs, which could be sewing needles, jewellers' broaches or, simply, thin wires with tip diameters 0.05 to 0.3 mm. Bare wires attached to the prongs at their two ends are common, but are subject to aerodynamic and thermal prong interference effects. In an alternate configuration, the sensing element often forms the central segment of a thicker wire, which in turn is attached to the prongs. To accomplish this, the sensing wire is either gold plated at its two ends or etched from Wollaston wire (Wollaston wire has a thin core of platinum or platinum alloy and a thicker sleeve of silver, which can be removed by a stream of nitric acid). Usual thicknesses of hot-wire sensors are in the range 1 to 5 μm, although Pt wires as thin as 0.2 μm as well as thicker than 10 μm have been used in special studies. Typical lengths for hot-wires sensors are between 0.5 and 4 μm.

Hot-films are manufactured by depositing a thin platinum film (about 0.1 μm thickness) on a quartz substrate and coating it with a layer of quartz or alumina, which offers electrical isolation as well as mechanical and chemical protection. Among the common sensor shapes are cylindrical (with external diameter 20 to 100 μm), conical, wedge-type hemispherical and flush-mounted sensors.

Heat Transfer Characteristics

Consider a metallic sensor with purely ohmic resistance R, crossed by a current I and placed in a flow with velocity U and temperature T_a. Energy conservation requires that the rate of heat stored in the sensor must be equal to the balance between the supplied power, in this case the Joule power RI^2, and the total heat flux Q from the sensor to the flow and to supports in contact with the sensor. For a sensor having mass m, specific heat c and uniform temperature T, heat balance is expressed as

$$mc\frac{dT}{dt} = RI^2 - Q \qquad (21)$$

Resistance and temperature of usual hot sensor materials are related by an essentially linear relationship over their operating range.

$$R = R_0[1 + \alpha_0(T - T_0)] \tag{22}$$

Heat transfer from the heated sensor is taking place mostly by forced convection. Heat conduction to the fluid is always negligible; that to the end supports is negligible for sensor aspect-ratio above a hundred or so; although heat conduction from a hot-film to the insulating substrate appears to have some contribution to the overall heat transfer, its effect is usually considered only implicitly through calibration. Heat radiation is also negligible except when the sensor temperature is particularly high, such as in combustion research. Heat convection to a stream generally depends on a large number of geometrical, material, and flow properties. More than seventy years of continuous investigations and use have not yet established a generally acceptable relationship for the thermal sensor response that would be valid in a wide range of flow conditions and fluid properties. It appears that, even if such a relationship were established, it would contain adjustable coefficients to account for individual sensor differences as well as for material property changes due to aging and exposure to dirt and contaminants. Usual practice is to utilize available semiempirical or purely empirical relationships and evaluate the adjustable constants by calibrating the instrument under flow conditions similar to the experimental ones. King's [79] potential flow solution with various modifications [80] remains a popular choice, providing the Nusselt number of the sensor ($Nu = hd/k$; h is heat transfer coefficient, d is the sensor characteristic dimension and k is the thermal conductivity of the fluid) in terms of the sensor Reynolds number ($Re = Ud/\nu$) and overheat ratio $a = (T - T_r)/T_r$ (T_r is a reference temperature, for example the flow temperature during calibration).

$$Nu = (A_1 + B_1 Re^n)(1 + \tfrac{1}{2}a)^m \tag{23}$$

The coefficients A_1, B_1, n, m are, presumably, weak functions of other parameters and should be determined by calibration as constants over a narrow operating range. In any case, for given sensor and fluid properties as well as calibration parameters, the heat balance law cannot be solved because it contains three variables, T, I, and U. The required additional constraint is introduced by using special circuitry: either I or T are held constant ("constant current" and "constant temperature" anemometry, respectively) and, thus, the above law permits a relationship between flow velocity U and some circuit voltage, E.

Constant Current Anemometry

Constant current can be easily achieved by connecting the sensor in series with a large ballast resistor to a constant voltage source. Various bridge configurations (e.g. Figure 16a) have also been used. Due to the circuit simplicity, constant current anemometry was the first to develop and played an important role in turbulence research. However, it has several disadvantages:

1. The frequency response of the system is severely limited, due to the inability of the sensor to adjust its temperature to rapid velocity fluctuations ("thermal inertia" of the sensor). Various electronic compensation circuits have been added to the system to extend the frequency range sufficiently for many turbulence applications.
2. The sensor performance (e.g. the "time constant", according to first-order models) varies with flow velocity, so that continuous adjustments of the circuit are necessary if the flow velocity range is relatively wide.
3. The overheat ratio depends on the flow velocity. At low velocities, the sensor temperature might increase to melting point, while, at high velocities, the temperature difference might decrease causing poor sensitivity.

For these reasons, constant current anemometry is now used only in very few situations, if ever. Compared to the widely used constant temperature anemometry, it is still reputed to provide higher signal to noise ratio in extremely low turbulent intensity flow (less than 0.1%) and to perform more reliably in supersonic and hypersonic flows [81].

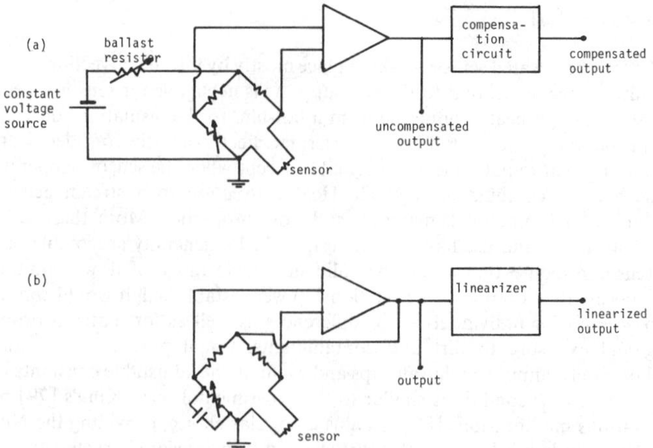

Figure 16. Sketch of hot-wire anemometer circuits: (a) constant current anemometer, (b) constant temperature anemometer.

Constant Temperature Anemometry

This technique utilizes a high gain differential amplifier in a feedback configuration (Figure 16b) that senses differences between the sensor resistance and a reference resistance and adjusts the current to restore the sensor resistance to a preset value. Thus, resistance and, through Equation 22, temperature remain constant independent of the flow velocity. This bypasses the sensor thermal inertia effect and simplifies the heat balance equation into

$$I^2R = Q \tag{24}$$

Using further a simplified expression for the heat convection from the sensor, one can derive a particularly simple relation for constant temperature anemometry as (E is a suitable bridge output voltage)

$$\frac{E^2}{T - T_a} = A + BU^n \tag{25}$$

where n is usually between 0.35 and 0.55 and A, B, n are calibration constants computed for example by linear least squares fitting to data within a given velocity range. More involved derivations and other related expressions can be found in the general references. Unlike other empirical expressions (for example polynomial type curves, used in conjunction with analog linearizers) Equation 25 resembles the theoretical solution of heat transfer from a cylindrical body and, more important, it provides a means for correcting hot-wire response for ambient temperature changes. The validity of this expression has been checked by several researchers. It is recommended for usual hot-wire operating conditions and with minor reservations for hot-film operation. The sensor temperature is selected as high as the sensor material permits. For hot-wires, common *resistance* overheats $(R - R_r)/R_r$ are between 1.5–1.8, while for hot-films in water between 1.05–1.2 and in gases up to 1.5.

In its early stages of development [82, 83] constant temperature anemometry was plagued by the inadequacy of electronic amplifiers, but modern integrated circuitry has provided this instrument with a dynamic range exceeding most turbulence requirements. Standard anemometer bridges are available commercially but can be also manufactured without particular difficulty. Home-made units are much more economical, especially for multi-sensor applications.

Sensor Calibration and Circuit Adjustments

The usual calibration procedure involves placing the heated sensor in a low turbulence stream, for example, the free-stream of a wind-tunnel or the potential core of a calibration jet, and collecting 6–10 pairs of values (E, U) while maintaining the flow temperature at a constant value T_r. Reference velocity is usually measured with pressure tubes or other standard pressure difference instruments. The sensor operating temperature, T, can be determined from the operating resistance, R, using Equation 22. One could easily devise a computational algorithm that would select the optimum exponent n by minimizing the total absolute or squared differences of calibration points from the resulting curve. Otherwise, one could plot $E^2/(T - T_r)$ vs U^n for various values of n between 0.35 and 0.55 and select the value that gives the "best" straight line fit to the data; an example is given in Figure 17.

The nonlinearity of Equation 25 introduces errors in turbulent velocity measurements, especially at large turbulence intensities. Furthermore, it implies that the instrument sensitivity to velocity and temperature changes is not constant but it depends on the operating point. Such errors can be removed by applying theoretical corrections to the results, or, more often, by passing the anemometer output through an analog linearizer, which provides an output, E_L, proportional to velocity U. Digital data acquisition systems permit the linearization of Equation 25 using software. This approach should be preferable to the use of analog linearizers, since it is more flexible, it permits straightforward temperature corrections and avoids the additional signal noise and drift possibly introduced by the linearizer.

Directional Effects

The previous relations assume that the sensor is perpendicular to the flow direction, or, at least, that the sensor-flow relative orientations during calibration and experiment are identical. In a turbulent flow, however, the velocity direction changes continually, affecting the heat transfer characteristics. Then, an "effective cooling velocity," U_{eff} rather than U must be used in the equations. The simplest relationship, assuming negligible heat transfer tangential to a long cylindrical

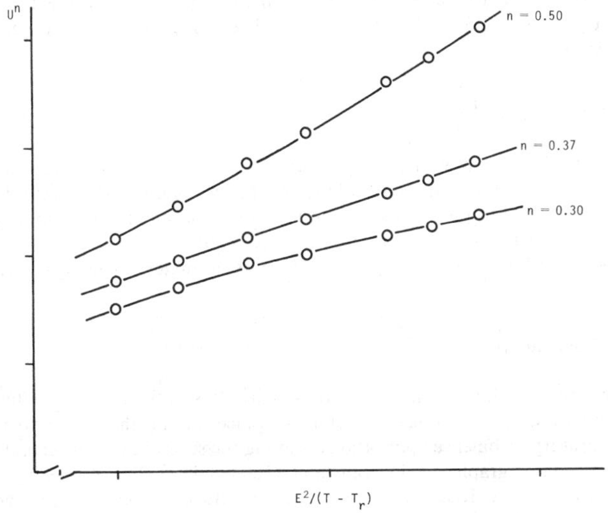

Figure 17. Example of hot-wire calibration data plots for the selection of optimum exponent in "King's law."

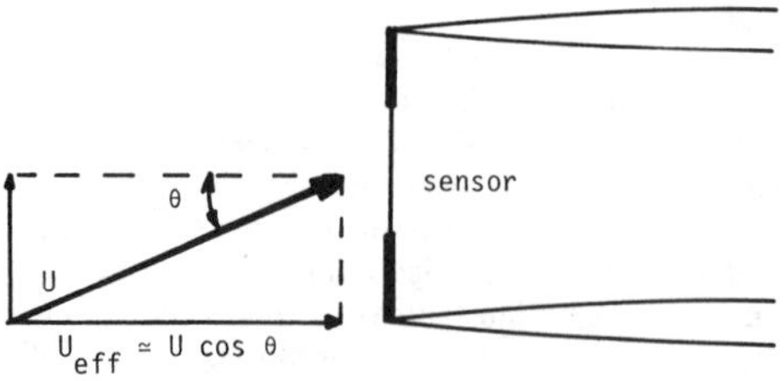

Figure 18. Sketch of directional dependence of heat transfer from a long hot-wire sensor; the velocities are projected on the wire-prong plane of symmetry.

sensor (see Figure 18), would be

$$U_{eff} = U \cos\theta \tag{26}$$

Further empirical corrections can be made by adding adjustable terms, such as [84]

$$U_{eff} = U \cos\theta \sqrt{1 + k^2 \tan^2\theta} \tag{27}$$

where k is found by calibration (usually between 0 and 0.2). Other empirical expressions may be more suitable for non-cylindrical sensors, since in addition to heat transfer changes, they must also account for probe interference effects.

The directional sensitivity of cylindrical sensors permits the measurement of two or three velocity components. A single sensor, operated at three different orientations in a stationary flow can resolve the three r.m.s. turbulent velocities but no cross-statistics. Two sensors at $\pm 45°$ angles with the mean flow direction (Figure 15i) can measure simultaneously the streamwise and one transverse velocity component. Three-sensor arrays (Figure 15l) can resolve the three-dimensional velocity vector. For details about the response characteristics, see the general references.

Temperature Sensitivity

Equation 25 can be used for estimating the temperature variation effects. A probe calibrated in a flow with temperature T_r but operating in a stream with mean temperature $\bar{T}_a$ would indicate a mean velocity, $\bar{U}_{ind}$, different from the actual one, $\bar{U}$. Their difference, to first-order approximation [85] (Figure 19a) is

$$\frac{\bar{U} - \bar{U}_{ind}}{\bar{U}_{ind}} \approx \frac{1}{n}\left(1 + \frac{A}{B\bar{U}_{ind}^{n}}\right)\frac{\bar{T}_a - T_r}{\bar{T} - T_r} \tag{28}$$

Mean temperature changes as well as temperature fluctuations introduce errors on the turbulence statistics, as shown, for example, in Figure 19b for the mean squared turbulence velocity. Although corrections can be applied to the results a posteriori, a simpler and more accurate procedure is to measure simultaneously the flow temperature (using cold-wires or other sensors) and substitute its instantaneous value in Equation 25 using analog circuits or, preferably using software, in digitized signal records.

Laser Doppler Anemometers

Since its conceptualization in the mid-sixties [86], this technique has witnessed intensive development and by now it occupies a prominent place among the flow velocity transducers, especially in laboratory turbulence applications. Among the several existing general references, one could mention the monographs by Durrani and Greated [87], Durst, Melling, and Whitelaw [88], and Watrasiewicz and Rudd [89], the articles by Buchhave et al. [90], Lading [91], and Riethmuller [92] as well as manufacturers' notes [93, 94].

A great variety of system configurations have already been proved suitable for fluid mechanical research and it seems certain that various other schemes and improvements will appear in the future. For economy of space, the present discussion will deal only with a very common system,

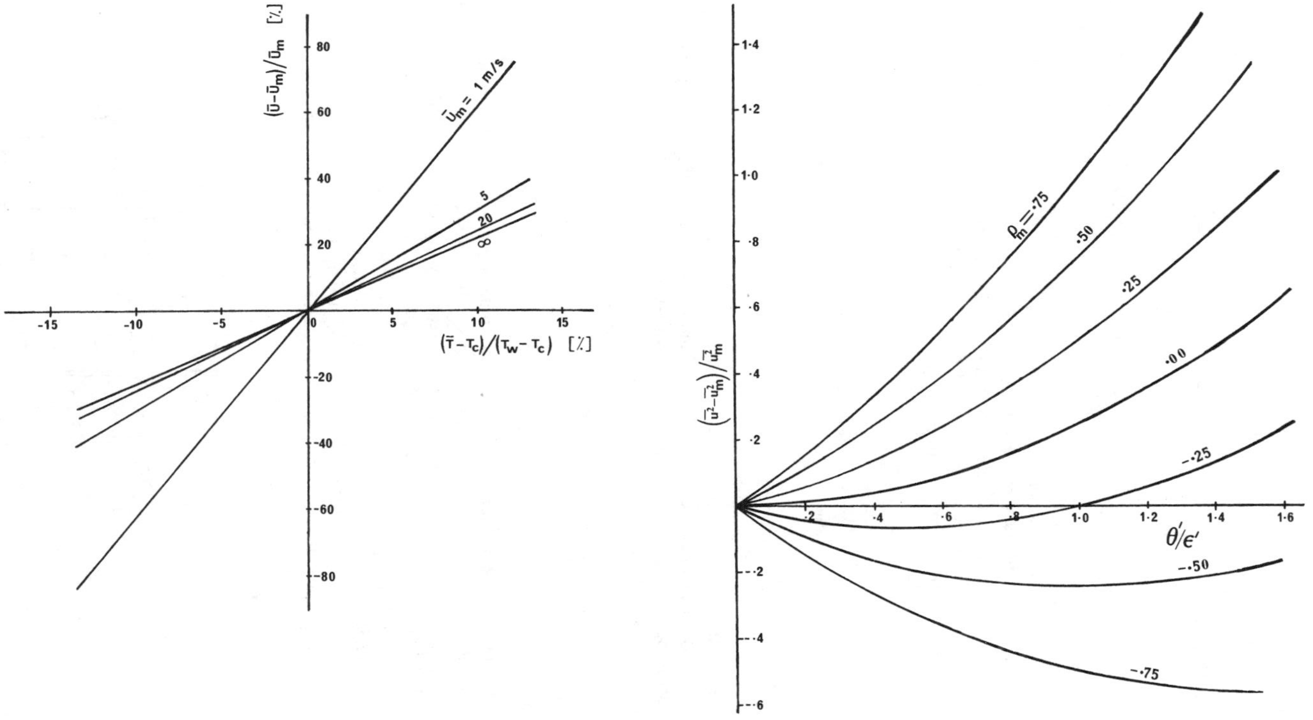

Figure 19. Examples of flow temperature variation effects on the hot-wire output: (a) effect on mean velocity measurement, (b) effect on mean-squared velocity measurement.

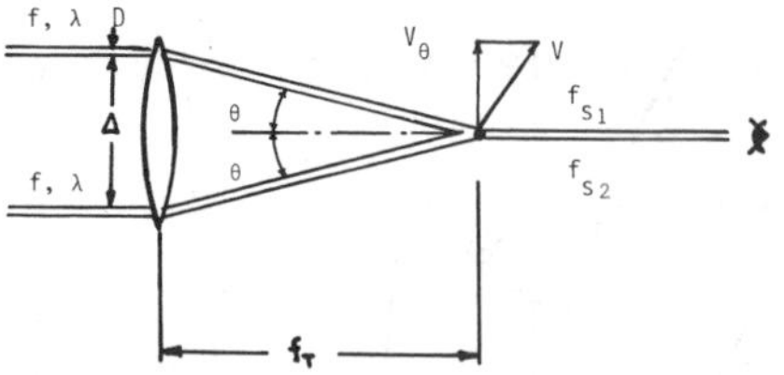

Figure 20. Sketch of a dual-beam system.

the "dual beam" anemometer. Its principle of operation is illustrated in Figure 20. Two identical laser beams with frequency f (wavelength λ) intersect at an angle 2θ. A small particle crossing the intersection volume with velocity $\vec{V}$ scatters the light of both beams, part of which is collected by a photodetector. The frequency of each collected beam is shifted from the incident light frequency by an amount proportional to the particle velocity (Doppler-Fizeau phenomenon). Optical mixing of the two collected beams on the photodetector produces a photocurrent, I,

$$I \sim E_{s_1}^2 + E_{s_2}^2 + 2E_{s_1}E_{s_2}\cos[2\pi(f_{s_2} - f_{s_1})t] \quad (29)$$

where E_{s_1}, E_{s_2} are the amplitudes of the scattered beams and f_{s_1}, f_{s_2} are the scattered beam frequencies. It turns out that the difference $f_{s_2} - f_{s_1}$ is equal to the "Doppler frequency"

$$f_D = \frac{2\sin\theta}{\lambda} V_\theta \quad (30)$$

which is independent of the observation angle. Thus, the flow velocity component V_θ perpendicular to the two-beam bisector can be measured directly from measurements of f_D, via Equations 29 and 30. A great advantage of the response Equation 30 is that it is linear and it contains no undetermined constants, thus eliminating the need for calibration.

A typical dual-beam, forward-scatter, laser Doppler anemometer is sketched in Figure 21. Briefly, its main components and their function are as follows.

The Laser

A laser (an acronym for Light Amplification by the Stimulated Emission of Radiation) light source consists of an optical cavity, which selectively amplifies radiation and produces a light beam that is monochromatic (i.e. its energy is concentrated in extremely narrow bandwidths) and coherent (i.e. all emitted radiation has the same phase, both in space and in time). Laser beams are highly columnated, having divergence angles of the order of a millirad and present a plane of minimal cross-section, called the "waist" of the beam. The light intensity across the beam has a roughly Gaussian distribution; the beam diameter is normally defined as the distance between points where the light intensity has dropped to $1/e^2$ of its maximum value. Most commonly used in fluid mechanics

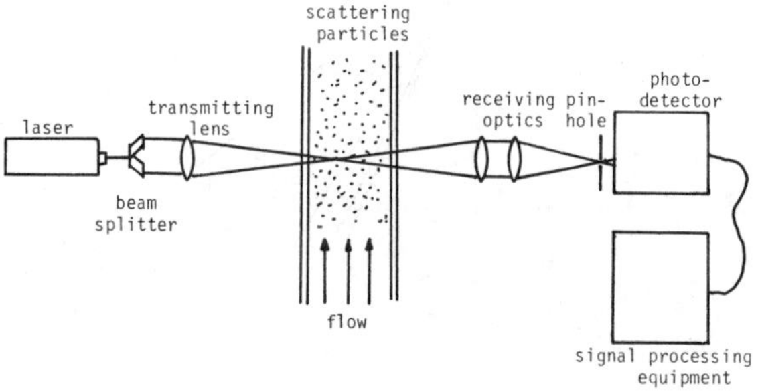

Figure 21. Typical dual-beam, laser Doppler system configuration (forward scattering).

Table 2
Some Properties of Lasers Used in Fluids Research

Laser	Main Wavelengths (μm)	Mode of Operation	Power Range	Typical Beam Waist (mm)
He-Ne	0.633 (red)	continuous	0.5–100 mW	0.5
Ar-Ion	0.515 (green) 0.488 (blue)	continuous	0.005–15 W	0.5
CO_2	about 10.6 (9.2 to 11.0)	continuous or pulsed	1–1,000 W	20

are the He-Ne and the Ar-Ion lasers and to a lesser degree the CO_2 laser. Their main features are shown in Table 2 [95].

Transmitting Optics and Measuring Volume

The laser beam is first split into two parallel beams by passing through a "beam splitter" (for example, a partially transparent mirror); then the two beams are brought to intersection within the flow region, with the use of a convergent lens or mirror. The transmitting system might also contain filters, polarizers, beam path equalizers, beam expanders, and other special components. The measuring volume, designed to contain the waists of the two focused beams, has an ellipsoidal shape, defined by the locus of $1/e^2$ intensity points (Figure 22), with width equal to the focussed beam waist

$$d = \frac{4f_T\lambda}{\pi D} \tag{31}$$

(f_T is the transmitting lens focal distance and D is the diameter of beam waist before the transmitting lens), length

$$l = \frac{d}{\sin\theta} \tag{32}$$

and height

$$h = \frac{d}{\cos\theta} \tag{33}$$

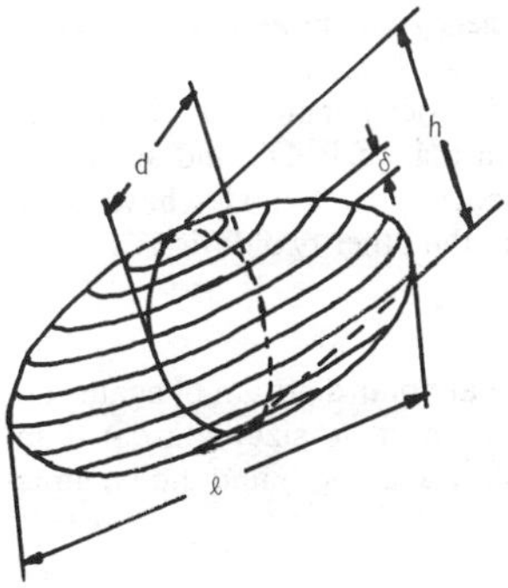

Figure 22. Measuring volume dimensions in a dual-beam system.

Interference of the two light beams in the measuring volume produces a number of nearly parallel fringes (i.e. bright and dark stripes) whose spacing is

$$\delta = \frac{\lambda}{2 \sin \theta} \tag{34}$$

and number is (Δ is the parallel beam separation)

$$n = \frac{4}{\pi} \frac{\Delta}{D} \tag{35}$$

The Doppler signal can be also interpreted by the fringe-crossing sequence of the moving particle.

Light Scattering Particles

The intensity of light scattered by a spherical particle depends on particle size, refractive index ratio, light wavelength and angle between incident and scattered light; it can be estimated using Mie scattering theory. The light intensity is maximum, when the scattering angle is zero (forward scattering), as illustrated in Figure 23. The total power of scattered light generally decreases with decreasing particle size, but not in a monotonic fashion. Naturally existing particles, such as dirt, lint etc. are sometimes sufficient to produce a detectable Doppler signal, especially in forward scattering in liquids. In general, however, addition of scattering particles ("seeding" of the flow) is necessary. Special devices for the generation and distribution of particles in gas and liquid flows are available [88, 93, 94]. Usual particle size is of the order of 1 μm.

Receiving Optics

The scattered light is collected by a convergent lens or mirror focussed on the measuring volume and "cleared" with the use of a pinhole and, occasionally, other filters.

Photodetectors

These are devices that utilize the "photoelectric effect," namely the absorption of photons and emission of electrons, which constitute the photoelectric current. Their quality is measured by the "quantum efficiency," which represents the relative number of emitted electrons with respect to collected photons. The photocurrent is subject to several sources of noise: "photon noise" or "shot noise" is due to random fluctuations in the rate of collected photons and to background illumination, while "electronic noise" or "thermal noise" is due to amplification of current within the photodetector or in external amplifiers. Two types of photodetectors are in common use:

1. The photomultiplier tubes (PMT) have a low quantum efficiency but have a large internal amplification, which provides a larger signal to noise ratio for weak signals.
2. The photodiodes or photoelectric cells have a high quantum efficiency but no internal gain and require external amplification. Their cost is much lower than that of PMTs and should be preferable when the signal is sufficiently strong. The "avalanche" photodiodes have some internal gain and exhibit a performance intermediate between the other two devices.

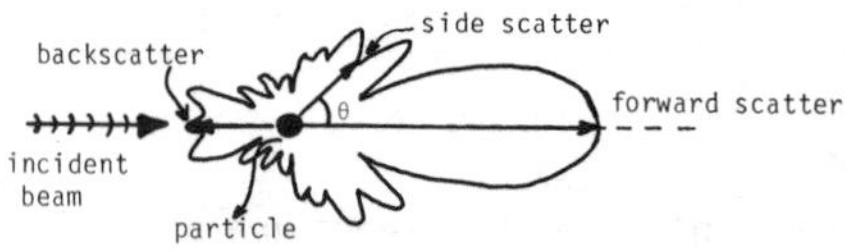

Figure 23. Typical polar diagram of scattered light intensity; the particle size is larger than the incident light wavelength and the scale is logarithmic.

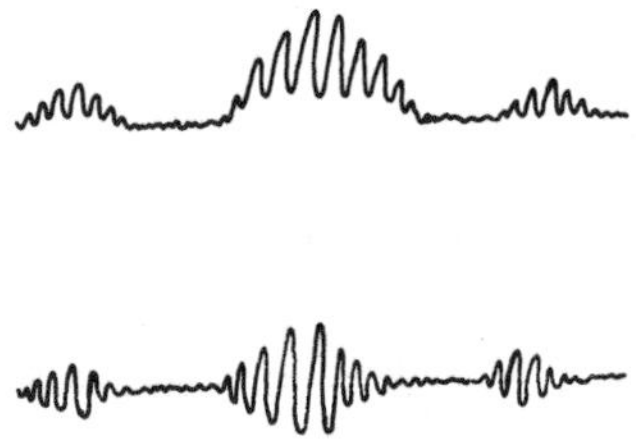

Figure 24. Sketch of a low particle density Doppler signal with (top) and without (bottom) the pedestal.

Doppler Signal Processing

The appearance of the photodetector output signal depends on the collected light intensity ("photon density") and on the number of particles crossing the measuring volume at any given time ("burst density"). At extremely low photon densities, the signal consists of a train of pulses corresponding to individual collected photons. Then special techniques, such as photon counting and correlation must be used to recover the Doppler frequency. In the following, it is assumed that the photon density is large enough to provide a continuous signal, according to Equation 29.

When the particle density is low enough for the measuring volume to contain at most one particle, the signal consists of a series of "bursts" (corresponding to particle crossings and called the "pedestal") each of which is an amplitude-modulated sinusoidal function with frequency f_D (Figure 24–top). The amplitude modulation reflects the light intensity variation within the measuring volume, while differences between bursts reflect differences between particle sizes and crossing paths. After removal of the pedestal and high frequency noise, the Doppler frequency can be found with the use of "Frequency Counters," which basically time a fixed number, N, of zero crossings. Then, the particle velocity is simply (δ is the fringe spacing)

$$V_\theta = \frac{N\delta}{\Delta t} \tag{36}$$

Frequency counters are very accurate and have a wide dynamic range, but they provide an output at irregular intervals and thus necessitate the use of special statistical procedures. Furthermore, they cannot perform at higher particle concentrations, when two or more particles are likely to coexist in the measuring volume.

On the other extreme, if particle concentration is high enough for many particles to occupy the measuring volume at all times, the Doppler signal (Figure 25) is continuous, with frequency f_D, but with phase and amplitude that vary randomly. This randomness introduces an additional error in the Doppler frequency measurement, called "ambiguity noise." Devices suitable for such signal types are "Frequency Trackers." In brief, they contain an electronic oscillator, which scans a frequency range and "locks" at the Doppler frequency providing an analog output proportional to f_D, and, thus, to particle velocity. As the particle velocity fluctuates, the tracker output is continually updated within a fixed number of Doppler cycles. When there are no particles in the measuring volume or when the particle velocity is outside the specified range of operation, the tracker "loses" the signal; commercial trackers usually display the last valid signal value until the tracker locks again. Trackers provide an analog output, which is readily suitable for statistical analysis and are preferable in flows where "heavy seeding" is possible. Their main disadvantage is limited dynamic range.

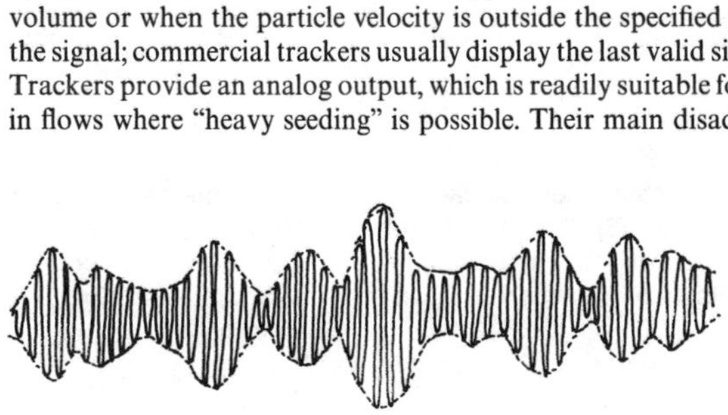

Figure 25. Sketch of a high particle density Doppler signal (pedestal removed).

Frequency Shifting

Doppler signals contain an ambiguity with respect to flow direction and would produce erroneous readings near regions of reverse flow. This ambiguity can be resolved by frequency shifting of one beam by an amount Δf. This produces a motion of the interference fringe pattern with velocity

$$V_f = \Delta f\, \delta \tag{37a}$$

so that the particle velocity would be given as

$$V = (f_D - \Delta f)\delta \tag{37b}$$

which can be zero, even negative. Frequency shifting also increases the dynamic range of the instrument by increasing the number of fringes intersected by each particle crossing the control volume. Frequency shifting is necessary for highly turbulent flows. It is usually accomplished by passing one laser beam through one or more acousto-optical cells, "Bragg cells"; each cell shifts the beam frequency by about 40 MHz.

Multicomponent and Special Systems

Since laser-Doppler systems do not interfere with the flow or with each other, multi-component and multi-point measurements can in principle be made by using a number of independent systems. For economic reasons, though, such systems are designed to utilize several common components. For example, the two main frequencies of a single Ar-Ion laser can be used for measuring two, even three, velocity components (Figure 26). Transmission of laser beams through optical fibers [96] has recently permitted the use of the system in flows with complex boundaries, such as rod bundles.

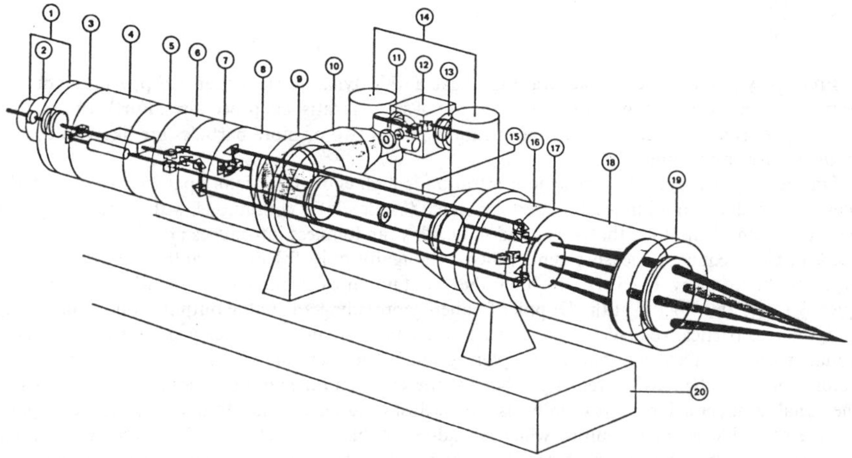

Figure 26. Diagram of a four-beam, two-component, back-scatter, laser Doppler system: (1) cover and retarder, (2) beam waist adjuster, (3), (5), (7) beamsplitters, (4) Bragg cell, (6) beam displacer, (8) backscatter section, (9) support, (10) photomultiplier optics, (11), (13) interference filters, (12) color separator, (14) photomultiplier section, (15) pinhole, (16) beam translator, (17) lens mounting ring, (18) beam expander, (19) front lens, (20) mounting bench; not shown are the Ar-ion laser and the signal processing equipment. (Courtesy of DANTEC)

TEMPERATURE AND CONCENTRATION MEASUREMENT

Temperature Measurement Techniques

Temperature monitoring in a flow is often necessary in order to prevent or correct temperature contamination of velocity measurement techniques, especially in hot-wire anemometry. If temperature changes are slow and non-localized, such as time of the day temperature variation, they can easily be monitored with slow-response instruments, such as liquid-in-glass thermometers and thermistors. Similar thermometers are used as references during static calibration of turbulence transducers.

The measurement of temperature variations, however, is often a main target of turbulence investigations, since it is related to heat transfer processes. Heat at small amounts, introduced locally by electrical or optical heating of bodies, is often used as a representative passive scalar contaminant, whose transport characteristics can be also applied to a wide class of other contaminants. In these situations, there is a need for measuring temperature fluctuations with spacial and temporal resolutions comparable to those required in velocity measurement.

Miniature thermocouple junctions that can be inserted in a flow are available with time constants in the range 2 to 10 ms [97]. Thermistor beads have time constants of the order of 1 s minimum [98] (miniature thermistor probes used as velocity transducers are reported to respond to up to 40 Hz fluctuations [99]). The only temperature instruments whose frequency response extends to several kHz appear to be thin resistance thermometers, similar in appearance to hot-wires, but operating at extremely low overheats (hence, the name, "cold-wires"). These are usually platinum or tungsten wires with diameters 0.25 to 2 μm and lengths between 0.5 and 2 mm, mounted on regular hot-wire probes. They operate in constant current mode, with current supplied by AC or DC bridges [100, 101]. For extremely low currents, less than about 0.5 mA, their velocity sensitivity becomes negligible, (see also Figure 27 [102]) while their temperature sensitivity is (E_T is the amplified voltage

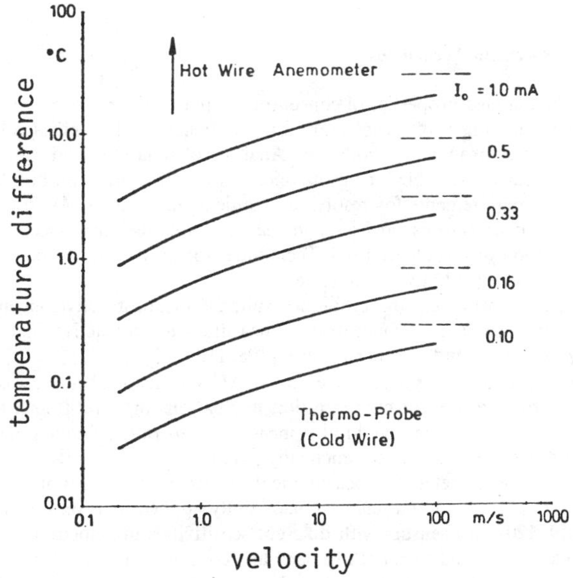

Figure 27. Diagram showing the equivalent velocity and temperature effects on the voltage across a 0.65 μm diameter platinum wire at small currents [102].

across the wire, g is the amplifier gain)

$$\frac{\partial E_T}{\partial T} \approx \alpha_0 R_0 I g \qquad (38a)$$

Since the current has to be kept low and the product $\alpha_0 R_0$ is determined by the wire material and size, a large gain-bandwidth product amplifier is required to detect small temperature fluctuations. Detailed studies of the resolution of cold-wires [100, 103, 104, 105] have established that its performance is suitable for usual gas turbulence applications. One must keep in mind, however, that cold-wire performance deteriorates rapidly with aging and dirt accumulation, especially when exposed to atmospheric conditions [106]. Electronic compensation circuits similar to those used in constant-current anemometry, are sometimes employed to improve the system frequency response [107].

Flow temperature measurement by inserted probes is accurate in incompressible flow, but becomes erroneous as the Mach number increases above about 0.2. Then, the probe reads the "total" or "stagnation temperature" (C_p is the gas specific heat under constant pressure).

$$T_s = T + \frac{1}{2C_p} U^2 \qquad (38b)$$

Consistent cold-wire calibrations have been performed in both subsonic and supersonic flows, while in the transonic range cold-wire response has not yet been totally understood. Other sources of errors in high-speed flows are due to radiation and free molecular flow effects [108, 109].

Various optical methods have been developed for high-temperature measurements in combustion and reactive processes. Such flows are outside the present scope, so only a few general references are provided here [108–111].

Concentration Measurement Techniques

Knowledge of the detailed properties of concentration (i.e. of the relative amount of a chemical species in a fluid) fluctuations is often necessary for the prediction of chemical reactions, industrial mixing processes, and environmental pollution. Analysis of removed fluid samples by chemical, physical or optical means is suitable for mean concentration measurement, but does not meet the spatial and temporal requirements for resolving turbulent fluctuations. As discussed earlier, additives such as smoke, dyes, various particles, and heat are sometimes introduced in a turbulent flow as markers for the study of turbulence itself. Techniques for measuring the concentration of such addities are as varied as the additives themselves.

The present discussion will focus on techniques, suitable for non-reactive, incompressible turbulence. Discussion and references for concentration measurements in reactive, high temperature and compressible flows can be found in other reviews [108, 112, 113].

A single electrode conductivity probe, driven by an AC Wheatstone bridge, was used by Gibson and Schwarz [114] to measure salt concentration fluctuations in water-tunnel turbulence. Their probe had a resolution 0.003% and could also measure temperature fluctuations with resolution 0.001°C. Similar probes are available commercially [115].

The use of thermal anemometers for measurement of species concentration in gas mixtures was suggested by Corrsin [116, 117] and has been sucessfully applied by several investigators. In one variation [118, 119, 120] two sensors with different sensitivities to velocity and composition are operated next to each other and then both velocity and concentration are calculated by combining the two outputs according to a two-dimensional calibration chart (e.g. Figure 28). The velocity sensitivity effect on thermal sensors can be eliminated by placing the sensor in a suction stream with constant velocity controlled by a sonic nozzle [121–124] (Figure 29). Then, a single sensor provides an output directly proportional to concentration. Such techniques are directly applicable to isothermal flows, otherwise temperature corrections become necessary [120].

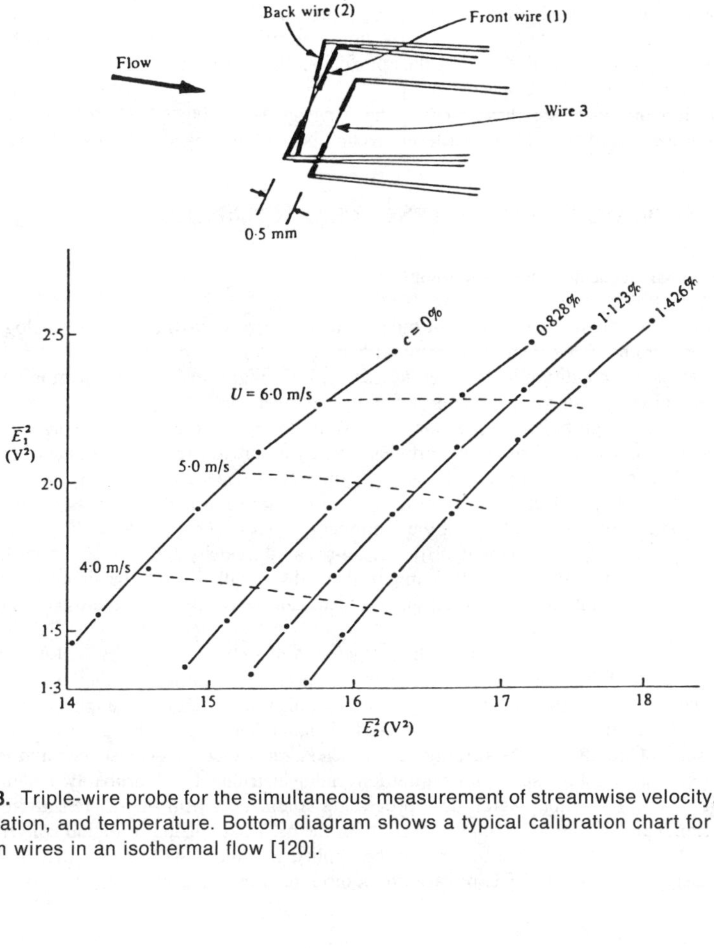

Figure 28. Triple-wire probe for the simultaneous measurement of streamwise velocity, helium concentration, and temperature. Bottom diagram shows a typical calibration chart for the two upstream wires in an isothermal flow [120].

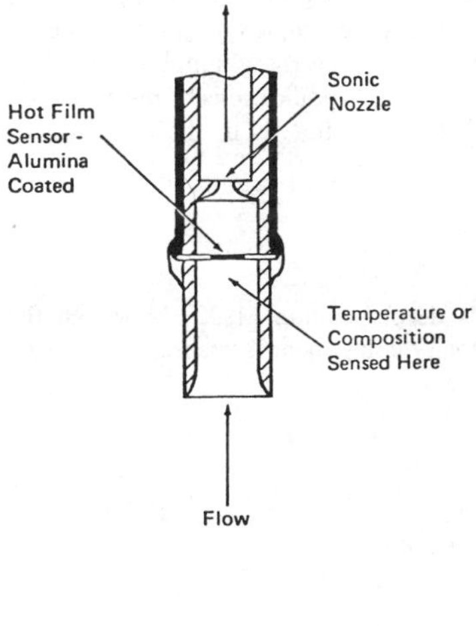

Figure 29. Suction probe tip for the measurement for gas concentration fluctuations. (Courtesy of Thermosystems Inc.)

Optical measurements of concentration fluctuations have been made by several investigators. As typical examples, mention can be made of the light-scatter technique for measuring concentration of suspended particles in gas flows [125], the optical fiber light-technique [126] and the laser-induced fluorescence technique [127]. Spectroscopic techniques have been developed for measuring species concentration in gas flows; these include the Rayleigh scattering technique, suitable for two-component gases, and the Raman scattering technique, suitable for multicomponent gases.

THE MEASUREMENT OF TURBULENCE CHARACTERISTICS

Random Signal Acquisition and Processing

In information provided by most turbulence transducers is in the form of analog electrical signals and permits convenient processing with both analog and digital systems. In most cases, some initial "signal conditioning" is often necessary. This might involve amplification, attenuation and/or DC offset of a signal in order to bring it to a level compatible with the other instrument requirements; and high-pass, low-pass, or band-pass filtering in order to remove noise, interfering inputs, or instrumentation drift. Since turbulent energy is spread over a wide range of frequencies, the choice of filter cut-off points must be carefully optimized, depending on the signal-to-noise ratio and on the type of turbulent quantity measured. For example, a high Reynolds number velocity signal, low-pass filtered to one third of the Kolmogoroff frequency $f_K = \bar{U}(\varepsilon/\nu^3)^{1/4}/2\pi$, will provide a reasonably accurate measurement of the mean-squared velocity but not of the mean-squared velocity derivative. Use of analog filters might also substantially distort the signal waveform due to non-linear phase shift characteristics; this problem can be avoided with the use of digital filters [128].

A variety of statistical information can be extracted from a random signal by analog instrumentation, designed specifically for providing correlations, spectra, intermittency indicators etc. However, current availability of low-cost, reliable digital data aquisition systems has simplified extremely the extraction of statistical information from electrical signals and has rendered most analog instruments obsolete. Only the simplest (averaging voltmeters, r.m.s. meters, oscilloscopes) and some other high-performance (e.g. high-speed differentiators) analog instruments will probably remain in use for turbulent signal processing. Digital data processing commences with the discretization of a conditional signal at fixed sampling intervals, Δt, with the use of an analog-to-digital converter, interfaced with the computer. Discretization can be applied to either "live" or recorded signals. Then an analog signal $S(t)$, $0 < t < T$ is replaced by a discrete "time history"

$$S_i = S(t_i) \qquad i = 1, 2, \ldots, N \tag{39}$$

where $t_i = (i - 1)\,\Delta t$ and $T = (N - 1)\,\Delta t$. To avoid biasing the sampling time Δt must be smaller than $1/2f_H$, where f_H is the highest frequency existing in the signal, for example the cut-off frequency of a low-pass filter must be set to at most f_H. A thorough discussion concerning sampling and time-series analysis can be found in standard references [129, 130, 131]. Modifications of the statistical analysis would be necessary, when the sampling interval Δt is variable as in the case of low-particle-density laser Doppler signals [132].

Moments and Correlations

A general approach in turbulence analysis is Reynolds decomposition [1, 33] by which the instantaneous value S of a random variable (such as a velocity component, pressure, transported scalar etc.) is decomposed into a "mean" $\bar{S}$ and a fluctuation s

$$S = \bar{S} + s \tag{40}$$

where, by definition, s has zero mean.

In general, all statistical properties of random processes must be calculated by "ensemble averaging", namely by averaging corresponding values measured over a number of repeated realizations of the same experiment. In many situations, however, the turbulent flow can be assumed stationary (i.e. with statistical properties which are independent of time origin shifts) and ergodic (i.e. with all realizations of the experiment having the same statistical properties). Then, it is possible to average over a finite sample of a single realization. For example, the mean $\bar{S}$ can be estimated from a discrete time history, as

$$\bar{S} = \frac{1}{N} \sum_{i=1}^{N} S_i \tag{41}$$

The number of points N must be sufficient large for the sample average to approximate the population average. Usually, N is of an order-of-magnitude between 10^3 and 10^6; the averaging time $T = (N - 1)\,\Delta t$ must be sufficiently larger than the largest characteristic time of the problem (e.g. 100 times the integral time scale). Special care in the selection of sample size and averaging time must be taken, when the process is not precisely stationary.

The mean (also called "first-order moment") is normally removed from the signals for further statistical analysis. Then one can estimate any central moment of n-th order as

$$\overline{s^n} = \frac{1}{N} \sum_{i=1}^{N} s_i^n \tag{42}$$

where

$$s_i = S_i - \bar{S} \qquad i = 1, 2, \ldots, N \tag{43}$$

When two or more random variables S_a, S_b, ... are measured, their covariances (or cross-correlations or double correlations) can be estimated as

$$\overline{s_a s_b} = \frac{1}{N} \sum_{i=1}^{N} (s_{ai} s_{bi}) \tag{44}$$

Similarly, triple and higher-order correlations can be estimated as

$$\overline{s_a s_b s_c \ldots} = \frac{1}{N} \sum_{i=1}^{N} (s_{ai} s_{bi} s_{ci} \ldots) \tag{45}$$

This procedure is suitable, for instance, for the measurement of Reynolds stresses $-\rho\overline{u_i u_j}$, the heat flux vector components $\overline{u_i \theta}$ etc. Improved accuracy of the estimates can be achieved by ensemble averaging of several "sample averages" obtained as above [132]. Cross-correlations are often presented in normalized form by dividing by the corresponding r.m.s. values. The resulting "correlation coefficient"

$$\rho_{ab} = \overline{s_a s_b}/(\overline{s_a^2}\,\overline{s_b^2})^{1/2} \tag{46}$$

takes values between -1 and 1 depending on the flow type and the coordinate axes orientation. For example, in grid-generated, nearly isotropic turbulence

$$\rho_{u_i u_j} \simeq 0, \qquad i \neq j$$

while in shear flows with mean velocity mainly in x_1-direction and mean shear mainly in the x_2-direction

$$|\rho_{u_1 u_2}| \approx 0.4 - 0.6$$

When the quantities S_a, S_b, ... are measured at the same point in the flow, $\overline{s_a s_b}$ etc. are called "one-point" correlations. If, however, S_a, S_b are measured at different points one gets "two-point," "three-point" etc. correlations.

The measurement of multi-point correlations, also known as spacial correlations, is very useful since they describe the average structure of the flow. In general, inhomogeneous turbulence, spacial correlations are functions of the positions of all measuring points. When, however, the flow can be assumed homogeneous (i.e. with statistical properties which are invariant in space) along one, two, or three axes, these correlations are only functions of the magnitudes are relative orientations of the separation vectors between measuring points. Specialization of the definitions and reviews of related theories are included in all turbulence text-books.

Temporal correlations are measured by introducing a time delay, τ, into a signal and then computing the averages of products of delayed and non-delayed quantities. Temporal correlations are also presented in normalized form: the "autocorrelation coefficient" is defined as

$$R_a(\tau) = \frac{\overline{s_a(0)s_a(\tau)}}{\overline{s_a^2}} \tag{47}$$

while temporal cross-correlation coefficients are defined as

$$R_{ab}(\tau) = \frac{\overline{s_a(0)s_b(\tau)}}{(\overline{s_a^2}\,\overline{s_b^2})^{1/2}} \tag{48}$$

Equations 47 and 48 assume stationarity of the flow. Correlations involving time-delayed quantities can be easily estimated from digital records of the signals, as, for example,

$$\overline{s(0)s(\tau)} = \frac{1}{N}\sum_{i=1}^{N} s_i s_{i+j} \tag{49}$$

where

$$j\,\Delta t = \tau \tag{50}$$

More general types of correlations involve both spacial and temporal separation of the measuring points and are known as space-time correlations.

For example, a two-point velocity, space-time correlation coefficient in a homogeneous, stationary turbulent flow is defined as

$$R_{ij}(\vec{r};\tau) = \frac{\overline{u_i(\vec{x};t)u_j(\vec{x}+\vec{r};t+\tau)}}{(\overline{u_i^2(\vec{x})}\,\overline{u_j^2(\vec{x}+\vec{r})})^{1/2}} \tag{51}$$

Typical examples of correlation coefficients are shown in Figures 30 and 31 [134]. Further information can be found in textbooks of turbulence and in specialized review articles [132, 135, 136].

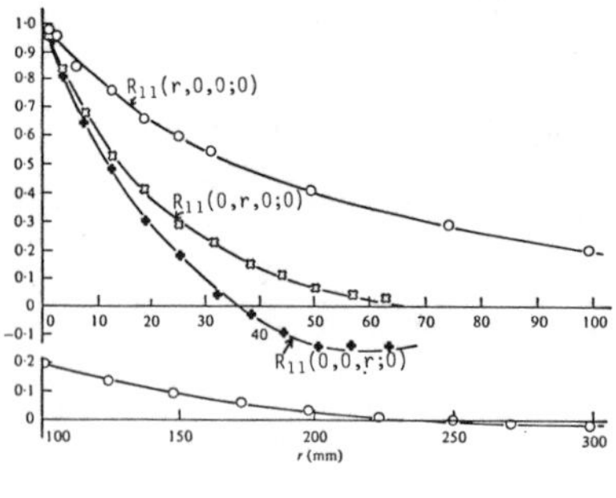

Figure 30. Streamwise turbulent velocity autocorrelation functions in a nearly homogeneous shear flow [134].

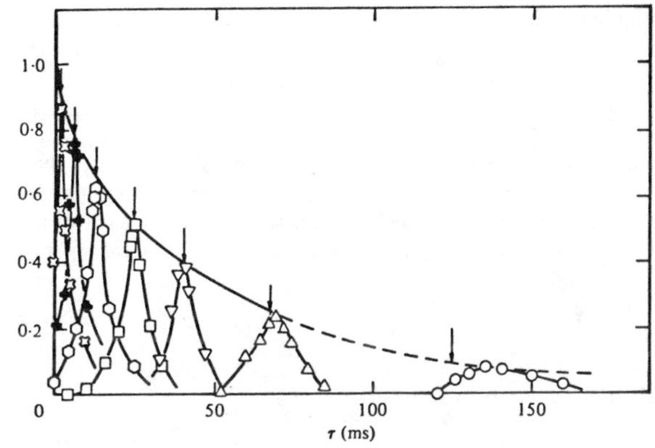

Figure 31. Temperature space-time correlations in a heated nearly homogeneous shear flow [134].

Spectra

Energy spectra play an important role in the theory of turbulence and, thus, considerable insight would result from their measurement. Unfortunately, "wave-number spectra," defined via the Fourier transform of spacial correlation functions, are not easily measurable and, in fact, consistent spectral theories have only been developed for exactly homogeneous and stationary turbulence, a rather unlikely experimental situation. On the other hand, one-dimensional "frequency spectra" are relatively easy to measure. They are defined as the Fourier transform of a time-correlation

$$F(f) = \int_{-\infty}^{\infty} R(\tau)e^{-j2\pi f\tau}\,dt, \qquad j = \sqrt{-1} \tag{52}$$

Early spectral measuring techniques involved the definition (52) or narrow band-pass filtering. An alternative approach, estimates the frequency spectrum F(f) of a discrete time history s_i, $i = 1, 2, \ldots, N$ as

$$F_k = \frac{1}{N}\sum_{i=1}^{N}\left|s_i e^{-j\frac{2\pi ik}{N}}\right|^2, \qquad k = 1, 2, \ldots, N/2 \tag{53}$$

which provides N/2 discrete values of the spectrum at frequencies

$$f_k = (k-1)/(N\,\Delta t), \qquad k = 1, 2, \ldots, N/2 \tag{54}$$

The application of highly efficient computational algorithms ("Fast Fourier transforms") has permitted the fast and inexpensive computation of Equation 53. Ensemble averaging and smoothing techniques are further applied to the raw spectral estimates to reduce scatter and statistical uncertainty.

The only component of the one-dimensional wave-number spectrum tensor that can be estimated easily from measured frequency spectra is the one corresponding to spacial separation in the streamwise direction, say x_1. When the turbulent intensity is sufficiently small for Taylor approximation to be valid[137,138], the streamwise one-dimensional, wave-number spectrum is

$$E_{11}(\kappa_1) = \frac{\bar{U}}{2\pi} F(f) \tag{55}$$

where the wave-number κ_1 is

$$\kappa_1 = \frac{2\pi f}{U} \tag{56}$$

These relations are correct only when the turbulent intensity is small such that Taylors' approximation is valid [137, 138]. An estimate off the "three-dimensional energy spectrum" (reminder: this is a strictly isotropic concept) $E(\kappa)$ is sometimes based on measurements of the one-dimensional spectrum of the streamwise velocity u_1 as

$$E(\kappa) = \frac{1}{2}\kappa^3 \frac{\partial}{\partial \kappa}\left[\frac{1}{\kappa}\frac{\partial}{\partial \kappa} E_{11}(\kappa)\right] \tag{57}$$

Then, the "dissipation spectrum" can be estimated as

$$D(\kappa) = 2\nu\kappa^2 E(\kappa) \tag{58}$$

For further information about spectral measurement techniques, one can consult special reviews [132, 139] and textbooks [130, 131].

Length and Time Scales

Turbulent phenomena are characterized by a continuous distribution of length and time scales. However, it is possible to define certain magnitudes, which are characteristic, on the average, of particular turbulent activities.

The integral length scales of the Eulerian velocity fluctuations represent the order of magnitude of relative motions contributing most to the turbulent kinetic energy ("energy containing eddies"). Their usual definition is (no summation over i, j, k)

$$L_{ij,k} = \int_0^{r_0} \frac{\overline{u_i u_j}(r_k)}{\overline{u_i u_j}(0)}\, dr_k \tag{59}$$

where r_k is point separation along the x_k axis. The integration limit should, in principle, approach infinity; in practice, r_0 is taken as the first zero crossing of the correlation coefficient, if there is one, or as a separation where the correlation coefficient becomes negligible. A discussion of related problems can be found in Reference 140.

Eulerian integral time scales represent the typical duration of turbulent activity of the energy containing eddies ("eddy turnover time") and are defined as

$$T_{ij} = \int_0^{T_0} R_{u_i u_j}(\tau)\, d\tau \tag{60}$$

where the limit T_0 is defined in a manner similar to r_0. Alternatively, integral time scales can be estimated from the low-frequency range of one-dimensional frequency spectra as

$$T_{ij} = \frac{1}{4\overline{u_i u_j}} \lim_{f \to 0} F_{u_i u_j}(f) \tag{61}$$

A different type of scales representing to a certain degree the relative intensity of kinetic energy dissipation are the Taylor microscales, usually defined as (δ_{ij} is Kronecker's delta, $\delta_{ij} = 0$ if $i \neq j$, $\delta_{ij} = 1$ if $i = j$).

$$\lambda_{ij} = \left(\frac{(2 - \delta_{ij})\overline{u_i^2}}{\overline{(\partial u_i/\partial x_j)^2}}\right)^{1/2} \tag{62}$$

The measurement of space-correlations and length scales requires simultaneous measurements at two points. This increases the complication and cost of experimental set up and introduces possible errors due to probe interference. Time correlations and scales, on the other hand, are computed from the generation of time delayed signals, a trivial matter when digitized records are available. Thus, it is quite common to estimate spacial correlations and length scales in the streamwise direction from corresponding temporal quantities. This can be justified only when the turbulence intensity is low enough (typically, less than 10%) so that the structure of turbulent eddies is nearly "frozen" during their convection past the measuring probe (Taylor's hypothesis, also known as Taylor's approximation). In such cases, the following approximations can be made [140, 141]:

$$\frac{\partial(\cdots)}{\partial x_1} \approx \frac{1}{\bar{U}}\frac{\partial(\cdots)}{\partial t} \tag{63}$$

$$L_{ij,1} = \bar{U}T_{ij} \tag{64}$$

$$\lambda_{11} = \left[\frac{\overline{u_1^2}}{\overline{(\partial u_1/\partial t)^2}}\right]^{1/2}\bar{U} \tag{65}$$

The smallest "eddies" present in a turbulent flow are characterized by viscous actions, by which turbulent kinetic energy is dissipated into heat. An estimate of the size of such "dissipation eddies" is the "Kolmogoroff microscale"

$$\eta = (\nu^3/\epsilon)^{1/4} \tag{66}$$

Related time and velocity scales can be determined as

$$\tau_K = (\nu/\epsilon)^{1/2} \tag{67}$$

$$v_K = (\nu\epsilon)^{1/4} \tag{68}$$

The kinematic viscosity ν is measured independently and can be found in tables for common fluids. The turbulent kinetic energy dissipation rate ϵ, however, defined as

$$\epsilon = \nu \overline{\sum_{i=1}^{3}\sum_{k=1}^{3}\frac{\partial u_i}{\partial x_k}\left(\frac{\partial u_i}{\partial x_k} + \frac{\partial u_k}{\partial x_i}\right)} \tag{69}$$

is extremely painstaking to measure directly. A usual approximation based on the assumption of local isotropy is

$$\epsilon \approx 15\nu\overline{\left(\frac{\partial u_1}{\partial x_1}\right)^2} \tag{70}$$

Another common procedure is to estimate ϵ as the balance of the other terms in the turbulent kinetic energy equation. Due to errors in the measurement of some terms and uncertainties in the contributions of non-measurable terms, such estimates of ϵ must be treated with caution.

Various scales can be combined to provide dimensionless measures of turbulent activity. A Reynolds number characteristic of the macroscopic flow condition can be defined as

$$R = \frac{\bar{U}L}{\nu} \tag{71}$$

where L is a selected (e.g. $L_{11.1}$) or "average" integral length scale. The "turbulent Reynolds number"

$$R_{\lambda_1} = \frac{\sqrt{\overline{u_1^2}}\,\lambda_{11}}{\nu} \tag{72}$$

represents the relative intensity of turbulent activity and is more relevant for classifying different turbulent flows. The Reynolds number characteristic of energy dissipation is, by definition, a constant:

$$R_K = \frac{v_K \eta}{\nu} = 1 \tag{73}$$

Definitions and properties of other Eulerian and Lagrangian scales and dimensionless parameters can be found in the general turbulence literature.

Probabilities

Measurements of probability density functions often provide considerable insight in the turbulent structure and dynamics. Complete probabilistic formulations of the turbulence euqations have been developed, and although also subject to the closure problem, they sometimes appear advantageous over the physical formulation, especially for reacting flows. Estimates of probabilities as histograms from digitized records of turbulent signals are straightforward, limited only by measuring resolution and computer capacity. Some initial normalization of the digitized signals is recommended, if the results are to be used for comparisons with measurements in other flows or with standard random processes. It is advisable not to analyze raw signals, but first to determine the quantity of interest S_i, $i = 1, 2, \ldots, N$ from the signal records, especially when the transducer output is non-linearly related to the measured quantity (e.g. hot-wire velocity sensitivity) or if it contains contributions from more than one variable (e.g. mixed velocity and temperature sensitivity of hot-wire output). Then, it is easy to transform the record into a normalized form, as

$$r_i = \frac{S_i - \bar{S}}{(\overline{s^2})^{1/2}} \qquad i = 1, 2, \ldots, N \tag{74}$$

Then, the range of r_i is divided into a number of narrow ranges, with width w (non-uniform width might be more economical in some cases) and centered at points $\zeta_k = kw$, $k = 0, \pm 1, \pm 2, \ldots, \pm K$. The probability density function of r_i can be estimated as

$$P_r(\zeta_k) = \frac{P\left(\zeta_k - \frac{w}{2} \le r_i \le \zeta_k + \frac{w}{2}\right)}{w} = \frac{M(\zeta_k, w)}{Nw} \tag{75}$$

where $M(\zeta_k, w)$ is the number of points r_i that are in the interval $[\zeta_k - w/2, \zeta_k + w/2]$. A common procedure is to compare this probability with the pdf of a Gaussian (normal) distribution

$$P(\zeta_k) \frac{1}{2\pi} e^{-1/2\zeta_k^2} \tag{76}$$

Deviations of measured probabilities from the Gaussian have often revealed important properties of the turbulence structure.

Joint statistics of two or more turbulent quantities can be computed, if simultaneously sampled records of these quantities are available. For example, the joint probability density function of two

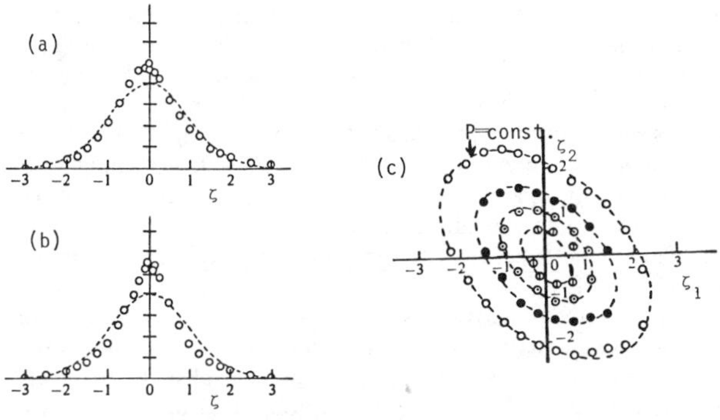

Figure 32. Examples of probability plots: (a) pdf of streamwise velocity derivative [128]; (b) pdf of transverse velocity derivative [128]; (c) joint pdf of streamwise and transverse velocities (isoprobability contours) [134]. Dashed lines indicate corresponding values of Gaussian random variables.

normalized discrete time histories

$$r_{ai}, r_{bi} \qquad i = 1, 2, \ldots, N$$

can be estimated as

$$P_{r_a r_b}(\zeta_{ak}, \zeta_{bk}) = \frac{P\left(\zeta_{ak} - \frac{w_a}{2} \le r_{ai} \le \zeta_{ak} + \frac{w_a}{2} \text{ and } \zeta_{bk} - \frac{w_b}{2} \le r_{bi} \le \zeta_{bk} + \frac{w_b}{2}\right)}{w_a w_b}$$

$$= \frac{M(\zeta_{ak}, w_a; \zeta_{bk}, w_b)}{N w_a w_b} \tag{77}$$

where M is the number of pairs r_{ai}, r_{bi} in the rectangle centered at (ζ_{ak}, ζ_{bk}) with sides (w_a, w_b). For comparison, the jointly normal probability density function is

$$P(\zeta_{ak}, \zeta_{bk}) = \frac{1}{2\pi(1 - \rho_{ab}^2)^{1/2}} e^{-(\zeta_{ak}^2 + \zeta_{bk}^2 - 2\rho_{ab}\zeta_{ak}\zeta_{ab})/2(1-\rho_{ab}^2)^{1/2}} \tag{78}$$

where ρ_{ab} is the correlation coefficient between r_a, r_b. Examples of probabilistic measurements in a turbulent shear flow [128, 134] are shown in Figure 32.

Derivative and Higher-Order Statistics

Statistics of various derivatives of velocity, pressure, scalars, etc. appear in many balance equations of turbulence (e.g. through dissipation rate, Equation 69), so that considerable benefits are obtained from their measurement. Spacial derivatives can be estimated, in principle, from differences of the outputs of two nearby probes, e.g.

$$\frac{\partial u_i}{\partial x_j} \simeq \frac{u_i(x_j + \Delta x_j) - u_i(x_j)}{\Delta x_j} \tag{79}$$

This procedure is commonly used to measure derivatives in directions perpendicular to the mean flow. Streamwise derivatives are usually estimated from temporal derivatives (obtained from differences of consecutive samples in digitized records) with the use of Taylor's approximation, Equation 63. Practical problems, such as probe cross-talk and low signal-to-noise ratio, make the measurement of derivative statistics significantly more difficult than that of straight quantities. Careful matching of sensors [103], reduction of noice and repeated trials with different sampling times, probe separations, and operating conditions are necessary for increasing the reliability of such measurements.

Derivatives accentuate the fine structure effects on a signal, and so derivative statistics, and especially skewness and flatness factors of velocity and scalar derivatives, are in common use as measures of the spacial distribution of regions of intense turbulent kinetic energy dissipation ("internal intermittency" [142, 143]). Velocity derivatives also appear in the definition of vorticity. Reflecting the importance of vorticity fluctuations in the dynamics of turbulence, special probes and measuring configurations have been developed for the measurement of both streamwise [144, 145] and transverse [146] vorticity fluctuations. Mechanical probes and multi-wire sensors are the most commonly used at present but optical techniques (optical tracing of miniature mirrors, multi-beam laser Doppler systems) are expected to find wider application in the future.

Due to their sensitivity, derivative statistics are normally used as sampling criteria in conditional sampling, as discussed in the next subsection.

The computation of higher moments ($\overline{u_i u_j u_k}$, $\overline{\theta^2 u_i}$, etc.) presents no particular difficulty if simultaneous digitized records are available. Of course, the reliability of such measurements decreases with increasing order of statistical quantity, as factors such as probe interference, filter distortion, and finite sample errors become more significant.

Conditional Sampling—Intermittency

Despite their apparent "randomness" many turbulent flows are known to exhibit some recurrent, "organized" phenomena. Indiscriminant averaging of turbulent signals tends to smear out information about such phenomena, which, however, can be recovered by proper discriminatory signal processing, termed as "conditional sampling" [147, 148]. One of the earlier forms of conditional sampling has been "zone averaging" near turbulent/non turbulent interfaces in shear layers and boundary layers. Signals are considered turbulent when the value of a criterion (e.g. velocity derivative, vorticity temperature) is above a preset threshold and non-turbulent otherwise. Thus, averages can be performed separately for the turbulent and for the non-turbulent zones. A useful indicator is the "intermittency factor," defined as the relative duration of turbulent vs. total signal.

A different, more sophisticated form of conditional sampling, termed as "point averaging," involves ensemble averaging of a signal at a specified time after a certain event has been detected. The event can be periodic, such as external pulsing of a flow or periodic heating, or random, such as detachment of "coherent structures" from a wall in turbulent boundary layers. "Pattern recognition" techniques [149], rescaling selected segments, permit the reconstruction of waveforms typical to organized events. Simultaneous visualization and local signal acquisition has greatly enhanced the confidence in the detection and description of such phenomena.

Early conditional sampling studies utilized analog detectors and discriminators and demanded lengthy adjustments and trials. Digital data processing has simplified the implementation of sampling criteria and further statistical computations. The remaining difficulties are conceptual: establishing a sampling criterion, setting a threshold, selecting scaling laws for grouping of detected events. The current trends in turbulence research indicate that future experiments will provide a wealth of conditional information, capable of resolving the space-time dependence of the flow structure.

SELECTING A MEASURING PROCEDURE

As discussed earlier, a variety of accurate transducers and measuring procedures have been developed for measurements in turbulent flows. Instrumentation can be purchased commercially or, sometimes, built locally with reasonable chance of success. As options increase, however, the inexperienced researcher might be faced with the tough question of choosing the optimum technique that is suitable to the particular flow conditions. Economy of time, effort, and cost dictate that one adopts not the most advanced technique available, but rather the simplest possible technique that can provide all the required information. Therefore, the first necessary step would be to prepare a list of the quantities that must be measured, separating the essential ones from those of secondary importance, and the level of required accuracy. Next to each desirable quantity, one should list the instruments and techniques that can be used for its measurement, including approximate cost, accuracy, and ease of operation. A comparative study of the entire list will reveal which procedure would be the optimum one.

For example, if one tries to detect only whether separation of a turbulent boundary layer occurs, one should employ simple flow visualization techniques and not a hot-wire or laser-Doppler anemometer. Similarly, if only average velocities are required, one should utilize a pressure tube instead of a fast response transducer. Simplicity of operation reduces significantly the chance for systematic errors.

On the other hand, once a sophisticated measuring system is available it would be inefficient to utilize it for mundane tasks alone. Extraction of as much information as the system permits may be rewarding to the immediate research objective, but also to future reconsideration and uses of the results by the same investigator or others. Careful documentation of experimental conditions and measurements, often in similar experimental setups by different groups, increases the confidence of the overall turbulence database and provides valuable input to theorists and modelists. A certain degree of redundancy in experimental conditions is also desirable.

When one tries to develop an instrumentation system that will be used in a variety of unforeseen turbulence applications, it is advisable to select one that has a wide range of operation and that permits upgrading and modification by the addition of other components. Modular systems that can be used alone or in combinations are best choice. Further, it is advisable to avoid overspecialized instruments, such as correlators, spectrum analyzers, etc., if a general use, programmable digital data acquisition and processing system could be acquired instead.

A final warning is due: The difficulties associated with setting up, debugging, and operating a turbulence measuring system should never be underestimated and sufficient development time expense must be allowed, especially when the flow under investigation is complex and outside the usual range of well-documented studies. Finally, one should keep in mind that, as a rule, turbulence experiments require much more time and effort during the development of flow facility, instrumentation, and measuring techniques, than for the collection of the desired measurements.

NOTATION

A	amplitude
A, B	calibration constants in Equation 25
c	specific heat
D(k)	dissipation spectrum
d	beam waist diameter
d_p	particle diameter
E(w)	wave-number spectrum of turbulence
E	voltage
E_{s1}, E_{s2}	amplitudes of scattered beams
E_T	amplified voltage
F(f)	frequency spectrum
f	frequency
f_D	Doppler frequency
f_{s1}, f_{s2}	beam frequencies
f_T	focal length
g	gravitational acceleration
g	amplifier gain
h	height
I	current
K	Gladstone-Dale constant
K	shear parameter
k	coefficient
$L_{ij,k}$	eddy length scale
l	length
M	Mach number
m	mass
N	number of samples
N	number of molecules per unit volume
N_s	Stokes number
Nu	Nusselt number
n	coefficient
P, p	pressure
P_r	probability density function
Q	heat flux
q^2	mean squared turbulent kinetic energy

R	universal gas constant	u_i	velocity fluctuation
Re	Reynolds number	u_p	particle velocity
R, R_0	resistance	u_F	local fluid velocity
$R_{i,j}$	correlation coefficient	V	volume
r_k	point separation along x_k axis	V_θ	particle velocity projection
T	absolute temperature	V_k	velocity scale
T_0	time limit	v_s	settling velocity
t	time	W	work
U	velocity	X	distance

Greek Symbols

α_0	coefficient in Equation 22	ν	kinematic viscosity
γ	ratio of specific heats	ν_F	fluid kinematic viscosity
δ	displacement	ρ	density
δ	fringe or beam spacing	ρ_p	particle density
δ_{ij}	Kronecher's delta	ρ_{ab}	correlation coefficient between r_a, r_b
ϵ	kinetic energy dissipation rate	ρ_F	fluid density
η	Kolmogoroff microscale	τ	time constant
θ	angle	σ_{ij}	normal stress
κ	wave-number	τ_k	time scale, Equation 67
λ_{ij}	Taylor microscales	ω	frequency
μ	viscosity		

REFERENCES

1. Monin, A. S., and Yaglom, A. M., *Statistical Fluid Mechanics; Mechanics of Turbulence*, MIT Press, Cambridge, Vol. 1, pp. 205–256 (1971).
2. Kollmann, W., (ed.), *Prediction Methods for Turbulent Flows*, Hemisphere, Washington (1980).
3. Rogallo, R. S., and Moin, P., "Numerical Simulation of Turbulent Flows," *Ann. Rev. Fluid Mech.*, Vol. 16, pp. 99–137 (1984).
4. Rodi, W., "Turbulence Models and Their *Application in Hydraulics–A State-of-the-Art Review*," *Intern. Assoc. for Hydraulic Res.*, Delft, The Netherlands (June 1980).
5. Hagen, G. H. L., "Über den Einfluss der Temperatur auf die Bewegung des Wassers in Rohren," Mathem. Abhandl. Akad. Wissensch., Berlin, p. 17 (1854).
6. Reynolds, O., "An Experimental Investigation of the Circumstances Which Determine Whether the Motion of Water Should be Direct or Sinuous and of the Law of Resistance in Parallel Channels," *Phil. Trans. Roy. Soc. London*, Vol. 174, pp. 935–982 (1883).
7. Burgers, J. M., "Experiments on the Fluctuations of the Velocity in a Current of Air," Proc. Koninklijke Akad. von Wetenschappen te Amsterdam, Vol. 29, pp. 547–558 (1926).
8. Dryden, H. L., and Kuethe, A. M., "The Measurement of Fluctuations of Air Speed by the Hot-wire Anemometer," NACA Tech. Rep. 320, (1929).
9. Bradshaw, P., *An Introduction to Turbulence and Its Measurement*, Pergamon Press (1971).
10. Corrsin, S., "Turbulence: Experimental Methods," *Encycl. of Physics* (S. Flügge, ed.), Vol. 8/2, pp. 524–590, Springer, Berlin (1963).
11. Reynolds, A. J., *Turbulent Flows in Engineering*, Ch. 2, John Wiley & Sons (1974).
12. Hinze, J. O., *Turbulence*, 2nd Edition, Ch. 2, McGraw-Hill (1975).
13. Frost, W., and Moulden, T. H., (ed.), *Handbook of Turbulence*, Vol. 1, Plenum, New York (1977).
14. Richards, B. E., (ed.), "*Measurement of Unsteady Fluid Dynamic Phenomena*, Hemisphere, Washington (1977).
15. Emrich, R. J., (ed.) *Methods of Experimental Physics: Fluid Dynamics*, Vol. 18, Parts A and B, Academic Press (1981).
16. Goldstein, R. J. (ed.), *Fluid Mechanics Measurements*, Hemisphere, Washington (1983)

17. Monro, P. A. G., *Adv. Opt. Electron Microsc.*, Vol. 1, p. 1 (1966).
18. Falco, R. E., "Combined Simultaneous Flow Visualization/Hot-wire Anemometry for the Study of Turbulent Flows," *J. Fluids Eng.*, Vol. 102, pp. 174–182 (1980).
19. Freymuth, P., Bank, W., and Palmer, M., "Flow Visualization and Hot-wire Anemometry," *TSI Quart.*, Vol. 9, Issue 4, pp. 11–14 (1983).
20. Merzkirch, W., *Flow Visualization*, Academic Press (1974).
21. Yang, W. J., et al. (eds.), *Flow Visualization* III, Hemisphere, Washington (1985).
22. Ref. 12, pp. 460–471.
23. Ref. 15, part A, p. 35.
24. Dimotakis, P. E., and Brown, G. L., "The Mixing Layer at High Reynolds Number: Large Structure Dynamics and Entrainment," *J. Fluid Mech.*, Vol. 78, pp. 535–560 (1976).
25. Ref. 20, pp. 45–52.
26. Corke, T., Koga, D., Drubka, R., and Nagib, H., "A New Technique for Introducing Controlled Sheets of Smoke Streaklines in Wind Tunnels," *Proc. Intern. Congress on Instrum. in Aerospace Simul. Facil.*, IEEE Publication 77 CH1251-8 AES, pp. 74–80 (1977).
27. Nagib, H., Corke, T., Hellard, K., and Way, J., "Computer Analysis of Flow Visualization Records Obtained by the Smoke-wire Technique," *Proc. Dynamic Flow Conf.* 1978, Skovlunde, Denmark, pp. 567–581 (1978).
28. Corrsin, S., and Karweit, M., "Fluid Line Growth in Grid-generated Isotropic Turbulence," *J. Fluid Mech.*, Vol. 39, pp. 87–96 (1969).
29. Schraub, F. A., Kline, S. J., Henry, J., Runstadler, P. W., and Littell, A., "Use of Hydrogen Bubbles for Quantitative Determination of Time-dependent Velocity Fields in Low-speed Water Flows," *J. Basic Eng.*, Vol. 87, pp. 429–444 (1965).
30. Thompson, D. H., "Flow Visualization Using the Hydrogen Bubble Technique," Aerodynamics Note 338, Australian Defence Scientific Service, Aeronautical Research Laboratory, Melbourne, Australia (1973).
31. Ref. 20, pp. 64–70.
32. Smith, C. R., and Paxson, R. D., "A Technique for Evaluation of Three-Dimensional Behavior in Turbulent Boundary Layers Using Computer Augmented Hydrogen Bubble-Wine Flow Visualization," *Exper. in Fluids*, Vol. 1, pp. 43–49 (1983).
33. Born, M., and Wolf, E., *Principles of Optics*, Pergamon, Oxford (1964).
34. Van deHulst, H. C., *Light Scattering by Small Particles*, Wiley, New York (1957).
35. Benedict, R. P., *Fundamentals of Temperature, Pressure, and Flow Measurements*, 2nd Edition, pp. 287–390, John Wiley and Sons, New York (1977).
36. Doebelin, E. O., *Measurement Systems Application and Design*, 3rd Edition, pp. 404–490, McGraw-Hill (1983).
37. Harvey, G. F., (ed.) *Transducer Compedium*, 2nd Edition, IFI/Plenum, New York (1969).
38. Bernstein, L., "Measurement of Unsteady Pressures, Forces and Accelerations," in Ref. 14, pp. 21–61 (1977).
39. Willmarth, W. W., "Unsteady Force and Pressure Measurements," *Ann. Rev. Fl. Mech.*, Vol. 3, pp. 147–170 (1971).
40. Bynum, D. S., Ledford, R. L., and Smotherman, W. E., "Wind-Tunnel Pressure Measuring Techniques," AGARDograph 145 (1970).
41. Corcos, G. M., "Pressure Measurements in Unsteady Flows," *ASME Symp. on Measurements in Unsteady Flow*, Worcester, Massachusetts, pp. 15–21 (1962).
42. Soloukhin, R. I., Curtis, C.W., and Emrich, R. J., "Measurement of Pressure," in Ref. 15, part B, pp. 499–610 (1981).
43. Remenyik, C. J., and Kovasznay, L. S. G., "The 'Orifice-Hot-Wire' Probe and Measurements of Wall Pressure Fluctuations," *Proc. Heat Transfer and Flow Measurement Inst.*, Seattle, Washington, pp. 76–88, Stanford University Press (1962).
44. Strasberg, M., and Cooper, R. D., "Measurements of Fluctuating Pressure and Velocity in the Wake Behind a Cylinder," *Proc. 9th Int. Cong. Appl. Mech.*, Brussels, Belgium, Vol. 2, pp. 384–393 (1965).
45. Kobashi, Y., "Measurements of Pressure Fluctuation in the Wake of a Cylinder," *J. Phys. Soc.* Japan, Vol. 12, pp. 533–543 (1957).

46. Strasberg, M., "Measurement of Fluctuating Static and Total-head Pressure in a Turbulent Wake," NATO AGARD Rept. 464 (1963).
47. Sami, S., Carmody, T., and Rouse, H., "Jet Diffusion in the Region of Flow Establishment," *J. Fluid Mech.*, Vol. 27, pp. 231–252 (1967).
48. Sami, S., "Balance of Turbulence Energy in the Region of Jet-flow Establishment," *J. Fluid Mech.*, Vol. 29, pp. 81–92 (1967).
49. Siddon, T. E., "On the Response of Pressure Measuring Instrumentation in Unsteady Flow," Univ. Toronto Inst. Aerosp. Stud., Rep. UTIAS 136 (1969).
50. Spencer, B. W., and Jones, B. G., "A Bleed-type Transducer for In-Stream Measurements of Static Pressure Fluctuations," *Rev. Scient. Instru.*, Vol. 42, pp. 450–454 (1971).
51. Jones, B. G., "A Bleed-type Pressure Transducer for In-stream Fluctuating Static Pressure Sensing," *TSI Quart.*, Vol. 7, Issue 2, pp. 5–11 (1981).
52. Taylor, R. J., "Thermal Structures in the Lowest Layers of the Atmosphere," *Australian J. Phys.*, Vol. 11, p. 168 (1958).
53. Bradbury, L. J. S., and Castro, I. P., "A Pulsed-Wire Technique for Velocity Measurements in Highly Turbulent Flows," *J. Fluid Mech.*, Vol. 49, pp. 657–691 (1971).
54. Bradbury, L. J. S., "Examples of the Use of the Pulsed Wire Anemometer in Highly Turbulent Flow," *Proc. Dyn. Flow Conf.* 1978, pp. 489–509, P.O. Box 121, DK-2740 Skovlunde, Denmark (1978).
55. Wyngaard, J. C., "Cup, Propeller, Vane, and Sonic Anemometers in Turbulence Research," *Ann. Rev. Fluid Mech.*, Vol. 13, pp. 399–423 (1981).
56. Zimmerman, R., and Deery, D., *Turbine Flow Meter Handbook*, Flow Technology Inc., Phoenix, Arizona (1977).
57. Stevens, G. H., "Dynamic Calibration of Turbine Flow Meters by Means of Frequency Response Tests," NASA TM X-1736 (1969).
58. Tritton, D. J., "The Use of a Fibre Anemometer in Turbulent Flows," *J. Fluid Mech.*, Vol. 16, p. 269 (1963).
59. Siddon, T. E., and Ribner, H. S., "An Aerofoil Probe for Measuring the Transverse Component of Turbulence," *AIAA J.*, Vol. 3, pp. 747–749 (1965).
60. Cheng, D. Y., "Introduction of the Viscous Force Sensing Fluctuating Probe Technique with Measurement in the Mixing Zone of a Circular Jet," AIAA Paper No. 73–1004 (1973).
61. Cliff, W. C., "Optical and Acoustical Measuring Techniques," in *Handbook of Turbulence*, Frost, W. and Moulden, T. H., (eds) Vol. 1, pp. 403–432, Plenum (1977).
62. Waller, J. M., "Guidelines for Applying Doppler Acoustic Flow Meters," *In Tech.*, pp. 55–57, October 1980.
63. Kaimal, J. C., "Sonic Anemometer Measurement of Atmospheric Turbulence," *Proc. Dyn. Flow Conf.* 1978, pp. 551–565, P. O. Box 121, DK-2740, Skovlunde, Denmark (1978).
64. Miyake, M., Stewart, R. W., Burling, R. W., Tsuang, L. R., Koprov, B. M., and Kuznetzov, O. A., "Comparison of Acoustic Instruments in an Atmospheric Turbulent Flow Over Water," *Boundary Layer Meteorol.*, Vol. 2, pp. 228–245 (1971).
65. Lynnworth, L. C., "Ultrasonic Flow Meters," Ch. 5, *Physical Acoustics*, Mason, W. P. and Thurston, R. N., (eds), Academic Press (1979).
66. Reid, A. M., "Turbulence Measurement Using a Glow Discharge as an Anemometer," *R. and D.*, No. 6, p. 69 (February 1962).
67. Werner, F. D., "An Investigation of the Possible Use of a Glow Discharge as a Means of Measuring Air Flow Characteristics," *Rev. Sci. Instrum.*, Vol. 21, pp. 61–68 (1950).
68. Franzen, B., Fucks, W., and Schmitz, G., "Koronaanemometer zur Messung von Turbulenzkomponenten," Zeit. für Flugwiss., Vol. 9, p. 347 (1961).
69. Bean, H. S., (ed.), *Fluid Meters—Their Theory and Application*, ASME Research Committee Report, 6th Edition, Amer. Soc. Mech. Eng., New York (1971).
70. *Handbook of Measurement and Control*, Instruments Publishing, Pittsburgh, Pennsylvania (1954).
71. Considine, D. M., *Encyclopedia of Instrumentation and Control*, McGraw-Hill, New York (1971).

72. Bryer, D. W., and Pankhurst, R. C., "Pressure-Probe Methods for Determining Wind Speed and Flow Direction," Nat. Phys. Lab., London (1971).
73. Benedict, R. P., *Fundamentals of Temperature, Pressure, and Flow Measurements*, 2nd Edition, pp. 339–374, Wiley (1977).
74. Sami, S., "The Pitot Tube in Turbulent Shear Flow," *Proc. 11th Midwestern Mech. Conf., Dev. in Mechanics*, Vol. 5, Paper 11, p. 171 (1967).
75. Freymuth, P., "A Bibliography of Thermal Anemometry," *TSI Quant.*, Vol. 4, Issue 4, pp. 3–26 (Nov./Dec. 1978). See also *J. Fluids Eng.*, Vol. 102, pp. 152–159 (1980).
76. Perry, A. E., *Hot-Wire Anemometry*, Clarendon Press, Oxford (1982).
77. Compte-Bellot, G., "Hot-Wire Anemometry," *Ann. Rev. Fluid Mech.*, Vol. 8, pp. 209–231 (1976).
78. Fingerson, M., and Freymuth, P., "Thermal Anemometry," *Fluid Mechanics Measurements*, Goldstein, R. J. (ed.) pp. 99–151, Hemisphere (1983).
79. King, L. V., "On the Convection of Heat from Small Cylinders in a Stream of Fluid; Determination of the Convection Constants of Small Platinum Wires with Applications to Hot-Wire Anemometry," *Phil. Trans. Roy. Soc. London*, Vol. A214, pp. 373–432 (1914).
80. Collis, D. C., and Williams, M. J., "Two-Dimensional Convection from Heated Wires at Low Reynolds Numbers," *J. Fluid Mech.*, Vol. 6, pp. 357–384 (1959).
81. Bestion, D., and Gaviglio, J., "Comparison Between Constant-Current and Constant-Temperature Hot-Wire Anemometers in High-Speed Flows," *Rev. Sci. Instrum.*, Vol. 54, pp. 1513–1524 (1983).
82. Ossofsky, E., "Constant Temperature Operation of the Hot-Wire Anemometer at High Frequency," *Rev. Sci. Instrum.*, Vol. 19, pp. 881–889 (1948).
83. Kovasznay, L. S. G., "Simple Analysis of the Constant Temperature Feedback Hot-Wire Anemometer," AERO/JHU CM-478, The Johns Hopkins University, Baltimore (1948).
84. Champagne, F. H., Sleicher, C. A., and Wehrmann, O. H., "Turbulence Measurements with Inclined Hot-Wires," *J. Fluid Mech.*, Vol. 28, pp. 153–182 (1967).
85. Tavoularis, S., "Simple Corrections for the Temperature Sensitivity of Hot-Wires," *Rev. Sci. Instrum.*, Vol. 54, pp. 741–743 (1983).
86. Yeh, H., and Cummins, H. Z., "Localized Fluid Flow Measurements with a He-Ne Spectrometer," *App. Phys. Lett.*, Vol. 4, p. 178 (1964).
87. Durrani, T. S., and Greated, C. A., *Laser Systems in Flow Measurement*, Plenum Press, New York (1977).
88. Durst, F., Melling, A., and Whitelaw, J. H., *Principles and Practice of Laser Doppler Anemometry*, Academic Press, New York (1976).
89. Watrasiewicz, B. M., and Rudd, M. J., *Laser Doppler Measurements*, Butterworths, London (1976).
90. Buchhave, P., George, W. K., and Lumley, J. L., "The Measurement of Turbulence with the Laser Doppler Anemometer," *Ann. Rev. Fluid Mech.*, Vol. 11, pp. 443–503 (1979).
91. Lading, L., "Processing of Laser Anemometry Signals," *Proc. Dyn. Flow Conf.* 1978, P.O. Box 121, DK-2740, Skovlunde, Denmark (1978).
92. Riethmuller, M., "Laser Doppler Velocimetry," *Measurement of Unsteady Fluid Dynamic Phenomena*, Richards, B. E., (ed.), pp. 163–187 (1977).
93. "Users Course Notes, CTA and LDA," DISA Electronics, Skovlunde, DK-2740, Denmark (1983).
94. Adrian, R. J., and Fingerson, L. M., *Laser Anemometry; Theory, Application, and Techniques*, Thermosystems Inc., St. Paul, Minnesota (1982).
95. Bloom, A. L., *Gas Lasers*, John Wiley and Sons, New York (1968).
96. "Laser Doppler Anemometry," DISA Publication No. 3205, pp. 46–49 (1983).
97. *Temperature Measurement Handbook*, Omega Engineering Co., Stamford, Connecticut, p. B-14 (1980).
98. *Thermistor Manual*, Fenwal Electronics, EMC-6A, Framingham, Massachusetts, pp. 16–17 (1974).
99. Ishizaki, H., *Wind Tunnel Modeling for Civil Engineering Applications*, Reinhold, T. A., (ed.),

pp. 567–572, Cambridge University Press (1982).
100. LaRue, J. C., Deaton, T., and Gibson, C. H., "Measurement of High-Frequency Turbulent Temperature," *Rev. Sci. Instrum.*, Vol. 46, pp. 757–764 (1975).
101. Tavoularis, S., "A Circuit for the Measurement of Instantaneous Temperature in Heated Turbulent Flows *J. Phys. E: Sci. Instrum.*, Vol. 11, pp. 21–23 (1978).
102. Fiedler, H., "On Data Acquisition in Heated Turbulent Flows," *Proc. Dyn. Flow Conf.*, P.O. Box 121, DK-2740 Skovlunde, Denmark, pp. 88–100 (1978).
103. Mestayer, P., and Chambaud, P., "Some Limitations to Measurements of Turbulence Micro-Structure with Hot and Cold Wires," *Bound. Layer Meteor.*, Vol. 16, pp. 311–329 (1979).
104. Hojstrup, J., Rasmussen, K., and Larsen, S. E., "Dynamic Calibration of Temperature Wires in Still Air," DISA Information, Vol. 20, pp. 22–30 (September 1976).
105. Wyngaard, J. C., "Spatial Resolution of a Resistance Wire Temperature Sensor," *Phys. Fluids*, Vol. 14, pp. 2052–2054 (1971).
106. Schacher, G. E., and Fairall, C. W., "Frequency Response of Cold Wires Used for Atmospheric Turbulence Measurements in the Marine Environment," *Rev. Sci. Instrum.*, Vol. 50, pp. 1463–1466 (1979).
107. Nieuwvelt, C., Bessem, J. M., and Trines, G. R. M., "A Rapid Thermometer for Measurement in Turbulent Flow," *Int. J. Heat Mass Transfer*, Vol. 19, pp. 975–980 (1976).
108. Jones, T. V., "Heat Transfer, Skin Friction, Total Temperature and Concentration Measurements," *Measurements of Unsteady Fluid Dynamic Phenomena*, Richards, B. E., (ed.), Hemisphere, pp. 63–102 (1977).
109. Sandborn, V. A., "Resistance Temperature Transducers," *Metrology Press*, Fort Collins, Colorado (1972).
110. Lapworth, K. C., "Spectroscopic Temperature Measurement in High Temperature Gases and Plasmas," *J. Phys. E: Sci Instrum.*, Vol. 7, p. 413 (1974).
111. Sandborn, V. A., "A Review of Turbulence Measurements in Compressible Flow," NASA Rep. No. TM X 62–337 (1974).
112. Dove, J. E., "Measurement of Composition," *Methods of Experimental Physics, Fluid Dynamics*, Emrich, R. J., (ed.), Vol. 18, Part B, pp. 611–661, Academic Press (1981).
113. Swithenbank, J., "Measurement in Combustion Processes," *Measurement of Unsteady Fluid Dynamic Phenomena*, Richards, B. E., (ed.), pp. 189–212, McGraw-Hill (1977).
114. Gibson, C. H., and Schwarz, W. H., "Detection of Conductivity Fluctuations in a Turbulent Flow Field," *J. Fluid Mech.*, Vol. 16, pp. 357–364 (1963).
115. Flow Industries 1100 Series Conductivity Gauges, Flow Research Company, Kent, Washington (1977).
116. Corrsin, S., "Extended Applications of the Hot-Wire Anemometer," *Rev. Sci. Instrum.*, Vol. 18, pp. 469–471 (1947).
117. Corrsin, S., "Extended Applications of the Hot-Wire Anemometer," NACA Tech. Note No. 1864 (1949).
118. Way, J., and Libby, P. A., "Application of Hot-Wire Anemometry and Digital Techniques to Measurements in a Turbulent Helium Jet," *AIAA* J., Vol. 9, pp. 1567–1573 (1971).
119. Stanford, R. A., and Libby, P. A., "Further Applications of Hot-Wire Anemometry to Turbulence Measurements in Helium-Air Mixtures," *Phys. Fluids*, Vol. 17, pp. 1353–1361 (1974).
120. Sirivat, A., and Warhaft, Z., "The Mixing of Passive Helium and Temperature Fluctuations in Grid Turbulence," *J. Fluid Mech.*, Vol. 120, pp. 475–504 (1982).
121. Brown, G. L., and Rebollo, M. R., "A Small, Fast-Response Probe to Measure Composition of a Binary Gas Mixture," *AIAA J.*, Vol. 10, pp. 649–652 (1972).
122. Adler, D., "A Hot Wire Technique for Continuous Measurement in Unsteady Concentration Fields of Binary Gaseous Mixtures," *J. Phys. E: Sci. Instrum.*, Vol. 5, pp. 163–169 (1972).
123. Hallett, W., and Lenz, W., "Eine Spurengasmethode zur Aufenthalt-szeitmessung in Gasstromungen," *Verweilzeit & Messtechnik*, Vol. 15, pp. 157–160 (1981).
124. "Hot Wire/Hot Film Anemometry, Probes and Accessories," Thermosystems Inc., HWACAT 184 ISM SM, p. 30, St. Paul, Minnesota (1984).
125. Becker, H. A., Hottel, H. C., and Williams, G. C., "On the Light-Scatter Technique for the Study of Turbulence and Mixing," *J. Fluid Mech.*, Vol. 30, pp. 259–284 (167).

126. Nye, J. O., and Brodkey, R. S., "Light Probe for Measurement of Turbulent Concentration Fluctuations," *Rev. Sci. Instrum.*, Vol. 38, pp. 26–28 (1967).
127. Dimotakis, P. E., Miake-Lye, R. C., and Papantoniou, D. A., "Structure and Dynamics of Round Turbulent Jets," *Phys. Fluids*, Vol. 26, pp. 3185–3192 (1983).
128. Tavoularis, S., and Corrsin, S., "Experiments in Nearly Homogeneous Turbulent Shear Flow with a Uniform Mean Temperature Gradient, Part 2. The Fine Structure," *J. Fluid Mech.*, Vol. 104, pp. 349–367 (1981).
129. Stearns, S. D., *Digital Signals Analysis*, Rochelle Park, Haydon, New Jersey (1975).
130. Otnes, R. K., and Enochson, L., *Digital Time Series Analysis*, John Wiley and Sons, New York (1972).
131. Bendat, J. S., and Piersol A. G., *Random Data: Analysis and Measurement Procedures*, Wiley-Interscience, New York (1971).
132. George, W. K., "Processing of Random Signals," *Proc. Dyn. Flow Conf.*, pp. 757–793, P.O. Box 121, DK-2740 Skovlunde, Denmark (1978).
133. Reynolds, O., "On the Dynamical Theory of Incompressible Fluids and the Determination of the Criterion," *Phil. Trans. Roy. Soc. London*, Vol. 186, pp. 123–161 (1894).
134. Tavoularis, S., and Corrsin, S., "Experiments in Nearly Homogeneous Turbulent Shear Flow with a Uniform Mean Temperature Gradient. Part 1." *J. Fluid Mech.*, Vol. 104, pp. 311–347 (1981).
135. Dumas, R., "Corrélation Spatiotemporelles d'Ordre Elevé," *Proc. Dyn. Flow Conf.*, pp. 963–981, P.O. Box 121, DK-2740 Skovlunde, Denmark (1978).
136. Comte-Bellot, G., and Sabot, J., "Space-Time Correlations," *Measurement of Unsteady Fluid Dynamic Phenomena*, Richards, B. E., (ed.) pp. 261–290 (1977).
137. Lumley, J. L., "Interpretation of Time Spectra Measured in High-Intensity Shear Flows," *Phys. Fluids*, Vol. 8, pp. 1056–1062 (1965).
138. Wyngaard, J. C., and Clifford, S. F., "Taylor's Hypothesis and High-Frequency Turbulence Spectra," *J. Atmosph. Sci.*, Vol. 34, pp. 922–929 (1977).
139. Mayo, W. T., "Spectrum Measurements with Laser Velocimeters," *Proc. Dyn. Flow Conf.*, pp. 851–868, P.O. Box 121, DK-2740 Skovlunde, Denmark (1978).
140. Comte-Bellot, G., and Corrsin, S., "Simple Eulerian Time Correlation of Full- and Narrow-Band Velocity Signals in Grid-Generated 'Isotropic' Turbulence," *J. Fluid Mech.*, Vol. 48, pp. 273–337 (1971).
141. Heskestad, G., "A Generalized Taylor Hypothesis with Application for High Reynolds Number Turbulent Shear Flows," *J. Appl. Mech.*, Paper No. 65-APM-G, pp. 1–5 (December 1965).
142. Kuo, A. Y. S., and Corrsin, S., "Experiments on Internal Intermittency and Fine Structure Distribution Functions in Fully Turbulent Fluid," *J. Fluid Mech.*, Vol. 50, pp. 285–319 (1971).
143. Gagne, Y., Hopfinger, E. J., and Marechal, J., "Measurements of Internal Intermittency and Dissipation Correlations in Fully Developed Turbulence," *Proc. Dyn. Flow Conf.*, pp. 271–277, P.O. Box 121, DK-2740, Skovlunde, Denmark (1978).
144. Kovasznay, L. S. G., *Physical Measurements in Gasdynamics and Combustion*, Princeton University Press (1954).
145. Willmarth, W. W., "Nonsteady Vorticity Measurements: Survey and New Results," *Proc. Dyn. Flow Conf.*, pp. 1003–1012, P.O. Box 121, DK-2740, Skovlunde, Denmark (1978).
146. Foss. J. F., "Transverse Vorticity Measurements," *Proc. Dyn. Flow* Conf., pp. 983–1001, P.O. Box 121, DK-2740, Skovlunde, Denmark (1978).
147. Dopazo, C., "On Conditioned Averages for Intermittent Turbulent Flows," *J. Fluid Mech.*, Vol. 81, pp. 433–438 (1977).
148. Kovasznay, L. S. G., "Measurement in Intermittent and Periodic Flow," *Proc. Dyn. Flow Conf.*, P.O. Box 121, DK-2740, Skovlunde, Denmark, pp. 133–159 (1978).
149. Eckelmann, H., Wallace, J.M., and Brodkey, R. S., "Pattern Recognition, a Means for Detection of Coherent Structures in Bounded Turbulent Shear Flows," *Proc. Dyn. Flow Conf.*, P.O. Box 121, DK-2740, Skovlunde, Denmark, pp. 161–172 (1978).

CHAPTER 37

THE MEASUREMENT OF REYNOLD'S STRESSES

I. P. Castro

Department of Mechanical Engineering
University of Surrey, U.K.

CONTENTS

INTRODUCTION

It is probably true to say that the great increase in understanding, and hence the power to predict, turbulent flows over the last half-century can be largely attributed to the still developing ability to make accurate measurements of the time averaged turbulence quantities. As discussed earlier in this volume, the time averaged momentum equations governing turbulent fluid flow contain terms that are essentially gradients of turbulent stresses, arising from the ability of the fluctuating velocity field to transport momentum. These are additional to the laminar stresses which arise because of the viscous nature of the fluid. Whilst time-averaging the Navier Stokes equations undoubtedly "hides" some of the important physical features of turbulent flows, it has provided the framework for the great majority of turbulent flow study so that, since the turbulent stress terms are additional unknowns in the equation set, their measurement has been of fundamental importance.

In the engineering context, knowledge of these stresses is usually not a requirement of industry in a particular design problem, but the ability to predict the quantities actually required—like mean velocities, mass flow rates, surface pressures, for example—often rests on the adequacy of a turbulence model. This, in turn, depends critically on the ability of the latter to deal satisfactorily with the turbulent, or Reynolds, stress terms in the equations of motion. Measurement of the

Reynolds stresses in turbulent flows is consequently almost always a requirement of the research community rather than the practicing engineer and is therefore often approached from a different point of view than would be the routine measurement of, say, a mass flow rate through a pipe.

It should also be noted that whereas engineering cases encompass a wide variety of different fluids, nearly all turbulent stress measurements made in the laboratory are in flows of either air or water. This is partly because measurements in more exotic fluids sometimes present particular difficulties, but more often because standard dimensional analysis methods often allow dynamically similar conditions to be set up in the laboratory using one or other of these two commonly available fluids. The present paper therefore concentrates entirely on Reynolds stress measurements in air or water although all the techniques can in principle be used in many other fluids.

Measurement of velocity parameters other than the mean value in non-Newtonian and/or two- or multi-phase fluids is exceedingly difficult. Measurement of Reynolds stresses in such situations are rare; useful backgrounds are contained in Hewitt [1] and Goldstein [2].

Until the last two decades, the only commercial instrument available to the experimenter for measurement of the fluctuating velocities in a turbulent flow was the hot-wire anemometer. The literature surrounding the principles and practice of its use is truly enormous; arguably, no other instrument in the engineering sciences has been the subject of so much scrutiny. Certainly, within the fluid dynamics research community, it has been used more than any other experimental device (except, perhaps, the wind tunnel itself!). However, it would not be appropriate here to attempt a review of what has been previously written—this has been successfully undertaken elsewhere [3, 4]. Rather, we give a summary of the almost inevitable limitations of the instrument arising from its mode of operation and the nature of turbulence, along with some indications of the current "state of the art". It is admitted from the outset that these, particularly the latter, are to some extent personal opinions, but they have been formulated after many years of "hands-on" experience.

Two of the classic limitations of hot-wire anemometry are its virtual inability to make measurements in highly turbulent or reversing flows (but see "Hot-Wire Measurements") and the fact that a piece of hardware has to be located at the point of measurement, thus affecting, to a greater or lesser degree, the flow being measured. This also prevents it being used to probe small scale flows or those with difficult geometry, for example, some regions within rotating machinery. Over the last 15–20 years, two quite distinct techniques have been developed that can in principle overcome one or both of these problems: laser Doppler and pulsed-wire anemometry. The former is now demonstrating its early promise and has the great advantage of being non-intrusive. It can also be used to make measurements in highly turbulent regions. The latter, while being intrusive, like the ordinary hot wire, can also make measurements in such regions. Along with hot-wire anemometry, these instruments will be discussed in later sections. There is one other technique that has been used to make mean and fluctuating velocity measurements, and which seems to have considerable potential yet to be fully exploited. This is the quantitative analysis of high-speed movie, or video, pictures of the motion of small particles or bubbles introduced into the flow. If the particles are small enough to follow the high-frequency flow fluctuations faithfully, then this technique is also, in one sense, non-intrusive, since the presence of the particles will not generally affect the flow, unless the seeding levels are extraordinarily high. The technique has the unique potential of being able, in principle, to yield quantitative information about the instantaneous state of a whole region of the flow: all the other methods make essentially one- (or, at most, a few) point measurements. Its great potential is really a result of the rapid advances in digital video photography and image processing, although impressive measurements have been made over a number of years using less sophisticated analysis methods (e.g. Kim et al. [5]). There are many variants of this general technique, but no commerical system for "routine" measurements is yet available and the method will not be discussed further in this report. A recent example of its use is the work of Paxton and Smith [6].

Various other assorted methods for turbulence measurements have been developed in the past, usually for special applications. Some of these are briefly outlined in Bradshaw [7] but, again, will not be discussed here. We concentrate on the three, major, commercially available instruments—hot-wire, pulsed-wire and laser anemometry.

It is important to emphasize that the limitations in the use of each of these instruments are a direct result not only of the inherent features of the instrument itself but also of the nature of turbulence. In a turbulent flow the velocity vector varies in magnitude and direction. Now in many practical cases it is sufficient to assume that variations in the angle of the instantaneous velocity vector are small—usually because the turbulent intensity, measured with respect to the local mean velocity, is small. The instrument is therefore required to operate in a known fashion only over a small range of flow angles. Even in these circumstances, accurate measurements ideally require an instrument that is sensitive to only one velocity component. (Multi-element wire probes, or multi-sets of fringes in laser anemometry, are often used to reduce experimental time, but in such systems the individual elements—wires or fringes—are oriented differently so as to obtain instantaneous information about more than one velocity component). However, the directional response is never perfect—the transducer signal always contains *some* information about all three components—so it is useful to have some understanding of the likely statistics of the velocity vector.

In flows having high turbulent intensities, the probability of instantaneous flow angles being greater than the limit of the instrument's defined directional response is not necessarily small and in this case there can be significant errors. Generally, a knowledge of the flow angle probability distribution would not in itself permit an accurate assessment of such errors but it is certainly helpful to know when the errors are likely to be large. The problem can be stated as "how high must the turbulent intensity be before errors caused by the instrument's inadequate or ill-defined yaw response become significant?"

There are, of course, many other sources of error inherent in anemometry of any kind and the more important of these will be mentioned in later sections. But there is no doubt that in many, perhaps even most, practical cases it is the actual nature of the flow being investigated, coupled with the directional response of the instrument, that is the major factor in determining measure ment errors. In the following section, therefore, some basic ideas about the statistics of the instantaneous velocity vector are outlined and it is shown how such ideas can be used in deciding which isntrument would be most suitable in a given situation and what the measurement errors are likely to be.

STATISTICS OF THE INSTANTANEOUS VELOCITY VECTOR

In most fully turbulent flows the point-velocity statistics are not very different from Gaussian. Usually only when higher order correlations or velocity derivatives are considered does the turbulence exhibit significant non-Gaussianity (e.g. Frenkiel [8]). There is, of course, significant turbulent diffusion in nearly all shear flows so that the skewness factors (third-order moments of the probability density function) are not zero, but for the present purpose the exact form of the amplitude distribution is unimportant so it is sufficient to assume normality of the velocity fluctuations.

Neglecting for the moment any correlation between the u or v and the w components of velocity and assuming that the mean lateral component, W, is zero, the joint probability function of the velocity vector is then:

$$F = [(\sqrt{2\pi})^3 \alpha_u \alpha_v \alpha_w \sqrt{(1 - r^2)}]^{-1} \exp\left[\frac{-(u - U)^2}{2\alpha_u^2(1 - r^2)} - \frac{(v - V)^2}{2\alpha_v^2(1 - r^2)} + \frac{2r(u - U)(v - V)}{2\alpha_u \alpha_v (1 - r^2)} - \frac{w^2}{2\alpha_w^2(1 - r^2)}\right] \quad (1)$$

where α_u, α_v and α_w are the square roots of the turbulent energies, $\sqrt{\overline{u^2}}$, etc., (standard deviations) in the x, y, and z directions, respectively, and r is the correlation coefficient, $\overline{uv}/(\overline{u^2} \cdot \overline{v^2})^{1/2}$. Overbars refer as usual to time averages. The above assumptions merely restrict the analysis to the common practical case of a two-dimensional mean flow, but lead to great simplification in the theoretical or numerical solutions. The basic problem is now to calculate the probability, P, that the instantaneous velocity vector lies outside a semi-infinite cone of included angle 2θ, aligned with its axis in the x-direction. This would be appropriate for an instrument aligned so that it responds essentially to the u-component of velocity for flow angles less than θ, but for greater flow angles becomes sensitive to the other components, see Figure 1. In general, we may require P for a cone aligned

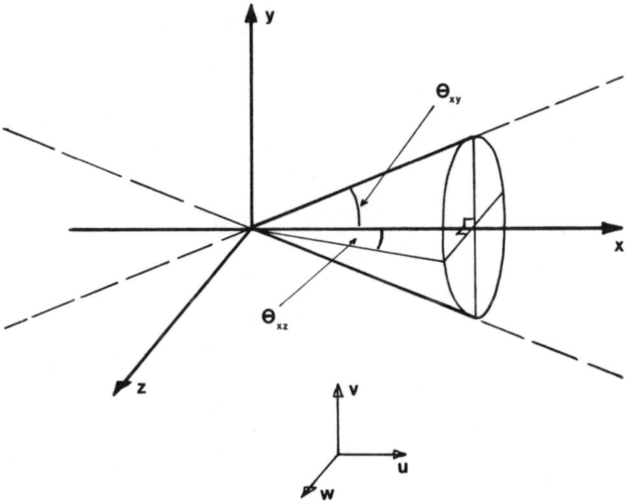

Figure 1. "Acceptance" cone for any instrument, oriented to measure the u-component with yaw response well defined for $\theta - \theta_{xy}, \theta_{xz}$.

in any direction, but this is most easily obtained by solving the case specified above and applying an axis transformation.

The required probability is simply:

$$P = 1 - \iiint_{\text{cone}} F \, du \, dv \, dw$$

There is no analytic solution to this general problem and therefore a numerical technique has to be used. However, it is instructive to investigate one or two limiting cases that can be tackled analytically. If we assume the turbulence to be isotropic, then $r = 0$ and $\alpha_u = \alpha_v = \alpha_w$. It can be shown that for this case

$$P = \tfrac{1}{2}[1 - \operatorname{erf}(U/\alpha\sqrt{2})] + \tfrac{1}{2}\cos\theta \cdot \left[1 + \operatorname{erf}\left(\frac{U\cos\theta}{\alpha\sqrt{2}}\right)\right]\exp\left[\frac{-U^2\sin^2\theta}{2\alpha^2}\right]$$

which clearly has the correct behavior in the limits: if $\theta = 0°$, $P = 1$ and if $\theta = 90°$, $P = \tfrac{1}{2}(1 - \operatorname{erf}(U/\alpha\sqrt{2}))$, which is simply the probability that the instantaneous velocity is negative. If the local turbulent intensity (α/U) is zero, P is zero for all θ (laminar flow) and if α/U is infinite (i.e. at a free stagnation point) $P = \tfrac{1}{2}(1 + \cos\theta)$, which is simply one minus the ratio of the surface area subtended on a sphere by the cone of semi-angle θ to the total surface area ($4\pi r^2$). Figure 2 shows how P varies with α/U and θ. It is obvious that for moderate turbulent intensities there are unlikely to be very significant measurement errors unless θ is particularly restricted, provided the instrument does respond only to the u-component for flow angles less than θ.

In most turbulent flows there is a significant correlation between the u- and v-components (and the $u - w$ and $v - w$ components if the mean flow is three dimensional). Figure 3 shows the variation of P in a case where $r = 0.5$ (and the turbulent intensity is high) compared with the $r = 0$ result; clearly the effect of even quite a large correlation on the probability variation is small. In this case the results had to be obtained numerically. The physically unrealistic but mathematically

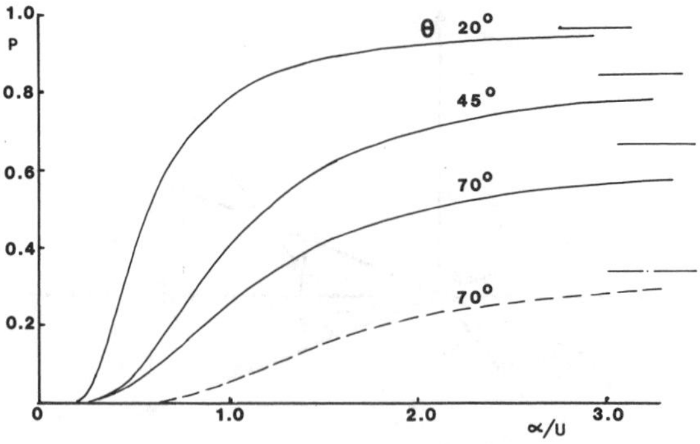

Figure 2. Probability that the velocity vector lies outside the acceptance cone of semi-angle θ: solid line, semi-infinite cone (+ve. velocities only); dashed lines, infinite cone, isotropic turbulence; asymptotic results for $\alpha/U \rightarrow \infty$ are shown by the short lines on the right.

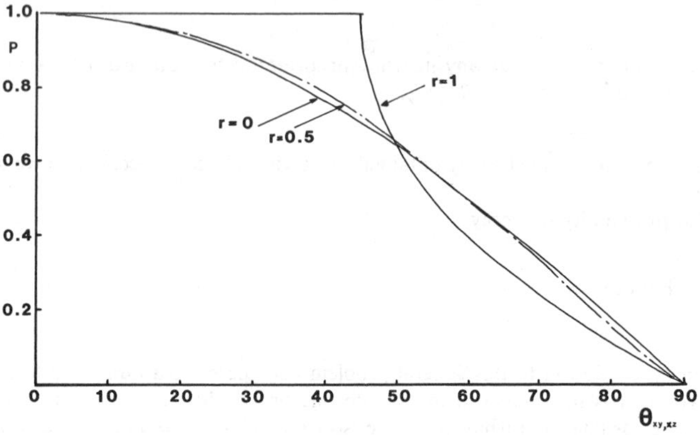

Figure 3. Probability of lying outside "acceptance" cone in the case $\alpha/U = \infty$, for various correlation coefficients, $r = \overline{uv}/(\overline{u}^2 \cdot \overline{v}^2)^{1/2}$.

interesting case of a perfect correlation, r = 1, can be analyzed by considering the degenerate two-dimensional problem. In this case it can be shown that

$$P = \frac{2}{\pi} \tan^{-1}\left\{\frac{\alpha_w}{(\alpha_u^2 \tan^2 \theta - \alpha_v^2)^{1/2}}\right\}, \qquad \tan\theta > (\alpha_v/\alpha_u)$$

$$= 1 \qquad \tan\theta < (\alpha_v/\alpha_u)$$

This is also plotted, for $\alpha_u = \alpha_v$, in Figure 3, although it should be noted that r rarely exceeds 0.5 in practical cases.

As an example of the application of this approach to a particular instrument, consider the case of a standard hot-wire anemometer. It is often assumed that as long as the velocity vector is not closer than 20° to the hot-wire axis, the wire responds *only* to the component of velocity normal to its axis. For a single wire with its axis in the y-direction, placed there to measure the axial Reynolds stress, $\overline{u^2}$, (strictly, the stress is density $\times$ $\overline{u^2}$), we might choose a cone of semi-angle 70° in the x-y

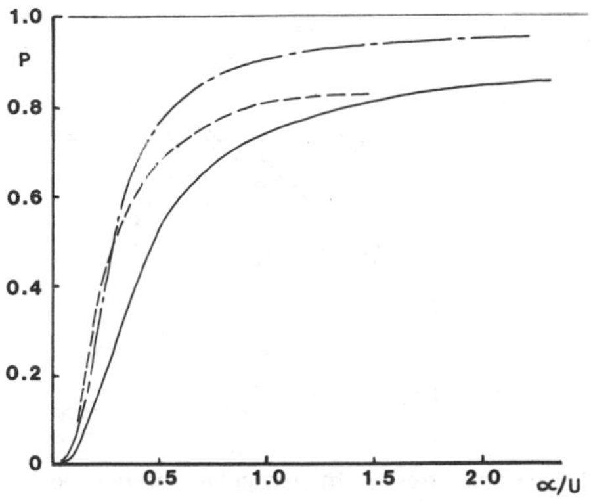

Figure 4. Variation of P for elliptic cones at various orientations to the x-axis (ψ): ______ $\theta_{xy} = 70°$, $\theta_{xz} = 20°$, $\psi = 0$; _ _ _ _ $\theta_{xy} = \theta_{xz} = 20°$, $\psi = 0$; _ _ _ _ $\theta_{xy} = 70°$, $\theta_{xz} = 20°$, $\psi = 45°$.

plane. There is, however, a more severe restriction in the x-z plane, since practically all representations of a hot-wire's response ignore the effect of the w-components of velocity (not unreasonably, since the errors can be shown to be often small, Guitton [9]). In principle, therefore, we should restrict the semi-angle in this plane to, say, 20°. These restrictions amount to assuming that providing the instantaneous values of w/U and v/U do not exceed about 0.35 and 2.7, respectively, the wire measures the u-component correctly. Many workers would consider these to be rather optimistic values.

Figure 4 shows the variation of P for this elliptic cone as the turbulent intensities (assumed equal) rise, for a typical correlation coefficient of 0.3. If a slant wire at 45° is used the cone of interest should also be inclined at 45° to the x-axis and Figure 4 includes the results for such a case. The worst case is shown ($\psi = -45°$): clearly P would be a little lower for $\psi = +45°$ since we chose $r > 0$. For turbulent intensities around 30% the results indicate that the velocity vector is only within the cone for about half the time and even for a normal wire P is 25%. Note that P rises very rapidly, in all cases, between intensities of about 10% and 100%. If a crossed wire probe is used (two orthogonal wires at $\pm 45°$) errors are likely to occur if the instantaneous velocity vector lies outside a circular cone of semi-angle 20° and this case is also shown in Figure 4. It is worth noting that for $\alpha/U = 0.5$ the probability of the velocity vector approaching at least one of the wires of a standard crossed wire probe from *behind* is about 20%!

This kind of information, along with a knowledge of the response characteristics of the instrument in question, allows one to make some rough estimates of the conditions in which the instrument is likely to suffer from significant measurement errors caused by inadequate yaw response. For instance, Figure 4 might suggest that for intensities greater than 20–30%, a crossed wire will give significant errors, since $P > 0.3$. Since the turbulent Reynolds stresses are *second* moments of the velocity probability distributions, errors in these quantities will usually be higher than errors in mean velocity measurements: inadequate or ill-defined yaw response leads to errors in the "tails" of the probability density functions and these contribute proportionally more to second and higher order moments.

In addition to qualitative guidelines that can be obtained in this way, it is obviously important in many instances to be able to quantify the errors. This in general requires detailed knowledge of the entire directional response characteristics of the instrument and will be discussed in the following sections.

HOT-WIRE MEASUREMENTS

Calibration and Linearization

A good introduction to the science (some would say, the art) of hot-wire anemometry is given by Bradshaw [7], with rather more detail being supplied by Comte-Bellot [3] in her 1976 review of the whole subject. It is our purpose here to emphasize the features of the technique that are

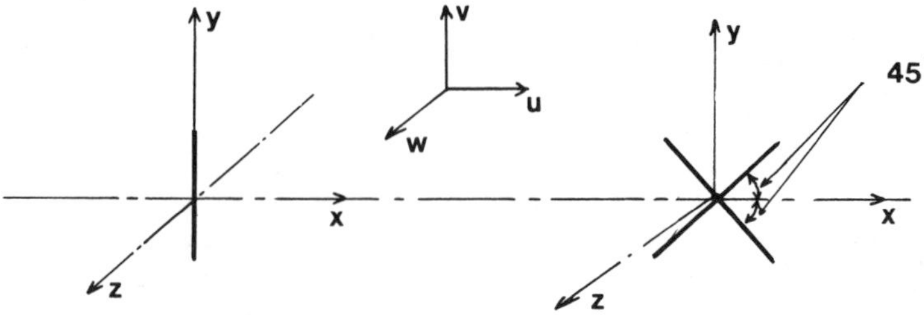

Figure 5. Common hot-wire configurations.

of particular relevance for the measurement of Reynolds stresses. In particular, the question of accuracy and what constitutes "good engineering practice" are addressed.

The usual arrangement of hot wires, or films (which are used mainly in fluids) is shown in Figure 5. For measurement of U and $\overline{u^2}$ a single wire normal to the flow is commonly used, whereas measurements of $\overline{v^2}$ and $\overline{uv}$ (and V, if the mean velocity is not in the x-direction) require either separate measurements with a single slanted wire at different angles, or simultaneous measurements from two such wires. The 45° x-wire arrangement shown in Figure 5 is the usual configuration.

In most practical cases, the rate of heat loss from the wire or film can be expressed as a function of wire Reynolds number and "overheat" ratio—essentially a measure of the ratio of the wire operating temperature to that of the surroundings. Whilst there have been a great many "universal" relationships proposed, it is always advisable and is common practice to calibrate each probe individually. The majority of workers use a relationship of the form:

$$E^2 = A + BU^n \tag{2}$$

where U is the velocity normal to the axis of the wire, E is the anemometer bridge output voltage, and A, B, and n are constants. For many years it was accepted that a static calibration, that is, one in which the velocity U is held constant while a measurement of E is obtained, would yield satisfactory measurements in turbulent (dynamic) situations, with n assumed constant throughout. Note that for measurement of turbulence quantities it is the sensitivity to velocity fluctuations, $\partial E/\partial U$, that is required. Changes of 1%, say, in the measured E values can lead to 10% changes in $\partial E/\partial U$ so quite accurate calibrations are required if accurate turbulence measurements are to be made. In the early 1970s Perry and his co-workers showed that the measured sensitivity seemed to be different if the wire was calibrated dynamically, by "shaking" it (Perry and Morrison [10], Morrison et al. [11]). It appeared that the prongs which support the sensing element were exerting some influence via their relatively low thermal inertia, a not unreasonable deduction. However, Bruun [12] suggested that most of the difference could be accounted for by recognizing that the response equation (2) does not provide a good fit with a constant value of n if the velocity range is too wide. Figure 6, taken directly from Bruun [12] shows the variation of sensitivity with velocity deduced from careful *local* fits to calibration data, compared to what would be deduced if the data were fitted to Equation 2 over the whole velocity range. Up to U = 20 m/s, n = 0.45 gives a good fit but fails at higher velocity. There is clearly no value of n that would be adequate over the whole range.

In boundary layers a typical ratio between the maximum and minimum instantaneous velocity to which the probe might be subjected would be about three or four, but in free shear flows it is often considerably higher. However, in the latter case local turbulent intensities would normally be high in the regions of lowest velocity and then errors caused by transverse velocity components and/or rectification would dominate. It now seems generally accepted that static calibrations are quite satisfactory provided they are carried out with sufficient care, enabling a "best-fit" n to be determined. n = 0.45 is a commonly used exponent since it has been found to give good fits for many commercially available probes at low velocity. This finding (e.g. Bruun [12]) agrees with the

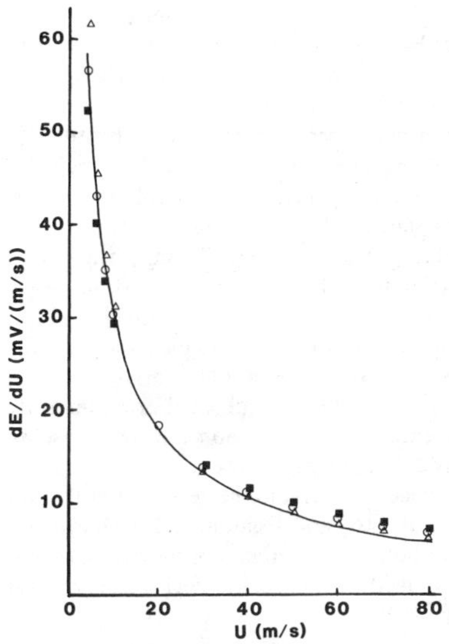

Figure 6. Hot wire sensitivities. ______, from local calibration fits; Δ, n = 0.4; 0, n = 0.45; ■, n = 0.5 (assumed constant throughout range). (Bruun, 1976).

results of careful fundamental experiments on convective heat transfer from fine wires by Collis and Williams [13] and Bradbury and Castro [14].

Turning now to the effects of the non-linear character of Equation 2, for small fluctuations it is clear that

$$2Ee' = nU^{n-1}u'$$

by differentiating Equation 2 so that

$$\overline{u^2} = 4E^2/(n^2B^{2/n}) \cdot (E^2 - A)^{2(1-n)/n}\overline{e^2} \tag{3}$$

A measurement of the mean and mean square of the bridge output will therefore enable the axial turbulent stress to be obtained (for a single wire normal to the u-direction). Higher order terms arising from the non-linearity of Equation 2 and ignored in Equation 3 will lead to increasingly large errors as the turbulence level rises. For measurements in flows where local turbulent intensities (u'/U) are below 10–15% such errors are small. It is common practice, however, to use a linearizer which, by applying electronically the appropriate transfer function to the bridge output, can be set-up to yield a voltage directly proportional to velocity. This non-linear source of error then disappears.

Nowadays, many users digitize the signal and perform the data analysis on a computer. In those circumstances an analogue linearizer is largely superfluous. Any analogue electronics additional to the bridge itself will inevitably increase the possibility of calibration drift and, certainly as far as linearizers are concerned, will considerably increase the calibration time. Digital analysis, including linearization, can easily be undertaken on-line, even with small micro-computers, and is not susceptible to drift! It also reduces experimental time and hence the likelihood of problems with temperature drift and dirt accumulation on the probe.

Temperature and Dirt Effects

In low-intensity flows slow ambient temperature drift or unwanted temperature fluctuations and/or gradual contamination of the heated wire or film are often the major sources of error. Dirt contamination is usually more serious in the case of measurements in liquids but it can cause problems in gas flows, particularly if no filtering is possible. The only really satisfactory way to minimize errors is to ensure as "clean" a flow as possible and to reduce the necessary experimental time between calibrations to a minimum. Any data for which an immediate calibration check reveals that drift has occurred during the run must be regarded as suspect. There is no substitute for frequent calibration.

If calibration drift has been caused by a drift in the ambient temperature, it is possible to correct the data to some extent. Various temperature correction schemes have been proposed and some of these can be relatively easily implemented (e.g. Bearman [15]). "Automatic" temperature compensation via a temperature sensitive element in the anemometer bridge is also sometimes used. Commercial temperature compensated probes are available. In air, a 1°C temperature change can produce an error of 2% in mean velocity, although errors in the stress measurements are usually lower. If large temperature drifts are unavoidable some correction is therefore essential.

If the flow contains significant temperature fluctuations, then rather complicated procedures are required to separate the effects of velocity and temperature fluctuations. Some workers use an additional very fine "thermometer" wire (running virtually cold), mounted close to the other wire(s), to monitor the instantaneous temperature directly and hence unscramble the response of the hot wires. Good examples of this are provided by Antonia et al. [16] and Dean and Bradshaw [17], although only relatively small fluctuations were present in both cases. Other workers use standard probes but make separate measurements at different wire overheat ratios; this technique can also be used in fluctuating density flows (e.g. Way and Libby [18]).

In most work of the sort just described, the temperature fluctuations were *deliberately* introduced as a means of "tagging" part of the flow. With care it is possible to make Reynolds stress measurements with an accuracy similar to what could be achieved in the equivalent, truly isothermal flow. Temperature effects will therefore not be considered further in this work. It is the writer's experience that in carefully controlled experiments unwanted temperature effects can almost always be made small enough to ensure that other sources of measurement error are dominant. These are discussed in the following sections, but the reader with special problems in the previously discussed areas should consult the literature for further information.

Yaw Response

In a turbulent flow, the instantaneous velocity will not normally be perpendicular to the wire axis. The question then arises, "what is the effective velocity, U_e, that should replace U in the response equation (2)?" Since hot-wires are not infinitely long, it is evident that the velocity component in the direction parallel to the wire will have some effect on the heat transfer from it. There have been very many attempts to include the effect of this axial component. Perhaps the most well known and commonly used approach is to express the effective normal velocity as

$$U_e = \{U_n^2 + k^2 U_a^2\}^{1/2},$$

where U_n and U_a are the velocity components normal and parallel to the wire and k is defined as the axial sensitivity. Champagne et al. [19] demonstrated how k depends on the length to diameter ratio (l/d) of the wire and Champagne and Sleicher [20] worked out the resulting corrections that must be applied to the measurements, although this was only done for low turbulent intensities. Typically, $k = 0.2$ for a wire with $l/d = 200$ and that would lead to errors of about 8% and 16% in measurements of $\overline{uv}$ and $\overline{v^2}$ with a standard crossed wire.

Many workers assume a value for k based on their wire l/d and simply apply the corrections after making the measurements, assuming that the geometric wire angles are $\pm 45°$. However, k values are also quite dependent on the precise geometry of the probe. Few workers would consider an ordinary U-component sensitivity calibration (e.g. Equation 2) to be unnecessary so it seems inconsistent not to undertake a yaw calibration to determine the V-component sensitivity directly (i.e. to find k). While k may well be fairly constant for a particular probe type—though it may well depend on the wire Reynolds number—it is difficult to measure accurately the geometric angle of the wire with respect to the flow and to ensure that it remains the same each time the probe is inserted into the test-rig.

An alternative to the k-factor approach is to assume that the effective velocity is:

$$U_e = U \cos \psi_e + V \sin \psi_e$$

where ψ_e is the "effective" wire angle, not necessarily equal to the actual geometric angle, θ. (Taking $\psi_e = \theta$ is equivalent, of course, to neglecting the axial velocity component). This is the "effective cosine law" approach discussed by Bradshaw [7] and amounts to curve fitting the yaw-response calibration data to a simple cosine law. For $\theta = 45^\circ$ and if ψ_e differs from this by a small quantity ϵ, it is easy to show that the effective cosine law response is equivalent to the k-factor response with $\epsilon = \frac{1}{2}\sin^{-1}(k^2)$.

The ability to achieve a good fit over a wide range of angles is dependent on probe geometry. If the sensing element is the central third, say, of the total wire length (achieved by using plated wire and etching away the plating material over only the central region) then good fits can be obtained for velocity vectors at angles up to at least 70° to the normal to the wire axis. Some commercially available probes have that geometry but many do not. Hot film probes generally have much lower 1/d values. In such cases the "effective cosine law" is rather less accurate. Note, however, that the k-factors quoted by Champagne and Sleicher [20] were only valid for angles up to 60° to the normal and, indeed, Tutu and Chevray [21] have pointed out that "local" k values change for higher angles. Both these approaches, therefore, become less accurate when the local turbulent intensity is relatively high, for then instantaneous velocity vectors may approach the wire at angles outside the range of validity of the yaw response fit. However, in those circumstances there will be significant errors arising from the velocity component normal to the calibration plane of the wire and also, perhaps, rectification. In practice, these are then the dominant sources of error.

Since it is difficult to be sure of the exact geometric angle of the wire and the end effects caused by the prongs, the present writer believes that a direct yaw response calibration is essential for accurate measurements and the "effective cosine law" approach is the simplest to implement. Provided good curve fits can be achieved up to angles within about 20–25° of the wire axis, the major sources of error in the measurement of turbulence quantities in flows of relatively high intensity will then be those discussed in the following section.

Effects of Cross-Stream Velocity Components and Rectification

For a single hot wire with its axis in the V-direction, the effective normal velocity "seen" by the wire is

$$U_n = (U^2 + W^2)^{1/2}$$

If the mean velocity is $\bar{U}$ and u' and w' denote fluctuating components as usual,

then $$U_n = [(\bar{U} + u')^2 + w'^2]^{1/2}$$

or $$U_n = \bar{U}[1 + u'/\bar{U} + O(u'^2, w'^2)]$$

where O(x, y) denotes terms of order x, y and higher. In the case of a wire inclined at θ to the u-direction in the u-v plane,

$$U_n = \{[(\bar{U} + u') \cos \theta + v' \sin \theta]^2 + w'^2\}^{1/2}$$

or $$U_n = \bar{U} \cos \theta[1 + u'/\bar{U} + (v'/\bar{U}) \tan \theta + O(u'^2, v'^2, w'^2)]$$

The "effective cosine law" approach would then put the effective velocity, U_e, equal to U_n in the above expression but with θ replaced by ψ_e, the effective wire angle. In both the previous cases it should be noted that if the fluctuations are small the wire responds essentially to the velocity vector normal to its axis and in the U,V plane, but larger fluctuations lead inevitably to some sensitivity to the cross-stream component, w, via the $O(w'^2)$ terms. So even if the yaw response in the U-V plane were accurately known over the whole angular range, measurements would become increasingly inaccurate as the turbulence levels rise because of the wire's inherent inability to separate the two velocity components in the plane normal to its axis.

It should also be noted that additional errors arise if the instantaneous velocity vector approaches the wire from behind—the positive square root was used in the above expressions for U_n. As pointed out earlier, this occurs over 20% of the time for a crossed wire probe in a flow whose local turbulent intensity is 50%. In principle it would be possible to square the instantaneous signal, making measurements with wires at six different angles to deduce all the mean and fluctuating components. Rodi [22] has employed this sort of technique but it is obviously extremely difficult to implement successfully. An example of the effects of this "rectification" is presented in Figure 7, where the measured probability distribution of the flow angle obtained from a crossed wire mounted conventionally in the outer low velocity region of a two-dimensional mixing layer is shown. Data is presented for two measurement locations where the (measured) axial turbulent intensity was 57% and 87%. It is clear that the distribution is distorted because the wires rectify all velocity vectors for which $\tan^{-1}(V/U) > 45°$. Since the turbulence is far from isotropic and the flow is intermittent in this region, the real $p(\theta)$ distribution is skewed and the mean θ has a significant negative value, so that the rectification effect is noticeable only on the $\theta < 0$ side. The errors could possibly be reduced a little by inclining the probe so that it faced the local mean velocity direction. Note that the cut-off at $\theta = -45°$ is not sharp, as it would be if there were no sensitivity to the velocity component along the wire axis. It is obvious that the measurements, of $\overline{u^2}$, $\overline{v^2}$ and $\overline{uv}$ are likely to be in considerable error—even without the influence of the w-component.

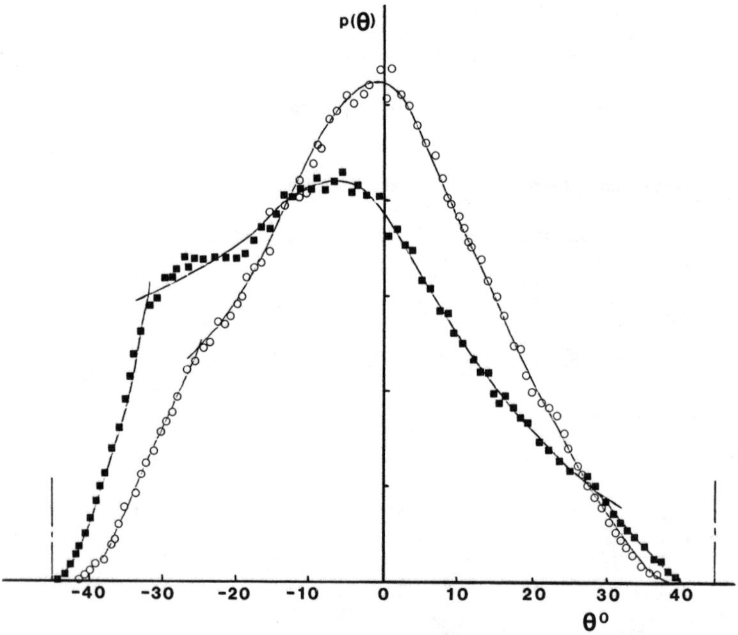

Figure 7. Probability distribution of instantaneous velocity vector direction, measured with a crossed hot-wire in a plane turbulent mixing layers. (O) $\alpha_u/U = 0.57$; (■) $\alpha_u/U = 0.87$.

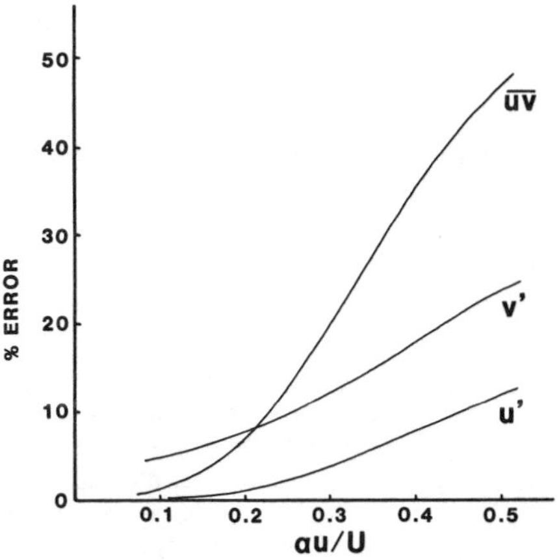

Figure 8. Errors as a function of turbulence intensity; (crossed hot-wire measurements). $\alpha_v = \alpha_w = 0.8\alpha_u$, r = 0.3. Note that for the normal stresses, errors in r.m.s. are shown. (Tutu and Chevray, 1975.)

There have been innumerable efforts to quantify the errors arising from the w-component and rectification. Perhaps the most thorough and certainly the most useful is the work described by Tutu and Chevray [21]. They assumed that the three velocity components were jointly normal and calculated the joint probability density functiion, p*(u, v), that would be measured with a standard crossed wire, for an actual joint function p(u, v, w). The integrations had to be performed numerically, and having found p*(u, v) further numerical integrations for the appropriate moments of this function yielded the measured values of the mean and turbulent quantities. Some of their tabulated results, expressed as errors in measured quantities, are replotted in Figure 8; these are the *total* errors and therefore include the effects of the transverse component and of rectification. They do not include errors due to uncertainty in the yaw response—Tutu and Chevray [21] assumed a "k-factor" response. The results shown in Figure 8 are for k = 0.15, $\alpha_v = \alpha_w = 0.8\alpha_u$, where α_i is the local turbulent intensity in the i-direction, and a correlation in the U-V plane defined by $\overline{uv}/(\overline{u^2} \cdot \overline{v^2})^{1/2} = 0.3$. These are all typical values. It is evident from Figure 8 that the errors become quite large once the local axial intensity exceeds about 15%, particularly in the case of the stresses involving the transverse component. Tutu and Chevray [21] were also able to separate the effects of the transverse w-component and rectification; they found that errors due to rectification are small at low intensities, as expected, but become comparable to those due to the w-component when the longitudinal intensity reaches about 40%.

The results in Figure 8 seem to provide a reasonable estimate of likely errors and can be used to deduce correction factors for measured values, although it should be noted that it is not, in principle, possible to deduce p(u, v, w) from p*(u, v) since this inverse transformation is multivalued. Figure 9 presents the data of Figure 8 in a rather more useful form, enabling a measured local longitudinal intensity, α_u^*/U^*, to be used to estimate the corrections that should be applied to the measured quantities. Now the actual measurement errors will depend on the form of the probability density function which, in most practical cases, will not be jointly normal. They will also depend on the magnitude of the mean V-component of velocity. Figure 9 may therefore often give overestimates of the errors actually occurring, but it does indicate the very rapid rise in the errors as the turbulent intensity rises.

These results demonstrate clearly that if accurate shear stress measurements are required it will often be necessary to apply corrections. There have been a number of attempts to circumvent some of the difficulties posed by high intensities by using probes with three or more wires. Sometimes, such probes are used solely to obtain simultaneous three-dimensional measurements, rather than explicitly in an attempt to avoid errors, since conventional single and crossed hot wires are tedious to use in fully three-dimensional flows. Recent examples are provided by Moffat et al. [23] and Acrivellis [24]. Calibration of such probes is arduous and usually on-line digital techniques are practically obligatory since the complete system of response equations is quite complex. If accurate

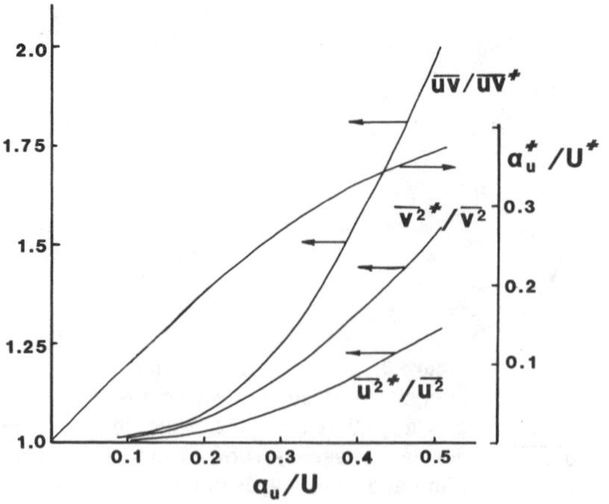

Figure 9. Correction factors required for measured values of Reynolds stresses, as a function of local turbulence intensity (from data of Figure 8).

measurements of Reynolds stresses are required in regions where local intensities can exceed about 40%, serious consideration should be given to one of the two major alternative instruments—pulsed-wire or laser anemometers.

In cases where the local turbulent intensity does not exceed about 10%, a useful guide to the accuracy of hot-wire measurements was given at the Stanford 1980/81 Conference on Complex Turbulent Flows. A committee of experienced hot-wire users agreed that the longitudinal turbulence energy, $\overline{u^2}$, could generally be obtained with an error no more than ± 5–8%; corresponding figures for the shear stress, $\overline{uv}$, and the lateral energy components, $\overline{v^2}$ and $\overline{w^2}$, were given as ± 5–15% and ± 15%, respectively. No consensus was reached for cases where local intensities were higher than 10%, except in so far that all were agreed that errors would be higher.

There is one variant of standard hot-wire anemometry that has been developed to cope with the problems posed by high intensities and, in particular, flow reversals. This is the so called "flying hot wire," which is simply an ordinary hot-wire probe arranged to traverse through the flow at relatively high speed so that the instantaneous velocity seen by the wire(s) is always within the calibrated yaw-response range. Accurate measurement of the instantaneous velocity and position of the probe simultaneously with measurement of the wire response is obviously necessary. In practice this means that relatively complex computer data acquisition and analysis is necessary, but the technique has been successfully used (Coles and Wadcock [25], Watmuff et al. [26]). It is not recommended for the faint-hearted and seems unlikely to become a standard technique.

PULSED WIRE ANEMOMETRY

General Response

The pulsed hot-wire anemometer is a device designed specifically to make measurements in flows where the turbulent intensities are so high that accurate hot-wire measurements are impossible. Initially, the principle was conceived independently by Bradbury [27] and Tombach [28] and consists essentially of measuring the time required by a heat tracer to traverse the distance between the heat generating wire and a downstream sensor wire. After early work describing the theoretical and actual response of the instrument (Bradbury and Castro [29]) there have been many demonstrations of its use in highly turbulent, and often separated, flows. Figure 10 shows the usual probe arrangement. The instrument is calibrated in a low intensity (usually laminar) flow whose velocity is perpendicular to the plane of the wires. In a turbulent flow each firing of the central "pulser" wire will then lead to a measurement of U cos θ, where U and θ are the magnitude and direction of the instantaneous velocity vector. Now U cos θ is just the component of velocity in the direction perpendicular to the plane of the wires so in principle, recognizing that the presence of a sensor

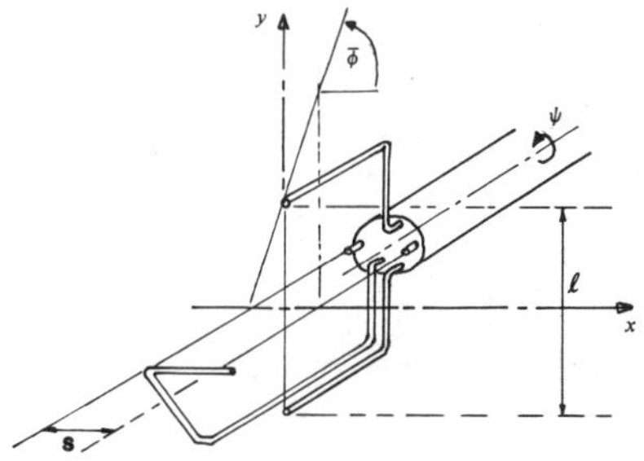

Figure 10. Typical pulsed wire probe geometry, $\tan \bar{\phi} = \ell/s$.

on both sides of the pulsed wire allows the *sign* of the instantaneous velocity to be distinguished, taking a sufficient number of samples of flight times will allow calculation of the statistics of that particular velocity component. The "ideal" response of the instrument is simply $U = s/2T$, where U is the component of velocity normal to the plane of the wires, s is the spacing between the two sensor wires, and T is the heat tracer flight time. Thermal diffusion and various features of the electronics required to measure the flight time lead to an actual response like $U = A/T + B/T^3$, where A and B are constants. (Earlier work used B/T^2, but a cubic has recently been found to give a better fit to the calibration data for wide velocity variations, 0.25–15 m/s, say). This non-linearity in the response causes no problems or errors in practice since, as in much hot-wire work, the measurements are invariably linearized in the computer used to control data acquisition and analysis.

The instrument cannot be used if there are temperature fluctuations in the flow. Under the usual operating conditions, fluctuations of fractions of 1°C would lead to sufficient additional "noise" on the sensor signals to prevent adequate discrimination between that and the signal caused by arrival of the heat tracer. It is also not generally possible to use the instrument at velocities exceeding about 20 m/s, nor in very small scale flows. However, it is far less susceptible to drift, caused by either dirt accumulation on the wires or slow ambient temperature variations, than is the ordinary hot-wire anemometer and has been successfully used in a variety of low-speed separated or high-intensity flows. Recent examples include Bradbury [30], Castro and Robins [31], Eaton et al. [32], Britter and Hunt [33], Ruderich and Fernholz [34]. Bradbury [30] is particularly relevant in this context since he calculated the expected response of a single hot wire in a highly turbulent flow on the basis of measured pulsed wire turbulence parameters. The hot-wire measurements were quite close to those expected and this approach seems a rather satisfactory, though indirect, way of checking the accuracy of the pulsed wire.

The Yaw Response

There are two features of the instrument that make its directional response less than ideal. Firstly, unless the wires are infinitely long, if the pulsed wire is fired when the instantaneous velocity is inclined at an angle greater than $\tan^{-1}(l/s)$ to the x-direction (see Figure 10), the heat tracer will miss the sensor wire altogether, so the instrument will effectively record a zero velocity. If the probe geometry is such that $\tan^{-1}(l/s) \simeq 80°$, say, the consequent errors in mean and turbulence quantities may be small, since cos 80° is only 0.17. Using the analysis outlined earlier, it is possible to quantify the errors since, unlike the case of the hot wire, the instrument's response for velocity vectors outside the $\tan^{-1}(l/s)$ limit is known exactly. We return to this point later.

Secondly, it has been assumed thus far that the probe has an exactly cosinusoidal response. Early calibrations suggested a yaw response behavior roughly like

$$U_\theta = U_0 \cos \theta (1 + \epsilon \tan \theta) \tag{4}$$

where ϵ is a small parameter, typically about 0.1. Thermal diffusion was thought to be responsible for the slight reduction in flight time for vectors inclined at θ to the probe (see Figure 10) from

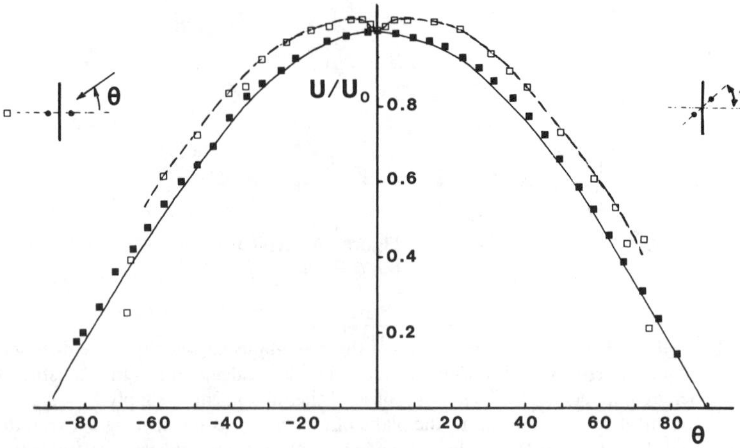

Figure 11. Yaw response characteristics of pulsed wire probes. □ standard probe; ■ offset probe; ______ cosine function.

the expected value based on the result for $\theta = 0°$ (Bradbury and Castro [29], Bradbury [30]). However, recent work has indicated that this effect is smaller than had been thought. Figure 11 shows results of a standard yaw response calibration, in which closely spaced data around $\theta = 0°$ has been obtained. It is clear that the viscous wake of the upstream sensor wire has a measurable effect and, since the normalizing velocity is usually taken as that measured at $\theta = 0°$, leads to measured velocities at non-zero angles apparently exceeding $U_0 \cos \theta$. A probe with the two sensor wires arranged so that they only lie directly upstream and downstream of each other when $\theta = 45°$ has a yaw response typified by the additional data shown in Figure 11. This is much closer to the ideal cosine law response, so that measurement errors arising from a non-ideal response are lower if such a probe is used.

The other obvious feature of Figure 11 is the probe's finite yaw response: $\tan^{-1}(1/s) \simeq 70°$ in the case of the normal probe but the offset probe has wires a little closer together so that velocity vectors at angles up to about 80° can be measured.

These two features of the directional response of a pulsed wire seem to be the major sources of error in measurement of mean and fluctuating velocities although they are not the only ones. If there is a substantial velocity gradient in the direction of the pulsed (or sensor) wire axis one might, for example, anticipate errors caused by this velocity shear. However, at low intensities, it is the wire *spacing*, rather than their length, which is the dominant parameter. This is typically the same order as the length of an ordinary hot wire, so shear related errors will only become relatively significant (compared to those occurring in hot wire measurements) at higher intensities, when the errors caused by the previously mentioned imperfections in directional response will be dominant.

Errors and Measurements

Using the ideas discussed earlier, it would seem that errors arising from the finite yaw response will occur once the intensity is high enough to give a significant probability that velocity vectors will lie outside a cone of semi-angle 70°, say. In this case, however, the presence of a sensor wire on the "upstream" side of the pulsed wire ensures that negative velocity vectors are correctly measured (provided $\theta \lesssim 110°$). The integrations can therefore be carried out for an *infinite* cone of semi-angle 70°. Assuming a jointly normal, isotropic turbulence field analytic results for this probability can be easily obtained and are included in Figure 4. Bradbury and Castro [29] derived expressions for the resulting error in mean velocity and longitudinal turbulent intensity measurements. They neglected the influence of the ϵ term in Equation 4 and assumed that the probe was aligned so that the plane of the wires was normal to the mean velocity. For the "standard" probe with a 70° response, the error in measurement of $\overline{u^2}$ reaches a peak of about 5% at a local turbulent intensity of around 50% and decreases as the intensity rises (because of the presence of the second sensor wire). Later calculations showed that the presence of imperfections for $\theta < 70°$ (such that $\epsilon = 0.1$) leads to errors in $\overline{u^2}$ rising to about 18% as $\alpha_u/U \rightarrow \infty$, but of opposite sign to the finite yaw response errors (Bradbury [30]). More recent numerical integrations for the case in which *both* features are included and a correlation of 0.4 between the u and v-components is assumed again gives an error in $\overline{u^2}$ of about 20% as $\alpha_u/U \rightarrow \infty$ (Castro and Cheun [35]). But this was for $\epsilon = 0.15$ and a 70° response so, since the error is dominated by the former, the offset probe geometry shown in Figure 11 would certainly lead to much smaller errors.

Transverse components of turbulence energy and the turbulent shear stress can be measured by using the probe like an ordinary single slant hot-wire. To obtain $\overline{u^2}$, $\overline{v^2}$, and $\overline{uv}$, for example, three measurements with (u-v) probe orientations of 0°, −45° and +45°, say, are required. Castro and Cheun [35] have analyzed this problem in a similar way in order to determine the likely measurement errors. They showed that "in flows of such high intensity that hot wires would be useless, measurements of all the Reynolds stresses can be made with an accuracy probably better than 30% (for $\overline{v^2}$) or even 15% (for $\overline{u^2}$ and $\overline{uv}$). In the medium intensity range (10–30%, say), provided the yaw response extends to large enough angles, pulsed-wire measurements can be as accurate as hot-wire measurements." Probes with a geometry like that shown in Figure 11 would now seem to enable even more accurate measurements (since ϵ is near zero). It should be emphasized that one of the features of the results of both Castro and Cheun [35] and Bradbury [30] is that the errors are quite sensitive to the particular distribution of turbulence energy between the three components and to the presence of a mean velocity component in the plane parallel to the wires. For example, a "70°" probe with $\epsilon = 0.15$ would measure $\overline{uv}$ with an error of about 20% in a flow for which $\alpha_w = \alpha_u$ and $r = 0$ when $\alpha_v = \alpha_u = 0.16$ or $\alpha_v = 0.2$, $\alpha_u = 0.4$, but with roughly no error when $\alpha_v = \alpha_u = 0.4$ or $\alpha_v = 0.08$, $\alpha_u = 0.16$. Similarly, in a location where $\alpha_u = \alpha_v = \alpha_w = 0.25$ and $r = 0$, the error in $\overline{uv}$ would be about 30% for $V/U = 0$, but −30% for $V/U = 0.1$. Note that in the latter case these correspond to the maximum errors (accuracy improves with increasing α for $\alpha >$ 0.25) and that in all cases the errors would be substantially lower for a probe with ϵ nearer zero.

Figure 12 (discussed later) shows that it is possible to make reasonable shear stress measurements with a pulsed wire in "classic" turbulent shear flows (see also Castro and Cheun [35]).

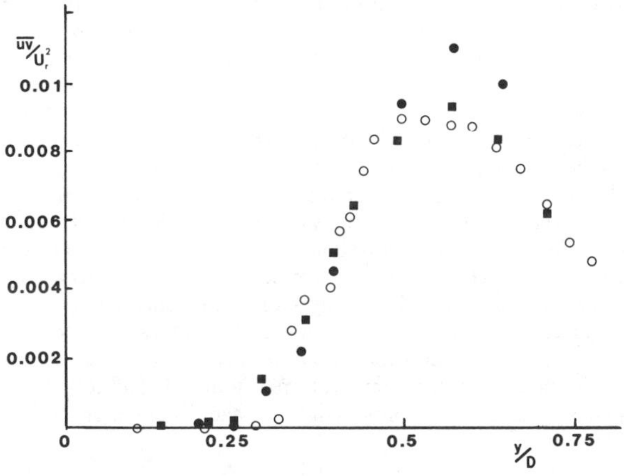

Figure 12. Shear stress measurements in an axisymmetric mixing layer. (O) pulsed wire; (■) crossed hot wire (corrected); (●) laser transit anemometer (Brown and Inman, 1983).

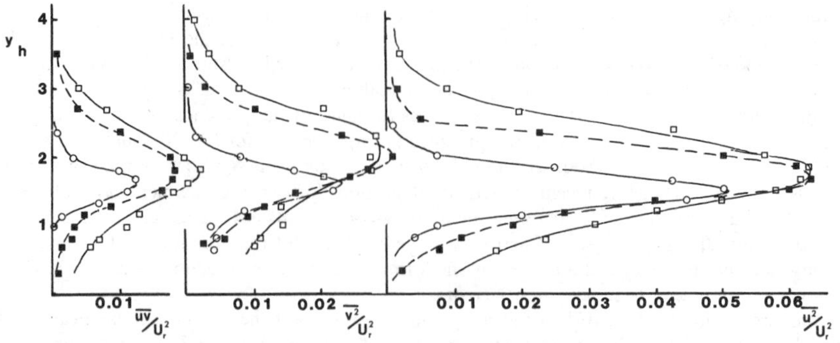

Figure 13. Reynolds stress data for flow shown in Figure 14. x/h = (O) 2; (■) 5; (□) 9 (mean flow reattachment is around x/h = 13). Pulsed-wire measurements, obtained by making measurements with probe at $\psi = 0^0$ and $\pm 45°$ (see Figure 10).

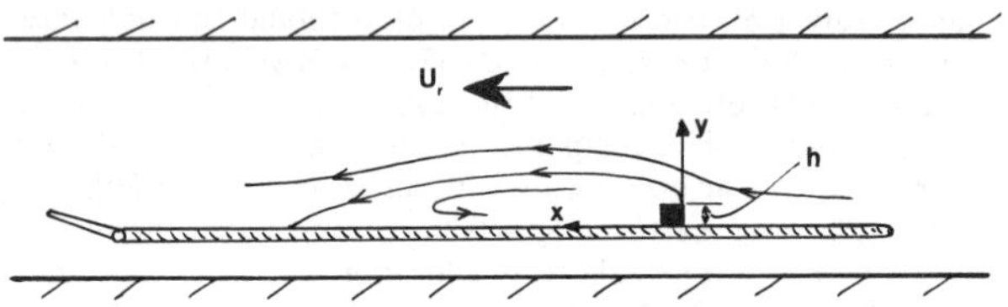

Figure 14. Sketch of surface-mounted two-dimensional bluff body flow.

Finally, some data are presented as an example of Reynolds stress measurements in a flow obviously not amenable to hot-wire anemometry. Figure 13* shows a few distributions of $\overline{u^2}$, $\overline{v^2}$ and $\overline{uv}$ in the near wake of a two-dimensional surface-mounted body. The details of the flow are unimportant here; Figure 14 is a sketch of the general set-up. A probe of the offset type shown in Figure 11* was used so that the errors are thought to be rather lower than those present in the similar measurements of Castro and Cheun [35]. The major point to note is that the maximum shear stress exceeds the maximum value in an ordinary plane mixing layer by a factor of about two. Crossed-wire measurements give much lower results, but the local intensities exceed 50% at the center of the separated shear layer so corrections of the sort outlined earlier are quite inadequate.

Before the pulsed-wire anemometer can be considered by the wider fluid mechanics community to be a well-tested and satisfactory device for measurements of transverse Reynolds stresses in highly turbulent flows, it is accepted that many more independent studies should be made. In particular, only one direct comparison between pulsed-wire and laser anemometry stress measurements has ever been made. There seems to be no other way of independently assessing the instrument's accuracy, since accuracy in the numerical prediction of highly turbulent flows is similarly uncertain. Corresponding remarks, of course, could be made about laser anemometry, as pointed out in the following section.

---

* The author is grateful to Dr. A. Haque for provision of this data.

LASER-DOPPLER ANEMOMETRY (LDA)

The first application in fluid mechanics of the long known Doppler shift effect was described by Yeh and Cummins [36] and over the last twenty years there has been an enormous effort to exploit the obvious potential of laser-Doppler anemometry. Much of the effort has inevitably been directed towards new and improved optical arrangements and the design of signal-processing equipment. By 1972, when the first European Mechanics Colloquium to be devoted to laser anemometry was held, there were already hundreds of published papers describing various aspects of the technique. It is worth noting immediately that in the published report on that Conference Durst et al. [37] stated that the reliability of laser anemometry "for measuring simple quantities, such as second order correlations, is not yet firmly established," despite some papers that suggested turbulence stresses could be obtained to precisions of 5% (e.g. Durst and Whitelaw [38], Bourke et al. [39]). However, this work was on relatively simple laboratory flows; there were very few measurements even of that sort available at the time.

Indeed, despite the burgeoning interest in the laser anemometer—it seems destined to overtake hot-wire anemometry as the most discussed tool for turbulence research—it remains true that there have been relatively few careful sets of turbulence measurements, particularly in air, that can be used to assess the accuracy of the technique. What laser anemometry has clearly demonstrated—and the importance of this should not be underestimated—is its ability to make measurements at all in flows that cannot be quantitatively studied in any other way. These include, for example, blade-to-blade flows in turbomachinery (e.g. Wisler and Mossey [40], Strazisar and Powell [41]), flows in piston-cylinder assemblies (e.g. Gosman et al. [42], Cole and Swords [43]), transonic flows and shock-wave/boundary layer interactions (e.g. East [44], Bachalo and Johnson [45]), combustion systems (e.g. Self and Whitelaw [46], Hutchinson et al. [47]) and capillary flows (Born et al. [48]).

This author has very little "hands on" experience of laser anemometry so any attempt here to discuss definitively the accuracy of laser anemometer measurements of turbulent stresses would, perhaps, be inappropriate. It is in any case a much more complex problem than in the case of wire anemometry. This is partly because of the bewildering variety of optical arrangements and signal processing techniques that have been used. Different flow situations often dictate quite different approaches to both these aspects of the technique. The variety arises partly of course just because the laser anemometer *can*, in principle, be employed in such a wide variety of different kinds of fluid mechanical situations. Each optical arrangement and processing method has its own sources of measurement error, although many of the basic ones are common to many different techniques. Excellent reviews of the whole subject are provided by Durst et al. [49] and Buchave et al. [50]. The latter attempts in particular to isolate and quantify theoretically the major sources of error, whereas the former contains much useful information on the different experimental arrangements that can be used, with their advantages and disadvantages. The authors of the former book have many years of practical experience of a wide variety of laser anemometry, which makes it a particularly valuable reference source.

In this paper, therefore, we attempt merely to point out the major common sources of measurement error referring the reader to particular experiments as examples which indicate the likely errors in practical turbulence measurements.

Sources of Error

The basic principle of the laser-Doppler anemometer is well-known. In terms of the usual fringe mode arrangement, two light beams are arranged to intersect and therefore interfere, in a "control volume." The waves scattered by the passage of a small particle through this volume will have slightly different Doppler-shifted frequencies, neither of which could be accurately measured directly because they are so high. However, the interference between them yields a beat frequency which, provided the particle velocity is much smaller than the velocity of light, is directly proportional to the velocity of the particle, and *can* be measured since it is very much smaller than the Doppler frequencies. In fact, the beat frequency depends only on the component of the velocity normal to the "fringes" in the control volume and is totally insensitive to the other two components. This insensitivity and the linear relationship between frequency and velocity makes calibration easy, since the constant of proportionality is fixed by the laser wavelength and the angle between the two intersecting beams.

However, the properties of the signal at the photodetector and the consequent accuracy of the measurement *do* depend on the total velocity vector, since that determines the "residence time" of the particle in the control volume. The geometry of the control volume is also clearly important in that respect. Proper design of the optical system appropriate in any particular situation will depend partly on how many particles are likely to be present in the control volume at any one time. Errors can be caused by the particles not faithfully following the flow, by detector shot noise or by variations in the refractive index of the fluid, but these can normally be made insignificant, though the last of them is more important in combusting flows. Apart from these, there are two major factors that determine the performance of laser-Doppler anemometry in turbulence measurements. These are, firstly, the type of signal generated by the photodetector and, secondly, the type of signal processor used. Very often, the broad approach to the latter will depend on the former, although details will usually depend on the particular properties to be measured and the required precision. If there are sufficient particles present in the flow and sufficient laser power, then the Doppler signal will often be essentially analogue; in those circumstances processing by "frequency tracking" is common, though other techniques are also used. Now even in laminar flow, the random arrival and departure of particles into and out of the scattering volume causes phase and frequency fluctuations in the Doppler signal. The consequent uncertainties in frequency measurement (whether by frequency tracking, spectrum analysis or any other "frequency domain" processing) are therefore linked closely with the transit time of the particles through the control volume. This effect is usually termed "transit time broadening." Similar broadening also occurs even if only one particle passes through the control volume, since the spectrum will be essentially the Fourier transform of the sinusoidal signal whose duration is finite (a delta function spectrum would only be possible for an infinite integration time of a continuous sinusoid). There are other sources of spectrum broadening; increasing the spatial resolution, for example, by reducing the scattering volume dimensions, will lead to increased "velocity gradient broadening" errors if the mean velocity varies significantly across the control volume. This effect is usually only important in very small scale flows or near a solid surface. In principle, these broadening effects make frequency domain processing less suitable for very low intensity flows, since they will contribute relatively more to the measured velocity fluctuations. However, there have been several demonstrations that frequency domain processing can be used to make quite accurate measurements of Reynolds stresses in low to moderate intensity flows. Bourke et al. [39] and Durst and Whitelaw [38], for example, compared laser-Doppler anemometer measurements of the turbulent stresses in a fully developed channel flow (of water) with classic hot-wire measurements (in air) at the same Reynolds number. They included broadening corrections and obtained results within the experimental uncertainty of the hot-wire data. Similar laser-Doppler anemometer measurements in air (usually more difficult because of seeding requirements) have also been made (Yanta and Smith [51]). One of the most recent examples is the work of Simpson et al. [52], who obtained good agreement between crossed-wire and laser-Doppler anemometer measurements of the Reynolds stresses in the regions of a separating boundary layer flow where crossed wire measurements could sensibly be made. They conducted an error analysis and deduced that $\overline{u^2}$ and $\overline{v^2}$ uncertainties were no greater than $\pm 4\%$ of their maximum values and $\overline{uv}$ errors were less than $\pm 6\%$ of the maximum value.

At higher turbulent intensities, it is difficult to maintain high particle densities throughout the flow so that the signal from the photodetector becomes very intermittent. The passage of individual particles lead to signal "bursts" and under these circumstances it is usual to employ counting methods to obtain a velocity measurement. The function of the signal processor is then essentially digital, producing a single velocity value—often as a digital word—at the (random) times when a particle traverse occurs. Such methods are also preferable when the intensities are very low, since they do not suffer from transit time broadening. The technique is effectively a time-domain measurement and the errors depend on how accurately the processor can deduce the frequency present in each single burst. This in turn obviously depends on the signal noise ratio within each burst.

Buchave et al. [50] have shown that the randomly arriving samples are capable of reproducing the statistics of the desired Eulerian velocity field, provided the burst signal is analyzed for the period of time that the particle is in the scattering volume. This implies that direct ensemble averaging of the velocity samples would lead to bias errors (since each particle spends a different amount of time in the sampling volume), so that "residence-time weighting" has to be employed, or suitable corrections made. There are also other sources of bias error in burst-type laser-Doppler anemometer processing. In general, bias errors become increasingly significant as the turbulent intensity rises. Various correction or weighting techniques have been suggested but as yet there seems to be no consensus about the most effective approach. Measurements by Karpuk and Tiederman [53] and Quigley and Tiederman [54] in the viscous sub-layer in pipe flow have shown good agreement between laser-Doppler anemometer and hot-wire turbulence data, so it does seem possible to make quite accurate measurements by burst-processing at intensities around 30%. On the other hand, Lau et al. [55] measured the fluctuating velocities in axi-symmetric subsonic and supersonic free jets, using a frequency shifted, two-color system with burst counting, and report discrepancies of up to 75% in local values of the normal turbulent stress ($\overline{v^2}$).

In cases where the scattered light intensity or signal noise ratio at the photodetector output is very low, photon correlation techniques are potentially very powerful. This method was originally suggested by Pike [56] and Jakeman [57] and consists essentially of digitally autocorrelating a sequence of logic "on/off" pulses generated from the photodetector response to the arrival of individual light photons from the scattering volume. In turbulent flow, the resulting autocorrelation is, in general, a decaying cosine wave superimposed on a decaying mean value. The latter is caused by the usual Gaussian variation of light intensity across the laser beam and the former contains all the mean and fluctuating velocity information. Various methods of deducing this information from the measured autocorrelation have been proposed but only in the last few years have reasonably accurate measurements of the turbulence stresses been obtained. Moore and Smart [58] compared laser-Doppler anemometer and hot-wire axial turbulence intensity measurements in an axisymmetric free jet and obtained discrepancies no worse than 10% (20% on the stress) on the flow centerline. Abbiss [59] made shear stress measurements in a similar flow, by combining two sets of measurements obtained with the fringe system at two orientations. His results agreed with the shear stress calculated from the momentum integral equation to within about 15% over much of the flow. Similar measurements, this time using a two-component system, were made in a supersonic boundary layer (see also East [44]).

While there have been some laser-Doppler anemometer measurements in separated flows (e.g. Simpson et al. [52], Crabb et al. [60], and Durst and Rastogi [61]) it is only very recently that direct comparisons have been made with data obtained by one of the other techniques capable of making such measurements—pulsed-wire anemometry. Brown [62] has used photon correlation techniques to make laser-Doppler anemometer measurements in the separated wake of a circular disc placed at the exit of an axisymmetric jet. Agreement between the turbulent stresses and those measured with a pulsed wire was encouragingly close, although in view of the possible errors in the latter it is at least possible that this was entirely fortuitous.

There seems no doubt, however, that if the optical system and signal processing technique are carefully chosen, measurements of the turbulence Reynolds stresses *can* be made in a wide variety of flows, with a precision certainly no worse than that obtainable with standard wire anemometry in flows appropriate to its use. That is a significant step forward in experimental fluid mechanics, although the difficulties in effective use of laser-Doppler anemometers (and the cost) should not be underestimated.

CONCLUSION

In this chapter we have tried to emphasize the major sources of error usually present in anemometry measurements of turbulent Reynolds stresses. Attention has been concentrated on wire anemometry, with which the author is most familiar. It has been argued that major sources of error in hot-wire measurements occur once local turbulent intensities exceed, say, 20–25% and that these are a direct and inevitable result of the inadequate or ill-defined yaw response of the instrument and the statistics of the instantaneous velocity vector. The work of Tutu and Chevray [21] allows estimation of the errors and some measure of correction is therefore possible. However, the adequacy of such corrections can depend on the details of the mean and fluctuating velocity field. While multi-wire probe and flying hot-wire techniques have been developed to overcome the major problems, these are difficult to implement successfully and, arguably, represent attempts to push the hot-wire technique beyond the bounds of what can sensibly be expected from it. If, in any particular measurement problem, a significant quantity of data is required from regions of flow where the turbulent intensities exceed about 30%, serious consideration should be given to the use of either laser-Doppler or pulsed-wire anemometry. Neither of these techniques are simple to use effectively, but there is now sufficient evidence to show that both can give reasonably accurate Reynolds stress data in isothermal gas flows. If local velocities exceed about 20 m/s, or the flow is more hostile or access is difficult then laser-Doppler techniques have to be used.

A final example illustrates the confidence that *can* be gained in all three measurement techniques. Figure 12 shows the variation of Reynolds shear stress across the axi-symmetric mixing layer formed downstream of the exit from a circular jet. This is, to the author's knowledge, the first direct comparison between all three techniques, in a flow where the intensity varies from very low (at the high velocity side of the layer, small y/D) to very high (at the low velocity side, high y/D). The results were obtained by Brown and Inman [63]. Note that the hot-wire data has been corrected using Tutu and Chevray's [21] results and the laser transit data has been corrected for bias effects. There is no more than about 20% scatter between all three sets of measurements, although the pulsed-wire data is clearly not very reliable in the low intensity region ($y/D < 0.3$). Detailed discussion of these results can be found in Brown and Inman [63] and Brown [62], but further work of a similar nature, particularly in flows where standard hot-wire techniques cannot be applied, would clearly be of great value.

NOTATION

A, B	empirical coefficients	P	probability
E	anemometer bridge output voltage	r	correlation coefficient
e'	voltage fluctuation	U	normal velocity
F	probability function	u, v, w	velocity components
n	coefficient		

Greek Symbols

$\alpha_u, \alpha_v, \alpha_w$	square roots of turbulent energies	ψ_e	effective wire angle
θ	angle		

REFERENCES

1. Hewett, G. F., *Measurement of Two-Phase Flow Parameters*, Academic, London (1978).
2. Goldstein, R. J., *Fluid Mechanics Measurements*, Hemisphere (1983).
3. Comte-Bellot, G., "Hot Wire Anemometry," *Ann. Rev. Fluid Mech.* 8, 209. (1976).
4. Vagt, J. D., "Hot-Wire Probes in Low-Speed Flow," *Prog. in Aero. Sci.* 18, (1978).
5. Kim, H. T., Kline, S. J., and Reynolds, W. C., "The Production of Turbulence Near a Smooth Wall in a Turbulent Boundary Layer," *J. Fluid Mech.*, 50, 133 (1971).

6. Smith, C. R., and Paxson, R. D., "A Technique for Evaluation of Three-Dimensional Behavior in Turbulent Boundary Layers Using Computer Augmented Hydrogen Bubble-Wire Flow Visualization" *Expt. in Fluids* 1, 43 (1983).
7. Bradshaw, P., *An Introduction to Turbulence and Its Measurement*, 2nd ed. Pergamon, Oxford (1975).
8. Frenkiel, P., "Higher order correlations in a Turbulent Jet", *Phys. Fluids* 10, 507 (1967).
9. Guitton, D. E., "Correction of Hot-Wire Data for High Intensity Turbulence, Longitudinal Cooling and Probe Interference," McGill Univ. Mech. Eng. Rpt No. 68-6 (1968).
10. Perry, A. E., and Morrison, G. L., "Static and Dynamic Calibrations of Constant Temperature Hot-Wire Systems," *J. Fluid Mech.* 47, 765 (1971).
11. Morrison, G. L., Perry, A. E., and Samuel, A. E., "Dynamic Calibration of Inclined and Crossed Hot Wires," *J. Fluid Mech.* 52, 465 (1972).
12. Bruun, H. H., "A Note on Static and Dynamic Calibration of Constant-Temperature Hot-Wire Probes," *J. Fluid Mech.* 76, 145 (1976).
13. Collis, D. C., and Williams, M. J., "Two-dimensional convection from heated wires at low Reynolds Numbers," *J. Fluid Mech.* 6, 357 (1959).
14. Bradbury, L. J. S., and Castro, I. P., "Some Comments on Heat Transfer Laws for Fine Wires," *J. Fluid Mech.* 51, 487 (1972).
15. Bearman, P. W., "Corrections for the Effect of Ambient Temperature Drift on Hot-Wire Measurements in Incompressible Flow," DISA Information 11 (1971).
16. Antonia, R. A., Prabhu, A., and Stephenson, S. E., "Conditionally Sampled Measurements in a Heated Turbulent Jet," *J. Fluid Mech.* 72, 455 (1975).
17. Dean, R. B., and Bradshaw, P., "Measurements of Interacting Turbulent Shear Layers in a Duct," *J. Fluid Mech.* 78, 641 (1976).
18. Way, J., and Libby, P. A., "Hot-Wire Probes for Measuring Velocity and Concentration in Helium-Air Mixtures," *AIAA J.* 8, 976, (1970).
19. Champagne, F. W., Sleicher, C. A., and Wehrmann, O. H., "Turbulence Measurements With Inclined Hot Wires. Part I. Heat Transfer Experiments." *J. Fluid Mech.* 28, 153 (1967).
20. Champagne, F. W., and Sleicher, C. A., "Turbulence Measurements with Inclined Hot-Wires. Part II. Hot-wire Response Equations," *J. Fluid Mech.* 28, 177 (1967).
21. Tutu, N. K., and Chevray, R., "Cross-Wire Anemometry in High Intensity Turbulence," *J. Fluid Mech.* 71, 785 (1975).
22. Rodi, W., "A New Method of Analyzing Hot-Wire Signals in Highly Turbulent Flow, and its Evaluation in a Round Jet," DISA Information 17, 9 (1974).
23. Moffat, R. J., Yavuzkurt, S., and Crawford, M. E., "Real-Time Measurements of Turbulence Quantities with a Triple Hot-Wire System," Flow Dynamics Conf., Marseille, France (1979).
24. Acrivellis, M., "Measurements by Means of Triple Sensor Probes," *J. Phys. E: Sci. Instrum.* 13, 986 (1980).
25. Coles, D. E., and Wadcock, J. A., "Flying Hot-Wire Study of Flow Past a NACA 4412 Airfoil at Maximum Lift," *J. AIAA*, 17, 321 (1979).
26. Watmuff, J. H., Perry, A. E., and Chong, M. S., "A Flying Hot-Wire System," *Exp. in Fluids*, 1, (1983).
27. Bradbury, L. J. S., "A Pulsed-Wire Technique for Velocity Measurements in Highly Turbulent Flows," NPL Aero Report 1284 (1969).
28. Tombach, I. H., "Velocity Measurements with a New Probe in Inhomogeneous Turbulent Jets," PhD Thesis, California Institute of Technology (1969).
29. Bradbury, L. J. S., and Castro, I. P., "A Pulsed-Wire Technique for Velocity Measurements in Highly Turbulent Flows," *J. Fluid Mech.*, 49, 657 (1971).
30. Bradbury, L. J. S., "Measurements with a Pulsed-Wire and a Hot-wire Anemometer in the Highly Turbulent Wake of a Flat Plate," *J. Fluid Mech.*, 77, 473 (1976).
31. Castro, I. P., and Robins, A. G., "The Flow Around a Surface-Mounted Cube in Uniform and Turbulent Streams," *J. Fluid Mech.*, 79, 307 (1977).
32. Eaton, J. K., Johnston, J. P., and Jeans, A. H., "Measurements in a Reattaching Turbulent Shear Layer," In *Proc. 2nd Int. Symp. on Turbulent Shear Flows*, London, Paper No. 16.7 (1979).

33. Britter, R. E., and Hunt, J. C. R., *J. Ind. Aero.*, 4, 165 (1979).
34. Ruderich, R., and Fernholz, H. H., "An Experimental Investigation of the Turbulent Shear Flow Downstream of a Normal Flat Plate with a Long Splitter Plate Modification of a Model." In *Proc. 4th Int. Symp. on Turbulent Shear Flows*, Karlsruhe (1983).
35. Castro, I. P., and Cheun, B. S., "The Measurement of Reynolds Stresses with a Pulsed-Wire Anemometer," *J. Fluid Mech.*, 118, 41 (1982).
36. Yeh, Y., and Cummins, H. Z., "Localized Flow Measurements with an He-Ne Laser Spectrometer," *Appl. Phs. Letters*, 4, 176 (1964).
37. Durst, F., Melling, A., and Whitelaw, J. H., "Laser Anemometry: A Report on Euromech 36," *J. Fluid Mech.*, 56, 143 (1972).
38. Durst, F., and Whitelaw, J. H., "Measurements of Reynolds Shear Stress in Water Using Laser Anemometry," DISA Inf. 12, 11 (1971).
39. Bourke, P. J., Brown, C. G., and Drain, L. E., "Measurements of Reynolds Shear Stress in Water Using Laser Anemometry," DISA INF. 12, 21 (1971).
40. Wisler, D. C., and Mossey, P. W., "Gas Velocity Measurements Within a Compressor Rotor Passage Using a Laser-Doppler Velocimeter." *Trans. ASME, J. Eng. Power*, 95, 91 (1973).
41. Strazisar, A. J., and Powell, J. A., "Laser Anemometer Measurements in a Transonic Axial Flow Compressor Rotor." Symp. on Measurement Methods in Rotating Components of Turbomachinery, ASME Gas Turbine Conference, New Orleans, 165 (1980).
42. Gosman, A. D., Melling, A., Watkins, P., and Whitelaw, J. H., "Axisymmetric Flow in a Motored Reciprocating Engine." *Proc. I. Mech. E.*, 192, 213 (1978).
43. Cole, J. P., and Swords, M. D., "Laser-Doppler Anemometry Measurements in an Engine," *Appl. Optics* 18, 1539 (1979).
44. East, L. F., "The Application of a Laser Anemometer to the Investigation of Shock-Wave Boundary Layer Interactions," AGARD-CP-193, Paper S (1976).
45. Bachalo, W. D., and Johnson, D. A., An Investigation of Transonic Turbulent Boundary Layer Separation Generated on an Axisymmetric Flow Model," AIAA Paper 79-1479 (1979).
46. Self, S. A., and Whitelaw, J. H., "Laser Anemometry for Combustion Research," *Comb. Sci. & Tech.* 13, 171 (1976).
47. Hutchinson, P., Khalil, E. E., and Whitelaw, J. H., "The Measurement and Calculation of Furnace-Flow Properties," *J. Energy* 1, 212 (1977).
48. Born, G. V. R., Melling, A., and Whitelaw, J. H., "Laser-Doppler Microscope for Blood Velocity Measurements," *Biorheology* 15, 163 (1978).
49. Durst, F., Melling, A., and Whitelaw, J. H., *Principles and Practice of Laser-Doppler Anemometry*, 2nd edition, Academic, London (1981).
50. Buchhave, P., George, W. K., and Lumley, J. L., "The Measurement of Turbulence with the Laser-Doppler Anemometer" *Ann. Rev. Fluid Mech.*, 11, 443 (1979).
51. Yanta, W. J., and Smith, R. A., "Measurements on Turbulence-transport Properties with a Laser-Doppler Anemometer," AIAA paper 73-169 (1973).
52. Simpson, R. L., Strickland, J. H., and Barr, P. W., "Features of a Separating Turbulent Boundary Layer in the Vicinity of Separation," *J. Fluid Mech.* 79, 553 (1977).
53. Karpuk, M. E., and Tiederman, W. G., "Effect of Finite Probe Volume Upon Laser-Doppler Anemometer Measurements," *AIAA J.* 14, 1105 (1976).
54. Quigley, M. S., and Tiederman, W. G., "Experimental Evaluation of Sampling Bias in Individual Realization Laser Anemometry," *AIAA J.* 15, 266 (1977).
55. Lau, J. C., Morris, P. J., and Fisher, M. J., "Measurements in Subsonic and Supersonic Free Jets Using a Laser Velocimeter," *J. Fluid Mech.* 93, 1 (1979).
56. Pike, E. R., "Photon Statistics" *Proc. Florence Inaugural Conf. European Phys. Soc.*, Rivista del Nuovo Climento, 1 (Numero speciale), 277 (1969).
57. Jakeman, E., "Theory of Optical Spectroscopy by Digital Autocorrelation of Photon-Counting Fluctuations," *J. Phys. A: Gen. Phys.* 3, 201 (1970).
58. Moore, C. J., and Smart, A. E., "Retrieval of Flow Statistics Derived from Laser Anemometry by Photon Correlation," *J. Phys. E: Sci. Instrum.* 9, 997 (1976).
59. Abbiss, J. B., "Development of Photon correlation Anemometry for Application to Supersonic flows," AGARD-CP-193, Paper 11 (1976).

60. Crabb, D., Durao, D. F. G., and Whitelaw, J. H., "A Jet Normal to a Cross Flow." *Trans. ASME J. Fluids Eng.* (1981).
61. Durst, F., and Rastogi, A. K., "Theoretical and Experimental Investigations of Turbulent Flows with Separation," In *Turbulent Shear Flows* I, p. 208, Springer Verlag, Berlin (1979).
62. Brown, R. G. W., "An Exploration of New Optical Techniques for Turbulence Measurements Using Photon Correlation Laser Anemometry," PhD Thesis, University of Surrey (1984).
63. Brown, R. G. W., and Inman, P. I., "Direct Comparisons of Laser-Doppler, Transit, Hot-Wire, and Pulsed-Wire Anemometer Measurements in an Axisymmetric Jet." In *Proc.* ICIASF-83, ISL, St. Louis, France (1983).

CHAPTER 38

MEASUREMENT TECHNIQUES FOR MULTIPHASE FLOWS

N. P. Cheremisinoff

Exxon Chemical Co.
Linden, New Jersey, USA

CONTENTS

INTRODUCTION

In subsequent volumes, complex flow phenomena involving two or more phases are addressed. Investigations of the hydrodynamics and transport mechanisms occurring in these systems have been conducted using a variety of experimental techniques and instrumentation. Studies are most often aimed at quantifying the behavior of the discrete phase; e.g. bubble dynamics in a gas-solid fluidized bed or in a gas-liquid bubble columns droplet formation and dispersion in a mixed reactors, particle behavior in various suspension or slurry flows, aerosol particle dynamics in the atmosphere, interphase and phase separation detection in multi-component flows, and many others.

This chapter describes specific techniques that fall under two general categories of flow behavior measurements, namely electrical sensing techniques and those methods based on light scattering and optics principles. In the first category, a variety of sensor configurations have been devised by investigators to quantify the actions of the discrete phase. Many of these methods base their measuring principles on electrical properties differences between the flowing phases. Interpretation of measured signals requires application of signal processing principles, namely, autocorrelation, cross-correlation and spectral density analyses.

The latter approach enables studies by the transmission of light waves and resultant scattering properties from light's interception with entities such as traveling particles and droplets. The physics of these methods are quite complex and some rather rigorous mathematics have been devised to formulate the theoretical bases for measurements. However, for the purposes of obtaining a working knowledge of the state-of-the art of light scattering instrumentation, only the simplified mathematics of ray optics is presented. Although modeling light as a ray with finite boundaries is not rigorous, it does provide instruction in the measurement principles.

Experimental arrangements for photographic and cinegraphic techniques are also described. These are perhaps the most widely used methods for flow visualization studies and are capable of providing valuable quantitative data on particle sizes in dispersed-type flows over both spacial and temporal distributions. Experimental set-ups for multiphase flow studies require that careful considerations be given to a variety of parameters to ensure high quality data. Generalizations on specific system arrangements are difficult and often require trial and error to determine optimum lighting, depth of field, and flash arrangements. Discussions should therefore be viewed only as examples and not recommended procedures except for specific flow systems studied.

ELECTRICAL SENSING TECHNIQUES

This first section describes various methods and probe configurations that measure flow regimes and flow discontinuities, such as bubbles based on electrical property differences between phases. The basic requirement for the use of most of the methods described this section, excluding transducers, is that the electrical conducting properties of the two phases be measurably different. Interpretation of measured signals requires application of auto-correlation, cross-correlation, and spectral density analysis. It is assumed that the reader has a fundamental knowledge of electricity and basic circuit theory and therefore, no review of such principles is given in the foregoing discussions.

Piezoelectric Transducers

The use of pressure transducers is a straightforward approach to inferring certain properties about two-phase flow systems. Local pressure fluctuations imply information about the voidage. In entrained flows, for example, sensitive transducers can be used to detect high concentrations of a discrete phase along the flow's axis. By positioning wall-tap transducers along the height of a fluidized column of either liquid or solids, and measuring the pressure differential with respect to some reference point (say the top of the column), the free surface of the bed can be readily detected. The pressure drop profile for a gas-solid fluidized bed is shown in Figure 1 for example. Local

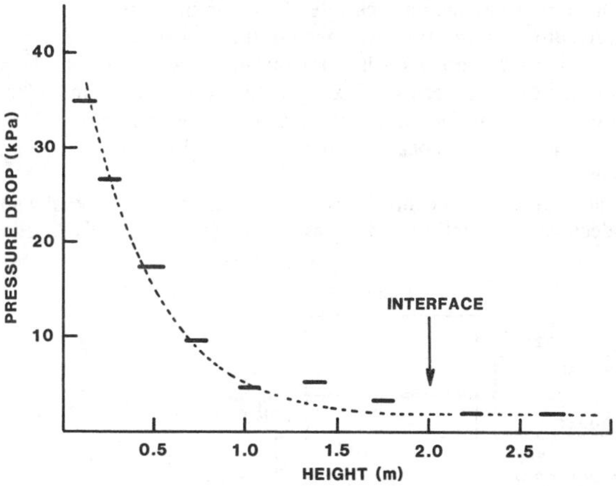

Figure 1. Pressure profiles along a fluidized bed of air and catalyst particles. The profile provides information on the bed height.

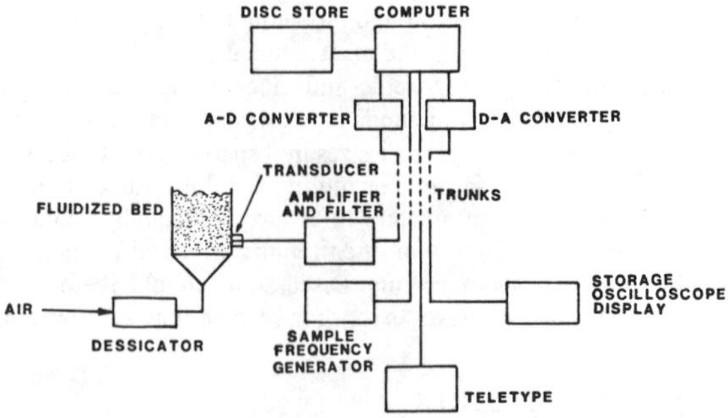

Figure 2. Schematic diagram of piezoelectric transducer technique employed by Taylor et al. [2].

pressure fluctuations provide other information about the average or statistical properties of flow behavior. A number of investigators [1–5] have used sensitive transducers to obtain more detailed information on the flow structure in bubbling systems by applying spectral density and correlation techniques. One of the more unique experimental set-ups is that of Taylor et al. [2] in which the power spectrum of the noise from a transducer was shown to relate information on the flow regime identification. The system used is shown in Figure 2.

A pressure transducer provides an electrical signal that is proportional to the pressure at the probe's tip. Such a signal can be amplified to remove the d.c. components and spurious high frequency noise. In the system shown in Figure 2, the signal is then converted to digital form using an A–D converter, set for sampling at prespecified intervals. The sampling interval is established by a sample frequency generator. This type of experiment requires large amounts of data be processed to infer meaningful statistical properties. Consequently, a computer was used to generate the power spectrum of a batch of samples from the signal and after conversion to analog form by the D–A converter, the spectrum was displayed on a storage oscilloscope. For tall beds, a specially designed piezoelectric transducer, shown in Figure 3 was employed. The transducer was capable of a resolution of 1 mm water gauge with a maximum design of 150 m water gauge. The design consisted of two piezoelectric ceramic discs mounted on each side of an aluminum plate. One plate was exposed to the fluidized bed pressure while the other was open to the atmosphere. The two elements were connected to the input of a charge amplifier, where the output is proportional to the pressure difference on the elements (in this case, between the frequency limits of 0.01 Hz and 1,000 Hz). The system devised by Taylor et al. [2] provided on-line analysis of the power spectrum. Their analysis scheme on the computer had three major blocks: a controller, a fast Fourier transform algorithm, and an allocator of storage.

The controller block organized a continual sampling of the pressure signal while a spectrum was simultaneously calculated, controlled the disc based storage area, and displayed the last spectrum.

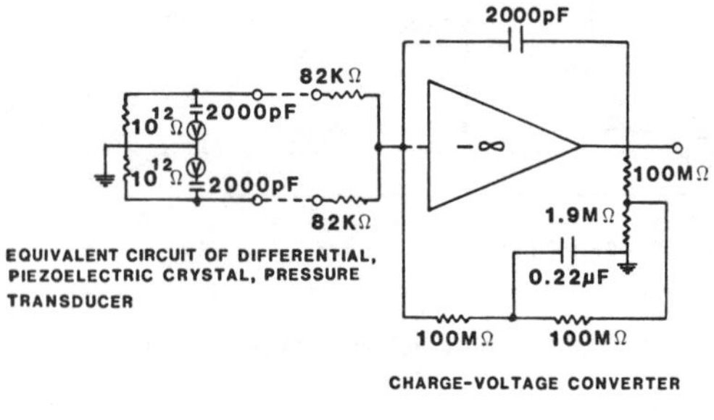

Figure 3. Pressure transducer circuit designed by Taylor et al. [2].

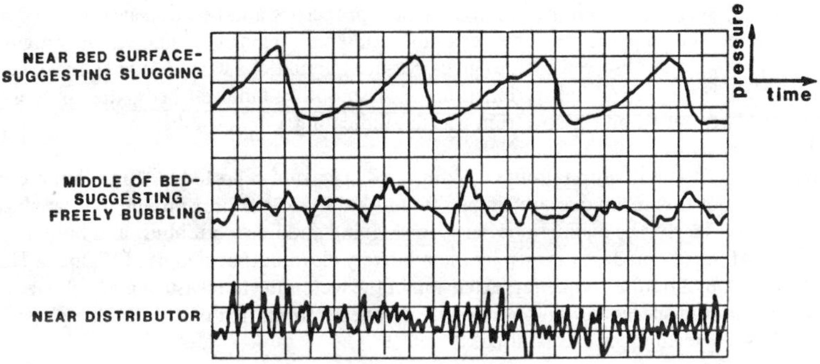

Figure 4. Pressure traces obtained at various heights in a 3-ft column.

The spectrum was formed using a "decimation in time fast Fourier transform algorithm" followed by a calculation of the modulus of the Fourier coefficients to form the power spectrum. An integer format was necessary since a floating point format would effectively half the storage capability while increasing computation time. The data was smoothed with a 10% Cosine Bell data window before transformation and the power spectrum smoothed by a 3-point weighted moving averaging.

Broadhurst and Becker [3] also used spectral analysis of bed pressure fluctuations for interpreting bubbling system behavior. The technique, in general, is relatively simple to apply and is often used in industrial pilot studies. However, interpretation of the statistical properties of the transducer output signal is not always as straightforward. Typical transducer output signals recorded on an oscillograph by the author for various heights in a three foot diameter column using char particles and air are shown in Figure 4. The signals clearly show different zones of fluidization.

The two most important statistical properties of the pressure fluctuations are the autocorrelation and the frequency spectrum. The first is a measure of the degree of correlation between neighboring values of the pressure fluctuations. The autocorrelation is also used as an intermediate step in the estimation of the spectrum, which measures the distribution of energy with frequency. Broadhurst et al. [3] present various correlations that enable estimation of the peak frequency and the magnitude

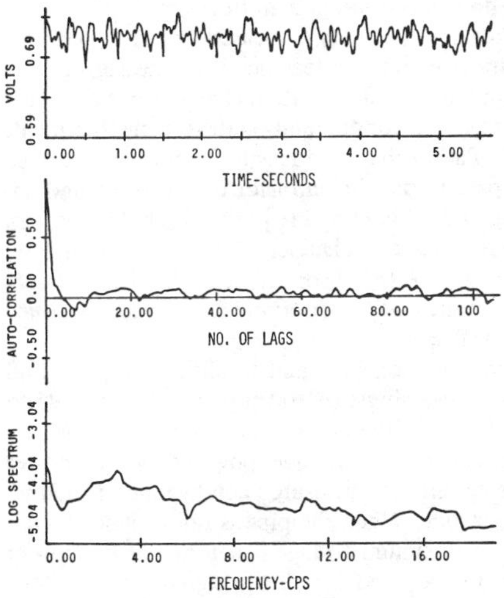

Figure 5. Shows typical spectral analysis of Pressure fluctuations reported by Broadhurst et al. [3].

of the frequency spectrum, from the gas and particle properties and bed dimensions, for fluid beds operating in the slugging regime. They examined different regimes of fludization ranging from smooth to slugging and observed each to give pressure fluctuations with characteristically different autocorrelation and spectral density functions. For smoothly fluidized beds, the autocorrelation damped quickly to zero, and the spectrum was flat indicating a uniform distribution of energy with frequency.

Figure 5 shows the autocorrelation and computed spectrum reported by Broadhurst et al. [3]. The advantages of this technique are simplicity and ability to perform on-line data analysis on a mini-computer. Clearly, the distinctions between slugging and freely bubbling, and bubbling versus nonbubbling are easily made, as shown by the relatively clean output signals of Figure 4. However, with the exception of bubble frequency, extension of the technique to measuring bubble size and rise velocity has not been successful. To obtain more detailed information on the discrete phase properties, the following techniques can be used.

Electroresistivity Probes

These methods include conductivity, capacitance, inductance, and impedance sensing elements. Flow discontinuties such as bubbles in fluidized system, or droplets in emulsions are detected from local measurements of the electrical properties of the continuum phase using these probes. Again, the most widely practiced application with these devices has been in studying fluidized systems, but also in investigating the mixing characteristics of immiscible fluids and/or efficiency of mixing systems in general. In fluidized systems, bubble properties have been measured from point resistivity measurements by various investigators [6–9]. Burgess et al. [8] employed a five-element probe to derive the shape and size of bubbles. Resistivity measurements to detect bubbles along with simultaneous transient gas measurements were made by Calderbank [9] to differentiate the compositions of the gas in the emulsion and bubble phases in gas-solid fluid beds.

In liquid systems, capacitance, conductance, impedance and resistivity probes are relatively straightforward to use. Any one of these properties can be monitored to detect local disturbances or flow regimes. Barnea et al. [10] have used conductance probes to identify flow patterns in two-phase (gas-liquid) horizontal, near horizontal and upward flows. Flow pattern information in gas liquid flows is most often obtained by visual observation, however, the designation of flow pattern has not yet been accurately standardized and depends largely upon individual interpretation of visual observations. The major difficulty of visual observation, even using high-speed photography is that the picture is often confusing and difficult to interpret, especially when dealing with high-velocity flows. Also, some systems are opaque and flow visualization is not possible.

Johnes and Delhaye [11] have reviewed and summarized a variety of measuring techniques used in two-phase flow. Hsu et al. [12], used a hot-wire anemometry technique for measuring void distribution for vertical flow and used the signal output also for flow pattern characterization. Jones and Zuber [13] developed an X-ray void measurement system for obtaining statistical measurements in vertical air water flow in a rectangular channel. The probability density function of the void fraction fluctuations was used as a quantitive flow pattern discriminator for bubbly, slug, and annular flows. Govier et al. [14], Chaudhry et al. (15), and Isbin et al. [16] related the flow pattern to the pressure gradient variation as described earlier. Hubbard and Dukler [17] describe the method by which the flow pattern can be determined from the spectral distribution of the wall pressure fluctuations, again, as described earlier. They could not however, discriminate between stratified and annular flows, or between the dispersed liquid or dispersed gas flow regimes.

The conductance probe method was employed by Solomon [18] and Griffith [19], Fiori and Bergles [20], and Bergles et al. [21]. These studies used a single central probe and were able to detect differences among bubbly, slug, and annular flows. In the more recent study by Barnea et. al. [10], the conductivity probes were composed of parallel probe electrodes powered by a d.c. power source with variable voltage from 0 to 10 volts. The circuit used measured conductance as a function of time between the probes and a flat large electrode. When the pipe is full of fluid (liquid) the electric current is closed through the liquid and a maximum voltage is detected at the output signal. When the pipe is empty of liquid the circuit is open and the output signal is zero. Three

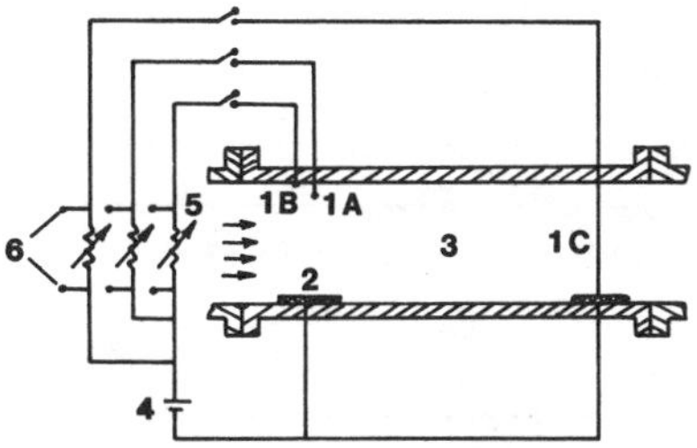

Figure 6. Scheme of probing system described by Barnea et al. [10]: (1A) probe-tip electrode, stainless steel wire, 0.25mm dia., teflon coated, 3mm down from the top; (1B) probe-tip electrode, stainless steel wire, 0.25mm dia., teflon coated, flat with tube surface; (1C) probe-metal needle, not insulated; (2) flat copper electrode; (3) two-phase flow; (4) voltage supply; (5) variable resistors; (6) output signal (to oscilloscope).

probes were used at various axial locations along a pipe with a flat large electrode located at the floor of the pipe.

One probe could detect bubbles just below the top of the pipe. Since the probe is quite small it can detect easily even small gas bubbles yielding zero voltage during a bubble passage. The fall, as well as the rise, of the voltage, was found immediately, and for bubble flow the voltage fluctuated from zero to a maximum. A second probe was flush with the upper part of the wall. It is designed to detect any surface wetness around the inner tube periphery. Even thin liquid films that are present in annual flow or the remaining film after the passage of the liquid slug in intermittent flow can cause a voltage output of this probe. Cheremisinoff [22] used a similar technique to study film thickness in annular flows. The last probe could detect the liquid level under stratified conditions. It is constructed of a non-insulated needle that is inserted vertically along the pipe diameter and almost reaches the bottom flat electrode. A change of the liquid level interface is easily detected by this arrangement. In the case of vertical upflow, the flow is symmetrical and two probes are sufficient to detect the flow patterns.

Figure 6 shows the probe scheme described by Barnea et al. [10] and Figure 7 shows typical output tracings obtained by the author using a similar arrangement for vertical pipe flow. The top

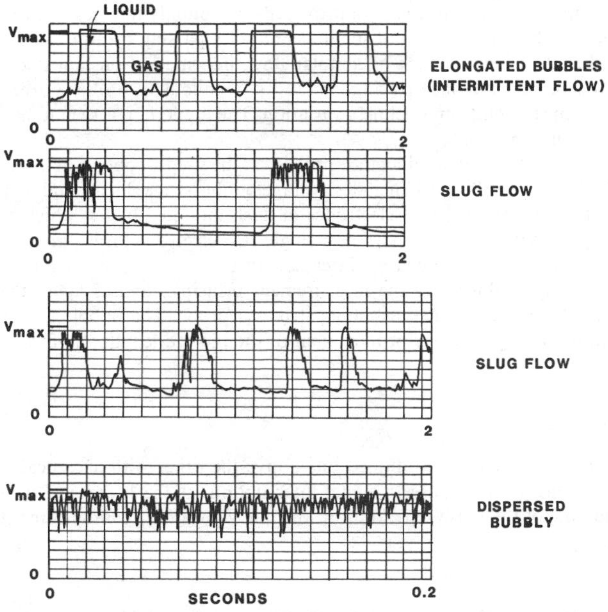

Figure 7. Typical oscillograph tracings obtained using a conductivity probe in air-water flow.

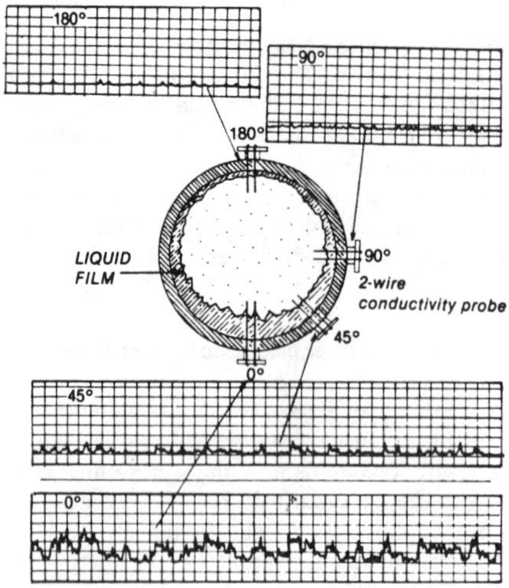

Figure 8. Wave height tracings obtained from conductivity probes in annular flow study [22].

tracing in Figure 7 shows the presence of elongated bubbles where the voltage fluctuates intermittently from maximum conductivity (corresponding to full liquid bridging) to zero conductivity which corresponds to the gas zone. Barnea et al. define elongated bubbles as an intermittent flow pattern in which the liquid slug is free of entrained gas bubbles. The second and third tracing in Figure 7 correspond to slug flow where adjacent gas zones are separated by a liquid zone with entrained gas bubbles. The entrained gas bubbles in the liquid slug correspond to the short pulses in the fully conducting slug zone. The bottom trace is a dispersed-type bubbly flow regime. In this regime the probe is exposed to a high population of small gas bubbles passing by at a fast rate.

To identify other flow regimes such as annular flow, information on film thickness is needed. Figure 8 shows gas-liquid interface tracings made by conductivity probes positioned about the periphery of a pipe in studying film thickness variation. The liquid film is seen to be fully supported around the entire inner pipe wall.

Resistivity and capacitance or conductance probes can be used to obtain point measurements of complete bubble characteristics which includes average bubble size, velocity, and frequency. A single probe can consist of two parallel wire electrodes of flush mounted plates, which are excited by a high-frequency signal. When a bubble strikes a single two electrode probe, an interruption in the electrical signal occurs due to the change in capacitance, resistance, or conductance of the surrounding media. Figure 9 illustrates the measurement principle. In this figure the signal U(t) is shown versus time for the case when a bubble strikes a probe. The bubble generates an electric impulse of duration time t_B, which is proportional to the pierced length (ℓ) of the bubble.

$$\ell = U_B t_B \tag{1}$$

where U_B is the rise velocity of the bubble.

By using two probes A and B, positioned one above the other with a vertical displacement, s, we note that the rising bubble causes a signal interruption, at the lower and then at the higher probe. From the time of the pulse separation t_s, the rise velocity U_B can be computed.

$$U_B = s/t_s \tag{2}$$

Hence, a single probe enables measurements of the local bubble frequencies (by simply counting the number of electric impulses), and two vertically aligned probes provide measurements of the bubble rise velocity and pierced length of the bubble.

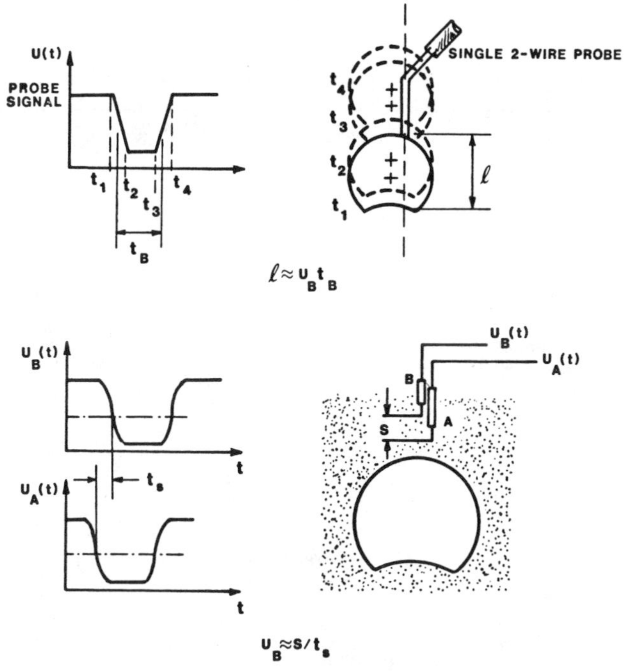

Figure 9. Illustration of measurement principle for rise velocity and pierced length of a single bubble.

The displacement height s is a critical parameter in the measurements because it is difficult to identify corresponding bubbles in the signals $U_A(t)$ and $U_B(t)$ due to splitting or coalescence of bubbles between probes. If displacement s is too small, it is difficult to measure individual pulse separations because small variations in the shape of the leading and trailing edges of the pulses can lead to errors in the magnitude of t_s. The instantaneous rise velocity of a bubble depends on several factors:

1. The rise velocity of an individual bubble is influence by the proximity and size of neighboring bubbles.
2. During the period prior to coalescence, bubbles about to coalesce influence each other's rise velocity.
3. The size and geometry of the probe can alter both the bubble shape and rise velocity.

The optimum separation between two probes is on the order of the radius of the maximum bubble size expected.

The geometric configuration of the sensor itself can play an important role. Various probe configurations have been employed, ranging from parallel plates to flush-mounted designs. Different geometries tend to give different responses. This is particularly true with capacitance probes. Capacitance probes are perhaps the most widely used probe-type in fluidized (gas-solid) bed studies because they work well with nonconducting mediums. The measurement of bubble properties is however, much more difficult than in the case of a gas-liquid flow both from the standpoints of optimum probe geometry and the effect of spurious capacitance signals due to the solids motion itself.

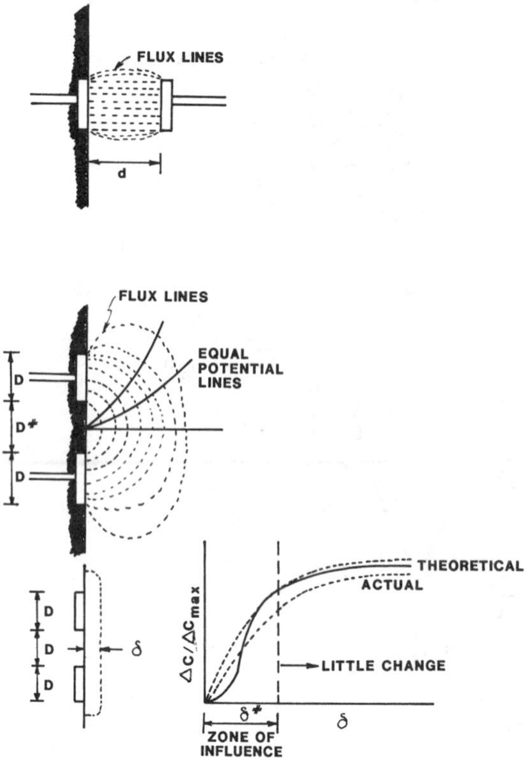

Figure 10. Fluxes generated by plate capacitance probes.

Figure 10 shows the lines of flux generated by a plate capacitance probe immersed in a continuous medium. Also shown in this figure is the relative change in capacitance as a function of distance from the electrode surface. As shown, there is a relatively insensitive region in which the relative difference between the capacitance of the gas and fluid cannot be readily distinguished. This means that an optimum size plate or electrode provides the greatest sensitivity. When two plates are immersed in a fluidized bed of solids, the change in capacitance between the plates provides a measure of the emulsion-phase density in the vicinity of the plates. The plates form an electric field that appears quite uniform in the presence of solids, but when a gas bubble passes the field becomes broken or distorted. The size of the plates and their separation are also critical from the point of view of the solids diameter. Directionally, wide size distribution particles as those typically found in fluid catalytic crackers improves the statistical averaging by the plates. As a rule of thumb, the optimum gap between the plates should be on the order of 10 times the mean particle size.

Parallel plate configurations are typically 10 mm in size on a side and typical measurements are in the range of tenths of picofarads. With such low sensitivity in solids, proper shielding to reduce spurious electronic noise becomes an important issue. Shielding should be of good quality coaxial cable and should be brought as close as possible to the plate. Also, in gas-solids applications, the problem of static charges can lead to difficulties, because they can alter the agglomerating properties of solids causing denser packing of solids in the vicinity of the plates. Furthermore, particles may at times discharge onto the probe, producing spurious electrical signals that might be mistaken as a property of the actual flow. Fortunately, these problems can be minimized with proper grounding of metal parts in the test section.

The sensitivity range for capacitance is favorable such that measurements can be made with simple bridge imbalance techniques. Investigators such as Morse and Ballon [23], Yoshida et al.

[24], and Matsen [25] employed parallel plate probes. Dotson [26] was among the first to recognize the importance of probe geometry in terms of disturbing the flow environment and therefore, modified the parallel plate configuration into thin parallel rods. This geometry however, gives a less sensitive response to capacitance changes.

Ormison et al. [27], Angelino et al. [28], Nguyen, et al. [29], Baskakov et al. [30], and Thiel and Potter [31], employed flush-mounted large-area capacitance probes. These large-area designs are favorable to large capacitance changes and hence, can be detected with conventional bridge circuitry

Tinly flush-mounted probes along with simultaneous heat transfer measurements in fluid beds were made by Catipovic et al. [32]. In this study, the measured capacitance change was on the order of 0.01 picofarads. This design resembles the radio frequency-diode probes developed by Lanneau [33]. Lanneau's probes were originally designed to be tiny pencil point type probes. These were later modified by Rooney and Harrison [34] to a multi-pin probe in order to measure bubble shape (described later). Werther and Molerous [35] and Werther [36] describe a tiny pencil point type probe to measure bubbles with minimal flow disturbances. In Werther's design, variations in the bed's capacitance are detected by a shift in the frequency of an oscillator. This type of circuitry is considerably more sensitive than a conventional bridge circuit, and as such, tends to measure the distributed capacitance of the leads along with the capacitance of the probe itself. Fortunately, this can be readily filtered out with an appropriate discriminator system.

The principle difference between measurements made in gas-liquid and gas-solid systems with any of the methods described in this section is that in the latter, the output signal of the probe contains two types of information: information on the disturbance of interest (i.e., a bubble), and on the local porosity variations of the particulate solids in the vicinity of the sensor. Hence, to understand the true nature of the flow and to obtain useful data that can be related to transport mechanisms, we must be able to separate these two types of signals.

Figure 11 shows how a capacitance probe signal can be decomposed into the two types of information (i.e., bubble pulses and those due to the variations of porosity of the dense phase). A discriminator circuit can be used to filter out fluctuations in bed porosity. Only that part of the signal

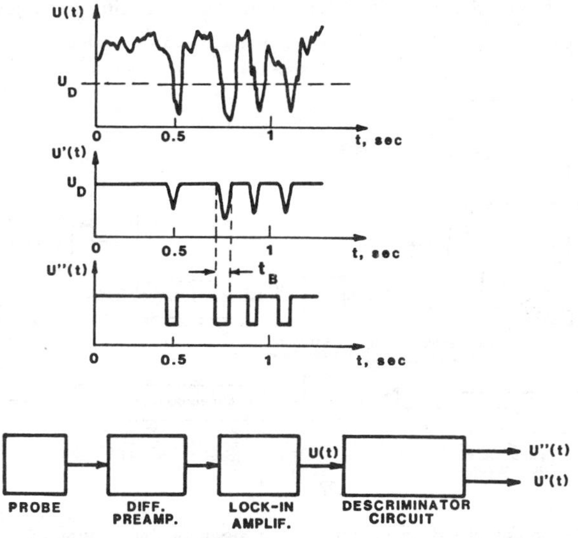

Figure 11. Decomposition of the probe signal into components U_2 due to bubbles and U_1 due to dense phase porosity fluctuations.

U(t) that lies below an adjustable reference value U_D is transmitted undistorted, while for other parts of U(t) the signal is blocked and the output is constant U_D. The discriminator circuit can be designed to produce a signal more suitable for further automatic processing U″(t) in the form of rectangular pulses of constant heights. The duration of such a rectangular pulse is identically t_B for which the discriminator circuit detects the corresponding pulse. Hence, once U_D is established (by comparing the circuit output to cinegraphic detection of the bubble frequency, say in a two-dimensional bed for calibration purposes), the output U(t) of the discriminator circuit contains all bubble pulses but no contributions due to the bed porosity fluctuations.

Since instantaneous bubble rise velocities and lengths pierced by the probe are virtually stochastic quantities, the measurement of individual bubbles is not meaningful. Instead, the local state of fluidization is best described by mean values as detected at the point of the probe. The bubble mean rise time is measured by means of correlation techniques. Referring back to Figure 9, the pierced length is affected by two probes A and B displaced a distance 's' apart in the direction of motion of the structure. These probes provide the electric signals $U_A'(t)$ and $U_B'(t)$ in response to the surroundings. It can be assumed that all elements move at the same velocity, hence, the two signals are identical except for a constant time displacement. This means that the mean rise velocity can be estimated by a cross-correlation of the duration signal times of the two probes and the mean pierced length obtained by an autocorrelation function of the displaced time t_B. A standard correlator can be used to process these mean values. The specific mathematics of the auto- and cross-correlation functions are described by Werther [35, 36].

Note that the output signal U′(t) of the discriminator circuit, assuming U_D is properly set, contains all bubbles and is suitable for triggering an electronic counter. The discriminator sends the signal to a counter for a preselected period T. The counter registers the number n of the bubble pulses within the set period. Hence, for sufficiently large T, the bubble frequency f is directly measured.

The author has combined both resistivity and capacitance rates into a single system for fluidized bed studies. Parallel electrode, flush-mounted probes provide good signal response and can be miniaturized to prevent flow distortion. By combining a capacitance and resistivity bridges in a single unit, the operator can switch the probe to either capacitance or resistivity measurements, depending on the properties of the material in the bed. The instrument setup is shown by the block diagram in Figure 12 and details of the discriminator circuit and bridge circuits are given in Figure 13.

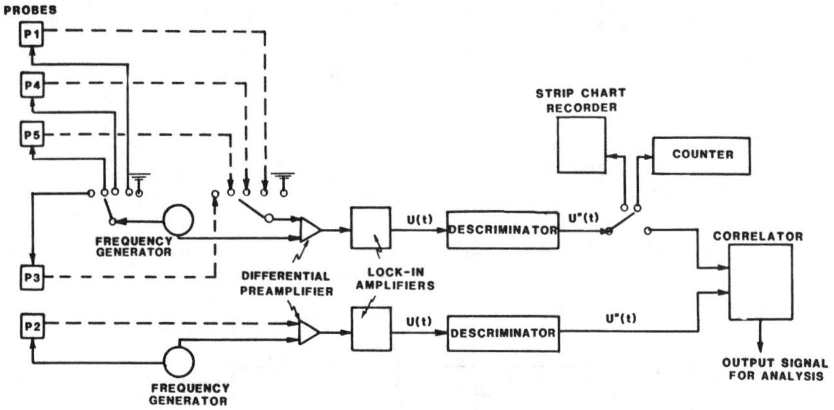

Figure 12. Instrumentation set up for measuring mean bubble properties. For simultaneous operation of probes P2 and P3, the correlator is operated in the cross-correlation mode. Probe P2 operation alone requires autocorrelation mode. Remaining probes provide bubble frequency measurements only.

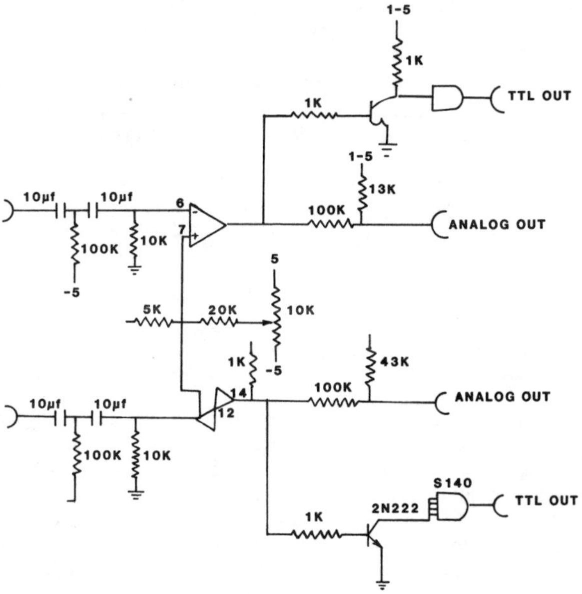

Figure 13. Details of discriminator circuit.

Two Model 125-A Princeton Applied Research Lock-In Amplifiers are used to excite the probes and amplify the probe output signals. Two Lock-In Amplifiers are required when probes are synchronized to obtain measurements of the rise velocity. Each Lock-In has a frequency generator that supplies a 10-kHz excitation signal to the probes. A differential preamplifier (diff. preamp.) is used to amplify the probe output signal. The lock-in provides real-time comparison of the probe distorted signal to the reference signal supplied for excitation. Hence, the difference between the probe input and output signals contains information on both bubbles and porosity fluctuations. The portion of the signal containing this information is then sent to the discriminator circuit to separate bubble and porosity information. The processed signal containing information on bubbles only is then sent to the correlator where cross- correlation of the double probe signal (rise velocity) and autocorrelation of the single probe signal (bubble pierced length) can be made. Capability for switching to an electronic counter (for local bubble frequency measurements) and a strip chart recorder (for discriminator base level monitoring) is included. Each probe has a specially designed d.c. bridge circuit enabling it to be used either as a capacitance probe or a resistivity probe.

Bubbles of considerably different sizes and shapes pass through a given horizontal control volume containing the probe as one of its points. To obtain a statistically representative mean bubble size an adequate description of the size distribution of the array of bubbles is needed. Because of the irregular shape of bubbles a measurement of the pierced length in itself does not define bubble size. Werther [36] provides a detailed statistical analysis of the bubble size distribution on the basis of the following model: A size distribution defined as either the cumulative number distribution $N_0(D_h)$ of the horizontal diameters D_h, $D_{h,min} < D_h < D_{h,max}$; or the cumulative number distribution $Q_0(D_v)$ of the vertical diameters D_v, with $D_{v,min} < D_v < D_{v,max}$.

Because a pierced length lies in the range $0 < \ell < D_{v,max}$, from probability theory:

$$g(\ell)\, d\ell = \sum_{i=1}^{n} P_{i,1} P_{i,2} \tag{3}$$

where 1 and n denote the first and nth size class corresponding to the minimum bubble diameter ($D_{v,min}$) and the maximum bubble diameter ($D_{v,max}$), respectively.

$g(\ell)d$ = probability that a measurement yields a pierced length in the interval $(\ell, \ell + d\ell)$

$P_{i,1}$ = probability that the struck bubble belongs to the ith class $(D_{v,i}, D_{v,i} + \Delta D_{v,i})$

$P_{i,2}$ = probability that the measured pierced length lies in the interval $(\ell, \ell + d\ell)$ if a bubble in the ith class $(D_{v,i}, D_{v,i} + \Delta D_{v,i})$ is struck

The criterion developed by Werther [36] for deriving the bubble size distribution $Q_0(D_v)$ from a measured distribution of pierced lengths $g(\ell)$ is

$$Q_0(\ell) = 1 - \left\{\frac{g(\ell)}{g(D_{v,min})}\right\}\left\{\frac{D_{v,min}}{\ell}\right\}, \qquad D_{v,min} < \ell < D_{v,max} \tag{4}$$

where $Q_0(\ell)$ is the cumulative number density distribution of pierced lengths (cm^{-1}) and $g(\ell)$ is the number density distribution (cm^{-1}). Werther has shown that the distribution $g(\ell)$ is a linear relation passign through the origin, and hence, the lower limit, $D_{r,min}$ of the bubble size distribution can be found. Note that Equation 4 shows that only the ratio $g(\ell)/g(C_{v,min})$ is necessary to determine the bubble size distribution. To enable the measurement of the distribution of pulse durations to be automatic, the discriminator circuit is also designed to generate a signal $U'''(t)$ from the signal $U''(t)$, which consists of a series of rectangular pulses of constant duration but of varying heights. The height of an individual pulse of this last signal is always proportional to the duration of the corresponding bubble pulse.

Because of the uniform duration of pulses, the probability density distribution of the amplitudes of the signal $U'''(t)$, is proportional to the distribution of pulse duration t_g for $U''' > U_0'''$, or more specifically, the bubble pulse durations t_b. A calibration voltage U_e''' must be used to convert the measured signal directly to units of pulse duration, t_b. Typical oscillograph tracings of the unprocessed and processed signals from a capacitance probe are shown in Figure 14.

Another approach to discriminating signals to obtain a better representation of mean size and shape is the system described by Calderbank et al. [9]. This approach uses a three-dimensional probe having five channels, to sense the local bubble interface approach angle as well as measure bubble size and velocity. The system in this study was interfaced with a high-speed digital computer capable of rapid, accurate conversion of analog voltage signals to discrete binary numbers and with software to undertake logical decisions consequent on the spatial orientation of the bubble

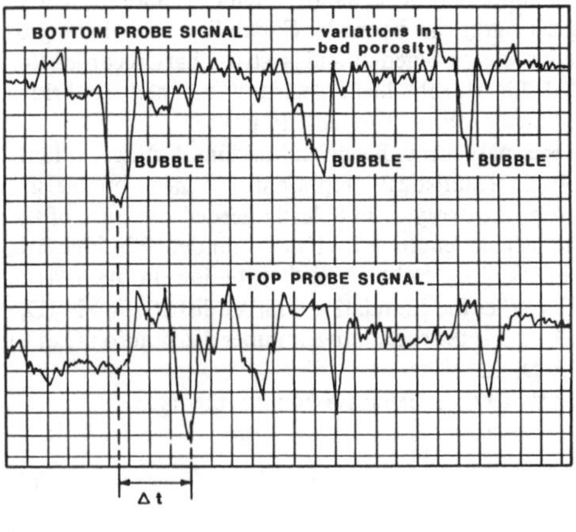

Figure 14. Typical oscillograph tracings made by capacitance probe in a fluidized bed of coal char.

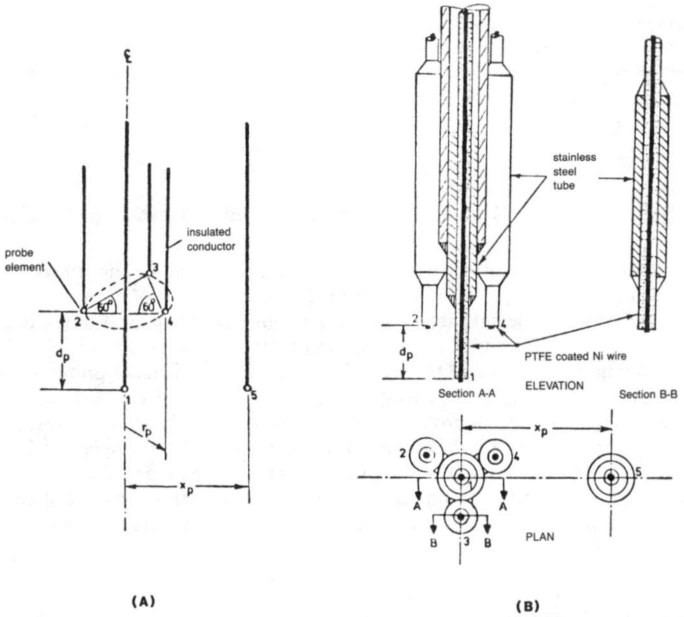

Figure 15. (A) Isometric projection of the spacial orientation of the probe contact elements, (B) details of the probe configuration [9].

with respect to the probe axis. The logic selects only those bubbles whose central axes are coincident with the probe axis and calculates their size, shape and velocity by correction of the resulting non-square pulses. It was assumed in deriving the characteristic bubble-size distribution function of a dispersion that bubbles rise into the probe in a random fashion and the measurements are therefore characteristic of unbiased sampling.

The probe arrangement consisted of three symmetrically oriented electrodes around and above the first electrode. Hence, all three electrodes were positioned in a horizontal plane at a specified distance above the central electrode and radially spaced from it. A fifth contact was positioned in the same horizontal plane as the central contact but somewhat distant from it. The arrangement and details of the probe array is shown in Figure 15.

The probe array was connected so that each contact formed part of an electrical resistivity circuit whereby current flows from an external d.c. power supply while that contact is resident in the conducting phase and thus develops a voltage across a load resistor, which is measured by the computer in digital form simultaneously for each channel [9].

The purpose of the fifth electrode (leg 5 in Figure 15A) is to measure approximate bubble shapes for those bubbles whose horizontal axis exceeds length x_p. The probe array selects bubbles relatively close to their centerlines, so the outer electrode 5 measures the vertical distance, L_d, between the bubble leading surface at the centerline and at the radial distance, x_p (Figure 16). It also measures the bubble vertical length at this position, L_O, as follows:

$$L_d = U_B t_5 \tag{5}$$

and

$$L_O = U_B(T_t + t_c) \tag{6}$$

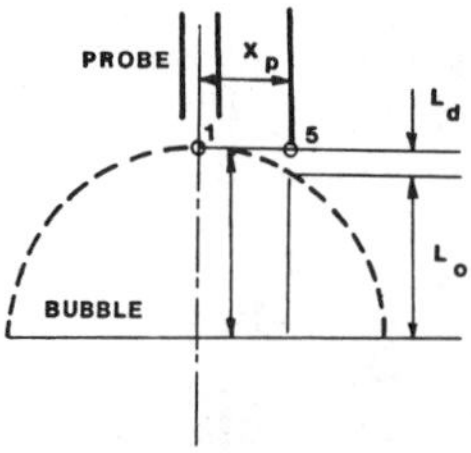

Figure 16. Five-contact probe for measuring bubble shape [9].

As a very crude discrimination scheme, a three-element probe can be constructed. Various statistical sizes can then be computed from the data. The 16K memory of an Apple II computer is more than sufficient to handle data acquisition and number crunching for this type of application.

Two additional types of probes that fall into the category of electroresistivity techniques are inductance and impedance devices. One example of the use of inductance probes is the work by Cranfield [37], in which a small probe having a tip that functions similar to the head on a magnetic tape recorder is used. This type of probe can be used to detect the presence of particles that have ferromagnetic properties. Hence, if the fluidized bed is composed of ferromagnetic solids, the signal at the proble provides an indication of whether the emulsion phase or bubbles are present. The bed does not have to be entirely composed of ferromagnetic particles; instead, it can be seeded with ferromagnetic tracer particles. This approach is useful in studying solids movement and mixing in fluidized beds.

Electrical Discharge Technique

Most measurements reported using capacitance and resistance probe techniques were made at room temperature. The few experiments that were made at high temperature reveal that the bubble characteristics are significantly affected by the change of temperature. Mii et al. [38] obtained bubble frequency measurements in a fluidized bed using a conductivity probe for graphite particles fluidized by nitrogen in the range of 20°C to 800°C. It was reported that the bubble frequency increased with a rise in temperature. Hoshida et al. [39] noted a decrease in bubble size with an increase in temperature from observations in a two-dimensional bed. Otake et al. [40], using a capacitance probe to measure bubble sizes during the catalytic cracking of cyclohexane, observed a temperature dependence for both bubble size and frequency. On study opposite to these findings is that of Whittmann et al [41], who measured bubble size and rise velocity during the thermal cracking of $NaHCO_3$ of bed temperatures ranging from 65 to 200°C; observing no temperature effect on bubble properties.

The capacitance probe does appear to have limitations in high-temperature measurements. The principle restriction is that the probe requires cooling as a whole. Any high-temperature measuring method must be heat-resistant. Yoshida et al. [42] describes an improved technique based on exploitation of the alteration in an electric discharge associated with the probe being in the bubble or emulsion phases in gas-solid fluid beds. The probe and circuitry are capable of measurements up to about 1,000°C. The probe geometry and circuitry are shown in Figure 17. The probe's operation is based on the fact that the voltage breakdown of dielectric particles is higher in the emulsion phase than in the bubble phase. The probe shown in Figure 17 consists of two needle-shaped electrodes with their tips facing each other at a gap of several millimeters. An electric discharge is generated across the electrodes when a bubble passes over the gap between probe tips and for a length of time where the impressed voltage across the probe is higher than the breakdown potential in the bubble phase and lower than in the emulsion phase. Hence, detection of rapid changes in electric current provides detection of the presence of bubbles. The electrode materials are an important consideration in proble construction as high temperatures can cause oxidation reactions. The probe material used by Yoshida et al. was Kanthal alloy.

Figure 18 shows the incipient discharge voltage in the bubble and emulsion phases for quartz sand particles. As shown, the incipient voltage decreases with an increase in bed temperature. The

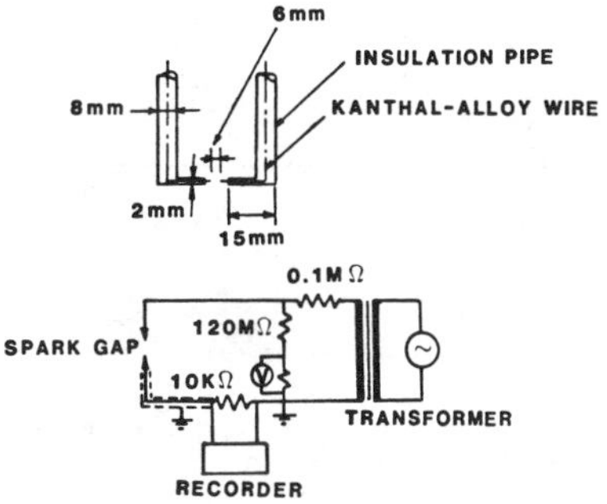

Figure 17. Probe configuration and circuitry for electric discharge probe [42].

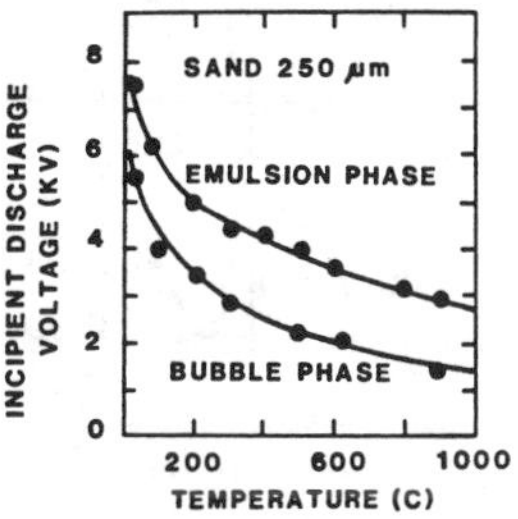

Figure 18. Relationship between incipient discharge voltage and temperature reported by Yoshida et al. [42].

circuitry employs a 50-Hz alternating current. The use of d.c. causes particle sticking to the electrode surface. Reported voltages range between 5 and 10 kV. With conductive particles, such as coal chars, the measuring voltage can be set at the incipient discharge voltage of the bubble phase. In this manner the probe can be employed as a conductivity type probe. The electric current will then flow only when a cloud of conductive particles passes between the pair of electrodes. A sufficiently low voltage must be employed to prevent electric discharge in the emulsion phase. It should be noted that although the characteristics of the detected signal may vary depending on the conductivity of the solids, the technique described by Yoshida et al. is always suitable provided that there is a difference in the electrical properties between the bubble and emulsion phases. Through the use of dual, double probe electrodes, complete bubble characteristics (i.e., size, rise velocity, as well as frequency), can be determined, as described earlier.

Thermal Methods

Constant temperature anemometry techniques have also been applied to studying bubbling phenomenon. One example of a gas solid fluidized bed study is that of Wen et al. [43]. The investigators constructed a miniature probe from a self-heating thermister. The measurement principle is based on the fact that the probe's heat transfer coefficient in the emulsion phase is significantly greater than in the bubble phase. Consequently, when the probe is in the emulsion phase, its temperature decreases, resulting in an increase in the resistivity of the probe. In Wen's design, the resistivity itself is measured and is used to detect the presence of bubbles. The technique is easily extended to gas-liquid flow studies.

Other studies [44–46] aimed at examining the heat transfer mechanisms in fluidized beds, describe methods that indirectly sense bubbling conditions. Basakakov et al. [44] describe a thermo-anamometer for directly measuring the instantaneous fluctuations of local heat transfer coefficients. The probe itself consisted of a 5 μm thick platinum foil glued on a surface immersed in the fluid bed. The temperature of the foil was observed to drop abruptly when a packet of cold particles approached the foil, then rose with the heating of particles. The temperature drop continued even further when gas bubbles passed over the foil surface. By simultaneous measurements using photo-transducer technique, the presence of bubbles were detected. Figure 19 shows oscillograph tracings of the simultaneous recording of signals from the photo-transducer and the thermo-anemometer, indicating, as in Wen's study, that temperature fluctuations provide a satisfactory approach to bubble detection. Details of the probe configuration employed by Basakakov et al. are shown in Figure 19. Similiar techniques are described by Bernis et al. [45] and Goosens and Hellinckx [46].

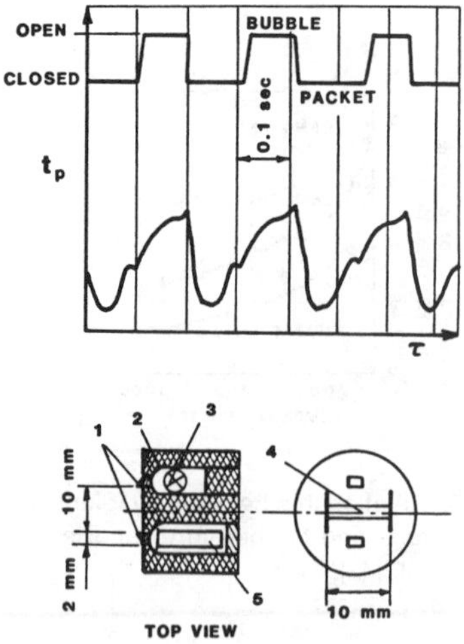

Figure 19. Oscillograph tracings of signals from a phototransducer and thermo-anemometer reported by Baskakov et al. [44].

Measurement Techniques for Three-Phase Systems

The main interest in three-phase systems over the past decade has been in potential applications of fluidized beds. There are a variety of reactions between a gas and a liquid on a solid surface in chemical processing. However, many of these reactions have traditionally been carried out in fixed bed reactors. Fixed beds have a number of limitations such as high pressure drops, fouling, and flow maldistributions. Consequently, interest has shifted towards the three-phase reactor, where the solid particles serve as catalyst material for reactions. This type of system is attractive to such applications as the hydrogenation of liquid petroleum fractions, hydrogenation of unsaturated fats, coal liquefaction processes, and a variety of biological reactors. In relatively dilute solid phase systems, the conductivity and resistivity probe methods can be used to derive information on the bubble phase (i.e., bubble frequency, bubble size and size distribution, rising velocity and local volume fraction occupied by gas bubbles).

The conventional approach to defining the various holdup fractions is through the following equations:

$$\epsilon_L + \epsilon_a + \epsilon_s = 1 \tag{7}$$

$$\frac{dP}{dh} = g(\rho_L \epsilon_L + \rho_G \epsilon_G + \rho_s \epsilon_s) \tag{8}$$

$$\epsilon_s = M_s/\rho_s AH \tag{9}$$

where ϵ is the phase holdup and subscripts L, a, and s refer to liquid, gas and solid, respectively; H = expanded bed height; h = position down the column; M_s = mass of solids; A = cross-sectional area of bed; g = gravitational acceleration; P = local pressure in column; ρ = density.

Begovich et al. [47] noted that the electroconductivity of a liquid system with a fixed ion concentration at constant conditions is proportional to the cross-sectional area of the conducting liquid and inversely proportional to the length of the path between two electrodes. If the tortuosity factor remains approximately constant, the conductivity is approximately proportional to the liquid holdup in the bed:

$$\epsilon_L = \gamma/\gamma_0 \tag{10}$$

Similar assumptions have been made by other investigators [48–51]. For nonconducting solids Turner [51] showed that:

$$\gamma/\gamma_0 = (1 - \epsilon_s)/(1 + \epsilon_s/2) \tag{11}$$

For two-phase fluidization, this can be expressed in terms of the liquid holdup as:

$$\epsilon_L = (3\gamma/\gamma_0)/(2 + \gamma/\gamma_0) \tag{12}$$

In the work of Begovich and Watson [47], electroconductivity probes were composed of two platinum electrodes, each approximately 1.4 cm^2, positioned 180° apart on the inside of a movable Plexiglas ring. The ring could be lowered or raised by two stainless steel tubes threaded into the ring. Insulated wires were passed through the tubing and soldered to the electrodes. The wires were connected by a coaxial cable to a conductivity meter. A digital millivoltmeter and a resistor-capacitor circuit connected to the conductivity meter permitted a time averaged digital read out. Saturated potassium chloride solution was added to the water in the feed tank to allow readings on the 5 mmho scale of the conductivity meter.

The conductivity was first measured above the bed in the liquid alone. After the liquid and gas velocities were adjusted to their desired flow rates, the liquid manometer heights were recorded. Then the conductivity between adjacent pressure taps was also recorded. Equations 7, 8, and 10 could then be solved to yield each of the three-phase holdups as a function of position in the column.

A more accurate approach to three-phase holdup measurements is the impedance probe devised by Linneweber and Blass [52]. This technique provides a direct measurement of the local gas and solids hold-up in three-phase systems. The technique employs the double needle probes described earlier to detect gas bubbles and thus obtain the gas phase holdup. The addition of an impedance probe obtains both the solids and gas holdup, and hence, comparison of the two measurements provides the solids holdup (refer to Equation 7). In the case of three phases, the capacitance or conductivity probe is unaffected by the solids, but rather is influenced by the continuous liquid phase. Figure 20 reviews the measuring principle of the needle probes previously described. Note

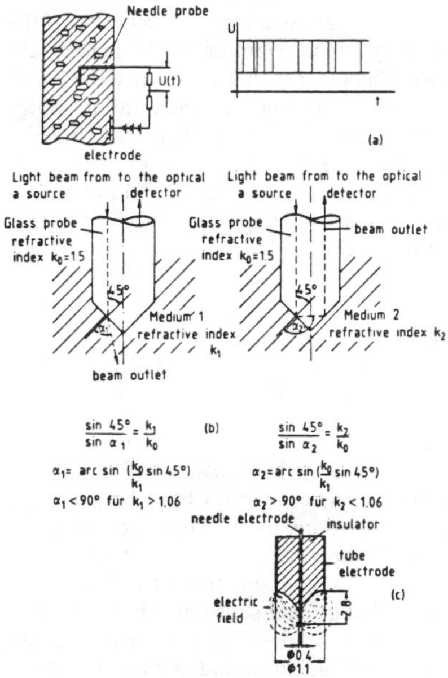

Figure 20. Summary of measuring principles behind the double conductivity probes as applied by Lenneweber et al. [52] to measure bubble properties in three-phases.

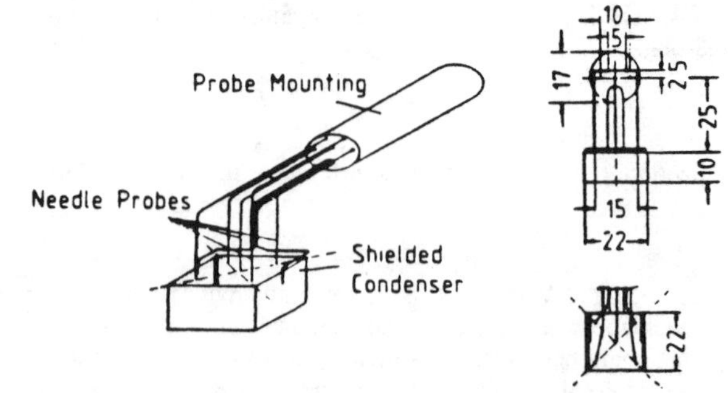

Figure 21. Details of impedance/conductivity probe designed by Lenneweber et al. [52].

that an electric circuit is completed over the conducting fluid suspension between the needle probe, which is insulated apart from its tip, and an electrode, sufficiently large so as not to be affected by gas bubbles. The probe's signal is disturbed when a gas bubble strikes the tip. The time intervals during which the probe tip is surrounded by gas bubbles can be summed to provide the mean value of the gas holdup. As emphasized earlier, this approach only works when the dipersed and continuous phases have significantly different conductivities. If the conductivities of these phases are similar, Linneweber, et al. recommend the use of an optical-type probe (Figure 20b). In this case, differences in the refractive indices between the dispersed and continuous phases provide an indirect measurement of holdup. The flush-mounted probe configuration and circuitry for the capacitance system described earlier are equally suited for the bubble holdup measurement in the former case.

The impedance probe method is based on differences in the dielectric constants and conductivities of the liquid, solids, and gas mediums. The capacity of a plate condenser, for example, is a function of the volume fractions of dispersed phases exposed to the electric field. The actual relationship between impedance and holdup is a function of the size, geometry, and distribution of the dispersed phase particles. The relationships for the particle system distribution can be mathematically described by standard models such as the well known Maxwell distribution. Linneweber, et al. have combined the two separate measurement principles into a single probe configuration shown in Figure 21. A relatively large condenser is employed to detect the bubbles. Hence, the bubbles are treated as small particles compared with the volume of the electric field. Note that the design includes five needle probes to detect the mean values of gas holdup in the electric field of a plate condenser, whose plates are surrounded by Faraday's cave. The resulting signal from the impedance probe is the sum of the gas and solids holdups in the electric field.

LIGHT SCATTERING AND OPTICAL METHODS

Light Transmission Techniques

Two techniques that have been successfully applied to studying the dynamics of multiphase flows in the laboratory are light scattering and fiber optics. These methods have been applied specifically to measuring interfaces between two phases by measuring local disturbances of either the continuum-phase or the presence of a discrete phase. These methods have produced useful correlations and understanding of bubbling states in which pockets of gas rise through a bed of aggregate solids or a column of liquid, where the liquid and aggregate solids are the continuum phases. The size distribution of bubbles in these systems and their degree of uniformity throughout the column or reactor volume plays an important role in establishing heat and mass exchanges as

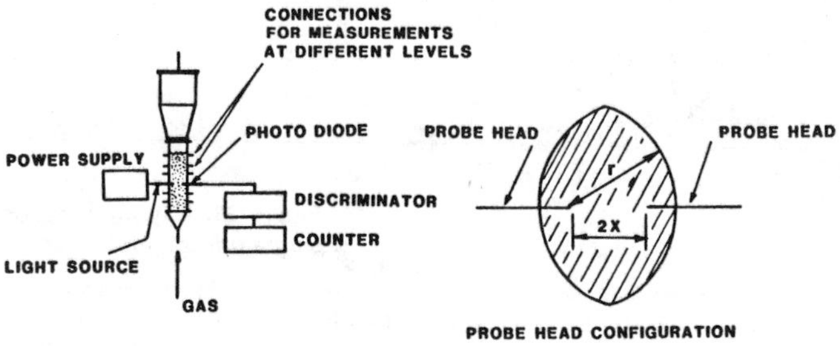

Figure 22. Bubble detection system used by Put et al. [53].

well as the rates of chemical reactions. Light scattering and fiber optics are methods that can detect and measure the size and local behavior of bubbles.

The local size distribution of bubbles in these systems are perceived as a probability density function of characteristic size, which reflects information about the average bubble radius. This can be described using Maxwell-Boltzmann statistics, which provides a reasonable approximation of the energy distribution associated with bubbles in a fluidized system. This energy in turn is proportional to the average surface area of the bubbles. Put et al. [53] employed this analogy with a light transmission technique to evaluating bubble sizes in a gas-solid fluidized system. A characteristic bubble size was directly sensed in this study by the interaction between a horizontal light beam and the surrounding fluidized medium. In the case of a fluidized bed, the transmitted light is scattered when solids obstruct the light path, causing attenuation of the transmitted signal. When gas is present there is no attenuation of the light beam at all.

Figure 22 shows a schematic of the experimental set-up used by Put et al. in the study along with the geometrical configuration of the probe. The small light source used generated a circular beam of parallel light rays 5 mm wide. The light sensing element consisted of a photodiode that faced the light source. This technique provides a local measurement of the number of bubbles passing the probe per unit time (i.e., bubble frequency). This is achieved by simply counting the non-attenuated signal interruptions over a known time interval. Counting is done by a discriminator and electronic counter. The discriminator level must be predetermined (for example, with the use of an oscilloscope and correlator) so that only the presence of a single bubble bridging the probe opening is detected. By varying the size of the probe opening, several points of the cumulative density function of bubble width, n(x), can be obtained. The function is calculated by accounting for the probability of detecting a certain bubble size in the probe opening:

$$n(x) = \int_x^\infty D(r, x)D(r)\, dr \tag{13}$$

where D(r, x) is a weight function defined by Put et al. [53].

The derived integral is:

$$n(x) = \frac{2\pi k r^2}{S} m(1 - \mathrm{erf}(x/\sqrt{2}\, r_m)) \tag{14}$$

where erf( ) denotes the error integral.

A weight function is introduced to the distribution to correct for the geometrical configuration of the light probe. The weight function, D(r, x) provides the probability that a spherical bubble with radius r is detected by the probe having a total opening of 2x. The critical assumption in this technique is that bubbles rise randomly and that any location in a column section has an identical probability for any bubble. Figure 22 shows that the probe head configuration is taken as small points lying along a horizontal distance 2x. The weight function D(r, x) can then be taken as being equal to the ratio of the shaded area (i.e., the area of the centers of the bubbles with radius r which

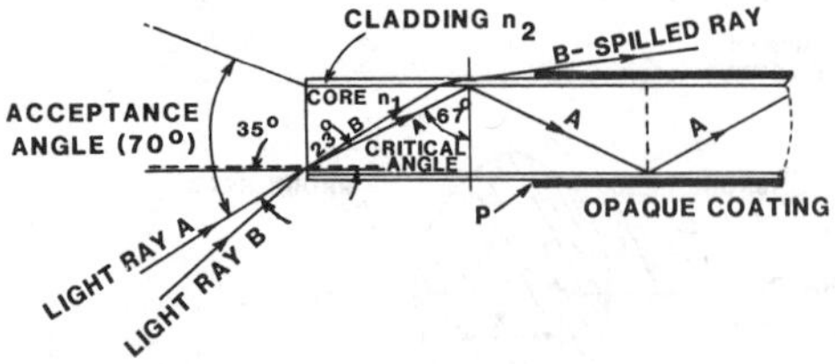

Figure 23. Ray propagation through a single optical fiber.

can be detected) and the total area S of the column. The following expression is reported in the study:

$$D(r, x) = \frac{2r^2}{S}\left(\frac{\pi}{2} - \arc \sin \frac{\pi}{r}\right) - \frac{2x}{S}\sqrt{r^2 - x^2} \tag{15}$$

Examples of other studies using this measurement principle are the investigations of Yasui and Johanson [54], Kilkis et al. [55] and Yoshida et al. [56]. Further discussions with reference to fluidized beds are given by Cheremisinoff et al. [57]. This method, as well as the fiber optics type probes to be described are also well adapted to studying discrete or entrained flows. For example, droplet entrainment and deposition from two-phase gas-liquid flows can be just as readily measured. In this case, a photodiode is again used but this time to translate the attenuated light signal corresponding to droplets into an electrical one. A discriminator and counter can then count droplets and distinguish sizes.

With the introduction of fiber optics bundles, the use of miniaturized light probes for localized measurements was adopted in a variety of two-phase flow investigations [58–61]. An optical fiber is a transparent linear element through which radiation propagates by total internal reflection. Figure 23 illustrates the physical principle by which a fiber optic functions. When light is transmitted through the front surface, all rays acquiring an inclination smaller than the critical angle become totally reflected inside the fiber core. The light rays will continue to travel in this manner until they reach the opposite end or are totally absorbed. Figure 23 gives the example of a fiber having a critical angle of 67°, in which case can acceptance angle of 70° is realized. This means that all rays that are incident onto the fiber's front surface at an angle of ≤35° with its axis become trapped inside the fiber because of total internal reflection.

Note that the critical angle is a function of the ratio of the refractive indexes of the glass of which the core is constructed of and the surrounding medium (n_1/n_2). The acceptance angle of a fiber optic element can therefore be varied by changing the n_1/n_2 ratio. A fixed n_1/n_2 is obtained by applying the technique of cladding which consists of coating the core with a layer of solid material of a desired index of refraction. The cladding is necessary especially when several fibers are arranged into a bundle. Without this coating adjacent fibers would "leak" light into each other since they all have the same index of refraction.

An additional coating using an absorbent layer of material is also applied onto the outer surface of the cladding (shown to the right of point P in Figure 23). This coating prevents stray rays such as those surpassing the critical angle from splitting out into adjacent fiber elements. Light rays can exceed the critical angle when the fiber makes a bend or curve, for example. In this case, the outer opaque coating will simply absorb the stray light. Also, at a bend in the element, radiation cannot enter into the fiber from the outside, thus eliminating any chance of interference with the rays traveling through the fiber. The ability to bend fiber optical elements is an important feature that enables access to an environment of interest that may be difficult to reach with other types of sensors. Transmission losses are generally negligible provided the ratio of the bend radius to the fiber diameter exceeds a value of around 40 [62]. This means that a fiber having a 25-micron core diameter for example can be wound about a 1-mm mandrel without experiencing significant transmission losses.

Optical fibers can receive wide acceptance angles of incident radiation, which makes them competitive with fast conventional optical systems. That is, they are comparable to optical systems having large numerical aperture (NA) numbers. The NA number is a measure of the ability of the

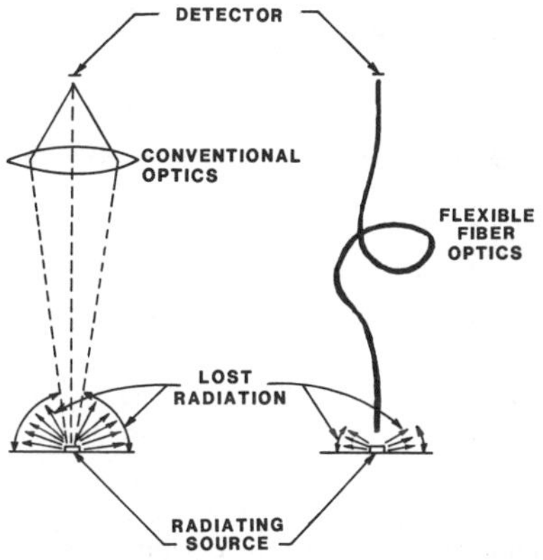

Figure 24. Comparison of the abilities of conventional optics and fiber optics in capturing energy.

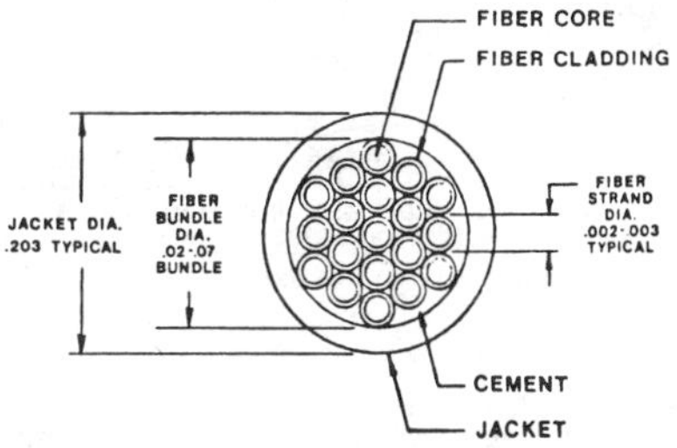

Figure 25. End-view of fiber optic bundle.

optics to accept incident light rays. This is a function of the limit angle of acceptance, and the larger the NA number is, the larger the cone of radiation entering the optical fiber. Optical fibers can capture a relatively high degree of radiating power. Figure 24 qualitatively compares the cones of radiation energy collected by a conventional lens and an optical fiber. As shown, an optical fiber can be more efficient.

Practical sensing of radiant energy can only be made using a bundle of fibers, which is an assembly of a number of single fiber elements that may range from a few to several hundred. The bundle is usually lined inside a containing sleeve that can be either rigid or flexible. The ends of these bundles are generally rigid so that they may be held firmly in place. A cementing compound such as epoxy resin is often used to create the rigid end points of the bundle. This enables optical finishing of the end surfaces to be performed. An end-view of a fiber bundle is shown in Figure 25. A typical outside diameter of a single fiber element is on the order of 25 microns.

There are basically two types of fiber bundles, namely, coherent and incoherent.
have an identical geometrical distribution of fibers at both ends. Consequently, the light distribution transmitted at the front of the bundle is, within practical limitations, duplicated at the output. Hence, an image is transmitted from one end of the bundle to the other with the only degradation in the image's resolution resulting from transmission losses and the resolution restrictions imposed by the size of and spacing between individual fibers.

In contrast, incoherent fiber bundles are arranged with a random distribution of fibers at the two ends. In this case, no image is transmitted through the bundle. This type of an arrangement provides only the radiation output level that reflects the strength of the input source. An example where this is of practical importance is in infrared fiber optics systems for temperature sensing and control. Optical fiber bundles can be used with infrared radiometers to accurately measure high temperatures. It should be noted however, that optical fibers in general have poor transmitivity in the infrared region since they are composed of glass (or in some cases, plastic). Radiation of intermediate and far infrared (wavelengths > 2.5 μm) is absorbed by the fibers within a few millimeters of travel. This property makes it difficult to accurately measure temperatures below 100°C (most of the radiation emitted by a surface below 100°C has a wavelength of under 7 μm). There is some radiation in the near infrared region at the lower temperatures, but its energy level is low and often masked by background noise.

A fiber optic probe can be viewed as a low-pass filter with gradual cutoff of transmissivity above the threshold wavelength of 2.5 μm. This characteristic is an advantage from the standpoint that it enhances thermal resolution. Radiation power increments in the near-infrared region exceed the

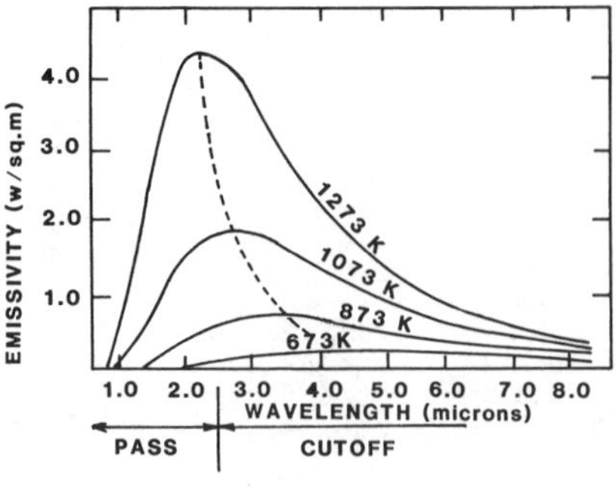

Figure 26. Plot of emissivity vs. wavelength for fiber optics.

corresponding power increments in the total radiation area for the same time period of a target area. Figure 26 is a plot of emissivity versus wavelength, illustrating that the area increments to the left of the vertical cutoff line are larger than the area increments of the total surface area subtended by the full radiation curves. Ohki et al. [61] used fiber optics in measuring both local particle movement and bubble characteristics in gas-solid fluidized beds. The components of the electronic system used for measuring bubbles is shown in Figure 27.

Simultaneous measurement of bubble rise velocity, fraction, frequency, and size of the bubble in the bed is possible with this set-up. Bubbles are detected by the probe (shown in Figure 28) based on the light transmission principle as follows: when a bubble wraps around the probe, the light from a stroboscope of flash frequency N_0 is transmitted to the light receiving fiber of the probe, where it is converted to an electric signal via a photomultiplier. The local bubble fraction ϵ_B can be computed from the ratio of the photomultiplier's output signal mean frequency and the stroboscope flash frequency, N_0:

$$\epsilon_B = N_c/N_0 \tag{16}$$

To obtain the mean rising velocity of bubbles, u_B is computed from the output signals of the frequency to voltage (F–V) converters by means of a cross-correlation method. Ohki observed from simultaneous measurements of bubble characteristics and particle movement that the rising velocity of particles is proportional to the local bubble velocity.

These methods can be combined with light scattering principles to obtain droplet size distributions in gas-liquid flows [63–68]. They have been combined rather successfully with the optical and photographic techniques described below.

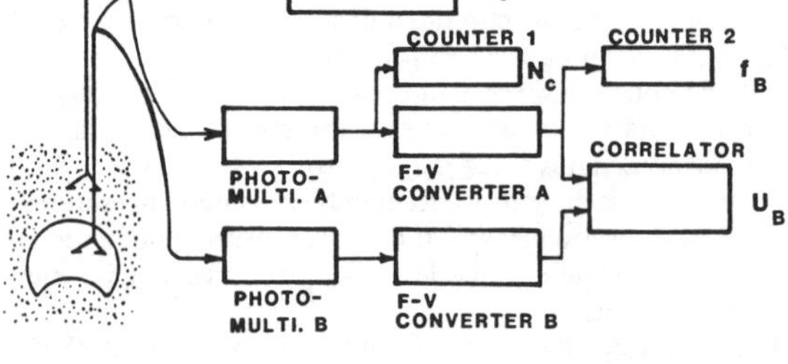

Figure 27. Block diagram for bubble characteristics detection using fiber optics probes [61].

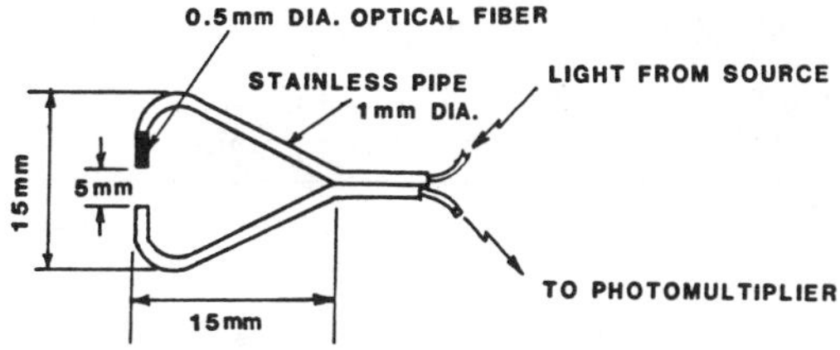

Figure 28. Details of optical bubble detector used by Ohki et al. [61].

Photographic and Optical Techniques

In entrained flows (gas liquid or gas-solid flows) one is faced with the problem of sensing small, fast-moving individual particles. The time response of the sensing circuitry needed to detect particles and droplets in the micron size range must be considerably faster than the case of a fluidized bed where bubbles may be on the order of inches. The same problem exists in studying emulsions in mixing systems where drops or particles may range from only a few microns to less than 100 μm in size.

In a fluidized system the assumptions that bubbles randomly rise through a column and that the probability of their existence is the same everywhere is reasonable for a symmetrical deep bed containing no internals such as baffles. In other systems, for example, atomized flows, the probability of particles being present everywhere at one time varies. In fact, most experiments must be geared towards obtaining spacial distributions of information in addition to temporal.

Photographic and optical methods provide a basis for obtaining both spacial and temporal distributions in that they are non-insitu techniques. That is, the sensing system itself can be moved with respect to a stationary test section to obtain spacial information on the flow phenomenon. Systems described in earlier chapters basically use a fixed probe that can only measure temporal information at one point and hence, many probe sensors must be used to obtain all information about the flow disturbances.

Small, fast-moving particles are difficult to photograph, however, when proper consideration is given to illumination and the limitations and requirements of photography, detailed information on entrained flows can be obtained. Equally important to obtaining reliable data from photography is the proper interpretation of the images produced.

When studying dispersed flows, the major considerations are the required illumination intensity, the expected range of particle sizes, the velocity range of travelling particles, and the angular variation of scattered light.

The degree of illumination required to reproduce a particle's image depends on the particle size and velocity. It is a fundamental axiom of geometric optics that the light scattered from a small sphere in the size range of 10–10^3 μm is:

$$I = I_0 F(\theta) d^2 \tag{17}$$

where I_0 = incident illumination

$F(\theta)$ = function dependent on the direction of observation with respect to that of illumination

d = particle size

The term *exposure time* refers to the time interval required for a particle to travel a small fraction of its own diameter. *Exposure* is defined as the total amount of light covering any portion of a photographic plate. The exposure for the portion of the plate onto which the image of a particle is focused is

$$E = (I_0 F(\theta) d^2)\left(\frac{K_E d}{V_p}\right) = \frac{K_E I_0 F(\theta) d^3}{V_p} \tag{18}$$

where K_E is the fraction of the diameter particle permitted to move during photography and V_p is the particle velocity.

The value of E must be set to a prescribed level to provide sufficient light to produce a visible image on the developing plate. Its value depends on the incident light intensity, which in turn is a function of particle size and velocity; i.e.:

$$I_0 \propto d_p^{-3} \qquad I_0 \propto V_p \tag{19}$$

From the theories of Fraunhofer diffraction and Mie scattering, $F(\theta)$ increases with incident angle θ, where θ is the angle included between the illuminating and observing light paths. Angle θ tends to go to 0°. It follows that illumination is also a function of particle size and velocity, and that it should be set as close to 0° to the observing direction as possible. At exactly 0°, the shadows of particles in motion will also be recorded on the photographic plate.

To produce contrast of photographed images, back illumination can be used in which recorded particles appear as black dots or dark dashes on a bright background. Kirkman et al. [69] give guidelines for photographing droplet sprays using such a technique.

To extract information on the direction of motion and velocity of particles, a double exposure technique outlined in the literature for studying sprays can be used [70, 71]. This involves the use of double flash exposures, where measurements are taken from the start of one streak to the start of the next. Multiple flashes can be provided by a stroboscopic light source. Note however, that these units do not always provide sufficient illumination intensity and/or a fast enough rate of interruption. To compensate for this, McCreath et al. [78] describes a double flash system, where the first spark focuses at the point of origin of the second spark by means of a lens system. A further lens focuses them both onto the camera shutter with a small region of the flow. A technique described by Finlay and Welsh [71] employs two flash units arranged such that their directions of illumination are perpendicular to each other, and one of these units faces the particle. A half-silvered mirror is positioned at the intersection of the light paths and at 45° to each other as shown in Figure 29.

If the particle flow is comprised of very small particles, magnification of the photographic stage may be necessary. This usually requires small image distances which can pose difficulties experi-

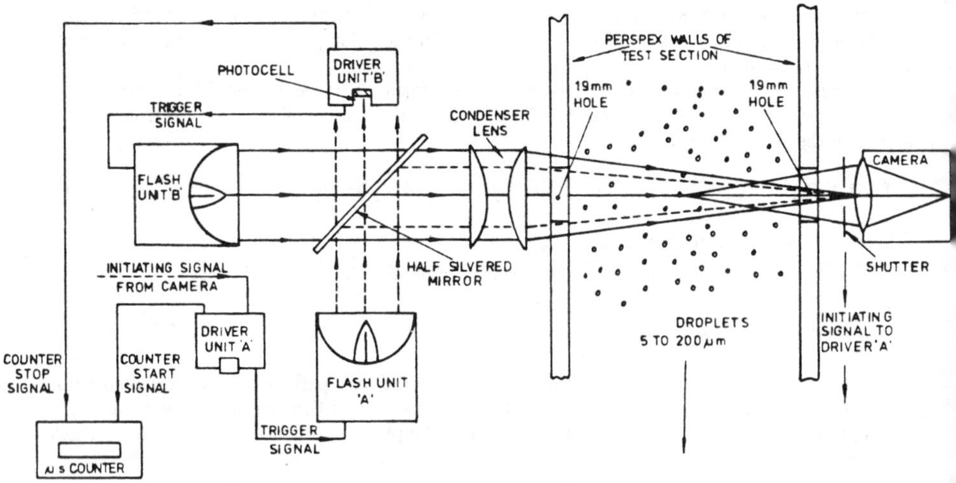

Figure 29. Illustrates the double flash technique employed by Finlay and Welsh [71].

mentally. Large magnifications are generally achieved using small depths of field. Hence, particles that fall outside small elements on either side of the plane of focus appear blurred. The limits of the depth of field are usually defined with respect to the smallest diffuse image of a point that is indistinguishable from a point. There are several criteria that relate the depth of field to various parameters such as aperture size, magnification, and focal length.

A small depth of field tends to mask some information of particle behavior and consequently is not useful in obtaining special size distributions of particles. When using large depths of field, the magnification error caused by the different particle paths and sizes can be significant. There are several methods to overcome this problem, such as steroscopy and simultaneous observation of several neighboring planes, each employing a small depth of field to provide exact image distances [72–75]. Azzopardi [76] provides an overview of these alternate methods.

Simmons [77] and Dix et al. [78] have replaced the photographic plate by photo-detector arrays. The photo-detector signals can be sorted in a minicomputer. In Dix's set-up, a focusing system is used that allows one to distinguish between particles in focus, nearly in focus and completely out of focus. This provides information on the longitudinal positions of particles. Outputs from the detector arrays are electronically formatted and punched onto tape for subsequent computer analysis. The analysis reveals information along three coordinates plus time.

Light scattering principles offer other alternatives to obtaining detailed information on discrete flows. The amount of light scattered by a particle is a function of the intensity of the illuminating radiation, the diameter and refractive index of particles, the wavelength and polarization of light, and the direction of observation relative to that of illumination. Mie [79] formalized the theory of scattering on the basis of electromagnetic theory, and reviews of the principles are given by van de Hulst [80] and Kirker [81]. The theory shows that the scattering caused by a spherical object varies in an oscillatory manner with the angle of observation.

The specific pattern produced by scattered light depends on the particle's relative refractive index and a dimensionless particle size defined as $\pi d/\lambda$ [82]. For single spherical particles larger or smaller than the wavelength of light, limiting but simpler theories apply. For particles smaller than the wavelength of light, the principles of geometric optics apply. That is, the angular distribution of scattered intensity can be used for sizing particles as can variations in polarization or color. The limiting range is for particles less than 1 μm in size [81]. For particles greater than 1 μm in size, the theory of Raleigh applies, where the intensity of the light scattered at a given angle is directly proportional to the diameter raised to the square power.

Miès theory [79] provides an exact solution of light scattered by spherical particles of arbitrary size. A formal analysis of theoretical optics for describing the interaction of a plane monochromatic wave with a homogeneous sphere of arbitrary radius suspended in a homogeneous medium involves solution of Maxwell's equations of electromagnetic waves. The classical solution of this problem was given by Mie [79]. The solution is expressed in terms of expansions with coefficients that are evaluated from appropriately defined boundary conditions. Second-order differential equations used in the derivation are the Legendre and Bessel equations and solutions may be found in any standard numerical analysis textbook [83].

In general, Mie's theory is sophisticated, often requiring complex solutions that are best handled by computers. In the case of particles that are large in comparison to the wavelength of light, computations become quite extensive.

For large particles a less rigorous theory can be applied to obtain approximate solutions to light scattering. The criterion for application of the simplified approach is:

$$\alpha = \frac{\pi d}{\lambda} \gg 1 \tag{20}$$

where d = particle diameter
λ = wavelength of light

That is, for particles much larger than the wavelength of light, the simplified analysis of diffraction and geometrical optics provide practical solutions. The phenomenon of light scattering in this case

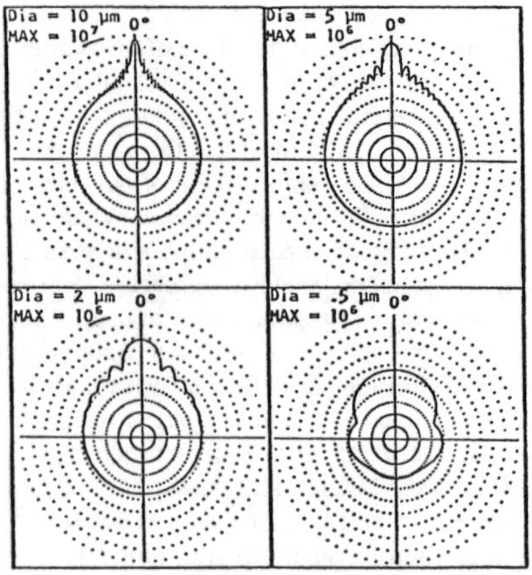

(A)

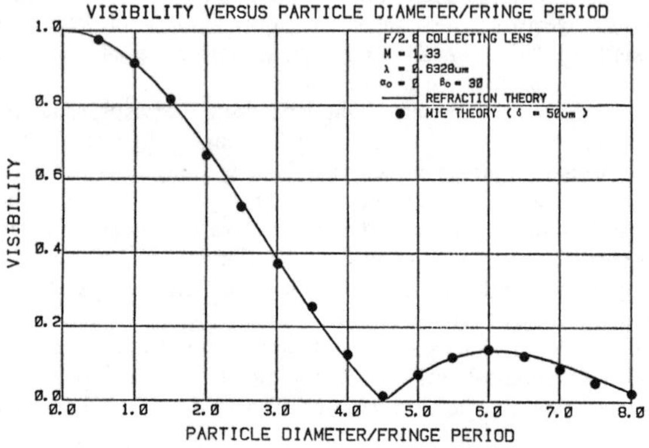

(B)

Figure 30. (A) Polar plots showing magnitudes of the Mie scattering coefficients for particles (relative refractive index ~ 1.5; $\lambda \sim 0.515$ μm); (B) plot of visibility vs. particle size/fringe period for comparison of Mie and refraction theories.

can be described by three distinct properties, namely, diffraction, reflection, and refraction. The criterion for this approximation is:

$$\alpha = \frac{\pi d}{\lambda} \simeq 10$$

or a particle size of about 2 μm.

A comparison of Mie theory predictions to those obtained from refraction theory are shown in Figure 30. Figure 30a gives polar plots of scattered light by different particle sizes. Figure 30b shows the visibility curves predicted from both theories. As shown, both theories are in excellent agreement for large particles. Because the range of experimental applications discussed in this chapter deal with large particulates (droplets or solid particles), light scattering techniques will be explained in terms of the simpler theory.

The first property, *diffraction*, refers to the behavior of light in the form of rectilinear propagation. It is a general property of wave phenomena that occurs whenever a wavefront becomes obstructed. For example, light will interact with an obstacle such that the light wave's amplitude and phase becomes distorted. This phenomenon occurs regardless whether the object is opaque or transparent, but obviously the degree of distortion varies. Basically, every point on a primary wavefront acts as a source of spherical secondary wavelets. In this manner, the primary wavefront at some time later becomes the envelope of these wavelets. The wavelets tend to advance with a speed and frequency equivalent to that of the primary wave and does so maintaining spacial similarity. This description is known as Huygen's principle, and is illustrated in Figure 31. The figure shows that a wavefront ϕ' is constructed from wavefront ϕ, and the envelope of the wavelets corresponds to the advancing primary wavefront ϕ'.

Diffraction occurs due to the interaction of electromagnetic waves with the electron cloud of atoms surrounding the obstruction. Electrons in the vicinity of the illuminated surface become excited by the impinging light and in turn, re-emit radiant energy. This energy discharge has been altered both in phase and amplitude. This interaction produces a so-called diffraction pattern.

The formation of a diffraction pattern is illustrated in Figure 32. Light is transmitted through a small aperture, which can be divided into equal elements. The individual elements are small in comparison to the wavelength of light; consequently, each element can be considered as a source of a secondary Huygen's wave. A lens behind the aperture is employed to produce the pattern at the barrier without having to be at a large distance in comparison to the aperture size.

The light rays that reach point P exit the slit at angle θ. In terms of ray optics, we may consider four rays positioned at the top of four equal zones. For a θ value corresponding to one-half wavelength (bb'), rays r_1 and r_2 cancel out at point P_2 (i.e., they are out of phase by $1/2\lambda$). Furthermore,

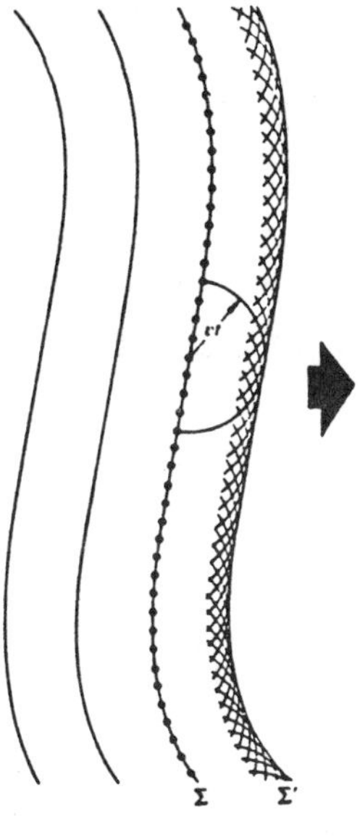

Figure 31. Propagation of a wavefront of light

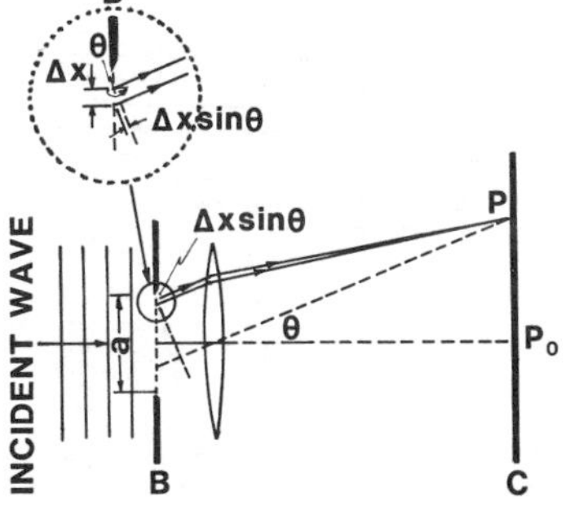

Figure 32. Formation of diffraction patterns.

rays r_3 and r_4 also cancel out along with other rays at angle θ. Hence, no light reaches point P_2 (a point of zero intensity). A simple experiment to demonstrate this principle is to shine a laser beam through a small pin hole (e.g., 1 mm diameter). Diffraction patterns, called Airy rings will be observed on a screen intersecting the light after it passes through the hole. The scattered light intensity for particles that appear for opaque disks produce patterns that are identical to those generated by orifices (known as Babinet's principle).

The intensity of diffracted light is independent of the obstruction's index of refraction. It may be described mathematically by:

$$I_0(\theta, d) = \frac{\alpha^2}{4\pi}\left\{\frac{2J_1(\alpha \sin\theta)}{\alpha \sin\theta}\right\}^2 \tag{21}$$

where d is a characteristic size of the obstruction (assume diameter of a sphere), and

λ = wavelength of light

J_1 = first-order Bessel function of the first kind

α = scattering coefficient

We may now direct attention to the properties of refraction and reflection. Consider a single, homogeneous spherical particle suspended in a homogeneous medium. When a beam of light is transmitted onto that particle, approximately 50% of the light is scattered due to diffraction, and the other half is scattered due to reflection and refraction. Both reflection and refraction are functions of the particle's index of refraction.

The description of light scattering by reflection and refraction given in the previous chapter is straightforward using geometrical optics theory. When a ray strikes the surface of the sphere, it produces a reflected and refracted ray as shown in Figure 33. The direction of these rays can be described by the laws of reflection and Snell's law for the refracted ray. The relative intensity of the

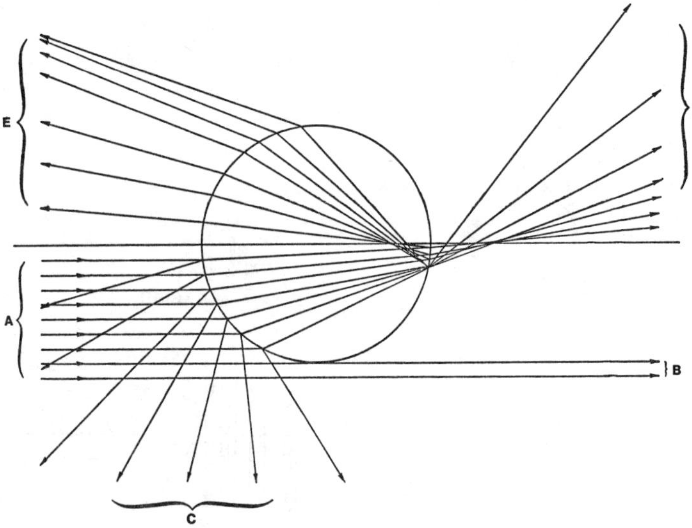

Figure 33. Geometrical optics of a normal cylinder.

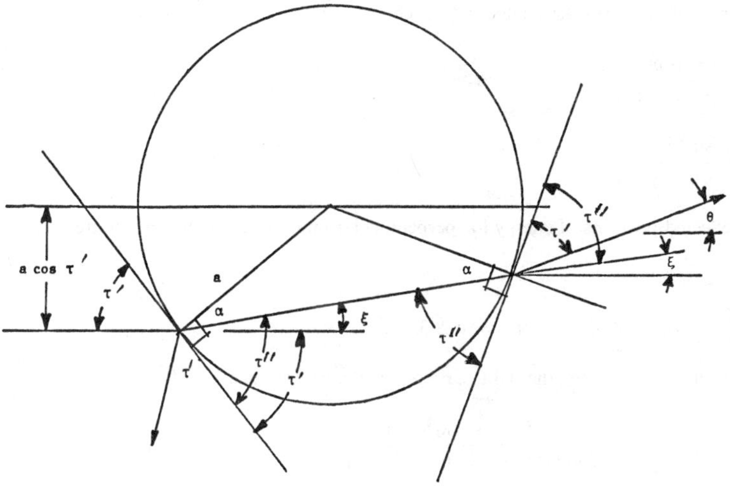

Figure 34. Ray trace through a spherical droplet.

rays may be described by the Fresnel coefficients. The light scattered by a sphere having a real refractive index m can be expressed in terms of simple mathematics. Figure 34 shows a ray characterized by its distance from the axis gives as (a cos τ') where a is the sphere radius and τ' is the angle between the incident ray and the surface. If the light is a cylindrical beam of intensity I_0, then its energy flux is:

$$F = I_0 a^2 \cos \tau' \sin \tau' \, d\tau' \, d\phi \tag{22}$$

Note that Equation 22 is simply the product of the light's intensity and the derivative for the area of a circle of radius a cos τ'. As the light beam emerges from the sphere it spreads as a result of refraction and reflection into a solid angle $\sin \theta \, d\theta \, d\phi$ over an area:

$$As = r^2 \sin \theta \, d\theta \, d\phi \tag{23}$$

Equation 23 applies to large distances r from the sphere. The angle of the emergent light ray is given by:

$$\theta = 2\tau' - 2p\tau'' \tag{24}$$

where τ'' is obtained from Snell's law:

$$\cos \tau' = \frac{1}{m} \cos \tau \tag{25}$$

Parameter p in Equation 12 denotes the successive reflections and refractions. That is, p = 0 for the first surface reflection; p = 1 for a single pass through the sphere, etc.

Dividing the emergent flux by this area gives an expression for the intensity of the scattered light:

$$I(p_1, \tau) = \frac{\epsilon_1^2 I_0 a^2 \cos \tau' \sin \tau' \, d\tau' \, d\phi}{r^2 \sin \theta \, d\theta \, d\phi} \tag{26}$$

The amount of energy associated with each ray is given by the Fresnel reflection coefficients:

$$r_1 = \frac{\sin\tau' - m\sin\tau''}{\sin\tau' + m\sin\tau''}$$

$$r_2 = \frac{m\sin\tau' - \sin\tau''}{m\sin\tau' + \sin\tau''} \quad (27)$$

The reflected portions of energy for perpendicular and parallel polarization are:

$$\epsilon_1 = r,\ p = 0$$

$$\epsilon_1 = (1 - r_1^2)(-r_1)^{p-1} \qquad \text{for } p = 1, 2, 3, \ldots \quad (28)$$

The exact expression for the light scattered at the first surface by reflection is:

$$I_1(\theta) = I_0 \frac{a^2}{8}\left\{\frac{\sin(\theta/2) - [m^2 - 1 + \sin^2(\theta/2)]^{1/2}}{\sin(\theta/2) + [m^2 - 1 + \sin^2(\theta/2)]^{1/2}}\right\}^2 \quad (29)$$

For scattering due to transmission with refraction, the exact formula for the transmitted intensity is:

$$I(\theta) = 2a^2 I_0\left(\frac{m}{m^2 - 1}\right)^4 \frac{(m\cos\theta/2 - 1)^3(m - \cos\theta/2)^3}{\cos\theta/2\,(m^2 + 1 - 2m\cos\theta/2)^2} \cdot 1 + \sec^4\frac{\theta}{2} \quad (30)$$

These expressions surprisingly show that the size of the spherical particle plays no role in the angular distribution of the scattered light. This means that the angular distribution of the geometrical components is the same for all particle sizes. The magnitude of the light scattered, however, is proportional to the particle diameter squared.

These expressions can be used to compute the angular distribution of light scattered by a sphere for a given index of refraction. Figure 35 shows a splot of the intensity coefficient versus angle θ. The curves represent light scattered by a single transmission through a spherical particle.

In the case of multiple spheres it is generally assumed that the light scattered by an array of particles is incoherent and as such the scattering functions for single particles apply. The sum of the intensity scattered by individual particles is therefore assumed to be cumulative. On this basis, it is further assumed that there is no rescattering of scattered light. However, these assumptions are only justified for a sufficiently dilute system. A plot of the scattering coefficient, K, is given in Figure 36. As shown, the total amount of light scattered only departs from the square law dependence for very small particle sizes. Below 1 μm in size the function is monotonically increasing, and between 1 and 10 μm, the function is shown to be oscillated. At $d > 10\ \mu$m, K is independent of particle size. It can be concluded that a method based on total light scattered is feasible only for particles smaller than 1 μm in size.

Limiting discussions to the principles of light scattering for single particles, a region approached by dilute two-phase flows, a narrow parallel beam of illumination in conjuction with a similar observation beam can be used to sense particle flows. This arrangement permits a small probe volume to be established that minimizes the opportunity of particle collisions. In other words, a very small probe volume increases the probability of detecting a single particle. This, in turn, extends the range of particle concentrations in flows that can be measured. A discriminator circuit is required in order to reject any signal arising from the coincidence of two particles (a coincidence which thus produces a doublet pulse). To avoid detecting a particle that is only partially within the probe measuring volume, an annular beam of light of a different color can be used to form a ring around the volume. Through the use of beam splitters and filters, the signals from each color can be separately monitored. Using appropriate circuitry, particles only partially in the detection volume can be identified as they

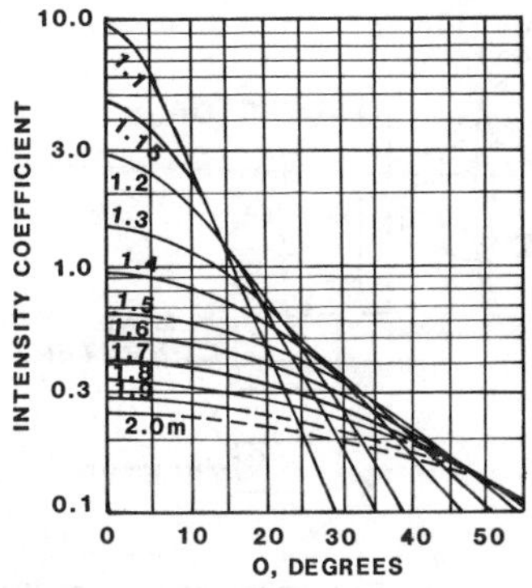

Figure 35. Angular distribution of light scattered by transmission through a particle, sum of both polarizations for refractive indices 1.1 to 2.0.

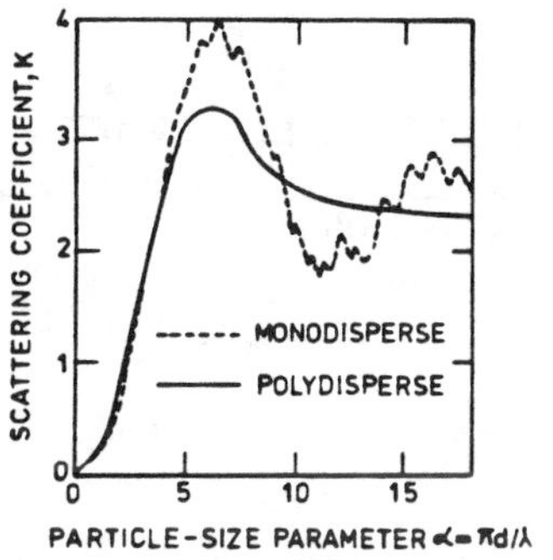

Figure 36. Plot of the scattering coefficient.

produce a simultaneous signal in each color, and hence, are removed during the data analysis or acquisition stages. The optical techniques based on light scattering principles from single particles are described in References 82 and 84–87.

Now consider a cluster of spherical particles of the same diameter. The additivity rule applies to light scattering from a monosized group of particles. To evalute the particle size from this summing, the concentration must be known. The same principle can be applied to a distribution of particle sizes, where the scatter is summed with respect to a weighted average with the fraction of each size present in the sample. For particles in the size range of 10 to 1,000 μm, the forward direction must be used to attain a sufficiently sensitive intensity variation to different particle sizes.

Several investigators have used light scattering techniques to obtain droplet size distributions in gas-liquid flows [63–68] Dobbins et al. [63] studied the small angle forward scatter for a polydispersion fitted by an upper limit log normal distribution. This particular study revealed the scattered light distribution to be relatively insensitive to parameters that characterize the particle size distribution. This method is restricted to obtaining only the Sauter mean diameter. Deich, et al. [64, 65] and Chin et al. [66] observed the opposite; that is, the size/light distribution was found to be sensitive to the method employed for particle size determination.

Regardless of the actual method used, there are many difficulties in experiments of this nature. One problem is that the light that is diffracted at a fixed angle from particles located at different distances from a detector arrive at different areas of the detector, thus causing extraneous or confusing signals. Azzopardi [76] describes a solution to this by positioning a lens between the object and the detector so that the latter is at the focal plane of the lens. In this manner, the lens serves as a Fourier transformer, where the light diffracted by particles in any given direction is collected at a single point, irrespective of the particle's position. Figure 37 shows the diffraction system and optical geometry of a Fourier transform lens. The displacement of a point from the optical axis will only depend on the focal length of the lens and the angle at which light is diffracted (i.e., $r = \theta f$; r is the displacement from the optical axis, f is the len's focal length and θ is the angle of the diffracted beam). By traversing the detector across the focal plane, the light diffracted at different angles can then be sensed.

The angular distribution of diffracted light varies according to the Airy function:

$$I \propto \left[\frac{2J_1(\alpha \sin\theta)}{\alpha \sin\theta}\right]^2 \tfrac{1}{2}(1 + \cos^2\theta) \tag{31}$$

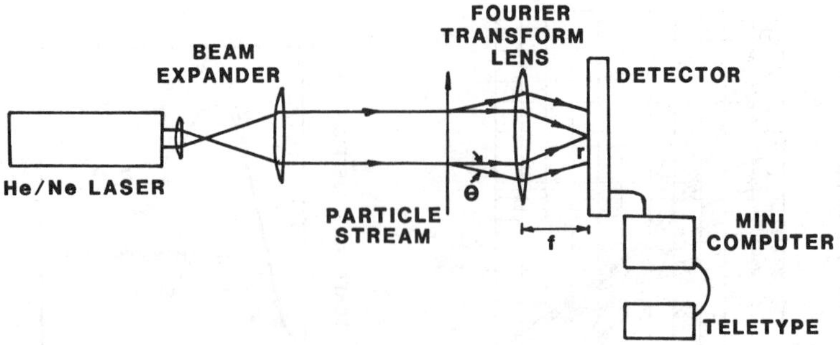

Figure 37. Diffraction equipment and optical geometry for a Fourier transform lens [24].

For small angles this function reduces to:

$$I \propto \left[\frac{2J_1(\alpha\theta)}{\alpha\theta}\right]^2 \tag{32}$$

where α is a dimensionless particle size ($= \pi d/\lambda$); J_1 is the first order Bessel function of the first kind; and I is the illumination. The Airy function is shown plotted in Figure 38 for the distribution of light diffracted by an opaque disc. This poses a very practical problem in that the range of intensities to be detected drops off sharply, thus requiring a detector having a very large range. Chin, et al. [66] have proposed the use of photographic plates and a microdensitometer as one solution. Density is a linear function of log intensity over a wide range of intensity and hence, can be compressed into a small density range. An alternative method is that of Cornillault [67], who employed a screen having various size apertures for a photodetector. In this approach, as the detector moves from the optical axis it is covered by a screen having an aperture of increasing size. Annular ring detectors described in References 68 and 88 perform a similar function to the variable area screen. These latter arrangements are used in conjunction with a Fourier transform lens.

A backscattering device for measuring mass flux data is described by Shofner, et al. [89]. The measurement depends on the light scattered per unit volume of particle. The system employs a large probe volume in which the total light backscattered from all the particles within this volume is collected. The optical systems described thus far are reasonably sophisticated, and yet measurement interpretation is still subject to interpretation by the investigator.

A relatively simple approach that could be adapted to test environments in actual plant equipment, provided viewing ports are permissible, is based on the principle of light obscuration. As a particle flows across a light beam, the amount of light that emerges from the control volume along

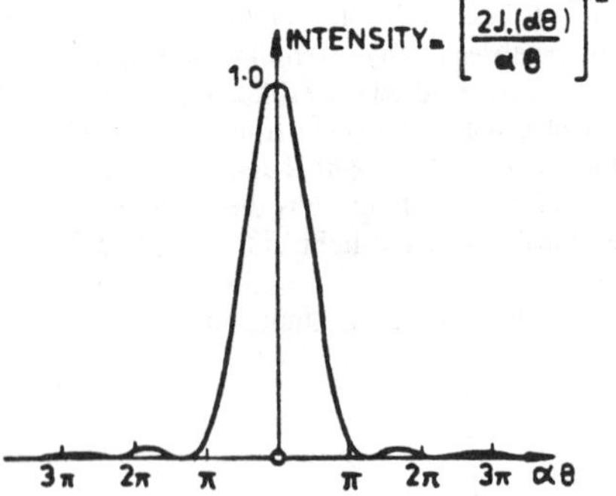

Figure 38. Plot of Airy function using Equation 32.

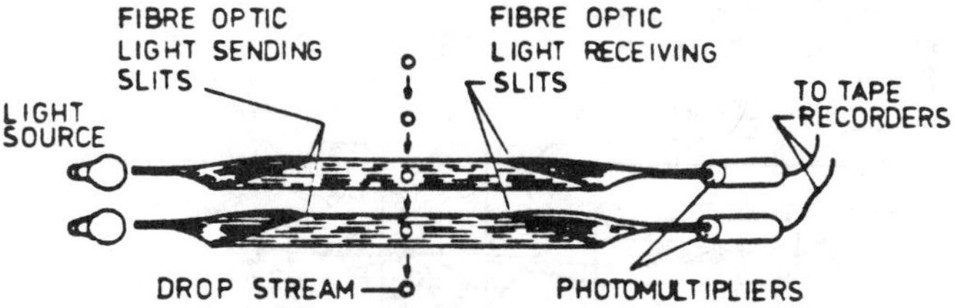

Figure 39. Setup for particle detection using ight obscuration principle [90].

the irradiation direction diminishes. The diminution of light can be described by:

$$I_{out} = I_{in} - (I_{scA} + I_{ABs}) \tag{33}$$

where I_{scA} = amount of light scattered in directions other than the irradiation direction
I_{ABs} = amount of light absorbed by the particle

For large, single particles, the last two terms are indistinguishable and considered as one group, which is proportional to the projected area of the particle or that portion of it in the control volume. Various techniques based on this principle are described in the literature [90–95]. The general approach involves the use of a light beam having a rectangular cross-section with the particle moving across the longer side. The area involved in this case is proportional to d^2 where d must be smaller than the shortest beam width. If the particle is larger than the width, then the linear relationship turns out to be a good approximation [76]. Fiber optics probes can be used in this approach, where a rectangular beam is produced by illuminating one end of the fiber bundle while the other end is splayed out flat and a reverse fiber bundle is used for recording the observation. The technique used by Ritter, et al. [90] is illustrated in Figure 39. By employing two such detectors, one positioned directly below the other, complete information on particle dynamics such as diameter, flow velocity, and frequency or concentration can be obtained. In this case, cross-correlation of the signals between the two probes must be performed to obtain particle velocity.

Laser Doppler Anemometry

A laser-Doppler anemometer measures the velocities of flowing particles from frequency information contained in light scattered by the particles as they pass through a fringe or interference pattern. The intensity of the scattered light also contains information; specifically, it is proportional to particle size. Several investigations directed at studying the influence of size on velocity measurements, as well as in obtaining particle size information in atomized and dispersed flows, have been done [96–102].

Several theories have been postulated to provide a basis of an accurate technique for measuring the velocity-size dependency in discrete flows. Among the most recent is an attempt to define a practical relationship between the signal visibility (i.e., the ratio of a.c. and d.c. components of the burst of scattered light), and the ratio of particle size to fringe spacing. Fistrom, et al. [98] provides one relationship from geometric optics. Hong and Jones [100] and others [103–105] provides more rigorous calculation approaches. One difficulty with this technique is that the observation angle and solid collection angle must be carefully chosen to optimize the signal strength. The limited range of drop sizes for which the visibility is an unambiguous function (i.e., constant fringe spacing) is a disadvantage of the method. According to Farmer's relationship, a necessary condition to avoid ambiguity is:

$$\frac{d_{max}}{\lambda^*} < 1.22\pi \tag{34}$$

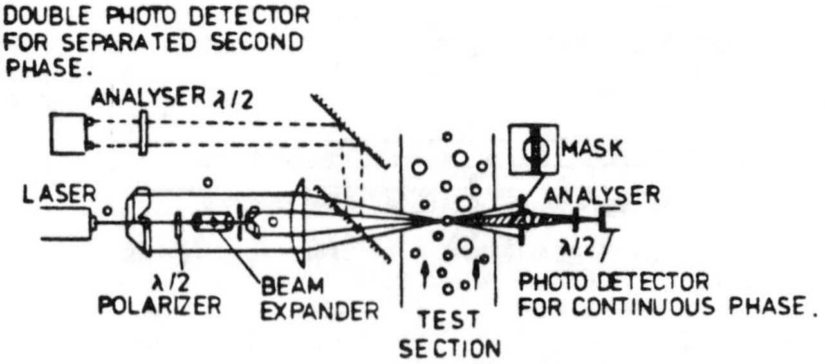

Figure 40. Laser-Doppler scheme posed by Durst et al. [99].

This criterion stipulates that a relatively large probe volume is required since a minimum of 8–10 fringes are usually needed to produce a discernable signal. Another disadvantage is that particles that do no pass through the center of the probe volume yield Doppler bursts that are different from those from the center of the probe volume [96]. Thus, the method is difficult to use when "off center" particles are encountered. Careful focusing of the cross beams and the observing optics can minimize this problem to an extent. A discriminator module in the output circuitry is an essential component in the system that minimizes the problem. The addition of a gate photomultiplier at 90° to the input light beams permits the selection of those particles passing through the center of the probe volume [106]. The coincidence of two particles within the probe volume can result in a spurious signal and hence, the distance between particles is a controlling variable. The method described by Ungut et al. [106] requires only one measurement detector at a fixed position. The Doppler burst is the measured signal, and hence, allows extraction of information on both velocity and size on a particle by particle basis.

Durst, et al. [99] have conducted extensive tests to assess the suitability of different illumination/detection arrangements. One of their configurations that employs dual photodetectors is shown in Figure 40. When a particle travels through the probe volume, the outputs of the photodetectors are Doppler burst that are out of phase with one another. The phase difference is a function of particle size, and will decrease along with size. Various other configurations and techniques using laser anemometry principles are described by Chigier, et al. [107], Yule et al. [108], and Wigley [109]. All of these systems have limitations, a principle one being saturation of the scattered light signal when applied to high number density particle systems.

Particle sizing interferometric techniques have greatly extended the particle number density range over which measurements can be made and are commercially available as packaged systems such as the Spectron Development Labs laser-Doppler interferometer [110, 111]. These systems have the advantage of being insensitive to the absolute intensity or scattered light and light absorbed by the particle, and provide information on both size and velocity, simultaneously.

In the previous section, light scattered by spherical particles of diameters much larger than the wavelength of the incident light was described by a simple mathematical model using geometrical optics theory. Van de Hulst [80] notes that when $\alpha > 10$ (where $\alpha = \pi d/\lambda$), scattered electromagnetic radiation can be described using the theories of diffraction, refraction and reflection. One half of the scattered light is due to diffraction which is concentrated in the forward direction about the laser beams while the other is due to refraction and reflection that are scattered in all directions about the particle. With the appropriate selection of off-axis angles for the size and index of refraction of the particles to be measured, the diffracted light can be ignored in the analysis.

Bachalo, et al. [110] describe a technique for sizing spherical particles based on the measurement of the relative phase shift that occurs when two light waves pass through transparent particles on different paths. Figures 41a illustrates the principle in which two rays (which are normal to the waves) pass through a droplet and arrive at a common point on the plane of detection. The interference pattern of the scattered light can be measured and related to the particle size. The optical arrangement used in the analysis is shown in Figure 41b. By measuring the visibility or amplitude

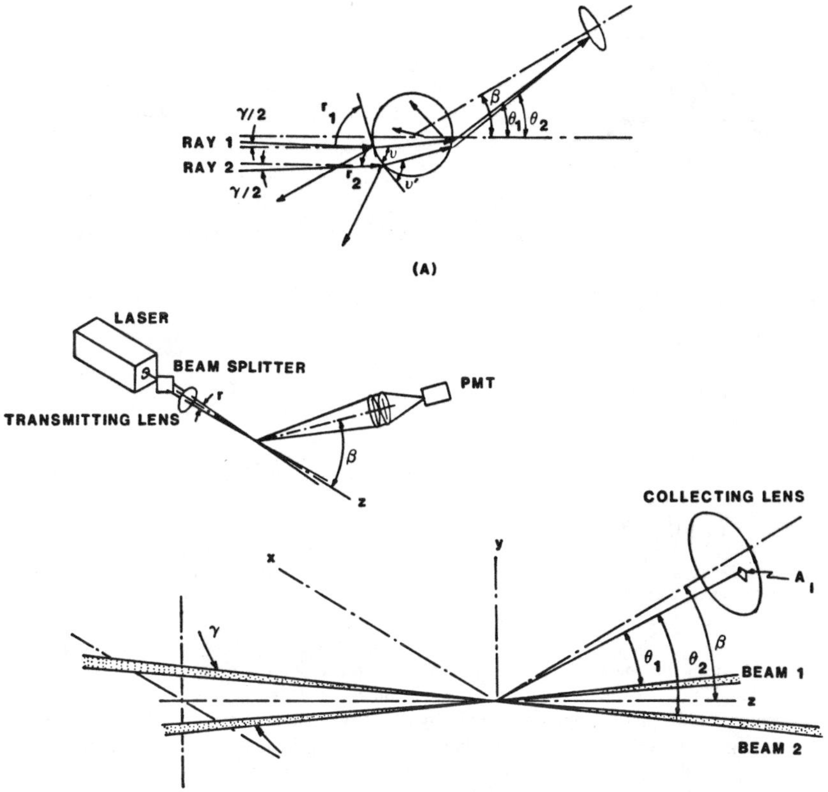

Figure 41. (A) Ray trace diagram illustrating the angles used in light scattering analysis; (B) interferometer setup described by Bachalo et al. [110].

modulation of the interference pattern formed by the scattered light and collected over a finite collecting aperture, information to size the droplets is obtained.

The intensity of the scattered light by a droplet located at the intersection of two laser beams is:

$$I = \frac{1}{n}\{|E_{s1}|^2 + |E_{s2}|^2 + 2|E_{s1}||E_{s2}| \cos \sigma\} \tag{35}$$

where $|E_s|$ is the magnitude of the complex field function, σ is the phase angle between the scattered fields E_{s1} and E_{s2}, and n is the wave impedence. The visibility that is the ratio of the a.c. or Doppler component of the signal to the d.c. or pedestal component is:

$$\tilde{v} = 2\left|\frac{\int A_{lens} \int |E_{s1}||E_{s2}| \cos \sigma \, dA}{\int A_{lens} \int \{|E_{s1}|^2 + |E_{s2}|^2\} \, dA}\right| \tag{36}$$

The Doppler burst signal and its components are shown in Figure 42.

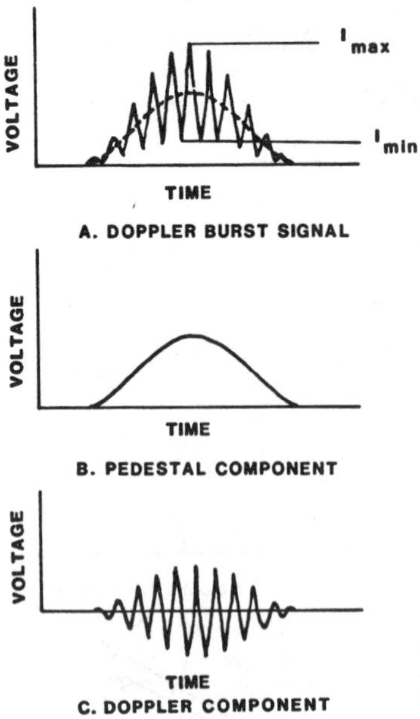

Figure 42. Doppler burst signal and Doppler and pedestal components.

Information on the size of the droplet is contained in the phase shift of the light waves passing through the spherical droplet. Two planes waves incident on the droplet, one from each beam, pass through the droplet at angles that differ by the beam intersection angle. The waves scattered from each beam arrive at the detection plane with a phase difference which forms the interface pattern. Relationships describing the phase shift of the light waves passing through the droplet (or reflected from it) assume that the particles are spherical.

There are angles at which the light reflected and refracted by particles are of the same order of magnitude but will be out of phase. Light collection at these angles should be avoided in order to prevent ambiguity in the measurements.

With high density particle flows, a 90° forward-scattering mode is used. The light scattered by the particles in this case, is dominated by the laws of reflection. Figure 43 shows the schematic of the measurement principle for this case. The 90° cone of light collection intersecting the focused laser beams forms a small measurement volume. The focused beam diameters can be as small as 50 μm and the aperture on the photomultiplier is set to admit the image of the beam.

Particles passing through the probe volume scatter light at an intensity that is proportional to their size. The pedestal component of the signal will grow with the size of the particles while the Doppler modulation grows as the product of the pedestal and the visibility [110, 111]. This product is zero when the diameter is very small, reaches a maximum for intermediate size particles, and decreases to zero again for larger particles.

The magnitude of the pedestal and the Doppler component thus establishes the size of the sampling volume. Particles with a large magnitude of the Doppler component are detected over a large volume than those with a smaller modulated signal.

The system described by Bachalo et al. provides a number count of particles over different size ranges. Because of the probability of detecting particles that produce a small modulated signal is smaller than the probability of detecting particles with a large modulated signal, the system displays a biased histogram of particle size versus the number of particles. To correct for this biasing, the detectable cross-sectional area of each particle size must be known. The number of particles detected at a given size range is then divided by the normalized cross-sectional area to obtain the proper distribution.

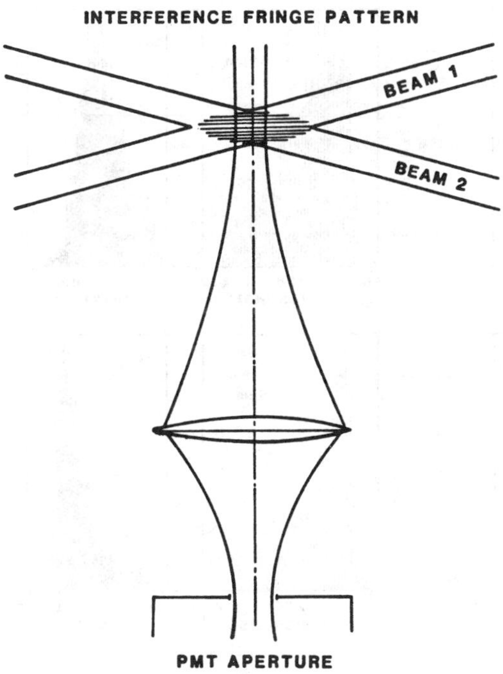

Figure 43. 90° light scatter detection scheme.

The light source used in the Spectron Labs system is a 15 mW helium-neon laser. Several optical configurations are required to provide the probe volumes and interference fringe spacings appropriate to the size ranges to be measured. With the combination of the collimating beam expander and transmitting lens that can be inserted or removed from the optical path, a large range of probe diameters is attainable. Figure 44 shows typical size and velocity histograms obtained from a pneumatic type nozzle. Measurements were made using a 30° forward scattering mode.

As noted earlier, the projected interference pattern received by the photomultiplier tube is a function of the ratio of the droplet size to the fringe spacing. When a droplet is much smaller than the fringe spacing, only a part of a fringe is projected. In contrast with particle that is larger than the fringe spacing, several fringes are observed to be projected.

There are three scattering modes that contribute to the projection of the interference pattern:

1. Diffraction, which is concentrated in the forward direction
2. Reflection, which is characteristic of backscattering
3. Refraction, where the particle is transparent and acts as a spherical lens

The photomultiplier tube converts the scattered light into an electrical signal with the same time and amplitude variation as the modulated light. In other words, the intensity of the signal is high when a particle is on the middle of a bright fringe, and low when it is in the middle of a dark fringe. The Doppler bursts of a small and large particle when observed on an oscilloscope appear as in Figure 45. The intensity distribution of the smaller particle is more modulated than the larger one because it can scatter light from one fringe at a time resulting in the production of either low or high intensity. A large particle scatters light from several fringes at a time and hence, there is insufficient time for the intensity level to drop to ground. It should also be noted that the intensity maximum is greater for larger particles.

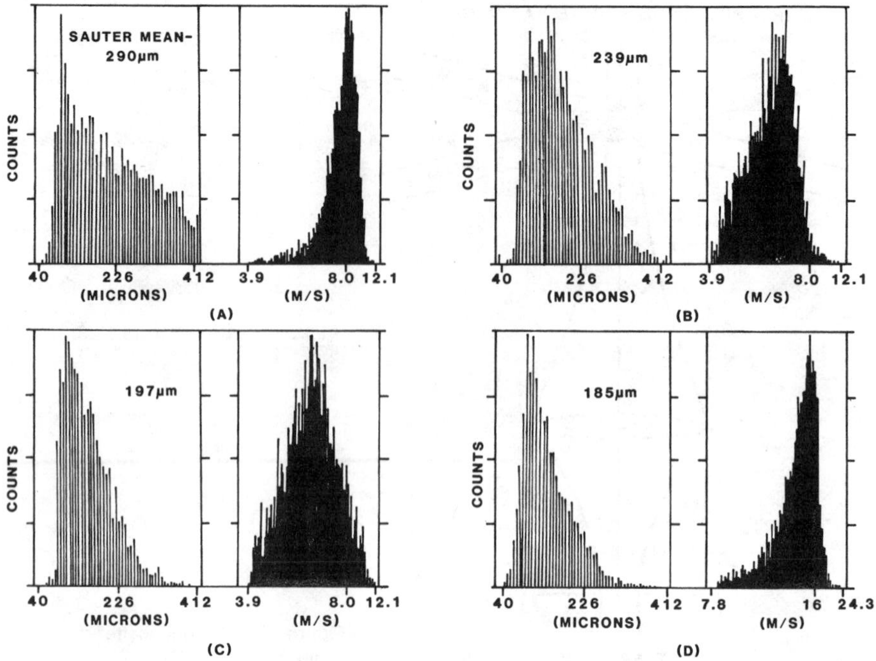

Figure 44. Typical histograms for water sprays obtained using 30° forward scattering. The plots are in order of increasing liquid throughput.

Mathematically, the light waves forming the interference pattern in the probe volume are [111]:

$$\bar{E}_1 = \sqrt{I_0}\,\bar{P}_1 \exp i\left[\phi_0 + W_0 t - \frac{2\pi}{\lambda}\left(x\cos\frac{\gamma}{2} + z\sin\frac{\gamma}{2}\right)\right] \tag{37}$$

$$\bar{E}_2 = \sqrt{I_0}\,\bar{P}_2 \exp i\left[\phi_0 + W_0 t - \frac{2\pi}{\lambda}\left(-x\cos\frac{\gamma}{2} + z\sin\frac{\gamma}{2}\right)\right] \tag{38}$$

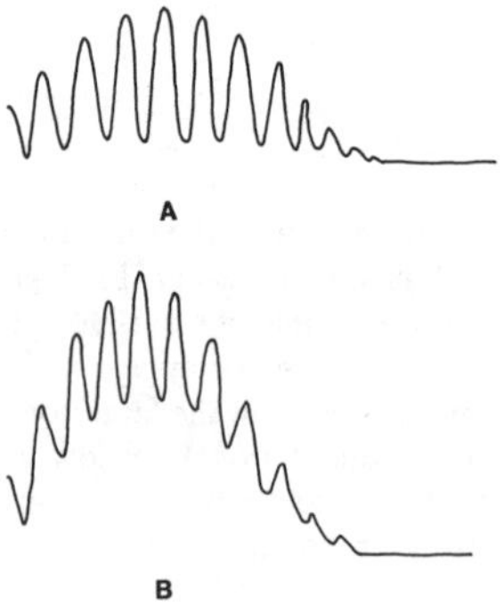

Figure 45. Doppler burst signals for large and small particles.

And the total vector resulting from the interference of two light waves is:

$$\bar{E} = \bar{E}_1 + \bar{E}_2 \tag{39}$$

where I_0 = intensity of the laser beam
$\bar{p}$ = polarization vector
$i = \sqrt{-1}$
ϕ_0 = phase of laser beam
W_0 = frequency of laser beam
λ = wavelength of laser beam
γ = angle of intersection between the two beams
x, z = coordinates

The total intensity is the product of $\bar{E}$ and its complement:

$$I = 2I_0\left(1 + \bar{P}_1 \cdot \bar{P}_2 \cos\frac{4\pi \times \sin\gamma/2}{\lambda}\right) \tag{40}$$

The fringe spacing can be determined by evaluating the x-value that results in the maximum I. This expression is

$$\delta = \frac{\lambda}{2\sin(\gamma/2)} \tag{41}$$

Particle velocity is determined in this technique through the relationship between fringe spacing and particle velocity. Particle size is determined from the relationship between fringe spacing and the signal modulation or visibility. The term visibility is actually an intensity ratio defined as [111]:

$$\tilde{V} = \frac{I_{max} - I_{min}}{I_{max} + I_{min}} \tag{42}$$

The Doppler period, τ, is the time that it takes a particle to travel between fringes. Hence, for particles traveling normal to the fringes, particle velocity is:

$$V_p = \frac{\delta}{\tau} = \delta f_D \tag{43}$$

where $f_D = 1/\tau$ is the Doppler frequency.

Note that the technique only measures the velocity of particles that cross the fringes at a normal direction, which is not necessarily that of the carrying fluid. For very small particles it may be assumed that the particles are traveling at the same speed as the carrying fluid. However, if the size distribution is broad, this can be a poor assumption. A reasonably accurate measurement of the carrier fluid's velocity can be obtained by seeding the flow with tiny particles and then measuring the flow velocity.

A major limitation with the technique as previously described is that measurements are flow direction dependent. Recall that since frequency is a positive parameter, no information on the flow direction is derived, and hence, the optics must be aligned with respect to the test section so that flow is normal to the fringe patterns. There are, however, many situations in which drop dynamics in recirculating flows are to be studied. Recirculating lows lend themselves to ambiguous signals. This problem is overcome by imposing a frequency shift in one of the laser beams so that fringes formed move with a shifted frequency f_s. A stationary particle will scatter light with a frequency of f_s, since the frequency is established by the relative motion between the particle and

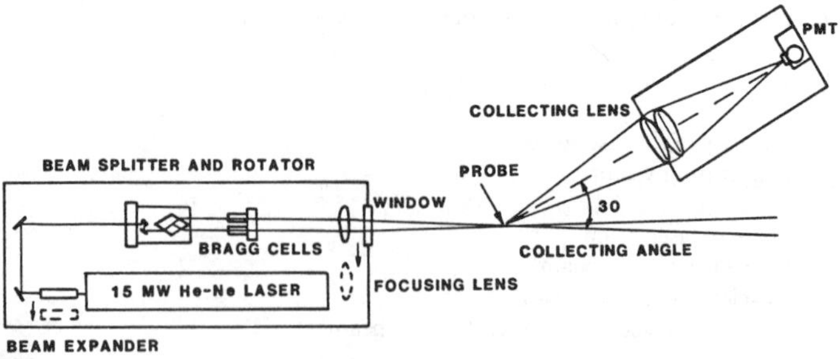

Figure 46. Optical configuration for interferometer system described by Bachalo et al. [110].

fringes. The Doppler frequency in this case is:

$$f_D = f_s \pm U/\delta \tag{44}$$

The governing criterion is that $f_s > U/\delta$.

The frequency shift can be achieved using either a rotating diffraction grating or by fitting the transmitting system with Bragg cell frequency shifting to control the relative frequency bandwidth and to resolve any flow direction ambiguity. With the latter, the collected scattered light is focused by a second lens onto an aperture on the photomultiplier tube. The size of the aperture is variable to accommodate the particular probe diameter being used and still remain small enough to block light scattered from outside of the probe volume. The optical configuration for such a system is shown in Figure 46.

Modarress and Tan [112] have addressed one serious consideration concerning the accuracy of velocity measurements in two-phase flows using laser-Doppler anemometry. The problem concerns the bias of the data by the "cross talk" between the phases. The accuracy of data depends on the ability to discriminate the signals originating from large particles from those of smaller particles that more closely follow the carrier fluid's behavior. The problem is of most concern when the population density of the discrete phase is high. Modarress et al. note that simple amplitude discriminators that identify the size of light scattering particles by the amplitude of their signals are not always reliable. To better understand this argument, we should note that the probe volume has a spheroid shape with a Gaussian intensity distribution. A large particle passing through the edge of the probe volume can produce a signal amplitude of the same order as the signal amplitude of seeding particles passing through the center of the probe volume. This is the definition of the "cross-talk" error. Discrimination based on the frequency of the signal is not a reliable approach since it only applies when the slip velocity between phases is large enough such that the probability distribution functions (Pdf) for each phase do not overlap. Lee et al. [113] reduced the cross-talk error by using a combination of amplitude discriminator, filter banks and photodiodes. Yeoman et al. [114] have simultaneously measured drop size and velocity in a two-phase flow using two-color coincident beams to create two parallel sets of fringes. The visibility and maximum intensity techniques were used concurrently to make simultaneous size and velocity measurements. In principal, this method forms the basis of a discriminator for two-phase flows.

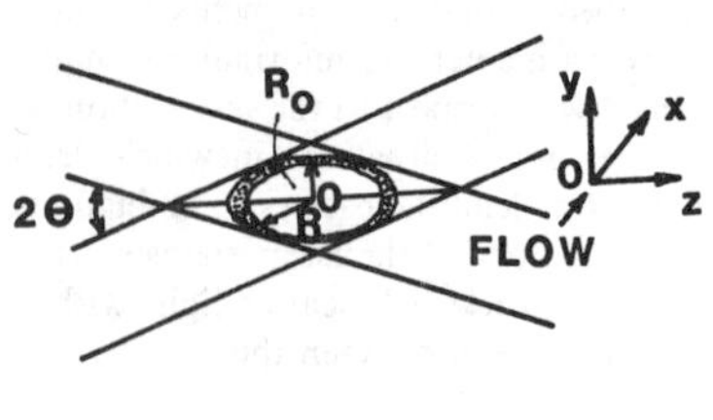

Figure 47. Probe volume region created by two intersecting beams.

Modarress Tan [112] provide a method for estimating the cross talk error for a typical laser anemometry system in two-phase flow. In the analysis, particles are assumed to be of two uniform sizes, where the seeding particle diameter d_g (representing the gas-phase) is much smaller than the discrete-phase diameter, d_s. A simple amplitude discriminator is used to distinguish between the size of the light scattering particle from the amplitude of the pedestal of their signal.

For a scattering particle located at a distance R from the center of the probe volume, Durrani and Greated [115] provide the following expression for the photodetector current:

$$i = 2(\tfrac{1}{2}MC_{sc})^2\epsilon_0 \exp[-2R^2/R_0^2]\left[\cosh\left(\frac{2XZ}{R_0^2}\sin 2\theta\right) + \cos\left(\frac{4\pi}{\lambda}X\sin\theta\right)\right] \tag{45}$$

The locus of the particle is on an ellipsoid as shown in Figure 47, and is given by:

$$R^2 = X^2\cos^2\theta + Y^2 + Z^2\sin^2\theta \tag{46}$$

For small θ and spherical scattering particles, the pedestal amplitude of the signal can be derived from Equation 45:

$$V(R) = C\sigma^2 \exp(-2R^2/R_0^2) \tag{47}$$

Following the analysis of Madarress and Tan, V_0 is defined as the signal amplitude of a seeding particle at R_0, from whence Equation 47 is:

$$V_g(R) = V_0 \exp(2 - 2R^2/R_0^2) \tag{48}$$

For the discrete-phase Equation 47 becomes:

$$V_s(R) = V_0 \exp(2n^2 - 2R^2/R_0^2) \tag{49}$$

For specified lower threshold limit V_l and upper threshold limit V_u, data will be accepted if $V_l \leqq V \leqq V_u$. Relative to the position of the particles in the probe volume, only the signals from particles passing at a distance from the probe volume axis of between R_l and R_u will then be accepted. The data rate is expressed by:

$$N = \pi\lambda(R_l^2 - R_u^2) \tag{50}$$

This expression provides the data rate as a function of threshold voltages:

$$N = \tfrac{1}{2}\pi\lambda R_0^2 \ln(V_u/V_l) \qquad \text{for } V_l \geq V_0 \quad \text{and} \quad V_u \leq V_{max} \tag{51}$$

Note that when $V_1 = V_0$, and $N = 0$ for $V_u \leq V_0$, and $N = \text{Max}\{N\}$ for $V_u \geq V_{max}$. This establishes the following criteria

$$\text{Max}\{N_g\} = \pi\lambda_g R_0^2,$$

$$\text{Max}\{N_s\} = \pi n^2 \lambda_s R_0^2 \tag{52}$$

The velocity measurement for the solid-phase can be made by using the following criteria:

$$V_l \geq (V_g)_{max} = V_0 e^2$$

$$V_u \geq (V_s)_{max} = V_0 \exp(2n^2) \tag{53}$$

For the data rate, at $N_g = 0$,

$$N_2 = \pi(n^2 - 1)\lambda_s R_0^2 \tag{54}$$

This analysis shows the cross-talk error for the discrete-phase measurement can be eliminated by setting the lower threshold at a sufficiently high value. This is accomplished by lowering the gain on the photomultiplier tube.

For the measurement of the gas phase in the presence of the discrete-phase, the lower and upper values of the voltage are set to accept signals from the seeding particles. In this case, $V_l > V_0$ and $V_u = V_0 e^2$. The data rates in this case will be:

$$\begin{aligned} N_g &= \tfrac{1}{2}\pi\lambda_g R_0^2 \ln(V_u/V_l) \\ N_s &= \tfrac{1}{2}\pi\lambda_s R_0^2 \ln(V_u/V_l) \end{aligned} \tag{55}$$

The cross-talk error can be defined as the ratio of the solid-phase samples to that of the gas phase ($N_s/N_g = \psi = \lambda_s/\lambda_g$, where λ is the particle arrival rate per unit area).

Provide N_s and N_g do not reach maximas (see Equation 52), the cross-talk error will be independent of the threshold voltages. Modarress et al. note that for a given probe volume geometry and discrete-phase mass loading, the cross-talk error can be expressed by:

$$\psi = \varphi(d_g/d_s)^3(\rho_g/\rho_s)(SU_g/\lambda_g v_g) \tag{56}$$

where U_g = mean velocity of gas
v_g = volume of seeding particle
S = probe volume cross-sectional area
φ = discrete-phase to gas-phase mass ratio

Figure 48 shows estimates of the cross-talk error reported by Modarress and Tan.

As shown in Figure 48, cross-talk error becomes most significant for particle sizes under 100 μm. Modarress and co-workers [112, 116, 117] devised a simple approach to discrimination against the large particles involving the use of an auxiliary probe volume with different polarization and having a cross-section larger than the measurement probe volume. In this arrangement, when a large particle passes along the edge of the measurement probe volume, it is still within the central region of the auxiliary probe. In this manner, monitoring of the pedestal signal of the auxiliary probe and the signal from the measurement probe volume are correctly identified. The auxiliary probe volume size is designed such that when a solid particle is at nR_0 position, the solid signal of the auxiliary probe is always greater than the maximum auxiliary signal from the seeding particle.

The system used consisted of a two-component frequency shifted laser-Doppler anemometer. By proper expansion of the beams, the axial (green) and the radial (blue) probe sizes were set in which the blue signal had a dual role of measuring the radial component of the velocity and also of providing the pedestal signal for discrimination of the axial signal.

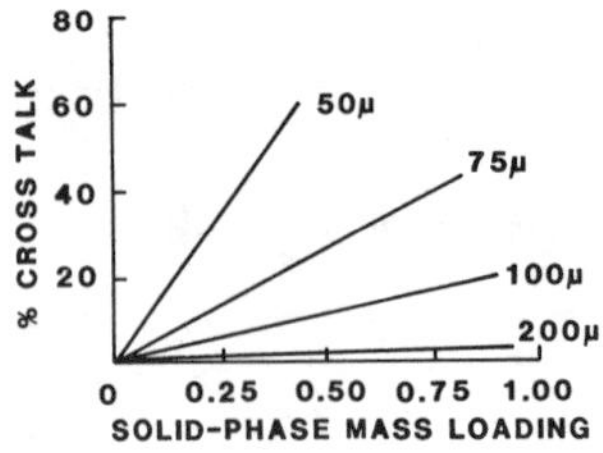

Figure 48. Variation of cross-talk with solid-phase mass loading [112].

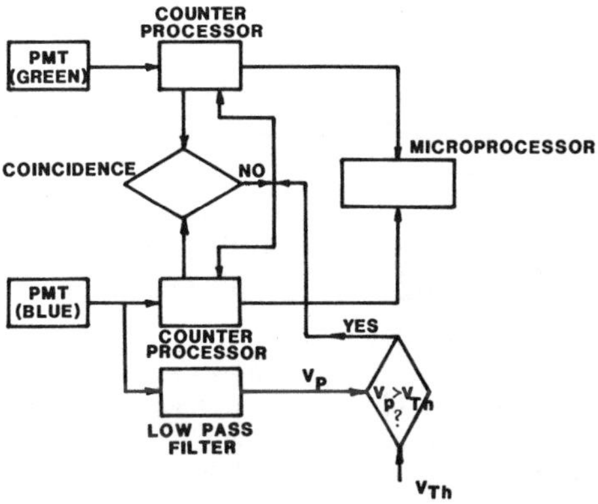

Figure 49. Logic diagram of auxiliary pedestal discriminator reported by Modarress and Tan [112].

Their system was comprised of two receiving optics packages for both components of the velocity. Solid-phase data were collected when the system's sensitivity to the seeding particles was reduced below noise level (lower laser power and amplification gain were used and power threshold level was set at $V_1 > (V_g)_{max}$). The gas-phase velocity measurements were made with $V_1 \approx V_0$. A block diagram of the auxiliary pedestal discriminator developed by Modarress et al. [112] for eliminating the cross-talk is shown in Figure 49.

Time of Flight Measurements

Essentially all the optical techniques described for studying particle dynamics in two-phase flows can be extended to measure transit times between two measurement points. From a measurement of the residence time within one of the dual probe volumes, particle sizes can be obtained. Recall from the discussions on cross-correlation techniques, the relationship between velocity, residence time and particle size is:

$$V_p = \frac{d + \ell}{t_R} \tag{57}$$

where V is the particle velocity, ℓ is a characteristic width of the probe volume, t_R is the residence time, and d is the particle diameter or cord length. Distance ℓ must obviously be minimized otherwise it will obscure the accuracy of d. It should also be minimized to prevent coincidence.

These techniques appear well developed, and there are many successful examples in the literature. By miniaturizing the optical probe used for studying bubble rise velocity in a fluid bed, Oki et al. [118] records the transit time between two pairs of fiber optics transmitters/receivers as illustrated in Figure 50. From the time of flight between the pairs, the velocity can be obtained, and then from the residence time within the vision of one pair the diameter is computed as given by Equation 57. The time of flight (i.e., transit time) depends on several factors—velocity and particle size, but also on the distance between the particle and the probe(z), and the eccentricity of the particle from the probe centerline (y). These last two factors call for the use of three observation fibers surrounding a transmitter to provide sufficient information on transit and residence times to enable calculation of diameters for any y and z coordinates. Again, correlation techniques can be used to evaluate a mean particle size. A distribution of sizes can also be obtained by computing the size of each particle and sorting the information. This approach will also provide a distribution of particle velocity. An inherent assumption with this approach is that particle motion is unidirectional.

Another example is the work of Lading [119] who employed two parallel beams a short distance apart in the flow direction and detected the obscuration of the beams. Correlation techniques were employed to evaluate mean velocities and a method for obtaining the particle size versus particle velocity spectrum is given.

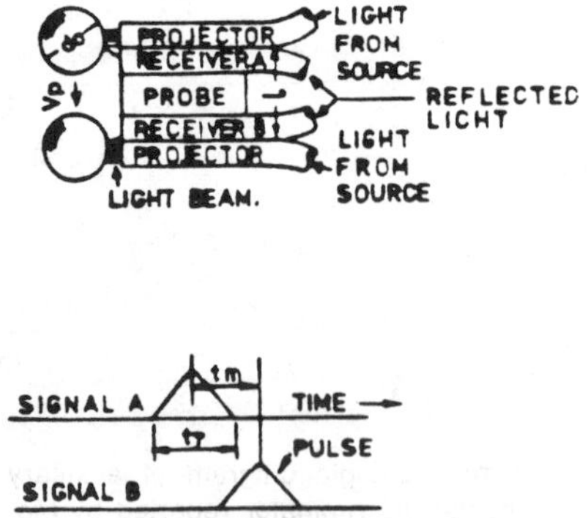

Figure 50. Time of flight technique used by Ohki et al. [118].

Techniques described earlier (in particular, that of Ritter el al. [90] and Lafrance et al. [91], can be extended to provide time of flight measurements. These can be arranged to detect transit times between beams, which will thus yield velocities, and from residence times within one beam particle sizes are computed. Color edging techniques described earlier can be used to eliminate particles not totally in the beam for cases where a particle chord and not a diameter is measured.

Wigley [109] devised an elaborate time of flight measurement technique using forward and backscatter signals from a laser anemometry system shown in Figure 51. The system provides a probe volume having a small dimension in the transverse direction of the particles. By means of a backscatter signal, the velocity is measured in the usual manner. In the forward direction, when a particle enters the probe volume of beam A, it is reflected at the glancing angle through the slit into the photo multiplier. As the same particle leaves the probe volume, the event reoccurs at beam B, but at a displaced time. The two separate events produce photo multiplier signals that detect the start and finish of the time of the residence. The slit aperture is set so as to ensure that signals are only generated by particles passing through the center of the probe volume. In applying

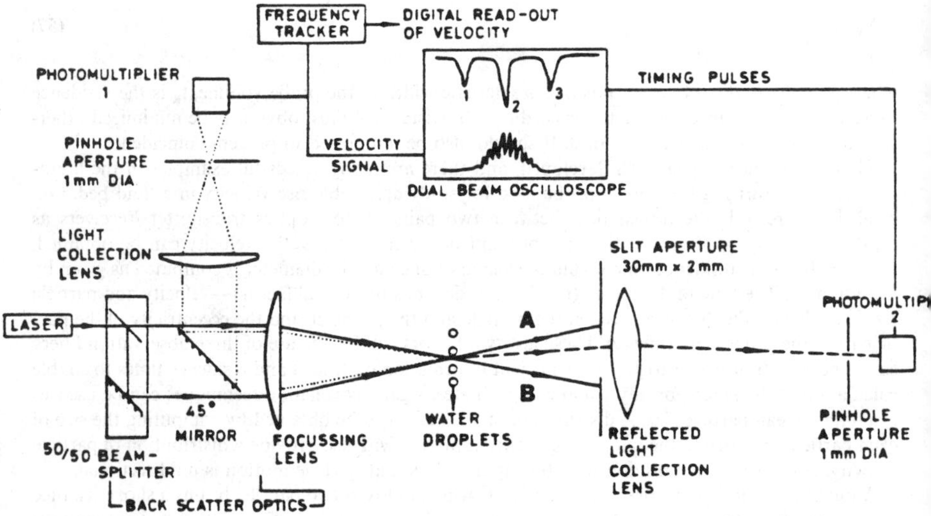

Figure 51. Laser anemometer system described by Wighley [109].

Equation 57, ℓ is considered negligible, and the relevant particle size thus obtained from the measured velocity and residence times. Wigley [109] applied the technique to studying droplet sprays and found it a suitable method for particle sizes in the range of 100 $\simeq$ 1,000 μm.

Specialized Photographic Techniques

Photography is perhaps the most widely used analytical tool in the laboratory for gaining insight and understanding of complex flow phenomena. The specific arrangements and detailed set-up required to obtain useful data are, however, unique to the system under investigation. In the previous section various photographic methods for studying dispersed flows were described and some generalizations in setting up a measuring system given. However, methods of high-speed photography require a fair amount of experimentation before a reliable measuring system can produce information of high quality. There are specialized cameras and techniques that can provide very detailed information on flow structure, however, the capital investment for such elaborate systems may go well beyond the practical budgets of industrial research laboratories. Fortunately, some of the techniques developed for specialized cameras have been successfully adapted to conventional cameras. The following techniques seem well adapted to studying interfacial phenomenon in stratified type flows and jet formation and breakup. High-speed photography provides a means of obtaining useful information on the interfacial turbulent structure leading to entrainment and deposition of droplets and the formation of atomized flows from nozzles.

To appreciate the problems associated with photography consider a water jet emerging from a nozzle. Typical discharge velocities can be on the order of 30 m/sec. To photograph the jet, the system can be illuminated with an electronic flash, which typically has a duration of 10 μsec. However, during the time of the flash, the jet has travelled a distance of 0.3 mm. To obtain greater resolution of the flow structure on the negative, a 0.75 mm distortion or blur would be observed using a 2.5 × magnification. This image blur would obscure a large portion of the fine detail. Even with faster flashes the basic problem still remains.

Taylor et al. [120] describe an approach to overcome distorted images by compensating for the predominantly uni-direction of the jet by moving either the film or the image at a speed which makes the image appear stationary on the film at the instant of exposure. The principle behind this method is illustrated in Figure 52 and is described in detail in References 120 through 122. The technique of image-motion compensation is shown to produce a substantial increase in image quality.

The moving-film camera described by Taylor and Hoyt [120] used commercial cut film. The system employed individual 57 mm × 82 mm (2° × 3° inch) cut film pieces accelerated to the image speed by rubber rollers. The film was passed behind the lens at a constant speed and the rollers were driven by a variable-speed d.c. motor with speed control and tachometer readout. The peripheral speed of the rollers could be matched to the image speed corresponding to the known jet speed and the magnification in the camera. Rollers must be operated at the predetermined and measured speed before the exposure.

When the moving-film speed is well matched to the image speed, and ample light is provided, a very short exposure is obtained by using a slot or slit in a plate covering the film plane. This slot is then analogous to the slit in a focal-plane shutter typically used in commercial single-lens reflex cameras. With a moving-film camera, however, much higher linear speeds can be achieved since the rollers are up to speed, and only a small piece of film need be accelerated.

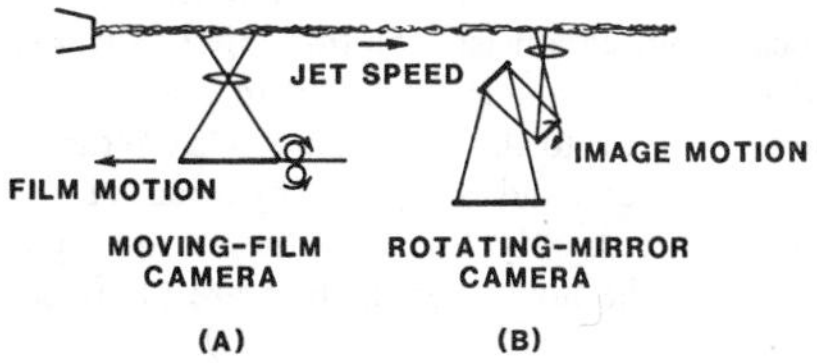

Figure 52. Principle of image-motion compensation in photography [120].

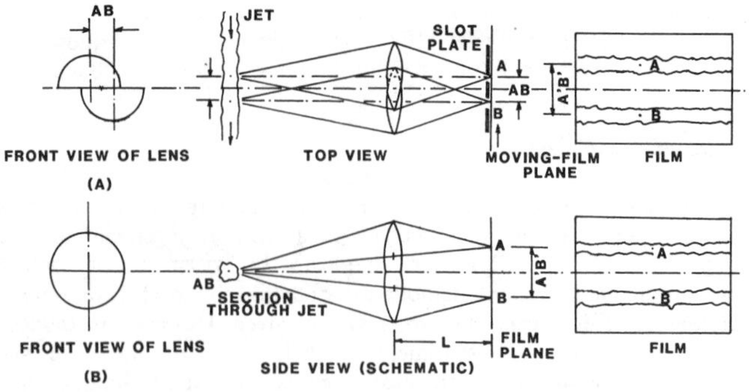

Figure 53. Split lens configurations for obtaining two exposures on the same film [120].

The rotating-mirror camera system shown in Figure 52b moves the image instead of the film. This technique provides good definition on stationary and axially moving objects. The arrangement requires the rotating mirror be in dynamic mechanical balance to avoid vibration. A stationary second mirror is used to correct the image and to bring the film plane away from the object being photographed. The flash is activated by a rotary switch driven by the rotating-mirror shaft at the exact instant when the image is formed on the film plane. Results obtained by analyzing photographs of jets using the rotating-mirror camera technique are given in References 123 through 125.

In many dynamic or transient type flows, it is useful to compare photographs taken in critical flow regimes at a few milliseconds to microseconds apart. When such comparisons are necessary, it is best to position two photos on the same piece of film. Consecutive exposures can be made by either following the phenomenon as it moves through the test section or by recording events at the same location separated in time. Consecutive exposures of the same objects as they move can be accomplished in the moving-film slot camera method. This can be done by using two lenses and two slots positioned one above the other. In this manner, two images will appear on the film, each occupying one half of the film height. Each lens is focused on the phenomenon, a slight distance apart, and its field covers one of the two slots. Physically, this arrangement is awkward since the lenses may interfere with one another. It is therefore necessary to trim the vertical dimension of each lens as recommended by Taylor [120]. This is illustrated in Figure 53a. Lenses can be split in half using diamond optical saws. The optical quality of the image and lens focal length are unaffected when they are cut in half. The f-number is however reduced by half. Figure 53a shows the two-slot, moving-film camera in which photos are taken of two side-by-side locations of the flow phenomenon. Figure 53b shows the rotating mirror system using a conventional camera. In this case, photos are taken of the same location but at two different times. It should be noted that distance L is 2 $\sim$ 4 times the object-to-lens distance and the vertical distance between lens centers is small. Special lighting is required for image separation using the latter approach. The lighting must be "compartmented," either physically or optically, so that light from one flash enters one lens and exposes one view of the flow scene, while the second flash, occurring at a specified time later, enters the second lens only and exposes the second view.

A dual-flash lighting scheme for obtaining time-separated photos by color differences is also possible. This is shown in Figure 54 where the center beam splitter consists of a thin mirror silvered on both sides with approximately 50% of the silvered area on both sides removed in a pattern of 4 mm diameter dots. The filter numbers shown in the figure refer to the Eastman Kodak Wratten series. In this arrangement, a red filter covers half of the film plane and is exposed by a red-filtered flash. The other half of the film is exposed by a blue-filtered flash only. Since the flashes are separated in time, time-separated exposures of the same area of the flow phenomenon are obtained. These can be done on black and white film.

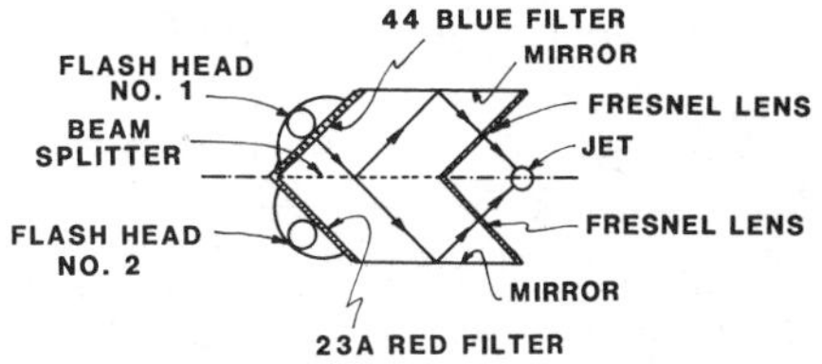

Figure 54. Dual-flash lighting scheme described by Taylor and Hoyt [120].

The importance of proper lighting to obtaining good quality pictures of high-speed events cannot be underemphasized. This often requires a great deal of experimentation before lighting conditions are optimized. In general, frontal lighting produces a more three-dimensional appearance to the object and is especially good in delineating the interfacial wave structure in flows and jets. Back lighting tends to illuminate objects in a more visible manner.

To produce the effect of a very bright object contrasted against a dark background, an opaque strip can be placed in the flow plane so that most of the light comes from the sides. Electronic-flash photography has been found to be very effective when simultaneously lighting areas 45° above and below the centerline of flow [120].

In many cases, the flow situation may be comprised of a transparent fluid. One example is a water jet discharging into a column of water. If the interest is to photograph the cavitation phenomenon accompanying the jet discharge, the method based on Kohler illumination is employed. Kohler illumination is used in microscopy and is also referred to as transillumination in high-contrast enlargers [126]. A dual transillumination setup for jet cavitation photography is shown in Figure 55. In this figure, A and B are the electronic flash units, C is a condenser, A′/B′ is camera lens.

Cavitation bubbles tend to interfere with the light transmission by refracting and reflecting the light so that they appear as black photographs. Flashes from the two flashheads can be separated in time, electronically, so that two images with a known time separation are formed on the same film frame. The electronic flash units can be operated at low power because of the high illumination efficiency and thus the flash duration can be limited to about 4 μsec. The camera lens is focused on the cavitation bubbles to form the image. A tracer dye can be injected in with the jet to reveal the upper boundary of the jet for reference purposes. Further details are given in Reference 127.

Another useful approach to obtaining clear photographs is to make the flow part of the optical system itself. Taylor has done this with jets using a tracer injection. Backlighting causes reflections from the cylindrical jet surface but no jet penetrates the jet. Figure 56 shows the optical arrangement in which an auxiliary half-cylinder lens is used whereby an intense line of light is focused near the edge of the jet on the plane of its axis. The jet itself actually becomes a cylindrical lens. Most of the light is refracted through the jet revealing the path of the tracer particles within.

When photographing through cylindrical test sections, the curvature of the test vessel wall can distort the images viewed by the camera. One approach to overcoming this problem is to construct a box around the viewing system and filling it with a fluid whose index of refraction will enable

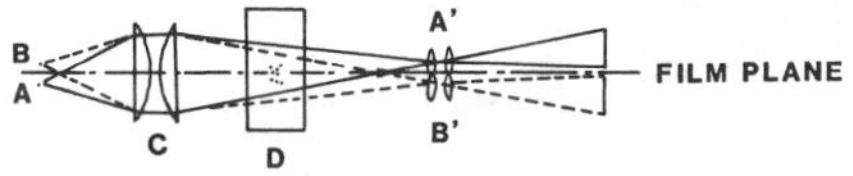

Figure 55. Dual transillumination setup for jet cavitation photography.

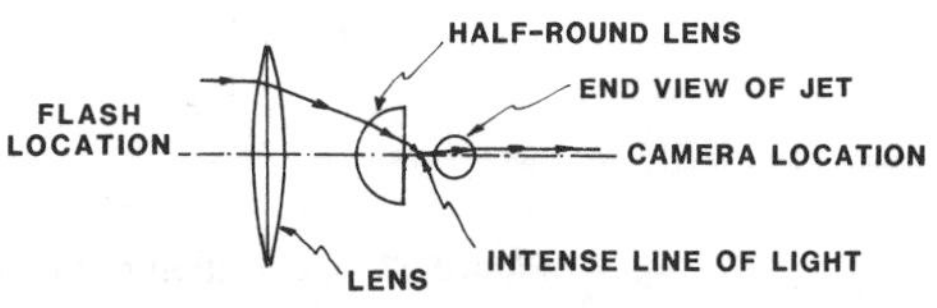

Figure 56. Light path arrangement for flow visualization studies inside a jet [120].

most of the light to be refracted through to the flow. Again, tracer particles can be used to highlight the phenomenon's recording.

Holographic Techniques

Whereas photography provides a record of events and details of the flow structure in two-dimensions, holography freezes three-dimensional objects and scenes. A hologram is created by the light scattered by an object and light that is unaffected by the object. The principle behind the formation of a hologram is shown in Figure 57. A beam of coherent light is split into two parts and illuminates the object in its path. The light is reflected by the object (a droplet for example) onto a photographic plate on which the hologram is to be produced, and arrives at the same instant as a reference beam that is deflected around the particle unchanged in phase. A phase difference between the two beams causes interference from whence the recorded hologram is composed of zones of high and low intensity. To examine the hologram, reconstruction is needed, which is done by illuminating the hologram with the reference beam from the same direction as during the recording experiment. The hologram itself serves as a diffraction barrier or screen for the reference beam, thus producing a wave pattern. The end result is a three-dimensional picture of the original object.

There are several disadvantages with holography. The principal limitations are:

1. The lack of sensitivity in low density flow fields,
2. Time-varying distortions produced by gaseous boundary layers on walls, windows, etc. in the optical paths and their deleterious effects on time-averaged or multiple-exposure holography,
3. Susceptibility to mechanical vibrations in the test facility or its environment.
4. The amount of light required for large facilities.
5. The required light source coherence lengths (i.e., beam purity).

Holography has been used successfully in single fluid flow studies such as wind tunnel experiments. In this case, real-time flow visualization studies are achieved using holographic moire patterns. The moire principle involves the production of a very fine system of fringes that is magnified to produce real-time visible interferometric-type fringes that can be readily recorded. The production of moire fringes is illustrated in Figure 58. As shown, two sets of straight-line interference patterns or grids are superimposed. This arrangement resembles the interferometer in that two coherent beams are superimposed to produce an interference pattern. The first set of fringes is recorded on a photographic plate with no disturbance in either beam. The hologram is then repositioned in its original spot and a second set of live straight-line fringes from the system is

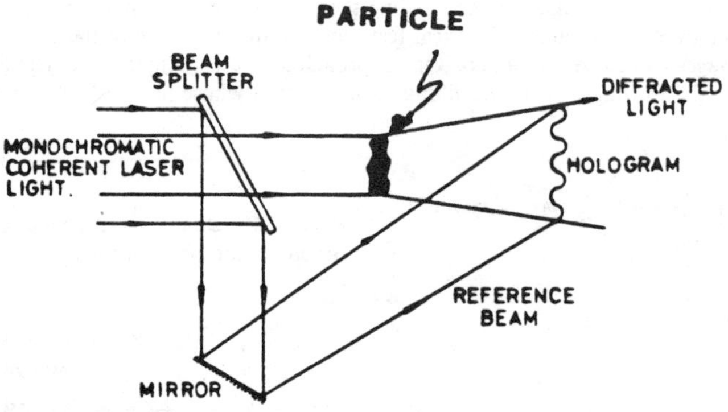

Figure 57. Illustration of the formation of holograms.

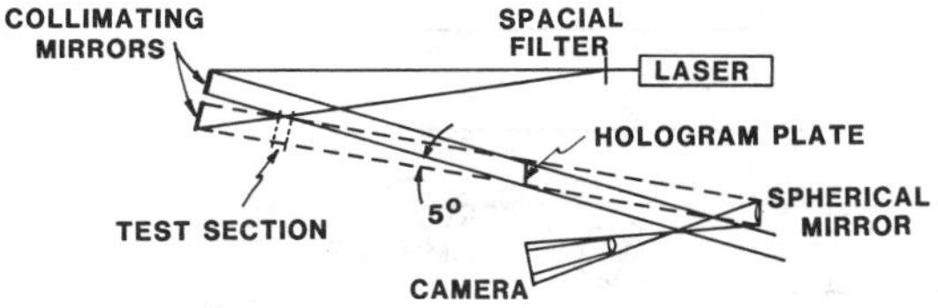

Figure 58. Formation of holographic Moire fringes using interferometer arrangement.

superimposed over the first set. A disturbance in either beam causes the Moire fringes to be displaced as with an interferometer and hence, an indication of the real-time flow phenomenon is obtained.

Another application of three-dimensional holographic interferograms is in studying turbulent boundary layers on the windows of test sections. The ground glass used to illuminate the subject provides light traversing the subject from all different directions. If the subject possesses sufficient axial symmetry such that the interference patterns for different directions are basically the same, then the interference patterns appear to focus at the subject's center plane when viewed with a large aperture lens. Contributions from turbulence in the boundary layer can however affect such direction of view differently. When a sufficiently large aperture is employed (of a size such that a large sample of the boundary layer is averaged), an accurate view of the subject interferogram is obtained. Turbulence has the effect of reducing the contrast of the subject fringes, however it does not displace them.

Holographic techniques for studying dispersed and reacting-atomized flows are possible but experimentation requires an elaborate set-up. One area of practical interest is in combustion research where fuel and an oxidizer are brought into intimate contact through an injector (nozzle) and then sprayed into the combustion chamber. The chemical reaction is coupled within the chamber producing high-temperature product gas. Information on reacting-spray particle-size distributions defines the rates at which reactions take place and hence, is fundamental design data. The problem crosses many practical processes such as liquid rocket propulsion, fuel-oil furnaces, recovery furnaces. Combustion rates in these examples are closely related to the initial atomization processes. An understanding of these processes and the ability to characterize the mass and particle distributions are prerequisites to specifying engineering design criteria for nozzle injectors. Such systems are extremely difficult to scale-up from cold model studies and hence, the most reliable information must be obtained from experiments in environments that most closely approximate the real system. Most studies have employed non-reacting fluids, however, combustion tests alone do not always correlate with nonreactive results. Hence, there is a real need to correlate nonreacting- and reacting-spray properties. Holography is one approach to obtaining this type of information because of its ability to record a three-dimensional scene depicting complete particle size and spacial distribution. This type of study requires the use of a high-power, pulsed-laser light source. The high light intensity of the laser can penetrate a reasonably dense optical path through a combustion environment. One example of this type of study is that of Clayton et al. [128].

Perhaps the greatest advantage of holography (and interferometry) is that it provides information on both flow phenomena and heat transfer. Examples of areas where holography has been used are in the investigation of convective gas paths and turbulence in flowing gases and in bubble formation in incipient nucleate boiling at water-cooled surfaces (in this case, cooling is by liuid-to-gas phase change). Early studies were based on flash holography to record heat fluxes from surfaces [129].

Other examples of holographic applications include flow field studies of small free jets and supersonic nozzles. These flow systems require the use of double-exposure holograms and provide such information as pressure ratios in the nozzle throat. One optical arrangement suitable for flow visualization studies is illustrated in Figure 59. The set-up employs a servocontrol system that stabilizes the holographic fringe pattern, and a variable ratio beam splitter. The light is split into the signal and reference beams before being passed through the test section. The signal beam is sent through a prism for spacial matching and then through a polarization rotator, which alters the polarization of the signal beam light to the same plane as that of the reference beam. A telescope

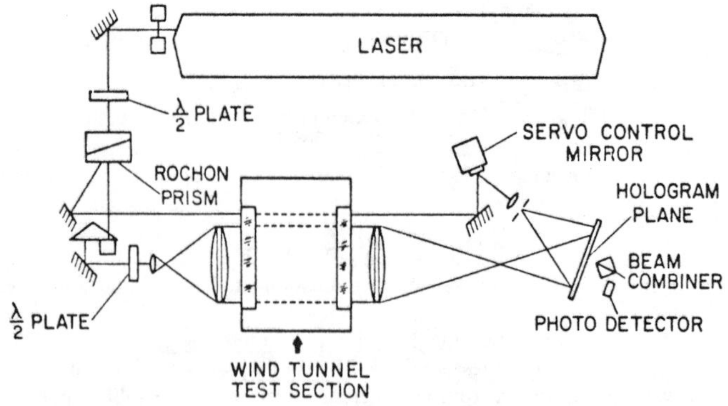

Figure 59. Schematic of holographic flow visualization system [126].

arrangement is used to enlarge the beam diameter. The enlarged beam then passes through the test section after being refocused, and intersects onto the hologram plane. The other beam simultaneously passes through the test section above the flow disturbances. It is reflected from a piezoelectric servo-control mirror and passes through a spacial filter and forms the reference beam.

Although holography has a number of limitations it appears well suited for two-phase flow studies within the range of industrial applications. The technique allows sensing of discrete particles in flow to sizes below 10 μm. In these applications there appears to be little restriction on the depth of field and a uniform magnification of particles longitudinally can be achieved when the reconstructed hologram is projected. As a rule of thumb, the depth of field should be 100 d^2/λ, where d is the particle size and λ the wavelength of the illumination [76].

In order to ensure that the particles appear as stationary objects, movement during the formation of the hologram should be restricted to 10% of the fringe spacing. Pulsed ruby lasers are capable of producing an exposure of approximately 30 ms. This essentially renders motionless particles moving with velocities in the upper range of 100 ms^{-1}.

Holography has not been extensively applied to studying multiphase flows, however, the technique may become more widely used as it evolves, especially in studies aimed at coupling heat and mass exchanges with system hydrodynamics.

A summary of the particle flow measurement techniques and range of measurements for light scattering techniques is given in Table 1.

NOTATION

A	area
a	sphere radius
C	constant
C_{so}	particle scattering cross section
D	plate diameter
D(r, x)	weight function
d, d_p	particle diameter or size
E	exposure—amount of light covering photographic plate; or normalized threshold voltage
$\lvert E_s \rvert$	magnitude of complex field function
F(θ)	illumination function
f_D	Doppler frequency
g	gravitational acceleration
g(l)	probability that measurement yields pierced length
H	expanded bed height
h	vertical position
I	light intensity
I_0	incident illumination
i	imaginary number ($\sqrt{-1}$)
J_1	first-order Bessel function
K_E	fraction of particle permitted to move in photography
k	coefficient
L	vertical length

Notation continued on page 1334.

Table 1
Summary of Light Scattering Techniques for Two-Phase Flow Investigations

General Basis of Method	Detailed Basis of Method	Number or Mass Flux Data	Single Particles Counted	Distribution of Mean	Type of Distribution	Size Range μm	Disadvantages/ Difficulties in Use	Advantages
Photography	Photography	Number and mass flux	Yes	Distribution	Spatial	5 → mm	1. Errors in data abstraction 2. Data abstraction tedious 3. Subjective judgement involved in selection of 'in-focus' drops 4. Very high quality photographs needed for automatic or semi-automatic data abstraction 5. Depth of field problems often considered	1. Very simple equipment required 2. Non disturbing to to some flow fields (if unenclosed)
	Photography related methods	Number and mass flux	Yes	Distribution	Spatial/ Temporal			
	Holography	Number and mass flux	Yes	Distribution	Spatial	2 → mm	1. Same as photography 2. Requires more equipment than photography	1. Freeze 3-D scenes which can be analyzed at leisure 2. No depth of field problems
Optical	Scattering by single particle at one angle	Number and mass flux	Yes	Distribution	Temporal	500–2,000	1. Problems of coincidence edge effects 2. Uses absolute intensity measurement	

(Continued)

Table 1 (Continued)

General Basis of Method	Detailed Basis of Method	Number or Mass Flux Data	Single Particles Counted	Distribution of Mean	Type of Distribution	Size Range μm	Disadvantages/ Difficulties in Use	Advantages
	Scattering by single particle-ratio methods	Number and mass flux	Yes	Distribution	Temporal	<1	1. Method insensitive above 1 μm	
	Multiple particle scattering—diffraction	—	No	Distribution	Spatial/ Temporal	2–1,000		1. Sensitive in changes in mean width of distribution
	Obscuration	Number and mass flux	Yes	Distribution	Spatial/ Temporal	50–1,000	1. Uses absolute intensity measurements 2. Not space specific	1. Signal/diameter relation linear for large particles
	Obscuration—detector counting	Number and mass flux	Yes	Distribution	Spatial/ Temporal	10–100	1. Not space specific	1. Does not use absolute intensity measurements
	Laser doppler—Visibility	Number and mass flux	Yes	Distribution	Temporal	<10	1. Ambiguity if fringe spacing $\nless Kd_{max}$ 2. Simple relationships only exist for limited conditions, e.g. forward scatter 3. Analysis and relationship very complex in backscatter	1. Also provides velocity
	Laser doppler—Variable fringe spacing	Number and mass	Yes	Distribution	Temporal	Not known	1. Variable fringe spacing difficult to produce 2. Point of equivalent size difficult to determine	
	Laser doppler—phase lag	Number and mass flux	Yes	Distribution	Temporal	Not known	1. Very small phase differences to be measured	1. Also provides velocity

	Laser doppler—envelope modulation	Number and mass flux	Yes	Distribution	Temporal	Not known	1. Expected modulations not seen in experiments	
	Sampling followed by optical analysis	Number and mass flux	Yes	Distribution		1–80	1. Sampling disturbs flow structure	
	Fibers	Number and mass flux	Yes	Distribution	Temporal	100–400	1. Probe in flow stream 2. Slow removal of particles dense sprays 3. Data abstraction method incomplete 4. Lower limit set by noise	
Time of Residence	Fibers	Number and mass flux	Yes	Distribution	Temporal	200–1,000	1. Probe in flow 2. Distance of particles from probe and from probe centre line important but not allowed for (can be overcome)	1. Also provides velocities
	Laser anemometer	Number and mass flux	Yes	Distribution	Temporal	100–1,000	1. Needs good access	1. Also provides velocities 2. Can identify non-spherical particles
	Scanning beam	Number and mass flux	Yes	Distribution	Temporal	10–1,000	1. Limited to flow through narrow cell 2. Measures chord length 3. Not space specific	
	Beam interrupt	Number and mass flux	Yes	Distribution	Spatial/Temporal		1. Not space specific 2. Coincidence and edge effects 3. Can measure chord	

ℓ	separation between dual probes
M	magnification factor of receiving optics
M_s	mass of solids in bed
m	variable or index of refraction
N	data rate
N_0	cummulative number size distribution
N_0, N_c	flash and photomultiplier signal frequency, respectively
n	wave impedance
n(x)	cumulative density function of bubble width
P	pressure
$\bar{P}$	polarization vector
p	parameter in Equation 12
Q_0	bubble size distribution
R	radial position of particle center
R_l, R_v	position R corresponding to V_l, V_u
r	radius
r_1, r_2	Fresnel reflection coefficients
S	area
s	probe separation distance
T_t	bubble rise time
t_B	bubble duration time
t_R	residence time
$U_{A,B}(t)$	probe output signal
u_B	bubble rise velocity
V	signal amplitude
$\tilde{V}$	visibility
V_p	particle velocity
V_0	signal amplitude for seeding particle
V_l, V_u	lower and upper threshold limits
W_0	frequency of laser beam
x, z	coordinates

Greek Symbols

α	scattering coefficient
γ	angle; fluidized bed density
δ	fringe spacing; plate separation
ϵ_0	optical wave amplitude
ϵ_B	bubble voidage
ρ	density
θ	angle
λ	wavelength; also particle arrival rate per unit area
v_g	volume of seeding particle
ρ	density
σ	phase angle between E_{s1} and E_{s2}; also particle scattering amplitude function
τ	time
τ', τ''	angle
ϕ	phase angle
ψ	discrete-phase to gas-phase mass ratio
ψ	ratio of discrete-phase to gas-phase data rate

REFERENCES

1. Gerald, C. F., *Chem. Eng. Prog.*, 47, 199 (1951).
2. Taylor, P. A., Lorenz, M. M., and Sweet, M. R., in *International Congress on Fluidization and Its Applications*, Toulouse, France, (1973).
3. Broadhurst, T. E., and Becker, H. A., in *Fluidization Technology*, Vol. 1, Cambridge University Press, London (1978).
4. Kang, W. K., Sutherland, J. P., and Osberg, G. L., *I&EC Fundamentals*, 6, 499 (1967).
5. Littman, H., and Homolka, G. A. J., *Chem. Eng. Progr. Symp. Series*, 105, Vol. 66, 37 (1970).
6. Park, W. H., Kang, W. K., Capes, C. E., and Osberg, G. L., *Chem. Eng. Sci.*, 24 (1969).
7. Rigby, G. R., Van Blockland, G. P., Park, W. H., and Capes, C. E., *Chem. Eng. Sci.*, 25, 1729 (1970).
8. Burgess, J. M., and Calderbank, P. H., *Chem. Engr. Sci.*, 30, 1511 (1975).
9. Calderbank, P. H., Pereira, J., and Burgess, J. M., *Fluidization Technology*, Vol. I, McGraw-Hill Pub., NY (1976).
10. Barnea, D., Shoham, O., and Taitel, Y., *Int. J. Multiphase Flow*, Vol. 6, pp. 387–397(1980).
11. Jones, O. C., Jr., and Delhaye, J. M., *Int. J. Multiphase Flow*, 3, pp. 89–116 (1976).
12. Hsu, Y. Y., Simons, F. F., and Graham, R. W., "Application of Hot Wire Anemometry for Two-Phase Flow Measurements," ASME Winter Meeting, Philadelphia, PA (1963).
13. Jones, O. C., Jr., and Zuber, N., *Int. J. Multiphase Flow*, 2, pp. 273–306 (1975).
14. Govier, G. W., Radford, B. A., and Dun, J. S. C., *Can. J. Chem. Engrg.*, 35, pp. 58–70 (1957).

15. Chaudry, A. B., Emerton, A. C., and Jackson, R., "Flow Regimes in the Concurrent Upwards Flow of Water and Air," Paper presented at the Symp. Two-Phase Flow, Exeter, pp. 21–23 (1965).
16. Isbin, H. S., Moen, R. H., Wickey, R. O., Mosher, D. R., and Larson, H. C., *Chem. Engrg. Series*, 23, 55, pp. 75–84 (1959).
17. Hubbard, M. G., and Dukler, A. E., *Proc. 1966 Heat Transfer and Fluid Mechanics Institute*, M. A. Saad and J. A. Moller (eds.), pp. 100–121, Stanford Univ. Press, CA (1966).
18. Solomon, J. V., M. Sc. thesis, Mech. Eng. Dept., MIT (1962).
19. Griffith, P., AMSE Paper 64-WA/MT-43 (1964).
20. Fiori, M. P., and Bergles, A. E., "A Study of Boiling Water Flow Regimes at Low Pressure," Rep. 5382–40, Dept. of Mech. Engrg., MIT (1966).
21. Bergles, A. E., Lopina, R. F., and Fiori, M. P., *J. Heat Transfer*, 89, pp. 69–74 (1967).
22. Cheremisinoff, N. P., "An Experimental and Theoretical Investigation of Horizontal Stratified and Annular Two-Phase Flow with Heat Transfer," Ph.D. thesis, Clarkson College of Technology, Potsdam, NY (1977).
23. Morse, R. D., and Ballou, C. O., *Chem. Engr. Prog.*, *47*, 99 (1951)
24. Yoshida, K., Ueno, T., and Kunii, D., *Chem. Engr. Sci.*, *29*, 77 (1974).
25. Matsen, J. M., *AIChE Symp. Series*, 128, Vol. 69, 31 (1973).
26. Datson, M., *AIChE J.*, 5, 169 (1959).
27. Ormiston, R. M., Mitchell, F. R. G., and Davidson, J. F., *Trans. Inst. Chem. Engrs.*, 43, T209 (1965).
28. Angelino, H., Charzat, C., and Williams, R., *Chem. Engr. Sci.*, 19, 289 (1964).
29. Nguyen, X. T., Leung, L. S., and Weiland, R. H., *Chem. Engr. Sci.*, 30, 1187 (1975).
30. Baskakov, A. P., Viit, O. K., Kirakosyan, V. A., Makayev, V. K., and Filippovsky, N. F., *Int'l. Congress on Fluidication and Its Applications*, Toulouse, France (1973).
31. Thiel, W. J., and Potter, O. E., *I&EC Fundamentals*, 16, 242 (1977).
32. Catipovic, N. M., Fitzgerald, T., Javanovic, G., and Levenspiel, O., Heat Transfer to Horizontal Tubes in Fluidized Beds, EPRI Report, Palo Alto, CA (1979).
33. Lanneau, K. P., *Trans. Inst. Chem. Engrs.*, 38, 125 (1960).
34. Rooney, N. M., and Harrison, D., *Fluidization Technology*, Vol. II, McGraw-Hill Pub., NY (1976).
35. Werther, J., and Molerous, O., *Int. J. Multiphase Flow* 1, 103 (1973).
36. Werther, J., *Fluidization Technology*, Vol. I, McGraw-Hill Pub., NY (1976).
37. Cranfield, R. R., *Chem. Engr. Sci.*, 27, 239 (1972).
38. Mii, T., Yoshida, K., and Kunii, D., *J. Chem. Eng. Japan*, 6, 196, (1973).
39. Yoshida, K., Fujii, S., and Kunii, D., "Characteristics of Fluidized Beds at High Temperatures," in *Fluidization Technology*, Vol. 1, D. L. Keains (ed.), Hemisphere Pub. Corp., Washington (1976).
40. Otake, T., Tone, S., Kawashima, M., and Shibata, T., *J. Chem. Eng.* Japan, 8, 388 (1975).
41. Whittmann, K., Wippern, D., Hermrich, H., and Schugerl, K., *Fluidization*, J. F. Davidson (ed.), Cambridge Univ. Press (1978).
42. Yoshida, K., Sakane, J., and Shimizu, F., *Ind. Eng. Chem. Fundam.*, 21, 83–85 (1982).
43. Wen, C. Y., Krishnan, R., Khosravi, R., and Dutta, S., *Fluidization*, Cambridge Univ. Press, 32 (1978).
44. Baskakov, A. P., Vitt, O. K., Kirakosyan, V. A., Maskayer, V. K., and Filippovsky, N. F., *Int'l Congress on Fluidization and Its Applications*, Toulouse, France (1973).
45. Bernis, A., Coeuret, F., Vergnes, F., and Le Goff, P., *Int'l. Congress on Fluidization and Its Applications*, Toulouse, France (1973).
46. Goosens, W. R. A., and Hellinckx, L., *Int'l Congress on Fluidization and Its Applications*, Toulouse, France (1973).
47. Begovich, J. M., and Watson, J. S., *AIChE Journal*, Vol. 24, No. 2, pp. 351–354 (March 1978).
48. Achwal, S. K., and Stepanek, J. B., *Chem. Engrg. Sci.*, 30, 1443 (1975).
49. Achwal, S. K., and Stepanek, J. B., *Chem. Engrg. Journ.*, 12, 69 (1976).
50. Rigby, G. R., Van Blockland, G. P., Park, W. H., and Capes, C. E., *Chem. Engrg. Sci.*, 25, pp. 1729–1741 (1970).

51. Turner, J. C. R., *Chem. Eng. Sci.*, 31, 487 (1976).
52. Linneweber, K. W., and Blass, E., *Germ. Chem. Engr.*, 6, 28–33 (1983).
53. Put, M., Francesconi, A., and Goossens, W., *International Congress on Fluidization and Its Applications*, Toulouse, France (1973).
54. Yasui, G., and Johanson, L. N., *A.I.Ch.E. J.*, 4, 446 (1958).
55. Kilkis, B., DeGeyter, F. M., and Ginoux, J. J., *International Congress on Fluidization and Its Applications*, Toulouse, France (1973).
56. Yoshida, K., Nakajima, K., Hamatani, N., and Shimizu, F., *Fluidization*, 13, Cambridge University Press, London (1978).
57. Cheremisinoff, N. P., and Cheremisinoff, P. N., *Hydrodynamics of Gas-Solid Flows*, Gulf Publishing Co., Houston, TX (1984).
58. Lockett, M. J., and Safekourdi, A. A., *A.I.Ch.E. J.*, Vol. 23, No. 3, 395–398 (1977).
59. Cheremisinoff, N. P., "An Experimental Investigation of Pressure Drop and Heat Transfer in Two-Phase Flows," m.s. thesis Clarkson College of Technology, Potsdam, NY (1975).
60. Coulaloglou, C. A., and Tavlarides, L. L., *A.I.Ch.E. J.*, Vol. 22, No. 2, 289–297 (March 1976).
61. Ohki, K. and Shirai, T., "Particle Velocity in Fluidized Beds," in *Fluidization Technology*, Vol. 1, Cambridge University Press, London (1978).
62. Vanzetti, R., "Fiber Optics in Infrared Instrumentation and Control," in *Analytical Measurements and Instrumentation for Process and Pollution Control*, pp. 395–441, P. N. Cheremisinoff and H. J. Perlis (eds.) Ann Arbor Science Pub., Ann Arbor, MI (1981).
63. Dobbins, R. A., Crocco, L., and I. Glassman, *Am. Inst. Aeronaut. Astronaut. J.*, 1, 1982 (1963).
64. Deich, M. E., Saltanov, G. A., and Kurshakov, A. V., *Thermal Engng.*, 18, 127 (1971).
65. Deich, M. E., Tsiklauri, G. W., and Shanin, V. K., *High Temp.*, 16, 102 (1972).
66. Chin, J. H., Sliepcevich, C. M., and Trius, M., *J. Phys. Chem.*, Ithaca, 5, 841 (1955).
67. Cornillault, J., *Appl. Optics* 11, 265 (1972).
68. Switherbank, J., Beer, J. M., Taylor, D. S., Abbot, D., and McCreath, G. C., *Prog. Astronaut. Aeronaut.* 53, 421 (1976).
69. Kirkman, G. A., and Ryley, C. J., "The Use of Laser Photography for Measuring the Diameters of Entrained Droplets in Two-Phase Flow," Liverpool Univ., Dept. Mech. Engrg., Report (1969).
70. McCreath, C. G., Roett, M. F., and Chigier, N. A., *J. Phys. E. Sci. Instrum.*, 5, 60 (1972).
71. Findlay, I. C., and Welsh, N., *J. Photographic Sci.*, 16, 70 (1968).
72. Cox, A., *Photographic Optics*, Focal Press, London (1966).
73. Reddy, K. V. S., van Wijk, M. C., and Pei, D. C. T., *Can. J. Chem. Engrg.*, 47, 85 (1969).
74. Whalley, P. B., Azzopardi, B. J., Pshyk, L., and Hewitt, G. F., UKAEA Report, AERE-R-8787 (1977).
75. Dombrowski, N. and Weston, J. A., *J. Photographic Sci.*, 14, 215 (1966).
76. Azzopardi, B. J., *Intl. J. Heat & Mass Transfer*, 22, 1245–1279, (Sept. 1979).
77. Simmons, H. C., and Gaag, H. H., U.S. Patent 3609043 (1971).
78. Dix, M. J., Sawistowski, H., and Tyley, L. R. T., *Proc. 11th Inst. Congress High Speed Photography*; P. J. Rolls (ed.), Chapman and Hall Pub., London (1975).
79. Mie, G., Annln. *Physik*, 25, 377 (1908).
80. van de Hulst, H. C., *Light Scattering by Small Particles*, Wiley, New York (1957).
81. Kerker, M., *The Scattering of Light and Other Electromagnetic Radiation*, Academic Press, NY (1969).
82. Blau, H. H., McCleese, D. J., and Watson, D., *Appl. Optics*, 9, 2522 (1970).
83. Jacquez, J. A., *A First Course in Computing and Numerical Methods*, Addison-Wesley Publishing Co., Reading, MA (1970).
84. Landa, I., and Tebay, E. S., I.E.E.E., *Trans. Instrum. Measurements*, 21, 516 (1972).
85. Keller, A., Trans. A.S.M.E., *J. Basic Engrg.*, 94, 917 (1972).
86. Mason, B. J., and Ramanadham, R., *Q.J.R. Meteorol. Soc.*, 79, 490 (1953).
87. Shofner, F. M., Kreikebaum, G., Schmitt, H. W., and Barnhart, B. E., Fifth Annual Industrial Air Pollution Control Conference, Knoxville, TN (1975).
88. McSweency, A., and Rivers, W., *Appl. Optics*, 11, 2101 (1972).

89. Shafner, F. M., Kreikebaum, G., and Schmitt, H. W., 68th Air Pollution Control Assoc. Meeting, Boston (1975).
90. Ritter, R. C., Zinner, N. R., and Sterling, A. M., *Phys. Med. Biol.*, 19, 161 (1974).
91. Lafrance, P., Aiello, G., Ritter, R. C., and Trefil, J. S., *Physics Fluids*, 17, 1469 (1974).
92. Schleusener, S. A., and Reed, A. A., *Rev. Scient. Instrum.*, 38, 1152 (1967).
93. Schleusener, S. A., *Powder Technol.*, 1, 364 (1968).
94. Shuster, B. G., and Knollenberg, R., *Appl. Optics*, 11, 1515 (1972).
95. Rhodes, C. A., Stowers, I. F., Hawkins, L., Bonnell, R. D., and Raines, W., *Powder Technol.* 14, 203 (1976).
96. Farmer, W. M., *Appl. Optics*, 11, 2603 (1972).
97. Farmer, W. M., *Appl. Optics*, 13, 610 (1974).
98. Fristrom, R. M., Jones, A. R., Schwar, M. J. R., and Weinberg, F. J., Faraday Symp. Chem. Soc., 7, 183 (1973).
99. Durst, F., Zare, M., and Karlsruhe, U., SFB-80/TM63 (1975).
100. Hong, N. S., and Jones, A. R., *J. Phys. D: Appl. Phys.*, 9, 1839 (1976).
101. Jones, A. R., *J. Phys. D: Appl. Phys.*, 6, 417 (1973).
102. Jones, A. R., *J. Phys. D: Appl. Phys.*, 7, 1369 (1974).
103. Robards, D. W., *Appl. Optics*, 16, 1861 (1977).
104. Adrain, R. J., and Orloff, K. L., *Appl. Optics*, 16, 677 (1977).
105. Chu, W. P., and Robinson, D. M., *Appl. Optics*, 16, 619 (1977).
106. Ungut, A., Yule, A. J., Taylor, D. A., and Chigier, N. A., A.I.A.A. 16th Aerospace Sciences Meeting, Huntsville, AL (1978).
107. Chigier, N. A., and Yule, A. J., Conf. on Physical Chemistry and Hydrodynamics, Oxford (1977).
108. Yule, A. J., Chigier, N. A., Atakan, S., and Ungut, A., Energy, J., 1, 220 (1977).
109. Wigley, G., UKAEA Report, AERA-R-8771 (1977).
110. Bachalo, W. D., Hess C. F., and Hartwell, C. A., Winter Annual Meeting of ASME, NY (Dec. 2–7, 1979).
111. Hess, C., Spectron Development Laboratories, 330 Harbor Blvd., Costa Mesa, CA, personal correspondence (1983).
112. Modarress, D., and Tan, H., *Experiments in Fluids*, Vol. 1, No. 3, Springer International Pub., London, pp. 129–134 (1983).
113. Lee, S. L., and Durst, F., *Int. J. Multiphase Flow*, 8, 125–146 (1982).
114. Yoeman, M. L., Azzopardi, B. J., White, H. J., Bates, C. J., and Roberts, P. J., "Optical Development and Application of a Two-Colour LDA System for the Simultaneous Measurement of Particle Size and Particle Velocity," Paper presented at the Winter Annual Meeting of Amer. Soc. Mech. Eng., Phoenix, AZ (Nov. 14–19, 1982).
115. Durrani, T. S., and Greated, C. A., *Laser Systems in Flow Measurement*, Plenum Press, London, pp. 118–129 (1977).
116. Modarress, D., Wuerer, J., and Elghobashi, S., "An Experimental Study of a Turbulent Round Two-Phase Jet," AIAA paper 82-0964 (1982).
117. Modarress, D. H., Tan, and Elghobashi, S., "Two Component LDA Measurement in a Two-Phase Turbulent Jet," AIAA paper 83-0052 (1983).
118. Oki, K., Akehata, T., and Shirai, T., *Powder Technol.*, 11, 51 (1975).
119. Lading, L., Third Biennial Symposium on Turbulence in Liquids, U. Missouri-Rolla, MO (1973).
120. Taylor, J. J., and Hoyt, J. W., *Experiments in Fluids*, Vol. 1, No. 3, Springer International Pub., pp. 113–120 (1983).
121. Taylor, J. J., Camera Apparatus for Making Photographic Images on Moving Cut Film Pieces, U.S. Patent 3, 925, 796 (1975).
122. Hoyt, J. W., Taylor, J. J., and Runge, C. D., *J. Fluid Mech.*, 63, 635–640 (1974).
123. Hoyt, J. W., Taylor, J. J., and Altman, R. L., *J. Rheology*, 24, 685–699 (1980).
124. Hoyt, J. W., and Taylor, J. J., *J. Fluid Mech.*, 83, 119–127 (1977).
125. Hoyt, J. W., and Taylor, J. J., *Physics Fluids*, 20, S253–S257 (1977).

126. Slater, E. M., *Optical Methods in Biology*, John Wiley & Sons Inc., New York, pp. 268–269 (1970).
127. Hoyt, J. W., and Taylor, J. J., *ASME Trans., J. Fluids Eng.*, 103, 14–18 (1981).
128. Clayton, R. M., and Wuerker, R. F., Paper 11 (pp. 117–127) in Holographic Instrumentation Applications, NASA SP-248, at Ames Res. Center, Moffett Field, CA (Jan. 13–14, 1970).
129. Miller, C. G., and Stephens, J. B., Paper 18 (pp. 205–212) in Holographic Instrumentation Applications, NASA 2 SP-248, Conf. at Ames Res. Center, Moffett Field, CA (Jan. 13–14, 1970).
130. Brown, R. M., paper 19 (pp. 213–219) in Holographic Instrumentation Applications, NASA SP-248, Conf. at Ames Res. Center, Moffett Field, CA (Jan. 13–14, 1970).

CHAPTER 39

INDUSTRIAL FLOW MEASURING DEVICES

N. P. Cheremisinoff

Exxon Chemical Co.
Linden, New Jersey, USA

CONTENTS

INTRODUCTION

Flow measuring devices are employed in a wide range of industrial applications. Under laboratory conditions, high precision is often required. In process operations, accurate measurements may be necessary for evaluating materials-handling equipment. Less precise measurements may be used for material balances around various unit operations such as distillation, extraction, polymerization, or in waste stream processing. Regardless of the intended application, it is essential that all the variables be considered in selecting the proper flow measuring device. Variables that should be considered are meter accuracy, maintenance requirements, type of indication and flow control units, flexibility, resistance to materials being processed, properties of the fluid, and system economy.

The major types of flow measuring devices discussed in this chapter are as follows:

1. Differential pressure meters
   - Venturi
   - Flow nozzles
   - Orifice plates
   - Pitot tube
2. Positive displacement meters
   - Reciprocating piston
   - Nutating disk
   - Rotary piston
   - Rotary vane
3. Mechanical meters
   - Rotameter
   - Turbine
4. Acoustic
   - Travel time difference method
   - Beam deflection method
   - Doppler method
5. Thermal
   - Hot wire/hot film
6. Open channel techniques
   - Weirs
   - Flumes

DIFFERENTIAL FLOWMETERS

Differential pressure flow measuring instruments are referred to as head meters or rate meters. Their primary characteristics are that they measure the flow rate without sectioning the fluid into isolated quantities. The primary device creates a differential head that is a function of the fluid velocity and density. Many devices are based on the differential head method—Venturi meters, flow nozzles, orifice meters, and pitot tubes.

Venturi Meter

Figure 1 illustrates the basic design of a Venturi meter. As fluid passes through the reduced area of the Venturi throat, its velocity increases, resulting in a pressure differential between the inlet and throat regions. That portion immediately following the throat gradually increases in flow area; consequently the fluid's velocity decreases, causing pressure recovery. The differential pressure across the Venturi's throat can be recorded directly or translated into actual flow units (e.g., gallons per minute (GPM) or liters per second) by employing various types of differential pressure meters and capacity curves.

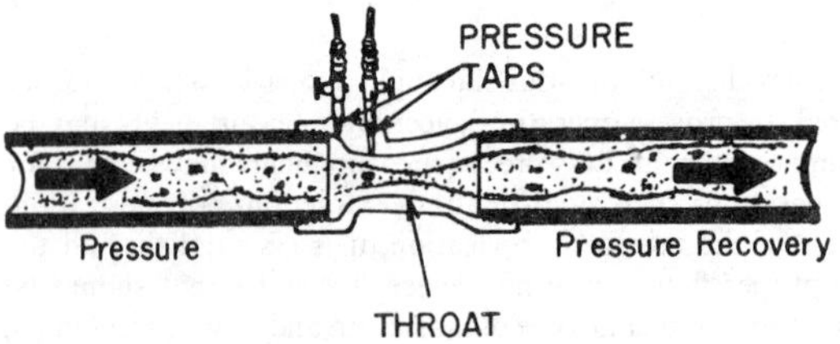

Figure 1. Operation of the Venturi meter.

Venturi tubes are employed whenever the permanent loss of pressure is to be reduced to a minimum or in cases where fluids handled contain sufficient amounts of materials in suspension that other devices such as orifice plates or flow nozzles are not effective. These systems are recommended for use where metering conditions require relatively low pressure loss. Widest application has traditionally been in low-pressure gas lines and water mains. Its streamlined design makes the Venturi suitable for metering liquids with solids in suspension. Venturi tubes are employed in systems that use hot and chilled water for heating and air conditioning. They are used in office buildings, manufacturing plants, hotels, or other edifices where comfort is an important criterion. Units are used in the processing industries where piping systems transport chemicals.

In many fan systems, Venturi tubes are positioned in return lines for the purpose of reheating coils. Usually a balancing valve is installed downstream to allow accurate balancing [1]. Venturi meters have long been used for measuring sewage flow. At each piezometer or annular chamber on the units, valves are positioned so that pressure openings can be closed. The valves are designed so that, upon closing, a rod is forced through the opening to clean out any debris that clogs. When all the valves are closed, plates covering the hand-holds in the pressure chamber can be dismantled and the chamber flushed.

Whenever a finite length exists between two pressure taps, the differential developed is the sum of the energy-transfer effect and the frictional effect. For a Venturi tube there is an appreciable distance between the taps and a persistent decrease in diameter. Both these factors exert a significant impact on the differential developed in cases where fluid-friction values are high. For solids in suspension, fluid friction and the differential pressure are sensitive to the solids concentration.

The American Society of Mechanical Engineers (ASME) [2, 3] has standardized the construction of these types of head meters. Figure 2 provides the ASME-recommended specifications fo Venturi meters. The design calls for pressure taps to be connected to manifolds that encompass the upstream and throat portions of the tube. The manifolds receive a portion of the pressure all around the sections so that an average differential can be obtained. A piezometer ring is inserted at the large end of the inlet section, and the determination of fluid flowing is based on the difference in pressures indicated at this point and at the throat.

In general, variations from these specifications in the outlet cone have a small effect on the discharge coefficient, but they do impact on the pressure recovery. The outlet cone may, however, be truncated considerably on low diameter ratios without seriously impacting on the pressure recovery (refer to Figure 3). In applications where there are space limitations a truncated cone Venturi tube is to be preferred over the full-length exit cone. (See Figure 4 for other Venturi designs.)

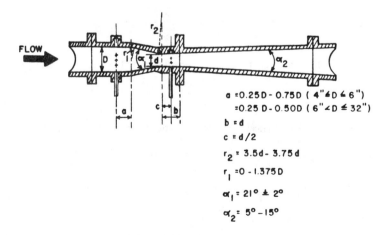

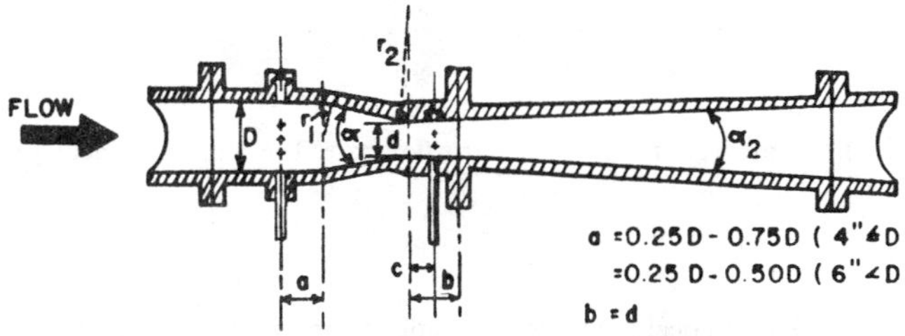

$a = 0.25D - 0.75D$ ($4'' \leq D \leq 6''$)
$= 0.25D - 0.50D$ ($6'' < D \leq 32''$)
$b = d$
$c = d/2$
$r_2 = 3.5d - 3.75d$
$r_1 = 0 - 1.375D$
$\alpha_1 = 21° \pm 2°$
$\alpha_2 = 5° - 15°$

Figure 2. Recommended specifications for Venturi tubes [2].

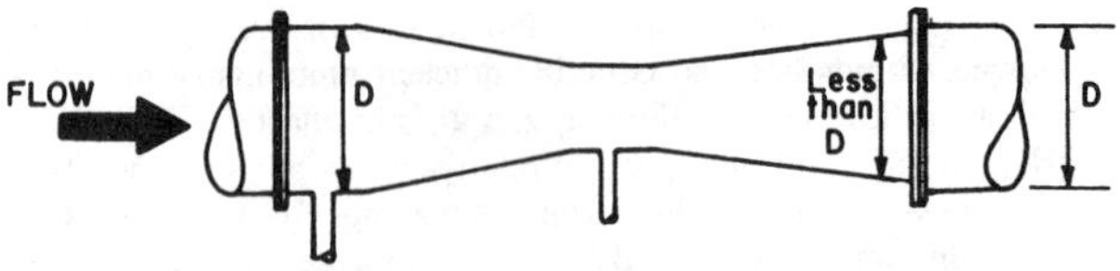

Figure 3. Truncated exit cone Venturi tube.

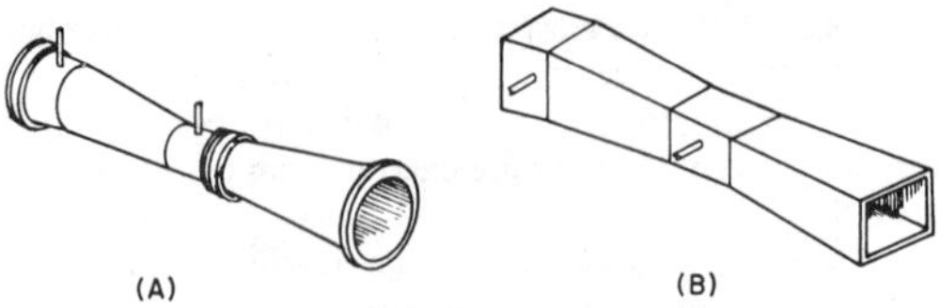

Figure 4. (A) Eccentric Venturi tube; (B) rectangular design Venturi tube.

If the fluid is incompressible and the flow is adiabatic and frictionless, then the Bernoulli expression can be written as:

$$\frac{P_1}{\rho_1} + \frac{u_1^2}{2g_c} = \frac{P_2}{\rho_2} + \frac{u_2^2}{2g_c} \tag{1}$$

where subscript 1 refers to the inlet of the Venturi and 2 refers to the throat. Combining Equation 1 with the continuity equation, with $\rho_1 = \rho_2$, an expression for the pressure drop can be derived:

$$\Delta P = \frac{u_2^2 \rho}{2g_c}[1 - (A_2/A_1)^2] \tag{2}$$

and for the volumetric flow rate

$$Q_i = u_2A_2 = u_1A_1 = \frac{A_2}{\sqrt{1-(A_2/A_1)^2}}\sqrt{\frac{2g_c}{\rho}\Delta P} \tag{3}$$

Equation 3 gives the theoretical or ideal volumetric flow rate. The ratio of the actual volumetric flow rate to the ideal value is the discharge coefficient.

$$C = Q_a/Q_i \tag{4}$$

It should be noted that the empirical coefficient may not be constant but rather a strong function of the channel geometry and Reynolds number.

The following semiempirical equation can be used to compute the actual volumetric flow rate for a Venturi:

$$Q_a = CMA_2\sqrt{\frac{2g_c}{\rho}(P_1 - P_2)} \tag{5}$$

M is referred to as the velocity of approach factor. This empirical constant is:

$$M = [1 - (A_2/A_1)^2]^{-1/2} \tag{6}$$

Now consider the case of an ideal gas, whereby the equation of state applies:

$$P = \rho RT \tag{7}$$

If the flow is reversible adiabatic, the steady-flow energy equation can be written as:

$$C_P T_1 + \frac{u_1^2}{2g_c} = C_P T_2 + \frac{u_2^2}{2g_c} \tag{8}$$

where C_P is the specific heat of the fluid (for an ideal gas, C_P is constant). Combining the previous expressions with the continuity equation gives:

$$\dot{m}_i^2 = 2g_c A_2^2 \frac{\gamma}{\gamma - 1} \frac{P_1^2}{RT_1} \left[\left(\frac{P_2}{P_1} \right)^{2/\gamma} - \left(\frac{P_2}{P_1} \right)^{(\gamma+1)/\gamma} \right] \tag{9}$$

where $\dot{m}_i = \rho Q_i$ = the ideal mass flow rate (kg/s)

γ = ratio of the specific heat of the gas at constant pressure (C_P) to the specific heat at constant volume (C_V)

Equation 9 can be simplified (for the case where $\Delta P < P_1/10$) to the following expression for the actual mass flow rate ($\dot{m}_a$):

$$m_a = YCMA_2\sqrt{2g_c\rho_1(P_1 - P_2)} \tag{10}$$

the additional parameter Y is known as the expression factor defined by:

$$Y = \left[\left(\frac{P_2}{P_1} \right)^{2/\gamma} \frac{\gamma}{\gamma - 1} \frac{1 - (P_2/P_1)^{(\gamma-1)/\gamma}}{1 - (P_2/P_1)} \frac{1 - \beta^4}{1 - \beta^4 (P_2/P_1)^{2/\gamma}} \right]^{1/2} \tag{11}$$

The parameter β is the diameter ratio (i.e., the Venturi's throat diameter d divided by the inlet diameter D):

$$\beta = \frac{d}{D} = \sqrt{\frac{A_2}{A_1}} \tag{12}$$

In specifying a Venturi for a particular application, the first step should be to determine the flow requirements and establish all line sizes in the piping system. The semi-theoretical formulas just presented can be used to size the units, or more often the manufacturer's capacity curves will allow fast estimates. Figure 5 shows a typical capacity curve for one manufacturer's design. Venturis are normally available in a range of β ratios for each line size.

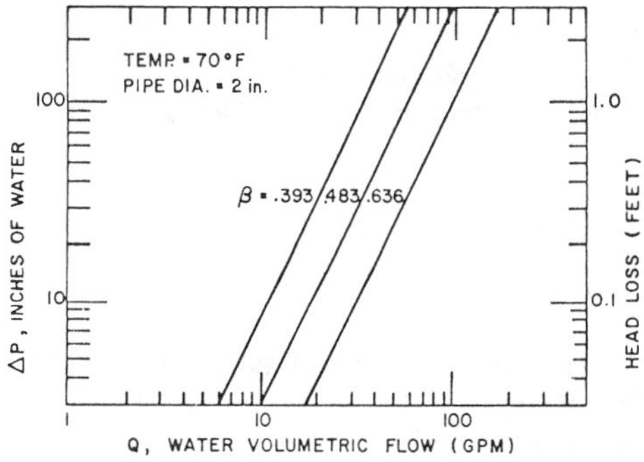

Figure 5. Manufacturer's flow curves for Venturi in a 2-in. pipeline. (Courtesy of Aeroquip Corp., subsidiary of Libbey-Owens-Ford Co., Jackson, MI.)

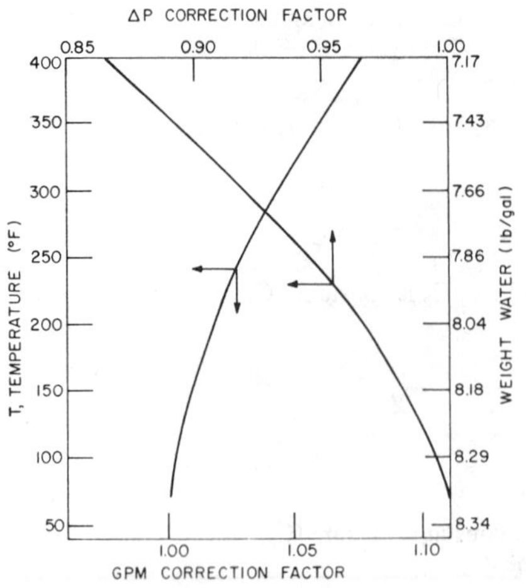

Figure 6. Differential pressure and volumetric flow corrections for water at elevated temperatures. To use, multiply ΔP specified for flow rate with 70°F water by the correction factor to obtain the ΔP for a specified flow at another temperature. To obtain the corrected volumetric flow, multiply gpm by the correction factor to obtain the actual gpm at the indicated temperature. (Courtesy of Aeroquip Corp., subsidiary of Libbey-Owens-Ford Co., Jackson, MI.)

Each β ratio will have a separate capacity or flow curve. As shown, capacity curves provide information on the differential pressure at various volumetric flow rates. Note that since the plot is on the log-log scale, the flow curve is linear.

Temperature corrections and conversion factors for fluids other than water are usually provided with the capacity curves such as Figure 5. Figure 6 provides volumetric flow and differential pressure corrections for water flows at elevated temperatures.

For other fluids, capacities can be converted to equivalent GPM water at 70°F by the following set of equations [1].

For liquids other than water,

$$Q = Q_f \sqrt{S_g} \tag{13}$$

where Q = sizing quantity equivalent GPM at 70°F, water
Q_f = given flow of fluid
S_g = fluid's specific gravity

Note that there are no specific guidelines for viscous fluids; normally the manufacturer must be consulted for recommendations.

For air flow, [1]

$$Q = \frac{Q_f}{C_A} F_{pa} F_{ta} \tag{14}$$

where C_A = 3.8 scfm at 0 psig and 70°F (equivalent to 1.0 gpm of water at 70°F); F_{pa} is the pressure correction factor for air; and F_{ta} is the temperature corrections for air. Values for F_{pa} and T_{ta} are given in Table 1.

For saturated steam flow,

$$Q = \frac{Q_f}{C_s} F_{ps} \tag{15}$$

where C_s = 12.25 lb/h at 0 psig saturated (212°F), equivalent to 1.0 gpm of water at 70°F, and F_{ps} is the pressure correction factor (see Table 1). For gas flow other than air, Equation 14 is applicable for rough approximations with $C_A = 3.8/\sqrt{S_g}$ scfm at 0 psig and 70°F.

Table 1
Recommended Values for Pressure and Temperature Correction Factors for Air and Steam Flow*

Temperature		Pressure		
°F	Air/gas temp. (F_{ta})	psig	Air/gas pressure (F_{pa})	Sat. steam pressure (F_{ps})
0	0.932	0	1.000	1.000
2	0.933	2	0.938	0.934
4	0.936	4	0.886	0.887
6	0.938	6	0.843	0.846
8	0.940	8	0.805	0.811
10	0.942	10	0.771	0.780
12	0.944	12	0.742	0.752
14	0.946	14	0.716	0.727
16	0.948	16	0.692	0.705
18	0.950	18	0.670	0.685
20	0.952	20	0.651	0.666
25	0.956	25	0.608	0.626
30	0.961	30	0.573	0.592
35	0.966	35	0.544	0.564
40	0.971	40	0.518	0.539
50	0.981	50	0.477	0.498
60	0.990	60	0.443	0.466
70	1.000	70	0.416	0.439
80	1.009	80	0.394	0.416
90	1.019	90	0.375	0.397
100	1.028	100	0.358	0.380
120	1.046	120	0.330	0.352
140	1.064	140	0.308	0.331
160	1.081	160	0.290	0.312
180	1.099	180	0.275	0.296
200	1.116	200	0.261	0.282
225	1.137	225	0.247	0.267
250	1.157	250	0.235	0.255
275	1.177	275	0.225	0.244
300	1.197	300	0.216	0.234
325	1.217	325	0.208	0.226
350	1.236	350	0.201	0.218
375	1.255	375	0.194	0.211
400	1.274	400	0.188	0.204
425	1.292	425	0.183	0.198
450	1.310	450	0.178	0.193
475	1.328	475	0.173	0.188
500	1.346	500	0.169	0.183

* *Courtesy of Aeroquip Corp., subsidiary of Libbey-Owens-Ford Co., Jackson, Mich.*

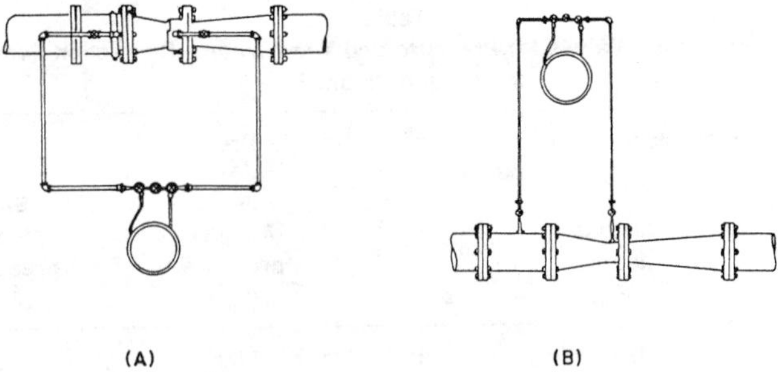

Figure 7. (A) Proper installation for liquid flow; (B) proper installation for gas flow.

Installing a Venturi consists of setting the unit into the line as another section of the pipe. The shorter cone forms the inlet or upstream end. The meter can be installed in the horizontal, vertical, or inclined position. It is good practice to install the Venturi as far downstream as possible from the source(s) of flow disturbances (e.g., reducers, valves, or combinations of fittings). Figure 7 illustrates the proper hookups of a Venturi for liquid and gas flow.

When the Venturi is installed in a vertical line, pressure connections can obviously be made to any part of the tube. In horizontal or inclined geometries, care should be taken to ensure that pressure connections are installed in proper locations, or else faulty measurements will result.

For gas flow applications, locate pressure connections at the top of the tube.

For liquid flow applications, make connections at the side of the tube.

For steam installations where the meter is above the line, connections should be made at the top of the tube; and when the meter is below the line, to the side of the tube [5].

Dall Flow Tube

This tube is a modified Venturi tube that was developed in England. It has found wide use in gas, water, and steam flow applications and fluids which do not contain settleable solids. The basic design of this device is similar to the classical Venturi; however, it is considerably shorter in length. Figure 8 illustrates the device.

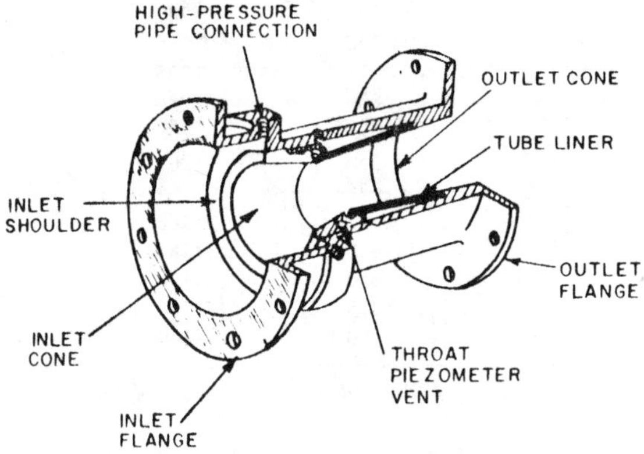

Figure 8. The Dall tube.

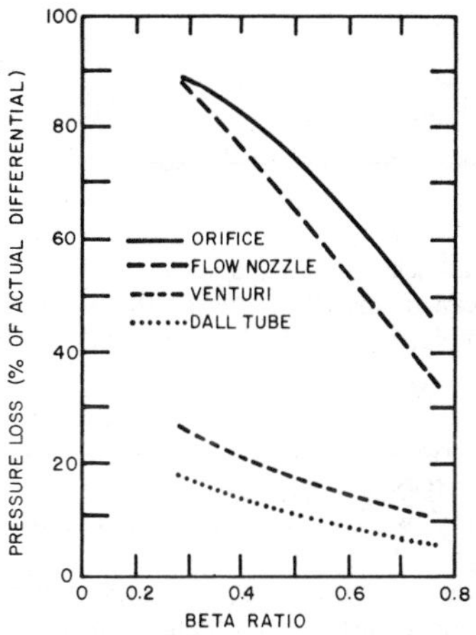

Figure 9. Comparison of the pressure loss in a Venturi to other head devices.

The Dall tube has a lower permanent pressure loss than the classical Venturi for a given flow rate and differential. It does, however, have a higher differential than a Venturi with the same β ratio. Its differential pressure drop is nearly twice that of the Venturi (at the same β and flow rate).

As shown in Figure 8, the device has a short, straight inlet section, the end of which decreases in diameter to the inlet shoulder. It has a converging cone section, a narrow annular gap or slotted throat annulus and a diverging outlet cone. The throat tap is located over the annulus, and the inlet-pressure tap is positioned directly upstream of the inlet shoulder [5].

Accuracy and Operational Characteristics

In general, the Venturi can supply extremely good accuracy in flow measurement. A properly calibrated unit can provide accurate measurements within $\pm 0.5\%$ for nearly all range of sizes. Venturis will maintain their accuracy over relatively long periods of time. The Venturi is a self-cleaning device in most applications as its internal configuration allows smooth flow and efficient pressure recovery and minimizes erosion and clogging.

For the most part, Venturi tubes are maintenance free. They have no moving parts nor mechanical features or glass that can undergo fatigue, strain, or breakage.

Probably the greatest advantage of a Venturi is its low pressure loss as compared to other types of heat meters. Figure 9 shows a comparison of the pressure loss in a Venturi and Dall tube to other head meters. Pressure recovery is generally smooth and gradual within a minimum length of pipe after the fluid has passed through the throat area [1].

Orifice Plate

Figure 10 shows the basic design of the orifice plate and the resultant flow patterns of a fluid in motion. The differential pressure between the upstream and downstream sides of the unit are measured with pressure taps located on either side of the orifice plate. The most common design for the orifice is a circular hole in a metal diaphragm that is mounted concentrically between the flanges

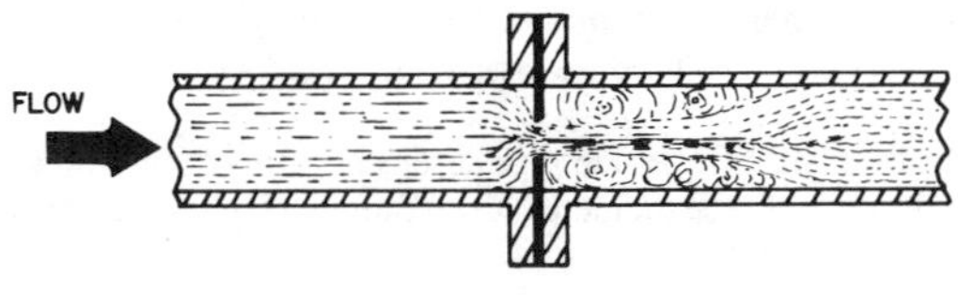

Figure 10. Basic design of the orifice plate and the flow patterns of a fluid passing through it.

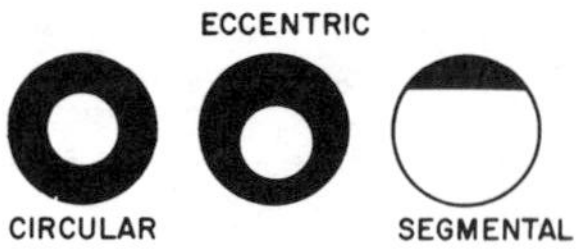

Figure 11. Various types of orifice shapes.

in the flow pipe. Other common designs are the eccentric and segmental types shown in Figure 11. Most orifice meters are of the concentric type. Orifices can be installed between the existing flanges in a piping system. On new installations, however, manufactured orifice flanges unions with built-in pressure taps are recommended.

Flange-mounted orifice plates offer the advantage of flexibility. They may be inserted and removed from a line without disturbing the piping or the differential pressure connections. Note that any increase in the flow rate of a fluid under pressure will cause a decrease in the pressure. The difference in pressures caused by changes in flow rates is a measure of the flow. With an orifice plate, a highly accurate prediction of the contraction of the flow stream (Figure 10) can be made.

The orifice plate's upstream edge should be sharp and squared. The sharp edge should always be installed upstream as there are no appropriate correction factors if the beveled side faces upstream.

Orifice plates cannot be used for two-directional flow. There are no coefficients or operating characteristics that can be applied if the device is improperly installed.

An important design consideration is the position of the orifice opening, which should be accurately centered in the conduit. Improper location can again cause erroneous ΔP measurements.

Disadvantages of the sharp-edged, concentric design include inability to handle dirty fluids and viscous flows, inadequate disposal of condensate in flowing steam and vapors, and higher pressure loss than in the Venturi (refer to Figure 9).

The eccentric design shown in Figure 11 is bored tangent to a circle concentric with the pipe and having a diameter roughly 98% that of the pipe. It is important that no portion of the hole be covered by the flange or gasket upon installation.

The segmental orifice design consists of a segment of a circle located in an arc concentric with the pipe and with a diameter approximately 98% that of the pipe. The diameter ratio for this design is defined as the square root of the area ratio and is therefore a fictitious value.

The design has been used for metering wet stream flow, liquids containing particulates, and oils containing water.

For horizontal lines containing vapor or gas flows, the unit is used with the opening at the top of the pipe. It is not recommended for use with highly viscous fluids or liquids containing sticky solids as deposits build up on the edge of the orifice.

Pressure taps must be installed in such a manner as to prevent debris and sediment from clogging lines and to maintain the device sealed with the fluid. Furthermore, provision should be made for allowing gases or vapors to return to the line. Gas taps should always be positioned at the top of the unit or pipe. Measurement taps for steam and gas applications should be mounted in the side of the unit or pipe.

The most widely used taps for orifices are of the flange type. They are positioned 25.4 mm from both the upstream and downstream faces of the plate.

The location of the upstream tap is somewhat arbitrary, except on high diameter ratios; however, the downstream tap location is critical. The most reliable measurements are obtained when the tap is located at the vena contracta, where the pressure profile is flat. For most orifice diameter ratios, the location can be roughly one-half pipe diameter downstream from the inlet face. Downstream from the vena contracta a highly unstable flow regime exists, and pressure taps in this region should be avoided [6].

Vena contracta taps are located one pipe diameter upstream and approximately one-half pipe diameter downstream. Usually these types of taps are not used on pipe diameters less than 10.2 cm [5].

Taps are made through the pipe wall and flush with the inside pipe surface. For maximum pressure drops, pipe taps are positioned two and one-half pipe diameters upstream and eight diameters downstream [5].

In a manner similar to the development of the expressions for Venturis, semiempirical flow equations can be obtained for the orifice plate. For incompressible flow the actual volumetric flow can be written as follows:

$$Q_s = kA_2\sqrt{\frac{2g_c}{\rho}(P_1 - P_2)} \tag{16}$$

where k is known as the flow coefficient given by

$$k = CM \tag{17}$$

Here C and M are defined as before.

The expansion factor for an orifice employing flange taps or vena contracta taps is given by the following semiempirical expression (for compressible flows):

$$Y = 1 - \left[0.41 + 0.35\left(\frac{A_2}{A_1}\right)^2\right]\frac{P_1 - P_2}{g\rho P_1} \tag{18}$$

(note A_2 is the area of the hole).

For orifices with pipe taps, the expansion factor is

$$Y = 1 - \left[0.333 + 1.145\left(\frac{A_2}{A_1}\right) + 0.7\left(\frac{A_2}{A_1}\right)^{5/2} + 12\left(\frac{A_2}{A_1}\right)^{13/2}\right]\frac{P_1 - P_2}{g\rho P_1} \tag{19}$$

or in terms of the ratio:

$$Y = 1 - [0.333 + 1.145(\beta^2 + 0.7\beta^5 + 12\beta^{13})]\frac{P_1 - P_2}{g\rho P_1} \tag{20}$$

And the actual mass flow rate is

$$\dot{m}_a = YkA_2\sqrt{2g_c\rho_1(P_1 - P_2)} \tag{21}$$

For convenience Table 2 tabulates values of the velocity of approach factor, M (Equation 16) and orifice ratios (i.e., β values; also applicable to Venturi calculations).

Table 2
Tabulated Values of the Velocity-of-Approach Factor (M)

β	M	β	M
1	1.0000	0.55	1.0492
0.05	1.0000	0.60	1.0719
0.10	1.0001	0.65	1.1033
0.15	1.0003	0.70	1.1472
0.20	1.0008	0.75	1.2095
0.25	1.0020	0.80	1.3014
0.30	1.0041	0.85	1.4464
0.35	1.0076	0.90	1.7052
0.40	1.0131	0.95	2.3219
0.45	1.0212	0.99	5.0377
0.50	1.0328		

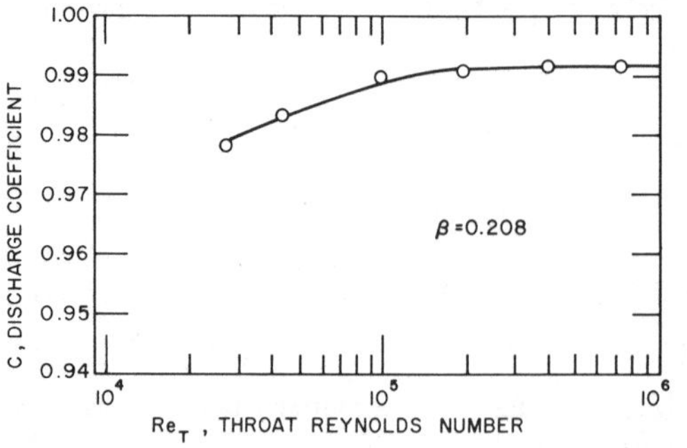

Figure 12. Illustration of how the discharge coefficient in a homemade Venturi varies with the Reynolds number [7].

The discharge coefficient, k, varies with fluid properties and flow conditions and is graphically related to the Reynolds number with the β ratio as a parameter. As an example, Figure 12 shows the variation of the flow coefficient, C, as a function of the throat Reynolds number for air flow through a Venturi, fabricated in the laboratory from fiberglass-reinforced mat.

This type of curve can be based on a Reynolds number definition at the throat conditions as shown (or based on upstream conditions if so desired).

$$Re_T = \frac{\rho u_m d}{\mu} \tag{22}$$

where Re_T is the throat Reynolds number and u_m the mean flow velocity. The mass flow is:

$$\dot{m} = \rho u_m A_m \tag{23}$$

Here A_m is the cross-sectional area for the flow where u_m is measured.

For throat Reynolds numbers above 200,000 the coefficient for a Venturi tube can be assumed to be 0.99 to 1.00.

For an orifice plate, an approximation of the apparent flow can be made by assuming $C = 0.62$.

For more precise flow calculations, it will be necessary to account for fluid properties and flow conditions through the Reynolds number. Flow coefficients for orifice plates, Venturis, and nozzles can be found in two ASME publications [2, 3], and in other reference books, e.g., see Tuve [8].

Orifice plates have found wide acceptance in automatic flow control systems. One such design combines an adjustable orifice and an automatic internal regulating valve. The regulating valve is positioned by a force balance of the differential pressure across the orifice. The force balancing action maintains constant differential pressure across the orifice, thus maintaining the flow rate as determined by the size of the orifice [9].

The principle of operation is illustrated in Figure 13. The orifice is rectangular, narrow, long, and adjustable in length. This allows a uniform linear scale to indicate area and hence flow rate.

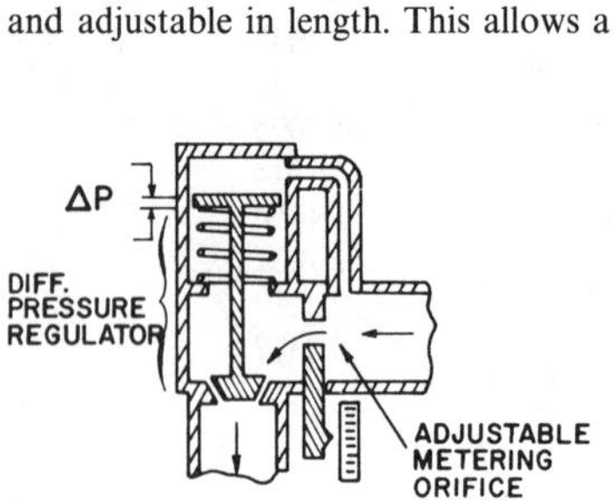

Figure 13. Illustration of the principle behind the automatic flow rate controller with adjustable orifice. (Courtesy of the W. A. Kates Co., Deerfield, IL.)

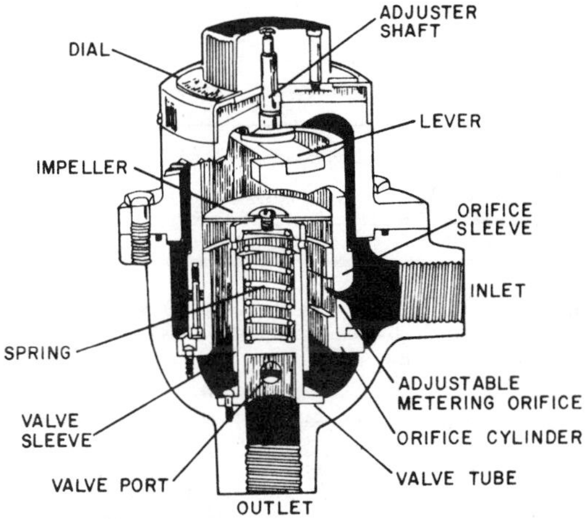

Figure 14. One manufacturer's design of the automatic flow rate controller. (Courtesy of the W. A. Kates Co., Deerfield, IL.)

A manufacturer's specific design is illustrated in Figure 14. The orifice is an accurate slot in a vertical cylindrical sleeve. The sleeve encompasses a second cylinder, which is slotted similarly. The outer sleeve can be rotated about the inner one. When the two slots coincide, the metering orifice opening is maximum. When the slots are in another position, the measuring orifice area is directly proportional to the angle of rotation [9].

The regulating valve is located in the orifice sleeve-cylinder mechanism. The valve includes a valve sleeve that is capable of sliding on a valve tube and can close or open valve ports. An impeller disk is driven downwards by the pressure differential across the orifice, causing the valve sleeve to close the ports. The force of a spring resists this closing action. The sleeve's position results from the direct balance of the two forces, and it sizes the valve port openings to cause the force balance.

The force balance maintains the orifice pressure differential constant regardless of upstream or downstream pressures or the size of the orifice opening. Complete corrective action has extremely fast response as control elements are direct-connected and directly actuated by the pressures of the fluid medium being controlled. Lag times are generally on the order of 1 to 2 seconds.

For remote stepless adjustment of controller flow rate set point, units can be fabricated with pneumatic or electric actuators. The actuator operation can be from a manual loading station or can be integrated into other process controls (e.g., temperature, level, velocity).

This type of design has been used in a variety of applications; some are illustrated in Figures 15 and 16. Figure 15a shows one application: a cascaded final control element of a complete pneumatic or electric control system employing a throttling actuator. The set point is changed by a system controller. In Figure 15b the unit is applied to a pressure filter where the drop across the filter increases as the filter cake builds up. A controller in the outlet line maintains the filter effluent constant at the allowable or safe rate. The controller can also be reset to suit downstream process requirements as warranted. As shown in the figure, a low-pressure alarm system can be used to signify the end of a filtration run.

Figure 15c illustrates another application: the blending of several liquids. Continuous proportionate blending of liquids can be achieved despite large, erratic variations in line pressures by the use of a controller in each line. An operator can preset each controller for the desired flow rates.

Figure 16a illustrates the metering of liquids in a batch operated process. The scheme shown employs a controller, a timer, and an automatic stop valve. Figure 16b illustrates a water-flooding oil recovery operation. Controllers maintain the drive rate for any injection station, regardless of the water supply pressure or formation backpressure [9].

Other typical applications include corn processing, oil production, the pulp and paper industry, sewage treatment, the pharmaceutical industry, water treatment, the plastics industry, and various other industrial chemical processes.

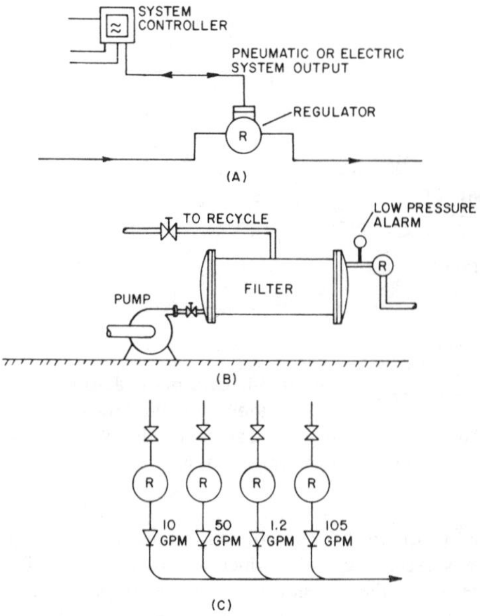

Figure 15. Various applications of the automatic flow rate controller: (A) cascaded final control element; (B) application to pressure filtration. (C) application to proportionate blending. (Courtesy of the W. A. Kates Co., Deerfield, IL.)

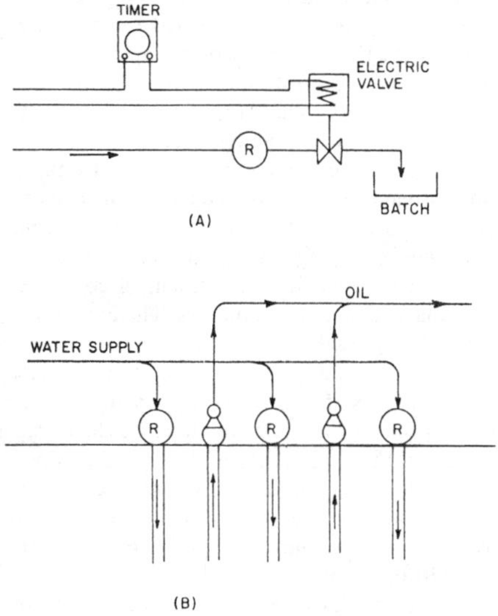

Figure 16. Various applications of the automatic flow rate controller: (A) application to a batch-operated process; (B) application to a water-flooding oil recovery operation. (Courtesy of the W. A. Kates Co., Deerfield, IL.)

Flow Nozzles

The basic design of the standard flow nozzle is illustrated in Figure 17. As shown, the rounded approach has a curvature that is equivalent to the quadrant of an ellipse. The distance between the front face and tip is roughly one-half the pipe diameter (where the throat ratio of the nozzle is between 0.4 and 0.8). For long radius nozzles, throat ratios are usually less than 0.4. The discharge coefficient for a flow nozzle generally lies between that of an orifice plate and a Venturi. For a rough

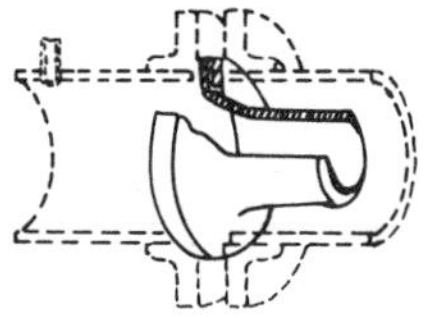

PLAN VIEW **Figure 17.** Basic design of the flow nozzle.

approximation of the apparent flow, the discharge coefficient can be assumed to have a value of 0.98. The effect of Reynolds number on flow nozzle discharge coefficient is affected primarily by the type finish of the inlet and throat regions. Rough finishes produce a flatter coefficient over a wider range of Reynolds numbers. Special calibration of the device is however, required.

The incompressible and compressible flow equations presented for orifice plates are directly applicable to nozzles as well. For more accurate calculations, detailed computations and values of discharge coefficients for flow nozzles can be found in References 2, 3, 10–13.

Flow nozzles have found wide use in the measurement of wet gases (e.g., saturated steam with condensate in suspension). Entrainment in gas or vapor streams can cause excessive erosion problems on some types of head meters; however, the nozzle's curved, surface face guards tend to minimize such action.

Flow nozzles have been used in the metering of high-velocity fluids. This application is most often used when plant capacities undergo an increase with no changes in piping. In the production of high-temperature steam, for example, pipe sizes are usually maintained at a minimum due to the high accelerating costs of high-temperature piping as pipe diameter increases. Flow nozzles are used in these cases to meter resultant high-velocity high-temperature steam flow [14].

Flow nozzles are more efficient than orifice plates. In general, they can handle roughly 60% more fluid than an orifice plate at the same pressure drop [5]. Another advantage over orifice plates is that nozzles are capable of handling liquids with suspended solids. When metering liquids with suspended solids, it is generally recommended that the flow should be vertically downward. This allows some suspended matter to drop out of suspension and will limit variations in the approach conditions as there will be no place for matter to lodge. The streamlined design of nozzles causes solids to be swept through the throat.

Flow nozzles are not suited for highly viscous fluids or fluids containing large amounts of sticky solids (generally magnetic flowmeters are preferred for the latter, if the fluid is conductive).

Table 3 provides a rough guide to the range of operating characteristics of flow nozzles and other head meters discussed thus far.

Sonic flow nozzles have been the subject of considerable investigation for a number of years [15]. Sonic flow conditions can exist at minimum flow area when the flow rate and pressure differential become sufficiently high. The flow is described as being "choked," and flow rates achieve maximum values at inlet conditions [16]. For isentropic flow and assuming an ideal gas, the pressure ratio at the choked or critical condition can be expressed as:

$$\left(\frac{P_2}{P_1}\right)_{crit} = \left(\frac{2}{\gamma - 1}\right)^{\gamma/(\gamma-1)} \tag{24}$$

where $\gamma = C_P/C_V$

An expression for the mass flow rate can be obtained by substituting Equation 24 into the mass flow expression given by Equation 9:

$$\dot{m} = A_2 P_1 \sqrt{\frac{2g_c}{RT_1}} \left[\frac{\gamma}{\gamma + 1}\left(\frac{2}{\gamma + 1}\right)^{2/(\gamma-1)}\right]^{1/2} \tag{25}$$

Table 3
Characteristics of Flowmeters

Meter	Range of max. flow (GPM)	Max. pressure (psig)	Temp. range (°F)	Max. viscosity (cP)	Construction materials	Tolerance (%)
Orifice	0.2–3,500	6,000	−455 to 2,000	4,000	Most metals	1–2
Flow nozzle	0.5–15,000	1,500	−60 to 1,500	4,000	Bronze, iron, steel	>2
Venturi	0.5–1,500	1,500	−60 to 1,500	4,000	Bronze, iron, steel, plastics	>1

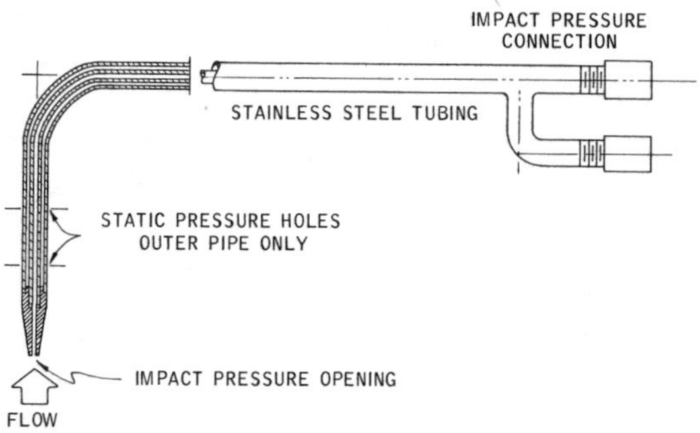

Figure 18. L-type pitot tube designs.

Equation 25 allows calculation of the ideal sonic-nozzle mass flow rate. To compute the actual flow rate, a discharge coefficient should be included in the expression. A value of 0.97 for C is applicable for most engineering calculations. The criterion for Equation 25 is that only upstream stagnation conditions can be used for values of P_1 and T_1.

Pitot Tubes

The pitot tube is one of the oldest basic devices used for measuring velocity and is used to measure liquid and gas flows. The basic design is shown in Figure 18; that of a single-tip type that incorporates static-pressure connections. In general, they consist of two concentric tubes; the inner tube transmits the impact pressure while the annular space between the two tubes transmits static pressure. The static pressure taps shown are located well back from the tube's front face to prevent interaction from eddy currents. Common designs are generally of the sharp- or round-nose fashions.

Manometers are generally used as the pressure-indicating device, as in the case of the Venturi meter. The range of rates that pitot tubes are capable of measuring (using standard secondary differential measuring devices) is narrow and hence has limited their industrial usage. In addition, the ratio of average-to-center velocities in pipes varies through wide limits; thus, it is necessary to perform a velocity traverse over the pipe diameter to obtain a meaningful average velocity. Figure 19 shows the data obtained from a 10-point traverse over a 63.5-mm I.D. pipe for air flow. The plot shown is in terms of the square root of the impact head. Readings obtained from a pitot tube are a direct measure of the velocity head ΔH. For ΔH expressed in millimeters of fluid flowing, the velocity at any point along the diameter is given by:

$$u_0 = C\sqrt{2g_c\,\Delta H} = C\sqrt{2g_c(P_1 - P_0)/\rho_0} \tag{26}$$

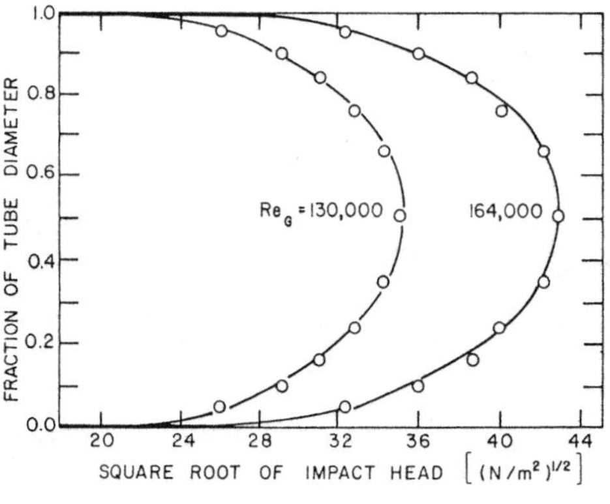

Figure 19. Typical velocity head profiles for a vertical traverse across a 63.5-mm I.D. pipe with air flow.

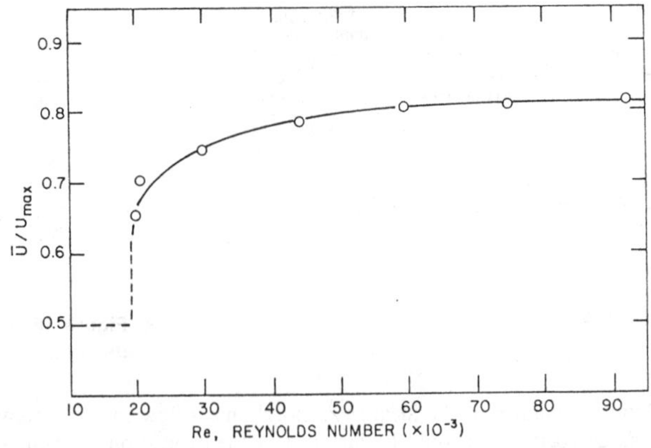

Figure 20. Ratio of average velocity ($\bar{u}$) to center or maximum velocity (u_{max}) plotted as a function of the Reynolds number (where Re is based on $\bar{u}$). Data obtained for air flow in a 63.5-mm I.D. smooth pipe.

A proper traverse is made by dividing the cross-sectional area of flow into a number of equal subareas and taking a local measurement at a representative point within each subarea. Numerically integrating over the measured velocity profile gives the mean velocity (integration can be done by Simpson's rule). The greater the number of subdivisions of flow area, the greater precision achieved in the average or mean tube velocity. For a 20-point traverse, the average velocity is good to within $\pm 0.1\%$. For a rough approximation (good to within $\pm 5\%$) a single measurement at the center of the tube times 0.9 is sufficient.

If quantity rate measurements are needed, values can be computed from the ratio of average velocity to the velocity at the point of measurement. Figure 20 shows the ratio of average-to-center velocity for air flow through a smooth acrylic pipe. At high Reynolds numbers, pipe roughness causes an increase in the relative center velocity.

The primary disadvantage of the pitot tube is that it is limited to only point measurements (although this is an advantage in some applications). In addition, pitot tubes are very susceptible to fouling by foreign matter in the fluid medium. Another shortcoming is that the pressure drops they create are often too small to be detected accurately by standard differential pressure meters.

In order to overcome the differential pressure effects, the pitot-Venturi design, shown in Figure 21, has evolved. Single- or double-Venturi sections may be added to a pitot tube to increase the pressure differential. The Venturi head essentially decreases the static pressure, thus creating a magnified differential. No measurable pressure loss results from the use of this design.

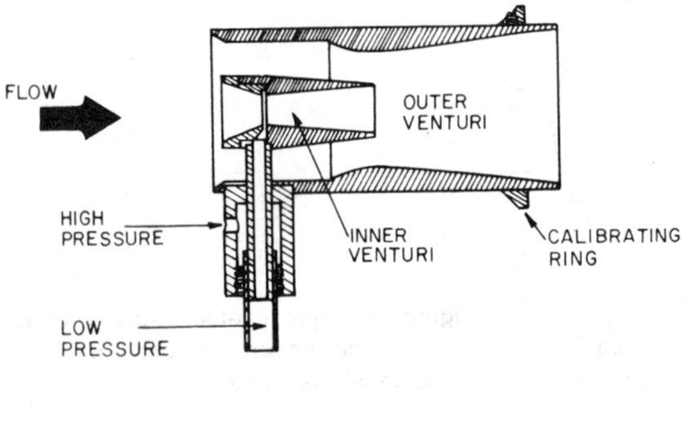

Figure 21. Pitot-Venturi design.

POSITIVE DISPLACEMENT FLOWMETERS

Positive displacement type of flowmeters are commonly employed where a consistently high degree of accuracy is required under steady flow conditions. These devices are accurate with well-defined tolerances, normally $\pm 1\%$ over flow ranges of up to 20:1. They display relatively low pressure drops and are well suited to batch operations and mixing or blending systems.

Positive displacement meters function by channeling portions of the flow into separate volumes, according to the meter's physical dimensions, and measures the flow by counting or totalizing these volumes.

Positive displacement devices are further characterized by one or more moving parts, positioned in the flow stream. These components physically separate the fluid into volumetric increments. Mechanical parts are driven by energy supplied from the fluid motion, which produces a pressure loss between the inlet and outlet of the device. The accuracy of these systems largely depends on clearances between moving and stationary components. As such, these meters require precision-machined parts for the small clearances necessary. In general, this class of flowmeters is not adaptable to metering slurries or fluids with appreciable amounts of suspended particles.

The positive displacement meters described in this chapter include reciprocating piston, nutating disk, rotary piston, and rotary vane meters. Positive displacement flowmeters have been used in a variety of liquid handling operations ranging from the household water meter to a spectrum of industrial applications. (See Table 4.)

Reciprocating Piston Meters

These meters are of the single and multipiston types—the specific choice of design depends on the range of flow rates experienced in an application. With a suitable choice of construction materials, the reciprocating piston meter can be employed to meter practically all liquids within ± 1 accuracy. (Extreme accuracy is achievable to 0.1% for these designs). Periodic lubrication is required.

The piston meter is essentially a piston pump operating backwards. Figure 22 illustrates the operating principle of the reciprocating single-piston meter. The piston undergoes back and forth motion from the fluid under pressure. As the piston moves, reaching the end of its stroke, it shifts

Table 4
Examples of Specific Liquids Handled by Positive Displacement Meters

Diethanolamine	Acetates
Phosphoric acid	Cyclohexanone
Methylene chloride	Liquefied petroleum gas (LPG)
Gasoline	Glycols
Liquid nitrogen	Honey
Toluene	Latex
Vinegar	Fuel oil
Water	Liquid oxygen
Syrup	Caustics
Asphalt	Liquid sugar
Xylene	Butadiene
Fruit juices	Acetone
Crude oil	Liquid lard
Cyclohexylamine	Whiskey
Jet fuel	Fish oil
Alcohols	Ammonia
Styrene	Acrylonitrile
Cresylic acid	Formalin

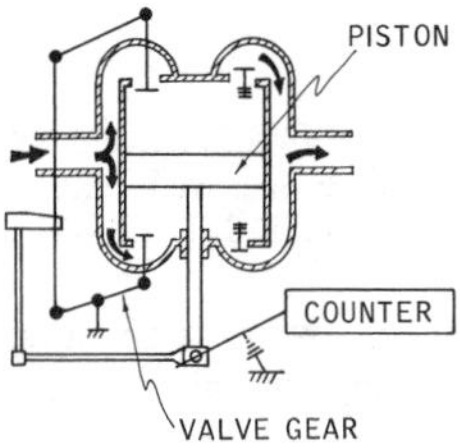

Figure 22. Operating principle of the reciprocating piston meter.

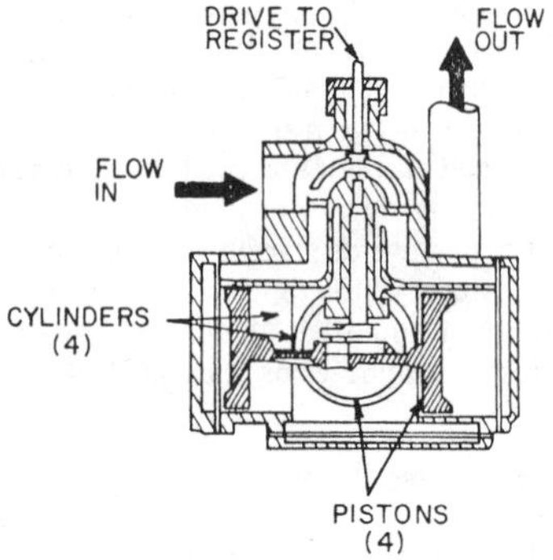

Figure 23. Reciprocating multipiston meter design.

the intake and discharge valve. This in turn operates a counter, which adds a fixed volume increment of fluid with each cycle or stroke of the piston. Figure 23 illustrates a typical multipiston design.

Nutating Disk Meters

This type of meter consists of a movable disk mounted on a concentric sphere. The disk is contained in a working chamber with spherical side-walls and top and bottom surfaces that extend conically inward. It is restricted from rotating about its own axis by a radial partition which extends across the entire height of the working chamber. The disk is slotted to fit over this partition. Figure 24 shows a cutaway view of the meter.

In the figure, the liquid enters the left side of the meter and strikes the disk, forcing it to rock (nutate) in a circular path without rotating about its own axis. A pin extending out from the inner sphere, perpendicular to the disk, traces a circular path as it is driven by the nutating motion. This pin drives the undergear that controls the meter's register [4].

It is important to note that the nutating disk is the only moving part in the measuring chamber. There are several disk materials available for use with a broad range of liquids: measuring chambers are available in bronze, iron and steel; Ni-Resist and stainless steel chambers are available in a range of sizes [16, 17]. Bronze meters are employed for use with most liquids that are noncorrosive to bronze, for example, water, glycol, brine, liquid sugar, and fish solubles. Iron meters are used with most liquids that are noncorrosive to iron. Examples are water, animal fats, fuel oils, vegetable oils, molasses, benzene, gasoline, and coal tar distillates.

Stainless steel meters can service most liquids, such as nitric acid, fruit juices, deionized water, phenol, formaldehyde, and oleum. The direct-drive mechanical shaft couples the meter to the register through a stuffing box (refer to Figure 24). Models are also available with magnetic drive.

The accuracy rating of nutating disk liquid meters is normally expressed as a percentage variation over the full recommended flow range (note: not as a percent of the full flow value). This means that the percentage variation is improved as the actual operating flow range is reduced.

Figure 25 shows the performance curve for one manufacturer's meter. By changing the ratio of the calibration gears, the meter's performance curve can be vertically shifted so that the midpoint

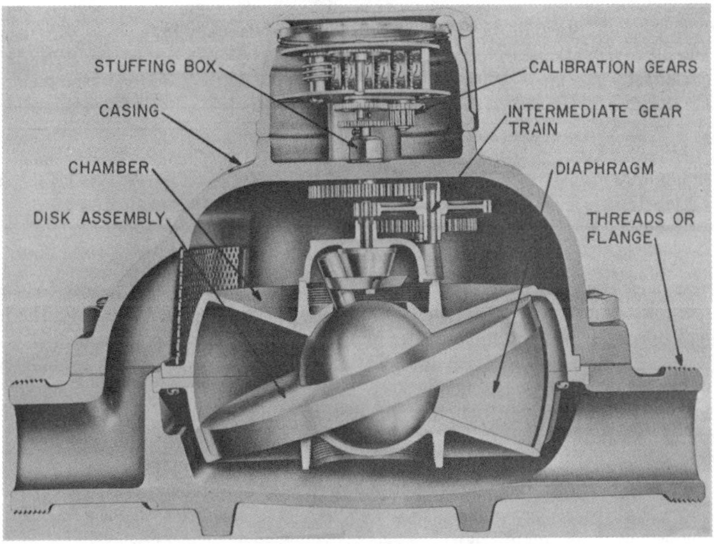

Figure 24. Cutaway view of a nutating disk meter. (Courtesy of Hersey Products Inc., Spartanburg, SC).

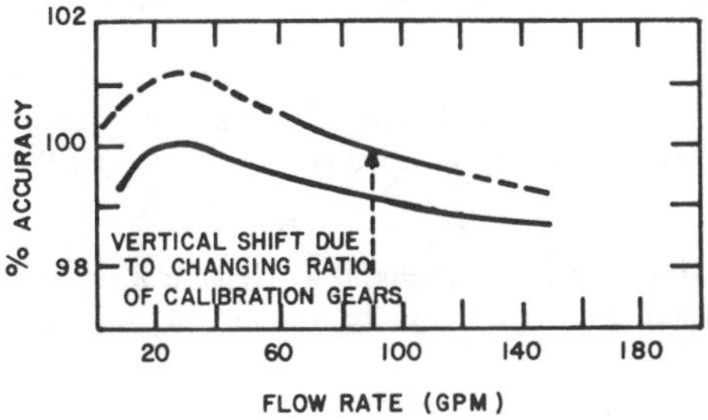

Figure 25. Performance curve of a nutating disk meter. The bottom curve represents the accuracy over the total flow range. By changing the accuracy of the calibration gears, the accuracy curve shifts upward. (Courtesy of Hersey Products Inc., Spartanburg, SC.)

of the actual flow range is 100% accurate. (This, of course, largely depends on the fluid being handled).

The performance curves for all types of displacement meters will change form, with changes in fluid viscosity. Variations in viscosity affect the meter's slip characteristic, which in turn causes variations in flow accuracy. The slip factor in these types of meters are a result of the clearances. Fixed manufacturing tolerances for the measuring chamber and nutating disk provide the clearances between moving and stationary components.

Manufacturing clearances are usually scaled to match a specific viscosity range. Additional accuracy adjustment is provided by the meter calibration gears.

Increases in viscosity causes a proportional increase in the pressure drop. As such, the recommended flow range is reduced with an increase in viscosity in order to maintain a realistic service life factor. Table 5 shows the accuracy variation of one manufacturer's nutating disk meter over various viscosity ranges.

Temperature can also play a key role in meter accuracy. A significant increase in fluid temperature will result in predictable changes in the dimensions of the internal meter parts. Meter parts are machined to operate within a certain temperature range. For example, a meter that has been built to operate in a temperature range of 350–500°F will have excessive clearances and consequently

Table 5
Variation of Meter Accuracy with Viscosity [17]

Viscosity (SSU)	Variation (%)	Viscosity (SSU)	Variation (%)
30–75	1.0	125–150	1.0
75–100	1.0	150–200	1.5
100–125	1.0	200–10,000	~3.0

proportionate inaccuracy when operated at 90°F. Conversely, a meter that is specified for a temperature operating range of 50–80°F cannot provide satisfactory service at 250°F. The following are guidelines to selecting the optimum meter for a specific fluid flow application:

1. Specify the materials of construction.
2. Verify temperature range or limitation for operating conditions.
3. Verify the operating pressure.
4. Specify the meter size.
5. Select the meter register most suitable to the particular application.

A brief discussion of each step follows.

Materials of Construction. The selection of meter materials should be given careful consideration, especially when the metering corrosive liquid is extended. Chemical suppliers should be consulted for specific recommendations on materials selection. Tables 6 and 7 can be used as a rough guide to material selection for nutating disk meters. The viscosity groups given in Table 6 are explained in Table 8.

Temperature Limitations. Verifying operating temperatures will help to establish a range of viscosities, meter construction materials, and degree of flow accuracy that can be achieved by the meter.

Operating Pressure. All meters will have a maximum pressure rating at a specific temperature or over a range of temperatures. In general, steel and bronze meters can be used for relatively high pressure applications.

Meter Size. Manufacturers will supply tables and/or charts that will provide information on the meter's flow capacity range from which a specific size meter can be selected to meet the flow requirements Meter capacities are normally tabulated along with viscosity groups (as in Table 8) and meter size.

In specifying meter size, the anticipated pressure drop through the meter must be determined in order to confirm the meter's compatibility with flow conditions required. Figure 26 is a nomograph for estimating the pressure drop across a meter. The nomograph is based on tests performed by one manufacturer on their stock nutating disk meters; however, it is useful for rough estimates.

Meter Register. There are a variety of meter registers available each suited for a particular aplication. Registers fall under the following general classifications according to application groupings:

1. Totalizing registers
2. Flow rate indicators
3. Manual batch registers
4. Electricontact batch registers
5. Batch controllers

Reference 7 gives examples of totalizing registers and batch controllers.

Table 6
Rough Guide to Meter Material Selection[a,b]

Liquid[c]	Meter construction	Disk/ball material	Viscosity group
* Acetaldehyde	Iron	M	1
‡* Acetic acid	Stainless steel	M	2
* Acetic anhydride	Stainless steel	D	2
Acetone	Bronze	C	1
Alcohol (ethyl or methyl)	Bronze	A	1
Alcohol (denatured)	Bronze or iron	C	1
* Alum solution	Stainless steel	M	2
* Aluminum nitrate 5%	Stainless steel	M	<2
* Aluminum sulfate 50%	Stainless steel	M	<2
* Ammonia (anhydrous cold)	Iron or steel	M	2
* Ammonia (aqueous)	Iron	A or M	2
* Ammonium hydroxide	Iron	A or M	2
‡* Ammonium nitrate	Iron or stainless steel	A or M	2
* Ammonium phosphate	Stainless steel	M	2
Amyl acetate	Bronze	C	2
* Analine	Iron (black)	M	2
‡* Animal fat	Iron	D	3
* Apple juice	Stainless steel	H or M	2
Asphalt (mastic)	Iron	D	6
‡* Benzine	Iron	D	1
Benzol	Bronze	D	1
* Black liquor soap	Stainless steel	M	2
Brine (sodium)	Bronze	A	2
Bunker C oil	Iron	D	5
Butadiene	Iron	C	1
Butadyne	Bronze	C	1
Buttermilk (for cattle feed)	Bronze or iron	A or H	2
* Butylamine	Iron	C	3
* Calcium chloride 30%	Iron	A or H	<2
* Carbon bisulfide	Iron	M	1
Carbon tetrachloride	Iron	C	1
Casein	Iron	H	2
‡* Caustic soda	Iron or stainless steel	A or M	2
* Chloroform	Iron	C	1
* Chocolate liquor	Iron (black)	D	2
* Chrome liquor	Stainless steel	M	2
* Citrus fruit juices	Stainless steel	M	2
* Coal tar distillate	Iron	D	6
* Cocoa butter	Iron (black)	D	4
* Coconut oil	Iron	D	3
Coffee (hot)	Bronze	C	1
Core oil	Iron	D	3

Table 6 (Continued)

Liquid[c]	Meter construction	Disk/ball material	Viscosity group
‡* Corn oil	Iron	D	3
* Cottonseed oil	Iron (black)	D	3
Corn syrup	Iron	D	6
Creosote	Bronze or iron	D	4
Cutting oil	Bronze or iron	H	4
* Cyanide solution	Iron	A	2
DDT solution	Iron	H	1
* Dibutyl phthalate	Iron	M	2
* Diethylamine	Iron	C	1
* Distilled water	Stainless steel	C or M	1
Emulsion oil and water	Bronze or iron	C or M	5
Ester (aromatic)	Bronze	C or D	2
* Ether	Iron	C	1
* Ether (ethyl)	Steel or stainless steel	C	1
Ethyl acetate	Iron	C	1
* Ethylene diamine	Stainless steel	C	3
Ethylene glycol	Iron	D	2
* Ferric sulfate solution	Stainless steel	A	1
* Formaldehyde	Stainless steel	M	1
Fuel oils Nos. 1 and 2	Iron	D	2
Nos. 3 and 4	Iron	D	3
Nos. 5 and 6	Iron	D	4
Fish oil	Iron	D	3
Fish solubles	Bronze or iron	C	2
Freon	Bronze or iron	C	1
Gasoline	Bronze or iron	D	1
Glue	Iron	D	6
Glycerine	Bronze or iron	D	2
Grease	Iron	D	6
* Hydrogen peroxide	Stainless steel	F	2
Isobutylene	Iron	C	1
* Isopropylamine	Iron	D	1
Kerosene	Bronze or iron	D	1
Ketones	Iron	C	1
Lacquer	Iron	D	3
* Lactic acid	Stainless steel	M	2
‡* Lard (molten)	Iron (black)	D	3
Latex solution	Iron	D or H	6
Lecithin	Iron	A	1
* Lime sulfur solution	Iron	M	1
* Liquid soap solution	Iron	H	2
Malt syrup	Bronze	H	3
* Methyl or ethyl acrylate	Iron	D	1

Table 6 (Continued)

Liquid[c]	Meter construction	Disk/ball material	Viscosity group
* Methyl ethyl ketone	Iron	C	1
* Methyl formate	Steel	C	1
Mineral oil	Iron	D	2
Mineral spirits	Iron	D	1
Molasses (cold)	Iron	H	6
(hot)	Iron	D or H	5
* Monochlorobenzene	Iron (black)	C	2
Monochlorobenzol	Bronze	D	1
Naphtha	Iron	D	1
* Naphthenic acids	Stainless steel	C	5
* Nitrogen solutions	Iron or stainless steel	A or M	2
* Nitric acid	Stainless steel	F	2
Oil (soluble cutting)	Iron	H	2
‡* Oleic acid (red oil)	Iron	D	2
* Oleum	Stainless steel	F	2
* Organic acid	Stainless steel	M	2
Paracol wax	Iron	D	3
Paraffin (molten)	Iron	I	3
Paint (oil base)	Iron	D	3
Pentachlorophenol	Bronze	C	3
Perchloroethylene	Iron	C	2
* Phenol	Stainless steel	M	2
Phenolic resin	Iron	D	6
* Pineapple juice	Stainless steel	M	2
Plasticizer	Iron	C or D	3
Printing ink	Iron	D	6
* Polyvinyl chloride resin	Stainless steel	M	2
* Potassium chloride 30% (cold)	Iron	A, H, M	<2
* Potassium aluminum sulfate	Iron	A or M	2
* Potassium hydroxide 50%	Iron or steel	M	<2
Resin emulsion	Iron	H	4
Resin polyester	Iron	C	3
Resin size	Bronze	H	3
Rubber cement	Iron	H	6
‡* Salad oil	Iron (black)	D	3
‡* Soap	Iron	M	2
Soap (resin)	Iron	A or H	3
* Soda ash	Iron	H or M	2
‡* Sodium carbonate	Iron	A or M	2
‡* Sodium hydroxide	Iron or steel	C or M	2
Sodium silicate	Bronze or iron	A or D	6
‡* Soya oil	Iron	D	3
Soy bean oil	Iron	D	3

Table 6 (Continued)

Liquid[c]	Meter construction	Disk/ball material	Viscosity group
* Stearic acid (fatty acid)	Stainless steel	D or M	3
Stoddart solvent	Iron	D	1
Sugar cane juice	Bronze	I	2
Sugar (liquid)	Bronze or iron	H	2
* Sulfide liquor	Stainless steel	M	2
* Sulfuric acid 94%	Iron or stainless steel	F	>2
* Tall oil	Stainless steel	C	3
‡* Tallow (molten)	Iron	D	3
* Tanning liquor	Iron	M	2
Tar	Iron	D	5
Tetrachloroethane	Iron	D	1
Thinners	Bronze or iron	D	1
Toluene	Bronze or iron	D	1
Toxaphene and DDT	Iron	C	2
Trichlorethylene	Iron	C	1
Turpentine	Iron	D	2
Vanilla extract	Iron	A	2
Varsol	Iron	D	1
‡* Vegetable fat or oil	Iron	D	2
‡* Vinegar	Stainless steel	A or M	2
Vinsol	Iron	C	1
* Vinyl acetate	Stainless steel	M	1
‡* Vinyl chloride	Iron	D	2
Vinyl plastics paint	Iron	D	4
* Viscose	Iron or stainless steel	M	4
Water, cold to 100°F	Bronze or iron	A	2
Water, 100–180°F	Bronze or iron	A	2
above 180°F	Bronze or iron	C or H	4
* Water deionized, cold	Stainless steel	M	2
hot	Stainless steel	M	1
Water gas tar	Bronze or iron	D	4
Wax emulsion	Iron	H	2
Wax (hot)	Iron	D	1
Weed killer	Bronze or iron	C	2
Whey	Bronze or iron	A or H	2
Whiskey	Stainless steel	F	2
* Wine (cold)	Stainless steel	A or H	2
Xylol (xylene)	Iron	C or D	1

[a] *Courtesy of Hersey Products Inc., Spartanburg, S.C.*
[b] *Use in conjunction with Tables 7 and 8.*
[c] *Key: *Indicates stainless steel chamber and gear train should be used.*
‡ *Indicates Ni-Resist chamber and gear train may be used.*

Table 7
Material Selection Guide for Disk and Ball

Disk and ball	Letter code	Maximum temperature rating
Hard rubber	A	100°F
Carbon	C	400°F
Aluminum	D	400°F*
Kel-F	F	200°F
Disk/ball	**Letter code**	**Maximum temperature rating**
Spauldite/carbon	H	225°F
Aluminum/carbon	I	Same as aluminum
Kel-F/carbon	M	200°F

Courtesy of Hersey Products Inc., Spartanburg, S.C.
* *For liquids between 200°F and 300°F, disk diameter is as follows:*
undersized 0.010 in. on meter sizes 0.75 through 2.50 in.
undersized 0.020 in. on meter sizes 3 and 4 in.
undersized 0.030 in. on meter size 6 in.
For higher temperatures consult the manufacturer.

Table 8
Viscosity Groups in Table 6 [17]

Group	Viscosity range		Examples
	SSU	cP	
1	30	0.20–0.75	Freon 12, dipropyl ketone, ether, toluene, MEK (methyl ethyl ketone), carbon tetrachloride
2	31–450	1–90	Water (<80°F), olive oil, ammonia, blood, sulfuric acid, turpentine, SEA 10 oil (80°F)
3	450–1,000	90–220	SAE 30 oil (100°F), Grade 4 road tar (120°F), sucrose 64 to 68 Brix (70°F)
4	1,000–5,000	220–1,100	Paints, castor oil, ketchup, glycerine (70°F), black liquor (122°F)
5	5,000–20,000	1,100–4,400	Table molasses, wood resin (200°F), bunker C oil (100°F)
6	20,000–50,000	4,400–11,000	Honey (70°F), blackstrap molasses (70°F), Grade 8 road tar (>100°F), corn syrup 41°B at 80°F

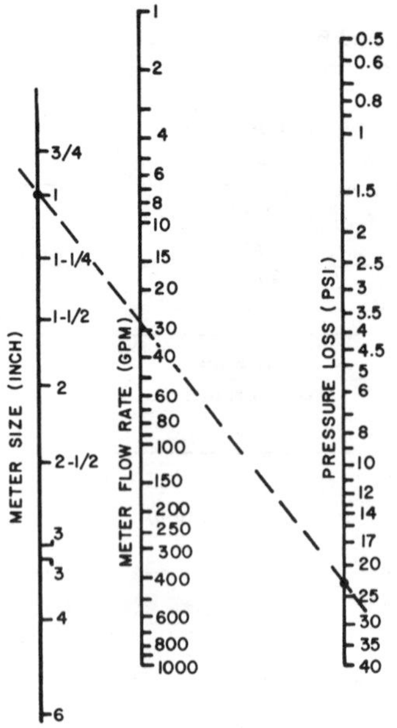

Figure 26. Nomograph to obtain pressure drop. Locate meter size value and flow rate on left and middle columns, respectively. Draw a straight line through the two points, intersecting the right-hand column, ΔP in psi.

Rotating Mechanism Designs

Rotary Piston Meters

The basic design of the internal workings of a rotary piston meter is illustrated in Figure 27. A cylindrical working chamber houses a hollow cylindrical piston of equal length. The central hub of the piston is guided in a circular motion by two short inner cylinders. The piston and cylinder are alternately filled and emptied by the fluid flowing through the meter. A slot in the sidewall of the piston is removed such that a partition extending inward from the bore of the working chamber can be inserted. This restricts the piston to a sliding motion along the partition.

Rotary Vane Meters

Figure 28 illustrates a rotating vane type rotary meter. To ensure that the vanes are in continual contact with the meter casing, they are spring loaded. As the eccentric drum rotates, a fixed quan-

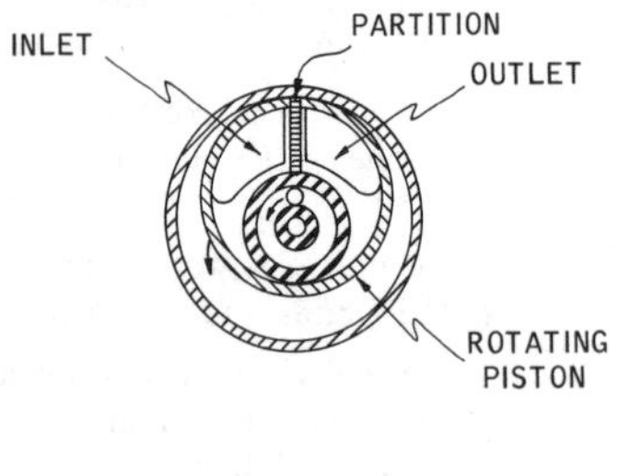

Figure 27. Internals of a rotary piston meter.

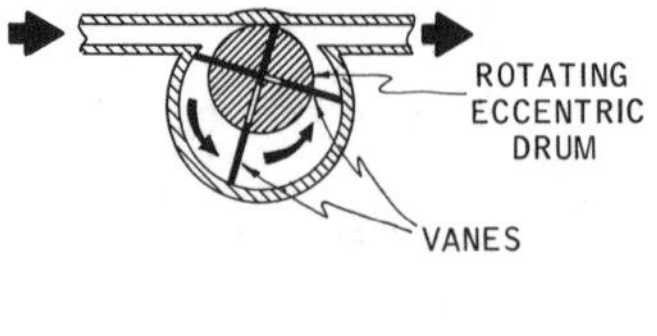

Figure 28. Basic design of a rotary vane flowmeter.

tity of liquid is swept through each section to the outlet. The volume of the displaced fluid is measured by a register attached to the shaft of the eccentric drum. In general, these are highly reliable meters and relatively insensitive to fluid viscosity changes.

The rotating vane type rotary meter is a lightweight unit that can be easily installed in the field. Flow accuracy is established by machined casings that form the measurement compartments. These meters are employed in gas and liquid applications for distribution measurement in commercial and industrial areas and in-plant metering.

For gas applications, it is generally good practice to purge the gas line prior to installation of the meter. In lines that contain high amounts of pipe scale or other particulates, the use of filters is recommended to extend meter service life.

Other Positive Displacement Flowmeters

An additional differential pressure flowmeter worth noting is a calibrated flow nozzle design. The flow through a calibrated nozzle is sensed by an arrangement of opposed bellows. Displacement of the bellows is transferred by a low-friction cam and lever to a rotary geared movement to indicate flow rate. The operation of this design is entirely mechanical [18].

Designs are self-contained and do not require separate orifices, blocking, purging, or equalizing valves. This type of flowmeter is well suited for applications where compactness, low cost, and shock or impact resistance are priority factors. Typical applications include metering of fuel oil to burners, purging of instrument lines, injection of chemicals into process streams, and monitoring flow of inert shielding gases.

MECHANICAL FLOWMETERS

The rotameter and turbine meter are the most common of this type of flowmeter. Both devices accomplish flow measurement by drag effects. The drag coefficient is a function of the Reynolds number and hence depends on the fluid viscosity. As such, both instruments must be recalibrated for each new fluid handled, with calibrations periodically checked.

Rotameter

Rotameters are often referred to as variable-area meters. They are widely employed both in laboratory and industrial applications. The basic design, illustrated in Figure 29, consists of a tapered vertical tube in which the fluid enters through the bottom, causing a float or bob to ascend. The tube is narrow at the bottom where the fluid velocity is the greatest, and widest at the top, where the velocity is at a minimum.

The float, used to indicate the flow rate, is slightly heavier than the fluid measured, and as such will sink to the bottom of the tapered tube when there is no flow. As fluid flows through the tube, the float continues to rise until it reaches a point where the drag forces are balanced by the weight and buoyancy forces. The float's position in the tube, measured by a linear scale, is taken as an indication of the flow rate.

A force balance about the float in Figure 29 gives the following equation:

$$F_D + \rho_f \phi \frac{g}{g_c} = \rho_B \phi \frac{g}{g_c} \tag{27}$$

where ϕ represents the volume of the float, ρ the density, F_D the drag force, g the acceleration of gravity (32.2 ft/s^2), g_c a conversion factor, and subscripts f and B indicate fluid and bob (float), respectively.

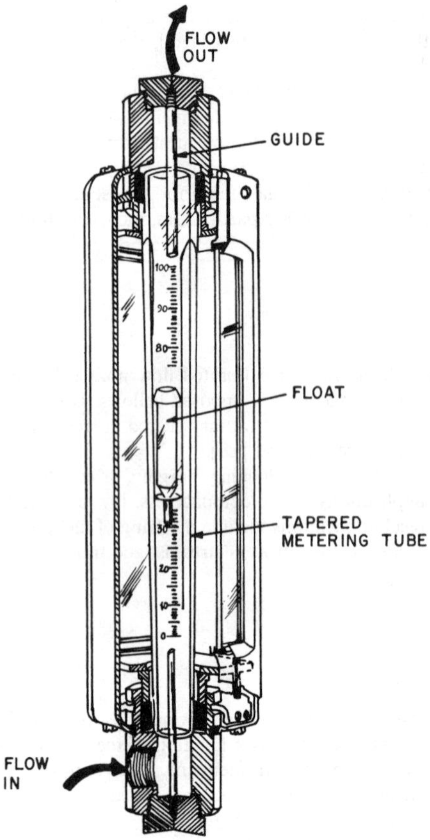

Figure 29. Basic design of the rotameter.

The drag force, F_D, is defined by:

$$F_D = C_D A_B \frac{\rho_f U_m^2}{2g_c} \tag{28}$$

where C_D = drag coefficient
A_B = area of the front face of the float
U_m = mean fluid flow velocity in the annular space between the float and tube wall

Substituting Equation 28 into 27 and rearranging, an expression for the mean fluid velocity is obtained.

$$U_m = \left[\frac{2g\phi}{C_D A_B}\left(\frac{\rho_B - \rho_f}{\rho_f}\right)\right]^{1/2} \tag{29}$$

From this expression the volumetric flow rate can be obtained by multiplying U_m by the annular flow area ($Q = U_m A$), where

$$A = \frac{\pi}{4}[(D + \alpha z)^2 - d^2] \tag{30}$$

Here D = diameter of the tube at the inlet (cm)
d = maximum bob diameter (cm)
z = vertical distance from the entrance (cm)
α = constant related to the tube taper

To estimate the mass flow rate, the area relationship given by Equation 28 can be assumed linear for the actual tube and float dimensions. This is useful in that an expression for mass flow rate can be written simply as:

$$\dot{m} = C'z[(\rho_B - \rho_f)\rho_f]^{1/2} \tag{31}$$

where C′ is an appropriate meter constant obtained through calibration.

It is advantageous to use a flowmeter that indicates measurements that are independent of fluid density variations. Special float construction can allow rotameters to be used under fluid density changes. In such cases the mass flow rate is expressed by:

$$\dot{m} = \frac{C'z\rho_B}{2} \tag{32}$$

where the float's density is approximated by:

$$\rho_B = 2\rho_f \tag{33}$$

In general, rotameters measure small flow rates relative to the flow capacities of other methods already discussed. An operational problem frequently encountered with rotameters is that the float has a tendency to stick to the tube wall upon contact. In unstable flows, the float is often pushed against the tube wall. This problem is normally eliminated by the use of a guide wire that passes up through the center of the float and tube. Still another solution to this problem involves the use of glass ribs on the inside of the tube.

Rotameters are primarily limited to relative clear fluids although there are methods that facilitate measurement of dark or opaque fluids. For example, rib walls can be fabricated into the tube to guide the float so as to allow such a small clearance that the float will be visible through the film between it and the tube wall. This particular design has the disadvantage of blockage or plugging from suspended particles in the fluid.

Another approach to handling dark fluids is to employ a transparent disk that is illuminated from the back wall of the tube, creating a line of light that allows reading the measurement.

Unstable or pulsating flow will cause the float to undergo vertical oscillations. The mean measurement of these oscillations does not necessarily represent the true flow through the rotameter. Only under ideal conditions, where the oscillations of the float correspond to the amplitude of the flow pulsation, does the average measurement provide the true flow.

Rotameters can be fabricated from a variety of materials. There is a broad line of acrylic plastic rotameters useful for measuring low flow rates of both liquids and gases. The meter body for these is usually supported in an extruded aluminum frame which can accommodate process connectors and absorb connection strain.

A metal-tube rotameter is illustrated in Figure 30. Steel or other metal can be used to house the float. The float's position can be detected by observing an indicating magnet that moves up and down with a mating magnet located within the metal extension tube.

There are a large number of float designs from which to choose. Two main classifications of floats are the extension type and spool type. Figure 31 illustrates several designs and indicates the proper manner in which they are read.

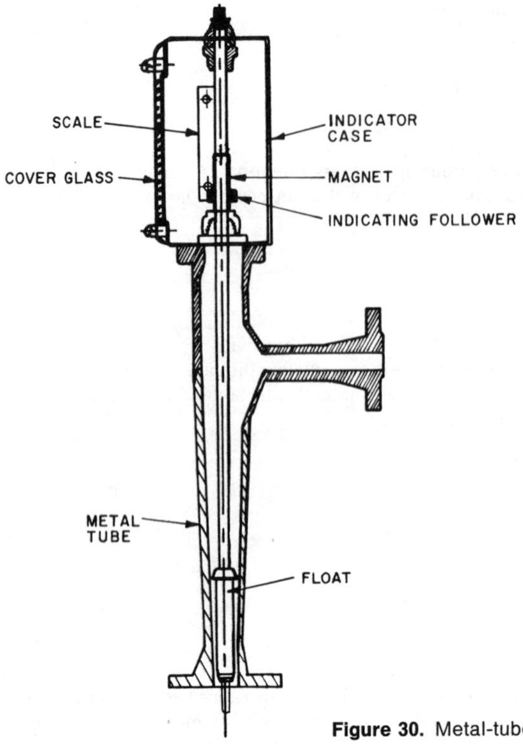

Figure 30. Metal-tube rotameter.

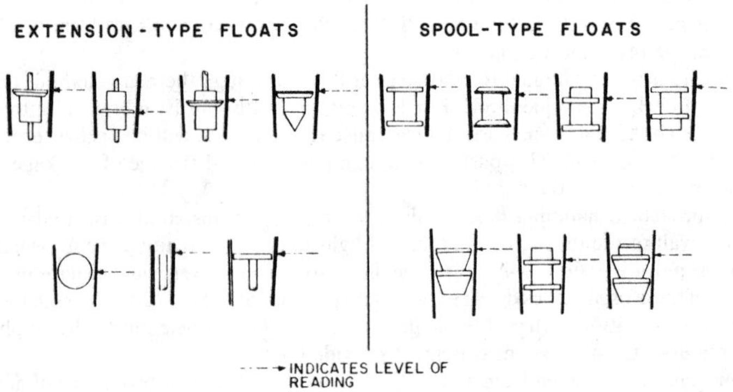

Figure 31. Various types of rotameter floats: extension-type floats and spool-type floats.

Turbine Meter

In principle, the turbine meter operates similarly to an hydraulic turbine. The turbine is positioned within a designed passage through which the fluid is directed. The fluid's impact on the blade surfaces causes rotation. The rotor can be positioned such that it can be driven by radial or axial flow, or a combination of both. The rotor motion may directly drive a register. Frequently, magnets

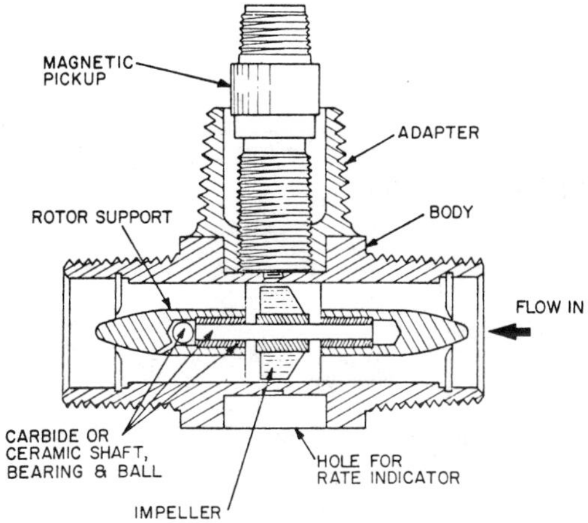

Figure 32. Major components of the turbine meter.

on the rotor are employed to generate a rotating field and thus produce a current indicative of the flow rate. Figure 32 illustrates the basic design of the turbine meter.

In the figure, the fluid to be metered enters the flowmeter body chamber at the inlet port. The rotor assembly contained in the chamber consists of the turbine and hover plate elements mounted at each end of the shaft. The rotor assembly rests on the top hover seat in a no-flow condition. When flow begins, a pressure drop occurs across both hover plates and corresponding seats. This pressure drop lifts the rotor assembly from the top seat. The fluid element is split within the body chamber, and the flow through the two turbines causes the rotor assembly to rotate with stability at a speed that is directly proportional to the flow velocity. The fluid leaves the meter body chamber through two tapered seating ports that lead the flow to the exit port [7, 19]. A magnetic pickup is employed (see Figure 32) to sense the rotational speed of the rotor, and an a.c. sinusoidal signal is generated. The frequency of these signals is directly proportional to the fluid flow rate. The a.c. signal is transmitted by means of a conductor-shielded cable to an appropriate readout instrument.

Turbine meters provide an accurate and relatively economical method of metering to a wide variety of liquids. In the food industry, for example, turbine meters have been employed in metering milk, cheese, whey, cream, syrups, vegetable oils, vinegar, etc. [20]. Examples of sanitary turbine flowmeters widely used throughout the beverage and food processing industries are given in Reference 7.

Another application where turbine meters have found widespread use is in oil field automation. During the 1960s, oil companies began automating their production facilities employing computers for data storage of production statistics [21]. Computer utilization expedited data retrieval for oil, water production reports, gas produced, as well as oil and gas sold and water injected for water-flooding or pressure-maintenance purposes. Turbine flowmeters have been incorporated into these liquid flow systems. The wide range and variety of output signals has allowed these flowmeters to be interfaced with almost all data acquisition instrumentation.

In Figure 33 the turbine flowmeter is employed in an oil/water separator operation. These flowmeters can be used on two-phase and three-phase high- and low-pressure separators. Flowmeters are normally installed roughly ten pipe diameters of straight pipe upstream and five pipe diameters downstream. Valves are usually of the snap-acting type so that a constant flow rate can be provided each time the separator dumps. Pilot valves can be lever operated or actuated by liquid head provided that the pilot valves are pressure balanced against the static pressure head of the vessel [21].

Another oil production application is in water injection projects. Oil that is produced from formation pressure (primary production) results in the recovery of roughly 30–35% crude oil. When the crude is pumped out with nothing to replace it, surface subsidence often results. This surface subsidence can be raised by means of injecting water back into the formation to replace the removed oil. The water injection method is also used to force more crude oil out of a given formation. The

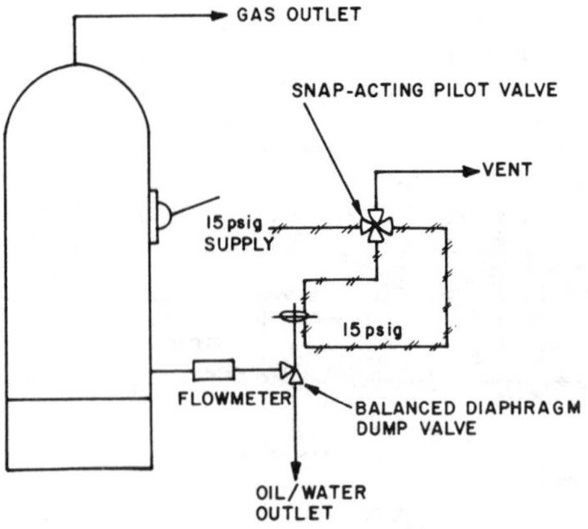

(A)

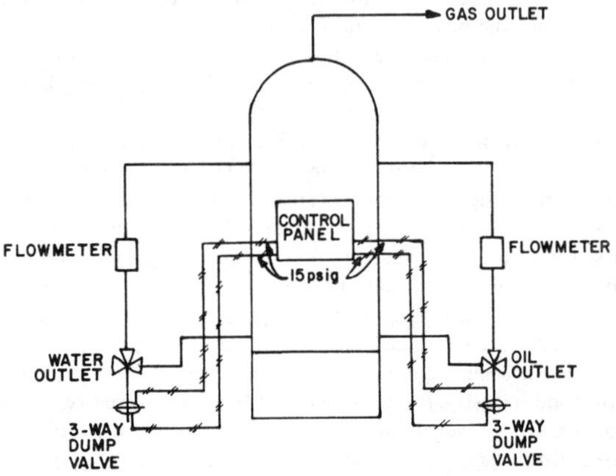

(B)

Figure 33. Turbine meter application to oil production: (A) application to a two-phase separator, (B) application to a three-phase separator. (Courtesy of Flow Technology, Inc., Phoenix, AZ.)

injected water usually consists of a produced water plus a certain quantity of makeup. Produced water is obtained from the crude oil because wells tend to capture salt water with the crude. Most injection water used is salt water, which presents a serious corrosion problem. Turbine flowmeters are different from other differential flow measuring devices in that they are capable of handling liquids that are corrosive or abrasive, as well as liquids containing appreciable solids and dissolved gases. Turbine flowmeters for metering injected water are tailored specifically to handle problem liquids.

There are many designs suited to particular liquid handling applications. Typical capacity ranges for flanged-connections turbine meters are given in Table 9. Table 10 is a partial listing of various liquids measured with turbine meters.

Table 9
Flow Range Capabilities and Performance Characteristics of the Flanged Model Turbine Meter

Flange size (in.)	Flow range			Frequency output (Hz)
	GPM	BPH	BPD	
3/4	1.3–13	1.9–19	45–450	100–1,000
3/4	3.2–23	4.6–46	109.7–1,097	100–1,000
1	6.4–64	9–90	219–2,190	100–1,000
1½	17.4–174	25–250	600–6,000	100–1,000
2	33–290	48–480	1141.7 11,417	100–1,000
3	60–600	85–850	2,057–20,570	50–500
4	107–1,071	153–1,530	3,672–36,720	50–500
6	300–3,000	429–4,290	10,286–102,860	50–500
8	789–7,890	1,128–11,280	27,068–270,680	50–500

Repeatability (±% of rate)	Accuracy (±% actual flow)	Temp. range (°F)	Coil inductance (mH)	Coil resistance (ohms)
0.1	0.5	−100 to +225 Std. −100 to +450 Opt. −300 to +800 Opt.	400–605	1,190–1,450

Courtesy of C-E Invalco, Combustion Engineering Co., Tulsa, Okla.

Readout Instrumentation

There are two major components in any flow metering system: the meter and the readout instrument. In the case of the turbine meter, the unit generates electrical pulses that are proportional to the flow. The readout instrument accepts these pulses and displays its reading directly in units such as gallons or barrels, in the form of totalized flow on a counter and/or flow rate on an indicating meter.

Readout instruments for turbine meters fall under two basic categories: totalizers and rate indicators. Rate indicators are basically frequency-to-d.c. converters. They accept the turbine meter output frequency and convert the signal to d.c. to drive an electrical meter. The electrical meter can then be calibrated in any usable units (e.g., barrels per day (bpd) or gallons per minute (gpm)). Figure 34 shows a basic instrumentation hookup for a flow rate indicator system [34].

The totalizer system (illustrated in Figure 35) accepts the pulses from the pickup, amplifies and squares the pulses, and through a dividing network allows a certain percentage of the pulses to pass to the counter to readout measurements directly in gallons, barrels, etc. The dividing network is established for the turbine meter being used. These models display continuous totalization of flow and instantaneous flow rate from a turbine meter. The counter is mounted in a plug-in module that contains the necessary circuitry for totalization and rate indication of one turbine flowmeter.

Turbine flowmeters have been miniaturized to allow traversing or point measurements as in the case of pitot tubes. One design involves an axial turbine flowmeter element that is mounted on the end of a strut, which allows the flowmeter element to be positioned as desired in a large pipe or channel [21, 23].

With this design, the flow velocity is measured by a turbine rotor. The flow velocity profile in a pipe can, for example, be determined by traversing over the pipe diameter with a retractable or hand-held turbine probe. The flow velocity can be directly related to the total flow volume. These

Table 10
Partial Listing of Liquids Measured with Turbine Flowmeters

Anhydrous ammonia	Casillion acid	Fuel oil	Maleic acid	Rosin
Acetic acid	Chlorine propellant	Fish oil	Maleic anhydride	Rust inhibitor
Amines	Coffee extract		Methyl amines	
API condensate	Chocolate slurry	Gasoline	Milk	Silicone tetrachloride
Acetic anhydride	Corn oil	Glacial acetic acid	Milk products	Salt water
Actinium hydroxide	Cellulose slurry	Glycol	Molasses (hot)	Sugar
Arnel	Cyclohexanone		Methanol formaldehyde	Soft drink syrup
Ammonium sulfate	Caustics	Hydraulic oil		Sodium hydroxide
Aromatic naphtha	Chicken fat (hot)	Heat transfer oil	Nitrogen tetraoxide	Sulfuric acid
Aromatic hydrocarbons		Heptane	Nitrogen	Slurries
Apple juice	DMT (dimethyltryptamine)	Hydrogen peroxide	Nitric acid	Synthetic fibers
Adipic acid	Dimethylaniline	Hydrocarbon condensate	Naphtha	
Amyl	Debutanized isopentane	HCN (acid)		Tetraethyl lead
Animal fat (hot)	Detergent		Oil	Transformer oil
Alcohol	Demineralized water	Isocyanate	Oil dispersant	Transformer coolant
Ammonium phosphate	Deionized water	Isopropyl alcohol		Tea extract
Amino phosphonic acid	Distilled water	Isobutanol acetate	p-Dioxane	Toluolene diisocyanate
	Dimethyl ether	Isobutyl	Phthalic anhydride	Toluolene
Benzoic acid	Diesel fuel	Ice cream mix	Phosphoric acid	Trioxane
Brine			Pickle brine	Transmission fluid
Benzene		Jet fuel	Phenol	Tetrachloroethylene
Brackish sulfate	Ethylacetate		Propylene oxide	Toluene
Beer	Ethylene glycol	Kerosene	Propane	
Butanol	Ethyl toluene		Pentane	Urea
Bisphor A	Egg yolk	Lard (hot)	Paint	Vinyl chloride
Butane A		LPG (liquefied petroleum gas)	Polymers	
	Fluorocarbon R-13		Pluoronic L62	Water
Caprolactan	Fluorocarbon R-12			Wax (hot)
Crude oil	Fatty acid	Methanol	Resin	Whiskey
Coconut oil	Fabric dyes	Methylene chloride	Rum	Whey

Courtesy of C-E Invalco, Combustion Engineering Co., Tulsa, Okla.

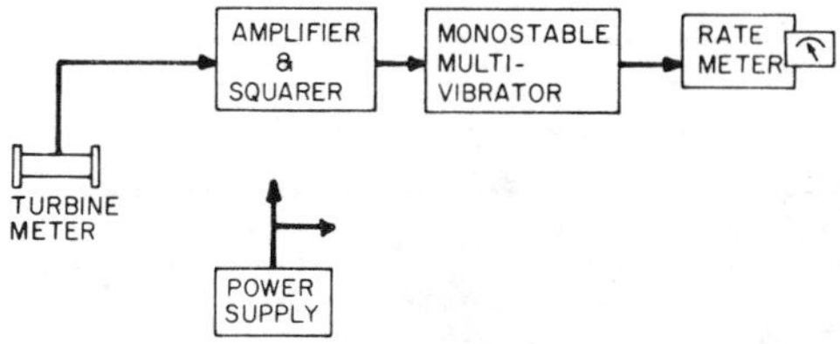

Figure 34. Basic setup for the flow rate indicator. (Courtesy of C-E Invalco, Combustion Engineering Co., Tulsa, OK.)

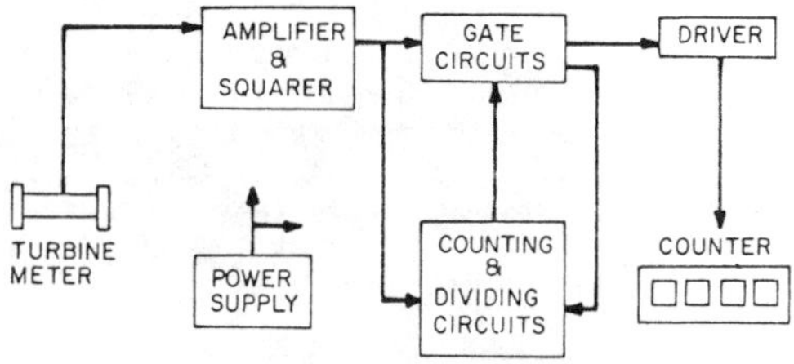

Figure 35. Basic setup for flow totalizer system. (Courtesy of C-E Invalco, Combustion Engineering Co., Tulsa, OK.)

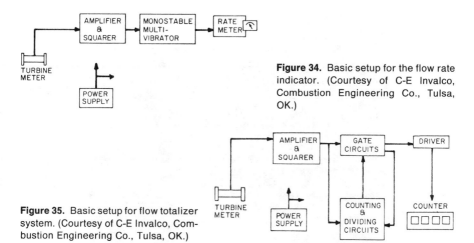

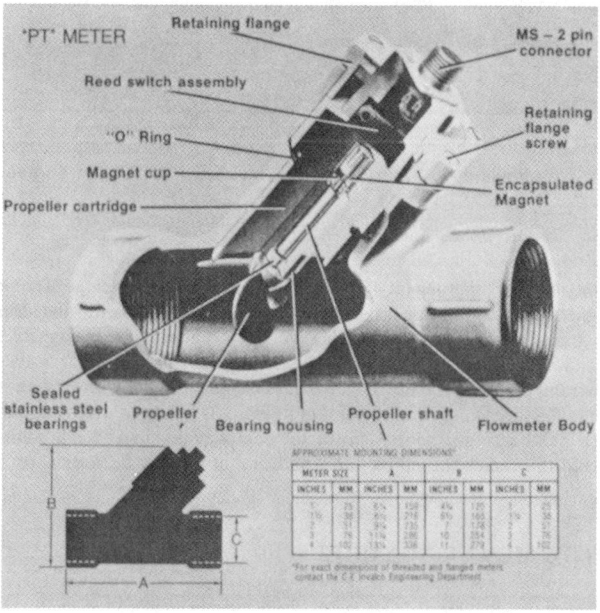

Figure 36. Propeller-type flowmeter. (Courtesy of C-E Invalco, Combustion Engineering Co., Inc., Tulsa, OK.)

systems are specifically designed for large pipe or channel applications and can be coupled with standard readout instrumentation.

There are many flowmeters on the market that are similar to the basic turbine meter. Figures 36 and 37 show two meters that are specifically designed for metering low liquid flow rates.

Figure 36 shows a compact propeller-type flowmeter, designed with a Y-type body to position all components, except a propeller, out of the liquid stream. The main leg of the body forms part

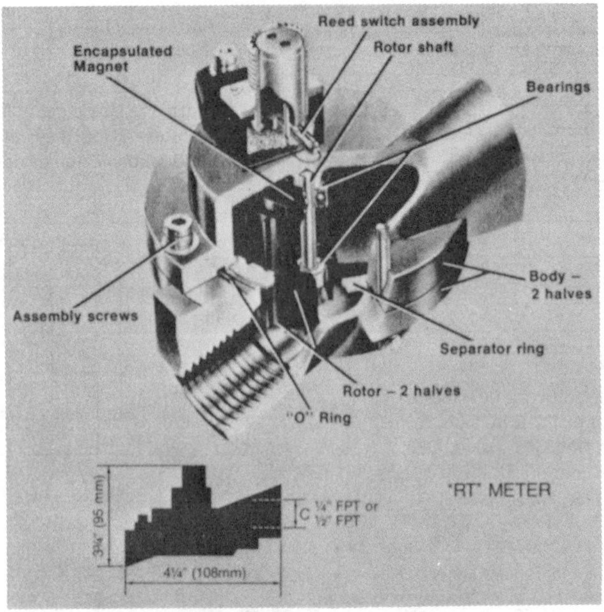

Figure 37. Rotor-type flowmeter. (Courtesy of C-E Invalco, Combustion Engineering Co., Inc., Tulsa, OK.)

of the flow stream, and the other leg (at 45°) contains the propeller cartridge. The propeller is three bladed and designed for maximum clearance between each blade, thus allowing suspended particles to pass. As the propeller goes through each revolution, encapsulated magnets generate pulses through the pickup device (and the number of pulses is directly proportional to the flow rate) [24].

Figure 37 illustrates a rotor-type flowmeter that is designed for flow lines with less than 2.5-cm diameter. The manufacturer notes that maximum linearity is achieved at low liquid flow rates. The design includes a ball-bearing-supported rotor for compact design and minimum restrictions to flow. Flow through this meter is unidirectional and mounting may be vertical or horizontal.

MAGNETIC FLOW MEASURING SYSTEMS

The operation of a magnetic flowmeter is based on Faraday's law of electromagnetic induction, which states that the relative motion, at right angles between a conductor and magnetic field, induces a voltage within the conductor. The voltage induced is proportional to the relative velocity of the conductor and the magnetic field. The operating principle is illustrated in Figure 38. The flow metering system consists of a primary device, which is the magnetic flowtube, and a secondary device, which is the magnetic flow transmitter. The magnetic flowtube is mounted in the process pipeline, whereas the secondary device can be remote from the flowtube.

The basic components of the flowtube consists of an electrically insulated lined metal or unlined fiberglass-reinforced plastic metering tube with an opposed pair of metal electrodes mounted in the tube wall. A pair of electromagnetic coils are mounted external to the metering tube in a protective casing [25].

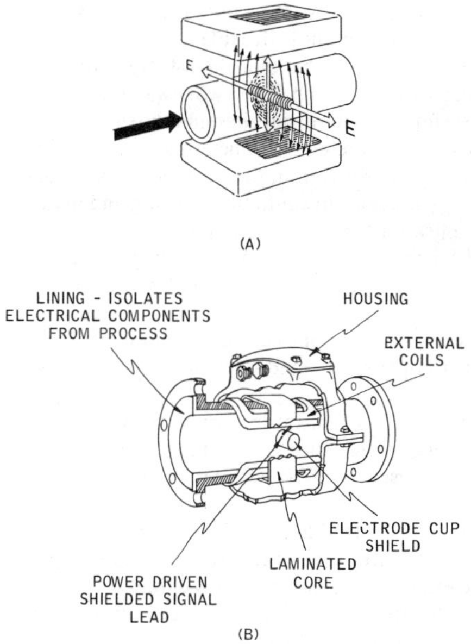

Figure 38. (A) Basic operational concept of the magnetic flowmeter, (B) major components of the magnetic flowmeter.

Electric current (a.c.) is used to excite the coils and generate a magnetic field at right angles to the axis of the process fluid passing through the metering tube.

The liquid being metered must be conductive. As the fluid passes through the magnetic field, a voltage (electrical potential) develops across the electrodes. This voltage is directly proportional to the volumetric flow rate.

The voltage signal generated can then be converted in the magnetic flow transmitter to either a standardized d.c. current signal, which can be transmitted to a suitable receiver unit, or a 0–10 V (d.c.) output signal [25].

Magnetic flowmeters were introduced commercially in the 1950s by Foxboro Corp. They have been applied to metering the flow of a wide variety of conductive fluids, ranging from as low as 200 μS (microsiemens)/m (or 2 μmho/cm) [25].

Magnetic flowmeters fall into the classification of obstructionless meters; that is, there is no restriction to the flow of the process fluid. Hence, there is no pressure loss across the meter other than that experienced by an equivalent length of pipe. This is an important advantage to low-pressure applications (e.g., fluid systems that are gravity fed and not pumped).

A distinct advantage of this class of flowmeters is their minimum exposure of system components to the actual fluid. As opposed to mechanical meters, which have metallic parts exposed to the process fluid, the magnetic flowmeter's only exposed parts are the electrodes and the liner or unlined plastic tube. This minimizes corrosion and/or erosion from problem fluids and liquids containing appreciable suspended solids.

Unlike the Venturi meter or differential flow metering devices, the magnetic flowmeter is unaffected by variations in fluid viscosity, temperature, and pressure. It is, however, sensitive to large changes in fluid conductivity and slightly affected by supply frequency and voltage variations. In general, magnetic flow systems feature low power consumption.

Magnetic flowmeters have been applied to a wide variety of process industries. In the chemical process industries, these flowmeters have been widely employed in control loops where reproducibility and flow stability are important. These meters are well suited to operations where flashing is a problem, since no pressure drop occurs across these systems. Magnetic flowmeters are used for metering all types of corrosives, strong acids and bases, and lumpy and viscous materials. Specific metering applications include nitric acid, aqua regia, caustics, polystyrene, butadiene latexes, aluminum chloride, various inorganic salts, ferric acids, and ammonium hydroxide.

In the pulp and paper industry, magnetic flowmeters are used in handling high-temperature black liquors. Examples of applications include measuring stock and pulp flows, whitewater, dyes, additives, bleaching chemicals, lime sludge, clay slurries, liquors, alum, and titanium dioxide slurries.

In the metals mining and refining industries, these meters are used for metering iron ore, copper and aluminum slurries, grinding circuits, and basic mineral slurries. In the steel industry, magnetic flowmeters are employed in a variety of water flow applications in continuous casting and influent/effluent waste treatment systems. Most of these applications involve abrasive slurries.

In the power industry, these systems are useful for measuring the flow of cooling tower water, recirculating water, carbon slurries, chemical wash, and in SO_2 scrubbing and waste treatment systems.

In the food industry, magnetic flowmeters offer the advantage of clean-in-place design. They are employed in all segments of the food industry, for example, milk standardization, corn processing, aeration control, brewing, and sugar refining.

In the textile industry, magnetic flowmeters are used in chemical metering and waste stream monitoring applications and in measuring the flow of dyes, caustics, resins, and water.

These meters are also widely used in virtually all aspects of wastewater treatment systems, including dredging and mass flow metering.

Magnetic flowmeters are capable of measuring most industrial liquids. Water, acids, bases, slurries, liquids with suspended solids, and industrial wastes are some of the examples already given. The primary limitation of this type of meter is the electrical conductivity of the fluid.

Meter size, liner material, and materials for the electrodes as well as the specific arrangements for installation will largely depend on the specific fluid to be metered and the process temperature. If the liquid to be metered contains solids that may coat the liner, the meter's transmitter should be sized for normal operating velocities at a minimum of 1.5 m/s. This tends to minimize buildup and allows the coils to be wired in series, thus minimizing the temperature rise [26].

The transmitter should be installed vertically with the flow upward (this again will minimize coating from solids settling out). The line should be arranged so that when no flow conditions exist the transmitter remains full and thus at a lower temperature than if it were empty.

Liquids that are highly viscous will leave a coating on the tube walls, even after drainage. In such cases, the transmitter should be arranged in the line so that it can be conveniently flushed with a suitable solvent. Also, during lengthy shutdown periods, the tube should be left full of water in order to minimize baking of solids onto the liner surface and/or electrodes.

It should be noted that magnetic flowmeters are far from being trouble-free. Periodic maintenance and cleaning are necessary. For this reason it is often advisable to install the transmitter in a bypass loop or with a "Y" or "T" adjacent to it with removable plug for accessibility [26]. The flow transmitter should be installed with the flow upward. This arrangement tends to equally distribute liner wear.

For liquids that are relatively noncorrosive but are abrasive, an abrasion-resistant material should be selected for the liner. The manufacturer should be consulted for recommendations. The transmitter should be sized for a maximum fluid velocity of 3 m/s-again, to minimize wear [26]. The unit should be installed sufficiently downstream of any flow disturbances in order to prevent the fluid stream from impinging on or causing excessive wear to any portion of the liner.

For liquid/liquid flows, where the fluids are immiscible, the transmitter should be sized for normal operating velocities, no less than 1.5 m/s. The unit should be installed relatively close to a pump or other point where the stream has undergone mixing and the flow is close to being homogeneous. An elbow, for example, will often provide sufficient mixing action to the stream.

For applications handling liquids in the form of raw sludge, the unit should be installed in such a manner as to provide easy accessibility for cleaning and repairs. The transmitter should be sized for normal operating velocities in excess of 1.8 m/s.

Excessive or rapid temperature rises in the fluid can adversely effect the meter's accuracy. Permanent damage to the meter and recording instrumentation can occur when excessive temperatures are reached, particularly under no-flow conditions. The maximum temperature reached in a transmitter under stagnant conditions depends on the initial fluid temperature and the contributed rise

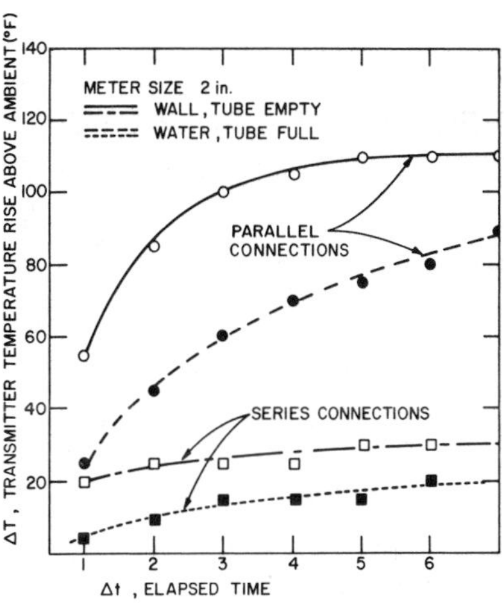

Figure 39. Transmitter temperature rise above ambient conditions. Note that for liquids other than water, the ΔT can be estimated by dividing the temperature rise for water by the product of the specific heat and the specific gravity of the liquid under consideration. The ΔT for a liquid, however, will never exceed that for air [27].

due to the energy dissipated by the field coils. The contributed temperature use depends on a number of factors, including the meter size, the field coils connection arrangement (i.e., series or parallel), whether the transmitter is in contact with liquid or not, and the thermal conductivity of the fluid in contact with the tube wall.

Figure 39 illustrates the temperature rise with time for a 2-in. magnetic meter. The temperature used above ambient conditions is reported for stagnant conditions, with the transmitter full of water and with air. Figure 39 also illustrates the magnitude of the temperature rise experienced for both parallel and series coil connections. As shown, the temperature rise experienced by the parallel-connected coils is substantially greater than with the series-connected arrangements.

The maximum temperature experienced by the transmitter will impact on the maximum permissible temperature for any given liner material as well as the maximum temperature to which the process liquid can be subjected. The selection of materials for tube and electrodes will primarily depend on bulk fluid properties, process temperature, and trace constituents in the fluid stream. The user must employ considerable discretion when selecting materials for a specific fluid metering application. Meter manufacturers and/or materials experts should always be consulted for troublesome applications or where there is a problem in selecting the right materials.

The selection of the metal for the electrodes can be a critical decision because of the small size of the electrode and the integrity of sealing that is required. A minimal degree of corrosion is often tolerable for massive structures; however, it can be intolerable on small electrodes. Corrosion rates in excess of 0.0051 cm/year are generally considered unacceptable for magnetic flowmeter electrodes [27]. Temperature considerations must also be weighed in selecting the proper electrode to ensure that it is compatible with the maximum temperature limitation on the liner material.

With regard to liner materials, the manufacturer will supply information on the chemical and abrasion resistance and temperature limitations of various materials. It is important to note that it is not often possible to predict or assess a particular liner materials' usefulness in service. Properties are normally reported at one temperature. Properties will, however, vary considerably with temperature (for example, chemical resistance will often decrease with increasing temperature).

The amount and types of contamination that may be present in a fluid stream will also influence liner material selection. The liner material can often be easily selected to meet compatibility requirements with the main constituent in the stream; however, it is usually much more difficult to select a liner that is compatible with a mixture of fluids.

Table 11 contains information on the compatibility of various materials for magnetic flowmeter applications. The tabulated data are based on secondary information compiled by one manufacturer. Table 11 is useful, however, as a rough guide to materials selection for magnetic flow transmitters.

Table 11
Rough Guide to Materials Selection for Magnetic Flowmeters*

Process liquid	Flowtube i.d. Material†							Electrode Metal			
	TFE (T)	Kynar (X)	Kel-F (X)	Polyurethane (A)	Epoxy (Fibercast RB 2530)	Polyester (Atlac 382)	Neoprene (N)	Platinum ±10% iridium (P)	Tantalum (B)	316 SST (S)	Hastelloy C (H)
Acetaldehyde	300	—	250	X	NR	NR	X	P	P	S[1]	P
Acetic acid 10%	300	230	200	X	150	*200	0	P	P	S[1]	P
Acetic acid 25%	300	230	180	X	100	200	0	P	P	S[1]	P
Acetic acid 50%	300	230	180	X	100	160	—	P	P	NR	P
Acetic acid 75%	300	—	120	X	100	160	—	P	P	NR	P
*Glacial 100%	300	150	120	X	100	—	X	P	P	NR	P
Alum	*350	*250	220	—	200	*200	0	P	P	NR	NR
Aluminum chloride	*350	*250	220	—	*200	*200	0	P	P	NR	NR
Aluminum chlorohydrate	*350	*250	—	—	—	200	—	P	P	NR	NR
Aluminum fluoride	*350	*250	—	—	150	75	0	P	NR	NR	NR
Aluminum hydroxide	*350	*250	220	—	150	—	—	P	P	S	NR
Aluminum nitrate	*350	*250	220	—	*200	160	0	P	NR	NR	NR
Aluminum sulfate	*350	*250	220	0	*200	*200	0	P	P	S	S
Ammonium bicarbonate	*350	*250	*250	—	200	160	—	P	P	NR	NR
Ammonium carbonate	*350	*250	*250	—	200	75	0	P	P	NR	NR
Ammonium chloride	*350	*250	*250	—	200	*200	0	P	P	NR	S
Ammonium bifluoride	*350	*250	—	—	—	—	—	P	P	NR	NR
Ammonium fluoride	*350	*250	—	—	150	—	—	P	NR	NR	S
Ammonium hydroxide 10%	*350	*250	*250	0	200	140	0	P	S	S	S
Ammonium hydroxide 20%	*350	*250	*250	0	150	140	0	P	S	S	S
Ammonium hydroxide 30%	*350	*250	*250	0	100	100	0	P	S	S	S
Ammonium nitrate	*350	*250	*250	X	*200	*200	LT	P	P	NR	S

Ammonium persulfate	*350	*250	*250	X	NR	180	0	P	P	NR	NR
Ammonium phosphate	*350	*250	*250	—	150	75	0	P	P	NR	NR
Ammonium sulfate	*350	*250	*250	0	*200	*200	—	P	P	NR	NR
Ammonium sulfide	*350	*250	*250	—	—	—	—	P	P	NR	NR
Antimony pentachloride	*350	*250	—	—	—	70	—	P	P	NR	NR
Antimony trichloride	*350	70	—	—	150	*200	—	P	P	NR	NR
Aqua regia	—	70	120	X	—	NR	X	NR	P	NR	NR
Arsenic acid	*350	*250	*250	X	—	70	0	P	P	NR	NR
Arsenious acid	*350	*250	*250	—	—	180	—	P	P	NR	NR
Barium acetate	*350	*250	*250	—	—	140	—	P	P	NR	NR
Barium carbonate	*350	*250	*250	—	*200	*200	—	P	P	NR	NR
Barium chloride	*350	*250	*250	0	*200	*200	0	P	P	NR	S
Barium hydroxide	*350	*250	*250	0	200	70	0	P	NR	S	NR
Barium sulfate	*350	*250	*250	0	*200	70	0	P	P	NR	NR
Barium sulfide	*350	*250	*250	0	*200	140	0	P	P	NR	NR
Benzene sulfonic acid	—	70	160	—	NR	200	0	P	P	NR	NR
Borax	*350	*250	*250	0	*200	70	0	P	P	NR	NR
Cadmium chloride	*350	*250	*250	—	*200	70	—	S	P	NR	NR
Calcium bisulfite	*350	*250	175	0	200	180	0	P	P	NR	NR
Calcium carbonate	*350	*250	175	—	*200	*200	—	P	P	S	S
Calcium chlorate	*350	*250	175	—	*200	*200	—	P	P	NR	NR
Calcium chloride	*350	*250	175	0	*200	*200	0	P	P	NR	NR
Calcium hydroxide 15%	*350	*250	175	0	200	160	0	P	P	NR	S
Calcium hydroxide 20%	*350	*250	175	0	200	160	0	P	P	NR	S
Calcium hydroxide 25%	*350	*250	175	0	200	140	0	P	P	NR	S
Calcium hypochlorite	*350	*250	85	—	NR	85	X	P	P	NR	S
Calcium nitrate	*350	*250	175	0	*200	*200	0	P	P	NR	NR
Calcium sulfate	*350	*250	175	0	*200	*200	—	P	P	NR	NR
Chloroacetic acid 25%	300	212	200	—	100	200	—	P	P	NR	S
Chloroacetic acid 50%	300	212	—	—	—	140	—	P	P	NR	S
Choroacetic acid 100%	300	—	—	—	—	—	—	P	P	NR	S
Chlorosulfonic acid	200	NR	*250	X	—	NR	X	P	P	NR	S
Chromic acid 50%	300	*240	*250	X	150	70	X	P	P	NR	NR
Chromic acid 10%	300	*250	*250	X	150	70	X	P	P	NR	NR

Table 11 (Continued)

Process liquid	Flowtube i.d. Material[†]							Electrode Metal			
	TFE (T)	Kynar (X)	Kel-F (X)	Polyurethane (A)	Epoxy (Fibercast RB 2530)	Polyester (Atlac 382)	Neoprene (N)	Platinum ±10% iridium (P)	Tantalum (B)	316 SST (S)	Hastelloy C (H)
Chromic acid 30%	300	150	*250	X	75	NR	X	P	P	NR	NR
Chromic acid 100%	300	—	—	X	NR	NR	X	P	P	NR	NR
Chromic fluoride	*350	—	—	X	75	—	X				
Chromium sulfate	*350	*250	*250	X	—	140	X	P	P	NR	S
Copper chloride	*350	*250	*250	0	*200	*200	0	NR	P	NR	NR
Copper cyanide	*350	*250	*250	0	—	*200	0	P	P	S	S
Copper fluoride	*350	*250	—	—	*200	—	—	P	NR	NR	NR
Copper nitrate	*350	*250	*250	—	*200	*200	—	P	P	NR	S
Copper oxychloride	*350	*250	—	—	—	—	—	P	NR	NR	NR
Copper sulfate	*350	*250	*250	0	200	*200	0	P	P	S	S
Ferric chloride	*350	*250	*250	0	*200	*200	0	NR	P	NR	S
Ferric nitrate	*350	*250	*250	—	*200	*200	0	P	P	NR	S
Ferric sulfate	*350	*250	*250	—	200	*200	0	P	P	NR	S
Ferrous chloride	*350	*250	*250	—	*200	*200	—	P	P	NR	NR
Ferrous nitrate	*350	*250	*250	—	—	*200	—	P	P	NR	S
Ferrous sulfate	*350	*250	*250	—	200	*200	—	P	P	NR	NR
Fluoboric acid	200	—	—	—	NR	*200	0	P	NR	NR	NR
Fluosilici acid 10%	200	—	—	—	200	150	0	P	NR	NR	NR
Fluosilicic acid 30%	200	—	—	—	—	70	0	P	NR	NR	NR
Fluosilicic acid 40%	200	—	—	—	—	—	0	P	NR	NR	NR
Fluosulfonic acid	200	—	—	—	—	—	—	—	—	—	—
Formaldehyde	*350	*250	*250	X	150	—	0	P	P	NR	S

Formic acid 10%	*350	*250	180	X	—	150	0	P	P	NR	S
Formic acid 25%	*350	250	180	X	100	—	0	P	P	NR	S
Formic acid 50%	*350	250	180	X	—	70	0	P	P	NR	S
Formic acid 100%	*350	250	—	X	—	—	0	P	P	NR	S
Glycerin	*350	*250	*250	0	*200	*200	0	P	P	P	P
Hydrobromic acid 50%	*350	*250	200	X	150	160	0	P	P	NR	NR
Hydrochloric (cone)	*350	*250	180	X	200	160	LT	X	P	NR	NR
Hydrocyanic (all)	*350	*250	*250	—	NR	200	LT	P	P	S	S
Hydrofluoric acid 10%	*350	250	120	X	NR	180	LT	P	NR	NR	NR
Hydrofluoric acid 20%	*350	250	120	X	NR	120	LT	P	NR	NR	NR
Hydrofluosilicic 10%	*350	*250	—	—	—	150	LT	P	NR	NR	S
Hydrofluosilicic 35%	*350	*250	—	—	—	70	LT	P	NR	NR	S
Hydrofluosilicic 100%	100	*250	—	—	—	—	LT	P	NR	NR	S
Hyrdofluoric 100%	78	212	120	X	NR	—	LT				
Hydrogen peroxide 5%	*350	*250	150	—	150	150	—	P	P	S	S
Hydrogen peroxide 30%	*350	*250	150	—	75	70	—	P	P	S	S
Hydroxy acetic acid 35%	200	—	—	—	—	—	—	P	P	S	S
Hyrdoxy acetic acid 70%	150	—	—	—	—	—	—	P	P	S	S
Hypochlorous acid 10%	300	—	—	—	200	—	—	NR	P	NR	S
Hypochlorous acid 20%	300	*250	—	—	—	160	—	NR	P	NR	S
Lead acetate	*350	*240	*250	—	*200	*200	LT	NR	P	NR	NR
Lithium chloride	*350	*250	*250	—	—	—	—	P	P	NR	S
Magnesium bisulfite	*350	*250	*250	—	*200	175	—	—	—	S	S
Magnesium carbonate	*350	*250	*250	—	*200	175	—	P	P	S	S
Magnesium chloride	*350	*250	*250	0	*200	*200	0	P	P	NR	S
Magnesium hydroxide	*350	*250	*250	0	*200	—	0	P	NR	NR	NR
Magnesium nitrate	*350	*250	*250	—	*200	—	—	P	P	NR	NR
Magnesium sulfate	*350	*250	*250	—	*200	*200	0	P	P	S	NR
Mercuric chloride	*350	250	150	—	—	*200	0	P	P	NR	NR
Mercurous chloride	*350	250	150	—	—	*200	—	—	—	—	—
Nickel chloride	*350	*250	*250	—	*200	*200	0	P	P	NR	S
Nickel nitrate	*350	*250	*250	—	200	*200	—	P	P	NR	NR
Nickel sulfate	*350	*250	*250	0	—	*200	0	P	P	NR	NR
Nitric acid 10%	*350	*250	*250	X	NR	160	X	P	P	S	S

Table 11 (Continued)

Process liquid	Flowtube i.d. Material[†]							Electrode Metal			
	TFE (T)	Kynar (X)	Kel-F (X)	Polyurethane (A)	Epoxy (Fibercast RB 2530)	Polyester (Atlac 382)	Neoprene (N)	Platinum ±10% iridium (P)	Tantalum (B)	316 SST (S)	Hastelloy C (H)
Nitric acid 15%	*350	*250	*250	X	NR	—	X	P	P	NR	NR
Nitric acid 20%	*350	200	*250	X	NR	—	X	P	P	NR	NR
Nitric acid 40%	*350	200	*250	X	NR	70	X	P	P	NR	NR
Nitric acid 70%	200	120	*250	X	NR	NR	X	P	P	NR	NR
Oxalic acid (saturated)	*350	120	*250	—	*200	*200	LT	P	P	NR	NR
Perchloric acid 70%	300	120	75	—	75	—	0	P	P	NR	NR
Phenol 10%	BP	212	175	X	150	NR	X	P	P	S	S
Phosphoric acid 75%	300	*250	*250	—	*200	*200	—	P	P	NR	NR
Phosphoric acid 85%	300	230	*250	—	NR	*200	—	P	P	NR	NR
Phosphoric oxychloride	300	—	—	—	—	—	—	—	—	—	—
Potassium aluminum sulfate	*350	*250	*250	—	—	*200	—	P	P	NR	NR
Potassium bicarbonate	*350	*250	*250	—	*200	70	—	P	P	S	S
Potassium carbonate	*350	*250	*250	—	*200	70	—	P	P	S	S
Potassium chloride	*350	*250	*250	0	*200	*200	0	P	P	NR	S
Potassium dichromate	*350	*250	*250	0	*200	*200	0	P	P	S	NR
Potassium ferricyanide	*350	*250	*250	—	—	*200	0	NR	P	NR	NR
Potassium ferrocyanide	*350	*250	*250	—	—	*200	—	NR	P	NR	NR
Potassium hydroxide 10%	200	200	250	LT	200	150	0	P	NR	NR	NR
Potassium hydroxide 20%	200	200	250	LT	—	70	0	P	NR	NR	NR
Potassium hydroxide 45%	200	212	250	LT	—	—	0	P	NR	NR	NR
Potassium nitrate	*350	*250	*250	0	*200	*200	0	P	P	NR	NR
Potassium permanganate	200	*250	250	—	150	*200	—	P	P	NR	NR

Potassium persulfate	*350	*250	250	—	—	*200	—	P	P	S	S
Potassium sulfate	*350	*250	250	0	150	*200	0	P	P	S	NR
Silver nitrate	*350	*250	*250	0	*200	*200	0	P	P	NR	S
Sodium acetate	*350	*250	*250	X	200	*200	LT	P	P	NR	NR
Sodium bicarbonate	*350	*250	*250	—	*200	140	0	P	P	S	S
Sodium bisulfate	*350	*250	*250	—	*200	*200	—	P	P	NR	NR
Sodium bisulfide	*350	*250	*250	—	—	—	—	P	P	S	NR
Sodium bisulfite	*350	*250	*250	—	—	*200	0	P	P	NR	S
Sodium borate	*350	*250	*250	—	—	—	0	P	P	NR	NR
Sodium bromide	*350	*250	*250	—	200	*200	—	P	P	NR	NR
Sodium carbonate	*350	*250	*250	—	*200	160	—	P	P	S	S
Sodium chlorate	*350	*250	*250	—	—	*200	—	P	P	NR	NR
Sodium chloride	*350	*250	*250	0	*200	*200	0	P	P	NR	NR
Sodium chlorite	*350	*250	*250	—	—	150	—	S	P	NR	NR
Sodium chromate	*350	*250	*250	—	—	*200	—	P	P	NR	NR
Sodium cyanide	*350	*250	*250	—	*200	*200	0	NR	P	S	—
Sodium dichromate	*350	*250	*250	—	*200	*200	—	—	—	—	—
Sodium ferricyanide	*350	*250	*250	—	—	*200	—	NR	S	NR	—
Sodium ferrocyanide	*350	*250	*250	—	*200	—	—	NR	S	NR	—
Sodium fluoride	*350	*250	—	—	*200	—	—	P	NR	NR	NR
Sodium hexametaphosphate	*350	*250	*250	—	—	—	—	—	—	—	—
Sodium hydrosulfide	*350	*250	*250	—	—	—	—	P	P	S	S
Sodium hydroxide 5%	*350	212	*250	LT	200	200	0	P	NR	S	NR
Sodium hydroxide 10%	*350	212	*250	LT	200	160	0	P	NR	S	NR
Sodium hydroxide 25%	*350	212	*250	LT	200	140	0	P	NR	S	NR
Sodium hydroxide 50%	*350	212	*250	LT	200	70	0	P	NR	S	NR
Sodium hypochlorite 5%	BP	*250	75	X	NR	—	LT	P	P	NR	S
Sodium hypochlorite 10%	—	*250	—	X	NR	—	LT	P	P	NR	S
Sodium hypochlorite 15%	—	*250	—	X	NR	—	LT	P	P	NR	S
Sodium hypochlorite 20%	200	*250	—	X	NR	—	LT	P	P	NR	S
Sodium nitrate	*350	*250	*250	—	*200	*200	0	P	P	S	NR
Sodium nitrite	*350	*250	*250	—	—	*200	—	P	P	NR	NR
Sodium silicate	*350	*250	*250	—	150	*200	0	P	P	NR	NR
Sodium sulfate	*350	*250	*250	0	*200	*200	0	P	P	NR	NR

Table 11 (Continued)

Process liquid	Flowtube i.d. Material†							Electrode Metal			
	TFE (T)	Kynar (X)	Kel-F (X)	Polyurethane (A)	Epoxy (Fibercast RB 2530)	Polyester (Atlac 382)	Neoprene (N)	Platinum ±10% iridium (P)	Tantalum (B)	316 SST (S)	Hastelloy C (H)
Sodium sulfide	*350	*250	*250	—	—	*200	0	P	P	NR	NR
Sodium sulfite	*350	*250	*250	—	200	*200	0	P	P	S	NR
Sodium tetraborate	*350	*250	*250	—	—	—	—	P	P	S	S
Sodium thiosulfate	*350	*250	*250	0	150	—	0	NR	NR	NR	P
Stannic chloride	*350	*250	*250	—	200	*200	0	P	P	NR	NR
Stannous chloride	*350	*250	*250	—	—	*200	0	P	P	NR	NR
Sulfamic acid	300	*250	—	—	—	*200	—	P	P	—	—
Sulfuric acid 10%	*350	*250	*250	LT	*200	*200	LT	P	P	NR	NR
Sulfuric acid 25%	*350	*250	*250	0	150	*200	0	P	P	NR	NR
Sulfuric acid 50%	*350	*250	*250	0	100	*200	0	P	P	NR	NR
Sulfuric acid 80%	*350	212	*250	X	NR	70	X	P	P	NR	NR
Sulfuric acid 100%	300	150	*250	X	NR	—	X	NR	NR	NR	NR
Sulfurous acid 10%	*350	212	—	0	200	—	LT	P	P	NR	NR
Trisodium phosphate	*350	*250	250	—	150	70	—	P	P	S	S
Zinc chloride	*350	*250	*250	—	*200	*200	0	P	P	NR	NR

Courtesy of Foxboro Corp., Foxboro, Mass.

* *Key: X, NR, not recommended; P, preferred material (virtually unlimited life); S, satisfactory material (reasonable service life); LT, low temperature use only (~100°F); 0, compatible to rated tube temperature.*

† *A maximum recommended temperature marked with an asterisk (*) indicates that the transmitter construction is the limiting factor as opposed to the tube i.d. material in the particular liquid being the limiting factor. Maximum recommended temperatures of process liquids in transmitters, unless otherwise noted, are as follows: Kynar or Kel-F, 250°F; urethane, 160°F; neoprene, 150°F; tetrafluoroethylene (TFE), 350°F; and epoxy- or polyester-reinforced fiberglass (unlined), 200°F.*

ULTRASONIC FLOW MEASUREMENT

A Doppler transducer or ultrasonic flow measuring systems meter fluid flow by employing the Doppler frequency shift of ultrasonic signals reflected from discontinuities in the flowing liquid. This class of flowmeters offers the following advantages:

1. Nonintrusive flow measurement
2. No moving parts
3. No pressure loss

Flow measurement is achieved by using the Doppler frequency shift of ultrasonic signals that are reflected from various forms of discontinuities in the liquid stream. These discontinuities can be in the form of suspended solids, bubbles, or interfaces generated by turbulent eddies in the flow stream. Figure 40 illustrates the principle of operation. The meter or "sensor" is mounted on the outside of the pipe, and an ultrasonic beam from a piezoelectric crystal is transmitted through the pipe wall into the fluid at an angle to the flow stream.

Signals reflected off of flow disturbances are detected by a second piezoelectric crystal located in the same sensor. Transmitted and reflected signals are compared in an electric circuit. The frequency shift is proportional to the flow velocity. The electronic circuit is employed for the addition of factors to take into account pipe and liquid constants prior to generating an analog signal to feed a meter and totalizer or batch counter (refer to Figure 41) [28].

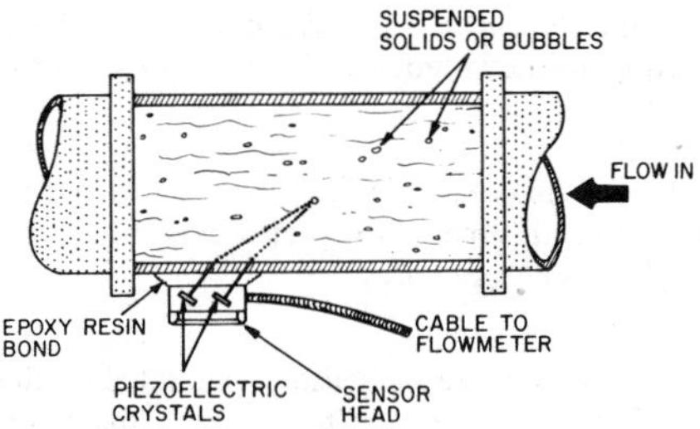

Figure 40. Application of the Doppler flowmeter to metering liquids in pipe flow. (Courtesy of Bestobell Mobrey Limited, Slouth, Bucks, England.)

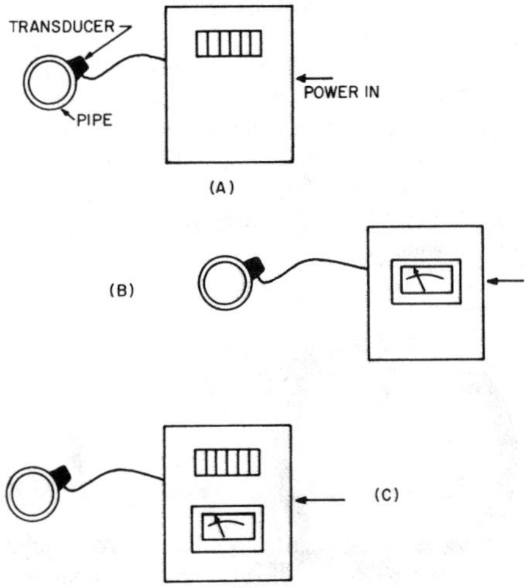

Figure 41. The Doppler transducer flowmeter can be coupled with: (A) a controller that consists of a totalizer which indicates the volume metered; (B) a flow indicator that records the rate of mass flow or velocity; or (C) a flowmeter totalizer that indicates the rate of flow or velocity.

Electronic units are factory calibrated to provide readouts of mean velocity (i.e., meters per second). Internal adjustments can be made to accommodate variations generated from different flow profiles or site conditions. Usually other internal adjustments enable flow speed readings to be translated into suitable volume units for meter or totalizer readouts, calculated in accordance with pipe sizes.

Factory calibration is generally good to within 5%, based on information supplied on pipe material and size, site conditions, fluid composition, and temperature. On-site flow calibration can, however, provide measurement accuracy to within 1% of the actual flow [28]. The amplitude of reflected signals will largely depend on the number of particles or discontinuities present in the flow stream. Doppler flowmeters have several advantages. They allow noncontact flow measurements by means of an externally mounted transducer. The unit is relatively easy to install and align, permitting flow measurement with no alterations to piping, no downtime, and no pressure loss.

There is no upper size limit on pipe diameters; hence, capital costs are the same regardless of the application. In addition, this flow measuring system is adaptable to a fairly wide range of piping materials. The system works well with metal, plastic, and other homogeneous piping materials.

The Doppler system is applicable to many problem liquids, including slurries, acids, viscous fluids, and various liquids containing suspended solids.

The Doppler system does have some limitations, however. The principal limitations are:

1. Liquids to be metered must have an excess of approximately 2% suspended solids by volume.
2. Liquid linear velocities must exceed 0.15 m/s.
3. The transducer head must be maintained at a temperature below 83°C.
4. Piping material must be of a homogeneous composition (i.e., materials composed of more than one substance, such as concrete, reinforced concrete, vitreous clay, or asbestos cement cannot be used with this method).
5. Pipe wall thickness cannot be greater than 1.91 cm. The Doppler system is only applicable to full pipe flow. Multiphase and some types of multicomponent flows cannot be metered by this method. Probably the most serious limitation on this method is that it can only meter liquids flows. The Doppler method will not work on solids or gases.

Doppler flowmeters are generally low-maintenance systems. Since the sensor is mounted on the outside of an existing standard plant pipework (i.e., there is no intrusion to the flow), the meter cannot become fouled by dirt, corrosion, or suspended solids. In fact, the presence of dirt or suspended solids provides excellent reflectors for the ultrasonic signal to give a high-amplitude Doppler signal.

The ideal location for the transducer, as recommended by manufacturers, is on a vertical pipe [28, 29]. If the sensor has to be mounted on a horizontal pipeline, a 45° or 135° position from the pipe bottom is recommended (refer to Figure 42).

The transducer should be located roughly ten diameters of straight pipe downstream of any flow disturbances (i.e., pumps, tees, elbows, reducers, etc.). If the sensor is located near such disturbances, calibration adjustments are often necessary, which usually affect the linearity of the instrument over a wide flow span.

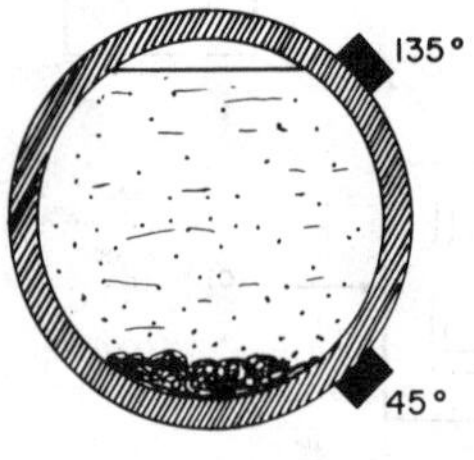

Figure 42. Proper location of Doppler transducer for a horizontal flow application.

The sensor is normally bonded to the outside pipe wall with a slow hard-setting epoxy resin. The surface to which the transducer is to be bonded should be thoroughly cleaned of any oil, dirt, paint, rust, etc. Epoxy is normally applied to the transducer face and the unit firmly pressed onto the pipe surface. Care must be taken to avoid cracks or air holes in the epoxy. The sensor should be evenly attached to the pipe surface as a wedge-shaped bond or excessive epoxy will cause the signal to be dispersed, causing erroneous measurements. During the bonding process the sensor is positioned parallel to the pipe axis and maintained stationary by a mechanical clamp.

Note that it is often advisable to select the permanent installation site after site testing is performed. Initial site testing can be done by using silicon grease as a couplant between the pipe wall and sensor.

Ultrasonic flow measuring techniques can be used in open channel flow applications. The principle of operation is different from that of the Doppler flowmeter. Ultrasonic open channel flow measuring systems translate ultrasonic water depth measurements to a linear readout of flow rate. A computer integrator is usually used to translate depth measurements to an indication of flow rate, to totalize flow, and to control flow-proportional composite sampling [30]. Figure 43 illustrates one type of installation to metering the flow in open channels. The unit shown can be programmed to match the characteristics of all standard flumes, weirs, and nonrestricted free-flowing pipes and channels where flow rate is always the same at a given depth.

The front face of the sensor (see Figure 43) must be positioned so that the signal beam is a 7° cone projected from the edges of the front face, encompassing the area of the flow being monitored. The sensor can be mounted up to 152 m from the electronic system (this varies with the model). Special cable connections between the sensor and recording instrument are necessary for temperature compensation [31]. System calibration is accomplished by positioning targets to represent maximum and zero flow with reference to the ultrasonic sensor face. Span and zero potentiometers are adjusted until calibration has been achieved.

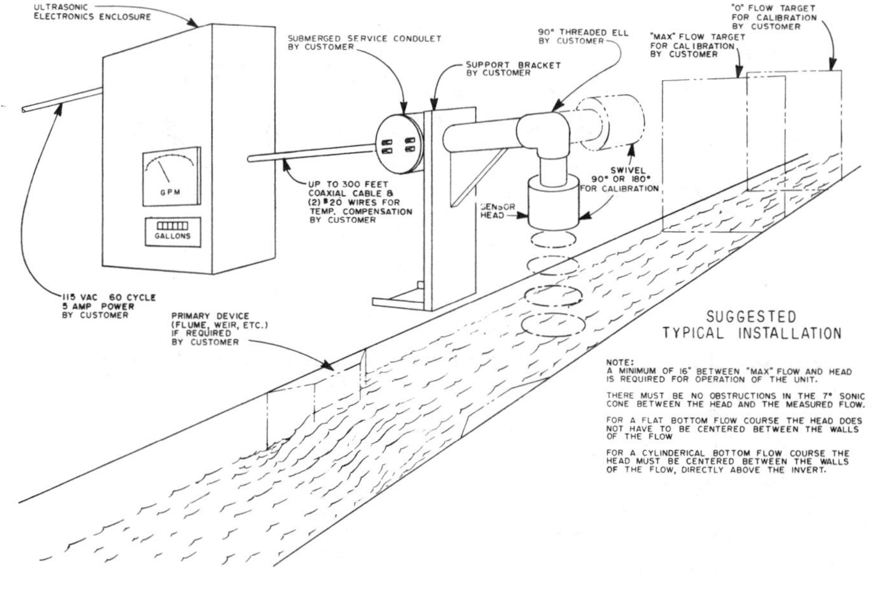

Figure 43. Ultrasonic flow measuring system as applied to open channel flow. Such an arrangement can be applied to industrial and municipal flow applications. (Courtesy of Cochrane Environmental Systems, PA.)

MASS FLOW MEASUREMENT

The need for highly accurate flow measurement becomes very important when processes must be optimized and waste streams minimized. The need is especially acute when dealing with costly chemicals or valuable petrochemicals and petroleum products. Mass flow measurement techniques allow this close control. In addition, because energy utilization is critical, mass flow measurement techniques have been found to be one of the most effective methods of optimizing fuel consumption.

There are a number of advantages to using mass flow measurement. One is the significant reduction in the number of measurements that may be required and their associated error. Process parameters such as pressure, temperature, and specific gravity need not be considered in the measurements; hence, data-handling is often simplified.

In principle, linear mass flowmeters can be considered hydraulic equivalents of the electrical Wheatstone bridge. The meter is schematically shown in Figure 44. As shown, the meter consists of four matched orifices arranged in a hydraulic bridge. The pump shown across the bridge serves to create a constant recirculating reference flow or hydraulic potential within the bridge. Since the orifices are matched, the recirculating flow q becomes equally distributed in the arms of the bridge.

Linear mass flowmeters have been widely used in metering relatively low flow rates. In Figure 45 a fluid stream is imposed across the bridge. If q > Q, then the flow through orifices b and d will be:

$$q_{bd} = \frac{q + Q}{2} \tag{34}$$

and the flow through orifices a and c will be:

$$q_{ac} = \frac{q - Q}{2} \tag{35}$$

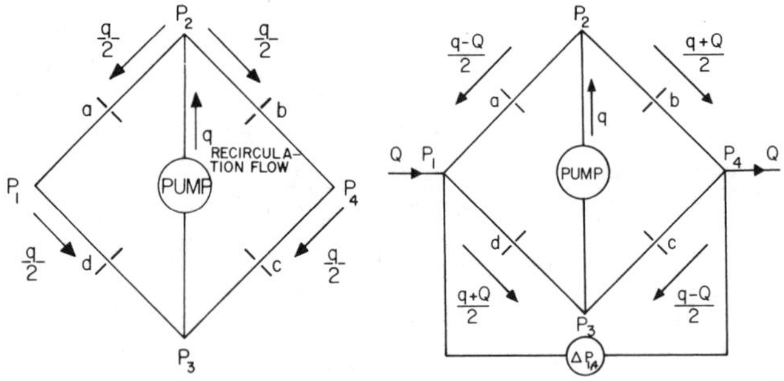

Figure 44. Operation of the linear mass flowmeter. (Courtesy of Flo-Tron, Inc., Paterson, NJ.)

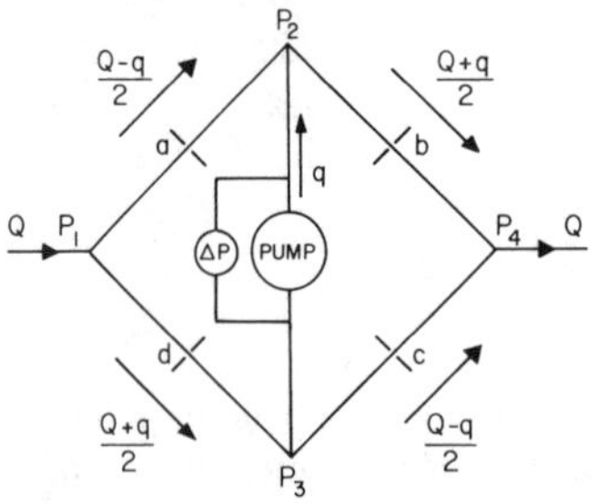

Figure 45. Hydraulic bridge with output signal for Q < q. (Courtesy of Flo-Tron, Inc., Paterson, NJ.)

Recall that the equation for flow through a single orifice is:

$$Q = CA\sqrt{\frac{\Delta P}{\rho}} \tag{36}$$

where C is the orifice discharge coefficient; A is the orifice area; ΔP is the pressure drop across the orifice; and ρ is the fluid density.

Hence, the flow versus pressure drop relationship for orifice b can be written as:

$$\frac{q + Q}{2} = C_b A_b \sqrt{\frac{P_2 - P_4}{\rho}} \tag{37}$$

and for orifice a:

$$\frac{q - Q}{2} = C_a A_a \sqrt{\frac{P_2 - P_1}{\rho}} \tag{38}$$

By squaring Equations 37 and 38, the following is obtained:

$$\frac{q^2 - 2Qq + Q^2}{4} = C_b^2 A_b^2 \frac{P_2 - P_4}{\rho} \tag{39}$$

$$\frac{q^2 - 2Qq + Q^2}{4} = C_a^2 A_a^2 \frac{P_2 - P_1}{\rho} \tag{40}$$

Since it has been assumed earlier that the orifices are balanced or matched, then:

$$C_b A = C_a A_a = CA \tag{41}$$

Then, combining Equations 39 and 40, the following expression is derived:

$$Qq = C^2 A^2 \frac{P_1 - P_4}{\rho} \tag{42}$$

Equation 42 can be rearranged in terms of the pressure differential across the hydraulic bridge:

$$\Delta P_{1,4} = K\dot{m} \tag{43}$$

where $K = q(CA)^2$ = a constant, and $\dot{m}$ is the process stream mass flow rate ($Q\rho$).

The principle of the hydraulic Wheatstone bridge has also been applied to handling fairly large ranges of flow conditions. As in the previous case, the output signal is a differential pressure signal, linear and proportional to the true mass flow.

For the case where $q < Q$, the flow through orifices b and d and through a and c will be given by Equations 34 and 35, respectively (refer to Figure 45, which is the normal arrangement for high flow application).

The flow-pressure drop expression for orifice d is given as:

$$\frac{Q + q}{2} = C_d A_d \sqrt{\frac{P_1 - P_3}{\rho}} \tag{44}$$

and for orifice a

$$\frac{Q - q}{2} = C_a A_a \sqrt{\frac{P_1 - P_2}{\rho}} \tag{45}$$

Following the derivation for low mass flow,

$$Qq = C^2 A^2 \frac{P_2 - P_3}{\rho} \tag{46}$$

$$\Delta P_{2,3} = KQ\rho \tag{47}$$

where K is defined as before.

Equation 47 allows the mass flow rate to be computed as a function of the pressure rise across the recirculating pump for the case where $Q > q$.

The selection of whether the recirculation rate q is greater or less than the process stream flow depends on the magnitude of Q, desired rangeability, and other factors [32, 33].

Typical mass flowmeter units are capable of measuring extremely low flow rates down to 0.1 lb/h with extreme accuracy (to within ±0.5% of the reading (±0.03% of full scale)). Models are often used to assist in work on automobile engine emission control and fuel economy.

Integrated with the flowmeter is a small heat exchanger. Fuel normally comes from a temperature-controlled supply. Measured quantities of fuel pass through the flowmeter while an additional amount of fuel is passed through the heat exchanger and serves as a coolant. The recirculating pump is located on the other side of the heat exchanger. This arrangement allows the orifice housing to be maintained at a constant temperature, independent of the surroundings [33].

Table 12 gives typical operating ranges and illustrates the versatility of these systems.

The four orifices are contained in an orifice housing that has the differential pressure ports and influent and effluent ports. Standard differential pressure transducers can be employed to sense the meter signal. The recirculating pump is usually a gear pump being driven at constant speed by a synchronous motor. This provides the necessary constant volume reference flow through the bridge arms.

Units having low flow capability are generally plagued with a series of problems. Therefore, care should be taken in selecting a particular system. Major problems include signal noise, signal drift due to temperature changes, and sensitivity.

Table 12
Specifications of a Low Flow Rate Meter (Model 10M Microflow Meter)

Flow range (lb/h)	0–10
Accuracy	±0.5% (±0.03% of full scale)
Repeatability	±0.25% (±0.03% of full scale)
Accuracy and repeatability span	50/1 with 0.1 lb/h minimum flow
Liquids	Gasoline, diesel, kerosene, similar hydrocarbons
Meter pressure drop (in H_2O/lb/h)	1.2
Operating pressure, maximum (psig)	200
Hazardous location	Explosion proof, Class 1, Group D
Heat exchanger fuel coolant flow (lb/h)	150 (minimum)
Materials of construction	Aluminum, steel, bronze, stainless, Buna N
Size (in.)	28(L) × 10-3/16(W) × 11-7/16(H)
Weight (lb)	68
Internal fluid volume (points)	2
Response	1 Hz or 1/16-s step response

Courtesy of Flo-Tron, Inc., Paterson, N.J.

In general, signals from all systems have some noise associated with them. Noise reduction is usually achieved by filtering the signal or by down response time, which is not desirable when metering low flows. Another approach to noise reduction is through the use of the low flow recirculating pump on these systems. This reduces turbulent flow noise in the meter, a primary source. Systems are usually equipped with special screened baffles within the orifice housing to further dampen turbulent noise.

Drift resulting from transient fluid temperature variations is another critical parameter affecting low flow measurement. Temperature changes can be attributed to variations in the process stream or ambient conditions. Changes in temperature will consequently impact on fluid density, causing transient flow conditions to arise. That is, temperature changes can cause a flow transient, which is directly proportional to the magnitude of the time rate of temperature change and the volume of fluid affected [33, 34]. Although for large flows this effect is often neglected, the impact on microflow measurements can be great. Because of this, the fluid temperature must be stabilized so that the flow transient is minimized. This is done by the use of the aforementioned heat exchanger.

For most applications the coolant can be water or in some cases fuel. The important criterion for the specific coolant used is its own temperature stability.

The weigh scale and timer method can be used to calibrate this type of flowmeter. The time of flow of a known amount of fluid can be measured by employing a balanced scale to start and stop an electronic timer. The weight of fluid collected divided by the recorded time interval gives the mass flow rate. High-precision scales can be employed for extreme accuracy.

For volatile liquids (for example, gasoline), the amount of evaporation from the collection vessel can cause significant errors in the calibration data. Evaporation may be controlled by enclosing the collection vessel so that the volume above the liquid surface does not circulate to the surroundings. The enclosed volume will eventually reach a saturated condition and usually variations in the vapor content during a run can be considered negligible. Note that the flow tube feeding the collection tank or beaker must discharge within the enclosed vessel to minimize vaporization losses.

The weigh scale and timer can be mounted on a calibration stand along with a reservoir. The reservoir contains the liquid for calibrating the meter as well as the coolant for the heat exchanger.

Typical output data are shown in Figure 46. Recall that the hydraulic bridge produces an output of differential pressure, which is both linear and proportional to the true mass liquid flow rate. The primary advantage with this type of flowmeter i- that measurements are essentially unaffected by fluid temperature variations and (consequently) density. In addition, these meters are unaffected by changes in viscosity. The meter can be set to measure flow rate in direct units of mass flow (e.g., lb/h, kg/s).

Figure 47 shows a typical calibration curve. The error band shown is $\pm 0.5\%$ of the reading (with $\pm 0.03\%$ of full scale).

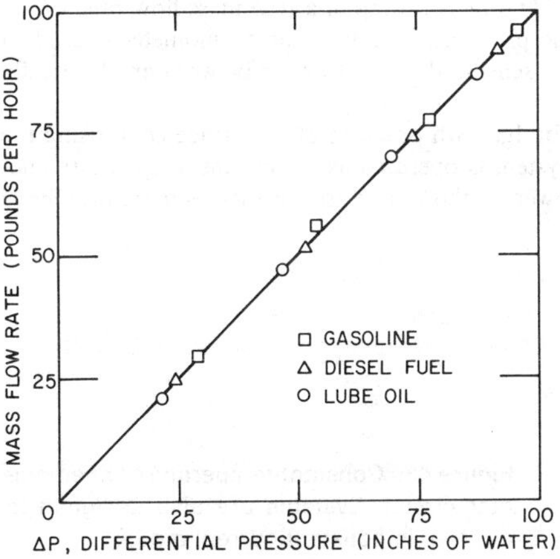

Figure 46. Typical output signal and calibration curve from mass flowmeters. (Courtesy of Flo-Tron, Inc., Paterson, NJ.)

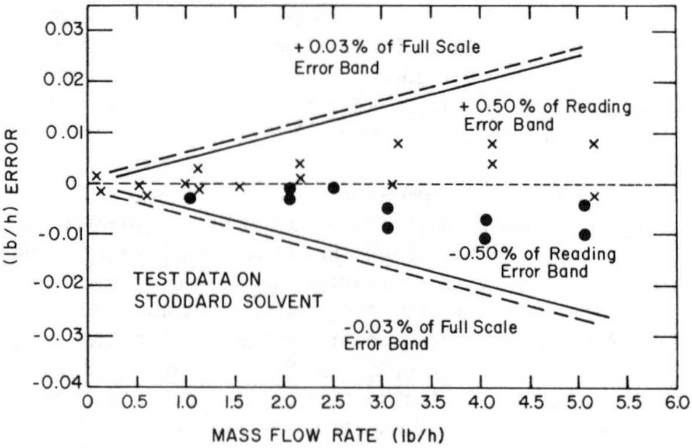

Figure 47. Typical calibration curve for a low mass flow linear flowmeter. The plot shows the error band associated with the recorded data. (Courtesy of Flo-Tron, Inc., Paterson, NJ.)

It should be noted that the systems described are true mass flowmeters. That is, this class of meters gives the mass rate of flow regardless of the fluid (see data in Figure 46). In the automotive industry, mass flow is important because fuel energy is related to the mass of the fuel (i.e., Btu/lb). As such, if volume flow were the measuring parameter, specific gravity would also have to be measured in order to convert volume flow to mass flow. Specific gravity measurements introduce additional error into flow calculations. Furthermore, fluid specific gravity may vary from batch to batch for the same fluid as well as with temperature. As an example, Masnik [33] notes that a 2°F change in fuel temperature can cause a 0.1% specific gravity variation in gasoline. Under constant volumetric flow conditions with this temperature variation, the mass flow would experience a 0.1% change. In some applications with flow rates as low as 0.1 lb/h, this error can be significant. Direct measurement of mass flow is preferable to measurement of volumetric flow in order to minimize all possible sources of measuring error.

Mass Flow Transducers

Mass flow transducers have been applied to mass flow measurements of gases and liquids. These systems employ a heat transfer principle to provide direct reading of a true mass flow rate.

A heated-sensor technique is used as the basic principle. A hot sensor anemometer is used to instantaneously measure fluid-flow parameters by sensing the heat transfer between an electrically heated sensor and the flow medium.

The anemometer is essentially a Wheatstone bridge with one arm of the bridge comprising the probe or heated sensor (Figure 48). When the system is operated as a constant temperature anemometer, the bridge circuit supplies heating power to the sensor to raise its temperature above

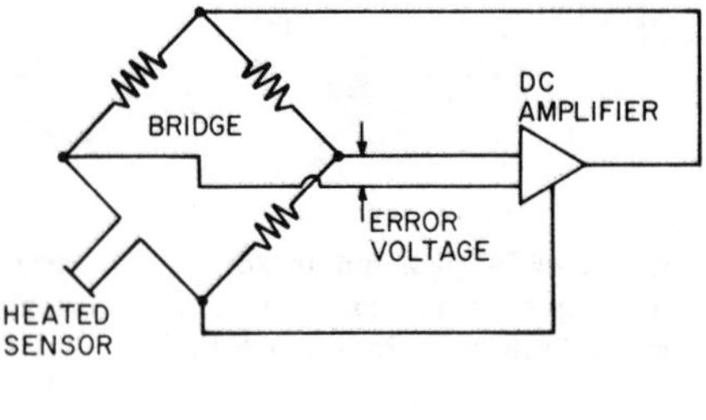

Figure 48. Constant-temperature anemometer circuit. Systems are also designed to operate in constant current mode.

ambient conditions. When the sensor is exposed to a flow stream (which can be either a liquid or gas) the flow conducts heat away from the sensor. The rate at which heat is lost from the sensor is a direct measure of the fluid velocity. The rate of heat removed is indirectly measured by the instantaneous power surge from the circuit that is required to maintain the probe at one temperature. Mathematically, the heat transfer from the heated element (in this case let us say a wire) as a function of the fluid velocity can be expressed as:

$$I^2R = (R - R_e)[\alpha + \beta(\rho u)^{1/n}] \tag{48}$$

Here, α and β are constants that depend on the sensor dimensions and on fluid properties such as thermal conductivity, viscosity, and specific heat. The other parameters in Equation 48 are:

ρ = density of fluid

u = fluid velocity

n = empirical constant that is dependent on the velocity range and fluid

R = resistance of the hot element

R_e = resistance of the element under ambient conditions

I = electrical current in the heated sensor

Equation 48 is a nonlinear function. The output signal can be in terms of either I or power (I^2R). The basic variable that is measured is the mass flow (ρu) per unit area. From this measurement mass flow can be obtained by ρuA, where A is the area of the flow path.

Figure 49 shows a typical output curve for a heated sensor exposed to a moving fluid stream. Linearization circuits have been designed to obtain linear output and to suppress zero flow offset. Linearization curves can be adjusted to match the calibration curve for a specific type of sensor, fluid, and flow range. Note that this type of flow measurement system can be applied to turbulent studies as well as full channel flow measurement.

Application to measuring total flow in a channel or pipe can follow either of two approaches:

- One approach is to divide the cross-section of flow into subareas and obtain measurements by traversing the sensor or probe over the cross-section. This is the same approach that is used with a pitot tube traverse. The flow profile can be controlled so that only a single point measurement is necessary. Measurements can then be appropriately averaged to obtain an indication of the total mass flow. The total mass flow is the product of the single-point (averaged) measurement times the cross-sectional area of the pipe and an appropriate proportionality constant.

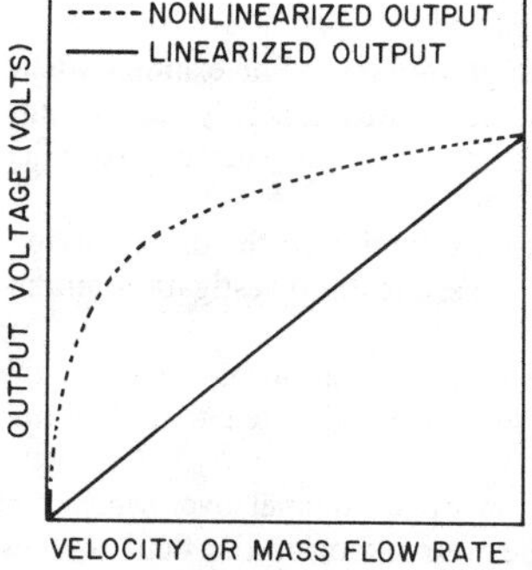

Figure 49. Typical nonlinear output from heated sensor exposed to fluid flow.

- The other approach is to use an in-line transducer, which maintains a flat flow profile at the sensor location. These units can be calibrated to measure total mass flow. The sensor is electrically heated as before and generates a d.c. voltage that is proportional to the rate of mass flow. Note that both systems described are unaffected by pressure and temperature changes.

In general, mass flow transducers are intended for use in relatively clean fluids. Dirty fluids (i.e., fluids with suspended particles) or highly viscous materials can cause fouling problems and impair probe sensitivity. In liquid flow applications, probes are normally coated with a thin film (for example, sputtered quartz). This can deter excessive fouling or damage to the actual sensor and prevent electrolysis from occurring with certain types of probes.

Calibration will normally change with time on miniaturized probes. Dirt or suspended particles will bake onto the sensor. Thus, new probes undergo an aging process. Calibration curves should be periodically checked to ensure no drastic changes. It is good practice to expose a new probe or transducer to actual flow conditions for several hours to a day and then check the factory calibration before use. Note that some probes are designed with a self-cleaning feature that allows burn-off contamination.

Fluid temperature variations will affect the signal output and as such must be corrected for. Corrections can be made by measuring the ambient temperature of the fluid and correcting output data via a calculation procedure. Another approach is to use a second sensor in the bridge circuit, so that the circuit maintains a fixed resistance ratio in the two sensors. The heated or flow sensor carries the higher current, whereas the unheated or temperature sensor carries a lower current. The unheated sensor acts like a compensating sensor, which automatically corrects for errors caused by variations in the fluid's ambient or bulk temperature. If the unheated sensor is properly trimmed with balanced fixed resistors, it corrects for virtually all temperature effects [35]. In general, these systems offer the advantages of no moving parts, high sensitivity, and high accuracy in flow measurement. Air flows less than 10^{-4} scfm can be detected with probes. In standard applications, accuracy is within $\pm 2\%$ of reading ($\pm 0.2\%$ or better of full scale) for the linearized signal.

It is advisable to calibrate mass transducers in actual flow streams where possible, but this is not essential. For gas applications, these flowmeters are factory calibrated in air for standard conditions (i.e., 70°F (21.2°C) and 14.7 psi (101.3 kPa)). Standard flowmeters can be arranged to readout flow in SCFM, SCCM, SFPM, SM/S, etc., depending on the user's requirements. For corrosive fluids, the transducer material must be checked for compatibility. Some gases such as chlorine can greatly shorten transducer life.

OPEN CHANNEL AND SEWER FLOW MEASUREMENT

Open channel flow is usually measured by constructing an obstruction across the flow path and metering a parameter or characteristic variable resulting from the fluid flowing over or under the obstruction. Open channel flow is important to many industrial applications. Effluent streams are generated by nearly every industry. Examples include breweries, slaughter houses, the pulp and paper industry, the chemical industry, and various other kinds of processing plants. The handling and cleaning of effluents is one of industry's major concerns. Accurate flow measurement has therefore become an essential operation prior to final effluent treatment.

Obtaining open channel flow measurement is not always straight-forward. For example, there are numerous instances where an open sewage flow must be monitored with the only access being through a manhole. Often space limitations prevent the construction of a primary device (i.e., sluice gate or weir).

The use of sluice gates is not discussed in this section. A good introduction to the theory of flow measurement with sluice gates is given by Benedict [36] and an experimental investigation on discharge coefficients for sluice gates was carried out by Henry [37].

To evaluate open channel flows requires establishment of a definite relationship between the level of liquid and discharge. This can be accomplished by inserting a calibrated weir or flume into the flow channel.

A weir is essentially some type of obstruction built across an open channel over which the liquid must flow. Usually there is an opening or notch on the weir plate for the fluid to flow

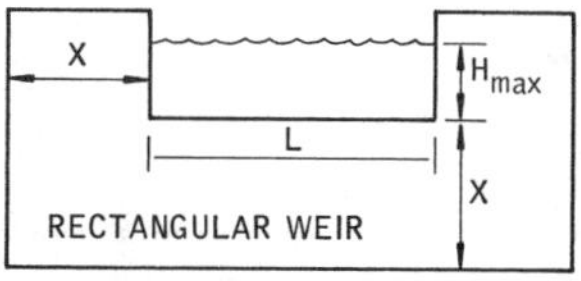

Figure 50. Ideal dimensions for a fully contracted weir. The liquid flow is contracted as it passes through the rectangular opening since the ends of the weir notch are a distance from the sides of the weir pool.

through. There are basically three configurations of sharp-crested weirs: rectangular weirs, V-notch weirs, and Cipolletti weirs.

Figure 50 illustrates a rectangular weir with end contractions. The sheet or layer of fluid flowing over the weir is referred to as the "nappe." To determine the ideal flow rate over this configuration weir, one must assume approach velocities are negligible, and the nappe is under atmospheric pressure. The second assumption imposes the condition that the nappe be treated as a free-falling body, in which case the velocity of liquid flowing over the notch can be expressed as:

$$u_h = (2gh)^{1/2} \tag{49}$$

where h is an incremental height of liquid (head) over the notch (refer to Figure 50b).

The ideal volumetric flow rate can be expressed as:

$$dQ_i = u_h dA \tag{50}$$

where dA = Ldh.

Equations 49 and 50 can be combined to give:

$$dQ_i = 2gh(Ldh) \tag{51}$$

and, integrating between the limits of 0 and H,

$$Q_i = \tfrac{2}{3}L(2g)^{1/2}H^{3/2} \tag{52}$$

Equation 52 states that the ideal volumetric flow rate over a rectangular weir is proportional to the $\frac{3}{2}$ power of the liquid head.

In practice, empirical or semiempirical correlations are used to compute actual flow conditions, one of the reasons being that the first assumption used in the above derivation often has a significant effect on flow rate and cannot be neglected. The Francis formula given by Equation 53 is one of the most widely used empirical expressions for flow over rectangular weirs:

$$Q_a = 3.33\left(L - \frac{nH}{10}\right)\left[\left(H - \frac{u_a^2}{2g}\right)^{3/2} - \left(\frac{u_a^2}{2g}\right)^{3/2}\right] \tag{53}$$

where u_a is the fluid's approach velocity and n is the number of lateral contractions of the weir. Lateral contractions arise when the weir plate is not constructed over the entire width of the approach channel. For the case illustrated in Figure 50, n = 0 and assumption 1 becomes valid; hence Equation 53 reduces to:

$$Q_a = 3.33LH^{3/2} \tag{54}$$

Thin plastic or metal strips are normally secured along the edge of the weir notch, or the entire weir is sometimes cut from a sheet of 1/8-in. aluminum or stainless steel. Care is usually taken in selecting a weir so that at average flow conditions, the minimum head obtained is 0.2 ft. Under conditions where the overflow nappe does not clear the weir crest, accuracy in flow measurement can be significantly reduced. For relatively small flows, a V-notch weir is preferred over the rectangular design [38].

Figure 51 illustrates the V-notch (sometimes called triangular) weir design. By applying the same assumptions used for the flow over the rectangular weir to the geometry shown in Figure 51, an

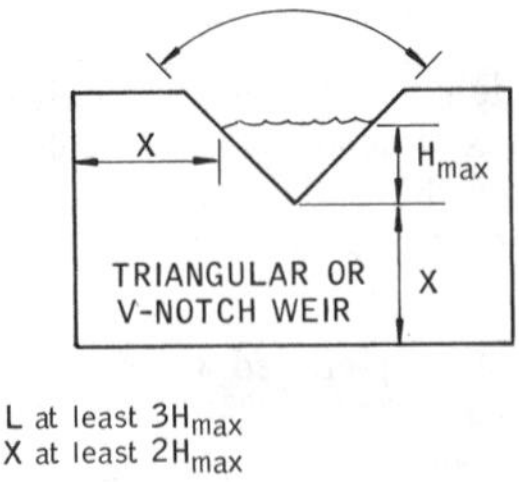

Figure 51. V-notch weir.

expression for the ideal volumetric flow rate is obtained:

$$Q_i = \frac{4L}{15H}(2g)^{1/2}H^{5/2} \tag{55}$$

There are a variety of empirical expressions that compensate for losses and allow reliable computations for the volumetric flow rate. The reader should refer to References 39–41 for details.

Flumes are specially designed open channel flow sections that provide a restriction in area that results in an increase in fluid velocity. They can be designed to be installed in a circular pipe section which has applications to open channel sewer work. Figure 52 is a cross-sectional drawing of a flume installed in a manhole.

Flumes have several advantages over weirs. The flow velocity through flumes is high, and so they tend to be self-cleaning systems, which minimizes deposition of sediment or solids. Flumes can operate with much smaller head losses than experienced with weirs, which can be a decided advantage for irrigation and existing sewer applications.

Flumes generally require relatively smooth, near-laminar flow for proper operation. As such, they should not be installed near a sharp change in slope or near obstructions in the channel. The user must determine whether the flow is laminar at least ten pipe diameters upstream of the proposed location in sewer applications.

A Parshall flume, illustrated in Figure 52, is employed in applications where it is important to maintain a low head loss, or if the liquid contains a large amount of suspended solids.

Once the weir or flume has been installed in the flow path, the flow rate can be determined. The simplest and least sophisticated method of accomplishing this in open channels with low volume flow is by the sandbag method. This involves damming the open channel with sandbags so that

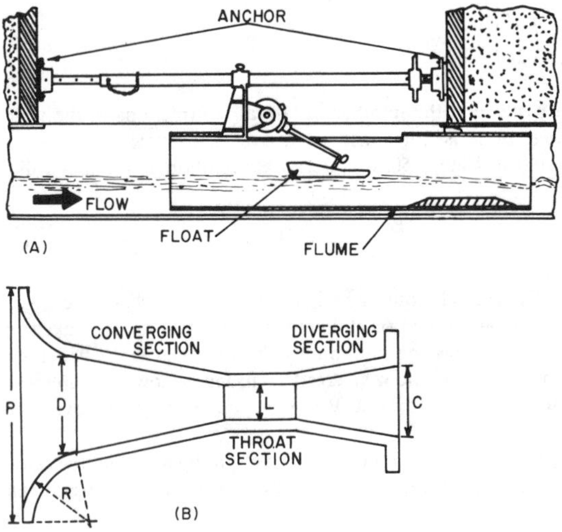

Figure 52. (A) Flume installed in a manhole cover. (B) Parshall flume, often employed when liquids contain a large amount of settleable solids.

Figure 53. Float operated liquid level/flow rate recorder. The unit shown is used for metering open channel flow of sewage, industrial wastes, municipal water supplies, irrigation water, etc. The liquid volume is continuously totalized on a seven-digit counter, and the chart can provide up to 180 days of unattended record depending on the chart drive speed used. (Courtesy of Leupold & Stevens, Inc., Beaverton, OR.)

the water will rise and overflow at one point. A small container is then used to collect the overflow, and a stopwatch records the collection time. From the volume of liquid collected over a recorded time interval, the volumetric flow rate can be computed. A staff gauge located on the influent side of the weir records the liquid level, which can then be correlated with the measured flow rate. If sufficient data have been obtained, a correlation can be developed that allows one to compute the volumetric flow rate through the flume or weir from a measurement of the liquid level.

For large flows, the sandbag and stopwatch method is not practical. For large flows mechanical recorders are to be preferred. A float operated liquid level recorder is shown in Figure 53. The unit operates as follows: A float follows the liquid level as it rises and falls. The float's pulley, while turning, rotates a drum containing a chart. A clock monitor is employed to drive a pen from left to right on the instrument, producing a graphic record of the changing water levels. Depending on the gearing mechanism and type of clock drive, recording times can be varied over several hours to many days.

The recorders are normally equipped with several scales on the same chart so that flow rate (e.g., mgd (million gallons per day)) as well as level can be read directly. Other units have been designed with built-in totalizing features [42].

As the instrument is a float-operated mechanism, a stilling well must be used along with it. The stilling well can range from a simple metal cylinder to a concrete shaft anchored in the floor of a large instrument shelter. For flumes, stilling wells are normally located on the outside wall adjacent to the inlet area. A crossover pipe can be used to connect the bottom of the well to the side of the flume.

It should be noted that these recorders are not universal. That is, the instrument used for recording flows is based on the particular flow characteristics of the primary measuring device, and, as such, should be selected on that basis. Table 13 gives one manufacturer's partial listing of some of the more common devices and flow ranges experienced by their recorders.

Table 13
Partial List of One Manufacturer's Total Flowmeters—Measuring Devices and Flow Ranges

Weir or flume	Max. flow (MGD)	Totalizer multiplier (to determine) vol. in gallons)	Sampling rate of optional actuating switch (gallons through weir of flume between samples)			Recommended float diameter (in.)
			10-lobe cam	20-lobe cam	25-lobe cam	
$22\frac{1}{2}°$ V-notch weir	0.350	100	1,000	500	400	7
	0.140	40	400	200	160	9
	0.070	20	200	100	80	10
	0.035	10	100	50	40	12
	0.0175	5	50	25	20	14
	0.014	4	40	20	16	14
45° V-notch weir	1.750	500	5,000	2,500	2,000	6
	0.700	200	2,000	1,000	800	8
	0.350	100	1,000	500	400	9
	0.175	50	500	250	200	10
	0.140	40	400	200	160	10
	0.070	20	200	100	80	12
60° V-notch weir	1.750	500	5,000	2,500	2,000	7
	1.400	400	4,000	2,000	1,600	7
	0.700	200	2,000	1,000	800	8
	0.350	100	1,000	500	400	9
	0.175	50	500	250	200	10
	0.140	40	400	200	160	12
	0.070	20	200	100	80	14
90° V-notch weir	7.000	2,000	20,000	10,000	8,000	6
	3.500	1,000	10,000	5,000	4,000	7
	1.750	500	5,000	2,500	2,000	8
	1.400	400	4,000	2,000	1,600	8
	0.700	200	2,000	1,000	800	9
	0.350	100	1,000	500	400	10
	0.175	50	500	250	200	12

120° V-notch weir	35.000	10,000	100,000	50,000	40,000	6
	17.500	5,000	50,000	25,000	20,000	6
	7.000	2,000	20,000	10,000	8,000	6
	3.500	1,000	10,000	5,000	4,000	7
12-in. Cipolletti weir	0.700	200	2,000	1,000	800	12
	0.350	100	1,000	500	400	14
2-ft Cipolletti weir	1.750	500	5,000	2,500	2,000	10
12-in. rectangular weir without end contraction	3.500	1,000	10,000	5,000	4,000	6
	1.750	500	5,000	2,500	2,000	8
	1.400	400	4,000	2,000	1,600	9
2-ft rectangular weir without end contraction	3.500	1,000	10,000	5,000	4,000	8
	1.400	400	4,000	2,000	1,600	12
12-in. rectangular weir with end contraction	0.700	200	2,000	1,000	800	12
	0.350	100	1,000	500	400	14
2-ft rectangular weir with end contraction	3.500	1,000	10,000	5,000	4,000	8
	1.750	500	5,000	2,500	2,000	10
	1.400	400	4,000	2,000	1,600	12
1-in. Parshall flume	0.140	40	400	200	160	9
	0.070	20	200	100	80	12
	0.035	10	100	50	40	14
2-in. Parshall flume	0.350	100	1,000	500	400	9
	0.175	50	500	250	200	10
	0.070	20	200	100	80	14
3-in. Parshall flume	0.700	200	2,000	1,000	800	8
	0.350	100	1,000	500	400	9
	0.175	50	500	250	200	12
	0.140	40	400	200	160	14
6-in. Parshall flume	3.500	1,000	10,000	5,000	4,000	6
	1.750	500	5,000	2,500	2,000	7
	1.400	400	4,000	2,000	1,600	8
	0.700	200	2,000	1,000	800	9
	0.350	100	1,000	500	400	12

Table 13 (Continued)

Weir or flume	Max. flow (MGD)	Totalizer multiplier (to determine) vol. in gallons)	Sampling rate of optional actuating switch (gallons through weir of flume between samples) 10-lobe cam	20-lobe cam	25-lobe cam	Recommended float diameter (in.)
9-in. Parshall flume	7.000	2,000	20,000	10,000	8,000	6
	3.500	1,000	10,000	5,000	4,000	6
	1.750	500	5,000	2,500	2,000	8
	1.400	400	4,000	2,000	1,600	8
	0.700	200	2,000	1,000	800	10
	0.350	100	1,000	500	400	14
12-in. Parshall flume	7.000	2,000	20,000	10,000	8,000	6
	3.500	1,000	10,000	5,000	4,000	7
	1.750	500	5,000	2,500	2,000	8
	0.700	200	2,000	2,000	800	12
18-in. Parshall flume	7.000	2,000	20,000	10,000	8,000	6
	3.500	1,000	10,000	5,000	4,000	8
	1.750	500	5,000	2,500	2,000	10
	0.700	200	2,000	1,000	800	14
2-ft Parshall flume	17.500	5,000	50,000	25,000	20,000	6
	14.000	4,000	40,000	20,000	16,000	6
	7.000	2,000	20,000	10,000	8,000	7
	3.500	1,000	10,000	5,000	4,000	9
	1.400	400	4,000	2,000	1,600	12
3-ft Parshall flume	35.000	10,000	100,000	50,000	40,000	6
	14.000	4,000	40,000	20,000	16,000	6
5-ft Parshall flume	70.000	20,000	200,000	100,000	80,000	6
	35.000	10,000	100,000	50,000	40,000	6
6-in. Palmer-Bowlus flume	0.350	100	1,000	500	400	12
	0.175	50	500	250	200	14
	0.140	40	400	200	160	14

8-in. Palmer-Bowlus flume	0.700	200	2,000	1,000	800	10
	0.350	100	1,000	500	400	12
	0.175	50	500	250	200	14
10-in. Palmer-Bowlus flume	1.400	400	4,000	2,000	1,600	9
	0.700	200	2,000	1,000	800	10
	0.350	100	1,000	500	400	12
12-in. Palmer-Bowlus flume	1.750	500	5,000	2,500	2,000	9
	1.400	400	4,000	2,000	1,600	9
	0.700	200	2,000	1,000	800	10
	0.350	100	1,000	500	400	12
15-in. Palmer-Bowlus flume	3.500	1,000	10,000	5,000	4,000	6
	1.750	500	5,000	2,500	2,000	8
	1.400	400	4,000	2,000	1,600	9
6-in. Leopold-Lagco flume	0.140	40	400	200	160	14
8-in. Leopold-Lagco flume	0.350	100	1,000	500	400	12
	0.175	50	500	250	200	14
10-in. Leopold-Lagco flume	0.700	200	2,000	1,000	800	9
	0.350	100	1,000	500	400	12
12-in. Leopold-Lagco flume	1.400	400	4,000	2,000	1,600	8
	0.700	200	2,000	1,000	800	10
15-in. Leopold-Lagco flume	1.750	500	5,000	2,500	2,000	8
	1.400	400	4,000	2,000	1,600	8
	0.700	200	2,000	1,000	800	10
	0.350	100	1,000	500	400	14
18-in. Leopold-Lagco flume	3.500	1,000	10,000	5,000	4,000	7
	1.750	500	5,000	2,500	2,000	8
	1.400	400	4,000	2,000	1,600	9
	0.700	200	2,000	1,000	800	12
	0.350	100	1,000	500	400	14

Courtesy of Leupold & Stevens, Inc., Beaverton, Oreg.

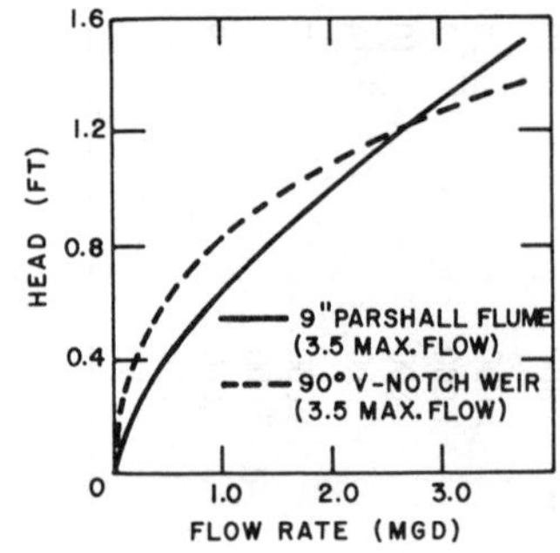

Figure 54. Typical rating curves. Shown are the rating curves for a 90° V-notch weir and a 9-in. (23-cm) Parshall flume. (Courtesy of Leupold & Stevens, Inc., Beaverton, OR.)

Manhole flow measuring installations are usually temporary systems. For this reason, instrumentation is portable and preferably operated independent of an external power source. Weirs and flumes are used in sewer flow metering applications. As indicated earlier, flow through the primary measuring device is a function of the liquid level and, as such, the discharge at any given instant can be determined from a single level measurement. The relationship between level and fluid discharge is termed "rating." Rating curves (refer to Figure 54) and tabulated rating values for standard weirs and flumes are available from manufacturers.

Flumes are generally selected for use with untreated sewage as they tend to be self-cleaning systems and operate at a lower head loss than weirs do. Figure 52 illustrates the use of a flume in this type of application.

Weirs are often considered the least expensive and simplest devices to install in a manhole for flow measurement. One type of design is illustrated in Figure 55. The arrangement shown is useful for situations where portable flumes face difficult installation problems; for example, in manholes with a junction or a manhole where the incoming or outgoing pipes, or both, are not in line with the channel [43]. The unit shown consists of an arrangement of sliding plastic bars that can be positioned to dam or block the open channel under the bulkhead. Slotted mounting holes and an integral bubble level allow adjustment of the weir to match flow conditions.

In the upper corners of the bulkhead are pointed threaded rods which, when tightened against the manhole walls, support the bulkhead until a head of fluid builds up behind the weir. A watertight seal between the bulkhead and manhole walls can be achieved with a temporary sealing compound.

If maximum and minimum flow conditions are known, the primary unit can be sized based on empirical or theoretical expressions using a safety factor. For weirs the Francis expression (Equation 53) along with the recommended dimensions in Figure 50 can be used to size the unit.

The Foxboro Co. [44] has prepared graphs that can be used in the selection of size and type of weir. Figure 56 can be used for rectangular and V-notch weirs. The range of head, H, and construction details can be obtained from Figure 50.

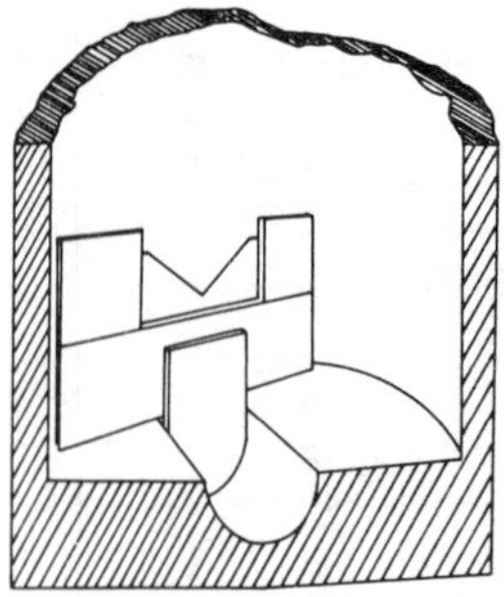

Figure 55. V-notch, bulkhead weir for measuring flows in manholes. (Courtesy of NB Instruments, Inc., Horsham, PA.)

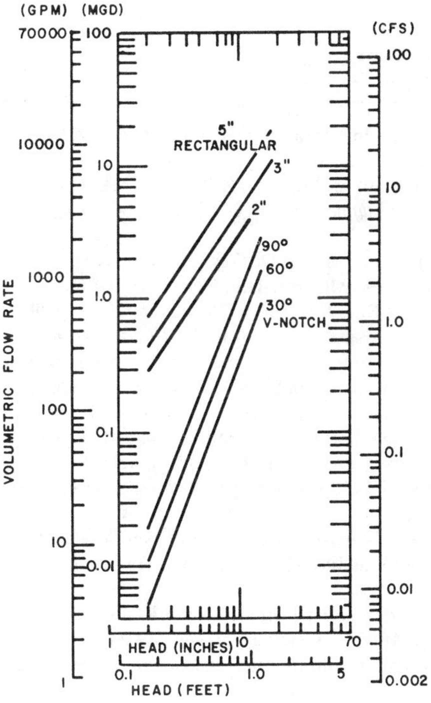

Figure 56. Graph for obtaining size and type of weir. (Courtesy of The Foxboro Co., Foxboro, MA.)

NOTATION

A	area
C	discharge coefficient
C_A	correction factor
C_P	specific heat at constant pressure
C_S	correction factor
C_V	specific heat at constant volume
D	diameter
d	throat diameter
F_{pa}	pressure correction factor for air
F_{ps}	pressure correction factor for steam flow
F_{ta}	temperature correction factor for air
g_c	conversion factor (32.174 lb_m ft/lb_f s^2 or 4.17×10^8 lb_m ft/lb_f hr^2)
ΔH	velocity head (mm)
k	flow coefficient
L	length or weir width
M	velocity-of-approach factor
$\dot{m}$	mass flow rate
P	pressure
ΔP	pressure differential
Q	volumetric flow rate
R	universal gas constant (1,544 ft lb_f/lb mol °R, 1.986 Btu/lb mol °R
Re_G	gas phase Reynolds number
Re_T	throat Reynolds number
S_g	specific gravity
T	temperature
u	velocity
u_m	mean flow velocity at throat
Y	expansion factor

Greek Symbols

β	ratio of throat to tube diameter
γ	C_P/C_V
ρ	density

REFERENCES

1. Aeroquip Corp., Jackson, MI, "Barco-Venturi Flow Measurement System," (1975).
2. Fluid Meters, *Their Theory and Application*, 5th ed., ASME, New York, (1959).
3. *Flowmeter Computation Handbook*, ASME, New York, (1961).
4. Cheremisinoff, N. P., and Niles, R., "Survey of Fluid Flow Measurement Techniques and Fundamentals," *Water & Sewage Works*, 122, No. 12, (1975).
5. Fribance, A. E., *Industrial Instrumentation Fundamentals*, McGraw-Hill, New York, (1962).
6. Spink, L. K., *Principles and Practice of Flow Meter Engineering*, 8th ed., The Foxboro Co., Foxboro, MA, (1958).
7. Cheremisinoff, N. P., *Applied Fluid Flow Measurement*, Marcel Dekker Inc., New York, (1979).
8. Tuve, G. L., *Mechanical Engineering Experimentation*, McGraw-Hill, New York, (1961).
9. Kates, W. A., Co., Deerfield, IL, "Automatic Flow Rate Controllers," Brochure 701, (1978).
10. Froude, W., "Discharge of Elastic Fluids Under Pressure," *Proc. Inst. Civil Engrs.*, (London), *6* (1947).
11. Bean, H. S., and Beitler, S. R., "Some Results from Research on Flow Nozzles," *Trans. A.S.M.E.*, (April 1938).
12. Bean, H. S., Beitler, S. R., and Sprenkle, R. E., "Discharge Coefficients of Long Radius Flow Nozzles When Used with Pipe Wall Pressure Taps," *Trans. A.S.M.E.*, (July 1941).
13. Folsom, R. G., "Nozzle Coefficients for Free and Submerged Discharge," *Trans. A.S.M.E.*, (April 1939).
14. Kallen, H. P., (ed.), *Handbook of Instrumentation and Controls*, McGraw-Hill, New York, (1961).
15. Arnberg, B. T., "Review of Critical Flowmeters for Gas Flow Measurements," *Trans. A.S.M.E.*, 84D (1962).
16. Holman, J. P., *Experimental Methods for Engineers*, McGraw-Hill, New York, (1971).
17. Hersey Products Inc., Spartanburg, SC, "*Liquid Meters*," Catalog 67710M, (1978).
18. RCM Industries, Orinda, CA, "Product Specification Sheet-Series 7000 Flow Gage," Sheet S-700-677 (1978).
19. C-E Invalco, Combustion Engineering Inc., Tulsa, OK, "Hoverflow Bearingless Turbine Flowmeter," Brochure IVC-251-A4, (1978).
20. C-E Invalco, Combustion Engineering Inc., Tulsa, OK, "Sanitary Turbine Flowmeters," Brochure IVC-255-A1, (1978).
21. Flow Technology, Inc., Phoenix, AZ, "Turbine Flow Transducers," Bulletin PB-753, (1978).
22. Flow Technology, Inc., Phoenix, AZ, "Turbine Flow Transducers," Bulletin TP-771, (1978).
23. Flow Technology, Inc., Phoenix, AZ, "Retractable Turbo-Probes," Bulletin TR-761, (1978).
24. C-E Invalco, Combustion Engineering Co., Tulsa, OK, "Flowmeters: PT-Series, RT-Series," Bulletin IVC-252, (1978).
25. Foxboro Corp., Foxboro, MA, "Magnetic Flow Measurement Systems," Bulletin E10-D, (July 1975).
26. Foxboro Corp., Foxboro, MA, "Magnetic Flow Meter Application Information," Technical Information T127-71e, (May 1968).
27. Foxboro Corp., Foxboro, MA, "Magnetic Flow Transmitter Materials Guide," Technical Information T1 27-71f, (July 1973).
28. Bestobell Mobrey Limited, Slough, Bucks., England, "Doppler Flowmeter: Ultrasonic Flow Measurement of Liquids or Slurries in Pipes," Bulletin MF1, (May 1977).
29. Hersey Products Inc., Spartanburg, SC, "Doppler Flow Measurement Systems," Bulletin DFT-1, (March 1978).
30. Cochrane Environmental Systems, PA, Specifications sheet on "Uni Pac Monitor-Model M-1," (1978).
31. Cochrane Environmental Systems, PA, Specifications sheet on "Non-contacting Characterized Flow Transmitter," Model M-3, (1978).
32. Flo-Tron, Inc., Paterson, NJ, "Model 10 Low Flow Linear Mass Flowmeter," Bulletin F-10, (1978).
33. Masnik, W., "Flo-Tron Microflow Meter," paper presented at International Symposium on Automotive Technology and Automation, Wolfsburg, W. Germany, (Sept. 12–15, 1977).

34. Flo-Tron, Inc., Paterson, NJ, "Model 10M-Microflow Meters," Bulletin F-15, (1978).
35. Datametrics Inc., Wilmington, MA, "Gould Datametrics, Series-1000, Mass Flow Transducers," Bulletin 1050B, (May 1977).
36. Benedict, R. P., *Fundamentals of Temperature, Pressure, and Flow Measurements*, Wiley, New York, (1969).
37. Henry, H. R., "A Study of Flow from a Submerged Sluice Gate," M. S. thesis, State University of Iowa, Ames, 10, (Feb. 1950).
38. Leupold & Stevens, Inc., Beaverton, OR, "Measuring Open Channel Waste-water Flows," Stevens Application Notes, No. 3 (1977).
39. Bakhmeteff, B. A., *Hydraulics of Open Channels*, McGraw-Hill, New York, (1932).
40. Chow, V. T., *Open Channel Hydraulics*, McGraw-Hill, New York, (1959).
41. Rouse, H., *Fluid Mechanics for Hydraulic Engineers*, Dover, New York, (1961).
42. Leupold & Stevens, Inc., Beaverton, OR, "One Recorder-Many Uses," Stevens Application Notes, No. 2, (1978).
43. NB Instruments, Inc., Horsham, PA, "Bulkhead Weir for Manholes," Bulletin 708R, (1976).
44. The Foxboro Co., Foxboro, MA, "Open Channel Flow Measurement with Parshall Flumes and Weirs," Technical Information Sheet T19-12a, (Feb. 1974).

CHAPTER 40

DESIGN AND OPERATION OF TARGET FLOWMETERS

W. Tom Clark

Ramapo Instrument Co., Inc.
Subsidiary of Hersey Products Inc.
Montville, New Jersey, USA

CONTENTS

INTRODUCTION

Flow holds a unique position among the process variables in that there is a wide choice of available flow transducers. Selecting a specific flowmeter presents a formidable task for the instrument engineer. The basis for selection must include some knowledge of the general types of flowmeters. The most common types are head, area, and velocity meters. The head class meter is, by far, the most common of the flowmeters in use today.

In the head type flowmeter a restriction is placed inside the conduit where the flow measurement is to be made. The flow through this restriction creates a pressure differential, which is measured. The conventional orifice plate is a good example of such a flowmeter. In spite of the general acceptance of the orifice plate approach, it is quite vulnerable to dirt build up in front of the orifice plate. To overcome this problem the annular orifice was developed.

In the annular orifice approach, a circular disk supported in the center of a pipe acts as the restricting element and forms an annulus between the disk and pipe wall. The annulus provides a space for the dirt to pass freely beneath the disk and gas bubbles above it. A refinement of this approach is to suspend the disk in the center of the pipe and measure the force on the disk instead of the pressure differential. This force is transmitted through a bar to a force measuring device and the flowrate is calculated by taking the square root of the output. The target flowmeter offers the advantage of the annular orifice and at the same time eliminates the need for impulse lines.

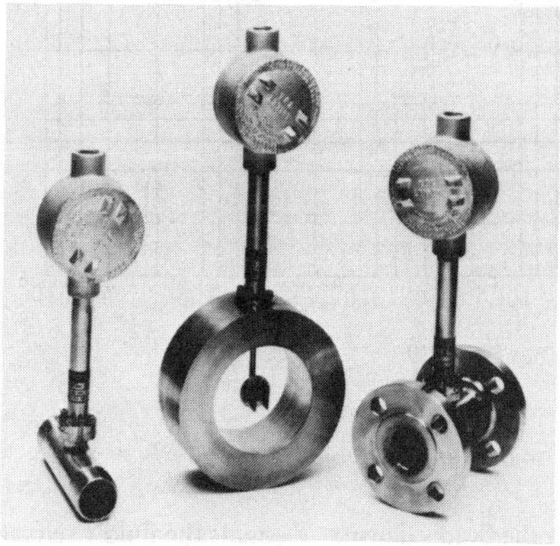

Figure 1A. Typical target flowmeters.

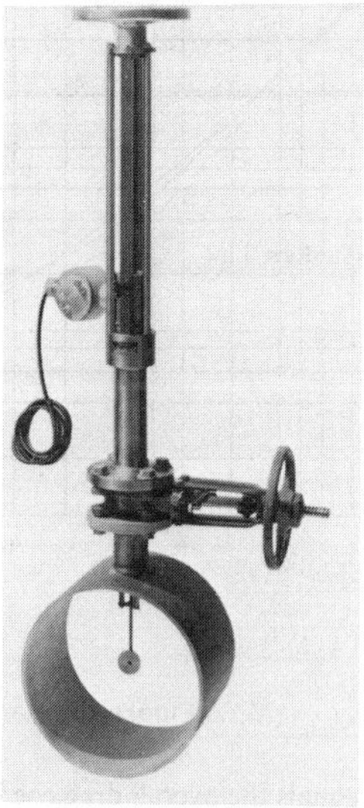

Figure 1B. Retractable probe.

Since its introduction more than thirty years ago, the target flowmeter (see Figure 1) has proven to be one of the most reliable, versatile, and easy to use flowmeters. The inherent flexibility of the target flowmeter provides many advantages to the user when compared to the other types of flowmeters. Certain types of target flowmeters can handle wide temperature and pressure ranges. They also can use special alloy materials compatible with almost any fluid, and can be easily field calibrated.

PRINCIPLE OF OPERATION

The target flowmeter utilizes the principle of fluid drag on a three-dimensional body. A basic review of the theory is included in this chapter. For a complete analysis of fluid drag, refer to References 1–3.

The total drag on any three-dimensional body suspended in a fluid stream, gas or liquid, is the sum of the friction drag and the pressure drag.

$$F_D = F_F + F_P$$

The friction drag is equal to the integration of the shear stresses along the boundary of the body in the direction of the general motion of the fluid stream. The pressure drag is equal to the integration of the components in the direction of motion of all the pressure forces acting on the body's surface. Pressure drag is the dynamic component of the stagnation pressure acting on the projected area of the immersed body normal to flow.

When considering an equation to describe the response of the target flowmeter, only the total drag is of interest, and the equation becomes,

$$F_D = C_D \rho A \frac{V^2}{2} \quad (1)$$

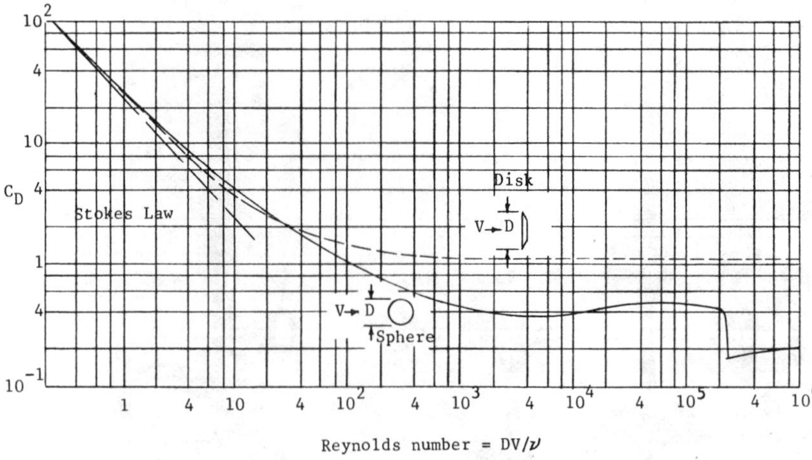

Figure 2. Drag coefficients for bodies of revolution.

where C_D equals the overall drag coefficient, ρ equals the fluid's density, V equals the fluid's velocity at the point of measurement, and A is the projected area of the body normal to flow. The overall drag coefficient C_D is determined by empirical data.

A flat circular disk whose principal area is normal to the direction of flow will generate a wake behind the disk that extends over the entire diameter. This geometry gives a constant drag coefficient over a broad range of flow rates. A sphere or streamlined body produces a wake that varies with the separation point which is a function of the fluid's velocity. The result is a varying drag coefficient that greatly limits the dynamic flow range of the flowmeter.

For very low flow rates, the drag becomes a function of viscosity and is described by Stoke's law,

$$F_D = 3\pi\mu DV \tag{2}$$

where μ is the absolute viscosity and D the body's diameter.

As can be seen in Figure 2, as the fluid velocity increases and the pressure drag becomes more dominant, the drag coefficient levels off and the drag becomes proportional to the velocity squared. These relationships shall be expanded upon in the discussion of the calibration of the target flowmeter.

DESIGN AND CONSTRUCTION

It should be noted that drag is a force and that the target flowmeter is a force transducer. Three primary transducers have been used to measure the force or the deflection that occurs to the drag disk during flow. They are measuring the balancing force, the displacement of the drag disk, and the strain produced in the drag disk support structure.

As the target flowmeter is subjected to flow, the drag disk and support structure deflect (Figure 3). In the force-balance approach, the disk is returned to its original position. This restoring force is the measurand of the force-balance system. A feedback force is applied and senses the equilibrium position. It is this feedback mechanical system that limits the turndown ratio of the force-balance method.

The second technique measures the physical displacement that the drag disk undergoes. A linear variable differential transformer (LVDT) or strain gage flexure is used to measure this small deflection. The limitation of this approach is the required sensitivity to very small displacements, and for

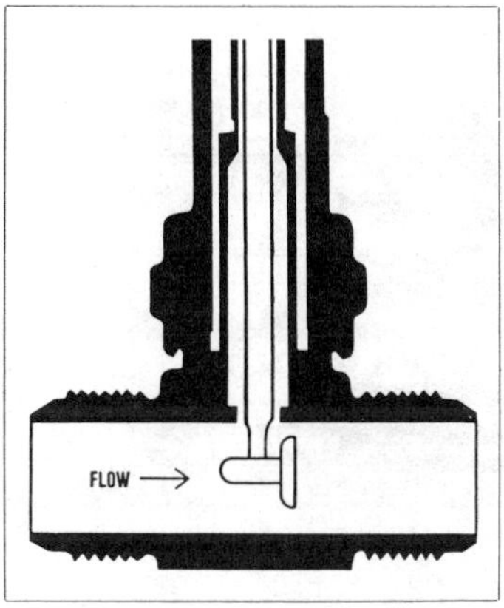

Figure 3. Operating principle of the target flowmeter.

the LVDT, the requirement for AC excitation, which does not lend itself to two-wire direct current loops.

Although there are many applications where the force-balance and LVDT approaches will work satisfactorily, the strain measuring approach is by far the most versatile and reliable.

Figure 4 shows the typical construction of a strain measuring target flowmeter. High strength materials are used for the target rod, base, and sensing tube. These pieces are autogenously welded together forming a complete seal from the process fluid. Strain gages are bonded to the non-wetted exterior surface of the sensing tube. The strain gages are bonded to this section in such a way as to form a four active arm strain gage bridge.

There are several pertinent facts pertaining to the sensing tube. First, the drag or force generated by flow can be transferred to this section as a couple (bending moment and force). These moments produce a stress/strain field in the thin walled section of the sensing tube that is uniform and directly proportional to the applied force. Second, if the strain gage bridge is positioned on the sensing tube to take advantage of the principle strains, two gages on the compression side and two on the tension side, an electrical signal is generated that is sensitive only to the general fluid motion down the conduit. Any stresses or strains induced in other directions such as by flow turbulence are not sensed. Third, by nature of the sensing tube geometry and the strain gage bridge circuitry, pressure induced strains electrically cancel out. The temperature induced strains also tend to cancel.

The capability to handle high-pressure applications with little effect on the output signal is one of the major features of the strain gage target flowmeter. The sensing tube, which is the critical portion of the strain gage target flowmeter, is a thin-walled tubular section. The wall thickness determines the maximum operating pressure and the sensitivity, the electrical output for a given force. Material properties, such as corrosion resistance, yield strength, and the minimum acceptable sensitivity, have limited the operating pressure to 15,000 psi (100,000 kN/m^2). By controlling the tolerances and concentricity of the thin-walled section, the target flowmeter becomes insensitive to pressure variations. The stress-induced strains from internal pressure will be the same magnitude and positive throughout the thin-walled section. Strain gage bridge theory states that if all four arms of the bridge experience the same strain, the net effect on the output will be zero. Empirical data show that the pressure effect will be less than 0.10% from atmospheric to rated pressure and is linear. That is, at 50% rated pressure, the pressure effect would be 0.050%.

For basically the same reasons, the thermal effects tend to reduce. However, these are more complex than the pressure effect. A change in temperature changes: (a) the resistance of each gage in the bridge, (b) the gage factor or the strain sensitivity of each gage, and (c) Young's modulus (MOE) as it pertains to the thin-wall section of the sensing tube. Ideally, as with the pressure effect, the gage resistance change with temperature is identical for each gage and should cancel. From a practical point of view, the gage-to-gage connections also contribute an effect. The net effect is an off-set of the bridge output, which is termed the bridge zero temperature coefficient.

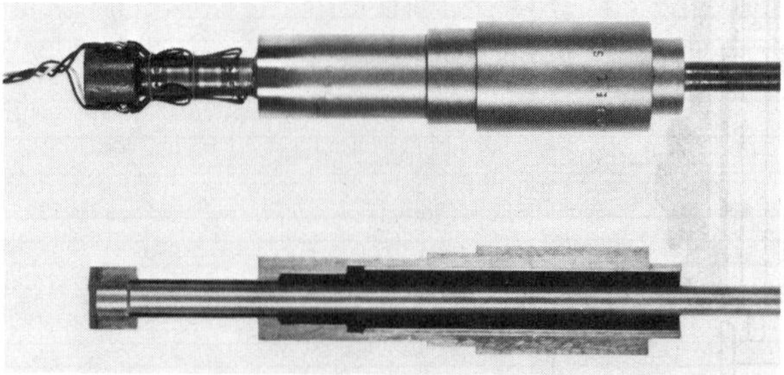

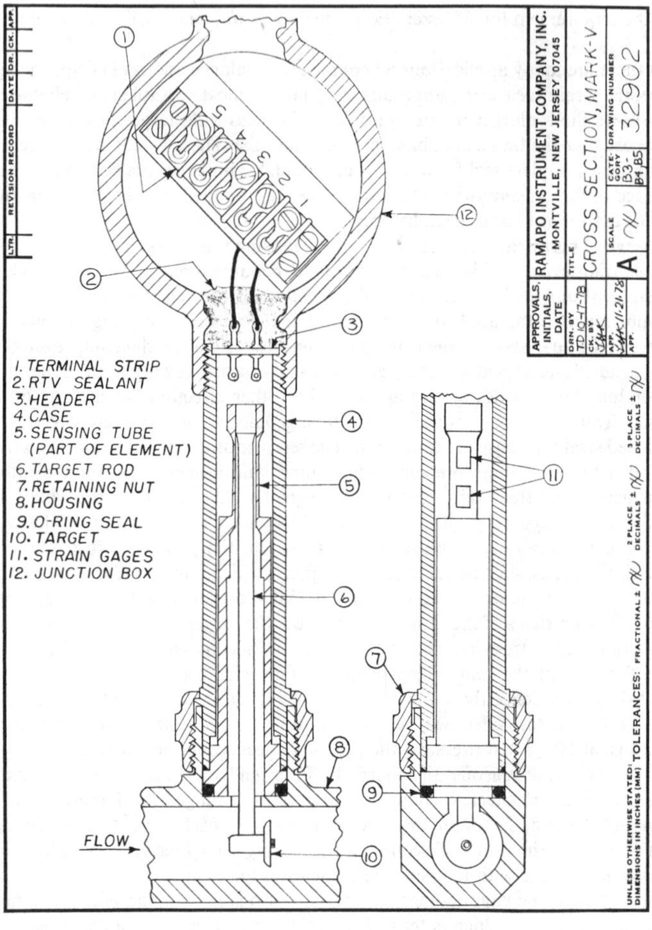

Figure 4A. Typical construction features of a strain measuring target flowmeter; (B) cross-sectional view of flowmeter.

The other two effects are combined in the characteristic termed span temperature coefficient. Young's modulus is defined as the relationship of stress to strain. Ideally, the strain gage material is chosen so that its gage factor thermal coefficient is matched to the modulus coefficient. Generally, for the materials used, the modulus coefficient is linear and the gage factor is not. The resulting span coefficient is small and almost linear.

The temperature limitation of the target flowmeter is a function of the strain gage bonding materials and the temperature characteristics of the base metal's yield strength. Bonded strain gage bridges like the ones used in the target flowmeter have been used in temperatures ranging from −420°F (−250°C) to 1,000°F (538°C). The strain gage target flowmeter has limited its temperature range to −320°F (−196°C) to 650°F (343°C). At the elevated temperatures, both the yield strength and creep characteristics of the base metal deteriorate. Super alloys such as Inconel X750 and 718 are substituted for the standard material of 17–4 PH and have successfully extended the upper temperature limit for short duration to 1,000°F.

Bonded strain gage bridges are generally applied with epoxy cements. These cements can handle temperature ranges of −320°F to 650°F with a high degree of reliability. For higher temperatures the bonding is done with a ceramic cement. Because of the flow turbulence and the brittle nature of the ceramic material, the temperature and output level are limited to 750°F and 1 millivolt/volt.

RELIABILITY

Strain gage force transducers such as load cells and pressure transducers have been manufactured for many decades. The technology used for these transducers is the same as that for the strain gage target flowmeter. The following examples of long-term reliability and stability are from a paper by Shapiro [4].

"In each instance, the transducers utilized bonded strain gage systems developed and manufactured by Dynisco or its sister company, BLH Electronics.

"*Third-year recalibration:* This unit was shipped to the customer on September 20, 1977 and returned to Dynisco for evaluation after the user's yearly check. The unit was tested by Dynisco as received on July 1, 1980.

"Three consecutive runs were made without any prior exercise and without touching the zero or span adjustments. Neither adjustment had been touched after the transducer left the factory.

Parameter	9/30/77	7/1/80
Zero Balance	0.000 VDC	0.000 VDC
F.S. Sensitivity	5.000 VDC	4.985 VDC
Linearity	+0.08%	+0.08%
Hysteresis	−0.02%	−0.04%
Repeatability	0.02%	0.02%
Non-Return to Zero	0.00%	0.00%

"Over the 33-month period that the unit was out in the field, all performance parameters showed outstanding stability. The F.S. Sensitivity showed an average change of −0.055%/6 months. The combined error (linearity, hysteresis, and repeatability), zero balance and non-return to zero showed no measurable changes over the 2–3/4-year time span. This clearly demonstrates the outstanding stability and accuracy achievable in the bonded, strain gage transducer.

"*Sixteen-month continuous study:* A major university evaluated three types of pressure transducers for long term drift. One of the units tested was a model DHF, bonded foil strain gage pressure transducer with a full scale range of 50 psia.

"The test schedule was such that the pressure was maintained continuously. On weekdays the nominal pressure of 40 psia was lowered to 35 psia, assured by a dead weight tester. With increments of 2.5 psia, the pressure was increased to 45 psia, and at each point the system outputs were recorded. Upon completing the test pressures, the transducer was brought back to 40 psia and maintained by a nitrogen bottle and regulator.

Drift Period	Drift in Percent of Full Scale 35 psia	40 psia	45 psia
Day 1 to day 7	+0.028%	+0.028%	+0.028%
Day 7 to day 14	+0.014%	+0.0085%	+0.0085%
Day 100 to day 130	+0.0057%	+0.011%	+0.011%
Day 130 to day 474	−0.0085%	+0.0057%	−0.0028%
Net Drift, 474 days	+0.039%	+0.042%	+0.045%

"The 474-day test period (approximately 16 months), showed an average 6-month drift rate for each pressure point, as follows:

(a) 35 psia	+0.015%
(b) 40 psia	+0.016%
(c) 45 psia	+0.017%

"The data shows excellent long-term stability and the characteristic signature of a transducer that changed very little over time. The small positive and negative changes in drift between time intervals is the net stability demonstrated by the complete measuring system. This includes the transducer, the read out instrument, the enviroment, and the operator. It is impossible to isolate the contribution of each to the net drift value.

"*An 800,000 pound capacity force transducer* used as a shop standard by BLH Electronics was initially calibrated by the National Bureau of Standards in 1968 and checked again in 1980 to confirm its accuracy. Both calibration runs were performed at the National Bureau of Standards in Gaithersburg, Maryland. All loadings were done in compression using dead weights and following standard N.B.S. procedures.

Applied Force Pounds	Calibration Error, Percent F.S. 3/29/68	5/9/80
0	0	0
100,000	0.07%	0.09%
200,000	0.13%	0.15%
300,000	0.16%	0.18%
400,000	0.165%	0.185%
500,000	0.155%	0.17%
600,000	0.12%	0.135%
700,000	0.07%	0.08%
800,000	0	0
600,000	0.095%	0.10%
400,000	0.135%	0.14%
200,000	0.11%	0.11%
0	0	0

"Over the twelve-year time period (144 months), the calibration change was a maximum of +0.02% F.S. This represents an average 6-month calibration stability of +0.0008%.

"*After 29 years of service in the field*, a force transducer purchased in 1952 was returned for recalibration in March, 1981. A comparison of the specification and recalibration data shows remarkable stability and performance over the time period.

Parameter	Specification 1952	Recalibration March 24, 1981
Bridge Resistance	120 ±0.2 ohms	120.1 ohms
F.S. Output	2.4 ±0.004 mV/V	2.3977 mV/V
Zero Balance	1.00% F.S.	0.72% F.S.
Linearity	Not Specified	−0.72% F.S.
Hysteresis	Not Specified	+0.04% F.S.
Repeatability	Not Specified	0.012% F.S.

"Assuming a worst case situation, the change in the first three parameters shows the following:

Bridge Resistance:	+0.3 ohms	(120.1–119.8)
F.S. Output:	−0.11% F.S.	(2.3977–2.4004)
Zero Balance:	−0.28% F.S.	(0.72–1.00)

"The transducer displays not only rock solid stability over three decades, but also ultra-precision performance. The unit was subsequently returned to the customer and is still in use. A large number and variety of bonded strain gage transducers are out in the field with decades of service; attesting to their long-term stability."

The Ramapo strain gage target flowmeter has been used to meter the production of heavy water. Because of the critical nature of the application, the user developed a test procedure to check the reliability and stability of the flowmeter. The test used a reciprocating driving force. The zero-unbalance and element sensitivity were monitored periodically. The driving force was a sine wave with an amplitude approximately ±70% of full-scale output and a frequency of 30 Hertz. After 101×10^6 cycles, the test was terminated with no measurable effect in the mentioned parameters.

Ramapo Instrument Company ran a similar test and the data is shown in Figure 5. The loading was from 0 to 90% full scale at a frequency of 28.75 Hertz. The test was also terminated after 50×10^6 cycles. No significant effect can be seen in the parameters listed.

PERFORMANCE EVALUATION

The major performance parameters of an instrument of this type are accuracy, repeatability, hysteresis, linearity, conformance, sensitivity, rangeability, and response time.

The accuracy of any instrument is defined as the ability of an instrument to produce an output reading equal to the actual input value. The accuracy is generally expressed as a percent of full scale. If, for instance, an actual flow of 100 gpm caused the flowmeter to read 99 gpm, a one percent error would have occurred. However, the device is sometimes referred to as having a "one percent accuracy." This accuracy can be expressed as percent of full scale or as a percent of the actual value of the instantaneous rate. In the first approach, a flowmeter ranged for 100 gpm full scale with an accuracy of ±0.5% would have an output accurate to ±0.5 gpm over the entire range. In the second approach, an accuracy of ±0.5% would mean that the same flowmeter would have an output accurate to ±0.5 gpm at 100 gpm and this error would linearly decrease to ±0.05 gpm at 10 gpm.

The rated accuracy of the target flowmeter is determined by the statistical summation of all precision and bias errors. Examples of bias errors are temperature and pressure effects, and the calibration of the flowmeter itself. Examples of precision errors are repeatability, hysteresis, linearity, and conformance.

The bias errors can generally be numerically corrected. That is, if temperature and pressure effect data are recorded for each flowmeter, the data can be used to correct the electrical output. With today's technology these characteristics can be programmed into a real-time computer and the output corrected immediately. The calibration accuracy is the most significant contribution to the overall accuracy of the instrument. The calibration is dependent upon the basic accuracy of

LIFE TEST

Mark V Flow Meter Probe Sensing Element No. 52236
Output Cycling Range: .07 to 1.8 MV/V ①
Cycling Rate: 1725 Cycles/Minute

Date	Number of Cycles Completed	Force Factor MV/V/KG	Zero Unbalance MV/V	Bridge Resistance Ohms		Insulation Resistance to Ground Megohms	Pressure Effect MV/V/5000 psi	Natural Frequency Hz ②
				Input	Output			
3/28/74	0	1.214	+0.0010	120.3	120.3	>250	-0.002	27.8
3/29/74	2,935,950	1.214	+0.0010					
3/30/74	4,798,950	1.214	+0.0010	120.3	120.3	>250		
3/31/74	7,481,325	1.214	+0.0020	120.3	120.3	>250	-0.002	
4/1/74	9,301,200	1.213	+0.0010	120.3	120.3	>250	-0.002	
4/2/74	11,707,575	1.212	+0.0020					
4/3/74	14,165,700	1.215	+0.0020					
4/4/74	16,623,825	1.214	+0.0030					
4/8/74	26,542,575	1.214	+0.0035					
4/9/74	28,983,450	1.213	+0.0035					
4/10/74	31,441,575	1.215	+0.0020	120.3	120.3	>250	-0.002	27.8
4/11/74	33,822,075							
4/12/74	36,254,325	1.215	+0.0035					
4/15/74	43,654,575	1.215	+0.0030					
4/17/74	46,138,575	1.214	+0.004	120.3	120.3	>250	-0.0035	27.8
4/19/74	51,020,325	1.215	+0.003	120.3	120.3	>250	-0.0035	27.8

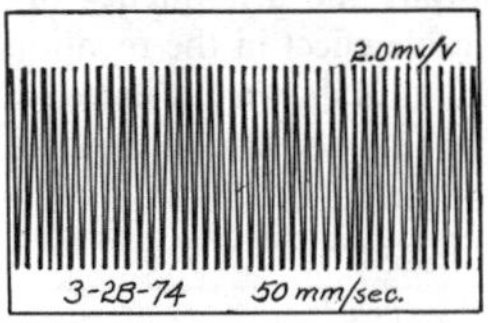

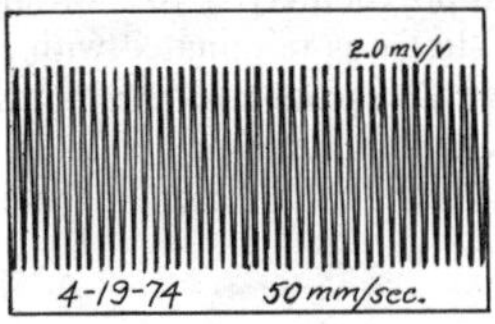

① Maximum rated output for AECL and CGE units: 1.5 MV/V

② With 1.350" diameter target.

Figure 5. Life test data of flowmeter.

the calibration facility itself. An elaboration on the calibration of the strain gage target flowmeter is included later in this text.

Linearity, repeatability, hysteresis, and conformance are examples of precision errors. Linearity is the ability of the transducer to produce a calibration curve that is a straight line. The target flowmeter is designed to give a linear output as a function of force. The non-linearity is no greater than $\pm 0.10\%$ full scale.

Repeatability is the ability of a transducer to reproduce an output signal when the measurand (force) is applied consecutively and from the same direction.

Hysteresis is the inability to give the same output for increasing measurand values and decreasing measurand values. The maximum difference in any pair of identical measurand readings during one complete calibration cycle is the hysteresis of the transducer, and is usually given as a percent of full-scale output. The strain gage target has a combined effect of repeatability and hysteresis of $\pm 0.15\%$ of reading.

The target flowmeter output signal is proportional to flowrate squared. The transfer characteristic is a parabola described by the equation:

$$E = KV^2 \tag{3}$$

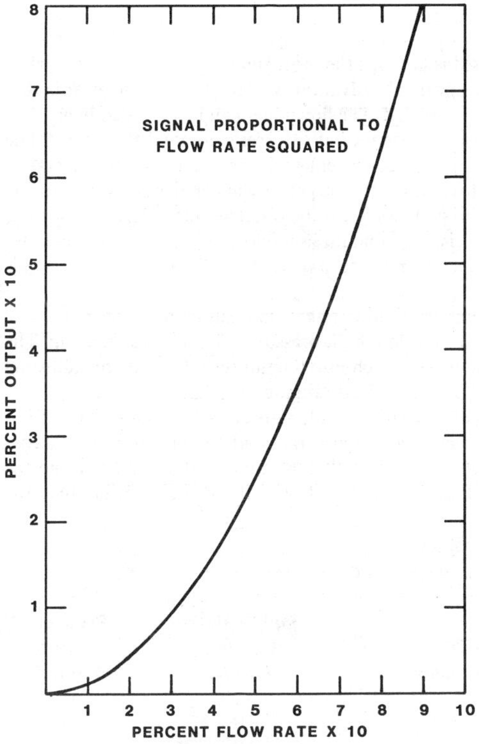

Figure 6. Typical calibration plot for a target type flowmeter.

Since the transfer characteristic is non-linear, the usual linearity specification is replaced by a conformity specification generally referred to as square law conformity. For Reynolds numbers in the turbulent region, the calibration data will conform to the curve generated by Equation 3. Since the net strain signal is zero when the force or flowrate is zero, the curve's intercept is the origin. A typical calibration plot is shown in Figure 6.

The sensitivity of a transducer is the ratio of the change in output to a change in the value of the measurand. For the strain gage target flowmeter, the sensitivity is given as millivolts/volt per kilogram. The sensitivity is a function of the wall thickness of the sensing tube or the pressure rating of the element and the gage factor of strain gages.

One of the more useful performance features of the strain gage target flowmeter is its rapid response. As a consequence, this target flowmeter has been used extensively for transient response studies. The natural frequency of the sensing element is a good indicator for determining the response time. Testing for the exact response time has been difficult due to the limitations of the test facility. The tests have indicated that the movement or stroke time of the controlling valve limits the response time tests.

The most comprehensive test of the response characteristics of the strain gage target flowmeter were done in connection with the measurement of thrust and propellent flowrates for evaluating the performance of rocket thrusters. The quick response of the flowmeter made it possible to throttle a pulse as small as 10 milliseconds. For additional information, refer to the paper by Ginsburg [5]. Other testing has indicated a response time of 1 millisecond for a 9·% full scale step function. and that the flowmeter's response, magnitude ratio, is flat to 100 Hertz for continuously changing sinusoidal flow.

RANGING AND PRESSURE DROP

The range of flow that can be accommodated is perhaps the most significant consideration when specifying a flow measuring device. One of the principle advantages of the target flowmeter is that almost any desired full-scale flow range can be provided. The flowmeter can be provided in any line size to insure the maximum accuracy and best performance at the user's nominal flowrate. A line size is chosen that will accommodate the flows to be consistent with the size piping in the system and the minimum pressure drop or energy loss. Since the target flowmeter is a force measuring device, by adjusting the drag disk diameter the same force can be obtained for different flow rates (Equation 1). That is, for high velocities the disk is small and for low velocities the disk is large. Thus, a 2 inch pipe unit can have the same full scale electrical output whether the full scale flowrate is 25 gpm or 250 gpm.

Most flowmeter manufacturers limit the number of flow ranges that can be handled in any one line size. This can result in reduced accuracy, resolution, high pressure drop, or over-ranging. The strain gage target flowmeter gives the user the option to choose the full-scale flowrate by adjusting the target size, which greatly reduces the bad effects of fixed ranging.

The strain gage target flowmeter operates at a fixed full scale output and therefore has a fixed pressure drop for each line size. Bernoulli's energy equation states that the pressure drop across the target flowmeter is a function of the annulus created by the disk and conduit diameter and the fluid velocity. By maintaining a constant full scale output or force, the relationship between the

Table 1
Flow Range and Pressure Drop Data

Nominal Pipe Size	Flow Ranges (in GPM)* Minimum	Maximum	Full-Scale† Output mV/V	Pressure** Drop PSI (F.S.)
$\frac{3}{4}$	1.0–10	6–60	2.0	8
1	1.5–15	8–80	2.0	4
$1\frac{1}{4}$	2.5–25	10–100	2.0	3
$1\frac{1}{2}$	3–30	15–150	2.0	2
2	4–40	20–200	2.0	1.5
$2\frac{1}{2}$	5–50	25–250	2.0	1.0
3	7–70	35–350	2.0	0.6
4	10–100	55–550	2.0	0.4
6	40–400	120–1,200	2.0	0.2
8	120–1,200	200–2,000	2.0	0.1
10	190–1,900	300–3,000	2.0	0.1
12	280–2,800	450–4,500	2.0	0.1
14	330–3,300	550–5,500	2.0	0.1
16	420–4,200	700–7,000	2.0	0.1
18	550–5,500	900–9,000	2.0	0.1
20	650–6,500	1,100–11,000	2.0	0.1
24	950–9,500	1,600–16,000	2.0	0.1
30	1,550–15,500	2,600–26,000	2.0	0.1
36	2,250–22,500	3,900–39,000	2.0	0.1
LARGER LINE SIZES AVAILABLE				

Notes

* *Customer may select any range desired between the above limits, with a nominal range ratio of 10 : 1. Ranges are given in GPM of water.*

† *Full scale output is nominal value. Different flow ranges and/or pressure drop may be obtained by selecting a full scale output voltage of either 1.0 or 2.0 mV/V. (Applicable to any line size.)*

** *Maximum pressure drop occurs at full scale (F.S.) output. For line sizes >6 in., the pressure drop has been rounded up to 0.1 psi. Actuaal drop is lower.*

annulus and the fluid velocity remains constant. Hence, in any line size, the full-scale pressure drop is constant, no matter what full-scale flow range has been chosen.

The pressure drop is linear with output and directly proportional to the percent flowrate squared. By increasing the flowmeter's sensitivity, the annulus increases reducing the pressure drop. This makes it possible to operate at very low system pressures. Because of the high sensitivity of the strain gage target flowmeter, the pressure drop will generally be lower than most other insertion type flowmeters. Some typical full scale pressure drops for standard strain gage target flowmeters are shown in Table 1.

RANGEABILITY AND CALIBRATION

Prior to actually calibrating a measurement device one must determine the range over which the device will be used. The rangeability or turn-down ratio of a flowmeter is generally defined as the ratio of the maximum to the minimum recommended flowrate. For example, a flowmeter that is to be used for measuring flowrates between 5 and 50 gpm has a 10:1 turn-down.

The strain gage target flowmeter has a much better than average rangeability. Recall that the strain gage target flowmeter is really a strain gage force transducer, analogous to a load cell. The load cell's high accuracy and almost infinite resolution are well documented. These strain gage force transducers are regularly used in precision measurements over ranges of from 100:1 up to 5,000:1.

The strain gage target flowmeter is capable of measuring forces over a range of 2,500:1. Because of the velocity squared relationship to force, the effective flow turn-down for a force range of 2,500:1 is 50:1. A calibration curve showing the actual calibration points is shown in Table 2 and Figure 7. The data presented in Table 2 were collected for a critical government test program, and experiments were conducted at a flow calibration facility.

In actual industrial practice, the rangeability specification has been reduced to 20:1 or even 10:1. This is done because of practical considerations of piping configurations that can generate excessive flow noise and limitations in some analog signal conditioning. These turn-down ratios compare rather well with other head type flowmeters whose turn-down ratios are generally 4:1. With the range of the flowmeter established, the calibration procedure can be started.

The purpose of calibrating any flowmeter is to determine the relationship between the measurand, flowrate, and the electrical output of the transducer.

There are several methods of calibration in use today. They are weight versus time, volume versus time, and the use of a secondary standard.

The weight versus time approach utilizes a large catch tank, weigh scale, and timer. The fluid, usually water, passes through the flowmeter test section and is directed into a catch tank at the same instant the timer is started. At the end of the calibration run, the fluid is diverted away from the tank and the timer stopped. The collected water is weighed and the elapsed time recorded. The weight is then converted into a volumetric value and divided by the elapsed time.

$$Q = \frac{W}{(t)\rho}(448.83)$$

where W = accumulated weight (lb)
t = elapsed time (s)
ρ = density of fluid (lb/ft^3)
Q = volumetric flowrate (gpm)

The conversion of weight to volume is accomplished by dividing the weight by the fluid's density. The density is determined by making a density measurement with a hydrometer or by measuring the temperature and referencing a temperature versus density curve.

The volume versus time approach is quite similar to the weight versus time. The only difference is that the water is diverted into a tank of known volume and the measurement is made directly.

The secondary standard method is nothing more than the use of a highly accurately calibrated flowmeter used to calibrate another flowmeter. This approach is generally used where time and

Table 2
Calibration Data for Plot Shown in Figure 7

Calibration Test	Calibration Data
Flow range: 1 to 50 gpm	Model: Mark V-1-SRBD
Fluid: Water	Serial no.: 3451
Specific gravity: 1.001*	Bridge resistance: 120.1 ohms
Fluid temperature: 62°F*	Force factor: 0.628 mV/V/Kg
Fluid pressure: 40 psi	Sensor type: SPIR-B-654
Viscosity: 1.0 cps	

Flow Rate (gpm)	Electrical Output (mV/V)	Accumulated Weight (lbs)	Elapsed Time (sec)	Direction of Flow
45.236	1.800	300	47.75	Forward
41.182	1.500	300	52.45	Forward
35.008	1.090	300	61.70	Forward
24.914	0.560	200	57.80	Forward
14.999	0.206	200	96.00	Forward
9.877	0.090	100	72.90	Forward
8.126	0.060	100	88.60	Forward
5.023	0.0234	100	143.35	Forward
4.018	0.015	100	179.20	Forward
2.990	0.0085	100	240.80	Forward
2.019	0.0038	60	214.00	Forward
1.019	0.0010	40	282.40	Forward
44.906	1.803	300	48.10	Reverse
41.143	1.500	300	52.50	Reverse
34.727	1.091	300	62.20	Reverse
24.785	0.561	200	58.10	Reverse
14.938	0.206	200	96.40	Reverse
9.890	0.090	100	72.80	Reverse
8.090	0.0601	100	89.00	Reverse
4.950	0.0234	100	145.40	Reverse
3.962	0.0150	75	136.30	Reverse
2.987	0.0085	75	180.80	Reverse
1.976	0.0038	50	182.20	Reverse
1.014	0.0010	25	177.55	Reverse

* *Water temperature 62°F. Rust inhibitors and other additives result in specific gravity shown.*

cost are a factor. Use of a secondary standard will increase the overall error band of the calibrated flowmeter. However, the repeatability of the instrument is unaffected.

The strain gage target flowmeter, by nature of the construction of the sensing element, inherently provides the same sensitivity to force in either direction. Changing the direction of the applied force simply reverses the polarity of the output signal. Using a symmetrical drag disk that is matched to provide the same output versus force relationship from flow in either direction, the strain gage target flowmeter provides both magnitude and direction of flow. One set of calibration data is generally matched to within 0.50% for the same flow in either direction. Refer to Table 2 and Figure 7.

One of the most significant advantages of the target flowmeter is its ability to perform with extreme accuracy and repeatability over large turn-down ratios and wide flow ranges. The ability of the target flowmeter to perform so well under such wide ranges is best explained by reviewing Reynolds numbers and their effect on the response of the target flowmeter.

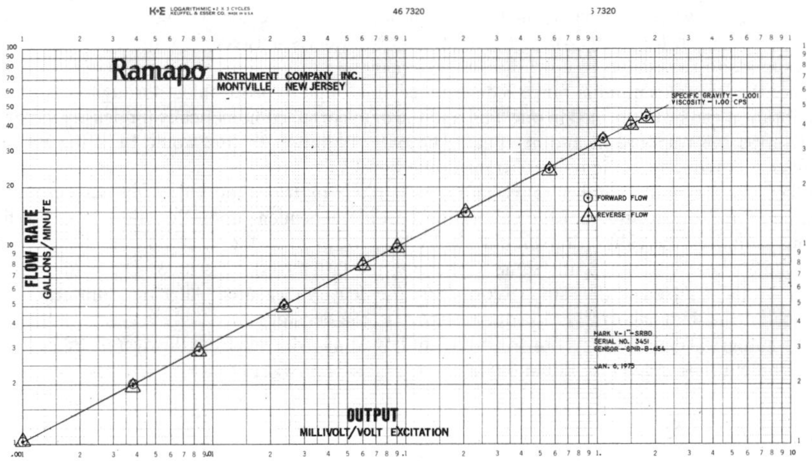

Figure 7. Calibration curve for strain gage target flowmeter.

There does not appear to be one single equation that adequately relates all of the fluid flow variables that determine the response of the target flowmeter. This chapter, so far, has referred to the response as being proportional to the density times the velocity squared. This does indeed describe the response of the flowmeter in most cases and is true for moderate to high velocities. This flow regime is characterized as turbulent. Very low velocity flows are referred to as laminar and the velocities between laminar and turbulent are called transitional.

During the late 1800s, Osborne Reynolds developed a relationship that can be used to predict the type of flow characteristic. The relationship is a dimensionless number named after him, the Reynolds number. The Reynolds number relates the inertial forces to the viscous forces. The equation for Reynolds number is:

$$R_D = \frac{DV\rho}{\mu}$$

where D = the characteristic length, for a circular pipe D equals the inside diameter
V = average fluid velocity
ρ = fluid density
μ = absolute viscosity

Reynolds found that for numbers less than 2,000 the pipe flow was laminar; for Reynolds numbers greater than 4,000 the flow was turbulent. The area in between, 2,000 to 4,000, was considered the transitional zone.

Reynolds conducted an experiment to detect the fluid motion. A dye was injected into a glass pipe and traveled down the length of pipe. At low velocities, or Reynolds numbers less than 2,000 (laminar flow), the injected dye would travel down the pipe in a nice straight line. In turbulent flow, the dye would disperse throughout the pipe. Thus, laminar flow can be visualized as cylindrical layers of fluid that glide smoothly past each other, the fastest moving in the center. In turbulent flow, each fluid particle moves randomly when compared to an adjacent particle, but the general motion of the fluid, as a whole, continues to move down the pipe. The boundaries set by Reynolds' experiment, 2,000 and 4,000, are based upon a non-obstructed length of pipe. Data show that obstructions placed inside the pipe will induce turbulence. The very presence of the drag disk in the target flowmeter makes the flow turbulent at lower Reynolds numbers. A measure of the turbulence with a target flowmeter is the flowmeter's conformance to square law. The conformance to square law is excellent in flowmeters with Reynolds number ranges between 500 and 5,000. This range corresponds to the 10:1 turn-down for the flowmeter.

Once the Reynolds number range is determined and found to be turbulent throughout the range, a calibration can be performed in one fluid and the data corrected numerically for density variations to describe the response in another fluid. The only limitation to this would be if the Reynolds number was reduced to the point of transitional flow.

When the Reynolds number range indicates laminar or transitional flow the response of the flowmeter and the accuracy of the calibration become dependent upon the viscosity of the fluid. In this situation, the calibration should ideally be performed in the same fluid as the end user's, or a fluid whose viscosity is the same. Generally, the latter method is used. The fluid, usually a blend of oils, is mixed to the same viscosity and the density measured. After the calibration of the flowmeter, the data is corrected for the difference in the densities.

The numerical density correction for turbulent flow can be derived from the overall drag equation for turbulent flow, reference, Equation 1.

$$F = C_D \rho A \frac{V^2}{2}$$

For a given flowmeter and providing the Reynolds number is high enough, the C_D drag coefficient and the drag disk area are constant when correcting to another fluid.

What is to be determined is the full-scale velocity of the new fluid or when the force is the same for both fluids. Let subscript 1 equal the conditions of the calibration and subscript 2 equal to the conditions of the new fluid.

$$F_1 = K_1 \rho_1 V_1^2 \tag{4}$$

where $K_1 = \frac{C_D A}{2}$

$$F_2 = K_2 \rho_2 V_2^2 \tag{5}$$

where $K_2 = \frac{C_D A}{2}$

Since $K_1 = K_2$ and $F_1 = F_2$ by definition one can equate Equations 4 and 5.

$$\rho_1 V_1^2 = \rho_2 V_2^2$$

Solving for V_2 yields,

$$V_2 = V_1 \left(\frac{\rho_1}{\rho_2} \right)^{1/2} \tag{6}$$

Take the equation for volumetric flowrate,

$$Q = VA_C$$

where Q = volumetric flowrate
V = fluid velocity
A_C = conduit cross-sectional area

and substitute the solution for velocity into Equation 5. Note that since the cross-sectional area will be the same for both the calibration conditions and the new conditions it can be omitted from the equation. The density correction equation for volumetric flowrates then becomes,

$$Q_2 = Q_1\left(\frac{\rho_1}{\rho_2}\right)^{1/2} \tag{7}$$

When one is interested in gravimetric units such as pounds per minute, a substitution of the mass flow equation into Equation 6 is made. The mass flow equation is defined by,

$$W = \rho A_C V$$

where W = mass flowrate
ρ = fluid's density
A_C = cross sectional area of the conduit
V = fluid's velocity

solving for V, the equation becomes,

$$V = \frac{W}{\rho A_C} \tag{8}$$

One can then substitute this equation into Equation 6 and the following correction term results:

$$\frac{W_2}{\rho_2 A_2} = \frac{W_1}{\rho_1 A_1}\left(\frac{\rho_1}{\rho_2}\right)^{1/2}$$

Note: $A_1 = A_2$

$$W_2 = W_1\frac{\rho_2}{\rho_1}\left(\frac{\rho_1}{\rho_2}\right)^{1/2}$$

$$W_2 = W_1\left(\frac{\rho_2}{\rho_1}\right)^{1/2} \tag{9}$$

The density correction term is determined by the relationship of the output of the flowmeter to the flowrate (see Figure 7). If the slope of the curve is determined, it can be seen that the curve has excellent square law conformity and is described by,

$$Q = K(E)^{1/2}$$

where Q = flowrate
E = electrical output
K = proportionality constant

Recall that if an equation of this form, $y = ax^n$, is plotted on log-log paper, the slope is the exponent (n) in the equation. The density correction term can then be determined by the slope of the log-log plot of the calibration data. Put another way, the slope of the log-log plot yields the inverse of the proportionality exponent of the velocity term in the overall drag equation. For turbulent flow, the exponent is $\frac{1}{2}$, and for laminar flow the exponent is 1. The proportionality exponent then has the boundary conditions of $\frac{1}{2}$ and 1. When the flow range is laminar or transitional, a general flow equation can be developed.

$$Q = K(E)^n$$

where $\frac{1}{2} < n < 1$

The exponent (n) can be substituted into Equations 6, 7, and 9. In these equations, the correction term will be the same for both the calibration and the application because the Reynolds number range is the same.

A target flowmeter can also be calibrated in a liquid, usually water, and corrected to give the electrical output relationship for a gas flowrate application. This is based on the assumption that the equivalent water flow range will produce a turbulent flow Reynolds number. The Reynolds number range for the gas application is almost always turbulent because of the very low viscosities of gases.

The process for determining the equivalent water flow rate for gas applications is quite complex so a basic review of some volumetric gas measurements is included.

Since a cubic foot of gas at one set of temperature and pressure conditions can have a different volume at another set of conditions, measurement in actual cubic feet per minute (acfm) may not always be meaningful. The economic value of a gas is frequently related to its mass; pounds per hour may be the desirable flow rate unit. In SI units kg/s is used.

The use of standard cubic feet per minute (scfm) retains the volumetric unit but makes it independent of temperature and pressure variations. A standard cubic foot is that volume of gas flowing at the operating conditions of temperature and pressure, which would occupy one cubic foot if the gases were at standard conditions (14.7 psia and 70°F). A shorthand notation for standard temperature and pressure is STP.

There is one caution to be observed, however, in specifying the flow range in scfm units. The correct way to specify the flow would be, for example, "90–900 scfm flowing at 15.7 psi and 300°F." The operating conditions must be stated since the signal from the flowmeter will be dependent upon the actual flow velocity and density at the operating conditions. Therefore, the operating conditions as well as the flow range in scfm must be specified.

If these conditions are specified, a correction can be made to an air flowrate at standard conditions, 14.7 and 70°F, and this flowrate can then easily be changed to an equivalent water flowrate by the following equation:

$$Q_{gpm} = 7.48052 \frac{gal}{ft^3} \, scfm_{stp} \frac{ft^3}{min} \left(\frac{\rho_1}{\rho_2}\right)^{1/2}$$

where $\rho_1 = 0.0745$ lb/ft
$\rho_2 = 62.43$ lb/ft (water)

$$Q_{gpm\,(water)} = \frac{1}{3.87} \, scfm_{stp} \tag{10}$$

In order for Equation 10 to be applied properly, the scfm flow rate must be expressed as scfm flowing at 14.7 psia and 70°F (STP). As mentioned earlier, flowrates can be expressed in terms of acfm or scfm at any set of conditions. The following derivation shows how one can correct one gas flowrate at one set of conditions to another set of conditions. That is acfm to scfm, scfm to scfm or acfm to acfm.

Again referring to Equation 1, the response of the target flowmeter is proportional to ρV^2, all other parameters being constant. When one is determining a water flow equivalent for a gas calibration, one should correct the gas flow from acfm or scfm (at some specified temperature and pressure) to scfm at stp.

It may also be necessary to change the flowing conditions for a particular application, and in doing so these same corrections are used to reflect these changes.

The following potential situations for corrections can occur. For a complete derivation, refer to Appendix 1.

Case 1. If a unit is calibrated to indicate acfm at one set of conditions, P_1, T_1, Z_1, and G_1 and the conditions change to P_2, T_2, Z_2, and G_2, the following correction equation should be used:

$$VA_2 = VA_1 \left(\frac{G_1}{G_2} \frac{P_1}{P_2} \frac{T_2}{T_1} \frac{Z_2}{Z_1}\right)^{1/2}$$

Case 2. If a flowmeter is calibrated in scfm at conditions P_1, T_1, Z_1, and G_1 and the conditions change to P_2, T_2, Z_2, and G_2, the following correction equation should be used.

$$VS_2 = VS_1\left(\frac{G_1}{G_2}\frac{P_2}{P_1}\frac{T_1}{T_2}\frac{Z_1}{Z_2}\right)^{1/2} \tag{11}$$

Case 3. If a flowmeter was calibrated in acfm at P_1, T_1, Z_1, and G_1 and one wanted to correct the indicated reading to scfm at P_2, T_2, Z_2, and G_2, the following equation should be used:

$$VS_2 = VA_1\frac{530}{14.7}\left(\frac{G_1}{G_2}\frac{P_1}{(T_1)(Z_1)}\frac{P_2}{(T_2)(Z_2)}\right)^{1/2} \tag{12}$$

Case 4. If a flowmeter was calibrated in scfm at P_1, T_1, Z_1, and G_1 and one wanted to correct the indicated flow rate to acfm flow at P_2, T_2, Z_2, and G_2 the following equation should be used:

$$VA_2 = VS_1\frac{14.7}{530}\left(\frac{G_1}{G_2}\frac{(T_1)(Z_1)}{P_1}\frac{(T_2)(Z_2)}{P_2}\right)^{1/2}$$

In the four preceding equations, the subscript 1 refers to the calibration conditions and subscript 2 refers to the new set of conditions:

VA( ) = volumetric flow rate in acfm
VS( ) = volumetric flow rate in scfm
P( ) = the absolute pressure
T( ) = the absolute temperature
Z( ) = the super compressibility factor for the particular gas
G( ) = the specific gravity of the particular gas

When determining the water flow equivalent, only Equations 11 and 12 are of interest. Since the correction is performed to bring the gas to standard conditions all the variables pertaining to subscript 2 are those of air at STP, $P_2 = 14.7$, $T_2 = 530°R$, $Z_2 = 1.0$, and $G_2 = 1.0$. Values for temperature and pressure must be absolute and the units consistent in the equations, SI units can be substituted for the British engineering units.

The ability to calibrate the target flowmeter in a safe and easy-to-handle fluid like water is one of the major benefits of this calibration approach. It allows the manufacturer and the instrumentation engineer the ability to provide a solution with high accuracy, for hazardous applications, where a calibration with the actual fluid may not be possible.

An alternative to the actual flow calibration is a theoretical calibration. The drag disk diameter can be determined by substituting $\pi D^2/4$ for the drag disk area term in Equation 1 and solving for D.

The only variable not defined numerically by the application is the overall drag coefficient C_D, which is determined empirically. The statistical uncertainty of the drag coefficient has to be incorporated into the overall accuracy evaluation of the flowmeter. This method of calibration is generally referred to as a force calibration and the system accuracy is reduced by about one percent full scale.

SIGNAL CONDITIONING AND APPLICATION

The electrical output of the strain gage target flowmeter is a low level d.c. voltage. It is measured directly from the sensing element and is proportional to the flow rate squared. The signal can be processed in a variety of ways and provide the user with a system capable of indicating, totalizing, batching, controlling and proportioning. The indicating system is the most common. The indicator can be digital or analog meter scaled to read as percent full scale or directly in engineering units (see Figure 8).

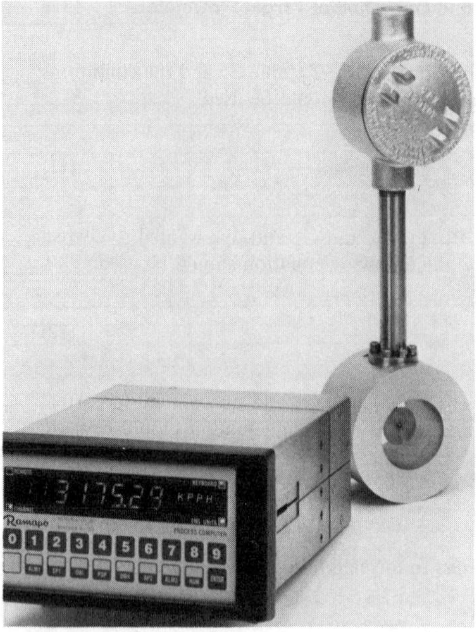

Figure 8. Shows target flowmeter equipped with a digital readout.

The signal conditioning can be quite simple or very complex. A simple two-wire transmitter is an example of the first. A microprocessor based transmitter is an example of the latter. The two-wire transmitter provides an output signal capable of being transmitted long distances. The output is a 4 to 20 mADC signal proportional to the flow rate squared. Because of the transmitter's simplicity, it can be incorporated as an integral part of the flowmeter. With the microprocessor based transmitter, intricate relationships between a number of sensed variables such as pressure, temperature, flow and density can be readily converted into precise information in the desired format. The computer (microprocessor) is used to calculate mass flow rate of saturated or superheated steam and energy flow rate, such as in Btu per hour of steam, heated or chilled water. The computer extends the range of the usefulness and limits the errors introduced by the sensing system. It can be programmed to automatically correct for the bias errors such as the thermal characteristics of the flowmeter.

Unlike many other types of flowmeters, the target flowmeter has been applied successfully in virtually every type of application. The only two applications considered unacceptable are sewage and paper pulp. The following are examples of the wide range of applications handled by the strain gage target flowmeter.

- Measurement of saturated or superheated steam is critical in today's energy conscious environment. Conventional flowmeters are not practical due to the damaging effects of condensate that can occur in saturated steam. The design of the target flowmeter permits the addition of an over-range stop to limit the possibility of damage to the primary sensing element. Many steam metering applications are provided with integral temperature and pressure sensors for use with a microprocessor for calculating mass flow.
- Two-phase flow measurement has been handled successfully using a target flowmeter in conjunction with a turbine flowmeter (see Chapter 39 for a discussion of turbine flowmeters). Screen-like targets or drag disks are used sometimes to sample the entire cross-sectional area of the flowstream. The target flowmeter is calibrated in terms of ρV^2. In the calculation of mass flow rate the velocity term is obtained from the turbine flowmeter, which is insensitive to density variations, and the equation is then solved for density.
- As a result of the successful use of the strain gage target flowmeter by NASA, the government specified the flowmeter for all the liquid systems in the National Solar Heating and Cooling

Demonstration Program [6]. Its use was based on ease of installation and high reliability. A large quantity of flowmeters have been used successfully in this program.

- The automotive industry uses the strain gage target flowmeter extensively in the construction of new test stands. Flows of diesel fuel, gasoline, air, and transmission fluid are measured. Turbine flowmeters had previously been used, but did not have the response and viscous immunity at low flows as did the target flowmeter.
- Pollution control requires the measurement of gas flows in large ducts and stacks. Because of the critical nature of this measurement, the government sponsored a testing program to evaluate available flowmeters for particulate-laden gas flows [7]. The test found the strain gage target flowmeter the most accurate and to have the best characteristics of the point sensors tested.

Advances in strain gage technology and application procedures, the use of new materials with improved physical properties, and the introduction of new corrosion resistant materials offer future application opportunities. It should be noted, however, that the application potential of the target flowmeter as it is presently configured has not been fully explored.

CONCLUSION

There are many different techniques of measuring flow rate and not one will offer a solution to every flowmetering problem. The strain gage target flowmeter has proven itself as one of the most versatile and reliable flow measurement devices available today. An inspection of this ten point summary of the strain gage flowmeter should provide the instrumentation engineer the confidence to consider and then specify the strain gage target flowmeter:

1. *Line sizes* of 3/8″ and larger (with no upper limit) can be handled easily and economically with the target flowmeter. Applications have been handled in pipe sizes to 96 inches and ducts 10 feet square.
2. *Rangeability:* a 10:1 range ratio is standard; ranges to 20:1 and higher are available. Target size selection makes it possible to provide almost any reasonable flow range compatible with the designated line size. The range of the flowmeter can be changed at any time (even in the field) simply by replacing the target.
3. *Bi-directional flow:* the construction of the sensing element allows with the use of a symmetrical target the capability of measuring flow in either direction. The change in direction is indicated by the polarity change in the flowmeter's electrical output.
4. *Low pressure drop:* typical pressure drops can range from 2 psi at maximum flow for a 1″ pipe size to a few inches of water for pipe sizes of 8″ and larger.
5. *Materials:* the simplicity of construction permits use of a wide range of materials such as stainless steel, Hastelloy C, Inconel X-750, and tantalum to handle most applications.
6. *Pressure:* ratings to 10,000 psi and higher are available.
7. *Temperature:* ratings from −320°F to +750°F are standard; temperatures to 1,000°F can be handled short-term, or if the service is not continuous.
8. *Reliability:* target flowmeters are in service after 20 years of service. Life tests have been performed by independent users covering 100,000,000 forward and reverse cycles to approximately 70% of full-scale flowrate. Similar life tests have also been performed covering 51,000,000 cycles from zero to approximately 90% of full scale. There were no significant changes in any of the important measured variables following these tests.
9. *Field calibration:* a check of the original calibration accuracy can be made easily in the field. The calibration is a function of the target diameter and the force factor or sensitivity of the sensing element. This information, provided with every flowmeter, can be verified in the field by a dimensional measurement and simple force check.
10. *Simplicity:* the target flowmeter, even in large pipe sizes, can be easily and quickly installed by one person. Skilled instrument personnel are not required either for installation or use.

The reader should refer to References 4–7 for more in-depth discussions.

APPENDIX: GAS FLOW MEASUREMENT WITH THE TARGET FLOWMETER

Basic Concept of Measurement

The target flowmeter measures fluid flow as a function of the drag force exerted on a disc suspended in the fluid stream. The force is related to the flow rate and also to the fluid density. When measuring the flow of a gas, a knowledge of the effect of density changes is important when considering the use of the flow meter in various applications.

The drag force on a disc is given by:

$$F_D = C_D \rho V_f^2 \frac{(A)}{2} \tag{13}$$

This equation is taken from Marks' *Standard Handbook for Mechanical Engineers*, eighth edition, page 3–52 (McGraw-Hill Book Company, New York), which also defines the following terms:

	Customary Units	SI Units
F_D, drag force	lb f	N
C_D, drag coefficient	dimensionless	
ρ, density	slug/ft^3	kg/m^3
V_f, velocity of fluid	ft/s	m/s
A, area of target	ft^2	m^2

Gas flow rate is frequently expressed in volumetric units such as ft^3/min or m^3/s, rather than velocity. In Equation 14, V is the volumetric flow rate and A_p is the cross-section area of the pipe.

$$V = V_f A_p \tag{14}$$

The drag force on the target is measured by a strain gage bridge having an electrical output directly proportional to the drag force. Consequently, the flowmeter output signal, E, for a given flowmeter becomes (from Equations 13 and 14).

$$E = \rho V^2 K_1 \tag{15}$$

All of the parameters that do not vary, target area, pipe area, and drag coefficient, have been included in the one constant K_1.

The target flowmeter can be used in a wide variety of gas flow measurement systems. In the simplest system, where the gas density is known and does not vary, there is no need to monitor temperature and pressure. The flowmeter can be sized at the factory so that the relationship between output signal and flow rate is known.

Even if small density changes are expected, the system without pressure and temperature monitoring may be useful. Because of the ρV^2 relationship, Equation 15, the error in flow reading, is about half the percent change in density. For example, a +2.5% variation in density would result in an error of +1.25% in the indicated flow rate.

If the gas density varies but the gas line is equipped with temperature and pressure indicators, the flow rate can be calculated at any time by taking the indicated flow rate reading and using correction factor equations to correct for the temperature and pressure changes.

In some applications the gas density does not change during a particular run, but can be changed to another predictable density for another run. The same equations can be used to find the new full scale flow rate for the new density.

Totalizing the flow is practical in a system without pressure and temperature monitoring if the density remains constant. It is also practical when the density changes from run-to-run (but not during the run), but then the total at the end of each run must also be corrected by the same factor used to correct the full scale flow rate.

When the density varies continuously over a range that would introduce flow errors that are not acceptable without correction, then a gas flow computer must be used. In such a system pressure, temperature and flow are continuously monitored and the signals from the three transducers are used to produce, in real time, a corrected gas flow rate signal. This corrected flow signal can be used for totalizing.

Customary Units for Gas Flow Rate

Since a cubic foot of gas at one set of temperature and pressure conditions can have a different volume at another set of conditions, measurement in actual cubic feet per minute (acfm) may not always be meaningful. The economic value of a gas is frequently related to its mass; pounds per hour may be the desirable flow rate unit; in SI units, kg/s is used.

The use of standard cubic feet per minute (scfm) retains the volumetric unit but make it independent of temperature and pressure variations. A standard cubic foot is that volume of gas flowing at the operating conditions of temperature and pressure, which would occupy one cubic foot if the gas were at "standard" conditions (14.7 psia and 70.0°F).

There is one caution to be observed, however, in specifying the flow range in scfm units. The correct way to specify the flow would be, for example, "90–900 scfm flowing at 15.7 psia and 300°F." The operating conditions must be stated since the signal from the flowmeter will be dependent on the actual flow velocity and density at the operating conditions. Therefore the operating conditions as well as the flow range in scfm must be specified.

Derivation of Correction Factor Equations

The correction factor equations shown can be derived from the equations already given and the applicable gas laws. As shown earlier, the density ρ is the ratio of mass to unit volume.

$$\rho_g = \frac{\text{mass}}{\text{Vol}} \tag{16}$$

For a given gas, the density ρ_g is not fixed but varies with pressure and temperature. To refer to a particular density, the pressure and temperature must be stipulated (for instance, the standard conditions, 14.7 psia and 70°F). As an alternate, the specific gravity at standard conditions can be stated. The specific gravity of a gas is the ratio of the density of that gas to the density of a reference gas (air or nitrogen) when both gases are at the same pressure and temperature.

$$G = \frac{\rho_{gc}}{\rho_{ac}} \tag{17}$$

G is the specific gravity of the gas. The densities, ρ_{gc} and ρ_{ac} are of the gas and air, respectively, at the same conditions. This can be written as:

$$\rho_{gc} = G\rho_{ac} \tag{18}$$

The perfect or ideal gas fulfills this relationship:

$$P\ \text{Vol} = RTK_2 \tag{19}$$

This equation adapted from *Fluid Mechanics*. V. L. Streeter and E. B. Wylie, sixth edition, page 14 (McGraw-Hill Book Co., New York) holds true in practical applications where super compressibility is negligible (i.e. 100 psig and below). P is the absolute pressure in psi, Vol is the volume in ft^3 and T is the temperature in degrees Rankine (°F + 460). The term R is a constant for a given gas and K_2 is a factor related to the particular units used.

Equation 19 rewritten:

$$\text{Vol} = \text{RK}_2 \frac{\text{T}}{\text{P}} \text{Z} \tag{20}$$

Z equals the super-compressibility factor and is used to correct the ideal gas relationship. Referring back to Equation 15 for the signal output from the flow meter:

$$\text{E} = \rho \text{V}^2 \text{K}_1 \tag{15}$$

The value of ρ is actually ρ_{gc} in Equation 18. Substituting that value from Equation 18:

$$\text{E} = \text{G}\rho_{ac}\text{V}^2\text{K}_1 \tag{21}$$

From Equation 16, substituting for ρ_{ac}

$$\text{E} = \text{G}\frac{\text{mass}_{ac}}{\text{Vol}_{ac}}\text{V}^2\text{K}_1 \tag{22}$$

The value of Vol_{ac}, the volume of air at the same temperature as the gas, can be obtained from Equation 20 and substituted into Equation 22.

$$\text{E} = \text{G}\frac{\text{mass}_{ac}}{\text{RK}_2\text{T/P}}\text{V}^2\text{K}_1 = \text{G}\left(\frac{\text{mass}_{ac}}{\text{RK}_2}\right)\frac{\text{P}}{\text{T}}\text{V}^2\text{K}_1 \tag{23}$$

The expression in parenthesis is a constant. The value of mass_{ac} is considered constant by definition and the value of R is that for air. The constants are then combined and re-written

$$\text{E} = \text{GV}^2\frac{\text{P}}{\text{TZ}}\text{K} \tag{24}$$

Flowmeter Calibrated in scfm

Let P_1, T_1 be the operating conditions for the flow rate V_{1S} which is given in scfm. Let V_{1A} be the acfm flowing at the operating conditions and therefore the flow rate which is actually measured by the flow meter. From Equation 24, the signal produced by the flowmeter under test:

$$\text{E}_1 = \text{G}_1\text{V}_{1A}^2\frac{\text{P}_1}{\text{T}_1\text{Z}_1}\text{K} \tag{25}$$

The actual gas flow rate can be equated to the flow rate at standard conditions by the use of Boyle's and Charles' Laws:

$$\frac{\text{V}_m\text{P}_m}{\text{T}_m\text{Z}_m} = \frac{\text{V}_n\text{P}_n}{\text{T}_n\text{Z}_n} \tag{26}$$

In this equation, V refers to volume. It is also applicable to velocity, in this case, since velocity is proportional to volume per unit time. Rewritting Equation 26 with appropriate subscripts and the values standard and operating conditions:

$$\frac{\text{V}_{1A}\text{P}_{1A}}{\text{T}_{1A}\text{Z}_{1A}} = \frac{\text{V}_{1S}\text{P}_{1S}}{\text{T}_{1S}\text{Z}_{1S}} \tag{27}$$

$$\frac{V_{1A}P_1}{T_1Z_1} = \frac{V_{1S}\,14.7}{(1)530} Z_{1S} = 1.0 \tag{28}$$

$$\therefore V_{1A} = V_{1S}\frac{14.7}{530}\frac{T_1}{P_1} Z_1 \tag{29}$$

Substituting Equation 29 in 25:

$$E_1 = G_1\left(V_{1S}\frac{14.7}{530}\frac{Z_1T_1}{P_1}\right)^2 \frac{P_1}{Z_1T_1} K = G_1V_{1S}^2\left(\frac{14.7}{530}\right)^2 \frac{T_1}{P_1} Z_1K \tag{30}$$

This equation is the signal produced by a flowmeter calibrated in scfm.

Flowmeter Calibrated in acfm

Let P_2 and T_2 be the operating conditions for the flow rate V_{2A} which is given in acfm. From Equation 24 the signal produced by the flowmeter under test is:

$$E_2 = G_2V_{2A}^2 \frac{P_2}{T_2Z_2} K \tag{31}$$

Change in Conditions: Original in acfm, New in acfm

Premise: A flowmeter originally calibrated in acfm is to be used under new conditions. With the new conditions it is also desired to read acfm. With the new flow conditions the output signal and/or meter reading will still be in the range of the original calibration. When the gas is flowing the reading will be V_{1A}. It must be corrected by Equation 36 below to obtain the flow rate V_{2A}.

Equation 31 applies to both the original and new conditions, since they are both in acfm. The equation is shown twice below, with the appropriate subscripts.

$$E_1 = G_1V_{1A}^2 \frac{P_1}{T_1Z_1} K \tag{32}$$

$$E_2 = G_2V_{2A}^2 \frac{P_2}{T_2Z_2} K \tag{33}$$

The two equations will be made equal to each other, since it is desired to find what flow rate under the new conditions would produce the same signal as the original conditions: $E_2 = E_1$.

$$G_2V_{2A}^2 \frac{P_2}{T_2Z_2} K = G_1V_{1A}^2 \frac{P_1}{T_1Z_1} K \tag{34}$$

$$\frac{V_{2A}^2}{V_{1A}^2} = \frac{G_1}{G_2}\frac{P_1}{T_1}\frac{T_2}{P_2}\frac{Z_2}{Z_1} \tag{35}$$

$$V_{2A} = V_{1A}\sqrt{\frac{G_1}{G_2}\frac{P_1}{T_1}\frac{T_2}{P_2}\frac{Z_2}{Z_1}} \tag{36}$$

Change in Conditions: Original in scfm, New in acfm

Premise: A flowmeter originally calibrated in scfm is to be used under new conditions where it is desired to read in acfm. The procedure is the same as for the derivation of Equation 36. The two equations to be made equal to each other are, for V_{1S}, Equation 30 and for V_{2A}, Equation 31. With appropriate subscripts, the resultant equation is:

$$G_1 V_{1S}^2 \left(\frac{14.7}{530}\right)^2 \frac{T_1}{P_1} Z_1 K = G_2 V_{2A}^2 \frac{P_2}{T_2} \frac{1}{Z_2} K \tag{37}$$

$$\frac{V_{2A}^2}{V_{1S}^2} = \frac{G_1}{G_2} \left(\frac{14.7}{530}\right)^2 \frac{Z_1 T_1}{P_1} \frac{T_2 Z_2}{P_2}$$

$$V_{2A} = V_{1S} \frac{14.7}{530} \sqrt{\frac{G_1}{G_2} \frac{T_1}{P_1} \frac{T_2}{P_2} \frac{Z_1 Z_2}{1}} \tag{38}$$

Change in Conditions: Original in acfm, New in scfm

Premise: A flowmeter originally calibrated in acfm is to be used under new conditions where it is desired to read in scfm. The procedure is the same as previous derivations except that Equation 31 is used for V_{1A} and Equation 30 for V_{2S}.

$$G_1 V_{1A}^2 \frac{P_1}{T_1 Z_1} K = G_2 V_{2S}^2 \left(\frac{14.7}{530}\right)^2 \frac{T_2 Z_2}{P_2} K \tag{39}$$

$$\frac{V_{2S}^2}{V_{1A}^2} = \frac{G_1}{G_2} \left(\frac{530}{14.7}\right)^2 \frac{P_1}{T_1} \frac{P_2}{T_2} \frac{1}{Z_1 Z_2} \tag{40}$$

$$V_{2S} = V_{1A} \frac{530}{14.7} \sqrt{\frac{G_1}{G_2} \frac{P_1}{T_1} \frac{P_2}{T_2} \frac{1}{Z_1 Z_2}} \tag{41}$$

Change in Conditions: Original in scfm, New in scfm

Premise: A flowmeter originally calibrated in scfm is to be used under new conditions where it is desired to read in scfm. The same procedure is used. Equation 30 applies for both sets of conditions, original and now.

$$G_2 V_{2S}^2 \left(\frac{14.7}{530}\right)^2 \frac{T_2}{P_2} Z_2 K = G_1 V_{1S}^2 \left(\frac{14.7}{530}\right)^2 \frac{T_1}{P_1} Z_2 K \tag{42}$$

$$\frac{V_{2S}^2}{V_{1S}^2} = \frac{G_1}{G_2} \frac{T_1}{P_1} \frac{P_2}{T_2} \frac{Z_1}{Z_2} \tag{43}$$

$$V_{2S} = V_{1S} \sqrt{\frac{G_1}{G_2} \frac{T_1}{P_1} \frac{P_2}{T_2} \frac{Z_1}{Z_2}} \tag{44}$$

NOTATION

A	area
C_D	drag coefficient
D	diameter
E	voltage signal
F	frictional force
G	specific gravity
K	calibration coefficient
n	empirical exponent
P	pressure
Q	volumetric flowrate
R	universal gas law constant
R_D	Reynolds number
T	temperature
t	time

V	velocity
V_A	volumetric flowrate in ACFM
V_S	volumetric flowrate in SCFM
W	weight
Z	compressibility factor

Greek Symbols

ρ	fluid density
μ	viscosity

REFERENCES

1. Cheremisinoff, N. P., *Applied Fluid Flow Measurement—Fundamentals and Technology*, Marcel Dekker, Inc., New York (1979).
2. Cheremisinoff, N. P., *Fluid Flow: Pumps, Pipes and Channels*, Ann Arbor Science Pub., Ann Arbor, MI (1981).
3. Azbel, D. S., and Cheremisinoff, N. P., *Fluid Mechanics and Units Operations*, Ann Arbor Science Pub., Ann Arbor, MI (1983).
4. Shapiro, B. H., "Putting Strain Gage Pressure Transducer Stability in Perspective," Dynisco Co., Norwood, MA (1983 reprint).
5. Ginsburg, B. R., "Flow Rate Measurements for Spacecraft Thrusters," Rocketdyne Division of North American Aviation (1983).
6. "Instrumentation Installation Guidelines for National Solar Heating and Cooling Demonstration Program," SCH-1006 (Aug. 4, 1976), Washington, DC.
7. Brooks, E. F., Beder, E. C., Flegal, C. H., Luciani, D. J., and Williams, R., "Continuous Measurement of Total Gas Flowrate from Stationary Sources," EPA-650/2-75-020 (Feb. 1975).

CHAPTER 41

OPTIMAL DIMENSIONS OF WEIRS

Devendra Kumar

Superintending Engineer
Pragatipuram,
Near-Rishikesh, India

CONTENTS

INTRODUCTION

A weir founded on porous medium is designed for surface and subsurface flow conditions. The design from surface flow consideration is primarily concerned with the fixation of pond level, waterway, and effective dissipation of excess energy of water flowing over the structure. The stilling basin forms an integral part of the weir apron for dissipation of energy. Its elements are proportioned in such a way that most of the turbulence of flow dies out on the apron. However, scour on the downstream of the stilling basin due to some turbulence passing on to the unprotected river bed is a common phenomenon.

The weir apron is also subjected to forces due to seepage through the porous foundation underneath. The seepage flow causes uplift pressure under the apron and tends to lift the soil particles at the exit. The weir apron is, therefore, designed to ensure safety against uplift pressures and undermining.

POND LEVEL

The pond level of a weir is determined by the functional requirements of the development project of which the weir is a component. When the structure is used to divert the river flow into an open channel for irrigation purposes, the pond level is determined by the full supply level of the channel and the head of water needed to push the stipulated discharge into the channel. If water is being ponded to mitigate the variation inflow rate in the stream due to some structure located upstream, the pond level may depend upon the volume of water to be stored and variation in capacity of the reservoir with pond level. In case the weir is to impound river flow for industrial or domestic consumption, the pond level will be governed by requirements of the intake structure.

WATERWAYS

Prior to construction of a weir, the water surface slope and cross section of the stream varies with the flow rate of the stream. During floods the water surface slope, water depth, and width of flow increase. The depth of flow in the stream increases due to a rise in water level. In alluvial plains the depth of flow is also increased because of scour during floods.

Generally, the waterway of weirs is kept smaller than the river width during floods. This results in constriction of the river section in the horizontal direction. The apron of the weir does not permit erosion of the river bed at the site of the structure. This leads to constriction of the waterway in vertical direction as well. To push the flood discharge through the limited passage over a weir, afflux is caused on the upstream of the structure. The extent of rise of water level on the upstream side depends mainly on the waterway of the weir.

The permissible afflux, A_{f_p}, governs the minimum waterway of a weir. The rise in flood level due to afflux upstream of a weir inundates additional area. In the upper reaches of the stream, where the bed slope is steep, the rise in flood level due to afflux does not extend for long. This permits higher afflux and hence more constriction. In alluvial plains, where the river bed slope is less and the country is flatter, the rise in water level due to afflux may travel long distances and may inundate large areas. This necessitates more waterways in lower reaches of rivers.

The other aspect that may govern the waterways of weirs is the type of land getting submerged due to afflux. If the area is fertile or habitated, higher afflux may not be permissible. In case the land is not so fertile, a higher value of afflux can be adopted depending on economics. The cost of protecting the area getting submerged may also at some places govern the waterway.

STILLING BASIN

Due to afflux, the water flowing over a weir carries extra energy. If the river bed is exposed to the turbulent flow passing over the weir, it causes erosion on the downstream of the weir apron. The

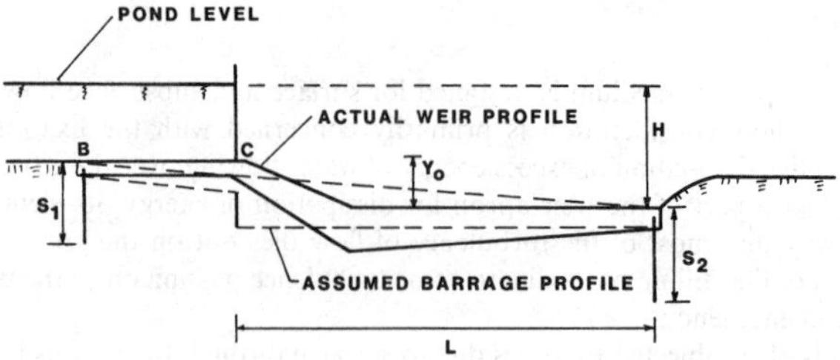

Figure 1. Profile of a weir.

tail erosion may lead to piping. To dissipate the energy of water flowing over the structure, the stilling basin forms an integral part of weir apron.

Hydraulic jump is most commonly used to dissipate the excess energy of water flowing over a weir. The water is made to flow over a glacis (Figure 1), where it becomes super-critical. The super-critical flow is allowed to impinge against sub-critical flow in the stilling basin. If the depth of flow in the stilling is adequate, a hydraulic jump is formed on the glacis. Let H_L be the head loss and q the discharge intensity over the weir. The pre- and post-jump depths are given by the equation

$$d_2 = -\frac{d_1}{2} \pm \left(\frac{d_1^2}{4} + \frac{2q^2}{gd_1}\right)^{1/2} \tag{1}$$

$$H_L = (d_2 - d_1)^3/4d_1d_2 \tag{2}$$

where d_1 = prejump depth of flow over the glacis
d_2 = post-jump depth of flow in the still basin

The depth of flow in a stream at any site is a function of flow rate, water surface slope, cross section of the river, and value of the rugosity coefficient. Downstream of a weir the stage-discharge relationship of the river will be unique and it can be represented by a curve (Figure 2). If the design discharge, waterway and site of a weir are decided, the discharge intensity q, head loss H_L, and the stage of the river are uniquely defined. The pre- and post-jump depths d_1 and d_2 can be obtained

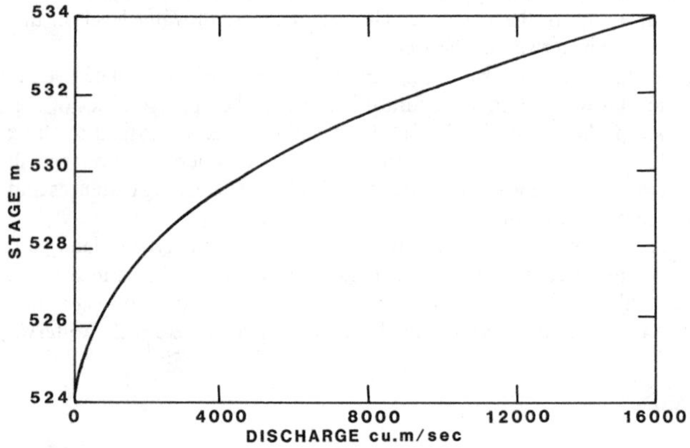

Figure 2. State-discharge curve.

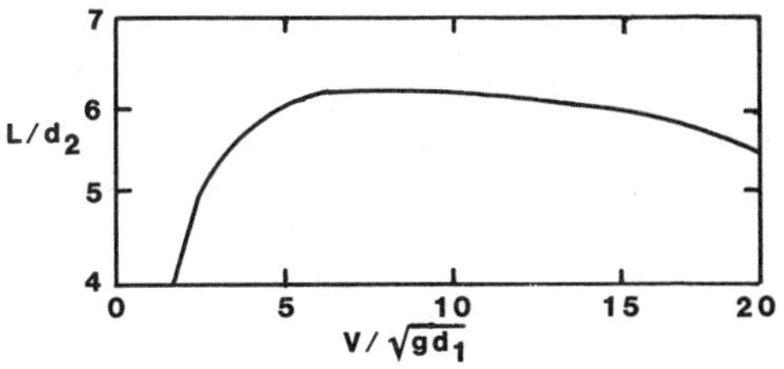

Figure 3. Variation of L/d_2 with Froude Number (Based on Recommendations of U.S. Bureau of Reclamation).

using Equations 1 and 2. The level of stilling basin to attain the post jump depth d_2 can be obtained using the relation.

$$\text{Stilling basin level} = \text{Downstream flood level} - d_2 \quad (3)$$

The downstream flood level corresponding to design discharge can be read from the stage discharge curve of the river. It is evident that for a particular site and design discharge, as the waterway is reduced, the head loss H_L and discharge intensity q increase. Consequently, post-jump depth d_2 increases and the stilling basin becomes deeper.

The length of the stilling basin is kept equal to the jump length so that most of the turbulence dies out on the apron. The length of the stilling basin is a function of Froude number of flow, F, and post jump depth d_2. The value of Froude number of weirs generally varies between 1.7 to 4.5. The length of jump for the different values of Froude number between 1.7 and 20 can be obtained from the curve shown in Figure 3. As the waterway is reduced the length of jump also increases.

EXIT GRADIENT

The safe exit gradient, adopted for design purposes, depends upon the nature of the foundation soil. The value of safe exit gradient governs the depth of downstream sheet pile—the lesser the value of safe exit gradient adopted the greater is the depth of sheet pile for the same floor length. The soil particles at the boundary of flow domain get moved as the exit gradient approaches the critical value. For ordinary soil met in practice the value of the critical gradient is about 1.0. However, the recommended values of safe exit gradient are as follows [5]:

Gravel	1/4 to 1/5
Sand	1/5 to 1/6
Silt	1/6 to 1/7

Such low values of exit gradient have been recommended due to ignorance about distribution of true exit gradient in the stream bed, which has been subjected to scour. The shape of a scoured bed due to surface flow (Figure 4), as reported by various investigators, resemble the arc of a circle or an aerofoil. (Bilashevsky [1], Colaric et al. [2], Doddiah [3], Tsuchiya [4], Khosla [5], Sharma [6], Leliavsky [7]).

EXIT GRADIENT WITH CIRCULAR SCOUR

In the z-plane, where $z = x + iy$, the profile of a flat bottom weir with a vertical sheet pile under its heal is shown in Figure 5a. The weir is resting on a porous medium of infinite depth. In the figure, b is the width of the apron and s is the depth of the vertical sheet pile. The scoured portion is circular in nature and it is described by

$$\left| z - \left(-\ell + i\frac{\ell^2 - h^2}{2h} \right) \right| = R \quad (3)$$

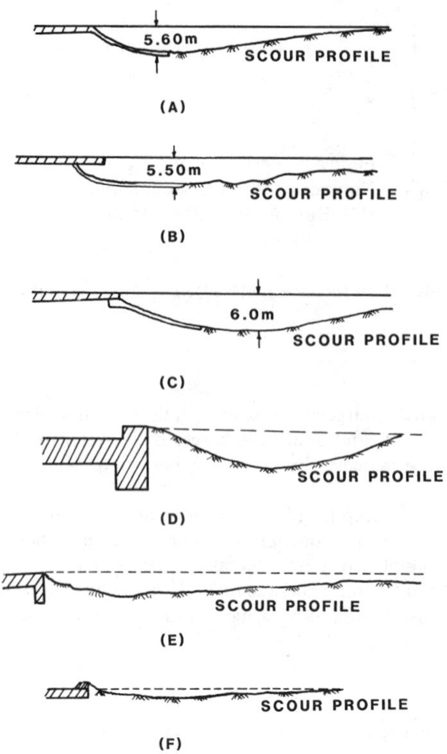

Figure 4. Scour observed in the field and on models. (A) Scour profile observed on Sambesk Barrage (after Leliavsky). (B) Observed on a Bye Pass (after Leliavsky). (C) Observed on Grave Barrage (after Leliavsky). (D) Typical profile observed on models (after Leliavsky). (E) Observed on a model of a barrage. (F) Observed on a model of a barrage.

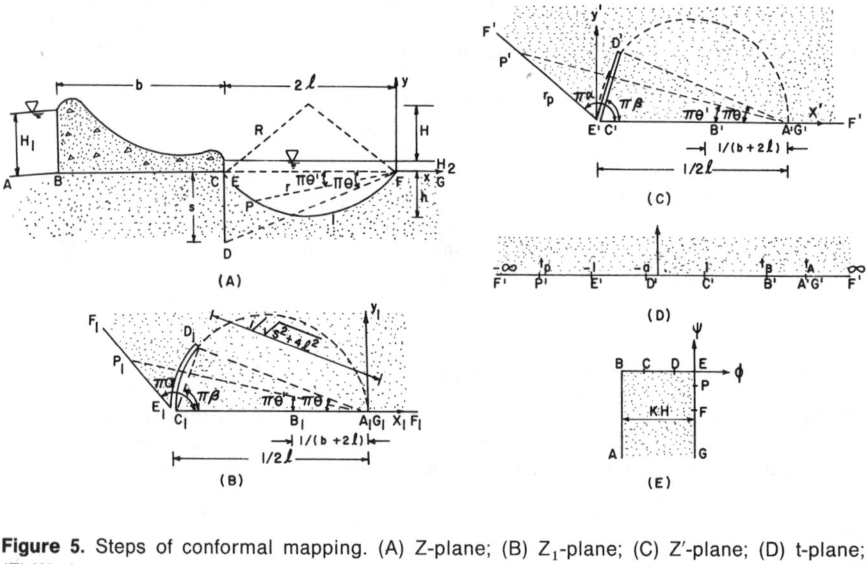

Figure 5. Steps of conformal mapping. (A) Z-plane; (B) Z_1-plane; (C) Z'-plane; (D) t-plane; (E) W-plane.

where, R is the radius of curvature, h is the depth, and 2ℓ is the length of scour profile. The depth of water on the upstream and downstream sides are H_1 and H_2, respectively, and the difference in head is H. The porous medium under the apron extends to infinite depth.

Approximate Analysis

The flow domain ABCDEIFG, comprising a circular boundary, can be converted into a polygon with straightline boundaries by its inversion about E (Figure 5a). But with this polygon it has not been possible to apply conformal mapping technique. In view of this limitation point F has been taken as the origin and the inverse flow domain 1/z is shown in Figure 5b. The circular profile in the z-plane can be described by

$$x^2 + y^2 + 2\ell x - \left(\frac{\ell^2 - h^2}{h}\right) y = 0 \tag{4}$$

Under the transformation $z_1 = 1/z$, where $z_1 = x_1 + iy_1$, the circle in the z-plane can be mapped on to the straight line given by the equation

$$y_1 = -\frac{2\ell h}{\ell^2 - h^2} x_1 - \frac{h}{\ell^2 - h^2} \tag{5}$$

The vertical sheet pile CDE, being parallel to the y-axis, maps onto the arc of a circle passing through the origin in the inverse plane. As an approximation, the image of the sheet pile, $C_1D_1E_1$ in the inverse plane can be replaced by an inclined straight slit joining C_1D_1.

For the purpose of this analysis, the inverse plane z_1 has been mapped onto the z′-plane shown in Figure 5c where $z' = z_1 + 1/2\ell$. From Figure 5c the modified inverse length of the sheet pile is given by

$$S = \frac{s}{2\ell(4\ell^2 + s^2)^{1/2}} \tag{6}$$

and the angle D′C′B′ is given by

$$\beta\pi = \pi/2 - \tan^{-1}\left(\frac{s}{2\ell}\right) \tag{7}$$

Considering Equation 5, the angle $\alpha\pi$ made by the straight line E′F′ with X′ axis is found to be,

$$\alpha\pi = \tan^{-1}\left(\frac{-2\ell h}{\ell^2 - h^2}\right) \tag{8}$$

According to the Schwarz-Christoffel transformation, the conformal mapping of the inverse flow domain in z′-plane onto the upper half of the auxiliary t-plane, shown in Figure 5d, is given by

$$z' = M \int \frac{(t - a)\, dt}{(1 - t)^{1-\beta}(1 + t)^{1+\beta-\alpha}} + N \tag{9}$$

the vertices E′, D′, and C′ of the polygon in z′-plane being mapped at −1, a, and 1, respectively, on the real axis of the t-plane. In Equation 9, M and N are constants.

Since the domain in z′-plane comprises a radial slit alone, Verigin's transformation is also applicable to this domain.

According to Verigin's transformation, the mapping of the flow domain with radial slit onto the upper half of the t-plane is given by

$$z' = C(1+t)^{\alpha-\beta}(1-t)^{\beta} \tag{10}$$

the points E′, D′, and C′ going onto the points −1, a and 1, respectively, on the t-plane.
Since Equations 9 and 10 must be equal, it follows after differentiation,

$$\frac{dz'}{dt} = M(1-t)^{\beta-1}(1+t)^{\alpha-\beta-1}(t-a)$$

$$= -C\alpha(1+t)^{\beta-1}(1+t)^{\alpha-\beta-1}\left(t+\frac{2\beta}{\alpha}-1\right) \tag{11}$$

Hence, $M = -C\alpha$ (12)

$$a = 1 - \frac{2\beta}{\alpha} \tag{13}$$

At $t = a$, $z' = Se^{i\beta\pi}$. Making use of this relationship, Equation 10 gives,

$$C = \frac{Se^{i\beta\pi}}{(1+a)^{\alpha-\beta}(1-a)^{\beta}} \tag{14}$$

Using Equations 6, 7, 8, 13, and 14, the unknowns S, β, α, a, and C can be evaluated respectively for given values of s, ℓ and h.

Determination of t_A and t_B

Considering point A′ G′ for which $z' = 1/2\ell$ and $t = t_A$, and substituting these values in Equation 10

$$(1+t_A)^{\alpha-\beta}(t_A-1)^{\beta} = \frac{(1+a)^{\alpha-\beta}(1-a)^{\beta}}{2\ell S} \tag{15}$$

Similarly, considering point B′ for which $z' = \frac{1}{2\ell} - \frac{1}{b+2\ell}$ and $t = t_B$,

$$(1+t_B)^{\alpha-\beta}(t_B-1)^{\beta} = \frac{b(1+a)^{\alpha-\beta}(1-a)^{\beta}}{2\ell S(b+2\ell)} \tag{16}$$

t_A and t_B can be evaluated from Equations 15 and 16 respectively by an iteration procedure.

Relationship Between t, z′ and z for the Circular Portion EIF

Let P be any point on the circle for which $z = z_p$. In polar coordinate system $z_p = re^{i\pi(\theta'+1)}$, r and $(\theta'+1)\pi$ being modulus and argument of point P, respectively. The corresponding point z_{1p} in z_1-plane is given by $z_{1p} = 1/re^{i\pi(1-\theta')}$. In z′-plane the location of the point P is given by the relation.

$$z'_p = \left[\left(-\frac{\cos\theta'\pi}{r}+\frac{1}{2\ell}\right)^2+\left(\frac{\sin\theta'\pi}{r}\right)^2\right]^{1/2} e^{i\pi\alpha}$$

or

$$z'_p = \left[\frac{1}{r^2} + \frac{1}{4\ell^2} - \frac{\cos \theta'\pi}{\ell r}\right]^{1/2} e^{i\pi\alpha} \tag{17}$$

Let t_p be the location of point P on t-plane. Using Equations 10, and 17 and simplifying

$$S\left(\frac{-t_p - 1}{1 + a}\right)^{\alpha-\beta}\left(\frac{1 - t_p}{1 - a}\right)^{\beta} = \left(\frac{1}{r^2} + \frac{1}{4\ell^2} - \frac{\cos \theta'\pi}{\ell r}\right)^{1/2} \tag{18}$$

From Equation 18 the value of t_p for any point P on the circle can be obtained by an iteration procedure.

Mapping of Complex Potential Plane

The complex potential w is defined as $w = \phi + i\psi$ in which the velocity potential function

$$\phi = -k\left(\frac{p'}{\gamma_w} + y\right) + c' \tag{19}$$

In Equation 19 k = coefficient of permeability, c′ = constant, p′ = pressure, ϕ = potential function and γ_w = unit weight of water. The w-plane for the flow domain is shown in Figure 5e. The transformation of w-plane to the upper half of t-plane is given by

$$w = M'\int \frac{dt}{(t - t_A)\sqrt{(t - t_B)}\sqrt{(t + 1)}} + N' \tag{20}$$

in which M′ and N′ are constants. Integrating and simplifying [8]

$$w = \frac{M'}{\sqrt{(t_A + 1)}\sqrt{(t_B - t_A)}} \sin^{-1}\left[\frac{(2t_A + 1 - t_B)(t - t_A) + 2(t_A + 1)(t_A - t_B)}{(t - t_A)(t_B + 1)}\right] + N' \tag{21}$$

For the point E, t = −1 and w = 0, Therefore,

$$\frac{\pi M'}{2\sqrt{(t_A + 1)}\sqrt{(t_B - t_A)}} + N' = 0 \tag{22}$$

For point B, $t = t_B$ and w = −kH. Hence,

$$-\frac{\pi M'}{2\sqrt{(t_A + 1)}\sqrt{(t_B - t_A)}} + N' = -kH \tag{23}$$

Solving for M′ and N′

$$M' = \frac{kH}{\pi}\sqrt{(t_A + 1)}\sqrt{(t_B - t_A)} \tag{24}$$

and

$$N' = -\frac{Kh}{2} \tag{25}$$

Pressure Distribution Along Base

The complex potential w for a point t', $-1 \le t' \le t_B$ is given by

$$w = \frac{kH}{\pi} \sin^{-1}\left[\frac{(2t_A + 1 - t_B)(t' - t_A) + 2(t_A + 1)(t_A - t_B)}{(t' - t_A)(t_B + 1)}\right] - \frac{kH}{2} \tag{26}$$

Since $\psi = 0$ along the base of the structure, $w = \Phi$ in Equation 26. The pressure can be determined by substituting this relation for Φ into Equation 19. The value of z corresponding to point t', $-1 \le t' \le t_B$ is given by

$$z_t' = \left[S\theta^{i\beta\pi}\left(\frac{1 + t'}{1 + a}\right)^{\alpha-\beta}\left(\frac{1 - t'}{1 - a}\right)^{\beta} - \frac{1}{2\ell}\right]^{-1} \tag{27}$$

Exit Gradient Beyond Point E

The exit gradient, I_E, is given by

$$I_E = \frac{i}{k}\left(\frac{dw}{dt} \cdot \frac{dt}{dz}\right) \tag{28}$$

$$\frac{dt}{dz} = \frac{dt}{dz'}\frac{dz'}{dz_1}\frac{dz_1}{dz}$$

or

$$\frac{dt}{dz} = \frac{dt}{dz'} \cdot 1 \cdot \left(-\frac{1}{z^2}\right) = -\frac{(1 - t)^{1-\beta}(1 + t)^{1+\beta-\alpha}}{M(t - a)}\frac{1}{z^2} \tag{29}$$

Substituting the expression for dw/dt and dt/dz from Equations 20 and 29, respectively, in Equation 28.

$$I_E = -\frac{i}{k}\frac{M'}{(t - t_A)\sqrt{(t - t_B)}\sqrt{(t + 1)}}\frac{(1 - t)^{1-\beta}(1 + t)^{1+\beta-\alpha}}{M(t - a)z^2} \tag{30}$$

For a given value of r and θ' the exit gradient can be evaluated using Equation 30 after determining the corresponding value of t from Equation 18 for the circular portion. Beyond the circular portion the exit gradient can be evaluated as follows. For an assumed value of $t > t_A$, the corresponding value of z can be evaluated using equation,

$$z = \left[S\left(\frac{1 + t}{1 + a}\right)^{\alpha-\beta}\left(\frac{t - 1}{1 - a}\right)^{\beta} - \frac{1}{2\ell}\right]^{-1} \tag{31}$$

After determining value of z, the value of exit gradient beyond point F can be evaluated using Equation 30.

Exact Solution for Semi-Circular Scour

Figure 6a shows the flow domain ABCDEIFG and profile of the weir in the z-plane. The scour on the downstream side of the weir is semi-circular in shape. It is described by

$$|z - \ell| = \ell \tag{32}$$

where ℓ is the radius of scour profile.

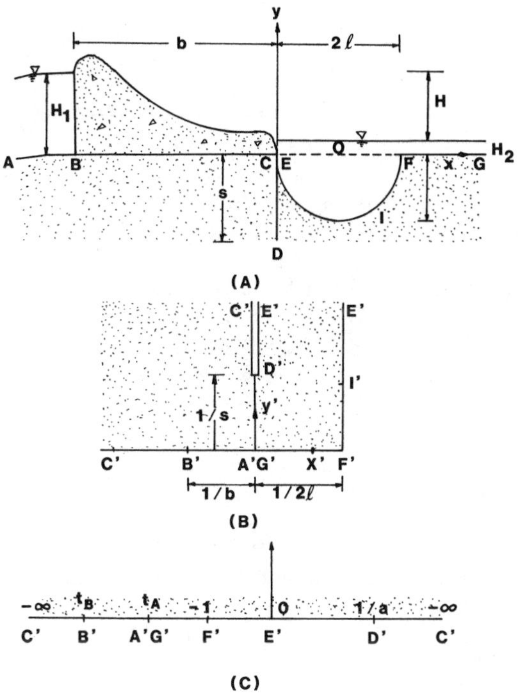

Figure 6. Steps of conformal mapping for semi-circular scour. (A) Z-plane; (B) Z′-plane; (C) t-plane.

The semi-circular profile in the z-plane can be described by

$$x^2 + y^2 - 2\ell x = 0 \tag{33}$$

The inverse flow domain, 1/z, is shown in Figure 6b. Since the flow domain comprises a circle and straight lines passing through the origin, the inverse of the flow domain has straight line boundaries.

Under the transformation 1/z, the circle maps onto the straight line given by

$$x' = 1/2\ell \tag{34}$$

where $1/z = x' + iy'$ (35)

According to the Schwarz-Christoffel transformation, the conformal mapping of the inverse flow domain in z′-plane onto the upper half of the auxiliary t-plane, shown in Figure 6c, is given by,

$$z' = M\int \frac{dt}{(t - 1/a)^{-1}\sqrt{(t+1)t}} + N \tag{36}$$

the vertices F′, E′ and D′ of the polygon in z′-plane being mapped at −1, 0 and 1/a, respectively, on the real axis of the t-plane. In Equation 36, M and N are constants.

Performing the integration [8]

$$z' = 2M(1+t)^{1/2} - \frac{M}{a}\log\left[\frac{\sqrt{(t+1)} - 1}{\sqrt{(t+1)} + 1}\right] + N \tag{37}$$

For point F′, t = −1 and z′ = 1/2ℓ. Therefore,

$$N = (1/2\ell) + \frac{iM\pi}{a} \tag{38}$$

From Equation 36

$$dz' = \frac{M\,dt}{t\sqrt{(t+1)}(t-1/a)^{-1}} \tag{39}$$

Putting $t = re^{i\theta}$, $dt = r \cdot e^{i\theta}\, i\, d\theta$. As t passes around a semi-circle of small radius r at E′ the corresponding change in z′ is 1/2ℓ. Thus,

$$1/2\ell = M\int_0^{\pi} \frac{re^{i\theta} i\, d\theta}{re^{i\theta}(re^{i\theta}+1)^{1/2}(re^{i\theta}-1/a)^{-1}}, \; r \to 0 \tag{40}$$

Simplifying

$$M = \frac{ia}{2\ell\pi} \tag{41}$$

Considering Equations 38 and 41, the constant N = 0. For point D′, t = 1/a, z′ = i/s. Using this condition in Equation 37 and substituting the value of M and N.

$$1 = \frac{as}{\ell\pi}\frac{\sqrt{(a-1)}}{\sqrt{a}} - \frac{s}{2\pi\ell}\log\left[\frac{\sqrt{(a+1)}-\sqrt{a}}{\sqrt{(a+1)}+\sqrt{a}}\right] \tag{42}$$

For a given value of ℓ and s, the value of a can be evaluated from Equation 42 by an iteration procedure.

Determination of t_A and t_B

Considering point A′ G′ for which z′ = 0 and $t = t_A$ and substituting these values in Equation 37 and simplifying

$$a\sqrt{(1+t_A)} - \frac{1}{2}\log\left[\frac{\sqrt{(t_A+1)}-1}{\sqrt{(t_A+1)}+1}\right] = 0$$

or

$$a\sqrt{(1+t_A)} - \frac{1}{2}\log\left[\frac{(\sqrt{(t_A+1)}-1)^2}{t_A}\right] = 0$$

or

$$a\sqrt{(1+t_A)} - \log[i\sqrt{(-t_A-1)}-1] + \frac{1}{2}\log[(-1)(-t_A)] = 0$$

or

$$a\sqrt{(1+t_A)} - \log\left[\sqrt{(-t_A)}e^{i}\, e^{i\tan^{-1}\left\{\frac{\sqrt{-t_A-1}}{-1}\right\}}\right] + \frac{1}{2}\log[(-t_A) + \frac{i\pi}{2} = 0$$

or

$$a\sqrt{(-1-t_A)} - \tan^{-1}\left[\frac{\sqrt{(-t_A-1)}}{-1}\right] + \pi/2 = 0 \tag{43}$$

Similarly, considering point B for which $z' = 1/b$ and $t = t_B$

$$a\sqrt{(-1-t_B)} - \tan^{-1}\left[\frac{\sqrt{(-t_B-1)}}{-1}\right] + \frac{\pi}{2} = \frac{\pi\ell}{b} \tag{44}$$

t_A and t_B can be evaluated from Equations 43 and 44, respectively, by an iteration procedure.

Relationship Between t and z for the Portion FG

Considering any point t' in between F and G, $t_A < t' < -1$, using Equation 37 and simplifying

$$z = \left[\frac{1}{\pi\ell}\tan^{-1}\left\{\frac{\sqrt{(-t'-1)}}{-1}\right\} - \frac{a}{\pi\ell}\sqrt{(-t'-1)} - \frac{1}{2\ell}\right]^{-1} \tag{45}$$

Mapping of Complex Potential Plane

The transformation of w-plane to upper half of t-plane is given by

$$w = M'\int\frac{dt}{\sqrt{(t-t_B)}\sqrt{t(t-t_A)}} + N' \tag{46}$$

where, M' and N' are constants. Integrating [8]

$$w = -\frac{M'}{\sqrt{t_A}\sqrt{(t_B-t_A)}}\sin^{-1}\left\{\frac{(2t_A-t_B)(t-t_A)+2t_A(t_A-t_B)}{t_B(t-t_A)}\right\} + N' \tag{47}$$

For point E, $t = 0$, $w = 0$. Therefore,

$$-\frac{M'}{\sqrt{t_A}\sqrt{(t_B-t_A)}}\cdot\frac{\pi}{2} + N' = 0 \tag{48}$$

For point B, $t = t_B$, $w = -kH$. Hence,

$$\frac{M'}{\sqrt{t_A}\sqrt{(t_B-t_A)}}\cdot\frac{\pi}{2} + N' = -kH \tag{49}$$

Solving for M' and N':

$$N' = -\frac{kH}{2}$$

and

$$M' = -\frac{kH}{\pi}\sqrt{t_A}\sqrt{(t_B-t_A)} \tag{50}$$

Pressure Distribution Under Base

The complex potential w for a point t', $-\infty < t' < t_B$ and $0 < t' < \infty$ is given by

$$w = \frac{kH}{\pi}\sin^{-1}\left[\frac{(2t_A - t_B)(t - t_A) + 2t_A(t_A - t_B)}{t_B(t - t_A)}\right] - \frac{kH}{2} \tag{51}$$

Since $\psi = 0$ along the base of the structure, $w = \phi$ in Equation 51. The pressure can be determined by substituting this relation for ϕ into Equation 19. The value of z corresponding to point t' is given by

$$z_{t'} = \left[-\frac{ia}{\pi\ell}(1 + t')^{1/2} + \frac{i}{2\pi\ell}\log\left(\frac{\sqrt{(t' + 1)} - 1}{\sqrt{(t' + 1)} + 1}\right)\right]^{-1} \tag{52}$$

Exit Gradient Beyond E

The exit gradient is given by

$$I_E = \frac{i}{k}\left(\frac{dw}{dt}\cdot\frac{dt}{dz}\right) \tag{53}$$

$$\frac{dt}{dz} = \frac{dt}{dz'}\frac{dz'}{dz}$$

or

$$\frac{dt}{dz} = \frac{dt}{dz'}\left(-\frac{1}{z^2}\right) = -\frac{t(t + 1)^{1/2}(t - 1/a)^{-1}}{M}\frac{1}{z^2} \tag{54}$$

Substituting the expression for $\frac{dw}{dt}$ and $\frac{dt}{dz}$ from Equations 46 and 54, respectively, in Equation 53.

$$I_E = -\frac{i}{k}\frac{M}{\sqrt{t}\sqrt{(t - t_B)(t - t_A)}}\frac{t\sqrt{(t + 1)}}{Mz^2(t - 1/a)} \tag{55}$$

For a given value of t, $-1 < t < 0$, the exit gradient can be evaluated using Equation 55 after determining the corresponding value of z from Equation 52.

For an assumed value of t, $t_A < t < -1$, the exit gradient can be evaluated using Equation 55 after determining the corresponding value of z using Equation 45.

Factor of Safety Against Piping

Water flows through porous medium under the weir. While moving through soil it imparts energy to soil particles through friction. The direction of the seepage force, acting on soil particles, will be along the gradient direction. The seepage force on the soil particles, located at the downstream boundary of the flow domain, which is an equipotential line, will act normal to the downstream surface. The other force acting on the soil particles, located at the downstream boundary of the flow domain, is gravitational force due to the effective weight of the soil particles.

Considering a volume V of soil mass, the effective weight of the soil is given by

$$F_g = V(G - 1)(1 - n)\gamma_w \tag{56}$$

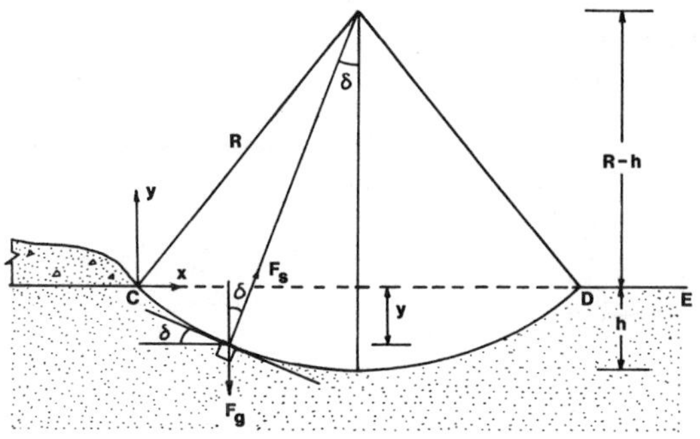

Figure 7. Forces acting on a soil particle at the exit.

where F_g is the gravitational force acting vertically downwards, G is the specific gravity of the soil, n is porosity of the soil, and γ_w is the unit weight of water.

The seepage force exerted on the soil mass of volume V, by the hydraulic gradient I_E, is given by [9]

$$F_s = -I_E\gamma_w V \tag{57}$$

When the boundary of the flow domain is horizontal, the seepage force F_s acts vertically upwards. Thus, the factor of safety against piping, neglecting cohesion forces, is given by [5]

$$\text{Factor of safety} = \frac{F_g}{F_s} = -\frac{(G-1)(1-n)}{I_E} \tag{58}$$

However, with a curved flow domain boundary, the direction of the hydraulic gradient at the exit also changes with the location of the point under consideration along the curved surface. Figure 7 shows a soil particle at the boundary of the flow domain. The tangent to the curved surface at the point under consideration makes an angle δ with the horizontal plane. The seepage force, being normal to the curved surface, will be at an angle δ with the vertical. Therefore, the factor of safety against piping for the soil particle at this point is given by

$$\text{Factor of safety} = \frac{(G-1)(1-n)\cos\delta}{-I_E} \tag{59}$$

For a circular scour profile the stream lines at the exit, being normal to the curved surface, will be radial. Therefore, from Figure 7

$$\cos\delta = \frac{R-h+y}{R} \tag{60}$$

where, y is the ordinate of the point under consideration on the circular boundary in the z-plane.

The value of the factor of safety against piping at any point on the circular scour profile can be evaluated using Equation 59. The values of $\cos\delta$ and I_E can be determined using Equations 60 and 55, respectively.

The variation of $I_E s/H$ with x/b for $h/\ell = 0.10$ and $\ell/b = 0.0625$, 0.0833 and 0.125 is shown in Figures 8–10 for various values of s/b. The exit gradient at point E is zero, since the angle between the streamline DE and the equipotential line EF at point E is less than $\pi/2$. The exit gradient beyond point E increases to a maximum value and then decreases to become zero at point F. Since the angle between FIE and FG at point F is less than π, and EIF and FG are part of the same equipotential line, the exit gradient at point F is also zero. Beyond point F the exit gradient increases

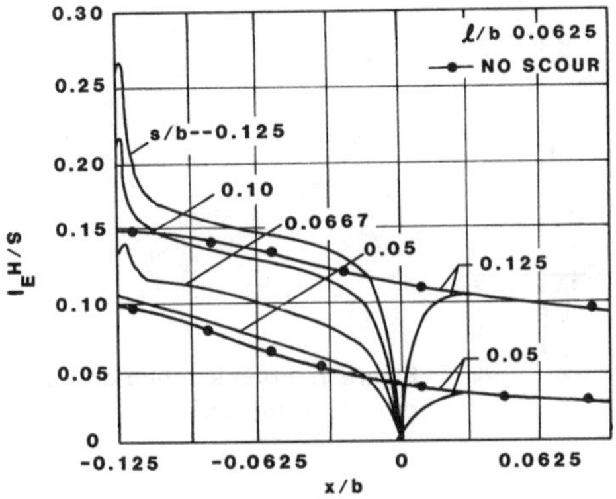

Figure 8. Variation of I_Es/H with x/b for h/ℓ = 0.10.

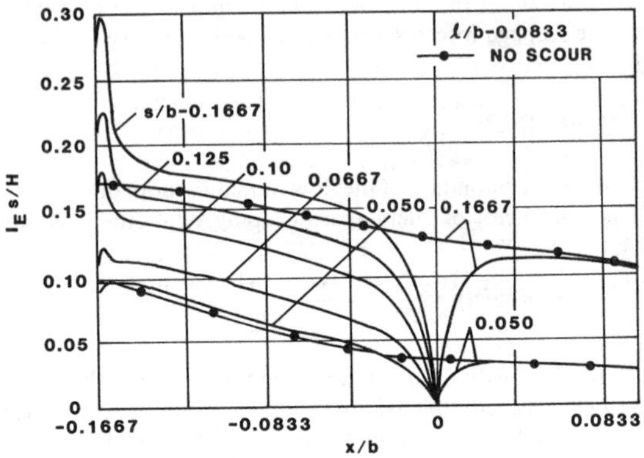

Figure 9. Variation of I_Es/H with x/b for h/ℓ = 0.10.

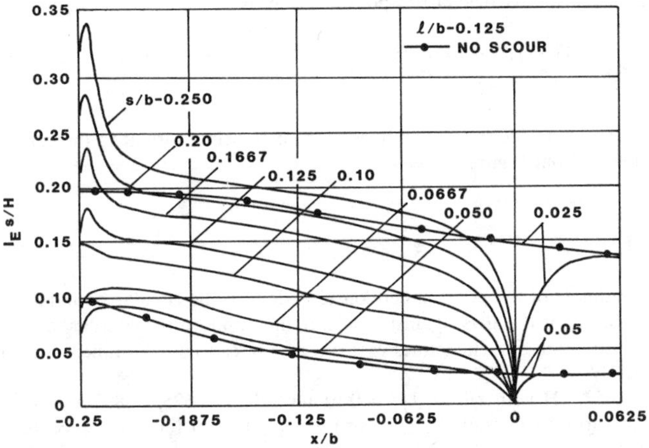

Figure 10. Variation of I_Es/H with x/b for h/ℓ = 0.10.

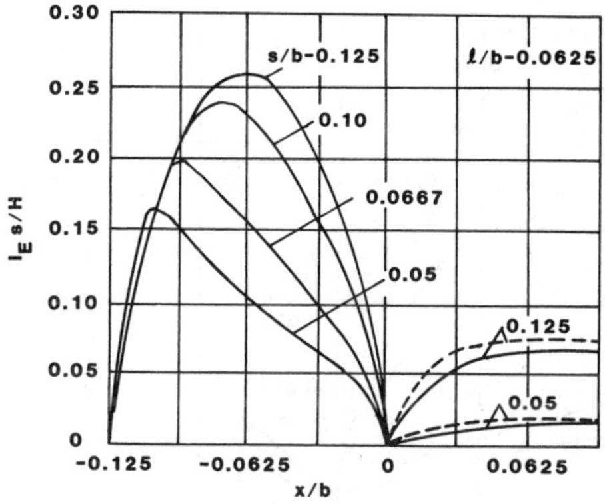

Figure 11. Variation of $I_E s/H$ with x/b for $h/\ell = 1.0$.

to a maximum value and then decreases monotonically with distance. For no scour condition, the exit gradient, which has been obtained from Khosla's analysis is also shown in Figures 8–10. For s/b = 0.125, when there is no scour, the maximum value of $I_E s/H$ is 0.1495 [5]. When the downstream bed is scoured and the circular profile of the scoured bed is described by $\ell/b = 0.0625$ and $h/\ell = 0.10$, the maximum $I_E s/H = 0.2651$. Thus, the exit gradient is increased by 67.32% because of scour. For $\ell/b = 0.833$ and 0.125 the corresponding increase is 50.97% and 20.84%, respectively.

In Figures 11–13 the exit gradient distribution for semi-circular scour has been plotted for $\ell/b =$ 0.0625, 0.0833, and 0.125 for various values of s/b. These results have been obtained making use of the approximate analysis. As seen from the figure, for the same scour profile, as s/b increases, the point of maximum exit gradient shifts away from the sheet pile. For a scour profile described by $\ell/b = 0.0625$ and $h/\ell = 1$, the maximum exit gradient for a weir of dimension s/b = 0.125, is increased by 73%. For $\ell/b = 0.0833$ and 0.125, the corresponding increases are 77.3% and 78.5%, respectively.

The exit gradient for semi-circular scour obtained from exact analysis are shown in Figures 14–16. For s/b = 0.125 and $\ell/b = 0.0625$, the difference in the maximum value of $I_E s/H$ obtained from approximate and exact analysis is 0.0201. The error in the approximate analysis is 7.2% and the maximum exit gradient is under-estimated. For $\ell/b = 0.0833$ and 0.125, the corresponding errors are 3.76% and 2.02%, respectively. The results of a approximate solution for the exit gradient are thus quite near to the true value, and for all practical purposes the exit gradient for circular scour, obtained by the approximate analysis, can be used without involving significant error.

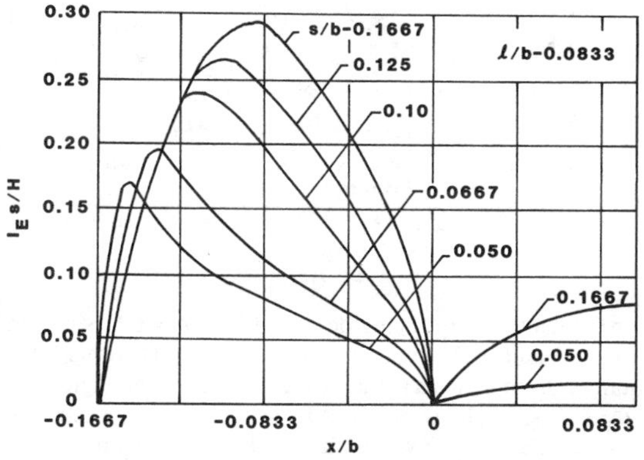

Figure 12. Variation of $I_E s/H$ with x/b for $h/\ell = 1.0$.

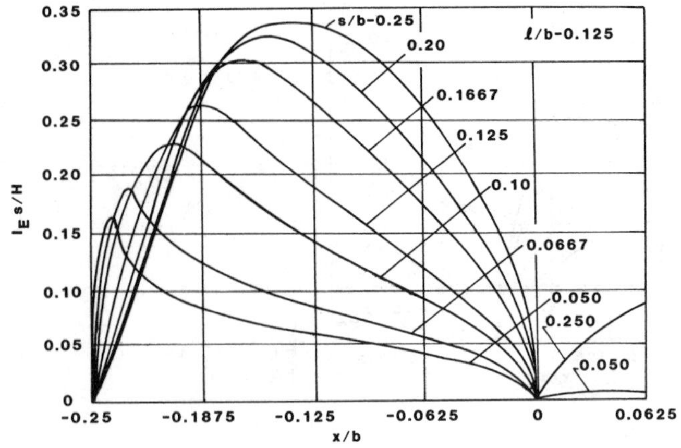

Figure 13. Variation of $I_E s/H$ with x/b for $h/\ell = 1.0$.

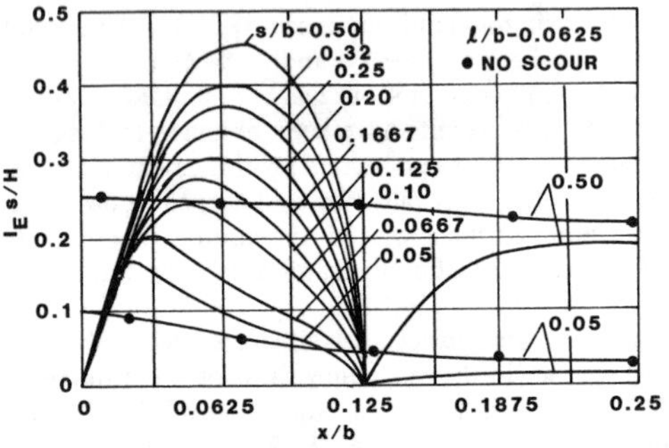

Figure 14. Variation of $I_E s/H$ with x/b for $h/\ell = 1.0$.

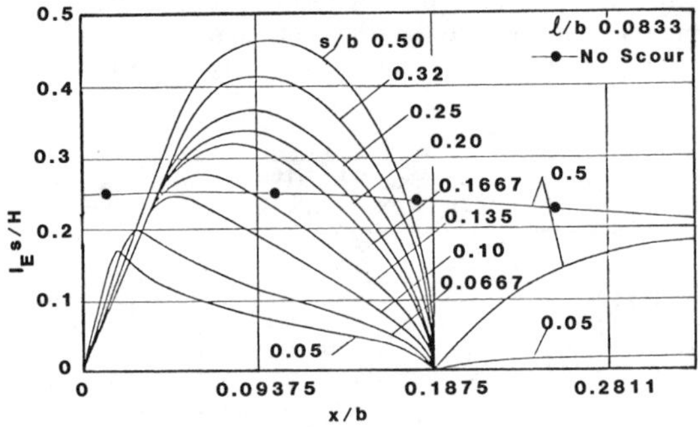

Figure 15. Variation of $I_E s/H$ with x/b for $h/\ell = 1.0$.

The variation of $-\Phi_D/kH$ with s/b for segmental scour, being described by $h/\ell = 0.1$, is shown in Figure 17 for values of $\ell/b = 0.0625$, 0.0833 and 0.125. Also variation of $-\Phi_D/kH$ with s/b for no scour situation, which has been obtained from Khosla's analysis, is shown in this figure. In Figure 18 the results of variation of $-\Phi_D/kH$ with s/b, as obtained from exact and approximate analyses for semi-circular scour are shown. Besides this the variation of $-\Phi_D/kH$ with s/b with no scour situation has been plotted in this figure. As seen from this figure the potential at point

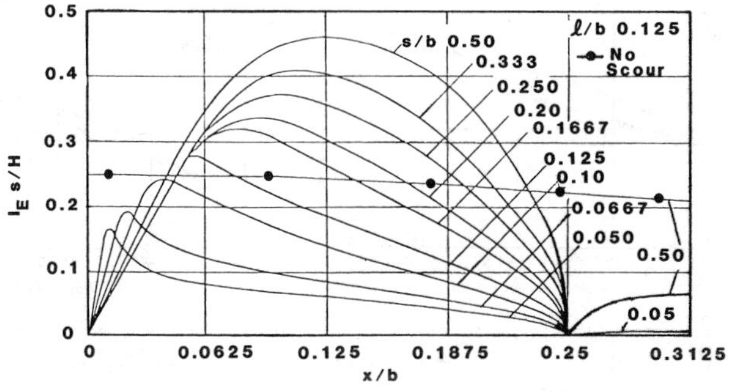

Figure 16. Variation of I_Es/H with x/b for h/ℓ = 1.0.

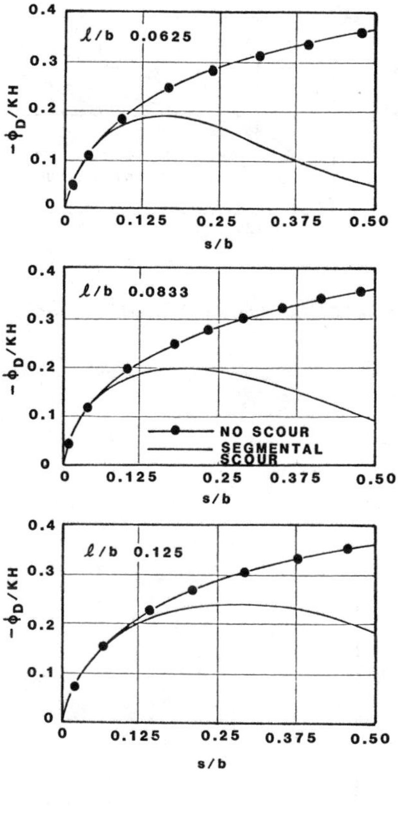

Figure 17. Variation of $-\phi_D$/kH with s/b for h/ℓ = 0.10.

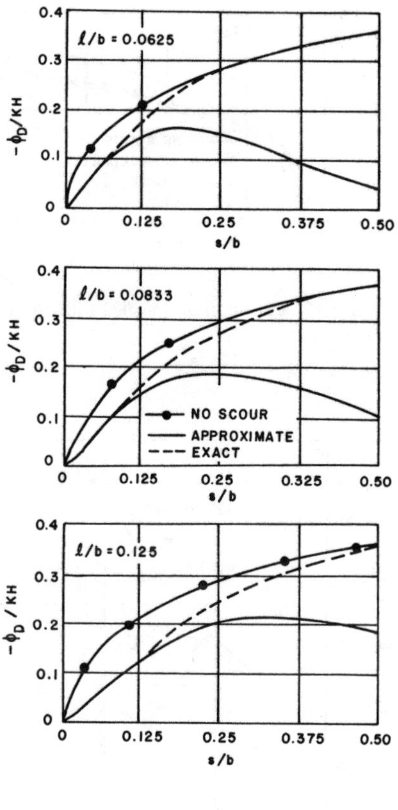

Figure 18. Variation of $-\phi_D$/kH with s/b for h/ℓ = 1.0.

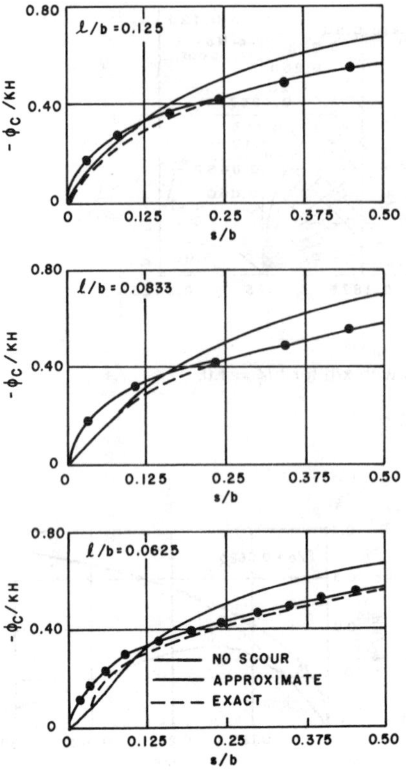

Figure 19. Variation of $-\phi_C$/kH with s/b for h/ℓ = 1.0.

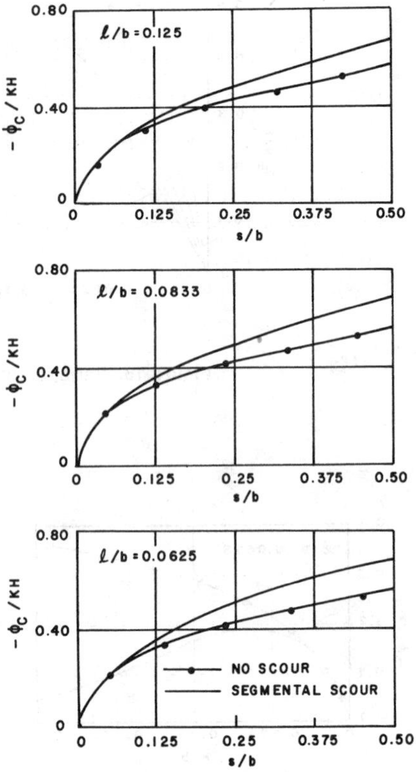

Figure 20. Variation of $-\phi_C$/kH with s/b for h/ℓ = 0.10.

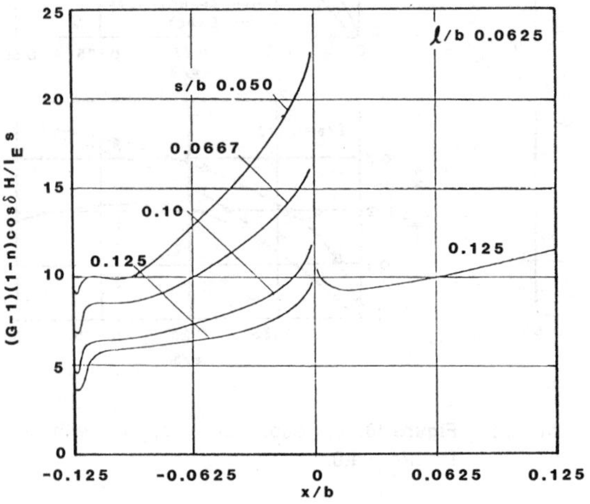

Figure 21. Variation of (G − 1) × (1 − n) cos $\delta H/I_E s$ with s/b for h/ℓ = 0.10.

D decreases because of scour. The figure also indicates that in case of semi-circular scour for s/b = 0.1, there is no appreciable difference in the values of $-\Phi_D/kH$ obtained from the exact and approximate analyses. Hence, up to s/b = 0.1 the approximate analysis will not introduce any significant error in estimating the potential under the weir for segmental scour condition.

The variation of $-\Phi_D/kH$ with s/b for the above scour condition is shown in Figures 19 and 20. Because of scour the potential at point C also decreases up to s/b = 0.1. This condition gets violated for higher value of s/b.

The variation of $H(G-1)(1-n)\cos\delta/I_E s$ with x/b for G = 2.65, n = 0.4, $h/\ell = 0.1$ and ℓ/b = 0.0625, 0.0833 and 0.1250 is shown in Figures 21–23 for various values of s/b. The factor of safety

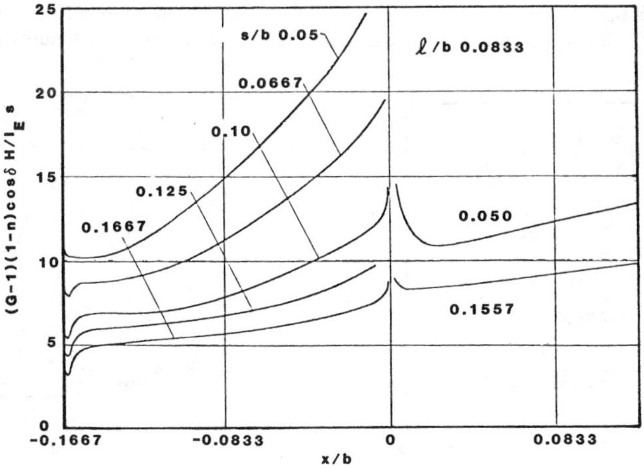

Figure 22. Variation of $(G-1)(1-n)\cos dH/I_E s$ with s/b for $h/\ell = 0.10$.

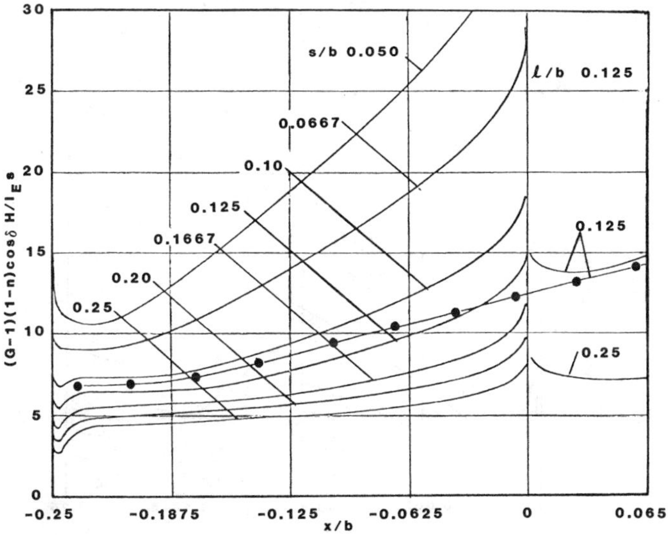

Figure 23. Variation of $(G-1)(1-n)\cos dH/I_E s$ with s/b for $h/\ell = 0.10$.

at point E is infinite, since the exit gradient at this point is zero. The factor of safety beyond point E decreases rapidly to a minimum value and then increases to become infinite at point F, where exit gradient is again zero. Beyond point F it again decreases to another minimum value and then increases monotonically with distance.

For no scour condition, the factor of safety, which has been obtained from Khosla's analysis, is also shown in Figure 23. For s/b = 0.1250, when there is no scour, the minimum value of factor of safety, which occurs at E, is 6.69 H/s. When downstream bed is scoured and the circular profile is described by ℓ/b = 0.0625 and h/ℓ = 0.1, the minimum factor of safety is equal to 3.66 H/s. Thus, the minimum factor of safety is reduced by about 50 percent, because of scour. For ℓ/b = 0.0833 and 0.125, the corresponding reduction is 36% and 19.6% respectively. However, the point where minimum factor of safety occurs, shifts away from the sheet pile. The variation of H(G − 1)(1 − n) cos δ/I_Es with x/b for h/ℓ = 1.0 and ℓ/b = 0.0625, 0.0833 and 0.125 is shown in Figures 24–26.

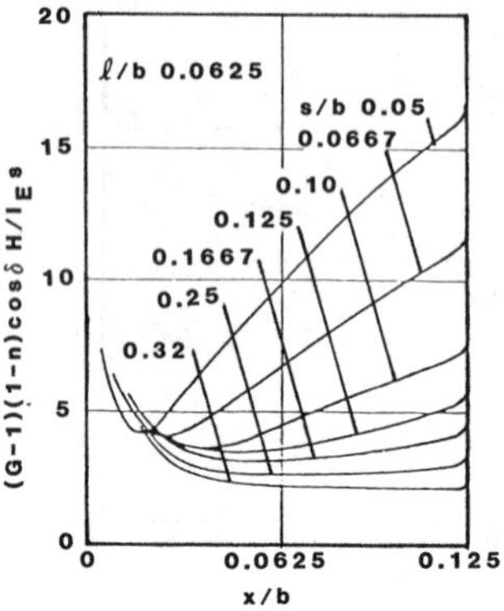

Figure 24. Variation of (G − 1)(1 − n) cos δH/I_Es with s/b for h/ℓ = 1.0.

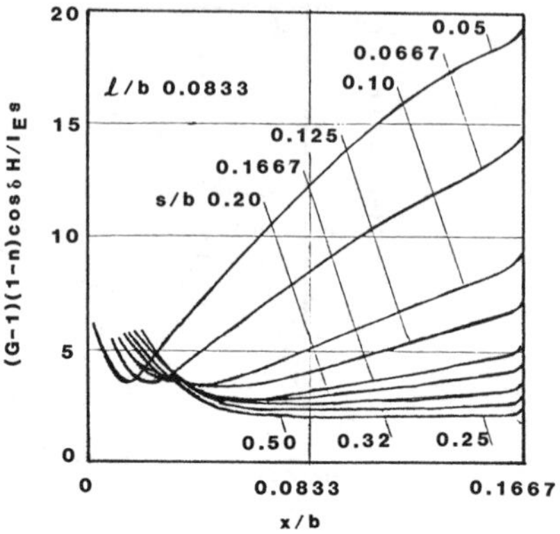

Figure 25. Variation of (G − 1)(1 − n) cos δH/I_Es with s/b for h/ℓ = 1.0.

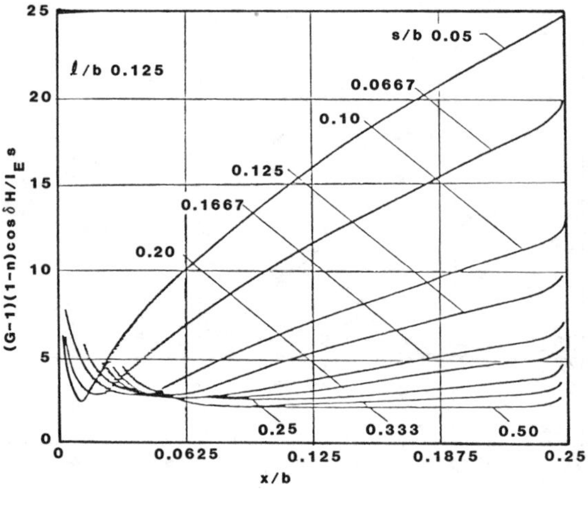

Figure 26. Variation of (G − 1)(1 − n) cos cH/I_Es with s/b for h/ℓ = 1.0.

EXIT GRADIENT WITH AEROFOIL SCOUR

In the z-plane, where $z = x + iy$, the profile of a weir with a vertical sheet pile under its heal, with aerofoil scour, is shown in Figure 27a. The weir is resting on porous medium of infinite depth. In the figure, b is the width of the apron and s is the depth of vertical sheet pile. The chord length of the aerofoil is L. The maximum depth of scour is Y_m, occurring at a distance τ from one end of aerofoil as shown in the figure. The area of the aerofoil is A_r. The parameters L, Y_m, τ, and A_r completely define an unique aerofoil. The porous medium under the weir extends to an infinite depth.

Mapping of an Aerofoil on to a Circle

Let ℓ be the radius of a circle, with its center located on the real axis, at a distance c from the origin, in T_1-plane (Figure 27b), where, $T_1 = R_1 + iS_1$. Let the dimension B_1C_1 be b_1 and C_1D_1 be s_1. Once ℓ and c are fixed, the configuration of the domain in T_1-plane is also fixed. Let under the transformation

$$z = T_1 + \frac{J}{T_1} \tag{61}$$

where, J is a real constant, the domain $A_1B_1C_1D_1E_1P_1F_1G_1$ in T_1-plane transform to the physical flow domain ABCDEPFG in z-plane. The origin 0_1 in the T_1-plane maps onto 0 in the z-plane. The chord length, L, of the aerofoil is EF. Since, EF = E0 +.0F, therefore,

$$L = \left| -(\ell + c) + \frac{J}{-(\ell + c)} \right| + (\ell - c) + \frac{J}{(\ell - c)}$$

or

$$L = 2\ell \left\{ 1 + \frac{J}{\ell^2 - c^2} \right\} \tag{62}$$

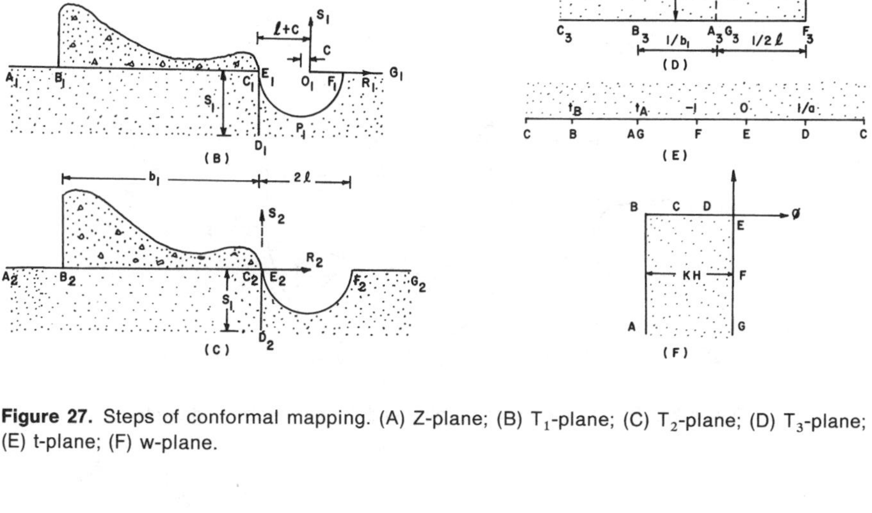

Figure 27. Steps of conformal mapping. (A) Z-plane; (B) T_1-plane; (C) T_2-plane; (D) T_3-plane; (E) t-plane; (F) w-plane.

Equation 61 can be rewritten as

$$z = x + iy = R_1 + iS_1 + \frac{J(R_1 - iS_1)}{R_1^2 + S_1^2} \tag{63}$$

Equating real and imaginary parts

$$x = R_1 + \frac{JR_1}{R_1^2 + S_1^2} \tag{64}$$

and

$$y = S_1 - \frac{JS_1}{R_1^2 + S_1^2} \tag{65}$$

Taking any point on the circle in T_1-plane, such that the radial line to that point makes an angle ϵ with the real axis, (Figure 28)

$$R_1 = -c + \ell \cos \epsilon \tag{66}$$

and

$$S_1 = \ell \sin \epsilon \tag{67}$$

On substituting value of R_1 and S_1 from Equations 66 and 67, respectively, in Equations 64 and 65 and simplifying

$$x = (-c + \ell \cos \epsilon)\left\{1 + \frac{J}{c^2 + \ell^2 - 2c\ell \cos \epsilon}\right\} \tag{68}$$

$$y = \ell \sin \epsilon \left\{1 - \frac{J}{c^2 + \ell^2 - 2c\ell \cos \epsilon}\right\} \tag{69}$$

The area of the aerofoil is given by

$$(\ell - c) + \frac{J}{\ell - c}$$

$$A_r = \int y \, dx - (\ell + c) + \frac{J}{\ell + c} \tag{70}$$

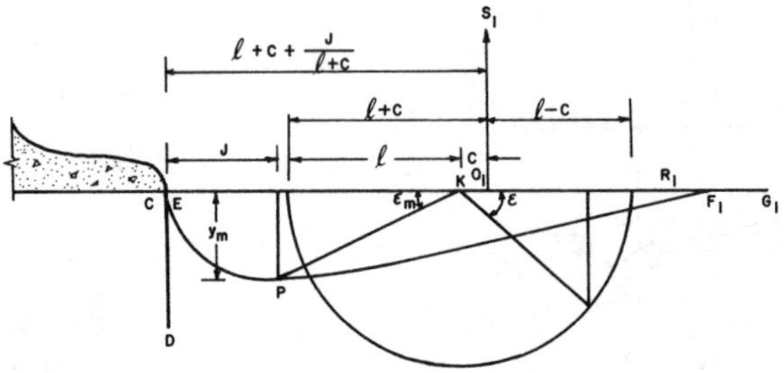

Figure 28. The Aerofoil scour profile and the corresponding circle.

On differentiation with respect to ϵ and simplification Equation 68 gives,

$$dx = -\ell \sin \epsilon \left\{ 1 + \frac{J}{c^2 + \ell^2 - 2c\ell \cos \epsilon} + \frac{2cJ(-c + \ell \cos \epsilon)}{(c^2 + \ell^2 - 2c\ell \cos \epsilon)^2} \right\} d\epsilon \tag{71}$$

Substituting values of y and dx from Equations 69 and 71 respectively, in Equation 70 and simplifying

$$\begin{aligned} A_r = {} & J\ell^2(2c^2 + J)\int_\pi^{2\pi} \frac{\sin^2 \epsilon \, d\epsilon}{(c^2 + \ell^2 - 2c\ell \cos \epsilon)^2} - 2cJ\ell^3 \int_\pi^{2\pi} \frac{\sin^2 \epsilon \cos \epsilon \, d\epsilon}{(c^2 + \ell^2 - 2c\ell \cos \epsilon)^2} \\ & - 2c^2\ell^2J^2 \int_\pi^{2\pi} \frac{\sin^2 \epsilon \, d\epsilon}{(c^2 + \ell^2 - 2c\ell \cos \epsilon)^3} + 2cJ^2\ell^3 \int_\pi^{2\pi} \frac{\sin^2 \epsilon \cos \epsilon \, d\epsilon}{(c^2 + \ell^2 - 2c\ell \cos \epsilon)^3} \\ & - \ell^2 \int_\pi^{2\pi} \sin^2 \epsilon \, d\epsilon \end{aligned} \tag{72}$$

Substituting

$$c^2 + \ell^2 - 2c\ell \cos \epsilon = p$$

$$d\epsilon = \frac{dp}{2c\ell \sin \epsilon}$$

$$\cos \epsilon = \frac{\ell^2 + c^2 - p}{2c\ell}$$

and

$$\sin \epsilon = \frac{\sqrt{-p^2 + 2(\ell^2 + c^2)p - (\ell^2 - c^2)^2}}{2c\ell}$$

$$\begin{aligned} A_r = {} & \frac{J(2c^2 + J)}{4c^2} \int_{(\ell+c)^2}^{(\ell-c)^2} \frac{\sqrt{-p^2 + 2p(\ell^2 + c^2) - (\ell^2 - c^2)^2} \, dp}{p^2} \\ & - \frac{J(\ell^2 + c^2)}{4c^2} \int_{(\ell+c)^2}^{(\ell-c)^2} \frac{\sqrt{-p^2 + 2p(\ell^2 + c^2) - (\ell^2 - c^2)^2} \, dp}{p^2} \\ & + \frac{J}{4c^2} \int_{(\ell+c)^2}^{(\ell-c)^2} \frac{\sqrt{-p^2 + 2p(\ell^2 + c^2) - (\ell^2 - c^2)^2} \, dp}{p} \\ & - \frac{J^2}{2} \int_{(\ell+c)^2}^{(\ell-c)^2} \frac{\sqrt{-p^2 + 2p(\ell^2 + c^2) - (\ell^2 - c^2)^2} \, dp}{p^3} \\ & + \frac{J^2(\ell^2 + c^2)}{4c^2} \int_{(\ell+c)^2}^{(\ell-c)^2} \frac{\sqrt{-p^2 + 2p(\ell^2 + c^2) - (\ell^2 - c^2)^2} \, dp}{p^3} \\ & - \frac{J^2}{4c^2} \int_{(\ell+c)^2}^{(\ell-c)^2} \frac{\sqrt{-p^2 + 2p(\ell^2 + c^2) - (\ell^2 - c^2)^2} \, dp}{p^2} + \frac{\pi\ell^2}{2} \end{aligned} \tag{73}$$

On integration Equation 73 gives

$$A_r = \frac{\pi\ell^2}{2} - \frac{\pi J^2 \ell^2}{2(\ell^2 - c^2)^2} \tag{74}$$

Recalling Equation 69 and differentiating with respect to ϵ

$$\frac{dy}{d\epsilon} = \ell \cos\epsilon\left(1 - \frac{J}{c^2 + \ell^2 - 2c\ell\cos\epsilon}\right) + \ell\sin\epsilon\left\{\frac{2c\ell J\sin\epsilon}{(c^2 + \ell^2 - 2c\ell\cos\epsilon)^2}\right\} \tag{75}$$

$$\text{At } y = y_m,\ \frac{dy}{d\epsilon} = 0$$

Equating $dy/d\epsilon$ given by Equation 75 to zero and simplifying

$$4c^2\ell^2\cos^3\epsilon_m - 4c\ell(c^2 + \ell^2)\cos^2\epsilon_m + (\ell^2 + c^2)(\ell^2 + c^2 - J)\cos\epsilon_m + 2c\ell J = 0 \tag{76}$$

where, ϵ_m is the angle corresponding to the point of maximum scour. Also from Figure 28

$$\cos\epsilon_m = \frac{1 + \dfrac{J}{\ell + c} - \tau}{\left\{y_m^2 + \left(\ell + \dfrac{J}{\ell + c} - \tau\right)^2\right\}^{1/2}} \tag{77}$$

Substituting the value of $\cos\epsilon_m$ from Equation 77 in Equation 76, the values of ℓ, c and J can be evaluated by iterative procedure using Equations 62, 74, and 76 for known values of L, A_r, y_m and τ for the given aerofoil.

Mapping of salient points in the z-plane onto the T_1-plane: In the z-plane the vertical sheet pile CD is described by the equation

$$x = -\left(\ell + c + \frac{J}{\ell + c}\right) \tag{78}$$

For any point $(-\ell - c - J/(\ell + c), iy)$ in z-plane on the line CD, let (R_1, iS_1) be the corresponding point in T_1-plane. Then, using Equation 61

$$-\left(\ell + c + \frac{J}{\ell + c}\right) + iy = R_1 + iS_1 + \frac{J}{R_1 + iS_1} \tag{79}$$

Equating real and imaginary parts Equation 79 gives

$$-\ell - c - \frac{J}{\ell + c} = R_1\left(1 + \frac{J}{R_1^2 + S_1^2}\right) \tag{80}$$

and

$$y = S_1\left(1 - \frac{J}{R_1^2 + S_1^2}\right) \tag{81}$$

From Equation 81

$$R_1 = \left(\frac{JS_1}{S_1 - y} - S_1^2\right)^{\frac{1}{2}} \tag{82}$$

Substituting for R_1 in Equation 80 and simplifying

$$(2S_1 - y)^2\left\{\frac{J}{S_1(S_1 - y)} - 1\right\} - \left(c + \ell + \frac{J}{\ell + c}\right)^2 = 0 \tag{83}$$

For any value of y on the sheet pile, the corresponding S_1 can be known from Equation 83 by an iteration procedure.

The vertical sheet pile CD in z-plane gets mapped onto a curve in T_1-plane described by the Equation 80.

Let R_1' and S_1' be the co-ordinates of the tip of the sheet pile in T_1-plane.

Using the condition at point D, for which, $y = -s$, the ordinate S_1' of point D_1 can be obtained by an iterative procedure, from Equation 83. Knowing the value of S_1' the value of R_1' can be obtained from Equation 80.

Considering Figure 27c, S_1, the length of sheet pile in T_1-plane can be approximated by

$$s_1 = \sqrt{S_2'^2 + (R_1' - \ell - c)^2} \tag{84}$$

In z-plane for point B

$$z_B = -\left(\ell + c + \frac{J}{\ell + c}\right) - b \tag{85}$$

The corresponding point B_1 in T_1-plane is given by

$$T_1 = -(\ell + c) - b_1 \tag{86}$$

where, b_1 is the apron length in T_1-plane. Again making use of Equation 61, the mapping of point B_1 onto z-plane is given by

$$z_B = -(\ell + c + b_1) - \frac{J}{(\ell + c + b_1)} \tag{87}$$

Since Equations 85 and 87 are equal,

$$b_1 + \frac{J}{\ell + c + b_1} = b + \frac{J}{\ell + c} \tag{88}$$

The unknown s_1 and b_1 can be evaluated using Equations 84 and 88 by an iterative procedure.

Analysis

The mapping of the flow domain in T_1-plane is now defined. The scour profile $E_1P_1F_1$ is described by semi-circle with radius ℓ and the sheet pile C_1D_1 by the curve described by Equation 80.

Using Equation 80 for point C_1, where, $S_1 = 0$, $R_1 = -(\ell + c)$, and making use of the same equation, when $S_1 = \infty$, for a hypothetical case, $R_1 = -\ell - c - J/\ell + c$. Thus, the maximum variation in $|R_1|$ along the sheet pile CD is less than $J/\ell + c$. As an approximation, the image of the sheet pile $C_1D_1E_1$ in the T_1-plane has been replaced by a vertical straight slit of the length s_1.

For purpose of analysis, the T_1-plane has been mapped on to T_2-plane, shown in Figure 27c according to the relation given by

$$T_2 = T_1 + (\ell + c) \tag{89}$$

where

$$T_2 = R_2 + iS_2 \tag{90}$$

The flow domain in T_2-plane can be converted into a polygon with straightline boundaries by its inversion about E_2. The inverse of the flow domain in T_2-plane is shown in Figure 27d.

Under the transformation $1/T_2$ the circle maps on to the straightline given by

$$R_3 = \frac{1}{2\ell} \tag{91}$$

where

$$\frac{1}{T_2} = T_3 = R_3 + iS_3 \tag{92}$$

Using Schwarz-Christoffel transformation, the conformal mapping of the inverse flow domain in T_3-plane onto the upper half of the auxiliary t-plane, shown in Figure 27e, is given by

$$T_3 = M \int \frac{dt}{(t - 1/a)^{-1}(t + 1)^{\frac{1}{2}}t} + N \tag{93}$$

the vertices F_3, E_3, and D_3 of the polygon in T_3-plane being mapped at -1, 0 and $1/a$, respectively, on the real axis of the t-plane. In Equation 93 M and N are constants. Performing the integration [8]

$$T_3 = 2M(1 + t)^{\frac{1}{2}} - \frac{M}{a} \log\left(\frac{\sqrt{(t + 1)} - 1}{\sqrt{(t + 1)} + 1}\right) + N \tag{94}$$

For point F_3, $t = -1$ and $T_3 = 1/2\ell$. Therefore,

$$N = \frac{1}{2\ell} + \frac{iM\pi}{a} \tag{95}$$

From Equation 93

$$dT_3 = \frac{M\,dt}{t(t + 1)^{\frac{1}{2}}(t - 1/a)^{-1}} \tag{96}$$

Putting $t = re^{i\theta}$, $dt = re^{i\theta}i\,d\theta$. As t passes around a semi-circle of small radius r at E_3, the corresponding T_3 is $1/2\ell$. Thus,

$$1/2\ell = M \int_{0,\; r \to 0}^{\pi} \frac{re^{i\theta}i\,d\theta}{re^{i\theta}(re^{i\theta} + 1)^{\frac{1}{2}}(re^{i\theta} - 1/a)^{-1}} \tag{97}$$

simplifying

$$M = \frac{ia}{2\pi\ell} \tag{98}$$

Considering Equations 95 and 98 the constant $N = 0$. For point D_3, $t = 1/a$, $T_3 = i/s_1$. Using this condition in Equation 94 and substituting the values of M and N

$$1 = \frac{as_1}{\ell\pi}\frac{\sqrt{a + 1}}{\sqrt{a}} - \frac{s_1}{2\pi\ell} \log\left(\frac{\sqrt{a + 1} - \sqrt{a}}{\sqrt{a + 1} + \sqrt{a}}\right) \tag{99}$$

For the known values of ℓ and s_1, the values of a can be evaluated from Equation 99 by an iteration procedure.

Determination of t_A and t_B: Considering point A_3G_3 for which $T_3 = 0$ and $t = t_A$, and substituting these values in Equation 94 and simplifying

$$a\sqrt{1 + t_A} - 1/2 \log\left(\frac{\sqrt{t_A + 1} - 1}{\sqrt{t_A + 1} + 1}\right) = 0$$

or

$$a\sqrt{-1 - t_A} - \tan^{-1}\left(\frac{\sqrt{-t_A - 1}}{-1}\right) + \pi/2 = 0 \tag{100}$$

Similarly considering point B_3 for which $T_3 = -1/b_1$ and $t = t_B$

$$a\sqrt{-t_B - 1} - \tan^{-1}\left(\frac{\sqrt{-t_B - 1}}{-1}\right) + \pi/2 = \frac{\pi\ell}{b_1} \tag{101}$$

t_A and t_B can be evaluated from Equations 100 and 101, respectively, by an iteration procedure.

Relationship between t and T_3 for the portion F_3G_3: Considering any point t′ in between F_3 and G_3, $t_A < t' < -1$, using Equation 94 and simplifying

$$T_3 = \frac{1}{\pi\ell} \tan^{-1}\left[\left(\frac{\sqrt{-t' - 1}}{-1}\right) - \frac{a}{\pi\ell}\sqrt{-t' - 1} - \frac{1}{2\ell}\right] \tag{102}$$

Mapping of Complex Potential Plane

The complex potential w is defined as $w = \Phi + i\psi$ in which the velocity potential function.

$$\phi = -k\left(\frac{p'}{\gamma_w} + y\right) + c' \tag{103}$$

in which, k = coefficient of permeability, p′ = pressure, γ_w = unit weight of water, y = ordinate, c′ = constant, and ψ = stream function. The w-plane for the flow domain is shown in Figure 27f. The transformation of w-plane on to the upper half of t-plane is given by

$$w = M' \int \left(\frac{dt}{(t - t_B)^{\frac{1}{2}} t^{\frac{1}{2}} (t - t_A)}\right) + N' \tag{104}$$

where, M′ and N′ are constants. On integration Equation 104 gives [8]

$$w = -\frac{M'}{\sqrt{t_A(t_B - t_A)}} \sin^{-1}\left[\frac{(2t_A - t_B)(t - t_A) + 2t_A(t_A - t_B)}{t_B(t - t_A)}\right] + N' \tag{105}$$

For point E_2, t = 0, w = 0, Therefore,

$$-\frac{M'}{\sqrt{t_A(t_B - t_A)}} \cdot \frac{\pi}{2} + N' = 0 \tag{106}$$

For point B, $t = t_B$, $w = -kh$. Hence,

$$\frac{M'}{\sqrt{t_A(t_B - t_A)}} \cdot \frac{\pi}{2} + N' = -kH \tag{107}$$

Solving for M′ and N′

$$N' = -\frac{kH}{2} \tag{108}$$

and

$$M' = -\frac{kH}{\pi}\sqrt{t_A(t_B - t_A)} \tag{109}$$

Pressure distribution under base: The complex potential w for a point t′, $-\infty < t' < t_B$ and $0 < t' < \infty$ is given by

$$w = -\frac{kH}{\pi}\sin^{-1}\left[\frac{(2t_A - t_B)(t - t_A) + 2t_A(t_A - t_B)}{t_B(t - t_A)}\right] - \frac{kH}{2} \tag{110}$$

Since $\psi = 0$ along the base of the structure $w = \Phi$ in Equation 110. The pressure can be determined by substituting this relation for Φ into Equation 103. The value of T_3 corresponding to point t′ is given by

$$T_3 = \left[-\frac{ia}{\pi\ell}(1 + t')^{1/2} + \frac{i}{2\pi\ell}\log\left\{\frac{\sqrt{t' + 1} - 1}{\sqrt{t' + 1} + 1}\right\}\right] \tag{111}$$

Exit Gradient Between E and F

The exit gradient, I_E, is given by

$$I_E = \frac{i}{k}\left(\frac{dw}{dt} \cdot \frac{dt}{dz}\right) \tag{112}$$

$$\frac{dt}{dz} = \frac{dt}{dT_3} \cdot \frac{dT_3}{dT_2} \cdot \frac{dT_2}{dT_1} \cdot \frac{dT_1}{dz} \tag{113}$$

Obtaining dT_1/dz, dT_2/dT_1 and dT_3/dT_2 from differentiation of Equations 61, 89, and 92, respectively, and substituting in Equation 113,

$$\frac{dt}{dz} = \frac{dt}{dT_3}\left(-\frac{1}{T_2^2}\right) \cdot 1 \cdot \frac{1}{\left(1 - \frac{J}{T_1^2}\right)} \tag{114}$$

Substituting value of dt/dT_3 from Equation 96 in Equation 114

$$\frac{dt}{dz} = -\frac{T_1^2}{T_2^2(T_1^2 - J)} \frac{t(t + 1)^{1/2}(t - 1/a)^{-1}}{M} \tag{115}$$

Substituting the expression for dw/dt and dt/dz from Equations 104 and 115 in Equation 112, the expression for exit gradient I_E is given by

$$I_E = -\frac{i}{k}\frac{M'}{(t - t_A)\sqrt{t(t - t_B)}}\frac{t\sqrt{t + 1}}{M(t - 1/a)}\frac{T_1^2}{T_2^2(T_1^2 - J)} \tag{116}$$

Obtaining T_1 and T_2 from Equations 61 and 89 and substituting in Equation 116.

$$I_E = -\frac{i}{k}\frac{M'}{M}\frac{1}{(t - t_A)(t - 1/a)}\frac{\sqrt{t(t + 1)}}{(t - t_B)}$$

$$\frac{4(z - \sqrt{z^2 - 4J})^2}{[z - \sqrt{z^2 - 4J} + 2(\ell + c)]^2[(z - \sqrt{z^2 - 4J})^2 - 4J]} \tag{117}$$

For a given value of t, $-1 < t < 0$, the corresponding value of z can be obtained making use of Equations 111, 89, and 61. After determining the value of z, the exit gradient at z can be evaluated from Equation 117.

For $t_A < t < -1$, the corresponding value of z can be obtained from Equations 102, 89, and 61 and the exit gradient can be found out from Equation 117.

Factor of Safety Against Piping

The factor of safety against piping in the scoured portion can be evaluated using Equation 59. In this equation δ is the angle the tangent to the curved scour profile at the point under consideration makes with the horizontal plane. The angle δ is also the angle the stream line at the exit makes with the vertical.

The slope of the tangent to the curved scour profile at any point is given by

$$\tan \delta = \frac{dy}{dx}$$

$$\tan \delta = \frac{dy}{d\epsilon}\cdot\frac{d\epsilon}{dx}$$

On differentiation with respect to ϵ and simplification Equation 69 gives

$$\frac{dy}{d\epsilon} = \frac{\ell \cos \epsilon(\ell^2 + c^2 - 2c\ell \cos \epsilon)(\ell^2 + c^2 - 2c\ell \cos \epsilon - J) + 2J\ell^2 c \sin^2 \epsilon}{(\ell^2 + c^2 - 2c\ell \cos \epsilon)^2}$$

Substituting the expression for $d\epsilon/dx$ from Equation 71 and $dy/d\epsilon$ from the equation just given, respectively, and simplifying one gets

$$\delta = \tan^{-1}\left[\frac{2c\ell J \sin \epsilon + (c^2 + \ell^2 - 2c\ell \cos \epsilon)(c^2 + \ell^2 - 2c\ell \cos \epsilon - J) \cot \epsilon}{J(c^2 - \ell^2) - (c^2 + \ell^2 - 2c\ell \cos \epsilon)^2}\right] \tag{118}$$

The value of factor of safety against piping can be evaluated using Equation 59 for known values of G and n, after determining value of I_E and δ from Equations 117 and 118, respectively.

The variation of I_{E_s}/H with x/b, for $Y_{m/L} = 0.125$, 0.25, and 0.45 and L/b 0.25, for aerofoil scour shown in Figure 29, is given in Figures 30-32, for s/b = 0.5, 0.25, 0.2, 0.125, 0.10, and 0.0667. The exit gradient at point E is zero, since the angle between streamline DE and equipotential line FE at point E is less than $\pi/2$. The exit gradient beyond point E increases to a maximum value and

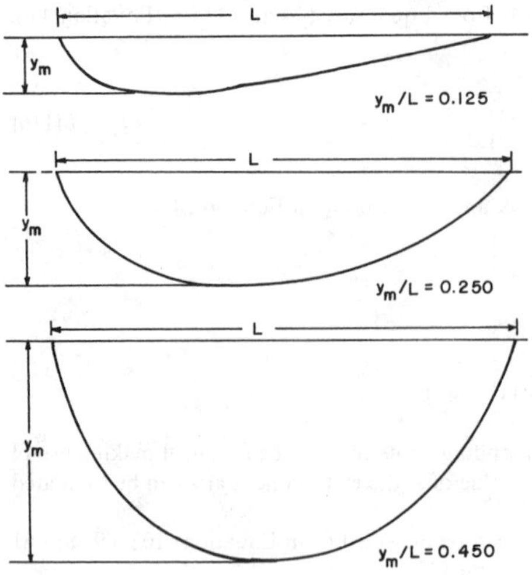

Figure 29. Assumed aerofoil scour profiles.

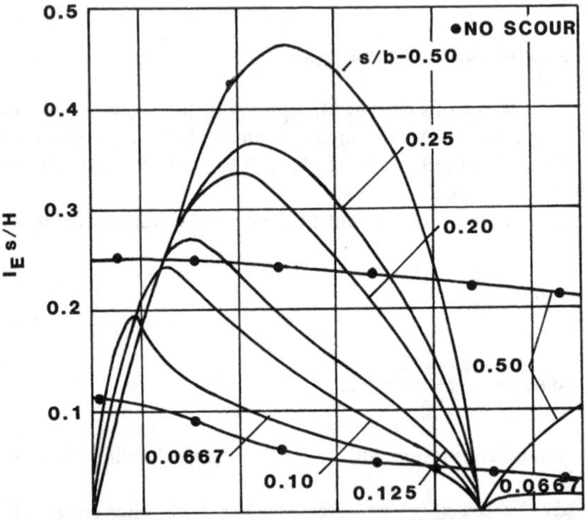

Figure 30. Variation of $I_E s/H$ with x/b for $y_m/L = 0.45$, $L/b = 0.25$.

then decreases to become zero again at point F. As the angle between FIE and FG at point F is less than π, and FIE and FG are the parts of the same equipotential surface, the exit gradient at point F is also zero. Beyond point F exit gradient again increases, to reach a maximum value and then decreases monotomically with distance. For no scour condition the exit gradient, which has been obtained form Khosla's analysis, is also shown in Figures 30–32. For $s/b = 0.125$ when there is no scour, the maximum value of exit gradient is 0.1495. When the downstream bed is scoured and the aerofoil profile of scoured bed is described by $L/b = 0.25$, $Y_m/L = 0.125$, for $s/b = 0.125$ the maximum $I_E s/H = 0.2756$. Thus, the exit gradient is increased by 84.4% because of scour. For $Y_m/L = 0.25$ and 0.45 the corresponding increase is 83.6% and 82.7%, respectively.

The variation of $-\Phi_D/KH$ with s/b for $Y_m/L = 0.125$, 0.25 and 0.45 is shown in Figure 33. Also variation of $-\Phi_D/kH$ with s/b for no scour situation, which has been obtained from Khosla's

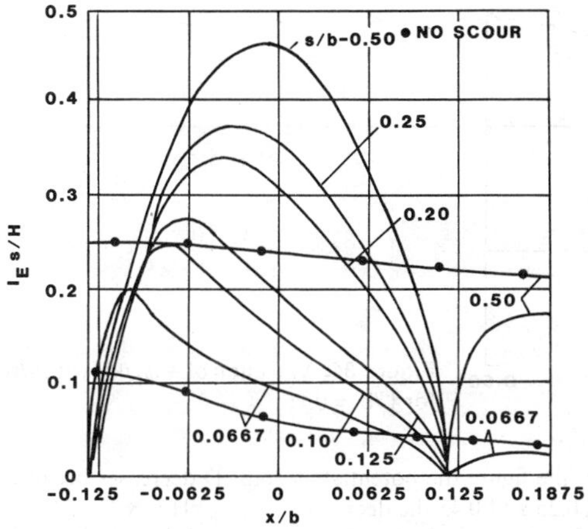

Figure 31. Variation of I_Es/H with x/b for y_m/L = 0.25, L/b = 0.25.

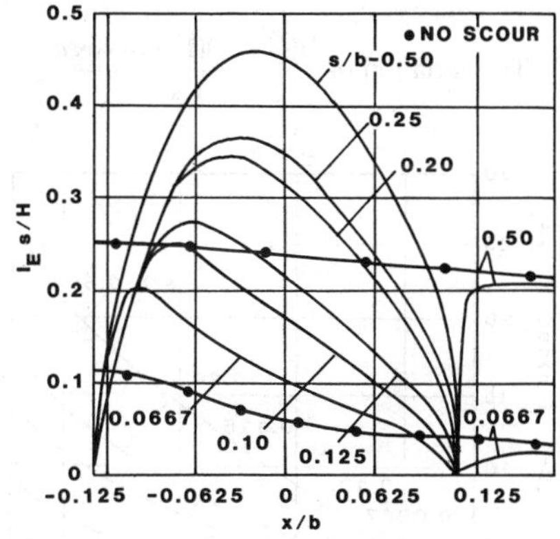

Figure 32. Variation of I_Es/H with x/b for y_m/L = 0.125, L/b = 0.25.

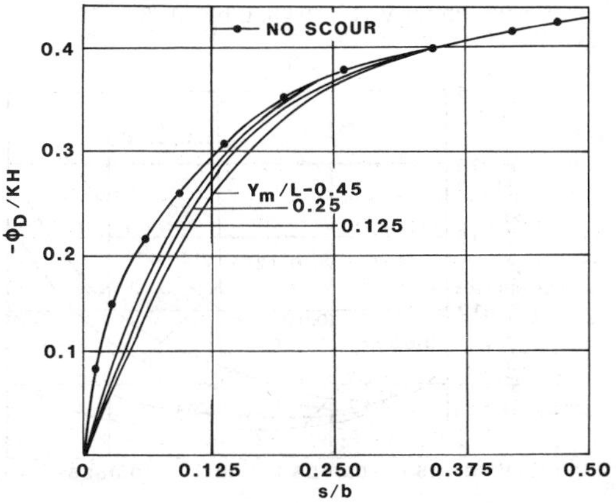

Figure 33. Variation of $-\phi_D$/kH with s/b for L/b = 0.25.

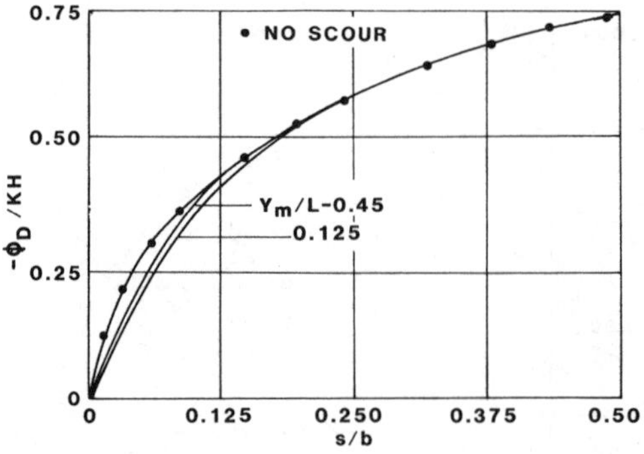

Figure 34. Variation of $-\phi_C$/kH with s/b for L/b = 0.25.

analysis is shown in this figure. As seen from the figure the potential at point D decreases because of scour. For s/b = 0.125 and Y_m/L = 0.125, 0.25 and 0.45 the decrease in $-\Phi_D$/kH is 8.9%, 24.5%, and 32.7%, respectively. The variation of $-\Phi_C$/kH with s/b for above scour condition is shown in Figure 34. Because of scour the potential at point C also decreases. The decrease in $-\Phi_C$/kH for Y_m/L = 0.125, 0.25 and 0.45 is 9.9%, 13.6%, 17.6%, respectively.

The variation of $\dfrac{H(G-1)(1-n)\cos\delta}{I_E s}$ for G = 2.65,

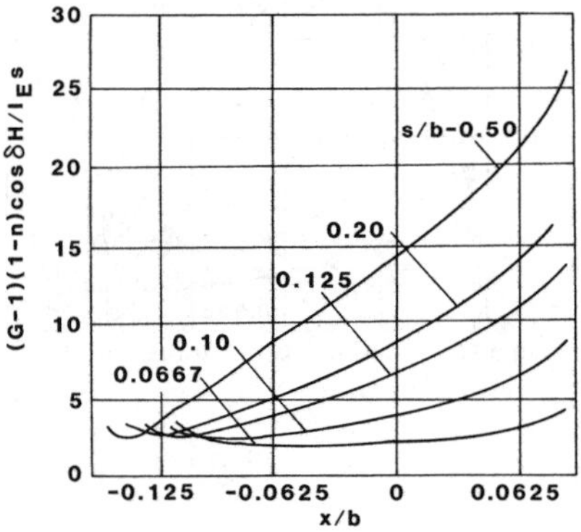

Figure 35. Variation of (G − 1)(1 − n) cos δH/I_Es with x/b for y_m/L = 0.45, L/b = 0.25.

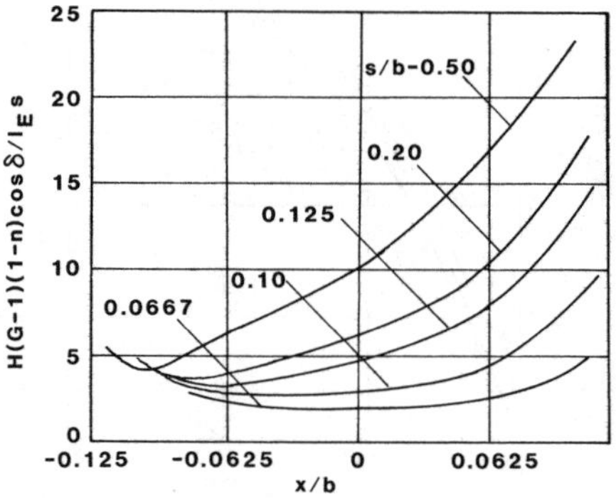

Figure 36. Variation of (G − 1)(1 − n) and H/I_Es with x/b for y_m/L = 0.25, L/b = 0.25.

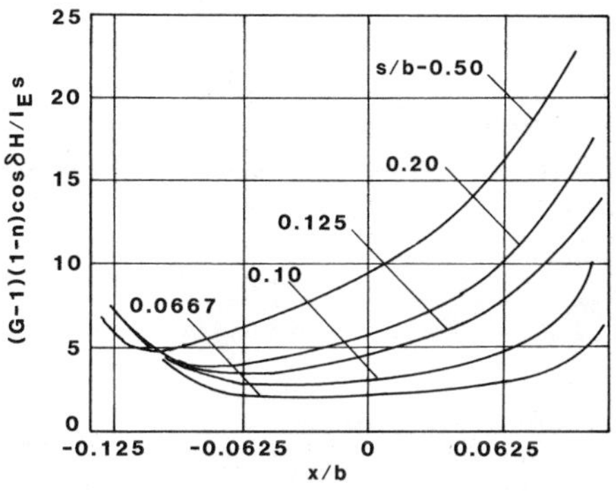

Figure 37. Variation of $(G - 1)(1 - n)$ and $H/l_E s$ with x/b for $y_m/L = 0.125$, $L/b = 0.25$.

$n = 0.4$ with aerofoil scour, has been shown in Figures 35–37, for $Y_m/L = 0.125$, 0.25 and 0.45, for $s/b = 0.50$, 0.20, 0.125, 0.10 and 0.0667. As seen from the figure for the same scour, as s/b increases, the point of minimum factor of safety shifts away from the sheet pile. For $s/b = 0.125$, when there is no scour, the minimum value of factor of safety against piping, as obtained from Equation 59 for $\delta = 0$, is 6.63 H/s. When the downstream bed is subjected to scour and the aerofoil profile of scour is described by $L/b = 0.25$, $Y_m/L = 0.125$ for $s/b = 0.125$ the minimum value of factor of safety is reduced to 3.48 H/s. Thus, the factor of safety is reduced by about 46% because of scour. For $Y_m/L = 0.25$ and 0.45 the corresponding reduction in factor of safety is about 48 percent and 50 percent respectively.

DISTRIBUTION OF PRESSURE UNDER WEIR APRON

The profile of a weir generally comprises a horizontal apron with a sloping step and two vertical sheet piles at the toe and heal, as shown in Figure 1. However, the exact analysis of seepage flow under such a profile is not available. For assessing the pattern of uplift pressure under the apron and the value of the exit gradient on the downstream side various approximate methods are used. According to Khosla's principle of independent variables, the complex weir profile is split into simple profiles comprising a flat floor and a vertical sheet pile. The pressures at salient points under a complex weir profile are determined by superimposing the values of pressure at such points obtained for simpler profiles using the analysis given by Khosla. The electrical anology models are also used to assess the seepage flow characteristics under a complex weir profile.

A sloping apron with vertical sheet piles under its toe and heal, as shown in Figure 38a, is a very close approximation to the actual weir profile. In view of this, an analytical method for estimating the pressures under a sloping apron with two vertical sheet piles and exit gradient on the downstream side of such a weir follows.

The flow domain is z-plane, where, $z = x + iy$ is shown in Figure 38a. In the Figure 38 b is the sloping floor length of the weir. It is inclined at an angle of $\alpha\pi$ with the upstream vertical sheet pile. The length of upstream and downstream sheet piles is s_1 and s_2, respectively. H_1 and H_2 are the depth of water on the upstream and downstream boundaries. The head causing seepage is represented by H.

The Schwarz-Christoffel conformal mapping of the flow domain in z-plane on to lower half of the auxiliary t-plane, shown in Figure 38b, is given by

$$z = \int_0^t \frac{M(t - t_C)(t - t_F)\,dt}{t^{1/2}(t - t_D)^{1-\alpha}(t - t_E)^{\alpha}(t - 1)^{1/2}} + N \tag{119}$$

the vertices A, B, C, D, E, F, G and H being mapped on to $-\infty$, 0, t_C, t_D, t_E, t_F, 1, and ∞, respectively, on real axis of t-plane, where, M and N are constants.

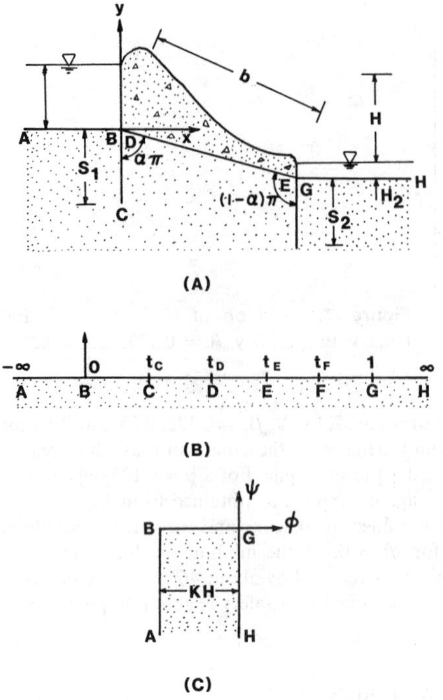

Figure 38. Steps of conformal mapping. (A) Z-plane; (B) t-plane; (C) W-plane.

The Equation 119 can be rewritten as

$$
\begin{aligned}
z = {} & M \int_0^t t^{3/2}(t - t_D)^{\alpha-1}(t - t_E)^{-\alpha}(t - 1)^{-1/2}\,dt \\
& - M(t_C + t_F) \int_0^t t^{1/2}(t - t_D)^{\alpha-1}(t - t_E)^{-\alpha}(t - 1)^{-1/2}\,dt \\
& + M t_C t_{F_0} \int_0^t t^{-\frac{1}{2}}(t - t_D)^{\alpha-1}(t - t_E)^{-\alpha}(t - 1)^{-1/2}\,dt + N_1
\end{aligned}
\tag{120}
$$

Since at $t = 0$, $z = 0$, therefore, $N_1 = 0$. For $0 \le t \le t_D$, substituting

$$t/t_D = r$$

in Equation 120 and simplifying

$$
\begin{aligned}
z = {} & M(-1)^{-3/2} t_D^{\alpha-1/2} t_E^{-\alpha} \Bigg[t_D^2 \int_0^r r^{5/2-1}(1 - r)^{\alpha+5/2-5/2-1}(1 - x_1 r)^{-\alpha}(1 - y_1 r)^{-1/2}\,dr \\
& (t_C + t_F) t_D \int_0^r r^{3/2-1}(1 - r)^{\alpha+3/2-3/2-1}(1 - x_1 r)^{-\alpha}(1 - y_1 r)^{-1/2}\,dr \\
& + t_C t_F \int_0^r r^{1/2-1}(1 - r)^{\alpha+1/2-1/2-1}(1 - x_1 r)^{-\alpha}(1 - y_1 r)^{-1/2}\,dr \Bigg]
\end{aligned}
\tag{121}
$$

where $x_1 = t_D/t_E$
$y_1 = t_D$

On integration Equation 121 gives [10]

$$
\begin{aligned}
z = {} & M(-1)^{-3/2} t_D^{3/2+\alpha} t_E^{-\alpha}\Bigg[\beta_r\left(\frac{5}{2}, \alpha\right) F_1\left(\frac{5}{2}, \alpha, \frac{1}{2}; \frac{5}{2} + \alpha; x_1, y_1\right) \\
& - r^{5/2}(1-r)^{\alpha}\Bigg\{\frac{2}{5} F_1\left(\frac{5}{2}, \alpha, \frac{1}{2}; \frac{5}{2} + \alpha, x_1, y_1\right) {}_2F_1\left(\frac{5}{2} + \alpha, 1; 1 + \frac{5}{2}; r\right) \\
& - \sum_{n=0}^{n=\infty} \frac{C_{1,n}}{\frac{5}{2} + n} r_2^n F_1\left(\frac{5}{2} + \alpha + n, 1; 1 + \frac{5}{2} + n; r\right)\Bigg\}\Bigg] \\
& - M(-1)^{-3/2}(t_A + t_F) t_0^{1/2} t^{\alpha} t^{-\alpha}\Bigg[Br\left(\frac{3}{2}, \alpha\right) F_1\left(\frac{3}{2}, \alpha, \frac{1}{2}; \frac{3}{2} + \alpha; x_{1E}, y_1\right) \\
& - r^{3/2}(1-r)^{\alpha}\Bigg\{\frac{2}{3} F_1\left(\frac{3}{2}, \alpha, \frac{1}{2}; \frac{3}{2} + \alpha; x_1, y_1\right) {}_2F_1\left(\frac{3}{2} + \alpha, 1; 1 + \frac{3}{2}; r\right) \\
& - \sum_{n=0}^{n=\infty} \frac{C_{1,n}}{\frac{3}{2} + n} r_2^n F_1\left(\frac{3}{2} + \alpha + n, 1; 1 + \frac{3}{2} + n; r\right)\Bigg\}\Bigg] \\
& + M(-1)^{-3/2} t_C t_F t_D^{-1/2+\alpha} t_E^{-\alpha}\Bigg[\beta_r\left(\frac{1}{2}, \alpha\right) F_1\left(\frac{1}{2}, \alpha, \frac{1}{2}; \frac{1}{2} + \alpha; x_1, y_1\right) \\
& - r^{1/2}(1-r)^{\alpha}\Bigg\{\frac{2}{1} F_1\left(\frac{1}{2}, \alpha, \frac{1}{2}; \frac{1}{2} + \alpha; x_1, y_1\right) {}_2F_1\left(\frac{1}{2} + \alpha, 1; 1 + \frac{1}{2}; r\right) \\
& - \sum_{n=0}^{n=\infty} \frac{C_{1,n}}{\frac{1}{2} + n} r_2^n F_1\left(\frac{1}{2} + \alpha + n, 1; 1 + \frac{1}{2} + n; r\right)\Bigg\}\Bigg]
\end{aligned} \tag{122}
$$

where $F_1(\ ) =$ hypergeometric series of two variables
$_2F_1(\ ) =$ Gauss hypergeometric function
$\beta_r(\ ) =$ incomplete beta function

$$C_{1,n} = A_{1,n}B_{1,0} + A_{1,n-1}B_{1,1} + A_{1,n-2}B_{1,2} + \ldots + A_{1,1}B_{1,n-1} + A_{1,0}B_{1,n}$$
$n = 0, 1, 2, 3, \ldots$

$$A_{1,n} = \frac{\alpha(\alpha+1)(\alpha+2)\cdots(\alpha+\overline{n-1})x_1^n}{n!}; \qquad n = 1, 2, 3, \cdots$$

$$B_{1,n} = \frac{0.5(0.5+1)(0.5+2)\cdots(0.5+\overline{n-1})y_1^n}{n!}; \qquad n = 1, 2, 3, \ldots$$

$$A_{1,0} = B_{1,0} = 1$$

For point D, $t = t_D$, hence $r = 1$. Using the condition for point D, for which $z = 0$, and $r = 1$, Equation 122 gives

$$
\begin{aligned}
& (-1)^{-3/2} M t_D^{\alpha-1/2} t_E^{-\alpha}\Bigg[t_D^2 \beta\left(\frac{5}{2}, \alpha\right) F_1\left(\frac{5}{2}, \alpha, \frac{1}{2}; \frac{5}{2} + \alpha; x_1, y_1\right) \\
& - t_D(t_C + t_F)\beta\left(\frac{3}{2}, \alpha\right) F_1\left(\frac{3}{2}, \alpha, \frac{1}{2}; \frac{3}{2} + \alpha; x_1, y_1\right) + t_C t_F \beta\left(\frac{1}{2}, \alpha\right) F_1\left(\frac{1}{2}, \alpha, \frac{1}{2}; \frac{1}{2} + \alpha; x_1; y_1\right)\Bigg] = 0
\end{aligned} \tag{123}
$$

where $\beta(\ ) =$ complete beta function

Using the condition at point C, for which $z = -is_1$, and $r = t_C/t_D$ in Equation 122, the following equation is obtained:

$$
\begin{aligned}
-is_1 = {} & M(-1)^{-3/2} t_D^{3/2+\alpha} t_E^{-\alpha}\Bigg[\beta_{t_C/t_D}\left(\frac{5}{2},\alpha\right)F_1\left(\frac{5}{2},\alpha;\frac{1}{2};\frac{5}{2}+\alpha;x_1,y_1\right) \\
& -\left(\frac{t_C}{t_D}\right)^{5/2}\left(1-\frac{t_C}{t_D}\right)^{\alpha}\Bigg\{\frac{2}{5}F_1\left(\frac{5}{2},\alpha,\frac{1}{2};\frac{5}{2}+\alpha;x_1\,y_1\right){}_2F_1\left(\frac{5}{2}+\alpha,1;1+\frac{5}{2};\frac{t_C}{t_D}\right) \\
& -\sum_{n=0}^{n=\infty}\frac{C_{1,n}}{\frac{5}{2}+n}\left(\frac{t_C}{t_D}\right)^{n}{}_2F_1\left(\frac{5}{2}+\alpha+n,1;1+\frac{5}{2}+n;\frac{t_C}{t_D}\right)\Bigg\}\Bigg] \\
& -(t_C+t_F)M(-1)^{-3/2} t_D^{1/2} t_E^{-\alpha}\Bigg[\beta_{t_C/t_D}\left(\frac{3}{2},\alpha\right)F_1\left(\frac{3}{2},\alpha,\frac{1}{2};\frac{3}{2}+\alpha;x_1,y_1\right) \\
& -\left(\frac{t_C}{t_D}\right)^{3/2}\left(1-\frac{t_C}{t_D}\right)^{\alpha}\Bigg\{\frac{2}{3}F_1\left(\frac{3}{2},\alpha,\frac{1}{2};\frac{1}{2};\frac{3}{2}+\alpha;x_1,y_1\right){}_2F_1\left(\frac{3}{2}+\alpha,1;1+\frac{3}{2};\frac{t_C}{t_D}\right) \\
& -\sum_{n=0}^{n=\infty}\frac{C_{1,n}}{\frac{3}{2}+n}\left(\frac{t_C}{t_D}\right)^{n}{}_2F_1\left(\frac{3}{2}+\alpha+n,1;1+\frac{3}{2}+n;\frac{t_C}{t_D}\right)\Bigg\}\Bigg] \\
& +t_C t_F M(-1)^{-3/2} t_D^{-1/2+\alpha} t_E^{-\alpha}\Bigg[\beta_{t_C/t_D}\left(\frac{1}{2},\alpha\right)F_1\left(\frac{1}{2},\alpha,\frac{1}{2};\frac{1}{2}+\alpha;x_1,y_1\right) \\
& -\left(\frac{t_C}{t_D}\right)^{1/2}\left(1-\frac{t_C}{t_D}\right)^{\alpha}\Bigg\{\frac{2}{1}F_1\left(\frac{1}{2},\alpha,\frac{1}{2};\frac{1}{2}+\alpha;x_1,y_1\right){}_2F_1\left(\frac{1}{2}+\alpha;1;1+\frac{1}{2};\frac{t_C}{t_D}\right) \\
& -\sum_{n=0}^{n=\infty}\frac{C_{1,n}}{\frac{1}{2}+n}\left(\frac{t_C}{t_D}\right)^{n}{}_2F_1\left(\frac{1}{2}+\alpha+n,1;1+\frac{1}{2}+n;\frac{t_C}{t_D}\right)\Bigg\}\Bigg] \qquad (124)
\end{aligned}
$$

For $t_D \le t \le t_E$, Equation 119 can be rewritten as

$$
\begin{aligned}
z = {} & M\int_{t_D}^{t}[1-(1-t)]^{3/2}(t-t_D)^{\alpha-1}(t-t_E)^{-\alpha}(t-1)^{-1/2}\,dt \\
& - M(t_C+t_F)\int_{t_D}^{t}[1-(1-t)]^{1/2}(t-t_D)^{\alpha-1}(t-t_E)^{-\alpha}(t-1)^{-1/2}\,dt \\
& + Mt_C t_F\int_{t_D}^{t}[1-(1-t)]^{-1/2}(t-t_D)^{\alpha-1}(t-t_E)^{-\alpha}(t-1)^{-1/2}\,dt + N_2\ldots \qquad (125)
\end{aligned}
$$

Since for $t = t_D$, $z = z_D = 0$, therefore, $N_2 = 0$. Expanding the terms $[1-(1-t)]^{3/2}$, $[1-(1-t)]^{1/2}$, and $[1-(1-t)]^{-1/2}$ binomially, as $(1-t)$ is less than 1, and substituting their binomial expansion in Equation 125.

$$
\begin{aligned}
z = {} & M\int_{t_D}^{t}\sum_{p=0}^{p=\infty}U_{1,p}(1-t)^{p}(t-t_D)^{\alpha-1}(t-t_E)^{-\alpha}(t-1)^{-1/2}\,dt \\
& - M(t_C+t_F)\int_{t_D}^{t}\sum_{p=0}^{p=\infty}U_{2,p}(1-t)^{p}(t-t_D)^{\alpha-1}(t-t_E)^{-\alpha}(t-1)^{-1/2}\,dt \\
& + Mt_C t_F\int_{t_D}^{t}\sum_{p=0}^{p=\infty}U_{3,p}(1-t)^{p}(t-t_D)^{\alpha-1}(t-t_E)^{-\alpha}(t-1)^{-1/2}\,dt\ldots \qquad (126)
\end{aligned}
$$

where

$$U_{1,p} = \frac{\frac{3}{2}(\frac{3}{2}-1)\cdots(\frac{3}{2}-\overline{p-1})(-1)^{p}}{p!}; \qquad p = 1, 2, 3\ldots$$

$$U_{1,0} = 1$$

$$U_{2,p} = \frac{\frac{1}{2}(\frac{1}{2}-1)\cdots(\frac{1}{2}-\overline{p-1})(-1)^p}{p!}; \qquad p = 1, 2, 3 \ldots$$

$$U_{2,0} = 1$$

$$U_{3,p} = \frac{-\frac{1}{2}(\frac{1}{2}-1)\cdots(-\frac{1}{2}-\overline{p-1})(-1)^p}{p!}; \qquad p = 1, 2, 3 \ldots$$

and

$$U_{3,0} = 1$$

Substituting

$$r = \frac{t - t_D}{t_E - t_D} \tag{127}$$

and simplifying Equation 126 gives

$$\begin{aligned} z &= (-1)^{-\alpha-1/2} M \sum_{p=0}^{p=\infty} U_{1,p}(1-t_D)^{p-1/2} \int_0^r r^{\alpha-1}(1-r)^{1-\alpha-1}(1-x_2 r)^{-(1/2-p)}\, dr \\ &\quad - (t_C + t_F)(-1)^{-\alpha-1/2} M \sum_{p=0}^{p=\infty} U_{2,p}(1-t_D)^{p-1/2} \int_0^r r^{\alpha-1}(1-r)^{1-\alpha-1}(1-x_2 r)^{-(1/2-p)}\, dr \\ &\quad + t_C t_F(-1)^{-\alpha-1/2} M \sum_{p=0}^{p=\infty} U_{3,p}(1-t_D)^{p-1/2} \int_0^r r^{\alpha-1}(1-r)^{1-\alpha-1}(1-x_2 r)^{-(1/2-p)}\, dr \ldots \end{aligned} \tag{128}$$

where $x_2 = \dfrac{t_E - t_D}{1 - t_D}$

On term by term integration, for $p = 0, 1, 2, 3 \ldots$ Equation 128 reduces to [10]

$$\begin{aligned} z &= (-1)^{-\alpha-1/2} M \sum_{p=0}^{p=\infty} U_{1,p}(1-t_D)^{p-1/2} \Bigg[\beta_r(\alpha, 1-\alpha)\, {}_2F_1\left(\alpha, \frac{1}{2} - p; 1; x_2\right) \\ &\quad - r^\alpha (1-r)^{1-\alpha} \Bigg\{ \frac{1}{\alpha}\, {}_2F_1\left(\alpha, \frac{1}{2} - p; 1; x_2\right) {}_2F_1(1, 1; \alpha+1; r) \\ &\quad - \sum_{n=0}^{n=\infty} \frac{A_n r^n}{\alpha+n}\, {}_2F_1(1+n, 1; \alpha+1; r) \Bigg\} \Bigg] \\ &\quad - (-1)^{-\alpha-1/2}(t_C + t_F) M \sum_{p=0}^{p=\infty} U_{2,p}(1-t_D)^{p-1/2} \Bigg[\beta_r(\alpha, 1-\alpha)\, {}_2F_1\left(\alpha, \frac{1}{2} - p; 1; x_2\right) \\ &\quad - r^\alpha (1-r)^{1-\alpha} \Bigg\{ \frac{1}{\alpha}\, {}_2F_1\left(\alpha, \frac{1}{2} - p; 1; x_2\right) {}_2F_1(1, 1; \alpha+1; r) \\ &\quad - \sum_{n=0}^{n=\infty} \frac{A_n r^n}{\alpha+n}\, {}_2F_1(1+n, 1; \alpha+1; r) \Bigg\} \Bigg] \\ &\quad + (-1)^{-\alpha-1/2} t_C t_F M \sum_{p=0}^{p=\infty} U_{3,p}(1-t_D)^{p-1/2} \Bigg[\beta_r(\alpha, 1-\alpha)\, {}_2F_1\left(\alpha, \frac{1}{2} - p; 1; x_2\right) \\ &\quad - r^\alpha (1-r)^{1-\alpha} \Bigg\{ \frac{1}{\alpha}\, {}_2F_1\left(\alpha, \frac{1}{2} - p; 1; x_2\right) {}_2F_1(1, 1; \alpha+1; r) \\ &\quad - \sum_{p=0}^{p=\infty} \frac{A_n r^n}{\alpha+n}\, {}_2F_1(1+n, 1; \alpha+1; r) \Bigg\} \Bigg] \ldots \end{aligned} \tag{129}$$

where $A_n = \dfrac{(\frac{1}{2}-p)(\frac{1}{2}-p+1)\cdots(\frac{1}{2}-p+\overline{n-1})x_2^n}{n!}$

$n = 1, 2, 3, \ldots$

For point E, $z = b(\sin \pi\alpha - i \cos \pi\alpha)$, $t = t_E$, and $r = 1$.
Making use of this condition, Equation 129 gives

$$b(\sin \pi\alpha - i\cos \pi\alpha) = (-1)^{-\alpha-1/2}M\left[\sum_{p=0}^{p=\infty} U_{1,p}(1-t_D)^{p-1/2}\beta(\alpha, 1-\alpha)_2F_1(\alpha, \tfrac{1}{2}-p; 1; x_2)\right.$$
$$- (t_c + t_F)\sum_{p=0}^{p=\infty} U_{2,p}(1-t_D)^{p-1/2}\beta(\alpha, 1-\alpha)_2F_1(\alpha, \tfrac{1}{2}-p; 1; x_2)$$
$$\left. + t_c t_F \sum_{p=0}^{p=\infty} U_{3,p}(1-t_D)^{p-1/2}\beta(\alpha, 1-\alpha)_2F_1(\alpha, \tfrac{1}{2}-p; 1; x_2)\right] \tag{130}$$

For $t_E \le t \le 1$, Equation 119 can be rewritten as

$$z = M\int_{t_E}^{1} t^{3/2}(t-t_D)^{\alpha-1}(t-t_E)^{-\alpha}(t-1)^{-1/2}\,dt$$
$$- M(t_c + t_F)\int_{t_E}^{1} t^{1/2}(t-t_D)^{\alpha-1}(t-t_E)^{-\alpha}(t-1)^{-1/2}\,dt$$
$$+ Mt_c t_F \int_{t_E}^{t} t^{-1/2}(t-t_D)^{\alpha-1}(t-t_E)^{-\alpha}(t-1)^{-1/2}\,dt + N_3 \tag{131}$$

Since, for $t = t_E$, $z = b(\sin \pi\alpha - i\cos \pi\alpha)$, therefore,

$$N_3 = b(\sin \pi\alpha - i\cos \pi\alpha)$$

On substituting

$$r = \frac{t - t_E}{1 - t_E} \tag{132}$$

in Equation 131 and simplifying

$$z = (-1)^{-1/2}(1-t_D)^{\alpha-1}(1-t_E)^{-\alpha+1/2}M\left[\int_0^r r^{-\alpha}(1-\pi)^{-1/2}\{1-(1-\pi)x_3\}^{3/2}\{1-(1-r)y_3\}^{\alpha-1}\,dr\right.$$
$$- (t_c + t_F)\int_0^r r^{-\alpha}(1-r)^{-1/2}\{1-(1-r)x_3\}^{1/2}\{1-(1-r)y_3\}^{\alpha-1}\,dr$$
$$\left. + t_c t_F \int_0^r r^{-\alpha}(1-r)^{-1/2}\{1-(1-r)x_3\}^{-1/2}\{1-(1-r)y_3\}^{\alpha-1}\,dr\right] + N_3 \tag{133}$$

where $x_3 = (1 - t_E)$

$$y_3 = \frac{t - t_E}{1 - t_D}$$

On substituting $R = 1 - r$ in Equation 133 and simplifying

$$
\begin{aligned}
z = {} & (-1)^{-1/2}(1-t_D)^{\alpha-1}(1-t_E)^{-\alpha+1/2}M \\
& \times \Bigg[\int_0^1 R^{1/2-1}(1-R)^{3/2-\alpha-1/2-1}(1-x_3R)^{-(-3/2)}(1-y_3R)^{-(1-\alpha)}\,dR \\
& - (t_C+t_F)\int_0^1 R^{1/2-1}(1-R)^{3/2-\alpha-1/2-1}(1-x_3R)^{-(-1/2)}(1-y_3R)^{-(1-\alpha)}\,dR \\
& + t_Ct_F\int_0^1 R^{1/2-1}(1-R)^{3/2-\alpha-1/2-1}(1-x_3R)^{-1/2}(1-y_3R)^{-(1-\alpha)}\,dR \\
& - \int_0^R R^{1/2-1}(1-R)^{3/2-\alpha-1/2-1}(1-x_3R)^{-(-3/2)}(1-y_3R)^{-(1-\alpha)}\,dR \\
& + (t_C+t_F)\int_0^R R^{1/2-1}(1-R)^{3/2-\alpha-1/2-1}(1-x_3R)^{-(-1/2)}(1-y_3R)^{-(1-\alpha)}\,dR \\
& - t_Ct_F\int_0^R R^{1/2-1}(1-R)^{3/2-\alpha-1/2-1}(1-x_3R)^{-1/2}(1-y_3R)^{-(1-\alpha)}\,dR \Bigg] + N_3 \ldots
\end{aligned}
\tag{134}
$$

On integration Equation 134 gives

$$
\begin{aligned}
z = {} & (-1)^{-1/2}(1-t_D)^{\alpha-1}(1-t_E)^{-\alpha+1/2}M\left[\left\{\beta\left(\frac{1}{2}, 1-\alpha\right) - \beta_R\left(\frac{1}{2}, 1-\alpha\right)\right\}\right. \\
& \times \left\{F_1\left(\frac{1}{2}, -\frac{3}{2}, 1-\alpha; \frac{3}{2}-\alpha; x_3, y_3\right) - (t_C+t_F)F_1\left(\frac{1}{2}, -\frac{1}{2}, 1-\alpha; \frac{3}{2}-\alpha; x_3, y_3\right)\right. \\
& \left.\left. + t_Ct_FF_1\left(\frac{1}{2}, \frac{1}{2}, 1-\alpha; \frac{3}{2}-\alpha; x_3, y_3\right)\right\}\right] + (-1)^{-1/2}(1-t_D)^{\alpha-1}(1-t_E)^{-\alpha+1/2}2MR^{1/2} \\
& \times (1-R)^{1-\alpha}{}_2F_1\left(\frac{3}{2}-\alpha, 1; 1+\frac{1}{2}; R\right)\left[F_1\left(\frac{1}{2}, -\frac{3}{2}, 1-\alpha; \frac{3}{2}-\alpha; x_3, y_3\right)\right. \\
& \left. - (t_C+t_F)F_1\left(\frac{1}{2}, -\frac{1}{2}, 1-\alpha; \frac{3}{2}-\alpha; x_3, y_3\right) + t_Ct_FF_1\left(\frac{1}{2}, \frac{1}{2}, 1-\alpha; \frac{3}{2}-\alpha; x_3, y_3\right)\right] \\
& -(-1)^{-1/2}(1-t_D)^{\alpha-1}(1-t_E)^{-\alpha+1/2}MR^{1/2}(1-R)^{1-\alpha}\sum_{n=0}^{n=\infty}\left[\left\{\frac{C_{2,n}}{\frac{1}{2}+n} - (t_C+t_F)\frac{C_{3,n}}{\frac{1}{2}+n}\right.\right. \\
& \left.\left. + t_Ct_F\frac{C_{4,n}}{\frac{1}{2}+n}\right\}R^n{}_2F_1\left(\frac{3}{2}-\alpha+n, 1; 1+\frac{1}{2}+n; R\right)\right] + N_3 \ldots
\end{aligned}
\tag{135}
$$

where $C_{i,n} = A_{i,n}B_{i,0} + A_{i,n-1}B_{i,1} + \cdots + A_{i,1}B_{i,n-1} + A_{i,0}B_{i,n}$
$n = 0, 1, 2, 3, \ldots$
$i = 2, 3$, and 4

$$A_{2,n} = \frac{(-\frac{3}{2})(-\frac{3}{2}+1)(-\frac{3}{2}+2)\cdots(-\frac{3}{2}+\overline{n-1})x_3^n}{n!}$$

$$A_{3,n} = \frac{(-\frac{1}{2})(-\frac{1}{2}+1)(-\frac{1}{2}+2)\cdots(-\frac{1}{2}+\overline{n-1})x_3^n}{n!}$$

$$A_{4,n} = \frac{(\frac{1}{2})(\frac{1}{2}+1)(\frac{1}{2}+2)\cdots(\frac{1}{2}+\overline{n-1})x_3^n}{n!}$$

$$B_{i,n} = \frac{(1-\alpha)(1-\alpha+1)(1-\alpha+2)\cdots(1-\alpha+\overline{n-1})x_3^n}{n!}$$

n = 1, 2, 3, ...

i = 2, 3, and 4

At point F where, $Z = b \sin \alpha\pi - i(b \cos \alpha\pi + s_2)$, $t = t_F$,

$$r = \frac{t_F - t_E}{1 - t_E} \quad \text{and} \quad R = \frac{1 - t_F}{1 - t_E}$$

Making use of this condition Equation 135 gives

$$\begin{aligned}
& b \sin \alpha\pi - i(b \cos \alpha\pi + s_2) \\
&= (-1)^{-1/2}(1 - t_D)^{\alpha-1}(1 - t_E)^{-\alpha+1/2} M\left[\left\{\beta\left(\frac{1}{2}, 1-\alpha\right) - \beta\frac{1-t_F}{1-t_E}\left(\frac{1}{2}, 1-\alpha\right)\right.\right. \\
&\quad \left.+ 2\left(\frac{1-t_F}{1-t_E}\right)^{1/2}\left(\frac{t_F-t_E}{1-t_E}\right)^{-\alpha} {}_2F_1\left(\frac{3}{2}-\alpha, 1; 1+\frac{1}{2}; R\right)\right\}\left\{F_1\left(\frac{1}{2}, -\frac{3}{2}, 1-\alpha; \frac{3}{2}-\alpha; x_3, y_3\right)\right. \\
&\quad \left. - (t_C + t_F)F_1\left(\frac{1}{2}, -\frac{1}{2}, 1-\alpha; \frac{3}{2}-\alpha; x_3, y_3\right) + t_C t_F F_1\left(\frac{1}{2}, \frac{1}{2}, 1-\alpha; \frac{3}{2}-\alpha; x_3, y_3\right)\right\} \\
&\quad - \left(\frac{1-t_F}{1-t_E}\right)^{1/2}\left(\frac{t_F-t_E}{1-t_E}\right)^{1-\alpha} \sum_{n=0}^{n=\infty}\left\{\frac{C_{2,n}}{\frac{1}{2}+n} - (t_C + t_F)\frac{C_{3,n}}{\frac{1}{2}+n} + t_C t_F \frac{C_{4,n}}{\frac{1}{2}+n}\right\} \\
&\quad \left.\times R^n {}_2F_1\left(\frac{3}{2} - \alpha + n, 1; 1 + \frac{1}{2} + n; R\right)\right] + N_3 \ldots
\end{aligned} \tag{136}$$

At point G, where, $z = b(\sin \pi\alpha - i \cos \pi\alpha)$, $t = 1$, $r = 1$ and $R = 0$. Making use of this condition Equation 135 gives

$$\begin{aligned}
& b(\sin \alpha\pi - i \cos \alpha\pi) = (-1)^{-1/2}(1 - t_D)^{\alpha-1}(1 - t_E)^{-\alpha+1/2} M\beta(\tfrac{1}{2}, 1 - \alpha) \\
& [F_1(\tfrac{1}{2}, -\tfrac{3}{2}, 1 - \alpha; \tfrac{3}{2} - \alpha; x_3, y_3) - (t_C + t_F)F_1(\tfrac{1}{2}, -\tfrac{1}{2}, 1 - \alpha, \tfrac{3}{2} - \alpha; x_3, y_3) \\
& + t_C t_F F_1(\tfrac{1}{2}, \tfrac{1}{2}, 1 - \alpha; \tfrac{3}{2} - \alpha; x_3, y_3)] + b(\sin \alpha\pi - i \cos \alpha\pi) \ldots
\end{aligned} \tag{137}$$

For given values of α, b, s_1, and s_2, the unknowns M, t_c, t_D, t_E, and t_F can be evaluated by solving Equations 123, 124, 130, 136, and 137 by an interative procedure.

Mapping of the Complex Potential Plane

The complex potential w is defined as $w = \Phi + i\psi$, in which the velocity potential function

$$\Phi = -k\left(\frac{p'}{\gamma_w} + y\right) + c' \ldots \tag{138}$$

in which k = coefficient of permeability, c′ = constant; γ_w = unit weight of water, y = the ordinate and ψ = the stream function. The w-plane for this problem is shown in Figure 38c. The transformation of w-plane onto the lower half of t-plane is given by

$$w = M' \int \frac{dt}{\sqrt{t(t-1)}} + N' \cdots \tag{139}$$

where M′ and N′ are constants.
On integration Equation 139 gives

$$w = M' \sin^{-1}(2t-1) + N' \ldots \tag{140}$$

Making use of condition at point G where $t = 1$; $w = 0$. Equation 140 gives

$$M' \frac{\pi}{2} + N' = 0 \ldots \tag{141}$$

Using condition at point B where $w = -kH$, and $t = 0$. Equation 140 gives

$$M' = \frac{kH}{\pi} \ldots \tag{142}$$

Thus,

$$N' = -\frac{kH}{2} \ldots \tag{143}$$

Pressure Distribution Along Base

The complex potential w for a point $0 \le t \le 1$ is given by

$$w = \frac{kH}{\pi} \sin^{-1}(2t-1) - \frac{kH}{2} \ldots \tag{144}$$

Since $\psi = 0$ along the base of the structure, $w = \phi$ in Equation 144. The pressure can be determined by subsituting this relationship for ϕ into Equation 138. The value of z corresponding to a point $0 \le t \le 1$ can be evaluated by using Equation 122, 129, or 135 depending upon the location of the point.

Exit Gradient

The exit gradient I_E is given by

$$I_E = \frac{dw}{dz} = \frac{dw}{dt} \cdot \frac{dt}{dz} \ldots \tag{145}$$

Substituting the expressions for dw/dt and dt/dz from Equations 139 and 119 Equation in 145 and simplifying

$$I_E = \frac{M'(t - t_D)^{1-\alpha}(t - t_E)^{\alpha}}{M(t - t_c)(t - t_F)} \ldots \tag{146}$$

The value of I_E can be evaluated using Equation 146 for any assumed value of t, $t \geq 1$. The corresponding value of z can be evaluated by numerical integration of Equation 119.

The variation of $-\Phi_C/kH$, $-\Phi_D/kH$, $-\Phi_E/kH$, and $-\Phi_F/kH$ with s_1/b for $\alpha = 0.495$, 0.4167, and 0.3333 and $s_2/b = 0.05$, 0.1 and 0.2 is shown in Figures 39–50. The figures show that the value of $-\Phi/kH$ for points C, D, E, and F under the weir increases as the value of α is reduced. For $\alpha = 0.495$, 0.4167 and 0.3333, $s_1/b = s_2/b = 0.2$, the value of $-\Phi D/kH$ are 0.652, 0.701 and 0.755, respectively, and the values of $-\Phi E/kH$ are 0.3525, 0.396, and 0.435, respectively.

The variation of maximum value of $I_E s_2/H$ at the toe of the weir with s_1/b for $\alpha = 0.495$, 0.4167, and 0.333, and $s_1/b = 0.2$ is shown in Figures 51–53. The figures show that the value of $I_E s/H$ increases as the value of α is reduced. For $\alpha = 0.495$, 0.4167, and 0.3333, $s_1/b = s_2/b = 0.2$, the values of $I_E s/H$ are 0.106, 0.11 and 0.114, respectively.

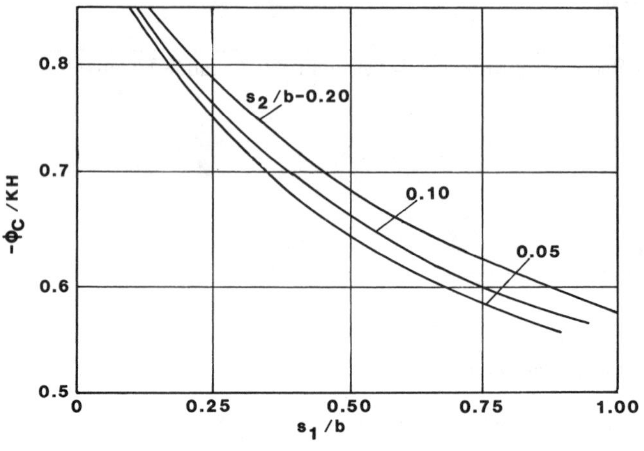

Figure 39. Variation of $-\phi_C/kH$ with s_1/b for $\alpha = 0.333$.

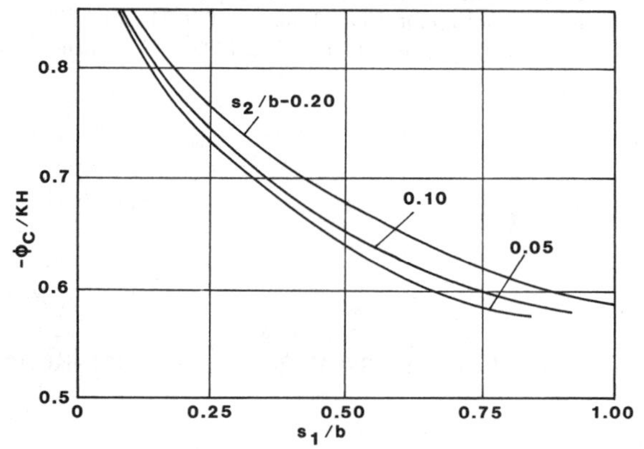

Figure 40. Variation of $-\phi_C/kH$ with s_1/b for $\alpha = 0.4167$.

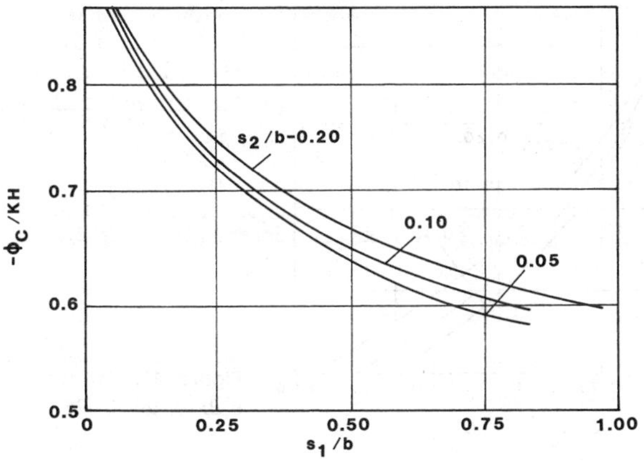

Figure 41. Variation of $-\phi_C/kH$ with s_1/b for $\alpha = 0.495$.

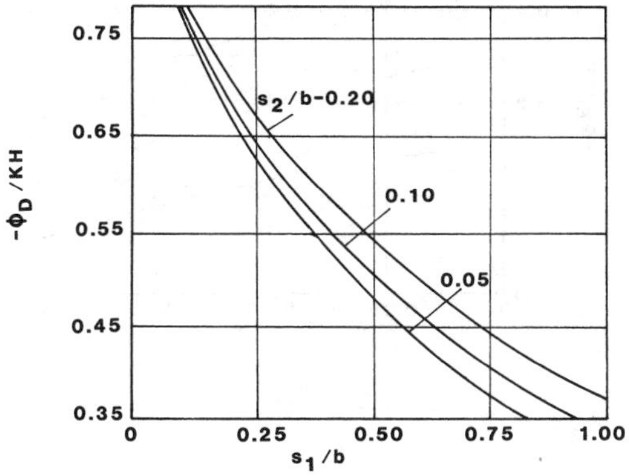

Figure 42. Variation of $-\phi_D/kH$ with s_1/b for $\alpha = 0.333$.

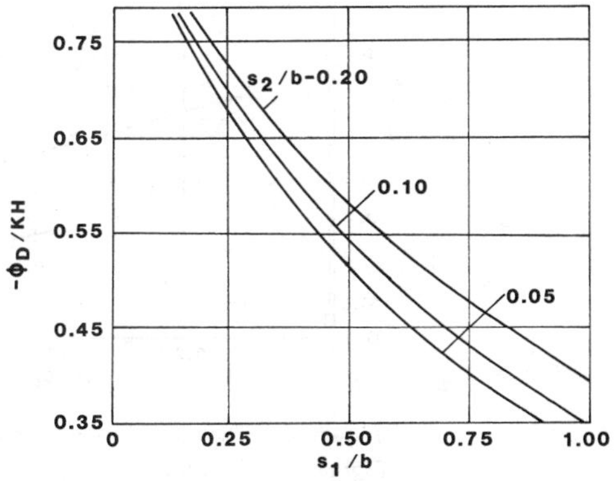

Figure 43. Variation of $-\phi_D/kH$ with s_1/b for $\alpha = 0.4167$.

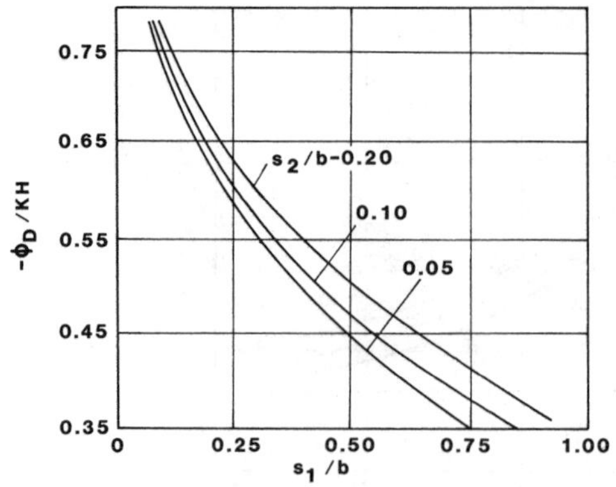

Figure 44. Variation of $-\phi_D$/kH with s_1/b for $\alpha = 0.495$.

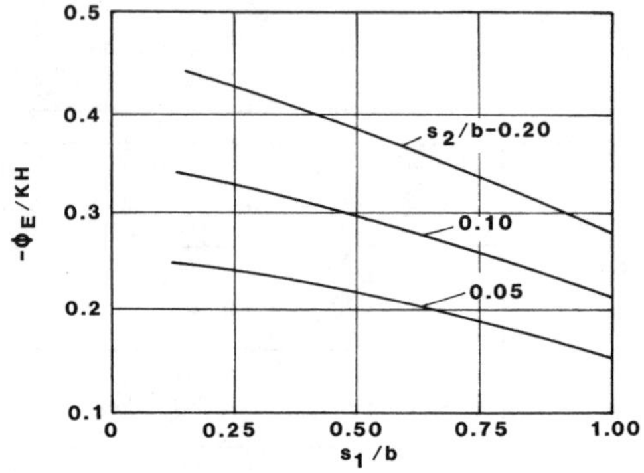

Figure 45. Variation of $-\phi_E$/kH with s_1/b for $\alpha = 0.333$.

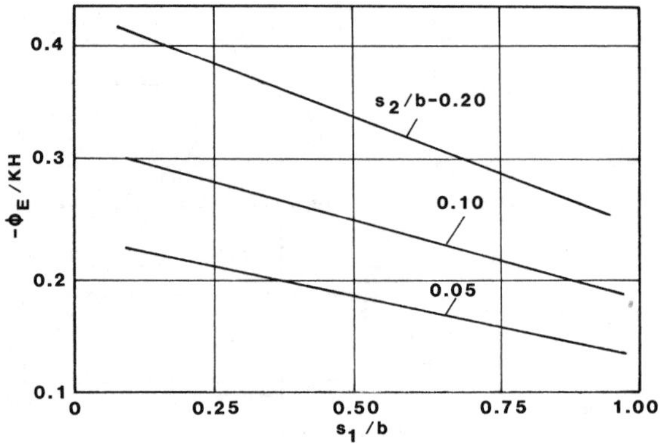

Figure 46. Variation of $-\phi_E$/kH with s_1/b for $\alpha = 0.4167$.

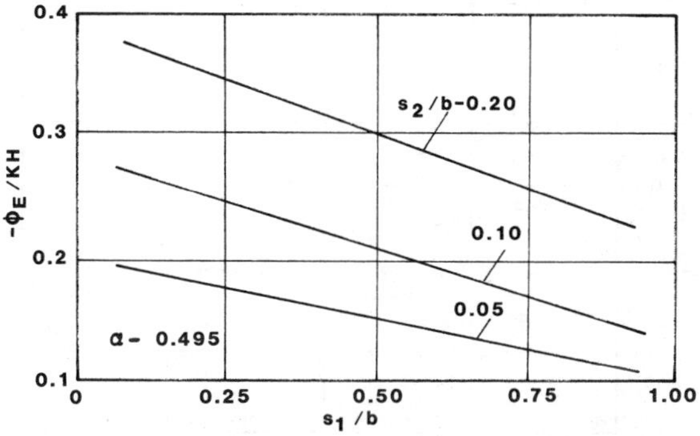

Figure 47. Variation of $-\phi_E/kH$ with s_1/b for $\alpha = 0.495$.

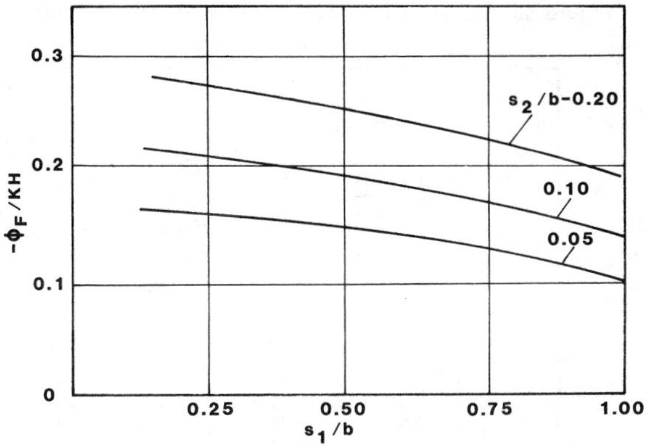

Figure 48. Variation of $-\phi_F/kH$ with s_1/b for $\alpha = 0.333$.

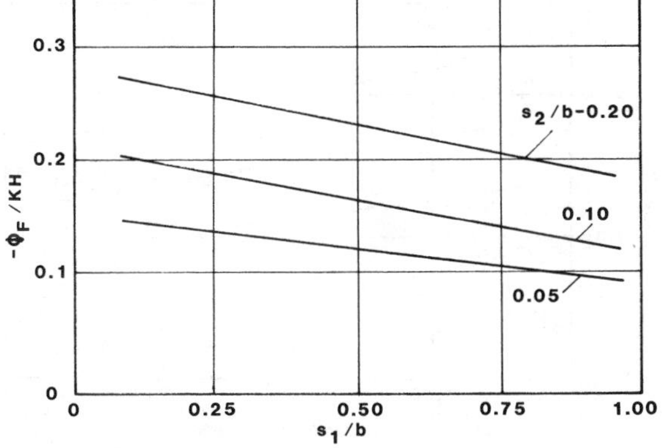

Figure 49. Variation of $-\phi_F/kH$ with s_1/b for $a = 0.4167$.

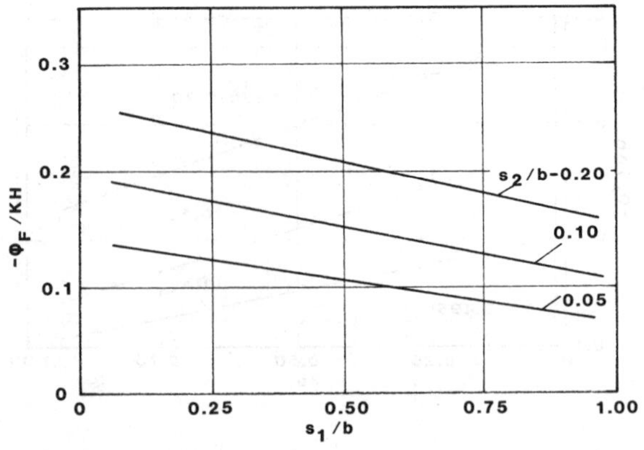

Figure 50. Variation of $-\phi_F$/kH with s_1/b for $\alpha = 0.495$.

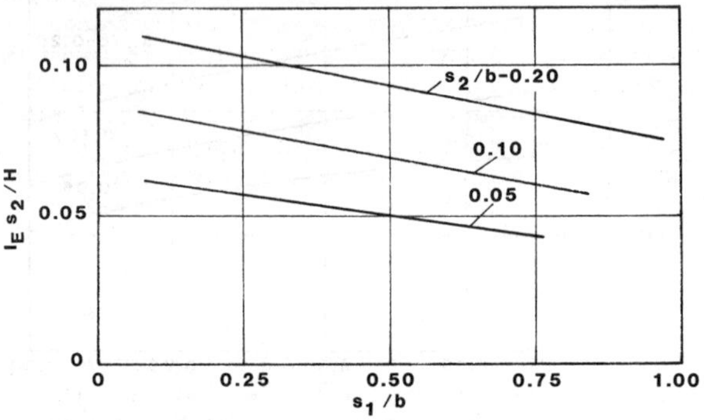

Figure 51. Variation of I_Es/H with s_1/b for $\alpha = 0.495$.

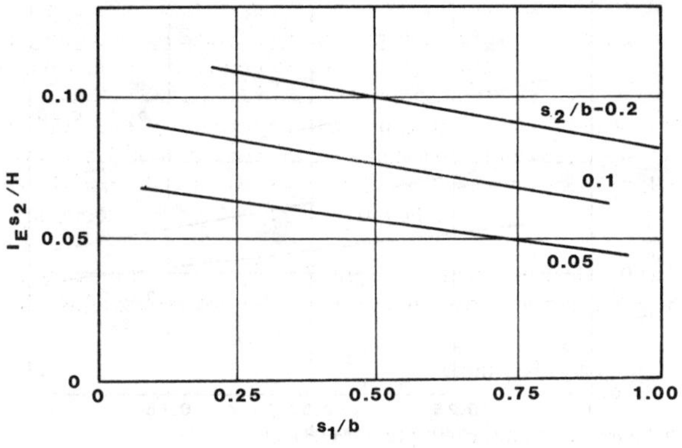

Figure 52. Variation of I_Es/H with s_1/b for $\alpha = 0.4167$.

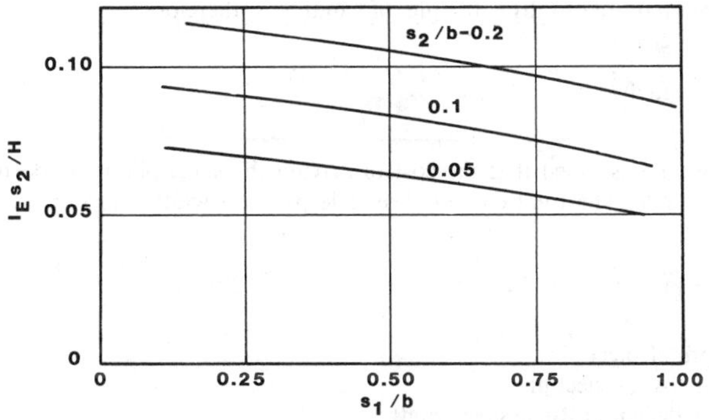

Figure 53. Variation of $I_E s/H$ with s_1/b for $\alpha = 0.333$.

COST OF WEIR

Having fixed the site of a weir, stage discharge curve, design flood discharge, pond level, and waterway, the design of the structure from surface flow consideration is uniquely defined. The value of the safe exit gradient can be decided using the results of analysis given earlier. The cost of weir can now be minimized with respect to surface and sub-surface flow characteristics.

The cost of a weir comprises the cost of the upstream and downstream sheet piles, impervious apron, flank walls, gates and their hoisting arrangement, earth work and dewatering for laying the apron in the stream bed. The cost of each component of a weir can be expressed as a function of the waterway, W, overall length of impervious floor, b, depth of sheet piles s_1 and s_2 and cost coefficients, as discussed in the following.

Cost of Sheet Piles

The sheet piles located at upstream and downstream end of the impervious apron are provided to guard the apron against scour. The minimum depth of sheet piles is, therefore, designed from the consideration of scour. When the weir is founded on permeable soil, seepage flow under the apron causes uplift pressure and tends to move the soil particles located on the boundary of the flow domain on the downstream side. The sheet piles help to reduce exit gradient. The upstream sheet pile also reduces the pressure under the apron. The value of exit gradient depends upon the head causing seepage, H, the overall floor length b and depth of sheet piles s_1 and s_2. For the same value of head causing seepage and safe exit gradient G_E, several combinations of b, s_1, and s_2 are possible. But values of s_1 and s_2 cannot be less than those required from scour consideration.

The head causing seepage, H, is a function of waterway of the weir. Hence, for a given value of G_E and b the depth of sheet piles can be expressed as a function of waterway, W.

The total cost of sheet pile can be considered in two parts, cost of pile and cost of driving. The cost of the sheet pile will vary linearly with its length, while cost of driving will increase nonlinearly with depth of piling. The cost of driving the pile can be evaluated from the energy required to drive it.

Let ℓ be the intermediate drive length of the pile. The energy, E_D, required to drive the pile a unit distance, when the pile is already driven to a length ℓ, may be expressed as

$$E_D = \int_0^{\ell} 2K_n \gamma z \, dz \tan \phi = K_n \gamma \tan \Phi \ell^2$$

where K_n = coefficient of lateral earth pressure at rest
γ = unit weight of soil
Φ = angle of internal friction

The total energy required to drive the pile of length s is, therefore,

$$\int_0^s E_D \, d\ell = K_n \gamma \tan \Phi \frac{s^3}{3}$$

It can, therefore, be assumed that the cost of driving the sheet pile is proportional to cube of the depth of sheet pile. The total cost of sheet pile per unit width and of depth, s, can now be expressed as

$$Y_{US} = C_1 s + C_2 s^3$$

where s = depth of sheet piles
C_1 = unit cost of sheet pile
C_2 = a coefficient in Rupees per length^3

If W is the waterway of the weir, the cost of both upstream and downstream sheep piles, Y_s, can be written as

$$Y_s = W[C_1(s_1 + s_2) + c_2(s_1^3 + s_2^3)] \ldots \tag{147}$$

Cost of Weir Apron

The weir apron comprises the stilling basin, sloping glacis, and upstream floor. The stilling basin is provided to protect the stream bed against scour.Its dimensions and level are designed to effectively dissipate the excess energy of flow over the weir. Generally, hydraulic jump is used to destroy the excess energy of water. The flow is made to run over the glacis. The supercritical flow so generated is made to impinge upon the sub-critical flow in the stilling basin, where adequate depth is available for jump formation.

The level of the basin is decided on the consideration that post jump depth, d_2, is available for jump formation. The level of stilling basin is, given by the equation,

Level of stilling basin = Post-construction high flood level on the downstream of the weir − d_2

The length of the jump depends upon the Froude number of flow and post-jump depth d_2. It can be obtained from the curve shown in Figure 3. The length of the basin is kept equal to length of the jump.

The difference in elevation of the upstream apron and stilling basin, Y_0 is negotiated through a sloping glacis. The water flowing over the glacis develops supercritical velocities. This ensures formation of jump over the glacis itself. The length of glacis depends upon the difference in elevation of upstream floor and stilling basin and the slope of the glacis. The level of the upstream floor is generally kept at the river bed level.

To increase overall length of the apron it is sometimes extended on the upstream side of the weir crest. The overall length of the apron depends upon the value of safe exit gradient, G_E, head causing seepage, H, and depth of upstream and downstream sheet piles.

The weight of the impervious apron counters the uplift pressures due to seepage flow underneath. The quantity of concrete in the apron varies with the pattern of uplift pressure distribution, which depends upon the seepage head, geometry and overall length of the apron, and depth of upstream and downstream sheet piles. Generally, a weir apron has a horizontal floor with an inclined step and two vertical sheet piles (Figure 1). But solution of seepage flow under such a profile is not available. Therefore, as an approximation the weir profile has been assumed to be a sloping apron with two vertical sheet piles (Figure 54). The pressures at the salient points under the apron and value of exit gradient for such profile can be obtained using analysis given earlier. Let the pressure at points A and B be P_A and P_B (Figure 54). Assuming the variation of pressure under the apron to be linear

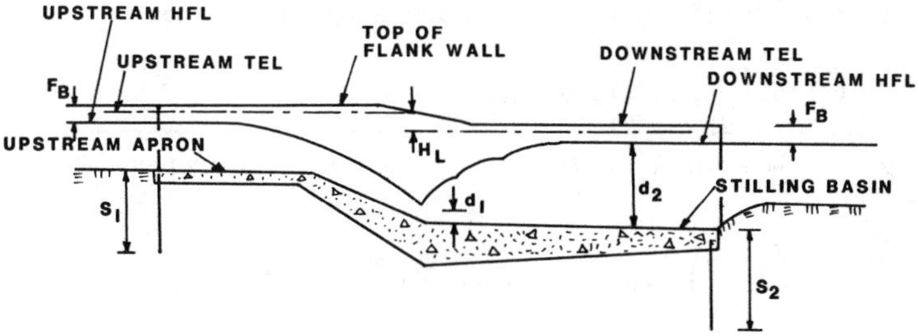

Figure 54. Actual and assumed weir profiles.

the pressure at point C under the apron is given by

$$P_C = P_A + (P_B - P_A)\frac{L}{b} \tag{148}$$

where L = length of impervious floor downstream of crest.

The depth of concrete required at point A to counteract the unlift pressure is given by

$$d_A = \frac{P_A}{\gamma_{cs}} \tag{149}$$

where γ_{cs} = submerged unit weight of concrete
d_A = thickness of concrete floor at A

The thickness of floor at C is given by

$$d_c = \frac{1}{\gamma_{cs}}\left[P_A + (P_B - P_A)\frac{L}{b}\right] \tag{150}$$

where d_c = thickness of concrete floor at C.

As the variation of pressure under apron between A and B has been assumed to be linear, the variation of thickness of concrete floor, to provide adequate dead weight to counteract uplift pressure, is linear.

Hence, the volume of concrete, Q_s, in the floor of the stilling basin can be expressed as

$$Q_s = \frac{(d_A + d_C)}{2} WL \tag{151}$$

Substituting the values of d_A and d_C from Equations 149 and 150 and simplifying Equation 151 gives

$$Q_s = \frac{WL}{2\gamma_{cs}}\left[P_A\left(2 - \frac{L}{b}\right) + P_B\frac{L}{b}\right] \tag{152}$$

The part of weir floor on upstream of gate line remains loaded with water. Therefore, the thickness of this portion of the floor is generally kept nominal and uniform as it has not to withstand any resultant pressure. If d_u is the depth of upstream floor, the quantity of concrete in upstream floor, Q_u, can be expressed as

$$Q_u = d_u W(b - L) \tag{153}$$

Using Equations 152 and 153, the volume of concrete in weir apron is given by

$$Q_A = \frac{WL}{2\gamma_{cs}}\left[P_A\left(2 - \frac{L}{b}\right) + P_B\frac{L}{b}\right] + Wd_u(b - L) \tag{154}$$

and cost of concrete in weir apron

$$Y_A = \frac{C_3WL}{2\gamma_{cs}}\left[P_A\left(2 - \frac{L}{b}\right) + P_B\frac{L}{b}\right] + C_3Wd_u(b - L) \tag{155}$$

where C_3 = unit cost of concrete.

For a given design flood discharge Q the length and level of stilling basin depend upon the waterway of the weir. When there is no water on the downstream side and pond level is being maintained, the head causing seepage, H, is given by (Figure 54).

H = Pond level − stilling basin level

The pond level is determined on the basis of functional requirements of the weir and is independent of the waterway, while level of stilling basin depends upon the waterway.

Thus, for the same value of design discharge the length of the apron downstream of the crest and the head causing seepage, both can be expressed as function of the waterway. The value of pressure under the apron, P_A and P_B varies with the head causing seepage, H, the overall length of apron, b, and depth of sheet piles s_1 and s_2. Hence

Cost of weir apron, $Y_A = f(W, C_3, b, s_1, s_2)$

Cost of Flank Walls

The flank walls on both sides retain the approaches to the weir. The elevation of the top of the flank walls is so fixed that certain minimum freeboard over the flood level both upstream and downstream of the weir crest is available.

Let H_{DF} and H_{UF} be the height of downstream and upstream flank walls, then (Figure 1)

H_{DF} = elevation of top of downstream flankwall-level of stilling basin

$= F_B + d_2$

where F_B = the free board. The value of F_B is generally constant, while d_2 is a function of the waterway. Let C_4 be the unit cost of the material of the flank walls. The total cost of the down stream flank walls, Y_{DF}, can be written as

$$Y_{DF} = C_4K_{DF}H_{DF}^2 \cdot L \tag{156}$$

where K_{DF} = a coefficient, depending up the type of soil. It will be constant for a specific sites. As values of L and H_{DF} both depend upon waterway. Therefore,

$Y_{DF} = f(C_4, W)$

The height of upstream flank walls (Figure 1)

H_{UF} = Elevation of top of upstream flankwall-level of upstream floor.

$= F_B$ + Afflux + (HFL-level of upstream floor)

The total cost of upstream flank walls Y_{UF} can be written as

$$Y_{UF} = C_4 K_{UF} H_{UF}^2 \cdot (b - L) \quad (157)$$

where K_{UF} = a coefficient depending upon type of soil.

The value of afflux varies with waterway while the values of HFL, level of upstream floor and F_B are constant for a site. Therefore, Y_{UF} can be expressed as

$$Y_{UF} = f(C_4, W, b)$$

Cost of Gates and Hoisting Arrangement

The gates resting over the crest of the weir help to maintain the pond level during low water season. During floods the gates can be opened to allow the excess flood water to flow down. To operate the gates suitable arrangement is made. The gates can be vertically lifting or radial type. The height of the gates is the difference in elevation of the pond level and the crest of the weir. The total length of gates, and peirs supporting the gates is equal to the waterway of the weir. If C_5 is the cost of the gates per unit length including cost of peirs and the hoisting arrangement, the total cost of gates,

$$Y_G = C_5 W \quad (158)$$

The unit cost of gates depends upon their height. Having decided the pond level and crest level the height of gates becomes constant. Therefore, the cost of gates is a function of waterway only and can be written as

$$Y_G = f(C_5, W)$$

Cost of Earth Work

The quantity of earth work for laying the downstream apron is given by

$$Q_{ES} = W.L.(\text{RBL} - \text{stilling basin level}) + Q_s$$

where Q_{ES} = volume of earthwork
RBL = general river bed level
Q_s = volume of concrete in stilling basin (Equation 152)

The volume of earthwork for upstream floor,

$$Q_{EU} = W(b - L)(\text{RBL-Upstream floor level}) + W(b - L)d_u$$
$$= W(b - L)(\text{RBL-upstream floor level} + d_u)$$

The cost of earthwork, Y_E, is given by,

$$Y_E = C_6(Q_{ES} + Q_{EU})$$

where C_6 is the unit cost of earthwork. The river bed level being constant Y_E can be expressed as

$$Y_E = f(C_6, W, b, s_1, s_2)$$

Cost of Dewatering

The cost of dewatering depends upon waterway, site conditions, sequence of construction, ground water level, coefficient of permeability of the soil, depth of dewatering, and pattern and means of dewatering. In view of the many variables that many change with the site and sequence of construction, the cost of dewatering has not been considered.

The total cost of a weir, Y, can now be written as

$$Y = Y_S + Y_A + Y_{DF} + Y_{UF} + Y_G + Y_E \tag{160}$$

The values of Y_S, Y_A, Y_F, Y_G and Y_E can be obtained using Equations 155–159. The total cost of barrage can be written as

$$Y = f(C_1, C_2, C_3, C_4, C_5, C_6, W, b, s_1, s_2)$$

OPTIMAL DIMENSIONS OF WEIR

The value of cost function of a weir depends upon the cost coefficient of various items of works and the values of W, b, s_1 and s_2. In case the cost coefficients are determined for a work, the optimization problem can be written as

$$\text{Min } Y = f(W, b, s_1, s_2)$$

Subject to the constraints

$$\begin{aligned}
&g_1(W) && \le A_{fp} \\
&g_2(b, s_1, s_2) && \le G_E \\
&b && \ge L \\
&s_1 && \ge s_{1c} \\
&s_2 && \ge s_{2c} \\
&s_1 && \le s_{1m} \\
&s_2 && \le s_{2m} \\
&W, b, s_1, s_2 \ge 0
\end{aligned}$$

where A_{fp} is the permissible afflux, G_E is the permissible saft exit gadient, s_{1c} and s_{2c} are the minimum length of upstream and downstream sheet piles from scour consideration, and s_{1m} and s_{2m} are the maximum length of sheet piles from practical consideration.

The cost function, Y, depends upon the value of the uplift pressures under the apron at points A and B. For the configuration of the floor system the values of these pressures can be determined using the analysis given earlier. But the cost function cannot be expressed as an explicit expression of the variables W, b, s_1, and s_2. Therefore, it is not amenable to calculus methods and direct search method has been used for optimization of design of the weir [11].

The objective function in the present case is a function of the four design vectors W, b, s_1, s_2. The range of variation of these vectors generally met in practice is also known and the field of search is not very wide. In view of this, the univariate search technique has been used. The starting value of cost function has been evaluated for a set of assumed values of design vectors W, b, s_1 and s_2, known as initial solution. The value of only one design vector is then varied to generate a sequence of improved approximations to the optimal [12]. In the process, the value of the other three decision variables is kept fixed. The problem is thus reduced to that of one-dimensional minimization.

Initial Solution

To begin with the value of waterway, W, of the weir is assumed for which the afflux is equal to the permissible afflux A_{fp}. The initial solution thus lies on the boundary of the feasible region. Using the stage-discharge curve, the level and length of stilling basin are computed for the given design discharge, Q. The pond level, upstream floor level and length of sloping glacis are decided. Now the length of apron downstream of the crest, L, and head causing seepage H are computed. The value of safe exit gradient G_E is decided for the soil type at the site.

The weir apron is assumed to be horizontal and $s_1 = 0.0$. The weir now is comprised a horizontal floor of width b and a downstream sheet pile of length s_2. Assuming the water depth on the downstream side to be zero, the pressures P_A and P_B at points A and B respectively are given by [5]

$$P_B = H\gamma_w \tag{161}$$

$$P_A = \frac{H}{\pi}\gamma_w \cos^{-1}\left(\frac{\lambda - 2}{\lambda}\right) \tag{162}$$

where $$\lambda = \frac{1 + \sqrt{1 + b^2/s_2^2}}{2} \tag{163}$$

For the assumed waterway, the value of H and L, the cost of downstream flank walls, earth work for downstream floor, and gates and hoisting arrangement becomes constant. Using Equations 147, 155, and 157 and considering the ocst of the sheet pile, apron and upstream flank walls the cost function becomes

$$Y_{AS} = W(C_1 s_2 + C_2 s_2^3) + \frac{WC_3 L\gamma_w}{2\gamma_{cs}}\left[P_A\left(2 - \frac{L}{b}\right) + P_B\frac{L}{b}\right] + C_3 W d_u(b - L) + Y_{UF}$$

If the flank walls are also of concrete and Q_{UF} is the quantity of concrete per unit length, the cost of upstream flank walls can be written as

$$Y_{UF} = C_3 \cdot Q_{UF} \cdot (b - L)$$

and the cost function is

$$Y_{AS} = W(C_1 s_2 + C_2 s_2^3) + \frac{WC_3 L\gamma_w}{2\gamma_{cs}}\left[P_A\left(2 - \frac{L}{b}\right) + P_B\frac{L}{b}\right] + WC_3 d_u(b - L) + C_3 Q_{UF}(b - L) \tag{164}$$

Substituting value of P_A and P_B from Equations 161 and 162, respectively, Equation 164 gives

$$\frac{1}{W}Y_{AS} = C_1 s_2 + C_2 s_2^3 + \frac{C_3 L\gamma_w H}{2\pi\gamma_{cs}}\left[\left(2 - \frac{L}{b}\right)\cos^{-1}\left(\frac{\lambda - 2}{\lambda}\right) + \frac{\pi L}{b}\right] + C_3(b - L)\left(d_u + \frac{Q_{UF}}{W}\right) \tag{165}$$

The maximum exit gradient, I_E, for horizontal weir with downstream sheet pile is given by [5]

$$I_E = \frac{H}{\pi s_2}\frac{1}{\sqrt{\lambda}} \tag{166}$$

Since the value of maximum exit gradient I_E should not be greater than the safe exit gradient, G_E, substituting G_E for I_E in Equation 166 and simplifying

$$\lambda = \frac{H^2}{\pi^2 s_2^2 G_E^2} \tag{167}$$

Substituting value of λ from Equation 163 in Equation 167

$$\frac{1 + \sqrt{1 + \frac{b^2}{s_2^2}}}{2} = \frac{H^2}{G_E^2 \pi^2 s_2^2}$$

or

$$b = \frac{2H}{\pi G_E} \sqrt{\frac{H^2}{G_E^2 \pi^2 s_2^2} - 1} \tag{168}$$

Substituting values of λ and b from Equations 167 and 168 on simplification Equation 165 gives

$$\frac{1}{W} Y_{AS} = C_1 s_2 + C_2 s_2^3 + \frac{C_3 L \gamma_w H}{2\gamma_{cs}\pi}\left[\left(2 - \frac{L\pi^2 G_E^2 s_2}{2H\sqrt{H^2 - G_E^2\pi^2 s_2^2}}\right)\cos^{-1}\left(1 - \frac{2\pi^2 G_E^2 s_2^2}{H^2}\right) + \frac{L\pi^3 G_E^3 s_2}{2H\sqrt{H^2 - G_E^2\pi^2 s_2^2}}\right] + C_3\left[\frac{2H\sqrt{H^2 - G_E^2\pi^2 s_2^2}}{\pi^2 G_E^2 s_2} - L\right]\left(d_u + \frac{Q_{UF}}{W}\right) \tag{169}$$

Differentiating Equation 169 with respect to s_2 and simplifying

$$\frac{1}{W}\frac{\partial Y_{AS}}{\partial s_2} = C_1 + 3C_2 s_2^2 + \frac{C_3 L \gamma_w H}{2\pi\gamma_{cs}}\left[\left\{\frac{L\pi^2 G_E^3 H^2}{2(H^2 - G_E^2\pi^2 s_2^2)^{3/2}}\right\}\left\{\pi - \cos^{-1}\left(1 - \frac{2^2\pi^2 G_E^2 s_2^2}{H^2}\right)\right\} + \left\{2 - \frac{L\pi^2 G_E^2 s_2}{2H(H^2 - G_E^2\pi^2 s_2^2)^{1/2}}\right\}\frac{2\pi G_E}{(H^2 - \pi^2 G_E^2 s_2^2)^{1/2}}\right] - C_3\left(d_u + \frac{Q_{UF}}{W}\right)\frac{2H^3}{\pi^2 G_E^2 s_2^2 (H^2 - G_E^2\pi^2 s_2^2)^{1/2}}$$

The second order derivative is given by

$$\frac{1}{W}\frac{\partial^2 \gamma_{AS}}{\partial s_2^2} = 6C_2 s_2 + \frac{C_3 L\gamma_w H}{2\pi\gamma_{cs}}\left[\frac{3\pi^4 G_E^4 s_2 HL}{2(H^2 - \pi^2 G_E^2 s_2^2)^{5/2}}\left\{\pi - \cos^{-1}\left(1 - \frac{2\pi^2 G_E^2 s_2^2}{H^2}\right)\right\} - \frac{\pi^3 G_E^3 HL}{(H^2 - \pi^2 G_E^2 s_2^2)^2} + \frac{4\pi^3 G_E^3 s_2}{(H^2 - \pi^2 G_E^2 s_2^2)^{3/2}} \frac{L\pi^3 G_E^3}{H(H^2 - \pi^2 G_E^2 s_2^2)} - \frac{2L\pi^5 G_E^5 s_2^2}{H(H^2 - \pi^2 G_E^2 s_2^2)^2}\right] - C_3\left(d_u + \frac{Q_{UF}}{W}\right)\frac{2H^3}{\pi^2 G_E^2 s_2^3 (H^2 - \pi^2 G_E^2 s_2^2)^{3/2}}(3\pi^2 G_E^2 s_2^2 - 2H^2) \tag{170}$$

For minimum value of Y_{AS} the necessary and sufficient conditions are

$$\left.\frac{\partial Y_{AS}}{\partial s_2}\right|_{s_2 = s_2^*} = 0; \quad \text{and} \quad \left.\frac{\partial^2 Y_{AS}}{\partial s_2^2}\right|_{s_2 = s_2^*} > 0$$

where, s_2^* is the value of s_2 at the optimal point.

At the optimal point the following relation is, therefore, satisfied.

$$C_1 + 3C_2 s_2^{*2} + \frac{C_3 L \gamma_w H}{2\pi \gamma_{cs}} \left[\left\{ \pi - \cos^{-1}\left(1 - \frac{2\pi^2 G_E^2 s_2^{*2}}{H^2}\right) \right\} \left\{ \frac{\pi^2 G_E^2 H L}{2(H^2 - G_E^2 \pi^2 s_2^{*2})^{3/2}} \right\} \right.$$
$$\left. + \left\{ 2 - \frac{L \pi^2 G_E^2 s_2^{*2}}{2H(H^2 - G_E^2 \pi^2 s_2^{*2})^{1/2}} \right\} \frac{2\pi G_E}{(H^2 - \pi^2 G_E^2 s_2^{*2})^{1/2}} \right] - C_3\left(d_u + \frac{Q_{UF}}{W}\right)$$
$$- \frac{2H^3}{\pi^2 G_E^2 s_2^{*2} (H^2 - G_E^2 \pi^2 s_2^{*2})^{1/2}} = 0 \qquad (171)$$

For given value of H, L, C_1, C_2, C_3 and G_E the value of s_2 can be obtained using Equation 171 by an iterative procedure. This gives a combination of values of W, b and s_2 for a hypothetical weir profile, when $s_1 = 0$, the weir apron is horizontal and the value of waterway is such that the afflux is equal to the allowable afflux A_{fp}.

Direct Search Algorithm

To initiate the search, the value of cost function Y for the weir dimensions W, b, and s_2 as obtained for a horizontal weir (Equation 160) is evaluated. The slope of the apron for computed values of Y_0 and b is evaluated and value of $\alpha\pi$ (Figure 54) determined. The length of downstream sheet pile s_2 is now reduced and the s_1 increased, and a new combination of values of s_1 and s_2 is reached so that the value of maximum exit gradient reaches the specified safe exit gradient G_E. The pressures under the sloping apron with two vertical sheet piles is obtained using analysis given earlier. The value of cost function is now computed. This process of reducing the length of downstream pile and increasing length of upstream sheet pile, for the same values of W and b, is repeated till the cost function is increased with decrease in s_2 value.

The length of downstream sheet pile, s_2, is then increased by smaller magnitude, and that of s_1 is reduced till the minimum value of the cost function is reached, satisfying the exit gradient constraint. This search gives the minimum value of cost function value of b. The pattern of search is illustrated in Figure 55. In this figure the numerical value of Y for b = 60.0 m and for various combinations of s_1 and s_2, satisfying the exit gradient constraint are indicated to draw the contours of equal cost.

The search is repeated for changed values of b to find the new minimum value of cost function until the desired optimal combination of b, s_1 and s_2 is reached.

Having obtained the optimal values of b, s_1 and s_2 corresponding to the initially assumed value of the waterway, the value of W is now increased and the search for the new optimal combination of b, s_1 and s_2 is repeated. This is continued until the optimal combination of W, b, s_1 and s_2 is reached.

For an assumed value of waterway and corresponding values of L = 40.0 m, H = 15.0 m, Y_0 = 2.0 m and various value of b, the minimum value of the cost function, Y, and corresponding values of s_1 and s_2 have been obtained for d_u = 1.0 m, Q_{UF}/W = 0.20 m^2, C_1 = Rs. 250.0/m^2, C_2 = Rs. 2.0/m^3, C_3 = Rs. 200.0/m^3, b > L and G_E < 0.1666, 0.25 and 0.5. The minimum of the minimum values of Y for L = 40.0 m can be obtained from Figure 56 for various values of G_E. For G_E = 0.1666, 0.25, and 0.50 the optimal values of b are 140.0 m, 100.0 m, and 80.0 m, respectively. The corresponding optimal values of s_1 and s_2 are found to be 3.2 m, 11.2 m; 4.6 m, 6.6 m and 5.1 m, 2.2 m, respectively (Figures 57 and 58). The variation of the optimal value of the cost function Y with G_E is shown in Figure 59. The figure shows that the optimal value of the cost function is influenced by the value of the specified safe exit gradient. When the specified safe exit gradient is reduced from 0.25 to 0.1666 the optimal value of Y_W is increased from Rs. 47,634.0 to Rs. 59,447.0. Thus, the optimal cost is increased by 24.8%. The increase in optimal cost, when the safe exit gradient is reduced from 0.5 to 0.1666 is 51.0%.

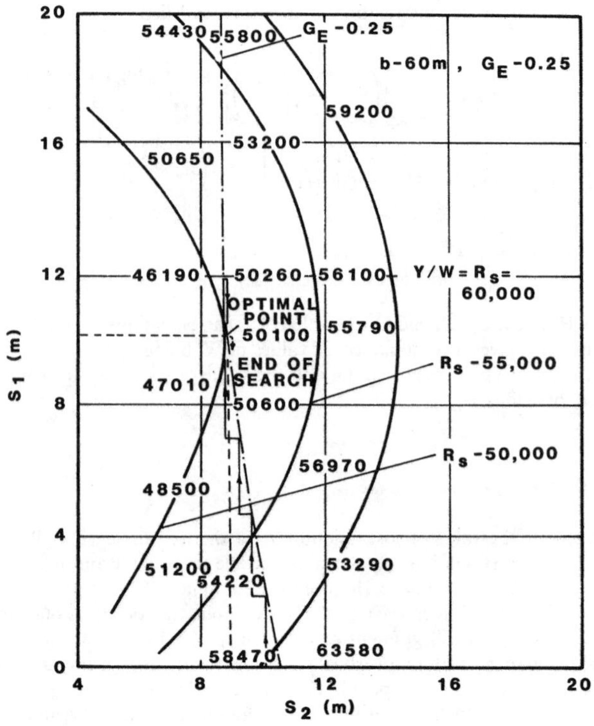

Figure 55. Cost contours and pattern of search.

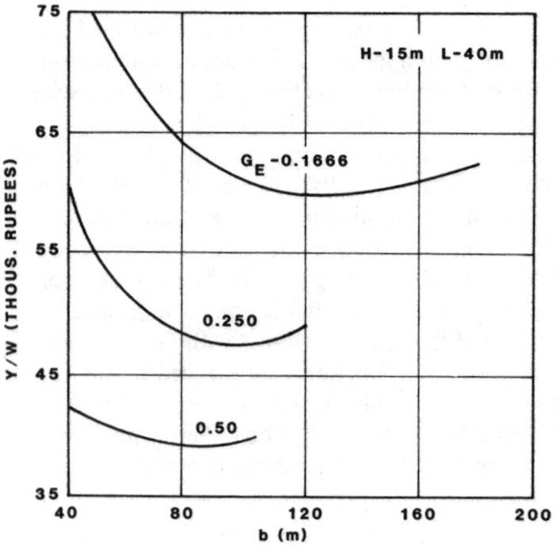

Figure 56. Variation of minimum value of Y/W with b for $G_E = 0.1666$, 0.25 and 0.500.

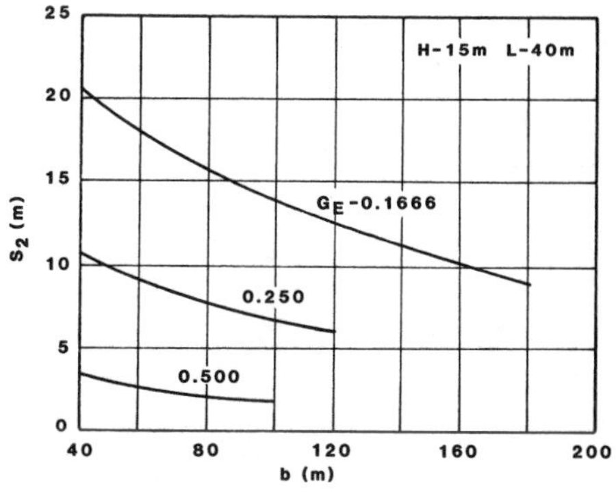

Figure 57. Variation of optimal value of s_2 with b for $G_E = 0.1666$, 0.25 and 0.500.

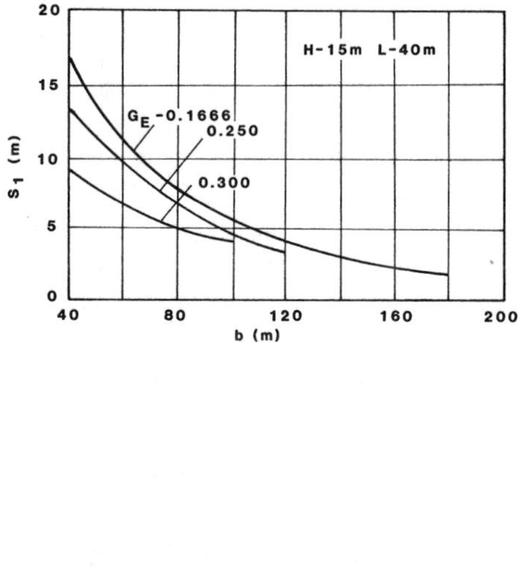

Figure 58. Variation of optimal value of s_1 with b for $G_E = 0.1666$, 0.25 and 0.500.

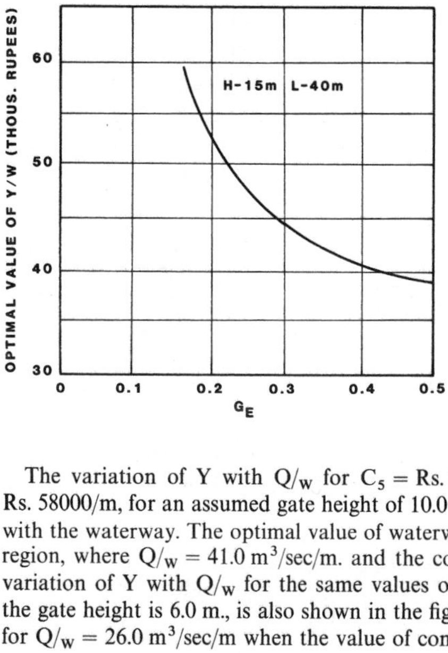

Figure 59. Variation of optimal value of Y/W with G_E.

The variation of Y with $Q/_W$ for $C_5 =$ Rs. 200.0/m^3, $C_6 =$ Rs. 30.0/m^3, $G_E = 0.2$, and $C_4 =$ Rs. 58000/m, for an assumed gate height of 10.0 m, is shown in Figure 60. The value of Y increases with the waterway. The optimal value of waterway of the weir lies on the boundary of the feasible region, where $Q/_W = 41.0$ m^3/sec/m. and the computed afflux is equal to permissible afflux. The variation of Y with $Q/_W$ for the same values of C_5, C_6 and G_E and $C_4 =$ Rs. 28,000.0/m, where the gate height is 6.0 m., is also shown in the figure. In this case the optimal waterway is obtained for $Q/_W = 26.0$ m^3/sec/m when the value of computed afflux is less than the permissible afflux.

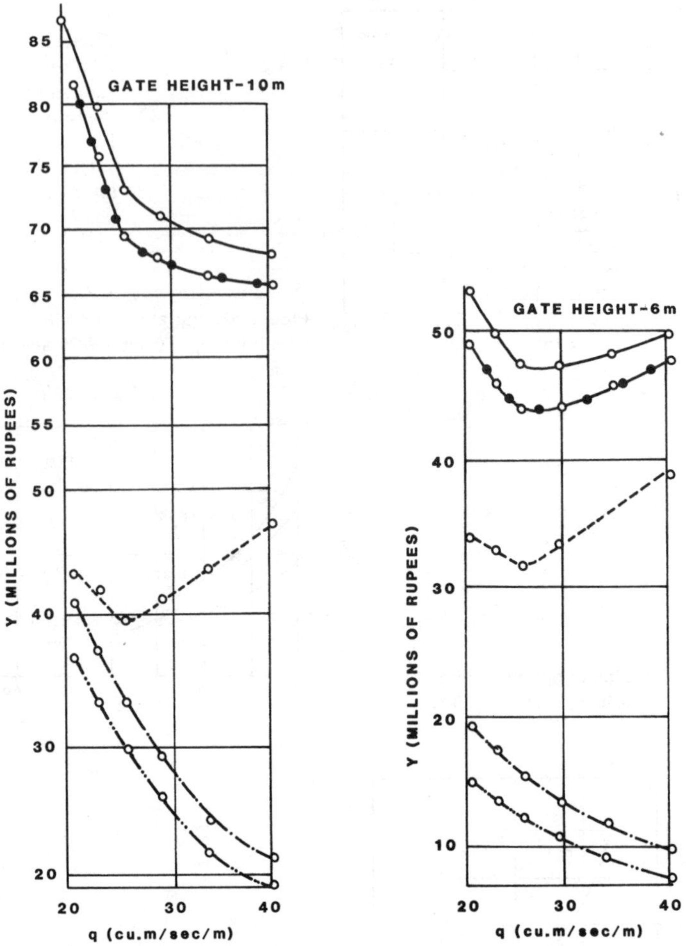

Figure 60. Variation of optimal value of Y with q. —··— Cost of gates vs. discharge intensity. —·— Cost of gates and bridge vs. discharge intensity. ------ Cost of concrete earth work etc. vs. discharge intensity. —●— Cost of barrage excluding bridge vs. q.——Total cost of barrage vs. discharge intensity.

NOTATION

A_f	afflux
A_{fp}	permissible afflux
A_r	area of aerofoil
$A_{1,n}, A_{2,n}, A_{3,n}, A_{4,n}$	coefficients
a	value of t corresponding to point D; inverse of the value of t corresponding to point D
B( )	complete beta function
B_r( )	incomplete beta function
$B_{1,m}, B_{2,m}, B_{3,m}, B_{4,m}$	coefficients
b	width of apron
b_1	width of apron in fictitious plane
C	Constant

C_1 cost of sheet pile/m^2
C_2 cost of driving 1.0 m wide and 1.0 m deep pile
C_3 cost of concrete/m^3
C_4 cost of material of flank walls/m^3
C_5 cost of gates and hoisting arrangement/m
C_6 cost of earth work/m^3
$C_{1,n}, C_{2,n}, C_{3,n}, C_{4,n}$ coefficient
c distance of centre of the circle in T_1-plane from the origin
c′ constant
d_A, d_C depth of floor at points A and C
d_1 prejump depth
d_2 post jump depth
d_u depth of upstream floor
E_D energy required to drive unit length of pile
F_B value of free board
F_g gravitational force
F_s seepage force
$F_1(\ )$ hypergeometric series of two variables
Q_U volume of concrete in upstream floor
R radius of curvature of circular scour
RBL general river bed level
r dummy variable
$R_1, S_1, R_2, S_2, R_3, S_3$ Cartesian coordinates
S length of sheet pile in modified inverse plane
S_{1c} minimum permissible depth of upstream sheet pile
S_{1m} maximum permissible depth of upstream sheet pile
S_{2c} minimum permissible depth of downstream sheet pile
S_{2m} maximum permissible depth of downstream sheet pile
s length of sheet pile
s_1, s_2 length of upstream and downstream sheet piles
T_1 $R_1 + iS_1$ = complex variable
T_2 $R_2 + iS_2$ = complex variable
T_3 $R_3 + iS_3$ = complex variable
t parametric plane
t_A, t_B, T_P values of t corresponding to points A, B, and P
t_C, t_D, t_E, t_F values of t corresponding to points C, D, E and F
$U_{1,P}, U_{2,P}, U_{3,P}, U_{4,P}$ coefficient
V volume of soil mass
w $\Phi + i\psi$ = complex potential
W waterway of barrage
x, y coordinates
x', x_1, y', y_1 coordinates
x_1, x_2, x_3 dummy variables
Y_0 difference in levels of upstream and downstream ends of floor
${}_2F_1(\ )$ Gauss hypergeometric series
G specific gravity of soil
G_E specified safe exit gradient
H head causing seepage
H_1 head of water on upstream side
H_2 head of water on downstream side
H_L head loss in hydraulic jump
H_{DF}, H_{UF} height of downstream and upstream flank walls
h depth of scour
I_E exit gradient
i $\sqrt{-1}$
J constant
K coefficient of permeability
K_{DF}, K_{UF} coefficient pertaining to quantity of materials in downstream and upstream flank walls
L Length of scour length of floor downstream of gate line
ℓ half length of scour half length of scour in fictitious T_1-plane
M, M′, N, N′ constants
N_2, N_3 constants
n porosity of soil
p a dummy variable
p′ pressure

P_A, P_B, P_C	pressure at points A, B, and C
Q_A	volume of concrete in weir apron
Q_{ES}, Q_{EU}	volume of earthwork in downstream and upstream apron
Q_{UF}	volume of concrete per unit length of upstream flank walls
Q_s	volume of concrete in stilling basin floor
Y	cost of weir floor, sheet piles, earthwork and gates
Y_A	cost of concrete
Y_{AS}	cost of assumed weir apron
Y_E	cost of earthwork
Y_{DF}, Y_{UF}	cost of downstream and upstream flank walls
Y_G	cost of gates and hoisting arrangement
Y_m	maximum depth of scour
Y_s	cost of sheet piles
y_1, y_2, y_3	dummy variables
z	$x + iy$ = complex variable
z'	$x' + iy'$ = complex variable
z_1	$x_1 + iy_1$ = complex variable
z_P	value of z for point P
z_B	value of z for point B

Greek Symbols

α	angle in units of π
β	angle in units of π
γ_w	unit weight of water
γ_{cs}	submerged unit weight of concrete
δ	angle
θ	angle in units of π
θ'	angle in units of π
λ	$\frac{1 + \sqrt{1 + (b/s)^2}}{2}$
ϵ	angle
ϵ_m	angle
τ	distance of point of maximum scour from one end of scour
Φ	velocity potential function
$\Phi_C, \Phi_D, \Phi_E, \Phi_F$,	value of velocity potential function at points C, D, E, F, respectively
ψ	stream function

REFERENCES

1. Bilashevsky, N. N., "The Mechanism of the Local Scour Behind the Spillway Structures with the Aprons and Influence of the Macro Turbulence Upon the Scour," *Proceedings Twelfth Congress of the International Association for Hydraulic Research*, Vol. 3, pp. 260–265, (1967).
2. Colar, et al., "Etude des Affouillements a L'Aval d' un Seuil Deversant," Proceedings Twelfth Congress of The International Association for Hydraulic Research, Vol. 3, pp. 322–329, (1967).
3. Doddiah, D., "Scour Below Submerged Solid Bucket Type Energy Dissipators," *Proceedings Twelfth Congress of The International Association for Hydraulic Research*, Vol. 3, pp. 105–116, (1967).
4. Tsuchiya, Y., and Iwagaki, Y., (1967), "On the Mechanism of the Local Scour From Flows Downstream of an Outlet," *Proceedings Twelfth Congress of The International Association for Hydraulic Research*, Vol. 3, pp. 55–64.
5. Khosla, A. N., Bose, N. K., and Taylor, E. M., "Design of Weirs on Permeable Foundations," Central Board of Irrigation and Power, New Delhi, India, (1936).
6. Sharma, H. D., et al., "Hydraulic Design of Undersluice Bays, the Barrage Bays and Head Regulator for Kosi Barrange near Ramnagar—A Model Study," Technical Memorandum No. 42-RR. (H 1–5) Irrigation Research Institute, Roorkee, (1972).

7. Leliavsky, S., *Irrigation and Hydraulic Design*, Vol. I, Chapman and Hall, Ltd., London, (1955).
8. Peirce, B. O., *A Short Table of Integrals*, Oxford and IBH Publishing Co., Calcutta, (1956).
9. Cedergren, H. R., *Seepage, Drainage, and Flow Nets*, John Wiley and Sons, Inc., New York, (1967).
10. Basu, U., "Steady State Confined and Unconfined Flow Through Anistropic Soils," Ph.D. Thesis, Faculty of Engineering, Indian Institute of Sciences, Bangalore, (1976).
11. Himmelblau, David, M., *Applied Non-Linear Programming*, McGraw-Hill Book Company, Inc., New York, NY, (1972).
12. Fox, R. L., *Optimization Methods for Engineering Design*, Addison Wesley Publishing Company, Inc., California, (1970).

INDEX